The Hvac/r Professional's Field Guide to

Medium & High Efficiency Gas Furnaces

Richard Jazwin

BNP
Business News Publishing Company
Troy, Michigan

Library of Congress Cataloging-in-Publication Data

Jazwin, Richard.
    The Hvac/r professional's field guide to medium
& high efficiency gas furnaces/Richard Jazwin.
       p. cm.
    ISBN 0-912524-83-9
     1. Gas--Heating and cooking--Handbooks, manuals,
etc. 2. Furnaces--Handbooks, manuals, etc. I. Title.
II. Title: Medium & high efficiency gas furnaces. III. Title:
Medium and high efficiency gas furnaces
TH7466.G3J38   1993   93-8584
697'. 07--dc20   CIP

Administrative Editor: Joanna Turpin
Technical Editor: Barbara A. Checket-Hanks
Art Director: Mark Leibold
Copy Editor: Carolyn Thompson
Cover Illustration: TechWorld

Printed in United States of America
7 6 5 4 3

Dedication

To all of the heating and air conditioning technicians who bear the unbearable to make our lives better.

Disclaimer

This book was written as a general guide. The author and publisher have neither liability nor can they be responsible to any person or entity for any misunderstanding, misuse or misapplication that would cause loss or damage of any kind, including loss of rights, material or personal injury, or alleged to be caused directly or indirectly by the information contained in this book. The service manuals are reprinted unedited and the author and publisher are not responsible for their content.

Table of Contents

American Standard

> Installer's Guide: Super Efficiency Gas-Fired Condensing Furnace, Models TUC-B; TDC-B
> User's Information Manual: Super Efficiency Condensing Gas Furnace Models TUC, TDC and TUX, TDX
> REDDI FACTS: Gas Furnace Model TUC100B960A0

Arcoaire/Comfortmaker

> Service Manual: Deluxe Gas-Fired Forced-Draft Furnace, GUA/GDA Series
> User's Information Manual: GUA Series Deluxe High Efficiency Forced Draft Upflow Gas-Fired Furnace
> Service Manual: Super High Efficiency Condensing Gas Furnace, GUH Series
> Installation Instructions and Operating Procedures: Forced Draft Upflow Gas-Fired Furnaces
> Installation Instructions and Operating Procedures: Super High Efficiency Upflow Gas-Fired Furnaces, GUH Series
> User's Information Manual: Super High Efficiency Upflow Gas-Fired Furnaces, GUH Series

Armstrong

> Installation and Maintenance Instructions/User's Information Manual: Gas-Fired High Efficiency Furnace,
> Model Series EG6E and EG7E

Bard

> Installation Instructions/User's Information Manual: High Efficiency Gas Furnace, Models: HI-BOY
> DCH050D30A, DCH065D36A, DCH080D48A, DCH095D60A

Introduction

Heat is a necessity to humanity. Without an adequate heating system, many of our endeavors would be curtailed. Nothing can distract a person faster than being cold.

In physics, there is no such thing as *cold*. Everything contains heat energy; cold is merely the absence of heat. While the "non existence" of cold may be a valid principle, it will not help you feel better when you're freezing to death. If you are exposed to extreme cold, you first feel distracted; then feel extreme discomfort; and then enter a potentially permanent sleep.

Early pictures show cave dwellers huddled around a fire trying to keep warm. Fire, one of the first heat sources, has been, and still is, an excellent heat source.

The fireplace was a stock fixture in years past. Many homes still use the fireplace as a primary heat source. Fireplaces cause concern because of factors such as pollution, lack of efficiency, and the user's labor. However, the main disadvantage of the fireplace is its efficiency factor. Early inventors, such as Ben Franklin, developed methods that improved fireplace efficiency. Later inventors developed heat exchangers that substantially improved fireplace efficiency. Fireplaces have been replaced by central heating systems, which have become the basic heat source in most homes.

This book deals with medium- and high-efficiency gas furnaces in central heating applications. Understanding basic furnace design and operation allows us to see the significant advances that have been made in furnace design. These advances have led to today's high-efficiency furnaces.

What characterizes a high-efficiency furnace? It utilizes less fuel to produce more heat. There are systems in place to measure furnace efficiency. A primary efficiency measurement is the annual fuel utilization efficiency (AFUE) rating.

AFUE is a ratio that compares annual output energy to the annual input of energy. If a quantity of gas with a known Btu value is sent to a furnace, some of that Btu value does not end up heating the space; some is lost up the vent. AFUE rating provides a measure of that loss.

For our purposes, we will use a guideline that categorizes furnaces into three AFUE ranges:

- Low-efficiency furnace — AFUE rating of 71% or less.

- Middle-efficiency furnace — AFUE rating of 71% to 83%.

- High-efficiency furnace — AFUE of 84% and up.

Some sources state that any furnace over 80% AFUE can be considered a high-efficiency furnace; others state that 90% is a minimum AFUE rating. What the reader must understand is that the higher the AFUE number, the less fuel the furnace uses to heat the space. Less fuel means fewer dollars spent by the consumer each year. From the consumer's standpoint, the real measure of efficiency is how much it costs to operate the unit.

An act passed in 1976 requires that all furnaces sold from 1992 on must have an AFUE percentage of no less than 78%. The purpose of this act is to encourage conservation of fuels, for the benefit of future generations. The consumer switching to a high-efficiency furnace helps future generations and, at the same time, saves money.

From the technician's standpoint, a basic understanding of AFUE is essential. A furnace is a major investment for most homeowners. High-efficiency units cost more to purchase and install. There will come a time when repairing the old furnace is just not cost effective. At that time, the technician should make the consumer aware of high-efficiency alternatives. Furnaces offering an AFUE greater than 90% are readily available. If the old furnace is in the 60% range, replacing the unit with a mid- or high-efficiency unit gives the user a long-term payback, provided a high-efficiency furnace can be installed.

This book is divided into two parts. The first covers basic theories and operations of furnaces, with particular emphasis placed on what makes a furnace a mid- or high-efficiency unit. The second part reprints actual service and installation manuals for many mid- and high-efficiency furnaces, provided by the manufacturers.

The inclusion of a manual is not an endorsement of that manufacturer's product. The choice of manuals was the author's. The decision to reprint a manual indicates the author's desire to demonstrate specific features that can differ in construction, components, and terminology. A manufacturer whose manuals do not appear in this book does not manufacture an inferior product. More than likely, the lack of the manual indicates that the manufacturer chose not to have a particular manual reprinted.

The reader who studies the first part of this book will develop a solid understanding of today's furnaces. Using the second part of the book on a service call will assist the technician in solving a specific problem by understanding how a particular manufacturer approaches the combustion process.

To compete successfully in service today, a new tool must be added to the toolbox: **Information.** Today's equipment is becoming more complex, and without proper information, the possibility of successful service decreases. This book was written with that thought in mind. The author wishes to thank all of the furnace manufacturers, without whose cooperation this book could not have been possible.

Combustion

Heat is energy. As such, it cannot be created or destroyed. A gas furnace burns natural gas or liquefied petroleum and oxygen to create heat, which is then transferred to the space that needs it. Combustion is the process that allows all of this to occur.

Combustion

Combustion can be defined as the process of burning. When something is combustible, it catches fire easily. To create heat, we burn a combustible fuel. When oxygen combines with a combustible fuel, such as natural gas, and is heated to its ignition temperature — the lowest temperature at which a fuel burns — heat and light are produced. The ignition process shows itself as flame. The flame itself gives off heat and light.

Combustion continues as heat from the burning mixture fires the remaining unburned mixture of air (oxygen) and the combustible gas. The combustion process liberates the heat energy in the oxygen-gas mixture. In a furnace application, this heat energy is used to heat space.

In order to have proper combustion, the correct amount of oxygen must be introduced with the correct amount of gas. Different fuels require different amounts of oxygen to achieve a satisfactory combustion process.

Complete combustion occurs when all of the oxygen-gas mixture is burned. The amount of oxygen introduced into the combustion process must be controlled, so that complete combustion will result. When all of the air-gas mixture does not burn, combustion remains incomplete. If combustion is not completed, carbon monoxide, aldehydes, and soot (unburned carbon) result. These byproducts affect furnace efficiency and, more importantly, are dangerous to the home's inhabitants.

Technicians must strive to ensure that combustion is complete.

Now that we have a basic understanding of combustion, let's examine the factors necessary to achieve it:

Fuel + Oxygen + Ignition = Combustion

FUEL

Throughout history, many substances have been used for fuel. This book will deal with two primary sources in use today: natural gas and liquefied petroleum gas (LPG).

Natural gas

Natural gas is used extensively in the heating market because of its widespread availability. If we burn 1 cu ft of natural gas it will yield, on the average, 1,000 to 1,050 British Thermal Units of heat (Btu). A Btu is the amount of heat necessary to raise the temperature of 1 lb of water 1°F. Natural gas can yield from 900 to 1,400 Btu per cu ft. The Btu yield varies according to the source of the original gas and the area in which the gas is used. When servicing a natural gas furnace, the technician should check the Btu value for the area with the local gas utility.

Natural gas is lighter than air. It also has an odorant added to assist in leak checking. Because natural gas is lighter than air, leaking natural gas dissipates easier than LPG; LPG, which is heavier than air, will "puddle" on the ground in the event of a leak. Gas in any form can be dangerous, but a puddle of LPG at ground level has an extremely high accident potential.

Natural gas is a hydrocarbon. It is composed of hydrogen and carbon atoms. To achieve proper combustion, 15 cu ft of air is required for every cubic foot of natural gas. The boiling point of natural gas is low, thus allowing it to remain in a gaseous state for use in residential heating. Under ideal conditions, natural gas has an ignition temperature of 1,100°F, burning temperature of 3,500°F, and a specific gravity of 0.65.

Liquefied petroleum gas

For those markets where natural gas isn't readily available, LPG is an acceptable alternative. Because LPG can be stored in a tank, piping and distribution lines from a central utility aren't needed. This accounts for the popularity of LPG as a heating fuel in remote areas.

LPG is typically contained in a dedicated storage tank. Because of its high boiling point, LPG in a storage tank will be in liquid form, with some evaporating gas present on top of the liquid. When a pressure drop occurs, such as the gas valve opening, LPG evaporates into a gaseous state. An additional benefit of liquid storage is the concentration of Btu potential: 1 gal of LPG evaporates into more than 36 cu ft of gas.

LPG is composed of either propane, butane, or a combination of the two. Propane and butane have different boiling points, so storage can be a problem. Butane, at about 40°F, has a vapor pressure in the tank of only about 3 psig. Propane, at the same temperature, has a pressure of 78 psig. To use LPG, 11 in. of water column (wc) are necessary as a minimum pressure for the fuel to enter the heating unit.

The boiling point differences account for the necessity of mixing propane and butane. Because of its low boiling point, butane that is stored without a heat source surrounding the tank would not, on a cold day, be able to maintain sufficient vapor pressure in order to flow. Mixing butane and propane, however, creates a fuel that will flow in gaseous form on a cold day.

Both butane and propane are heavier than air. The specifications for butane are 3,267 Btu per cu ft of vapor, a specific gravity of 2, approximately 1,100°F ignition temperature, and 3,300°F burning temperature. The specifications for propane are 2,521 Btu per cu ft of vapor, a specific gravity of 2, approximately 1,100°F ignition temperature, and 2,975°F burning temperature.

These specifications are for the individual fuels and do not apply to a mixture. As indicated earlier, a mixture may be necessary for heating purposes. Consult the fuel supplier for specifications on their product.

With certain changes, a furnace that burned natural gas can be converted to LPG.

OXYGEN

In order to have oxygen for combustion, we must have air. We already know that oxygen combined with fuel and heated to the appropriate temperature creates a flame. To achieve this flame, we use oxygen present in the air. Air is composed of 21% oxygen, 79% nitrogen, and a sprinkling of some other gases.

Complete combustion can be defined as burning all of the fuel present, while supplying more than enough oxygen to ensure combustion. In order to achieve complete combustion, sufficient air must be introduced into the combustion process, Table 1-1.

COMPLETE COMBUSTION NEEDS	COMPLETE COMBUSTION RESULTS
15 Cubic Feet of Air	12 Cubic Feet of Nitrogen
1 Cubic Foot of Natural Gas	2 Cubic Feet of Water Vapor
	1 Cubic Foot of Carbon Dioxide
	1 Cubic Foot of Oxygen
	HEAT

Table 1-1. Complete Combustion

The quantity of air varies from fuel to fuel. The variance exists because of the grade and quality of available fuel. In order to have complete combustion, we must have 10 cu ft of air for each cubic foot of natural gas. If less than 10 cu ft of air is supplied, incomplete combustion results. Actually, about 5 cu ft of "extra air" is supplied for the combustion process, to ensure that there is enough air to achieve complete combustion. The presence of this excess air is also a negative, since it absorbs heat by lowering the unit's temperature.

If we achieve complete combustion, *carbon dioxide* is a byproduct of the process. Carbon dioxide is harmless. If we do not have enough oxygen and incomplete combustion results, one of the byproducts is *carbon monoxide*. This can be dangerous.

Incomplete combustion is the opposite of complete combustion. If all of the fuel is not consumed, the result is incomplete combustion. Incomplete combustion results in carbon monoxide, gray smoke, and soot buildup. The products of incomplete combustion can be dangerous to building occupants, as well as to the furnace itself.

To achieve complete combustion, we employ the principle of *fuel oxidation*. When fuel is oxidized, chemicals in the fuel combine with oxygen, resulting in burning. It is the introduction and mixing of air with fuel that causes the burning process. Air is introduced into the furnace from two sources: primary and secondary air.

Primary air is mixed with the fuel before the fuel reaches the burner for the combustion process. Primary air controls the rate at which combustion takes place.

Secondary air mixes with fuel during the combustion process. Secondary air taken from the atmosphere helps ensure complete combustion.

It is extremely important for a service technician to have a basic understanding of the combustion process. Tuning a furnace to achieve complete combustion is a skill. The first requirement is understanding that the proper amount of oxygen, mixed with the proper amount of fuel, results in the desired combustion rate. A lack of understanding of the combustion process can result in serious injury, perhaps death.

There is no acceptable trial-and-error approach in furnace service. Understand what you are going to do before you do it — or don't service the unit.

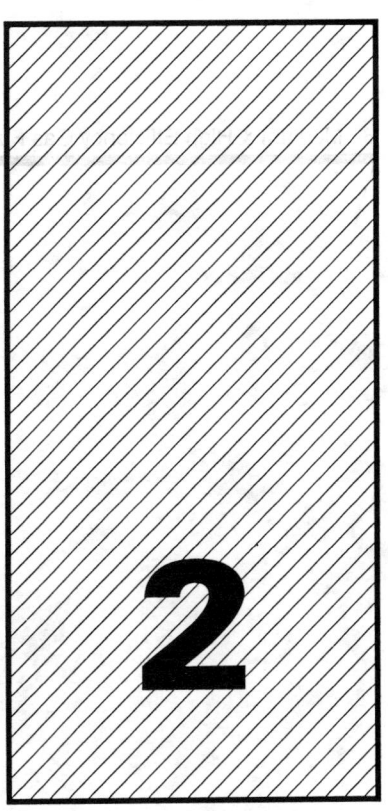

Furnace Construction

In order to heat a space, air from the space must be brought into contact with a heat source hot enough to heat the air. At the same time, the air that is heated must not become contaminated with any byproducts of the combustion process. A furnace's construction helps create these conditions.

Furnace Types

Return air (air supplied to the furnace) flows across the furnace heat exchanger in one of three ways:

- **Upflow** — Flow travels up and through the heat exchanger.

- **Downflow** — Flow travels down and through the heat exchanger.

- **Horizontal** — Flow travels across the heat exchanger.

Modern furnaces are configured in four basic designs: upflow, downflow, horizontal, and lowboy. The different configurations help reconcile the space in which the furnace will be installed, as well as its airflow requirements.

Upflow furnaces are the most popular. Air enters from the bottom of the furnace, moves across the heat exchanger and leaves through the ductwork. These furnaces can be installed in a closet or basement, as long as safety codes are followed. The blower motor is mounted in the bottom of the cabinet.

To provide cooling as well as heating, an evaporator coil, connected to an outdoor condensing coil, is added to the top of the furnace, and a multi-speed blower motor is installed within the furnace.

Downflow furnaces, sometimes called counterflow furnaces, have an airflow pattern that is the reverse of an upflow furnace. In a downflow furnace, the return air enters at the top and leaves at the bottom. Typical installations would be a mobile home or other residence where the ductwork is in a slab or crawl space. Warm supply air is supplied to the space from floor vents. The blower motor is mounted in the top of the cabinet.

Horizontal furnaces are best used in an attic location (in homes with no utility area in which to install a furnace). The airflow in this type of furnace travels across the heat exchanger.

Lowboy furnaces are essentially smaller models that can be configured for add-on air conditioning (low enough to be mounted in a service closet, while maintaining sufficient ceiling clearance for adding central air at a later date).

Basic Components

Regardless of a furnace's efficiency, it must have certain basic components. The following would be found in a typical upflow furnace, Figure 2-1.

AIR FILTER

Most furnaces come with a washable filter media, allowing the user to clean the filter on a regular basis. Disposable filters can also be used.

It is extremely important to point out the filter's location to the user and to establish a filter changeout or cleaning schedule. Explain that a dirty filter can create temperatures detrimental to furnace operation. And if the system includes add-on air conditioning, a dirty filter can decrease operational pressures to the point where floodback damages the compressor.

Filters also add resistance to airflow. Piling extra filters into the filter chamber can seriously impair furnace operation. Airflow must be considered if the thickness of the filter is changed from original equipment specifications.

Electronic air cleaners as an add-on feature are growing more popular. They have the advantage of conditioning the air beyond the capabilities of media-type filters.

BLOWER MOTOR

An electric motor, generally in the ¼- to ½-hp range, can be a split-phase, capacitor-start, shaded-pole, or permanent-split-capacitor model. These motors direct a centrifugal blower to make air flow across the heat exchanger and evaporator coil. In a direct-drive application, the blower and the motor turn at the same number of rpm.

In some larger installations, a belt-driven centrifugal blower wheel is used; an adjustable pulley creates speed changes. Blower motors driving a centrifugal blower can be multi-speed models. When replacing a blower motor, cfm requirements and rotation must be considered. Refer to manufacturer's data on recommended blower speed for a particular model furnace. Remember — a blower operating in heating mode generally moves less air than a blower operating in cooling mode.

In some of today's furnaces, new technology causes the blower motor to vary speed according to the space's heating requirements. The furnace's heat output is constantly monitored by solid-state controls. Firing rate and blower speed are monitored and adjusted to meet the space needs. This approach reduces the furnace from cycling on and off — not the most efficient method of heating a space.

BURNER

A typical burner is made out of steel; ports can be single, slotted, or multiple. Burners mix the metered fuel and air to the furnace. Fuel arrives from the supply line through an orifice and mixes with primary air in the burner tube. Secondary air enters from the atmosphere. The secondary air, fuel, and primary air create the necessary mix for combustion. Most furnaces have a primary air shutter to control the fuel-primary air mix. Primary air constitutes about 50% of the air necessary to achieve combustion. Secondary accounts for 50% more air than is necessary to achieve combustion.

CABINET

A typical furnace has a steel cabinet. The blower compartment is lined with insulation to reduce Btu loss. Service access panels are built in, and a compartment for an air filter is provided (filter location depends

B. IDENTIFICATION OF COMPONENTS

① Wrap Casing Assembly
② Control Louverd Door
③ Blower Door
④ Blower Door Interlock Switch
⑤ Burner Door
⑤Ⓐ Gasket
⑥ Flash Shield
⑦ Burner Assembly
⑧ Cross-over On Burners
⑨ Flue Box (Porcelain)
⑨Ⓐ Gasket
⑩ Collector Box
⑩Ⓐ Gasket
⑪ Equalizer Plate
⑪Ⓐ Gasket (2)
⑫ Heat Exchanger Flue Baffles
⑬ Heat Exchanger (RPJ)
⑭ Pressure Switch
⑮ Fan & Limit Switch
⑯ Forced Draft Motor
⑰ Gas Valve (Redundant)
   See Page 5
⑱ Manifold Connection-Dell Fitting
   See Figure 14

⑲ Manifold
⑳ Burner Orifice and Spring
   See Page 4
㉑ Ignitor Box
   See Page 12
㉒ Blower Assembly with Motor
㉓ Transformer, Relay Package
㉔ Capacitor
㉕ Burner Box
㉖ Burner Inspection Window
㉗ Ignitor
㉘ Sensor
㉙ Aux. Limit Control
㉚ Flame Roll-out Switch

HAND TOOLS

The hand tools carried by a qualified service mechanic will normally be sufficient to do any service work upon this type furnace.

Recommended tools needed:

1. Thermometers
2. U-Tube Manometer with fittings
3. Clamp-on Amp Meter
4. Watch with second hand or stopwatch
5. Soap bubbles for checking leaks
6. Incline Manometer or Magnahelic with Pitot tube or static pressure tip
7. V.O.M. Meter

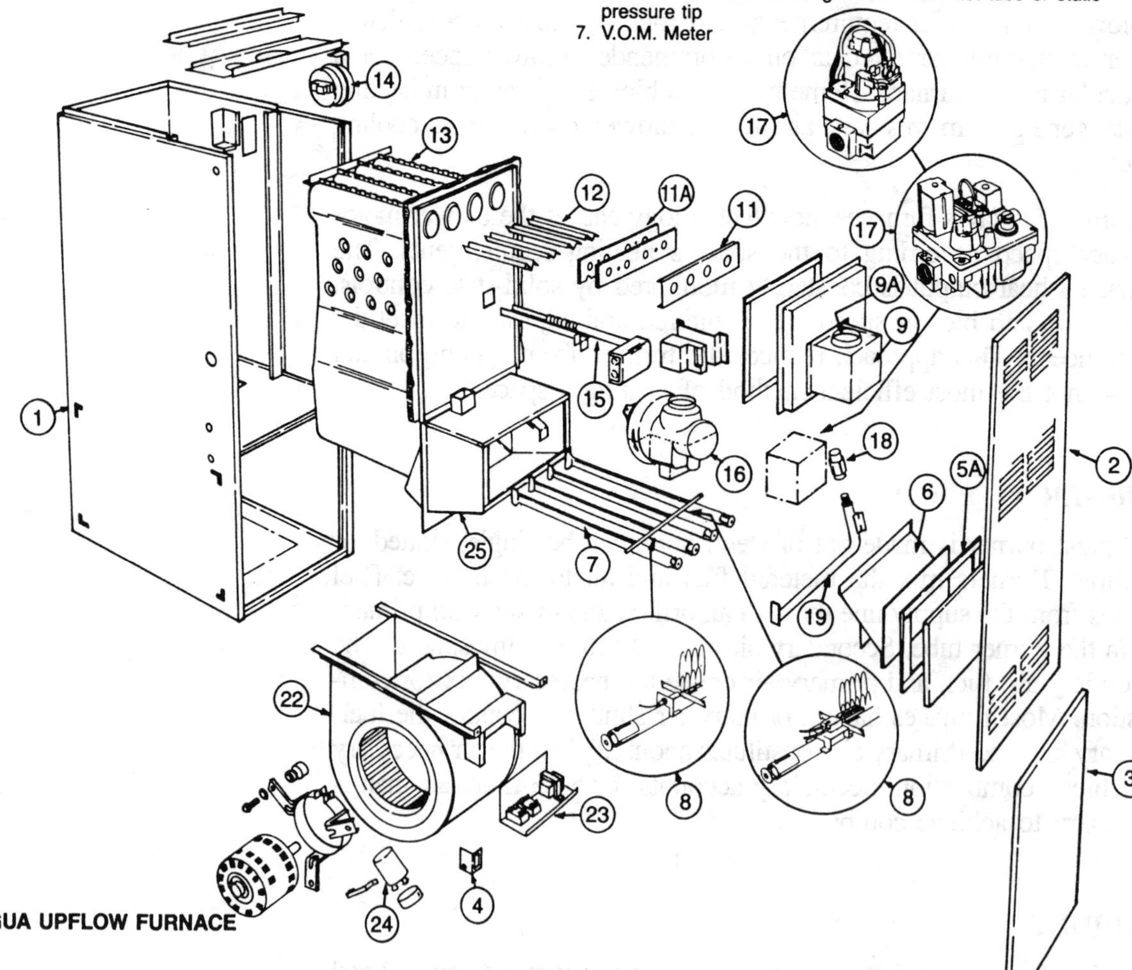

GUA UPFLOW FURNACE

Figure 2-1. Components in an Upflow Furnace (Courtesy, Inter-City Products Corp.)

on the furnace's configuration). A safety interlock is mounted on the filter access door to de-energize the electrical circuits when the user changes the filter.

COMBUSTION BLOWER

Burners in older furnaces relied on atmospheric conditions to deliver combustion air. Today's high-efficiency furnaces use combustion blowers to boost air volume for combustion. The combustion blower can be an induced-draft type, which pulls the products of combustion from the heat exchanger and discharges up the vent; or a forced-draft type, which drives air into the combustion chamber to improve combustion.

HEAT EXCHANGER

The heat exchanger allows heat to be transferred from its surface to the air that will heat the space. (For those readers familiar with the refrigeration process, an evaporator or condenser is a heat exchanger.)

Heat exchangers are constructed from a variety of materials. Each manufacturer has its own opinion as to the best substance. Some common materials include stainless steel, glass-lined cells, multi-cell, ceramic-coated, gauge steel, cast iron, and aluminized steel. The variety of materials can be accounted for by the different heat transfer rates and longevity of the materials.

It is extremely important that a heat exchanger be as strong as possible. A crack in the body of a heat exchanger can allow combustion gases to mix with the space air, which can be fatal to occupants.

Many high-efficiency furnaces use a secondary heat exchanger, allowing for additional heat transfer to the space air. These units, generally referred to as condensing furnaces, are designed to remove as much heat as possible from the air that has been heated. Extremely high efficiencies have been achieved using the condensing furnace because the heat ends up in the space rather than going up the vent.

IGNITION SYSTEMS

There are many different forms of ignition systems in use. The old thermocouple standing-pilot system is being replaced by everything from automotive spark plugs to hot surface ignition. (Ignition systems will be covered in more detail in a later chapter.)

LOW-VOLTAGE CONTROL SYSTEM

With rare exceptions, all furnaces have low-voltage control systems, a prime example being the thermostat circuit. In addition to low-voltage components, some safety devices also control on-line voltage.

TRANSITION TO SPACE VIA DUCTWORK

Heated air is moved by the blower across the heat exchanger, into the ductwork, until it finally reaches the space. Unless the heat exchanger is cracked, there will be no byproducts of combustion in the space air.

VENTING SYSTEMS FOR COMBUSTION BYPRODUCTS

Furnaces may vent using either a natural-draft or a power-venting system.

In a natural-draft (atmospheric) venting system, combustion byproducts are vented to the atmosphere, due to the difference in density between vent gases in the flue and air outside the flue. More than 15 cu ft of combustion gases go up the flue pipe. Another 15 cu ft of air (dilution air) is mixed at the draft diverter, making a total of 30 cu ft of flue gas going up the vent per cubic foot of burned gas.

In power venting, gases are vented due to the positive pressure created by the blower.

> *Note — Not all blower systems produce positive pressure in the vent system.*

Furnaces are rated into "categories" based on their venting characteristics. The furnace data plate lists the category to which a furnace belongs. Listed here are the different furnace categories and their characteristics:

Category I

Natural or induced blower; vent gas at 275°F or higher; low efficiency with a draft hood, or mid efficiency with an induced blower that does not pressurize the vent; can be common vented. A Type B gas vent can be used to vertically vent a Category I fan-induced installation; single-wall vent-pipe connectors are not recommended, due to low vent gas temperature and increased heat loss, thus creating the potential for condensation.

Category II

Currently no furnaces are manufactured in this category.

Category III

Mid efficiency with flue gas in the 275° to 360°F range; positive pressure in the vent system. Condensation may occur in the venting system; this must be accounted for by using special venting materials. Some applications can use a common vent.

Category IV

The highest-efficiency furnaces are in this category. Flue gases are in the 100° to 275°F range. Condensation occurs in the heat exchanger; heat-resistant plastic venting materials must be used.

Furnace venting requirements are governed by building codes and the particular unit's design category. Five-in. vent pipe to the atmosphere has been standard; however, high-efficiency furnaces expel a much cooler vent gas, so ductwork can be made of products such as PVC, CPVC, PEI, ABS, and others. Use the type of piping recommended by the furnace manufacturer and that conforms to code requirements.

High-efficiency furnaces utilize improved design features to ensure better operating efficiency. For example:

- A high-efficiency furnace may have two heat exchangers, to "wring out" every last Btu.

- Significant advances in ignition; the constant standing pilot of the past is outmoded.

- Airflow is now forced, so proper combustion can be achieved.

The goal of any high-efficiency furnace is to reduce energy consumption on the inlet side and squeeze every possible Btu out of the fuel before exhausting the byproducts of combustion.

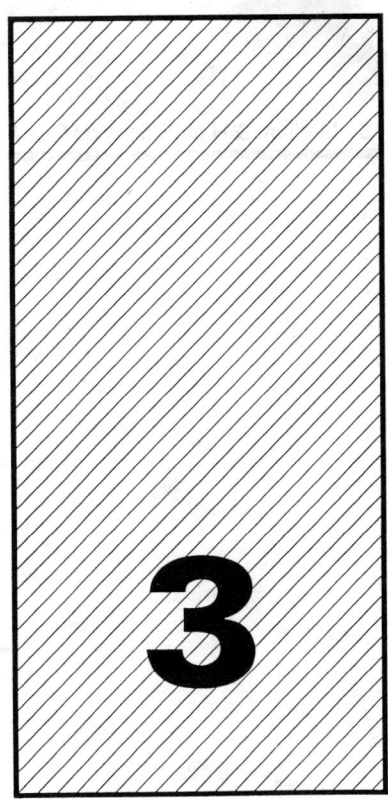

3

Furnace Controls and Components

In the last chapter we viewed some basic furnace components. This chapter will expand that view and explore changes that have been made in controls and components to improve efficiency.

There are many methods of increasing furnace efficiency. A manufacturer may modify a control or component, causing it to function differently than it did before. Currently, manufacturers are making modifications in eight different areas to improve efficiency. These eight areas are by no means a complete listing but merely focus on trends that will result in higher operational efficiencies. Current approaches to improve efficiency are as follows:

- Eliminating the free-standing pilot to conserve fuel. Free-standing pilots are being replaced by electronic ignition systems. Electronics are also being used to monitor furnace operation to ensure maximum efficiency.

- Adding vent dampers to reduce cycle air loss.

- Utilizing secondary condensing heat exchangers to capture heat that would otherwise be lost.

- Using new materials in heat exchanger construction to improve heat transfer. This includes redesign of heat exchangers to allow for methods of combustion and ignition that are different from what has been used in the past.

- Controlling the blower motor's run time to allow heated air (normally lost in the ducts) to reach the space. This includes controlling fan speed to match the load requirements of the space.

- Creating zoned systems, so each area of the space controls its own heating needs.

- Installing automatic setback thermostats, which match the heating requirements of the space to the appropriate time of day or day of the week;

- Using forced- and induced-draft blowers to ensure proper fuel mix and venting.

Controls and Components

For a furnace to operate, it must have controls and components that deliver the correct amount of fuel and air. The fuel and air must then be mixed in a proper ratio, ignited, and burned. Finally, heated air must be sent to the space and byproducts of the combustion process exhausted. While all of these events are occurring, the furnace must operate in a safe, efficient manner.

REGULATOR

A regulator reduces the supply line pressure to a constant working pressure, which a furnace requires. Gas appliances use a stand-alone regulator, while most of today's natural gas furnaces incorporate the regulator into the gas valve. The regulator portion of the gas valve is set to provide the necessary pressure of 3½ in. water column (wc) for a natural gas furnace.

> *Note — Water column is a measurement of gas and vapor pressure, made by balancing a column of liquid against fuel pressure to see how high the pressure will cause the liquid to rise. Water column is measured using a manometer.*

Liquid propane systems may use more than one regulator. A regulator can be located at the fuel source and also at the building. The pressure setting for a furnace using liquid propane is 11 inches water column (in. wc).

GAS VALVE

The gas valve turns gas flow to the burners off and on. A gas valve can use manual or automatic operation. Several manufacturers produce the gas valves used in furnaces today. Each approaches operation from a slightly different standpoint. It is not uncommon to have one furnace manufacturer use different brands of gas valves on different models of their equipment. The different types of gas valves are discussed below.

Solenoid valve

Used with a stand-alone regulator, the electric solenoid valve opens and allows gas to flow. Construction varies, but solenoid valves always open or close based on an electrical signal sent from a control.

Diaphragm valve

Magnetic. Gas pressure opens and closes the valve. For the valve to open, a coil must be energized to allow gas pressure on one side of a diaphragm to bleed off. When pressures are equal on the diaphragm, gas cannot flow. Bleeding creates a pressure difference on one side of a diaphragm, which allows gas to flow through the valve.

The bleed feature of a diaphragm valve is significant to system operation. When the energized coil allows bleeding to occur, the gas that is bled is piped through a line to the furnace for combustion. If the bleed line is plugged, the diaphragm valve can't operate.

Bimetal diaphragm. A bimetal is created when two pieces of dissimilar metal are bonded together, creating a straight piece of metal that can look like a stick of chewing gum, Figure 3-1. When heat is applied to a bimetal, it warps, forming an arc. Warping occurs because the expansion and contraction rates of the two metals are different.

A heater is mounted on top of the bimetal, which is attached to a bleed valve mounted on the top side of the diaphragm. When the heater is energized, it warps the bimetal, opening a bleed port through which gas flows. The bleeding gas creates a pressure difference on the diaphragm, which then allows gas to flow to the burner. When the heater is de-energized, the bimetal cools and returns to its original position; pressures are equalized; and gas flow ceases.

Figure 3-1. Bimetal Diaphragm

Bimetallic. This type of valve uses a bimetal mounted to a disc. The disc keeps a port closed, halting gas flow. When the bimetal heats, the warping action exposes enough of the port to allow gas flow.

Heat motor. This type of valve uses wire wrapped around a rod. When the wire is energized, heat is given off. The heat causes the rod to expand and move a disc, allowing gas to flow through a port.

Combination gas valve

The pressure regulator and gas valve, which control fuel flow to the burner, are incorporated in one unit. A combination gas valve uses a manual on-off control. Gas flow is controlled electrically; shutdown occurs when the pilot flame is out or is not large enough to maintain sufficient flame at the thermocouple.

Redundant gas valve

A redundant gas valve is designed with safety as a criteria. The solenoids in the valve are wired so that a failure on the part of one solenoid causes the other to close and stop gas flow, Figure 3-2. As many as three solenoid valves may exist to control and, if necessary, shut down gas flow to pilot and burners.

Two stage

A two-stage gas valve is used with a two-stage thermostat to supply fuel in quantities necessary to meet staged demand. The valves are built to supply less fuel on a first-stage heating call. When the second stage is energized, 100% fuel flow is sent to the burner. Two-stage valves can be constructed using magnetic diaphragms and bimetals as the flow fuel control.

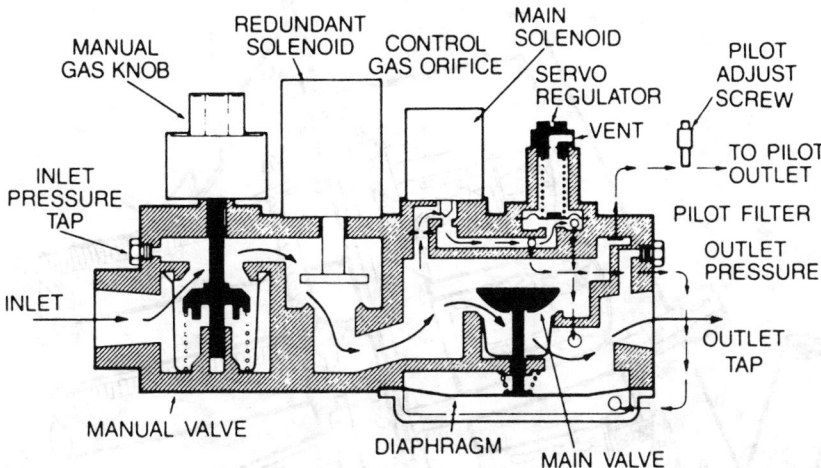

Figure 3-2. Redundant Gas Valve (Courtesy, Inter-City Products Corp.)

Valve Shutoff

100% shutoff. As the name implies, a 100% shutoff valve will stop all gas flow to the burner and to the pilot. A 100% shutoff valve uses an internal safety to shut down fuel flow to the burners and the pilot when the pilot is not lit. In order to light the pilot, a switch must be manually depressed to override the internal safety feature.

Non-100% shutoff. A non-100% shutoff does *not* have any internal safety shutoff in the pilot circuit. In a system using ignition such as direct spark, fuel will flow to the pilot without depressing a pilot switch. When the gas valve is on, fuel flows to the pilot. There is a safety circuit in the pilot wiring that will prevent fuel flow to the burner unless the pilot is lit.

MANIFOLD

The manifold is the pathway for fuel to reach the burner. Fuel flows through the manifold and into the burner through the orifice, or "spud," Figure 3-3. Fuel pressure in the manifold varies depending on the type of fuel that is used. Natural gas pressure is 3½ to 4 in. wc. Liquid propane requires a pressure of 11 in. wc.

BURNERS

The burner supplies, meters, and mixes fuel and air to the head of the burner for combustion. The burner is designed to produce a burning pattern that will result in efficient combustion, Figure 3-4. Most burners are made of cast iron or formed steel.

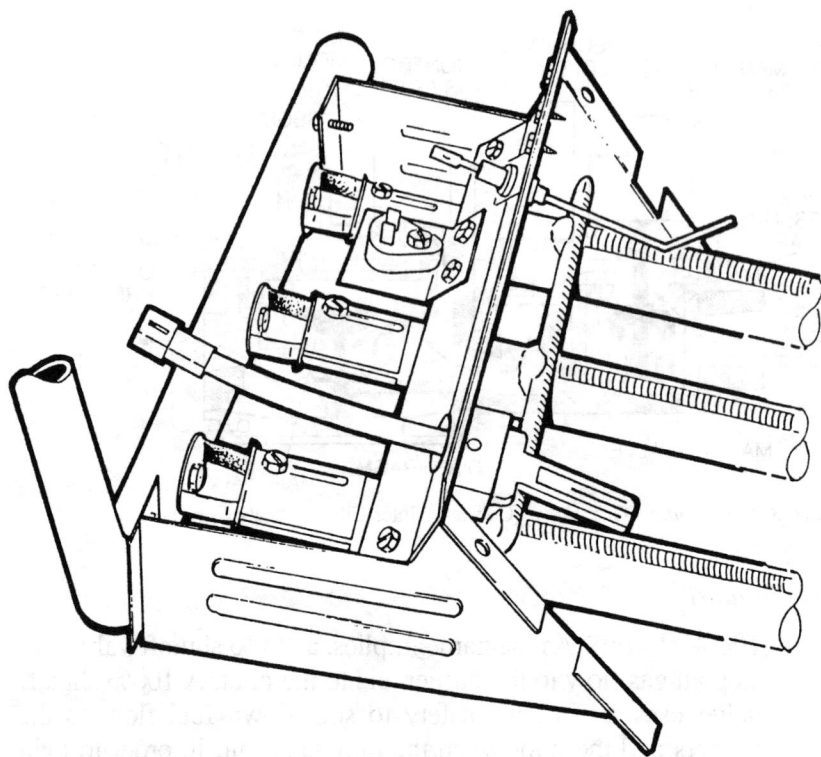

Figure 3-3. Manifold (Courtesy, American Standard, Inc.)

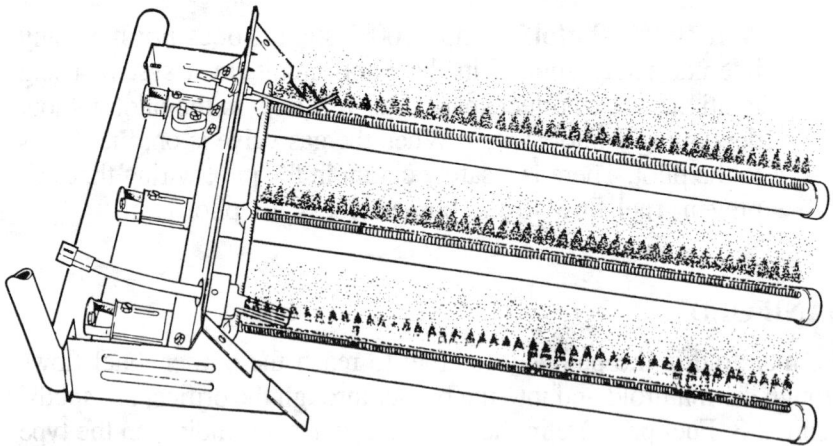

Figure 3-4. Burner (Courtesy, American Standard, Inc.)

The burner is mounted in the combustion chamber. Primary air enters through the face of the burner and continues through the venturi, or mixing tube, to the head of the burner, where the ports of the burner are located. Burner ports are designed by manufacturers for their particular furnace. Burner ports can be configured as single, multiple, inshot, or slotted.

When the burner is atmospheric (no combustion blower is present), the fuel goes through the orifice mounted in the face of the burner. After going through the orifice, the burner is restricted in an area referred to as the venturi or mixing tube. When the gas enters the venturi it speeds up, and this increase in speed creates low pressure in the venturi.

Primary air rushes through shutters in the face of the burner into a low-pressure area and mixes with the fuel. If the burner has adjustable shutters, Figure 3-5, the amount of primary air can be controlled by adjusting the opening of the shutters. If the shutters are not adjustable, the mixture is set by the manufacturer.

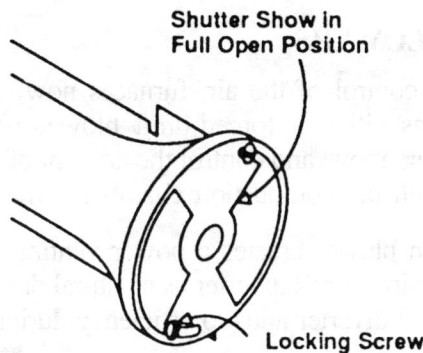

Shutter Show in Full Open Position

Locking Screw

Figure 3-5. Adjustable-Shutter Burners (Courtesy, Inter-City Products Corp.)

Secondary air is the additional air necessary to achieve combustion and ensure that the byproducts of combustion are exhausted. Secondary air enters from outside of the burner and mixes with the fuel at the burner head. The heat exchanger has built-in baffles which act to control secondary airflow.

ORIFICE (SPUD)

An orifice is mounted in the manifold protruding into the center of the burner face. The orifice causes fuel to flow down the center of the burner body. The velocity of the fuel flowing down the body of the burner draws the primary air into the burner where the air mixes with the fuel. The mixing of the primary air and the fuel creates a positive pressure, which is exhausted through the burner ports.

The orifice is drilled to size in order to meter the proper amount of gas for combustion. The size of the orifice and the pressure in the manifold affect fuel and primary air flow. For example, if the fuel velocity is too slow because of a smaller sized orifice, then there will be insufficient primary air mixing in the burner. Orifice sizing charts in Part Two of this book show the necessary orifice size for a particular application and altitude. In order to size an orifice, the Btuh/burner, Btu/cu ft, and the specific gravity of the gas must be known.

A furnace that operates at an altitude higher than 2,000 ft must be *derated*. Derated means that the Btu output of the furnace must be adjusted downward when a furnace is operating at a higher altitude. Derating is based on a formula that calls for a deration of 4% per 1,000 ft above sea level. A furnace operating at a higher altitude has less air available for combustion because there is less oxygen available in the air at higher altitude. Therefore, as altitude increases, the fuel input should be decreased. Sizing the orifice smaller will allow this to occur.

COMBUSTION BLOWERS

In order to achieve control of the air, furnaces now employ a small blower designated as either a forced-draft blower or induced-draft blower. These blowers move and control the amount of combustion air or byproducts through the combustion chamber or heat exchanger.

Use of a combustion blower creates a power venting system. Power venting does not require a draft diverter as a natural draft furnace does. The absence of a draft diverter adds to efficiency during the off cycle.

Forced-draft type

Forced draft uses a small blower to force air into the combustion chamber. The air is under a positive pressure which, after combustion occurs, allows proper venting to occur.

Induced-draft type

Induced draft is essentially the opposite of forced draft. Air is pulled into the heat exchanger and discharged into the vent, Figure 3-6. Pressure in the combustion chamber is negative, and the venting system can be at atmospheric or slightly positive pressure.

Combustion air requirements for a high-efficiency furnace are stricter than those for earlier furnaces. In particular, the air must not contain chemicals that could be acidic (e.g., chlorine, cleaning solvents, varnishes, etc.). These chemicals will have a negative effect on any furnace, but in high-efficiency furnaces, the problem is greater. When burned in the combustion process, these chemicals create acids that, when condensed, will destroy furnace components and create unsafe operating conditions. When an installation is encountered that main-

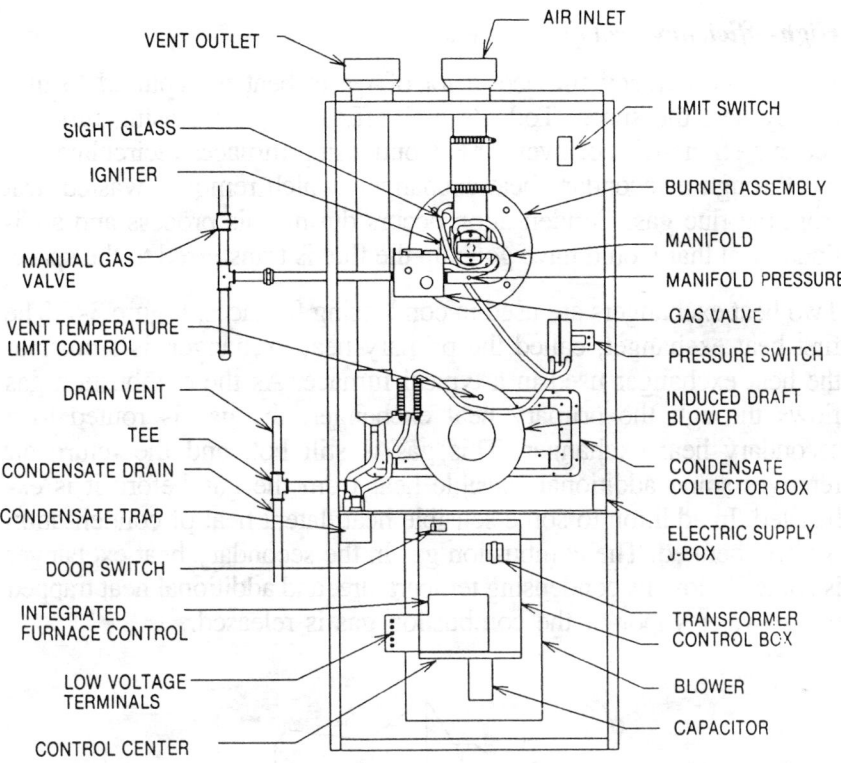

Figure 3-6. Induced-Draft Furnace Layout (Courtesy, Rheem Manufacturing Company)

tains a constant presence of these types of chemicals (such as an artist's studio or beauty shop), consideration must be given to using outside air for combustion and ventilation.

For proper service to occur, the technician must understand that the air requirements for combustion are established by the manufacturer. Combustion blower systems are designed as a sealed system, thereby controlling air requirements. Safety controls exist to shut down the system when motor failure, leaks, or abnormal conditions occur that will affect pressure inside the sealed system.

HEAT EXCHANGERS

Heat exchangers have undergone the greatest change in order to achieve higher efficiencies. In early furnaces, efficiencies of 68% were common. Combustion gases, after transferring some heat to the space air, passed up and out the vent. Anyone who has ever touched a vent pipe knows that there is still a considerable amount of heat left in the vented gases, and in older furnaces, that heat was headed for the atmosphere. With many of today's high-efficiency furnaces, that is no longer the case.

High-efficiency heat exchangers

In a high-efficiency furnace, a lot of waste heat is captured to ultimately heat the space. Today's high-efficiency, condensing furnaces achieve efficiencies of over 90%. Condensing furnaces recirculate flue gas through a secondary heat exchanger, which removes wasted heat from the flue gas. Condensation occurs during this process and additional heat that would have gone up the flue is transferred to the space.

Two heat exchangers are used in condensing furnaces, Figure 3-7. The first heat exchanger, called the primary heat exchanger, is similar to the heat exchanger used in a typical furnace. As the combustion gas flows through the primary heat exchanger, the gas is routed to a secondary heat exchanger. This gas is still hot, and the return air removes some additional sensible heat from the gas before it is exhausted. In addition to some sensible heat, latent heat of condensation is also liberated. The combustion gas in the secondary heat exchanger is cooled below its condensing temperature, and additional heat trapped in the water vapor of the combustion gas is released.

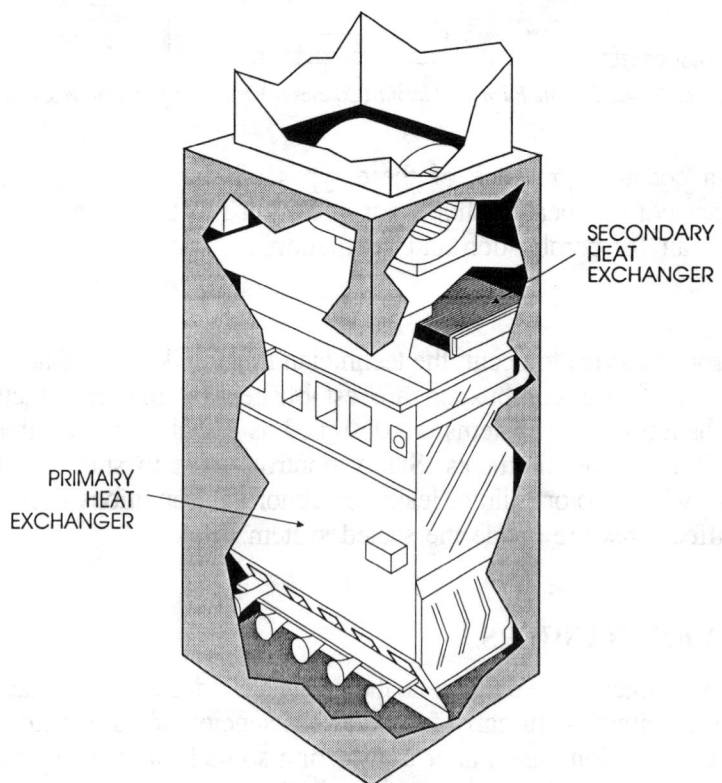

Figure 3-7. Gas Flow: Primary-to-Secondary Heat Exchangers

Remember, the process of condensation is essentially that of releasing heat in order for a gas to condense into a liquid. A secondary heat exchanger allows the combustion gas to stay in contact longer with the return air, which allows condensation to occur. The end result is that a cooler vent gas goes to exhaust and more heat is retained by the supply air.

In designing a high-efficiency furnace, the secondary heat exchanger is positioned so that return air comes in contact with the secondary heat exchanger first. The cooler return air is warmed *before* entering the primary heat exchanger. Condensation must take place only in the secondary heat exchanger. The furnace was designed that way to ensure a longer life for the primary heat exchanger.

> *Note — Manufacturers' guidelines must be adhered to when servicing a high-efficiency furnace. This is not to imply that manufacturers' guidelines should ever be ignored. In a condensing furnace, the venting and drain processes are different than in standard furnaces.*

VENT PIPING

In high-efficiency furnaces, the vent gas is cooler and most of the water vapor has been removed, so venting requirements are different from those used for other furnaces. The use of plastic vent piping is allowed, Figure 3-8, and the piping will be smaller in size. The correct choice of which plastic pipe can be used is governed by the manufacturer and local building codes. ABS-DVW, PVC-40, or PVC-DWV are just a few of the different types of plastics available.

CONDENSATE DRAIN

A condensing furnace creates condensation in the process of transferring heat. This condensation must be drained from the unit.

Due to the condensing process and the increased exposure time of the combustion gases to furnace materials, there are many corrosive products present in the condensate. High-efficiency furnaces are constructed to handle these corrosive products; however, piping a drain is a serious matter. Local codes must be adhered to and materials used in the piping must be as specified by the manufacturer and local building code. An installer cannot, under any circumstances, allow condensate products to flow out of the drain pipe and onto the ground.

> *Note — As with most products, there are variations on the design and construction of high-efficiency heat exchangers. For example, one manufacturer installs the secondary heat*

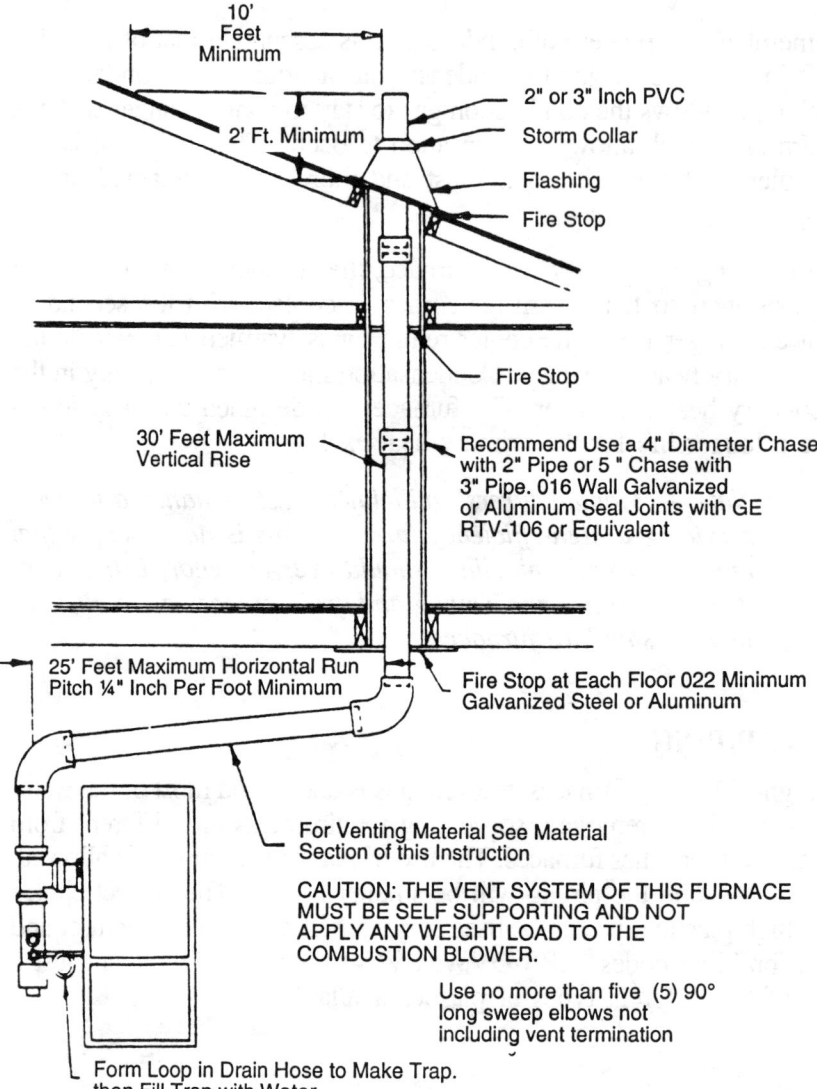

Figure 3-8. High-Efficiency Furnace Venting (Courtesy, Armstrong Air)

exchanger next to the primary heat exchanger as an additional heat source for space air. In this type of installation, additional space air comes in contact with the secondary heat exchanger but not the primary heat exchanger.

PILOT IGNITION SYSTEMS

Chapter Four explains pilot ignition systems in detail.

HIGH-TEMPERATURE LIMIT (HTL) SYSTEMS

The HTL is a safety control that is designed to shut down the furnace in case it overheats. HTL switches can be found in more than one configuration.

Limit switch

One manufacturer will use a limit switch that serves as a safety, shutting the furnace down when the furnace overheats. The limit switch is normally closed and is designed to open and shut down the furnace when the plenum (heated air chamber) temperature reaches approximately 200°F.

If the heated air temperature reaches 200°F, the possible cause could either be a lack of airflow or the furnace itself is overfiring. The 200°F temperature is an average and the manufacturer's setting must be used. When a stand-alone, high-temperature limit switch is used, a second switch (blower switch) is used to control blower motor operation.

Combination fan and limit switch

A second configuration employs a combination blower limit switch. The limit switch functions the same as it does in a single limit system. The blower switch controls the blower, and the limit portion acts as a safety.

Limit switches rarely fail on their own. When a limit switch opens, find the cause — don't just replace the switch.

Fan switch:

> **Single unit blower only.** This switch allows the heat exchanger to warm up before the blower motor is started, so cold air is not sent to the space. After the space temperature has been satisfied, the blower switch allows the blower to continue running, which allows the heat exchanger to cool down.

> The blower will normally be energized around 130° to 150°F and will turn off at about 100° to 105°F. *(Always use manufacturers' specifications.)* Allowing the fan to operate after burner shutdown results in heat ending up in the space that would have otherwise been lost. If the installation had the blower set for continuous duty, then the delay function on start up and shut down would not occur.

Fan switches that operate on a time delay relay (TDR) are also used. TDRs have an adjustable time delay to cycle the fan on and off based on a time setting established by the installer.

Fan switch as part of combination fan limit. In this configuration, one temperature-sensing bimetal, spiral or flat bimetal, senses temperature for both the limit and the fan switch, Figure 3-9. The temperature settings for the limit and the fan are different.

Separate switches control the limit or fan switch even though the same temperature sensing device is used. The switches can be mounted in the furnace plenum or in the duct itself, depending on design. In both the single and combination switch configuration, failure of the blower would result in higher air temperatures shutting the unit down on the limit switch.

If the unit functions both for heating and air conditioning, a fan relay is normally used to select blower motor speed for heating and cooling. Lower fan speeds are normally used in heating operation.

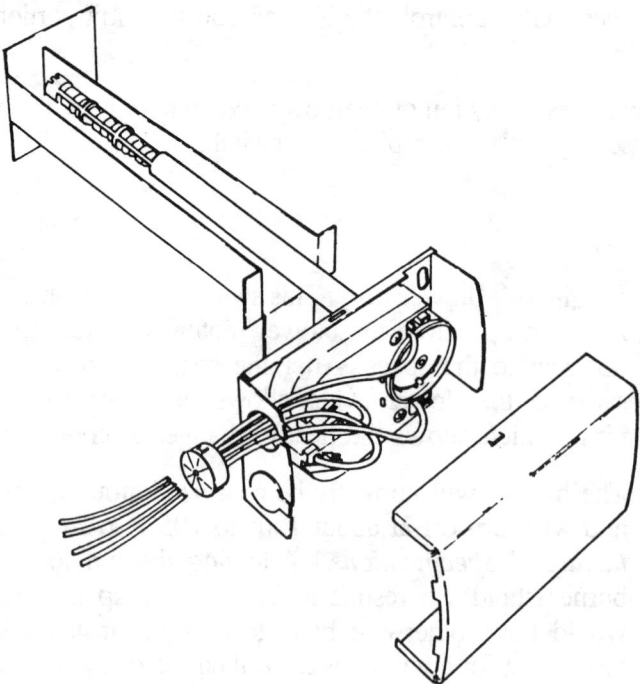

Figure 3-9. Combination Fan Limit (Courtesy, Inter-City Products Corp.)

DRAFT DIVERTER, HOOD

The products that result from combustion must be exhausted to the outside. In an atmospheric vent furnace, the draft diverter or draft hood serves this function. The draft diverter helps create a draft, which causes diluted airflow into the furnace area.

When operating, the draft diverter supplies air, which mixes with the combustion gases that are vent bound. The draft hood allows combustion gases to escape, if there is no draft, backdraft, or stoppage. A draft diverter helps prevent a backdraft from entering the furnace. It can also prevent a free-standing pilot from being blown out, in the event of a down draft, by diverting the down draft away from the pilot and burner.

Furnaces that use power vent combustion (forced or induced) do not generally use a draft hood. Increased efficiency results from not using a draft hood because off cycle heat loss is reduced.

New styles of high-efficiency furnaces may employ a direct vent, Figure 3-10. A direct vent furnace can be natural draft or power assisted. Combustion air is obtained from the outside by means of a combustion air pipe. In a direct vent furnace, no draft hood is used. Mobile homes must use a direct vent type of furnace.

VENT SAFETY SWITCH

If the furnace vent is blocked, a dangerous condition exists. The vent safety switch is designed to shut down the furnace in the event of a vent blockage. The switch can be temperature-sensitive, mounted in the draft hood or pressure-sensitive, mounted in the vent, Figure 3-11.

VENT DAMPER

A vent damper is an energy-saving accessory designed to increase efficiency. The vent damper, when used with *electronic ignition*, will close the vent pipe when the furnace stops. The damper uses electric power to operate.

THERMOSTAT

The thermostat is the switch designed to turn the system on and off. Thermostats are generally 24-volt or millivolt systems.

Inside a low-voltage thermostat is a heat anticipator. This anticipator adds heat to the thermostat bimetal to achieve better temperature con-

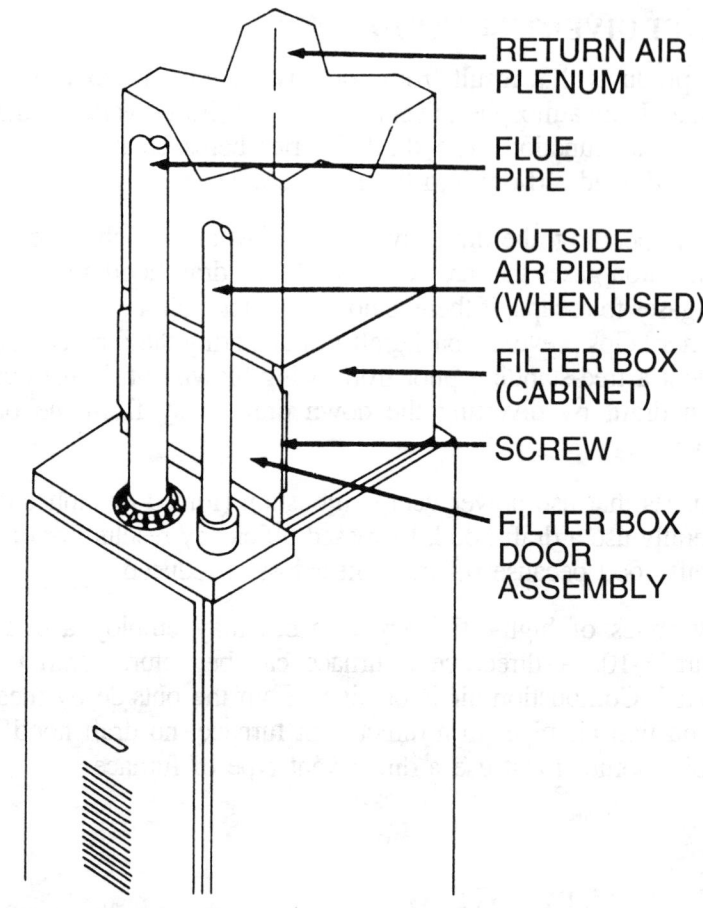

RETURN AIR PLENUM

FLUE PIPE

OUTSIDE AIR PIPE (WHEN USED)

FILTER BOX (CABINET)

SCREW

FILTER BOX DOOR ASSEMBLY

STANDARD ARRANGEMENT OF DOWNFLOW FURNACE AIR FILTERS

Figure 3-10. Direct-Vent Furnace (Courtesy, American Standard, Inc.)

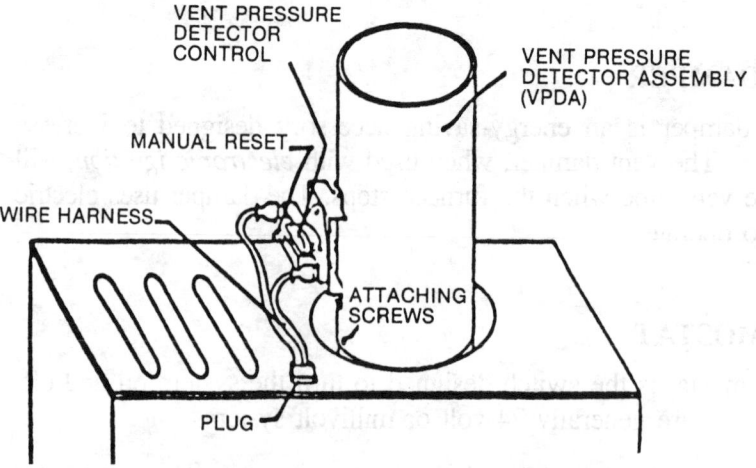

VENT PRESSURE DETECTOR CONTROL

VENT PRESSURE DETECTOR ASSEMBLY (VPDA)

MANUAL RESET

WIRE HARNESS

ATTACHING SCREWS

PLUG

Figure 3-11. Vent Safety Switch (Courtesy, Heat Controller, Inc.)

trol by allowing less swing during the heating cycle. The anticipator is wired in series with the contacts of the thermostat, Figure 3-12.

The anticipator must be set to the amp rating of the gas valve, which is marked on the valve. For example, if the valve is marked 0.4 amps, then the movable slide of the anticipator should be set to 0.4. If the valve is not marked, an amp reading on the heating circuit should be taken and the anticipator set to correspond. Millivolt thermostats do not use heat anticipators.

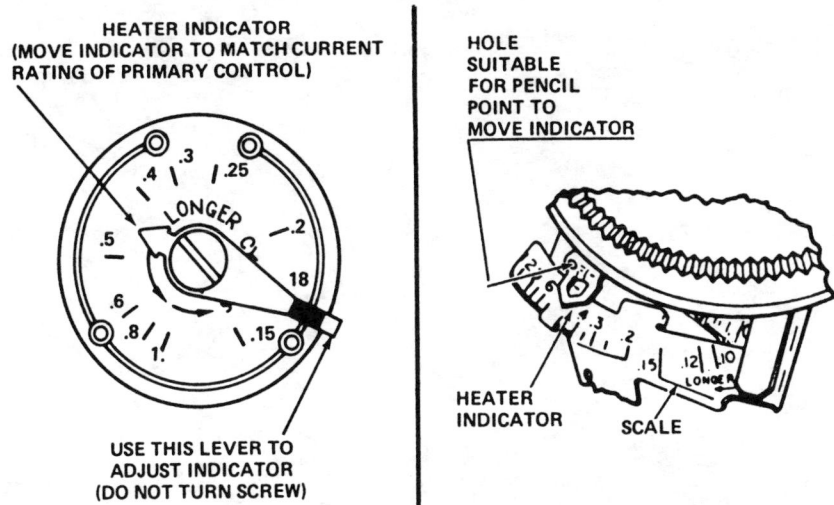

Figure 3-12. Typical Heat Anticipator (Courtesy, Inter-City Products Corp.)

Many new installations use solid-state thermostats. The solid-state thermostat is similar to a regular thermostat but is programmable to allow the setpoint to be adjusted for different conditions. A solid-state thermostat has a microprocessor, which controls thermostat functions.

Most, if not all, solid-state thermostats do not use a heat anticipator. This type of thermostat relies on cycle time. The thermostat allows the furnace to run a fixed number of cycles every 60 minutes. For example, the thermostat will allow the furnace to maintain a cycle rate of three times per hour. The cycle rate is set by the furnace manufacturer, and the method for adjusting the rate can be found in the literature provided by the thermostat manufacturer.

> *Warning — Never short the terminals of a gas valve when performing service. The anticipator is wired in series and will burn out.*

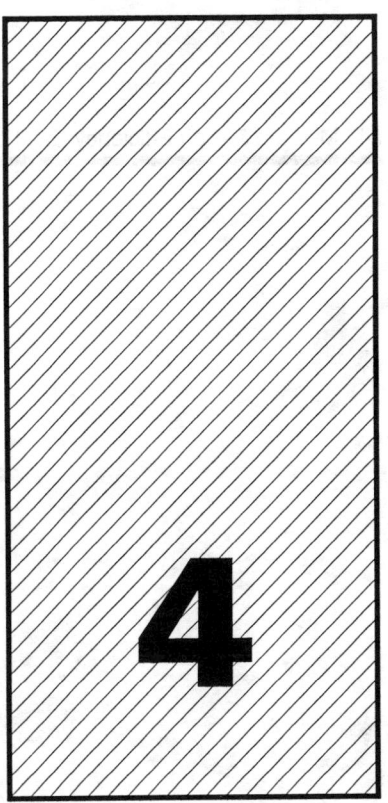

Ignition Systems

An ignition system provides the flame or spark to fire the burners. To ensure safe operation, the ignition system can also monitor furnace operation by locking out gas flow to the burner in case of flame loss.

Furnace manufacturers have modified the ignition system in order to obtain higher efficiencies. New types of ignition systems were developed to meet the following needs:

- Higher efficiency requirements mandated by government

- Conservation of fossil fuels

- Greater pilot-relighting convenience in remote locations

- Fewer annoying shutdowns caused by such incidents as freestanding pilots blowing out

- Greater use of electronics to control furnace operation, thus improving both efficiency and safety

Ignition System Types

There are three basic types of ignition systems:

- Continuous ignition

• Intermittent or automatic pilot

• Direct ignition

Manufacturers often configure these ignition systems according to their own engineering requirements. Because of these different configurations, certain operational concepts vary from manufacturer to manufacturer. However, the basic operational principles are consistent.

CONTINUOUS IGNITION

The most widely used ignition system in the past was the continuous-ignition system. With the increased availability of electronics, the use of intermittent- and direct-ignition systems has become prevalent in new, high-efficiency systems. However, many furnaces still in operation employ continuous ignition. They may not have the highest efficiency, but they still need service.

One type of continuous-ignition system is the *standing pilot*. When the standing pilot is lit, it is always available for burner ignition.

An important component of the continuous ignition system is the *thermocouple*. A thermocouple is a device that generates electricity when heated. It is composed of two dissimilar metals, with one of the metals wrapped around the core of the other, and the two are welded together at either end, Figure 4-1. The end of one metal is placed in the pilot flame (hot junction) and the other end is kept cooler by combustion airflow (cold junction).

The temperature difference between hot and cold junctions causes a dc voltage to be generated, ranging from 15 to 30 mV. This small amount of voltage is enough to allow the thermocouple to act as a safety device. As long as the hot junction is kept hot, sufficient current is generated to keep an electromagnetic relay open in the gas valve, allowing gas flow. If the thermocouple's hot junction cools due to flame loss, the thermocouple's electrical output drops. This, in turn, causes the electromagnet in the gas valve to be overcome by spring pressure. Gas flow ceases.

Thermocouples are used on 100%-shutoff systems. A manual reset button is used to start a thermocouple system. Depressing and holding the manual reset 30 to 45 seconds allows sufficient time for the current flow necessary to operate the electromagnetic relay in the gas valve.

A variation on the thermocouple is the *thermopile*. A thermopile is composed of thermocouples wired in series to produce a higher output voltage. The increased voltage operates the gas valve directly, rather than through the electromagnetic relay. A thermopile eliminates the

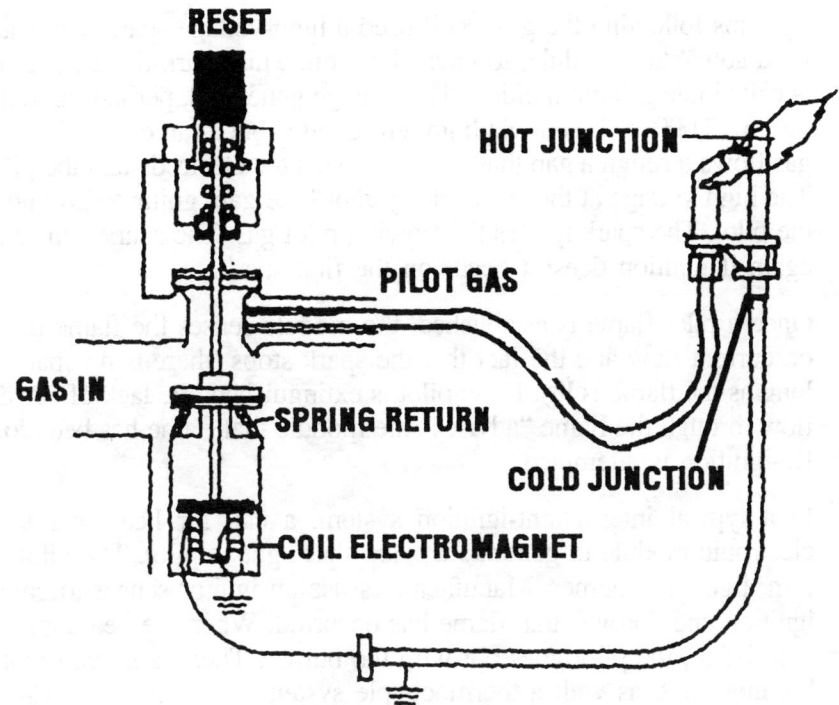

Figure 4-1. Thermocouple

need for an outside power source; the thermostat uses thermopile power to operate the gas valve.

You can replace a thermocouple system with an intermittent-ignition system. Kits are available that will add the module and replace the thermocouple with an ignitor sensor. Keep in mind, though, that following the manufacturer's instructions is necessary if you want to install one of these kits successfully and *safely*.

INTERMITTENT OR AUTOMATIC PILOT

A system that uses a thermocouple must be manually reset each time the pilot goes out. This isn't always possible, so automatic pilot systems were developed to overcome this problem.

A glow coil pilot system was one of the first to create ignition without requiring manual reset. The glow coil consists of a tightly wound wire conducting current flow. The current flow creates heat, causing gas passing over the hot glow coil to ignite. The glow coil is then de-energized by a pilot safety switch. The pilot safety switch also makes the contact necessary to energize the gas valve. If the glow coil is not de-energized, it will burn out. Glow coil systems operate on about 12 V of input. The input power is obtained by a center tap on a 24-V transformer or by using fixed resistance to drop the voltage.

Systems following the glow coil used a high-voltage spark, generated by a solid-state module, to cause ignition. This spark-driven ignition is called intermittent ignition. The voltage generated, perhaps amounting to 15,000 V, is sent via high-tensioned cable to an electrode. Pilot gas flows through a gap that exists between the electrode and the pilot. The high voltage at the electrode "jumps" the gap, going to ground at the pilot. The spark ignites the flowing pilot gas. The spark will jump again if ignition doesn't occur on the first spark.

Once a pilot flame is established, the module senses the flame based on current flow and the fact that the spark stops (there is no spark as long as the flame is lit). If the pilot is extinguished, the lack of current flow through the flame "advises" the module that flame has been lost. Re-ignition is attempted.

In a typical intermittent-ignition system, a call for heat causes an electronic module to generate a spark that lights a pilot. The pilot, in turn, lights the burner. Manufacturers use an ignitor sensor to create ignition and "prove" that flame has occurred. When the heating need is satisfied, the pilot goes out with the burner. There is no constantly burning pilot, as with a thermocouple system.

The flow chart shown in Figure 4-2 is a representation of how a typical intermittent system operates. Areas such as the amount of time dedicated to trial ignition or prepurge, vary from manufacturer to manufacturer. Consult the individual service manual for the system you are servicing, for information on that system's sequence.

The following are components of a typical intermittent-ignition system:

Electronic ignition module

- Checks to see that it's all right to fire the furnace

- Reacts to sensor input

- Powers the ignitor

- Operates the gas valve

- Monitors the pilot and burner flame while the unit is heating

- Shuts down the heating system in case of problems

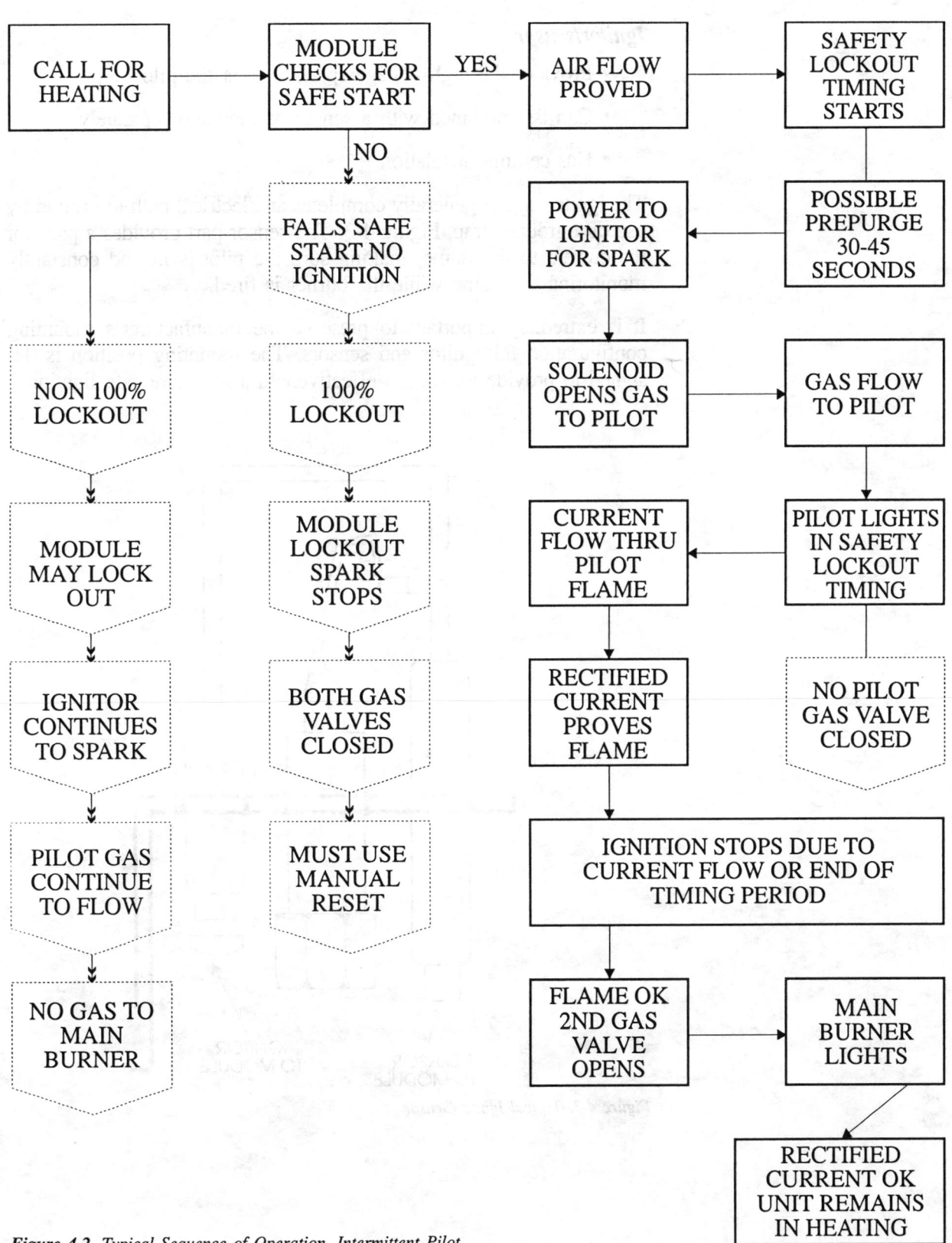

Figure 4-2. *Typical Sequence of Operation, Intermittent Pilot*

Ignitor/sensor

- Provides the high-voltage spark to light the pilot

- Can be combined with a sensor or mounted separately

- Has ceramic insulation

The ignitor circuit generally completes an electrical path to ground by use of a ground strap, Figure 4-3. The sensor part provides a path for the current to the flame, making sure the pilot is lit and constantly monitoring the flame while the burner is fired.

It is extremely important to preserve the manufacturer's mounting configuration for ignitor and sensors. The mounting position is designed to provide maximum effectiveness and ensure safe lighting.

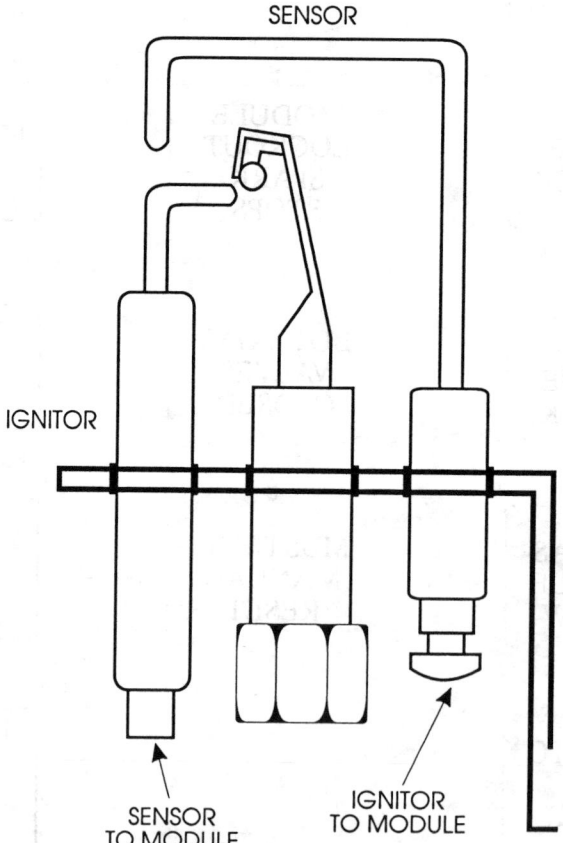

Figure 4-3. Typical Ignitor/Sensor

Pilot burner

- Lights the main burner

- Maintains the ignitor/sensor in the right position to allow ignition

- Maintains the proper flame pattern for monitoring

The pilot burner used in electronic ignition is similar to the free-standing pilot used in continuous-ignition systems.

Gas valve

Gas valves may be a combination valve or two single valves mounted in the gas service. In a continuous system, one main gas valve may be used as long as there is a temperature-operated, manual, external shutoff, which shuts down pilot thermocouple operation in case of system failure. A single main gas valve is not allowed in an intermittent-ignition system.

A dual valve for an intermittent-ignition system is necessary; in the event of problems, there is no free-standing pilot to shut off gas flow. With electronic ignition, the pilot safety is activated by an electronic signal rather than a thermocouple.

High-tension cabling and additional controls

High-tension cabling provides the path from the module to the ignitor for the high voltage. In addition, every system has safety controls, such as high limit switches and components such as vent dampers, installed by the manufacturer.

DIRECT IGNITION

When a system uses direct ignition, there is no pilot to light. The burner is lit by a direct-spark ignition (DI) or hot-surface ignitor (HSI). The primary difference between a direct-ignition system and an intermittent system is the absence of a pilot flame in any form. Ignition is always direct to the burner.

In direct ignition, a spark lights the main burner directly. In hot-surface ignition, current flow heats a silicon carbide element to a red-hot condition. The element then lights the main burner. Using either method, the electronic module will cease ignition either after a preset time period or when the main burner is lit.

A typical direct-ignition system is composed of the following components:

Electronic ignition module

- Checks to see that it's all right to fire the furnace
- Reacts to sensor input
- Powers the ignitor
- Operates the gas valve
- Monitors the pilot and burner flame while the unit is heating
- Shuts down the heating system in case of problems

Ignitor

Direct-spark ignitors provide a spark directly to the burner, causing ignition, Figure 4-4. There is no pilot flame to light.

Hot-surface ignitors become red hot and cause the burner to light. Hot-surface ignitors are made out of silicon carbide, Figure 4-5. The placement of the hot-surface ignitor is important, Figure 4-6.

Sensor

A sensor can be mounted with the ignitor or as a separate component. As with an intermittent system, the sensor proves the burner flame and monitors the heating operation.

Gas valve

The gas valve functions the same as in an intermittent system. It must be a dual (combination) valve and be specified for use in a direct system.

Note — It is important to understand that in intermittent and direct systems, components such as gas valves, modules, and sensors may appear to perform the same function. They are not identical and should not be interchanged. Use only manufacturer-recommended replacements.

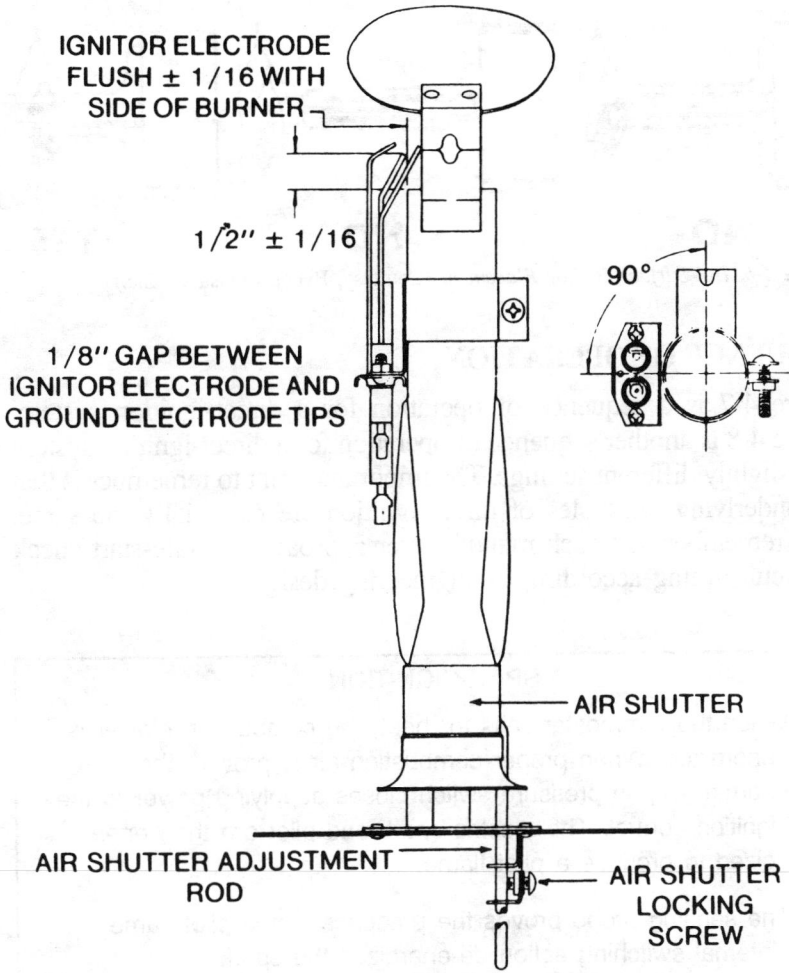

IGNITOR ELECTRODE
FLUSH ± 1/16 WITH
SIDE OF BURNER

1/2″ ± 1/16

1/8″ GAP BETWEEN
IGNITOR ELECTRODE AND
GROUND ELECTRODE TIPS

90°

AIR SHUTTER

AIR SHUTTER ADJUSTMENT
ROD

AIR SHUTTER
LOCKING
SCREW

Figure 4-4. *Direct-Spark Ignitor* (Courtesy, Evcon Industries, Inc.)

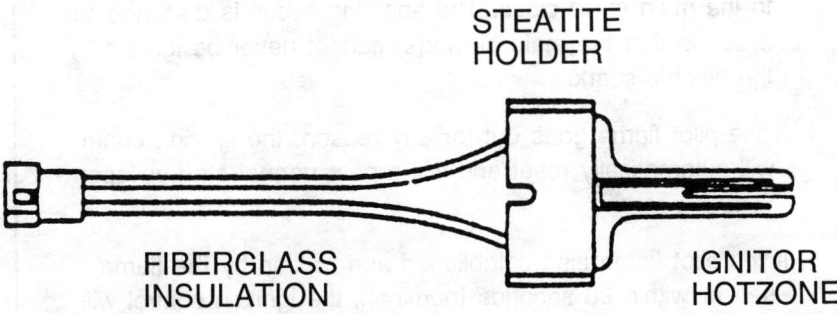

STEATITE
HOLDER

FIBERGLASS
INSULATION

IGNITOR
HOTZONE

Figure 4-5. *Typical Hot-Surface Element* (Courtesy, Evcon Industries, Inc.)

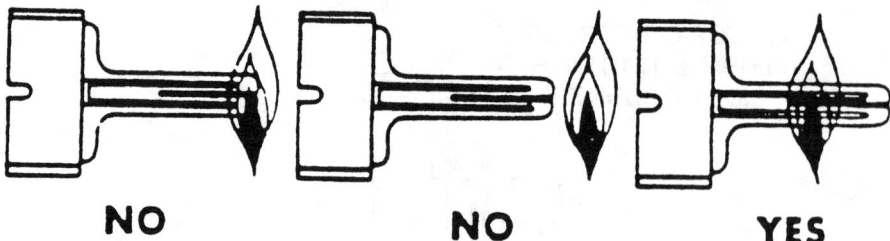

NO   **NO**   **YES**

Figure 4-6. Hot-Surface Ignitor Placement (Courtesy, Evcon Industries, Inc.)

SEQUENCE OF OPERATION

Figure 4-7 is a sequence of operation for a direct-ignition system. Figure 4-8 is another sequence of operation for a direct-ignition system with slightly different settings. The important point to remember is that the underlying principles of direct ignition are essentially the same. Also remember that each manufacturer approaches a safe-start check or circuit timing according to engineering design.

SPARK IGNITION

1. When the thermostat calls for heat, the combustion blower is energized. When proper combustion air is proven, the normally open pressure switch closes supplying power to the ignition control. The electric spark and pilot are then energized to produce a pilot flame.

2. The sensing probe proves the presence of the pilot flame internal switching action de-energizes the spark.

3. The sensor will cause contacts to close energizing the main valve solenoid; the main burners will light when the contacts to the main valve close. The sparking circuit is disconnected, assuring that the main burner(s) cannot never be ignited by the electric spark.

4. If the pilot flame goes out for any reason, the ignition control will automatically reset and will repeat normal start-up operation.

5. If the pilot flame isn't established and proven by the flame sensor within 90 seconds (nominal), the ignition control will lock out, shutting off the spark and pilot gas flow. The control can be reset by momentarily setting the room temperature to its lowest setting, then setting it to the desired temperature.

Figure 4-7. Sequence of Operation for an ULTRAII Model Series EGGE and EG7E Furnace (Courtesy, Armstrong Air Conditioning)

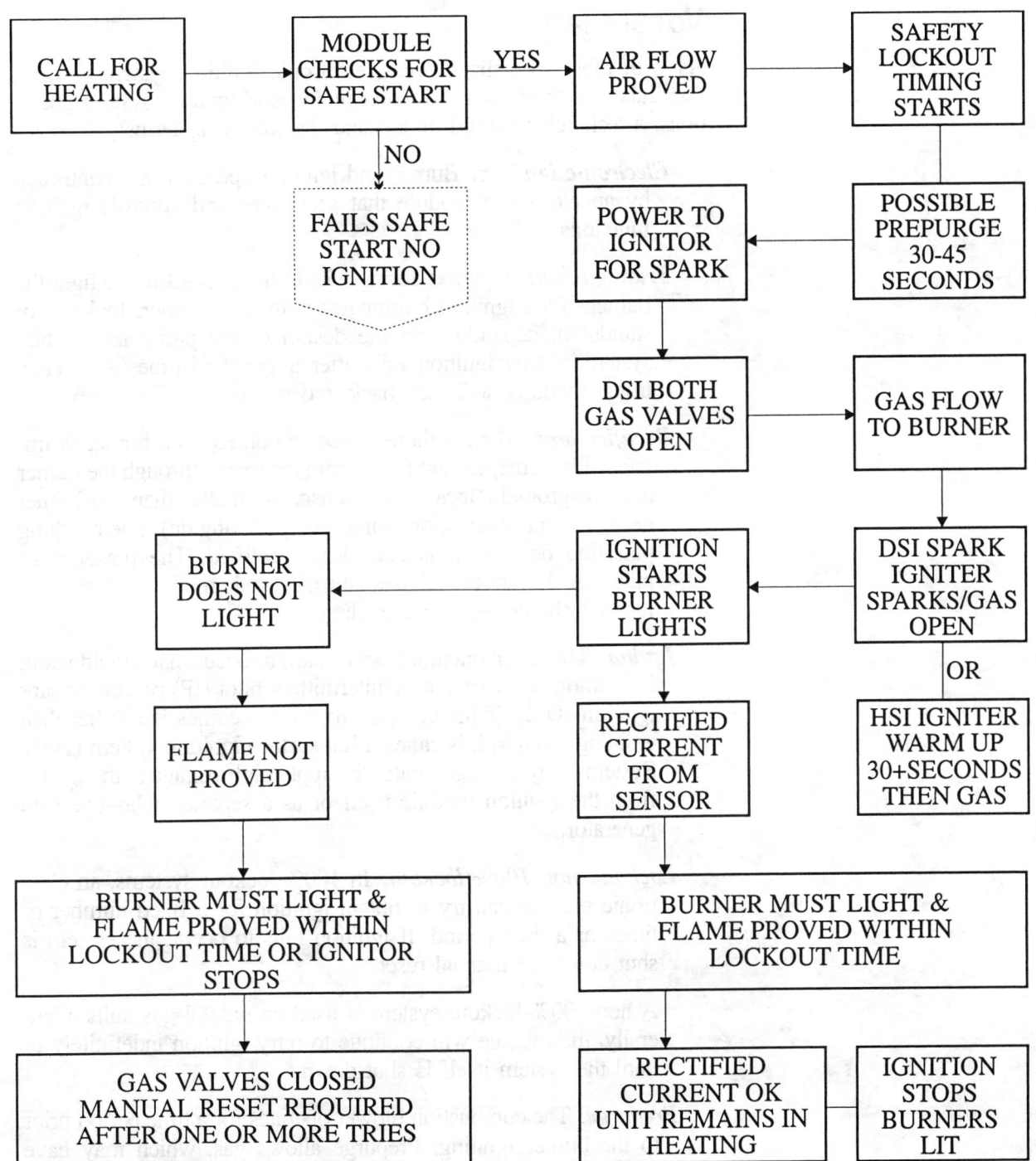

Figure 4-8. Typical Sequence of Operation, Direct Ignition

Vocabulary

As with most specialized areas, electronic ignition systems have a vocabulary of their own. To better understand ignition system operations, a technician should understand the following terms:

Electronic ignition: Burner and ignition operation are controlled by an electronic module that sequences and controls ignition functions.

Trial ignition: A period during which the ignitor tries to light the burner. Trial ignition culminates with a lit burner, lockout, or shutdown, depending on the design of the particular ignition system. If trial ignition fails after a specified time, most electronic modules will kick back into retrial.

Rectification: When a flame sensor mounted on a burner is immersed in flame, current flows from the sensor through the burner head to ground. Because the sensor is smaller than the burner head, current flows in one direction (pulsating dc). The resulting pulsating dc current is considered rectified. The presence of pulsating dc current "advises" the electronic module that the trial ignition should terminate in flame.

Ignitor: This component allows a spark to occur that should result in ignition. Ignitors can be intermittent pilot (IP) or direct-spark ignition (DSI). If the component itself becomes hot rather than emitting a spark, it is called a hot-surface ignition system (HSI). Systems may use a generator to supply high voltage to the ignitor from the ignition module itself or as a separate relay-operated generator.

Lockout, non-100% lockout: In 100%-lockout systems, an electronic module can try to obtain ignition for a fixed number of times or a time period. If ignition fails to occur, the system is shut down for manual reset.

A non-100%-lockout system is used on natural gas units. Generally, the module will continue to retry ignition indefinitely or until the system itself is shut down.

Prepurge: The combustion blower operates for a time period prior to the burner igniting. Prepurge allows gas, which may have been left in the combustion chamber to be purged out. The time period, usually 30 to 45 seconds, is set by the electronic module. Sparking is not allowed to occur until prepurge is complete.

Spark generator: This device provides the high-voltage spark for ignition. The generator may be in the module or relay, controlled as a separate component.

Combination gas valve: This has a manual on-off position, single or double automatic valve, and pressure regulator mounted in one housing.

Sensor or rod for flame: The sensor sends a signal to the electronic module that flame has been established. This signal is based on current flow through the sensor when immersed in flame.

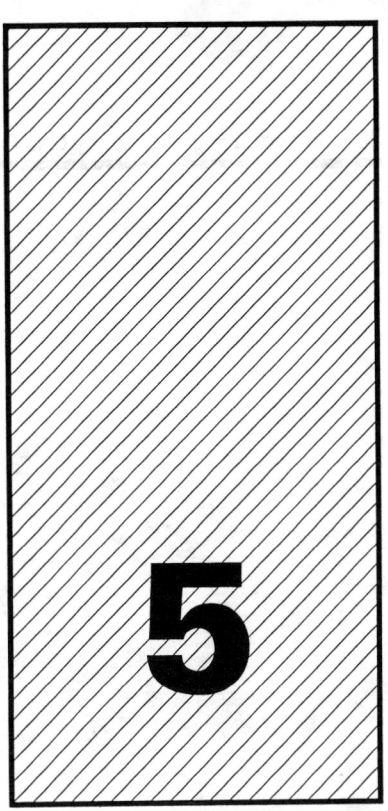

Sequences of Operation

A sequence of operation is the series of steps a furnace goes through when it receives a call for heating. To successfully repair furnaces, you must first know how they should operate. Manufacturers configure their furnaces differently. Differences can be as simple as a different time frame before an event occurs, to a totally different use of a component.

With the increased use of electronics, it is becoming more difficult to check a system's operation. For example, in the past, when a call for heating was made through the thermostat, the furnace would respond immediately. In today's units, a call for heat may be postponed by a time-delay circuit in the thermostat, which indicates a 3-min start-up delay. Nothing is wrong with the unit — the time delay is built into the control circuit. But, not understanding this, the technician might try to repair something that isn't broken.

The following sequences have been reprinted, with permission, in their entirety from manufacturers' service manuals. They represent a sampling of furnaces in service today. (Additional sequences can be found in Part Two of this book.) Reading through these sequences will help you understand how manufacturers expect the furnace to fire. It is one of the best methods of learning furnace operation.

American Standard, Inc.

Sequence of Operation for Super-efficiency, Gas-fired, Condensing Furnace TUC-B; TDC-B

Thermostat calls for heat:

R and W Thermostat contacts close to supply power, through safety limit switches, to control circuit which starts the induced draft blower. When the required combustion air is established, the pressure switch allows power to flow through the safety controls to the flame control module and the ignitor.

The ignitor will heat for approx. 15 seconds, then the gas valve is energized to permit gas flow to the burners. The flame sensor confirms that ignition has been achieved within the 7 seconds ignition trial period. All models utilize a remote sensor.

If the sensor does not confirm ignition in the 7 second time period, the ignition control will go into a re-try mode. During the second trial for ignition there is a 60 second pre-purge time for the induced draft motor, then the ignitor will energize for approx. 25 seconds before the gas valve is energized. If the flame sensor does not confirm ignition during this second trial, the ignition control will go into a re-try mode again. The sequence of operation will be repeated. If the ignition has not been confirmed at the end of the 3rd trial, the ignition control will go into lock out mode.

When ignition and flame is established, the flame control module monitors the flame and supplies power to the gas valve until the thermostat is satisfied. When the thermostat is satisfied, R and W contracts open to deenergize the control circuit.

Comfort-Aire, Heat Controller, Inc.

Sequence of Operation for Upflow/Downflow and Horizontal Induced Draft Furnaces

This furnace is equipped with a Honeywell ST9201A or Hamilton Standard 1012-800 integrated ignition and blower control board. This control combines functions of the hot surface ignition 100% lockout safety control and fixed time on/time off blower controls. It also provides a low voltage heat/cool thermostat control terminal board and connection points for field installed humidifier and electronic air cleaner optional accessories. Two indicator lights are also provided to aid the service technician.

When the heating thermostat closes (connection of R and W terminals), the induced draft blower starts and runs through a 30 second prepurge cycle. After the induced draft blower starts, the air proving differential negative pressure switch closes and starts the main burner ignition cycle. The hot surface ignitor is energized for 36 seconds to heat up, then the gas valve is energized to start gas flow to the main burner for ignition. The main gas burner flame is sensed by the de-energized hot surface ignitor within 0.8 seconds. If main burner flame is not sensed within the six second maximum trial for ignition time, the control will repeat the prepurge and ignition cycle for four additional retries. After a total of five cycles without sensing main burner flame, the system will then go into a 100% lockout mode. During the lockout mode neither the hot surface ignitor or the gas valve will be energized until the system is reset by opening the thermostat (disconnecting the R and W terminals) or interrupting the electrical power for ten seconds or longer. The induced draft blower and main burner will shut off when the thermostat is satisfied (R and W open).

The fixed time blower control will start the circulating air blower on heat speed thirty seconds after the main burner is ignited. The circulating air blower will continue to run during burner operation then shut down at a preset time after the burner shuts off. The circulating air blower will start and run on heating speeds if the thermostat fan switch is in the "On" position and the thermostat mode switch is in the "Heat" position. When the thermostat closes while in this mode, the blower will stop and go through a delay until 30 seconds after the burner lights.

When the thermostat is in the cooling mode, the blower control will start and stop the circulating air blower on cooling speed when the cooling thermostat contacts close or open respectively. The circulating air blower will start and run continuously at heating speed with the thermostat fan switch in the "On" position and the thermostat in the cooling mode. The blower will step up to cooling speed when both terminals G and Y are energized.

Note — The heating blower speed and the heating off delay come from the factory set for cooling applications. A lower heating speed and shorter off delay may be more desirable for heating only applications.

Coleman Evcon

Sequence of Operation for 90 Series

These furnaces are equipped with an electric hot surface burner ignition system. In response to a call for heat by the room thermostat, the burner is lighted by a hot glowing ignitor at the beginning of each operation cycle. The burner will continue to operate until the thermostat is satisfied at which time all burner flame is extinguished. During the off cycle no gas is consumed. With the room thermostat set below room temperature, and with the electrical power and gas supply to the furnace on, the normal sequence of operation is as follows:

1. When the room temperature falls below the setting of the room thermostat, the thermostat energizes the heating relay.

2. When the heating relay closes, a circuit is made starting the vent blower. A circuit is also made through the heating relay to the normally open vent air pressure switch contacts.

3. As the vent blower increases in speed, a negative pressure is developed in the vent blower. When sufficient negative pressure has developed, the contacts of the vent air pressure switch will close and complete the electrical circuit through the normally closed limit switch to the electronic ignition module.

4. During the next 40 to 50 seconds, the vent blower will bring fresh air into the heat exchanger and the ignitor will begin to glow. At the end of this period, the ignition module will open the gas valve and energize a safety lock-out circuit.

5. When the burner lights, the ignitor then acts as a flame probe which checks for the presence of a flame. As long as flame is present, the system will monitor it and hold the gas valve open.

6. If the burner fails to light within 6-8 seconds after the gas valve opens, the ignition module will close the gas valve and de-energize the ignitor. After a short pause, the system will recycle and try again for ignition. If the burner fails to light after three tries, the ignition module will lock off the gas valve and the ignitor. The system will remain in lock-out mode until the room thermostat is set below room temperature. The lock-out circuit will then be released and setting the thermostat to above room temperature will cause the system to try for ignition again.

7. The lapsed time from the moment the room thermostat closes to when the burner lights may be 50-60 seconds. This delay is caused by:(1)the time required for the vent blower to come to full speed, (2) the 40 to 50 seconds required for the ignitor to heat up, and (3) the

time required for the vent blower to bring fresh air into the heat exchanger.

8. Thirty to fifty seconds after the burner has lighted, the fan switch will close and the furnace air circulation blower will run.

NOTE - If a heating/cooling thermostat is being used and the fan switch is set in the continuous blower position, the furnace circulating blower will run at the air conditioning speed. If the room thermostat calls for heat, the furnace air circulating blower will shut off for 50-60 seconds and then the burner will light. One to two minutes after the burner has lighted, the furnace circulating blower will begin running at heating speed. There is no pause in the blower operation when the thermostat is satisfied; the furnace circulating air blower just changes back over to cooling speed.

9. When room thermostat is satisfied the circuit to the heating relay is broken and the relay contacts return to the normally open position. The circuit to the vent blower, the blower sequencer, and the ignition module is broken and the burner is extinguished. The contacts of the blower sequencer return to the normally open position within 30 seconds after the burner extinguishes. Then as heat is drawn from the heat exchanger and the air temperature is reduced to below the fan switch setting, the fan switch will open which stops the furnace air circulation blower.

Lennox Industries

Sequence of Operation for G21V Pulse 21™ Series Upflow Gas Furnace (See Figure 5-1)

Room thermostat, on a demand for heat, will initiate purge blower operation for a pre-purge cycle (30 seconds) followed by energizing and opening of the gas valve. As ignition occurs, the flame sensor reacts to proof of ignition and de-energizes the spark plug igniter and purge blower. Furnace blower operation is initiated 45 seconds after combustion ignition. When thermostat is satisfied, gas valve is closed and purge blower is re-energized for a post-purge cycle (34 seconds). Furnace blower will remain in operation until "fan off" factory setting of 330 seconds (adjustable from 90 to 330 seconds) is reached. Should loss of flame occur before thermostat is satisfied, flame sensor controls will initiate 5 attempts at re-ignition before locking out unit operation. Additionally, loss of either combustion intake air or flue exhaust will automatically terminate system operation. If unit becomes locked out, Watchguard control automatically resets ignition controls after one hour.

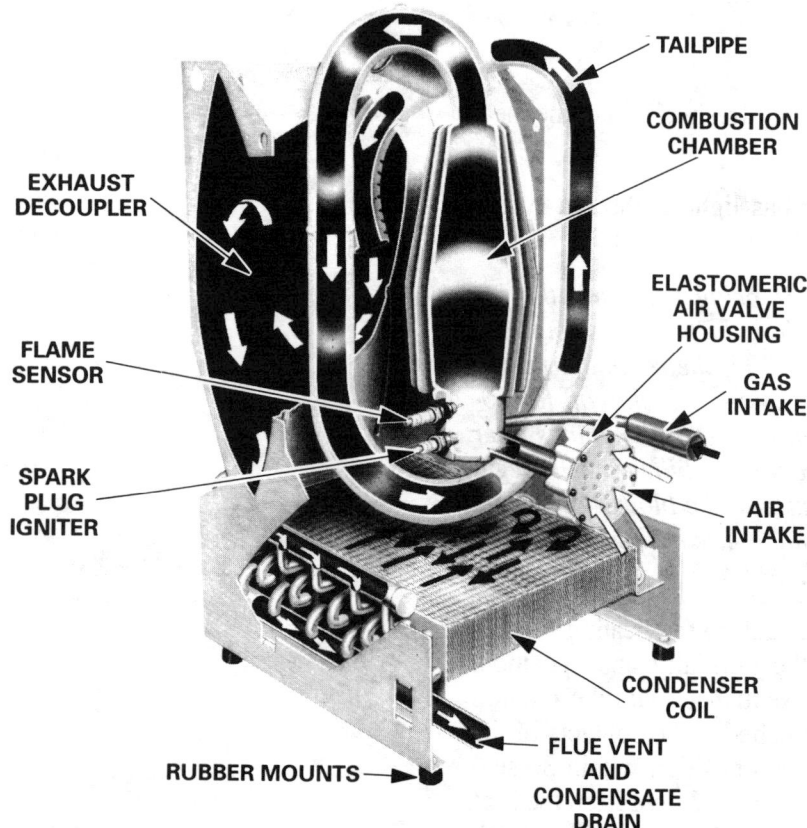

EXHAUST
DECOUPLER

FLAME
SENSOR

SPARK
PLUG
IGNITER

RUBBER MOUNTS

TAILPIPE

COMBUSTION
CHAMBER

ELASTOMERIC
AIR VALVE
HOUSING

GAS
INTAKE

AIR
INTAKE

CONDENSER
COIL

FLUE VENT
AND
CONDENSATE
DRAIN

Figure 5-1. Process of Combustion (Courtesy, Lennox Industries)

PROCESS OF COMBUSTION

The process of combustion begins as gas and air are introduced into the sealed combustion chamber with the spark plug igniter. Spark from the plug ignites the gas/air mixture, which in turn causes a positive pressure buildup that closes the gas and air inlets. This pressure relieves itself by forcing the products of combustion out of the combustion chamber through the tailpipe into the heat exchanger exhaust decoupler and on into the heat exchanger coil. As the combustion chamber empties, its pressure becomes negative, drawing in air and gas for the next pulse of combustion. At the same instant, part of the pressure pulse is reflected back from the tailpipe at the top of the combustion chamber. The flame remnants of the previous pulse of combustion ignites the new gas/air mixture in the chamber, continuing the cycle. Once combustion is started, it feeds upon itself allowing the purge blower and spark plug igniter to be turned off. Each pulse of gas/air mixture is ignited at a rate of 60 to 70 times per second, producing from one-fourth to one-half of a Btu per pulse of combustion. Almost complete combustion occurs with each pulse. The force of these series of ignitions creates great turbulence which forces the products of combustion through the entire heat exchanger assembly resulting in maximum heat transfer.

York International Corporation

Sequences of Operation for Hot-Surface Ignition Stellar™ Models P3UC (Style A) Upflow, 45- through 106-MBH Output; P3CC (Style A) Downflow, 45- through 106-MBH Output; P9UC (Style A) Upflow, 45- through 106-MBH Output; P9CC (Style A) Downflow, 45- through 106-MBH Output (See Figure 5-2)

> *Warning — Do not attempt to light this furnace by hand (with a match or any other means). There may be a potential shock hazard from the components of the hot surface ignition system. The furnace can only be lit automatically by its hot surface ignition system.*

The following describes the sequence of operation of the furnace. Refer to Figure 5-2 for component location.

CONTINUOUS BLOWER

On the cooling/heating units with fan switch, when the fan switch is set in the ON position, a circuit is completed between terminals R and

G of the thermostat. This energizes the 1R relay. Contact 1R-1 closes and contact 1R-2 opens. The motor is energized through the black, high speed tap. The blower then operates on high speed.

INTERMITTENT BLOWER

When the system switch is set on HEAT and the fan switch is set on AUTO, and the room thermostat calls for heat, a circuit is completed between terminals R and W of the thermostat. This energizes the venter relay 3R, which energizes the venter. When the proper amount of combustion air is being provided, a pressure switch (1LP) activates the 50E47 ignition control. The rollout switch control, primary limit and auxiliary limit are also in this circuit and must be in the closed position for the ignition control to be activated.

The 50E47 ignition control provides a 45-second warm-up period. The gas valve then opens for four seconds.

As gas starts to flow and ignition occurs, the flame sensor begins its sensing function. If a flame is detected within four seconds after ignition, normal furnace operation continues until the thermostat circuit between R and W is opened. After approximately 60 seconds (or the supply air temperature reaches 155° to 125°F), the fan switch closes.

When the thermostat opens, the venter is de-energized, along with the ignition control. With the ignition control de-energized, the gas flow stops and the burner flames are extinguished.

The blower motor continues to operate until the supply air temperature drops to between 85° and 100°F. When this occurs, the fan switch opens, de-energizing the blower motor. The heating cycle is then complete, and the unit is ready for the start of the next heating cycle.

If flame is not detected in the four second sensing period, the gas valve is de-energized. The 50E47 control is equipped with a re-try option. This provides a 60-second wait following an unsuccessful ignition attempt (flame not detected). After the 60 second wait, the ignition sequence is restarted with an additional 10 seconds of igniter warm-up time. If this ignition attempt is unsuccessful, one more re-try will be made before lockout.

50E47 SERIES HOT SURFACE IGNITION CONTROL

All White-Rogers 50E47 controls will repeat the ignition sequence for a total of five recycles if flame is lost within the first 10 seconds of establishment.

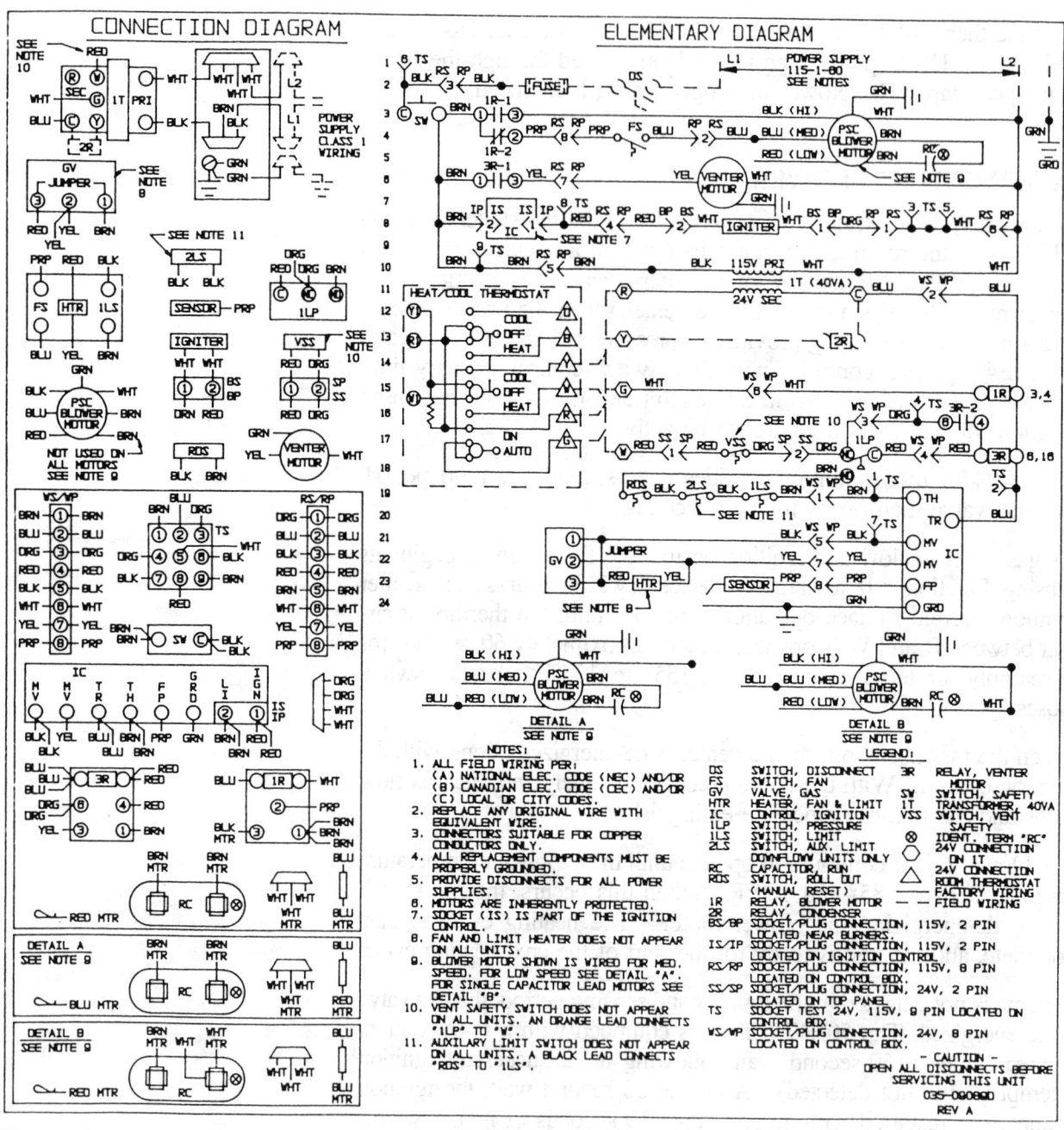

Figure 5-2 Unit Wiring Diagram (Courtesy, York Heating and Air Conditioning)

If flame is established for more than 10 seconds after ignition, the controller will clear the ignition attempt (re-try) counter. If flame is lost after 10 seconds, it will restart the ignition sequence. This can occur a maximum of five times.

During burner operation, a momentary loss of power of 50 milliseconds or longer will drop out the main gas valve. When power is restored, the gas valve will remain de-energized, and a restart of the ignition sequence will begin immediately.

A momentary loss of gas supply, flame blowout, or a shorted or open condition in the flame probe circuit will be sensed with 0.8 seconds. The gas valve will de-energize and the control will restart the ignition sequence after waiting 60 seconds. Recycles will begin and the burner will operate normally if the gas supply returns, or the fault condition is connected prior to the last ignition attempt. Otherwise, the control will lock out.

If the control is locked out, it may be reset by momentary power interruption of 1/20 second or longer. Either the 24-volt thermostat or line voltage may be interrupted.

> *Author's Note* — *Each manufacturer approaches the sequence from a slightly different position. It is important for the technician to be aware of the time delays that exist before a furnace will begin operating. These time delays are based on both component function and safety features. The fact that something is not happening a given point in time does not indicate failure. The delay may be nothing more than a system analyzing itself.*

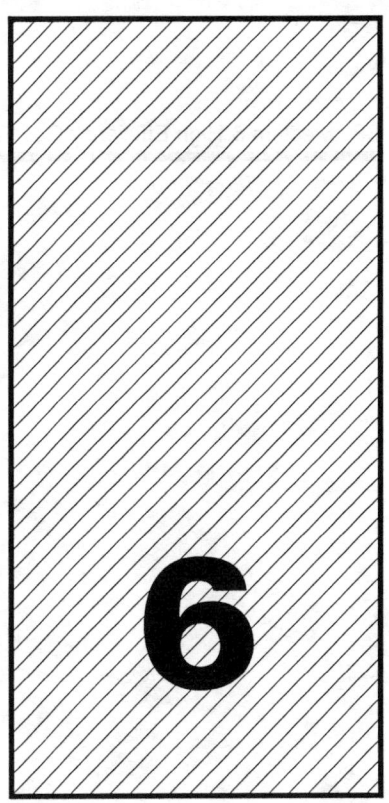

Basic Service Procedures

This chapter explores basic service procedures as they apply to most furnaces, in high-, mid-, and low-efficiency ranges.

On January 1, 1992, the National Appliance Energy Act became law. Under this act, the minimum-efficiency standard for the installation of a new furnace is 78% AFUE (annual fuel utilization efficiency). The days of furnaces with 60%-plus efficiency are gone. The furnace of the future will be 90%-plus efficient; it may use two-stage burners with variable-speed blower motors constantly adjusting to load.

A service call means repairing the existing unit or, if repair cannot be effected, replacing the unit. Consumer resistance to purchasing a high-efficiency furnace is still great. When you quote the price of a 90%-plus, high-efficiency furnace, make sure you give the consumer alternatives. Furnaces in the 80%-plus, mid-efficiency range are still acceptable from a cost standpoint.

Contractors bidding on tract homes need to consider the price of the furnace versus the overall salability of the home. The long-term payback of a high-efficiency furnace is valid, but sometimes long-term payback considerations just don't matter enough.

With those thoughts in mind, furnace service will continue to be profitable, as long as the technician can fix the existing unit in a reasonable amount of time. A furnace is capable of lasting 20-plus years; therefore, a number of service calls will involve repairing older units. In the future, medium- and high-efficiency service calls will become the norm.

Note — Not all service procedures apply to all units. Manufacturers compete by making changes in their products. If you unsure about a service procedure, obtain the manufacturer's literature and double-check.

Safety

Before we discuss furnace service, let's talk about safety. Fossil fuels are used throughout the U.S., and the incidence of accidents is extremely low. When accidents do occur during service, it is generally because the technician failed to use procedures that ensure a safe working environment.

Servicing a furnace involves both electrical and fuel dangers, which can easily become life threatening. Technicians should be aware of the following points before servicing any furnace:

- Know where the gas shut-off is located before you start the job. Make sure the shut-off is not frozen open.

- Carry an approved fire extinguisher as a tool, or at least know where the nearest extinguisher is.

- If you smell gas, danger is imminent. Shut the system down, let the gas disperse, then find and fix the leak.

- Don't use a match to check a joint for a leak. Use soap bubbles.

- Remember: **sparks + gas = ignition**.

- Use a pilot-lighting tool to light a standing pilot. A flame rollout can cause havoc.

- If you lose a pilot, let the gas disperse before trying again.

- Replace a part with its *exact* replacement.

- Check local codes for the proper type of piping and the permitted use of flex piping.

- Find and lock the fuse or circuit boxes when you turn them off. Someone can throw a switch and energize you.

- The customer is *not* always right: If you think the furnace should not be repaired, don't do it! A cracked heat exchanger cannot work for a little while longer.

- A jumper wire can help make a repair easier or destroy a component and the user. Know what you are doing, or don't do it.

Service Procedures

Good service procedures for start-up and repair require that the fuel-air mixture, gas pressure, and flame be correct. The following procedures are the tools a technician uses to ensure that the furnace operates properly.

AIR ADJUSTMENT

Primary air — the air mixing in the burner — should be adjusted to ensure proper-colored flame and complete combustion. Primary air controlled by a shutter can be increased by opening the shutter wider and decreased by narrowing the shutter opening, Figure 6-1. Some burners do not have a shutter-adjustment feature. If this is the case, check the manufacturer's literature for primary air-adjustment procedures.

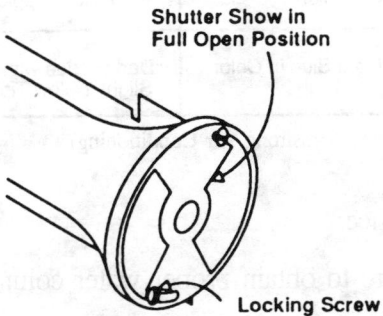

Figure 6-1. Air Shutter Adjustment (Courtesy, Inter-City Products Corp.)

When primary air adjustment is required on a shuttered burner, use the following procedure to set the flame:

1. Open the shutter full.

2. Let the furnace run 10 to 15 min, then check the flame.

3. If the flame requires adjustment, loosen the shutter screw.

4. Close the shutter enough for the flame to develop a yellow tip.

5. Reopen the shutter enough to lose the yellow tip; lock the shutter.

6. Check the flame. Is it correct?

7. Cycle the furnace several times to ensure that the flame is correct.

8. Adjust primary air as required by flame condition.

FLAME DIAGNOSIS

A powerful tool in furnace service is the technician's vision and hearing. The flame for a properly-fired furnace should have a soft blue color, without yellow tipping or lifting. If the flame doesn't look like this, you must make certain adjustments, Figure 6-2.

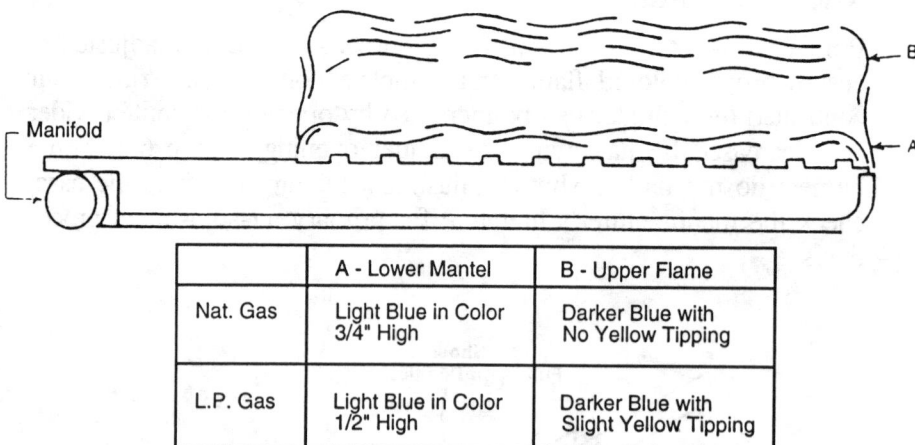

	A - Lower Mantel	B - Upper Flame
Nat. Gas	Light Blue in Color 3/4" High	Darker Blue with No Yellow Tipping
L.P. Gas	Light Blue in Color 1/2" High	Darker Blue with Slight Yellow Tipping

Figure 6-2. Main Burner (Courtesy, Armstrong Air Conditioning)

To correct flame appearance:

1. Adjust the gas pressure to obtain proper water column (in. wc).

2. Time the meter for gas flow. The flow must be within 5% of manufacturer's specifications. (In some systems, only 2% is allowable.)

3. Check the combustion air side, including blowers and sealing.

4. Check for blocked vents.

5. Check the burners for soot, dirt, improper mount, and other obstructions.

6. Check the flue for blockage, distortion, warping, or poor piping.

7. Check the heat exchanger for cracks, rust-through, or burn-out.

The flame should *not* be: yellow, lifting, flashing-back, popping, floating, or rolling-out.

Yellow flame

A flame should not be yellow; it should be soft blue, ascending solidly from the burners.

> *Possible cause* — Incomplete combustion is occurring. Soot may be created, which further hampers burner operation. Lack of primary air is the main cause.

> *Probable correction* — Reset the primary air; check for blockage, such as dirt in the primary air openings; and check whether the orifice or burner port are improperly aligned or the draft hood is clogged.

Lifting flame

A portion of the flame at the burner head rises off the burner head and may drop back to the head. Lifting flame may be limited to a few flames present on the burner head.

> *Possible cause* — The gas velocity is faster than the speed at which the gas can burn.

> *Probable correction* — Input gas or primary air may need to be reduced. If the lifting flame is limited to one burner, its orifice size should be checked against the other orifices.

Flashback

The air and gas are igniting inside the burner and burning near the orifice. Listen for a loud sound associated with this.

> *Possible cause* — The gas is flowing slower than the speed at which burn is occurring.

Probable correction — Reduce the primary air, check the burner rating and orifice size; the input rate or spud size may need to be increased. The gas valve may be leaking, or the burner itself may need to replaced, if the corrections listed earlier don't work.

Pop (Extinction)

The technical name for pop is extinction. Under the burner head, a small explosion occurs that creates a popping sound.

Possible cause — Flashback burning occurred when the burner was shut off, and burning continued after the gas was shut off. The pilot light also could have been blown out by the force of the explosion.

Probable correction — Reduce primary air if possible, increase gas pressure, and compensate by reducing the orifice. Also, check the gas valve and burner — if necessary, replace.

Floating flame

A floating flame doesn't appear to have a shape — it looks like it's all over the burner area — and is quiet. An aldehyde odor is generally present.

Probable cause — Incomplete combustion is the cause. Floating flames *must* be corrected. Carbon monoxide is loose in the burner area, and this can create serious health and safety risks.

Probable correction — Check the gas flow, adjust the primary air, check every opening and space that involves secondary air, and check flues and burners.

Rollout

When the flame leaves the combustion chamber on igniting, rollout is occurring. The flame is trying to find air outside of the combustion area. This can be dangerous!

Possible cause — Look for poor draft, blocked flue, not enough supply air, and burner overfiring.

Probable correction — Check gas flow, inspect flues, adjust primary air, clean the burner and all secondary air sources.

Other flame considerations

Other variations may exist for an improper flame. The flame shouldn't vary in length over a given time. If the flame is changing size, check the gas pressure (it may be fluctuating). The flame should never appear as if it is being blown all over the place. If this is the case, look for drafts across the burner (they shouldn't be there).

Are the panels in place or missing? If the pilot flame is moving, baffle it. Do not rule out a cracked heat exchanger. If the heat exchanger is cracked, replace it or the furnace. *Do not leave a unit operating with a cracked heat exchanger; it is deadly.*

PILOT FLAMES

The proper pilot flame is crucial to a unit's operation, *from both operational and safety standpoints.* New units have different approaches to achieve heating, but in older units, pilot flame still controls many functions.

Thermocouple

The thermocouple is the primary pilot device on lower-efficiency furnaces. There are still many furnaces requiring thermocouple service.

A thermocouple creates a small dc current, which powers an electromagnet, which opens the gas valve. The current is created by the application of heat to the thermocouple. The heat, in the form of flame, must be applied at the proper location between the hot and cold junction.

If the pilot flame (heat source) covers the thermocouple at the wrong location, this creates current flow problems. The pilot flame should be $3/8$ to $1/2$ in. from the tip, and it must cover the hot junction end of the thermocouple to be effective.

Pilot flames that float, lift, etc., at the thermocouple must be corrected, or the furnace will operate poorly — or not operate at all.

Intermittent-pilot (IP) system

There is no free-standing pilot during shutdown on an IP system. There is a flame during heating operation that is similar to a flame on a thermocouple system.

An intermittent-pilot system uses a flame rod-sensor to allow rectified current flow to the ignition module when a flame is present at the rod,

Figure 6-3. This configuration allows you to check that a pilot flame is present after spark has occurred and establishes conditions necessary for turning off the spark. The pilot flame also provides for main burner ignition.

Most manufacturers call for the flame to cover $^3/_8$ to $^1/_2$ in. of the flame rod-sensor.

Flame current flow can be measured and adjusted on many units by amperage check. For maximum amperage readings, to ensure proper system operation, check the manufacturer's specifications.

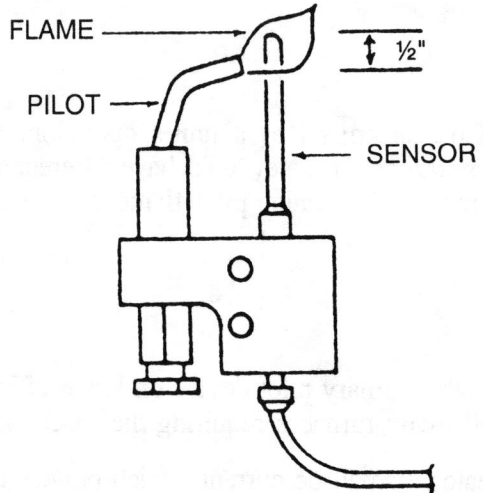

Figure 6-3. *Pilot Burner* (Courtesy, Armstrong Air Conditioning)

Direct-surface ignition

There is no pilot flame in a DSI system, Figure 6-4. The burner flame allows the sensor or sensor/ignitor to provide the necessary flame signal.

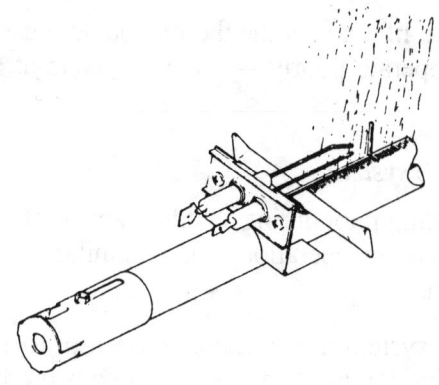

Figure 6-4. *DSI System* (Courtesy, Inter-City Products Corp.)

First, check the burner flame: Is it a healthy, solid, blue flame? The sensor itself should be immersed about 1 in. into the burner flame for a correct signal to be sent. Sensor flame coverage can be adjusted on some units by bending or relocating the sensor bracket when the sensor is alone. Ignitors or combination ignitor/sensors *must not* be adjusted using this method.

Ceramic insulator temperature problems and excessive amperage can be caused by improper flame coverage at the sensor rod. These conditions create problems and must be corrected to conform with manufacturers' specifications.

Hot-surface ignition

In an HSI system, the flame sensor is immersed about $3/4$ to 1 in. in the burner flame, Figure 6-5. Again, the burner flame must be correct.

If the immersion rate of the sensor isn't correct, the bracket can be bent or relocated. Ignitors or combination ignitor/sensors *must not* be adjusted using this method.

Ceramic insulator temperature concerns and excessive amperage can be caused by improper flame coverage at the sensor rod. This condition must be corrected.

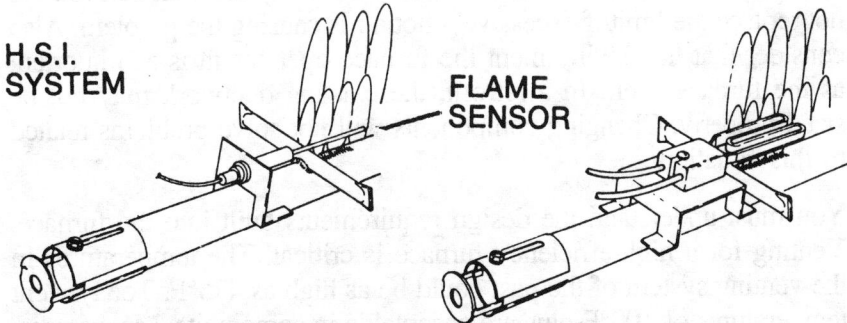

Figure 6-5. HSI System (Courtesy, Inter-City Products Corp.)

> *Note — The above techniques do not apply to all modules or makes. Always check the manufacturers' specifications, as that is the only way to ensure correct, safe performance. The days of routine quick fixes using rules of thumb are gone!*

BLOWER MOTORS

In the past, blower service was relatively simple. Temperature rise and rules of thumb were all you needed to set blower speed. For example, if the temperature rise was lower than 80°F, too much air was being moved and the blower speed was reduced. If temperature rise was greater than 80°F, the blower was moving too little air and should be sped up.

With the entrance of high-efficiency furnaces and blower motors, the rules have changed. The new motors are more powerful and use less energy than in the past. Technicians who install and service today's high-efficiency furnaces must start by using the manufacturer's recommended setting for blower speed. If the manufacturer's data on speed settings does not create the desired temperature rise, correction is necessary. The old rule of high speed for cooling and low for heating may not work.

The starting point should be the manufacturer's recommendation for the proper cfm airflow in cooling. The rules for air conditioning airflow also have changed.

Fan speed should be set by the temperature rise called for on the data plate. Temperature rise is simply the difference in degrees between supply and return air. It is possible to have a higher fan-speed flow in heating than in cooling operation. This is the way the manufacturer built the unit, and this is the way the unit should run.

From the service standpoint, remember that a high temperature rise and lower blower speed can cause the unit to trip on limit control. Do not replace the limit if excessively hot air is causing the problem. Also consider that the environment the furnace operates in is as important as the furnace itself. In a retrofit, the existing ductwork may not be sized properly. Changing components will not solve problems related to this situation.

You must understand the design requirements built into the furnace. Venting for a high-efficiency furnace is critical. The temperatures in the venting system of the past could be as high as 475°F. Today, vent temperatures of 100°F-plus are acceptable in some units. Manufacturers' installation instructions provide the specific data necessary to ensure efficient operation. There is no longer a single method of venting a furnace; nor is there a single manufacturer who sets the standard for the industry.

The rule for blower speed setting is:

Increasing speed = decreasing temperature rise

Decreasing speed = increasing temperature rise

Some installations in the field are now using variable-speed blower motors. A microprocessor unit controls the blower speed to maintain proper airflow and temperature as determined by the manufacturer. The motor itself uses a speed controller governed by the micro-processor. The draft-inducing motor can also be controlled by the microprocessor. This type of system gives the furnace the ability to operate under maximum-efficiency conditions and may also allow the use of staged burners.

TIMING A GAS METER

For service calls that involve procedures such as derating a furnace, you need to check gas flow to the unit. Gas flow to a natural gas furnace can be checked using the gas meter installed at the site. (The Btu content of natural gas in the area must be known. Check with the local gas company for this value.)

For this exercise, we will assume 1,000 Btu per cu ft of gas.

Time the meter:

1. Turn off all gas appliances at the site.

2. Call for heat and allow the furnace to operate. Make sure that any combustion air box is in place.

3. Look at the gas meter; locate the dials labeled *cubic foot* or *two cubic feet*, Figure 6-6.

4. Using a stopwatch, determine the length of time it takes for two cubic feet to pass through the meter. If the dial completes one full revolution and is labeled two cubic feet, with a Btu value of 1,000, then the gas Btu to the furnace content is 2,000 Btu. Let's assume the stopwatch showed 48 sec.

Apply formula:

Btu Gas Content = 2 cu ft x 1,000 Btu/cu ft = 2,000 Btu to Furnace

Time to Pass 2 cu ft = 48 Sec

Seconds in Hour x Btu Content of Gas ÷ Watch Time = Btuh Input

3,600 x 2,000 ÷ 48 = 150,000 Btuh

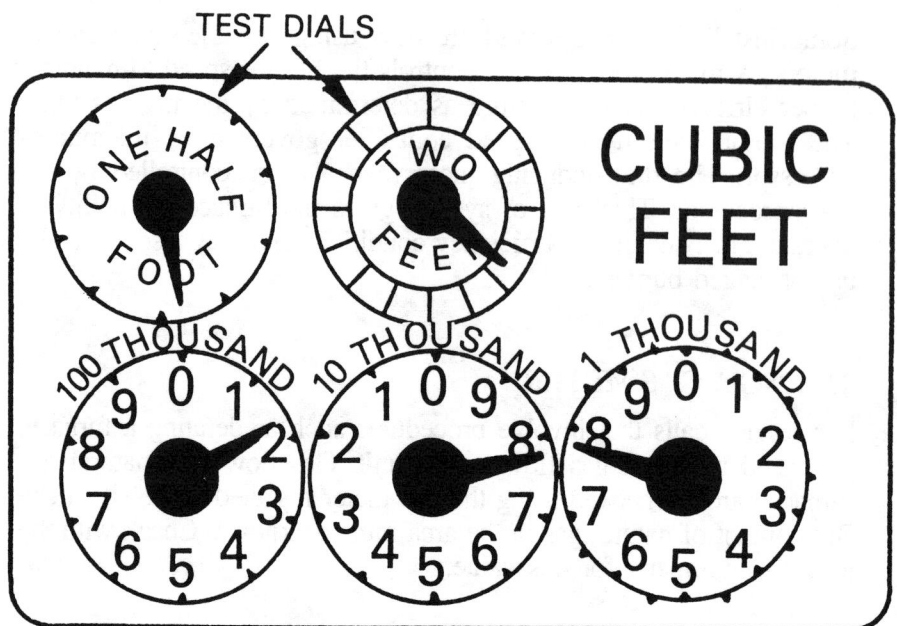

Figure 6-6. *Test Dials* (Courtesy, Lennox Industries)

The furnace times out at 150,000 Btu per hour. Check the rating plate to determine what the furnace is rated for, and adjust gas flow in the regulator.

> *Warning* — *Do not allow the furnace to either over- or under-fire.*

MANOMETER PRESSURE CHECK

The simplest way to measure gas pressure is with a U-tube manometer, Figure 6-7. The U tube converts pressure placed on a column of liquid in the tube into a reading measured as inches of water column (in. wc). The pressure reading is taken from an inch scale mounted on the U tube. If you blow in one side of the U tube, the liquid column moves proportionately to the pressure that your breath has exerted on the liquid.

Manifold pressure is measured in in. wc because the pressure in a typical gas system is less than 1 lb per sq in; 1 lb per sq in. of pressure equals 27.71 in. wc; 1 oz of pressure, therefore, equals 1.732 in. wc. The pressure in a typical gas furnace is 3.5 in. wc or a little over 2 oz.

The U tube is attached to the appropriate pressure tap at the gas valve or at a tap in the manifold pipe. The main burner and pilot light should be operating. Gas pressure should meet the manufacturer's specifications for the gas service provided.

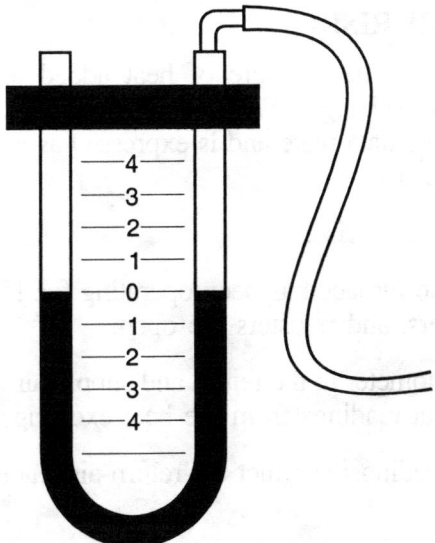

Figure 6-7. U-Tube Manometer

If the pressure is lower or higher than called for, the gas valve pressure regulator should be adjusted.

> *Note — The adjustment of manifold pressure at the regulator is for minor pressure adjustments only. For example, if the required pressure is 3.5 in. wc and there is a larger pressure difference, regulator adjustment is not the cure. If a large difference exists, the burner orifice should be changed to meet the unit's needs.*

Don't rely on regulator adjustment to compensate for major pressure deficiencies. It's accepted common practice that a pressure difference greater or less than 0.03 in. wc should not be compensated for by regulator screw adjustment.

> *Note — When a furnace uses liquid propane (LP), the gas pressure is controlled by a regulator at the LP tank. Don't adjust this regulator; call the gas supplier to handle this adjustment.*

Another area of pressure checks in newer furnaces is for flue vent pressures. Newer furnaces are using power venting and flue vent pressures as a means of establishing draft control. Check manufacturers' information in Part Two for specific applications on troubleshooting pressure switches, pilot tubes (J tubes), and flue pressures.

TEMPERATURE RISE

Temperature rise is the measure of heat added to supply air after crossing the heat exchanger, Figure 6-8. This value can generally be found on the unit's data plate and is expressed as a range (i.e., 40° to 70°F temperature rise).

To obtain temperature rise:

1. Make sure the furnace has been operating for 15 min and that all ducts, dampers, and registers are open.

2. Place a thermometer in the return and supply air ducts. (Be careful of radiant heat readings from the heat exchanger.)

3. Record the readings; subtract the return-air reading from the supply-air reading.

4. The resulting number should be in the mid range of the data plate requirement for temperature rise. If not, assuming that all other conditions of operation are being met, then the blower speed should be adjusted.

5. Blower speed is reduced for low temperature rise and increased for high temperature rise correction.

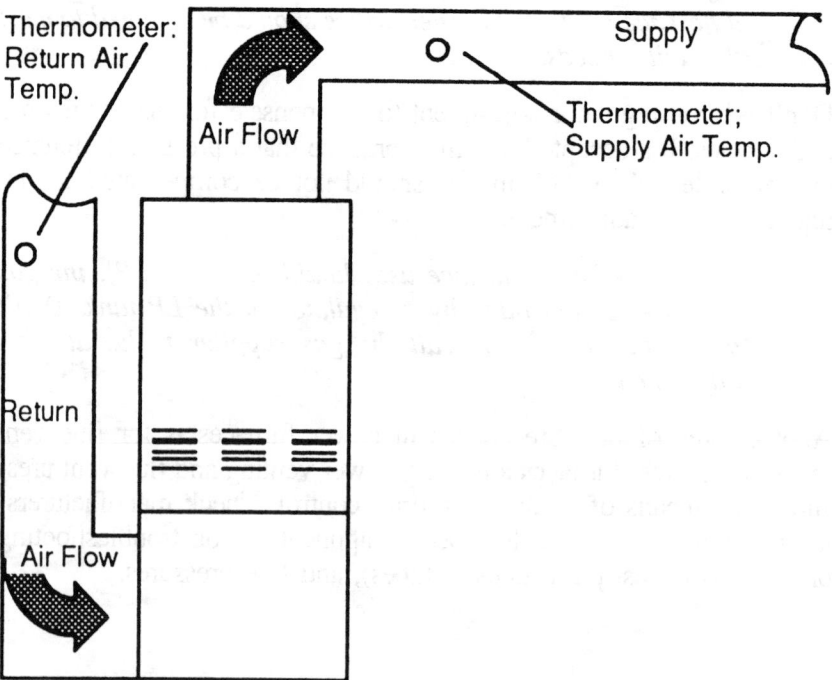

Figure 6-8. Temperature Rise Diagram (Courtesy, Inter-City Products Corp.)

ANTICIPATOR SETTING

The heat anticipator helps prevent a furnace from short-cycling, Figure 6-9.

To set an anticipator on a mechanical thermostat, wrap 10 turns of wire, connected between R and W, around the jaws of an induction ammeter, Figure 6-10. Take the reading from the meter, and divide by 10 to obtain the amp draw. This number is the setting. With a direct-reading digital voltmeter that has an amp induction clamp, this process is not necessary. Just use the amp reading from R to W as your setting.

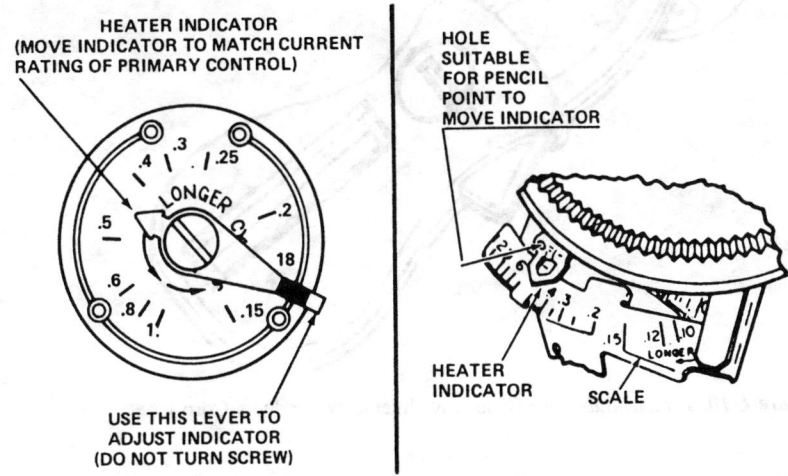

Figure 6-9. *Typical Heat Anticipator* (Courtesy, Inter-City Products Corp.)

Many mid- and high-efficiency furnaces are being installed with electronic thermostats. These do not use an amp draw as the anticipator setting. Cycle time, defined as the number of times the unit is allowed to run in an hour, is the basis for anticipation control. Under normal conditions, most modern units use a cycle rate of three per hour. This rate is adjustable, depending on local conditions. Use manufacturer's literature to set the rate.

COMBUSTION TESTING

Measuring exhaust gases for carbon dioxide (CO_2), carbon monoxide (CO), oxygen (O_2), and gas temperature are excellent methods of checking furnace efficiency.

CO_2

To test for CO_2, follow the procedure outlined on the next pages (reprinted with permission from Lennox Industries, Inc. *Service Unit Information* on the *G14 Series* furnace).

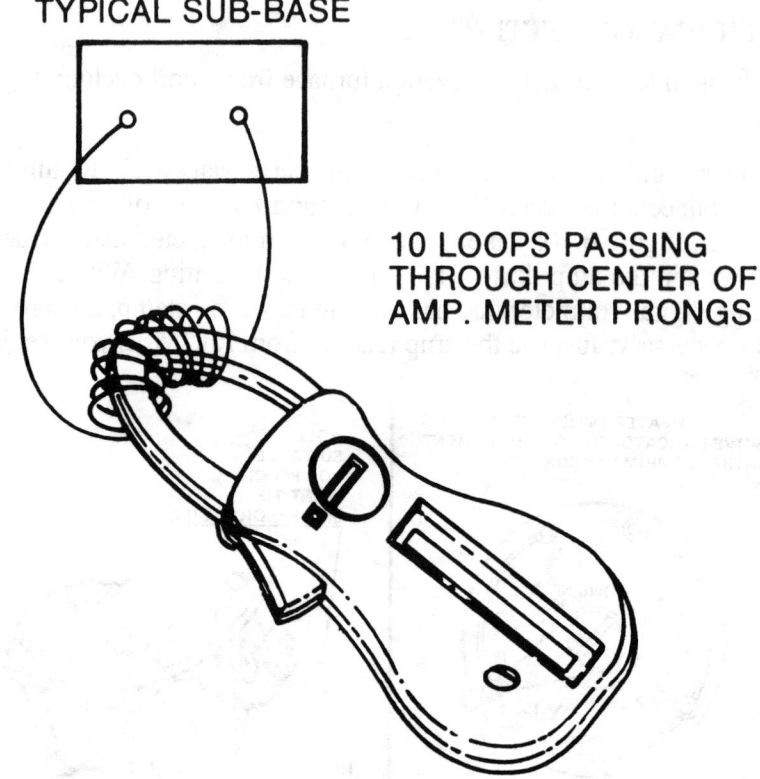

TYPICAL SUB-BASE

10 LOOPS PASSING THROUGH CENTER OF AMP. METER PRONGS

Figure 6-10. Typical Sub-Base (Courtesy, Inter-City Products Corp.)

Exhaust CO_2 Content

When the unit is properly installed and operating normally the CO_2 content of the exhaust gases is 8.5 to 10% for natural gas and 11 to 12% for L.P. gases.

The CO_2 content can be measured using the Bacharach CO_2 test with a Fyrite CO_2 indicator. Other testers are available and instructions packaged with the tester should be used.

To Measure:

1. Drill size "R" or $^{11}/_{32}$ in. hole on top of the exhaust outlet PVC elbow (inside unit cabinet) and tap 1/8-27 NPT, Figure 6-11. This hole is used as a CO_2 test port.

2. Install a hose barb connector into the CO_2 test port, Figure 6-11.

3. Attach end of Fyrite sampling tube to hose barb on exhaust outlet elbow.

4. Set thermostat to highest setting and allow unit to run for 15 minutes.

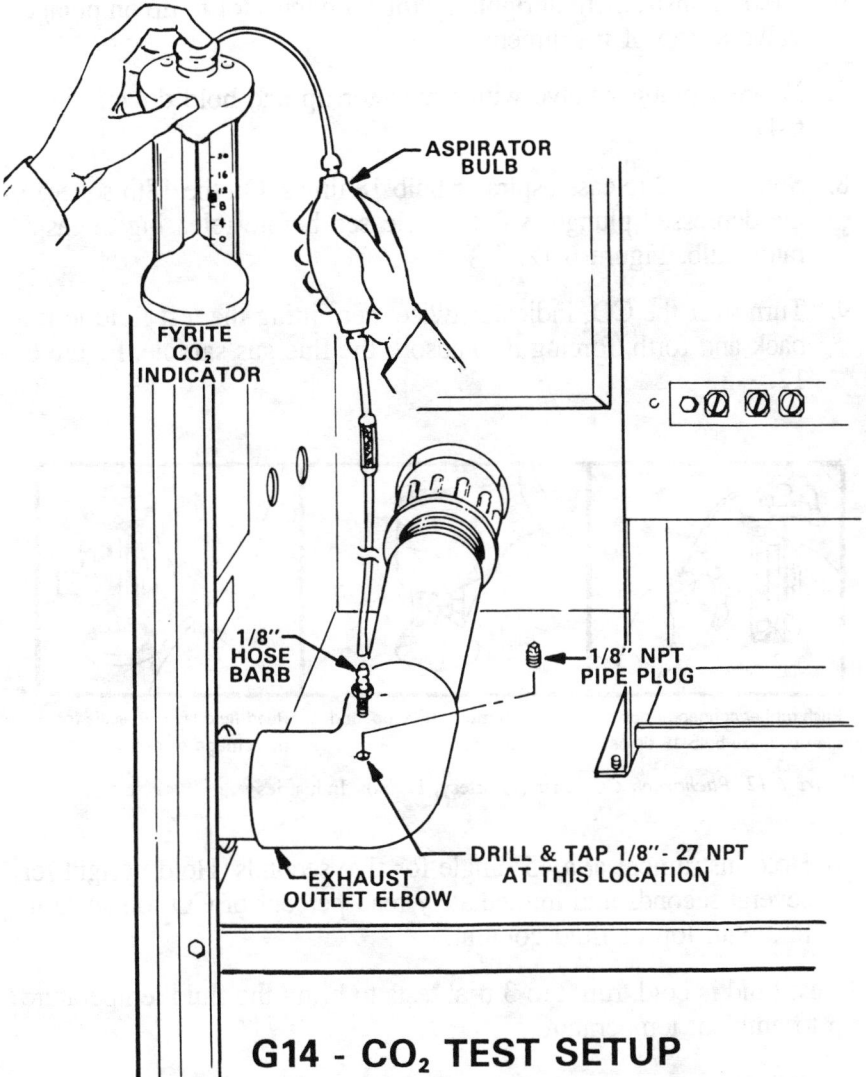

ASPIRATOR BULB

FYRITE CO₂ INDICATOR

1/8" HOSE BARB

1/8" NPT PIPE PLUG

DRILL & TAP 1/8" - 27 NPT AT THIS LOCATION

EXHAUST OUTLET ELBOW

G14 - CO₂ TEST SETUP

Figure 6-11. CO₂ Setup (Courtesy, Lennox Industries, Inc.)

5. Adjust the Fyrite indicator scale to zero:

a. Invert Fyrite until all the fluid has run into top reservoir.

b. Turn Fyrite upright until all fluid has run into bottom reservoir.

c. Hold Fyrite upright at a 45° angle for five seconds to drain excess fluid droplets from inside surfaces.

d. Hold Fyrite upright and depress plunger valve and release.

e. Loosen lock nut in rear of scale. Slide scale until zero percent CO₂ scale division lines up with top of fluid column. Tighten scale lock nut.

6. Holding instrument upright, lay rubber connector to tip on plunger valve at top of instrument.

7. Depress plunger valve with connector tip and hold down, Figure 6-12.

8. Squeeze and release aspirator bulb 18 times. On the 18th squeeze, the depressed plunger valve is released before releasing the aspirator bulb, Figure 6-12.

9. Turn over the CO_2 indicator twice, permitting the test fluid to run back and forth, forcing it to absorb the flue gas sample, Figure 6-12.

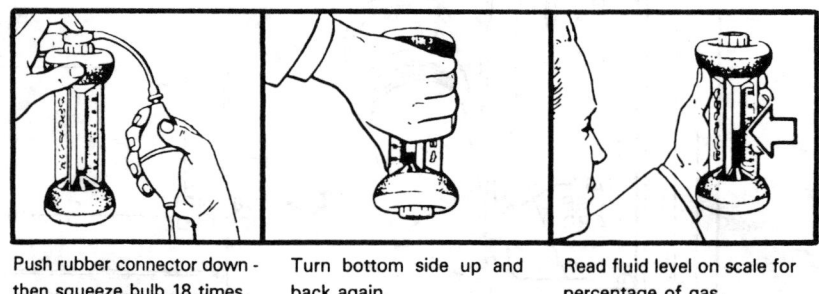

Push rubber connector down - then squeeze bulb 18 times. Turn bottom side up and back again. Read fluid level on scale for percentage of gas.

Figure 6-12. Bacharach CO₂ Test (Courtesy, Lennox Industries)

10. Hold instrument at a 45° angle for five seconds. Hold upright for several seconds and immediately read percent of CO_2 on scale in line with top of fluid column.

If test fluid is cold, run 2 to 3 trial tests to bring the fluid temperature up to ambient temperature.

Fluid Strength Check

When a sample has been absorbed and CO_2 percent read, without venting to atmosphere, turn instrument over and back two more times and take a second reading on the same sample. If the CO_2 reading has increased by more than ½ percent, the fluid needs replacing. Refer to the instructions packaged with the Fyrite CO_2 indicator for fluid replacement and maintenance.

Fresh Fyrite fluid will absorb CO_2 from about 400 samples.

> ***Caution*** *— Fyrite fluid is corrosive to skin, clothing, some metals, painted and lacquered surfaces. Care should be taken when turning instrument and when changing fluid. Do not get fluid in eyes.*

11. When CO_2 test is completed, turn off the unit, remove hose barb from exhaust outlet elbow and use a $^1/_8$ inch plastic pipe plug to close off the test port.

> *Caution — The exhaust vent pipe operates under positive pressure and must be completely sealed to prevent leakage of combustion products into the living space.*

CO

The content of CO must be under 0.04% to be operating normally and safely. If the level exceeds 0.04%, something is wrong. Pipes may be blocked, improper gas usage exists, or atmospheric conditions at the site may be creating discharge problems. In any case, the CO reading must be brought into a normal range.

To test for CO, follow the same procedure that was outlined for CO_2.

A high CO rating can be fatal. Shut the unit down and find the source.

O_2

If the O_2 count is high — normally no more than 1% O_2 count is allowed — there is probably excess air. An adjustment of the air-fuel ratio should reduce primary air and solve this concern.

Exhaust gas temperature

On older units, exhaust gas temperatures greater than 400°F were permissible. In high-efficiency units, the exhaust gas is substantially cooler, since additional heat has been removed from the gas. A new, high-efficiency furnace that is operating efficiently may have normal exhaust gas temperatures between 110° and 125°F. Temperatures in excess of 135°F indicate problems.

Understand that these temperatures are not for every furnace. In some cases, the actual operating range may be different among particular models within the same product line. Additionally, lower temperature ranges are possible and quite normal in some furnaces. To make an accurate diagnosis, you must have the manufacturer's data.

GAS VALVE SERVICE

Checking a gas valve is not a difficult task. As a general rule, a gas valve receives an electrical signal from a source, opens, and allows gas to flow.

With HSI and other types of ignition, different valves are used. Figure 6-13 shows a valve used on an HSI system that is referred to as a step valve. Consult manufacturers' data for the type of valve used on the system requiring service.

To check a gas valve:

1. Call for heat at the thermostat.

2. If the burner comes on, the valve opened.

3. If the burner did not come on, an electrical or mechanical problem exists.

4. To check for electrical failure, see if there is voltage at the valve. If there is no voltage, check circuitry for an open switch or failed sources, such as the transformer.

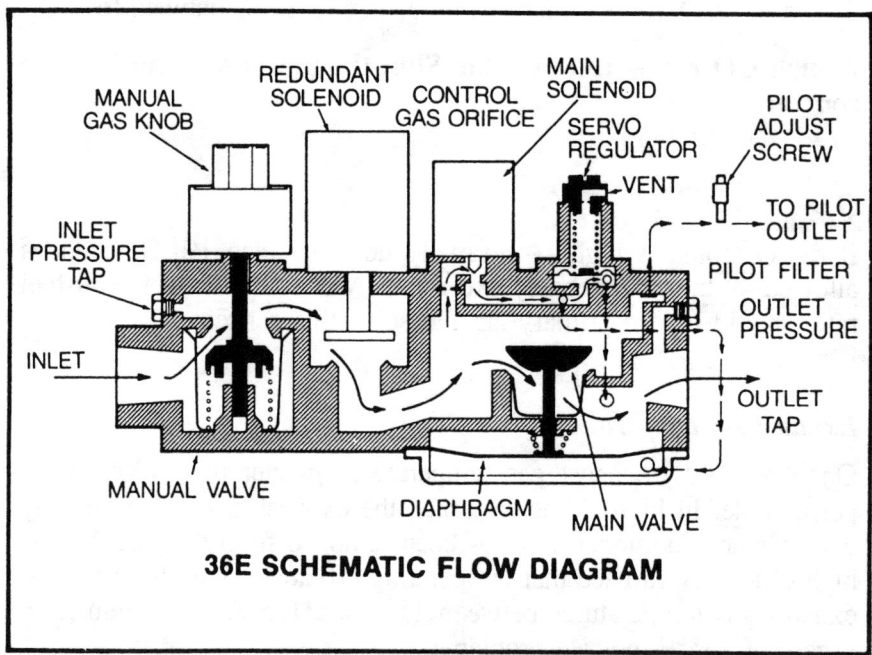

36E SCHEMATIC FLOW DIAGRAM

Upon a call for heat, the valve is energized through the closure of the thermostat contacts, centrifugal switch and ignition system module circuitry. Both the main gas valve and redundant gas valve operators are energized simultaneously, opening both valves and permitting gas flow to the burners. (First at low pressure then stepping up to full input.) Upon interruption of power to the gas valve operators, both valves will close immediately.

Figure 6-13. Gas Valve Opening (Courtesy, Inter-City Products Corp.)

5. If there is voltage, disconnect at terminals and "ohm-out" the valve, checking for resistance. This should identify an open coil by showing "infinite." Readable resistance indicates that the coil or heater is all right and the problem is probably mechanical.

6. For mechanical checks:

   • Cycle the valve; if the burner fires, the valve is sticking.

   • If the valve is stuck open, the burners stay lit after the valve is turned off. *This is an extremely dangerous condition — correct it!*

   • For both electrical and mechanical failures, replace the valve immediately.

   Note — When checking the valve, a call for heating may not produce an immediate burner flame. There may be a time delay at the valve. Check the manufacturer's data when in doubt.

7

Electric/Electronic Troubleshooting and Repair

Many of the furnaces sold today employ solid-state electronic modules, which govern operation. The heart of these modules is the microprocessor. The modules are not designed to be serviced in the field, but diagnostics are built into some of them to allow the technician to determine whether or not a module has failed.

Safety

In the previous chapter we discussed safety primarily from the fuel point of view. Furnaces also rely on electricity, which can be dangerous. Safety should be a prime concern of every technician. Always consider the following points:

- Lock the fuse or circuit box when you turn them off. Someone can throw a switch and energize you.

- A jumper wire can help make a repair easier — or destroy a component and the user. Know what you're doing, or don't do it.

- Do not work on live units unless the problem requires a check on a live circuit.

- Technicians conduct electricity as effectively as any wire. The results, however, are quite different. Use good tools with properly insulated handles.

- Many times, the work environment lends itself to situations where even the smallest shock can cause a loss of balance and a fall.

- Approach every unit as if it were live. Remember, the fact that the unit isn't running doesn't necessarily mean it's without power.

Service Charts, Trouble Trees

To assist technicians, manufacturers have developed flow charts that simplify diagnoses and repairs. These charts are called by many names; two of the most popular are *service charts* or *trouble trees*. These charts provide a troubleshooting road map. Following the chart generally leads the technician to the problem by proceeding down a logical path.

Charts are necessary because of the complexity of new units, but technicians must understand how and when to use them. In addition, a technician must have a good working knowledge of schematic reading and know how to properly use a digital volt ohmmeter (DVOM).

Many technicians are afraid of servicing units because of the black box that controls the unit. This fear is needless. The control module (black box) sends power to the proper component at the proper time. Its switches open and close, and its timing circuits allow a proper sequence to take place. What the technician is looking for is power in and power out to the appropriate component at the correct time.

To diagnose a system, the technician must know what the unit is supposed to be doing and when the unit is supposed to do it. Time delays, common in today's units, are necessitated by safety and system check concerns. When servicing a unit, remember that lack of power to a component may not be a failure, but rather the clock running on a time-delay circuit.

Figures 7-1 through 7-5 are excellent examples of flow charts and how they are used in diagnostics. The figures represent the approach to troubleshooting used by Coleman Company, Inc., on its *2800 Series* furnace. There are four problems cited in the charts, with the appropriate diagnostic path. All of the figures represent possible service calls on a furnace.

In other furnaces, the wiring and some of the components will be different. In Part Two of this book, many manufacturers have contributed service and installation information for the technician's use. The next time a service call arises, use the appropriate flow chart to trouble-shoot the problem. If the proper chart is not available, adapt one of the other charts, recognizing that, while there are individual differences, the principles are the same.

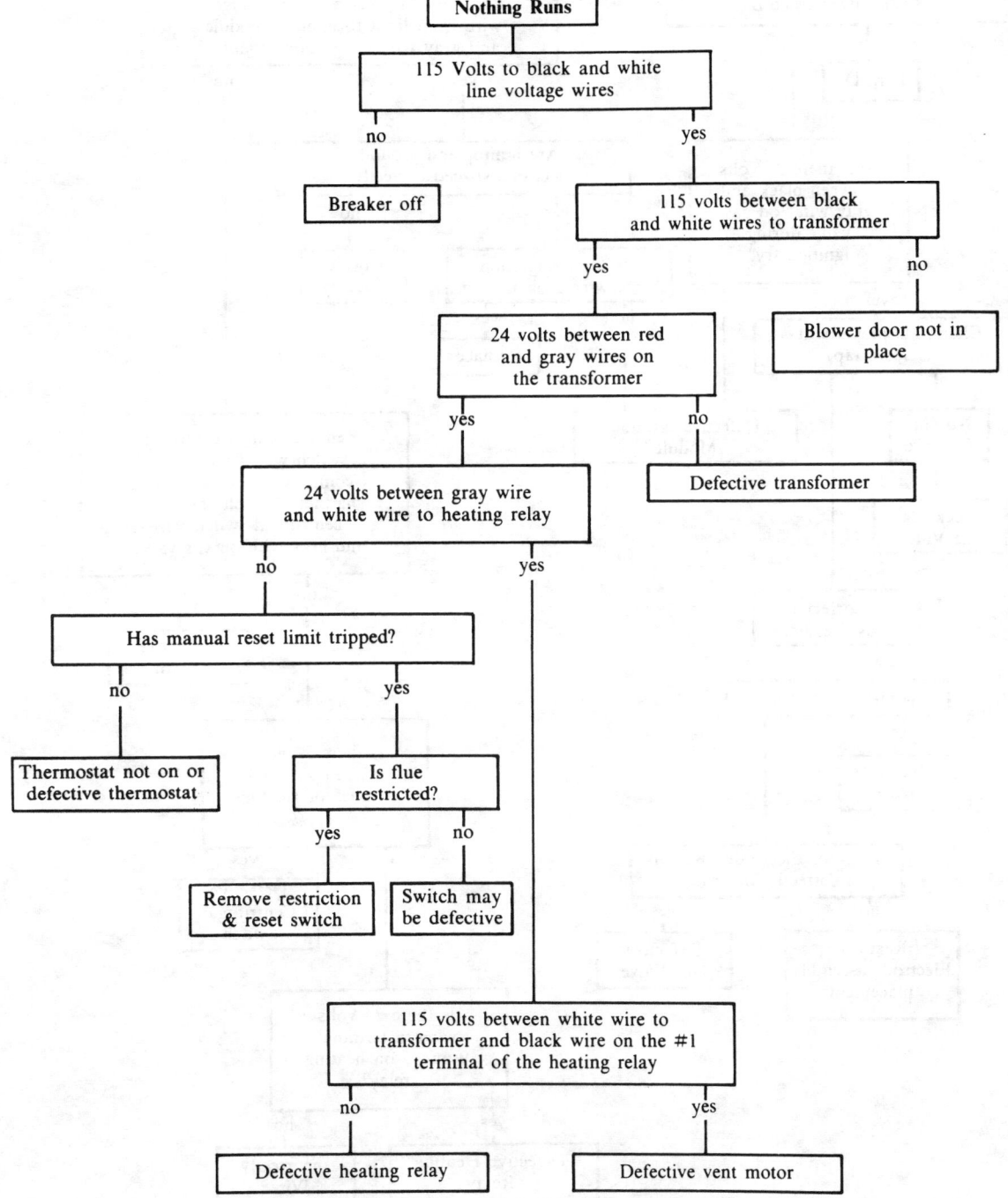

Figure 7-1. Nothing Runs (Courtesy, Coleman Company, Inc.)

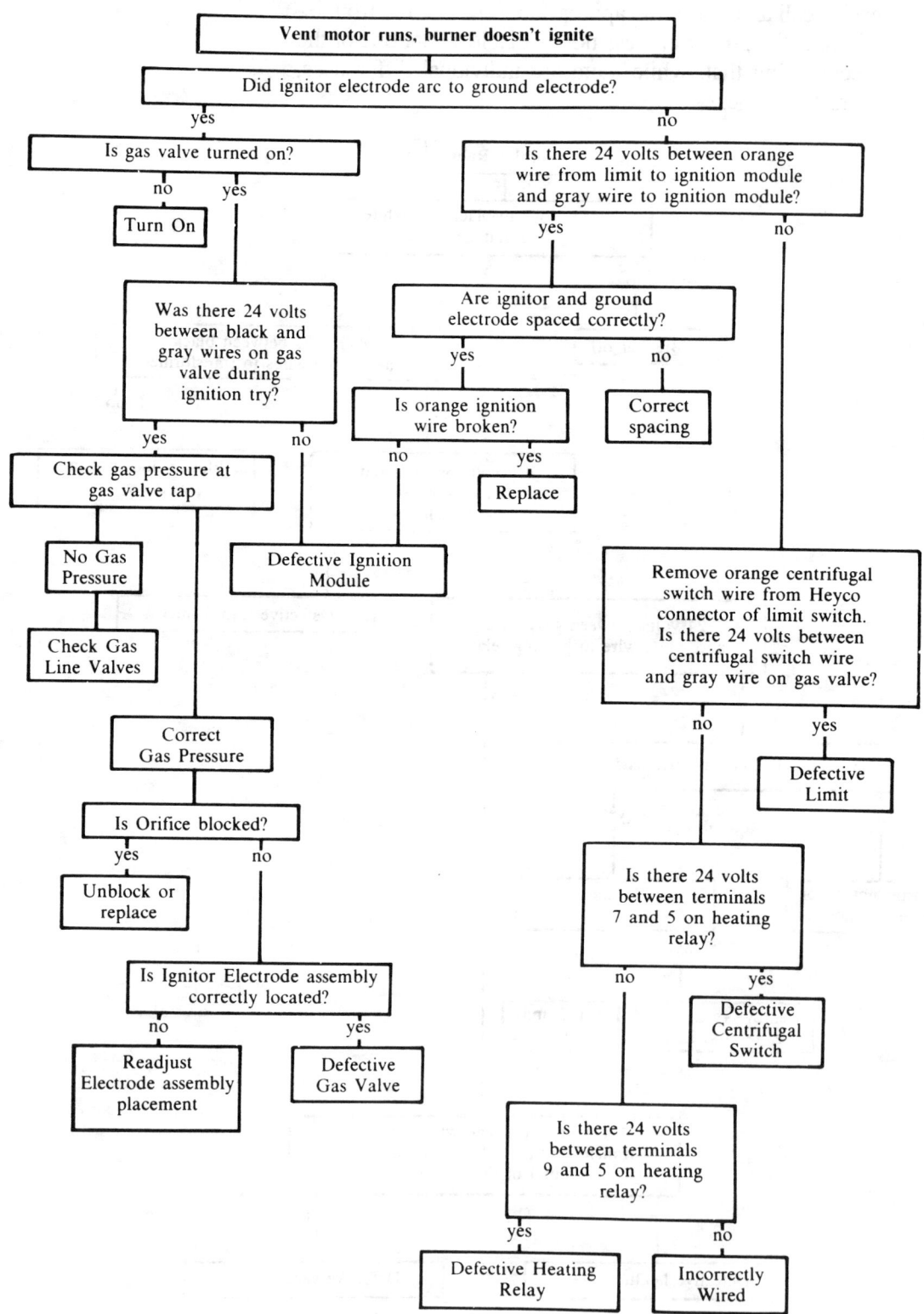

Figure 7-2. Vent Motor Runs, Burner Doesn't Ignite (Courtesy, Coleman Company, Inc.)

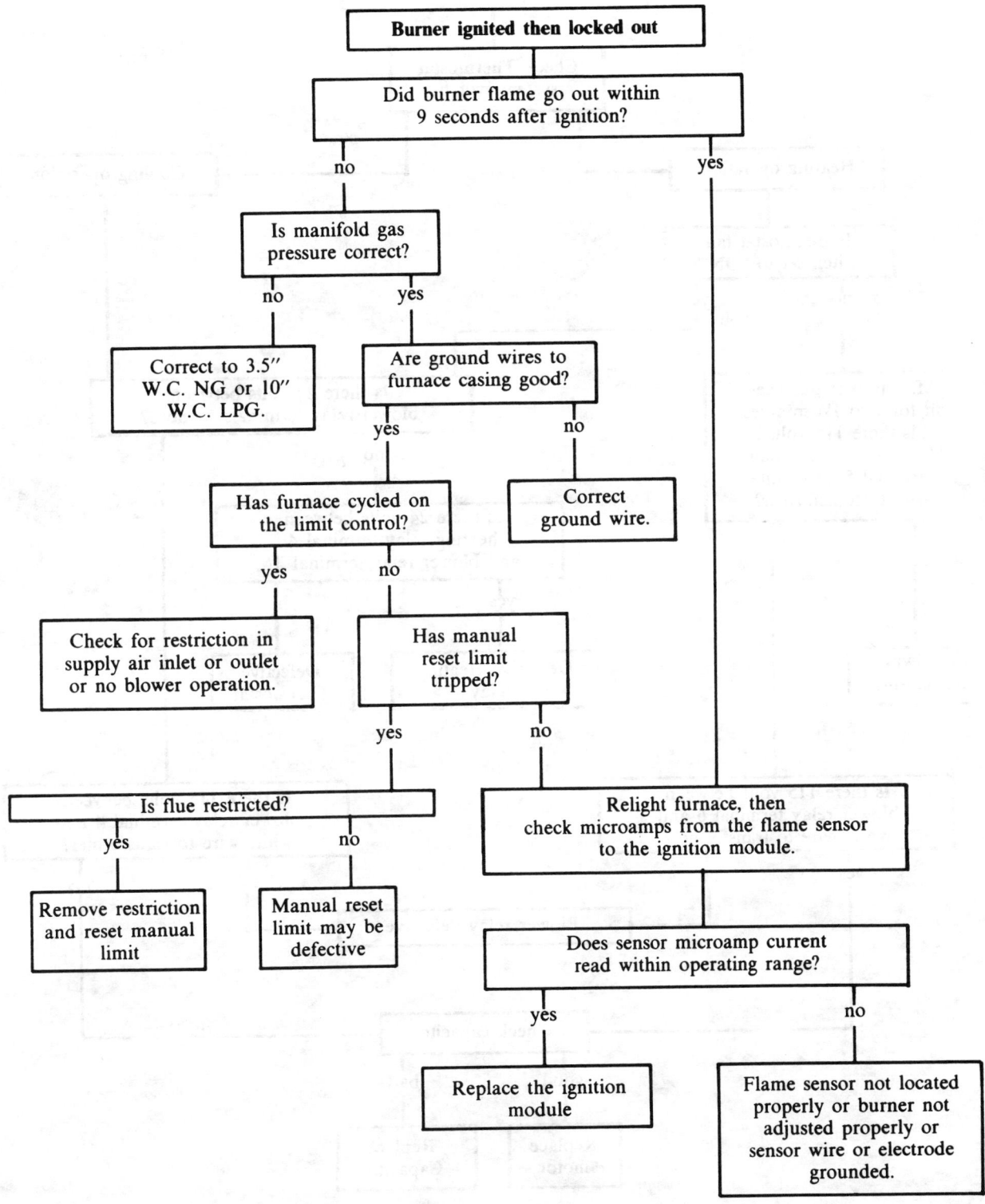

Figure 7-3. Burner Ignited Then Locked Out (Courtesy, Coleman Company, Inc.)

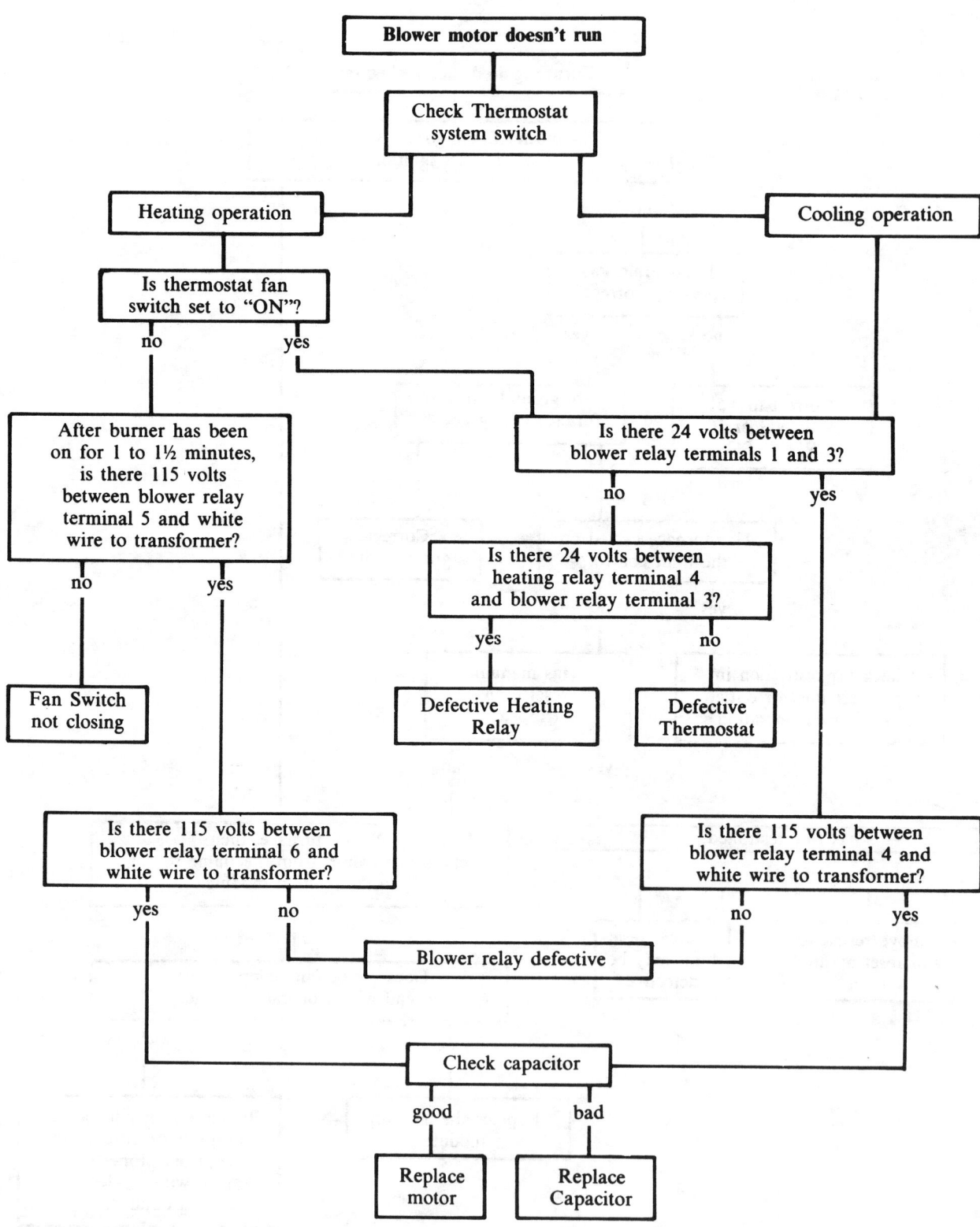

Figure 7-4. Blower Motor Doesn't Run (Courtesy, Coleman Company, Inc.)

80 SERIES
COLEMAN GAS FORCED AIR FURNACE
USE ONLY 115 VAC 60 HZ 1 PH
LESS THAN 12 AMPS   MAX. OVERCURRENT PROTECTION 15 AMPS

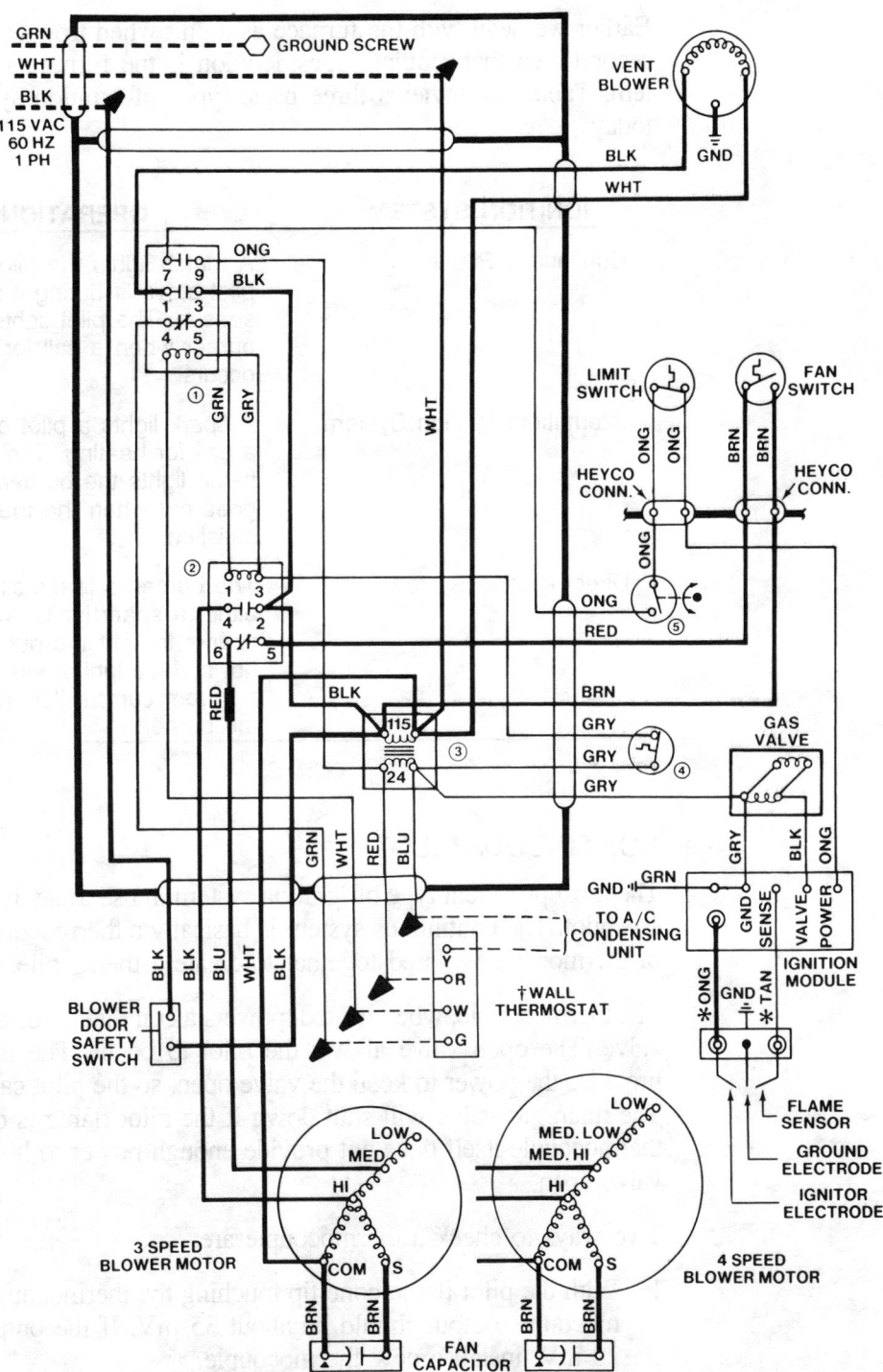

① HEATING RELAY, ② BLOWER RELAY, ③ TRANSFORMER, ④ MANUAL RESET LIMIT SWITCH, ⑤ CENTRIFUGAL SWITCH

FACTORY INTERNAL WIRING SHOWN SOLID.

Figure 7-5. Wiring Diagram (Courtesy, Coleman Company, Inc.)

Ignition Systems

Earlier we dealt with the furnace as a unit when troubleshooting. The second area that requires consideration is the furnace's ignition system. Table 7-1 reviews three basic types of ignition systems in use today.

IGNITION SYSTEM	OPERATION
Continuous Pilot	A match lights the pilot and the pilot stays lit during the heating season. The pilot lights the main burner when a call for heating occurs.
Intermittent Ignition System	A spark lights a pilot burner on a call for heating. The pilot flame lights the burner. The pilot goes out when the thermostat is satisfied.
Direct Ignition System	The burner is lit directly by either a spark ignitor which sparks to light the burner or a hot surface ignitor which gets hot from current flow and lights the burner.

Table 7-1. Ignition Types

CONTINUOUS PILOT

The most prevalent type of ignition system in use today is the continuous pilot. The continuous system is basically a thermocouple or group of thermocouples wired together to create a thermopile.

The thermocouple, when heated, powers a coil that holds open a safety valve. The open valve allows the pilot to be lit. The thermocouple provides the power to keep the valve open, so the pilot can remain lit. The main gas valve will shut down if the pilot flame is out, or if the thermocouple itself does not provide enough power to hold the safety valve open.

Two ways to check a thermocouple are:

1. With the pilot flame cone tip touching the thermocouple, the thermocouple output should be about 35 mV. If the output is below 20 mV, install a new thermocouple.

2. There is an adapter available that allows the thermocouple to be tested under load. Using the adapter, the thermocouple should test output greater than 9 mV. If not, replace the thermocouple.

INTERMITTENT-IGNITION SYSTEMS

An intermittent system uses a spark to light a pilot. Figure 7-6 is a service chart for a Honeywell S8600M spark-to-pilot ignition system. It provides an excellent reference for troubleshooting an intermittent system.

DIRECT-IGNITION SYSTEMS

With direct ignition, the burner is lit directly by either a spark or a hot-surface ignitor. On the following pages, we will discuss operational characteristics of each system.

Sequence of operation for spark-generated direct-ignition control

The following text and flow chart, Figure 7-7, are for the Arcoaire/Comfortmaker Forced-Draft Furnace GUA/GDA Series (reprinted with permission).

1. When the thermostat calls for heat, the spark generator and gas valve are energized simultaneously after pre-purge cycle. Ignition sparks are generated and the gas valve remains open for ignition trial period of up to 5 seconds maximum. If the main burner flame is not established, the system goes into "lockout" and the thermostat must be reset for another ignition attempt. This can be accomplished either by breaking line voltage or control circuit.

2. When the main burner flame is established the spark generator is put on standby and the flame-sensor monitors the main burner flame until the thermostat is satisfied.

3. If there is a loss of flame during normal burner operation, the spark generator is activated in 0.8 sec. or less and attempts to re-establish the main burner flame for up to 5.0 seconds maximum. If the flame is not re-established, the system goes into "Lock-out" and the thermostat must be reset for another ignition attempt.

4. If power interruption occurs during normal burner operation, the system goes thru a normal start-up sequence when the power is restored.

WIRING: DO NOT APPLY POWER TO THE CONTROL MODULE UNTIL WIRING IS COMPLETED AND THE ELECTRODE ASSEMBLY IS PROPERLY INSTALLED.

Attach the wiring harness by inserting the connector and pushing it firmly onto the printed circuit board. Connect the lead wires to the valve, flame-sensing electrode and 24 VAC power source. Push the high-voltage lead wire onto the terminal located on top of the control module.

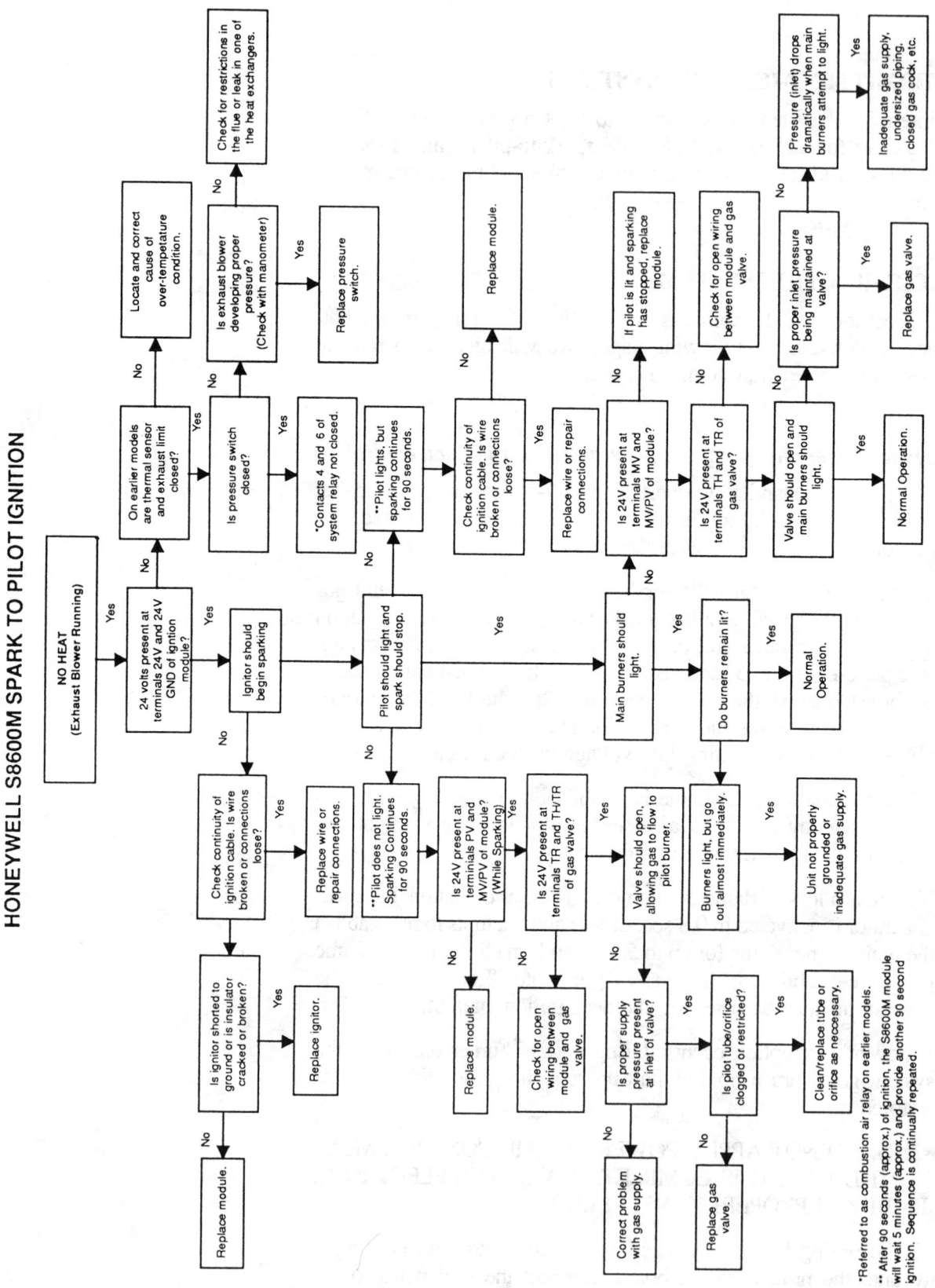

Figure 7-6. Honeywell S8600M Spark to Pilot Ignition (Courtesy, Inter-City Products Corp. from information on their Heil-Quaker products)

Turn Thermostat up to Call for Heat

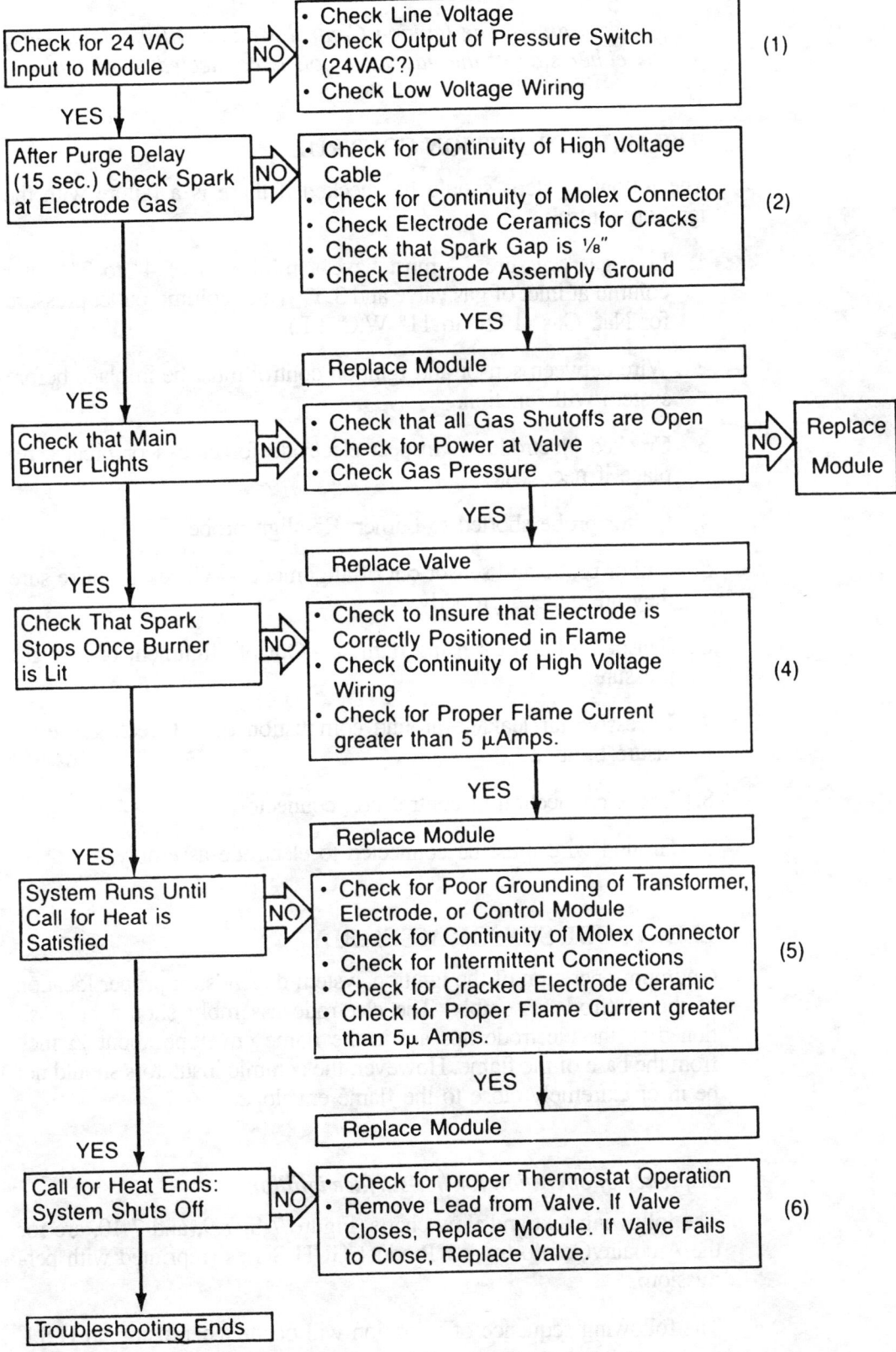

Figure 7-7. DSI Troubleshooting Table (Courtesy, Inter-City Products Corp.)

Caution — The Ignition Control Module will be damaged if either side of the gas valve coil is connected to ground.

FAILURE OF IGNITION CONTROL

The following steps should be checked if there is a failure with the ignition control:

1. Low gas pressure — must have a minimum of 4" to 7" water column at inlet of gas valve and 3.5" Water column outlet pressure for Nat. Gas (10.5" to 11" W.C. LP).

2. Wire between sensor and Ignition control must be in place before system will function.

3. Cracked electrode — Inspect electrode for cracks or breaks, replace if necessary.

4. Ignitor probe shorted to burner. Re-align probe.

5. Ignitor probe and sensor out of alignment — Check to make sure they are gapped properly.

6. Sensor Probe not sensing flame — Out of alignment or low gas pressure.

7. Excessive air leakage around combustion door Check gasket to insure tight fit.

8. Loose connection at control box connector.

9. Ground wire must be connected to electrode assembly.

ELECTRODE SENSOR ASSEMBLY

Optimum operation of the ignition system depends on proper location of the electrode assembly. The electrode assembly should be positioned so the electrode tips are in the flame envelope about ½ inch from the base of the flame. However, the ceramic insulators should not be in or extremely close to the flame envelope.

Sequence of operation for hot-surface ignition

The following text and flow charts, Figure 7-8, 7-9, and 7-10, are for the Arcoaire/Comfortmaker Product GUH Series (reprinted with permission).

The following sequence of operation will occur when using an H.S.I. system on a gas furnace:

The ignition control must have 115 Volts available to the L1 terminal to operate the ignitor.

The pre-purge mode in the ignition control will delay thirty seconds before applying power to the silicone carbide ignitor. The ignitor is allowed to warm-up for approximately 15 seconds. At the end of the warm-up time, the valve is energized.

The main burner flame must be detected within 7 seconds or the valve is de-energized and the ignitor turned off.

If there is a failure to sense flame on the first attempt, there will be a 90 second delay. The second attempt at lighting the burner flame will duplicate the first attempt with pre-purge, warm-up times and lock-out times identical to those in the first try. Failure to prove on the second attempt will allow yet another 90 second delay and one more complete ignition attempt before complete system lock-out.

Lock-out of the 50E47 control after the final ignition attempt, will require interruption of the power to the ignition control (either the 24V thermostat or line voltage to the furnace may be interrupted) for 1/20 second or longer to reset the system.

All 50E47 Controls will repeat the ignition sequence for a total of three recycles if flame is lost within the first thirty seconds of establishment. The total of ignition attempts plus the number of recycles cannot exceed three.

If flame is established for more than thirty seconds after ignition, the 50E47 will clear the recycle counter. If flame is lost after 30 seconds, there will be a 60 second wait and restart of the ignition sequence.

A momentary loss of power or flame to the sensing circuit will cause the valve to be de-energized and a delay of 60 seconds before restarting the ignition sequence. Recycles will begin and the burner will operate normally if the gas supply returns or the fault condition is corrected prior to the last ignition attempt. Otherwise the control will lock-out.

With the ignitor warmed-up, gas flow to the burners will allow ignition. The burners will stay lit as long as the thermostat calls for heat.

The ignitor is switched off at the end of the flame detection period (7 sec).

The heat exchanger heats up the combination fan and limit control. When the temperature reaches the "on" set point, the blower will be energized. There is also a safety limit combined in this control. If for some reason the blower does not start, the temperature will increase until the limit set point is reached. At this time, the 24 volt secondary of the transformer is broken and the gas valve will close.

FAILURE TO IGNITE

If 24 Volts are available to Terminals "TH" and "TR" of the ignition module, refer to the HOT SURFACE TROUBLE-SHOOTING CHART on the following page.

⚠ CAUTION:

The Ignition Control Module will be damaged if either side of the gas valve coil is connected to ground or the gas valve leads are shorted together.

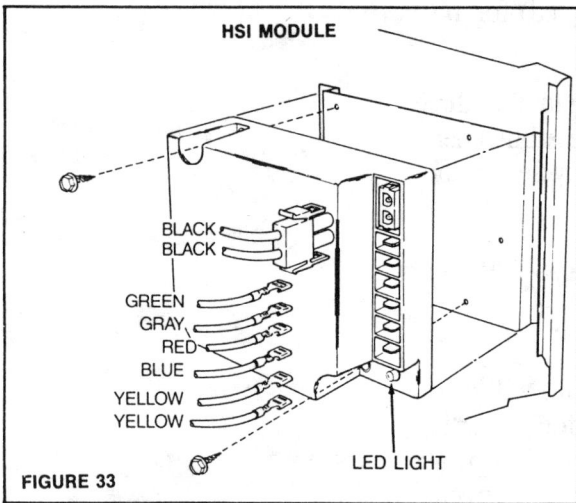

FIGURE 33

BLACK: (MODULAR PLUG)
       LINE VOLTAGE FOR IGNITOR

GREEN: CHASSIS GROUND

GREY: FLAME SENSOR

RED: 24 VOLT FROM SAFETY CIRCUITS

BLUE: TRANSFORMER COMMON (24V)

YELLOW: GAS VALVE LEADS

ORANGE: CONDENSATE LEVEL
        SAFETY SENSOR

Figure 7-8. White-Rodgers Hot-Surface Ignition System (Courtesy, Inter-City Products Corp.)

THE WHITE-RODGERS HOT SURFACE IGNITION SYSTEM

DANGER

DO NOT OMIT THIS STEP!
LINE VOLTAGE (120 VAC) COULD BE PRESENT ON THE SURFACE OF THE IGNITOR, IF THE SYSTEM IS NOT CORRECTLY WIRED. SUCH VOLTAGE CAN CAUSE SERIOUS INJURY OR DEATH.

1. Disconnect electric power to system at main fuse or circuit breaker.

2. Remove burner access shield (if necessary) to gain access to the ignitor. See Furnace Installation Instructions.

3. Connect an AC voltmeter from the surface of the ignitor to chassis ground, and then reconnect electric power to system.

4. If voltage exists between the surface of the ignitor and the chassis ground, the main power supply lines are improperly connected to the furnace.

   Reverse incoming line voltage leads.

In an effort to minimize the time required to troubleshoot this system:

1. Turn off the gas supply at the main gas valve.

2. Disconnect electric power to system at main fuse or circuit breaker if connected.

3. Visually inspect equipment for apparent damage. Check wiring for loose connections.

4. With burner access shield removed (if necessary), inspect ignitor for visible cracking or scale deposits and flame sensor for position or deposits shorting sensor to burner.

After performing the above inspections, restore gas supply and electric power to equipment. Close thermostat contacts to cycle the system. If a "no heat" condition persists, the three visual indicators (below) will help determine if the system is operating properly.

1. The ignitor will warm up and glow red.

2. The main burner flame will ignite.

3. The main burner flame will continue to burn after the ignitor is turned off.

Troubleshooting the system consists of checking for these three visual indications. The attached charts define the proper action if any of the indications do not occur.

A diamond on the chart encloses a question that must be answered yes or no. A rectangle on the chart encloses a statement or instruction.

FIRST VISUAL CHECK

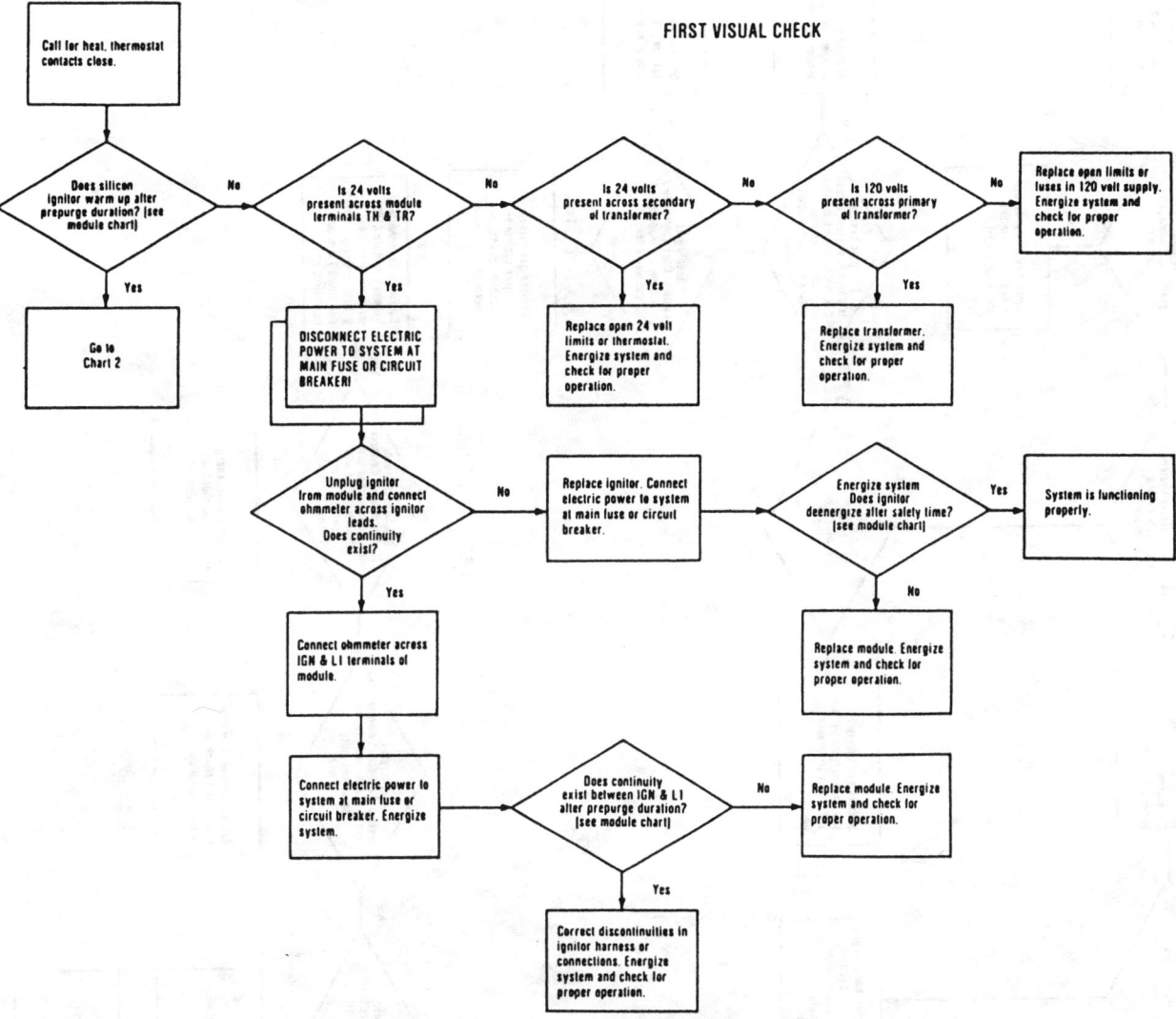

Figure 7-9. Operation Flow Chart, First Visual Check (Courtesy, Inter-City Products Corp.)

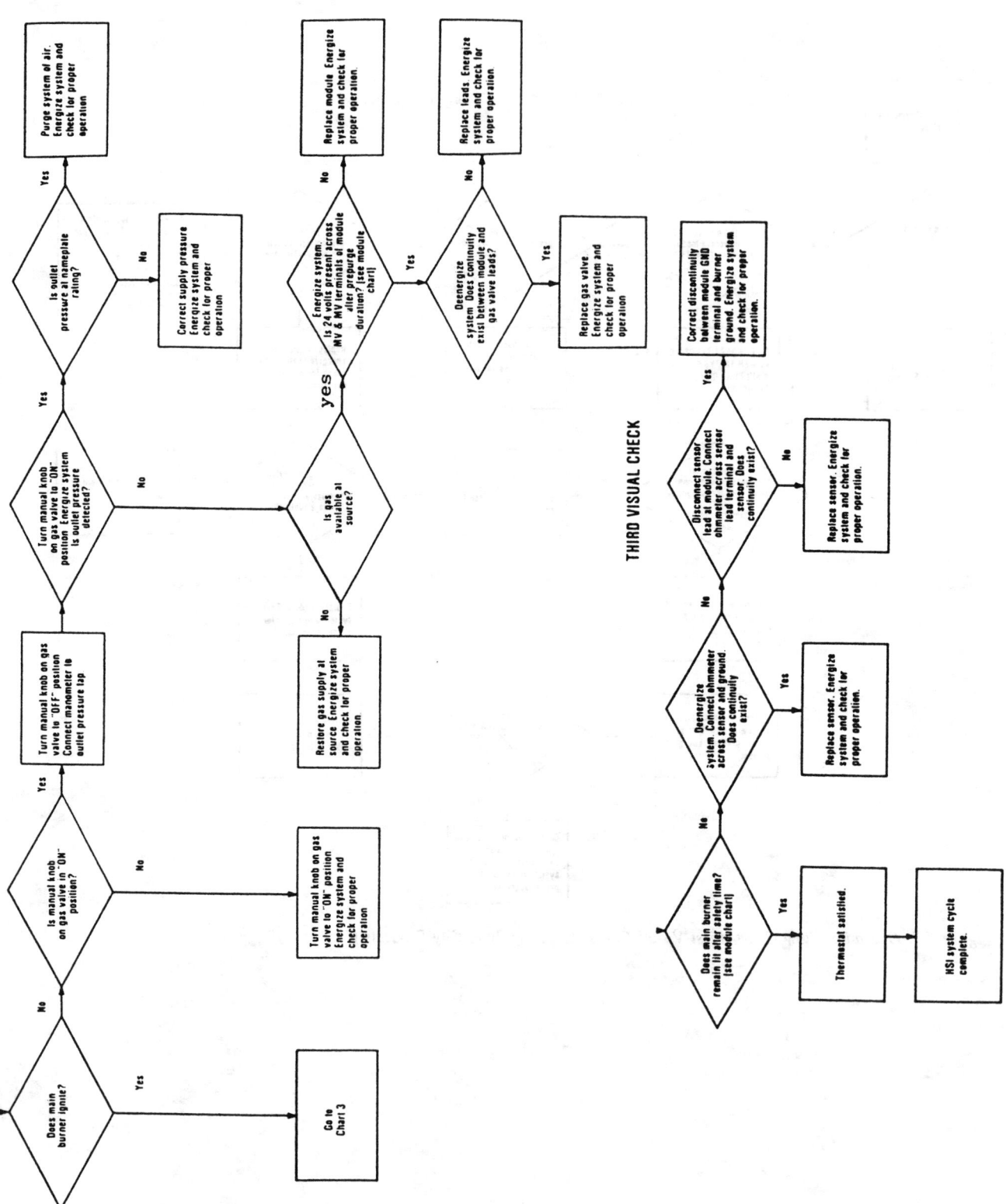

Figure 7-10. Operation Flow Chart, Second Visual Check (Courtesy, Inter-City Products Corp.)

When the thermostat is satisfied, the gas valve will close, shutting down the burners. At this time, the blower will continue to operate until the temperature cools down below the "off" set point of the fan control. When this point is reached, the blower will shut off.

Now the furnace is ready to repeat the same cycle when the thermostat calls for heat.

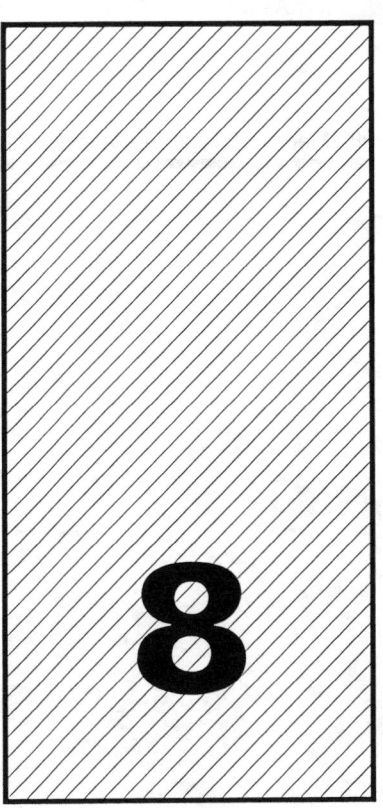

Venting

No book on medium- to high-efficiency furnaces would be complete without looking at venting. The introduction of new furnaces with forced or induced draft has changed the requirements for properly venting a system.

Proper venting is important, because it causes the products of combustion to exit the system. Venting also affects the quantity of combustion supply air that enters the furnace for combustion. Furnaces can be vented in two different configurations:

- *Atmospheric venting*, which utilizes a draft hood connected to a heat exchanger. The products of combustion flow through at what is essentially atmospheric pressure. This arrangement also allows constant pressure to be maintained in the combustion chamber. Safety controls now exist to shut the furnace down in case the vent system becomes blocked.

- *Power venting*, which uses a small blower to force or induce movement of the flue gases through the furnace, generally under positive pressure. Power venting was one of the first changes made to improve furnace efficiency.

Atmospheric pressure, building pressure, and gas temperature all have an effect on the air and flue gas flow in a unit. Proper air flow for combustion depends on the pressure difference between the air entering the furnace and the flue gas leaving the furnace. If a unit is improperly vented, overall air flow and operation are affected.

To understand the importance of venting, a brief review of basics is in order. In older-style furnaces (60%-plus efficiency), the products of combustion (vent gases) are vented to the atmosphere. The force moving these gases is the difference in density (in this case, defined as weight) between the vent gases and the surrounding air. As the temperature difference between the surrounding air and the vent gases increases, the force causing gas movement increases.

For our purposes, older-style furnaces can be considered natural-draft or atmospheric-venting units. In atmospheric venting, there is no positive pressure. Older-style furnaces employ a draft hood, which utilizes the temperature difference between the surrounding air and the vent gases to allow vent gas movement.

With the introduction of higher-efficiency furnaces, blower-moved gas has become standard. Blowers are configured as *forced* or *induced*. **Forced-draft units** use a blower to move air into the combustion chamber and allow byproducts to flow out through the venting system. **Induced-draft units** utilize a blower to pull air through the combustion area and move the gases out the vent. The blower approach creates positive pressure, resulting in improved furnace efficiency. *Note — Increased efficiency causes furnaces to operate with slightly cooler flue gases than-older style furnaces. As furnace efficiencies reach higher numbers, flue gases become even cooler.*

Efficiency is a relative value. When furnace efficiency increases from 68% to 78%, it is a higher-efficiency furnace than its predecessor but not high efficiency per se. Early high-efficiency units operate above dewpoint, so the condensation process doesn't take place. The lack of condensation causes the latent heat present in the gases to be lost to the atmosphere, rather than used in space heating.

Furnaces operating at higher-efficiency levels come in different configurations. *Condensing furnaces* (furnaces employing a medium such as glycol to circulate through the system) and *pulse furnaces* (which employ only outside air, brought into an enclosed chamber and ignited initially by a spark) are both approaches used to achieve the highest-possible efficiency. Both are designed to obtain as much heat as possible for the space before exhausting the byproducts of combustion.

Capturing and using latent heat to increase efficiency has resulted in 90%-plus-efficiency furnaces. These allow condensation to occur in a secondary heat exchanger or downstream of the primary heat-exchange process. Allowing condensation to occur means latent heat, which normally would be vented to the atmosphere, is actually sent to the space.

Dewpoint and Condensation

DEWPOINT

Water vapor is a combustion byproduct when natural gas is burned. More than 2 lb of water vapor is produced for every pound of natural gas burned. This moisture, which is in a vapor state, contains heat. If we can change the vapor to liquid, usable heat is liberated. Dewpoint is a temperature at which vapor begins to condense into liquid.

Examples of dewpoint are everywhere. Dew on a rooftop is water vapor present in the air that has condensed into liquid because the rooftop was at its dewpoint. In flue gases, the actual dewpoint temperature varies, depending on the composition of the vent gas and the amount of excess combustion and dilution air.

Excess air causes dewpoint temperatures to be lowered. Lowering the dewpoint temperature in the flue system creates the potential for condensation to occur. The dewpoint temperature for undiluted flue gas is around 140°F. In diluted gas, the temperature will drop to about 105°F.

Dewpoint can create concerns about the temperature of the flue gas leaving the flue pipe. When water vapor in flue gas comes into contact with the wall of a flue whose temperature is below dewpoint and the flue gas is at its condensing temperature, condensation occurs. Condensation can also occur if flue gas traveling from the furnace to the atmosphere loses too much heat. For example, if the flue gas lost excessive heat due to poor piping practice, the flue pipe's wall at the end of the flue run could be below the flue gas dewpoint. When this occurs, condensation may follow.

A second concern arises because the excessively cooled gas may lose its venting power due to draft reduction. If the manufacturer requires a double-wall flue and single wall was used incorrectly, watch out; double-wall vent pipe slows down vent gas heat loss, thereby eliminating condensation and draft-reduction concerns. It is important that manufacturers' data be used, regardless of what type of piping exists in the original installation.

CONDENSATION

Condensation occurs when a vapor changes state to liquid and, in the process, gives up heat. This can be a problem when water vapor condenses in the flue pipe and begins to corrode the pipe itself and furnace components. If the temperature within the flue pipe walls is below the dewpoint, condensation will occur.

Condensation must be controlled. In a medium-efficiency furnace, condensation occurs because the flue vent pipe has lost too much heat and cooled below the vent gas condensing temperature. This situation can occur when the system is improperly vented.

The higher efficiencies achieved by some medium-efficiency furnaces are created by efficient design and forced-air movement — not by release of latent heat. Condensation and latent heat capture are not functions of all higher-efficiency furnaces. Know what type of furnace you are installing before you vent the furnace.

Heat released during condensation can be captured and utilized in the space. In a 90%-plus furnace, condensation is desirable and controlled. It occurs because we are trying to move the latent heat present in the water vapor into the space.

VENTING CATEGORIES

Furnaces are classified into two separate ranges for purposes of venting discussion: medium and high efficiency. When higher-efficiency furnaces were introduced, categorizing became necessary.

Chapter Two describes the categories from an operational standpoint. Table 8-1 is more technical in nature. It is based on ANSI Standard Z21.47A-1990 and sets the venting standards for temperature and pressures required by proper venting.

Category I	Furnace has a non-positive vent pressure and operates with a vent gas temperature at least 140°F above its dew point.
Category II	Furnace has a non-positive vent pressure and operates with a vent gas temperature less than 140°F above its dew point.
Category III	Furnace has a positive vent pressure and operates with a vent gas temperature at least 140°F above its dew point.
Category IV	Furnace has a positive vent pressure and operates with a vent gas temperature less than 140°F above its dew point.

Table 8-1. ANSI Standard Venting Classifications

Medium-Efficiency Furnace Venting

Medium-efficiency furnaces (80%-plus) rely on forced-air movement either from the inlet (forced-air) or outlet (induced-air) side of the furnace. Industry has not totally agreed on the definition of medium efficiency. A furnace operating in the 76%-efficiency range with a draft hood or 78% with an induced blower, can still be marketed as a Category I furnace. The presence of a blower does not automatically move the furnace up in category. Check the furnace's category. This is identified by the manufacturer and listed on a plate attached to the unit. The category sets the basic standard for venting the system.

Condensation is a concern for Category I, induced-draft furnaces, and it must be considered when retrofitting. Always consider the following points:

- The blower may not be powerful enough to create positive pressure in the vent. The name plate venting requirement for the unit is still Category I, so install the furnace as such.

- There is less dilution air, if any. Excess combustion air also is reduced. A lower volume of gas at a cooler temperature increases chances of condensation.

- Type B vent piping is allowed for vertical venting; however, there is less volume of flue gas to move. Double-wall venting should be considered for *at least* a portion of the run. A rule of thumb is, no more than 5 ft of single wall on a fan-induced, Category I furnace.

- Type B venting is not airtight; avoid taping the joints, since allowing even a little dilution air through the joints impedes condensation.

- Sizing is critical. Incorrect sizing, particularly oversizing, leads to condensation.

More lower-, but still higher-than-original-efficiency furnaces, will be retrofitted in the future. Read and follow the manufacturer's information. Heat load, insulation values, capacity, length of run, type of piping, temperature rise, air flow, and vent-size tables are critical with today's equipment. It isn't as complicated as it sounds. Just follow the manufacturer's recommendations and conform to local codes.

High-Efficiency Furnace Venting (Categories II, III, IV)

Categories II, III, and IV deal with high-efficiency furnaces. (As indicated in an earlier chapter, Category II furnaces are not currently being manufactured.)

Category III. In this category, the furnace has positive vent pressure and operates with a vent gas temperature of at least 140°F above its dewpoint. Category III is primarily used by mid-efficiency furnaces operating with a positive vent pressure. Flue gas temperatures are greater than 275°F. Always consider the following points:

- Vent connections must be leak tight.

- Furnaces may be, in special circumstances, vented with another appliance.

- Type B vent piping is not allowed.

- Can be horizontally vented when approved by manufacturer and codes.

- Can use thermoplastic vent piping, such as polytherimid.

- Uses a condensate drain fitting to ensure that possible condensation in the vent does not reach furnace.

Category IV. In this category, the furnace has positive vent pressure and operates with a vent gas temperature less than 140°F above its dewpoint. Category IV is primarily used by high-efficiency condensing furnaces with positive vent pressure. Flue gas temperatures are under 275°F. Always consider the following points:

- Vent connections must be leak tight.

- Furnace may never be vented with another appliance.

- Type B vent piping is not allowed.

- Generally uses a special venting kit.

- Can be horizontally vented.

- Can use thermoplastic vent piping, such as PVC or CPVC.

- Can use outside air for combustion air that is direct vented. No inside air is used.

- A piped condensate drain fitting is installed to ensure draining of all condensation that forms in the heat exchanger.

In a high-efficiency furnace (90%-plus), condensation generally occurs in the secondary heat exchanger. Combustion products are purposely cooled below dewpoint to promote latent heat release. A high-efficiency furnace must dispose of condensate as well as combustion byproducts. Condensate must be disposed of by meeting code requirements. You cannot just allow condensate to run freely across the nearest sidewalk.

Part Two of this book contains service manuals that cover, in detail, the specific venting practices required by manufacturers in order for their units to operate properly. It is important for technicians to understand that not all high-efficiency furnaces are equal; nor is lack of equality an indicator of poorer quality. Different furnaces operate at different levels, because the manufacturer felt that a particular design was necessary for the marketplace. Don't carry the misconception that if the furnace is high efficiency, it automatically uses PVC pipe for venting.

The high-efficiency market is new; the lines between medium- and high-efficiency equipment are still not crystal clear. In the end, a unit that operates at a higher efficiency than the unit it replaces is a high-efficiency unit from the user's standpoint.

Part Two

Manufacturers' Service Data

This section contains service and installation manuals on different furnaces in the medium- and high-efficiency operating range.

These unedited manuals are furnished by the manufacturers and are reproduced as originally published by them. The manuals are arranged in alphabetical order by brand name; no order of preference is expressed, implied, or intended by the author or publisher.

Not every unit manufactured today is included in this appendix. The author attempted to show a sampling of furnaces in service. Some manufacturers chose not to allow publication of their manuals; others desired that only certain models be included in this book. If the unit you are installing isn't covered in this book, **obtain a manufacturer's service manual**.

The author and publisher wish to thank the manufacturers for their cooperation in granting permission to reprint the manuals.

American Standard

Super Efficiency Gas-fired Condensing Furnaces
TUC—B; TDC—B

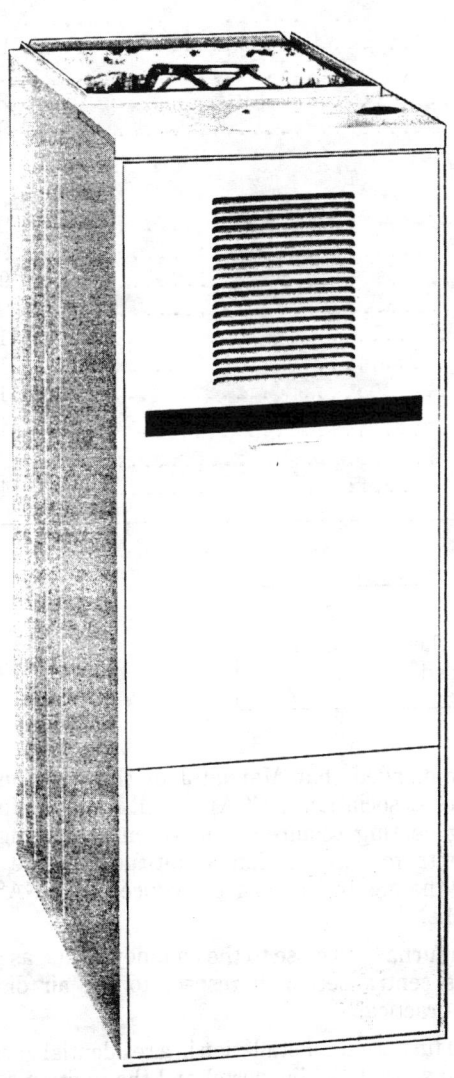

FOR YOUR SAFETY

Do not store or use gasoline or other flammable vapors and liquids in the vicinity of this or any other appliance.

WARNING
FOR YOUR SAFETY

If you smell gas:
1. Open windows
2. Don't touch electrical switches
3. Extinguish any open flame
4. Immediately call your gas supplier.

IMPORTANT MESSAGE TO OWNER: CONGRATULATIONS on your purchase of this warm air furnace. Your warm air furnace is the product of years of engineering and modern designing techniques. Proper care and periodic attention to this unit will result in years of troublefree-service and comfort for you. It is Suggested That An Annual Check-up Of This Heating Unit Be Made By A Qualified Service Technician. These instructions should be carefully read and kept for future reference. It is suggested that this booklet be affixed on or adjacent to the furnace in the envelope provided.

CONTENTS

FURNACE EFFICIENCY

The furnace efficiency rating is a product thermal efficiency rating determined independent of any installed system.

GENERAL INFORMATION NOTES

The TUC and TDC series of central furnaces has been classified as CATEGORY IV furnaces in accordance with ANSI Z21.47-1987 standards. Category IV furnaces operate with a positive vent pressure and with a vent gas temperature less than 140°F above its dewpoint. These conditions require special venting systems, which must be gastight and watertight. The special venting system requirements are outlined in the INSTALLER'S GUIDE (Pub. No. 18-CH15D2), Venting and Condensate Drain Instructions for these Gas Furnaces.

GENERAL INSTALLATION INSTRUCTIONS

The manufacturer assumes no responsibility for equipment installed in violation of any code or regulation. The Home Owner's Manual will instruct you in the service and care of your unit. Have your installer go over the operational portion of that manual with you so that you fully understand your warm air furnace and how it is intended to function.

It is recommended that Manual J of the Air Conditioning Contractors Association (ACCA) or A.R.I. 230 be followed in estimating heating requirements. When estimating heating requirements for installation at altitudes above 2000 ft. remember the gas input must be reduced (See GAS INPUT ADJUSTMENT).

Locate the furnace as close to the chimney or flue as permissible and as centralized with respect to the air distribution system as practical.

A gas-fired furnace for installation in a residential garage must be installed so that the burner(s) and the ignition source are located not less than 18 inches above the floor, and that the furnace be located or protected to avoid physical damage by vehicles.

Take into consideration all clearances that are required

A. Clearance from combustible materials and for service accessibility (See INSTALLATION CLEARANCES)

B. Clearance for combustion and ventilation (see AIR FOR COMBUSTION AND VENTILATION)

C. Clearance for attaching duct work (See DUCT CONNECTIONS)

D. Clearance for proper installation of venting and for running condensate drain to the floor drain (See VENTING and CONDENSATE DRAIN INSTRUCTION, Pub. No. 18-CH15D2.)

E. Clearances for any external filter racks or sub-bases to be used.

F. Clearances for gas piping (See GAS PIPING).

The installation must conform with local gas company regulations, national and local electrical codes, local building, plumbing or, in the absence of local codes, with the latest National Fuel Gas Code ANSI Standard Z223.1-1988.

INSPECTION AND HANDLING

Material in this shipment has been inspected at the factory and released to transportation agency without known damage. Inspect exterior of carton for evidence of rough handling in shipment. Unpack carefully after moving equipment to approximate location. If damage to contents is found, report damage immediately to delivering agency.

ORDERING PARTS

When ordering shortages or damaged parts, or when ordering repair parts, give the complete unit model and serial number as stamped on the unit nameplate. Order all parts through your local contractor.

INSTALLATION, CLEARANCES AND SUPPORT

UPFLOW models are A.G.A. design certified for installation on combustible floors and for closet installation.

For installation on combustible flooring, the furnace shall not be installed directly on carpeting, tile, or other combustible material other than wood flooring.

DOWNFLOW (counterflow) models are A.G.A. design certified for installation on noncombustible floors and for closet installation. For installation on combustible floor use combustible floor subbase. Combustible floor subbase must also be used when cooling coil and jacket are applied in installation on combustible floor.

The following lists minimum permissible clearances to combustible material, for all models of recuperative gas furnaces.

BACK	FRONT	TOP OF PLENUM	SIDE
0″	6″	1″	0″

ACCESSIBILITY CLEARANCES MUST TAKE PRECEDENCE OVER FIRE PROTECTION CLEARANCES.

Minimum accessibility clearances are as follows:

1. 24 inches at side where access is required for servicing and cleaning.

2. 18 inches at a side where passage is required to another side requiring servicing or cleaning, or at a side requiring inspection or replacement of the flue connector.

3. In utility room installations, the door shall be wide enough to allow the largest furnace part to enter or to permit the replacement of another appliance, such as a gas water heater.

4. Where external filter racks are used, clearance must be provided for removal of filters.

The furnace must be installed on a solid surface and must be level front-to-back and side-to-side.

If this furnace is installed in an attic or other insulated space it must be kept free and clear of all insulating materials as some insulating materials are combustible.

If additional insulation is added after the furnace is installed, the area around the furnace must be inspected to insure it is free and clear of insulation.

AIR FOR COMBUSTION AND VENTILATION

OUTDOOR AIR IS RECOMMENDED.

When installing the TUC—B and TDC—B gas furnace, it is recommended to use the outdoor combustion air option. The use of indoor air for most applications is acceptable, unless there is the presence of corrosive chemicals or contamination in the air supply. Certain types of installations will require the use of outdoor air for combustion. For further information regarding the combustion air option, refer to Installer's Guide (Pub. No. 18-CH15D2), Venting and Condensate Drain Instructions for these Gas Furnaces.

When the furnace is installed with the indoor air option, the following guidelines must be followed:

1. If the furnace is installed in a confined space, the necessary combustion air must come from the outdoors by way of attic, crawl space, air duct, or direct opening.

2. **There must be no exposure to the corrosive chemicals and substances identified in the Venting and Condensate Drain Installer's Guide.**

3. All provisions for indoor combustion air must meet the requirements for combustion air supply indicated in the National Fuel Gas Code, ANSI Z223.1-1988, section 5.3, and/ or any applicable local codes.

AIR FOR COMBUSTION AND VENTILATION:
(Extracted from National Fuel Gas Code ANSI Z223.1-1988)

5.3 AIR FOR COMBUSTION AND VENTILATION

5.3.1 GENERAL
a. The provisions of 5.3 apply to gas utilization equipment installed in buildings and which require air for combustion, ventilation and dilution of flue gases from within the building. They do not apply to (1) direct vent equipment which is constructed and installed so that all air for combustion is obtained from the outside atmosphere and all flue gases are discharged to the outside atmosphere, or (2) enclosed furnaces which incorporate an integral total enclosure and use only outside air for combustion and dilution of flue gases.

b. Equipment shall be installed in a location in which the facilities for ventilation permit satisfactory combustion of gas, proper venting and the maintenance of ambient temperature at safe limits under normal conditions of use. Equipment shall be located so as not to interfere with proper circulation of air. When normal infiltration does not provide the necessary air, outside air shall be introduced.

c. In addition to air needed for combustion, process air shall be provided as required for: cooling of equipment or material, controlling dew point, heating, drying, oxidation or dilution, safety exhaust, odor control.

d. In addition to air needed for combustion, air shall be supplied for ventilation, including all air required for comfort and proper working conditions for all personnel.

5.3.3 EQUIPMENT LOCATED IN CONFINED SPACES:

All air from outdoors: The confined space shall be provided with two permanent openings, one commencing within 12 inches of the top and one commencing within 12 inches of the bottom of the enclosure. The openings shall communicate directly, or by ducts, with the outdoors or spaces (crawl or attic) that freely communicate with the outdoors.

1. When directly communicating with the outdoors, each opening shall have a minimum free area of 1 square inch per 4,000 Btu per hour of total input rating of all equipment in the enclosure.

2. When communicating with the outdoors through vertical ducts, each opening shall have a minimum free area of 1 square inch per 4,000 Btu per hour of the total input rating of all equipment in the enclosure.

3. When communicating with the outdoors through horizontal ducts, each opening shall have a minimum free area of 1 square inch per 2,000 Btu per hour of total input rating of all equipment in the enclosure.

4. When ducts are used, they shall be of the same cross sectional area as the free area of the openings to which they connect. The minimum dimension of rectangular air ducts shall be not less than 3 inches.

5.3.4 SPECIALLY ENGINEERED INSTALLATIONS:

The requirements of 5.3.3 shall not necessarily govern when special engineering, approved by the authority having jurisdiction, provides an adequate supply of air for combustion, ventilation, and dilution of flue gases.

5.3.5 LOUVERS AND GRILLES:

In calculating free area in 5.3.3, consideration shall be given to the blocking effect of louvers, grilles or screens protecting openings. Screens used shall not be smaller than 1/4 inch mesh. If the free area through a design of louver or grille is known. It should be used in calculating the size opening required to provide the free area specified. If the design and free area is not known, it may be assumed that wood louvers will have 20-25 percent free area and metal louvers and grilles will have 60-75 percent free area. Louvers and grilles shall be fixed in the open position or interlocked with the equipment so that they are opened automatically during equipment operation.

5.3.6 SPECIAL CONDITIONS CREATED BY MECHANICAL EXHAUSTING OR FIREPLACES:

Operation of exhaust fans, ventilation systems, clothes dryers, or fireplaces may create conditions requiring special attention to avoid unsatisfactory operation of installed gas utilization equipment.

In especially cold climates it is possible that use of openings of the sizes specified above in outside walls may result in over ventilation of excessive cooling of the utility room or other confined space furnace locations. Under certain conditions, this might introduce the hazard of freezing water lines or water heater storage tanks. In this case, a supply air duct should be used to heat this space.

Air openings in the casing front, return grille and warm air diffusers or registers must be kept free of obstructions restricting airflow.

COIL ENCLOSURE INSTALLATION (UPFLOW UNITS ONLY)

1. Assemble coil enclosure per Installation Instruction 18-AH10D1.

2. Flatten front flange on TUC — B Furnace.

3. Cut side flanges on TUC — B Furnace 1-3/4″ from the front of flange. Then fold 1-3/4″ flange flat, see below.

4. Coil enclosure will now fit snugly on the TUC — B Gas Furnace.

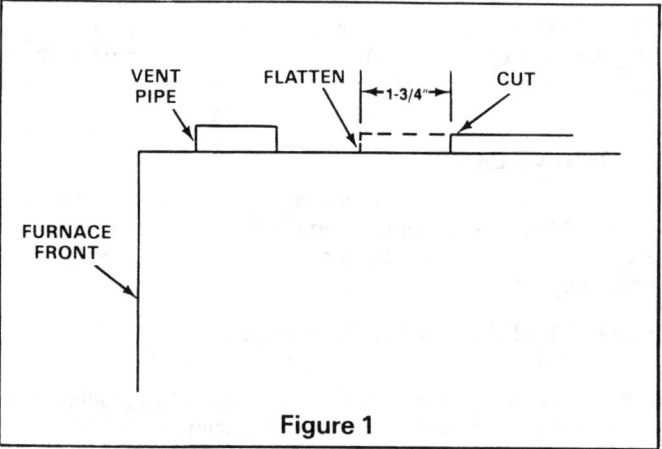

Figure 1

This procedure is not required for the TX series furnace coils, as those coils will fit onto the furnace without additional modification.

TABLE 1
MINIMUM EXTERNAL FILTER AREA (SPECIFIED IN SQUARE INCHES)

Filter Type	Air Temp. Rise °F.	Input (BTUH)				
		40	60	80	100	120
300 FPM Low Velocity	25-55	333	—	—	—	—
	30-60	—	444	—	—	—
	35-65	—	—	533	—	—
	40-70	—	—	—	606	—
	50-80	—	—	—	—	615
500 FPM High Velocity	25-55	200	—	—	—	—
	30-60	—	267	—	—	—
	35-65	—	—	320	—	—
	40-70	—	—	—	364	—
	50-80	—	—	—	—	369
520 FPM	25-55	192	—	—	—	—
	30-60	—	256	—	—	—
	35-65	—	—	308	—	—
	40-70	—	—	—	350	—
	50-80	—	—	—	—	355

OUTSIDE AIR SUPPLY INSTRUCTIONS FOR UPFLOW AND DOWNFLOW FURNACES

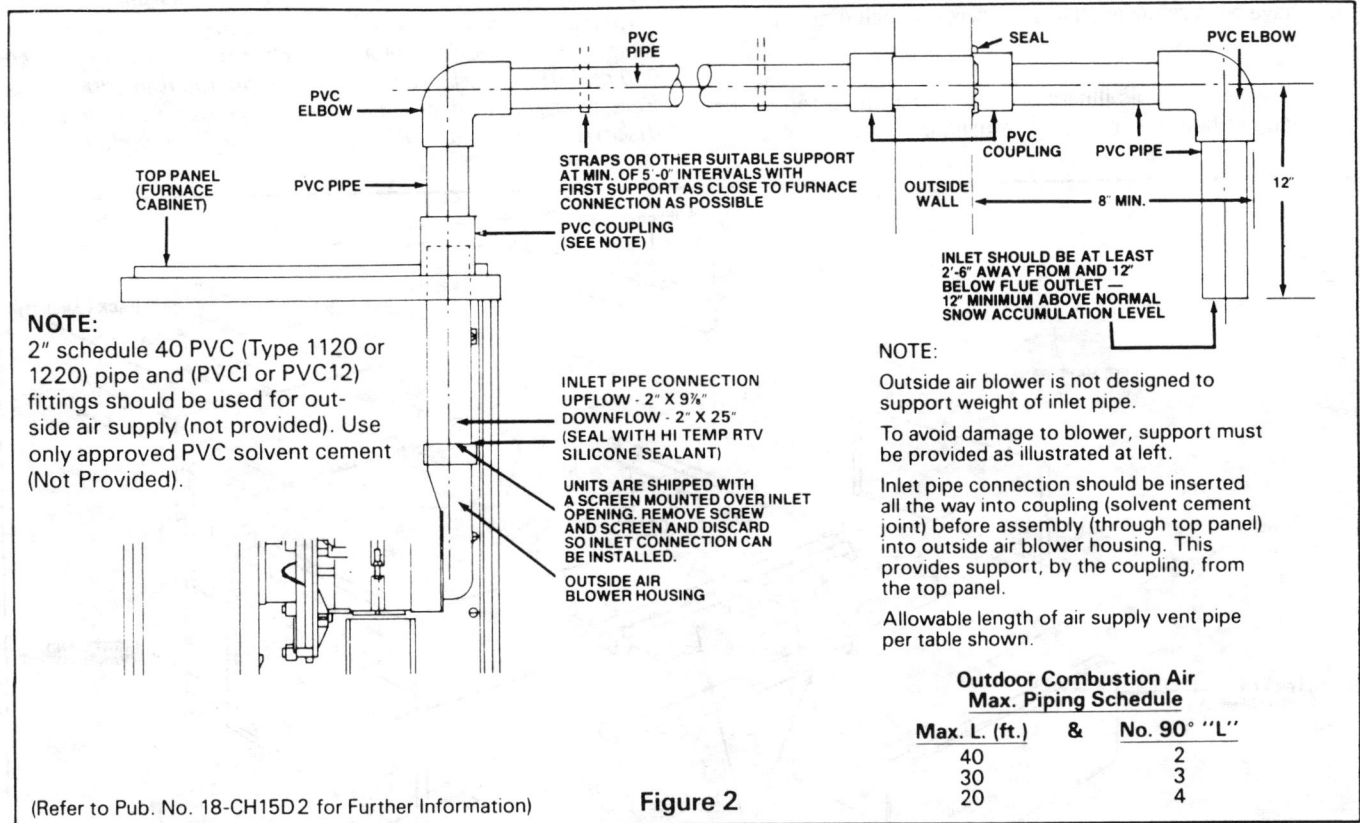

NOTE:
2" schedule 40 PVC (Type 1120 or 1220) pipe and (PVCI or PVC12) fittings should be used for outside air supply (not provided). Use only approved PVC solvent cement (Not Provided).

PVC PIPE
PVC ELBOW
TOP PANEL (FURNACE CABINET)
PVC PIPE
PVC COUPLING (SEE NOTE)
STRAPS OR OTHER SUITABLE SUPPORT AT MIN. OF 5 -0" INTERVALS WITH FIRST SUPPORT AS CLOSE TO FURNACE CONNECTION AS POSSIBLE

INLET PIPE CONNECTION
UPFLOW - 2" X 9⅞"
DOWNFLOW - 2" X 25"
(SEAL WITH HI TEMP RTV SILICONE SEALANT)

UNITS ARE SHIPPED WITH A SCREEN MOUNTED OVER INLET OPENING. REMOVE SCREW AND SCREEN AND DISCARD SO INLET CONNECTION CAN BE INSTALLED.

OUTSIDE AIR BLOWER HOUSING

SEAL
PVC ELBOW
PVC COUPLING
PVC PIPE
OUTSIDE WALL
8" MIN.
12"

INLET SHOULD BE AT LEAST 2'-6" AWAY FROM AND 12" BELOW FLUE OUTLET — 12" MINIMUM ABOVE NORMAL SNOW ACCUMULATION LEVEL

NOTE:

Outside air blower is not designed to support weight of inlet pipe.

To avoid damage to blower, support must be provided as illustrated at left.

Inlet pipe connection should be inserted all the way into coupling (solvent cement joint) before assembly (through top panel) into outside air blower housing. This provides support, by the coupling, from the top panel.

Allowable length of air supply vent pipe per table shown.

Outdoor Combustion Air Max. Piping Schedule

Max. L. (ft.)	&	No. 90° "L"
40		2
30		3
20		4

(Refer to Pub. No. 18-CH15D2 for Further Information)   **Figure 2**

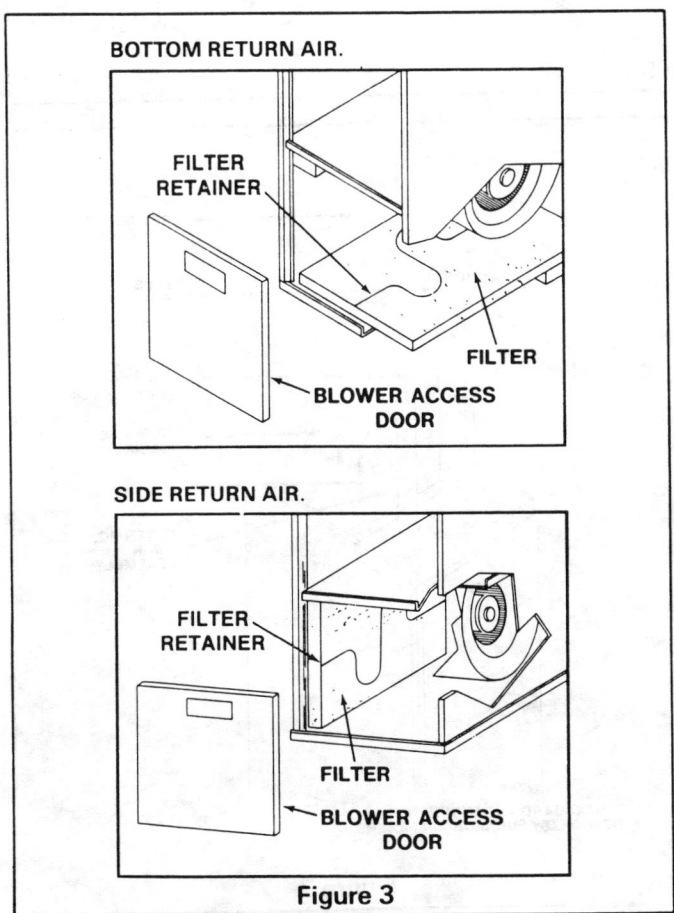

BOTTOM RETURN AIR.

FILTER RETAINER
FILTER
BLOWER ACCESS DOOR

SIDE RETURN AIR.

FILTER RETAINER
FILTER
BLOWER ACCESS DOOR

Figure 3

RETURN AIR-FILTERS — UPFLOW MODELS

Upflow models are supplied with standard size permanent type filter one inch in thickness. Filters may be located within unit cabinet for either bottom or side return air inlet. Cut to fit as required.

To replace filters, remove blower access door, loosen the filter retaining wire at the front of the unit. Replace the filter in the same manner, making sure that filter retaining wire is secured in place in both front and back of the unit. Replace blower access door.

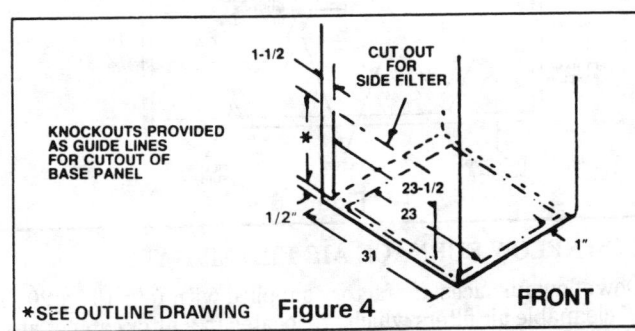

KNOCKOUTS PROVIDED AS GUIDE LINES FOR CUTOUT OF BASE PANEL

1-1/2
CUT OUT FOR SIDE FILTER
23-1/2
23
1/2"
31
1"
FRONT

*SEE OUTLINE DRAWING   **Figure 4**

RETURN AIR FILTERS — DOWNFLOW MODELS: ASSEMBLY OF AIR FILTER SUPPORT PARTS

Install blower compartment door assembly when carton and all parts have been removed, using six screws, including one used for clip.

Assemble filter box as illustrated. Then install on top of furnace cabinet before connecting return air plenum.

NOTE: Filter box parts are packed in carton and secured for shipment on right side of air circulating blower. Blower compartment door assembly is packed inside carton for 040, 060 and 080 models - outside carton for 100 and 120 models. Remove screw and clip used to secure carton, save screw, discard clip and carefully remove parts from carton.

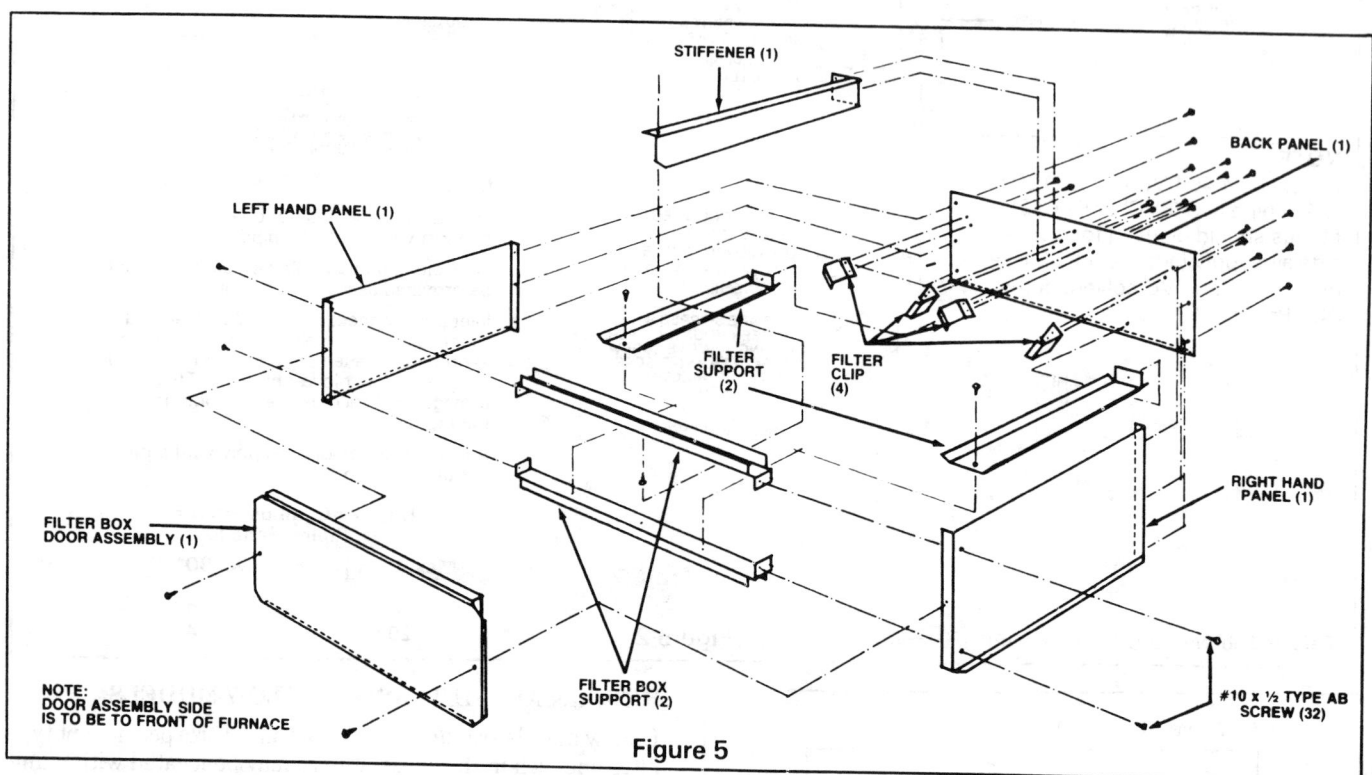

Figure 5

FILTER ARRANGEMENT IS SHOWN IN VIEW BELOW

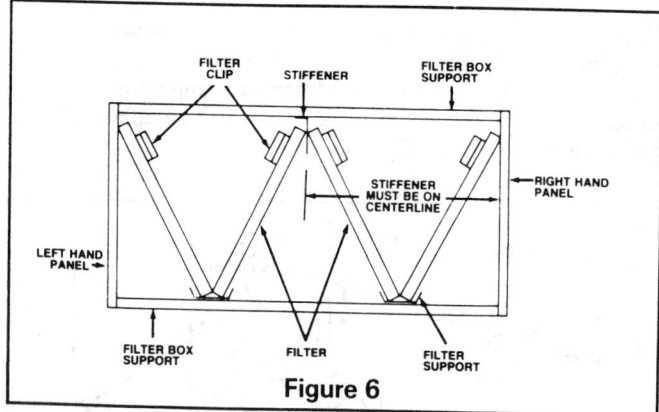

Figure 6

DOWNFLOW FURNACE AIR FILTER DATA

Downflow furnaces are factory supplied with four 10″ x 20″ x 1″ cleanable air filters which are located in a filter cabinet atop the furnace.

Access to filters requires removal of two screws and filter box door assembly (See Illustration). Filters are removed by pulling out of box, guiding as required to pass pipes. Replacement filters are inserted in a reverse order, making sure they are properly positioned in retaining clips on back panel. Then re-install filter box door assembly and secure with two screws previously removed.

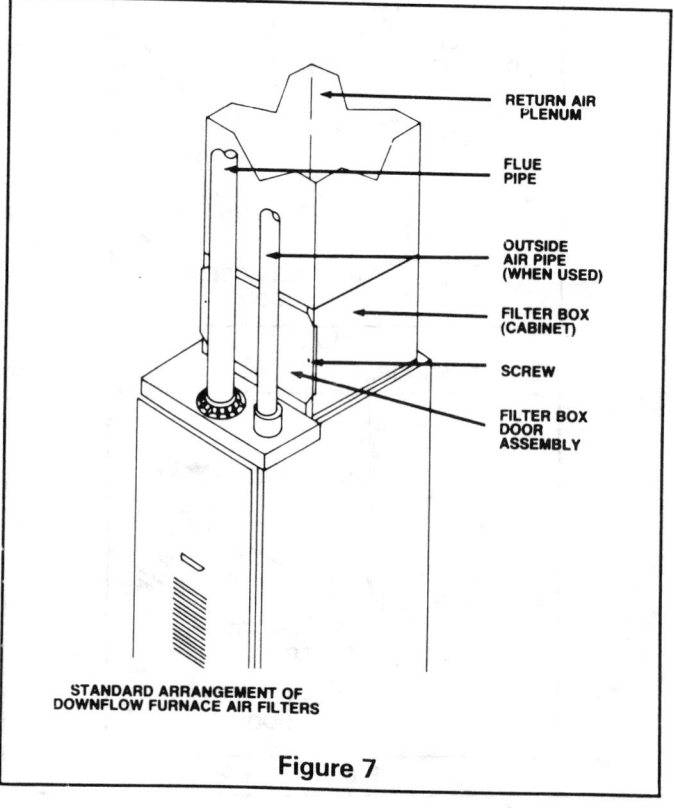

STANDARD ARRANGEMENT OF
DOWNFLOW FURNACE AIR FILTERS

Figure 7

dwg. no. 21X145100 P01

ACCESSORIES

If installing furnace on combustible floor, sub-base (BAY-BASE201 or BAYBASE202) must be used depending on furnace width. If cooling unit is to be installed at this time, or later, coil enclosure should be installed with sub-base. Installation instructions are packed with the accessories. If a 30″ wide coil is to be used with a 24″ wide furnace, sub-base BAYBASE 203 must be used.

In year-round air conditioning installations where the coil enclosure is used with the downflow furnace — the sub-base

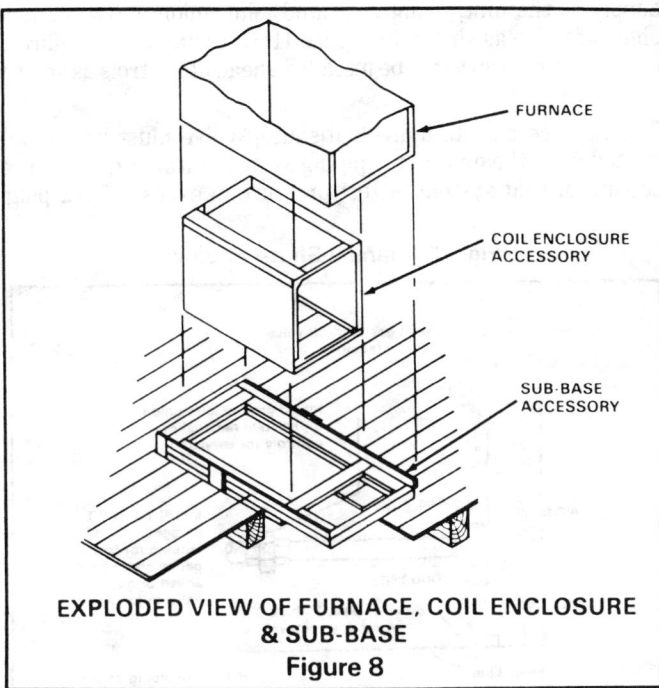

EXPLODED VIEW OF FURNACE, COIL ENCLOSURE & SUB-BASE

Figure 8

must be installed beneath the coil enclosure on combustible flooring.

DUCT CONNECTIONS

Air duct systems should be installed in accordance with standards for air conditioning systems, National Fire Protection Association Pamphlet No. 90. They should be sized in accordance with ACCA Manual D or whichever is applicable. Check on controls to make certain they are correct for the electrical supply.

Central furnaces, when used in connection with cooling units, shall be installed in parallel or on the upstream side of the cooling units to avoid condensation in the heating element, unless the furnace has been specifically approved for downstream installation. With a parallel flow arrangement, the dampers or other means used to control flow of air shall be adequate to prevent chilled air from entering the furnace, and if manually operated, must be equipped with means to prevent operation of either unit unless the damper is in full heat or cool position.

The installer must make the plenum chamber at least 8 inches in depth and sufficiently wide to take the largest duct to be attached to it.

Though these units have been specifically designed for quiet, vibration free operation, air ducts can act as sounding boards and could, if poorly installed, amplify the slightest vibration to the annoyance level. If this is the case, a flex connector of non-flammable material may be installed in the return and supply ducts.

Where the furnace is located in a utility room adjacent to the living area, the system should be carefully designed with returns to minimize noise transmission through the return air grille. Although these winter air conditioners are designed with large blowers operating at moderate speeds, any blower

SUB-BASE FOR DOWNFLOW MODELS (REQUIRED FOR INSTALLATION ON COMBUSTIBLE FLOOR)

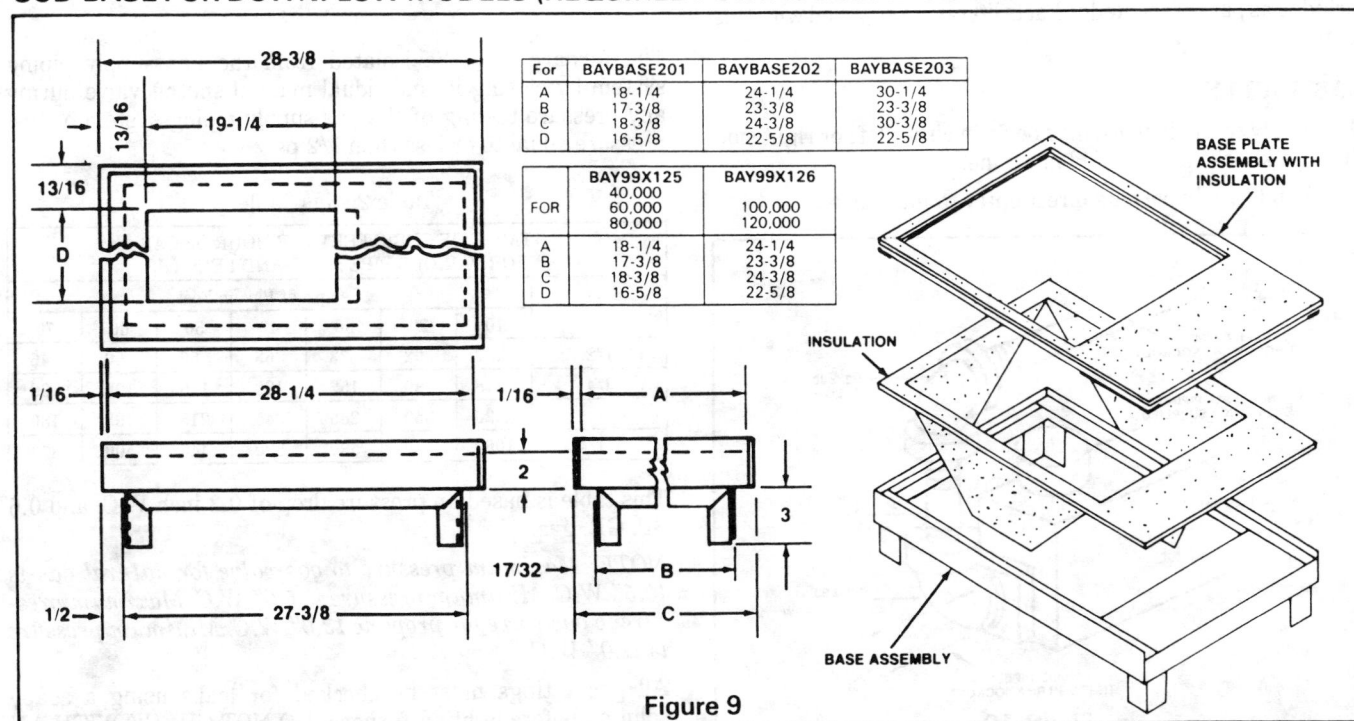

For	BAYBASE201	BAYBASE202	BAYBASE203
A	18-1/4	24-1/4	30-1/4
B	17-3/8	23-3/8	23-3/8
C	18-3/8	24-3/8	30-3/8
D	16-5/8	22-5/8	22-5/8

FOR	BAY99X125 40,000 60,000 80,000	BAY99X126 100,000 120,000
A	18-1/4	24-1/4
B	17-3/8	23-3/8
C	18-3/8	24-3/8
D	16-5/8	22-5/8

Figure 9

DUCT CONNECTIONS (cont.)

moving a high volume of air will produce audible noise which could be objectionable when unit is located very close to a living area. It is often advisable to carry the return air ducts under the floor or through the attic. Such design permits the installation of air return remote from the living area (i.e. central hall).

When furnace is installed so that supply ducts carry air circulated by the furnace to areas outside the space containing the furnace the return air shall also be handled by a duct(s) sealed to the furnace and terminating outside the space containing the furnace.

WHERE THERE IS NO COMPLETE RETURN DUCT SYSTEM THE RETURN CONNECTION MUST BE RUN FULL SIZE FROM THE FURNACE TO A LOCATION OUTSIDE THE UTILITY ROOM, BASEMENT, ATTIC OR CRAWL SPACE

IMPORTANT: Do not take return air through back of furnace cabinet.

ELECTRICAL CONNECTIONS

Make wiring connections to unit as indicated on enclosed wiring diagram. As with all gas appliances using electrical power, this furnace shall be connected into a permanently live electric circuit. It is recommended that it be provided with a separate "circuit protection device" electric circuit. Furnace must be electrically grounded in accordance with local codes or in the absence of local codes with the National Electrical Code. ANSI/NFPA No. 70-1984, if an external electrical source is utilized.

Wiring to be done in the field between the furnace and devices not attached to the furnace or between separate devices which are field installed and located, shall conform with the temperature limitation for Type T wire (63° F. rise [35° C.]) when installed in accordance with the manufacturer's instructions.

Wiring diagrams and schematic (ladder style) diagrams are included as part of the Reddi-Facts literature supplied with this furnace.

GAS SUPPLY

Gas supply to the furnace may be from either left or right side through knockouts provided in cabinet.

Left hand supply permits direct entry to gas valve.

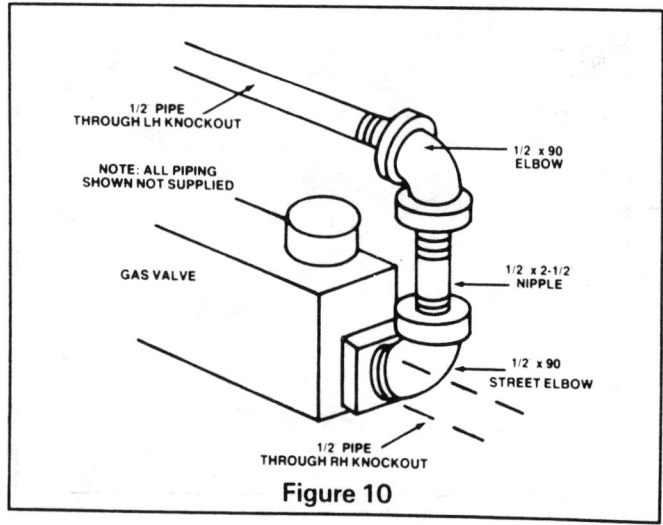

Figure 10

Right hand supply requires elbows and nipple (shown below) to complete supply connection to gas valve.

GAS PIPING

The installation of piping shall be in accordance with piping codes and the regulations of the local gas company. The gas supply shall be a separate line. Pipe joint compound must be resistant to the chemical reaction with liquified petroleum gases.

Refer to piping table, Table 2, for delivery sizes. Connect gas supply to the unit, using a ground joint union and a manual shut-off valve as shown in Figure 11. National codes require a condensation drip leg to be installed ahead of controls as shown in Figure 11.

The furnace and its individual shutoff valve must be disconnected from the gas supply piping system during any pressure testing of that system or test pressure in excess of 1/2 psig.

Fig. 11, Manual Shut-Off Valve

Main gas shut-off valve (required)

Install ground joint union to facilitate removal of controls for servicing

40" to 60"   Tee

Install 1/8" N.P.T. plugged tapping for test gauge connection ahead of gas valve in furnace

Pipe Cap   Drip Leg

Block at least 1" thick Allow for drip removal

Floor Line

NOTE: All piping shown not supplied

SUPPLY PIPING

The furnace must be isolated from the gas supply piping system by closing its individual manual shutoff valve during any pressure testing of the gas supply piping system or test pressure equal to or less than 1/2 psig.

Table 2, Gas Table

TABLE OF CUBIC FEET PER HOUR OF GAS FOR VARIOUS PIPE SIZES AND LENGTHS							
Pipe Size	Length of Pipe in Feet						
	10	20	30	40	50	60	70
1/2	132	92	73	63	56	50	46
3/4	278	190	152	130	115	105	96
1	520	350	285	245	215	195	180
1-1/4	1050	730	590	520	440	400	370

This table is based on pressure drop of 0.3 inch W.C. and 0.6 SP. GR. gas.

NOTE: *Maximum pressure to gas valve for natural gas is 10.5" W.C. Minimum pressure is 5.0" W.C. Maximum pressure to gas valve for propane 13.0" W.C. Minimum pressure is 11.0" W.C.*

All gas fittings must be checked for leaks using a soapy solution before lighting furnace. DO NOT CHECK WITH AN OPEN FLAME.

START UP & ADJUSTMENT

PRELIMINARY INSPECTIONS

With gas and electrical power "OFF"

1. Duct connections properly sealed
2. Filters in place
3. Venting properly assembled
4. Condensate drains properly installed
5. Blower door in place

Adjust Heat Anticipator on Thermostat, for Natural or Propane Gas, as follows:
0.85 — White Rodgers Gas Valve
0.75 — Robertshaw Gas Valve
0.65 — Johnson Gas Valve

Turn knob on main gas valve within unit to "OFF". Turn external gas valve to "ON". Purge air from gas lines. After purging, check all gas connections with soapy solution. DO NOT CHECK WITH AN OPEN FLAME. Allow 5 minutes for any gas that might have escaped to dissipate. LP Gas being heavier than air may require forced ventilation. Turn knob on gas valve in unit to "ON".

REVIEW SEQUENCE OF OPERATION

THERMOSTAT CALLS FOR HEAT

R and W Thermostat contacts close to supply power, through safety limit switches, to control circuit which starts the induced draft blower. When the required combustion air is established, the pressure switch allows power to flow through the safety controls to the flame control module and the ignitor.

The ignitor will heat for approx. 15 seconds, then the gas valve is energized to permit gas flow to the burners. The flame sensor confirms that ignition has been achieved within the 7 seconds ignition trial period. All models utilize a remote sensor.

If the sensor does not conform ignition in the 7 second time period, the ignition control will go into a re-try mode. During the second trial for ignition there is a 60 second pre-purge time for the induced draft motor, then the ignitor will energize for approx. 25 seconds before the gas valve is energized. If the flame sensor does not confirm ignition during this second trial, the ignition control will go into a re-try mode again. The sequence of operation will be repeated. If the ignition has not been confirmed at the end of the 3rd trial, the ignition control will go into lock out mode.

When ignition and flame is established, the flame control module monitors the flame and supplies power to the gas valve until the thermostat is satisfied. When thermostat is satisfied, R and W contracts open to deenergize the control circuit.

LIGHTING INSTRUCTIONS

WARNING: DO NOT ATTEMPT TO MANUALLY LIGHT THE BURNER.

Lighting instructions appear on each unit. Each installation must be checked out at the time of initial start up to insure proper operation of all components. Check out should include putting the unit through one complete cycle as outlined below.

Turn on main electrical supply and set thermostat above indicated temperature. The ignitor will automatically heat for approximately 15 seconds, then the gas valve is energized to permit gas flow to burners. After ignition and flame is estab-lished, flame control module monitors flame and supplies power to gas valve until thermostat is satisfied.

If burner fails to ignite, lower thermostat setting or disconnect electrical supply, wait 5 minutes, raise thermostat setting above indicated temperature, turn electrical supply on. Unit will repeat lighting sequence.

Table 3
GAS FLOW IN CFH

GAS FLOW IN CUBIC FEET PER HOUR							
2 CUBIC FOOT DIAL							
Sec.	Flow	Sec.	Flow	Sec.	Flow	Sec.	Flow
8	900	29	248	50	144	82	88
9	800	30	240	51	141	84	86
10	720	31	232	52	138	86	84
11	655	32	225	53	136	88	82
12	600	33	218	54	133	90	80
13	555	34	212	55	131	92	78
14	514	35	206	56	129	94	76
15	480	36	200	57	126	96	75
16	450	37	195	58	124	98	73
17	424	38	189	59	122	100	72
18	400	39	185	60	120	104	69
19	379	40	180	62	116	108	67
20	360	41	176	64	112	112	64
21	343	42	172	66	109	116	62
22	327	43	167	68	106	120	60
23	313	44	164	70	103	124	58
24	300	45	160	72	100	128	56
25	288	46	157	74	97	132	54
26	277	47	153	76	95	136	53
27	267	48	150	78	92	140	51
28	257	49	147	80	90	144	50

For 1 Cu. Ft. Dial Gas Flow CFH = Chart Flow Reading ÷ 2
For 1/2 Cu. Ft. Dial Gas Flow CFH = Chart Flow Reading ÷ 4
For 5 Cu. Ft. Dial Gas Flow CFH = 10X Chart Flow Reading ÷ 4

COMBUSTION AND INPUT CHECK

1. Make sure all gas appliances are off except furnace.

2. Clock meter with furnace operating (Determine dial rating of meter) for one revolution.

3. Match "Sec" column in the gas flow (in cfh) table with the time clocked.

4. Read "Flow" column opposite the number of seconds clocked.

5. Use factors under table if necessary.

6. Multiply the final figure by the heating value of the gas obtained from utility company and compare to nameplate rating. This must not exceed nameplate rating.

7. Changes can be made by adjusting manifold pressure and orifices (orifice change may not always be required).

A. Attach manifold pressure gauge.

B. Remove cap on top of gas valve.

C. Turn adjustment screw in to increase gas rate, and out to decrease gas rate. Replace cap.

D. The final setting shall be at a manifold pressure no less than 3.0" W.C. and no more than 3.5" W.C. with an input of no more than nameplate rating and no less than 93% of nameplate rating.

Example: If manifold pressure is above 3.5" and rate is less than 93%, a larger orifice is needed. If manifold pressure is less than 3.0" and rate is more than 100%, a smaller orifice is needed.

MAIN BURNER ORIFICE SELECTION

1. ALL furnaces are factory equipped with main burner orifices to provide the nameplate input rating using natural gas (having a heating value of 1100 Btu/cu. ft.). To adjust to nameplate input rating of natural gases having a different heating value refer to the National Fuel Gas code or the following tables.

Table 4
ORIFICES SHIPPED WITH FURNACES

ORIFICE DRILL NUMBERS	
Models	Natural Gas 1100 BTU/cu. ft. .60 Sp. Grav.
TUC040,080,120B TDC080,120B	31
TUC-TDC060B	37
TUC-TDC100B	35

FLOW OF GAS THROUGH FIXED ORIFICES

Table 5 Utility Gases
(Cubic feet per hour at sea level)

Specific Gravity = 0.60
Orifice Coefficient = 0.90
For utility gases of another specific gravity, select multiplier from Table F-3.
For altitudes above 2,000 feet, first select the equivalent orifice size at sea level from Table 9.

Orifice or Drill Size	Pressure at Orifice — Inches Water Column	
	3	3.5
56	5.68	6.13
55	7.11	7.68
54	7.95	8.59
53	9.30	10.04
52	10.61	11.46
51	11.82	12.77
50	12.89	13.92
49	14.07	15.20
48	15.15	16.36
47	16.22	17.52
46	17.19	18.57
45	17.73	19.15
44	19.45	21.01
43	20.73	22.39
42	23.10	24.95
41	24.06	25.98
40	25.03	27.03
39	26.11	28.20
38	27.08	29.25
37	28.36	30.63
36	29.76	32.14
35	32.36	34.95
34	32.45	35.05
33	33.41	36.08
32	35.46	38.30
31	37.82	40.85
30	43.40	46.87
29	48.45	52.33

Table 6
LP-Gases
(Btu per hour at sea level)

	Propane	Butane
Btu per Cubic Foot —	2,516	3,280
Specific Gravity —	1.52	2.01
Pressure at Orifice, Inches Water Column —	11	11
Orifice Coefficient —	0.9	0.9

For altitudes above 2,000 feet, first select the equivalent orifice size at sea level from Table 9.

BTUH Flow Rates		
Orifice or Drill Size	Propane	Butane
57	15,026	17,035
56	17,572	19,921
55	21,939	24,872
54	24,630	27,922
53	28,769	32,615
52	32,805	37,190
51	36,531	41,414
50	39,842	45,168

Table 7
Multipliers for Utility Gases of Another Specific Gravity

Specific Gravity	Multiplier	Specific Gravity	Multiplier
0.45	1.155	0.95	0.795
0.50	1.095	1.00	0.775
0.55	1.045	1.05	0.756
0.60	1.000	1.10	0.739
0.65	0.961	1.15	0.722
0.70	0.926	1.20	0.707
0.75	0.894	1.25	0.693
0.80	0.866	1.30	0.679
0.85	0.840	1.35	0.667
0.90	0.817	1.40	0.655

Divide rated burner input by Table 7 factor.

Table 8
CATALOG NUMBERS FOR REPLACEMENT ORIFICES

Drill Size	Cat. No.	Drill Size	Cat. No.
31	WG16X187	41	WG16X0217
32	WG16X0213	42	WG16X191
33	WG16X204	43	WG16X190
34	WG16X205	44	WG16X0215
35	WG16X186	49	WG16X0218
36	WG16X0214	50	WG16X194
37	WG16X188	51	WG16X195
38	WG16X206	52	WG16X196
39	N/A	53	WG16X192
40	WG16X0216	54	WG16X193

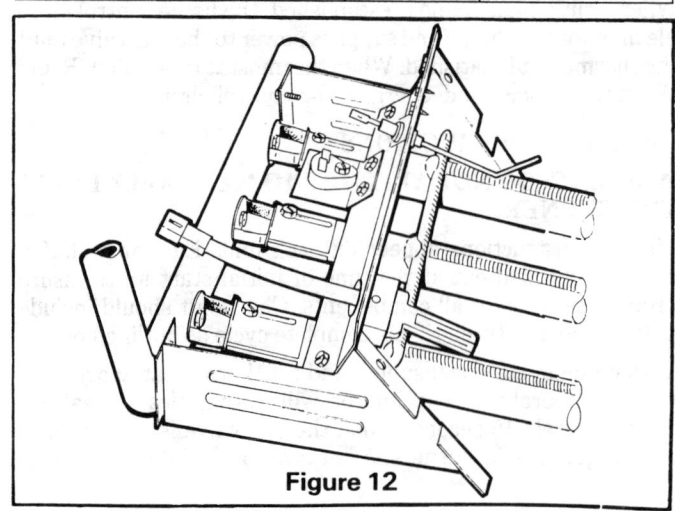

Figure 12

BURNER ADJUSTMENT

NOTE: On LP (propane) units, some light yellow tipping of the outer mantle is normal. Inner mantle should be bright blue. Natural gas unit should not have any yellow tipped flames. See Figure 12.

Table 9
Correction Table for Burner Orifice Drill Sizes for furnaces Installed at Altitudes 2000 Feet and More Above Sea Level

Orifice Twist Drill Size If Installed At Sea Level	ALTITUDE ABOVE SEA LEVEL						
	2000	3000	4000	5000	6000	7000	8000
31	32	32	32	33	34	35	36
32	33	34	35	35	36	36	37
33	35	7/64	36	36	37	38	38
34	35	36	36	37	37	38	39
35	36	36	37	37	38	39	40
36	37	38	38	39	40	41	3/32
37	38	39	39	40	41	3/32	42
38	39	40	41	41	3/32	42	43
39	40	41	3/32	42	42	43	43
40	3/32	3/32	42	42	43	43	44
41	42	42	42	43	43	44	44
42	42	43	43	43	44	44	45
43	44	44	44	45	45	46	47
44	45	45	45	5/64	47	47	48
45	46	5/64	47	47	48	48	49
50	51	51	51	51	52	52	1/16
51	51	52	52	1/16	52	53	53
52	1/16	53	53	53	53	53	54
53	54	54	54	54	54	54	55
54	54	55	55	55	55	3/64	56
55	55	3/64	3/64	3/64	3/64	3/64	56

HIGH ALTITUDE INSTALLATIONS

IMPORTANT: The sea level rated input of the furnace installed at elevations above 2,000 feet should be reduced 4% for each 1,000 feet above sea level. For example, for elevations of 4,350 feet, the correct input would be the sea level rated input less 17%. Refer to table 9 for guide to orifice selection.

High altitude application (4,000 ft. and above elevations) of the new TUC,TDC-B Gas Furnaces necessitates a change in the air pressure differential switch which controls the ignition, to compensate for the less dense air. Due to the lower density, there would be insufficient negative developed by the induced draft blower resulting in nuisance shut down. Please note, with the reduced firing ratings attendant with high altitude application, the induced draft blower has the capacity to provide sufficient air for proper, safe combustion and operation. Refer to replacement parts literature for information concerning correct high altitude switch for each model.

CAUTION:

1. Pressure switches are factory calibrated and sealed. Field adjustment is not permitted. If seal is broken the product warranty is voided.

2. The use of high altitude calibrated switches in sea level application is not permitted.

CONTROLS AND SAFETY SWITCH ADJUSTMENT

ON AN INITIAL START UP, THE FAN SWITCH AND LIMIT SWITCH ADJUSTMENT MUST BE CHECKED.

The fan switch is factory set for the blower to come on at 120°F.* and to shut off at 100°F. These settings are satisfactory in most applications. If necessary to field adjust, disconnect power and proceed as follows:

1. Adjust fan-off pointer to shut blower off when air at the farthest register begins to feel cool, or to setting that will not cause blower to recycle.

2. Fan-on differential is adjustable from 25°F. to 45°F. above fan-off setting.

NOTE.: Do not exceed 125°F. fan-on setting for downflow units.

LIMIT SWITCH CHECK OUT

The limit switch is a safety device designed to close the gas valve should the furnace become overheated. Since proper operation of this switch is important to the safety of the unit, it **must** be checked out on initial start up by the installer.

To check for proper operation of the limit switches, set thermostat to temperature higher than room temperature to bring on the gas valve. Restrict airflow by blocking the return air. When the furnace reaches the maximum outlet temperature as shown on the rating plate, the burners must shut off. If they do not shut off after a reasonable time and over-heating is evident, a faulty limit is probable and the limit switch must be replaced. After checking the operation of the fan and limit control, be sure to remove the paper or cardboard from the return air inlet.

AIRFLOW ADJUSTMENT

Check inlet and outlet air temperatures to make sure they are within the range specified on the furnace rating plate. If airflow needs to be increased or decreased, see wiring diagram for information on changing the wiring to blower motor. Speed changes can be made **at molex connector in bottom left vestibule.**

NOTE TO INSTALLER

Review warnings below with owner. Review contents of "HOME OWNER'S MANUAL" with owner.

WARNING: DISCONNECT POWER TO UNIT BEFORE REMOVING BLOWER DOOR.

Unit is equipped with a blower door switch which cuts power to blower and gas valve causing shutdown when door is removed. Unit must not be altered to allow operation with the blower door removed. Operation with doors removed or ajar can permit the escape of dangerous fumes. All panels must be securely closed at all times for safe operation of the furnace.

INSTRUCTIONS TO THE OWNERS

IN THE EVENT THAT ELECTRICAL, FUEL OR MECHANICAL FAILURES OCCUR, THE OWNER SHOULD IMMEDIATELY TURN OFF THE GAS SUPPLY AT THE MANUAL GAS VALVE LOCATED IN THE BURNER COMPARTMENT, THE ELECTRIC POWER TO THE FURNACE AND CONTACT THE SERVICE AGENCY DESIGNATED BY YOUR DEALER.

OPERATING INFORMATION
FLAME ROLL-OUT DEVICE

All models are equipped with a fusible link on burner cover. In case of flame roll-out, link will open and close off flow of all gas. See instruction label on front panel of heat exchanger.

ABNORMAL CONDITIONS

1. EXCESSIVE COMBUSTION PRESSURE (WIND IN EXCESS OF 40 MPH), VENT OR FLUE BLOCKAGE

If pressure against induced draft blower outlet becomes excessive, pressure switch will react and shut off gas valve until acceptable combustion pressure is again available.

2. LOSS OF FLAME

If loss of flame occurs during heating cycle, when flame is not present at the sensor, flame control module will close gas valve. The flame control module will then re-cycle ignition sequence, then if ignition is not achieved, it will shut off gas valve and lock out system.

3. RESET AFTER LOCK OUT

To reset after lock out, set thermostat to lowest possible setting or turn thermostat to "OFF" for approx. 30 seconds, then return thermostat to desired setting or turn thermostat to "ON".

4. POWER FAILURE

If there is a power failure during heating cycle, system will re-start ignition sequence automatically when power is restored, if thermostat still calls for heat.

5. GAS SUPPLY FAILURE

If loss of flame occurs during heating cycle, system flame control module will re-cycle ignition sequence, then if ignition is not achieved, flame control module will shut off gas valve and lock out system.

6. INDUCED DRAFT BLOWER FAILURE

If pressure is not sensed by pressure switch, contacts will remain open and not allow gas valve to open, therefore unit will not start. If failure occurs during running cycle, pressure switch contacts will open and gas valve will close to shut unit down.

7. CONDENSATE DRAIN BLOCKAGE

If the condensate drain is blocked, either by debris, improper draining, or by freezing condensate, the pressure switch will sense the accumulation of condensate in the furnace drain pan. The pressure switch contacts will open and remain open, not allowing unit operation. The unit will not operate until the condensate drain has been cleared, and the condensate flows freely.

REMOVAL OF BLOWER DOWNFLOW MODELS

If the blower should have to be removed for servicing, turn off electricity to unit. Disconnect electrical lead in junction box or at the motor. You must disconnect the flue pipe to remove the blower completely. Remove panels by removing screws, remove flue pipe in blower compartment.

NOTE: Care must be taken in removing the blower to prevent damage to the auxiliary limit control mounted on the blower housing.

RELAY INSTRUCTIONS

All models are factory equipped with a transformer and relay.

MAINTENANCE

This maintenance information has been prepared so that you may better understand and care for your warm air furnace.

1. THERMOSTAT — The thermostat is the heart of the warm air furnace control center. Its operation depends on the surrounding air temperature, therefore, it should be mounted on a draft-free inside wall for best operation. Because the thermostat is sensitive to heat, devices such as radios, televisions, or lamps should not be placed near it. The thermostat also accumulates lint which affects its accuracy. For best operation, the thermostat should be cleaned annually.

2. FILTERS — High efficiency warm air furnaces are equipped with appropriate air filters. The filter removes dust and debris from the air before it is heated and circulated to the living spaces. Your filters must be changed or cleaned when dirty. Inspection of the filters must be made on a monthly basis. The use of continuous air circulation (CAC) or unusually dusty conditions may decrease this time interval. Cleanable filters should be washed with a household detergent and replaced only when dry. It is a good idea to keep several filters on hand when disposable filters are used. NEVER OPERATE YOUR WARM AIR FURNACE WITH FILTERS REMOVED.

3. BLOWERS — The blower size and speed determine the air volume delivered by the furnace. Annual cleaning of the blower wheel and housing is recommended for maximum air output.

4. MOTORS — Direct drive motors have bearings which are permanently lubricated and do not under normal use require lubrication.

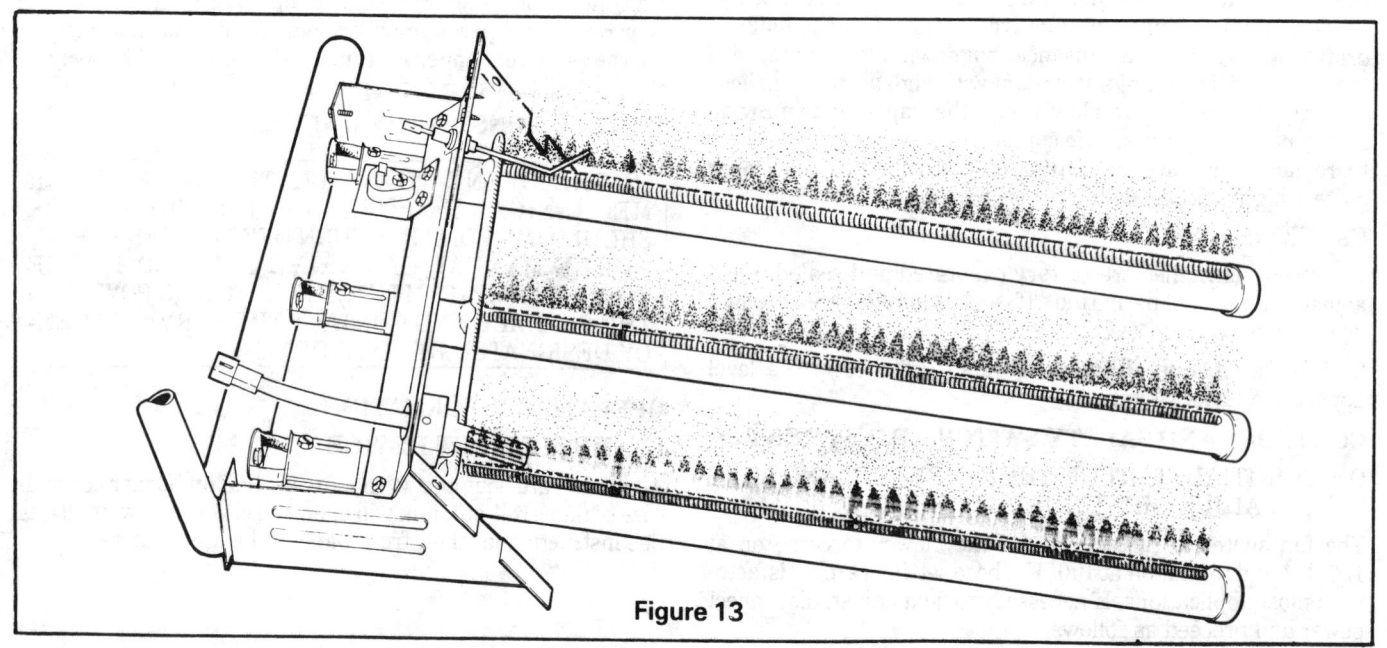

Figure 13

dwg. no. 21X145100 P01

5. BURNER — Gas burners do not normally require scheduled servicing, however, accumulation of lint may cause a yellowing flame or delayed ignition. Either condition indicates that a service call is required. For best operation burners must be cleaned annually using brushes and vacuum cleaner.

To clean burners, remove burner cover, lift front of burner from orifice, then slide burner assembly forward and out of furnace. NOTE: Be careful not to break ignitor when removing burner assembly. Clean burners with brush and/or vacuum cleaner. Reassemble the parts by reversal of the procedure just described.

For additional information refer to main burner adjustment.

NOTE: On LP (propane) units, due to variations in BTU content and altitude, servicing may be required at shorter intervals.

6. HEAT EXCHANGER/FLUE PIPE — These items must be inspected for signs of corrosion, and/or deterioration at the beginning of each heating season by a qualified service technician and cleaned annually for best operation. To clean flue gas passages, follow recommendation steps as outlined.

NOTE: Before cleaning is started make sure you have these parts on hand:
Gasket — Access cover
Gasket — Flue collection box

a. Turn off gas and electric power supply.

b. Remove screws, retaining clips and flue collector box insulation sections.

c. Remove screws fastening flue collector box and remove collector box.

d. Remove flue baffles.

e. Remove burners.

f. Clean heat exchangers using wire brush and vacuum cleaner.

g. Replace parts removed in reverse order. Inspect burners. Clean if necessary.

h. Restore gas supply. Check for leaks. Restore electrical supply. Check unit for normal operation.

7. RECUPERATIVE CELL (EXTERNAL)

a. Turn off gas and electric power supply.

b. Remove wires from pressure switch.

c. Remove screws which secure control box to side support. Remove 2 screws which secure ignitor control and relay to side support. Carefully lower control box and ignitor control assembly out of the way.

d. Remove 2 screws and cover from partition to provide access to upstream side of cell for inspection and cleaning.

e. Fins may be cleaned using a suitable brush and vacuum cleaner attachment.

f. Install new gasket on access cover, if needed, and replace parts removed in reverse order.

g. Restore gas supply. Check for leaks. Restore electrical supply. Check unit for normal operation.

8. RECUPERATIVE CELL (INTERNAL)

a. With flue collector box removed (See Item 6) inspect inside upper header cover for soot and/or dirt accumulation.

b. If cleaning is needed, fashion an extension tube of 3/8 or 1/4 O.D. copper tube (for connection to a garden hose) with a 90° curved end.

c. Insert curved end of adapter tube into each tube in cell and flush with water, using a suitable container to collect water flushed through cell and/or any spillage during cleansing operation.

d. Install new gasket on flue collector box, if needed, and replace parts removed in reverse order.

9. CIRCUIT BREAKERS OR FUSES — If blower or gas valve fails to operate, the cause could be a tripped circuit breaker or a loose or blown fuse. When fused circuits are used, replacement fuses should always be kept on hand.

10. OPERATION — Your warm air furnace should not be operated in a corrosive atmosphere. Paint solvents, cleaning chemicals, spray propellants, and bleaches should not be used in the vicinity of the furnace during normal operation.

11. CONDENSATE DRAIN SYSTEM — Condensate drain system must be checked every three months to assure that condensate can flow freely from unit to drain.

A cap is installed in drain system inside cabinet for inspection and cleaning as needed. When cap is removed, clamp must be in place when re-installed. If a drain problem cannot be corrected, call your service company.

IMPORTANT NOTE: Additional service information, replacement parts listing, wiring diagrams, and troubleshooting charts are included in the REDDI-FACTS literature provided with these instructions. The REDDI-FACTS literature is customer property and is to remain with the unit when installed.

FIELD WIRING DIAGRAM FOR HEATING/COOLING
(OUTDOOR SECTION WITHOUT TRANSFORMER)

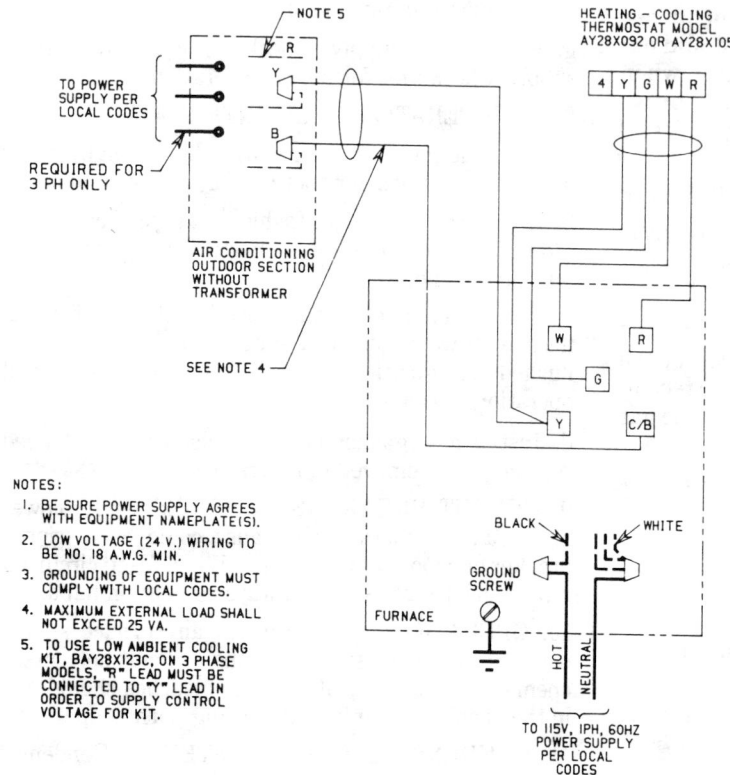

HEATING – COOLING THERMOSTAT MODEL AY28X092 OR AY28X105

INTER-COMPONENT WIRING

- – – – – 24 V. } FACTORY
- – – – – LINE V. } WIRING
- ———— 24 V. } FIELD
- ———— LINE V. } WIRING

NOTES:
1. BE SURE POWER SUPPLY AGREES WITH EQUIPMENT NAMEPLATE(S).
2. LOW VOLTAGE (24 V.) WIRING TO BE NO. 18 A.W.G. MIN.
3. GROUNDING OF EQUIPMENT MUST COMPLY WITH LOCAL CODES.
4. MAXIMUM EXTERNAL LOAD SHALL NOT EXCEED 25 VA.
5. TO USE LOW AMBIENT COOLING KIT, BAY28X123C, ON 3 PHASE MODELS, "R" LEAD MUST BE CONNECTED TO "Y" LEAD IN ORDER TO SUPPLY CONTROL VOLTAGE FOR KIT.

TO 115V, 1PH, 60HZ POWER SUPPLY PER LOCAL CODES

From Dwg. 21B134671 Rev. 2

FIELD WIRING DIAGRAM FOR HEATING ONLY

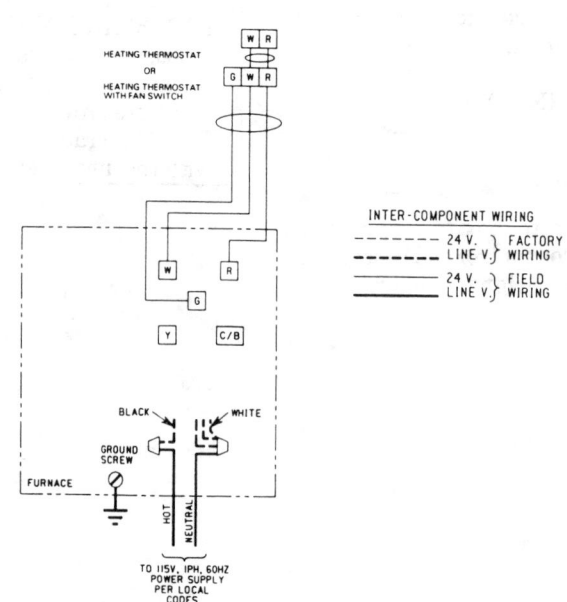

INTER-COMPONENT WIRING

- – – – – 24 V. } FACTORY
- – – – – LINE V. } WIRING
- ———— 24 V. } FIELD
- ———— LINE V. } WIRING

NOTES
1 BE SURE POWER SUPPLY AGREES WITH EQUIPMENT NAMEPLATE(S)
2 LOW VOLTAGE (24 V) FIELD WIRING TO BE NO 18 A W G MIN
3 GROUNDING OF EQUIPMENT MUST COMPLY WITH LOCAL CODES

TO 115V, 1PH, 60HZ POWER SUPPLY PER LOCAL CODES

From Dwg. 21B134672 Rev. 1

TUC — B OUTLINE DRAWING
(ALL DIMENSIONS ARE IN INCHES)

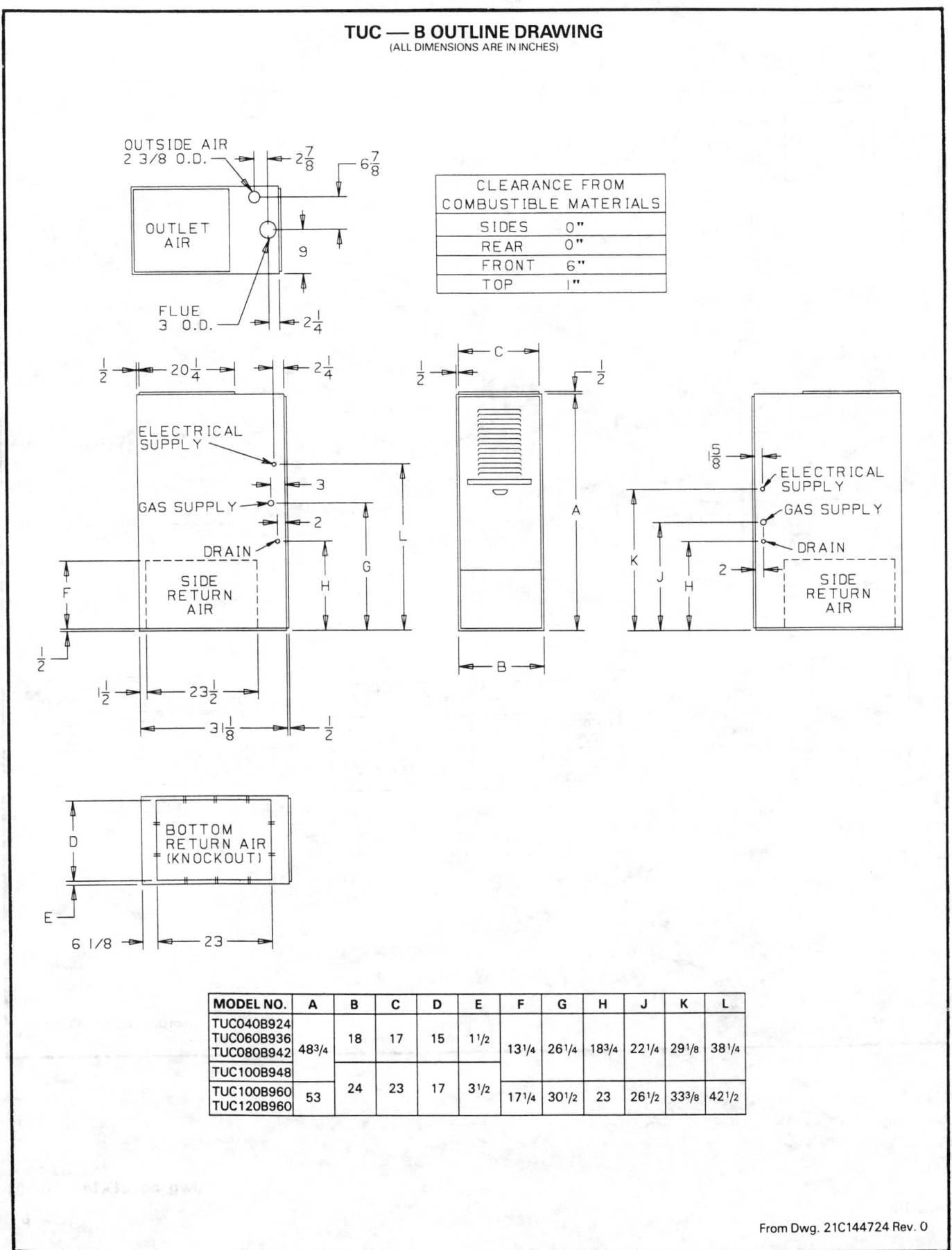

CLEARANCE FROM COMBUSTIBLE MATERIALS	
SIDES	0"
REAR	0"
FRONT	6"
TOP	1"

MODEL NO.	A	B	C	D	E	F	G	H	J	K	L
TUC040B924 TUC060B936 TUC080B942	48³/₄	18	17	15	1¹/₂	13¹/₄	26¹/₄	18³/₄	22¹/₄	29¹/₈	38¹/₄
TUC100B948											
TUC100B960 TUC120B960	53	24	23	17	3¹/₂	17¹/₄	30¹/₂	23	26¹/₂	33³/₈	42¹/₂

From Dwg. 21C144724 Rev. 0

TDC — B OUTLINE DRAWING
(ALL DIMENSIONS ARE IN INCHES)

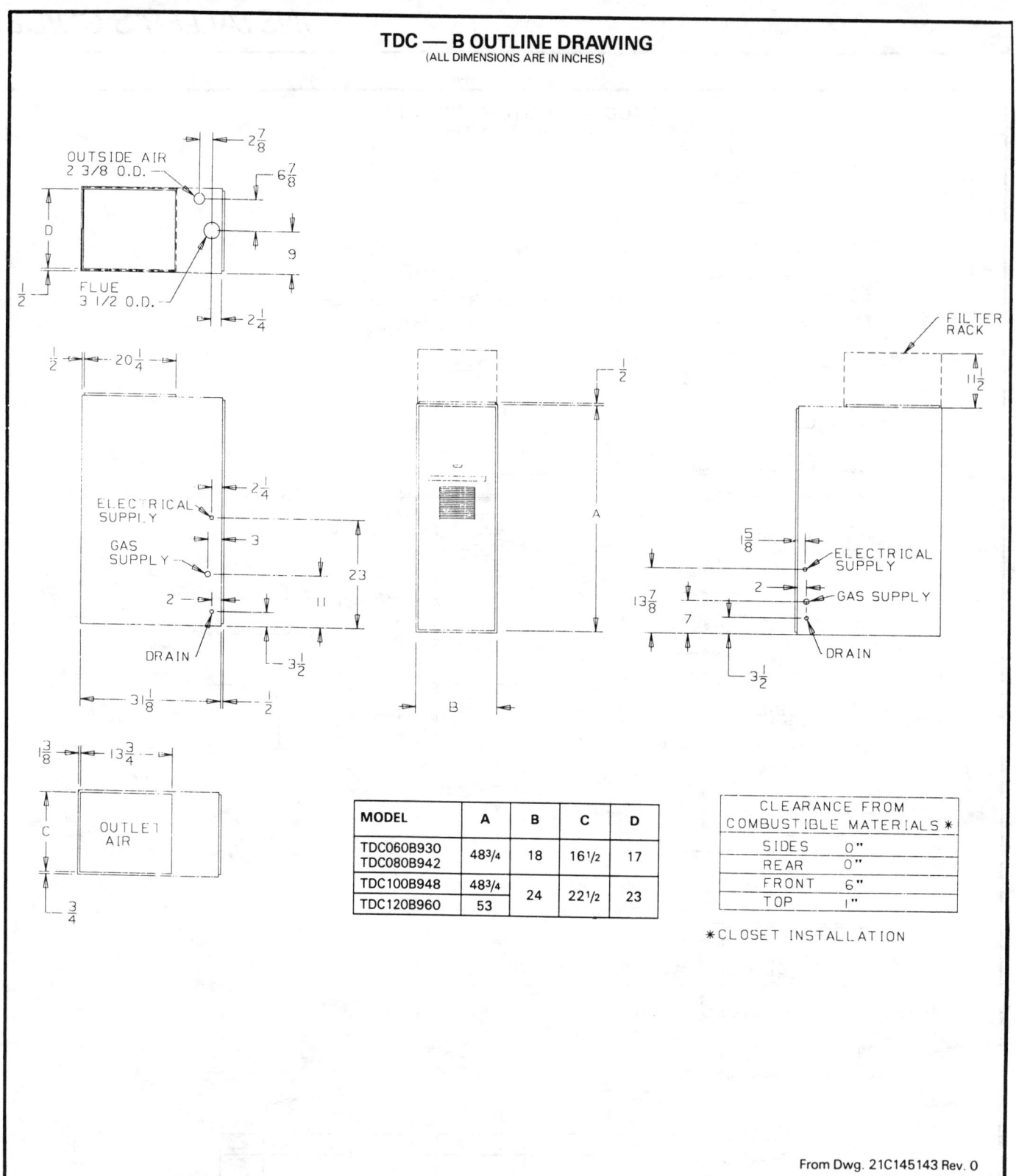

MODEL	A	B	C	D
TDC060B930 TDC080B942	48³/4	18	16¹/2	17
TDC100B948	48³/4	24	22¹/2	23
TDC120B960	53			

CLEARANCE FROM COMBUSTIBLE MATERIALS *	
SIDES	0"
REAR	0"
FRONT	6"
TOP	1"

*CLOSET INSTALLATION

From Dwg. 21C145143 Rev. 0

dwg. no. 21X145100 P01

P.I.

User's Information Manual
Super efficiency condensing gas furnace TUC, TDC and TUX, TDX Models

FOR YOUR SAFETY: WHAT TO DO IF YOU SMELL GAS	**WARNING:**	**FOR YOUR SAFETY**

FOR YOUR SAFETY: WHAT TO DO IF YOU SMELL GAS

- Do not try to light any appliance.
- Do not touch any electrical switch; do not use any phone in your building
- Immediately call your gas supplier from a neighbor's phone. Follow the gas supplier's instructions.
- If you cannot reach your gas supplier, call the fire department.

WARNING:

Improper installation, adjustment, alteration, service, or maintenance can cause injury or property damage. Refer to this manual. For assistance or additional information, consult a qualified installer, service agency, or the gas supplier.

FOR YOUR SAFETY

Do not store or use gasoline or other flammable vapors and liquids in the vicinity of this or any other appliance.

Safety notice.

WARNING: Improper installation, adjustment, alteration, service or maintenance can cause injury or property damage. Refer to the installation instructions provided with the furnace and this manual. For assistance or additional information consult a qualified installer, service agency or the gas supplier.

This information is intended for use by individuals possessing adequate backgrounds of electrical and mechanical experience. Any attempt to repair a central air conditioning product may result in personal injury and/or property damage. American Standard Inc. or seller cannot be responsible for the interpretation of this information, nor can it assume any liability in connection with its use.

Do not use this furnace if any part has been under water. Immediately call a qualified service technician to inspect the furnace, and to replace any part of the control system and any gas control which has been under water.

Filter maintenance reduces energy use.

A clean filter saves money.

When the furnace circulates and filters the air in your home, dust and dirt particles build up on the filter. Excessive accumulation can block the airflow, forcing the unit to work harder to maintain desired temperatures.

The harder your unit has to work, the more energy it uses. So you pay more any time your system is running with a dirty filter.

CAUTION: Never operate your unit for either heating or cooling with filters removed.

Help ensure top efficiency by cleaning the filter once a month.

Clean it twice a month during seasons when the unit runs more often.

You can leave the filter in the frame and vacuum it. Or you can take it out of the furnace and wash it with a household detergent.

Both methods are quick and easy — and guaranteed to improve the performance of your system.

Replacing your filter.

When replacing your furnace filters, always use the same size and type that was originally supplied. Filters are available from your dealer.

Where disposable filters are used, they must be replaced every month with the same size as originally supplied.

How to remove your filter.

WARNING: Disconnect power to unit before removing blower door.

Downflow furnace.

Downflow furnaces are factory supplied with two or four 10" x 20" x 1" disposable air filters which are located in a filter cabinet atop the furnace.

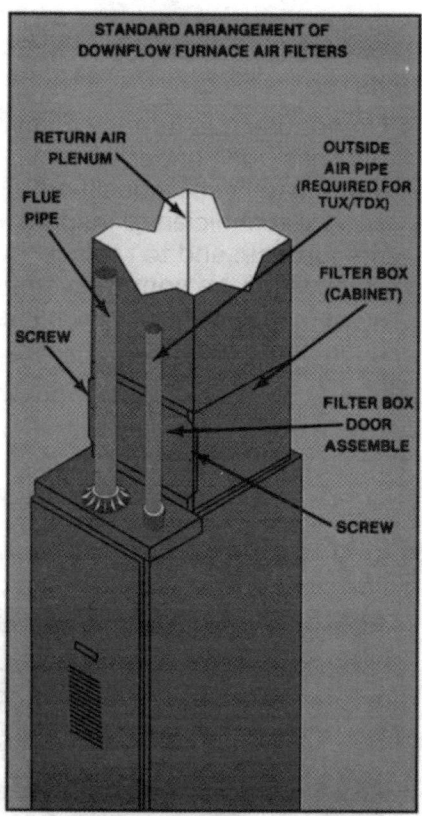

STANDARD ARRANGEMENT OF DOWNFLOW FURNACE AIR FILTERS

RETURN AIR PLENUM
FLUE PIPE
OUTSIDE AIR PIPE (REQUIRED FOR TUX/TDX)
FILTER BOX (CABINET)
SCREW
FILTER BOX DOOR ASSEMBLE
SCREW

Access to filters requires removal of two screws and filter box door assembly (see illustration). Filters are removed by pulling out of box, flexing as required to pass pipes. Replacement filters are inserted in a reverse order making sure they

are properly positioned in retaining clips on back panel. Then re-install filter box door assembly and secure with two screws previously removed.

Upflow furnace.

Upflow furnaces are factory supplied with a standard size permanent type air filter which may be located within the furnace blower compartment in either a BOTTOM or SIDE (left or right) return air inlet.

To replace filters, remove blower access door, loosen the filter retaining wire at the front of the unit. Replace the filter in the same manner, making sure that filter retaining wire is secured in place in both front and back of the unit. Replace blower access door.

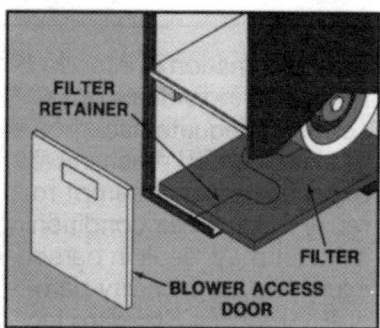

FILTER RETAINER
FILTER
BLOWER ACCESS DOOR

A bottom return air inlet as above features a 16" x 25" x 1" filter in the 18" wide furnaces, a 20" x 25" x 1" filter in the 24" wide furnace cabinets.

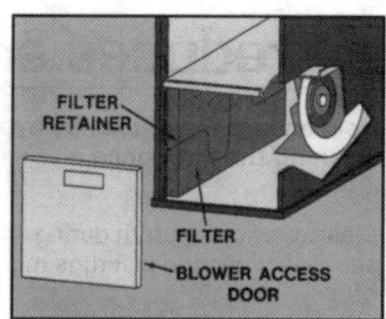

FILTER RETAINER
FILTER
BLOWER ACCESS DOOR

A left or right return air inlet as above (left side shown) requires trimming of factory supplied filter to 14½" x 25" x 1" for both the 18" and 24" wide furnaces.

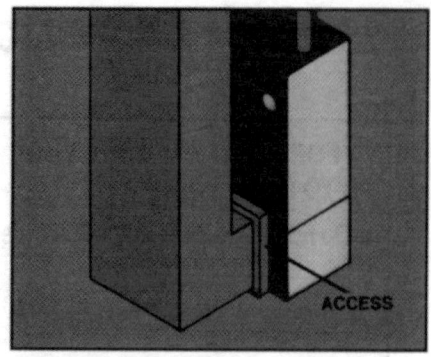

ACCESS

Air filters may also be located outside of the furnace using a SIDE FILTER FRAME.

A furnace is not a household appliance. It is complex and requires professional maintenance and repair.

That's why attempts at "do-it-yourself" repairs on an in-warranty unit may void the remainder of your warranty.

Other than performing the simple maintenance recommended in this manual, you should not attempt to make any adjustments to your furnace. Your dealer will be able to take care of any questions or problems you may have. A periodic inspection of your furnace should be made by a qualified service agency at the start of each heating season.

The Problem Solver.

Save time and money.
Before calling for service, check the following:

Problem	Possible Trouble	Possible Remedy
No Heating - Blower Does Not Operate	1. Thermostat set incorrectly. 2. Blown fuse or tripped circuit breaker. 3. Defective component. 4. Burner may not ignite. 5. Main gas line turned off. 6. Blower door removed or ajar. 7. Combustion Air Inlet or Vent Gas Outlet blocked.	1. Adjust thermostat - See operating instructions. 2. Replace or reset protective device or call for servicer. 3. Most controls are automatic and will recycle. If your unit still does not operate, call for servicer. 4. Call servicer. 5. Have gas company check. 6. Close door securely to restore power to blower and gas valve. 7. Call servicer.
Insufficient Heating - Blower Operates Continuously	1. Dirty air filters. 2. Blocked supply or return registers.	1. Clean or replace filters. 2. Make sure registers are open and no obstacles blocking off the air.
No Heat - Vent motor is running.	Restricted or plugged furnace condensate drain.	1. Remove drain clamp and cap to access drain pan outlet. 2. Flush or clear drain blockage. 3. Install cap and clamp.
Unusual Noise		Call your servicer.

Keep your furnace looking like new for years.

Clean the enamel finish of your furnace with ordinary soap and water. For stubborn grease spots, use a household detergent. Lacquer thinner or other synthetic solvents may damage the finish.

Do away with surprise repair bills with a Service Agreement.

Service Agreements may be available from your Dealer or installer. The agreement has the following advantages:
1. Established cost for service resulting from normal usage . . . no need for an unexpected service cost to "upset" budgeted expenses.
2. Includes both parts and labor for the duration of the Agreement. **Be certain you read the Agreement for complete details and exclusions.**
3. Service is performed by servicers knowledgeable of the operation of this equipment.

Maintenance information for the owner.

Stopping the system by shutting off the main power —

If the main power to your air conditioner is ever disconnected for more than three hours, turn off the thermostat. Then wait for at least three more hours after the power has been restored before turning the thermostat back on. Failure to follow this procedure could result in damage to your air conditioning system.

WARNING —

Hazardous Voltage. Disconnect power before servicing.

1. **General Inspection —**
Examine the furnace installation for the following items:
a. All flue product and combustion air inlet areas external to furnace (i.e. chimney, vent connector, inlet air termination) are clear and free of obstruction.
b. The vent connector is in place, slopes upward and is physically sound without holes or excessive corrosion.
c. The return air duct connection(s) is physically sound, is sealed to furnace and

terminates outside space containing the furnace.
d. The physical support of the furnace should be sound without sagging, cracks, gaps, etc., around the base so as to provide a seal between the support and the base.
e. There are no obvious signs of deterioration of furnace.

2. **Blowers —** The blower size and speed determine the air volume delivered by the furnace. The blower motor bearings are factory lubricated and under normal operating

conditions usually do not require servicing. Your Servicer can advise you if oiling is required. Annual cleaning of the blower wheel and housing is recommended for maximum air output and combustion air requirements.

3. **Igniter** — This unit has a special hot surface direct ignition device that automatically lights the burners. Please note that it is very fragile and should be handled with care.

CAUTION: Do not touch igniter. It is extremely hot.

*4. **Burner** — Gas burners do not normally require scheduled servicing, however, accumulation of foreign material may cause a yellowing flame or delayed ignition. On LP (propane) units, some light yellow tipping of the outer mantle is normal. Inner mantle should be bright blue. Natural gas units should not have any yellow tipped flames. This condition indicates that a service call is required. For best operation, burners must be cleaned annually using brushes and vacuum cleaner.

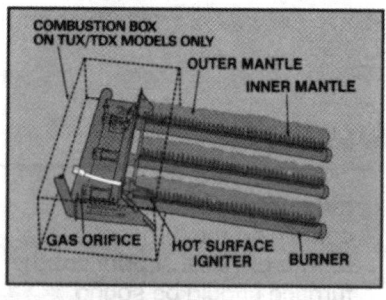

CAUTION: TUX, TDX only. If sealed combustion box top or front cover must be removed, contact a qualified servicer. Gasket material must be replaced if torn or shredded to maintain seal.

NOTE: On LP (propane) units, due to variations in BTU content and altitude, servicing may be required at shorter intervals.

5. **Heat Exchanger/Flue Pipe** — These items must be inspected

for signs of corrosion and/or deterioration at the beginnning of each heating season by a qualified service technician and cleaned annually for best operation. To clean flue gas passages, follow these recommendations.

a. Inspect flue pipe exterior for cracks, leaks, holes or leaky joints. Some discoloration of PVC pipe is normal.

b. Turn off gas and electrical power to furnace.

c. Remove burner compartment door from furnace.

d. Inspect around insulation covering the flue collector box. Inspect induced draft blower connections from recuperative cell and to the flue pipe connection.

e. Remove burner cover (front cover on combustion box on TUX, TDX must be removed first), but do not remove burners. Use a mirror and flashlight to inspect interior of heat exchanger by looking past burners. Be careful not to damage the igniter, flame sensor or other components.

f. If any corrosion is present, contact a service agency. Heat exchangers should be cleaned by a qualified service technician.

g. After inspection is completed replace the burner cover (combustion box front cover of TUX, TDX models) and furnace door.

h. Restore gas and electrical power to furnace. Check unit for normal operation.

6. **Circuit Protection** — If blower or gas valve fail to operate, the cause could be the circuit breaker or a loose or blown fuse. Replace fuse or reset circuit breaker.

7. **Operation** — Your warm air furnace should not be operated in a corrosive atmosphere. Paint solvents, cleaning chemicals, spray propellents, and bleaches should not be used in the vicinity of the furnace during normal operation.

8. **Furnace Condensate Drain Tubes** — Condensate drain tubes must be checked periodically to assure that condensate can flow freely from unit to drain. Drain tubes should not be installed where freeze up is a possibility.

A cap is installed in drain system inside cabinet for inspection and cleaning as needed. When cap is removed, clamp must be in place when reinstalled. If a drain problem cannot be corrected, call your service company.

9. **Cooling Coil Condensate Drain** — If you have a cooling coil installed with your furnace, condensate drains should be checked and cleaned periodically to assure that condensate can drain freely from coil to drain. If condensate cannot drain freely water damage could occur.

WARNING —

Unit is equipped with a blower door switch which cuts power to blower and gas valve causing shutdown when door is removed. Unit must not be altered to allow operation with the blower door removed. Operation with doors removed or ajar can permit the escape of dangerous fumes. All panels must be securely closed at all times for safe operation of furnace.

WARNING: Should overheating occur, or gas supply fail to shut off, shut off the manual gas valve to the furnace before shutting off the electrical supply.

In the event that electrical, fuel or mechanical failures occur, the owner should immediately turn off the gas supply at the manual gas valve located in the burner compartment and electrical power to the furnace and contact your servicer.

For your safety —

Furnace area must be kept clear and free of combustible materials, gasoline, and other flammable vapors and liquids.

Air for combustion and ventilation.

The flow of combustion and ventilating air must not be obstructed from reaching the furnace. Air openings provided in the casing of furnace must be kept free of obstructions which would restrict airflow, thereby affecting efficiency and safe operation of your furnace. Also, air openings provided to the area in which the furnace is installed and the space around the furnace shall not be blocked or obstructed. Keep this in mind should you choose to remodel the area which contains your furnace.

Furnaces must have air for proper performance. If this furnace is installed in an attic or other insulated space it must be kept free and clear of all insulating materials as some insulating materials are combustible.

If additional insulation is added after the furnace is installed, the area around furnace must be inspected to ensure it is free and clear of insulation.

Condensate drain.

Provisions must be made to prevent winter freeze-up of condensate drain tubes from furnace. Frozen condensate tubes will result in furnace shutdown. Condensate drain tubes should not be installed where freeze up is a possibility.

To light furnace.

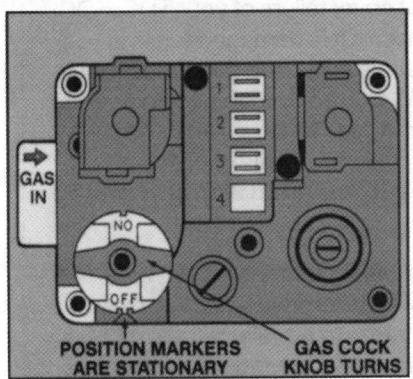

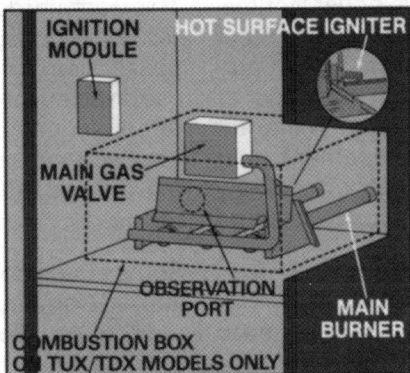

Lighting instructions.

Your furnace is equipped with a hot surface direct ignition device.

WARNING: Do not attempt to manually light the furnace.

1. **Please read all safety information in this book before operating furnace.**

2. Set thermostat to lowest setting. Turn off all electric power to furnace.

3. Remove control access panel.

4. Turn gas cock knob on main gas valve within unit clockwise to "OFF" position. If external gas cock is used, turn to "OFF" position. Allow 5 minutes for any gas within unit to escape. LP gas being heavier than air may require forced ventilation. If you smell gas STOP! Follow the "What To Do If You Smell Gas" instructions on the front cover of this book. If you don't smell gas, go to next step.

5. Turn gas cock knob counterclockwise to "ON" marker.

6. Replace control access panel.

7. Turn on main electrical supply and set thermostat to desired setting. Combustion blower will start and ignition device will start to heat up. After several seconds main gas valve will open and burners will ignite. You can confirm ignition on TUX/TDX models by looking through the observation port in combustion box.

8. When thermostat is satisfied, main burners will extinguish.

9. If main burners fail to ignite lower thermostat setting or disconnect electrical supply, wait 5 minutes, raise thermostat setting above indicated temperature.

10. If furnace will not light, turn "OFF" all gas and electricity to unit and call Servicer or gas supplier.

For complete shut-down.

Turn gas cock knob on main gas valve to "OFF" position. Disconnect electrical supply to unit.

CAUTION: If this is done during the cold weather months, provisions must be taken to prevent freeze-up of all water pipes and water receptacles.

Whenever your house is to be vacant, arrange to have someone inspect your house for proper temperature. If your furnace should fail to operate, damage could result, such as frozen water pipes.

Safety cutoff device. (Thermal limit)

All models are equipped with a fusible link behind gas valve. In case of main gas valve malfunction and consequent overheating, link will open and close off flow of all gas. See instruction label on front panel of heat exchanger.

Limited Lifetime Warranty
High Efficiency Condensing Gas Furnace
TUC, TDC, TUX and TDX Models (Parts Only)

This warranty is extended by American Standard Inc., to the original purchaser for lifetime under the conditions as defined below or for a period of twenty years from the date of installation for any succeeding owner.

If any part of your TUC, TDC, TUX or TDX Gas-fired Furnace fails because of a manufacturing defect within one year from the date of original purchase, Warrantor will furnish without charge the required replacement part. **Any local transportation, related service labor, air filters and diagnosis calls are not included.**

Lifetime Warranty
The heat exchanger of your TUC, TDC, TUX or TDX Gas Furnace when installed to serve a single family residence or single condominium is warranted to the original purchaser for use during his or her lifetime, provided the dwelling is the original purchaser's primary, uninterrupted residence from the date of purchase until a defect in the primary heat exchanger or secondary heat exchanger is discovered. Warrantor will furnish without charge a replacement heat exchanger F.O.B. nearest Parts Distribution point. **Any local transportation, related service labor, and diagnosis calls are not included.**

Twenty Year Warranty
The heat exchanger of the Gas Furnace is warranted to any purchaser other than the original purchaser for a period of twenty (20) years from the date of original installation of the furnace, subject to proof of original purchase. Warrantor will furnish without charge a replacement heat exchanger F.O.B. nearest Parts Distribution point. **Any local transportation, related service labor, and diagnosis calls are not included.**

Extended Warranties
First Through Tenth Year — type AL 29-4C™ Vent Pipe — (Applies only to installations requiring type AL 29-4C™ Vent Pipe). If the external type AL 29-4C™ vent pipe fails due to perforation caused by corrosion within ten years from the date of original installation of the TUC, TDC, TUX or TDX model gas-fired Furnace and vent pipe, Warrantor will furnish without charge a replacement external vent pipe F.O.B. nearest Parts Distribution point. **Any local transportation, related service labor and diagnosis calls are not included. This warranty** covers only the type AL 29-4C™ external vent pipe specified by Warrantor for use with your Gas Furnace.

During the first year following installation, Warrantor will furnish a replacement for a covered failed component, F.O.B. nearest Parts Distribution point. Thereafter in order to fill any Extended Warranty, **WARRANTOR WILL AT ITS OPTION** provide a heat exchanger including secondary heat exchanger without charge F.O.B. nearest Parts Distribution point or allow a credit (in the amount of the then current wholesale price) of an equivalent heat exchanger toward the purchase price of a comparable heating unit. **Any local transportation, related service labor and diagnosis calls are not included.**

This warranty does not cover failure of your TUC, TDC, TUX or TDX Gas Furnace if it is damaged while in your possession or if the failure is caused by unreasonable use. In no event shall Warrantor be liable for incidental or consequential damages. **In no event shall any implied warranty of merchantability or fitness for use exceed the term of the limited warranty stated above.**

Some states do not allow limitations on how long an implied warranty lasts, so the above limitation may not apply to you. Some states do not allow the exclusion or limitation of incidental or consequential damages, so the above limitation or exclusion may not apply to you. This warranty gives you specific legal rights, and you may also have other rights which vary from state to state.

Parts will be provided by our factory organization or an authorized service organization in your area. All you need do is look us up in the yellow pages or write to the address given below. If you wish further help or information concerning this warranty, contact:

Manager — Product Service
American Standard Inc.
Troup Highway
Tyler, Texas 75711-9010

American Standard Inc.,
Troup Highway, Tyler, Texas 75711-9010
Warrantor

GW-509-4090

All model designs are certified by the American Gas Association Laboratories to comply with national standards for safety performance and durability.

Pub No 32-5002-05
© American Standard Inc. 1991

P.I. 3/91
21X330081 P02

Gas Furnace
Model: TUC100B960A0

B - FURNACES

IMPORTANT — This document is customer property and is to remain with this unit. Please return to service information pack upon completion of work.

Library	
Product Section	
Product	
Model	
Literature Type	
Sequence	
Date	
File No.	
Supersedes	

PRODUCT SPECIFICATIONS

MODEL	TUC100B960A0
TYPE	UPFLOW, INTERMITTENT ELECTRONIC IGNITION
RATINGS① Input, BTUH② Temp. Rise (Min. — Max.) °F.	100,000 35 — 60
BLOWER DRIVE Dia. — Width (in.) No. Used Speeds (No.) CFM vs. in. w.g. Motor HP R.P.M. Volts/Ph/Hz	DIRECT 12 x 11 1 4 SEE FAN PERFORMANCE TABLE 3/4 1075 115/1/60
FILTER — Furnished? Type Recommended Lo Vel. (No. - Size - Thk.) Hi Vel. (No. - Size - Thk.)	YES 1 - 20 x 25 - 1 in.
VENT — Size (in.)	3.0 ROUND
HEAT EXCHANGER Type — Fired — Unfired Gauge (Fired)	ALUMINIZED STEEL TYPE 1 29-4C 20
ORIFICES — Main Nat. Gas Qty. — Drill Size L.P. Gas Qty. — Drill Size	3 — 35 3 — 52
GAS VALVE	REDUNDANT — SINGLE STAGE
DIRECT IGNITION DEVICE Type	HOT SURFACE
BURNERS — Type Number	LINEAR 3
POWER CONN. — V/Ph/Hz③ Ampacity (In Amps) Fuse Size — Max. (Amps.)	115/1/60 19 25
PIPE CONN. SIZE (IN.)	1/2
DIMENSIONS Crated (in.)	H x W x D 54.75 x 26 x 33½
WEIGHT Shipping (lbs.) / Net (lbs.)	235 / 217

① Central furnace heating designs are certified by the American Gas Association Inc. Laboratories.
② Ratings shown are for elevations up to 2000 feet. For elevations above 2000 feet; ratings should be reduced at the rate of 4% for each 1000 feet above sea level.
③ The above wiring specifications are in accordance with the National Electrical Code; however installations must comply with local codes.

OPTIONAL EQUIPMENT

ELECTRONIC AIR CLEANER BEF140C100A
AIR CLEANER RELAY KIT BAY24X043
LP CONVERSION KIT BAYLPKT208
HIGH ALTITUDE PRESSURE SWITCH BAYHALT209

EMERGENCY SHUT-OFF INSTRUCTIONS

IF IT IS SUSPECTED THAT A FAILURE OF THE ELECTRICAL, FUEL, OR MECHANICAL SYSTEMS WITHIN THIS FURNACE HAS OCCURRED, THE GAS SUPPLY SHOULD IMMEDIATELY BE TURNED OFF AT THE MANUAL GAS VALVE, LOCATED IN THE BURNER COMPARTMENT AND/OR AT LEVER-HANDLED COCK, AND ELECTRICAL POWER TO THE FURNACE SHOULD BE DISCONNECTED. THE FAILURE MUST BE CORRECTED BY A QUALIFIED SERVICER BEFORE OPERATING THE FURNACE.

WARNING — DO NOT ATTEMPT TO MANUALLY LIGHT THE GAS BURNERS.

Lighting instructions appear on each unit. Each installation must be checked out at the time of initial start up to insure proper operation of all components. Check out should include putting the unit through one complete cycle as outlined below.

Turn on main electrical supply and set thermostat above indicated temperature. The ignitor will automatically heat for approx. 15 seconds, then the gas valve is energized to permit gas flow to burners. After ignition and flame is established, flame control module monitors flame and supplies power to gas valve until thermostat is satisfied. Ignition sequence will cycle 3 times before lockout occurs.

If burner fails to ignite, lower thermostat setting or disconnect electrical supply, wait 5 minutes, raise thermostat setting above indicated temperature, turn electrical supply on. Unit will repeat lighting sequence.

SAFETY NOTICE

THIS INFORMATION IS INTENDED FOR USE BY INDIVIDUALS POSSESSING ADEQUATE BACKGROUNDS OF ELECTRICAL AND MECHANICAL EXPERIENCE. ANY ATTEMPT TO REPAIR A CENTRAL AIR CONDITIONING PRODUCT MAY RESULT IN PERSONAL INJURY AND OR PROPERTY DAMAGE. THE MANUFACTURER OR SELLER CANNOT BE RESPONSIBLE FOR THE INTERPRETATION OF THIS INFORMATION, NOR CAN IT ASSUME ANY LIABILITY IN CONNECTION WITH ITS USE.

RECONNECT ALL GROUNDING DEVICES

ALL PARTS OF THIS PRODUCT CAPABLE OF CONDUCTING ELECTRICAL CURRENT ARE GROUNDED. IF GROUNDING WIRES, SCREWS, STRAPS, CLIPS, NUTS OR WASHERS USED TO COMPLETE A PATH TO GROUND ARE REMOVED FOR SERVICE, THEY MUST BE RETURNED TO THEIR ORIGINAL POSITION AND PROPERLY FASTENED.

DISCONNECT POWER BEFORE SERVICING

REDDI PARTS

COMPONENT	QTY.	DESCRIPTION	CAT. #
Blower Wheel	1	12"D x 11"W, CW, Concave	WG74X0068
Burner Asm.	4	Replacement	WW54X0112
Capacitor	1	15 MFD, 440V	WW20X0126
Fan - Induced Draft	1	Replacement	WW73X0128
Fan & Limit Control (Fan/Limit)	1	Limit 190°F, 8" Probe Honeywell # L4064A2675	WW29X0902
Filter	1	25" x 20"	WG85X0051
Flame Sensor	1	Fenwal # 22-100000-6H	WW37X0058
Fusible Link (Flame Roll-Out)	1	333°F./167°C Rating, Micro Devices # 404333A	WG09X0033
Gas Valve	1	White Rodgers #36E01-221	WG19X0231
Heat Exchanger Asm.	1	Replacement	WW93X0081
Ignitor Asm.	1	Norton #271N	WW37X0057
Ignitor Control	1	24 VAC, 60 Hz., 3-Try Board White Rodgers #50E47-160	WW37X0056
Motor - Blower	1	3/4 HP, 1075 RPM, 115V, 60 Hz., 1 Phase, 4 Speed, 8.8 FLA, PSC, Sleeve Bearings	WG94X0169
Motor - Comb. Air/I.D. Blower (Vent)	1	Replacement	WW94X0059
Orifice	3	# 35 Drill	WG16X0186
Recup Cell Asm.	1	Replacement	WW93X0096
Relay - Blower	1	DPDT, 16FLA, 48LRA @ 120V, 24V Coil, P & B #KUM1078	WW24X0231
Relay - Combustion Blower	1	SPST, 12FLA, 60LRA @ 125V, 24V Coil, P & B #S87R1A2B1D-24V	WG24X0119
Relay - Time Delay	1	13.8RLA, 82.8LRA @ 120V, 40-80 Sec. Time Delay, ThermO-Disc	WW24X0232
Switch - Aux. Limit	1	Open 155° ± 5°F, Close 125° + 7°F T.I. # 1NT01L-0388	WG23X0090
Switch - Blower Door	1	SPST, N.O., 3/4 H.P. @ 125VAC	WG23X0073
Switch - Pressure	1	1.08" ± .06" Neg. W.C., 28VA Pilot Duty @ 24V, Tridelta #FS6399-615	WW26X0110
Transformer	1	115V Pri., 24V Sec., 35VA Jard # TF-351124-B11A	WW32X0092
Wheel - Combustion Blower	1	Vernco # AL-30-101	WG92X0162

SEQUENCE OF OPERATION

With gas and electrical power "OFF"

1. Duct connections properly sealed
2. Filters in place
3. Venting properly assembled
4. Condensate drains properly installed
5. Blower door in place on upflow models

Adjust Heat Anticipator on Thermostat, for Natural or Propane Gas, as follows:
0.85 — White Rodgers Gas Valve
0.75 — Robertshaw Gas Valve
0.65 — Johnson Gas Valve

Turn knob on main gas valve within unit to "OFF". Turn external gas valve to "ON". Purge air from gas lines. After purging, check all gas connections with soapy solution-**DO NOT CHECK WITH AN OPEN FLAME.** Allow 5 minutes for any gas that might have escaped to dissipate. LP Gas being heavier than air may require forced ventilation. Turn knob on gas valve in unit to "ON".

THERMOSTAT CALLS FOR HEAT

R and W Thermostat contacts close to supply power, through safety limit switches, to control circuit which starts the induced draft blower. When the required combustion air is established, the pressure switch allows power to flow through the safety controls to the flame control module and the ignitor.

The ignitor will heat for approx. 15 seconds, then the gas valve is energized to permit gas flow to the burners. The flame sensor confirms that ignition has been achieved within the 7 second ignition trial period. All models utilize a remote sensor.

If the sensor does not confirm ignition in the 7 second time period, the ignition control will go into a re-try mode. During the second trial for ignition there is a 60 second prepurge time for the induced draft motor, then the ignitor will energize for approx. 45 seconds before the gas valve is energized. If the flame sensor does not confirm ignition during this second trial, the ignition control will go into a re-try mode again. The sequence of operation will be repeated. If the ignition has not been confirmed at the end of the 3rd trial, the ignition control will go into lock out mode.

When ignition and flame is established, the flame control module monitors the flame and supplies power to the gas valve until the thermostat is satisfied. When the thermostat is satisfied, R and W contacts open to de-energize the control circuit.

MAIN BURNER ADJUSTMENT

The regulator on the unit's gas valve is set for a manifold pressure of 3.5" W.C. for natural gas. This can be checked by turning off the gas valve and removing the plug on the gas valve labeled "Outlet Pressure" and installing a water manometer. Turn the gas valve to ON. Put furnace in operation and read the manometer. If it is necessary to adjust the gas consumption, turn the adjustment screw (beneath cap on pressure regulator) clockwise or counterclockwise to 3.5" W.C. After checking pressure, turn gas valve OFF. Remove manometer. Replace plug. Put furnace into operation and leak check plug with soapy solution for leaks.

For LP Models, the regulator is set at 10.5" W.C. If the gas consumption needs to be increased, do not exceed 11.0" W.C.

FLAME CHARACTERISTICS

On LP (propane) units, some light yellow tipping of the outer mantle is normal. Inner mantle should be bright blue. Natural gas units should not have any yellow tipped flames.

Another method for checking input is to clock the gas meter with **all** other gas appliances turned OFF.

HIGH ALTITUDE INSTALLATIONS:

IMPORTANT: The sea level rated input of the furnace installed at elevations above 2,000 feet should be reduced 4% for each 1,000 feet above sea level. For example, for elevations of 4,350 feet, the correct input would be the sea level rated input less 17%.

Check orifice size. The drill size is stamped on the gas manifold or see Table below.

Input Rating BTUH (000)	No. of Burners	Main Burner Orifice Drill Size	
		Nat. Gas	Propane
100	3	35	52

CONTROLS AND SAFETY SWITCH ADJUSTMENT

The fan switch is factory set for the blower to come on at 120°F.* and to shut off at 100°F. These settings are satisfactory in most applications. If necessary to field adjust, disconnect power and proceed as follows:

1. Adjust fan-off pointer to shut blower off when air at the farthest register begins to feel cool, or to setting that will not cause blower to recycle.

2. Fan-on differential is adjustable from 25°F. to 45°F. above fan-off setting.

*NOTE: Do not exceed 125°F. fan-on setting for counterflow units.

BLOWER

All models are factory equipped with a transformer and relay. Additional relay is not required when adding air conditioning.

REMOVAL OF BLOWER:

If the blower should have to be removed for servicing, **TURN OFF ELECTRICITY TO UNIT** and disconnect electrical leads at control box.

AIRFLOW ADJUSTMENT

Check inlet and outlet air temperatures to make sure they are within the ranges specified on the furnace rating plate. If airflow needs to be increased or decreased, see wiring diagram for information on changing the wiring to blower motor. Speed changes can be made at molex plug in blower compartment.

LIMIT SWITCH CHECK OUT

The limit switch is a safety device designed to electrically de-energize the gas valve should the furnace become overheated.

To check for proper operation of the limit switches, set thermostat to temperature higher than room temperature to energize the gas valve. Restrict airflow by blocking the return air or by interrupting power to the blower. When the furnace reaches the maximum outlet temperature as shown on the rating plate, the burners must shut off. If they do not shut off after a reasonable time and over-heating is evident, a faulty limit is probable and the limit switch must be replaced. After checking the operation of the fan and limit control, be sure to remove the paper or cardboard restriction from the filter.

FLAME ROLL-OUT DEVICE

All models are equipped with a fusible link on burner cover. In case of flame roll-out, link will open and close off flow of all gas. See instruction label on front panel of heat exchanger.

FAN PERFORMANCE
FURNACE AIRFLOW (CFM) VS EXTERNAL STATIC PRESSURE (m. w.g.)

MODEL	SPEED TAP	— FILTER IN PLACE —								
		.10	.20	.30	.40	.50	.60	.70	.80	.90
TUC100B960A0	4	—	2086	2050	2016	1979	1926	1868	1808	1729
	3	—	1894	1870	1839	1811	1773	1722	1684	1611
	2	1768	1749	1727	1697	1668	1637	1598	1554	1478
	1	1516	1497	1473	1457	1433	1409	1380	1341	1274

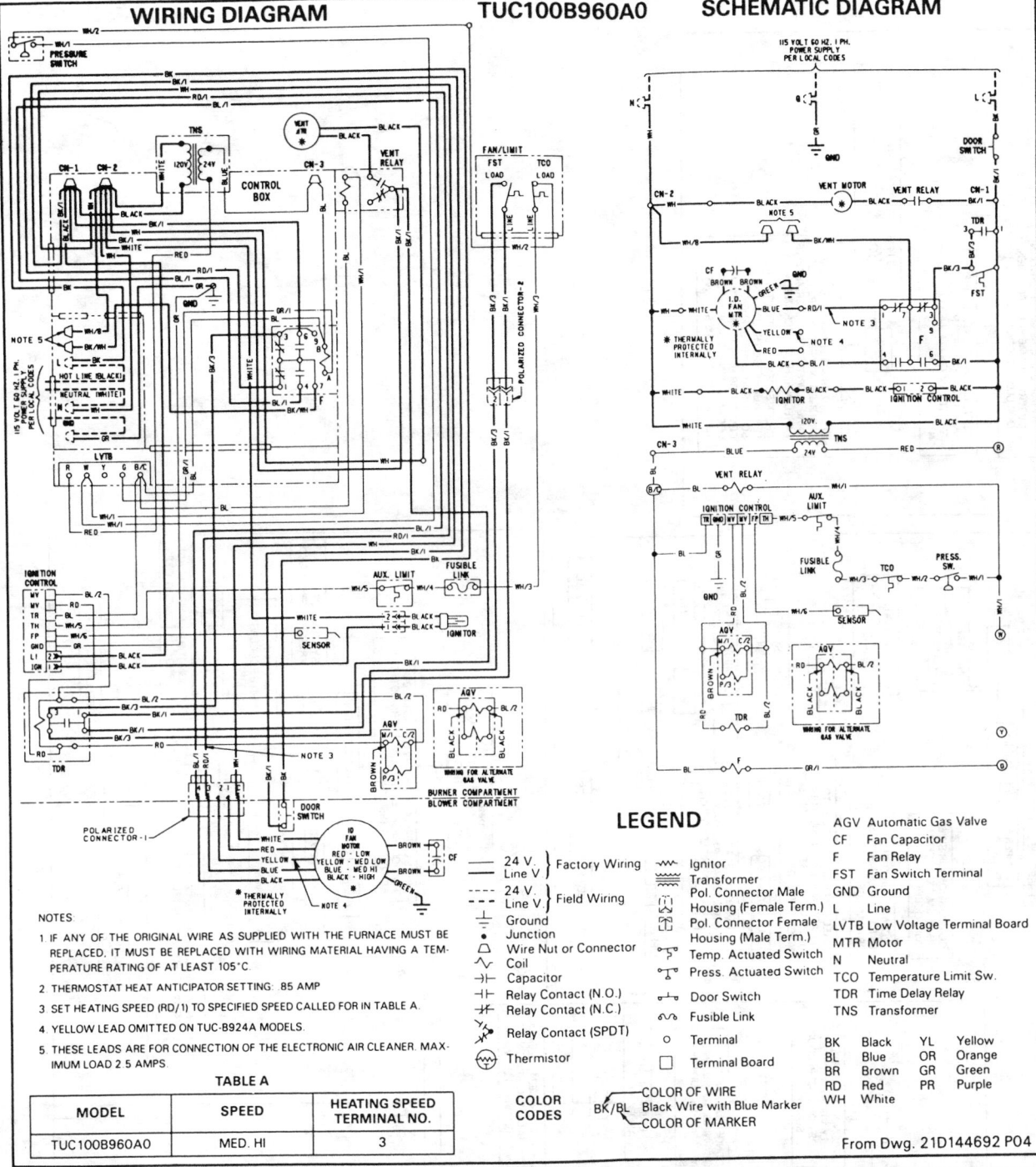

WIRING DIAGRAM — TUC100B960A0 — SCHEMATIC DIAGRAM

NOTES:

1. IF ANY OF THE ORIGINAL WIRE AS SUPPLIED WITH THE FURNACE MUST BE REPLACED, IT MUST BE REPLACED WITH WIRING MATERIAL HAVING A TEMPERATURE RATING OF AT LEAST 105°C.

2. THERMOSTAT HEAT ANTICIPATOR SETTING: .85 AMP

3. SET HEATING SPEED (RD/1) TO SPECIFIED SPEED CALLED FOR IN TABLE A.

4. YELLOW LEAD OMITTED ON TUC-B924A MODELS.

5. THESE LEADS ARE FOR CONNECTION OF THE ELECTRONIC AIR CLEANER. MAXIMUM LOAD 2.5 AMPS.

TABLE A

MODEL	SPEED	HEATING SPEED TERMINAL NO.
TUC100B960A0	MED. HI	3

LEGEND

24 V.	Factory Wiring
Line V.	
24 V.	Field Wiring
Line V.	
	Ground
•	Junction
	Wire Nut or Connector
	Coil
	Capacitor
	Relay Contact (N.O.)
	Relay Contact (N.C.)
	Relay Contact (SPDT)
	Thermistor
	Ignitor
	Transformer
	Pol. Connector Male Housing (Female Term.)
	Pol. Connector Female Housing (Male Term.)
	Temp. Actuated Switch
	Press. Actuated Switch
	Door Switch
	Fusible Link
o	Terminal
	Terminal Board

AGV	Automatic Gas Valve
CF	Fan Capacitor
F	Fan Relay
FST	Fan Switch Terminal
GND	Ground
L	Line
LVTB	Low Voltage Terminal Board
MTR	Motor
N	Neutral
TCO	Temperature Limit Sw.
TDR	Time Delay Relay
TNS	Transformer

COLOR CODES

BK	Black	YL	Yellow
BL	Blue	OR	Orange
BR	Brown	GR	Green
RD	Red	PR	Purple
WH	White		

BK/BL — Black Wire with Blue Marker / COLOR OF WIRE / COLOR OF MARKER

From Dwg. 21D144692 P04

WARNING

The furnace must be isolated from the gas supply piping system by closing its individual manual shutoff valve during any pressure testing of the gas supply piping system at test pressures equal to or greater than 1/2 psig.

All gas fittings must be checked for leaks using a soapy solution before lighting furnace. **DO NOT CHECK WITH AN OPEN FLAME.**

WARNING: DISCONNECT POWER TO UNIT BEFORE REMOVING BLOWER DOOR.

Unit is equipped with a blower door switch which cuts power to blower and gas valve causing shutdown when door is removed. Unit must not be altered to allow operation with the blower door removed. Operation with doors removed or ajar can permit the escape of dangerous fumes. All panels must be securely closed at all times for safe operation of the furnace.

PRINCIPLE OF OPERATION

The system utilizes a silicone carbide element for ignition. The ignitor is an electrically heated resistance element which thermally ignites the gas. The flame detector circuit utilizes flame rectification for monitoring the gas flame.

Upon a call for heat, the element is powered from the 120 VAC line and allowed to heat for 15 seconds (typical). Then the main valve is powered, permitting gas flow to the burner for the trial-for-ignition period. At the end of this period, the flame sensor checks for the presence of flame. If flame is present, the system will monitor it and hold the main valve open. If flame is not established within the trial-for-ignition period, the system will recycle through the complete ignition sequence. If after three cycles combustion has not been established, the system will lockout. If lockout does occur the combustion vent motor will continue to run as long as the thermostat is calling for heat.

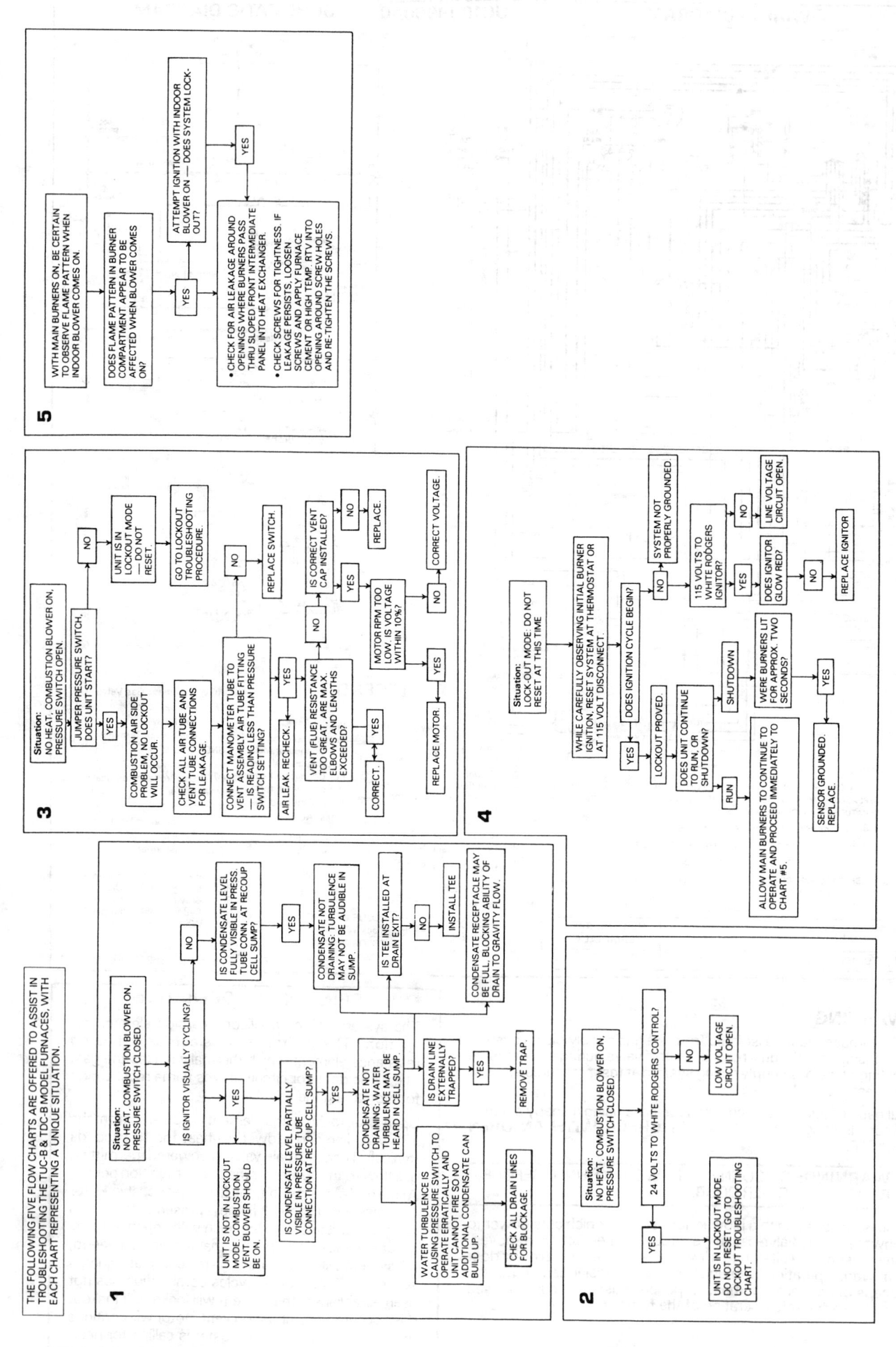

THE FOLLOWING FIVE FLOW CHARTS ARE OFFERED TO ASSIST IN TROUBLESHOOTING THE TUC-B & TDC-B MODEL FURNACES, WITH EACH CHART REPRESENTING A UNIQUE SITUATION.

1

Situation:
NO HEAT. COMBUSTION BLOWER ON, PRESSURE SWITCH CLOSED.

IS IGNITOR VISUALLY CYCLING?

YES

UNIT IS NOT IN LOCKOUT MODE. COMBUSTION VENT BLOWER SHOULD BE ON.

IS CONDENSATE LEVEL PARTIALLY VISIBLE IN PRESSURE TUBE CONNECTION AT RECOUP CELL SUMP?

NO — IS CONDENSATE LEVEL FULLY VISIBLE IN PRESS. TUBE CONN. AT RECOUP CELL SUMP?

YES — CONDENSATE NOT DRAINING. TURBULENCE MAY NOT BE AUDIBLE IN SUMP.

CONDENSATE NOT DRAINING. WATER TURBULENCE MAY BE HEARD IN CELL SUMP.

IS DRAIN LINE EXTERNALLY TRAPPED?

YES — REMOVE TRAP.

IS TEE INSTALLED AT DRAIN EXIT?

NO — INSTALL TEE

CONDENSATE RECEPTACLE MAY BE FULL, BLOCKING ABILITY OF DRAIN TO GRAVITY FLOW.

WATER TURBULENCE IS CAUSING PRESSURE SWITCH TO OPERATE ERRATICALLY AND UNIT CANNOT FIRE SO NO ADDITIONAL CONDENSATE CAN BUILD UP.

CHECK ALL DRAIN LINES FOR BLOCKAGE.

2

Situation:
NO HEAT. COMBUSTION BLOWER ON, PRESSURE SWITCH CLOSED.

24 VOLTS TO WHITE RODGERS CONTROL?

NO — LOW VOLTAGE CIRCUIT OPEN

YES

UNIT IS IN LOCKOUT MODE. DO NOT RESET. GO TO LOCKOUT TROUBLESHOOTING CHART.

3

Situation:
NO HEAT. COMBUSTION BLOWER ON, PRESSURE SWITCH OPEN.

JUMPER PRESSURE SWITCH. DOES UNIT START?

NO — UNIT IS IN LOCKOUT MODE — DO NOT RESET.

GO TO LOCKOUT TROUBLESHOOTING PROCEDURE.

YES

COMBUSTION AIR SIDE PROBLEM. NO LOCKOUT WILL OCCUR.

CHECK ALL AIR TUBE AND VENT TUBE CONNECTIONS FOR LEAKAGE.

CONNECT MANOMETER TUBE TO VENT ASSEMBLY AIR TUBE FITTING — IS READING LESS THAN PRESSURE SWITCH SETTING?

NO — REPLACE SWITCH

YES

AIR LEAK. RECHECK.

YES

VENT (FLUE) RESISTANCE TOO GREAT. ARE MAX. ELBOWS AND LENGTHS EXCEEDED?

NO — IS CORRECT VENT CAP INSTALLED?

NO — REPLACE.

YES — MOTOR RPM TOO LOW. IS VOLTAGE WITHIN 10%?

NO — CORRECT VOLTAGE.

YES — REPLACE MOTOR.

CORRECT. YES

4

Situation:
LOCK-OUT MODE. DO NOT RESET AT THIS TIME

WHILE CAREFULLY OBSERVING INITIAL BURNER IGNITION, RESET SYSTEM AT THERMOSTAT OR AT 115 VOLT DISCONNECT.

DOES IGNITION CYCLE BEGIN?

NO — SYSTEM NOT PROPERLY GROUNDED.

115 VOLTS TO WHITE RODGERS IGNITOR?

NO — LINE VOLTAGE CIRCUIT OPEN

YES — DOES IGNITOR GLOW RED?

YES

NO — REPLACE IGNITOR

YES — LOCKOUT PROVED.

DOES UNIT CONTINUE TO RUN, OR SHUTDOWN?

SHUTDOWN — WERE BURNERS LIT FOR APPROX. TWO SECONDS?

YES — SENSOR GROUNDED. REPLACE.

RUN — ALLOW MAIN BURNERS TO CONTINUE TO OPERATE AND PROCEED IMMEDIATELY TO CHART #5.

5

WITH MAIN BURNERS ON, BE CERTAIN TO OBSERVE FLAME PATTERN WHEN INDOOR BLOWER COMES ON

DOES FLAME PATTERN IN BURNER COMPARTMENT APPEAR TO BE AFFECTED WHEN BLOWER COMES ON?

YES

ATTEMPT IGNITION WITH INDOOR BLOWER ON — DOES SYSTEM LOCK-OUT?

YES

• CHECK FOR AIR LEAKAGE AROUND OPENINGS WHERE BURNERS PASS THRU SLOPED FRONT INTERMEDIATE PANEL INTO HEAT EXCHANGER.

• CHECK SCREWS FOR TIGHTNESS. IF LEAKAGE PERSISTS, LOOSEN SCREWS AND APPLY FURNACE CEMENT OR HIGH TEMP. RTV INTO OPENING AROUND SCREW HOLES AND RE-TIGHTEN THE SCREWS.

Arcoaire/Comfortmaker

Deluxe Gas Fired
Forced Draft Furnace
GUA/GDA Series
Service Manual

7508-640

REVISED: FEBRUARY, 1990
SUPERSEDES: MAY, 1989

SERVICE CHECK SHEET

OWNER NAME: _____

ADDRESS: _____

CITY: _____ STATE: _____ ZIP: _____

MODEL NUMBER: _____ SERIAL NUMBER _____

TYPE OF GAS - NAT: _____ LPG: _____

GAS VALVE: _____ LINE PRESSURE: _____ MANIFOLD PRESSURE: _____

THERMOSTAT: _____ SUBBASE: _____ HEAT ANTICIPATOR SETTING: _____

BLOWER MOTOR H.P. _____ VOLTAGE: _____

FAN OPENS: _____ °F   FAN CLOSES: _____ °F

FAN AND LIMIT
MODEL NUMBER: _____ LIMIT OPENS: _____ °F   LIMIT CLOSES _____ °F

TEMPERATURE RISE: RETURN AIR: _____ °F   SUPPLY AIR: _____ °F

FILTER SIZE AND TYPE: _____

BURNER FLAME PROPERLY ADJUSTED? _____

DRIP-LEG INSTALLED PRIOR TO GAS VALVE? _____

EVAPORATOR BLOWER SPEED CHECKED? _____

EVAPORATOR BLOWER PROPERLY LUBRICATED? _____

FORCED DRAFT MOTOR SPEED CHECKED? _____

FORCED DRAFT MOTOR PROPERLY LUBRICATED? _____

ALL ELECTRICAL CONNECTIONS, MOUNTING SCREWS, ETC. TIGHT? _____

CHECK SEALED COMBUSTION CHAMBER FOR AIR LEAKS (GASKETS) _____

_____

_____

_____

_____

_____

_____

SERVICEMAN _____

INDEX

I. INTRODUCTION

INTRODUCTION TO THE
FORCED DRAFT FURNACE SERVICE MANUAL

This Service Manual is designed to be used in conjunction with the Installation Manual and Operating Procedure furnished with each furnace.

Because of the features present within the operation and servicing of the Forced Draft furnace, a thorough understanding of the sequence of operation is a must.

This Manual is divided into several sections. **The First Section** deals with general operation of the furnace, pointing out the different components within the furnace and how they function, including the Forced Draft Blower.

Most furnaces use the conventional type venting system which draws the combustion air through the furnace. With our Forced Draft Furnace, we push the air through the heat exchanger and out the vent.

The Forced Draft furnace is manufactured in input ranges of 40,000 to 120,000 BTUH. Each burner fires at a range of 20,000 BTUH each. In order to control combustion within this furnace, we use a sealed and pressurized combustion chamber. A sealed chamber means that all air entering the furnace is provided by the Forced Draft motor and not by conventional methods. If for any reason the seal is broken, the furnace will not operate properly.

A direct spark ignition or hot surface ignition is used for main burner lighting, instead of a pilot relight system. This simply means that the pilot has been eliminated in this furnace and direct spark (or hot surface) is used for burner ignition.

The Second Section deals with troubleshooting, heating operation and electrical controls. This section is to be used by a qualified service technician. The technician must have a thorough understanding of the combustion process and electrical controls operation, before trying to service this equipment.

Three things are important to remember when servicing the electrical controls in this furnace:

1. Make sure that furnace is properly grounded.

2. Within the ignition control there is a 7 second "lockout" function. This is a safety device that gives the furnace 100% shut-off capabilities in case the gas entering the burner fails to ignite.

3. The Blocked Flue Pressure Switch—This is another safety device that operates on a positive pressure principle. It is normally closed and will remain closed and allow the furnace to operate until the vent system is blocked. When blockage occurs, (.35″ W.C. set Point) the switch will sense too much positive pressure and will open up, breaking control voltage and preventing further furnace operation.

The Third Section includes a step-by-step description of how to clean, inspect and maintain the furnace. Also included, if it should become necessary, a step-by-step sequence of how to remove and replace the heat exchanger.

A final word of caution; although we have mentioned some new features that have been incorporated within the Forced Draft Furnace, standard operation of the furnace is still the same as in conventional furnaces in many ways. It still requires the following steps when servicing:

1. The furnace must be isolated from the gas supply piping system by closing individual shut-off valves during any pressure testing of the gas supply piping system at test pressures equal to or more than ½ PSIG (13.86″ Water Column).

2. Use pipe compounds on gas pipe threads that are resistant to all gases.

3. Do not use open flame to check for gas leaks. Use a soapy solution or a commercial leak tester.

4. Keep instruments and tools in good repair and calibrated.

5. Instruct the owner or operator of the furnace in the proper procedure to follow for shutting off the furnace in case of an emergency. Should overheating occur or the gas supply fail to shut-off, instruct the owner or user to turn off the manual gas valve to the appliance before turning off the electrical supply. Explain how to change, service or clean filters and setting of controls for best comfort conditions.

   Refer the owner to the user's manual 7112-012 for future reference and further information.

6. Explain to the owner or operator that the furnace has lock-out safety devices within its operation, and that if it should fail to operate, a qualified service technician should be called to service the unit.

A. FURNACE

FIGURE 1

SHOWING UPFLOW FURNACE

FIGURE 2

SHOWING DOWNFLOW FURNACE

B. IDENTIFICATION OF COMPONENTS

① Wrap Casing Assembly
② Control Louverd Door
③ Blower Door
④ Blower Door Interlock Switch
⑤ Burner Door
⑤A Gasket
⑥ Flash Shield
⑦ Burner Assembly
⑧ Cross-over On Burners
⑨ Flue Box (Porcelain)
⑨A Gasket
⑩ Collector Box
⑩A Gasket
⑪ Equalizer Plate
⑪A Gasket (2)
⑫ Heat Exchanger Flue Baffles
⑬ Heat Exchanger (RPJ)
⑭ Pressure Switch
⑮ Fan & Limit Switch
⑯ Forced Draft Motor
⑰ Gas Valve (Redundant)
　　See Page 5
⑱ Manifold Connection-Dell Fitting
　　See Figure 14

⑲ Manifold
⑳ Burner Orifice and Spring
　　See Page 4
㉑ Ignitor Box
　　See Page 12
㉒ Blower Assembly with Motor
㉓ Transformer, Relay Package
㉔ Capacitor
㉕ Burner Box
㉖ Burner Inspection Window
㉗ Ignitor
㉘ Sensor
㉙ Aux. Limit Control
㉚ Flame Roll-out Switch

HAND TOOLS

The hand tools carried by a qualified service mechanic will normally be sufficient to do any service work upon this type furnace.

Recommended tools needed:

1. Thermometers
2. U-Tube Manometer with fittings
3. Clamp-on Amp Meter
4. Watch with second hand or stopwatch
5. Soap bubbles for checking leaks
6. Incline Manometer or Magnahelic with Pitot tube or static pressure tip
7. V.O.M. Meter

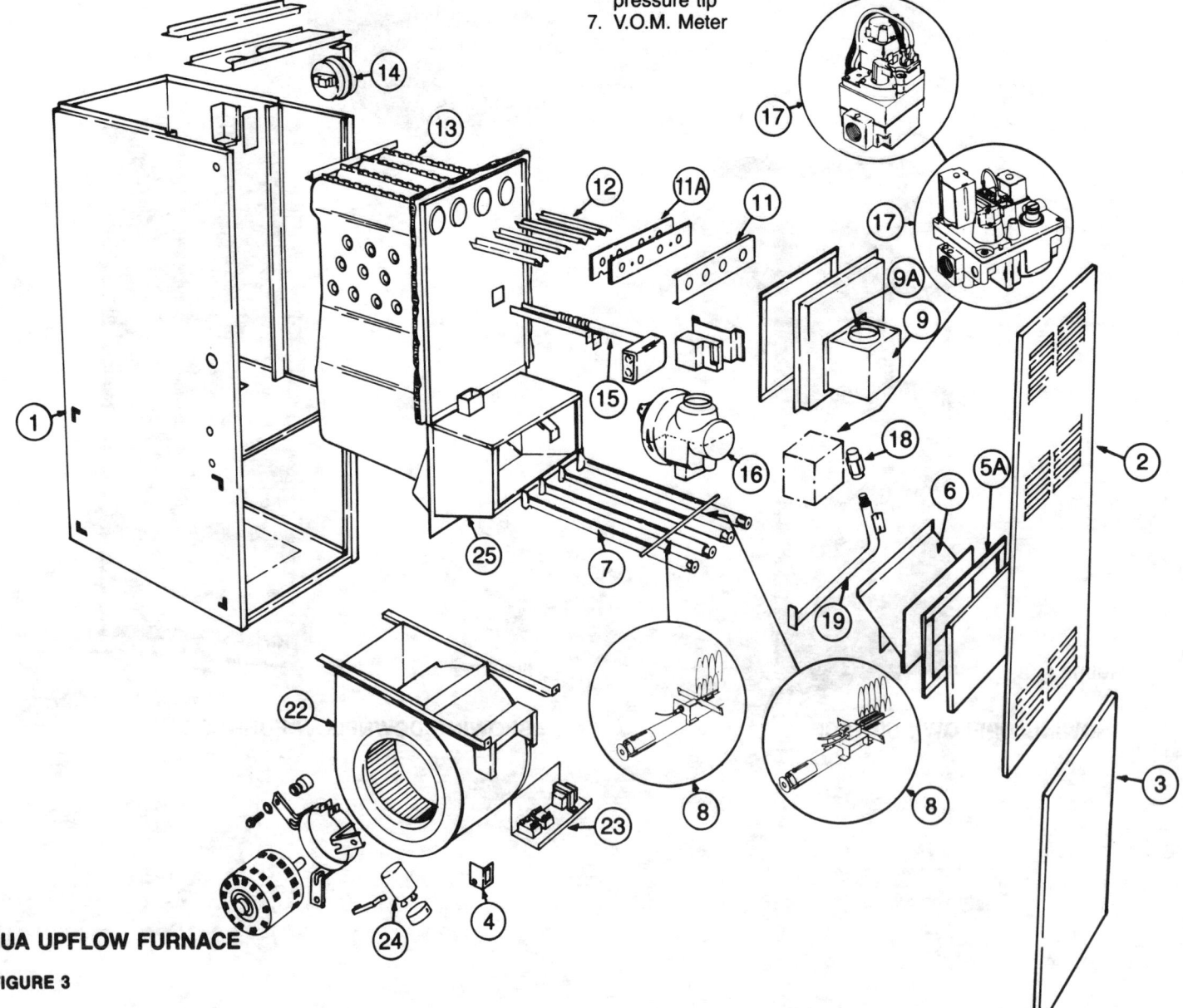

GUA UPFLOW FURNACE

FIGURE 3

3

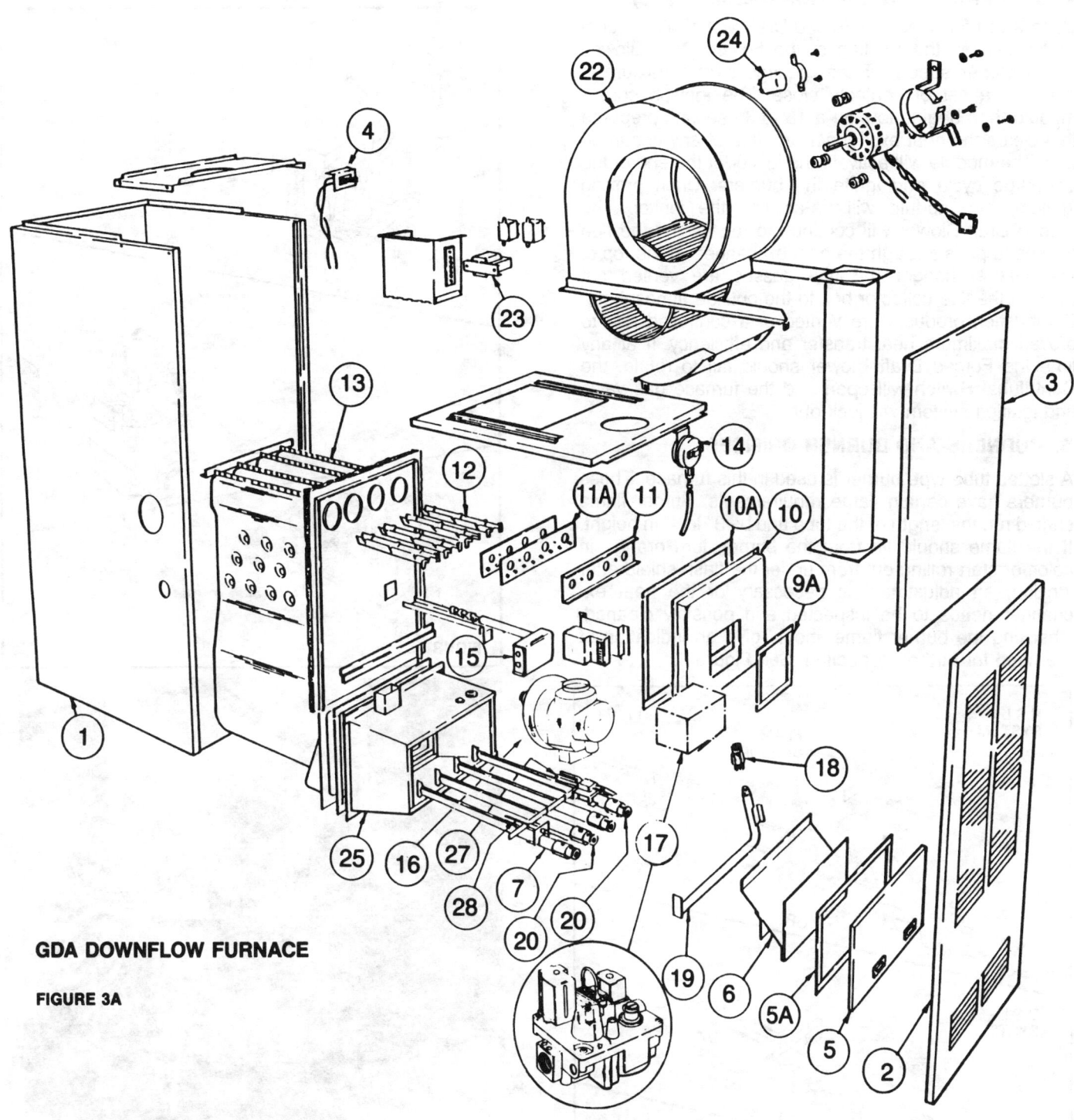

GDA DOWNFLOW FURNACE

FIGURE 3A

4

II. FURNACE OPERATION

A. SEQUENCE OF OPERATION (Heating Cycle)

Upon a call for heat, the Forced Draft Motor's centrifugal switch senses the rotation of the Forced Draft Blower. When proper speed for the required air/gas mixture is reached, a set of contacts close. The ignition control module then begins to time a 15 to 30 second prepurge that clears the heat exchanger and flue of any unburned gas. The module will activate the ignitor at the end of the pre-purge cycle and ignite the burners. Upon proving ignition, the module will de-activate the ignitor. The Forced Draft Blower will continue to run forcing the flue gasses to pass through the heat exchanger. At the top of the heat exchanger the flue gasses are vented out through the flue collector box to the outside atmosphere. Combustion products are vented at a controlled rate to proved maximum heat transfer and efficiency. If at any time the Forced Draft Blower should fail to rotate, the Centrifugal Switch will open and the furnace gas valve and ignition system will lock-out.

B. BURNERS AND BURNER ORIFICES

A slotted tube type burner is used in this furnace. These burners have certain flame requirements. Burner flame should run the length of the tube and be 3″ to 4″ in height. If the flame should lift from the burner, turn orange in color, or start rolling out from under the flash shield after ignition, an adjustment is necessary or the heat exchanger needs to be inspected and possibly cleaned. Checking the burner flame should give an indication of the need for further inspection. See Figure 4.

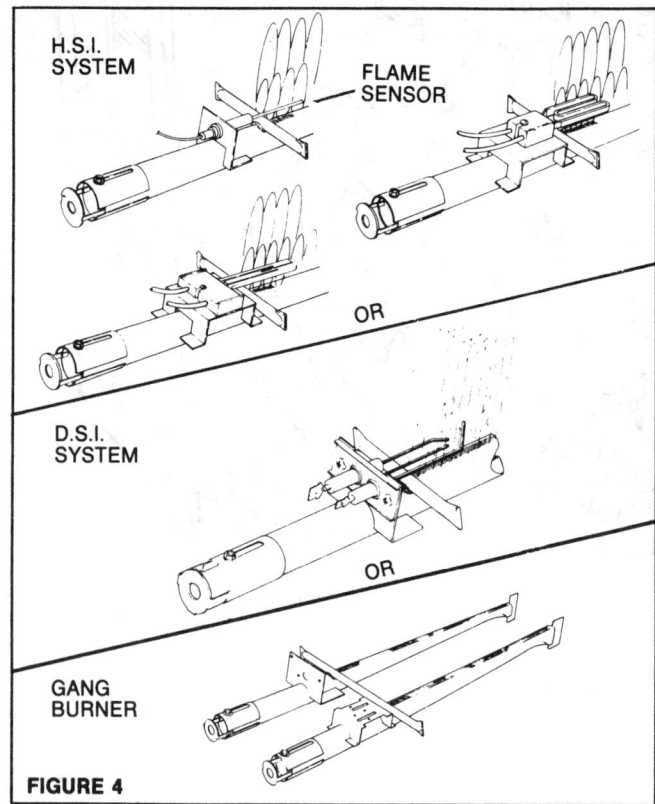

FIGURE 4

C. GAS VALVE

The control used on this furnace is a redundant combination gas valve. This design provides manual main valve, screw regulation automatic main diaphragm valve, plus the added safety of an additional in-line electromechanical operated redundant valve. See Figure 5.

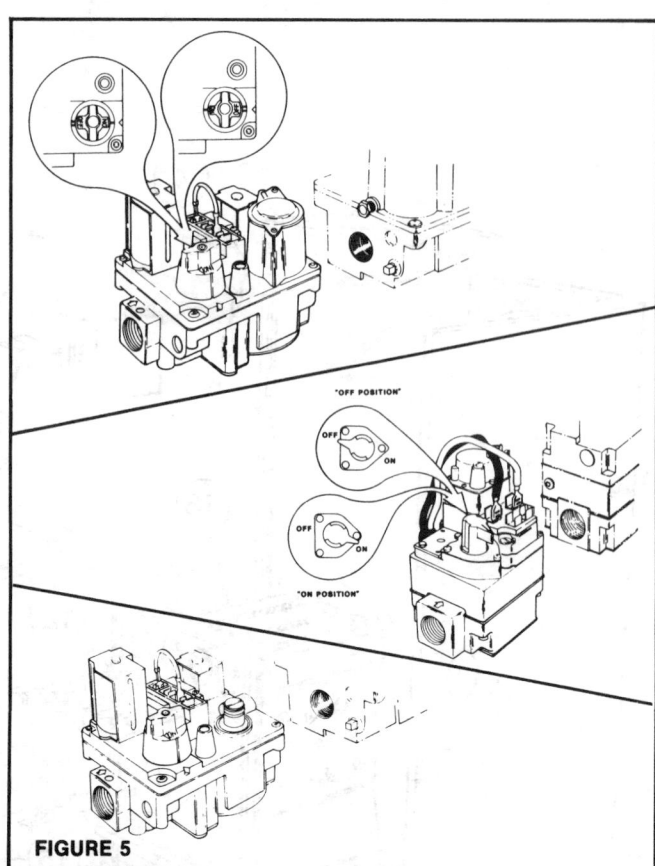

FIGURE 5

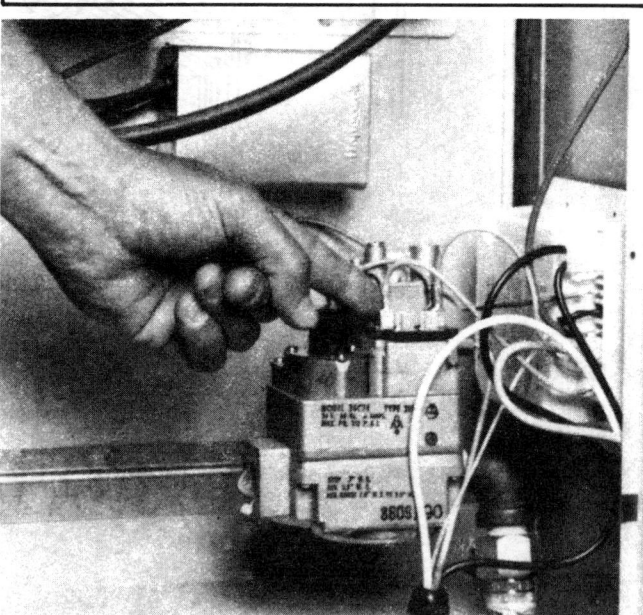

FIGURE 6

D. PRESSURE REGULATOR

A Pressure tap has been provided on the gas valve. See Figure 5. Attach a pressure gauge fitting to this tap and check the outlet pressure. The pressure should be set as indicated on the rating plate. To adjust the pressure regulator, remove the cap covering the Pressure regulator adjusting screw. On valves with more than one adjusting screw, adjust only the screw marked "HI". Turn the adjusting screw out (counter-clockwise) to decrease the pressure, and in (Clockwise) to increase the pressure. Reinstall the cap securely.

III. FURNACE OPERATION (ELECTRICAL)

A. SEQUENCE OF OPERATION

Heating Cycle—Direct Spark Ignition System

a) HEATING CYCLE (D.S.I)

1. With blower door in place, the N.O. Switch Ⓔ closes, allowing 115 to 120VAC to feed to transformer Ⓐ. When the wall thermostat is placed in heating position and mercury bulb or contacts are made, 24 VAC from Ⓡ will feed to Ⓦ on furnace strip Ⓜ. This will energize coil Ⓒ which will close contacts ⓒ to energize the Forced Draft Blower Ⓖ.

Direct Spark Ignition System

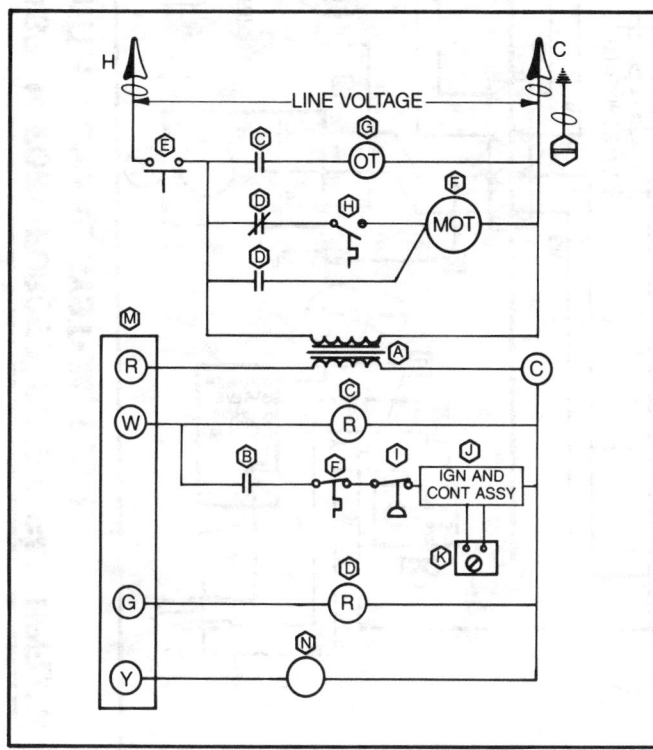

FIGURE 7

LEGEND

- Ⓐ Transformer 120VAC/24VAC
- Ⓑ Switch (Centrifugal)
- Ⓒ Relay (Combustion Blower)
- ⓒ Contact
- Ⓓ Relay (Cooling)
- Ⓔ Switch (Blower Door)
- Ⓕ Motor (Blower)
- Ⓖ Motor (Combustion Blower)
- Ⓗ Control (Fan/Limit)
- Ⓘ Pressure Switch (Block Flue)
- Ⓙ Control (Ignition with Pre-Purge)
- Ⓚ Gas Valve
- Ⓛ Screw (Grounding)
- Ⓜ Switch (Thermostat)
- Ⓝ Contactor (Compressor)

2. Rotation of the Forced Draft Blower Ⓖ will close centrifugal switch Ⓑ within the motor to prove blower operation and send power through the fan/limit Ⓗ N.C. set of contacts—and pressure switch Ⓘ to the ignition control Ⓙ.

3. The ignition control Ⓙ allows a pre-purge period of between 15 to 30 seconds before trial ignition.

4. After pre-purge, the ignition control Ⓙ simultaneously powers the gas valve Ⓚ and provides the spark for ignition of main burner flames. Should the main flame fail to ignite after a trial period of five (5) seconds, from the initiation of spark and gas flow, the system will automatically LOCK-OUT. To reset the system, you must interrupt electrical power (at the thermostat or main power supply) for 30 seconds before you can call for the heat cycle again.

5. Power will feed through N.C. Contact ⓓ of the blower relay to the fan/limit control Ⓗ. As the heat exchanger heats up, between 120° and 140°F, the fan switch will close and allow blower motor Ⓕ to start.

6. After the thermostat is satisfied, the ignition system Ⓙ will cut power to the gas valve to drop out the main burner flame.

7. Blower motor Ⓕ will continue operation until heat exchanger cools off between 95° and 105°F. The fan switch is adjustable up to a maximum fan "ON" setting of 130°F, for this heating up and cooling off period. See Section V for adjustment.

8. Furnace is ready to repeat the same cycle as above when thermostat calls for heat.

b) APPROXIMATE TIMING FOR DIFFERENT COMPONENTS WHEN CALLING FOR HEAT:

1. Centrifugal Switch—5 Seconds
2. Pre-Purge Cycle—30 Seconds
3. Ignition Lock-Out—5 Seconds (Burner Gas Fails to Ignite)
4. Burner Ignition—5 Seconds
5. Blower (Furnace)—60 to 90 Seconds

c) SEQUENCE OF OPERATION—COOLING CYCLE:

1. Place thermostat system switch in "Cool" position. Fan Switch can either be set on "AUTO" or "ON" position. If fan switch is in the "ON" position, current will flow from the transformer Ⓐ through the Ⓡ leg directly to Ⓖ on therminal strip Ⓜ and energize cooling relay Ⓓ, closing contact ⓒ to energize the blower motor.

2. If Fan Switch is in the "AUTO" position and cooling is called for by the thermostat, power will feed from Ⓡ to Ⓨ on the sub-base and from Ⓨ to Ⓖ and bring on the fan. It will also energize the contactor through Ⓡ to Ⓨ and bring on the compressor for cooling.

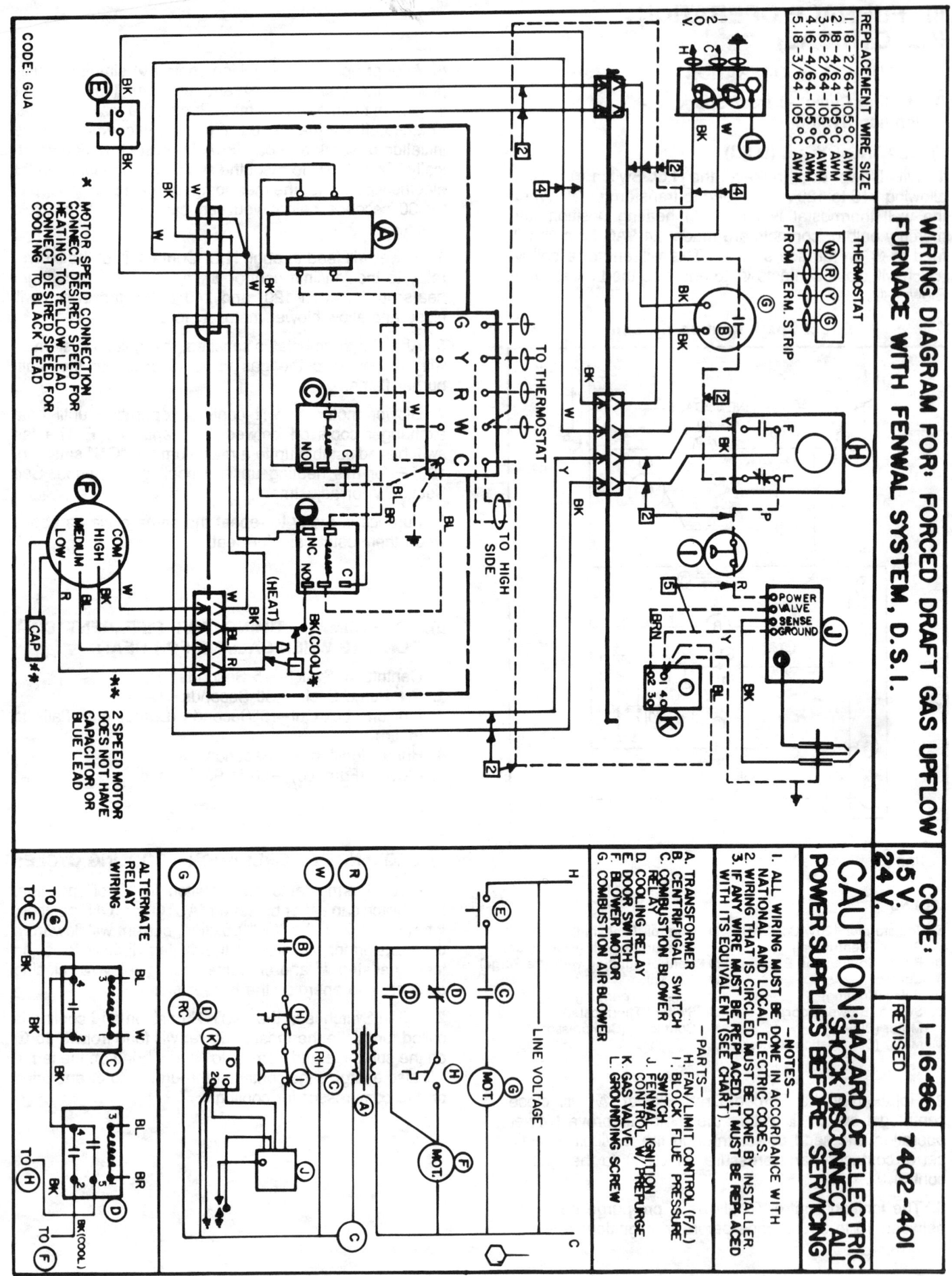

WIRING DIAGRAM FOR: FORCED DRAFT GAS UPFLOW
FURNACE WITH FENWAL SYSTEM, D.S.I.

CODE: 1-16486 REVISED 7402-401

115 V.
24 V.

CAUTION: HAZARD OF ELECTRIC SHOCK DISCONNECT ALL POWER SUPPLIES BEFORE SERVICING

—NOTES—
1. ALL WIRING MUST BE DONE IN ACCORDANCE WITH NATIONAL AND LOCAL ELECTRIC CODES.
2. WIRING THAT IS CIRCLED MUST BE DONE BY INSTALLER.
3. IF ANY WIRE MUST BE REPLACED, IT MUST BE REPLACED WITH ITS EQUIVALENT (SEE CHART).

—PARTS—
A. TRANSFORMER
B. CENTRIFUGAL SWITCH
C. COMBUSTION BLOWER RELAY
D. COOLING RELAY
E. DOOR SWITCH
F. BLOWER MOTOR
G. COMBUSTION AIR BLOWER
H. FAN/LIMIT CONTROL (F/L)
I. BLOCK FLUE PRESSURE SWITCH
J. FENWAL IGNITION CONTROL W/PREPURGE
K. GAS VALVE
L. GROUNDING SCREW

REPLACEMENT WIRE SIZE
1. 18-2/64-105°C AWM
2. 18-4/64-105°C AWM
3. 16-2/64-105°C AWM
4. 16-4/64-105°C AWM
5. 18-3/64-105°C AWM

CODE: GUA

THERMOSTAT

FROM TERM. STRIP

* MOTOR SPEED CONNECTION
CONNECT DESIRED SPEED FOR HEATING TO YELLOW LEAD
CONNECT DESIRED SPEED FOR COOLING TO BLACK LEAD

** 2 SPEED MOTOR DOES NOT HAVE CAPACITOR OR BLUE LEAD

ALTERNATE RELAY WIRING

LINE VOLTAGE

7

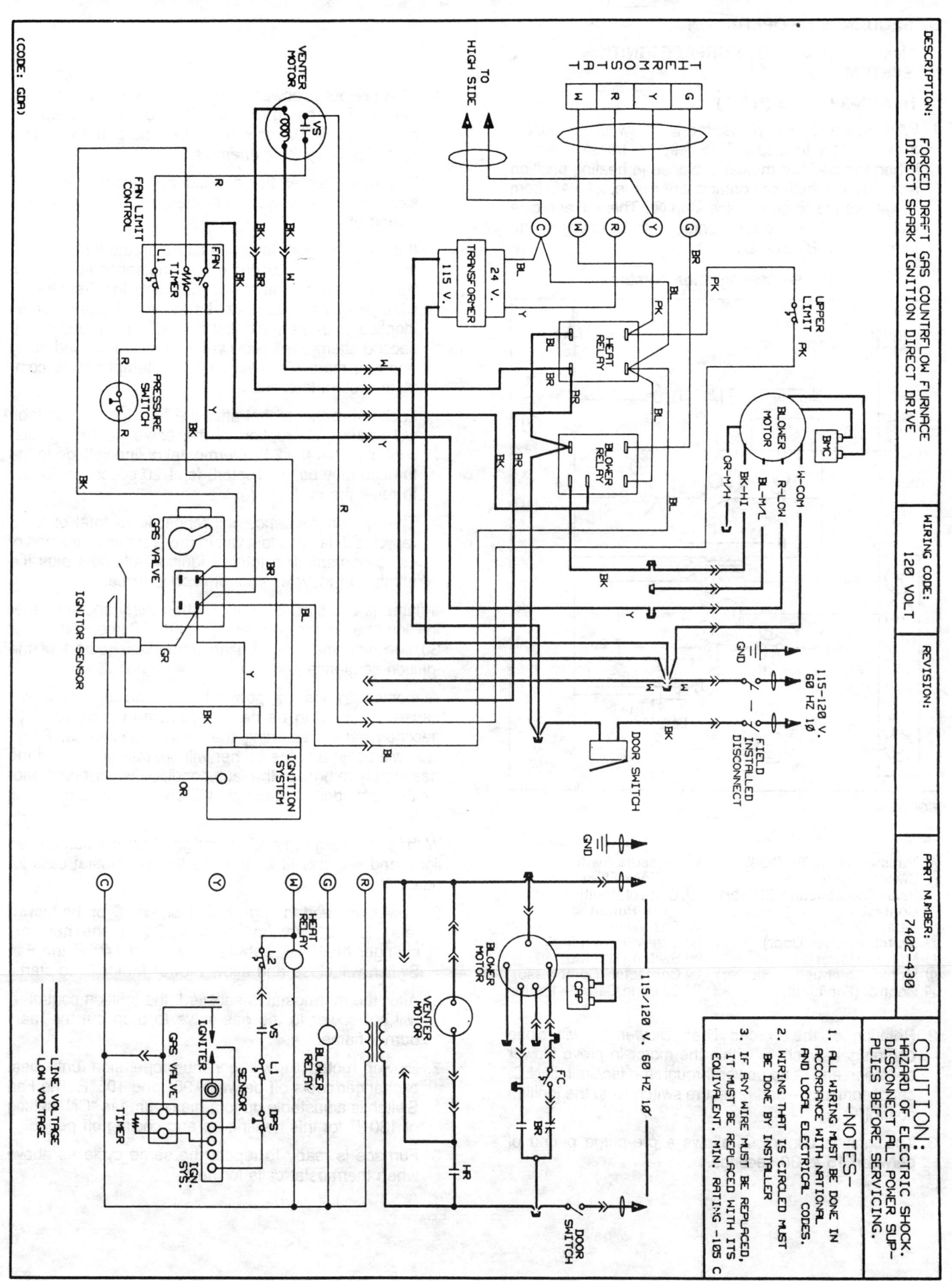

8

B. SEQUENCE OF OPERATION

Heating Cycle—HOT SURFACE IGNITION SYSTEM

a) HEATING CYCLE (H.S.I.)

1. With blower door in place, the N.O. Switch Ⓔ closes, allowing 115 to 120VAC to feed to transformer Ⓐ. When the wall thermostat is placed in heating position and mercury bulb or contacts are made, 24 VAC from Ⓡ will feed to Ⓦ on furnace strip Ⓜ. This will energize coil Ⓒ which will close contacts Ⓒ to energize the Forced Draft Blower Ⓖ.

Hot Surface Ignition System

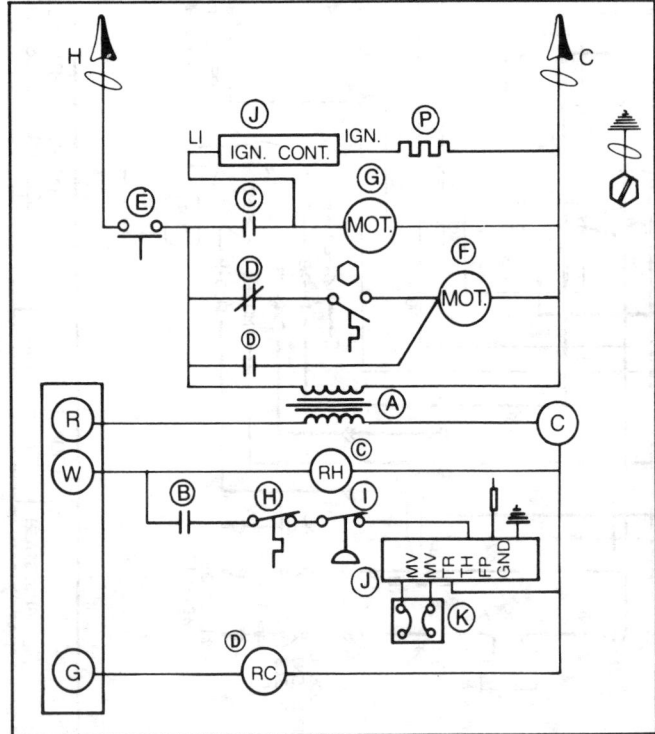

FIGURE 8

LEGEND

Ⓐ Transformer 120VAC/24VAC
Ⓑ Switch (Centrifugal)
Ⓒ Relay (Combustion Blower)
Ⓒ¹ Contact
Ⓓ Relay (Cooling)
Ⓔ Switch (Blower Door)
Ⓕ Motor (Blower)
Ⓖ Motor (Combustion Blower)
Ⓗ Control (Fan/Limit)
Ⓘ Pressure Switch (Block Flue)
Ⓙ Control (Ignition with Pre-Purge)
Ⓚ Gas Valve
Ⓛ Screw (Grounding)
Ⓜ Switch (Thermostat)
Ⓝ Contactor (Compressor)
Ⓟ Silicone Carbide Ignitor

2. Rotation of the Forced Draft Blower Ⓖ will close centrifugal switch Ⓑ within the motor to prove blower operation and send power through the fan/limit Ⓗ N.C. set of contacts—and pressure switch Ⓘ to the ignition control Ⓙ.

3. The ignition control Ⓙ allows a pre-purge period of between 15 to 30 seconds.

4. After pre-purge power is applied to the silicone carbide ignitor Ⓟ. The ignitor is allowed to warm-up for approximately 15 seconds. At the end of the warm-up time, the valve Ⓚ is energized.

The main burner flame must be detected within 7 seconds or the valve is de-energized and the ignitor turned off.

If there is a failure to sense flame on the first attempt, there will be a 60 second delay. The second attempt at lighting the burner flame will duplicate the first attempt with pre-purge, warm-up times and lock-out times identical to those in the first try. Failure to prove on the second attempt will allow yet another 60 second delay and one more complete ignition attempt before complete system lock-out.

Lock-out of the control after the final ignition attempt, will require interruption of the power to the ignition control (either the 24V thermostat or line voltage to the furnace may be interrupted) for 1/20 second or longer to reset the system.

The ignition sequence will repeat for a total of three recycles if flame is lost within the first thirty seconds of establishment. The total of ignition attempts plus the number of recycles cannot exceed three.

If flame is established for more than thirty seconds after ignition, the recycle counter will clear. If flame is lost after 30 seconds, there will be an immediate restart of the ignition sequence.

A momentary loss of power to the sensing circuit will cause the valve to be de-energized and a delay of 15 seconds before restarting the ignition sequence. Recycles will begin and the burner will operate normally if the gas supply returns or the fault condition is corrected prior to the last ignition attempt. Otherwise the control will lock-out.

With ignitor warm-up and proper gas flow, the burners will light, and will stay lit as long as the thermostat calls for heat.

5. Power will feed through N.C. Contact Ⓓ of the blower relay to the Fan/Limit Control Ⓗ. As the heat exchanger heats up, between 120° and 140°F, the Fan Switch will close and allow Blower Motor Ⓕ to start.

6. After the thermostat is satisfied, the ignition control Ⓙ will cut power to the gas valve to drop out the main burner flame.

7. Blower motor Ⓕ will continue operation until heat exchanger cools off between 95° and 105°F. The Fan Switch is adjustable up to a maximum fan "ON" setting of 130°F, for this heating up and cooling off period.

8. Furnace is ready to repeat the same cycle as above when thermostat calls for heat.

b) APPROXIMATE TIMING FOR DIFFERENT COMPONENTS WHEN CALLING FOR HEAT:

1. Centrifugal Switch—5 Seconds
2. Pre-Purge Cycle—15 or 30 Seconds
3. Silicone Carbide Ignitor Warm-Up—12 Seconds
4. Ignition Lock-Out—7 Seconds (Burner Gas Fails to Ignite)
5. Second Re-Try—90 Seconds
6. Third Re-Try—90 Seconds
7. Burner ignition—7 Seconds
8. Blower (Furnace)—60 to 90 Seconds

c) SEQUENCE OF OPERATION—COOLING CYCLE:

1. Place thermostat system switch in "Cool" position. Fan Switch can either be set on "AUTO" or "ON" position. If fan switch is in the "ON" position, current will flow from the transformer Ⓐ through the Ⓡ leg directly to Ⓖ on thermical strip Ⓜ and energize cooling relay Ⓓ, closing contact Ⓞ2 to energize the blower motor.

2. If Fan Switch is in the "AUTO" position and cooling is called for by the thermostat, power will feed from Ⓡ to Ⓨ on the sub-base and from Ⓨ to Ⓖ and bring on the fan. It will also energize the contactor through Ⓡ to Ⓨ and bring on the compressor for cooling.

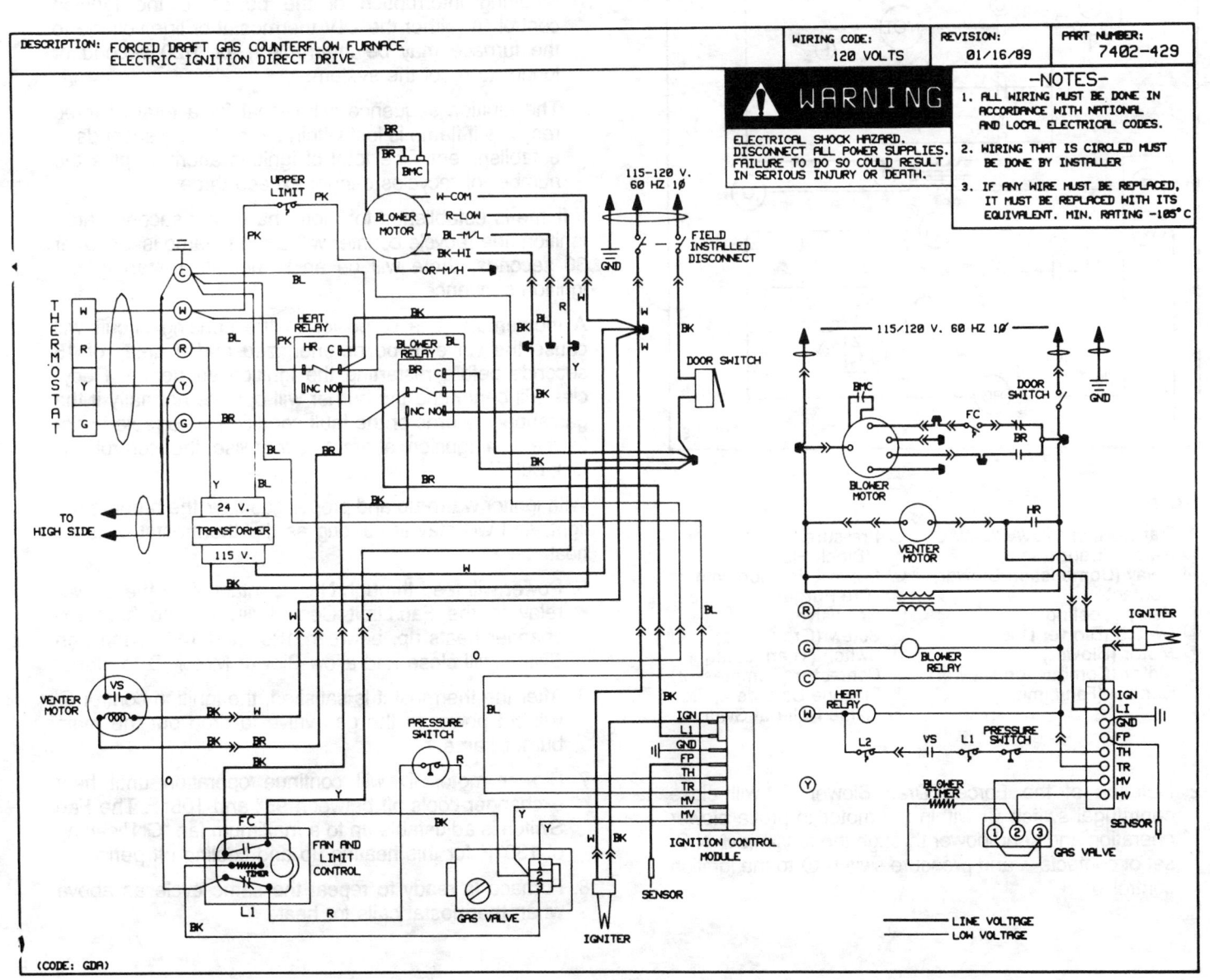

C. SEQUENCE OF OPERATION—WITH LED CONTROL

Heating Cycle—HOT SURFACE IGNITION SYSTEM

a) HEATING CYCLE (H.S.I.)

1. With blower door in place, the N.O. Switch Ⓔ closes, allowing 115 to 120VAC to feed to transformer Ⓐ. When the wall thermostat is placed in heating position and mercury bulb or contacts are made, 24 VAC from Ⓡ will feed to Ⓦ on furnace strip Ⓜ. This will energize coil Ⓒ which will close contacts Ⓒ to energize the Forced Draft Blower Ⓖ.

Hot Surface Ignition System

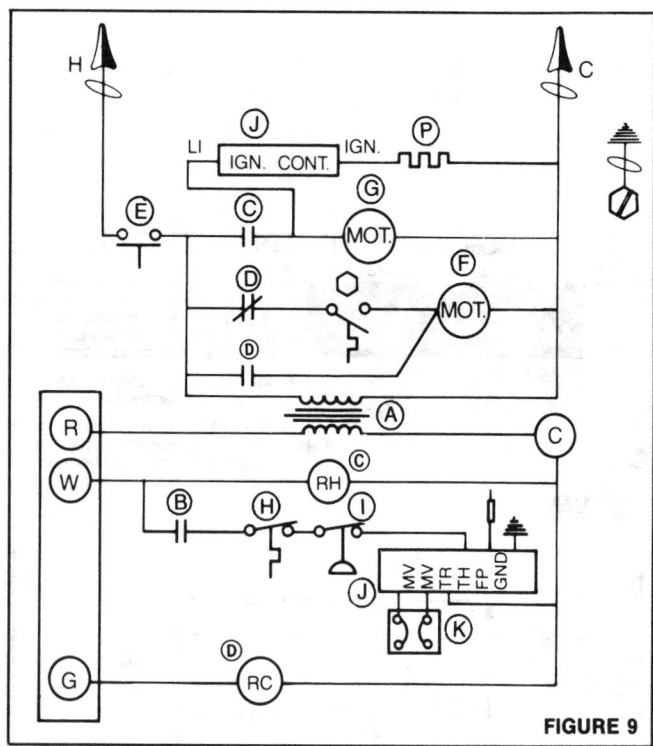

FIGURE 9

LEGEND

- Ⓐ Transformer 120VAC/24VAC
- Ⓑ Switch (Centrifugal)
- Ⓒ Relay (Combustion Blower)
- Ⓒ¹ Contact
- Ⓓ Relay (Cooling)
- Ⓔ Switch (Blower Door)
- Ⓕ Motor (Blower)
- Ⓖ Motor (Combustion Blower)
- Ⓗ Control (Fan/Limit)
- Ⓘ Pressure Switch (Block Flue)
- Ⓙ Control (Ignition with Pre-Purge)
- Ⓚ Gas Valve
- Ⓛ Screw (Grounding)
- Ⓜ Switch (Thermostat)
- Ⓝ Contactor (Compressor)
- Ⓟ Silicone Carbide Ignitor
- Ⓡ Flame Roll-out Switch

2. Rotation of the Forced Draft Blower Ⓖ will close centrifugal switch Ⓑ within the motor to prove blower operation and send power through the fan/limit Ⓗ N.C. set of contacts—and pressure switch Ⓘ to the ignition control Ⓙ.

3. The "LED" indicator light on the ignition control Ⓙ will flash once to indicate operation. There will be a 30 second delay for pre-purge of the burner compartment.

4. After pre-purge power is applied to the silicone carbide ignitor Ⓟ. The ignitor is allowed to warm-up for approximately 15 seconds. At the end of the warm-up time, the valve Ⓚ is energized.

The main burner flame must be detected within 7 seconds or the valve is de-energized and the ignitor turned off.

If there is a failure to sense flame on the first attempt, there will be a 60 second delay. The second attempt at lighting the burner flame will duplicate the first attempt with pre-purge, warm-up times and lock-out times identical to those in the first try. Failure to prove on the second attempt will allow yet another 60 second delay and one more complete ignition attempt before complete system lock-out.

Lock-out of the ignition control Ⓙ after the final ignition attempt will display a flashing "LED" indicator light requiring interruption of the power to the ignition control Ⓙ (either the 24V thermostat or line voltage to the furnace may be interrupted) for 1/20 second or longer to reset the system.

The ignition sequence will repeat for a total of three recycles if flame is lost within the first thirty seconds of establishment. The total of ignition attempts plus the number of recycles cannot exceed three.

If flame is established for more than thirty seconds after ignition, the recycle counter will clear. If flame is lost after 30 seconds, there will be an immediate restart of the ignition sequence.

A momentary loss of power to the sensing circuit will cause the valve to be de-energized and a delay of 15 seconds before restarting the ignition sequence. Recycles will begin and the burner will operate normally if the gas supply returns or the fault condition is corrected prior to the last ignition attempt. Otherwise the control will lock-out.

With ignitor warm-up and proper gas flow, the burners will light, and will stay lit as long as the thermostat calls for heat.

5. Power will feed through N.C. Contact Ⓓ of the blower relay to the Fan/Limit Control Ⓗ. As the heat exchanger heats up, between 120° and 140°F, the Fan Switch will close and allow Blower Motor Ⓕ to start.

6. After the thermostat is satisfied, the ignition control Ⓙ will cut power to the gas valve to drop out the main burner flame.

7. Blower motor Ⓕ will continue operation until heat exchanger cools off between 95° and 105°F. The Fan Switch is adjustable up to a maximum fan "ON" setting of 130°F, for this heating up and cooling off period.

8. Furnace is ready to repeat the same cycle as above when thermostat calls for heat.

III. FURNACE OPERATION (Cont.)

D. THERMOSTAT ANTICIPATOR SETTING

1. Proper control of the indoor area temperature can only be achieved if the thermostat is calibrated to the heating and/or cooling cycle. A vital consideration of this calibration is related to the thermostat heat anticipator.

2. The proper thermostat heat anticipator setting is 0.8 AMPS. for furnace operation only. To increase length of cycle, increase setting of heat scale; to decrease length of cycle, decrease setting of heat scale.

Anticipators for the cooling operation are generally pre-set by the thermostat manufacturer and require no adjustment.

Anticipators for the heating operation are of two types, pre-set and adjustable. Those that are pre-set will not have an adjustment scale and are generally marked accordingly.

Thermostat models having a scale as shown in Figure 10 must be adjusted to each application.

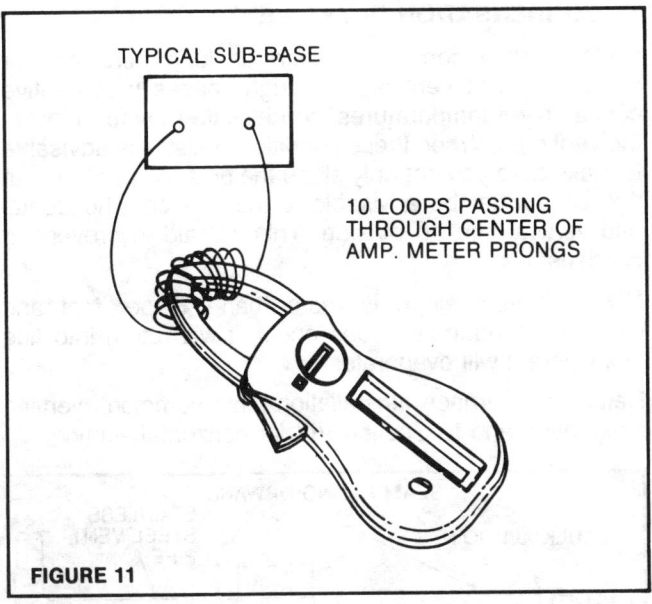

FIGURE 11

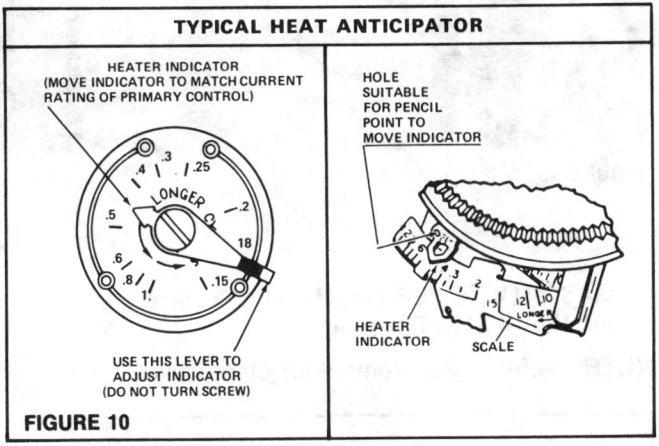

FIGURE 10

In many cases, this adjustment setting can be found in the thermostat Installation Instructions. If this information is not available, or if the correct setting is questioned, the following procedures should be followed:

PREFERRED METHOD:
Use a low scale ammeter such as an ampcheck. Connect the meter across terminals "R" and "W" on the sub-base ("RH" & "W" on multi-stage thermostat sub-base) or

STEP 1. Wrap 10 loops of single strand insulated thermostat wire around the prongs of an ampmeter. Set the scale to the 1 to 5 or 1 to 6 amp. scale.

STEP 2. Connect the uninsulated ends of this wire jumper across terminals "R" and "W" on the sub-base ("RH" and "W" on multi-stage thermostat sub-base). See Figure 11. This test must be performed without the thermostat attached to the sub-base.

STEP 3. Let the heating system operate in this position for about one minute. Read the ampmeter scale. Whatever reading is indicated must be divided by 10 (for 10 loops of wire).

This is the setting at which the adjustable heat anticipator should be set.

$$\text{FORMULA:} \quad \frac{\text{Ampmeter reading}}{10 \text{ loops}} = \text{Anticipator setting}$$

$$\text{OR:} \quad \frac{2.5 \text{ AMPS}}{10} = .25 \text{ AMPS Setting}$$

STEP 4. If a slightly longer cycle is desired, the pointer should be moved to a higher setting. Slightly shorter cycles can be achieved by moving to a lower setting.

STEP 5. Remove the meter jumper wire and reconnect the thermostat. Check the thermostat in the heating mode for proper operation.

NOTE: The length of the heating cycle can also be affected by the fan limit control settings (if applicable). The fan "ON" and "OFF" settings should be checked at this point.

For thermostats having 2 stage heat, Step 1, 2 and 3 must be repeated. Second stage heat is controlled through terminals "RH" and "W2" on the sub-base.

If Digital Amp Probe is used, read amp draw direct from meter, then Step #1 is not required.

IV. VENTING

A. CONDENSATION IN THE VENT

Under certain conditions, such as long horizontal or vertical runs of vent pipe through spaces with relative cool ambient temperatures, condensate may form inside the vent pipe. When these conditions exist, it is advisable to make sure you not only slope the pipe ¼" per foot, but it would further be advisable to insulate both horizontal and vertical vent to outside. This will aid in preventing condensation.

NOTE: If vent slope is more than ¼" per foot and moderate condensate does occur, it will return into flue box where it will evaporate.

Refer to Furnace Installation for common venting information and to Section VIII for horizontal venting.

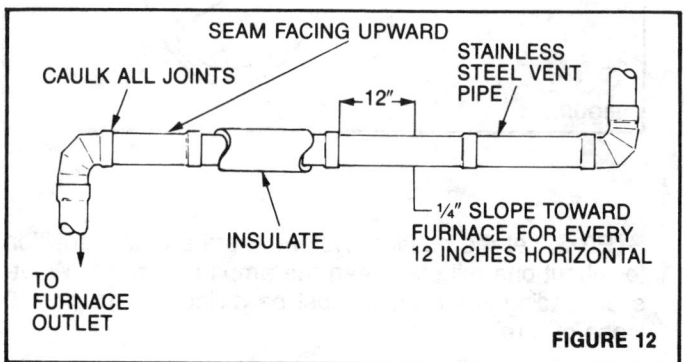

FIGURE 12

The furnace should not be vented into unlined masonry chimneys that are exposed to outdoor ambient conditions.

V. SERVICE AND TROUBLE SHOOTING.

A. FURNACE—OPERATION

1. Gas Pressure

Check to insure all connections have been made properly, then preceed as follows:

The gas supply line coming in may not have been purged of air. Properly purge all the air out of the incoming gas line before trying to test the furnace.

GAS VALVE—Inlet Pressure

The maximum inlet pressure on the gas valve is 0.5 PSIG (14.0 Inches of water column) for natural and LP gases. To allow inlet pressure to exceed this value may cause the regulator, solenoid and valve to fail in an unsafe mode. This will damage gas valve and require replacement. The incoming pressure to gas valve inlet side should be 4" to 7" for Nat. gas, and 11" for LP gas.

GAS VALVE—Outlet Pressure

The pressure on the outlet side of the gas valve should read 3.5" Water Column for Natural and 10" Water Column for LP gas. Minor changes in manifold pressure may be made by regulator adjustments. Refer to Figure 5 for adjustments.

Figure 13 shows U-Tube Manometer being used to check outlet pressure.

FIGURE 13

B. GAS VALVE OPERATION—Hot Surface Ignition—Step Opening

NOTE: Refer to Supplement for other valves.

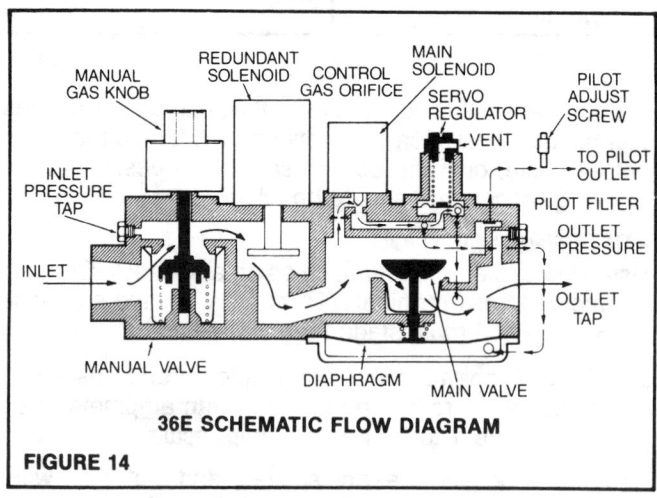

36E SCHEMATIC FLOW DIAGRAM

FIGURE 14

Upon a call for heat, the valve is energized through the closure of the thermostat contacts, centrifugal switch and ignition system module circuitry. Both the main gas valve and redundant gas valve operators are energized simultaneously, opening both valves and permitting gas flow to the burners. (First at low pressure then stepping up to full input.) Upon interruption of power to the gas valve operators, both valves will close immediately. See Figure 14.

C. BURNER INSPECTION DOOR

Figure 15 is the inspection door for the burner compartment. If for any reason this door must be removed, care must be taken to prevent gasket from being torn or destroyed. If gasket is torn or frayed, it should be replaced.

An inspection door that does not have a good air seal, will cause improper combustion air control and possible furnace failure by lock-out or flame instability.

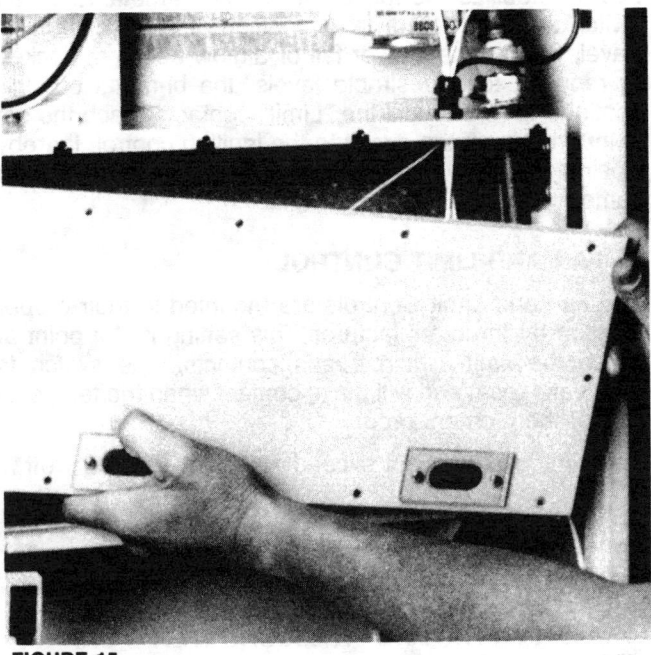

FIGURE 15

D. BURNER COMPARTMENT—Flame Instability

The main cause of flame instability is lack of combustion air. The Forced Draft Furnace combustion air is circulated by the Forced Draft Blower. See Figure 16.

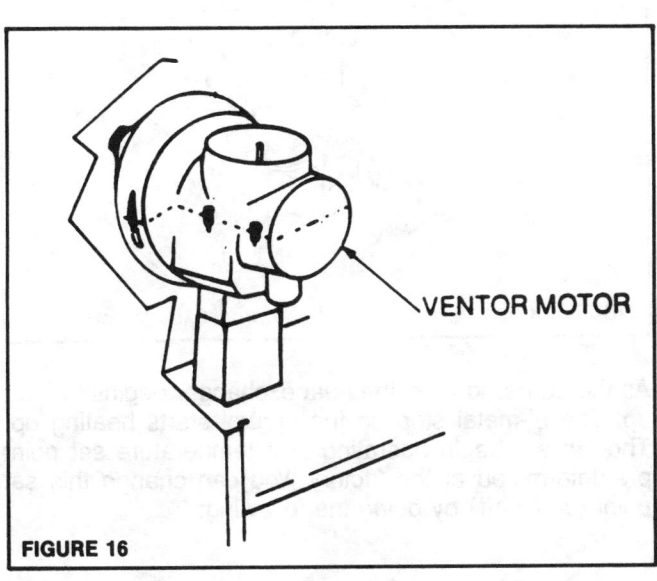

VENTOR MOTOR

FIGURE 16

Each basic furnace model has a different size air restrictor used in conjunction with the Forced Draft Blower.

DO NOT INTER-CHANGE OR REMOVE COMBUSTION AIR RESTRICTORS.

MODEL	RESTRICTOR PART NUMBER
040	4020-007
060	4020-008
080	4020-009
100	4020-010
120	4020-011

Causes of Flame Instability:

1. Open Burner Compartment Panel or inadequate seal.
2. Severe blockage of heat exchanger
3. Heat Exchanger Leaks
4. Excessive manifold gas pressure

NOTE: Pressure at gas valve outlet should not exceed 3.5 inches of water column.

5. Burner orifices oversized resulting in an over-fired condition within the furnace.

E. HIGH ALTITUDE

Ratings are approved for altitudes up to 2000 Feet for all gases. Ratings for altitudes over 2000 feet are decreased by 4% for each 1000 feet above sea level. See Table below. Check and note the orifice size required at that altitude. If the orifices shipped with the furnace are not the recommended size, remove them and install the proper size.

EQUIVALENT ORIFICE SIZES AT HIGH ALTITUDES
(Includes 4% Input Reduction For Each 1000 Feet)

ORIFICE SIZE AT SEA LEVEL	ORIFICE SIZE REQUIRED AT OTHER ELEVATIONS								
	2,000	3,000	4,000	5,000	6,000	7,000	8,000	9,000	10,000
21	23	23	24	25	26	27	28	28	29
22	23	24	25	26	27	27	28	29	29
23	25	25	26	27	27	28	29	29	30
24	25	26	27	27	28	28	29	29	30
25	26	27	27	28	28	29	29	30	30
26	27	28	28	28	29	29	30	30	30
27	28	28	29	29	29	30	30	30	31
28	29	29	29	30	30	30	30	31	31
29	29	30	30	30	30	31	31	31	32
30	30	31	31	31	31	32	32	33	35
31	32	32	32	33	34	35	36	37	38
32	33	34	35	35	36	36	37	38	40
33	35	35	36	36	37	38	38	40	41
34	35	36	36	37	37	38	39	40	42
35	36	36	37	37	38	39	40	41	42
36	37	38	38	39	40	41	41	42	43
37	38	39	39	40	41	42	42	43	43
38	39	40	41	41	42	42	43	43	44
39	40	41	41	42	42	43	43	44	44
40	41	42	42	42	43	43	44	44	45
41	42	42	42	43	43	44	44	45	46
42	42	43	43	43	44	44	45	46	47
43	44	44	44	45	45	46	47	47	48
44	45	45	45	46	47	47	48	48	49
45	46	47	47	47	48	48	49	49	50
46	47	47	47	48	48	49	49	50	50
47	48	48	49	49	49	50	50	51	51
48	49	49	49	50	50	51	51	51	52
49	50	50	50	51	51	51	52	52	52
50	51	51	51	51	52	52	52	53	53
51	51	52	52	52	52	53	53	53	54
52	52	53	53	53	53	53	54	54	54
53	54	54	54	54	54	54	55	55	55
54	54	55	55	55	55	55	56	56	56
55	55	55	55	56	55	56	56	56	57
56	56	56	57	57	57	58	59	59	60
57	58	59	59	60	60	61	62	63	63
58	59	60	60	61	62	62	63	63	64
59	60	61	61	62	62	63	64	64	65
60	61	61	62	63	63	64	64	65	65

V. SERVICE AND TROUBLESHOOTING (Cont.)

F. FURNACE—Electrical

A thorough understanding of the sequence of the electrical operation for this furnace is a must before you can service the electrical components. See Sequence of Operation for details.

Check all electrical connections to insure they are tight before proceeding:

1. FORCED DRAFT MOTOR AND RELAY

See enclosed Wiring Diagrams.

When the thermostat calls for "HEAT" the relay © is energized by 24 VAC. When the relay coil is energized, the normally open (N.O.) contacts will close between terminals #4 and #2 on Relay. This allows 120 VAC to feed to the Forced Draft Blower Motor.

When the blower motor starts rotation, the normally open (N.O.) End Switch ® on blower motor will close allowing 24V power to feed through normally closed (N.C.) contact on Limit.

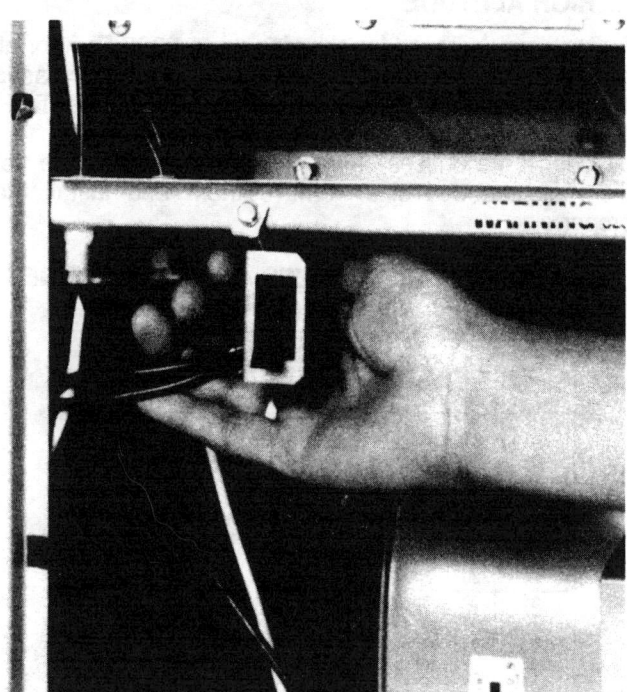

FIGURE 17

The following is a check list of steps to follow if Forced Draft Motor does not function:

NOTE: See individual items for further trouble shooting.

1. Check power to furnace—120 VAC
2. Is door switch in blower compartment closed?
3. Is transformer secondary side 24 VAC output?
4. Is thermostat calling for Heat?
5. Is Forced Draft Relay coil energized?
6. Are Forced Draft Relay contacts closed?
7. Is Forced Draft Motor operational?

2. HIGH LIMIT

The High Limit is part of the combination Fan and Limit Control. Both the fan circuit and the limit circuit are activated by the action of a calibrated bi-metal coil which is sensitive to heat. The fan/limit bi-metal probe containing this coil is inserted into the heat exchanger compartment. As the flame heats the heat exchanger, the bi-metal reacts to the heat and rotates a shaft connected to a switch mechanism. The first switch is the normally open fan contacts which upon closing energize the Evaporator blower motor. When the Blower is energized the airflow over the heat exchanger removes sufficient heat to stabilize the movement of the bi-metal actuated switch and maintain a fixed position of the switch in its travel. Should the blower fail or airflow become blocked or reduced to undesirable levels, the bi-metal coil will continue to rotate until the "Limit" contacts reach the set point and break the circuit to the ignition control, thereby closing the gas valve and extinguishing the main burner flame.

3. FAN AND LIMIT CONTROL

The Fan and Limit Controls are mounted in their proper location by the manufacturer. This setting is the point at which the limit switch opens contacts. The switch is automatic reset and will make contact when the temperature of the element drops.

The setting should not exceed 130° "on" and 90° "off".

⚠ CAUTION:
DO NOT TAMPER WITH THE INTERNAL FAN OR LIMIT CONTROL MECHANISM. TO DO SO CAN CAUSE INJURY AND WILL SHORTEN FURNACE LIFE.

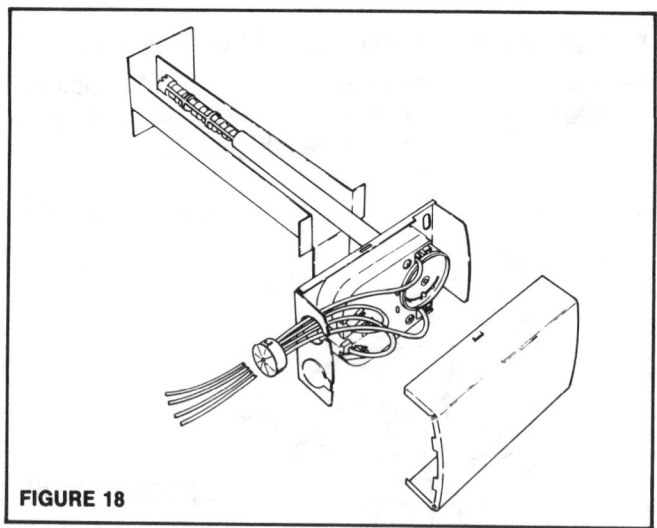

FIGURE 18

As the burner ignites, the heat exchanger begins to heat up. The bi-metal strip on the control starts heating up. The fan will begin operating at a temperature set point pre-determined at the factory. You can change this set point (on or off) by doing the following:

V. SERVICE AND TROUBLESHOOTING (Cont.)

1. Hold the scaleplate dial to keep it from turning and straining the sensing element.

2. Move the setting levers to the control point recommended by the furnace or burner manufacturer. Use gentle finger pressure.

3. The readings at the pointed ends of the setting levers indicate the control points.

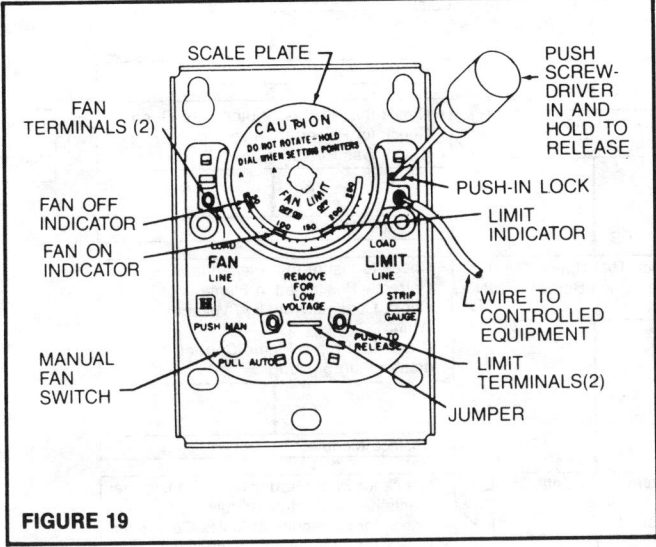

FIGURE 19

4. SAFETY CHECK LIMIT

The recommended method for checking the Limit Control is to block off the return air after furnace blower has been operating for 5 minutes. Blocking off return air will make the furnace over heat and cut out on high limit control. The temperature of cut-out should be no more than 200°F or as set by manufacturer. As soon as the limit control has been proven safe, the return air opening should be unblocked to permit air circulation. When you use this method, you have made sure that the furnace limit safety operation is functioning properly in case of blower failure, blocked filters or excessive outlet restriction.

5. FLUE SAFETY PRESSURE SWITCH

A Pressure switch is incorporated in the thermostat circuit to the ignition control and will break the circuit if too much pressure is sensed at the flue outlet (.35″ W.C.). Too much pressure may be the result of a blockage in the vent to the outside atmosphere or a severe downdraft condition. See Figures 20 & 21.

When checking with AC Volt Meter, if you get voltage across the two points on the pressure switch, this means the switch is open. Possible causes as stated above, is flue blockage or possible downdraft conditions not allowing the Pressure Switch to remain closed.

The switch can be temporarily "jumped" to test a suspect switch. Under no circumstances should the switch be jumped other than to test.

The Switch Pressure setting is factory set and cannot be adjusted.

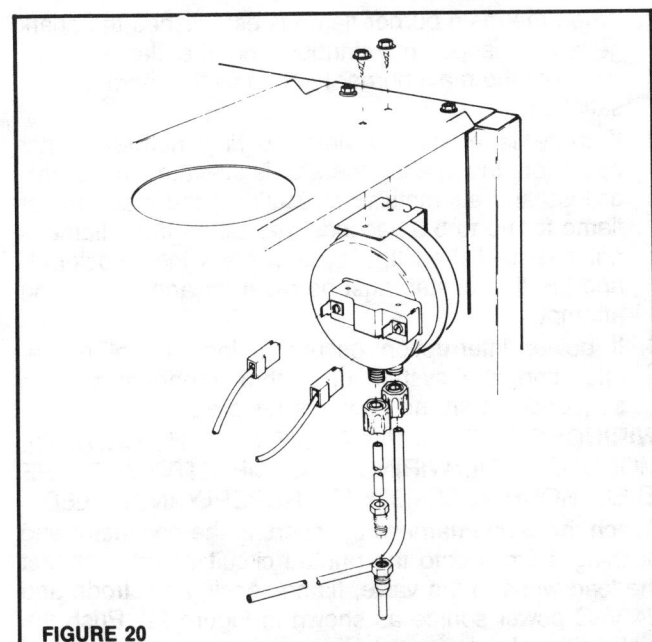

FIGURE 20

FIGURE 21

6. DIRECT IGNITION CONTROL—Sequence of Operation:

1) When the thermostat calls for heat, the spark generator and gas valve are energized simultaneously after pre-purge cycle. Ignition sparks are generated and the gas valve remains open for ignition trial period of up to 5 seconds maximum. If the main burner flame is not established, the system goes into "lockout" and the thermostat must be reset for another ignition attempt. This can be accomplished either by breaking line voltage or control circuit.

2. When the main burner flame is established the spark generator is put on standby and the flame-sensor monitors the main burner flame until the thermostat is satisfied.

3. If there is a loss of flame during normal burner operation, the spark generator is activated in 0.8 sec. or less and attempts to re-establish the main burner flame for up to 5.0 seconds maximum. If the flame is not re-established, the system goes into "Lock-out" and the thermostat must be reset for another ignition attempt.

4. If power interruption occurs during normal burner operation, the system goes thru a normal start-up sequence when the power is restored.

WIRING: DO NOT APPLY POWER TO THE CONTROL MODULE UNTIL WIRING IS COMPLETED AND THE ELECTRODE ASSEMBLY IS PROPERLY INSTALLED.

Attach the wiring harness by inserting the connector and pushing it firmly onto the printed circuit board. Connect the lead wires to the valve, flame-sensing electrode and 24 VAC power source as shown in Figure 22. Push the high-voltage lead wire onto the terminal located on top of the control module.

CAUTION: The Ignition Control Module will be damaged if either side of the gas valve coil is connected to ground.

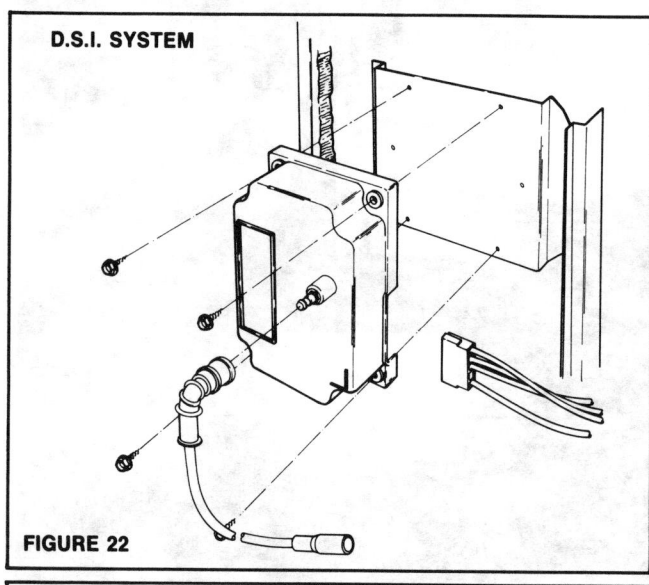

D.S.I. SYSTEM

FIGURE 22

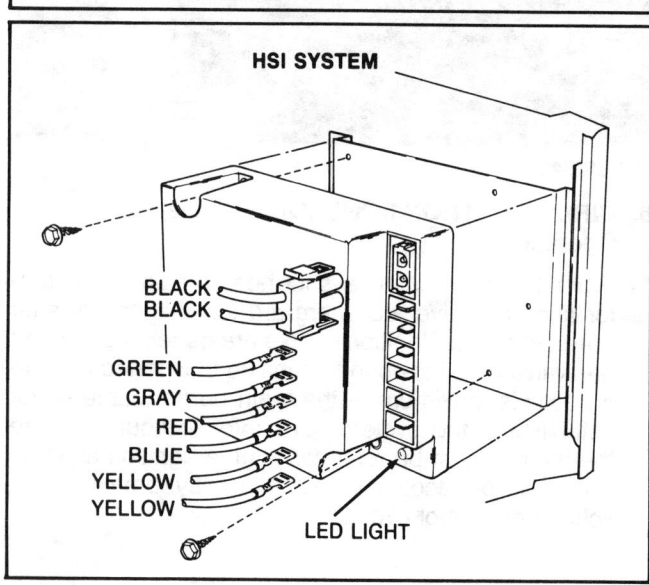

HSI SYSTEM

BLACK
BLACK
GREEN
GRAY
RED
BLUE
YELLOW
YELLOW

LED LIGHT

D.S.I. TROUBLESHOOTING TABLE

Turn Thermostat up to Call for Heat

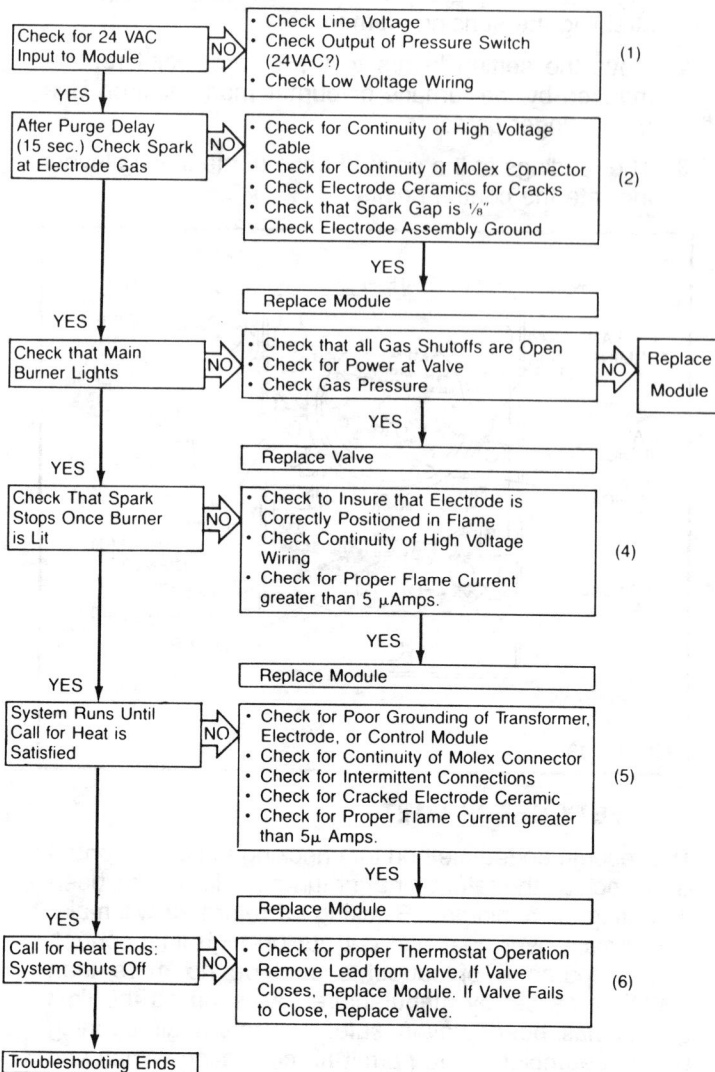

7. FAILURE OF IGNITION CONTROL

The following steps should be checked if there is a failure with the Ignition control:

1. Low gas pressure—must have a minimum of 4″ to 7″ water column at inlet of gas valve and 3.5″ Water column outlet pressure for Nat. gas (10.5″ to 11″ W.C. LP).

2. Wire between sensor and Ignition control must be in place before system will function.

3. Cracked electrode—Inspect electrode for cracks or breaks, replace if necessary.

4. Ignitor probe shorted to burner Re-align probe.

5. Ignitor probe and sensor out of alignment—Check to make sure they are gapped properly.

6. Sensor Probe not sensing flame—Out of alignment or low gas pressure.

7. Excessive air leakage around combustion door—Check gasket to insure tight fit.

8. Loose connection at control box connector.

9. Ground wire must be connected to electrode assembly.

8. ELECTRODE SENSOR ASSEMBLY

Optimum operation of the ignition system depends on proper location of the electrode assembly. The electrode assembly should be positioned so the electrode tips are in the flame envelope about ½ inch from the base of the flame. However, the ceramic insulators should not be in or extremely close to the flame envelope.

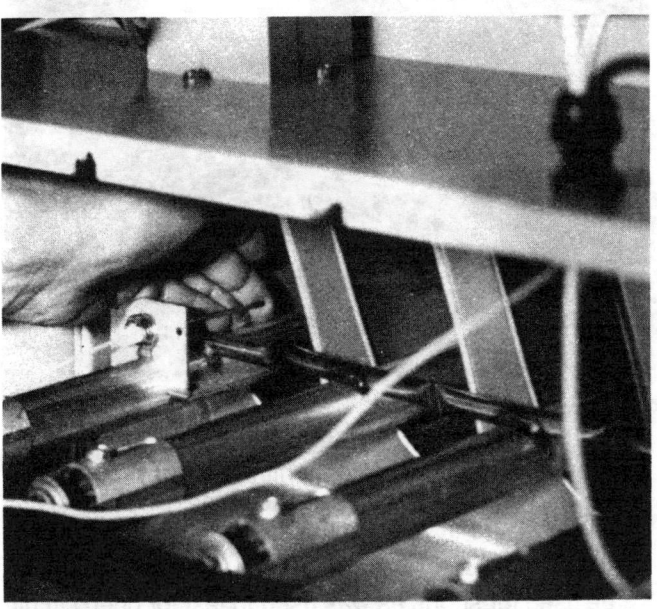

FIGURE 23

SEQUENCE OF OPERATION FOR: HOT SURFACE IGNITION SYSTEM

The following sequence of operation will occur when using an H.S.I. system on a gas furnace.

When the thermostat switch is set on "HEAT" and desired temperature is selected, the thermostat control make, completing the electrical circuit to the furnace ignition control.

The pre-purge mode in the ignition control will delay thirty seconds before applying power to the silicone carbide ignitor. The ignitor is allowed to warm-up for approximately 10 seconds. At the end of the warm-up time, the valve is energized.

The main burner flame must be detected within 4 seconds or the valve is de-energized and the ignitor turned off.

If there is a failure to sense flame on the first attempt, there will be a 60 second delay. The second attempt at lighting the burner flame will duplicate the first attempt with pre-purge, warm-up times and lock-out times identical to those in the first try. Failure to prove on the second attempt will allow yet another 60 second delay and one more complete ignition attempt before complete system lock-out.

Lock-out of the 50D47 control after the final ignition attempt, will require interruption of the power to the ignition control (either the 24V thermostat or line voltage to the furnace may be interrupted) for 1/20 second or longer to reset the system.

All 50D47 Controls will repeat the ignition sequence for a total of three recycles if flame is lost within the first thirty seconds of establishment. The total of ignition attempts plus the number of recycles cannot exceed three.

If flame is established for more than thirty seconds after ignition, the 50D47 will clear the recycle counter. If flame is lost after 30 seconds, there will be a 60 second wait and restart of the ignition sequence.

A momentary loss of power or flame to the sensing circuit will cause the valve to be de-energized and a delay of 60 seconds before restarting the ignition sequence. Recycles will begin and the burner will operate normally if the gas supply returns or the fault condition is corrected prior to the last ignition attempt. Otherwise the control will lock-out.

With ignitor warm-up and gas flow in the burners will light, and will stay lit as long as the thermostat calls for heat.

The heat exchanger heats up the combination fan and limit control. When the temperature reaches the "on" set point, the blower will be energized. There is also a safety limit combined in this control. If for some reason the blower does not start, the temperature will increase until the limit set point is reached. At this time, the 24 volt secondary of the transformer is broken and the gas valve will close.

When the thermostat is satisfied, the gas valve will close, shutting down the ignitor and burners. At this time, the blower will continue to operate until the temperature cools down below the "off" set point of the fan control. When this point is reached, the blower will shut off.

Now the furnace is ready to repeat the same cycle as above when the thermostat calls for heat.

FAILURE TO IGNITE

PRELIMINARY CHECKS:

Before gas supply is turned on re-check the following:

1. System properly grounded?
2. High voltage connection secured at both ends?
3. Flame sensing wire connected?
4. Electrode assembly adjusted and secured?
5. System properly vented?

9. BLOWER MOTOR AND RELAY—Blower Speed Adjustment:

The multi-speed direct drive motor is wired to the hi-side tap for cooling operation and low speed for heating operation. A change to a lower speed may be made in the lower electrical box. See Figure 24.

Heating and cooling models equipped with a blower relay, change speeds automatically. Instructions on checking CFM can be found in the Installation Manual.

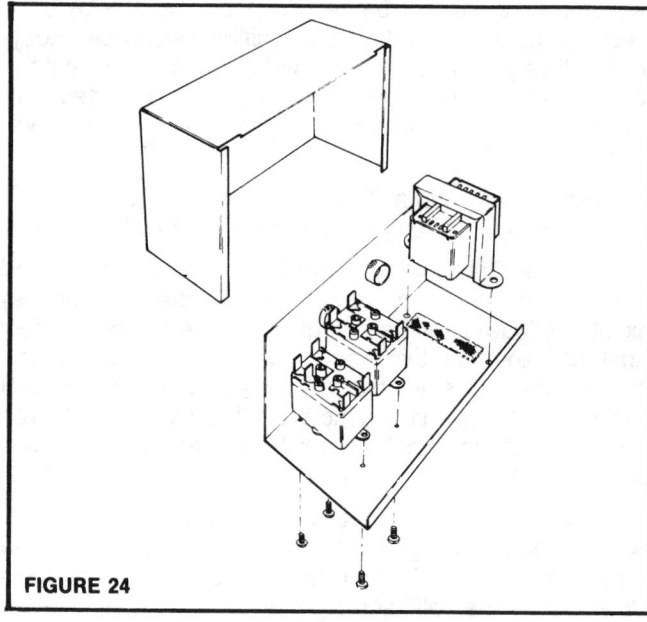

FIGURE 24

In case motor does not run, the following steps can be used to check for malfunction:

CAUTION: DISCONNECT ELECTRICAL POWER BEFORE PROCEEDING: DISCONNECT MOTOR LEADS FROM THE RELAY AND FROM THE CAPACITOR TERMINALS:

1. Grounded—Using a V.O.M., set an ohmmeter scale for highest resistance reading zero to meter. Check continuity between each motor lead and a bare metal section on motor frame. A grounded motor will show continuity to ground.

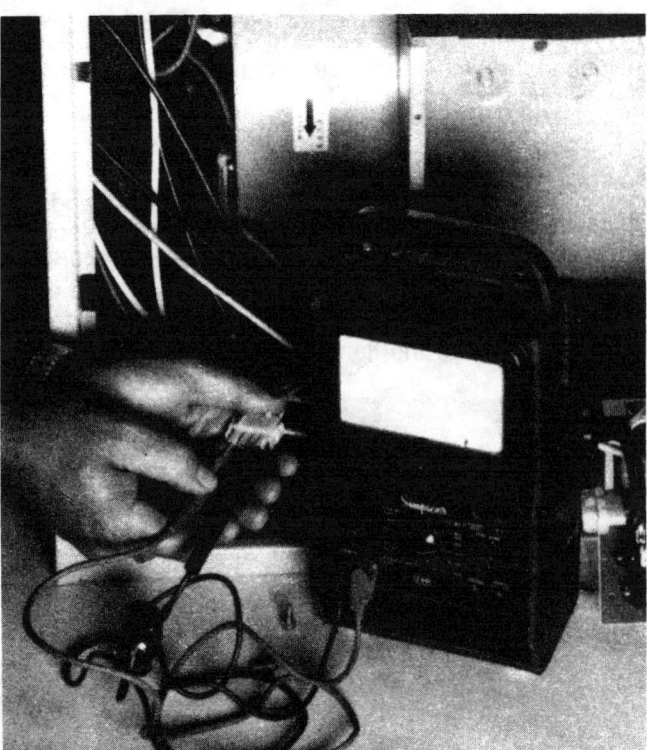

FIGURE 25

2. Open Circuit—Check for continuity between common wire from motor and each speed lead. An open circuit will measure infinity.

3. Shorts between windings—The same procedure can be used for shorts as was used to test for open circuit. But, a shorted winding will indicate zero Ohms. In most cases you will either see or smell the shorted windings.

4. RUN CAPACITOR—Place Ohmmeter on highest scale and zero adjust. Touch meter leads to the two terminals on capacitor. Meter should move toward zero Ohms and return slowly to infinity. Reverse lead and repeat.

   SHORT—Place Ohmmeter on highest scale and touch on lead to bare metal of capacitor and the other lead to each terminal. If meter defects from infinity, replace capacitor.

TROUBLESHOOTING

THE WHITE-RODGERS HOT SURFACE IGNITION SYSTEM

DANGER! DO NOT OMIT THIS STEP!

LINE VOLTAGE (120 VAC) COULD BE PRESENT ON THE SURFACE OF THE IGNITOR, IF THE SYSTEM IS NOT CORRECTLY WIRED. SUCH VOLTAGE CAN CAUSE SERIOUS INJURY OR DEATH.

1. Disconnect electric power to system at main fuse or circuit breaker.

2. Remove draft shield (if necessary) to gain access to the ignitor. See Furnace Installation Instructions.

3. Connect an AC voltmeter from the surface of the ignitor to chassis ground, and then reconnect electric power to system.

4. If voltage exists between the surface of the ignitor and the chassis ground, the main power supply lines are improperly connected to the furnace.

   Reverse incoming line voltage leads.

In an effort to minimize the time required to troubleshoot this system:

1. Turn off the gas supply at the main gas valve.

2. Disconnect electric power to system at main fuse or circuit breaker if connected.

3. Visually inspect equipment for apparent damage. Check wiring for loose connections.

4. With draft shield removed (if necessary), inspect ignitor for visible cracking or scale deposits and flame sensor for position or deposits shorting sensor to burner.

After performing the above inspections, restore gas supply and electric power to equipment. Close thermostat contacts to cycle the system. If a "no heat" condition persists, the three visual indicators (below) will help determine if the system is operating properly.

1. The ignitor will warm up and glow red.

2. The main burner flame will ignite.

3. The main burner flame will continue to burn after the ignitor is turned off.

Troubleshooting the system consists of checking for these three visual indications. The attached charts define the proper action if any of the indications do not occur.

A diamond on the chart encloses a question that must be answered yes or no. A rectangle on the chart encloses a statement or instruction.

FIRST VISUAL CHECK

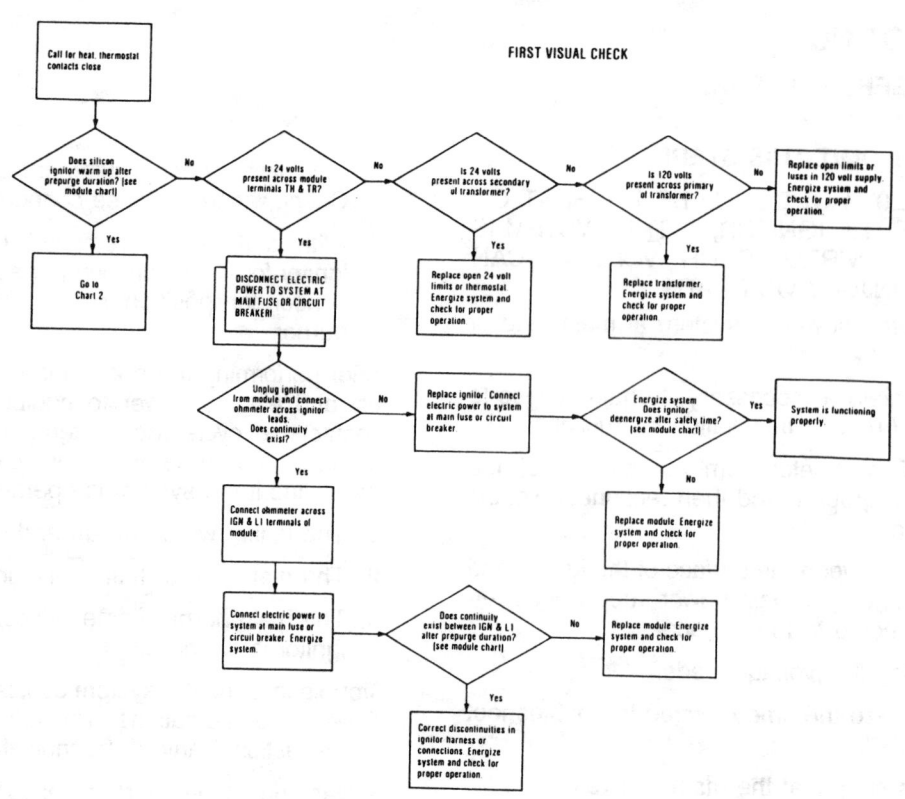

SECOND VISUAL CHECK

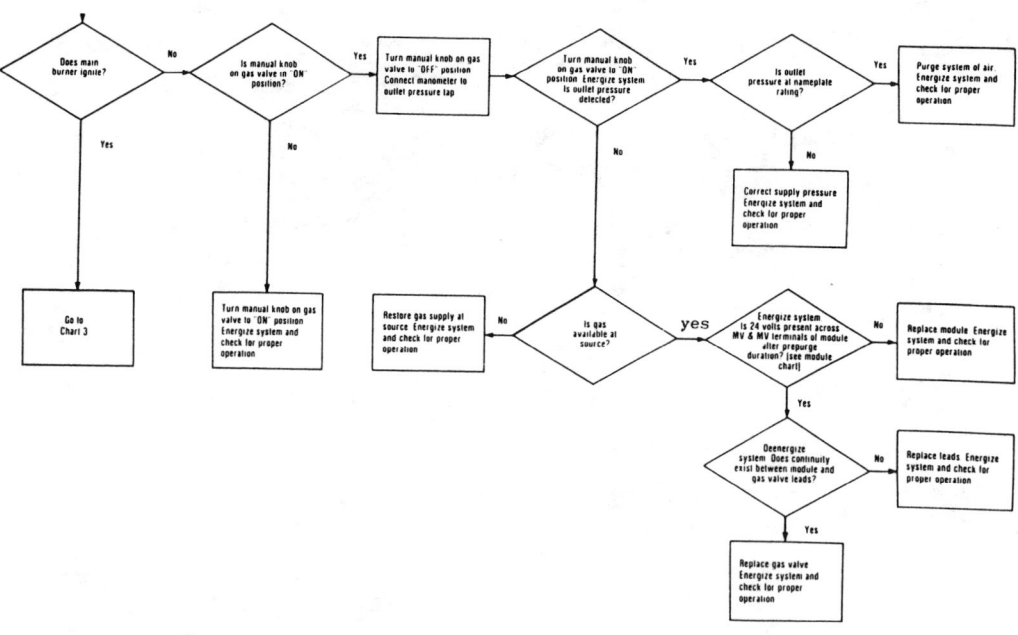

THIRD VISUAL CHECK

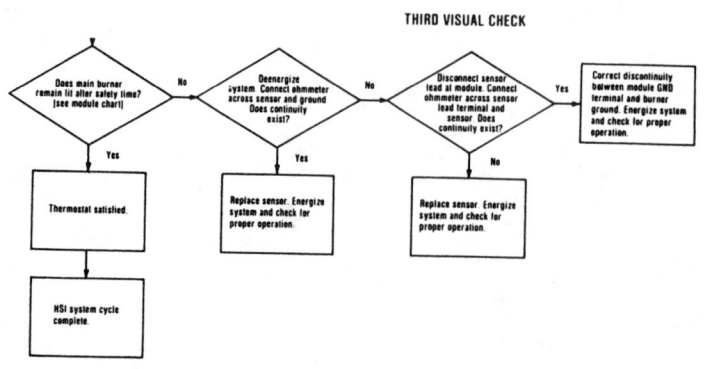

HSI MODULE WITH LED LIGHT

FLOW CHART

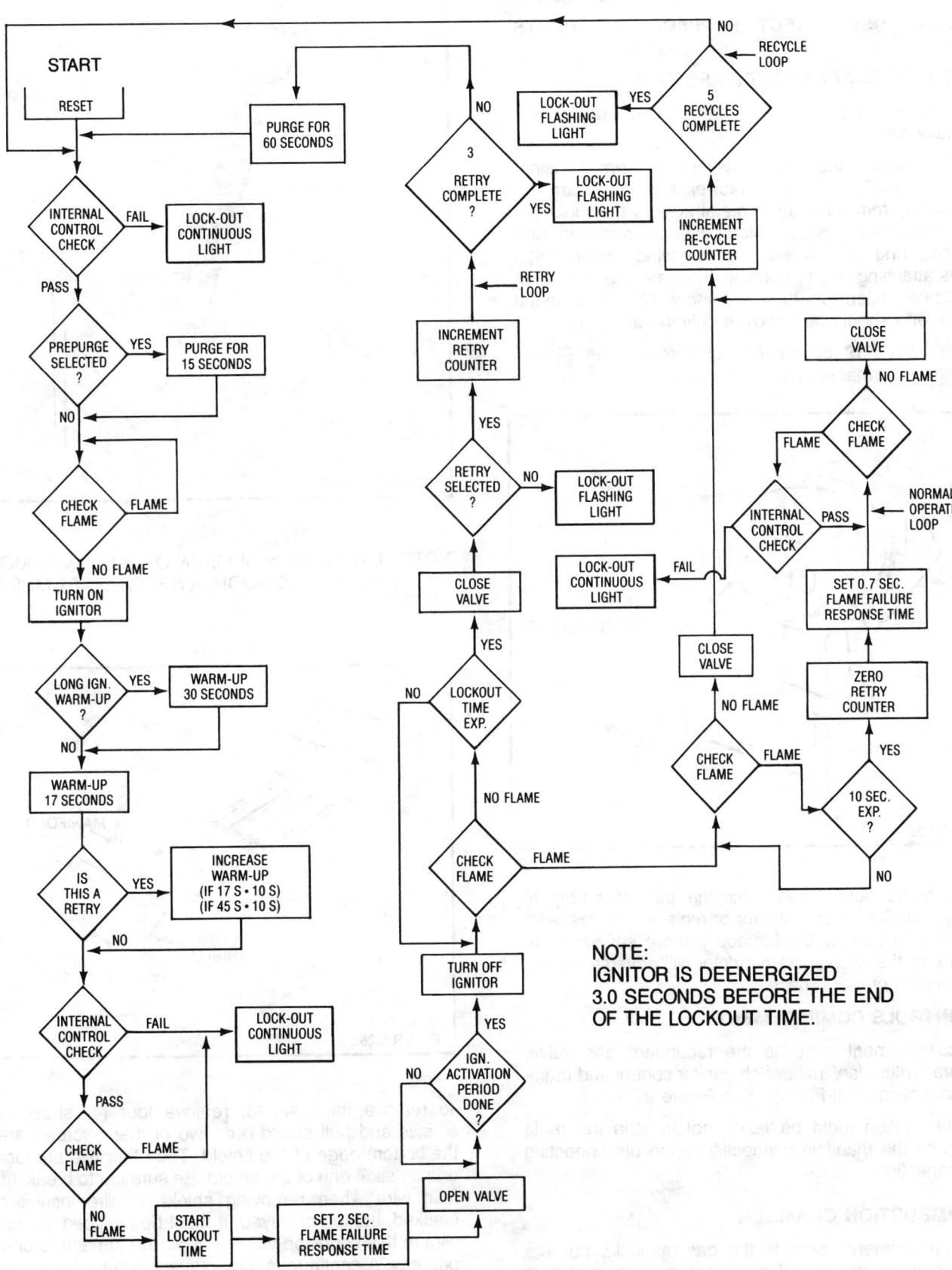

NOTE:
IGNITOR IS DEENERGIZED
3.0 SECONDS BEFORE THE END
OF THE LOCKOUT TIME.

VI. REPAIR AND REPLACEMENT PROCEDURE

Figure 3 gives an overview of the complete furnace. If need arises for disassembly of furnace, refer to Figure 3.

CAUTION: USE DIRECT REPLACEMENT PARTS ONLY.

A. FORCED DRAFT BLOWER ASSEMBLY

Parts that can be replaced in the Forced Draft blower assembly are:

Wheel, mounting ring, motor, gasket and blower housing. See Figure 26. The total blower assembly can be removed by removing sheet metal screws that hold the assembly to the furnace. Motor can be removed with mounting ring and wheel, by removing sheet metal screws attaching ring to housing. The restrictor is held in the blower extension tube with four (4) sheet metal screws, and should be removed only if damaged.

NOTE: With P19 production, all Forced Draft Blower motors are permanently lubricated.

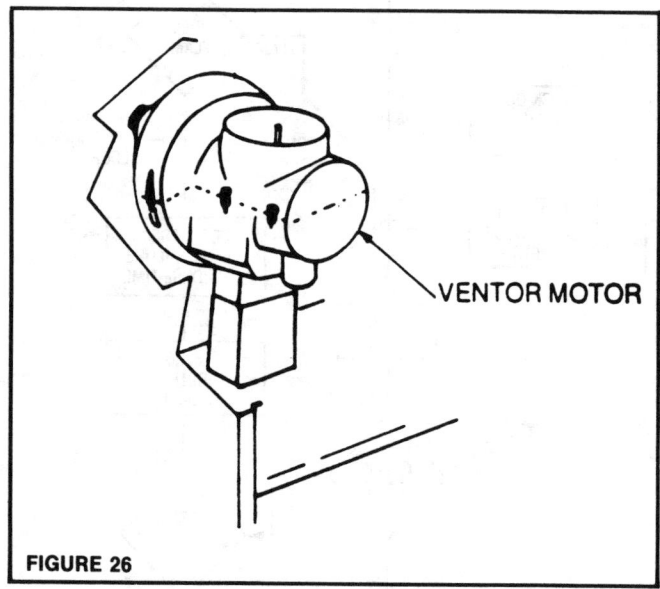

FIGURE 26

NOTE: Make sure when removing this assembly to service, that the exact restrictor or replacement restrictor is the correct part for the furnace you are servicing. No restrictor or the wrong size restrictor will adversely affect the operation of the furnace.

B. CONTROLS COMPARTMENT

This compartment contains the redundant gas valve, pressure switch, fan/limit switch, ignitor control and manifold connection (Dell-Fitting). See Figure 27.

Extreme caution must be taken not to strip the male threads on the manifold connection when disconnecting or reconnecting.

C. COMBUSTION CHAMBER

This compartment contains the gas manifold, burners, burner orifices, ignitor or ignitor/sensor probe, and flash shield. Care must be taken when removing the burner combustion door as not to destroy the gasket.

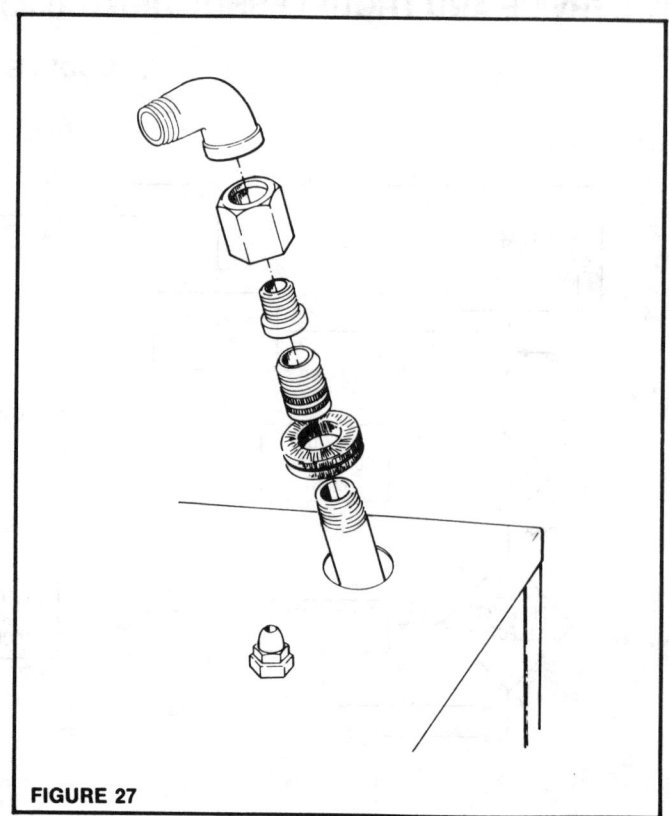

FIGURE 27

NOTE: IF GASKET IS BROKEN OR LOST IN REMOVAL PROCESS OF THIS DOOR, A NEW GASKET MUST BE INSTALLED.

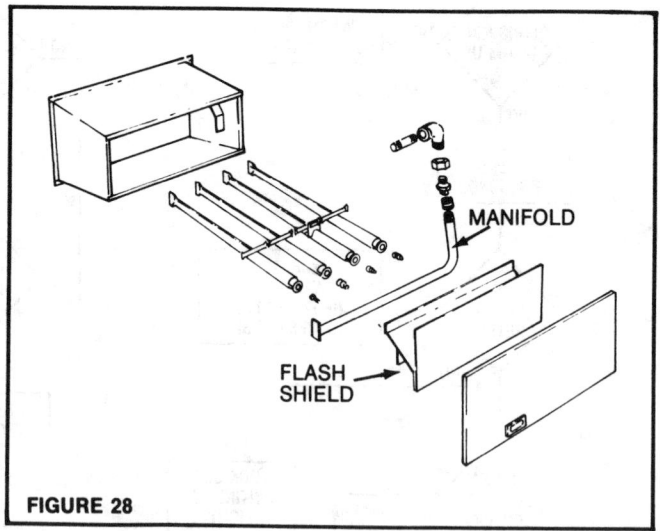

FIGURE 28

To remove flash shield, remove four (4) sheet metal screws and pull shield out. Two of these screws are on the bottom edge of the shield. The other two are located one on each end of the shield. Be sure not to break ignitor lead wire when removing shield. If wire insulation is cracked, broken or frayed, it must be replaced. An ignitor wire in this condition will not allow the furnace to operate properly. A continuous lock-out will occur.

D. EVAPORATOR BLOWER COMPARTMENT

This compartment contains the blower housing, motor, wheel, capacitor, door safety switch, transformer relay box and low voltage terminal strip, filter and retainer spring. See Figure 29.

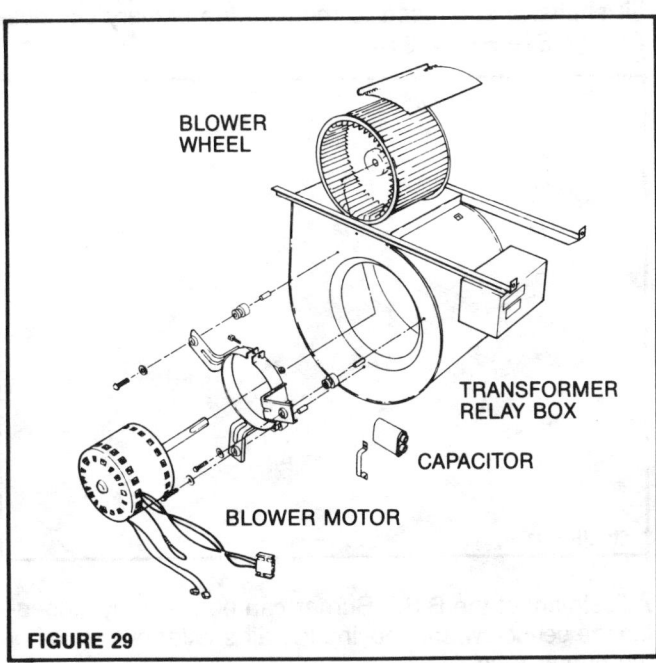

FIGURE 29

To remove blower housing from furnace, unplug both AMP connectors, remove low voltage wire connections from terminal strip and remove connector from door safety switch. Remove the two (2) sheet metal screws holding blower assembly to furnace and blower assembly can be pulled toward the front of the furnace then drops for easy removal.

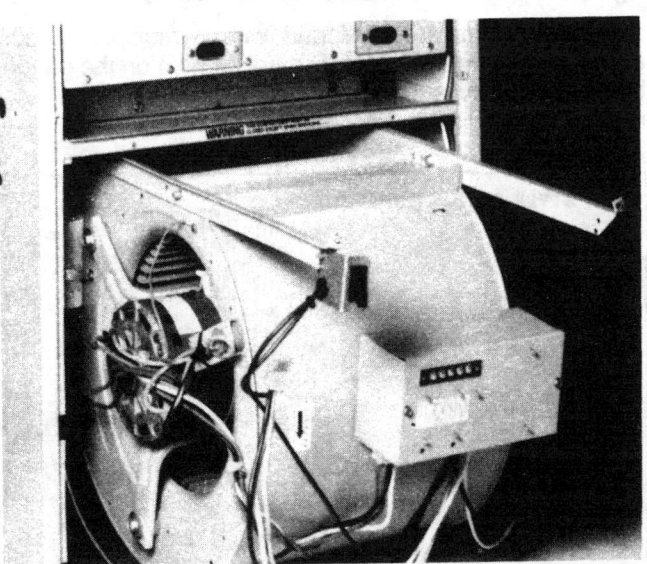

FIGURE 30

E. HEAT EXCHANGER REPLACEMENT

CAUTION: Both electrical and gas supply must be turned OFF before replacement of heat exchanger. Failure to do this could result in property damage or personal injury.

NOTE: When ordering heat exchanger, a new gasket kit must be ordered with the heat exchanger.

1. Remove the front louvered door.

2. Disconnect the flue pipe from the flue box.

3. Loosen the fitting on the Pressure Switch and disconnect the tubing at the switch. Take care not to bend or kink the metal tubing. Remove wire leads.

4. Remove the access cover to the field wired junction box and disconnect all field wiring connections.

5. Remove the screws holding the junction box to the top panel and let the box hang loosely and out of the way of further disassembly.

6. Remove screws holding the top panel. Remove top panel (with Pressure Switch attached).

7. Disconnect the access plug and ignition wire from the ignition control module.

8. Remove the screws attaching the ignition control bracket to the furnace casing, remove the control and control bracket from the furnace.

9. Remove the wiring assembly connectors for the control circuitry from their positions in the blower partition by squeezing the snap tabs on the connector and pulling the connector free of the partition. DO NOT pull on the wires to accomplish this or the connection pins may be damaged or displaced.

10. Disconnect the gas supply piping from the furnace at the gas valve.

NOTE: At this point it should be possible to remove the entire heat exchanger assembly, burner enclosure and controls from the furnace casing.

11. Remove the two (2) screws at the top corner of the furnace front partition.

12. Remove the large hex-head screws on each side of the front partition just above the upper outside corners of the burner enclosure.

13. Remove the large hex-head screws at the bottom under side of the blower enclosure.

14. Remove the screw on each side of the furnace at the blower partition. This will allow some spreading of the casing front to allow clearance for removal of the heat exchanger.

15. Carefully pull the entire assembly forward and free of the casing support brackets.

16. With the assembly as far forward as possible, lift slightly and turn it to one side to clear the front channel of the casing. As the edge clears the channel, shift the assembly over to clear the far side and pull straight out of the furnace.

17. With the entire assembly clear of the furnace, the various component parts may now be removed for installation on the replacement heat exchanger.

    Begin at the top and work down.

18. Remove the hex-head screws which attach the collector assembly to the heat exchanger front partition. Remove the collector assembly.

    DO NOT remove the porcelainized flue box from the collector unless it is damaged or otherwise in need of replacement.

19. Remove the screws attaching the flue equalizer baffle to the front partition. Remove the equalizer baffle and gaskets.

20. Remove the screws securing the flue "V" baffles in the flue outlet passages and slide the baffles out the front of the assembly.

21. Remove the Fan/Limit Control and baffle.

22. Remove the Forced Draft Blower by removing the screw from the bracket at the extension tube restrictor port.

23. Remove the burner enclosure front cover.

24. Remove the flash shield attached to the interior brackets with two (2) screws.

25. Disconnect leads from ignitor or ignitor/sensor electrode.

26. Remove the burners and burner retention springs.

27. Remove the screws in the side flanges of the burner enclosure that secure the enclosure to the pouch.

28. Remove the burner enclosure top bracket. This is attached with screws at the enclosure top flange.

29. Pull the burner enclosure away from the heat exchanger pouch. The gas manifold and valve assembly may remain attached to the enclosure for this disassembly.

REPLACEMENT HEAT EXCHANGER—INSTALLATION

Upon arrival of new heat exchanger inspect for damage. If no damages are found and new gasket kit is available proceed with replacement as follows:

1. Clean all old gasket residue off of components requiring new gaskets.

2. Cut gaskets to size for the furnace you are servicing.

3. Steps 1 thru 29 on "HEAT EXCHANGER REPLACEMENT" can be followed in reverse order for reinstalling the new heat exchanger.

F. BURNERS

During the design process of the Forced Draft Furnace, we tested two different burners, to assure an adequate parts inventory.

Older units were manufactured with either a set of Lincoln Brass Burners Part number 2145-702 or a set of B.S.I. burners part number 2145-700. These burners are interchangeable *as a set*. However, the two different brands CANNOT be intermixed in the same furnace.

Although both operate the same, there is a physical difference between the two burners. Should there be a need for burner replacement on a furnace, it will be necessary to order the correct burner, or order a complete set of burners.

Newer Production Uses B.S.I. Gang Burner Assemblies. The operation and adjustment are the same as the B.S.I. single burner.

G. PRIMARY COMBUSTION AIR

Since we have two burners that have been approved and cannot be interchanged, proper adjustment of primary air will depend on which burner you are dealing with.

The B.S.I. Burner is equipped with a slide type primary air shutter that is used to regulate the primary combustion air. See Figure 31A.

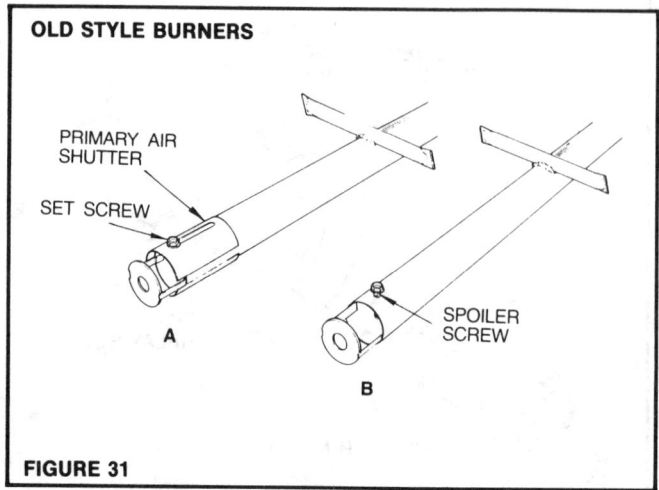

OLD STYLE BURNERS

PRIMARY AIR SHUTTER

SET SCREW

SPOILER SCREW

A

B

FIGURE 31

Adjustment of the B.S.I. Burner can be made by loosening the set screw and moving the air shutter onto or off of the burner tube.

The Lincoln Brass Burner does not have a primary air shutter. See Figure 31B. Burner primary combustion air is adjusted by inserting or retracting the gas "spoiler" screw on the top of the burner tube.

The gas "spoiler" screw regulates the velocity of gas flow through the burner venturi, thereby increasing or decreasing primary combustion air to the burner.

The newer units are equipped with a multiple burner assembly. See FIGURE 32. The adjustment on the gange burner assembly is done by loosening the set screw and moving the air shutter onto or off of the tube.

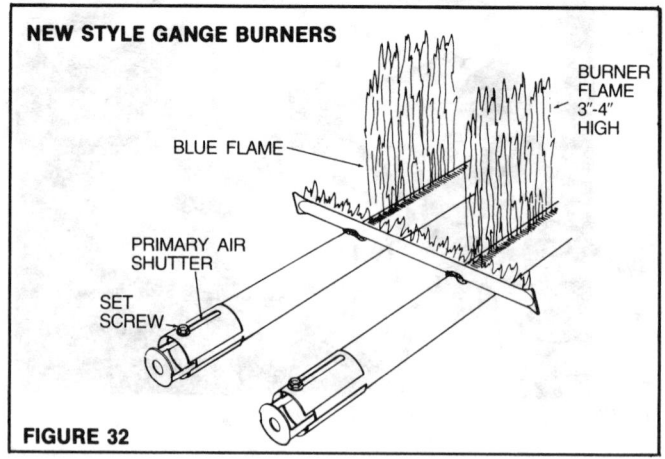

NEW STYLE GANGE BURNERS

BURNER FLAME 3"-4" HIGH

BLUE FLAME

PRIMARY AIR SHUTTER

SET SCREW

FIGURE 32

VII. FURNACE CLEANING— INSPECTION AND CLEANING PROCEDURE

NOTE: This furnace is equipped with a pressurized combustion chamber. It is mandatory to replace all fiber gaskets that are removed. Replacement gasket parts should be on hand prior to starting service procedure.

A. HEAT EXCHANGER AND BURNER CLEANING

1. Disconnect electrical sources to the unit.
2. Remove louver panel from the unit and set aside.
3. Turn gas supply "OFF" at the gas valve. Allow a few minutes for unit to cool down.
4. Remove the screw from the front panel of the burner enclosure.
5. Remove the front panel. **NOTE:** Discard old gasket.
6. Remove the two screws holding the flash shield in place.
7. Remove the flash shield. See Figure 33.
8. Disconnect the ignitor and sensor leads from the ignitor/sensor.

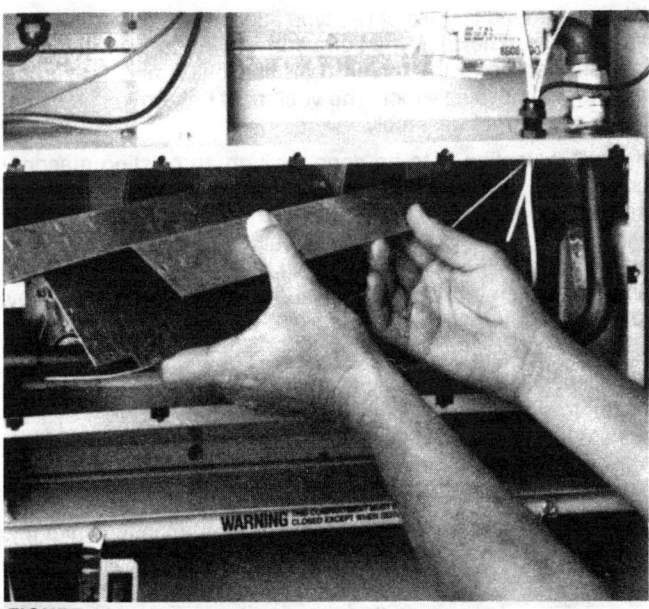

FIGURE 33

NOTE: At this point inspect for any loose particles or carbon deposits in the burner area. Using a flashlight and a small mirror check the upper portions of the heat exchanger for any signs of blockage or dirt accumulation. If any of these conditions are apparent proceed with the following steps:

9. Remove all burners from the unit. See Figure 34.

Removal of the burners is accomplished by pulling back on the burner so that it compresses the spring that is located between the manifold and the burner. Tilt the burner up into the heat exchanger tube. Slide the burner off the orifice and remove from the unit. MAKE NOTE OF LOCATION OF THE BURNER WITH THE IGNITOR FOR REASSEMBLY IN THE CORRECT POSITION.

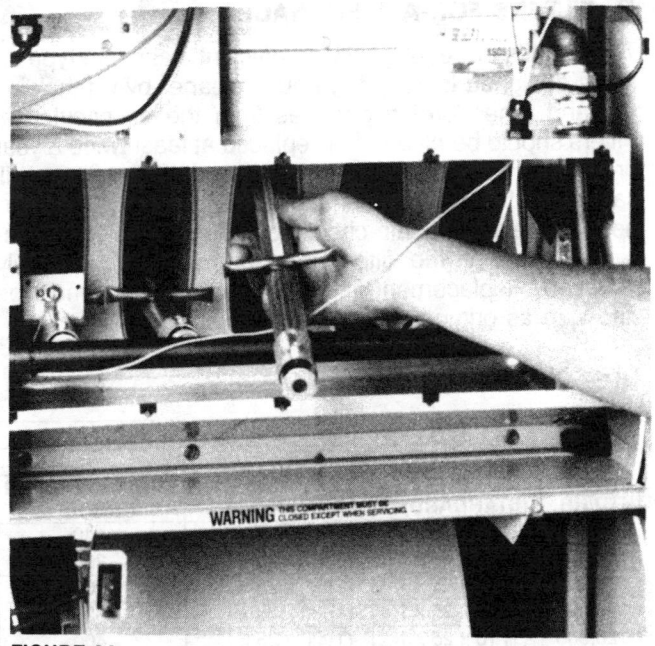

FIGURE 34

10. After each burner has been taken from the unit, tap the end of the burner that slips over the orifice *lightly,* thereby removing any residue from the inside of the burner, then run a vacuum hose across the top of the burner to remove any foreign material that might be lodged between the ports.
11. Disconnect the flue pipe from the flue box collar.
12. Attach the exhaust end of a high pressure vacuum cleaner (shop vacuum) to the furnace flue outlet to force any loose particles from the upper heat exchanger into the bottom burner area.
13. With a stiff bristled brush, clean the bottom area of the heat exchanger, brushing into the burner enclosure where the vacuum can easily be used to pick up the residue. See Figure 35.
14. Reassemble the unit by reversing the above procedure.

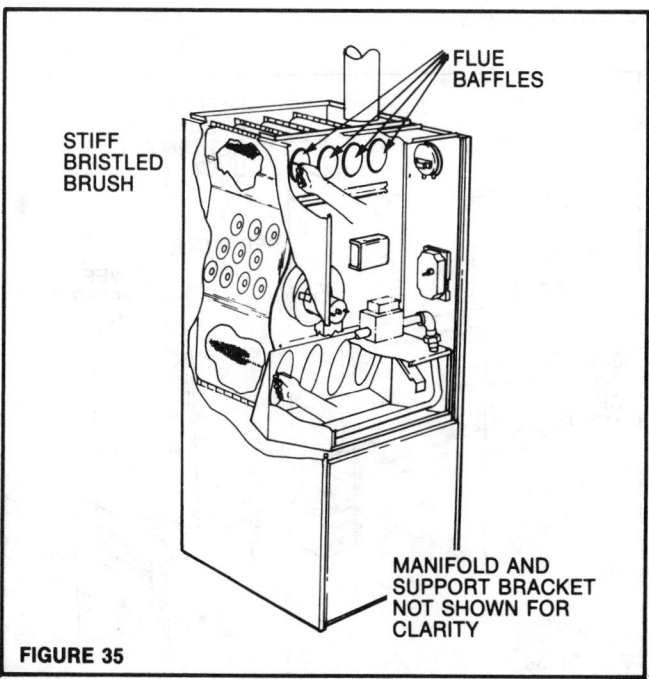

FIGURE 35

B. FILTERS FOR ALL FURNACES

These units are equipped with efficient washable type air filters designed to keep the house cleaner by withholding dirt and other foreign particles from the air circulation. Filters should be cleaned or replaced at least twice a year or at any time they become clogged. Dirty filters will cause the furnace to heat improperly and possibly overheat by restricting air circulation. These units are designed so that the filters may be quickly and easily replaced. Replacement filters must be of the same type and size as originally furnished with the furnace.

UPFLOW FURNACE

FILTER REQUIREMENTS (HEATING)*			
FILTER NUMBER AND SIZE			
INPUT BTU/HR.	INTERNAL (STANDARD)**	EXTERNAL (OPTIONAL)***	MINIMUM FILTER AREA(SQ.INCH)**
40,000	(1) 16 x 25 x 1	(1) 20 x 25 x 1	140
60,000	(1) 16 x 25 x 1	(1) 20 x 25 x 1	200
80,000	(1) 16 x 25 x 1	(1) 20 x 25 x 1	260
100,000	(1) 16 x 25 x 1	(1) 20 x 25 x 1	320
120,000	(1) 16 x 25 x 1	(1) 20 x 25 x 1****	400

*For cooling requirements consult specification sheet.
**Filters rated at 520 FPM or more, permanent type.
***See price book or specification sheet for external filter rack options.
****For quantities over 1700 CFM requires (2) external filters.

COUNTERFLOW FURNACE

The filters supplied with the furnace may be installed in the return air plenum above the furnace as shown in Figure 36. A filter rack is supplied with each furnace.

INPUT BTUH	FILTER SIZE	QUANTITY	"A" TOP OF FILTER ABOVE UNIT
60	16 x 16 x 1	2	14½
80	16 x 16 x 1	2	13⁹⁄₁₆
100	16 x 16 x 1	2	12¼
120	16 x 16 x 1	2	10⅝

NOTE: The return air plenum must extend a sufficient height above dimension "A" as shown in Figure 36, to provide for the attachment of a return air duct or grille above the filters.

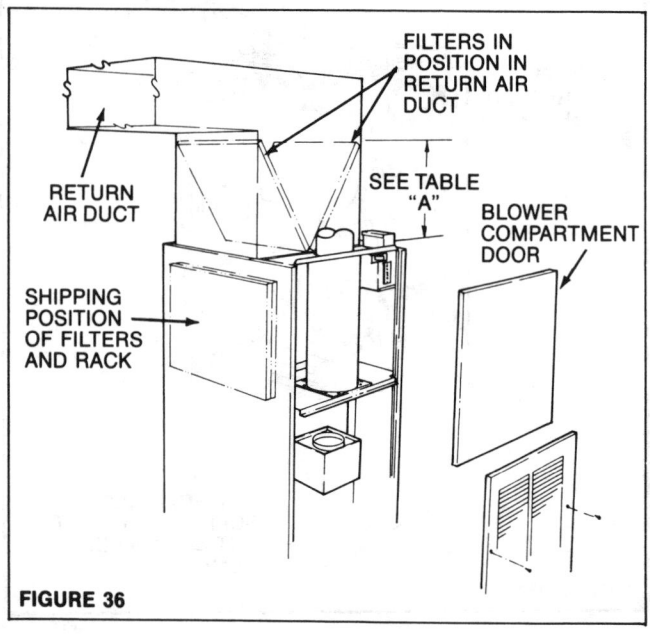

RETURN AIR DUCT

FILTERS IN POSITION IN RETURN AIR DUCT

SEE TABLE "A"

BLOWER COMPARTMENT DOOR

SHIPPING POSITION OF FILTERS AND RACK

FIGURE 36

C. VENTILATION

The flow of combustion and ventilating air must NOT BE obstructed from reaching the furnace. Air openings provided in the casing of the furnace must be kept free of obstructions which would restrict air flow, thereby affecting efficiency and safe operation of the furnace. Furnaces MUST have air for proper performance.

WARNING: The air for combustion and ventilation must NOT come from a corrosive atmosphere.

Combustion air must be free of acid forming chemicals; such as sulphur, fluorine and chlorine. These elements are found in aerosol sprays, detergents, bleaches, cleaning solvents, air fresheners, paint and varnish removers, refrigerants and many other commercial and household products. Vapors from these products when burned in a gas flame, form acid compounds. The acid compounds increase the dew point temperature of the flue products and become highly corrosive after they condense.

NEW APPLICATION VENTING—VERTICAL

Flue vent installations for this series shall be in accordance with Part 7, "Venting of Equipment", of the National Fuel Gas Code, ANSI Z223.1-1988 or application provisions of local building codes.

We recommend the use of B-1 type flue vent for *all* vertical vent applications. The B-1 type flue vent referred to is *vertical flue vent* that starts at the forced draft flue outlet and terminates through the building's roof. The vent must be terminated with a listed cap or roof assembly.

If the flue venting is to be done through an existing masonry chimney, it is strongly recommended that when local experience indicates that flue gases may condense to liquid within the chimney, provisions should be made to provide a suitable metal liner, chimney should be lined with an approved flue pipe which terminates at or above the top of the chimney, and must be terminated with a listed cap or roof assembly.

Proper vent termination is required so that wind from any direction will not create positive pressure in the vicinity of the outlet, to prevent entry of rain water into the flue vent system and to prevent flue restriction by ice build-up during subfreezing weather conditions.

Additionally, provisions must be made, by the installer, for the safe removal of any condensate that might form in the chimney.

> ### ⚠ CAUTION:
> Failure to provide for safe removal of condensate, could result in water damage to the building.

If the flue venting is to be done through an existing chimney, we recommend the following:

1. The chimney should be inspected for conformance with local and national codes (refer to Article X of the National Building Code or the Standard for Chimney and Vents NFPA 211).

2. If the chimney is of outside construction (not enclosed by the building) the chimney should be lined or relined with a suitable metal liner:

   a. B-1 type vent with proper terminals, or

   b. With single wall metal pipe having resistance to heat and corrosion not less than that of galvanized sheet steel or aluminum not less than 0.016 inch thick and with proper ends or terminations, or

   c. If permitted by local codes, with stainless steel vent pipe not less than 0.016 inch thick.

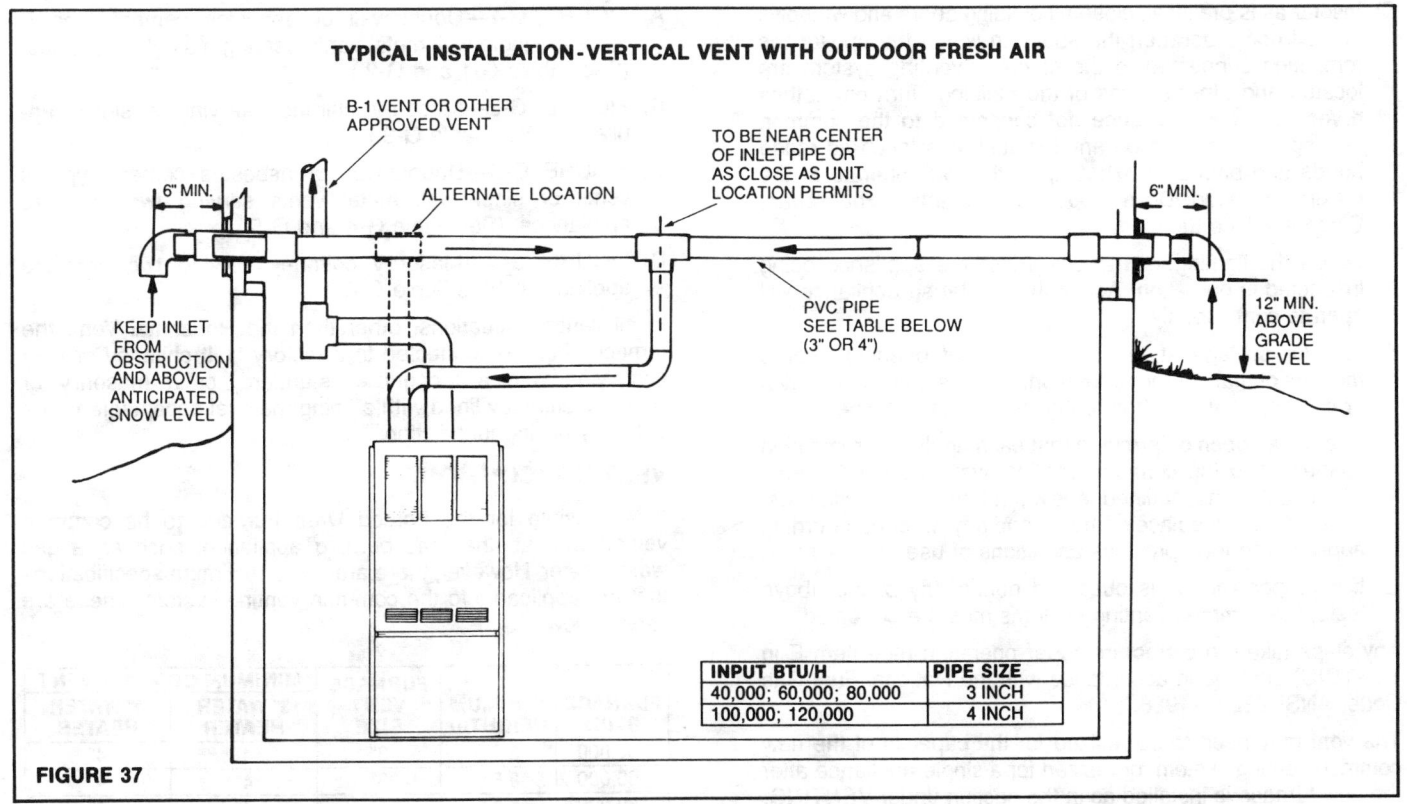

TYPICAL INSTALLATION-VERTICAL VENT WITH OUTDOOR FRESH AIR

B-1 VENT OR OTHER APPROVED VENT

ALTERNATE LOCATION

6" MIN.

KEEP INLET FROM OBSTRUCTION AND ABOVE ANTICIPATED SNOW LEVEL

TO BE NEAR CENTER OF INLET PIPE OR AS CLOSE AS UNIT LOCATION PERMITS

PVC PIPE SEE TABLE BELOW (3" OR 4")

6" MIN.

12" MIN. ABOVE GRADE LEVEL

INPUT BTU/H	PIPE SIZE
40,000; 60,000; 80,000	3 INCH
100,000; 120,000	4 INCH

FIGURE 37

The chimney liner or flue vent pipe must be joined together in accordance with the vent pipe manufacturer's instructions, but in any event all joints must be assembled to provide a good seal for the entire flue system.

d. It is recommended that the vent from the furnace and any other vented appliance be attached to the liner, and that the liner be terminated with an elbow or a closed bottom "T" to trap any condensate that may form.

3. If the chimney is located within the building and extends through less than five (5) feet of uninsulated attic space, then the flue vent connector may connect directly to the chimney and the chimney used without a metal liner.

The following additional recommendations are made for flue vent systems for Forced Draft furnaces:

1. Select flue connection material that is satisfactory for installation and that meets requirements of local codes.

   In applications where the flue vent or chimney is not directly above the flue opening of the furnace, a vent connector of single wall galvanized steel or aluminum of not less than 0.016 inch wall thickness (28 gauge) may be used, where clearances to combustible materials is not less than 6 inches.

2. Flue connection pipe must be at least as large as the diameter of the outlet collar on the furnace. Do not reduce the diameter of the vent connector between the furnace outlet collar and the flue vent.

3. Run flue vent pipe from the vent system or chimney directly to the furnace outlet collar, utilizing the shortest route possible.

4. Frequently, the job requires horizontal vent pipe runs and elbows between the furnace flue collar and the vertical vent. We recommend no more than 4 elbows in vent system.

   In applications where the horizontal run of B-1 type vent passes through an unheated space, the pipe should be insulated with R-7 or equal insulation.

5. Horizontal runs should slope upward from the furnace not less than 1/4 inch per foot (21 mm/m).

6. Horizontal portions of the vent piping should be supported with hangers or straps every 3 feet to prevent sagging and to insure that there will be no movement after installation.

7. Vertical flue vents should extend high enough above the roof or neighboring structures so that wind from any direction will not create positive pressure in the vicinity of the outlet (refer to ANSI Z223.1-1988 for recommended dimensions).

8. The furnace shall not be connected to a chimney flue serving a separate appliance designed to burn solid fuel.

9. The furnace shall not be connected into any portion of mechanical draft systems operating under positive pressure, such as Category III and IV furnaces.

It is always advisable to consult local codes, or in the absence of local codes the National Fuel Gas Code when installing a flue vent system.

VENTING—VERTICALLY EXISTING APPLICATION

In many venting installations, the furnace may be installed and connected to a common vent system and vented in conjunction with other gas burning appliances. When an existing furnace is removed from a venting system serving other appliances, the venting system is likely to be too large to properly vent the remaining attached appliances.

A test procedure has been developed under ANSI Standards to determine the acceptable operation of the vent system.

The following steps shall be followed with each appliance remaining connected to the common venting system placed in operation, while the other appliances remaining connected to the common venting system are not in operation.

A) Seal any unused opening in the common venting system.

B) Visually inspect the venting system for proper size and horizontal pitch and determine there is no blockage or restriction, leakage, corrosion and other deficiencies which could cause an unsafe condition.

28

C) Insofar as is practical, close all building doors and windows and all doors between the space in which the appliances remaining connected to the common venting system are located and other spaces of the building. Turn on clothes dryers and any appliance not connected to the common venting system. Turn on any exhaust fans, such as range hoods and bathroom exhaust, so they will operate at the maximum speed. Do not operate a summer exhaust fan. Close fireplace dampers.

D) Follow the lighting instructions. Place the appliance being inspected in operation. Adjust thermostat so appliance will operate continuously.

E) Test for spillage at the drafthood relief opening after 5 minutes of main burner operation. Use the flame of a match or candle, or smoke from a cigarette, cigar, or pipe.

F) After it has been determined that each appliance remaining connected to the common venting system properly vents when tested as outlined above, return doors, windows, exhaust fans, fireplace dampers and any other gas-burning appliance to their previous conditions of use.

G) If improper venting is observed during any of the above tests, the common venting systems must be corrected.

Any steps taken to correct improper operation (see Item E in VENTING) shall be in accordance with the National Fuel Gas Code, ANSI Z223.1-1988.

The vent may need to be resized for the capacity of the new common venting system, or resized for a single appliance after the new furnace is installed as in the section under VENTING, HORIZONTAL. When resizing any portion of the venting system, common or single vent, the system should be resized to approach the minimum size as determined using the appropriate tables in Appendix G in the National Fuel Gas Code.

A) FIGURE G-1—Doublewall or asbestos cement type B vents or single-wall metal vents serving a single appliance. (See Tables G-1 and G-2.)

B) FIGURE G-2—Masonry chimney serving a single appliance. (See Table G-3.)

C) FIGURE G-3—Double-wall or asbestos cement type B vents or single-wall metal vents serving two or more appliances. (See Table G-4 and G-5.)

D) FIGURE G-4—Masonry chimney serving two or more appliances. (See Table G-6.)

In all venting situations, other than the Horizontal Vent, the furnace shall be connected to a factory built chimney or vent complying with a recognized standard, or a masonry or concrete chimney lined with a lining material acceptable to the authority having jurisdiction.

VENTING (COMMON)

It is possible for the Forced Draft Furnace to be common vented with another gas burning appliance, such as a gas water heater. However, there are some minimum specifications that are applicable to the common venting system. These are listed below.

FURNACE BTUH	MINIMUM HEIGHT①	FURNACE VENT SIZE	MINIMUM COMMON VENT 3" WATER HEATER	4" WATER HEATER
40,000	8 FT	3"	4"	4"
60,000	8 FT	3"	4"	4"
80,000	8 FT	3"	4"	4"
100,000	8 FT	4"	5"	5"
120,000	8 FT	4"	5"	5"

① MINIMUM VENT HEIGHT MEASURED FROM WATER HEATER OR 9 FT. FROM FURNACE WHICHEVER IS LOWEST.

ADDENDUM A

INSTALLATION INSTRUCTIONS FOR DOWNFLOW
FORCED DRAFT FURNACE HORIZONTAL VENTING
(See Furnace Installation and Category Label)

<u>INSTALLED AS A DIRECT VENT FURNACE</u>

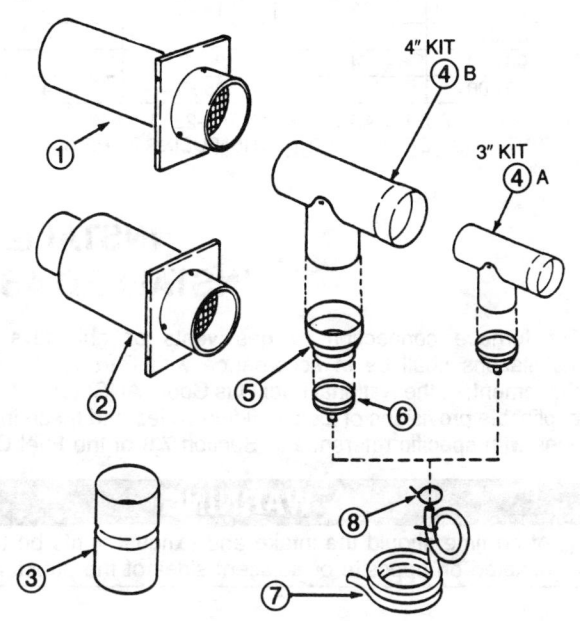

3" KIT NO. 8G03004—4" KIT NO. 8G03005

ITEM NUMBER	DESCRIPTION	8G03004 3" KIT NO.	8G03005 4" KIT NO.
①	INTAKE TERMINAL	0548010001	0548010002
②	VENT TERMINAL WITH CLEARANCE SHIELD	0548010003	0548010004
③	VENT ADAPTOR FITTING	2311-170	2311-171
④A	CONDENSATE TRAP TEE WITH CAP & DRAIN	2311-182	—
④B	CONDENSATE TRAP TEE	—	2311-181
⑤	REDUCER 4" TO 3"	—	2311-172
⑥	CAP WITH DRAIN	—	2311-192
⑦	8 FT. x ⅜" DRAIN HOSE	2939-156	2939-156
⑧	HOSE CLAMP	2344-433	2344-433
⑨	INLET ADAPTER 4" TO 3"	—	1708-605

INSTALLER SUPPLIED MATERIALS

		TYPE 29-4C STAINLESS VENT		TYPE 304 STAINLESS VENT	
		3 INCH	4 INCH	3 INCH	4 INCH
⑧	Straight sections— Stainless Steel Pipe	SNAPPY NO. S-905-3	SNAPPY NO. S-905-4	Type 304 Stainless Pipe	
⑨	Elbows—Stainless Steel	SNAPPY NO. S-900-3	SNAPPY NO. S-900-4	Type 304 Stainless Fittings	
⑩	Intake—Lightweight Plastic (Drain Pipe)	Type ASTM-D 335-OMS (Conform to ASTMD2564)	Type ASTM-D 335-OMS	Type ASTM-D 335-OMS (Conform to ASTMD2564)	Type ASTM-D 335-OMS
⑪	PVC Bonding Cement	—	—	—	—
⑫	Silicone Sealant (Dow Corning TRV 732 or the equivalent)	—	—	—	—
⑬	Pipe Hangers (Perforated Metal Strapping, Etc.)	—	—	—	—
⑭	Stainless Steel Sheet Metal Screws	—	—	Type 304	Type 304
⑮	Outlet Pipe Insulation (Optional) Any heavy duty fiberglass pipe insulation type ASJ may be used, if this option is desired.			Outlet Insulation Required 1 inch ASJ Fiberglass shall be used	

For the location of the Snappy Pipe Distributor closest to your area call:

SNAPPY
Air Distribution Products
A Division of Standex Energy Systems
1090 Legion Road (P.O. Box 1168)
Detroit Lakes, MN 56501
(218) 847-9258

SEE TABLE FOR VENT LENGTHS

For purposes of Horizontal Venting these units may be vented up to 30 feet with a maximum of (4) four ninety degree elbows. However, to minimize condensate in the vent pipe on individual units follow the guidelines in the following chart.

MAXIMUM LENGTH OF VENT (HORIZONTAL APPLICATION)

INPUT BTUH	VENT OR INTAKE DIA.	INSULATED VENT PIPE②			UNINSULATED VENT PIPE		
		MAX. LENGTH	MAX. ELBOWS	CLEARANCE TO COMBUSTIBLE MATERIALS①	MAX. LENGTH	MAX. ELBOWS	CLEARANCE TO COMBUSTIBLE MATERIALS
40,000	3"	10 FT.	4	1'	—	—	6"
60,000	3"	15 FT.	4	1"	10 FT.	3	6"
80,000	3"	25 FT.	4	1"	15 FT.	3	6"
100,000	4"	25 FT.	4	1"	15 FT.	3	6"
120,000	4"	25 FT.	4	1"	15 FT.	3	6"

① NATIONAL FUEL GAS CODE, ANSI Z-223.1-1988
② VENT <u>MUST BE INSULATED</u> WHEN USING TYPE 304 STAINLESS STEEL PIPE AND FITTINGS.

INSTALLATION INSTRUCTIONS
INSTALLED AS A DIRECT VENT FURNACE

For furnace connection to gas vents or chimneys, vent installations shall be in accordance with Part 7, Venting of Equipment, of the National Fuel Gas Code, ANSI Z223.1-1988, applicable provisions of local building codes and these instructions with specific reference to Section 7.8 of the Fuel Code.

> ## ⚠ WARNING
> At no time should the intake and exhaust vents be terminated on opposite or adjacent sides of the structure.

1. DIRECTION OF VENT FITTINGS

All Vent fittings and pipe **must** be installed with male ends toward the furnace.

Longitudinal Vent pipe seams **must** be on top.

All horizontal vent pipe **must** be pitched up (Downhill toward furnace) at least ¼" per foot.

All horizontal pipe should be supported to prevent sagging. Place pipe hangers approximately five feet apart for both vent and intake pipe runs. Greater distance apart may be used if pipe lengths are 10 feet or more between joints.

2. CONDENSATE IN THE VENT

Under certain conditions, such as long runs of vent pipe through spaces with relatively cool ambient temperatures, condensate may form inside the flue vent pipe. The prescribed pitch of the pipe and the supplied condensate trap/drain connected in the horizontal pipe close to the furnace outlet will allow removal of any accumulation that may occur.

3. SEALING THE PIPE JOINTS

> ## ⚠ WARNING
> Solvent cements are combustible liquids and should be kept away from all ignition sources, (ie. sparks, open flames and excessive heat). Avoid breathing cement vapors or contact with skin and eyes.

Because the forced draft unit produces positive vent pressures when the vent arrangement is in the horizontal position, it is necessary to seal all joints and connections to prevent the escape of flue products into the surrounding area. Sealing of the pipe joints of both intake and outlet is necessary to prevent pressure losses in the pipe under wind conditions. These losses can produce an imbalance between the intake and exhaust, at the furnace, which could shut down the system. This would include the horizontal pipe connections, pipe longitudinal seams, seams in elbows, tees, tee reducers and caps.

4. SEALING OF PLASTIC INTAKE PIPING

Use a good quality cement on clean, oil free joints.

5. THE STAINLESS STEEL VENT PIPING

Use a high temperature silicone sealant (such as Dow-Corning RTV 732—400°F or equivalent). All joints to be sealed should be oil and dirt free.

Where it is necessary to provide extra strength to a joint in the stainless steel exhaust pipe, the use of stainless steel sheet metal screws is recommended (stainless steel type

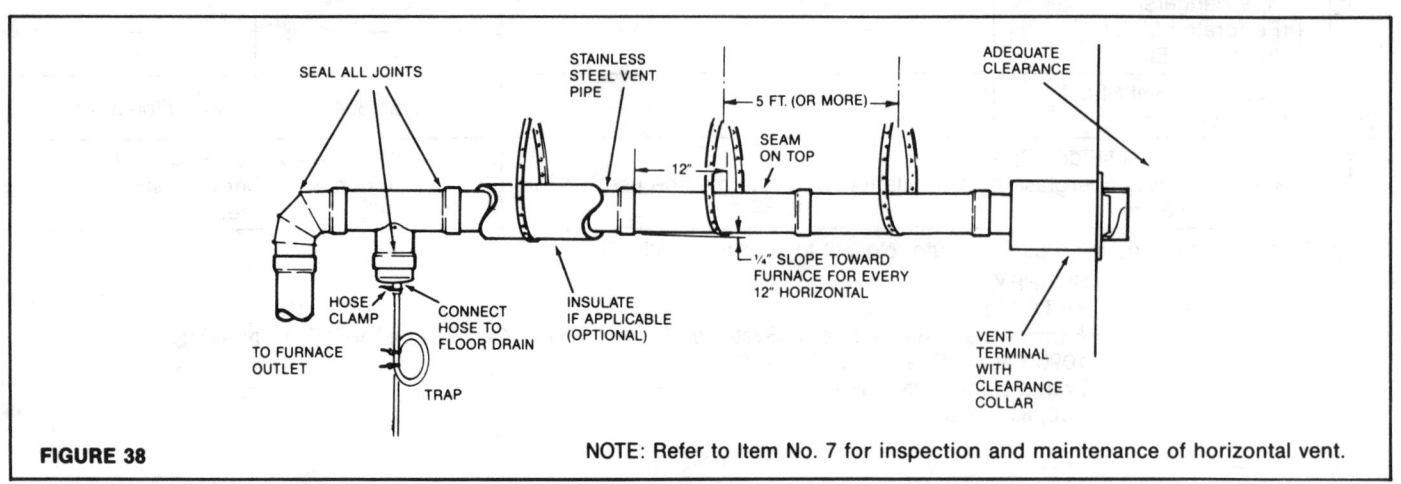

FIGURE 38   NOTE: Refer to Item No. 7 for inspection and maintenance of horizontal vent.

304). DO NOT PUT A SCREW IN THE BOTTOM OF THE PIPE. Dab the screw with silicone sealant to prevent leakage around the threads. **Screwing the joint together is not a substitute for sealing the connection with the silicone sealant.**

6. LOCATION OF THE DIRECT VENT TERMINAL

The terminations supplied in the kit must be used for all horizontal installations to ensure the proper performance of the appliance and to prohibit the entrance of birds or debris. The location of terminals is important for performance as well as safety:

1. The distances from adjacent public walkway; adjacent buildings, open windows and building openings shall be consistent with the National Fuel Gas Code Z223.1 (current edition).

2. The venting system terminal must be at least three (3) feet above any forced air inlet located within 10 feet.

   Exception: This provision shall not apply to the **combustion air intake.**

3. The vent terminal of a direct vent appliance with an input of 50,000 Btu per hour or less shall be located at least 9 inches from any opening through which flue gases could enter a building, and such an appliance with an input over 50,000 Btu per hour shall require a 12-inch vent termination clearance.

4. The vent terminal must be located at least one foot above grade to avoid blockage by snow or as required to be above snow level.

5. A clearance of four (4) feet is required to combustible materials opposite the outlet terminal.

6. Vent and intake must be on the same face of the building in order to be in the same pressure zone. The centerline of terminals must be a minimum 16 to a maximum of 24 inches apart on a horizontal line. Never install the intake terminal higher than the vent terminal.

7. The vent terminal of the appliance shall not terminate over public walkways or over an area where condensate or vapor could create a nuisance or hazard or could be detrimental to the operation of regulators, relief valves, or other equipment.

> ### ⚠ CAUTION:
> At no time should the intake and exhaust vents be terminated on opposite or adjacent sides of structure.

For typical vent-inlet air installations, see FIGURES 39 and 40.

8. Do not mix the venting portion of this system with components or materials from other systems made by other manufacturers.

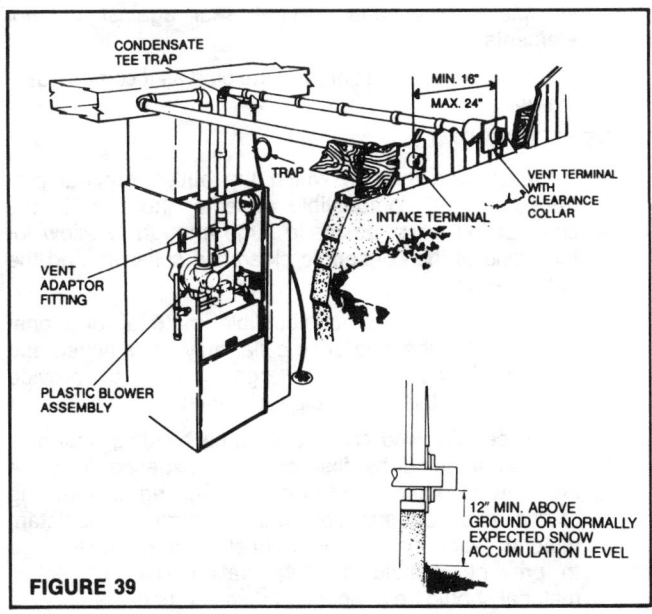

FIGURE 39

9. Do not use Type B gas vent components anywhere in this inlet-vent system.

10. Do not connect more than one appliance to the inlet-vent system.

11. Do not connect equipment vented by natural draft (units with draft hoods or draft diverters) into any portion of this positive pressure system.

7. PERIODIC INSPECTION AND MAINTENANCE

Terminals

Inspect the intake and outlet terminals at the beginning and periodically throughout the heating season to prevent the accumulation of debris at the terminal barrier screens.

Vent Pipe

Inspect the vent pipe (Stainless Steel Pipe) at the beginning and periodically throughout the heating season to insure that the joints are leak tight and that the pipe is in proper condition for safe operation. At the same time check the hose/trap from the tee connector to ensure that it is intact and unrestricted.

Also check that there is water in the tube loop. Some water should be in the trap to prevent leakage of a small amount of flue products thru the drain tube. Check for water by removing tube clamp, removing tube and pouring water into tube till it flows out end of tube. Replace tube and clamp. This procedure required only if original installation includes tee, plug and drain hose.

READ ASSEMBLY PROCESS THOROUGHLY BEFORE PROCEEDING:

ASSEMBLY INSTRUCTIONS:

> ### ⚠ CAUTION:
> SHUT OFF ALL ELECTRICAL POWER TO THE FURNACE.

1. ASSEMBLY OF OUTSIDE INTAKE/VENT TERMINALS

Follow all the conditions described earlier in these instructions with regard to clearances and positioning when determining the location of the intake and outlet vent terminals.

INTAKE TERMINAL:

a. The hole made in the wall through which the terminal is installed, should be one inch (1″) larger than the outside diameter of the terminal pipe. Four inch for the three (3) inch terminal and 5 inch for the four (4) inch terminal.

b. Insert the terminal through the hole from the outside and secure it in place with the proper type of anchor for the type of construction to which it is being attached.

c. From the outside, caulk around the edge of the mounting plate to provide a tight seal against exterior elements.

d. From the inside, caulk around the pipe as it comes out of the wall.

VENT TERMINAL:

a. If the surface through which the outlet terminal protrudes is of a combustible material, the hole in the construction **must** be made large enough to allow for the fitting of the concentric clearance collar around the outlet terminal.

If the wall is a non-combustible material of stone, concrete etc., the clearance collar may be removed and a hole should be made large enough to provide clearance for the plastic pipe of 4 inches.

Under certain wind conditions some building materials may be affected by flue products expelled in close proximity to unprotected surfaces. Sealing or shielding of the exposed surfaces with a corrosion resistant material (such as aluminum sheeting) may be required to prevent staining or deterioration. The protective material should be attached and sealed (if necessary) to the house before attaching the vent terminal.

b. Slip the terminal in place from the outside and secure it with the appropriate anchoring device.

c. Caulk around the perimeter of the mounting plate.

d. From the inside, caulk around the clearance collar if combustible construction or around vent pipe (with plastic pipe in place), if the opening is through non-combustible material of concrete, stone, etc.

2. PIPE ASSEMBLY

Before hanging the intake and outlet pipe permanently, assemble the pipe from the terminal to the furnace and cut the lengths necessary to give the positioning required in final assembly. See FIGURES 39 and 40 for examples.

INTAKE PIPE:

There are two main types of PVC pipe that will work on the combustion air intake of the forced draft furnace. These are specified as thin wall PVC sewer pipe (ASTMD2729) or scheduled 40 PVC standard wall pipe (ASTMD2665). Refer to FIGURE 40 for proper application of the pipe.

Continue with straight pipe and elbows of intake design configuration to the intake terminal. Do not cement connections at this time. At the inlet air terminal the pipe should overlap the metal connection for at least 3 inches for a good seal.

VENT PIPE:

Start the vent system assembly by placing the vent adaptor fitting on the furnace. Assemble the stainless steel vent pipe and fittings with the male ends toward the vent adaptor fitting. All other pipe sections and fittings must be oriented in the same direction to insure condensate drainage toward the vent drain tee. The drain tee should be installed as close to the furnace as possible (**within 18″ laterally**).

DO NOT allow the metal outlet pipe to come within six inches of combustible construction. See National Fuel Gas Code, ANSI Z-223.1-1988. (One inch clearance is required from the outer surface of an insulation wrapped pipe.) Allow ¼″ per foot drop toward furnace, for the vent pipe.

To obtain desired piping length:

Final trim on the pipe may be accomplished at either the furnace end or the terminal end. If it is necessary to trim the pipe at points between the furnace and terminal ends, use the following procedure:

1. Trim only the unflared end at these points.

2. Do not trim drain tee.

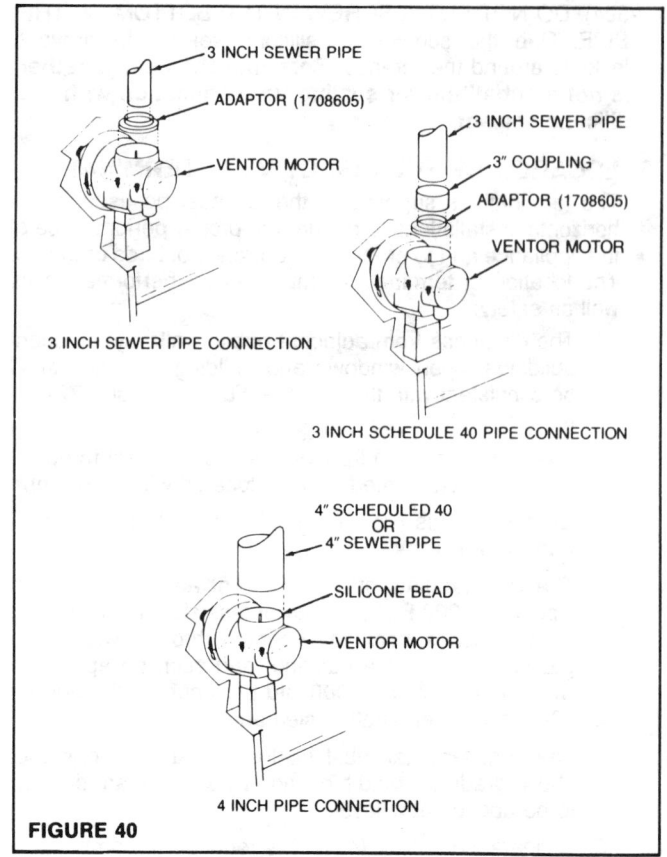

FIGURE 40

DRAIN TEE:

A drain tee assembly using a tee (#3UT) and drain plug (3UDP) should be used in those pipe assemblies that exceed a horizontal pipe length of 15 feet. Locate the drain tee as close to the appliance outlet collar as possible using as few offsets as possible to reduce back pressure. See FIGURES 38 and 39 for example. Connect ⅜ I.D. plastic tubing to the drain plus hose connection. If not provided, make a permanent 3-inch loop in the tubing for a condensate trap.

At final assembly, before connecting the tube, "prime" the trap with water. Pour a small amount of water into the tube to act as a water seal to prevent any flue products from escaping out of the drain tube. Run the loose end of the tube to a drain. See FIGURE 41.

3. FINAL ASSEMBLY

VENT PIPE:

a. Begin at the vent terminal—**NOTE: LONGITUDINAL SEAM OF THE VENT PIPE MUST BE ON TOP WHEN INSTALLED.** See FIGURE 39.

b. Seal each piping joint as you assemble the connections by spreading at least a ⅛ inch thick bead of silicone sealant one (1) inch in from the end of the male connection. After the connection is pushed together, check that there is no gapping in the seal. Reseal any gaps or open seams.

Sealing of the piping would include the horizontal pipe connections, pipe longitudinal seams, seams in elbows, tees, tee reducers and caps.

c. Support the pipe as you go along to prevent excess movement at the joints.

d. Where screws are needed for rigidity (stainless steel screws) be sure to seal around the screw to prevent leakage.

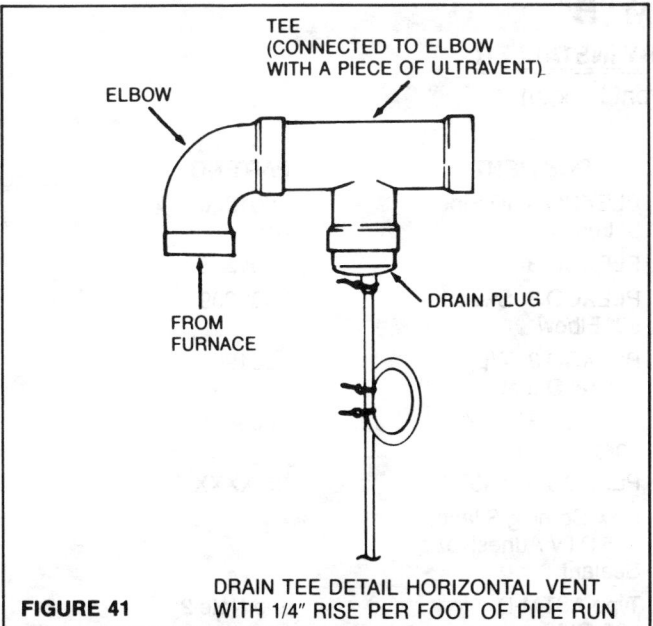

FIGURE 41 — DRAIN TEE DETAIL HORIZONTAL VENT WITH 1/4" RISE PER FOOT OF PIPE RUN

ELBOW

TEE (CONNECTED TO ELBOW WITH A PIECE OF ULTRAVENT)

FROM FURNACE

DRAIN PLUG

e. At the furnace connection, caulk around the base of the collar at the vent box collar with silicone. Then install the Vent adaptor fitting into the silicone caulking and check for a good perimeter seal.

f. Seal the joint at the vent terminal with silicone and make one final inspection of each pipe joint seal.

g. Wrap the areas requiring insulation.

h. Attach the condensate drain hose to the fitting on the drain tee cap and secure with the plastic tie provided. The looped end of the hose, which will act as a trap in the line, must be attached at the tee drain. The free end should then be run to, or as close as possible to a floor drain. At final assembly, before connecting the tube, "prime" the trap with water. Pour a small amount of water into the tube to act as a water seal to prevent any flue products from escaping out of the drain tube.

⚠ CAUTION:

THE CONDENSATE DRAIN HOSE MUST BE ROUTED INSIDE THE HOUSEHOLD DRAIN SO THAT THE CONDENSATE PRODUCED BY THE FURNACE CANNOT BE ACCESSED BY SMALL CHILDREN OR PETS.

4. FRESH AIR INTAKE PIPE ASSEMBLY

a. Beginning at the intake terminal, slip the pipe end over the terminal pipe after placing the bead of sealant over the pipe as in the vent terminal assembly.

b. Connect the pipe back toward the furnace, cementing each joint with standard plastic cement and support the structure with pipe hangers as needed. **BE SURE JOINTS ARE CLEAN AND DRY BEFORE CEMENTING.**

c. At the furnace connection, lay a bead of silicone at the base of the collar on the intake transition box. Then slip the final connection into place to complete the joint. If the unit is equipped with the plastic forced draft blower, use silicone to attach the inlet blower adaptor for three (3) inch inlet pipe and then add the bead of silicone to attach inlet pipe. With four (4) inch pipe to the blower, place at least 1/4 inch bead about 1/4 inch outside and below the collar inlet.

d. Caulk the joint at the terminal connection with silicone, forcing the silicone into the gap between the pipes.

e. Check the entire inlet assembly for integrity of joint seal.

⚠ CAUTION:

The Horizontal Vent Kit is designed for use with a Forced Draft Gas Furnace. As a natural part of the unit's operation, normal products of combustion, including water vapor are vented to the atmosphere. Since the outside air temperature can be well below 32°F., it is possible that the water vapor in the exhaust will freeze, causing an ice buildup around the discharge opening of the pipe. During periods of extremely cold weather and prolonged operation of the furnace, this ice build-up could become quite large. The manufacturer does not recommend the installation of these units in locations above frequent vehicular and/or pedestrian traffic. The ice build-up could present a potentially hazardous situation if it becomes dislodged. The manufacturer will NOT be held responsible for any injury or property damage resulting from any improper installation.

FIGURES 42 and 43 are recommended installations to prevent property or personal injury.

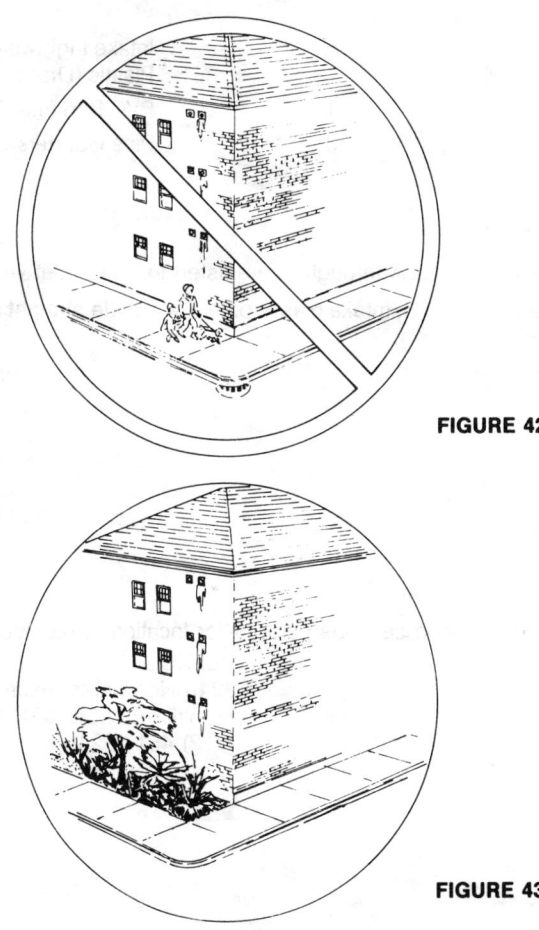

FIGURE 42

FIGURE 43

FINAL INSPECTION QUESTIONNAIRE

System now should be operated and the following items checked:

1. All vent fittings MUST be installed with the horizontal pipe to furnace pitched at least 1/4 inch per foot.

2. Have all joints from furnace to terminals been adequately sealed?

3. Make sure that hose with condensate trap is properly attached to tee with water seal.

4. Have terminals been adequately caulked to prevent leakage?

5. Make sure these instructions are kept available for your periodic inspection and maintenance of the system.

ADDENDUM "B"

MATERIAL SUPPLIED BY INSTALLER

(Refer to Tables for Vent Length)

ITEM NO.	DESCRIPTION	COMMENT	PART NO.
A	Straight Sections 3" Plastic Pipe	PLEXCO 3" ID Pipe 5' Lengths	901220
B	Coupling	PLEXCO 3"	901218
C	90° Elbow	PLEXCO 3" Sweep 90° Elbow	902299
D	Drain Tee Assembly	PLEXCO 3" T/D Lateral Outlet	901962
E	Long Nipple	PLEXCO 3" x 30" Long Nipple	903421
F	45° Elbow	PLEXCO 3" x 45°	90XXXX
G	Sealant	Dow Corning Silastic 736 RTV Adhesive/ Sealant	—
H	Intake Lightweight Plastic (Drain Pipe)	Type ASTM-D 335-OMS	See Note 2
I	PVC Bonding Cement	Any Good Grade	—
J	Pipe Hangers	Perforated Metal Strapping, ETC.	—

Note 1: Items A through G are listed for the outlet vent connections.

Note 2: Item H, intake plastic pipe will include straight sections, elbows 90° or 45°, and couplings for installation of inlet air to furnace as follows:

Furnace Input	Intake Pipe
40,000 BTU/HR	3 inch
60,000 BTU/HR	3 inch
80,000 BTU/HR	3 inch
100,000 BTU/HR	4 inch
120,000 BTU/HR	4 inch

*For ordering catalogs parts or for location of distributor closest to your area call or write:

PLEXCO
3240 North Mannheim Road
Franklin Park, IL 60131
(312) 451-2924

ADDENDUM "B-1"

INSTALLATION INSTRUCTIONS FOR
FORCED DRAFT FURNACE HORIZONTAL VENTING
INSTALLED AS A DIRECT VENT FURNACE

These kits contain parts and information for the installation of the high temperature thermo-plastic venting systems. It is essential to follow these instructions for proper venting, to obtain maximum efficiency of the unit, and safe disposal of flue gases.

Kit Numbers	Unit Heat Input BTU/HR
3 Inch Kit No. 8G03024	40,000, 60,000, 80,000
3 Inch-4 Inch Kit No. 8G03025	100,000, 120,000

CONTENTS OF KIT

ITEM NUMBER	DESCRIPTION	8G03024 3" KIT NO.	8G03025 3"-4" KIT NO.
1	INTAKE TERMINAL	0548010001 (3")	0548010002 (4")
2	VENT TERMINAL WITH CLEARANCE SHIELD	0548010003 (3")	0548010003 (3")
3	INLET AIR BLOWER ADAPTER	1708605	
4	8' x ⅜" DRAIN HOSE	2939156	2939156
5	HOSE CLAMP	2344433	2344433
6	INSTALLATION INSTRUCTIONS	7502127	7502127

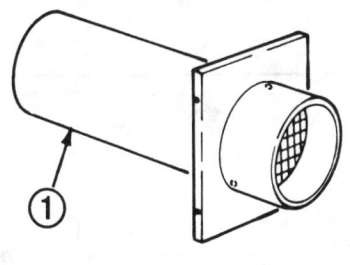

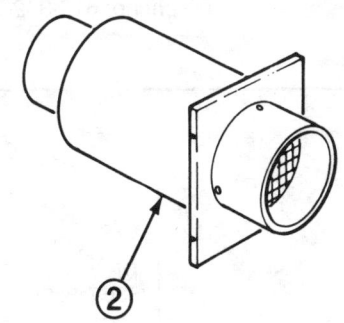

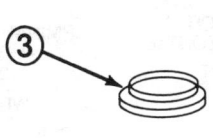

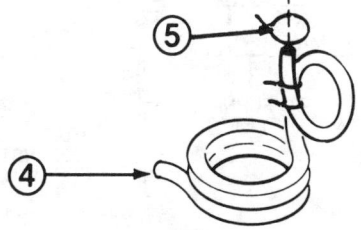

NOTE: The outlet vent size for furnaces of heat input 100,000 and 120,000 BTU/HR is reduced to 3 inch when using only the Hart and Cooley venting system.

⚠ CAUTION:

EMPLOY ONLY THE COMPONENTS LISTED FOR USE IN THIS VENTING SYSTEM. USE OF OTHER COMPONENTS MAY CAUSE NUISANCE SHUT-DOWN OR LOCK-OUT.

INSTALLER SUPPLIED MATERIALS
SEE TABLE FOR VENT LENGTHS AND CLEARANCES

ITEM NO.	DESCRIPTION	COMMENT	PART NO.①
A	STRAIGHT SECTIONS 3" PLASTIC PIPE	HART & COOLEY #3VP10 10 FT. PSC	17850
B	90 DEGREE ELBOW	HART & COOLEY #3UES90	17864
C	45 DEGREE ELBOW	HART & COOLEY #3UE45	17852
D	VENT REDUCER 4" TO 3"	HART & COOLEY #4UR3	17863
E	DRAIN TEE (3 x 3 x 3)	HART & COOLEY #3UT	17854
F	COUPLING	HART & COOLEY #3UC	17855
G	DRAIN PLUG	HART & COOLEY #UDP	17858
H	SEALANT	10.3 FL. OZ. #URTV	17861
I	INTAKE LIGHTWEIGHT② PLASTIC (DRAIN PIPE)	TYPE ASTM D2729 (OR EQUIVALENT)	—
J	PVC BONDING CEMENT	ANY GOOD GRADE	—
K	PIPE HANGERS	PERFORATED METAL STRAPPING, ETC.	—

①HART AND COOLEY PART NO.
②IF LOCAL CODES REQUIRE, SCHEDULE 40 PIPE MAY BE USED.

For location of the Hart and Cooley distributor closest to your area call:

Hart and Cooley, Inc.
500 East Eight Street
Holland, MI 49423
Telephone: 616-392-7855

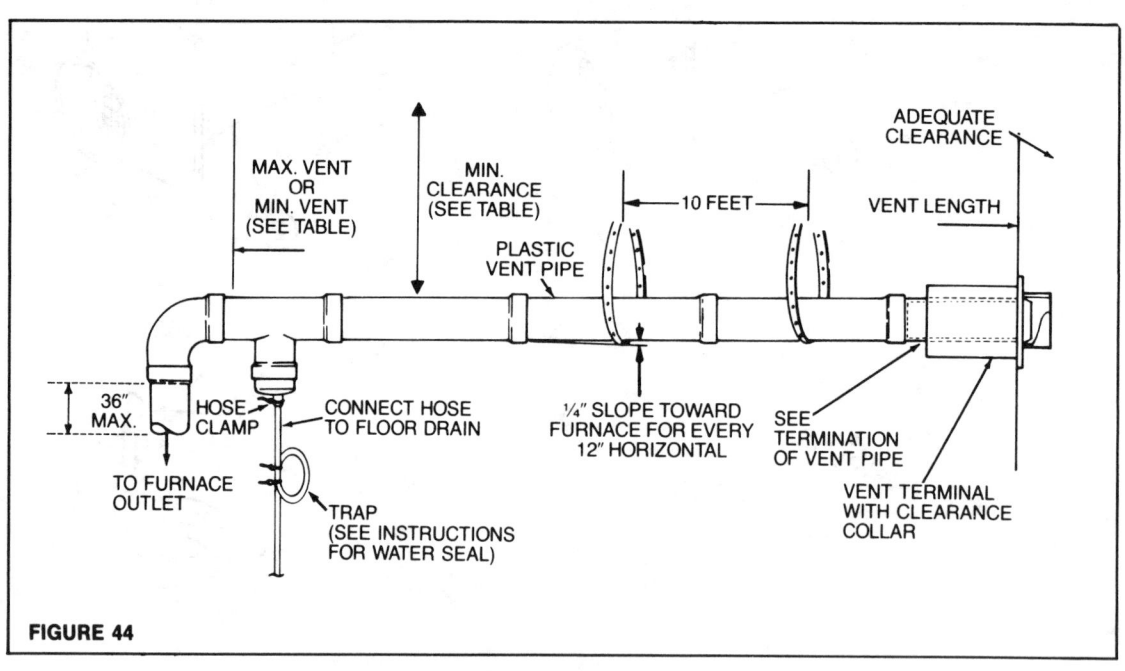

FIGURE 44

MIN.-MAX. LENGTH OF VENT (HORIZONTAL APPLICATION)
(See FIGURE 44)

INPUT BTU/HR	VENT DIA.	INLET DIA.	MAX. LENGTH	MIN. LENGTH	ELBOWS (MAX NO)	CLEARANCE TO COMBUSTIBLE MATERIALS (IN.)
40,000	3"	3"	30'	3'	4	3"
60,000	3"	3"	30'	3'	4	3"
80,000	3"	3"	30'	3'	4	3"
100,000	3"	4"	30'	3'	4	3"
120,000	3"	4"	30'	3'	4	3"

For the purposes of horizontal venting, as a direct vent unit, these furnaces may be vented up to a maximum of 30 feet with a maximum of four (4) ninety degree elbows. The minimum horizontal vent length is three (3) feet. See FIGURE 44.

The dimensions shown in the TABLE for min. max length of vent, are those allowed by test. For other allowable reduced clearances to combustible materials, refer to National Fuel Gas Code ANSI Z223.1 current edition and applicable provisions of local building codes. Maximum wall thickness at vent terminal is 10 inches. See FIGURE 44.

INSTALLATION INSTRUCTIONS
INSTALLED AS A DIRECT VENT FURNACE

GENERAL INSTRUCTIONS

A. For furnace connection to gas vents or chimneys, the vent installations shall be in accordance with Part 7, **Venting of Equipment** of the National Fuel Gas Code, ANSI Z223.1-1988, applicable provisions of local building codes and these instructions, with specific reference to section 7.8 of the Fuel Gas Code.

B. Vent assembly requires special attention for gas and liquid tight joints and proper removal of any condensate that can accumulate.

1. It is not necessary to drill holes in the pipe or fittings. Do **not** use sheet metal or other type screws to mount pipe.

2. If the vent system extends 15 feet horizontally, the system should include a tee and drain plug for collection and disposal of condensate.

   Under certain conditions, such as long runs of vent pipe through spaces with relatively cool ambient temperatures, condensate may form inside the flue vent pipe. The prescribed pitch of the pipe and the supplied condensate trap/drain connected in the horizontal pipe close to the furnace outlet will allow removal of any accumulation that may occur. See instructions for drain tee installation.

3. All horizontal sections in the venting must have a slope upward toward the terminal end of not less than ¼ inch per foot. This will prevent collection of any possible condensate. See FIGURE 44.

4. Horizontal runs must be supported with ¾ inch or equivalent perforated pipe strap at a maximum of 10 foot intervals and at each point where an elbow is used.

5. Maintain a minimum air space to combustibles from all sections of the vent pipe as outlined in Table No. 11.

6. All joints must be sealed correctly using the Ultravent sealant (3URTV) since this is the method of sealing the joints together for structural integrity and to prevent the escape of products of combustion into the occupied zone. Pipe ends must be smooth and clean before applying sealant especially if the pipe has been cut. The sealer is applied by running a minimum ¼ inch bead entirely around the pipe end no more than ⅛ inch back from the edge. Push the fitting and pipe together fully with a twisting motion to create a good seal. Allow a minimum of 24 hours for a functional cure. After sealant has cured, inspect each joint to determine that flue gas will not escape. If necessary, reapply sealant to any joint that is suspect.

7. Foil tape or temporary supports may be used ONLY to assist in holding joints together during installation and cure but tape may not be used for sealing purposes.

C. The plastic inlet piping may be lightweight plastic drain pipe preferably type ASTM D2729 (or equivalent). If local codes require, schedule 40 pipe may be used.

1. A good quality cement, specified for the type pipe used, must be applied on clean, oil free joints. The pipe ends must be smooth and square before applying sealant, especially if the pipe has been cut.

2. The horizontal runs must be adequately supported. Follow instructions for vent piping outlined above.

D. Vent and Intake Terminals

The terminations supplied in the kit must be used for all horizontal installations to ensure the proper performance of the appliance and to prohibit the entrance of birds or debris. The location of terminals is important for performance as well as safety:

1. The distances from adjacent public walkway; adjacent buildings, open windows and building openings shall be consistent with the National Fuel Gas Code Z223.1 (current edition).

2. The venting system terminal must be at least three (3) feet above any forced air inlet located within 10 feet.

   Exception: This provision shall not apply to the **combustion air intake.**

3. The vent terminal of a direct vent appliance with an input of 50,000 Btu per hour or less shall be located at least 9 inches from any opening through which flue gases could enter a building, and such an appliance with an input over 50,000 Btu per hour shall require a 12-inch vent termination clearance.

4. The vent terminal must be located at least one foot above grade to avoid blockage by snow or as required to be above snow level.

5. A clearance of four (4) feet is required to combustible materials opposite the outlet terminal.

6. Vent and intake must be on the same face of the building in order to be in the same pressure zone. The centerline of terminals must be a minimum 16 to a maximum of 24 inches apart on a horizontal line. Never install the intake terminal higher than the vent terminal.

7. The vent terminal of the appliance shall not terminate over public walkways or over an area where condensate or vapor could create a nuisance or hazard or could be detrimental to the operation of regulators, relief valves, or other equipment.

⚠ CAUTION:
At no time should the intake and exhaust vents be terminated on opposite or adjacent sides of the structure.

For typical vent-inlet air installations, see FIGURE 45.

8. Do not mix the venting portion of this system with components or materials from other systems made by other manufacturers.

9 Do not use Type B gas vent components anywhere in this inlet-vent system.

10. Do not connect more than one appliance to the inlet-vent system.

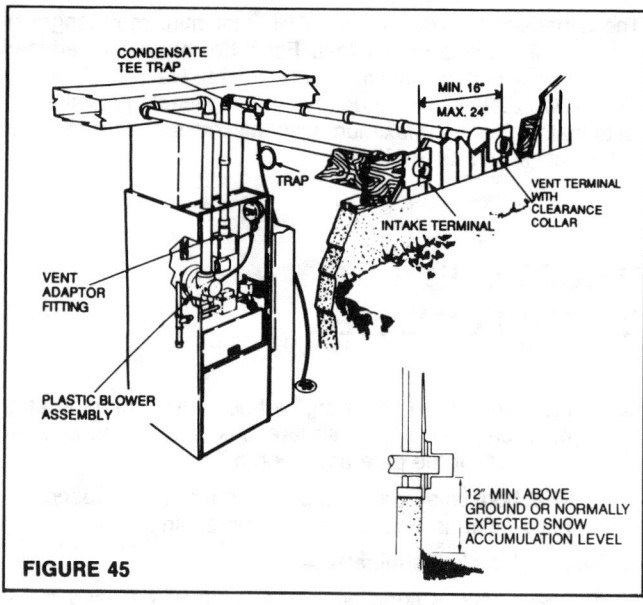

FIGURE 45

Labels in figure:
CONDENSATE TEE TRAP
TRAP
VENT TERMINAL WITH CLEARANCE COLLAR
INTAKE TERMINAL
MIN. 16"
MAX. 24"
VENT ADAPTOR FITTING
PLASTIC BLOWER ASSEMBLY
12" MIN. ABOVE GROUND OR NORMALLY EXPECTED SNOW ACCUMULATION LEVEL

11. Do not connect equipment vented by natural draft (units with draft hoods or draft diverters) into any portion of this positive pressure system.

E. Periodic Inspection and Maintenance

1. Terminals

   Inspect the intake and outlet terminals at the beginning and periodically throughout the heating season to prevent the accumulation of debris at the terminal screens.

2. Vent Pipe

   Inspect the vent pipe at the beginning and periodically throughout the heating season to insure that the joints are leak tight and that the pipe is in proper condition for safe operation. At the same time check the hose/trap from the tee connector to ensure that it is intact and the tube end is unrestricted. Also check that there is water in the tube loop. Some water should be in the trap to prevent leakage of a small amount of flue products thru the drain tube. Check for water by removing tube clamp, remove tube and pour water into the tube until it flows out the end of the tube. Replace tube and clamp. This procedure is required only if original installation includes tee, plug and drain hose.

ASSEMBLY INSTRUCTIONS:

Read assembly process thoroughly before proceeding.

⚠ CAUTION:
SHUT OFF ALL ELECTRICAL POWER TO THE FURNACE.

1. ASSEMBLY OF OUTSIDE INTAKE/VENT TERMINALS

Follow all the conditions described earlier in these instructions with regard to clearances and positioning when determining the location of the intake and outlet vent terminals.

INTAKE TERMINAL:

a. The hole made in the wall through which the terminal is installed should be one inch (1") larger than the outside diameter of the terminal pipe. Four inch for the three (3) inch terminal and 5 inch for the four (4) inch terminal.

b. Insert the terminal through the hole from the outside and secure it in place with the proper type of anchor for the type of construction to which it is being attached.

c. From the outside, caulk around the edge of the mounting plate to provide a tight seal against exterior elements.

d. From the inside, caulk around the pipe as it comes out of the wall.

VENT TERMINAL:

a. If the surface through which the outlet terminal protrudes is of a combustible material, the hole in the construction **must** be made large enough to allow for the fitting of the concentric clearance collar around the outlet terminal.

   If the wall is a non-combustible material of stone, concrete etc., the clearance collar may be removed and a hole should be made large enough to provide clearance for the plastic pipe of 4 inches.

   Under certain wind conditions some building materials may be affected by flue products expelled in close proximity to unprotected surfaces. Sealing or shielding of the exposed surfaces with a corrosion resistant material (such as aluminum sheeting) may be required to prevent staining or deterioration. The protective material should be attached and sealed (if necessary) to the house before attaching the vent terminal.

b. Slip the terminal in place from the outside and secure it with the appropriate anchoring device.

c. Caulk around the perimeter of the mounting plate.

d. From the inside, caulk around the clearance collar if combustible construction or around vent pipe (with plastic pipe in place), if the opening is through non-combustible material of concrete, stone, etc.

2. PIPE ASSEMBLY

Before hanging the intake and outlet pipe permanently, assemble the pipe from the terminal to the furnace and cut the lengths necessary to give the positioning required in final assembly. See FIGURE 45 for examples.

INTAKE PIPE:

There are two main types of PVC pipe that will work on the combustion air intake of the forced draft furnace. These are specified as thin wall PVC sewer pipe (ASTMD2729) or scheduled 40 PVC standard wall pipe (ASTMD2665). Refer to FIGURE 46 for proper application of the pipe.

Start the intake plastic pipe so the unflared (male) end is at the furnace connection (inlet of the blower). This pipe will be four (4) inches for the 100 and 120M units without any reductions and three (3) inches with the kit reducer for the 40, 60, and 80M units. See FIGURE 47.

Continue with straight pipe and elbows of intake design configuration to the intake terminal. Do not cement connections at this time. At the inlet air terminal the pipe should overlap the metal connection for at least 3 inches for a good seal.

VENT PIPE:

Start the vent plastic pipe assembly by placing the plastic pipe to the three (3) inch pipe connection at the outlet collar

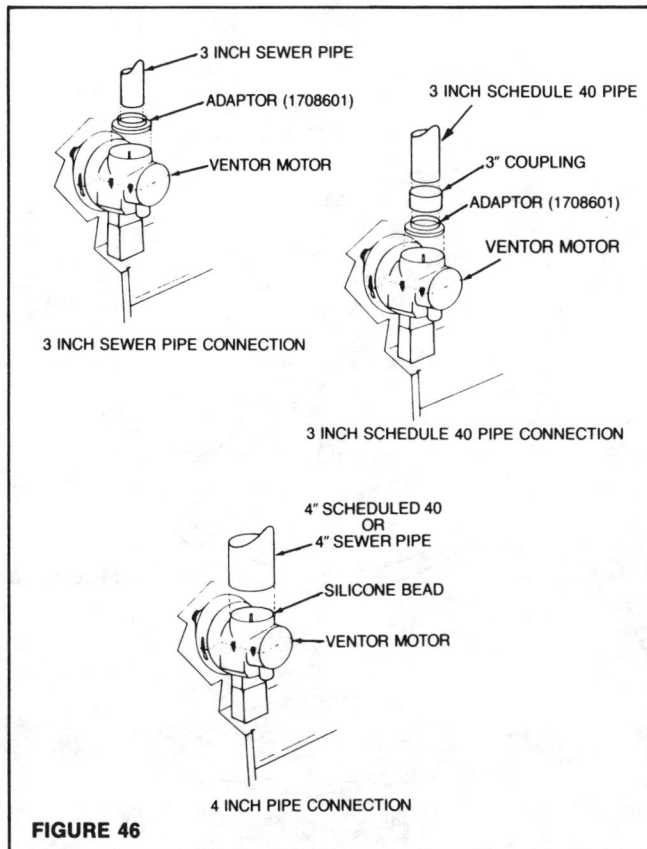

FIGURE 46

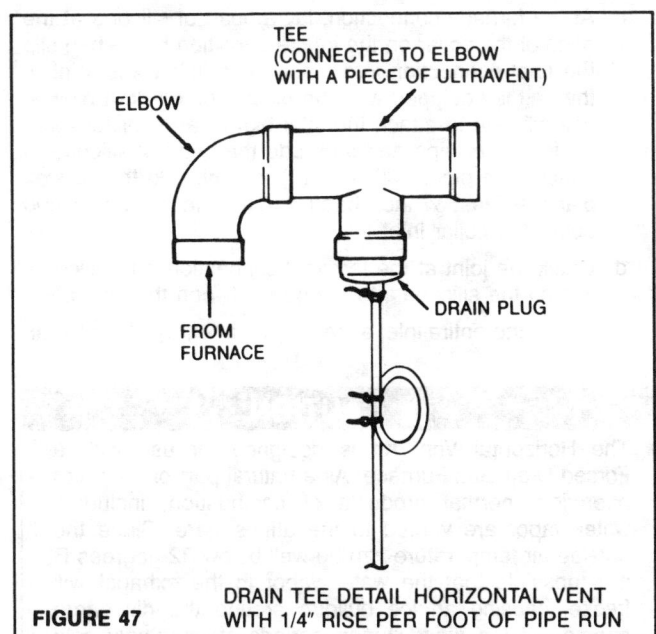

FIGURE 47 DRAIN TEE DETAIL HORIZONTAL VENT WITH 1/4" RISE PER FOOT OF PIPE RUN

of the collector box. NOTE: The 100 and 120M units require the 4 to 3 inch reducer #4UR3. Use only this part for a good assembly. The three smaller units do not require the reducer. The three (3) inch pipe connects directly to their 3 inch outlet. (See FIGURE 47 for Drain Tee details.) Continue with straight pipe couplings and elbows to the vent terminal. The pipe (3UP10) may be cut to length using a hand-held hacksaw with a blade having 18 to 24 teeth per inch. Scrape the burrs off with a sharp edged tool.

The cut must be square with the pipe. Wipe clean the pipe ends and fitting sockets before sealing joints. Use couplings (3UC) to extend pipe runs. At the terminal, the last pipe section must extend the full length of the metal vent pipe tube. See FIGURE 47. Do not cement at this time.

DRAIN TEE:

A drain tee assembly using a tee (#3UT) and drain plug (3UDP) should be used in those pipe assemblies that exceed a horizontal pipe length of 15 feet. Locate the drain tee as close to the appliance outlet collar as possible using as few offsets as possible to reduce back pressure. See FIGURES 44 and 45 for example. Connect 3/8 I.D. plastic tubing to the drain plug hose connection. If not provided, make a permanent 3-inch loop in the tubing for a condensate trap.

At final assembly, before connecting the tube, "prime" the trap with water. Pour a small amount of water into the tube to act as a water seal to prevent any flue products from escaping out of the drain tube. Run the loose end of the tube to a drain. See FIGURE 47.

3. FINAL ASSEMBLY

VENT PIPE:

a. Begin at the vent terminal. Put a large bead of sealant on the metal pipe and a small bead on the very edge of the plastic pipe. With a twisting motion put the plastic pipe on the terminal pipe. For all pipe connections, use twisting motion at assembly.

b. Seal each piping joint as you assemble the connections by spreading at least 1/4 inch thick bead of Ultravent sealant one (3URTV). After the connection is pushed together, check that there is no gaps in the seal. Reseal any gaps or open seams.

c. Support the pipe as you go along to prevent excess movement at the joints.

d. At the furnace connection, caulk around the base of the collar at the collector box with sealant. Then install the vent adaptor fitting if required or straight pipe (three smaller units), into the caulking and check for a good perimeter seal.

e. Make one final inspection of each pipe joint seal.

⚠ CAUTION:

THE CONDENSATE DRAIN HOSE MUST BE ROUTED INSIDE THE HOUSEHOLD DRAIN SO THAT THE CONDENSATE PRODUCED BY THE FURNACE CANNOT BE ACCESSED BY SMALL CHILDREN OR PETS.

---

f. Attach the condensate drain hose to the fitting on the drain tee cap and secure with the plastic tie provided. The looped end of the hose, which will act as a trap in the line, must be attached at the tee drain. The free end should then be run to, or as close as possible to a floor drain. Remember to add water to the loop before connection.

4. FRESH AIR INTAKE PIPE ASSEMBLY

a. Beginning at the intake terminal, slip the pipe end over the terminal pipe after placing the bead of sealant over the pipe as in the vent terminal assembly.

b. Connect the pipe back toward the furnace, cementing each joint with standard plastic cement and support the structure with pipe hangers as needed. **BE SURE JOINTS ARE CLEAN AND DRY BEFORE CEMENTING.**

⚠ WARNING

Solvent cements are combustible liquids and should be kept away from all ignition sources, (ie. sparks, open flames, and excessive heat). Avoid breathing cement vapors or contact with skin and eyes.

---

c. At the furnace connection, lay a bead of silicone at the base of the collar on the intake transition box. Then slip the final connection into place to complete the joint. If the unit is equipped with the plastic forced draft blower, use silicone to attach the inlet blower adaptor for three (3) inch inlet pipe and then add the bead of silicone to attach inlet pipe. With four (4) inch pipe to the blower, place at least ¼ inch bead about ¼ inch outside and below the collar inlet.

d. Caulk the joint at the terminal connection with silicone, forcing the silicone into the gap between the pipes.

e. Check the entire inlet assembly for integrity of joint seal.

⚠ CAUTION:

The Horizontal Vent Kit is designed for use with a Forced Draft Gas Furnace. As a natural part of the unit's operation, normal products of combustion, including water vapor are vented to the atmosphere. Since the outside air temperature can be well below 32 degrees F, it is possible that the water vapor in the exhaust will freeze, causing an ice buildup around the discharge opening of the pipe. During periods of extremely cold weather and the prolonged operation of the furnace, this ice build-up could become quite large. The manufacturer does not recommend the installation of these units in locations above frequent vehicular and/or pedestrian traffic. The ice build-up could present a potentially hazardous situation if it becomes dislodged. The manufacturer will NOT be held responsible for any injury or property damage resulting from any improper installation.

FIGURES 48 and 49 are recommended installations to prevent property or personal injury.

FINAL INSPECTION QUESTIONNAIRE

System now should be operated and the following items checked.

1. All vent fittings MUST be installed with the horizontal pipe to furnace pitched at least ¼ inch per foot.

2. Have all joints from furnace to terminals been adequately sealed?

3. Make sure that hose with condensate trap is properly attached to tee with water seal.

4. Have terminals been adequately caulked to prevent leakage?

5. Make sure these instructions are kept available for your periodic inspection and maintenance of the system.

FIGURE 48

FIGURE 49

GLOSSARY

AIR-GAS RATIO - The ratio of combustion air supply flow rate to the fuel gas supply flow rate.

AIR SHUTTER - An adjustable shutter on the primary air openings of a burner, which is used to control the amount of combustion air introduced into the burner body.

ATMOSPHERIC PRESSURE - The pressure exerted upon the earth's surface by the weight of atmosphere above it.

AUTOMATIC GAS PILOT DEVICE - A gas pilot incorporating a device, which acts to automatically shut off the gas supply to the appliance burner if the pilot flame is extinguished.

BREAKER BOLT - (See Spoiler Screw).

BRITISH THERMAL UNIT - (Btu) - The quantity of heat required to raise the temperature of one pound of pure water one degree F.

BURNER - A device for the final conveyance of gas, or a mixture of gas and air, to the combustion zone (See also specific type of burner).

1. *Injection Burner.* A burner employing the energy of a jet of gas to inject air for combustion into the burner and mix it with gas.

   a. *Atmospheric Injection Burner.* A burner in which the air injected into the burner by a jet of gas is supplied to the burner at atmospheric pressure.

2. *Power Burner.* (See also Forced Draft Burner, Induced Draft Burner, Premixing Burner, and Pressure Burner). A burner in which either gas or air or both are supplied at pressures exceeding, for gas, the line pressure, and for air, atmospheric pressure.

3. *Yellow-Flame Burner.* A burner in which secondary air only is depended on for the combustion of the gas.

BURNER FLEXIBILITY - The degree to which a burner can operate with reasonable characteristics with a variety of fuel gases and/or variations in input rate (gas pressure).

COMBUSTION - The rapid oxidation of fuel gases accompanied by the production of heat or heat and light.

COMBUSTION AIR - Air supplied in an appliance specifically for the combustion of a fuel gas.

COMBUSTION CHAMBER - The portion of an appliance within which combustion normally occurs.

COMBUSTION PRODUCTS - Constituents resulting from the combustion of a fuel gas with the oxygen in air, including the inert gases, but excluding excess air.

CONDENSABLE - A gas which can be easily converted to liquid form, usually by lowering the temperature and/or increasing pressure.

CONTROLS - Devices designed to regulate the gas, air, water or electricity supplied to a gas appliance. They may be manual, semi- automatic or automatic.

CUBIC FOOT OF GAS - (Standard Conditions). The amount of gas which will occupy 1 cubic foot when at a temperature of 60F., and under a pressure equivalent to that of 30 inches of mercury.

DEWPOINT - The temperature at which a vapor will start to condense into its liquid form.

DRAFT HOOD - (Draft Diverter) - A device built into an appliance, or made part of a vent connector from an appliance, which is designed to: (1) assure the ready escape of the products of combustion in the event of no draft, backdraft, or stoppage beyond the draft hood; (2) prevent a backdraft from entering the appliance; and (3) neutralize the effect of stack action of a chimney or gas vent upon the operation of the appliance.

DOWNDRAFT - Excessive high air pressure existing at the outlet of chimney or stack which tends to make gases flow downward in the stack.

DRILLED PORT BURNER - A burner in which the ports have been formed by drilled holes in a thick section in the burner head or by a manufacturing method which results in holes similar in size, shape and depth.

EXCESS AIR - Air which passes through an appliance and the appliance flues in excess of that which is required for complete combustion of the gas. Usually expressed as a percentage of the air required for complete combustion of the gas.

EXTINCTION POP - (See Flashback).

FAHRENHEIT - The common scale of temperature measurement in the English system of units. It is based on the freezing point of water being 32 F. and the boiling point of water being 212 F. at standard pressure conditions.

FIXED ORIFICE - (See Orifice Spud).

FLAME ROLLOUT - A condition where flame rolls out of a combustion chamber when the burner is turned on.

FLAME VELOCITY - The speed at which a flame moves through a fuel-air mixture.

FLAMMABILITY LIMITS - The maximum percentages of a fuel in an air- fuel mixture which will burn.

FLASHBACK - An undesirable flame characteristic in which burner flames strike back into a burner to burn there or to create a pop after the gas supply has been turned off.

FLUE GASES, FLUE PRODUCTS - Products of combustion and excess air in appliance flues or heat exchangers before the draft hood.

FLUE LOSS - The heat lost in flue products leaving from the flue outlet of an appliance.

FLUE OUTLET - The opening provided in an appliance for the escape of flue gases.

FLUID - A gas or liquid, as opposed to a solid.

FORCED DRAFT BURNER - A burner in which combustion air is supplied by a fan or blower.

FUEL - Any substance used for combustion.

FUEL GAS - Any substance in a gaseous form when used for combustion.

HARD FLAME - A flame with a hot, tight, well-defined inner cone.

HEAT EXCHANGER - Any device for transferring heat from one fluid to another.

HEATING SURFACE - All surfaces which transmit heat from flames or flue gases to the medium being heated.

HEATING VALUE - The number of British thermal units produced by the complete combustion at constant pressure of one cubic foot of gas. Total heating value includes heat obtained from cooling the products to the initial temperature of the gas and air and condensing the water vapor formed during combustion.

IGNITION - The act of starting combustion.

IGNITION TEMPERATURE - The minimum temperature at which combustion can be started.

IGNITION VELOCITY - (See Flame Velocity).

IMPINGEMENT TARGET BURNER - A burner consisting simply of a gas orifice and a target, with the gas jet from the orifice entraining combustion air in the open and the mixture striking and burning on the target surface. No usual burner body is used.

INCHES OF MERCURY COLUMN - A unit used in measuring pressures. One inch of mercury column equals a pressure of 0.491 pounds per square inch.

INCHES OF WATER COLUMN - A unit used in measuring pressures. One inch of water column equals a pressure of 0.578 ounces per square inch. One inch mercury column equals about 13.6 inches water column.

INCOMPLETE COMBUSTION - Combustion in which the fuel is only partially burned.

INDUCED DRAFT BURNER - A burner which depends on draft induced by a fan or blower at the flue outlet to draw in combustion air and vent flue gases.

INPUT RATING - The gas-burning capacity of an appliance in Btu per hour as specified by the manufacturer. Appliance input ratings are based on sea level operation and need not be changed for operation up to 2,000 feet elevation. For operation at elevations above 2,000 feet, input ratings should be reduced at the rate of 4 percent for each 1,000 feet above sea level.

LIFTING FLAMES - An unstable burner flame condition in which flames lift or blow off the burner port (s).

LIQUEFIED PETROLEUM GASES - The terms "Liquefied Petroleum Gases", "LPG", and "LP Gas" mean and include any fuel gas which is composed predominantly of any of the following hydrocarbons, or mixtures of them: propane, pro-pylene, normal butane or isobutane and butylenes.

LNG - Liquefied natural gas. Natural gas which has been cooled until it becomes a liquid.

LP GAS-AIR MIXTURES - Liquefied petroleum gases distributed at relatively low pressures and normal atmospheric temperatures which have been diluted with air to produce desired heating value and utilization characteristics.

LUMINOUS FLAME BURNER - (See Burner, Yellow Flame).

MANIFOLD - The conduit of an appliance which supplies gas to the individual burners.

MANIFOLD PRESSURE - The gas pressure in an appliance manifold, upstream of burner orifices.

MANUFACTURED GAS - A fuel gas which is artificially produced by some process, as opposed to natural gas, which is found in the earth. Sometimes called town gas.

NATURAL DRAFT - The motion of flue products through an appliance generated by hot flue gases rising in a vent connected to the furnace flue outlet.

NATURAL GAS - Any gas found in the earth, as opposed to gases which are manufactured.

ODORANT - A substance added to an otherwise odorless, colorless and tasteless gas to give warning of gas leakage and to aid in leak detection.

ORIFICE - An opening in an orifice cap (hood), orifice spud or other device through which gas is discharged, and whereby the flow of gas is limited and/or controlled. (See the also Universal Orifice).

Orifice Cap (Hood). - A movable fitting having an orifice which permits adjustment of the flow of gas by changing its position with respect to a fixed needle or other device extending into the orifice.

Orifice Discharge Coefficient - (See Discharge Coefficient).

Orifice Spud - A removable plug or cap containing an orifice which permits adjustment of the gas flow either by substitution with a spud having different sized orifices (fixed orifice) or by motion of an adjustable needle into or out of the orifice (adjustable orifice).

OVERRATING - Operation of a gas burner at a greater rate than it was designed for.

PILOT - A small flame which is used to ignite the gas at the main burner.

PORT - Any opening in a burner head through which gas or an air-mixture is discharged for ignition.

PRESSURE BURNER - A burner in which an air and gas mixture under pressure is supplied, usually at 0.5 to 14 inches water column.

PRESSURE REGULATOR - A device for controlling and maintaining a uniform outlet gas pressure.

PRIMARY AIR - The combustion air introduced into a burner which mixes with the gas before it reaches the port. Usually expressed as a percentage of air required for complete combustion of the gas.

PRIMARY AIR INLET - The opening or openings through which primary air is admitted into a burner.

PROPANE - A Hydrocarbon gas heavier than methane but lighter than butane. It is used as a fuel gas alone, mixed with air or as a major constituent of liquefied petroleum gases (See Liquefied Petroleum Gases).

SECONDARY AIR - Combustion air externally supplied to a burner flame at the point of combustion.

SINGLE PORT BURNER - A burner in which the entire air-gas mixture issues from a single port.

SPECIFIC GRAVITY - Specific gravity is the ratio of the weight of a given volume of gas to that of the same volume of air, both measured at the same temperature and pressure.

SPOILER SCREW - (Breaker Bolt) - A screw or bolt moved in or out of the gas jet in a burner to control primary air injection.

SPUD - (See Orifice).

STANDARD CONDITIONS - Pressure and temperature conditions selected for expressing properties of gases on a common basis. In gas appliance work, these are normally 30 inches of mercury and 60 F.

STATIC PRESSURE - The pressure exerted by a motionless gas.

TARGET BURNER - (See Impingement Burner).

THERM - A unit of heat energy equal to 100,000 Btu.

THROAT - (See Venturi).

TOTAL AIR - The total amount of air supplied to a burner. It is the sum of primary air, secondary air and excess air.

TOTAL PRESSURE - Also called impact pressure. The pressure measured in a moving fluid by an impact tube. It is the sum of velocity pressure and the static pressure.

UPDRAFT - Excessively low air pressure existing at the outlet of a chimney or stack which tends to increase the velocity and volume of gases passing up the stack.

UTILITY GASES - Natural gas, manufactured gas, liquefied petroleum gas-air mixtures or mixtures of any of these gases.

VELOCITY PRESSURE - Pressure exerted by a flowing gas by virtue of its movement in the direction of its motion. It is the difference between total pressure and static pressure.

VENT - A device, such as a pipe or chimney, to transmit flue products from an appliance to the outdoors. This term also is used to designate a small hole or opening for the escape of a fluid (such as in a gas control).

VENT GASES - Products of combustion from gas appliances plus excess air, plus dilution air in the venting system above a draft hood.

VENTURI - A section in a pipe or a burner body that narrows down and then flares out again.

VISCOSITY - The property of a fluid to resist flow, ie. thickness.

WATER COLUMN - Abbreviated as W.C. A unit used for expressing pressure. One inch of water column equals a pressure of 0.578 ounces per square inch.

YELLOW FLAME BURNER - (See Burner).

YELLOW TIPS - (Yellow Tipping) - The appearance of yellow tips in an otherwise blue flame, indicating the need for additional primary air.

ELECTRICAL TERMS

ALTERNATING CURRENT (A.C.) - A flow of electricity which rises from zero to some maximum value in one direction, falls off to zero, then reverses and reaches a maximum value in the other direction, then again falls off to zero. This cycle is repeated continuously at a fixed frequency.

AMPERE OR AMP. - A unit of low electrical current (flow of electrons). One ampere represents the flow of 6.25×10^{18} electrons per second past a given point in the circuit. One volt (potential difference) across a resistance of one ohm will cause one ampere of current to flow.

AUTOMATIC PILOT DEVICE (SAFETY SHUTOFF DEVICE) - A device which is designed to automatically shut off the gas supply to the controlled burner(s) in the event the source of ignition fails. This device may interrupt gas flow to the main burner(s) only, or to pilot(s) and main burner(s) under its supervision.

BI-METAL - A device consisting of two strips of different materials fastened firmly together in the form of a strip, coil or dished disc. When the bi-metal is heated, one material will try to expand (lengthen) more than the other, but is restricted from doing so by the other material. As a result, the strip will bend, the coil twist or the disc invert producing a motion which can be used, for example in opening or closing a switch.

CAPACITOR - An electrical device consisting basically of two or more conducting surfaces separated by an insulating material. It acts to store electrical energy, to block flow of DC current, and to act as a conductor of AC current.

CIRCUIT - An arrangement of conductors and devices connected together for the purpose of carrying an electrical current.

CIRCUIT BREAKER - A device designed to open and close a circuit by nonautomatic means, and to open the circuit automatically on a predetermined overload of current, without injury to itself when properly applied within its rating.

CURRENT (Electrical) - The rate of movement of free electrons through an electrical conductor. It is usually expressed in amperes and is designated by the symbol I.

DIRECT CURRENT (DC) ELECTRICITY - A form of electricity in which current flows at an essentially constant value in one direction only in a circuit.

IGNITOR - Any device used to light gas. A spark ignitor uses an electric spark generated across an air gap for this purpose.

INSULATOR - A material which offers very high resistance to flow of electrical current, so that essentially no current flows through it. Used to isolate and separate conductors in a circuit so that electricity only flows through those conductors in the intended manner.

JUMPER WIRE - A length of wire used to temporarily complete a circuit, specifically by-passing a device (such as a switch) in that circuit.

KILOWATT (KW) - Unit of electrical power, equal to 1000 watts. Equivalent to 3413 BTU per hour.

KILOWATT HOUR - Unit of electrical energy supplied by 1000 watts (1 KW) for one hour. Equivalent to 3413 BTU of heat.

LOAD (ELECTRICAL) - The power consumed by components in an electrical circuit. Sometimes all elements which consume electrical power are referred to as the load in the circuit.

OHM - The basic unit of electrical resistance. It is the resistance which will allow one ampere of current flow when the voltage across the resistance is one volt.

RESISTANCE (ELECTRICAL) - The property of a material to resist the flow of electric current through it. Designated as R and is expressed in units of ohms.

RESISTOR - A device which acts to limit current flow in a circuit by virtue of its electrical resistance.

1. COMPOSITION RESISTOR - A resistor made of carbon particles mixed with a binder and provided with terminal leads to connect the resistor (usually cylindrical in shape) into a circuit. The resistance value is determined by the relative amounts of carbon and binder used.

2. FIXED RESISTOR - A resistor with one essentially constant resistance value (except changes due to temperature variations.

3. WIRE WOUND RESISTOR - A resistor made by winding wire around a non-conducting cylinder. The resistance value is determined by the diameter and the number of turns of the wire.

4. VARIABLE RESISTOR - A resistor constructed so that the value of resistance may be varied over a range usually by mechanical means.

SERIES CIRCUIT - A circuit in which only one continuous path is provided for current flow and so the same current flows through all components in the circuit.

SHORT CIRCUIT - Also called a short. An abnormal, low resistance connection between two points in a circuit, such as created when two conductors of different polarity inadvertently touch each other. This condition can create a relatively large flow of current between the points, sometimes causing damage.

SPARK GAP - A gap between two electrodes. When a high potential difference is imposed across the electrodes, current will jump the gap, creating an electrical spark for such purposes as igniting gas.

SWITCH - A device which completes or breaks the path of current flow in a circuit or which sends it into a different path.

1. NORMALLY OPEN SWITCH - A switch is in the open position, interrupting current flow in the circuit which it controls, when the switch is not activated. Designated as N.O.

2. NORMALLY CLOSED SWITCH - A switch is in the closed position, allowing current to flow in the circuit it controls when this switch is not activated. Designated as N.C.

THERMISTOR - A device whose resistance varies with temperature to such a degree that it can be used as a control element.

THERMOCOUPLE - A device consisting of two wires or strips of dissimilar materials which are joined together at one end (the hot junction). When this hot junction is heated, the thermocouple produces a DC voltage across the other two ends (cold junction).

TRANSFORMER - An electrical device which, by electromagnetic induction, transforms AC power in one circuit to another circuit(s), usually at different voltage and current values.

VOLT - The unit of voltage (potential difference, emf). One volt will produce a flow of one ampere through a one ohm resistance.

VOLTAGE - The relative amount of electric charge at one point in an electric circuit compared with that at another point in the circuit, which causes a current flow through a continuous path between the two points. Also referred to as electromotive force and potential difference.

VOLTMETER - An instrument for measuring voltage.

WATT - Unit of electrical power. In a purely resistive AC circuit or DC circuit, one watt of power is consumed by a load when one volt causes a flow of one ampere through the load.

WATTMETER - An instrument which measures true power directly.

USER'S
INFORMATION MANUAL

GUA SERIES DELUXE HIGH EFFICIENCY

FORCED DRAFT UPFLOW GAS FIRED FURNACE

REVISED:
SEPTEMBER, 1989
SUPERSEDES:
SEPTEMBER, 1988

FOR YOUR SAFETY
WHAT TO DO IF YOU SMELL GAS

- DO NOT TRY TO LIGHT ANY APPLIANCE.
- DO NOT TOUCH ANY ELECTRICAL SWITCH; DO NOT USE ANY PHONE IN YOUR BUILDING.
- IMMEDIATELY CALL YOUR GAS SUPPLIER FROM A NEIGHBOR'S PHONE. FOLLOW THE GAS SUPPLIER'S INSTRUCTIONS.
- IF YOU CANNOT REACH YOUR GAS SUPPLIER, CALL THE FIRE DEPARTMENT.

FOR YOUR SAFETY

Do not store or use gasoline or other flammable vapors and liquids in the vicinity of this or any other appliance.

⚠ WARNING

Improper installation, adjustment, alteration, service or maintenance can cause injury or property damage. Refer to this manual. For assistance or additional information, consult a qualified installer, service agency or the gas supplier.

SECTION I

DO'S AND DO NOT'S

The information contained in this manual is designed to identify any hazards and to serve as a guide to the user of any malfunctions.

- DO READ THE USER'S MANUAL AND INSTALLATION INSTRUCTIONS.

- DO HAVE YOUR FURNACE SERVICED ONCE A YEAR. Proper care of your furnace will assure many years of economical comfort. We strongly recommend that your furnace be inspected and serviced by a qualified serviceman at the beginning of each heating season.

- DO INSPECT THE INSTALLATION OF YOUR FURNACE.

> ⚠ **WARNING**
>
> It is mandatory that the area around the furnace be kept clear and free of any combustible materials (cardboard boxes, etc.). Gasoline and other volatile liquids or flammable vapors. DO NOT work with volatile materials near an open flame.

- DO REVIEW THE LIGHTING INSTRUCTIONS BEFORE OPERATING THE UNIT.

- DO SHUT OFF MANUAL GAS VALVE BEFORE ELECTRICAL SUPPLY IN AN EMERGENCY.

> ⚠ **WARNING**
>
> Should over heating occur or the gas supply fail to shut off, turn off the manual gas valve to the appliance before shutting off the electrical supply.

- DO NOT ATTEMPT TO ALTER YOUR FURNACE.

> ⚠ **WARNING**
>
> Improper installation, adjustment, alteration, service or maintenance can cause injury or property damage.

- DO NOT ATTEMPT TO LIGHT AN ELECTRIC IGNITION FURNACE WITH A MATCH.

> ⚠ **CAUTION:**
>
> Your furnace is equipped with an electric ignition. DO NOT attempt to light the pilot manually with a match or other lighting means.

- DO NOT BLOCK OR OBSTRUCT THE AIR INLETS OR OUTLETS TO THE FURNACE.

- DO NOT TAMPER WITH THE INTERNAL FAN AND LIMIT CONTROL MECHANISM.

- DO NOT INSTALL FURNACE IN A SPACE THAT CONTAINS COMBUSTIBLE INSULATION.

- DO NOT INSULATE THE FURNACE CABINET.

SECTION II

FACTS ABOUT YOUR FURNACE

UNIT DESCRIPTION

This furnace is a gas Forced Draft Upflow furnace. It is designed for indoor installation.

The nameplate label on the furnace shows the model number of the furnace. This label is located on the left hand side of the casing inside unit. See FIGURE 1 for the location of the nameplate label.

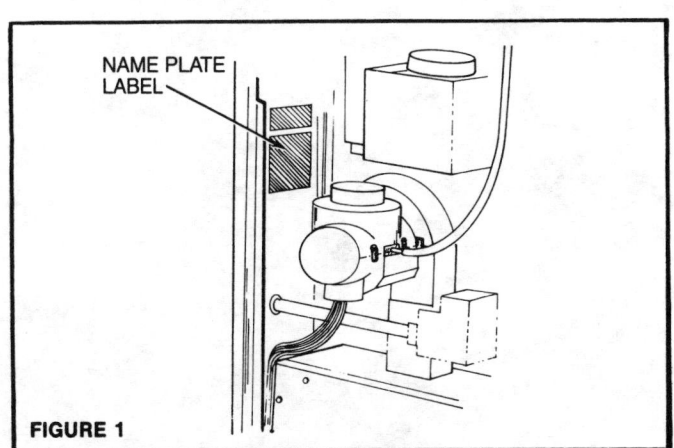

FIGURE 1

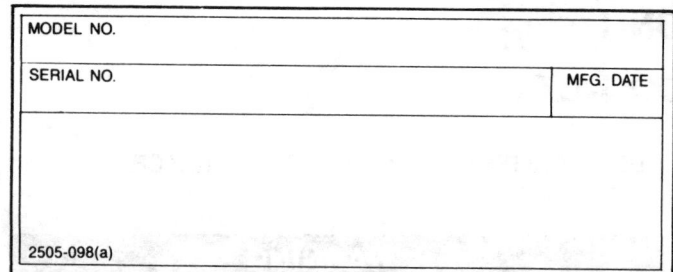

MODEL NO.

SERIAL NO.   MFG. DATE

2505-098(a)

Check the furnace model number on the nameplate label. The suffix of the model number indicates the type of fuel and the ignition system.

Please note that each digit of the model number has a special significance:

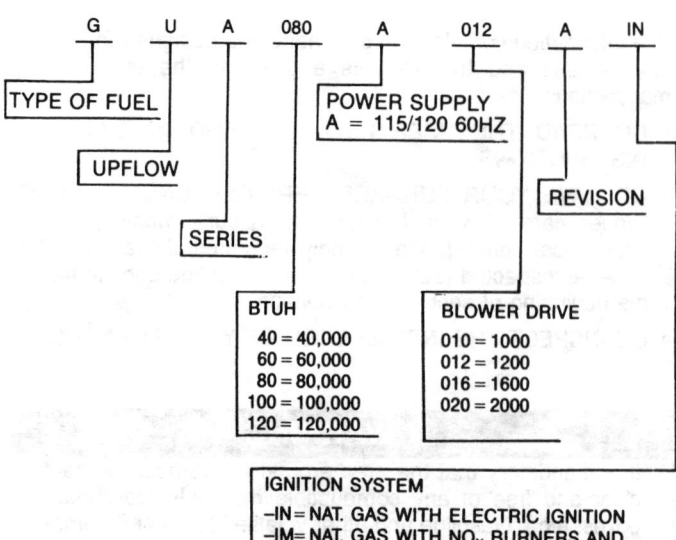

G   U   A   080   A   012   A   IN

TYPE OF FUEL

UPFLOW

SERIES

POWER SUPPLY
A = 115/120 60HZ

REVISION

BTUH

40 = 40,000
60 = 60,000
80 = 80,000
100 = 100,000
120 = 120,000

BLOWER DRIVE

010 = 1000
012 = 1200
016 = 1600
020 = 2000

IGNITION SYSTEM

–IN = NAT. GAS WITH ELECTRIC IGNITION
–IM = NAT. GAS WITH NO_x BURNERS AND
         ELECTRIC IGNITION
–IL = LPG. GAS WITH ELECTRIC IGNITION

SECTION III
PROPER INSTALLATION

This section serves as a guide line to the user for recognizing unsafe operating conditions.

INSULATION AND COMBUSTIBLES

The user is advised to examine the furnace area when the furnace is installed in an attic or when insulation is added to the attic. A furnace installed in an attic or other insulated space must be kept clear of the insulating material. DO NOT install the furnace in a space that is insulated with combustible insulation.

⚠ WARNING

It is mandatory that the area around the furnace be kept clear and free of any combustible materials (cardboard boxes, etc.). Gasoline and other volatile liquids or flammable vapors. DO NOT work with volatile materials near an open flame.

AIR REQUIREMENTS

This furnace needs air for combustion of the gas and for distribution into the living space. Therefore, DO NOT block or obstruct:

A) The air openings on the furnace.

B) The air path into the area where the furnace is installed.

C) The space around the furnace.

CLEARANCES

If the storage shelves, a furnace room, a recreation area, etc. is to be constructed around the furnace area, the following criteria must take preference:

A minimum of 36" must be allowed in front of the furnace to permit an ample amount of space for service. SEE FIGURE 2.

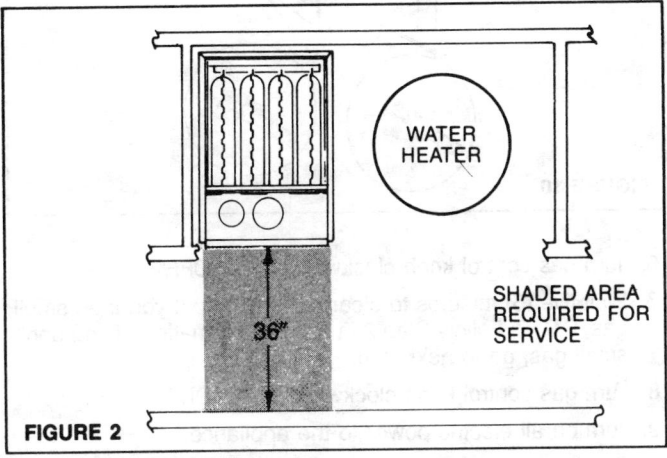

FIGURE 2

WATER HEATER

SHADED AREA REQUIRED FOR SERVICE

36"

Any enclosed installation (i.e. furnace room, recreation area, etc.) requires two openings in the door or one opening in the wall of the enclosure. One opening should be located near the floor and the other near the ceiling. SEE FIGURE 3.

Another alternative to two (2) grilles is a fully louvered door at the entrance to the furnace room.

PROPER VENTING

> ### ⚠ CAUTION:
> It is vitally important that the furnace have proper venting. Never attach another appliance or heater to the furnace chimney. Any alteration to the flue pipe or chimney must be in accordance with local and national codes.

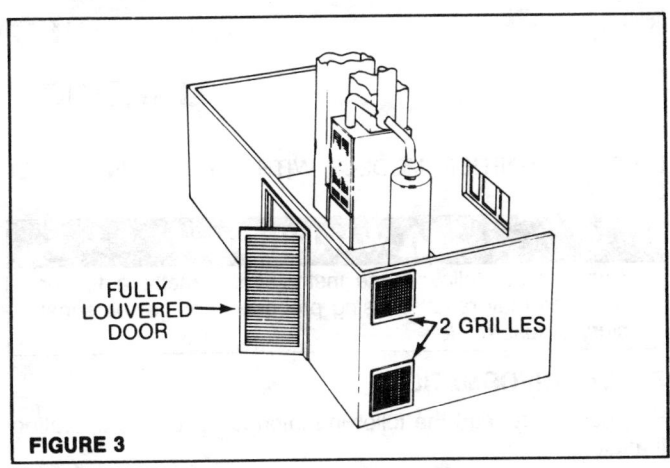

FULLY LOUVERED DOOR

2 GRILLES

FIGURE 3

The furnace shall not be connected to a chimney flue serving a separate appliance designed to burn any fuel other than gas.

INSTRUCTIONS FOR EXAMINING THE FURNACE AND ALL FLUE PRODUCT CARRYING AREAS

For safe operation, the flue product carrying areas of the furnace, including the associated vent system, and main burners, should be checked at the beginning of the heating season and periodically during the season by the owner. The following items should be checked and even if no deterioration is observed, the furnace should be examined periodically by a qualified service agency.

1. All flue product carrying areas external to the furnace (i.e., chimney, vent connector) are clear and free of obstructions.

2. The vent connector is in place, slopes upward and is physically sound without holes or excessive corrosion.

3. The return air duct connection(s) is physically sound, is sealed to the furnace casing, and terminates outside the space containing the furnace.

4. The physical support of the furnace is sound without sagging, cracks, gaps, etc., around the base so as to provide a seal between the support and the base.

5. There are no obvious signs of deterioration of the furnace.

6. The burner flames are in good adjustment (by comparison with pictorial sketches or drawings, as provided in this manual, of the main burner flame).

SECTION IV
LIGHTING INSTRUCTIONS

ELECTRIC IGNITION MODELS WITH "–IN" or "IM" or "IL" SUFFIX

> ## ⚠ WARNING
>
> If you do not follow these instructions exactly, a fire or explosion may result causing property damage, personal injury or loss of life.

SAFETY INFORMATION

For your safety read the following information before operating furnace:

1. This appliance does not have a pilot. It is equipped with an ignition device which automatically lights the burner. Do not try to light the burner by hand.

2. BEFORE LIGHTING smell all around the appliance area for gas. Be sure to smell next to the floor because some gas is heavier than air and will settle on the floor.

WHAT TO DO IF YOU SMELL GAS

* Do not try to light any appliance.
* Do not touch any electric switch; do not use any phone in your building
* Immediately call the gas supplier from a neighbor's phone. Follow the gas supplier's instructions
* If you cannot reach your gas supplier, call the fire department

3. Use only your hand to push in or turn the gas control knob. Never use tools. If the knob will not push in or turn by hand, don't try to repair it, call a qualified service technician. Force or attempted repair may result in a fire or explosion.

4. Do not use this appliance if any part has been under water. Immediately call a qualified service technician to inspect the appliance and to replace any part of the control system and any gas control which has been under water.

OPERATING INSTRUCTIONS

1. STOP! Read the safety information above.
2. Set the thermostat or heat control to lowest setting.
3. Turn off all electric power to the appliance.
4. This appliance is equipped with an ignition device which automatically lights the burner. **DO NOT TRY TO LIGHT THE BURNER BY HAND.**

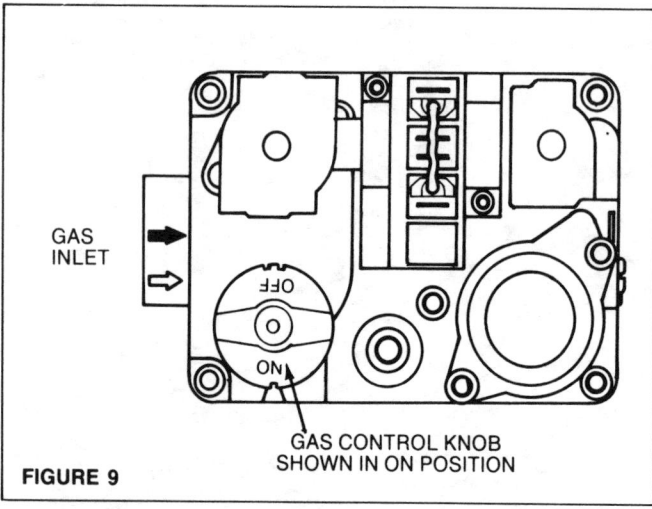

GAS INLET

OFF

ON

GAS CONTROL KNOB SHOWN IN ON POSITION

FIGURE 9

5. Remove louvered access panel. SEE FIGURE 10.

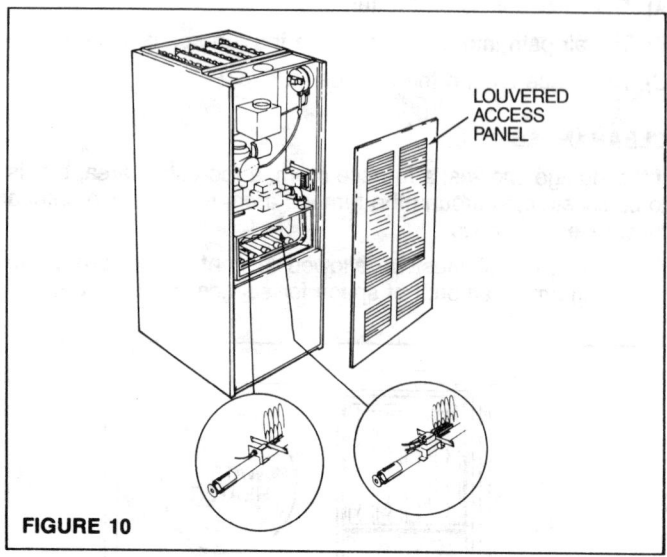

LOUVERED ACCESS PANEL

FIGURE 10

6. Turn gas control knob clockwise ⟳ to "OFF."

7. Wait five (5) minutes to clear out any gas. If you then smell gas, STOP! Follow Step 2 in Safety Information. If you don't smell gas, go to next step.

8. Turn gas control knob clockwise ⟳ to "ON."

9. Turn on all electric power to the appliance.

10. Set thermostat to desired setting.

11. If the appliance will not operate, follow the instructions "To Turn Off Gas to Appliance" and call your service technician or gas supplier.

12. Replace louvered access panel.

BURNER FLAME REQUIREMENTS

Burner flame should run the length of the burner ports. The flame should be blue in color. If for any reason the flame should lift from the burner, change color to orange or start rolling out from under the burner access shield after ignition, an adjustment to the burner is necessary or the heat exchanger needs to be inspected and cleaned by a qualified service technician.

These furnaces are equipped with a multiple burner assembly. SEE FIGURE 11.

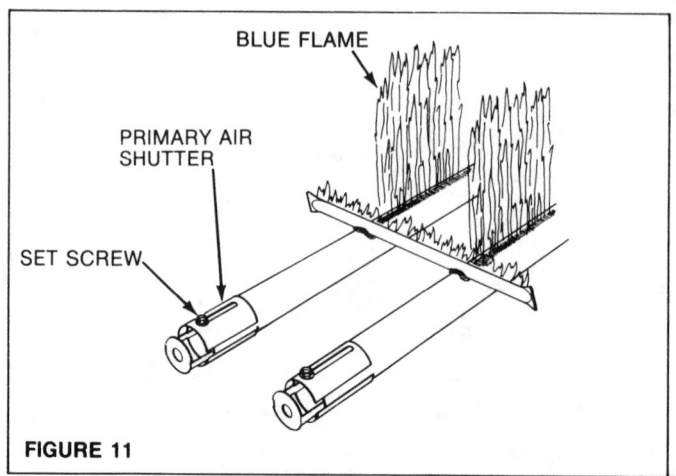

BLUE FLAME

PRIMARY AIR SHUTTER

SET SCREW

FIGURE 11

TO TURN OFF GAS TO APPLIANCE

1. Set the thermostat to lowest setting.

2. Turn off all electric power to the appliance if service is to be performed.

3. Remove louvered access panel.

4. Turn gas control knob clockwise ⌒ to "OFF."

5. Replace louvered access panel.

> ### ⚠ WARNING
> Should overheating occur or the gas supply fail to shut off, turn off the manual gas valve to the appliance before shutting off the electrical supply.

SEQUENCE OF OPERATION—ELECTRIC IGNITION

1. Thermostat calls for heat, energizing the Forced Draft Blower Relay (HR) which closes contacts that power the Forced Draft Blower.

2. The rotation of the Forced Draft Combustion Blower will make a centrifugal switch (CS) within the motor to prove blower operation and send power from the ignition circuit to the ignition control (IC). When the ignition control receives the call for heat signal from the closure of contacts of the centrifugal switch, the LED on the ignition module will flash once indicating the module is functioning properly.

3. There will be a 30 second delay before power is applied to the silicone carbide ignitor.

4. The ignitor is allowed to warm-up for approximately 15 seconds.

5. At the end of the warm-up time the gas valve is energized.

6. The ignitor provides the heat to light the main burner which lights the remaining burners via the carryovers. The main burner flame must be detected within 7 seconds or the gas valve is de-energized and the ignitor is turned off.

7. If there is a failure to sense flame on the first attempt, there will be a 60 second delay. The second attempt at lighting the burner flame will duplicate the first attempt with pre-purge, warm-up times and lock-out times identical to those in the first try. Failure to prove on the second attempt will allow yet another complete ignition attempt before complete system

lockout. If ignition module lock-out occurs after the third attempt, the LED on the module will flash continually indicating an external cause for lock-out. (i.e. ignitor failure, low voltage, low gas pressure, etc.)

8. Flashing "LED" indicator light will result in interruption of the power to the ignition control (either the 24V thermostat or line voltage to the furnace may be interrupted) for 1/20 second or longer to reset the system.

9. The ignition sequence will repeat for a total of two recycles. If flame is lost within the first thirty seconds of establishment, the total of ignition attempts plus the number of recycles will not exceed three.

10. If flame is established for more than thirty seconds after ignition, the recycle counter will clear.

11. A momentary loss of power or flame to the sensing circuit will cause the gas valve to be de-energize and a delay of 15 seconds before restarting the ignition sequence will occur. Recycles will begin and the burner will operate normally if the gas supply returns or fault condition is corrected prior to the last ignition attempt. Otherwise the control will lock-out.

12. Typical conditions are that the burner will light and stay lit as long as the thermostat calls for heat.

13. Heat from the burner flames warm the heat exchanger surface, the flue products pass through the baffles and vent outdoors through the flue pipe.

14. The fan and limit control senses the supply air temperature adjacent to the heat exchanger and will signal the blower to operate when the air temperature reaches the "ON" fan set point. The fan & limit control will shut the gas off if the supply air temperature reaches the "High Temperature" set point. The blower continues to operate and conditioned air is supplied to the space through the duct system.

15. When the thermostat set point temperature is reached, the gas valve is de-energized which stops the flow of gas to the burners. The blower continues to operate until the air temperature falls below the "OFF" blower set point.

16. The furnace repeats this same cycle each time the thermostat calls for heat.

NOTE: If ignition module lock-out occurs and is indicated by continuous illumination of the LED, the module has internally failed and must be replaced.

SECTION V
MAINTENANCE

It is very important that this appliance be maintained according to the following operation & maintenance procedures. Failure to do so could jeopardize the warranty coverage.

This section gives basic guidelines for maintenance of the furnace. The owner/operator of the unit can perform these tasks.

> ### ⚠ WARNING
> DISCONNECT ELECTRIC POWER TO THIS FURNACE BEFORE ANY MAINTENANCE.

OILING INSTRUCTIONS

BLOWER MOTOR—DIRECT DRIVE (MULTI-SPEED)

1. Disconnect power to the furnace before oiling.

2. Remove bottom door (blower door). This is done by lifting up on the door and pulling forward. SEE FIGURE 5.

3. Disconnect the wires from the low voltage terminal strip located on the transformer relay box.

   NOTE: When removing low voltage wiring, mark wires with appropriate letters so replacement of wires can be reconnected correctly. SEE FIGURE 5.

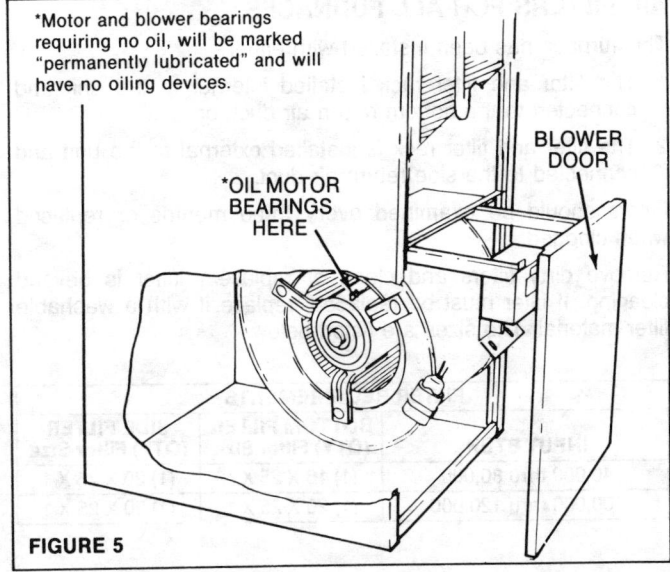

*Motor and blower bearings requiring no oil, will be marked "permanently lubricated" and will have no oiling devices.

BLOWER DOOR

*OIL MOTOR BEARINGS HERE

FIGURE 5

4. Unplug line voltage wiring from transformer relay box to receptical in blower partition.

5. Remove the 2 screws holding the blower assembly in the furnace and pull the assembly out of the casing.

6. Remove the screws or rubber plugs from the oil holes at the front and back of the motor and relubricate the motor with 5 drops of #10 weight motor oil, or any light weight oil. Replace the screws or plugs.

7. Slide blower assembly back into position and secure with 2 screws previously removed.

8. Reconnect the line voltage and low voltage wiring making sure all wires are connected to the proper points.

9. Turn the power on and check the operation of the furnace blower.

FORCED DRAFT BLOWER MOTOR

PLEASE NOTE:

The Forced Draft blower motor (shown below) is permanently lubricated. **DO NOT** attempt to add oil to this motor.

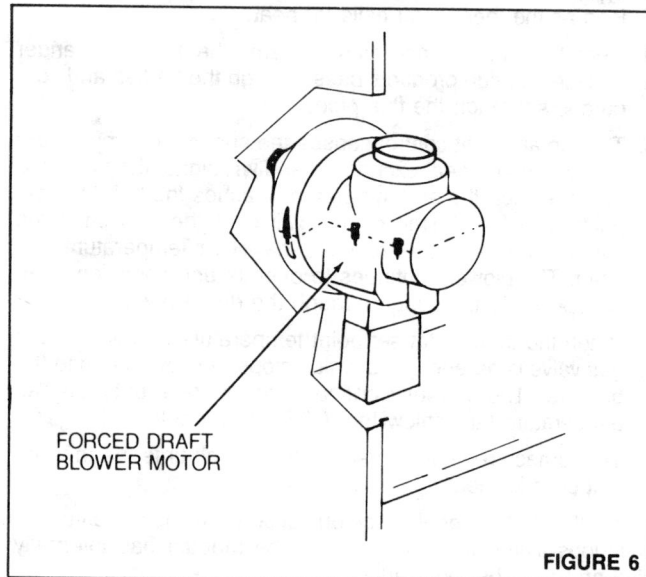

FORCED DRAFT BLOWER MOTOR

FIGURE 6

⚠ WARNING

DISCONNECT ELECTRICAL POWER TO THE FURNACE BEFORE MAINTENANCE.

AIR FILTERS FOR ALL FURNACES

This furnace has been installed with either:

1. The filter and filter rack installed internal to the unit and connected to the bottom return air duct; or

2. The filter and filter rack is installed external to the unit and connected to the side return air duct.

Filters should be examined every three months or replaced when clogged.

Remove dirty filters and clean or replace if filter is beyond cleaning. If filter must be replaced, replace it with a washable filter material. The sizes are listed below.

FILTER REQUIREMENTS		
INPUT BTUH	BOTTOM FILTER (QTY) Filter Size	SIDE FILTER (QTY) Filter Size
40,000 thru 80,000	(1) 16 X 25 X 1	(1) 20 X 25 X 1
100,000 thru 120,000	(1) 20 X 25 X 1	(1) 20 X 25 X 1

BOTTOM RETURN

To clean filter for bottom return application:

1. Disconnect power supply to the unit.

2. Remove louvered door and blower door.

3. Lift filter rack straight up and slide filter out of the rack. SEE FIGURE 7.

4. Clean or replace filter and reinstall the filter reversing the procedure described above.

5. Reconnect power supply to the unit.

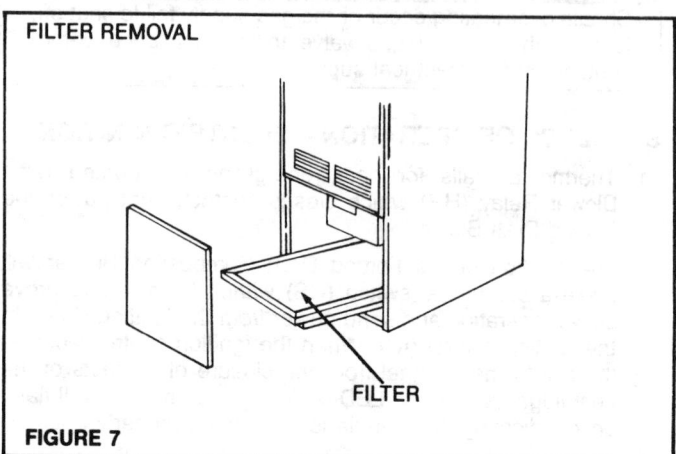

FILTER REMOVAL

FILTER

FIGURE 7

SIDE RETURN

To clean filter for side return application:

1. Disconnect power supply to furnace.

2. Slide filter out of the rack. SEE FIGURE 8.

3. Clean or replace filter.

4. Slide filter back into the filter rack.

5. Reconnect power supply.

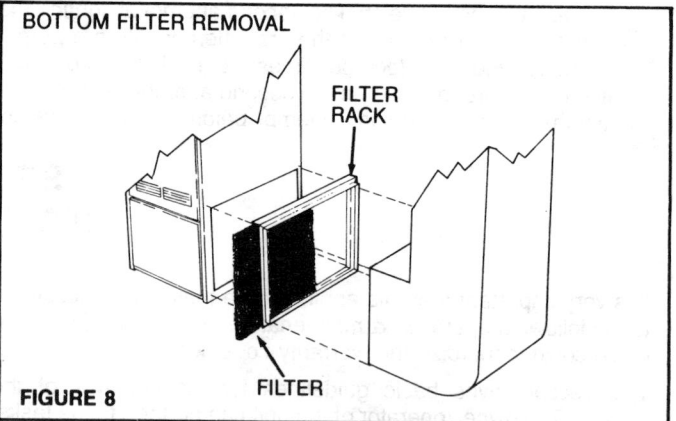

BOTTOM FILTER REMOVAL

FILTER RACK

FILTER

FIGURE 8

OPERATIONAL SAFETY FEATURES

1. FLUE SAFETY PRESSURE SWITCH

A Pressure Switch is incorporated in the thermostat circuit to the ignition control and will break the circuit if too much pressure is sensed at the flue outlet, as in the event of a blockage in the vent to the outside atmosphere or a severe downdraft condition.

2. IGNITION CONTROL LOCKOUT FEATURE

This feature of the ignition system automatically de-energizes the ignitor and shuts off power to the gas valve in the event of of the loss or absence of main burner flame. Should the energized ignition circuit not sense the presence of main burner flame on the flame sensor within 7 seconds the system will try again for ignition for a total of 3 tries. If the system fails to sense burner flame, the system will lockout.

3. HIGH LIMIT

The high limit is part of the Combination Fan and Limit Control. Both the fan circuit and the limit circuit are activated by the action of a calibrated bi-metal coil which is sensitive to heat. The probe containing this coil is inserted into the heat exchanger compartment in proximity to an area which is heated by the main burner flame and cooled by the flow of air from the Blower. As the flame heats the heat exchanger, the bi-metal reacts to the heat and rotates a shaft connected to a switch mechanism. The first switch is the normally open fan contacts which upon closing, energize the Blower Motor. When the blower is energized the airflow over the heat exchanger removes sufficient heat to stabilize the movement of the bi-metal actuated switch and maintain a fixed position of the switch in its travel. Should the blower fail or airflow become blocked or reduced to undesirable levels, the bi-metal coil will continue to rotate until the "limit" contacts reach the set point and break the circuit to the ignition control, thereby closing the gas valve and extinguishing the main burner flame.

4. CONTROL MODULE

Diagnostic LED indicators to help the serviceman indicate whether a failure do to the ignition control or something external in the system.

A. Initial Start Up—On call for heat, the LED indicator will flash once indicating normal operation.

B. Internal Check Failure—The enhanced control will immediately lock-out and indicate the failure by a continuous LED indicator light.

C. External Check Failure—The control will retry and ultimately lock-out; it will also indicate the failure by a flashing LED indicator light.

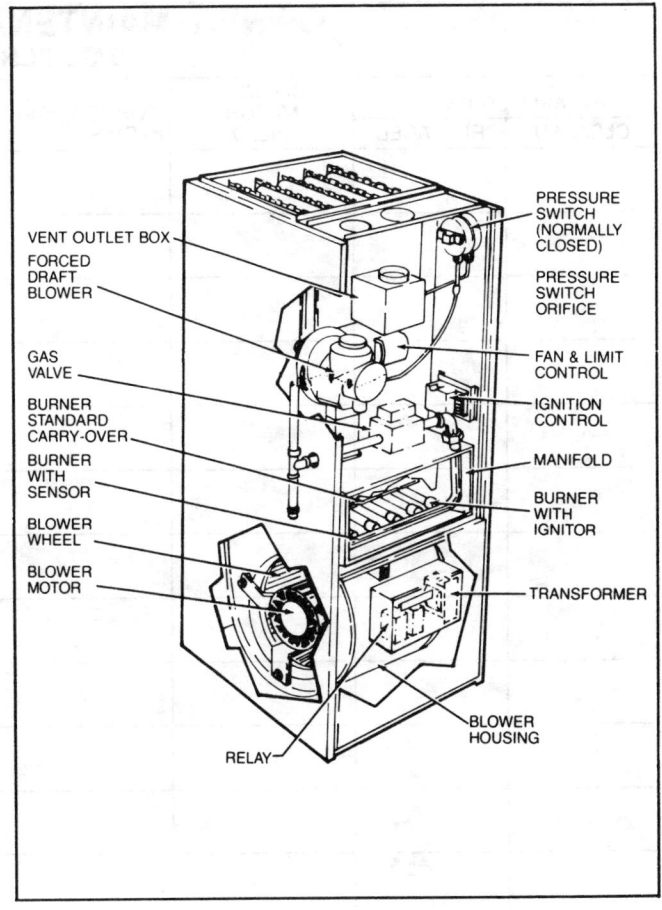

7

OWNER MAINTENANCE SCHEDULE
(DATE PERFORMED)

AIR FILTERS		BLOWER MOTOR OILED	FORCED DRAFT BLOWER MOTOR	COMPLETE PREVENTIVE EXAMINATION BY A SERVICING CONTRACTOR	OTHER
CLEANED	REPLACED				

SUPER HIGH EFFICIENCY CONDENSING GAS FURNACE

GUH SERIES

Service Manual

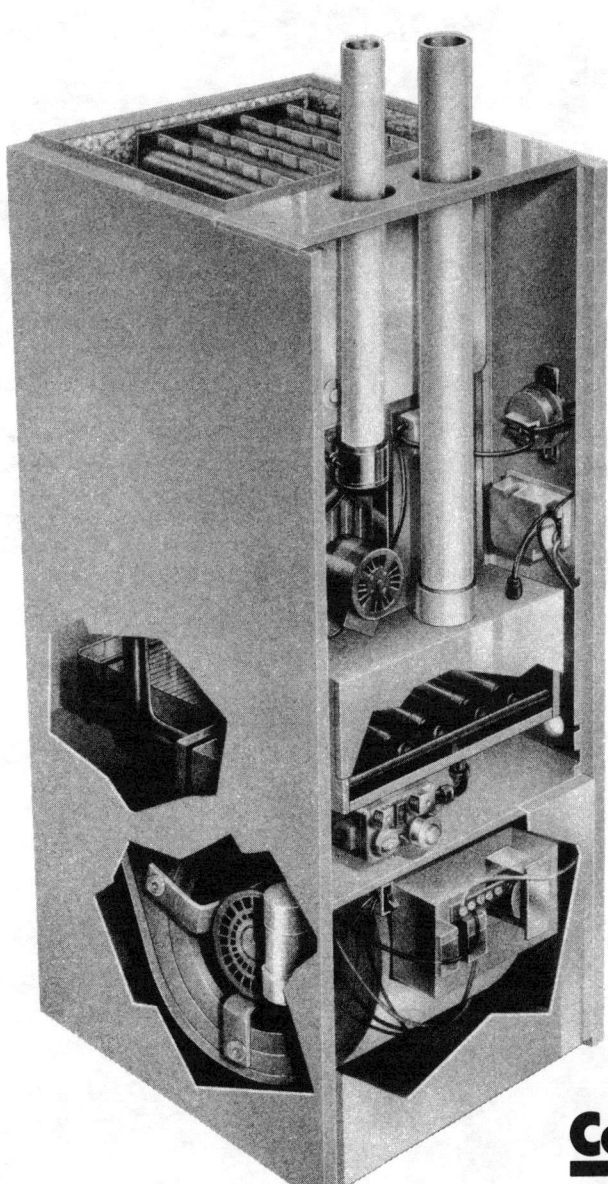

Comfortmaker®
Air Conditioning & Heating

Arcoaire®
Air Conditioning & Heating

REVISED: FEBRUARY, 1990
SUPERSEDES: APRIL, 1989

INDEX

IDENTIFYING THE UNIT

UNIT DESCRIPTION

The nameplate label and serial number label on the furnace shows the model number of the furnace. This label is located on the right hand side inside the casing. See Figure 1 for the location of these labels.

Check the furnace model number on the nameplate label. The suffix of the model number indicates the type of fuel and the ignition system.

Please note that each digit of the model number has a special significance:

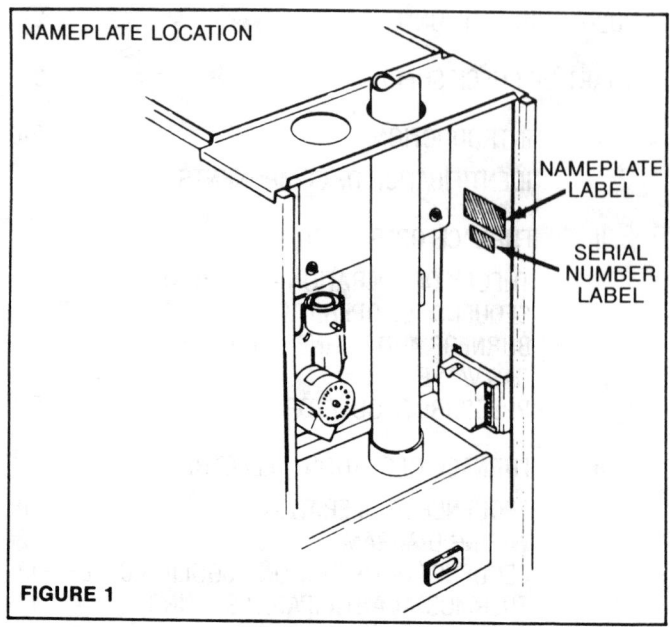

NAMEPLATE LOCATION

NAMEPLATE LABEL

SERIAL NUMBER LABEL

FIGURE 1

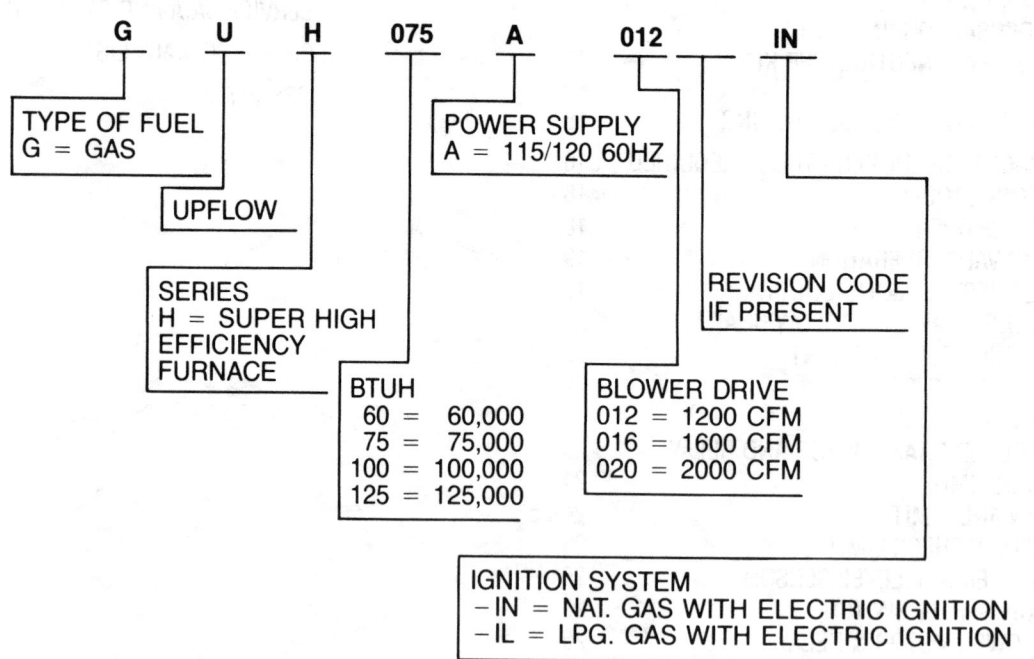

G U H 075 A 012 IN

TYPE OF FUEL
G = GAS

UPFLOW

SERIES
H = SUPER HIGH EFFICIENCY FURNACE

BTUH
60 = 60,000
75 = 75,000
100 = 100,000
125 = 125,000

POWER SUPPLY
A = 115/120 60HZ

REVISION CODE IF PRESENT

BLOWER DRIVE
012 = 1200 CFM
016 = 1600 CFM
020 = 2000 CFM

IGNITION SYSTEM
– IN = NAT. GAS WITH ELECTRIC IGNITION
– IL = LPG. GAS WITH ELECTRIC IGNITION

START-UP CHECK SHEET

OWNER NAME: _____ DEALER NAME: _____

ADDRESS: _____ ADDRESS: _____

CITY, STATE, ZIP: _____ CITY, STATE, ZIP: _____

MODEL NUMBER: _____ SERIAL NUMBER: _____

TYPE OF GAS: NAT: _____ LPG: _____ CONV. STICKER APPLIED: _____ CONV. KIT NUMBER: _____

GAS
VALVE _____ LINE
PRESSURE _____ MANIFOLD
PRESSURE _____ INPUT _____

THERMOSTAT: _____ SUBBASE: _____ HEAT ANTICIPATOR SETTING: _____

BLOWER MOTOR H.P.: _____ VOLTAGE: _____

FAN OPENS: _____ °F FAN CLOSES: _____ °F

FAN AND LIMIT
MODEL NUMBER: _____ LIMIT OPENS: _____ °F LIMIT CLOSES: _____ °F

TEMPERATURE RISE: RETURN AIR: _____ °F SUPPLY AIR: _____ °F

FILTER SIZE AND TYPE: _____

BURNER FLAME PROPERLY ADJUSTED? _____

DRIP-LEG INSTALLED PRIOR TO GAS VALVE? _____

CONDENSATE DRAIN CONNECTED? _____

BLOWER SPEED CHECKED? _____ BLOWER PROPERLY LUBRICATED? _____

ALL ELECTRICAL CONNECTIONS, MOUNTING SCREWS, ETC. TIGHT? _____

ADJUSTMENT REQUIRED: _____

DATE OF INSTALLATION? _____

DATE OF START-UP: _____

_____

_____

_____

_____

_____

_____

_____

_____

_____

SERVICE AND MAINTENANCE LOG

AIR FILTERS		BLOWER MOTOR OILED	COMPLETE PREVENTIVE EXAMINATION BY A SERVICING CONTRACTOR	OTHER
CLEANED	REPLACED			

I. INTRODUCTION

INTRODUCTION TO THE
CONDENSING FURNACE SERVICE MANUAL

This Service Manual is designed to be used in conjunction with the installation Manual and Operating Procedure furnished with each furnace.

WARNING:

The Condensing Gas Furnace is a modern, complex product, built of materials and ideas based on the leading edge of gas appliance technology. Therefore, it is not only recommended, but imperative that those installing, servicing and maintaining this product are experienced, qualified, professional HVAC Technicians. This Service Manual was written to assist those professionals to do their job quickly, more accurately, and more efficiently. **It is not intended for the do-it-yourselfer, nor is it a substitute for the Owner's Manual.** The manufacturer will not be responsible for injury or damage caused by misapplication or misuse of this manual, or of the product itself.

This furnace uses an induced draft blower to draw combustion air through the furnace.

This furnace is manufactured in input ranges of 60,000 to 125,000 BTU/H. Each burner fires at a range of 20,000 or 25,000 BTUH each. In order to control combustion within this furnace, we use a sealed combustion chamber. A sealed combustion chamber means that all air entering the furnace is drawn in by the induced draft motor and not by conventional methods. If for any reason the seal is broken around the combustion chamber, the furnace will not operate properly.

A Hot Surface Ignition System is used for main burner ignition, instead of a pilot relight system. This simply means that the pilot has been eliminated in this furnace and a hot surface ignitor is used for burner ignition.

A final word of **CAUTION:** although we have mentioned some new features that have been incorporated within the condensing furnace, standard operation of the furnace is still the same as conventional furnaces in many ways. It still requires the following steps when servicing:

1. The furnace must be isolated from the gas supply piping system by closing individual shut-off valves during any pressure testing of the gas supply piping system at test pressures equal to or more than ½ PSIG (13.86″ Water Column).

2. Use pipe compounds on gas pipe threads that are resistant to all gases.

3. Do not use open flame to check for gas leaks. Use a soapy solution or a commercial leak tester.

4. Keep instruments and tools in good repair and calibration.

5. Instruct the owner or operator of the furnace in the proper procedure to follow for shutting off the furnace in case of an emergency. Should overheating occur or the gas supply fail to shut-off, instruct the owner or user to turn off the manual gas valve to the appliance before turning off the electrical supply. Explain how to change, service or clean filters and setting of controls for best comfort conditions.

   Refer to the Installation Instructions and Operation Procedure 7212-056 for specific information on items not covered in this Manual. Also refer the owner to the User's Manual 7112-031 for future reference and further information.

6. Explain to the owner or operator that the furnace has lock-out safety devices within its operation, and that if it should fail to operate, a qualified service technician should be called to service the unit.

I. INTRODUCTION (Cont.)

FIGURE 1

I. INTRODUCTION (Cont.)

B IDENTIFICATION OF KEY COMPONENTS

① WRAP CASING ASSEMBLY
② LOUVER DOOR
③ BLOWER COMPARTMENT DOOR
④ BLOWER DOOR INTERLOCK SWITCH
⑤ COMBUSTION DOOR
5A GASKET
⑥ BURNER ACCESS SHIELD
⑦ BURNER ASSEMBLY
⑧ CROSSOVER ON BURNERS
⑨ COLLECTOR
9A GASKET
9B COLLECTOR INSULATION
⑩ EQUALIZER PLATE
10A GASKET (2)
⑪ HEAT EXCHANGER FLUE BAFFLES
⑫ HEAT EXCHANGER (RPJ) PRIMARY
⑬ HEAT EXCHANGER 29-4C STAINLESS
SECONDARY

⑭ VENT PRESSURE SWITCH
⑮ CONDENSATE LEVEL SENSOR
⑯ FAN AND LIMIT CONTROL
⑰ COMBUSTION BLOWER MOTOR
⑱ GAS VALVE
⑲ MANIFOLD
⑳ BURNER ORIFICE AND SPRING
㉑ IGNITOR MODULE
㉒ BLOWER ASSEMBLY
㉓ MULTI-SPEED P.S.C. BLOWER
㉔ BLOWER RUN CAPACITOR
㉕ BURNER COMPARTMENT
㉖ BURNER INSPECTION WINDOW
㉗ IGNITOR
㉘ SENSOR

㉙ DRAIN TRAP/COLLECTOR
㉚ CONTROL CENTER
㉛ TRANSFORMER
㉜ HEAT RELAY
㉝ BLOWER RELAY
㉞ COMBUSTION BLOWER
㉟ AIR INTAKE
㊱ TRANSFER TUBE
㊲ DISCHARGE TUBE
㊳ FLAME SENSOR

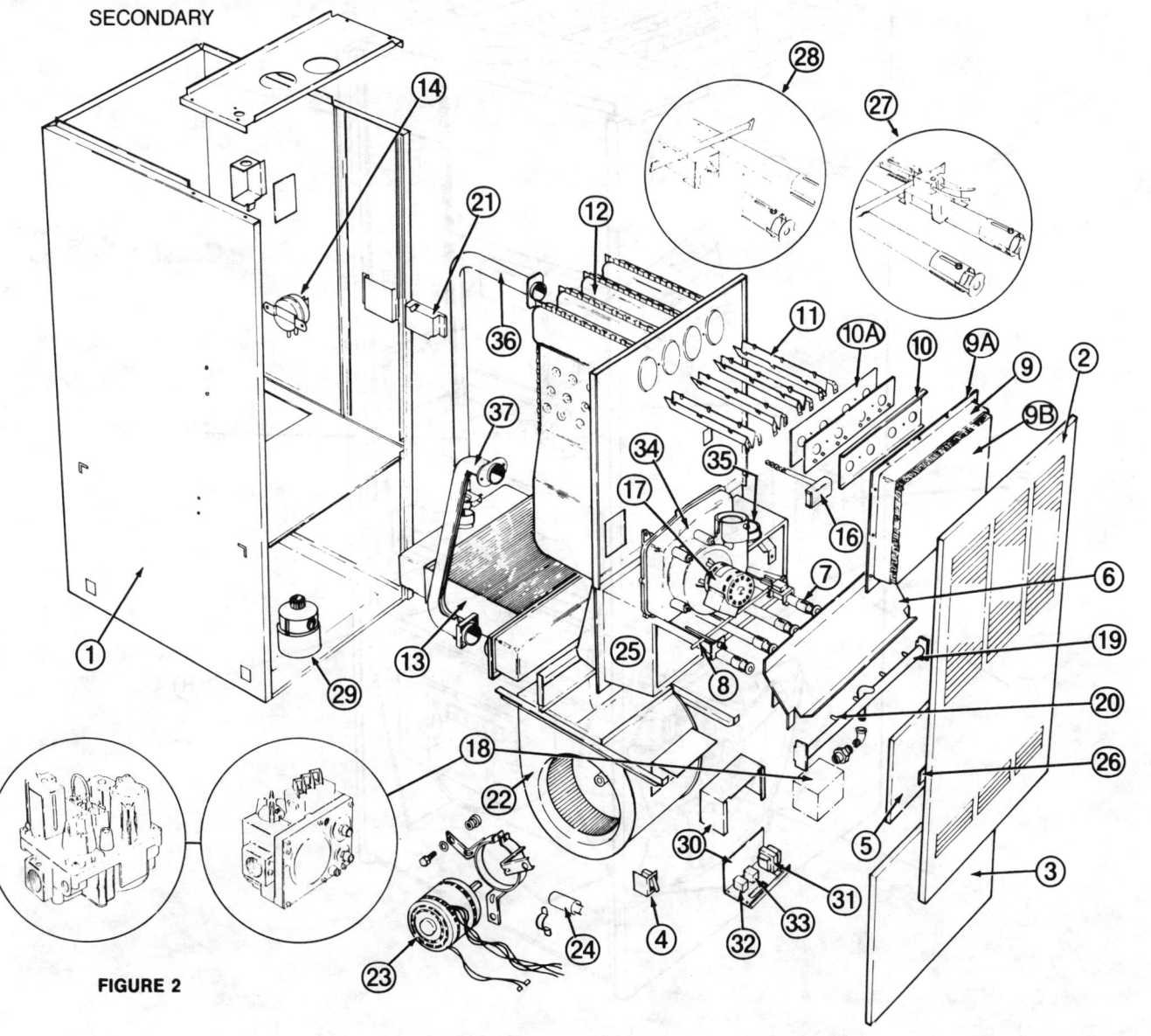

FIGURE 2

7

II. FURNACE OPERATION

A. Theory of Operation (Combustion)

The efficiency of this furnace is obtained through three basic processes that are not found on standard efficiency gas furnaces. These are:

1. Very accurate control of the combustion process by metering the volume of gas **and** fresh air entering the combustion chamber.

2. The use of a **secondary** heat exchanger to scour the residual heat from the flue gases that are leaving the primary exchanger.

3. The introduction of air from **outdoors** to the combustion area of the furnace, rather than from the immediate area of the furnace's installation.

Let's look closely at each one of these processes and see how they work closely in unison to achieve the highest possible efficiencies.

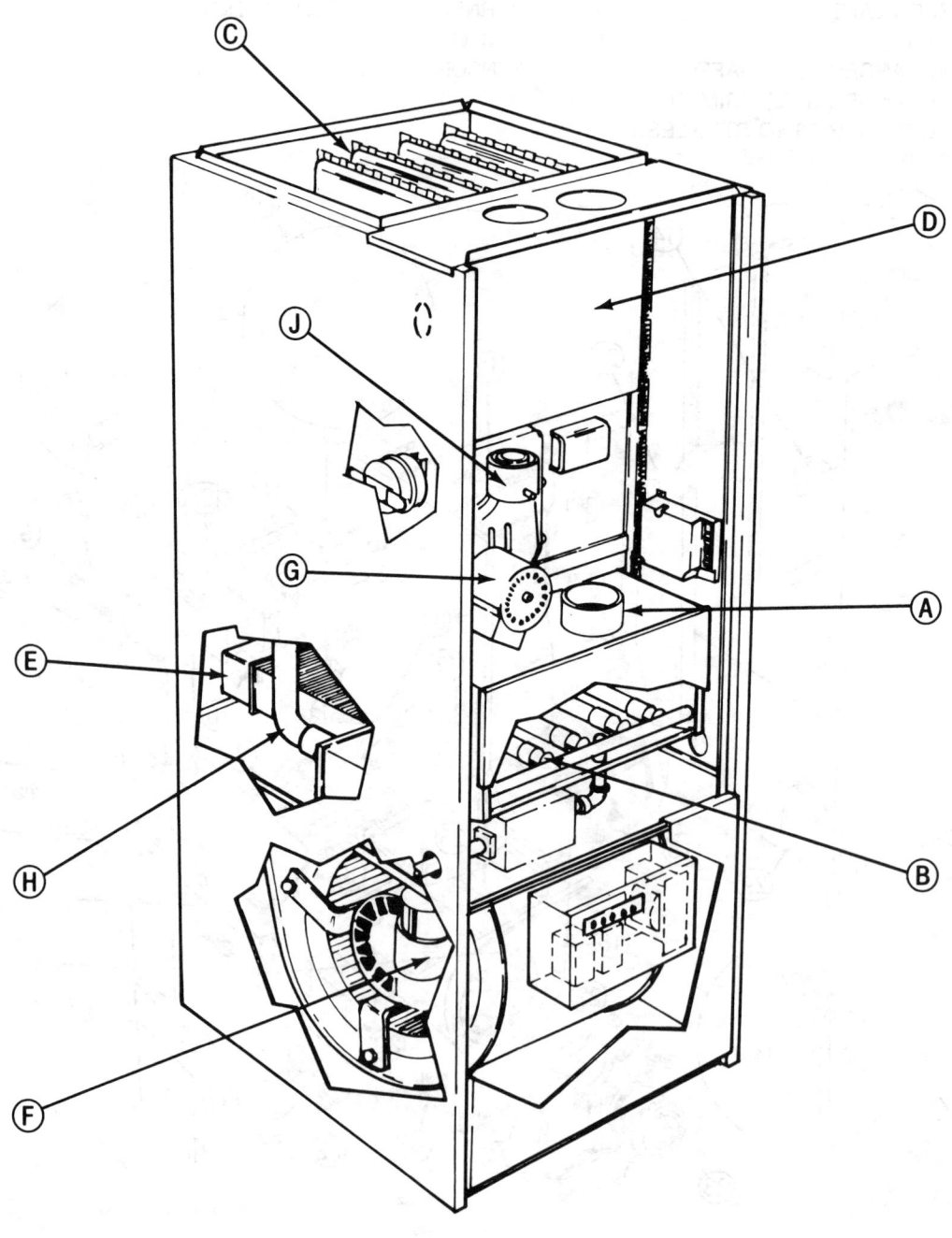

FIGURE 3

8

II. FURNACE OPERATION (Cont.)

SEQUENCE OF OPERATION (COMBUSTION)

Refer to Figure 3

As the furnace is firing, air is drawn through the combustion supply air tube Ⓐ into the combustion chamber. The chamber is sealed (in fact, **must** be sealed) and operates at a slight negative pressure. A baffle insures that each burner receives an adequate share of both primary and secondary combustion air.

The ratio of gas to primary air is controlled by the conventional method of slotted shutters Ⓑ between the gas orifice and burner tube venturi. The greater the shutter is opened, the greater the proportion of the mixture will be air.

Each burner fires into a cell of the heat exchanger at an input of 20,000 or 25,000 BTUH, depending on the capacity of the furnace. This is the primary heat exchanger Ⓒ and is fabricated of aluminized steel, utilizing the weldless, one piece RPJ process of construction. The operation of the furnace at this point is essentially the same as a conventional furnace. Air flow across the exchanger takes heat to the conditioned space and the combustion products exit the main exchanger into the collector Ⓓ through a series of baffles. These baffles slow the exit of the combustion gases so that more heat is drawn from them before they leave the primary heat exchanger.

The similarity in operation between a standard furnace and a condensing furnace end here. The exiting flue gases would leave the conventional furnace and go up the flue at a temperature of 350° to 550° F., wasting a significant amount of heat. The condensing gas furnace utilizes this heat by routing the combustion gases to the **secondary** heat exchanger. This secondary exchanger Ⓔ removes heat from the gases, but also takes latent heat from the water vapor in the combustion products. This causes the water vapor to **condense**; hence the term "Condensing Furnace," and the need for a drainage system. The liquids that condense in the secondary exchanger are not all so harmless as water. Several dilute but powerful acids are also produced. This is the reason for the use of 29-4C stainless steel and high temperature plastics in all areas of the furnace that are exposed to condensed combustion products. Even though these materials **are** expensive to buy and manufacture, they are **essential** to long product life and to the furnace's safety and reliability.

The combustion products now have little heat left in them. The remaining liquids from condensation are drained from the secondary exchanger to the drain trap/collector Ⓕ. The cooled (140° to 200° F.) gases are drawn to the induced draft combustion blower (Power Ventor) Ⓖ via a discharge tube Ⓗ and through a restrictor. The restrictor controls the volume of which gases drawn through the entire combustion system. Any remaining liquids are routed from the combustion blower to the drain trap/collector and drained from there to an external drain. The remaining gases exit the furnace through the vent pipe Ⓙ to the outdoors.

BURNERS AND BURNER ORIFICES

A slotted tube type burner is used in this furnace. These burners have certain flame requirements. Burner flame should run the length of the tube and be 3" to 4" in height. If the flame should lift from the burner, turn orange in color, or start rolling out from under the flash burner access shield after ignition, an adjustment is necessary or the heat exchanger needs to be inspected and possibly cleaned. Checking the burner flame should give an indication of the need for further inspection. See Figures 4 & 5.

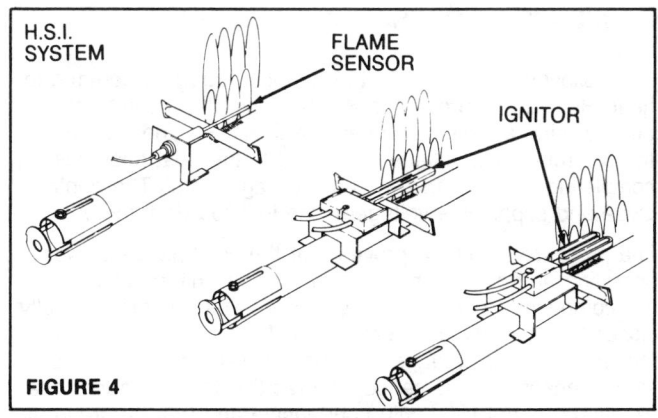

FIGURE 4

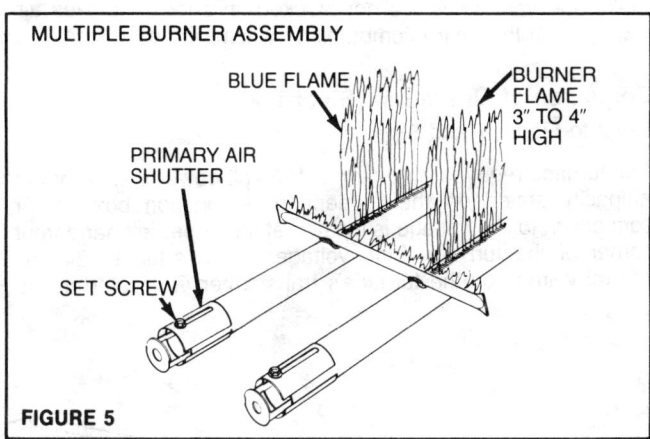

FIGURE 5

GAS VALVE

The control used on this furnace is a redundant combination gas valve. This design provides a manual main valve, servo regulated automatic main diaphragm valve, plus the added safety of an additional in-line electromechanical operated redundant valve. See Figure 6.

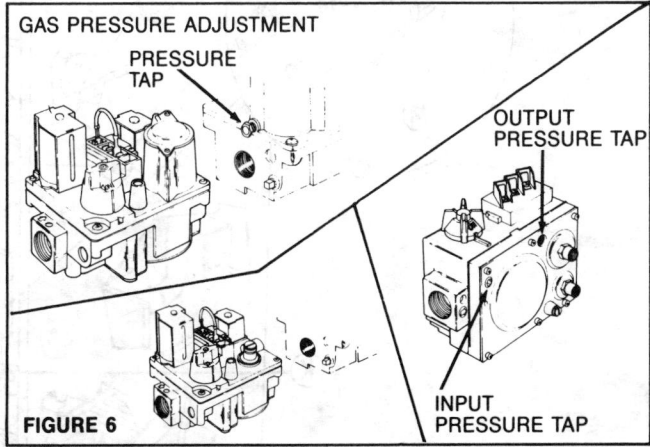

FIGURE 6

PRESSURE REGULATOR

A pressure tap has been provided on the gas valve. See Figure 6. Attach a pressure gauge fitting to this tap and check the outlet pressure. The pressure should be set as indicated on the rating plate. To adjust the pressure regulator, remove the cap covering the pressure regulator adjusting screw. On valves with more than one adjusting screw, adjust only the screw marked "HI". Turn the adjusting screw out (counter-clockwise) to decrease the pressure, and in (clockwise) to increase the pressure. Reinstall the cap securely.

III. FURNACE OPERATION

SEQUENCE OF OPERATION (GENERAL)

Let's follow the operation of the furnace through a demand for heat. Refer to Figure 7 and 8. The room thermostat calls for heat, closing a circuit between terminals "R" and "W". Inside the control center, voltage is applied to the heat relay coil ℗, completing the "W" and "C" Low Voltage circuit. The contacts close and supply line voltage to the induced draft motor.

The pressure switch ⒟ proves that the combustion blower is operating and closes on a rise in pressure differential across the combustion blower. This sends 24 volts to the normally closed limit switch of the combination fan and limit control Ⓔ. If the limit is closed, voltage becomes available to a condensate safety sensor Ⓖ. This sensor insures that the secondary heat exchanger is not filled with condensate and that the fresh air intake and vent pipes are not blocked. In effect, it proves air flow through the entire combustion system of the furnace.

Sequence of Operation (Electrical)

Refer to Figure 7 and 8.

The furnace requires an external 115 volt power source of the ampacity stated on the nameplate. A junction box Ⓕ for connection to line voltage is located at the upper left hand front corner of the furnace. Line voltage is converted to 24 volt control voltage by the furnace's transformer Ⓐ located in the control center in the blower compartment. Neither line or control voltage will be available unless the door inter-lock switch Ⓑ is closed.

To operate the furnace as a heating unit only, a 25 volt thermostat is the only required external control. This thermostat would be connected to the "R" and "W" terminals of the terminal strip Ⓒ. For operation with a heating and cooling unit or if continuous indoor blower operation is required, a combination heating/cooling thermostat will be required. In this application "R" to "G" will energize the blower at high speed.

Terminal "C" is 24 volt common; clock type thermostats and an air conditioning unit would require use of "C" terminal. The "Y" terminal is simply a binding post to connect the "Y" lead from the thermostat to the "Y" lead of the condensing unit and is not electrically connected to the furnace controls in any way.

Under no circumstances should the "R" and "C" terminals be connected together. Such action constitutes abuse and **will** damage the furnace. This damage will render the furnace transformer inoperable and will **not** be covered under warranty! The furnace transformer is capable of handling: all internal furnace functions, any recommended thermostat and any control voltage requirements of any matching approved split system air conditioning unit or heat pump. Heat pumps require the application of an "Add-On" Kit to interlock controls.

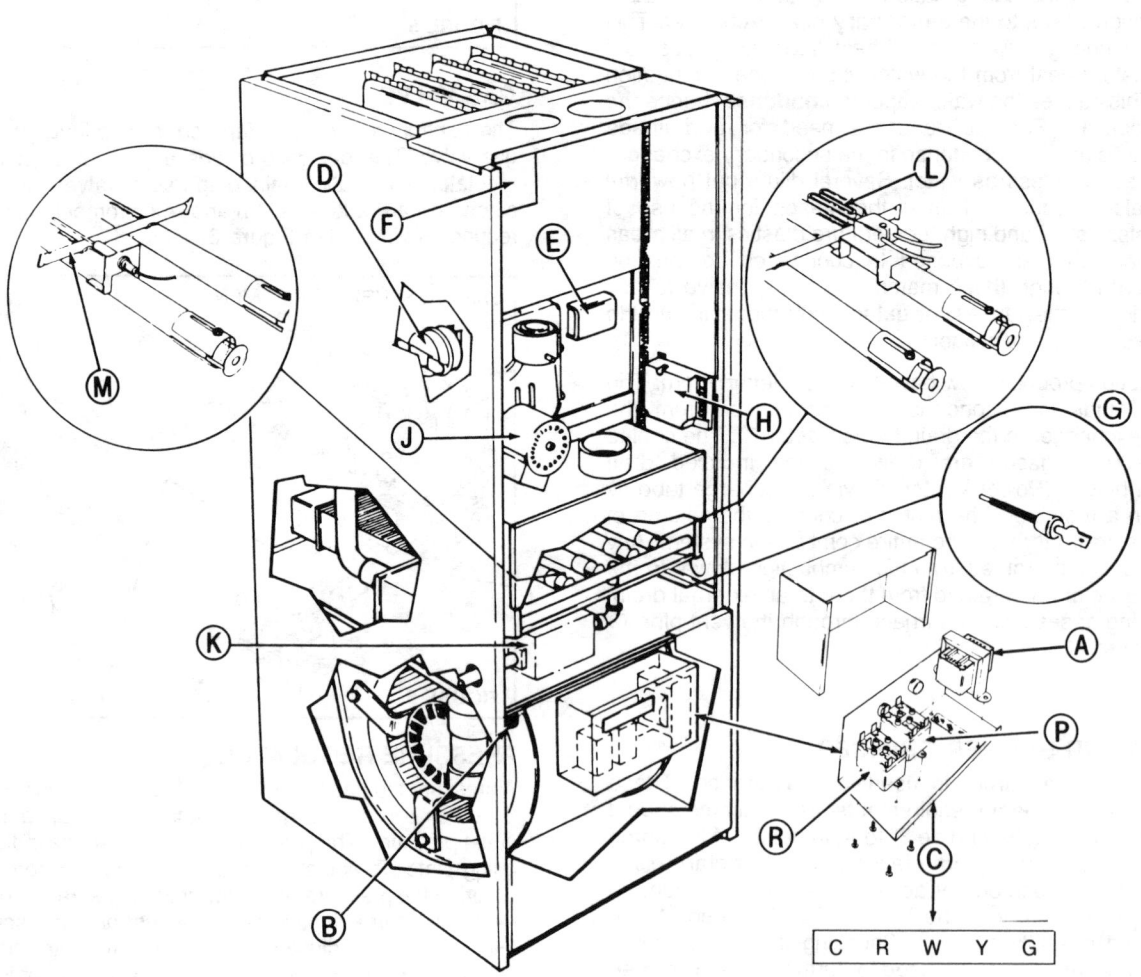

FIGURE 7

III. FURNACE OPERATION (Cont.)

When this series of controls have all closed (the entire sequence occurs in a few seconds), 24 volts is available to the ignition module ⓗ. At this time the module will start the ignition sequence. The module will require a 30 second "Pre-Purge" from the time it receives power. This means that the combustion blower ⓙ will clear both heat exchangers as well as vent pipes of all gases (other than fresh air) before ignition can take place. This is extremely important. Igniting residual combustible vapors in a sealed combustion system can cause severe equipment damage.

When the pre-purge cycle is complete, the ignition module ⓗ closes a 115 volt circuit to the igniter ⓛ. The module will allow approximately 15 seconds for the igniter to warm up. The igniter will now be glowing almost white-hot over the right hand burner. The gas valve ⓚ will receive 24 volts from the module.

Ignition Occurs

The sensor ⓜ, located over the left hand burner, must sense flame in 7 seconds or less; if it does not, the module will shut the gas valve off and re-try the ignition sequence twice more before locking out. See the section on hot surface ignition for more detail.

As the heat exchangers begin to heat up, the fan switch of the Fan and Limit Control ⓝ will close, causing the main blower to run at heating speed, (selected by the normally closed contacts of the blower relay), bringing air from the conditioned space across the secondary heat exchanger, primary heat exchanger, then returning it to the conditioned space.

When the space thermostat is satisfied and the call for heat ends, the 24 Volt supply will be interrupted to terminal "W". The combustion blower will stop and the gas valve closes; the burner will extinguish. The main blower will continue to run until the heat exchangers cool off and the fan switch of the fan and limit control opens. The furnace remains in this condition until there is another demand for heat.

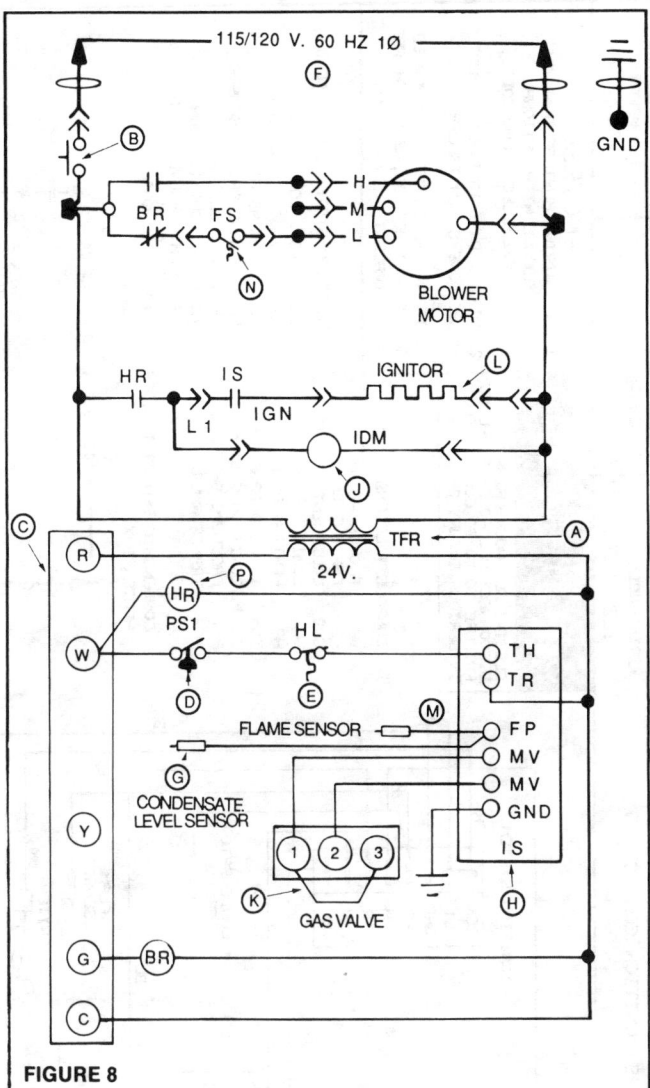

FIGURE 8

LEGEND

- ⓐ TRANSFORMER 120V/24V
- ⓑ DOOR INTERLOCK SWITCH
- ⓒ TERMINAL STRIP
- ⓓ PRESSURE SWITCH (ONE)
- ⓔ FAN LIMIT CONTROL
- ⓕ JUNCTION BOX
- ⓖ CONDENSATE LEVEL SENSOR
- ⓗ IGNITION MODULE
- ⓙ INDUCED DRAFT MOTOR
- ⓚ GAS VALVE
- ⓛ IGNITOR
- ⓜ SENSOR
- ⓝ FAN SWITCH
- ⓟ HEAT RELAY
- ⓠ BLOWER MOTOR (CONDITIONED SPACE BLOWER)
- ⓡ BLOWER RELAY

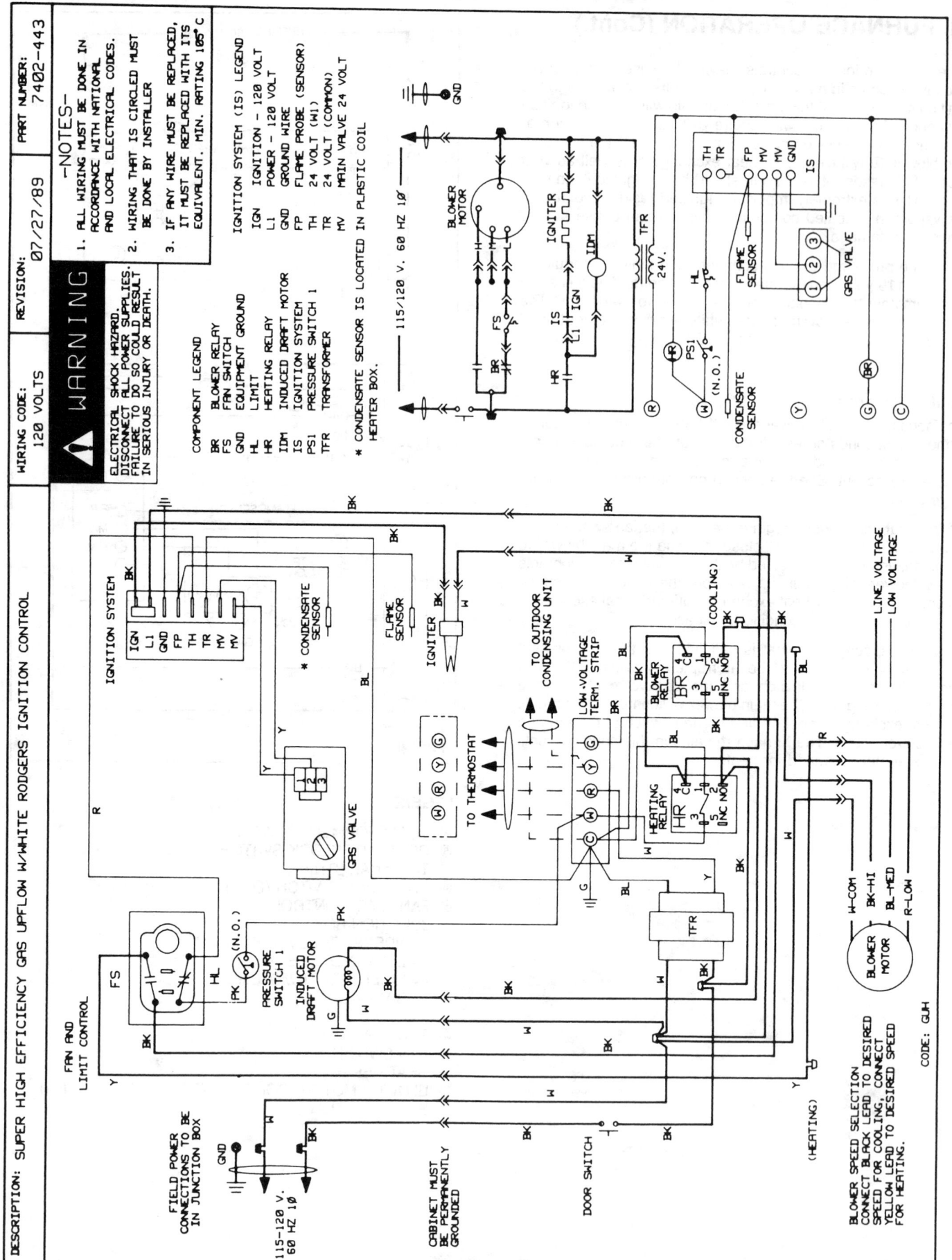

FIGURE 9

III. FURNACE OPERATION (Cont.)

SEQUENCE OF OPERATION—COOLING CYCLE:

1. Placing the thermostat subbase Fan Switch in the "ON" position will take 24 volts from the furnace "R" terminal and complete the circuit to Terminal "G". This energizes the blower relay, causing the normally open relay contacts to close and power the blower motor at cooling speed.

2. If the Fan Switch is in the "AUTO" position, the "G" terminal will only be energized when the "Y" terminal is energized. This occurs only if the system switch is set for "COOL" and the thermostat calls for cooling. In essence, this will cause the furnace blower and outdoor unit to cycle together.

THERMOSTAT ANTICIPATOR SETTING

Proper control of the indoor area temperature can only be achieved if the thermostat is calibrated to the heating and/or cooling cycle. A vital consideration of this calibration is related to the thermostat heat anticipator.

Anticipators for the cooling operation are generally pre-set by the thermostat manufacturer and require no adjustment.

Anticipators for the heating operation are of two types, pre-set and adjustable. Those that are pre-set will not have an adjustment scale and are generally marked accordingly.

Thermostat models having a scale as shown in Figure 10 must be adjusted to each application.

The proper thermostat heat anticipator setting is a MINIMUM of 0.8 Amps for furnace operation only. A lower setting will result in short cycling of the furnace and in extreme cases will cause a complaint of "no heat."

To increase length of cycle, increase setting of heat scale; to decrease length of cycle, decrease setting of heat scale.

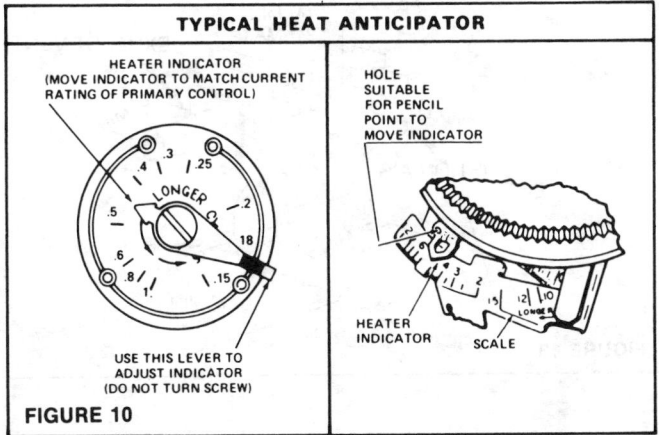

TYPICAL HEAT ANTICIPATOR

HEATER INDICATOR
(MOVE INDICATOR TO MATCH CURRENT RATING OF PRIMARY CONTROL)

USE THIS LEVER TO ADJUST INDICATOR (DO NOT TURN SCREW)

HOLE SUITABLE FOR PENCIL POINT TO MOVE INDICATOR

HEATER INDICATOR   SCALE

FIGURE 10

A third type uses a "cycle rate" adjustment. See Figure 11. Generally the rate is factory set for the forced air furnaces. Consult the thermostat instruction sheet for details.

In many cases, this adjustment setting can be found in the thermostat instruction sheet. If this information is not available, or if the correct setting is questioned, the following procedures should be followed:

PREFERRED METHOD:

Use a low scale ammeter such as an ampcheck or milliameter. Connect the meter across terminals "R" and "W" on the sub-base ("RH" & "W1" on multi-stage thermostat sub-base).

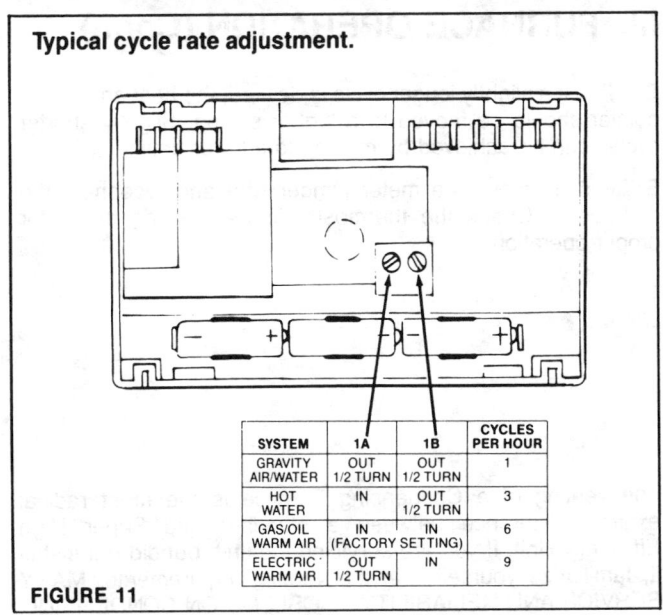

SYSTEM	1A	1B	CYCLES PER HOUR
GRAVITY AIR/WATER	OUT 1/2 TURN	OUT 1/2 TURN	1
HOT WATER	IN	OUT 1/2 TURN	3
GAS/OIL WARM AIR	IN	IN (FACTORY SETTING)	6
ELECTRIC WARM AIR	OUT 1/2 TURN	IN	9

FIGURE 11

STEP 1. Wrap 10 loops of single strand insulated thermostat wire around the prongs of an ampmeter. Set the scale to the lowest amp scale.

STEP 2. Connect the uninsulated ends of this wire jumper across terminals "R" and "W" on the sub-base ("RH" and "W1" on multi-stage thermostat sub-base). See Figure 12. This test must be performed without the thermostat attached to the sub-base.

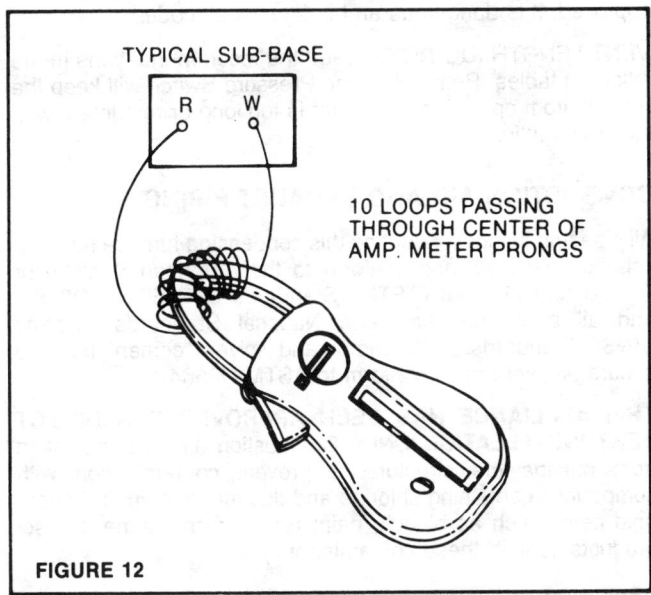

TYPICAL SUB-BASE

R   W

10 LOOPS PASSING THROUGH CENTER OF AMP. METER PRONGS

FIGURE 12

STEP 3. Let the heating system operate in this position for about one minute. Read the ampmeter scale. Whatever reading is indicated must be divided by 10 (for 10 loops of wire).

This is the setting at which the adjustable heat anticipator should be set.

FORMULA: $\dfrac{\text{Ampmeter reading}}{10 \text{ loops}} = $ Anticipator setting

OR: $\dfrac{8.5 \text{ AMPS}}{10} = .85$ AMPS Setting

III. FURNACE OPERATION (Cont.)

STEP 4. If a slightly longer cycle is desired, the heat anticipator pointer should be moved to a higher setting. Slightly shorter cycles can be achieved by moving to a lower setting.

STEP 5. Remove the meter jumper wire and reconnect the thermostat. Check the thermostat in the heating mode for proper operation.

NOTE: The length of the heating cycle can also be affected by the fan control settings (of the combination fan and limit control). The fan "ON" and "OFF" settings should be checked at this point.

If Digital Ampmeter is used, read the amp draw direct from the meter, then Step #1 and #3 are not required.

IV. VENTING

The venting of a Condensing Furnace is the most radical external difference between a standard and Super High Efficiency Unit. If you are servicing the unit, but did not install it, familiarize yourself with the venting requirements. MANY SERVICE AND RELIABILITY PROBLEMS ON CONDENSING FURNACES CAN BE TRACED TO INCORRECT VENTING PROCEDURES AND MATERIALS!

The most notable are:

MATERIAL: Schedule 40 PVC is the ONLY vent material approved for use with this furnace.

COMBUSTION AIR: MUST be taken from outdoors. This prevents heat exchanger damage from corrosion.

COMMON VENTING: DO NOT EVEN THINK ABOUT VENTING THIS UNIT WITH ANOTHER APPLIANCE! It is NOT approved; it is dangerous and is against all codes.

VENT LENGTH: DO NOT exceed the recommendations in the following tables. Remember, the Pressure Switch will keep the furnace from operating if the vent is too long or restricted, with excessive fittings.

COMBUSTION AIR AND EXHAUST PIPING

All pipe and fittings for venting this condensing furnace must be schedule 40 PVC and conform to the American Society for Testing and Material (ASTM) Standards D1785 and D2665, and all applicable American National Standards Institute (ANSI) Standards. PVC primer and solvent cement used to secure all joints must conform to ASTM D2564.

THIS APPLIANCE HAS BEEN APPROVED FOR DIRECT VENT INSTALLATION ONLY. Combustion air must be taken from outside the structure to prevent contamination with compounds containing chlorine and fluorine. (Common household items such as bleach, paint remover and some aerosol products contain these contaminants.)

⚠ CAUTION:

Terminate the combustion air intake as far as possible from the air conditioning unit or heat pump, swimming pools, swimming pool pumping units, and dryer vents.

All combustion air and exhaust piping must be installed in accordance with local codes and these instructions.

When the exhaust vent passes through an unconditioned space or raceway, it must be insulated with ½" wall, closed cell, neoprene insulation or equivalent.

If the original furnace was common vented with another gas fired appliance, the vent may need to be resized for the capacity of the other appliance. Consult the National Fuel Gas Code (ANSI Z223.1a-1987) for proper sizing and revise as required. **COMMON VENTING OF THIS FURNACE WITH ANOTHER GAS FIRED APPLIANCE (WATER HEATER, ETC.) IS NOT ALLOWED.**

VENT TERMINAL INSTALLATION

1. This furnace can be installed with either a vertical or horizontal direct vent. In either case, the exhaust vent and combustion air intake pipe must be located on the same side of the structure and separated by no less than 14" and no more than 24". See Figure 13.

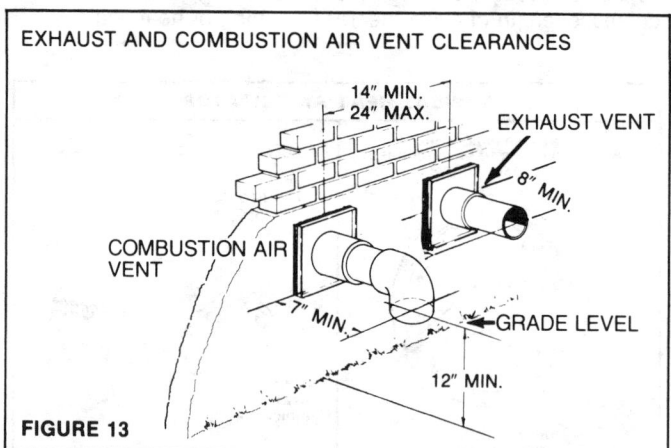

EXHAUST AND COMBUSTION AIR VENT CLEARANCES

14" MIN.
24" MAX.
EXHAUST VENT
8" MIN.
COMBUSTION AIR VENT
7" MIN.
GRADE LEVEL
12" MIN.

FIGURE 13

⚠ WARNING

At no time should the intake and exhaust vents be terminated on opposite or adjacent sides of the structure. Also, combination vertical and horizontal venting is NOT permitted.

---

2. Position the exhaust vent and combustion intake terminals away from obstructions and above anticipated snow accumulation or grade level as shown in Figures 13-14.

⚠ CAUTION:

Always determine anticipated snow accumulation level and install vent terminal accordingly to prevent vent/intake blockage.

IV. VENTING (Cont.)

If installation above snow accumulation level is chosen as shown in Figure 14 replace the exterior coupling with a 90° elbow pointing up to secure the termination assembly. The elbow must be secured with a stainless steel screw as shown in Figure 14.

DO NOT point terminals into window wells, stairwells, alcoves or other recessed areas. Maintain a 3 foot clearance from plumbing vent stacks. It is preferable for the vent intake to terminate on opposite or adjacent sides of the structure than the dryer vent. If this is not possible, maintain a 10 foot clearance.

Maintain a 1 foot clearance from any opening through which flue gas could enter the building and a 7 foot clearance above public walkways.

3. Consult Tables 1-3 to select the proper diameter exhaust and combustion air piping. Exhaust and combustion air piping is sized for each unit model based on total lineal vent length and number of 90° elbows required. Two 45° elbows can be substituted for one 90° elbow. The elbow used for vent termination outside the structure is **not** counted when using Tables 1 thru 3. The additional two 90° elbows shown in Figure 14 should be counted however. **When the vent system required is borderline with the next size combination category, always use the next larger size.**

EXAMPLE: Refer to Table 1; A vent system utilizing 20 feet of pipe and 2 elbows could use either a 2" intake and 2" exhaust or a 3" intake and 2" exhaust.—The 3" intake and 2" exhaust should be used.

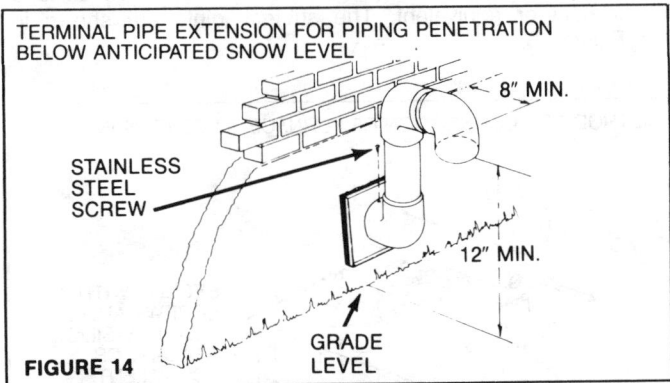

TERMINAL PIPE EXTENSION FOR PIPING PENETRATION BELOW ANTICIPATED SNOW LEVEL

8" MIN.

STAINLESS STEEL SCREW

12" MIN.

GRADE LEVEL

FIGURE 14

TABLE 1—FOR 60,000-75,000 BTUH FURNACE

NO. OF ELBOWS	VENT TYPE	FEET OF PIPE					
		0-10'	10-15'	15-20'	20-25'	25-30'	30-35'
1	INLET	2	2	2	2	3	3
	EXHAUST	2	2	2	2	2	2
2	INLET	2	2	2	3	3	3
	EXHAUST	2	2	2	2	2	2
3	INLET	2	2	3	3	3	3
	EXHAUST	2	2	2	2	2	2
4	INLET	2	3	3	3	3	3
	EXHAUST	2	2	2	2	2	3
5	INLET	3	3	3	3	3	3
	EXHAUST	2	2	2	2	3	3

TABLE 2—FOR 100,000 BTUH FURNACE

NO. OF ELBOWS	VENT TYPE	FEET OF PIPE					
		0-10'	10-15'	15-20'	20-25'	25-30'	30-35'
1	INLET	2	2	2	3	3	3
	EXHAUST	2	2	2	2	2	2
2	INLET	2	2	3	3	3	3
	EXHAUST	2	2	2	2	2	2
3	INLET	2	3	3	3	3	3
	EXHAUST	2	2	2	2	3	3
4	INLET	3	3	3	3	3	3
	EXHAUST	2	2	2	2	3	3
5	INLET	3	3	3	3	3	3
	EXHAUST	2	2	2	3	3	3

TABLE 3—FOR 125,000 BTUH FURNACE

NO. OF ELBOWS	VENT TYPE	FEET OF PIPE					
		0-10'	10-15'	15-20'	20-25'	25-30'	30-35'
1	INLET	2	3	3	3	3	3
	EXHAUST	2	2	2	2	3	3
2	INLET	3	3	3	3	3	N.A.
	EXHAUST	2	2	2	3	3	N.A.
3	INLET	3	3	3	3	N.A.	N.A.
	EXHAUST	2	2	2	3	N.A.	N.A.
4	INLET	3	3	3	3	N.A.	N.A.
	EXHAUST	2	3	3	3	N.A.	N.A.
5	INLET	3	3	3	N.A.	N.A.	N.A.
	EXHAUST	3	3	3	N.A.	N.A.	N.A.

NOTE: Vent pipe diameters apply for installation from sea level to 2000 ft. For installation above 2000 ft. reduce the maximum pipe length as follows:
2000-5000 ft. above sea level—Reduce max. length allowable by 5 ft.
5001 7000 ft. above sea level—Reduce max. length allowable by 10 ft.

SIDE WALL VENT TERMINATION

NOTE: When horizontal venting application is used, a Horizontal Vent Kit (#8G03022) **MUST** be ordered. The following is the instructions for installing the Horizontal Vent Kit.

1. Figure 15 shows the proper means for a horizontal side wall vent termination.

2. The optional Horizontal Vent Kit consists of the following:

Two stainless steel face plates; two galvanized face plates and two ½ inch foam gaskets. One stainless steel plate, galvanized plate and one ½" gasket is used on the flue exhaust and the other set is used for the combustion air intake.

3. The face plates are adaptable to 2 or 3 inch PVC pipe. If 3 inch PVC pipe is used, cut the 3 tabs on the radial slot and remove the center disc. If 2 inch PVC pipe is used, use the plates as they are shipped.

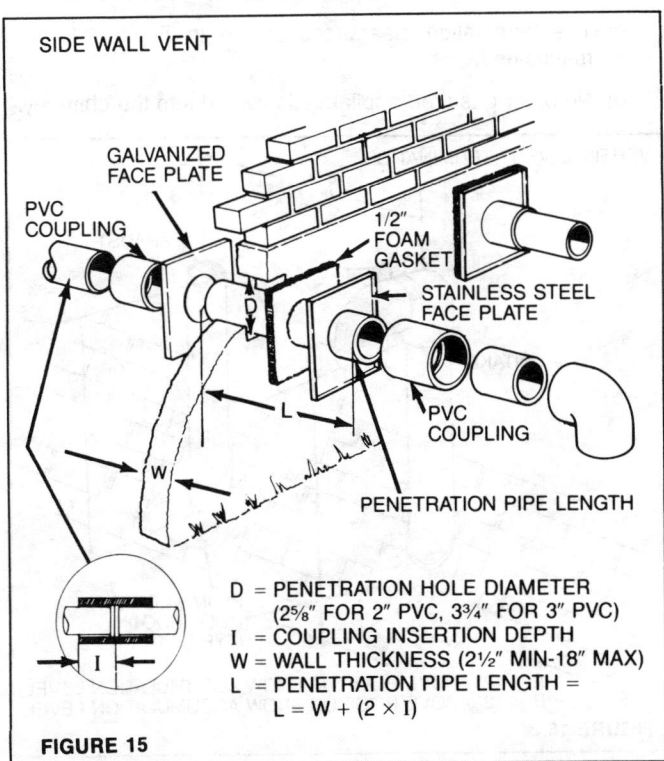

SIDE WALL VENT

GALVANIZED FACE PLATE

PVC COUPLING

1/2" FOAM GASKET

STAINLESS STEEL FACE PLATE

PVC COUPLING

PENETRATION PIPE LENGTH

D = PENETRATION HOLE DIAMETER (2⅝" FOR 2" PVC, 3¾" FOR 3" PVC)
I = COUPLING INSERTION DEPTH
W = WALL THICKNESS (2½" MIN-18" MAX)
L = PENETRATION PIPE LENGTH =
   L = W + (2 × I)

FIGURE 15

IV. VENTING (Cont.)

4. Drill or saw 2 wall penetration holes spaced as shown in Figure 15. The penetration hole diameter should be 2⅝" for 2" diameter pipe and 3¾" for 3" diameter pipe.

5. Cut wall penetration pipe to a length equal to the wall thickness plus 2 times the coupling insertion depth. Remove all shavings from inside pipe. Debur and chamfer both ends of the pipe. See Figure 15.

6. Apply PVC cement primer to one end of the pipe and the coupling elbow or socket. Apply cement liberally to **both** pipe end and coupling elbow or socket. Insert pipe into coupling or elbow with ¼ turn twist. Slip **stainless steel** face plate and gasket over end of pipe.

NOTE: Flange on face plate should be facing out.

From the outside, insert pipe assembly into the wall penetration hole. Insure that the gasket is located between the structure and the face plate. From the inside, slip the **galvanized** face plate over the pipe end and cement coupling to secure the assembly.

NOTE: The outside gasket should be slightly compressed.

Repeat this procedure for the other vent termination. Complete outside termination as shown in Figure 14 thru 16.

VERTICAL VENT TERMINATION

1. Figure 17 shows the proper installation and clearances for vertical vent termination. The vertical roof termination should be sealed with a plumbing roof boot or equivalent flashing. The inlet of the intake pipe and end of the exhaust vent must be terminated no less than 12" above the roof or snow accumulation level, and 12" away from a vertical wall or other protrusion.

2. In no case should the air intake or exhaust vent extend more than 24" above the roof penetration. The vent system can be installed in a existing chimney provided that:

   a) Both the exhaust vent and air intake run the length of the chimney.

   b) The top of the chimney is sealed and weather proofed.

   c) The termination clearances shown in Figure 16 are maintained.

   d) No other gas fired appliance is vented into the chimney.

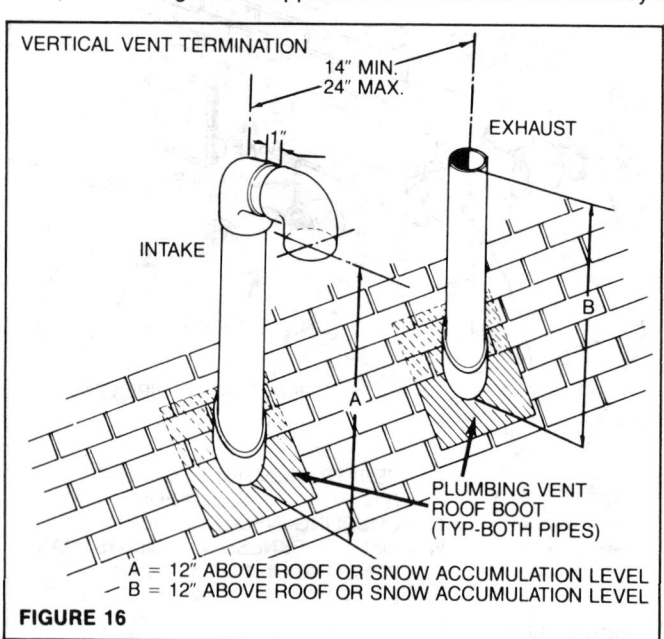

FIGURE 16

EXHAUST AND COMBUSTION AIR VENT INSTALLATION

1. Preassemble the exhaust and combustion air piping from the furnace to the vent termination. DO NOT cement any joints together until the preassembly process is complete.

   NOTE: If 2" combustion air piping is used, a 3" to 2" bell reducer fitting is required. 3" schedule 40 PVC should be run from the connection collar on the burner box to the outside of the casing before reducing to 2".

   The reducing section must be after the 90° elbow in a horizontal section. See Figure 17.

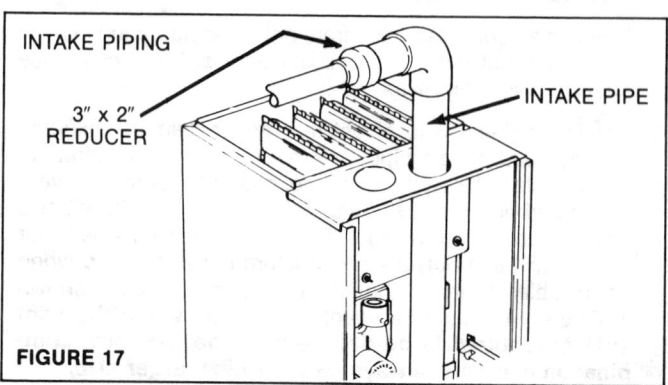

FIGURE 17

2. For serviceability the section of intake air and exhaust vent connecting the intake collar and the flue outlet of the appliance to the vent system should be secured with 2 screws and sealed air and water tight with RTV-Silicone sealant or equivalent. The subject joints are shown in Figure 18.

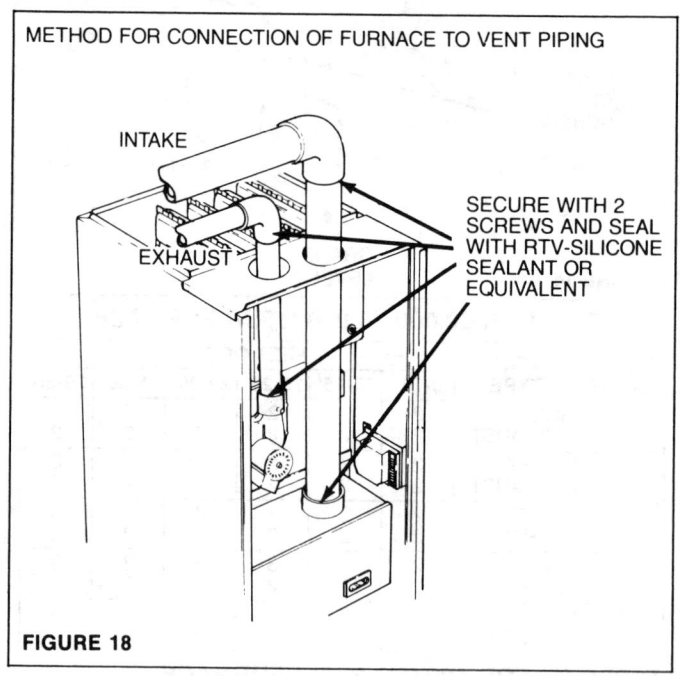

FIGURE 18

3. Debur and chamfer all pipe ends and cement joints. Remove all shavings from inside of pipe. A continuous bead of cement will be visible around the perimeter of a properly made joint.

4. Support horizontal runs of exhaust and combustion air piping every 5 feet using perforated metal hanger strips. Slope piping ¼" per lineal foot back toward the furnace. All piping should be supported to prevent sagging. See Figure 19.

IV. VENTING (Cont.)

CONDENSATE DRAIN PIPING

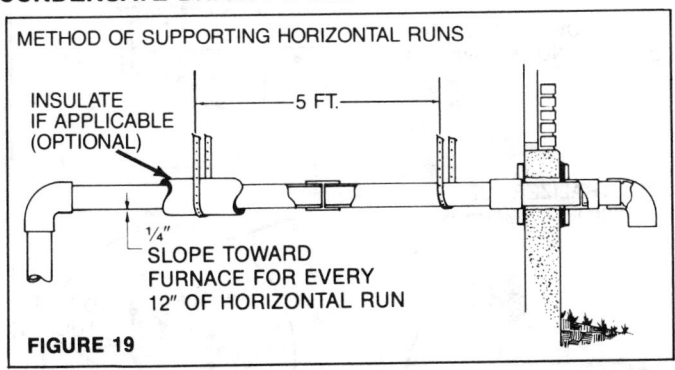

METHOD OF SUPPORTING HORIZONTAL RUNS

INSULATE IF APPLICABLE (OPTIONAL)

5 FT.

¼" SLOPE TOWARD FURNACE FOR EVERY 12" OF HORIZONTAL RUN

FIGURE 19

⚠ CAUTION:

Condensate drain tubing must be routed inside the household drain so that the condensate produced by the furnace cannot be accessed by small children or household pets.

---

1. Use ½" CPVC pipe per ASTM D2946 and CPVC cement per ASTM F493 for condensate drain piping.

2. A condensate collection trap is located in the return air section of the furnace. **No additional collection trap is required.** A female, ½" solvent joint connection is provided on the outlet of the collection trap.

3. For proper condensate drainage, the furnace must be within ½" of level.

4. A ½" compression fitting is provided for a right discharge condensate drain line. Figure 19B shows proper condensate drain installation.

5. If both left and right side return air openings are used, route condensate drain line through the back of the cabinet as shown in Figure 19A.

6. Preassemble the complete condensate drain line prior to cementing any joint. Insure that the drain line has continuous slope to the floor drain without any low spots.

7. Debur and chamfer both ends of piping and remove all shavings from inside of pipe. Cement to secure joints.

8. Condensate drain routed through unconditioned spaces must be insulated with closed cell, neoprene tubing insulation or equivalent.

9. The condensate drain must not be exposed to freezing temperatures.

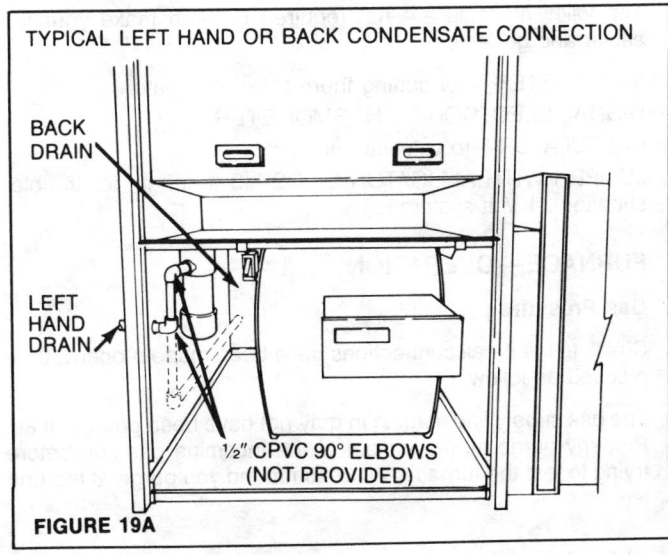

TYPICAL LEFT HAND OR BACK CONDENSATE CONNECTION

BACK DRAIN

LEFT HAND DRAIN

½" CPVC 90° ELBOWS (NOT PROVIDED)

FIGURE 19A

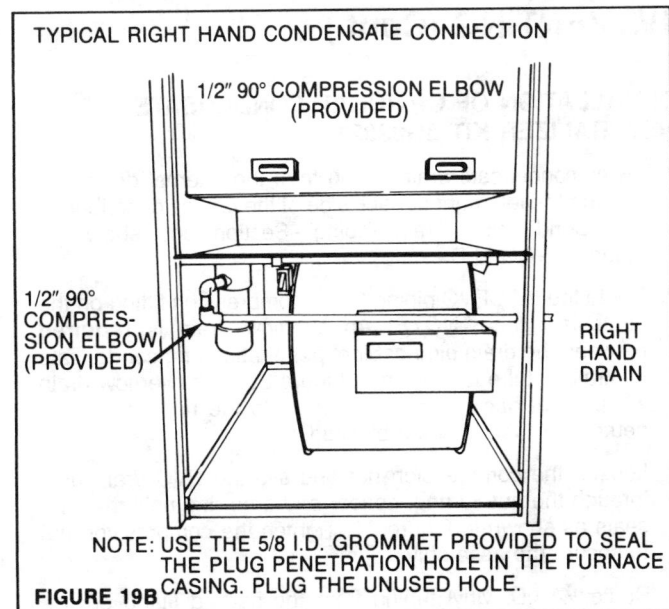

TYPICAL RIGHT HAND CONDENSATE CONNECTION

½" 90° COMPRESSION ELBOW (PROVIDED)

½" 90° COMPRESSION ELBOW (PROVIDED)

RIGHT HAND DRAIN

NOTE: USE THE 5/8 I.D. GROMMET PROVIDED TO SEAL THE PLUG PENETRATION HOLE IN THE FURNACE CASING. PLUG THE UNUSED HOLE.

FIGURE 19B

CONDENSATE SAFETY LEVEL SENSOR

This furnace is equipped with a condensate level sensor that will detect the back-up of condensate into the furnace due to a blocked condensate drain line. If condensate should back-up into the front manifold of the furnace to the level of the sensor, the sensor will ground the flame sensing circuit of the ignition module terminating unit operation. See Figure 19C. The unit will retry until lock-out of the ignition module occurs. To put the furnace back into operation, the blockage in the condensate line must be removed and the ignition module must be reset by interrupting 24 VAC power to the unit.

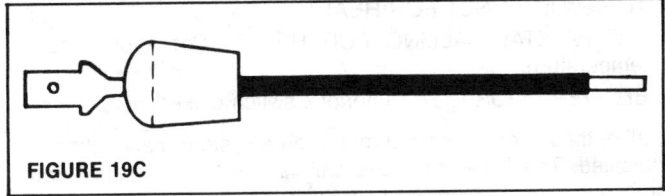

FIGURE 19C

CONDENSATE NEUTRALIZER KIT

Some locales have code requirements that prohibit the introduction of the acidic condensate produced by the furnace into the sewer system. Additionally, furnaces installed in homes that are using septic tanks should not have condensate dumped directly into the sewer system. The acidic condensate can destroy the bacteria necessary for septic tank operation.

In these and any other cases that require the acid content of the condensate to be rendered harmless, a Condensate Neutralizer Kit should be installed in the condensate drain line. The effectiveness of a questionable condensate neutralizer tube can be tested by sampling the condensate after it leaves the tube. Common acid tests such as litmus paper are available locally. See Figure 20.

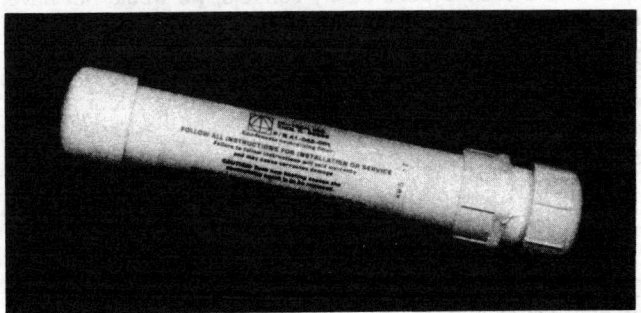

FIGURE 20

IV. VENTING (Cont.)

INSTALLATION OF OPTIONAL CONDENSATE NEUTRALIZER KIT 8G03021

1. Install condensate drain piping from the internal drain trap through the left, right or back side of the unit as described in the "Condensate Drain Piping" Section and shown in Figures 19A and B.

2. Route the ½" CPVC piping to the compression fitting on the neutralizer tube. **NOTE:** An overflow drain is **required** between the drain pipe cabinet penetration and condensate neutralizer tube as shown in Figure 21. The overflow drain will prevent back-up of condensate to the unit should the neutralizer tube become blocked.

3. Loosen the compression nut and slip the PVC drain pipe through the nut, O-ring, spacer and grab ring until the pipe seats as shown in Figure 21. Tighten the compression nut to secure the assembly.

4. Route ½" I.D. vinyl tubing from the barbed fitting on the neutralizer tube to a household plumbing drain.

Replacement Rock can be purchased from the Parts Dept. at Red Bud. Part. No. 1690999 (one lb. bags).

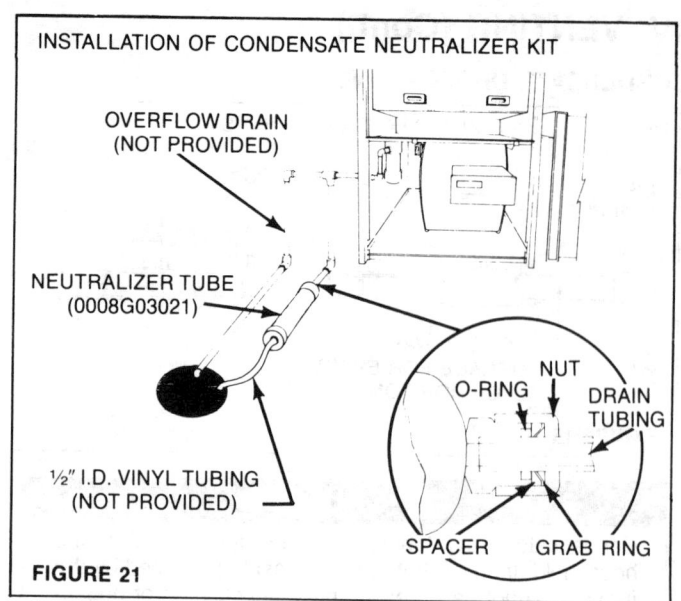

INSTALLATION OF CONDENSATE NEUTRALIZER KIT

OVERFLOW DRAIN (NOT PROVIDED)

NEUTRALIZER TUBE (0008G03021)

½" I.D. VINYL TUBING (NOT PROVIDED)

O-RING   NUT   DRAIN TUBING

SPACER   GRAB RING

FIGURE 21

V. SERVICE AND TROUBLE SHOOTING

Before troubleshooting the furnace make sure the furnace has all requirements met for normal operation. Always check the following items FIRST!

- GAS ON?—Check inlet pressure.
- POWER ON?—Check switch, fuses, circuit breakers.
- THERMOSTAT SET FOR HEAT?
- THERMOSTAT CALLING FOR HEAT?—Set above room temperature.
- BLOWER DOOR ON?—Interlock switch closed.

If all of these conditions are met, then a system malfunction is indicated. The following pages will list the tools required and procedures recommended to service the furnace. A troubleshooting guide has been developed for your convenience. The guide for the furnace follows the TOOL LIST. A separate guide for the Hot Surface Ignition System is found on page 28.

NOTE: If 24 Volts is available to terminals "TH" and "TR" of the ignition module but the furnace will not light, the problem is mostly likely in the ignition system!

CAUTION

Work Safely! Turn off power and gas before servicing furnace. Remember to allow time for blowers to coast to a stop after power is removed. Likewise, allow hot surfaces time to cool before you touch them. **DO NOT** "temporarily fix" or "rig" the furnace to operate when components are needed. Any jumpers used in troubleshooting must be removed before you leave the job site. **DO NOT** bypass safety controls; we would not have installed them in manufacturing if they were not important. **UNDER NO CIRCUMSTANCES SHOULD YOU ATTEMPT TO FIRE THE FURNACE WITH THE COMBUSTION DOOR OFF!** Viewport windows are installed in the combustion door so that you may safely view the ignition and firing of the furnace.

SERVICE PROCEDURES—INSTRUMENTS REQUIRED

MULTIMETER—capable of reading resistances of 0-100 OHMS, Voltages of 0-125 volts A.C.
CLAMP ON AMMETER—Capable of reading amperages of 0-60 Amps

MANOMETER, PRESSURE GAUGE OR HYDRO-GAUGE—capable of reading gas pressures of 0-12" W.C.
SLOPE GAUGE
DIAL TYPE POCKET THERMOMETER

TOOLS REQUIRED
WRENCHES: ¼", ½", and ⁹⁄₁₆" combination box-open end.
PIPE WRENCH—8"-12"
ADJUSTABLE WRENCH—6"-12"
NUT DRIVER—¼", ⁵⁄₁₆", and ⅜"
SCREWDRIVERS—#2 Flat and #2 Phillips
ASSORTED ALLEN WRENCHES

Most of the required tools will be found in any technicians tool kit. However, if you are to service the furnace, we strongly recommend you acquire **ALL** these tools before starting. Remember, doing a good job with the correct tools and instruments assures rapid and accurate troubleshooting, effective service and most importantly, customer satisfaction.

OPTIONAL INSTRUMENTS AND TOOLS

The following tools are not required but can make your job easier and go faster.

MILLIAMETER—for setting thermostat anticipator
DIGITAL ELECTRONIC THERMOMETER
CALCULATOR—to calculate firing rate
JUMPER WITH ALLIGATOR CLIPS—3 ft. length for troubleshooting 24 Volt system.

FURNACE—OPERATION

Gas Pressure

Check to insure all connections have been made properly, then proceed as follows:

The gas supply line coming in may not have been purged of air. Properly purge all the air out of the incoming gas line before trying to test the furnace. We recommend you purge at the drip leg.

V. SERVICE AND TROUBLESHOOTING (Cont.)

GAS VALVE—Inlet Pressure

The maximum inlet pressure on the gas valve is 0.5 PSIG (14.0 Inches of water column) for natural and LP gases. To allow inlet pressure to exceed this value may cause the regulator, solenoid and valve to fail in an unsafe mode. This will damage the gas valve and require replacement. The inlet pressure to the gas valve should be 4" to 7" for Nat. gas, and 13.0" for LP gas.

GAS VALVE—Outlet Pressure

The pressure on the outlet side of the gas valve should read 3.5" Water Column for Natural and 10" Water Column for LP gas. Minor changes in manifold pressure may be made by regulator adjustments. Refer to Figure 6 for adjustments.

Figure 22 shows Hydro Gauge Manometer being used to check outlet pressure.

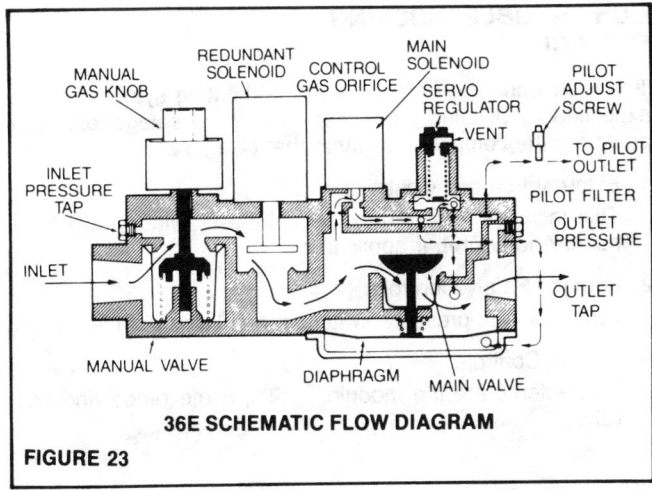

36E SCHEMATIC FLOW DIAGRAM

FIGURE 23

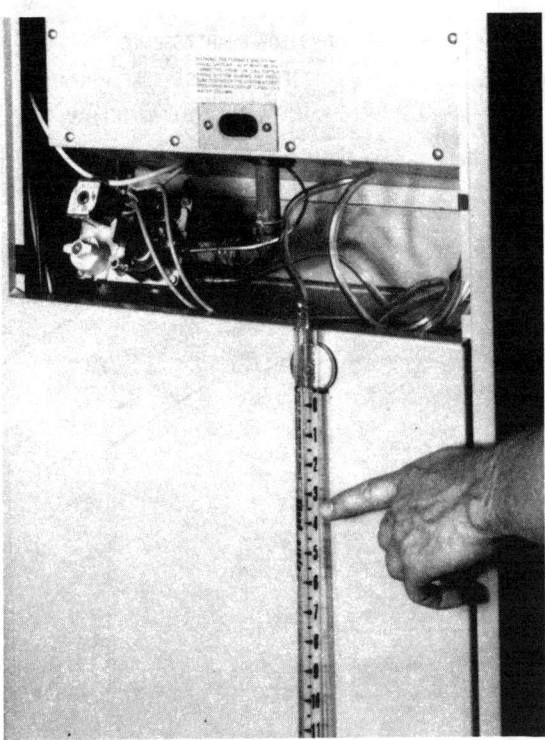

FIGURE 22

BURNER INSPECTION DOOR

Figure 24 shows the inspection door for the burner compartment. If for any reason this door must be removed, care must be taken to prevent the gasket from being torn or destroyed. If the gasket is torn or frayed, it should be replaced.

An inspection door that does not have a good air seal, will cause improper combustion air control and possible furnace failure by lock-out or flame instability.

DO NOT ATTEMPT TO FIRE THE FURNACE WITH THIS DOOR REMOVED!

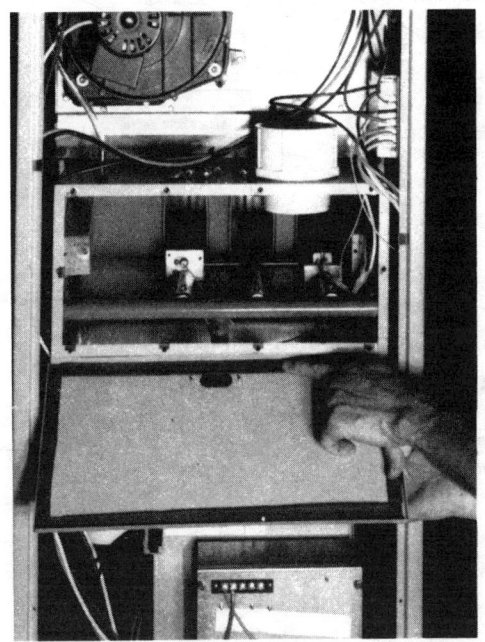

FIGURE 24

GAS VALVE OPERATION—Hot Surface Ignition—Step Opening

NOTE: Refer to Supplement for other valves.

Upon a call for heat, the valve is energized through the closure of the thermostat contacts, safety switches and ignition system module circuitry. Both the main gas valve and redundant gas valve operators are energized simultaneously, opening both valves and permitting gas flow to the burners. (First at low pressure then stepping up to full input.) Upon interruption of power to the gas valve operators, both valves will close immediately. See Figure 23.

V. SERVICE AND TROUBLESHOOTING (Cont.)

GUH TROUBLESHOOTING FLOW-CHART

Troubleshooting the furnace at a total heating system can be expedited by dividing the system into three categories, then analyzing the components within that category.

1. External Furnace Controls;

   This includes the thermostat, sub-base, wiring, and heat pump Add-On Kit, if applicable.

2. Furnace Safety Controls.

   Items such as pressure switches, and limit controls.

3. Ignition Controls.

   These items are the module, ignitor, flame probe and gas valve.

Briefly, if the unit does not operate at all, start with Category 1. If the combustion blower runs, but no heat is being produced, start with Category 2. If 24 Volts is available to terminals "TH" and "TR" of the ignition module, then start with Category 3.

The following flow chart will take you through all three categories of components.

NOTE: THIS FLOW CHART ASSUMES THAT ALL WIRING IS CORRECT AND TERMINATIONS SECURE, VENT PIPE IS OF CORRECT LENGTH AND DIAMETER, AND CONDENSATE DRAIN IS OPEN.

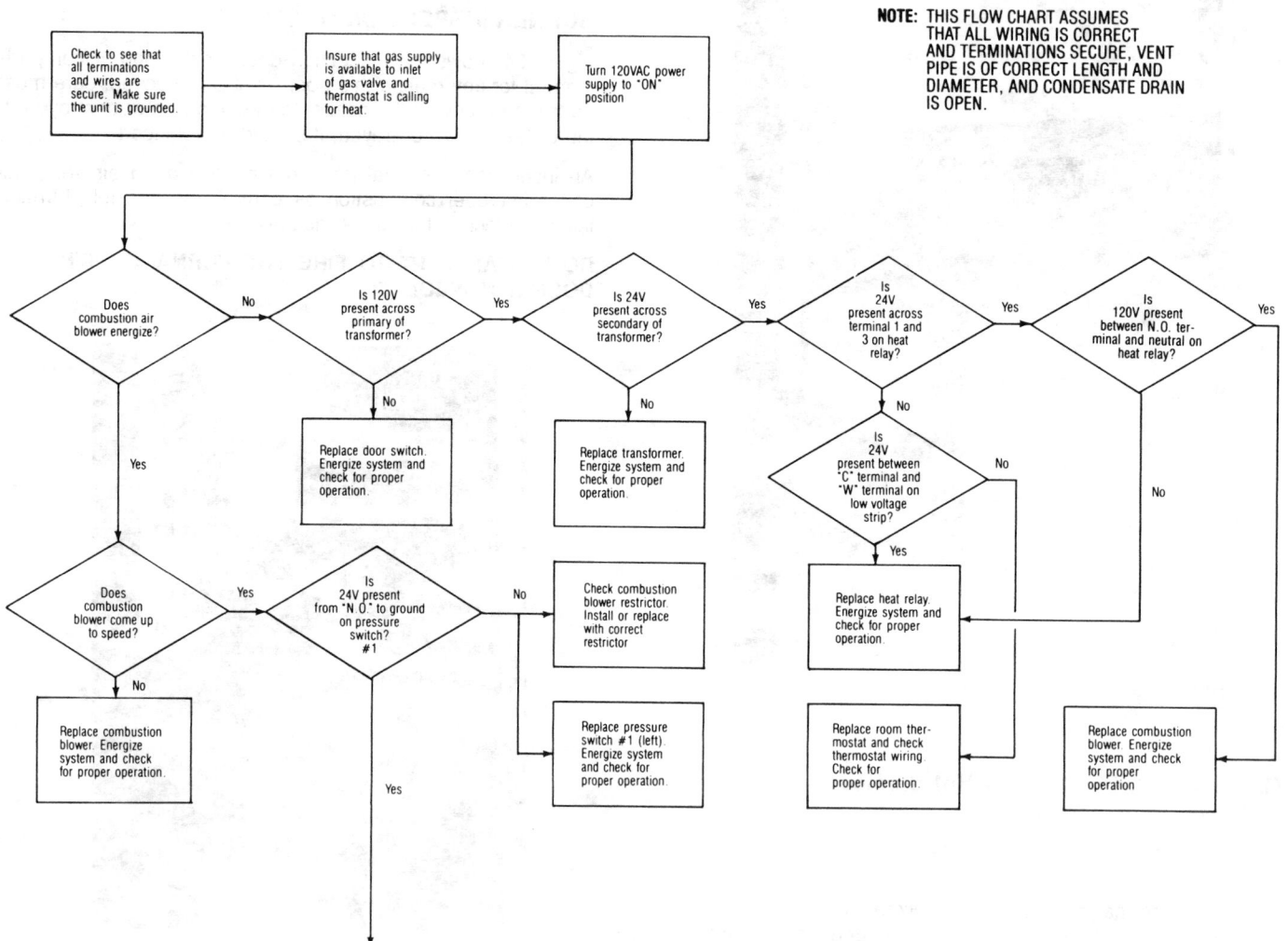

Continued on next page

V. SERVICE AND TROUBLESHOOTING (Cont.)

Check high limit control. Continuity should exist across limit terminals.

Does pressure switch #2 have 24V from N.O. to ground?

No →

Pressure switch #2 open. Replace and energize system and check for proper operation.

Yes ↓

FIRST VISUAL CHECK

Does silicon ignitor warm up after prepurge duration?

No → Is 24 volts present across module terminals TH & TR?

No → Repeat previous troubleshooting steps carefully. Energize system and check for proper operation

Yes ↓

Refer to Hot Surface Ignition Trouble-Shooting Chart Pg 28. First Visual Check

Yes ↓

SECOND VISUAL CHECK

Does main burner ignite?

No → Is manual knob on gas valve in "ON" position?

Yes → Refer to Hot Surface Ignition Troubleshooting Chart Page 29. Second Visual Check

No ↓

Turn manual knob on gas valve to "ON" position. Energize system and check for proper operation.

Yes ↓

THIRD VISUAL CHECK

Does main burner remain lit after safety time?

No → Refer to Hot Surface Ignition Trouble-Shooting Chart, Pg 29. Third Visual Check

Yes ↓

Does F & L dial indicator rotate as unit heats up?

No → Replace F & L control. Energize system and check for proper operation.

Yes ↓

Check supply air blower capacitor. Replace if shorted or open

With blower plug connected, does supply air blower start as indicated on F & L?

No → Is 120V present from yellow wire terminal to ground on F & L control?

Yes → Is 120V present across "N.C." terminal and ground on blower relay?

Yes → Replace supply air blower. Energize system and check for proper operation.

No ↓ (from yellow wire)

Replace F & L control energize system and check for proper operation.

No ↓ (from N.C. terminal)

Is 120V present across "C" terminal and neutral on blower relay?

No → Check wiring. Energize system and check for proper operation.

Yes ↓

Replace blower relay. Energize system and check for proper operation

Yes ↓

Does supply air blower shut off as indicated on F & L dial on cool down?

No → Is 120V present across "ON" terminal on F & L control?

Yes → Replace F & L control. Energize system and check for proper operation

Yes ↓

Complete

21

HSI MODULE WITH LED LIGHT

FLOW CHART

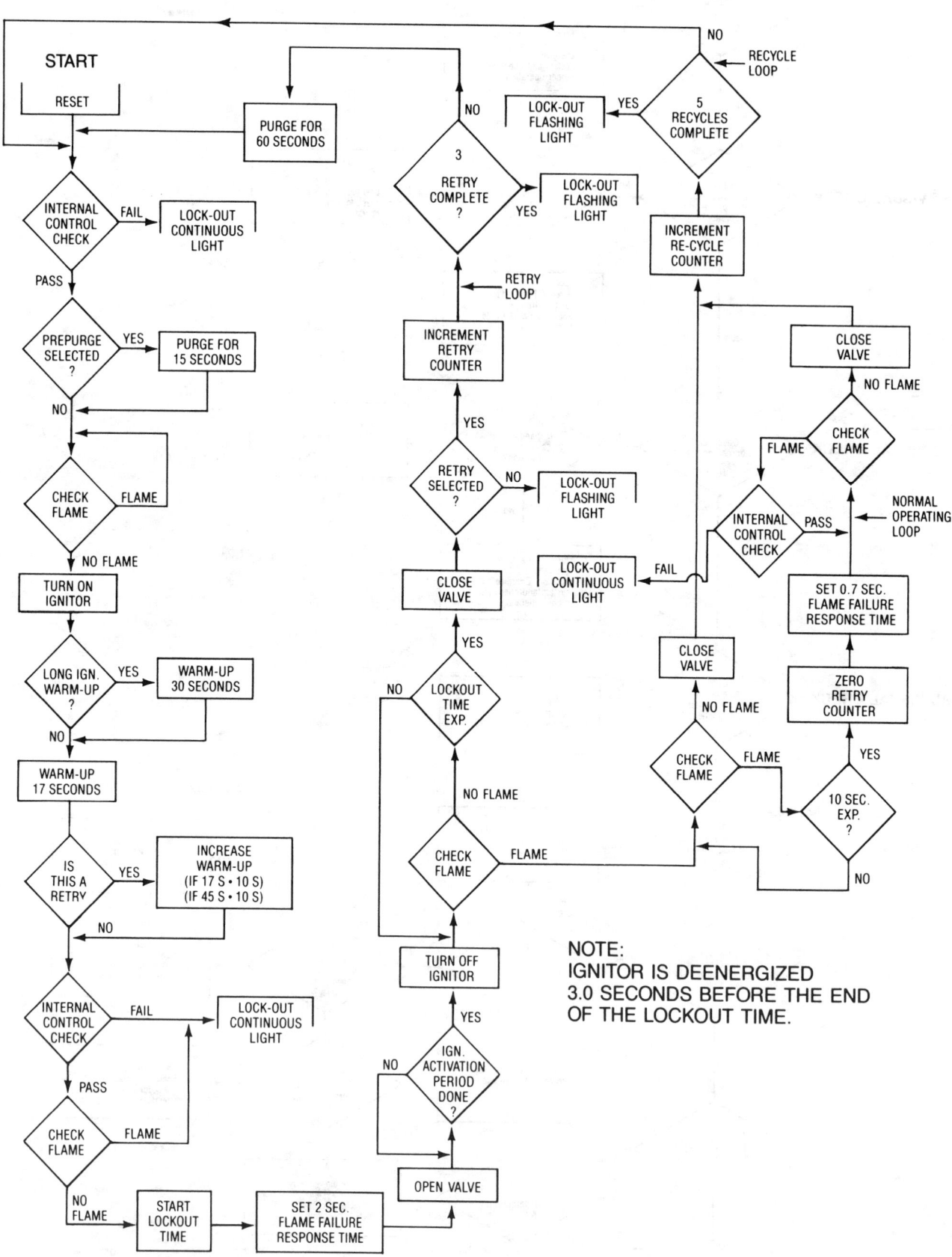

NOTE:
IGNITOR IS DEENERGIZED
3.0 SECONDS BEFORE THE END
OF THE LOCKOUT TIME.

V. SERVICE AND TROUBLESHOOTING (Cont.)

BURNER COMPARTMENT—Flame Instability

The main cause of flame instability is lack of combustion air. The condensing furnace combustion air is circulated by the induced draft blower. See Figure 25.

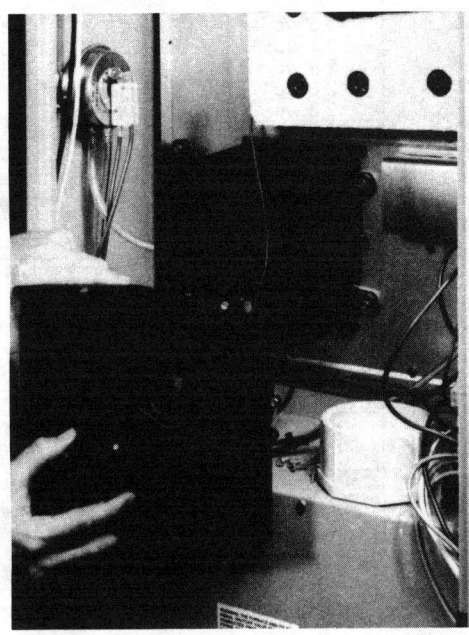

FIGURE 25

Each basic furnace model has a different size air restrictor used in conjunction with the induced draft blower.

DO NOT INTER-CHANGE OR REMOVE COMBUSTION AIR RESTRICTORS.

MODEL	RESTRICTOR COLOR	PART NUMBER
060	RED	2972-010
075	BLACK	2972-011
100	NATURAL	2972-012
125	NONE USED	—

CAUSES OF FLAME INSTABILITY:

1. Open burner compartment panel or inadequate seal.
2. Blockage of heat exchanger.
3. Heat exchanger leaks.
4. Excessive or insufficient manifold gas pressure.

NOTE: Pressure at gas valve outlet should not exceed 3.5 inches of water column.

5. Primary air shutter incorrectly adjusted.
6. Burner orifices incorrectly sized, cross threaded or mismatched resulting in mis-fired condition within the furnace.
7. Incorrectly sized or excessive length and restriction within the venting system.

HIGH ALTITUDE

Ratings are approved for altitudes up to 2000 Feet for all gases. Ratings for altitudes over 2000 feet are decreased by 4% for each 1000 feet above sea level. See Table 1. Check and note the orifice size required at that altitude. If the orifices shipped with the furnace are not the recommended size, remove them and install the proper size.

To use the natural gas table, first consult your local gas utility for the heating value of the gas supply. Select the heating value on the vertical border and follow across the table until the appropriate elevation for the installation is reached. The first value in the box at the intersection of the heating value and elevation will be the manifold pressure required. If a gas orifice change is also required, the box is shaded. The required orifice size is shown in the table. **NOTE:** The manifold pressure may still need to be reset after a gas orifice change. The high altitude label supplied with the unit and also in the required orifice kit must be completed and affixed to the front of the burner enclosure box. Fill in the manifold pressure, orifice size and revised input rate. The revised input rate is determined in the following manner:

$$\text{High Altitude Input Rate} = \begin{array}{c} \text{Sea Level} \\ \text{Nameplate} \\ \text{Input Rate} \end{array} \times \text{(Multiplier)}$$

Elevation	High Altitude Multiplier
2000' - 2999'	0.92
3000' - 3999'	0.88
4000' - 4999'	0.84
5000' - 5999'	0.80
6000' - 6999'	0.76
7000' - 7999'	0.72

Example:

For a GUH075A installed at an altitude of 5280' with a gas supply heating value of 950 BTU/Ft3, a manifold pressure of 2.5" WC is required. The revised altitude input is:

High Altitude Input Rate = 75,000 × 0.80 = 60,000 BTU/HR

NOTE: A gas orifice change was not required in this example.

V. SERVICE AND TROUBLESHOOTING (Cont.)

TABLE 1

MANIFOLD PRESSURE AND ORIFICE SIZE FOR HIGH ALTITUDE APPLICATIONS

NATURAL GAS

HEATING VALUE BTU/CU FT		MEAN ELEVATION FEET ABOVE SEA LEVEL					
		2000 to 3000	3000 to 4000	4000 to 5000	5000 to 6000	6000 to 7000	7000 to 8000
800		3.5"wc	3.5"wc	3.5"wc	3.5"wc	3.2"wc	2.8"wc
850		3.5"wc	3.5"wc	3.5"wc	3.1"wc	2.8"wc	2.5"wc
900		3.5"wc	3.3"wc	3.0"wc	2.8"wc	2.5"wc	2.3"wc
950		3.3"wc	3.0"wc	2.7"wc	2.5"wc	2.3"wc	3.5"wc
1000		2.9"wc	2.7"wc	2.5"wc	2.3"wc	3.5"wc	3.1"wc
1050		2.7"wc	2.4"wc	2.3"wc	3.5"wc	3.2"wc	2.8"wc
1100		2.4"wc	2.3"wc	3.5"wc	3.2"wc	2.8"wc	2.5"wc
GUH060	ORIFICE SIZE	STANDARD FACTORY ORIFICE		#48	#48	#48	#48
GUH075				#44	#44	#44	#44
GUH100				#44	#44	#44	#44
GUH125				#44	#44	#44	#44

▨ SHADED AREA REQUIRES ORIFICE CHANGE
NO SHADING INDICATES MANIFOLD PRESSURE CHANGE ONLY.

PROPANE

HEATING VALUE BTU/CU FT		MEAN ELEVATION FEET ABOVE SEA LEVEL					
		2000 to 3000	3000 to 4000	4000 to 5000	5000 to 6000	6000 to 7000	7000 to 8000
2500		8.5"wc	7.8"wc	7.0"wc	10.0"wc	9.0"wc	8.1"wc
GUH060	ORIFICE SIZE	STANDARD FACTORY ORIFICE			3/64	3/64	3/64
GUH075					#55	#55	#55
GUH100					#55	#55	#55
GUH125					#55	#55	#55

NOTE: NATURAL GAS DATA BASED ON 0.60 SPECIFIC GRAVITY.
PROPANE DATA BASED ON 1.53 SPECIFIC GRAVITY.
FOR FUELS WITH DIFFERENT SPECIFIC GRAVITY CONSULT THE NATIONAL FUEL GAS CODE ANSI Z223.1a-1987.
IMPORTANT: FILL OUT HIGH ALTITUDE LABEL SUPPLIED WITH UNIT OR HIGH ALTITUDE KIT AND AFFIX TO COVER OF BURNER BOX.

FURNACE—Electrical

A thorough understanding of the sequence of the electrical operation for this furnace is a must before you can service the electrical components. See Sequence of Operation for details.

Check all electrical connections to insure they are tight before proceeding:

INDUCED DRAFT MOTOR AND RELAY

See enclosed Wiring Diagram.

When the thermostat calls for "HEAT" the relay is energized by 24 VAC. When the relay coil is energized, the normally open (N.O.) contacts will close between the Terminals #4 and #2 on Relay. This allows 120 VAC to feed to the Induced Draft Blower Motor and L1 Terminal of the ignition module. See Figure 30.

When the combustion blower motor reaches full speed, the normally open (N.O.) Pressure Switch connected to the blower will close allowing 24V power to feed through normally closed (N.C.) contact on the Limit Switch.

The following is a check list of steps to follow if the Induced Draft Motor does not function:

NOTE: See individual items for further trouble shooting.

1. Check power to furnace—120 VAC
2. Is door switch in blower compartment closed?
3. Is transformer secondary side 24 VAC output?—Check Terminals "R" to "C".

4. Is thermostat calling for "Heat"?—Check Terminals "R" to "W".
5. Is Heating Relay coil energized?—Check Terminals "1" and "3" on Heating relay. See Figure 26.

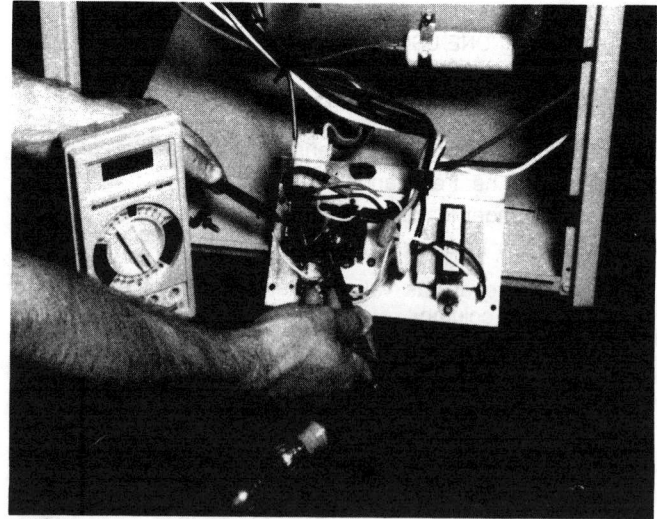

FIGURE 26

6. Are Heating Relay contacts closed?—Check Terminals "2" and "4" on Heating Relay.
7. Is Induced Draft Motor operational?

V. SERVICE AND TROUBLESHOOTING (Cont.)

HIGH LIMIT

The High Limit is part of the combination Fan and Limit Control. Both the fan circuit and the limit circuit are activated by the action of a calibrated bi-metal coil which is sensitive to heat. The fan/limit bi-metal probe containing this coil is inserted into the heat exchanger compartment. As the flame heats the heat exchanger, the bi-metal reacts to the heat and rotates a shaft connected to a switch mechanism. The first switch is the normally open fan contacts which upon closing energize the Indoor blower motor. When the Blower is energized the airflow over the heat exchanger removes sufficient heat to stabilize the movement of the bi-metal actuated switch and maintain a fixed position of the switch in its travel. Should the blower fail or airflow become blocked or reduced to undesirable levels, the bi-metal coil will continue to rotate until the "Limit" contacts reach the set point and break the circuit to the ignition control, thereby closing the gas valve and extinguishing the main burner flame. Refer to Specifications on page 37 for correct Limit Set Point.

FAN CONTROL ADJUSTMENT

In the event that factory set points require adjustment for the fan cycle (during the heating cycle), the following procedure is recommended:

Locate a standard duct thermometer in the supply duct. SEE FIGURE 27. With the furnace in a normal heating mode (burner and blower on) wait until the thermometer stabilizes. Then turn the thermostat "OFF" so the gas valve closes. The indoor blower should continue to operate until the supply air drops to around 85° to 95°. If the blower turns off at a higher temperature, lower the "fan off" setting approximately 5° to 10° and repeat the heat cycle again. If the fan runs too long (blowing cold air), raise the "fan off" setting 5° to 10° and recycle again.

The "fan on" temperature is normally set above 35° to 40° above the "fan off" setting.

FAN AND LIMIT CONTROL

The Fan and Limit Controls are mounted in their proper location by the manufacturer. This setting is the point at which the limit switch opens contacts and fan contacts close. The control is automatic reset and will actuate when the temperature of the element drops.

The setting should not exceed 130° "on" and 90° "off".

⚠ CAUTION:
DO NOT TAMPER WITH THE INTERNAL FAN OR LIMIT CONTROL MECHANISM. TO DO SO CAN CAUSE INJURY AND WILL SHORTEN FURNACE LIFE.

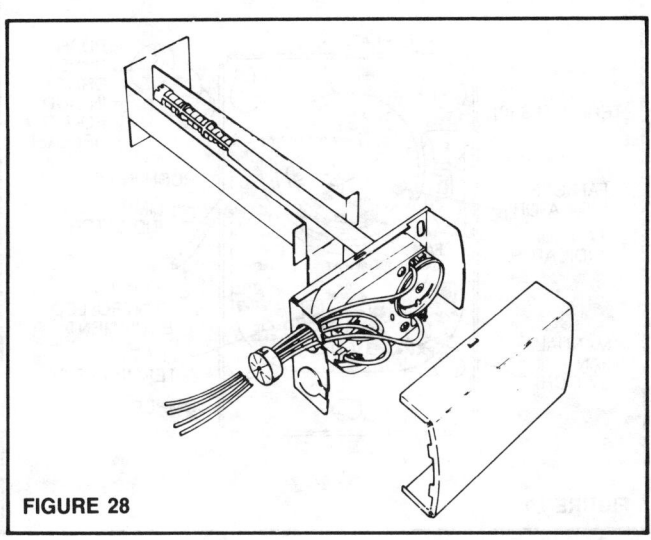

FIGURE 28

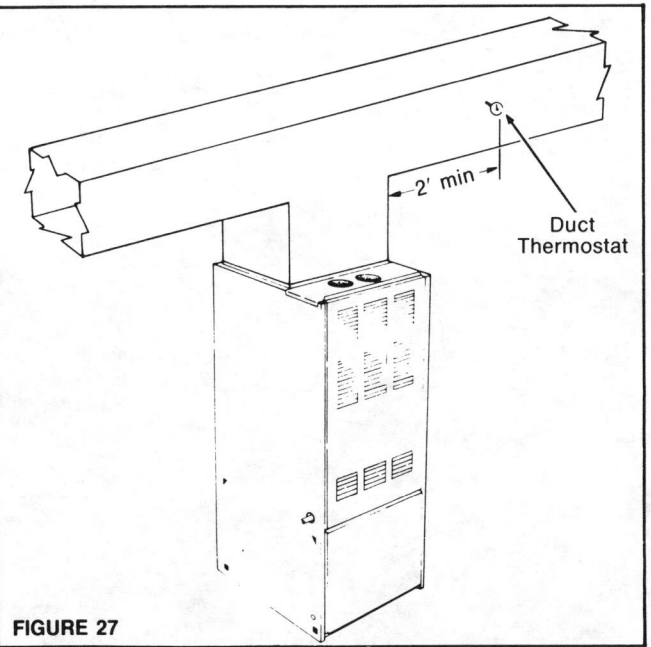

FIGURE 27

25

V. SERVICE AND TROUBLESHOOTING (Cont.)

After the burners ignite, the heat exchanger begins to heat up. The bi-metal strip on the control starts heating up. The fan will begin operating at a temperature set point pre-determined at the factory. You can change this set point (on or off) by doing the following:

1. Hold the scale plate dial to keep it from turning and straining the sensing element.

2. Move the setting levers to the control point recommended by the furnace or burner manufacturer. Use gentle finger pressure.

3. The readings at the pointed ends of the setting levers indicate the approximate control points.

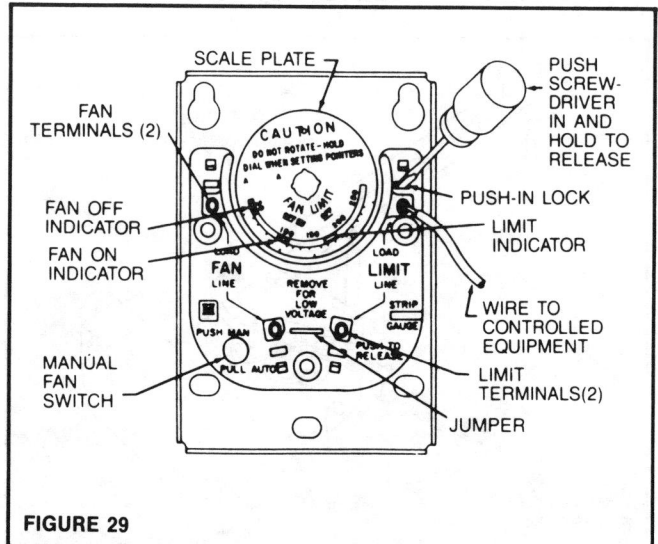

FIGURE 29

SAFETY LIMIT CHECK

The recommended method for checking the Limit Control is to block off the return air after furnace blower has been operating for 5 minutes. Blocking off return air will make the furnace over heat and cut out on high limit control. The temperature of cut-out should be no more than manufacturer recommendations. See specifications Page 39. As soon as the limit control has terminated unit operation, (the gas valve and ignitor should shut down), the return air opening should be unblocked to permit air circulation. When you use this method, you have made sure that the furnace limit safety operation is functioning properly in case of blower failure, blocked filters or excessive outlet restriction.

COMBUSTION DRAFT PROVING PRESSURE SWITCH

A pressure switch is located directly above the combustion blower on the left side of the furnace. The Pressure Switch is open and will not allow the ignition module to receive control voltage until the switch closes. Two conditions must be met to "prove" or close the switch.

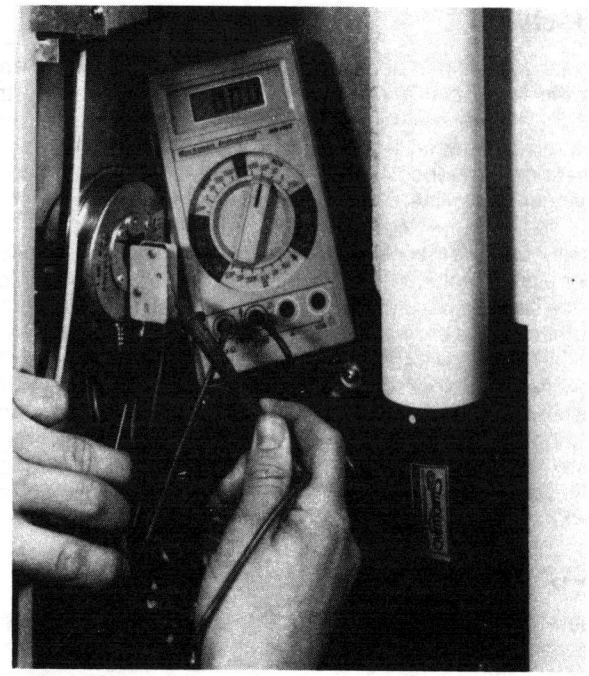

FIGURE 30

These are:

1. The combustion blower MUST be up to speed.

2. Air must be moving through the combustion chamber in quantities sufficient to safely fire the furnace. Figure 30 shows the voltage method of checking the pressure switch(es).

V. SERVICE AND TROUBLESHOOTING (Cont.)

BLOWER MOTOR AND RELAY—
Blower Speed Adjustment

The multi-speed direct drive motor is wired for high speed for cooling operation and low speed for heating operation. A change to a lower speed may be made in the control center.

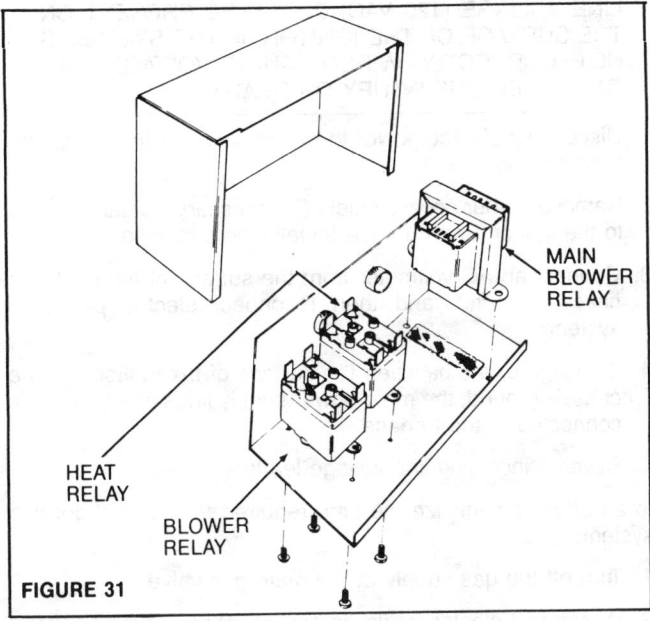

HEAT RELAY

BLOWER RELAY

MAIN BLOWER RELAY

FIGURE 31

This furnace is equipped with a blower relay, changing speeds automatically. Instructions on checking and changing airflow can be found in the Installation Manual. See Figure 31 for relay location.

In the event the motor does not run, the following steps can be used to check for malfunction:

⚠ CAUTION:

DISCONNECT ELECTRICAL POWER BEFORE PRO-CEEDING: DISCONNECT MOTOR LEADS FROM THE RELAY AND FROM THE CAPACITOR TERMINALS.

---

1. Grounded—Using a V.O.M., set an ohmmeter scale for highest resistance reading zero to meter. Check continuity between each motor lead and a bare metal section on motor frame. A grounded motor will show continuity to ground.

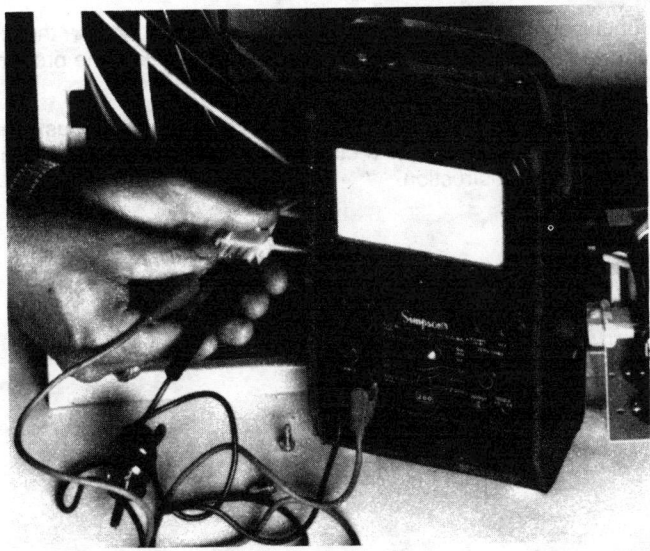

FIGURE 32

2. Open Circuit—Check for continuity between common wire from motor and each speed lead. An open circuit will measure infinity.

3. Shorts between windings—The same procedure can be used for shorts as was used to test for open circuit. But, a shorted winding will indicate zero Ohms. In most cases you will either see or smell the shorted windings.

4. RUN CAPACITOR—Place Ohmmeter on highest scale and zero adjust. Touch meter leads to the two terminals on capacitor. Meter should move toward zero Ohms and return slowly to infinity. Reverse leads and repeat. If the needle does not move the capacitor is open; it must be replaced. If the needle goes to zero and does not return to infinity, the capacitor is shorted and must be replaced.

To check for a grounded capacitor (metal cased capacitors only)—Place Ohmmeter on highest scale and touch one lead to bare metal of capacitor and the other lead to each terminal. If meter deflects from infinity, replace capacitor.

SEQUENCE OF OPERATION FOR:
HOT SURFACE IGNITION SYSTEM

The following sequence of operation will occur when using an H.S.I. system on a gas furnace.

All of the previously mentioned controls must be closed to start the ignition sequence. The ignition control must have 115 Volts available to the L1 terminal to operate the ignitor.

The pre-purge mode in the ignition control will delay thirty seconds before applying power to the silicone carbide ignitor. The ignitor is allowed to warm-up for approximately 15 seconds. At the end of the warm-up time, the valve is energized.

The main burner flame must be detected within 7 seconds or the valve is de-energized and the ignitor turned off.

If there is a failure to sense flame on the first attempt, there will be a 90 second delay. The second attempt at lighting the burner flame will duplicate the first attempt with pre-purge, warm-up times and lock-out times identical to those in the first try. Failure to prove on the second attempt will allow yet another 90 second delay and one more complete ignition attempt before complete system lock-out.

Lock-out of the 50E47 control after the final ignition attempt, will require interruption of the power to the ignition control (either the 24V thermostat or line voltage to the furnace may be interrupted) for 1/20 second or longer to reset the system.

All 50E47 Controls will repeat the ignition sequence for a total of three recycles if flame is lost within the first thirty seconds of establishment. The total of ignition attempts plus the number of recycles cannot exceed three.

If flame is established for more than thirty seconds after ignition, the 50E47 will clear the recycle counter. If flame is lost after 30 seconds, there will be a 60 second wait and restart of the ignition sequence.

A momentary loss of power or flame to the sensing circuit will cause the valve to be de-energized and a delay of 60 seconds before restarting the ignition sequence. Recycles will begin and the burner will operate normally if the gas supply returns or the fault condition is corrected prior to the last ignition attempt. Otherwise the control will lock-out.

With the ignitor warmed-up, gas flow to the burners will allow ignition. The burners will stay lit as long as the thermostat calls for heat.

The ignitor is switched off at the end of the flame detection period (7 sec.).

V. SERVICE AND TROUBLESHOOTING (Cont.)

The heat exchanger heats up the combination fan and limit control. When the temperature reaches the "on" set point, the blower will be energized. There is also a safety limit combined in this control. If for some reason the blower does not start, the temperature will increase until the limit set point is reached. At this time, the 24 volt secondary of the transformer is broken and the gas valve will close.

When the thermostat is satisfied, the gas valve will close, shutting down the burners. At this time, the blower will continue to operate until the temperature cools down below the "off" set point of the fan control. When this point is reached, the blower will shut off.

Now the furnace is ready to repeat the same cycle when the thermostat calls for heat.

FAILURE TO IGNITE

If 24 Volts are available to Terminals "TH" and "TR" of the ignition module, refer to the HOT SURFACE TROUBLE-SHOOTING CHART on the following page.

> ### ⚠ CAUTION:
>
> The Ignition Control Module will be damaged if either side of the gas valve coil is connected to ground or the gas valve leads are shorted together.

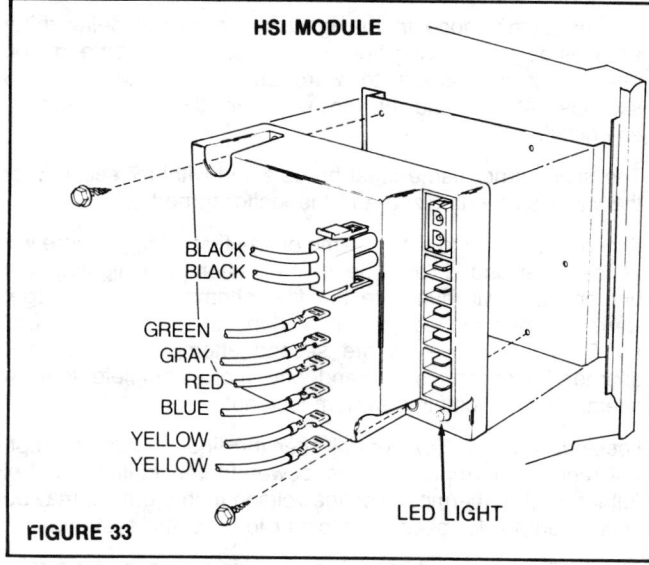

HSI MODULE

BLACK
BLACK
GREEN
GRAY
RED
BLUE
YELLOW
YELLOW

LED LIGHT

FIGURE 33

BLACK: (MODULAR PLUG)
    LINE VOLTAGE FOR IGNITOR

GREEN: CHASSIS GROUND

GREY: FLAME SENSOR

RED: 24 VOLT FROM SAFETY CIRCUITS

BLUE: TRANSFORMER COMMON (24V)

YELLOW: GAS VALVE LEADS

ORANGE: CONDENSATE LEVEL
    SAFETY SENSOR

THE WHITE-RODGERS HOT SURFACE IGNITION SYSTEM

> ### DANGER
>
> **DO NOT OMIT THIS STEP!**
> LINE VOLTAGE (120 VAC) COULD BE PRESENT ON THE SURFACE OF THE IGNITOR, IF THE SYSTEM IS NOT CORRECTLY WIRED. SUCH VOLTAGE CAN CAUSE SERIOUS INJURY OR DEATH.

1. Disconnect electric power to system at main fuse or circuit breaker.

2. Remove burner access shield (if necessary) to gain access to the ignitor. See Furnace Installation Instructions.

3. Connect an AC voltmeter from the surface of the ignitor to chassis ground, and then reconnect electric power to system.

4. If voltage exists between the surface of the ignitor and the chassis ground, the main power supply lines are improperly connected to the furnace.

   Reverse incoming line voltage leads.

In an effort to minimize the time required to troubleshoot this system:

1. Turn off the gas supply at the main gas valve.

2. Disconnect electric power to system at main fuse or circuit breaker if connected.

3. Visually inspect equipment for apparent damage. Check wiring for loose connections.

4. With burner access shield removed (if necessary), inspect ignitor for visible cracking or scale deposits and flame sensor for position or deposits shorting sensor to burner.

After performing the above inspections, restore gas supply and electric power to equipment. Close thermostat contacts to cycle the system. If a "no heat" condition persists, the three visual indicators (below) will help determine if the system is operating properly.

1. The ignitor will warm up and glow red.

2. The main burner flame will ignite.

3. The main burner flame will continue to burn after the ignitor is turned off.

Troubleshooting the system consists of checking for these three visual indications. The attached charts define the proper action if any of the indications do not occur.

A diamond on the chart encloses a question that must be answered yes or no. A rectangle on the chart encloses a statement or instruction.

FIRST VISUAL CHECK

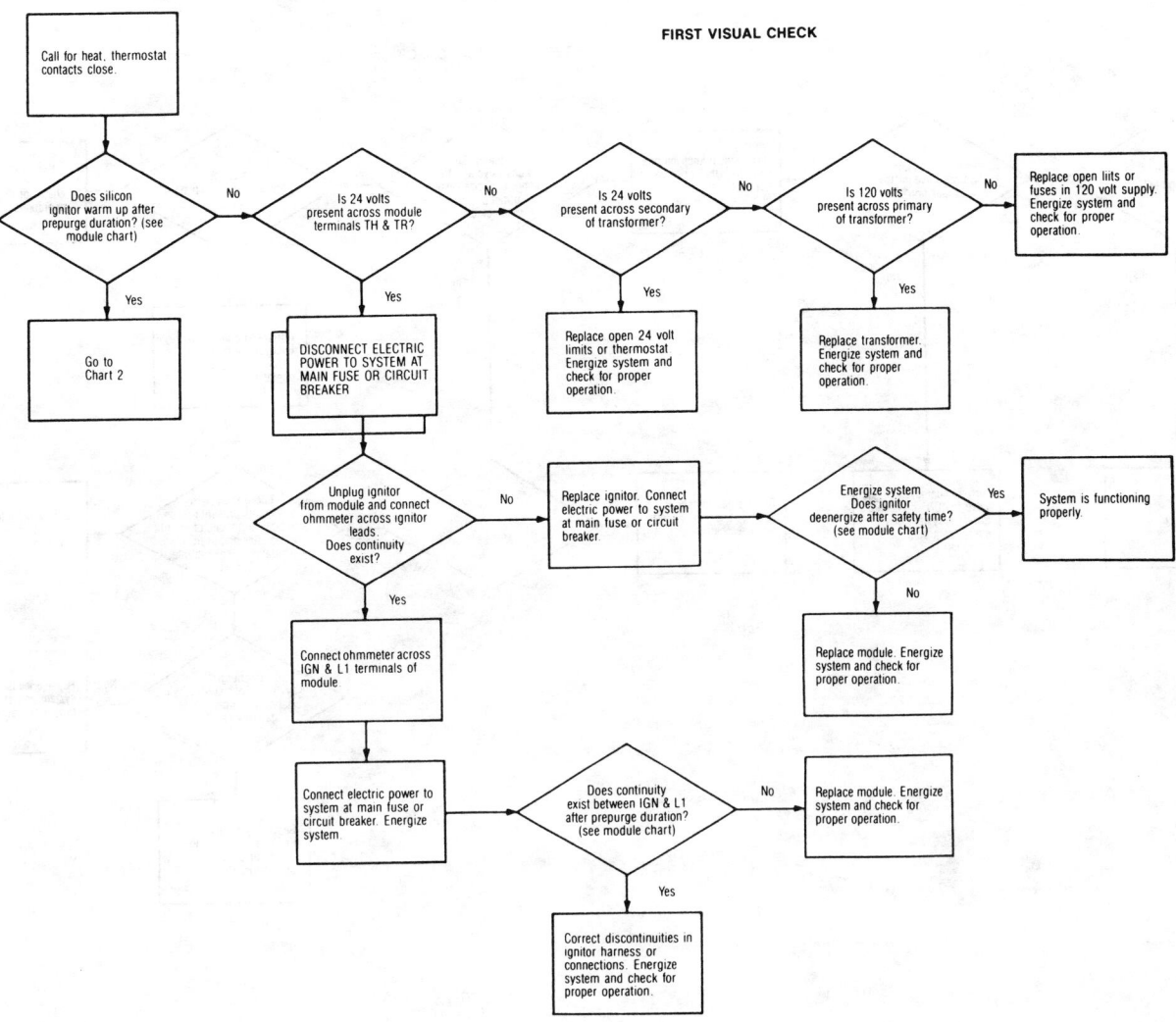

Call for heat, thermostat contacts close.

Does silicon ignitor warm up after prepurge duration? (see module chart)

Yes → Go to Chart 2

No → Is 24 volts present across module terminals TH & TR?

No → Is 24 volts present across secondary of transformer?

No → Is 120 volts present across primary of transformer?

No → Replace open liits or fuses in 120 volt supply. Energize system and check for proper operation.

Yes (TH & TR) → DISCONNECT ELECTRIC POWER TO SYSTEM AT MAIN FUSE OR CIRCUIT BREAKER

Yes (secondary) → Replace open 24 volt limits or thermostat. Energize system and check for proper operation.

Yes (primary) → Replace transformer. Energize system and check for proper operation.

Unplug ignitor from module and connect ohmmeter across ignitor leads. Does continuity exist?

No → Replace ignitor. Connect electric power to system at main fuse or circuit breaker.

Energize system. Does ignitor deenergize after safety time? (see module chart)

Yes → System is functioning properly.

No → Replace module. Energize system and check for proper operation.

Yes (continuity) → Connect ohmmeter across IGN & L1 terminals of module.

Connect electric power to system at main fuse or circuit breaker. Energize system.

Does continuity exist between IGN & L1 after prepurge duration? (see module chart)

No → Replace module. Energize system and check for proper operation.

Yes → Correct discontinuities in ignitor harness or connections. Energize system and check for proper operation.

V. SERVICE AND TROUBLESHOOTING (Cont.)

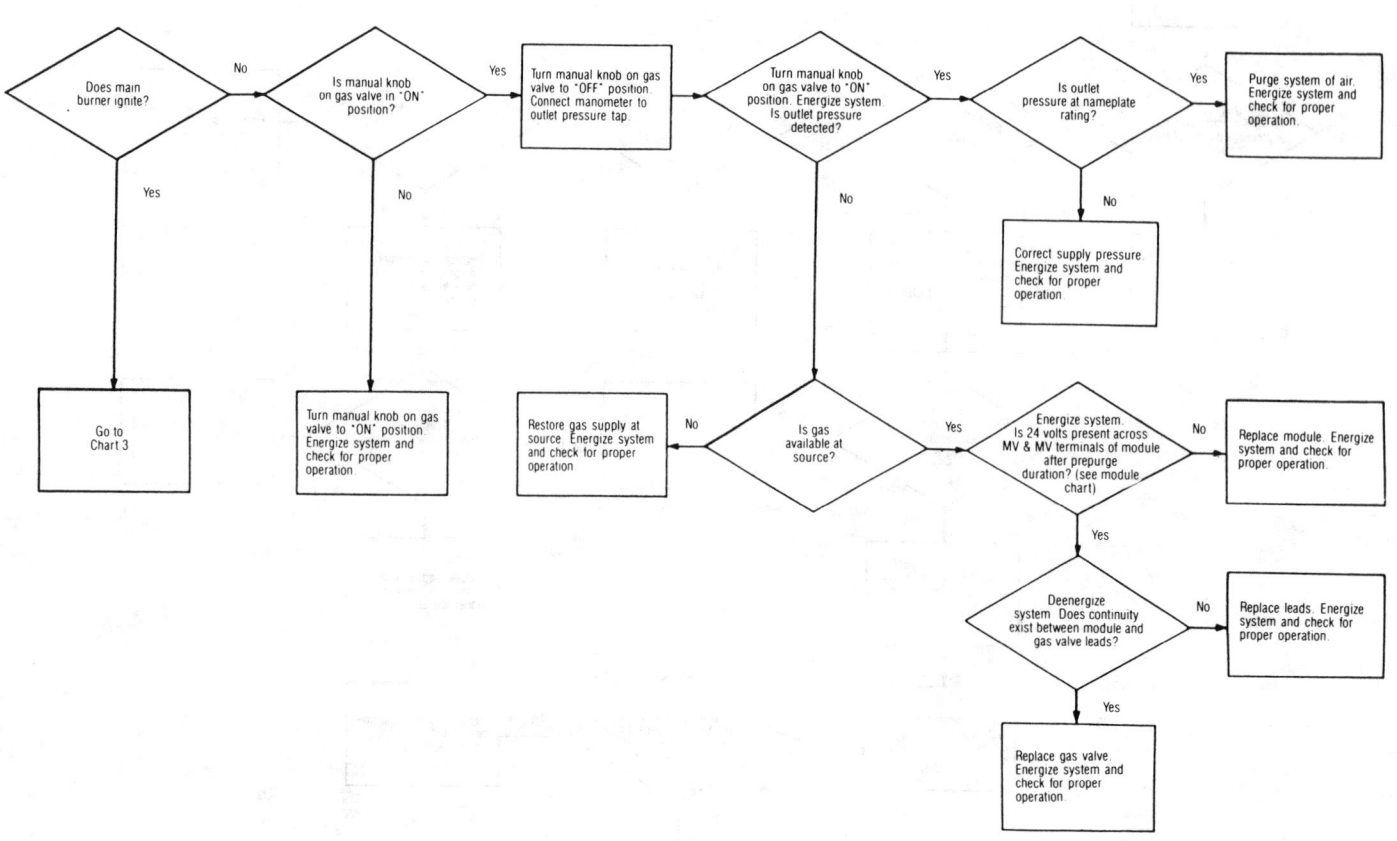

SECOND VISUAL CHECK

Does main burner ignite?

- No → Is manual knob on gas valve in "ON" position?
  - Yes → Turn manual knob on gas valve to "OFF" position. Connect manometer to outlet pressure tap. → Turn manual knob on gas valve to "ON" position. Energize system. Is outlet pressure detected?
    - Yes → Is outlet pressure at nameplate rating?
      - Yes → Purge system of air. Energize system and check for proper operation.
      - No → Correct supply pressure. Energize system and check for proper operation.
    - No → Is gas available at source?
      - No → Restore gas supply at source. Energize system and check for proper operation.
      - Yes → Energize system. Is 24 volts present across MV & MV terminals of module after prepurge duration? (see module chart)
        - No → Replace module. Energize system and check for proper operation.
        - Yes → Deenergize system. Does continuity exist between module and gas valve leads?
          - No → Replace leads. Energize system and check for proper operation.
          - Yes → Replace gas valve. Energize system and check for proper operation.
  - No → Turn manual knob on gas valve to "ON" position. Energize system and check for proper operation.
- Yes → Go to Chart 3

THIRD VISUAL CHECK

Does main burner remain lit after safety time? (see module chart)

- No → Deenergize system. Connect ohmmeter across sensor and ground. Does continuity exist?
  - Yes → Replace sensor. Energize system and check for proper operation.
  - No → Disconnect sensor lead at module. Connect ohmmeter across sensor lead terminal and sensor. Does continuity exist?
    - Yes → Correct discontinuity between module GND terminal and burner ground. Energize system and check for proper operation.
    - No → Replace sensor. Energize system and check for proper operation.
- Yes → Thermostat satisfied. → HSI system cycle complete.

VI. REPAIR AND REPLACEMENT PROCEDURE

Figure 2 gives an overview of the complete furnace. If need arises for disassembly of the furnace, refer to Page 7.

⚠ CAUTION:
USE DIRECT REPLACEMENT PARTS ONLY. DO NOT FIELD MODIFY THIS FURNACE.

INDUCED DRAFT BLOWER ASSEMBLY

The induced Draft Combustion Blower is constructed of Polypropelene, a man-made high temperature plastic type material. Polypropelene is impervious to the corrosive products of combustion condensed in the Secondary Heat Exchanger. However, it is not as forgiving of mishandling as steel components would be. Should any part of the blower fail, including the motor, it will be necessary to replace the entire assembly.

INSTALLATION INSTRUCTIONS FOR:
REPLACEMENT VENTOR MOTOR ASSEMBLY
KIT 8G19005

BILL OF MATERIAL

DESCRIPTION	PART NO.	QUANTITY
VENTOR MOTOR AND HOUSING ASSEMBLY	1708603	1
PLASTIC FASTENER PIN	2344276	3
SCREWS	2360950	8
NEOPRENE ROPE GASKET	—	1 piece
AMP. RECEPTACLE	1822001	1
INSTRUCTIONS	7502-126	1

READ THESE INSTRUCTIONS CAREFULLY BEFORE INSTALLING THE VENTOR MOTOR ASSEMBLY TO THE FURNACE.

Follow these steps for proper installation of the Ventor motor Replacement Kit.

1. Disconnect the power supply to the furnace.

2. Remove the Ventor Motor and housing assembly as follows:

   a) Disconnect the plug from the receptacle in the blower compartment and remove the receptacle from the blower partition.

   b) Remove the ground wire attached to the collector box.

   c) Remove the eight (8) screws holding the ventor motor assembly to the back plate. See Figure 34.

NOTE: The Neoprene Rope Gasket may be reused if it is NOT DAMAGED. If it is damaged, use the replacement gasket supplied in this Kit.

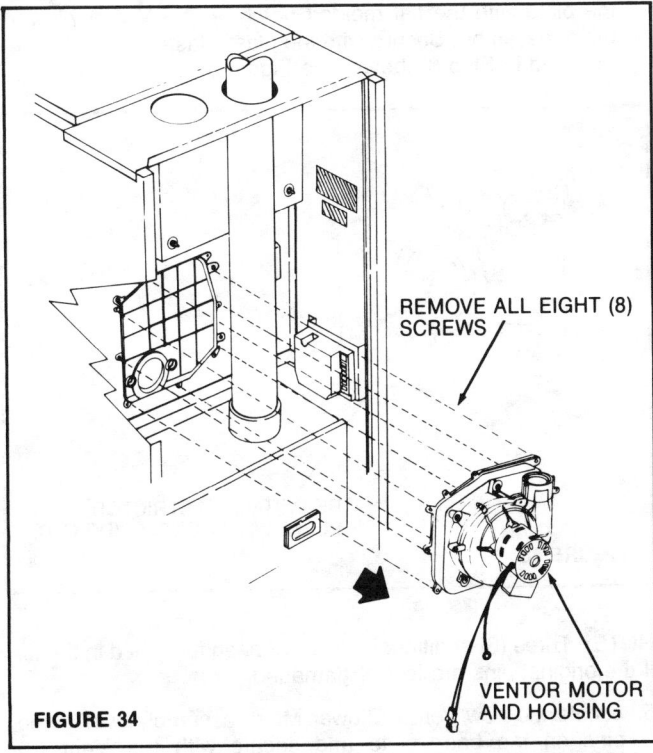

REMOVE ALL EIGHT (8) SCREWS

VENTOR MOTOR AND HOUSING

FIGURE 34

WARNING: The Neoprene Gasket must be installed to prevent condensate leakage.

3. Using a slot-headed screw driver, pry out the plastic fastener pins holding the restrictor plate, and remove plate from the defective blower assembly. See Figure 35.

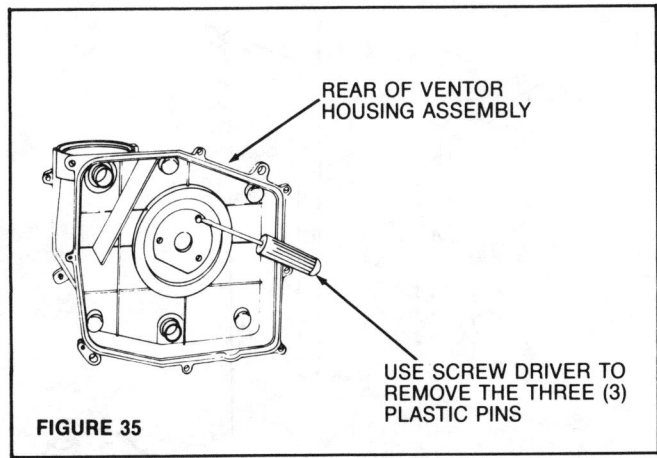

REAR OF VENTOR HOUSING ASSEMBLY

USE SCREW DRIVER TO REMOVE THE THREE (3) PLASTIC PINS

FIGURE 35

4. Install the restrictor plate with the *rough side out* on the new Ventor Blower motor assembly by aligning the flat edge of the plate with the flat molded notch on the Ventor blower motor assembly. Secure with the plastic fasteners that were removed in Step 3 above. See Figure 36.

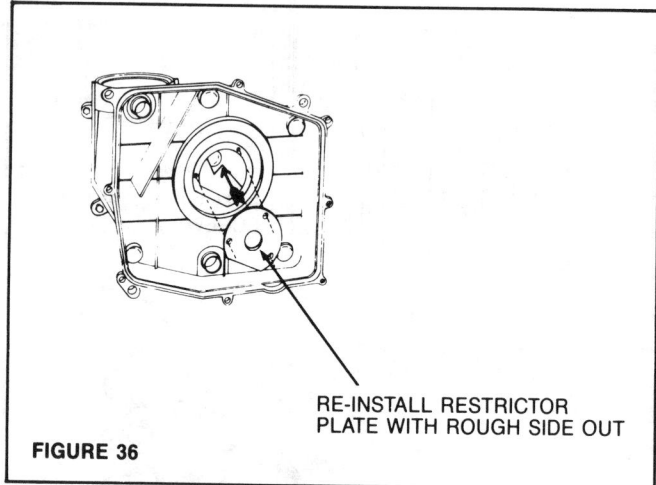

RE-INSTALL RESTRICTOR PLATE WITH ROUGH SIDE OUT

FIGURE 36

NOTE: Three (3) additional pins have been furnished in the kit if the original pins are lost or damaged.

5. Locate the new Ventor Blower Motor assembly over guide pins on the back plate and secure with the eight (8) mounting screws. Use screws removed in Step 2 c, or use the screws in the Kit. See Figure 37.

NOTE: Make sure the Neoprene Rope is in place.

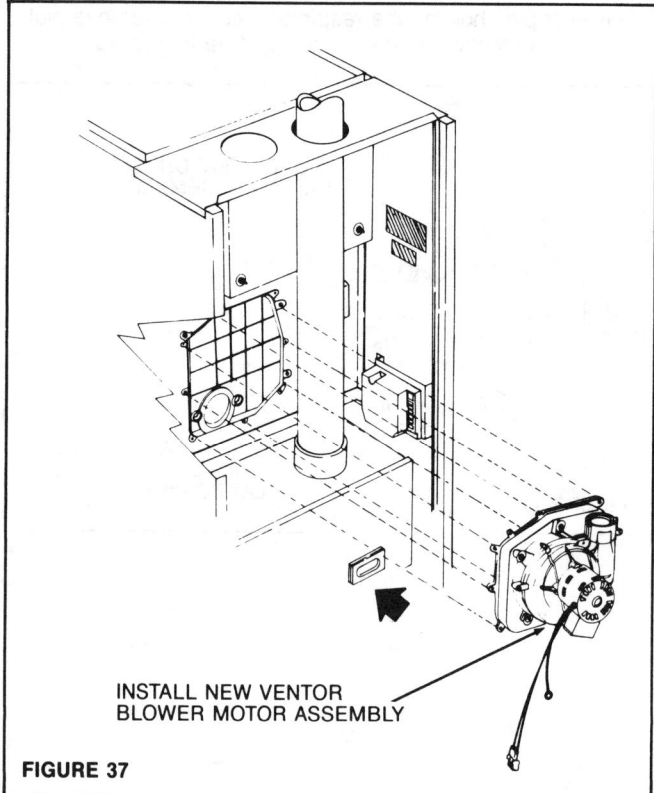

INSTALL NEW VENTOR BLOWER MOTOR ASSEMBLY

FIGURE 37

6. Rewire the motor by snapping the receptacle into the blower partition and plugging the power wire from the blower compartment into the receptacle. See Figure 38.

7. Reconnect the ground wire to the collector box.

Reconnect power supply to the furnace.

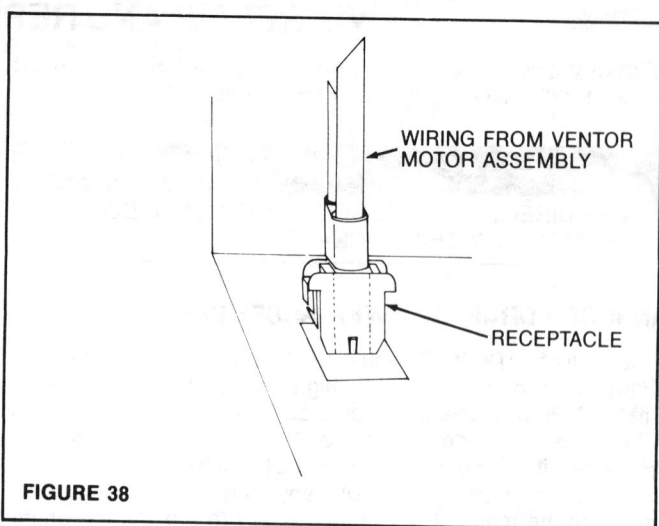

WIRING FROM VENTOR MOTOR ASSEMBLY

RECEPTACLE

FIGURE 38

NOTE that the assembly is the same for all units. But that each one has a different restrictor except the 125 MBTUH unit. Refer to the Restrictor Chart on Page 22

NOTE: All Induced Draft Blower motors are permanently lubricated.

RESTRICTOR

NOTE: Make sure when removing the Induced Draft blower assembly to service, that the exact restrictor or replacement restrictor is the correct part for the furnace you are servicing. The wrong size restrictor will adversely affect the operation of the furnace.

CONTROLS COMPARTMENT

This compartment contains the redundant gas valve, pressure switches, fan/limit switch, ignition control, manifold connection (Dell-Fitting), line voltage junction box and the combustion blower.

Extreme caution must be taken not to strip the male threads on the manifold connection when disconnecting or reconnecting.

COMBUSTION CHAMBER

This compartment contains the gas manifold, burners, burner orifices, ignitor, sensor probe, and flash shield. Care must be taken when removing the burner combustion door as not to destroy the gasket.

FIGURE 39

VI. REPAIR AND REPLACEMENT PROCEDURE (Cont.)

⚠ CAUTION:

If gasket is broken or damaged in the removal process of this door, a new gasket must be installed.

To remove the burner access shield, remove two (2) sheet metal screws and pull shield out. The screws are located one on each end of the shield. Be sure not to break ignitor or sensor lead wire when removing shield. If wire insulation is cracked, broken or frayed, it must be replaced. An ignitor or sensor wire in this condition will not allow the furnace to operate properly. A continuous lock-out will occur.

INDOOR BLOWER COMPARTMENT

This compartment contains the blower housing, motor, wheel capacitor, door safety switch, control center, low voltage terminal strip, wiring diagram, and condensate trap assembly.

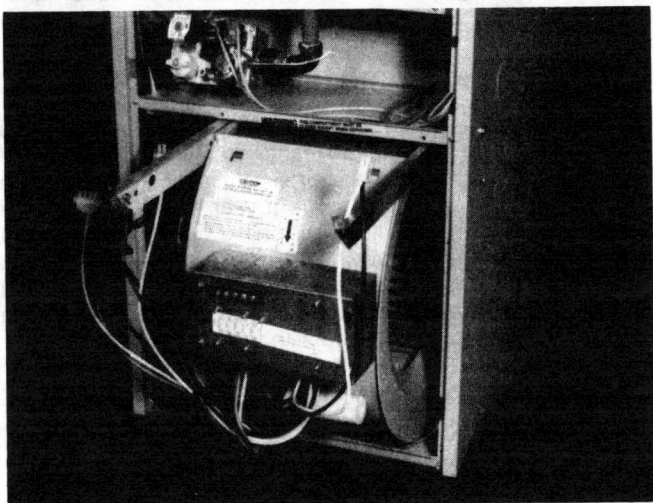

FIGURE 40

To remove blower housing from the furnace, first unplug both AMP connectors; one on the left side of the blower partition and the other one on the right side. Remove the low voltage wire connections from the terminal strip. Remove the two (2) sheet metal screws holding the blower assembly to the furnace. The blower assembly can be pulled toward the front of the furnace for easy removal. See Figure 40.

PRESSURE SWITCH REPLACEMENT

The Pressure Switch is NOT identical on all furnaces, even though physical appearance is the same. For identification purposes, both in manufacturing, and field replacement, the LABEL on the Pressure Switch is color coded. The Switch should bear the same color label. The following table illustrates the color code.

FURNACE MODEL	BTUH INPUT	LABEL COLOR
GUH060	60,000	White
GUH075	75,000	Yellow
GUH100	100,000	Red
GUH125	125,000	Red

In the event there is a mis-match on the furnace you are servicing, replace the Switch with the correct part before attempting any further repairs or troubleshooting.

The Pressure Switch is mounted on screws which protrude from outside the furnace case inward. Removal of a ⅜" nut from each end of the Pressure Switch bracket releases the Switch from the furnace casing. Nut clips have been installed behind the Pressure Switch bracket to hold the screws in place in close or zero clearance installations.

The Draft Proving Switch has two hose connections. Make sure the hoses go to the correct side of the Switch. The short hose (from the combustion blower) should connect to the left side of the Switch. See Figure 41. The long hose (from the Combustion Chamber) should connect to the right side of the Switch.

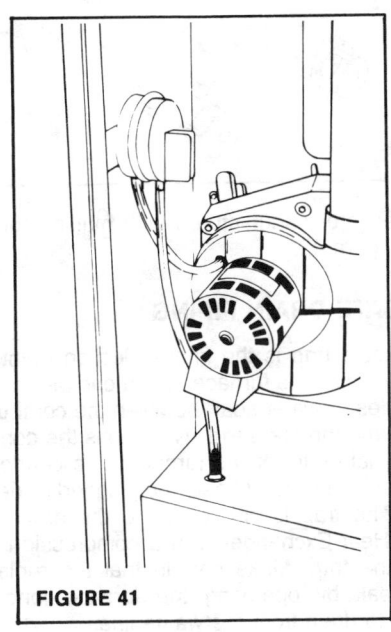

FIGURE 41

The Pressure Switch has three terminals, but only two leads connected to them. Insure that on BOTH switches one lead is attached to Common "C" and the other lead is attached to the Normally Open (N.O.) Terminal. Test run the furnace after any Pressure Switch replacement.

REPLACEMENT HEAT EXCHANGER—

A faulty heat exchanger requires furnace replacement. Consult your Tech Rep. See Page 40.

BURNERS

The Condensing furnace uses 1¼" slotted gang burners. DO NOT attempt to substitute smaller burners. The Burners should be replaced as a complete set. See below for burner adjustment.

VI. REPAIR AND REPLACEMENT PROCEDURE (Cont.)

PRIMARY AIR ADJUSTMENT (Air Shutter)

These units have individual primary air shutters on each burner. To adjust, loosen the lock screw and close the air shutter until yellow tips appear on the flames, then open the air shutter until the yellow tips just disappear, and tighten the lock screw. Follow this procedure on each burner.

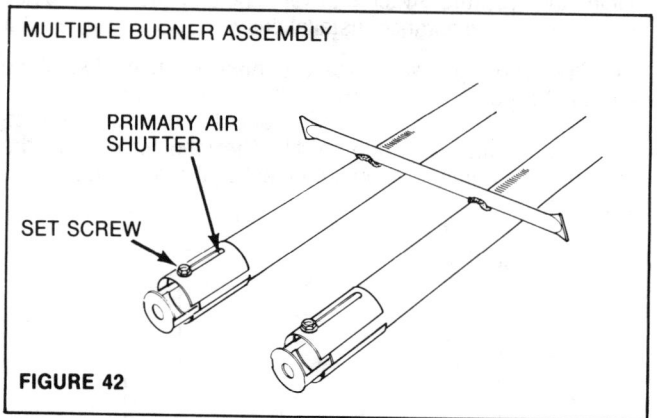

MULTIPLE BURNER ASSEMBLY

PRIMARY AIR SHUTTER

SET SCREW

FIGURE 42

The furnaces are manufactured with multiple burner assemblies. See Figure 42.

CONDENSATE DRAIN PIPING

The condensate trap is the final collection point for all liquid waste produced by the furnace. It is a one piece plastic device which provides a "water seal" between the combustion area of the furnace and the drain exit. The trap is the connection point for the external drain for the furnace. The condensate trap is NOT servicable, if it should become plugged or develop a leak, replace it. The trap is connected to the drain outlet of the Secondary Heat Exchanger with a compression fitting that is integral to the trap. Make certain that the replacement trap does not leak by operating the furnace long enough for condensate to drain from the waste line.

VENTILATION

The flow of ventilating air must NOT BE obstructed from reaching the furnace. Air openings provided in the casing of the furnace must be kept free of obstructions which would restrict air flow, thereby affecting efficiency and safe operation of the furnace. Furnaces MUST have air for proper performance.

WARNING: The air for combustion and ventilation must NOT come from a corrosive atmosphere.

Combustion air must be free of acid forming chemicals; such as sulphur, fluorine and chlorine. These elements are found in aerosol sprays, detergents, bleaches, cleaning solvents, air fresheners, paint and varnish removers, refrigerants and many other commercial and household products. Vapors from these products when burned in a gas flame, form acid compounds. The acid compounds increase the dew point temperature of the flue products and become highly corrosive after they condense.

FILTERS FOR ALL FURNACES

These units are equipped with efficient washable type air filters designed to keep the house cleaner by withholding dirt and other foreign particles from the air circulation. Filters should be cleaned or replaced at least twice a year or at any time they become clogged. Dirty filters will cause the furnace to heat improperly and possibly overheat by restricting air circulation. These units are designed so that the filters may be quickly and easily replaced. Replacement filters must be of the same type and size as originally furnished with the furnace.

FILTER REQUIREMENTS (HEATING AND COOLING)②

| INPUT BTU/H | STANDARD SIDE EXTERNAL FILTER AND RACK | OPTIONAL FILTER KITS | | MINIMUM FILTER AREA (IN²) |
		EXTERNAL SET-AWAY FILTER RACK KIT①	INTERNAL BOTTOM FILTER KIT②	
60,000	(1) 16 x 25 x 1	NOT APPLICABLE	(1) 16 x 25 x ½	400
75,000	(1) 16 x 25 x 1	NOT APPLICABLE	(1) 16 x 25 x ½	400
100,000	(2) 16 x 25 x 1	(1) 20 x 25 x 1	(1) 20 x 25 x ½	500
125,000	(2) 16 x 25 x 1	(1) 20 x 25 x 1	(1) 25 x 25 x ½	500

①SEE OPTIONAL EQUIPMENT FOR KIT PART NUMBER.
②FILTERS ARE RATED AT 520 FPM OR MORE (WASHABLE TYPE).

VII. FURNACE CLEANING—
INSPECTION AND CLEANING PROCEDURE

> ### ⚠ CAUTION:
> NOTE: THIS FURNACE IS EQUIPPED WITH A PRESSURIZED COMBUSTION CHAMBER. IT IS MANDATORY TO REPLACE ALL GASKETS THAT ARE REMOVED, ORDER REPLACEMENT GASKET PRIOR TO STARTING SERVICE.

1. Disconnect power source to the unit.

2. Remove louver panel from the unit and set aside.

3. Turn gas supply "off" at the gas valve. Allow a few minutes for unit to cool down.

4. Remove the screws from the front panel of the burner enclosure and set aside.

5. Remove the front panel. **NOTE:** Discard old gasket.

6. Remove the two screws holding the burner access shield in place and set aside.

7. Remove the burner access shield.

8. On the H.S.I. System, disconnect the sensor lead from the remote sensor (far left burner) and the ignitor leads at the ignitor disconnect.

NOTE: At this point inspect for any loose particles or carbon deposits in the burner area. Using a flashlight and a small mirror, check the upper portions of the heat exchanger for any signs of blockage or dirt accumulation. If any of these conditions are apparent, proceed with the following steps.

9. Remove all burners from the unit. This is accomplished by pulling back on the burner assembly so that it compresses the springs that are located between the manifold and the burners.

   Tilt the burners up into the heat exchanger tube. Slide the burner assembly off the orifices and remove from the unit.

   Note the location of the burner assembly. The burner assembly with the ignitor is on the right hand side and the sensor burner is positioned on the left.

10. After the burner assemblies have been taken from the unit, tap the end of the burners (with the primary air shutter attached) lightly on the ground, thereby removing any residue from inside the burners. Then run a vacuum hose across the top of the burner assembly removing any foreign material that might be lodged between the ports.

11. Remove the collector box at the top of the unit. Using the exhaust end of a high pressure vacuum (shop vacuum) blow any loose particles from the upper heat exchanger to the bottom burner area.

12. With a stiff bristled brush, clean the bottom area of the heat exchanger, brushing into the burner enclosure where the vacuum can easily be used to pick up the residue.

13. Flush the secondary heat exchanger with fresh water through the transfer tube inlet in the collector area. The water flush will drain through the drain trap in the blower compartment.

14. Reassemble the unit by reversing the above procedure. Make sure all old gaskets are discarded and replaced with new gaskets.

NOTE: Extreme care must be taken to protect all new seals and gaskets to assure tightness of the system. Failure to provide tight seals can result in faulty or improper system operation.

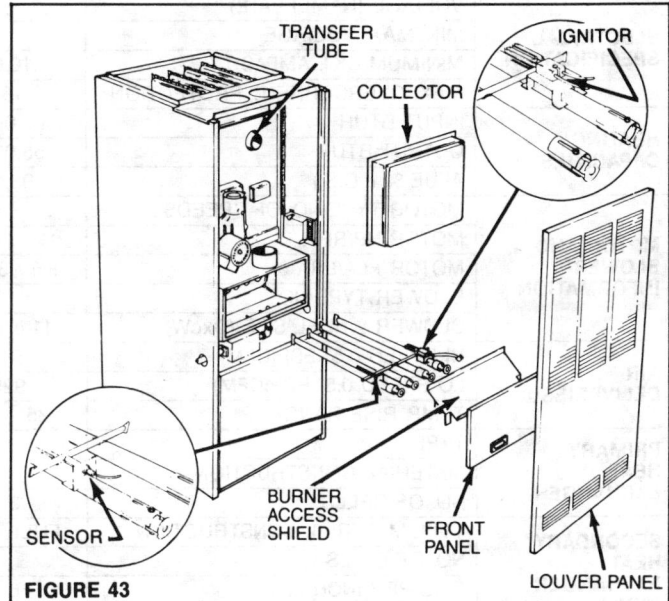

FIGURE 43

VIII. SPECIFICATIONS

The Specification Tables are divided into three parts. Section One is general information. Section Two would be of interest to installers and applications personnel. Section Three is of primary interest to the serviceman.

SECTION 1 GENERAL SPECIFICATIONS AND INFORMATION

SPECIFICATION AND INFORMATION

MODEL		GUH060A012	GUH075A012	GUH100A016	GUH125A016	GUH125A020
ELECTRICAL SPECIFICATION	VOLTAGE (NAMEPLATE)	115 / 120				
	MIN./MAX. VOLTAGE	102 / 132				
	MINIMUM CKT. AMPACITY	10.6	10.6	15.0	15.0	15.0
	MAX. OVERCURRENT PROTECTION	15	15	15	15	15
HEATING① CAPACITIES	INPUT BTUH	60,000	75,000	100,000	125,000	125,000
	OUTPUT BTUH	56,000	70,000	92,000	115,000	113,250
	AFUE % (I.C.S.)	93.1	91.5	91.4	90.8	90.0
MOTOR AND BLOWER INFORMATION	MOTOR TYPE/NO. OF SPEEDS	PSC-DIRECT DRIVE / 1				
	MOTOR HP/RPM	½ / 1100	½ / 1100	¾ / 1100	¾ / 1100	¾ / 1100
	MOTOR FLA/LRA②	8.5 / 34.5	8.5 / 34.5	12.0 / 48	12.0 / 48	12.0 / 48
	BLOWER TYPE/NO.	CENTRIFUGAL / 1				
	BLOWER WHEEL SIZE (DxcW)	11.8 x 8	11.8 x 8	11.8 x 10.6	11.8 x 10.6	11.8 x 10.6
AIR DELIVERIES③	HI SPEED 0.5" ESP(CFM)	1365	1365	1359	1577	1820
	LO SPEED 0.5" ESP(CFM)	925	925	1152	1313	1448
	TEMP. RISE RANGE °F	35 - 65	45 - 75	45 - 75	55 - 85	45 - 75
PRIMARY HEAT EXCHANGER	TYPE	RPJ® WELDLESS SECTIONAL				
	MATERIAL CONSTRUCTION	ALUMINIZED STEEL				
	NO. OF CELLS	3	3	4	5	5
SECONDARY HEAT EXCHANGER (COIL)	TYPE/MATERIAL CONSTRUCTION	FULL COLLAR ALUMINUM FIN AND 29-4C STAINLESS STEEL TUBE				
	NO. OF COILS	1				
	FINS PER INCH	6	6	6	6	6
	TUBE SIZE	5.505" O.D. x 0.020" WALL				
PIPING INFORMATION	GAS CONNECTION (IN.)	½"				
	VENT SIZE DIA. (IN.)	SEE CAPACITY SIZING CHART ON INSIDE PAGE				
	CONDENSATE CONN. (IN.) TYPE	½" / COMPRESSION FITTING				
IGNITION SYSTEM	TYPE/IGNITOR MATERIAL	ELECTRIC (HOT SURFACE) / SILICONE CARBIDE				
	IGNITION SEQUENCE	THREE TRY				
	FLAME PROVING INTERVAL	7 SECONDS				
	SENSOR TYPE	FLAME RECTIFICATION				
BURNER INFORMATION	TYPE/MAT. CONSTRUCTION	SLOTTED PORT / ALUMINIZED STEEL MULTIPLE BURNER				
	SIZE	1¼" DIA.				
	NO. OF BURNERS	3	3	4	5	5
GENERAL INFORMATION	AIR FILTER TYPE	WASHABLE				
	AIR FILTER SIZE/QUANTITY④	16 x 25 x 1 (1)		16 x 25 x 1 (2)		
	SHIPPING WEIGHT LBS.	192	192	223	242	242
	NET WEIGHT LBS.	182	182	213	232	232

①CAPACITY AND EFFICIENCY RATINGS IN ACCORDANCE WITH D.O.E. TEST PROCEDURES.
②FULL LOAD AMPS DEPEND ON INSTALLATION CONDITIONS AND HEATING OR COOLING OPERATION.
③REFER TO BLOWER CHART ON PAGE 5 FOR COMPLETE BLOWER INFORMATION.
④REFER TO FILTER REQUIREMENT CHART FOR PROPER FILTER AND NUMBER OF FILTERS.

VIII. SPECIFICATIONS (Cont.)

SECTION 2

INSTALLATION AND APPLICATION INFORMATION

RECOMMENDED CLEARANCE GUIDE (FIRE PROTECTION CLEARANCE)

(Dimensions shown in inches are the minimum clearances to combustible materials for which the furnace design has been certified.)

TABLE 1

INPUT BTU/HR.	FRONT*	REAR	SIDES	TOP	VENT CONNECTION	COMBUSTIBLE FLOOR**
60,000	6	0	0	1	0	0
75,000	6	0	0	1	0	0
100,000	6	0	0	1	0	0
125,000	6	0	0	1	0	0

*A MINIMUM OF 36" CLEARANCE IN FRONT OF THE FURNACE SHOULD BE AVAILABLE FOR SERVICING.
**SHALL NOT BE INSTALLED DIRECTLY ON CARPETING, TILE OR OTHER COMBUSTIBLE MATERIALS OTHER THAN WOOD FLOORING.

TABLE 2—FILTER REQUIREMENTS (HEATING AND COOLING)**

INPUT BTU/H	STANDARD SIDE EXTERNAL FILTER AND RACK	OPTIONAL FILTER KITS		MINIMUM FILTER AREA (IN²)
		EXTERNAL SET-AWAY FILTER RACK KIT*	INTERNAL BOTTOM FILTER KIT**	
60,000	(1) 16 x 25 x 1	NOT APPLICABLE	(1) 16 x 25 x ½	400
75,000	(1) 16 x 25 x 1	NOT APPLICABLE	(1) 16 x 25 x ½	400
100,000	(2) 16 x 25 x 1	(1) 20 x 25 x 1	(1) 20 x 25 x ½	500
125,000	(2) 16 x 25 x 1	(1) 20 x 25 x 1	(1) 25 x 25 x ½	500

*SEE OPTIONAL EQUIPMENT FOR KIT PART NUMBER.
**FILTERS ARE RATED AT 520 FPM OR MORE (WASHABLE TYPE).

GAS PIPING TABLES

TABLE 6
Maximum capacity of pipe in cubic feet of gas per hour for gas pressures of 0.5 PSIG or less and a pressure drop of 0.5 inch water column (based on 0.60 specific gravity gas)

NOMINAL IRON PIPE SIZE INCH	INTERNAL DIAMETER INCHES	LENGTH OF PIPE, FEET													
		10	20	30	40	50	60	70	80	90	100	125	150	175	200
⅜	.493	95	65	52	45	40	36	33	31	29	27	24	22	20	19
½	.622	175	120	97	82	73	66	61	57	53	50	44	40	37	35
¾	.824	360	250	200	170	151	138	125	118	110	103	93	84	77	72
1	1,049	680	465	375	320	285	260	240	220	205	195	175	160	145	135
1¼	1,380	1,400	950	770	660	580	530	490	460	430	400	360	325	300	280
1½	1,610	2,100	1,460	1,180	990	900	810	750	690	650	620	550	500	460	430

TABLE 7 Multipliers to be used when Specific Gravity is Other than 0.60

SPEC. GRAV.	MULT.	SPEC.GRAV.	MULT.	SPEC. GRAV.	MULT.	SPEC. GRAV.	MULT.
.35	1.31	.65	.962	1.00	.775	1.60	.612
.40	1.23	.70	.926	1.10	.740	1.70	.594
.45	1.16	.75	.895	1.20	.707	1.80	.577
.50	1.10	.80	.867	1.30	.680	1.90	.565
.55	1.04	.85	.841	1.40	.655	2.00	.547
.60	1.00	.90	.817	1.50	.633	2.10	.535

TABLE 8
Capacity of Semi-Rigid Tubing in 1000 BTU Per Hour of Undiluted Liquefied Petroleum Gases of 0.50 Pressure Drop

OUTSIDE DIA. INCHES	LENGTH OF TUBING (FEET)									
	10	20	30	40	50	60	70	80	90	100
⅜	39	26	21	19	—	—	—	—	—	—
½	92	62	50	41	37	35	31	29	27	26
⅝	199	131	107	90	79	72	67	62	59	55
¾	329	216	181	145	131	121	112	104	95	90
⅞	501	346	277	233	198	187	164	155	146	138

NOTES:
1. ALLOWANCE HAS BEEN MADE FOR AN AVERAGE NUMBER OF FITTINGS.
2. WHEN USING A GAS WITH SPECIFIC GRAVITY OTHER THAN 0.60, MULTIPLY THE CAPACITY BY THE FACTOR GIVEN IN TABLE 6.
3. TO CONVERT BTU INPUT PER HOUR TO CUBIC FEET PER HOUR, DIVIDE BTU PER HOUR BY THE HEATING VALUE OF THE GAS TO BE USED. THIS CAN BE OBTAINED FROM YOUR LOCAL UTILITY.
4. PIPING SYSTEMS OF SEMI-RIGID TUBING (TABLE 8) THAT ARE TO BE SUPPLIED WITH GAS OF A SPECIFIC GRAVITY OF 1.53 OR LESS CAN BE SIZED DIRECTLY FROM TABLE 6 UNLESS THE AUTHORITY HAVING JURISDICTION SPECIFIES THAT A GRAVITY FACTOR BE APPLIED. WHEN THE SPECIFIC GRAVITY OF THE GAS IS GREATER THAN 1.53, THE GRAVITY FACTOR SHALL BE APPLIED.

VIII. SPECIFICATIONS (Cont.)

SECTION 2 (Cont.)

VENT PIPING REQUIREMENTS

ALTITUDE DERATE

High Altitude Input Rate = $\dfrac{\text{Sea Level}}{\text{Nameplate}} \times$ (Multiplier)
Input Rate

Elevation	High Altitude Multiplier
2000' - 2999'	0.92
3000' - 3999'	0.88
4000' - 4999'	0.84
5000' - 5999'	0.80
6000' - 6999'	0.76
7000' - 7999'	0.72

MANIFOLD PRESSURE AND ORIFICE SIZE FOR HIGH ALTITUDE APPLICATIONS

NATURAL GAS

HEATING VALUE BTU/CU FT		MEAN ELEVATION FEET ABOVE SEA LEVEL					
		2000 to 3000	3000 to 4000	4000 to 5000	5000 to 6000	6000 to 7000	7000 to 8000
800		3.5"wc	3.5"wc	3.5"wc	3.5"wc	3.2"wc	2.8"wc
850		3.5"wc	3.5"wc	3.5"wc	3.1"wc	2.8"wc	2.5"wc
900		3.5"wc	3.3"wc	3.0"wc	2.8"wc	2.5"wc	2.3"wc
950		3.3"wc	3.0"wc	2.7"wc	2.5"wc	2.3"wc	3.5"wc
1000		2.9"wc	2.7"wc	2.5"wc	2.3"wc	3.5"wc	3.1"wc
1050		2.7"wc	2.4"wc	2.3"wc	3.5"wc	3.2"wc	2.8"wc
1100		2.4"wc	2.3"wc	3.5"wc	3.2"wc	2.8"wc	2.5"wc
GUH060	ORIFICE SIZE	STANDARD FACTORY ORIFICE		#48	#48	#48	#48
GUH075				#44	#44	#44	#44
GUH100				#44	#44	#44	#44
GUH125				#44	#44	#44	#44

▨ SHADED AREA REQUIRES ORIFICE CHANGE
NO SHADING INDICATES MANIFOLD PRESSURE CHANGE ONLY.

PROPANE

HEATING VALUE BTU/CU FT		MEAN ELEVATION FEET ABOVE SEA LEVEL					
		2000 to 3000	3000 to 4000	4000 to 5000	5000 to 6000	6000 to 7000	7000 to 8000
2500		8.5"wc	7.8"wc	7.0"wc	10.0"wc	9.0"wc	8.1"wc
GUH060	ORIFICE SIZE	STANDARD FACTORY ORIFICE			3/64	3/64	3/64
GUH075					#55	#55	#55
GUH100					#55	#55	#55
GUH125					#55	#55	#55

NOTE: NATURAL GAS DATA BASED ON 0.60 SPECIFIC GRAVITY.
PROPANE DATA BASED ON 1.53 SPECIFIC GRAVITY.
FOR FUELS WITH DIFFERENT SPECIFIC GRAVITY CONSULT THE NATIONAL FUEL GAS CODE ANSI Z223.1a-1987.
IMPORTANT: FILL OUT HIGH ALTITUDE LABEL SUPPLIED WITH UNIT OR HIGH ALTITUDE KIT AND AFFIX TO COVER OF BURNER BOX.

VIII. SPECIFICATIONS (Cont.)

SECTION 2 (Cont.)
VENT PIPING REQUIREMENTS

FOR 60,000-75,000 BTUH FURNACE

NO. OF ELBOWS	VENT TYPE	FEET OF PIPE					
		0-10'	10-15'	15-20'	20-25'	25-30'	30-35'
1	INLET	2	2	2	2	3	3
	EXHAUST	2	2	2	2	2	2
2	INLET	2	2	2	3	3	3
	EXHAUST	2	2	2	2	2	2
3	INLET	2	2	3	3	3	3
	EXHAUST	2	2	2	2	2	2
4	INLET	2	3	3	3	3	3
	EXHAUST	2	2	2	2	2	3
5	INLET	3	3	3	3	3	3
	EXHAUST	2	2	2	2	3	3

FOR 100,000 BTUH FURNACE

NO. OF ELBOWS	VENT TYPE	FEET OF PIPE					
		0-10'	10-15'	15-20'	20-25'	25-30'	30-35'
1	INLET	2	2	2	3	3	3
	EXHAUST	2	2	2	2	2	2
2	INLET	2	2	3	3	3	3
	EXHAUST	2	2	2	2	2	2
3	INLET	2	3	3	3	3	3
	EXHAUST	2	2	2	2	2	3
4	INLET	3	3	3	3	3	3
	EXHAUST	2	2	2	2	3	3
5	INLET	3	3	3	3	3	3
	EXHAUST	2	2	2	3	3	3

FOR 125,000 BTUH FURNACE

NO. OF ELBOWS	VENT TYPE	FEET OF PIPE					
		0-10'	10-15'	15-20'	20-25'	25-30'	30-35'
1	INLET	2	3	3	3	3	3
	EXHAUST	2	2	2	2	3	3
2	INLET	3	3	3	3	3	N.A.
	EXHAUST	2	2	2	3	3	N.A.
3	INLET	3	3	3	3	N.A.	N.A.
	EXHAUST	2	2	3	3	N.A.	N.A.
4	INLET	3	3	3	3	N.A.	N.A.
	EXHAUST	2	3	3	3	N.A.	N.A.
5	INLET	3	3	3	N.A.	N.A.	N.A.
	EXHAUST	3	3	3	N.A.	N.A.	N.A.

NOTE: Vent pipe diameters apply for installation from sea level to 2000 ft. For installation above 2000 ft. reduce the maximum pipe length as follows:

2000-5000 ft. above sea level—Reduce max. length allowable by 5 ft.

5001 7000 ft. above sea level—Reduce max. length allowable by 10 ft.

AIR DELIVERY—(CFM) FOR DIRECT DRIVE BLOWERS

MODEL NUMBER	BLOWER SIZE	MOTOR H.P.	BLOWER SPEED	EXTERNAL STATIC PRESSURE—INCHES WATER COLUMN													
				0.1		0.2		0.3		0.4		0.5		0.6		0.7	
				CFM	TEMP. RISE	CFM	TEMP. RISE	CFM	TEMP. RISE	CFM	TEMP. RISE	CFM	TEMP. RISE	CFM	TEMP. RISE	CFM	TEMP. RISE
GUH060A012	11.8x8	½	HIGH	1609	32.4	1550	33.6	1491	35.0	1431	36.4	1365	38.2	1280	40.7	1202	43.4
			MED.	1294	40.3	1268	41.1	1231	42.4	1188	43.9	1131	46.1	1070	48.9	992	52.6
			LOW	1045	49.9	1021	51.1	995	52.4	960	54.3	924	56.5	875	59.6	820	63.6
GUH075A012	11.8x8	½	HIGH	1609	40.1	1550	41.6	1491	43.3	1431	45.1	1365	47.3	1280	50.4	1202	53.7
			MED.	1294	49.9	1268	50.9	1231	52.4	1188	54.3	1131	57.1	1070	60.3	992	65.1
			LOW	1045	61.8	1021	63.2	995	64.9	960	67.2	924	69.8	875	73.8	820	78.7
GUH100A016	11.8x10.6	¾	HIGH	1642	52.4	1568	54.9	1505	57.2	1441	59.7	1359	63.3	1270	67.8	1170	78.5
			MED. HI	1621	53.1	1557	55.3	1487	57.9	1415	60.8	1357	63.4	1252	68.7	1152	74.7
			MED. LOW	1465	58.7	1399	61.5	1343	64.1	1285	67.0	1197	71.9	1122	76.7	1030	83.6
			LOW	1335	64.5	1295	66.4	1249	68.9	1207	71.3	1152	74.7	1084	79.4	1010	85.2
GUH125A016	11.8x10.6	¾	HIGH	1810	58.8	1750	60.8	1675	63.5	1628	65.4	1577	67.5	1470	72.4	1371	77.6
			MED. HI	1802	59.0	1730	61.5	1655	64.3	1575	67.6	1478	72.0	1380	77.1	1262	84.3
			MED. LOW	1685	63.1	1625	65.5	1565	68.0	1490	71.4	1403	75.8	1308	81.4	1188	89.6
			LOW	1562	68.1	1515	70.2	1460	72.9	1390	76.6	1313	81.0	1220	87.2	1103	96.5
GUH125A020	11.8x10.6	¾	HIGH	2160	48.5	2085	50.0	2010	52.1	1955	53.6	1890	55.4	1820	57.6	1680	62.4
			MED. HI	1891	55.4	1837	57.0	1784	58.7	1729	60.6	1675	62.6	1582	66.2	1490	70.3
			MED. LOW	1777	59.0	1728	60.6	1680	62.4	1614	64.9	1549	67.6	1495	70.1	1441	72.7
			LOW	1631	64.2	1696	65.7	1562	67.1	1505	69.6	1448	72.4	1389	75.4	1330	78.8

FIELD INSTALLED KITS AND ACCESSORIES

OPTIONAL EQUIPMENT			
DESCRIPTION	PART NUMBER	SHIP WT.	WHERE USED:
CONDENSATE NEUTRALIZER KIT	8G03021	2	ALL UNITS
HORIZONTAL VENT KIT	8G03022	10	ALL UNITS
COIL ADAPTER	8G11008	8	GUH060, 075
COIL ADAPTER	8G11009	8	GUH100
COIL ADAPTER	8G11010	8	GUH125
HIGH ALTITUDE KIT (NAT. GAS)	8G07903	1	ALL UNITS
HIGH ALTITUDE KIT (LPG)	8G07905	1	ALL LPG UNITS
LPG TO NAT. CONV. KIT	8G02028	5	ALL LPG UNITS
BOTTOM FILTER KIT WITH FILTER	8G05009	10	GUH060, 075
	8G05010	10	GUH100
	8G05011	10	GUH125
STAND-OFF FILTER RETURN KIT WITH FILTER	8G05008	10	GUH100, 125

FURNACE IS **NOT** FIELD CONVERTIBLE FROM NATURAL TO LP GAS.

VIII. SPECIFICATIONS (Cont.)

SECTION 3

Service data specifications and parts substitutions.

FAN AND LIMIT CONTROL

FURNACE MODEL	FAN ON	FAN OFF	LIMIT (f10 F)	MFG. NAME/ NUMBER	PART NUMBER
GUH060	130°F	90°F	250°F	HONEYWELL	1300-163
GUH075	130°F	90°F	250°F	L4064T-1996	1300-163
GUH100	130°F	90°F	170°F		1300-163
GUH125A016	130°F	90°F	160°F		1300-163
GUH125A020	130°F	90°F	200°F		1300-163

PRESSURE SWITCHES

FURNACE	COLOR CODE	COMBUSTION BLOWER (LH)
GUH060	White	1445-587
GUH075	Yellow	1445-589
GUH100	Red	1445-591
GUH125	Red	1445-591

PRESSURE SWITCH SPECIFICATIONS

PART NUMBER	COLOR CODE	SET POINT (IN. W.C.)	DIFFERENTIAL (IN. W.C.)
1445-587	White	1.78 ± 0.08	0.15
1445-589	Yellow	1.60 ± 0.12	0.15
1445-591	Red	0.90 ± 0.09	0.12

NOTE: Vent switch pressures are shown as positive values. Both sides of switch operate a negative pressure.

COMBUSTION BLOWER RESTRICTOR

FURNACE MODEL	COLOR CODE	OPENING DIAMETER	PART NUMBER
GUH060	Red	¾"	2972-010
GUH075	Black	1"	2972-011
GUH100	White	1⅜"	2972-012
GUH125	—	—	None

CRITICAL TORQUE REQUIREMENTS

AREA	TORQUE
Transfer Tube to Secondary Heat Exchanger clamp	18 ft./lb.
Transfer Tube to Front Partition	30 in./lb.
Discharge Tube to Secondary Heat Exchanger	50 in./lb.
Discharge Tube to Combustion Blower Backplate	30 in./lb.
Combustion Blower Assembly to Combustion Blower Backplate	7 in./lb.

APPROVED PARTS SUBSTITUTION

ITEM	FURNACE MODEL	ORIGINAL PART NO.	MANUFACTURERS PART NUMBER	SECOND ALT. PART NO. AND MFG. NOTE	
IGNITOR	ALL	1380-654	WHITE-RODGERS 767A-353	1380-672	NORTON MODEL 271
IGNITION MODULE	ALL	1380-669	WHITE-RODGERS 50F47-140	1380-671	ROBERTSHAW HS 780/17PR308-02
GAS VALVE (NAT.)	ALL "IN" SUFFIX	1585-975	WHITE-RODGERS 36E01-244	1585-973	ROBERTSHAW 7100 DERSO
GAS VALVE (LPG)	ALL "IL" SUFFIX	1585-976	WHITE-RODGERS 36E01-245	1585-974	ROBERTSHAW 7100 DERSO-LP
BLOWER MOTOR	GUH060 GUH075	1068-690 1068-690	FASCO 7126-1469	1068-621 1068-621	EMERSON KA55HXHBH2088
	GUH100 GUH125	1069-694 1069-694	FASCO 7126-2036	1069-640 1069-640	G.E. 5KCP39RGK627AS

TECHNICAL REPRESENTATIVES

If you encounter difficulties not covered in this Manual, or have further questions, you may wish to call your nearest Technical Representative. The list below relates the name and phone number of technical representatives available at the time of this publication. SnyderGeneral Corporation also offers factory and regional service training Seminars on this and other products. Contact your Technical Representative for details.

Steve Allemore Manager of Service and Training	Hutchins, Texas (214) 225-7351
Ron Froning	Anaheim, California (714) 978-8989
Shawn Hill	Atlanta, Georgia (404) 452-1699
Tom Hughes	Charlotte, N.C. (704) 588-5050
Gordon Nelson	Orlando, Florida (305) 855-6500
Open	Baltimore, Maryland (301) 792-0301
Open	Washington, D.C. (301) 953-9469
Bill Taylor	Chicago, Illinois (312) 832-0300
Charles Whitaker	St. Louis, Missouri (314) 991-1033
Larry Sutton Trainer	Hutchins, Texas (214) 225-7351

GLOSSARY

AIR-GAS RATIO - The ratio of combustion air supply flow rate to the fuel gas supply flow rate.

AIR SHUTTER - An adjustable shutter on the primary air openings of a burner, which is used to control the amount of combustion air introduced into the burner body.

ATMOSPHERIC PRESSURE - The pressure exerted upon the earth's surface by the weight of atmosphere above it.

AUTOMATIC GAS PILOT DEVICE - A gas pilot incorporating a device, which acts to automatically shut off the gas supply to the appliance burner if the pilot flame is extinguished.

BREAKER BOLT - (See Spoiler Screw).

BRITISH THERMAL UNIT - (Btu) - The quantity of heat required to raise the temperature of one pound of pure water one degree F.

BURNER - A device for the final conveyance of gas, or a mixture of gas and air, to the combustion zone (See also specific type of burner).

1. *Injection Burner.* A burner employing the energy of a jet of gas to inspire air for combustion into the burner and mix it with gas.

   a. *Atmospheric Injection Burner.* A burner in which the air inspired into the burner by a jet of gas is supplied to the burner at atmospheric pressure.

2. *Power Burner.* (See also Forced Draft Burner, Induced Draft Burner, Premixing Burner, and Pressure Burner). A burner in which either gas or air or both are supplied at pressures exceeding, for gas, the line pressure, and for air, atmospheric pressure.

3. *Yellow-Flame Burner.* A burner in which secondary air only is depended on for the combustion of the gas.

BURNER FLEXIBILITY - The degree to which a burner can operate with reasonable characteristics with a variety of fuel gases and/or variations in input rate (gas pressure).

COMBUSTION - The rapid oxidation of fuel gases accompanied by the production of heat or heat and light.

COMBUSTION AIR - Air supplied in an appliance specifically for the combustion of a fuel gas.

COMBUSTION CHAMBER - The portion of an appliance within which combustion normally occurs.

COMBUSTION PRODUCTS - Constituents resulting from the combustion of a fuel gas with the oxygen in air, including the inert gases, but excluding excess air.

CONDENSABLE - A gas which can be easily converted to liquid form, usually by lowering the temperature and/or increasing pressure.

CONTROLS - Devices designed to regulate the gas, air, water or electricity supplied to a gas appliance. They may be manual, semi- automatic or automatic.

CUBIC FOOT OF GAS - (Standard Conditions). The amount of gas which will occupy 1 cubic foot when at a temperature of 60F., and under a pressure equivalent to that of 30 inches of mercury.

DEWPOINT - The temperature at which a vapor will start to condense into its liquid form.

DRAFT HOOD - (Draft Diverter) - A device built into an appliance, or made part of a vent connector from an appliance, which is designed to: (1) assure the ready escape of the products of combustion in the event of no draft, backdraft, or stoppage beyond the draft hood; (2) prevent a backdraft from entering the appliance; and (3) neutralize the effect of stack action of a chimney or gas vent upon the operation of the appliance.

DOWNDRAFT - Excessive high air pressure existing at the outlet of chimney or stack which tends to make gases flow downward in the stack.

DRILLED PORT BURNER - A burner in which the ports have been formed by drilled holes in a thick section in the burner head or by a manufacturing method which results in holes similar in size, shape and depth.

EXCESS AIR - Air which passes through an appliance and the appliance flues in excess of that which is required for complete combustion of the gas. Usually expressed as a percentage of the air required for complete combustion of the gas.

EXTINCTION POP - (See Flashback).

FAHRENHEIT - The common scale of temperature measurement in the English system of units. It is based on the freezing point of water being 32 F. and the boiling point of water being 212 F. at standard pressure conditions.

FIXED ORIFICE - (See Orifice Spud).

FLAME ROLLOUT - A condition where flame rolls out of a combustion chamber when the burner is ignited.

FLAME VELOCITY - The speed at which a flame moves through a fuel-air mixture.

FLAMMABILITY LIMITS - The maximum percentages of a fuel in an air- fuel mixture which will burn.

FLASHBACK - An undesirable flame characteristic in which burner flames strike back into a burner to burn there or to create a pop after the gas supply has been turned off.

FLUE GASES, FLUE PRODUCTS - Products of combustion and excess air in appliance flues or heat exchangers before the vent exit.

FLUE LOSS - The heat lost in flue products leaving from the flue outlet of an appliance.

FLUE OUTLET - The opening provided in an appliance for the escape of flue gases.

FLUID - A gas or liquid, as opposed to a solid.

FORCED DRAFT BURNER - A burner in which combustion air is supplied by a fan or blower at a positive pressure.

FUEL - Any substance used for combustion.

FUEL GAS - Any substance in a gaseous form when used for combustion.

HARD FLAME - A flame with a hot, tight, well-defined inner cone.

HEAT EXCHANGER - Any device for transferring heat from one fluid to another.

HEATING SURFACE - All surfaces which transmit heat from flames or flue gases to the medium being heated.

HEATING VALUE - The number of British thermal units produced by the complete combustion at constant pressure of one cubic foot of gas. Total heating value includes heat obtained from cooling the products to the initial temperature of the gas and air and condensing the water vapor formed during combustion.

IGNITION - The act of starting combustion.

IGNITION TEMPERATURE - The minimum temperature at which combustion can be started.

IGNITION VELOCITY - (See Flame Velocity).

IMPINGEMENT TARGET BURNER - A burner consisting simply of a gas orifice and a target, with the gas jet from the orifice entraining combustion air in the open and the mixture striking and burning on the target surface. No usual burner body is used.

INCHES OF MERCURY COLUMN - A unit used in measuring pressures. One inch of mercury column equals a pressure of 0.491 pounds per square inch.

INCHES OF WATER COLUMN - A unit used in measuring pressures. One inch of water column equals a pressure of 0.578 ounces per square inch. One inch mercury column equals about 13.6 inches water column. 27.7 IM = 1 P.S.I.

INCOMPLETE COMBUSTION - Combustion in which the fuel is only partially oxidized.

INDUCED DRAFT BURNER - A burner which depends on draft induced by a fan or blower at the flue outlet to draw in combustion air and vent flue gases, at a negative pressure.

INPUT RATING - The gas-burning capacity of an appliance in Btu per hour as specified by the manufacturer. Appliance input ratings are based on sea level operation and need not be changed for operation up to 2,000 feet elevation. For operation at elevations above 2,000 feet, input ratings should be reduced at the rate of 4 percent for each 1,000 feet above sea level.

LIFTING FLAMES - An unstable burner flame condition in which flames lift or blow off the burner port (s).

LIQUEFIED PETROLEUM GASES - The terms "Liquefied Petroleum Gases", "LPG", and "LP Gas" mean and include any fuel gas which is composed predominantly of any of the following hydrocarbons, or mixtures of them: propane, propylene, normal butane or isobutane and butylenes.

LNG - Liquefied natural gas. Natural gas which has been cooled until it becomes a liquid.

LP GAS-AIR MIXTURES - Liquefied petroleum gases distributed at relatively low pressures and normal atmospheric temperatures which have been diluted with air to produce desired heating value and utilization characteristics.

LUMINOUS FLAME BURNER - (See Burner, Yellow Flame).

MANIFOLD - The conduit of an appliance which supplies gas to the individual burners.

MANIFOLD PRESSURE - The gas pressure in an appliance manifold, upstream of burner orifices.

MANUFACTURED GAS - A fuel gas which is artificially produced by some process, as opposed to natural gas, which is found in the earth. Sometimes called town gas.

NATURAL DRAFT - The motion of flue products through an appliance generated by hot flue gases rising in a vent connected to the furnace flue outlet.

NATURAL GAS - Any gas found in the earth, as opposed to gases which are manufactured.

ODORANT - A substance added to an otherwise odorless, colorless and tasteless gas to give warning of gas leakage and to aid in leak detection.

ORIFICE - An opening in an orifice cap (hood), orifice spud or other device through which gas is discharged, and whereby the flow of gas is limited and/or controlled. (See the also Universal Orifice).

Orifice Cap (Hood). - A movable fitting having an orifice which permits adjustment of the flow of gas by changing its position with respect to a fixed needle or other device extending into the orifice.

Orifice Discharge Coefficient - (See Discharge Coefficient).

Orifice Spud - A removable plug or cap containing an orifice which permits adjustment of the gas flow either by substitution with a spud having different sized orifices (fixed orifice) or by motion of an adjustable needle into or out of the orifice (adjustable orifice).

OVERRATING - Operation of a gas burner at a greater rate than it was designed for.

PILOT - A small flame which is used to ignite the gas at the main burner.

PORT - Any opening in a burner head through which gas or an air-mixture is discharged for ignition.

PRESSURE BURNER - A burner in which an air and gas mixture under pressure is supplied, usually at 0.5 to 14 inches water column.

PRESSURE REGULATOR - A device for controlling and maintaining a uniform outlet gas pressure.

PRIMARY AIR - The combustion air introduced into a burner which mixes with the gas before it reaches the port. Usually expressed as a percentage of air required for complete combustion of the gas.

PRIMARY AIR INLET - The opening or openings through which primary air is admitted into a burner.

PROPANE - A Hydrocarbon gas heavier than methane but lighter than butane. It is used as a fuel gas alone, mixed with air or as a major constituent of liquefied petroleum gases (See Liquefied Petroleum Gases).

SECONDARY AIR - Combustion air externally supplied to a burner flame at the point of combustion.

SINGLE PORT BURNER - A burner in which the entire air-gas mixture issues from a single port.

SPECIFIC GRAVITY - Specific gravity is the ratio of the weight of a given volume of gas to that of the same volume of air, both measured at the same temperature and pressure.

SPOILER SCREW - (Breaker Bolt) - A screw or bolt moved in or out of the gas jet in a burner to control primary air injection.

SPUD - (See Orifice).

STANDARD CONDITIONS - Pressure and temperature conditions selected for expressing properties of gases on a common basis. In gas appliance work, these are normally 30 inches of mercury and 60 F.

STATIC PRESSURE - The pressure exerted by a motionless gas.

TARGET BURNER - (See Impingement Burner).

THERM - A unit of heat energy equal to 100,000 Btu.

THROAT - (See Venturi).

TOTAL AIR - The total amount of air supplied to a burner. It is the sum of primary air, secondary air and excess air.

TOTAL PRESSURE - Also called impact pressure. The pressure measured in a moving fluid by an impact tube. It is the sum of velocity pressure and the static pressure.

UPDRAFT - Excessively low air pressure existing at the outlet of a chimney or stack which tends to increase the velocity and volume of gases passing up the stack.

UTILITY GASES - Natural gas, manufactured gas, liquefied petroleum gas-air mixtures or mixtures of any of these gases.

VELOCITY PRESSURE - Pressure exerted by a flowing gas by virtue of its movement in the direction of its motion. It is the difference between total pressure and static pressure.

VENT - A device, such as a pipe or chimney, to transmit flue products from an appliance to the outdoors. This term also is used to designate a small hole or opening for the escape of a fluid (such as in a gas control).

VENT GASES - Products of combustion from gas appliances plus excess air, plus dilution air in the venting system above a draft hood.

VENTURI - A section in a pipe or a burner body that narrows down and then flares out again.

VISCOSITY - The property of a fluid to resist flow, ie. thickness.

WATER COLUMN - Abbreviated as W.C. A unit used for expressing pressure. One inch of water column equals a pressure of 0.578 ounces per square inch. One pound per square inch equals 27.7" W.C.

YELLOW FLAME BURNER - (See Burner).

YELLOW TIPS - (Yellow Tipping) - The appearance of yellow tips in an otherwise blue flame, indicating the need for additional primary air.

ELECTRICAL TERMS

ALTERNATING CURRENT (A.C.) - A flow of electricity which rises from zero to some maximum value in one direction, falls off to zero, then reverses and reaches a maximum value in the other direction, then again falls off to zero. This cycle is repeated continuously at a fixed frequency.

AMPERE OR AMP. - A unit of low electrical current (flow of electrons). One ampere represents the flow of 6.25×10^{18} electrons per second past a given point in the circuit. One volt (potential difference) across a resistance of one ohm will cause one ampere of current to flow.

AUTOMATIC PILOT DEVICE (SAFETY SHUTOFF DEVICE) - A device which is designed to automatically shut off the gas supply to the controlled burner(s) in the event the source of ignition fails. This device may interrupt gas flow to the main burner(s) only, or to pilot(s) and main burner(s) under its supervision.

BI-METAL - A device consisting of two strips of different materials fastened firmly together in the form of a strip, coil or dished disc. When the bi-metal is heated, one material will try to expand (lengthen) more than the other, but is restricted from doing so by the other material. As a result, the strip will bend, the coil twist or the disc invert producing a motion which can be used, for example in opening or closing a switch.

CAPACITOR - An electrical device consisting basically of two or more conducting surfaces separated by an insulating material. It acts to store electrical energy, to block flow of DC current, and to act as a conductor of AC current.

CIRCUIT - An arrangement of conductors and devices connected together for the purpose of carrying an electrical current, and done work.

CIRCUIT BREAKER - A device designed to open and close a circuit by nonautomatic means, and to open the circuit automatically on a predetermined overload of current, without injury to itself when properly applied within its rating.

CURRENT (Electrical) - The rate of movement of free electrons through an electrical conductor. It is usually expressed in amperes and is designated by the symbol I.

DIRECT CURRENT (DC) ELECTRICITY - A form of electricity in which current flows at an essentially constant value in one direction only in a circuit.

IGNITOR - Any device used to light gas. A spark ignitor uses an electric spark generated across an air gap for this purpose.

INSULATOR - A material which offers very high resistance to flow of electrical current, so that essentially no current flows through it. Used to isolate and separate conductors in a circuit so that electricity only flows through those conductors in the intended manner.

JUMPER WIRE - A length of wire used to temporarily complete a circuit, specifically by-passing a device (such as a switch) in that circuit.

KILOWATT (KW) - Unit of electrical power, equal to 1000 watts. Equivalent to 3413 BTU per hour, when converted to heat.

KILOWATT HOUR - Unit of electrical energy supplied by 1000 watts (1 KW) for one hour. Equivalent to 3413 BTU of heat.

LOAD (ELECTRICAL) - The power consumed by components in an electrical circuit. Sometimes all elements which consume electrical power are referred to as the load in the circuit.

OHM - The basic unit of electrical resistance. It is the resistance which will allow one ampere of current flow when the voltage across the resistance is one volt.

RESISTANCE (ELECTRICAL) - The property of a material to resist the flow of electric current through it. Designated as R and is expressed in units of ohms.

RESISTOR - A device which acts to limit current flow in a circuit by virtue of its electrical resistance.

1. COMPOSITION RESISTOR - A resistor made of carbon particles mixed with a binder and provided with terminal leads to connect the resistor (usually cylindrical in shape) into a circuit. The resistance value is determined by the relative amounts of carbon and binder used.

2. FIXED RESISTOR - A resistor with one essentially constant resistance value (except changes due to temperature variations.

3. WIRE WOUND RESISTOR - A resistor made by winding wire around a non-conducting cylinder. The resistance value is determined by the diameter and the number of turns of the wire.

4. VARIABLE RESISTOR - A resistor constructed so that the value of resistance may be varied over a range usually by mechanical means.

SERIES CIRCUIT - A circuit in which only one continuous path is provided for current flow and so the same current flows through all components in the circuit.

SHORT CIRCUIT - Also called a short. An abnormal, low resistance connection between two points in a circuit, such as created when two conductors of different polarity inadvertently touch each other. This condition can create a relatively large flow of current between the points, sometimes causing damage.

SPARK GAP - A gap between two electrodes. When a high potential difference is imposed across the electrodes, current will jump the gap, creating an electrical spark for such purposes as igniting gas.

SWITCH - A device which completes or breaks the path of current flow in a circuit or which sends it into a different path.

1. NORMALLY OPEN SWITCH - A switch is in the open position, interrupting current flow in the circuit which it controls, when the switch is not activated. Designated as N.O.

2. NORMALLY CLOSED SWITCH - A switch is in the closed position, allowing current to flow in the circuit it controls when this switch is not activated. Designated as N.C.

THERMISTOR - A device whose resistance varies with temperature to such a degree that it can be used as a control element.

THERMOCOUPLE - A device consisting of two wires or strips of dissimilar materials which are joined together at one end (the hot junction). When this hot junction is heated, the thermocouple produces a DC voltage across the other two ends (cold junction).

TRANSFORMER - An electrical device which, by electromagnetic induction, transforms AC power in one circuit to another circuit(s), usually at different voltage and current values.

VOLT - The unit of voltage (potential difference, emf). One volt will produce a flow of one ampere through a one ohm resistance.

VOLTAGE - The relative amount of electric charge at one point in an electric circuit compared with that at another point in the circuit, which causes a current flow through a continuous path between the two points. Also referred to as electromotive force and potential difference.

VOLTMETER - An instrument for measuring voltage.

WATT - Unit of electrical power. In a purely resistive AC circuit or DC circuit, one watt of power is consumed by a load when one volt causes a flow of one ampere through the load.

WATTMETER - An instrument which measures true power directly.

INSTALLATION
INSTRUCTIONS
AND
OPERATING
PROCEDURE

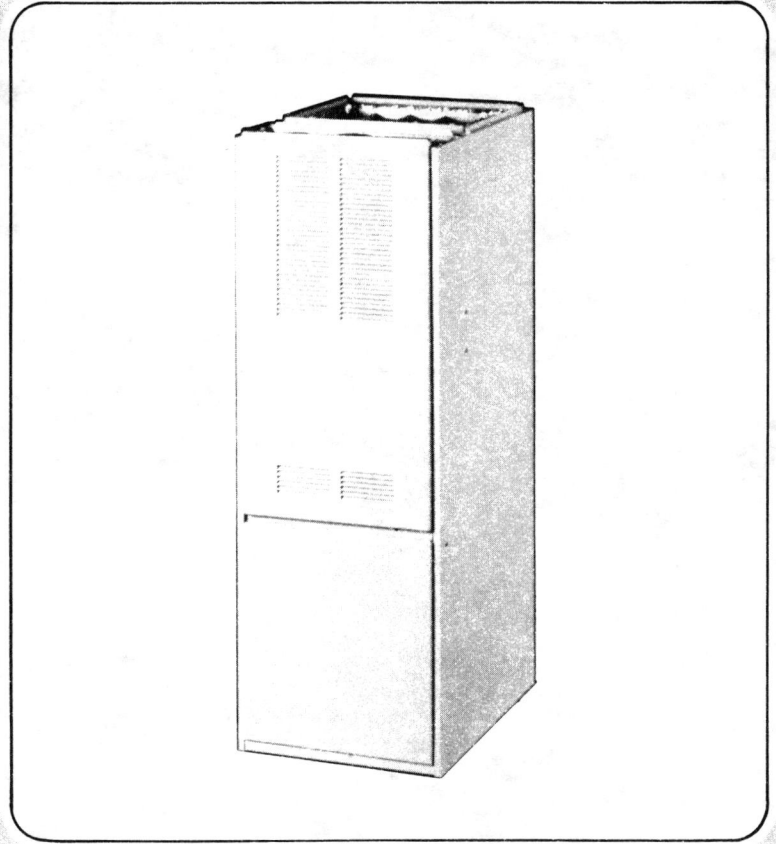

DELUXE
ENERGY SAVER

FORCED
DRAFT
UPFLOW
GAS FIRED
FURNACE

REVISED: OCTOBER, 1987
SUPERSEDES:
NOVEMBER, 1986

START-UP CHECK SHEET

OWNER NAME:_____ DEALER NAME:_____

ADDRESS:_____ ADDRESS:_____

CITY, STATE, ZIP:_____ CITY, STATE, ZIP:_____

MODEL NUMBER:_____ SERIAL NUMBER:_____

TYPE OF GAS: NAT:_____ LPG:_____ LP STICKER APPLIED:_____ LP KIT NUMBER_____

GAS
VALVE_____ LINE
PRESSURE_____ MANIFOLD
PRESSURE_____ INPUT_____

THERMOSTAT:_____ SUBBASE:_____ HEAT ANTICIPATOR SETTING:_____

BLOWER MOTOR H.P._____ VOLTAGE:_____

FAN OPENS:_____ °F   FAN CLOSES:_____ °F

FAN AND LIMIT
MODEL NUMBER:_____ LIMIT OPENS:_____ °F   LIMIT CLOSES:_____ °F

TEMPERATURE RISE: RETURN AIR:_____ °F   SUPPLY AIR:_____ °F

FILTER SIZE AND TYPE:_____

BURNER FLAME PROPERLY ADJUSTED?_____

DRIP-LEG INSTALLED PRIOR TO GAS VALVE?_____

BLOWER SPEED CHECKED?_____ BLOWER PROPERLY LUBRICATED?_____

ALL ELECTRICAL CONNECTIONS, MOUNTING SCREWS, ETC. TIGHT?_____

SHIPPING BLOCKS AND SCREWS REMOVED?_____

ADJUSTMENT REQUIRED:_____

DATE OF INSTALLATION?_____

DATE OF START-UP:_____

_____

_____

_____

_____

_____

_____

_____

_____

INDEX

SPECIFICATIONS AND INFORMATION										
MODELS			040A	060A	080A		100A		120A	
HEATING CAPACITIES	INPUT (BTUH)*		40,000	60,000	80,000	80,000	100,000	100,000	120,000	120,000
	OUTPUT (BTUH)**		32,000	48,000	64,000	64,000	80,000	80,000	96,000	96,000
	AFUE %		83.0	83.0	83.0	83.0	83.0	83.0	83.0	83.0
ELECTRICAL SPECIFICATIONS	UNIT VOLTAGE		120V 60 HZ				120V 60 HZ			
	MIN./MAX. VOLTAGE RANGE		115/120				115/120			
	DIRECT DRIVE	BLOWER DRIVE	010	012	012	016	016	020	016	020
		RATED LOAD ① CURRENT	5.4	5.8	6.5	9.4	8.2	11.7	8.2	11.7
		AMPACITY	7.1	7.6	8.5	12.1	10.6	15.0	10.6	15.0
		MAX. OVERCURRENT PROTECTION	10	15	15	20	15	20	15	20
AIR DELIVERIES①	MAX. CFM @ 0.5 ESP	DIRECT DRIVE	1100	1195	1325	1660	1590	2025	1520	2180
STANDARD CONTROLS AND PIPING INFORMATION	ADJ. FAN & FIXED LIMIT CONTROL		STANDARD							
	TRANSFORMER (120V/24V)		STANDARD							
	COOLING BLOWER RELAY		STANDARD							
	GAS VALVE		STANDARD ½" INLET							
	MANIFOLD SIZE (INLET)		STANDARD ½" N.P.T.							
GENERAL INFORMATION	AIR FILTER	FILTER TYPE	WASHABLE							
		FILTER SIZE (QTY)	16 x 25 x 1 (1)							
	UNIT VENT SIZE (INCHES)		3	3	3	3	4	4	4	4
	UNIT SHIP WEIGHT LBS. (APPROX.)		176	191	191	191	220	220	225	225

① NOMINAL AMP. LOAD DEPENDING ON INSTALLATION CONDITIONS AND HEATING OR COOLING OPERATIONS.
* RATINGS SHOWN ARE FOR ELEVATIONS UP TO 2000 FT. RATINGS SHOULD BE REDUCED 4% PER 1000 FT. FOR EACH 1000 FT. ABOVE SEA LEVEL.
** RATED IN ACCORDANCE WITH D.O.E. TEST PROCEDURES.

CODE: GUA 72120351087

GAS FIRED—FORCED DRAFT FURNACE—PREWIRED & PREASSEMBLED UPFLOW SERIES

INFORMATION CONTAINED IN THIS MANUAL IS DESIGNED TO ACQUAINT YOU WITH THE MECHANICAL FEATURES, AND TO SERVE AS AN EMERGENCY REPAIR GUIDE IN CASE OF HEAT FAILURE.

PROPER CARE OF YOUR GAS FURNACE WILL ASSURE YOU OF MANY YEARS OF ECONOMICAL COMFORT AND SERVICE. LIKE EVERY PIECE OF FINE MECHANICAL EQUIPMENT, YOUR FURNACE SHOULD BE SERVICED PERIODICALLY BY AN EXPERT. WE RECOMMEND A COMPLETE INSPECTION BY AN AUTHORIZED SERVICEMAN AT LEAST ONCE EACH YEAR—USUALLY AT THE BEGINNING OF THE HEATING SEASON.

SECTION I

FACTS ABOUT THE FURNACE

UNIT DESCRIPTION

This furnace is a gas Forced Draft Upflow furnace. It is designed for indoor installation, and is listed as a Category I furnace.

The nameplate label on the furnace shows the model number of the furnace. This label is located on the left hand side of the casing inside unit. See FIGURE 1 for the location of the nameplate label.

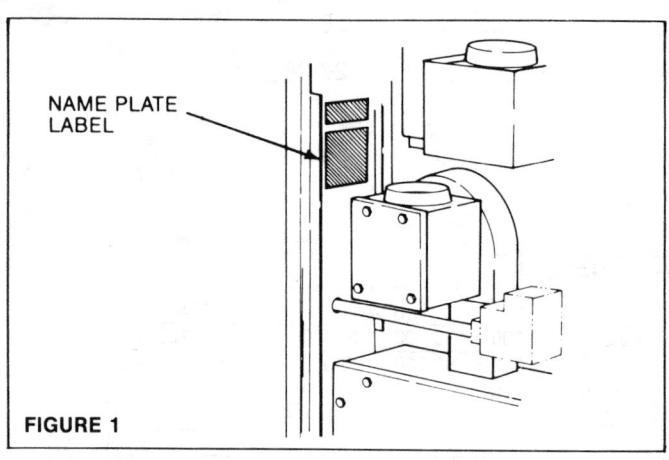

FIGURE 1

MODEL NO.	
SERIAL NO.	MFG. DATE
2505-098(a)	

Please note that each digit of the model number has a special significance:

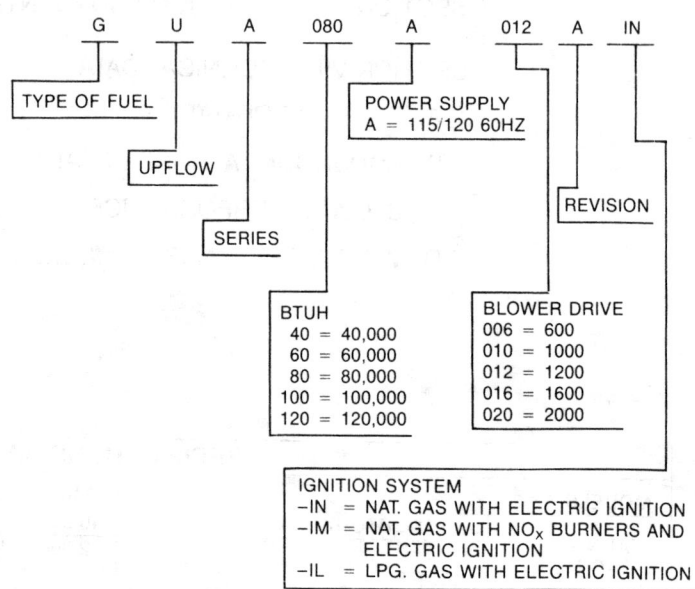

WARNING

Before operating, smell all around the appliance area for gas. Be sure to smell next to the floor because some gas is heavier than air and will settle on the floor.

Check the furnace model number on the nameplate label. The suffix of the model number indicates the type of fuel and the ignition system.

GENERAL INSTALLATION REQUIREMENTS

The information in this manual is designed to provide the installer with adequate information to install this furnace properly.

> ### ⚠ WARNING
> Improper installation, adjustment, alteration, service or maintenance can cause injury or property damage. Refer to the installation instructions and user's manual provided with the furnace. For assistance or additional information, consult a qualified installer, service agency or the gas supplier.

UNCRATING AND INSPECTING

After uncrating the unit, inspect thoroughly for concealed damage, if found, notify transportation company immediately and file concealed damage claim.

GENERAL REQUIREMENTS

All installations should be made as outlined in the latest edition of National Fuel Gas Code ANSI-Z223.1-1984, and other applicable National and Local codes, including requirements of Local Utilities. In addition, the unit must be electrically grounded in accordance with the latest edition of National Electric Code ANSI-NFPA-#70-1987. Current Code Edition(s) are available for a fee from the American Gas Association, 1515 Wilson Blvd. Arlington, Virginia 22209.

Select a level spot for the installation of the unit. Install as near to the flue or chimney as possible and install the vent pipe with a minimum number of elbows. It is also recommended that if possible, the unit be centralized with relation to the distribution system.

LOCATION OF THERMOSTAT

Thermostat should be mounted 4 to 5 feet above the floor, on an inside wall of the living room, dining room, or a hallway that has good air circulation from the other rooms being controlled by the thermostat. It is essential that there be free circulation of air at this location, of the same average temperature as other rooms being controlled. Movement of air should not be obstructed by furniture, doors, draperies, etc. The thermostat should not be mounted where it will be affected by drafts, hot or cold water pipes or air ducts in walls, radiant heat from a fireplace, lamps, the sun, T.V., etc., or on an outside wall. Consult Instruction Sheet packed with thermostat for mounting instructions.

ACCESSIBILITY CLEARANCES

All servicing and cleaning of these units can be done from the front, and a minimum of 36″ horizontal clearance should be allowed. If necessary, these units can be installed at the clearances listed in TABLE 1. However, if access is required to some other piece of equipment, a minimum of 18″ should be allowed for passage.

In a closet or utility room installation, the door must be large enough to allow replacement of the unit if necessary or to permit replacement of any other appliance within the confined space. A minimum of 36″ clearance should be allowed at the front of the unit for servicing and cleaning when the closet door is open. See FIGURE 2.

In addition adequate clearances must be provided around air openings into the combustion chamber. The furnace should not be installed directly on carpeting, tile or other combustible materials other than wood flooring.

If the furnace is installed in a residential garage the unit must be installed so that the burners and the ignition source are located not less than 18 (457 mm) inches above the floor. The furnace must also be located or protected to avoid physical damage by vehicles.

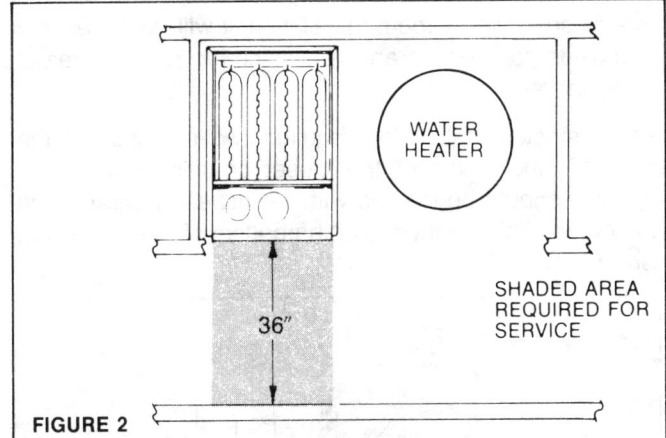

WATER HEATER

SHADED AREA REQUIRED FOR SERVICE

36″

FIGURE 2

PROVISIONS FOR COMBUSTION AIR AND VENTILATION

The flow of combustion and ventilating air must not be obstructed from reaching the furnace. Air openings provided in the casing of the furnace must be kept free of obstructions which would restrict air flow, thereby affecting efficiency and safe operation of the furnace. Keep this in mind should you choose to remodel the area which contains the furnace. Furnaces must have air for proper performance.

RECOMMENDED CLEARANCE GUIDE

(Dimensions shown in inches are the minimum clearances to combustible materials for which the furnace design has been certified.)

TABLE 1

INPUT BTU/HR.	FRONT	REAR	SIDES	TOP	VENT CONNECTION*	FURNACE FLUE PIPE SIZE	COMBUSTIBLE FLOOR**
40,000	6	1	1	1	6	3 Dia.	0
60,000	6	1	1	1	6	3 Dia.	0
80,000	6	1	1	1	6	3 Dia.	0
100,000	6	1	1	1	6	4 Dia.	0
120,000	6	1	1	1	6	4 Dia.	0

*May be 1 inch when listed Type B-1 Vent is used. This is a Category I Furnace.
**Shall not be installed directly on carpeting, tile or other combustible materials, other than wood flooring.

> ### ⚠ WARNING
>
> The air for combustion and ventilation must not come from a corrosive atmosphere.

Combustion air must be free of acid forming chemicals; such as sulphur, fluorine and chlorine. These elements are found in aerosol sprays, detergents, bleaches, cleaning solvents, air fresheners, paint and varnish removers, refrigerants and many other commercial and household products. Vapors from these products when burned in a gas flame form acid compounds. The acid compounds increase the dew point temperature of the flue products and are highly corrosive after they condense.

In open basements of normal construction, an adequate supply of air is obtained from infiltration.

In a closet or utility room installation, it will be necessary to provide combustion and ventilation air from an area of adequate air supply.

Any restricted installation requires two openings in the door or single wall of the closet or utility room. One opening should be located with 12 inches of the top and the other within 12 inches of the bottom of the enclosure. See FIGURE 3.

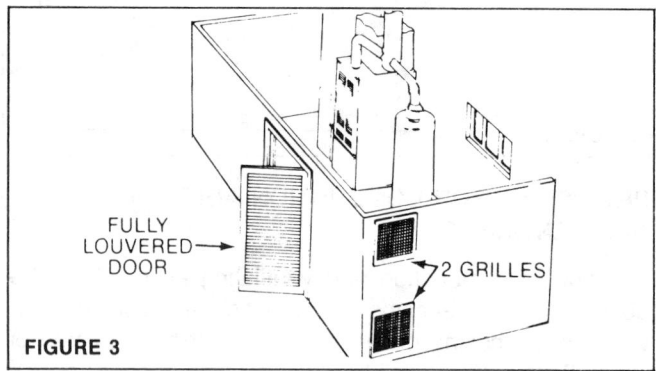

FULLY LOUVERED DOOR

2 GRILLES

FIGURE 3

The following minimum free area is required for these openings:

a) Openings to interior space—Each opening to interior space must have a free area of at least 1 square inch per 1,000 BTU per hour input rating, but not less than 100 square inches.

b) Openings to outdoors—Each opening to the outdoors must have a free area of at least 1 square inch per 2,000 BTU per hour input rating if horizontal air ducts are used.

Air openings in the casing front, return air grilles, and warm air registers must not be obstructed.

In all installations provisions for adequate combustion air must be made in accordance with Section 5.3, Air for Combustion and Ventilation, of the National Fuel Gas Code, ANSI Z223.1-1984, or applicable provisions of the local building codes.

DUCT CONNECTIONS

It is recommended that a flexible duct connection of a non-flammable material be used for the return air and supply air connections to prevent transmission of vibration.

Never make an installation without a ducted return air system to the unit. The return connection must be made full size to a location outside the utility room. RETURN AIR MUST NEVER BE DRAWN FROM THE INSIDE OF A CLOSET, OR A UTILITY ROOM.

In addition, when the furnace is located in an area near or adjacent to the living area, the system should be carefully designed with returns to minimize noise transmission through the return air grille. Any blower moving a high volume of air will produce audible noise which could be objectionable when the unit is located very close to a living area. It is often advisable to route the return air ducts under the floor or through the attic.

When a furnace is installed so that supply ducts carry air circulated by the furnace to areas outside the space containing the furnace, the return air shall also be handled by a duct(s) sealed to the furnace casing and terminating outside the space containing the furnace.

ADDITION OF AIR CONDITIONING

When a refrigeration coil is used in conjunction with this unit, it must be installed on the discharge side of the unit to avoid condensation on the heat exchanger. With a parallel flow arrangement, dampers must be installed to prevent chilled air from entering the furnace. If manually operated dampers are used, they must be equipped with a means to prevent operation of either unit unless the damper is in full heat or full cool position.

HIGH STATIC OPERATION

These blowers can be operated against external static pressures up to 0.5" water column, or as indicated on rating plate.

ELECTRICAL CONNECTIONS

These units require 115 VOLT SINGLE PHASE POWER SUPPLY. Connect 115 Volt power supply and 24 Volt thermostat wires as indicated in the Wiring Diagram, attached to the inside of the blower compartment door, or refer to the diagram inserted in this book. Optional electrical power connections are also shown on the diagram supplied with the furnace.

⚠ WARNING

In order to prevent the possibility of electrical shock, this equipment must be electrically grounded in accordance with the latest edition of the National Electrical Code ANSI/NFPA 70-1984.

VENTING

Please refer to Figure 20 Parts List for difference between inlet air box and vent outlet box.

Flue vent installations for this series shall be in accordance with Part 7, "Venting of Equipment", of the National Fuel Gas Code, ANSI Z223.1-1984 or applicable provisions of local building codes.

Horizontal flue vent requires the use of Stainless Steel vent pipe and our approved *horizontal vent kit* (4" P/N 8G03005, & 3" P/N 8G03004). A horizontal flue vent system also requires a piped intake combustion air vent from the outside (refer to Installation Instructions for Upflow Forced Draft Furnace Horizontal Venting P/N 7502-084).

We recommend the use of B-1 type flue vent for *all* vertical vent applications. The B-1 type flue vent referred to is *vertical flue vent* that starts at the forced draft flue outlet and terminates through the building's roof. The vent must be terminated with a listed cap or roof assembly.

If the flue venting is to be done through an existing masonry chimney, it is strongly recommended that when local experience indicates that flue gases may condense to liquid within the chimney, provisions should be made to provide a suitable metal liner, the chimney should be lined with an approved flue pipe which terminates at or above the top of the chimney, and must be terminated with a listed cap or roof assembly.

Proper vent termination is required so that wind from any direction will not create positive pressure in the vicinity of the outlet, to prevent entry of rain water into the flue vent system and to prevent flue restriction by ice build-up during sub-freezing weather conditions.

Additionally, provisions must be made, by the installer, for the safe removal of any condensate that might form in the chimney.

⚠ CAUTION:

Failure to provide for safe removal of condensate, could result in water damage to the building.

If the flue venting is to be done through an existing chimney and in the installers judgement, it would be impractical or impossible to line the chimney with new B-1 type vent, we recommend the following:

1. The chimney should be inspected for conformance with local and national codes (refer to Article X of the National Building Code or the Standard for Chimney and Vents NFPA 211).

2. If the chimney is of outside construction (not enclosed by the building) the chimney should be lined or relined with a suitable metal liner:

    a. B-1 type vent with proper terminals, or

    b. With single wall metal pipe having resistance to heat and corrosion not less than that of galvanized sheet steel or aluminum not less than 0.016 inch thick and with proper ends or terminations, or

    c. If permitted by local codes, with stainless steel vent pipe not less than 0.016 inch thick.

    The chimney liner or flue vent pipe must be joined together in accordance with the vent pipe manufacturer's instructions, but in any event all joints must be assembled to provide a tight seal for the entire flue system.

    d. It is recommended that the vent from the furnace and any other vented appliance be attached to the liner, and that the liner be terminated with an elbow or a closed bottom "T" to trap any condensate that may form.

3. If the chimney is located within the building and extends through less than five (5) feet of uninsulated attic space, then the flue vent connector may connect directly to the chimney and the chimney used without a metal liner.

The following additional recommendations are made for flue vent systems for Forced Draft furnaces:

1. Select flue connection material that is satisfactory for installation and that meets requirements of local codes.

    In applications where the flue vent or chimney is not directly above the flue opening of the furnace, a vent connector of single wall galvanized steel or aluminum of not less than 0.016 inch wall thickness (28 gauge) may be used, where clearances to combustible materials is not less than 6 inches.

2. Flue connection pipe must be at least as large as the diameter of the outlet collar on the furnace. Do not reduce the diameter of the vent connector between the furnace outlet collar and the flue vent.

3. Run flue vent pipe from the vent system or chimney directly to the furnace outlet collar, utilizing the shortest route possible.

4. Frequently, the job requires horizontal vent pipe runs and elbows between the furnace flue collar and the vertical vent. We recommend no more than 4 elbows in vent systems with insulated vent pipe or 3 elbows in vent system with uninsulated vent pipe.

   In applications where the horizontal run of B-1 type vent passes through an unheated space, the pipe should be insulated with R-7 or equal insulation.

5. Horizontal runs should slope upward from the furnace not less than 1/4 inch per foot (21 mm/m).

6. Horizontal portions of the vent piping should be supported with hangers or straps every 3 feet to prevent sagging and to insure that there will be no movement after installation.

7. Vertical flue vents should extend high enough above the roof or neighboring structures so that wind from any direction will not create positive pressure in the vicinity of the outlet (refer to ANSI Z223.1-1984 for recommended dimensions).

It is always advisable to consult local codes, or in the absence of local codes the National Fuel Gas Code when installing a flue vent system.

VENTING (HORIZONTAL)

For Horizontal venting, special intake and vent terminals with stainless steel pipe connections must be used. Follow instructions furnished with the Horizontal Venting Kit, or as outlined in Addendum A in this manual.

VENTING (COMMON)

It is possible for the Forced Draft Furnace to be common vented with another gas burning appliance, such as a gas water heater. However, there are some minimum specifications that are applicable to the common venting system. These are listed below with additional references made to FIGURES 4 thru 7 as shown:

TABLE 2

FURNACE BTUH	MINIMUM HEIGHT①	FURNACE VENT SIZE	MINIMUM COMMON VENT	
			3" WATER HEATER	4" WATER HEATER
40,000	8 FT	3"	4"	4"
60,000	8 FT	3"	4"	4"
80,000	8 FT	3"	4"	4"
100,000	8 FT	4"	5"	5"
120,000	8 FT	4"	5"	5"

① MINIMUM VENT HEIGHT MEASURED FROM WATER HEATER OR 9 FT. FROM FURNACE WHICHEVER IS LOWEST.

REFER TO THE FOLLOWING FIGURES 4 thru 7 ILLUSTRATING MINIMUM COMMON VENTING ARRANGEMENTS:

FURNACE BTUH	VENTED WITH 3" WATER HEATER D.H.	VENTED WITH 4" WATER HEATER D.H.
40,000	FIGURE 4 & 5	FIGURE 4 & 5
60,000	FIGURE 4 & 5	FIGURE 4 & 5
80,000	FIGURE 4 & 5	FIGURE 4 & 5
100,000	FIGURE 6 & 7	FIGURE 6 & 7
120,000	FIGURE 6 & 7	FIGURE 6 & 7

These illustrations show the minimum pipe sizes and dimensions. Larger pipe sizes and dimensions may be used to accomodate the various furnace and water heater installations.

The height of the connector above the furnace will vary from the dimensions shown. Connector height will depend on the rise of the water heater connection pipe due to the length of pipe used, height of the water heater and height of the draft hood vent collar above the top of the water heater.

For the use of Type B-1 gas vent or single wall metal pipe and installation of these vents, refer to Part No. 7, Venting of Equipment, of the National Fuel Gas Code ANSI Z223.1 and these instructions. OTHER CONFIGURATIONS MAY BE USED ESPECIALLY THOSE FOUND IN PUBLICATIONS BY VARIOUS GAS VENT (B-TYPE VENT) MANUFACTURERS AND AS PERMITTED BY LOCAL CODES OR UTILITY COMPANIES.

TYPICAL INSTALLATION FOR COMMON VENTING REQUIREMENTS
FOR FORCED DRAFT UPFLOW FURNACE 40,000 - 80,000 BTUH

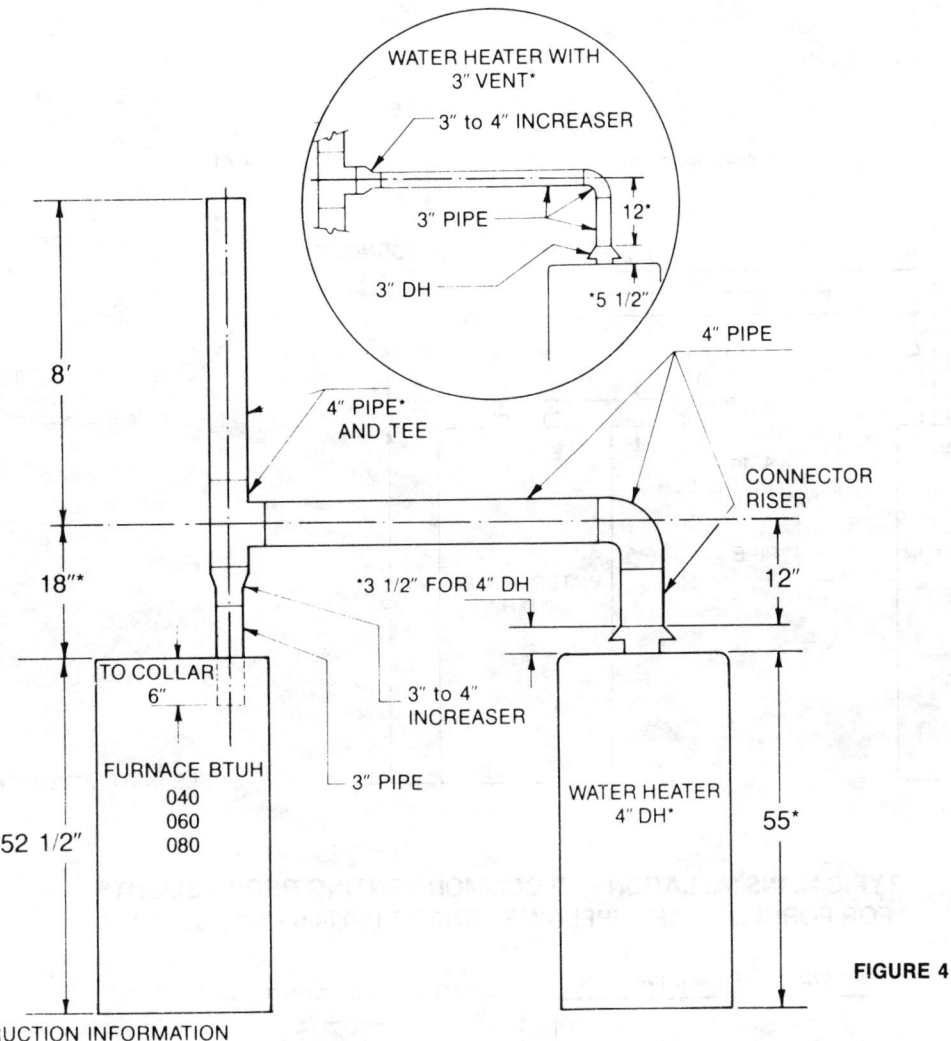

FIGURE 4

*SEE INSTRUCTION INFORMATION

TYPICAL INSTALLATION FOR COMMON VENTING REQUIREMENTS
FOR FORCED DRAFT UPFLOW FURNACE 40,000 - 80,000 BTUH

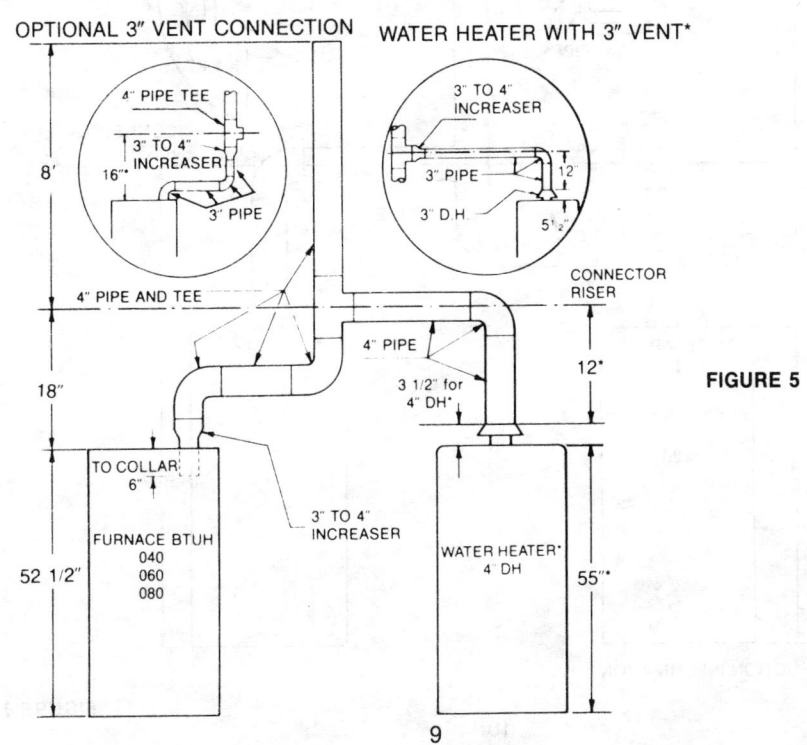

FIGURE 5

9

TYPICAL INSTALLATION FOR COMMON VENTING REQUIREMENTS
FOR FORCED DRAFT UPFLOW FURNACE 100,000 - 120,000 BTUH

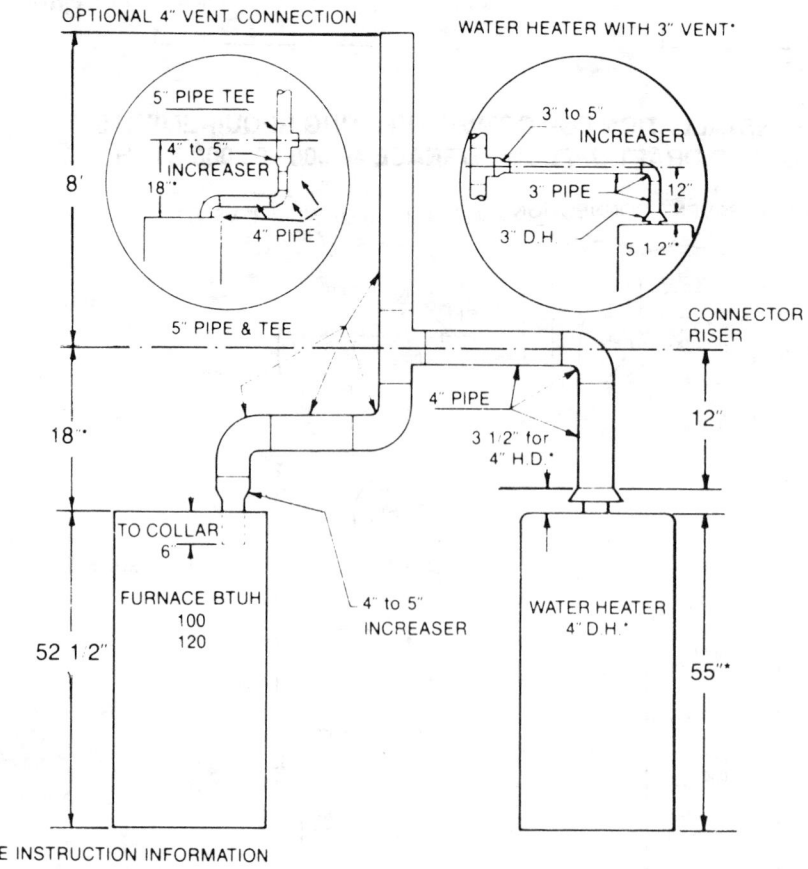

8'

18"*

52 1/2"

5" PIPE AND TEE

4" PIPE

CONNECTOR RISER

12"

3 1/2" for 4" DH*

4" to 5" INCREASER

4" PIPE

TO COLLAR
6"

FURNACE BTUH
100
120

WATER HEATER
4" D.H.*

55"*

WATER HEATER WITH 3" VENT*

3" TO 5" INCREASER

3" PIPE

3" DH

12"*

5 1/2"*

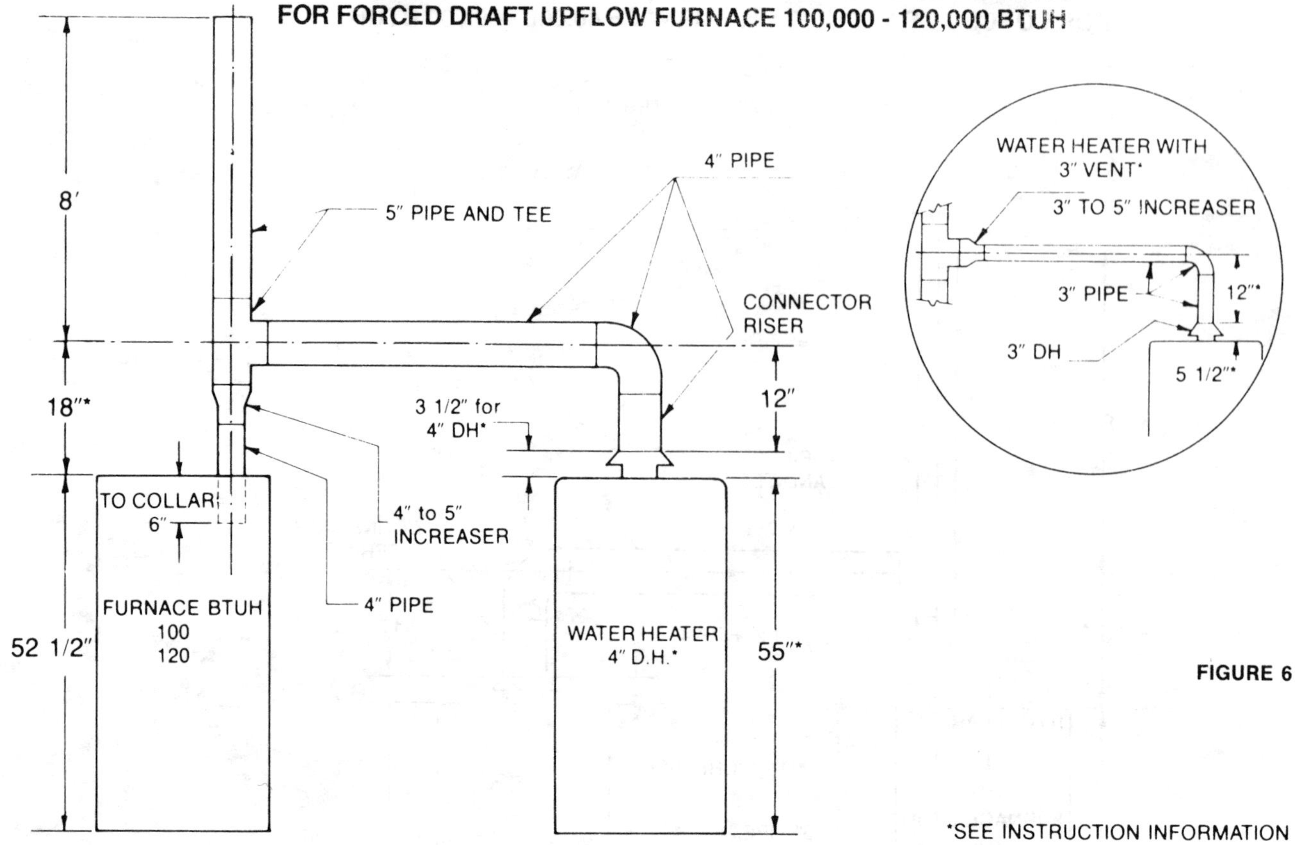

FIGURE 6

*SEE INSTRUCTION INFORMATION

TYPICAL INSTALLATION FOR COMMON VENTING REQUIREMENTS
FOR FORCED DRAFT UPFLOW FURNACE 100,000 - 120,000 BTUH

OPTIONAL 4" VENT CONNECTION

WATER HEATER WITH 3" VENT*

5" PIPE TEE

4" to 5" INCREASER

18"

4" PIPE

3" to 5" INCREASER

3" PIPE

3" D.H

12"

5 1 2"

8'

5" PIPE & TEE

CONNECTOR RISER

4" PIPE

12"

3 1/2" for 4" H.D.

18"

TO COLLAR
6"

FURNACE BTUH
100
120

4" to 5" INCREASER

WATER HEATER
4" D.H.*

55"*

52 1 2"

*SEE INSTRUCTION INFORMATION

FIGURE 7

10

VENT PIPE JOINTS

It is recommended that all the pipe joints be caulked with a high temperature sealant (such as RTV Silicone) to assure a tight seal for the flue system from the furnace to chimney/vent.

⚠ CAUTION:

It is vitally important that the furnace have proper venting. Never attach another appliance or heater to the furnace chimney without proper approval of the installer/servicing mechanic. Any alteration to the flue pipe or chimney must be in accordance with all local and national codes.

GAS PIPING

⚠ CAUTION:

PRESSURE TESTING: The furnace and its individual shutoff valve must be disconnected from the gas supply piping system during any pressure testing of that system at test pressures in excess of ½ PSIG (3.48 kPa.) (13.87″ Water Column.)

The furnace must be isolated from the gas supply piping system by closing its individual manual shut-off valve during any pressure testing of the gas supply piping system at test pressures equal to or less than ½ PSIG (3.48 kPa) (13.87″ Water Column).

⚠ WARNING

Do not use open flame for leak testing gas piping system.

⚠ CAUTION:

Use pipe compounds on gas pipe threads that are resistant to all gases.

An opening has been provided on both sides of the furnace for the entry of the gas piping.

For right hand gas pipe entry, remove the 2″ metal cap and install cap on the unused left side opening.

Install the gas piping from the meter to the unit. Size the pipe as recommended in TABLES 3, 4 & 5. It is recommended, and many local codes require a manual shutoff valve to be located externally to the unit and within 4 feet of the unit.

TABLE 3

Maximum capacity of pipe in cubic feet of gas per hour for gas pressures of 0.5 PSIG or less and a pressure drop of 0.5 inch water column (based on 0.60 specific gravity gas)

Nominal Iron Pipe Size Inch	Internal Diameter Inches	Length of Pipe, Feet													
		10	20	30	40	50	60	70	80	90	100	125	150	175	200
3/8	.493	95	65	52	45	40	36	33	31	29	27	24	22	20	19
1/2	.622	175	120	97	82	73	66	61	57	53	50	44	40	37	35
3/4	.824	360	250	200	170	151	138	125	118	110	103	93	84	77	72
1	1.049	680	465	375	320	285	260	240	220	205	195	175	160	145	135
1 1/4	1.380	1,400	950	770	660	580	530	490	460	430	400	360	325	300	280
1 1/2	1.610	2,100	1,460	1,180	990	900	810	750	690	650	620	550	500	460	430

TABLE 4

Multipliers to be used when Specific Gravity is Other than 0.60

Spec. Grav.	Mult.	Spec. Grav.	Mult.	Spec. Grav.	Mult.	Spec. Grav.	Mult.
.35	1.31	.65	.962	1.00	.775	1.60	.612
.40	1.23	.70	.926	1.10	.740	1.70	.594
.45	1.16	.75	.895	1.20	.707	1.80	.577
.50	1.10	.80	.867	1.30	.680	1.90	.565
.55	1.04	.85	.841	1.40	.655	2.00	.547
.60	1.00	.90	.817	1.50	.633	2.10	.535

TABLE 5

Capacity of Semi-Rigid Tubing in 1000 BTU Per Hour of Undiluted Liquefied Petroleum Gases of 0.50 Pressure Drop

Outside Dia. Inches	Length of Tubing (Feet)									
	10	20	30	40	50	60	70	80	90	100
3/8	39	26	21	19	—	—	—	—	—	—
1/2	92	62	50	41	37	35	31	29	27	26
5/8	199	131	107	90	79	72	67	62	59	55
3/4	329	216	181	145	131	121	112	104	95	90
7/8	501	346	277	233	198	187	164	155	146	138

NOTES
1. Allowance has been made for an average number of fittings.
2. When using a gas with a specific gravity other than 0.60, multiply the capacity by the factor given in Table 2.
3. To convert BTU input per hour to cubic feet per hour, divide BTU per hour by the heating value of the gas to be used. This can be obtained from your local utility.
4. Piping systems of semi-rigid tubing (Table 3) that are to be supplied with gas of a specific gravity of 1.53 or less can be sized directly from Table 3 unless the authority having jurisdiction specifies that a gravity factor be applied. When the specific gravity of the gas is greater than 1.53, the gravity factor shall be applied.

11

GAS PIPING (Cont.)

Install a "Tee" in the gas pipe at the same elevations as the gas inlet connection to the unit. At the bottom of the "Tee" install a "drip-trap" (sediment trap). Install a ground joint union at the top of the "Tee" or between "Tee" and furnace as shown in FIGURE 8.

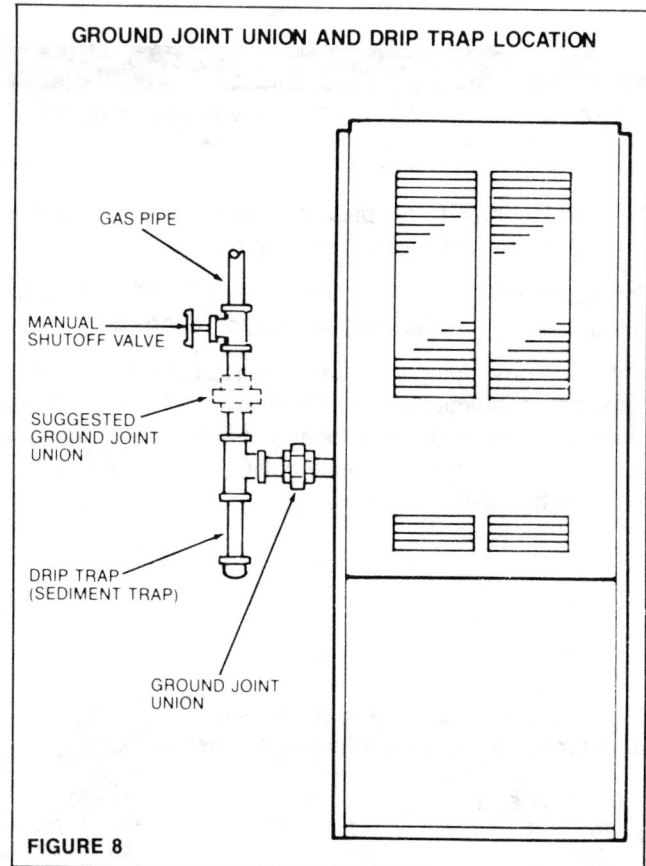

GROUND JOINT UNION AND DRIP TRAP LOCATION

GAS PIPE

MANUAL SHUTOFF VALVE

SUGGESTED GROUND JOINT UNION

DRIP TRAP (SEDIMENT TRAP)

GROUND JOINT UNION

FIGURE 8

A ⅛" N.P.T. plugged tap, accessible for test gauge connection, must be installed immediately upstream of the gas supply connection to the furnace.

All gas piping must conform with National and Local Codes and Local Utility requirements. Upon completion of piping, check for leaks with a soapy solution.

⚠ WARNING

Do not use an open flame such as a candle or match for leak testing the gas piping system.

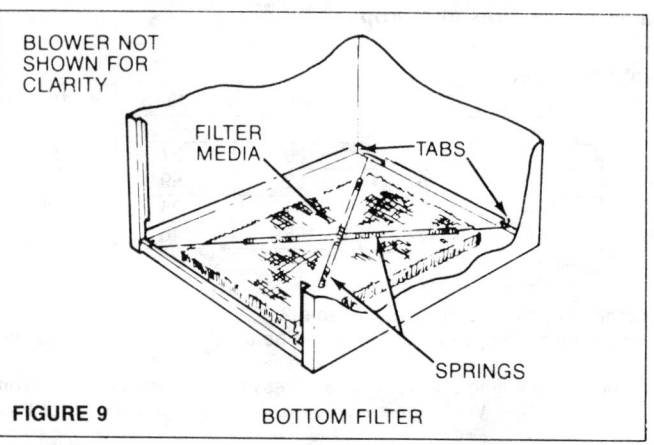

BLOWER NOT SHOWN FOR CLARITY

FILTER MEDIA

TABS

SPRINGS

FIGURE 9   BOTTOM FILTER

BOTTOM AIR RETURN

If return air is to enter the unit from the bottom, cut opening in bottom, center and level the unit over the return air opening. If necessary, grout around the base to seal leaks between the unit and the floor. Insert air filters into guides cut into the furnace base. Proceed with duct work. See FIGURE 9.

SIDE AIR RETURN

If the return air is to enter the unit from the side, it will be necessary to cut the return air opening in the side casing panel. Mark the opening to be cut, using the square embossed knockouts on the casing, as guides. Proceed with duct work. See FIGURE 10.

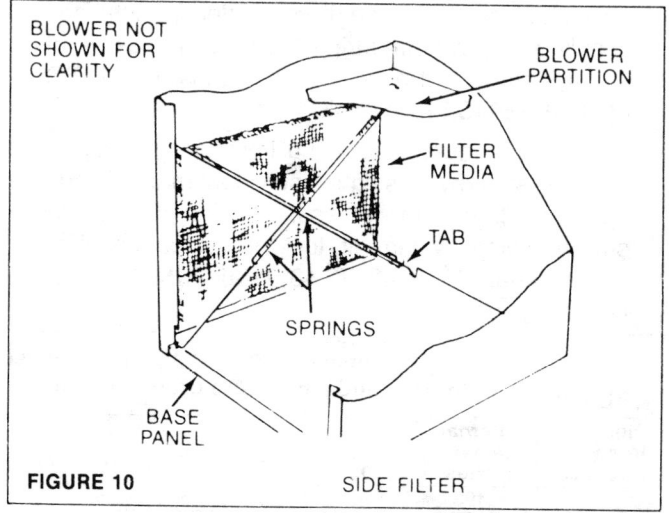

BLOWER NOT SHOWN FOR CLARITY

BLOWER PARTITION

FILTER MEDIA

TAB

SPRINGS

BASE PANEL

FIGURE 10   SIDE FILTER

When using the external filter rack, center the filter rack on the side panel, flush with bottom edge of the furnace. Mark the fastening holes. Drill the fastening holes in the side panel, and fasten the filter rack in place with sheet metal screws. Proceed with duct work. See FIGURE 11.

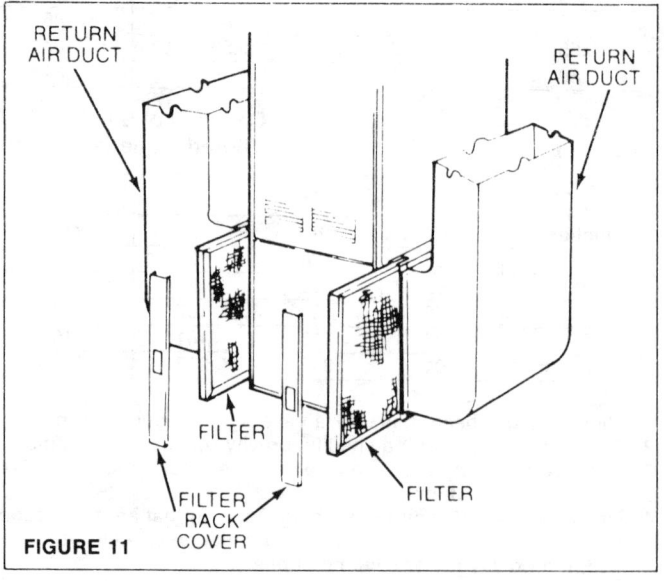

RETURN AIR DUCT

RETURN AIR DUCT

FILTER

FILTER

FILTER RACK COVER

FILTER

FIGURE 11

12

AIR FILTERS FOR ALL FURNACES:

Filters supplied with the external filter rack are of the permanent type. Replacement of either the internal or external air filters should be of the same size and capacity as the original. Failure to provide adequate filter media, can cause equipment malfunction, uneven room temperature, and excessive fuel usage. Refer to the following table for filter size and minimum filter area required for all furnace sizes. See TABLE 6.

TABLE 6	FILTER REQUIREMENTS (HEATING)*		
	FILTER NUMBER AND SIZE		
INPUT BTU/HR.	INTERNAL (STANDARD)**	EXTERNAL (OPTIONAL)***	MINIMUM FILTER AREA (SQ. INCH)**
40,000	(1) 16 x 25 x 1	(1) 20 x 25 x 1	140
60,000	(1) 16 x 25 x 1	(1) 20 x 25 x 1	200
80,000	(1) 16 x 25 x 1	(1) 20 x 25 x 1	260
100,000	(1) 16 x 25 x 1	(1) 20 x 25 x 1	320
120,000	(1) 16 x 25 x 1	(1) 20 x 25 x 1****	400

*For cooling requirements consult specification sheet.
**Filters rated at 520 FPM or more, permanent type.
***See price book or specification sheet for external filter rack options.
****For quantities over 1700 CFM requires (2) external filters.

SECTION IV

LIGHTING INSTRUCTIONS

ELECTRIC IGNITION MODELS WITH "-IN" or "-IM" or "-IL" SUFFIX

⚠ WARNING

If you do not follow these instructions exactly, a fire or explosion may result casing property damage, personal injury or loss of life.

SAFETY INFORMATION

For your safety read the following information before operating furnace:

1. This appliance does not have a pilot. It is equipped with an ignition device which automatically lights the burner. **DO NOT TRY TO LIGHT THE BURNER BY HAND.**

2. BEFORE LIGHTING smell all around the appliance area for gas. Be sure to smell next to the floor because some gas is heavier than air and will settle on the floor.

   WHAT TO DO IF YOU SMELL GAS

   • Do not try to light any appliance

   • Do not touch any electric switch; do not use any phone in your building

   • Immediately call your gas supplier from a neighbor's phone. Follow the gas supplier's instructions.

   • If you cannot reach your gas supplier, call the fire department.

3. Use only your hand to push in or turn the gas control knob. Never use tools. If the knob will not push in or turn by hand, don't try to repair it, call a qualified service technician. Force or attempted repair may result in a fire or explosion.

4. Do not use this appliance if any part has been under water. immediately call a qualified service technician to inspect the appliance and to replace any part of the control system and any gas control which has been under water.

OPERATING INSTRUCTIONS

1. STOP! Read the safety information above.

2. Set the thermostat to lowest setting.

3. Turn off all electric power to the appliance.

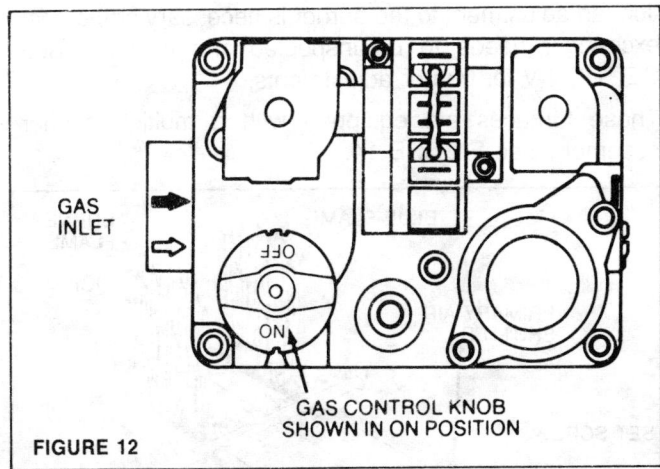

GAS INLET

GAS CONTROL KNOB SHOWN IN ON POSITION

FIGURE 12

4. This appliance is equipped with an ignition device which automatically lights the burner. **DO NOT TRY TO LIGHT THE BURNER BY HAND.**

5. Remove louvered access panel. See FIGURE 13

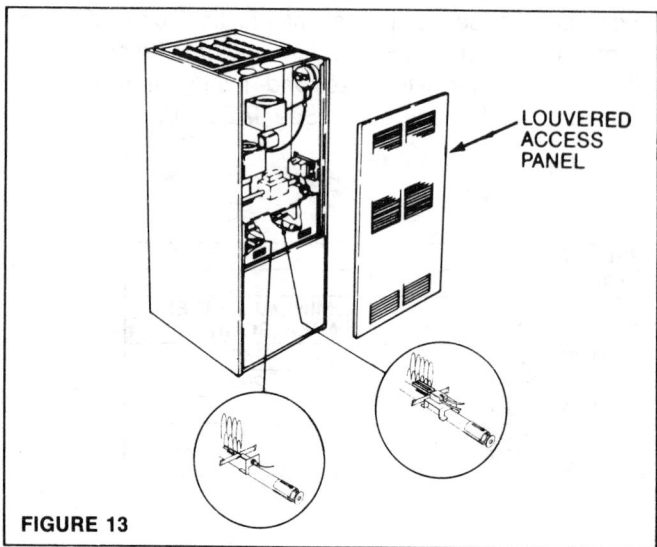

FIGURE 13

6. Turn gas control knob clockwise ⌒ to "OFF".
7. Wait five (5) minutes to clear out any gas. If you then smell gas, STOP! Follow step 2 in Safety Information. If you don't smell gas, go to next step.
8. Turn gas control knob clockwise ⌒ to "ON."
9. Turn on all electric power to the appliance.
10. Set thermostat to desired setting.
11. If the appliance will not operate, follow the instructions "To Turn Off Gas To Appliance" and call your service technician or gas supplier.
12. Replace louvered access panel.

BURNER FLAME REQUIREMENTS

Burner flame should run the length of the burner ports and be 3" to 4" in height. The flame should be blue in color. If for any reason the flame should lift from the burner, change color to orange or start rolling out from under the burner access shield/vertical shield after ignition, an adjustment to the burner is necessary or the heat exchanger needs to be inspected and cleaned. See SECTION V for burner adjustments.

These furnaces are equipped with a multiple burner assembly. See FIGURE 14.

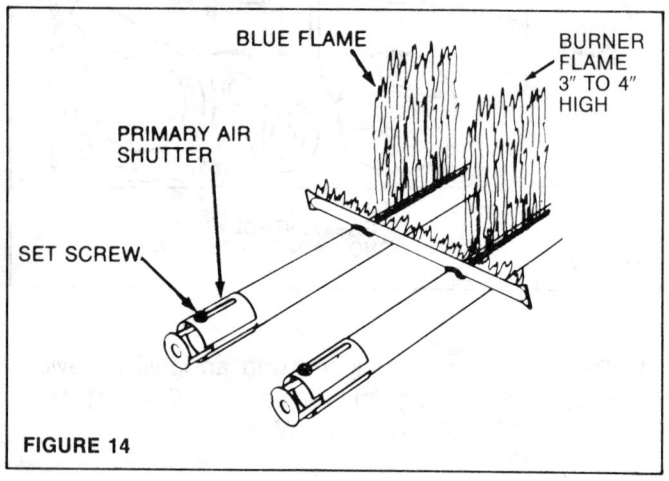

FIGURE 14

TO TURN OFF GAS TO APPLIANCE

1. Set the thermostat to lowest setting.
2. Turn off all electric power to the appliance if service is to be performed.
3. Remove louvered access panel.
4. Turn gas control knob clockwise ⌒ to "OFF."
5. Replace louvered access panel.

⚠ **WARNING**

Should over heating occur or the gas supply fail to shut off, turn off the manual gas valve to the appliance before shutting off the electrical supply.

SEQUENCE OF OPERATION—ELECTRIC IGNITION

1. Thermostat calls for heat, energizing the Forced Draft Blower Relay whilch closes contacts that power the Forced Draft Blower.
2. The rotation of the Forced Draft Combustion Blower will make a centrifugal switch within the motor to prove blower operation and send power from the ignition circuit to the ignition control.
3. The pre-purge mode will cause a (30) thirty second delay before energizing the silicone carbide ignitor.
4. The ignitor is energized for approximately a 10 second warm-up time before the gas valve receives the signal to open.
5. The ignitor provides the heat to light the main burner which lights the remaining burners via the carry-overs. The control system must detect the main burner flame within 7 seconds or the gas valve and the ignitor are de-energized.
6. If the main burner flame is not sensed on the first attempt, the control system de-energizes the gas valve and causes a 90 second delay before attempting re-ignition.
7. The second attempt at lighting the burner flame will duplicate the first attempt with pre-purge, warm-up times and lock-out times identical to those in the first try. Failure to prove on the second attempt will allow yet another 90 second delay and one more complete ignition attempt before complete system lock-out.
8. The 24 Volt circuit or the line voltage circuit must be interrupted for 1/20th of a second or longer in order to reset the system.
9. The ignition sequence will repeat for a total of three re-cycles if the flame is extinguished within 30 seconds of ignition. The total of ignition attempts plus re-cycles must not exceed 3 or the control system will lock-out.

10. If flame is established for more than 30 seconds after ignition, the recycle counter will clear. If flame is lost after 30 seconds, there will be a 60 second wait and restart of the ignition sequence.

11. A momentary loss of power or flame to the sensing circuit will cause the valve to be de-energized and a delay of 60 seconds before restarting the ignition sequence. Re-cycles will begin and the burner will operate normally if the gas supply returns or fault condition is corrected prior to the last ignition attempt. Otherwise the control will lock-out.

12. Typical conditions are that the burner will light and stay lit as long as the thermostat calls for heat.

13. Heat from the burner flames warm the heat exchanger surface, the flue products pass through the baffles and vent outdoors through the flue pipe.

14. The fan and limit control senses the supply air temperature adjacent to the heat exchanger and will signal the blower to operate when the air temperature reaches the "ON" fan set point. The fan & limit control will shut the gas off if the supply air temperature reaches the "High Temperature" set point. The blower continues to operate and conditioned air is supplied to the space through the duct system.

15. When the thermostat set point temperature is reached, the gas valve is de-energized which stops the flow of gas to the burners. The blower continues to operate until the air temperature falls below the "OFF" blower set point.

16. The furnace repeats this same cycle each time the thermostat calls for heat.

SECTION V
FINAL ADJUSTMENTS

TEMPERATURE RISE

The furnace installation is to be adjusted to obtain a temperature rise within the range specified on the rating plate.

PRESSURE REGULATOR ADJUSTMENT

A pressure tap has been provided on the gas valve. See FIGURE 15. Attach a pressure gauge fitting to this tap and check the outlet pressure. The pressure should be set as indicated on the rating plate. To adjust the pressure regulator, remove the cap covering the Pressure regulator adjusting screw. On valves with more than one adjusting screw, adjust only the screw marked "HI". Turn the adjusting screw out (counter-clockwise) to decrease the pressure, and in (Clockwise) to increase the pressure. Reinstall the cap securely.

After use, replace securely the 1/8" pipe plug. For most models the pressure rating is normally 3.5 inches W.C. for Nat. Gas, and 11.0 inches W.C. for LP Gas.

After the pressure regulator has been adjusted to its proper value, check the input by the following formula:

INPUT IN BTU/HR =

$$\frac{①\text{Heating value of gas in BTU-Per-CU. ft. x } 3600 \times \text{CU. ft. gas measured}②}{③\text{Time in seconds for CU. ft. of gas measured}}$$

To obtain energy input:
① Heating value of gas (BTU/CU. Ft.) is obtained from the Local Gas Utility.
② To calculate CU. FT. Gas Measured: turn off all other gas appliances and pilots. REMEMBER to relight all pilots after test is completed.
③ Measure Time (in seconds) for gas meter test dial to complete one revolution (local Utility will provide size of meter).
④ Calculated energy input must not exceed gas input on unit rating plate. If the input does not correspond to that shown on the rating plate, adjust the regulator screw of the pressure regulator until the rating plate input is obtained. The pressure in the manifold should not vary more than plus or minus 0.3 inches of water column as specified on rating plate. Any major changes in the gasflow should be made by changing the size of the burner orifices.

HIGH ALTITUDE INSTALLATION

These units may be used at full input rating when installed at altitudes up to 2000 feet. When installed above 2000 feet, the input must be decreased 4% for each 1000 feet above sea level.

PRIMARY AIR ADJUSTMENT (Air Shutter)

These units have individual primary shutters on each burner. To adjust, loosen the lock screw and close the air shutter until yellow tips appear on the flames, then open the air shutter until the yellow tips just disappear, and tighten the lock screw. Follow this procedure on each burner.

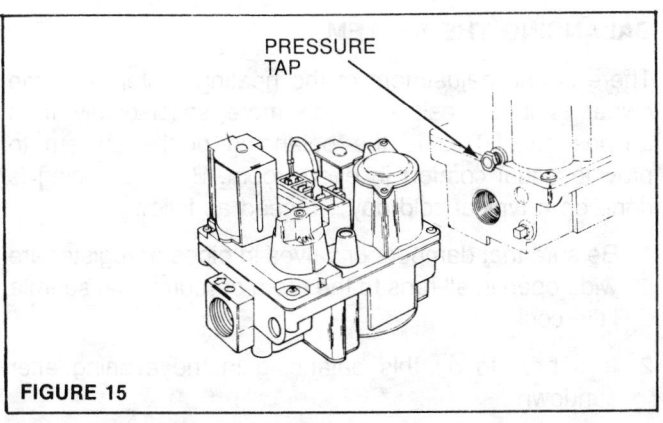

PRESSURE TAP

FIGURE 15

Should there be a need for burner replacement on a furnace with an RPJ heat exchanger, it will be necessary to order the complete set of burners.

The furnaces are manufactured with multiple burner assemblies. See FIGURE 16.

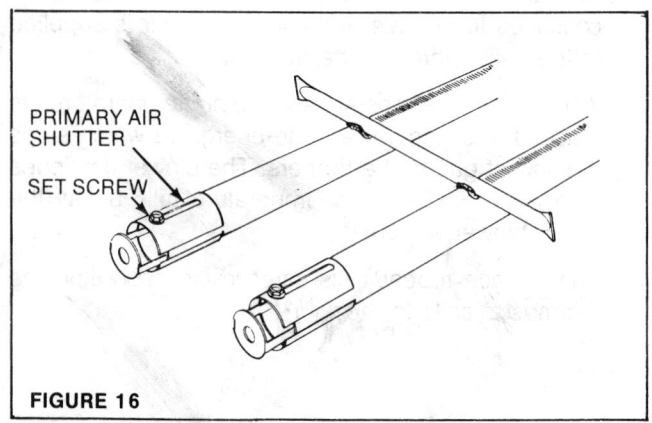

FIGURE 16

THERMOSTAT ANTICIPATOR SETTING

The proper thermostat heat anticipator setting is the amp. rating on the gas control valve and is indicated on a label attached to the furnace. To increase length of cycle increase setting of heat scale. To decrease length of cycle, decrease setting of heater scale.

Proper control of the indoor air temperature can only be achieved if the thermostat is calibrated to the heating and/or cooling system. A vital consideration of this calibration is related to the thermostat heat anticipator.

Anticipators for the cooling operation are generally preset by the thermostat manufacturer and require no adjustment. Anticipators for the heating operation are of two types: preset or adjustable. Those that are pre-set will not have an adjustable scale and are generally marked accordingly.

Thermostat models having a scale as shown in FIGURE 17 *must be* adjusted to each application.

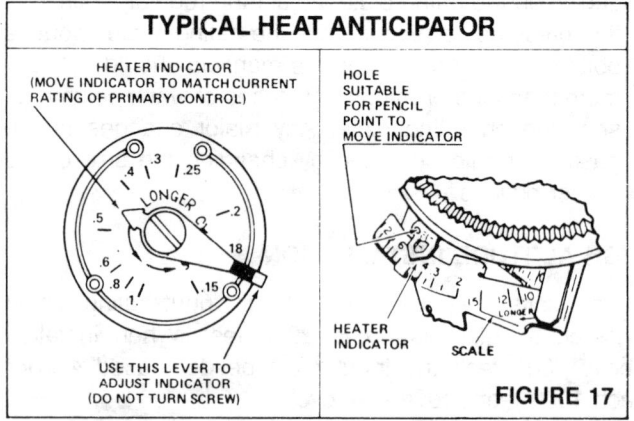

TYPICAL HEAT ANTICIPATOR

HEATER INDICATOR
(MOVE INDICATOR TO MATCH CURRENT RATING OF PRIMARY CONTROL)

LONGER

USE THIS LEVER TO ADJUST INDICATOR (DO NOT TURN SCREW)

HOLE SUITABLE FOR PENCIL POINT TO MOVE INDICATOR

HEATER INDICATOR

SCALE

LONGER

FIGURE 17

In many cases, this adjustment setting can be found in the thermostat instructions. If this information is not available, or if the correct setting is questioned, the below procedure should be followed:

STEP 1. Wrap 10 loops of single strand, insulated thermostat wire around the prongs of an ampmeter. Set the scale to the 1 to 5 or 1 to 6 amp. scale.

STEP 2. Connect the uninsulated ends of this wire jumper across terminals "R" and "W" on the sub-base ("RH" and "W", on multi-stage thermostat sub-base). See FIGURE 18.

This test must be performed without the thermostat attached to the sub-base.

STEP 3. Let the heating system operate in this position for about one minute. Read the ampmeter scale. Whatever reading is indicated must be divided by 10 (for 10 loops of wire).

This is the setting at which the adjustable heat anticipator should be set.

$$\text{FORMULA: } \frac{\text{Ampmeter reading}}{10 \text{ loops}} = \text{Anticipator Setting}$$

$$\text{OR: } \frac{2.5 \text{ Amps}}{10} = .25 \text{ AMPS Setting}$$

STEP 4. If a slightly longer cycle is desired, the pointer should be moved to a higher setting. Slightly shorter cycles can be achieved by moving to a lower setting.

STEP 5. Remove the meter jumper wire and reconnect the thermostat. Check the thermostat in the heating mode for proper operation.

NOTE: The length of the heating cycle can also be affected by the fan limit control settings (if applicable). The fan "on and off" settings should be checked at this point.

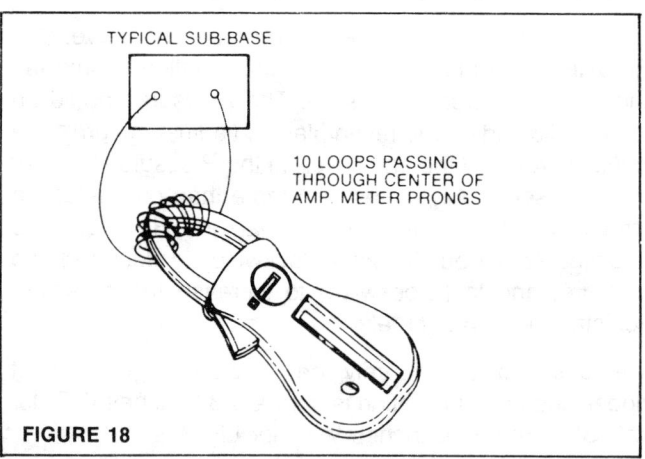

TYPICAL SUB-BASE

10 LOOPS PASSING THROUGH CENTER OF AMP. METER PRONGS

FIGURE 18

For thermostats having 2 stage heat, step 1, 2, and 3 must be repeated. Second stage heat is controlled through terminals "RH" and "W$_2$" on the sub-base.

BALANCING THE SYSTEM

There is one aajustment of the heating system that the owner is in a position to do more satisfactorily than anyone else. This is the balancing of the system to provide equal comfort in every room. Best balancing is done on a typical cold day. Proceed as follows:

1. Be sure that dampers or valves in pipes or register are wide open in all runs to the room or rooms that seem a little cool.

2. It is best to do this balancing in the evening after sundown.

3. The room thermostat should be left at one setting for several hours before you attempt to do any balancing.

4. Check the temperature in all rooms, you can do this with a thermometer.

5. In the room or rooms that are too warm, turn down the register valve.

6. Wait several hours until room comfort balances out at

this new setting and then recheck temperature.

7. By doing the adjusting a little at a time, you will accomplish the temperature balance you want. After a pipe damper or register valve has been adjusted it takes time for the temperature to conform with the new setting. By adjusting a little at a time, you will soon accomplish better system balance than can be done by anyone else who does not live in the house.

SECTION VI

OPERATION AND MAINTENANCE

THERMOSTAT

The thermostat is designed to provide the utmost in heating comfort, and is also designed to be attractive in appearance.

Please remember that these controls are precision devices, and therefore should be handled accordingly. Also, remember that the Thermostat will be affected by any heat source, so do not place lamps, radio, TV, etc., so that heat from these sources can affect the Thermostat operation.

TEMPERATURE SELECTION

To select the temperature control point, move the temperature selection dial until the pointer is in line with the desired point on the temperature scale. The thermometer pointer will not change position on the temperature scale when the temperature selection lever is moved, but will change position as room temperature varies.

FAN AND LIMIT CONTROL

The Fan and Limit Controls are mounted in their proper location. The limit is set at the correct setting. This setting is the point at which the Limit Switch opens contact. The switch is automatic-reset and will make contact when the temperature of the element drops. On installations with adjustable fan controls, the settings should not exceed 140° on and 90° off.

⚠ CAUTION:

Do not tamper with the internal fan or limit control mechanism.

BLOWER SPEED ADJUSTMENT
(DIRECT DRIVE BLOWER)

The multi-speed direct drive motor is wired to the Hi-speed tap for cooling operation and low speed for heating

operation. A change to a lower speed may be made at the motor. Heating and cooling models equipped with a blower relay, change speeds automatically. Refer to wiring diagram attached to the furnace.

LUBRICATION INSTRUCTIONS

For Lubrication Instructions for motor or blower bearings refer to labels attached to blower housing.

FILTERS

If filters are provided as standard equipment, the forced air units are then equipped with efficient washable type air filters designed to keep house cleaner by withholding dirt and other foreign particles from the air circulation. The filters should be cleaned or replaced at least twice a year or at any time they become clogged. Dirty filters will cause the furnace to heat improperly and possibly overheat by holding back air circulation. These furnaces are designed so that the filters may be quickly and easily replaced. Replacement filters must be of the same type and size.

HEAT EXCHANGER AND BURNER CLEANING PROCEDURE:

⚠ CAUTION:

On Units with suffix "-IN", "-IM", or "-IL" a silicon carbide ignition is used. Extreme care must be exercised in handling the assembly due to the fragile nature of the ignition assembly.

⚠ CAUTION:

NOTE: THIS FURNACE IS EQUIPPED WITH A PRESSURED COMBUSTION CHAMBER. IT IS MANDATORY TO REPLACE ALL GASKETS THAT ARE REMOVED, ORDER REPLACEMENT GASKET PRIOR TO STARTING SERVICE.

---

1. Disconnect power source to the unit.

2. Remove louver panel from the unit and set aside. See FIGURE 19.

3. Turn gas supply "off" at the gas valve. Allow a few minutes for unit to cool down.

4. Remove the screws from the front panel of the burner enclosure and set aside.

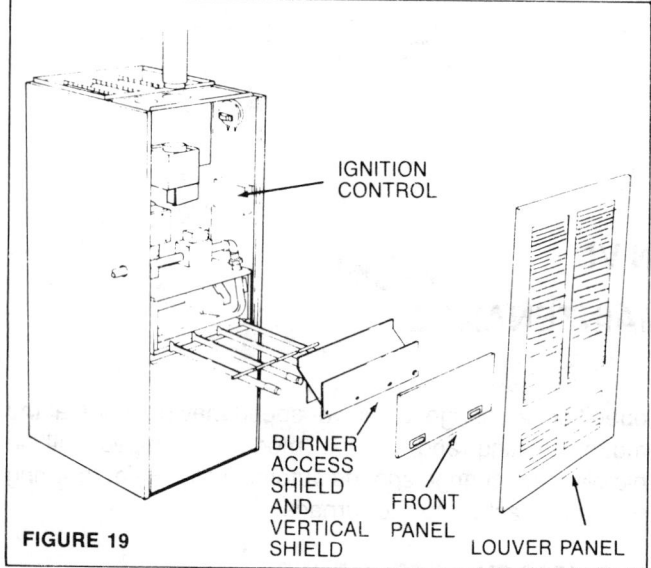

IGNITION CONTROL

BURNER ACCESS SHIELD AND VERTICAL SHIELD

FRONT PANEL

LOUVER PANEL

FIGURE 19

5. Remove the front panel

6. Remove the four screws holding the burner access shield and vertical shield in place and set aside.

7. On the H.S.I. System, disconnect the sensor lead from the remote sensor (far left burner) and the ignitor leads at the ignitor disconnect.

8. Remove the burner access shield, and vertical shield.

NOTE: At this point inspect for any loose particles or carbon deposits in the burner area. Using a flashlight and a small mirror, check the upper portions of the heat exchanger for any signs of blockage or dirt accumulation. If any of these conditions are apparent, proceed with the following steps.

9. Remove all burners from the unit. This is accomplished by pulling back on the burner assembly so that it compresses the springs that are located between the manifold and the burners.

   Tilt the burners up into the heat exchanger tube. Slide the burner assembly off the orifices and remove from the unit.

   Note the location of the burner assembly. The burner assembly with the ignitor is on the right hand side and the sensor burner is positioned on the left.

10. After the burner assemblies have been taken from the unit, tap the end of the burners (with the primary air shutter attached) lightly on the ground, thereby removing any residue from inside the burners. Then run the vacuum hose across the top of the burner assembly removing any foreign material that might be lodged between the ports.

11. Disconnect the flue pipe from the flue outlet box collar.

12. Attach the exhaust end of a high pressure vacuum cleaner (shop vacuum) to the furnace flue outlet to force any loose particles from the upper heat exchanger into the bottom burner area.

13. With a stiff bristled brush, clean the bottom area of the heat exchanger, brushing into the burner enclosure where the vacuum can easily be used to pick up the residue.

14. Reassemble the unit by reversing the above procedure. Make sure all old gaskets are discarded and replaced with new gaskets.

NOTE: Extreme care must be taken to protect all new seals and gaskets to assure tightness of the system. Failure to provide tight seals can result in faulty or improper system operation.

EMERGENCY SERVICE

In case of operating difficulty, check the following items before calling for service:

1. Make sure thermostat is set above room temperature.

2. Make sure line switch is "ON" and that fuse is not blown.

3. Make sure air filter is clean.

4. Make sure the Blower Door is properly in place.

5. Make sure there is no blockage of the flue to the outside atmosphere.

6. Make sure connections on the safety pressure switch are secure.

7. Make sure all compartments are securely closed and all fasteners are in place.

If difficulty still exists, call your Installer.

SECTION VII
PART LIST

REPAIR PARTS ARE AVAILABLE FROM THE FURNACE MANUFACTURER, DEALER OR DISTRIBUTOR. WHEN ORDERING PARTS, REFER TO THAT PART OF THE RATING PLATE, ATTACHED TO THE FURNACE THAT INCLUDES THE COMPLETE FURNACE MODEL NUMBER, SERIAL NUMBER AND LISTED NAME AND ADDRESS WHERE PARTS ARE AVAILABLE.

BURNER GROUP
GAS MANIFOLD
MAIN BURNER ORIFICES
BURNER ASSEMBLY WITH IGNITOR BRACKET
BURNER ASSEMBLY WITH SENSOR BRACKET

BLOWER GROUP
BLOWER WHEEL
BLOWER MOTOR
BLOWER MOTOR CAPACITOR
BLOWER MOTOR SUPPORT (MOUNT)
FORCED DRAFT BLOWER
BLOWER MOTOR (COMBUSTION)
BLOWER WHEEL (COMBUSTION)

CONTROL GROUP
GAS VALVE
IGNITION CONTROL (ELECTRIC IGNITION)
IGNITOR (ELECTRIC IGNITOR)
SENSOR-REMOTE (ELECTRIC IGNITION)
FAN & LIMIT CONTROL
PRESSURE SWITCH
PRESSURE SWITCH ORIFICE
TRANSFORMER
RELAY-MAIN BLOWER
RELAY-COMBUSTION BLOWER
BLOWER DOOR INTERLOCK SWITCH

AN EXPANDED PARTS LIST, FOR THE PARTICULAR FURNACE MODEL MAY BE FOUND IN THE FURNACE INSTALLATION INSTRUCTION AND OWNER'S MANUAL ENVELOPE.

⚠ WARNING

IMPROPER INSTALLATION, ADJUSTMENT, SERVICE OR MAINTENANCE CAN CAUSE INJURY OR PROPERTY DAMAGE. CONSULT A QUALIFIED INSTALLER, SERVICE AGENCY, DEALER, DISTRIBUTOR OR THE GAS SUPPLIER FOR INFORMATION OR ASSISTANCE.

GAS FIRED FORCED DRAFT FURNACE—UPFLOW SERIES

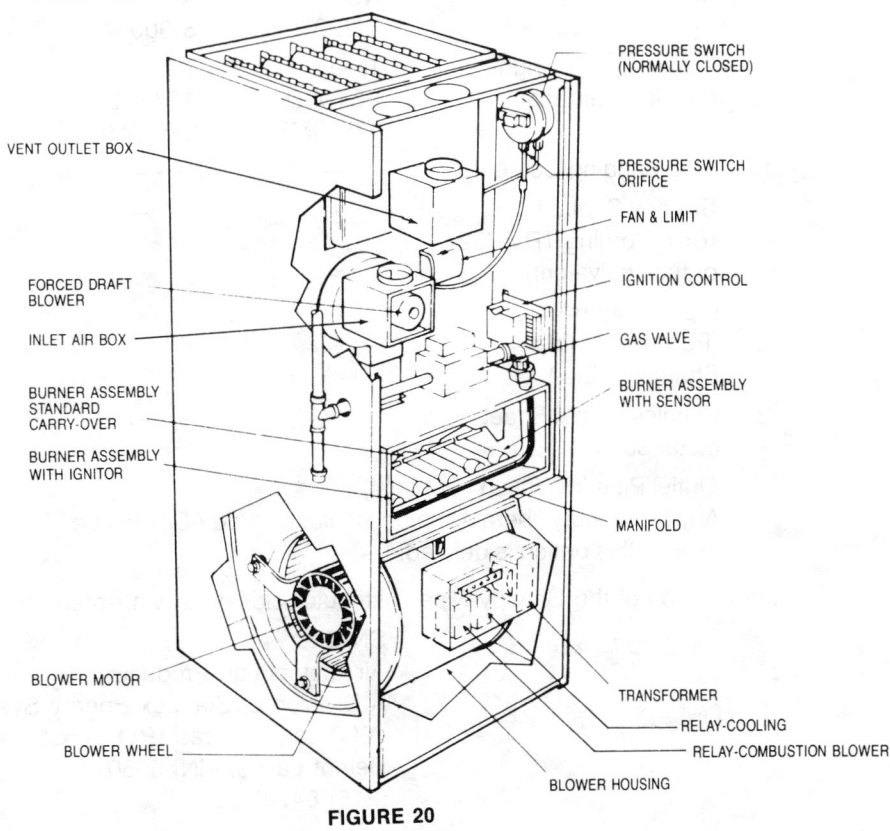

FIGURE 20

19

ADDENDUM A

INSTALLATION INSTRUCTIONS FOR UPFLOW FORCED DRAFT FURNACE HORIZONTAL VENTING

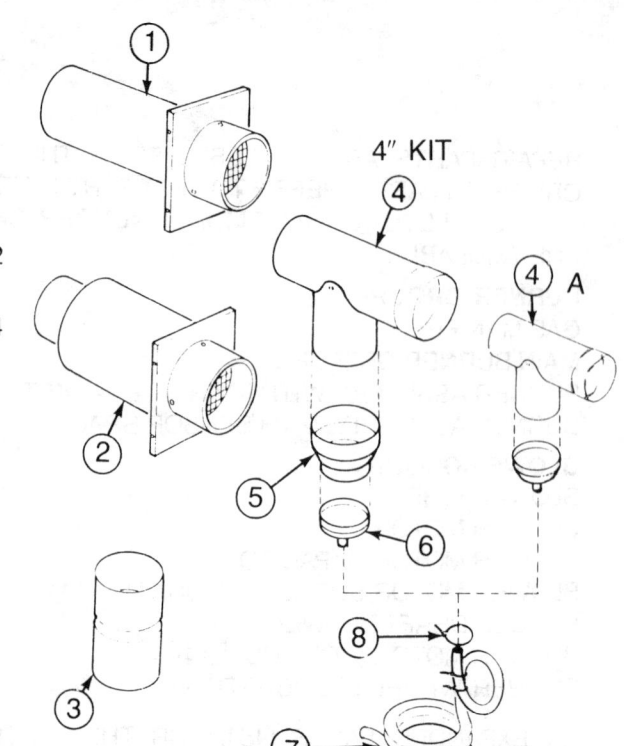

4" KIT

ITEM NUMBER	DESCRIPTION	8G03004 3" KIT NO.	8G03005 4" KIT NO.
①	INTAKE TERMINAL	0548010001	0548010002
②	VENT TERMINAL WITH CLEARANCE SHIELD	0548010003	0548010004
③	VENT ADAPTOR FITTING	2311-170	2311-171
④A	CONDENSATE TRAP TEE WITH CAP & DRAIN	2311-182	—
④B	CONDENSATE TRAP TEE	—	2311-181
⑤	REDUCER 4" TO 3"	—	2311-172
⑥	CAP WITH DRAIN	—	2311-192
⑦	8 FT. x ⅜" DRAIN HOSE	2939-156	2939-156
⑧	HOSE CLAMP	2344-433	2344-433
⑨	INSTALLATION INSTRUCTIONS	7502-084	7502-084

INSTALLER SUPPLIED MATERIALS

SEE TABLE 1 FOR VENT LENGTHS

⑧	Straight sections— Stainless Steel Pipe	SNAPPY NO. S-905-3	SNAPPY NO. S-905-4
⑨	Elbows—Stainless Steel	SNAPPY NO. S-900-3	SNAPPY NO. S-900-4
⑩	Intake—Lightweight Plastic (Drain Pipe)	Type ASTM-D 335-OMS	Type ASTM-D 355-OMS
⑪	PVC Bonding Cement	—	—
⑫	Silicone Sealant (Dow Corning TRV 732 or the equivalent)	—	—
⑬	Pipe Hangers (Perforated Metal Strapping, Etc.)	—	—
⑭	Stainless Steel Sheet Metal Screws	—	—
⑮	Outlet Pipe Insulation (Optional) Any heavy duty fiberglass pipe insulation type ASJ may be used, if this option is desired.		

For the location of the Snappy Pipe Distributor closest to your area call:

SNAPPY
Air Distribution Products
A Division of Standex Energy Systems
1090 Legion Road (P.O. Box 1168)
Detriot Lakes, MN. 56501
(218) 847-9258

20

For purposes of Horizontal Venting these units may be vented up to 30 feet with a maximum of (4) four ninety degree elbows. However, to minimize condensate in the vent pipe on individual units follow the guidelines in the following chart.

MAXIMUM LENGTH OF VENT (HORIZONTAL APPLICATION)

INPUT BTUH	VENT OR INTAKE DIA.	INSULATED VENT PIPE			UNINSULATED VENT PIPE		
		MAX. LENGTH	MAX. ELBOWS	CLEARANCE TO COMBUSTIBLE MATERIALS*	MAX. LENGTH	MAX. ELBOWS	CLEARANCE TO COMBUSTIBLE MATERIALS
40,000	3"	10 FT.	4	1"	—	—	6"
60,000	3"	15 FT.	4	1"	10 FT.	3	6"
80,000	3"	25 FT.	4	1"	15 FT.	3	6"
100,000	4"	25 FT.	4	1"	15 FT.	3	6"
120,000	4"	25 FT.	4	1"	15 FT.	3	6"

*NATIONAL FUEL GAS CODE, ANSI Z-223.1-1984

INSTALLATION INSTRUCTIONS
FOR FURNACE HORIZONTAL VENT WITH OUTSIDE AIR INTAKE

For furnace connection to gas vents or chimneys, vent installations shall be in accordance with Part 7, Venting of Equipment, of the National Fuel Gas Code, ANSI Z223.1-1984, applicable provisions of local building codes and these instructions.

1. DIRECTION OF VENT FITTINGS

All Vent fittings and pipe **must** be installed with male ends toward the furnace.

Longitudinal Vent pipe seams **must** be on top.

All horizontal vent pipe **must** be pitched up (Downhill toward furnace) at least ¼" per foot.

All horizontal pipe should be supported to prevent sagging. Place pipe hangers approximately five feet apart for both vent and intake pipe runs.

2. CONDENSATE IN THE VENT

Under certain conditions, such as long runs of vent pipe through spaces with relatively cool ambient temperatures, condensate may form inside the flue vent pipe. The prescribed pitch of the pipe and the supplied condensate trap/drain connected in the horizontal pipe close to the furnace outlet will allow removal of any accumulation that may occur.

3. SEALING THE PIPE JOINTS

Because the forced draft unit produces positive vent pressures when the vent arrangement is in the horizontal position, it is necessary to seal all joints and connections to prevent the escape of flue products into the surrounding area. Sealing of the pipe joints of both intake and outlet is necessary to prevent pressure losses in the pipe under wind conditions. These losses can produce an imbalance between the intake and exhaust, at the furnace, which could shut down the system. This would include the horizontal pipe connections, pipe longitudinal seams, seams in elbows, tees, tee reducers and caps.

4. SEALING OF PLASTIC INTAKE PIPING

Use a good quality cement on clean, oil free joints.

5. THE STAINLESS STEEL VENT PIPING

Use a high temperature silicone sealant (such as Dow-Corning RTV 732—400°F or equivalent). All joints to be sealed should be oil and dirt free.

Where it is necessary to provide extra strength to a joint in the stainless steel exhaust pipe, the use of stainless steel sheet metal screws is recommended (stainless steel type 304). DO NOT PUT A SCREW IN THE BOTTOM OF THE PIPE. Dab the screw with

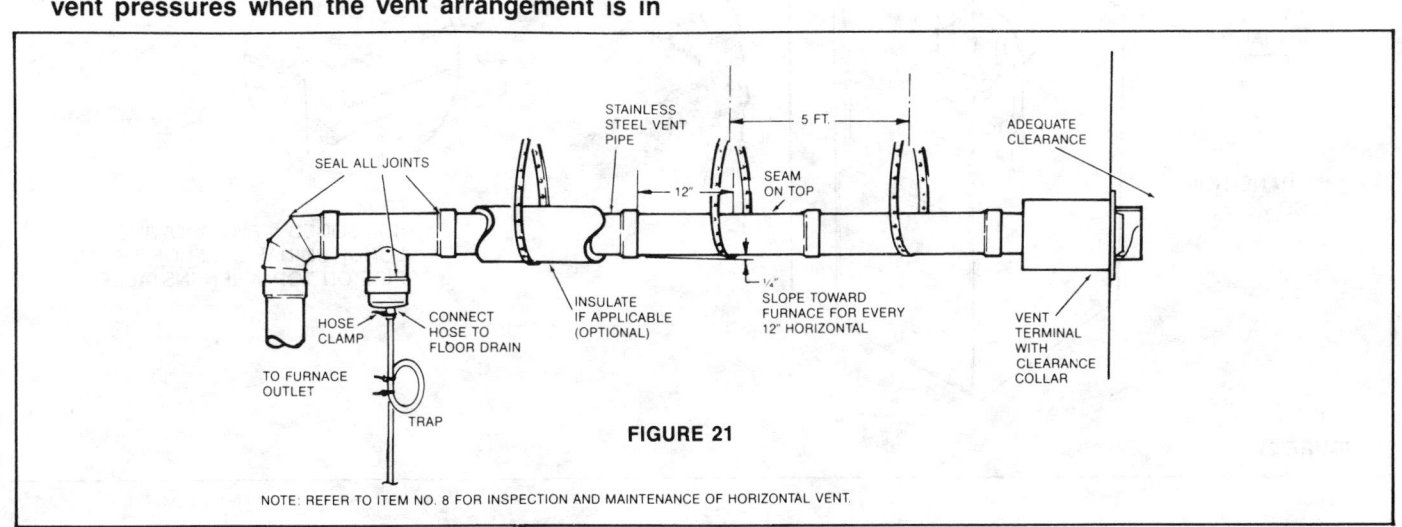

FIGURE 21

NOTE: REFER TO ITEM NO. 8 FOR INSPECTION AND MAINTENANCE OF HORIZONTAL VENT.

silicone sealant to prevent leakage around the threads. **Screwing the joint together is not a substitute for sealing the connection with the silicone sealant.**

6. LOCATION OF THE VENT TERMINAL

The vent terminal must be located at least one foot above grade to avoid blockage by snow or as required to be above snow level.

The terminal must be at least three (3) feet above any forced air inlet terminal servicing any other appliance located within 10 feet horizontally.

The vent terminal shall be located at least four (4) feet below, four (4) feet horizontally from, or one (1) foot above any door, window or gravity air inlet into any building.

A clearance of four (4) feet is required to combustible materials opposite the outlet terminal.

Should the outlet terminal be located adjacent to a public walkway, it must be not less than seven (7) feet above grade.

7. Vent and intake must be on the same face of the building in order to be in the same pressure zone. The centerline of the terminals must be a minimum 16 inches apart on a horizontal line. Never install the intake terminal higher than the vent terminal.

8. PERIODIC INSPECTION AND MAINTENANCE

Under certain wind conditions some building materials may be affected by flue products expelled in close proximity to unprotected surfaces. Sealing or shielding of the exposed surfaces with a corrosion resistant material (such as aluminum sheeting) may be required to prevent staining or deterioration.

TERMINALS

Inspect the intake and outlet terminals at the beginning and periodically throughout the heating season to prevent the accumulation of debris at the terminal barrier screens.

VENT PIPE

Inspect the vent pipe (stainless Steel Pipe) at the beginning and periodically throughout the heating season to insure that the joints are leak tight and that the pipe is in proper condition for safe operation. At the same time check the hose/trap from the tee connector to ensure that it is intact and unrestricted.

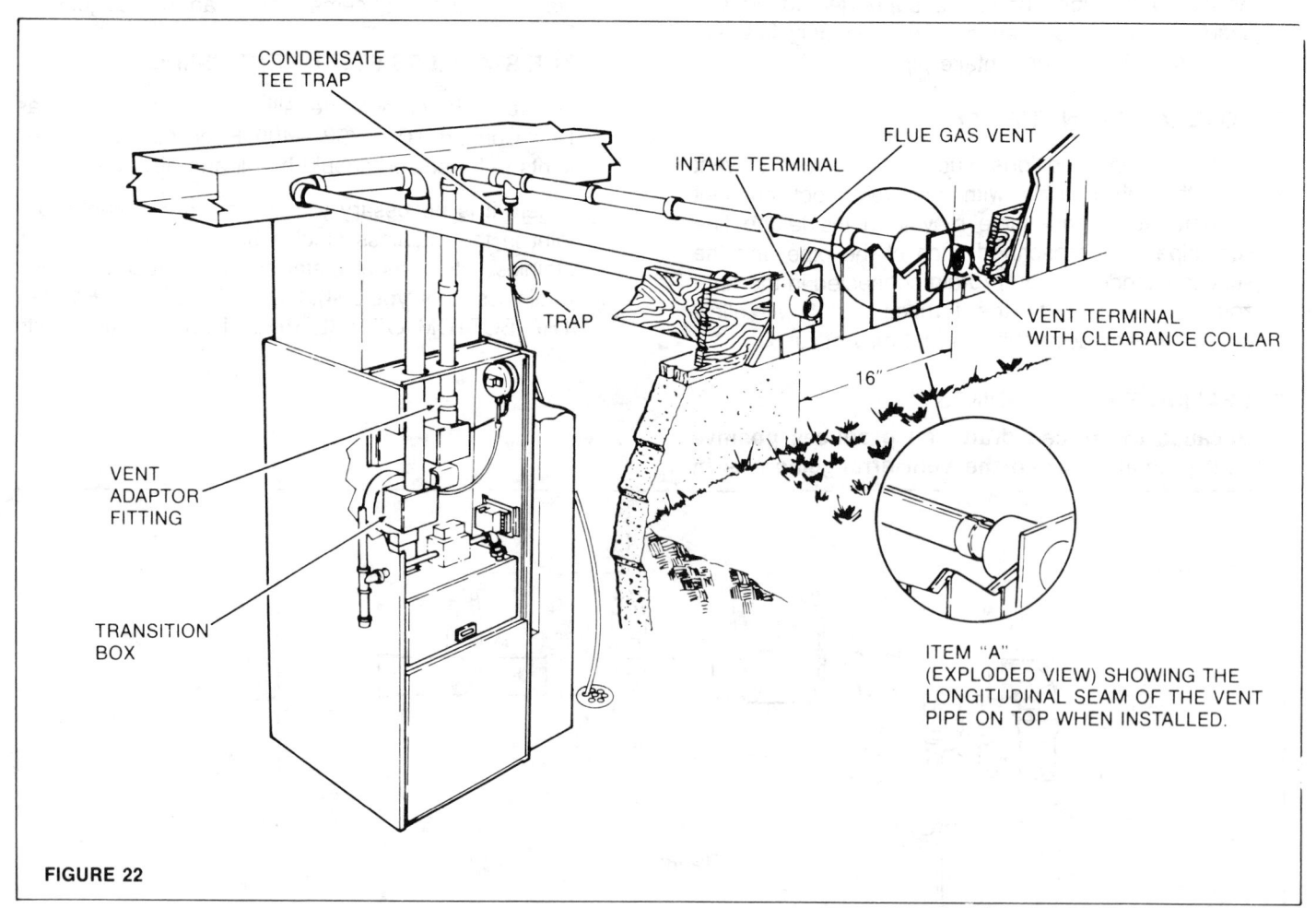

CONDENSATE TEE TRAP

INTAKE TERMINAL

FLUE GAS VENT

TRAP

VENT TERMINAL WITH CLEARANCE COLLAR

16"

VENT ADAPTOR FITTING

TRANSITION BOX

ITEM "A" (EXPLODED VIEW) SHOWING THE LONGITUDINAL SEAM OF THE VENT PIPE ON TOP WHEN INSTALLED.

FIGURE 22

ASSEMBLY INSTRUCTIONS:

CAUTION: SHUT OFF ALL ELECTRICAL POWER TO THE FURNACE.

1. Assembly of Outside Intake/Vent Terminals

FOLLOW ALL THE CONDITIONS DESCRIBED EARLIER IN THESE INSTRUCTIONS WITH REGARD TO CLEARANCES AND POSITIONING WHEN DETERMINING THE LOCATION OF THE INTAKE AND OUTLET VENT TERMINALS.

INTAKE TERMINAL:

a. The hole made in the wall through which the terminal is installed, should be one inch (1″) larger than the outside diameter of the terminal pipe.

b. Insert the terminal through the hole from the outside and secure it in place with the proper type of anchor for the type of construction to which it is being attached.

c. From the outside, caulk around the edge of the mounting plate to provide a tight seal against exterior elements.

d. From the inside, caulk around the pipe as it comes out of the wall.

VENT TERMINAL:

a. If the surface through which the outlet terminal is protruding is of combustible material, the hole in the construction must be made large enough to allow the fitting of the concentric clearance collar around the outlet terminal.

b. Slip the terminal in place from the outside and secure it with the appropriate anchoring device.

c. Caulk around the perimeter of the mounting plate.

d. From the inside, caulk around the pipe, (or clearance collar if used).

2. PIPE ASSEMBLY

Before hanging the intake and outlet pipe permanently, assemble the pipe from the terminal to the furnace and cut the lengths necessary to give the positioning required in final assembly.

Start the intake plastic pipe so the unflared (male) end is at the furnace connection. Start the vent system assembly by placing the vent adaptor fitting on the furnace. Assemble the stainless steel vent pipe and fittings with the male ends toward the vent adaptor fitting. All other pipe sections and fittings must be oriented in the same direction to insure condensate drainage toward the vent drain tee. The drain tee should be installed as close to the furnace as possible, **(within 18″ laterally).**

DO NOT allow the metal outlet pipe to come within six inches of combustible construction. See National Fuel Gas Code, ANSI Z-223.1-1984. (One inch clearance is required from the outer surface of an insulation wrapped pipe.) Allow ¼″ per foot drop toward furnace, for the vent pipe.

To obtain desired piping length:

Final trim on the pipe may be accomplished at either the furnace end or the terminal end. If it is necessary to trim the pipe at points between the furnace and terminal ends, use the following procedure:

1. Trim only the unflared end at these points.

2. Do not trim drain tee.

3. FINAL ASSEMBLY

Vent Pipe

a. Begin at the vent terminal—**NOTE: LONGITUDINAL SEAM OF THE VENT PIPE MUST BE ON TOP WHEN INSTALLED.** (See FIGURE 22 INSERT A.)

b. Seal each piping joint as you assemble the connections by spreading at least a ⅛ inch thick bead of silicone sealant one (1) inch in from the end of the male connection. After the connection is pushed together, check that there is no gapping in the seal. Reseal any gaps or open seams.

c. Support the pipe as you go along to prevent excess movement at the joints.

d. Where screws are needed for rigidity (stainless steel screws) be sure to seal around the screw to prevent leakage.

e. At the furnace connection, caulk around the base of the collar at the vent box collar with silicone. Then install the Vent adaptor fitting into the silicone caulking and check for a good perimeter seal.

f. Seal the joint at the vent terminal with silicone and make one final inspection of each pipe joint seal.

g. Wrap the areas requiring insulation.

h. Attach the condensate drain hose to the fitting on the drain tee cap and secure with the plastic tie provided. The looped end of the hose, which will act as a trap in the line, must be attached at the tee drain. The free end should then be run to, or as close as possible to a floor drain.

4. FRESH AIR INTAKE PIPE ASSEMBLY

a. Beginning at the intake terminal, slip the pipe end over the terminal pipe.

b. Connect the pipe back toward the furnace, cementing each joint with standard plastic cement and support the structure with pipe hangers as needed. BE SURE JOINTS ARE CLEAN AND DRY BEFORE CEMENTING.

c. At the furnace connection, lay a bead of silicone at the base of the collar on the intake transition box. Then slip the final connection into place to complete the joint.

d. Caulk the joint at the terminal connection with silicone, forcing the silicone into the gap between the pipes. Check the assembly for integrity of seal.

CAUTION: The Horizontal Vent Kit is designed for use with a Forced Draft Gas Furnace. As a natural part of the unit's operation, normal products of combustion, including water vapor are vented to the atmosphere. Since the outside air temperature can be well below 32°F., it is possible that the water vapor in the exhaust will freeze, causing an ice buildup around the discharge opening of the pipe. During periods of extremely cold weather and prolonged operation of the furnace, this ice build-up could become quite large. The manufacturer does not recommend the installation of these units in locations above frequent vehicular and/or pedestrian traffic. The ice build-up could present a potentially hazardous situation if it becomes dislodged. The manufacturer will NOT be held responsible for any injury or property damage resulting from any improper installation.

FIGURE 24 is a recommended installation to prevent property or personal injury.

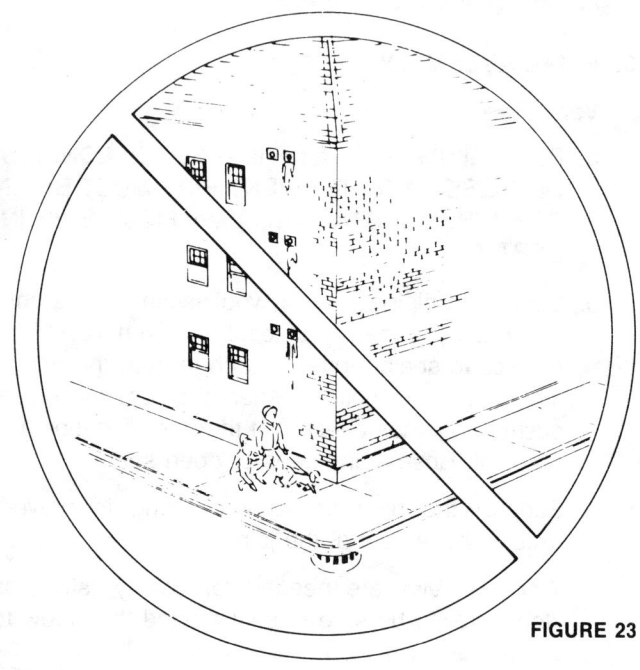

FIGURE 23

FIGURE 24

FINAL INSPECTION QUESTIONNAIRE

System now should be operated and the following items checked:

1. All vent fittings MUST be installed with male ends toward furnace (pitched at least ¼ inch per foot).

2. Have Vent Pipe and Vent Terminals been installed with seams on top?

3. Have all joints from furnace to terminals been adequately sealed?

4. Make sure that hose with condensate trap is properly attached to tee and unrestricted.

5. Have Terminals been adequately caulked to prevent leakage?

6. Make sure these instructions are kept available for your periodic inspection and maintenance of the system.

BLOWER PERFORMANCE DATA—CFM (WITH FILTER)

UNIT INPUT	WHEEL SIZE	MOTOR H.P. (RPM)	MOTOR SPEED	EXTERNAL STATIC PRESSURE INCHES WATER COLUMN						
				0.1	0.2	0.3	0.4	0.5	0.6	0.7
040A010	10 x 6	⅕ (1100)	LOW	425	450	475	490	500	510	505
			MED.	795	810	820	825	830	820	800
			HIGH	1220	1190	1160	1135	1100	1065	1020
060A012	10 x 6	¼ (1100)	LOW	950	945	940	935	930	925	920
			MED.	1090	1075	1050	1035	1020	1000	980
			HIGH	1335	1300	1260	1230	1195	1160	1130
080A012	10 x 8	¼ (1100)	LOW	900	895	885	880	865	840	800
			MED.	1090	1075	1060	1045	1025	990	945
			HIGH	1440	1425	1400	1365	1325	1270	1195
080A016	11.8 x 8	½ (1100)	LOW	1060	1080	1090	1100	1090	1085	1150
			MED.	1385	1385	1380	1380	1370	1360	1345
			HIGH	1800	1765	1735	1695	1660	1620	1580
100A016	10 x 10	⅓ (1100)	LOW	1010	1060	1085	1075	1065	1035	995
			MED.	1350	1325	1300	1275	1250	1225	1200
			HIGH	1795	1745	1690	1640	1590	1540	1490
100A020	11.8 x 10.6	¾ (1100)	LOW	1675	1650	1620	1585	1530	1465	1375
			MED. LOW	1920	1870	1820	1760	1705	1625	1535
			MED. HI	2050	2005	1945	1885	1820	1745	1650
			HIGH	2270	2210	2150	2090	2025	1970	1910
120A016	10 x 10	⅓ (1100)	LOW	1150	1145	1130	1100	1045	950	810
			MED.	1340	1335	1330	1305	1275	1220	1145
			HIGH	1650	1630	1600	1570	1520	1460	1380
120A020	11.8 x 10.6	¾ (1100)	LOW	1725	1720	1705	1675	1635	1580	1515
			MED. LOW	1980	1950	1915	1870	1875	1770	1690
			MED. HI	2235	2190	2140	2085	2025	1970	1890
			HIGH	2400	2355	2305	2245	2180	2105	2025

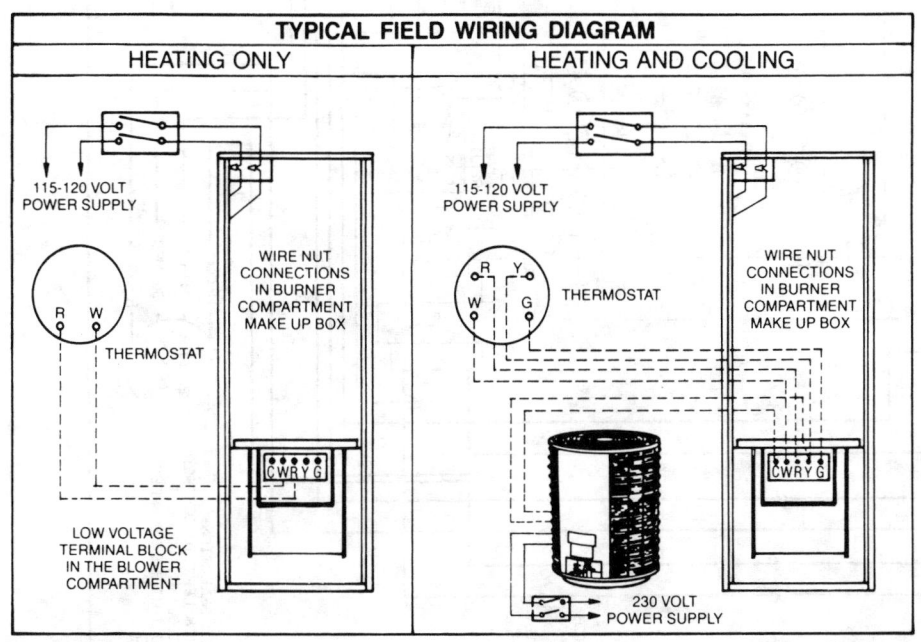

TYPICAL FIELD WIRING DIAGRAM

HEATING ONLY

115-120 VOLT POWER SUPPLY

WIRE NUT CONNECTIONS IN BURNER COMPARTMENT MAKE UP BOX

R   W

THERMOSTAT

LOW VOLTAGE TERMINAL BLOCK IN THE BLOWER COMPARTMENT

C W R Y G

HEATING AND COOLING

115-120 VOLT POWER SUPPLY

R   Y
W   G

THERMOSTAT

WIRE NUT CONNECTIONS IN BURNER COMPARTMENT MAKE UP BOX

230 VOLT POWER SUPPLY

C W R Y G

SUGGESTED THERMOSTAT WIRE SIZE
CONTROL CIRCUIT WIRE SIZE
(FROM UNIT TO THERMOSTAT)

COPPER WIRE ONLY

0'-75' .... 18 GA. A.W.G.
76'-150' .... 16 GA. A.W.G.

ALL WIRING MUST BE IN ACCORDANCE WITH THE N.E.C. AND ALL LOCAL CODES

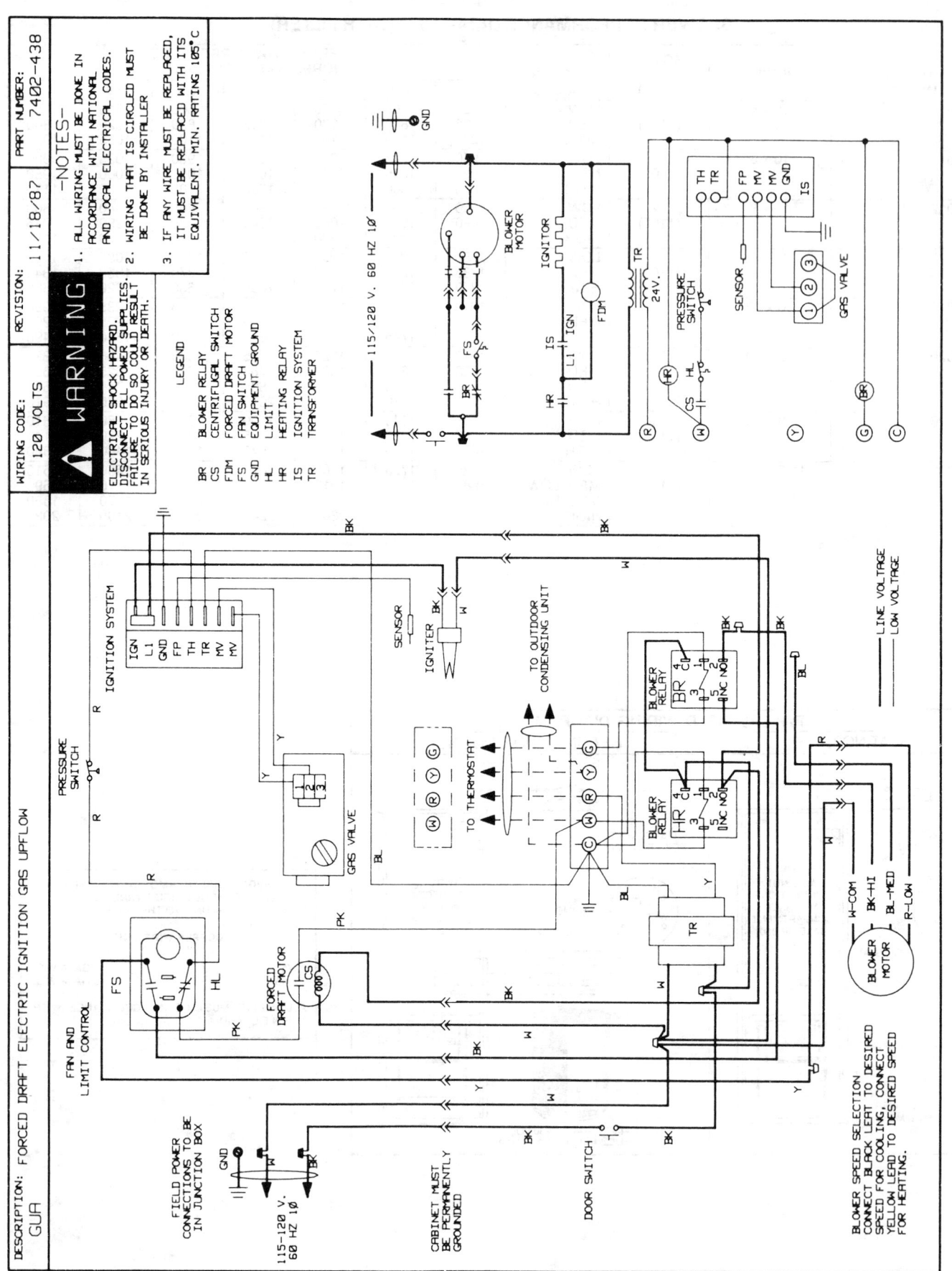

WIRING CODE: 120 VOLTS | REVISION: 11/18/87 | PART NUMBER: 7402-438

DESCRIPTION: FORCED DRAFT ELECTRIC IGNITION GAS UPFLOW
GUA

⚠ WARNING

ELECTRICAL SHOCK HAZARD. DISCONNECT ALL POWER SUPPLIES. FAILURE TO DO SO COULD RESULT IN SERIOUS INJURY OR DEATH.

—NOTES—

1. ALL WIRING MUST BE DONE IN ACCORDANCE WITH NATIONAL AND LOCAL ELECTRICAL CODES.
2. WIRING THAT IS CIRCLED MUST BE DONE BY INSTALLER.
3. IF ANY WIRE MUST BE REPLACED, IT MUST BE REPLACED WITH ITS EQUIVALENT. MIN. RATING 105°C.

LEGEND

BR — BLOWER RELAY
CS — CENTRIFUGAL SWITCH
FDM — FORCED DRAFT MOTOR
FS — FAN SWITCH
GND — EQUIPMENT GROUND
HL — LIMIT
HR — HEATING RELAY
IS — IGNITION SYSTEM
TR — TRANSFORMER

115/120 V. 60 HZ 1Ø

BLOWER MOTOR

IGNITOR

PRESSURE SWITCH

SENSOR

GAS VALVE

IGNITION SYSTEM

PRESSURE SWITCH

FAN AND LIMIT CONTROL

FORCED DRAFT MOTOR

GAS VALVE

TO THERMOSTAT

TO OUTDOOR CONDENSING UNIT

IGNITER

SENSOR

BLOWER RELAY BR

BLOWER RELAY HR

TR

BLOWER MOTOR

FIELD POWER CONNECTIONS TO BE IN JUNCTION BOX

115-120 V. 60 HZ 1Ø

CABINET MUST BE PERMANENTLY GROUNDED

DOOR SWITCH

BLOWER SPEED SELECTION
CONNECT BLACK LEAD TO DESIRED SPEED FOR COOLING, CONNECT YELLOW LEAD TO DESIRED SPEED FOR HEATING.

W-COM
BK-HI
BL-MED
R-LOW

LINE VOLTAGE
LOW VOLTAGE

INSTALLATION
INSTRUCTIONS
AND
OPERATING
PROCEDURE

**SUPER
HIGH
EFFICIENCY
UPFLOW
GAS FIRED
FURNACE**

GUH SERIES

REVISED: JULY, 1989
SUPERSEDES:
NOVEMBER, 1988

START-UP CHECK SHEET

OWNER NAME: _____ DEALER NAME: _____

ADDRESS: _____ ADDRESS: _____

CITY, STATE, ZIP: _____ CITY, STATE, ZIP: _____

MODEL NUMBER: _____ SERIAL NUMBER: _____

TYPE OF GAS: NAT: _____ LPG: _____ LP STICKER APPLIED: _____ LP KIT NUMBER: _____

GAS
VALVE _____ LINE
PRESSURE _____ MANIFOLD
PRESSURE _____ INPUT _____

THERMOSTAT: _____ SUBBASE: _____ HEAT ANTICIPATOR SETTING: _____

BLOWER MOTOR H.P.: _____ VOLTAGE: _____

FAN OPENS: _____ °F FAN CLOSES: _____ °F

FAN AND LIMIT
MODEL NUMBER: _____ LIMIT OPENS: _____ °F LIMIT CLOSES: _____ °F

TEMPERATURE RISE: RETURN AIR: _____ °F SUPPLY AIR: _____ °F

FILTER SIZE AND TYPE: _____

BURNER FLAME PROPERLY ADJUSTED? _____

DRIP-LEG INSTALLED PRIOR TO GAS VALVE? _____

CONDENSATE DRAIN CONNECTED? _____

BLOWER SPEED CHECKED? _____ BLOWER PROPERLY LUBRICATED? _____

ALL ELECTRICAL CONNECTIONS, MOUNTING SCREWS, ETC. TIGHT? _____

ADJUSTMENT REQUIRED: _____

DATE OF INSTALLATION? _____

DATE OF START-UP: _____

_____

_____

_____

_____

_____

_____

_____

_____

INDEX

SPECIFICATION AND INFORMATION

MODEL		GUH060A012	GUH075A012	GUH100A016	GUH125A016	GUH125A020
ELECTRICAL SPECIFICATION	VOLTAGE (NAMEPLATE)	115 / 120				
	MIN./MAX. VOLTAGE	102 / 132				
	MINIMUM CKT. AMPACITY	10.6	10.6	15.0	15.0	15.0
	MAX. OVER CURRENT PROTECTION	15	15	15	15	15
HEATING① CAPACITIES	INPUT BTUH	60,000	75,000	100,000	125,000	125,000
	OUTPUT BTUH	56,400	69,750	93,000	115,000	
	AFUE %	94	93	93	92	91.0
	AFUE % (I.C.S.)	93.1	92.2	92.1	91.2	90.1
MOTOR AND BLOWER INFORMATION	MOTOR TYPE/NO. OF SPEEDS	PSC-DIRECT DRIVE / 1				
	MOTOR HP/RPM	½ / 1100	½ / 1100	¾ / 1100	¾ / 1100	¾ / 1100
	MOTOR FLA/LRA②	8.5 / 34.5	8.5 / 34.5	12.0 / 48	12.0 / 48	12.0 / 48
	BLOWER TYPE/NO.	CENTRIFUGAL / 1				
	BLOWER WHEEL SIZE (DxW)	11.8 x 8	11.8 x 8	11.8 x 10.6	11.8 x 10.6	11.8 x 10.6
AIR DELIVERIES③	HI SPEED 0.5" ESP (CFM)	1365	1365	1359	1577	1820
	LO SPEED 0.5" ESP (CFM)	925	925	1152	1313	1448
	TEMP. RISE RANGE °F	35 - 65	45 - 75	45 - 75	55 - 85	45 - 75
PRIMARY HEAT EXCHANGER	TYPE	RPJ® WELDLESS SECTIONAL				
	MATERIAL CONSTRUCTION	ALUMINIZED STEEL				
	NO. OF CELLS	3	3	4	5	5
SECONDARY HEAT EXCHANGER (COIL)	TYPE/MATERIAL CONSTRUCTION	FULL COLLAR ALUMINUM FIN AND 29-4C STAINLESS STEEL TUBE				
	NO. OF COILS	1				
	FINS PER INCH	6	6	6	6	6
	TUBE SIZE	5.505" O.D. x 0.020" WALL				
PIPING INFORMATION	GAS CONNECTION (IN.)	½"				
	VENT SIZE DIA. (IN.)	SEE CAPACITY SIZING CHART ON INSIDE PAGE				
	CONDENSATE CONN. (IN.) TYPE	½" / COMPRESSION FITTING				
IGNITION SYSTEM	TYPE/IGNITOR MATERIAL	ELECTRIC (HOT SURFACE) / SILICONE CARBIDE				
	IGNITION SEQUENCE	THREE TRY				
	FLAME PROVING INTERVAL	7 SECONDS				
	SENSOR TYPE	FLAME RECTIFICATION				
BURNER INFORMATION	TYPE/MAT. CONSTRUCTION	SLOTTED PORT / ALUMINIZED STEEL MULTIPLE BURNER				
	SIZE	1¼" DIA.				
	NO. OF BURNERS	3	3	4	5	5
GENERAL INFORMATION	AIR FILTER TYPE	WASHABLE				
	AIR FILTER SIZE/QUANTITY④	16 x 25 x 1 (1)		16 x 25 x 1 (2)		
	SHIPPING WEIGHT LBS.	192	192	223	242	242
	NET WEIGHT LBS.	182	182	213	232	232

①CAPACITY AND EFFICIENCY RATINGS IN ACCORDANCE WITH D.O.E. TEST PROCEDURES.
②FULL LOAD AMPS DEPEND ON INSTALLATION CONDITIONS AND HEATING OR COOLING OPERATION.
③REFER TO BLOWER CHART ON PAGE 5 FOR COMPLETE BLOWER INFORMATION.
④REFER TO FILTER REQUIREMENT CHART FOR PROPER FILTER AND NUMBER OF FILTERS.

CODE: GUH

SUPER HIGH EFFICIENCY GAS FIRED UPFLOW FURNACE

INFORMATION CONTAINED IN THIS MANUAL IS DESIGNED TO ACQUAINT YOU WITH THE MECHANICAL FEATURES, AND TO SERVE AS AN EMERGENCY REPAIR GUIDE IN CASE OF HEAT FAILURE.

PROPER CARE OF YOUR GAS FURNACE WILL ASSURE YOU OF MANY YEARS OF ECONOMICAL COMFORT AND SERVICE. LIKE EVERY PIECE OF FINE MECHANICAL EQUIPMENT, YOUR FURNACE SHOULD BE SERVICED PERIODICALLY BY AN EXPERT. WE RECOMMEND A COMPLETE SERVICING BY AN AUTHORIZED SERVICEMAN AT LEAST ONCE EACH YEAR—USUALLY AT THE BEGINNING OF THE HEATING SEASON.

SECTION I
FACTS ABOUT THE FURNACE

UNIT DESCRIPTION

The nameplate label and serial number label on the furnace shows the model number of the furnace. This label is located on the right hand side inside the casing. SEE FIGURE 1 for the location of these labels.

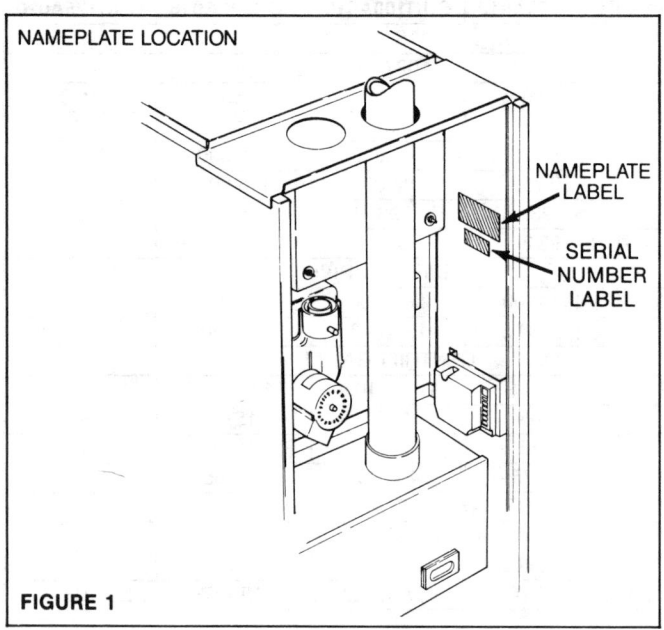

FIGURE 1

Check the furnace model number on the nameplate label. The suffix of the model number indicates the type of fuel and the ignition system.

Please note that each digit of the model number has a special significance:

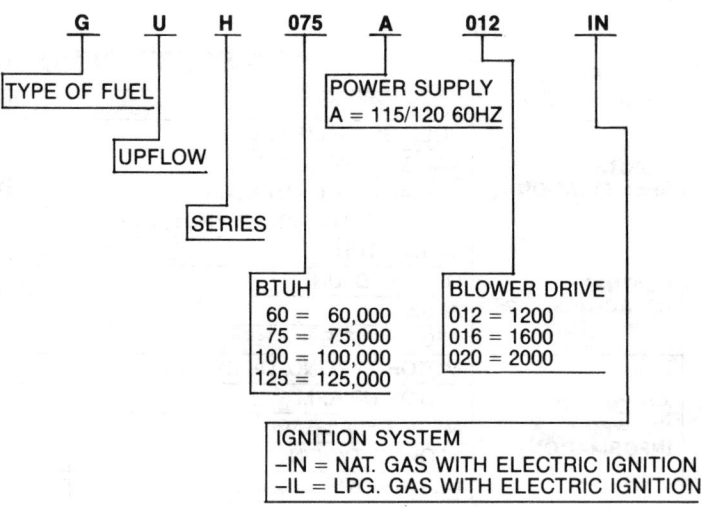

⚠ WARNING

Before operating, smell all around the appliance area for gas. Be sure to smell next to the floor because some gas is heavier than air and will settle on the floor.

NOTICE TO INSTALLER:

Instruct the owner or operator of the furnace:

1. The proper operating and shut-down procedure to follow for shutting off the furnace in case of an emergency.
2. How to change, service or clean the filter.
3. Setting of controls for best comfort conditions.
4. Where the lighting instruction label is located.
5. Explain the lighting procedure.

SECTION II
GENERAL INSTALLATION REQUIREMENTS

The information in this manual will provide the installer with adequate information to install this furnace properly.

> ## ⚠ WARNING
>
> Improper installation, adjustment, alteration, service or maintenance can cause injury or property damage. Refer to the installation instructions and user's manual provided with the furnace. For assistance or additional information, consult a qualified installer, service agency or the gas supplier.

UNCRATING AND INSPECTING

After uncrating the unit, inspect thoroughly for concealed damage. If found, notify transportation company immediately and file a concealed damage claim.

GENERAL REQUIREMENTS

All installations should be made as outlined in the latest edition of National Fuel Gas Code ANSI-Z223.1a-1987, and other applicable National and Local codes, including requirements of Local Utilities. In addition, **the unit must be electrically grounded** in accordance with the latest edition of National Electric Code ANSI-NFPA-#70-1987. Current Code Edition(s) are available for a fee from the American Gas Association, 1515 Wilson Blvd. Arlington, Virginia 22209.

Select a level spot for the installation of the unit. Install as near to the vent termination point as possible and install the vent pipe with a minimum number of elbows. It is also recommended that the unit be centralized with relation to the distribution system.

LOCATION OF THERMOSTAT

Thermostat should be mounted 4 to 5 feet above the floor, on an inside wall of the living room, dining room, or a hallway that has good air circulation from the other rooms being controlled by the thermostat.

Movement of air should not be obstructed by furniture, doors, draperies, etc. The thermostat should not be mounted where it will be affected by drafts, hot or cold water pipes or air ducts in walls, radiant heat from a fireplace, lamps, the sun, T.V., etc., or on an outside wall. Consult the Instruction Sheet packed with thermostat for mounting instructions.

ACCESSIBILITY CLEARANCES

All servicing and cleaning of these units can be done from the front, and a minimum of 36″ horizontal clearance should be allowed. If necessary, these units can be installed at the clearances listed in TABLE 1. However, if access is required to some other piece of equipment, a minimum of 18″ should be allowed for passage.

In a closet or utility room installation, the door must be large enough to allow replacement of the unit if necessary or to permit replacement of any other appliance within the confined space. A minimum of 36″ clearance should be allowed at the front of the unit for servicing and cleaning when the closet door is open. SEE FIGURE 2.

In addition adequate clearances must be provided around air openings. The furnace should not be installed directly on carpeting, tile or other combustible materials other than wood.

If the furnace is installed in a residential garage, the unit must be installed so that the burners and the ignition source are located not less than 18 inches (457 MM) above the floor. The furnace must also be protected to avoid physical damage by vehicles.

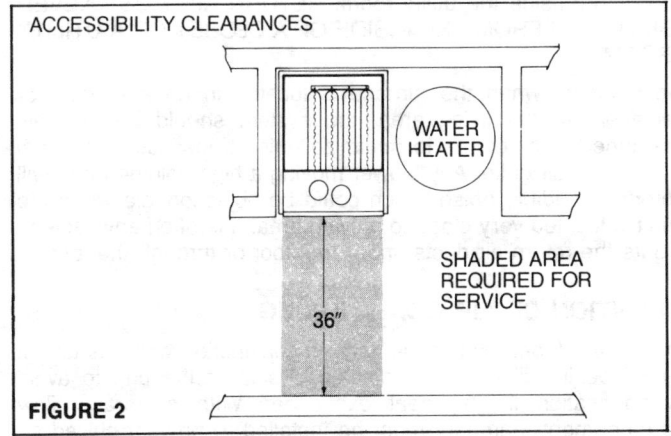

FIGURE 2

RECOMMENDED CLEARANCE GUIDE (FIRE PROTECTION CLEARANCE)

(Dimensions shown in inches are the minimum clearances to combustible materials for which the furnace design has been certified.)

TABLE 1

INPUT BTU/HR.	FRONT	REAR	SIDES	TOP	VENT CONNECTION	COMBUSTIBLE FLOOR**
60,000	6	0	0	1	0	0
75,000	6	0	0	1	0	0
100,000	6	0	0	1	0	0
125,000	6	0	0	1	0	0

**Shall not be installed directly on carpeting, tile or other combustible materials, other than wood flooring.

IMPORTANT NOTE:

THIS FURNACE IS A DIRECT VENT APPLIANCE AND MUST BE INSTALLED IN ACCORDANCE WITH THE "VENTING AND CONDENSATE DRAIN INSTRUCTION" INCLUDED AS A PART OF THESE INSTALLATION INSTRUCTIONS. SEE SECTION 3.

⚠ WARNING

Combustion and ventilation air must come from a non-corrosive atmosphere.

Combustion air must be free of acid forming chemicals; such as sulphur, fluorine and chlorine. These elements are found in aerosol sprays, detergents, bleaches, cleaning solvents, air fresheners, paint and varnish removers, refrigerants and many other commercial and household products. Vapors from these products when burned in a gas flame form acid compounds which are highly corrosive when they condense.

PROVISIONS FOR VENTILATION

Air openings provided in the casing of the furnace must be kept free of obstructions.

It is recommended (but not required) that an adequate amount of ventilation air to the furnace enclosure be provided to avoid excessive heat build-up.

HIGH STATIC OPERATION

The supply air blower can be operated against external static pressures up to 0.5″ water column, or as indicated on the rating plate.

SECTION III
INSTALLATION PROCEDURE

DUCT CONNECTIONS

It is recommended that a flexible duct connection of a non-flammable material be used for the return air and supply air connections to prevent transmission of vibration.

A ducted return air system to the unit must be used in all installations. The return connection must be made full size to a location outside the utility room. RETURN AIR MUST NEVER BE DRAWN FROM THE INSIDE OF A CLOSET OR A UTILITY ROOM.

In addition, when the furnace is located in an area near or adjacent to the living area, the system should be carefully designed with returns to minimize noise transmission through the return air grille. Any blower moving a high volume of air will produce audible noise which could be objectionable when the unit is located very close to a living area. It is often advisable to route the return air ducts under the floor or through the attic.

ADDITION OF AIR CONDITIONING

When a refrigeration coil is used in conjunction with this unit, it must be installed on the discharge side of the unit to avoid condensation in the heat exchanger. With a parallel flow arrangement, dampers must be installed to prevent chilled air from entering the furnace. If manually operated dampers are used, they must be equipped with a means to prevent operation of the unit unless the damper is in full heat or full cool position.

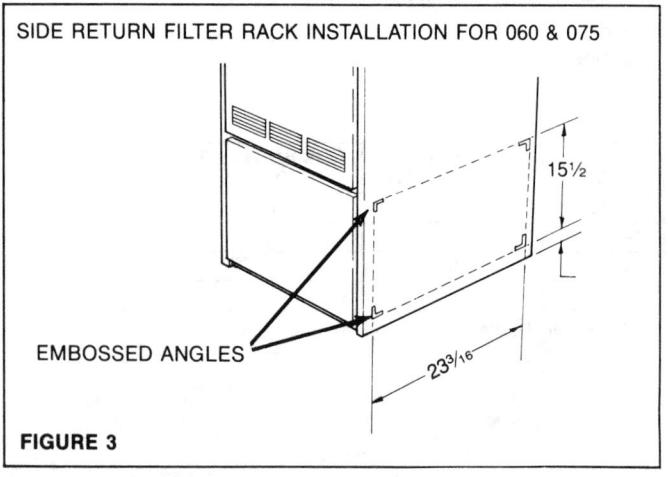

SIDE RETURN FILTER RACK INSTALLATION FOR 060 & 075

15½

EMBOSSED ANGLES

23³/₁₆

FIGURE 3

SIDE AIR RETURN

The 60,000 & 75,000 BTUH furnace comes with (1) one 16 x 25 x 1 washable filter and filter rack for side return. The 100,000 and 125,000 BTUH furnace comes with (2) two 16 x 25 x 1 washable filters and filter racks for side return.

To install the filter rack(s) cut the return opening in the side casing panel (right or left) using the embossed angles as guides on the casing. SEE FIGURE 3.

Mount the filter rack over the opening on the side of the furnace. SEE FIGURE 4. Proceed with duct work.

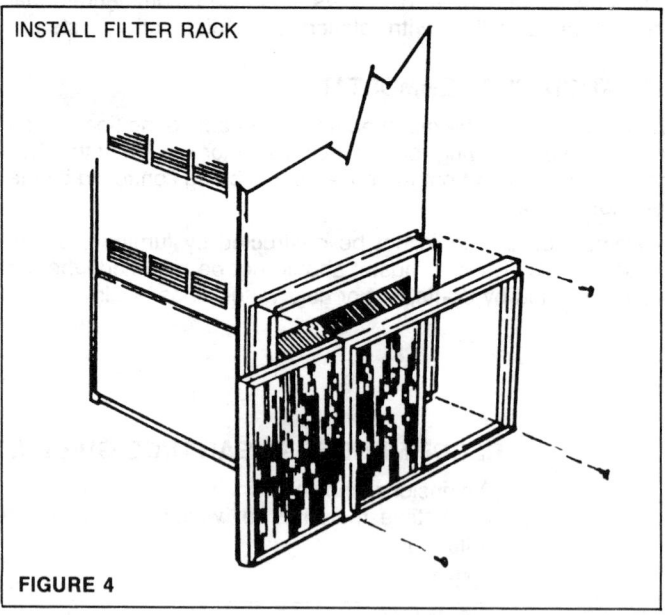

INSTALL FILTER RACK

FIGURE 4

The 100,000 BTUH and 125,000 BTUH furnace will require (2) two 16 x 25 x 1 filters and racks or the optional Stand-Off Filter Rack. SEE FIGURE 5.

OPTIONAL BOTTOM AIR RETURN FILTER KIT

A bottom return air filter is NOT provided with this furnace. Either an external filter rack must be built between the base of the unit and the return air plenum (furnished by the installer) or the Optional Bottom Return Air Filter Kit can be used. See Instructions in Kit for installing filter in bottom of unit.

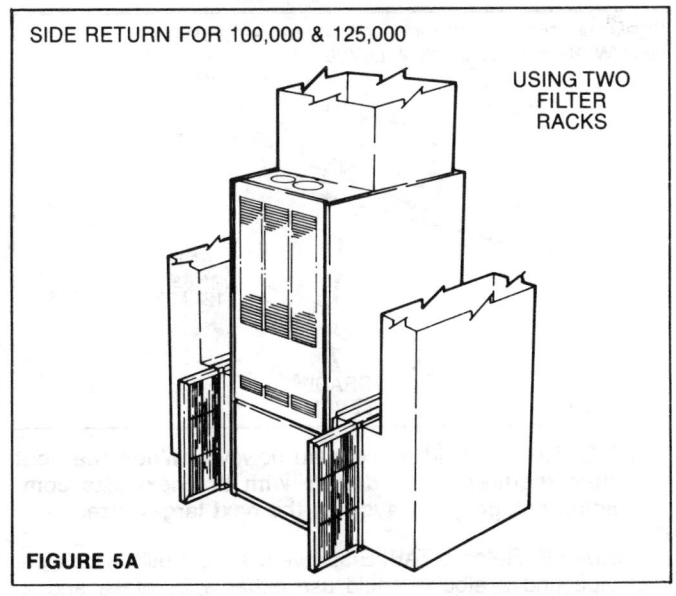

SIDE RETURN FOR 100,000 & 125,000

USING TWO FILTER RACKS

FIGURE 5A

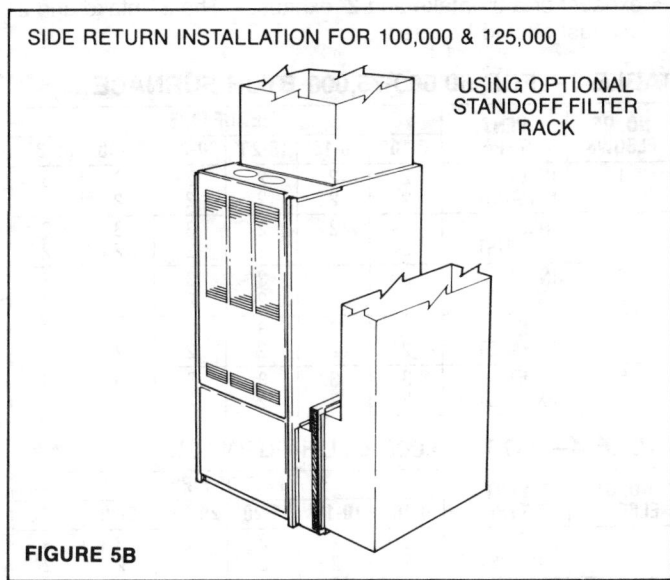

SIDE RETURN INSTALLATION FOR 100,000 & 125,000

USING OPTIONAL STANDOFF FILTER RACK

FIGURE 5B

BOTTOM AIR RETURN FILTER KITS

KIT NUMBER	USED ON:
8G05009	GUH060 & 070
8G05010	GUH100
8G05011	GUH125

OPTIONAL STAND-OFF SIDE FILTER RETURN KIT

This Kit is available if only one filter rack is to be used on the 100,000 and 125,000 BTUH furnace. The Kit contains a special stand-off filter rack and (1) one 20 x 25 x 1 filter. Refer to Instructions in the Kit for attaching the filter rack to the furnace side.

STAND-OFF FILTER KIT

KIT NUMBER	USED ON:
8G05008	GUH100 & 125

AIR FILTERS FOR ALL FURNACES:

Filters supplied with the external filter rack are of the washable type. Replacement of either the internal or external air filters should be of the same size and capacity as the original. Failure to provide adequate filter media, can cause equipment malfunction, uneven room temperature and excessive fuel usage. Refer to the chart for filter size and minimum filter area required for all furnace sizes. SEE TABLE 2.

VENTING AND CONDENSATE DRAIN INSTRUCTIONS

COMBUSTION AIR AND EXHAUST PIPING

A. All pipe and fittings used for venting this condensing furnace must be schedule 40 PVC and conform to the American Society for Testing and Material (ASTM) Standards D1785 and D2665, and all applicable American National Standards Institute (ANSI) Standards. PVC primer and solvent cement used to secure all joints must conform to ASTM D2564.

> ### ⚠ WARNING
>
> Solvent cements are combustible liquids and should be kept away from all ignition sources, (ie. sparks, open flames and excessive heat). Avoid breathing cement vapors or contact with skin and eyes.

B. **THIS APPLIANCE HAS BEEN APPROVED FOR DIRECT VENT INSTALLATION ONLY.** Combustion air must be taken from outside the structure to prevent contamination with compounds containing chlorine and fluorine. (Common household items such as bleach, paint remover and some aerosol products contain these contaminants.)

> ### ⚠ CAUTION:
>
> Terminate the combustion air intake as far as possible from the air conditioning unit or heat pump, swimming pools, swimming pool pumping units, and dryer vents.

C. All combustion air and exhaust piping must be installed in accordance with local codes and these instructions.

D. When the exhaust vent passes through an unconditioned space or raceway, it must be insulated with ½″ wall, closed cell, neoprene insulation or equivalent.

TABLE 2—FILTER REQUIREMENTS (HEATING AND COOLING)①

	STANDARD FILTER RACK	OPTIONAL FILTER KITS		
INPUT BTU/H	SIDE EXTERNAL FILTER AND RACK	EXTERNAL STAND OFF FILTER RACK KIT②	BOTTOM FILTER KIT②	MINIMUM FILTER AREA (IN.²)
60,000	(1) 16 x 25 x 1	NOT APPLICABLE	(1) 16 x 25 x ½	400
75,000	(1) 16 x 25 x 1	NOT APPLICABLE	(1) 16 x 25 x ½	400
100,000	(2) 16 x 25 x 1	(1) 20 x 25 x 1	(1) 20 x 25 x ½	500
125,000	(2) 16 x 25 x 1	(1) 20 x 25 x 1	(1) 25 x 25 x ½	500

①FILTERS ARE RATED AT 520 FPM OR MORE (WASHABLE TYPE).
②SEE OPTIONAL EQUIPMENT FOR KIT PART NUMBER.

E. If the original furnace was common vented with another gas fired appliance, the vent may need to be resized for the capacity of the other appliance. Consult the National Fuel Gas Code (ANSI Z223.1a-1987) for proper sizing and revise as required. **COMMON VENTING OF THIS FURNACE WITH ANOTHER GAS FIRED APPLIANCE (WATER HEATER, ETC.) IS NOT ALLOWED.**

VENT TERMINAL INSTALLATION

1. This furnace can be installed with either a vertical or horizontal direct vent. In either case, the exhaust vent and combustion air intake pipe must be located on the same side of the structure and separated by no less than 14" and no more than 24". SEE FIGURE 6.

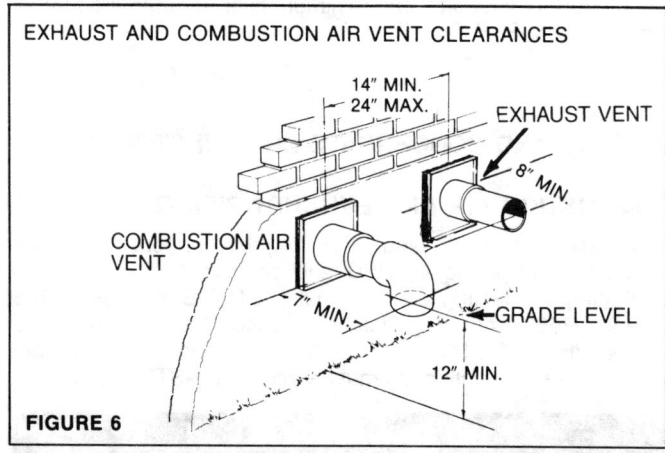

EXHAUST AND COMBUSTION AIR VENT CLEARANCES

14" MIN.
24" MAX.

EXHAUST VENT

8" MIN.

COMBUSTION AIR VENT

7" MIN.

GRADE LEVEL

12" MIN.

FIGURE 6

⚠ WARNING

At no time should the intake and exhaust vents be terminated on opposite or adjacent sides of the structure. Also, combination vertical and horizontal venting is NOT permitted.

2. Position the exhaust vent and combustion intake terminals away from obstructions and above anticipated snow accumulation or grade level as shown in FIGURE 6 & 7.

⚠ CAUTION:

Always determine anticipated snow accumulation level and install vent terminal accordingly to prevent vent/intake blockage.

If installation above snow accumulation level is chosen as shown in FIGURE 7, replace the exterior coupling with a 90° elbow pointing up to secure the termination assembly. The elbow must be secured with a stainless steel screw as shown in FIGURE 7.

DO NOT point terminals into window wells, stairwells, alcoves or other recessed areas. Maintain a 3 foot clearance from plumbing vent stacks. It is preferable for the vent intake to terminate on opposite or adjacent sides of the structure than the dryer vent. If this is not possible, maintain a 10 foot clearance.

Maintain a 1 foot clearance from any opening through which flue gas could enter the building and a 7 foot clearance above public walkways.

3. Consult TABLES 3-5 to select the proper diameter exhaust and combustion air piping. Exhaust and combustion air piping is sized for each unit model number based on total lineal vent length and number of 90° elbows required. Two 45° elbows can be substituted for one 90° elbow. The elbow used for vent termination outside the structure is **not** counted when using TABLES 3 thru 5. The additional two 90° elbows shown

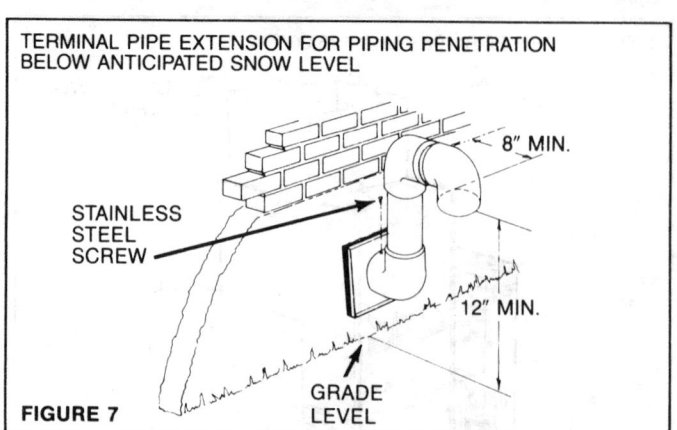

TERMINAL PIPE EXTENSION FOR PIPING PENETRATION BELOW ANTICIPATED SNOW LEVEL

8" MIN.

STAINLESS STEEL SCREW

12" MIN.

GRADE LEVEL

FIGURE 7

in FIGURE 7 should be counted however. **When the vent system required is borderline with the next size combination category, always use the next larger size.**

EXAMPLE: Refer to TABLE 3; A vent system utilizing 20 feet of pipe and 2 elbows could use either a 2" intake and 2" exhaust or a 3" intake and 2" exhaust.—The 3" intake and 2" exhaust should be used.

TABLE 3—FOR 60,000-75,000 BTUH FURNACE

NO. OF ELBOWS	VENT TYPE	0-10'	10-15'	15-20'	20-25'	25-30'	30-35'
1	INLET	2	2	2	2	3	3
	EXHAUST	2	2	2	2	2	2
2	INLET	2	2	2	3	3	3
	EXHAUST	2	2	2	2	2	2
3	INLET	2	2	3	3	3	3
	EXHAUST	2	2	2	2	2	2
4	INLET	2	3	3	3	3	3
	EXHAUST	2	2	2	2	2	3
5	INLET	3	3	3	3	3	3
	EXHAUST	2	2	2	2	3	3

TABLE 4—FOR 100,000 BTUH FURNACE

NO. OF ELBOWS	VENT TYPE	0-10'	10-15'	15-20'	20-25'	25-30'	30-35'
1	INLET	2	2	2	3	3	3
	EXHAUST	2	2	2	2	2	2
2	INLET	2	2	3	3	3	3
	EXHAUST	2	2	2	2	2	2
3	INLET	2	3	3	3	3	3
	EXHAUST	2	2	2	2	2	3
4	INLET	3	3	3	3	3	3
	EXHAUST	2	2	2	2	3	3
5	INLET	3	3	3	3	3	3
	EXHAUST	2	2	2	3	3	3

TABLE 5—FOR 125,000 BTUH FURNACE

NO. OF ELBOWS	VENT TYPE	0-10'	10-15'	15-20'	20-25'	25-30'	30-35'
1	INLET	2	3	3	3	3	3
	EXHAUST	2	2	2	2	3	3
2	INLET	3	3	3	3	3	N.A.
	EXHAUST	2	2	2	3	3	N.A.
3	INLET	3	3	3	3	N.A.	N.A.
	EXHAUST	2	2	3	3	N.A.	N.A.
4	INLET	3	3	3	3	N.A.	N.A.
	EXHAUST	2	3	3	3	N.A.	N.A.
5	INLET	3	3	3	N.A.	N.A.	N.A.
	EXHAUST	3	3	3	N.A.	N.A.	N.A.

NOTE: Vent pipe diameters apply for installation from sea level to 2000 ft. For installation above 2000 ft. reduce the maximum pipe length as follows:

2000-5000 ft. above sea level—Reduce max. length allowable by 5 ft.

5001-7000 ft. above sea level—Reduce max. length allowable by 10 ft.

7

SIDE WALL VENT TERMINATION

NOTE: When horizontal venting application is used, a Horizontal Vent Kit (#8G03022) **MUST** be ordered. The following is the instructions for installing the Horizontal Vent Kit.

1. FIGURE 8 shows the proper means for a horizontal side wall vent termination.

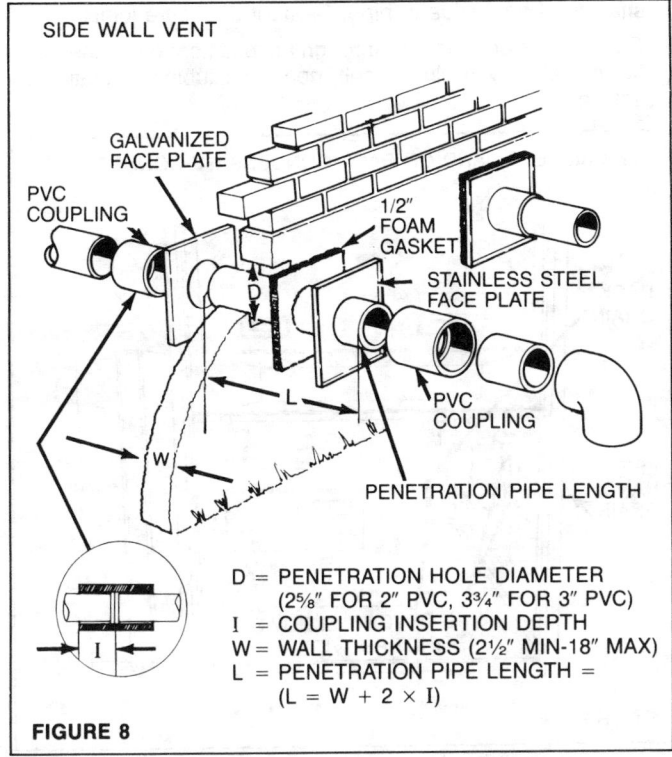

D = PENETRATION HOLE DIAMETER
(2⅝" FOR 2" PVC, 3¾" FOR 3" PVC)
I = COUPLING INSERTION DEPTH
W = WALL THICKNESS (2½" MIN-18" MAX)
L = PENETRATION PIPE LENGTH =
(L = W + 2 × I)

FIGURE 8

2. The optional Horizontal Vent Kit consists of the following:

Two stainless steel face plates; two galvanized face plates and two ½ inch foam gaskets. One stainless steel plate, galvanized plate and one ½" gasket is used on the flue exhaust and the other set is used for the combustion air intake.

3. The face plates are adaptable to 2 or 3 inch PVC pipe. If 3 inch PVC pipe is used, cut the 3 tabs on the radial slot and remove the center disc. If 2 inch PVC pipe is used, use the plates as they are shipped.

4. Drill or saw 2 wall penetration holes spaced as shown in FIGURE 6. The penetration hole diameter should be 2⅝" for 2" diameter pipe and 3¾" for 3" diameter pipe.

5. Cut wall penetration pipe to a length equal to the wall thickness plus 2 times the coupling insertion depth. Remove all shavings from inside pipe. Debur and chamfer both ends of the pipe. SEE FIGURE 8.

6. Apply PVC cement primer to one end of the pipe and the coupling elbow or socket. Apply cement liberally to **both** pipe end and coupling elbow or socket. Insert pipe into coupling or elbow with ¼ turn twist. Slip **stainless steel** face plate and gasket over end of pipe.

NOTE: Flange on face plate should be facing out.

From the outside, insert pipe assembly into the wall penetration hole. Insure that the gasket is located between the structure and the face plate. From the inside, slip the **galvanized** face plate over the pipe end and cement coupling to secure the assembly.

NOTE: The outside gasket should be slightly compressed.

Repeat this procedure for the other vent termination. Complete outside termination as shown in FIGURE 6 thru 8.

VERTICAL VENT TERMINATION

1. FIGURE 9 shows the proper installation and clearances for vertical vent termination. The vertical roof termination should be sealed with a plumbing roof boot or equivalent flashing. The inlet of the intake pipe and end of the exhaust vent must be terminated no less than 12" above the roof or snow accumulation level, and 12" away from a vertical wall or other protrusion.

2. In no case should the air intake or exhaust vent extend more than 24" above the roof penetration. The vent system can be installed in an existing chimney provided that:
   a) Both the exhaust vent and air intake run the length of the chimney.
   b) The top of the chimney is sealed and weather proofed.
   c) The termination clearances shown in FIGURE 9 are maintained.
   d) No other gas fired appliance is vented into the chimney.

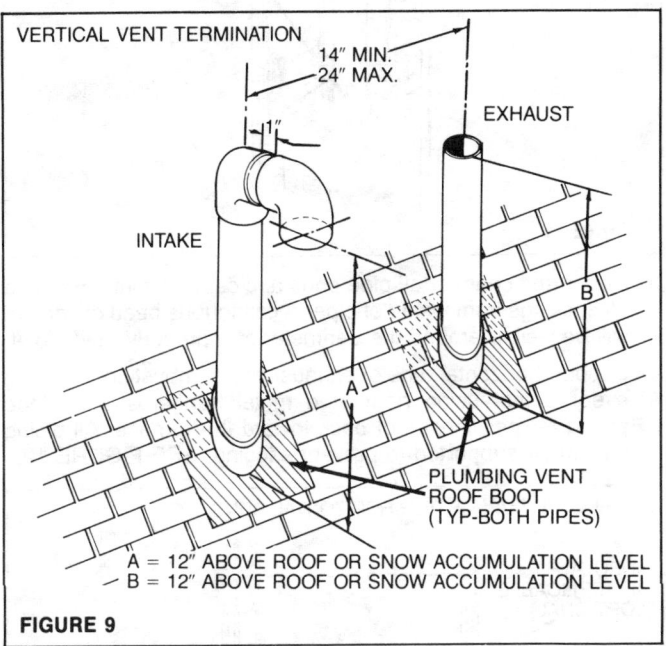

A = 12" ABOVE ROOF OR SNOW ACCUMULATION LEVEL
B = 12" ABOVE ROOF OR SNOW ACCUMULATION LEVEL

FIGURE 9

EXHAUST AND COMBUSTION AIR VENT INSTALLATION

1. Preassemble the exhaust and combustion air piping from the furnace to the vent termination. DO NOT cement any joints together until the preassembly process is complete.

NOTE: If 2" combustion air piping is used, a 3" to 2" bell reducer fitting is required. 3" schedule 40 PVC should be run from the connection collar on the burner box to the outside of the casing before reducing to 2".

The reducing section must be after the 90° elbow in a horizonal section. SEE FIGURE 10.

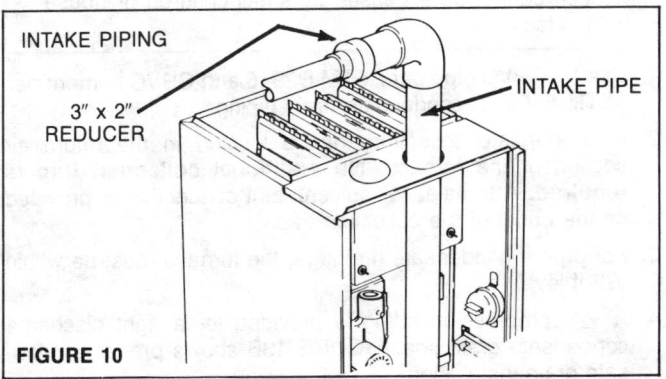

FIGURE 10

2. For serviceability the section of intake air and exhaust vent connecting the intake collar and the flue outlet of the appliance to the vent system should be secured with 2 screws and sealed air and water tight with RTV-Silicone sealant or equivalent. The subject joints are shown in FIGURE 11.

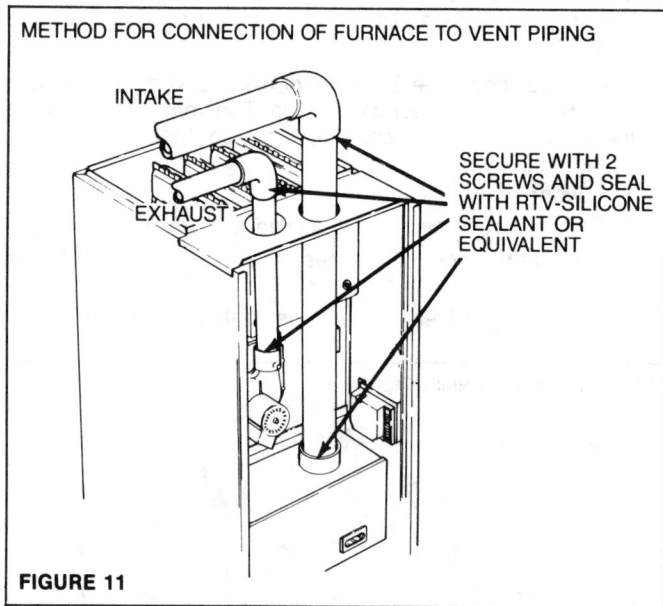

METHOD FOR CONNECTION OF FURNACE TO VENT PIPING

INTAKE

EXHAUST

SECURE WITH 2 SCREWS AND SEAL WITH RTV-SILICONE SEALANT OR EQUIVALENT

FIGURE 11

3. Debur and chamfer all pipe ends and cement joints. Remove all shavings from inside of pipe. A continuous bead of cement will be visible around the perimeter of a properly made joint.

4. Support horizontal runs of exhaust and combustion air piping every 5 feet using perforated metal hanger strips. Slope piping ¼" per lineal foot back toward the furnace. All piping should be supported to prevent sagging. SEE FIGURE 12.

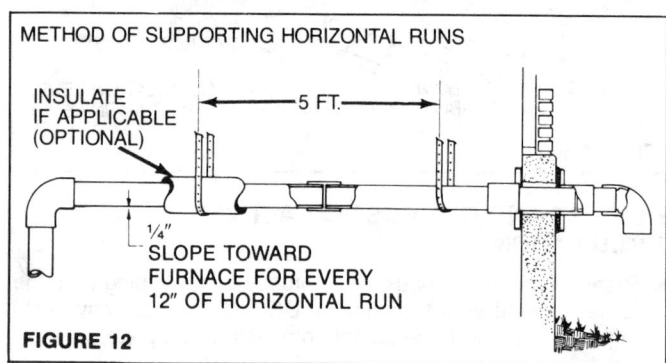

METHOD OF SUPPORTING HORIZONTAL RUNS

INSULATE IF APPLICABLE (OPTIONAL)

5 FT.

¼"
SLOPE TOWARD FURNACE FOR EVERY 12" OF HORIZONTAL RUN

FIGURE 12

CONDENSATE DRAIN PIPING

⚠ CAUTION:

Condensate drain tubing must be routed inside the household drain so that the condensate produced by the furnace cannot be accessed by small children or household pets.

1. Use ½" CPVC pipe per ASTM D2946 and CPVC cement per ASTM F493 for condensate drain piping.

2. A condensate collection trap is located in the return air section of the furnace. **No additional collection trap is required.** A female, ½" solvent joint connection is provided on the outlet of the collection trap.

3. For proper condensate drainage, the furnace must be within ½" of level.

4. A ½" compression fitting is provided for a right discharge condensate drain line. FIGURE 13B shows proper condensate drain installation.

5. If both left and right side return air openings are used, route condensate drain line through the back of the cabinet as shown in FIGURE 13A.

6. Preassemble the complete condensate drain line prior to cementing any joint. Insure that the drain line has continuous slope to the floor drain without any low spots.

7. Debur and chamfer both ends of piping and remove all shavings from inside of pipe. Cement to secure joints.

8. Condensate drain routed through unconditioned spaces must be insulated with closed cell, neoprene tubing insulation or equivalent.

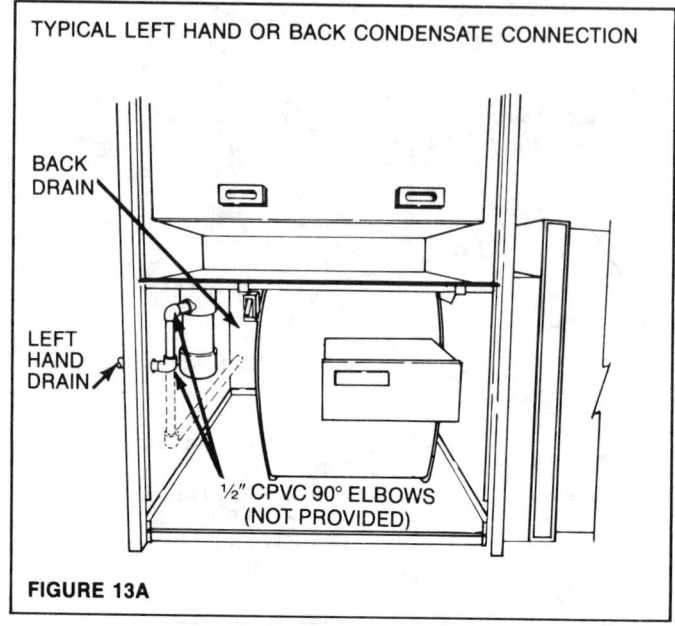

TYPICAL LEFT HAND OR BACK CONDENSATE CONNECTION

BACK DRAIN

LEFT HAND DRAIN

½" CPVC 90° ELBOWS (NOT PROVIDED)

FIGURE 13A

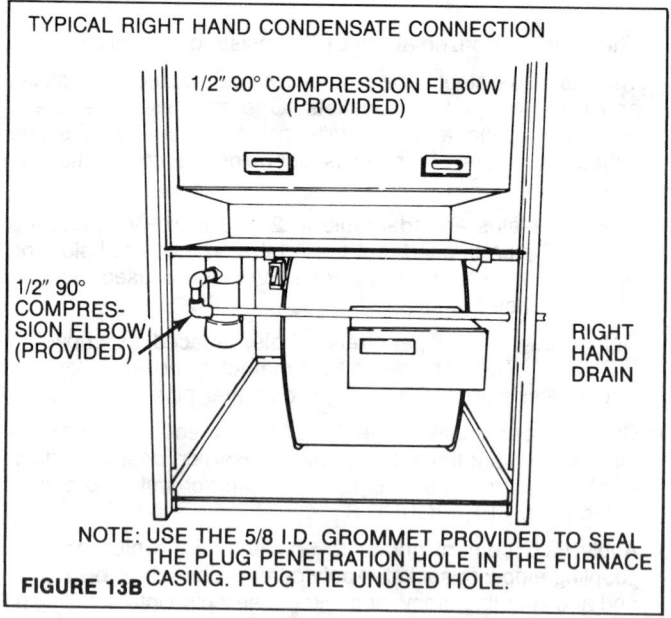

TYPICAL RIGHT HAND CONDENSATE CONNECTION

1/2" 90° COMPRESSION ELBOW (PROVIDED)

1/2" 90° COMPRESSION ELBOW (PROVIDED)

RIGHT HAND DRAIN

NOTE: USE THE 5/8 I.D. GROMMET PROVIDED TO SEAL THE PLUG PENETRATION HOLE IN THE FURNACE CASING. PLUG THE UNUSED HOLE.
FIGURE 13B

CONDENSATE SAFETY LEVEL SENSOR

This furnace is equipped with a condensate level sensor that will detect the back-up of condensate into the furnace due to a blocked condensate drain line. If condensate should back-up into the front manifold of the furnace to the level of the sensor, the sensor will ground the flame sensing circuit of the ignition module terminating unit operation. SEE FIGURE 13C. The unit will retry until lock-out of the ignition module occurs. To put the furnace back into operation, the blockage in the condensate line must be removed and the ignition module must be reset by interrupting 24 VAC power to the unit.

FIGURE 13C

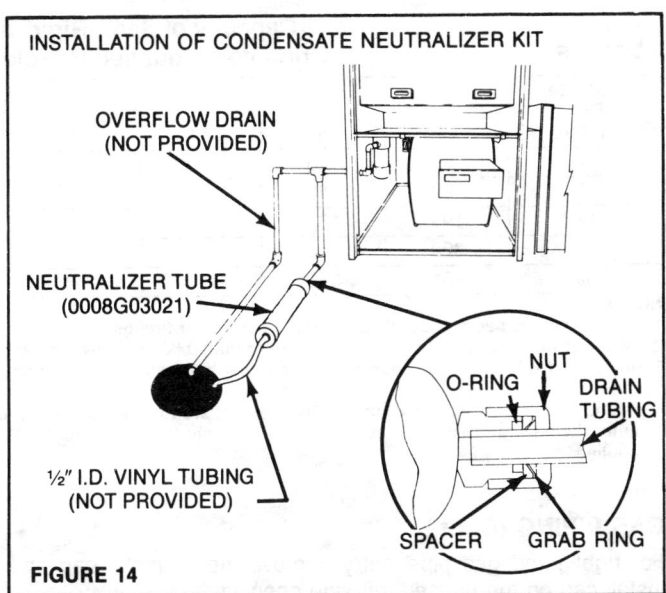

INSTALLATION OF CONDENSATE NEUTRALIZER KIT

OVERFLOW DRAIN
(NOT PROVIDED)

NEUTRALIZER TUBE
(0008G03021)

½" I.D. VINYL TUBING
(NOT PROVIDED)

O-RING — NUT — DRAIN TUBING

SPACER — GRAB RING

FIGURE 14

INSTALLATION OF OPTIONAL CONDENSATE NEUTRALIZER KIT 8G03021

1. Install condensate drain piping from the internal drain trap through the left, right or back side of the unit as described in the "Condensate Drain Piping" Section and shown in FIGURE 13.

2. Route the ½" CPVC piping to the compression fitting on the neutralizer tube. **NOTE:** An overflow drain is **required** between the drain pipe cabinet penetration and condensate neutralizer tube as shown in FIGURE 14. The overflow drain will prevent back-up of condensate to the unit should the neutralizer tube become blocked.

3. Loosen the compression nut and slip the PVC drain pipe through the nut, O-ring, spacer and grab ring until the pipe seats as shown in FIGURE 14. Tighten the compression nut to secure the assembly.

4. Route ½" I.D. vinyl tubing from the barbed fitting on the neutralizer tube to a household plumbing drain.

GAS PIPING

⚠ CAUTION:

PRESSURE TESTING: The furnace and its individual shutoff valve must be disconnected from the gas supply piping system during any pressure testing of that system at test pressures in excess of ½ PSIG (3.48 kPa) (13.87" Water Column.)

The furnace must be isolated from the gas supply piping system by closing its individual manual shutoff valve during any pressure testing of the gas supply piping system at test pressures equal to or less than ½ PSIG (3.48 kPa) (13.87" Water Column).

⚠ WARNING

Do not use open flame for leak testing gas piping system.

⚠ CAUTION:

Use pipe compounds on gas pipe threads that are resistant to all gases.

An opening has been provided on both sides of the furnace for the entry of the gas piping.

TABLE 6 — Maximum capacity of pipe in cubic feet of gas per hour for gas pressures of 0.5 PSIG or less and a pressure drop of 0.5 inch water column (based on 0.60 specific gravity gas)

Nominal Iron Pipe Size Inch	Internal Diameter Inches	Length of Pipe, Feet													
		10	20	30	40	50	60	70	80	90	100	125	150	175	200
⅜	.493	95	65	52	45	40	36	33	31	29	27	24	22	20	19
½	.622	175	120	97	82	73	66	61	57	53	50	44	40	37	35
¾	.824	360	250	200	170	151	138	125	118	110	103	93	84	77	72
1	1.049	680	465	375	320	285	260	240	220	205	195	175	160	145	135
1¼	1.380	1,400	950	770	660	580	530	490	460	430	400	360	325	300	280
1½	1.610	2,100	1,460	1,180	990	900	810	750	690	650	620	550	500	460	430

TABLE 7 — Multipliers to be used when Specific Gravity is Other than 0.60

Spec. Grav.	Mult.	Spec. Grav.	Mult.	Spec. Grav.	Mult.	Spec. Grav.	Mult.
.35	1.31	.65	.962	1.00	.775	1.60	.612
.40	1.23	.70	.926	1.10	.740	1.70	.594
.45	1.16	.75	.895	1.20	.707	1.80	.577
.50	1.10	.80	.867	1.30	.680	1.90	.565
.55	1.04	.85	.841	1.40	.655	2.00	.547
.60	1.00	.90	.817	1.50	.633	2.10	.535

TABLE 8

**Capacity of Semi-Rigid Tubing in 1000 BTU Per Hour of
Undiluted Liquefied Petroleum Gases of 0.50 Pressure Drop**

Outside Dia. Inches	Length of Tubing (Feet)									
	10	20	30	40	50	60	70	80	90	100
³/₈	39	26	21	19	—	—	—	—	—	—
¹/₂	92	62	50	41	37	35	31	29	27	26
⁵/₈	199	131	107	90	79	72	67	62	59	55
³/₄	329	216	181	145	131	121	112	104	95	90
⁷/₈	501	346	277	233	198	187	164	155	146	138

NOTES
1. Allowance has been made for an average number of fittings.
2. When using a gas with specific gravity other than 0.60, multiply the capacity by the factor given in Table 6.
3. To convert BTU input per hour to cubic feet per hour, divide BTU per hour by the heating value of the gas to be used. This can be obtained from your local utility.
4. Piping systems of semi-rigid tubing (Table 8) that are to be supplied with gas of a specific gravity of 1.53 or less can be sized directly from Table 6 unless the authority having jurisdiction specifies that a gravity factor be applied. When the specific gravity of the gas is greater than 1.53, the gravity factor shall be applied.

GAS PIPING (Cont.)

For right hand gas pipe entry, remove the 2″ metal cap and install cap on the unused left side opening.

Install the gas piping from the meter to the unit. Size the pipe as recommended in TABLES 6, 7, and 8. Install a manual shutoff valve external to the unit (max. 4 ft. from furnace).

Install a "Tee" in the gas pipe at the same elevation as the gas inlet connection to the unit. At the bottom of the "Tee" install a "drip-trap" (sediment trap). Install a ground joint union at the top of the "Tee" or between "Tee" and furnace as shown in FIGURE 15.

A ¹/₈″ N.P.T. plugged tap, accessible for test gauge connection, must be installed immediately upstream of the gas supply connection to the furnace.

All gas piping must conform with National and Local Codes and Local Utility requirements. Upon completion of piping, check for leaks with a soapy solution.

⚠ WARNING

Do not use an open flame such as a candle or match for leak testing the gas piping system.

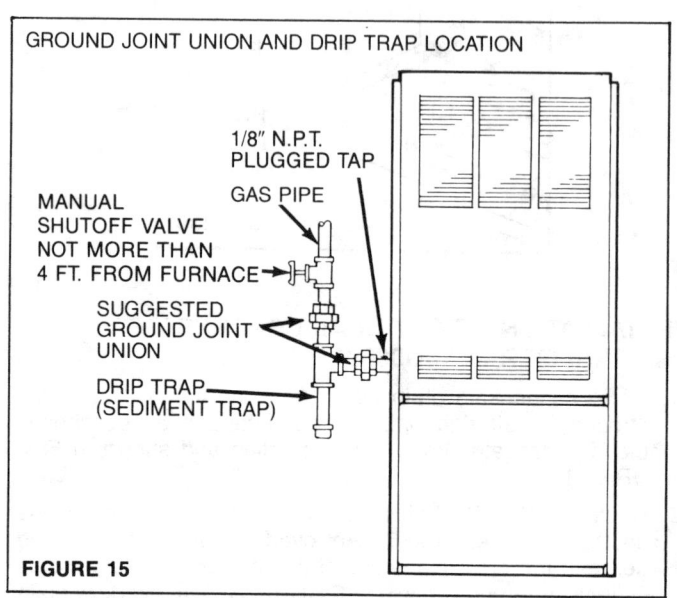

GROUND JOINT UNION AND DRIP TRAP LOCATION

1/8″ N.P.T. PLUGGED TAP

GAS PIPE

MANUAL SHUTOFF VALVE NOT MORE THAN 4 FT. FROM FURNACE

SUGGESTED GROUND JOINT UNION

DRIP TRAP (SEDIMENT TRAP)

FIGURE 15

SECTION IV
ELECTRICAL CONNECTIONS

These units require 115 VOLT SINGLE PHASE POWER SUPPLY. Connect 115 Volt power supply and 24 Volt thermostat wires as indicated in the Wiring Diagram attached to the inside of the blower compartment door, as shown in FIGURE 16.

⚠ WARNING

In order to prevent the possibility of electrical shock, this equipment must be electrically grounded in accordance with the latest edition of the National Electrical Code ANSI/NFPA 70-1987.

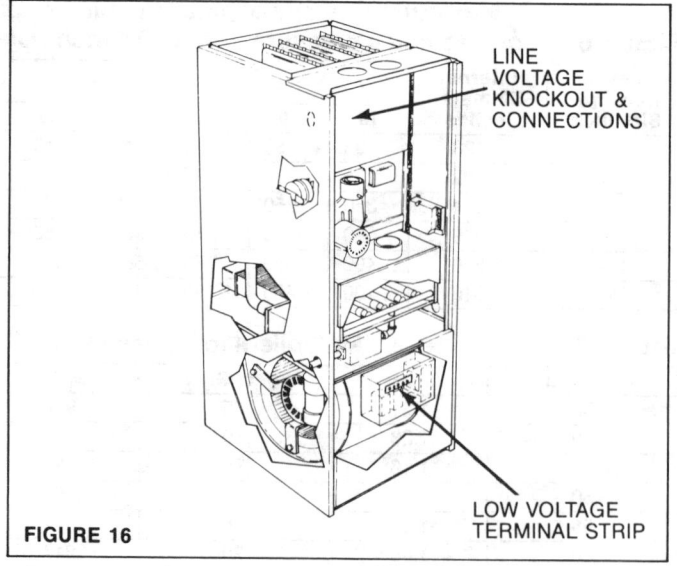

LINE VOLTAGE KNOCKOUT & CONNECTIONS

LOW VOLTAGE TERMINAL STRIP

FIGURE 16

SECTION V
LIGHTING INSTRUCTIONS

ELECTRIC IGNITION MODELS WITH "-IN" OR "-IL" SUFFIX

⚠ WARNING

If you do not follow these instructions exactly, a fire or explosion may result causing property damage, personal injury or loss of life.

SAFETY INSTRUCTIONS

For your safety read the following information before operating furnace:

1. This appliance does not have a pilot. It is equipped with an ignition device which automatically lights the burners. **DO NOT TRY TO LIGHT THE BURNERS BY HAND.**

2. BEFORE LIGHTING, smell all around the appliance area for gas. Be sure to smell next to the floor because some gas is heavier than air and will settle on the floor.

   WHAT TO DO IF YOU SMELL GAS

   • Do not try to light any appliance.

   • Do not touch any electric switch; do not use any phone in your building.

   • Immediately call your gas supplier from a neighbor's phone. Follow the gas supplier's instructions.

   • If you cannot reach your gas supplier, call the fire department.

3. Use only your hand to push in or turn the gas control knob. See FIGURE 17. Never use tools. If the knob will not push in or turn by hand, don't try to repair it, call a qualified service technician. Force or attempted repair may result in a fire or explosion.

4. Do not use this appliance if any part has been under water. Immediately call a qualified service technician to inspect the appliance and to replace any part of the control system and any gas control which has been under water.

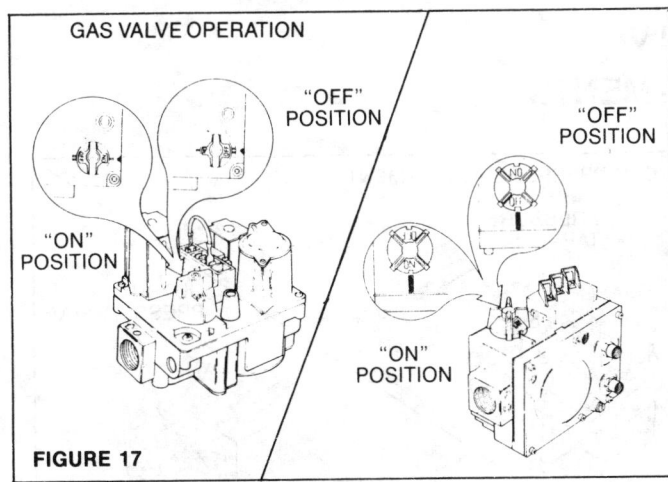

GAS VALVE OPERATION

"OFF" POSITION

"OFF" POSITION

"ON" POSITION

"ON" POSITION

FIGURE 17

OPERATING INSTRUCTIONS

1. STOP! Read the safety information above.

2. Set the thermostat to lowest setting.

3. Turn off all electric power to the appliance.

4. This appliance is equipped with an ignition device which automatically lights the burner. **DO NOT TRY TO LIGHT THE BURNER BY HAND.**

5. Remove louvered access panel. SEE FIGURE 18.

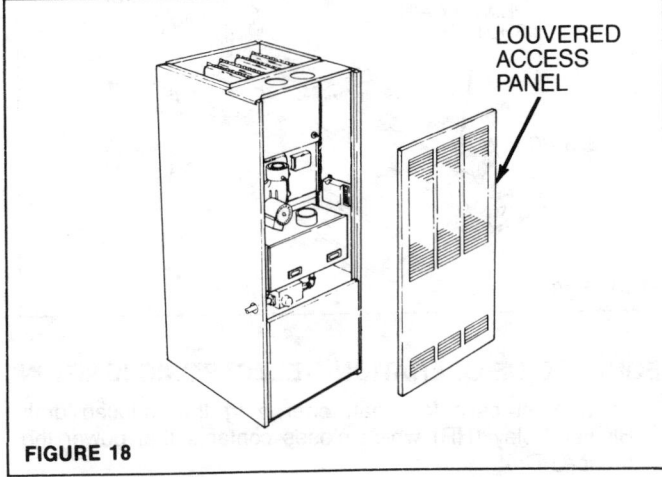

LOUVERED ACCESS PANEL

FIGURE 18

6. Turn gas control knob clockwise ↻ to "OFF." SEE FIGURE 17.

7. Wait five (5) minutes to clear out any gas. If you then smell gas. STOP! Follow step 2 in Safety Information. If you don't smell gas, go to next step.

8. Turn gas control knob clockwise ↻ to "ON." SEE FIGURE 17.

9. Turn on all electric power to the appliance.

10. Set thermostat to desired setting.

11. If the appliance will not operate, follow the instructions "To Turn Off Gas To Appliance" below and call your service technician or gas supplier.

12. Replace louvered access panel.

TO TURN OFF GAS TO APPLIANCE

1. Set the thermostat to lowest setting.

2. Turn off all electric power to the appliance if service is to be performed.

3. Remove louvered access panel.

4. Turn gas control knob clockwise ↻ to "OFF."

5. Replace louvered access panel.

⚠ WARNING

Should overheating occur or the gas supply fail to shut off, turn off the manual gas valve to the appliance before shutting off the electrical supply.

BURNER FLAME REQUIREMENTS

Burner flame should run the length of the burner ports and be 3" to 4" in height. The flame should be blue in color. If for any reason the flame should lift from the burner, change color to orange or start rolling out from under the burner access shield/vertical shield after ignition, an adjustment to the burner is necessary or the heat exchanger needs to be inspected and cleaned. SEE SECTION VI for burner adjustments.

These furnaces are equipped with a multiple burner assembly. SEE FIGURE 19.

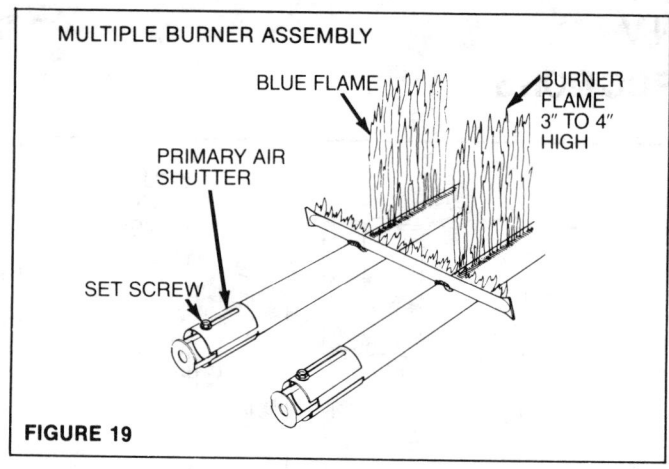

FIGURE 19

SEQUENCE OF OPERATION—ELECTRONIC IGNITION

1. Thermostat calls for heat, energizing the Induced draft Blower Relay (HR) which closes contacts that power the Induced Draft Blower.

2. The rotation of the Induced Draft Combustion Blower will make the vent/intake pressure switch (PS#1) to prove blower operation and send power from the ignition circuit to the ignition control.

3. The pre-purge mode will delay 30 seconds before applying power to the silicone carbide ignitor.

4. The ignitor is allowed to warm up for approximately 15 seconds.

5. At the end of the warm-up time the valve is energized.

6. The main burner flame must be detected within 7 seconds or the valve is de-energized and the ignitor turned off.

7. If there is a failure to sense flame on the first attempt, there will be a 60 second delay. The second attempt at lighting the burner flame will duplicate the first attempt with pre-purge, warm-up times and lock-out times identical to those in the first try. Failure to prove on the second attempt will allow yet another complete ignition attempt before complete system lockout.

8. Lock-out of the control after the final ignition attempt, will require interruption of the power to the ignition control (either the 24V thermostat or line voltage to the furnace may be interrupted) for 1/20 second or longer to reset the system.

9. The ignition sequence will repeat for a total of two recycles. If flame is lost within the first thirty seconds of establishment, the total of ignition attempts plus the number of recycles will not exceed three.

10. If flame is established for more than thirty seconds after ignition, the recycle counter will clear.

11. A momentary loss of power or flame to the sensing circuit will cause the valve to de-energize and a delay of 15 seconds before restarting the ignition sequence will occur. Recycles will begin and the burner will operate normally if the gas supply returns or the fault condition is corrected prior to the last ignition attempt. Otherwise the control will lock-out.

12. Typical conditions are that the burner will light and stay lit as long as the thermostat calls for heat.

13. Heat from the burner flames warm the heat exchanger surface, the flue products pass through the baffles and vent outdoors through the flue pipe.

14. The fan and limit control senses the supply air temperature adjacent to the heat exchanger and will signal the blower to operate when the air temperature reaches the "ON" fan set point. The fan and limit control will shut the gas off if the supply air temperature reaches the "High Temperature" set point. The blower continues to operate and conditioned air is supplied to the space through the duct system.

15. When the thermostat set point temperature is reached, the gas valve is de-energized which stops the flow of gas to the burners. The blower continues to operate until the air temperature falls below the "OFF" blower set point.

16. The furnace repeats this same cycle each time the thermostat calls for heat.

SECTION VI
FINAL ADJUSTMENTS

TEMPERATURE RISE

The furnace installation is to be adjusted to obtain a temperature rise within the range specified on the rating plate.

PRESSURE REGULATOR ADJUSTMENT

A pressure tap has been provided on the gas valve. SEE FIGURE 20. Attach a pressure gauge fitting to this tap and check the outlet pressure. The pressure should be set as indicated on the rating plate. To adjust the pressure regulator, remove the cap covering the Pressure regulator adjusting screw. On valves with more than one adjusting screw, adjust only the screw marked "HI." Turn the adjusting screw out (counterclockwise) to decrease the pressure, and in (Clockwise) to increase the pressure. Reinstall the cap securely.

After use, replace the 1/8" pipe plug. **NOTE:** The pressure rating is normally 3.5 inches W.C. for Nat. Gas, and 10.0 inches W.C. for LP Gas.

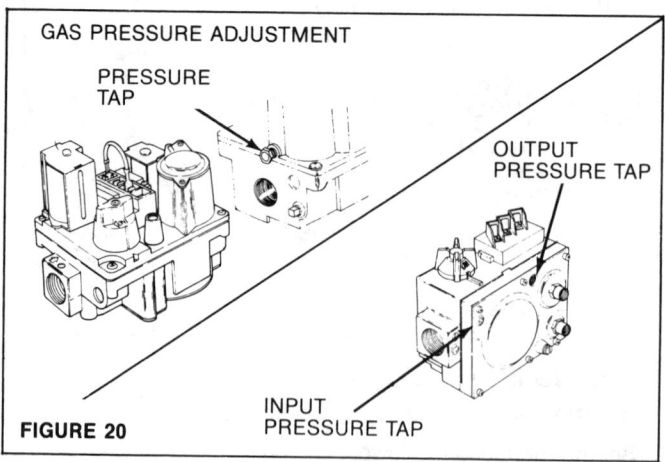

FIGURE 20

After the pressure regulator has been adjusted to its proper value, check the input by the following formula:

INPUT IN BTU/HR =

$$\text{INPUT IN BTU/HR} = \frac{①\text{Heating value of gas in BTU-Per-CU. ft} \times 3600 \times \text{CU. ft. gas measured②}}{③\text{Time in seconds for CU. ft. of gas measured}}$$

To obtain energy input:

① Heating value of gas (BTU/CU. Ft.) is obtained from the Local Gas Utility.

② To calculate CU. FT. Gas Measured: turn off all other gas appliances and pilots. REMEMBER to relight all pilots after test is completed.

③ Measure Time (in seconds) for gas meter test dial to complete one revolution (local Utility will provide size of meter).

④ Calculated energy input must not exceed gas input on unit rating plate. If the input does not correspond to that shown on the rating plate, adjust the regulator screw of the pressure regulator until the rating plate input is obtained. The pressure in the manifold should not vary more than plus or minus 0.3 inches of water column of the value specified on rating plate. Any major changes in the gasflow should be made by changing the size of the burner orifices.

HIGH ALTITUDE INSTALLATION

These units may be used at full input rating when installed at altitudes up to 2000 feet. When installed above 2000 feet, the input must be decreased 4% for each 1000 feet above sea level. This may be accomplished by a simple adjustment of manifold pressure or an orifice change, or a combination of a pressure adjustment and an orifice change. The changes required depend on the installation altitude and the heating value of the fuel. TABLE 9 shows the proper furnace manifold pressure and gas orifice size to achieve proper performance based on elevation above sea level for both natural gas and propane.

TABLE 9

MANIFOLD PRESSURE AND ORIFICE SIZE FOR HIGH ALTITUDE APPLICATIONS

NATURAL GAS

HEATING VALUE BTU/CU FT		MEAN ELEVATION FEET ABOVE SEA LEVEL					
		2000 to 3000	3000 to 4000	4000 to 5000	5000 to 6000	6000 to 7000	7000 to 8000
800		3.5"wc	3.5"wc	3.5"wc	3.5"wc	3.2"wc	2.8"wc
850		3.5"wc	3.5"wc	3.5"wc	3.1"wc	2.8"wc	2.5"wc
900		3.5"wc	3.3"wc	3.0"wc	2.8"wc	2.5"wc	2.3"wc
950		3.3"wc	3.0"wc	2.7"wc	2.5"wc	2.3"wc	3.5"wc
1000		2.9"wc	2.7"wc	2.5"wc	2.3"wc	3.5"wc	3.1"wc
1050		2.7"wc	2.4"wc	2.3"wc	3.5"wc	3.2"wc	2.8"wc
1100		2.4"wc	2.3"wc	3.5"wc	3.2"wc	2.8"wc	2.5"wc
GUH060	ORIFICE SIZE	STANDARD FACTORY ORIFICE		#48	#48	#48	#48
GUH075				#44	#44	#44	#44
GUH100				#44	#44	#44	#44
GUH125				#44	#44	#44	#44

▨ SHADED AREA REQUIRES ORIFICE CHANGE
NO SHADING INDICATES MANIFOLD PRESSURE CHANGE ONLY.

PROPANE

HEATING VALUE BTU/CU FT		MEAN ELEVATION FEET ABOVE SEA LEVEL					
		2000 to 3000	3000 to 4000	4000 to 5000	5000 to 6000	6000 to 7000	7000 to 8000
2500		8.5"wc	7.8"wc	7.0"wc	10.0"wc	9.0"wc	8.1"wc
GUH060	ORIFICE SIZE	STANDARD FACTORY ORIFICE			3/64	3/64	3/64
GUH075					#55	#55	#55
GUH100					#55	#55	#55
GUH125					#55	#55	#55

NOTE: NATURAL GAS DATA BASED ON 0.60 SPECIFIC GRAVITY.
   PROPANE DATA BASED ON 1.53 SPECIFIC GRAVITY.
   FOR FUELS WITH DIFFERENT SPECIFIC GRAVITY CONSULT THE NATIONAL FUEL GAS CODE ANSI Z223.1a-1987.
IMPORTANT: FILL OUT HIGH ALTITUDE LABEL SUPPLIED WITH UNIT OR HIGH ALTITUDE KIT AND AFFIX TO COVER OF BURNER BOX.

To use the natural gas table, first consult your local gas utility for the heating value of the gas supply. Select the heating value on the vertical border and follow across the table until the appropriate elevation for the installation is reached. The first value in the box at the intersection of the heating value and elevation will be the manifold pressure required. If a gas orifice change is also required, the box is shaded. The required orifice size is shown in the table. **NOTE:** The manifold pressure may still need to be reset after a gas orifice change. The high altitude label supplied with the unit and also in the required orifice kit must be completed and affixed to the front of the burner enclosure box. Fill in the manifold pressure, orifice size and revised input rate. The revised input rate is determined in the following manner:

$$\text{High Altitude Input Rate} = \frac{\text{Sea Level}}{\text{Nameplate}} \times (\text{Multiplier})$$
$$\text{Input Rate}$$

Elevation	High Altitude Multiplier
2000' - 2999'	0.92
3000' - 3999'	0.88
4000' - 4999'	0.84
5000' - 5999'	0.80
6000' - 6999'	0.76
7000' - 7999'	0.72

Example:

For a GUH075A installed at an altitude of 5280' with a gas supply heating value of 950 BTU/Ft3, a manifold pressure of 2.5" WC is required. The revised high altitude input is:

High Altitude Input Rate = 75,000 × 0.80 = 60,000 BTU/HR

NOTE: A gas orifice change was not required in this example.

PRIMARY AIR ADJUSTMENT (Air Shutter)

These units have individual primary air shutters on each burner. To adjust, loosen the lock screw and close the air shutter until yellow tips appear on the flames, then open the air shutter until the yellow tips just disappear, and tighten the lock screw. Follow this procedure on each burner.

Should there be a need for burner replacement on a furnace, it will be necessary to order the complete set of burners.

The furnaces are manufactured with multiple burner assemblies. SEE FIGURE 21.

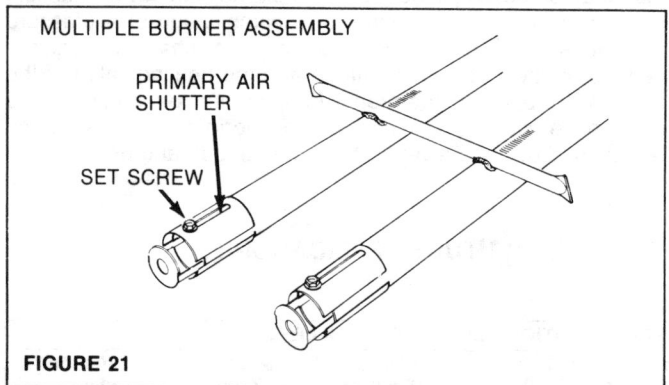

MULTIPLE BURNER ASSEMBLY

PRIMARY AIR SHUTTER

SET SCREW

FIGURE 21

THERMOSTAT ANTICIPATOR SETTING

The proper thermostat heat anticipator setting is the amp rating on the gas control valve and is indicated on a label attached to the furnace. To increase length of cycle increase setting of heat scale. To decrease length of cycle decrease setting of heater scale.

Proper control of the indoor air temperature can only be achieved if the thermostat is calibrated to the heating and/or cooling system. A vital consideration of this calibration is related to the thermostat heat anticipator.

Anticipators for the cooling operation are generally preset by the thermostat manufacturer and require no adjustment. Anticipators for the heating operation are of two types: preset or adjustable. Those that are pre-set will not have an adjustable scale and are generally marked accordingly.

Thermostat models having a scale as shown in FIGURE 22 *must be* adjusted to each application.

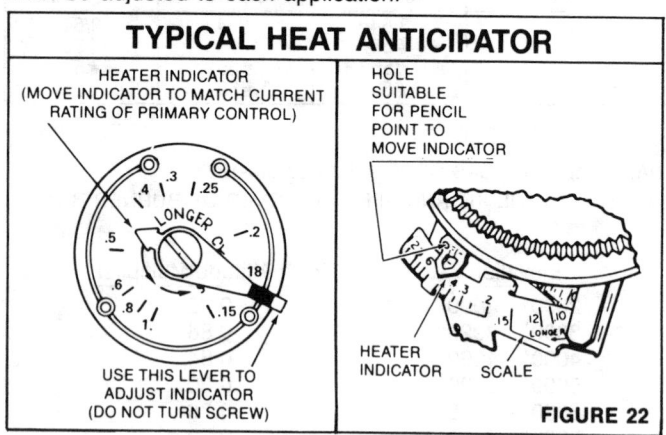

TYPICAL HEAT ANTICIPATOR

HEATER INDICATOR
(MOVE INDICATOR TO MATCH CURRENT RATING OF PRIMARY CONTROL)

LONGER

USE THIS LEVER TO ADJUST INDICATOR
(DO NOT TURN SCREW)

HOLE SUITABLE FOR PENCIL POINT TO MOVE INDICATOR

HEATER INDICATOR   SCALE

FIGURE 22

In many cases, this adjustment setting can be found in the thermostat instructions. If this information is not available, or if the correct setting is questioned, the following procedure should be followed:

STEP 1. Wrap 10 loops of single strand, insulated thermostat wire around the prongs of an ampmeter. Set the scale to the 1 to 5 or 1 to 6 amp. scale.

STEP 2. Connect the uninsulated ends of this wire jumper across terminals "R" and "W" on the sub-base ("RH" and "W", on

multi-stage thermostat sub-base). SEE FIGURE 23.

This test must be performed without the thermostat attached to the sub-base.

STEP 3. Let the heating system operate in this position for about one minute. Read the ampmeter scale. Whatever reading is indicated must be divided by 10 (for 10 loops of wire).

This is the setting at which the adjustable heat anticipator should be set.

$$\text{FORMULA:} \quad \frac{\text{Ampmeter reading}}{10 \text{ loops}} = \text{Anticipator Setting}$$

$$\text{OR:} \quad \frac{2.5 \text{ Amps}}{10} = .25 \text{ AMPS Setting}$$

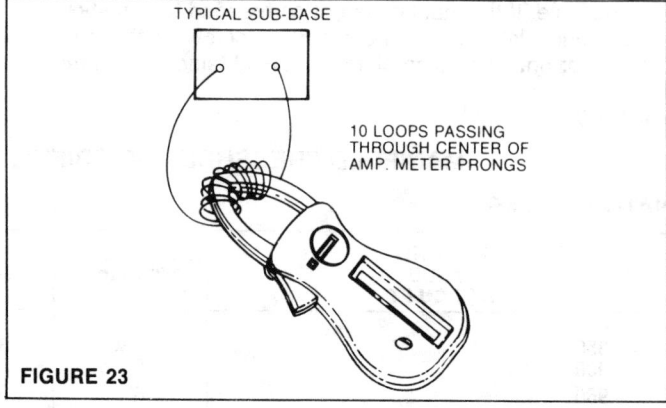

TYPICAL SUB-BASE

10 LOOPS PASSING THROUGH CENTER OF AMP. METER PRONGS

FIGURE 23

STEP 4. If a slightly longer cycle is desired, the pointer should be moved to a higher setting. Slightly shorter cycles can be achieved by moving to a lower setting.

STEP 5. Remove the meter jumper wire and reconnect the thermostat. Check the thermostat in the heating mode for proper operation.

NOTE: The length of the heating cycle can also be affected by the fan limit control settings (if applicable). The fan "on and off" settings should be checked at this point.

For thermostats having 2 stage heat, step 1, 2, and 3 must be repeated. Second stage heat is controlled through terminals "RH" and "W_2" on the sub-base.

BALANCING THE SYSTEM

There is one adjustment of the heating system that the owner is in a position to do more satisfactorily than anyone else. This is the balancing of the system to provide equal comfort in every room. Best balancing is done on a typical cold day. Proceed as follows:

1. Be sure that dampers or valves in pipes or register are wide open in all runs to the room or rooms that seem a little cool.

2. It is best to do this balancing in the evening after sundown.

3. The room thermostat should be left at one setting for several hours before you attempt to do any balancing.

4. Check the temperature in all rooms, you can do this with a thermometer.

5. In the room or rooms that are too warm, turn down the register valve.

6. Wait several hours until room comfort balances out at this new setting and then recheck temperature.

7. By doing the adjusting a little at a time, you will accomplish the temperature balance you want. After a pipe damper or register valve has been adjusted it takes time for the temperature to conform with the new setting. By adjusting a little at a time, you will soon accomplish better system balance than can be done by anyone else who does not live in the house.

NOTE: IT IS VERY IMPORTANT THAT THIS APPLIANCE BE MAINTAINED ACCORDING TO THE FOLLOWING OPERATION & MAINTENANCE PROCEDURES. FAILURE TO DO SO COULD JEOPARDIZE THE WARRANTY CONVERAGE FOR THIS UNIT.

THERMOSTAT

The thermostat is designed to provide the utmost in heating comfort, and is also designed to be attractive in appearance.

Please remember that these controls are precision devices, and therefore should be handled accordingly. Also, remember that the Thermostat will be affected by any heat source, so do not place lamps, radio, TV, etc., so that heat from these sources can affect the Thermostat operation.

TEMPERATURE SELECTION

To select the temperature control point, move the temperature selection dial until the pointer is in line with the desired point on the temperature scale. The thermometer pointer will not change position on the temperature scale when the temperature selection lever is moved, but will change position as room temperature varies.

FAN AND LIMIT CONTROL

The Fan and Limit Control is mounted in its proper location. The limit is set at the correct setting. This setting is the point at which the Limit Switch contacts open. The switch is automatic-reset and will make contact when the temperature of the element drops. On installations with adjustable fan controls, the settings should not exceed 140° on and 90° off.

⚠ **CAUTION:**

Do not tamper with the internal fan or limit control mechanism.

BLOWER SPEED ADJUSTMENT
(DIRECT DRIVE BLOWER)

The multi-speed direct drive motor is wired to the Hi-speed tap for cooling operation and low speed for heating operation. A change to a lower speed may be made in the transformer box. Heating and cooling models with a blower relay, change automatically. Refer to wiring diagram attached to the furnace.

LUBRICATION INSTRUCTIONS

For Lubrication Instructions for motor or blower refer to labels attached to blower housing.

FILTERS

If filters are provided as standard equipment, the forced air units are then equipped with efficient washable type air filters designed to keep house cleaner by withholding dirt and other foreign particles from the air circulation. The filters should be cleaned or replaced at least twice a year or at any time they become clogged. Dirty filters will cause the furnace to heat improperly and possibly overheat by holding back air circulation. These furnaces are designed so that the filters may be quickly and easily replaced. Replacement filters must be of the same type and size.

HEAT EXCHANGER AND BURNER CLEANING PROCEDURE:

⚠ **CAUTION:**

On units with suffix "-IN," or "-IL" a silicon carbide ignition is used. Extreme care must be exercised in handling the assembly due to the fragile nature of the ignition assembly.

⚠ **CAUTION:**

NOTE: THIS FURNACE IS EQUIPPED WITH A PRESSURIZED COMBUSTION CHAMBER. IT IS MANDATORY TO REPLACE ALL GASKETS THAT ARE REMOVED, ORDER REPLACEMENT GASKET PRIOR TO STARTING SERVICE.

---

1. Disconnect power source to the unit.
2. Remove louver panel from the unit and set aside. SEE FIGURE 24.

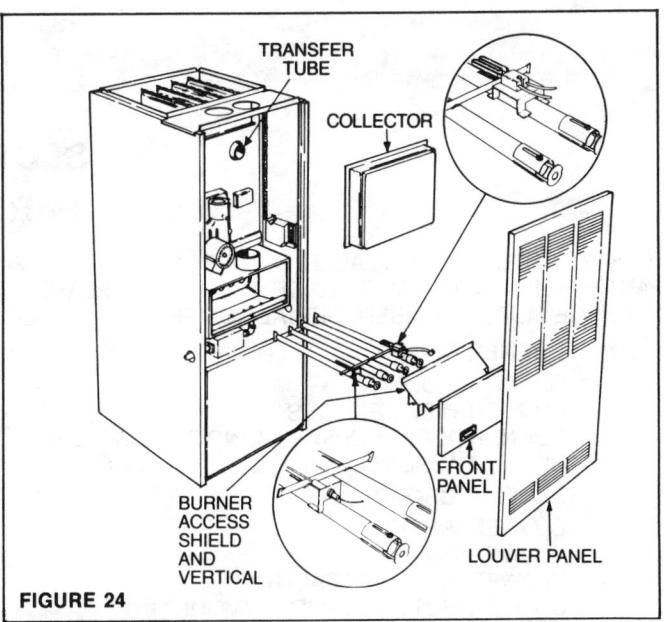

FIGURE 24

3. Turn gas supply "off" at the gas valve. Allow a few minutes for unit to cool down.
4. Remove the screws from the front panel of the burner enclosure and set aside.
5. Remove the front panel.
6. Remove the two screws holding the burner access shield and vertical shield in place and set aside.
7. Remove the burner access shield, and vertical shield.
8. On the H.S.I. System, disconnect the sensor lead from the remote sensor (far left burner) and the ignitor leads at the ignitor disconnect.

NOTE: At this point inspect for any loose particles or carbon deposits in the burner area. Using a flashlight and a small mirror, check the upper portions of the heat exchanger for any signs of blockage or dirt accumulation. If any of these conditions are apparent, proceed with the following steps.

9. Remove all burners from the unit. This is accomplished by pulling back on the burner assembly so that it compresses the springs that are located between the manifold and the burners.

Tilt the burners up into the heat exchanger tube. Slide the burner assembly off the orifices and remove from the unit.

Note the location of the burner assembly. The burner assembly with the ignitor is on the right hand side and the sensor burner is positioned on the left.

10. After the burner assemblies have been taken from the unit, tap the end of the burners (with the primary air shutter attached) lightly on the ground, thereby removing any residue from inside the burners. Then run the vacuum hose across the top of the burner assembly removing any foreign material that might be lodged between the ports.

11. Remove the flue transition box at the top of the unit. Using the exhaust end of a high pressure vacuum (shop vacuum) blow any loose particles from the upper heat exchanger to the bottom burner area.

12. With a stiff bristled brush, clean the bottom area of the heat exchanger, brushing into the burner enclosure where the vacuum can easily be used to pick up the residue.

13. Flush the secondary heat exchanger with fresh water through the transfer tube inlet in the flue transition area. The water flush will drain through the drain trap in the blower compartment.

14. Reassemble the unit by reversing the above procedure. Make sure all old gaskets are discarded and replaced with new gaskets.

NOTE: Extreme care must be taken to protect all new seals and gaskets to assure tightness of the system. Failure to provide tight seals can result in faulty or improper system operation.

EMERGENCY SERVICE

In case of operating difficulty, check the following items before calling for service:

1. Make sure thermostat is set above room temperature.

2. Make sure line switch is "ON" and that fuse is not blown.

3. Make sure air filter is clean.

4. Make sure the Blower Door is properly in place.

5. Make sure there is no blockage of the flue or intake to the outside atmosphere.

6. Make sure connections on the safety pressure switch, both tubing and wiring are secure.

7. Make sure all compartments are securely closed and all fasteners are in place.

If difficulty still exists, call your Installer.

SECTION VIII
PARTS LIST

REPAIR PARTS ARE AVAILABLE FROM THE FURNACE MANUFACTURER, DEALER OR DISTRIBUTOR. WHEN ORDERING PARTS, REFER TO THAT PART OF THE RATING PLATE, ATTACHED TO THE FURNACE THAT INCLUDES THE COMPLETE FURNACE MODEL NUMBER, SERIAL NUMBER AND LISTED NAME AND ADDRESS WHERE PARTS ARE AVAILABLE.

BURNER GROUP
GAS MANIFOLD
MAIN BURNER ORIFICES
BURNER ASSEMBLY WITH IGNITOR BRACKET
BURNER ASSEMBLY WITH SENSOR BRACKET

BLOWER GROUP
BLOWER WHEEL
BLOWER MOTOR
BLOWER MOTOR CAPACITOR
BLOWER MOTOR SUPPORT (MOUNT)
INDUCED DRAFT BLOWER/COLLECTOR PACKAGE

CONDENSING HEAT EXCHANGER GROUP
CONDENSING COIL
DISCHARGE TUBE
DRAIN TRAP
TRANSFER TUBE

CONTROL GROUP
GAS VALVE
IGNITION CONTROL (ELECTRIC IGNITION)
IGNITOR (ELECTRIC IGNITOR)
SENSOR-REMOTE (ELECTRIC IGNITION)
FAN & LIMIT CONTROL
PRESSURE SWITCH (VENT)
CONDENSATE LEVEL SENSOR
TRANSFORMER
RELAY-MAIN BLOWER
RELAY-COMBUSTION BLOWER
BLOWER DOOR INTERLOCK SWITCH

AN EXPANDED PARTS LIST, FOR THE PARTICULAR FURNACE MODEL MAY BE FOUND IN THE FURNACE INSTALLATION INSTRUCTION AND USER'S MANUAL ENVELOPE.

⚠ WARNING

IMPROPER INSTALLATION, ADJUSTMENT, SERVICE OR MAINTENANCE CAN CAUSE INJURY OR PROPERTY DAMAGE. CONSULT A QUALIFIED INSTALLER, SERVICE AGENCY, DEALER, DISTRIBUTOR OR THE GAS SUPPLIER FOR INFORMATION OR ASSISTANCE.

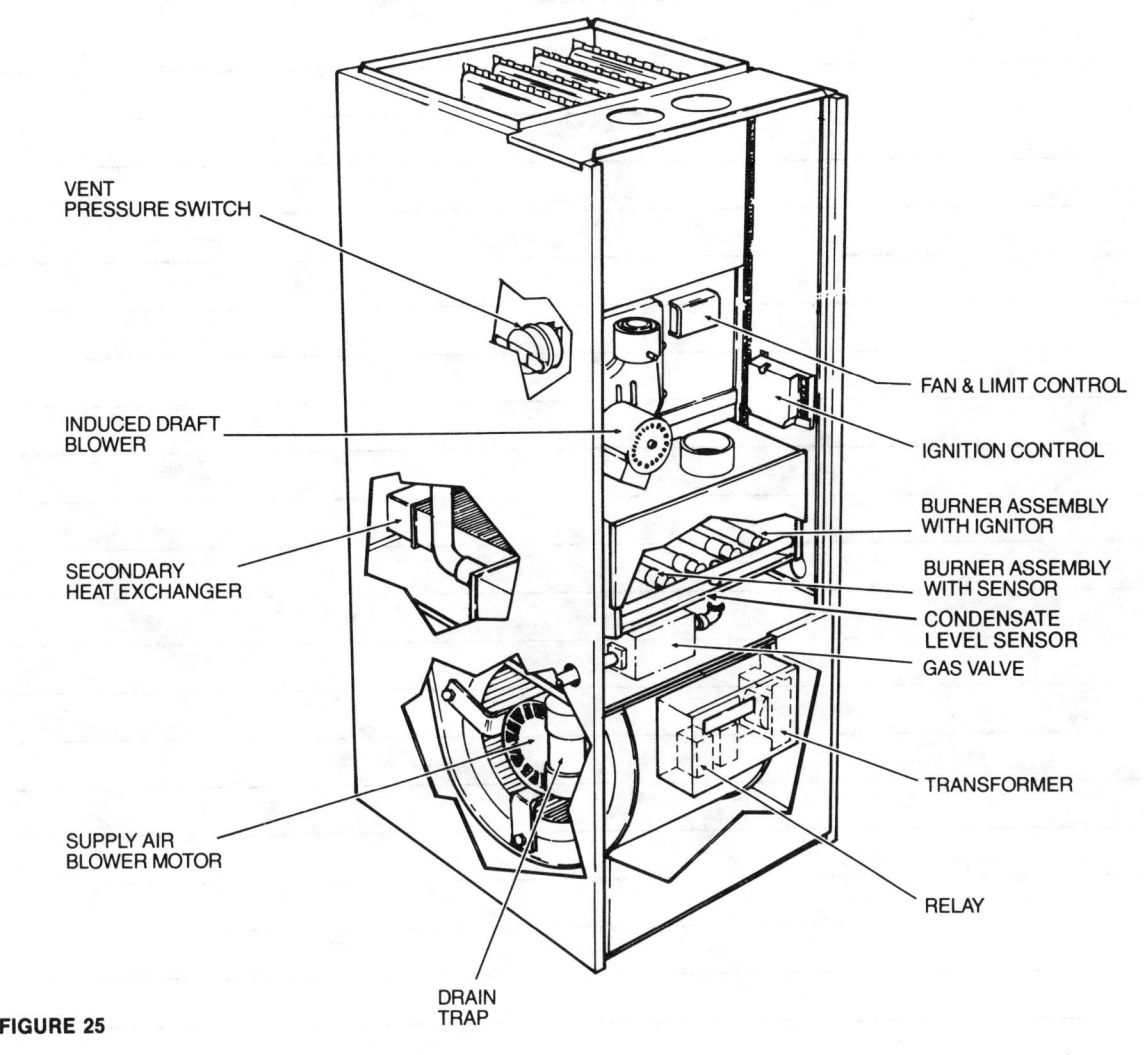

VENT PRESSURE SWITCH

INDUCED DRAFT BLOWER

SECONDARY HEAT EXCHANGER

SUPPLY AIR BLOWER MOTOR

DRAIN TRAP

FAN & LIMIT CONTROL

IGNITION CONTROL

BURNER ASSEMBLY WITH IGNITOR

BURNER ASSEMBLY WITH SENSOR

CONDENSATE LEVEL SENSOR

GAS VALVE

TRANSFORMER

RELAY

FIGURE 25

AIR DELIVERY—(CFM) FOR DIRECT DRIVE BLOWERS

MODEL NUMBER	BLOWER SIZE	MOTOR H.P.	BLOWER SPEED	EXTERNAL STATIC PRESSURE—INCHES WATER COLUMN													
				0.1		0.2		0.3		0.4		0.5		0.6		0.7	
				CFM	TEMP. RISE	CFM	TEMP. RISE	CFM	TEMP. RISE	CFM	TEMP. RISE	CFM	TEMP. RISE	CFM	TEMP. RISE	CFM	TEMP. RISE
GUH060A012	11.8x8	½	HIGH	1609	32.4	1550	33.6	1491	35.0	1431	36.4	1365	38.2	1280	40.7	1202	43.4
			MED.	1294	40.3	1268	41.1	1231	42.4	1188	43.9	1131	46.1	1070	48.9	992	52.6
			LOW	1045	49.9	1021	51.1	995	52.4	960	54.3	924	56.5	875	59.6	820	63.6
GUH075A012	11.8x8	½	HIGH	1609	40.1	1550	41.6	1491	43.3	1431	45.1	1365	47.3	1280	50.4	1202	53.7
			MED.	1294	49.9	1268	50.9	1231	52.4	1188	54.3	1131	57.1	1070	60.3	992	65.1
			LOW	1045	61.8	1021	63.2	995	64.9	960	67.2	924	69.8	875	73.8	820	78.7
GUH100A016	11.8x10.6	¾	HIGH	1642	52.4	1568	54.9	1505	57.2	1441	59.7	1359	63.3	1270	67.8	1170	78.5
			MED. HI	1621	53.1	1557	55.3	1487	57.9	1415	60.8	1357	63.4	1252	68.7	1152	74.7
			MED. LOW	1465	58.7	1399	61.5	1343	64.1	1285	67.0	1197	71.9	1122	76.7	1030	83.6
			LOW	1335	64.5	1295	66.4	1249	68.9	1207	71.3	1152	74.7	1084	79.4	1010	85.2
GUH125A016	11.8x10.6	¾	HIGH	1810	58.8	1750	60.8	1675	63.5	1628	65.4	1577	67.5	1470	72.4	1371	77.6
			MED. HIGH	1802	59.0	1730	61.5	1655	64.3	1575	67.6	1478	72.0	1380	77.1	1262	84.3
			MED. LOW	1685	63.1	1625	65.5	1565	68.0	1490	71.4	1403	75.8	1308	81.4	1188	89.6
			LOW	1562	68.1	1515	70.2	1460	72.9	1390	76.6	1313	81.0	1220	87.2	1103	96.5
GUH125A020	11.8x10.6	¾	HIGH	2160	48.5	2085	50.0	2010	52.1	1955	53.6	1890	55.4	1820	57.6	1680	62.4
			MED. HIGH	1891	55.4	1837	57.0	1784	58.7	1729	60.6	1675	62.6	1582	66.2	1490	70.3
			MED. LOW	1777	59.0	1728	60.6	1680	62.4	1614	64.9	1549	67.6	1495	70.1	1441	72.7
			LOW	1631	64.2	1596	65.7	1562	67.1	1505	69.6	1448	72.4	1389	75.4	1330	78.8

NOTES

NOTES

USER'S
INFORMATION MANUAL

SUPER HIGH EFFICIENCY UPFLOW GAS FIRED FURNACE

GUH SERIES

REVISED: JULY, 1989
SUPERSEDES: JUNE, 1988

FOR YOUR SAFETY
WHAT TO DO IF YOU SMELL GAS

- DO NOT TRY TO LIGHT ANY APPLIANCE.
- DO NOT TOUCH ANY ELECTRICAL SWITCH; DO NOT USE ANY PHONE IN YOUR BUILDING.
- IMMEDIATELY CALL YOUR GAS SUPPLIER FROM A NEIGHBOR'S PHONE. FOLLOW THE GAS SUPPLIER'S INSTRUCTIONS.
- IF YOU CANNOT REACH YOUR GAS SUPPLIER, CALL THE FIRE DEPARTMENT.

FOR YOUR SAFETY

Do not store or use gasoline or other flammable vapors and liquids in the vicinity of this or any other appliance.

⚠ WARNING

Improper installation, adjustment, alteration, service or maintenance can cause injury or property damage. Refer to this manual. For assistance or additional information consult a qualified installer, service agency or the gas supplier.

SECTION I
DO'S AND DO NOT'S

The information contained in this manual is designed to identify any hazards and to serve as a guide to the user of any malfunctions.

- DO READ THE USER'S MANUAL AND INSTALLATION INSTRUCTIONS.
- DO HAVE YOUR FURNACE SERVICED ONCE A YEAR. Proper care of your furnace will assure many years of economical comfort. We strongly recommend that your furnace be inspected and serviced by a qualified serviceman at the beginning of each heating season.
- DO INSPECT THE INSTALLATION OF YOUR FURNACE.

> **⚠ WARNING**
>
> It is mandatory that the area around the furnace be kept clear and free of any combustible materials (cardboard boxes, etc.). Gasoline and other volatile liquids or flammable vapors. DO NOT store or use volatile materials near an open flame.

- DO REVIEW THE LIGHTING INSTRUCTIONS BEFORE OPERATING THE UNIT.
- DO SHUT OFF MANUAL GAS VALVE BEFORE ELECTRICAL SUPPLY IN AN EMERGENCY.

> **⚠ WARNING**
>
> Should over heating occur or the gas supply fail to shut off, turn off the manual gas valve to the appliance before shutting off the electrical supply.

- DO NOT ATTEMPT TO ALTER YOUR FURNACE.

> **⚠ WARNING**
>
> Improper installation, adjustment, alteration, service or maintenance can cause injury or property damage.

- DO NOT ATTEMPT TO LIGHT AN ELECTRIC IGNITION FURNACE WITH A MATCH.

> **⚠ CAUTION:**
>
> Your furnace is equipped with an electric ignition. DO NOT attempt to light the furnace manually with a match or other lighting means.

- DO NOT BLOCK OR OBSTRUCT THE AIR INLETS OR OUTLETS TO THE FURNACE.
- DO NOT TAMPER WITH THE INTERNAL FAN AND LIMIT CONTROL MECHANISM.
- DO NOT INSTALL FURNACE IN A SPACE THAT CONTAINS COMBUSTIBLE INSULATION.
- DO NOT INSULATE THE FURNACE CABINET.

SECTION II
FACTS ABOUT YOUR FURNACE

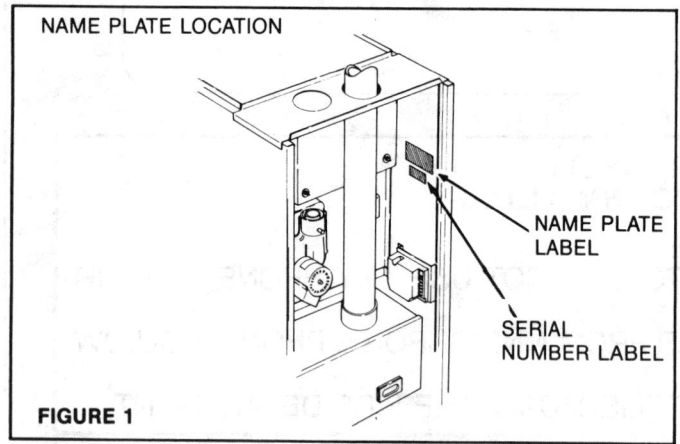

NAME PLATE LOCATION

NAME PLATE LABEL

SERIAL NUMBER LABEL

FIGURE 1

UNIT DESCRIPTION

The nameplate label and serial number label shows the model number of the furnace. These labels are located on the right hand side, inside the casing. SEE FIGURE 1 for the location of labels.

Please note that each digit of the model number has a special significance:

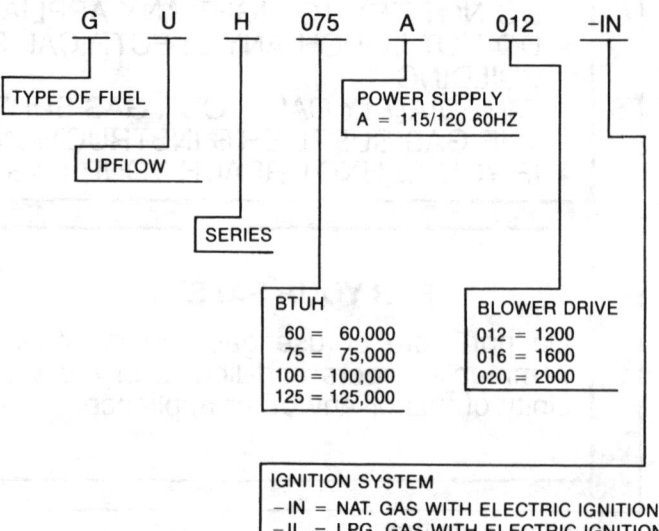

| G | U | H | 075 | A | 012 | -IN |

TYPE OF FUEL

UPFLOW

SERIES

POWER SUPPLY
A = 115/120 60HZ

BTUH

60 =	60,000
75 =	75,000
100 =	100,000
125 =	125,000

BLOWER DRIVE

012 =	1200
016 =	1600
020 =	2000

IGNITION SYSTEM

-IN = NAT. GAS WITH ELECTRIC IGNITION
-IL = LPG. GAS WITH ELECTRIC IGNITION

SECTION III
PROPER INSTALLATION

This section serves as a guide line to the user for recognizing unsafe operating conditions.

INSULATION AND COMBUSTIBLES

The user is advised to examine the furnace area when the furnace is installed in an attic or when insulation is added to the attic. A furnace installed in an attic or other insulated space must be kept clear of the insulating material. DO NOT install the furnace in a space that is insulated with combustible insulation.

⚠ WARNING

It is mandatory that the area around the furnace be kept clear and free of any combustible materials (cardboard boxes, etc.). Gasoline and other volatile liquids or flammable vapors. DO NOT work with volatile materials near an open flame.

AIR REQUIREMENTS

Combustion air must be free of acid forming chemicals; such as sulphur, fluorine and chlorine. These elements are found in aerosol sprays, detergents, bleaches, cleaning solvents, air fresheners, paint and varnish removers, refrigerants and many other commercial and household products. Vapors from these products when burned in a gas flame form acid compounds which are highly corrosive when they condense.

PROVISIONS FOR VENTILATION

Air openings provided in the casing of the furnace must be kept free of obstructions.

It is recommended (but not required) that an adequate amount of ventilation air be provided to the furnace enclosure to avoid excessive heat build-up.

⚠ WARNING

The air for combustion and ventilation must not come from a corrosive atmosphere.

⚠ CAUTION:

It is vitally important that the furnace have proper venting. This furnace is a direct vent appliance. NEVER common vent with another gas fired appliance. Any alteration to the venting system must be in accordance with local and national codes, and the manufacturers installation instructions.

CLEARANCES

If storage shelves, a furnace room, a recreation area, etc. is to be constructed around the furnace area, the following criteria must take preference:

A minimum of 36" must be allowed in front of the furnace to permit an ample amount of space for service. SEE FIGURE 2.

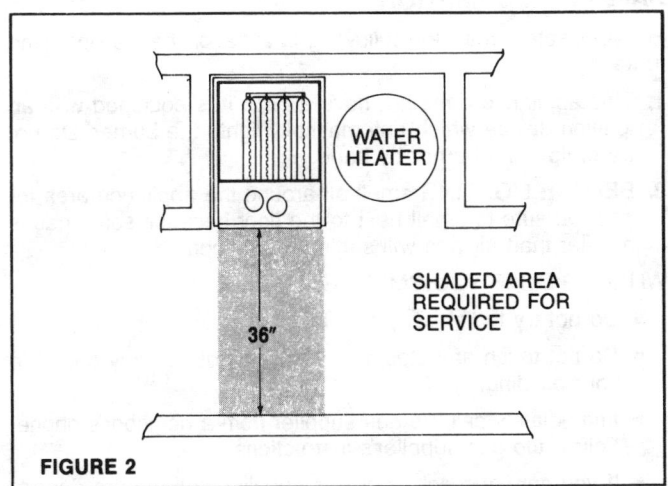

FIGURE 2

INSTRUCTIONS FOR EXAMINING ALL FLUE PRODUCT CARRYING AREAS

For safe operation, the Vent/Air Intake areas of the furnace, including the associated vent system should be checked at the beginning of the heating season and periodically during the season by the owner. This check is accomplished in the following manner:

1. All Vent and Air Intake areas external to the furnace (i.e., Vent, Intake and terminal) are clear and free of obstructions.

2. The vent and intake piping is properly supported without sags and, sloped back toward the furnace.

3. The return-air duct connection(s) is physically sound, is sealed to the furnace casing, and terminates outside the space containing the furnace.

4. The physical support of the furnace is sound without sagging, cracks, gaps, etc. around the base so as to provide a seal between the support and the base.

5. There are no obvious signs of deterioration of the furnace.

6. The burner flames are in good adjustment.

SECTION IV
LIGHTING INSTRUCTIONS

ELECTRIC IGNITION MODELS WITH "–IN" or "–IL" SUFFIX

⚠ WARNING

If you do not follow these instructions exactly, a fire or explosion may result causing property damage, personal injury or loss of life.

SAFETY INFORMATION

For your safety read the following information before operating furnace:

1. This appliance does not have a pilot. It is equipped with an ignition device which automatically lights the burner. Do not try to light the burner by hand.

2. BEFORE LIGHTING smell all around the appliance area for gas. Be sure to smell next to the floor because some gas is heavier than air and will settle on the floor.

WHAT TO DO IF YOU SMELL GAS

- Do not try to light any appliance.
- Do not touch any electric switch; do not use any phone in your building.
- Immediately call the gas supplier from a neighbor's phone. Follow the gas supplier's instructions.
- If you cannot reach your gas supplier, call the fire department.

3. Use only your hand to turn the gas control knob. See Figure 3. Never use tools. If the knob will not turn by hand, don't try to repair it, call a qualified service technician. Force or attempted repair may result in a fire or explosion.

4. Do not use this appliance if any part has been under water. Immediately call a qualified service technician to inspect the appliance and to replace any part of the control system and any gas control which has been under water.

OPERATING INSTRUCTIONS

1. STOP! Read the safety information above.
2. Set the thermostat to lowest setting.
3. Turn off all electric power to the appliance.

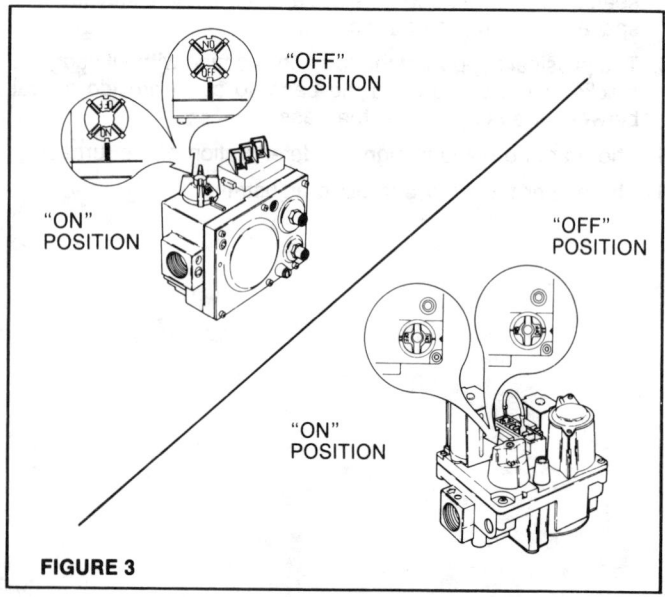

FIGURE 3

4. This appliance is equipped with an ignition device which automatically lights the burners. **DO NOT TRY TO LIGHT THE BURNER BY HAND.**

5. Remove louvered access panel. See FIGURE 4.

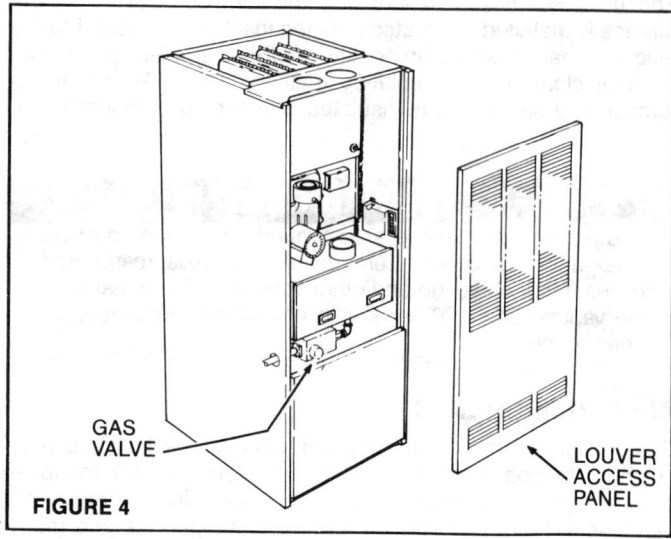

GAS VALVE

LOUVER ACCESS PANEL

FIGURE 4

6. Turn gas control knob clockwise ⌐ to "OFF."

7. Wait five (5) minutes to clear out any gas. If you then smell gas, STOP! Follow Step 2 in Safety Information. If you don't smell gas, go to next step.

8. Turn gas control knob clockwise ⌐ to "ON."

9. Turn on all electric power to the appliance.

10. Set thermostat to desired setting.

11. If the appliance will not operate, follow the instructions "To Turn Off Gas To Appliance" and call your service technician or gas supplier.

12. Replace louvered access panel.

BURNER FLAME REQUIREMENTS

Burner flame should run the length of the burner ports. The flame should be blue in color. If for any reason the flame should lift from the burner, change color to orange or start rolling out from under the burner access shield after ignition, an adjustment to the burner is necessary or the heat exchanger needs to be inspected and cleaned by a qualified service technician.

These furnaces are equipped with a multiple burner assembly. SEE FIGURE 5.

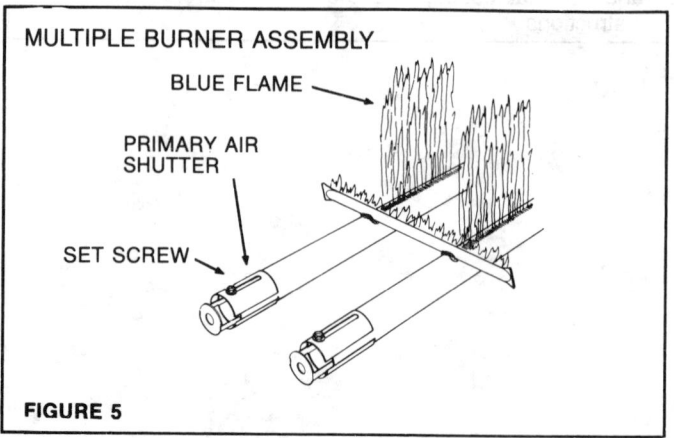

MULTIPLE BURNER ASSEMBLY

BLUE FLAME

PRIMARY AIR SHUTTER

SET SCREW

FIGURE 5

TO TURN OFF GAS TO APPLIANCE

1. Set the thermostat to lowest setting.
2. Turn off all electric power to the appliance if service is to be performed.
3. Remove louvered access panel.
4. Turn gas control knob clockwise ↷ to "OFF."
5. Replace louvered access panel.

⚠ WARNING

Should overheating occur or the gas supply fail to shut off, turn off the manual gas valve to the appliance before shutting off the electrical supply.

SEQUENCE OF OPERATION—ELECTRONIC IGNITION

1. Thermostat calls for heat, energizing the Induced Draft Blower Relay (HR) which closes contacts that power the Induced Draft Blower.
2. The rotation of the Induced Draft Combustion Blower will make the vent/intake pressure switch (PSI) prove blower operation and send power from the ignition circuit to the ignition control.
3. The pre-purge mode will delay 30 seconds before applying power to the silicone carbide ignitor.
4. The ignitor is allowed to warm up for approximately 15 seconds.
5. At the end of the warm-up time the valve is energized.
6. The main burner flame must be detected within 7 seconds or the valve is de-energized and the ignitor turned off.
7. If there is a failure to sense flame on the first attempt, there will be a 60 second delay. The second attempt at lighting the burner flame will duplicate the first attempt with pre-purge, warm-up times and lock-out times identical to those in the first try. Failure to prove on the second attempt will allow yet

another complete ignition attempt before complete system lockout.

8. Lock-out of the control after the final ignition attempt, will require interruption of the power to the ignition control (either the 24V thermostat or line voltage to the furnace may be interrupted) for 1/20 second or longer to reset the system.
9. The ignition sequence will repeat for a total of two recycles. If flame is lost within the first thirty seconds, the total number of ignition attempts will not exceed three.
10. If flame is established for more than thirty seconds after ignition, the recycle counter will clear.
11. A momentary loss of power or flame to the sensing circuit will cause the valve to de-energize and a delay of 15 seconds before restarting the ignition sequence will occur. Recycles will begin and the burner will operate normally if the gas supply returns or the fault condition is corrected prior to the last ignition attempt. Otherwise the control will lock-out.
12. Typical conditions are that the burner will light and stay lit as long as the thermostat calls for heat.
13. Heat from the burner flames warm the heat exchanger surface, the flue products pass through the baffles and vent outdoors through the flue pipe.
14. The fan and limit control senses the supply air temperature adjacent to the heat exchanger and will signal the blower to operate when the air temperature reaches the "ON" fan set point. The fan & limit control will shut the gas off if the supply air temperature reaches the "High Temperature" set point. The blower continues to operate and conditioned air is supplied to the space through the duct system.
15. When the thermostat set point temperature is reached, the gas valve is de-energized which stops the flow of gas to the burners. The blower continues to operate until the air temperature falls below the "OFF" blower set point.
16. The furnace repeats this same cycle each time the thermostat calls for heat.

SECTION V
MAINTENANCE

NOTE: IT IS VERY IMPORTANT THAT THIS APPLIANCE BE MAINTAINED ACCORDING TO THE FOLLOWING OPERATION & MAINTENANCE PROCEDURES. FAILURE TO DO SO COULD JEOPARDIZE THE WARRANTY COVERAGE FOR THIS UNIT.

This section gives basic guidelines for maintenance of the furnace. The owner/operator of the unit can perform these tasks.

⚠ WARNING

DISCONNECT ELECTRIC POWER TO THIS FURNACE BEFORE ANY MAINTENANCE.

OILING INSTRUCTIONS

BLOWER MOTOR—DIRECT DRIVE (MULTI-SPEED)

1. Disconnect power to the furnace before oiling.
2. Remove bottom door (blower panel). This is done by lifting up on the door and pulling forward. SEE FIGURE 6.
3. It may be necessary to remove the two screws holding the blower assembly in the furnace and pull the assembly forward, about 4 inches.

DO NOT REMOVE BLOWER FROM FURNACE COMPLETELY.

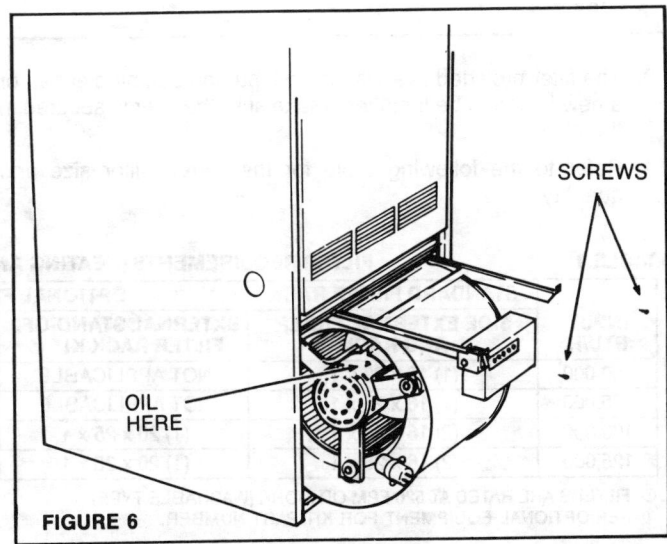

SCREWS

OIL HERE

FIGURE 6

4. The motor (located on the left hand side of the blower) has a small white plastic oiling device inserted in both end bells of the motor.

5. Apply five (5) drops of #10 weight lubricating oil or any light weight lubricating oil, at the beginning of the heating season.

6. Slide blower assembly back into position and secure with two (2) screws previously removed.

7. After the oiling, replace blower door and energize power source.

INDUCED DRAFT BLOWER MOTOR

PLEASE NOTE:

The Induced Draft blower motor is permanently lubricated. **DO NOT** attempt to add oil to this motor.

FILTER CARE—INTERNAL FILTERS

⚠ WARNING
DISCONNECT ELECTRIC POWER TO THIS FURNACE BEFORE ANY MAINTENANCE.

1. Remove bottom blower door from furnace.

2. The filter is flexible. Reach in and remove the filter from the clips and pull out through the front of the furnace. SEE FIGURE 7.

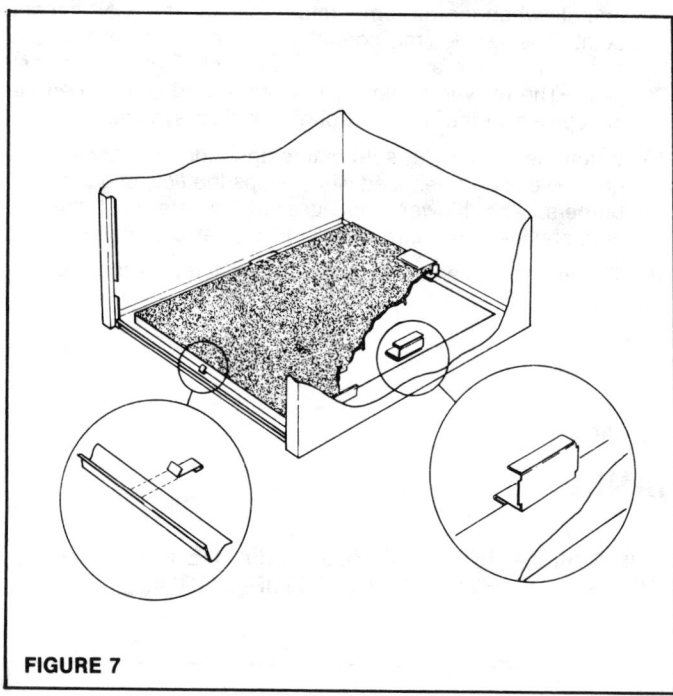

FIGURE 7

3. The filter provided is a washable type and can be cleaned or a new filter can be installed. Make sure the filter is secured in the clips.

4. Refer to the following table for the correct filter size and quantity.

FILTER CARE FOR UNITS USING EXTERNAL FILTER RACK

1. Grip filter with finger tips and slide filter out through opening. SEE FIGURE 8.

2. The filter is a washable type, clean and reinstall. If filter must be replaced, replace with the size listed in TABLE 1.

3. This should be done every 3 months or when filter becomes clogged.

4. After filter is returned to the rack (make sure arrow printed on side of filter points in the direction of the air flow).

Two filters are required with the 100,000 and 125,000 BTUH unit.

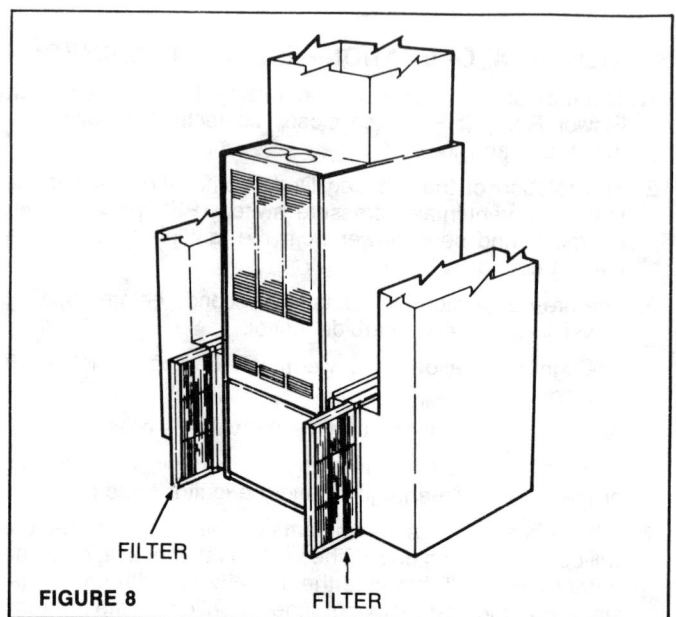

FILTER

FIGURE 8   FILTER

TABLE 1   **FILTER REQUIREMENTS (HEATING AND COOLING)①**

INPUT BTU/H	STANDARD FILTER RACK	OPTIONAL FILTER KITS		MINIMUM FILTER AREA (IN²)
	SIDE EXTERNAL FILTER AND RACK	EXTERNAL STAND-OFF FILTER RACK KIT ②	INTERNAL BOTTOM FILTER KIT ②	
60,000	(1) 16 x 25 x 1	NOT APPLICABLE	(1) 16 X 25 X ½	400
75,000	(1) 16 x 25 x 1	NOT APPLICABLE	(1) 16 X 25 X ½	400
100,000	(2) 16 x 25 x 1	(1) 20 x 25 x 1	(1) 20 x 25 x ½	500
125,000	(2) 16 x 25 x 1	(1) 20 x 25 x 1	(1) 25 x 25 x ½	500

① FILTERS ARE RATED AT 520 FPM OR MORE (WASHABLE TYPE).
② SEE OPTIONAL EQUIPMENT FOR KIT PART NUMBER.

OPERATIONAL SAFETY FEATURES

1. VENT AND HEAT EXCHANGER SAFETY PRESSURE SWITCH

A) A pressure switch to detect vent or air intake blockage is incorporated in the thermostat circuit to the ignition control and will break the circuit if excessive pressure due to blockage in the venting system is sensed.

B) A Condensate Safety Level Sensor, to detect secondary heat exchanger blockage, is incorporated into the flame sensing circuit and will ground the sensing circuit of the ignition module and terminate unit operation.

2. IGNITION CONTROL LOCKOUT FEATURE

This feature of the ignition system automatically de-energizes the ignitor and shuts off power to the gas valve in the event of the loss or absence of main burner flame. Should the energized ignition circuit not sense the presence of main burner flame on the flame sensor within 7 seconds the system will try again for ignition for a total of 3 tries. If the system fails to sense burner flame, the system will lockout.

3. HIGH LIMIT

The high limit is part of the Combination Fan and Limit Control. Both the fan circuit and the limit circuit are activated by the action of a calibrated bi-metal coil which is sensitive to heat. The probe containing this coil is inserted into the heat exchanger compartment in proximity to an area which is heated by the main burner flame and cooled by the flow of air from the Supply Air Blower. As the flame heats the heat exchanger, the bi-metal reacts to the heat and rotates a shaft connected to a switch mechanism. The first switch is the normally open fan contacts which upon closing, energize the Supply Air Blower Motor. When the blower is energized the airflow over the heat exchanger removes sufficient heat to stabilize the movement of the bi-metal actuated switch and maintain a fixed position of the switch in its travel. Should the blower fail or airflow become blocked or reduced to undesirable levels, the bi-metal coil will continue to rotate until the "limit" contacts reach the set point and break the circuit to the ignition control, thereby closing the gas valve and extinguishing the main burner flame.

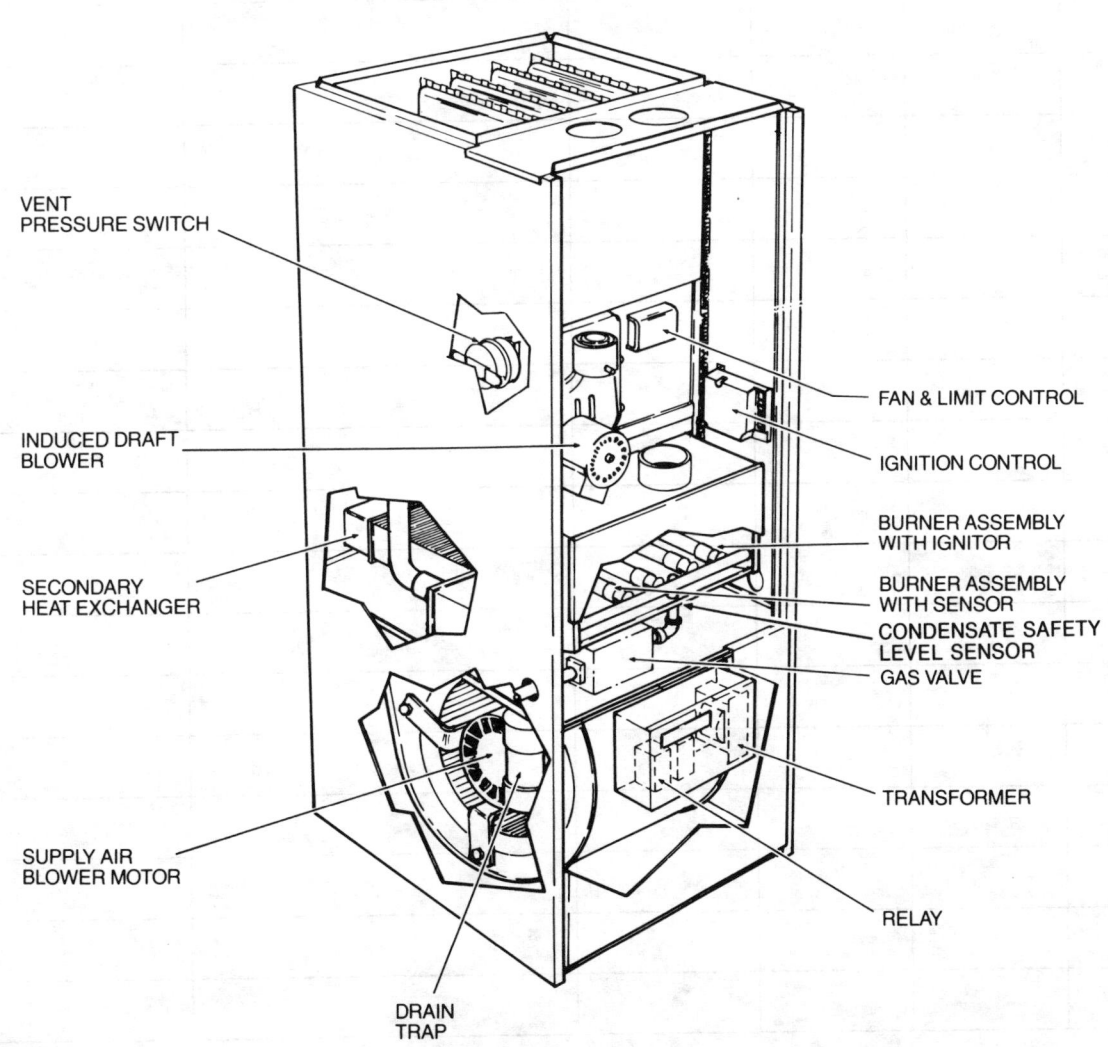

VENT PRESSURE SWITCH

INDUCED DRAFT BLOWER

SECONDARY HEAT EXCHANGER

SUPPLY AIR BLOWER MOTOR

DRAIN TRAP

FAN & LIMIT CONTROL

IGNITION CONTROL

BURNER ASSEMBLY WITH IGNITOR

BURNER ASSEMBLY WITH SENSOR

CONDENSATE SAFETY LEVEL SENSOR

GAS VALVE

TRANSFORMER

RELAY

FIGURE 9

MAINTENANCE SCHEDULE
(DATE PERFORMED)

AIR FILTERS		BLOWER MOTOR OILED	FORCED DRAFT BLOWER MOTOR	COMPLETE PREVENTIVE EXAMINATION BY A SERVICING CONTRACTOR	OTHER
CLEANED	REPLACED				

Armstrong

Installation and Maintenance Instructions

ARMSTRONG
AIR CONDITIONING INC.
Bellevue, Ohio 44811

INSTALLATION SHALL BE MADE IN ACCORDANCE WITH THE RE-QUIREMENTS OF THE LOCAL UTILITY, ANY OTHER AUTHORITIES HAVING JURSIDICTION, OR WITH THE NATIONAL FUEL GAS CODE, ANSIZ223.1 (LATEST EDITION). THESE UNITS ARE NOT CERTIFIED FOR USE IN MANUFACTURED HOMES (MOBILE HOMES) OR TRAVEL TRAILERS.

THESE INSTRUCTIONS MUST BE HUNG ON OR NEAR THE FURNACE IN A CONSPICUOUS PLACE.

GENERAL

All models are suitable for closet or utility room installation. The furnace design is certified by the American Gas Association in compliance with the latest edition of American National Standard Z21.47 Central Furnaces for operation with natural gas or L.P. gas. Consult the rating plate on your furnace for gas type before installing.

The maximum hourly heat loss of each heated space shall be calculated in accordance with the procedure described in the current manuals of Air Conditioning Contractors of America, or by any other recognized method which is suitable for local conditions, provided the results obtained are in substantial agreement with, and not less than, those obtained using the procedure described in the manuals.

If the furnace is used in conjunction with a cooling unit, the furnace must be installed in parallel with, or on the upstream side of the evaporator coil to avoid condensation on the heating element. In a parallel installation, dampers or other means must be provided to prevent chilled air from entering the furnace. If the air control is manually operated, an interlock must be provided to prevent operation of either unit unless the dampers are in full heat or full cool position.

UNPACKING

FURNACE IS SHIPPED ON ONE PACKAGE COMPLETELY ASSEMBLED AND WIRED EXCEPT FOR THE DRAIN TRAP ASSEMBLY. See Figure 1 for installation. The thermostat is shipped in a separate carton when ordered. If any damage is found at time of delivery, proper notation should be made on the carrier's freight bill. Damage claims should be filed with the carrier at once. Claims of shortages should be filed with the seller within five (5) days.

Check the rating plate to ascertain correct model has been received.

LOCATION

To provide proper operation and satisfactory performance, care must be taken in choosing the location for this furnace. The atmosphere in which the furnace operates must be free of contaminants such as chlorides and sulfates.

The furnace should be well centralized with respect to the duct distribution system and set on a firm, level foundation.

A gas-fired furnace for installation in a residential garage must be installed so that the burner(s) and the ignition source are located not less than 18 inches above the floor and that the furnace is located or protected to avoid physical damage by vehicles. **CARE MUST ALSO BE GIVEN IN CHOOSING A LOCATION SO THAT CONDENSATE TRAP AND DRAIN LINE DO NOT FREEZE.**

All normal servicing of the furnace can be performed from the front. If installed in a closet or utility room provide 27 inches clearance in front for service if the door to the room is not in line with the front of the furnace. The opening to the room must be wide enough to allow the furnace to enter and to permit the replacement of any other appliance in the room.

The furnace shall be installed in a location where all the electrical components are not exposed to water.

The furnace shall not be installed directly on carpeting, tile or combustible material other than wood flooring.

The Furnace may be installed in a closet provided the following minimum clearances to combustible materials are held. Clearances for the cabinet are as follows:

Rear/Sides—0 inches; Front—4 inches; Flue Pipe—0 inches; Plenum Top—1 inch.

The counterflow furnace is certified for installation on combustible flooring provided a special base assembly is used. The part number for the special base is noted on the rating plate, located in the burner compartment. Check the part number on the base to make sure the proper base is available.

When counterflow unit is installed on a combustible floor, one (1) inch clearance must be provided between supply duct and floor.

ACCESSIBILITY CLEARANCES MUST TAKE PRECEDENCE OVER FIRE PROTECTION CLEARANCES, PROVIDE SUFFICIENT CLEARANCE FOR CONDENSATE DRAIN TRAP.

CAUTION: DO NOT STORE COMBUSTIBLE MATERIALS NEAR FURNACE OR WARM AIR DUCTS. THE MATERIAL MAY IGNITE BY SPONTANEOUS COMBUSTION CREATING A FIRE HAZARD.

DUCT CONNECTION—COUNTERFLOW MODELS

If a unit is installed on a non-combustible floor, unit may be installed directly over supply duct or plenum.

If installation is made on a combustible floor, the special base assembly must be used. Cut hole on floor two (2) inches larger in each direction than duct size.

The four angles on the base assembly should recess into floor openings and base should rest on all four outside flanges. Construct duct connections with 1½ inch - 2 inch right angle flanges and long enough to extend below floor joists. Drop duct connections through top of base assembly with right angle flanges in good contact with glass tape on top of base assembly. Carefully position furnace over right angle duct flanges. See View A.

A return air duct system is recommended. If the unit is installed in a confined space or closet, a return connection must be run, full size, to a location outside the closet. The air duct in the closet must be tight to prevent any entrance of air from the closet into the circulating air.

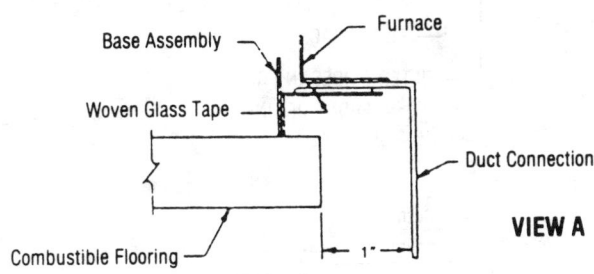

VIEW A

COMBUSTION AIR & VENTILATION

Adequate provisions for combustion air, ventilation of furnace and dilution of the gases must be made. See Section 5.3, Air for Combustion and Ventilation of the National Fuel Gas Code, ANSI Z223.1 (latest edition), or applicable provisions of the local building code. When a furnace is installed in an unconfined space in a building, it can be assumed that infiltration will be sufficient to supply the required air. If the furnace is installed in a confined space within the building and combustion air is taken from the heated space, the combustion air and ventilating air must enter and leave the space through two (2) permanent openings of equal area. One (1) opening shall be located within twelve (12) inches of the ceiling and the other within twelve (12) inches of the floor, each having a free area of one (1) square inch per 1,000 BTU/HR of total input rating of all appliances within the space and not less than 100 square inches each. If the furnace is installed in a space within a building of tight

construction, make-up air must be supplied from outdoors. In this case one (1) opening shall be within twelve (12) inches of the ceiling and one (1) opening within twelve (12) inches of the floor. Each opening shall have a free area of one (1) square inch per 4,000 BTU/HR if combustion air ducts are vertical. If horizontal combustion air ducts are run, one (1) square inch per 2,000 BTU/HR of the total input rating of all appliances within the enclosure is required.

WARNING—COMBUSTION AIR OPENING MUST BE KEPT FREE OF OBSTRUCTIONS. ANY OBSTRUCTION MAY CAUSE IMPROPER BURNER OPERATION AND MAY RESULT IN A FIRE HAZARD OR PERSONAL INJURY.

Additional combustion air may be delivered to the furnace by piping directly to the outdoors. (See Figure 1) Adding combustion air in this manner may be necessary if the furnace is installed in an unusually tight house. To determine whether the addition of combustion air from the outside is required, use the ''Recommended Procedure for Safety Inspection of an Existing Appliance Installation'' as described in the National Fuel Gas Code, ANSI Z223.1 (latest edition) or an equivalent method.

If the furnace is installed in a ''confined space'', the use of the outside air connection is not intended to eliminate the need for permanent openings that communicate with additional combustion and ventilation air.

CONTAMINATED COMBUSTION AIR

The recommended source of combustion air is to use the outdoor air supply option. However, the use of indoor air in most applications is acceptable if these guidelines are followed:

1. If the furnace is installed in a confined space, the necessary combustion air must come from the outdoors by way of attic, crawl space, air duct, or direct opening.
2. If indoor combustion air option is used, there must be no exposure to the installations or substances listed below.
3. All provisions for indoor combustion air must meet the requirements for combustion air supply indicated in the National Fuel Gas Code, Z223.1 latest edition, Section 5.3, and/or any applicable local codes.
4. The following types of installation will require OUTDOOR AIR for combustion, due to chemical exposures:
   ° Commercial buildings
   ° Buildings with indoor pools
   ° Furnaces installed in laundry rooms
   ° Furnaces installed in hobby or craft rooms.
   ° Furnaces installed near chemical storage areas.

Exposure to the following substances in the combustion air supply will also require OUTDOOR AIR for combustion:
   ° Permanent wave solutions
   ° Chlorinated waxes and cleaners
   ° Chlorine based swimming pool chemicals
   ° Water softening chemicals

- ° De-icing salts or chemicals
- ° Carbon tetrachloride
- ° Halogen type refrigerants
- ° Cleaning solvents (such as perchloroethylene)
- ° Printing inks, paint removers, varnishes, etc.
- ° Hydrochloric acid
- ° Cements and glues
- ° Antistatic fabric softeners for clothes dryers
- ° Masonry acid washing materials

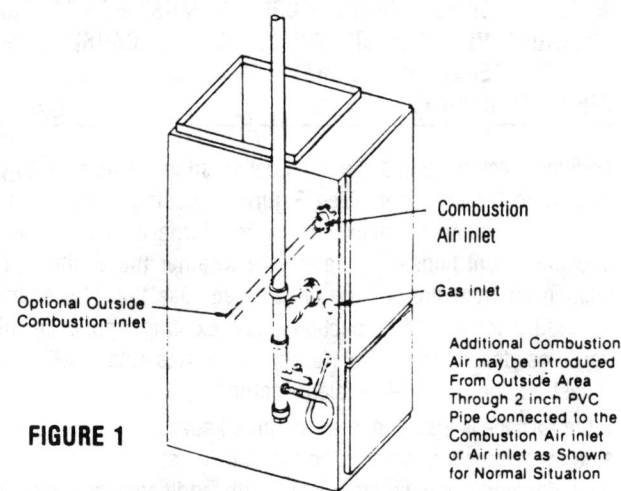

FIGURE 1

Combustion Air inlet

Optional Outside Combustion inlet

Gas inlet

Additional Combustion Air may be introduced From Outside Area Through 2 inch PVC Pipe Connected to the Combustion Air inlet or Air inlet as Shown for Normal Situation

RETURN AIR AND FILTER RACK (See Figure 2)

Where the furnace is installed in a confined space and there is not a complete return duct system, a return duct must run to an unconfined area. The air duct must be tight to prevent any entrance of air from the confined space into the circulating air stream.

For upflow furnaces return air connection can be made on either side or the bottom of the furnace. A filter and an external filter rack is provided for side return (see instructions below for installation). If bottom return is used, a filter and rack of the proper size should be provided. Models with higher CFM requiring filter areas over 400 square inches, may be installed with return air connections on both sides or an external filter rack, with the required filter area, must be used. A side return air cabinet is also available, the use of which can improve the appearance of the installation.

INSTALLATION EXTERNAL SIDE FILTER RACK

The upflow comes equipped with a filter rack or racks which are located in blower compartment. To install filter assemblies, the following steps must be taken.

1. Cut a 14 inches high by 22 inches wide opening in side or sides of cabinet, using the lanced corners as locating points.
2. Loosen the four (4) screws at the base of the side panel.
3. Slide the outer flange of filter frame into base and rear panel flange.
4. Add two (2) #8 × ½ sheet metal screws to top flange of rack into engagement holes found in side panels. Tighten screws at base.

FILTER - COUNTERFLOW

The filter can be installed either on the left or right side of the unit depending on which side the return air duct is installed.

To install the filters, remove the two (2) upper blower compartment panels and install the filter as shown in figure 2 if the return air duct is installed coming off the return air plenum on the right side.

If the return duct is installed on the left side, remove the two (2) screws holding the side filter rack and install the side filter rack on the left side of the furnace in holes provided in the unit. Install filter opposite than what is shown in Figure 2.

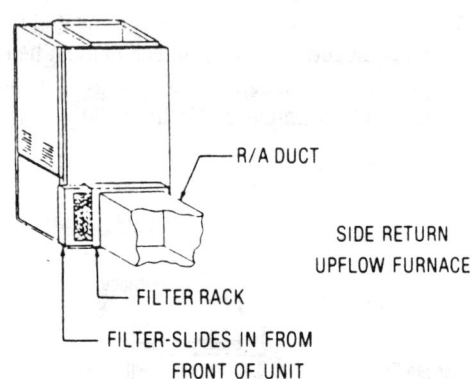

R/A DUCT

SIDE RETURN UPFLOW FURNACE

FILTER RACK

FILTER-SLIDES IN FROM FRONT OF UNIT

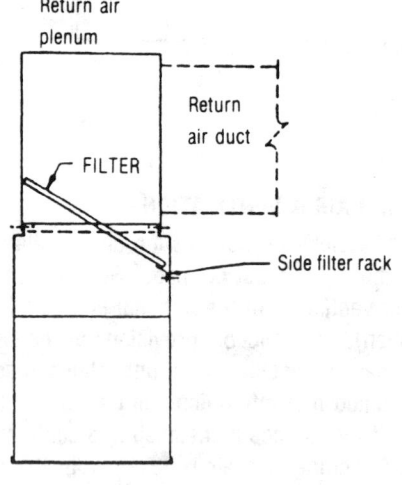

Return air plenum

Return air duct

FILTER

Side filter rack

COUNTERFLOW FURNACE

FIGURE 2

continued on following pg.

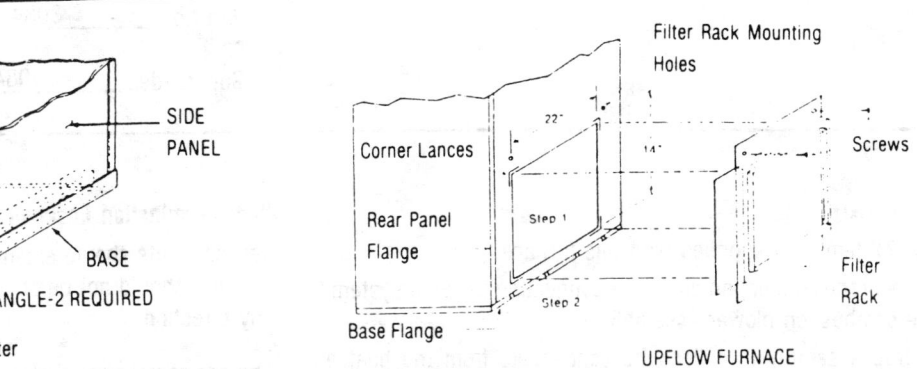

FIGURE 2 *(continued)*

VENTING

General

The high efficiency of this furnace is accomplished by the removal of both sensible and latent heat from the combustion flue gases. The removal of the latent heat results in the condensation of moisture from the flue gases. This condensation occurs in the secondary heat exchanger and also in the vent system; therefore the following special considerations and instructions must be followed to assure the proper venting of this furnace and other appliances. All venting must be in accordance with codes having jurisdiction and these installation instructions.

This is a Category IV Furnace

Do not use more than five (5) 90° elbows excluding vent termination. Two (2) 45° elbows are equal to one (1) 90° elbow. Horizontal runs must not exceed 25 feet.

Do not vent this appliance into a masonry chimney or any other type of vent passage not designed or intended for use with condensating appliances.

Do not connect the flue system of this appliance to a common vent with any other appliance.

When replacing a furnace which has been connected to a common vent with other appliances the remaining vent system and appliances should be inspected and evaluated for proper operation.

When an existing furnace is removed from a venting system serving other appliances, the venting system is likely to be too large to properly vent the remaining attached appliances.

The following steps shall be followed with each appliance remaining connected to the common venting system placed in operation, while the other appliances remaining connected to the common venting system are not in operation.

(a) Seal any unused openings in the common venting system.

(b) Visually inspect the venting system for proper size and horizontal pitch and determine there is no blockage or restriction, leakage, corrosion and other deficiencies which could cause an unsafe condition.

(c) Insofar as is practical, close all building doors and windows and all doors between the space in which the appliances remaining connected to the common venting system are located and other spaces of the building. Turn on clothes dryers and any appliance not connected to the common venting system. Turn on any exhaust fans, such as range hoods and bathroom exhausts, so they will operate at a maximum speed. Do not operate a summer exhaust fan. Close fireplace dampers.

(d) Follow the lighting instructions. Place the appliance being inspected in operation. Adjust thermostat so appliance will operate continuously.

(e) Test for spillage at the draft hood relief opening after 5 minutes of main burner operation. Use the flame of a match or candle.

(f) After it has been determined that each appliance remaining connected to the common venting system properly vents when tested as outlined above, return doors, windows, exhaust fans, fireplace dampers and any other gas-burning appliance to their previous conditions of use.

(g) If improper venting is observed during any of the above tests, the common venting system must be corrected.

See National Fuel Gas Code ANSI Z223.1 (latest edition) to correct improper operation of common venting system.

Multistory venting is allowed as permitted by the NFGC or local codes and within the manufacturer's guidelines.

Vent Materials

Seven (7) items are supplied with this furnace.

1. A neoprene connecting clamp for connecting the vent system to the combustion blower assembly.

2. A trap assembly to collect the condensate from the heat exchanger and vent system.

3. A combustion air coupling is provided and is to be connected to the left side panel.

4. A nut is provided to secure the combustion air coupling to the left side panel.

5. A two (2) inch vent tee is provided to be connected to the drain trap assembly.

6. A screen is provided to be inserted inside the termination elbow.

7. A hose clamp to be used on the end of the condensate drain tubing.

In addition to the above listed items, on the 100,000 BTUH Model a 3 to 2 inch reducing coupling is furnished to be located on top of the drain trap tee for a 3 inch vent line.

All other venting materials, pipe and fittings, for this furnace are to be two (2) inch PVC (Poly-Vinyl Chloride) or CPVC (Chlorinated Poly-Vinyl Chloride) schedule 40 minimum. NOTE: INSTALLATION OF 100,000 BTU FURNACE MUST BE DONE WITH THREE (3) INCH VENT PIPE OF THE SAME RATING. Elbows used to connect vertical runs to horizontal runs should be the type DWV to provide the correct slope for horizontal runs. Horizontal runs should not be allowed to sag as water may collect in low spots and cause blocked flue switch to open. In areas where flue pipe may be subject to abnormal stress or damage, schedule 80 PVC pipe should be used or protection must be provided for the flue pipe.

Vent termination for horizontal venting through side wall should be a two (2) inch PVC elbow oriented with the opening to the bottom. USE A THREE (3) INCH ELBOW FOR THE 100,000 BTU FURNACE. The opening in the vent termination should be protected with ½ inch mesh screen. See Figure 5 for further description.

All joints are to be sealed using a PVC sealer-primer and a solvent cement such as Genova 15015 or equivalent. Follow all instructions and recommendations of the cleaner-primer and solvent cement manufacturer when making joints.

All pipe, fittings, primers, solvent cement and procedures must conform to the following ANSI-ASTM standards.

Pipe and Fittings ASTM D1785, D2466, and D2665 Primer and Solvent Cement ASTM D2564 Procedure REF. ASTM D2855.

Vent Termination Location

Vertical vents should extend through the roof a minimum of two (2) feet and should not be obstructed for a minimum of ten (10) feet in any direction.

The horizontal vent system shall terminate at least 4 feet below, 4 feet horizontally from, or 1 foot above any door, window, or gravity air inlet into the building. The vent system shall terminate at least 3 feet above any forced air inlet located within 10 feet. The bottom of the vent terminal shall be located at least 12 inches above grade.

Minimum horizontal clearance of 4 feet from electric meters, gas meters, regulators, and relief equipment.

In addition to the above requirements, consideration must be given to prevent unwanted ice buildup from the vent condensate. The vent should not be located on the side of a building where the prevailing winter winds could trap the moisture causing it to freeze on the walls or on overhangs (under eaves). The vent location should not discharge over a sidewalk, patio or other walkway, where the condensate could cause surface to become slippery.

Venting of the furnace and vent terminal locations must be installed in accordance with the National Fuel Gas Code and/or local codes.

Installation

Connect the vent system and drain trap assembly to the outlet of the combustion air blower using the supplied neoprene connection clamp and two (2) inch tee. Under no circumstances should the trap assembly be cutoff or altered in any way. The trap assembly must be readily accessible for cleaning and in the same confined space with the furnace. For additional clearance to cold air return or other obstacles a two (2) inch street elbow may be used to move drain assembly toward the front of the unit. See Figure 3.

Route the condesate drain tubing in a manner so that the tubing won't kink.

The evaporator drain **must not** be connected to the overflow opening of the tee on the furnace drain trap assembly. If this were done and the drain line downstream of this fitting became blocked, the furnace heat exchanger could fill with water draining from the evaporator coil.

THE OVERFLOW MUST BE LEFT OPEN

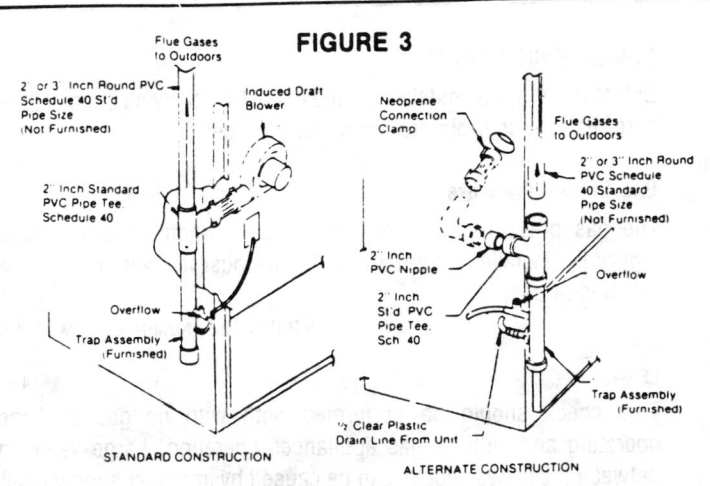

FIGURE 3

STANDARD CONSTRUCTION

ALTERNATE CONSTRUCTION

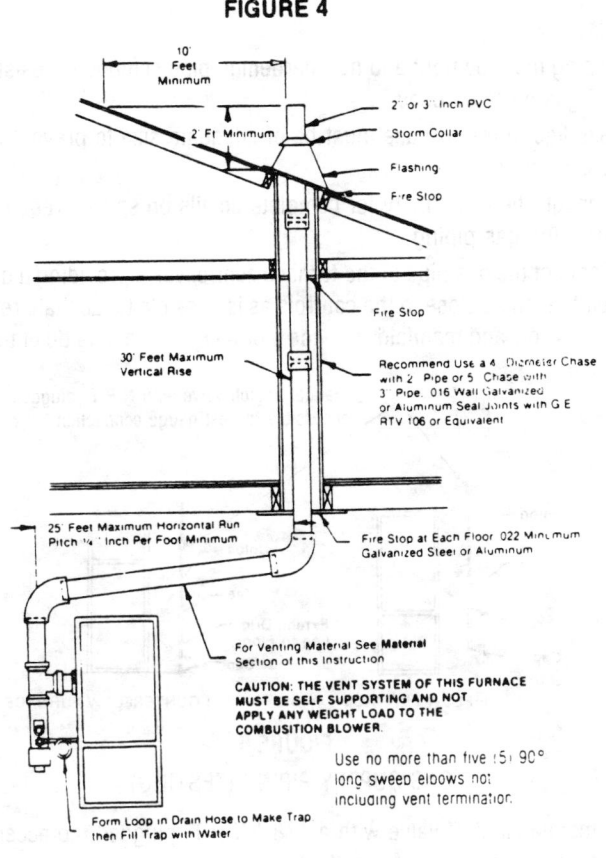

FIGURE 4

CAUTION: THE VENT SYSTEM OF THIS FURNACE MUST BE SELF SUPPORTING AND NOT APPLY ANY WEIGHT LOAD TO THE COMBUSTION BLOWER.

Use no more than five (5) 90° long sweep elbows not including vent termination.

The vent system must be supported with mounting straps to prevent any weight load from being applied to the vent blower. Horizontal vent pipe must be supported every five (5) feet and vertical pipe should be supported every ten (10) feet to prevent sagging and provide rigid support.

For vertical runs through confined space where flue pipe cannot be inspected, it is recommended the flue pipe be installed in four (4) inch diameter chase with proper fire stops to prevent damage to flue pipe. Use five (5) inch diameter chase when three (3) inch PVC pipe is used.

Refer to Figure 4 and 5 for recommended venting installations.

ELECTRICAL CONNECTIONS

ALL THE WIRING SHOULD BE DONE IN ACCORDANCE WITH THE NATIONAL ELECTRICAL CODE, ANSI/NFPA NO. 70 (LATEST EDITION), OR WITH LOCAL CODES, WHERE THEY PREVAIL.

Using wiring with a temperature limitation for type T wire (63°F rise), (36°C); run the 115 volt, 60 hertz electric power supply through fused disconnect switch to the junction box of furnace and connect as shown in the wiring diagram, located on the inside of the control-access panel.

Install thermostat according to directions furnished with it. Select a location away from drafts and direct sunshine. Set heat anticipator to amperage draw of complete system, as measured or as shown on wiring diagram.

Furnaces must be electrically grounded in accordance with local codes or in the absence of local codes, with the National Electric Code, ANSI/NFPA No. 70 (latest edition).

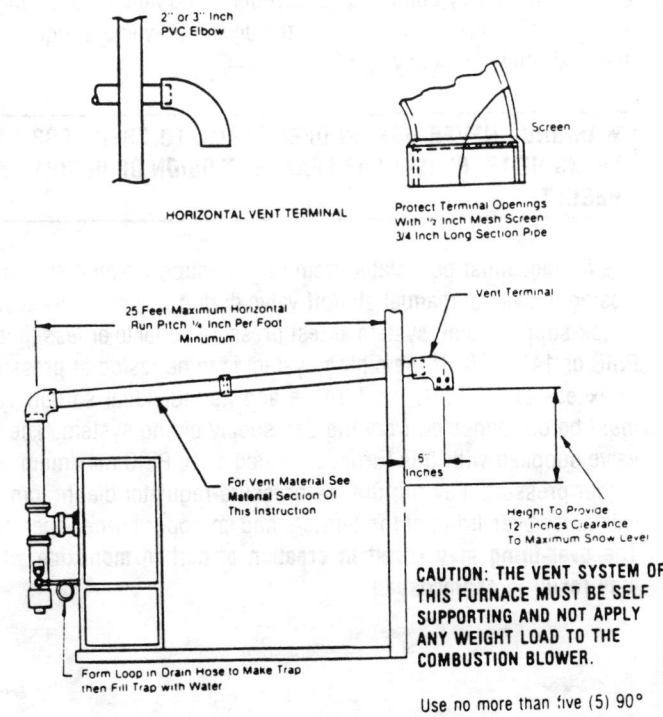

FIGURE 5

HORIZONTAL VENT TERMINAL

Protect Terminal Openings With ½ Inch Mesh Screen ¾ Inch Long Section Pipe

CAUTION: THE VENT SYSTEM OF THIS FURNACE MUST BE SELF SUPPORTING AND NOT APPLY ANY WEIGHT LOAD TO THE COMBUSTION BLOWER.

Use no more than five (5) 90° long sweep elbows not including vent termination.

Gas

Piping must be tight and non-hardening pipe compound resistant to LP gas must be used.

Gas line to the furnace must be of adequate size to prevent undue pressure drop.

Consult the local utility for complete details on special requirement in sizing gas piping.

Connect the gas pipe to the furnace control valve providing a ground joint union as close to the controls as is possible to facilitate removal of controls and manifold. Provide a drip leg on the outside of the furnace.

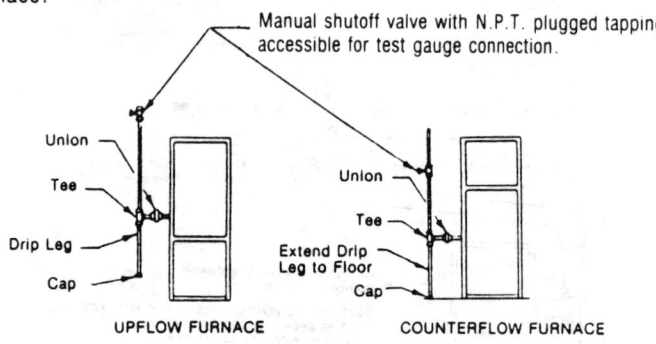

FIGURE 6

GAS SUPPLY PIPING (TESTING)

A manual shutoff valve with a 1/8'' N.P.T. plugged tap accessible for test-gauge connections shall be installed in the gas line, outside the unit, five (5) feet above the floor, or in accordance with any local codes.

The pilot is factory connected and requires no additional piping. A pilot gum filter may be installed in the pilot supply line if required by the local utility company.

WARNING—NEVER USE AN OPEN FLAME TO CHECK FOR GAS LEAKS. IF THERE IS A GAS LEAK, EXPLOSION OR INJURY CAN RESULT.

The furnace must be isolated from the gas-supply piping system by closing individual manual shutoff valve during any pressure testing of gas-supply piping system at test pressure equal to or *less* than ½ PSIG or 14'' W.C. If the piping system is to be tested at pressures in excess of ½ PSIG, the furnace and its individual shutoff valve *must* be disconnected from the gas-supply piping system. The gas valve supplied with this furnace is rated at ½ PSIG maximum. Any higher pressure may rupture the pressure-regulator diaphragm and will cause over-firing of the burners and improper burner operation. **The over-firing may result in creation of carbon monoxide which may result in asphyxiation.**

FURNACE CHECK-OUT

Before leaving, the installer should make the following checks to ensure that the controls are functioning properly.

Gas Supply Pressure

The gas pressure measured at manual shutoff valve (reference Figure 6), installed external to the unit, must fall within the following standards.

	MINIMUM	NOMINAL	MAXIMUM
Natural Gas	5	7	9
LP Gas	11	12	14

This check should be performed both with no gas appliance operating and with all gas appliances operating. Large variations between these two checks can be caused by improper supply piping or a defective regulator.

Gas Regulator

Gas input must exceed the input capacity shown on the rating plate. The furnace is equipped for rated inputs with manifold pressure as follows:

NATURAL GAS 3.5″ W.C.

L.P. GAS 10″ W.C.

The manifold pressure can be measured by removing the piping plug in the downstream side of the gas valve and connecting a water manometer or gauge.

Turn gas valve ON. To adjust the regulator, turn the adjusting screw on the regulator clockwise to increase pressure and input, counterclockwise to decrease pressure and input.

Only small variations in gas input may be made by adjusting the regulator. In no case should the final manifold pressure vary more than 0.3″ W.C. from the above specified pressures.

For natural gas installations, check furnace rate by observing gas meter, making sure all other gas appliances are turned off. The test hand on the meter should be timed for at least one revolution. **Note the number of seconds for one revolution.**

BTU/HR. INPUT EQUALS

$$\frac{\text{Cubic feet per revolution}}{\text{No. seconds per revolution}} \times 3600 \times \text{Heating Value}$$

The heating value of your gas can be obtained from your local utility company.

NOTE: BTU RATINGS SHOWN ON THE RATING PLATE AREA FOR ELEVATIONS UP TO 2,000 FEET. FOR ELEVATIONS ABOVE 2,000 FEET, RATING SHOULD BE REDUCED AT THE RATE OF FOUR (4) PERCENT FOR EACH 1,000 FEET ABOVE SEA LEVEL.

BURNER AND PILOT:

See Figure 7 and 8 For Visual View

FIGURE 7

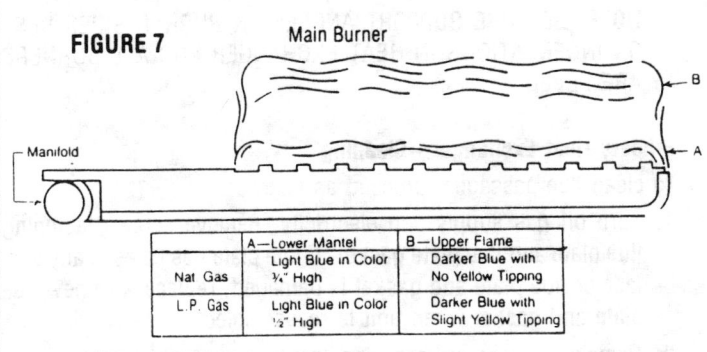

Main Burner

	A—Lower Mantel	B—Upper Flame
Nat Gas	Light Blue in Color ¼" High	Darker Blue with No Yellow Tipping
L.P. Gas	Light Blue in Color ½" High	Darker Blue with Slight Yellow Tipping

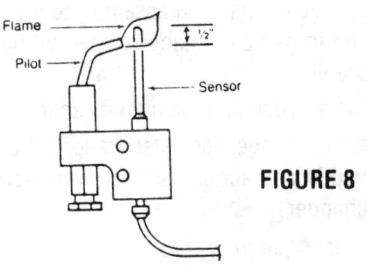

Pilot Burner

FIGURE 8

FOR NATURAL GAS

Before making burner adjustments, permit furnace to burn 5-10 minutes. The air shutters are factory set with 1/8 inch clearance from the manifold pipe, however, if yellow tipping occurs, remove the air shutter locking screws and push shutter forward until yellow disappears. Install locking screws.

FOR L.P. GAS

Before making burner adjustments, permit furnace to burn 5-10 minutes. Remove air shutter locking screw. Close air shutter until you have a distinct yellow flame then push air shutter forward until flame has approximately 1/8 inch yellow tip. Install locking screw.

Fan & Limit Controls

To obtain maximum comfort, low settings on the fan control are recommended to obtain nearly continuous air circulation as recommended by the *Air Conditioning Contractors of America*. A 90°F degree fan off position will usually be satisfactory.

Test the limit switch by blocking air inlet, turn on unit and wait until automatic gas valve shuts off. Before starting test, remove cover from limit switch and note position of dial. The reference point for the Honeywell switch is on the vertical centerline point. If valve does not shut off when reference point is reached by limit stop, replace switch and repeat test.

CAUTION—DO NOT OVERFIRE IF UNIT SWITCH IS DEFECTIVE AND WILL NOT BREAK CIRCUIT. FIRE UNIT ONLY LONG ENOUGH TO ROTATE LIMIT STOP ON DIAL PAST THE REFERENCE POINT.

Interlock (Blower Door) Switch

When blower door is removed, the interlock switch breaks the power supply to burner controls and blower motor. The switch operation *must* be checked.

Roll-Out Switch

This furnace is equipped with a manual reset roll-out switch for the purpose of preventing furnace operation with a blocked heat exchanger, or under other abnormal conditions that would cause burner flames to burn outside the combustion chamber. To manually reset this switch, push and release the button on top. If the furnace does not operate, or if this switch must repeatedly be reset, call a serviceman.

MAINTENANCE

It is recommended that this furnace be inspected by a qualified Service Technician at the beginning of each heating season.

Disconnect electrical power to the unit before doing any service. Turn off main gas valve before breaking gas connections.

Furnace Start-Up and Check-Out Procedure

After maintenance has been performed or the heat exchanger cleaned and re-assembled, the following checks should be performed. FILL TRAP WITH WATER, turn unit on, check the control operating pressure and flame appearance when main blower is cycled on and off. Any fluctuation in this pressure or variation in flame appearance are an indication of air leakage between the circulating air compartment and heat exchangers. The source of the leakage must be found and repaired. Control operating pressure is measured at the pressure tap directly above the coil drain connection. Install a tee to allow the pressure switches to remain connected. Normal pressure is negative 1.40 to negative 1.60 inch water column. The normally open pressure switch will remain open if pressure is excessively low (excessive leakage or blocked flue). If pressure is excessively high (restricted coil or plugged coil drain) flame roll-out will occur and roll-out switch will open.

Filter

The air filter should be cleaned, at least, every three (3) months. More frequent service is required if unusually dirty conditions are encountered.

This unit is supplied with a permanent filter. To clean this filter, shake filter to remove excess dirt, use vacuum cleaner or wash in soapy or detergent water and replace after filter is dry. Filter should not be oiled after washing.

If electronic air cleaner or other filter media has been installed, clean per the manufacturers instructions.

Blower

Installation is to be adjusted to obtain a temperature rise within the range specified on the appliance rating plate.

The direct drive blower in this unit will provide an air temperature rise at a static pressure as noted on the rating plate. If a duct system with less resistance is used, a lower air temperature will result. To increase the air temperature rise, add restriction to the return air system. The direct drive motors have permanently lubricated bearings and therefore do not require oiling. On the upflow furnace, the blower can be removed for servicing by removing screws under burners in the blower partition panel. The blower will slide out on side supporting rails. On the counterflow furnace access to the blower is made by removing the two blower access panels at the top of the furnace after the upper front panel is removed.

Main Burners: See Figure #7 For Visual View

Distorted flame, yellow tipping of the natural gas main burner flame or more than 1/8" high yellow tips on LPG may be caused by lint accumulation or dirt inside the burner or burner ports, lint at air inlet between air shutter and manifold pipe or obstruction over main burner orifice. Use soft brush or vacuum to clean affected areas. If it is necessary to remove burners:

1. Remove the screws securing the burner shield and set it aside.
2. Remove hexagon nut that secures burners to manifold.
3. Mark air shutter setting, remove air shutter locking screw, then push air shutter to full open position.

4. Remove screws that secure pilot to main burner.
5. After burners have been removed and cleaned, re-assemble burners in unit by reversing procedure Steps #4 through #1 above. Then adjust air shutters to original location before installing locking screws.
6. Check unit for normal operation per **Furnace Start-Up and Check-Out Procedure** on page 9.

NOTE: BE SURE SUPPORT ANGLES ON BURNER SIDES REST ON INDENTATIONS IN HEAT EXCHANGER AND ALL BURNERS ARE LEVEL.

Primary Heat Exchanger—Cleaning

To clean flue passages, proceed as follows:

1. Turn off gas supply and electricity. Remove screws fastening flue plate and flue plate gasket. If flue plate has holes in any surface or flue plate and gasket is damaged, replace with new flue plate and gasket when unit is reassembled.
2. Remove the screws securing the burner shield located above burners.
3. Remove four (4) screws that secure manifold assembly to the inner front panel. Break union in gas piping and remove burners and manifold assembly as unit.
4. Clean flue passages with wire brush and vacuum cleaner.
5. After cleaning, inspect heat exchanger for deterioration. If any holes or cracks are observed in any surface of heat exchanger, replace with new heat exchanger.
6. Inspect burners and clean, if required.
7. Reassemble parts removed in reverse order.
8. Turn on gas. Check for leaks. Turn on electrical supply.
9. Check unit for normal operation per **Furnace Start-Up and Check-Out Procedure** on page 9.

TO CLEAN

Secondary Heat Exchanger (Exterior)

Remove rear panel to clean bottom side of fins.

If rear panel cannot be readily removed, remove blower assembly for accessibility to coil (applies to upflow only).

Clean fins with suitable brush or vacuum cleaner attachment. If any fins are damaged, use a fin comb to straighten.

Secondary Heat Exchanger Internal Cleaning and Removal.

1. Remove top front collector plate cover.

2. Inspect inside of upper collector box and tube inlets for soot and/or dirt accumulation.

3. If cleaning is needed, fashion a 1/4 inch or 3/8 inch tube extension with a 90° curved end. Adapt other end for connection to a garden hose or garden sprayer.

4. Insert curved end of adapter tube, center over each tube and flush with water. Remove bottom cap from trap assembly and place a suitable container under the trap to collect water flushed through cells.

If further cleaning is required, the secondary heat exchanger must be removed from the furnace. Proceed as follows:

5. Remove rear panel.

6. Remove the combustion air blower assembly, drain tubing and connecting tubing for pressure switches.

7. Remove four (4) nuts that secure top coil box to primary heat exchanger collector box and the screw in the embossed hole above flue outlet in the inner-front panel.

8. Remove four (4) screws from left side panel that secure lower section of secondary heat exchanger to this panel.

9. Remove secondary heat exchanger out the rear of the unit.

10. Remove screws that secure top box to coil.

11. Remove turbulators from each tube and any contaminants from the turbulators.

12. Use a small diameter bottle brush to clean the inside of each tube.

13. Rinse each tube with clear water.

14. If bottom coil box drain becomes blocked, remove screws that secure box to bottom coil. Clean pan and replace. To assure gas tight seal, add a 1/4 diameter bead of G.E. RTV106 sealant around flange of pan before assembly to coil.

15. Install turbulators in each tube.

16. Apply 1/4 diameter bead of G.E. RTV106 high temperature sealant around top flange of coil plate. Add top coil box and secure box to coil plate.

17. If silicone gasket on front of top coil box is damaged, add a new gasket or apply a ring of 1/4 diameter bead of RTV106 before attaching coil box to heat exchanger collector box using four nuts previously removed.

18. Install combustion air blower, tubing and collector cover and re-assemble furnace.

19. Check unit for normal unit operation per **Furnace Start-Up and Check-Out Procedures** on page 9.

Removal of Primary Heat Exchanger

Should it be necessary to remove primary heat exchanger proceed as follows:

1. Remove secondary heat exchanger as described above.

2. Remove manifold and burners as described under cleaning burners.

3. Remove fan limit control.

4. Remove screws that secure inner front panel to heat exchanger and slide out heat exchanger carefully to avoid damage to insulation.

Before replacing heat exchanger, repair any damaged insulation with aluminum foil tape.

Sequence of Operation—Spark Ignition

1. When thermostat calls for heat, the combustion blower is energized. When proper combustion air is proven, the normally open pressure switch closes supplying power to the ignition control. The electric spark and pilot valve are then energized to produce a pilot flame.

2. The sensing probe proves the presence of the pilot flame and internal switching action de-energizes the spark.

3. The sensor will cause contacts to close energizing the main valve solenoid, and the main burners will light when the contacts to the main valve close. The sparking circuit is disconnected assuring that the main burner(s) can never be ignited by the electric spark.

4. If the pilot flame goes out for any reason, the ignition control will automatically reset and will repeat normal start-up operation.

5. If pilot flame is not established and proven by the flame sensor within 90 seconds (nominal), the ignition control will lock out, shutting off the spark and pilot gas flow. The control can be reset by momentarily setting the room thermostat to its lowest setting, then setting it to the desired temperature.

PARTS LIST

The following repair parts are available from your local ARMSTRONG AIR CONDITIONING DEALER.

When ordering parts, include the complete furnace model number and serial number which are printed on the rating plate located on the furnace.

Transformer Relay
Fan and Limit Control
Gas Valve
Ignition Control
Flame Sensor
Blower Door Interlock Switch
Pressure Switch—Normally Open
Combustion Blower Relay
Flame Roll-Out Protector Switch
Drain Trap Assembly

Combustion Blower Assembly
Aux. Limit Switch (Counterflow Only)

Blower Group

Blower Housing Assembly
Blower Wheel
Blower Motor
Blower Motor Mount
Blower Motor Capacitor

Burner Group

Gas Manifold
Main Burner Orifice
Pilot Burner Assembly
Pilot Orifice
Main Burners
    Burners with Pilot Bracket
    Burner with Carryover
    Burner Right Side

Heat Exchanger Group

Heat Exchanger—Primary
Heat Exchanger—Secondary

WARNING—IMPROPER INSTALLATION, ADJUSTMENT, SERVICE OR MAINTENANCE CAN CAUSE INJURY OR PROPERTY DAMAGE. CONSULT A QUALIFIED INSTALLER, SERVICE AGENCY OR THE GAS SUPPLIER FOR INFORMATION OR ASSISTANCE.

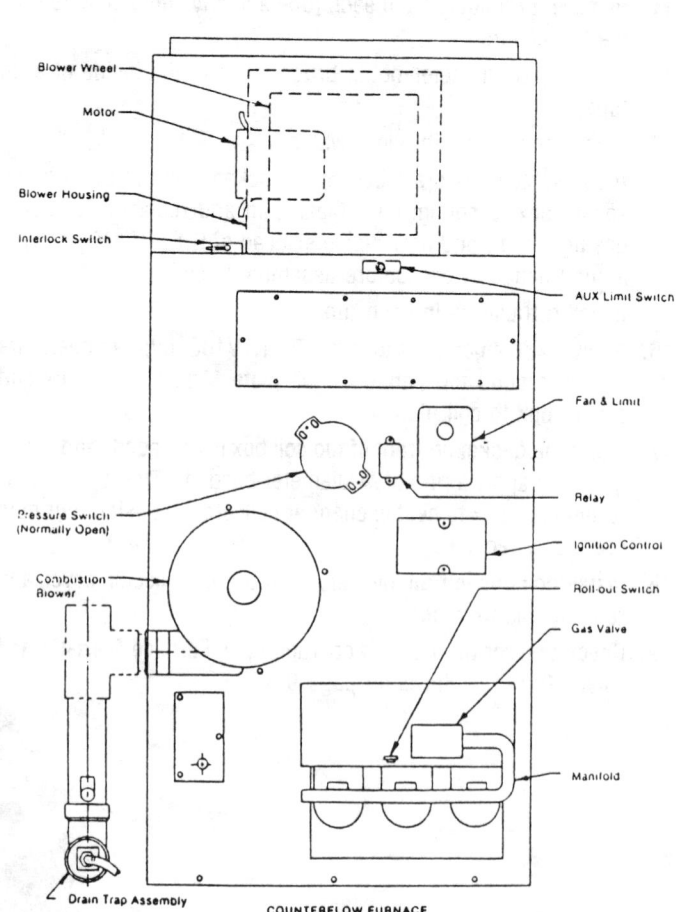

UPFLOW FURNACE

COUNTERFLOW FURNACE

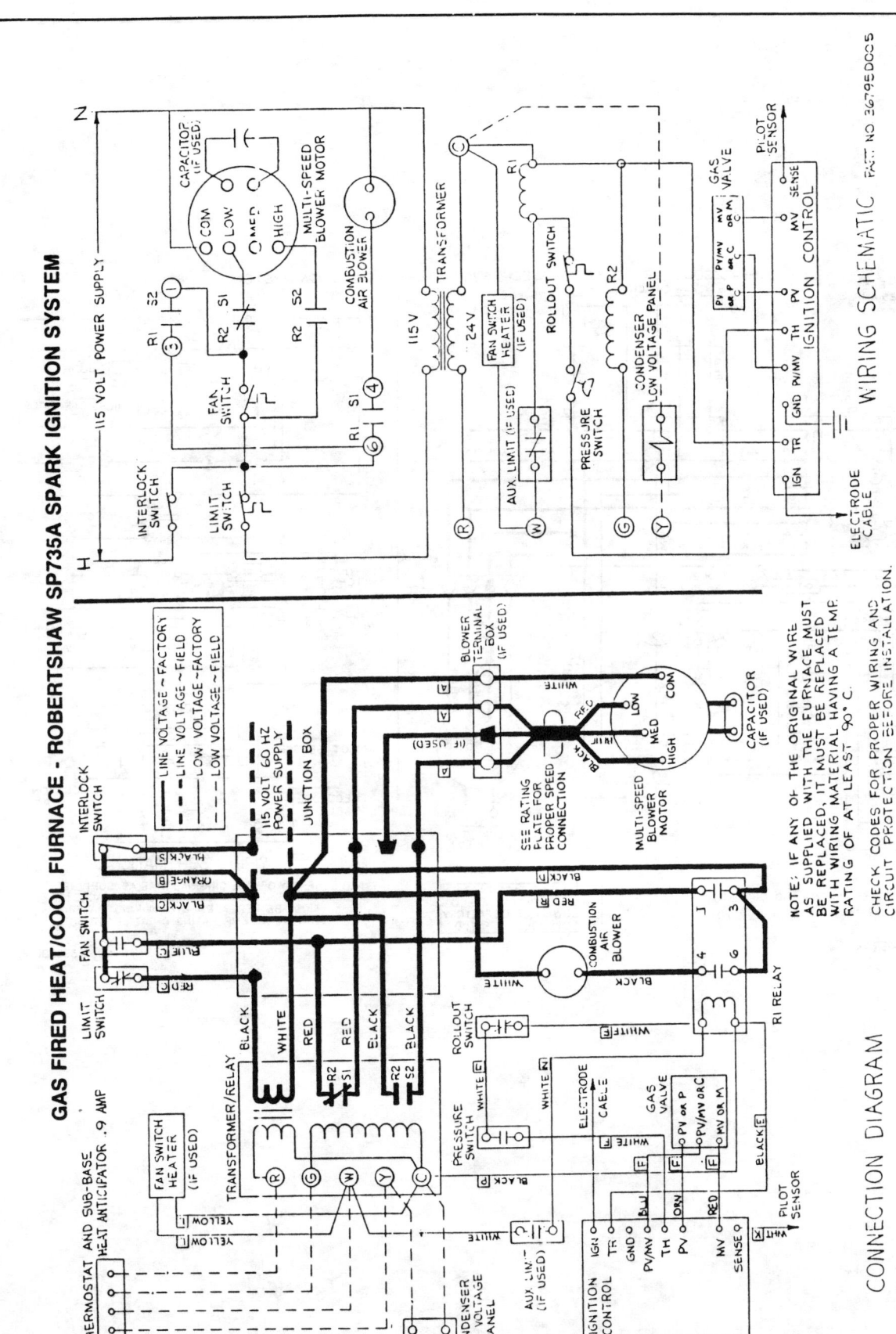

GAS FIRED HEAT/COOL FURNACE - ROBERTSHAW SP735A SPARK IGNITION SYSTEM

WIRING SCHEMATIC

CONNECTION DIAGRAM

EG6E & EG7E MODELS

GAS FIRED HEAT/COOL FURNACE – HONEYWELL S8600C SPARK IGNITION SYSTEM

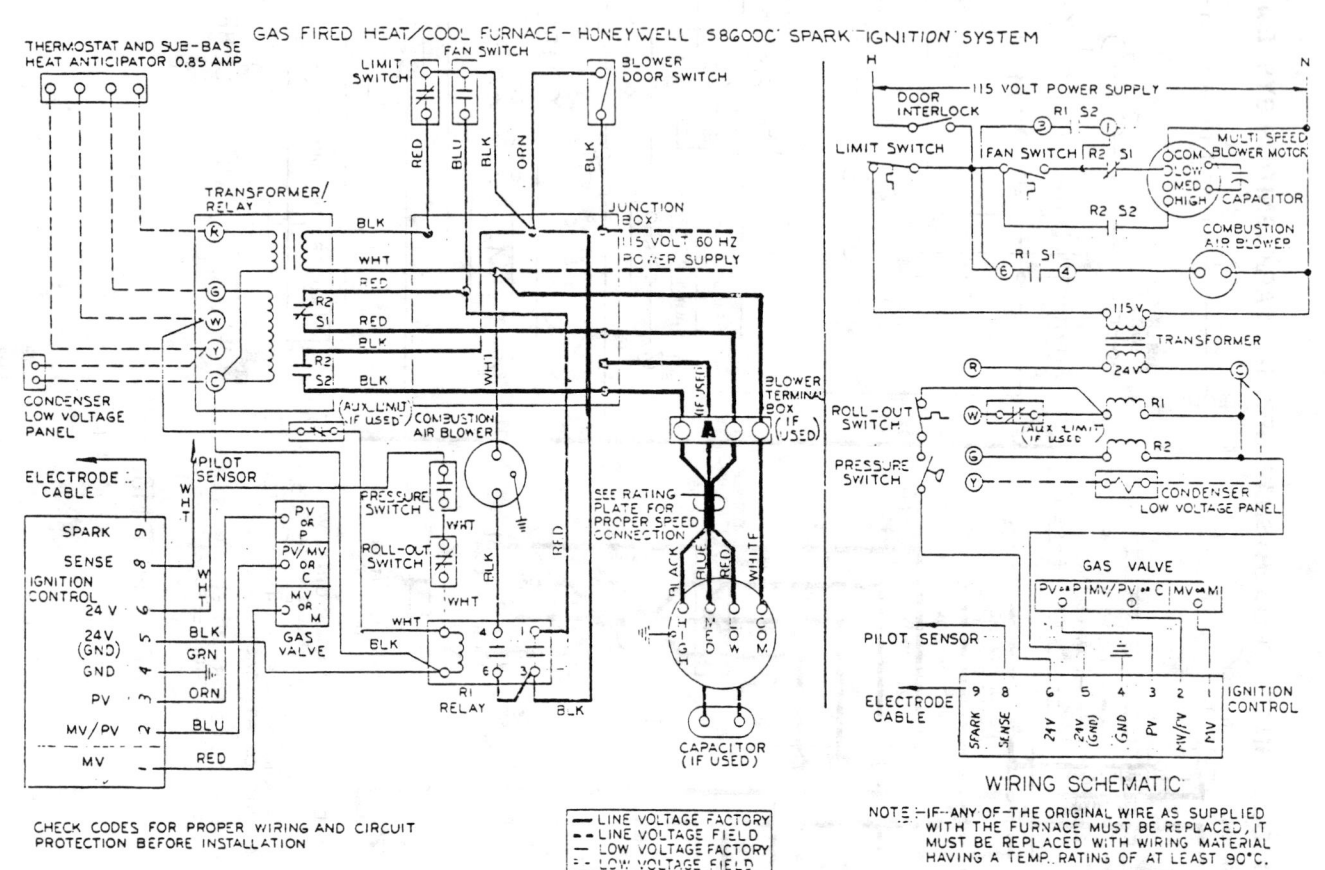

CHECK CODES FOR PROPER WIRING AND CIRCUIT
PROTECTION BEFORE INSTALLATION

CONNECTION DIAGRAM

WIRING SCHEMATIC

NOTE: IF ANY OF THE ORIGINAL WIRE AS SUPPLIED
WITH THE FURNACE MUST BE REPLACED, IT
MUST BE REPLACED WITH WIRING MATERIAL
HAVING A TEMP. RATING OF AT LEAST 90°C.

PART NO. 39969D001

Ultra II HIGH EFFICIENCY
Gas-fired Furnace

USER'S INFORMATION MANUAL

WARNING: If the information in this manual is not followed exactly, a fire or explosion may result causing property damage, personal injury or loss of life.

— Do not store or use gasoline or other flammable vapors and liquids in the vicinity of this or any other appliance.

—WHAT TO DO IF YOU SMELL GAS

- Do not try to light any appliance.
- Do not touch any electrical switch; do not use any phone in your building.
- Immediately call your gas supplier from a neighbor's phone. Follow the gas supplier's instructions.
- If you cannot reach your gas supplier, call the fire department.

— Installation and service must be performed by a qualified installer, service agency or the gas supplier.

designed and manufactured by

ARMSTRONG
AIR CONDITIONING INC.
Bellevue, Ohio 44811

Your unit

☐ **Upflow** ☐ **Counterflow**

TEMPERATURE CONTROL

THERMOSTAT

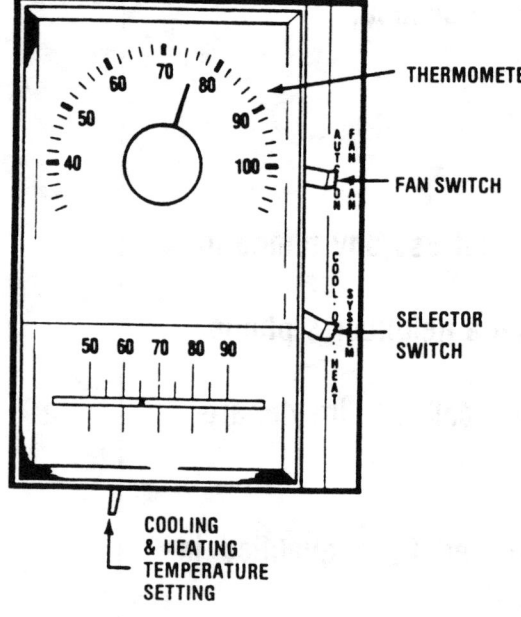

THERMOMETER

FAN SWITCH

SELECTOR SWITCH

COOLING & HEATING TEMPERATURE SETTING

Ultra II

HIGH EFFICIENCY Gas Fired Furnace
Congratulations...

...you have one of the most efficient gas furnaces made. Your unit has been carefully selected to keep you warm and comfortable during the winter months. It will deliver superb performance with only minimal help from you.

To keep your operating costs low and to eliminate unnecessary service calls, we have provided a few guidelines. The few minutes spent reading this manual will be well worth your while.

Your high efficiency gas furnace is designed to extract more usable heat from the burning of natural gas than that obtained from a standard design gas furnace. In this high efficiency heat transfer process the water vapor in the products of combustion are cooled to form condensation inside the furnace secondary heat exchanger.

This is normal operation for your furnace. You can observe this water being drained away by a drain line connected from the furnace to an adjacent sanitary sewer opening.

How To Operate Your System

THERMOSTAT OPERATION WHEN YOU HAVE HEATING ONLY
The thermostat is the only control with which to concern yourself. The furnace is completely automatic. Simply set the temperature setting at the desired comfort level and allow automatic operation.

THERMOSTAT OPERATION WHEN YOU HAVE HEATING AND COOLING
There are two switches located on the thermostat. One switch is for ''COOLING-HEATING-OFF''. The other switch is for ''FAN'' operation, either continuous or automatic. On the thermostat is the temperature selection range for the heating temperature and the cooling temperature desired.

To put the system into operation, push the switch to either: ''HEAT'' or ''COOL'' position.

In the cooling season, you will set the switch for straight cooling. The same for heating when you get into the colder heating season.

After you have selected the type of operation you desire, you then select the temperature you would like the system to maintain.

FAN OPERATION
You may wish to increase your comfort by setting your system for continuous air circulation of the indoor air. The fan switch on the thermostat permits you to do this.

With the switch in the ''ON'' position, the fan will operate continuously. ''AUTO'' position gives fan operation only when the unit is in either the heating or cooling cycle.

OWNER RECORD

Model _____ Serial No. _____

Installation Date _____

INSTALLED BY: _____

Dealer _____

Address _____

_____

Telephone No. _____

For Your Safety
Read Before Operating

> **WARNING: IF YOU DO NOT FOLLOW THESE INSTRUCTIONS EXACTLY, A FIRE OR EXPLOSION MAY RESULT CAUSING PROPERTY DAMAGE, PERSONAL INJURY OR LOSS OF LIFE.**

A. This appliance is equipped with an ignition device which automatically lights the pilot. Do **not** try to light the pilot by hand.

B. **BEFORE OPERATING** smell all around the appliance area for gas. Be sure to smell next to the floor because some gas is heavier than air and will settle on the floor.

WHAT TO DO IF YOU SMELL GAS

- Do not try to light any appliance.
- Do not touch any electric switch; do not use any phone in your building.
- Immediately call your gas supplier from a neighbor's phone. Follow the gas supplier's instructions.
- If you cannot reach your gas supplier, call the fire department.

C. Use only your hand to turn the gas control knob. Never use tools. If the knob will not turn by hand, don't try to force or repair it, call a qualified service technician. Force or attempted repair may result in a fire or explosion.

D. Do not use this appliance if any part has been under water. Immediately call a qualified service technician to inspect the appliance and to replace any part of the control system and any gas control which has been under water.

Operating Instructions

1. **STOP!** Read safety information above.
2. Set the thermostat to lowest setting.
3. Turn off all electric power to the appliance.
4. This appliance is equipped with an ignition device which automatically lights the pilot. Do **not** try to light the pilot by hand.

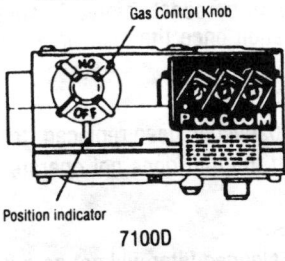

7100D
ROBERTSHAW

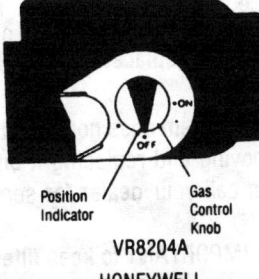

VR8204A
HONEYWELL

5. Remove control access panel.
6. Turn the gas control knob clockwise ↻ to "OFF".
7. Wait five (5) minutes to clear out any gas. Then smell for gas, including near the floor. If you smell gas, **STOP!** Follow "B" in the safety information above. If you don't smell gas, go to next step.
8. Turn gas control knob counterclockwise ↺ to "ON".
9. Replace control access panel.
10. Turn on all electric power to the appliance.
11. Set thermostat to desired setting.
12. If the appliance will not operate, follow the instructions "To Turn Off Gas To Appliance" and call your service technician or gas supplier.

To Turn Off Gas To Appliance

1. Set the thermostat to lowest setting.
2. Turn off all electric power to the appliance if service is to be performed.
3. Remove control access panel.
4. Turn the gas control knob clockwise ↻ to "OFF". Do not force.
5. Replace control access panel.

Maintenance of Your Unit

CAUTION: There are routine maintenance steps you should take to keep your unit operating efficiently. This will assure longer life, lower operating costs and fewer service calls. The steps given in this publication are easy to follow and are not time consuming. Certain service and maintenance procedures require the skill of a trained service person who has specialized tools and training for their use. Please call your dealer for service. Personal injury can result if you are not qualified to do this work.

CLEANING

Your unit's cabinet has an acrylic baked enamel finish. This finish can be cleaned with soap and water. Grease spots can be removed with any good household cleaning agent. The cabinet can be kept attractive indefinitely by polishing with automobile wax at least twice a year.

CLEAN FURNACE AREA

Insulation materials may be combustible. Therefore, a furnace installed in an attic or other insulated space must be kept free and clear of insulating materials. Make sure to examine the furnace area when a furnace or additional insulation has been added.

WHAT TO DO IF YOUR UNIT IS NOT WORKING CORRECTLY

If your unit is operating but fails to provide complete comfort, check the following before calling for service.

1. Be sure that the thermostat setting is correct.
2. Check to see if the filter is clean.
3. Be sure air can circulate freely throughout your home. Do not block supply registers or return grilles with furniture or rugs.

And if you also have cooling ...

4. Keep surface of the outdoor coil free from dirt, lint, paper or leaves.
5. Check and clean indoor coil if necessary. (This check should be made at the start of each cooling season by your service technician.)

WHAT TO DO IF YOUR UNIT FAILS TO OPERATE

1. Be sure the main switch that supplies power to the unit is in the "ON" position.
2. Replace any burned-out fuses or reset circuit breakers.
3. Be sure the thermostat is properly set.

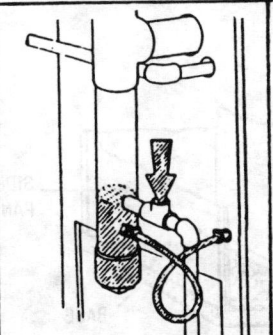

4. **IMPORTANT:** For proper operation of your furnace, the condensate trap and plastic drain tube must be kept full of liquid. If plastic tube does not show full, then fill trap with clean water by adding at the open top of the small tee. Fill until plastic drain tube is full or until liquid becomes visible in bottom of small tee.

Opening in small tee must not be restricted or closed.

5. If the unit still does not start, call your service technician.

Specific Filter Changing Instructions

VERY IMPORTANT! Cleaning or replacing the filter.

A dirty air filter can sharply increase the operational costs of your unit ... in some cases it can double the cost. It is a good idea to check the filters at least every six weeks. If they are dirty they should be replaced, or cleaned if you have a cleanable filter.

The unit may contain either a disposable filter or a permanent filter. The type of filter may be indicated on a label attached to the filter. If a disposable filter is provided, replace with a filter the same size. If a permanent filter is provided, clean filter and replace. To clean permanent filter, shake filter to remove excess dirt and/or use a vacuum cleaner. Wash filter in soap or detergent water and replace after filter is dry. Permanent filters supplied need not be oiled after washing. If your air distribution system has a central return air filter-grille, you do not need a filter in your furnace. Be sure to clean the filter-grille filter as recommended above.

Upflow Models

1. If filter is located in an external filter rack or cabinet on either side of the furnace cabinet, it may be easily removed, cleaned or changed, and re-inserted in its proper location.

2. If the furnace has a bottom return air, remove the lower front panel by lifting up and pulling out. The filter(s) may be removed by sliding toward the front. Replace in a reverse procedure.

Filter sizes Upflow Models

Side return, all models — 16" × 25"

Bottom return

CABINET WIDTH	FILTER SIZE
17½"	15½" × 25"
21½"	19½" × 27"
25½"	23½" × 27"

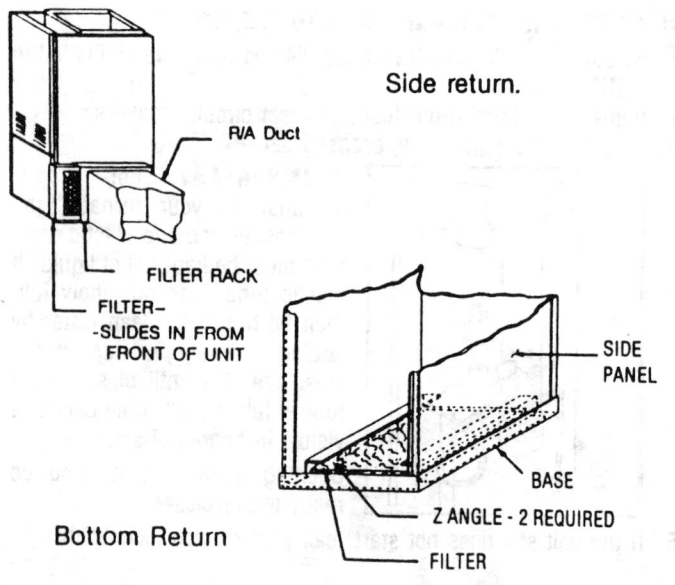

Side return.

R/A Duct

FILTER RACK

FILTER—
—SLIDES IN FROM
FRONT OF UNIT

SIDE PANEL

BASE

Z ANGLE - 2 REQUIRED

FILTER

Bottom Return

Counterflow Models

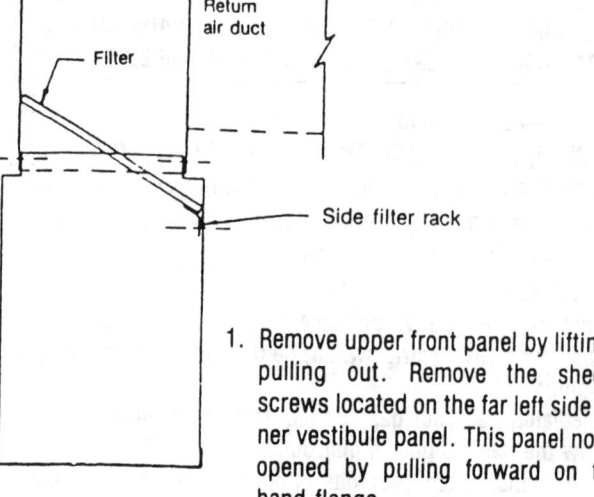

Return air plenum

Filter

Return air duct

Side filter rack

1. Remove upper front panel by lifting up and pulling out. Remove the sheet metal screws located on the far left side of the inner vestibule panel. This panel now can be opened by pulling forward on the right hand flange.

2. The two filters will be found at the top of the blower compartment. By moving each filter up on one side into the return air duct, it can be removed through the left panel opening.

3. Replace filters using a reverse procedure.

Filter sizes Counterflow Models

CABINET WIDTH	FILTER SIZE
17½"	(2) 9" × 18"
21½"	(2) 9" × 22"
25½"	(2) 9" × 26"

CAUTION: The blower compartment door on your high efficiency gas furnace is equipped with a safety interlock switch that will automatically shut off your complete system (including the blower) once this door is removed. This is for your personal safety. Be sure to check your furnace for proper operation once the door or panel has been replaced.

If the system does not operate once the panel has been replaced, try removing and replacing it once again. If the unit does not operate, then call your dealer for service.

It is IMPORTANT to keep filters clean. A clogged filter will not permit adequate airflow. This can cause heat exchanger failure or cooling coil freeze-up.

Lubrication

Direct drive blower and power venter contain motors which do not require oiling.

For Your Safety

The furnace area must be kept clear and free of combustible materials, gasoline and other flammable vapors and liquids.

> **WARNING: Should overheating occur, or the gas supply fail to shut off, shut off the manual gas valve to the appliance before shutting off the electrical supply.**

COMBUSTION AND VENTILATION AIR

> **WARNING: ADEQUATE AIR MUST REACH YOUR GAS FURNACE TO PROVIDE FOR PROPER AND SAFE OPERATION. ANY OBSTRUCTION OF THIS AIRFLOW CAN CAUSE AN UNSAFE CONDITION WHICH MAY RESULT IN DEATH OR PERMANENT INJURY.**
>
> **LETHAL CARBON MONOXIDE GAS CAN BE PRODUCED IF COMBUSTION AIR IS RESTRICTED.**

Additional combustion and ventilation air may be brought to your furnace at the side of the cabinet between the front door and the supply air plenum. Your furnace may have outside air brought directly by means of a plastic (PVC) pipe. Air movement through this pipe must not be blocked or restricted, or your furnace will not operate due to the functioning of the pressure sensing control located on the furnace.

THE OUTSIDE AIR PIPE MUST BE FREE FROM ANY RESTRICTIONS INTERNALLY OR AT ITS INTAKE AT ALL TIMES.

If outside air is not piped directly to your furnace, the top front area must be kept open.

DO NOT PLACE ANY OBJECTS OR RESTRICTIONS OVER OR WITHIN THIS OPENING AT ANY TIME.

Furnaces located in a closet, alcove or utility room must have provision for adequate air supply by means of upper and lower grilles in the door, or by the introduction of outside air, or both. National Fuel Gas Code ANSI Z223.1 (latest edition) and local code requirements are generally alike. However, local codes take precedence.

Venting

Venting of this furnace **must** comply with our published instructions. Be sure the installer has followed these requirements. If they have not been, you should request the installer to comply as soon as possible. For your safety please note the following.

1. This furnace **must not be vented with any other appliance.** The flue (vent) system is under positive pressure from the power venter. Connection of any other appliance to the furnace flue may create a hazardous condition that could cause either appliance to malfunction.
2. An automatic vent damper **must not be used** with this furnace. The furnace is not designed for use with a vent damper. Use of such a device will not improve the efficiency of this furnace.

The vent from your furnace may rise vertically and terminate above the roof. Or, it may be run horizontally to an exterior foundation wall. If this is the case, your dealer or installer will have terminated the vent, outdoors, facing down and covered with a ½" mesh screen — all according to our published installation instructions.

It is therefore, important that the outside area where the vent terminates be kept clear of any obstructions which might block or impede the venting of the furnace. Should venting become blocked at any time, your furnace is equipped with a special safety control to prevent operation of the furnace until the condition has been corrected. Your dealer will be glad to give you additional information about this.

Should any unusual conditions be observed during your inspections, be sure to call your authorized service dealer at once.

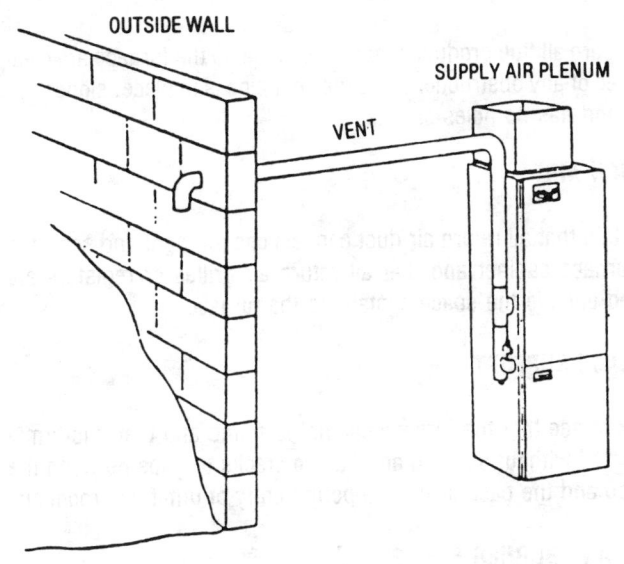

Note: Counterflow models will be connected similar to arrangement shown.

Annual Inspection

The following inspection procedure should be conducted each year prior to the beginning of the heating season. Regular inspections and planned maintenance from your authorized dealer will assure many years of economical performance from your new gas furnace.

CONDENSATE COLLECTION AND DISPOSAL SYSTEM

Be sure the condensate drain line does not become blocked or plugged. Visual inspection of condensate flow can easily be made while the furnace is in operation. A flashlight can be used to illuminate the discharge end placed in the sewer opening.

Clean the condensate trap and flush the condensate tube to make sure condensate flows freely while the furnace is in operation. Refill with clean water as shown on page 2.

VENT

Make sure all flue product materials external to the furnace are clear and free of any obstruction; that the vent pipe is in place, slopes upward and has no holes or leaks.

RETURN AIR

Ascertain that all return air duct connections are tight and sealed to the furnace cabinet and that all return air grilles or registers are located outside the space containing the furnace.

FURNACE SUPPORT

Check to see that the furnace cabinet is sound and that it is firmly supported without sagging and has no cracks or gaps between the furnace and the base or floor to permit entry of unfiltered room air.

PILOT AND BURNER FLAME

By removal of the front door (lift up, then off) and while the furnace is in operation, observe the main burner and the pilot flame. Compare these observations to the diagrams as shown on this page to determine if proper flame adjustment is present. If your observations indicate improper flame adjustment, then call your authorized dealer for service.

DO NOT ATTEMPT TO ADJUST EITHER FLAME! Your service representative will perform this adjustment correctly.

IMPORTANT!

Your gas furnace is designed to give many years of efficient, satisfactory service. However, the varied air pollutants commonly found in most areas can affect longevity and safety. Chemicals contained in everyday household items such as laundry detergents, cleaning sprays, hair sprays, deodorizers and other products which produce airborne residuals may have an adverse effect upon the metals used to construct your furnace. It is important that you visually inspect the conditions of the gas burners and the gas vent from the furnace. All of these locations are found at the front of the furnace. A flashlight will be useful for these inspections. Make one inspection prior to the beginning of the heating season and another during the middle. SHOULD YOU OBSERVE UNUSUAL AMOUNTS OF RUST, FLAKES OR OTHER DEPOSITS, COATINGS OR CORROSION, IT IS IMPORTANT THAT YOU CALL YOUR AUTHORIZED DEALER AT ONCE TO OBTAIN A QUALIFIED SERVICE INSPECTION.

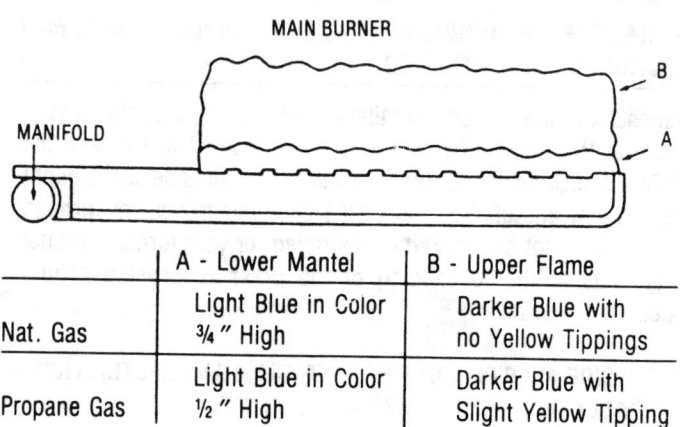

MAIN BURNER

	A - Lower Mantel	B - Upper Flame
Nat. Gas	Light Blue in Color ¾" High	Darker Blue with no Yellow Tippings
Propane Gas	Light Blue in Color ½" High	Darker Blue with Slight Yellow Tipping

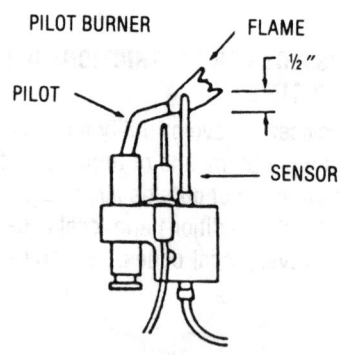

A.G.A. DESIGN CERTIFIED

The American Gas Association symbol on each nameplate is your assurance that your furnace design meets nationally recognized standards for safety and performance.

CERTIFIED EFFICIENCY RATINGS

The Gas Appliance Manufacturers Association (GAMA) symbol verifies that Annual Fuel Utilization Efficiency (AFUE) ratings for our gas furnaces have been derived from U.S. Government standard tests.

37638L091
Printed in U.S.A.

Ultra II 97
ADVANCED HIGH EFFICIENCY

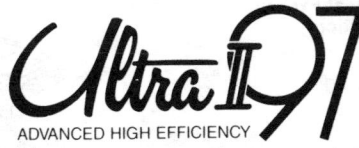

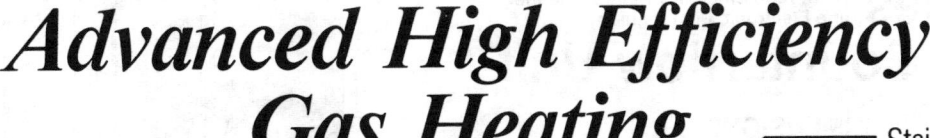

Advanced High Efficiency Gas Heating

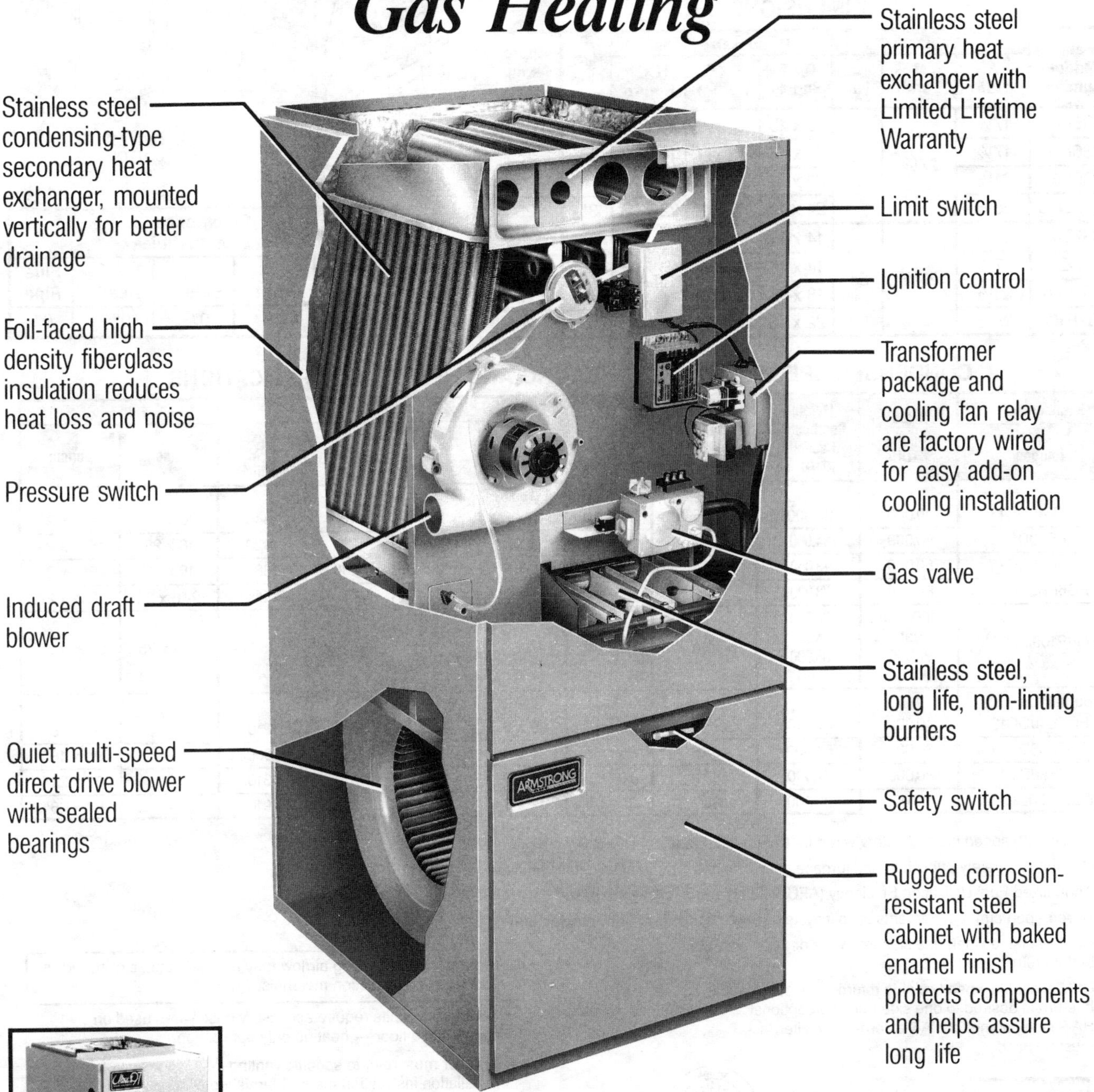

Stainless steel condensing-type secondary heat exchanger, mounted vertically for better drainage
Foil-faced high density fiberglass insulation reduces heat loss and noise
Pressure switch
Induced draft blower
Quiet multi-speed direct drive blower with sealed bearings
Stainless steel primary heat exchanger with Limited Lifetime Warranty
Limit switch
Ignition control
Transformer package and cooling fan relay are factory wired for easy add-on cooling installation
Gas valve
Stainless steel, long life, non-linting burners
Safety switch
Rugged corrosion-resistant steel cabinet with baked enamel finish protects components and helps assure long life
ARMSTRONG

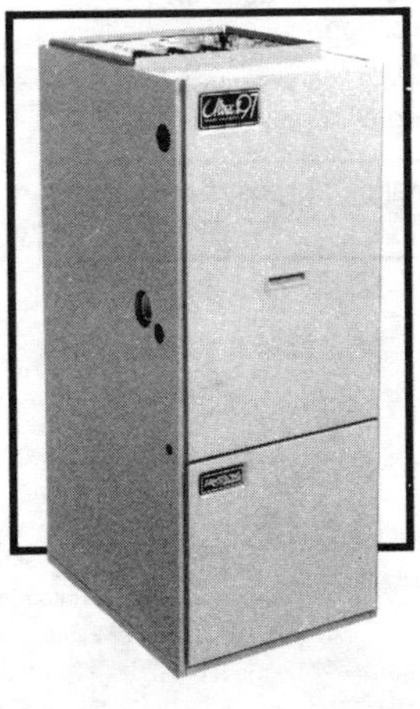

Ultra II 97 offers homeowners fuel efficiencies up to 94.5% AFUE (isolated combustion) with a Limited Lifetime Heat Exchanger Warranty.

ARMSTRONG AIR

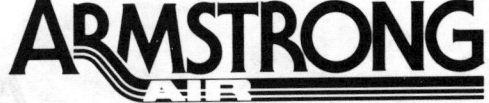

EG6E Series (With High Efficiency Heat Exchanger and Electronic Pilot Ignition)
Gas Fired Hi-Boy Unit

EG7E Series (With High Efficiency Heat Exchanger and Electronic Pilot Ignition)
Gas Fired Counterflow Unit

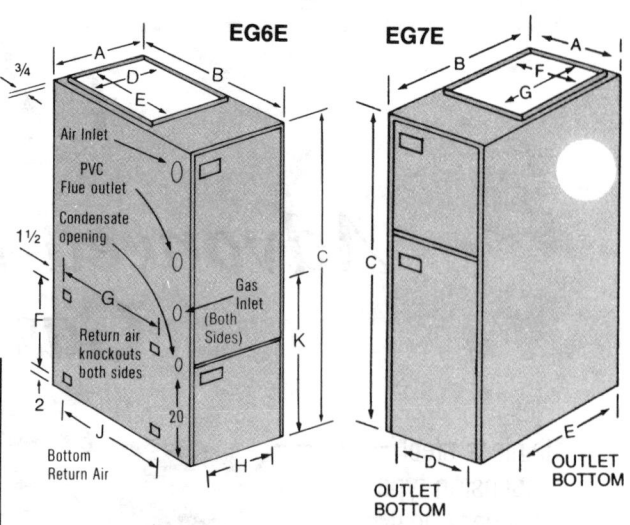

DIMENSIONS

Basic Model Number	Cabinet			Plenum Openings			► PVC Flue Dia.	Approx. Ship Wt.
	A Width	B Depth	C Ht.	D x E Supply	F x G Return	H x J Return		
EG6E40	17½		51	16 x 18		14 x 23		180
EG6E60	17½	27½	51	16 x 18	14 x 22	14 x 23	2"	185
EG6E80	21½		51	20 x 18		18 x 23		230
EG6E100	25½			24 x 18		22 x 23	3"	260
EG7E40	17½			14 x 16	16 x 18			185
EG7E60	17½	27½	54	14 x 16	16 x 18	N.A.	2"	190
EG7E80	21½			18 x 16	20 x 18			235
EG7E100	25½			22 x 16	24 x 18		3"	265

Clearances To Combustibles				
Plenum Top	Front	Sides	Rear	Flue Pipe
1"	4"	0"	0"	0"

CONDENSED SPECIFICATIONS — HEATING AND HEAT/COOL APPLICATIONS

Model No. Electronic Ignition	Rated Input BTUH	Isolated Combustion Capacity BTUH ●	Isolated Combustion Fuel Eff'y. AFUE ●	■ Nominal Cooling Capacity	Airflow				Filters ②	
					Blower Size①	H.P.	Min. CFM 0.20"	Max. CFM 0.50"	Side	Bottom
Upflow										
EG6E40DC13	40,000	38,000	94.5	2.5 - 3.0	10 x 8	1/3	985	1340	16 x 25	15½ x 25
EG6E60DC14	60,000	54,000	91.2	3.0 - 3.5	10 x 8	1/2	1210	1420	16 x 25	15½ x 25
EG6E80DC16	80,000	73,000	90.5	3.0 - 4.0	10 x 9	1/2	1230	1680	16 x 25	19½ x 27
EG6E80DC19	80,000	73,000	90.7	3.0 - 5.0	12 x 9	3/4	1230③	2025④	2-16 x 25	19½ x 27
EG6E100DC20A	100,000 Nat. 97,500 LPG	90,000 Nat. 88,000 LPG	90.5	3.0 - 5.0	12 x 9	3/4	1210③	2050④	2-16 x 25	23½ x 27
Counterflow										
EG7E40DC13	40,000	37,000	93.2	2.5 - 3.0	10 x 8	1/3	865	1210	—	9" x 18"
EG7E60DC14	60,000	54,000	88.5	3.0 - 3.5	10 x 8	1/2	1220	1300	—	9" x 18"
EG7E80DC16	80,000	71,000	89.0	3.0 - 4.0	10 x 9	1/2	1310	1310	—	9" x 22"
EG7E100DC20	100,000	88,000	88.7	3.0 - 5.0	12 x 9	3/4	1375	1910	—	9" x 26"

① Multiple (3) speed motor. Factory wired for two-speed operation. Direct Drive blower — all models.

② Consult separately published gas furnace specifications for complete details on filter use and return air application.

● Annualized Fuel Utilization Efficiency (AFUE) rating and capacity are based on U.S. Government Standard Tests.

■ Based upon cfm requirements of split system evaporator coils of our manufacture.

③ CFM available when one side return air or bottom return used.

④ CFM available when two side return air used. If return is desired to one side only then optional HA-9 side cabinet or HA-7 external side filter rack must be used.

Not approved for use in mobile home applications.

High efficiency heating airflow may necessitate use of sound and vibration reduction materials.

Counterflow units require accessory base when used on combustible floor — heating only application.

► Installer must refer to specific venting instructions given in installation instruction manual furnished with each unit.

Data shown herein provided for information purposes only and are subject to revision.

Bellevue, Ohio 44811

Form No. AU-297 11/91

Printed in U.S.A.

Bard

INSTALLATION INSTRUCTIONS
HIGH EFFICIENCY GAS FURNACE

IMPORTANT
READ ALL INSTRUCTIONS CAREFULLY BEFORE BEGINNING THE INSTALLATION.

MODELS:

HI-BOY

DCH050D30A
DCH065D36A
DCH080D48A
DCH095D60A

IMPORTANT NOTICE
THIS FURNACE IS NOT INTENDED FOR USE AS A CONSTRUCTION HEATER.
USE OF THIS FURNACE DURING CONSTRUCTION AND FINISHING PHASES OF A STRUCTURE IS CONSIDERED AS "OPERATION IN A CORROSIVE ATMOSPHERE" AND "UNUSUAL, NEGLIGENT OR IMPROPER USE" AND AS SUCH ARE CONSIDERED EXCLUSIONS BY THE BARD MANUFACTURING COMPANY LIMITED WARRANTY.

MANUAL 2100-187 REV.
SUPERSEDES REV.
FILE VOL. I, TAB 2

Section	TABLE OF CONTENTS	Page

SECTION 1 --GETTING OTHER INFORMATION AND PUBLICATIONS

These publications can help you install the furnace. You can usually find these at your local library or purchase them directly from the publisher. Be sure to consult current edition of each standard.

National Fuel Gas Code	-ANSI Z223.1/NFPA 54
National Electrical Code	-ANSI/NFPA 70
Standard For The Installation Of Air Conditioning and Ventilating Systems	-ANSI/NFPA 90A
Standard For Warm Air Heating and Air Conditioning Systems	-ANSI/NFPA 90B
Standard For Chimneys, Fireplaces, Vents, and Solid Fuel Burning Appliances	-NFPA 211
Load Calculation For Residential Winter and Summer Air Conditioning	-ACCA Manual J
Duct Design For Residential Winter and Summer Air Conditioning and Equipment Selection	-ACCA Manual D

FOR MORE INFORMATION, CONTACT THESE PUBLISHERS

ACCA: AIR CONDITIONING CONTRACTORS OF AMERICA
1513 16th Street NW
Washington, DC 20036
Telephone: (202) 483-9370   Fax: (202) 234-4721

ANSI: AMERICAN NATIONAL STANDARDS INSTITUTE
1430 Broadway
New York, NY 10018
Telephone: (212) 354-3300   Fax: (212) 302-1286

ASHRAE: AMERICAN SOCIETY OF HEATING REFRIGERATING AND
AIR CONDITIONING ENGINEERS, INCORPORATED
1791 Tullie Circle, N.E.
Atlanta, GA 30329-2305
Telephone: (404) 636-8400   Fax: (404) 321-5478

NFPA: NATIONAL FIRE PROTECTION ASSOCIATION
Batterymarch Park
P. O. Box 9101
Quincy, MA 02269-9901
Telephone: (800) 344-3555   Fax: (617) 984-7057

SECTION 2 -- IMPORTANT SAFETY RULES

WARNING: Read and exactly follow these rules. Failure to do so could cause improper furnace operation, resulting in damage, injury or death.

A. Signal words.

To alert you to potential hazards, we use the signal words **"WARNING"** and **"CAUTION"** throughout this manual. **"WARNING"** alerts you to situations that could cause serious injury or death. **"CAUTION"** alerts you to situations that could cause minor or moderate injury or property damage. To help you, we use the words "must" and "should" in this manual. "Must" is mandatory. "Should" is advisory.

"IMPORTANT" is used to draw your attention to information that is high priority.

B. Use only the type of gas approved for this furnace; refer to furnace rating plate.

WARNING

Only use natural gas in furnaces designed for natural gas.
Only use Propane (LP) gas for furnaces designed for Propane (LP) gas.
Make sure furnace will operate properly on gas type available to user.
Do not use this furnace with butane. Using wrong gas could create a hazard, resulting in damage, injury, or death.

C. **DO NOT** install this furnace outdoors or in a mobile home, trailer, or recreational vehicle. It is not AGA design-certified for these installations. This furnace is suitable for a home built on site or manufactured home completed at final site. See Section 4 for more information.

D. Carefully choose furnace installation site. **DO NOT** directly expose furnace to drafts, wind or other outdoor conditions.

E. **DO NOT** install furnace in a corrosive or contaminated atmosphere. See Section 4 for more information.

F. **DO NOT** use this furnace during construction when adhesives, sealers, and/or new carpets are being installed. See Section 4 for more information.

G. Provide adequate ventilation air to space where furnace is being installed. See Section 9 for more information.

H. Connect this furnace to an approved vent and combustion air intake system. See Section 10 for more information.

I. Never test for gas leaks with an open flame. Use a commercial soap made specifically for leak detection to check all connections. See Section 17 for more information.

J. Always install duct system with furnace. Be sure duct system has external static pressure within allowable furnace range. See Section 7 for more information.

K. Completely seal supply and return air ducts to furnace casing. Duct work must run to an area outside furnace air space. Seal duct work wherever it runs through walls, ceilings or floors. See Section 7 for more information.

SECTION 3 -- MEETING CODES

Before installing furnace, make sure you know all applicable codes. National, state and local codes may take precedence over any instructions in this manual. Be sure to consult:

- Authorities having jurisdiction over furnaces;
- Local code authorities for information on electrical wiring, gas piping and vent pipe;
- Current National Fuel Gas Code ANSI/NFPA 54;
- Current National Electrical Code ANSI/NFPA 70.

See Section 1 for information on getting copies of these codes.

SECTION 4 -- APPLICATION

This is a fan-assisted Category IV forced air direct vent gas furnace for indoor installation in building constructed on site. The furnace installation must conform with local building codes and ordinances or, in their absence with the National Fuel Gas Code, ANSI Z223.1-latest edition, and the National Electrical Code, ANSI/NFPA 70-latest edition. It is the personal responsibility and obligation of the purchaser to contract a qualified installer to assure that installation is adequate and is in conformance with governing codes and ordinances.

```
***** IMPORTANT NOTICE *****
_____

THIS FURNACE IS NOT INTENDED FOR USE AS A
CONSTRUCTION HEATER.

USE OF THIS FURNACE DURING CONSTRUCTION
AND FINISHING PHASES OF A STRUCTURE IS
CONSIDERED AS "OPERATION IN A CORROSIVE
ATMOSPHERE" AND "UNUSUAL, NEGLIGENT OR
IMPROPER USE" AND AS SUCH ARE CONSIDERED
EXCLUSIONS BY THE BARD MANUFACTURING
COMPANY LIMITED WARRANTY.
```

When a furnace is used as a construction heater, it is operated under unusual and abnormal conditions that can cause condensation to occur in some portions of the DCH-series furnaces that will not condense under normal operation conditions when properly sized, installed and set-up for operation.

Combined with condensation problems are the many sources of chloride that are present in high concentrations during construction phases of a structure. These chloride sources are either very reduced or the vapors have left once the construction activities are completed and the structure is ready to occupy.

Many of the more common construction sources of chlorides are listed below, and they are usually present in large quantities.

Cement/Concrete Mixtures	Tile/Counter Cements
Paint	Adhesives
Stain, Varnish	Cements and Glues
Solvents	Dust Particles
Wood Preservatives	Foam Insulations
Floor Sealers	

SECTION 5 -- SIZING OF FURNACE

The sizing of high efficiency gas furnaces for both new and replacement installations is critical due to condensing design of the appliances. Oversizing of the furnace for any application can cause short cycling (short on-time) conditions, and this in turn permits condensate to occur in locations within the furnace where it was not intended to be.

It is a normal occurrence for some condensate to form in the upper areas of the primary heat exchanger and in the combustion air blower section of the furnace on start-up. As the furnace attains normal operating conditions, these areas dry up, and all condensing takes place in the secondary stainless steel heat exchanger.

Short-cycling of the furnace permits condensate to form in some areas and never dry up. This continuously wet condition can lead to corrosion and metal deterioration causing premature failure of those components.

The following guidelines must be used to properly size and select the furnace for all applications:

1. Always conduct an accurate heating load calculation using appropriate methods, typically Air Conditioning Contractors of America (ACCA) Manual J.

2. Always use the correct outdoor and indoor design temperatures for the area; the actual dimensional information for the structure; the correct insulation values for windows, doors, walls, and ceilings; make sure that the correct values are used for the tightness of the building; and all other characteristics are input correctly into the calculation. There is no need to inflate or adjust any of these values just to make sure the furnace will be large enough. There is adequate safety factor built into Manual J, and any deviation from real or actual values will cause a potentially gross oversizing.

3. Always make the furnace selection based upon the useful heat rating of the furnace expressed as heating capacity or output Btuh. Never use input capacity.

4. Never make a furnace selection based upon the size of the previous furnace. It was undoubtedly grossly oversized to begin with, and very probably energy efficiency improvements were done to the structure over time further reducing the Btu requirement.

5. The DCH series 90+ gas furnaces are available in six heating output capacities. A nominal 15 percent oversizing and 10 percent undersizing rule should be used in determining which furnace should be installed. The following chart will assist in furnace selection:

Calculated Heat Loss Range(Btuh)	28,000-36,000	36,001-50,000	50,001-63,000	63,001-75,000	75,001-92,000	92,001-114,000
Use Furnace Output Rating	DCH036D30A	DCH050D30A	DCH065D36A	DCH080D48A	DCH095D60A	DCH115D60A

It is important to remember that furnace selections are made based upon the winter outdoor design conditions, and that very few hours of operation per year are at or near that temperature. And when the outdoor temperature is at the winter design condition, a correctly sized furnace will be operating the majority of the time with few and/or short off cycles.

Approximately 85 percent of the operating time of a typical furnace is when the building heat loss is only 10-80 percent of the winter design heat loss rating of the building, so it is essential not to oversize the furnace by either having an inflated calculated heat loss and/or picking one size larger furnace than calculated just to be sure.

6

SECTION 6 -- LOCATING THE FURNACE

When selecting a location for the furnace, observe the following rules.

1. The furnace should be set on a level floor. If the floor may become damp or wet at times, the furnace should be supported above the floor using a concrete base, bricks, patio blocks, etc., making sure adequate support is available for the furnace. Furnace approved for installation on combustible flooring shall not be installed directly on carpeting, tile or other combustible material other than wood flooring.

2. The furnace should be as centralized as practical with respect to the air distribution system.

3. The vent and air intake pipe should be as short as practical but must be at least 10 feet and no more than 65 feet in total equivalent length from furnace to the outside termination point. See Section 10 for more information.

4. Provide at least the minimum clearances specified in Table 1 for fire protection, proper operation and service access. These clearances must be permanently maintained. The ventilating air openings in the front of the furnace must never be obstructed.

5. All models are approved for either closet or alcove installation. An alcove is similar to a closet except there is no door permitted to enclose the front of furnace. See Table 1 for approved installation.

6. Fresh air for combustion must be piped from the outside of the building to the connection on the burner enclosure. See Section 10 for more information.

7. Minimum service clearances must take precedence over fire protection clearances (minimum installation clearances).

8. A gas-fired furnace installed in a residential garage must be installed so that the burners and ignition source are located not less than 18 inches above the floor, and the furnace must be located or protected to avoid physical damage by vehicles.

9. This furnace must be installed as to protect all electrical components from exposure to condensation and/or water.

CAUTION	WARNING
DO NOT locate furnace where temperature may drop below freezing as condensate may freeze resulting in improper operation or furnace damage.	DO NOT store combustible materials near furnace or warm air ducts. The material may ignite by spontaneous combustion creating a fire hazard.

TABLE 1   MINIMUM CLEARANCES (INCHES)

Model	Type Of Installation	Furnace Front	Back	Left Side	Right Side	Plenum Top	Sides	(1) Duct	Vent Pipe	Floor	Front	Back	Sides	Minimum Ventilation Openings For Confined Spaces-- Square Inches Of Free Area (2)
DCH050D30A	Closet	4	0	0	0	1	1	1	0	C	24	0	0	100 (2 req'd)
DCH065D36A	Closet	4	0	0	0	1	1	1	0	C	24	0	0	100 (2 req'd)
DCH080D48A	Closet	4	0	0	0	1	1	1	0	C	24	0	0	100 (2 req'd)
DCH095D60A	Closet	4	0	0	0	1	1	1	0	C	24	0	0	100 (2 req'd)

(1) For the first three (3) feet from plenum. After the first three (3) feet, no clearance required.
(2) See Section 9--Ventilation Air--for additional information.
C Floor may be combustible material.

SECTION 7 -- DUCT WORK

<u>Inadequate Supply Air and/or Return Air Duct Systems</u>. Short cycling because of limit control operation can be created by incorrectly designed or installed supply and/or return air duct systems.

The duct systems must be designed using ASHRAE or ACCA design manuals and the equipment cfm and external static pressure ratings to insure proper air delivery capabilities.

On replacement installations, particularly if equipment is oversized, the duct systems can easily be undersized. Modifications may be required to assure that the equipment is <u>operating within the approved temperature rise range when under full input conditions,</u> and that no short cycling on limit controls is occurring.

WARNING

When a furnace is installed so that supply ducts carry air circulated by the furnace to areas outside the space containing the furnace, the return air must also be handled by a duct(s) sealed to the furnace casing and terminating outside the space containing the furnace.

This is to prevent drawing possibly hazardous combustion products into the circulated air which can result in bodily injury or death.

When the furnace is used in connection with a cooling unit*, the furnace shall be installed parallel with or on the upstream side of the cooling unit to avoid condensation in the heating element. With a parallel flow arrangement, the dampers or other means used to control flow of air shall be adequate to prevent chilled air from entering the furnace and, if manually operated, must be equipped with means to prevent operation of either unit, unless the damper is in the full heat or cool position.

*A cooling unit is an air conditioning coil, heat pump coil or chilled water coil.

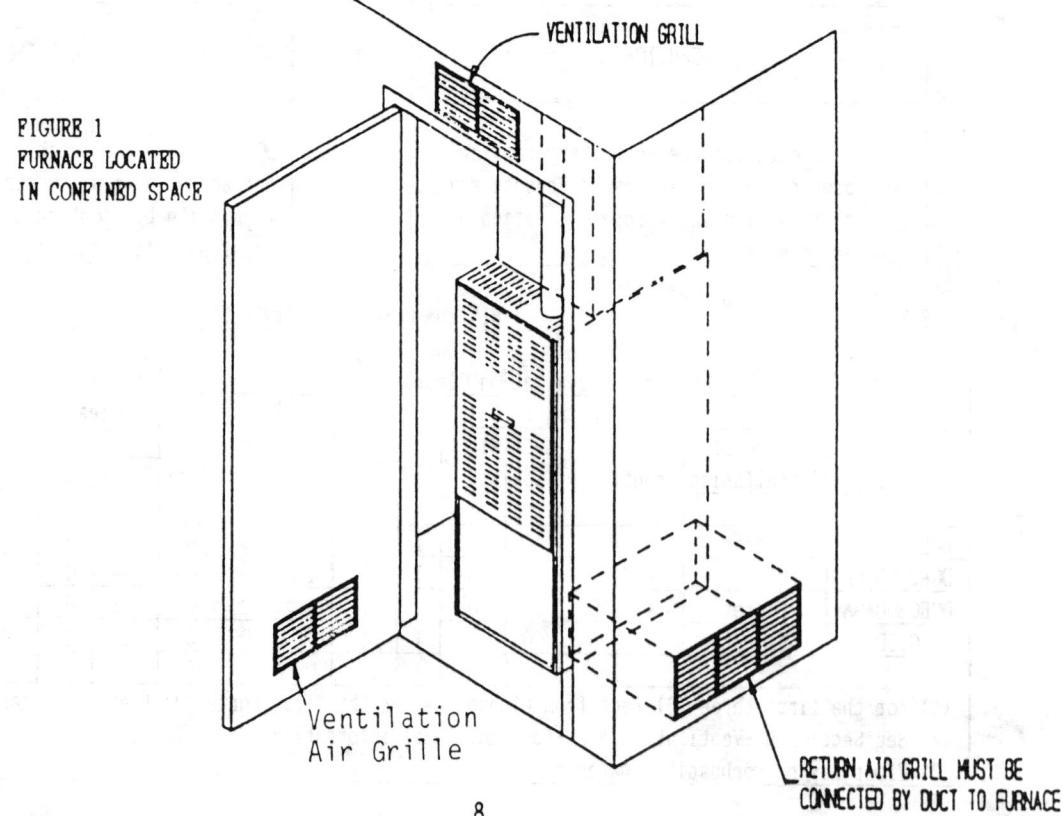

FIGURE 1
FURNACE LOCATED
IN CONFINED SPACE

VENTILATION GRILL

Ventilation
Air Grille

RETURN AIR GRILL MUST BE
CONNECTED BY DUCT TO FURNACE

8

SECTION 8 -- ELECTRICAL, GAS, DRAIN, AND SUPPLY AIR DUCT CONNECTIONS

Shown below are the electrical, gas and drain entrance/exit locations available on the furnace. There is a single fixed connection point for both the 115V electrical supply and 24V thermostat wiring. There is an optional left or right side entrance/exit point for the gas piping, condensate drain tubing, and optional accessory wiring harness for field supplied and installed electronic air cleaner and humidifier.

For attachment of supply air duct system, there are two methods provided:

1. If optional Cooling Coil Cabinet is used, it will fit over the 20" connection depicted in Figure 1A. The coil cabinets are designed to accomodate the standard A-coil evaporator sections.

2. If the optional coil cabinet is not used, installing a plenum 21" deep x required width and using the forward plenum flange 21" from back flange will better facilitate later installtaion of the A-coil evaporator sections which have a 20-1/2" pan dimension.

   The 20" dimension can be used but would require an offset cover door be fabricated to accomodate the 20-1/2" A-coil dimension if decided to install at a later date.

TABLE 1A

	SUPPLY AIR DUCT CONNECTIONS	
Model	If Coil Cabinet Is Used W x D	If No Coil Cabinet Is Used W x D
DCH050D30A DCH065D36A	18 x 20	18 x 21
DCH080D48A DCH095D60A	22 x 20	22 x 21

FIGURE 1A

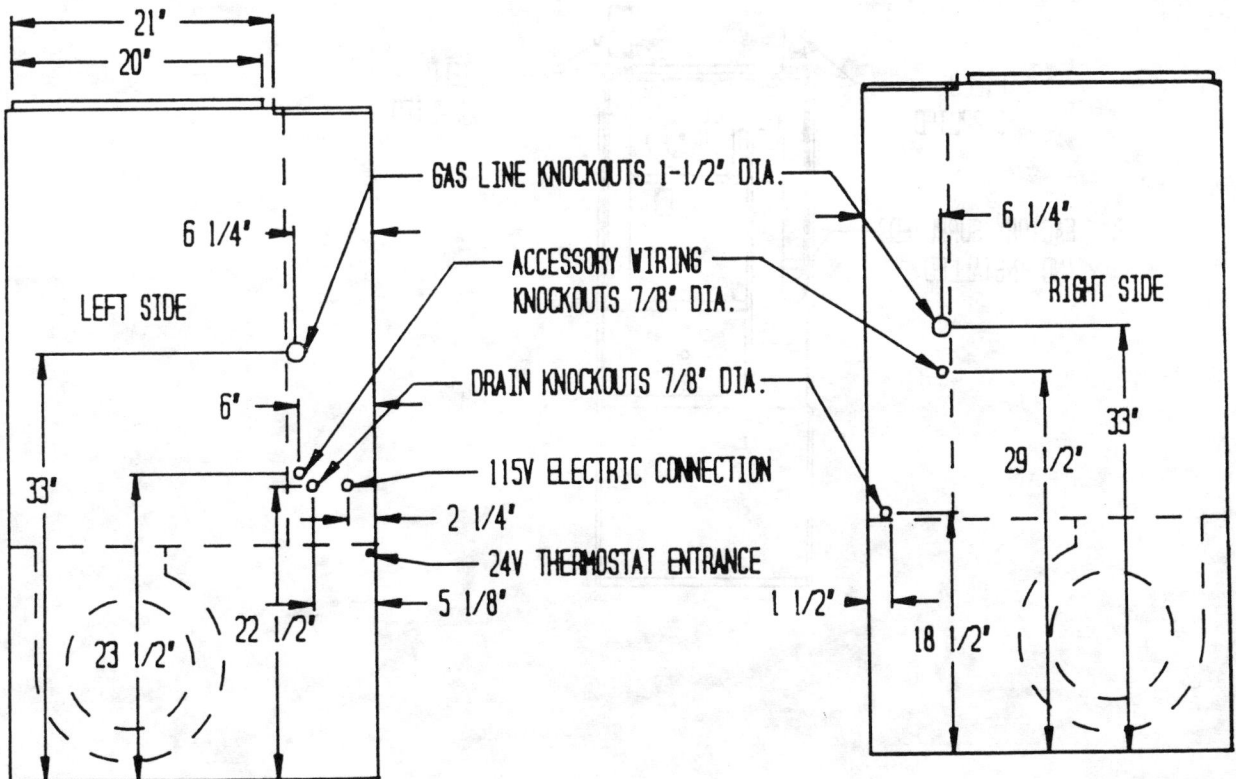

9

SECTION 9 -- VENTILATION AIR

If the furnace is installed in a closet or utility room, ventilating air must be allowed to enter the room through two permanent openings of equal area. One opening shall be located within twelve (12) inches of the ceiling and one opening within twelve (12) inches of the floor. Each opening shall have a face area of 100 square inches minimum of free area. If grilles are used, the opening will need to be larger to provide the 100 square inches of free area due to the grille restriction. Consult grille catalog to determine free area of various grilles. See Figure 2 for more information.

SECTION 10 -- VENT AND COMBUSTION AIR INTAKE PIPING SYSTEM DESIGN

This furnace must be vented to the outdoors with round PVC (poly-vinyl chloride) or ABS (accylonoitrile--butadiene-styrene) Schedule 40 pipe fittings. See Table 2 for vent and air intake pipe size selection. Both vent and air intake pipes must be the same size and approximately the same total length in equivalent feet, and neither is to exceed the total length shown in Table 2.

This furnace must use outside air for combustion. Failure to connect the air intake to the outside may result in premature failure of the heat exchanger(s), and will affect operation whenever there are any wind conditions. Connection of the air intake is a requirement for limited lifetime warranty of primary and secondary heat exchanger to be in effect.

This furnace removes both sensible and latent heat from the combustion flue gases. Removal of latent heat results in condensation of flue gas water vapor. This condensed water vapor drains from the secondary heat exchanger into a 29-4C stainless steel drain pan. The condensate exits the drain pan by means of 3/8 I.D. vinyl hose. See Section 15 for drain information.

FIGURE 2

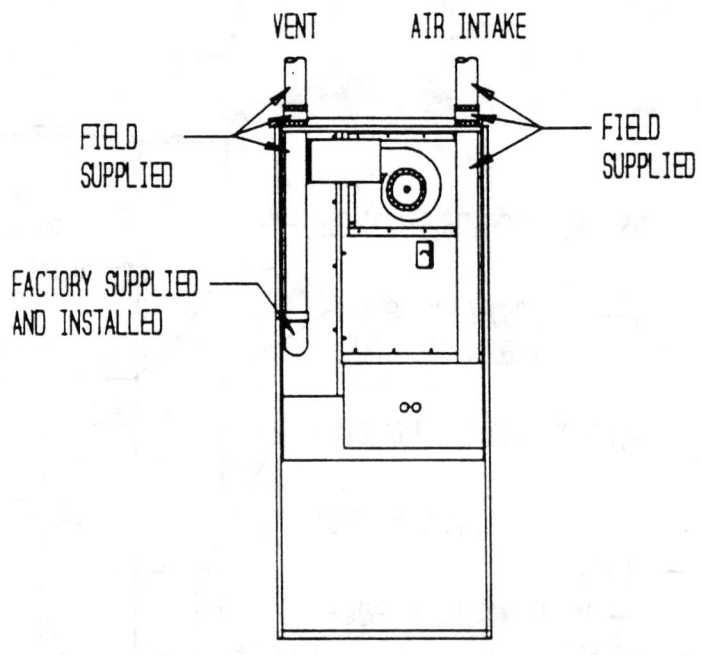

TABLE 2 PROPER VENT AND AIR INTAKE SELECTION

Furnace Models	Vent & Air Intake Length and Size *
DCH050D30A, DCH065D36A	8 - 30 ft. x 2", OR 8 - 65 ft. x 2-1/2", OR 8 - 65 ft. x 3"
DCH080D48A, DCH095D60A	8 - 65 ft. x 3"

*Vent length is in equivalent feet measurement. Refer to Item 1, General Instructions below.

IMPORTANT
For horizontal vent systems see special vent termination information in Section 13.

This vent must be in accordance with Part 7, Venting of Equipment, of the National Fuel Gas Code, ANSI Z223.1-latest edition, and Addenda Z223.1a--latest edition, or applicable provisions of the local building codes.

CAUTION	WARNING
Each vent must serve only one furnace. Connecting more than one furnace to a vent system will cause one or both furnaces to malfunction.	Do not connect vent to an existing masonry chimney or vent. Failure to adhere to this warning can result in property damage, bodily injury, or death.

General Instructions

1. The maximum length in equivalent feet for vent and air intake piping system is shown in Table 2 with each 45° elbow counting as 2-1/2 feet and each 90° elbow counting as 5 feet. Do not count the elbow within the furnace cabinet in this measurement. Drainage type (long radius) elbows should be used. Pressure type (short radius 90° elbows count as 10 equivalent feet.

    Example: 20 feet of straight vent pipe with 3 - 90° long radius elbows equals 35 equivalent feet.

2. Minimum horizontal piping length is 3 feet and 1 elbow.

3. A maximum of 5 elbows are permitted in the piping system.

4. Pipe diameter must not be reduced.

5. All horizontal runs must slope upwards not less than 1/4 inch per foot from the furnace to the vent and air intake terminal.

6. All horizontal pipe runs must be supported at least every 4 feet with metal pipe strapping. No sags or dips or low spots are permitted.

7. Do not install the piping system in the same chase with a vent from another gas or other fuel burning appliance.

8. For any sections of the piping system in free air, do not install the pipe within 6 inches of a single wall vent pipe (2 inch B-vent) from another gas or other fuel burning appliance.

9. The piping system can be run in the same chase or adjacent to supply or vent pipe for water supply or waste plumbing.

10. The vent pipe must be insulated if there is any chance of condensate freezing inside the pipe. This can occur if the vent pipe passes through an unconditioned space such as attic, crawl, uninsulated chase or a masonry chimney. It can also occur where the vent terminates above the roof or if an exterior vertical riser (Figure 3) is used to get above snow levels. Local climatic conditions and vent length must be considered. If vent height above roof exceeds 12 inches because of snow accumulation, it must be insulated.

FIGURE 3

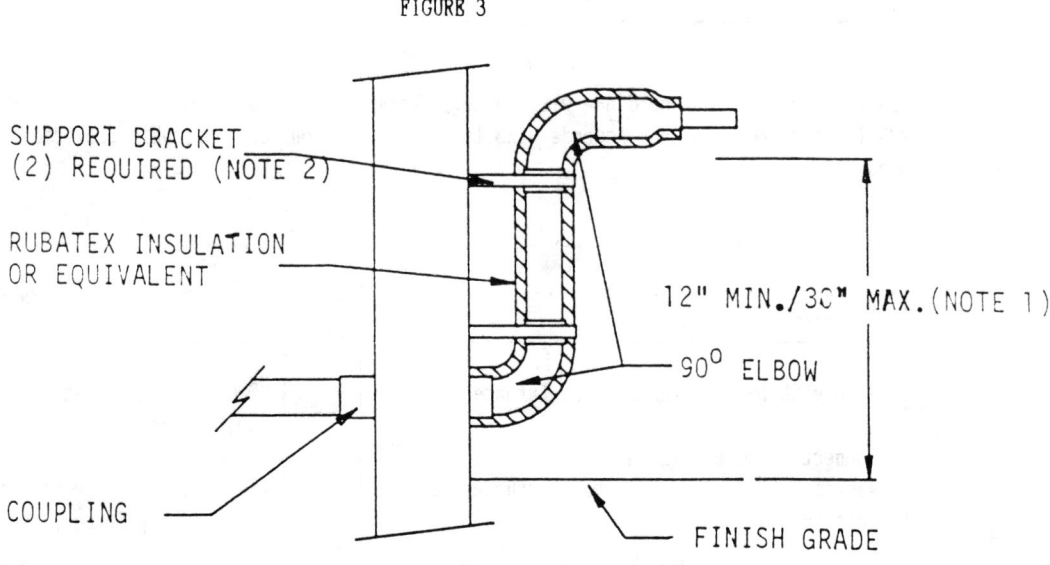

(NOTE 1) If installation requires rise greater than 30" above grade, an alternate exit location or interior vertical vent/air intake system should be used.

(NOTE 2) Brackets should be designed to securely hold and position riser piping to prevent movement in any direction, including back towards wall.

IMPORTANT: Short cycling conditions by pressure switch operation can be created by vent system installations that have too much restriction (pressure drop) because of too many elbows, too much equivalent feet length, not the proper slope on horizontal runs, sag or low spots in horizontal piping sections creating condensation collection points, or incorrect vent terminal application or location.

Types Of Insulation

FOR INDOOR OR OUTDOOR USE
Armaflex closed cell foam or equal. Recommended thickness is one inch. Additional layers may be required for extreme cold climate conditions.

FOR INDOOR USE ONLY
Fiberglass insulation with vapor barrier, or equal. Recommended thickness of 1 inch up to 10 feet. 2 inch thickness if unconditioned exposure exceeds 10 feet.

Connecting Piping System To The Furnace

The furnace is shipped from the factory with a 2" PVC street ell installed in the furnace vestibule.

The vent is constructed in the field by installing the 2" PVC or ABS pipe into the 2" PVC street ell connected to furnace. No-hub connectors are recommended at top of furnace for ease of any service requirements at a later date.

The air intake pipe connects directly to the burner box by being placed over a short section of aluminum tubing crimped to the burner box.

See Figure 4 for details on making both connections to the furnace.

See Figure 5 for details on upsizing the piping system as required by Table 3.

See Figure 6 for details on how to use two 45° elbows to offset air intake pipe to provide clear access to add-on air conditioning coil refrigerant lines and condensate drain connections.

FIGURE 4

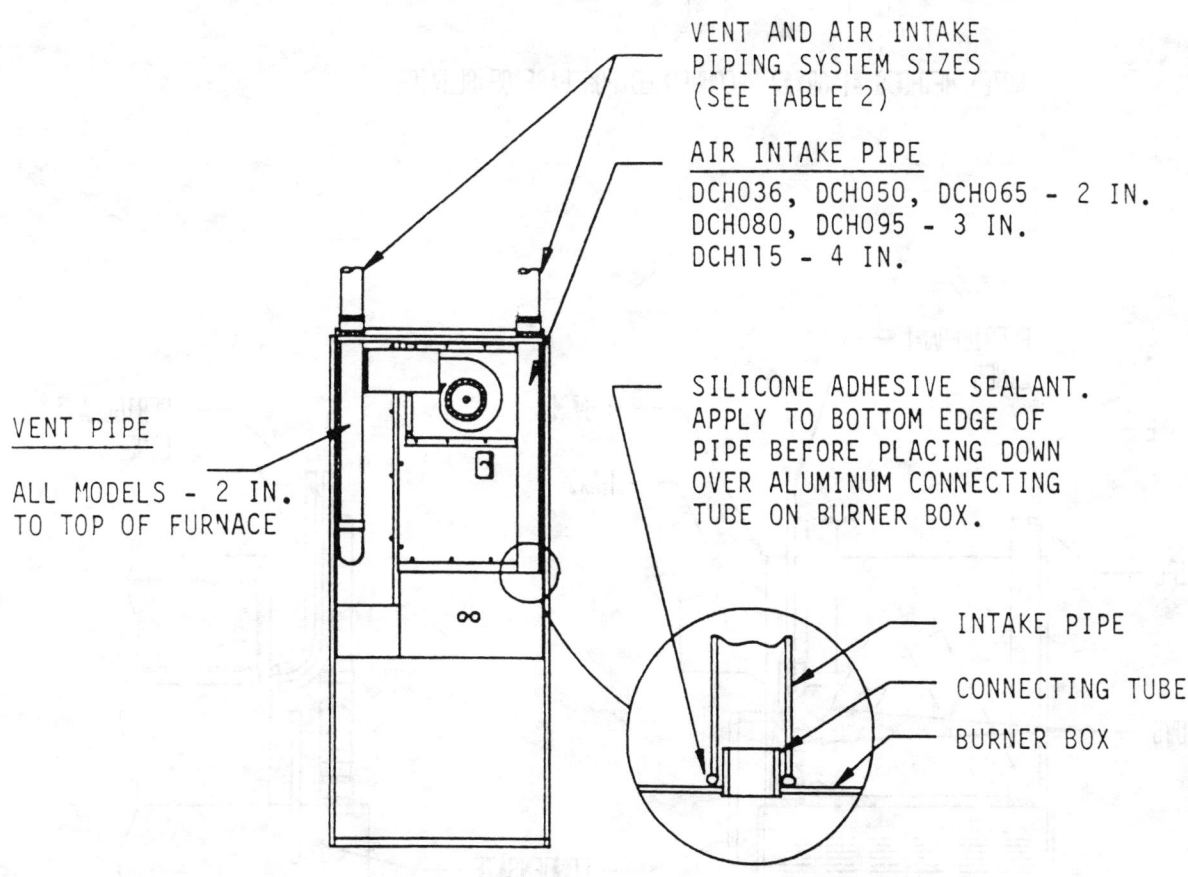

VENT AND AIR INTAKE
PIPING SYSTEM SIZES
(SEE TABLE 2)

AIR INTAKE PIPE
DCH036, DCH050, DCH065 - 2 IN.
DCH080, DCH095 - 3 IN.
DCH115 - 4 IN.

SILICONE ADHESIVE SEALANT.
APPLY TO BOTTOM EDGE OF
PIPE BEFORE PLACING DOWN
OVER ALUMINUM CONNECTING
TUBE ON BURNER BOX.

VENT PIPE

ALL MODELS - 2 IN.
TO TOP OF FURNACE

INTAKE PIPE

CONNECTING TUBE

BURNER BOX

FIGURE 5

RECOMMENDED METHODS FOR VENT AND AIR INTAKE
PIPE UPSIZING AT TOP OF FURNACE

METHOD #1   METHOD #2   METHOD #3

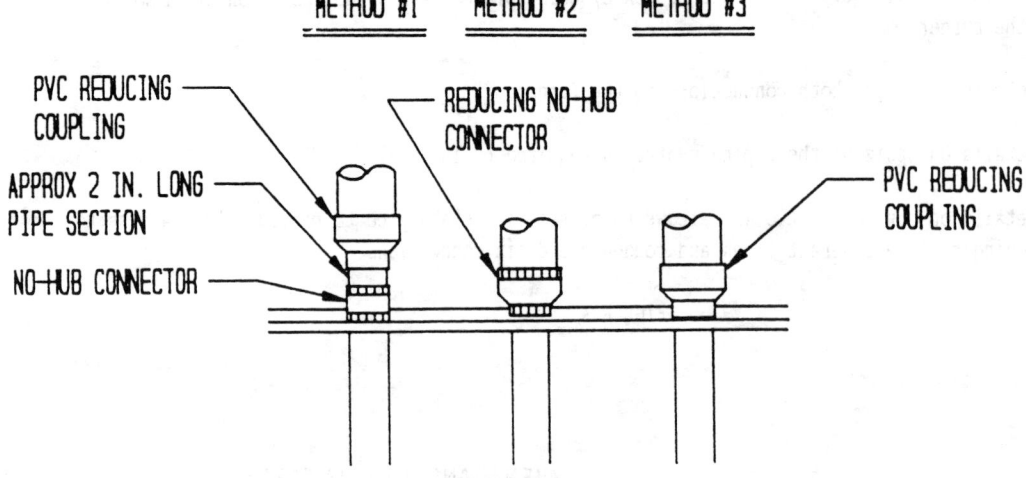

NOTE: METHODS #1 OR #2 RECOMMENDED FOR EASE OF SERVICE

FIGURE 6

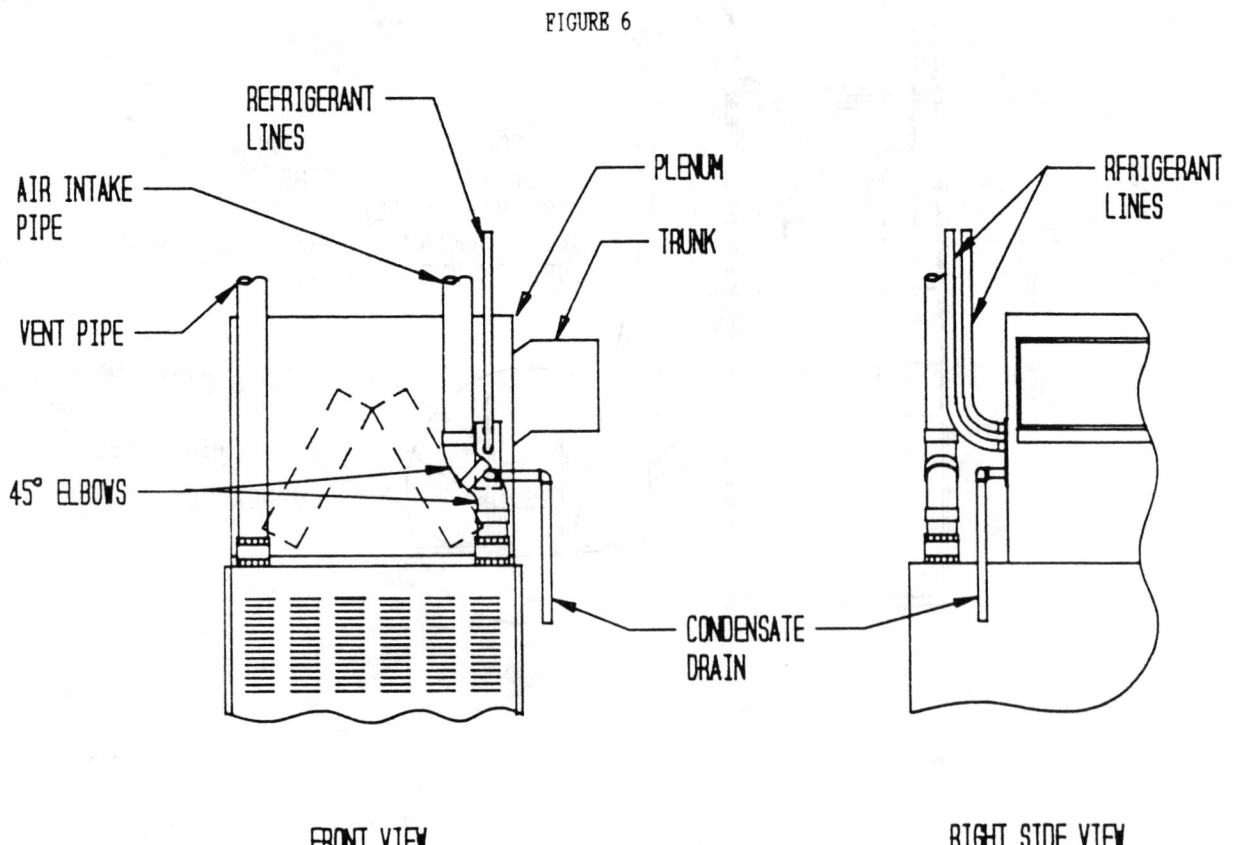

FRONT VIEW   RIGHT SIDE VIEW

SECTION 11 -- JOINING PIPE AND FITTINGS

All pipe, fittings, solvent cement, primers and procedures must conform to American National Standard Institute and American Society for Testing and materials (ANSI/ASTM) standards.

Pipe and Fittings--All connections must be made using cleanr-primer and all purpose solvent cement rated to ASTM Standards D-2564, D-2846, and D-2235.

WARNING

- Danger of fire or bodily injury.

- Solvent cements and primers are highly flammable.

- Provide adequate ventilation and DO NOT assemble near heat source or open flame. DO NOT smoke.

- Avoid skin or eye contact. Observe all cautions and warnings printed on material containers.

All joints in the piping system must be properly sealed using the following material and procedure.

IMPORTANT

FOR PROPER INSTALLATION:

DO NOT use solvent cement that has become curled, lumpy or thickened.

DO NOT thin. Observe shelf precautions printed on containers.

For applications below 40° F use only low temperature type solvent cement.

---

1. Cut pipe end square, remove ragged edges and burrs. Chamfer end of pipe, then clean fitting socket and pipe joint area of all dirt, grease or moisture.

2. After checking pipe and socket for proper fit, wipe socket and pipe with cleaner-primer. Apply a liberal coat of primer to inside surface of socket and outside of pipe. DO NOT ALLOW PRIMER TO DRY BEFORE APPLYING CEMENT.

3. Apply a thin coat of cement evenly in the socket. Quickly apply a heavy coat of cement to the pipe end and insert pipe into fitting with a slight twisting movement until it bottoms out.

   NOTE: Cement must be fluid, if not, recoat.

4. Hold the pipe in the fitting for 30 seconds to prevent the tapered socket from pushing the pipe out of the fitting.

5. Wipe all excess cement from the joint with a rag. Allow 15 minutes before handling. Cure time varies accordingly to fit, temperature and humidity.

    NOTE: Stir the solvent cement frequently while using. Use a natural bristle brush or the dauber supplied with the can. The proper size is one inch.

SECTION 12 -- VERTICAL VENTING

A typical vent installation is shown in Figure 7.

1. When vent penetrates through the roof and is brought above anticipated snow level, the pipe should be cut off on a 45° angle. This will help prevent freezing and blockage of the vent system.

2. The air intake pipe should have two 90° elbows installed with opening on bottom to prevent water entrance. A bird screen is provided to insert into open end of elbow.

3. Vent pipe running through an unconditioned space must be insulated. See "Types of Insulation" in Section 10.

4. An interior masonry chimney can be used as a chase as long as steps 8 and 9 under General Instructions are followed.

5. If an exterior masonry chimney is used as a raceway or chase, the vent piping must be insulated. See "Types of Insulation" in Section 10. If a B-vent for a water heater is also run up through the same chimney, a minimum of one inch clearance from the insulation to the B-vent must be maintained at all points.

6. For either an interior or exterior chimney, the top of the chimney must be sealed with a metal cap to prevent cold air from blowing into the chimney and surrounding the vent pipe(s).

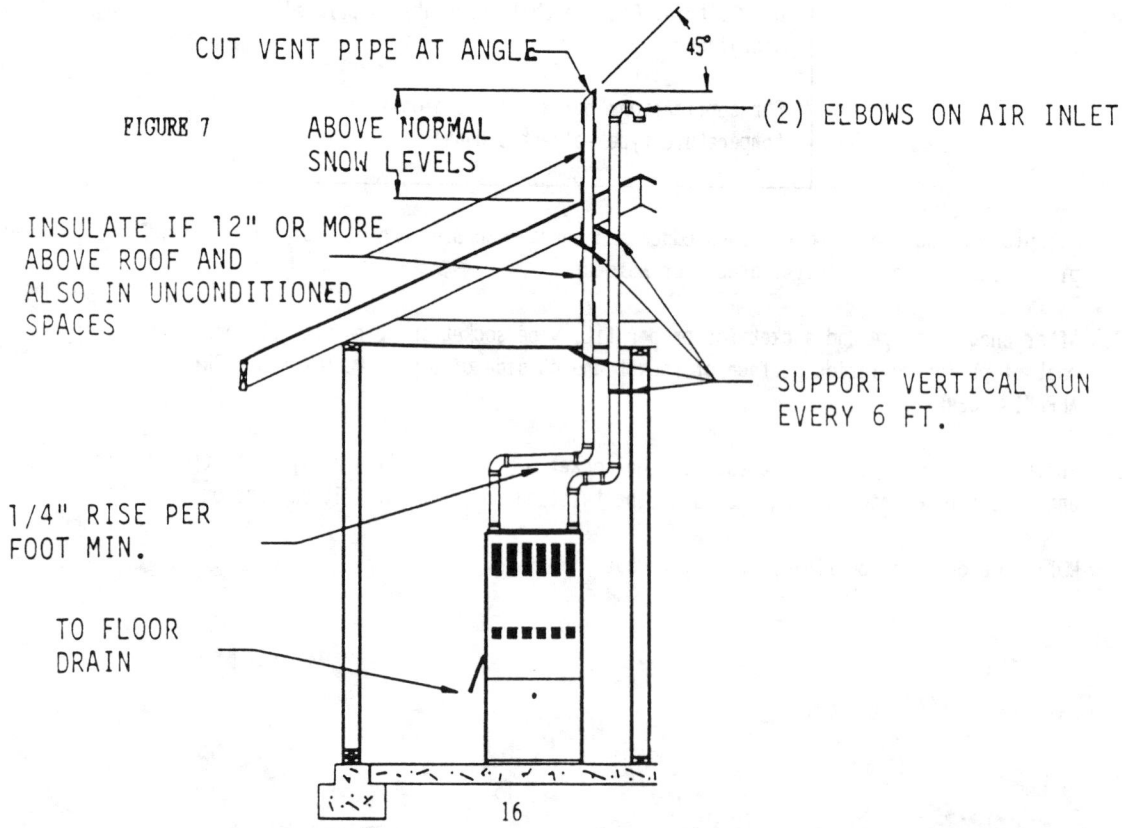

CUT VENT PIPE AT ANGLE — 45°

FIGURE 7   ABOVE NORMAL SNOW LEVELS

(2) ELBOWS ON AIR INLET

INSULATE IF 12" OR MORE ABOVE ROOF AND ALSO IN UNCONDITIONED SPACES

SUPPORT VERTICAL RUN EVERY 6 FT.

1/4" RISE PER FOOT MIN.

TO FLOOR DRAIN

16

SECTION 13 -- HORIZONTAL VENTING

The furnace may be vented horizontally through an outside wall, using all of the applicable instructions under Vent Pipe Installation with these additional requirements. The requirements and limitations for Horizontal Venting are very strict. ALL HORIZONTAL VENT INSTALLATIONS MUST BE MADE IN ACCORDANCE WITH THESE INSTRUCTIONS.

Vent Air Intake Terminal Location

The vent and air intake terminal location must meet the requirements listed in the following instructions or applicable codes, whichever specifies the most clearance or strictest limitations. See Table 3 for sizing requirements.

IMPORTANT	CAUTION
The combustion products and moisture in the flue gases may condense as they leave the terminal fitting. The condensate may freeze on the exterior wall, under the eaves and on surrounding objects. Some discoloration to the exterior of the building may occur.	As a natural part of the unit's operation, normal products of combustion, including water vapor are vented to the atmosphere. Since the outside air temperature can be well below 32°F, it is possible that the water vapor in the exhaust will freeze, causing an ice buildup around the discharge opening of the pipe. During periods of extremely cold weather and prolonged operation of the furnace, this ice build-up could become quite large. The manufacturer does not recommend the installation of these units in locations above frequent vehicular and/or pedestrian traffic. The ice build-up could present a potentially hazardous situation if it becomes dislodged. The manufacturer will NOT be held responsible for any injury or property damage resulting from any improper installation.

Location Requirements--Horizontal (Sidewall) Installation

The vent and air intake terminal must be installed with the following minimum clearances and requirements:

1. The vent and air intake terminals must be located not less than 10 inches or more than 20 inches on centerline. The two terminals must be in a side by side relationship to one another within this requirement. See Figure 8.

2. Bird screens are provided for vent and air intake terminals. See Figure 9.

3. 12 inches minimum above ground level, and above anticipated normal snow levels. See Figure 9 and 16.

   NOTE: Ice or snow may cause the furnace to shut down if the vent or air intake becomes obstructed. If required, use a vertical riser or shield vent and air intake to prevent blockage from drifting snow. See Figure 3. All fittings and pipe for riser must be included in total equivalent feet calculation. Refer to Section 10, General Instructions, Item 1.

17

4. Not above the walkway or area that may cause a hazard or nuisance or be detrimental to the operation of other equipment.

5. 4 feet minimum from and not above or below any door, window, gravity inlet or forced air inlet for the building.

6. 4 feet minimum horizontal clearance from electric meters, gas meters, regulators, and relief equipment.

7. At least 4 feet from any soffit or under eave vent.

8. Do not vent under any kind of patio or deck.

9. Locate vent and air intake terminal on the side of the building away from prevailing winter winds when practical.

10. Do not locate too close to shrubbery as condensate may stunt or kill them.

11. Caulk all cracks, seams, and joints within 3 feet of vent.

TABLE 3 — VENT AND AIR INTAKE TERMINALS (See Figure 9)

| Pipe Size (A) Note 1 | Air Intake Terminal | | Vent Terminal | | |
	Tee (B)	Bird Screen (C)	Reducing Coupler (D)	Nozzle Tube (E)	Bird Screen (F)
2"	2"	2-3/8" O.D.	2" 1-1/2"	1-1/2" x 4"	2-3/8" O.D.
2-1/2"	2-1/2"	2-7/8" O.D.	2-1/2" x 2"	2" x 4"	2-1/2" O.D.
3"	3"	3-1/2" O.D.	3" x 2"	2" x 4"	3-1/2" O.D.
4"	4"	4-1/2" O.D.	4" x 3"	3" x 4"	4-1/2" O.D.

Note 1: Pipe size determined from Table 2.

FIGURE 8
VENT AND AIR INTAKE TERMINAL ORIENTATION

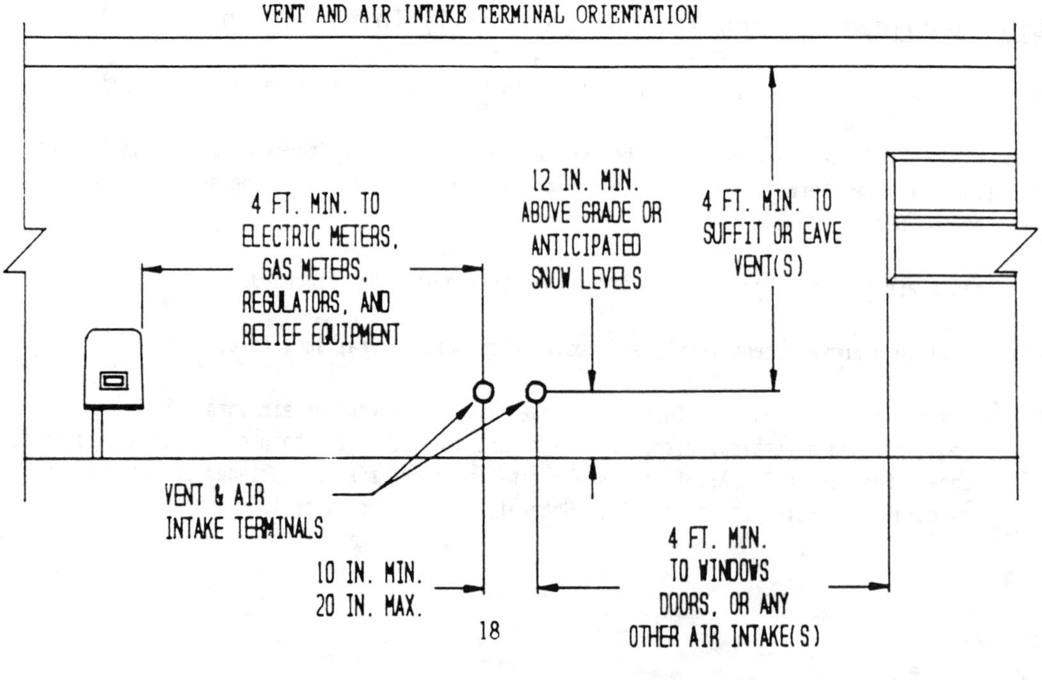

4 FT. MIN. TO ELECTRIC METERS, GAS METERS, REGULATORS, AND RELIEF EQUIPMENT

12 IN. MIN. ABOVE GRADE OR ANTICIPATED SNOW LEVELS

4 FT. MIN. TO SOFFIT OR EAVE VENT(S)

VENT & AIR INTAKE TERMINALS

10 IN. MIN. 20 IN. MAX.

4 FT. MIN. TO WINDOWS DOORS, OR ANY OTHER AIR INTAKE(S)

FIGURE 9

AIR INTAKE TERMINAL

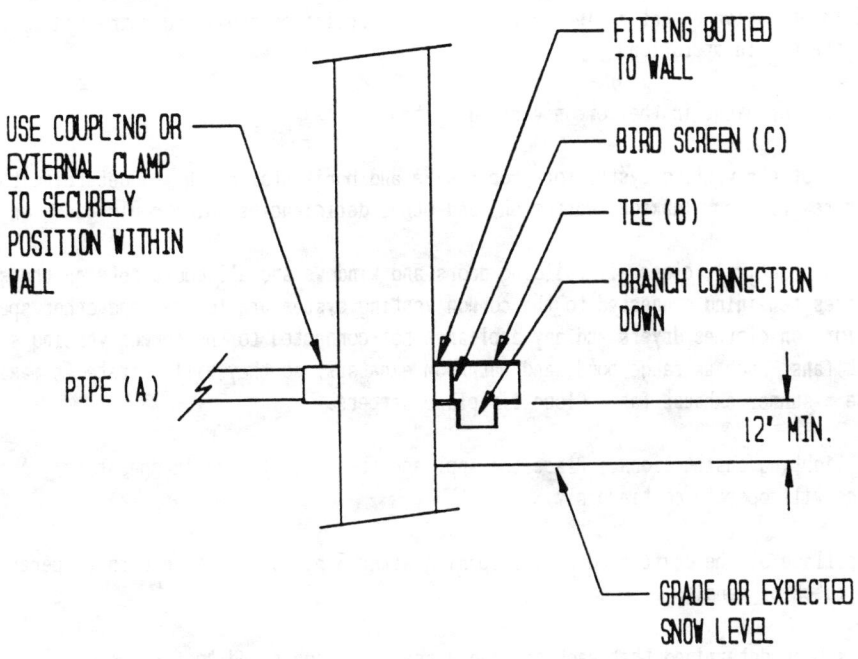

USE COUPLING OR
EXTERNAL CLAMP
TO SECURELY
POSITION WITHIN
WALL

PIPE (A)

FITTING BUTTED
TO WALL

BIRD SCREEN (C)

TEE (B)

BRANCH CONNECTION
DOWN

12" MIN.

GRADE OR EXPECTED
SNOW LEVEL

VENT TERMINAL

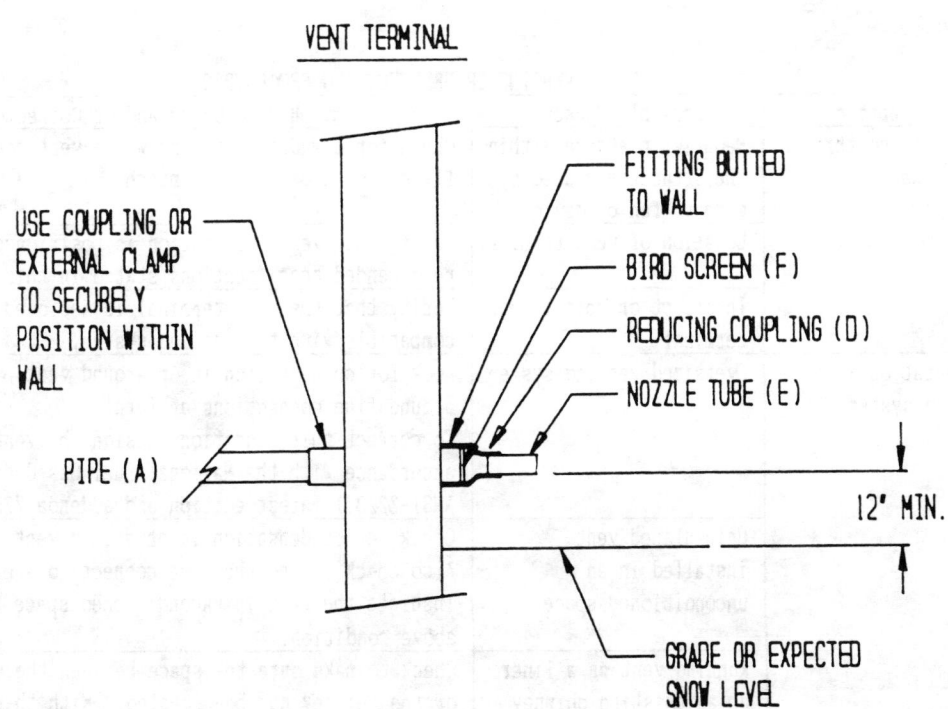

USE COUPLING OR
EXTERNAL CLAMP
TO SECURELY
POSITION WITHIN
WALL

PIPE (A)

FITTING BUTTED
TO WALL

BIRD SCREEN (F)

REDUCING COUPLING (D)

NOZZLE TUBE (E)

12" MIN.

GRADE OR EXPECTED
SNOW LEVEL

SECTION 14 -- VENTING RESIZING INSTRUCTIONS (REPLACEMENT FURNACES ONLY)

When an existing furnace is removed from a venting system serving other appliances, the venting system is likely to be too large to properly vent the remaining attached appliances.

The following steps shall be followed with each of the appliances remaining connected to the common venting system, placed in operation one at a time while the other appliances remaining connected to the common venting system are not in operation.

1. Seal any unused openings in the common venting system.

2. Visually inspect the venting system for proper size and horizontal pitch and determine there is no blockage or restriction, leakage, corrosion, and other deficiencies which could cause an unsafe condition.

3. Insofar as is practical, close all building doors and windows and all doors between the space in which the appliances remaining connected to the common venting system are located and other spaces of the building. Turn on clothes dryers and any appliance not connected to the common venting system. Turn on any exhaust fans, such as range hoods and bathroom exhausts, so they will operate at maximum speed. Do not operate a summer exhaust fan. Close fireplace dampers.

4. Follow the lighting instructions. Place the appliance being inspected in operation. Adjust thermostat so appliance will operate continuously.

5. Test for spillage at the draft hood relief opening after 5 minutes of main burner operation. Use the flame of a match or candle.

6. After it has been determined that each appliance remaining connected to the common venting system properly vents when tested as outlined above, return doors, windows, exhaust fans, fireplace dampers and any other gas-burning appliance to their previous conditions of use.

7. If improper venting is observed during any of the above tests, the common venting system must be corrected.

VENTING TROUBLESHOOTING PROCEDURES

Symptoms	Possible Causes	How To Check And/Or Correct
Downdrafting through the furnace	Negative pressure within the structure caused by exhaust fan of device	Check for downdraft in vent where vent connects to unit with flame from a candle or a match.
	Location of vent terminal	Verify that vent termination is positioned with the recommended specifications stated in this manual.
	Incorrect or absent vent terminal	Verify that the vent terminal is designed for and is compatible with the venting system.
Condensation in venting system	Oversized venting system	Look for condensation in or around vent pipe joints or around flue connections at furnace. To correct this condition, design the venting system in accordance with the National Fuel Gas Code, ANSI-Z223.1-latest edition and addenda Z223.1a latest edition
	Uninsulated vent installed in an unconditioned space	Check for condensation in or around vent pipe joints. Also check around the flue connecting areas on the unit. Insulate the vent in unconditioned space to prevent the above condition.
	Running vent as a liner up an existing chimney without capping off the chimney	Check to make sure the space between the vent and the inside of the chimney has been sealed. With this space being open, it may allow too much cold air in and around the vent pipe allowing the flue products to condense causing condensation back at the unit.

20

SECTION 15 -- CONDENSATE DRAIN

The drain tubing that is provided with this unit is for the purpose of removing condensation from the furnace. A condensate trap is required for operation and is easily obtained when installed as shown in Figure 10. The drain line should slope "downhill" to the drain after exiting the furnace cabinet. Excessive condensate trap (long uphill and/or level runs) can cause the furnace to malfunction.

CAUTION	CAUTION
Do not run drain to an area where temperature may drop below freezing point 32° F. Freezing of condensate could result in property damager or furnace malfunction.	Do not connect drain tube into a drainage system that may become pressurized. Terminate furnace drain tubing into a drainage system that has some type of relief opening to prevent airlocking of furnace drain system.

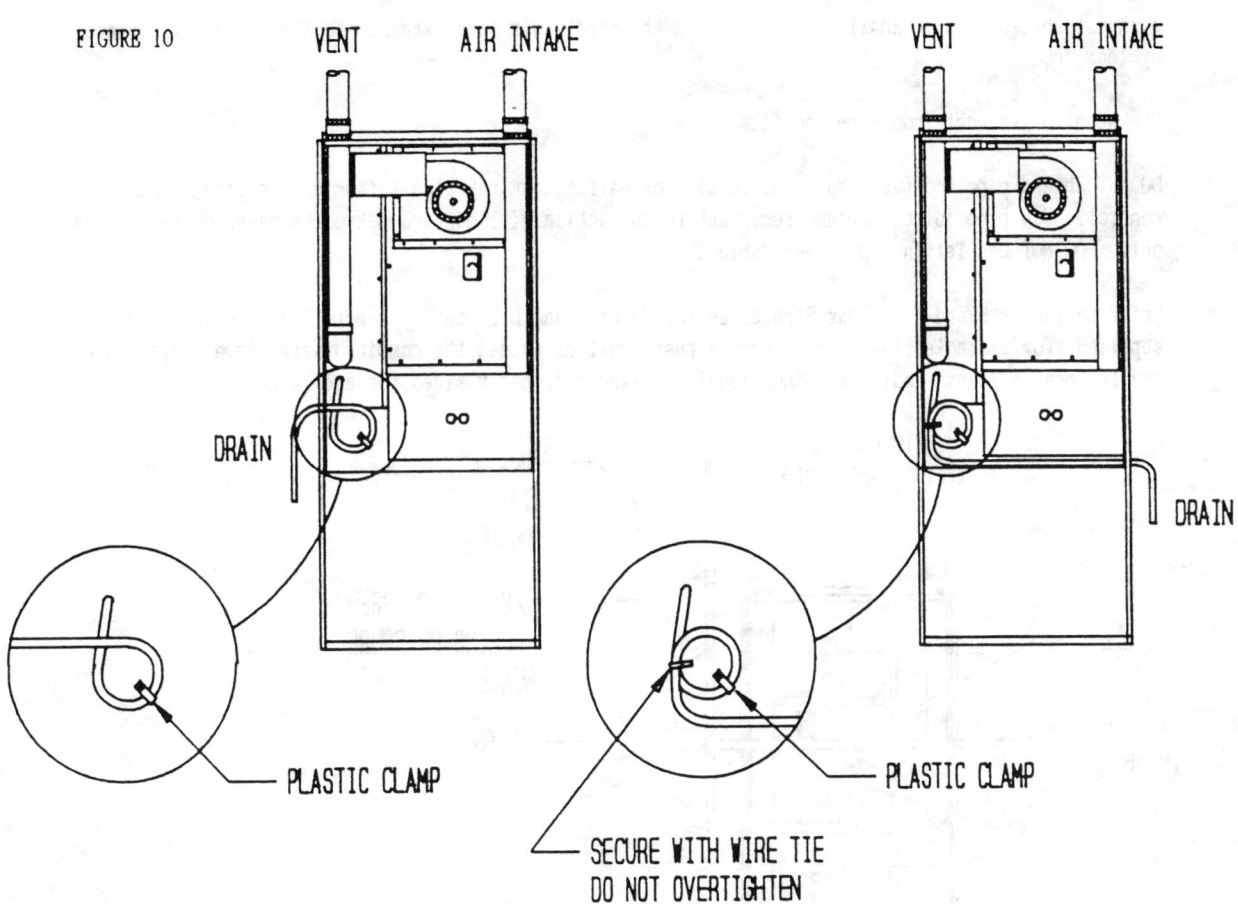

FIGURE 10

SECTION 16 -- CONDENSATE PUMP AND NEUTRALIZER

If no floor drain is available, a condensate pump or sump pump must be used for pumping condensate to the nearest drain. A condensate neutralizer cartridge may be required depending on the type of pump and/or local codes. If a condensate pump is used or if local codes require, install a condensate neutralizer cartridge in the drain line as it exists the furnace.

Neutralizer part No. 8620-031 can be ordered separately as an option. It has a barbed fitting on both ends sized for 3/8" I.D. tubing. It can be installed either vertically or horizontally and should be located somewhere in drain line after it exits the furnace and ahead of condensate pump (if used) or drainage system.

SECTION 17 -- GAS SUPPLY AND PIPING

General Recommendations

1. Be sure the gas line complies with the local codes and ordinances, or in their absence with National Fuel Gas Code, ANSI Z223.1-latest edition.

2. The gas line can be piped to the gas valve from either the left or right side of the furnace through the knockout opening provided. See Figure 11 for typical installation.

3. A sediment trap or drip leg must be installed in the supply line to the furnace.

4. A ground joint union shall be installed in the gas line adjacent to the upstream from the gas valve and downstream from the manual main shut-off valve.

5. A 1/8" N.P.T. plugged tapping accessible for test gauge connection shall be installed immediately upstream of the gas supply connection to the furnace for the purpose of determining the supply gas pressure.

6. A manual shut-off valve shall be installed in the supply gas line external to the furnace when required by local code.

7. Use steel or wrought iron pipe and fittings.

8. DO NOT thread pipe too far. Valve distortion or malfunction may result from excess pipe within the control. Use pipe joint compound resistant to the action of liquefied petroleum gases on male threads only. DO NOT use Teflon tape. See Table 4.

9. Refer to Tables 5 and 6 for Gas Pipe Sizes for Natural and L.P. gas. If more than one appliance is supplied from a single line size, capacity must equal or exceed the combined input to all appliances, and the branch lines feeding the individual appliances properly sized for each input.

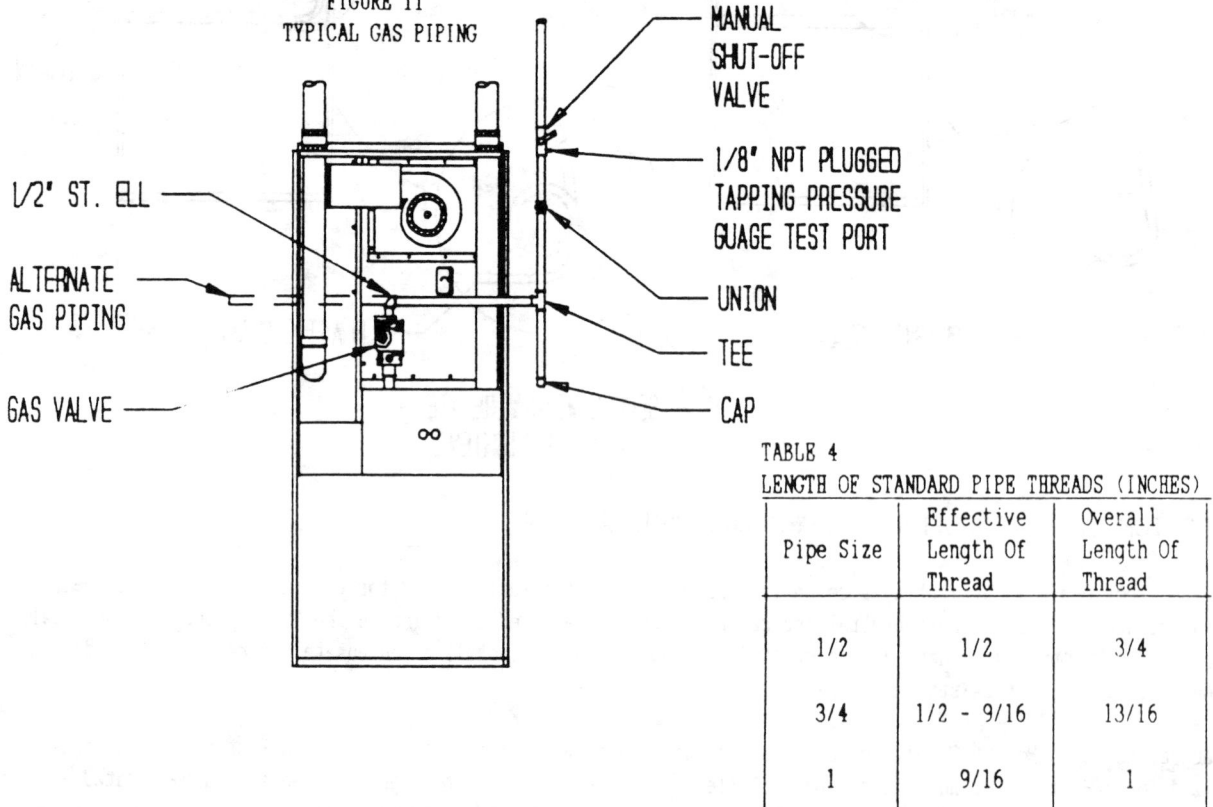

FIGURE 11
TYPICAL GAS PIPING

MANUAL SHUT-OFF VALVE

1/8" NPT PLUGGED TAPPING PRESSURE GUAGE TEST PORT

UNION

TEE

CAP

1/2" ST. ELL

ALTERNATE GAS PIPING

GAS VALVE

TABLE 4
LENGTH OF STANDARD PIPE THREADS (INCHES)

Pipe Size	Effective Length Of Thread	Overall Length Of Thread
1/2	1/2	3/4
3/4	1/2 - 9/16	13/16
1	9/16	1

Checking The Gas Piping

Before turning gas under pressure into piping, all openings from which gas can escape should be closed. Immediately after turning on gas, the system should be checked for leaks. This can be done by watching the 1/2 cubic foot test dial and allowing 5 minutes to show any movement, and by soaping each pipe connection and watching for bubbles. If a leak is found, make the necessary repairs immediately and repeat the above test. The furnace must be isolated from the gas supply piping system by closing the manual shutoff valve on the combination gas control valve during pressure testing of the gas supply piping system at pressures up to 1/2 psig. The furnace must be disconnected from supply piping and supply piping capped during any pressure testing of supply piping system at test pressure in excess of 1/2 psig.

Defective pipes or fittings should be replaced and not repaired. Never use a flame or fire in any form to locate gas leaks, use a soap solution.

After the piping and meter have been checked completely, purge the system of air. DO NOT bleed the air inside the furnace. Be sure to relight all the gas pilots on other appliances that may have been extinguished because of interrupted gas supply.

TABLE 5
GAS PIPE SIZES/CAPACITY
NATURAL GAS

| Length of Pipe--Ft. | Capacity--Btuh Per Hour Input | | |
| | Pipe Size | | |
	1/2"	3/4"	1"
20'	92,000	190,000	350,000
40'	63,000	130,000	245,000
60'	50,000	105,000	195,000

TABLE 6
GAS TUBING AND PIPE SIZES
LP GAS

| Length In Feet | Capacity--BTU Per Hour Input | | | |
| | Copper Tubing* | | Iron Pipe | |
	1/2"**	3/4"**	1/2"	3/4"
20'	62,000	216,000	189,000	393,000
40'	41,000	145,000	129,000	267,000
60'	35,000	121,000	103,000	217,000

*Copper tubing for gas supply must comply with limitation in National Fuel Gas Code, reference "2.6.3 Metallic Tubing".
**Outside diameter.

SECTION 18 -- CHECKING GAS INPUT

NATURAL GAS

The following is a procedure in which to measure gas input:

1. Turn off all gas appliances other than the furnace.

2. From local gas supplier, obtain the average heating value in BTU/CU FT of gas supplied to the installation site.

3. Light furnace following the lighting and operating instructions label.

4. With a stop watch, measure the amount of time, in seconds, it takes to consume two (2) cubic feet of gas.

5. Use the following formula to calculate the gas input of the furnace.

$$\text{Gas input rating in BTU/HR} = \frac{\text{BTU/CU FT} \times 7200}{\text{Time in seconds for two cubic feet of gas flow}}$$

Example: Assume a time of 60 seconds for two cubic feet has been determined with a heating value of 1,000 BTU/CU FT.

$$\text{Gas input rating} = \frac{1000 \times 7200}{60}$$

$$\text{Gas input rating} = 120,000 \text{ BTU/HR}$$

6. If the input rate is not within two percent of the rated input, it may be achieved by adjusting the manifold pressure. If the specified input cannot be obtained, the furnace must be reorificed.

   CAUTION: For operation at elevations above 2,000 feet, the installer must comply with the "Standard Orifice Size and High Altitude Derate" specifications in Section 20.

LP Gas

The pressure regulator on the storage tank must be adjusted to maintain a 11.0" - 13.0" W.C. line supply pressure to the furnace, and the manifold pressure set at 10.0" W.C. on the exit side of the furnace gas valve. This pressure will result in correct input when the burners are orificed properly (based on 2500 Btu/cu. ft. heat content for LP gas).

WARNING
These furnaces must be equipped and operated with the correct orifice sizes and manifold pressures as specified in Sections 19 and 20. Failure to do so could result in property damage, bodily injury, or death.

SECTION 19 -- GAS PRESSURE SPECIFICATIONS

This furnace is equipped with fixed orifices for rated input with the following gas pressures:

TABLE 7

	Supply		Manifold
	Minimum	Maximum	
Natural @ 1,000 BTU/CU FT	4.5" WC	11.0" WC	3.5" WC
LP @ 2,550 BTU/CU FT	11.0" WC	13.0" WC	10.0" WC

The supply pressure may be measured by attaching a water column gauge to a pressure tap adjacent to and up stream from the gas valve. The manifold pressure may be measured by removing the small pipe plug located in the top of the main burner manifold, and inserting a pressure tap. Attach water column gauge to tap and ignite main burner. Small variations in the gas pressure may be made, not to exceed $\pm$ 0.3" W.C., to achieve rated input conditions based on local gas supply Btu conditions for natural gas. See Section 18. This may be achieved by turning the gas valve regulator adjusting screw clockwise to increase pressure or counterclockwise to decrease pressure. Major changes in flow rate must be made by changing the size of the main burner orifices. See Section 20.

Underfire Conditions. Underfire conditions (not setting up furnace installation for full input operation) can also cause condensation and corrosion problems. Natural gas models are to be operated at 3.5 inches W.C. manifold pressure with standard factory supplied orifice sizes. Models converted to L.P. gas are to be operated at 10.0 inch W.C. using the designated orifices from the approved L.P. conversion kit based on 2500 Btu/cu. ft.

It is not acceptable to either reduce manifold pressure or to reduce orifice size to compensate for oversized equipment. It is mandatory that manifold pressure be checked at time of installation/start-up of the furnace. Adjustments, if necessary, must be made to achieve correct manifold pressure as stated above and also on the rating plate on the furnace.

WARNING
Before changing orifices, turn off electrical power and gas. Failure to do so could result in property damage, bodily injury, or death.

25

SECTION 20 -- STANDARD ORIFICE SIZING AND HIGH ALTITUDE DERATE

Rating of gas utilization equipment are based on sea level operation and need not be changed for operation at elevations up to 2,000 feet. For operation at elevations above 2,000 feet and, in the absence of specific recommendations from the local authority having jurisdiction, equipment ratings shall be reduced at the rate of 4 percent for each 1,000 feet above sea level before selecting appropriately sized equipment. (Ref. National Fuel Gas Code ANSI Z223.1 (NFPA 54), latest edition).

These furnaces are shipped with fixed gas orifices for use with Natural Gas and sized for 1000 Btu/cubic feet of gas. Make sure actual furnace gas input does not exceed furnace rating plate input. You may need to change orifices to get correct gas input. Whether you do or not depends on furnace input, your gas heat value at standard conditions and elevation. Consult your local gas supplier for gas heat value and any special derating requirements. Table 8 below gives normal orifice specifications based upon standard conditions as shown. See Table 9 for decimal equivalent for all orifice sizes.

For propane (LP) gas operation, the furnace must be converted using authorized LP Kit sold separately.

EQUIVALENT ORIFICE SIZES AT HIGH ALTITUDES
TABLE 8 — (INCLUDES 4% INPUT REDUCTION FOR EACH 1,000 FEET)

Orifice Size Chart

Fuel Gas Type	Gas Heat Value Btu/Cu. Ft. *	0 to 2000 Feet	2001 to 3000 Feet	3001 to 4000 Feet	4001 to 5000 Feet	5001 to 6000 Feet	6001 to 7000 Feet	7001 to 8000 Feet	8001 to 9000 Feet	9001 to 10,000 Feet
Natural	800 - 849									
	850 - 899									
	900 - 949									
	950 - 999									
	1000 - 1049**									
	1050 - 1100									
Propane (LP)	2500***									

MODEL DCH050D30A — Orifice Size Chart

Fuel Gas Type	Gas Heat Value Btu/Cu. Ft. *	0 to 2000 Feet	2001 to 3000 Feet	3001 to 4000 Feet	4001 to 5000 Feet	5001 to 6000 Feet	6001 to 7000 Feet	7001 to 8000 Feet	8001 to 9000 Feet	9001 to 10,000 Feet
Natural	800 - 849	#37	#39	#40	2.45mm	2.40mm	#42	2.35mm	2.30mm	2.25mm
	850 - 899	2.60mm	2.50mm	2.45mm	#41	2.40mm	2.35mm	2.30mm	#43	2.20mm
	900 - 949	2.50mm	2.40mm	#42	2.35mm	2.30mm	#43	2.25mm	#44	2.15mm
	950 - 999	2.45mm	#42	2.35mm	2.30mm	#43	2.20mm	#44	2.10mm	#45
	1000 - 1049**	3/32"	2.30mm	#43	2.25mm	2.20mm	2.15mm	2.10mm	#45	2.05mm
	1050 - 1100	2.30mm	2.25mm	2.20mm	#44	2.15mm	2.10mm	#45	2.05mm	5/64"
Propane (LP)	2500***	1.45mm	1.40mm	1.40mm	1.35mm	1.35mm	#55	1.30mm	1.30mm	1.25mm

MODEL DCH065D36A — Orifice Size Chart

Fuel Gas Type	Gas Heat Value Btu/Cu. Ft. *	0 to 2000 Feet	2001 to 3000 Feet	3001 to 4000 Feet	4001 to 5000 Feet	5001 to 6000 Feet	6001 to 7000 Feet	7001 to 8000 Feet	8001 to 9000 Feet	9001 to 10,000 Feet
Natural	800 - 849	3.00mm	#32	2.90mm	#34	2.80mm	2.75mm	#36	#37	#38
	850 - 899	#32	#34	2.80mm	2.75mm	#36	#37	2.60mm	#38	2.50mm
	900 - 949	#33	#35	2.75mm	#36	#37	2.60mm	2.55mm	2.50mm	2.45mm
	950 - 999	#35	2.70mm	2.65mm	2.60mm	#38	#39	#40	#41	#42
	1000 - 1049**	#36	2.60mm	#38	2.55mm	2.50mm	2.45mm	2.40mm	2.35mm	2.30mm
	1050 - 1100	#37	#39	#40	#40	#41	#42	2.35mm	2.30mm	2.25mm
Propane (LP)	2500***	1.65mm	1/16"	1.55mm	1.55mm	#53	1.50mm	1.45mm	1.45mm	1.40mm

TABLE 8 (Continued)

MODEL DCHO80D48A — Orifice Size Chart

Fuel Gas Type	Gas Heat Value Btu/Cu. Ft. *	0 to 2000 Feet	2001 to 3000 Feet	3001 to 4000 Feet	4001 to 5000 Feet	5001 to 6000 Feet	6001 to 7000 Feet	7001 to 8000 Feet	8001 to 9000 Feet	9001 to 10,000 Feet
Natural	800 - 849	2.75mm	#37	2.60mm	#38	#39	2.50mm	2.45mm	2.40mm	2.35mm
	850 - 899	#37	#39	#40	2.45mm	2.40mm	#42	2.35mm	2.30mm	2.25mm
	900 - 949	2.60mm	2.50mm	2.45mm	#41	2.40mm	2.35mm	2.30mm	#43	2.20mm
	950 - 999	#39	#41	2.40mm	#42	2.30mm	2.30mm	2.25mm	2.20mm	2.15mm
	1000 - 1049**	2.45mm	#42	2.35mm	2.30mm	#43	2.20mm	#44	2.10mm	#45
	1050 - 1100	2.40mm	2.30mm	#43	2.25mm	2.20mm	#44	2.10mm	2.10mm	2.05mm
Propane (LP)	2500***	#53	1.45mm	1.45mm	1.40mm	1.40mm	1.35mm	1.35mm	1.30mm	1.30mm

MODEL DCHO95D60A — Orifice Size Chart

Fuel Gas Type	Gas Heat Value Btu/Cu. Ft. *	0 to 2000 Feet	2001 to 3000 Feet	3001 to 4000 Feet	4001 to 5000 Feet	5001 to 6000 Feet	6001 to 7000 Feet	7001 to 8000 Feet	8001 to 9000 Feet	9001 to 10,000 Feet
Natural	800 - 849	3.00mm	#32	2.90mm	#34	2.80mm	2.75mm	#36	#37	#38
	850 - 899	2.90mm	2.80mm	7/64"	2.75mm	2.70mm	#37	#38	#39	#40
	900 - 949	#34	2.75mm	2.70mm	#37	2.60mm	#38	#39	#40	#41
	950 - 999	2.75mm	#37	2.60mm	#38	#39	2.50mm	2.45mm	2.40mm	2.35mm
	1000 - 1049**	2.70mm	2.60mm	#38	#39	#40	#41	2.40mm	2.35mm	2.30mm
	1050 - 1100	2.60mm	2.50mm	2.45mm	#41	2.40mm	2.35mm	2.30mm	#43	2.20mm
Propane (LP)	2500***	1.65mm	1/16"	1.55mm	1.55mm	#53	1.50mm	1.45mm	1.45mm	1.40mm

Orifice Size Chart

Fuel Gas Type	Gas Heat Value Btu/Cu. Ft. *	0 to 2000 Feet	2001 to 3000 Feet	3001 to 4000 Feet	4001 to 5000 Feet	5001 to 6000 Feet	6001 to 7000 Feet	7001 to 8000 Feet	8001 to 9000 Feet	9001 to 10,000 Feet
Natural	800 - 849									
	850 - 899									
	900 - 949									
	950 - 999									
	1000 - 1049**									
	1050 - 1100									
Propane (LP)	2500***									

* At standard conditions: Sea level pressure and 60°F temperature.

** Standard factory supplied orifice size.

*** BTU/cu. ft. at 60°F temperature.

TABLE 9 — ORIFICE DRILL SIZE DECIMAL EQUIVALENTS

Drill No.	3.50mm	#29	3.40mm	3.30mm	#30	3.25mm	3.20mm	1/8"	3.10mm	#31	3.00mm	#32	2.90mm	#33
Decimal	.1378	.1360	.1339	.1299	.1285	.1279	.1260	.1250	.1221	.1200	.1181	.1160	.1142	.1130
Drill No.	#34	2.80mm	#35	7/64"	2.75mm	#36	2.70mm	#37	2.60mm	#38	#39	2.50mm	#40	2.45mm
Decimal	.1110	.1102	.1100	.1094	.1082	.1065	.1063	.1040	.1024	.1015	.0995	.0984	.0980	.0964
Drill No.	#41	2.40mm	3/32"	#42	2.35mm	2.30mm	#43	2.25mm	2.20mm	#44	2.15mm	2.10mm	#45	#46
Decimal	.0960	.0945	.0938	.0935	.0925	.0906	.0890	.0885	.0866	.0860	.0846	.0827	.0820	.0810
Drill No.	2.05mm	2.00mm	#47	5/64"	1.95mm	#48	1.90mm	#49	1.85mm	1.80mm	#50	1.75mm	#51	1.70mm
Decimal	.0807	.0787	.0785	.0781	.0767	.0760	.0748	.0730	.0728	.0709	.0700	.0688	.0670	.0669
Drill No.	1.65mm	#52	1.60mm	1/16"	1.55mm	#53	1.50mm	1.45mm	1.40mm	#54	1.35mm	#55	1.30mm	1.25mm
Decimal	.0649	.0635	.0630	.0625	.0610	.0595	.0590	.0570	.0551	.0550	.0531	.0520	.0512	.0492

SECTION 21 -- WIRING SPECIFICATIONS

```
┌─────────────────────────────────────────────┐
│                    WARNING                    │
├─────────────────────────────────────────────┤
│ For your personal safety, turn off electric  │
│ power at service entrance panel before        │
│ making any electrical connections.            │
│                                               │
│ Failure to do so could result in property     │
│ damage, bodily injury, death.                 │
└─────────────────────────────────────────────┘
```

All electrical work must conform with local codes and ordinances, or in their absence, with the National Electrical Code, ANSI/NFPA 70-latest edition.

Electrical Power Supply

Run a separate 120 volt, AC circuit from a separate fuse or circuit breaker in the service entrance panel with an ampacity rating as shown in Table 10. Locate a shut off switch at the furnace. Make connections from this switch to the furnace junction box as shown in the furnace wiring diagram.

TABLE 10

Model	Volts/ HZ/PH	Total Amps	Blower Motor HP	Blower Motor FLA	Inducer Motor HP	Inducer Motor FLA	Minimum Circuit Ampacity	Minimum Time Delay Fuse OR HACR Circuit Breaker
DCH050D30A	115-60-1	6.0	1/3	4.5	1/35	1.4	15	15
DCH065D36A	115-60-1	6.5	1/3	6.5	1/35	1.4	15	15
DCH080D48A	115-60-1	10.5	1/2	10.5	1/35	1.4	15	20
DCH095D60A	115-60-1	12.5	3/4	12.5	1/35	1.4	16	20

Electrical Grounding

RECOMMENDED GROUNDING METHOD

When installed, the furnace must be electrically grounded in accordance with local codes or in the absence of local codes, with the National Electrical Code, ANSI/NFPA No. 70-latest edition. Use #14 AWG copper wire from green ground wire in the field wiring junction box to a grounded connection in the service panel or a properly driven and electrically grounded ground rod.

Field Installed Equipment

Wiring to be done in the field between the furnace and devices not attached to the furnace, or between separate devices which are field installed and located, shall conform with the temperature limitation for Type T wire (63°F rise (36°C)) when installed in accordance with the manufacturer's instructions. Refer to wiring diagrams.

Electronic Air Cleaner and Humidifier Accessories

See Section 29--Accessory Wiring, for specific information on connecting the accessories.

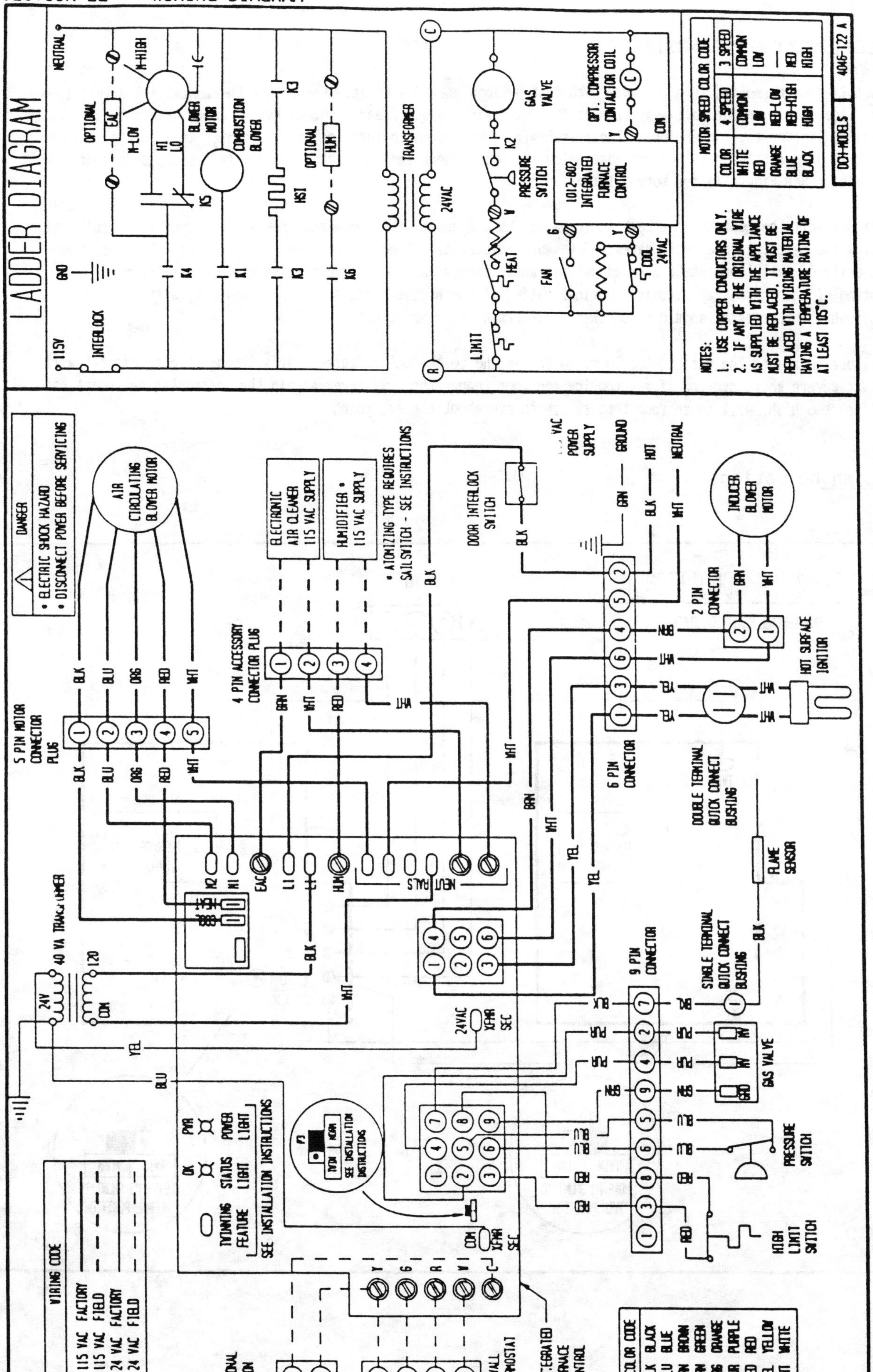
LADDER DIAGRAM
115V
INTERLOCK
GND
NEUTRAL
OPTIONAL
EAC
M-LOW
M-HIGH
HI
LO
BLOWER MOTOR
K5
K4
COMBUSTION BLOWER
K1
K3
HSI
OPTIONAL
HUM
K6
TRANSFORMER
24VAC
GAS VALVE
K2
PRESSURE SWITCH
1012-802 INTEGRATED FURNACE CONTROL
OPT. COMPRESSOR CONTACTOR COIL
COM
HEAT
FAN
COOL
24VAC
LIMIT
NOTES:
1. USE COPPER CONDUCTORS ONLY.
2. IF ANY OF THE ORIGINAL WIRE AS SUPPLIED WITH THE APPLIANCE MUST BE REPLACED, IT MUST BE REPLACED WITH WIRING MATERIAL HAVING A TEMPERATURE RATING OF AT LEAST 105°C.
MOTOR SPEED COLOR CODE
COLOR | 4 SPEED | 3 SPEED
WHITE | COMMON | COMMON
RED | LOW | LOW
ORANGE | MED-LOW | —
BLUE | MED-HIGH | MED
BLACK | HIGH | HIGH
DCH-MODELS
4046-122 A
DANGER
• ELECTRIC SHOCK HAZARD
• DISCONNECT POWER BEFORE SERVICING
AIR CIRCULATING BLOWER MOTOR
5 PIN MOTOR CONNECTOR PLUG
4 PIN ACCESSORY CONNECTOR PLUG
ELECTRONIC AIR CLEANER 115 VAC SUPPLY
HUMIDIFIER • 115 VAC SUPPLY
• ATOMIZING TYPE REQUIRES SAILSWITCH - SEE INSTRUCTIONS
DOOR INTERLOCK SWITCH
115 VAC POWER SUPPLY
GROUND
HOT
NEUTRAL
INDUCER BLOWER MOTOR
2 PIN CONNECTOR
HOT SURFACE IGNITOR
6 PIN CONNECTOR
DOUBLE TERMINAL QUICK CONNECT BUSHING
FLAME SENSOR
SINGLE TERMINAL QUICK CONNECT BUSHING
9 PIN CONNECTOR
GAS VALVE
PRESSURE SWITCH
HIGH LIMIT SWITCH
40 VA TRANSFORMER
120
24V
COM
24VAC XFMR SEC
EAC
L1
HUM
NEUTRALS
OK
STATUS LIGHT
PWR
POWER LIGHT
TWINNING FEATURE
SEE INSTALLATION INSTRUCTIONS
SEE INSTALLATION INSTRUCTIONS
COM XFMR SEC
WIRING CODE
115 VAC FACTORY
115 VAC FIELD
24 VAC FACTORY
24 VAC FIELD
OPTIONAL ADD-ON A/C
WALL THERMOSTAT
INTEGRATED FURNACE CONTROL
COLOR CODE
BLK BLACK
BLU BLUE
BRN BROWN
GRN GREEN
ORG ORANGE
PUR PURPLE
RED RED
YEL YELLOW
WHT WHITE

SECTION 23 --THERMOSTAT

Install the thermostat in accordance with instructions packed with it. Locate the thermostat 4-1/3 feet from the floor on an inside wall away from drafts, warm air registers and floor or table lamps. Refer to furnace wiring diagrams for connections. Thermostat wiring is routed to furnace control through an insulated bushing installed in left side of furnace just below the blower deck and 1/2 inch back from left front corner of furnace. See Figure 1A for more information.

All 24V wall thermostats have heat anticipators to compensate the thermostat for various system controls and allow the best possible cycle rates. Some anticipators are fixed and require no adjustment. However, the majority of wall thermostats have adjustable anticipators and do require adjustment to match the current rating of the thermostat circuit. Nominal rating of thermostat circuit is .35A , however, actual amp draw through the thermostat should be checked to determine heat anticipator setting.

Failure to adjust the anticipator lever to correspond to the actual current draw through the thermostat will cause severe short cycling if set too low and room temperature may never attain the thermostat set point, and if set too high, will cause room temperature to overshoot the set point.

Twinning Options

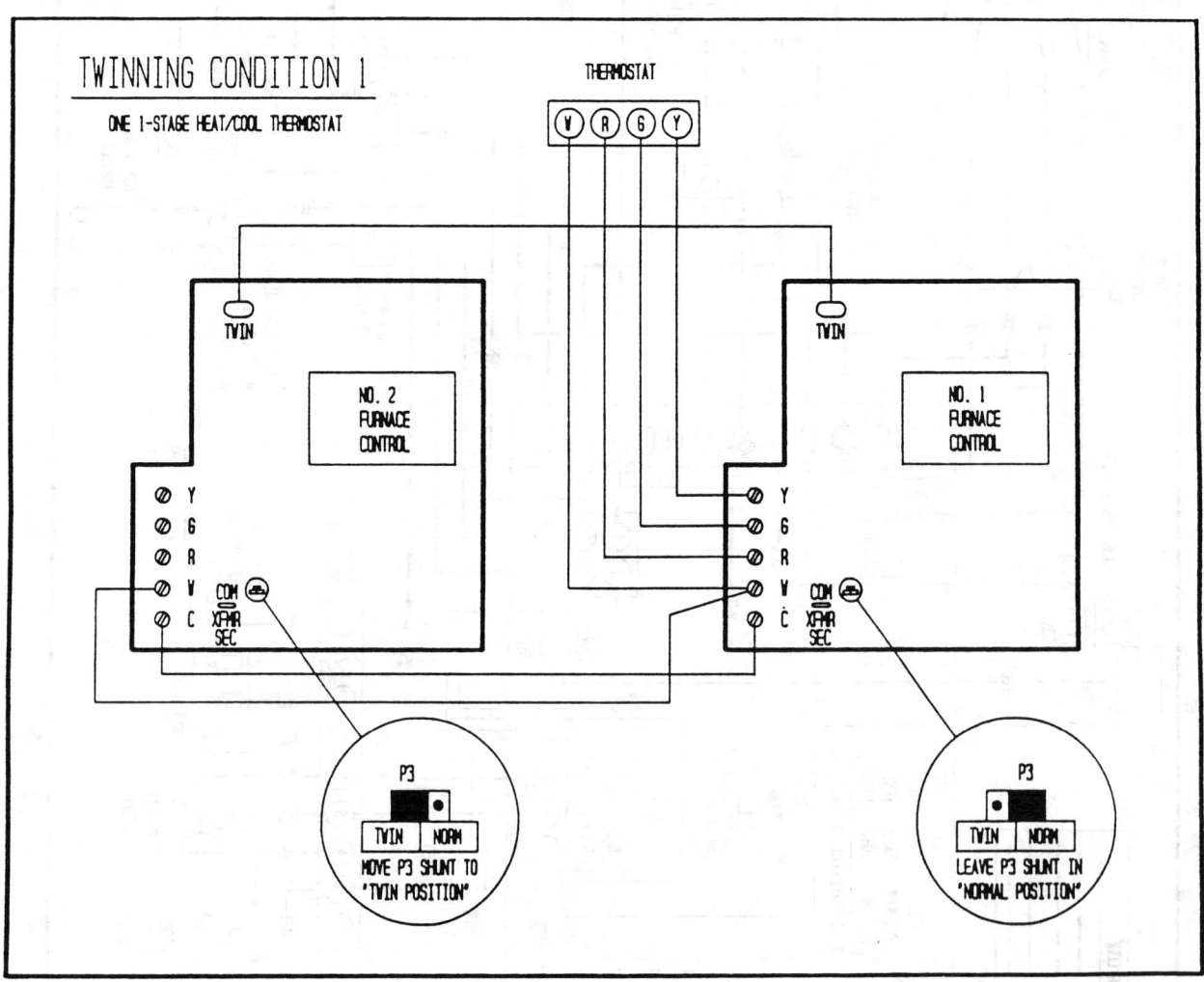

30

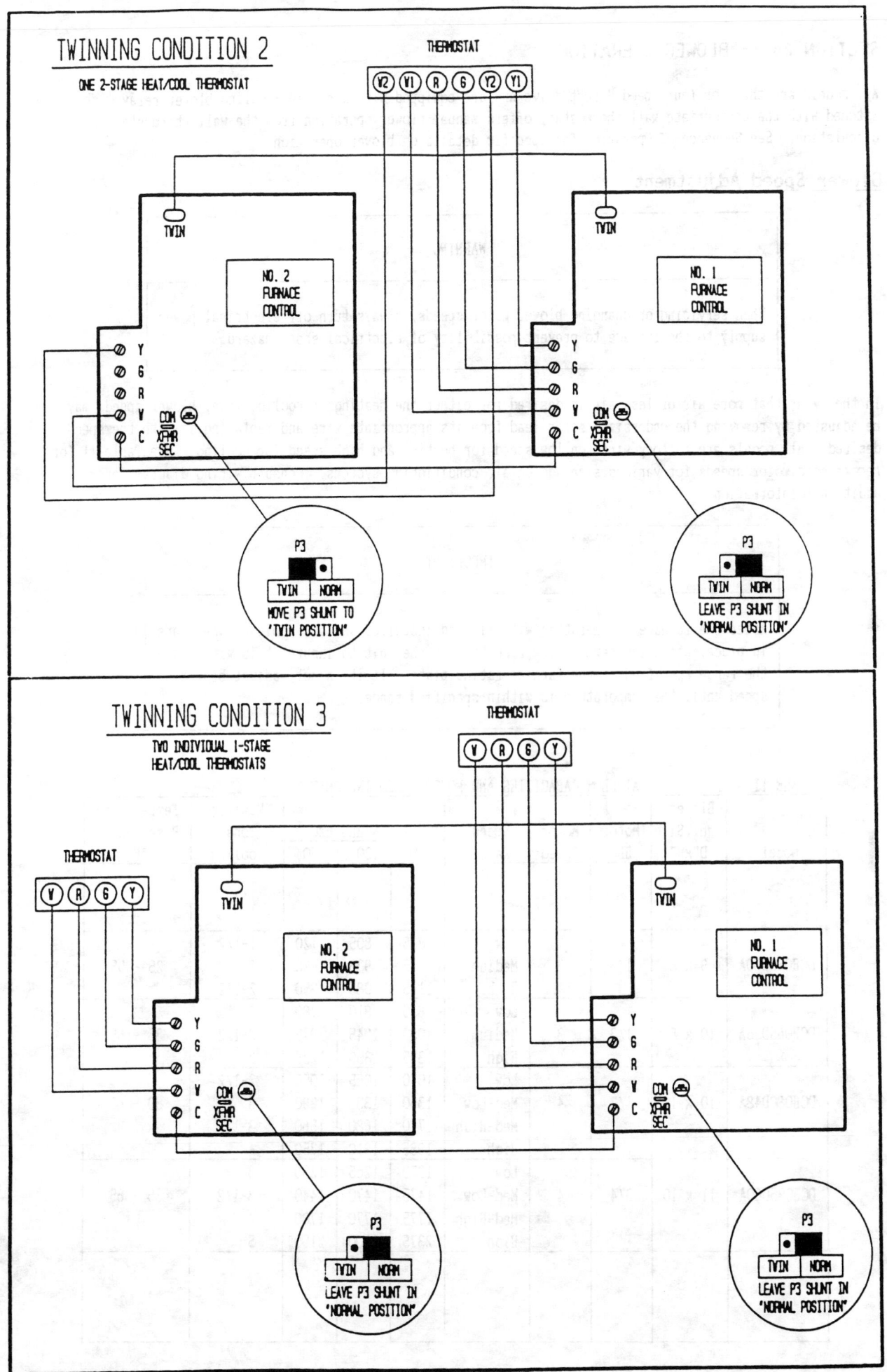

TWINNING CONDITION 2

ONE 2-STAGE HEAT/COOL THERMOSTAT

THERMOSTAT

Y2 Y1 R G Y2 Y1

TWIN

TWIN

NO. 2
FURNACE
CONTROL

NO. 1
FURNACE
CONTROL

Y
G
R
V
C

COM
XFMR
SEC

Y
G
R
V
C

COM
XFMR
SEC

P3

TWIN NORM

MOVE P3 SHUNT TO
'TWIN POSITION'

P3

TWIN NORM

LEAVE P3 SHUNT IN
'NORMAL POSITION'

TWINNING CONDITION 3

TWO INDIVIDUAL 1-STAGE
HEAT/COOL THERMOSTATS

THERMOSTAT

V R G Y

THERMOSTAT

V R G Y

TWIN

TWIN

NO. 2
FURNACE
CONTROL

NO. 1
FURNACE
CONTROL

Y
G
R
V
C

COM
XFMR
SEC

Y
G
R
V
C

COM
XFMR
SEC

P3

TWIN NORM

LEAVE P3 SHUNT IN
'NORMAL POSITION'

P3

TWIN NORM

LEAVE P3 SHUNT IN
'NORMAL POSITION'

31

SECTION 24 -- BLOWER OPERATION

All models are three or four speed direct drive and are equipped with a heating-cooling blower relay. When matched with the appropriate wall thermostat, offers manual blower operation from the wall thermostat for air circulation. See Sequence of Operation for specific details on blower operation.

Blower Speed Adjustment

WARNING
When servicing or changing blower motor speeds, always turn off electrical power supply to the furnace to prevent possibility of electrical shock hazard.

In the event that more air or less air is desired for either the heating or cooling mode, blower speeds may be adjusted by removing the undersized motor lead from its appropriate wire and replacing it with the speed desired. All models are factory wired on low speed for heating and high speed for cooling. See Table 11 for recommended motor speeds for various size add-on air conditioning systems. Consult wiring diagram for additional information.

IMPORTANT
After the furnace is operating with filters installed and all cabinet panels are in place, check the temperature rise through the unit to insure it is within the the range specified on the furnace rating plate. If it is not, adjust blower speed until the temperature is within specified range.

TABLE 11   AIRFLOW CAPACITIES AND MOTOR SPEED INFORMATION

Model	Blower Whl.Size D"xW"	Motor HP	Motor Speed	Speed Range	CFM--INCHES H2O .20	.30	.50	Maximum Tons Cooling	Temperature Rise Range °F
DCH050D30A	9 x 8	1/3	3	Low	835	805	720	1-1/2	25 - 55
				Medium	985	935	820	2	
				High	1145	1085	950	2-1/2	
DCH065D36A	10 x 8	1/3	3	Low	830	810	780	2	35 - 65
				Medium	1060	1045	975	2-1/2	
				High	1395	1350	1235	3	
DCH080D48A	10 x 10	1/2	4	Low	1070	1065	1050	2-1/2	30 - 60
				Med-Low	1350	1335	1290	3	
				Med-High	1720	1680	1560	3-1/2	
				High	1985	1910	1750	4	
DCH095D60A	11 x 10	3/4	4	Low	1270	1265	1250	3	35 - 65
				Med-Low	1475	1470	1440	3-1/2	
				Med-High	1775	1770	1700	4	
				High	2375	2300	2150	5	

SECTION 25 -- AIR FILTERS

All models are shipped with filters and external filter racks that can be installed on either side of the furnace. See Table 12 for sizes and Figures 12 and 13 for installation details.

TABLE 12

FILTER SIZE FOR GAS FURNACE		
Model	Standard Size*	Optional Size*
DCH050D30A		
DCH065D36A	1 - 16x25x1	
DCH080D48A		
DCH095D60A	1 - 16x25x1	1 - 20x25x1 Model FR22

* Shipped with all furnaces.

** Optional filter rack. Must be used on all D60 versions when set-up for 5 ton add-on air conditioning system.

Filter Locations--Removal and Replacement Procedures--All Hi-Boy Furnaces

16 x 25 x 1 external filter racks are supplied as standard equipment for all DCH models. A 14 x 23 opening is to be cut into either the left or right side of the furnace depending upon installation requirements.

See Figure 12 for typical installation of a 16 x 25 x 1 filter rack centered over the 14 x 23 cutout.

Figure 13 shows a typical installation of optional 20 x 25 x 1 filter rack. The same 14 x 23 cutout is required in the furnace side, and the bottom of the filter rack is aligned over the bottom of the 14 x 23 cutout. The top of the filter rack rises approximately 6 inches above the top of the cutout. The 3 inch depth of the filter rack provides ample spacing between furnace side and leaving edge of filter for the entire filter surface to be effective.

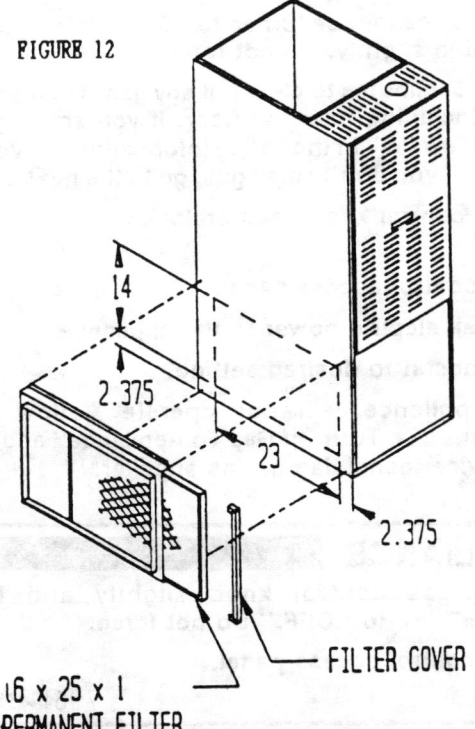

FIGURE 12

16 x 25 x 1
PERMANENT FILTER

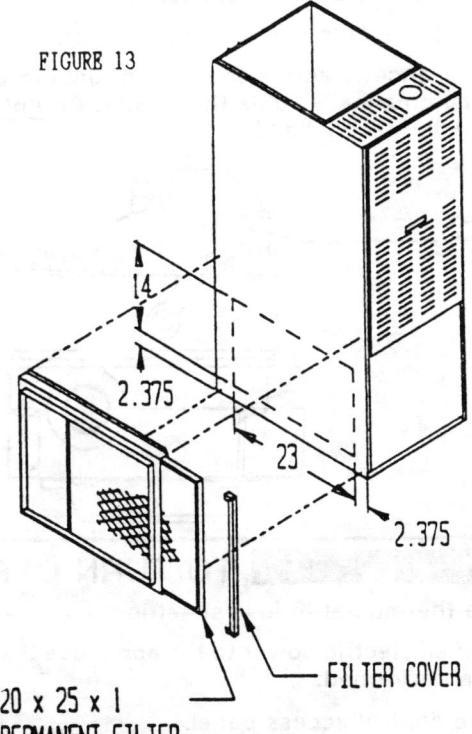

FIGURE 13

20 x 25 x 1
PERMANENT FILTER

To remove filter from the filter rack , remove the filter access cover and grasp the end of the filter. This filter is the permanent high velocity type foam filter which may be washed and used over and over. DO NOT replace it with a fiberglass disposable filter of the same size.

To replace filter, reverse the above procedure.

FOR YOUR SAFETY READ BEFORE OPERATING

> **WARNING:** If you do not follow these instructions exactly, a fire or explosion may result causing property damage, personal injury or loss of life.

A. This appliance does not have a pilot. It is equipped with an ignition device which automatically lights the burner. Do <u>not</u> try to light the burner by hand.

B. BEFORE OPERATING smell all around the appliance area for gas. Be sure to smell next to the floor because some gas is heavier than air and will settle on the floor.

   WHAT TO DO IF YOU SMELL GAS
   • Do not try to light any appliance.
   • Do not touch any electric switch; do not use any phone in your building.
   • Immediately call your gas supplier from a neighbor's phone. Follow the gas supplier's instructions.

   • If you cannot reach your gas supplier, call the fire department.

C. Use only your hand to push in or turn the gas control knob. Never use tools. If the knob will not push in or turn by hand, don't try to repair it, call a qualified service technician. Force or attempted repair may result in a fire or explosion.

D. Do not use this appliance if any part has been under water. Immediately call a qualified service technician to inspect the appliance and to replace any part of the control system and any gas control which has been under water.

OPERATING INSTRUCTIONS

1. STOP! Read the safety information above on this label.

2. Set the thermostat to lowest setting.

3. Turn off all electric power to the appliance.

4. This appliance is equipped with an ignition device which automatically lights the burner. Do <u>not</u> try to light the burner by hand.

GAS CONTROL KNOB SHOWN IN OFF POSITION

5. Remove control access panel.

6. Push in gas control knob slightly and turn clockwise to "OFF."

   NOTE: Knob cannot be turned to "OFF" unless knob is pushed in slightly. Do not force.

7. Wait five (5) minutes to clear out any gas. Then smell for gas, including near the floor. If you smell gas, STOP! Follow "B" in the safety information above on this label. If you don't smell gas, go to the next step.

8. Turn gas control knob counterclockwise to "ON."

9. Replace control access panel.

10. Turn on all electric power to the appliance.

11. Set thermostat to desired setting.

12. If the appliance will not operate, follow the instructions "To Turn Off Gas To Appliance" and call your service technician or gas supplier.

TO TURN OFF GAS TO APPLIANCE

1. Set the thermostat to lowest setting.

2. Turn off all electric power to the appliance if service is to be performed.

3. Remove control access panel.

4. Push in gas control knob slightly and turn clockwise to "OFF." Do not force.

5. Replace control access panel.

7961-366

34

SECTION 27 -- SEQUENCE OF OPERATION

The operation sequence shown below describes normal operation. See Section 28 for detailed explanation of Integrated Furnace Control (IFC), and explanations of abnormal occurences.

1. **BASIC FURNACE--HEATING CYCLE**

   This furnace is operated by an Integrated Furnace Control (IFC) which controls all functions of the furnace.

   On a call for heat from the thermostat, the IFC first checks the pressure switch to verify contacts are open, then turns on the inducer motor. The pressure switch then closes signaling the IFC to proceed with ignition function.

   There is a 36 second ignitor heat-up period, then the gas valve opens introducing main burner gas to be ignited by the hot surface ignitor. The flame sensor proves ignition and carryover across all burners.

   There is a 60 second delay after main burner is on until the comfort air blower starts on heating speed. After the thermostat is satisfied, the burners will go off as gas valve closes. The inducer will continue to run for 30 seconds, and the comfort air blower will continue to run for 120 seconds standard. Can be increased to 180 seconds, see Section 28.

   ### IMPORTANT

   The furnace cannot be recycled from the wall thermostat during the heating blower off delay period.

   A new call for heat can be initiated during this period but the system will "wait" until the blower off delay times out.

   ---

2. **ACCESSORIES (OPTIONAL, FIELD INSTALLED)**

   The furnace and IFC are designed to accommodate both electronic air cleaner and humidifier. The electronic air cleaner circuit is powered any time the comfort air blower runs on either heating or cooling speed. The humidifier circuit is powered anytime the inducer motor is operating. See Section 27 for more information.

3. **COOLING CYCLE OPERATION**

   If optional add-on air conditioning system is installed, the comfort air blower starts on cooling speed immediately on call for cool, will run continuously during the call for cool, and will stop 60 seconds after the thermostat is satisfied.

4. **MANUAL FAN (CONTINUOUS AIRFLOW) OPERATION**

   If wall thermostat is set to MANUAL (ON) position to operate comfort air blower continuously to provide air circulation throughout the building, the blower will operate on the heating speed rather than the cooling as is typical with most systems. This permits the air to circulate as desired but helps keep the operating noise level down as well as conserving energy.

   When all call for heat occurs, the blower will continue to run during the ignitor heat up period, but shut off for 60 seconds after gas valve opens and burners ignite to accelerate the heat exchanger heat-up process. The blower then restarts and runs continuously until the next burner cycle as described above.

   During a call for cooling, the blower automatically shifts up to cooling speed, and remains there until 60 seconds after thermostat is satisfied then drops back to heating speed.

SECTION 28 -- INTEGRATED FURNACE CONTROL (IFC)

The Integrated Furnace Control (IFC) controls all aspects of the furnace operation. Shown below are the IFC specifications covering Ignition Sequence Control, Combustion Blower Control, Comfort Fan Control, Accessory Control, and Diagnostics.

Specific information on TWINNING of furnaces (2 furnaces tied into a large common duct system installation) is detailed under THERMOSTATS--Section 23. This twinning feature assures simultaneous starting and stopping of the comfort air blowers in each furnace which is essential to prevent air short circuiting through one of the furnaces if the blower is only running in the other furnace.

DIAGNOSTICS are provided through 2 light emitting diodes (LED's) mounted on the IFC. A hole is punched in the control box cover, and a clear sight glass is located in the blower access panel lining up with the LED's. Therefore, the LED's can be observed without removing any furnace panels to determine operating status or problem condition that might be present.

IMPORTANT

Be sure to observe the LED signals on the IFC <u>before</u> removing the blower access panel.

Removing the blower access panel opens the blower door interlock safety switch which interrupts power to the furnace, thus removing all LED signaling.

There is no memory retention capability and any problem condition would have to be redetermined by the IFC in due course of time.

Ignition Sequence Control

Ignition Source	115 VAC HSI (Norton 201)
Flame Sensing	Remote
Prepurge	0 seconds
Total Ignition Trials	3 per call for heat
Ignitor Heat Up	1st trial 36 seconds, 2-3 trial 46 seconds
Ignitor De-energized	6 seconds after gas valve opens
Ignition Trial Period	9 seconds from when gas valve opens
Inter Trial Purge	30 seconds
Lockout Purge	30 seconds after 3rd trial lockout
Post Purge	30 seconds
Ignition Sequence Lockout	Reset by cycling thermostat or by internal timer 1 hour from end of lockout purge
Limit Operation Lockout	Forces ignition system to 1 hour lockout w/auto reset. After 2nd automatic reset, control will require manual reset by recycling main power. One hour timing starts when limit closes.
Thermostat Recycle Period	Normal minimum recycle timing is equal to comfort fan OFF delay (120 or 180 seconds). If high limit control has operated the minimum recycle time is 5 minutes after limit resets automatically on cool down (this is a variable time).

Combustion Blower Control

Normal Operation	ON with call for heat (after pressure switch check)
	OFF after postpurge
Limit Action	ON when limit OPEN and flame is sensed
	OFF 10 seconds after limit opens and no flame is sensed
Flame Sense	ON whenever any flame is sensed

Safety Inputs

High Limit/Rollout	SPST in 24 volt circuit
Pressure Switch	SPST, safe start check (60 second proving time)

Comfort Fan Control

HEATING SPEED FAN

Normal Operation

- On Delay — 60 seconds fixed. Timing starts when ignitor de-energized.
- OFF delay — 2 selectable timings -- 120 seconds standard, can be changed to 180 seconds with slide switch located upper left corner of IFC. "ON" = 120, "OFF = 180.

Limit Operation — On when limit OPEN
OFF after OFF delay when limit CLOSES

Flame Sense — ON if flame is sensed <u>and</u> there is no call for heat

COOLING SPEED FAN

- ON Delay — None
- OFF Delay — 60 Seconds

MANUAL FAN — ON continuously on HEATING speed. When call for heat the fan stops when gas valve opens for selected heating ON delay. When call for cool, the fan switches to COOLING speed. Then when thermostat satisfied, the fan switches back to HEATING speed after COOLING OFF delay.

Accessory Control

Electronic Air Cleaner	ON/OFF with comfort fan
Humidifier	ON/OFF with combustion blower*

*Atomizing type humidifier requires use of sail switch

Diagnostics

Green LED
- ON when 24V present
- OFF when no 24V present

Amber LED
- HEARTBEAT when waiting for thermostat call for heat
- ON when call for heat and all functions normal
- 1 FLASH when control cycles through all 3 ignition trials and goes into 1 hour SOFT LOCKOUT(1) (problem could be gas valve, ignitor, flame sensor or fuel supply).
- 2 FLASH when pressure switch does not close or fails safe start test (problem could be inducer, pressure switch, or connecting tubing).
- 3 FLASH when control goes into 1 hour SOFT LOCKOUT(1) due to high limit opening. Can occur 2 times before going into HARD LOCKOUT(2) at 3rd consecutive high limit operation (thermostat not satisfied or manually recycled).
- 4 FLASH if there have been 3 consecutive high-limit cycles and control has gone into HARD LOCKOUT(2).
- 5 FLASH if there is any flame sensed out of normal sequence.
- OFF when there is an internal problem with integrated furnace control.

(1) SOFT LOCKOUT will auto reset 1 hour after lockout occurs or can be manual reset by recycling wall thermostat.
(2) HARD LOCKOUT can only be reset by interrupting 115V main power supply to furnace.

SECTION 29 -- ACCESSORY WIRING

General Information

An optional accessory wiring harness, part No. 8620-044, is available to simplify wiring connections for installation of an electronic air cleaner and/or a humidifier. A 115V power supply is output to these two accessory wiring terminals under the following conditions of unit operation:

115V Electronic Air Cleaner -- Powered whenever the comfort air blower motor is operating on either heating or cooling speed.

115V Humidifier -- Powered whenever the combustion air blower is operating.

Humidifier Notes:

1. Atomizing type humidifiers also require a soil switch to be installed in the comfort air duct system to permit humidifier operation only when comfort air fan is on. The soil switch is wired in series with the RED wire on the accessory wire harness.

2. A 24V humidifier motor or solenoid will require a 115 x 24V transformer. If not included with humidifier, it will need to be supplied separately.

Accessory Wiring Installation

The 8620-044 accessory wiring harness consists of a polarized 4 pin connector with 4 color-coded wires 58 inches long. The 4 pin connector plugs into the back of control box located behind the comfort air blower access panel. The wires are routed through a knockout in left side of blower deck near the 6 pin and 9 pin connectors installed in blower deck, and are long enough to go to either the left or right side of the furnace. See Figure 14.

A junction box should be field supplied and installed on outside of furnace cabinet, and wiring from one or both accessories routed to this junction box for connection to the accessory wiring harness.

The BROWN and one WHITE wires are for the electronic air cleaner, and the RED and one WHITE wire are for the humidifier.

A stain relief is provided to be used where all 4 wires pass through the knockout in the blower deck. Two snap bushings are provided to protect wiring as it passes through the knockout(s) in side panels at junction box location. Two are provided in case humidifier wiring is needed on one side of furnace and air cleaner on the other side. Four (4) wire nuts are provided for connecting to the accessory wiring.

WARNING
Hazard of electric shock. All unused wires must be terminated with wire nuts as they will be energized as described in General Information Section above.

See Figure 14 for pictorial detail of wire routing. Also see Section 8 for additional detail to assure correct knockouts are used.

FIGURE 14

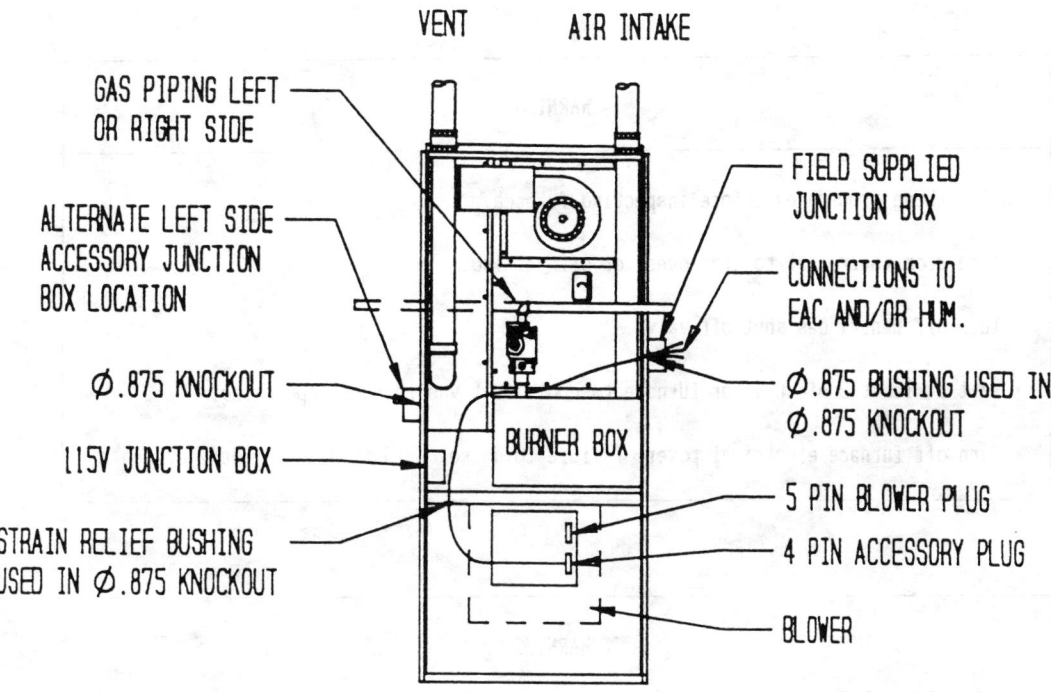

VENT AIR INTAKE

GAS PIPING LEFT OR RIGHT SIDE

ALTERNATE LEFT SIDE ACCESSORY JUNCTION BOX LOCATION

Ø.875 KNOCKOUT

115V JUNCTION BOX

STRAIN RELIEF BUSHING USED IN Ø.875 KNOCKOUT

FIELD SUPPLIED JUNCTION BOX

CONNECTIONS TO EAC AND/OR HUM.

Ø.875 BUSHING USED IN Ø.875 KNOCKOUT

BURNER BOX

5 PIN BLOWER PLUG

4 PIN ACCESSORY PLUG

BLOWER

SECTION 30 -- ROUTINE MAINTENANCE INSTRUCTIONS

Routine maintenance procedures are the responsibility of the user and are contained in the User's Information Manual. These are briefly outlined below.

A detailed inspection of the furnace and its vent/air intake system should be conducted annually by a qualified service agency, generally prior to the heating season. See Section 31.

Routine Maintenance

1. Air Filters. Check the condition on at least a monthly basis when the furnace is in use and clean or replace whenever it is necessary. Permanent filters must not be replaced with disposable fiberglass filters.

2. Lubrication Requirements. The main blower motor and induced draft blower motor is permanently lubricated, no maintenance required.

3. Periodic Inspection of the Vent and Air Intake. Visual inspection of the vent and air intake for any leaking, sags, dips or defective parts. The vent and intake should also be inspected outside of the structure for any blockage in the openings. If blockage is present, the screens should be cleaned with a vacuum cleaner.

4. Periodic Inspection of Drain Line. The drain line on a condensing furnace should be checked monthly for blockage or freezing of drain condensate. Blockage of drain will prevent furnace operation.

SECTION 31 -- SERVICING THE FURNACE

The following procedures should be performed by a qualified dealer serviceman.

WARNING
Follow these procedures before inspecting furnace. • Turn room thermostat to its lowest or off setting. • Turn off manual gas shut-off valve. • Wait at least 5 minutes for furnace to cool if it was recently operating. • Turn off furnace electrical power; failure to do so could result in injury or death.

WARNING
Use replacement parts listed in parts list. Failure to do so could cause improper furnace operation, resulting in damage, injury or death.

Perform periodic preventative maintenance once before heating season begins and once during heating season, Inspect, clean, and repair as needed following items:

Checking the Circulating Air Blower Assembly

1. Remove blower service panel.

2. Remove 2 screws securing front edge of blower support angles to blower base.

3. Disconnect the 6 pin and 9 pin connectors from blower base on left side of blower.

4. If installed, disconnect 4 pin accessory wiring plug from back of control box.

5. If thermostat wires are not sufficiently long enough inside blower compartment, disconnect from the 24V screw terminal on integrated furnace control.

6. The blower assembly is now free to be pulled from the furnace.

7. Reinstallation is completed by reversing all of the above steps.

Checking the Hot Surface Ignitor

1. Disconnect the 2 YELLOW leads connected to the double 3/16" quick connect terminal bushing on left side of burner box.

2. Using an ohmmeter, check the resistance of the ignitor. Cold resistance should be 45 - 250 ohm. Resistance above 250 ohms could indicate a fracture or hairline crack and ignitor should be replaced.

3. To replace ignitor, remove burner box cover, and then disconnect the 2 ignitor leads from the double 3/16" terminal bushing on inside of burner box.

4. Carefully remove the two screws securing ignitor mounting bracket, and extract ignitor through the open space to left of gas manifold vertical riser.

5. Replace ignitor by removing nut and bolt from bracket, placing new ignitor in bracket, and fastening with nut and bolt.

6. Reverse above steps to install ignitor/bracket assembly back into furnace. The 1/4" tab section on top of ignitor bracket must fit into 1/8" slot of ignitor bracket mounting panel, then secure bracket with the 2 screws.

Checking the Burners

1. Inspect the burner assembly for any scaling, sooting or blockage of ports. If there are signs of this, the burner should be removed and cleaned with a wire brush until all soot and scale is removed and burner ports are not obstructed.

CAUTION
Before removing burners for cleaning, the hot surface ignitor. MUST BE removed first to avoid damaging the ignitor. Failure to do so can result in breakage of the ignitor requiring its replacement.

2. The gas piping to gas valve will usually require disconnection to permit valve and manifold assembly to be unfastened and moved out of the way to allow burner assembly to be extracted from furnace. The split grommet around manifold vertical riser will slide out of the slotted opening on top of burner box.

3. Remove 2 screws on both left and right side burner brackets securing brackets to the interior support plate.

4. The burner is now free to be pulled out and examined and cleaned as outlined in Step 1 above.

5. Reinstall all parts for reversing above steps.

WARNING
The grommet that seals the gas manifold vertical riser to the top panel of the burner box is a critical seal. It must be reinstalled exactly as it was removed to assure proper operation of the pressure switch supervising vent/air intake piping systems. Failure to install properly could result in property damage, bodily injury, or death.

41

Checking the Heat Exchanger and Flue Gas Passageways

Furnaces that are properly installed and maintained will normally not require cleaning of the heat exchangers.

THE ONLY TIME it should be necessary to disassemble and clean the interior of both the Primary and Secondary Heat Exchangers would be due to a sooting conditioning caused by abnormal combustion.

The inside of the primary heat exchanger can be examined for scale and soot using a light and a mirror on an extension handle. If soot and/or scale is evident, the heat exchanger must be cleaned as follows:

1. Remove burner assembly from inlet to heat exchanger. See above section.

2. Disconnect the 2 pin connector for inducer motor, unplug pressure switch tubing from inducer, and carefully remove the 4 screws securing inducer housing to the flue collector box.

3. Carefully work the inducer housing and the 3-inch plastic elbow fitting connected between blower outlet and secondary heat exchanger inlet loose. Both ends of the 3-inch elbow are sealed with a high temperature silicone sealant, and by carefully working the inducer assembly up and away from the flue collector box one or both of the silicone sealant connections will break free.

4. Now the flue collector box can be removed, along with the internal flue baffles.

5. Using a small wire brush on extension handle, brush inside walls of the heat exchanger until soot and/or scale is removed. The loose scale and/or soot is easily removed using a vacuum cleaner at the bottom of the heat exchanger.

6. To clean secondary heat exchanger requires removing the section of vent piping within the furnace. Disconnect the 2-inch no-hub connector securing 2-inch PVC street elbow to outlet of secondary heat exchanger inside the furnace. Take whatever action necessary to then disconnect and remove the vertical 2-inch pipe section rising of the furnace. This is necessary to permit secondary heat exchanger to be pulled out the front opening of furnace.

7. Disconnect wires to pressure switch, and unplug tubing from pressure switch to burner box (unplug at the burner box).

8. Disconnect the drain hose from the drain tube projecting through the cover panel.

9. Remove all the screws around the edge of the sheet metal panel that covers the secondary heat exchanger-- do not remove the 4 screws securing secondary heat exchanger to panel.

10. Gently work the secondary heat exchanger assembly forward out of the furnace being careful not to damage the fins.

11. Once the coil is removed from the furnace, take it to a place where the following steps can be performed.

    A. Set coil on its back so the inlet, outlet and drain openings are pointed up. Pour 2 quarts of hot water into the coil and cap off openings. Shake coil vigorously and pout out water. Repeat this procedure until the water being poured from the coil is clear.

    B. Thoroughly wash off the exterior of the heat exchanger using a soft brush and a mild stream of water. DO NOT use a hard stream of water as this may damage the fins.

12. Reinstallation of all parts is done by reversing all of the above steps.

WARNING	WARNING
High temperature silicone sealant/adhesive rated at least 450°F continuous must be used to reseal the 3-inch elbow fitting to both the inducer blower housing and the secondary heat exchanger. Failure to seal properly could result in improper furnace operation resulting in property damage, bodily injury, or death.	The gas supply piping must be reconnected and checked for leaks. Never use an open flame when testing for gas leaks. Use of an open flame could cause a fire or explosion.

Checking the Heat Exchanger and Flue Gas Passageways

After all inspections are made and the furnace is completely reassembled and all electrical, gas, drain, air intake and vent connections are reconnected and verified, the furnace should be checked out for proper operation.

The general operation should be in accordance with that outlined in Sequence of Operation--Section 27.

Also, observe the main burners in operation. The flame should be mostly "blue" with possibly a little orange (not yellow) at the tips of the flame. The flames should be in the center of the heat exchanger compartments and not impinging on the heat exchanger surfaces themselves. See Figure 15.

FIGURE 15

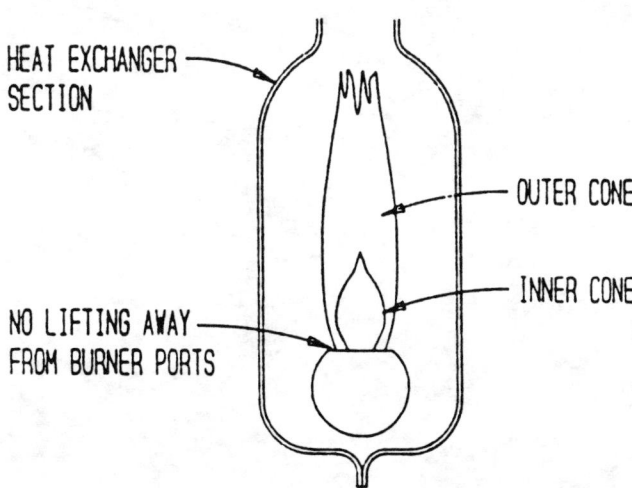

Observe the fire until the blower starts (there is a normal delay period until the heat exchanger warms up). There should be no change in the size or shape of the flame. If there is any wavering or blowing of the flame on blower start-up, it is an indication of a possible leak in the heat exchanger.

SECTION 32 -- REPLACEMENT PARTS

Replacement parts for the gas furnaces are available through local distributor.

Parts list covering all furnace components is shown in the Replacement Parts Manual. When ordering parts or making inquiries pertaining to any of the furnaces covered by these instructions, it is very important to always supply the COMPLETE model number and serial number of the furnace. This is necessary to assure that the correct parts (or an approved alternate part) are issued to the service agency.

USER'S INFORMATION MANUAL
HIGH EFFICIENCY GAS FURNACE

We're pleased you've chosen our gas furnace to supply your heating needs. Please keep this manual in a safe, yet readily available place. It contains important and useful information.

ATTENTION, INSTALLER! After installing furnace, give the user:

--User's Information Manual   --Parts List
--Installation Instructions   --Warranty Information

ATTENTION, USER! Your furnace installer should give you the above four important documents relating to your furnace. Keep these as long as you do your furnace. Pass these documents on to later purchasers or furnace users. If any of the four documents are missing or damaged, contact your installer or furnace manufacturer for replacement. For efficient service, please give your furnace model and serial number, from Section 1 or from your furnace rating plate.

WARNING: If the information in this manual is not followed exactly, a fire or explosion may result causing property damage, personal injury or loss of life.

— Do not store or use gasoline or other flammable vapors and liquids in the vicinity of this or any other appliance.

— WHAT TO DO IF YOU SMELL GAS

  • Do not try to light any appliance.
  • Do not touch any electrical switch; do not use any phone in your building.
  • Immediately call your gas supplier from a neighbor's phone. Follow the gas supplier's instructions.
  • If you cannot reach your gas supplier, call the fire department.

— Installation and service must be performed by a qualified installer, service agency or the gas supplier.

WARNING: Read and follow all safety information in this manual, operating instructions and furnace safety labels. Failure to follow safety precautions could result in damage, injury, or death.

IMPORTANT SAFETY NOTE: You must know how to turn off gas and electricity to furnace. Your qualified installer, service agency or gas supplier can teach you to use controls and switches.

WARNING: Do not use this furnace if any part has been under water. Corrosion can start if electrical and gas control systems become wet. Corrosion can cause gas to leak, which could result in damage, injury or death. Consult a qualified installer, service agency or gas supplier to inspect furnace. Instruct them to replace any part which has been under water.

Thank you for reading these safety statements. Please read on so you will know how to maintain your furnace for years of dependable service.

MANUAL 2100-188	REV.	SUPERSEDES	FILE VOL. 1, TAB 2

Section	TABLE OF CONTENTS	Page

COPYRIGHT MAY, 1992
BARD MANUFACTURING COMPANY
BRYAN, OHIO USA 43506

SECTION 1 -- RATING PLATE INFORMATION

Record the manufacturer's name, furnace model number and serial number below. These are on your furnace rating plate located on blower door. Record installation date, which is important for warranty purposes.

Also fill in the installer's name, address and telephone number. This will be handy if you have questions later. Some companies install an identification tag on furnaces they install or service. If not, ask for the information.

YOUR FURNACE INFORMATION

Furnace Type _____

Manufacturer's Name _____

Model Number _____

Serial Number _____

Date Installed _____

Installer/Servicer _____

Address _____

City/State/Zip Code _____

Telephone Number _____

SECTION 2 -- IMPORTANT SAFETY PRECAUTIONS

A. SIGNAL WORDS

Years of safe, dependable service, are assured when you understand and follow all safety precautions. REMEMBER: Your furnace contains flames, gas, electricity, rotating parts and metal edges.

Signal words **"WARNING"** and **"CAUTION"** alert you to potential hazards.

"WARNING" alerts you to situations that could cause serious injury or property damage.

"CAUTION" alerts you to situations that could cause minor or moderate injury or property damage

B. SAFETY PRECAUTIONS

These are some of our most important safety precautions; others are throughout this manual. Please read and follow them.

1. Gas and Combustion Products.

WARNING: Any condition that will allow gas or combustion products to enter furnace area can cause nausea, asphyxiation or fire resulting in damage, injury or death.

Natural gas and propane (LP) gas have characteristic odors. When your furnace is operating correctly, you should not smell any unfamiliar odor. Normally, burning gas with air produces combustion products which contain carbon dioxide, oxygen and water vapor. Under abnormal conditions, combustion products can contain aldehydes and carbon monoxide.

3

--Aldehydes have a strong pungent, acrid smell that can cause nausea.

--Carbon monoxide is tasteless, colorless and odorless. It can cause headaches, flu-like symptoms or nausea. We refer to all these symptoms as nausea in this manual. It can also cause death by asphyxiation.

WARNING: Any unfamiliar smell can alert you to presence of gas or aldehydes. If you detect any unfamiliar odor, follow instructions in Section 4.B.1, Otherwise, nausea, asphyxiation or fire could occur, resulting in damage, injury or death.

WARNING: Do not block or cover combustion openings in the furnace door or closet door. Blocking or covering these openings could cause nausea, asphyxiation or fire resulting in damage, injury or death.

WARNING: Do not block or cover any openings from outside the furnace area which supply combustion and ventilation air to your furnace. Keep insulation away from these openings. Blocking or covering these openings could cause nausea, asphyxiation or fire, resulting in damage, injury or death.

WARNING: A loud noise may mean faulty burner ignition. If your furnace makes a loud noise, turn it off. Follow instructions in Section 4.B.2. If you don't turn your furnace off, it could cause fire or an explosion, resulting in damage, injury or death.

WARNING: If your furnace is in an attic or other insulated space, keep all insulating materials at least 12 inches away from its burner combustion air openings. Blocking or covering these openings could cause nausea, asphyxiation or fire resulting in damage, injury or death.

WARNING: Do not operate furnace with blower door open or removed. Do not alter furnace to allow operation with blower door removed. Doing either could allow combustion products to circulate throughout the furnace area, causing nausea, asphyxiation or fire resulting in damage, injury or death.

WARNING: Front door must be in place during furnace operation. Hot surfaces behind front door could cause damage or injury.

2. Storage and Use of Flammable, Corrosive and Combustible Products Near Your Furnace.

WARNING: Never store or use flammable liquids or vapors near or on your furnace. These include gasoline, kerosene, cigarette lighter fluid, cleaning fluids, solvents, paint thinners or painting compounds. Flammable vapors can travel great distances before igniting. Flammables could cause fires or explosions and result in damage, injury or death.

WARNING: Never store or use anything near or on your furnace that can produce vapors that are corrosive to gas-fired furnaces. Vapors from products containing chlorines, fluorines, bromines and iodines can cause vent system or heat exchanger failure. Examples of such products are spray or aerosol containers, detergents, bleaches, cat litter, waxes, adhesives, solvents and other cleaning compounds. Vent system or heat exchanger failure could cause nausea, asphyxiation or fire, resulting in damage, injury or death.

WARNING: Never store anything combustible near or on your furnace. These include brooms, dustmops, vacuum cleaners, other cleaning tools or items, plastic or plastic containers, paper bags or other paper products. A fire could occur, resulting in damage, injury or death.

3. Alteration of Furnace Controls.

WARNING: Do not alter any gas or electrical controls (gas control, pilot or safety controls) in any manner. Altering them could cause furnace to operate unsafely, resulting in damage, injury or death.

4

SECTION 3 -- UNDERSTANDING HOW YOUR FURNACE WORKS

Your installer should have given you a detailed explanation of how the furnace operates. Shown below are the basic operating characteristics and sequence of operation. If you have any questions, consult your installer/service agency.

IMPORTANT

There are many types of thermostats compatible with this furnace. Make sure you understand the specific type installed. Ask installer for detailed explanation, and retain thermostat instruction/manual for reference.

A. BASIC FURNACE -- HEATING CYCLE

This furnace is operated by an Integrated Furnace Control (IFC) which controls all functions of the furnace.

On a call for heat from the thermostat, the IFC first checks the pressure switch to verify contacts are open, then turns on the inducer motor. The pressure switch then closes signaling the IFC to proceed with ignition function.

There is a 36 second ignitor heat-up period, then the gas valve opens introducing main burner gas to be ignited by the hot surface ignitor. The flame sensor proves ignition and carryover across all burners.

There is a 60 second delay after main burner is on until the comfort air blower starts on heating speed. After the thermostat is satisfied, the burners will go off as gas valve closes. The inducer will continue to run for 30 seconds, and the comfort air blower will continue to run approximately two minutes.

B. ACCESSORIES (OPTIONAL, FIELD INSTALLED)

The furnace and IFC are designed to accommodate both electronic air cleaner and humidifier. The electronic air cleaner circuit is powered any time the comfort air blower runs on either heating or cooling speed. The humidifier circuit is powered anytime the inducer motor is operating.

C. COOLING CYCLE OPERATION

If optional add-on air conditioning system is installed, the comfort air blower starts on cooling speed immediately on call for cool, will run continuously during the call for cool, and will stop 60 seconds after the thermostat is satisfied.

D. MANUAL FAN (CONTINUOUS AIRFLOW) OPERATION

If wall thermostat is set to MANUAL (ON) position to operate comfort air blower continuously to provide air circulation throughout the building, the blower will operate on the <u>heating</u> speed rather than the cooling as is typical with most systems. This permits the air to circulate as desired but helps keep the operating noise level down as well as conserving energy.

When all call for heat occurs, the blower will continue to run during the ignitor heat up period, but shut off for 60 seconds after gas valve opens and burners ignite to accelerate the heat exchanger heat-up process. The blower then restarts and runs continuously until the next burner cycle as described above.

During a call for cooling, the blower automatically shifts up to cooling speed, and remains there until 60 seconds after thermostat is satisfied then drops back to heating speed.

SECTION 4 -- TURNING OFF FURNACE IN AN EMERGENCY

WARNING: Have a qualified installer, service agency or gas supplier teach you location and operation of gas and electrical shutoff devices. Ask them any questions you have about this section. If you don't turn off your furnace in an emergency, damage, injury or death could result.

In an emergency you MUST know how to turn off gas and electricity. Find out how BEFORE THE EMERGENCY.

WARNING: Should overheating occur or the gas supply fail to shut off, shut off the manual gas control to furnace before shutting off the electrical supply, failure to do so can cause a fire or explosion which could result in damage, injury or death.

A. GAS AND ELECTRICAL SHUTOFF DEVICES.

1. Gas Shutoff Devices.

    In an emergency, you may not be able to reach all the gas shutoff devices. You must know how to turn off gas using any one of the three manual types:

    A. Manual Shutoff Knob on Gas Control

        Gas control location is behind door.

        See Lighting and Shutdown Instructions in Section 5 for more information.

        To turn gas control furnace knob OFF, turn it clockwise >. Use this same procedure when you leave a vacation home vacant and do not want the furnace to operate.

    B. Manual In-Line Shutoff Valve In Gas Supply Line

        This valve is next to furnace. Figure 1 shows a typical installation.

        Normally, gas is ON when you turn the shutoff valve handle parallel to gas pipe. Gas is OFF when turn handle 90° from gas pipe.

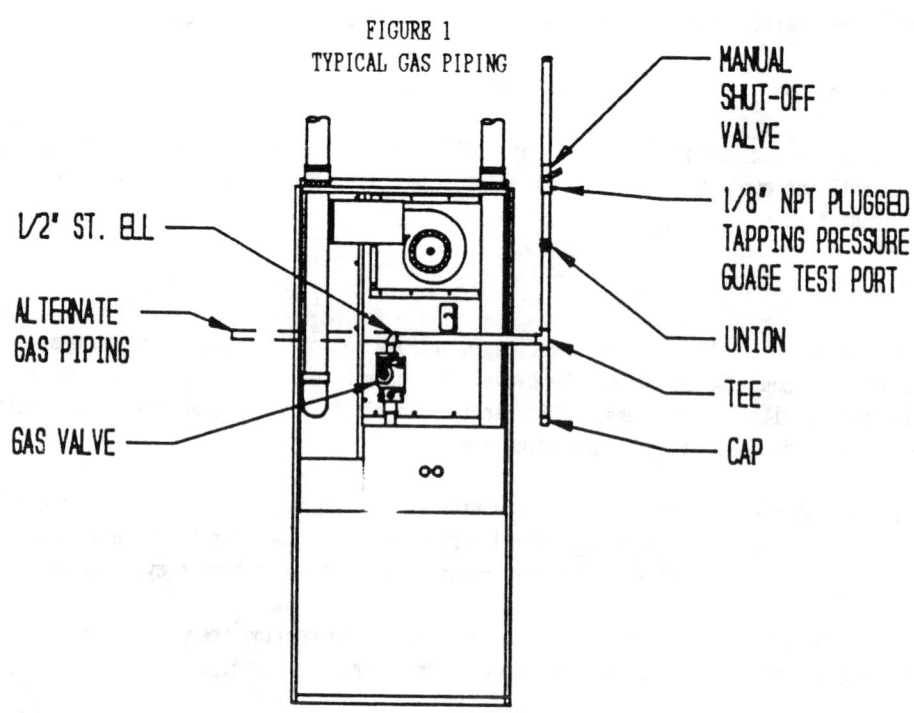

FIGURE 1
TYPICAL GAS PIPING

1/2" ST. ELL

ALTERNATE GAS PIPING

GAS VALVE

MANUAL SHUT-OFF VALVE

1/8" NPT PLUGGED TAPPING PRESSURE GUAGE TEST PORT

UNION

TEE

CAP

6

C. Manual Shutoff Valve at Natural Gas Meter or Propane (LP) Gas Tank

Normally, natural gas is ON when you turn shutoff parallel to gas pipe. Gas is off when you turn shutoff 90° from gas pipe. Some valves require a wrench or other tools.

2. Electrical Shutoff Devices

In an emergency, you may not be able to reach both of your electrical shutoff devices. Therefore, you must know how to turn off electricity using either one of them. Here are two types of electrical shutoff devices:

A. There should be an electrical shutoff device located on or immediately adjacent to the furnace.

B. There should be a separate circuit breaker or fuse serving only the furnace located in the main circuit breaker or fuse panel. Know its location and make sure this device is clearly identified.

B. POSSIBLE EMERGENCIES AND RECOMMENDED ACTIONS.

WARNING: If gas or electricity is off due to an emergency, only a qualified installer, service agency or gas supplier should turn it back on. Doing it yourself could result in damage, injury or death.

1. Possible emergency: Smelling gas or other unfamiliar smell; or not knowing what may be wrong or what to do about it.

ACTION: For your safety:

A. Leave your house or building immediately.

B. Go to a neighbor's or another building.

C. Use their telephone.

D. Call your gas supplier; tell them you smell gas; give them your name and address.

E. If you cannot reach gas supplier, call fire department.

CAUTION: Three important things not to do:

1. Don't try to light any gas appliances.

2. Don't touch any electrical switches.

3. Don't use the telephone in your house or building.

2. Possible emergency: Your thermostat is set below room temperature: yet, whether the blower is off or on, the air coming from your room registers continually gets hotter.

ACTION:

A. Turn room thermostat to its lowest or OFF setting.

B. If you can do so safely, turn gas off. Use manual shutoff valve at gas meter or on propane (LP gas) tank (you may need wrench or (tools). If you can safely turn off electricity at the main circuit panel, do so. If you cannot do these safely, leave your home or building immediately. Call your gas supplier or fire department from a neighbor's phone for help.

C. Your furnace should now be off. If it is, call your service technician or gas supplier.

7

D. If your furnace continues to run, leave your home or building immediately. Call your gas supplier or fire department from a neighbor's phone for help.

3. Possible Emergency: Your thermostat is set above room temperature. The blower is on, but the air coming from your room registers is hot, then cold, then hot, then cold in a continuing cycle. This condition indicates lack of airflow through furnace.

ACTION:

A. Make sure air filters are clean and installed correctly.

B. Check that registers and return air grilles are open and unobstructed.

C. If condition continues, call your local qualified service technician or gas supplier.

4. Possible Emergency: While furnace is operating, you smell unfamiliar odors that go away when furnace is off.

WARNING: Unfamiliar odors may mean gas or aldehydes are present which could result in damage, injury or death.

ACTION:

A. Turn thermostat to its lowest or OFF setting.

B. Turn gas control knob clockwise > to OFF.

C. If blower is not operating, immediately turn off electricity to furnace using shut off device near furnace or at main circuit panel.

D. If blower is operating, wait five minutes for furnace to cool down and then turn off electricity to furnace using shutoff device near furnace or at main circuit panel.

E. Call your local qualified service technician or gas supplier.

5. POSSIBLE EMERGENCY: Main electrical circuit breaker for furnace cannot be reset without tripping again or new fuses continue to blow.

ACTION:

A. Turn gas control know clockwise > to OFF.

B. Call your local qualified service technician or gas supplier.

SECTION 5 -- OPERATING YOUR FURNACE

After reading the Safety Information and Precautions, follow Operating Instructions on front door of furnace and instructions repeated here.

WARNING: If you do not exactly follow these instructions, a fire or explosion could occur, resulting in damage, injury or death.

WARNING: Never use tools to turn gas control knob. Only use your hand. If gas control knob will not turn by hand, do not force it or try to repair it. Call a qualified installer, service agency or gas supplier. Forcing knob can cause gas to leak which could result in fire or explosion.

Properly operating your furnace requires certain abilities, mechanical skills and tools. If you are uncertain about your abilities or if you lack proper skills or tools, do not proceed. Instead, contact a qualified installer, service agency or gas supplier.

An automatic ignition device lights the burners. Do not try to light manually. See Figure 2 for step by step instructions.

WARNING: If you do not exactly follow these instructions, a fire or explosion could occur, resulting in damage, injury or death.

FIGURE 2

FOR YOUR SAFETY READ BEFORE OPERATING

WARNING: If you do not follow these instructions exactly, a fire or explosion may result causing property damage, personal injury or loss of life.

A. This appliance does not have a pilot. It is equipped with an ignition device which automatically lights the burner. Do **not** try to light the burner by hand.

B. BEFORE OPERATING smell all around the appliance area for gas. Be sure to smell next to the floor because some gas is heavier than air and will settle on the floor.

   WHAT TO DO IF YOU SMELL GAS
   • Do not try to light any appliance.
   • Do not touch any electric switch; do not use any phone in your building.
   • Immediately call your gas supplier from a neighbor's phone. Follow the gas supplier's instructions.

   • If you cannot reach your gas supplier, call the fire department.

C. Use only your hand to push in or turn the gas control knob. Never use tools. If the knob will not push in or turn by hand, don't try to repair it, call a qualified service technician. Force or attempted repair may result in a fire or explosion.

D. Do not use this appliance if any part has been under water. Immediately call a qualified service technician to inspect the appliance and to replace any part of the control system and any gas control which has been under water.

OPERATING INSTRUCTIONS

1. STOP! Read the safety information above on this label.

2. Set the thermostat to lowest setting.

3. Turn off all electric power to the appliance.

4. This appliance is equipped with an ignition device which automatically lights the burner. Do **not** try to light the burner by hand.

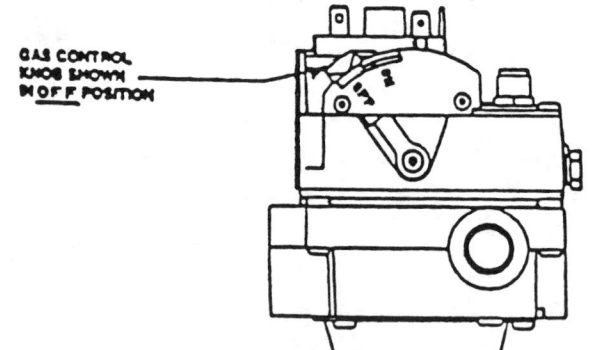

GAS CONTROL KNOB SHOWN IN OFF POSITION

5. Remove control access panel.

6. Push in gas control knob slightly and turn clockwise to "OFF."

   NOTE: Knob cannot be turned to "OFF" unless knob is pushed in slightly. Do not force.

7. Wait five (5) minutes to clear out any gas. Then smell for gas, including near the floor. If you smell gas, STOP! Follow "B" in the safety information above on this label. If you don't smell gas, go to the next step.

8. Turn gas control knob counterclockwise to "ON."

9. Replace control access panel.

10. Turn on all electric power to the appliance.

11. Set thermostat to desired setting.

12. If the appliance will not operate, follow the instructions "To Turn Off Gas To Appliance" and call your service technician or gas supplier.

TO TURN OFF GAS TO APPLIANCE

1. Set the thermostat to lowest setting.

2. Turn off all electric power to the appliance if service is to be performed.

3. Remove control access panel.

4. Push in gas control knob slightly and turn clockwise to "OFF." Do not force.

5. Replace control access panel.

7961-366

SECTION 6 -- PROPER MAINTENANCE OF YOUR FURNACE

You need special abilities, mechanical skills and tools to maintain your furnace properly. If you are uncertain about your abilities or if you lack proper skills or tools, do not try to maintain or repair your furnace yourself. Instead, contact a qualified installer, service agency or gas supplier.

A. IF YOU SMELL GAS OR ANY UNFAMILIAR SMELL WHILE WORKING ON YOUR FURNACE:

1. Do not try to light main burners.

2. Do not touch or turn on any electrical switch.

3. Do not use any phone in your building.

4. Immediately call your gas supplier from a neighbor's phone. Follow gas supplier's instructions.

5. If you cannot reach your gas supplier, call fire department.

B. LUBRICATION REQUIREMENTS

The main blower motor and the induced draft blower motor are permanently lubricated, and no maintenance is required.

C. MAKE SURE AIR FILTERS ARE IN PLACE

Ask your installer, local qualified service technician or gas supplier to make sure your filters are in place properly. Become familiar with their location and procedures for removing, cleaning and replacing them.

CAUTION: Operating furnace without clean air filters can damage blower motor, heat exchanger, or air conditioning system components. This can cause system failure which could result in damage or injury.

D. USE THE CORRECT SIZE AIR FILTERS.

All models are shipped with filters and external filter racks that can be installed on either side of the furnace. See Table 1 for sizes and Figures 3 and 4 for installation details.

TABLE 1

FILTER SIZE FOR GAS FURNACE		
Model	Standard Size*	Optional Size*
DCH050D30A DCH065D36A DCH080D48A	1 - 16x25x1	
DCH095D60A	1 - 16x25x1	1 - 20x25x1 Model FR22

* Shipped with all furnaces.
** Optional filter rack. Must be used on all D60 versions when set-up for 5 ton add-on air conditioning system.

E. KEEP AIR FILTERS CLEAN.

As a user, your personal responsibility is to keep air filters clean.

CAUTION: Dirty air filters reduce system efficiency and can cause erratic control performance. These could result in damage to blower motor or heat exchanger.

WARNING: To prevent electric shock, turn off electricity to furnace before removing, cleaning or replacing air filters. Failure to do so could result in injury or death.

WARNING: To prevent possibility of electrical shock or touching rotating parts, do not operate furnace with blower door removed. Doing so could result in injury or death.

1. During the first four weeks after your furnace is installed, inspect your air filters for dirt every week. Then check them monthly.

2. Permanent type foam filters are washable. Use a solution of soapy water, followed by a rinse in clear water and then dried by tapping the frame against a solid object to remove excess water.

WARNING: After cleaning or changing filter, filter access cover(s) must be replaced. Failure to do so could cause nausea, asphyxiation, or fire, resulting in damage, injury or death.

F. Filter Locations--Removal and Replacement Procedures--All Hi-Boy Furnaces

16 x 25 x 1 external filter racks are supplied as standard equipment for all DCH models. A 14 x 23 opening is to be cut into either the left or right side of the furnace depending upon installation requirements.

See Figure 3 for typical installation of a 16 x 25 x 1 filter rack centered over the 14 x 23 cutout.

Figure 4 shows a typical installation of optional 20 x 25 x 1 filter rack. The same 14 x 23 cutout is required in the furnace side, and the bottom of the filter rack is aligned over the bottom of the 14 x 23 cutout. The top of the filter rack rises approximately 6 inches above the top of the cutout. The 3 inch depth of the filter rack provides ample spacing between furnace side and leaving edge of filter for the entire filter surface to be effective.

FIGURE 3

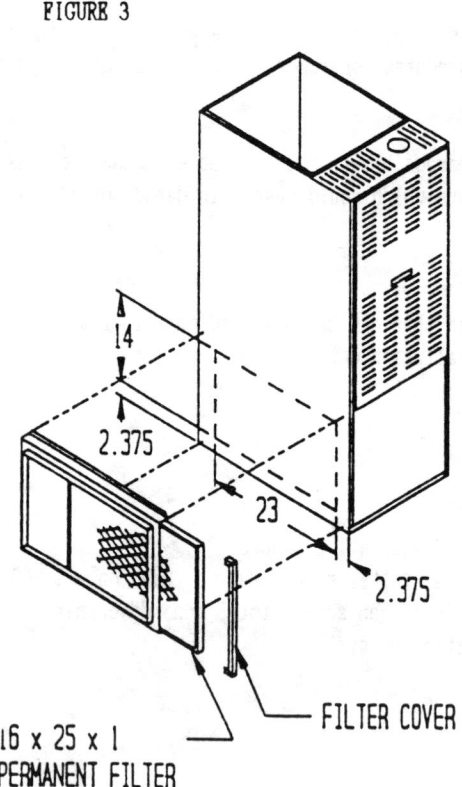

16 x 25 x 1
PERMANENT FILTER

FIGURE 4

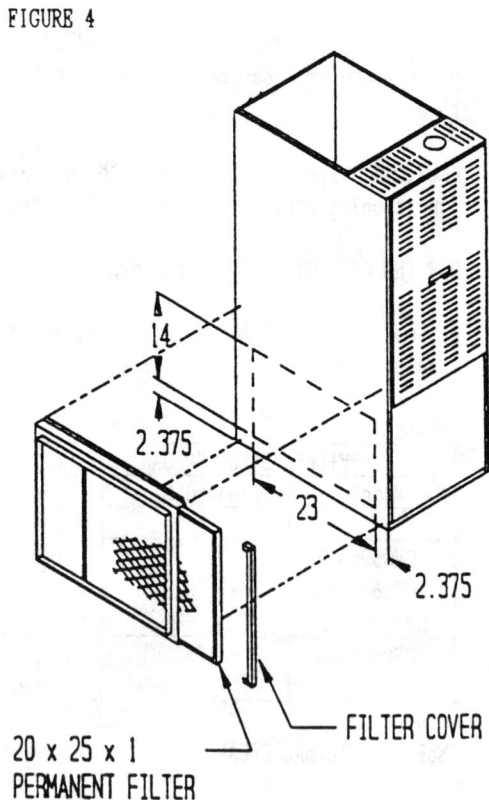

20 x 25 x 1
PERMANENT FILTER

To remove filter from the filter rack, remove the filter access cover and grasp the end of the filter. This filter is the permanent high velocity type foam filter which may be washed and used over and over. DO NOT replace it with a fiberglass disposable filter of the same size.

To replace filter, reverse the above procedure.

12

G. DO NOT OBSTRUCT DUCT WORK

For proper operation, keep registers and return air grilles open. Do not cover them with rugs, carpets, drapes or furniture.

H. HAVE YOUR FURNACE CHECKED ANNUALLY

The furnace and the vent and combustion air intake piping systems should be inspected yearly by a qualified service agency, generally prior to the heating season. Detailed procedures for this inspection are contained in the instructions booklet and should be handled by the qualified service agency only.

A general inspection of the furnace, the furnace area, and the vent and air intake piping systems, should be conducted on a regular basis by the owner/occupant. This review should include:

1. Make sure the furnace always has the minimum clearance as detailed on the furnace rating plate. Special attention must be given to these items if any remodeling is done.

2. Make sure the vent and air intake piping system is in place, slopes upward and is physically sound without holes, sags or dips.

3. Reviewing that the return air duct connection(s) is physically sound, is sealed to the furnace casing, and terminates outside the space containing the furnace.

4. The physical support of the furnace is sound without sagging, cracks, gaps, etc. around the base.

5. Inspect for any obvious signs of deterioration of the furnace.

6. The condensate drain must be inspected monthly for any blockage. If drain appears dirty or clogged, it must be removed and thoroughly washed out with warm water. Blockage of the drain can result in furnace malfunction.

   Periodic examinations of the vent and air intake piping system should also be conducted by the owner on a regular basis, preferably every month but at least every two months, during the heating season.

7. Check both the inlet and outlet terminals for any blockage. If any debris is present on the screens, use a small brush or vacuum to remove.

The following procedure should be followed for the periodic inspection as conducted by the owner/occupant:

1. Set the wall thermostat to the "off" position or lower the set point lever to a temperature well below the existing room temperature. Shut off electric power to the furnace. A switch should be mounted either on the outside of the furnace or adjacent to the furnace for this purpose.

2. Remove the outer panel that has ventilating slots in it, and then the inner cover over the burners (has two clean plastic windows) exposing the burner compartment.

3. Use flashlight or troublelight, observe the burner compartment and where the burner(s) extend into the heat exchanger. There should be very minimal scaling or sooting in this area. Some loose debris may have fallen down on to the floor of the heat exchanger from the upper flue passageways, and this may be vacuumed out. Also observe the sides of the heat exchanger for "hot spots" due to improper burner alignment or overfiring and give particular attention to any area where it looks like there may be any deterioration from corrosion or rusting. Observe for any corrosion on the burner themselves. Should anything appear questionable, contact your service agency.

   Replace the burner compartment cover immediately in front of the burners (this cover has two clear plastic windows) before continuing the inspection. Make sure all gaskets and seals are intact.

4. Inspect the vent pipe the full distance from the furnace to the exit point from the building, observing for any possible leakage that might result in products of combustion entering the living area.

```
┌─────────────────────────────────────────┐
│                                           │
│                 WARNING                   │
│                                           │
├─────────────────────────────────────────┤
│                                           │
│  Leakage of products of combustion into the │
│  living area may result in asphyxiation.  │
│                                           │
└─────────────────────────────────────────┘
```

Any questionable vent pipe should be replaced. Considering its importance, it is relatively cheap insurance to protect your family. Consult your service agency.

5. Restore the electrical power to the furnace by turning the switch back on. Adjust the thermostat to call for heating operation.

6. Observe the main burners in operation. The flame should be mostly "blue" with possibly a little "orange" at the tips of the flames. The flames should be in the center of the heat exchanger compartments and not impinging on the heat exchanger surfaces themselves. See Figure 5.

7. Observe the fire until the blower starts (there is a normal delay period until the heat exchanger warms up). There should be no change in the size or shape of the flame. If there is any wavering or blowing of the flame on the blower start-up, it is an indication of a possible leak in the heat exchanger. Turn off the gas valve in the gas line leading to the furnace, and then the main electrical switch to the furnace and call your service agency.

8. Replace the slotted outer furnace doors by reversing the procedure as outlined under Step 2 above.

FIGURE 5

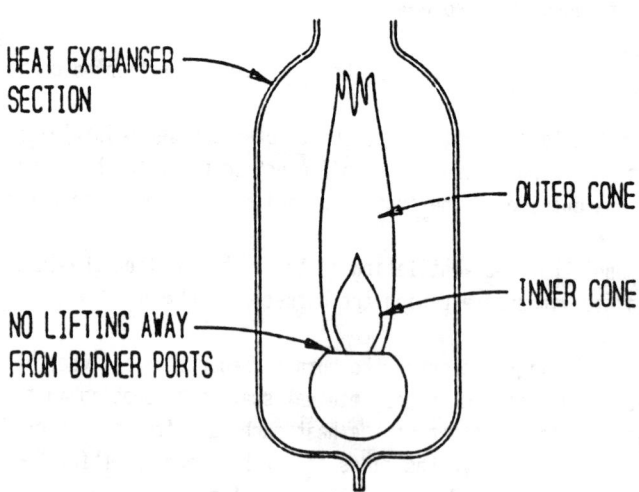

HEAT EXCHANGER SECTION

OUTER CONE

INNER CONE

NO LIFTING AWAY FROM BURNER PORTS

SECTION 7 -- CHECKING FURNACE BEFORE REQUESTING A SERVICE CALL

Before you call a local qualified service technician or gas supplier, check these items:

A. IF YOUR FURNACE IS OPERATING BUT NOT HEATING YOUR HOUSE TO DESIRED TEMPERATURE:

    1. Check to see if air filter is clean. See Section 6.D for checking instructions.

    2. Check both supply air register and return air grilles. They should be open, clean and unobstructed by rugs, carpets, drapes or furniture.

    3. A. Make sure room thermostat is in HEAT position and gas is ON, see Section 5.

        B. Set room thermostat above current room temperature.

        C. Make sure room thermostat is not near a heat source, such as a lamp, television, radio, computer, direct sunlight or fireplace. These can make your thermostat sense that the room is warmer than it is. Move heat source away from room thermostat.

    4. If furnace now provides sufficient heat, set room thermostat to desired temperature. If your furnace still fails to provide sufficient heat, call your local qualified service technician or gas supplier for repairs. Give furnace model and serial numbers, recorded in Section 1 or from the furnace rating plate.

B. IF YOUR FURNACE IS NOT OPERATING AT ALL.

    1. Make sure room thermostat is in HEAT position.

    2. Set room thermostat above current room temperature.

    3. Make sure electrical disconnect switch for furnace is ON.

    4. Check to see if main fuses have blown or main circuit breaker has tripped.

    5. Make sure blower door is securely in place. The blower door interlock switch prevents furnace operation if the door is not secured.

    6. Make sure gas is ON. If gas is OFF because of an emergency or unsafe condition, DO NOT turn gas ON. Call your local qualified technician.

Carrier

58DXC, 58SXC
Deluxe Gas-Fired Condensing Furnaces

Service and Maintenance Instructions
For Sizes 040-100, Series 100

NOTE: Read the entire instruction manual before starting the installation.

SAFETY CONSIDERATIONS

Installing and servicing heating equipment can be hazardous due to gas and electrical components. Only trained and qualified personnel should install, repair, or service heating equipment.

Untrained personnel can perform basic maintenance functions such as cleaning and replacing air filters. All other operations must be performed by trained service personnel. When working on heating equipment, observe precautions in the literature, tags, and labels attached to or shipped with the unit and other safety precautions that may apply.

Follow all safety codes, including NFPA 54/ANSI Z223.1-1988, National Fuel Gas Code. Wear safety glasses and work gloves. Have a fire extinguisher available during start-up and adjustment procedures and service calls.

Recognize safety information. This is the safety-alert symbol ⚠ . When you see this symbol on the unit and in instructions or manuals, be alert to the potential for personal injury.

Understand the signal word—DANGER, WARNING, or CAUTION. These words are used with the safety-alert symbol. DANGER identifies the most serious hazards which **will** result in severe personal injury or death. WARNING signifies hazards that **could** result in personal injury or death. CAUTION is used to identify unsafe practices, which **would** result in minor personal injury or product and property damage.

⚠ WARNING

The ability to properly perform maintenance on this equipment requires certain expertise, mechanical skills, tools, and equipment. If you do not possess these, do not attempt to perform any maintenance on this equipment other than those procedures recommended in the User's Manual. FAILURE TO FOLLOW THIS WARNING COULD RESULT IN POSSIBLE DAMAGE TO THIS EQUIPMENT, SERIOUS PERSONAL INJURY, OR DEATH.

⚠ WARNING

Never store anything on, near, or in contact with, the furnace, such as:
1. Spray or aerosol cans, rags, brooms, dust mops, vacuum cleaners, or other cleaning tools.
2. Soap powders, bleaches, waxes or other cleaning compounds, plastic or plastic containers, gasoline, kerosene, cigarette lighter fluid, dry cleaning fluids, or other volatile fluids.
3. Paint thinners and other painting compounds, paper bags or other paper products.

Failure to follow this warning can cause corrosion of the heat exchanger, fire, personal injury, or death.

Fig. 1—Model 58SXC Upflow Furnace A91098

A91103

Fig. 2—Model 58DXC Downflow Furnace

Manufacturer reserves the right to discontinue, or change at any time, specifications or designs without notice and without incurring obligations.

Book 1 4   PC 101   Catalog No. 535-832   Printed in U.S.A.   Form 58D,S-8SM   Pg 1   7-92   Replaces: 58D,S-6SM
Tab 6a 8a

CARE AND MAINTENANCE

For continuing high performance and to minimize possible equipment failure, it is essential that maintenance be performed annually on this equipment. Consult your local dealer for maintenance and the availability of a maintenance contract.

> **⚠ WARNING**
>
> Turn OFF the gas and electrical supplies to the unit before performing any maintenance or service. Follow the operating instructions on the label attached to the furnace. Failure to follow this warning could result in personal injury.

The minimum maintenance that should be performed on this equipment is as follows:

1. Check and clean or replace air filter each month as required.
2. Check blower motor and wheel for cleanliness and lubrication each heating and cooling season. Clean and lubricate as necessary. (See Step 2.)
3. Check electrical connections for tightness, and controls for proper operation each heating season. Service as necessary.
4. Check for proper condensate drainage; clean as necessary.
5. Check for blockages of combustion-air and vent pipes.

> **⚠ CAUTION**
>
> As with any mechanical equipment, personal injury could result from sharp metal edges, etc. Be careful when removing parts.

Step 1—Air Filter Cleaning and Replacement

The air filter arrangement may vary depending on the application.

> **⚠ CAUTION**
>
> Never operate unit without a filter or with filter access door removed. Failure to follow this warning could result in a fire or personal injury.

DOWNFLOW FURNACES ONLY — Each furnace accommodates 2 filters which are installed in the return-air duct. (See Fig. 3.) To clean or replace the filters, proceed as follows:

1. Turn OFF electrical supply to unit.
2. Remove blower access door.

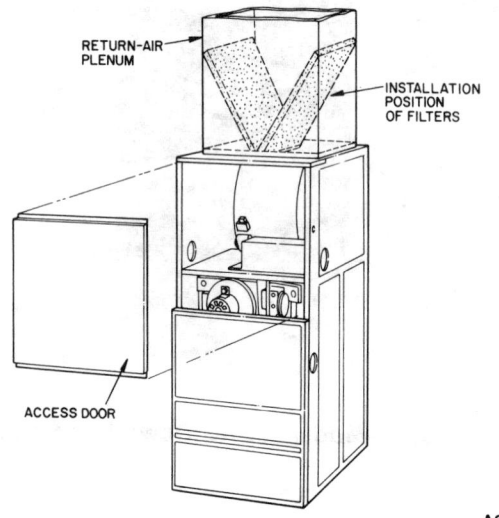

Fig. 3—Position of Filters in Downflow Furnace

A87300

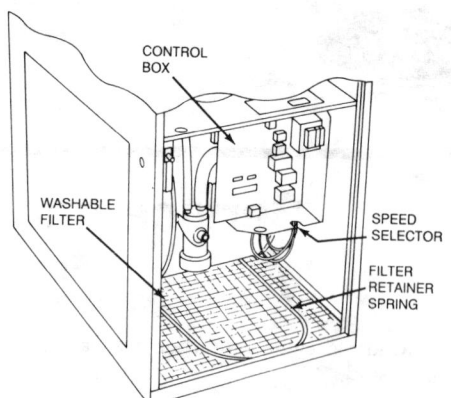

A91108

Fig. 4—Filter Installed for Bottom Inlet in Upflow Furnace

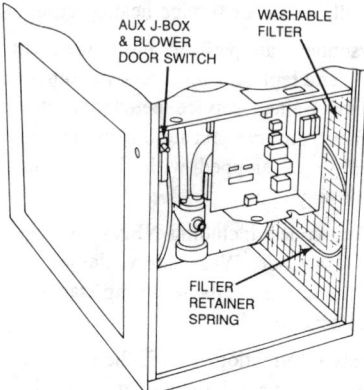

A91109

Fig. 5—Filter Installed for Side Inlet in Upflow Furnace

3. Reaching up behind top plate, tilt filters toward center of return-air plenum, remove filters, and replace or clean as needed.
4. Furnaces are equipped with permanent, washable filters. Clean these filters by spraying cold tap water through filter in opposite direction of airflow.
5. Rinse filters and let dry. Oiling or coating of filters is not recommended.
6. Reinstall filters with cross-mesh binding facing blower.
7. Replace access door.
8. Turn ON electrical supply to furnace.

UPFLOW FURNACES ONLY — To clean or replace the air filter, proceed as follows:

1. Turn OFF electrical supply to unit.
2. Remove access doors.
3. Release filter retainer spring from behind flange of furnace casing. (See Fig. 4 and 5.)
4. Slide filter out.
5. Furnaces are equipped with permanent, washable filters. Clean filter by spraying cold tap water through filter in opposite direction of airflow.
6. Rinse filter and let dry. Oiling or coating of filter is not recommended.
7. Place filter in furnace with cross-mesh binding either up or facing blower.
8. Replace access doors.
9. Turn ON electrical supply to furnace.

Step 2—Blower Motor and Wheel Maintenance

For long life, economy, and high efficiency clean accumulated dirt and grease from blower wheel and motor annually.

The following items should be performed by a qualified service technician:

Some motors have prelubricated, sealed bearings and require no lubrication. These motors can be identified by the absence of oil ports on each end of the motor. For motors with oil ports, lubricate as follows:

Lubricate motor every 5 years if motor is used for intermittent operation (thermostat FAN switch in AUTO position), or every 2 years if motor is in continuous operation (thermostat FAN switch in ON position).

Clean and lubricate as follows:

1. Turn OFF electrical supply to unit.

2. Remove access doors.

3. Downflow furnace only—disconnect vent pipe, elbow, and auxiliary limit switch. (See Fig. 6.)

   a. Remove vent pipe enclosure from top side of blower shelf and position to 1 side.

   b. Loosen hose clamps on outlet elbow and remove elbow.

   c. Loosen hose clamp on extension pipe outside of furnace and remove pipe.

   d. Disconnect wires from auxiliary limit on blower housing.

4. Note location of wires for reassembly, then remove electrical leads from numbered side of blower speed selector. (See Fig. 4 and 6.)

5. Upflow furnaces only—remove drain trap and control box.

   a. Remove control box from bottom side of blower shelf and position to 1 side.

   b. Using backup wrench, disconnect drain pipe at coupling in blower compartment.

   c. Loosen hose clamp and remove 7/8-in. diameter drain hose from drain trap.

   d. Loosen hose clamp and disconnect 5/8-in. diameter drain hose at bottom of inducer housing located under blower shelf.

   e. Remove screw securing drain trap assembly.

6. Remove screws securing blower assembly to blower shelf and slide blower assembly out of furnace.

7. Squeeze side tabs of blower speed selector and pull from blower housing bracket.

8. Loosen screw in strap holding motor capacitor to blower housing and slide capacitor from strap.

9. Mark blower wheel location on shaft before disassembly to insure proper reassembly.

10. Loosen setscrew holding blower wheel on motor shaft.

NOTE: Mark blower mounting arms, and blower housing so each arm is positioned at the same hole location during reassembly. This will insure that oilers point up.

11. Remove bolts holding motor mount to blower housing and slide motor and mounts out of housing. Disconnect ground wire attached to blower housing before removing motor.

12. Lubricate motor (when oil ports are provided).

   a. Remove dust caps or plugs from oil ports located at each end of motor. If motor does not have these caps or plugs, bearings are sealed and need no further lubrication.

   b. Use a good grade of SAE 20 nondetergent motor oil and add 1 teaspoon (5 cc, 3/16 oz, or 16 to 25 drops) in each oil port. The use of other types or grades of oil will damage the motor. Excessive oiling can cause premature bearing failures.

   c. Allow time for total quantity of oil to be absorbed by each bearing.

   d. After oiling motor, wipe excess oil from motor housing.

   e. Replace dust caps or plugs on oil ports.

13. Remove blower wheel from housing.

   a. Mark blower wheel orientation and cutoff plate location to insure proper reassembly.

   b. Remove screws securing cutoff plate and remove cutoff plate from housing.

   c. Remove blower wheel from housing.

14. Clean blower wheel and motor by using a vacuum with soft brush attachment. Be careful not to disturb balance weights (clips) on blower wheel vanes. Do not drop or bend wheel as balance will be affected.

15. Reassemble blower by reversing items 13.a. through 13.c. Ensure wheel is positioned for proper rotation.

16. Reassemble motor and blower by reversing items 6 through 11. If motor has ground wire, be sure it is reconnected.

⚠ CAUTION

Ensure the motor is properly positioned in the blower housing. The motor oil ports must be at a minimum of 30° above the horizontal centerline of the motor after the blower assembly has been reinstalled in the furnace.

---

17. Reinstall blower assembly in furnace.

18. Upflow furnace only—reinstall drain trap and control box.

   a. Inspect drain trap and hoses to ensure they are not blocked or restricted. Reinstall drain trap and hoses. Be sure to tighten hose clamps.

   b. Using backup wrench, attach drain pipe and tighten compression coupling.

   c. Reinstall control box on bottom side of blower shelf. Be sure edge connector is connected through top of blower shelf.

19. Downflow furnace only—reconnect vent pipe, elbow, and auxiliary limit switch.

   a. Reinstall outlet elbow and extension pipe. Be sure connections are tight and leak proof.

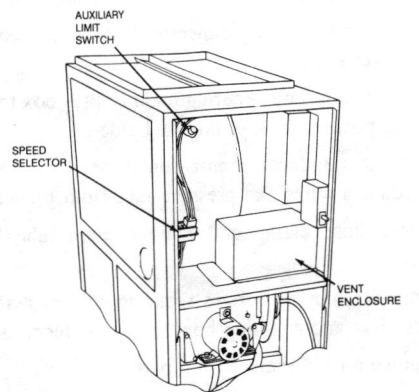

A91110

Fig. 6—Model 58DXC Downflow Furnace

3

b. Reinstall vent pipe enclosure.

c. Reconnect red wires to auxiliary limit switch.

20. Connect electrical leads to blower speed selector. Note that connections are polarized for correct assembly – **do not force.**

21. Turn ON electrical supply and check for proper rotation and speed changes between heating and cooling; operate unit 5 minutes and carefully check for condensate leaks.

Step 3—Cleaning Heat Exchangers

The following items should be performed by a qualified service technician:

If it becomes necessary to clean the heat exchanger because of carbon deposits, soot, etc., proceed as follows:

NOTE: Deposits of soot and carbon indicate a problem exists that needs to be corrected. Action must be taken to correct the problem.

1. Turn OFF gas and electrical supplies to furnace.

2. Remove control and blower access doors.

3. Loosen hose clamps on combustion-air pipe and move air pipe aside.

4. Using backup wrench, disconnect gas supply at ground joint union. Remove gas pipe from valve.

5. Disconnect hot surface ignitor and flame sensor leads at 3-circuit connector outside of burner enclosure.

6. Disconnect electrical wires from gas valve.

7. Disconnect pressure tubing from right side of burner enclosure and outlet end of gas valve.

8. Remove burner enclosure front.

9. Remove diffuser from inside top of burner enclosure. Remove screws that secure burner enclosure to cell panel. These screws are located inside the burner enclosure.

10. Using care not to damage cell inlet panel gasket, remove gas control assembly from furnace.

11. Remove vent pipe and drain.

   a. Upflow furnace only:

   (1.) Loosen hose clamps at vent pipe connection; disconnect vent pipe and position to 1 side.

   (2.) Loosen hose clamp and remove drain tube from inducer outlet box.

   b. Downflow furnace only:

   (1.) Remove vent pipe enclosure.

   (2.) Loosen hose clamps at vent pipe connection.

   (3.) Loosen hose clamp and remove drain tube from inducer outlet elbow.

12. Upflow furnace only — remove main control box.

   a. Disconnect edge connector from main control box at blower shelf.

   b. Remove screws securing main control box to blower shelf and position control box to 1 side.

13. Disconnect inducer motor connector from wiring harness. Disconnect wires and pressure tube from pressure switch.

14. Loosen hose clamp and remove drain tube from inducer housing.

15. Remove mounting screws securing inducer assembly to collector box and coupling box; remove inducer assembly.

16. Remove all old sealant from parts.

17. Remove coupling box(es).

   a. Upflow furnace only:

   (1.) Remove screws securing coupling box and remove from furnace. Remove all old sealant from parts.

   (2.) Remove choke plate (when used) from primary heat exchanger outlet.

   b. Downflow furnace only:

   (1.) Remove screws securing intake (upper) coupling box and remove from furnace. Remove all old sealant from parts.

   (2.) Remove screws securing primary (lower) coupling box and remove box. Clean old sealant from parts.

18. Loosen hose clamp and remove 7/8-in. drain tube from trap.

19. Hold bucket under 7/8-in. drain tube.

20. Using garden hose, flush each cell of the condensing heat exchanger with water. Use care not to spray water onto interior surfaces of control compartment. Dry all surfaces. Be careful **not** to remove sealant around cell openings in cell panel.

21. Using field-provided small wire brush, steel spring cable, reversible electric drill, and vacuum cleaner, clean primary heat exchanger cells. **Do not use wire brush or other sharp object to clean condensing heat exchanger.** Failure of the condensing heat exchanger will occur — flush with water only.

   a. Assemble wire brush and steel spring cable.

   (1.) Use 4 ft of 1/4-in. diameter high-grade steel spring cable (commonly known as drain cleaning or Roto-Rooter cable).

   (2.) Use 1/4-in. diameter wire brush (commonly known as 25-caliber rifle cleaning brush).

NOTE: The materials required in items (1.) and (2.) can be purchased at local hardware stores.

   (3.) Insert twisted wire end of brush into end of spring cable, and crimp tight with crimping tool or strike with ball-peen hammer. *Tightness is very important.*

   (4.) Remove metal screw fitting from wire brush to allow insertion into cable.

   b. Clean each primary heat exchanger cell.

   (1.) Attach variable-speed, reversible drill to end of spring cable (end opposite brush).

   (2.) Insert brush end of cable into upper opening of cell and slowly rotate with drill. Do **not** force cable. Gradually insert at least 3 ft of cable into 2 upper passes of cell. (See Fig. 7.)

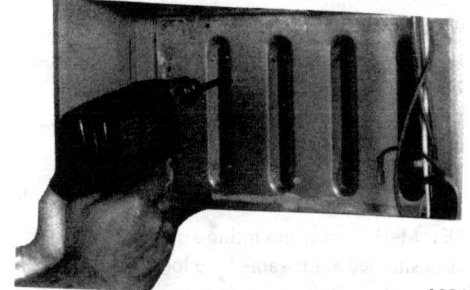

A88489

Fig. 7—Cleaning Primary Heat Exchanger Cell

   (3.) Work cable in and out of cell 3 or 4 times to obtain sufficient cleaning. Do **not** pull cable with great force. Reverse drill and gradually work cable out.

   (4.) Insert brush end of cable in lower opening of cell, and proceed to clean 2 lower passes of cell in same manner as 2 upper passes.

(5.) Repeat procedures (above) until each furnace cell has been cleaned.

(6.) Using vacuum cleaner, remove residue from each cell.

(7.) Using vacuum cleaner with soft brush attachment, clean burner assembly.

Step 4—Reassemble Furnace (After Cleaning Heat Exchangers)

1. Install choke plate (when used). Be sure choke plate bottom conforms to top flange of condensing heat exchanger.

2. Reinstall coupling box(es):

   a. Apply sealant releasing agent (Pam) to cell panel where coupling box flange matches. (See Fig. 8.)

   b. Apply a generous bead 3/16-in. dia) of G.E. RTV 122, 162, or Dow-Corning RTV 738 sealant (**NO substitute is permissible**) to flange of coupling box. Your distributor should have G.E. RTV 122, 162, or Dow-Corning RTV 738 sealants in stock.

   c. Being careful not to smear sealant, position coupling box so that slot in insulation is on left side and install coupling box.

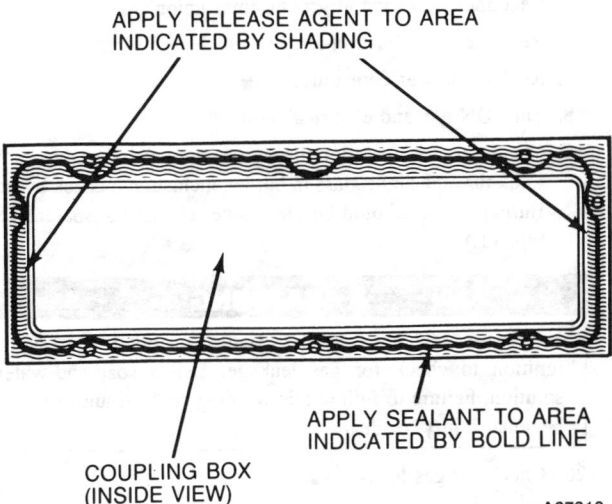

COUPLING BOX (INSIDE VIEW)

A87318

Fig. 8—Inside View of Coupling Box

NOTE: Downflow furnace only—position primary (lower) coupling box so that tallest end is on right side.

3. Reinstall inducer assembly.

   a. Upflow furnace only—Be sure small round gasket(s) is in place between blower shelf and inducer housing.

   b. Apply sealant releasing agent (Pam) to collector box. (See Fig. 9.)

   c. Apply 1/8-in. diameter bead of G.E. RTV 122, 162, or Dow-Corning RTV 738 sealant to back of inducer housing. Apply sealant around inlet air opening. (The sealant should be about 1/4 in. from the edge of the inlet air opening.)

   d. Install inducer assembly on collector box and support bracket to coupling box.

   e. Connect inducer motor plug-in connector to wiring harness. Reconnect wires to pressure switch using furnace wiring diagram. (See Fig. 17.)

   f. Reconnect pressure tubes to pressure switch. (See Fig. 13 or 14.)

4. Connect small drain tube from top of trap to fitting on bottom of inducer housing. (See Fig. 13 or 14.)

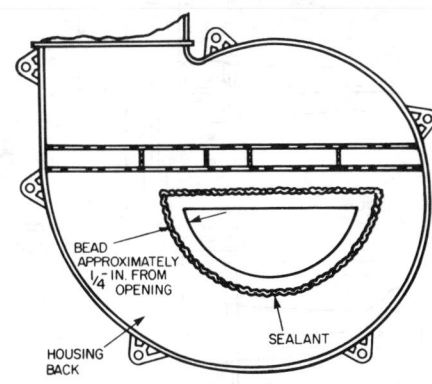

A86100

Fig. 9—Backside of Inducer Assembly Housing

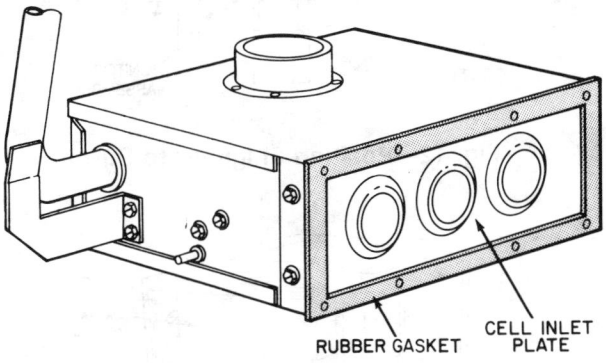

A87301

Fig. 10—Burner Enclosure

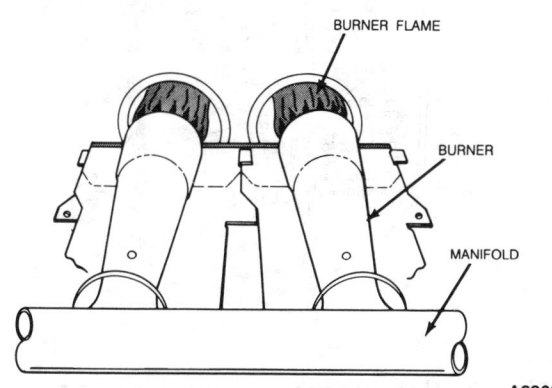

A89020

Fig. 11—Burner Flame

5. Connect 7/8-in. drain tube to trap and collector box, and tighten hose clamps. (See Fig. 13 or 14.)

6. Reinstall vent pipe and drain tube.

   a. Upflow furnace only:

      (1.) Reconnect vent pipe. Be sure clamps are tight.

      (2.) Connect drain tube from collector box to inducer outlet box.

   b. Downflow furnace only:

      (1.) Reconnect vent pipe. Be sure clamps are tight.

      (2.) Reinstall vent pipe enclosure.

      (3.) Connect drain tube from collector box to inducer outlet elbow.

7. Upflow furnace only—reinstall main control box.

   a. Reinstall main control box on blower shelf.

5

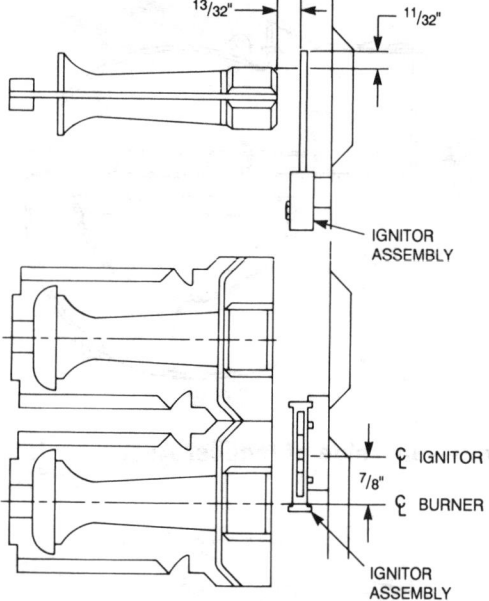

Fig. 12—Position of Ignitor to Burner

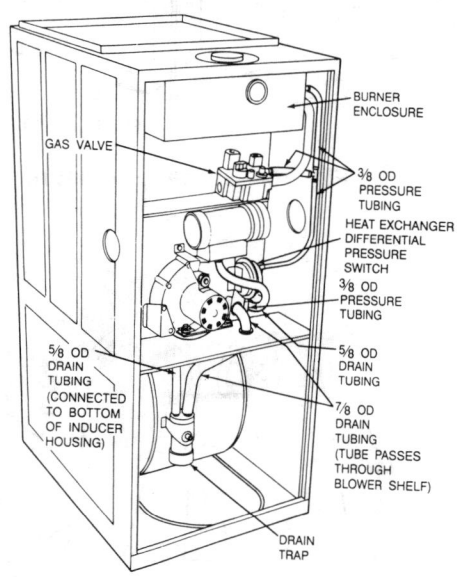

Fig. 13—Upflow Furnace Pressure and Drain Tubing Diagram

b. Reconnect edge connector at main control box on blower shelf.

8. Check condition of gasket on cell inlet panel of burner enclosure. Replace gasket if necessary. (See Fig. 10.)

9. Install gas control assembly in furnace.

10. Install diffuser and burner enclosure front.

11. Reconnect hot surface ignitor and flame sensor leads at 3-circuit connector.

12. Refer to furnace wiring diagram and connect wires to gas valve. (See Fig. 17.)

13. Reconnect pressure tubes to gas valve and burner enclosure. Be sure tubes are not kinked.

14. Using backup wrench, install gas pipe in gas valve.

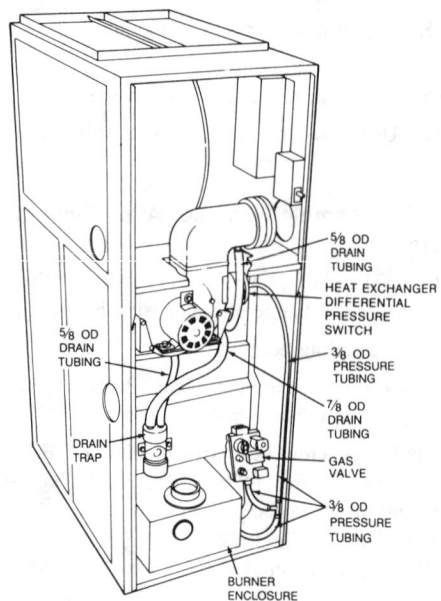

Fig. 14—Downflow Furnace Pressure and Drain Tubing Diagram

15. Reconnect gas pipe at ground joint union.

16. Reconnect combustion-air pipe. Tighten hose clamps.

17. Replace blower door only.

18. Turn ON gas and electrical supplies.

19. Check furnace operation through 2 complete operating cycles. Look through sight-glass in burner enclosure to check burners. Burner flames should be clear blue, almost transparent. (See Fig. 11.)

> ### ⚠ WARNING
> Never use matches, candles, flame, or other sources of ignition to check for gas leakage. Use a soap-and-water solution. Failure to follow this warning could result in a fire, personal injury, or death.

20. Check for gas leaks.

21. After condensate starts to drain, check for condensate leaks.

22. Replace control door.

Step 5—Clean Condensate Drainage System

1. Disconnect 5/8-in. drain tube from bottom of inducer housing. (See Fig. 13 or 14.)

2. Disconnect 7/8-in. drain tube from collector box. (See Fig. 13 or 14.)

3. Disconnect condensate drain line from drain trap at compression fitting.

4. Remove two 1/4-in. screws securing strap on drain trap to:

    a. blower housing (upflow furnaces only).

    b. bracket from cell panel (downflow furnaces only).

5. Remove drain trap/hose assembly from furnace and flush with water until clean.

6. Flush external condensate drain line with water until clean.

7. Reassemble condensate drainage system by reversing items 1. through 5.

Step 6—Hot Surface Ignitor

When removing the burner assembly, **use care to avoid breaking the hot surface ignitor**. See Fig. 12 for the correct ignitor

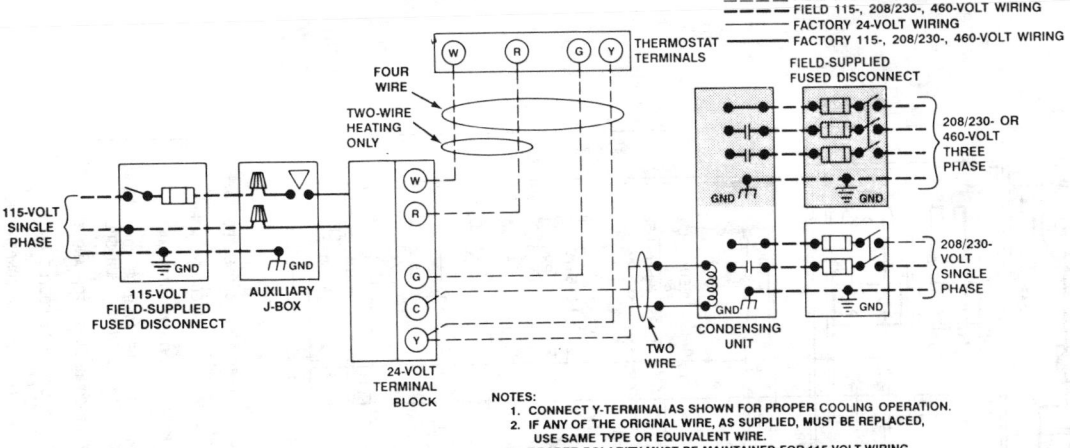

A87502

Fig. 15—Heating and Cooling Application Wiring Diagram

location. When reinstalling ignitor, use care to insure all wiring is away from the burners and is not touching the bottom of the sheet metal enclosure.

Step 7—Electrical Controls and Wiring

NOTE: There may be more than 1 electrical supply to the unit.

The electrical ground and polarity for 115-v wiring must be maintained properly. Refer to Fig. 15 for field wiring information and to Fig. 17 for unit wiring information. If the polarity is **not** correct, the microprocessor control will shut off the gas flow shortly after completion of the ignition trial period. The control system also requires an earth ground for proper operation of the microprocessor.

The 24-v circuit contains an automotive-type, 3-amp fuse (FU1) located on the main control board. (See Fig. 16.) Any direct shorts during installation, service or maintenance may cause this fuse to "blow." If fuse replacement is required, use **only** a fuse of identical size.

With power disconnected to the unit, check all electrical connections for tightness. Tighten all screws on electrical connections. If any smokey or burned connections are found, disassemble the connection, clean all parts, strip wire, and reassemble properly and securely.

Reconnect electrical power to the unit and observe unit through 1 complete operating cycle. Electrical controls are difficult to check without proper instrumentation; if there are any discrepancies in the operating cycle, contact your dealer and request service.

Step 8—Winterizing

> **⚠ CAUTION**
>
> The unit must not be installed, operated, and then turned off and left off in an unoccupied structure during cold weather when the temperature drops to 32° F and below. Freezing condensate left in the furnace will damage the equipment.

If the furnace will be off for an extended period of time in a structure where the temperature will drop to 32° F or below, winterize as follows:

1. Mix a solution of equal amounts of ethylene glycol (Prestone II antifreeze/coolant or equivalent) and water.
2. Turn OFF electrical supply to furnace.
3. Remove control access door.
4. Disconnect drain tube from bottom of inducer outlet box/elbow.
5. Insert funnel in drain tube and pour antifreeze/water solution into furnace until it is visible at point where condensate enters open drain.
6. Reconnect drain tube to outlet box/elbow.
7. Replace control access door.

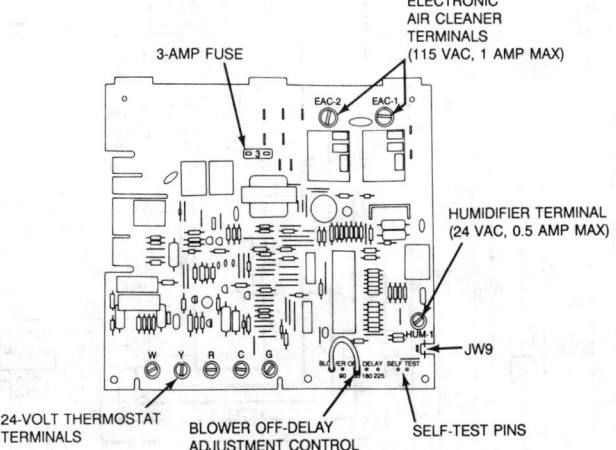

A91060

Fig. 16—Control Center

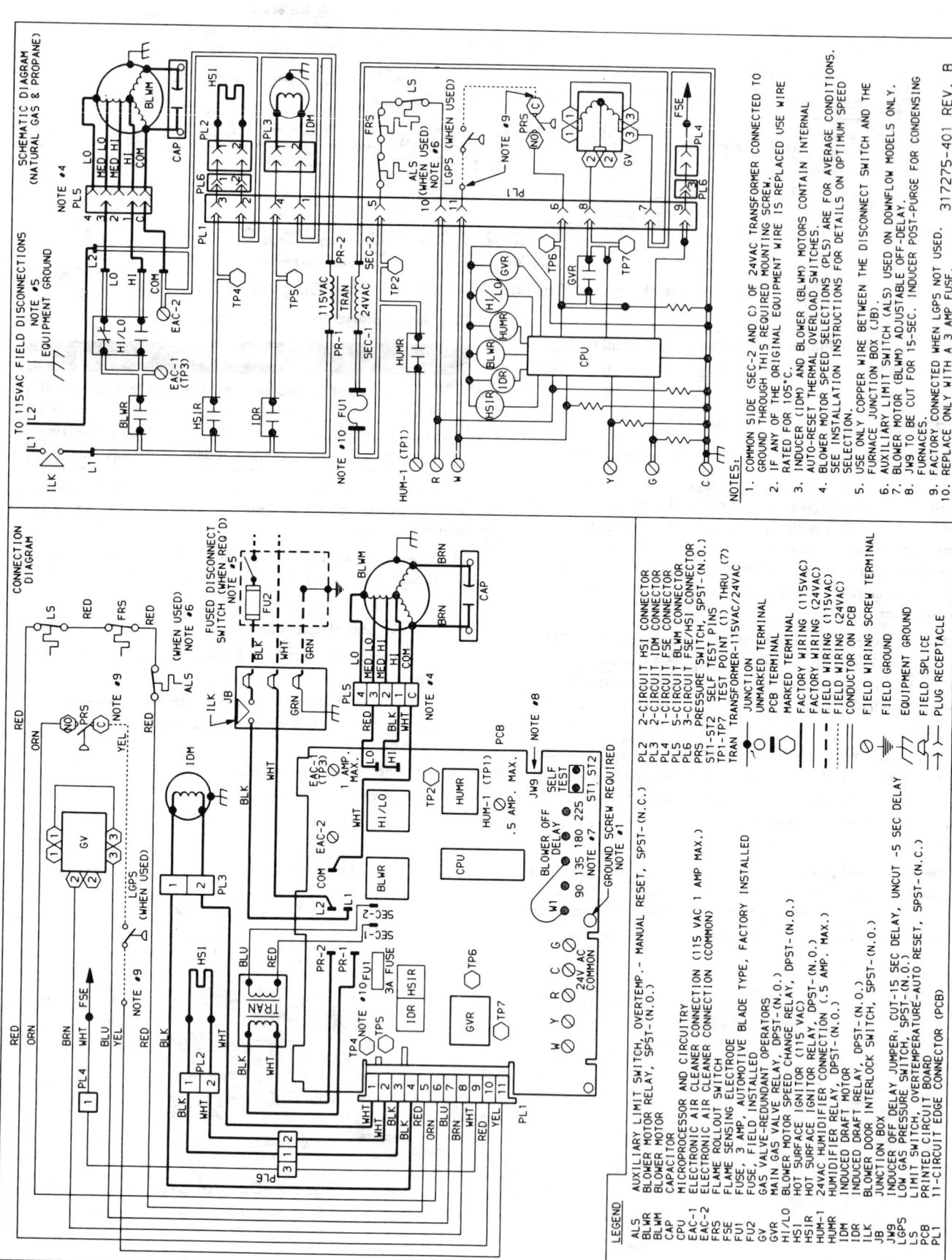

Fig. 17—Wiring Diagram

8

TROUBLESHOOTING GUIDE
HSI CONDENSING FURNACES

START

NOTES:

1. If 115-VAC power is energized or interrupted during a call for heat, the indoor blower will run for 90 sec before a heating cycle begins.

2. After replacing any component, verify correct operating sequence.

WARNING
ELECTRIC SHOCK HAZARD
ONLY QUALIFIED AND TRAINED SERVICE PERSONNEL SHOULD PERFORM THIS PROCEDURE

Turn thermostat OFF.

Cycle 115-v power OFF for 3 sec, then ON.

Set thermostat to call for heat. Set FAN switch to AUTO.

Draft inducer motor starts — No → Is indoor blower **ON**? — Yes →
- 24v should be present across R and C. If not, check for:
  1. Open flame rollout switch (FRS)
  2. Open limit switch (LS)
  3. Open Auxiliary Limit Switch (ALS)
  4. Check all low-voltage wiring connections

— No → Replace control only if all checks are OK.

Yes

Is indoor blower **ON**? — No →
- 24v should be present across C and W. If not, check for:
  1. Open thermostat
  2. Open 24-v fuse (FU1)
  3. Failed transformer
- If 24v are present across TP6 and C, the pressure switch is stuck closed. Replace switch and/or check pressure tube for blockage.
- 115v should be present at the inducer motor. If so, replace inducer motor. If not, check 115-v wiring.

— No → Replace control only if all checks are OK.

15-sec inducer prepurge

Yes

Ignitor warms up and glows orange/yellow; 17*-sec warm-up
*45 sec for subsequent ignition trials and twinned furnaces

— No →
- 24v should be present across TP6 and C. If not, check for:
  1. Open gas inlet pressure switch (when used)
  2. Open pressure switch (PRS) and/or tube
  3. Check all low-voltage wiring connections.
- If pressure switch is open, check for:
  1. Proper vent and air intake pipe sizing
  2. Proper vent condensate drainage
  3. Vent/air intake ice buildup
  4. Blocked condensate drain
- 115v should be present at the ignitor. If so, replace the ignitor; if not, check 115-vac wiring to ignitor.

— No → Replace control only if all checks are OK.

Yes

Main burners ignite

— No →
Control will attempt to light burners 4 times (approximately 1 minute between attempts). Voltage is present at the gas valve for 7 sec during each ignition trial. System will lockout after 4 attempts.
- Is the gas valve control knob in the **OPEN** or **ON** position?
- 24v should be present across the gas valve terminals 1 and 2 during the 7-sec ignition trial. If not:
  1. Check all low-voltage wiring connections to valve.
- If 24v are present, and main gas does not flow, replace gas valve.
- Check ignitor position.
- Check burner carryover gap.
- Check gas supply pressure (4.5-in. wc minimum).
- Check manifold pressure (3.2- to 3.8-in. wc).
- Check for proper orifice size.

— No → Replace control only if all electrical checks are OK.

Yes

Main burners stay on

— No →
If control goes into lockout, wait 5 sec then reset 115-v power.
- Check polarity of 115-v power at J-box and control. Twinned furnace polarities must match.
- Check ground continuity from J-box to control.
- Check flame sensor microamps (4.0 nominal; 0.5 for control to recognize flame).

— No → Replace control only if all checks are OK. Clean flame sensor if microamps are below nominal.

Yes

Indoor blower motor starts on heating speed after 66*-sec warm-up period
*45 sec delay if jumper JW9 is not cut

— No →
- 115v should be present at the blower motor. If so, check capacitor. If capacitor is OK, replace blower motor. If 115v are not present at the blower motor, check all 115-v wiring to motor.

— No → Replace control only if all checks are OK.

Yes

Furnace runs until call for heat ends

— No →
- 24v should be present across TP6 and C. If not, check for:
  1. Satisfied thermostat
  2. Open inlet gas pressure switch (when used)
  3. Open pressure switch (PRS)
  4. Open auxiliary limit (ALS)
  5. Open 24-v fuse (FU1)
  6. Open limit switch (LS)
  7. Open flame rollout switch (FRS)
  8. Check 115-v line voltage.
- Check for sources of electrical noise interference (electronic air cleaners, nearby TV, or radio antennas).

— No → Replace control only if all checks are OK.

Yes

Turn thermostat to OFF; gas valve shuts burners off; 15*-sec inducer post purge
* 5 sec post purge if jumper JW9 is not cut

— No →
- If inducer and burners continue to operate, check for 24v at the gas valve. If 24v are present, verify that the thermostat is open across R and W. If no voltage is present, turn the gas valve control knob to the **OFF** position. Replace gas valve.

— No → Replace control only if all checks are OK.

Yes

Indoor blower motor stops after 90, 135, 180, or 225 sec

— No →
- 24v should be present across R and C. If not, check for:
  1. Open limit switch (LS)
  2. Open flame rollout switch (FRS)
  3. Open auxiliary limit switch (ALS)
- 24v should not be present across R and G. If so, turn thermostat **FAN** switch to **AUTO.**

— No → Replace control only if all checks are OK.

Yes

Heating sequence of operation complete

A92218

9

Carrier
HEATING & COOLING

Maintenance and Service Instructions
For Sizes 040—140, Series 110

NOTE: Read the entire instructions before starting the installation.

SAFETY CONSIDERATIONS

Installation and servicing of heating equipment can be hazardous due to gas and electrical components. Only trained and qualified personnel should install, repair, or service heating equipment.

Untrained personnel can perform basic maintenance functions such as cleaning and replacing air filters. All other operations must be performed by trained service personnel. When working on heating equipment, observe precautions in the literature, tags, and labels attached to or shipped with the unit and other safety precautions that may apply.

Follow all safety codes. In the United States, follow all safety codes including the National Fuel Gas Code NFPA No. 54-1988/ANSI Z223.1-1988. In Canada, refer to the current edition of the National Standard of Canada CAN/CGA- B149.1- and .2-M86 Natural Gas and Propane Gas Installation Codes. Wear safety glasses and work gloves. Have fire extinguisher available during start-up and adjustment procedures and service calls.

Recognize safety information: This is the safety-alert symbol ⚠ . When you see this symbol on the furnace and in instructions or manuals, be alert to the potential for personal injury.

Understand the signal word DANGER, WARNING, or CAUTION. These words are used with the safety-alert symbol. DANGER identifies the most serious hazards which **will** result in severe personal injury or death. WARNING signifies a hazard that **could** result in personal injury or death. CAUTION is used to identify unsafe practices, which **would** result in minor personal injury or product and property damage.

⚠ WARNING

The ability to properly perform maintenance on this equipment requires certain expertise, mechanical skills, tools, and equipment. If you do not possess these, do not attempt to perform any maintenance on this equipment other than those procedures recommended in the Users Manual. A FAILURE TO FOLLOW THIS WARNING COULD RESULT IN POSSIBLE DAMAGE TO THIS EQUIPMENT, SERIOUS PERSONAL INJURY, OR DEATH.

CARE AND MAINTENANCE

For continuing high performance, and to minimize possible equipment failure, it is essential that periodic maintenance be performed on this equipment. Consult your local dealer as to the proper frequency of maintenance and the availability of a maintenance contract.

A88323

Fig. 1—Model 58DHC Horizontal

A88315

Fig. 2—Model 58DHC Downflow

A88319

Fig. 3—Model 58SSC Upflow

⚠ WARNING

Never store anything on, near, or in contact with, the furnace, such as:

1. Spray or aerosol cans, rags, brooms, dust mops, vacuum cleaners, or other cleaning tools.
2. Soap powders, bleaches, waxes or other cleaning compounds, plastic or plastic containers, gasoline, kerosene, cigarette lighter fluid, dry cleaning fluids, or other volatile fluids.
3. Paint thinners and other painting compounds, paper bags or other paper products. A failure to follow this warning could result in corrosion of the heat exchanger, fire, personal injury, or death.

⚠ WARNING

Turn OFF the gas and electrical supplies to the unit before performing any maintenance or service on it. Follow the operating instructions on the label attached to the furnace. A failure to follow this warning could result in personal injury.

The minimum maintenance that should be performed on this equipment is as follows:

1. Check and clean or replace air filter each month or more frequently if required.
2. Check blower motor and wheel for cleanliness and lubrication each heating and cooling season. Clean and lubricate as necessary.
3. Check electrical connections for tightness and controls for proper operation each heating season. Service as necessary.

⚠ CAUTION

As with any mechanical equipment, personal injury can result from sharp metal edges, etc.; therefore, be careful when removing parts.

AIR FILTER ARRANGEMENT — The air filter arrangement may vary depending on the application. Refer to Table 1 or 2 for filter size information.

Table 1—Filter Size Information (Downflow/Horizontal)

FURNACE CASING WIDTH	FILTER SIZE	FILTER TYPE
14-3/16	(2) 16 x 20 x 1	Cleanable
17-1/2	(2) 16 x 20 x 1	Cleanable
21	(2) 16 x 20 x 1	Cleanable
24-1/2	(2) 16 x 20 x 1	Cleanable

Table 2—Filter Size Information (Upflow)

FURNACE CASING WIDTH	FILTER SIZE		FILTER TYPE
	Side Return	Bottom Return	
14-3/16	(1) 16 x 25 x 1*	(1) 14 x 25 x 1	Cleanable
17-1/2	(1) 16 x 25 x 1*	(1) 16 x 25 x 1	Cleanable
21	(1) 16 x 25 x 1	(1) 20 x 25 x 1*	Cleanable
24-1/2	(2) 16 x 25 x 1*	(1) 24 x 25 x 1	Cleanable

* Factory provided with the furnace. Filters may be field modified by cutting as required.

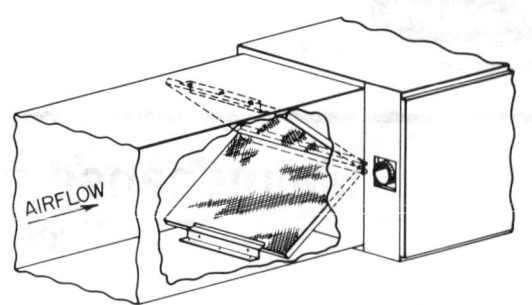

Fig. 4—Horizontal Filter Arrangement

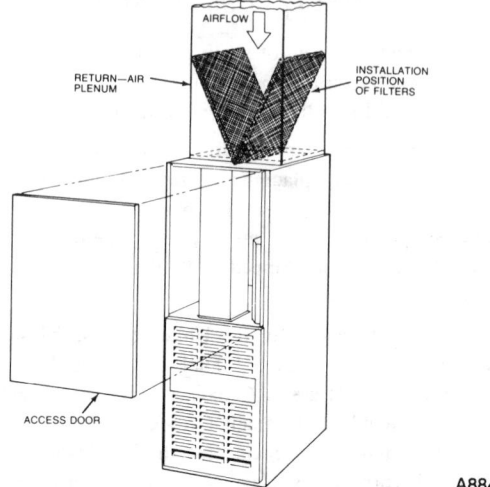

Fig. 5—Position of Filters

⚠ WARNING

Never operate unit without a filter or with filter access door removed. A failure to follow this warning could result in fire, personal injury, or death.

1. Horizontal and Downflow.

Each furnace requires two filters which are installed in the return-air duct. (See Fig. 4 and 5.)

To remove filters for cleaning or replacement, proceed as follows:

    a. Disconnect electrical power before removing access door.

    b. Remove blower access door.

    c. Reach up behind top plate, tilt filters toward center of return-air plenum, remove filters, and replace or clean as needed.

    d. Furnaces are equipped with permanent, washable filters. Clean those filters by spraying tap water through filter in opposite direction of airflow.

    e. Rinse and let dry. Oiling or coating of filters is not recommended.

    f. Reinstall filters with cross-mesh binding facing blower. Replace access door and restore electrical power to furnace.

2. Upflow.

Each furnace requires one or two filters which are installed in the blower compartment. (See Fig. 6.) To remove filters for cleaning or replacement, proceed as follows:

    a. Disconnect electrical power before removing access doors.

    b. Release filter retainer from clip at front of furnace casing. (See Fig. 6.) For side return, clips may be used on either (or both) sides of the furnace.

2

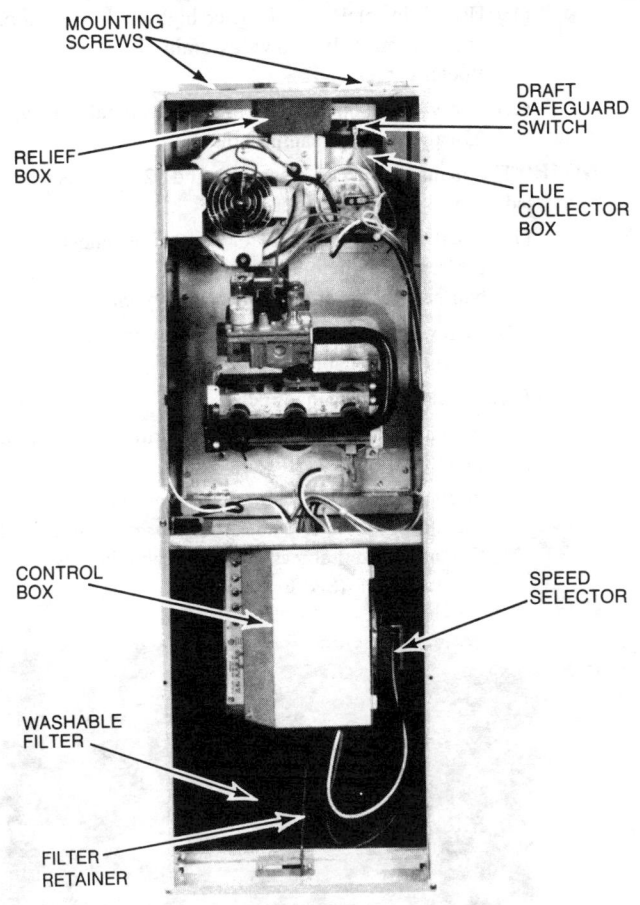

Fig. 6—Model 58SSC Upflow

A88487

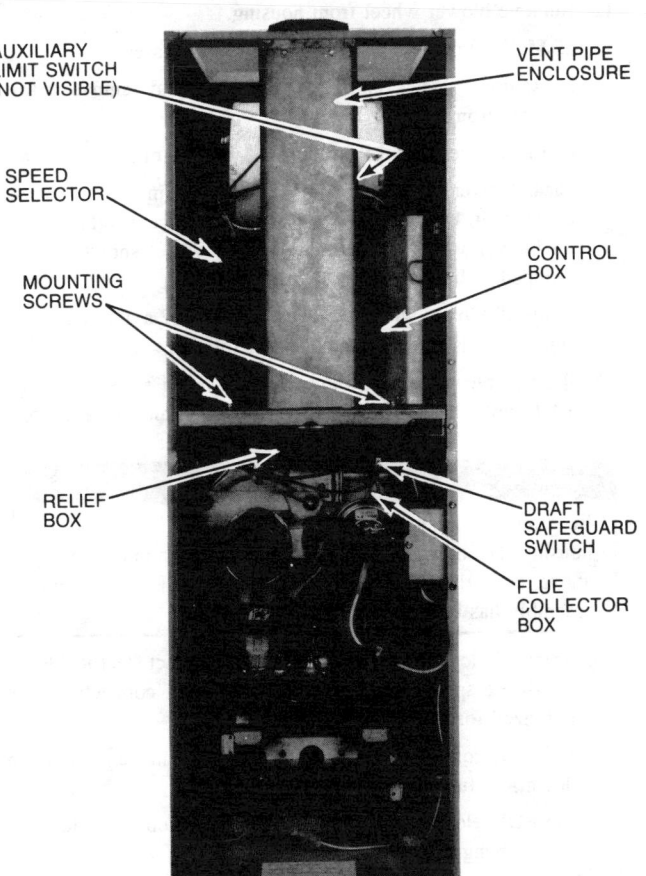

Fig. 7—Model 58RAV Downflow

A88488

c. Slide filter out.

d. Clean filter with tap water. Spray in opposite direction of airflow.

e. Rinse and let dry. Oiling or coating of filter is not recommended.

f. Place filter in furnace with cross-mesh binding up or facing blower. Replace access doors and restore electrical power to furnace.

BLOWER MOTOR AND WHEEL — For long life, economy, and high efficiency; clean accumulated dirt and grease from the blower wheel and motor annually.

The following steps should be performed by a qualified service technician:

Some motors have prelubricated sealed bearings and require no lubrication. These motors can be identified by the absence of oil ports on each end of the motor. For those motors with oil ports lubricate as follows:

Lubricate motor every 5 years if motor is used on intermittent operation (thermostat FAN switch in AUTO position), or every 2 years if motor is in continuous operation (thermostat FAN switch in ON position).

Remember to disconnect the electrical supply before removing access doors.

Clean and lubricate as follows:

1. Remove blower access door.

2. Disconnect vent pipe on downflow furnace only.

   a. Remove vent pipe enclosure.

   b. Disconnect vent pipe and remove short piece of pipe from furnace.

c. Disconnect wires from auxiliary limit on blower housing.

3. Remove control box.

4. Remove electrical leads from numbered side of Molex speed selector. (See Fig. 6 and 7.) Note location of wires for reassembly.

5. Remove screws holding blower assembly to blower deck and slide blower assembly out of furnace.

6. Squeeze side tabs of Molex speed selector and pull it from blower housing.

7. Loosen a screw in strap holding motor capacitor to blower housing and slide capacitor out from under strap.

8. Mark blower wheel, motor, and motor support in relation to blower housing before disassembly, to insure proper reassembly.

9. Loosen setscrew holding blower wheel on motor shaft.

10. Remove bolts holding motor mount to blower housing and slide motor and mount out of housing. Disconnect ground wire attached to blower housing before removing motor.

11. Lubricate motor (when oil ports are provided).

    a. Remove dust caps or plugs from oil ports located at each end of motor.

    b. Use a good grade of SAE 20 nondetergent motor oil and put one teaspoon, 5 cc, 3/16 oz, or 16 to 25 drops in each oil port. Do not over-oil.

    c. Allow time for total quantity of oil to be absorbed by each bearing.

    d. After oiling motor, be sure to wipe excess oil from motor housing.

    e. Replace dust cap or plugs on oil ports.

3

12. Remove blower wheel from housing.

   a. Mark cutoff location to insure proper reassembly.

   b. Remove screws holding cutoff plate and remove cutoff plate from housing.

   c. Lift blower wheel from housing through opening.

13. Clean blower wheel and motor using a vacuum with soft brush attachment. Do not remove or disturb balance weights (clips) on blower wheel blades. The blower wheel should not be dropped or bent as balance will be affected.

14. Reinstall blower wheel by reversing steps 12 a. through c. Be sure wheel is positioned for proper rotation.

15. Reassemble motor and blower by reversing steps 5 through 10. If motor has ground wire, be sure it is connected as before.

⚠ CAUTION

Be sure the motor is properly positioned in the blower housing. The motor oil ports must be at a minimum of 45 deg above the horizontal centerline of the motor after the blower assembly has been reinstalled in the furnace.

16. Reinstall blower assembly in furnace. Connect electrical leads to Molex speed selector. Please note that connections are polarized for assembly. **Do not force.**

17. Reinstall control box (reinstall vent pipe and enclosure on downflow furnace).

18. Turn ON electrical power and check for proper rotation and speed changes between heating and cooling.

CLEANING HEAT EXCHANGER

The following steps should be performed by a qualified service technician:

NOTE: Deposits of soot and carbon indicate the existance of a problem which needs to be corrected. Take action to correct the problem.

If it becomes necessary to clean the heat exchanger because of carbon deposits, soot, etc., proceed as follows:

1. Turn OFF gas and power to furnace.

2. Remove control and blower access doors.

3. Remove vent pipe enclosure (downflow/horizontal furnace only) and disconnect vent pipe from relief box.

4. Remove two screws that secure relief box. (See Fig. 6 or 7.)

5. Disconnect wires to the following components:

   a. Draft safeguard switch.

   b. Inducer motor.

   c. Pressure switch.

   d. Limit overtemperature switch(es).

   e. Gas valve.

   f. Hot surface ignitor.

   g. Flame-sensing electrode.

   h. Two wiring edge connectors leading to control box.

6. Remove eight screws that secure flue collector box to center panel. Be careful not to damage sealant.

7. Remove complete inducer assembly from furnace, exposing flue openings.

8. Using field-provided small wire brush, steel spring cable, reversible electric drill, and vacuum cleaner; clean cells as follows:

   a. Assemble wire brush and steel spring cable.

     (1.) Use 48 in. of ¼-in. diameter high-grade steel spring cable (commonly known as drain clean-out or Roto-Rooter cable).

     (2.) Use ¼-in. diameter wire brush (commonly known as 25-caliber rifle cleaning brush).

NOTE: The items needed in steps (1.) and (2.) can usually be purchased at local hardware stores.

     (1.) Insert twisted wire end of brush into end of spring cable, and crimp tight with crimping tool or strike with ball-peen hammer. *Tightness is very important.*

     (2.) Remove metal screw fitting from wire brush to allow insertion into cable.

   b. Clean each heat exchanger cell.

     (1.) Attach variable-speed, reversible drill to end of spring cable (end opposite brush).

     (2.) Insert brush end of cable into upper opening of cell and slowly rotate with drill. *Do not* force cable. Gradually insert at least 36 in. of cable into two upper passes of cell. (See Fig. 8.)

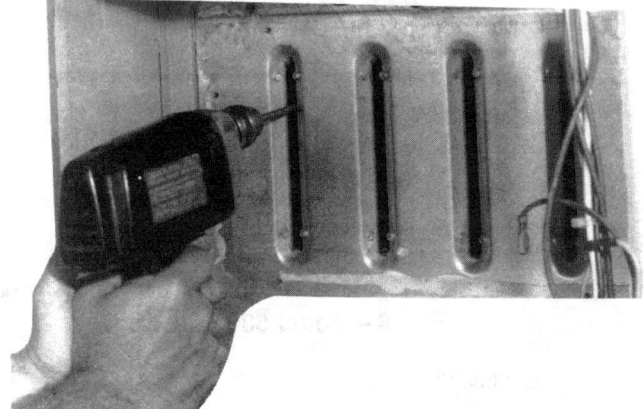

A88489

Fig. 8—Cleaning Heat Exchanger Cell

     (3.) Work cable in and out of cell three or four times to obtain sufficient cleaning. *Do not* pull cable with great force. Reverse drill and gradually work cable out.

     (4.) Remove burner assembly and cell inlet plates.

NOTE: Be very careful when removing the burner assembly to avoid breaking the ignitor. See Fig. 9 for the correct ignitor location

     (5.) Replace screws in center panel and cells before cleaning.

     (6.) Insert brush end of cable in lower opening of cell, and proceed to clean two lower passes of cell in same manner as two upper passes.

     (7.) Repeat foregoing procedures until each cell in furnace has been cleaned.

     (8.) Using vacuum cleaner, remove residue from each cell.

     (9.) Using vacuum cleaner with soft brush attachment, clean burner assembly.

     (10.) Reinstall cell inlet plates and burner assembly. Care must be exercised to center the burners in the cell openings.

9. After cleaning flue openings, check sealant on flue collector to ensure that it has not been damaged. If new sealant is needed, contact your dealer or distributor.

10. Clean and replace flue collector assembly, making sure all eight screws are secure.

12. Reinstall relief box.

13. Reconnect wires to the following components:

    a. Draft safeguard switch.

    b. Inducer motor.

    c. Pressure switch.

    d. Limit overtemperature switch(es).

    e. Gas valve.

    f. Hot surface ignitor.

    g. Flame sensing electrode.

    h. Two wiring edge connectors leading to control box.

⚠ WARNING

Never use a match or other open flame to check for gas leaks. Use a soap-and-water solution. A failure to follow this warning could result in fire, personal injury, or death.

14. Reconnect vent pipe to relief box. Replace vent pipe enclosure (when applicable).

15. Replace blower door only.

16. Turn on power and gas.

17. Set thermostat and check furnace for proper operation.

18. Check for gas leaks.

19. Replace control door.

HOT SURFACE IGNITOR

NOTE: Be very careful when removing the burner assembly to avoid breaking the ignitor. See Fig. 9 for the correct ignitor location.

ELECTRICAL CONTROLS AND WIRING

NOTE: There may be more than one electrical supply to unit.

The electrical ground and polarity for 115-volt wiring must be maintained properly. Refer to Fig. 10 for field wiring information and to Fig. 11 for unit wiring information. If the polarity is **not** correct, the microprocessor control will shut OFF the gas flow shortly after completion of the ignition trial period. The control system also requires an earth ground for proper operation of the microprocessor.

With power disconnected to unit, check all electrical connections for tightness. Tighten all screws on electrical connections. If any smoky or burned connections are noticed, disassemble the connection, clean all parts and stripped wire, and reassemble properly and securely. Electrical controls are difficult to check without proper instrumentation; therefore, reconnect electrical power to unit and observe unit through one complete operating cycle. If there are any discrepancies in the operating cycle, contact your dealer and request service.

→ The 24-volt circuit contains an automotive-type, 3-amp fuse located on the main control board. Any direct shorts during installation, service, or maintenance could cause this fuse to blow. If fuse replacement is required, use **only** a 3-amp fuse of identical size.

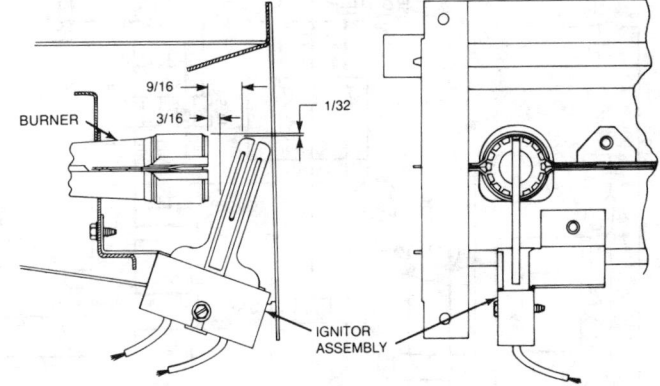

Fig. 9—Position of Ignitor to Burner

A88490

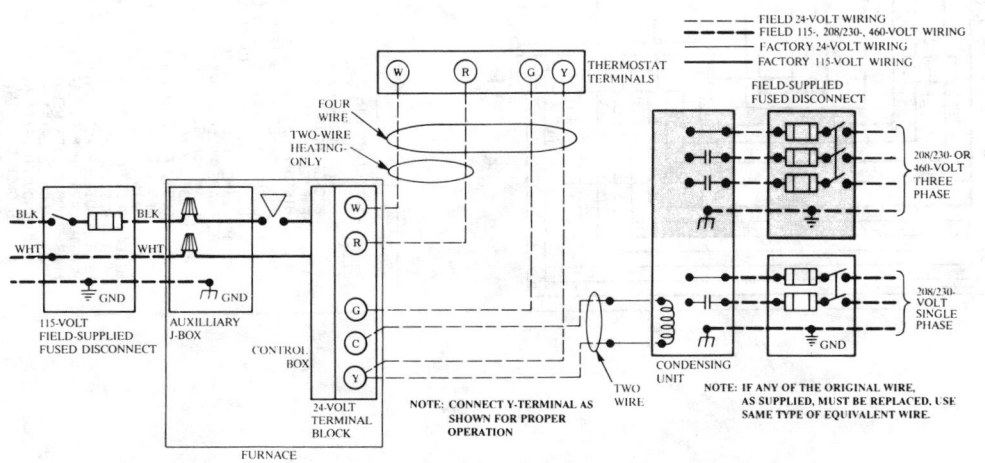

A78461

Fig. 10—Heating and Cooling Application Wiring Diagram

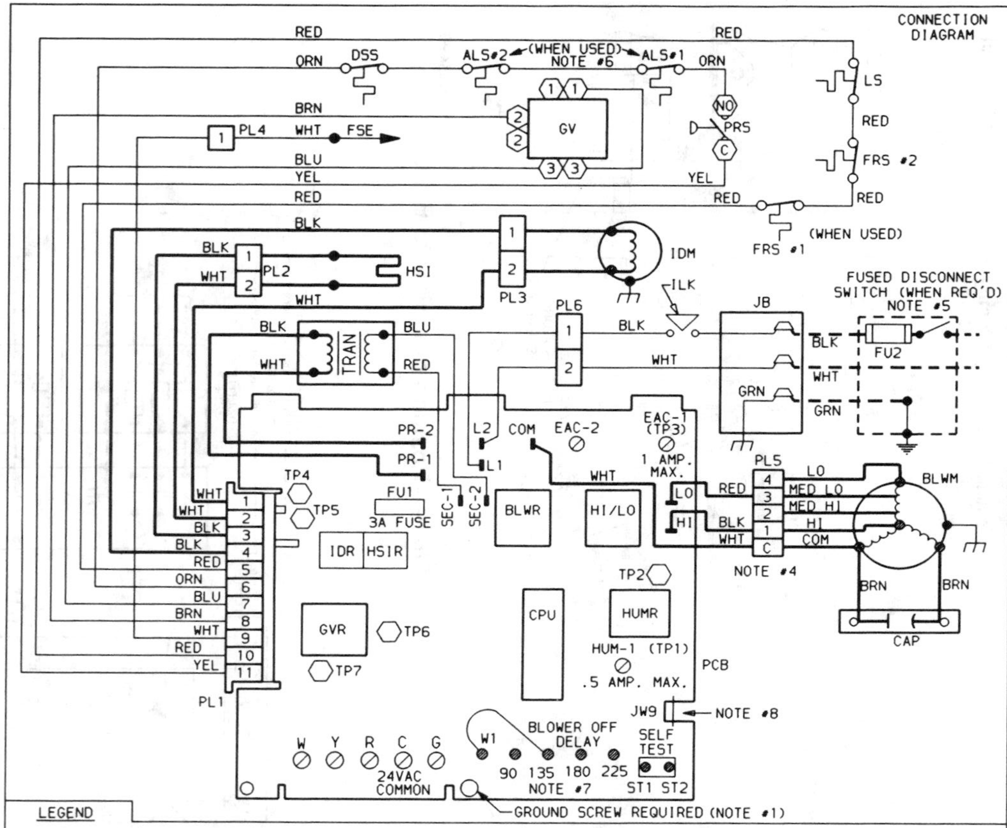

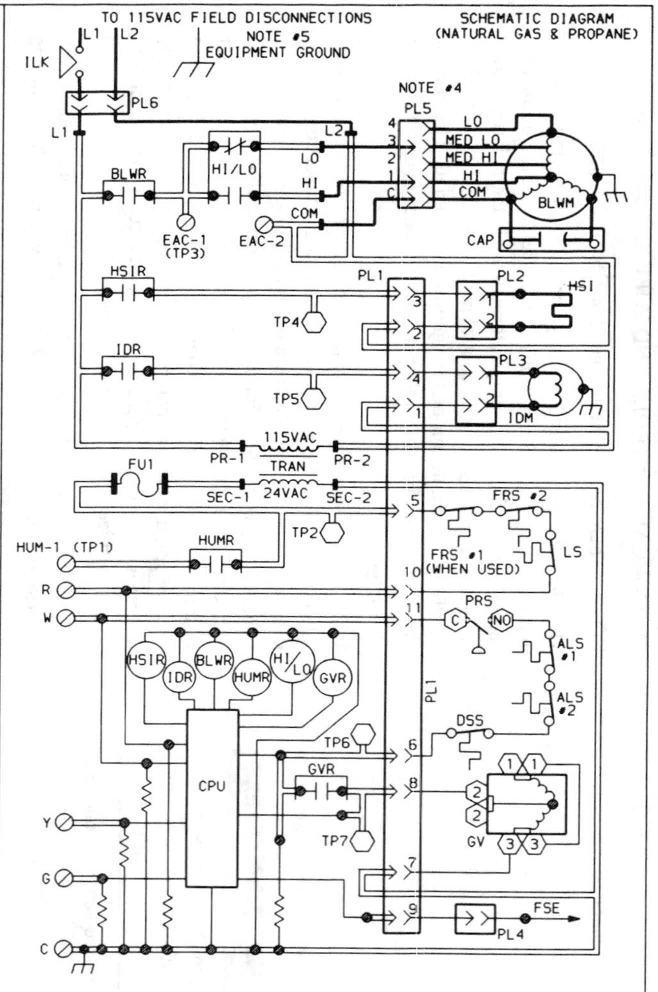

→ Fig. 11—Unit Wiring Diagram

A91059

318078-401 REV. A

15011

Model 58PAV & 58RAV
Induced-Combustion Furnaces

NOTE TO INSTALLER:
This manual must be left with the equipment user.

FOR YOUR SAFETY
Do not store or use gasoline or other flammable vapors and liquids in the vicinity of this or any other gas appliance.

FOR YOUR SAFETY
What to do if you smell gas:

1. Do not try to light any appliance.
2. Do not touch any electrical switch.
3. Do not use any phone in your building. Immediately call your gas supplier from a neighbor's phone. Follow the gas supplier's instructions.
4. If you cannot reach your gas supplier, call the fire department.

⚠**WARNING:** Improper installation, adjustment, alteration, service, maintenance, or use can cause carbon monoxide poisoning, explosion, fire, electrical shock, or other conditions which could result in personal injury, property damage, or death. Consult a qualified installer, service agency, or your local gas supplier for information or assistance.

USER'S INFORMATION MANUAL FOR THE OPERATION AND MAINTENANCE OF YOUR NEW GAS-FIRED FURNACE

GAS FURNACES

WELCOME TO A NEW GENERATION OF COMFORT

Congratulations! Your new, higher efficiency gas furnace is a sound investment which will reward you and your family with years of "warm memories" winter after winter.

Not only is your new furnace energy-efficient, it is also one of the most reliable. Spend just a few minutes with this booklet to learn about the operation of your new furnace—and the small amount of maintenance it takes to keep it operating at peak efficiency. Years went into the development of your new furnace. Take a little time now to assure its most efficient operation for years to come.

A90110

1

**MODEL 58PAV
UPFLOW FURNACE**

A90109

2

**MODEL 58RAV
DOWNFLOW/HORIZONTAL FURNACE**

FURNACE IDENTIFICATION

For your convenience, record the product and serial numbers of your new furnace on the form below. Should you ever require service, you will have ready access to the information needed by the service representative.

Product No. _____

Serial No. _____

Date Installed _____

Dealer Name _____

Address _____

City _____

State _____ Zip _____

Telephone _____

2

UPFLOW FURNACE COMPONENTS

1. Relief Box
2. Gas Valve Control Knob (On/Off)
3. Gas Valve
4. Gas Burner
5. Hot Surface Ignitor
6. Blower Door Safety Switch
7. Blower and Blower Motor
8. Draft Safeguard Tube & Switch
9. Rating Plate (Behind Junction Box)
10. Gas Manifold
11. Filter Retainer
12. Air Filter
13. Flame Sensor
14. Manual Reset Limit Switch

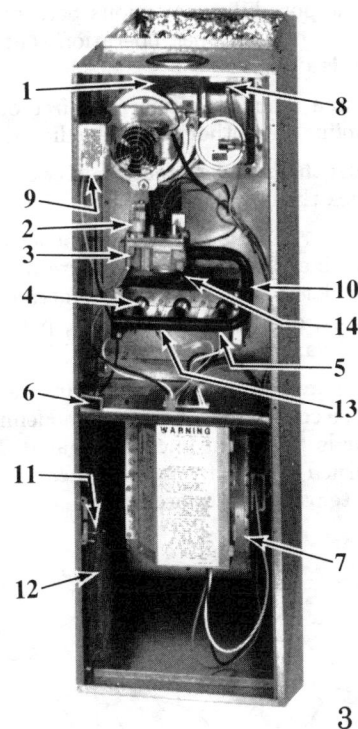

A90162

3

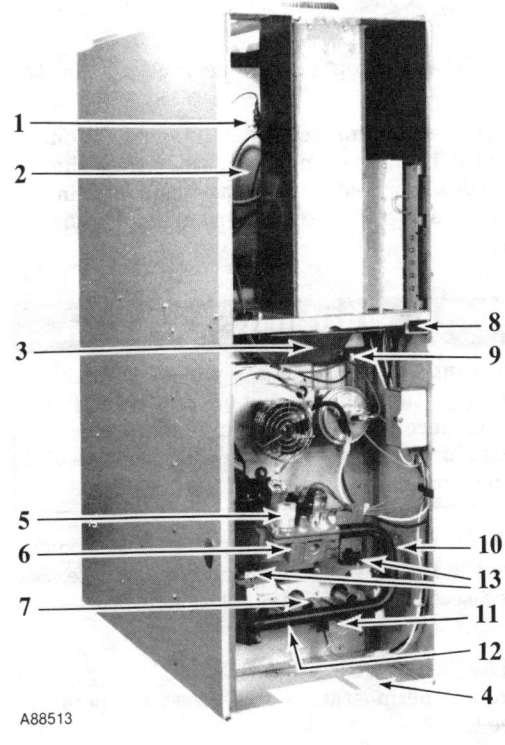

A88513

4

DOWNFLOW/HORIZONTAL FURNACE COMPONENTS

1. Manual-Reset Auxiliary Limit Switch
2. Blower and Blower Motor
3. Relief Box
4. Rating Plate
5. Gas Valve Control Knob (On/Off)
6. Gas Valve
7. Gas Burner
8. Blower Door Safety Switch
9. Draft Safeguard Tube & Switch
10. Gas Manifold
11. Hot Surface Ignitor
12. Flame Sensor
13. Manual Reset Limit Switch (2)

IMPORTANT FACTS

Your furnace must have adequate airflow for efficient combustion and safe ventilation. Do not enclose it in an airtight room or "seal" it behind solid doors.

To minimize the possibility of serious personal injury, fire, furnace damage, or improper operation; **carefully follow these safety rules:**

● Keep the area around your furnace free of combustible materials, gasoline, and other flammable liquids and vapors.

● Do not cover the furnace, store trash or debris near it, or in any way block the flow of fresh air to the unit.

● Combustion air must be clean and uncontaminated with chlorine or fluorine. These compounds are present in many products around the home, such as: water softener salts, laundry bleaches, detergents, adhesives, paints, varnishes, paint strippers, waxes, and plastics.

Make sure the combustion air for your furnace does not contain any of these compounds. During remodeling be sure the combustion air is fresh and uncontaminated. If these compounds are burned in your furnace, the heat exchangers and metal vent system may deteriorate.

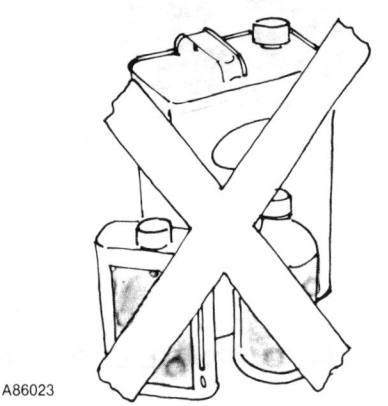

A86023   5

● A furnace installed in an attic or other insulated space must be kept free and clear of insulating material. Examine the furnace area when installing the furnace or adding more insulation. Some materials may be combustible.

NOTE: Do not use this furnace if any part has been under water. Immediately call a qualified service technician to inspect the furnace and to replace any part of the control system and any gas control which has been under water.

NOTE: The qualified installer or agency must use only factory-authorized replacement parts, kits, and accessories when modifying this product.

This furnace contains SAFETY DEVICES which must be MANUALLY RESET. If the furnace is left unattended for an extended period of time, have it checked periodically for proper operation. This precaution will prevent problems associated with NO HEAT, such as frozen water pipes, etc. See Before You Request a "Service Call" section in this manual.

SAFETY CONSIDERATIONS

Installation and servicing of heating equipment can be hazardous due to gas and electrical components. Only trained and qualified personnel should install, repair, or service heating equipment.

Untrained personnel can perform basic maintenance functions such as cleaning and replacing air filters. All other operations must be performed by trained service personnel. Observe safety precautions in this manual, on tags, and labels attached to the furnace and other safety precautions that may apply.

Recognize safety information: This is the safety-alert symbol △. When you see this symbol on the furnace and in instructions or manuals, be alert to the potential for personal injury.

Understand the signal word—DANGER, WARNING, or CAUTION. These words are used with the safety-alert symbol. DANGER identifies the most serious hazards which **will** result in severe personal injury or death. WARNING signifies hazards that **could** result in personal injury or death. CAUTION is used to identify unsafe practices, which would result in minor personal injury or product and property damage.

STARTING YOUR FURNACE

Instead of a continuously burning pilot flame which wastes valuable energy, your furnace uses an automatic hot surface ignition system to light the pilot each time the thermostat turns your furnace on. **Follow these important safeguards:**

● Never attempt to manually light the pilot with a match or other source of flame.

A86024   6

● Read and follow the operating instructions on the furnace.

● If a suspected malfunction occurs with your gas control system, such as the burners do not light when they should, refer to the shutdown procedures on the furnace or in the next section to turn off your system, then call your dealer as soon as possible.

> ### ⚠ WARNING
>
> Should overheating occur, or the gas valve fail to shut off the gas supply, turn off the manual gas valve (See Fig. 8) to the furnace BEFORE turning off the electrical supply. A failure to follow this warning could result in a fire or explosion and personal injury or death.

● **CHECK AIR FILTER:** Before attempting to start your furnace, be sure the furnace filter is clean and in place. See the maintenance section of this manual.) Then proceed as follows:

STEPS FOR STARTING YOUR FURNACE

1. Set your room thermostat to the lowest temperature setting. See Fig. 7.
2. Close the external manual gas valve. See Fig. 8.
3. Turn OFF the electrical supply to the furnace. See Fig. 9.

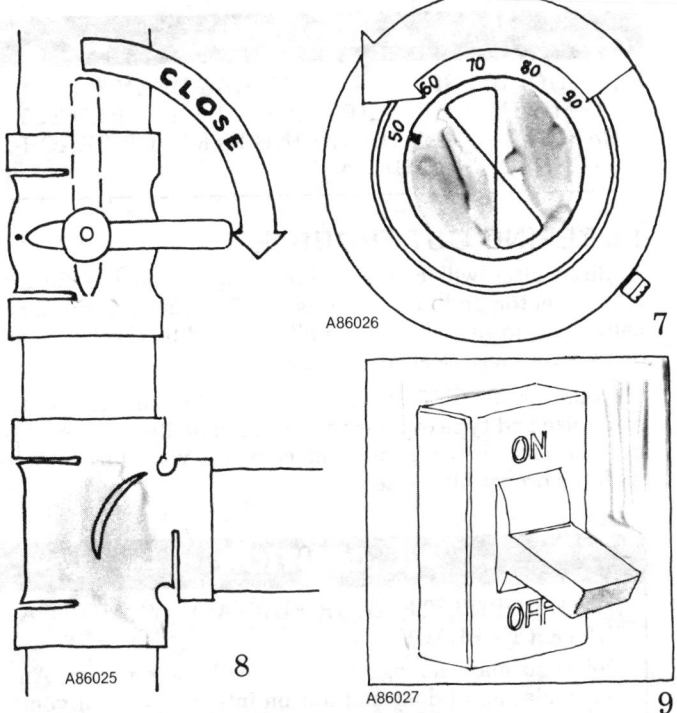

A86026
7

A86025
8

A86027
9

4. Remove the furnace access door(s).

   a. Upflow—remove control door. See Fig. 10.

   b. Downflow/Horizontal—remove blower door first, then remove the control door. See Fig. 11.

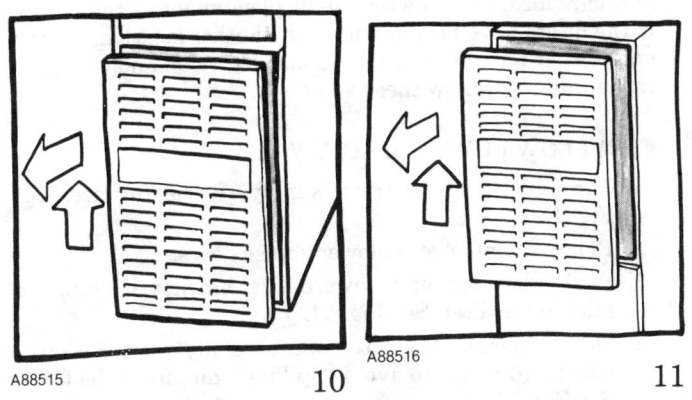

A88515
10

A88516
11

5. Turn the control knob on the gas valve to the OFF position and wait 5 minutes. See Fig. 12.

6. After waiting 5 minutes, turn the control knob on the gas valve to the ON position. See Fig. 13.

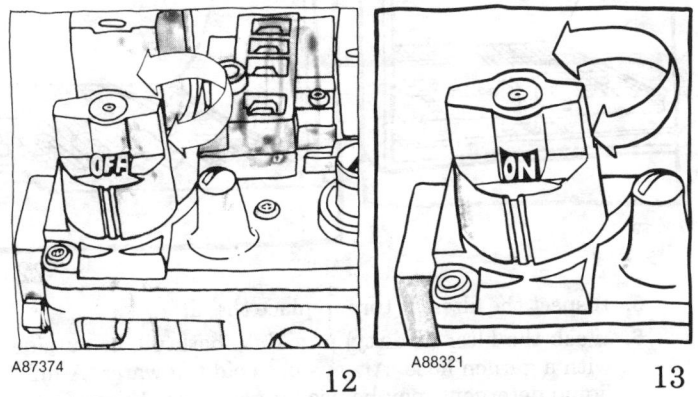

A87374
12

A88321
13

7. Replace the access door(s). See Fig. 14 for upflow and Fig. 15 for downflow. Replace control door first on downflow furnaces. Then replace blower door.

8. Turn ON the electrical supply to the furnace. See Fig. 16.

A88518
14

A88517
15

9. Open the external manual gas valve. See Fig. 17.

10. Set the room thermostat to a temperature slightly above the room temperature. This will automatically signal the furnace to start. The inducer motor will start, and the hot surface ignitor will energize. When hot, the ignitor will have an orange glow.

11. After 15 to 70 seconds, the gas valve permits gas to flow to the main burners where it is ignited. Hot flames begin to warm the furnace's heat exchanger. After a time delay of approximately 60 seconds, the furnace blower is switched on.

NOTE: If the main burners fail to ignite, the furnace control system will go thru three more ignition cycles. Then, if burners fail to ignite, the system will lockout. If lockout occurs, or the blower doesn't come on—shut down your furnace and call your dealer for service.

12. Set your thermostat to the temperature that satisfies your comfort requirements.

   SUGGESTION: Setting the thermostat back a few degrees—and compensating for the difference with warmer clothing—can make a big difference in your fuel consumption on extremely cold days. The few degrees at the top of your thermostat "comfort level" are the most costly degrees to obtain.

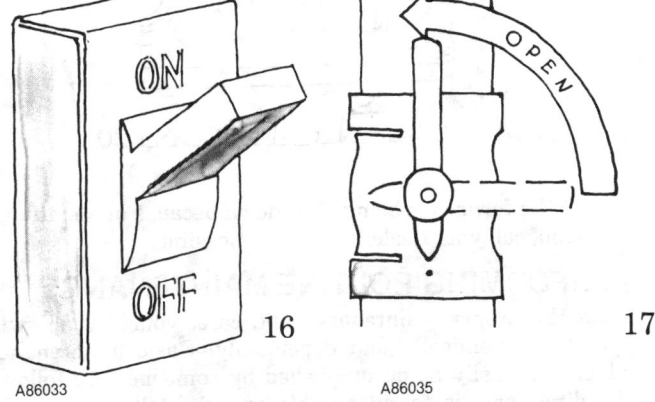

A86033
16

A86035
17

When the room temperature drops below the temperature selected on the thermostat, the furnace will be switched on automatically. When the room temperature reaches the degree selected on the thermostat, the furnace will be automatically switched off.

Some thermostats have a "fan" switch with two selections: AUTO or ON. When set on AUTO, the furnace blower cycles on and off, controlled by the thermostat. In the ON position, the furnace blower runs continuously except for a 60 second delay at the "call for heat." This keeps the temperature level in your home more evenly balanced. It also continuously filters the indoor air.

SHUTTING DOWN YOUR FURNACE

Should you ever suspect a malfunction in your furnace, you will need to turn the furnace off. The following procedures must be followed:

1. Set your room thermostat to the lowest temperature setting. See Fig. 18.
2. Close the external manual gas valve. See Fig. 8 on page 5.
3. Turn OFF the electrical power to your furnace. See Fig. 19.

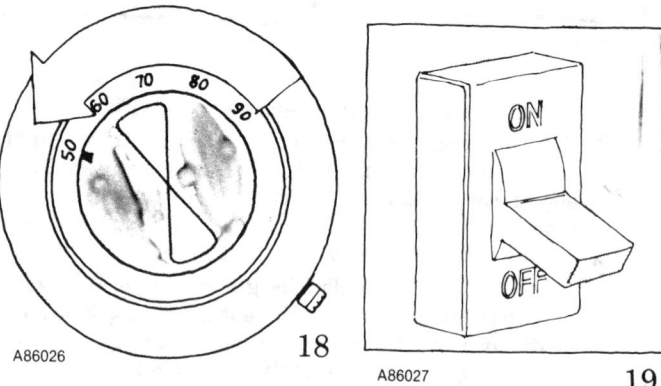

A86026 **18**   A86027 **19**

4. Remove the control access door on your furnace. See Fig. 10 or 11. On upflow furnaces, removing the blower access door is not required.
5. Turn the control knob on the gas valve to the OFF position. See Fig. 20.
6. Replace the control access door. See Fig. 14 or 15.

A87374 **20**

7. If the furnace is being shut down because of a malfunction, call your dealer as soon as possible.

PERFORMING ROUTINE MAINTENANCE

With the proper maintenance and care, your furnace will operate economically and dependably. Basic maintenance which can easily be accomplished by someone who follows the directions, is found on this and the following pages. However, before beginning maintenance, follow these safety precautions:

⚠ WARNING

TURN OFF ELECTRICAL POWER SUPPLY TO YOUR FURNACE BEFORE REMOVING THE ACCESS DOORS TO SERVICE OR PERFORM MAINTENANCE. A FAILURE TO FOLLOW THIS WARNING COULD RESULT IN PERSONAL INJURY OR DEATH.

⚠ CAUTION

ALTHOUGH SPECIAL CARE HAS BEEN TAKEN TO MINIMIZE SHARP EDGES, BE EXTREMELY CAREFUL WHEN HANDLING PARTS OR REACHING INTO THE FURNACE.

FILTERING OUT TROUBLE

A dirty filter will cause excessive stress on the furnace blower motor and can cause it to overheat and automatically shut down. The furnace filter should be checked every 3 or 4 weeks and cleaned if necessary.

If your furnace filter needs replacing, be sure to use the same size and type of filter that was originally supplied. Use the Furnace Filter Table and compare your furnace size with the proper filter size.

⚠ CAUTION

NEVER OPERATE YOUR FURNACE WITHOUT A FILTER IN PLACE.

Doing so may damage the furnace blower motor. An accumulation of dust and lint on internal parts of your furnace can cause a loss of efficiency.

The air filter for upflow furnaces is normally located in the blower compartment. Filters for the downflow furnaces are normally located in the return air plenum above the blower. If the filters have been installed in another location, contact your dealer for instructions. To inspect, clean and/or replace the air filter(s), follow these steps:

● UPFLOW FURNACES ONLY:

1. Turn OFF the electrical supply to the furnace. See Fig. 19.
2. Remove control and blower access doors.
3. Push filter retainer toward the bracket opening to release the filter. See Fig. 21.
4. Gently remove the filter and carefully turn the dirty side up (if dirty) to avoid "spilling" dirt from the filter. See Fig. 22.

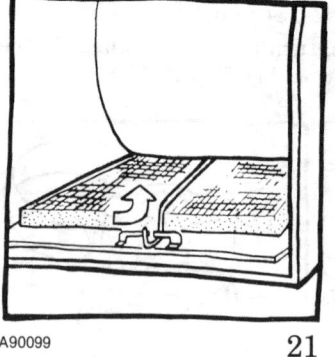

A90099 **21**   A90100 **22**

5. Inspect the filter. If torn, replace the filter.
6. Wash the filter (if dirty) in a sink, bathtub, or outside with a garden hose. Always use cold tap water. A mild liquid detergent may be used if necessary. Spray water through the filter in the opposite direction of airflow (through the cross-mesh binding side). Allow filter to dry.
7. Reinstall the clean filter with the cross-mesh binding side facing the furnace blower.

8. Put filter retainer back in the bracket opening and lock it in place.

9. Replace blower and control access doors and turn ON electrical power to your furnace. See Figs. 14, 16, and 23.

NOTE: For upflow models only—if side return ducts are used, 2 filters may be required in some models. The procedure listed above may be used to remove side filters.

23

● DOWNFLOW/HORIZONTAL FURNACES ONLY: Two filters are located in the return-air plenum above the blower (above line-of-sight) resting in the V-shaped channel on top of the furnace. See Fig. 24.

1. Turn OFF electrical supply to the furnace. See Fig. 19.

2. Remove blower access doors.

3. Remove the left side filter by tipping the filter toward the center—raise it from the V-shaped channel in which it rests. See Figs. 24 and 25.

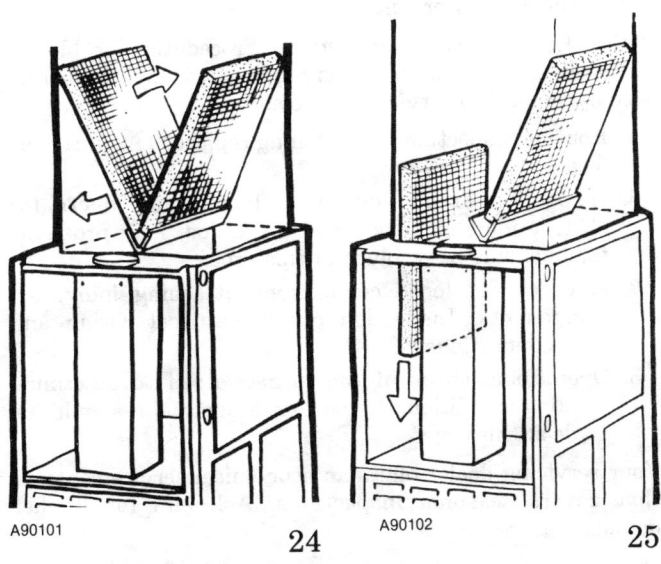

A90101   24   A90102   25

4. Lower filter down along side of the blower and remove from the furnace.

5. To remove the second filter, lift from V-shaped channel and remove the same way as left side filter.

6. Inspect the filter. If torn, replace the filter.

7. Wash the filter (if dirty) in a sink, bathtub, or outside with a garden hose. Always use cold tap water. A mild liquid detergent may be used if necessary. Spray water through the filter in the opposite direction of airflow through the cross-mesh binding side. Allow filter to dry.

8. Reinstall the clean filters with the cross-mesh binding side facing the furnace blower.

UPFLOW FURNACE FILTER TABLE

Furnace Casing Width	Filter Size		Filter Type
	Side Return	Bottom Return	
14-3/16	(1) 16 x 25 x 1*	(1) 14 x 25 x 1	Cleanable
17-1/2	(1) 16 x 25 x 1*	(1) 16 x 25 x 1	Cleanable
21	(1) 16 x 25 x 1	(1) 20 x 25 x 1*	Cleanable
24-1/2	(2) 16 x 25 x 1	(1) 24 x 25 x 1*	Cleanable

DOWNFLOW/HORIZONTAL FURNACE FILTER TABLE

Furnace Casing Width	Filter Size	Filter Type
14-3/16	(2) 16 x 20 x 1*	Cleanable
17-1/2	(2) 16 x 20 x 1*	Cleanable
21	(2) 16 x 20 x 1*	Cleanable
24-1/2	(2) 16 x 20 x 1*	Cleanable

*Factory provided with the furnace. Filter may be field modified by cutting as required. Alternate sizes and additional filters may be ordered from your dealer.

9. Replace blower door and turn ON electrical power to your furnace.

COMBUSTION AREA AND VENT SYSTEM

Inspect the combustion area and vent system before each heating season. An accumulation of dirt, soot, or rust can mean a loss of efficiency and improper performance. Build-ups on the main burners can cause faulty firing. This "delayed ignition" is characterized by an alarmingly loud sound. If your furnace makes a loud noise when the main burners are ignited, shut down the furnace—call your servicing dealer.

Use your flashlight and follow these steps for inspecting the combustion area and vent system of your furnace:

1. Turn off the electrical supply to the furnace and remove the access doors. See Figs. 9 and 10, or 11.

2. Carefully inspect the gas burner (see Fig. 26) for dirt, rust, or scale. Then, inspect the relief box, flue connection area, and the vent pipe for rust.

NOTE: If dirt, rust, soot, or scale accumulations are found, call your servicing dealer. DO NOT OPERATE THE FURNACE.

3. Inspect the vent pipe for a sag, holes, or a disconnection. A horizontal vent pipe must slope upward. If rusty joints or seams, or signs of water leakages are found call your dealer for service.

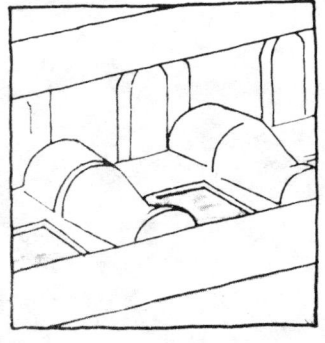

A86041   26

4. Replace the access doors and restore electrical power to the furnace. Be sure bottom door flange is inside of the furnace casing. See Figs. 14, 15, and 23.

5. Start the furnace and observe its operation. If possible, watch the burner flames. Are they burning bright blue? If not (or if you suspect some other malfunction), call your servicing dealer.

BEFORE YOU REQUEST A "SERVICE CALL"

BEFORE YOU CALL FOR SERVICE, CHECK FOR SEVERAL EASILY SOLVED PROBLEMS:

☐ Check for sufficient airflow. Check the air filter for dirt. Check for blocked return-air or supply-air grilles. Be sure they are open and unobstructed. If this isn't the cause, call your servicing dealer.

If your furnace isn't operating at all, check the following list for easily solved problems.

☐ Is your thermostat set above room temperature? Is the switch in the HEAT position?

☐ Is the electrical power supply switch ON? Is the blower access door firmly in place? Are any fuses blown—has a circuit breaker tripped?

☐ Is the manual shut-off valve in the gas supply pipe leading to the furnace open? Does the lever point in the same direction that the pipe runs (open)? Or is it at right angles (closed)?

NOTE: Before proceeding with the next checks, turn OFF the electrical power supply to the furnace. Remove the access doors.

☐ Is the control knob on the gas valve turned to the ON position? If this or the preceding check shows an interruption in the gas supply, make sure the gas has not been shut off for safety reasons. If nothing else seems to be wrong, follow the startup procedures found on pages 4 and 5 of this booklet.

☐ If for some reason the vent is blocked, the draft safeguard switch will shut off the furnace. Reset the switch by pushing the button located on top of the switch (see page 3 for switch location).

If the switch trips a second time, turn off the furnace and call for service.

☐ DOWNFLOW/HORIZONTAL ONLY—Check the manual-reset auxiliary limit switch located on the blower housing. If the blower motor fails, this switch will shut off the furnace. Reset it by pushing the button on the switch. If it trips again, turn off the furnace and call for service.

☐ Check the manual-reset limit switch(es) located near the burners. If the furnace has experienced a high-temperature condition, due to inadequate combustion air, these switches will shut off the furnace. Reset the switch(es) by pushing the button on the switch. If the switch trips a second time, turn off the furnace and call for service.

☐ If your furnace still fails to operate, call your servicing dealer for troubleshooting and repairs. Tell him the model and serial numbers for your furnace. (You should have them recorded on page 2 of this booklet.) If the dealer knows exactly which furnace you have, he may be able to offer suggestions over the phone, or save valuable time through knowledgeable preparation for the service call.

REGULAR DEALER MAINTENANCE

● In addition to the type of routine maintenance you might be willing to do, your furnace should be inspected regularly by a properly trained service technician. An annual inspection (or semiannual inspection, at least) should include the following:

1. Inspection of all flue product passages—including the burners, heat exchanger, relief box, and vent pipe.
2. Inspection of all combustion and ventilation air passages and openings.
3. Close check of all gas pipes leading to (and inside of) your furnace.
4. Inspection, cleaning, and lubrication (when required) of the blower motor, and wheel.

NOTE: Refer to the unit Service Procedures for blower motor oiling information. When required, the motor must be oiled by a qualified service technician.

5. Routine inspection and cleaning/replacement of the air filter.
6. Inspection of all supply- and return-air ducts for obstructions, air leaks, and insulation. Any problems found should be resolved at this time.
7. A check for loose connections attaching individual components. Inspection of all electrical wiring and their connections.
8. Operational check of the furnace itself to determine working condition. Repair or adjustment should be made at this time.

Your servicing dealer offers an economical service contract that covers seasonal inspections. Ask him for further details.

Detach & Mail Pro

Registration Card

STAPLE OR TAPE HERE

Carrier

IMPORTANT: Complete, detach and mail immediately for . . .

PRODUCT REGISTRATION

THE FEDERAL CONSUMER PRODUCT SAFETY ACT REQUIRES THAT YOU BE NOTIFIED OF ANY RECALLS INVOLVING THIS PRODUCT. YOUR NAME AND ADDRESS AND THE MODEL AND SERIAL NUMBERS OF YOUR PRODUCT WILL ASSIST US IN NOTIFYING YOU SHOULD THE NEED ARISE.

Your warranty coverage is not dependent upon the return of this card.

89A

1. 1. ☐ Mr. 2. ☐ Mrs. 3. ☐ Ms. 4. ☐ Miss

Name (First/Initial/Last)

Street

City _____ State _____ Zip

MODEL NO. (Copy from computer card in packet, or from name plate on unit.)

SERIAL NO.

2. DATE INSTALLED _____
   Mo. Day Yr.

3. **What Carrier product did you purchase?**
   1. ☐ Central air conditioner
   2. ☐ Gas furnace
   3. ☐ Oil furnace
   4. ☐ Electric furnace
   5. ☐ Heat pump

4. **When did you acquire your Carrier product?**
   1. ☐ Upon purchase of new dwelling
   2. ☐ To replace an older Carrier system
   3. ☐ To replace an older non-Carrier system
   4. ☐ Within a year of buying new dwelling with no central air system
   5. ☐ 2-4 years after buying new dwelling with no central air system
   6. ☐ Over 4 years after buying dwelling with no central air system

5. **If this was a replacement product, how old was the original?**
   1. ☐ 1-5 years
   2. ☐ 6-8 years
   3. ☐ 9-11 years
   4. ☐ 12-14 years
   5. ☐ 15-17 years
   6. ☐ 18 & over years

6. **Your Carrier product is installed in?**
   1. ☐ Single family/Townhouse
   2. ☐ Multi-family: 2-4 units
   3. ☐ Mobile home
   4. ☐ Apartment: 1-3 floors
   5. ☐ Apartment: 4 + floors
   6. ☐ Office/Bank
   7. ☐ Store
   8. ☐ Hospital/School
   9. ☐ Manufacturing building
   10. ☐ Other

7. **What 2 factors most influenced your purchase?**
   1. ☐ Carrier reputation
   2. ☐ Friend's recommendation
   3. ☐ Contractor/Dealer's recommendation
   4. ☐ Price
   5. ☐ Energy efficiency
   6. ☐ Ready availability
   7. ☐ Radio ads
   8. ☐ T.V. ads
   9. ☐ Newspaper ads
   10. ☐ Magazine ads
   11. ☐ Dealer display

8. **Which of the following have you done in the past 6 months? (check all that apply)**
   1. ☐ Redeemed a product coupon
   2. ☐ Ordered an item from mail order catalog
   3. ☐ Sent in product inquiry card from magazine
   4. ☐ Bought an item from offer received in mail
   5. ☐ Entered sweepstakes/contest

9. **In which age group are you?**
   1. ☐ 18-24
   2. ☐ 25-34
   3. ☐ 35-44
   4. ☐ 45-54
   5. ☐ 55-64
   6. ☐ 65 & over

10. **Marital status:**
    1. ☐ Married 2. ☐ Unmarried

11. **Which group best describes your family income?**
    1. ☐ Under $10,000
    2. ☐ $10,000-$14,999
    3. ☐ $15,000-$19,999
    4. ☐ $20,000-$24,999
    5. ☐ $25,000-$29,999
    6. ☐ $30,000-$34,999
    7. ☐ $35,000-$39,999
    8. ☐ $40,000-$44,999
    9. ☐ $45,000-$49,999
    10. ☐ $50,000 & over

12. **Do you have any children in any of the following age groups who are living at home?**
    1. ☐ Under age 2
    2. ☐ Age 2-4
    3. ☐ Age 5-7
    4. ☐ Age 8-10
    5. ☐ Age 11-12
    6. ☐ Age 13-15
    7. ☐ Age 16-18

13. **For your primary residence, do you:**
    1. ☐ Own a house?
    2. ☐ Rent a house?
    3. ☐ Own a townhouse/condominium?
    4. ☐ Rent an apartment?

14. **Which of the following types of credit cards do you use?**
    1. ☐ Travel/Entertainment (American Express, Diners Club, Carte Blanche)
    2. ☐ Bank (Master Charge, Visa)
    3. ☐ Gas, department store, etc.

15. **What is your occupation? (check one)**
    1. ☐ Professional/Technical
    2. ☐ Upper Mgt./Administrator
    3. ☐ Sales/Service/Middle Mgt.
    4. ☐ Clerical/White Collar
    5. ☐ Craftsman/Blue Collar
    6. ☐ Student
    7. ☐ Housewife
    8. ☐ Retired

16. **Which of the following interests and hobbies do you and your family enjoy?**
    1. ☐ Tennis
    2. ☐ Golf
    3. ☐ Snow Skiing
    4. ☐ Running/Jogging
    5. ☐ Camping/Hiking
    6. ☐ Hunting/Shooting
    7. ☐ Fishing
    8. ☐ Bicycling
    9. ☐ Racquetball
    10. ☐ Sailing/Boating
    11. ☐ Stamp/Coin Collecting
    12. ☐ Motorbiking/Motorcycling
    13. ☐ Home Video Games
    14. ☐ Physical Fitness/Exercise
    15. ☐ Home Video Recording
    16. ☐ Recreational Vehicle/4-WD
    17. ☐ Photography
    18. ☐ CB Radio
    19. ☐ Home Workshop/Do-It-Yourself
    20. ☐ Gardening/Plants
    21. ☐ Electronics
    22. ☐ Automotive Work
    23. ☐ Sewing/Needlework
    24. ☐ Crafts
    25. ☐ Collectibles/Collections
    26. ☐ Art & Antiques
    27. ☐ Stereo Music Equipment
    28. ☐ Foreign Travel
    29. ☐ Attending Cultural/Arts Events
    30. ☐ Gourmet Foods/Cooking
    31. ☐ Health/Natural Foods
    32. ☐ Wines
    33. ☐ Fashion Clothing
    34. ☐ Home Furnishings/Decorating
    35. ☐ Records & Tapes
    36. ☐ Avid Book Reading
    37. ☐ Science Fiction
    38. ☐ Astrology/Occult
    39. ☐ Stock/Bond Investments
    40. ☐ Real Estate Investments
    41. ☐ Self Improvement Programs
    42. ☐ Community/Civic Activities

We appreciate your taking the time to complete this card; the information provided will help us serve you better in the future. We participate in a multi-company program whereby you can receive information about new products, developments, trends, etc. related to the interest areas and other information you have indicated above. Please check here if you would prefer not to learn about such products and services. ☐

Other comments & suggestions about our product:

_____

Please Fold Here

PLACE
FIRST CLASS
STAMP
HERE

CARRIER CORPORATION
Product Registration Center
P.O. Box 17685
Denver, Colorado 80217

CARRIER CORPORATION

IF YOUR UNIT DOES NOT WORK, FOLLOW THESE STEPS IN ORDER:

FIRST: Contact the installer. You may find his name on the product or in your Homeowner's Packet. If his name is not known, call your builder if yours is a new residence.

SECOND: Contact the nearest CARRIER distributor. (See telephone yellow pages.)

THIRD: Contact:
CARRIER CORPORATION
Consumer Relations Department
Carrier Parkway
P.O. Box 4808
Syracuse, New York 13221
Telephone: 1-800-CARRIER (227-7437)
From Canada: 1-315-432-7885

Unit Model No. _____   Unit Serial No. _____

Date of Installation _____   Installed By _____

Name of Owner _____   Address of Installation _____

Carrier Corporation
Indoor Gas-Fired Furnace Limited Warranty

LIMITED ONE-YEAR WARRANTY — This CARRIER CORPORATION product is warranted to be free from defects in material and workmanship under normal use and maintenance for a period of one year from the date of original installation, whether or not actual use begins on that date. A new or remanufactured part to replace any defective part will be provided at CARRIER CORPORATION'S sole option without charge for the part itself, PROVIDED the defective part is returned to our distributor. This warranty applies only to the product in its original installation location and is voided if the product is reinstalled elsewhere.

THIS WARRANTY DOES NOT INCLUDE LABOR or other costs incurred for diagnosing, repairing, removing, installing, shipping, servicing or handling of either defective parts or replacement parts. SUCH COSTS MAY BE COVERED by a separate warranty or service agreement provided by the installer which is separate and distinct from this factory warranty.

EXTENDED NINE-YEAR LIMITED WARRANTY ON HEAT EXCHANGER ONLY — During the second through tenth years after the date of original installation, CARRIER CORPORATION further warranty the heat exchanger against defects in material and workmanship under normal use and maintenance.

LIMITATION OF WARRANTIES — ALL IMPLIED WARRANTIES (INCLUDING IMPLIED WARRANTIES OF MERCHANTABILITY AND FITNESS FOR A PARTICULAR PURPOSE) ARE HEREBY LIMITED IN DURATION TO THE PERIOD FOR WHICH EACH LIMITED WARRANTY IS GIVEN AND APPLIES. SOME STATES DO NOT ALLOW LIMITATIONS ON HOW LONG AN IMPLIED WARRANTY LASTS, SO THE ABOVE LIMITATIONS MAY NOT APPLY TO YOU. THE EXPRESSED WARRANTIES MADE IN THIS WARRANTY ARE EXCLUSIVE AND MAY NOT BE ALTERED, ENLARGED, OR CHANGED BY ANY DISTRIBUTOR, DEALER, OR OTHER PERSON WHATSOEVER. ALL WORK UNDER

THE TERMS OF THIS WARRANTY SHALL BE PERFORMED DURING NORMAL WORKING HOURS. ALL REPLACEMENT PARTS, WHETHER NEW OR REMANUFACTURED, ASSUME AS THEIR WARRANTY PERIOD ONLY THE REMAINING TIME PERIOD OF THIS WARRANTY.

CARRIER CORPORATION WILL NOT BE RESPONSIBLE FOR:
1. Normal maintenance as outlined in the installation and servicing instructions or owner's manual including filter cleaning and/or replacement and lubrication.
2. Damage or repairs as a consequence of faulty installation or application by others.
3. Failure to start due to voltage conditions, blown fuses, open circuit breakers or other damages due to the inadequacy or interruption of electrical service.
4. Damage or repairs needed as a consequence of any misapplication, abuse, improper servicing, unauthorized alteration or improper operation.
5. Damage as a result of floods, winds, fires, lightning, accidents, corrosive environments or other conditions beyond the control of CARRIER CORPORATION.
6. Costs for replacement parts or repair services which are not supplied or designated by CARRIER and which are specifically covered under this Warranty.
7. CARRIER CORPORATION products installed outside the continental U.S.A., Alaska, Hawaii, and Canada.
8. Electricity or fuel costs or increases in electricity or fuel costs for any reason whatsoever, including additional or unusual use of supplemental electric heat.
9. ANY SPECIAL, INDIRECT, OR CONSEQUENTIAL PROPERTY OR COMMERCIAL DAMAGE OF ANY NATURE WHATSOEVER. Some states do not allow the exclusion of incidental or consequential damages, so the above limitation may not apply to you.

This warranty gives you specific legal rights, and you may also have other rights which vary from state to state.

Form No. 530-059 (New 5-90)

HEATING & COOLING

Carrier Corporation ● Syracuse, New York 13221

Service and Maintenance Instructions

For Sizes 060-100, Series 100

NOTE: Read the entire instruction manual before starting the installation.

SAFETY CONSIDERATIONS

Installing and servicing heating equipment can be hazardous due to gas and electrical components. Only trained and qualified personnel should install, repair, or service heating equipment.

Untrained personnel can perform basic maintenance functions such as cleaning and replacing air filters. All other operations must be performed by trained service personnel. When working on heating equipment, observe precautions in the literature, tags, and labels attached to or shipped with the unit and other safety precautions that may apply.

Follow all safety codes, including NFPA 54/ANSI Z223.1-1988, National Fuel Gas Code. Wear safety glasses and work gloves. Have a fire extinguisher available during start-up and adjustment procedures and service calls.

Recognize safety information. This is the safety-alert symbol ⚠. When you see this symbol on the unit and in instructions or manuals, be alert to the potential for personal injury.

Understand the signal word—DANGER, WARNING, or CAUTION. These words are used with the safety-alert symbol. DANGER identifies the most serious hazards which **will** result in severe personal injury or death. WARNING signifies hazards that **could** result in personal injury or death. CAUTION is used to identify unsafe practices, which **would** result in minor personal injury or product and property damage.

Fig. 1—Model 58VUA Upflow Furnace A91128

⚠ WARNING

Never store anything on, near, or in contact with the furnace, such as:

1. Spray or aerosol cans, rags, brooms, dust mops, vacuum cleaners, or other cleaning tools.
2. Soap powders, bleaches, waxes or other cleaning compounds, plastic or plastic containers, gasoline, kerosene, cigarette lighter fluid, dry cleaning fluids, or other volatile fluids.
3. Paint thinners and other painting compounds, paper bags or other paper products.

Failure to follow this warning can cause corrosion of the heat exchanger, fire, personal injury, or death.

CARE AND MAINTENANCE

For continuing high performance and to minimize possible equipment failure, it is essential that maintenance be performed annually on this equipment. Consult your local dealer for maintenance and the availability of a maintenance contract.

A92095

Fig. 2—Model 58VCA Downflow Furnace

Manufacturer reserves the right to discontinue, or change at any time, specifications or designs without notice and without incurring obligations.

Book 1 4   PC 101   Catalog No. 565-997   Printed in U.S.A.   Form 58V-1SM   Pg 1   6-92   Replaces: 58SXB-1SM
Tab 6a 8a

The minimum maintenance that should be performed on this equipment is as follows:

1. Check and clean or replace air filter each month or as required.

2. Check blower motor and wheel for cleanliness and lubrication each heating and cooling season. Clean and lubricate as necessary. (See Step 2.)

3. Check electrical connections for tightness, and controls for proper operation each heating season. Service as necessary.

4. Check for proper condensate drainage; clean as necessary.

5. Check for blockages of combustion-air and vent pipes.

Step 1—Air Filter Cleaning and Replacement

The air filter arrangement may vary depending on the application.

DOWNFLOW FURNACES ONLY — Each furnace accommodates 2 filters which are installed in the return-air duct. (See Fig. 3.) To clean or replace the filters, proceed as follows:

1. Turn OFF electrical supply to unit.

2. Remove blower access door.

3. Reaching up behind top plate, tilt filters toward center of return-air plenum, remove filters, and replace or clean as needed.

4. Furnaces are equipped with permanent, washable filters. Clean these filters by spraying cold tap water through filter in opposite direction of airflow.

5. Rinse filters and let dry. Oiling or coating of filters is not recommended.

6. Reinstall filters with cross-mesh binding facing blower.

7. Replace access door.

8. Turn ON electrical supply to furnace.

UPFLOW FURNACES ONLY — To clean or replace the air filter, proceed as follows:

1. Turn OFF electrical supply to unit.

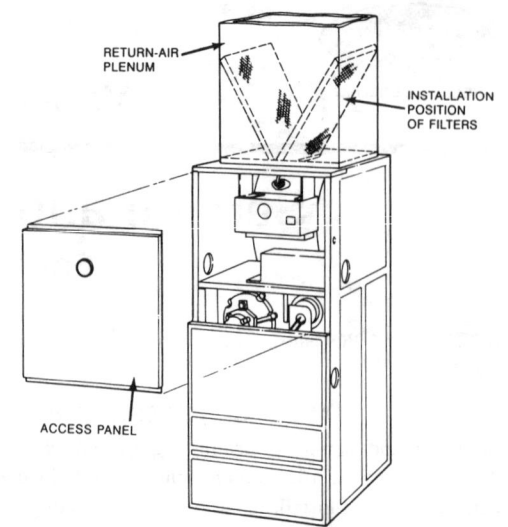

Fig. 3—Position of Filters in Downflow Furnace

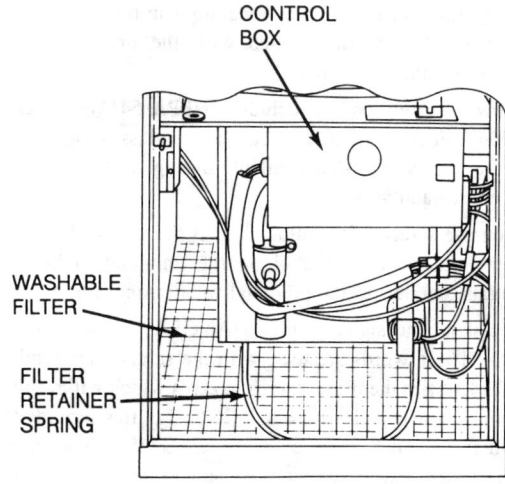

Fig. 4—Filter Installed for Bottom Inlet

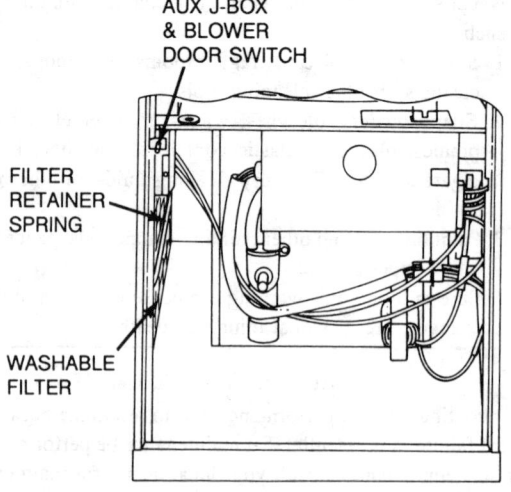

Fig. 5—Filter Installed for Side Inlet

2. Remove access doors.

3. Release filter retainer spring from behind flange of furnace casing. (See Fig. 4 and 5.)

4. Slide filter out.

WA

2

5. Furnaces are equipped with permanent, washable filters. Clean filter by spraying cold tap water through filter in opposite direction of airflow.

6. Rinse filter and let dry. Oiling or coating of filter is not recommended.

7. Place filter in furnace with cross-mesh binding either up or facing blower.

8. Replace access doors.

9. Turn ON electrical supply to furnace.

Step 2—Blower Motor and Wheel Maintenance

For long life, economy, and high efficiency, clean accumulated dirt and grease from blower wheel and motor annually.

The following items should be performed by a qualified service technician:

Some motors have prelubricated, sealed bearings and require no lubrication. These motors can be identified by the absence of oil ports on each end of the motor. For motors with oil ports, lubricate as follows:

Lubricate motor every 5 years if motor is used for intermittent operation (thermostat FAN switch in AUTO position), or every 2 years if motor is in continuous operation (thermostat FAN switch in ON position).

Clean and lubricate as follows:

1. Turn OFF electrical supply to unit.

2. Remove access doors.

3. Upflow furnaces only—remove drain trap and control box:

   a. Remove control box from bottom side of blower shelf and position to 1 side.

   b. Disconnect 9-circuit connector PL-13 from blower housing.

   c. Using backup wrench, disconnect drain pipe at coupling in blower compartment.

   d. Loosen hose clamp and remove 7/8-in. diameter drain hose from drain trap.

   e. Loosen hose clamp and disconnect 5/8-in. diameter drain hose at bottom of inducer housing located under blower shelf.

   f. Remove screw securing drain trap assembly.

4. Downflow furnaces only—disconnect vent pipe, elbow, and auxiliary limit switch. (See Fig. 6.)

   a. Remove control box from top plate and position to 1 side.

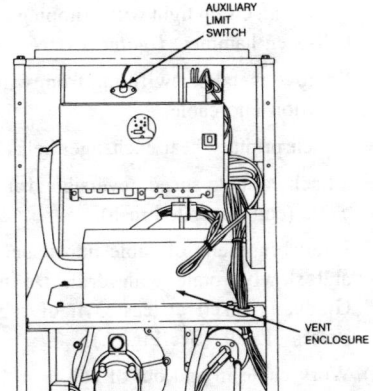

Fig. 6—Downflow Furnace Blower Compartment

   b. Disconnect 9-circuit connector PL-13 from blower housing.

   c. Disconnect wires from auxiliary limit on blower housing.

   d. Remove vent pipe enclosure from top side of blower shelf and position to 1 side.

   e. Loosen hose clamps on outlet elbow and remove elbow.

   f. Loosen hose clamp on extension pipe outside of furnace and remove pipe.

5. Remove screws securing blower assembly to blower shelf and slide blower assembly out of furnace.

6. Squeeze side tabs of connector PL-13 and pull from blower housing bracket.

7. Mark blower wheel location on shaft before disassembly to insure proper reassembly.

8. Loosen setscrew holding blower wheel on motor shaft.

NOTE: Mark blower mounting arms and blower housing so each arm is positioned at the same hole location during reassembly. This will insure that oilers point up.

9. Remove bolts holding motor mount to blower housing and slide motor and mounts out of housing.

10. Lubricate motor (when oil ports are provided).

    a. Remove dust caps or plugs from oil ports located at each end of motor. If motor does not have these caps or plugs, bearings are sealed and need no further lubrication.

    b. Use a good grade of SAE 20 nondetergent motor oil and add 1 teaspoon (5 cc, 3/16 oz, or 16 to 25 drops) in each oil port. The use of other types or grades of oil will damage the motor. Excessive oiling can cause premature bearing failures.

    c. Allow time for total quantity of oil to be absorbed by each bearing.

    d. After oiling motor, wipe excess oil from motor housing.

    e. Replace dust caps or plugs on oil ports.

11. Remove blower wheel from housing:

    a. Mark blower wheel orientation and cutoff plate location to insure proper reassembly.

    b. Remove screws securing cutoff plate and remove cutoff plate from housing.

    c. Remove blower wheel from housing.

12. Clean blower wheel and motor using a vacuum with soft brush attachment. Be careful not to disturb balance weights (clips) on blower wheel vanes. Do not drop or bend wheel, as balance will be affected.

13. Reassemble blower by reversing items 11.a. through 11.c. Ensure wheel is positioned for proper rotation.

14. Reassemble motor and blower by reversing items 6 through 9. If motor has ground wire, be sure it is reconnected.

⚠ CAUTION

Ensure the motor is properly positioned in the blower housing. The motor oil ports must be at a minimum of 30° above the horizontal centerline of the motor after the blower assembly has been reinstalled in the furnace.

15. Reinstall blower assembly in furnace.

16. Upflow furnaces only—reinstall drain trap and control box:

    a. Inspect drain trap and hoses to ensure they are not blocked or restricted. Reinstall drain trap and hoses. Be sure to tighten hose clamps.

    b. Using backup wrench, attach drain pipe and tighten compression coupling.

c. Reinstall control box on bottom side of blower shelf.

17. Downflow furnace only—reconnect vent pipe, elbow, and auxiliary limit switch.

   a. Reinstall outlet elbow and extension pipe. Be sure connections are tight and leak proof.

   b. Reinstall vent pipe enclosure.

   c. Reconnect red wires to auxiliary limit switch.

   d. Reinstall control box on top plate.

18. Connect 9-circuit connector PL-13 to blower harness. Note that connections are polarized for correct assembly—**do not force.**

19. Turn ON electrical supply and check for proper rotation and speed changes between low- and high-heat and cooling. Operate unit 10 minutes and carefully check for condensate leaks.

Step 3—Cleaning Heat Exchangers

The following items should be performed by a qualified service technician:

If it becomes necessary to clean the heat exchanger because of carbon deposits, soot, etc., proceed as follows:

NOTE: Deposits of soot and carbon indicate a problem exists that needs to be corrected. Action must be taken to correct the problem.

1. Turn OFF gas and electrical supplies to furnace.

2. Remove control and blower access doors.

3. Loosen hose clamps on combustion-air pipe and move air pipe aside.

4. Using backup wrench, disconnect gas supply at ground joint union. Remove gas pipe from valve.

5. Disconnect pilot leads at 3-circuit connector outside of burner enclosure.

6. Disconnect high-voltage lead at spark generator.

7. Disconnect gas valve leads at 6-circuit connector on top of valve.

8. Disconnect pressure tubing from right side of burner enclosure and outlet end of gas valve.

9. Remove burner enclosure front.

10. Remove diffuser from inside top of burner enclosure. Remove screws that secure burner enclosure to cell panel. These screws are located inside the burner enclosure.

11. Using care not to damage cell inlet panel gasket, remove gas control assembly from furnace.

12. Remove vent pipe and drain.

   a. Upflow furnace only:

      (1.) Loosen hose clamps at vent pipe connection; disconnect vent pipe and position to 1 side.

      (2.) Loosen hose clamp and remove drain tube from inducer outlet box.

   b. Downflow furnace only:

      (1.) Remove vent pipe enclosure.

      (2.) Loosen hose clamps at vent pipe connection.

      (3.) Loosen hose clamp and remove drain tube from inducer outlet elbow.

13. Upflow furnace only—remove main control box.

   a. Disconnect 15-circuit connector from main control box at blower shelf.

   b. Remove screws securing main control box to blower shelf and position control box to 1 side.

14. Loosen hose clamp and remove drain tube from inducer housing.

15. Disconnect both 6-circuit connectors from electronically commutated motor (ECM) inducer controller mounted on left side of furnace.

16. Remove screws securing ECM inducer controller to mounting plate attached to left side of furnace.

17. Disconnect 6-circuit connector from pressure switches.

18. Remove mounting screws securing inducer assembly to collector box and coupling box; remove inducer assembly and remove all old sealant from parts.

19. Remove coupling box(es).

   a. Upflow furnace only:

      (1.) Remove screws securing coupling box and remove from furnace. Remove all old sealant from parts.

   b. Downflow furnace only:

      (1.) Remove screws securing intake (upper) coupling box and remove box from furnace. Remove all old sealant from parts.

      (2.) Remove screws securing primary (lower) coupling box and remove box. Clean old sealant from parts.

20. Loosen hose clamp and remove 7/8-in. drain tube from trap.

21. Place bucket under 7/8-in. drain tube.

22. Using garden hose, flush each cell of the condensing heat exchanger with water. Use care not to spray water onto interior surfaces of control compartment. Dry all surfaces. Be careful **not** to remove sealant around cell openings in cell panel.

23. Using field-provided small wire brush, steel spring cable, reversible electric drill, and vacuum cleaner, clean primary heat exchanger cells. **Do not use wire brush or other sharp object to clean condensing heat exchanger.** Failure of the condensing heat exchanger will occur—flush with water only.

   a. Assemble wire brush and steel spring cable.

      (1.) Use 4 ft of 1/4-in. diameter high-grade steel spring cable (commonly known as drain cleaning or Roto-Rooter cable).

      (2.) Use 1/4-in. diameter wire brush (commonly known as 25-caliber rifle cleaning brush).

NOTE: The materials required above can be purchased at local hardware stores.

      (3.) Insert twisted wire end of brush into end of spring cable, and crimp tight with crimping tool or strike with ball-peen hammer. *Tightness is very important.*

      (4.) Remove metal screw fitting from wire brush to allow insertion into cable.

   b. Clean each primary heat exchanger cell:

      (1.) Attach variable-speed, reversible drill to end of spring cable (end opposite brush).

      (2.) Insert brush end of cable into upper opening of cell and slowly rotate with drill. **Do not** force cable. Gradually insert at least 3 ft of cable into 2 upper passes of cell. (See Fig. 7.)

      (3.) Work cable in and out of cell 3 or 4 times to obtain sufficient cleaning. **Do not** pull cable with great force. Reverse drill and gradually work cable out.

      (4.) Insert brush end of cable in lower opening of cell, and proceed to clean 2 lower passes of cell in same manner as 2 upper passes.

4

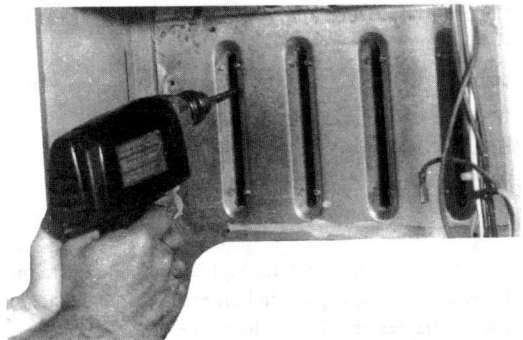

A88489

Fig. 7—Cleaning Primary Heat Exchanger Cell

(5.) Repeat procedures (previous) until each furnace cell has been cleaned.

(6.) Using vacuum cleaner, remove residue from each cell.

(7.) Using vacuum cleaner with soft brush attachment, clean burner assembly.

Step 4—Reassemble Furnace (After Cleaning Heat Exchangers)

1. Reinstall coupling box(es):

   c. Apply sealant releasing agent (Pam) to coupling box flange and cell panel where coupling box flange matches. (See Fig. 8.)

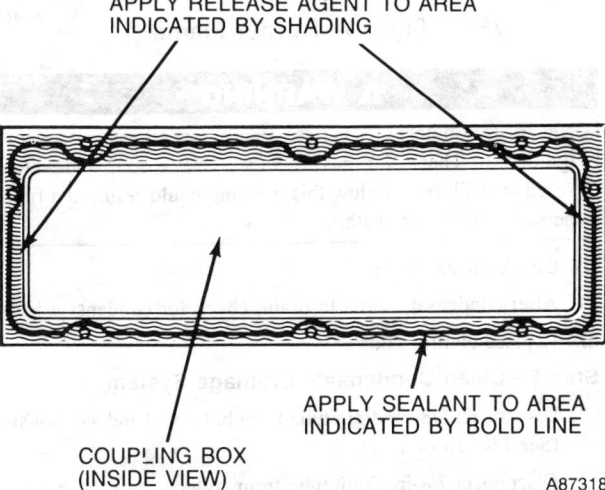

APPLY RELEASE AGENT TO AREA INDICATED BY SHADING

APPLY SEALANT TO AREA INDICATED BY BOLD LINE

COUPLING BOX (INSIDE VIEW)

A87318

Fig. 8—Inside View of Coupling Box

   d. Apply a generous bead (3/16-in. diameter) of G.E. RTV 122, 162, or Dow-Corning RTV 738 sealant (**NO substitute is permissible**) to flange of coupling box. (See Fig. 8.) Your distributor/dealer should have G.E. RTV 122, 162, or Dow-Corning RTV 738 sealants in stock.

   e. Being careful not to smear sealant, position coupling box so that slot in insulation is on left side and install coupling box.

2. Reinstall inducer assembly.

   a. Upflow furnace only—Be sure small round gasket(s) is in place between blower shelf and inducer housing.

   b. Apply sealant releasing agent (Pam) to collector box.

   c. Apply 1/8-in. diameter bead of G.E. RTV 122, 162, or Dow-Corning RTV 738 sealant to back of inducer housing. Apply sealant around inlet air opening. (The sealant should be about 1/4 in. from the edge of the inlet air opening.) (See Fig. 9.)

   d. Install inducer assembly on collector box and support bracket to coupling box.

   e. Connect 6-circuit inducer motor connector to inducer controller. Reconnect 6-circuit connector from pressure switches to main harness. (See Fig. 18.)

   f. Reconnect pressure tubes to pressure switch. (See Fig. 10 or 11.)

3. Connect small drain tube from top of trap to fitting on bottom of inducer housing. (See Fig. 10 or 11.)

4. Connect 7/8-in. drain tube to trap and collector box; tighten hose clamps. (See Fig. 10 or 11.)

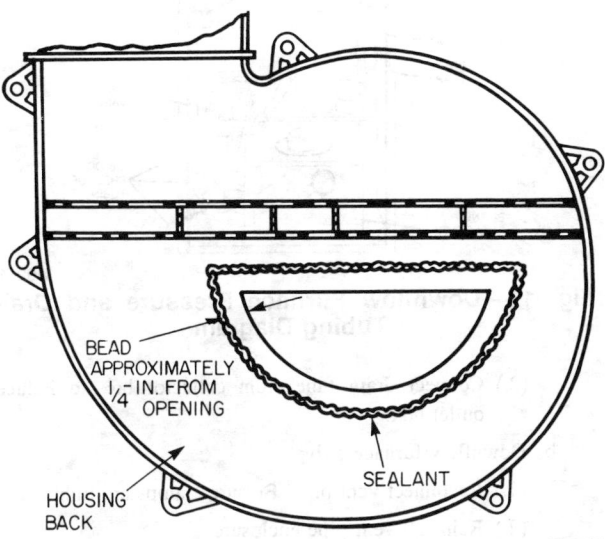

BEAD APPROXIMATELY 1/4-IN. FROM OPENING

SEALANT

HOUSING BACK

A86100

Fig. 9—Back of Inducer Assembly Housing

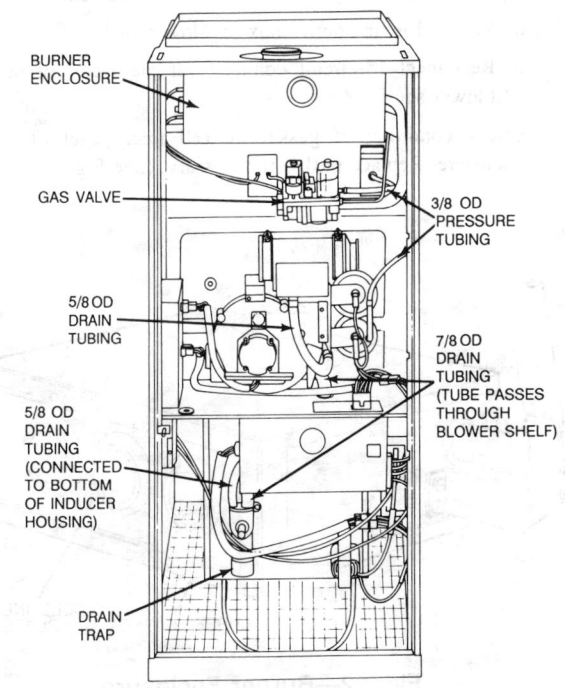

BURNER ENCLOSURE

GAS VALVE

5/8 OD DRAIN TUBING

5/8 OD DRAIN TUBING (CONNECTED TO BOTTOM OF INDUCER HOUSING)

DRAIN TRAP

3/8 OD PRESSURE TUBING

7/8 OD DRAIN TUBING (TUBE PASSES THROUGH BLOWER SHELF)

A91259

Fig. 10—Upflow Furnace Pressure and Drain Tubing Diagram

5. Reinstall vent pipe and drain tube.

   a. Upflow furnace only:

      (1.) Reconnect vent pipe. Be sure clamps are tight.

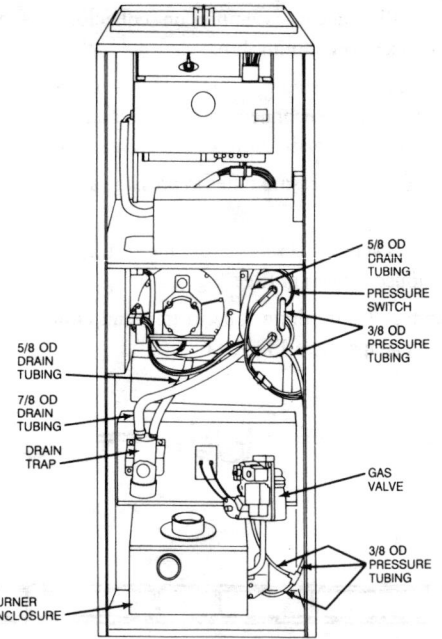

Fig. 11—Downflow Furnace Pressure and Drain Tubing Diagram

   (2.) Connect drain tube from collector box to inducer outlet box.

   b. Downflow furnace only:

   (1.) Reconnect vent pipe. Be sure clamps are tight.

   (2.) Reinstall vent pipe enclosure.

   (3.) Connect drain tube from collector box to inducer outlet elbow.

6. Upflow furnace only — reinstall main control box.

   a. Reinstall main control box on blower shelf.

   b. Reconnect 15-circuit connector at main control box on blower shelf.

7. Check condition of gasket on cell inlet panel of burner enclosure. Replace gasket if necessary. (See Fig. 12.)

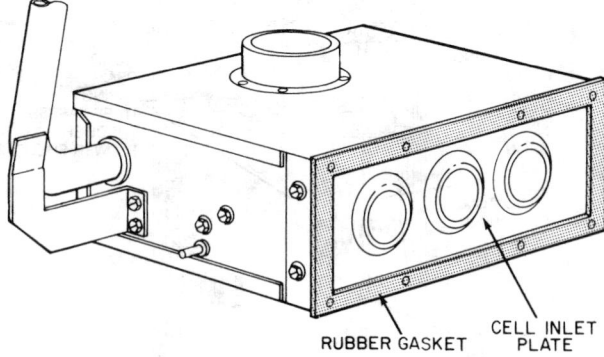

Fig. 12—Burner Enclosure

8. Install gas control assembly in furnace.

9. Install diffuser and burner enclosure front.

10. Reconnect pilot leads at 3-circuit connector.

11. Reconnect high-voltage lead to spark generator.

12. Reconnect gas valve leads at 6-circuit connector.

13. Reconnect pressure tubes to gas valve and burner enclosure. Be sure tubes are not kinked.

14. Using a backup wrench, install gas pipe in gas valve.

15. Reconnect gas pipe at ground joint union.

16. Reconnect combustion-air pipe. Tighten hose clamps.

17. Replace blower door only.

18. Turn ON gas and electrical supplies.

19. Check furnace operation through 2 complete operating cycles. Look through sight-glass in burner enclosure to check burners. Burner flames should be clear blue, almost transparent. (See Fig. 13.)

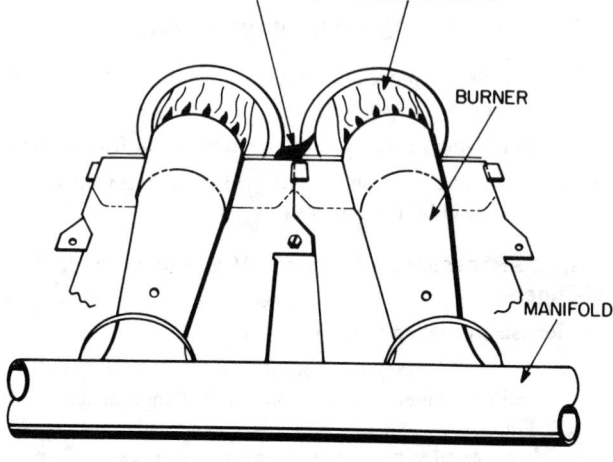

Fig. 13—Burner Flame

> ⚠ **WARNING**
>
> Never use matches, candles, flame, or other sources of ignition to check for gas leakage. Use a soap-and-water solution. Failure to follow this warning could result in a fire, personal injury, or death.

20. Check for gas leaks.

21. After condensate starts to drain, check for condensate leaks.

22. Replace control door.

Step 5—Clean Condensate Drainage System

1. Disconnect 5/8-in. drain tube from bottom of inducer housing. (See Fig. 10 or 11.)

2. Disconnect 7/8-in. drain tube from collector box. (See Fig. 10 or 11.)

3. Disconnect condensate drain line from drain trap at compression fitting.

4. Remove 1/4-in. screw(s) securing strap on drain trap.

5. Remove drain trap/hose assembly from furnace and flush with water until clean.

6. Flush external condensate drain line with water until clean.

7. Reassemble condensate drainage system by reversing items 1. through 5.

Step 6—Pilot Assembly

Check the pilot assembly and clean if necessary at the beginning of each heating season. The pilot flame should be high enough for proper impingement of the safety element and to light the burners. Remove any accumulation of soot and carbon from the safety element. Check spark electrode gap. (See Fig. 14 for proper spark gap and Fig. 15 for correct pilot location.)

Step 7—Electrical Controls and Wiring

NOTE: There may be more than 1 electrical supply to the unit.

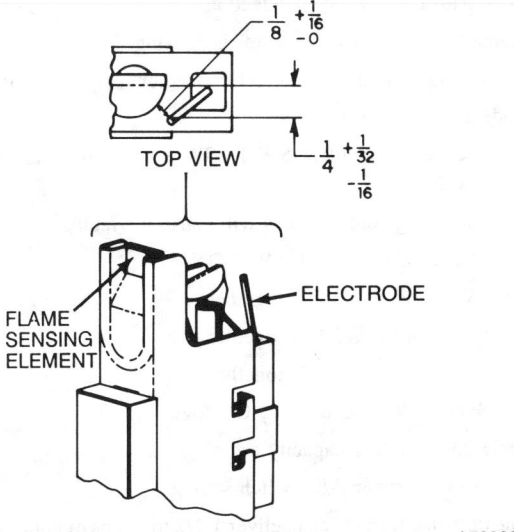

Fig. 14—Position of Electrode to Pilot

A79080

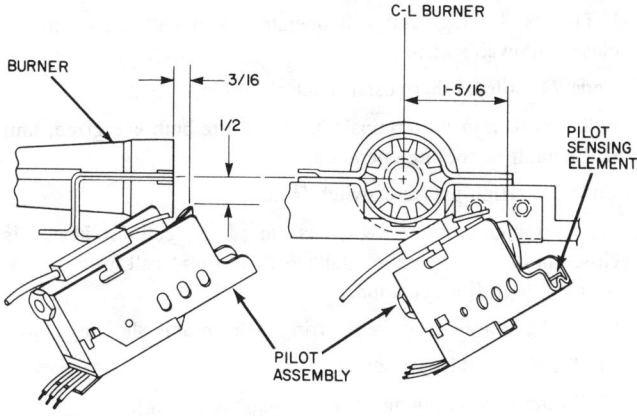

Fig. 15—Pilot/Burner Positions

A88221

With power disconnected to the unit, check all electrical connections for tightness. Tighten all screws on electrical connections. If any smokey or burned connections are found, disassemble the connection, clean all parts, strip wire, and reassemble properly and securely.

Reconnect electrical power to the unit and observe unit through 1 complete operating cycle. Electrical controls are difficult to check without proper instrumentation.

ELECTRICAL NOISE AND INTERFERENCE — This equipment generates and uses radio frequency energy and, if not installed and used properly (in strict accordance with the manufacturer's instructions), may cause interference with radio and television reception. The unit has been tested and found to comply with the limits for a Class B computing device in accordance with the specifications in Subpart J of Part 15 of Federal Communications Commission (FCC) Rules, which are designed to provide reasonable protection against such interference in a residential installation. However, there is no guarantee that interference will not occur in an installation. If this equipment does cause interference to radio or television reception (which may be determined by turning the equipment off and on), the user is encouraged to try correcting the interference by 1 or more of the following measures:

    (1.) Reorient receiving antenna.

    (2.) Relocate receiver with respect to equipment.

    (3.) Move receiver away from equipment.

    (4.) Plug receiver into different outlet so that equipment and receiver are on different branch circuits.

    (5.) Slide Ferrite core electrical noise suppressor over thermostat wire.

If necessary, the user should consult the dealer or an experienced radio/television technician for additional suggestions. The following booklet prepared by the FCC may be helpful: *"How to Identify and Resolve Radio-TV Interference Problems."* This booklet is available from the U.S. Government Printing Office, Washington, DC 20402, Stock No. 004-000-00345-4.

SERVICE DIAGNOSTICS — This furnace has a light emitting diode (LED) display to aid the installer, homeowner, or service technician in installing or servicing the unit. (See Fig. 17.) The display can be seen through the view port provided in the blower door. To decipher the meaning of the display, refer to Fig. 17 or the fault code label inside the control access door.

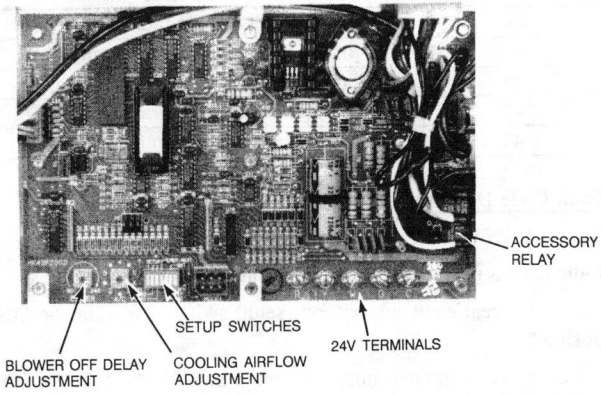

Fig. 16—Microprocessor Control Center

A92099

View the display through the port provided in the blower door. If YEL LED 3 is lit continuously, the furnace is operating in the high-heat mode. If GRN LED 4 is lit continuously, the furnace is operating in the low-heat mode. If the lower RED LED 1 is lit continuously, the furnace is operating in the emergency-heat mode. If the upper RED LED 2 is lit continuously, the microprocessor has malfunctioned.

DO NOT REMOVE THE BLOWER DOOR IF LED's ARE FLASHING; the fault code will be lost.

Alternate flashing of the YEL and GRN LED's indicates that a fault has occurred during operation. Count the number of times the YEL LED flashes and then count the number of times the GRN LED flashes. Once the fault code has been determined, refer to Table 1 or the fault code label inside the control access door to decipher the meaning of the display.

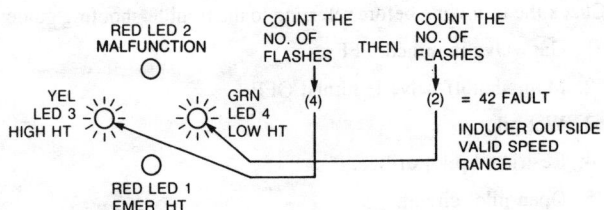

Fig. 17—Fault Code LED's

A87348

When power is turned ON at the main disconnect, a microprocessor self-test sequence will be completed in approximately 20 sec. During this period the GRN LED will light for 12 sec, followed by lighting of the YEL and GRN LED's for 1 sec. After this period, the unit will operate if a thermostat signal is initiated.

Table 1—Fault Code Descriptions

CODE	DESCRIPTION
11	No faults in history display
12	Blower calibration lockout
13	High limit lockout
14	Pilot proving lockout
21	Invalid model selection
22	Set-up error
23	Cooling capacity error
24	Illegal thermostat input
31	High-pressure switch fault
32	Low-pressure switch fault
33	High limit fault
34	Pilot proving fault
41	Blower outside valid speed range
42	Inducer outside valid speed range
43	Rpm ratio out of range
44	Blower calibration fault

Fault Code Descriptions:

Code 11 — No fault in recent history display.

Indicates no faults have occurred within last 5 cycles.

To read recent fault history, put setup switch SW-1 in the ON position.

To clear recent fault history, put setup switch SW-1 in the ON position and jumper R, W, and Y simultaneously until a code 11 is flashed.

Return setup switch SW-1 to the OFF position when complete.

Code 12 — Blower calibration lockout.

Indicates rpm calculated for low heat is less than 300 rpm or greater than 1300 rpm on 2 successive attempts.

Check the following before referring to the trouble-shooting guide:

1. Excessive high- or low-static pressure could be caused by dirty filters or undersized ductwork.

Code 13 — High-limit lockout.

Indicates the occurrence of 10 successive limit trips during high fire or 3 successive limit trips during low fire.

Check the following before referring to the trouble-shooting guide:

1. Improper or misaligned limit and/or limit shield.
2. Improper high- or low-gas input adjustment.
3. Stuck high-fire solenoid in gas valve.

Code 14 — Pilot-proving lockout.

Indicates pilot failed to prove in 5 minutes on 2 successive cycles. It can also indicate the pilot was proven at the *start* of 2 successive cycles.

Check the following before referring to the trouble-shooting guide:

1. Gas valve is turned OFF.
2. Main shutoff valve is turned OFF.
3. Wet pilot.
4. Restricted pilot orifice.
5. Open pilot circuit.
6. No spark at pilot.
7. Stuck pilot solenoid in gas valve.

Code 21 — Invalid model selection.

Indicates personality connector is missing or incorrect.

See wiring diagram for correct connector jumper location.

Code 22 — Setup error.

Indicates setup switch SW-1, SW-2, or SW-3 is positioned improperly.

The following combinations will cause the fault:

1. Thermostat call with SW-1 ON.
2. SW-2 and SW-3 ON together.
3. SW-1 and SW-2 ON together.
4. SW-1 and SW-3 ON together.
5. SW-1, SW-2, and SW-3 ON together.

Code 23 — Cooling capacity error.

Indicates improper A/C switch setting.

The 060-size furnace can deliver 1-1/2 to 3 tons of cooling airflow. The 080-size furnace can deliver 1-1/2 to 3-1/2 tons of cooling airflow. The 100-size furnace can deliver 2 to 5 tons of cooling airflow.

If fault is flashing, unit will operate, but it will default to the closest allowable airflow.

Code 24 — Illegal thermostat input.

Indicates thermostat terminals Y and W are both energized; unit will default to cooling operation.

Code 31 — High-pressure switch fault.

Indicates high-pressure switch is closed at "call for heat," is closed in low-fire operation, fails to close after "call for heat," or opens in high-fire operation.

Check the following before referring to the trouble-shooting guide:

1. Plugged condensate drain.
2. Water in vent piping (possibly sagging piping).
3. Pressure switch wiring or tubing connections incorrect.
4. Failed or out-of-calibration pressure switches.
5. Pilot flame adjustment.
6. Failed or out-of-calibration pilot.

Code 32 — Low-pressure switch fault.

Indicates low-pressure switch is closed at "call for heat," fails to close after "call for heat," or opens during operation.

Check the following before referring to the trouble-shooting guide:

1. Plugged condensate drain.
2. Water in vent piping (possibly sagging piping).
3. Pressure switch wiring or tubing connections incorrect.
4. Failed or out-of-calibration pressure switches.

Code 33 — High-limit fault.

Indicates the high limit is open or the unit is operating in high- heat only mode due to 2 successive low-fire limit trips.

Check the following before referring to the trouble-shooting guide:

1. Improper or misaligned limit and/or limit shield.
2. Improper low-gas input adjustment.
3. Stuck high-fire solenoid in gas valve.

Code 34 — Pilot-proving fault.

Indicates pilot failed to prove within 5 minutes, the pilot opened during the cycle, or the pilot was proven at the *start* of the cycle.

If this fault does not progress to a fault code 14, check the following. Otherwise, refer to fault code 14:

1. Combustion box diffuser plate missing or backwards.

2. Pilot flame adjustment.

3. Recirculation of combustion products at vent termination.

Code 41 — Blower outside valid speed range.

Indicates blower is not operating at the calculated or default rpm.

If this fault occurs in conjunction with fault code 44, check wiring to motor — otherwise refer to the troubleshooting guide.

If this fault occurs by itself, check torque taps on motor. Normal settings are White on pin-1, Black on pin-10, and Red on pin-11 unless used with a variable-speed cooling system.

Code 42 — Inducer outside valid speed range.

Indicates inducer is not operating at the calculated rpm or has not started within 10 sec after a "call for heat."

Check the following before referring to the troubleshooting guide:

1. Continuous pilot spark.

2. High-tension lead too close to wiring harness.

Code 43 — Inducer rpm ratio outside valid range.

Indicates the low- and high-pressure switch "make" points during purge are not within the calibration range.

Check the following before referring to the troubleshooting guide:

1. Plugged condensate drain.

2. Water in vent piping (possibly sagging piping).

3. Pressure switch wiring or tubing connections.

4. Failed or out-of-calibration pressure switches.

Code 44 — Blower calibration fault.

Indicates calculated blower speed is below 300 or above 1300 rpm. Unit will default to either mode if possible.

If this fault occurs in conjunction with fault 41, check wiring to motor. Otherwise, refer to troubleshooting guide.

If this fault occurs by itself, check for excessive static pressure caused by dirty filters or undersized ductwork.

Using these fault codes, the owner may save the expense of a service call by following the procedures provided in the User's Manual. A service technician can follow the steps furnished in the appropriate section of the trouble-shooting guide when correcting the problem.

NOTE: If the fault history is not cleared during servicing, the microprocessor will clear it internally after 5 heating cycles have been successfully completed without the fault occurring. This is done to prevent the storage of useless fault codes in the service history.

Step 8—Winterizing

⚠ **CAUTION**

The unit must not be installed, operated, and then turned off and left off in an unoccupied structure during cold weather when the temperature drops to 32° F and below. Freezing condensate left in the furnace will damage the equipment.

If the furnace will be off for an extended period of time in a structure where the temperature will drop to 32° F or below, winterize as follows:

1. Mix a solution of equal amounts of ethylene glycol (Prestone II antifreeze/coolant or equivalent) and water.

2. Turn OFF electrical supply to furnace.

3. Remove control access door.

4. Disconnect drain tube from bottom of inducer outlet box/elbow.

5. Insert funnel in drain tube and pour antifreeze/water solution into furnace until it is visible at point where condensate enters open drain.

6. Reconnect drain tube to outlet box.

7. Replace control access door.

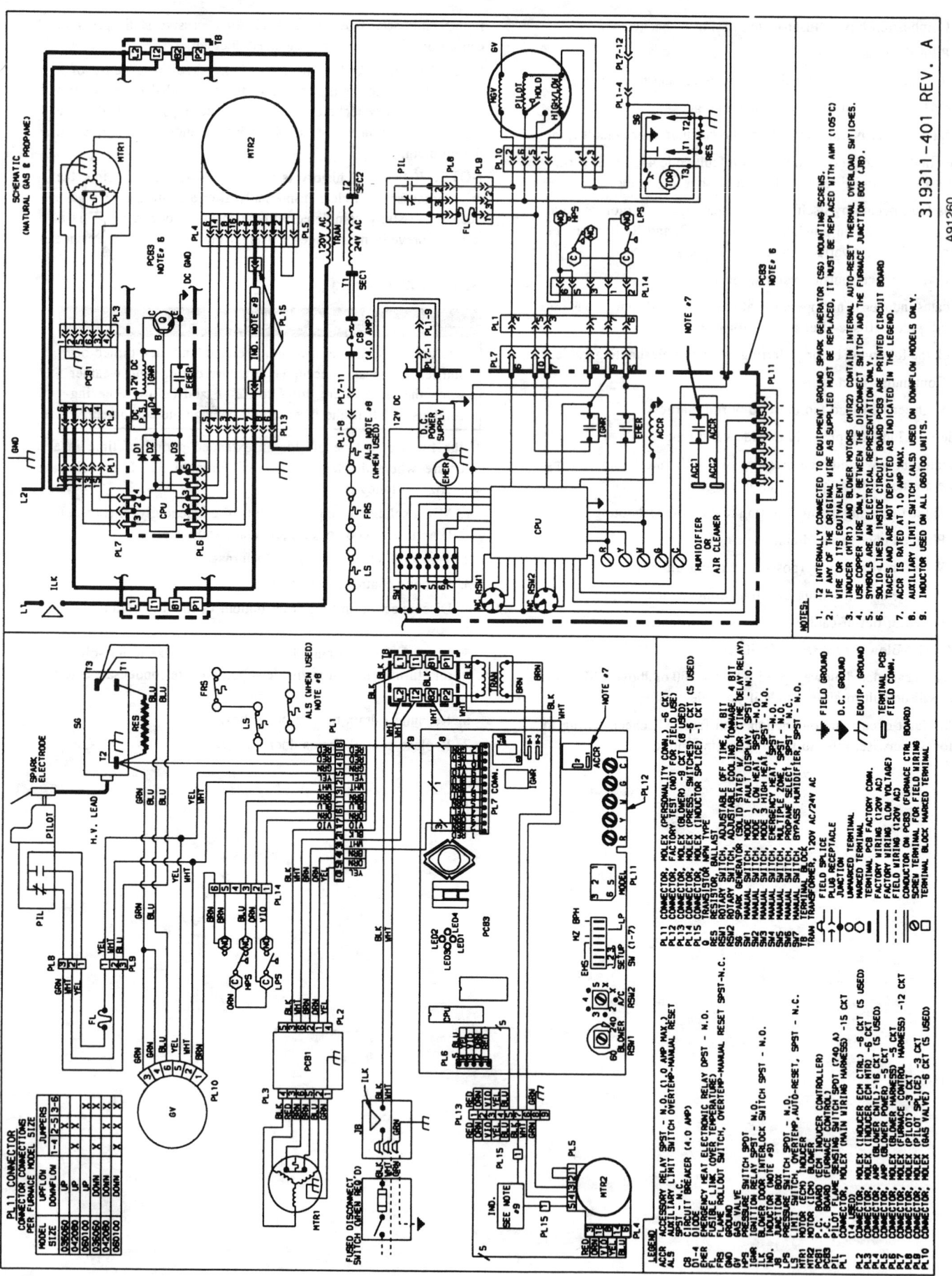

Fig. 18—Wiring Diagram

319311-401 REV. A

A91260

15055

Service and Maintenance Instructions
For Sizes 040—130, Series 100

NOTE: Read the entire instruction manual before starting the installation.

SAFETY CONSIDERATIONS

Installation and servicing of heating equipment can be hazardous due to gas and electrical components. Only trained and qualified personnel should install, repair, or service heating equipment.

Untrained personnel can perform basic maintenance functions such as cleaning and replacing air filters. All other operations must be performed by trained service personnel. When working on heating equipment, observe precautions in the literature, tags, and labels attached to or shipped with the unit and other safety precautions that may apply.

Follow all safety codes. In the United States, follow all safety codes including the National Fuel Gas Code NFPA No. 54-1988/ANSI Z223.1-1988. In Canada, refer to the current edition of the National Standard of Canada CAN/CGA- B149.1- and .2-M91 Natural Gas and Propane Gas Installation Codes. Wear safety glasses and work gloves. Have fire extinguisher available during start-up and adjustment procedures and service calls.

Recognize safety information: This is the safety-alert symbol ⚠. When you see this symbol on the furnace and in instructions or manuals, be alert to the potential for personal injury.

Understand the signal word DANGER, WARNING, or CAUTION. These words are used with the safety-alert symbol. DANGER identifies the most serious hazards which **will** result in severe personal injury or death. WARNING signifies a hazard that **could** result in personal injury or death. CAUTION is used to identify unsafe practices, which **would** result in minor personal injury or product and property damage.

⚠ WARNING
The ability to properly perform maintenance on this equipment requires certain expertise, mechanical skills, tools, and equipment. If you do not possess these, do not attempt to perform any maintenance on this equipment other than those procedures recommended in the User's Manual. A FAILURE TO FOLLOW THIS WARNING COULD RESULT IN POSSIBLE DAMAGE TO THIS EQUIPMENT, SERIOUS PERSONAL INJURY, OR DEATH.

CARE AND MAINTENANCE

For continuing high performance, and to minimize possible equipment failure, it is essential that periodic maintenance be performed on this equipment. Consult your local dealer as to the proper frequency of maintenance and the availability of a maintenance contract.

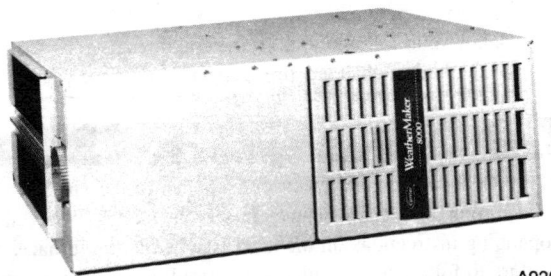

A92062

Fig. 1—Model 58ZAV Horizontal

A92061

Fig. 2—Model 58ZAV Downflow

A92060

Fig. 3—Model 58WAV Upflow

Manufacturer reserves the right to discontinue, or change at any time, specifications or designs without notice and without incurring obligations.

| Book | 1 | 4 | PC 101 | Catalog No. 535-801 | Printed in U.S.A. | Form 58W,Z-1SM | Pg 1 | 2-92 | Replaces: 58D,S-7SM |
| Tab | 6a | 8a | | | | | | | |

The minimum maintenance that should be performed on this equipment is as follows:

1. Check and clean air filter each month or more frequently if required. Replace if torn.
2. Check blower motor and wheel for cleanliness each heating and cooling season. Clean and lubricate as necessary.
3. Check electrical connections for tightness and controls for proper operation each heating season. Service as necessary.

AIR FILTER ARRANGEMENT — The air filter arrangement may vary depending on the application. Refer to Table 1 or 2 for filter size information.

Table 1—Filter Size Information (Downflow/Horizontal)

FURNACE CASING WIDTH	FILTER SIZE	FILTER TYPE
14-3/16	(2) 14 x 20 x 1	Cleanable
17-1/2	(2) 14 x 20 x 1	Cleanable
21	(2) 16 x 20 x 1	Cleanable
24-1/2	(2) 16 x 20 x 1	Cleanable

Table 2—Filter Size Information (Upflow)

| FURNACE CASING WIDTH | FILTER SIZE | | FILTER TYPE |
	Side Return	Bottom Return	
14-3/16	(1) 16 x 25 x 1*	(1) 14 x 25 x 1	Cleanable
17-1/2	(1) 16 x 25 x 1*	(1) 16 x 25 x 1	Cleanable
21	(1) 16 x 25 x 1	(1) 20 x 25 x 1*	Cleanable
24-1/2	(2) 16 x 25 x 1*	(1) 24 x 25 x 1	Cleanable

* Factory provided with the furnace. Filters may be field modified as required by cutting and folding the frame as indicated on the filter.

1. Horizontal and Downflow.

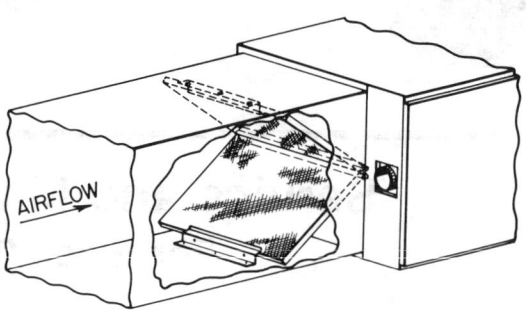

Fig. 4—Horizontal Filter Arrangement

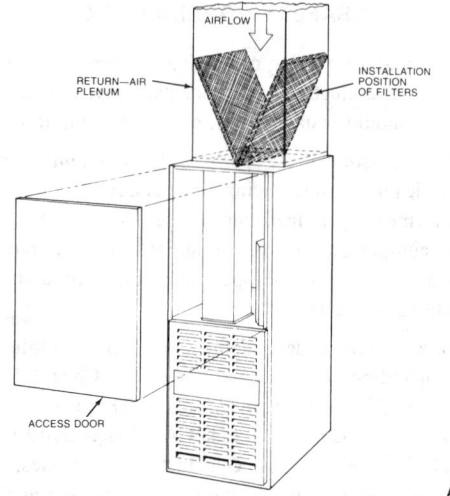

Fig. 5—Position of Filters

Each furnace requires two filters which are installed in the return-air duct. (See Fig. 4 and 5.)

To remove filters for cleaning or replacement, proceed as follows:

   a. Disconnect electrical power before removing access door.

   b. Remove blower access door.

   c. Reach up behind top plate, tilt filters toward center of return-air plenum, remove filters, and clean as needed. Replace if torn.

   d. Furnaces are equipped with permanent, washable filters. Clean those filters by spraying tap water through filter (from cross-mesh binding side) in opposite direction of airflow.

   e. Rinse and let dry. Oiling or coating of filters is not recommended or required.

   f. Reinstall filters with cross-mesh binding facing blower. Replace access door and restore electrical power to furnace.

2. Upflow.

Each furnace requires one or two filters which are installed in the blower compartment. (See Fig. 6.) To remove filters for cleaning or replacement, proceed as follows:

   a. Disconnect electrical power before removing access doors.

   b. Release filter retainer from clip at front of furnace casing. (See Fig. 6.) For side return, clips may be used on either (or both) sides of the furnace.

   c. Slide filter out.

   d. Clean filter with tap water. Spray water through cross-mesh binding side, in opposite direction of airflow.

   e. Rinse and let dry. Oiling or coating of filter is not recommended or required.

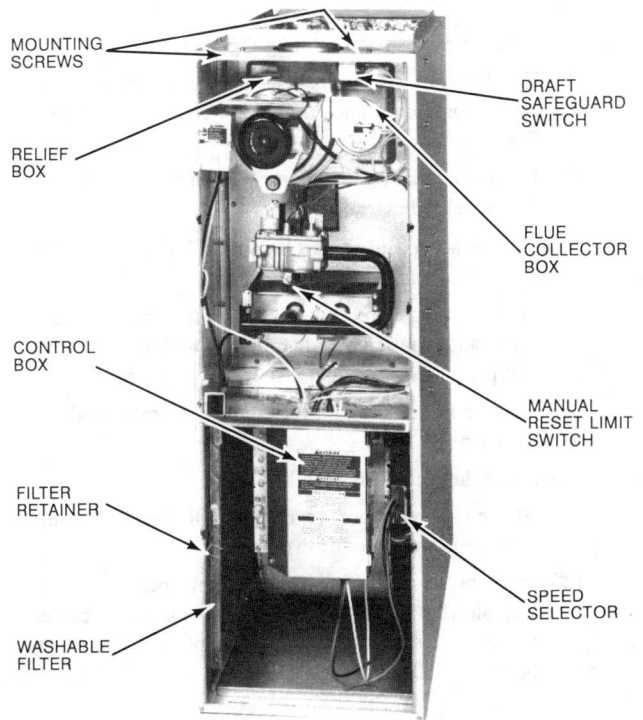

A92063

Fig. 6—Model 58WAV Upflow

f. Place filter in furnace with cross-mesh binding up or facing blower. Replace access doors and restore electrical power to furnace.

BLOWER MOTOR AND WHEEL — For long life, economy, and high efficiency; clean accumulated dirt and grease from the blower wheel and motor annually.

The following steps should be performed by a qualified service technician:

Some motors have prelubricated sealed bearings and require no lubrication. These motors can be identified by the absence of oil ports on each end of the motor. For those motors with oil ports lubricate as follows:

Lubricate motor every 5 years if motor is used on intermittent operation (thermostat FAN switch in AUTO position), or every 2 years if motor is in continuous operation (thermostat FAN switch in ON position).

Remember to disconnect the electrical supply before removing access doors.

Clean and lubricate as follows:

1. Remove blower access door.

2. Disconnect vent pipe on downflow/horizontal furnace only.

   a. Remove vent pipe enclosure.

   b. Disconnect vent pipe and remove short piece of pipe from furnace.

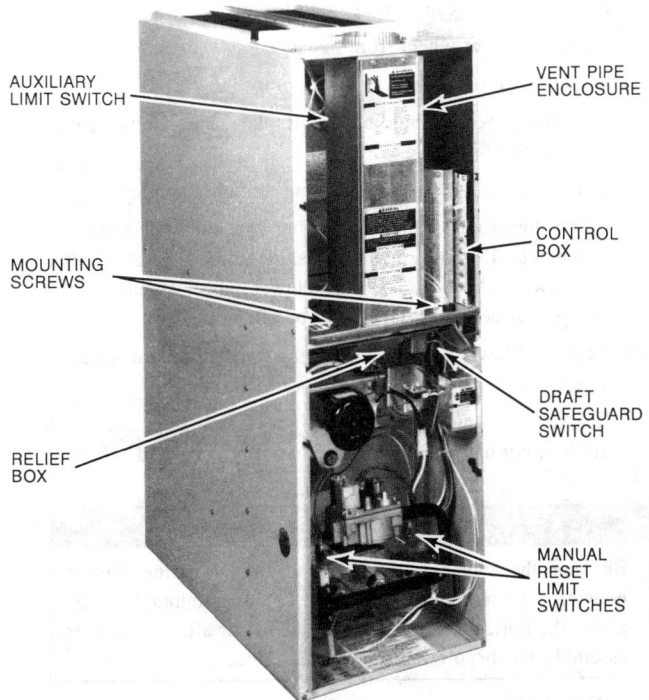

A92064

Fig. 7—Model 58ZAV Downflow

   c. Disconnect wires from auxiliary limit on blower housing.

3. Remove control box.

4. Remove electrical leads from numbered side of Molex speed selector. (See Fig. 6 and 7.) Note location of wires for reassembly.

5. Remove screws holding blower assembly to blower deck and slide blower assembly out of furnace.

6. Squeeze side tabs of Molex speed selector and pull it from blower housing.

7. Loosen a screw in strap holding motor capacitor to blower housing and slide capacitor out from under strap.

8. Mark blower wheel, motor, and motor support in relation to blower housing before disassembly to insure proper reassembly.

9. Loosen setscrew holding blower wheel on motor shaft.

10. Remove bolts holding motor mount to blower housing and slide motor and mount out of housing. Disconnect ground wire attached to blower housing before removing motor.

11. Lubricate motor (when oil ports are provided).

    a. Remove dust caps or plugs from oil ports located at each end of motor.

    b. Use a good grade of SAE 20 nondetergent motor oil and put one teaspoon, 5 cc, 3/16 oz, or 16 to 25 drops in each oil port. Do not over-oil.

    c. Allow time for total quantity of oil to be absorbed by each bearing.

d. After oiling motor, be sure to wipe excess oil from motor housing.

e. Replace dust cap or plugs on oil ports.

12. Remove blower wheel from housing.

   a. Mark cutoff location to insure proper reassembly.

   b. Remove screws holding cutoff plate and remove cutoff plate from housing.

   c. Lift blower wheel from housing through opening.

13. Clean blower wheel and motor using a vacuum with soft brush attachment. Do not remove or disturb balance weights (clips) on blower wheel blades. The blower wheel should not be dropped or bent as balance will be affected.

14. Reinstall blower wheel by reversing steps 12 a. through c. Be sure wheel is positioned for proper rotation.

15. Reassemble motor and blower by reversing steps 5 through 10. If motor has ground wire, be sure it is connected as before.

⚠ CAUTION

Be sure the motor is properly positioned in the blower housing. The motor oil ports must be at a minimum of 45° above the horizontal centerline of the motor after the blower assembly has been reinstalled in the furnace.

---

16. Reinstall blower assembly in furnace. Connect electrical leads to Molex speed selector. Please note that connections are polarized for assembly. **Do not force.**

17. Reinstall control box (reinstall vent pipe and enclosure on downflow/horizontal furnace).

18. Turn ON electrical power and check for proper rotation and speed changes between heating and cooling.

CLEANING HEAT EXCHANGER

The following steps should be performed by a qualified service technician:

NOTE: Deposits of soot and carbon indicate the existence of a problem which needs to be corrected. Take action to correct the problem.

If it becomes necessary to clean the heat exchanger because of carbon deposits, soot, etc., proceed as follows:

1. Turn OFF gas and power to furnace.

2. Remove control and blower access doors.

3. Remove vent pipe enclosure (downflow/horizontal furnace only) and disconnect vent pipe from relief box.

4. Remove two screws that secure relief box. (See Fig. 6 or 7.)

5. Disconnect wires to the following components:

   a. Draft safeguard switch.

   b. Inducer motor.

   c. Pressure switch.

   d. Limit overtemperature switch(es).

   e. Gas valve.

   f. Hot surface ignitor.

   g. Flame-sensing electrode.

   h. Two wiring connectors leading to control box.

6. Remove eight screws that secure flue collector box to center panel. Be careful not to damage sealant.

7. Remove complete inducer assembly from furnace, exposing flue openings.

8. Using field-provided small wire brush, steel spring cable, reversible electric drill, and vacuum cleaner; clean cells as follows:

   a. Assemble wire brush and steel spring cable.

      (1.) Use 48 in. of 1/4-in. diameter high-grade steel spring cable (commonly known as drain clean-out or Roto-Rooter cable).

      (2.) Use 1/4-in. diameter wire brush (commonly known as 25-caliber rifle cleaning brush).

NOTE: The items needed in steps (1.) and (2.) can usually be purchased at local hardware stores.

      (3.) Insert twisted wire end of brush into end of spring cable, and crimp tight with crimping tool or strike with ball-peen hammer. *Tightness is very important.*

      (4.) Remove metal screw fitting from wire brush to allow insertion into cable.

   b. Clean each heat exchanger cell.

      (1.) Attach variable-speed, reversible drill to end of spring cable (end opposite brush).

      (2.) Insert brush end of cable into upper opening of cell and slowly rotate with drill. *Do not* force cable. Gradually insert at least 36 in. of cable into two upper passes of cell. (See Fig. 8.)

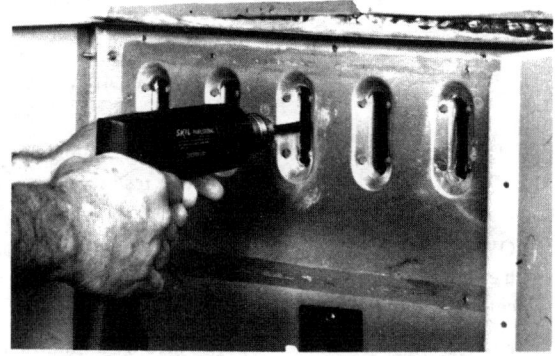

A91252

Fig. 8—Cleaning Heat Exchanger Cell

      (3.) Work cable in and out of cell three or four times to obtain sufficient cleaning. *Do not* pull cable with great force. Reverse drill and gradually work cable out.

      (4.) Remove burner assembly and cell inlet plates.

NOTE: Be very careful when removing the burner assembly to avoid breaking the ignitor. See Figure 9 for the correct ignitor location.

      (5.) Replace screws in center panel and cells before cleaning.

      (6.) Insert brush end of cable in lower opening of cell, and proceed to clean two lower passes of cell in same manner as two upper passes.

      (7.) Repeat foregoing procedures until each cell in furnace has been cleaned.

      (8.) Using vacuum cleaner, remove residue from each cell.

      (9.) Using vacuum cleaner with soft brush attachment, clean burner assembly.

      (10.) Reinstall cell inlet plates and burner assembly. Care must be exercised to center the burners in the cell openings.

9. After cleaning flue openings, check sealant on flue collector to ensure that it has not been damaged. If new sealant is needed, contact your dealer or distributor.

10. Clean and replace flue collector assembly, making sure all eight screws are secure.

11. Reinstall relief box.

12. Reconnect wires to the following components:

   a. Draft safeguard switch.

   b. Inducer motor.

   c. Pressure switch.

   d. Limit overtemperature switch(es).

   e. Gas valve.

   f. Hot surface ignitor.

   g. Flame sensing electrode.

   h. Two wiring connectors leading to control box.

13. Reconnect vent pipe to relief box. Replace vent pipe enclosure (when applicable).

14. Replace blower door only.

15. Turn on power and gas.

16. Set thermostat and check furnace for proper operation.

17. Check for gas leaks.

18. Replace control door.

⚠ WARNING

Never use a match or other open flame to check for gas leaks. Use a soap-and-water solution. A failure to follow this warning could result in fire, personal injury, or death.

ELECTRICAL CONTROLS AND WIRING

NOTE: There may be more than one electrical supply to unit.

The electrical ground and polarity for 115-volt wiring must be maintained properly. Refer to Fig. 10 for field wiring information and to Fig. 11 for unit wiring information. If the polarity is **not** correct, the microprocessor control will shut OFF the gas flow shortly after completion of the ignition trial period. The control system also requires an earth ground for proper operation of the microprocessor.

With power disconnected to unit, check all electrical connections for tightness. Tighten all screws on electrical connections. If any smoky or burned connections are noticed, disassemble the connection, clean all parts and stripped wire, and reassemble properly and securely. Electrical controls are difficult to check without proper instrumentation; therefore, reconnect electrical power to unit and observe unit through one complete operating cycle.

The 24-volt circuit contains an automotive-type, 3-amp fuse located on the main control board. Any direct shorts during installation, service, or maintenance could cause this fuse to blow. If fuse replacement is required, use **only** a 3-amp fuse of identical size.

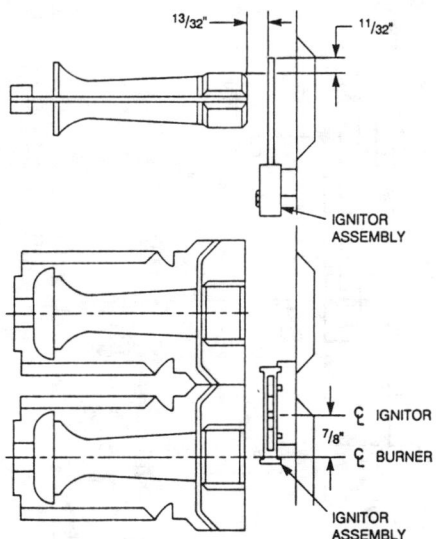

Fig. 9—Position of Ignitor to Burner

A91064

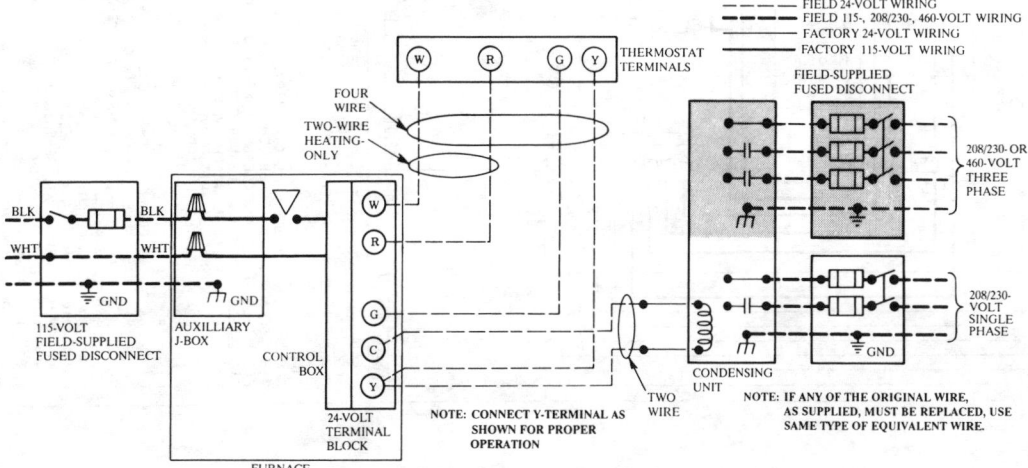

Fig. 10—Heating and Cooilng Application Wiring Diagram

A78461

5

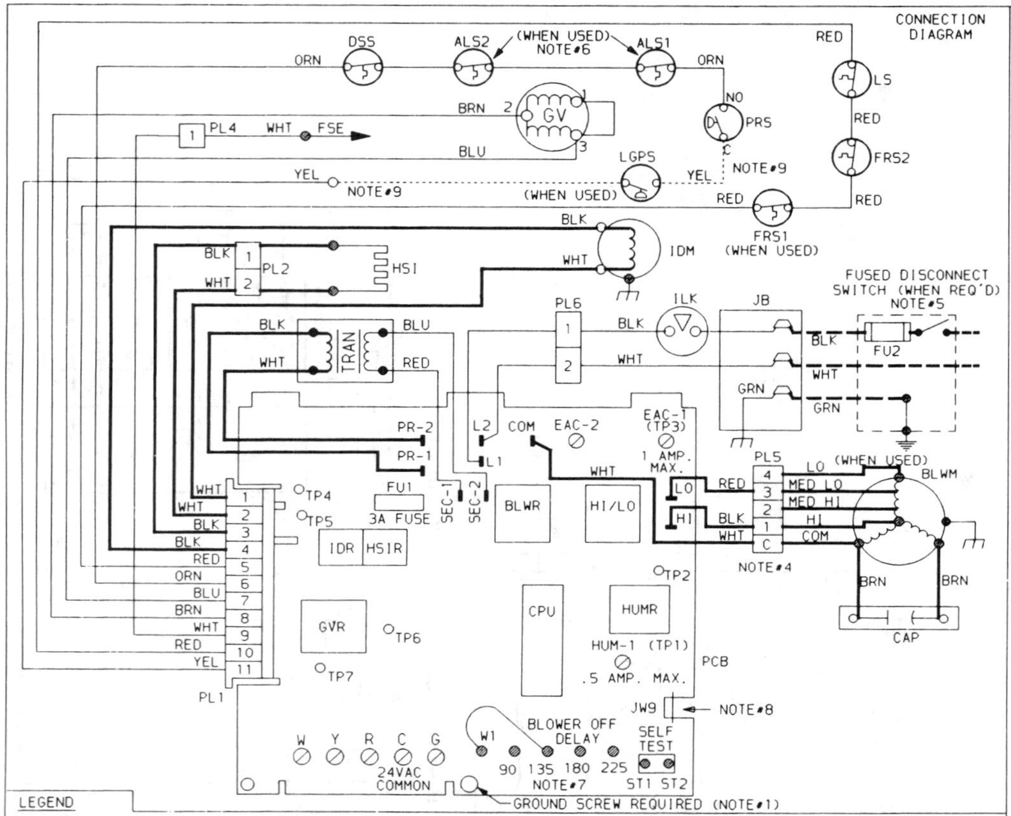

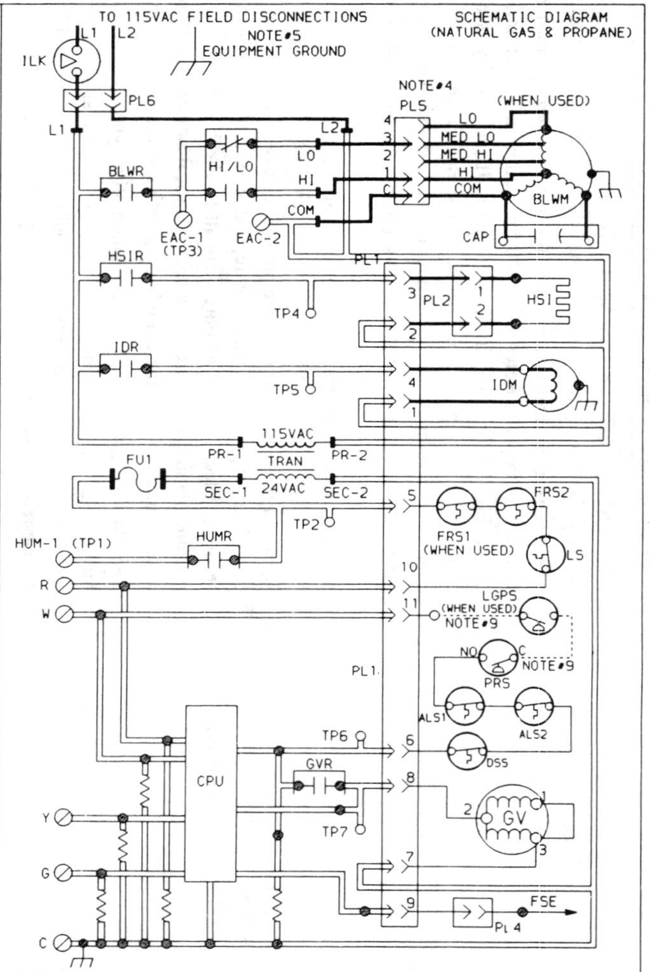

LEGEND

ALS1	AUXILIARY LIMIT SWITCH, OVERTEMP.- MANUAL RESET, SPST-(N.C.)
ALS2	AUXILIARY LIMIT SWITCH, OVERTEMP.- AUTO. RESET, SPST-(N.C.)
BLWR	BLOWER MOTOR RELAY, SPST-(N.O.)
BLWM	BLOWER MOTOR
CAP	CAPACITOR
CPU	MICROPROCESSOR AND CIRCUITRY
DSS	DRAFT SAFEGUARD SWITCH
EAC-1	ELECTRONIC AIR CLEANER CONNECTION (115 VAC 1 AMP MAX.)
EAC-2	ELECTRONIC AIR CLEANER CONNECTION (COMMON)
FRS1	FLAME ROLLOUT SWITCH OVERTEMP.-MANUAL RESET, SPST-(N.C.)
FRS2	FLAME ROLLOUT SWITCH OVERTEMP.-MANUAL RESET, SPST-(N.C.)
FSE	FLAME SENSING ELECTRODE
FU1	FUSE, 3 AMP, AUTOMOTIVE BLADE TYPE, FACTORY INSTALLED
FU2	FUSE, FIELD INSTALLED
GV	GAS VALVE-REDUNDANT OPERATORS
GVR	MAIN GAS VALVE RELAY, DPST-(N.O.)
HI/LO	BLOWER MOTOR SPEED CHANGE RELAY, DPST-(N.O.)
HSI	HOT SURFACE IGNITER (115 VAC)
HSIR	HOT SURFACE IGNITER RELAY, DPST-(N.O.)
HUM-1	24VAC HUMIDIFIER CONNECTION (.5 AMP. MAX.)
HUMR	HUMIDIFIER RELAY, DPST-(N.O.)
IDM	INDUCED DRAFT MOTOR
IDR	INDUCED DRAFT RELAY, DPST-(N.O.)
ILK	BLOWER DOOR INTERLOCK SWITCH, SPST-(N.O.)
JB	JUNCTION BOX
JW9	INDUCER OFF DELAY JUMPER: CUT-15 SEC DELAY, UNCUT-5 SEC DELAY
LGPS	LOW GAS PRESSURE SWITCH, SPST-(N.O.)
LS	LIMIT SWITCH, OVERTEMPERATURE-AUTO RESET, SPST-(N.C.)
PCB	PRINTED CIRCUIT BOARD
PL1	11-CIRCUIT EDGE CONNECTOR (PCB)
PL2	2-CIRCUIT HSI CONNECTOR
PL4	1-CIRCUIT FSE CONNECTOR
PL5	5-CIRCUIT BLWM CONNECTOR
PL6	2-CIRCUIT 115VAC CONNECTOR
PRS	PRESSURE SWITCH, SPST-(N.O.)
ST1-ST2	SELF TEST PINS
TP1-TP7	TEST POINT (1) THRU (7)
TRAN	TRANSFORMER-115VAC/24VAC

JUNCTION
TERMINAL
PCB TERMINAL
FACTORY POWER WIRING (115VAC)
FACTORY CONTROL WIRING (24VAC)
FIELD POWER WIRING (115VAC)
CONDUCTOR ON PCB
FIELD WIRING SCREW TERMINAL
FIELD GROUND
EQUIPMENT GROUND
FIELD SPLICE
PLUG RECEPTACLE

NOTES:

1. COMMON SIDE (SEC-2 AND C) OF 24VAC TRANSFORMER CONNECTED TO GROUND THROUGH THIS REQUIRED MOUNTING SCREW.
2. IF ANY OF THE ORIGINAL WIRE, AS SUPPLIED, MUST BE REPLACED, USE THE SAME OR EQUIVALENT TYPE WIRE.
3. INDUCER (IDM) AND BLOWER (BLWM) MOTORS CONTAIN INTERNAL AUTO-RESET THERMAL OVERLOAD SWITCHES.
4. BLOWER MOTOR SPEED SELECTIONS (PL5) ARE FOR AVERAGE CONDITIONS. SEE INSTALLATION INSTRUCTIONS FOR DETAILS ON OPTIMUM SPEED SELECTION.
5. USE ONLY COPPER WIRE BETWEEN THE DISCONNECT SWITCH AND THE UNIT.
6. AUXILIARY LIMIT SWITCHES (ALS1 & 2) USED ON DOWNFLOW MODELS ONLY.
7. BLOWER MOTOR (BLWM) ADJUSTABLE OFF-DELAY.
8. JW9 SHOULD BE LEFT UNCUT.
9. FACTORY CONNECTED WHEN LGPS NOT USED.
10. SYMBOLS ARE AN ELECTRICAL REPRESENTATION ONLY.

318975-401 REV. 1

A92048

Fig. 11—Unit Wiring Diagram

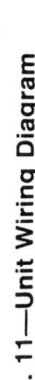

TROUBLESHOOTING GUIDE

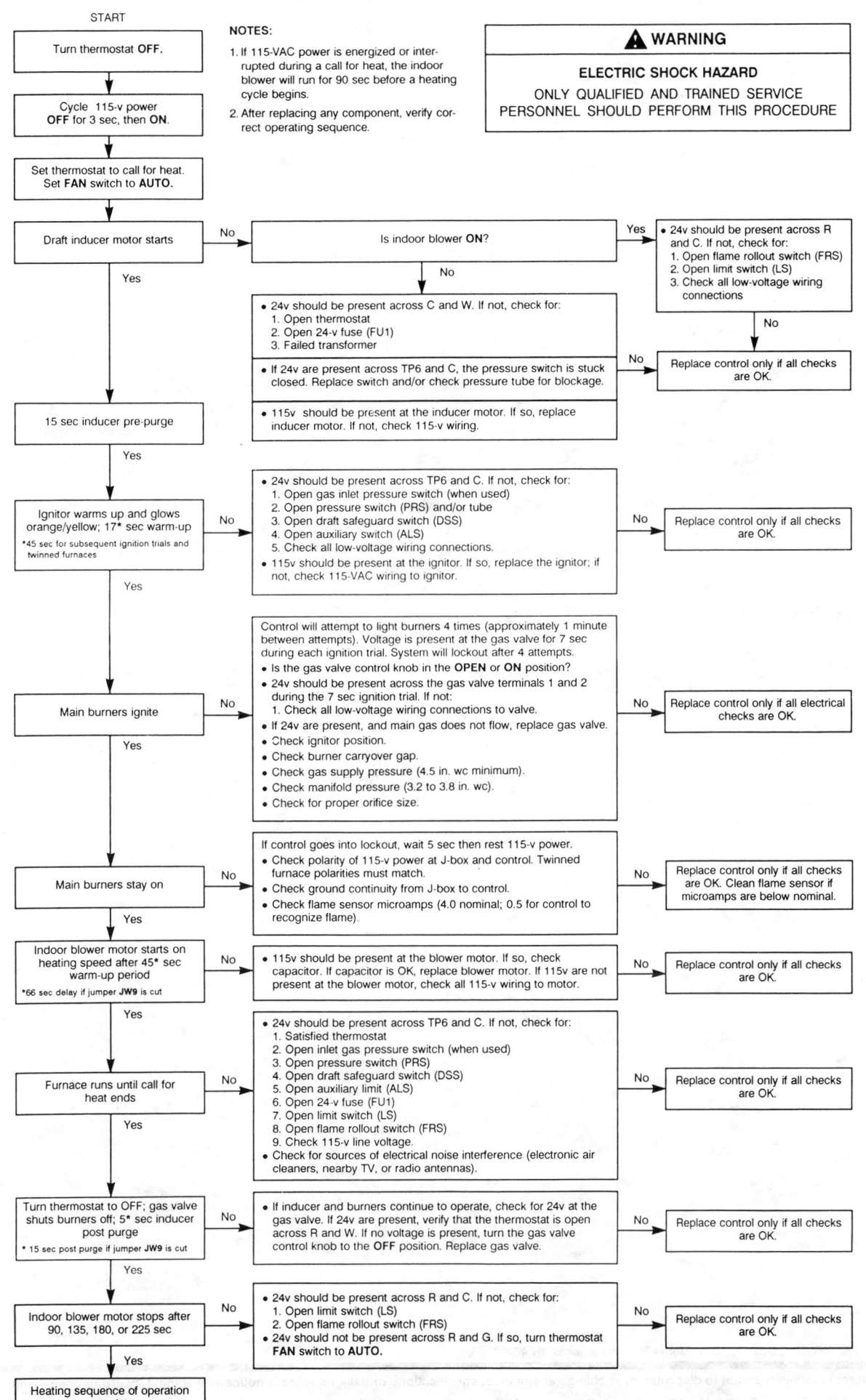

START

Turn thermostat **OFF.**

Cycle 115-v power **OFF** for 3 sec, then **ON.**

Set thermostat to call for heat. Set **FAN** switch to **AUTO.**

Draft inducer motor starts — No →

15 sec inducer pre-purge

Ignitor warms up and glows orange/yellow; 17* sec warm-up
*45 sec for subsequent ignition trials and twinned furnaces — No →

Main burners ignite — No →

Main burners stay on — No →

Indoor blower motor starts on heating speed after 45* sec warm-up period
*66 sec delay if jumper **JW9** is cut — No →

Furnace runs until call for heat ends — No →

Turn thermostat to **OFF**; gas valve shuts burners off; 5* sec inducer post purge
* 15 sec post purge if jumper **JW9** is cut — No →

Indoor blower motor stops after 90, 135, 180, or 225 sec — No →

Heating sequence of operation complete

NOTES:

1. If 115-VAC power is energized or interrupted during a call for heat, the indoor blower will run for 90 sec before a heating cycle begins.

2. After replacing any component, verify correct operating sequence.

⚠ **WARNING**

ELECTRIC SHOCK HAZARD

ONLY QUALIFIED AND TRAINED SERVICE PERSONNEL SHOULD PERFORM THIS PROCEDURE

Is indoor blower **ON?** — Yes →

- 24v should be present across R and C. If not, check for:
  1. Open flame rollout switch (FRS)
  2. Open limit switch (LS)
  3. Check all low-voltage wiring connections
  — No →

Replace control only if all checks are OK.

(No ↓)

- 24v should be present across C and W. If not, check for:
  1. Open thermostat
  2. Open 24-v fuse (FU1)
  3. Failed transformer

- If 24v are present across TP6 and C, the pressure switch is stuck closed. Replace switch and/or check pressure tube for blockage. — No →
Replace control only if all checks are OK.

- 115v should be present at the inducer motor. If so, replace inducer motor. If not, check 115-v wiring.

- 24v should be present across TP6 and C. If not, check for:
  1. Open gas inlet pressure switch (when used)
  2. Open pressure switch (PRS) and/or tube
  3. Open draft safeguard switch (DSS)
  4. Open auxiliary switch (ALS)
  5. Check all low-voltage wiring connections.
- 115v should be present at the ignitor. If so, replace the ignitor; if not, check 115-VAC wiring to ignitor. — No →
Replace control only if all checks are OK.

Control will attempt to light burners 4 times (approximately 1 minute between attempts). Voltage is present at the gas valve for 7 sec during each ignition trial. System will lockout after 4 attempts.
- Is the gas valve control knob in the **OPEN** or **ON** position?
- 24v should be present across the gas valve terminals 1 and 2 during the 7 sec ignition trial. If not:
  1. Check all low-voltage wiring connections to valve.
- If 24v are present, and main gas does not flow, replace gas valve.
- Check ignitor position.
- Check burner carryover gap.
- Check gas supply pressure (4.5 in. wc minimum).
- Check manifold pressure (3.2 to 3.8 in. wc).
- Check for proper orifice size. — No →
Replace control only if all electrical checks are OK.

If control goes into lockout, wait 5 sec then rest 115-v power.
- Check polarity of 115-v power at J-box and control. Twinned furnace polarities must match.
- Check ground continuity from J-box to control.
- Check flame sensor microamps (4.0 nominal; 0.5 for control to recognize flame). — No →
Replace control only if all checks are OK. Clean flame sensor if microamps are below nominal.

- 115v should be present at the blower motor. If so, check capacitor. If capacitor is OK, replace blower motor. If 115v are not present at the blower motor, check all 115-v wiring to motor. — No →
Replace control only if all checks are OK.

- 24v should be present across TP6 and C. If not, check for:
  1. Satisfied thermostat
  2. Open inlet gas pressure switch (when used)
  3. Open pressure switch (PRS)
  4. Open draft safeguard switch (DSS)
  5. Open auxiliary limit (ALS)
  6. Open 24-v fuse (FU1)
  7. Open limit switch (LS)
  8. Open flame rollout switch (FRS)
  9. Check 115-v line voltage.
- Check for sources of electrical noise interference (electronic air cleaners, nearby TV, or radio antennas). — No →
Replace control only if all checks are OK.

- If inducer and burners continue to operate, check for 24v at the gas valve. If 24v are present, verify that the thermostat is open across R and W. If no voltage is present, turn the gas valve control knob to the **OFF** position. Replace gas valve. — No →
Replace control only if all checks are OK.

- 24v should be present across R and C. If not, check for:
  1. Open limit switch (LS)
  2. Open flame rollout switch (FRS)
- 24v should not be present across R and G. If so, turn thermostat **FAN** switch to **AUTO.** — No →
Replace control only if all checks are OK.

7

Coleman Evcon

90 SERIES
UPFLOW FURNACES

RES 90

INSTALLATION INSTRUCTIONS
UPFLOW HI-EFFICIENCY *CONDENSING* FURNACE
WITH HOT SURFACE IGNITION

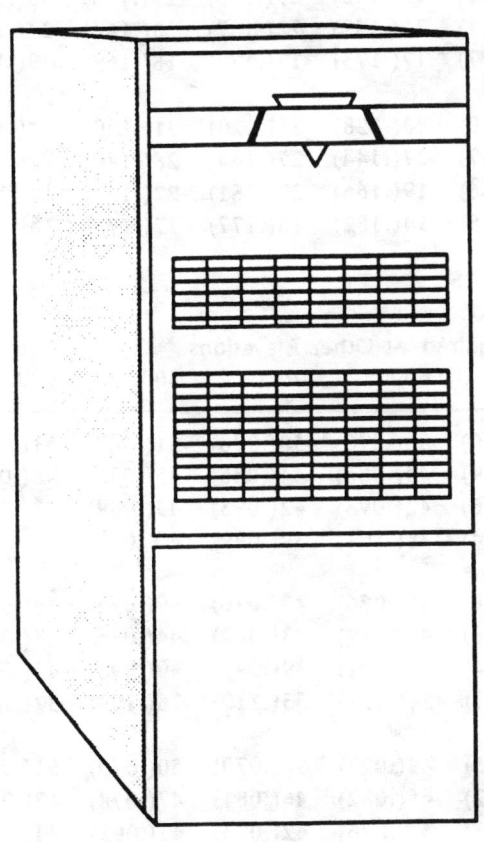

MODELS:
2940
2960
2970
2985

ORIFICE SIZES REQUIRED FOR HIGH ALTITUDES

NATURAL GAS

Model No.	Orifice at Sea Level	Orifice Size Required At Other Elevations								
		2000	3000	4000	5000	6000	7000	8000	9000	10000
2940	31(.120)	32(.116)	32(.116)	32(.116)	33(.113)	34(.111)	35(.110)	36(.106)	37(.104)	38(.101)
2960	27(.144)	28(.140)	28(.140)	29(.136)	29(.136)	29(.136)	30(.128)	30(.128)	30(.128)	31(.120)
2970	20(.161)	22(.157)	22(.157)	23(.154)	24(.152)	25(.149)	26(.147)	27(.144)	28(.140)	29(.136)
2985	17(.173)	18(.169)	19(.166)	19(.166)	20(.161)	21(.159)	22(.157)	23(.154)	24(.152)	26(.147)
2845	29(.136)	29(.136)	30(.128)	30(.128)	30(.128)	30(.128)	31(.120)	31(.120)	31(.120)	32(.116)
2865	20(.161)	22(.157)	22(.157)	23(.154)	24(.152)	25(.149)	26(.147)	27(.144)	28(.140)	29(.136)
2880	15(.180)	16(.177)	17(.173)	17(.173)	18(.169)	19(.166)	20(.161)	20(.161)	22(.157)	24(.152)
2895	8(.199)	10(.193)	11(.191)	12(.189)	13(.185)	13(.185)	15(.180)	16(.177)	17(.173)	18(.169)
2840	29(.136)	29(.136)	30(.128)	30(.128)	30(.128)	30(.128)	31(.120)	31(.120)	31(.120)	32(.116)
2855	25(.149)	26(.147)	27(.144)	27(.144)	28(.140)	28(.140)	29(.136)	29(.136)	30(.128)	30(.128)
2870	17(.173)	18(.169)	19(.166)	19(.166)	20(.161)	21(.159)	22(.157)	23(.154)	24(.152)	26(.147)
2890	12(.189)	13(.185)	14(.182)	15(.180)	16(.177)	17(.173)	17(.173)	18(.169)	19(.166)	20(.161)
2844	29(.136)	29(.136)	30(.128)	30(.128)	30(.128)	30(.128)	31(.120)	31(.120)	31(.120)	32(.116)
2860	22(.157)	23(.154)	24(.152)	25(.149)	26(.147)	27(.144)	27(.144)	28(.140)	29(.136)	29(.136)
2875	16(.177)	17(.173)	18(.169)	18(.169)	19(.166)	19(.166)	20(.161)	22(.157)	23(.154)	25(.149)
2892	9(.196)	11(.191)	12(.189)	12(.189)	13(.185)	14(.182)	16(.177)	17(.173)	18(.169)	19(.166)

PROPANE GAS

Model No.	Orifice at Sea Level	Orifice Size Required At Other Elevations								
		2000	3000	4000	5000	6000	7000	8000	9000	10000
2940	48(.076)	49(.073)	49(.073)	49(.073)	50(.070)	50(.070)	50(.070)	51(.067)	51(.067)	52(.063)
2960	42(.093)	42(.093)	43(.089)	43(.089)	43(.089)	44(.086)	44(.086)	45(.082)	46(.081)	47(.078)
2970	38(.101)	39(.099)	40(.098)	41(.096)	41(.096)	42(.093)	42(.093)	43(.089)	43(.089)	44(.086)
2985	35(.110)	36(.106)	36(.106)	37(.104)	37(.104)	38(.101)	39(.099)	40(.098)	41(.096)	42(.093)
2845	45(.082)	46(.081)	47(.078)	47(.078)	47(.078)	48(.076)	48(.076)	49(.073)	49(.073)	50(.070)
2865	40(.098)	41(.096)	42(.093)	42(.093)	42(.093)	43(.089)	43(.089)	44(.086)	44(.086)	45(.082)
2880	35(.110)	36(.106)	36(.106)	37(.104)	37(.104)	38(.101)	39(.099)	40(.098)	41(.096)	42(.093)
2895	31(.120)	32(.116)	32(.116)	32(.116)	33(.113)	34(.111)	35(.110)	36(.106)	37(.104)	38(.101)
2840	47(.078)	48(.076)	48(.076)	49(.073)	49(.073)	49(.073)	50(.070)	50(.070)	51(.067)	51(.067)
2855	43(.089)	44(.086)	44(.086)	44(.086)	45(.082)	45(.082)	46(.081)	47(.078)	47(.078)	48(.076)
2870	37(.104)	38(.101)	39(.099)	39(.099)	40(.098)	41(.096)	42(.093)	42(.093)	43(.089)	43(.089)
2890	33(.113)	35(.110)	35(.110)	36(.106)	36(.106)	37(.104)	38(.101)	38(.101)	40(.098)	41(.096)
2844	45(.082)	46(.081)	47(.078)	47(.078)	47(.078)	48(.076)	48(.076)	49(.073)	49(.073)	50(.070)
2860	42(.093)	42(.093)	43(.089)	43(.089)	43(.089)	44(.086)	44(.086)	45(.082)	46(.081)	47(.078)
2875	36(.106)	37(.104)	38(.101)	38(.101)	39(.099)	40(.098)	41(.096)	41(.096)	42(.093)	43(.089)
2892	32(.116)	33(.113)	34(.111)	35(.110)	35(.110)	36(.106)	36(.106)	37(.104)	38(.101)	40(.098)

The above is adapted from The National Fuel Gas Code, ANSI Z223.1-1988.

DIMENSIONS

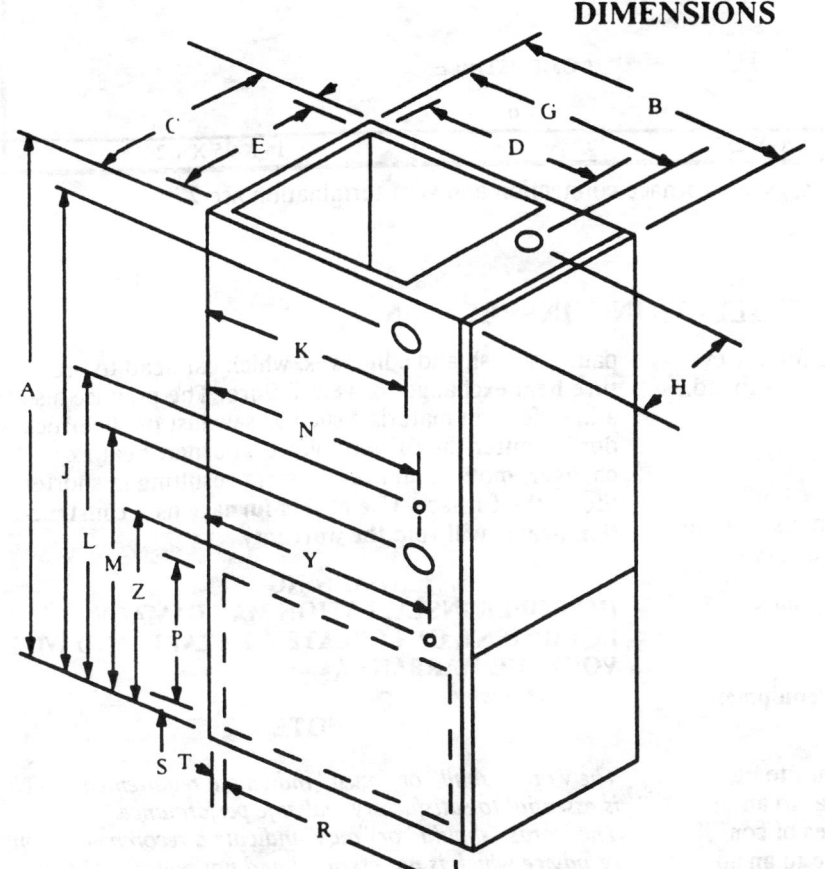

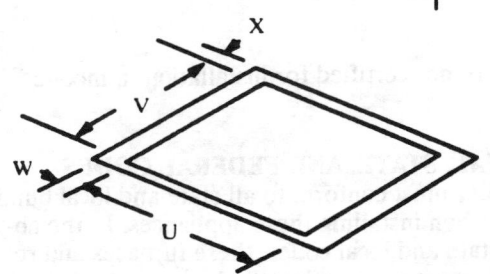

	2940E666 2960E666	2960E766 2970E766 2985E766
A	46	46
B	28	28
C	16–7/8	20
D	20–3/8	20–3/8
E	15–3/8	18–1/2
F	3/4	3/4
G	23–3/16	23–3/16
H	11–5/16	12–7/8
J	43–1/8	43–1/8
K	23–1/16	23–1/16
L	28–5/8	28–5/8
M	25–3/16	25–3/16
N	24–1/2	24–1/2
P	14–1/2	14–1/2
R	23	23
S	1	1
T	3–1/2	3–1/2
U	23–3/4	23–3/4
V	14–7/8	18
W	1	1
X	1	1
Y	26	26
Z	17–15/16	17–15/16

DIMENSIONS IN INCHES

Item	Dimension Location
Flue	G&H
Electrical Supply Entrance	J&K
Low Voltage Wire Entrance (either side)	L&N
Gas Inlet (either side)	M&N
Optional Side Return Air Inlet (either side)	P&R
Optional Bottom Return Air Inlet	U&V
Supply Air Outlet	D&E
Condensate Drain (either side)	Y&Z

SPECIFICATIONS

MODEL	NAT.	2940E666	2960E666	2960E766	2970E766	2985E766
UNIT RATING BTU/Hr.	Input:0-2000 Ft. Elevation	45,000	65,000	65,000	80,000	95,000
	High Altitude	FOR ELEVATIONS ABOVE 2000 FEET, REDUCE INPUT 4% FOR EACH 1000 FEET OF ELEVATION ABOVE SEA LEVEL.				
AIR TEMPERATURE RISE RANGE F		20-50	25-55	25-55	30-60	35-65
DESIGNED MAXIMUM OUTLET AIR TEMP. F		145	150	150	155	160
MAX. EXT. STATIC PRESSURE IN. W.C.		.5				
VENT PIPE		2 in. DIAMETER* SCHEDULE 40 PVC OR CPVC#				
VENT PIPE FITTINGS		2 in. DIAMETER SCHEDULE 40 OR 80 PVC OR CPVC#				
GAS CONNECTION		1/2 in. FPT				
ELECTRICAL SERVICE		115 VAC 60 HZ 1 PH				
MINIMUM DISTANCE TO COMBUSTIBLE MATERIALS IN INCHES						
FRONT		6				
BACK		0				
PLENUM-TOP		1				
FLOOR		COMBUSTIBLE				
SIDES		0				
VENT		0				
FILTER(Furnished)		20 X 25 X 1/2			16 X 25 X 1/2	

*3 in. diameter must be used with model 2985; except furnace connection and vent termination are 2 in.
First two feet of vent pipe must be CPVC

PRE-INSTALLATION INSPECTION

Inspect the shipping container and furnace for any evidence of shipping damage. If furnace damage is found, notify freight carrier and file claim.

IMPORTANT NOTICE

These instructions are for the use of qualified individuals specially trained and experienced in installation of this type equipment and related system components.

Installation and service personnel are required by some states to be licensed.

Persons not qualified shall not install this equipment nor interpret these instructions.

IMPORTANT NOTICE – This furnace is not to be used as a construction heater to supply heat to an unfinished building during the finishing phases of construction. This practice exposes the furnace to an abnormally corrosive atmosphere from sources such as paint, varnish and adhesives, which can lead to premature heat exchanger or vent failure. The practice also allows foreign materials such as sawdust or sheetrock dust to enter the furnace blower, burner, heat exchanger, motors, and vent system resulting in shorter life of the furnace. **Use of this furnace as a construction heater will void the warranty.**

WARNING
IMPROPER INSTALLATION MAY DAMAGE EQUIPMENT, CAN CREATE A HAZARD, AND WILL VOID THE WARRANTY.

NOTE

The words "shall" or "must" indicate a requirement which is essential to satisfactory and safe performance.
The words "should" or "may" indicate a recommendation or advice which is not essential and not required but which may be useful or helpful.

APPLICATION

FURNACE CERTIFICATION AND USAGE
The furnace described in these instructions are design certified by the American Gas Association to be in compliance with American National Standard Z21.47b-1989.
These furnaces are forced air type and may be utilized for indoor installation in manufactured buildings (modular only), or buildings constructed on site. These furnaces are not certified for installation in mobile homes.

MUNICIPAL, STATE, AND FEDERAL CODES
The installer must conform to all state and local building codes when installing these appliances. In the absence of state and local codes, these furnaces and related equipment must be installed in accordance with

the latest issue of the following:
NATIONAL FUEL GAS CODE - ANSI Z223.1-1988
NATIONAL ELECTRICAL CODE, ANSI/NFPA No. 70-1987
Applicable codes take precedence over any recommendation made in these instructions.

FURNACE SIZING AND DUCT SYSTEM DESIGN

Consideration should be given to the heating capacity required and also the air quantity (CFM) required if A/C is to be installed along with the furnace or at some future time. These factors can be determined by calculating the heat loss and heat gain of the home or structure.

If these calculations are not performed and the furnace is oversized, the following may result:
1. Short cycling of the furnace.
2. Wide temperature fluctuations from the thermostat setting.
3. Reduced overall operating efficiency of the furnace.

The supply and return duct system must be of adequate size and designed such that the furnace will operate within the designed air temperature rise range and **not exceed the maximum designed static pressure.** These values are listed in the specification table of these instructions and on the furnace rating plate.

Information, values and data necessary for heat loss, heat gain and duct system design may be found in ASHRAE HANDBOOK OF FUNDAMENTALS-1989 EDITION or in other nationally recognized publications recognized by municipal, state, and federal code authorities.

A/C USAGE DUCT SYSTEMS

1. When a single (common) duct system is used one of the following methods shall be used:
   a. A plenum type cooling coil must be installed on the air discharge side downstream from the furnace, or
   b. A Coil-Blower type cooling coil must be installed in parallel with and isolated from the furnace, or
   c. A self-contained A/C unit must be in parallel with and isolated from the furnace.

WARNING
DAMPERS MUST BE INSTALLED WHEN A COIL-BLOWER OR SELF-CONTAINED UNIT IS EMPLOYED TO PREVENT CONDITIONED COOL AIR FROM COMING IN CONTACT WITH THE HEAT EXCHANGER TO AVOID MOISTURE CONDENSATION AND RUST OUT, WHICH CAN ALLOW PRODUCTS OF COMBUSTION TO BE CIRCULATED INTO THE LIVING AREA BY THE FURNACE BLOWER RESULTING IN POSSIBLE ASPHYXIATION. IF DAMPERS ARE MANUALLY OPERATED TYPE, A MEANS MUST BE PROVIDED TO PREVENT EITHER THE FURNACE OR A/C UNIT FROM OPERATING UNLESS DAMPERS ARE IN FULL HEAT OR COOL POSITION.

2. If two duct systems are used as could be the case with a coil-blower or a self-contained A/C unit, the furnace and A/C unit should be controlled by a single combination heating and cooling thermostat which will prevent the furnace and A/C unit from operating simultaneously.

CAUTION
If a separate heating and separate cooling thermostat is used, a manually operated electrical interlock switch must be installed to prevent simultaneous operation of both systems and avoid a possible hazardous condition due to overheating of the conditioned space.

INSTALLATION PROCEDURE

NOTICE TO INSTALLER

This furnace is equipped with a metal blower brace which supports the blower during shipping. If the return air for the furnace is to be brought in the side, the blower brace may be left in place. However, if bottom return is to be used, the brace must be removed by removing the screw on each side of the casing which holds it in place. This must be done before the furnace is installed.

Models 29XXE766 are also equipped with a shipping strap on the blower motor shaft which supports the blower motor during shipping. This strap must be removed before the furnace is operated for the first time. It can be removed by removing the two fastening screws with a short screwdriver.

LOCATIONS AND CLEARANCES

The minimum clearances between the furnace, vent system, etc., and combustible materials are listed in the specification tables of these instructions and on the unit rating plate.

Installations on Combustible Flooring
This furnace may be installed directly on bare wood floors. It shall not be installed directly on carpeting, tile, or other type combustible flooring.

Clearance for Lighting and Service
Adequate clearance must be provided for lighting and maintenance. A minimum of 24" clearance should be provided between the front of the furnace and any opposite wall. If the furnace is in line with a door, you must provide a minimum of 6" clearance from the front of the furnace to the door.

If the furnace is to be installed in a close clearance closet, the door should be of adequate size to allow for removal of the furnace should it become necessary.

Installation in Residential Garages
When furnace is installed in a residential garage it must be located and installed such that it will be protected against vehicular damage. The furnace must be installed so that the burner is a minimum of 18" above the floor.

Installation in an Unconditioned Space

This is a condensing furnace. Therefore, if the temperature in an unconditioned space ever reaches freezing, 32 degrees F, or below, the condensate will freeze in the condensate tubes. If this happens, then condensate will not drain properly and will fill the condensing radiator until it blocks the flow of flue gases which in turn will shut down the furnace. The furnace will not operate until the condensate thaws and unblocks the flow of flue gases through the furnace. **DO NOT INSTALL THIS FURNACE IN ANY UNCONDITIONED SPACE THAT MAY HAVE TEMPERATURES THAT FALL BELOW THE FREEZING POINT.**

RETURN AIR AND FILTERS

Return Air Connection – The return air connection to the furnace may be through the bottom, through one side, or through both sides. Lances on casing sides locate the return air openings. The return air ducts to the furnace must have a total cross sectional area of not less than two square inches per 1000 BTUH of furnace input rating.

When bottom return air inlet is employed, filter is installed in the blower compartment, over the air inlet opening and secured in place with the retaining clip provided. When a side inlet return is employed, filters cannot be mounted inside the casing. Filters must be mounted in an external frame on the side of the furnace or in the return duct and be accessible for cleaning and replacement. One or more return grill-filter frames may be employed when side inlet returns are used.

IMPORTANT NOTICE – A solid metal block-off panel must be in place to block the bottom opening in the furnace when side return air ducts are used. Failure to block this opening could cause products of combustion to be circulated into the living space and create a potentially hazardous condition. A block-off panel has been provided with the furnace for this purpose.

WARNING – WHEN THE FURNACE IS INSTALLED IN A CLOSET OR OTHER CONFINED SPACE AND A SIDE INLET DUCT EMPLOYED, THE DUCT MUST BE SEALED TO THE FURNACE AND EXTEND TO THE CONDITIONED SPACE TO PREVENT ANY COMMUNICATION BETWEEN THE SPACE OR ROOM IN WHICH THE FURNACE IS INSTALLED, WHICH COULD RESULT IN ASPHYXIATION.

Do not exceed .5" W. C. total static pressure. See duct system design on page 4.

VENTILATION AND COMBUSTION AIR

Provide ventilation and combustion air in accordance with section 5.3, Air for Combustion and Ventilation, of the NATIONAL FUEL GAS CODE, ANSI Z223.1-1988, or applicable provisions of the local building codes. It is recommended that combustion air be obtained from outdoors as described in the section on confined spaces below.

WARNING – ADEQUATE VENTILATION AND COMBUSTION AIR MUST BE PROVIDED TO INSURE SATISFACTORY AND SAFE OPERATION OF THE FURNACE. AIR OPENINGS IN CASING FRONT PANEL AND VESTIBULE TOP PANEL MUST NOT BE OBSTRUCTED. FAILURE TO OBSERVE THIS RECOMMENDATION COULD RESULT IN ASPHYXIATION.

DO NOT STORE OR USE HALOGEN EMITTING SUBSTANCES IN THE VICINITY OF THIS APPLIANCE. SUCH SUBSTANCES INCLUDE CHLORINE BASED CLEANERS AND SWIMMING POOL CHEMICALS, WATER SOFTENING CHEMICALS, DE-ICING SALTS AND CHEMICALS, CLEANING SOLVENTS SUCH AS CARBON TETRACHLORIDE OR PERCHLORETHYLENE, HALOGEN TYPE REFRIGERANTS, PRINTING INKS, PAINT AND PAINT REMOVERS, VARNISHES, HYDROCHLORIC ACID, CEMENTS AND GLUES, AND MASONRY ACID WASHING MATERIALS. THE AIR USED BY THE BURNER FOR COMBUSTION MUST BE FREE OF HALOGENS TO AVOID POSSIBLE CORROSION TO THE HEATING SURFACES WHICH COULD RESULT IN ASPHYXIATION. EXPOSURE TO ANY OF THESE SUBSTANCES IN THE COMBUSTION AIR SUPPLY WILL REQUIRE THAT AIR FOR COMBUSTION BE BROUGHT IN FROM OUTDOORS AS DESCRIBED BELOW.

Installations in a Confined Space

If the unit is to be installed in a confined space such as a small closet or room, provisions must be made for supplying combustion and ventilation air to the space surrounding the furnace. This air must come from the outdoors by way of attic, crawl space, air duct, or direct opening. (See Fig. 1) Two openings of equal area must be provided; one starting within twelve inches of the ceiling and one starting within twelve inches of the floor of the confined space. The upper opening must always be above the top of the furnace casing. The lower opening, if in the side wall, floor or door, must be located below the level of the burner in the furnace.

When communicating directly with the outdoors through vertical ducts, the total free area of each opening must be at least one square inch for each 4,000 BTUH of furnace input, and when communicating directly with the outdoors through horizontal ducts, the total free area of each duct must be at least one square inch for each 2,000 BTUH of furnace input.

When ducts are used, they must be of the same cross-sectional area as the free area of the openings to which they connect. The minimum dimension of rectangular air ducts must not be less then three inches.

Installations in an Unconfined Space

In unconfined spaces in a building of conventional frame, masonry, or metal construction, infiltration is normally adequate to provide air for combustion and ventilation.

In buildings of tight construction, all air shall be obtained from outdoors or from spaces communicating freely with outdoors. A permanent opening or openings having a total free area of not less than one square inch for each 5000 BTUH of furnace input shall be provided.

If the furnace is to be installed in a commercial build-ing, a building with an indoor pool, a laundry room, hobby or craft room, or chemical storage area , all air must be brought in from outside as described above. Further details on supplying outdoor air for combustion may be obtained from Section 5.3.3 of the National Fuel Gas Code ANSI Z223.1-1988.

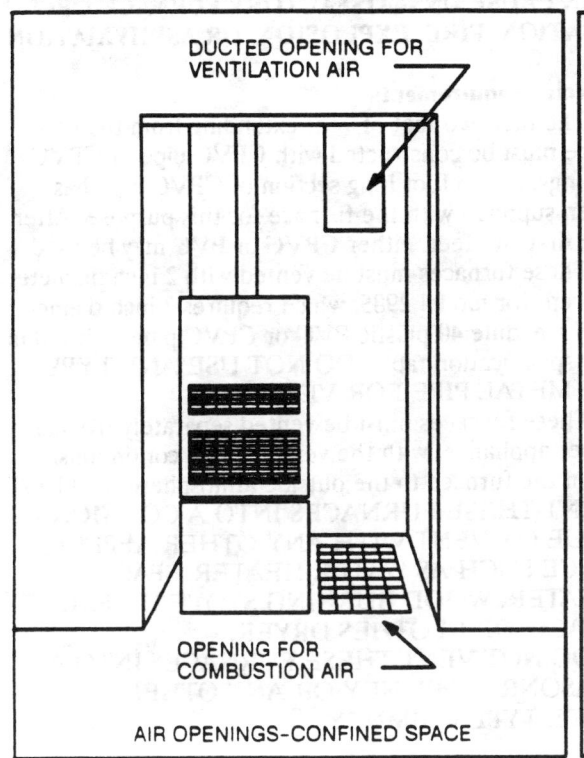

DUCTED OPENING FOR VENTILATION AIR

OPENING FOR COMBUSTION AIR

AIR OPENINGS-CONFINED SPACE

Figure 1

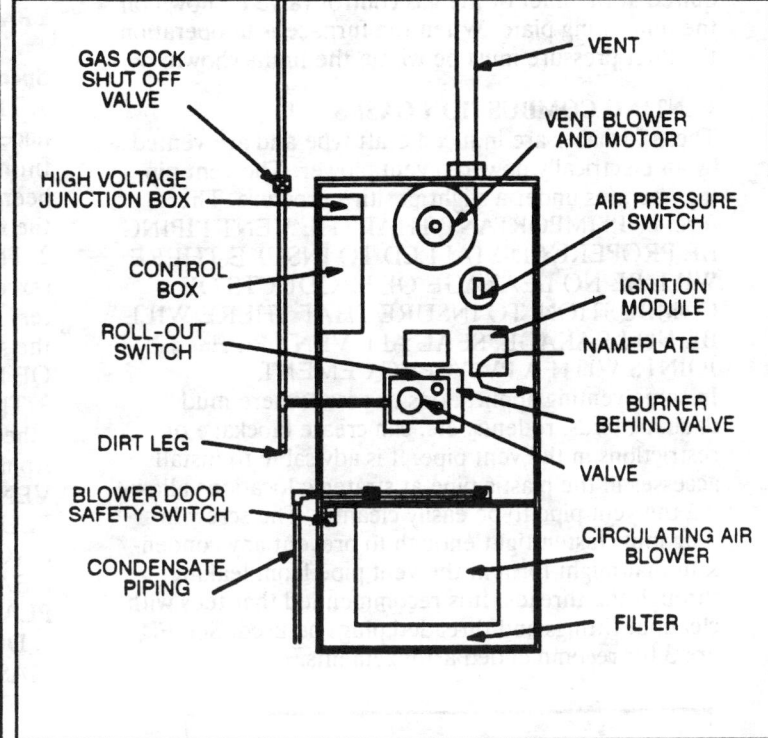

GAS COCK SHUT OFF VALVE

HIGH VOLTAGE JUNCTION BOX

CONTROL BOX

ROLL-OUT SWITCH

DIRT LEG

BLOWER DOOR SAFETY SWITCH

CONDENSATE PIPING

VENT

VENT BLOWER AND MOTOR

AIR PRESSURE SWITCH

IGNITION MODULE

NAMEPLATE

BURNER BEHIND VALVE

VALVE

CIRCULATING AIR BLOWER

FILTER

Figure 2

INSTALLATION OF PLENUM OR COIL CABINET

Because it may be necessary to remove the plate covering the top of the furnace vestibule compartment for a periodic maintenance inspection or service, observe the following procedure. Remove the plate if the plenum has a horizontal flange at its bottom and reinstall the plate over the top of the plenum flange. Drill the necessary holes and attach the plate to the plenum flange and furnace. If an A/C coil cabinet is being installed, remove and discard the back screws securing the plate to the vestibule. Install the coil cabinet over the top of the plate.

GAS PIPING AND SUPPLY PRESSURES

Before installing gas piping, check with local code authorities for requirements concerning gas piping. In the absence of local codes, follow the recommendations contained in NATIONAL FUEL GAS CODE ANSI Z223.1-1988 for gas piping materials, pipe sizing, and the requirements for installation. It is recommended that a gas cock shutoff valve be installed in the gas supply line **outside the casing** where it is readily accessible, as close to the furnace as is practicable. An 1/8 in. NPT (plugged) pressure tap for test gauge connection must be installed in the gas supply line immediately upstream from the furnace.

Install a dirt leg at the bottom of any vertical riser or

drop, as close to the furnace as possible, to collect moisture and foreign material. Install a ground joint union just ahead of the gas control valve. A typical arrangement is shown in Figure 2.

When making the connection at the gas control valve, use a wrench on the inlet side of the valve to prevent any possible twisting of the valve body which could cause damage and leaks. Connection sizes are shown in the specifications table. When making up pipe joints use pipe thread compound which is resistant to natural and LP gases.

CAUTION – During pressure testing of the gas supply piping system, observe the following to avoid fire, explosion, asphyxiation, or damage to the appliance.

a. If test pressure is less than or equal to 1/2 psig (3.48 kPa), isolate the furnace by closing its individual manual shutoff valve.

b. If test pressure is greater than 1/2 psig (3.48 kPa), the furnace and its individual shutoff valve must be disconnected from the gas supply system.

Following installation of the piping, first ensure that the gas control knob on the gas valve is in the off position and then pressurize the system with gas. Thoroughly check the piping system for leaks.

CAUTION – NEVER USE AN OPEN FLAME TO

7

CHECK FOR LEAKS. FIRE OR EXPLOSION COULD OCCUR. SINCE SOME LEAK SOLUTIONS INCLUDING SOAP AND WATER MAY CAUSE CORROSION OR STRESS CRACKING, THE PIPING MUST BE RINSED WITH WATER AFTER TESTING UNLESS IT HAS BEEN DETERMINED THAT THE LEAK SOLUTION IS NONCORROSIVE.

The maximum and minimum gas supply pressure required at the inlet of the gas control valve is shown on the unit rating plate. When the furnace is in operation, the inlet pressure must be within the limits shown.

VENTING COMBUSTION GASES

These furnaces are induced draft type and are vented by an electrically powered vent blower. The vent piping operates under a slight positive pressure. Therefore, IT IS IMPORTANT THAT THE VENT PIPING BE PROPERLY INSTALLED TO INSURE THERE WILL BE NO LEAKAGE OF PRODUCTS OF COMBUSTION. TO INSURE THAT THERE WILL BE NO LEAKAGE, SEAL ALL VENT PIPE JOINTS WITH A PVC/CPVC CEMENT.

In some venting applications or areas where mud daubers, birds, rodents, etc. can create blockage or restrictions in the vent pipe, it is advisable to install accesses in the plastic pipe at strategic locations allowing the vent pipe to be easily cleaned. The screw in plug must fasten tight enough to prevent any condensate that might form in the vent pipe from leaking through the threads. It is recommended that tees with cleanout fittings and threaded plugs be used. See Figure 3 for recommended arrangements.

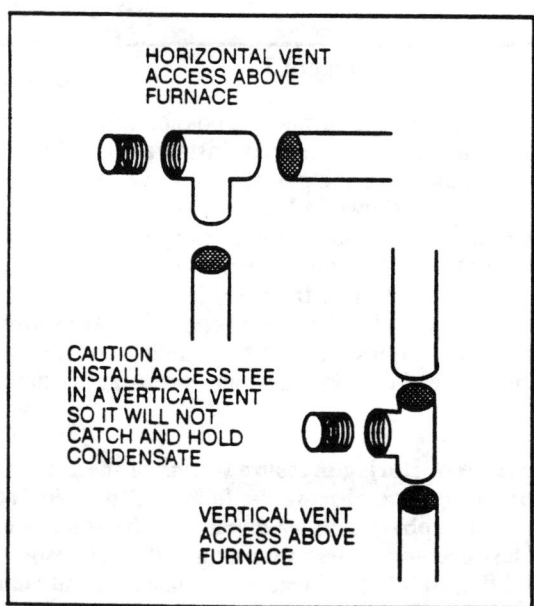

HORIZONTAL VENT ACCESS ABOVE FURNACE

CAUTION
INSTALL ACCESS TEE IN A VERTICAL VENT SO IT WILL NOT CATCH AND HOLD CONDENSATE

VERTICAL VENT ACCESS ABOVE FURNACE

Figure 3

Selecting the Best Vent Arrangement

The vent system may be one of two basic types: (1) A horizontal vent that terminates at the outer surface of an outside wall which is above grade. This type of venting must terminate with a 2 inch diameter schedule 40 or 80 plastic PVC or CPVC tee for the vent terminal. (2) A vertical vent which consists of a vertical riser or horizontal lateral which passes through the interior of the structure and terminates above the roof with a 2 inch diameter schedule 40 or 80 plastic PVC or CPVC return bend (two 90 degree elbows) or tee.

WARNING – FAILURE TO FOLLOW THE FOLLOWING LISTED REQUIREMENTS AND RULES CAN CAUSE UNSATISFACTORY FURNACE OPERATION, FIRE, EXPLOSION, OR ASPHYXIATION.

Special Requirements

1. The first two feet of vent extending from the furnace must be constructed with CPVC pipe or CPVC fittings. A two foot long section of CPVC pipe has been supplied with the furnace for this purpose. After the first two feet, either CPVC or PVC may be used.
2. These furnaces must be vented with 2 inch diameter (except for model 2985, which requires 3 inch diameter) schedule 40 plastic PVC or CPVC pipe as listed in the specification table. DO NOT USE ANY TYPE OF METAL PIPE FOR VENTING.
3. These furnaces must be vented separately from any other appliance, with the vent running continuously from the furnace to the outside atmosphere. DO NOT VENT THESE FURNACES INTO A COMMON FLUE OR VENT WITH ANY OTHER APPLIANCE SUCH AS WATER HEATER, SPACE HEATER, WOOD BURNING STOVE OR FIREPLACE, OR CLOTHES DRYER.
4. DO NOT VENT THESE FURNACES INTO A MASONRY CHIMNEY OR ANY OTHER ALL-FUEL TYPE CHIMNEY.

Observe the Following Rules:

1. Any horizontal run of the vent must be pitched at least 1/4 inch per foot upward so that water that condenses in the vent will run toward the furnace.
2. Joints in the vent pipe must be securely made and any horizontal run of the pipe supported no less than one support every three feet to prevent sagging or displacement of the pipe. This will prevent stressing of the joints and will keep water from collecting and restricting the vent pipe.
3. The vent piping route must be planned to minimize the number of elbows. Short offsets made by joining two 90 degree elbows together should be avoided. (Two 45 degree elbows are equal to one 90 degree elbow in flow resistance.) Each five foot reduction in vent pipe length will allow an additional 90 degree elbow to be used if necessary.
4. Before attaching the vent pipe, manually spin the vent blower by turning the plastic fan blade in the motor with a small screwdriver. This is to ensure that the blower will spin freely with no interference between the blower wheel and housing. If there is interference, see the service guide for these furnaces on making the proper adjustments.
5. The two foot section of CPVC pipe supplied with the furnace must be inserted into the vent blower outlet and, to provide a good mechanical connection to the furnace, securely connected by tightening the

bracket supplied on the furnace around the vent pipe. See Figure 4.

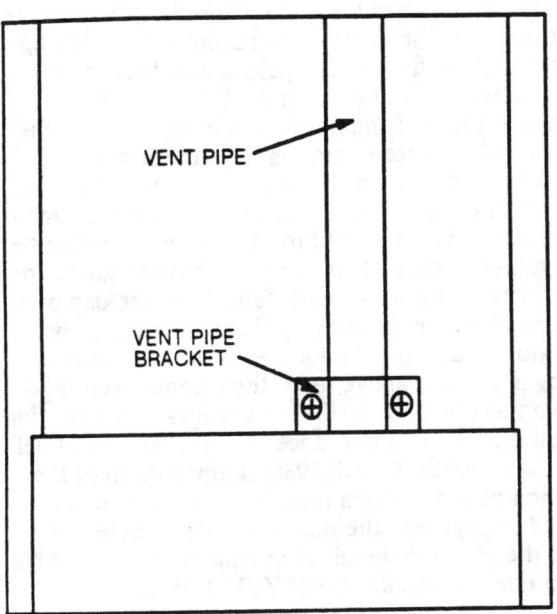

Figure 4

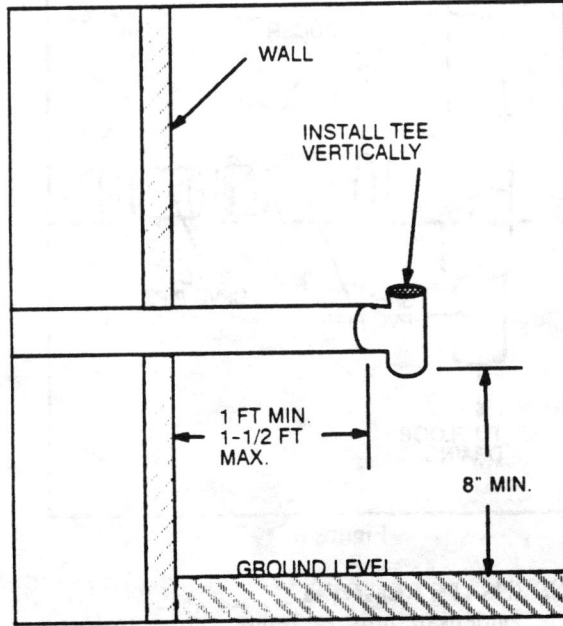

Figure 5

Horizontal Vents

1. The total run of the vent must not exceed 30 feet with a maximum of two 90 degree elbows. **Keep the vent as short and direct as practical.**
2. When the pipe exits the building structure, its exit must be at least three feet above any forced air inlet located within ten feet. Also, the exit must be at least four feet below, four feet horizontally from, or one foot above any door, window, or gravity air inlet. There must also be a minimum horizontal clearance of four feet from any electric meter, gas meter, regulator, and relief equipment.

3. The vent pipe must extend at least one foot but no more than one and one-half feet past the outside wall and must terminate with a two inch diameter schedule 40 or 80 plastic PVC or CPVC tee. The termination should be made so as to prevent possible blockage of the vent with snow and to protect any building materials from degradation from flue gases. See Figure 5.

WARNING – USE OF A NON-APPROVED TERMINAL COULD CAUSE FIRE OR ASPHYXIATION

4. FOR BEST RESULTS, EXIT THE HORIZONTAL VENT THROUGH A SIDE OF THE HOME THAT DOES NOT FACE THE PREVAILING WINTER WIND.

5. Vent pipe used on model 2985 furnaces must be three inch diameter schedule 40 plastic PVC or CPVC pipe. The pipe that connects to the furnace and that which extends through and past the outside wall must be two inch diameter schedule 40 plastic PVC or CPVC pipe. A suitable three to two inch reducer must be used when going from two inch to three inch diameter pipe. See Figure 6.

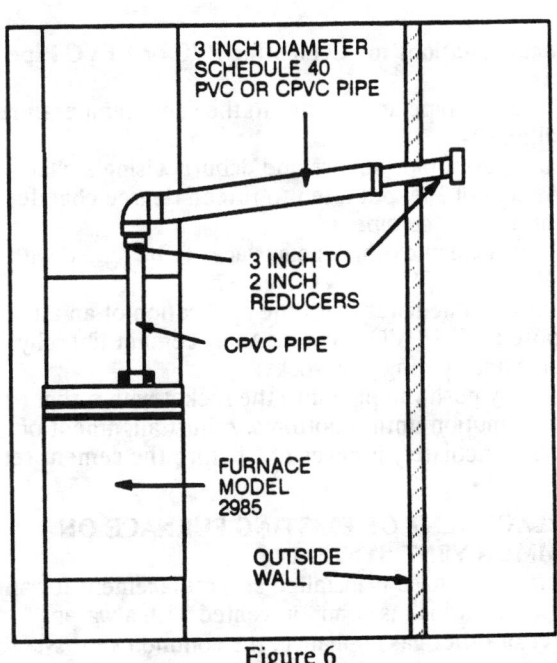

Figure 6

Vertical Vents

1. The total run of the vent must not exceed 30 feet with a maximum of two 90 degree elbows. **Keep the vent as short as practical.** The furnace may be vented vertically by running the vent pipe directly above the vent connection or with a horizontal lateral connected to a vertical riser.
2. Vertical vent risers must not be installed where they will be exposed to the outdoors.
3. The opening where the vent pipe penetrates the roof must be sealed with a plastic flashing.
4. The vent pipe must terminate with a two inch diameter schedule 40 or 80 plastic tee or a return bend made with two 90 degree elbows. The opening of the return bend or tee must be at least one foot but no

more than one and one-half feet above the roof. See Figure 7.

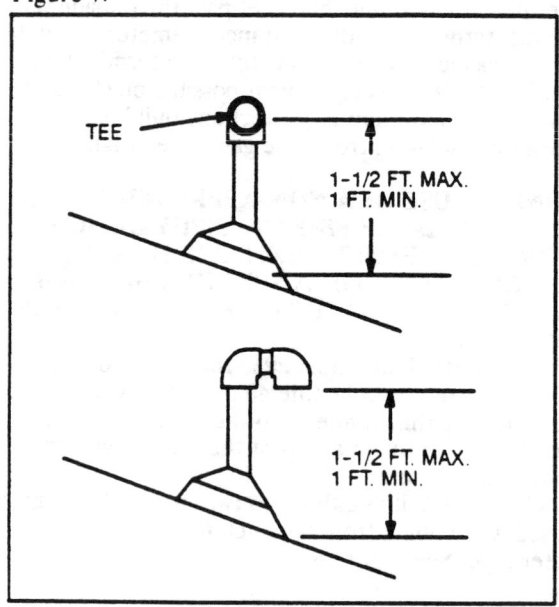

TEE

1-1/2 FT. MAX.
1 FT. MIN.

1-1/2 FT. MAX.
1 FT. MIN.

Figure 7

Recommendations for Cementing PVC or CPVC Pipe Joints

1. Condition pipe and fittings to the same temperature conditions.
2. Cut ends of pipe square and deburr. Using a chamfering tool or file, put a ten to fifteen degree chamfer on the end of the pipe.
3. Using a clean cloth, wipe surfaces to be joined with an all purpose PVC/CPVC cleaner.
4. Follow immediately with the application of an all purpose PVC/CPVC cement. Apply cement liberally on pipe and sparingly on socket.
5. Quickly push the pipe into the socket with a slight twisting motion until it bottoms. Adjust alignment of fitting immediately if necessary, before the cement sets up.

REPLACEMENT OF EXISTING FURNACE ON COMMON VENT SYSTEM

When this furnace is installed as a replacement for an old furnace which is common vented with a water heater or other gas appliance, the common vent system may be too large for the appliances remaining on the vent system after the old furnace is removed. To test for an oversized vent system, the following steps shall be followed with each appliance remaining connected to the common venting system placed in operation, while the other appliances remaining connected to the common venting system are not in operation: (a) Seal any unused openings in the common venting system. (b) Visually inspect the venting system for proper size and horizontal pitch and determine that there is no blockage or restriction, leakage, corrosion or other deficiencies which could cause an unsafe condition. (c) Insofar as is practical, close all building doors and windows and all doors between the space in which the appliances remaining connected to the common venting system are located and other spaces of the building. Turn on clothes dryers and any appliance not con-

nected to the common venting system. Turn on any exhaust fans, such as range hoods and bathroom exhausts, so they will operate at maximum speed. Do not operate a summer exhaust fan. Close fireplace dampers. (d) Follow the lighting instructions. Place the appliance being inspected in operation. Adjust thermostat so appliance will operate continuously. (e) Test for spillage at the draft hood relief opening after 5 minutes of main burner operation. Use the flame of a match or candle, or smoke from a cigarette, cigar, or pipe. (f) After it has been determined that each appliance remaining connected to the common venting system properly vents when tested as outlined above, return doors, windows, exhaust fans, fireplace dampers, and any other gas-burning appliance to their previous condition of use. (g) If improper venting is observed during any of the above tests, the common venting system must be corrected. Any changes to the venting system must be in accordance with the National Fuel Gas Code, ANSI Z223.1-1988. If any portion of the common venting system must be resized, it should be resized to approach the minimum size as determined using the appropriate tables in Appendix G in the National Fuel Gas Code, ANSI Z223.1-1988.

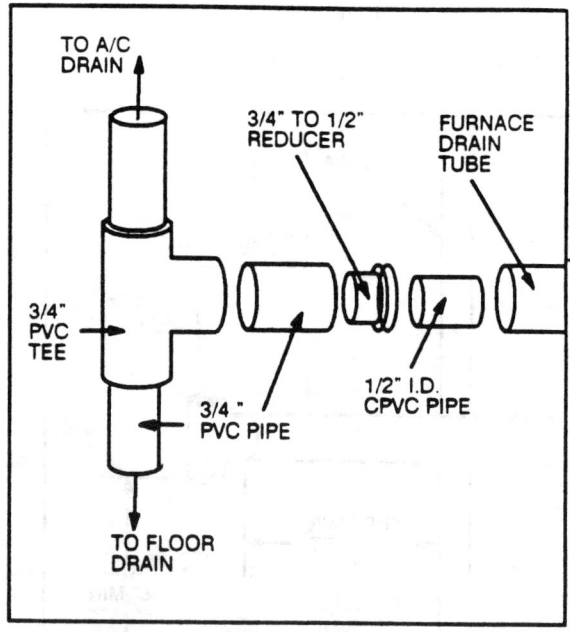

TO A/C DRAIN

3/4" TO 1/2" REDUCER

FURNACE DRAIN TUBE

3/4" PVC TEE

3/4" PVC PIPE

1/2" I.D. CPVC PIPE

TO FLOOR DRAIN

Figure 8

CONDENSATE REMOVAL

1. The condensate must be removed from the furnace with no less than 1/2 inch PVC or CPVC pipe. See Figures 2 and 8. DO NOT USE METAL PIPE FOR REMOVAL OF CONDENSATE FROM THE FURNACE. Make sure that the CPVC tee provided in the furnace condensate line is held securely in the metal clip to the side of the furnace casing.
2. The condensate from the furnace may be drained with the condensate from the air conditioning system using a common PVC or CPVC pipe. See Figure 8. In this application, the condensation drain pipe must be no less than 3/4 inch I.D.

CAUTION – The draining of other components not tested in combination with these furnaces may make

the equipment in violation of local codes, may create a hazard, and may ruin the equipment and void the warranty.

3. The condensate pipe from the furnace condensate exit to the drain must always be lower than the condensate exit point of the furnace.

4. Seal all condensate pipe joints with a PVC/CPVC cement to insure that there will be no leakage of condensate.

ELECTRICAL WIRING

All internal wiring has been made at the factory. Field wiring requires only the connection of line voltage supply wiring and low voltage thermostat wiring.
Refer to the unit rating plate and specification tables found in these instructions for applicable electrical characteristics and requirements. A complete wiring diagram is supplied on page 16.

Service wiring

Field wiring and electrical grounding of the unit should conform to local codes or in the absence of local codes with the National Electrical Code ANSI/NFPA 70-1987.
A separate fused circuit from the main electrical panel should serve only the furnace. Connect the power supply leads in the furnace junction box, providing an approved strain relief, as shown in the wiring diagram on the unit and in these instructions.

Accessories

A humidifier or electronic air cleaner may be used with this furnace. All accessories should be wired according to the manufacturer's instructions.
In most cases, the humidifier is intended to operate only when the room thermostat is calling for heat. Attach the humidifier control wires to the red and white thermostat wires to supply 24 volts to the humidifier only during the heating cycle. **Make sure that the total load on the furnace transformer does not exceed 40 VA, including gas valve, furnace relays, accessories, and air conditioner loads.**

In most cases, an electronic air cleaner will be provided with a pressure switch or sail switch which senses airflow and energizes the air cleaner. If not so equipped, a sail switch in the air duct may be used to switch the air cleaner on and off as needed. **Do not attempt to wire an electronic air cleaner into the furnace blower relay.** Damage to the furnace blower motor may result.

Control Wiring

The thermostat should be installed in accordance with the manufacturer's instructions, furnished with the thermostat, and make connections to the unit as shown on the unit wiring diagram. It is recommended that size 18 AWG wire be used.
If the thermostat has an adjustable heat anticipator, the setting should be .13 amps.

Blower Motor Speed Selection

These furnaces are equipped with blowers which have multi-speed direct drive motors.
The blower speed selected is dependent upon the design and static pressure loss of the duct system. The duct system external static pressure includes the combined total of the supply and return ducts and any plenum type air conditioning coil if used.
The furnace must be adjusted to operate at or below the maximum external static (in. W.C.) and within the air temperature rise range as shown on the unit rating plate and in the specification table.
These furnaces are equipped with a blower relay which will change blower speeds automatically when the furnace is properly connected to a heating and cooling type wall thermostat. The blower motors are factory connected to operate on high speed for heating operation and medium or medium high for cooling operation. Dependent upon the conditions in a particular installation, the blower speeds may be changed. However, ONLY THE HIGH OR MEDIUM HIGH SPEEDS ARE TO BE USED FOR HEATING OPERATION. Table I may be used for selection of motor speed tap for cooling operation.

TABLE I
MOTOR SPEED TAP SELECTION FOR COOLING

FURNACE MODELS	COOLING SYSTEM CAPACITY, BTUH							
	18,000	24,000	30,000	36,000	42,000	48,000	54,000	60,000
2940E666	Low	Med.	High					
2960E666		Low	Med.	High				
2970E766		Low	Low	Med.Low	Med. High	High	High	
2985E766		Low	Low	Med.Low	Med. High	High	High	
2960E766				Low	Med. Low	Med. High	High	High

DANGER – SHOCK HAZARD
BE SURE ELECTRICAL POWER TO FURNACE IS TURNED OFF BEFORE CHANGING MOTOR SPEEDS.

Motor Terminal Identification

In all cases the white wire from the furnace control box is the common circuit and is fitted with a 3/16

quick connect terminal. This wire must always be connected to the motor terminal marked White, Common, or 1.
The black wire from the furnace control box is controlled by the furnace fan switch and is the hot wire for heating operation.
The blue wire is controlled by the blower relay and is the hot wire for cooling.

The motor terminals of various motors which may be used are identified by one of the methods shown in Table II.

TABLE II

Motor Speed	Motor Terminal Identification
High Med. High Med. Med. Low Low	Common, White Hi, Black Med. High, Blue Med.,Blue Med.Low, Yellow Low, Red

PRE-OPERATIONAL CHECKS

DANGER – SHOCK HAZARD – BE SURE ELECTRICAL POWER TO FURNACE IS TURNED OFF BEFORE PERFORMING THE PRE-OPERATIONAL CHECKS.

1. Be sure that the furnace is equipped for the type of gas being supplied to the furnace. See unit rating plate.
2. Make sure that shipping strap was removed from blower housing of furnace models 29XXE766.
3. Manually spin circulating air blower wheel to ensure that it turns freely and does not strike the blower housing.
4. Manually spin the vent blower by turning the heat dissipation fan blades between the vent motor and the furnace vestibule. The blower wheel should spin freely with no interference between the blower wheel and the blower housing.
5. Was the gas piping tested and/or purged of air then checked for leaks? See instructions for gas piping. Even the smallest leak must be eliminated before attempting to light the furnace.

SEQUENCE OF OPERATION

These furnaces are equipped with an electric hot surface burner ignition system. In response to a call for heat by the room thermostat, the burner is lighted by a hot glowing ignitor at the beginning of each operation cycle. The burner will continue to operate until the thermostat is satisfied at which time all burner flame is extinguished. During the off cycle no gas is consumed. With the room thermostat set below room temperature, and with the electrical power and gas supply to the furnace on, the normal sequence of operation is as follows:

1. When the room temperature falls below the setting of the room thermostat, the thermostat energizes the heating relay.
2. When the heating relay closes, a circuit is made starting the vent blower. A circuit is also made through the heating relay to the normally open vent air pressure switch contacts.
3. As the vent blower increases in speed, a negative pressure is developed in the vent blower. When sufficient negative pressure has developed, the contacts of the vent air pressure switch will close and complete the electrical circuit through the normally closed limit switch to the electronic ignition module.
4. During the next 40 to 50 seconds, the vent blower will bring fresh air into the heat exchanger and the ignitor will begin to glow. At the end of this period, the ignition module will open the gas valve and energize a safety lock-out circuit.
5. When the burner lights, the ignitor then acts as a flame probe which checks for the presence of a flame. As long as flame is present, the system will monitor it and hold the gas valve open.
6. If the burner fails to light within 6-8 seconds after the gas valve opens, the ignition module will close the gas valve and de-energize the ignitor. After a short pause, the system will recycle and try again for ignition. If the burner fails to light after three tries, the ignition module will lock off the gas valve and the ignitor. The system will remain in lock-out mode until the room thermostat is set below room temperature. The lock-out circuit will then be released and setting the thermostat to above room temperature will cause the system to try for ignition again.
7. The lapsed time from the moment the room thermostat closes to when the burner lights may be 50-60 seconds. This delay is caused by:(1) the time required for the vent blower to come to full speed, (2) the 40 to 50 seconds required for the ignitor to heat up, and (3) the time required for the vent blower to bring fresh air into the heat exchanger.
8. Thirty to fifty seconds after the burner has lighted, the fan switch will close and the furnace air circulation blower will run.

NOTE – If a heating/cooling thermostat is being used and the fan switch is set in the continuous blower position, the furnace circulating blower will run at the air conditioning speed. If the room thermostat calls for heat, the furnace air circulating blower will shut off for 50-60 seconds and then the burner will light. One to two minutes after the burner has lighted, the furnace circulating blower will begin running at heating speed. There is no pause in the blower operation when the thermostat is satisfied; the furnace circulating air blower just changes back over to cooling speed.

9. When room thermostat is satisfied the circuit to the heating relay is broken and the relay contacts return to the normally open position. The circuit to the vent blower, the blower sequencer, and the ignition module is broken and the burner is extinguished. The contacts

of the blower sequencer return to the normally open position within 30 seconds after the burner extinguishes. Then as heat is drawn from the heat exchanger and the air temperature is reduced to below the fan switch setting, the fan switch will open which stops the furnace air circulation blower.

FURNACE OPERATION

GAS VALVE

The gas control valve is multi-function with two operating valves in line, a pressure regulator and a manual gas cock. See Figure 9.

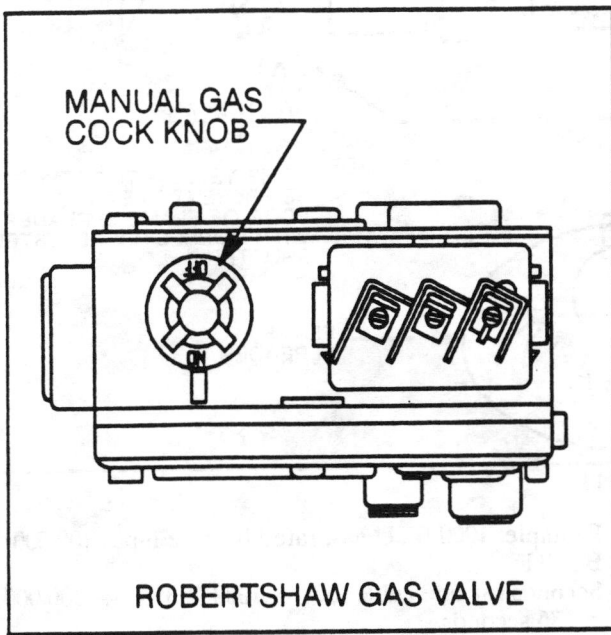

ROBERTSHAW GAS VALVE

Figure 9

The pressure regulator is factory set to provide an operating pressure of 3.5" W.C. on models equipped for natural gas and 10" W.C. on models equipped for LP gas.

The pressure regulator is a limited adjustment type. Refer to the section titled "Minor Input Adjustment" under "Furnace Input Capacity".

LIGHTING INSTRUCTIONS

THIS FURNACE IS EQUIPPED WITH AN AUTOMATIC ELECTRIC HOT SURFACE BURNER IGNITION SYSTEM WHICH LIGHTS THE BURNER EACH TIME THE THERMOSTAT CALLS FOR HEAT. THIS FURNACE CANNOT BE LIGHTED WITH A MATCH.

WARNING – FAILURE TO FOLLOW THESE INSTRUCTIONS MAY RESULT IN AN EXPLOSION AND POSSIBLE DAMAGE TO THE FURNACE AND INJURY TO THE OPERATOR.

1. Set room thermostat to lowest setting – or OFF.
2. Turn knob on gas control valve to OFF.
3. WAIT FIVE MINUTES.
4. Turn knob on gas control valve to ON.
5. Set room thermostat to desired setting above room temperature. Burner will light – which may take 50–60 seconds.
6. If burner fails to light, the furnace will try again for ignition. If after three tries and burner fails to light, go to complete shutdown and determine cause for failure to light. Be sure gas supply to furnace is on and supply piping has been purged of air. Be sure electric supply to furnace is on.
7. For complete shutdown – set thermostat to lowest setting or OFF and turn knob on gas valve to OFF.

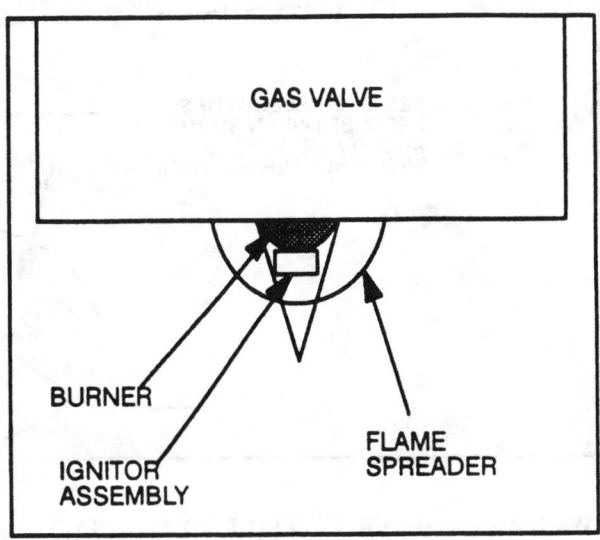

Figure 10

BURNER ADJUSTMENT

After lighting the furnace, allow the furnace to operate for 15 minutes and then adjust burner primary air shutter as follows:

1. Loosen air shutter adjustment rod locking screw. See Figure 10.
2. Close primary air shutter by pulling on adjustment rod until yellow tips appear in the flame at the end of the burner. See Figure 11.
3. Slowly open the primary air shutter by pushing adjustment rod in until yellow flame tips at the end of the burner disappear; then push rod in another 1/8 inch.
4. Secure primary air adjustment rod by tightening locking screw against it.

FURNACE INPUT CAPACITY

The maximum BTUH input capacity for each model is shown on the furnace rating plate and in the specification table. This input must not be exceeded.

The input shown may be used in geographic areas where the elevation is from 0 to 2000 feet. In areas above 2000 feet elevation, the furnace BTU input must be reduced 4% for each 1000 feet of elevation above sea level. The BTU input depends on the calorific heating value of the gas, orifice size, and manifold pressure. Orifice sizes are based on gas values of 1050 BTU/cu. ft. for natural gas and 2500 BTU/cu.ft. for LP (propane) gas. The orifice sizes supplied with the furnace should provide satisfactory input capacity for installations in most areas, except at high altitude.

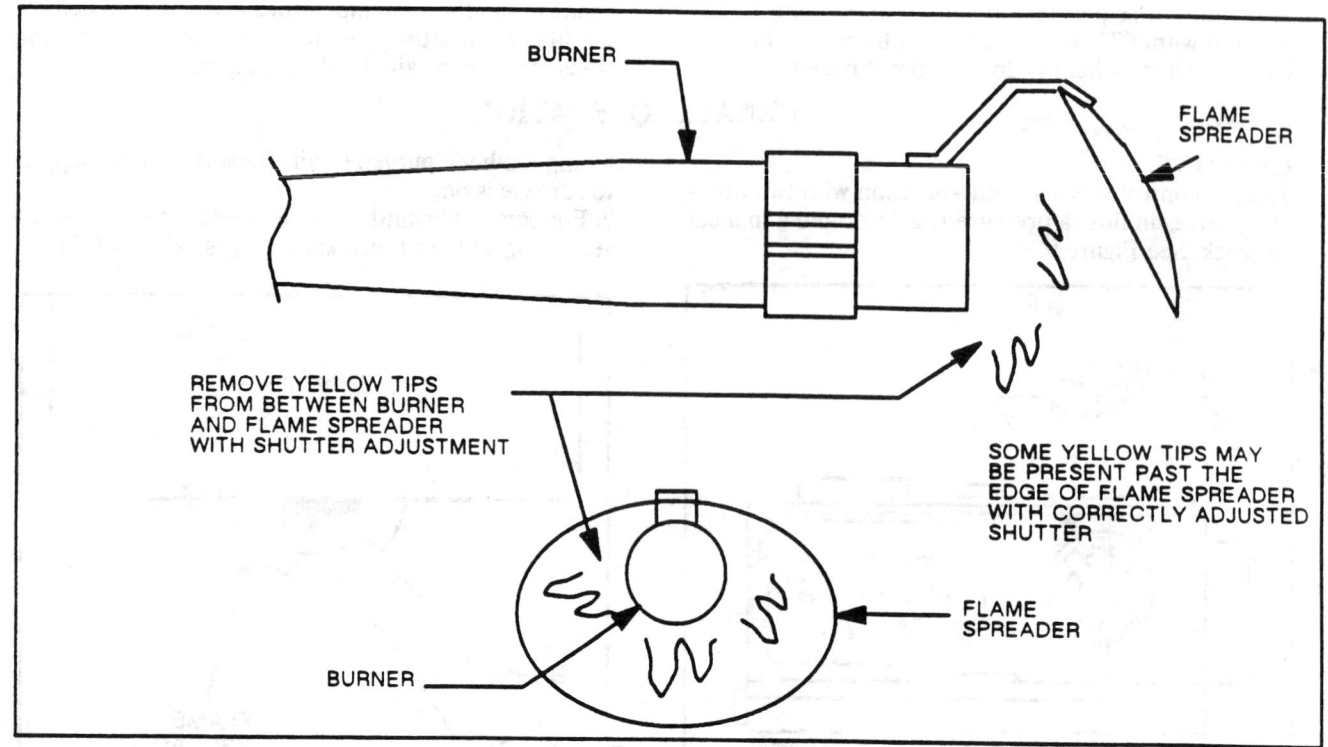

Figure 11

WARNING – NEVER ATTEMPT TO MODIFY THIS FURNACE – FIRE, EXPLOSION, OR ASPHYXI-ATION MAY RESULT. **If malfunction is apparent, contact qualified service agency and/or gas utility for assistance.**

Minor Input Adjustment

The input may be adjusted slightly by adjusting the pressure in the gas valve in order to change manifold pressure.

To adjust pressure regulator, remove cover screw (see location on Figure 9) on gas valve. Turn adjusting screw counter-clockwise to decrease pressure, turn clockwise to increase pressure. IN NO CASE SHOULD THE FINAL MANIFOLD PRESSURE VARY MORE THAN ± .3" W.C. FROM THE SPECIFIED REGULATOR PRESSURE SETTINGS (3.5" FOR NATURAL GAS AND 10" FOR LP GAS.)

Determining Gas Input Rate

Where gas is metered, the input rate may be determined by the following method:

Contact the gas supplier, public utility company or LP gas distributor to obtain the calorific gas value of the gas being used. When checking the gas input rate, any other gas burning appliances connected to the same meter should be completely off. The furnace should be allowed to operate for 15 minutes before attempting to check the gas input rate.

To check flow rate, observe the one cubic foot dial on the gas meter and determine the number of seconds required for the dial hand to complete one revolution (seconds to flow one cubic foot.)

To determine the number of seconds per cubic foot that is necessary to achieve the correct input rate, use the following formula: GAS VALUE X 3600 ÷ DESIRED INPUT = SECONDS NEEDED

Example: 1000 BTU gas, rated furnace input 100,000 BTUH

Seconds for one cubic foot = 1000 X 3600 ÷ 100,000 = 36 seconds.

If when clocking the meter, the one cubic foot dial makes a complete revolution in less time than was calculated that it should, the furnace is overfired and should be derated. If it takes more time for the meter to make one revolution than was calculated, the furnace is underfired and the rate should be increased. The orifice size must be changed to correct an over-fired or underfired condition. If it is determined that a different orifice is needed, please contact your distributor for assistance in selecting the correct replacement.

BALANCE THE SYSTEM – The air distribution system should be balanced, using in-line duct dampers if employed, to provide for satisfactory air delivery room to room. It is recommended that dampers in registers not be used for balancing the system.

CYCLE FURNACE FOR CORRECT OPERATION

After all parts of the installation have been completed, the furnace should be checked for normal operation prior to the time the system will be used by the home-owner/user.

1. Unplug the ignitor and check its resistance with a suitable ohmmeter. The cold resistance should be between 50 and 400 ohms. A reading outside of this range indicates a cracked or otherwise defective ignitor which should be replaced. Be sure to plug ignitor back in before continuing.

2. Place the heating system in the operating mode by following the lighting instructions outlined previously

14

in these instructions and on the furnace lighting instructions label.

3. Cycle the furnace on and off a number of times, allowing the furnace to light and the circulating air blower to come on. After thermostat is turned off each time, allow time for the circulating air blower to stop before starting the next operating cycle.

4. While unit is in operation check to determine that the circulating blower is operating smoothly with no undue vibration or noise. Check to insure that main burner has been adjusted and main burner flame is normal.

NOTICE TO INSTALLER – PROVIDE A HOOK, or other suitable means, and mount the envelope pucket supplied with the furnace in a conspicuous place near the furnace. Place all printed literature supplied with the furnace in the packet. Advise the owner/user of the location of the literature packet if possible. If installed system is not to be placed in operation immediately by the owner/user, it is recommended that the system be left in a shut down condition by setting the thermostat to off or lowest temperature setting. Turn gas control knob to off position and turn off electrical power to furnace.

SERVICE MAINTENANCE

A program of periodic inspection and preventative maintenance can help to insure trouble-free operation, provide more efficient operation and extend the life of the furnace.

DANGER
To avoid the possibility of electric shock, turn off electric power supply to furnace before any disassembly for inspection or maintenance work is performed.

HOME OWNER'S MAINTENANCE
Circulating Air Blower Assembly

1. Check to insure that blower wheel turns freely.
2. Check for build-up of foreign material on blades. Clean off any accumulation with stiff brush or by scraping with a suitable tool.
3. MOTOR LUBRICATION – If motor is provided with oil ports, lubricate motor with SAE 20 non-detergent type oil. Slowly add 5 to 10 drops to each oil port – DO NOT OVER OIL! Motors which are not provided with oil ports require no periodic lubrication.

SERVICE INSPECTION
Service inspection and repair of the furnace, and any related equipment which may be installed with the furnace, shall be done only by persons who are qualified and trained with this type of equipment. Some state and municipal code authorities require that persons servicing this type equipment be licensed. Persons who are not qualified or licensed shall not attempt to service this equipment.

It is recommended that the furnace be inspected by a qualified service company, such as your Coleman dealer, shortly before the beginning of each heating season. The furnace should be inspected prior to the beginning of the air conditioning season if air conditioning equipment is installed.

When required, adjustments should be made to insure that the operational performance of the furnace is in accordance with the specifications shown on the UNIT RATING PLATE and in these INSTALLATION INSTRUCTIONS.

Before removing components or assemblies for inspection, carefully observe how these components are installed, electrical connections made, and attachment means employed. All components and assemblies must be reinstalled in the same manner and position as they were originally.

WARNING – NEVER ATTEMPT TO MODIFY THIS FURNACE OR REPAIR DAMAGED OR DEFECTIVE COMPONENTS. SUCH ACTION COULD CAUSE UNSAFE OPERATION, FIRE, EXPLOSION, OR ASPHYXIATION

Damaged or defective components must be replaced. Only original equipment components, or substitute components authorized by Evcon Industries, Inc. may be used to repair this furnace.

Burner Assembly – The burner should be inspected for any accumulation of corrosion (rust) in the burner port. The burner should be blown out with the pressure side of a vacuum cleaner. Take care not to damage or displace the ignition assembly.

NOTE – Ignitor assembly is very fragile and should be handled with care. A cracked or broken ignitor will not function properly and must be replaced.

CAUTION – Ignitor operates on line voltage. Turn off electrical power before servicing to prevent shock hazard which could damage equipment or cause personal injury.

Heat Exchanger – The heat exchanger should be inspected for possible material corrosion, scaling, damage, or deposits of soot. (1) Remove the gas valve and mounting bracket and the burner assembly. (2) Place a suitable light through the combustion air tube and into the heat exchanger drum. Using a mirror, observe the heat exchanger inner surfaces and look for any deposit of black soot or any apparent scaling on the metal surfaces. The normal color of the heat exchanger may have a red color in some areas. (3) If soot deposits are found in the heat exchanger drum and sooting is also found in the vent box, the heat exchanger must be replaced and the cause of the soot determined and necessary corrective action taken. (4) If any light sooting is found, it and any debris from the bottom of the heat exchanger must be removed using a suitable vacuum cleaner. If sooting is noticed, its cause must be determined and necessary corrective action taken.

REPLACEMENT PARTS

Should it be necessary to replace any component parts, these may be obtained through a Coleman dealer, who is experienced and can be of assistance, or information on the nearest Distributor may be obtained directly from Evcon Industries, Inc., 3110 North Mead, P.O. Box 19014, Wichita, KS 67204-9014. Phone 316-832-6448.

CAUTION

Only genuine Coleman replacement parts should be used. Substitute parts should not be used as they may not be the same in operational and safety characteristics.

The parts listed are only those that are functional parts and those that might require replacement. The parts listed are by part name (description) only and CODE reference NUMBER. The CODE number is used to determine the location of the part in the generalized, pictorial illustration provided. See page 17. WHEN ORDERING PARTS - ORDER BY NAME AND DESCRIPTION. DO NOT ORDER BY CODE NUMBER. ALSO WHEN ORDERING PARTS please provide: (a) Complete furnace Model Number and Serial Number. These may be found on the rating plate located behind the upper front panel (door) of the furnace. (b) Type of gas being used. (c) Is furnace being used in conjunction with air conditioning or heat pump?

OVERALL OPERATION

Following major service or replacement of functional parts it is recommended that the furnace be operated in the various modes to insure that performance is normal and control components are functioning properly.

PARTS LIST

1. Thermostat - accessory
2. Casing Assembly
3. Shield
4. Vent Clamp
5.
6. Heat Exchanger
7. Vestibule
8. Limit Switch
9. Fan Switch
10. Switch Bracket
11. Terminal Bushing
12. Vent Blower Assembly
13. Pressure Tube
14. Pressure Switch
15. Logo

16. Front Panel
17. Blower Panel
18. Burner Assembly
19. Orifice
20. Valve Bracket
21. Gas Valve
22. Ignitor Bracket
23. Ignitor
24.
25. Ignition Module
26. Customer Envelope
27. Control Box Assembly
28. Control Box Cover
29. Heating Relay
30. Blower Relay

31. Transformer
32. Sequencer
33. Blower Assembly
34. Capacitor
35. Blower Scroll
36. Blower Wheel
37. Blower Motor
38. Motor Clamp
39. Rubber Mounts
40. Filter
41. Filter Rod
42. Safety Switch
43. Roll-out Switch
44. Terminal Board

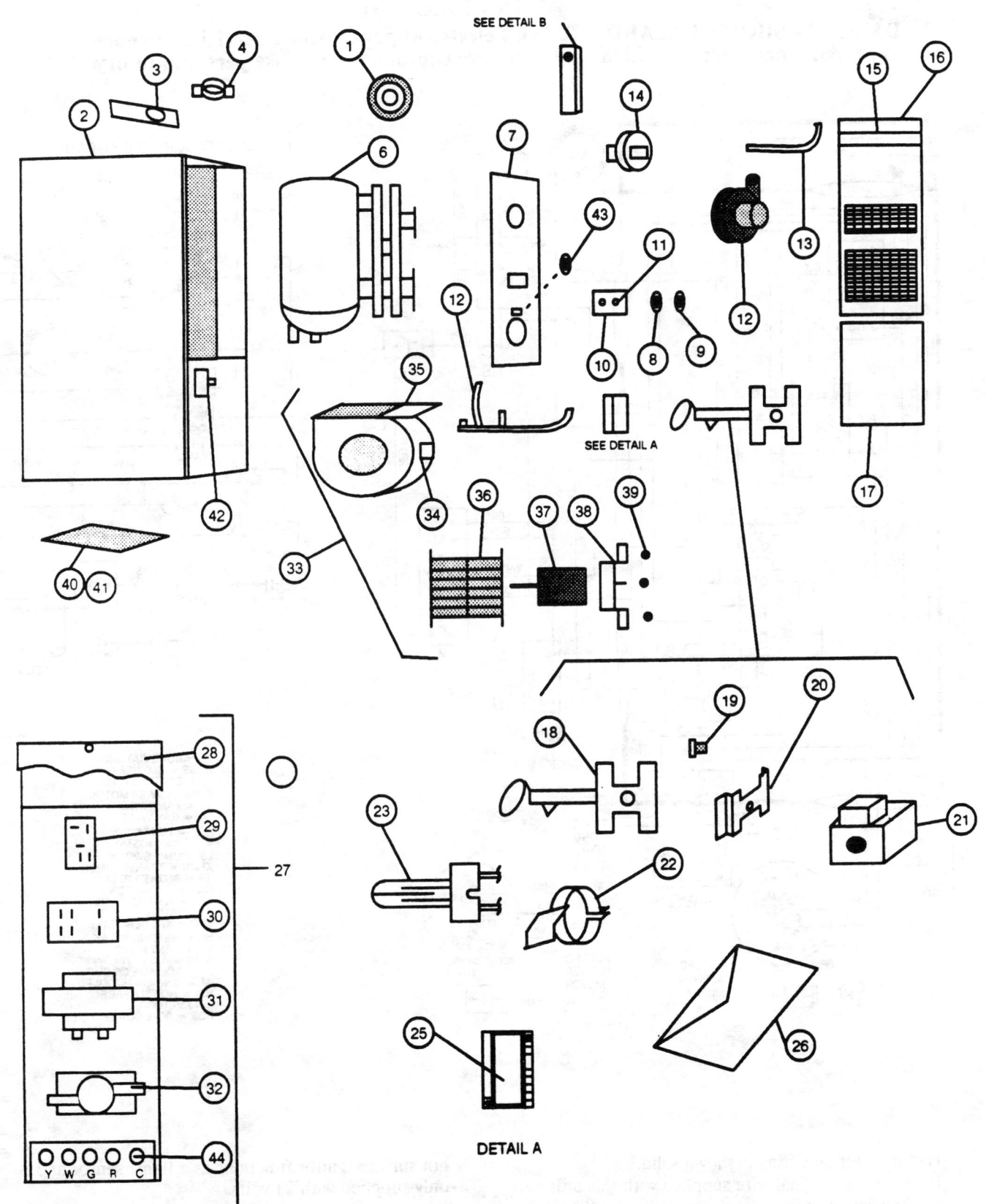

SEE DETAIL B

SEE DETAIL A

DETAIL A

Y W G R C

17

WIRING DIAGRAM
DANGER- SHOCK HAZARD – Turn off electrical power before servicing furnace to prevent shock hazard which could damage equipment or cause personal injury.

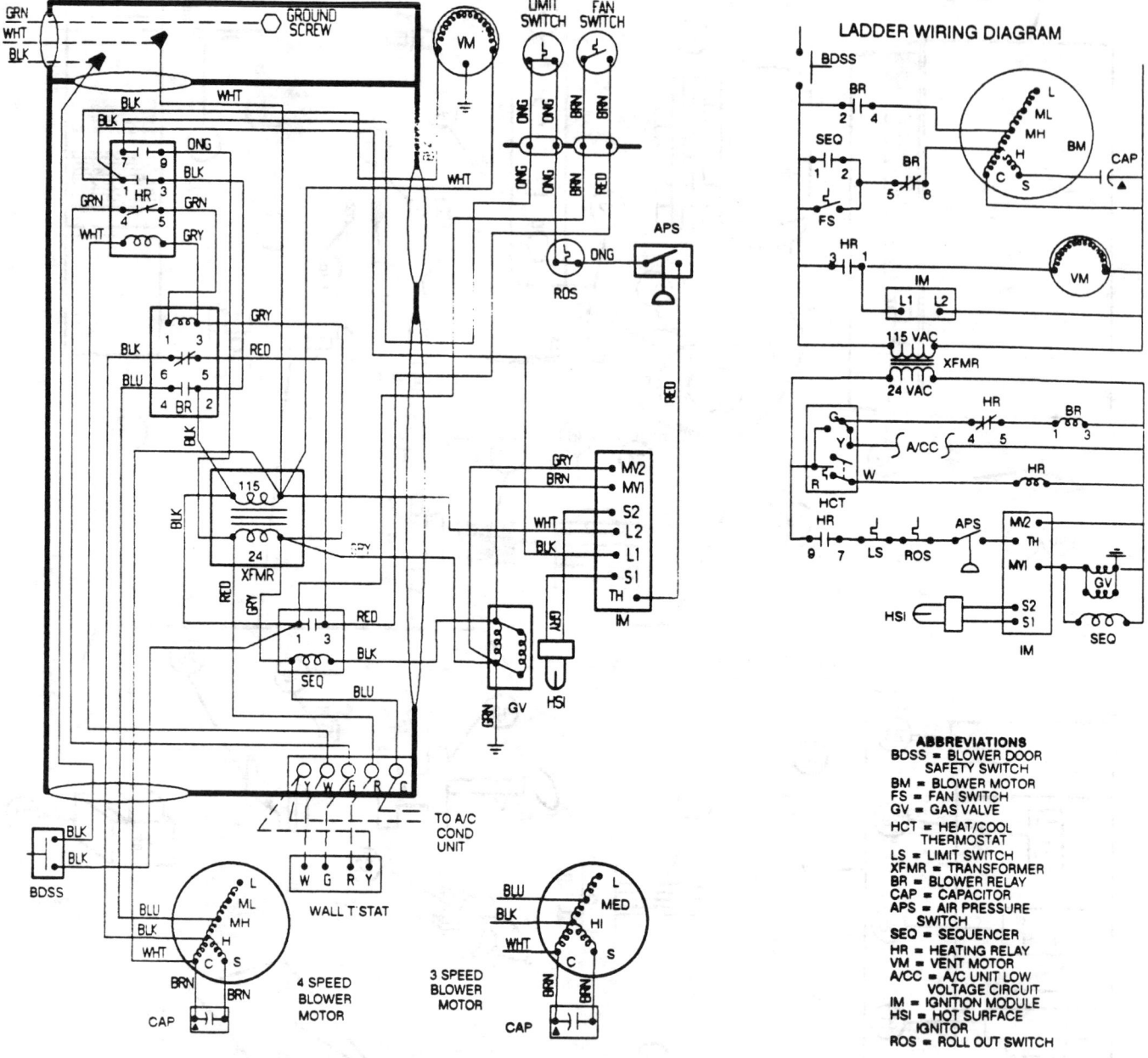

Factory internal wiring is shown solid.
If any of the original wire supplied with this unit must be replaced, it must be replaced with type 105°C thermoplastic wire or its equivalent.

IMPORTANT NOTICE – Prior to and during ignition, the ignition module supplies 120 volts to the hot surface ignitor. After ignition has been completed, the

hot surface ignitor functions as a flame sensor and is only supplied with 24 volts.

T.H.E.® UPFLOW HI-EFFICIENCY CONDENSING FURNACES

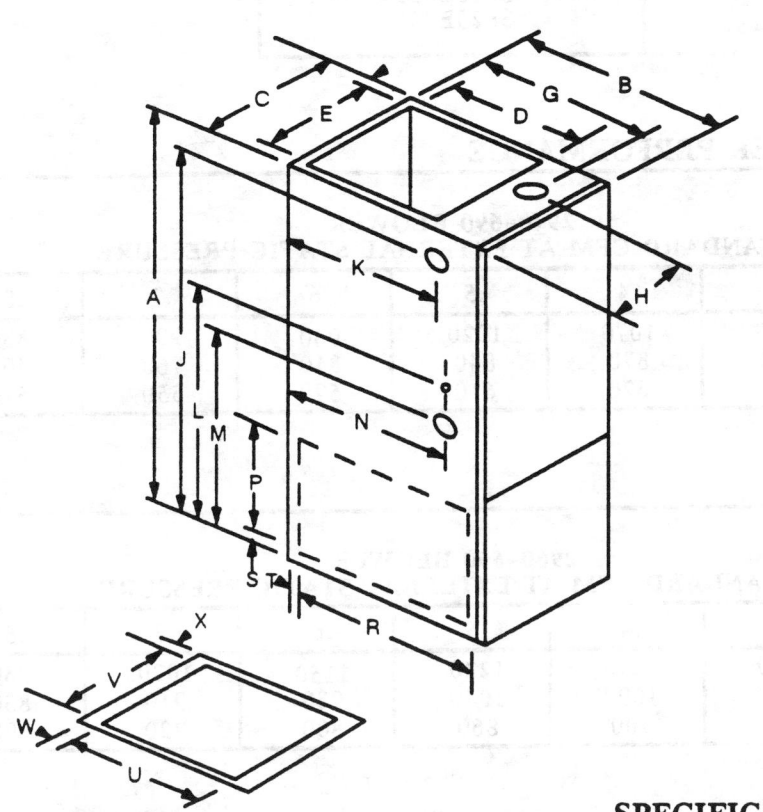

DIMENSIONS IN INCHES

FURNACE MODELS

	2940E666 2960E666	2960E766 2970E766 2985E766
A	46	46
B	28	28
C	16-7/8	20
D	20-3/8	20-3/8
E	15-3/8	18-1/2
F	3/4	3/4
G	23-3/16	23-3/16
H	11-5/16	12-7/8
J	43-1/8	43-1/8
K	23-1/16	23-1/16
L	28-5/8	28-5/8
M	25-3/16	25-3/16
N	24-1/2	24-1/2
P	23	23
R	1	1
S	3-1/2	3-1/2
T	23-3/4	23-3/4
U	14-7/8	18
V	1-3/4	1-3/4
W	1	1
X	1	1

SPECIFICATIONS

MODEL	NAT.	2940E666	2960E666	2960E766	2970E766	2985E766
	L.P. KIT	NONE	2960B5991	2960B5991	2970B5991	2985B5991
UNIT RATING BTU/Hr.	Input	45,000	65,000	65,000	80,000	95,000
	Output	41,000	58,000	58,000	70,000	84,000
	High Altitude	FOR ELEVATIONS ABOVE 2000 FEET, REDUCE INPUT 4% FOR EACH 1000 FEET OF ELEVATION ABOVE SEA LEVEL.				
AFUE %		91.5	90.0	90.0	90.0	90.0
AIR TEMPERATURE RISE RANGE F		20-50	25-55	25-55	30-60	35-65
DESIGNED MAXIMUM OUTLET AIR TEMP. F		145	150	150	155	160
MAX. EXT. STATIC PRESSURE IN. W.C.		.5				
VENT PIPE		2 in.* DIAMETER SCHEDULE 40 CPVC OR PVC#				
VENT PIPE FITTINGS		2 IN. DIAMETER SCHEDULE 40 OR 80 CPVC OR PVC				
CONDENSATE PIPE & FITTINGS		1/2 IN. CPVC				
GAS CONNECTION		1/2 in. FPT				
ELECTRICAL SERVICE		115 VAC 60 HZ 1 PH				
MINIMUM DISTANCE TO COMBUSTIBLE MATERIALS		FRONT - 6" BACK - 0 PLENUM TOP - 1" FLOOR - COMBUSTIBLE VENT - 0 SIDES - 0				
FILTER(Furnished)		16 X 25 X 1/2		20 X 25 X 1/2		
SHIPPING WEIGHTS		135		145	150	

Annual Fuel Utilization Efficiency is based on DOE isolated combustion system test procedure.
For LP use, optional kit is required. Model 2940 is not LP convertible.
* - 3 in. diameter must be used with model 2985. See installation instructions for details.
# - First two feet of vent pipe must be CPVC. (Supplied with unit)

COIL CABINET FURNACE MATCH-UP

FURNACE MODEL	CABINET
2940E666 2960E666	3617E835
2960E766 2970E766 2985E766	3620E835 3625E835

BLOWER PERFORMANCE

Furnace # 2940E666			2940-690 BLOWER STANDARD CFM AT EXTERNAL STATIC PRESSURE					
Blower Speed	.1	.2	.3	.4	.5	.6	.7	.8
High	1220	1170	1120	1070	1020	960	900	830
Medium	940	910	900	870	840	810	760	700
Low	630	630	630	620	610	590	550	510

Motor HP:1/3 FLA: 6.6 Impellor Size: 10-3/4 x 7-3/16

Furnace # 2960E666			2960-690 BLOWER STANDARD CFM AT EXTERNAL STATIC PRESSURE					
Blower Speed	.1	.2	.3	.4	.5	.6	.7	.8
High	1490	1420	1360	1290	1220	1150	1070	980
Medium	1210	1180	1140	1090	1040	980	910	830
Low	950	940	920	900	860	800	720	690

Motor HP:1/3 FLA: 9.2 Impellor Size: 10-5/8 x 8

Furnace # 2960E766			2970-690 BLOWER STANDARD CFM AT EXTERNAL STATIC PRESSURE					
Blower Speed	.1	.2	.3	.4	.5	.6	.7	.8
High	1870	1820	1760	1700	1650	1590	1510	1460
Medium High	1520	1500	1490	1460	1420	1390	1350	1300
Medium Low	1300	1290	1280	1260	1240	1220	1160	1120
Low	1080	1060	1040	1010	1000	960	910	870

Motor HP: 3/4 , FLA: 10.0 Impellor Size: 11-7/8 x 9-1/2

Furnace # 2970E766, 2985E766			2970-690 BLOWER STANDARD CFM AT EXTERNAL STATIC PRESSURE					
Blower Speed	.1	.2	.3	.4	.5	.6	.7	.8
High	1800	1760	1710	1660	1600	1540	1450	1400
Medium High	1480	1460	1440	1410	1400	1360	1310	1250
Medium Low	1280	1270	1260	1230	1210	1180	1140	1070
Low	1080	1050	1030	1010	990	960	920	880

Motor HP: 3/4 , FLA: 10.0 Impellor Size: 11-7/8 x 9-1/2
All motors are 1100 RPM and have 1/2" shafts.

T.H.E. is a registered trademark of Evcon Industries, Inc.

*Coleman is a registered trademark of The Coleman Co., Inc. used under license.

Evcon Industries, Inc.
3110 North Mead
P.O. Box 19014
Wichita, KS 67204-9014

LITHO IN U.S.A.

INSTALLATION INSTRUCTIONS

NATURAL TO LP CONVERSION KITS

For converting 2700, 2800, and 2900 Series automatic ignition natural gas furnaces to LP gas operation.

WARNING

This conversion kit is to be installed by a Coleman distributor (or other qualified agency) in accordance with the manufacturer's instructions and all codes and requirements of the authority having jurisdiction. Failure to follow instructions could result in serious injury or property damage. The qualified agency performing this work assumes responsibility for this conversion.

WARNING – IMPROPER INSTALLATION MAY DAMAGE EQUIPMENT, CAN CREATE A HAZARD, AND WILL VOID THE WARRANTY.

NOTE – The words "shall" or "must" indicate a requirement which is essential to satisfactory and safe product performance. The words "should" or "may" indicate a recommendation or advice which is not essential and not required but which may be useful or helpful.

CONTENTS:

A. Burner orifice for L.P. gas (size based on 2500

BTU/Cu. ft. propane gas and furnace located below 2,000 foot elevation). Size is marked on orifice.

B. Conversion plate sticker (silver color)

C. Honeywell valve conversion (unmarked envelope), which contains: (1) pressure regulator, (2) gasket, (3) two screws, (4) label, (5) Honeywell valve conversion instructions. (Not included in 2900 Series kits.)

D. Robertshaw valve conversion #78245, which contains: (1) yellow L.P. step spring, (2) black L.P. regulator spring, (3) two red slotted caps, (4) two adjusting screws, (5) label, and (6) Robertshaw valve conversion instructions. This conversion kit is used to convert either the 7100 or the 7200 Robertshaw valve.

TABLE I – APPLICATION TABLE

Conversion Kit Number	Furnace Model	Input (BTUH)	LP Orifice Size
2745-6991	2745	60,000	.086 (#44)
2755-6991	2755	70,000	.093 (#42)
2765-6991	2765	80,000	.101 (#38)
2775-6991	2775, 2780	100,000	.116 (#32)
2790-6991	2790	120,000	.128 (#30)
2840-5991	2840 Horiz.	50,000	.078 (#47)
2855-5991	2855 Horiz.	65,000	.089 (#43)
2870-5991	2870 Horiz.	85,000	.104 (#37)
2890-5991	2890 Horiz.	105,000	.113 (#33)
2845A5991	2845 Upflow	55,000	.082 (#45)
2865A5991	2865 Upflow	75,000	.098 (#40)
2880A5991	2880 Upflow	95,000	.110 (#35)
2895A5991	2895 Upflow	115,000	.120 (#31)
2844-5991	2844 Downflow	55,000	.082 (#45)
2860-5991	2860 Downflow	70,000	.093 (#42)
2875-5991	2875 Downflow	90,000	.106 (#36)
2892-5991	2892 Downflow	110,000	.116 (#32)
2960B5991	2960 Upflow	65,000	.093 (#42)
2970B5991	2970 Upflow	80,000	.101 (#38)
2985B5991	2985 Upflow	95,000	.110 (#35)

21

CONVERSION PROCEDURE

1. **WARNING: SHOCK HAZARD.** Turn off electrical supply to furnace.

2. Shut off gas supply at valve upstream from furnace or at meter as required.

3. Disconnect gas supply piping from gas valve on furnace.

4. Disconnect electrical wires from gas valve, noting which wires are connected to which terminals.

5. Remove the four sheet metal screws that attach the gas valve bracket to the furnace. See Figure 1.

6. Remove the gas valve and bracket from the furnace.

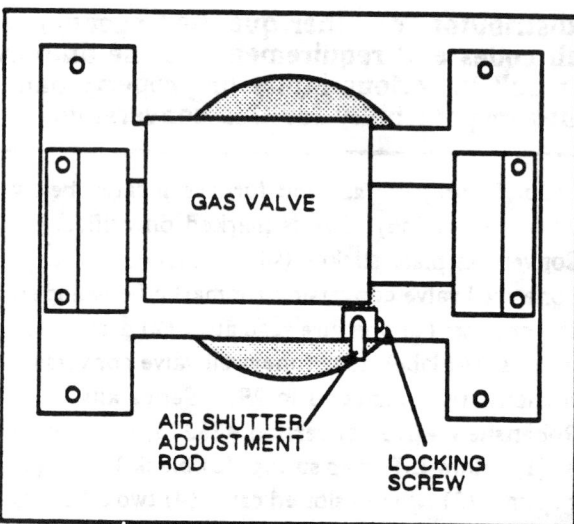

Figure 1

7. Remove and discard natural gas orifice.

8. Install the L.P. gas orifice supplied with the kit.

9. Reattach the gas valve and bracket to the furnace.

10. Convert the gas valve for L.P. gas operation by following the instructions and using the components supplied in the appropriate envelope for the gas valve on the furnace. See Figures 2, 3, and 4 for gas valve identification. Apply the label supplied in the kit to the gas valve to show that it has been converted.

11. Attach silver colored conversion plate sticker to furnace vestibule and next to the rating plate where it will be clearly visible. The name and address of the company making the conversion and the date of conversion must be written on the plate.

12. Reconnect the electrical wires to the gas valve using the wiring diagram as a guide.

13. Reconnect the gas supply piping to the gas valve and insure that all gas connections are tight.

14. Remove pressure tap plugs from gas valve and connect water gauge to the pressure tap ports. See Figures 2,3 and 4 for location of the gas valve pressure taps

and pressure regulator adjustment.

15. Turn on gas supply to furnace and check all gas connections with suitable leak detector.

CAUTION – Never use an open flame to check for leaks. Fire or explosion could occur. Since some leak solutions including soap and water may cause corrosion or stress cracking, the piping must be rinsed with water after testing unless it has been determined that the leak test solution is noncorrosive.

16. After assuring that there is no gas leakage, light the furnace using the lighting instructions shown on the furnace rating plate or on the label on the back of the door panel.

17. Verify that the gas supply pressure is between 11 and 14 inches water column with the furnace operating. Adjust gas valve pressure regulator to obtain gas pressure reading of 10 ± 0.3" W.C.

NOTE – On the Robertshaw gas valves, the step regulator must be adjusted separately to a pressure of between 4.5 and 5.0 inches water column. The step regulator adjustment screw is beneath the screw cap with the two small holes, and the main pressure adjustment screw is beneath the slotted screw cap. Make sure that these caps are replaced in their proper places after the pressures are set.

18. Turn gas supply off, remove pressure gauge, and replace pressure tap plugs.

19. Turn gas supply on, re-light the furnace, and check for gas leak at the pressure taps.

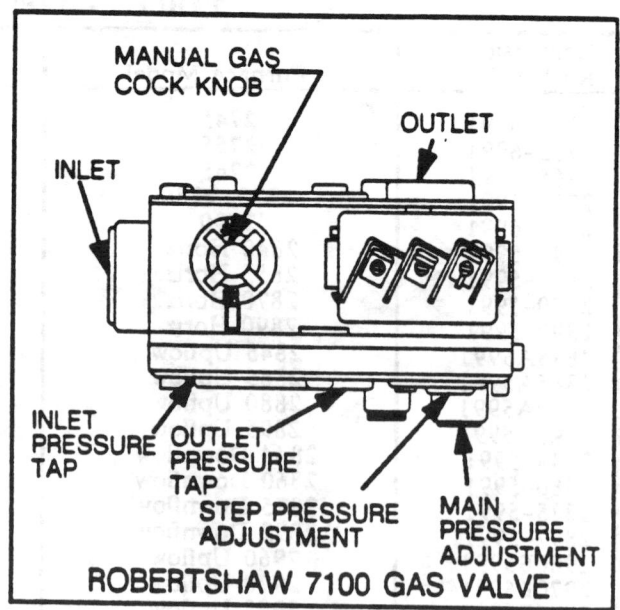

Figure 2

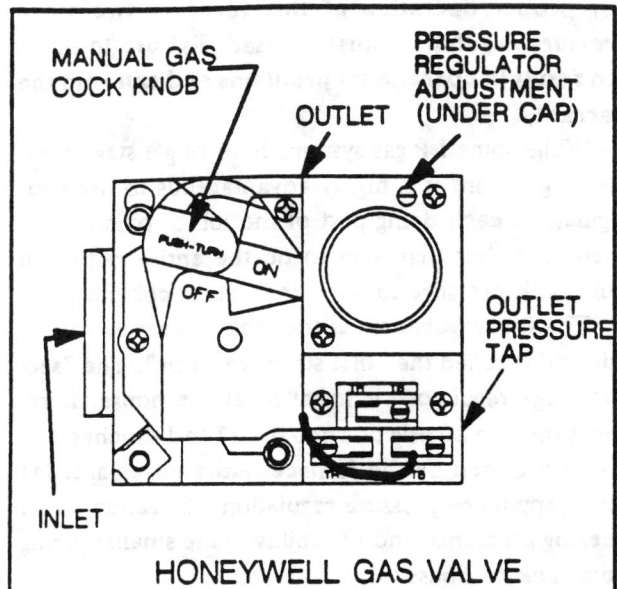

MANUAL GAS COCK KNOB

PRESSURE REGULATOR ADJUSTMENT (UNDER CAP)

OUTLET

OUTLET PRESSURE TAP

INLET

PUSH-TURN ON OFF

HONEYWELL GAS VALVE

Figure 3

20. After re-lighting the furnace, allow the furnace to operate for about 15 minutes and the adjust burner primary air as follows: (a) Loosen air adjustment rod locking screw. See Figure 1, (b) Close primary air shutter by pulling on adjustment rod until yellow tips appear in the flame at the end of the burner. See Figure 5. (c) Slowly open the primary air shutter by push-

ing adjustment rod in until yellow flame tips at the end of the burner disappear, and then push adjustment rod in another 1/8 inch, (d) Secure primary air adjustment rod by tightening screw against it.

21. Cycle furnace with thermostat a few times to insure that everything is working properly.

22. Check gas input rate as described below.

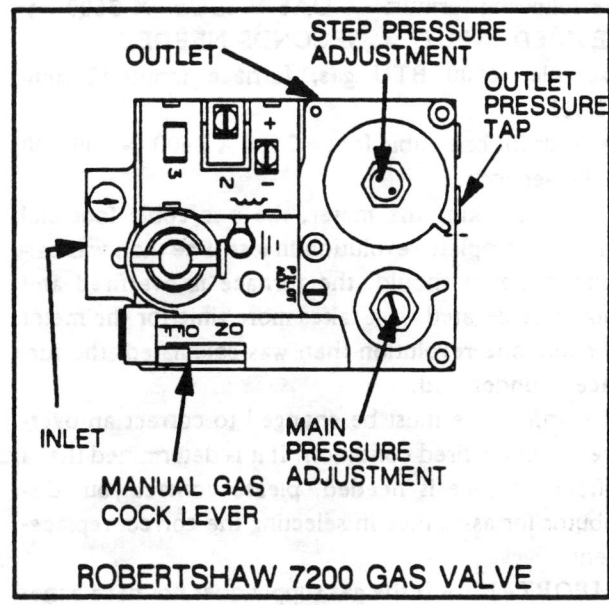

OUTLET

STEP PRESSURE ADJUSTMENT

OUTLET PRESSURE TAP

INLET

MAIN PRESSURE ADJUSTMENT

MANUAL GAS COCK LEVER

ROBERTSHAW 7200 GAS VALVE

Figure 4

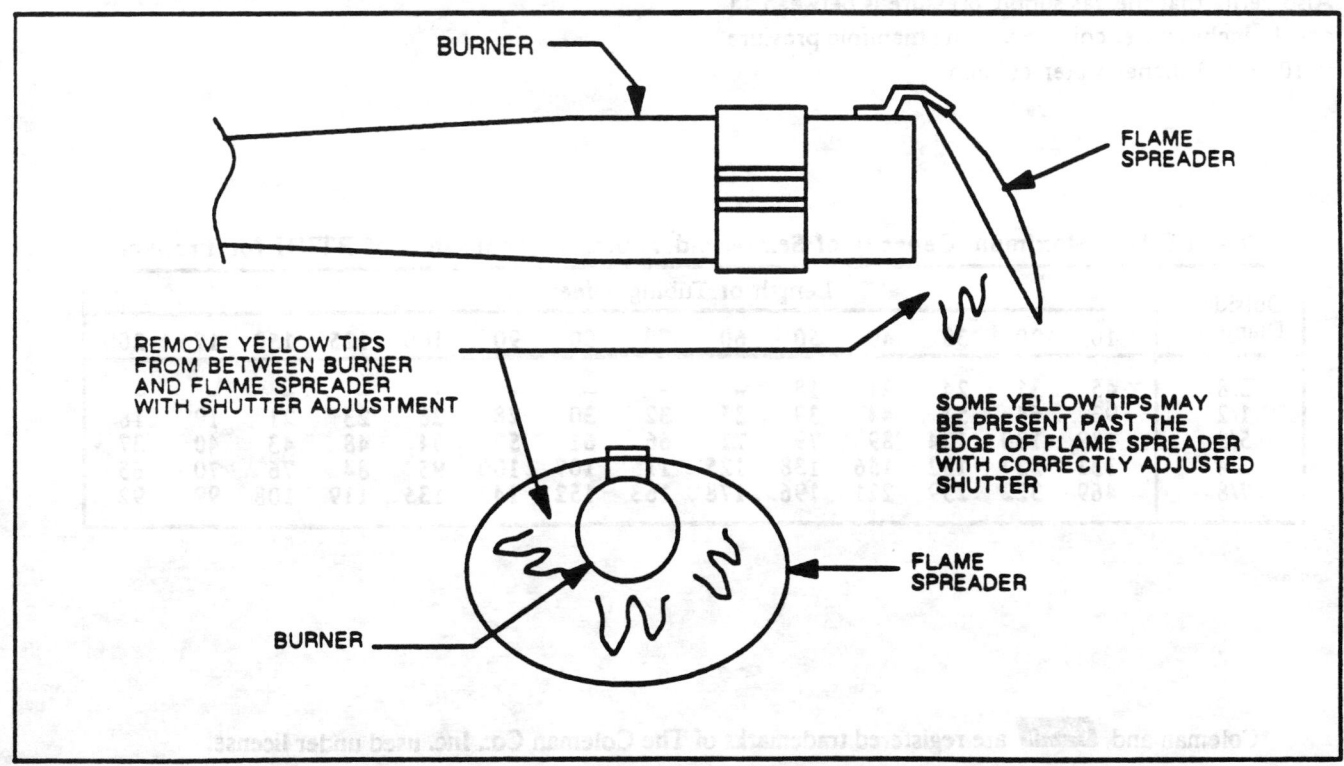

BURNER

FLAME SPREADER

REMOVE YELLOW TIPS FROM BETWEEN BURNER AND FLAME SPREADER WITH SHUTTER ADJUSTMENT

SOME YELLOW TIPS MAY BE PRESENT PAST THE EDGE OF FLAME SPREADER WITH CORRECTLY ADJUSTED SHUTTER

FLAME SPREADER

BURNER

Figure 5

DETERMINING GAS INPUT RATE

Where gas is metered, the input rate may be determined by the following method.

Contact the gas supplier, public utility company or LP gas distributor to obtain the calorific gas value of the gas being used. When checking the gas input rate, any other gas burning appliances connected to the same meter should be completely off. The furnace should be allowed to operate for 15 minutes before attempting to check the gas input rate.

To check flow rate, observe the one cubic foot dial on the gas meter and determine the number of seconds required for the dial hand to complete one revolution (seconds to flow one cubic foot.)

To determine the number of seconds per cubic foot that is necessary to achieve the correct input rate, use the following formula: GAS VALUE X 3600 ÷ DESIRED INPUT = SECONDS NEEDED

Example: 2500 BTU gas, furnace input 100,000 BTUH

Seconds for one cubic foot = 2500 X 3600 ÷ 100,000 = 90 seconds.

If when clocking the meter, the one cubic foot dial makes a complete revolution in less time than was calculated that it should, the furnace is overfired and should be derated. If it takes more time for the meter to make one revolution than was calculated, the furnace is underfired.

The orifice size must be changed to correct an overfired or underfired condition. If it is determined that a different orifice is needed, please contact your distributor for assistance in selecting the correct replacement.

IMPORTANT – If the gas supply does not have a meter, then be sure the proper orifice has been installed. Also verify that the gas supply pressure is between 11 and 14 inches water column and the manifold pressure is 10 ± 0.3 inches water column.

LP GAS PRESSURE REGULATION

For proper operation of this furnace, two-stage pressure regulation must be used. Failure to do so can result in operational problems and can void the warranty.

While some LP gas systems have single stage pressure regulation, it is highly advantageous to use two regulators, each doing part of the job of pressure reduction, rather than one to do the entire reduction from tank pressure to 11 inches water column.

The first regulator – at the tank – is set at 5 to 20 psig and is called the "first stage regulator". The "second stage regulator" is installed at the house. It reduces the 5 to 20 psig down to the 11 to 14 inches w.c. that is required by the furnace. Advantages are: (1) better appliance pressure regulation, (2) reduction of freezing problems, and (3) ability to use smaller piping from tank to house.

PIPING PRACTICES

A piping or tubing system must be designed to prevent the loss in pressure between the last stage regulator and the furnace, from exceeding 0.5 inches w.c. Use Table II to determine the proper size of pipe and fittings required for the distance from tank to the appliance and for the input of the appliances.

TABLE II – Maximum Capacity of Semi-rigid Tubing in Thousands of BTUH for Propane

Outside Diameter	Length of Tubing – feet													
	10	20	30	40	50	60	70	80	90	100	125	150	175	200
3/8	45	31	24	21	18	–	–	–	–	–	–	–	–	–
1/2	93	64	51	44	39	35	32	30	28	26	23	21	19	18
5/8	189	130	104	89	79	71	66	61	57	54	48	43	40	37
3/4	331	227	182	156	138	125	115	107	100	95	84	76	70	65
7/8	469	322	259	221	196	178	163	152	143	135	119	108	99	92

*Coleman and [Coleman] are registered trademarks of The Coleman Co., Inc. used under license.

Evcon Industries, Inc.
3110 North Mead
P.O. Box 19014
Wichita, KS 67204-9014

1951A994 (7-90) P.I.   LITHO IN U.S.A.

FENWAL®

SERIES 05-32 HOT SURFACE IGNITION SYSTEM

WITH OPTIONAL MULTIPLE TRIAL-FOR-IGNITION PERIODS

FEATURES

- Compact Size
- Rapid Heat-Up Time
- Patented System Check Circuit
- Local or Remote Sensing Capability
- Digital Trial-for-Ignition Timings
- Patented Dual Purpose Ignitor/Sensor System

GENERAL INFORMATION

The Series 05-32 Hot Surface Ignition System is available in single or multiple trial-for-ignition models. These compact ignitors have a nominal heat-up period of 45 seconds (30 to 60 seconds). Flame sensing is achieved through the patented technique of using the hot surface element as the sensor or, if required, a separate sensing probe may be employed. All models utilize the principle of flame rectification to monitor burner flame.

PRINCIPLE OF OPERATION

Single-Try Models
On a call for heat, the hot surface element is energized (120VAC) and is allowed to reach ignition temperature. The gas valve is then powered (24VAC). establishing the flame at the burner. The 05-32 then switches the hot surface element from the ignition mode to the sensing mode to monitor the flame for the duration of the duty cycle.
If the flame is not established during the trial-for-ignition period, the system will go into lockout. After the flame has been established, should it be extinguished for any reason, the ignitor will shut off the gas valve and reactivate the hot surface element. One complete retry cycle will occur.

Three-Try Models
These models provide up to three complete trials-for-ignition before going into lockout. If the flame is established on any of the three tries, the 05-32 unit will then monitor the flame for the duration of the duty cycle.

MOUNTING

The Series 05-32 is not position-sensitive and may be mounted vertically or horizontally. using #8 hardware.

WIRING

CAUTIONS:
I. *Note that both 24VAC and 120VAC are required for the system. Interrupt both supplies before proceeding with wiring.*
II. *Do not apply power to the module until all wiring is connected and insure that the system is properly grounded.*
III. *System must be checked after installation and before gas supply is turned on (see INITIAL OPERATION).*

Use wiring diagram Figure 1 for local sensing applications. Use wiring diagram Figure 2 for optional remote sensing applications.

WIRING (Continued)

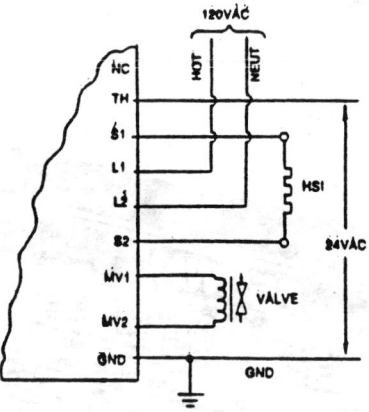

FIGURE 1 – Local Sensing
Wiring Diagram

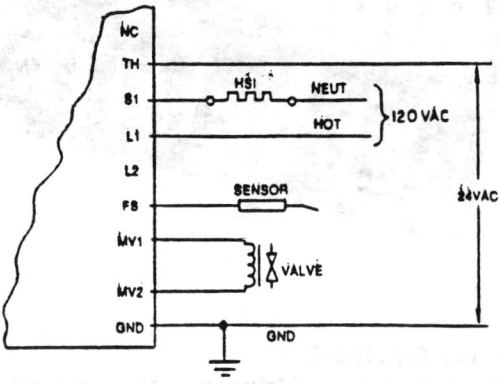

FIGURE 2 – Remote Sensing
Wiring Diagram

INITIAL OPERATION

1. Check installation. Mount and position the ignitor in accordance with the illustration in Figure 3.
2. With gas supply manually shut off, apply power to appliance and cycle thermostat above room temperature.
3. Insure that the ignitor "glows" during the heat-up and trial-for-ignition periods as noted in the Specification section. In multiple trial-for-ignition ignitors, the ignitor will automatically recycle for the specified number of times, then lockout.
4. Set the thermostat to the lowest setting.
5. Wait 15 seconds then manually open the gas supply and advance the thermostat above room temperature to recycle the system.
6. Check that ignition has been accomplished. The ignitor "glow" will diminish once the flame has been established. At this stage, the ignitor acts as the sensing element.

7. If the system ignites but fails to hold-in, check for proper grounding of the 24 volt circuit.

SILICON CARBIDE IGNITOR

Proper location of the silicon carbide ignitor is important to achieve optimum system performance for both ignition and flame sensing. See Figure 3.

NOTE: *The temperature of the steatite holder should not exceed the manufacturer's specifications.*

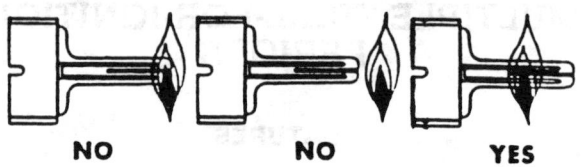

NO   NO   YES

FIGURE 3

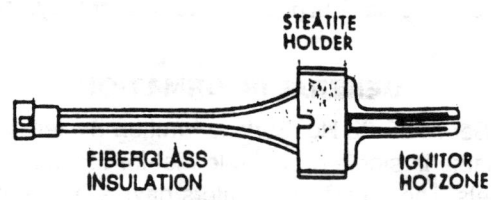

FIGURE 4 – Typical Hot Surface Element

SAFETY CHECKS

1. Manually shut off gas supply. Apply power to appliance and cycle thermostat above room temperature. After the system locks out, check that there is "0" voltage across the gas valve terminals with a suitable voltmeter. Next, set the thermostat to its lowest setting.
2. Wait 15 seconds, then manually open the gas supply and reactivate the system by advancing the thermostat to a setting above room temperature. The ignitor should glow brightly during the heat-up period, then diminish after the flame has been established. While the system is operating, manually shut off the gas supply. Following a short delay, the ignitor will re-energize and glow brightly. The ignition control will continue to cycle for the specified number of trial-for-ignition periods and then lockout. Check again that there is "0" voltage across the valve terminals after lockout.

SERIES 05-32 HOT SURFACE IGNITION SYSTEM

SERVICE CHECKS

Symptom	Cause/Cure
1) Dead	A) No 24 volt Input.
	B) Check system wiring.
	C) Check thermostat, transformer, limits, circuit breaker, etc.
2) Hot surface element does not heat - but unit cycles.	A) No 120 volt Input.
	B) Check system wiring.
	C) Check circuit breaker and 120 volt power source.
	D) Check for broken or cracked silicon carbide element.
3) Hot surface element heats up - but "0" voltage at valve during trial-for-ignition.	A) Check wiring between valve and module.
	B) Check power to valve.
4) Hot surface element heats. 24 volts to valve. Flame established but does not stay on.	A) Check system ground (24 volt supply).
	B) Hot surface element improperly located.
	C) Check all wiring connections.
	D) Burner out of adjustment.
5) Hot surface element heats. 24 volts to valve. System fails to ignite.	A) Turn gas supply off.
	B) Check gas valve.
	C) Burner out of adjustment (orifice plugged).
	D) Hot surface element incorrectly located.

REPAIRS

WARNING: The Fenwal Series 05-32 Hot Surface Ignition Module is NOT REPAIRABLE. Any modifications or repairs to this gas ignition module will invalidate Fenwal's standard warranty as well as agency certifications AND MAY CREATE HAZARDOUS CONDITIONS THAT COULD RESULT IN PROPERTY DAMAGE, PERSONAL INJURY OR EVEN DEATH FROM FIRE, EXPLOSION AND/OR TOXIC GASES. Faulty units should be replaced with a new unit.

SPECIFICATIONS

Size 3.234" × 4.234" × 1.593" (82.14 × 107.54 × 39.70mm)

Weight 5.11 ounces (145 grams) approx.

Input Voltage 24VAC 50/60 Hz (Operating range 20-28VAC)

Input Current Drain 80mA (Does not include valve power)

Valve Rating 24VAC, 1.0 A max.

Ignitor Heat-Up Time 45 seconds nominal (30-60 Seconds)

Ignitor Contact Rating 120VAC, 5.0 A max.

Ambient Temperature −40 to 160°F (−40 to 70°C)

System Check Circuit Checks valve control circuit and flame rectification circuit before ignition sequence is initiated.

Trial-For-Ignition Available in one or three try models.

Trial-For-Ignition Periods — (Flame establishing time)
- **-0X0 :** 1.5 seconds
- **-0X1 :** 3.0 seconds
- **-0X2 :** 5.0 seconds
- **-0X3 :** 7.0 seconds
- **-0X4 :** 10.0 seconds
- **-0X5 :** 15.0 seconds

Retry for Ignition Loss of flame will result in one retry for ignition. (Two retries on the three trial-for-ignition version.)

Vent Damper Connection (optional) Consult factory for details.

WARNING: Operation outside specifications could result in failure of the Fenwal product and other equipment with injury to people and property.

Specifications subject to change without notice!

HOW TO ORDER

1. Select Trial-For-Ignition mode option from Table 1.
2. Select required Ignition Time from Table 1.
3. Order by catalog number.

Example: **05-326266-052** Control Module with three tries for ignition and 5.0 second ignition period.

TABLE I

Catalog Number **05-3262X6-0XX**

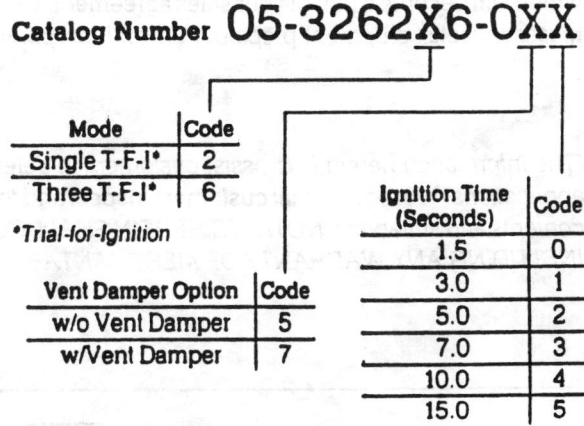

Mode	Code
Single T-F-I*	2
Three T-F-I*	6

*Trial-for-Ignition

Vent Damper Option	Code
w/o Vent Damper	5
w/Vent Damper	7

Ignition Time (Seconds)	Code
1.5	0
3.0	1
5.0	2
7.0	3
10.0	4
15.0	5

OUTLINE DIMENSIONS

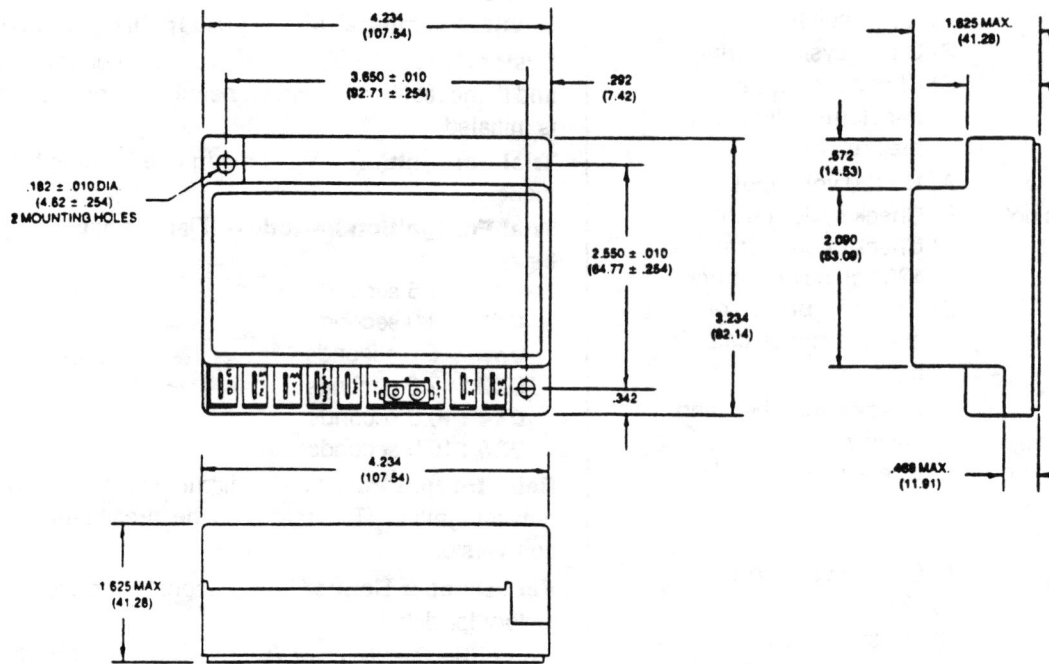

Metric dimensions are in parentheses.

LIMITED WARRANTY STATEMENT

Fenwal Incorporated represents that this product is free from defects in material and workmanship, and it will repair or replace any product or part thereof which proves to be defective in workmanship or material for a period of 18 months after delivery of the product to the buyer. For a full description of Fenwal's LIMITED WARRANTY, which, among other things, limits the duration of warranties of MERCHANTABILITY and FITNESS FOR A PARTICULAR PURPOSE and EXCLUDES liability for CONSEQUENTIAL DAMAGES, please read the entire LIMITED WARRANTY on the Fenwal Quotation, Acceptance of Order and/or Original Invoice which will become a part of your sales agreement. Defective units should be returned to the factory, Ashland, Massachusetts, shipment prepaid. Fenwal Incorporated will repair or replace and ship prepaid.

"The information herein is to assist customers in determining whether our products are suitable for their applications. We request that customers inspect and test our products before use and satisfy themselves as to contents and suitability. NOTHING HEREIN SHALL CONSTITUTE A WARRANTY, EXPRESSED OR IMPLIED, INCLUDING ANY WARRANTY OF MERCHANTABILITY OR FITNESS FOR A PARTICULAR PURPOSE."

FENWAL

400 Main Street Ashland, Massachusetts 01721 USA (508) 881-2000
Telex 948421 Fax: (508) 881-6729

5 032 1 2 5M 7/88

HOT SURFACE IGNITION MODULE KIT

For replacing the hot surface ignition module on any Coleman 2700, 2800, 2900, 8900, or 9100 Series furnace equipped with hot surface ignition.

WARNING - IMPROPER INSTALLATION MAY DAMAGE EQUIPMENT, CAN CREATE A HAZARD, AND WILL VOID THE WARRANTY.

NOTE - The words "shall" or "must" indicate a requirement which is essential to satisfactory and safe product performance. The words "should" or "may" indicate a recommendation or advice which is not essential and not required but which may be useful or helpful.

The ignition module supplied in this kit may be used to replace the Fenwal Series 05-32 hot surface ignition module found on Coleman furnaces built after August 15, 1990 or the Fenwal Series 05-21 or Channel Products hot surface ignition modules used on Coleman furnaces built before that date.

CONTENTS:
A. Fenwal 05-32 Series hot surface module.
B. Wiring adapter (used only when replacing older model Fenwal Series 05-21 hot surface ignition module.)

REPLACEMENT PROCEDURE:
If the furnace is equipped with a newer Series 05-32 ignition module:
1. **WARNING: SHOCK HAZARD.** Turn off electrical supply to furnace.
2. Remove the old ignition module.
3. Mount the new module in the same location with the same screws.
4. Reattach the wires as they were on the old module, as shown on the unit wiring diagram. The wiring adapter included in this kit will not be used and may be discarded.
5. Turn electrical power to furnace back on and cycle furnace with the thermostat a few times to insure that everything is working properly.

If the furnace is equipped with an older, Series 05-21 hot surface ignition module:
1. **WARNING: SHOCK HAZARD.** Turn off electrical supply to furnace.
2. Remove the old ignition module.
3. Mount the new module in the same location. It will be necessary to drill two .101 diameter holes. Use the old mounting screws.
4. Plug the edge connector on the wiring adapter included in this kit into the wiring harness plug on the furnace.
5. Plug the individual wires on the wiring adapter into the new ignition module as shown by the label on the adapter and as shown below.

Wire Color	Terminal
Yellow	GND
Brown	MV1
Blue	S2
White	L2
Black	L1
Gray	S1
Red	TH

6. Turn electrical power to furnace back on and cycle furnace with the thermostat a few times to insure that everything is working properly.

Evcon Industries, Inc.
3110 North Mead
P.O. Box 19014
Wichita, KS 67204-9014

1952-114 (9-90) P.I.

LITHO IN U.S.A.

SCHEMATIC DRAWING OF 7200DER-SO CONTROL

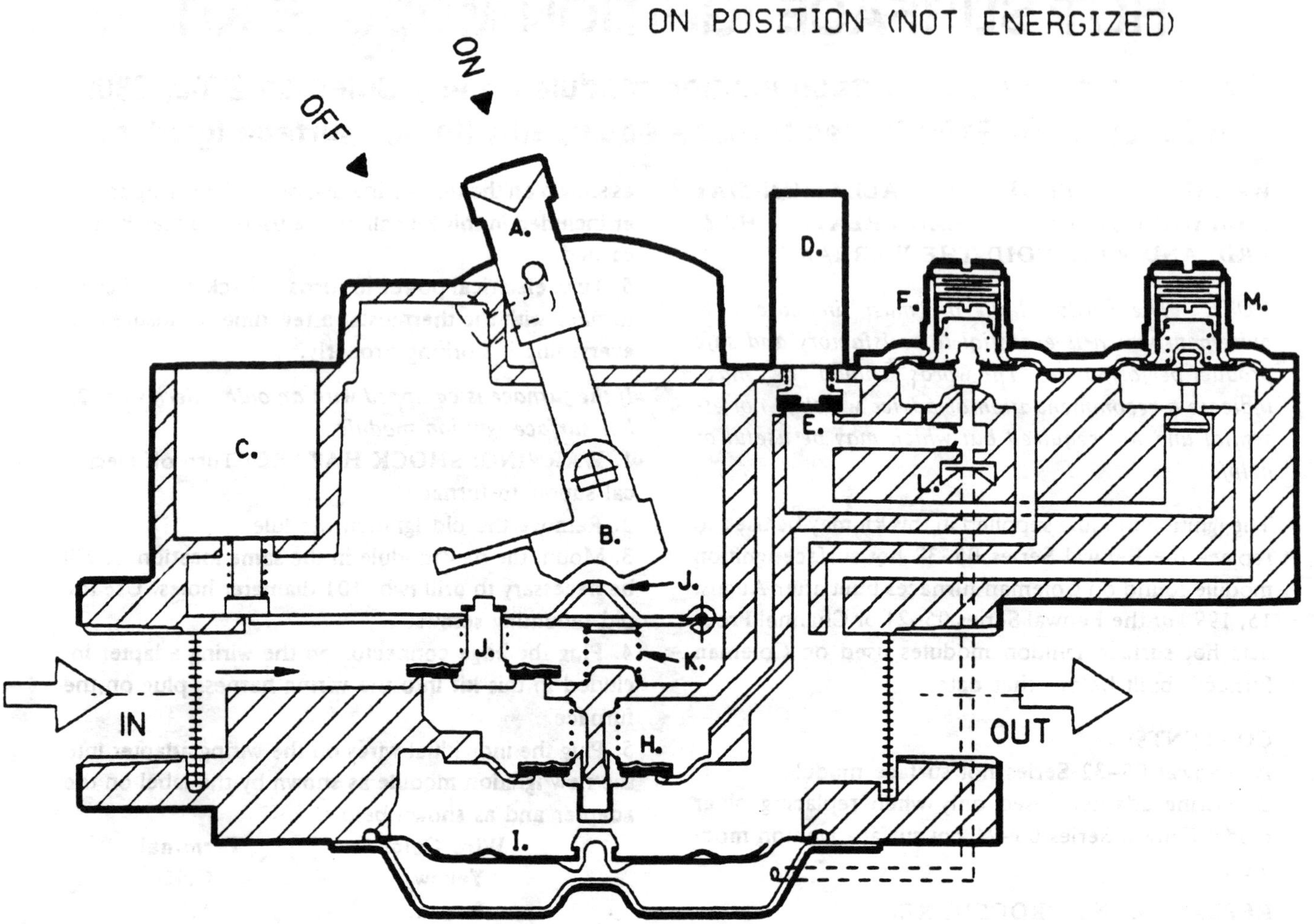

ON POSITION (NOT ENERGIZED)
ON
OFF
A.
B.
C.
D.
E.
F.
G.
H.
I.
J.
J.
K.
L.
M.
IN
OUT

A. MANUAL SELECTOR ARM
B. VALVE ACTUATING CAM
C. FIRST AUTOMATIC OPERATOR (ACTUATED BY ROOM THERMOSTAT)
D. SECOND AUTOMATIC OPERATOR (ALSO ACTUATED BY ROOM THERMOSTAT)
E. SERVO VALVE
F. SERVO REGULATOR
G. FIRST VALVE
H. DIAPHRAGM VALVE
I. MAIN DIAPHRAGM
J. FIRST VALVE LEVER
K. POWER SPRING
L. SLOW OPENER
M. STEP REGULATOR

II

DWG. *DE6156-9
DR: DEWEY
5-2-91

HEATING & AIR CONDITIONING

REPAIR
PARTS LIST

WITH HOT SURFACE IGNITION

MODELS

2940E
2960E
2970E
2985E

The High Efficiency

EVCON INDUSTRIES, INC.
P.O. Box 19014
Wichita, KS 67204-9014

R-912 (8/90) PP

#14 Catalog
Sec. 6-- Pg. 26.0

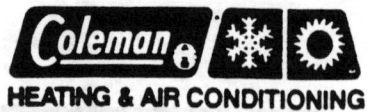

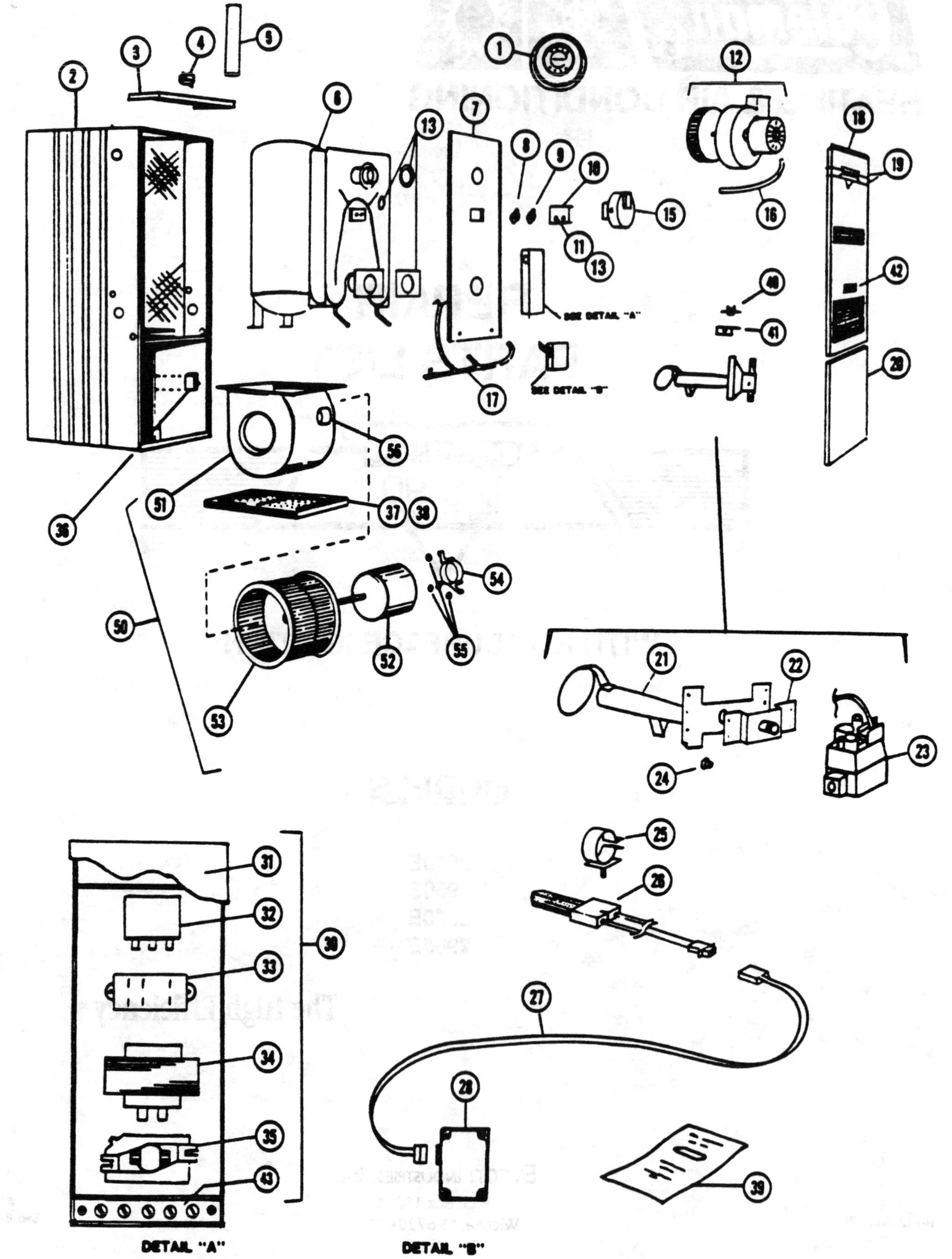
SEE DETAIL "A"
SEE DETAIL "B"
DETAIL "A"
DETAIL "B"

MODEL	2940E666	2960E666	2970E766	2985E766	2960E766
BTU/HR	45,000	65,000	80,000	95,000	65,000
A/C Capacity	See Blower Data on Tech spec. sheet – Form T-744				

CODE	MODEL	NO. REQ.	PART NUMBER	DESCRIPTION	VENDOR
-	All	X	8B246	Enamel (Spray 15 oz. Evcon Almond)	
★ 1	All	1	2661B2611	Thermostat (24V. Adj. Ant.) Accr.	HW#T87F
		1	2661B2601	Sub-Base	HW#HQ539A
2	2940E, 29605E666	1	2940B595	Casing Assy.	
	2960E, 2970E, 2985E766	1	2970B595	Casing Assy.	
3	2940E, 2960E666	1	2940A106	Shield (Vent Blower)	
	2960E, 2970E, 2985E766	1	2970A106	Shield (Vent Blower0	
4	All	1	2940B130	Vent Clamp	
5	All	1	2940-111	Vent Pipe (CPVC 2" Dia. x 24")	
6	2940E	1	2940A5751	Heat Exchanger (with Gaskets)	
	2960E	1	2960A5751	Heat Exchanger (with Gaskets)	
	2970E	1	2970A5751	Heat Exchanger (with Gaskets)	
	2985E	1	2985A5751	Heat Exchanger (with Gaskets)	
7	2940E, 29605E666	1	2940A544	Vestibule	
	2960E766	1	2960A544	Vestibule	
	2970E, 2985E766	1	2970A544	Vestibule	
★ 8	2940E	1	2845-3181	Limit Switch (O-170° - Close-140°)	TOD #60T11-312362 TI #20604L
	2960E, 2970E, 2985E	1	2985-3181	Limit Switch (O-190° - Close-160°)	TOD #60T11-312365 TI #20604L
★ 9	All	1	2940-3161	Fan Switch (C-120° - O-105°)	TOD #60T11-312361 TI #20604F
10	All	1	2940-115	Switch Bracket Assy. (less switches)	
11	All	2	1216-260	Terminal Bushing	
★ 12	All	1	2940B3901	Power Vent Blower Assy.	FASCO #7021-5478
13	All	1	2940A3011	Gasket Pkg. (5 Gaskets)	
★ 15	All	1	2940-3151	Pressure Switch (single pole- normally open — 24V.)	R-S #PD-50-1005 Tridelta # FS4109-95
16	All	1	2940-341	Pressure Tube (1/4 ID x 17)	
17	All	1	2970-5511	Drian Tube Kit	
18	2940E, 2960E666	1	2845-2111	Front Panel (Upper - 20" Wide)	
	2960E, 2970E, 2985E766	1	2895-2111	Front Panel (Upper - 20" Wide)	
19	All	1	1539-191	Evcon Logo (T.H.E. 90)	
20	2940E, 2960E666	1	2845-2131	Blower Panel (16-7/8" Wide)	
	2960E, 2970E, 2985E766	1	2895-2131	Blower Panel (20" Wide)	
★ 21	2940E, 2960E	1	2840A5201	Burner Assy. (30°) Includes Codes 25 & 26	
	2970E, 2985E	1	2870A5201	Burner Assy. (20°) Includes Codes 25 & 26	
22	All	1	2840-5451	Valve Bracket Assy.	
★ 23	All	1	2940-3271	Gas Valve (24V. - .6 Amp.)	R-S #7100-DERSO
24	2940E	1	9951-1201	Orifice (.120 Dia.-Nat)	
	2960E	1	9951-1441	Orifice (.144 Dia.-Nat.)	
		1	9951-0931	Orifice (.093 Dia.-LP)	
	2970E	1	9951-1611	Orifice (.161 Dia.-Nat)	
		1	9951-1011	Orifice (.101 Dia.-LP)	
	2985E	1	9951-1731	Orifice (.173 Dia.-Nat)	
		1	9951-1101	Orifice (.110 Dia.-LP)	
25	All	1	2735-5151	Ignitor Bracket	
★ 26	All	1	1474-0511	Ignitor	Norton #MB-255009
★ 27	All	1	2845-6611	Wire Bundle	
★ 28	All	1	1474-0021	Ignitor Module	Fenwall # 05-326265-052

Composit 2940E720

Coleman ❄ ☀
HEATING & AIR CONDITIONING

CODE	MODEL	NO. REQ.	PART NUMBER	DESCRIPTION	VENDOR
30	All	1	2940A535	Control Box Assy. (Complete)	
31	All	1	2940-182	Control Box Cover	
★32	All	1	2940-3551	Relay (Heating 24V-50/60 HZ)	Essex #93-232187-250
					Potter #KU 93-4221
★33	All	1	3110-3301	Relay (Blower-24V. 50/60 HZ)	RBM #184-20114-406
★34	All	1	2940A3541	Transformer (120/24V.-40VA)	Basler #BE17140-EEK
					Jard #TB401224-B11
★35	All	1	2940-3571	Sequencer (24V. _ 40VA)	TI # 60000A0-46
					TOD # 12520
★36	All	1	1462-1021	Safety Switch	Cherry #OE69-07C0
★37	2940E, 2960E666	1	2940A3391	Filter (Washable 25 x 14-1/2 - 12/Pkg.)	
	2960E, 2970E, 2985E766	1	2970A3391	Filter (Washable 25 x 18-1/2 - 12/Pkg.)	
38	All	2	2940-120	Filter Rod (3/32 Dia. x 25-1/8)	
39	All	1	1972Q221	Customer Envelope	
40	All	1	2940-3191	Rollout Switch (0 ~ ·Manual Reset)	TI #20611L
					TOD #60T14-206039
41	All	1	2840-110	Bracket (Rollout Switch)	
42	All	1	2845-350	Door Handle	
43	All	1	2845-300	Terminal Board	

	MODEL	2940E666	2960E666	2960E766 2970E766 2985E766
50	Blower Assy	▲2940A690	▲2960A690	▲2970-690
51	Scroll	2940-538	2960A538	2790A538
★52	Motor	2940-3239	2960-3239	2970A3249
		115V-FLA 6.6	115V-FLA 9.2	115V-FLA 8.4
		3 Speed 1100 RPM	3 Speed 1100 RPM	4 Speed 1120 RPM
		Fasco #7126-0994	Fasco #7126-0995	Fasco #7126-1023
		RMR # C628CW159	RMR # C640CW118	RMR # C640CW115
		Universal #HE3H133N	Universal #H3L056N	Universal #HE3T030N
		A O Smith #F48SQ6L28	A O Smith #F48SQ6L28	A O Smith #F48SQ6L14
★53	Blower Wheel	1472-2751	1472-2761	1472-2771
		10-3/4 Dia. x 7-3/16 x 1/2B	10-5/8 Dia. x 8 x 1/2B	11-7/8 Dia. x 9-1/2 x 1/2B
		Lau #027855-10	Lau #01332A-01	Lau #0127855-01
		Torin #CM1020-704-5	Torin #CM1020-800-5	Torin #CM11.5-9.5
		Morrison #03-42001-0	Morrison #03-43001-0	Morrison #03-54002-0
54	Motor Clamp	8680-5391	8680-6391	8680-6391
★55	Rubber Mounts (3/pkg)	7670-3551	7670-3551	7670-3551
★56	Run Capacitor (Mfd.-Volts)	1499-4441 (6-370)	1499-4491 (10-370)	1499-4491 (10-370)

NOTE: All parts with three digit suffix numbers as "Special Order" Parts. These parts are subject to factory availability and require extra time for delivery.

* Suggested Parts Inventory (2% of Units Installed - Minimum 1 each)

▲ Ref. Only - Not available for shipment. Order component parts.

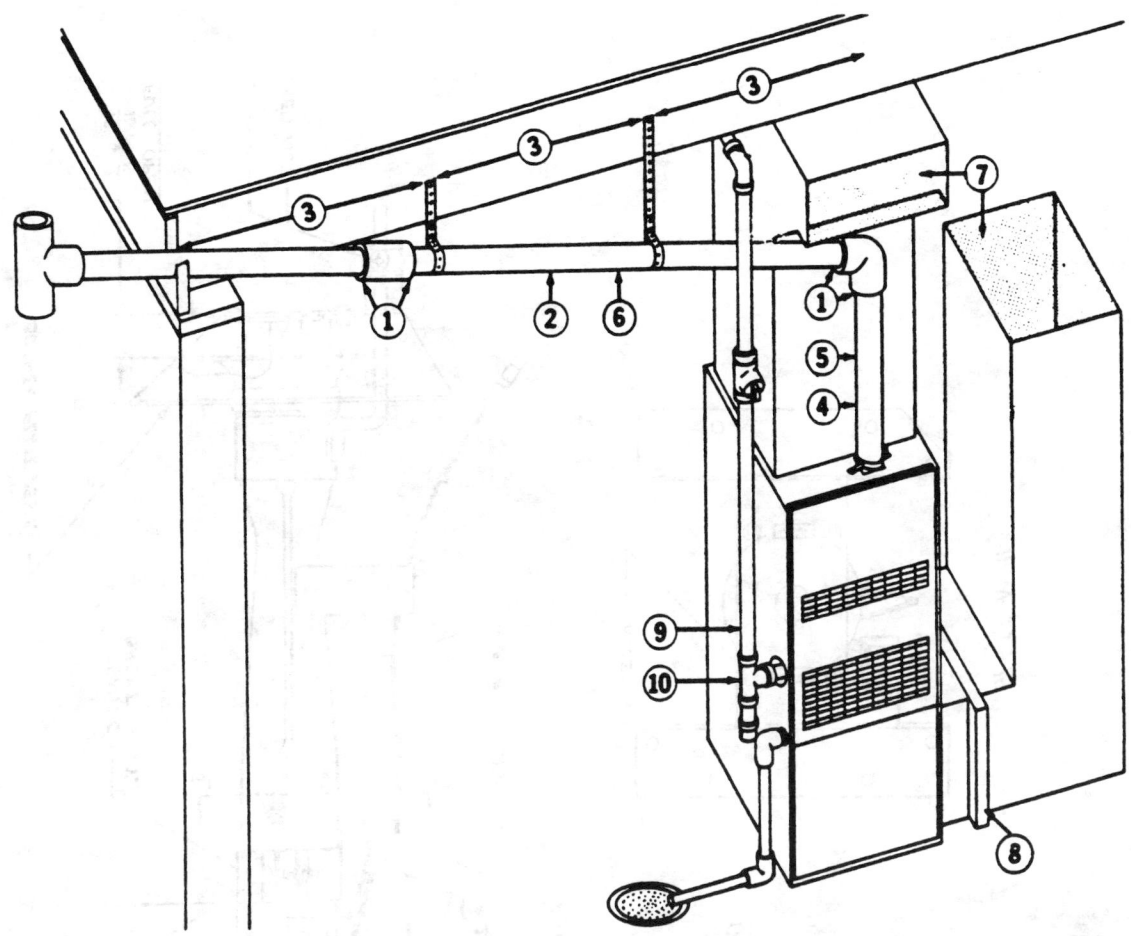

CHECK EACH INSTALLATION REQUIREMENT FOLLOWED	2900 SERIES FURNACE CORRECT INSTALLATION REQUIREMENTS (If these requirements are not followed, operational problems could result.)	ADDITIONAL INFORMATION FOUND IN INSTALLATION INSTRUCTIONS
	VENTING	
☐	1. Vent pipe must be tight and glued.	Page 9
☐	2. Horizontal vent pipe runs must have a minimum rise of ¼ inch per foot.	Page 7
☐	3. Horizontal vent pipe runs must have a minimum of one support every 3 feet.	Page 7
☐	4. Vertical section of vent pipe connected to furnace inducer must be straight and squared.	Page 7
☐	5. The first two feet of vent pipe, extending from the furnace, must be SCHEDULE 40 CPVC pipe.	Page 7
☐	6. Vent pipe must be SCHEDULE 40 PVC or CPVC and on model 2985, vent pipe must be 3 inch diameter	Page 7
	RETURN AND SUPPLY	
☐	7. Total system external static pressure must not exceed 0.5 inches water column.	Page 5
☐	8. Air filter must be clean and well supported.	Page 5
	GAS SUPPLY	
☐	9. Gas piping connections must be leak free.	Page 6
☐	10. Natural gas manifold pressure must be 3.5 ± 0.3 inches water column and inlet pressure between 5 inches minimum and 7 inches maximum water column; or for L.P. gas, manifold pressure must be 10 ± 0.3 inches water column and inlet pressure between 11 inches minimum and 14 inches maximum water column.	Page 6 & 13

1951A795 (8/90) PP   * Coleman and ▨▨▨ are registered trademarks of The Coleman Company, Inc. used under license.

35

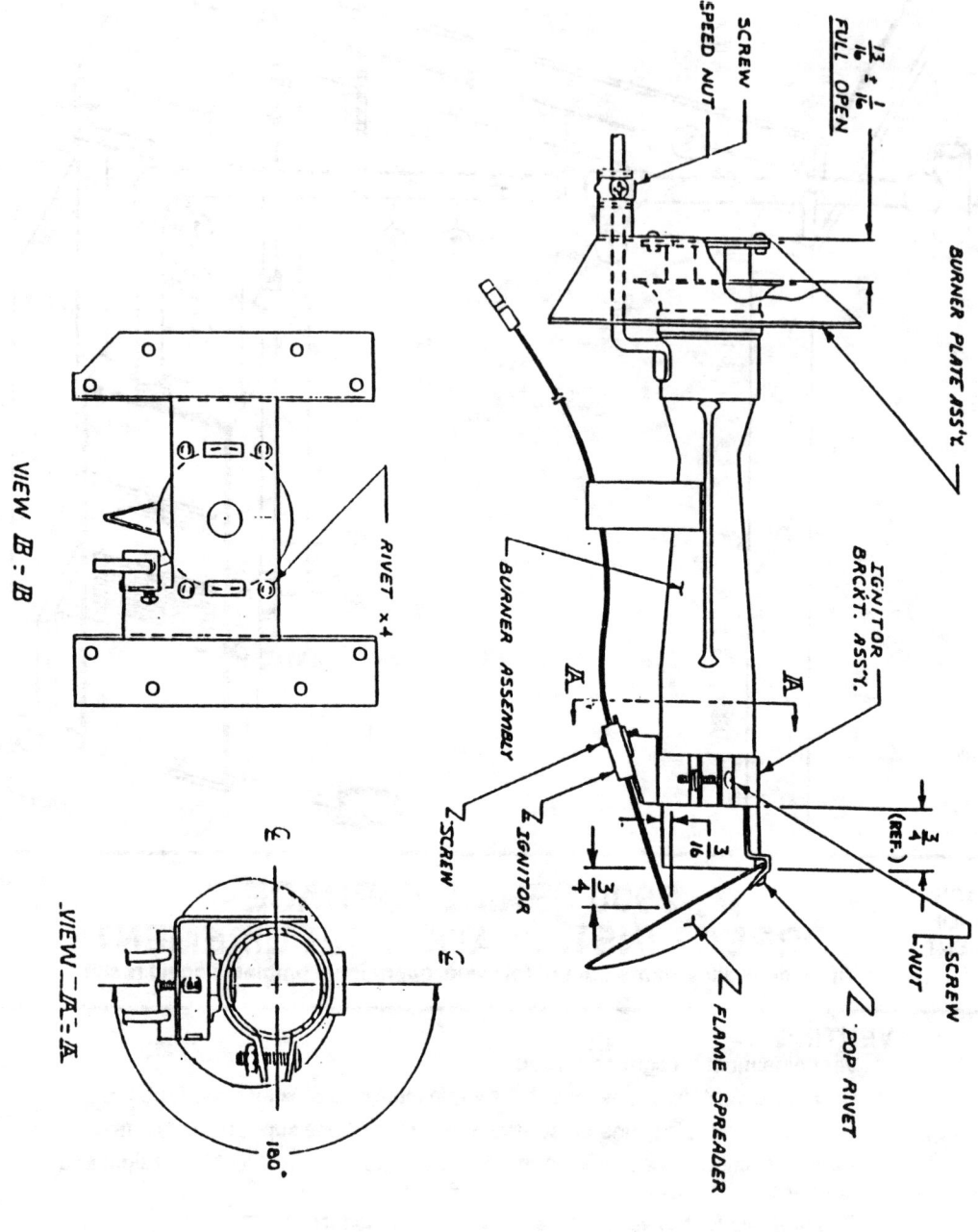

VIEW B-B

RIVET × 4

VIEW A-A

180°

FULL OPEN

$\frac{13}{16} \pm \frac{1}{16}$

SCREW

SPEED NUT

BURNER PLATE ASS'Y.

IGNITOR
BRCKT. ASS'Y.

BURNER ASSEMBLY

A

A

SCREW

IGNITOR

$\frac{3}{16}$

$\frac{3}{4}$

$\frac{3}{4}$
(REF.)

SCREW

NUT

POP RIVET

FLAME SPREADER

NEW FURNACE MODEL NUMBERS

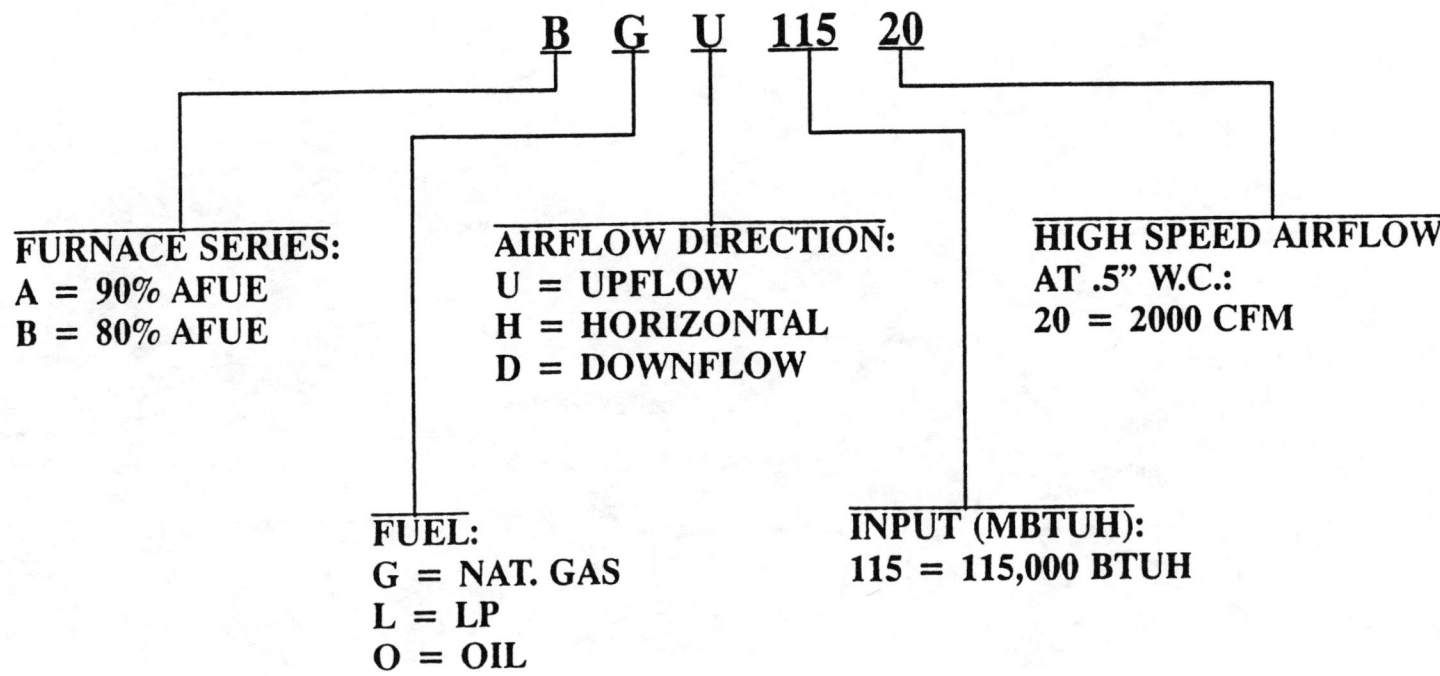

B G U 115 20

FURNACE SERIES:
A = 90% AFUE
B = 80% AFUE

AIRFLOW DIRECTION:
U = UPFLOW
H = HORIZONTAL
D = DOWNFLOW

HIGH SPEED AIRFLOW AT .5" W.C.:
20 = 2000 CFM

FUEL:
G = NAT. GAS
L = LP
O = OIL

INPUT (MBTUH):
115 = 115,000 BTUH

80 SERIES

UPFLOW, HORIZONTAL

& COUNTERFLOW

FURNACES

RES 80

HEATING AND AIR CONDITIONING PRODUCTS

INSTALLATION INSTRUCTIONS
UPFLOW HI-EFFICIENCY FURNACES
WITH HOT SURFACE IGNITION

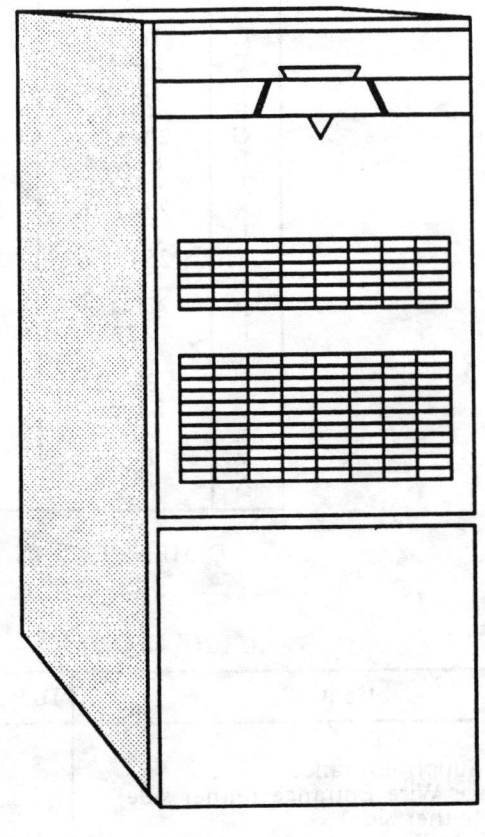

MODELS:
2845
2865
2880
2895

TABLE OF CONTENTS

DIMENSIONS

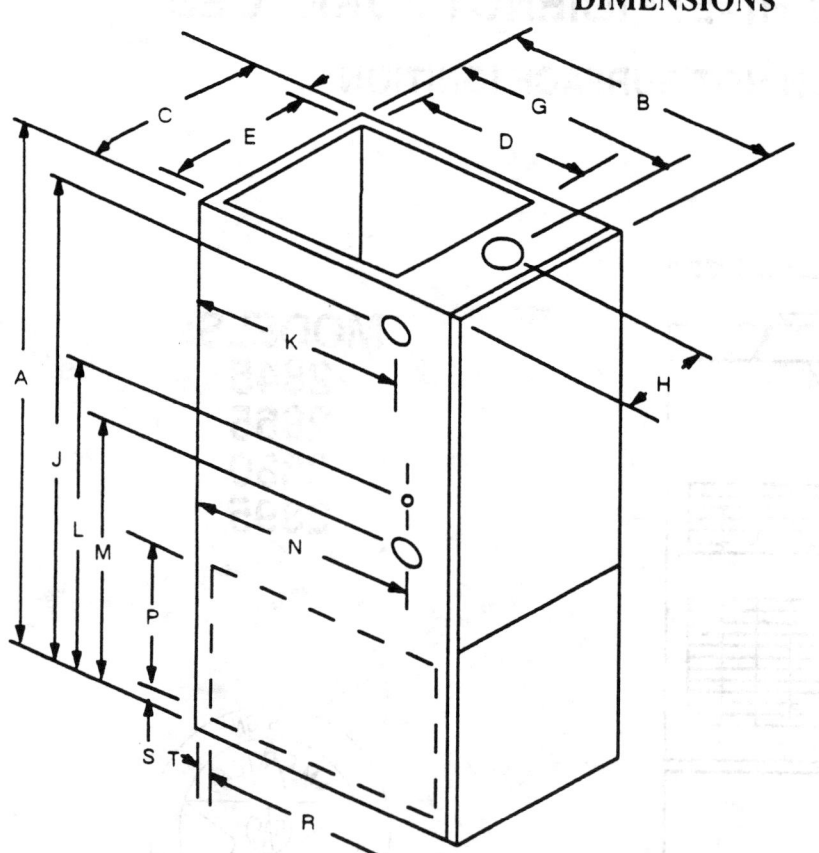

	2845E666 2865E666	2865E766 2880E766 2895E766
A	46	46
B	28	28
C	16-7/8	20
D	20-3/8	20-3/8
E	15-3/8	18-1/2
F	3/4	3/4
G	24-5/16	24-5/16
H	8-7/16	10
J	43-1/8	43-1/8
K	23-1/16	23-1/16
L	28-5/8	28-5/8
M	25-3/16	25-3/16
N	24-1/2	24-1/2
P	14-1/2	14-1/2
R	23	23
S	1	1
T	3-1/2	3-1/2
U	23-3/4	23-3/4
V	14-7/8	18
W	1-3/4	1-3/4
X	1-1/8	1-1/8

DIMENSIONS IN INCHES

Item	Dimension Location
Flue	G&H
Electrical Supply Entrance	J&K
Low Voltage Wire Entrance (either side)	L&N
Gas Inlet (either side)	M&N
Optional Side Return Air Inlet (either side)	P&R
Optional Bottom Return Air Inlet	U&V
Supply Air Outlet	D&E

MODEL	NAT.	2845E666	2865E666	2865E766	2880E766	2895E766
UNIT RATING BTU/Hr.	Input:0-2000 Ft. Elevation	55,000	75,000	75,000	95,000	115,000
	High Altitude	FOR ELEVATIONS ABOVE 2000 FEET, REDUCE INPUT 4% FOR EACH 1000 FEET OF ELEVATION ABOVE SEA LEVEL.				
AIR TEMPERATURE RISE RANGE F		30-60	40-70	40-70	45-75	50-80
DESIGNED MAXIMUM OUTLET AIR TEMP. F		155	165	165	170	175
MAX. EXT. STATIC PRESSURE IN. W.C.		.5				
FURNACE FLUE PIPE CONNECTION		4 in. ROUND			5 in. OVAL	
GAS CONNECTION		1/2 in. FPT				
ELECTRICAL SERVICE		115 VAC 60 HZ 1 PH				
MINIMUM DISTANCE TO COMBUSTIBLE MATERIALS IN INCHES						
FRONT		6				
BACK		0				
PLENUM-TOP		1				
FLOOR		COMBUSTIBLE				
SIDES		0				
FLUE Type B-1 Vent		1				
Single Wall Metal Pipe *		6				
FILTER(Furnished)		20 X 25 X 1/2			16 X 25 X 1/2	

PRE-INSTALLATION INSPECTION

Inspect the shipping container and furnace for any evidence of shipping damage. If furnace damage is found, notify freight carrier and file claim.

IMPORTANT NOTICE – These instructions are for the use of qualified individuals specially trained and experienced in installation of this type equipment and related system components.

Installation and service personnel are required by some states to be licensed. **Persons not qualified shall not install this equipment nor interpret these instructions.**

IMPORTANT NOTICE – This furnace is not to be used as a construction heater to supply heat to an unfinished building during the finishing phases of construction. This practice exposes the furnace to abnormally low return air temperatures which can cause condensation in the furnace or vent leading to premature failure. This practice also exposes the furnace to an abnormally corrosive atmosphere from sources such as paint, varnish and adhesives, which can lead to premature heat exchanger or vent failure. The practice also allows foreign materials such as sawdust or sheetrock dust to enter the furnace blower, burner, heat exchanger, motors, and vent system resulting in shorter life of the furnace. **Use of this furnace as a construction heater will void the warranty.**

APPLICATION

FURNACE CERTIFICATION AND USAGE
The furnace described in these instructions are design certified by the American Gas Association to be in compliance with American National Standard Z21.47b-1989.
These furnaces are forced air type and may be utilized for indoor installation in manufactured buildings (modular only), or buildings constructed on site. These furnaces are not certified for installation in mobile homes.

MUNICIPAL, STATE, AND FEDERAL CODES
The installer must conform to all state and local building codes when installing these appliances. In the absence of state and local codes, these furnaces and related equipment must be installed in accordance with the latest issue of the following:
NATIONAL FUEL GAS CODE - ANSI

Z223.1-1988
NATIONAL ELECTRICAL CODE, ANSI/NFPA No. 70-1987
Applicable codes take precedence over any recommendation made in these instructions.

FURNACE SIZING AND DUCT SYSTEM DESIGN
Consideration should be given to the heating capacity required and also the air quantity (CFM) required if A/C is to be installed along with the furnace or at some future time. These factors can be determined by calculating the heat loss and heat gain of the home or structure.
If these calculations are not performed and the furnace is oversized, the following may result:
1. Short cycling of the furnace.
2. Wide temperature fluctuations from the thermostat setting.

3. Reduced overall operating efficiency of the furnace.

The supply and return duct system must be of adequate size and designed such that the furnace will operate within the designed air temperature rise range and **not exceed the maximum designed static pressure**. These values are listed in the specification table of these instructions and on the furnace rating plate.

Information, values and data necessary for heat loss, heat gain and duct system design may be found in ASHRAE HANDBOOK OF FUNDAMENTALS-1989 EDITION or in other nationally recognized publications recognized by municipal, state, and federal code authorities.

A/C USAGE DUCT SYSTEMS

1. When a single (common) duct system is used one of the following methods shall be used:

   a. A plenum type cooling coil must be installed on the air discharge side downstream from the furnace, or

   b. A Coil-Blower type cooling coil must be installed in parallel with and isolated from the furnace, or

   c. A self-contained A/C unit must be in parallel with and isolated from the furnace.

WARNING
DAMPERS MUST BE INSTALLED WHEN A COIL-BLOWER OR SELF-CONTAINED UNIT IS EMPLOYED TO PREVENT CONDITIONED COOL AIR FROM COMING IN CONTACT WITH THE HEAT EXCHANGER TO AVOID MOISTURE CONDENSATION AND RUST OUT, WHICH CAN ALLOW PRODUCTS OF COMBUSTION TO BE CIRCULATED INTO THE LIVING AREA BY THE FURNACE BLOWER RESULTING IN POSSIBLE ASPHYXIATION. IF DAMPERS ARE MANUALLY OPERATED TYPE, A MEANS MUST BE PROVIDED TO PREVENT EITHER THE FURNACE OR A/C UNIT FROM OPERATING UNLESS DAMPERS ARE IN FULL HEAT OR COOL POSITION.

2. If two duct systems are used as could be the case with a coil-blower or a self-contained A/C unit,

the furnace and A/C unit should be controlled by a single combination heating and cooling thermostat which will prevent the furnace and A/C unit from operating simultaneously.

CAUTION
If a separate heating and separate cooling thermostat is used, a manually operated electrical interlock switch must be installed to prevent simultaneous operation of both systems and avoid a possible hazardous condition due to overheating of the conditioned space.

INSTALLATION PROCEDURE

NOTICE TO INSTALLER

This furnace is equipped with a metal blower brace which supports the blower during shipping. If the return air for the furnace is to be brought in the side, the blower brace may be left in place. However, if bottom return is to be used, the brace must be removed by removing the screw on each side of the casing which holds it in place. This must be done before the furnace is installed.

Models 28XXE766 are also equipped with a shipping strap on the blower motor shaft which supports the blower motor during shipping. This strap must be removed before the furnace is operated for the first time. It can be removed by removing the two fastening screws with a short screwdriver.

LOCATIONS AND CLEARANCES

The minimum clearances between the furnace, vent system, etc., and combustible materials are listed in the specification tables of these instructions and on the unit rating plate.

Installations on Combustible Flooring

This furnace may be installed directly on bare wood floors. It shall not be installed directly on carpeting, tile, or other type combustible flooring.

Clearance for Lighting and Service

Adequate clearance must be provided for lighting and maintenance. A minimum of 24" clearance should be provided between the front of the furnace and any opposite wall. If the furnace is in line with a door, you must provide a minimum of 6" clearance from the front of the furnace to the door.

If the furnace is to be installed in a close clearance closet, the door should be of adequate size to allow for removal of the furnace should it become necessary.

Installation in Residential Garages

When furnace is installed in a residential garage it must be located and installed such that it will be protected against vehicular damage. The furnace must be installed so that the burner is a minimum of 18" above the floor.

RETURN AIR AND FILTERS

Return Air Connection – The return air connection to the furnace may be through the bottom, through one side, or through both sides. Lances on casing sides locate the return air openings.

The return air ducts to the furnace must have a total cross sectional area of not less than two square inches per 1000 BTUH of furnace input rating.

When bottom return air inlet is employed, filter is installed in the blower compartment, over the air inlet opening and secured in place with the retaining clip provided. When a side inlet return is employed, filters cannot be mounted inside the casing. Filters must be mounted in an external frame on the side of the furnace or in the return duct and be accessible for cleaning and replacement. One or more return grill-filter frames may be employed when side inlet returns are used.

IMPORTANT NOTICE – A solid metal block-off panel must be in place to block the bottom opening in the furnace when side return air ducts are used. Failure to block this opening could cause products of combustion to be circulated into the living space and create a potentially hazardous condition. A block-off panel has been provided with the furnace for this purpose.

CAUTION
WHEN THE FURNACE IS INSTALLED IN A CLOSET OR OTHER CONFINED SPACE AND A SIDE INLET DUCT EMPLOYED, THE DUCT MUST BE SEALED TO THE FURNACE AND EXTEND TO THE CONDITIONED SPACE TO PREVENT ANY COMMUNICATION BETWEEN THE SPACE OR ROOM IN WHICH THE FURNACE IS INSTALLED, WHICH COULD RESULT IN ASPHYXIATION.

Do not exceed .5" W. C. total static pressure. See duct system design on page 3.

VENTILATION AND COMBUSTION AIR
Provide ventilation and combustion air in accordance with section 5.3, Air for Combustion and Ventilation, of the NATIONAL FUEL GAS CODE, ANSI Z223.1-1988, or applicable provisions of the local building codes.

WARNING
ADEQUATE VENTILATION AND COMBUSTION AIR MUST BE PROVIDED TO INSURE SATISFACTORY AND SAFE OPERATION OF THE FURNACE. AIR OPENINGS IN CASING FRONT PANEL AND VESTIBULE TOP PANEL MUST NOT BE OBSTRUCTED. FAILURE TO OBSERVE THIS RECOMMENDATION COULD RESULT IN ASPHYXIATION.

DO NOT STORE HALOGEN EMITTING SUBSTANCES SUCH AS LAUNDRY BLEACH AND DETERGENT, CLEANING FLUIDS, SPRAY CAN PROPELLENTS, AND SOLVENTS IN THE VICINITY OF THIS APPLIANCE. THE AIR USED BY THE BURNER FOR COMBUSTION MUST BE FREE OF HALOGENS TO AVOID POSSIBLE CORROSION TO THE HEATING SURFACES WHICH COULD RESULT IN ASPHYXIATION.

Installations in a Confined Space
If the unit is to be installed in a confined space such as a small closet or room, provisions must be made for supplying combustion and ventilation air to the space surrounding the furnace. (See Fig. 1)

Two openings of equal area must be provided; one starting within twelve inches of the ceiling and one starting within twelve inches of the floor of the confined space. The upper opening must always be above the top of the furnace casing. The lower opening, if in the side wall, floor or door, shall be located below the level of the burner in the furnace.

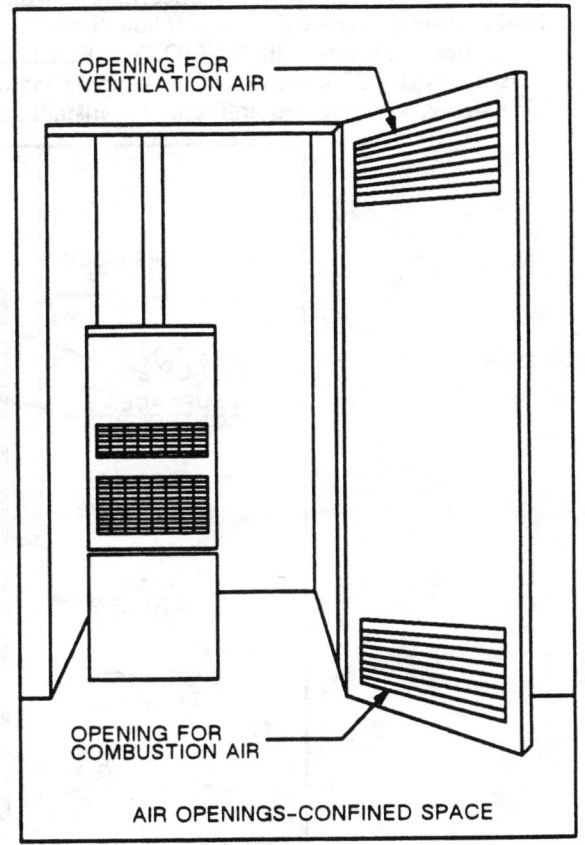

Figure 1

If all air is from inside building, the total free area of each opening must be at least one square inch for each 1,000 BTUH of furnace input but not less than 100 square inches.

If all air is from outdoors, when communicating directly with the outdoors through vertical ducts, the total free area of each opening must be at least one square inch for each 4,000 BTUH of furnace input, and when communicating directly with the outdoors through horizontal ducts, the total free area of each duct must be at least one square inch for each 2,000 BTUH of furnace input.

When ducts are used, they must be of the same cross-sectional area as the free area of the openings to which they connect. The minimum dimension of rectangular air ducts must not be less then three inches.

Installations in an Unconfined Space
In unconfined spaces in a building of conventional frame, masonry, or metal construction, infiltration is normally adequate to provide air for combustion and ventilation.

In buildings of tight construction, all air shall be

obtained from outdoors or from spaces communicating freely with outdoors. A permanent opening or openings having a total free area of not less than one square inch for each 5000 BTUH of furnace input shall be provided.

GAS PIPING AND SUPPLY PRESSURES

Before installing gas piping, check with local code authorities for requirements concerning gas piping. In the absence of local codes, follow the recommendations contained in NATIONAL FUEL GAS CODE ANSI Z223.1-1988 for gas piping materials, pipe sizing, and the requirements for installation. It

is recommended that a gas cock shutoff valve be installed in the gas supply line **outside the casing** where it is readily accessible, as close to the furnace as is practicable.

An 1/8 in. NPT (plugged) pressure tap for test gauge connection must be installed in the gas supply line immediately upstream from the furnace. Install a dirt leg at the bottom of any vertical riser or drop, as close to the furnace as possible, to collect moisture and foreign material. Install a ground joint union just ahead of the gas control valve. A typical arrangement is shown in Figure 2.

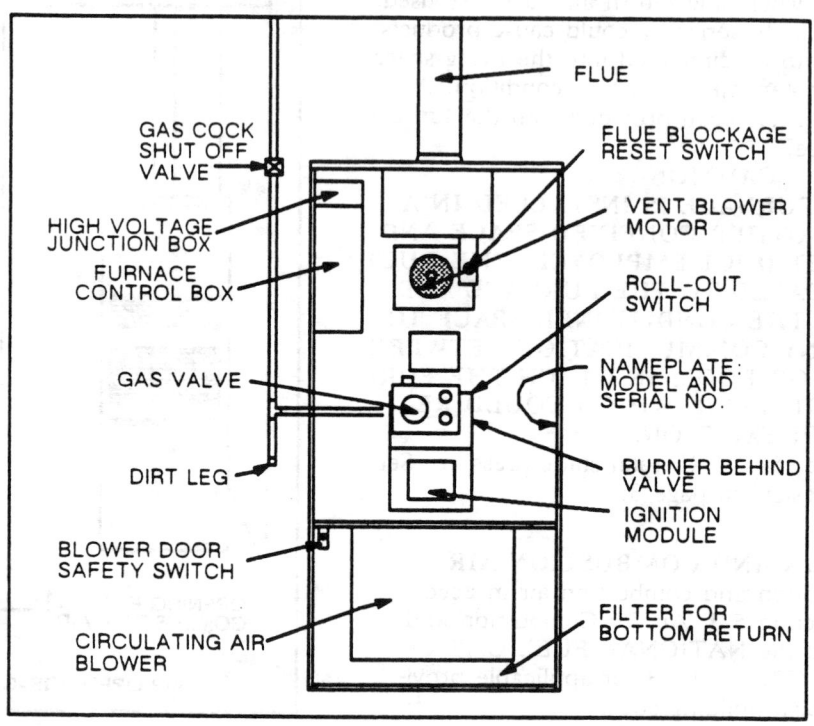

Figure 2

When making the connection at the gas control valve, use a wrench on the inlet side of the valve to prevent any possible twisting of the valve body which could cause damage and leaks. Connection sizes are shown in the specifications table. When making up pipe joints use pipe thread compound which is resistant to natural and LP gases.

CAUTION
During pressure testing of the gas supply piping system, observe the following to avoid fire, explosion, asphyxiation, or damage to the appliance.

a. **If test pressure is less than or equal to 1/2 psig (3.48 kPa), isolate the furnace by closing its individual manual shutoff valve.**

b. **If test pressure is greater than 1/2 psig (3.48 kPa), the furnace and its individual shutoff valve must be disconnected from the gas supply system.**

Following installation of the piping, first ensure that the gas control knob on the gas valve is in the off position and then pressurize the system with gas. Thoroughly check the piping system for leaks.

CAUTION
NEVER USE AN OPEN FLAME TO CHECK FOR LEAKS. FIRE OR EXPLOSION COULD OCCUR. SINCE SOME LEAK SOLUTIONS INCLUDING SOAP AND WATER MAY CAUSE CORROSION OR STRESS CRACKING, THE PIPING MUST BE RINSED WITH WATER AFTER TESTING UNLESS IT HAS BEEN DETERMINED THAT THE LEAK SOLUTION IS NON-CORROSIVE.

The maximum and minimum gas supply pressure required at the inlet of the gas control valve is shown on the unit rating plate. When the furnace is in operation, the inlet pressure must be within the limits shown.

VENTING COMBUSTION GASES

This furnace incorporates an induced draft blower. The vent box on the outlet of the inducer blower decouples the high pressure of the flue gas and allows natural venting of the hot flue gas. The pressure in the stack is slightly negative. A sensing tube and manual reset switch is located in the bottom of the vent box. In the event that the vent would become partially restricted or blocked, the manual reset switch will open and shut off both the inducer motor and the gas valve. After the blockage is removed, the manual reset switch may be reset by pushing the red button on the switch.

The vent installation shall be in accordance with Part 7, Venting of Equipment, of the NATIONAL FUEL GAS CODE, ANSI Z223.1-1988, or applicable provisions of the local building codes.

1. This furnace must be connected to a factory built chimney or vent complying with a recognized standard, or a masonry or concrete chimney lined with a lining material acceptable to the authority having jurisdiction.

2. The vent material can be single wall or B-1 (type B vent with one inch clearance) for the first five feet from the furnace. Beyond five feet it must be B-1 or a permanent chimney.

3. Maintain minimum six inch clearance to combustible material for single wall vent and one inch clearance for B-1 vent.

4. When using Listed B-1 vent materials and/or a listed Manufactured chimney, they must be installed in accordance with the manufacturer's installation instructions and the terms of the listing.

5. If horizontal pipe is required, it should be as short as possible and pitch upward toward the vertical riser or chimney inlet at least 1/4" per foot to help insure proper venting.

6. Do not vent the flue of these furnaces horizontally out through the side of the home.

7. The vent connector must be at least as large as the outlet on the furnace vent box.

8. Where two or more appliances are connected to a common vent or flue, use the vent tables in the separate booklet supplied with these instructions to determine the proper vent connector size.

9. **DO NOT USE** dampers or restrictors in vent piping or flue.

10. Vent pipe sections must be securely fastened together and fastened to the furnace flue collar using sheet metal screws where required.

11. Joints in the vent pipe must be securely made and any horizontal run of the vent pipe supported no less than one support every three feet to prevent sagging.

12. This furnace may be common vented with a water heater or other gas-fired naturally vented appliance.

CAUTION
THIS FURNACE UTILIZES NATURAL DRAFT VENTING AND MUST NOT BE CONNECTED INTO ANY PORTION OF A MECHANICAL DRAFT SYSTEM OPERATING UNDER POSITIVE PRESSURE.

CAUTION
THIS FURNACE MUST NOT BE CONNECTED TO A CHIMNEY FLUE SERVING A SEPARATE APPLIANCE DESIGNED TO BURN SOLID FUEL.

REPLACEMENT OF EXISTING FURNACE ON COMMON VENT SYSTEM
When this furnace is installed as a replacement for an old furnace which is common vented with a water heater or other gas appliance, and the new furnace is no longer connected to the common venting system, the common vent system may be too large for the appliances remaining on the vent system after the old furnace is removed. To test for an oversized vent system, the following steps shall be followed with each appliance remaining connected to the common venting system placed in operation, while the other appliances remaining connected to the common venting system are not in operation: (a) Seal any unused openings in the common venting system. (b) Visually inspect the venting system for proper size and horizontal pitch and determine that there is no blockage or restriction, leakage, corrosion or other deficiencies which could cause an unsafe condition. (c) Insofar as is practical, close all building doors and windows and all doors between the space in which the appliances remaining connected to the common venting system are located and other spaces of the building. Turn on clothes dryers and any appliance not connected to the common venting system. Turn on any exhaust fans, such as range hoods and bathroom exhausts, so they will operate at maximum speed. Do not operate a summer exhaust fan. Close fireplace dampers. (d) Follow the lighting instructions. Place the appliance being inspected in operation. Adjust thermostat so appliance will operate continuously. (e) Test for spillage at the draft hood relief opening after 5 minutes of main burner operation. Use the flame of a match or candle, or smoke from a cigarette, cigar, or pipe. (f) After it has been determined that each appliance remaining connected to the common venting system properly vents when tested as outlined above, return doors, windows, exhaust fans, fireplace dampers, and any other gas-burning appliance to their previous condition of use. (g) If improper venting is observed during any of the above tests, the common venting system must be corrected. Any changes to the venting system must be in accordance with the National Fuel Gas Code, ANSI Z223.1-1988. If any portion of the common venting system must be resized, it should be resized to approach the minimum size as determined using the appropriate tables in Appendix G in the National Fuel Gas Code, ANSI Z223.1-1988.

VENT SYSTEM CONSTRUCTION
Refer to the venting booklet supplied with these instructions for details on proper sizing of the vent system for this furnace.

Venting into an Existing Chimney
Whenever possible, B-1 metal pipe should be used for venting. Where use of an existing chimney is unavoidable, **the following rules must be followed:** (1) The masonry chimney must be built

and installed in accordance with nationally recognized building codes or standards and **must be lined with approved fire clay tile flue liners or other approved liner material** that will resist corrosion, softening, or cracking from flue gases. **THIS FURNACE IS NOT TO BE VENTED INTO AN UNLINED MASONRY CHIMNEY.** (2) This furnace may be vented into a fire clay tile lined masonry chimney only if a source of dilution air is provided, such as by common venting with a draft hood equipped water heater. If no such source of dilution air is available, Type B vent must be used. The existing chimney may be used as a chase for the Type B vent. (3) The chimney must extend at least three feet above the highest point where it passes through a roof of a building and at least two feet higher than any portion of the building within a horizontal distance of ten feet. (4) The chimney must extend at least five feet above the highest equipment draft hood or flue collar.

Inspection of Existing Chimney

(1) Before connecting the vent connector to a chimney, the chimney passageway must be examined to ascertain that it is clear and free of obstructions and must be cleaned if previously used for venting solid or liquid fuel-burning appliances or fireplaces. (2) Cleanouts must be examined to determine that they will remain tightly closed when not in use. (3) When inspection reveals that an existing chimney is not safe for this application, it must be rebuilt to conform to nationally recognized standards, lined or relined with a suitable liner, or replaced with a suitable vent or chimney.

Venting with Metal Pipe

Type B (double wall) vents must extend in a generally vertical direction with no offsets exceeding 45 degrees, except that one horizontal run may be allowed. Any angle greater than 45 degrees from the vertical is considered horizontal. The total horizontal run of a vent plus any horizontal vent connector must not be greater than 75% of the vertical height of the vent. A Type B vent must terminate at least five feet in vertical height above the highest connected equipment draft hood or flue collar.

Vent Connector

A vent connector must be sized properly for the equipment connected to it. If serving the furnace only, the vent connector area must not be less than the area of the furnace vent box outlet. If more than one appliance is connected to the vent connector, use the vent tables in the separate booklet supplied with these instructions to determine the proper vent connector size.

The vent connector must be as short as possible and the furnace should be located as close as practicable to the chimney or vent.

The horizontal run of a vent connector must not be more than 75% of the height of the vertical portion of the chimney or vent above the connector.

A vent connector must not pass through any ceiling, floor, fire wall, or fire partition. A single-wall metal pipe vent connector must not pass through any interior wall.

Condensation

These furnaces are not intended to have condensation occur in the furnace or in the venting system. Such condensation can cause corrosion and premature failure of the vent system, leading to possible asphyxiation. In most cases, condensation is a result of an oversized vent system. When sizing the vent system for this furnace, the vent pipe size should be kept to the minimum allowable according to the vent tables in the National Fuel Gas Code. Where local experience indicates that condensation may be a problem, the following steps may be taken: (1) Usage of single-wall vent pipe should kept to a minimum and should never be used in any unheated space. (2) As noted above, all masonry chimneys must be lined with the liner sized no larger than is necessary for the capacity of the attached appliances. (3) If local codes permit, the outside of metal vent pipe may be insulated with a noncombustible insulating material. (4) If necessary, the furnace circulating blower may be set to a lower speed, which will raise the flue temperature slightly. However, the temperature rise across the furnace must not exceed the maximum listed in the specification table in these instructions and on the furnace rating plate.

In certain conditions, condensation in a lined masonry chimney may be unavoidable. In such cases, provisions must be made to drain off and dispose of condensate to avoid damage to the chimney.

Further Instructions

For more details on venting or other aspects of gas appliance installation, consult the NATIONAL FUEL GAS CODE, ANSI Z223.1-1988. It is highly recommended that all gas appliance installers and servicemen have a copy of this manual.

ELECTRICAL WIRING

All internal wiring has been made at the factory. Field wiring requires only the connection of line voltage supply wiring and low voltage thermostat wiring.

Refer to the unit rating plate and specification tables found in these instructions for applicable electrical characteristics and requirements. A complete wiring diagram is supplied on page 15.

Service wiring

Field wiring and electrical grounding of the unit should conform to local codes or in the absence of local codes with the National Electrical Code ANSI/NFPA 70-1987.

A separate fused circuit from the main electrical panel should serve only the furnace. Connect the power supply leads in the furnace junction box, providing an approved strain relief, as shown in the wiring diagram on the unit and in these instructions.

Accessories

A humidifier or electronic air cleaner may be used with this furnace. All accessories should be wired according to the manufacturer's instructions.

In most cases, the humidifier is intended to operate only when the room thermostat is calling for heat. Attach the humidifier control wires to the red and white thermostat wires to supply 24 volts to the humidifier only during the heating cycle. **Make sure that the total load on the furnace transformer does not exceed 40 VA, including gas valve, furnace relays, accessories, and air conditioner loads.**

In most cases, an electronic air cleaner will be provided with a pressure switch or sail switch which senses airflow and energizes the air cleaner. If not so equipped, a sail switch in the air duct may be used to switch the air cleaner on and off as needed. **Do not attempt to wire an electronic air cleaner into the furnace blower relay.** Damage to the furnace blower motor may result.

Control Wiring

The thermostat should be installed in accordance with the manufacturer's instructions, furnished with the thermostat, and make connections to the unit as shown on the unit wiring diagram. It is recommended that size 18 AWG wire be used. If the thermostat has an adjustable heat anticipator, the setting should be .13 amps.

TABLE I
MOTOR SPEED TAP SELECTION FOR COOLING

FURNACE MODELS	COOLING SYSTEM CAPACITY, BTUH							
	18,000	24,000	30,000	36,000	42,000	48,000	54,000	60,000
2845E666	Low	Low	Med.	High				
2865E666	Low	Low	Med.	High				
2880E766		Low	Low	Med.Low	Med. High	High	High	
2895E766		Low	Low	Med.Low	Med. High	High	High	
2865E766				Low	Med. Low	Med. High	High	High

Blower Motor Speed Selection

These furnaces are equipped with blowers which have multi-speed direct drive motors.

The blower speed selected is dependent upon the design and static pressure loss of the duct system. The duct system external static pressure includes the combined total of the supply and return ducts and any plenum type air conditioning coil if used. The furnace must be adjusted to operate at or below the maximum external static (in. W.C.) and within the air temperature rise range as shown on the unit rating plate and in the specification table. These furnaces are equipped with a blower relay which will change blower speeds automatically when the furnace is properly connected to a heating and cooling type wall thermostat. The blower motors are factory connected to operate on high speed for cooling operation and medium or medium high for heating operation. Dependent upon the conditions in a particular installation, the blower speeds may be changed. Table I may be used for selection of motor speed tap for cooling operation.

DANGER – SHOCK HAZARD
BE SURE ELECTRICAL POWER TO FURNACE IS TURNED OFF BEFORE CHANGING MOTOR SPEEDS.

Motor Terminal Identification

In all cases the white wire from the furnace control box is the common circuit and is fitted with a 3/16 quick connect terminal. This wire must always be connected to the motor terminal marked White, Common, or 1.

The blue wire from the furnace control box is controlled by the furnace fan switch and is the hot wire for heating operation.

The black wire is controlled by the blower relay and is the hot wire for cooling.

The motor terminals of various motors which may be used are identified by one of the methods shown in Table II.

TABLE II

Motor Speed	Motor Terminal Identification
High Med. High Med. Med. Low Low	Common, White Hi, Black Med. High, Blue Med., Blue Med. Low, Yellow Low, Red

PRE-OPERATIONAL CHECKS

DANGER - SHOCK HAZARD - BE SURE ELECTRICAL POWER TO FURNACE IS TURNED OFF BEFORE PERFORMING THE PRE-OPERATIONAL CHECKS.

1. Be sure that the furnace is equipped for the type of gas being supplied to the furnace. See unit rating plate.

2. Make sure that shipping strap was removed from blower housing of furnace models 28XXE766.

3. Manually spin circulating air blower wheel to ensure that it turns freely and does not strike the blower housing.

4. Manually spin the vent blower by turning the heat dissipation fan blades between the vent motor and the furnace vestibule. The blower wheel should spin freely with no interference between the blower wheel and the blower housing.

5. Was the gas piping tested and/or purged of air then checked for leaks? See instructions for gas piping. Even the smallest leak must be eliminated before attempting to light the furnace.

SEQUENCE OF OPERATION

These furnaces are equipped with an electric hot surface burner ignition system. In response to a call for heat by the room thermostat, the burner is lighted by a hot glowing ignitor at the beginning of each operation cycle. The burner will continue to operate until the thermostat is satisfied at which time all burner flame is extinguished. During the off cycle no gas is consumed. With the room thermostat set below room temperature, and with the electrical power and gas supply to the furnace on, the normal sequence of operation is as follows:

1. When the room temperature falls below the setting of the room thermostat, the thermostat energizes the heating relay.

2. When the heating relay closes, a circuit is made starting the vent blower. A circuit is also made through the heating relay to the normally open vent blower centrifugal switch contacts.

3. As the vent blower increases in speed, the contacts of the centrifugal switch will close and complete the electrical circuit through the normally closed limit switch to the electronic ignition module.

4. During the next 40 to 50 seconds, the vent blower will bring fresh air into the heat exchanger and the ignitor will begin to glow. At the end of this period, the ignition module will open the gas valve and energize a safety lock-out circuit.

5. When the burner lights, the ignitor then acts as a flame probe which checks for the presence of a flame. As long as flame is present, the system will monitor it and hold the gas valve open.

6. If the burner fails to light within 6-8 seconds after the gas valve opens, the ignition module will close the gas valve and de-energize the ignitor. After a short pause, the system will recycle and try again for ignition. If the burner fails to light after three tries, the ignition module will lock off the gas valve and the ignitor. The system will remain in lock-out mode until the room thermostat is set below room temperature. The lock-out circuit will then be released and setting the thermostat to above room temperature will cause the system to try for ignition again.

7. The lapsed time from the moment the room thermostat closes to when the burner lights may be 50-60 seconds. This delay is caused by:(1) the time required for the vent blower to come to full speed, (2) the 40 to 50 seconds required for the ignitor to heat up, and (3) the time required for the vent blower to bring fresh air into the heat exchanger.

8. One to two minutes after the burner has lighted, the fan switch will close and the furnace air circulation blower will run.

NOTE – If a heating/cooling thermostat is being used and the fan switch is set in the continuous blower position, the furnace circulating blower will run at the air conditioning speed. If the room thermostat calls for heat, the furnace air circulating blower will shut off for 50-60 seconds and then the burner will light. One to two minutes after the burner has lighted, the furnace circulating blower will begin running at heating speed. There is no pause in the blower operation when the thermostat is satisfied; the furnace circulating air blower just changes back over to cooling speed.

9. When room thermostat is satisfied the circuit to the heating relay is broken and the relay contacts return to the normally open position. The circuit to the vent blower and the ignition module is broken and the burner is extinguished. Then as heat is drawn from the heat exchanger and the air temperature is reduced to below the fan switch setting, the fan switch will open which stops the furnace air circulation blower.

FURNACE OPERATION

GAS VALVE
The gas control valve is multi-function with two operating valves in line, a pressure regulator and a manual gas cock. See Figures 3 and 4.
The pressure regulator is factory set to provide an operating pressure of 3.5" W.C. on models equipped for natural gas and 10" W.C. on models equipped for LP gas.
The pressure regulator is a limited adjustment type.

Refer to the section titled "Minor Input Adjustment" under "Furnace Input Capacity".

LIGHTING INSTRUCTIONS
THIS FURNACE IS EQUIPPED WITH AN AUTOMATIC ELECTRIC HOT SURFACE BURNER IGNITION SYSTEM WHICH LIGHTS THE BURNER EACH TIME THE THERMOSTAT

CALLS FOR HEAT. THIS FURNACE CANNOT BE LIGHTED WITH A MATCH.

WARNING
FAILURE TO FOLLOW THESE INSTRUCTIONS MAY RESULT IN AN EXPLOSION AND POSSIBLE DAMAGE TO THE FURNACE AND INJURY TO THE OPERATOR.
1. Set room thermostat to lowest setting – or OFF.
2. Turn knob on gas control valve to OFF.
3. WAIT FIVE MINUTES.
4. Turn knob on gas control valve to ON.
5. Set room thermostat to desired setting above room temperature. Burner will light – which may take 50–60 seconds.

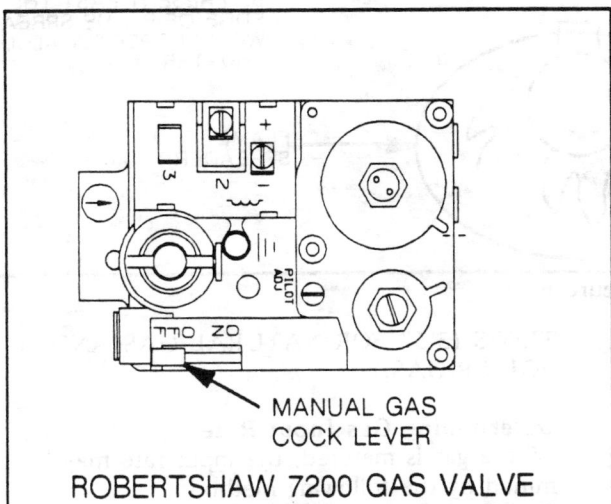

MANUAL GAS COCK LEVER

ROBERTSHAW 7200 GAS VALVE

Figure 3

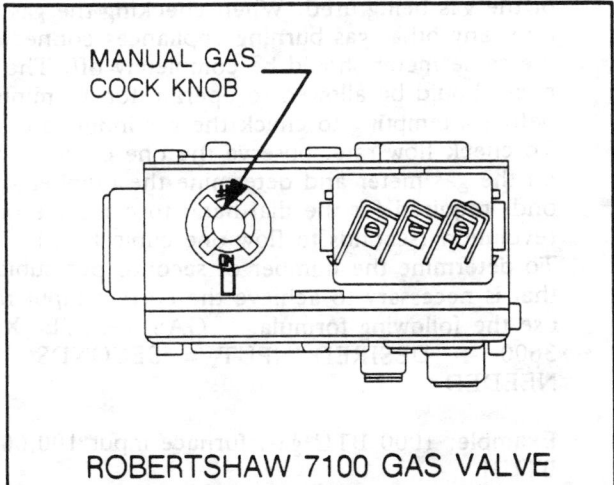

MANUAL GAS COCK KNOB

ROBERTSHAW 7100 GAS VALVE

Figure 4

6. If burner fails to light, set thermostat to OFF, then reset thermostat to ON and furnace will try again for ignition. If after three tries and burner fails to light, go to complete shutdown and determine cause for failure to light. Be sure gas supply to furnace is on and supply piping has been purged of air. Be sure electric supply to furnace is on.
7. For complete shutdown – set thermostat to lowest setting or OFF and turn knob on gas valve to OFF.

BURNER ADJUSTMENT
After lighting the furnace, allow the furnace to operate for 15 minutes and then adjust burner primary air shutter as follows:
1. Loosen air shutter adjustment rod locking screw. See Figure 5.
2. Close primary air shutter by pulling on adjustment rod until yellow tips appear in the flame at the end of the burner. See Figure 6.
3. Slowly open the primary air shutter by pushing adjustment rod in until yellow flame tips at the end of the burner disappear; then push rod in another 1/8 inch.
4. Secure primary air adjustment rod by tightening locking screw against it.

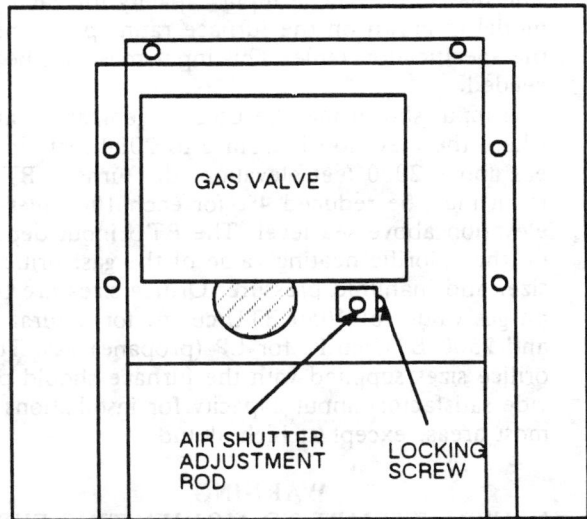

GAS VALVE

AIR SHUTTER ADJUSTMENT ROD

LOCKING SCREW

Figure 5

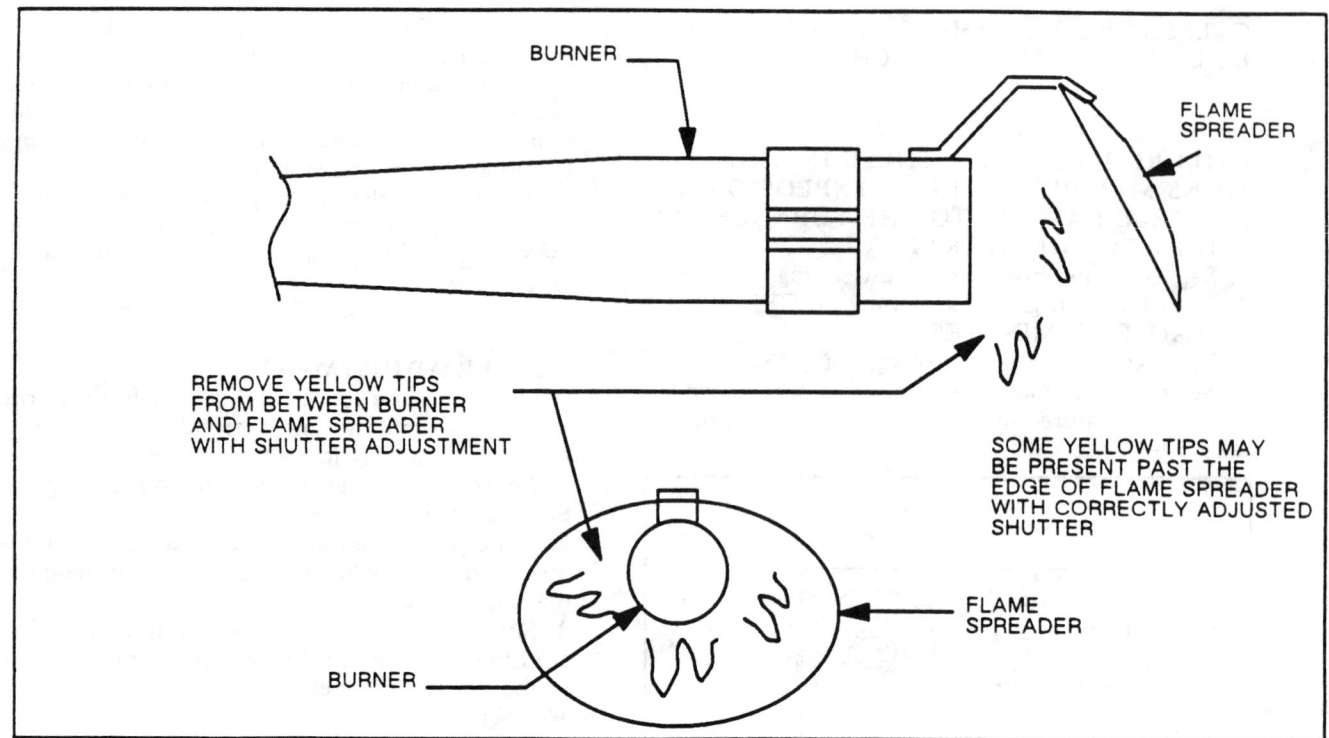

Figure 6

FURNACE INPUT CAPACITY

The maximum BTUH input capacity for each model is shown on the furnace rating plate and in the specification table. This input must not be exceeded.

The input shown may be used in geographic areas where the elevation is from 0 to 2000 feet. In areas above 2000 feet elevation, the furnace BTU input must be reduced 4% for each 1000 feet of elevation above sea level. The BTU input depends on the calorific heating value of the gas, orifice size, and manifold pressure. Orifice sizes are based on gas values of 1050 BTU/cu. ft. for natural gas and 2500 BTU/cu.ft. for LP (propane) gas. The orifice sizes supplied with the furnace should provide satisfactory input capacity for installations in most areas, except at high altitude.

WARNING
NEVER ATTEMPT TO MODIFY THIS FURNACE – FIRE, EXPLOSION, OR ASPHYXIATION MAY RESULT. If malfunction is apparent, contact qualified service agency and/or gas utility for assistance.

Minor Input Adjustment
The input may be adjusted slightly by adjusting the pressure in the gas valve in order to change manifold pressure.

To adjust pressure regulator, remove cover screw (see location on Figure 3, 4, or 5) on valve. Turn adjusting screw counter–clockwise to decrease pressure, turn clockwise to increase pressure. IN NO CASE SHOULD THE FINAL MANIFOLD PRESSURE VARY MORE THAN ± .3" W.C. FROM THE SPECIFIED REGULATOR PRESSURE SETTINGS (3.5" FOR NATURAL GAS AND 10" FOR LP GAS.)

Determining Gas Input Rate
Where gas is metered, the input rate may be determined by the following method:

Contact the gas supplier, public utility company or LP gas distributor to obtain the calorific gas value of the gas being used. When checking the gas input rate, any other gas burning appliances connected to the same meter should be completely off. The furnace should be allowed to operate for 15 minutes before attempting to check the gas input rate.

To check flow rate, observe the one cubic foot dial on the gas meter and determine the number of seconds required for the dial hand to complete one revolution (seconds to flow one cubic foot.)

To determine the number of seconds per cubic foot that is necessary to achieve the correct input rate, use the following formula: GAS VALUE X 3600 ÷ DESIRED INPUT = SECONDS NEEDED

Example: 1000 BTU gas, furnace input 100,000 BTUH

Seconds for one cubic foot = 1000 X 3600 ÷ 100,000 = 36 seconds.

If when clocking the meter, the one cubic foot dial makes a complete revolution in less time than was calculated that it should, the furnace is overfired and should be derated. If it takes more time for the meter to make one revolution than was calculated, the furnace is underfired.

The orifice size must be changed to correct an overfired or underfired condition. If it is determined that a different orifice is needed, please contact your distributor for assistance in selecting the correct replacement.

BALANCE THE SYSTEM - The air distribution system should be balanced, using in-line duct dampers if employed, to provide for satisfactory air delivery room to room. It is recommended that dampers in registers not be used for balancing the system.

CYCLE FURNACE FOR CORRECT OPERATION

After all parts of the installation have been completed, the furnace should be checked for normal operation prior to the time the system will be used by the homeowner/user.

1. Unplug the ignitor and check its resistance with a suitable ohmmeter. The resistance at room temperature should be between 50 and 400 ohms. A reading outside of this range indicates a cracked or otherwise defective ignitor which should be replaced. Be sure to plug ignitor back in before continuing.

2. Place the heating system in the operating mode by following the lighting instructions outlined previously in these instructions and on the furnace lighting instructions label.

3. Cycle the furnace on and off a number of times, allowing the furnace to light and the circulating air blower to come on. After thermostat is turned off each time, allow time for the circulating air blower to stop before starting the next operating cycle.

4. While unit is in operation check to determine that the circulating blower is operating smoothly with no undue vibration or noise. Check to insure that main burner has been adjusted and main burner flame is normal.

NOTICE TO INSTALLER - PROVIDE A HOOK, or other suitable means, and mount the envelope packet supplied with the furnace in a conspicuous place near the furnace. Place all printed literature supplied with the furnace in the packet. Advise the owner/user of the location of the literature packet if possible. If installed system is not to be placed in operation immediately by the owner/user, it is recommended that the system be left in a shut down condition by setting the thermostat to off or lowest temperature setting. Turn gas control knob to off position and turn off electrical power to furnace.

SERVICE MAINTENANCE

A program of periodic inspection and preventative maintenance can help to insure trouble-free operation, provide more efficient operation and extend the life of the furnace.

DANGER

To avoid the possibility of electric shock, turn off electric power supply to furnace before any disassembly for inspection or maintenance work is performed.

HOME OWNER'S MAINTENANCE
Circulating Air Blower Assembly

1. Check to insure that blower wheel turns freely.

2. Check for build-up of foreign material on blades. Clean off any accumulation with stiff brush or by scraping with a suitable tool.

3. MOTOR LUBRICATION - If motor is provided with oil ports, lubricate motor with SAE 20 non-detergent type oil. Slowly add 5 to 10 drops to each oil port - DO NOT OVER OIL! Motors which are not provided with oil ports require no periodic lubrication.

SERVICE INSPECTION

Service inspection and repair of the furnace, and any related equipment which may be installed with the furnace, shall be done only by persons who are qualified and trained with this type of equipment. Some state and municipal code authorities require that persons servicing this type equipment be licensed. Persons who are not qualified or licensed shall not attempt to service this equipment.

It is recommended that the furnace be inspected by a qualified service company, such as your Coleman dealer, shortly before the beginning of each heating season. The furnace should be inspected prior to the beginning of the air conditioning season if air conditioning equipment is installed.

When required, adjustments should be made to insure that the operational performance of the furnace is in accordance with the specifications shown on the UNIT RATING PLATE and in these INSTALLATION INSTRUCTIONS.

Before removing components or assemblies for inspection, carefully observe how these components are installed, electrical connections made, and attachment means employed. All components and assemblies must be reinstalled in the same manner and position as they were originally.

WARNING

NEVER ATTEMPT TO MODIFY THIS FURNACE OR REPAIR DAMAGED OR DEFECTIVE COMPONENTS. SUCH ACTION COULD CAUSE UNSAFE OPERATION, FIRE, EXPLOSION, OR ASPHYXIATION

Damaged or defective components must be replaced. Only original equipment components, or substitute components authorized by Evcon Industries, Inc. may be used to repair this furnace.

Burner Assembly - The burner should be inspected for any accumulation of corrosion (rust) in the burner port. The burner should be blown out with the pressure side of a vacuum cleaner. Take care not to damage or displace the ignition assembly.

NOTE – Ignitor assembly is very fragile and should be handled with care. A cracked or broken ignitor will not function properly and must be replaced.

CAUTION – Ignitor operates on line voltage. **Turn off electrical power before servicing to prevent shock hazard which could damage equipment or cause personal injury.**

Heat Exchanger – The heat exchanger should be inspected for possible material corrosion, scaling, damage, or deposits of soot. (1) Remove the gas valve and mounting bracket and the burner assembly. (2) Place a suitable light through the combustion air tube and into the heat exchanger drum.

Using a mirror, observe the heat exchanger inner surfaces and look for any deposit of black soot or any apparent scaling on the metal surfaces. The normal color of the heat exchanger may have a red color in some areas. (3) If soot deposits are found in the heat exchanger drum and sooting is also found in the vent box, the heat exchanger must be replaced and the cause of the soot determined and necessary corrective action taken. (4) If any light sooting is found, it and any debris from the bottom of the heat exchanger must be removed using a suitable vacuum cleaner. If sooting is noticed, its cause must be determined and necessary corrective action taken.

REPLACEMENT PARTS

Should it be necessary to replace any component parts, these may be obtained through a Coleman dealer, who is experienced and can be of assistance, or information on the nearest Distributor may be obtained directly from Evcon Industries, Inc., 3110 North Mead. P.O. Box 19014, Wichita, KS 67204-9014. Phone 316-832-6448.

CAUTION
Only genuine Coleman replacement parts should be used. Substitute parts should not be used as they may not be the same in operational and safety characteristics.

OVERALL OPERATION
Following major service or replacement of functional parts it is recommended that the furnace be operated in the various modes to insure that performance is normal and control components are functioning properly.

PARTS LIST
The parts listed are only those that are functional parts and those that might require replacement.
The parts listed are by part name (description) only and CODE reference NUMBER. The CODE number is used to determine the location of the part in the generalized, pictorial illustration provided. See page 16.
WHEN ORDERING PARTS – ORDER BY NAME AND DESCRIPTION. DO NOT ORDER BY CODE NUMBER. ALSO WHEN ORDERING PARTS please provide: (a) Complete furnace Model Number and Serial Number. These may be found on the rating plate located behind the upper front panel (door) of the furnace. (b) Type of gas being used. (c) Is furnace being used in conjunction with air conditioning or heat pump?

PARTS LIST
1. Thermostat – accessory
2. Casing Assembly
3. Shield
4.

5. Heat Exchanger
6. Power Vent Scroll
7. Vestibule
8. Limit Switch
9. Fan Switch
10. Switch Bracket
11. Terminal Bushing
12. Vent Motor Assembly
13. Vent Motor
14. Vent Blower Wheel
15. Vent Box Assembly
16. Limit Switch
17. Centrifugal Switch
18. Front Panel
19. Coleman Logo
20. Blower Panel
21. Burner Assembly
22. Valve Bracket Assembly
23. Gas Valve
24. Orifice
25. Ignitor Bracket
26. Ignitor
27.
28. Ignition Module
29. Control Box
30. Control Box Cover
31. Heating Relay
32. Blower Relay
33. Transformer
34. Safety Switch
35. Capacitor
36. Customer Envelope
37. Blower Assembly
38. Blower Scroll
39. Blower Motor
40. Blower Wheel
41. Motor Clamp
42. Rubber Mounts
43. Filter Box Assembly
44. Filter Rod
45. Roll-out Switch
46. Terminal Board

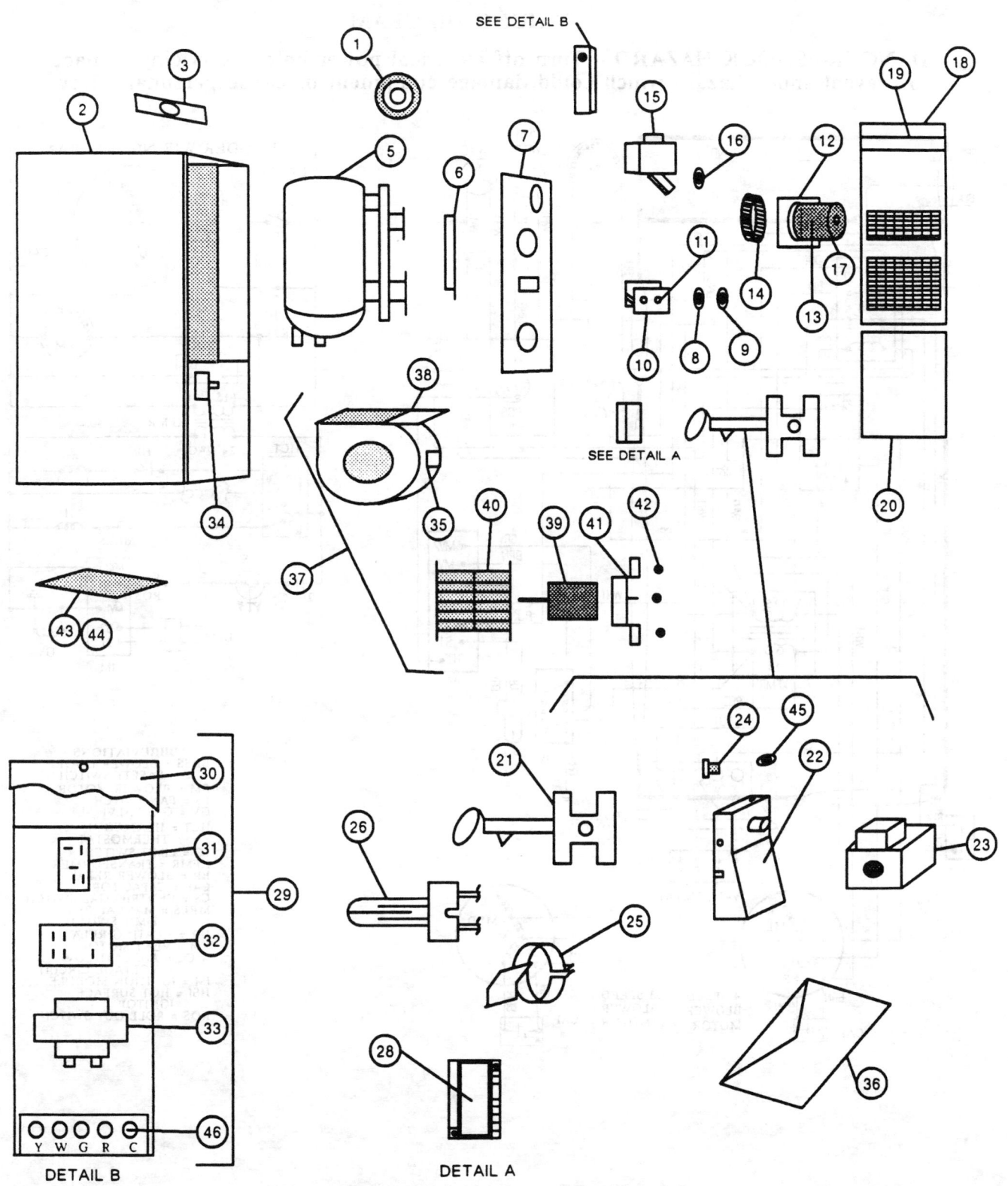

SEE DETAIL B

SEE DETAIL A

DETAIL B

Y W G R C

DETAIL A

Evcon Industries, Inc.
3110 North Mead
P.O. Box 19014
Wichita, KS 67204-9014

1972L214 (3-91) P.I.

LITHO IN U.S.A.

– 15 –

WIRING DIAGRAM

DANGER– SHOCK HAZARD – Turn off electrical power before servicing furnace to prevent shock hazard which could damage equipment or cause personal injury.

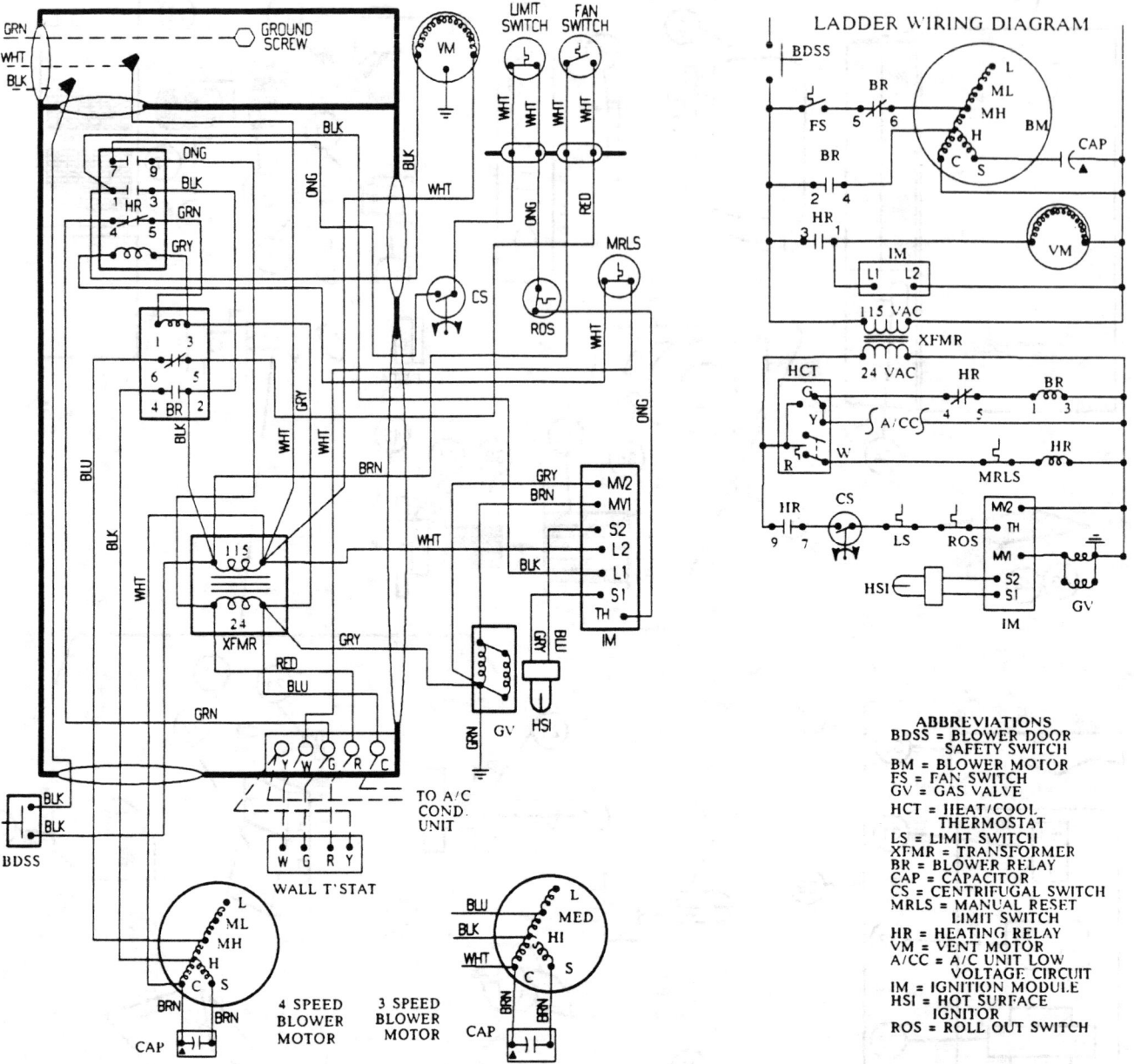

ABBREVIATIONS

BDSS = BLOWER DOOR
        SAFETY SWITCH
BM = BLOWER MOTOR
FS = FAN SWITCH
GV = GAS VALVE
HCT = HEAT/COOL
        THERMOSTAT
LS = LIMIT SWITCH
XFMR = TRANSFORMER
BR = BLOWER RELAY
CAP = CAPACITOR
CS = CENTRIFUGAL SWITCH
MRLS = MANUAL RESET
        LIMIT SWITCH
HR = HEATING RELAY
VM = VENT MOTOR
A/CC = A/C UNIT LOW
        VOLTAGE CIRCUIT
IM = IGNITION MODULE
HSI = HOT SURFACE
        IGNITOR
ROS = ROLL OUT SWITCH

Factory internal wiring is shown solid.
If any of the original wire supplied with this unit must be replaced, it must be replaced with type 105°C thermoplastic wire or its equivalent.
IMPORTANT NOTICE – Prior to and during igni-

tion, the ignition module supplies 120 volts to the hot surface ignitor. After ignition has been completed, the hot surface ignitor functions as a flame sensor and is only supplied with 24 volts.

DELUXE ENERGY SAVER 80 COUNTERFLOW GAS FURNACES

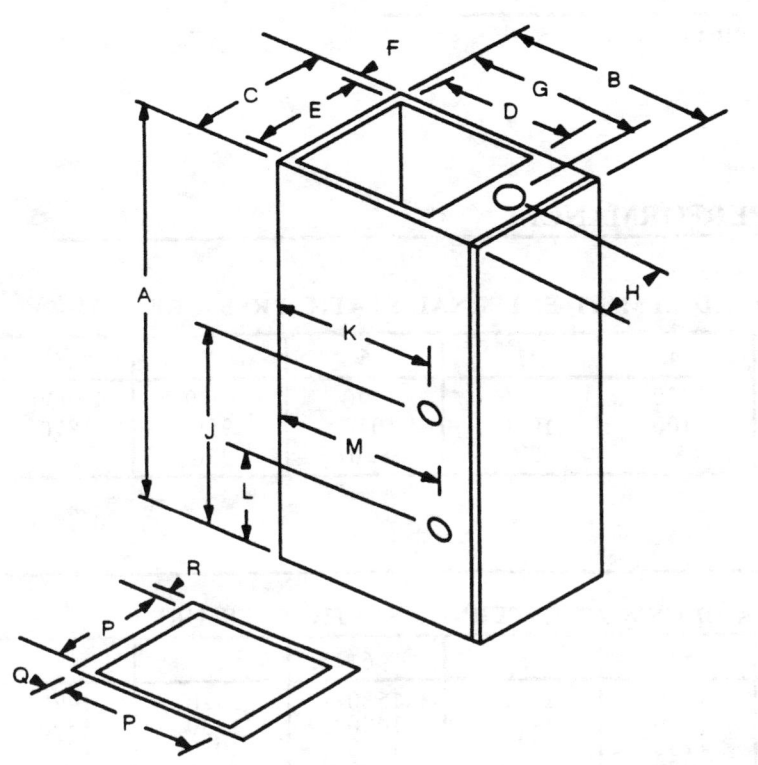

DIMENSIONS IN INCHES

FURNACE MODELS	
2844E866 2860E866	2860E966 2875E966 2892E966
A	49–5/8	49–5/8
B	28	28
C	16–7/8	20
D	20–3/8	20–3/8
E	15–3/8	18–1/2
F	3/4	3/4
G	24–5/16	24–5/16
H	8–7/16	10
J	14–1/4	14–1/4
K	23–1/16	23–1/16
L	9	9
M	24–1/2	24–1/2
N	18–1/2	18–1/2
P	14–7/8	18
Q	3–1/2	3–1/2
R	1	1

SPECIFICATIONS

MODEL	NAT.	2844E866	2860E866	2860E966	2875E966	2892E966
	L.P. KIT #	2844-5991	2860-5991	2860-5991	2875-5991	2892-5991
UNIT RATING BTU/Hr.	Input	55,000	70,000	70,000	90,000	110,000
	Output	45,000	57,000	57,000	74,000	90,000
	High Altitude	FOR ELEVATIONS ABOVE 2000 FEET, REDUCE INPUT 4% FOR EACH 1000 FEET OF ELEVATION ABOVE SEA LEVEL.				
AFUE % *		80.0	80.0	80.0	80.0	80.0
CALIFORNIA SEAS. EFFY.		76.0	76.0	76.0	76.0	76.0
AIR TEMPERATURE RISE RANGE F		30–60	40–70	35–65	35–65	45–75
DESIGNED MAXIMUM OUTLET AIR TEMP. F		160	170	165	165	175
MAX. EXT. STATIC PRESSURE IN. W.C.		.5				
FURNACE FLUE PIPE		Type B-1				
FLUE PIPE SIZE		4"			5"	
GAS CONNECTION		1/2 in. FPT				
ELECTRICAL SERVICE		115 VAC 60 HZ 1 PH				
MINIMUM DISTANCE TO COMBUSTIBLE MATERIALS		FRONT - 6" BACK - 1" SIDES - 1" FLOOR - NON-COMBUSTIBLE VENT - 6" (SINGLE WALL) 1" (B-1)				
FILTER(Furnished)		(2) 12 X 20		(2) 14 X 20		
SHIPPING WEIGHTS		130		140	145	

* – Annual Fuel Utilization Efficiency is based on DOE isolated combustion system test procedure.
\# – For LP use, optional kit is required.

COIL CABINET FURNACE MATCH-UP

FURNACE MODEL	CABINET
2844E866 2860E866	3617E835
2860E966 2875E966 2892E966	3620E835 3625E835*

* – Requires field fabricated block-off plates.

BLOWER PERFORMANCE

Furnace # 2844E866, 2860E866	STANDARD CFM AT EXTERNAL STATIC PRESSURE						
Blower Speed	.1	.2	.3	.4	.5	.6	.7	.8
High	1590	1530	1440	1350	1250	1120	1080	1030
Medium	1160	1120	1110	1100	1050	1010	950	870
Low	930	910	890	880	850	830	770	680

Motor HP: 1/3 FLA: 9.2 Impellor Size: 10-5/8 x 8

Furnace # 2860E966	STANDARD CFM AT EXTERNAL STATIC PRESSURE						
Blower Speed	.1	.2	.3	.4	.5	.6	.7	.8
High	1970	1900	1830	1760	1670	1580	1520	1490
Medium High	1620	1560	1520	1470	1430	1400	1330	1300
Medium Low	1300	1280	1260	1230	1210	1140	1080	1030
Low	1110	1100	1080	1070	1050	1030	990	950

Motor HP: 1/2, FLA: 10.0, Impellor Size: 11-7/8 x 9-1/2

Furnace # 2875E966, 2892E966	STANDARD CFM AT EXTERNAL STATIC PRESSURE						
Blower Speed	.1	.2	.3	.4	.5	.6	.7	.8
High	1940	1880	1810	1750	1650	1560	1500	1480
Medium High	1590	1530	1510	1450	1410	1390	1310	1290
Medium Low	1280	1260	1250	1210	1200	1120	1060	1020
Low	1090	1080	1060	1050	1030	1000	950	920

Motor HP: 1/2, FLA: 10.0, Impellor Size: 11-7/8 x 9-1/2

All motors are 1100 RPM and have 1/2" shafts.

Evcon Industries, Inc.

3110 North Mead
P.O. Box 19014
Wichita. KS 67204-9014

Manufacturers of Coleman Heating and Air Conditioning Products

DELUXE ENERGY SAVER 80 HORIZONTAL GAS FURNACES

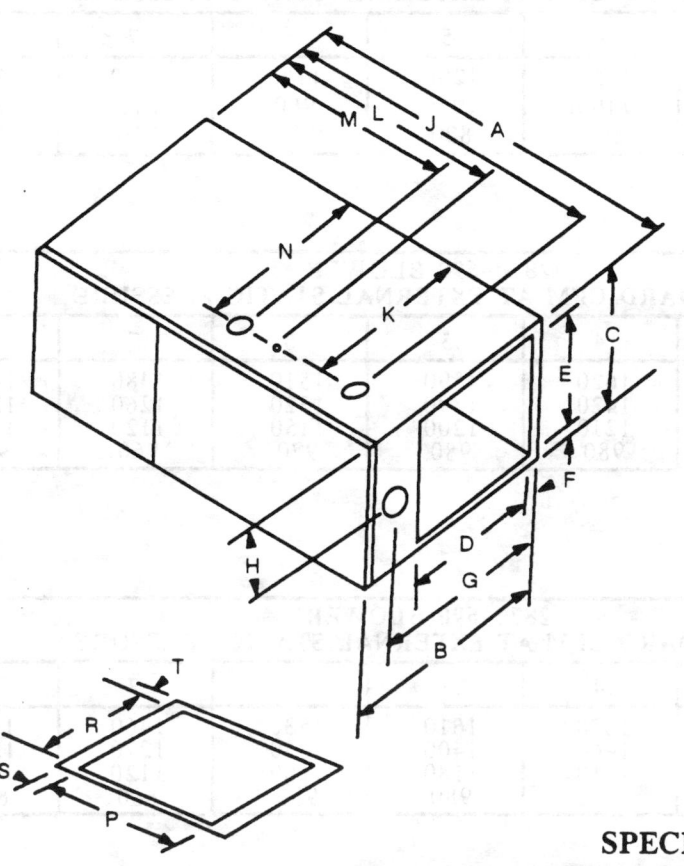

DIMENSIONS IN INCHES

FURNACE MODELS		
	2840E466 2855E466	2855E566 2870E566 2890E566
A	46	46
B	28	28
C	16-7/8	20
D	20-3/8	20-3/8
E	15-3/8	18-1/2
F	3/4	3/4
G	24-5/16	24-5/16
H	8-7/16	10
J	31-7/8	31-7/8
K	23-1/16	23-1/16
L	28-5/8	28-5/8
M	25-3/16	25-3/16
N	24-1/2	24-1/2
P	23-3/4	23-3/4
R	14-7/8	14-7/8
S	1-3/4	1-3/4
T	1	1

SPECIFICATIONS

MODEL	NAT.	2840E466	2855E466	2855E566	2870E566	2890E566
	L.P. KIT #	2840–5991	2855–5991	2855–5991	2870–5991	2890–5991
UNIT RATING BTU/Hr.	Input	50,000	65,000	65,000	85,000	105,000
	Output	41,000	54,000	54,000	70,000	86,000
	High Altitude	FOR ELEVATIONS ABOVE 2000 FEET, REDUCE INPUT 4% FOR EACH 1000 FEET OF ELEVATION ABOVE SEA LEVEL.				
AFUE % *		80.4	82.1	81.6	80.1	80.4
CALIFORNIA SEAS. EFFY.		74.0	77.7	75.4	75.4	76.8
AIR TEMPERATURE RISE RANGE F		30–60	40–70	35–65	45–75	50–80
DESIGNED MAXIMUM OUTLET AIR TEMP. F		160	170	165	175	180
MAX. EXT. STATIC PRESSURE IN. W.C.		.5				
FURNACE FLUE PIPE		Type B-1				
FLUE PIPE SIZE		4"			5"	
GAS CONNECTION		1/2 in. FPT				
ELECTRICAL SERVICE		115 VAC 60 HZ 1 PH				
MINIMUM DISTANCE TO COMBUSTIBLE MATERIALS		FRONT - 6" BACK - 1" PLENUM - 1" FLOOR - COMBUSTIBLE TOP - 2" VENT - 6" (SINGLE WALL) 1" (B-1)				
FILTER		EXTERNAL				
SHIPPING WEIGHTS		130		140		145

* – Annual Fuel Utilization Efficiency is based on DOE isolated combustion system test procedure.
# – For LP use, optional kit is required.

BLOWER PERFORMANCE

Furnace # 2840E466, 2855E466				2840-690 BLOWER STANDARD CFM AT EXTERNAL STATIC PRESSURE					
Blower Speed	.1	.2	.3	.4	.5	.6	.7	.8
High	1460	1400	1330	1250	1200	1120	1050	960
Medium	1220	1180	1140	1100	1030	960	870	820
Low	960	940	920	860	820	780	710	700

Motor HP:1/3 FLA: 9.2 Impellor Size: 10-5/8 x 8

Furnace # 2870E566, 2890E566				2870-690 BLOWER STANDARD CFM AT EXTERNAL STATIC PRESSURE					
Blower Speed	.1	.2	.3	.4	.5	.6	.7	.8
High	1780	1740	1680	1620	1590	1510	1480	1400
Medium High	1550	1530	1470	1420	1390	1320	1260	1240
Medium Low	1230	1220	1210	1210	1200	1150	1120	1100
Low	1000	1000	990	980	980	970	960	930

Motor HP: 3/4 , FLA: 10.0 Impellor Size: 11-7/8 x 9-1/2

Furnace # 2855E566				2870-690 BLOWER STANDARD CFM AT EXTERNAL STATIC PRESSURE					
Blower Speed	.1	.2	.3	.4	.5	.6	.7	.8
High	1800	1750	1710	1670	1610	1530	1460	1380
Medium High	1560	1550	1500	1460	1400	1350	1270	1260
Medium Low	1260	1230	1220	1210	1180	1160	1120	1090
Low	1050	1040	1030	1010	980	930	880	830

Motor HP: 3/4 , FLA: 10.0 Impellor Size: 11-7/8 x 9-1/2

All motors are 1100 RPM and have 1/2" shafts.

Evcon Industries, Inc.

3110 North Mead
P.O. Box 19014
Wichita. KS 67204-9014

Manufacturers of Coleman Heating and Air Conditioning Products

DELUXE ENERGY SAVER 80 UPFLOW GAS FURNACES

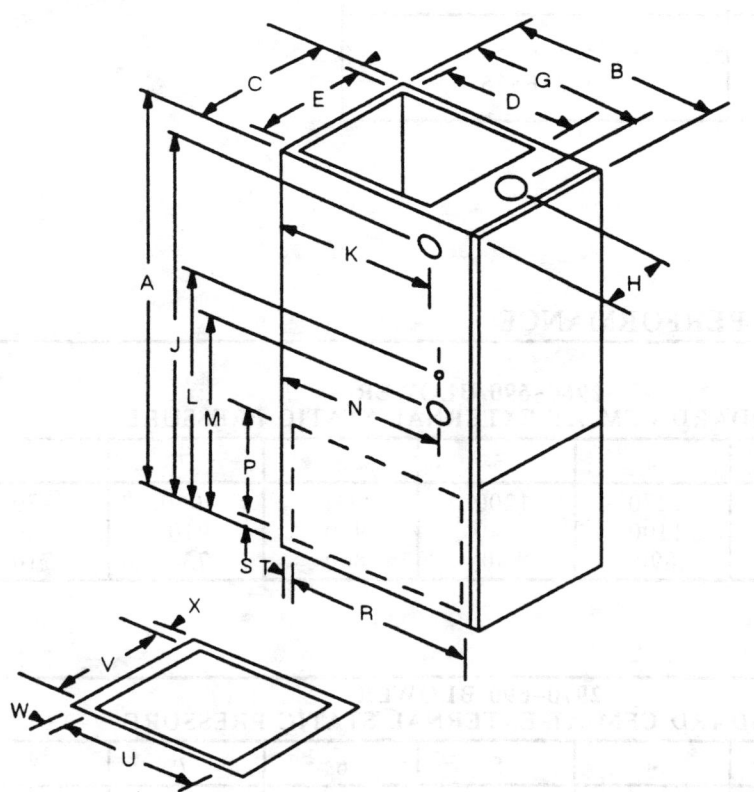

DIMENSIONS IN INCHES

	FURNACE MODELS	
	2845E666 2865E666	2865E766 2880E766 2895E766
A	46	46
B	28	28
C	16–7/8	20
D	20–3/8	20–3/8
E	15–3/8	18–1/2
F	3/4	3/4
G	24–5/16	24–5/16
H	8–7/16	10
J	43–1/8	43–1/8
K	23–1/16	23–1/16
L	28–5/8	28–5/8
M	25–3/16	25–3/16
N	24–1/2	24–1/2
P	14–1/2	14–1/2
R	23	23
S	1	1
T	3–1/2	3–1/2
U	23–3/4	23–3/4
V	14–7/8	18
W	1–3/4	1–3/4
X	1	1

SPECIFICATIONS

MODEL	NAT.	2845E666	2865E666	2865E766	2880E766	2895E766
	L.P. KIT #	2845A5991	2865A5991	2865A5991	2880A5991	2895A5991
UNIT RATING BTU/Hr.	Input	55,000	75,000	75,000	95,000	115,000
	Output	44,000	62,000	62,000	78,000	95,000
	High Altitude	FOR ELEVATIONS ABOVE 2000 FEET, REDUCE INPUT 4% FOR EACH 1000 FEET OF ELEVATION ABOVE SEA LEVEL.				
AFUE % *		80.0	81.5	81.6	80.0	80.5
CALIFORNIA SEAS. EFFY.		76.0	76.1	76.0	76.1	76.6
AIR TEMPERATURE RISE RANGE F		30–60	40–70	40–70	45–75	50–80
DESIGNED MAXIMUM OUTLET AIR TEMP. F		155	165	165	170	175
MAX. EXT. STATIC PRESSURE IN. W.C.		.5				
FURNACE FLUE PIPE		Type B–1				
FLUE PIPE SIZE		4"			5"	
GAS CONNECTION		1/2 in. FPT				
ELECTRICAL SERVICE		115 VAC 60 HZ 1 PH				
MINIMUM DISTANCE TO COMBUSTIBLE MATERIALS		FRONT – 6" BACK – 0 PLENUM TOP – 1" FLOOR – COMBUSTIBLE SIDES – 0 VENT – 6" (SINGLE WALL) 1" (B–1)				
FILTER(Furnished)		16 X 25 X 1/2			20 X 25 X 1/2	
SHIPPING WEIGHTS		130		140	145	

* – Annual Fuel Utilization Efficiency is based on DOE isolated combustion system test procedure.
# – For LP use, optional kit is required.

COIL CABINET FURNACE MATCH-UP

FURNACE MODEL	CABINET
2845E666 2865E666	3617E835
2865E766 2880E766 2895E766	3620E835 3625E835

BLOWER PERFORMANCE

Furnace # 2845E666, 2865E666				2960-690 BLOWER STANDARD CFM AT EXTERNAL STATIC PRESSURE				
Blower Speed	.1	.2	.3	.4	.5	.6	.7	.8
High	1480	1420	1340	1270	1200	1130	1060	970
Medium	1220	1180	1140	1100	1040	980	910	830
Low	960	950	930	890	840	800	730	710

Motor HP: 1/3 FLA: 9.2 Impellor Size: 10-5/8 x 8

Furnace # 2865E766				2970-690 BLOWER STANDARD CFM AT EXTERNAL STATIC PRESSURE				
Blower Speed	.1	.2	.3	.4	.5	.6	.7	.8
High	1810	1760	1720	1680	1620	1550	1480	1410
Medium High	1580	1560	1510	1470	1420	1370	1330	1270
Medium Low	1250	1240	1230	1220	1210	1170	1140	1120
Low	1030	1030	1020	1020	1020	1010	1000	970

Motor HP: 3/4 , FLA: 10.0 Impellor Size: 11-7/8 x 9-1/2

Furnace # 2880E766, 2895E766				2970-690 BLOWER STANDARD CFM AT EXTERNAL STATIC PRESSURE				
Blower Speed	.1	.2	.3	.4	.5	.6	.7	.8
High	1790	1750	1690	1640	1610	1530	1480	1410
Medium High	1580	1540	1490	1450	1400	1350	1300	1270
Medium Low	1240	1230	1230	1220	1210	1160	1140	1090
Low	1030	1030	1030	1020	1020	1010	990	960

Motor HP: 3/4 , FLA: 10.0 Impellor Size: 11-7/8 x 9-1/2

All motors are 1100 RPM and have 1/2" shafts.

Evcon Industries, Inc.
3110 North Mead
P.O. Box 19014
Wichita, KS 67204-9014

(8-90)

ORIFICE SIZES REQUIRED FOR HIGH ALTITUDES

NATURAL GAS

Model No.	Orifice at Sea Level	Orifice Size Required At Other Elevations								
		2000	3000	4000	5000	6000	7000	8000	9000	10000
2940	31(.120)	32(.116)	32(.116)	32(.116)	33(.113)	34(.111)	35(.110)	36(.106)	37(.104)	38(.101)
2960	27(.144)	28(.140)	28(.140)	29(.136)	29(.136)	29(.136)	30(.128)	30(.128)	30(.128)	31(.120)
2970	20(.161)	22(.157)	22(.157)	23(.154)	24(.152)	25(.149)	26(.147)	27(.144)	28(.140)	29(.136)
2985	17(.173)	18(.169)	19(.166)	19(.166)	20(.161)	21(.159)	22(.157)	23(.154)	24(.152)	26(.147)
2845	29(.136)	29(.136)	30(.128)	30(.128)	30(.128)	30(.128)	31(.120)	31(.120)	31(.120)	32(.116)
2865	20(.161)	22(.157)	22(.157)	23(.154)	24(.152)	25(.149)	26(.147)	27(.144)	28(.140)	29(.136)
2880	15(.180)	16(.177)	17(.173)	17(.173)	18(.169)	19(.166)	20(.161)	20(.161)	22(.157)	24(.152)
2895	8(.199)	10(.193)	11(.191)	12(.189)	13(.185)	13(.185)	15(.180)	16(.177)	17(.173)	18(.169)
2840	29(.136)	29(.136)	30(.128)	30(.128)	30(.128)	30(.128)	31(.120)	31(.120)	31(.120)	32(.116)
2855	25(.149)	26(.147)	27(.144)	27(.144)	28(.140)	28(.140)	29(.136)	29(.136)	30(.128)	30(.128)
2870	17(.173)	18(.169)	19(.166)	19(.166)	20(.161)	21(.159)	22(.157)	23(.154)	24(.152)	26(.147)
2890	12(.189)	13(.185)	14(.182)	15(.180)	16(.177)	17(.173)	17(.173)	18(.169)	19(.166)	20(.161)
2844	29(.136)	29(.136)	30(.128)	30(.128)	30(.128)	30(.128)	31(.120)	31(.120)	31(.120)	32(.116)
2860	22(.157)	23(.154)	24(.152)	25(.149)	26(.147)	27(.144)	27(.144)	28(.140)	29(.136)	29(.136)
2875	16(.177)	17(.173)	18(.169)	18(.169)	19(.166)	19(.166)	20(.161)	22(.157)	23(.154)	25(.149)
2892	9(.196)	11(.191)	12(.189)	12(.189)	13(.185)	14(.182)	16(.177)	17(.173)	18(.169)	19(.166)

PROPANE GAS

Model No.	Orifice at Sea Level	Orifice Size Required At Other Elevations								
		2000	3000	4000	5000	6000	7000	8000	9000	10000
2940	48(.076)	49(.073)	49(.073)	49(.073)	50(.070)	50(.070)	50(.070)	51(.067)	51(.067)	52(.063)
2960	42(.093)	42(.093)	43(.089)	43(.089)	43(.089)	44(.086)	44(.086)	45(.082)	46(.081)	47(.078)
2970	38(.101)	39(.099)	40(.098)	41(.096)	41(.096)	42(.093)	42(.093)	43(.089)	43(.089)	44(.086)
2985	35(.110)	36(.106)	36(.106)	37(.104)	37(.104)	38(.101)	39(.099)	40(.098)	41(.096)	42(.093)
2845	45(.082)	46(.081)	47(.078)	47(.078)	47(.078)	48(.076)	48(.076)	49(.073)	49(.073)	50(.070)
2865	40(.098)	41(.096)	42(.093)	42(.093)	42(.093)	43(.089)	43(.089)	44(.086)	44(.086)	45(.082)
2880	35(.110)	36(.106)	36(.106)	37(.104)	37(.104)	38(.101)	39(.099)	40(.098)	41(.096)	42(.093)
2895	31(.120)	32(.116)	32(.116)	32(.116)	33(.113)	34(.111)	35(.110)	36(.106)	37(.104)	38(.101)
2840	47(.078)	48(.076)	48(.076)	49(.073)	49(.073)	49(.073)	50(.070)	50(.070)	51(.067)	51(.067)
2855	43(.089)	44(.086)	44(.086)	44(.086)	45(.082)	45(.082)	46(.081)	47(.078)	47(.078)	48(.076)
2870	37(.104)	38(.101)	39(.099)	39(.099)	40(.098)	41(.096)	42(.093)	42(.093)	43(.089)	43(.089)
2890	33(.113)	35(.110)	35(.110)	36(.106)	36(.106)	37(.104)	38(.101)	38(.101)	40(.098)	41(.096)
2844	45(.082)	46(.081)	47(.078)	47(.078)	47(.078)	48(.076)	48(.076)	49(.073)	49(.073)	50(.070)
2860	42(.093)	42(.093)	43(.089)	43(.089)	43(.089)	44(.086)	44(.086)	45(.082)	46(.081)	47(.078)
2875	36(.106)	37(.104)	38(.101)	38(.101)	39(.099)	40(.098)	41(.096)	41(.096)	42(.093)	43(.089)
2892	32(.116)	33(.113)	34(.111)	35(.110)	35(.110)	36(.106)	36(.106)	37(.104)	38(.101)	40(.098)

The above is adapted from The National Fuel Gas Code, ANSI Z223.1-1988.

INSTALLATION INSTRUCTIONS

HORIZONTAL VENTING KIT

WARNING

This conversion kit shall be installed by a qualified service agency in accordance with the manufacturer's instructions and all applicable codes and requirements of the authority having jurisdiction. If the information in these instructions is not followed exactly, a fire or explosion may result causing property damage, personal injury or loss of life. The qualified service agency performing this work assumes responsibility for the proper conversion of this furnace with this kit.

APPLICATION

This kit is used to convert 2800 Series "E" model upflow and horizontal furnaces for horizontal through-the-wall venting. **When these models are horizontally vented, high temperature PLEX-VENT or Ultravent plastic pipe must be used.** The following table shows all 2800 Series upflow and horizontal furnace models along with the proper conversion kit number and the proper vent pipe diameter.

Furnace Model Number	Conversion Kit Number	Vent Pipe Diameter
2845E666	2845-5551	3"
2865E666	2845-5551	3"
2865E766	2845-5551	3"
2880E766	2845-5551	3"
2840E466	2845-5551	3"
2855E466	2845-5551	3"
2855E566	2845-5551	3"
2870E566	2845-5551	3"
2890E566	2895-5551	4"
2895E766	2895-5551	4"

CONTENTS

A. Pressure switch with bracket
B. Wire nut
C. Vent box
D. Orange wire
E. Louvered top panel (4" vent kit only)
F. Silicone rubber pressure hose
G. Wind shield
H. Two plastic straps
I. Venting label
J. Wiring diagram
K. Conversion instructions
L. Roll out switch

DANGER – SHOCK HAZARD

BE SURE ELECTRICAL POWER TO FURNACE

IS TURNED OFF BEFORE ANY DISASSEM-BLY WORK IS PERFORMED.

CONVERSION PROCEDURE

1. Remove louvered top panel from furnace.

2. Locate the two white wires going from the control box to the vent blockage switch mounted on the vent box. Cut off the terminals, strip the wires, and wire nut the ends together with the wire nut supplied in this kit. See Figure 1.

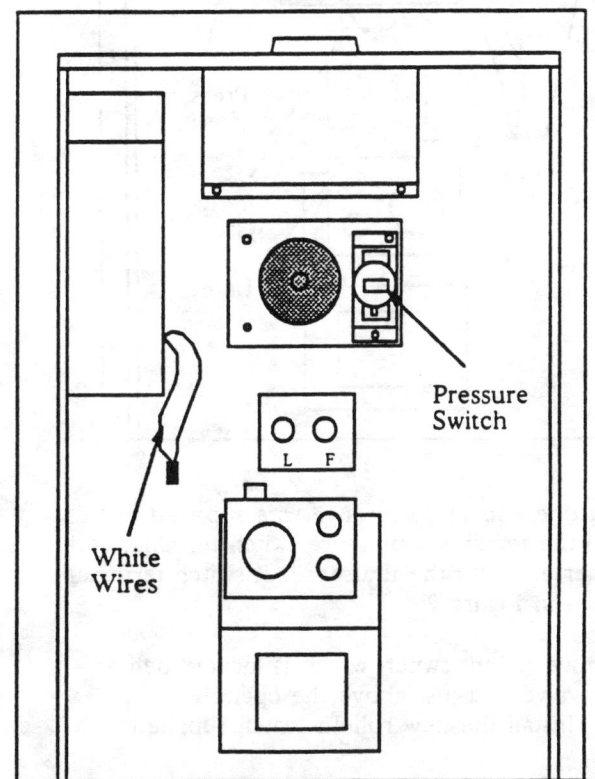

Figure 1

3. Remove old vent box from furnace. Save the two screws and two nuts for later use.

4. Install the new vent box using the two screws and two nuts saved in the previous step.

5. Remove the two right hand nuts which hold the vent motor assembly in place. Install the pressure switch from this kit on those two studs and hold in place with the same two nuts. The pressure switch should be installed with the pressure fitting pointing down as shown in Figure 1.

6. Connect the pressure hose in this kit from the pressure switch fitting to the pressure fitting in the right side of the new vent box. See Figure 2.

7. Locate the orange wire which goes from the vent blower centrifugal switch to the main furnace limit. Disconnect the orange wire from the main furnace limit switch and plug it in to one of the pressure switch terminals. See Figure 2.

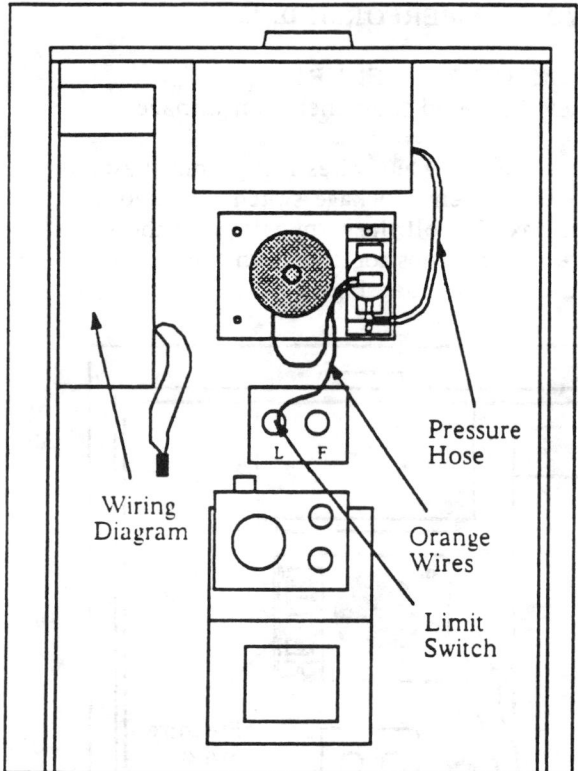

Figure 2

8. Plug one end of the orange wire supplied in this kit into the remaining pressure switch terminal and the other end into the unused limit switch terminal as shown in Figure 2.

9. Remove rollout switch which is located behind the gas valve and just above the opening to the burner. Install the new rollout switch supplied with this kit.

10. Reinstall top louvered panel. Kits for models which require 4" vent pipe contain a new louvered panel which should be installed at this time.

11. Apply the new wiring diagram supplied with this kit on top of the old wiring diagram located on the control box cover.

12. Apply the new venting label on top of the old venting label located either on the right inside casing side or on the vestibule next to the vent box.

VENTING INFORMATION

Horizontal venting of mid-efficiency furnaces increases the potential for condensation in the venting system. To avoid corrosion damage to the vent system, high temperature PLEXVENT or UltraVent plastic pipe MUST be used for venting. This plastic pipe must be installed according to the following requirements.

WARNING – FAILURE TO READ AND FOLLOW THE INSTRUCTIONS BELOW CAN CAUSE CORROSION AND PREMATURE FAILURE OF THE FURNACE AND VENT SYSTEM, LEADING TO POSSIBLE SERIOUS INJURY OR DEATH.

1. The maximum vent length is 30 feet of pipe plus three 90° elbows. Two 45° elbows are equivalent to one 90° elbow.

2. Since horizontal venting increases the potential for condensation in the vent system, **all horizontal vent installations must contain a drain tee in the vent as close as possible to the furnace** as shown in Figure 3. This will help to prevent condensate from running back into the furnace. Failure to install a drain tee will result in corrosion damage to the furnace. PLEXVENT and Ultravent plastic drain tees for this purpose are available from the manufacturers of the vent pipe. The drain hose should have a loop approximately 4" in diameter as shown in Figure 3. This loop acts as a trap to keep flue gases from escaping while still allowing the condensate to drain. The drain line should be routed to a suitable floor drain or other disposal drain. The condensate drain line must not be routed into or through any unconditioned area where the temperature may fall below freezing. This is to prevent the condensate from freezing in the line and blocking it.

3. Models 2890 and 2895 must use 4" PLEXVENT plastic vent pipe. All other models of may use 3" PLEXVENT or Ultravent pipe. Pipe and fittings in the 3" size are available in Hart & Cooley's Ultravent or Plexco's PLEXVENT. The 4" pipe and fittings are available only in PLEXVENT.

4. Horizontal venting creates positive pressure in the vent. Therefore, the entire vent system must be sealed with Dow-Corning RTV-732 or 736, GE

RTV-106 or equivalent silicone sealant rated to withstand 450° F temperatures. This sealant is commonly available from hardware stores and building materials stores. All pipe joints and the vent connection to the furnace must be sealed to prevent leakage of flue products into the home.

5. Pipe ends must be clean and smooth before applying sealant. Squeeze a generous bead of sealant entirely around the end of the pipe no more than 1/8" from the end. Push the fitting and the pipe together with a twisting motion to ensure a good seal. Full joint strength is developed after 24 hours curing time. Inspect each joint to make sure that a complete seal has been made. Apply additional sealant if required.

6. Do not drill holes through plastic pipe or fittings. Drilling may cause cracking around the hole. Do not use screws or rivets. They can cause stress around the hole and will cause cracking and leakage.

7. All vent systems must slope toward the furnace and must have at least 1/4" rise per foot of length to prevent collection of condensate at any point other than the drain tee. There must be no sags where condensate can collect. See Figure 3.

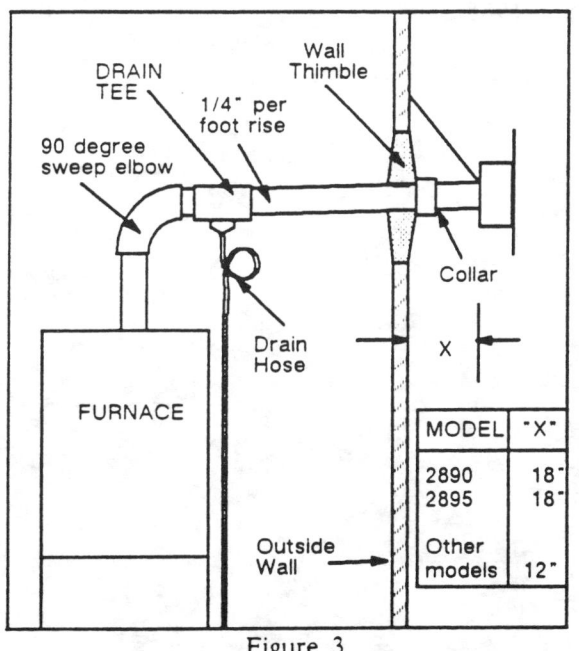

Figure 3

8. Horizontal runs must be supported every three feet and at every elbow.

9. A five inch minimum clearance to combustible material must be maintained at all points, except where a wall thimble is used to penetrate an outside wall.

10. If possible, the wall through which the vent

passes should not be exposed to the prevailing wind. If this is not possible, steps should be taken to protect the vent from strong winds, such as a fence or hedge. The wind shield provided with this unit will also help keep strong winds from affecting the furnace performance.

11. When penetrating a combustible wall, a wall thimble is required. The thimble must be sealed to the vent pipe and to the surrounding wall. For extra support of the vent, a wire support may be added from the tee to the wall.

12. The vent must be terminated with a vertical termination tee as shown in Figure 3. This type of tee has a screen installed to prevent the entry of animals, birds, etc. Manufacturers of the vent pipe also make termination tees for this purpose. The wind shield supplied in this kit must be installed on the termination tee.

13. The wind shield provided must be installed on the termination tee as shown in Figure 4. Two plastic straps are provided.

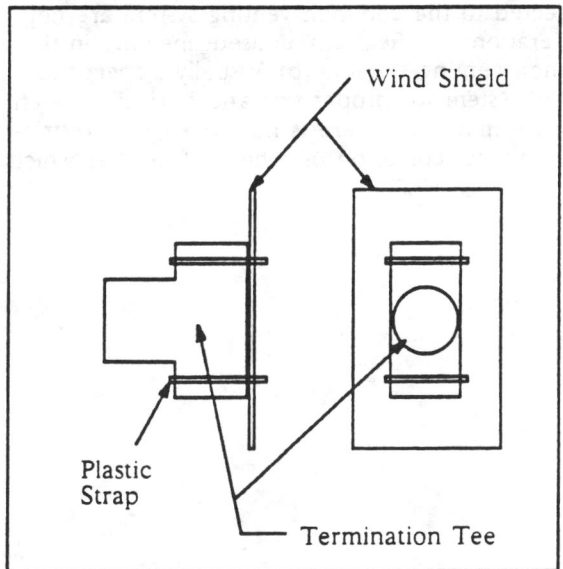

Figure 4

14. The termination of the vent system must be at least 12 inches above finished grade level, or at least 12 inches above the expected snow level when installed in areas where snow accumulates.

15. The termination of the vent must not be located in traffic areas such as walkways, unless the vent termination is at least seven feet above finished grade.

16. The vent termination must be at least four feet below or four feet horizontally from and not above any door or window. It must be at least three feet above any forced air inlet located within ten feet horizontally and at least four feet horizontally from any electric meter, gas meter, regulator, or relief

equipment.

17. Some building materials may be affected by flue products expelled in close proximity to unprotected surfaces. Sealing or shielding of the exposed surfaces with a corrosion resistant material may be required to prevent staining or deterioration.

18. The inside of the termination tee must extend beyond the outside wall surface at least the distance shown in Figure 3.

REPLACEMENT OF EXISTING FURNACE ON COMMON VENT SYSTEM

When this furnace is installed as a replacement for an old furnace which is common vented with a water heater or other gas appliance, and the new furnace is no longer connected to the common venting system, the common vent system may be too large for the appliances remaining on the vent system after the old furnace is removed. To test for an oversized vent system, the following steps shall be followed with each appliance remaining connected to the common venting system placed in operation, while the other appliances remaining connected to the common venting system are not in operation: (a) Seal any unused openings in the common venting system. (b) Visually inspect the venting system for proper size and horizontal pitch and determine that there is no blockage or restriction, leakage, corrosion or other deficiencies which could cause an unsafe condition. (c) Insofar as is practical, close all building doors and windows and all doors between the space in which the appliances remaining connected to the common venting system are located and other spaces of the building. Turn on clothes dryers and any appliance not connected to the common venting system. Turn on any exhaust fans, such as range hoods and bathroom exhausts, so they will operate at maximum speed. Do not operate a summer exhaust fan. Close fireplace dampers. (d) Follow the lighting instructions. Place the appliance being inspected in operation. Adjust thermostat so appliance will operate continuously. (e) Test for spillage at the draft hood relief opening after 5 minutes of main burner operation. Use the flame of a match or candle, or smoke from a cigarette, cigar, or pipe. (f) After it has been determined that each appliance remaining connected to the common venting system properly vents when tested as outlined above, return doors, windows, exhaust fans, fireplace dampers, and any other gas-burning appliance to their previous condition of use. (g) If improper venting is observed during any of the above tests, the common venting system must be corrected. Any changes to the venting system must be in accordance with the National Fuel Gas Code, ANSI Z223.1-1988. If any portion of the common venting system must be resized, it should be resized to approach the minimum size as determined using the appropriate tables in Appendix G in the National Fuel Gas Code, ANSI Z223.1-1988.

Evcon Industries, Inc.

3110 North Mead
P.O. Box 19014
Wichita. KS 67204-9014

Manufacturers of Coleman Heating and Air Conditioning Products

D. E. S. 80 UPFLOW GAS FURNACES

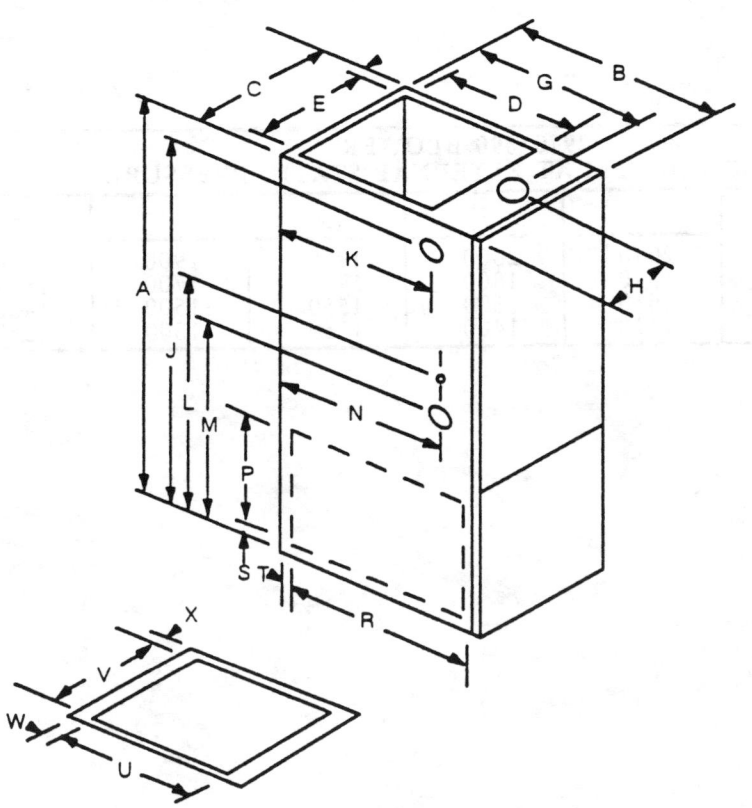

DIMENSIONS IN INCHES

	MODELS
	BGU11520 BGU13020
A	46
B	28
C	23
D	20-3/8
E	21-1/2
F	3/4
G	24-5/16
H	11-1/2
J	43-1/8
K	23-1/16
L	28-5/8
M	25-3/16
N	24-1/2
P	14-1/2
R	23
S	1
T	3-1/2
U	23-3/4
V	21
W	1-3/4
X	1

SPECIFICATIONS

MODEL	NAT.	BGU11520	BGU13020
	L.P. KIT #	2895A5991	2899-5991
UNIT RATING BTU/Hr.	Input	115,000	130,000
	Output	95,000	104,000
	High Altitude	FOR ELEVATIONS ABOVE 2000 FEET, REDUCE INPUT 4% FOR EACH 1000 FEET OF ELEVATION ABOVE SEA LEVEL.	
AFUE % *		80.0	79.0
AIR TEMPERATURE RISE RANGE F		40-70	50-80
DESIGNED MAXIMUM OUTLET AIR TEMP. F		170	180
MAX. EXT. STATIC PRESSURE IN. W.C.		.5	
FURNACE FLUE PIPE		Type B-1	
FLUE PIPE SIZE		5"	
GAS CONNECTION		1/2 in. FPT	
ELECTRICAL SERVICE		115 VAC 60 HZ 1 PH	
MINIMUM DISTANCE TO COMBUSTIBLE MATERIALS		FRONT - 6" BACK - 1" PLENUM TOP 1" FLOOR - COMBUSTIBLE SIDES - 1" VENT - 6" (SINGLE WALL) 1" (B-1)	
FILTER(Furnished)		23 x 25 x 1	
SHIPPING WEIGHTS		165	

* - Annual Fuel Utilization Efficiency is based on DOE isolated combustion system test procedure.
# - For LP use, optional kit is required.

COIL CABINET FURNACE MATCH-UP

FURNACE MODEL	CABINET
BGU11520 BGU13020	3625E835

Furnace # BGU11520, BGU13020					2970-690 BLOWER STANDARD CFM AT EXTERNAL STATIC PRESSURE			
Blower Speed	.1	.2	.3	.4	.5	.6	.7	.8
High	2040	2030	2020	2010	2000	1950	1900	1850
Medium High	1840	1830	1820	1810	1800	1750	1700	1650
Medium Low	1640	1630	1620	1610	1600	1550	1500	1450
Low	1440	1430	1420	1410	1400	1350	1300	1250

Motor HP: 3/4 , FLA: 10.0 Impellor Size: 10 X 10

All motors are 1100 RPM and have 1/2" shafts.

Evcon Industries, Inc.
3110 North Mead
P.O. Box 19014
Wichita. KS 67204-9014

FENWAL®

SERIES 05-32 HOT SURFACE IGNITION SYSTEM

WITH OPTIONAL MULTIPLE TRIAL-FOR-IGNITION PERIODS

FEATURES

- Compact Size
- Rapid Heat-Up Time
- Patented System Check Circuit
- Local or Remote Sensing Capability
- Digital Trial-for-Ignition Timings
- Patented Dual Purpose Ignitor/Sensor System

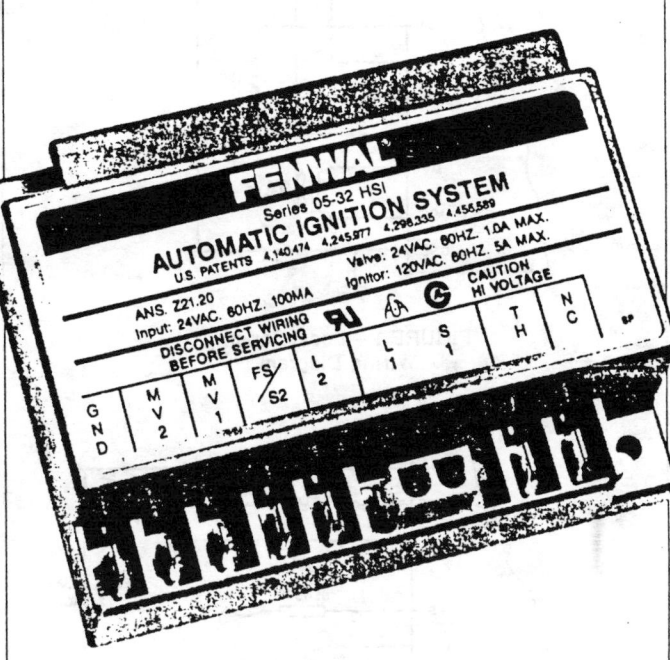

GENERAL INFORMATION

The Series 05-32 Hot Surface Ignition System is available in single or multiple trial-for-ignition models. These compact ignitors have a nominal heat-up period of 45 seconds (30 to 60 seconds). Flame sensing is achieved through the patented technique of using the hot surface element as the sensor or, if required, a separate sensing probe may be employed. All models utilize the principle of flame rectification to monitor burner flame.

PRINCIPLE OF OPERATION

Single-Try Models

On a call for heat, the hot surface element is energized (120VAC) and is allowed to reach ignition temperature. The gas valve is then powered (24VAC), establishing the flame at the burner. The 05-32 then switches the hot surface element from the ignition mode to the sensing mode to monitor the flame for the duration of the duty cycle.

If the flame is not established during the trial-for-ignition period, the system will go into lockout. After the flame has been established, should it be extinguished for any reason, the ignitor will shut off the gas valve and reactivate the hot surface element. One complete retry cycle will occur.

Three-Try Models

These models provide up to three complete trials-for-ignition before going into lockout. If the flame is established on any of the three tries, the 05-32 unit will then monitor the flame for the duration of the duty cycle.

MOUNTING

The Series 05-32 is not position-sensitive and may be mounted vertically or horizontally, using #8 hardware.

WIRING

CAUTIONS:
 I. *Note that both 24VAC and 120VAC are required for the system. Interrupt both supplies before proceeding with wiring.*
 II. *Do not apply power to the module until all wiring is connected and insure that the system is properly grounded.*
 III. *System must be checked after installation and before gas supply is turned on (see INITIAL OPERATION)*

Use wiring diagram Figure 1 for local sensing applications. Use wiring diagram Figure 2 for optional remote sensing applications.

WIRING (Continued)

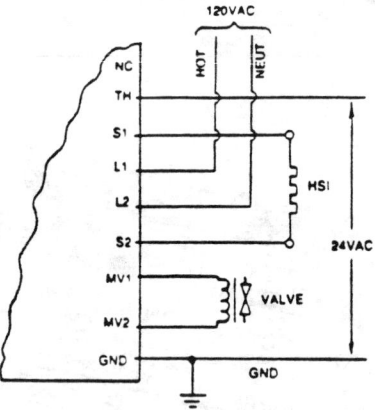

FIGURE 1 – Local Sensing
Wiring Diagram

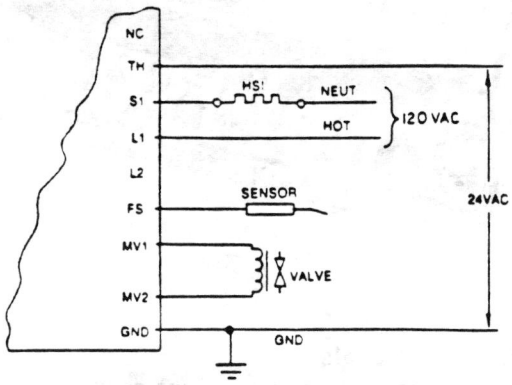

FIGURE 2 – Remote Sensing
Wiring Diagram

INITIAL OPERATION

1. Check installation. Mount and position the ignitor in accordance with the illustration in Figure 3.
2. With gas supply manually shut off, apply power to appliance and cycle thermostat above room temperature.
3. Insure that the ignitor "glows" during the heat-up and trial-for-ignition periods as noted in the Specification section. In multiple trial-for-ignition ignitors, the ignitor will automatically recycle for the specified number of times, then lockout.
4. Set the thermostat to the lowest setting.
5. Wait 15 seconds then manually open the gas supply and advance the thermostat above room temperature to recycle the system.
6. Check that ignition has been accomplished. The ignitor "glow" will diminish once the flame has been established. At this stage, the ignitor acts as the sensing element.

7. If the system ignites but fails to hold-in, check for proper grounding of the 24 volt circuit.

SILICON CARBIDE IGNITOR

Proper location of the silicon carbide ignitor is important to achieve optimum system performance for both ignition and flame sensing. See Figure 3.

NOTE: *The temperature of the steatite holder should not exceed the manufacturer's specifications.*

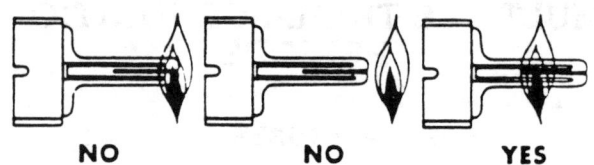

FIGURE 3

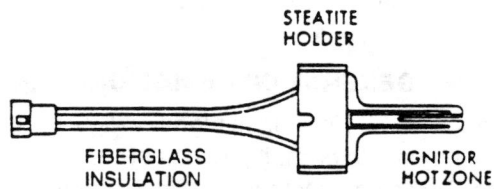

FIGURE 4 – Typical Hot Surface Element

SAFETY CHECKS

1. Manually shut off gas supply. Apply power to appliance and cycle thermostat above room temperature. After the system locks out, check that there is "0" voltage across the gas valve terminals with a suitable voltmeter. Next, set the thermostat to its lowest setting.
2. Wait 15 seconds, then manually open the gas supply and reactivate the system by advancing the thermostat to a setting above room temperature. The ignitor should glow brightly during the heat-up period, then diminish after the flame has been established. While the system is operating, manually shut off the gas supply. Following a short delay, the ignitor will re-energize and glow brightly. The ignition control will continue to cycle for the specified number of trial-for-ignition periods and then lockout. Check again that there is "0" voltage across the valve terminals after lockout.

SERIES 05-32 HOT SURFACE IGNITION SYSTEM

SERVICE CHECKS

Symptom	Cause/Cure
1) Dead	A) No 24 volt Input.
	B) Check system wiring.
	C) Check thermostat, transformer, limits, circuit breaker, etc.
2) Hot surface element does not heat - but unit cycles.	A) No 120 volt Input.
	B) Check system wiring.
	C) Check circuit breaker and 120 volt power source.
	D) Check for broken or cracked silicon carbide element.
3) Hot surface element heats up - but "0" voltage at valve during trial-for-ignition.	A) Check wiring between valve and module.
	B) Check power to valve.
4) Hot surface element heats. 24 volts to valve. Flame established but does not stay on.	A) Check system ground (24 volt supply).
	B) Hot surface element improperly located.
	C) Check all wiring connections.
	D) Burner out of adjustment.
5) Hot surface element heats. 24 volts to valve. System fails to ignite.	A) Turn gas supply off.
	B) Check gas valve.
	C) Burner out of adjustment (orifice plugged).
	D) Hot surface element incorrectly located.

REPAIRS

WARNING: *The Fenwal Series 05-32 Hot Surface Ignition Module is NOT REPAIRABLE Any modifications or repairs to this gas ignition module will invalidate Fenwal's standard warranty as well as agency certifications AND MAY CREATE HAZARDOUS CONDITIONS THAT COULD RESULT IN PROPERTY DAMAGE, PERSONAL INJURY OR EVEN DEATH FROM FIRE, EXPLOSION AND/OR TOXIC GASES Faulty units should be replaced with a new unit.*

SPECIFICATIONS

Size 3.234" × 4.234" × 1.593" (82.14 × 107.54 × 39.70mm)

Weight 5.11 ounces (145 grams) approx.

Input Voltage 24VAC 50/60 Hz (Operating range 20-28VAC)

Input Current Drain 80mA (Does not include valve power)

Valve Rating 24VAC, 1.0 A max.

Ignitor Heat-Up Time 45 seconds nominal (30-60 Seconds)

Ignitor Contact Rating 120VAC, 5.0 A max.

Ambient Temperature – 40 to 160°F (– 40 to 70°C)

System Check Circuit Checks valve control circuit and flame rectification circuit before ignition sequence is initiated.

Trial-For-Ignition Available in one or three try models.

Trial-For-Ignition Periods — (Flame establishing time)

- **-0X0 :** 1.5 seconds
- **-0X1 :** 3.0 seconds
- **-0X2 :** 5.0 seconds
- **-0X3 :** 7.0 seconds
- **-0X4 :** 10.0 seconds
- **-0X5 :** 15.0 seconds

Retry for Ignition Loss of flame will result in one retry for ignition. (Two retries on the three trial-for-ignition version.)

Vent Damper Connection (optional) Consult factory for details.

WARNING: *Operation outside specifications could result in failure of the Fenwal product and other equipment with injury to people and property.*

Specifications subject to change without notice!

HOW TO ORDER

1. Select Trial-For-Ignition mode option from Table 1.
2. Select required Ignition Time from Table 1.
3. Order by catalog number.
4. **Example: 05-326266-052** Control Module with three tries for ignition and 5.0 second ignition period.

TABLE I

Catalog Number 05-3262X6-0XX

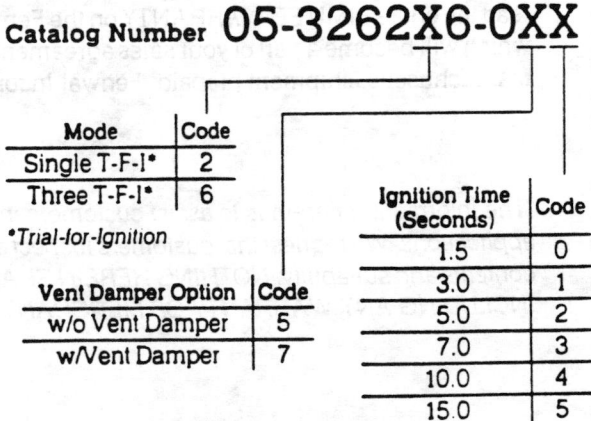

Mode	Code
Single T-F-I*	2
Three T-F-I*	6

*Trial-for-Ignition

Vent Damper Option	Code
w/o Vent Damper	5
w/Vent Damper	7

Ignition Time (Seconds)	Code
1.5	0
3.0	1
5.0	2
7.0	3
10.0	4
15.0	5

SERIES 05-32 HOT SURFACE IGNITION SYSTEM

OUTLINE DIMENSIONS

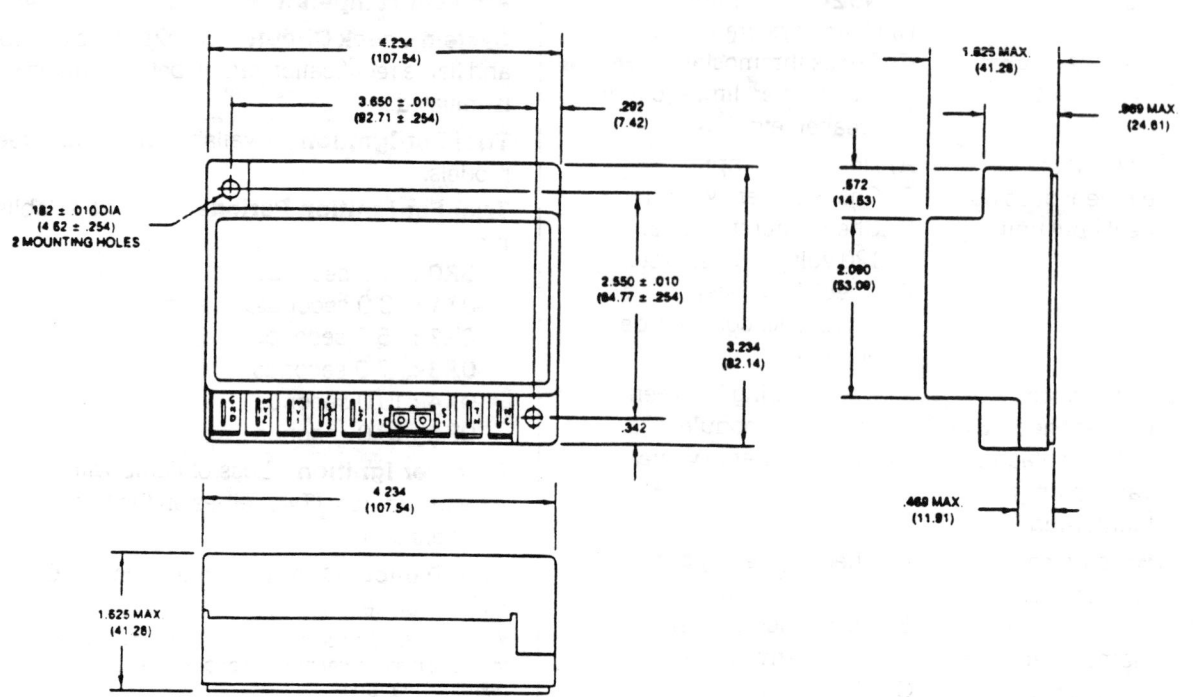

Metric dimensions are in parentheses.

LIMITED WARRANTY STATEMENT

Fenwal Incorporated represents that this product is free from defects in material and workmanship, and it will repair or replace any product or part thereof which proves to be defective in workmanship or material for a period of 18 months after delivery of the product to the buyer. For a full description of Fenwal's LIMITED WARRANTY, which, among other things, limits the duration of warranties of MERCHANTABILITY and FITNESS FOR A PARTICULAR PURPOSE and EXCLUDES liability for CONSEQUENTIAL DAMAGES, please read the entire LIMITED WARRANTY on the Fenwal Quotation, Acceptance of Order and/or Original Invoice which will become a part of your sales agreement. Defective units should be returned to the factory, Ashland, Massachusetts, shipment prepaid. Fenwal Incorporated will repair or replace and ship prepaid.

"The information herein is to assist customers in determining whether our products are suitable for their applications. We request that customers inspect and test our products before use and satisfy themselves as to contents and suitability. NOTHING HEREIN SHALL CONSTITUTE A WARRANTY, EXPRESSED OR IMPLIED, INCLUDING ANY WARRANTY OF MERCHANTABILITY OR FITNESS FOR A PARTICULAR PURPOSE."

FENWAL

400 Main Street   Ashland, Massachusetts 01721 USA   (508) 881-2000
Telex 948421   Fax (508) 881-6729

5 032 1   2 5M   7 88

HOT SURFACE IGNITION MODULE KIT

For replacing the hot surface ignition module on any Coleman 2700, 2800, 2900, 8900, or 9100 Series furnace equipped with hot surface ignition.

WARNING - IMPROPER INSTALLATION MAY DAMAGE EQUIPMENT, CAN CREATE A HAZARD, AND WILL VOID THE WARRANTY.

NOTE - The words "shall" or "must" indicate a requirement which is essential to satisfactory and safe product performance. The words "should" or "may" indicate a recommendation or advice which is not essential and not required but which may be useful or helpful.

The ignition module supplied in this kit may be used to replace the Fenwal Series 05-32 hot surface ignition module found on Coleman furnaces built after August 15, 1990 *or* the Fenwal Series 05-21 or Channel Products hot surface ignition modules used on Coleman furnaces built before that date.

CONTENTS:
A. Fenwal 05-32 Series hot surface module.
B. Wiring adapter (used only when replacing older model Fenwal Series 05-21 hot surface ignition module.)

REPLACEMENT PROCEDURE:
If the furnace is equipped with a newer Series 05-32 ignition module:
1. WARNING: SHOCK HAZARD. Turn off electrical supply to furnace.
2. Remove the old ignition module.
3. Mount the new module in the same location with the same screws.
4. Reattach the wires as they were on the old module,

as shown on the unit wiring diagram. The wiring adapter included in this kit will not be used and may be discarded.
5. Turn electrical power to furnace back on and cycle furnace with the thermostat a few times to insure that everything is working properly.

If the furnace is equipped with an older, Series 05-21 hot surface ignition module:
1. WARNING: SHOCK HAZARD. Turn off electrical supply to furnace.
2. Remove the old ignition module.
3. Mount the new module in the same location. It will be necessary to drill two .101 diameter holes. Use the old mounting screws.
4. Plug the edge connector on the wiring adapter included in this kit into the wiring harness plug on the furnace.
5. Plug the individual wires on the wiring adapter into the new ignition module as shown by the label on the adapter and as shown below.

Wire Color	Terminal
Yellow	GND
Brown	MV1
Blue	S2
White	L2
Black	L1
Gray	S1
Red	TH

6. Turn electrical power to furnace back on and cycle furnace with the thermostat a few times to insure that everything is working properly.

Evcon Industries, Inc.
3110 North Mead
P.O. Box 19014
Wichita, KS 67204-9014

1952-114 (9-90) P.I.

LITHO IN U.S.A.

NATURAL TO LP CONVERSION KITS

For converting 2700, 2800, and 2900 Series automatic ignition natural gas furnaces to LP gas operation.

WARNING

This conversion kit is to be installed by a Coleman distributor (or other qualified agency) in accordance with the manufacturer's instructions and all codes and requirements of the authority having jurisdiction. Failure to follow instructions could result in serious injury or property damage. The qualified agency performing this work assumes responsibility for this conversion.

WARNING - IMPROPER INSTALLATION MAY DAMAGE EQUIPMENT, CAN CREATE A HAZARD, AND WILL VOID THE WARRANTY.

NOTE - The words "shall" or "must" indicate a requirement which is essential to satisfactory and safe product performance. The words "should" or "may" indicate a recommendation or advice which is not essential and not required but which may be useful or helpful.

CONTENTS:

A. Burner orifice for L.P. gas (size based on 2500 BTU/Cu. ft. propane gas and furnace located below 2,000 foot elevation). Size is marked on orifice.

B. Conversion plate sticker (silver color)

C. Honeywell valve conversion (unmarked envelope), which contains: (1) pressure regulator, (2) gasket, (3) two screws, (4) label, (5) Honeywell valve conversion instructions. (Not included in 2900 Series kits.)

D. Robertshaw valve conversion #78245, which contains: (1) yellow L.P. step spring, (2) black L.P. regulator spring, (3) two red slotted caps, (4) two adjusting screws, (5) label, and (6) Robertshaw valve conversion instructions. This conversion kit is used to convert either the 7100 or the 7200 Robertshaw valve.

TABLE I - APPLICATION TABLE

Conversion Kit Number	Furnace Model	Input (BTUH)	LP Orifice Size
2745-6991	2745	60,000	.086 (#44)
2755-6991	2755	70,000	.093 (#42)
2765-6991	2765	80,000	.101 (#38)
2775-6991	2775, 2780	100,000	.116 (#32)
2790-6991	2790	120,000	.128 (#30)
2840-5991	2840 Horiz.	50,000	.078 (#47)
2855-5991	2855 Horiz.	65,000	.089 (#43)
2870-5991	2870 Horiz.	85,000	.104 (#37)
2890-5991	2890 Horiz.	105,000	.113 (#33)
2845A5991	2845 Upflow	55,000	.082 (#45)
2865A5991	2865 Upflow	75,000	.098 (#40)
2880A5991	2880 Upflow	95,000	.110 (#35)
2895A5991	2895 Upflow	115,000	.120 (#31)
2844-5991	2844 Downflow	55,000	.082 (#45)
2860-5991	2860 Downflow	70,000	.093 (#42)
2875-5991	2875 Downflow	90,000	.106 (#36)
2892-5991	2892 Downflow	110,000	.116 (#32)
2960B5991	2960 Upflow	65,000	.093 (#42)
2970B5991	2970 Upflow	80,000	.101 (#38)
2985B5991	2985 Upflow	95,000	.110 (#35)

CONVERSION PROCEDURE

1. WARNING: SHOCK HAZARD. Turn off electrical supply to furnace.
2. Shut off gas supply at valve upstream from furnace or at meter as required.
3. Disconnect gas supply piping from gas valve on furnace.
4. Disconnect electrical wires from gas valve, noting which wires are connected to which terminals.
5. Remove the four sheet metal screws that attach the gas valve bracket to the furnace. See Figure 1.
6. Remove the gas valve and bracket from the furnace.

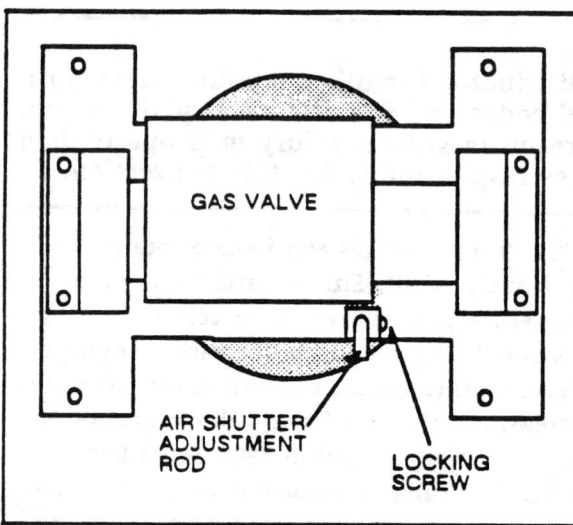

GAS VALVE

AIR SHUTTER ADJUSTMENT ROD

LOCKING SCREW

Figure 1

7. Remove and discard natural gas orifice.
8. Install the L.P. gas orifice supplied with the kit.
9. Reattach the gas valve and bracket to the furnace.
10. Convert the gas valve for L.P. gas operation by following the instructions and using the components supplied in the appropriate envelope for the gas valve on the furnace. See Figures 2, 3, and 4 for gas valve identification. Apply the label supplied in the kit to the gas valve to show that it has been converted.
11. Attach silver colored conversion plate sticker to furnace vestibule and next to the rating plate where it will be clearly visible. The name and address of the company making the conversion and the date of conversion must be written on the plate.
12. Reconnect the electrical wires to the gas valve using the wiring diagram as a guide.
13. Reconnect the gas supply piping to the gas valve and insure that all gas connections are tight.
14. Remove pressure tap plugs from gas valve and connect water gauge to the pressure tap ports. See Figures 2,3 and 4 for location of the gas valve pressure taps

and pressure regulator adjustment.

15. Turn on gas supply to furnace and check all gas connections with suitable leak detector.

CAUTION – Never use an open flame to check for leaks. Fire or explosion could occur. Since some leak solutions including soap and water may cause corrosion or stress cracking, the piping must be rinsed with water after testing unless it has been determined that the leak test solution is noncorrosive.

16. After assuring that there is no gas leakage, light the furnace using the lighting instructions shown on the furnace rating plate or on the label on the back of the door panel.
17. Verify that the gas supply pressure is between 11 and 14 inches water column with the furnace operating. Adjust gas valve pressure regulator to obtain gas pressure reading of 10 ± 0.3" W.C.

NOTE – On the Robertshaw gas valves, the step regulator must be adjusted separately to a pressure of between 4.5 and 5.0 inches water column. The step regulator adjustment screw is beneath the screw cap with the two small holes, and the main pressure adjustment screw is beneath the slotted screw cap. Make sure that these caps are replaced in their proper places after the pressures are set.

18. Turn gas supply off, remove pressure gauge, and replace pressure tap plugs.
19. Turn gas supply on, re-light the furnace, and check for gas leak at the pressure taps.

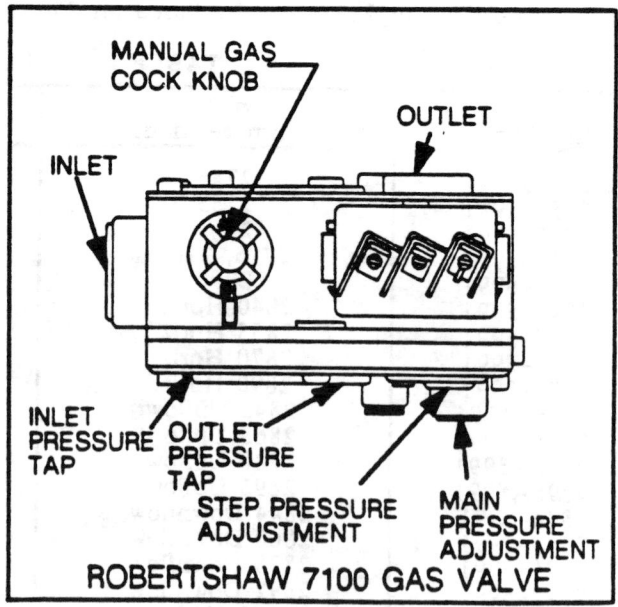

MANUAL GAS COCK KNOB

OUTLET

INLET

INLET PRESSURE TAP

OUTLET PRESSURE TAP

STEP PRESSURE ADJUSTMENT

MAIN PRESSURE ADJUSTMENT

ROBERTSHAW 7100 GAS VALVE

Figure 2

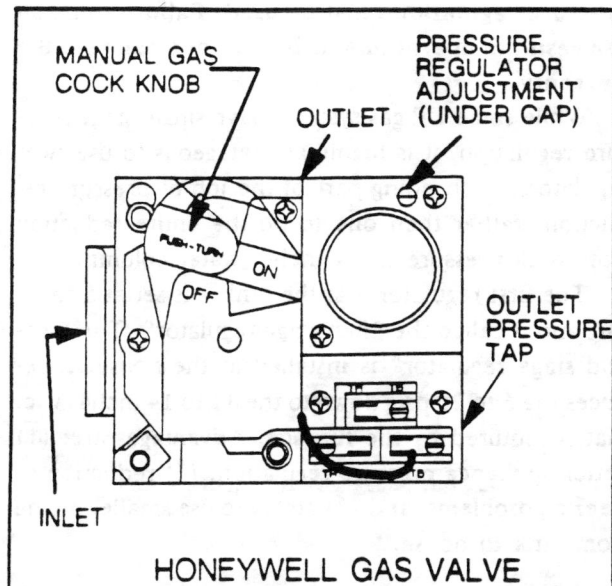

HONEYWELL GAS VALVE

Figure 3

20. After re-lighting the furnace, allow the furnace to operate for about 15 minutes and the adjust burner primary air as follows: (a) Loosen air adjustment rod locking screw. See Figure 1, (b) Close primary air shutter by pulling on adjustment rod until yellow tips appear in the flame at the end of the burner. See Figure 5. (c) Slowly open the primary air shutter by push-ing adjustment rod in until yellow flame tips at the end of the burner disappear, and then push adjustment rod in another 1/8 inch, (d) Secure primary air adjustment rod by tightening screw against it.

21. Cycle furnace with thermostat a few times to insure that everything is working properly.

22. Check gas input rate as described below.

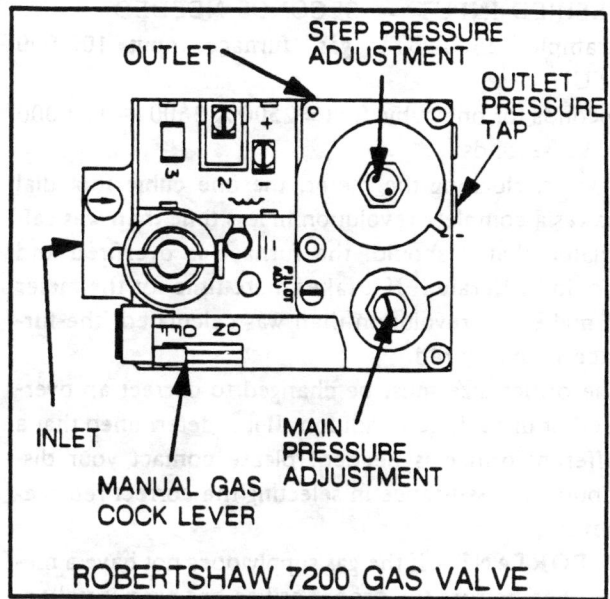

ROBERTSHAW 7200 GAS VALVE

Figure 4

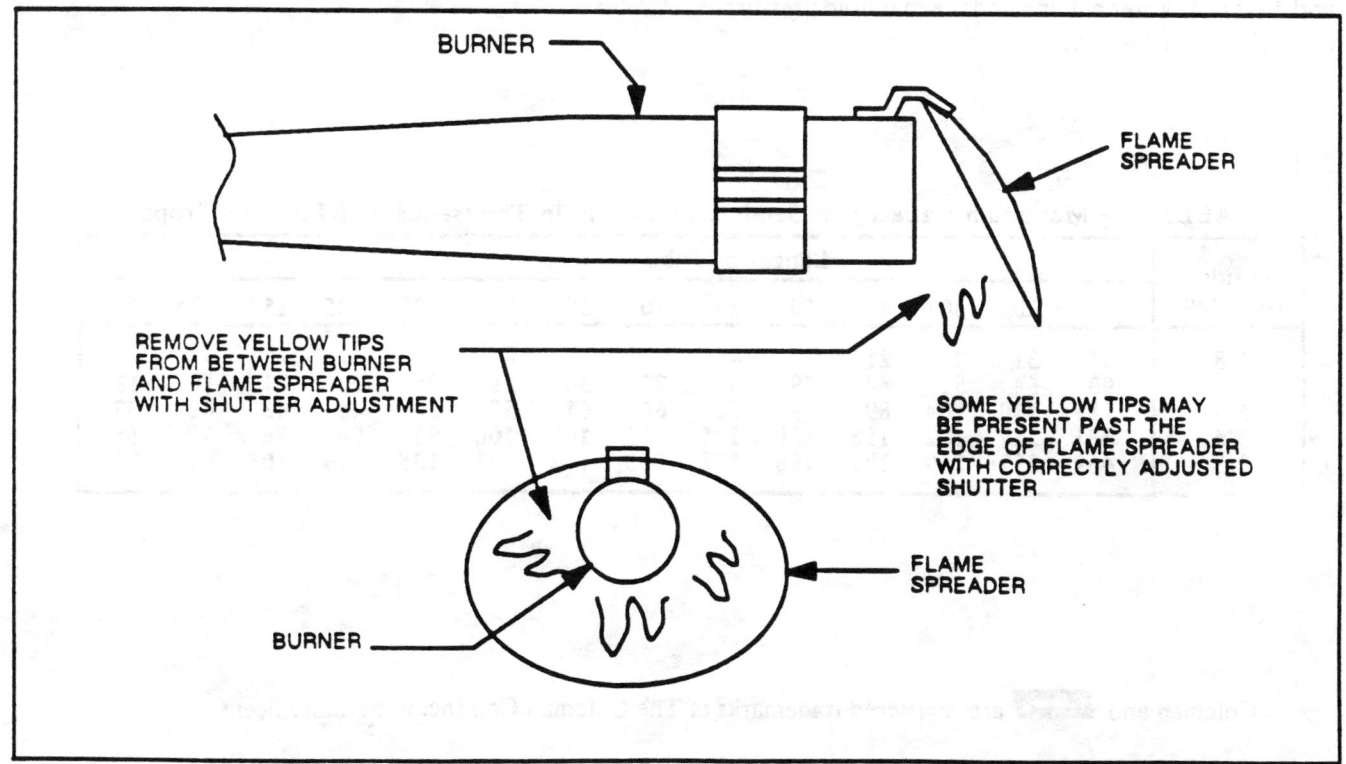

Figure 5

DETERMINING GAS INPUT RATE

Where gas is metered, the input rate may be determined by the following method.

Contact the gas supplier, public utility company or LP gas distributor to obtain the calorific gas value of the gas being used. When checking the gas input rate, any other gas burning appliances connected to the same meter should be completely off. The furnace should be allowed to operate for 15 minutes before attempting to check the gas input rate.

To check flow rate, observe the one cubic foot dial on the gas meter and determine the number of seconds required for the dial hand to complete one revolution (seconds to flow one cubic foot.)

To determine the number of seconds per cubic foot that is necessary to achieve the correct input rate, use the following formula: GAS VALUE X 3600 ÷ DESIRED INPUT = SECONDS NEEDED

Example: 2500 BTU gas, furnace input 100,000 BTUH

Seconds for one cubic foot = 2500 X 3600 ÷ 100,000 = 90 seconds.

If when clocking the meter, the one cubic foot dial makes a complete revolution in less time than was calculated that it should, the furnace is overfired and should be derated. If it takes more time for the meter to make one revolution than was calculated, the furnace is underfired.

The orifice size must be changed to correct an overfired or underfired condition. If it is determined that a different orifice is needed, please contact your distributor for assistance in selecting the correct replacement.

IMPORTANT - If the gas supply does not have a meter, then be sure the proper orifice has been installed. Also verify that the gas supply pressure is between 11 and 14 inches water column and the manifold pressure is 10 ± 0.3 inches water column.

LP GAS PRESSURE REGULATION

For proper operation of this furnace, two-stage pressure regulation must be used. Failure to do so can result in operational problems and can void the warranty.

While some LP gas systems have single stage pressure regulation, it is highly advantageous to use two regulators, each doing part of the job of pressure reduction, rather than one to do the entire reduction from tank pressure to 11 inches water column.

The first regulator - at the tank - is set at 5 to 20 psig and is called the "first stage regulator". The "second stage regulator" is installed at the house. It reduces the 5 to 20 psig down to the 11 to 14 inches w.c. that is required by the furnace. Advantages are: (1) better appliance pressure regulation, (2) reduction of freezing problems, and (3) ability to use smaller piping from tank to house.

PIPING PRACTICES

A piping or tubing system must be designed to prevent the loss in pressure between the last stage regulator and the furnace, from exceeding 0.5 inches w.c. Use Table II to determine the proper size of pipe and fittings required for the distance from tank to the appliance and for the input of the appliances.

TABLE II - Maximum Capacity of Semi-rigid Tubing in Thousands of BTUH for Propane

Outside Diameter	Length of Tubing - feet													
	10	20	30	40	50	60	70	80	90	100	125	150	175	200
3/8	45	31	24	21	18	–	–	–	–	–	–	–	–	–
1/2	93	64	51	44	39	35	32	30	28	26	23	21	19	18
5/8	189	130	104	89	79	71	66	61	57	54	48	43	40	37
3/4	331	227	182	156	138	125	115	107	100	95	84	76	70	65
7/8	469	322	259	221	196	178	163	152	143	135	119	108	99	92

Evcon Industries, Inc.
3110 North Mead
P.O. Box 19014
Wichita, KS 67204-9014

1951A994 (7-90) P.I.   LITHO IN U.S.A.

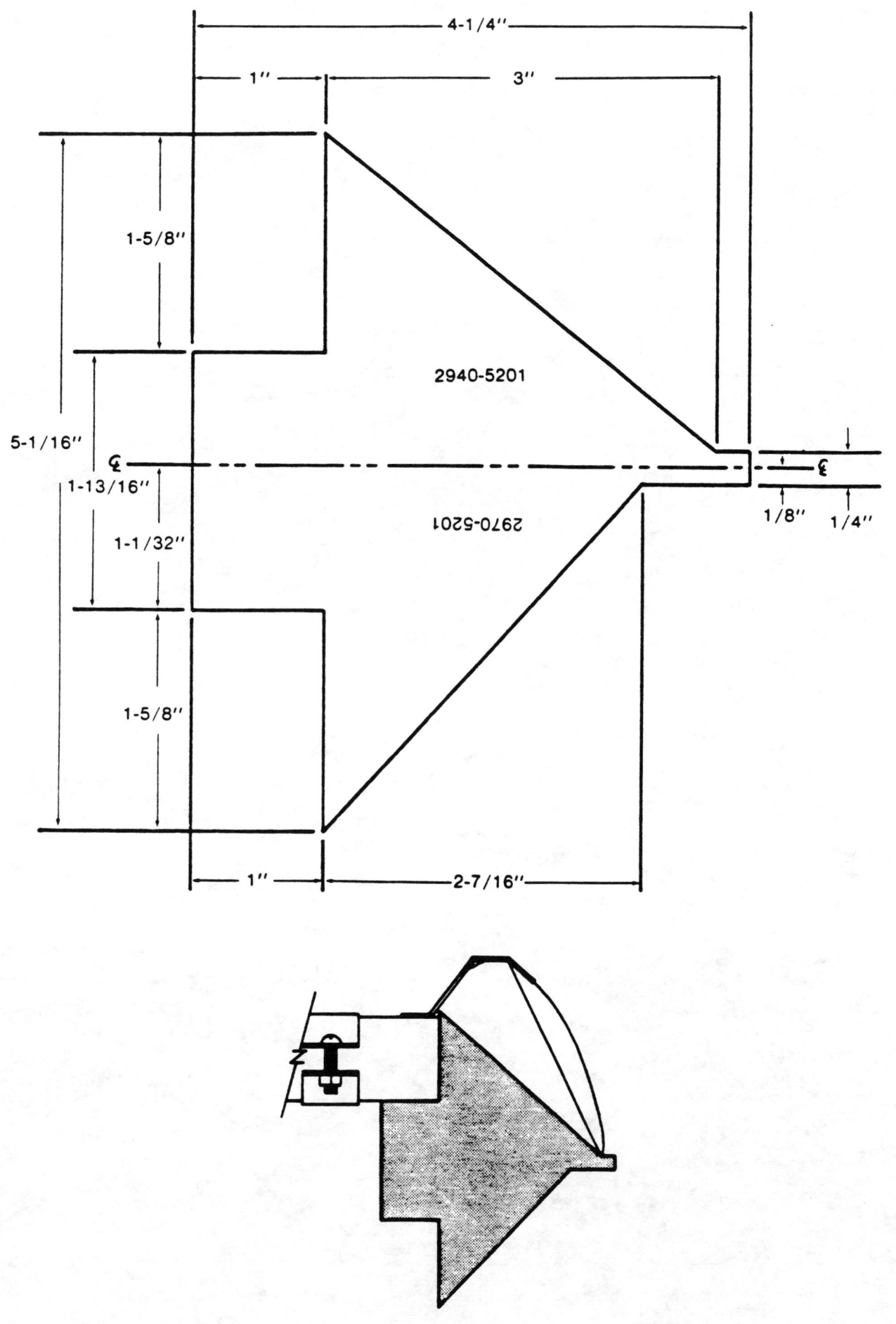

2940-5201

2970-5201

NEW FURNACE MODEL NUMBERS

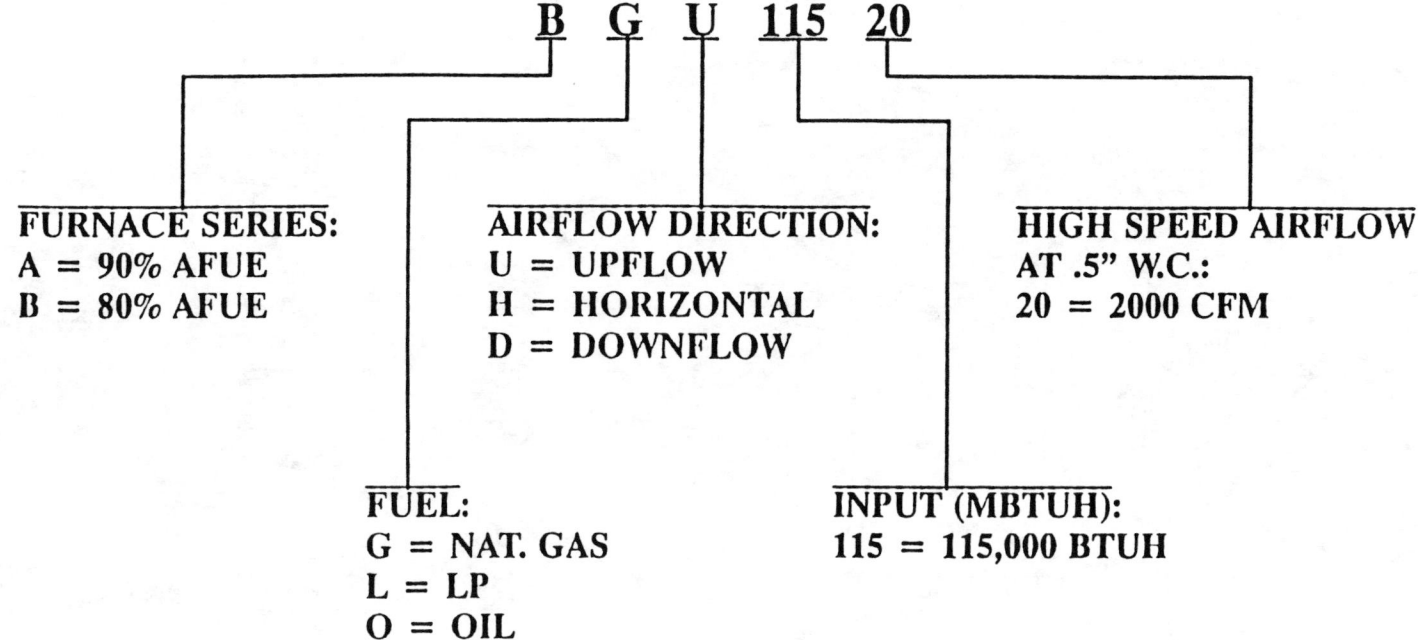

B G U 115 20

FURNACE SERIES:
A = 90% AFUE
B = 80% AFUE

AIRFLOW DIRECTION:
U = UPFLOW
H = HORIZONTAL
D = DOWNFLOW

HIGH SPEED AIRFLOW
AT .5" W.C.:
20 = 2000 CFM

FUEL:
G = NAT. GAS
L = LP
O = OIL

INPUT (MBTUH):
115 = 115,000 BTUH

HEATING & AIR CONDITIONING

UP FLOW HI-EFFICIENCY FURNACES
WITH DIRECT SPARK IGNITION

2800 SERIES

SERVICE GUIDE

FORM A-766

THE COLEMAN CO., INC.
HEATING AND AIR CONDITIONING
3050 N. ST. FRANCIS

NOTICE

These instructions are intended for the use of qualified individuals specially trained and experienced in servicing this type of equipment and related system components.

Service personnel are required by some states to be licensed. Persons not qualified should not attempt to install or service this equipment. Improper installation or service may damage the equipment, will void the warranty, and may create a hazard.

This is not a basic heating manual and does not, therefore, cover the basic principles of heating. The user of this manual should have already accomplished a thorough study of heating and should use this manual as an advanced text to apply only to Coleman Furnace Models listed in the specification table.

Being service oriented, this guide also does not cover all the details of the furnace installation. Installation instructions are packed with each furnace and copies are available upon request.

TABLE OF CONTENTS

INTRODUCTION

The furnaces covered in this guide are high efficiency models, featuring direct spark ignition (no pilot) and a vent blower to provide a constant supply of combustion air to the burners.

The direct spark ignition system is designed to provide a "purge" cycle so that any unburned gas which might have accumulated in the heat exchanger is dissipated before ignition occurs.

A centrifugal switch prevents furnace operation if for some reason the combustion air motor will not operate.

DIMENSIONS

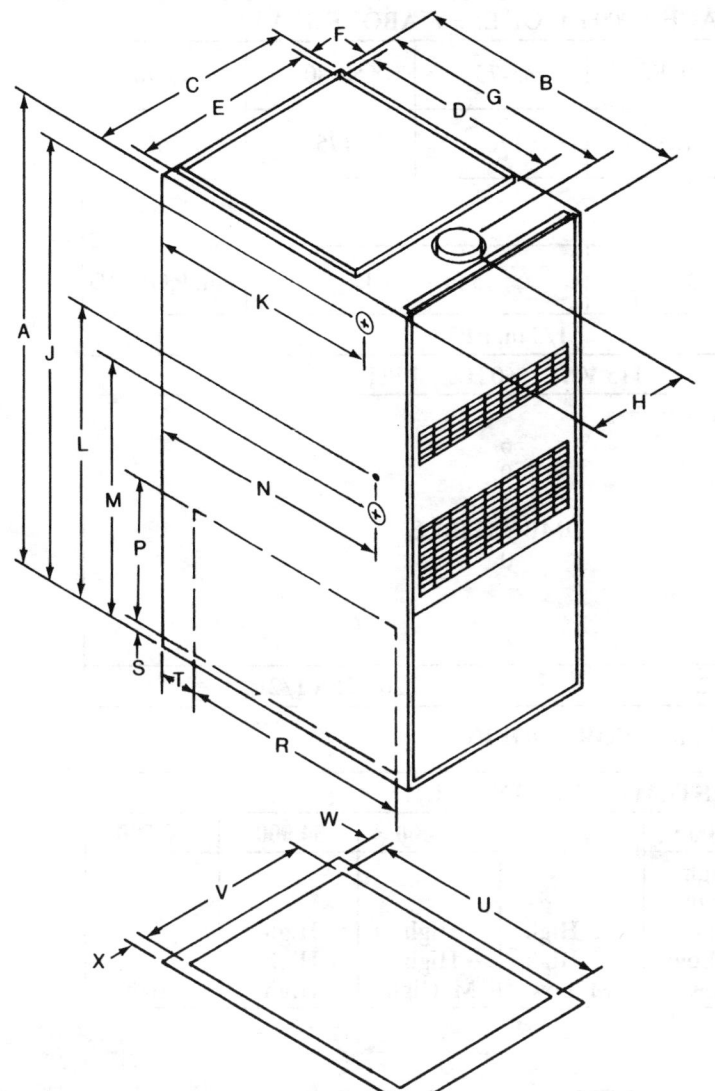

DIMENSIONS IN INCHES

	FURNACE MODELS		
	2845-666 2865-666	2865-766	2880-766 2895-766
A	46	46	46
B	28	28	28
C	16-7/8	20	20
D	20-3/8	20-3/8	20-3/8
E	15-3/8	18-1/2	18-1/2
F	3/4	3/4	3/4
G	24-5/16	24-5/16	24-5/16
H	8-7/16	10	10-11/32
J	43-1/8	43-1/8	43-1/8
K	23-1/16	23-1/16	23-1/16
L	28-5/8	28-5/8	28-5/8
M	25-3/16	25-3/16	25-3/16
N	24-1/2	24-1/2	24-1/2
P	14-1/2	14-1/2	14-1/2
R	23	23	23
S	1	1	1
T	3-1/2	3-1/2	3-1/2
U	23-3/4	23-3/4	23-3/4
V	14-7/8	18	18
W	1-3/4	1-3/4	1-3/4
X	1-1/8	1-1/8	1-1/8

Dimensional Differences

Item	Dimension Location
Flue	G&H
Electrical Supply Entrance	J&K
Low Voltage Wire Entrance (either side)	L&N
Gas Inlet (either side)	M&N
Optional Side Return Air Inlet (either side)	P&R
Optional Bottom Return Air Inlet	U&V
Supply Air Outlet	D&E

SPECIFICATIONS

MODEL	NAT.	2845-666	2865-666	2880-766	2895-766	2865-766
UNIT RATING Btu/Hr.	Input: 0-2,000 Ft. Elevation	55,000	75,000	95,000	115,000	75,000
	High Altitude	FOR ELEVATIONS ABOVE 2,000 FT. REDUCE INPUT 4% FOR EACH 1,000 FT. OF ELEV. ABOVE SEA LEVEL				
AIR TEMPERATURE RISE RANGE °F.		30-60	40-70	45-75	50-80	40-70
DESIGNED MAX. OUTLET AIR TEMP. °F.		155	165	170	175	165
MAX. EXTERNAL STATIC PRESSURE IN. W.C.		.5				
FURNACE FLUE PIPE CONNECTION		4 in. ROUND		5 in. OVAL		4 in. ROUND
GAS CONNECTION		1/2 in. FPT				
ELECTRICAL SERVICE		115 VAC 60 HZ 1 PH				
MINIMUM DISTANCE TO COMBUSTIBLE MATERIALS — inches FRONT BACK PLENUM-TOP FLOOR SIDES FLUE Type B-1 Vent Single Wall Metal Pipe		6 0 1 COMBUSTIBLE 0 1 6				
FILTER (Furnished)		16 x 25 x 1/2		20 x 25 x 1/2		

MOTOR SPEED TAP SELECTION FOR COOLING

FURNACE MODELS	COOLING SYSTEM CAPACITY, BTUH							
	18,000	24,000	30,000	36,000	42,000	48,000	54,000	60,000
2845-666	Low	Low	Med.	High				
2865-666	Low	Low	Med.	High				
2880-766		Low	Low	M. Low	M. High	High	High	
2895-766		Low	Low	M. Low	M. High	High	High	
2865-766				Low	M. Low	M. High	High	High

DANGER — SHOCK HAZARD

BE SURE ELECTRICAL POWER TO FURNACE IS TURNED OFF BEFORE CHANGING MOTOR SPEEDS.

Motor Terminal Identification

In all cases the **white wire** from the furnace control box is the **common** circuit and is fitted with a 3/16 quick connect terminal. This wire **must always** be connected to the motor terminal marked — White, Common or 1.

The **blue wire** from the furnace control box is controlled by the furnace fan switch and is the "hot" wire for **heating operation.**

The **black wire** is controlled by the blower relay and is the "hot" wire for **cooling operation.**

The motor terminals of various motors which may be used are identified by one of the following methods:

MOTOR TERMINAL IDENTIFICATION

	3-Speed Motor	4-Speed Motor
Motor Terminals And Speeds	Common, White or No. 1 High, Black or No. 2 Medium, Blue or No. 3 Low, Red or No. 4	Common, White or No. 1 High, Black or No. 2 Med. Hi, Blue or No. 3 Med. Low, Yellow or No. 4 Low, Red or No. 5

These furnaces are equipped with an electric spark, direct burner ignition system; therefore, in response to a call for heat by the room thermostat, the burner is lighted by an electric arc at the beginning of each operation cycle. The burner will continue to operate until the thermostat is satisfied at which time all burner flame is extinguished. During the off cycle, no gas energy is consumed. With the room thermostat set below room temperature and with the electrical power and gas supply to the furnace on, the normal sequence of operation is as follows:

1. When the room temperature falls below the setting of the room thermostat, the thermostat energizes the heating relay.

2. When the heating relay closes, a circuit is made starting the vent blower. A circuit is also made through the heating relay to the normally open vent blower centrifugal switch contacts.

3. As the vent blower increases in speed, the contacts of the centrifugal switch will close and complete the electrical circuit through the normally closed limit switch to the electronic ignition module.

4. After 15 to 20 seconds the electronic ignition module simultaneously energizes: (1) the electric ignition electrode, (2) the gas control valve, and (3) a safety lock out circuit.

5. When the burner lights, a flame sensor, which is part of the ignition electrode assembly, senses the presence of burner flame and causes the safety lock out circuit in the ignition module to be de-energized. This allows the gas valve to remain open and the burner to operate.

6. If the burner fails to light within 6-8 seconds from the time the ignition control is first energized by the centrifugal switch, the ignition module will de-energize; (1) the gas valve, and (2) the ignition electrode. The ignition control will again energize the gas valve, and the ignition electrode after a 15-20 second delay. If after three trials for ignition the burner still fails to light, the safety lock out circuit in the ignition module will de-energize and lock off: (1) the gas valve, and (2) the ignition electrode. The system will remain in a lockout mode until the room thermostat is set below room temperature causing the thermostat contacts to open and release the safety lock out circuit. Then setting the room thermostat above room temperature and causing the thermostat contacts to close will start the system to try for ignition again.

7. The lapsed time from the moment the wall thermostat closes to when the burner lights may be from 20 to 40 seconds. This delay in the ignition sequence is caused by: (1) the time required for the vent blower to develop sufficient speed to activate the centrifugal switch and (2) the 15 to 20 second delay designed into the ignition module. The lapsed time will also be effected by the temperature within the furnace and in the flue piping.

The 15 to 20 second delay designed into the ignition module is a purge cycle. This allows the vent blower time to replenish the heat exchanger with fresh air so ignition can occur safely.

8. One to two minutes after the burner has lighted, the normally open contacts of the fan switch close and the furnace air circulation blower runs.

NOTE

If a heating/cooling thermostat is being used and the fan switch is set in the "ON" (continuous blower) position, the furnace air circulation blower will run at the air conditioning speed. If the wall thermostat calls for heat, the furnace air circulation blower will shut off for 20 to 40 seconds then the burner will light. One to two minutes after the burner has lighted, the furnace air circulation blower will begin running again but at heating speed. There is no pause in the furnace air circulation blower operation when the wall thermostat is satisfied; the furnace air circulation blower just changes over to cooling (continuous blower) speed.

9. When the room thermostat is satisfied the circuit to the heating relay is broken and the heating relay contacts return to the normally open position. The circuit to the vent blower and the ignition module is broken and the burner is extinguished. Then as heat is drawn from the heat exchanger and the air temperature is reduced to below the fan switch setting, the fan switch will open which stops the furnace air circulation blower.

LIGHTING INSTRUCTIONS

This furnace is equipped with an automatic electric spark direct burner ignition system which lights main burner each time the thermostat calls for heat.

This furnace cannot be lighted with a match.

WARNING

FAILURE TO FOLLOW THESE INSTRUCTIONS MAY RESULT IN AN EXPLOSION AND POSSIBLE DAMAGE TO THE FURNACE AND INJURY TO THE OPERATOR.

1. Set room thermostat to lowest setting — or "off".

2. Turn knob on gas control valve to "off". The White Rodgers gas valve has a built in stop which prevents the knob from being turned directly to OFF. At this stop the knob must be depressed and then turned to OFF.

3. WAIT 5 MINUTES.

4. Turn knob on gas control valve to "on".

5. Set room thermostat to desired setting, above room

temperature, burner will light — which may take 20-40 seconds.

6. If after 3 trials for ignition, and burner fails to light, go to complete shutdown and determine cause for failure to light. Be sure gas supply to furnace is on and supply piping has been purged of air. Be sure electric power to furnace is on.

7. For complete shutdown — set thermostat to lowest setting, or "off" and turn knob on gas control valve to "off".

FURNACE COMPONENTS

Gas Valve — The gas valve is a redundant type valve. There are two operators which must open to allow gas to flow to the burner. Since furnace cannot be match lighted, there is no pilot position on the control knob. Operating voltage is 24 volts.

CAUTION

Never short the terminals on the gas valve. To do so may damage the valve or burn out the heat anticipator on the thermostat.

Figure 1

Gas Valve

Vent Motor Assembly — A 115 volt motor powers the assembly which creates the air draft through the burner and heat exchanger. Air for combustion is pulled through the front panel louvers, over the burner, through the heat exchanger and discharged into the vent box, then up through the flue pipe.

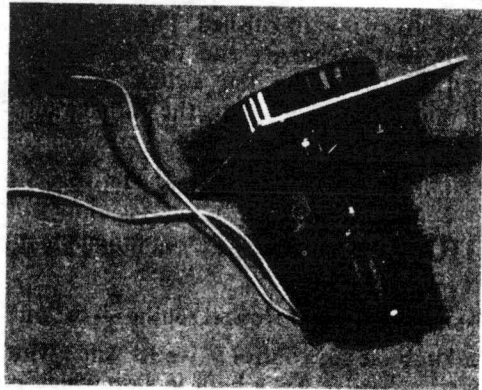

Figure 2

Vent Motor

Centrifugal Switch — A normally open switch attached to the vent motor and actuated by the vent motor r.p.m.'s. If the vent motor fails, the centrifugal switch will not close and the furnace will not operate in the heating mode.

Figure 3

Centrifugal Switch

Ignition Module — The ignition module contains the timing circuits which produces the high voltage spark to ignite the burner. At the same instant the spark starts, the gas valve is energized. After the third try for ignition, if the burner is not ignited, the module locks out the system and will stay locked out until the system is reset.

Figure 4

Ignition Module

Heating Relay — The heating relay has 3 sets of contacts — 2 normally open sets and 1 normally closed set. One set of normally open contacts control the power to the vent motor. The other set of normally open contacts control the power thru the normally closed limit switch and normally open air pressure switch to the ignition module. The set of normally closed contacts provides the power to the blower relay for blower operation in either the cooling or continuous mode.

Figure 5

Heating Relay

6

Blower Relay — Double pole double throw. One set of contacts, normally open, wired to cooling speed of blower motor. Normally closed contacts wired to heating speed of the blower motor.

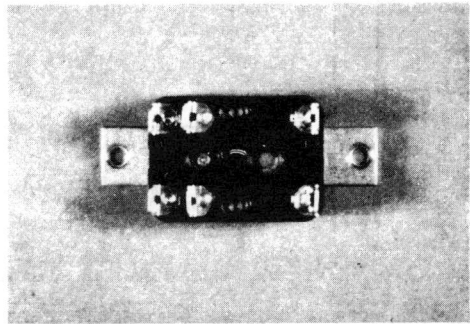

Figure 6
Blower Relay

Blower Motor — 115 volt, multi-speed permanent split capacitor type. Factory wired for medium high or medium speed in heating operation. Speed for cooling is optional and should be selected to match cooling load.

Transformer — The purpose of the transformer is to reduce the 115 volt primary voltage to 24 volts which the serviceman can work with easily and safely. The transformer is rated at 40 VA with sufficient capacity for add on air conditioning.

Fan Switch — Thermally actuated normally open switch — closes when heat exchanger temperature reaches switch temperature. Opens after burner is out and heat exchanger cools off.

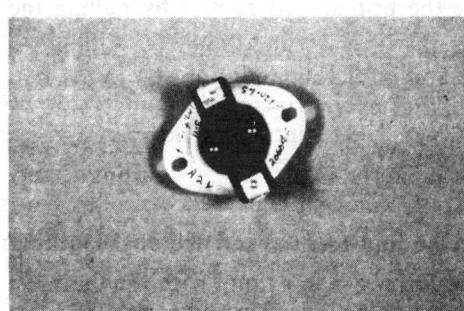

Figure 7
Fan Switch

Limit Control — Thermally actuated normally closed switch. Opens to de-energize the ignition module and close the gas valve when the heat exchanger temperature becomes excessive.

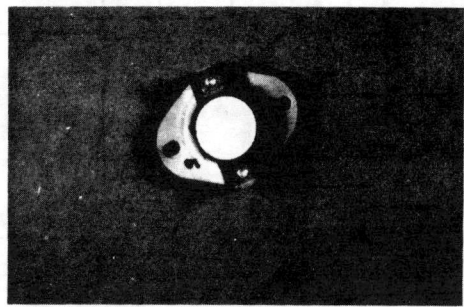

Figure 8
Limit Control

Manual Reset Limit Switch — A normally closed thermally actuated switch that does not reset automatically. The manual reset limit switch is mounted on the vent box sensor tube where it can sense the presence of flue gases if the flue is restricted.

If this switch trips, it will keep the two relays from functioning and that will shut off the operation of everything except the transformer and, if the fan switch is closed, the blower motor.

Figure 9
Manual Reset Limit Switch

Blower Door Safety Switch — This switch is located inside the blower compartment and controls the electrical supply to the furnace electrical circuits. When the blower door is put in place, the switch is activated allowing the furnace to operate. The switch is designed to prevent furnace operation if the blower door is removed and inadvertantly not re-installed, thus preventing the possibility of the blower creating a negative pressure in the furnace enclosure.

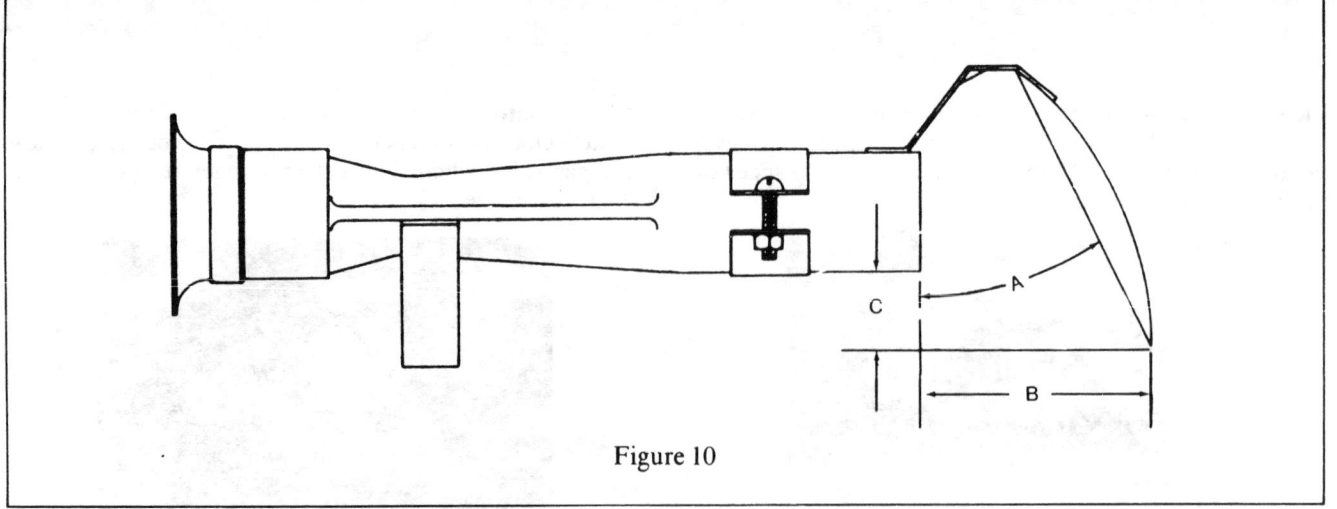

Figure 10

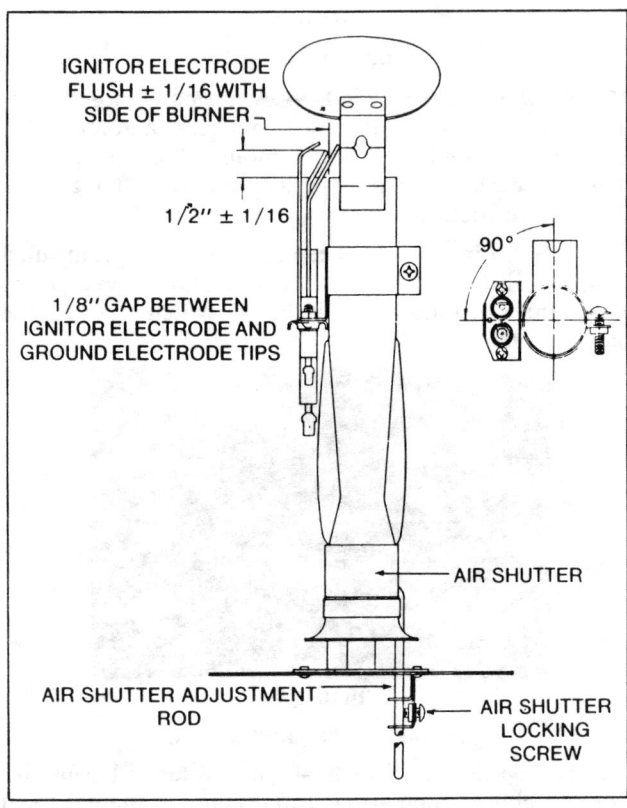

IGNITOR ELECTRODE
FLUSH ± 1/16 WITH
SIDE OF BURNER

1/2" ± 1/16

90°

1/8" GAP BETWEEN
IGNITOR ELECTRODE AND
GROUND ELECTRODE TIPS

AIR SHUTTER

AIR SHUTTER ADJUSTMENT
ROD

AIR SHUTTER
LOCKING
SCREW

Figure 11

Burner — Mono-Port with a stainless steel flame spreader and adjustable primary air.

Note "A"

Proper alignment of flame deflector is critical to the proper operation of the burner. Before burner is installed in furnace, check the alignment of the deflector as shown in figure 10. Constructing a template as shown on page 19 will help.

Furnace Model	Burner Number	Dimensions - Figure 10		
		A	B	C
2845, 2865	2940-5201	30°	3"	21/32"
2880, 2895	2970-5201	20°	2-7/16"	29/32"

To adjust burner flame.

1. Loosen locking screw — refer to figure 11.

2. Close the primary air shutter by pulling the adjustment rod until yellow tips appear in the flame at the end of the burner.

3. Slowly open the primary air shutter by pushing the adjustment rod in until the yellow tips disappear, then push in 1/8 inch more.

4. Hold the adjusting rod and tighten the locking screw.

5. Cycle furnace a few times with the thermostat to insure everything is working properly.

VENT SYSTEM

For more information regarding vent design and installation, refer to NATIONAL FUEL GAS CODE ANSI Z223.1-1980

1. The vent material can be class C or B1 for the first 5' from the furnace. Beyond 5', it must be B1, or a permanent chimney.

2. Maintain minimum 6" clearance to combustible material for class C vent and 1" clearance for B1 vent.

3. When using Listed Type B-1 vent materials and/or a Listed Manufactured chimney is used, they must be installed in accordance with the Manufacturer's Installation Instructions and the terms of the Listing.

4. If horizontal pipe is required, it should be as short as

possible and pitch upward toward the vertical riser or chimney inlet at least 1/4" per foot to help insure proper venting characteristics.

5. The vent pipe must be at least as large as outlet on vent box.

6. Where two or more appliances are connected to a common vent or flue, the area of the common vent or connector must be equal in area to the largest appliance vent size plus 50% of the area of the smaller vent size(s).

7. DO NOT USE dampers or restrictors in vent piping or flue.

8. Vent pipe sections must be securely fastened together and fastened to the furnace flue collar using sheet metal screws where required.

GAS CONVERSION

Furnaces can be converted to L.P. gas by using a conversion kit available from The Coleman Company, Inc. The conversion essentially consists of:

1. Instructions, burner orifice, two stickers and a spring for converting a White Rodgers gas valve.

2. Instructions, two screws, gasket, burner orifice, and pressure regulator, for converting a Honeywell gas valve.

Conversion parts for either valve are contained in one conversion kit. See the installation instructions packed with each kit for complete instructions.

SERVICE PROCEDURES

Before removing components for inspection or service, carefully observe how these components are installed, electrical connections made and attachment means employed.

All components and assemblies must be reinstalled in the same manner and position as they were originally.

Damaged or defective components must be replaced. Only original equipment components, or subsititute components authorized by The Coleman Company may be used in the repair of this furnace.

Vent Motor — If the vent motor does not run, disconnect the black and white wires from the vent motor and check for 115 volts between them. If voltage is present and the motor does not run, replace the motor. If voltage is not present use the service charts to determine the cause.

Burner Assembly — The burner should be inspected for any accumulation of dust or lint. Use the pressure side of a vacuum cleaner to blow out any dust or lint. The burner must be removed to inspect the ignition electrode assembly, and the heat exchanger.

To Remove the Burner Assembly —

1. Turn off the gas supply to the furnace.
2. Disconnect the gas line to the gas valve.
3. Disconnect low voltage wires to the gas valve.
4. Remove 4 screws holding gas valve mount to the furnace vestibule.
5. Remove 4 screws holding the burner assembly to the furnace vestibule.

CAUTION

Remove burner slowly, being careful not to damage electrode assembly.

Ignition Electrode Assembly — Check the condition of the electrode rods which should show no sign of serious deterioration, scaling, or carbon build up. Check the dimensional relationship between the rods and between the rods and burner. Refer to Fig. 11. Minor adjustments can be made by carefully bending the rods slightly. Take care not to damage the porcelain insulators.

Combustion Chamber — Using an inspection mirror and flashlight inspect the inside of the heat exchanger for any scaling, soot deposits, or metal fatigue. If any soot deposits are found the heat exchanger must be cleaned. Soot is caused by improper burner adjustment and must be corrected when the burner is re-installed. If any holes are found the heat exchanger must be replaced.

Following the inspection and cleaning re-install all components parts in the reverse order of their removal.

NOTE

If soot deposits are found in the heat exchanger drum, remove the vent motor assembly and inspect the vent blower wheel and housing for soot. If soot deposits are found here also, the heat exchanger will have to be replaced and the condition causing the sooting must be corrected.

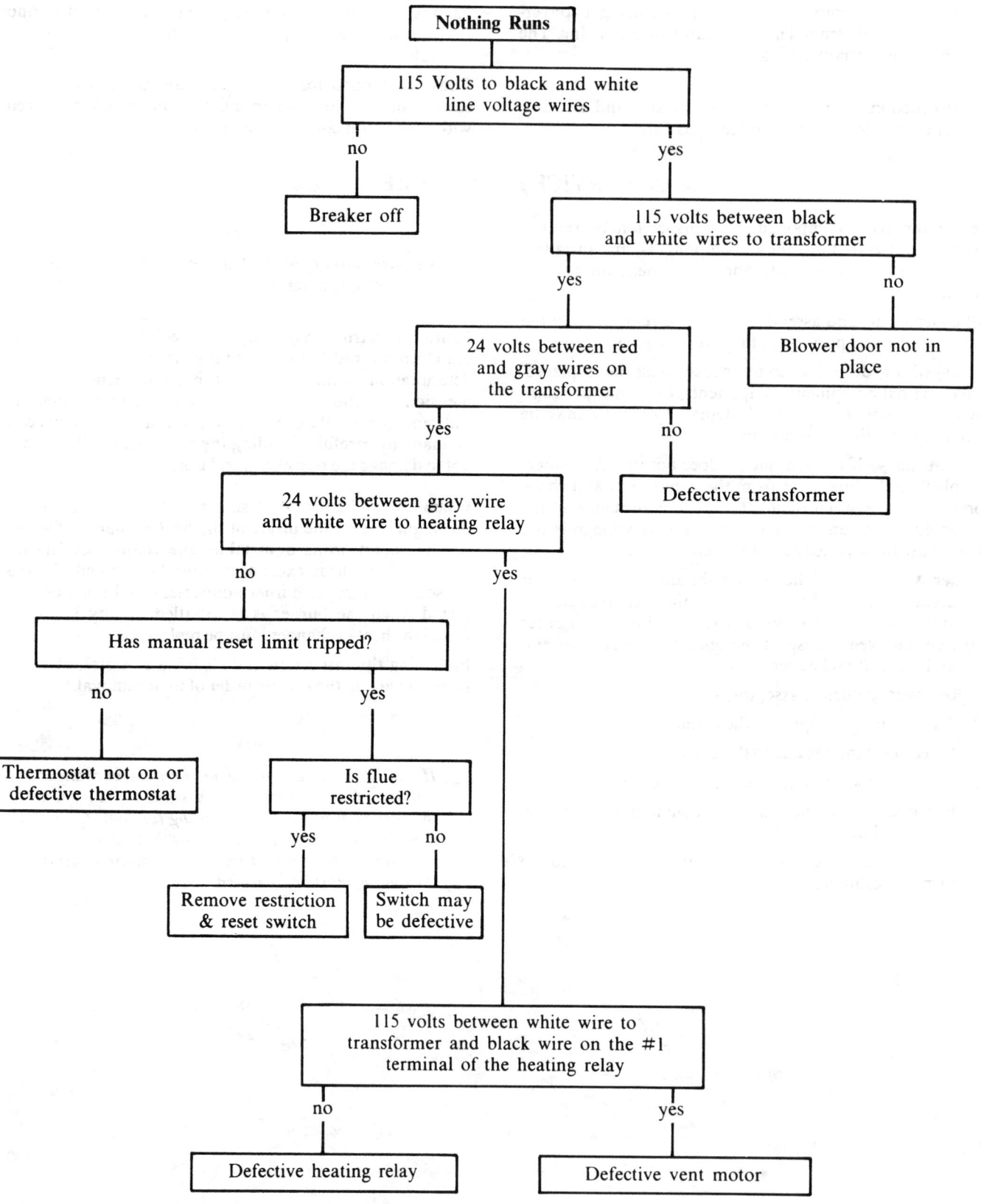

Nothing Runs

115 Volts to black and white line voltage wires

— no → Breaker off

— yes → 115 volts between black and white wires to transformer

- yes → 24 volts between red and gray wires on the transformer
  - yes → 24 volts between gray wire and white wire to heating relay
    - no → Has manual reset limit tripped?
      - no → Thermostat not on or defective thermostat
      - yes → Is flue restricted?
        - yes → Remove restriction & reset switch
        - no → Switch may be defective
    - yes → 115 volts between white wire to transformer and black wire on the #1 terminal of the heating relay
      - no → Defective heating relay
      - yes → Defective vent motor
  - no → Defective transformer
- no → Blower door not in place

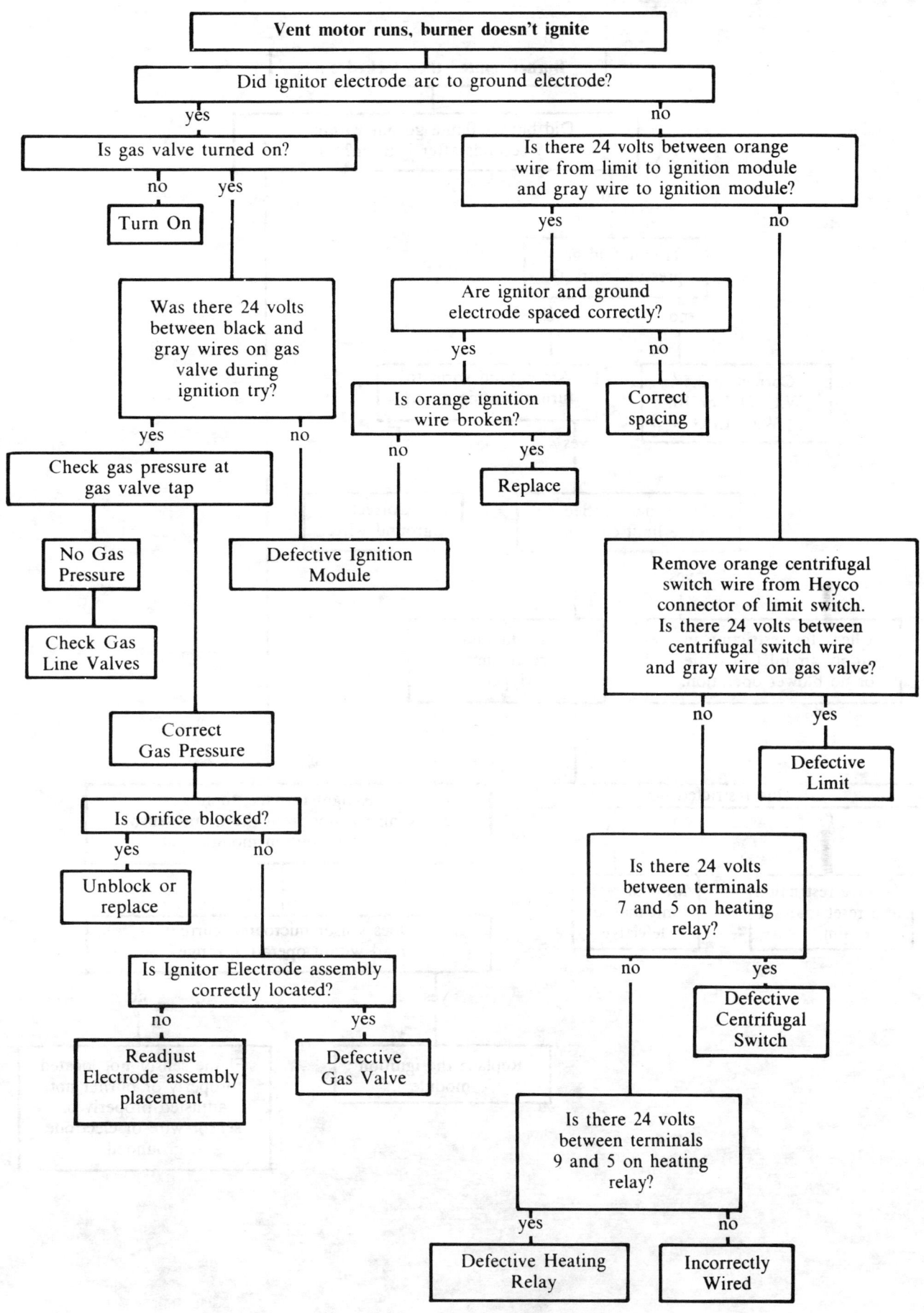

Vent motor runs, burner doesn't ignite

Did ignitor electrode arc to ground electrode?

- **yes** → Is gas valve turned on?
  - **no** → Turn On
  - **yes** → Was there 24 volts between black and gray wires on gas valve during ignition try?
    - **yes** → Check gas pressure at gas valve tap
      - No Gas Pressure → Check Gas Line Valves
      - Correct Gas Pressure → Is Orifice blocked?
        - **yes** → Unblock or replace
        - **no** → Is Ignitor Electrode assembly correctly located?
          - **no** → Readjust Electrode assembly placement
          - **yes** → Defective Gas Valve
    - **no** → Defective Ignition Module

- **no** → Is there 24 volts between orange wire from limit to ignition module and gray wire to ignition module?
  - **yes** → Are ignitor and ground electrode spaced correctly?
    - **yes** → Is orange ignition wire broken?
      - **no** → Defective Ignition Module
      - **yes** → Replace
    - **no** → Correct spacing
  - **no** → Remove orange centrifugal switch wire from Heyco connector of limit switch. Is there 24 volts between centrifugal switch wire and gray wire on gas valve?
    - **no** → Is there 24 volts between terminals 7 and 5 on heating relay?
      - **no** → Is there 24 volts between terminals 9 and 5 on heating relay?
        - **yes** → Defective Heating Relay
        - **no** → Incorrectly Wired
      - **yes** → Defective Centrifugal Switch
    - **yes** → Defective Limit

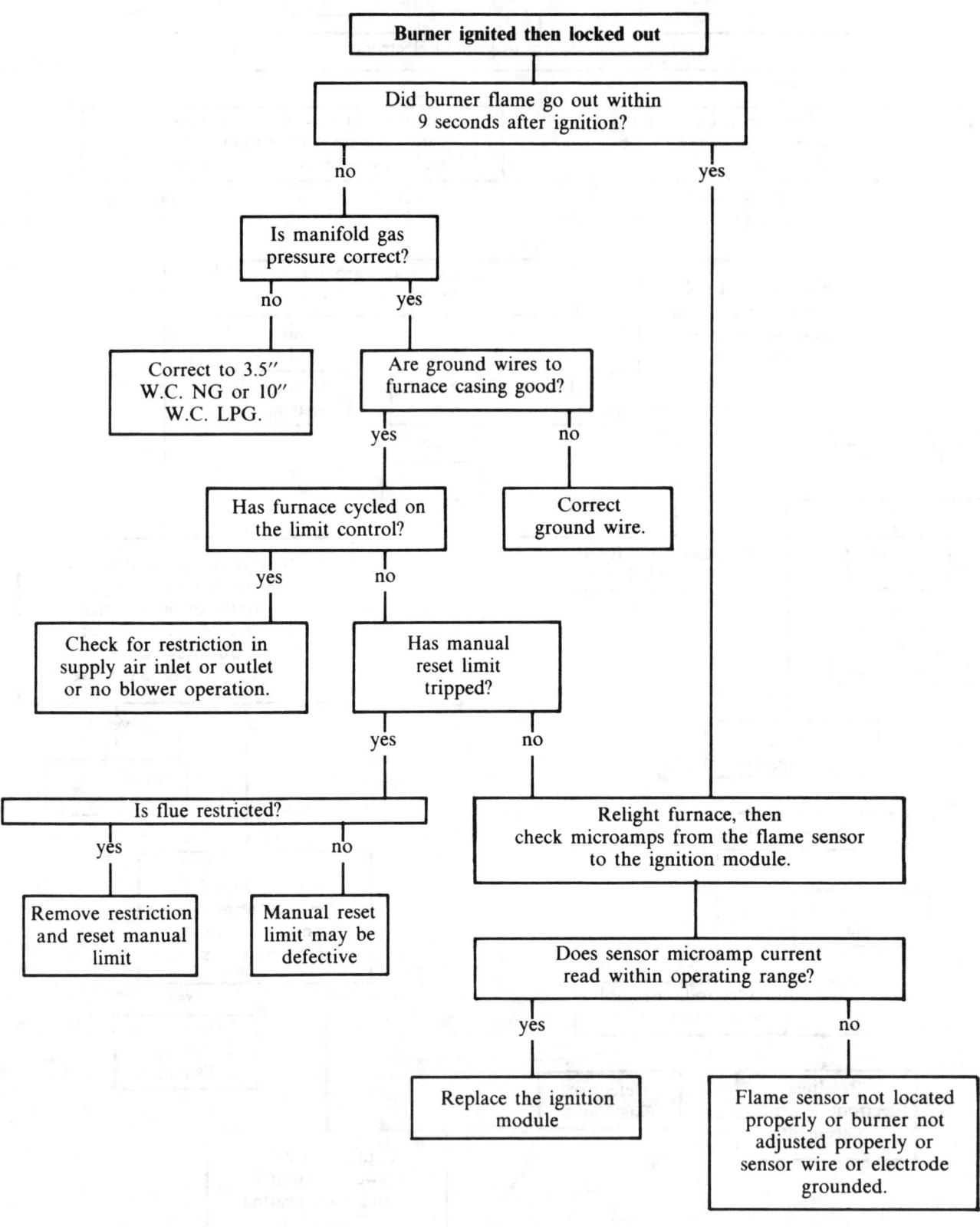

Burner ignited then locked out

Did burner flame go out within 9 seconds after ignition?

no — Is manifold gas pressure correct?

no — Correct to 3.5″ W.C. NG or 10″ W.C. LPG.

yes — Are ground wires to furnace casing good?

yes — Has furnace cycled on the limit control?

no — Correct ground wire.

yes — Check for restriction in supply air inlet or outlet or no blower operation.

no — Has manual reset limit tripped?

yes — Is flue restricted?

yes — Remove restriction and reset manual limit

no — Manual reset limit may be defective

no — Relight furnace, then check microamps from the flame sensor to the ignition module.

Does sensor microamp current read within operating range?

yes — Replace the ignition module

no — Flame sensor not located properly or burner not adjusted properly or sensor wire or electrode grounded.

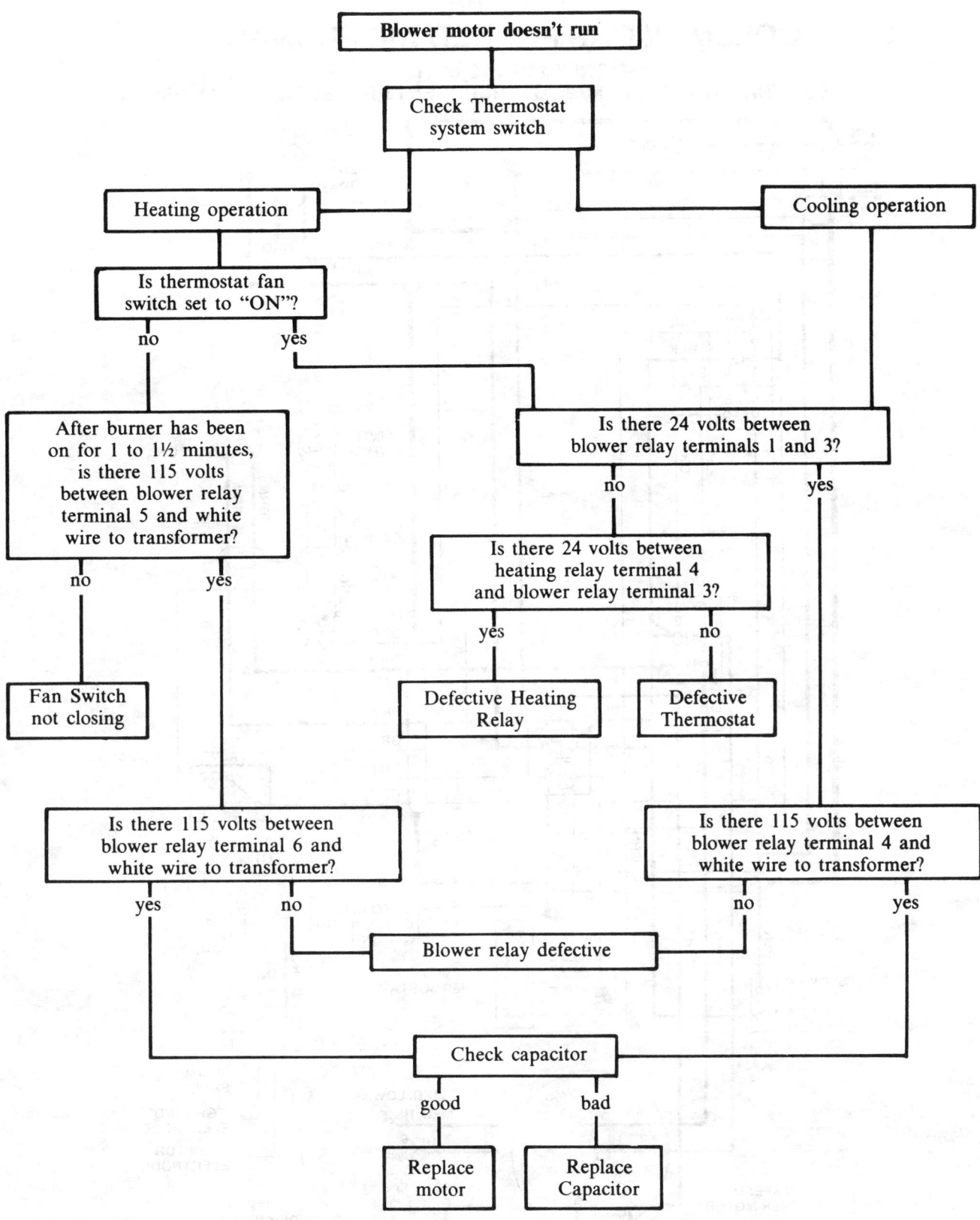

Blower motor doesn't run

Check Thermostat system switch

Heating operation

Cooling operation

Is thermostat fan switch set to "ON"?
— no
— yes

After burner has been on for 1 to 1½ minutes, is there 115 volts between blower relay terminal 5 and white wire to transformer?
— no
— yes

Is there 24 volts between blower relay terminals 1 and 3?
— no
— yes

Is there 24 volts between heating relay terminal 4 and blower relay terminal 3?
— yes
— no

Fan Switch not closing

Defective Heating Relay

Defective Thermostat

Is there 115 volts between blower relay terminal 6 and white wire to transformer?
— yes
— no

Is there 115 volts between blower relay terminal 4 and white wire to transformer?
— no
— yes

Blower relay defective

Check capacitor
— good
— bad

Replace motor

Replace Capacitor

13

80 SERIES
COLEMAN GAS FORCED AIR FURNACE
USE ONLY 115 VAC 60 HZ 1 PH
LESS THAN 12 AMPS   MAX. OVERCURRENT PROTECTION 15 AMPS

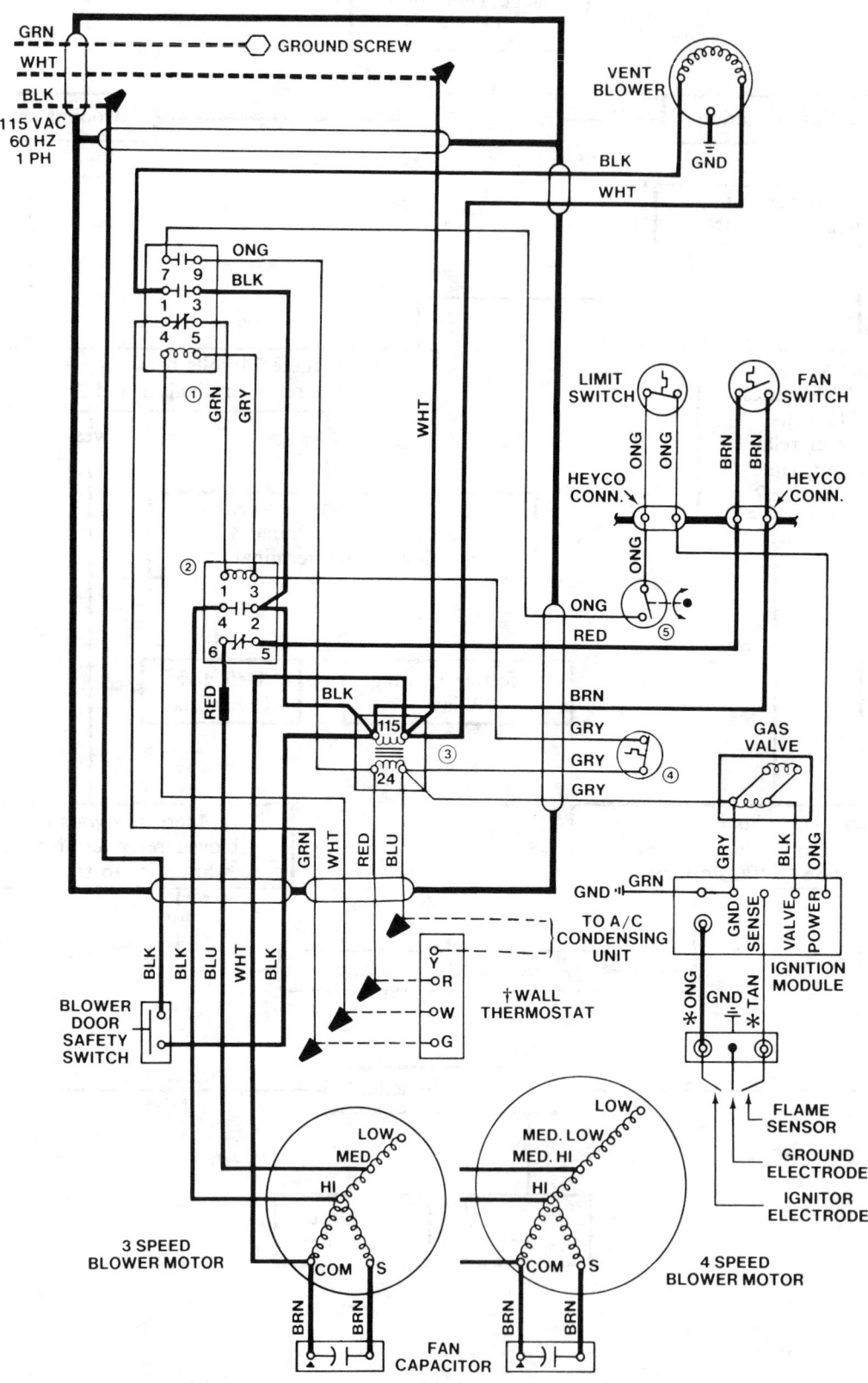

① HEATING RELAY, ② BLOWER RELAY, ③ TRANSFORMER, ④ MANUAL RESET LIMIT SWITCH,
⑤ CENTRIFUGAL SWITCH

FACTORY INTERNAL WIRING SHOWN SOLID.

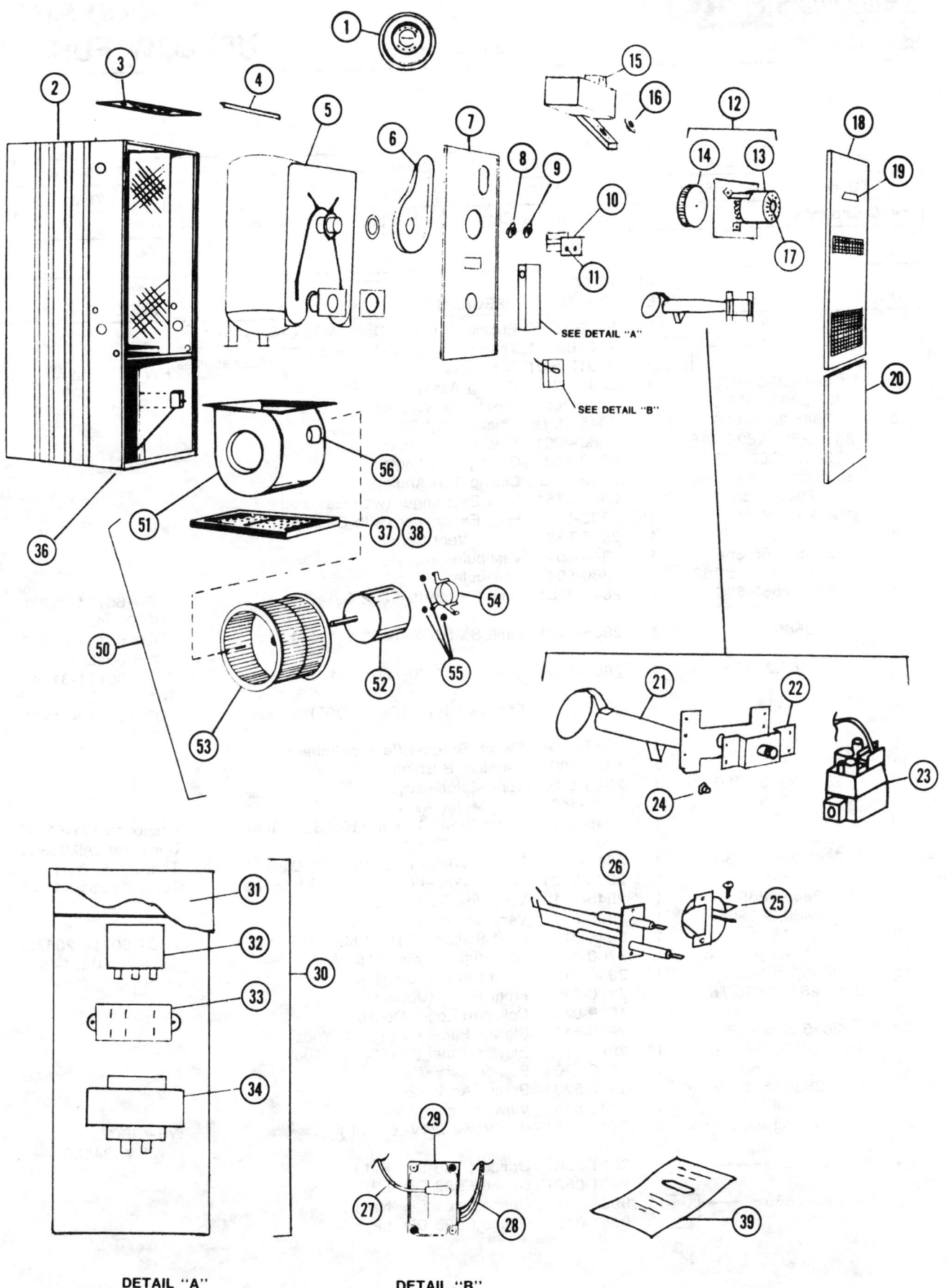

DETAIL "A"

DETAIL "B"

Model	2845-666	2865-666	2880-766	2895-766	2865-766
BTU/HR	55M	75M	95M	115M	75M
A/C Capacity	See Blower Data on Tech. Spec. sheet — Form T-744				

CODE	MODEL	NO. REQ.	PART NUMBER	DESCRIPTION	VENDOR
° -	All	X	8B190	Enamel (Spray - Dawn Mist Green)	
° 1	All	1	2661B2611	Thermostat (24V Adj. Ant.)	HW#T87F
		1	2661B2601	Sub-Base	HW #HQ539A
2	2845,2865-666	1	2845-585	Casing Assy.	
	2865,2880,2895-766	1	2880-585	Casing Assy.	
3	2845,2865-666	1	2845-2091	Shield (Vent Box)	
	2865,2880,2895-766	1	2880-2091	Shield (Vent Box)	
4	2845,2865-666	1	2940-103	Casing Top Angle	
	2865,2880,2895-766	1	2632A103	Casing Top Angle	
5	2845,2865	1	2845-5751	Heat Exchanger (with Gaskets)	
	2880,2895	1	2880-5751	Heat Exchanger (with Gaskets)	
6	All	1	2845-538	Power Vent Scroll	
7	2845,2865-666	1	2845-504	Vestibule	
	2865,2880-2895-766	1	2880-504	Vestibule	
° 8	2845,2865-666	1	2845-3181	Limit Switch (0-170° C-140°)	TOD#60T11-312362 TI#20604L
	2865-766	1	2865-3181	Limit Switch (0-145° C-115°)	TOD#60T11-312498 TI#20604L
	2880,2895	1	2880-3181	Limit Switch (0-160° C-130°)	TOD#60T11-312499 TI#20604L
° 9	All	1	2845-3171	Fan Switch (C-105° 0-95°)	TOD#60T12-312500 TI#20604F
10	All	1	2845-510	Switch Bracket (less switches)	
11	All	2	1216-260	Terminal Bushing	
° 12	2845,2865,2880	1	2845-540	Vent Motor Assy.	
	2895	1	2895-540	Vent Motor Assy.	
° 13	All	1	2845-3209	Motor (Vent Blower-115V-3200RPM)	Fasco #7121-5521 Universal JAIMI05NV
° 14	2845,2865,2880	1	2845-3401	Blower Wheel (1-9/16 x 5" Dia.)	Torin #FE500-108-1
	2895	1	2895-3401	Blower Wheel (1-11/16 x 5-5/8 Dia.)	Torin #FE519-119-1
15	2845,2865	1	2845-521	Vent Box Assy.	
	2880,2895	1	2895-521	Vent Box Assy.	
° 16	All	1	2845-3191	Limit Switch (C-195° Manual Reset)	TOD#60T14-205757
° 17	All	1	7700-2751	Centrifugal Switch (15 AMP)	R-S#243-0231-02
18	2845,2865-666	1	2940-2101	Front Panel (Upper)	
	2865,2880,2895-766	1	2970-2101	Front Panel (Upper)	
19	All	1	1539-0301	Coleman Logo (Decal)	
20	2845,2865-666	1	2940-5141	Blower Panel (15-11/16 Wide)	
	2865,2880,2895-766	1	2970-5141	Blower Panel (18-13/16 Wide)	
° 21	2845,2865	1	2940-5201	Burner Assy. (30°)	
	2880,2895	1	2970-5201	Burner Assy. (20°)	
22	All	1	2845-545	Valve Bracket Assy.	
23	All	1	2940-3261	Gas Valve (24V-.6 AMP)	WR#36C75-204 H-W#VR8450P
° 24	2845	1	9951-1361	Orifice (.136 Dia.-Nat.)	
		1	9951-0821	Orifice (.082 Dia.-LP)	
	2865	1	9951-1611	Orifice (.161 Dia.-Nat.)	
		1	9951-0981	Orifice (.098 Dia.-LP)	

Accessories

Composite 2845-720

Coleman ❄ ☀
HEATING & AIR CONDITIONING

CODE	MODEL	NO. REQ.	PART NUMBER	DESCRIPTION	VENDOR
	2880	1	9951-1801	Orifice (.180 Dia.-Nat.)	
		1	9951-1131	Orifice (.113 Dia.-LP)	
	2895	1	9951-1991	Orifice (.199 Dia.-Nat.)	
		1	9951-1201	Orifice (.120 Dia.-LP)	
25	All	1	2940-515	Electrode Mount Bracket	
○ 26	All	1	2940-3051	Electrode	Fenwal #22-100000-551
○ 27	All	1	7705-3071	High Tension Wire (Ignitor)	
○ 28	All	1	2845-6511	Wire Bundle	
○ 29	All	1	2940-3251	Ignitor Module (24V-Lockout 5-8 Sec.)	Fenwal #05-279000-023
30	All	1	2845-5301	Control Box Assy. Complete	
31	All	1	2940-182	Control Box Cover	
○ 32	All	1	2940-3551	Relay (Heating 24V-50/60 HZ)	Essex #93-232187-250 Potter #KU93-4221
○ 33	All	1	3110-3301	Relay (Blower-24V 50/60 HZ)	RBM #184-20114-406 Basler #BE-17140-GEK
○ 34	All	1	2940-3541	Transformer (120/24V-40VA.)	Essex #591-406800
○ 36	All	1	8632-3141	Safety Switch	Micro #6PL14 Cherry #2629-1010
○ 37	2845,2865-666	1	2940-3391	Filter (Washable 25 x 14-1/2 - 12/Pk'g.)	
	2865,2880,2895-766	1	2970-3391	Filter (Washable 25 x 18-1/2 - 12/Pk'g.)	
38	2845,2865-666	2	2940-120	Filter Rod (3/32 Dia. x 25-1/8")	
	2865,2880,2895-766	3			
39	All	1	1972-212	Customer Envelope	

	MODEL	2845-666 2865-666	2865-766 2880-766 2895-766
50	Blower Assy.	2960-690	2970-690
51	Scroll	2960-538	2970-538
○ 52	Motor	2960-3239 115V-FLA 9.2 3 Speed 1100 RPM Fasco#7126-0995 RMR#C640CW118 Universal#H3L056N A.O.Smith#F48SQ6L28	2970-3249 115V-FLA. 8.4 4 Speed 1120 RPM Fasco#7126-1023 Universal#HE3T030N A.O.Smith#F48SS6L14
○ 53	Blower Wheel	1462-2761 10-5/8 Dia. x 8 x 1/2B Lau#01332A-01 Torin#CM1020-800-5 Morrison#03-43001-0	1472-2771 11-7/8 Dia. x 9-1/2 x 1/2B Lau#027855-01 Torin#CM11.5-9.5 Morrison#03-54002-0
54	Motor Clamp	7700-5391	8680-6391
○ 55	Rubber Mounts (3/pkg)	7670-3551	7670-3551
○ 56	Run Capacitor (Mfd.-Volts)	1499-4491 (10-370)	1499-4491 (10-370)

Accessories

Conversion Kit (Nat. to LP)

Model	2845-666	2865-666 2865-766	2880-766	2895-766
Kit No.	2845-5991	2865-5991	2985-5991	2895-5991

NOTE: All parts with three digit suffix numbers are "Special Order" Parts. These parts are subject to factory availability and require extra time for delivery.
○ Suggested Parts Inventory (2% of Units Installed — Minimum 1 each)

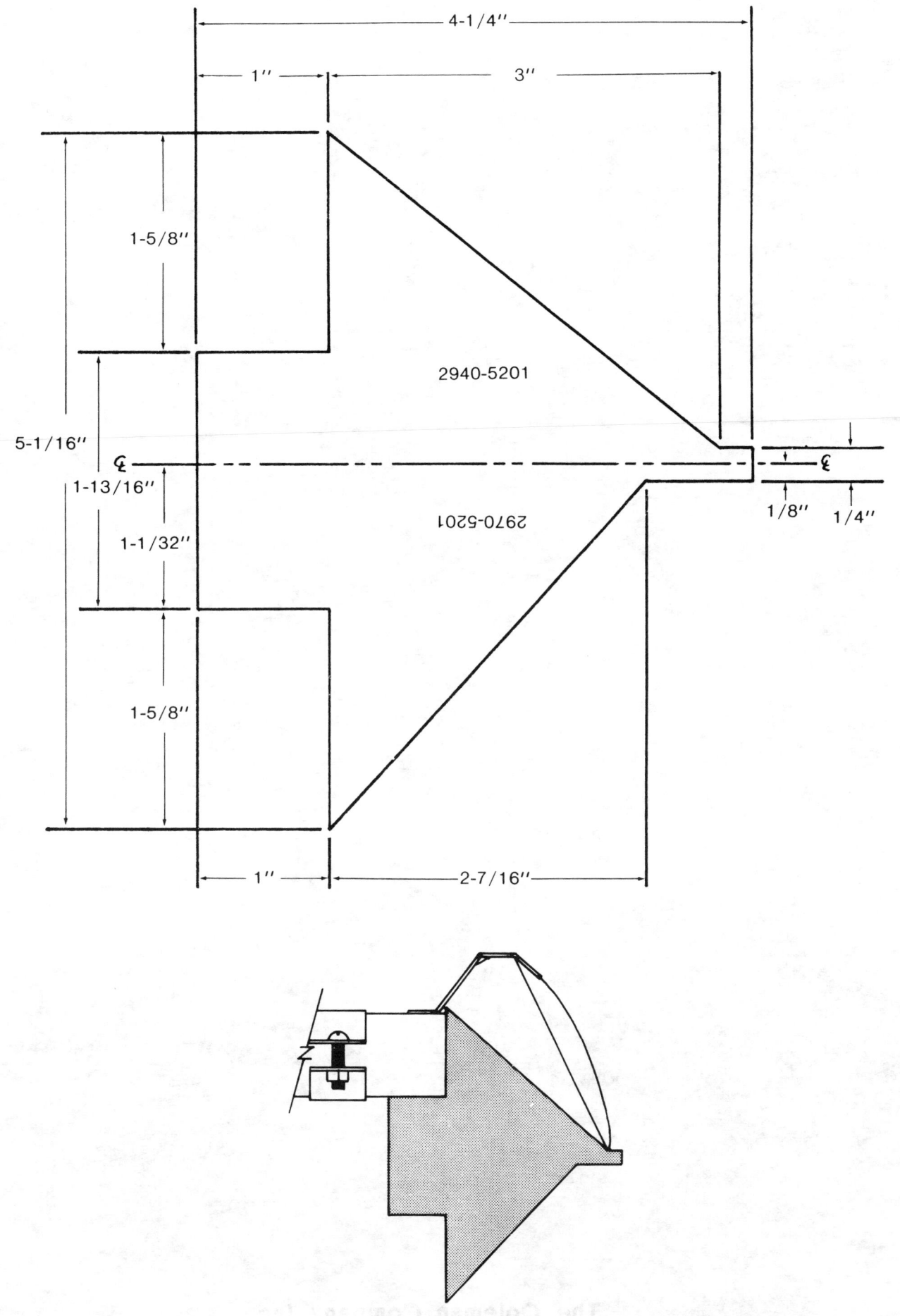

4-1/4"

1"   3"

1-5/8"

2940-5201

5-1/16"

1-13/16"

1-1/32"

2970-5201

1-5/8"

1/8"   1/4"

1"   2-7/16"

The Coleman Company, Inc.

HEATING AND AIR CONDITIONING GROUP
WICHITA, KANSAS 67201

HEATING & AIR CONDITIONING

OWNERS INFORMATION MANUAL
FOR
UP FLOW HI-EFFICIENCY FURNACES
WITH DIRECT SPARK IGNITION

FOR YOUR SAFETY

If you smell gas:

1. Open windows.
2. Don't touch electrical switches.
3. Extinguish any open flame.
4. Immediately call your gas supplier.

WARNING
FOR YOUR SAFETY

DO NOT STORE OR USE GASOLINE OR OTHER FLAMMABLE LIQUIDS, VAPORS OR MATERIALS IN THE VICINITY OF THIS FURNACE OR OTHER HEATING APPLIANCES, WHICH COULD CAUSE FIRE OR EXPLOSION.

WARNING: Improper installation, adjustment, alteration, service or maintenance can cause injury or property damage. Refer to this manual. For assistance or additional information consult a qualified installer, service agency or the gas supplier.

Keep This Booklet — You'll want to refer to it from time to time, as it contains important information about the operation and care of your Coleman Warm Air Furnace.

Your furnace will give you all the comforts of complete winter air conditioning — heating, circulation, filtering — you have control and yet the furnace is as automatic as you desire it to be.

If you will observe the few operating and maintenance instructions in this booklet the Coleman high efficiency forced warm-air furnace will give you many years of dependable service.

WARNING
FAILURE TO OBSERVE THE FOLLOWING SAFETY PRECAUTIONS COULD CAUSE FIRE, EXPLOSION, OR ASPHYXIATION.

FOR YOUR SAFETY

DO NOT STORE OR USE FLAMMABLE LIQUIDS, VAPORS OR MATERIALS IN THE IMMEDIATE AREA NEAR THE FURNACE OR OTHER HEATING APPLIANCE. DO NOT STORE BROOMS, MOPS, OR EQUIPMENT OR MATERIALS, NEAR THE FURNACE IN ANY CONFINED SPACE WHICH MAY BE AROUND OR IN FRONT OF THE FURNACE.

DO NOT STORE HALOGEN EMITTING SUBSTANCES SUCH AS LAUNDRY BLEACH AND DETERGENT, CLEANING FLUIDS, SPRAY CAN PROPELLENTS AND SOLVENTS IN THE VICINITY OF THIS APPLIANCE. THE AIR USED BY THE BURNER FOR COMBUSTION MUST BE FREE OF HALOGENS TO AVOID POSSIBLE CORROSION TO THE HEATING SURFACES WHICH COULD RESULT IN ASPHYXIATION.

A FURNACE INSTALLED IN AN ATTIC OR OTHER INSULATED SPACE MUST BE KEPT FREE AND CLEAR OF INSULATION MATERIALS. EXAMINE THE FURNACE AREA WHEN THE FURNACE IS INSTALLED, OR IF ADDITIONAL INSULATION IS ADDED. THE INSULATION MATERIAL MAY BE COMBUSTIBLE.

Louvered openings in the furnace front panel must be kept clean and unobstructed. Any openings into the furnace closet, or confined room in which the furnace is located, for the entrance of combustion and ventilation air must be kept open and unobstructed.

ADEQUATE VENTILATION AND COMBUSTION AIR MUST BE PROVIDED TO INSURE SATISFACTORY AND SAFE OPERATION OF THE FURNACE. AIR OPENINGS IN CASING FRONT PANEL AND VESTIBULE TOP PANEL MUST NOT BE OBSTRUCTED. FAILURE TO OBSERVE THIS RECOMMENDATION COULD RESULT IN ASPHYXIATION.

WARNING
SHOULD OVERHEATING OCCUR, OR THE GAS SUPPLY FAIL TO SHUT OFF, SHUT OFF THE MANUAL GAS VALVE TO THE APPLIANCE BEFORE SHUTTING OFF THE ELECTRICAL SUPPLY

DESCRIPTION

This furnace is equipped with an induced-draft vent blower and atmospheric burner. Combustion air is taken from the space or area in which the furnace is installed and drawn into the burner through the louvers in the front panel. Products of combustion are drawn from the furnace by the vent blower and discharged into the vent box then through the flue pipe to the outside atmosphere.

The furnace is equipped with controls necessary for prop-

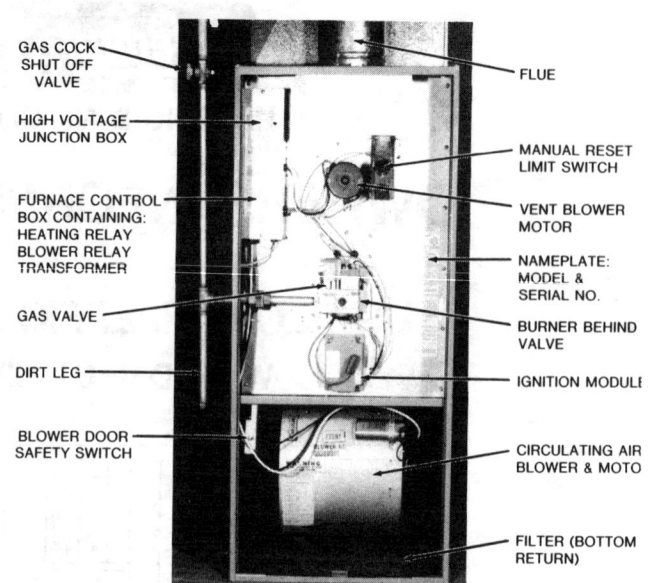

GAS COCK SHUT OFF VALVE
HIGH VOLTAGE JUNCTION BOX
FURNACE CONTROL BOX CONTAINING: HEATING RELAY BLOWER RELAY TRANSFORMER
GAS VALVE
DIRT LEG
BLOWER DOOR SAFETY SWITCH
FLUE
MANUAL RESET LIMIT SWITCH
VENT BLOWER MOTOR
NAMEPLATE: MODEL & SERIAL NO.
BURNER BEHIND VALVE
IGNITION MODULE
CIRCULATING AIR BLOWER & MOTO
FILTER (BOTTOM RETURN)

Figure 1

er operation. The burner ignition and control system incorporates a feature which senses the presence of burner flame. The ignition control will shut-off the gas valve in the event there is an interruption in the gas supply, during burner operation, or if the burner fails to light for any reason. A centrifugal switch prevents burner ignition until after the vent blower is in operation and a flue blockage manual reset limit switch will close the gas valve in the event the flue pipe becomes partially restricted or blocked.

Figure 1 pictures a typical model. The various components referred to in this manual and on the furnace Name Plate are identified.

SEQUENCE OF OPERATION

These furnaces are equipped with an electric spark, direct burner ignition system; therefore, in response to a call for heat by the room thermostat, the burner is lighted by an electric arc at the beginning of each operation cycle. The burner will continue to operate until the thermostat is satisfied at which time all burner flame is extinguished. During the off cycle, no gas energy is consumed. With the room thermostat set below room temperature and with the electrical power and gas supply to the furnace on, the normal sequence of operation is as follows:

1. When the room temperature falls below the setting of the room thermostat, the thermostat energizes the heating relay.

2. When the heating relay closes, a circuit is made starting the vent blower. A circuit is also made through the heating relay to the normally open vent blower centrifugal switch contacts.

3. As the vent blower increases in speed, the contacts of the centrifugal switch will close and complete the electrical circuit through the normally closed limit switch to the electronic ignition module.

4. After 15 to 20 seconds, the electronic ignition module simultaneously energizes: (1) the electric ignition electrode, (2) the gas control valve, and (3) a safety lock out circuit.

5. When the burner lights, a flame sensor, which is part of the ignition electrode assembly, senses the presence

of burner flame and causes the safety lock out circuit in the ignition module to be de-energized. This allows the gas valve to remain open and the burner to operate.

6. If the burner fails to light within 6-8 seconds from the time the ignition control is first energized by the centrifugal switch, the ignition module will de-energize; (1) the gas valve, and (2) the ignition electrode. The ignition control will again energize: the gas valve and the ignition electrode after a 15-20 second delay. If after three trials for ignition the burner still fails to light, the safety lock out circuit in the ignition module will de-energize and lock off: (1) the gas valve, and (2) the ignition electrode. The system will remain in a lock out mode until the room thermostat is set below room temperature causing the thermostat contacts to open and release the safety lock out circuit. Then setting the room thermostat above room temperature and causing the thermostat contacts to close will start the system to try for ignition again.

7. The lapsed time from the moment the wall thermostat closes to when the burner lights may be from 20 to 40 seconds. This delay in the ignition sequence is caused by: (1) the time required for the vent blower to develop sufficient speed to activate the centrifugal switch and (2) the 15 to 20 second delay designed into the ignition module. The lapsed time will also be effected by the temperature within the furnace and in the flue piping.

The 15 to 20 second delay designed into the ignition module is a purge cycle. This allows the vent blower time to replenish the heat exchanger with fresh air so ignition can occur safely.

8. One to two minutes after the burner has lighted, the normally open contacts of the fan switch close and the furnace air circulation blower runs.

NOTE

If a heating/cooling thermostat is being used and the fan switch is set in the "ON" (continuous blower) position, the furnace air circulation blower will run at the air conditioning speed. If the wall thermostat calls for heat, the furnace air circulation blower will shut off for 20 to 40 seconds then the burner will light. One to two minutes after the burner has lighted, the furnace air circulation blower will begin running again but at heating speed. There is no pause in the furnace air circulation blower operation when the wall thermostat is satisfied; the furnace air circulation blower just changes over to cooling (continuous blower) speed.

9. When the room thermostat is satisfied the circuit to the heating relay is broken and the heating relay contacts return to the normally open position. The circuit to the vent blower and the ignition module is broken and the burner is extinguished. Then as heat is drawn from the heat exchanger and the air temperature is reduced to below the fan switch setting, the fan switch will open which stops the furnace air circulation blower.

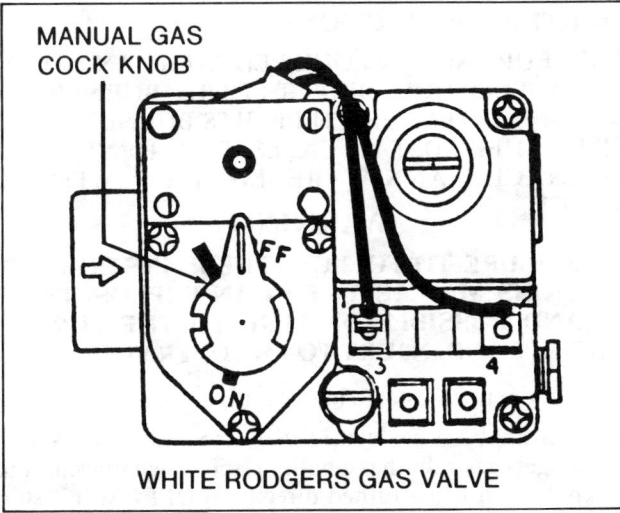

WHITE RODGERS GAS VALVE

Figure 2

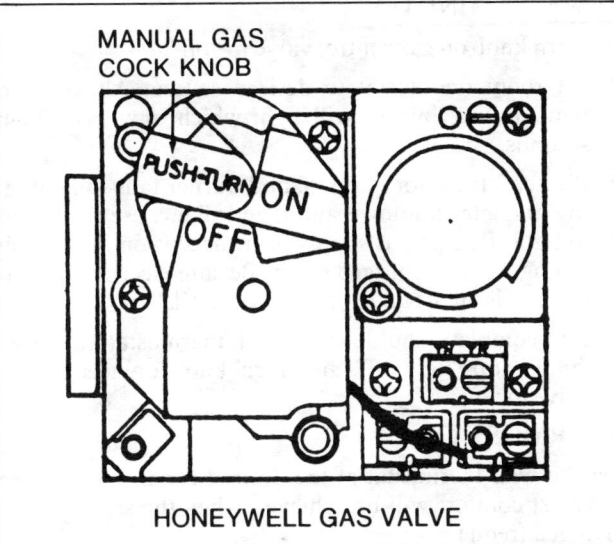

HONEYWELL GAS VALVE

Figure 3

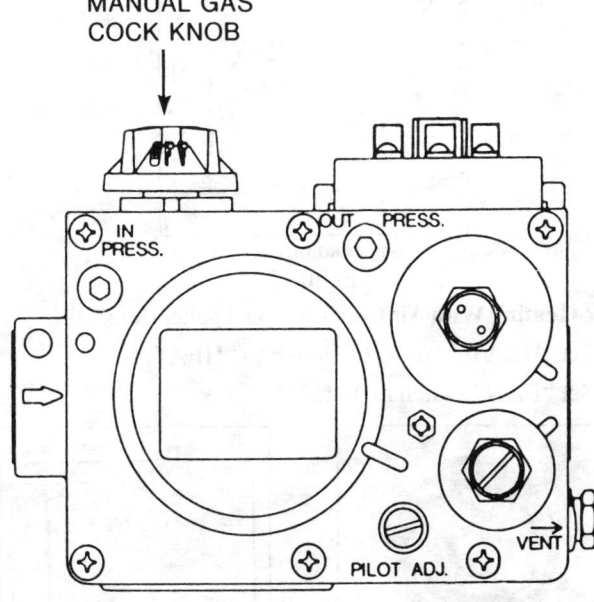

ROBERTSHAW GAS VALVE

Figure 4

3

LIGHTING INSTRUCTIONS

THIS FURNACE IS EQUIPPED WITH AN AUTO-MATIC ELECTRIC, DIRECT SPARK, BURNER IG-NITION SYSTEM WHICH LIGHTS BURNER EACH TIME THERMOSTAT CALLS FOR HEAT. THIS FURNACE CANNOT BE LIGHTED WITH A MATCH.

WARNING

FAILURE TO FOLLOW THESE INSTRUC-TIONS MAY RESULT IN AN EXPLOSION AND POSSIBLE DAMAGE TO THE FUR-NACE AND INJURY TO THE OPERATOR.

1. Set room thermostat to lowest setting — or "off".

2. Turn knob on gas control valve to "off". The White Rodgers gas valve has a built in stop which prevents the knob from being turned directly to OFF. At this stop the knob must be depressed and then turned to OFF. See figure 2.

3. WAIT 5 MINUTES.

4. Turn knob on gas control valve to "on".

5. Set room thermostat to desired setting. Above room temperature, burner will light-which may take 20-40 seconds.

6. If after 3 trials for ignition, and burner fails to light, go to complete shutdown and determine cause for failure to light. Be sure gas supply to furnace is on and supply piping has been purged of air. Be sure electric power to furnace is on. Refer to "If Furnace Fails".

7. For complete shutdown — set thermostat to lowest temperature or "off" and turn knob on gas control valve to "off".

THERMOSTAT

Set the wall thermostat at the desired room temperature. Greatest comfort will be achieved when the setting is not changed frequently.

For energy conservation and economy it is recommended the thermostat be set at 68° for heating and 80° for cooling.

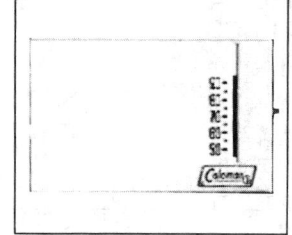

HEATING THERMOSTAT
Figure 5

For Heating With Air Conditioning Applications

1. Set "HEAT — COOL" switch to "HEAT."
2. Set "FAN" switch to "AUTO."

 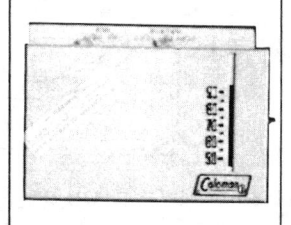

HEATING/AIR CONDITIONING THERMOSTATS
Figure 6

4

Energy Saver Thermostat

This thermostat will provide even greater fuel economy. The thermostat may be set to control the temperature at a "HI" setting during daytime hours and a "LO" setting during night time hours.

Follow the instructions supplied with the thermostat supplied by the manufacturer. If these are not available, check with the installing dealer or contractor for proper setting and operation of your specific thermostat.

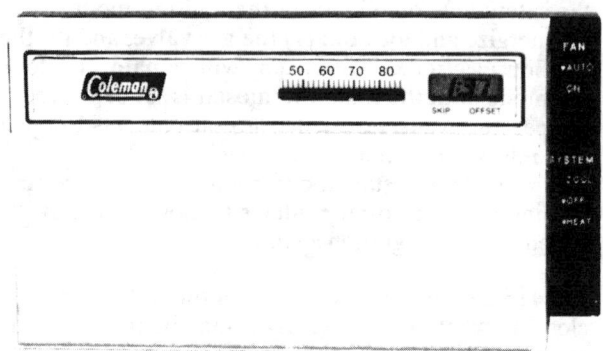

Figure 7

IF FURNACE FAILS TO OPERATE PROPERLY

1. Check setting of thermostat — and position of "Heat — Cool Switch" if air-conditioning is installed. If "Night-Setback" type thermostat is employed be sure thermostat is in the correct operating mode.

2. Check to see that electrical power is "on", or manual reset limit has not tripped.

3. Check to see that the knob on the gas control valve is in the full on position.

4. Make sure filter(s) is clean, return air grills are not obstructed and supply registers are open.

5. Be sure that furnace flue piping is open and unobstructed and flue blockage manual reset limit switch has not tripped.

6. This furnace is equipped with a BLOWER DOOR SAFETY SWITCH. It is located behind the blower door in the upper left hand corner. The switch controls the electrical power to the furnace and prevents the furnace from operating if the blower door is not in place or improperly installed. Door must be engaged at the bottom of the casing and hooked over the casing divider and pushed down firmly.

PERIODIC INSPECTION AND MAINTENANCE

It is recommended the home owner, or user, make an inspection of the furnace at least every 90 days, or more often if desired. It is also recommended a qualified service agency inspect the furnace before each operating season the furnace is used, both heating and air conditioning, and at anytime there is any indication of a malfunction. The owner/user should not attempt to disassemble the furnace unless they are experienced and qualified to do so.

CAUTION

For Safety — TURN OFF ELECTRICAL POWER TO FURNACE — Before performing service such as cleaning, replacing filters, oiling blower motor.

The furnace installation should be examined to determine that:

1. All flue product carrying areas external to the furnace (i.e., chimney, vent connector) are clear and free of obstructions.

2. The vent connector is in place, slopes upward and is physically sound without holes or excessive corrosion.

3. The return air connection is physically sound, is sealed to the furnace casing, and terminates outside the space containing the furnace.

4. The physical support of the furnace is sound without sagging, and the furnace is level.

5. There are no obvious signs of deterioration of the furnace.

Filters:

It is very important that filters in your furnace or air conditioning system be cleaned frequently and replaced when necessary. Clean filters not only provide added comfort and a more healthful environment, but also allow the system to operate more efficiently. Check filters every two or three weeks.

Your furnace is equipped with a permanent-type filter(s) which need not be replaced provided it is cleaned frequently. The permanent filter may be washed in a mild solution of detergent and water and then rinsed thoroughly with clear water. If the pores of the filter media become clogged with dirt or lint which cannot be washed out or the filter becomes damaged, it must be replaced. A replacement filter should be of the permanent type and be of the same dimensional size as the old filter.

In your installation the filter may be located in the bottom of the furnace as shown in Figure 1, or it may be located in the return air duct exterior to the furnace casing. If it is located in the bottom of the furnace casing, under the blower, it will be exposed with the blower door removed. To remove filter, disengage filter from retaining clip at the

Figure 8

front — holding one side down, raise the other side such the filter is at an angle — then pull filter forward to remove from casing. Reinstall the filter in the reverse manner.

If the filter is not located in the bottom of the furnace casing, but in the return duct, and the method of removal and replacement is not obvious, then contact the installing contractor of the furnace for assistance.

Burner Inspection (Figures 9 and 10)

With the furnace in operation, and the upper door of the furnace removed, observe the operation of the burner through the combustion air tube. A normal burner flame will appear as blue with slight yellow tips past the flame spreader. The flame may have intermittent orange tips, particularly when LP gas is being used, which is normal. A bright yellow, or illuminous flame is abnormal. Should this condition be observed, the furnace should be inspected by a qualified service agency and the cause for the abnormal operation determined and corrected.

Burner Ignition Electrode Assembly

The ignition electrode is located at the left side of the burner. The electrode assembly may be observed by looking under the gas valve and burner while the burner is in operation. The electrode rods will glow at a red heat, which is normal. See Figure 9.

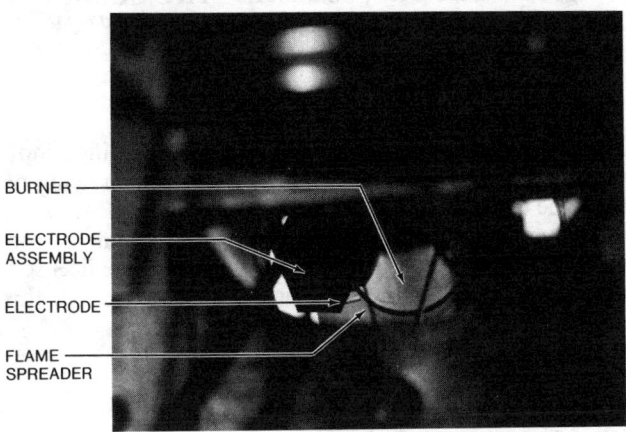

Figure 9

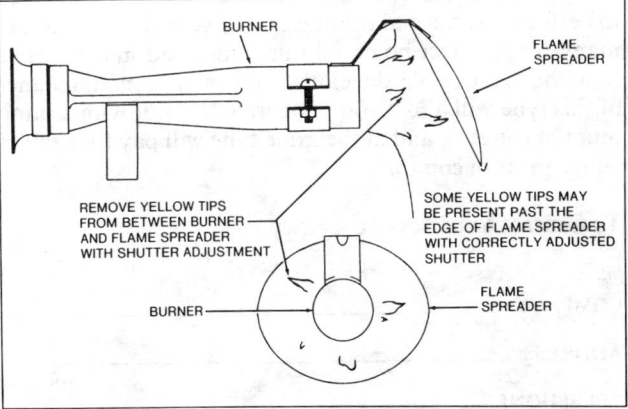

Figure 10

5

Inspection Of Flue Piping

Inspect the flue piping connection at the furnace and the connection of the individual sections to ensure that pipe joints have not become disengaged or there are any openings which could allow leakage of products of combustion. Inspect the termination of the flue outside the structure and look for any indication of carbon or soot streaks. The presence of any soot would indicate a malfunction — the cause must be determined and corrected.

Motor Lubrication

1. If circulating air blower motor is provided with oil ports, lubricate motor every 6 months or at the beginning of each operating season, heating and air conditioning. Use S.A.E. 20 non-detergent type oil. Slowly add 10 to 15 drops at each oil port — DO NOT OVER OIL. Motors which are not provided with oil ports do not require seasonal lubrication.

2. The vent blower motor is permanently lubricated and requires no lubrication.

DANGER

NEVER ATTEMPT TO MODIFY THIS FURNACE — FIRE, EXPLOSION, OR ASPHYXIATION MAY RESULT — IF MALFUNCTION OCCURS, OBTAIN THE ASSISTANCE OF A QUALIFIED SERVICE AGENCY.

IMPORTANT NOTICE

This equipment must be serviced only by qualified individuals specially trained and experienced in servicing of this type equipment and related system components, such as duct systems, air conditioning, etc. Installation and service personnel are required by some states to be licensed. Persons not qualified should not attempt to service this equipment.

YOUR COLEMAN SERVICE MAN

Your furnace's best friend is your Coleman Service Man. If the unit gives any indication of improper operation, call your Coleman Service Man. If the service man is allowed to perform the normal routine care of your furnace, he can many times detect potential difficulties and make corrections before trouble develops. Preventative maintenance of this type will allow you to operate the unit with a minimum of concern, and at the same time will pay for itself in added years of comfort.

To Contact Your Service Agency

(fill in)

COMPANY: _____

ADDRESS: _____

TELEPHONE: _____

When You Call For Service Assistance —

Very often time can be saved if you will give the service agency the MODEL and SERIAL NUMBER of your furnace. This will enable him to determine the specific components used, perhaps better identify the possible problem and be prepared if a service call is required.

This manual applies to models: 2845, 2865, 2880 and 2895.

See the nameplate on the furnace for model, series and serial number of your furnace. Figure 11 shows an example of the nameplate and figure 1 shows its location.

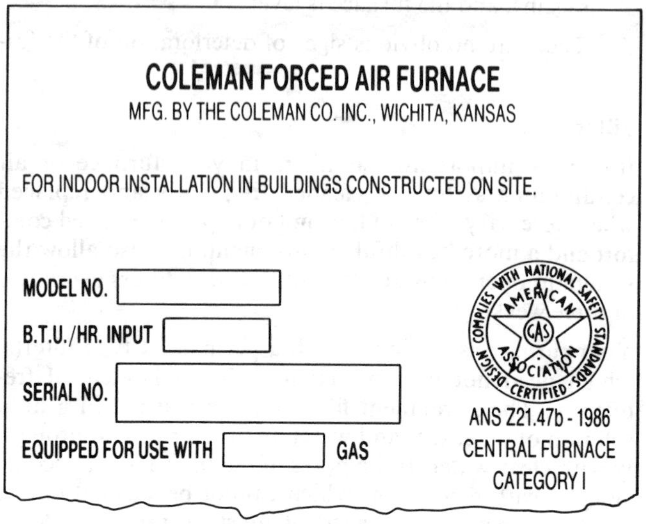

COLEMAN FORCED AIR FURNACE
MFG. BY THE COLEMAN CO. INC., WICHITA, KANSAS

FOR INDOOR INSTALLATION IN BUILDINGS CONSTRUCTED ON SITE.

MODEL NO. [   ]

B.T.U./HR. INPUT [   ]

SERIAL NO. [   ]

EQUIPPED FOR USE WITH [   ] GAS

ANS Z21.47b - 1986
CENTRAL FURNACE
CATEGORY I

Figure 11

Periodic Inspection Of Heat Exchanger

Prior to the heating season, when the service man is doing the periodic maintenance check of your furnace, have him inspect the inside surfaces of the heat exchanger drum for material corrosion, scaling, sooting, and damage. If any of these conditions are found, the cause of the condition must be determined and corrected. The heat exchanger must be replaced if it is damaged or if extensive soot is observed in the heat exchanger drum.

TRIPS AND VACATIONS

The furnace is equipped with controls which are designed to shut off the furnace burners should malfunction occur. However it is best never to assume that the furnace will operate unattended for long periods of time — especially if there is a possibility of damage to your property because of freezing. If you plan to be away for an extended period, it is suggested arrangements be made for someone to check the house and furnace operation frequently.

Comfort-Aire

INSTALLATION INSTRUCTIONS

FOR UPFLOW & DOWNFLOW / HORIZONTAL AND HORIZONTAL ONLY INDUCED DRAFT GAS FURNACES

WARNING: If the information in this manual is not followed exactly, a fire or explosion may result, causing property damage, personal injury or loss of life.

— Do not store or use gasoline or other flammable vapors and liquids, or other combustible materials in the vicinity of this or any other appliance.

— WHAT TO DO IF YOU SMELL GAS

- Do not try to light any appliance
- Do not touch any electrical switch; do not use any phone in your building.
- Immediately call your gas supplier from a neighbor's phone. Follow the gas supplier's instructions.
- If you cannot reach your gas supplier, call the fire department.

— Installation and service must be performed by a qualified installer, service agency or the gas supplier.

DO NOT DESTROY. PLEASE READ CAREFULLY AND KEEP IN A SAFE PLACE FOR FUTURE REFERENCE.

92-20552-89-02
SUPERSEDES 92-20552-89-01

CAUTION: *TO ENSURE PROPER INSTALLATION AND OPERATION OF THIS PRODUCT, COMPLETELY READ ALL INSTRUCTIONS PRIOR TO ATTEMPTING TO ASSEMBLE, INSTALL, OPERATE, MAINTAIN OR REPAIR THE PRODUCT.*

GENERAL INFORMATION

INSTALLATION

This furnace should be installed in accordance with the American National Standard Z223.1 - latest edition booklet entitled "National Fuel Gas Code" (NFPA 54) (in Canada, CAN/CGA B149.1 and .2 Installation Codes for gas burning appliances), and the requirements or codes of the local utility or other authority having jurisdiction including local plumbing or waste water codes.

Additional helpful publications available from the "National Fire Protection Association" are: NFPA-90A – Installation of Air Conditioning and Ventilating Systems 1985 or latest edition. NFPA-90B – Warm Air Heating and Air Conditioning Systems 1984.

These publications are available from:
National Fire Protection Association, Inc.
Batterymarch Park
Quincy, MA 02269

Canadian Gas Association
55 Scarsdale Road
Don Mills, Ontario, Canada M3B, 2R3

> **WARNING: DO NOT INSTALL A STANDING PILOT SYSTEM FURNACE IN A HORIZONTAL POSITION. IMPROPER INSTALLATION CAN CAUSE PROPERTY DAMAGE, BODILY INJURY, OR DEATH FROM SMOKE, FIRE OR CARBON MONOXIDE.**

CAUTION: Do not use this furnace as a heater in a building under construction. This furnace can be severely damaged due to the abnormal environment caused by construction. Chlorides from sources such as paint, stain or varnish, tile and counter cements, adhesives and foam insulations are abundant in a structure under construction and can be highly corrosive. Low return air temperature can cause condensation in the furnace and damage that can shorten the life of the furnace.

LOCATION

WARNING: THIS FURNACE IS NOT APPROVED FOR INSTALLATION IN A MOBILE HOME. DO NOT INSTALL THIS FURNACE IN A MOBILE HOME. INSTALLATION IN A MOBILE HOME COULD CAUSE FIRE, PROPERTY DAMAGE AND PERSONAL INJURY.

This furnace is suitable for installation in buildings constructed on-site. This heating unit should be located near the chimney and should be centralized with respect to the heat distribution system as much as practicable. When installed in a utility room, the door of the room should be wide enough to allow the largest part of the furnace to enter, or to permit the replacement of another appliance, such as a water heater.

CLEARANCE – ACCESSIBILITY

The design of forced air furnaces with input ratings as listed in table below is certified by A.G.A. Laboratories and CGA for the clearances to combustible materials shown in inches.

See name/rating plate and clearance label for specific model number and clearance information.

Service clearance of at least 24″ is recommended in front of all furnaces.

ACCESSIBILITY CLEARANCES WHERE GREATER MUST TAKE PRECEDENCE OVER FIRE PROTECTION CLEARANCES.

UPFLOW — Certified for use on combustible floor.

CAUTION: Furnaces must not be installed directly on carpeting, tile or other combustible material other than wood flooring.

DOWNFLOW WARNING: UNIT DESIGN IS CERTIFIED FOR INSTALLATION ON NON-COMBUSTIBLE FLOOR. A SPECIAL COMBUSTIBLE FLOOR SUB-BASE IS REQUIRED WHEN INSTALLING ON A COMBUSTIBLE FLOOR. FAILURE TO INSTALL THE SUB-BASE MAY RESULT IN FIRE, PROPERTY DAMAGE AND PERSONAL INJURY. THIS SPECIAL BASE IS OFFERED AS AN ACCESSORY FROM THE FACTORY.

UPFLOW WARNING: A SOLID METAL BASE PLATE (SEE TABLE BELOW OR FURNACE CLEARANCE LABEL FOR PART NUMBER) MUST BE IN PLACE WHEN THE FURNACE IS INSTALLED WITH SIDE OR REAR AIR RETURN DUCTS. FAILURE TO INSTALL A BASE PLATE COULD CAUSE PRODUCTS OF COMBUSTION TO BE CIRCULATED INTO THE LIVING SPACE AND CREATE POTENTIALLY HAZARDOUS CONDITIONS, INCLUDING CARBON MONOXIDE POISONING. REFER TO SECTION ON "CIRCULATING AIR SUPPLY" (PAGE 11) FOR RETURN AIR DUCTWORK INSTRUCTIONS.

FURNACE WIDTH	BASE PLATE NO.	BASE PLATE SIZE
14″	AE-61229-02	12¾″x24¹¹⁄₁₆″
17½″	AE-61229-03	16¼″x24¹¹⁄₁₆″
21″	AE-61229-04	19¾″x24¹¹⁄₁₆″
24½″	AE-61229-05	23¼″x24¹¹⁄₁₆″

A gas-fired furnace for installation in a residential garage must be installed so that the burner(s) and the ignition source are located not less than 18″ above the floor and the furnace is located or protected to avoid physical damage by vehicles.

WARNING: COMBUSTIBLE MATERIAL MUST NOT BE PLACED ON OR AGAINST THE FURNACE JACKET OR WITHIN 6″ OF THE VENT PIPE. THE AREA AROUND THE FURNACE MUST BE KEPT CLEAR AND FREE OF ALL COMBUSTIBLE MATERIALS INCLUDING GASOLINE AND OTHER FLAMMABLE VAPORS AND LIQUIDS. THE HOMEOWNER SHOULD BE CAUTIONED THAT THE FURNACE AREA MUST NOT BE USED AS A BROOM CLOSET OR FOR ANY OTHER STORAGE PURPOSES.

CLEARANCE TO COMBUSTIBLE MATERIAL (INCHES) UPFLOW MODELS

Model	A	B	C	D	E	Left Side	REDUCED CLEARANCE (IN.) Right Side	Back	Top	Front	Vent	Ship. Wgts.
05	14	$12\frac{27}{32}$	$10\frac{3}{8}$	①	$11\frac{1}{2}$	0	4②	0	1	6	6③	85 lbs.
07	17.5	$16\frac{11}{32}$	$12\frac{1}{8}$	①	15	0	3②	0	1	6	6③	105 lbs.
10(A)	17.5	$16\frac{11}{32}$	$12\frac{1}{8}$	①	15	0	3②	0	1	6	6③	115 lbs.
10(B)	21	$19\frac{27}{32}$	$13\frac{7}{8}$	①	$18\frac{1}{2}$	0	0	0	1	6	6③	120 lbs.
12	24.5	$23\frac{11}{32}$	$15\frac{5}{8}$	①	22	0	0	0	1	6	6③	140 lbs.
15	24.5	$23\frac{11}{32}$	$15\frac{5}{8}$	①	22	0	0	0	1	6	6③	150 lbs.

① May require 3″ to 4″ or 3″ or 5″ adapter.
② May be 0″ with type B vent.
③ May be 1″ with type B vent.

TOP   BOTTOM

$24\frac{1}{2}$

$24\frac{1}{2}$

C   E R.A.

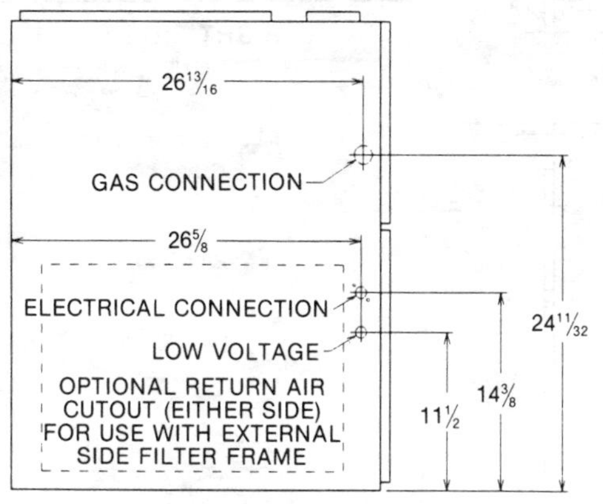

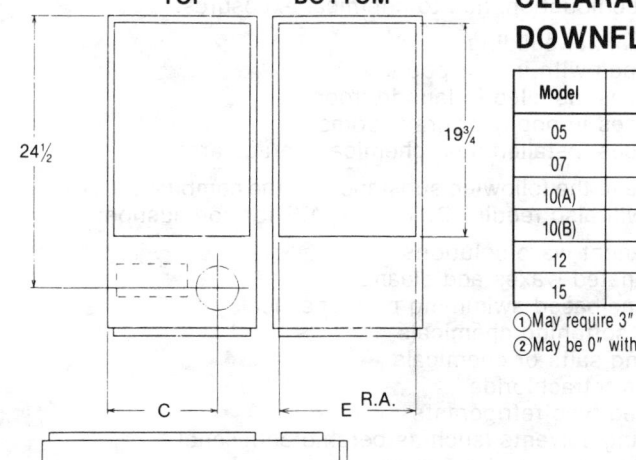

$26\frac{13}{16}$

GAS CONNECTION

$26\frac{5}{8}$

ELECTRICAL CONNECTION
LOW VOLTAGE
OPTIONAL RETURN AIR CUTOUT (EITHER SIDE) FOR USE WITH EXTERNAL SIDE FILTER FRAME

$24\frac{11}{32}$ $14\frac{3}{8}$ $11\frac{1}{2}$

LEFT SIDE

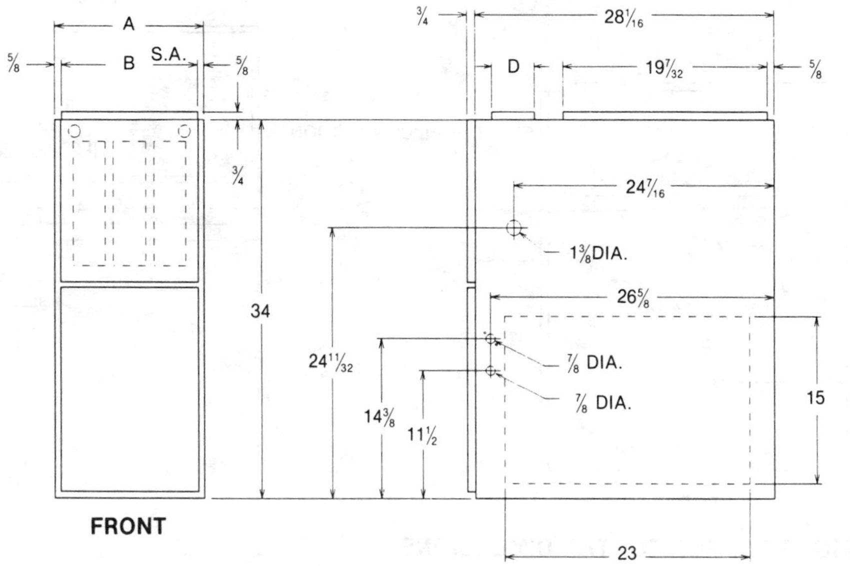

A   B S.A. $\frac{5}{8}$ $\frac{5}{8}$ $\frac{3}{4}$ $28\frac{1}{16}$

$\frac{3}{4}$ D $19\frac{7}{32}$ $\frac{5}{8}$

$\frac{3}{4}$

34

$24\frac{7}{16}$ $1\frac{3}{8}$ DIA.

$26\frac{5}{8}$

$24\frac{11}{32}$ $\frac{7}{8}$ DIA. $\frac{7}{8}$ DIA.

$14\frac{3}{8}$ $11\frac{1}{2}$ 15

23

FRONT   RIGHT SIDE

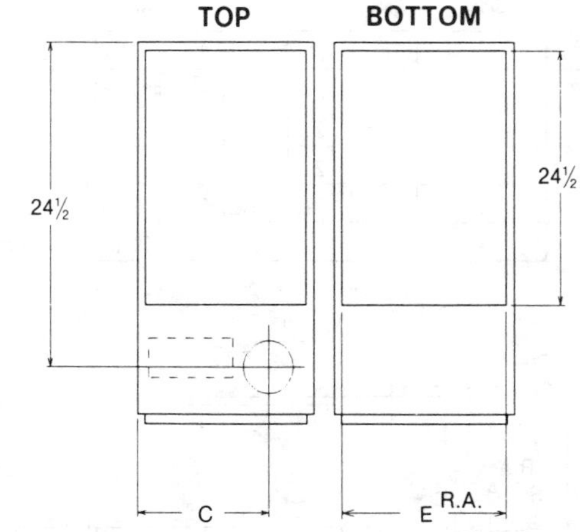

FIGURE 1. UPFLOW DIMENSIONS

CLEARANCE TO COMBUSTIBLE MATERIAL (INCHES) DOWNFLOW MODELS

Model	A	B	C	D	E	Left Side	REDUCED CLEARANCE (IN.) Right Side	Back	Top	Front	Vent	Ship. Wgts.
05	14	$12\frac{27}{32}$	$10\frac{3}{8}$	①	$13\frac{1}{8}$	0	4②	0	1	6	6③	85 lbs.
07	17.5	$16\frac{11}{32}$	$12\frac{1}{8}$	①	$16\frac{5}{8}$	0	3②	0	1	6	6③	105 lbs.
10(A)	17.5	$16\frac{11}{32}$	$12\frac{1}{8}$	①	$16\frac{5}{8}$	0	3②	0	1	6	6③	115 lbs.
10(B)	21	$19\frac{27}{32}$	$13\frac{7}{8}$	①	$20\frac{1}{2}$	0	0	0	1	6	6③	120 lbs.
12	24.5	$23\frac{11}{32}$	$15\frac{5}{8}$	①	$23\frac{5}{8}$	0	0	0	1	6	6③	140 lbs.
15	24.5	$23\frac{11}{32}$	$15\frac{5}{8}$	①	$23\frac{5}{8}$	0	0	0	1	6	6③	150 lbs.

① May require 3″ to 4″ or 3″ or 5″ adapter.   ③ May be 1″ with type B vent.
② May be 0″ with type B vent.

TOP   BOTTOM

$24\frac{1}{2}$

$19\frac{3}{4}$

C   E R.A.

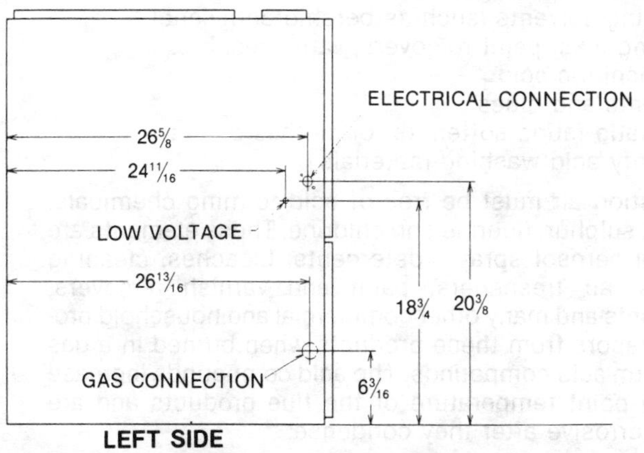

ELECTRICAL CONNECTION

$26\frac{5}{8}$

$24\frac{11}{16}$

LOW VOLTAGE

$26\frac{13}{16}$

$18\frac{3}{4}$ $20\frac{3}{8}$

GAS CONNECTION

$6\frac{3}{16}$

LEFT SIDE

FIGURE 2. DOWNFLOW DIMENSIONS

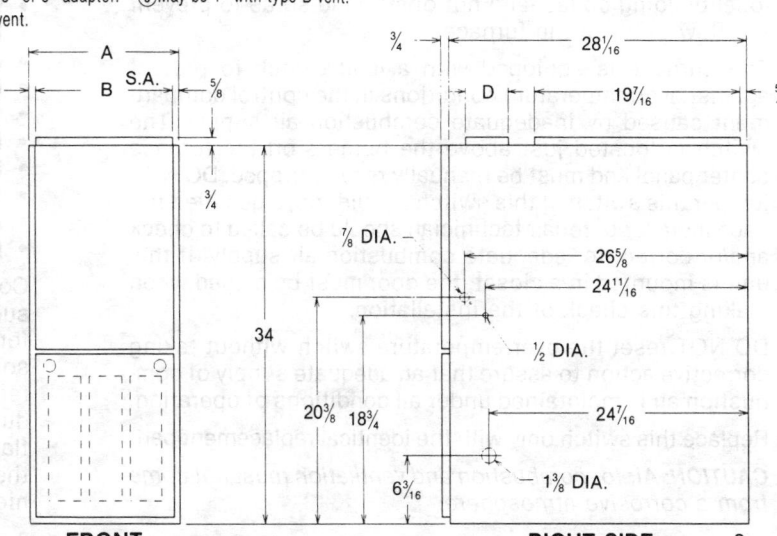

A   B S.A. $\frac{5}{8}$ $\frac{5}{8}$ $\frac{3}{4}$ $28\frac{1}{16}$

$\frac{3}{4}$ D $19\frac{7}{16}$ $\frac{5}{8}$

$\frac{3}{4}$

34

$\frac{7}{8}$ DIA.

$26\frac{5}{8}$

$24\frac{11}{16}$

$\frac{1}{2}$ DIA.

$20\frac{3}{8}$ $18\frac{3}{4}$

$24\frac{7}{16}$

$6\frac{3}{16}$

$1\frac{3}{8}$ DIA.

FRONT   RIGHT SIDE   3

CLEARANCE TO COMBUSTIBLE MATERIAL (INCHES) HORIZONTAL MODELS

LEFT SIDE

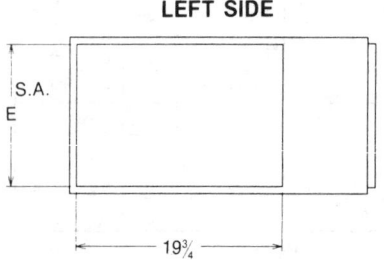

Model	A	B	C	D	E	Sides	REDUCED CLEARANCE (IN.)				Ship. Wgts.
							Back	Top	Front	Vent	
05	14	12²⁷⁄₃₂	10³⁄₈	①	13¹⁄₈	1	0	2	6	6②	85 lbs.
07	17.5	16¹¹⁄₃₂	12¹⁄₈	①	16⁵⁄₈	1	0	2	6	6②	105 lbs.
10(A)	17.5	16¹¹⁄₃₂	12¹⁄₈	①	16⁵⁄₈	1	0	2	6	6②	115 lbs.
10(B)	21	19²⁷⁄₃₂	13⁷⁄₈	①	20¹⁄₈	1	0	2	6	6②	120 lbs.
12	24.5	23¹¹⁄₃₂	15⁵⁄₈	①	23⁵⁄₈	1	0	2	6	6②	140 lbs.
15	24.5	23¹¹⁄₃₂	15⁵⁄₈	①	23⁵⁄₈	1	0	2	6	6②	150 lbs.

① May require 3" to 4" or 3" to 5" adapter.
② May be 1" with type B vent.

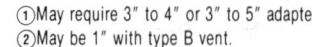

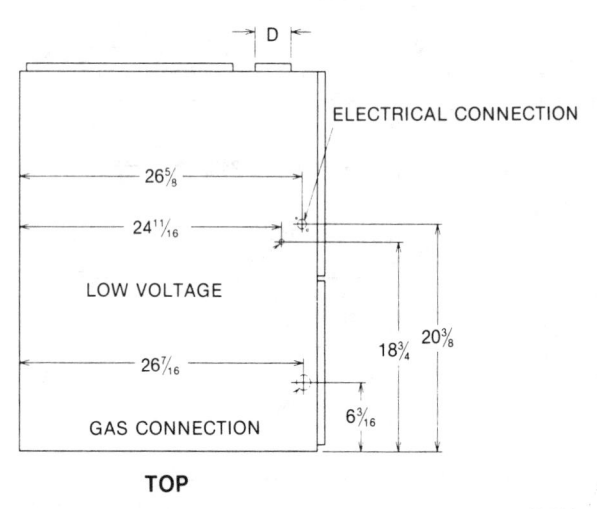

FRONT

RIGHT SIDE

TOP

FIGURE 3. HORIZONTAL DIMENSIONS

COMBUSTION & VENTILATION AIR

WARNING: THE FURNACE AND ANY OTHER FUEL-BURNING APPLIANCE MUST BE PROVIDED WITH ENOUGH FRESH AIR FOR PROPER COMBUSTION AND VENTILATION OF FLUE GASES. MOST HOMES WILL REQUIRE THAT OUTSIDE AIR BE SUPPLIED INTO THE FURNACE AREA.

Adequate facilities for providing air for combustion and ventilation must be provided in accordance with section 5.3, Air for Combustion and Ventilation, of the National Fuel Gas Code, ANSI Z223.1-1988, or applicable provisions for the local building codes, and not obstructed so as to prevent the flow of air to the furnace.

This furnace is equipped with a limit switch to protect against overtemperature conditions in the control compartment caused by inadequate combustion air supply. The switch is located just above the burners on the furnace center panel and must be manually reset if tripped. DO NOT jumper this switch. If this switch should trip, a qualified furnace installer or repair technician should be called to check and/or correct for adequate combustion air supply. If this unit is mounted in a closet, the door must be closed when making this check of the installation.

DO NOT reset the overtemperature switch without taking corrective action to assure that an adequate supply of combustion air is maintained under all conditions of operation.

Replace this switch only with the identical replacement part.

CAUTION: Air for combustion and ventilation must not come from a corrosive atmosphere.

The following types of installation will require OUTDOOR AIR for combustion, due to chemical exposures:

- Commercial buildings
- Buildings with indoor pools
- Furnaces installed in laundry rooms
- Furnaces in hobby or craft rooms
- Furnaces installed near chemical storage areas.

Exposure to the following substances in the combustion air supply will also require OUTDOOR AIR for combustion:

- Permanent wave solutions
- Chlorinated waxes and cleaners
- Chlorine-based swimming pool chemicals
- Water softening chemicals
- De-icing salts or chemicals
- Carbon tetrachloride
- Halogen type refrigerants
- Cleaning solvents (such as perchloroethylene)
- Printing inks, paint removers, varnishes, etc.
- Hydrochloric acid
- Cements and glues
- Antistatic fabric softeners for clothes dryers
- Masonry acid washing materials

Combustion air must be free of acid forming chemicals; such as, sulphur, fluorine and chlorine. These elements are found in aerosol sprays, detergents, bleaches, cleaning solvents, air fresheners, paint and varnish removers, refrigerants and many other commercial and household products. Vapors from these products when burned in a gas flame form acid compounds. The acid compounds increase the dew point temperature of the flue products and are highly corrosive after they condense.

WARNING: ALL FURNACE INSTALLATIONS MUST COMPLY WITH THE NATIONAL FUEL GAS CODE AND LOCAL CODES TO PROVIDE ADEQUATE COMBUSTION AND VENTILATION AIR FOR THE FURNACE.

Combustion air requirements are determined by whether the furnace is in an open (unconfined) area or in a confined space such as a closet or small room.

EXAMPLE 1.
FURNACE LOCATED IN UNCONFINED SPACE

Using indoor air for combustion.

An unconfined space must have at least 50 cubic feet for each 1,000 Btuh of the total input for all appliances in the space. Here are a few examples of the room sizes required for different inputs. The sizes are based on 8 foot ceilings.

Btuh Input	Minimum Sq. Feet With 8' Ceiling	Typical Room Size With 8' Ceiling
50,000	336	14'x24' or 18'x18'
75,000	469	15'x31' or 20'x24'
100,000	625	20'x31' or 25'x25'
125,000	833	23'x34' or 26'x30'
150,000	938	25'x38' or 30'x31'

If the open space containing the furnace is in a building with tight construction (contemporary construction), outside air may still be required for the furnace to burn and vent properly. Outside air openings should be sized the same as for a confined space.

EXAMPLE 2.
FURNACE LOCATED IN CONFINED SPACE

A confined space (any space smaller than shown above as "unconfined") must have two openings into the space. One opening must be within 12" of the ceiling and the other must be within 12" of the floor. The openings must be sized by how they are connected to the heated area or to the outside, and by the input of all appliances in the space.

If confined space is within a building with tight construction, combustion air must be taken from outdoors or area freely communicating with the outdoors.

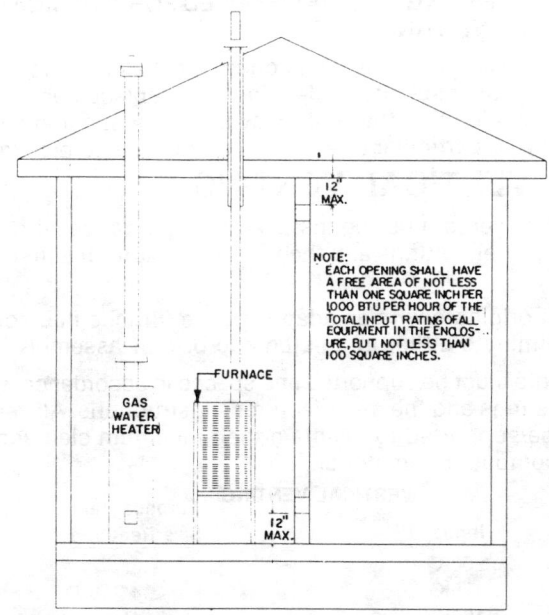

FIGURE 4. AIR FROM HEATED SPACE

A. USING INDOOR AIR FOR COMBUSTION

If combustion air is taken from the heated area, the openings must each have at least 100 square inches of free area. Each opening must have at least one square inch of free area for each 1,000 Btuh of total input in the space. Here are some examples of typical openings required.

Btuh Input	Free Area Each Opening
50,000	100 Square Inches
100,000	100 Square Inches
150,000	150 Square Inches

CAUTION: Air should not be taken from a heated space with a fireplace, exhaust fan or other device that may produce a negative pressure.

B. USING OUTDOOR AIR FOR COMBUSTION

If combustion air is taken from outdoors through vertical ducts, the openings and ducts must have at least one square inch of free area for each 4,000 Btuh of total appliance input. Here are some typical sizes.

Btuh Input	Free Area, Each Opening	Round Pipe Size
50,000	12.5 Square Inches	4"
75,000	18.75 Square Inches	5"
100,000	25 Square Inches	6"
125,000	31.25 Square Inches	7"
150,000	37.5 Square Inches	7"

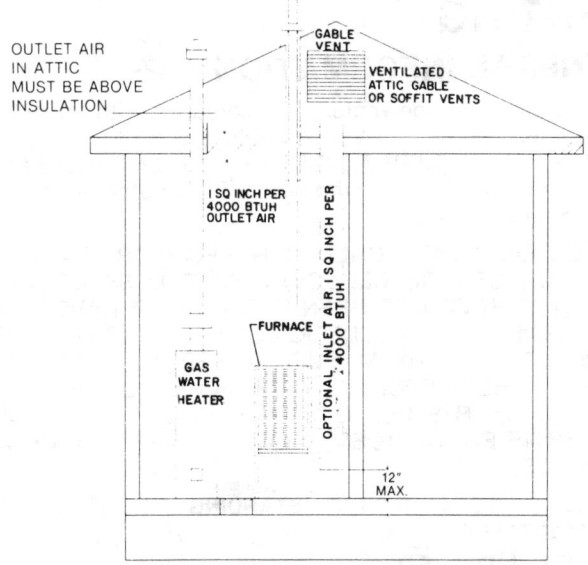

FIGURE 5. AIR FROM ATTIC

CAUTION: Do not take air from an attic space that is equipped with power ventilation.

If combustion air is taken from outdoors through horizontal ducts, the openings and ducts must have at least one square inch of free area for each 2,000 Btuh of total appliance input. Here are typical sizes.

Btuh Input	Free Area, Each Opening	Round Pipe Size
50,000	25 Square Inches	6"
75,000	37.5 Square Inches	7"
100,000	50 Square Inches	8"
125,000	62.5 Square Inches	9"
150,000	75 Square Inches	10"

If unit is installed where there is an exhaust fan, sufficient ventilation must be provided to prevent the exhaust fan from creating a negative pressure in the room.

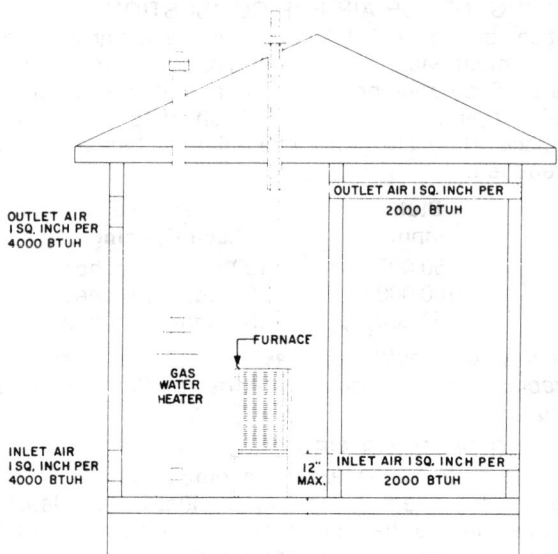

FIGURE 6. OUTSIDE AIR USING HORIZONTAL INLET & OUTLET

Combustion air openings must not be restricted in any manner.

CONSULT LOCAL CODES FOR SPECIAL REQUIREMENTS.

Air openings in furnace casing front, return air grilles, and warm air registers must not be obstructed.

VENTING
GENERAL INFORMATION

The furnace must be vented in accordance with these instructions, AGA-GAMA Venting Tables and the "National Fuel Gas Code" (NFPA No. 54-1988, ANSI Z223.1-1988 and requirements or codes of the local utility or other authority having jurisdiction.

NOTE: DEVICES ATTACHED TO THE FLUE OR VENT FOR THE PURPOSE OF REDUCING HEAT LOSS UP THE CHIMNEY HAVE NOT BEEN TESTED AND HAVE NOT BEEN INCLUDED IN THE DESIGN CERTIFICATION OF THIS FURNACE. WE, THE MANUFACTURER, CANNOT AND WILL NOT BE RESPONSIBLE FOR INJURY OR DAMAGE CAUSED BY THE USE OF SUCH UNTESTED AND/OR UNCERTIFIED DEVICES, ACCESSORIES OR COMPONENTS.

WARNING: DO NOT VENT STANDING PILOT MODELS HORIZONTALLY.

DRAFT INDUCER

CAUTION: Vent pipe attaching holes must be pre-drilled in draft inducer collar to prevent plastic material from crack-

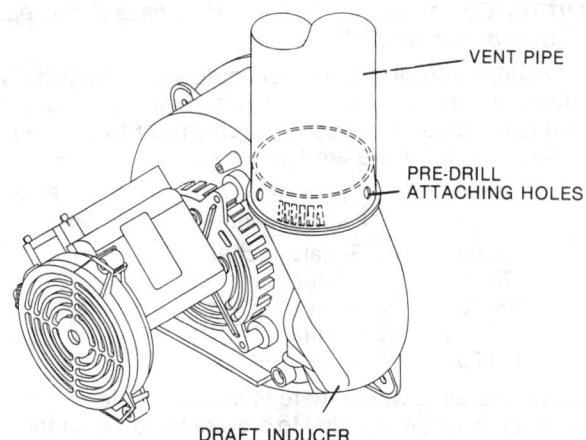

FIGURE 7. ATTACHING TO DRAFT INDUCER COLLAR

---

ing. Drill ⅛" diameter holes through vent pipe and collar and use #8 screws to attach. See Figure 7.

FURNACE CATEGORY INFORMATION

This furnace is shipped as a Category I type induced draft furnace. A Category I furnace operates with a nonpositive vent pressure and has a vent gas temperature at least 140°F above the dew point of the vent gases. A Category I type may be a draft hood equipped furnace or have a fan assisted combustion system (induced draft). The inducer is used to pull flue products through the combustion chamber and as they leave the furnace most of the energy has been dissipated. The buoyant effect of the flue gases provides venting to the outdoors.

During the off cycle, the inducer is off and there is very little flow through the vent, cooling the vent. During the on cycle there is no dilution airflow, as with a draft hood type furnace. Although the vent heats up rapidly without dilution air, the flue products contain more water vapor, which results in a higher dew point temperature. It is most important that you follow the guidelines in these instructions to prevent the possible formation of condensation in the venting system.

As a Category I furnace it may be vented vertically with type B vent pipe and also may be common vented, as described in these instructions.

IMPORTANT APPLICATION NOTES

When the furnace is used for a replacement, the existing vent system should be inspected for any obstructions and assurance that there is not an obstruction or blockage or any signs of corrosion.

VENT PIPE MAY BE TYPE "B" OR SPECIAL VENT SYSTEM (S.V.S.).

COMMON VENTING IS ALLOWED WITH VERTICAL B VENT SYSTEMS AND LINED INTERIOR MASONRY CHIMNEYS. FOLLOW THE AGA/GAMA VENTING TABLES FOR PROPER INSTALLATION PRACTICES.

SINGLE WALL VENT CONNECTORS TO "B VENT CHIMNEYS" MAY BE USED UNDER THE GUIDELINES OF THE AGA/GAMA VENTING TABLES.

THE SPECIAL VENT SYSTEM MUST BE USED FOR HORIZONTAL VENTING AND MAY BE USED FOR A DEDICATED VERTICAL VENTING SYSTEM.

The entire length of the vent connector shall be readily accessible for inspection, cleaning and replacement. It is recommended that the venting system be inspected annually, paying particular attention to corrosion and blockage.

"B" VERTICAL VENTING

Type "B" vents must be installed in accordance with the terms of their listings and the vent manufacturer's instructions.

It must originate with an adapter at the furnace flue collar and terminate either in a listed cap or roof assembly.

"B" vents must be supported and spaced in accordance with their listings and the manufacturer's instructions. All vents must be supported to maintain their minimum clearances from combustible material.

VERTICAL VENTING	
Input	Furnace Vent Size Required
50K	3"
75K	*4"
100K	*4"
125K	*4"
150K	*5"

*NOTE: All furnaces have a 3" vent connection as shipped from the factory. A 3" to 4" or 3" to 5" vent transition is required on all but the 50,000 BTUH model when vertically vented or common vented with metal vent pipe. **THE VENT TRANSITION CONNECTION MUST BE MADE AT THE FURNACE VENT EXIT.**

VERTICAL VENT SYSTEMS:

1. Must extend at least 5 feet above the furnace vent connection. (See A, Figure 8)

2. Must not have horizontal runs that total more than 75% of the vertical rise of the vent system. (See B, Figure 8)

3. Must not be smaller than the vent connection size on the furnace.

4. Must not be over 45 feet in total length.

5. Must not have more than 5 elbows. Do not increase pipe length for fewer elbows used. (See C, Figure 8)

6. Must extend at least 3 feet above the roof penetration and 2 feet higher than any object within 10 feet. If a UL listed vent cap is used, it may be installed in accordance with its listing terms and manufacturer's instructions.

7. Must rise ¼″ per foot away from the furnace on horizontal runs and be supported with straps or hangers so it has no sags or dips. Supports at 4 foot intervals and all elbows are recommended.

8. The vent connector must be mechanically fastened to the outlet collar of the furnace with at least (2) sheet metal screws except vent connectors of type B-1 material which shall be assembled in accordance with the manufacturer's instructions. See Figure 7.

NOTE: Refer to the AGA-GAMA venting tables for venting category I furnaces.

High efficiency furnaces should not have dedicated vents into any chimney totally exposed to the outdoors or any unlined masonry chimney. Chimneys with these conditions should be used as a chase to run B vent to the outside. See Figure 9.

If B vent is used with a chimney as a chase, do not vent another appliance into the chimney. Outside of B vent must not be exposed to flue products.

WARNING: DO NOT CONNECT THIS FURNACE TO A CHIMNEY USED TO VENT A SOLID FUEL APPLIANCE (WOOD OR COAL).

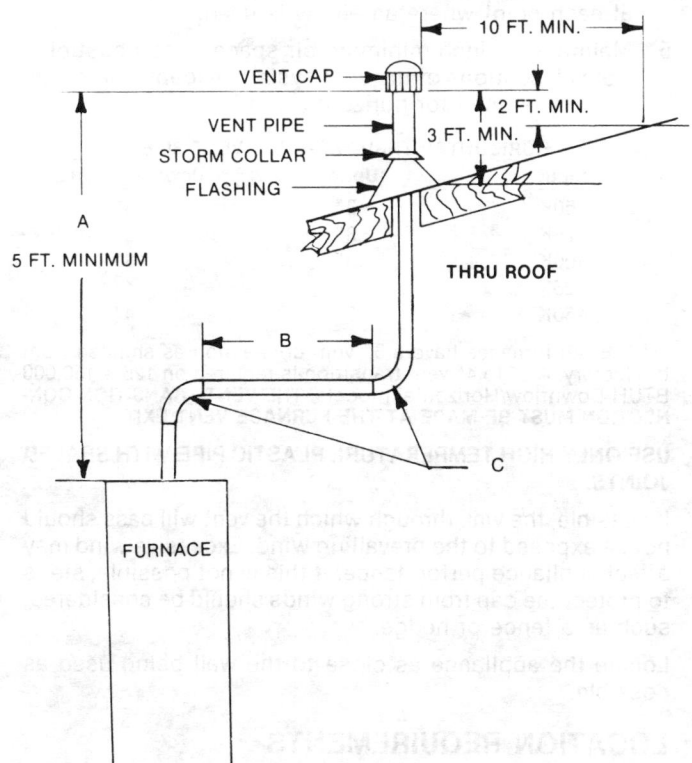

FIGURE 8. TYPICAL VENTING WITH "B" VENT

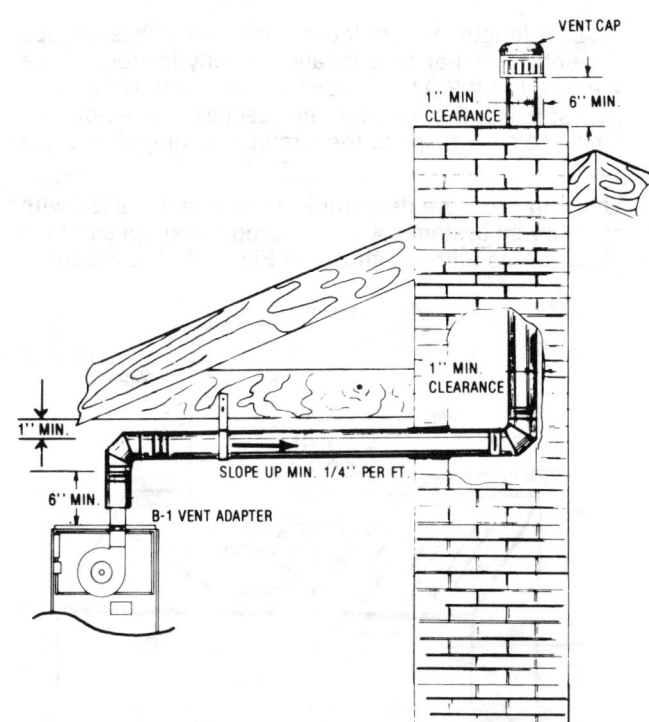

FIGURE 9. DEDICATED VENTING THROUGH CHIMNEY WITH "B" VENT

VENT PRESSURE DETECTOR ASSEMBLY

In Canada, the vent pressure detector assembly must be used with type "B-1" vent systems. It consists of the following components:

1. 7″ long pipe aluminized steel.

2. One standoff bracket secured with two sheet metal screws.

3. One vent pressure detector control (V.P.D.C.) secured with two sheet metal screws.

4. Pre-assembled wiring harness.

5. A plastic plug.

WARNING: THE VENT PRESSURE DETECTOR ASSEMBLY IS A PART OF THE FURNACE WHICH DOES REQUIRE FIELD INSTALLATION. THE FURNACE IS CERTIFIED BY CANADIAN GAS ASSOCIATION WITH THE V.P.D. ASSEMBLY INSTALLED. FAILURE TO INSTALL THE V.P.D. ASSEMBLY WILL VOID THE CGA APPROVAL.

The vent pressure detector assembly is factory assembled, electrical continuity checked, packaged in foam and located behind the upper front panel. Remove the assembly from the furnace and packaging. Check the components as listed above to be sure the assembly is complete. Place assembly on top of the flue box extension collar so the V.P.D.C. is located on the left or receptacle side. See Figure A. Drill two engaging holes for sheet metal screws, using the V.P.D. assembly clearance holes as locators. It is suggested that this joint be sealed with high temperature RTV. Secure the assembly with two (2) screws and insert the electrical plug terminal into the receptacle located on the top panel (front left of vent pipe when facing unit). Push manual reset button to insure furnace start-up.

The V.P.D. assembly, when installed as specified above, provides the user a safety feature for shutting off the gas valve if a positive pressure develops in the venting system. During normal operation, the venting system should operate with a negative internal pressure. If the venting system develops a positive pressure due to excessive length or restriction, the vent gases escape through the relief hole located directly in front of the V.P.D.C. The V.P.D.C. senses the temperature of the flue gases; and if the temperature reaches the factory set limit, the switch opens the circuit, shutting off the gas valve.

The vent pressure detector assembly is not used with special vent systems. A plug in jumper is supplied which replaces the harness shown in Figure A. See Figure B.

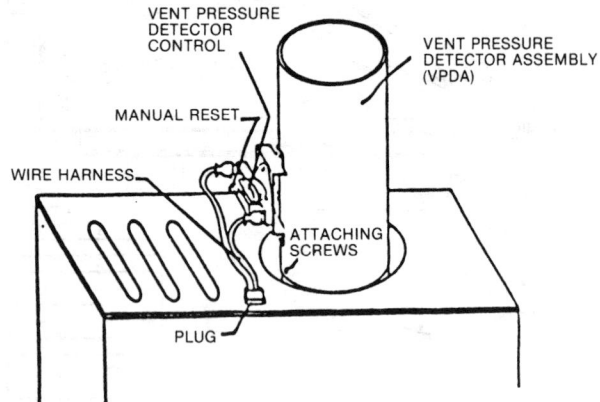

FIGURE A. VENT PRESSURE DETECTOR ASSEMBLY

NOTE: If the vent pressure detector control trips and shuts off the furnace, push the reset button. If the switch turns the furnace off a second time, check venting and combustion air provisions for cause of tripping.

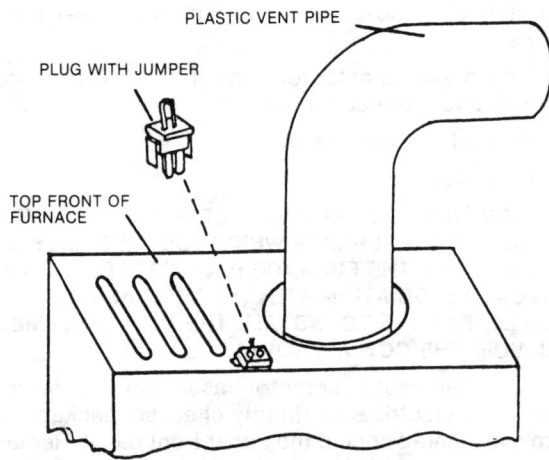

FIGURE B. JUMPER
The vent pressure detector assembly is not used with special vent systems. A plug-in jumper is supplied which replaces the harness shown in Figure A.

SPECIAL VENT SYSTEM (S.V.S.)
HIGH TEMPERATURE PLASTIC
GENERAL INFORMATION

This furnace is also certified with a special vent system using plastic vent material made from G.E.'s ULTEM resin which will withstand flue gas temperatures up to 480°F. This is a sealed system using single wall, 3 inch plastic pipe. The system components are Hart & Cooley's ULTRAVENT™ material or Plexco's PLEXVENT® material.

The maximum vent length is 30 feet plus three 90° sweep elbows. Part numbers are listed in Table 3. The minimum vent length is 5 feet plus one 90° sweep elbow.

The pipe may be cut to length using a hand-held hacksaw with a blade having 18 to 24 teeth per inch. Scrape the burrs off with a sharp-edged tool. The cut must be square with the pipe. Wipe clean the pipe ends and fitting sockets before sealing joints. Use coupling to extend pipe runs.

WARNING: ENTIRE VENT SYSTEM MUST BE SEALED WITH HIGH TEMPERATURE SEALANT WHICH WILL WITHSTAND TEMPERATURES TO 450°F, SUCH AS DOW-CORNING RTV-732 FOR HART AND COOLEY AND DOW CORNING RTV-738 FOR PLEXCO. SEAL THE VENT CONNECTION TO THE FURNACE, ALL PIPE JOINTS AND SEAMS. FAILURE TO SEAL VENT SYSTEM COULD ALLOW CARBON MONOXIDE LEAKAGE, RESULTING IN INJURY OR DEATH.

Pipe ends must be smooth and clean before applying sealant especially if the pipe has been cut. The sealer is applied by running a minimum ¼ bead entirely around the pipe end no more than ⅛ inch back from the edge. Push the fitting and pipe together fully with a twisting motion to create a good seal. Allow a minimum of 24 hours for a functional cure. After sealant has cured, inspect each joint to determine that flue gas will not escape. If necessary, reapply sealant to any joint that is suspect.

1. Do not drill holes in the pipe or fittings. Do not use sheet metal or other types of screws.

2. All vent systems must include a tee and drain plug for collection and disposal of condensate. The drain tee must be installed within the first 5 feet of vent run to protect the furnace.

3. All horizontal sections must have a slope toward the drain tee of not less than ¼ inch per foot to prevent collection of condensate at any location other than at the tee.

4. Horizontal runs must be supported with ¾ inch perforated pipe strap at a maximum of 3 foot intervals and at each point where an elbow is used.

5. Maintain a 5 inch minimum air space to combustibles from all sections of the vent system, except where wall thimble is used for horizontal venting.

HORIZONTAL THRU-THE-WALL VENT SIZE		
Input	Upflow	Downflow/Horiz.
50K	3″	3″
75K	3″	3″
100K	3″	3″
125K	3″	*4″
150K	3″	*4″

*NOTE: All furnaces have a 3″ vent connection as shipped from the factory. A 3″ to 4″ vent transition is required on 125 & 150,000 BTUH Downflow/Horizontal models. **THE VENT TRANSITION CONNECTION MUST BE MADE AT THE FURNACE VENT EXIT.**

USE ONLY HIGH TEMPERATURE PLASTIC PIPE WITH SEALED JOINTS.

If possible, the wall through which the vent will pass should not be exposed to the prevailing wind. Excessive wind may affect appliance performance. If this is not possible, steps to protect the cap from strong winds should be considered, such as a fence or hedge.

Locate the appliance as close to the wall being used as possible.

LOCATION REQUIREMENTS

The vent must be installed with the following minimum clearances. See Figure 10.

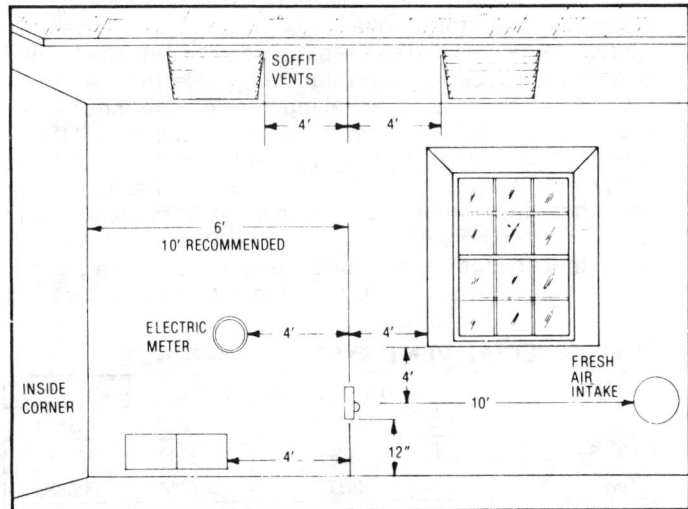

FIGURE 10. VENT TERMINAL CLEARANCES

1. 12 inches above grade level and above normal snow levels.

2. Not above any walkway.

3. 4 feet below, 4 feet horizontally from, or 1 foot above any door, window or gravity air inlet to the building or from gas or electric meters.

4. 6 feet from an inside corner formed by two exterior walls. 10 feet is recommended where possible.

5. At least 4 feet horizontally from any soffit or under eave vent.

6. 10 feet from any forced air inlet to the building. Any fresh air or make up air inlet such as for a dryer or furnace area is considered to be a forced air inlet.

7. Avoid areas where condensate drippage may cause problems such as above planters, patios, or adjacent to windows where steam may cause fogging.

Select the point of wall penetration where the minimum ¼ inch per foot of slope up can be maintained. The pipe can be located between joists spaced 16 inches on center.

When penetrating a non-combustible wall, the hole through the wall must be large enough to maintain the pitch, pipe clearance for passage, and provide proper sealing. (See Figure 8.)

Penetrating a combustible wall requires the use of a wall thimble. A 6½ inch square framed opening is required to insert the thimble halves. The thimble is adjustable to varying wall thickness and is held in place by applying sealant to the male sleeve before assembly. Also run a bead of sealant around the outer inside edge of the thimble halves before inserting in the wall thimble. Proceed with steps listed below.

1. The vent pipe must extend 1¼ inches through the outer thimble half. Be sure to check this carefully before cutting the vent pipe.

2. Attach a 3 inch coupling to the pipe that extends through the thimble. This prevents vent pipe from being pushed inward through the thimble.

Cut an 8¾ inch piece of vent pipe and connect the coupling to the termination tee. The inside of the tee must be a minimum of 13 inches from the outside wall. (See Figure 13.)

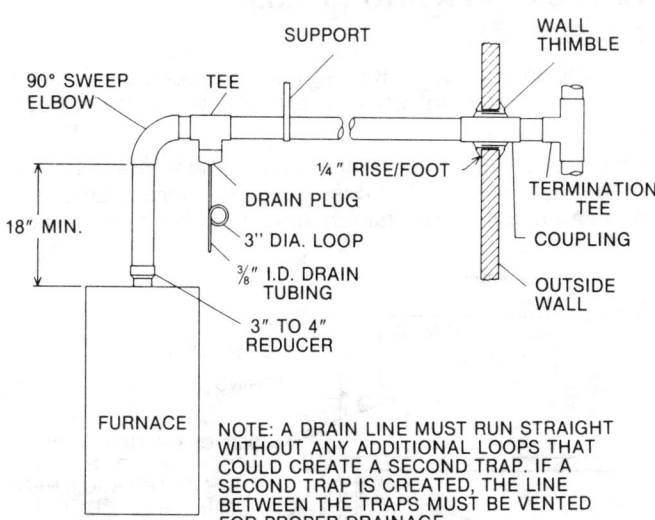

FIGURE 11. TYPICAL HORIZONTAL VENT SYSTEMS

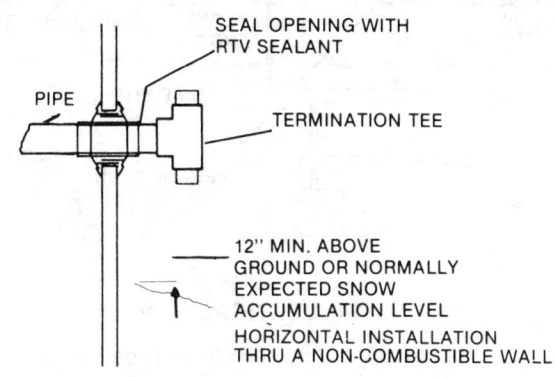

FIGURE 12. TERMINATION THROUGH WALL

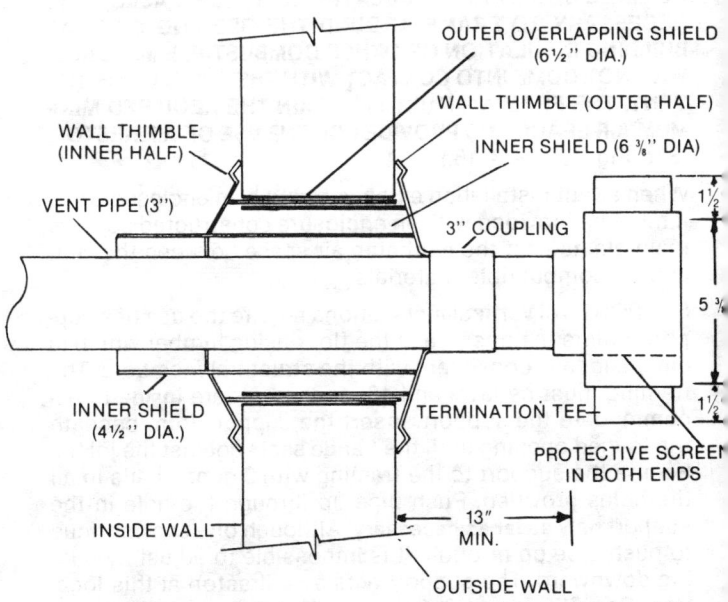

FIGURE 13. TERMINATION TEE

As shown in Figure 10, the termination tee must have a piece of straight pipe secured in each end of the tee. This shall protrude 1½″ as shown. This procedure must be followed to prevent high wind nuisance tripping of the pressure switch.

WARNING: STANDING PILOT MODELS CANNOT BE SAFELY VENTED HORIZONTALLY.

9

VERTICAL VENTING (S.V.S.)

(See Figure 14)

The appliance to be vented should be placed as close as possible to the point where the vent is to run up through the building.

Whenever possible, the vent should be continued straight up through to the roof. If it is necessary to make an offset in the attic, the horizontal run should slope upward ¼ inch

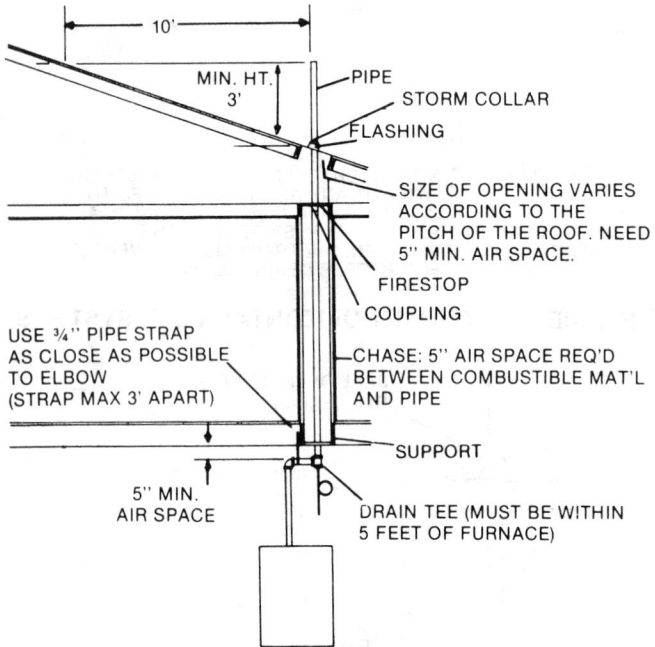

FIGURE 14. VERTICAL VENT

per foot going out and supported securely using ¾ inch perforated pipe strap at each elbow. (See Figure 12.)

WARNING: IN SOME INSTANCES, VENTS PASS THROUGH CEILINGS WHICH ARE INSULATED. IN THESE CASES, IT IS NECESSARY TO FRAME AROUND THE OPENING SO THAT BUILDING INSULATION OR OTHER COMBUSTIBLE MATERIAL WILL NOT COME INTO CONTACT WITH THE SURFACE OF THE VENT. THE FRAMING MUST MAINTAIN THE REQUIRED MINIMUM AIR SPACE AND PROVIDE FOR THE USE OF A FIRESTOP. (See Figures 14 & 15.)

When a vent installation extends through an enclosed zone, it must be provided with an enclosure constructed so as to maintain at least the minimum air space between the vent and all combustible materials.

SUPPORT: All vertical installations require the use of a support. Frame the opening in the floor using lumber which is dimensionally consistent with the structural members. The framing must be level and 13⅛ inches square **inside** to accommodate the support. Insert the support from beneath the framed opening until the flange seats against the joists. Secure the support to the framing with 8 penny nails in all the holes provided. Push pipe up through the hole in the support **only as far as necessary.** Although one can continue to push pipe up through, it is impossible to adjust by pulling downward. The support acts as a firestop at this location. See Figures 14 & 15.

FIRESTOP: A firestop must be provided whenever the vent passes through a combustible floor or ceiling. The opening must be framed as for the support (above) since the support also acts as a firestop. To install the firestop, extend a clear length of pipe (no couplings) up through the framed opening and then slide the firestop down over the pipe **flange side up.** Seat the flange in the opening and nail in place.

10

ROOF FLASHING: Where the vent passes through the roof, a flashing and storm collar must be used to maintain required clearances and to keep the weather out. The framed opening must be large enough to provide the necessary clearance to combustibles taking into account the slope of the roof. The flashing accommodates slopes from flat to a 6/12 pitch. Install the flashing against the roof while holding the pipe centered in the opening. Secure the flashing to the roof under the roofing material up-slope from the pipe and above the roofing below the pipe. Seal as required. A storm collar must be installed around the pipe immediately above the flashing and sealed to the pipe with RTV sealant.

SPECIAL VENT SYSTEM COMPONENTS

ITEM	3" Hart & Cooley	3" Plexco	4" Plexco
Pipe	3UP10	901220	903851
90° Sweep Elbow	3UES90	902299	905772
Tee	3UT	901214	906883
Drain Plug	3UDP	901462	903855
Coupling	3UC	901218	905807
Terminal Tee	3UTT	901971	906882
Wall Thimble	3UW	905295	907094
Support	3US	905655	907088
Roof Flashing	3UF	905337	907086
Storm Collar	3USC	905290	907085
Reducer	—	—	905744

FIGURE 15. FIRESTOP

EXISTING VENT SYSTEMS

When the existing furnace is removed from a venting system serving other appliances, the venting is likely to be too large to properly vent the remaining attached appliances.

The following steps shall be followed with each appliance remaining connected to the common venting system placed in operation, while the other appliances remaining connected to the common venting system are not in operation.

1. Seal any unused openings in the common venting system.

2. Visually inspect the venting system for proper size and horizontal pitch and determine there is no blockage or restriction, leakage, corrosion and other deficiencies which could cause an unsafe condition.

3. Insofar as is practical, close all building doors and windows and all doors between the space in which the appliances remaining connected to the common venting system are located and other spaces of the building. Turn on clothes dryers and any appliance not connected to the common venting system. Turn on any exhaust fans, such as range hoods and bathroom exhausts, so they will operate at maximum speed. Do not operate a summer exhaust fan. Close fireplace dampers.

4. Follow the lighting instructions. Place the appliance being inspected in operation. Adjust thermostat so appliance will operate continuously.

5. After it has been determined that each appliance remaining connected to the common venting system properly vents when tested as outlined above, return doors, windows, exhaust fans, fireplace dampers and any other gas-burning appliance to their previous conditions of use.

6. If improper venting is observed during any of the above tests, the common venting system must be resized. Refer to appendix G in the National Fuel Gas Code ANSI Z223.1, 1988 or the AGA-GAMA venting tables for category I furnaces.

GAS SUPPLY AND PIPING

Conversion

Any additions, changes or conversions required in order for the furnace to satisfactorily meet the application needs should be made by a qualified factory distributor or local service dealer, using factory specified or approved parts.

WARNING: THIS FURNACE WAS EQUIPPED AT THE FACTORY FOR USE ON NATURAL GAS ONLY. CONVERSION TO LP GAS REQUIRES A SPECIAL KIT SUPPLIED BY THE DISTRIBUTOR OR MANUFACTURER. MAILING ADDRESSES ARE LISTED ON THE FURNACE RATING PLATE, PARTS LIST AND WARRANTY. FAILURE TO USE THE PROPER CONVERSION KIT CAN CAUSE FIRE, CARBON MONOXIDE POISONING, EXPLOSION, PERSONAL INJURY OR PROPERTY DAMAGE.

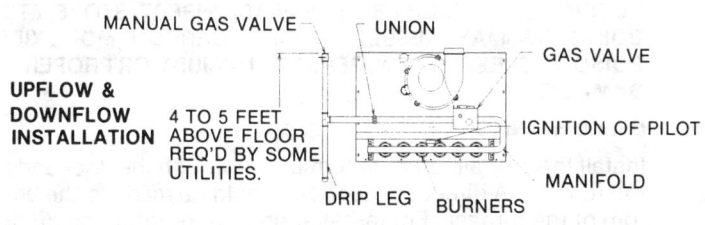

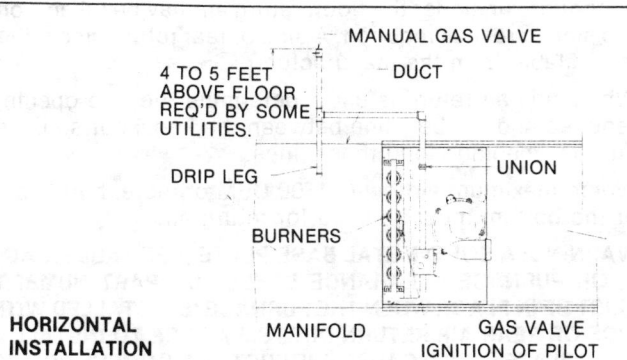

FIGURE 16. GAS PIPING

CAUTION: Check rating plate to make certain unit is equipped to burn the type of gas supplied. Care should be taken after installation of this equipment that gas control valve not be subjected to high gas supply line pressure. The furnace and its individual shut off valve must be disconnected from the gas supply piping system during any pressure testing of that system at test pressures in excess of 1/2 psig (3.48 kPa). The furnace must be isolated from the gas supply piping system by closing its individual manual shut off valve during any pressure testing of the gas supply piping system at test pressure equal to or less than 1/2 psig (3.48 kPa).

In making gas connections, avoid strains as they may cause noise and damage controls.

To check for leaks in piping, use a soap and water solution or other approved method. DO NOT USE AN OPEN FLAME.

CONNECT GAS SERVICE from meter to control assembly. See Figure 4 for typical hook-up. **A ground-joint union must be installed inside cabinet so the control assembly may be easily removed** and a 1/8″ NPT plug on the supply pipe to the valve for the purpose of making measurements of the inlet gas pressure. A manual shutoff valve or plugcock should be installed in the gas line outside of the furnace casing. Valve should be readily accessible for turning on or off. A drip leg should be installed in the gas supply line as close to the furnace as possible. **A pipe compound resistant to the action of liquefied petroleum gases must be used at all threaded pipe connections.**

Gas piping should be installed in accordance with local codes and regulations of the utility company. Consult local gas company for location of manual main valve. The gas line should be of adequate size to prevent undue pressure drop and never smaller than the pipe size to the combination gas valve. It is recommended that the size of pipe selected be in accordance with the gas pipe table for the length of pipe required and connected to the furnace as illustrated.

GAS PIPE CAPACITY TABLE (CU. FT./HR.)

Capacity of gas pipe of different diameters and lengths in cu. ft. per hr. with pressure drop of 0.3 in. and specific gravity of 0.60 (natural gas).

Nominal Iron Pipe Size, Inches	Length of Pipe, Feet							
	10	20	30	40	50	60	70	80
½	132	92	73	63	56	50	46	43
¾	278	190	152	130	115	105	96	90
1	520	350	285	245	215	195	180	170
1¼	1,050	730	590	500	440	400	370	350
1½	1,600	1,100	890	760	670	610	560	530

After the length of pipe has been determined, select the pipe size which will provide the minimum cubic feet per hour required for the gas input rating of the furnace. By formula:

$$\text{Cu. Ft. Per Hr. Required} = \frac{\text{Gas Input of Furnace (BTU/HR)}}{\text{Heating Value of Gas (BTU/FT}^3)}$$

The gas input of the furnace is marked on the furnace rating plate. The heating value of the gas (BTU/Ft³) may be determined by consulting the local natural gas utility or the LP gas supplier.

ELECTRICAL WIRING

WARNING: TURN OFF ELECTRIC POWER AT FUSE BOX OR SERVICE PANEL BEFORE MAKING ANY ELECTRICAL CONNECTIONS.

GROUND CONNECTION MUST BE COMPLETED BEFORE MAKING LINE VOLTAGE CONNECTIONS.

THE FURNACE MUST BE INSTALLED SO THAT THE ELECTRICAL COMPONENTS ARE PROTECTED FROM WATER (FURNACE CONDENSATE).

ELECTRICAL CONNECTIONS

The electrical supply requirements are listed on the furnace rating plate.

Use a separate fused branch electrical circuit containing a properly sized fuse or circuit breaker. Run this circuit directly from the main switch box to an electrical disconnect which must be readily accessible and located within sight of the furnace. Connect from the disconnect to the junction box on the left side of the furnace, inside the control compartment. See appropriate wiring diagram.

NOTE: L1 (hot) and L2 (neutral) polarity must be observed when making field connections to the furnace. The ignition module on electric ignition models will not sense flame if L1 and L2 are reversed.

WARNING: CABINET MUST BE PERMANENTLY GROUNDED. A GROUND SCREW IS PROVIDED IN THE JUNCTION BOX FOR THIS PURPOSE.

Installation of the electric supply line should be in accordance with the National Electric Code ANSI/NFPA No. 70, latest edition, or Canadian Electrical Code Part 1 - CSA Standard C22.1 and local building codes.

This can be obtained from:
National Fire Protection Association
Batterymarch Park
Quincy, MA 02269

Canadian Standards Association
178 Roxdale Boulevard
Roxdale, Ontario, Canada M9W 1R3

CIRCULATING AIR SUPPLY

The circulating air supply may be taken either: (1) from outside the building, (2) from return air ducts from several rooms, or (3) any combination of the two. When outside air is utilized, the system should be designed and adjusted such that the temperature of the supply air to the furnace will not be below 50°F during the heating season. When using a combination of outside air and return air, be sure the ducts are so designed and a diverting damper so installed that the volume of circulating air entering the furnace cannot be reduced or restricted below that which would normally enter through the circulating air intake of the furnace.

Plenum chambers and air ducts shall be installed in accordance with the Standard for the Installation of Air Conditioning and Ventilating Systems, NFPA No. 90A, or the Standard for the Installation of Warm Air Heating and Air Conditioning Systems, NFPA No. 90B.

When the furnace is installed so that the supply ducts carry air circulated by the furnace to areas outside the space containing the furnace, the return air shall be handled by a duct or ducts sealed to the furnace casing and terminated outside the space containing the furnace. If there is no complete return air duct system, the return air connection must be sealed to the furnace casing and run full size to a location outside the utility room or space housing the furnace to prevent a negative pressure on the venting systems.

If installed in parallel with a cooling unit, the damper, or other means used to control the flow of air, must be adequate to prevent chilled air from entering the furnace, and if manually operated must be equipped with means to prevent operation of the other unit unless the damper is in the full heat or cool position.

When a cooling coil is used in connection with a furnace, it must be installed downstream of the furnace (outlet end of furnace) or in a parallel with the furnace to avoid condensation in the heating element.

CAUTION: One of the most common causes of trouble in forced air heating systems is insufficient return air to the furnace. The return air system should be approximately equal to or greater than the area of the warm air discharge.

CONSULT LOCAL CODES FOR SPECIAL AND OTHER REQUIREMENTS.

Blower speed should be adjusted to maintain the air rise range shown on the rating plate.

NOTE: It is recommended that the outlet duct be provided with a removable access panel. It should be accessible after installation so that smoke or reflected light may be observed inside the casing to indicate the presence of leaks in the heat exchanger.

RETURN AIR

WARNING: NEVER ALLOW PRODUCTS OF COMBUSTION OR THE FLUE PRODUCTS TO ENTER THE RETURN AIR DUCTWORK, OR THE CIRCULATING AIR SUPPLY. ALL RETURN DUCTWORK MUST BE ADEQUATELY SEALED AND SECURED TO THE FURNACE WITH SHEET METAL SCREWS, AND JOINTS TAPED. ALL OTHER DUCT JOINTS MUST BE SECURED WITH APPROVED CONNECTIONS AND SEALED AIRTIGHT. WHEN A FURNACE IS MOUNTED ON A PLATFORM, WITH RETURN THROUGH THE BOTTOM, IT MUST BE SEALED AIRTIGHT BETWEEN THE FURNACE AND THE RETURN AIR PLENUM. THE RETURN AIR PLENUM MUST BE PERMANENTLY ENCLOSED. NEVER USE A DOOR AS A PART OF THE RETURN AIR PLENUM. THE FLOOR OR PLATFORM MUST PROVIDE SOUND PHYSICAL SUPPORT OF THE FURNACE, WITHOUT SAGGING, CRACKS, GAPS, ETC., AROUND THE BASE AS TO PROVIDE A SEAL BETWEEN THE SUPPORT AND THE BASE.

FAILURE TO PREVENT PRODUCTS OF COMBUSTION FROM BEING CIRCULATED INTO THE LIVING SPACE CAN CREATE POTENTIALLY HAZARDOUS CONDITIONS, INCLUDING CARBON MONOXIDE POISONING THAT COULD RESULT IN PERSONAL INJURY OR DEATH.

DO NOT, UNDER ANY CIRCUMSTANCES, CONNECT RETURN OR SUPPLY DUCTWORK TO OR FROM ANY OTHER HEAT PRODUCING DEVICE SUCH AS A FIREPLACE INSERT, STOVE, ETC. DOING SO MAY RESULT IN FIRE, CARBON MONOXIDE POISONING, EXPLOSION, PERSONAL INJURY OR PROPERTY DAMAGE.

UPFLOW UNITS

Install the cold air return to terminate through the floor under the furnace. A direct connection should be made to the bottom of the furnace. For installations where return air ducts cannot be run under the floor, return air may be taken from the side or rear as required. A side or rear return air cabinet is available from the manufacturer.

When side air return is used, determine the size opening required and scribe a line between the knockout squares, cut out opening along these lines.

Where maximum airflow is 1800 CFM or more, both sides or the bottom must be used for return air.

WARNING: A SOLID METAL BASE PLATE (SEE TABLE, PAGE 2, OR FURNACE CLEARANCE LABEL FOR PART NUMBER) MUST BE IN PLACE WHEN THE FURNACE IS INSTALLED WITH SIDE OR REAR AIR RETURN DUCTS. FAILURE TO INSTALL A BASE PLATE COULD CAUSE PRODUCTS OF COMBUSTION TO BE CIRCULATED INTO THE LIVING SPACE AND CREATE POTENTIALLY HAZARDOUS CONDITIONS, INCLUDING CARBON MONOXIDE POISONING.

WARNING: BLOWER AND BURNERS MUST NEVER BE OPERATED WITHOUT BLOWER DOOR IN PLACE. THIS IS TO PREVENT DRAWING GAS FUMES (WHICH COULD CONTAIN HAZARDOUS CARBON MONOXIDE) INTO THE HOME THAT COULD RESULT IN PERSONAL INJURY OR DEATH.

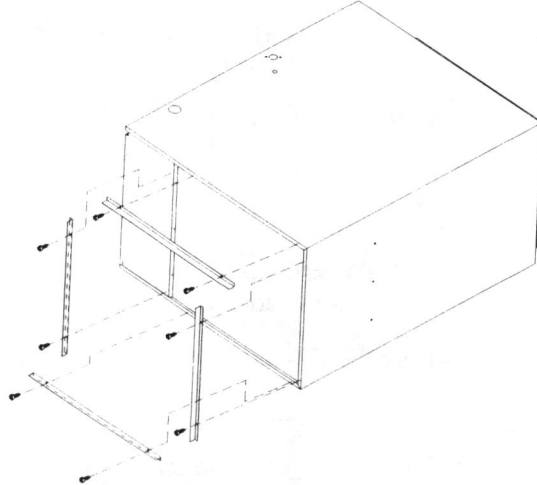

FOUR ANGLE BRACKETS ARE SHIPPED WITH EACH UNIT THAT CAN BE INSTALLED HORIZONTALLY. THESE BRACKETS MAY BE USED TO SECURE THE RETURN AIR DUCT TO A HORIZONTAL UNIT.

FIGURE 17. HORIZONTAL RETURN AIR DUCT

THERMOSTAT

Install room thermostat in accordance with instruction sheet in box with the thermostat. Run thermostat lead wires inside control compartment. Connect thermostat as shown on wiring diagram. Never install thermostat on an outside wall or where it will be influenced by drafts, concealed hot or cold water pipe or ducts, lighting fixtures, radiation from fireplace, rays of sun, lamps, television, radios or air streams from registers. Refer to instructions packed with thermostat for heat anticipator adjustment or selection.

HEAT ANTICIPATOR SETTINGS

For thermostat heat anticipator setting; (a) add the current draw of the various components in the system or (b) measure the current flow on either the R or W thermostat circuit and set the thermostat heat anticipator according to the current flow measured. Recommended setting on standing pilot and H.S.I. models is .75 amps.

START-UP PROCEDURE
LIGHTING INSTRUCTIONS (Standing Pilot Models)

Refer to instructions on furnace for specific cntrols used on that unit.

These instructions are for a standing pilot. If your unit is equipped with an automatic ignition device, do not attempt to light the furnace or burners using these instructions. Refer to Instructions for Hot Surface Ignition models.

TO START FURNACE

1. *CAUTION: Be sure that the manual gas control has been in the "Off" position for at least five minutes. Do not attempt to manually light the burner.*
2. Set room thermostat to the lowest setting.
3. HONEYWELL VALVE:
   a. Turn the gas control knob to the pilot position.
   b. Push down on red pilot button and hold while lighting pilot.

c. Allow pilot to burn approximately one-half minute before releasing red pilot button.
4. ROBERTSHAW VALVE:
   a. Move gas control lever to the pilot position.
   b. Push and hold gas control lever in "SET" position while lighting pilot burner.
   c. Allow pilot to burn approximately one-half minute before releasing lever to return to pilot position.
5. Turn gas control knob or lever to the "On" position.
6. Replace control access door.
7. Set room thermostat to a point above room temperature to light main burners. After main burners are lighted, set room thermostat to desired temperature.

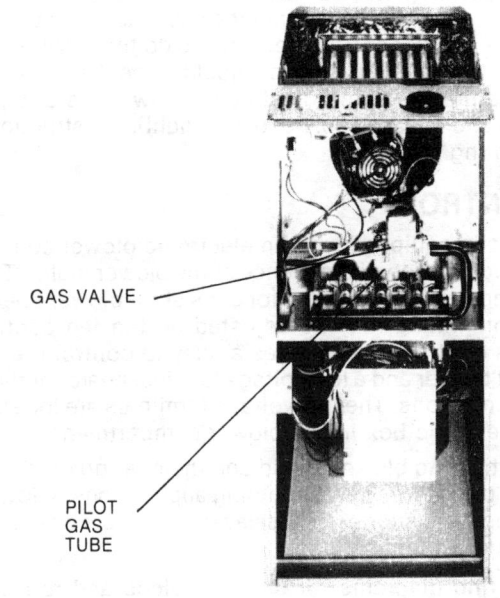

GAS VALVE

PILOT GAS TUBE

FIGURE 18. TYPICAL STANDING PILOT FURNACE

TO SHUT DOWN FURNACE

1. *Set thermostat to lowest setting.*
2. *Shut off gas to main burners and pilot by turning knob to "Off" position, or by depressing gas control lever and moving to the "Off" position.*

WARNING: SHOULD OVERHEATING OCCUR OR THE GAS SUPPLY FAIL TO SHUT OFF, SHUT OFF THE MANUAL GAS VALVE TO THE APPLIANCE BEFORE SHUTTING OFF THE ELECTRICAL SUPPLY.

PILOT ADJUSTMENT

The pilot flame should be adjusted to provide a soft flame that surrounds the tip of the thermocouple or flame sensor.

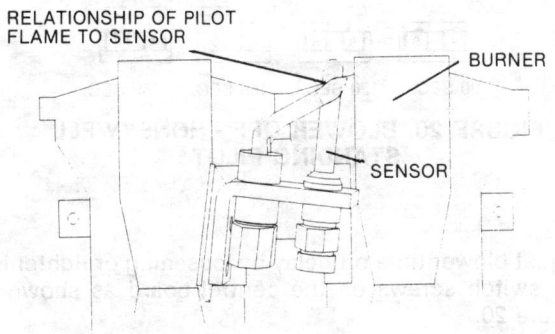

RELATIONSHIP OF PILOT FLAME TO SENSOR

BURNER

SENSOR

FIGURE 19. PILOT ADJUSTMENT

13

SEQUENCE OF OPERATION
(STANDING PILOT MODELS)

1. Lighted pilot remains on all of the time unless gas valve is turned off.

2. Each time the thermostat calls for heat, gas flows to the main burner and is ignited by the pilot flame.

3. The safety features of this system may be checked as follows:

   The pilot burner safety control does not permit gas to flow through the main gas burner unless there is a pilot flame present to ignite the gas. To check this function where combustion gas manifold controls are used: Remove cap from over pilot valve adjusting screw located on top of the manifold gas control. With the main gas burner operating, use a small scewdriver and turn the adjustment screw clockwise until the pilot gas is turned off. (Note the number of turns required to do this.) Within 90 seconds the main gas control should close the main gas valve. Return the pilot adjustment screw to its original position and replace cap. Follow the lighting instructions in relighting the pilot.

FAN CONTROL

The furnace is equipped with an electronic blower control which controls the on-off sequence of the blower motor. The on-off timings are factory fixed for consistent blower operation. The off timing may be adjusted on the fan control board. This control also includes a relay to control the induced draft blower and a low voltage terminal board for thermostat connections. The low voltage terminals are located inside the electric box in the blower compartment.

NOTE: The heating blower speed and the heating off delay come from the factory set for cooling applications. A lower heating speed and shorter OFF delay may be more desirable for heating only applications.

See the wiring diagrams for speed choices and refer to Figures 20, 21, 22, 23 for blower off delay choices.

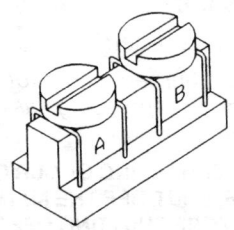

**HONEYWELL
DOWNFLOW MODELS**

MODEL ST9101A

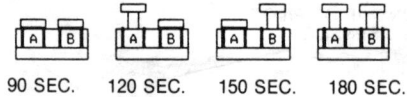

90 SEC.   120 SEC.   150 SEC.   180 SEC.

**FIGURE 20. BLOWER OFF - HONEYWELL
STANDING PILOT**

1. Adjust blower time off delay by loosening or tightening the switch screws on the control board as shown in Figure 20.

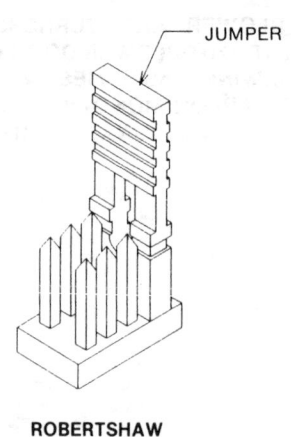

JUMPER

**ROBERTSHAW
UPFLOW MODELS**

MODEL RBC - 1

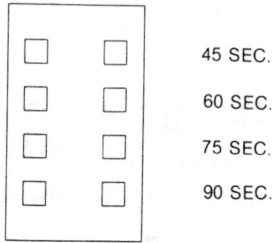

45 SEC.

60 SEC.

75 SEC.

90 SEC.

**FIGURE 21. BLOWER OFF - ROBERTSHAW
STANDING PILOT**

2. Adjust blower time off delay by moving the jumper on the control board to the pins required, as shown in Figure 21.

LIGHTING INSTRUCTIONS
(Hot Surface Ignition Models)

This appliance is equipped with a hot surface type ignition device. This device lights the main burners each time the room thermostat (closes) calls for heat. See lighting instructions in the furnace.

TO START FURNACE

1. *CAUTION: Be sure that the manual gas control has been in the "Off" position for at least five minutes. Do not attempt to manually light the burner.*

2. Set room thermostat to the lowest setting.

3. Turn gas control knob to the "On" position, or move gas control lever to the "ON" position.

4. Replace control access door.

5. Turn on electrical power.

6. Set room thermostat to a point above room temperature to light main burners. After burners are lighted, set room thermostat to desired temperature.

TO SHUT DOWN FURNACE

1. Set room thermostat to lowest setting.

2. Shut off gas to main burners and pilot by turning knob to "Off" position, or by depressing gas control lever and moving to the "Off" position.

WARNING: SHOULD OVERHEATING OCCUR OR THE GAS SUPPLY FAIL TO SHUT OFF, SHUT OFF THE MANUAL GAS VALVE TO THE APPLIANCE BEFORE SHUTTING OFF THE ELECTRICAL SUPPLY.

SEQUENCE OF OPERATION
(Honeywell or Hamilton Standard Integrated Control)

This furnace is equipped with a Honeywell ST9201A or Hamilton Standard 1012-800 integrated ignition and blower control board. This control combines functions of the hot surface ignition 100% lockout safety control and fixed time on/time off blower controls. It also provides a low voltage heat/cool thermostat control terminal board and connection points for field installed humidifier and electronic air cleaner optional accessories. Two indicator lights are also provided to aid the service technician.

When the heating thermostat closes (connection of R and W terminals), the induced draft blower starts and runs through a 30 second prepurge cycle. After the induced draft blower starts, the air proving differential negative pressure switch closes and starts the main burner ignition cycle. The hot surface ignitor is energized for 36 seconds to heat up, then the gas valve is energized to start gas flow to the main burner for ignition. The main gas burner flame is sensed by the de-energized hot surface ignitor within 0.8 seconds. If main burner flame is not sensed within the six second maximum trial for ignition time, the control will repeat the prepurge and ignition cycle for four additional retries. After a total of five cycles without sensing main burner flame, the system will then go into a 100% lockout mode. During the lockout mode neither the hot surface ignitor or the gas valve will be energized until the system is reset by opening the thermostat (disconnecting the R and W terminals) or interrupting the electrical power for ten seconds or longer. The induced draft blower and main burner will shut off when the thermostat is satisfied (R and W open).

The fixed time blower control will start the circulating air blower on heat speed thirty seconds after the main burner is ignited. The circulating air blower will continue to run during burner operation then shut down at a preset time after the burner shuts off. The circulating air blower will start and run on heating speeds if the thermostat fan switch is in the "On" position and the thermostat mode switch is in the "Heat" position. When the thermostat closes while in this mode, the blower will stop and go through a delay until 30 seconds after the burner lights.

When the thermostat is in the cooling mode, the blower control will start and stop the circulating air blower on cooling speed when the cooling thermostat contacts close or open respectively. The circulating air blower will start and run continuously at heating speed with the thermostat fan switch in the "On" position and the thermostat in the cooling mode. The blower will step up to cooling speed when both terminals G and Y are energized.

NOTE: The heating blower speed and the heating off delay come from the factory set for cooling applications. A lower heating speed and shorter off delay may be more desirable for heating only applications.

See the wiring diagrams for speed choices and refer to Figures 20, 21, 22, 23 for blower off delay choices.

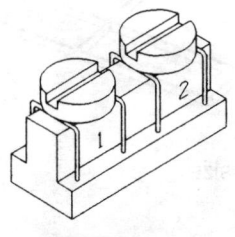

HONEYWELL
UPFLOW MODELS

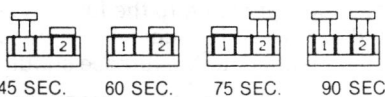

45 SEC.   60 SEC.   75 SEC.   90 SEC.

FIGURE 22. BLOWER OFF - HONEYWELL INTEGRATED CIRCUIT

1. Adjust blower time off delay by loosening or tightening the switch screws on the control board as shown in Figure 22.

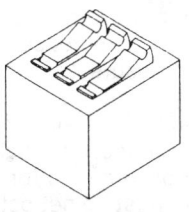

HAMILTON STANDARD
DOWNFLOW MODELS

MODEL 1012-800

OFF TIME	SWITCH 1	SWITCH 2	SWITCH 3
90 SEC.	ON	ON	ON
120 SEC.	OFF	OFF	OFF
150 SEC.	OFF	ON	OFF
180 SEC.	ON	OFF	OFF

FIGURE 23. BLOWER OFF - HAMILTON STANDARD INTEGRATED CIRCUIT

2. Adjust blower time off delay on the control board by setting the switches in the proper sequence.

FIELD INSTALLED OPTION ACCESSORIES

1. Electronic air cleaner line voltage power can be supplied from the screw terminal "EAC" and a line voltage neutral screw terminal on the control board. This will power the electronic air cleaner whenever the circulating air blower is in operation.

2. Humidifier line voltage power can be supplied from screw terminal "HUM" to a line voltage neutral screw terminal on the control board. This will power the humidifier whenever the burner is on and the circulating air blower is operating in the heating mode.

HELP GUIDES

A service trouble shooting guide is located on the outside of the control box cover. A wiring diagram is located on the inside of the control box cover. *CAUTION: The control box contains exposed line voltage connections; disconnect power before servicing. Service must be by a trained qualified service technician.*

Two green LED help lights are located on the control board. Both lights should be on when the board is operating. The first light marked PWR is lit when 24 volt power is available on the control board. The second light marked OK is lit when the board control circuit is operational.

15

TO RATE FURNACE

The maximum supply pressure to the furnace should be 7″ W.C. for natural gas. The minimum supply pressure for purposes of input adjustment to the furnace should be 5″ W.C.

A properly calibrated manometer or magnehelic gauge is required for accurate gas pressure readings.

Manifold pressure should be set at 3.5″ W.C. for natural gas. Only small variations in the gas flow should be made by means of the pressure regulator adjustment. In no case should the final manifold pressure vary more than plus or minus 0.3″ W.C. from the above specified pressures. To adjust pressure regulator, remove regulator cap and turn adjustment screw clockwise to increase pressure or counterclockwise to decrease pressure, then replace regulator cap securely. Any necessary major changes in the gas flow rate should be made by changing the size of the burner orifices. To change orifice spuds, shut off manual main gas valve and remove the manifold. For furnaces for use on LP gas, the LP gas supply pressure must be set between 11.0 and 14.0 W.C. by means of the tank or branch supply regulators. The furnace manifold pressure should be set at 10″ W.C. at the combination gas control valve. For elevations up to 2,000 feet, rating plate input ratings apply. For elevations over 2,000 feet, reduce input 4% for each 1,000 feet above sea level.

Check of input is important to prevent over firing of the furnace beyond its design-rated input. NEVER SET INPUT ABOVE THAT SHOWN ON THE RATING PLATE. Use the following table or formula to determine input rate. Start furnace and measure time required to burn one cubic foot of gas. Time the meter with only the furnace in operation.

Prior to checking furnace input, make certain that all other gas appliances are shut off, with the exception of pilot burners.

LIMIT CONTROL

The high limit cut-off is set at the factory and cannot be adjusted. It is calibrated to prevent the air temperature leaving the furnace from exceeding the maximum outlet air temperature.

METER TIME IN MINUTES AND SECONDS FOR NORMAL INPUT RATING OF FURNACES EQUIPPED FOR NATURAL OR LP GAS												
INPUT BTU/HR	METER SIZE CU. FT.	HEATING VALUE OF GAS BTU PER CU. FT.										
		900		1000		1040		1100		2500		
		MIN.	SEC.	MIN.	SEC.	MIN.	SEC.	MIN.	SEC.	MIN.	SEC.	
50,000	ONE	1	5	1	12	1	15	1	18	3	20	
	TEN	10	50	12	30	12	30	13	12	30	00	
75,000	ONE	0	44	0	48	0	50	0	53	2	0	
	TEN	7	12	8	0	8	19	8	48	20	0	
100,000	ONE	0	33	0	36	0	38	0	40	1	30	
	TEN	5	24	6	0	6	15	6	36	15	0	
125,000	ONE	0	26	0	29	0	30	0	32	1	12	
	TEN	4	19	4	48	5	0	5	17	12	0	
150,000	ONE	0	31	0	24	0	25	0	26	1	0	
	TEN	3	36	4	0	4	10	4	20	10	0	

$$\text{Input BTU/HR} = \frac{\text{Heating Value of Gas (BTU/Ft}^3\text{)} \times 3600}{\text{Time in Seconds (for 1 cu. ft.) of Gas}}$$

FLAME ROLL-OUT SAFETY SWITCHES

Furnaces are equipped with limit switches to protect against overtemperature conditions in the control compartment caused by inadequate combustion air supply. The switch for the UPFLOW FURNACE and DOWNFLOW STANDING PILOT SYSTEM is located just above the burners on the blower divider panel. Switches for the DOWNFLOW/HORIZONTAL HOT SURFACE IGNITION FURNACES are located on either side of the burner brackets and just above the burners on the blower divider panel. If a switch is tripped it must be manually reset. DO NOT jumper this switch. If this switch should trip, a qualified furnace installer or repair technician should be called to check and/or correct for adequate combustion air supply. If this unit is mounted in a closet, the door must be closed when making this check of the installation.

DO NOT reset the overtemperature switch without taking corrective action to assure that an adequate supply of combustion air is maintained under all conditions of operation. Replace this switch only with the identical replacement part.

PRESSURE SWITCH

This furnace has a pressure switch for sensing a blocked vent condition. It is normally open and closes when the induced draft blower starts, indicating air flow through the combustion chamber.

MAINTENANCE

WARNING: DISCONNECT MAIN ELECTRICAL POWER TO THE UNIT BEFORE ATTEMPTING ANY MAINTENANCE.

FILTERS

Keep air filters clean at all times. Vacuum dirt from filter, wash with detergent and water, air dry thoroughly and reinstall.

NOTE: Filters are not supplied with horizontal "only" models.

UPFLOW FILTER SIZES				
FURNACE WIDTH	INPUT BTUH	SIZE		QTY.
		BOTTOM	SIDE	
14″	50,000	12¼″x25″ *	15¾″x25″	1
17½″	75 & 100,000	15¾″x25″	15¾″x25″	1
21″	100,000	19¼″x25″	15¾″x25″ *	1
24½″	125 & 150,000	22¾″x25″	15¾″x25″ *	1

*NOTE: Some filters must be resized to fit certain units and applications.

1. 14″-50,000 btuh unit requires removal of 3½″ segment of filter and frame to get proper width for a bottom filter.

2. 21″-100,000 btuh unit requires removal of 3½″ segment of filter and frame to get proper width for a side filter.

3. 24½″-125,000 & 150,000 btuh units require removal of 7″ segment of filter and frame to get proper width for a side filter.

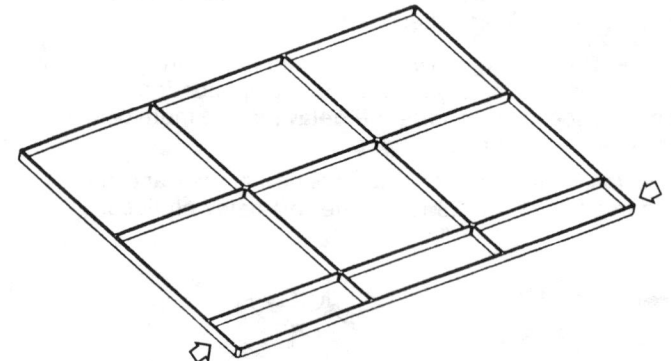

REMOVE SEGMENT TO SIZE AS REQUIRED

FIGURE 24. RESIZING FILTERS & FRAME

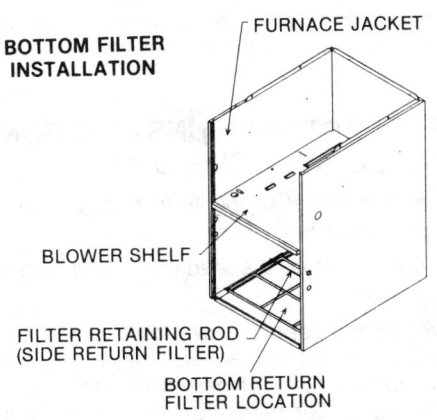

BOTTOM FILTER INSTALLATION

FURNACE JACKET

BLOWER SHELF

FILTER RETAINING ROD (SIDE RETURN FILTER)

BOTTOM RETURN FILTER LOCATION

SIDE FILTER INSTALLATION

FURNACE JACKET

BLOWER SHELF

SIDE RETURN FILTER LOCATION

FILTER RETAINING ROD (BOTTOM RETURN FILTER)

FIGURE 25. FILTER RETAINING RODS

UNIT SIZE	UNIT WIDTH	FILTER ROD LENGTH		APPLICATION	
		20¾ AE-61659-02	24¼ AE-61659-03	BOTTOM	SIDE
50,000	14″	1		Cut Off 7″	As Is
75,000	17½″	1		Cut Off 3½″	As Is
100,000	17½″	1		Cut Off 3½″	As Is
100,000	21″	1		As Is	As Is
125,000	24½″		1	As Is	Cut Off 3½″
150,000	24½″		1	As Is	Cut Off 3½″

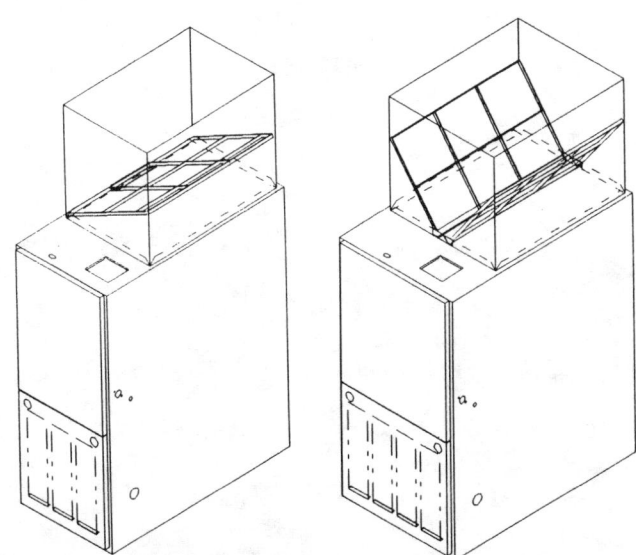

FIGURE 26. DOWNFLOW / HORIZONTAL FILTER SIZES

FURNACE WIDTH	INPUT BTUH	SIZE	QTY.
14″	50,000	14″x20″	1
17½″	75 & 100,000	12″x20″	2
21″	100,000	12″x20″	2
24½″	125 & 150,000	14″x20″	2

NOTE: Filters are not supplied with horizontal "only" models.

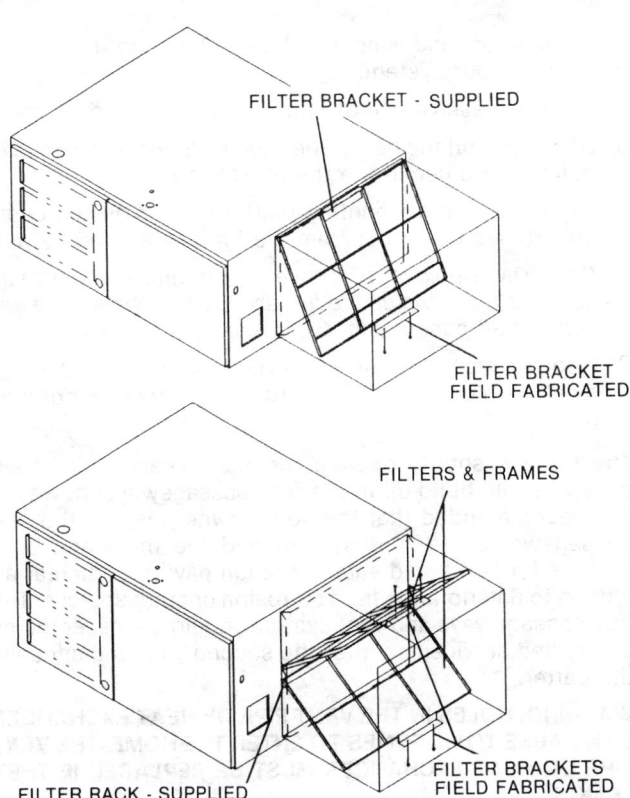

FILTER BRACKET - SUPPLIED

FILTER BRACKET FIELD FABRICATED

FILTERS & FRAMES

FILTER RACK - SUPPLIED

FILTER BRACKETS FIELD FABRICATED

FIGURE 27. HORIZONTAL FILTERS INSTALLATION

CAUTION: Do not operate your system for extended periods without filters. A portion of the dust entrained in the air may temporarily lodge in the air duct runs and at the supply registers. Any recirculated dust particles will be heated and charred by contact with the furnace heat exchanger. This residue will soil ceilings, walls, drapes, carpets, and other household articles.

LUBRICATION

The blower motor and induced draft motor are prelubricated by the manufacturer and do not require further attention.

Clean motor periodically to prevent the possibility of overheating due to an accumulation of dust and dirt on the windings or on the motor exterior. And, as suggested elsewhere in these instructions, the air filters should be kept clean because dirty filters can restrict airflow and the motor depends upon sufficient air flowing across and through it to keep from overheating.

HUMIDIFIER AND ELECTRONIC AIR CLEANER

Electric ignition models are equipped with connection points (screw terminals) on the control board for connecting a humidifier and electronic air cleaner. These points are labeled "HUM" and "EAC." They provide 115 volt output to each device. Connect to "HUM" and neutral and/or "EAC" and neutral to obtain 115 volts to your field added humidifier and electronic air cleaner. The air cleaner will operate any time the blower is running. The humidifier will operate only when there is a call for heat and the heating blower speed is on.

SYSTEM OPERATION INFORMATION

1. Keep the air filters clean. Your heating system will operate better, more efficiently and more economically.

2. Arrange your furniture and drapes so that the supply air registers and the return air grilles are unobstructed.

17

3. Close doors and windows. This will reduce the heating load on your system.

4. Avoid excessive use of kitchen exhaust fans.

5. Do not permit the heat generated by television, lamps or radios to influence the thermostat operation.

6. Exclusive of the mounting platform, keep all combustible articles three feet from the furnace and vent stack.

7. *CAUTION: Replace all blower doors and compartment covers after servicing the furnace. Do not operate the unit without all panels and doors securely in place.*

8. If you desire to operate your system with constant air circulation, please ask advice from your servicing contractor.

The furnace should operate for many years without excessive scale build-up in the flue passageways, however, it is recommended that the home owner inspect the flue passageways, the vent system and the main and pilot burners for continued safe operation paying particular attention to deterioration from corrosion or other sources. The flue passageways and vent system should be inspected by a qualified serviceman after the second year and annually thereafter.

WARNING: HOLES IN THE VENT PIPE OR HEAT EXCHANGER CAN CAUSE TOXIC FUMES TO ENTER THE HOME. THE VENT PIPE OR HEAT EXCHANGER MUST BE REPLACED IF THEY LEAK.

It is recommended that at the beginning of the heating season and approximately midway in the heating season a visual inspection be made of the main burner flames and pilot flame on standing pilot models for the desired flame appearance.

GAS FURNACE (DIRECT DRIVE) INSTRUCTIONS FOR CHANGING BLOWER SPEED

CAUTION: Disconnect electrical supply to furnace before attempting to change blower speed.

Blower motor is wired for blower speeds required for normal operation as shown.

If additional blower speed taps are available (leads connected to M1 and M2 on fan control), speeds may be changed if necessary to fit requirements of particular installation. Reconnect the unused motor leads to M1 or M2. Check motor lead color for speed designation.

Heating speeds should not be reduced where it could cause the furnace air temperature to rise to exceed the maximum outlet air temperature specified for the unit.

REPLACEMENT PARTS

Contact your local distributor for a complete parts list. See enclosed sheet.

PARALLEL FURNACE INSTALLATION

Two furnaces may be installed together for parallel operation by following the wiring diagrams included in these instructions. **NOTE:** Only furnaces with hot surface ignition systems may be used for parallel installation and the controls must be of the same manufacturer.

AIRFLOW PERFORMANCE

NOMINAL SIZE (BTUH)	BLOWER SIZE	MOTOR H.P.	BLOWER SPEED	CFM AIR DELIVERY EXTERNAL STATIC PRESSURE INCHES WATER COLUMN						
				.7	.6	.5	.4	.3	.2	.1
50,000	11x6	1/2	LOW	575	605	635	680	695	715	735
			MED-LO*	895	920	940	955	970	980	990
			MED-HI	1060	1085	1105	1125	1150	1170	1190
			HIGH	1250	1290	1320	1350	1390	1440	—
75,000	11x6	1/2	LOW	800	820	840	855	870	885	900
			MED*	995	1020	1045	1060	1080	1095	1115
			HIGH	1160	1200	1235	1265	1295	1320	1345
75,000	11x7	3/4	LOW	985	1010	1025	1035	1040	1045	1050
			MED-LO*	1105	1135	1155	1170	1185	1195	1205
			MED-HI	1385	1425	1450	1480	1510	1535	1565
			HIGH	1475	1525	1585	1640	1690	1740	—
100,000	11x6	1/2	LOW	770	795	810	830	845	860	875
			MED*	920	945	970	990	1010	1030	1050
			HIGH	1085	1120	1155	1190	1220	1250	1280
100,000	11x10	3/4	LOW	1095	1110	1120	1125	1130	1135	1140
			MED-LO*	1265	1285	1300	1310	1315	1320	1325
			MED-HI	1615	1655	1680	1705	1725	1735	1745
			HIGH	1865	1935	2010	2080	2145	2200	—
125,000	11x10	3/4	LOW	1130	1140	1150	1160	1170	1175	—
			MED-LO*	1300	1315	1330	1340	1350	1355	1360
			MED-HI	1665	1710	1740	1760	1775	1780	—
			HIGH	1895	1975	2050	2120	2195	2285	—
150,000	11x10	3/4	LOW	1150	1170	1175	1180	1185	1190	1195
			MED-LO*	1300	1330	1345	1360	1365	1375	1385
			MED-HI	1655	1695	1735	1775	1800	1830	1845
			HIGH	1775	1847	1915	1980	2045	2110	—

*HEATING SPEED COOLING APPLICATIONS
LOW SPEED MAY BE MORE DESIRABLE FOR HEATING ONLY APPLICATIONS.

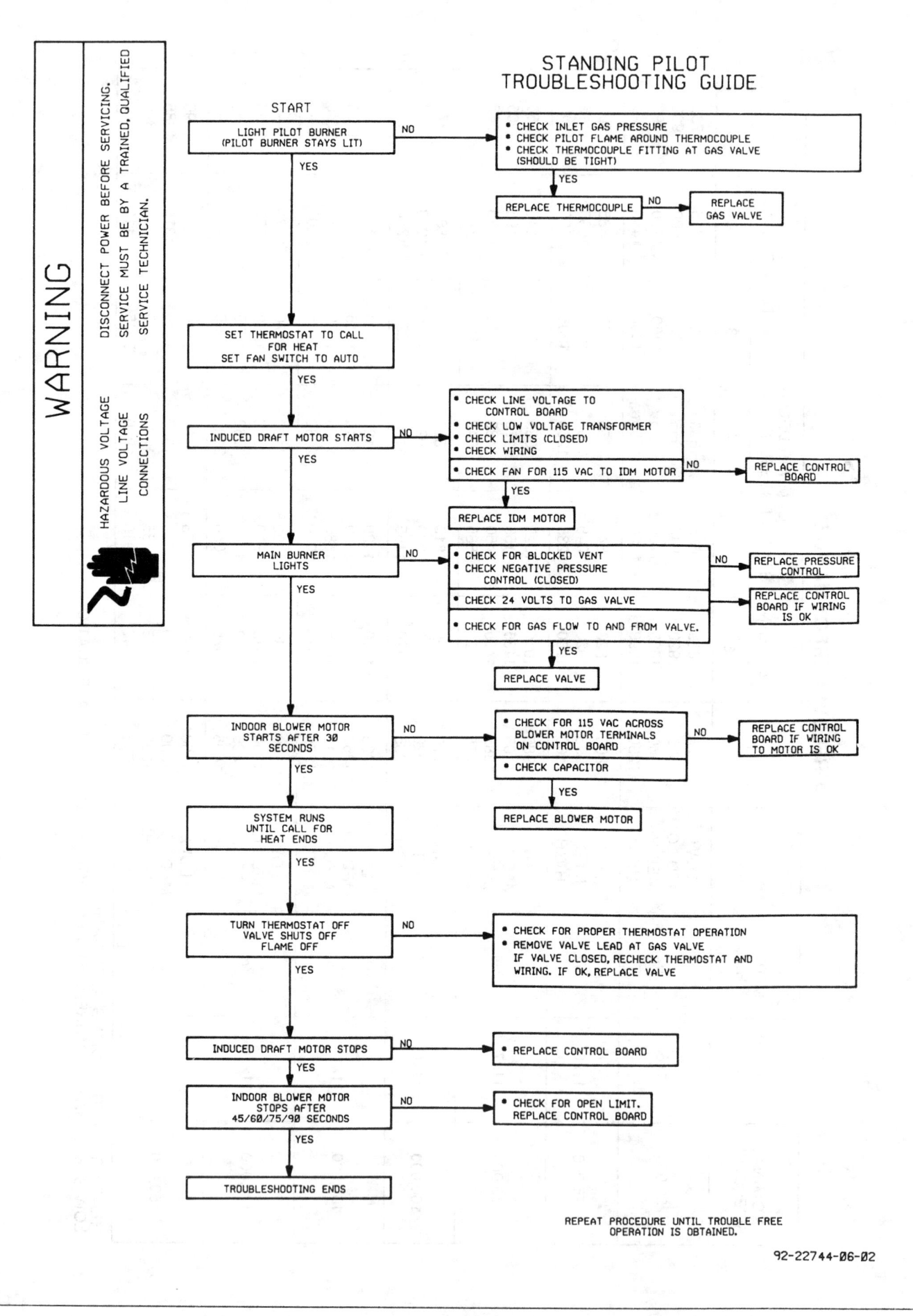

STANDING PILOT TROUBLESHOOTING GUIDE

WARNING

DISCONNECT POWER BEFORE SERVICING. SERVICE MUST BE BY A TRAINED, QUALIFIED SERVICE TECHNICIAN.

HAZARDOUS VOLTAGE
LINE VOLTAGE CONNECTIONS

START

LIGHT PILOT BURNER
(PILOT BURNER STAYS LIT) —NO→
- CHECK INLET GAS PRESSURE
- CHECK PILOT FLAME AROUND THERMOCOUPLE
- CHECK THERMOCOUPLE FITTING AT GAS VALVE
  (SHOULD BE TIGHT)

YES ↓

REPLACE THERMOCOUPLE —NO→ REPLACE GAS VALVE

YES ↓

SET THERMOSTAT TO CALL FOR HEAT
SET FAN SWITCH TO AUTO

YES ↓

INDUCED DRAFT MOTOR STARTS —NO→
- CHECK LINE VOLTAGE TO CONTROL BOARD
- CHECK LOW VOLTAGE TRANSFORMER
- CHECK LIMITS (CLOSED)
- CHECK WIRING
- CHECK FAN FOR 115 VAC TO IDM MOTOR —NO→ REPLACE CONTROL BOARD

YES ↓

REPLACE IDM MOTOR

YES ↓

MAIN BURNER LIGHTS —NO→
- CHECK FOR BLOCKED VENT
- CHECK NEGATIVE PRESSURE CONTROL (CLOSED) —NO→ REPLACE PRESSURE CONTROL
- CHECK 24 VOLTS TO GAS VALVE   REPLACE CONTROL BOARD IF WIRING IS OK
- CHECK FOR GAS FLOW TO AND FROM VALVE.

YES ↓

REPLACE VALVE

YES ↓

INDOOR BLOWER MOTOR STARTS AFTER 30 SECONDS —NO→
- CHECK FOR 115 VAC ACROSS BLOWER MOTOR TERMINALS ON CONTROL BOARD —NO→ REPLACE CONTROL BOARD IF WIRING TO MOTOR IS OK
- CHECK CAPACITOR

YES ↓

REPLACE BLOWER MOTOR

SYSTEM RUNS UNTIL CALL FOR HEAT ENDS

YES ↓

TURN THERMOSTAT OFF
VALVE SHUTS OFF
FLAME OFF —NO→
- CHECK FOR PROPER THERMOSTAT OPERATION
- REMOVE VALVE LEAD AT GAS VALVE IF VALVE CLOSED, RECHECK THERMOSTAT AND WIRING. IF OK, REPLACE VALVE

YES ↓

INDUCED DRAFT MOTOR STOPS —NO→ - REPLACE CONTROL BOARD

YES ↓

INDOOR BLOWER MOTOR STOPS AFTER 45/60/75/90 SECONDS —NO→ - CHECK FOR OPEN LIMIT. REPLACE CONTROL BOARD

YES ↓

TROUBLESHOOTING ENDS

REPEAT PROCEDURE UNTIL TROUBLE FREE OPERATION IS OBTAINED.

92-22744-06-02

20

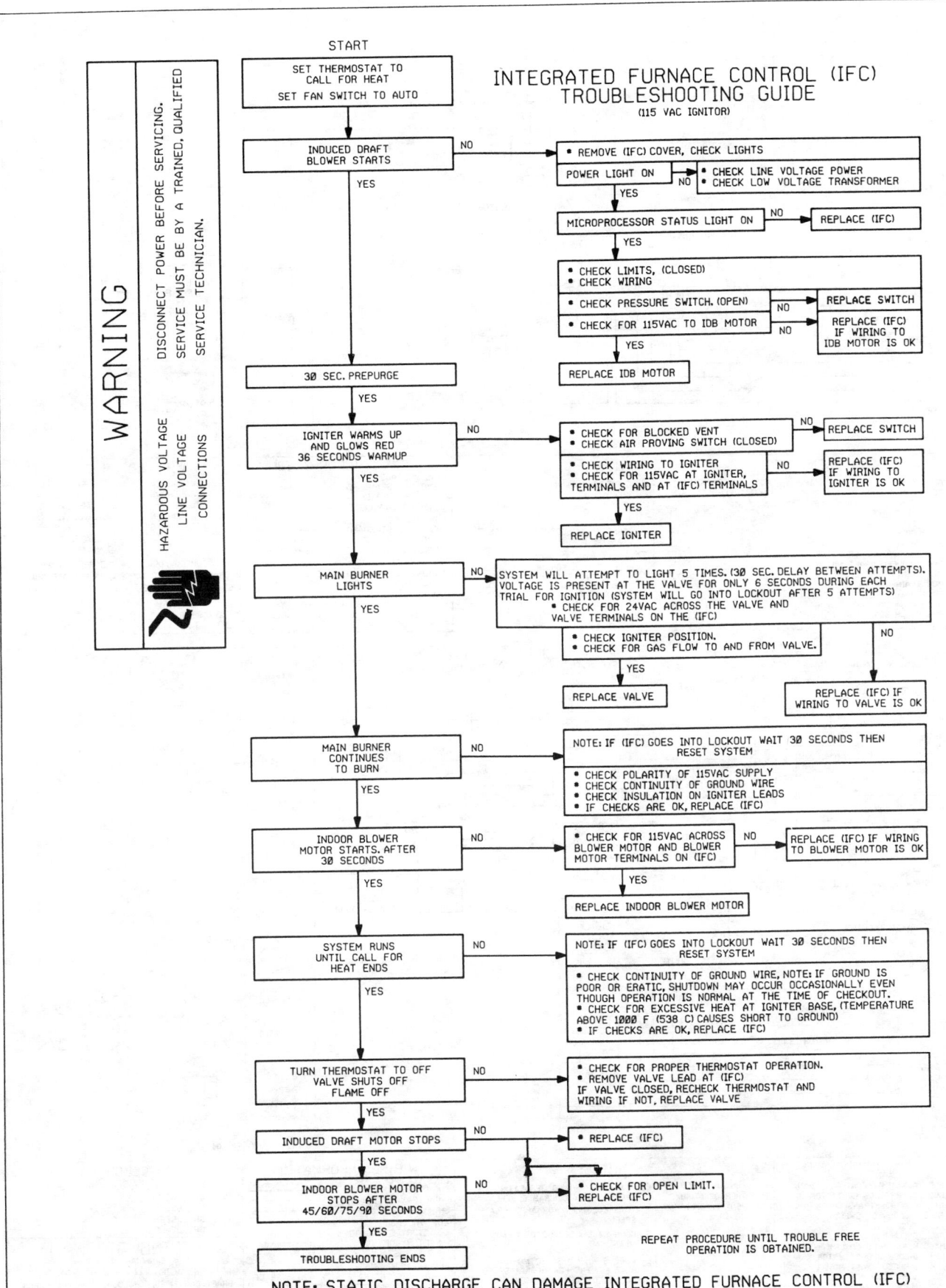

INTEGRATED FURNACE CONTROL (IFC) TROUBLESHOOTING GUIDE
(115 VAC IGNITOR)

START

SET THERMOSTAT TO CALL FOR HEAT
SET FAN SWITCH TO AUTO

INDUCED DRAFT BLOWER STARTS — NO →
- REMOVE (IFC) COVER, CHECK LIGHTS

POWER LIGHT ON — NO →
- CHECK LINE VOLTAGE POWER
- CHECK LOW VOLTAGE TRANSFORMER

YES

MICROPROCESSOR STATUS LIGHT ON — NO → REPLACE (IFC)

YES

- CHECK LIMITS, (CLOSED)
- CHECK WIRING
- CHECK PRESSURE SWITCH. (OPEN) — NO → REPLACE SWITCH
- CHECK FOR 115VAC TO IDB MOTOR — NO → REPLACE (IFC) IF WIRING TO IDB MOTOR IS OK

YES

REPLACE IDB MOTOR

30 SEC. PREPURGE

YES

IGNITER WARMS UP AND GLOWS RED 36 SECONDS WARMUP — NO →
- CHECK FOR BLOCKED VENT
- CHECK AIR PROVING SWITCH (CLOSED) — NO → REPLACE SWITCH

- CHECK WIRING TO IGNITER
- CHECK FOR 115VAC AT IGNITER, TERMINALS AND AT (IFC) TERMINALS — NO → REPLACE (IFC) IF WIRING TO IGNITER IS OK

YES

REPLACE IGNITER

MAIN BURNER LIGHTS — NO →
SYSTEM WILL ATTEMPT TO LIGHT 5 TIMES. (30 SEC. DELAY BETWEEN ATTEMPTS). VOLTAGE IS PRESENT AT THE VALVE FOR ONLY 6 SECONDS DURING EACH TRIAL FOR IGNITION (SYSTEM WILL GO INTO LOCKOUT AFTER 5 ATTEMPTS)
- CHECK FOR 24VAC ACROSS THE VALVE AND VALVE TERMINALS ON THE (IFC)

- CHECK IGNITER POSITION.
- CHECK FOR GAS FLOW TO AND FROM VALVE. — NO →

YES

REPLACE VALVE

REPLACE (IFC) IF WIRING TO VALVE IS OK

MAIN BURNER CONTINUES TO BURN — NO →
NOTE: IF (IFC) GOES INTO LOCKOUT WAIT 30 SECONDS THEN RESET SYSTEM
- CHECK POLARITY OF 115VAC SUPPLY
- CHECK CONTINUITY OF GROUND WIRE
- CHECK INSULATION ON IGNITER LEADS
- IF CHECKS ARE OK, REPLACE (IFC)

YES

INDOOR BLOWER MOTOR STARTS. AFTER 30 SECONDS — NO →
- CHECK FOR 115VAC ACROSS BLOWER MOTOR AND BLOWER MOTOR TERMINALS ON (IFC) — NO → REPLACE (IFC) IF WIRING TO BLOWER MOTOR IS OK

YES

REPLACE INDOOR BLOWER MOTOR

SYSTEM RUNS UNTIL CALL FOR HEAT ENDS — NO →
NOTE: IF (IFC) GOES INTO LOCKOUT WAIT 30 SECONDS THEN RESET SYSTEM
- CHECK CONTINUITY OF GROUND WIRE, NOTE: IF GROUND IS POOR OR ERATIC, SHUTDOWN MAY OCCUR OCCASIONALLY EVEN THOUGH OPERATION IS NORMAL AT THE TIME OF CHECKOUT.
- CHECK FOR EXCESSIVE HEAT AT IGNITER BASE, (TEMPERATURE ABOVE 1000 F (538 C) CAUSES SHORT TO GROUND)
- IF CHECKS ARE OK, REPLACE (IFC)

YES

TURN THERMOSTAT TO OFF VALVE SHUTS OFF FLAME OFF — NO →
- CHECK FOR PROPER THERMOSTAT OPERATION.
- REMOVE VALVE LEAD AT (IFC) IF VALVE CLOSED, RECHECK THERMOSTAT AND WIRING IF NOT, REPLACE VALVE

YES

INDUCED DRAFT MOTOR STOPS — NO →
- REPLACE (IFC)

YES

INDOOR BLOWER MOTOR STOPS AFTER 45/60/75/90 SECONDS — NO →
- CHECK FOR OPEN LIMIT. REPLACE (IFC)

YES

TROUBLESHOOTING ENDS

REPEAT PROCEDURE UNTIL TROUBLE FREE OPERATION IS OBTAINED.

NOTE: STATIC DISCHARGE CAN DAMAGE INTEGRATED FURNACE CONTROL (IFC)

WARNING

HAZARDOUS VOLTAGE LINE VOLTAGE CONNECTIONS

DISCONNECT POWER BEFORE SERVICING. SERVICE MUST BE BY A TRAINED, QUALIFIED SERVICE TECHNICIAN.

IDB - INDUCED DRAFT BLOWER
IFC - INTEGRATED FURNACE CONTROL

92-22744-07-02

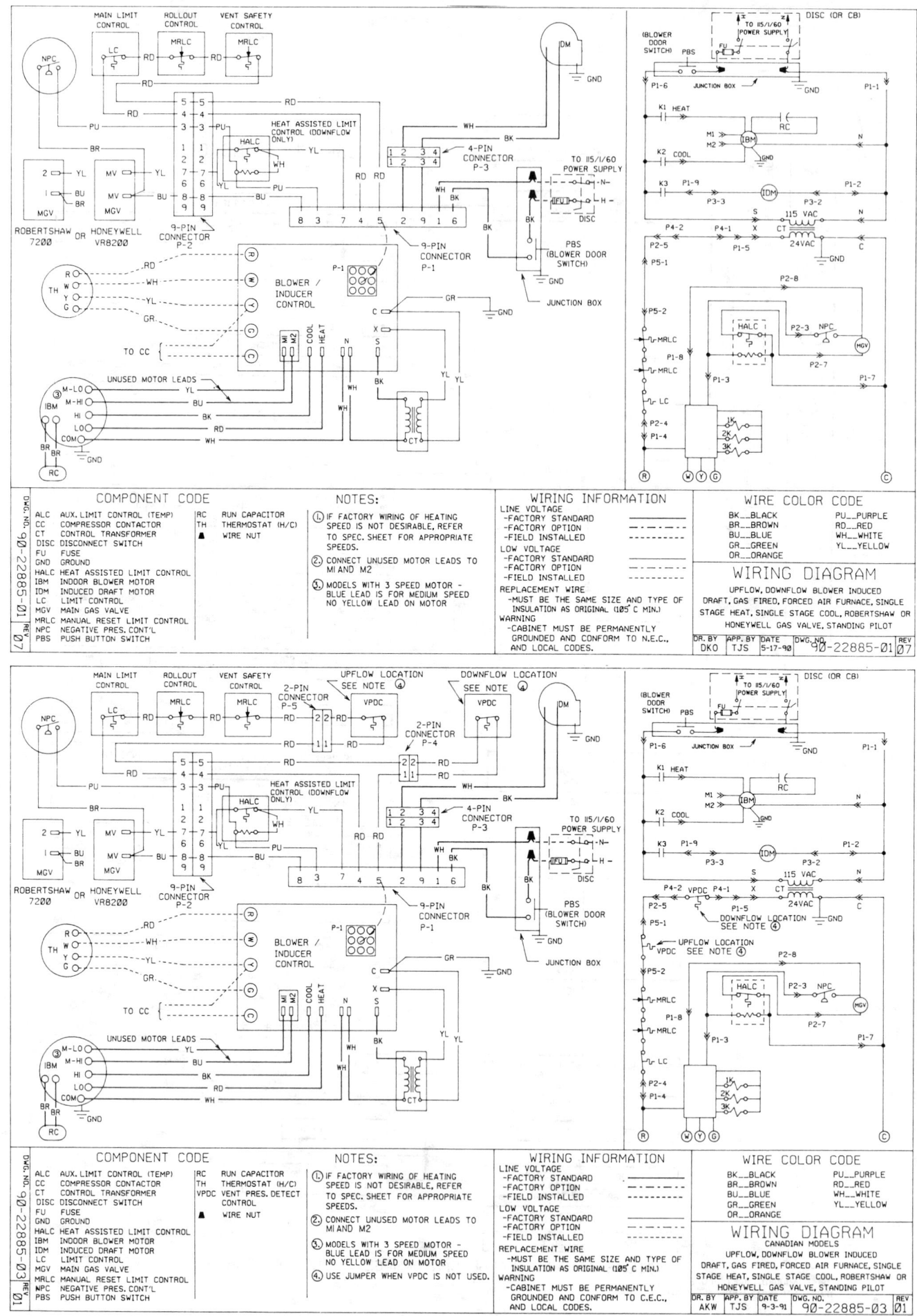

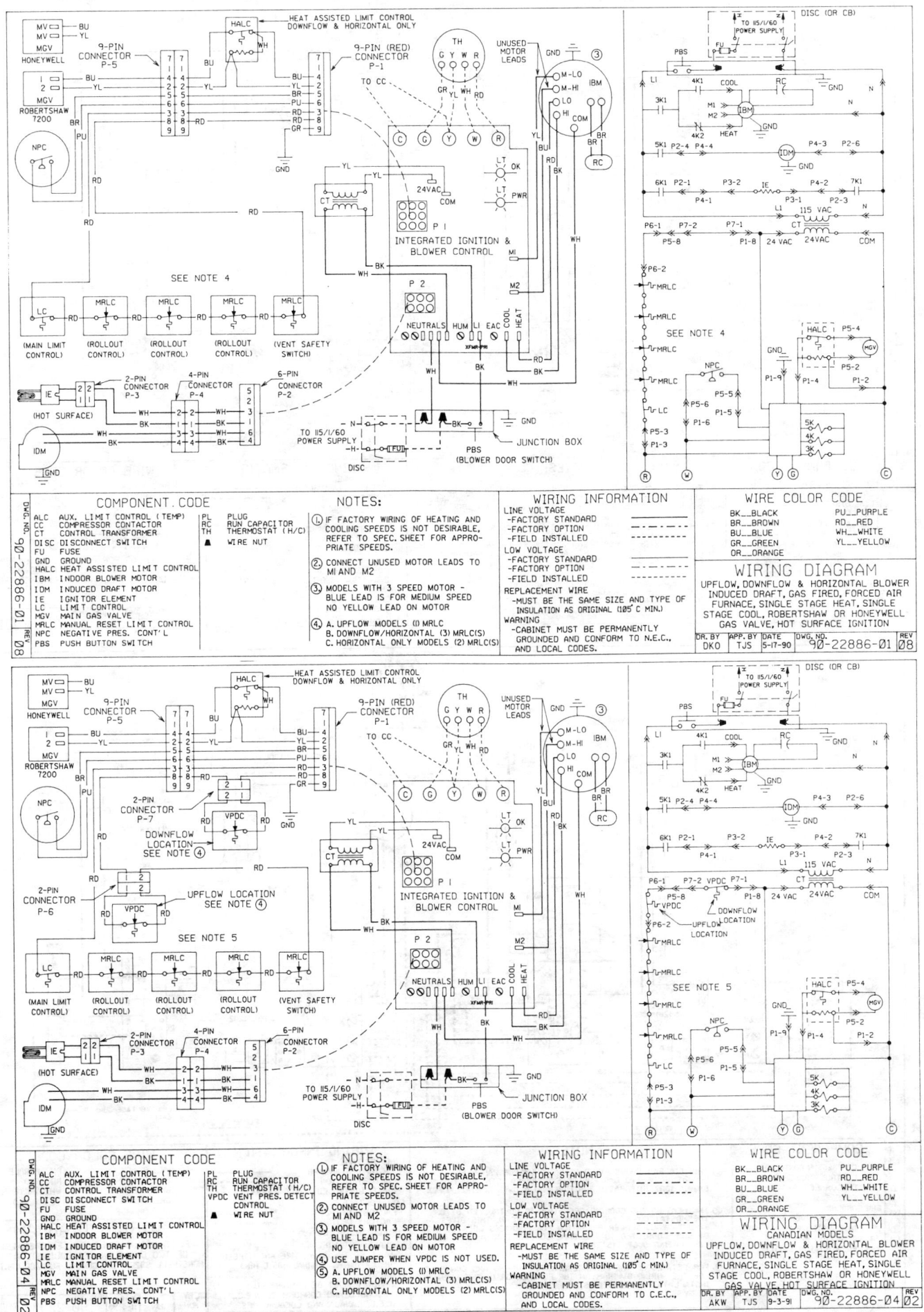

23

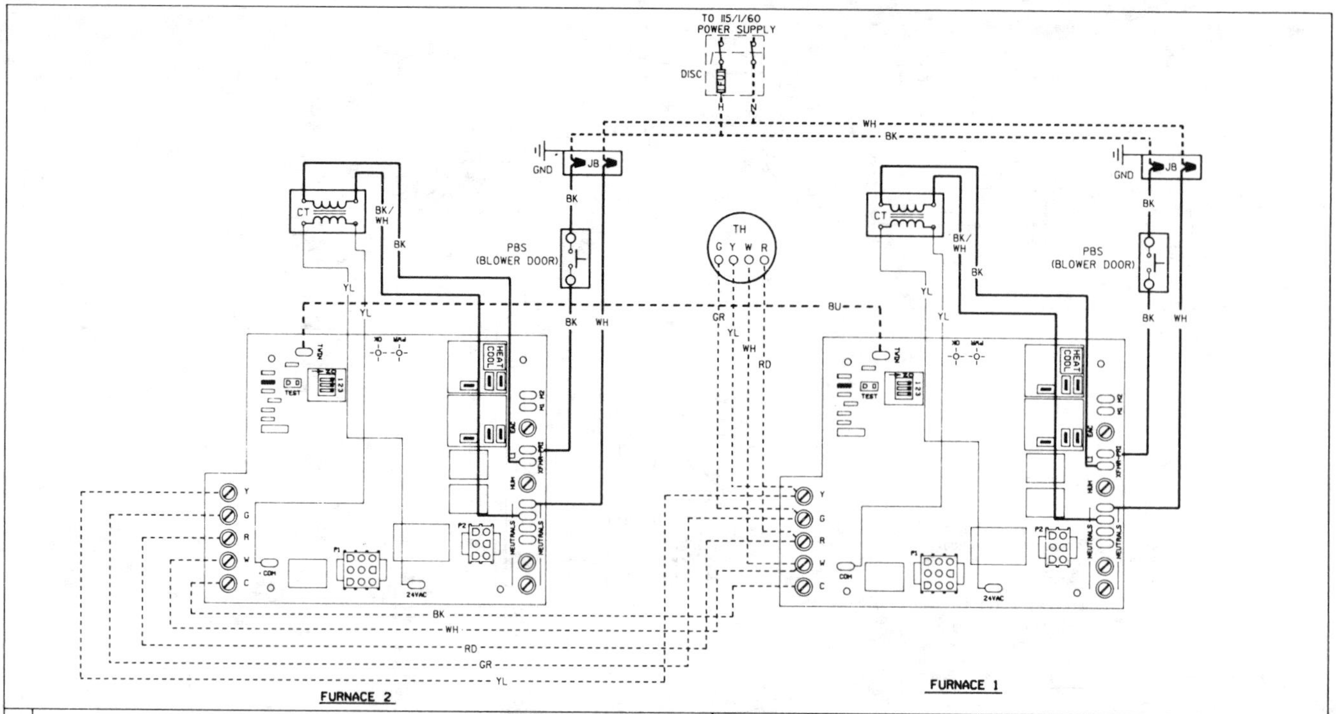

Top diagram (90-22775-01)

TO 115/1/60 POWER SUPPLY

DISC

FURNACE 2   FURNACE 1

TH — G Y W R

COMPONENT CODE

CC   COMPRESSOR CONTACTOR
CT   CONTROL TRANSFORMER
DISC   DISCONNECT SWITCH
FU   FUSE
GND   GROUND
IBM   INDOOR BLOWER MOTOR
IFC   INTEGRATED FURNACE CONT.
JB   JUNCTION BOX
PBS   PUSH BUTTON SWITCH
TH   THERMOSTAT (H/C)
▲   WIRE NUT

DWG. NO. 90-22775-01   REV 04

NOTES:

① THIS DIAGRAM SHOWS FIELD WIRING ONLY FOR PARALLEL OPERATION OF TWO GAS FURNACES EQUIPPED WITH HONEYWELL S9201A-1002 INTEGRATED IGNITION AND BLOWER CONTROL (IFC).

② THE HEATING AND COOLING OPERATION IS CONTROLLED BY ONE THERMOSTAT CONNECTED TO FURNACE 2 (IFC-2). THIS GIVES SIMULTANEOUS BLOWER OPERATION FOR FURNACES LOCATED SIDE BY SIDE CONNECTED TO THE SAME COOLING COIL AND/OR AIR CIRCULATING DUCT SYSTEM.

③ FOR DETAILS OF THE FURNACE INTERNAL WIRING AND CONTROLS, SEE THE WIRING DIAGRAM ON THE FURNACE OR IN THE FURNACE INSTALLATION INSTRUCTIONS.

WIRING INFORMATION

LINE VOLTAGE
- FACTORY STANDARD
- FACTORY OPTION
- FIELD INSTALLED

LOW VOLTAGE
- FACTORY STANDARD
- FACTORY OPTION
- FIELD INSTALLED

REPLACEMENT WIRE
- MUST BE THE SAME SIZE AND TYPE OF INSULATION AS ORIGINAL (105°C MIN.)

WARNING
- CABINET MUST BE PERMANENTLY GROUNDED AND CONFORM TO N.E.C., (C.E.C.-CANADA) AND LOCAL CODES.

WIRE COLOR CODE

BK	BLACK	PU	PURPLE
BR	BROWN	RD	RED
BU	BLUE	WH	WHITE
GR	GREEN	YL	YELLOW
OR	ORANGE		

WIRING DIAGRAM

FIELD WIRING FOR PARALLEL OPERATION OF TWO GAS FURNACES EQUIPPED WITH UTEC 1012-820 INTEGRATED FURNACE CONTROL

| DR. BY | APP. BY | DATE | DWG. NO. | REV |
| AKW | | 5-6-91 | 90-22775-01 | 04 |

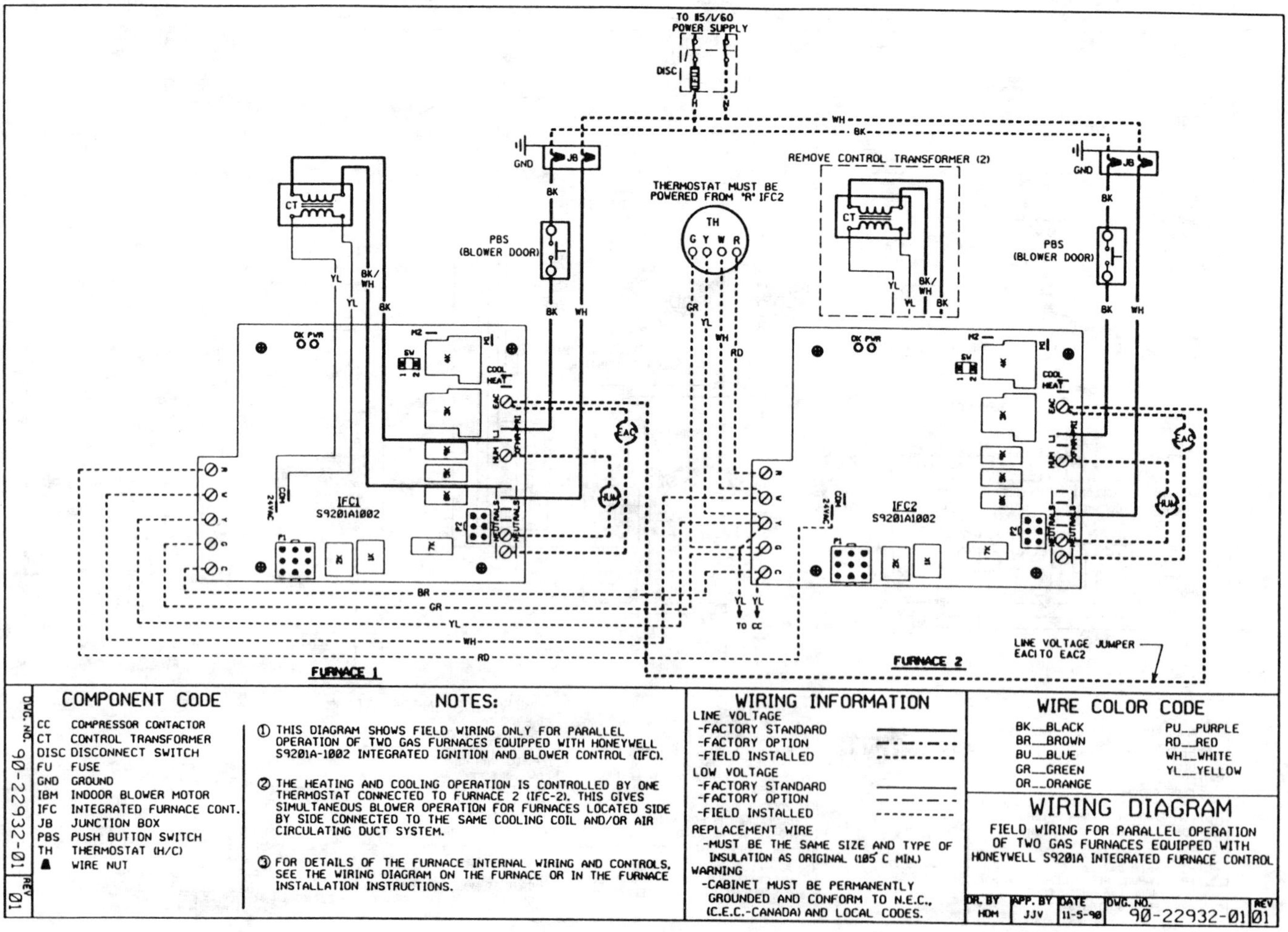

Bottom diagram (90-22932-01)

TO 115/1/60 POWER SUPPLY

DISC

REMOVE CONTROL TRANSFORMER (2)

THERMOSTAT MUST BE POWERED FROM 'R' IFC2

TH — G Y W R

IFC1 S9201A1002   IFC2 S9201A1002

FURNACE 1   FURNACE 2

LINE VOLTAGE JUMPER EAC1 TO EAC2

COMPONENT CODE

CC   COMPRESSOR CONTACTOR
CT   CONTROL TRANSFORMER
DISC   DISCONNECT SWITCH
FU   FUSE
GND   GROUND
IBM   INDOOR BLOWER MOTOR
IFC   INTEGRATED FURNACE CONT.
JB   JUNCTION BOX
PBS   PUSH BUTTON SWITCH
TH   THERMOSTAT (H/C)
▲   WIRE NUT

DWG. NO. 90-22932-01   REV 01

NOTES:

① THIS DIAGRAM SHOWS FIELD WIRING ONLY FOR PARALLEL OPERATION OF TWO GAS FURNACES EQUIPPED WITH HONEYWELL S9201A-1002 INTEGRATED IGNITION AND BLOWER CONTROL (IFC).

② THE HEATING AND COOLING OPERATION IS CONTROLLED BY ONE THERMOSTAT CONNECTED TO FURNACE 2 (IFC-2). THIS GIVES SIMULTANEOUS BLOWER OPERATION FOR FURNACES LOCATED SIDE BY SIDE CONNECTED TO THE SAME COOLING COIL AND/OR AIR CIRCULATING DUCT SYSTEM.

③ FOR DETAILS OF THE FURNACE INTERNAL WIRING AND CONTROLS, SEE THE WIRING DIAGRAM ON THE FURNACE OR IN THE FURNACE INSTALLATION INSTRUCTIONS.

WIRING INFORMATION

LINE VOLTAGE
- FACTORY STANDARD
- FACTORY OPTION
- FIELD INSTALLED

LOW VOLTAGE
- FACTORY STANDARD
- FACTORY OPTION
- FIELD INSTALLED

REPLACEMENT WIRE
- MUST BE THE SAME SIZE AND TYPE OF INSULATION AS ORIGINAL (105 C MIN.)

WARNING
- CABINET MUST BE PERMANENTLY GROUNDED AND CONFORM TO N.E.C., (C.E.C.-CANADA) AND LOCAL CODES.

WIRE COLOR CODE

BK	BLACK	PU	PURPLE
BR	BROWN	RD	RED
BU	BLUE	WH	WHITE
GR	GREEN	YL	YELLOW
OR	ORANGE		

WIRING DIAGRAM

FIELD WIRING FOR PARALLEL OPERATION OF TWO GAS FURNACES EQUIPPED WITH HONEYWELL S9201A INTEGRATED FURNACE CONTROL

| DR. BY | APP. BY | DATE | DWG. NO. | REV |
| HDM | JJV | 11-5-90 | 90-22932-01 | 01 |

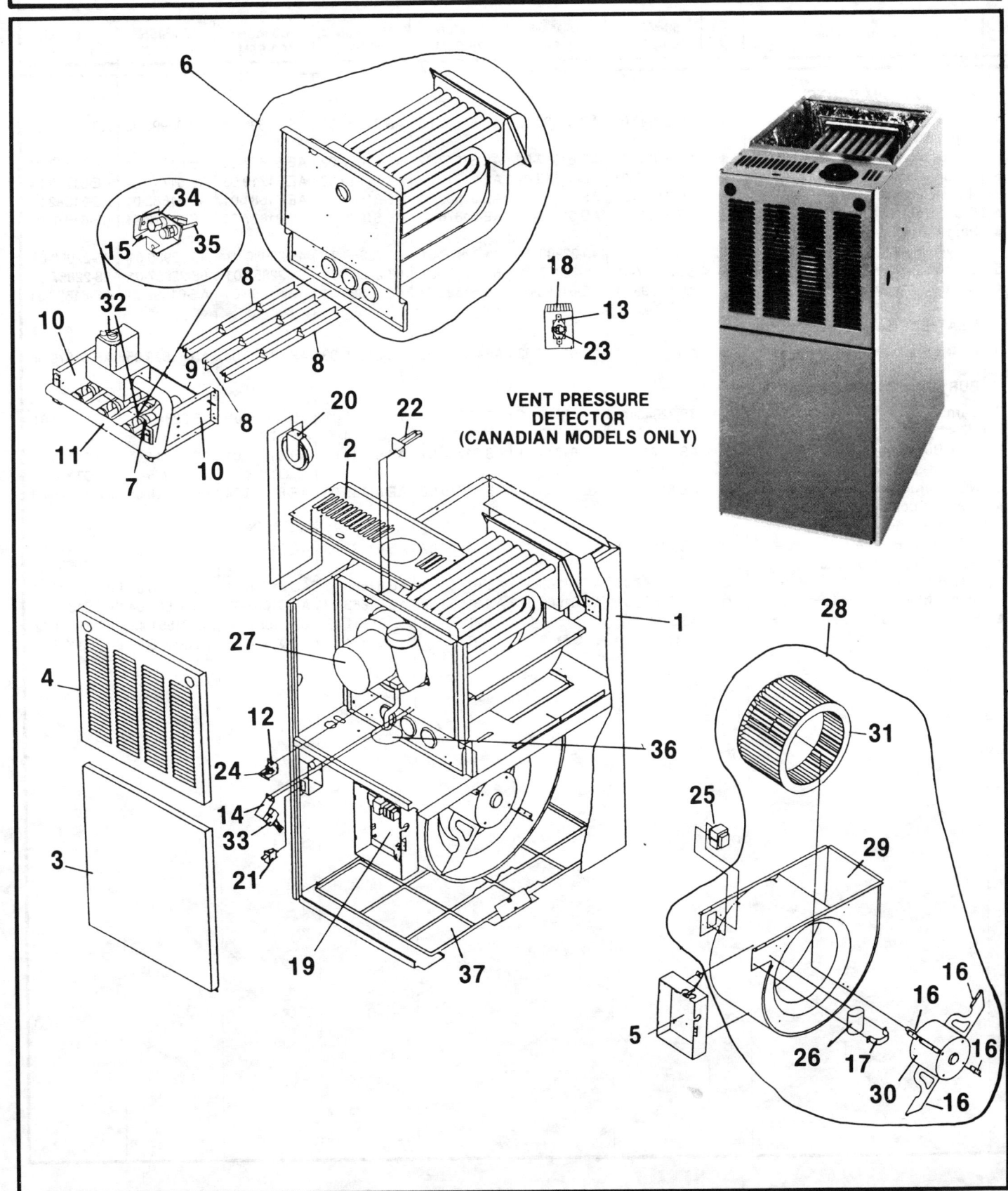

VENT PRESSURE DETECTOR (CANADIAN MODELS ONLY)

UPFLOW GAS FURNACES	GKUA SERIES	SIZES 50 THRU 150

Item No.	Part Description	Notes	50-M3N 50-E3N	75-M4N 75-E4N	75-M3N 75-E3N	100-M3N 100-E3N	100-M5N 100-E5N	125-M5N 125-E5N	150-M5N 150-E5N
	PANELS & S/M PARTS								
1	Jacket Assembly	☎	AS-61690-01	AS-61690-02	AS-61690-03	AS-61690-03	AS-61690-04	AS-66990-05	AS-66990-05
2	Top Plate-								
	U.S. Models	☎	AE-61657-01	AE-61657-02	AE-61657-02	AE-61657-02	AE-61657-03	AE-61657-04	AE-61657-04
	Canadian Models	☎	AE-61719-01	AE-61719-02	AE-61719-02	AE-61719-02	AE-61719-03	AE-61719-04	AE-61719-04
3	Blower Door	☎	AE-61679-01	AE-61680-01	AE-61680-01	AE-61680-01	AE-61681-01	AE-61682-01	AE-61682-01
4	Front Door W/O Plug Button	☎	AE-61675-01	AE-61676-01	AE-61676-01	AE-61676-01	AE-61677-01	AE-61678-01	AE-61678-01
▲	Plug Button W/Recessed Type Head (2 Required)	★	45-22980-01	45-22980-01	45-22980-01	45-22980-01	45-22980-01	45-22980-01	45-22980-01
5	Control Box W/O Components	☐	68-22857-01	68-22857-01	68-22857-01	68-22857-01	68-22857-01	68-22857-01	68-22857-01
▲	Control Box Cover	☎	AE-61699-01	AE-61699-01	AE-61699-01	AE-61699-01	AE-61699-01	AE-61699-01	AE-61699-01
	HEAT EXCHANGER GROUP								
6	Heat Exchanger Assy W/Insul.	☎	AS-61579-01	AS-61579-02	AS-61579-02	AS-61579-03	AS-61579-04	AS-61579-05	AS-61579-06
	BURNER / MANIFOLD GROUP								
7	Burner	☐	75-22840-01 QTY 2	75-22840-01 QTY 3	75-22840-01 QTY 3	75-22840-01 QTY 4	75-22840-01 QTY 4	75-22840-01 QTY 5	75-22840-01 QTY 6
8	NOx Burner Insert	☎	AS-61714-01 QTY 2	AS-61714-01 QTY 3	AS-61714-01 QTY 3	AS-61714-01 QTY 4	AS-61714-01 QTY 4	AS-61714-01 QTY 5	AS-61714-01 QTY 6
9	NOx Burner Retainer Rod	☎	AE-61715-01	AE-61715-02	AE-61715-02	AE-61715-03	AE-61715-04	AE-61715-05	AE-61715-06
▲	Burner Support-Holds the Burners Between the Burner Support Brackets	☎	AE-61577-01	AE-61577-02	AE-61577-02	AE-61577-03	AE-61577-03	AE-61577-04	AE-61577-05
10	Burner Support Bracket (2 Required)	☎	AE-61588-01	AE-61588-01	AE-61588-01	AE-61588-01	AE-61588-01	AE-61588-01	AE-61588-01
11	Manifold W/O Orifices	☎	AS-61663-01	AS-61663-02	AS-61663-02	AS-61663-03	AS-61663-03	AS-61663-04	AS-61663-05
▲	Burner Orifice-Natural	☐	62-21551-42 QTY 2	62-21551-42 QTY 3	62-21551-42 QTY 3	62-21551-42 QTY 4	62-21551-42 QTY 4	62-21551-42 QTY 5	62-21551-42 QTY 6

Item No.	Part Description	Notes	50-M3N 50-E3N	75-M4N 75-E4N	75-M3N 75-E3N	100-M3N 100-E3N	100-M5N 100-E5N	125-M5N 125-E5N	150-M5N 150-E5N
	BRACKETS / GASKETS / MISC.								
	Brackets								
12	Overtemperature Bracket-attached to Center Panel	☎	AE-61600-02	AE-61600-02	AE-61600-02	AE-61600-02	AE-61600-02	AE-61600-02	AE-61600-02
13	Vent Pressure Detector Control Bracket-Canadian Models Only	☎	AE-61508-01	AE-61508-01	AE-61508-01	AE-61508-01	AE-61508-01	AE-61508-01	AE-61508-01
14	Ignitor Bracket	☎	AE-61581-01	AE-61581-01	AE-61581-01	AE-61581-01	AE-61581-01	AE-61581-01	AE-61581-01
15	Pilot Bracket	☎	AE-61658-01	AE-61658-01	AE-61658-01	AE-61658-01	AE-61658-01	AE-61658-01	AE-61658-01
16	Motor Mounting Legs (4 Req.)-Mounting Screws Listed in Blower Assembly Group	☎1	AE-61698-01	AE-61698-01	AE-61698-01	AE-61698-01	AE-61698-01	AE-61698-01	AE-61698-01
17	Run Capacitor Strap	☎	AE-50892-01	AE-50892-01	AE-50892-01	AE-50892-01	AE-50892-01	AE-50892-01	AE-50892-01
	Gaskets								
▲	Induced Draft Motor Gasket	☐	68-22850-01	68-22850-01	68-22850-01	68-22850-01	68-22850-01	68-22850-01	68-22850-01
▲	Main Limit Gasket	☐	68-18724-04	68-18724-04	68-18724-04	68-18724-04	68-18724-04	68-18724-04	68-18724-04
	Miscellaneous								
▲	Jumper Plug-Canadian Models Only	☎	AS-60857-02	AS-60857-02	AS-60857-02	AS-60857-02	AS-60857-02	AS-60857-02	AS-60857-02
▲	Silicone Rubber Tubing-From the Induced Draft Blower Motor to the Pressure Switch	☎✔	79-21491-82	79-21491-82	79-21491-82	79-21491-82	79-21491-82	79-21491-82	79-21491-82
18	Vent Pressure Detector Collar-Canadian Models Only	☎	AE-61507-01	AE-61507-01	AE-61507-01	AE-61507-01	AE-61507-01	AE-61507-01	AE-61507-01
▲	Filter Retainer Rod	☐	45-22867-01	45-22867-01	45-22867-01	45-22867-01	45-22867-01	45-22867-02	45-22867-02
37	Filter-Permanent-Plastic	☐	54-22697-01	54-22697-01	54-22697-01	54-22697-01	54-22697-02	54-22697-03	54-22697-03

UPFLOW GAS FURNACES		GKUA SERIES		SIZES 50 THRU 150		

Item No.	Part Description	Notes	50-M3N 50-E3N	75-M4N 75-E4N	75-M3N 75-E3N	100-M3N 100-E3N	100-M5N 100-E5N	125-M5N 125-E5N	150-M5N 150-E5N
	ELECTRIC GROUP								
19	Integrated Ignition & Blower Control Board for Hot Surface Ignition Models	☐✔2	62-22694-82	62-22694-82	62-22694-82	62-22694-82	62-22694-82	62-22694-82	62-22694-82
19	Blower/Inducer Control Board for Standing Pilot Models	☐✔2	47-22827-82	47-22827-82	47-22827-82	47-22827-82	47-22827-82	47-22827-82	47-22827-82
20	Pressure Switch	☐	42-22682-02	42-22682-03	42-22682-03	42-22682-03	42-22682-03	42-22682-02	42-22682-02
21	Door Switch - Push Button	☐	42-22692-01	42-22692-01	42-22692-01	42-22692-01	42-22692-01	42-22692-01	42-22692-01
22	Main Limit	☐	47-22679-03	47-22679-01	47-22679-02	47-22679-02	47-22679-04	47-22679-03	47-22679-02
23	Vent Pressure Detector Control-Manual Reset Canadian Models Only								
	Hot Surface Ign. Models	☐✔	47-21900-21	47-21900-06	47-21900-06	47-21900-21	47-21900-21	47-21900-05	47-22383-11
	Standing Pilot Models	☐✔	47-21900-05	47-21900-06	47-21900-06	47-21900-05	47-21900-05	47-22383-11	47-22383-11
▲	Vent Safety Switch - Manual Reset Limit on Induced Draft Motor	☐	47-22862-01	47-22862-02	47-22862-02	47-22862-03	47-22862-03	47-22862-04	47-22862-03
24	Overtemperature Switch - Manual Reset Limit On Center Panel	☐	47-22861-01	47-22861-01	47-22861-01	47-22861-01	47-22861-01	47-22861-01	47-22861-01
25	Transformer-Hot Surface Ign.	☐	46-22863-01	46-22863-01	46-22863-01	46-22863-01	46-22863-01	46-22863-01	46-22863-01
25	Transformer-Standing Pilot	☐	46-22691-01	46-22691-01	46-22691-01	46-22691-01	46-22691-01	46-22691-01	46-22691-01
26	Run Capacitor 7.5/370	☐✔	43-20847-11		43-20847-11	43-20847-11			
26	Run Capacitor 15/370	☐		43-20847-17			43-20847-17	43-20847-17	43-20847-17
27	Induced Draft Blower Motor	☐	70-22838-02	70-22838-02	70-22838-02	70-22838-02	70-22838-02	70-22838-02	70-22838-02
	BLOWER GROUP								
28	Blower Assembly Including:	☎	AS-61697-07	AS-61697-02	AS-61697-01	AS-61697-01	AS-61697-03	AS-61697-03	AS-61697-03
29	Housing	☎	AS-61674-01	AS-61674-02	AS-61674-01	AS-61674-01	AS-61674-03	AS-61674-03	AS-61674-03
16	Motor Mounting Legs(4 Req)	☎1	AE-61698-01	AE-61698-01	AE-61698-01	AE-61698-01	AE-61698-01	AE-61698-01	AE-61698-01
▲	Motor Mounting Screws (8 Required)	☐3	63-22338-03	63-22338-03	63-22338-03	63-22338-03	63-22338-03	63-22338-03	63-22338-03
30	Motor 1/2 HP, 115V, 4 Spd.	★✔	51-22873-01		51-22873-01	51-22873-01			
30	Motor 3/4 HP, 115V, 4 Spd.	★		51-22700-01			51-22700-01	51-22700-01	51-22700-01
31	Wheel 11x6, CW, 1/2" Bore	☐	70-22683-01		70-22683-01	70-22683-01			
31	Wheel 11x7, CW, 1/2" Bore	☐✔		70-21476-01				70-21476-01	70-21476-01
31	Wheel 11x10, CW, 1/2" Bore	☐					70-20602-01		
26	Run Capacitor 7.5/370	☐✔	43-20847-11		43-20847-11	43-20847-11			
26	Run Capacitor 15/370	☐		43-20847-17			43-20847-17	43-20847-17	43-20847-17
17	Run Capacitor Strap	☎	AE-50892-01	AE-50892-01	AE-50892-01	AE-50892-01	AE-50892-01	AE-50892-01	AE-50892-01

	UPFLOW GAS FURNACES		GKUA SERIES		SIZES 50 THRU 150		

Item No.	Part Description	Notes	50-M3N 50-E3N	75-M4N 75-E4N	75-M3N 75-E3N	100-M3N 100-E3N	100-M5N 100-E5N	125-M5N 125-E5N	150-M5N 150-E5N
	HOT SURFACE IGNITION — GAS VALVE, GAS COMPONENTS AND LP (PROPANE) KIT								
	HONEYWELL GAS CONTROLS								
	SERIAL CONTROL CODES-BR, BS, CD, CE								
32	Gas Valve-HW VR8205H1060	□ △	60-22866-01	60-22866-01	60-22866-01	60-22866-01	60-22866-01	60-22866-01	60-22866-01
33	Ignitor Kit - Norton	□ ✔	62-22868-82	62-22868-82	62-22868-82	62-22868-82	62-22868-82	62-22868-82	62-22868-82
▲	Ignitor Shield	☎	AE-61627-01	AE-61627-01	AE-61627-01	AE-61627-01	AE-61627-01	AE-61627-01	AE-61627-01
14	Ignitor Bracket	☎	AE-61581-01	AE-61581-01	AE-61581-01	AE-61581-01	AE-61581-01	AE-61581-01	AE-61581-01
▲	**LP CONVERSION KIT - U.S. MODELS**	□ △	[EP-70H]	[EP-70H]	[EP-70H]	[EP-70H]	[EP-70H]	[EP-70H]	[EP-70H]
32	Gas Valve, For LP Gas, Use HW VR8205H1060, Nat. and	□ 4,5	60-22866-01	60-22866-01	60-22866-01	60-22866-01	60-22866-01	60-22866-01	60-22866-01
▲	Spring Conversion Assy.	□ 5	60-22513-01	60-22513-01	60-22513-01	60-22513-01	60-22513-01	60-22513-01	60-22513-01
▲	Burner Orifice - LP	□ 5	62-21551-54 QTY 2	62-21551-54 QTY 3	62-21551-54 QTY 3	62-21551-54 QTY 4	62-21551-54 QTY 4	62-21551-54 QTY 5	62-21551-54 QTY 6
▲	**PROPANE CONVERSION KIT - CANADIAN MODELS 0-4500'**	□ △	[EP-72H]	[EP-72H]	[EP-72H]	[EP-72H]	[EP-72H]	[EP-72H]	[EP-72H]
35	Gas Valve, For Propane Gas, Use HW VR8205H1060, Nat.	□ 4,5	60-22866-01	60-22866-01	60-22866-01	60-22866-01	60-22866-01	60-22866-01	60-22866-01
▲	and Spring Conversion Assy	□ 5	60-22513-01	60-22513-01	60-22513-01	60-22513-01	60-22513-01	60-22513-01	60-22513-01
▲	Burner Orifice - Propane 0-2000' Elevation	□ 5	62-21551-54 QTY 2	62-21551-54 QTY 3	62-21551-54 QTY 3	62-21551-54 QTY 4	62-21551-54 QTY 4	62-21551-54 QTY 5	62-21551-54 QTY 6
	2000'-4500' Elevation	★ 5	62-21551-55 QTY 2	62-21551-55 QTY 3	62-21551-55 QTY 3	62-21551-55 QTY 4	62-21551-55 QTY 4	62-21551-55 QTY 5	62-21551-55 QTY 6
	ROBERTSHAW GAS CONTROLS								
	SERIAL CONTROL CODES-BT								
32	Gas Valve-RS 7200DER-S7A	†	60-22525-11	60-22525-11	60-22525-11	60-22525-11	60-22525-11	60-22525-11	60-22525-11
33	Ignitor Kit - Norton	□ ✔	62-22868-82	62-22868-82	62-22868-82	62-22868-82	62-22868-82	62-22868-82	62-22868-82
▲	Ignitor Shield	☎	AE-61627-01	AE-61627-01	AE-61627-01	AE-61627-01	AE-61627-01	AE-61627-01	AE-61627-01
14	Ignitor Bracket	☎	AE-61581-01	AE-61581-01	AE-61581-01	AE-61581-01	AE-61581-01	AE-61581-01	AE-61581-01
▲	**LP CONVERSION KIT - U.S. MODELS**	⊘	[EP-69R]	[EP-69R]	[EP-69R]	[EP-69R]	[EP-69R]	[EP-69R]	[EP-69R]
32	Gas Valve, For LP Gas, Use RS 7200DER-S7A, Nat. and	† 4,5	60-22525-11	60-22525-11	60-22525-11	60-22525-11	60-22525-11	60-22525-11	60-22525-11
▲	Spring Conversion Assy.	† 5	60-18759-06	60-18759-06	60-18759-06	60-18759-06	60-18759-06	60-18759-06	60-18759-06
▲	Burner Orifice - LP	□ 5	62-21551-54 QTY 2	62-21551-54 QTY 3	62-21551-54 QTY 3	62-21551-54 QTY 4	62-21551-54 QTY 4	62-21551-54 QTY 5	62-21551-54 QTY 6
▲	**PROPANE CONVERSION KIT - CANADIAN MODELS 0-4500'**	⊘	[EP-71R]	[EP-71R]	[EP-71R]	[EP-71R]	[EP-71R]	[EP-71R]	[EP-71R]
32	Gas Valve, For Propane Gas, Use RS 7200DER-S7A, Nat.	† 4,5	60-22525-11	60-22525-11	60-22525-11	60-22525-11	60-22525-11	60-22525-11	60-22525-11
▲	and Spring Conversion Assy	† 5	60-18759-06	60-18759-06	60-18759-06	60-18759-06	60-18759-06	60-18759-06	60-18759-06
▲	Burner Orifice - Propane 0-2000' Elevation	□ 5	62-21551-54 QTY 2	62-21551-54 QTY 3	62-21551-54 QTY 3	62-21551-54 QTY 4	62-21551-54 QTY 4	62-21551-54 QTY 5	62-21551-54 QTY 6
	2000'-4500' Elevation	★ 5	62-21551-55 QTY 2	62-21551-55 QTY 3	62-21551-55 QTY 3	62-21551-55 QTY 4	62-21551-55 QTY 4	62-21551-55 QTY 5	62-21551-55 QTY 6

UPFLOW GAS FURNACES	GKUA SERIES	SIZES 50 THRU 150

Item No.	Part Description	Notes	50-M3N 50-E3N	75-M4N 75-E4N	75-M3N 75-E3N	100-M3N 100-E3N	100-M5N 100-E5N	125-M5N 125-E5N	150-M5N 150-E5N
	STANDING PILOT — GAS VALVE, GAS COMPONENTS AND LP (PROPANE) KIT								
	HONEYWELL GAS CONTROLS **SERIAL CONTROL CODES-** BA, BL, CB, CC								
32	Gas Valve-HW VR8200H1004	□△	60-22174-02	60-22174-02	60-22174-02	60-22174-02	60-22174-02	60-22174-02	60-22174-02
34	Pilot Assy-HW Q350A1065	□△6	62-22834-01	62-22834-01	62-22834-01	62-22834-01	62-22834-01	62-22834-01	62-22834-01
▲	Pilot Tubing	☎△	AE-60777-61	AE-60777-61	AE-60777-61	AE-60777-63	AE-60777-63	AE-60777-65	AE-60777-67
15	Pilot Bracket	☎∠	AE-61658-01	AE-61658-01	AE-61658-01	AE-61658-01	AE-61658-01	AE-61658-01	AE-61658-01
35	Thermocouple-HW Q309A2143	□△	62-22348-02	62-22348-02	62-22348-02	62-22348-02	62-22348-02	62-22348-02	62-22348-02
36	Exhaust Hood Assembly	☎	AS-61654-01	AS-61654-02	AS-61654-02	AS-61654-01	AS-61654-01	AS-61654-02	AS-61654-01
▲	**LP CONVERSION KIT -** **U.S. MODELS**	□	[EP-70H]	[EP-70H]	[EP-70H]	[EP-70H]	[EP-70H]	[EP-70H]	[EP-70H]
☞	Gas Valve, For LP Gas, Use								
32	HW VR8200H1004, Nat. and	□4,5	60-22174-02	60-22174-02	60-22174-02	60-22174-02	60-22174-02	60-22174-02	60-22174-02
▲	Spring Conversion Assy.	□5	60-22513-01	60-22513-01	60-22513-01	60-22513-01	60-22513-01	60-22513-01	60-22513-01
▲	Pilot Orifice - Kit - LP	□5,7	62-22834-83	62-22834-83	62-22834-83	62-22834-83	62-22834-83	62-22834-83	62-22834-83
▲	Burner Orifice - LP	□5	62-21551-54 QTY 2	62-21551-54 QTY 3	62-21551-54 QTY 3	62-21551-54 QTY 4	62-21551-54 QTY 4	62-21551-54 QTY 5	62-21551-54 QTY 6
▲	**PROPANE CONVERSION KIT -** **CANADIAN MODELS 0-4500'**	□∠	[EP-72H]	[EP-72H]	[EP-72H]	[EP-72H]	[EP-72H]	[EP-72H]	[EP-72H]
☞	Gas Valve, For Propane Gas,								
32	Use HW VR8200H1004, Nat.	□4,5	60-22174-02	60-22174-02	60-22174-02	60-22174-02	60-22174-02	60-22174-02	60-22174-02
▲	and Spring Conversion Assy	□5	60-22513-01	60-22513-01	60-22513-01	60-22513-01	60-22513-01	60-22513-01	60-22513-01
▲	Pilot Orifice - Kit - Propane	□5,7	62-22834-83	62-22834-83	62-22834-83	62-22834-83	62-22834-83	62-22834-83	62-22834-83
▲	Burner Orifice - Propane 0-2000' Elevation	□5	62-21551-54 QTY 2	62-21551-54 QTY 3	62-21551-54 QTY 3	62-21551-54 QTY 4	62-21551-54 QTY 4	62-21551-54 QTY 5	62-21551-54 QTY 6
	2000'-4500' Elevation	□5	62-21551-55 QTY 2	62-21551-55 QTY 3	62-21551-55 QTY 3	62-21551-55 QTY 4	62-21551-55 QTY 4	62-21551-55 QTY 5	62-21551-55 QTY 6
	ROBERTSHAW GAS CONTROLS **SERIAL CONTROL CODES-** BW, BZ								
35	Gas Valve-RS 7200ER-S7C	†	60-22525-10	60-22525-10	60-22525-10	60-22525-10	60-22525-10	60-22525-10	60-22525-10
37	Pilot Assy-RS 7CL-6	†6	62-22864-01	62-22864-01	62-22864-01	62-22864-01	62-22864-01	62-22864-01	62-22864-01
▲	Pilot Tubing	†	AE-60777-60	AE-60777-60	AE-60777-60	AE-60777-62	AE-60777-62	AE-60777-64	AE-60777-66
16	Pilot Bracket	☎	AE-61658-01	AE-61658-01	AE-61658-01	AE-61658-01	AE-61658-01	AE-61658-01	AE-61658-01
38	Thermocouple Kit - RS	†	62-22348-02	62-22348-02	62-22348-02	62-22348-02	62-22348-02	62-22348-02	62-22348-02
39	Exhaust Hood Assembly	☎	AS-61654-01	AS-61654-02	AS-61654-02	AS-61654-01	AS-61654-01	AS-61654-02	AS-61654-01
▲	**LP CONVERSION KIT -** **U.S. MODELS**	⊘	[EP-69R]	[EP-69R]	[EP-69R]	[EP-69R]	[EP-69R]	[EP-69R]	[EP-69R]
☞	Gas Valve, For LP Gas, Use								
35	RS 7200ER-S7C, Nat. and	†4,5	60-22525-10	60-22525-10	60-22525-10	60-22525-10	60-22525-10	60-22525-10	60-22525-10
▲	Spring Conversion Assy.	†5	60-18759-06	60-18759-06	60-18759-06	60-18759-06	60-18759-06	60-18759-06	60-18759-06
▲	Pilot Orifice - Kit - LP	†5,7	62-22834-83	62-22834-83	62-22834-83	62-22834-83	62-22834-83	62-22834-83	62-22834-83
▲	Burner Orifice - LP	□5	62-21551-54 QTY 2	62-21551-54 QTY 3	62-21551-54 QTY 3	62-21551-54 QTY 4	62-21551-54 QTY 4	62-21551-54 QTY 5	62-21551-54 QTY 6
▲	**PROPANE CONVERSION KIT -** **CANADIAN MODELS 0-4500'**	⊘	[EP-71R]	[EP-71R]	[EP-71R]	[EP-71R]	[EP-71R]	[EP-71R]	[EP-71R]
☞	Gas Valve, For Propane Gas,								
35	Use RS 7200ER-S7C, Nat.	†4,5	60-22525-10	60-22525-10	60-22525-10	60-22525-10	60-22525-10	60-22525-10	60-22525-10
▲	and Spring Conversion Assy	†5	60-18759-06	60-18759-06	60-18759-06	60-18759-06	60-18759-06	60-18759-06	60-18759-06
▲	Pilot Orifice - Kit - Propane	†5,7	62-22834-83	62-22834-83	62-22834-83	62-22834-83	62-22834-83	62-22834-83	62-22834-83
▲	Burner Orifice - Propane 0-2000' Elevation	□5	62-21551-54 QTY 2	62-21551-54 QTY 3	62-21551-54 QTY 3	62-21551-54 QTY 4	62-21551-54 QTY 4	62-21551-54 QTY 5	62-21551-54 QTY 6
	2000'-4500' Elevation	★5	62-21551-55 QTY 2	62-21551-55 QTY 3	62-21551-55 QTY 3	62-21551-55 QTY 4	62-21551-55 QTY 4	62-21551-55 QTY 5	62-21551-55 QTY 6

UPFLOW GAS FURNACES	GKUA SERIES	SIZES 50 THRU 150

EXPLANATION OF SERIAL NUMBER

④ ELEMENT FINISH ⑤ AREA REQ'T ⑥ MOTOR H.P. ⑩ PRODUCT NO.

① GAS CONTROL — **BS** **5D** **307** **M** **51** **91** **0000**

② LIMIT ③ BLOWER ⑦ PLANT ⑧ WEEK DATE CODE ⑨ YEAR

① TYPE OF CONTROL USED ON UNIT:

AU) (Field Test) ROBERTSHAW 7200DER W/HONEYWELL S9201A H.S.I.
BA) HONEYWELL VR8200H W/HONEYWELL ST9101A (STANDING PILOT)
BD) (Field Test) ROBERTSHAW 7200DER W/HONEYWELL ST9101A (STANDING PILOT)
BE) (Field Test) ROBERTSHAW 7200DER W/HAMILTON STANDARD 1033-1 (STANDING PILOT)
BH) (Field Test) ROBERTSHAW 7200DER W/WATSCO IBC (STANDING PILOT)
BK) (Field Test) ROBERTSHAW 7200DER W/HAMILTON STANDARD 1012-800
BL) HONEYWELL VRE8200H W/ROBERTSHAW RBC-1 (STANDING PILOT)
BM) (Field Test) ROBERTSHAW 7200ER W/ROBERTSHAW RBC-A (STANDING PILOT)
BR) HONEYWELL VR8205H W/HONEYWELL S9201A-1010 H.S.I. (9 SEC. IGN. TRIAL)
BS) HONEYWELL VR8205H W/HAMILTON STANDARD 1012-800
BT) ROBERTSHAW 7200DER-S7A W/HONEYWELL S9201A-1010 H.S.I. (9 SEC. IGN. TRIAL)
BW) ROBERTSHAW 7200ER-S7C W/HONEYWELL ST9101A (STANDING PILOT)
BZ) ROBERTSHAW 7200ER-S7C W/ROBERTSHAW RBC-1 (STANDING PILOT)
CB) HONEYWELL VR8200H W/HONEYWELL ST9101A (STANDING PILOT)
CC) HONEYWELL VR8200H W/ROBERTSHAW RBC-1 (STANDING PILOT)
CD) HONEYWELL VR8205H W/HONEYWELL S9201A-1010 H.S.I. (9 SEC. IGN. TRIAL)
CE) HONEYWELL VR8205H W/HAMILTON STANDARD 1012-822 OR 1012-823

② LIMIT CONTROL:
5 — T.O.D. OR SENSORS & SWITCH LIMIT CONTROL

③ BLOWER
D — DIRECT DRIVE

④ HEATING ELEMENT:
3 — ALUMINIZED

⑤ SPECIAL AREA REQUIREMENT:
O — NO AREA REQUIREMENT

⑥ MOTOR HORSEPOWER:
2 — 1/2 HP MOTOR
7 — 3/4 HP MOTOR

⑦ PLANT
F — FORT SMITH, AR
M — MILLEDGEVILLE, GA

⑧ WEEK

⑨ YEAR

⑩ STARTS WITH 0001 & RUNS THROUGH 9999 & THEN START WITH 0001 AGAIN.

NOTES:

WHEN A PARTICULAR COLOR IS REQUIRED IT MUST BE SPECIFIED WHEN ORDERING.

DNA DOES NOT APPLY

🐟 SPECIAL INFORMATION, PLEASE READ CAREFULLY AND FOLLOW APPLICABLE NOTES.

☐ THESE PARTS SHOULD BE STOCKED IN THE FIELD.

★ THIS IS A NEW REPLACEMENT PART AND SHOULD BE STOCKED IN THE FIELD.

☐ NOT A RECOMMENDED STOCKING ITEM AT TIME OF "PUBLICATION."

✔ THE REPLACEMENT PART IS EQUAL TO OR BETTER THAN THE ORIGINAL EQUIPMENT PART.

☎ COILS, PANELS AND OTHER MANUFACTURED PARTS ARE AVAILABLE WHILE UNIT IS IN PRODUCTION. AFTER UNIT IS OUT OF PRODUCTION ORDERING SUCH PARTS MUST BE CLEARED WITH THE PARTS DEPARTMENT AS PRICE AND AVAILABILITY WILL VARY. WHEN A PARTICULAR COLOR IS REQUIRED, IT MUST BE SPECIFIED WHEN ORDERING.

△ PRIMARY PART USED IN CURRENT PRODUCTION.

† ALTERNATE PART USED IN PRODUCTION.

▲ PARTS NOT SHOWN ON PICTORIAL COVER PAGE.

[ ] FACTORY APPROVED KITS MUST BE USED TO CONVERT FROM NATURAL TO LP(PROPANE) GAS AND MAY BE ORDERED AS AN OPTIONAL ACCESSORY.

1 THIS MOTOR IS SHELL MOUNTED. A SET OF SCREWS ARE SHIPPED WITH EACH REPLACEMENT MOTOR. NOTE: STANDARD MOTORS CAN BE USED WITH (2) SIX INCH RADIATOR CLAMPS IN PLACE OF SCREWS.

2 FOR ADJUSTING THE BLOWER CONTROL FUNCTION "ON-OFF." SEE THE MODEL INSTALLATION AND OPERATION MANUAL OR INSTRUCTIONS SHIPPED WITH THE REPAIR PART.

3 THESE ARE SPECIAL SCREWS FOR SHELL MOUNTED MOTORS - DO NOT SUBSTITUTE.

4 THIS PART IS A NATURAL GAS VALVE WHICH MUST BE CONVERTED TO LP(PROPANE) GAS BY USING THE REGULATOR SPRING ASSEMBLY LISTED BELOW THE GAS CONTROL.

5 THESE PARTS ARE TO BE USED ONLY AS REPAIR PARTS FOR UNITS WHICH HAVE BEEN CONVERTED TO LP(PROPANE) GAS BY THE APPROPRIATE CONVERSION KIT.

6 THE PILOT ORIFICE IS NOT AVAILABLE, ORDER PILOT ASSEMBLY.

7 PRODUCTION USES EITHER HONEYWELL OR ROBERTSHAW PILOTS, THIS KID INCLUDES BOTH ORIFICES.

Specifications and performance data subject to change without notice.

HEAT CONTROLLER, INC.

1900 WELLWORTH AVENUE • JACKSON, MICHIGAN 49203

THE QUALITY LEADER IN CONDITIONING AIR

Day & Night
Bryant/Payne

service and maintenance procedures
GAS-FIRED INDUCED-COMBUSTION FURNACE

376CAV
395CAV
Series B

Cancels: SP04-3   SP04-6
5/1/91

NOTE: Read the entire instructions before starting the installation. These procedures are for size 40,000- through 130,000-Btuh units.

SAFETY CONSIDERATIONS

Installation and servicing of heating equipment can be hazardous due to gas and electrical components. Only trained and qualified personnel should install, repair, or service heating equipment.

Untrained personnel can perform basic maintenance functions such as cleaning and replacing air filters. All other operations must be performed by trained service personnel. When working on heating equipment, observe precautions in the literature, tags, and labels attached to or shipped with the unit and other safety precautions that may apply.

Follow all safety codes. In the United States, follow all safety codes including the National Fuel Gas Code NFPA No. 54-1988/ANSI Z223.1-1988. In Canada, refer to the current edition of the National Standard of Canada CAN/CGA- B149.1- and .2-M86 Natural Gas and Propane Gas Installation Codes. Wear safety glasses and work gloves. Have fire extinguisher available during start-up and adjustment procedures and service calls.

Recognize safety information: This is the safety-alert symbol ⚠. When you see this symbol on the furnace and in instructions or manuals, be alert to the potential for personal injury.

Understand the signal word DANGER, WARNING, or CAUTION. These words are used with the safety-alert symbol. DANGER identifies the most serious hazards which **will** result in severe personal injury or death. WARNING signifies a hazard that **could** result in personal injury or death. CAUTION is used to identify unsafe practices, which **would** result in minor personal injury or product and property damage.

⚠ **WARNING:** The ability to properly perform maintenance on this equipment requires certain expertise, mechanical skills, tools, and equipment. If you do not possess these, do not attempt to perform any maintenance on this equipment other than those procedures recommended in the Users Manual. A FAILURE TO FOLLOW THIS WARNING COULD RESULT IN POSSIBLE DAMAGE TO THIS EQUIPMENT, SERIOUS PERSONAL INJURY, OR DEATH.

CARE AND MAINTENANCE

For continuing high performance, and to minimize possible equipment failure, it is essential that periodic maintenance be performed on this equipment. Consult your local dealer as to the proper frequency of maintenance and the availability of a maintenance contract.

A88317

Figure 1—Model 376CAV Horizontal

A88316

Figure 2—Model 376CAV Downflow

A88318

Figure 3—Model 395CAV Upflow

—1—

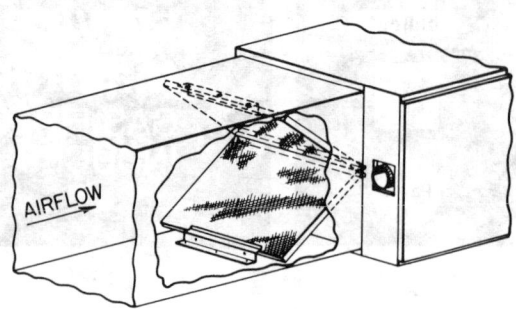

A82173

Figure 4—Horizontal Filter Arrangement

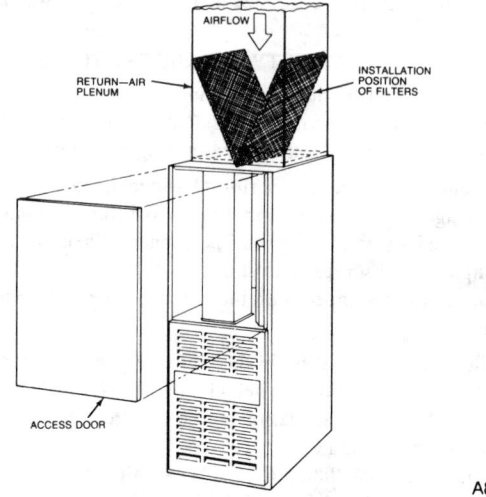

A88486

Figure 5—Position of Filters

> ⚠ **WARNING:** Never store anything on, near, or in contact with, the furnace, such as:
> 1. Spray or aerosol cans, rags, brooms, dust mops, vacuum cleaners, or other cleaning tools.
> 2. Soap powders, bleaches, waxes or other cleaning compounds, plastic or plastic containers, gasoline, kerosene, cigarette lighter fluid, dry cleaning fluids, or other volatile fluids.
> 3. Paint thinners and other painting compounds, paper bags or other paper products.
>
> A failure to follow this warning could result in corrosion of the heat exchanger, fire, personal injury, or death.

> ⚠ **WARNING:** Turn OFF the gas and electrical supplies to the unit before performing any maintenance or service on it. Follow the operating instructions on the label attached to the furnace. A failure to follow this warning could result in personal injury.

The minimum maintenance that should be performed on this equipment is as follows:

1. Check and clean or replace air filter each month or more frequently if required.

2. Check blower motor and wheel for cleanliness and lubrication each heating and cooling season. Clean and lubricate as necessary.

3. Check electrical connections for tightness and controls for proper operation each heating season. Service as necessary.

> ⚠ **CAUTION:** As with any mechanical equipment, personal injury can result from sharp metal edges, etc.; therefore, be careful when removing parts.

A. Air Filter Arrangement

The air filter arrangement may vary depending on the application. Refer to Table I or II for filter size information.

TABLE I—FILTER SIZE INFORMATION (DOWNFLOW/HORIZONTAL)

FURNACE CASING WIDTH	FILTER SIZE	FILTER TYPE
14-3/16	(2) 16 x 20 x 1	Cleanable
17-1/2	(2) 16 x 20 x 1	Cleanable
21	(2) 16 x 20 x 1	Cleanable
24-1/2	(2) 16 x 20 x 1	Cleanable

TABLE II—FILTER SIZE INFORMATION (UPFLOW)

| FURNACE CASING WIDTH | FILTER SIZE | | FILTER TYPE |
	Side Return	Bottom Return	
14-3/16	(1) 16 x 25 x 1*	(1) 14 x 25 x 1	Cleanable
17-1/2	(1) 16 x 25 x 1*	(1) 16 x 25 x 1	Cleanable
21	(1) 16 x 25 x 1	(1) 20 x 25 x 1*	Cleanable
24-1/2	(2) 16 x 25 x 1*	(1) 24 x 25 x 1	Cleanable

* Factory provided with the furnace. Filters may be field modified by cutting as required.

> ⚠ **WARNING:** Never operate unit without a filter or with filter access door removed. A failure to follow this warning could result in fire, personal injury, or death.

1. Horizontal and Downflow.

Each furnace requires two filters which are installed in the return-air duct. (See Figures 4 and 5.)

To remove filters for cleaning or replacement, proceed as follows:

  a. Disconnect electrical power before removing access door.

  b. Remove blower access door.

  c. Reach up behind top plate, tilt filters toward center of return-air plenum, remove filters, and replace or clean as needed.

  d. Furnaces are equipped with permanent, washable filters. Clean those filters by spraying tap water through filter in opposite direction of airflow.

  e. Rinse and let dry. Oiling or coating of filters is not recommended.

  f. Reinstall filters with cross-mesh binding facing blower. Replace access door and restore electrical power to furnace.

2. Upflow.

Each furnace requires one or two filters which are installed in the blower compartment. (See Figure 6.) To remove filters for cleaning or replacement, proceed as follows:

  a. Disconnect electrical power before removing access doors.

  b. Release filter retainer from clip at front of furnace casing. (See Figure 6.) For side return, clips may be used on either (or both) sides of the furnace.

  c. Slide filter out.

  d. Clean filter with tap water. Spray in opposite direction of airflow.

  e. Rinse and let dry. Oiling or coating of filter is not recommended.

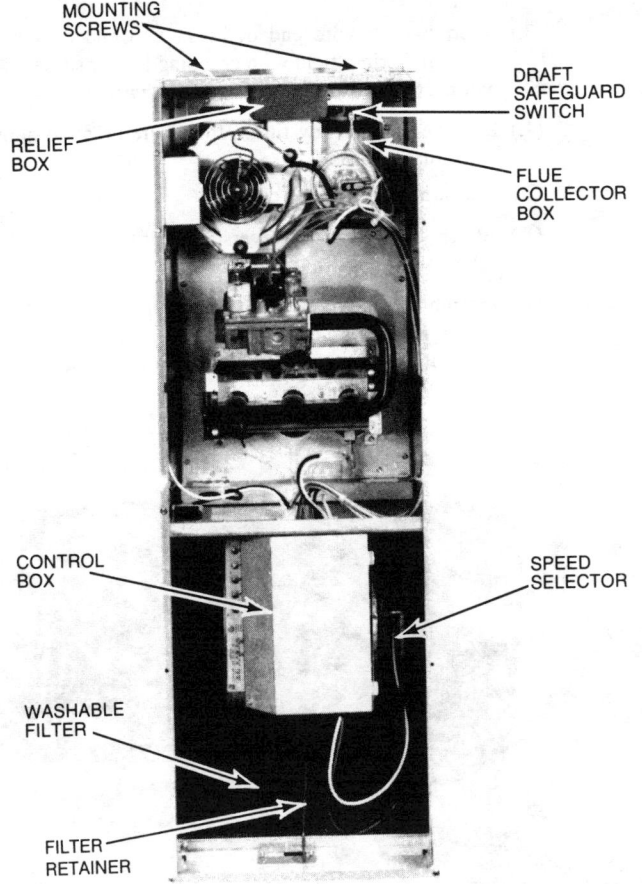

Figure 6—Model 395CAV Upflow

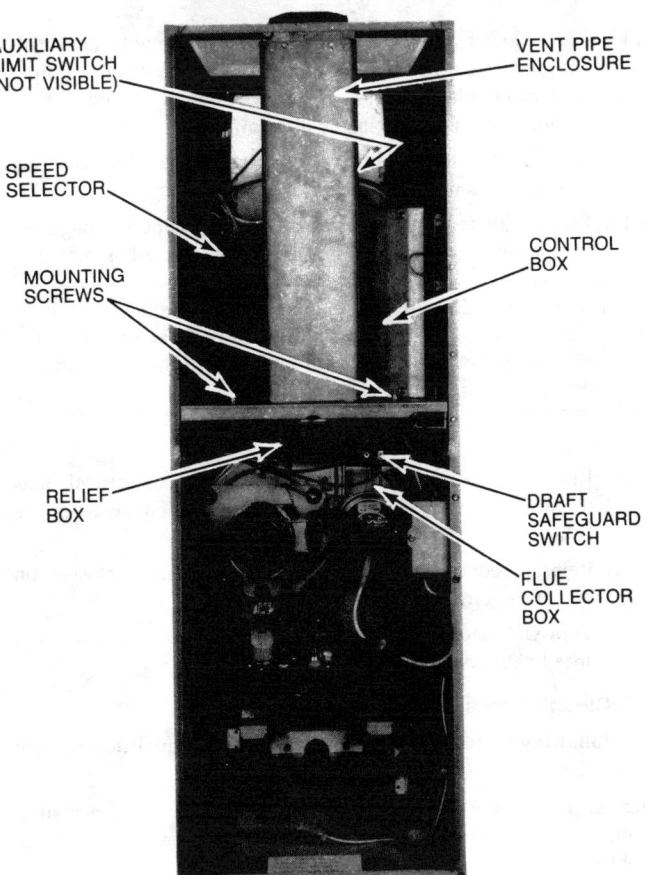

Figure 7—Model 376CAV Downflow

f. Place filter in furnace with cross-mesh binding up or facing blower. Replace access doors and restore electrical power to furnace.

B. Blower Motor and Wheel

For long life, economy, and high efficiency; clean accumulated dirt and grease from the blower wheel and motor annually.

The following steps should be performed by a qualified service technician:

Some motors have prelubricated sealed bearings and require no lubrication. These motors can be identified by the absence of oil ports on each end of the motor. For those motors with oil ports lubricate as follows:

Lubricate motor every 5 years if motor is used on intermittent operation (thermostat FAN switch in AUTO position), or every 2 years if motor is in continuous operation (thermostat FAN switch in ON position).

Remember to disconnect the electrical supply before removing access doors.

Clean and lubricate as follows:

1. Remove blower access door.
2. Disconnect vent pipe on downflow furnace only.
   a. Remove vent pipe enclosure.
   b. Disconnect vent pipe and remove short piece of pipe from furnace.
   c. Disconnect wires from auxiliary limit on blower housing.
3. Remove control box.
4. Remove electrical leads from numbered side of Molex speed selector. (See Figures 6 and 7.) Note location of wires for reassembly.

5. Remove screws holding blower assembly to blower deck and slide blower assembly out of furnace.
6. Squeeze side tabs of Molex speed selector and pull it from blower housing.
7. Loosen a screw in strap holding motor capacitor to blower housing and slide capacitor out from under strap.
8. Mark blower wheel, motor, and motor support in relation to blower housing before disassembly, to insure proper reassembly.
9. Loosen setscrew holding blower wheel on motor shaft.
10. Remove bolts holding motor mount to blower housing and slide motor and mount out of housing. Disconnect ground wire attached to blower housing before removing motor.
11. Lubricate motor (when oil ports are provided).
    a. Remove dust caps or plugs from oil ports located at each end of motor.
    b. Use a good grade of SAE 20 nondetergent motor oil and put one teaspoon, 5 cc, $\frac{3}{16}$ oz, or 16 to 25 drops in each oil port. Do not over-oil.
    c. Allow time for total quantity of oil to be absorbed by each bearing.
    d. After oiling motor, be sure to wipe excess oil from motor housing.
    e. Replace dust cap or plugs on oil ports.
12. Remove blower wheel from housing.
    a. Mark cutoff location to insure proper reassembly.
    b. Remove screws holding cutoff plate and remove cutoff plate from housing.
    c. Lift blower wheel from housing through opening.

—3—

13. Clean blower wheel and motor using a vacuum with soft brush attachment. Do not remove or disturb balance weights (clips) on blower wheel blades. The blower wheel should not be dropped or bent as balance will be affected.

14. Reinstall blower wheel by reversing steps 12 a. through c. Be sure wheel is positioned for proper rotation.

15. Reassemble motor and blower by reversing steps 5 through 10. If motor has ground wire, be sure it is connected as before.

⚠ **CAUTION:** Be sure the motor is properly positioned in the blower housing. The motor oil ports must be at a minimum of 45 deg above the horizontal centerline of the motor after the blower assembly has been reinstalled in the furnace.

16. Reinstall blower assembly in furnace. Connect electrical leads to Molex speed selector. Please note that connections are polarized for assembly: **Do not force.**

17. Reinstall control box (reinstall vent pipe and enclosure on downflow furnace).

18. Turn ON electrical power and check for proper rotation and speed changes between heating and cooling.

C. Cleaning Heat Exchanger

The following steps should be performed by a qualified service technician:

NOTE: Deposits of soot and carbon indicate the existence of a problem which needs to be corrected. Take action to correct the problem.

If it becomes necessary to clean the heat exchanger because of carbon deposits, soot, etc., proceed as follows:

1. Turn OFF gas and power to furnace.

2. Remove control and blower access doors.

3. Remove vent pipe enclosure (downflow/horizontal furnace only) and disconnect vent pipe from relief box.

4. Remove two screws that secure relief box. (See Figures 6 or 7.)

5. Disconnect wires to the following components:

   a. Draft safeguard switch.

   b. Inducer motor.

   c. Pressure switch.

   d. Limit overtemperature switch(es).

   e. Gas valve.

   f. Hot surface ignitor.

   g. Flame-sensing electrode.

   h. Two wiring edge connectors leading to control box.

6. Remove eight screws that secure flue collector box to center panel. Be careful not to damage sealant.

7. Remove complete inducer assembly from furnace, exposing flue openings.

8. Using field-provided small wire brush, steel spring cable, reversible electric drill, and vacuum cleaner; clean cells as follows:

   a. Assemble wire brush and steel spring cable.

   (1.) Use 48 inches of ¼-inch diameter high-grade steel spring cable (commonly known as drain clean-out or Roto-Rooter cable).

   (2.) Use ¼-inch diameter wire brush (commonly known as 25-caliber rifle cleaning brush).

NOTE: The items needed in steps (1.) and (2.) can usually be purchased at local hardware stores.

(3.) Insert twisted wire end of brush into end of spring cable, and crimp tight with crimping tool or strike with ball-peen hammer. *Tightness is very important.*

(4.) Remove metal screw fitting from wire brush to allow insertion into cable.

b. Clean each heat exchanger cell.

(1.) Attach variable-speed, reversible drill to end of spring cable (end opposite brush).

(2.) Insert brush end of cable into upper opening of cell and slowly rotate with drill. *Do not* force cable. Gradually insert at least 36 inches of cable into two upper passes of cell. (See Figure 8.)

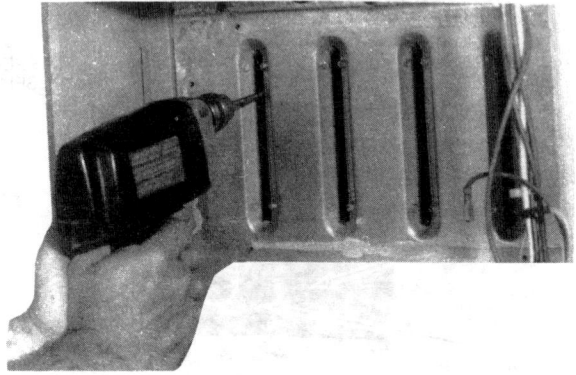

A88489

Figure 8—Cleaning Heat Exchanger Cell

(3.) Work cable in and out of cell three or four times to obtain sufficient cleaning. *Do not* pull cable with great force. Reverse drill and gradually work cable out.

(4.) Remove burner assembly and cell inlet plates.

(5.) Replace screws in center panel and cells before cleaning.

(6.) Insert brush end of cable in lower opening of cell, and proceed to clean two lower passes of cell in same manner as two upper passes.

(7.) Repeat foregoing procedures until each cell in furnace has been cleaned.

(8.) Using vacuum cleaner, remove residue from each cell.

(9.) Using vacuum cleaner with soft brush attachment, clean burner assembly.

(10.) Reinstall cell inlet plates and burner assembly. Care must be exercised to center the burners in the cell openings.

9. After cleaning flue openings, check sealant on flue collector to ensure that it has not been damaged. If new sealant is needed, contact your dealer or distributor.

10. Clean and replace flue collector assembly, making sure all eight screws are secure.

11. Reinstall relief box.

12. Reconnect wires to the following components:

    a. Draft safeguard switch.

    b. Inducer motor.

    c. Pressure switch.

    d. Limit overtemperature switch(es).

    e. Gas valve.

f. Hot surface ignitor.

g. Flame sensing electrode.

h. Two wiring edge connectors leading to control box.

13. Reconnect vent pipe to relief box. Replace vent pipe enclosure (when applicable).

14. Replace blower door only.

15. Turn on power and gas.

16. Set thermostat and check furnace for proper operation.

17. Check for gas leaks.

18. Replace control door.

 WARNING: Never use a match or other open flame to check for gas leaks. Use a soap-and-water solution. A failure to follow this warning could result in fire, personal injury, or death.

D. Hot Surface Ignitor

NOTE: Be very careful when removing the burner assembly to avoid breaking the ignitor. See Figure 9 for the correct ignitor location.

E. Electrical Controls and Wiring

NOTE: There may be more than one electrical supply to unit.

The electrical ground and polarity for 115-volt wiring must be maintained properly. Refer to Figure 10 for field wiring information and to Figure 11 for unit wiring information. If the polarity is **not** correct, the microprocessor control will shut OFF the gas flow shortly after completion of the ignition trial period. The control system also requires an earth ground for proper operation of the microprocessor.

With power disconnected to unit, check all electrical connections for tightness. Tighten all screws on electrical connections. If any smoky or burned connections are noticed, disassemble the connection, clean all parts and stripped wire, and reassemble properly and securely. Electrical controls are difficult to check without proper instrumentation; therefore, reconnect electrical power to unit and observe unit through one complete operating cycle. If there are any discrepancies in the operating cycle, contact your dealer and request service.

The 24-volt circuit contains an automotive-type, 3-amp fuse located on the main control board. Any direct shorts during installation, service, or maintenance could cause this fuse to blow. If fuse replacement is required, use **only** a 3-amp fuse of identical size.

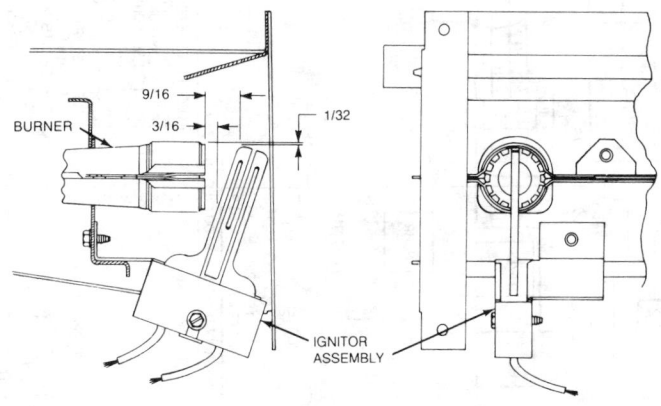

A88490

Figure 9—Position of Ignitor to Burner

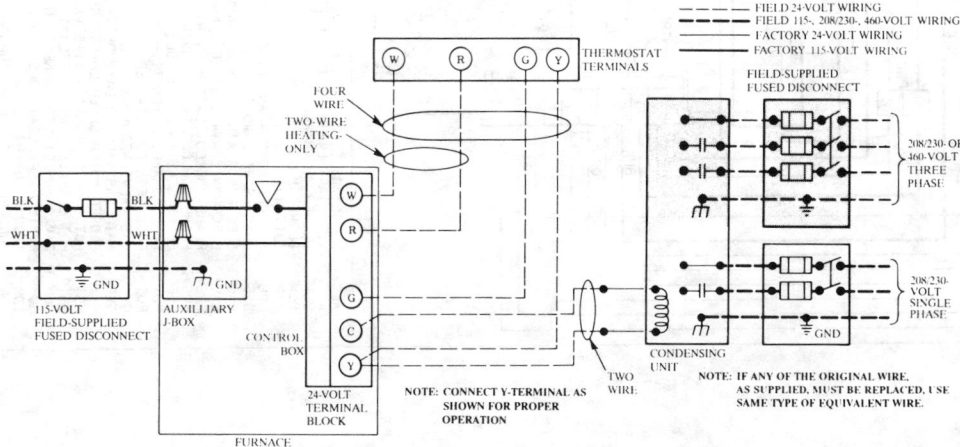

A78461

Figure 10—Heating and Cooilng Application Wiring Diagram

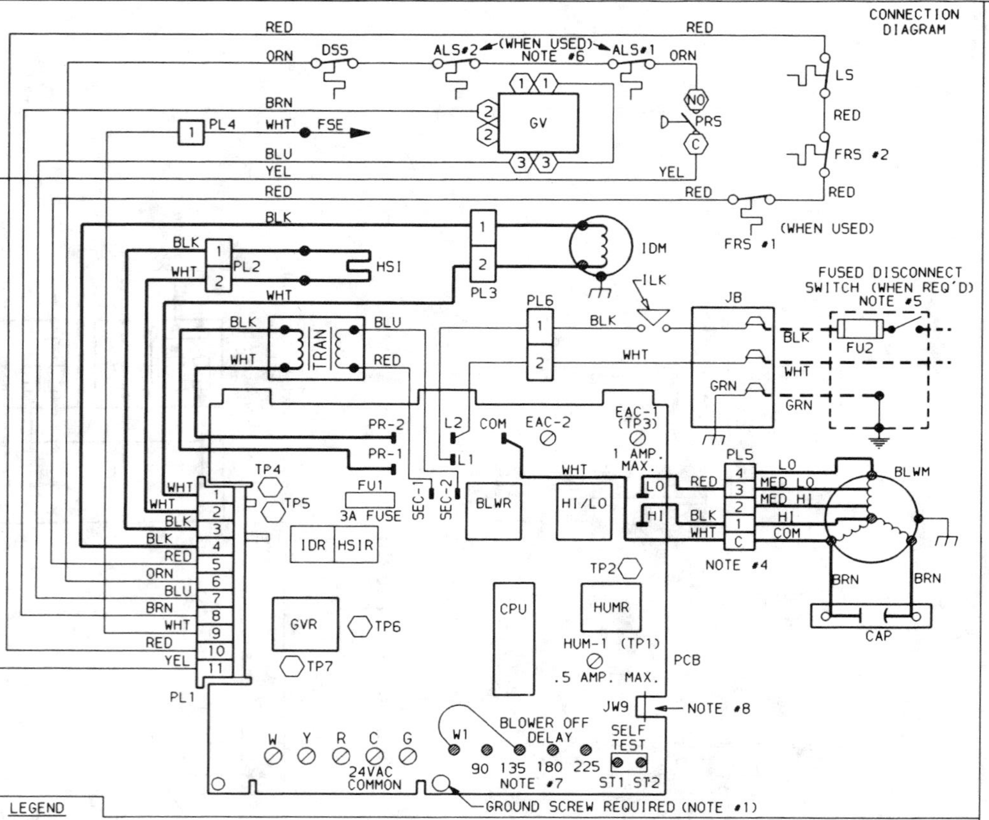

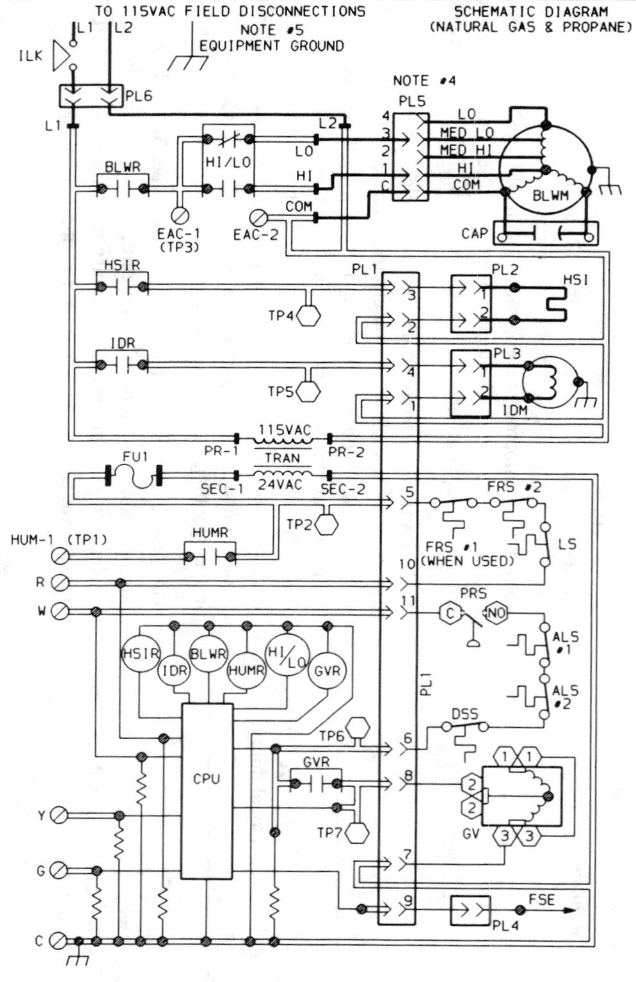

SCHEMATIC DIAGRAM
(NATURAL GAS & PROPANE)

318078-401 REV. A

A91059

NOTES:

1. COMMON SIDE (SEC-2 AND C) OF 24VAC TRANSFORMER CONNECTED TO GROUND THROUGH THIS REQUIRED MOUNTING SCREW.
2. IF ANY OF THE ORIGINAL EQUIPMENT WIRE IS REPLACED USE WIRE RATED FOR 105°C.
3. INDUCER (IDM) AND BLOWER (BLWM) MOTORS CONTAIN INTERNAL AUTO-RESET THERMAL OVERLOAD SWITCHES.
4. BLOWER MOTOR SPEED SELECTIONS (PL5) ARE FOR AVERAGE CONDITIONS. SEE INSTALLATION INSTRUCTIONS FOR DETAILS ON OPTIMUM SPEED SELECTION.
5. USE ONLY COPPER WIRE BETWEEN THE DISCONNECT SWITCH AND THE FURNACE JUNCTION BOX (JB).
6. AUXILIARY LIMIT SWITCHES (ALS#1,#2) USED ON DOWNFLOW MODELS ONLY.
7. BLOWER MOTOR (BLWM) ADJUSTABLE OFF-DELAY.
8. JW9 SHOULD BE LEFT UNCUT.

CONNECTION DIAGRAM

LEGEND

ALS#1	AUXILIARY LIMIT SWITCH, OVERTEMP.- MANUAL RESET, SPST-(N.C.)
ALS#2	AUXILIARY LIMIT SWITCH, OVERTEMP.- MANUAL RESET, SPST-(N.C.)
BLWM	BLOWER MOTOR RELAY, SPST-(N.O.)
BLWM	BLOWER MOTOR
CAP	CAPACITOR
CPU	MICROPROCESSOR AND CIRCUITRY
DSS	DRAFT SAFEGUARD SWITCH
EAC-1	ELECTRONIC AIR CLEANER CONNECTION (115 VAC 1 AMP MAX.)
EAC-2	ELECTRONIC AIR CLEANER CONNECTION (COMMON)
FRS#1	FLAME ROLLOUT SWITCH OVERTEMP.-MANUAL RESET, SPST-(N.C.)
FRS#2	FLAME ROLLOUT SWITCH OVERTEMP.-MANUAL RESET, SPST-(N.C.)
FSE	FLAME SENSING ELECTRODE
FU1	FUSE, 3 AMP, AUTOMOTIVE BLADE TYPE, FACTORY INSTALLED
FU2	FUSE, FIELD INSTALLED
GV	GAS VALVE-REDUNDANT OPERATORS
GVR	MAIN GAS VALVE RELAY, DPST-(N.O.)
HI/LO	BLOWER MOTOR SPEED CHANGE RELAY, DPST-(N.O.)
HS1	HOT SURFACE IGNITER (115 VAC)
HS1R	HOT SURFACE IGNITER RELAY, DPST-(N.O.)
HUM-1	24VAC HUMIDIFIER CONNECTION (.5 AMP. MAX.)
HUMR	HUMIDIFIER RELAY, DPST-(N.O.)
IDM	INDUCED DRAFT MOTOR
IDR	INDUCED DRAFT RELAY, DPST-(N.O.)
ILK	BLOWER DOOR INTERLOCK SWITCH, SPST-(N.O.)
JB	JUNCTION BOX
JW9	INDUCER OFF DELAY JUMPER: CUT-15 SEC DELAY, UNCUT-5 SEC DELAY
LS	LIMIT SWITCH, OVERTEMPERATURE-AUTO RESET, SPST-(N.C.)
PCB	PRINTED CIRCUIT BOARD

PL1	11-CIRCUIT EDGE CONNECTOR (PCB)
PL2	2-CIRCUIT HS1 CONNECTOR
PL3	2-CIRCUIT IDM CONNECTOR
PL4	1-CIRCUIT FSM CONNECTOR
PL5	5-CIRCUIT BLWM CONNECTOR
PL6	2-CIRCUIT 115VAC CONNECTOR
PRS	PRESSURE SWITCH, SPST-(N.O.)
ST1-ST2	SELF TEST PINS
TP1-TP7	TEST POINT (1) THRU (7)
TRAN	TRANSFORMER-115VAC/24VAC

●	JUNCTION
○	UNMARKED TERMINAL
▬	PCB TERMINAL
⬡	MARKED TERMINAL
──	FACTORY WIRING (115VAC)
──	FACTORY WIRING (24VAC)
─ ─	FIELD WIRING (115VAC)
⊘	CONDUCTOR ON PCB
⊘	FIELD WIRING SCREW TERMINAL
⏚	FIELD GROUND
🖧	EQUIPMENT GROUND
⌐	FIELD SPLICE
⊃─	PLUG RECEPTACLE

Figure 11—Unit Wiring Diagram

15009

—6—

Catalog No. BDP-3337-617

Models 376C & 395C
Induced-Combustion Furnaces

bryant

day & night

Payne

FOR YOUR SAFETY
What to do if you smell gas:

1. Do not try to light any appliance.
2. Do not touch any electrical switch.
3. Do not use any phone in your building. Immediately call your gas supplier from a neighbor's phone. Follow the gas supplier's instructions.
4. If you cannot reach your gas supplier, call the fire department.

FOR YOUR SAFETY
Do not store or use gasoline or other flammable vapors and liquids in the vicinity of this or any other gas appliance.

⚠**WARNING:** Improper installation, adjustment, alteration, service, maintenance, or use can cause carbon monoxide poisoning, explosion, fire, electrical shock, or other conditions which could result in personal injury, property damage, or death. Consult a qualified installer, service agency, or your local gas supplier for information or assistance.

USER'S INFORMATION MANUAL FOR THE OPERATION AND MAINTENANCE OF YOUR NEW GAS-FIRED FURNACE

NOTE TO INSTALLER: This manual must be left with the equipment user.

GAS FURNACE

WELCOME TO A NEW GENERATION OF COMFORT

Congratulations! Your new, higher efficiency gas furnace is a sound investment which will reward you and your family with years of "warm memories" winter after winter.

Not only is your new furnace energy-efficient, it is also one of the most reliable. Spend just a few minutes with this booklet to learn about the operation of your new furnace—and the small amount of maintenance it takes to keep it operating at peak efficiency. Years went into the development of your new furnace. Take a little time now to assure its most efficient operation for years to come.

A88318   1

**MODEL 395C
UPFLOW FURNACE**

A88316   2

**MODEL 376C
DOWNFLOW/HORIZONTAL FURNACE**

FURNACE IDENTIFICATION

For your convenience, record the product and serial numbers of your new furnace on the form below. Should you ever require service, you will have ready access to the information needed by the service representative.

Product No. _____

Serial No. _____

Date Installed _____

Dealer Name _____

Address _____

City _____

State _____ Zip _____

Telephone _____

UPFLOW FURNACE COMPONENTS

1. Relief Box
2. Gas Valve Control Knob (On/Off)
3. Gas Valve
4. Gas Burner
5. Hot Surface Ignitor
6. Blower Door Safety Switch
7. Blower and Blower Motor
8. Draft Safeguard Tube & Switch
9. Rating Plate (Behind Junction Box)
10. Gas Manifold
11. Filter Retainer
12. Air Filter
13. Flame Sensor
14. Manual Reset Limit Switch

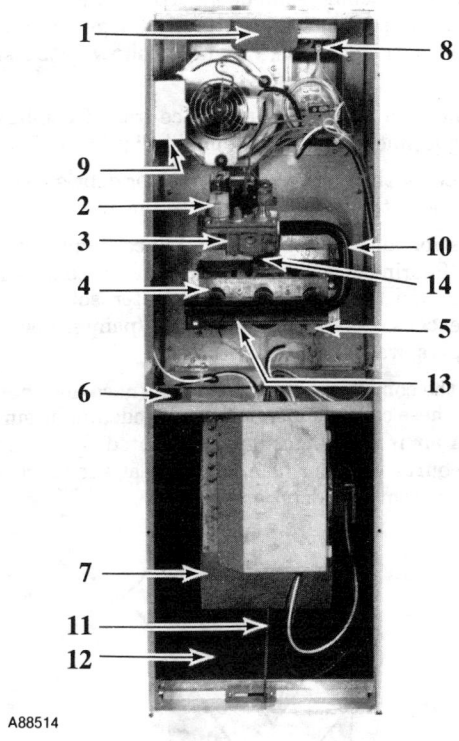

A88514

3

DOWNFLOW/HORIZONTAL FURNACE COMPONENTS

1. Manual-Reset Auxiliary Limit Switch
2. Blower & Blower Motor
3. Relief Box
4. Rating Plate
5. Gas Valve Control Knob (On/Off)
6. Gas Valve
7. Gas Burner
8. Blower Door Safety Switch
9. Draft Safeguard Tube & Switch
10. Gas Manifold
11. Hot Surface Ignitor
12. Flame Sensor
13. Manual Reset Limit Switch (2)

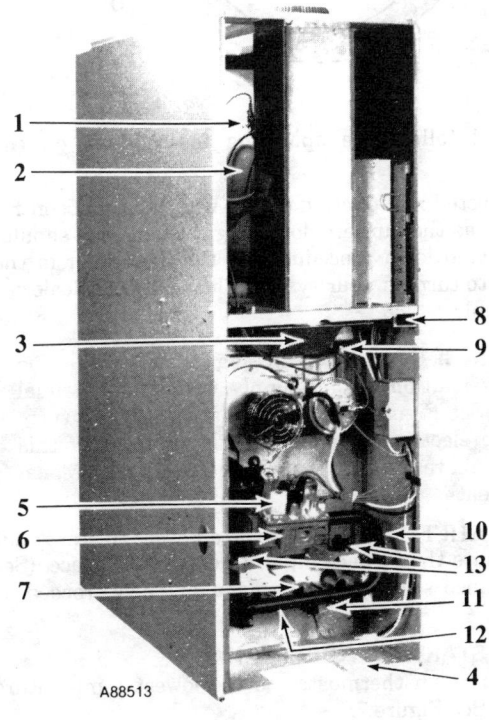

A88513

4

IMPORTANT FACTS

Your furnace must have adequate airflow for efficient combustion and safe ventilation. Do not enclose it in an airtight room or "seal" it behind solid doors. To minimize the possibility of serious personal injury, fire, damage to your furnace, or improper operation; **carefully follow these safety rules:**

● Keep the area around your furnace free of combustible materials, gasoline, and other flammable liquids and vapors.

● Do not cover the furnace, store trash or debris near it, or in any way block the flow of fresh air to the unit.

● Combustion air must be clean and uncontaminated with chlorine or fluorine. These compounds are present in many products around the home, such as: water softener salts, laundry bleaches, detergents, adhesives, paints, varnishes, paint strippers, waxes, and plastics.

Make sure the combustion air for your furnace does not contain any of these compounds. During remodeling be sure the combustion air is fresh and uncontaminated. If these compounds are burned in your furnace, the heat exchangers and metal vent system may deteriorate.

A86023
5

● A furnace installed in the attic or other insulated space must be kept free and clear of the insulating material. Examine the furnace area when installing the furnace or adding more insulation. Some materials may be combustible.

NOTE: Do not use this furnace if any part has been under water. Immediately call a qualified service technician to inspect the furnace and to replace any part of the control system and any gas control which has been under water.

NOTE: The qualified installer or agency must use only factory-authorized replacement parts, kits, and accessories when modifying or repairing this product.

This furnace contains SAFETY DEVICES which must be MANUALLY RESET. If the furnace is left unattended for an extended period of time, have it checked periodically for proper operation. This precaution will prevent problems associated with NO HEAT, such as frozen water pipes, etc. See Before You Request a "Service Call" section in this manual.

SAFETY CONSIDERATIONS

Installation and servicing of heating equipment can be hazardous due to gas and electrical components. Only trained and qualified personnel should install, repair, or service heating equipment.

Untrained personnel can perform basic maintenance functions such as cleaning and replacing air filters. All other operations must be performed by trained service personnel. Observe safety precautions in this manual, on tags, and labels attached to the furnace and other safety precautions that may apply.

Recognize safety information: This is the safety-alert symbol ⚠. When you see this symbol on the furnace and in instructions or manuals, be alert to the potential for personal injury.

Understand the signal word—DANGER, WARNING, or CAUTION. These words are used with the safety-alert symbol. DANGER identifies the most serious hazards which **will** result in severe personal injury or death. WARNING signifies hazards that **could** result in personal injury or death. CAUTION is used to identify unsafe practices, which would result in minor personal injury or product and property damage.

STARTING YOUR FURNACE

Instead of a continuously burning pilot flame which wastes valuable energy, your furnace uses an automatic hot surface ignition system to light the pilot each time the thermostat turns your furnace on. **Follow these important safeguards:**

● Never attempt to manually light the pilot with a match or other source of flame.

A86024
6

● Read and follow the operating instructions on the furnace.

● If a suspected malfunction occurs with your gas control system, such as the burners do not light when they should, refer to the shutdown procedures on the furnace or in the next section to turn off your system, then call your dealer as soon as possible.

⚠ **WARNING:** Should overheating occur, or the gas valve fail to shut off the gas supply, turn off the manual gas valve (See Figure 8) to the furnace BEFORE turning off the electrical supply. A failure to follow this warning could result in a fire or explosion, and personal injury or death.

● CHECK AIR FILTER: Before attempting to start your furnace, be sure the furnace filter is clean and in place. (See the maintenance section of this manual.) Then proceed as follows:

STEPS FOR STARTING YOUR FURNACE

1. Set your room thermostat to the lowest temperature setting. See Figure 7.
2. Close the external manual gas valve. See Figure 8.
3. Turn OFF the electrical supply to your furnace. See Figure 9.

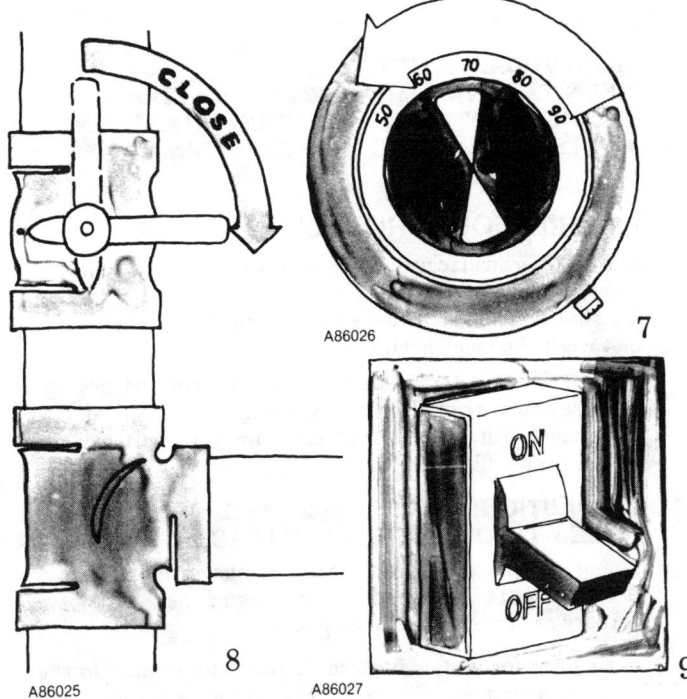

A86026 7

A86025 8   A86027 9

4. Remove the furnace access door(s).
    a. Upflow—remove control door. See Figure 10.
    b. Downflow/Horizontal—remove blower door first, then remove the control door. See Figure 11.

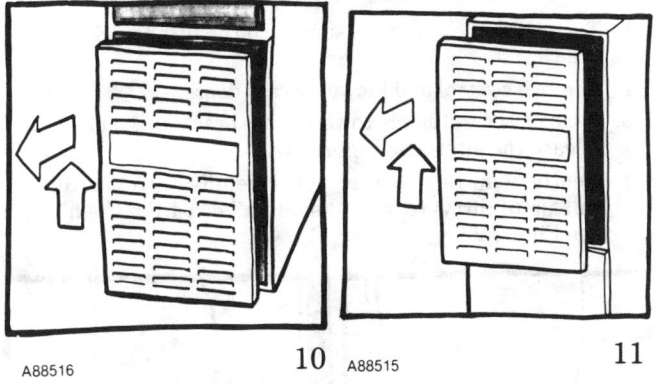

A88516   10   A88515   11

5. Turn the control knob on the gas valve to the OFF position and wait 5 minutes. See Figure 12.
6. After waiting 5 minutes, turn the control knob on the gas valve to the ON position. See Figure 13.

A87374   12   A88321   13

7. Replace the access door(s). See Figure 14 for upflow and Figure 15 for downflow. Replace control door first on downflow furnaces. Then replace blower door.
8. Turn ON the electrical supply to the furnace. See Figure 16.

A88518   14   A88517   15

9. Open the external manual gas valve. See Figure 17.
10. Set the room thermostat to a temperature slightly above the room temperature. This will automatically signal the furnace to start. The inducer motor will start, and the hot surface ignitor will energize. When hot, the ignitor will have an orange glow.
11. After 15 to 70 seconds, the gas valve permits gas to flow to the main burners where it is ignited. Hot flames begin to warm the furnace's heat exchanger. After a time delay of approximately 60 seconds, the furnace blower is switched on.

NOTE: If the main burners fail to ignite, the furnace control system will go thru three more ignition cycles. Then, if burners fail to ignite, the system will lockout. If lockout occurs, or the blower doesn't come on—shut down your furnace and call your dealer for service.

12. Set your thermostat to the temperature that satisfies your comfort requirements. SUGGESTION: Setting the thermostat back a few degrees—and compensating for the difference with warmer clothing—can make a big difference in your fuel consumption on extremely cold days. The few degrees at the top of your thermostat "comfort level" are the most costly degrees to obtain.

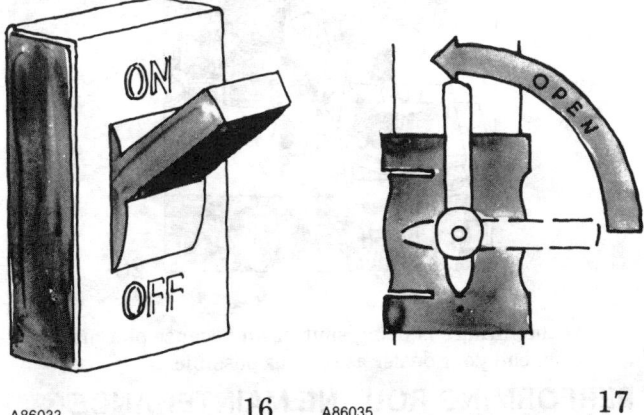

A86033   16   A86035   17

When the room temperature drops below the temperature selected on the thermostat, the furnace will be switched on automatically. When the room temperature reaches the degree selected on the thermostat, the furnace will be automatically switched off.

Some thermostats have a "fan" switch with two selections: AUTO or ON. When set on AUTO, the furnace blower cycles on and off, controlled by the thermostat. In the ON position, the furnace blower runs continuously except for a 60 second delay at the "call for heat." This keeps the temperature level in your home more evenly balanced. It also continuously filters the indoor air.

position, the furnace blower runs continuously except for a 60 seconds delay at the "call for heat." This keeps the temperature level in your home more evenly balanced. It also continuously filters the indoor air.

SHUTTING DOWN YOUR FURNACE

Should you ever suspect a malfunction in your furnace, you will need to turn the furnace off. The following procedures must be followed:

1. Set your room thermostat to the lowest temperature setting. See Figure 18.
2. Close the external manual gas valve. See Figure 8 on page 5.
3. Turn OFF the electrical supply to your furnace. See Figure 19.

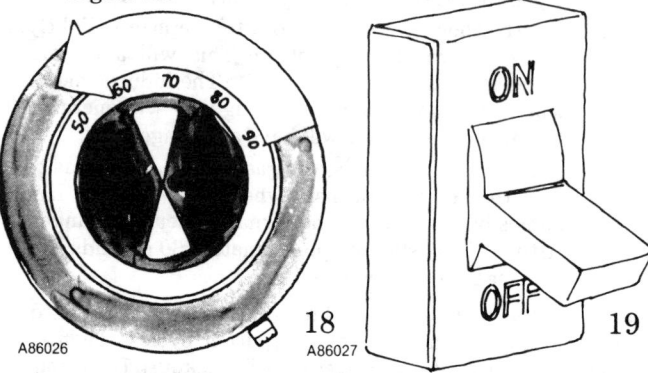

18
A86026

19
A86027

4. Remove the control access door on your furnace. See Figure 10 or 11. On upflow furnaces, removing the blower access door is not required.
5. Turn the control knob on the gas valve to the OFF position. See Figure 20.
6. Replace the control access door. See Figure 14 or 15.

A87374
20

7. If the furnace is being shut down because of a malfunction, call your dealer as soon as possible.

PERFORMING ROUTINE MAINTENANCE

With the proper maintenance and care, your furnace will operate economically and dependably. Basic maintenance, which can easily be accomplished by someone who follows the directions, is found on this and the following pages. However, before beginning maintenance, follow these safety precautions:

⚠ **WARNING:** TURN OFF ELECTRICAL POWER SUPPLY TO YOUR FURNACE BEFORE REMOVING THE ACCESS DOORS TO SERVICE OR PERFORM MAINTENANCE. A FAILURE TO FOLLOW THIS WARNING COULD RESULT IN PERSONAL INJURY OR DEATH.

⚠ **CAUTION:** ALTHOUGH SPECIAL CARE HAS BEEN TAKEN TO MINIMIZE SHARP EDGES, BE EXTREMELY CAREFUL WHEN HANDLING PARTS OR REACHING INTO THE FURNACE.

FILTERING OUT TROUBLE

A dirty filter will cause excessive stress on the furnace blower motor and can cause it to overheat and automatically shut down. The furnace filter should be checked every 3 or 4 weeks and cleaned if necessary.

If your furnace filter needs replacing, be sure to use the same size and type of filter that was originally supplied. Use the Furnace Filter Table and compare your furnace size with the proper filter size.

⚠ **CAUTION:** NEVER OPERATE YOUR FURNACE WITHOUT A FILTER IN PLACE.

Doing so may damage the furnace blower motor. An accumulation of dust and lint on internal parts of your furnace can cause a loss of efficiency.

The air filter for upflow furnaces is normally located in the blower compartment. Filters for the downflow furnaces are normally located in the return air plenum above the blower. If the filters have been installed in another location, contact your dealer for instructions. To inspect, clean and/or replace the air filter(s), follow these steps:

● UPFLOW FURNACES ONLY:

1. Turn OFF the electrical supply to the furnace. See Figure 19.
2. Remove control and blower access doors.
3. Push filter retainer toward the bracket opening to release the filter. See Figure 21.
4. Gently remove the filter and carefully turn the dirty side up (if dirty) to avoid "spilling" dirt from the filter. See Figure 22.

21
A88521

22
A88522

5. Inspect the filter. If torn, replace the filter.
6. Wash the filter (if dirty) in a sink, bathtub, or outside with a garden hose. Always use cold tap water. A mild liquid detergent may be used if necessary. Spray water through the filter in the opposite direction of airflow (through the cross-mesh binding side). Allow filter to dry.
7. Reinstall the clean filter with its cross-mesh binding side facing the furnace blower.
8. Put filter retainer back in the bracket opening and lock it in place.
9. Replace the blower and control access doors and turn ON electrical power to your furnace. See Figures 14, 16, and 23.

NOTE: For upflow models only—if side return ducts are used, 2 filters may be required in some models. The procedure listed above may be used to remove side filters.

A88523   23

• DOWNFLOW/HORIZONTAL FURNACES ONLY:
Two filters are located in the return-air plenum above the blower (above line-of-sight) resting in the V-shaped channel on top of the furnace. See Figure 24.

1. Turn OFF electrical supply to the furnace. See Figure 19.
2. Remove blower access door.
3. Remove the left side filter by tipping the filter toward the center—raise it from the V-shapped channel in which it rests. See Figures 24 and 25.

A88519   24   25
A88520

4. Lower filter down along side of the blower and remove from the furnace.
5. To remove the second filter, lift from V-shaped channel and remove the same way as left side filter.
6. Inspect the filters. If torn, replace the filter.
7. Wash the filters (if dirty) in a sink, bathtub, or outside with a garden hose. Always use cold tap water. A mild liquid detergent may be used if necessary. Spray water through the filter in the opposite direction of airflow through the cross-mesh binding side. Allow filter to dry.
8. Reinstall clean filters with the cross-mesh binding side facing the furnace blower.
9. Replace blower door and turn ON electrical power to your furnace.

UPFLOW FURNACE FILTER TABLE

Furnace Casing Width	Filter Size		Filter Type
	Side Return	Bottom Return	
14-3/16	(1) 16 x 25 x 1*	(1) 14 x 25 x 1	Cleanable
17-1/2	(1) 16 x 25 x 1*	(1) 16 x 25 x 1	Cleanable
21	(1) 16 x 25 x 1	(1) 20 x 25 x 1*	Cleanable
24-1/2	(2) 16 x 25 x 1*	(2) 12 x 25 x 1	Cleanable

DOWNFLOW/HORIZONTAL FURNACE FILTER TABLE

Furnace Casing Width	Filter Size	Filter Type
14-3/16	(2) 14 x 20 x 1*	Cleanable
17-1/2	(2) 14 x 20 x 1*	Cleanable
21	(2) 16 x 20 x 1*	Cleanable
24-1/2	(2) 16 x 20 x 1*	Cleanable

*Factory provided with the furnace. Filters may be field modified by cutting and folding the frame as indicated on the filter. Alternate sizes and additional filters may be ordered from your dealer.

COMBUSTION AREA AND VENT SYSTEM

Inspect the combustion area and vent system before each heating season. An accumulation of dirt, soot, or rust can mean a loss of efficiency and improper performance. Build-ups on the main burners can cause faulty firing. This "delayed ignition" is characterized by an alarmingly loud sound. If your furnace makes a loud noise when the main burners are ignited, shut down the furnace—call your servicing dealer.

Use your flashlight and follow these steps for inspecting the combustion area and vent system of your furnace:

1. Turn off the electrical supply to the furnace and remove the access doors. See Figures 9 and 10, or 11.
2. Carefully inspect the gas burner (see Figure 26) for dirt, rust, or scale. Then, inspect the relief box, flue connection area, and the vent pipe for rust.

NOTE: If dirt, rust, soot, or scale accumulations are found, call your servicing dealer. DO NOT OPERATE THE FURNACE.

3. Inspect the vent pipe for a sag, holes, or a disconnection. A horizontal vent pipe must slope upward. If rusty joints or seams, or signs of water leakages are found call your dealer for service.

A86041   26

4. Replace the access doors and restore electrical power to the furnace. Be sure bottom door flange is inside of the furnace casing. See Figures 14, 15, and 23.

5. Start the furnace and observe its operation. If possible, watch the burner flames. Are they burning bright blue? If not (or if you suspect some other malfunction), call your servicing dealer.

BEFORE YOU REQUEST A "SERVICE CALL"

BEFORE YOU CALL FOR SERVICE, CHECK FOR SEVERAL EASILY SOLVED PROBLEMS:

☐ Check for sufficient airflow. Check the air filter for dirt. Check for blocked return-air or supply-air grilles. Be sure they are open and unobstructed. If this isn't the cause, call your servicing dealer.

If your furnace isn't operating at all, check the following list for easily solved problems.

☐ Is your thermostat set above room temperature? Is the switch in the HEAT position?

☐ Is the electrical power supply switch ON? Is the blower access door firmly in place? Are any fuses blown—has a circuit breaker tripped?

☐ Is the manual shut-off valve in the gas supply pipe leading to the furnace open? Does the lever point in the same direction that the pipe runs (open)? Or is it at right angles (closed)?

NOTE: Before proceeding with the next checks, turn OFF the electrical power supply to the furnace. Remove the access doors.

☐ Is the control knob on the gas valve turned to the ON position? If this or the preceding check shows an interruption in the gas supply, make sure the gas has not been shut off for safety reasons. If nothing else seems to be wrong, follow the startup procedures found on pages 4 and 5 of this booklet.

☐ If for some reason the vent is blocked, the draft safeguard switch will shut off the furnace. Reset the switch by pushing the button located on top of the switch (see page 3 for switch location).

If the switch trips a second time, turn off the furnace and call for service.

☐ DOWNFLOW/HORIZONTAL ONLY—Check the manual-reset auxiliary limit switch located on the blower housing. If the blower motor fails, this switch will shut off the furnace. Reset it by pushing the button on the switch. If it trips again, turn off the furnace and call for service.

☐ Check the manual-reset limit switch(es) located near the burners. If the furnace has experienced a high-temperature condition, due to inadequate combustion air, these switches will shut off the furnace. Reset the switch(es) by pushing the button on the switch. If the switch trips a second time, turn off the furnace and call for service.

☐ If your furnace still fails to operate, call your servicing dealer for troubleshooting and repairs. Tell him the model and serial numbers for your furnace. (You should have them recorded on page 2 of this booklet.) If the dealer knows exactly which furnace you have, he may be able to offer suggestions over the phone, or save valuable time through knowledgeable preparation for the service call.

REGULAR DEALER MAINTENANCE

In addition to the type of routine maintenance you might be willing to do, your furnace should be inspected regularly by a properly trained service technician. An annual inspection (or semiannual inspection, at least) should include the following:

1. Inspection of all flue product passages—including the burners, heat exchanger, relief box, and vent pipe.
2. Inspection of all combustion and ventilation air passages and openings.
3. Close check of all gas pipes leading to (and inside of) your furnace.
4. Inspection, cleaning, and lubrication (when required) of the blower motor and wheel.

NOTE: Refer to the unit Service Procedures for blower motor oiling information. When required, the motor must be oiled by a qualified service technician.

5. Routine inspection and cleaning/replacement of the air filter.
6. Inspection of all supply- and return-air ducts for obstructions, air leaks, and insulation. Any problems found should be resolved at this time.
7. A check for loose connections attaching individual components. Inspection of all electrical wiring and their connections.
8. Operational check of the furnace itself to determine working condition. Repair or adjustment should be made at this time.

Your servicing dealer offers an economical service contract that covers seasonal inspections. Ask him for further details.

Detach & Mail Product Registration Card

IMPORTANT: Complete, detach and mail immediately for...
PRODUCT REGISTRATION

THE FEDERAL CONSUMER PRODUCT SAFETY ACT REQUIRES THAT YOU BE NOTIFIED OF ANY RECALLS INVOLVING THIS PRODUCT. YOUR NAME AND ADDRESS AND THE MODEL AND SERIAL NUMBERS OF YOUR PRODUCT WILL ASSIST US IN NOTIFYING YOU SHOULD THE NEED ARISE.

Your warranty coverage is not dependent upon the return of this card.

89B

1. 1. ☐ Mr. 2. ☐ Mrs. 3. ☐ Ms. 4. ☐ Miss

Name (First/Initial/Last)

Street

City ___ State ___ Zip

MODEL NO. ___ SERIAL NO.

2. DATE INSTALLED ___ Mo. Day Yr. (Copy from rating plate on unit.)

3. What brand did you purchase?
1. ☐ Bryant
2. ☐ Day & Night
3. ☐ Payne

4. What product did you purchase?
1. ☐ Central air conditioner
2. ☐ Gas furnace
3. ☐ Oil furnace
4. ☐ Electric furnace
5. ☐ Heat pump

5. When did you acquire your product?
1. ☐ Upon purchase of new dwelling
2. ☐ To replace an older system manufactured by our company
3. ☐ To replace an older system manufactured by a competitor
4. ☐ Within a year of buying new dwelling with no central air system
5. ☐ 2-4 years after buying dwelling with no central air system
6. ☐ Over 4 years after buying dwelling with no central air system

6. If this was a replacement product, how old was the original?
1. ☐ 1-5 years
2. ☐ 6-8 years
3. ☐ 9-11 years
4. ☐ 12-14 years
5. ☐ 15-17 years
6. ☐ 18 & over years

7. Your product is installed in?
1. ☐ Single family/Townhouse
2. ☐ Multi-family: 2-4
3. ☐ Mobile home
4. ☐ Apartment: 1-3 floors
5. ☐ Apartment: 4 + floors
6. ☐ Office/Bank
7. ☐ Store
8. ☐ Hospital/School
9. ☐ Manufacturing building
10. ☐ Other

8. What 2 factors meet influenced your purchase?
1. ☐ Our reputation
2. ☐ Friend's recommendation
3. ☐ Contractor/Dealer's recommendation
4. ☐ Price
5. ☐ Energy efficiency
6. ☐ Ready availability
7. ☐ Radio ads
8. ☐ T.V. ads
9. ☐ Newspaper ads
10. ☐ Magazine ads
11. ☐ Dealer display

9. Which of the following have you done in the past 6 months? (check all that apply)
1. ☐ Redeemed a product coupon
2. ☐ Ordered an item from mail order catalog
3. ☐ Sent in product inquiry card from magazine
4. ☐ Bought an item from offer received in mail
5. ☐ Entered sweepstakes/contest

10. In which age group are you?
1. ☐ 18-24
2. ☐ 25-34
3. ☐ 35-44
4. ☐ 45-54
5. ☐ 55-64
6. ☐ 65 & over

11. Marital status:
1. ☐ Married
2. ☐ Unmarried

12. Which group best describes your family income?
1. ☐ Under $10,000
2. ☐ $10,000-$14,999
3. ☐ $15,000-$19,999
4. ☐ $20,000-$24,999
5. ☐ $25,000-$29,999
6. ☐ $30,000-$34,999
7. ☐ $35,000-$39,999
8. ☐ $40,000-$44,999
9. ☐ $45,000-$49,999
10. ☐ $50,000 & over

13. Do you have any children in any of the following age groups who are living at home?
1. ☐ Under age 2
2. ☐ Age 2-4
3. ☐ Age 5-7
4. ☐ Age 8-10
5. ☐ Age 11-12
6. ☐ Age 13-15
7. ☐ Age 16-18

14. For your primary residence, do you:
1. ☐ Own a house?
2. ☐ Rent a house?
3. ☐ Own a townhouse/condominium?
4. ☐ Rent an apartment?

15. Which of the following types of credit cards do you use?
1. ☐ Travel/Entertainment (American Express, Diners Club, Carte Blanche)
2. ☐ Bank (Master Charge, Visa)
3. ☐ Gas, department store, etc.

16. What is your occupation? (check one)
1. ☐ Professional/Technical
2. ☐ Upper Mgt./Administrator
3. ☐ Sales/Service/Middle Mgt.
4. ☐ Clerical/White Collar
5. ☐ Craftsman/Blue Collar
6. ☐ Student
7. ☐ Housewife
8. ☐ Retired

Seal Here

Fold Here

17. Which of the following interests and hobbies do you and your family enjoy?
1. ☐ Tennis
2. ☐ Golf
3. ☐ Snow Skiing
4. ☐ Running/Jogging
5. ☐ Camping/Hiking
6. ☐ Hunting/Shooting
7. ☐ Fishing
8. ☐ Bicycling
9. ☐ Racquetball
10. ☐ Sailing/Boating
11. ☐ Stamp/Coin Collecting
12. ☐ Motorbiking/Motorcycling
13. ☐ Home Video Games
14. ☐ Physical Fitness/Exercise
15. ☐ Home Video Recording
16. ☐ Recreational Vehicle/4-WD
17. ☐ Photography
18. ☐ CB Radio
19. ☐ Home Workshop/Do-It-Yourself
20. ☐ Gardening/Plants
21. ☐ Electronics
22. ☐ Automotive Work
23. ☐ Sewing/Needlework
24. ☐ Crafts
25. ☐ Collectibles/Collections
26. ☐ Art & Antiques
27. ☐ Stereo Music Equipment
28. ☐ Foreign Travel
29. ☐ Attending Cultural/Arts Events
30. ☐ Gourmet Foods/Cooking
31. ☐ Health/Natural Foods
32. ☐ Wines
33. ☐ Fashion Clothing
34. ☐ Home Furnishings/Decorating
35. ☐ Records & Tapes
36. ☐ Avid Book Reading
37. ☐ Science Fiction
38. ☐ Astrology/Occult
39. ☐ Stock/Bond Investments
40. ☐ Real Estate Investments
41. ☐ Self Improvement Programs
42. ☐ Community/Civic Activities

We appreciate your taking the time to complete this card; the information provided will help us serve you better in the future. We participate in a multi-company program whereby you can receive information about new products, developments, trends, etc. related to the interest areas and other information you have indicated above. Please check here if you would prefer not to learn about such products and services. ☐

Other comments & suggestions about our product:

Fold Here

Carrier Corporation

PLACE FIRST CLASS STAMP HERE

BDP
Product Registration Center
P.O. Box 17686
Denver, Colorado 80217

Bryant, Day & Night, Payne Brands

FOR SERVICE OR REPAIR, FOLLOW THESE STEPS IN ORDER:

FIRST: Contact the installer. You may find his name on the product or in your User's Manual. If his name is not known, call your builder if yours is a new residence.

SECOND: Contact the nearest distributor. (See telephone yellow pages.)

THIRD: Contact:
BDP
Consumer Relations
P.O. Box 4952
Syracuse, New York 13221-4952
Phone: 1-800-428-4326 (TOLL FREE) from USA.
1-315-432-7885 from Canada.

Model No. _____   Unit Serial No. _____

Date of Installation _____   Installed by _____

Name of Owner _____   Address of Installation _____

Carrier Corporation

GAS-FIRED INDUCED-COMBUSTION FURNACE
LIMITED WARRANTY

ONE-YEAR WARRANTY—This CARRIER CORPORATION (herein after referred to as 'COMPANY') product is warranted to be free from defects in material and workmanship under normal use and maintenance for a period of one year from the date of original installation whether or not actual use begins on that date. A new or remanufactured part, at the COMPANY'S sole option, to replace any defective part will be provided within a reasonable time, without charge for the part itself: PROVIDED the defective part is returned to our distributor through a qualified servicing dealer. The replacement part assumes the unused portion of the warranty.

THIS WARRANTY DOES NOT INCLUDE LABOR OR OTHER COSTS incurred for diagnosing, repairing, removing, installing, shipping, servicing or handling of either defective parts or replacement parts, or complete unit. Such costs may be covered by a separate warranty provided by the installer.

EXTENDED 2-YEAR WARRANTY ON MICROPROCESSOR CONTROL CENTER ONLY—During the second through the third years after the date of original installation, the COMPANY further warrants the microprocessor control center against defects in material or workmanship under normal use and maintenance. A new or remanufactured part, at the COMPANY'S sole option, will be provided under the same conditions as stated in the ONE-YEAR WARRANTY.

EXTENDED 19-YEAR WARRANTY ON HEAT EXCHANGER ONLY—During the second through twentieth years after the date of original installation, the COMPANY further warrants the heat exchanger against defects in material or workmanship under normal use and maintenance. A new or remanufactured heat exchanger, at the COMPANY'S sole option, will be provided under the same conditions as stated in the ONE-YEAR WARRANTY or a credit will be allowed in the amount of the then current retail selling price of an equivalent heat exchanger toward the purchase of a new BRYANT, DAY & NIGHT or PAYNE gas furnace.

THESE EXTENDED WARRANTIES DO NOT INCLUDE LABOR OR OTHER COSTS incurred for diagnosing, repairing, removing, installing, shipping, servicing or handling of either defective parts or replacement parts.

THESE WARRANTIES APPLY ONLY TO PRODUCTS IN THEIR ORIGINAL INSTALLATION LOCATION AND BECOME VOID UPON REINSTALLATION.

LIMITATIONS OF WARRANTIES—ALL IMPLIED WARRANTIES (INCLUDING IMPLIED WARRANTIES OF MERCHANTABILITY AND FITNESS FOR A PARTICULAR PURPOSE) ARE HEREBY LIMITED IN DURATION TO THE PERIOD FOR WHICH THE LIMITED WARRANTY IS GIVEN. SOME STATES DO NOT ALLOW LIMITATIONS ON HOW LONG AN IMPLIED WARRANTY LASTS, SO THE ABOVE MAY NOT APPLY TO YOU. THE EXPRESS WARRANTIES MADE IN THIS WARRANTY ARE EXCLUSIVE AND MAY NOT BE ALTERED, ENLARGED, OR CHANGED BY ANY DISTRIBUTOR, DEALER, OR OTHER PERSON WHATSOEVER.

ALL WORK UNDER THE TERMS OF THIS WARRANTY SHALL BE PERFORMED DURING NORMAL WORKING HOURS. ALL REPLACEMENT PARTS, WHETHER NEW OR REMANUFACTURED, ASSUME AS THEIR WARRANTY PERIOD ONLY THE REMAINING TIME PERIOD OF THIS WARRANTY.

THE COMPANY WILL NOT BE RESPONSIBLE FOR:
1. Normal maintenance as outlined in the installation and servicing instructions or owner's manual including coil cleaning, filter cleaning and/or replacement and lubrication.
2. Damage or repairs required as a consequence of faulty installation, misapplication, abuse, improper servicing, unauthorized alteration or improper operation.
3. Failure to start due to voltage conditions, blown fuses, open circuit breakers or other damages due to the inadequacy or interruption of electrical service.
4. Damage as a result of floods, winds, fires, lightning, accidents, corrosive environments or other conditions beyond the control of the COMPANY.
5. Parts not supplied or designated by the COMPANY, or damages resulting from their use.
6. COMPANY products installed outside the continental U.S.A., Alaska, Hawaii and Canada.
7. Electricity or fuel costs or increases in electricity or fuel costs from any reason whatsoever including additional or unusual use of supplemental electric heat.
8. ANY SPECIAL INDIRECT OR CONSEQUENTIAL PROPERTY OR COMMERCIAL DAMAGE OF ANY NATURE WHATSOEVER. Some states do not allow the exclusion of incidental or consequential damages, so the above limitation may not apply to you.

This warranty gives you specific legal rights, and you may also have other rights which vary from state to state.

Effective on products manufactured after November 1, 1988.   Supersedes any other warranty certificates supplied with the product.   Catalog No. BDP-3339-501   39004DP81

Cancels: OM04-1   OM04-2

Carrier Corporation   Printed in U.S.A.   Catalog No. BDP-3339-505   10/15/89

Heil

CONDENSING GAS FURNACES

Service Manual

NUGK	NULS
NULK	NDGK
NUGS	NDLK

This manual supports condensing gas furnaces manufactured after 1988.

Manufactured by:

Inter-City Products
Corporation
Lavergne, TN USA 37086
Brantford, ONT. CANADA N3T 5P4

Part No.
2146635

Table of Contents

Table of Contents (Cont.)

Table of Contents (Cont.)

WARNING

The information contained in this manual is intended for use by a qualified service technician who is familiar with the safety procedures required in installation and repair and who is equipped with proper tools and testing instruments.

Installations and repairs made by unqualified persons can result in hazards subjecting the unqualified person making such repairs to the risk of injury or electrical shock which can be serious or even fatal not only to them, but also persons being served by the equipment.

If you install or perform service on equipment, you must assume responsibility for any bodily injury or property damage which may result to you or others. We will not be responsible for any injury or property damage arising from improper installation, service, and/or service procedures.

Furnace Troubleshooting Chart

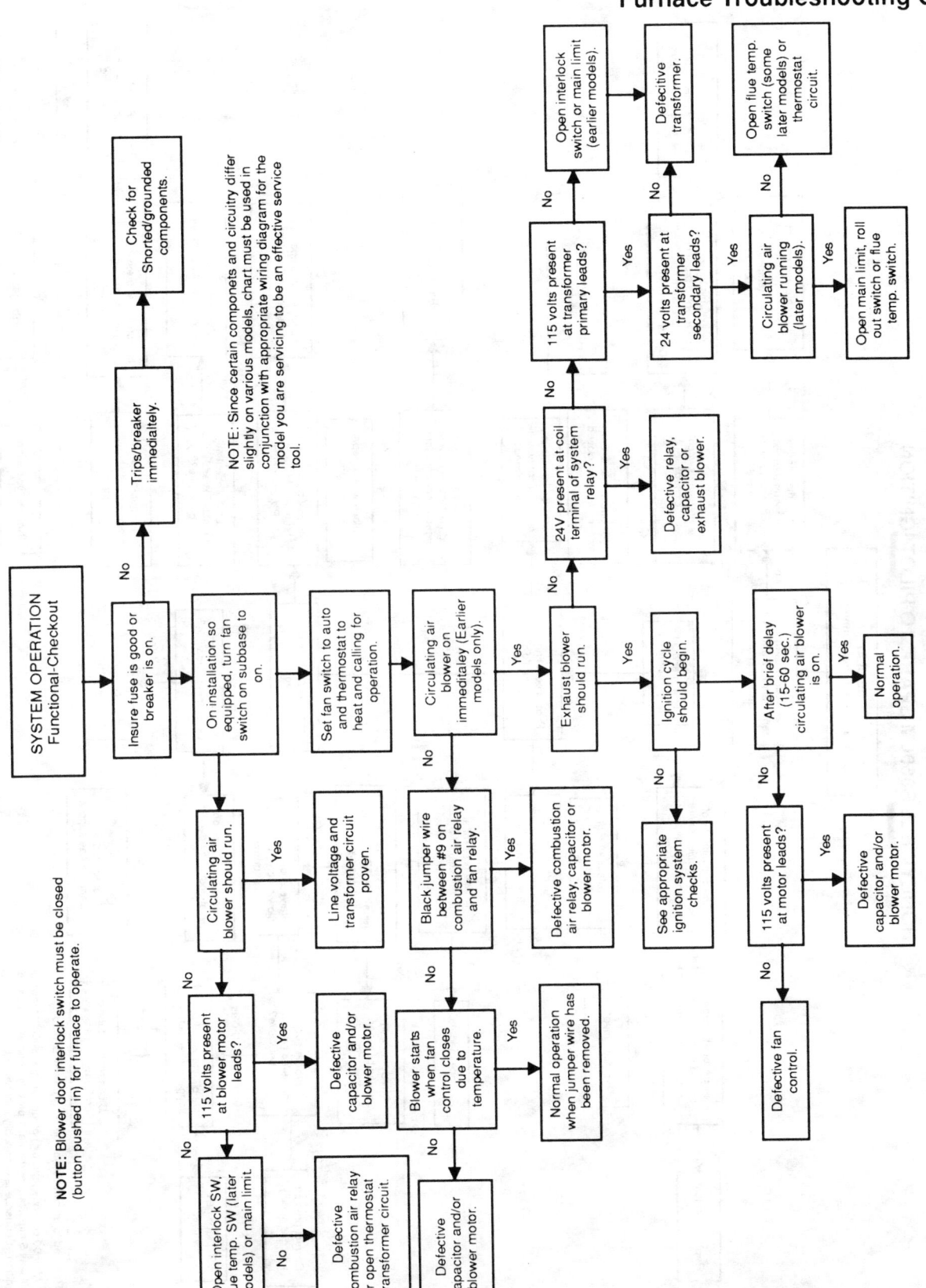

NOTE: Since certain componets and circuitry differ slightly on various models, chart must be used in conjunction with appropriate wiring diagram for the model you are servicing to be an effective service tool.

NOTE: Blower door interlock switch must be closed (button pushed in) for furnace to operate.

Honeywell S8600M Module Troubleshooting Chart

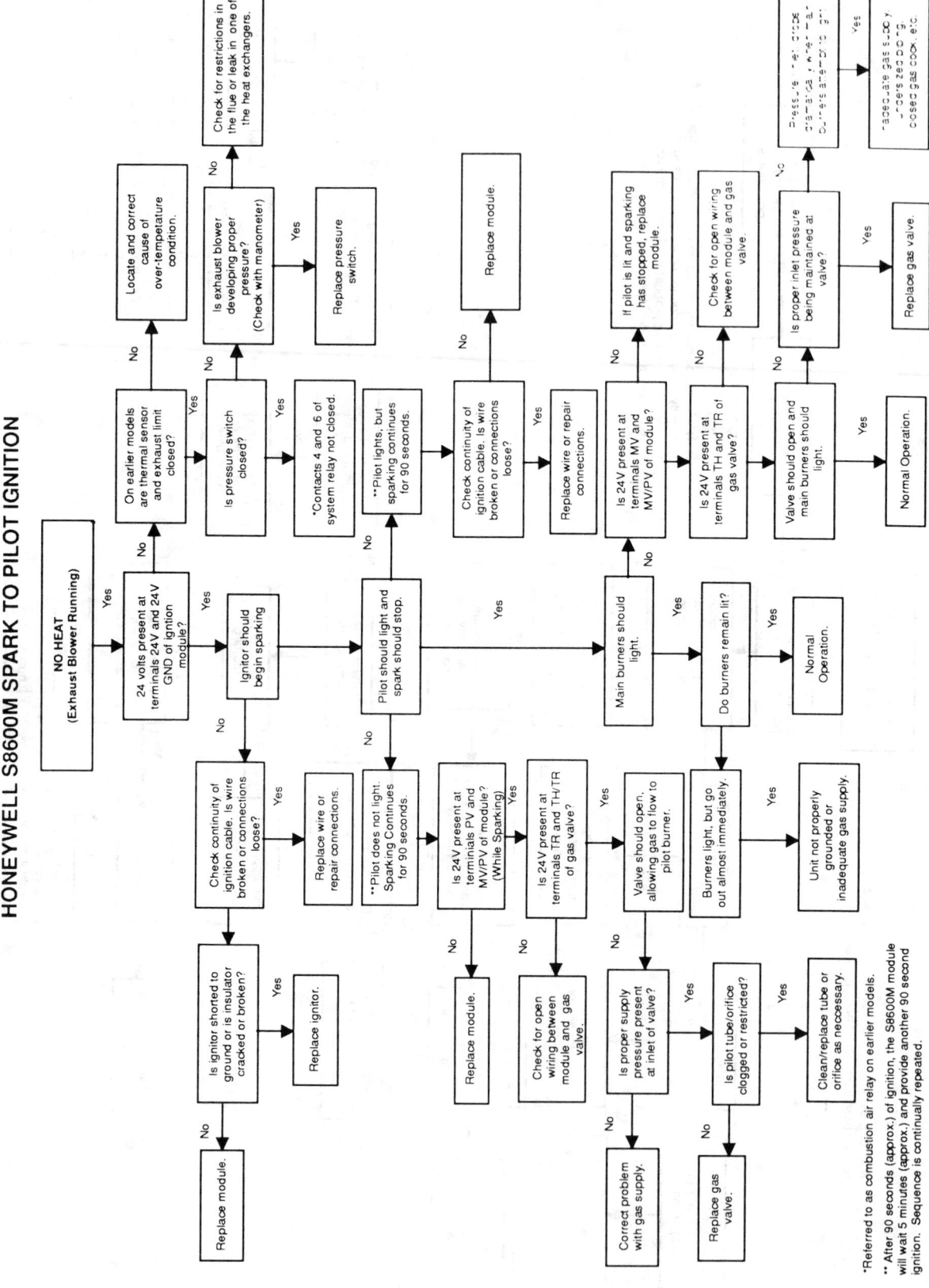

HONEYWELL S8600M SPARK TO PILOT IGNITION

* Referred to as combustion air relay on earlier models.

** After 90 seconds (approx.) of ignition, the S8600M module will wait 5 minutes (approx.) and provide another 90 second ignition. Sequence is continually repeated.

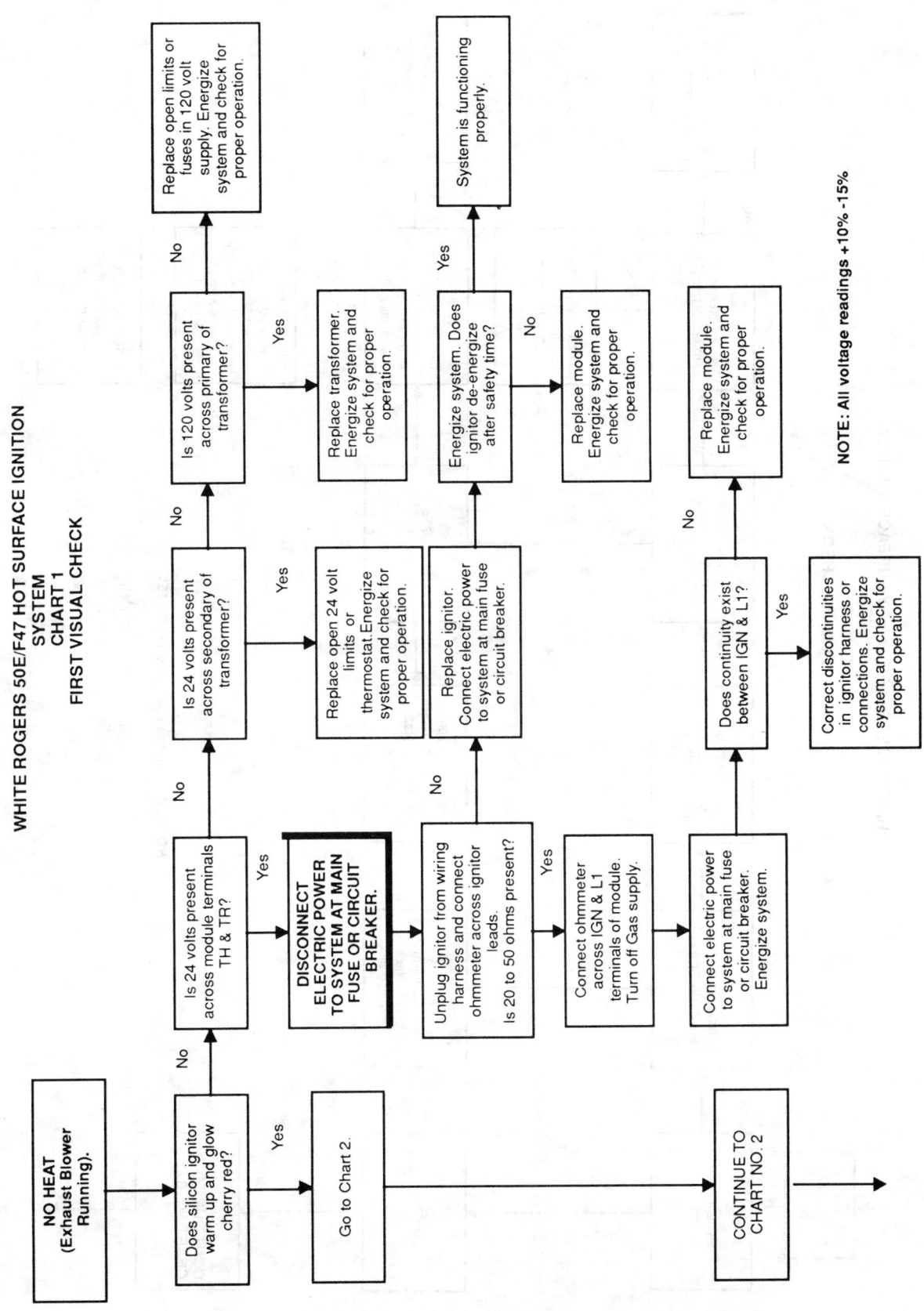

WHITE ROGERS 50E/F47 HOT SURFACE IGNITION SYSTEM
CHART 1
FIRST VISUAL CHECK

NO HEAT (Exhaust Blower Running).

Does silicon ignitor warm up and glow cherry red?

Yes → Go to Chart 2.

No → Is 24 volts present across module terminals TH & TR?

Yes → **DISCONNECT ELECTRIC POWER TO SYSTEM AT MAIN FUSE OR CIRCUIT BREAKER.**

No → Is 24 volts present across secondary of transformer?

Yes → Replace open 24 volt limits or thermostat. Energize system and check for proper operation.

No → Is 120 volts present across primary of transformer?

Yes → Replace transformer. Energize system and check for proper operation.

No → Replace open limits or fuses in 120 volt supply. Energize system and check for proper operation.

From DISCONNECT box → Unplug ignitor from wiring harness and connect ohmmeter across ignitor leads. Is 20 to 50 ohms present?

Yes → Connect ohmmeter across IGN & L1 terminals of module. Turn off Gas supply.

No → Replace ignitor. Connect electric power to system at main fuse or circuit breaker.

Connect ohmmeter across IGN & L1 → Connect electric power to system at main fuse or circuit breaker. Energize system.

Energize system. Does ignitor de-energize after safety time?

Yes → System is functioning properly.

No → Replace module. Energize system and check for proper operation.

Connect electric power... Energize system. → Does continuity exist between IGN & L1?

No → Replace module. Energize system and check for proper operation.

Yes → Correct discontinuities in ignitor harness or connections. Energize system and check for proper operation.

CONTINUE TO CHART NO. 2 →

NOTE: All voltage readings +10% -15%

White Rodgers 50E/F47 Module Chart2

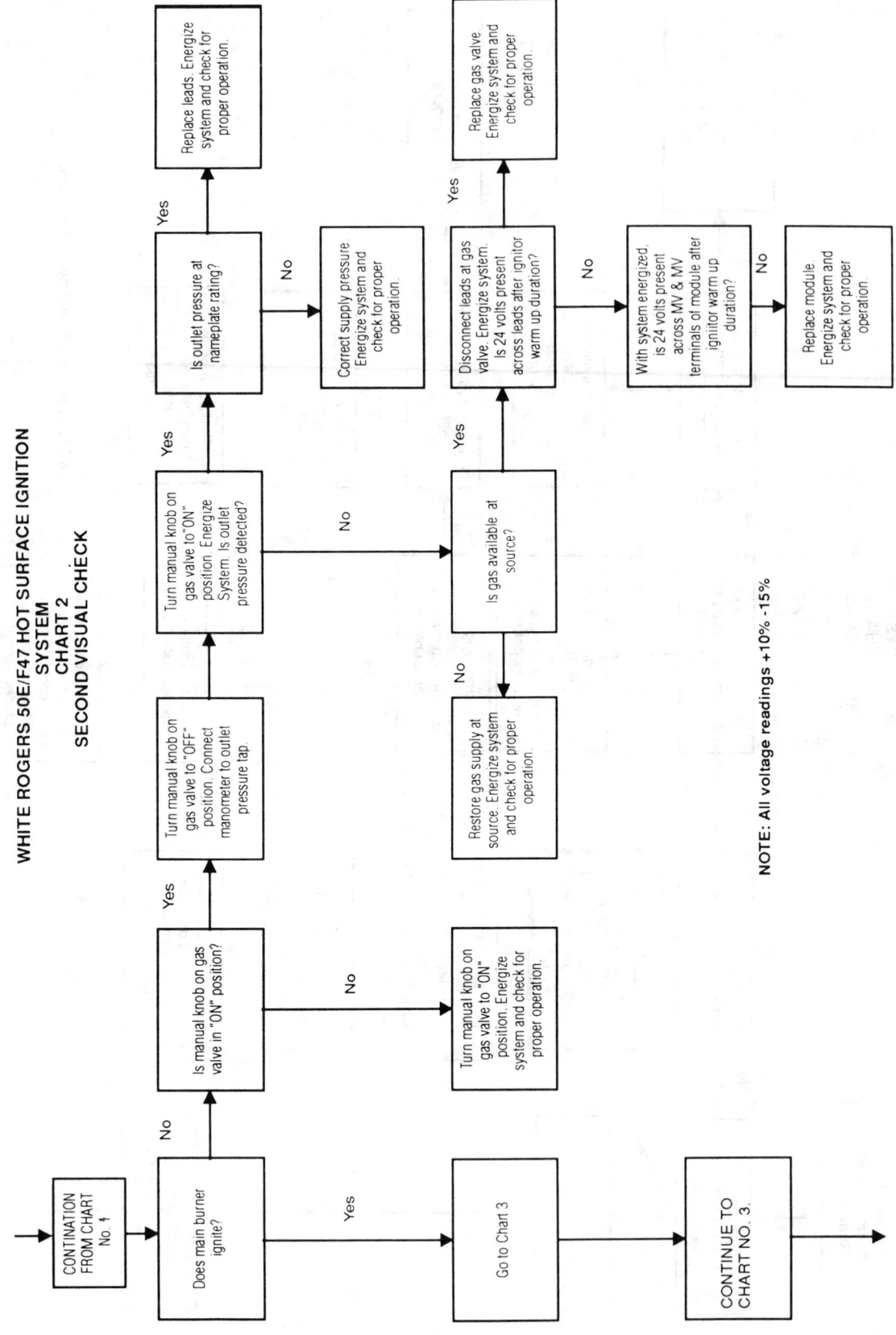

WHITE ROGERS 50E/F47 HOT SURFACE IGNITION
SYSTEM
CHART - 2
SECOND VISUAL CHECK

CONTINATION FROM CHART No. 1

Does main burner ignite?

Is manual knob on gas valve in "ON" position?

Turn manual knob on gas valve to "ON" position. Energize system and check for proper operation.

Turn manual knob on gas valve to "OFF" position. Connect manometer to outlet pressure tap.

Turn manual knob on gas valve to "ON" position. Energize System. Is outlet pressure detected?

Is outlet pressure at nameplate rating?

Replace leads. Energize system and check for proper operation.

Correct supply pressure. Energize system and check for proper operation.

Is gas available at source?

Restore gas supply at source. Energize system and check for proper operation.

Disconnect leads at gas valve. Energize system. Is 24 volts present across leads after ignitor warm up duration?

Replace gas valve. Energize system and check for proper operation.

With system energized is 24 volts present across MV & MV terminals of module after ignitor warm up duration?

Replace module. Energize system and check for proper operation.

Go to Chart 3

CONTINUE TO CHART NO. 3.

NOTE: All voltage readings +10% -15%

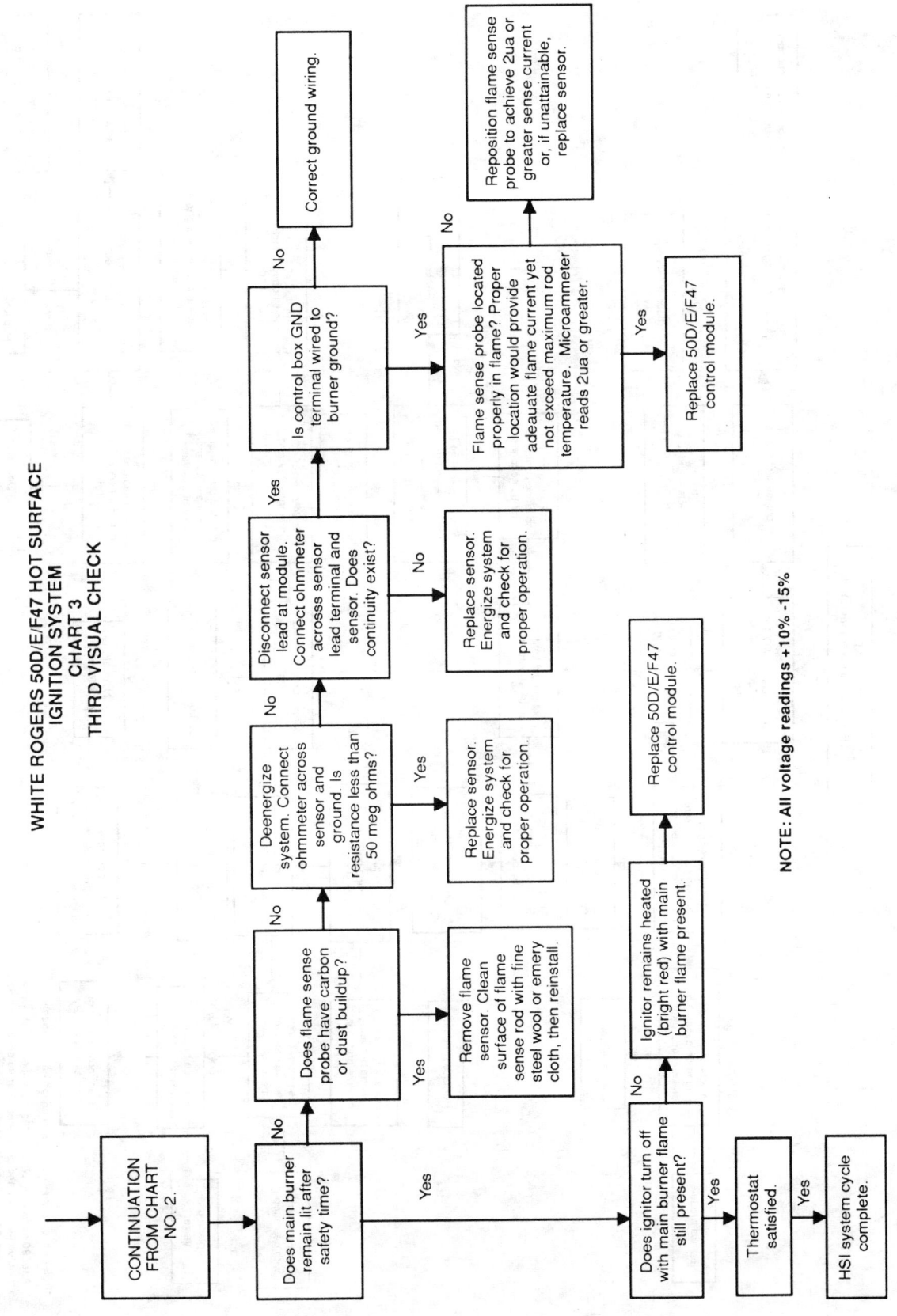

WHITE ROGERS 50D/E/F47 HOT SURFACE
IGNITION SYSTEM
CHART 3
THIRD VISUAL CHECK

NOTE: All voltage readings +10% -15%

Hamilton Standard Intermittent Pilot Ignition System

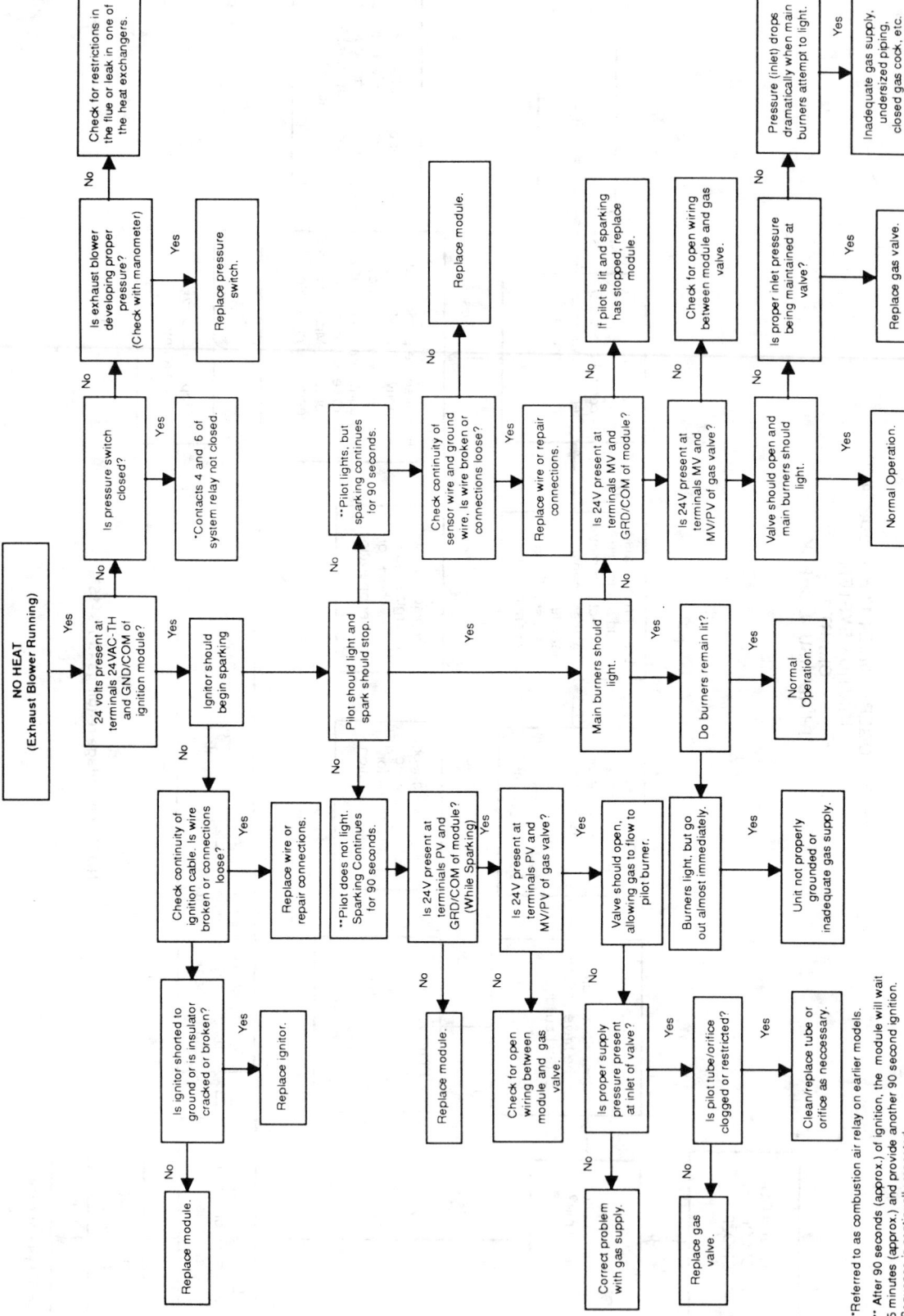

HAMILTON STANDARD INTERMITTENT PILOT IGNITION SYSTEM

NO HEAT (Exhaust Blower Running)

Check for restrictions in the flue or leak in one of the heat exchangers.

Is exhaust blower developing proper pressure? (Check with manometer)

Replace pressure switch.

Is pressure switch closed?

*Contacts 4 and 6 of system relay not closed.

24 volts present at terminals 24VAC-TH and GND/COM of ignition module?

Ignitor should begin sparking

Check continuity of ignition cable. Is wire broken or connections loose?

Replace wire or repair connections.

Is ignitor shorted to ground or is insulator cracked or broken?

Replace ignitor.

Replace module.

**Pilot does not light. Sparking Continues for 90 seconds.

Is 24V present at terminals PV and GRD/COM of module? (While Sparking)

Replace module.

Is 24V present at terminals PV and MV/PV of gas valve?

Check for open wiring between module and gas valve.

Valve should open, allowing gas to flow to pilot burner.

Is proper supply pressure present at inlet of valve?

Correct problem with gas supply.

Is pilot tube/orifice clogged or restricted?

Replace gas valve.

Clean/replace tube or orifice as necessary.

Pilot should light and spark should stop.

**Pilot lights, but sparking continues for 90 seconds.

Check continuity of sensor wire and ground wire. Is wire broken or connections loose?

Replace wire or repair connections.

Replace module.

Main burners should light.

Is 24V present at terminals MV and GRD/COM of module?

If pilot is lit and sparking has stopped, replace module.

Is 24V present at terminals MV and MV/PV of gas valve?

Check for open wiring between module and gas valve.

Do burners remain lit?

Burners light, but go out almost immediately.

Unit not properly grounded or inadequate gas supply.

Normal Operation.

Valve should open and main burners should light.

Is proper inlet pressure being maintained at valve?

Pressure (inlet) drops dramatically when main burners attempt to light.

Inadequate gas supply, undersized piping, closed gas cock, etc.

Replace gas valve.

Normal Operation.

*Referred to as combustion air relay on earlier models.

** After 90 seconds (approx.) of ignition, the module will wait 5 minutes (approx.) and provide another 90 second ignition. Sequence is continually repeated.

Figure 1.

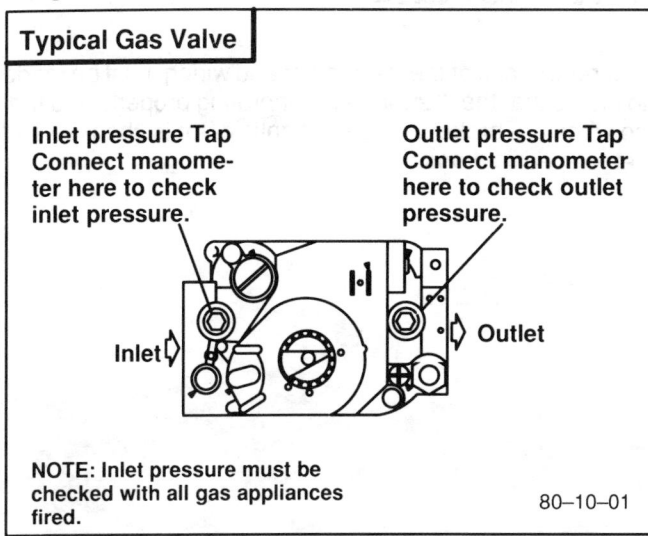

NOTE: Inlet pressure must be checked with all gas appliances fired.

80–10–01

Gas and Electrical Supply

Natural Gas

Check the gas inlet pressure to the gas valve. Inlet pressure to the valve must be a minimum 4.5 in W.C. with all other appliances fired.

L.P. Gas

Check the gas inlet pressure to the gas valve. Inlet pressure to the valve must be a minimum of 11.0 in. W.C. with all other gas appliances fired.

Figure 2.

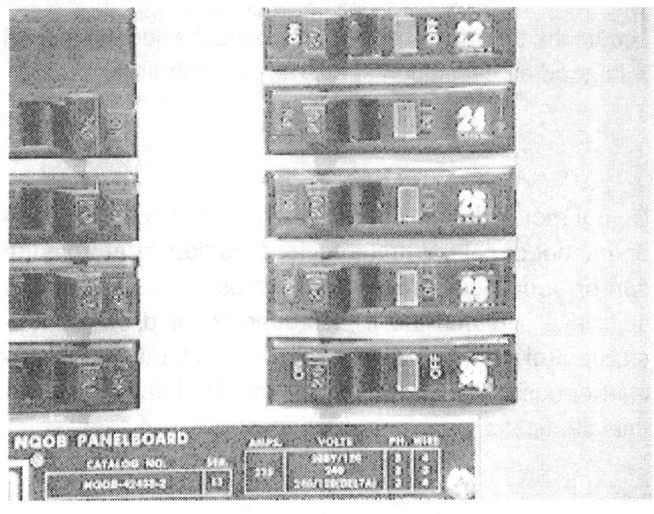

Electrical Power to Furnace

Check the fuse or circuit breakers to make sure they are not blown or tripped. Supply voltage to the unit should be within plus or minus 10% of the voltage shown on the rating plate with the unit operating normally.

Figure 3.

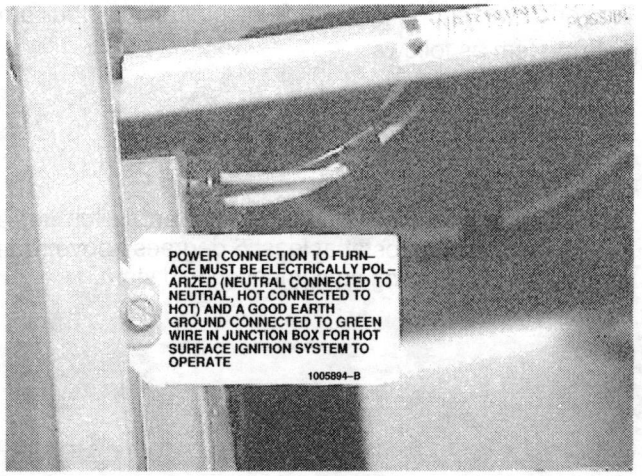

Furnace Electrical Ground

The furnace may not operate at all or may operate intermittently if the furnace is not properly grounded.

Figure 4.

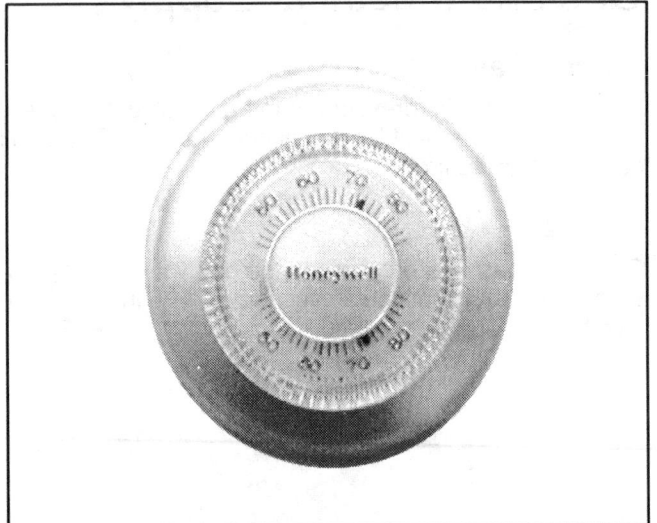

Figure 5.

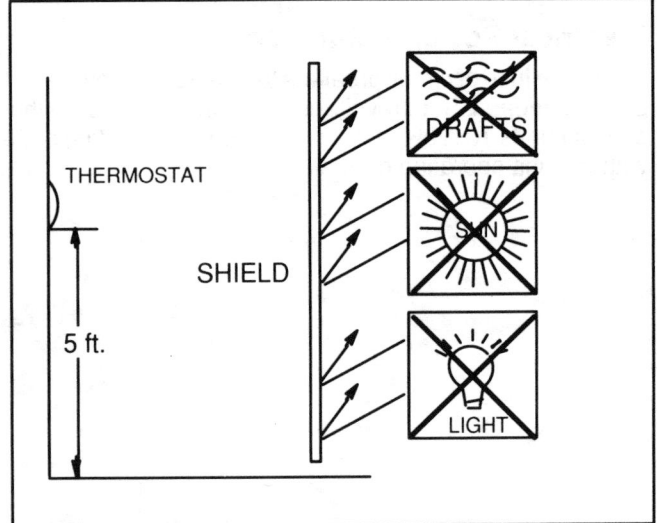

Figure 6.

Thermostats

A careful check of the thermostat and wiring must be made to insure that the thermostat is operating properly and that no wires are broken. The heat anticipator is checked and set to specifications.

LOCATION

Locate the thermostat about 5 ft. above the floor in an area with good air circulation at average temperature.

Do not mount the thermostat where it may be affected by drafts, hot or cold air from ducts, or radiant heat from the sun or appliances. Level the thermostat exactly using a spirit level or plumb line. If not properly leveled, the thermostat control point will deviate from set point. Electronic digital thermostats will operate properly if not level, however they should be level for appearance.

CHECKOUT

Turn on power supply and check the operation of the complete system as follows:

With system switch set at HEAT, and fan switch set at AUTO, set the thermostat at least 5 degrees above room temperature — heating equipment should start.

Figure 7.

With system switch set at COOL, and fan switch set at AUTO, set the thermostat at least 5 degrees below room temperature — cooling equipment should start.

Figure 8.

Set fan switch at ON — fan should run continuously with system changeover switch in either HEAT or COOL position.

Figure 9.

System Switch Off

Fan Switch to On

Set system changeover switch to OFF and fan switch to ON — fan should run continuously and neither heating or cooling equipment can be actuated by the thermostat.

Operate the entire system at least one complete cycle with switches in each position. Set thermostat at the desired setting and system and fan switches in proper positions.

Figure 10.

Heat Anticipators

The adjustable heat anticipator must be set to match the current draw from the heating control through the thermostat.

Check the tech service data section for the suggested heat anticipator setting for the furnace you are servicing. If setting is not available, connect an AC ammeter (about 0 to 2.0 amp. range) between the "W" and "R" terminals on the back of the room thermostat.

NOTE:

If a wallplate or subbase is used, connect ammeter between appropriate terminals on front of the wall plate or subbase.

Figure 11.

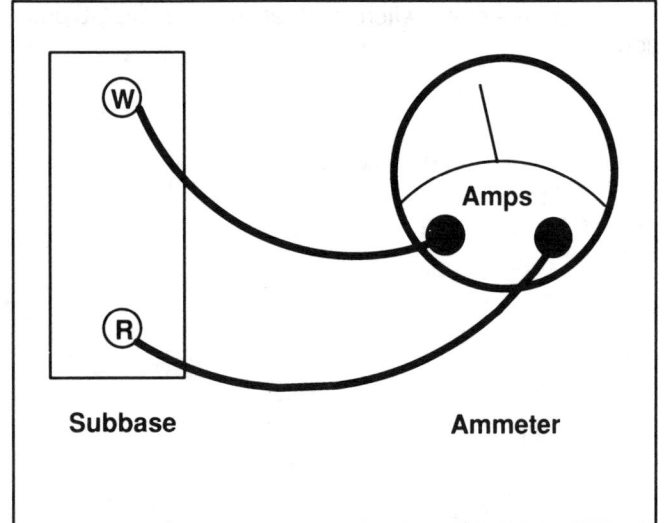

HEAT ANTICIPATORS

Move the thermostat or subbase system switch to OFF so no current passes through thermostat switch contacts. With the system operating through the ammeter, wait one minute and then read the ammeter. Use this reading to adjust the heat anticipator.

Figure 12.

The heat anticipator may require further adjustment for best performance. To lengthen burner–on–time, move the indicator in the direction of the "LONGER" arrows — not more than half a scale marking at a time. To shorten operation, move the indicator in the opposite direction.

Figure 13.

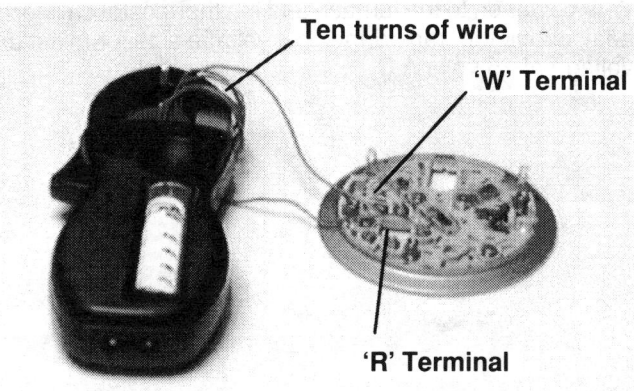

Ten turns of wire

'W' Terminal

'R' Terminal

If a low range ammeter (0 – 2.0 amps) is not available, a clamp on type ammeter may be used as follows:

Turn the power off to the system and remove the thermostat from the subbase or wallplate.

Wrap exactly 10 turns of wire around the jaw of a split jaw induction type current meter as illustrated.

Connect one end of wire to the R terminal and the other to the W terminal.

Figure 14.

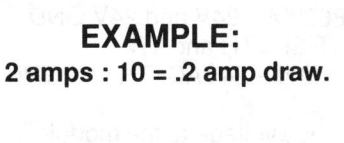

EXAMPLE:
2 amps : 10 = .2 amp draw.

Turn on power to the system, furnace should fire. Wait approximately 1 minute and read amp scale. Divide reading by 10 to obtain current draw.

Figure 15.

Turn off power to the system and remove the amp meter and wire coil. Set the anticipator to match the amp draw, and mount thermostat.

Figure 16.

Transformer

With power on to the furnace and the thermostat calling for heat, check the transformer low voltage side for minimum voltage of 25 volts A.C.

Figure 17.

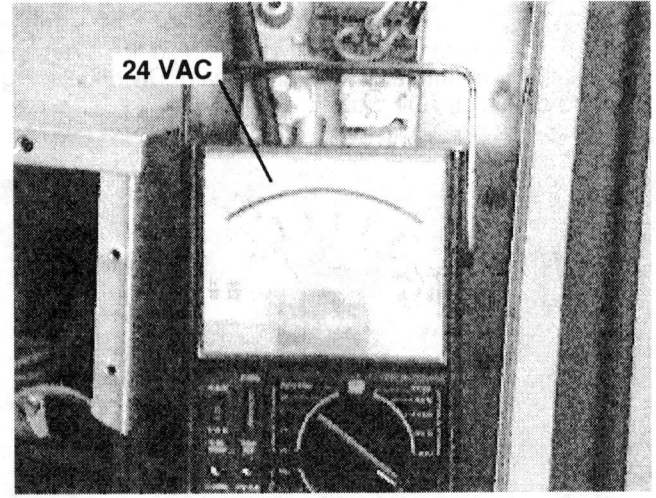

24 VAC

Voltage to Control Module (With Exhaust Blower Operating)

Using a volt meter, check for minimum voltage (24 volts) across the terminals of the control module as specified below.
Honeywell S8600M = 24V and 24V GND
White Rogers HSI = TH and TR
Hamilton Standard = 24VAC–TH and GND/COM

If you do not have voltage to the module, follow the wiring diagram for the unit you are servicing and check all relays, switches and wiring. If any parts prove to be defective they must be replaced. Electrical leads must be clean, tight and defective wiring replaced.

Figure 18.

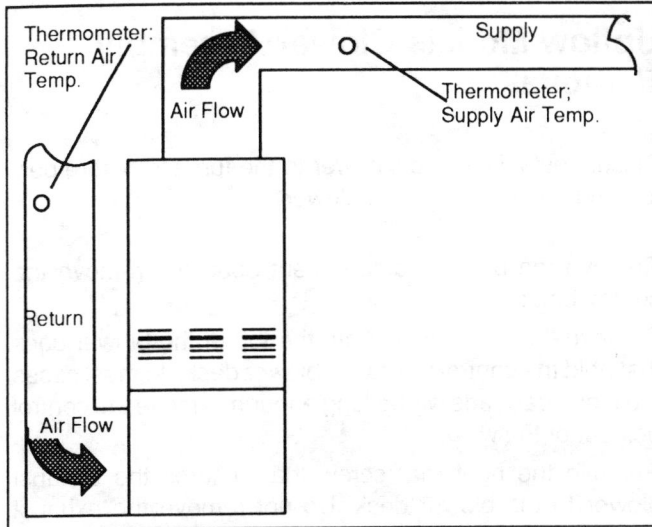

Figure 19.

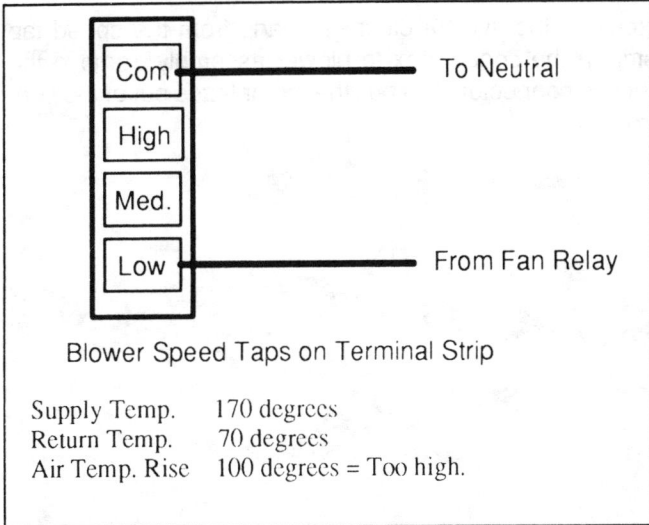

Blower Speed Taps on Terminal Strip

Supply Temp.   170 degrees
Return Temp.   70 degrees
Air Temp. Rise   100 degrees = Too high.

Figure 20.

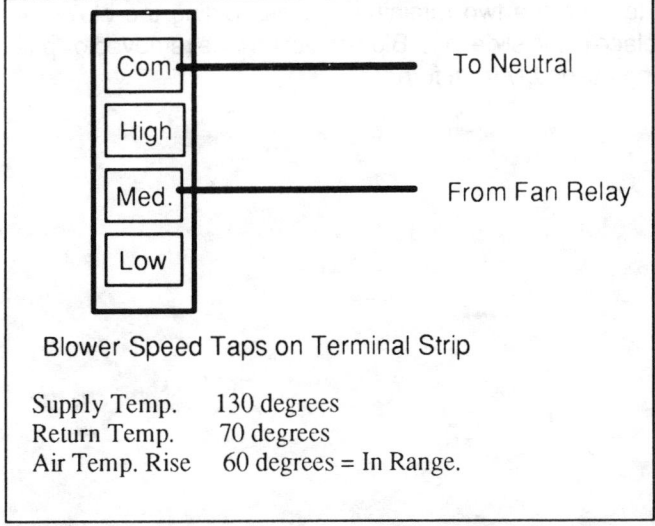

Blower Speed Taps on Terminal Strip

Supply Temp.   130 degrees
Return Temp.   70 degrees
Air Temp. Rise   60 degrees = In Range.

Temperature Rise

CHECK TEMPERATURE RISE

Check temperature rise through the unit by placing a thermometer in the return air duct as close to the unit as possible.   Place a second thermometer in the supply duct at least two (2) feet away from the furnace to prevent any false readings due to radiation from the furnace heat exchanger. All registers and duct dampers must be open and the unit should be operated at rated input for 15 minutes before taking readings.   Temperature rise must be within the range specified on the rating plate.

With a properly designed system, the proper amount of temperature rise will normally be obtained when the unit is operating at rated input with the recommended blower speed.

NOTE:
Air temperature rise is the temperature difference between supply and return air.

If the correct amount of temperature rise is not obtained, when operating on the recommended blower speed, it may be necessary to change the blower speed. A higher blower speed will lower the temperature rise.   A slower blower speed will increase the temperature rise.   See blower speed taps on page 16.

Example:

Supply Temp.   170 degrees
Return Temp.   70 degrees
Air Temp Rise   100 degrees = Too High

Solution:
Increase blower speed.

Re–check and readjust until proper temperature rise is obtained

Example:

Supply Temp.   130 degrees
Return Temp.   70 degrees
Air Temp Rise   60 degrees = In Range

Figure 21.

Remove Two Screws (Hidden)

Blower Removal

Upflow Models Blower Assembly Removal

Disconnect all electrical power to the furnace before performing any service to the blower.

Remove the blower compartment door and remove the control box cover.

Remove the two screws from the top of the blower deck that hold the control box to the blower deck. In most cases the electrical leads will be long enough to move the control box out of the way.

Remove the right rear screw that secures the exhaust blower to the blower deck. Do not remove the exhaust blower.

Figure 22.

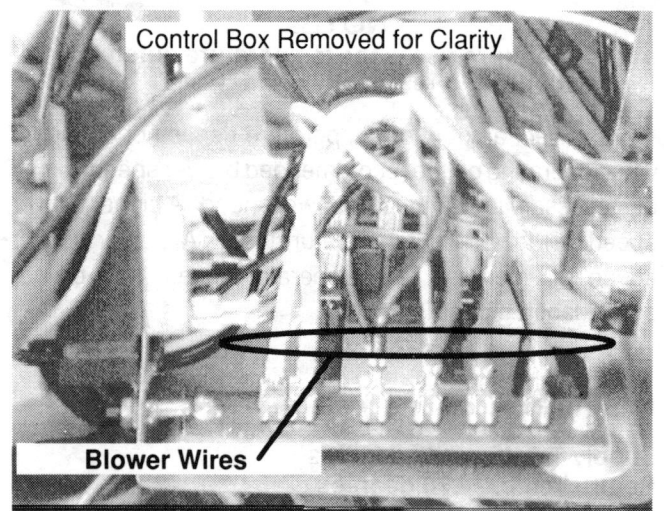

Control Box Removed for Clarity

Blower Wires

Remove the five (5) electrical leads from the speed tap strip in the control box to blower assembly, remove the "heyco connector" and pull the motor leads out of the control box.

Figure 23.

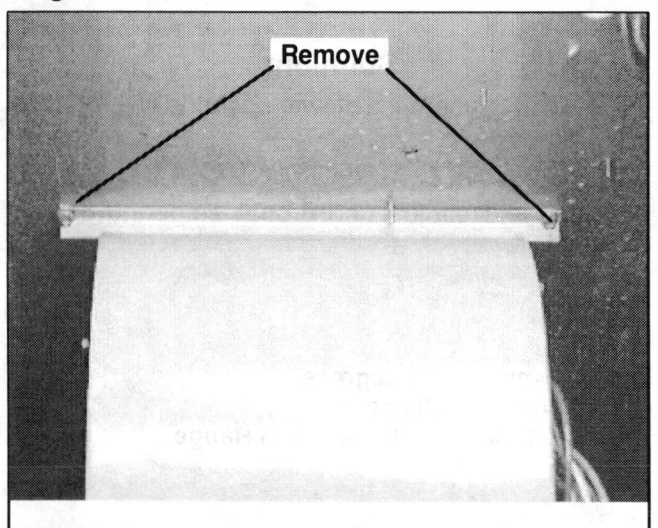

Remove

Remove the two retaining screws holding the blower in place in the slide rails. Blower can now be removed by pulling assembly from furnace.

Figure 24.

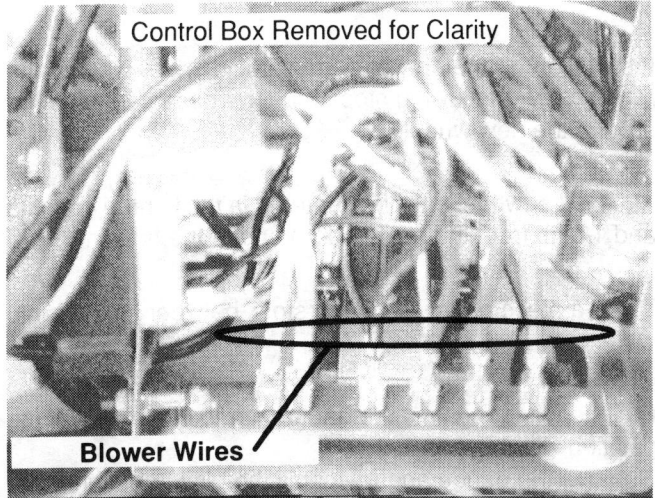

Control Box Removed for Clarity

Blower Wires

Counterflow Models: Blower Assembly Removal

Remove the electrical leads from the speed tap strip in the control box to blower assembly.

Figure 25.

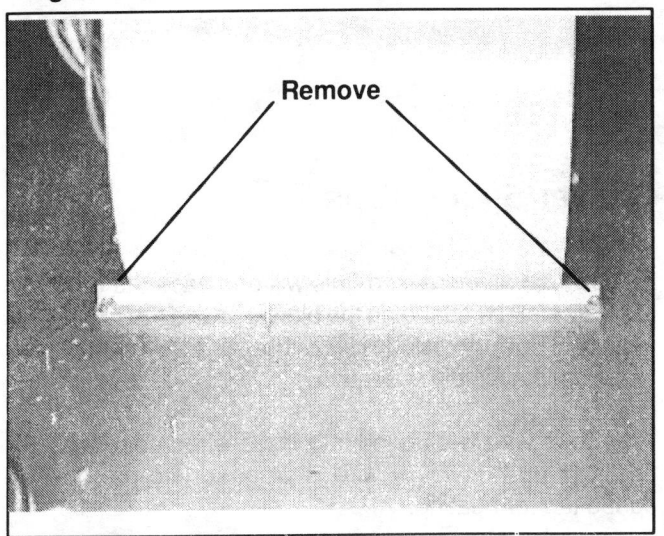

Remove

Remove the two retaining screws that hold the blower in position in the slide rails.

Note:

It is not necessary to remove the electrical control box before removing the blower. The blower can be removed by lifting up on the blower while pulling out and turning the blower to the right. Control box and junction box have been removed for clarity.

Figure 26.

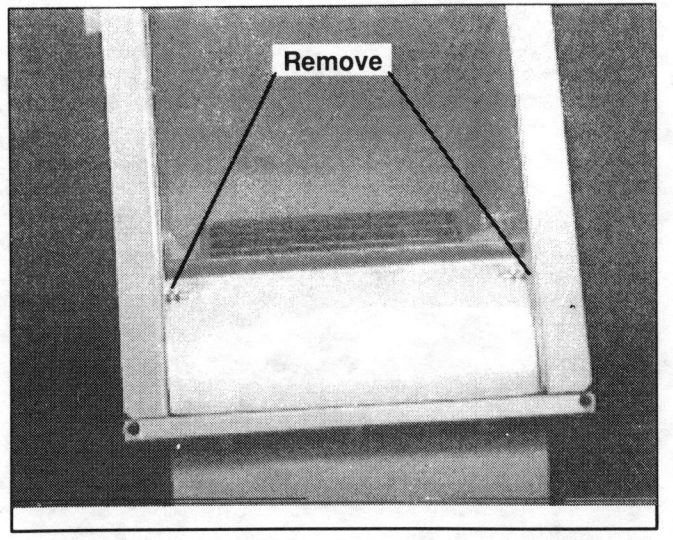

Remove

Blower Wheel Removal (All)

Remove the two screws securing the cut–off plate to the blower housing. Remove the blower cut–off plate from the throat of the blower scroll.

Loosen the set screw securing the blower wheel to the motor shaft.

Remove the three bolts securing the motor to blower housing and remove motor and bracket as an assembly. No further disassembly is necessary unless the motor is being replaced.

The blower wheel can now be removed from the scroll.

Figure 27.

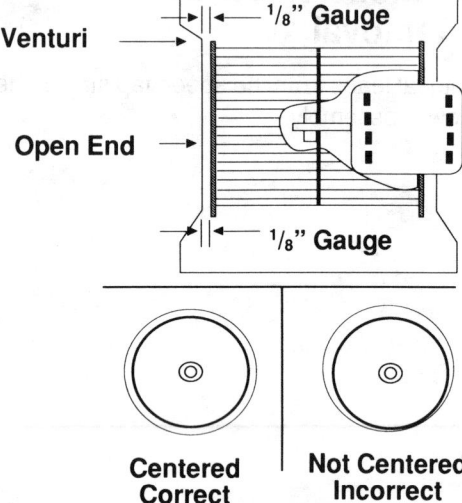

Venturi

Open End

1/8" Gauge

1/8" Gauge

Centered
Correct

Not Centered
Incorrect

Blower Wheel Installation

Installation of the new blower wheel is the reverse of the removal. The following steps must be observed when installing the new wheel.

The blower wheel must be centered in the venturi opening and 1/8 in. from the shaft (open) end of the housing.

Place a piece of 1/8 in. gauge stock between venturi and wheel.

Pull the wheel against the gauge and lock the wheel into position with the set screw.
Remove the gauge and re-install the blower assembly in reverse order.

Figure 28.

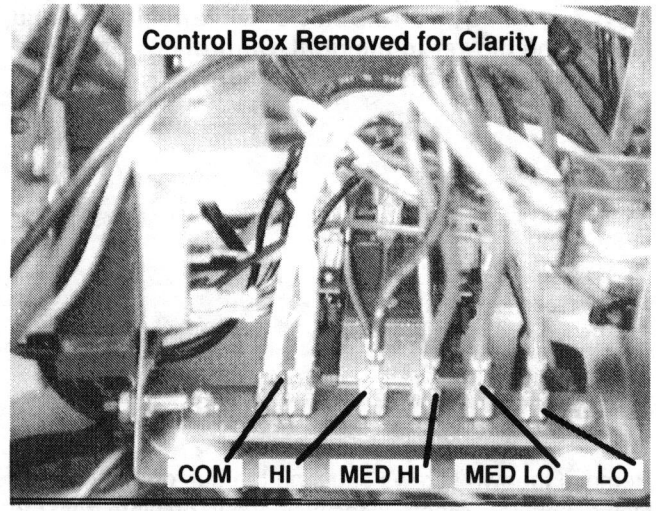

Control Box Removed for Clarity

COM   HI   MED HI   MED LO   LO

Blower Speed Taps

Air flow can be increased (this will decrease the outlet air temperature) by changing the blower speed tap to a higher setting. The terminal block in the electrical control box makes this a simple operation.

The violet wire is plugged into the desired speed tap for cooling and the yellow wire is plugged into the desired speed tap for heating.

Figure 29.

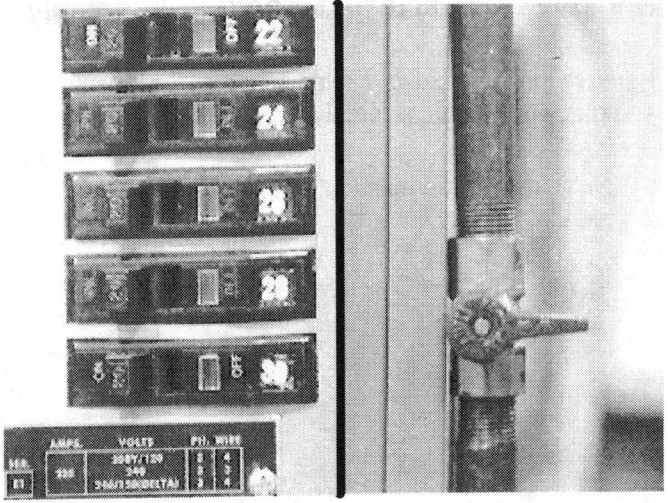

Burners

NUGK Series

Gas Valve and Manifold Removal

1. Turn off electrical power to the furnace at the disconnect, the circuit breakers, or remove the fuses.

2. Shut off the gas supply at the manual shut off valve. Disconnect the gas supply line to the gas valve at the union. After removing the gas valve, it is recommended that the gas line be capped if the line is to be open for an extended period of time.

Figure 30.

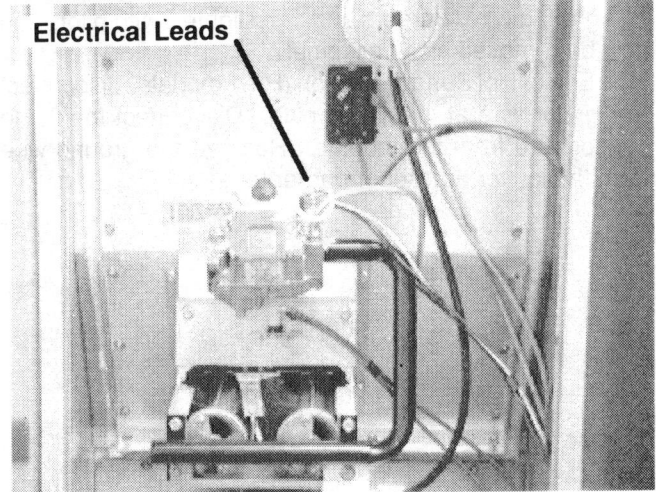

Electrical Leads

3. Disconnect the electrical leads to the gas valve.

Figure 31.

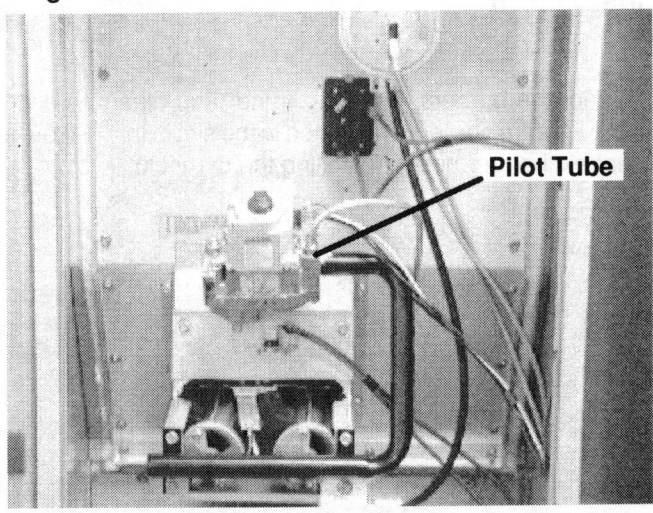

Pilot Tube

4. Disconnect the pilot tube from the gas valve, on those models equipped with spark to pilot ignition.

Figure 32.

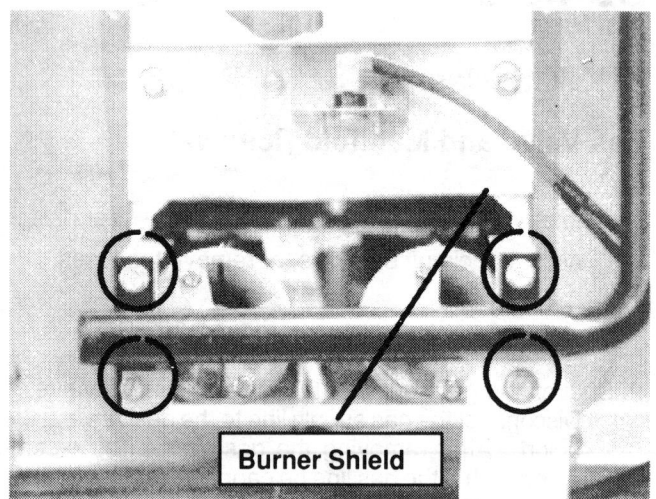

Burner Shield

Remove Gas Valve and Manifold Assembly

Remove the four screws, circled, and remove the manifold and gas valve as an assembly.

Carefully inspect the manifold and orifices to be sure that the manifold and orifices are clear.

Figure 33.

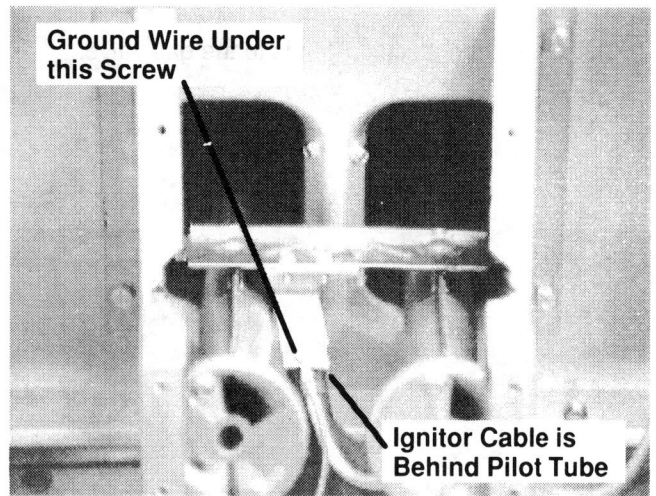

Ground Wire Under this Screw

Ignitor Cable is Behind Pilot Tube

Raise the burner shield and disconnect the ignitor cable and ground wire. Burners and crosslighter, with pilot, can now be removed as an assembly.
NOTE: For Hot Surface Ignition (HSI) models, disconnect the ignitor wire at the molex plug. Disconnect the flame sensor wire from the module. Remove the ground wire from the ignitor bracket or burner face.

Figure 34.

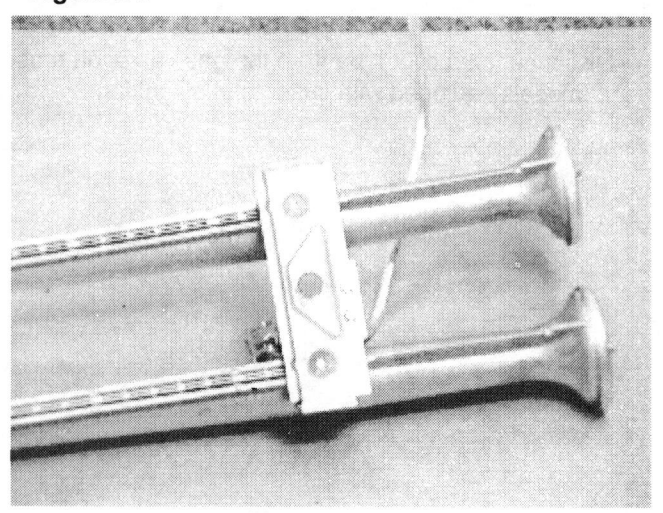

Inspection

Inspect the burners and crosslighter and clean if necessary. Be sure burners are seated in the slot in the rear of the heat exchanger before installing the manifold.

Figure 35.

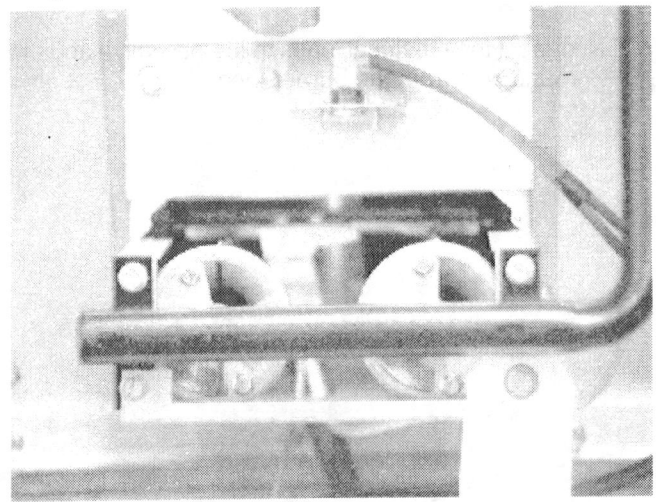

Reassembly

Reassemble all parts in reverse order of removal.

Insert burners into the heat exchanger. **Burners must be inserted into the slots at the back of the heat exchanger and leveled before installing the manifold.**

After reassembly and the gas has been turned on. All joints must be checked for gas leaks using a soapy solution. All leaks must be repaired immediately.

Burner shield MUST be replaced before returning furnace to normal operation.

Figure 36.

NUGS/NULS Series
Gas Valve and Manifold Removal

1. Turn off electrical power to the furnace at the disconnect, circuit breakers, or by removing fuses.

2. Shut off the gas supply at the manual shut off valve. Disconnect the gas supply line to the gas valve at the union. (After removing the gas valve, it is recommended that the gas line be capped if the line is to be open for an extended period of time.)

Figure 37.

Burner Compartment Cover   Burner Shield

3. Remove the burner compartment cover.

4. Remove the wires from the rollout switch on the burner shield and remove the burner shield.

Figure 38.

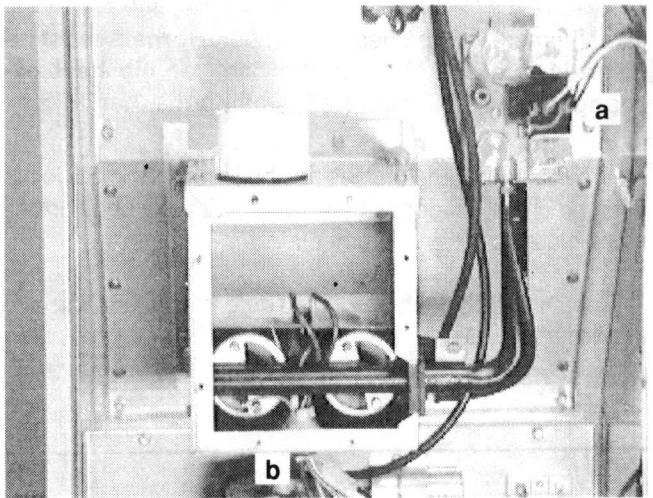

5. Disconnect the electrical leads to the gas valve and the ground wire from the burner face.

Figure 39.

Remove the screws securing the manifold to the burner box, the two screws shown and one screw from the opposite side of burner box. The manifold can now be removed. Be careful not to damage the rubber seal where the manifold enters burner compartment on the right side.

Figure 40.

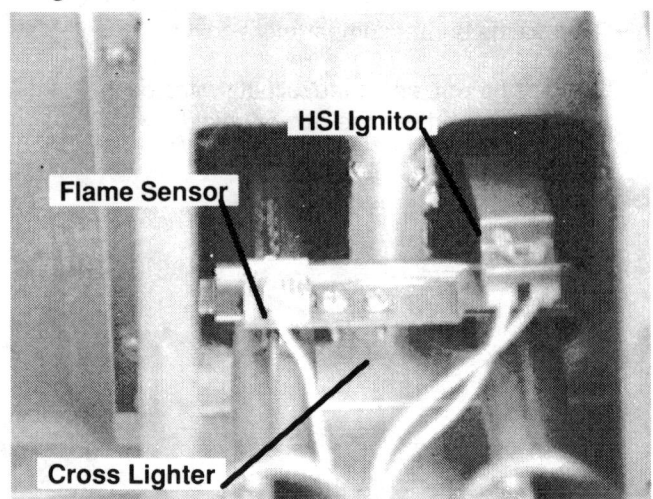

Remove the wing nut securing the ignitor to the crosslighter. Remove the shield and carefully remove the ignitor from the crosslighter laying in the bottom of the burner compartment. **EXTREME CAUTION MUST BE USED IN HANDLING THE HSI IGNITOR TO KEEP FROM DAMAGING THE IGNITOR.**

Remove the screw securing the flame sensor to the crosslighter and lay the flame sensor in the bottom of the burner compartment.

Figure 41.

This allows you to remove burners and crosslighter without disturbing the seal around the wiring.

The crosslighter and burners can now be inspected and cleaned if necessary.

Reassemble in reverse order, making sure the burners are seated in the rear of the heat exchanger and the seal around the manifold is tight.

Figure 42.

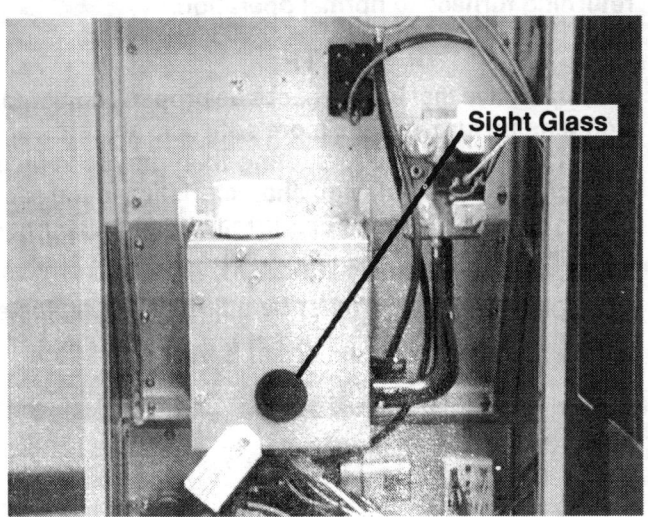

Sight Glass

Primary Air Adjustment (ALL)

If the burners are not equipped with Air Shutters (natural gas models), **NO** Adjustment is necessary. On L.P. models, adjustment of the air shutter may be necessary to obtain the correct flame characteristics and/or to minimize the resonance heat exchanger noise generated by the burner flame.

Resonance noise can be caused by:
Incorrect or misaligned orifices—Verify size and alignment
Improper gas pressure — check and adjust
Air shutters out of adjustment — adjust the air shutter (L.P. applications must have air shutters installed.)

Observe flame by looking through the inspection glass in cover. If adjustment is necessary, remove burner compartment cover and adjust air shutters as described below. Replace burner compartment cover and observe the flame through the inspection glass.

Figure 43.

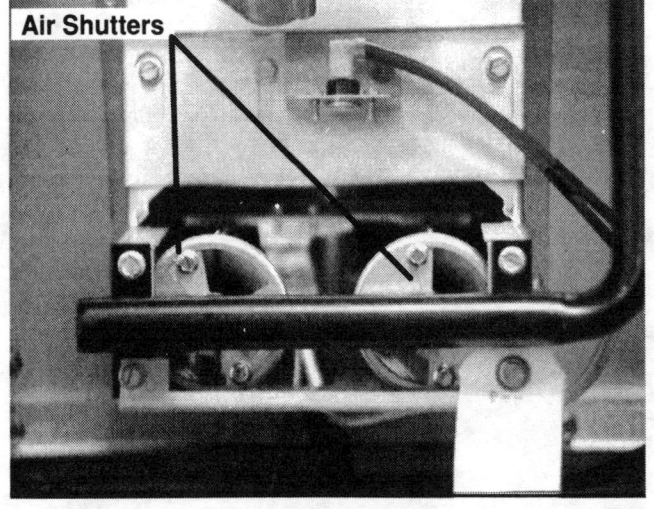

Air Shutters

Adjustment Procedures (NUGK Shown)

Adjustment procedure is the same for **DIRECT VENT MODELS**, except the burner compartment cover **must** be in place when checking flame characteristics.

Check air shutter position – should be full open.

1. Start the furnace; see lighting instructions on furnace or in the Owners Information Manual.

2. Allow furnace to run for 10 minutes then check flame characteristics.

Figure 44.

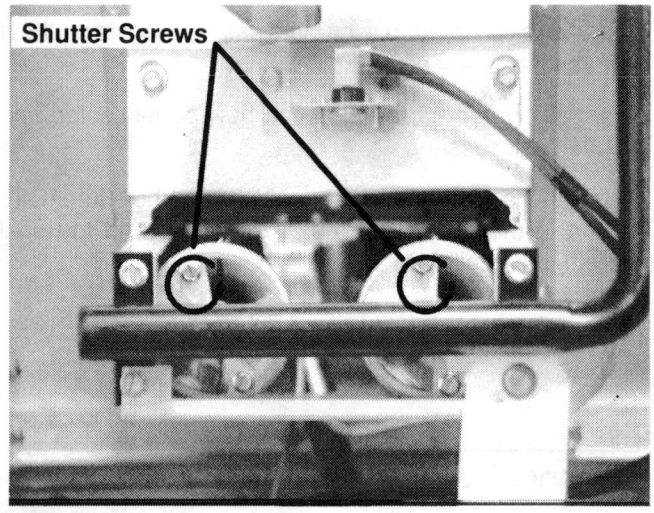

Shutter Screws

Figure 45.

Adjustment (Continued)

3. Loosen shutter screw(s) and close shutter until flame is yellow.

4. Open shutters just enough to eliminate yellow.

5. Tighten locking screws. (Flame will normally have slight yellow tip when adjusted properly).

The **DIRECT VENT** burner compartment cover **MUST** be in place when observing the flame.

Burners should fit loosely in the furnace and not bind against the manifold or heat exchanger.

Burner compartment cover MUST be in place before returning furnace to normal operation.

NOTE:

It is imperative that L.P. furnaces be properly adjusted. Condensing furnaces are more critical to adjustments and more susceptible to sooting than gravity vented furnaces due to the design characteristics in the secondary heat exchanger. For this reason, proper burner adjustment is extremely important.

Example: Air shutters must be adjusted to eliminate resonance and/or obtain correct burner operation.

Figure 46.

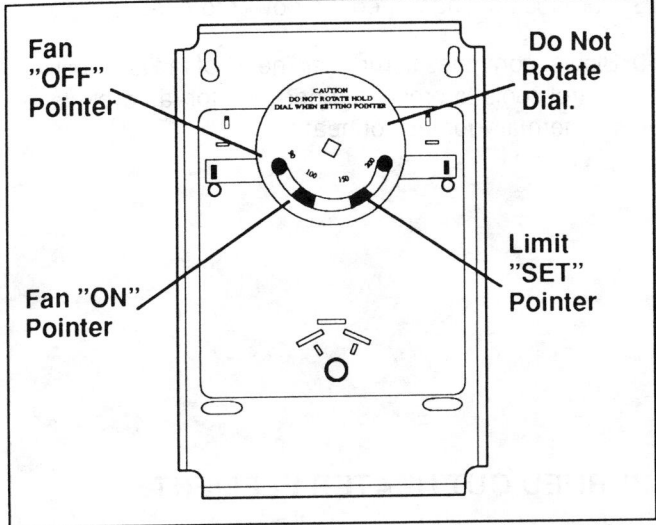

Fan "OFF" Pointer

Do Not Rotate Dial.

Fan "ON" Pointer

Limit "SET" Pointer

Fan and Limit Controls

ADJUSTMENT (The adjustment covers the Honeywell models discussed in this manual.)

NOTE:
The 'limit set pointer' on the limit control is factory preset and must not be adjusted.

1. Adjustment — **DO NOT** rotate the dial when setting pointers.

2. If necessary, adjust fan **ON–OFF** settings to obtain a satisfactory comfort level.

3. If the fan runs too long after furnace shutdown and blows cold air, adjust the fan **'OFF'** pointer up a few degrees.

4. If the fan goes off and then comes back on and off, adjust fan **'OFF'** pointer a few degrees lower.

Figure 47.

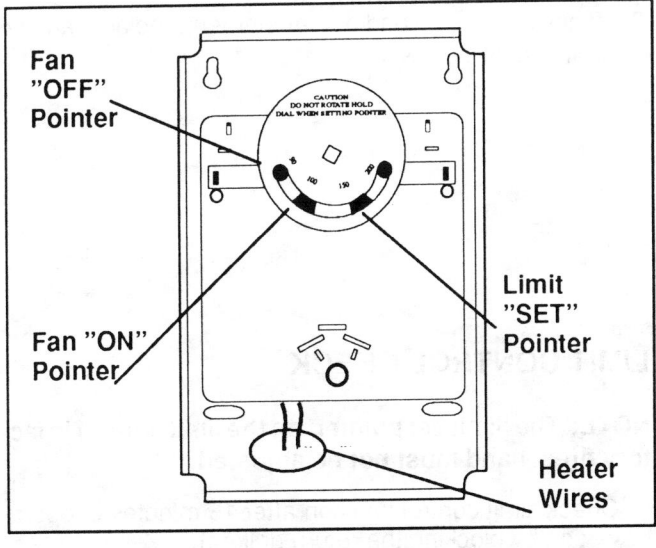

Fan "OFF" Pointer

Fan "ON" Pointer

Limit "SET" Pointer

Heater Wires

Fan and Limit Control (Honeywell–Timed On, Temperature Terminated)

The control is a timed on, temperature terminated switch. A 24 volt heater causes a 'warp switch' in the fan and limit control to turn on the fan after the furnace has fired.

How the Control Works

1. Start cycle.

2. Thermostat calls for heat.

3. Gas valve comes on; furnace fires.

4. Power goes to the fan limit control.

5. A 24 volt heater starts to heat a 'warp switch' and approximately 40 seconds later the blower starts.

Figure 48.

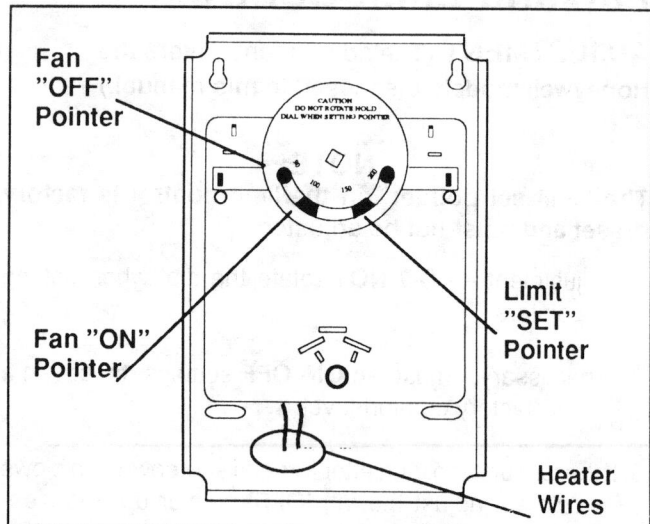

Fan "OFF" Pointer

Fan "ON" Pointer

Limit "SET" Pointer

Heater Wires

Figure 49.

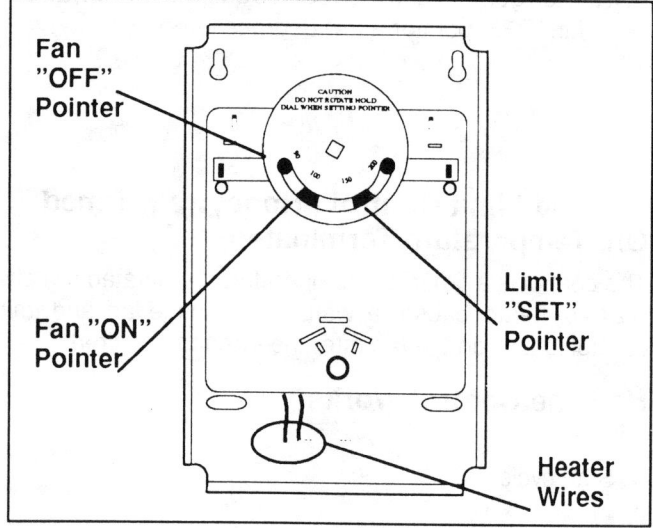

Fan "OFF" Pointer

Fan "ON" Pointer

Limit "SET" Pointer

Heater Wires

Figure 50.

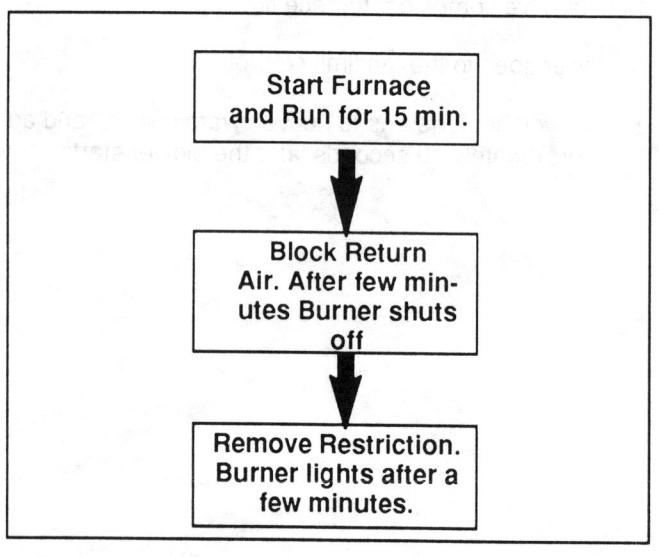

Start Furnace and Run for 15 min.

Block Return Air. After few minutes Burner shuts off

Remove Restriction. Burner lights after a few minutes.

6. Cycle Terminates.

7. Thermostat shuts off power to gas valve.

8. Furnace no longer fired, no power to heater.

9. Blower continues to run until "helix" of fan and limit control cools to preset temperature for fan cycle to end thermally for lack of heat.

BURNED OUT HEATER ELEMENT

1. Use an Ohmmeter to check the heater element using 1 x 10,000 scale.

2. Disconnect the heater wires and connect ohmmeter across wires going to heater in control.

3. Infinity reading — bad heater element, replace fan and limit control.

LIMIT CONTROL CHECK

NOTE: The 'limit set pointer' on the limit control is factory preset and must not be adjusted.

1. Check limit control function after 15 minutes of operation by blocking the return grille(s).

2. After several minutes the main burners should go **OFF**. Blower will continue to run.

3. Remove air restrictions and main burner will relight after a cool down period of a few minutes.

4. Fan and limit controls are preset at the factory.

5. Adjust the thermostat setting below room temperature.

6. Main burners should go off.

Figure 51.

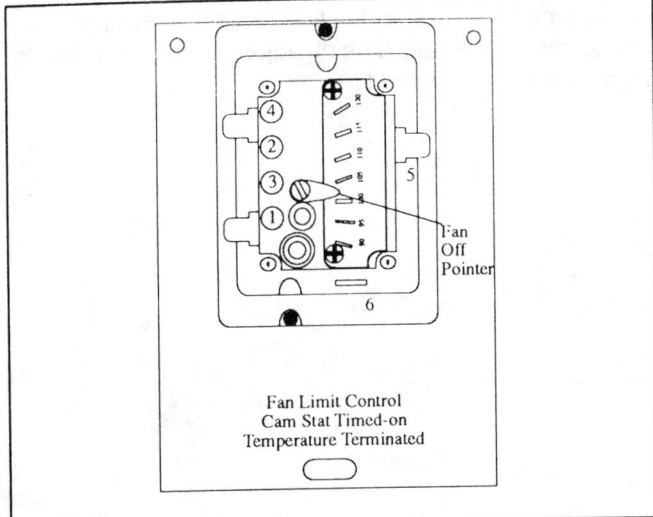

Fan Limit Control
Cam Stat Timed-on
Temperature Terminated

Fan and Limit Control (Cam–Stat Timed–on Temperature Terminated

The control is a timed–on temperature terminated switch. A 24VAC heater causes a 'warp switch' in the fan and limit control to turn on the fan after the furnace has fired.

The fan 'ON' timing is not adjustable. The fan 'OFF' setting may be adjusted with a small screwdriver. Rotating the pointer to a lower number will cause the fan to run longer.

Figure 52.

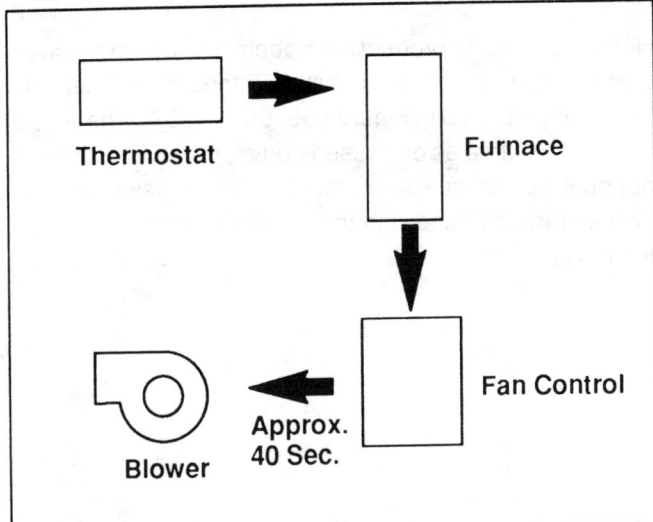

Thermostat

Furnace

Fan Control

Approx. 40 Sec.

Blower

HOW CONTROL WORKS

1. Start cycle.

2. Thermostat calls for heat.

3. Furnace fires.

4. Power goes to fan and limit control.

5. 24 volt heater starts to heat 'warp switch' and approximately 40 (nominal) seconds later blower starts.

6. Thermostat shuts off power to gas valve.

7. Furnace no longer fired — no power to heater.

8. Blower continues to run until fan/limit control cools to preset temperature for fan cycle to end thermally for lack of heat.

Figure 53.

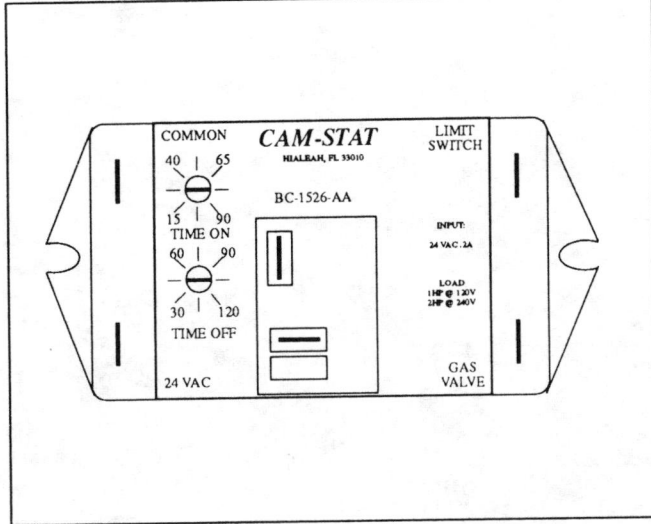

Electronic Fan Control (Timed–on, Timed Off)

How Control Works

1. Start cycle.

2. Thermostat calls for heat.

3. Timing on begins when gas valve is energized.

4. Fan comes on at end of timing on setting.

5. Fan runs until gas valve is deenergized at which time off timing begins. Fan continues to run until end of time off setting.

Figure 54.

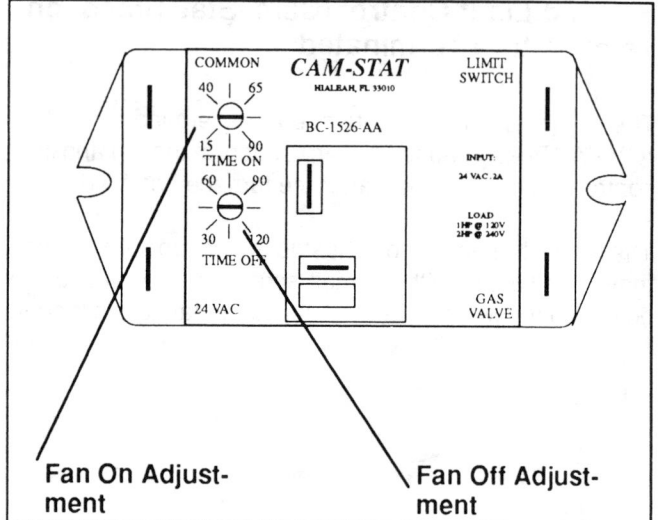

Fan On Adjustment

Fan Off Adjustment

ADJUST BLOWER ON/OFF

Rotate time on adjustment until screw slot lines up with the desired time on. Rotate time off adjustment until screw slot lines up with the desired time off.

Figure 55.

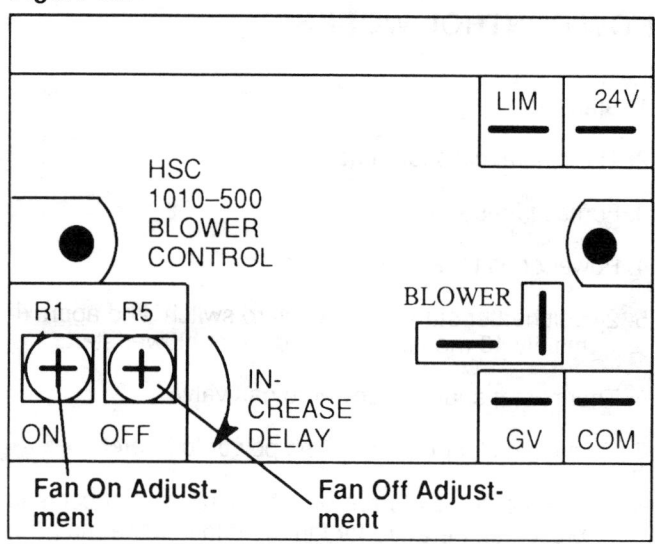

Fan On Adjustment

Fan Off Adjustment

NOTE: On recent production models, the unit may have a Cam–Stat, Heatcraft, or Hamilton Standard fan timer. The timers are pictured in Figures 55, 56, and 57. The adjustment is the same as discussed above, however times may not be indicated on some models. There may only be an indicator (arrow) showing the direction to turn to increase the delay.

Figure 56.

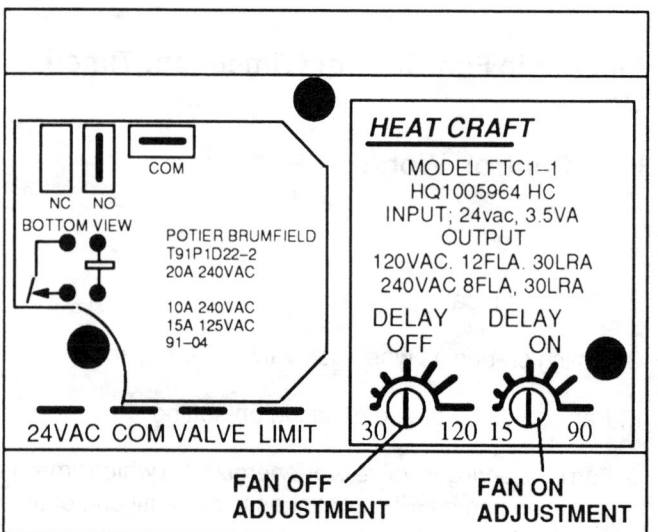

Figure 57.

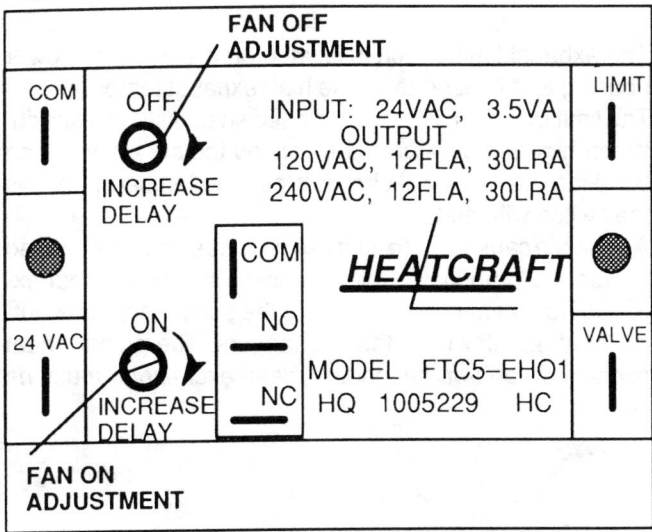

Figure 58.

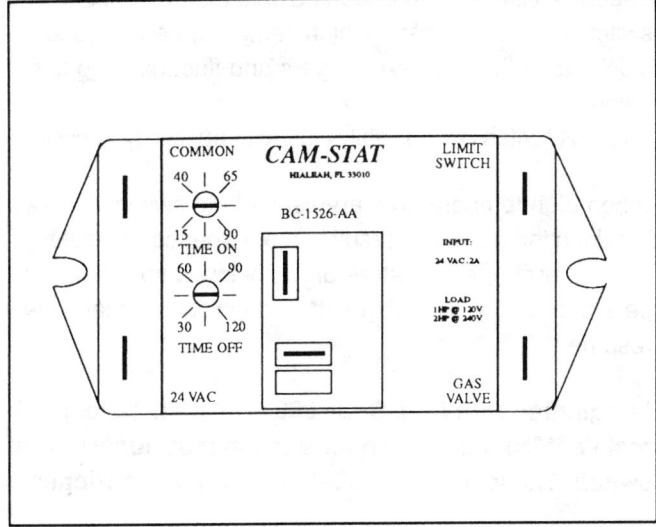

Checking for Defective Control

Any time there is power to the furnace, 24VAC should be present across 'COM' and 24VAC terminals and 'COM' and 'LIMIT' terminals. Once the control module has energized the gas valve, 24VAC should be present across the 'COM' and 'gas valve'. A quick check can be made as follows:

1. Connect a jumper across '24VAC' and the gas valve terminals. Blower should start after the selected time delay has expired.

2. Remove the jumper and the blower will turn off after the selected time delay has expired.

3. Disconnecting the wire at the limit terminal at any time should cause the blower to start immediately.

Limit Switch

A separate limit switch is used with the above control and is not field adjustable.

Checking Limit Control

Same as other controls.

Figure 59.

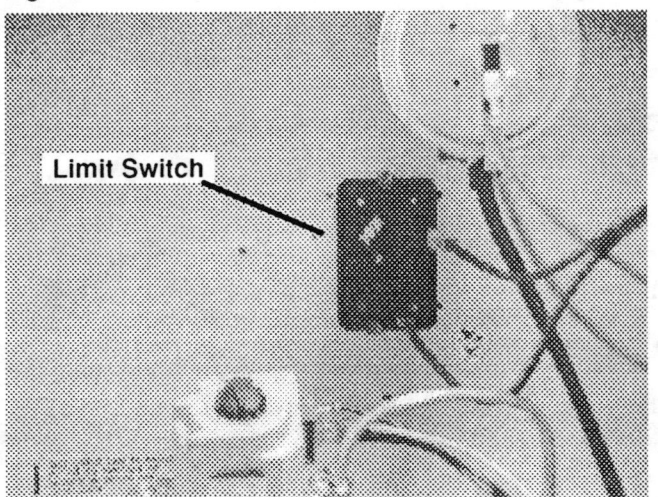

Figure 60.

Exhaust Limit Switch

The exhaust limit switch is located on the exhaust blower housing and is used to sense high exhaust temperature. The limit switch is a normally closed snap disc type switch. When high temperatures are sensed the switch will open breaking the electrical circuit. When the temperature falls the switch will reset.

A possible cause for the limit switch to open may be caused by soot build up inside the primary or secondary heat exchangers, or lint build up outside the secondary, preventing heat transfer from the exchangers. The primary heat exchanger and/or secondary heat exchanger must be cleaned.

Figure 61.

Rollout Limit Switch

The rollout limit switch is located on the burner shield. The switch is used to sense high temperatures caused by blockage in the heat exchangers and flue, causing flame rollout.

The limit switch is a normally closed snap disc type switch.

When high temperatures are sensed the switch will open breaking the electrical circuit to the burner control system. Whenever the temperature drops below a specified temperature, the switch will reset* and ignition sequence will resume.

* Beginning with Date Code L8944 all units (Except Direct Vent Models) were built with a manual reset rollout switch. These must be reset manually when tripped.

Figure 62.

White Rodgers 36E36 Gas Valve

Gas Valves
White–Rodgers 36E36

The manual gas knob of this valve differs from those of most other valves in that it does not have to be pushed in to be turned. The knob can also be turned a full 360 degrees in either direction. There are detents at the 'ON' and 'OFF' positions allowing it to remain at either position.

Figure 63.

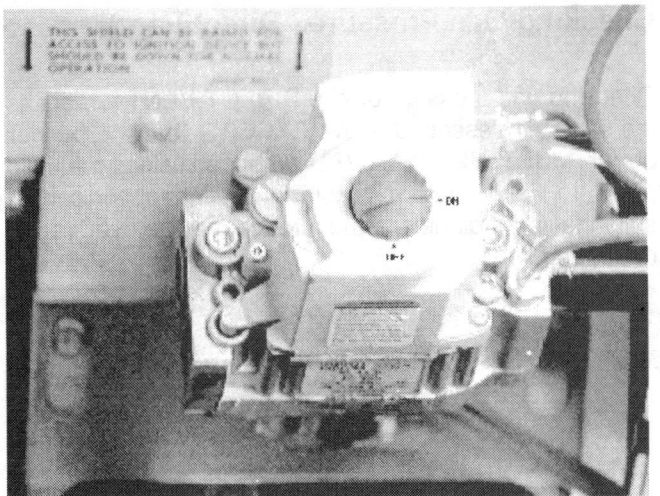

Honeywell VR8204A Gas Valve

Honeywell VR8204A Gas Valves

The VR8204A is a snap opening gas valve.

Figure 64.

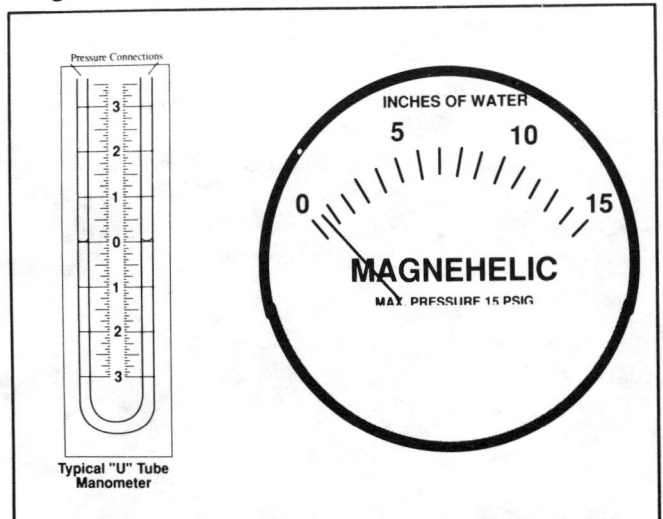

Checking Gas Pressure (Natural Gas)

Gas pressure is checked using a 'U' tube manometer or a magnehelic gauge. The instrument is connected to the inlet pressure tap on the gas valve to check gas line pressure and the outlet pressure tap to set manifold pressure. Incoming gas pressure to the gas valve, with all other gas appliances fired, is a minimum of 4.5 in. w.c. and a maximum of 11.0 in. w.c. The ideal input pressure to the gas valve should be 7.0 in. w.c. Outlet, or manifold, pressure should be adjusted to 3.5" w.c..

Figure 65.

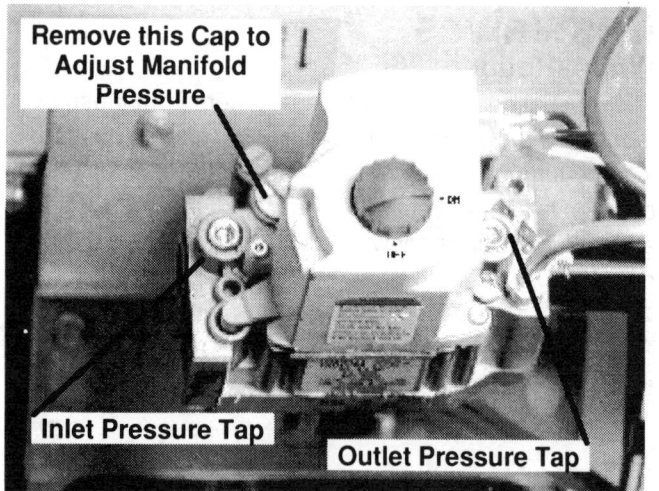

Remove this Cap to Adjust Manifold Pressure

Inlet Pressure Tap

Outlet Pressure Tap

Checking Gas Pressure To Valve (LP Gas)

Gas pressure is checked using a 'U' tube manometer or and magnehelic gauge. The instrument is connected to the inlet pressure tap on the gas valve to check gas line pressure and the outlet pressure tap to set manifold pressure. Incoming gas pressure to the gas valve, with all other gas appliances fired, is a minimum of 11.0" in. w.c. and a maximum of 14.0 in. w.c. The ideal input pressure to gas valve should be 12.0 in. w.c. Outlet, or manifold, pressure should be adjusted to 10.0" w.c..

Figure 66.

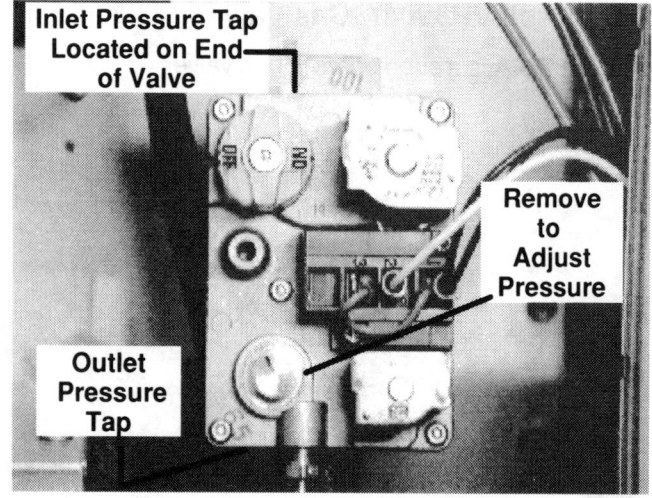

Inlet Pressure Tap Located on End of Valve

Remove to Adjust Pressure

Outlet Pressure Tap

Checking Gas Pressure (Direct Vent)

Checking gas pressure on the DIRECT VENT furnace is the same as described above **EXCEPT that the burner cover MUST BE REMOVED** when adjusting manifold pressure. Failure to have the burner cover removed can result in false readings and the furnace may be overfired or underfired.

Figure 67.

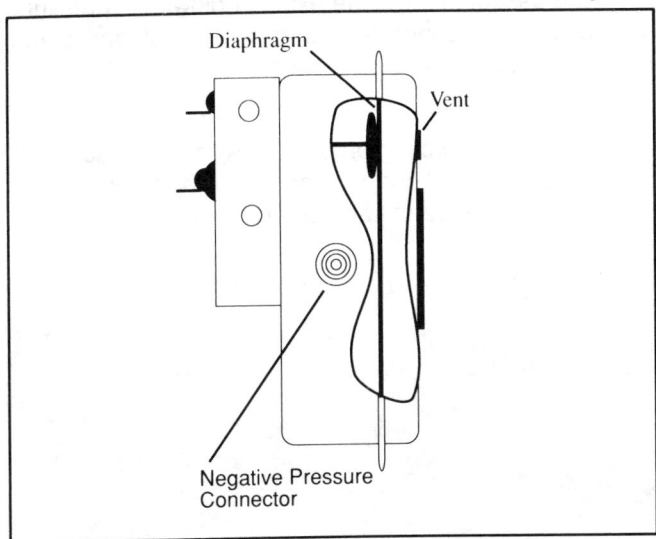

Figure 68.

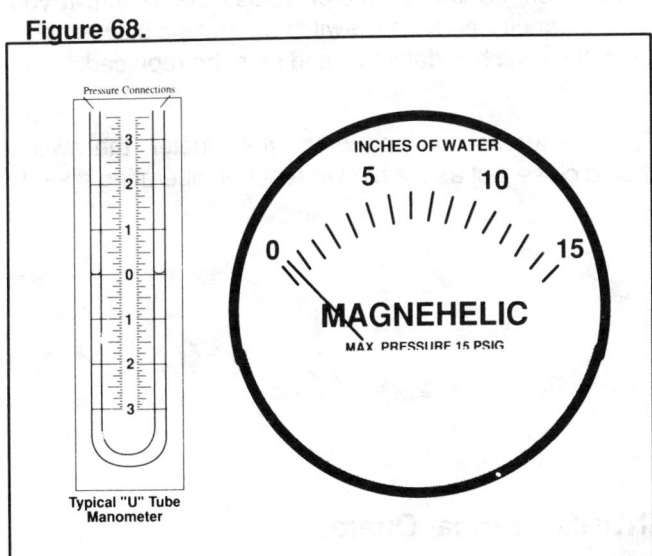

Figure 69.

Chart #1: Exhaust Blower Pressure Switch			
Switch Part #	Set in In. "WC"	Color	Units
			Units w/ date code H540 & earlier
611921	−2.5 ± .25	—	
1000743	−4.12 ± .175	Yellow	NDG/LK050
1000744	−3.25 ± .20	Orange	NDG/LK075
1000745	−2.36 ± .20	Red	NDG/LK100
1000746	−1.91 ± .15	Blue	NDG/LK125
1000747	−4.175 ± .175	Green	NUG/LK050
1000748	−3.2 ± .20	Gray	NUD/LK075
1000749	−1.95 ± .15	Purple	NUG/LK125
1001181	−2.3 ± .20	White	NUG/LK100

Pressure Switches

Exhaust Blower Pressure Switch

"NUGK" Normal Operation

The exhaust blower pressure switch is in the negative side of the exhaust blower system. The switch is used to sense a blockage in the exhaust system. On a call for heat from the thermostat, the exhaust blower starts to operate. (On one side of the blower, a negative (−) pressure is created and on the other side a positive (+) pressure is created. Combustion (− pressure) air is drawn through the primary heat exchanger, through the tube to the secondary heat exchanger, through the second heat exchanger, through he exhaust blower (+ pressure) and out the flue.

If the exhaust system is clear, the negative (−) pressure is felt by the diaphragm in the pressure switch. The normally open contact of the switch closes, completing the electrical circuit though the switch and the ignition cycle will begin. If the ignition cycle begins at the same time as the exhaust blower begins to run the switch is stuck closed.

Trouble Shooting

Manually turn off the gas supply to the furnace. Check for negative pressure using a "U" Tube manometer or a magnehelic gauge, by disconnecting the tube from the switch and connecting to the test instrument.

Turn the furnace on. The exhaust blower should start operate. Read the pressure just as the blower reaches operating speed and compare it to the pressures listed in Chart #1.

If the pressure comes up to specifications. reconnect the tube to the pressure switch.

Figure 70.

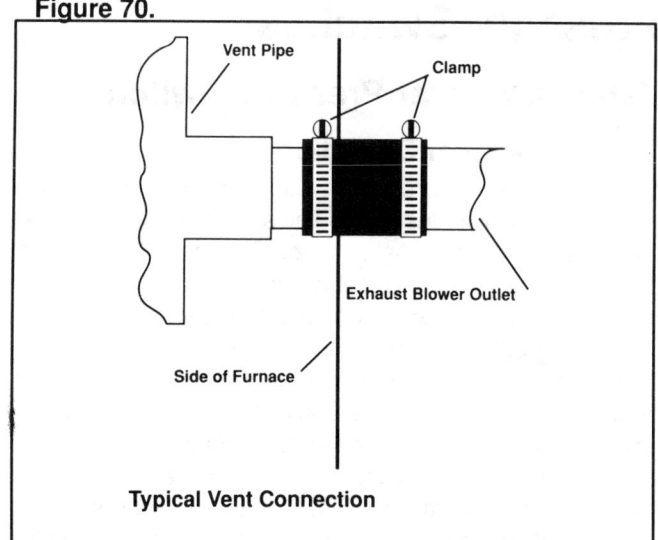

Typical Vent Connection

If pressure does not come up to specifications, remove the vent pipe assembly from the exhaust blower. If pressure now comes up to specifications you have problems in the vent system and it must be repaired or corrected.

If the blower is operating properly and pressure does not come up to specifications, you most likely have air leakage in the heat exchangers, transfer tube, or collector box.

Figure 71.

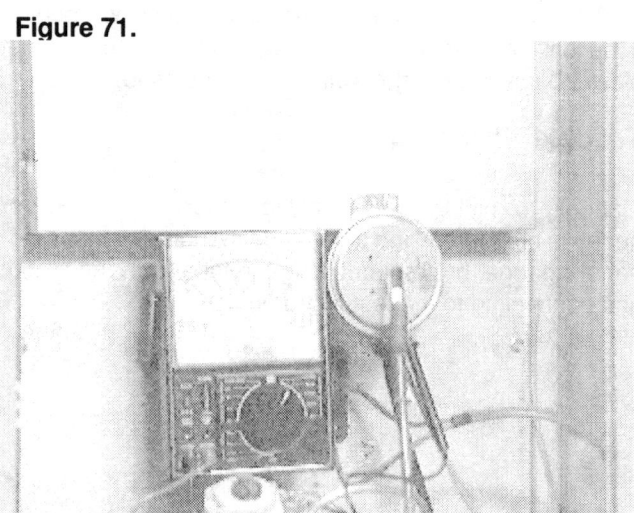

Turn off furnace and remove electrical leads from the switch. Connect an ohmmeter across the switch. If you have continuity across the switch without the blower operating, the switch is defective and must be replaced.

Turn the furnace on and watch the meter, the switch should close just as the blower reaches operating speed.

Figure 72.

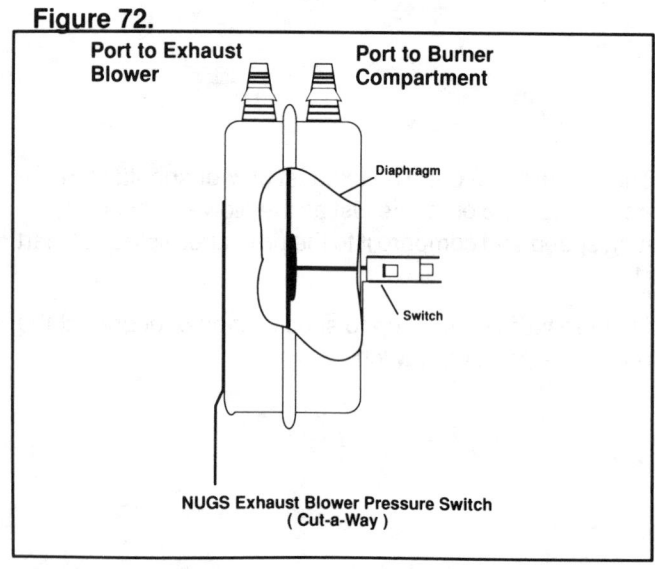

NUGS Exhaust Blower Pressure Switch
(Cut-a-Way)

"NUGS" Normal Operation

The exhaust blower pressure switch has two pressure taps. The switch senses the pressure differential between the negative side of the exhaust blower system and a lower negative pressure in the sealed burner compartment. The function of the exhaust blower pressure switch is to sense pressure differentials by means of a flexible diaphragm inside the pressure switch. As this diaphragm flexes in response to pressure, the movement is transmitted mechanically to a set of electrical contacts. If the switch fails to sense the proper pressure differential, it will shut down the system, preventing sooting of the primary and secondary heat exchangers.

Figure 73.

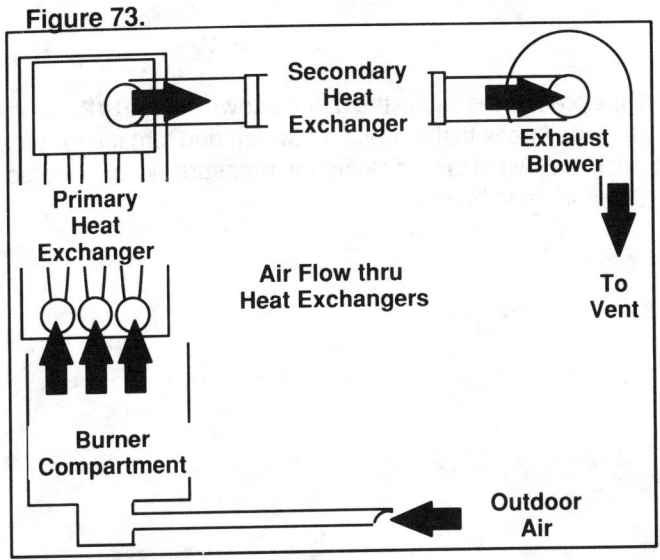

On a call for heat from the thermostat, the exhaust blower starts to operate. A negative pressure is created at the inlet side of the blower and this negative pressure is sensed by one side of the pressure switch. At the same time outside air is drawn through the combustion air inlet into the sealed burner compartment, creating a negative pressure of a lower value in the burner compartment. The pressure switch senses the difference between these two negative pressures. If the differential is within the range of the switch, the contacts on the pressure switch will close and complete the circuit through the switch.

Trouble Shooting

To troubleshoot the exhaust blower pressure switch, it must be determined that the pressure sensing part of the switch is seeing the proper pressures and that the electrical part of the switch is opening and closing. The tools needed to test the operation of the switch are the following:

To read pressures, either a "U" Tube or a magnehelic gauge.

An ohmmeter can be used to check continuity through the electrical part of the switch.

Turn off the gas supply at the shutoff valve in the gas supply line to the furnace.

Remove the hoses from the pressure switch and connect both sides of a "U" Tube manometer or magnehelic gauge manometer to the tubes.

Figure 74.

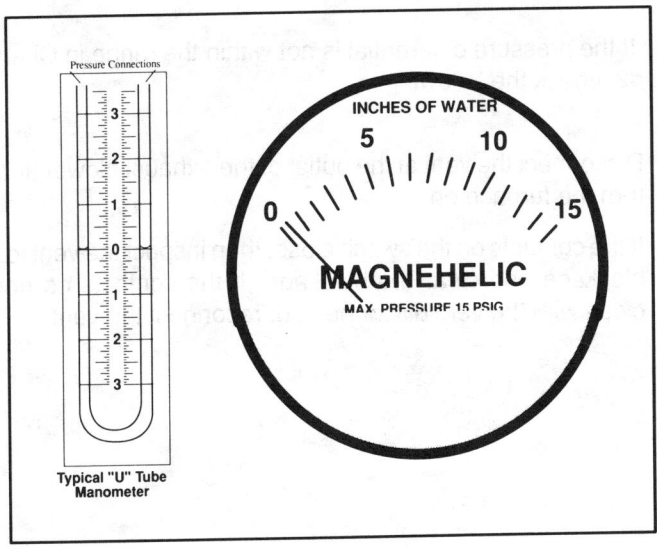

Figure 75.

Disconnect the wires to the pressure switch and connect an Ohmmeter across the terminals.

With the furnace off there should be no continuity across the terminals. If there is, the switch is defective and must be replaced. Turn the furnace on and read the pressure differential with the 'U' Tube or the magnehelic gauge.

Figure 76.

Switch Part #	Set in In. "WC"	Color	Units	Vent Size
Chart #2: Exhaust Blower Pressure Switch				
1006406	2.90 ± .10	Blue	NUGS100	3"
1006405	3.40 ± .10	Green	NUGS075	3"
1005605	2.60 ± .10	Yellow	NUGS125	3"
1005604	2.90 ± .10	Blue	NUGS100	2"
1005603	3.40 ± .10	Green	NUGS075	2"
1005602	3.60 ± .10	Red	NUGS050	2"
*1006410	2.35 ± .10	Yellow	NUGS125	3"
*1006409	2.60 ± .10	Yellow	NUGS100	2"or 3"
*1006408	3.05 ± .10	Yellow	NUGS075	2"or 3"
*1006407	3.20 ± .10	Yellow	NUGS050	2"

*High Altitude Switch, Available from Service Parts Only

If the pressure is within the range shown in Chart #2, reconnect the hoses to the pressure switch and turn the furnace on. If the switch fails to close, the pressure switch is defective and must be replaced.

Figure 77.

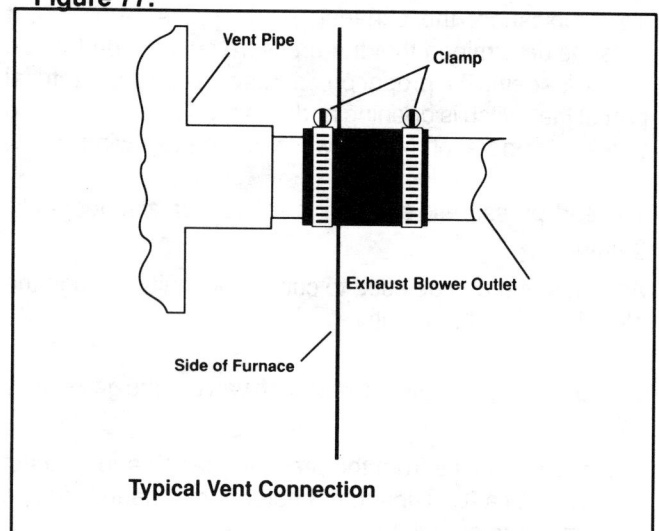

Typical Vent Connection

If the pressure differential is not within the range in Chart #2, check the following:

Disconnect the vent at the outlet of the exhaust blower and turn the furnace on.

If the contacts on the switch close, then inspect the vent for blockage and clear the blockage. If the contacts do not close with the vent disconnected, reconnect the vent.

Figure 78.

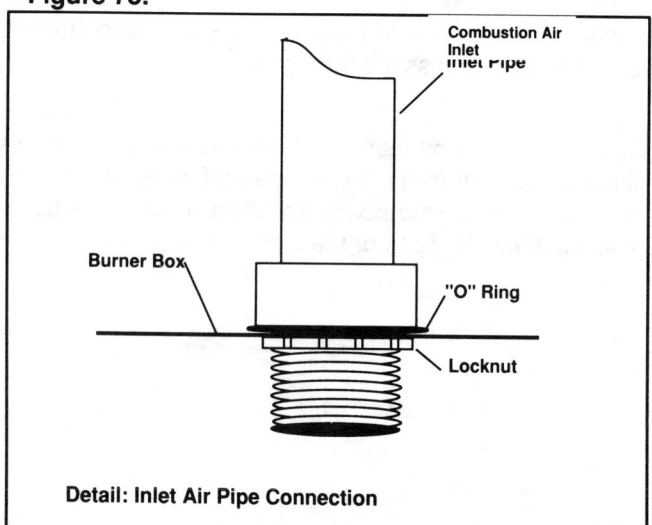

Detail: Inlet Air Pipe Connection

Remove the burner compartment cover and disconnect the inlet air pipe by removing the locknut securing it to the burner box

Replace the burner compartment cover and turn the furnace on. If the contacts close on the switch, inspect the inlet air system for blockage and clear the blockage. Reconnect all lines and replace the burner compartment cover.

Restore the gas supply to the furnace and place the furnace back in service.

Figure 79.

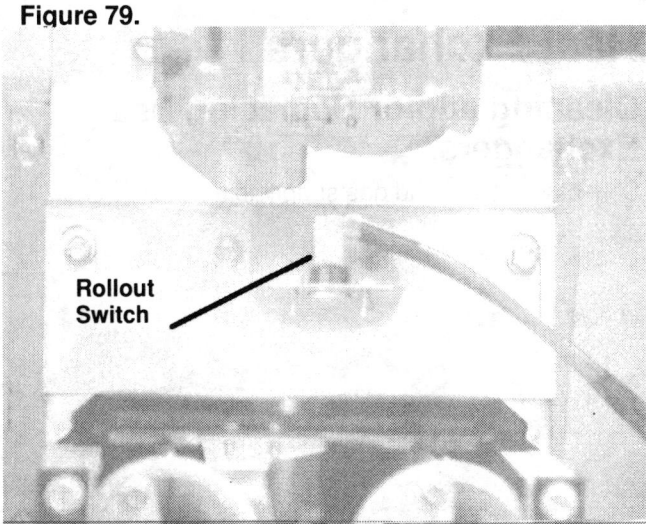

Rollout
Switch

NUGK & NUGS

Since the exhaust blower switch can only read a differential between the suction side of the exhaust blower and the burner compartment it may not detect an internal blockage in the heat exchanger. The roll–out switch provides this safety function. See limit switch section for operation and check out.

Figure 80.

Heat Exchangers
Cleaning and/or Replacing Heat Exchangers.
Shut off electrical and gas supply to furnace.

Figure 81.

PROPER FILTER MAINTENANCE
IS
THE KEY

Cleaning Secondary Heat Exchanger (Current Production) – Upflow Models

The exterior surface of the secondary heat exchanger may require cleaning if the furnace filters are not maintained properly. The inlet side surface of the secondary heat exchanger can be cleaned without removing it from the furnace.

Figure 82.

Remove two screws (Hidden)

Remove two screws from the burner side of the blower deck that secure the control box and remove the control box from the blower compartment.

Figure 83.

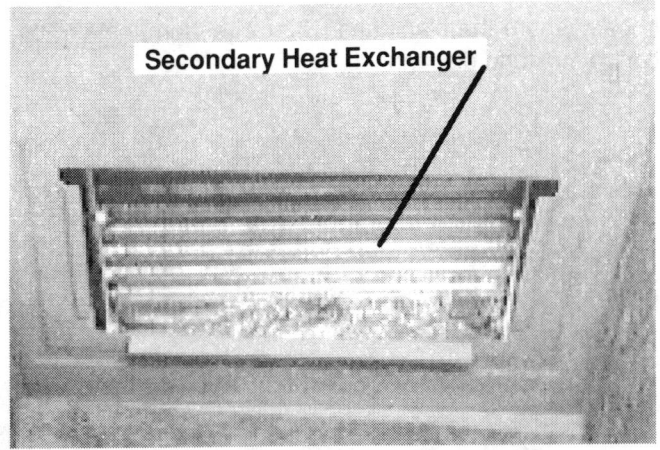

Secondary Heat Exchanger

Remove the two screws securing the blower and remove the blower from the furnace.

Once the blower has been removed the inlet side of the secondary heat exchanger can be accessed from the inside of the blower compartment for cleaning.

Figure 84.

**Clean with Soft Bristle Brush
&
Vacuum Cleaner**

Use a soft bristle brush and a vacuum cleaner to clean the surface. Do not use a wire brush or harsh chemicals to clean the surface as you may damage the fins. Inspect the blower wheel and, if necessary, clean before re–installing. Use caution when cleaning the blower wheel to keep from dislodging balance weights from the wheel.

If the secondary heat exchanger is being replaced, continue as follows:

Figure 85.

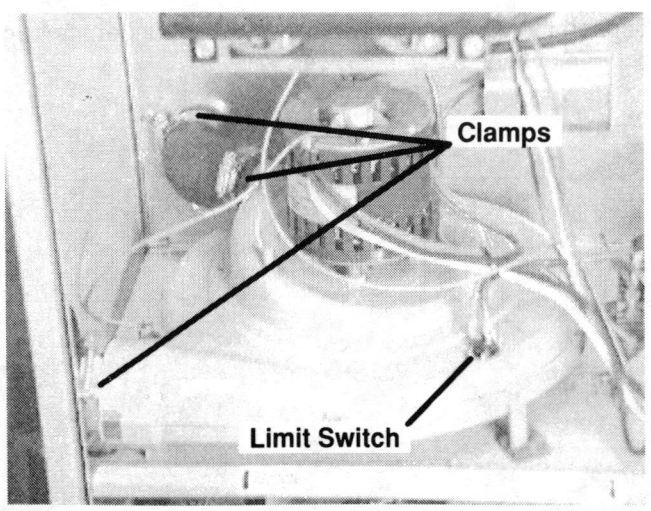

Clamps

Limit Switch

Secondary Heat Exchanger Removal/Replacement – Upflow Models

Loosen the clamps on the exhaust blower inlet and outlet. Disconnect the wires to the exhaust blower limit switch. Remove two screws securing exhaust blower to blower deck. **Do not remove exhaust blower at this time.**

Figure 86.

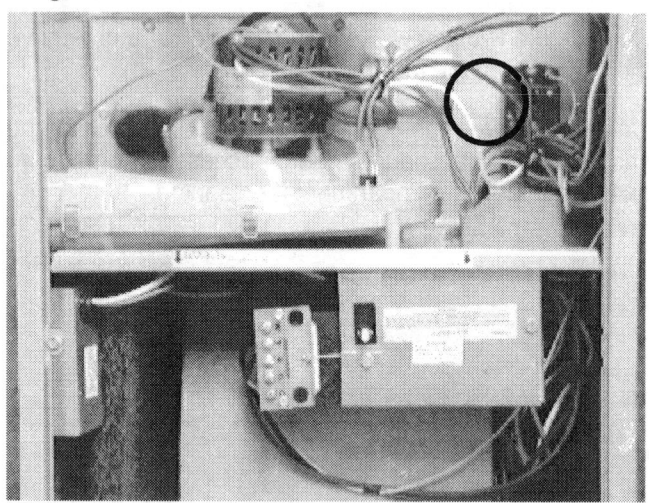

Locate where the black and white wires terminate in the control box and disconnect.

Figure 87.

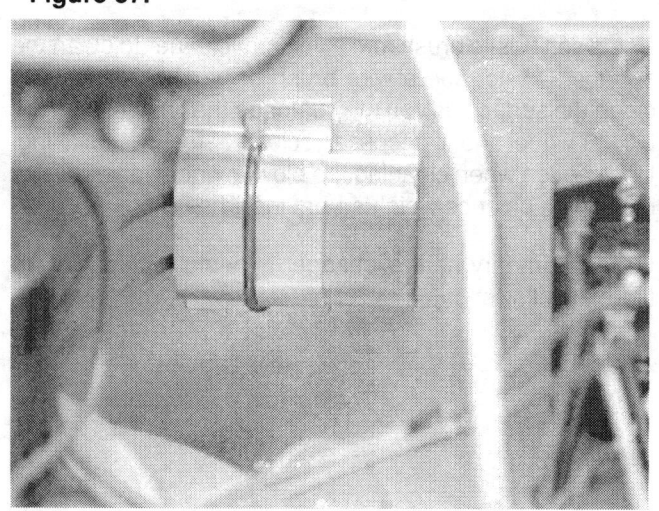

Remove the exhaust blower capacitor and strap from their mounting on the vestibule panel. (On earlier models the capacitor is mounted on the exhaust blower housing and do not have to be removed.)

Figure 88.

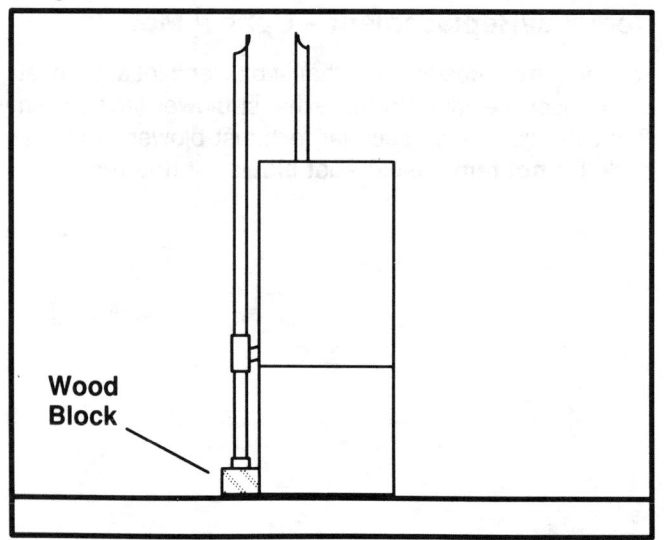

Support the vent pipe with a wooden block under the trap and remove the exhaust blower.

Figure 89.

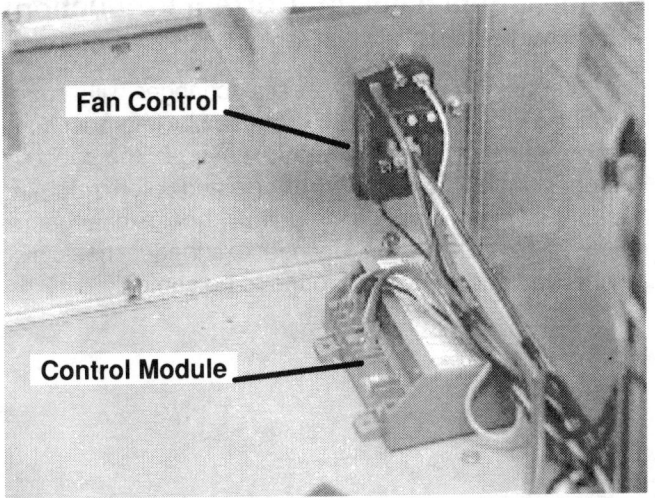

Fan Control

Control Module

Remove the electronic fan control, if equipped, and the control module.

Figure 90.

Secondary Heat Exchanger Panel

Remove the secondary heat exchanger panel in the vestibule.

Figure 91.

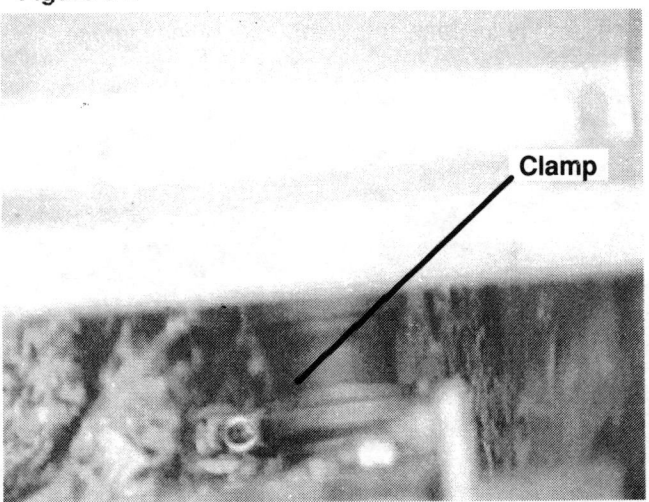

Clamp

Remove the muffler clamp securing the transfer tube to the secondary heat exchanger. Raise the transfer tube up and disengage from the secondary heat exchanger. The heat exchanger can now be removed from the furnace.

If the interior of the secondary heat exchanger is plugged with soot it should be considered defective and be replaced.

Figure 92.

Causes of Improper Combustion

Lack of Combustion Air

Improperly Adjusted
Burner Air Shutters

Improper
Inlet or Outlet
Gas Pressure

Dirty or Rusted Burners

Primary Heat Exchanger Removal/Replacement (Current Production) –Upflow Models

The interior of the primary heat exchanger will require cleaning as the result of sooting caused by improper combustion. After cleaning the heat exchanger, the cause of improper combustion should be corrected before returning the furnace to operation. If the primary heat exchanger requires cleaning the secondary heat exchanger should also be checked and replaced if found to be sooted.

Figure 93.

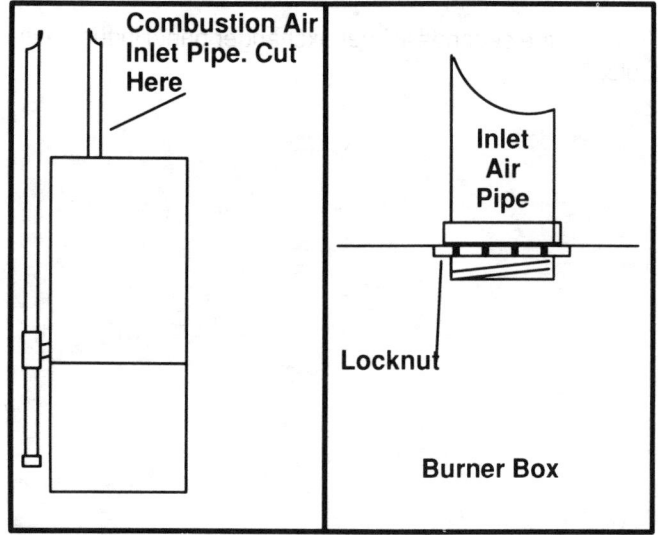

Direct Vent Models Only

Support the inlet air line. Cut the inlet air pipe outside the furnace cabinet. Remove the remaining inlet air section inside the furnace cabinet by removing the locknut inside the top of the burner box.

Figure 94.

Remove the tie panel at the top front of the furnace.

Figure 95.

Disconnect wires and tubing from the pressure switch. Remove the collector box cover.

Figure 96.

Remove the collector box from the flue outlet at the top of heat exchanger. Exercise care when removing collector box so as not to damage gasket. Should gasket be damaged in any way, a new gasket must be used when re-installing the collector box.

Figure 97.

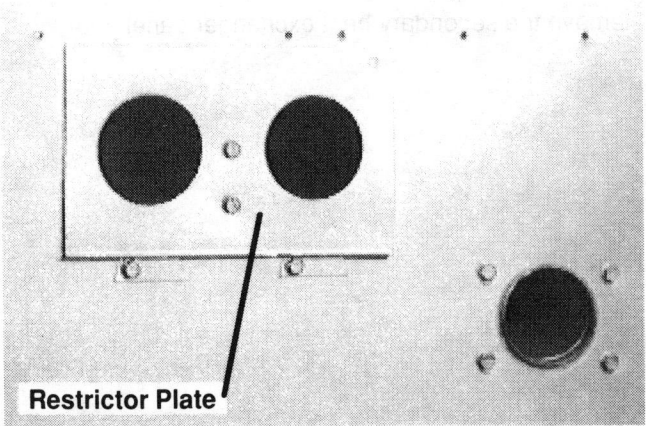

Remove flue restrictor plate.

Figure 98.

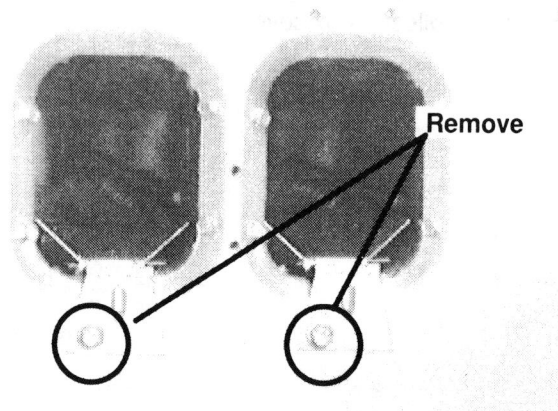

Remove the screws securing the flue baffles to the heat exchanger and remove the flue baffles.

Figure 99.

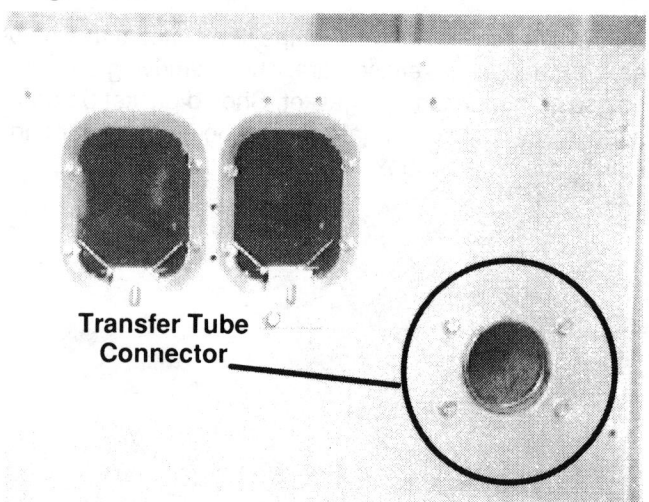

Remove the four screws securing the transfer tube connector to the heat exchanger front panel.

Figure 100.

Remove the secondary heat exchanger panel.

Figure 101.

Wood Blocks

Place two wooden blocks between the bottom of the primary heat exchanger and secondary heat exchanger to prevent the primary heat exchanger from dropping down and damaging the secondary heat exchanger.

Figure 102.

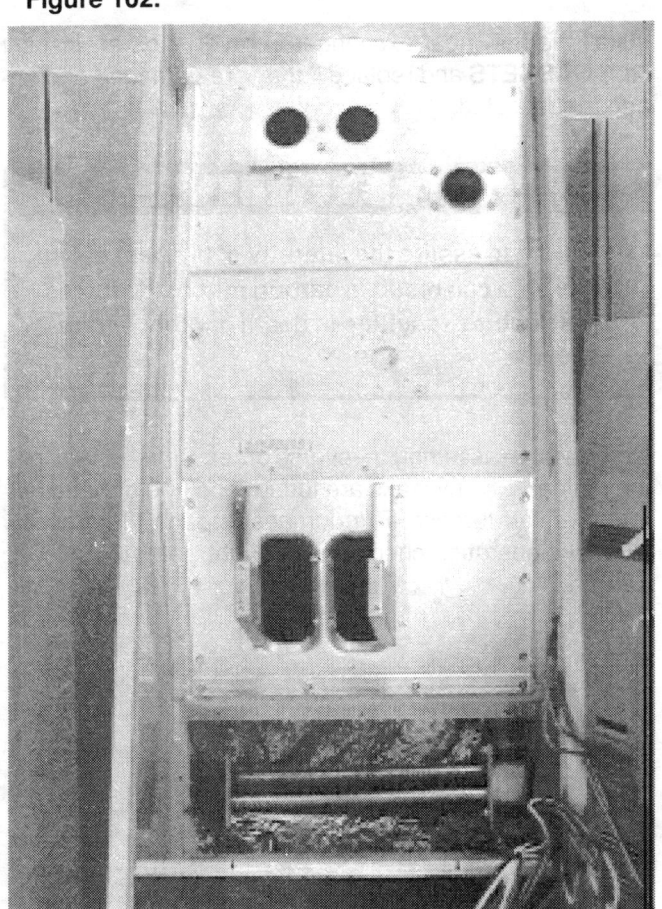

Remove the screws around the outer edge of the heat exchanger front panel and pull the heat exchanger out of the furnace cabinet.

Figure 103.

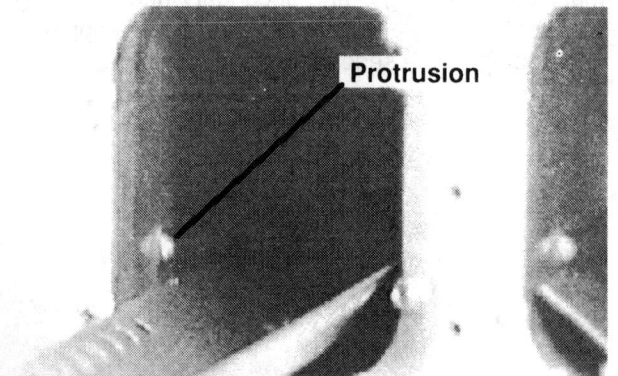

Install flue baffles in the new heat exchanger. Be sure that the top edge of the flue baffle is under the protrusions on either side of flue opening.

Tip: When installing flue baffle. Insert baffle into flue opening about 1". Get one edge of baffle under one of the protrusion on one side. Rotate baffle until the other edge snaps into place under the protrusion on the other side. Slide the baffle into place and secure with a screw.

Figure 104.

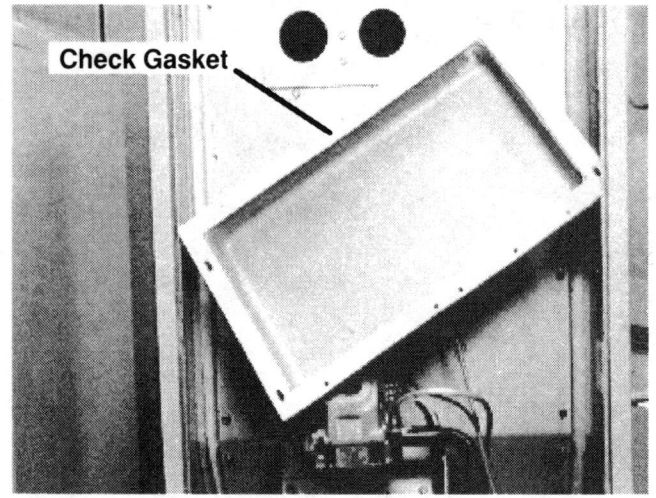

Install the new heat exchanger in reverse order. Inspect **ALL GASKETS** and replace if they are damaged or damage is suspected.

WARNING

Failure to assure the integrity of the gaskets in this area can result in carbon monoxide fumes in structure, resulting in death or other serious injury.

Complete re–assembly in reverse order. Check all gas piping for leaks. If any leaks are found repair before attempting to restart furnace. Run furnace through three or four cycles. If operation checks out all right, return furnace to service.

Figure 105.

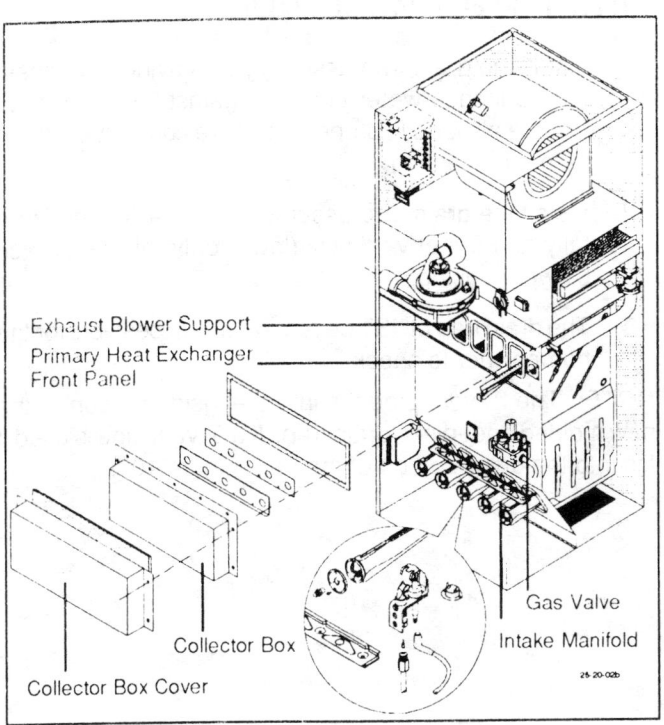

Exhaust Blower Support

Primary Heat Exchanger
Front Panel

Collector Box

Collector Box Cover

Gas Valve

Intake Manifold

Primary Heat Exchanger Removal/Replacement – Counterflow Models

1. Remove the exhaust blower support.

2. Remove the collector box cover and collector box.

3. Remove the screws to the transfer tube adapter.

4. Remove the control module.

5. Remove the gas valve, intake manifold, and any piping that will prevent primary heat exchangers from being removed.

6. Remove the burners and crosslighter from the burner ports.

7. Remove the burner shield and manifold support.

8. Remove all screws from the heat exchangers front panel. DO NOT REMOVE THE SCREWS FROM THE BURNER OR FLUE PORTS.

9. Remove the heat exchanger assembly.

10. Clean and inspect the heat exchanger using the procedures outlined for the Upflow Models.

11. Replace all parts and assemblies in reverse order as removed.

Figure 106.

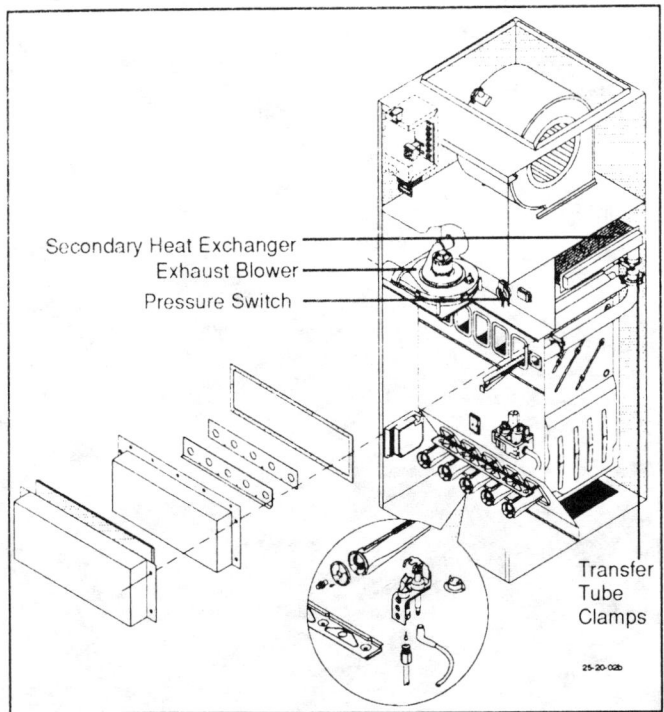

Secondary Heat Exchanger

Exhaust Blower

Pressure Switch

Transfer
Tube
Clamps

Secondary Heat Exchanger Removal/Replacement – Counterflow Models

1. Remove the exhaust blower.

2. Remove the pressure switch.

3. Remove the secondary heat exchanger panel.

4. Use a thin wall, deep well socket to remove the transfer tube clamps around the 'O' ring.

5. Remove the secondary heat exchanger by pulling the heat exchanger from the unit.

6. Clean and inspect the heat exchanger using the procedures outlined for the Upflow Models.

7. Replace all parts and assemblies in reverse order as removed.

Figure 107.

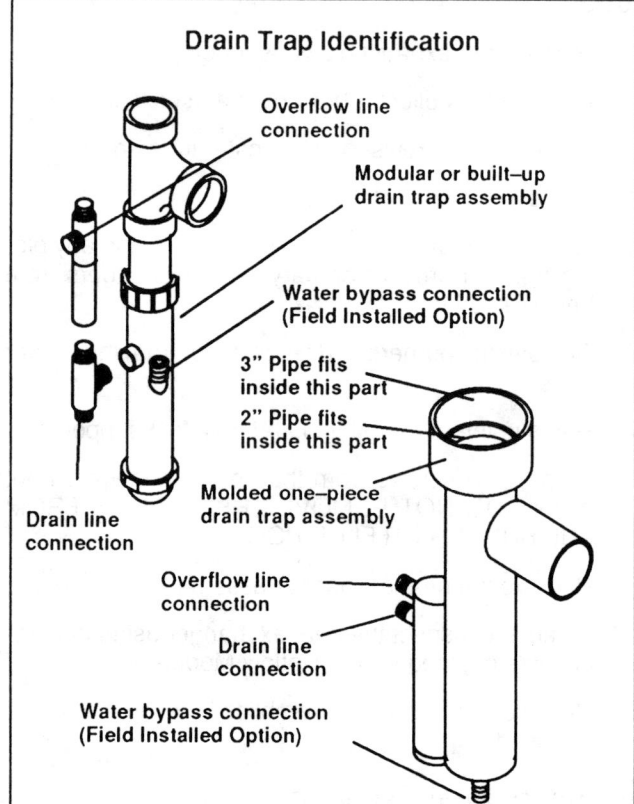

Drain Trap Identification

Overflow line connection

Modular or built-up drain trap assembly

Water bypass connection (Field Installed Option)

3" Pipe fits inside this part

2" Pipe fits inside this part

Molded one-piece drain trap assembly

Drain line connection

Overflow line connection

Drain line connection

Water bypass connection (Field Installed Option)

Drain Trap Assembly

1. Install the drain trap assembly to provide the necessary 5 inches water column against vent pressure. Ensure all parts fit properly and are correctly oriented before cementing.

2. Install the drain trap assembly within 4 feet horizontally and 5 feet vertically (lower only) of the furnace blower housing.

3. The drain trap MUST be reasonably accessible for the homeowner to check.

NOTE: The 2" vent pipe fits into the inside portion of the new molded one-piece drain trap. If a 3" vent pipe is used it fits into the outer portion of the trap, Figure 107.

THIS PAGE LEFT BLANK INTENTIONALLY

Specifications: Counterflow Natural Gas Models

MODEL/SERIES	NDGK040		NDGK050		NDGK075		NDGK100		NDGK125	
	STD	ALT	STD	ALT	STD	ALT	STD	ALT	STD	ALT
Input Rating (BTUH)	50,000	40,000	50,000	40,000	75,000	60,000	100,000	80,000	125,000	100,000
Output (BTUH)	47,000	37,500	47,000	37,500	69,000	–	89,500	–	109,500	–
Max. Ext. Static Press	.5		.5		.5		.5		.5	
Temperature Rise	20° – 50° F		20° – 50° F		40° – 70° F		50° – 80° F		35° – 65° F	
Volts/AMP	115/8.0		115/8.0		115/8.0		115/10.8		115/11.1	
Transformer Size (VA)	40		40		40		40		40	
Anticipator Setting	.15		.15		.15		.15		.15	
Limit Setting, MAX.	130		130		*200		**200		170	
Fan Switch Setting OFF	90		90		90		90		90	
***Gas Valve MFG/Type	WR36E		WR36E		WR36E		WR36E		WR36E	
Regulation Type	SNAP		SNAP		SNAP		SNAP		SNAP	
Manifold Pressure	3.5		3.5		3.5		3.5		3.5	
Orifice Sizes (Req'd)	#42(2) #45		#42(2) #45		#42(3) #44		#42(4) #44		#42(5) #44	
****Ignition Type (Hot Surface)	HSI/50F47		HSI/50F47		HSI/50F47		HSI/50F47		HSI/50F47	
Lock–Out	After 3 tries		After 3 tries		After 3 tries		After 3 tries		After 3 tries	
Filter Sq. In. HT/Cool	246/346		246/346		246/346		288/461		432/576	
Auxiliary Limit	120°F		120°F		130°F		130°F		150°F	
Std. Pressure Switch (Open)	–4.0 in.		–4.0 in.		–3.2 in		–2.2 in		–1.8 in	
Hi Alt. Press. Sw. (Open)	–3.82 in		–3.82 in		–2.8 in		–1.95 in		–1.65 in	

* NDGK075DF05 and NDGK075KF05 = 250
 NDGK075DF06, NDGK075KF06, NDGK075DF07, NDGK075KF07 = 170

** NDGK100DG05 and NDGK100KG05 = 250

*** Some models have a MH/VR8440P with STEP type regulation and a manifold pressure of 1.2/3.5.

**** Some models have spark to pilot ignitions, Model S86F or S8600M, with a spark gap of 1/8 in.

Specifications: Counterflow LP Gas Models

MODEL/SERIES	NDLK050		NDLK075		NDLK100		NDLK125	
	STD	ALT	STD	ALT	STD	ALT	STD	ALT
Input Rating (BTUH)	50,000	40,000	75,000	60,000	100,000	80,000	125,000	100,000
Output (BTUH)	47,000	37,500	69,000	–	89,500	–	109,500	–
Max. Ext. Static Press	.5		.5		.5		.5	
Temperature Rise	20° – 50° F		40° – 70° F		50° – 80° F		35° – 65° F	
Volts/AMP	115/8.0		115/8.0		115/10.8		115/11.1	
Transformer Size (VA)	40		40		40		40	
Anticipator Setting	.15		.15		.15		.15	
Limit Setting, MAX.	130		*200		**200		170	
Fan Switch Setting OFF	90		90		90		90	
***Gas Valve MFG/Type	WR36E36		WR36E36		WR36E36		WR36E36	
Regulation Type	SNAP		SNAP		SNAP		SNAP	
Manifold Pressure	10		10		10		10	
Orifice Sizes (Req'd)	#54(2) #55		#54(3) #55		#54(4) #55		#54(5) #55	
****Ignition Type (Hot Surface)	HSI/50F47		HSI/50F47		HSI/50F47		HSI/50F47	
Lock–Out	After 3 tries		After 3 tries		After 3 tries		After 3 tries	
Filter Sq. In. HT/Cool	246/346		246/346		288/461		432/576	
Auxiliary Limit	120°F		130°F		130°F		150°F	
Std. Pressure Switch (Open)	–4.0 in.		–3.2 in		–2.2 in		–1.8 in	
Hi Alt. Press. Sw. (Open)	–3.82 in		–2.8 in		–1.95 in		–1.65 in	

* NDLK075DF05 and NDLK075KF05 = 250.
 NDLK075DF06, NDLK075KF06, NDLK075DF07, NDLK075KF07 = 170.

** NDLK100DG05 and NDLK100KG05 = 250.

*** Some models have WR36E38 with STEP type regulation and a manifold pressure of 2.5/10.

**** Some models have a HSI/50E47 or spark to pilot Model S8600M with a spark gap of 1/8 inch.

Specifications: Upflow Natural Gas Models (NUGK)

1007516

MODEL/SERIES	NUGK040KF07		NUGK050MF07 NUGK050NF07		NUGK075DG08 NUGK075KG08		NUGK100DH11 NUGK100KH11		NUGK125DK08 NUGK125KK08	
	STD	ALT	STD	ALT	STD	ALT	STD	ALT	STD	ALT
Input Rating (BTUH)	40,000	50,000	50,000	40,000	75,000	60,000	100,000	80,000	125,000	100,000
Output (BTUH)	38,000	48,500	48,000	38,000	70,000	–	92,000	–	113,000	–
Temperature Rise	15° – 45° F		15° – 45° F		35° – 65° F		40° – 70° F		35° – 65° F	
Flue Size	2 in		2 in		2 in		2 in		2 in	
Elec.Volts/PH./F.L.A.	115/1/8.0		115/1/8.0		115/1/10.8		115/1/10.8		115/1/11.8	
Transformer Size (VA)	40		40		40		40		40	
Orifice Sizes (Req'd)	#45(2) #42		#42(2) #45		#42(3) #44		#42(4) #44		#42(5) #44	
Limit Setting	200		200		250		170		170	
Fan Setting Delay ON	15–90		15–90		15–90		15–90		15–90	
Fan Setting Delay OFF	30–120		30–120		30–120		30–120		30–120	
Exhaust Limit (Man.	120°F		120°F		130°F		130°F		150°F	
Thermal Sensor Reset)	300°F		300°F		300°F		300°F		300°F	
Gas Valve MFG/Type	WR36E		WR36E		WR36E		WR36E		WR36E	
Manifold Pressure	3.5		3.5		3.5		3.5		3.5	
Regulation Type	SNAP		SNAP		SNAP		SNAP		SNAP	
Ignition Type/Model	I.I.D./S8600M		I.I.D./S8600M		I.I.D./S8600M		I.I.D./S8600M		I.I.D.S8600M	
Pilot Orifice Size	.020		.020		.020		.020		.020	
Spark Gap	1/8 in.		1/8 in.		1/8 in		1/8 in		1/8 in	
Anticipator Setting	.55		.55		.55		.55		.55	
Cap. Rating MFD/Volts Combustion Air Blower	5/370		5/370		5/370		5/370		5/370	
Furnace Blower Rated Ext. Static Press	.10–.50		.10–.50		.12–.50		.15–50		.20–.50	

Specifications: Upflow LP Gas Models (NULK)

MODEL/SERIES	NULK050MF06		NULK075DG06		NULK100DH07		NULK125DK07	
	STD	ALT	STD	ALT	STD	ALT	STD	ALT
Input Rating (BTUH)	50,000	40,000	75,000	60,000	100,000	80,000	125,000	100,000
Output (BTUH)	48,000	38,000	70,000	–	92,000	–	113,000	–
Temperature Rise	15° – 45° F		35° – 65° F		40° – 70° F		35° – 65° F	
Flue Size	2 in		2 in		2 in		2 in	
Elec.Volts/PH./F.L.A.	115/1/8.0		115/1/10.8		115/1/10.8		115/1/11.1	
Transformer Size (VA)	40		40		40		40	
Orifice Sizes (Req'd)	#54(2) #55		#54(3) #55		#54(4) #55		#54(5) #55	
Limit Setting	200		250		170		170	
Fan Setting Delay ON	15–90		15–90		15–90		15–90	
Fan Setting Delay OFF	30–120		30–120		30–120		30–120	
Exhaust Limit (Man.	120°F		130°F		130°F		150°F	
Thermal Sensor Reset)	300°F		300°F		300°F		300°F	
Gas Valve MFG/Type	WR36E		WR36E		WR36E		WR36E	
Manifold Pressure	2.5/10		2.5/10		2.5/10		2.5/10	
Regulation Type	STEP		STEP		STEP		STEP	
Ignition Type/Model	I.I.D./S8600M		I.I.D./S8600M		I.I.D./S8600M		I.I.D.S8600M	
Pilot Orifice Size	.012		.012		.012		.012	
Spark Gap	1/8 in.		1/8 in		1/8 in		1/8 in	
Anticipator Setting	.55		.55		.55		.55	
Cap. Rating MFD/Volts Combustion Air Blower	5/370		5/370		5/370		5/370	
Furnace Blower Rated Ext. Static Press	.10–.50		.12–.50		.15–50		.20–.50	

1007200

Specifications: Upflow Natural Gas Models (NUGK)

MODEL/SERIES	NUGK040KF06		NUGK050MF06 NUGK050NF06		NUGK075DG07 NUGK075KG07		NUGK100DH09 NUGK100KH09		NUGK125DK07 NUGK125KK07	
	STD	ALT	STD	ALT	STD	ALT	STD	ALT	STD	ALT
Input Rating (BTUH)	40,000	50,000	50,000	40,000	75,000	60,000	100,000	80,000	125,000	100,000
Output (BTUH)	38,000	48,500	48,000	38,000	70,000	–	92,000	–	113,000	–
Temperature Rise	15° – 45° F		15° – 45° F		35° – 65° F		40° – 70° F		35° – 65° F	
Flue Size	2 in		2 in		2 in		2 in		2 in	
Elec.Volts/PH./F.L.A.	115/1/8.0		115/1/8.0		115/1/10.8		115/1/10.8		115/1/11.8	
Transformer Size (VA)	40		40		40		40		40	
Orifice Sizes (Req'd)	#45(2) #42		#42(2) #45		#42(3) #44		#42(4) #44		#42(5) #44	
Limit Setting	200		200		250		170		170	
Fan Setting Delay ON	15–90		15–90		15–90		15–90		15–90	
Fan Setting Delay OFF	30–120		30–120		30–120		30–120		30–120	
Exhaust Limit (Man.	120°F		120°F		130°F		130°F		150°F	
Thermal Sensor Reset)	300°F		300°F		300°F		300°F		300°F	
Gas Valve MFG/Type	MH/VR8204A		MH/VR8204A		MH/VR8204A		MH/VR8204A		MH/VR8204A	
Manifold Pressure	3.5		3.5		3.5		3.5		3.5	
Regulation Type	SNAP		SNAP		SNAP		SNAP		SNAP	
Ignition Type/Model	I.I.D./S8600M		I.I.D./S8600M		I.I.D./S8600M		I.I.D./S8600M		I.I.D.S8600M	
Pilot Orifice Size	.018		.018		.018		.018		.018	
Spark Gap	1/8 in.		1/8 in.		1/8 in		1/8 in		1/8 in	
Anticipator Setting	.75		.75		.75		.75		.75	
Cap. Rating MFD/Volts Combustion Air Blower	5/370		5/370		5/370		5/370		5/370	
Furnace Blower Rated Ext. Static Press	.10–.50		.10–.50		.12–.50		.15–50		.20–.50	

Specifications: Upflow LP Gas Models (NULK)

MODEL/SERIES	NULK050MF05		NULK075DG05		NULK100DH06		NULK125DK06	
	STD	ALT	STD	ALT	STD	ALT	STD	ALT
Input Rating (BTUH)	50,000	40,000	75,000	60,000	100,000	80,000	125,000	100,000
Output (BTUH)	48,000	38,000	70,000	–	92,000	–	113,000	–
Temperature Rise	15° – 45° F		35° – 65° F		40° – 70° F		35° – 65° F	
Flue Size	2 in		2 in		2 in		2 in	
Elec.Volts/PH./F.L.A.	115/1/8.0		115/1/10.8		115/1/10.8		115/1/11.1	
Transformer Size (VA)	40		40		40		40	
Orifice Sizes (Req'd)	#54(2) #55		#54(3) #55		#54(4) #55		#54(5) #55	
Limit Setting	200		250		170		170	
Fan Setting Delay ON	15–90		15–90		15–90		15–90	
Fan Setting Delay OFF	30–120		30–120		30–120		30–120	
Exhaust Limit (Man.	120°F		130°F		130°F		150°F	
Thermal Sensor Reset)	300°F		300°F		300°F		300°F	
Gas Valve MFG/Type	WR36E		WR36E		WR36E		WR36E	
Manifold Pressure	10		10		10		10	
Regulation Type	SNAP		SNAP		SNAP		SNAP	
Ignition Type/Model	HSI/50F47		HSI/50F47		HSI/50F47		HSI/50F47	
Lock–Out Timing	After 3 Tries		After 3 Tries		After 3 Tries		After 3 Tries	
Anticipator Setting	.55		.55		.55		.55	
Cap. Rating MFD/Volts Combustion Air Blower	5/370		5/370		5/370		5/370	
Furnace Blower Rated Ext. Static Press	.10–.50		.12–.50		.15–50		.20–.50	

Specifications: Upflow Natural Gas Models (NUGK)

1005218

MODEL/SERIES	NUGK040KF04 NUGK040KF05	NUGK050MF04 NUGK050MF05 NUGK050NF04 NUGK050NF05	NUGK075DG05 NUGK075DG06 NUGK075KG04 NUGK075KG05	NUGK100DH06 NUGK100DH07 NUGK100KH06 NUGK100KH07	NUGK125DK04 NUGK125DK05 NUGK125KK04 NUGK125KK05
	STD ALT	STD ALT	STD ALT	STD ALT	STD ALT
Input Rating (BTUH) Output (BTUH)	40,000 50,000 38,000 48,500	50,000 40,000 48,000 38,000	75,000 60,000 70,000 –	100,000 80,000 92,000 –	125,000 100,000 113,000 –
Temperature Rise	15° – 45° F	15° – 45° F	35° – 65° F	40° – 70° F	35° – 65° F
Flue Size Elec. Volts/PH./F.L.A. Transformer Size (VA)	2 in 115/1/8.0 40	2 in 115/1/8.0 40	2 in 115/1/10.8 40	2 in 115/1/10.8 40	2 in 115/1/11.8 40
Orifice Sizes (Req'd)	#45(2) #42	#42(2) #45	#42(3) #44	#42(4) #44	#42(5) #44
Limit Setting	170	170	170	170	170
Fan Switch Setting OFF	90	90	90	90	90
Exhaust Limit Thermal Sensor	120°F 300°F	120°F 300°F	130°F 300°F	130°F 300°F	150°F 300°F
Gas Valve MFG/Type Manifold Pressure Regulation Type	MH/VR8204C 1.2/3.5 STEP	MH/VR8204C 1.2/3.5 STEP	MH/VR8204C 1.2/3.5 STEP	MH/VR8204C 1.2/3.5 STEP	MH/VR8204C 1.2/3.5 STEP
Ignition Type/Model Pilot Orifice Size	I.I.D./S86F .018	I.I.D./S86F .018	I.I.D./S86F .018	I.I.D./S86F .018	I.I.D.S86F .018
Spark Gap	1/8 in.	1/8 in.	1/8 in	1/8 in	1/8 in
Anticipator Setting	.15	.15	.15	.15	.15
Cap. Rating MFD/Volts Combustion Air Blower	5/370	5/370	5/370	5/370	5/370
Furnace Blower Rated Ext. Static Press.	.10–.50	.10–.50	.12–.50	.15–.50	.20–.50

Specifications: Upflow LP Gas Models (NULK)

MODEL/SERIES	NULK050MF03 NULK050MF04	NULK075DG03 NULK075DG04	NULK100DH03 NULK100DH05	NULK125DK03 NULK125DK04
	STD ALT	STD ALT	STD ALT	STD ALT
Input Rating (BTUH) Output (BTUH)	50,000 40,000 48,000 38,000	75,000 60,000 70,000 –	100,000 80,000 92,000 –	125,000 100,000 113,000 –
Temperature Rise	15° – 45° F	35° – 65° F	40° – 70° F	35° – 65° F
Flue Size Elec. Volts/PH./F.L.A. Transformer Size (VA)	2 in 115/1/8.0 40	2 in 115/1/10.8 40	2 in 115/1/10.8 40	2 in 115/1/11.1 40
Orifice Sizes (Req'd)	#54(2) #55	#54(3) #55	#54(4) #55	#54(5) #55
Limit Setting	170	170	170	170
Fan Switch Setting OFF	90	90	90	90
Exhaust Limit Thermal Sensor	120°F 300°F	130°F 300°F	130°F 300°F	150°F 300°F
Gas Valve MFG/Type Manifold Pressure Regulation Type	WR36E 2.2/10 STEP	WR36E 2.2/10 STEP	WR36E 2.2/10 STEP	WR36E 2.2/10 STEP
Ignition Type/Model Lock–Out Timing	HSI/50E47 After 3 Tries	HSI/50E47 After 3 Tries	HSI/50E47 After 3 Tries	HSI/50E47 After 3 Tries
Anticipator Setting	.15	.15	.15	.15
Cap. Rating MFD/Volts Combustion Air Blower	5/370	5/370	5/370	5/370
Furnace Blower Rated Ext. Static Press.	.10–.50	.12–.50	.15–.50	.20–.50

Specifications: Upflow Natural Gas Models (NUGK)

1005997

MODEL/SERIES	NUGK040KF06	NUGK050MF06 NUGK050NF06	NUGK075DG07 NUGK075KG06	NUGK100DH08 NUGK100KH08	NUGK125DK06 NUGK125KK06
Input Rating (BTUH) Output (BTUH)	STD ALT 40,000 50,000 38,000 48,500	STD ALT 50,000 40,000 48,000 38,000	STD ALT 75,000 60,000 70,000	STD ALT 100,000 80,000 92,000 –	STD ALT 125,000 100,000 113,000 –
Temperature Rise	15° – 45° F	15° – 45° F	35° – 65° F	40° – 70° F	35° – 65° F
Flue Size Elec.Volts/PH./F.L.A. Transformer Size (VA)	2 in 115/1/8.0 40	2 in 115/1/8.0 40	2 in 115/1/10.8 40	2 in 115/1/10.8 40	2 in 115/1/11.8 40
Orifice Sizes (Req'd)	#45(2) #42	#42(2) #45	#42(3) #44	#42(4) #44	#42(5) #44
Limit Setting Fan Setting Delay ON Timer (Sec's) Delay OFF	200 15–90 30–120	200 15–90 30–120	250 15–90 30–120	170 15–90 30–120	170 15–90 30–120
Exhaust Limit (Man. Thermal Sensor Reset)	120°F 300°F	120°F 300°F	130°F 300°F	130°F 300°F	150°F 300°F
Gas Valve MFG/Type Manifold Pressure Regulation Type	MH/VR8204A 3.5 SNAP	MH/VR8204A 3.5 SNAP	MH/VR8204A 3.5 SNAP	MH/VR8204A 3.5 SNAP	MH/VR8204A 3.5 SNAP
Ignition Type/Model Pilot Orifice Size	I.I.D./S8600M .018	I.I.D./S8600M .018	I.I.D./S8600M .018	I.I.D./S8600M .018	I.I.D.S8600M .018
Spark Gap	1/8 in.	1/8 in.	1/8 in	1/8 in	1/8 in
Anticipator Setting	.75	.75	.75	.75	.75
Cap. Rating MFD/Volts Combustion Air Blower	5/370	5/370	5/370	5/370	5/370
Furnace Blower Rated Ext. Static Press	.10–.50	.10–.50	.12–.50	.15–50	.20–.50

Specifications: Upflow LP Gas Models (NULK)

MODEL/SERIES	NULK050MF05	NULK075DG05	NULK100DH06	NULK125DK05
Input Rating (BTUH) Output (BTUH)	STD ALT 50,000 40,000 48,000 38,000	STD ALT 75,000 60,000 70,000 –	STD ALT 100,000 80,000 92,000 –	STD ALT 125,000 100,000 113,000 –
Temperature Rise	15° – 45° F	35° – 65° F	40° – 70° F	35° – 65° F
Flue Size Elec.Volts/PH./F.L.A. Transformer Size (VA)	2 in 115/1/8.0 40	2 in 115/1/10.8 40	2 in 115/1/10.8 40	2 in 115/1/11.1 40
Orifice Sizes (Req'd)	#54(2) #55	#54(3) #55	#54(4) #55	#54(5) #55
Limit Setting Fan Setting Delay ON Timer (Sec's) Delay OFF	200 15–90 30–120	250 15–90 30–120	170 15–90 30–120	170 15–90 30–120
Exhaust Limit (Man. Thermal Sensor Reset)	120°F 300°F	130°F 300°F	130°F 300°F	150°F 300°F
Gas Valve MFG/Type Manifold Pressure Regulation Type	WR36E 10 SNAP	WR36E 10 SNAP	WR36E 10 SNAP	WR36E 10 SNAP
Ignition Type/Model Lock–Out Timing	HSI/50F47 After 3 Tries	HSI/50F47 After 3 Tries	HSI/50F47 After 3 Tries	HSI/50F47 After 3 Tries
Anticipator Setting	.85	.85	.85	.85
Cap. Rating MFD/Volts Combustion Air Blower	5/370	5/370	5/370	5/370
Furnace Blower Rated Ext. Static Press	.10–.50	.12–.50	.15–50	.20–.50

Specifications: Upflow Natural Gas Models (NUGK)

1005147

MODEL/SERIES	NUGK040KF03	NUGK050MF03 NUGK050NF03	NUGK075DG03 NUGK075DG04 NUGK075KG03	NUGK100DH03 NUGK100DH04 NUGK100DH05 NUGK100KH03 NUGK100KH04 NUGK100KH05	NUGK125DK03 NUGK125KK03
Input Rating (BTUH) Output (BTUH)	STD ALT 40,000 50,000 38,000 48,500	STD ALT 50,000 40,000 48,000 38,000	STD ALT 75,000 60,000 70,000 –	STD ALT 100,000 80,000 92,000 –	STD ALT 125,000 100,000 113,000 –
Temperature Rise	15° – 45° F	15° – 45° F	35° – 65° F	40° – 70° F	35° – 65° F
Flue Size Elec.Volts/PH./F.L.A. Transformer Size (VA)	2 in 115/1/8.0 40	2 in 115/1/8.0 40	2 in 115/1/10.8 40	2 in 115/1/10.8 40	2 in 115/1/11.1 40
Orifice Sizes (Req'd)	#45(2) #42	#42(2) #45	#42(3) #44	#42(4) #44	#42(5) #44
Limit Setting	170	170	170	170	170
Fan Switch Setting OFF	90	90	90	90	90
Exhaust Limit Thermal Sensor	120°F 300°F	120°F 300°F	130°F 300°F	130°F 300°F	150°F 300°F
Gas Valve MFG/Type Manifold Pressure Regulation Type	MH/VR8440P 1.2/3.5 STEP	MH/VR8440P 1.2/3.5 STEP	MH/VR8440P 1.2/3.5 STEP	MH/VR8440P 1.2/3.5 STEP	MH/VR8440P 1.2/3.5 STEP
Ignition Type/Model Pilot Orifice Size	I.I.D./S86F .018	I.I.D./S86F .018	I.I.D./S86F .018	I.I.D./S86F .018	I.I.D.S86F .018
Spark Gap	1/8 in.	1/8 in.	1/8 in	1/8 in	1/8 in
Anticipator Setting	.15	.15	.15	.15	.15
Cap. Rating MFD/Volts Combustion Air Blower	5/370	5/370	5/370	5/370	5/370
Furnace Blower Rated Ext. Static Press	.10–.50	.10–.50	.12–.50	.15–50	.20–.50

Specifications: Upflow LP Gas Models (NULK)

MODEL/SERIES	NULK050MF03	NULK075DG03 NULK075DG04	NULK100DH03 NULK100DH04 NULK100DH05	NULK125DK03
Input Rating (BTUH) Output (BTUH)	STD ALT 50,000 40,000 48,000 38,000	STD ALT 75,000 60,000 70,000 –	STD ALT 100,000 80,000 92,000 –	STD ALT 125,000 100,000 113,000 –
Temperature Rise	15° – 45° F	35° – 65° F	40° – 70° F	35° – 65° F
Flue Size Elec.Volts/PH./F.L.A. Transformer Size (VA)	2 in 115/1/8.0 40	2 in 115/1/10.8 40	2 in 115/1/10.8 40	2 in 115/1/11.1 40
Orifice Sizes (Req'd)	#54(2) #55	#54(3) #55	#54(4) #55	#54(5) #55
Limit Setting	170	170	170	170
Fan Switch Setting OFF	90	90	90	90
Exhaust Limit Thermal Sensor	120°F 300°F	130°F 300°F	130°F 300°F	150°F 300°F
Gas Valve MFG/Type Manifold Pressure Regulation Type	WR36E 2.2/10 STEP	WR36E 2.2/10 STEP	WR36E 2.2/10 STEP	WR36E 2.2/10 STEP
Ignition Type/Model Lock–Out Timing	HSI/50E47 After 3 Tries	HSI/50E47 After 3 Tries	HSI/50E47 After 3 Tries	HSI/50E47 After 3 Tries
Anticipator Setting	.15	.15	.15	.15
Cap. Rating MFD/Volts Combustion Air Blower	5/370	5/370	5/370	5/370
Furnace Blower Rated Ext. Static Press	.10–.50	.12–.50	.15–50	.20–.50

Specifications: Upflow Natural Gas Models (NUGS)

1007567

MODEL/SERIES	NUGS050AF03	NUGS075BG03	NUGS100BH03	NUGS125AK03
Gas Type–NAT	STD	STD	STD	STD
Input Rating (BTUH)	50,000	75.000	100,000	125,000
Output (BTUH)	48,000	70,000	92,000	113,000
Temperature Rise	15° – 45° F	35° – 65° F	40° – 70° F	35° – 65° F
Flue Size	2 in	3 in	3 in	3 in
Elec.Volts/PH./F.L.A.	115/1/8.0	115/1/10.8	115/1/10.8	115/1/11.8
Transformer Size (VA)	40	40	40	40
Orifice Sizes (Req'd)	#42(2)	#42(3)	#42(4)	#42(5)
Limit Setting	250	250	200	170
Fan Setting Delay ON	15–90	15–90	15–90	15–90
Timer (Sec's) Delay OFF	30–120	30–120	30–120	30–120
Exhaust Limit	150°F	150°F	150°F	180°F
Thermal Sensor	300°F	300°F	300°F	300°F
Gas Valve MFG/Type	WR36E	WR36E	WR36E	WR36E
Manifold Pressure	1.2/3.5	1.2/3.5	1.2/3.5	1.2/3.5
Regulation Type	STEP	STEP	STEP	STEP
Ignition Type/Model	IID/S8600M	IID/S8600M	IID/S8600M	IID/S8600M
Pilot Orifice Size	.20	.20	.20	.20
Anticipator Setting	.55	.55	.55	.55
Cap. Rating MFD/Volts Combustion Air Blower	5/370	5/370	5/370	5/370
Furnace Blower Rated Ext. Static Press	.10–.50	.12–.50	.15–50	.20–.50

Specifications: Upflow LP Gas Models (NULS)

MODEL/SERIES	NULS050AF03	NULS075BG03	NULS100BH03	NULS125AK03
Gas Type–NAT	STD	STD	STD	STD
Input Rating (BTUH)	50,000	75.000	100,000	125,000
Output (BTUH)	48,000	70,000	92,000	113,000
Temperature Rise	15° – 45° F	35° – 65° F	40° – 70° F	35° – 65° F
Flue Size	2 in	3 in	3 in	3 in
Elec.Volts/PH./F.L.A.	115/1/8.0	115/1/10.8	115/1/10.8	115/1/11.8
Transformer Size (VA)	40	40	40	40
Orifice Sizes (Req'd)	#54(2)	#54(3)	#54(4)	#54(5)
Limit Setting	250	250	200	170
Fan Setting Delay ON	15–90	15–90	15–90	15–90
Timer (Sec's) Delay OFF	30–120	30–120	30–120	30–120
Exhaust Limit	150°F	150°F	150°F	180°F
Thermal Sensor	300°F	300°F	300°F	300°F
Gas Valve MFG/Type	WR36E	WR36E	WR36E	WR36E
Manifold Pressure	2.5/10	2.5/10	2.5/10	2.5/10
Regulation Type	STEP	STEP	STEP	STEP
Ignition Type/Model	IID/S8600M	IID/S8600M	IID/S8600M	IID/S8600M
Pilot Orifice Size	.012	.012	.012	.012
Anticipator Setting	.55	.55	.55	.55
Cap. Rating MFD/Volts Combustion Air Blower	5/370	5/370	5/370	5/370
Furnace Blower Rated Ext. Static Press	.10–.50	.12–.50	.15–50	.20–.50

Specifications: Upflow Natural Gas Models (NUGS)

1007213

MODEL/SERIES	NUGS050AF02	NUGS075BG02	NUGS100BH02	NUGS125AK02
Gas Type–NAT Input Rating (BTUH) Output (BTUH)	STD 50,000 48,000	STD 75.000 70,000	STD 100,000 92,000	STD 125,000 113,000
Temperature Rise	15° – 45° F	35° – 65° F	40° – 70° F	35° – 65° F
Flue Size Elec.Volts/PH./F.L.A. Transformer Size (VA)	2 in 115/1/8.0 40	3 in 115/1/10.8 40	3 in 115/1/10.8 40	3 in 115/1/11.8 40
Orifice Sizes (Req'd)	#42(2)	#42(3)	#42(4)	#42(5)
Limit Setting Fan Timer Settings (Sec's) OFF	250 Timed	250 Timed	200 Timed	170 Timed
Exhaust Limit Thermal Sensor	150°F 300°F	150°F 300°F	150°F 300°F	180°F 300°F
Gas Valve MFG/Type Manifold Pressure Regulation Type	WR36E 3.5 SNAP	WR36E 3.5 SNAP	WR36E 3.5 SNAP	WR36E 3.5 SNAP
Ignition Type/Model	HSI/50F47	HSI/50F47	HSI/50F47	HSI/50F47
Anticipator Setting	.80	.80	.80	.80
Cap. Rating MFD/Volts Combustion Air Blower	5/370	5/370	5/370	5/370
Furnace Blower Rated Ext. Static Press	.10–.50	.12–.50	.15–50	.20–.50

Specifications: Upflow LP Gas Models (NULS)

MODEL/SERIES	NULS050AF02	NULS075BG02	NULS100BH02	NULS125AK02
Gas Type–NAT Input Rating (BTUH) Output (BTUH)	STD 50,000 48,000	STD 75.000 70,000	STD 100,000 92,000	STD 125,000 113,000
Temperature Rise	15° – 45° F	35° – 65° F	40° – 70° F	35° – 65° F
Flue Size Elec.Volts/PH./F.L.A. Transformer Size (VA)	2 in 115/1/8.0 40	3 in 115/1/10.8 40	3 in 115/1/10.8 40	3 in 115/1/11.8 40
Orifice Sizes (Req'd)	#54(2)	#54(3)	#54(4)	#54(5)
Limit Setting Fan Timer Settings (Sec's) OFF	250 Timed	250 Timed	200 Timed	170 Timed
Exhaust Limit Thermal Sensor	150°F 300°F	150°F 300°F	150°F 300°F	180°F 300°F
Gas Valve MFG/Type Manifold Pressure Regulation Type	WR36E 10 SNAP	WR36E 10 SNAP	WR36E 10 SNAP	WR36E 10 SNAP
Ignition Type/Model	HSI/50F47	HSI/50F47	HSI/50F47	HSI/50F47
Anticipator Setting	.80	.80	.80	.80
Cap. Rating MFD/Volts Combustion Air Blower	5/370	5/370	5/370	5/370
Furnace Blower Rated Ext. Static Press	.10–.50	.12–.50	..15–50	.20–.50

Specifications: Upflow Natural Gas Models (NUGS)

1006412

MODEL/SERIES	NUGS050AF01	NUGS075BG01	NUGS100BH01	NUGS125AK01
Gas Type–NAT Input Rating (BTUH) Output (BTUH)	STD 50,000 48,000	STD 75,000 70,000	STD 100,000 92,000	STD 125,000 113,000
Temperature Rise	15° – 45° F	35° – 65° F	40° – 70° F	35° – 65° F
Flue Size Elec.Volts/PH./F.L.A. Transformer Size (VA)	2 in 115/1/8.0 40	3 in 115/1/10.8 40	3 in 115/1/10.8 40	3 in 115/1/11.8 40
Orifice Sizes (Req'd)	#42(2)	#42(3)	#42(4)	#42(5)
Limit Setting Fan Timer Settings (Sec's) OFF	250 Timed	250 Timed	200 Timed	170 Timed
Exhaust Limit Thermal Sensor	120°F 300°F	130°F 300°F	130°F 300°F	150°F 300°F
Gas Valve MFG/Type Manifold Pressure Regulation Type	WR36E 3.5 SNAP	WR36E 3.5 SNAP	WR36E 3.5 SNAP	WR36E 3.5 SNAP
Ignition Type/Model Lock–Out Timing	HSI/50F47 After 3 Tries	HSI/50F47 After 3 Tries	HSI/50F47 After 3 Tries	HSI/50F47 After 3 Tries
Anticipator Setting	.80	.80	.80	.80
Cap. Rating MFD/Volts Combustion Air Blower	5/370	5/370	5/370	5/370
Furnace Blower Rated Ext. Static Press	.10–.50	.12–.50	..15–50	.20–.50

Specifications: Upflow LP Gas Models (NULS)

MODEL/SERIES	NULS050AF01	NULS075BG01	NULS100BH01	NULS125AK01
Gas Type–NAT Input Rating (BTUH) Output (BTUH)	STD 50,000 48,000	STD 75,000 70,000	STD 100,000 92,000	STD 125,000 113,000
Temperature Rise	15° – 45° F	35° – 65° F	40° – 70° F	35° – 65° F
Flue Size Elec.Volts/PH./F.L.A. Transformer Size (VA)	2 in 115/1/8.0 40	3 in 115/1/10.8 40	3 in 115/1/10.8 40	3 in 115/1/11.8 40
Orifice Sizes (Req'd)	#54(2)	#54(3)	#54(4)	#54(5)
Limit Setting Fan Timer Settings (Sec's) OFF	250 Timed	250 Timed	200 Timed	170 Timed
Exhaust Limit (Man. Thermal Sensor Reset)	120°F 300°F	130°F 300°F	130°F 300°F	150°F 300°F
Gas Valve MFG/Type Manifold Pressure Regulation Type	WR36E 10 SNAP	WR36E 10 SNAP	WR36E 10 SNAP	WR36E 10 SNAP
Ignition Type/Model Lock–Out Timing	HSI/50F47 After 3 Tries	HSI/50F47 After 3 Tries	HSI/50F47 After 3 Tries	HSI/50F47 After 3 Tries
Anticipator Setting	.80	.80	.80	.80
Cap. Rating MFD/Volts Combustion Air Blower	5/370	5/370	5/370	5/370
Furnace Blower Rated Ext. Static Press	.10–.50	.12–.50	..15–50	.20–.50

Specifications: Upflow Natural Gas Models (NUGS)

1006034

MODEL/SERIES	NUGS050AF01	NUGS075AG01	NUGS100AH01	NUGS125AK01
Gas Type–NAT	STD	STD	STD	STD
Input Rating (BTUH)	50,000	75.000	100,000	125,000
Output (BTUH)	48,000	70,000	92,000	113,000
Temperature Rise	15° – 45° F	35° – 65° F	40° – 70° F	35° – 65° F
Flue Size	2 in	2 in	2 in	3 in
Elec.Volts/PH./F.L.A.	115/1/8.0	115/1/10.8	115/1/10.8	115/1/11.8
Transformer Size (VA)	40	40	40	40
Orifice Sizes (Req'd)	#42(2)	#42(3)	#42(4)	#42(5)
Limit Setting	250	250	200	170
Fan Timer Settings (Sec's) OFF	Timed	Timed	Timed	Timed
Exhaust Limit	120°F	130°F	130°F	150°F
Thermal Sensor	300°F	300°F	300°F	300°F
Gas Valve MFG/Type	WR36E	WR36E	WR36E	WR36E
Manifold Pressure	3.5	3.5	3.5	3.5
Regulation Type	SNAP	SNAP	SNAP	SNAP
Ignition Type/Model	HSI/50F47	HSI/50F47	HSI/50F47	HSI/50F47
Lock–Out Timing	After 3 Tries	After 3 Tries	After 3 Tries	After 3 Tries
Anticipator Setting	.80	.80	.80	.80
Cap. Rating MFD/Volts Combustion Air Blower	5/370	5/370	5/370	5/370
Furnace Blower Rated Ext. Static Press	.10–.50	.12–.50	.15–50	.20–.50

Specifications: Upflow LP Gas Models (NULS)

MODEL/SERIES	NULS050AF01	NULS075AG01	NULS100AH01	NULS125AK01
Gas Type–NAT	STD	STD	STD	STD
Input Rating (BTUH)	50,000	75.000	100,000	125,000
Output (BTUH)	48,000	70,000	92,000	113,000
Temperature Rise	15° – 45° F	35° – 65° F	40° – 70° F	35° – 65° F
Flue Size	2 in	2 in	2 in	3 in
Elec.Volts/PH./F.L.A.	115/1/8.0	115/1/10.8	115/1/10.8	115/1/11.8
Transformer Size (VA)	40	40	40	40
Orifice Sizes (Req'd)	#54(2)	#54(3)	#54(4)	#54(5)
Limit Setting	250	250	200	170
Fan Timer Settings (Sec's) OFF	Timed	Timed	Timed	Timed
Exhaust Limit	120°F	130°F	130°F	150°F
Thermal Sensor	300°F	300°F	300°F	300°F
Gas Valve MFG/Type	WR36E	WR36E	WR36E	WR36E
Manifold Pressure	10	10	10	10
Regulation Type	SNAP	SNAP	SNAP	SNAP
Ignition Type/Model	HSI/50F47	HSI/50F47	HSI/50F47	HSI/50F47
Lock–Out Timing	After 3 Tries	After 3 Tries	After 3 Tries	After 3 Tries
Anticipator Setting	.80	.80	.80	.80
Cap. Rating MFD/Volts Combustion Air Blower	5/370	5/370	5/370	5/370
Furnace Blower Rated Ext. Static Press	.10–.50	.12–.50	.15–50	.20–.50

BLOWER PERFORMANCE DATA

NOTE: Use the Blower Performance Index to find the correct chart.

Model Number			NDGK040	NDGK050 NDLK050	NDGK075 NDLK075	NDGK100 NDLK100	NDGK125 NDLK125
Blower Type & Size			DD10–8A	DD10–8A	DD10–9A	DD10–9A	DD12–11AT
Motor Amps/RPM			8.0/1050	8.0/1050	8.0/1050	8.0/1050	11.8/1050
Nominal H.P./Type			1/2 PSC	1/2 PSC	3/4 PSC	3/4 PSC	1 PSC
Capacitor			7.5 MFD	7.5 MFD	10.0 MFD	10.0 MFD	15.0 MFD
Air Delivery in C.F.M. Varying Static Pressure (In WC.)	.10	LO MED. LO MED. HI HI	850 1090 1320 1460	850 1090 1320 1460	860 1100 1330 1470	1055 1330 1550 1690	1575 1720 1975 2210
	.20	LO MED. LO MED. HI HI	850 1080 1270 1400	850 1080 1270 1400	855 1080 1300 1415	1050 1295 1495 1625	1550 1695 1935 2135
	.30	LO MED. LO MED. HI HI	840 1050 1220 1340	840 1050 1220 1340	855 1060 1240 1360	1035 1245 1430 1550	1520 1665 1885 2075
	.40	LO MED. LO MED. HI HI	820 1010 1160 1275	820 1010 1160 1275	840 1030 1195 1295	1000 1190 1360 1470	1490 1620 1835 2020
	.50	LO MED. LO MED. HI HI	790 970 1105 1200	790 970 1105 1200	820 990 1140 1230	965 1140 1300 1380	1460 1580 1785 1965

BLOWER PERFORMANCE DATA (Cont.)

NOTE: Use the Blower Performance Index to find the correct chart.

Model Number			NDGK040	NDGK050 NDLK050	NDGK075 NDLK075	NDGK100 NDLK100	NDGK125 NDLK125
Blower Type & Size			DD10–8AT	DD10–8AT	DD10–9AT	DD10–9AT	DD12–11AT
Motor Amps/RPM			8.0/1050	8.0/1050	8.0/1050	10.8/1050	11.1/1050
Nominal H.P./Type			1/2 PSC	1/2 PSC	1/2 PSC	3/4 PSC	3/4 PSC
Capacitor			7.5 MFD	7.5 MFD	7.5 MFD	10.0 MFD	15.0 MFD
Air Delivery in C.F.M. Varying Static Pressure (In WC.)	.10	LO	850	850	860	1055	1575
		MED. LO	1090	1090	1100	1330	1720
		MED. HI	1320	1320	1330	1550	1975
		HI	1460	1460	1470	1690	2210
	.20	LO	850	850	855	1050	1550
		MED. LO	1080	1080	1080	1295	1695
		MED. HI	1270	1270	1300	1495	1935
		HI	1400	1400	1415	1625	2135
	.30	LO	840	840	855	1035	1520
		MED. LO	1050	1050	1060	1245	1665
		MED. HI	1220	1220	1240	1430	1885
		HI	1340	1340	1360	1550	2075
	.40	LO	820	820	840	1000	1490
		MED. LO	1010	1010	1030	1190	1620
		MED. HI	1160	1160	1195	1360	1835
		HI	1275	1275	1295	1470	2020
	.50	LO	790	790	820	965	1460
		MED. LO	970	970	990	1140	1580
		MED. HI	1105	1105	1140	1300	1785
		HI	1200	1200	1230	1380	1965

BLOWER PERFORMANCE DATA (Cont.)

NOTE: Use the Blower Performance Index to find the correct chart.

Model Number			NUGK040 NUGK050 NULK050	NUGK075 NULK075	NUGK100 NULK100	NUGK125 NULK125
Blower Type & Size			DD10–8AT	DD10–9AT	DD10–9AT	DD12–11AT
Motor Amps/RPM			8.0/1050	8.0/1050	10.8/1050	11.1/1050
Nominal H.P./Type			1/2 PSC	3/4 PSC	3/4 PSC	3/4 PSC
Capacitor			7.5 MFD	10.0 MFD	10.0 MFD	15.0 MFD
Air Delivery in C.F.M. Varying Static Pressure (In WC.)	.10	LO MED. LO MED. HI HI	800 1080 1350 1570	1035 1305 1545 1720	1050 1355 1660 1880	1460 1620 1950 2320
	.20	LO MED. LO MED. HI HI	850 1075 1320 1510	1030 1270 1490 1650	1040 1330 1600 1820	1455 1610 1910 2255
	.30	LO MED. LO MED. HI HI	855 1060 1280 1445	1020 1235 1430 1580	1035 1305 1545 1750	1460 1610 1885 2205
	.40	LO MED. LO MED. HI HI	855 1040 1230 1375	995 1195 1375 1510	1025 1280 1490 1675	1445 1590 1855 2145
	.50	LO MED. LO MED. HI HI	835 1010 1180 1300	960 1145 1300 1435	1010 1230 1425 1585	1430 1570 1825 2105

BLOWER PERFORMANCE DATA (Cont.)

NOTE: Use the Blower Performance Index to find the correct chart.

Model Number			NUGK040 NUGS050 NULK050 NULS050	NUGK075 NUGS075 NULK075 NULS075	NUGK100 NUGS100 NULK100 NULS100	NUGK125 NUGS125 NULK125 NULS125
Blower Type & Size			DD10–8A	DD10–9A	DD10–9A	DD12–11AT
Motor Amps/RPM			8.0/1050	10.8/1050	10.8/1050	11.8/1050
Nominal H.P./Type			1/2 PSC	3/4 PSC	3/4 PSC	1 PSC
Capacitor			7.5 MFD	10.0 MFD	10.0 MFD	15.0 MFD
Air Delivery in C.F.M. Varying Static Pressure (In WC.)	.10	LO MED. LO MED. HI HI	945 1190 1375 1625	1105 1360 1605 1775	1100 1370 1660 1900	1480 1565 1835 2135
	.20	LO MED. LO MED. HI HI	945 1180 1360 1570	1090 1335 1560 1705	1076 1350 1608 1830	1475 1560 1820 2090
	.30	LO MED. LO MED. HI HI	940 1160 1340 1500	1070 1360 1500 1630	1075 1325 1559 1752	1460 1545 1785 2050
	.40	LO MED. LO MED. HI HI	935 1140 1305 1445	1045 1255 1435 1560	1035 1280 1505 1675	1440 1525 1750 2005
	.50	LO MED. LO MED. HI HI	920 1110 1265 1385	1010 1205 1375 1485	1000 1245 1445 1605	1425 1500 1715 1965
	.60	LO MED. LO MED. HI HI	905 1080 1205 1320	960 1140 1295 1395	965 1195 1370 1520	1385 1475 1680 1900
	.70	LO MED. LO MED. HI HI	870 1030 1145 1245	885 1050 1205 1305	905 1125 1290 1430	1340 1430 1625 1855
	.80	LO MED. LO MED. HI HI	820 965 1070 1160	815 940 1080 1200	830 1040 1200 1335	1275 1360 1575 1790

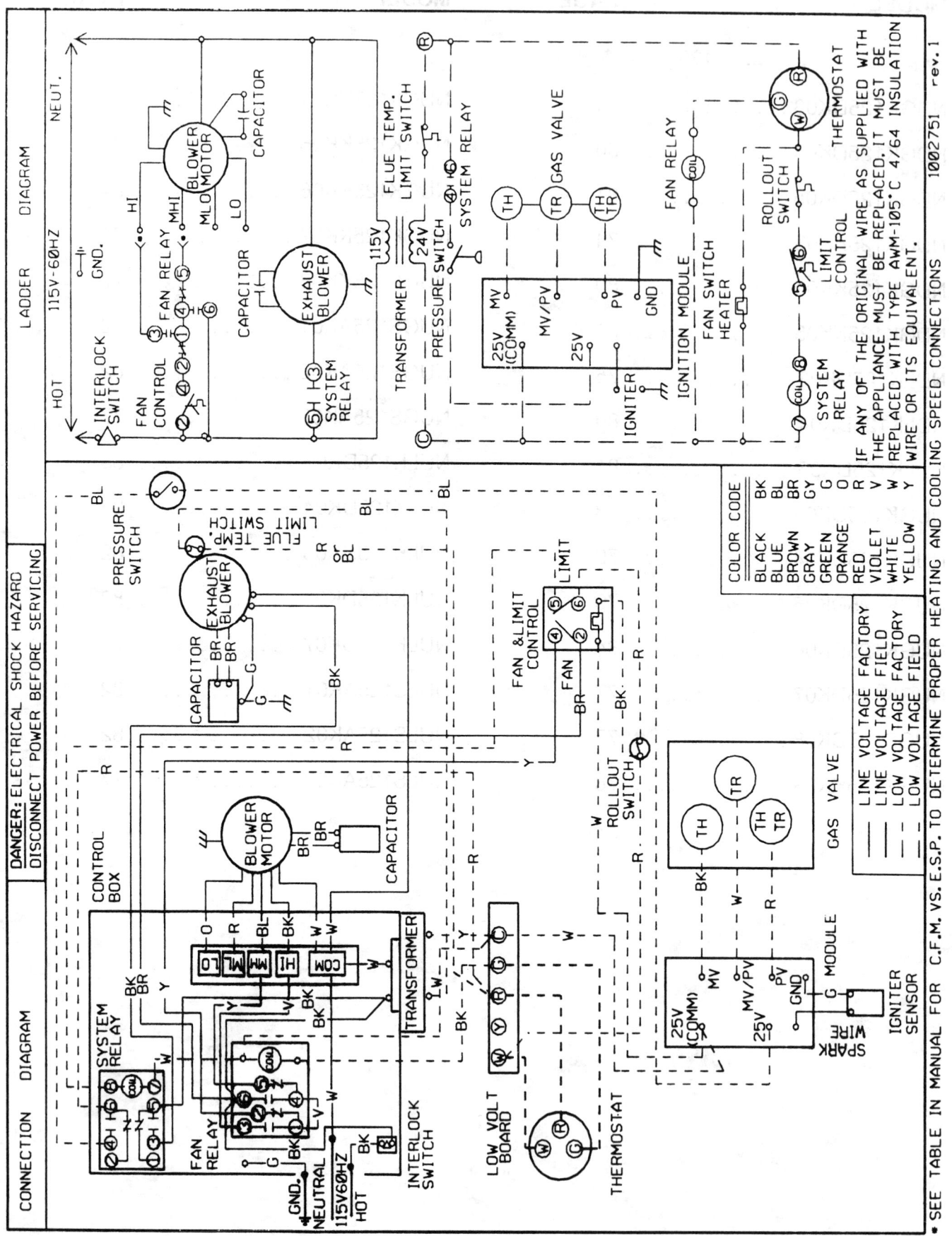

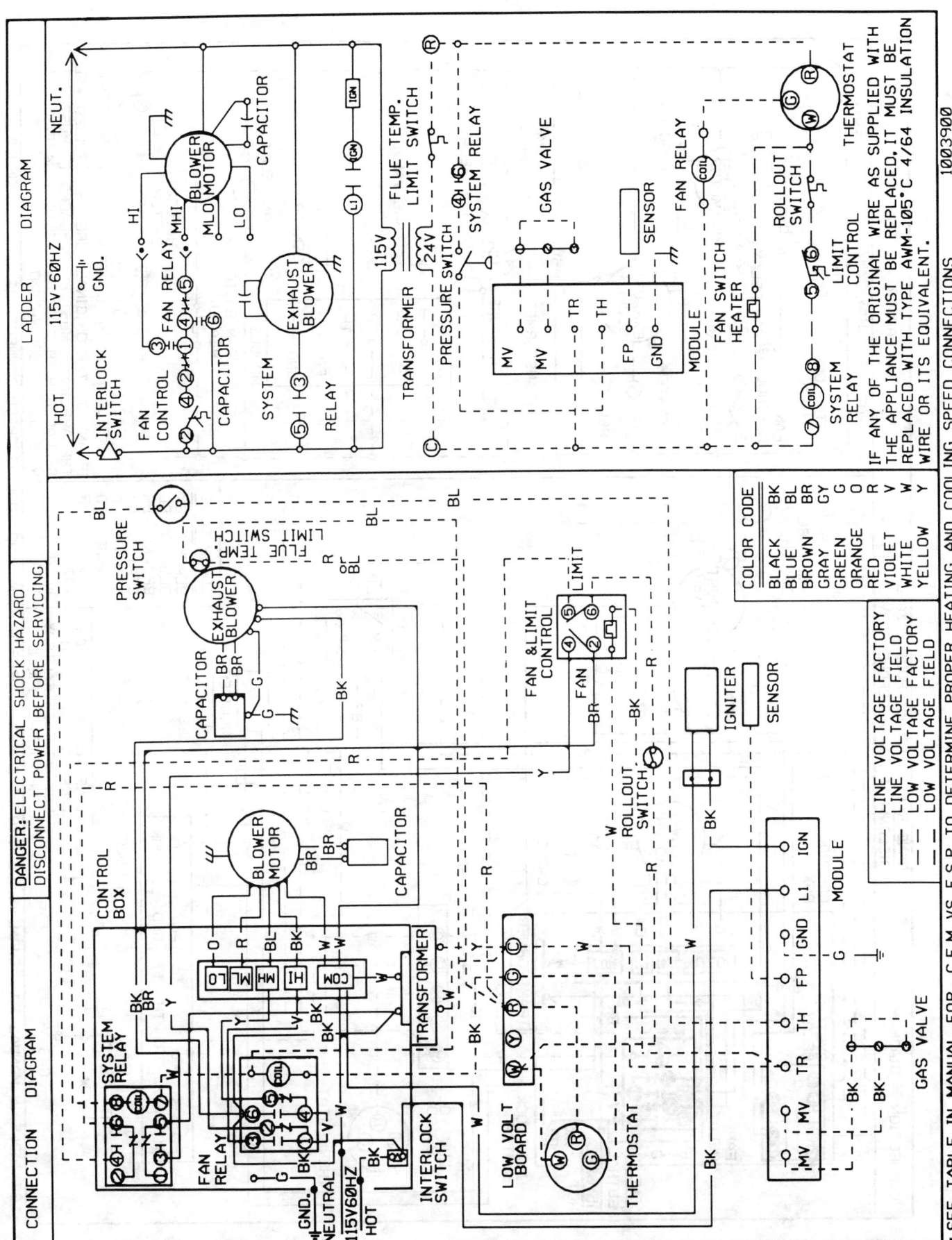

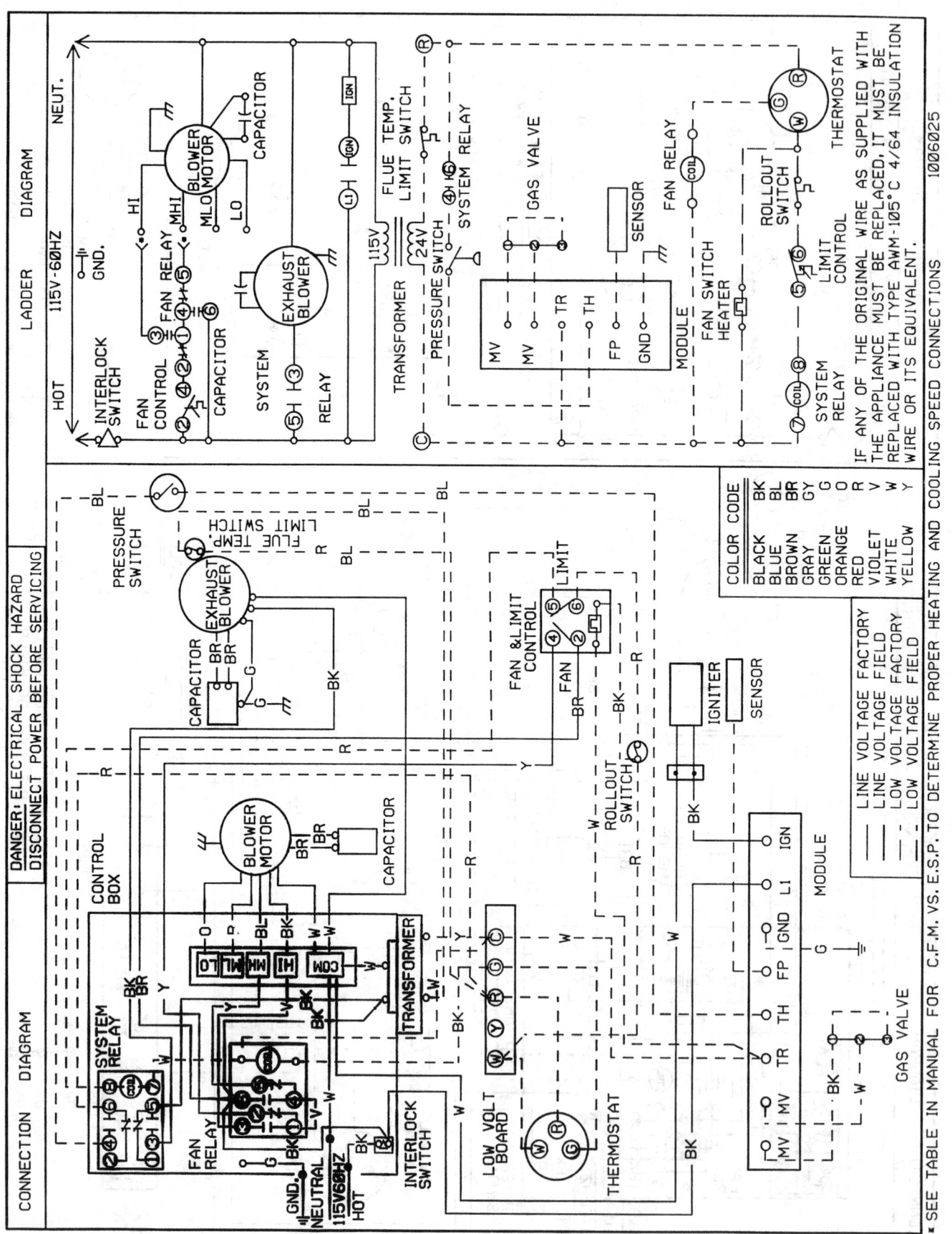

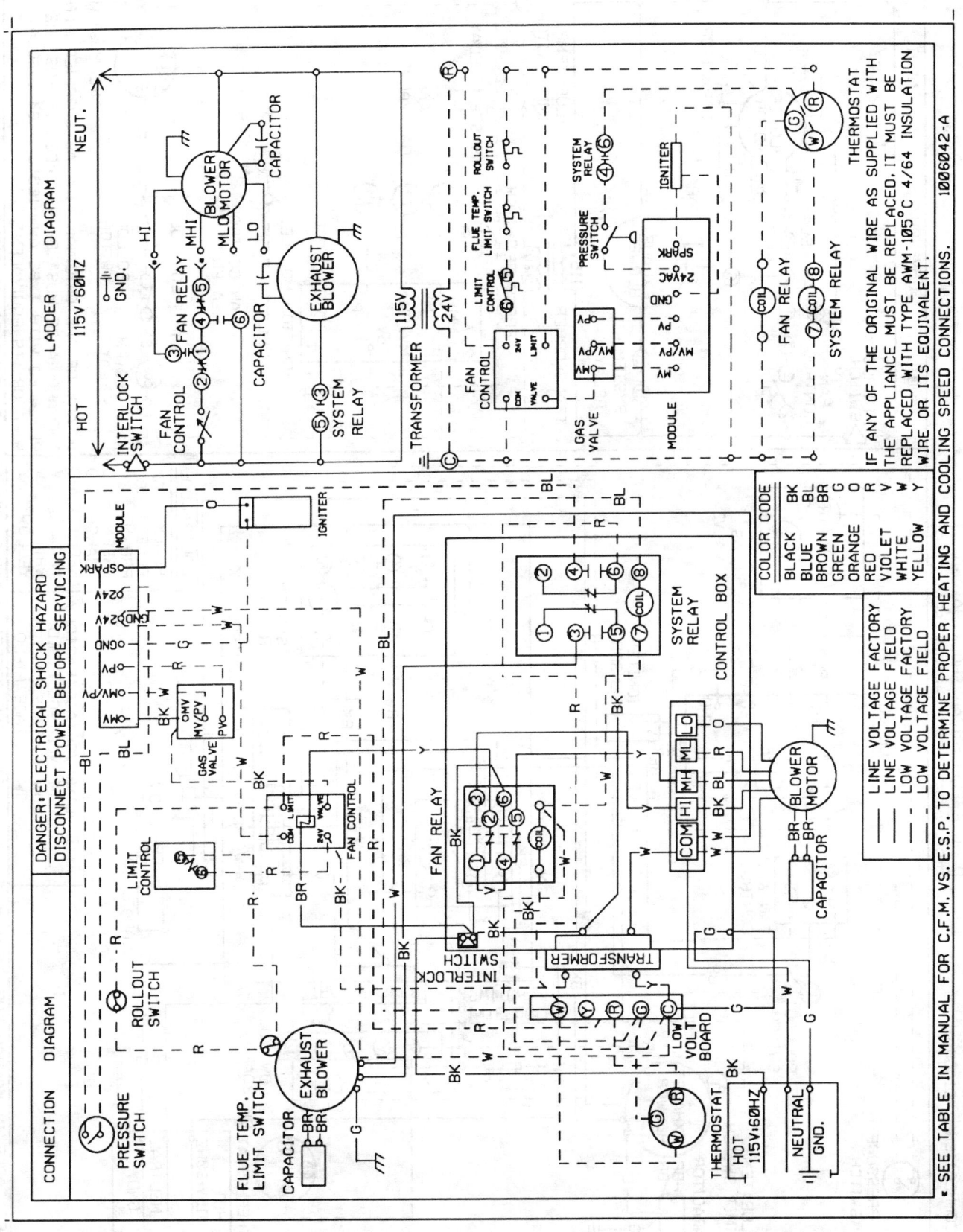

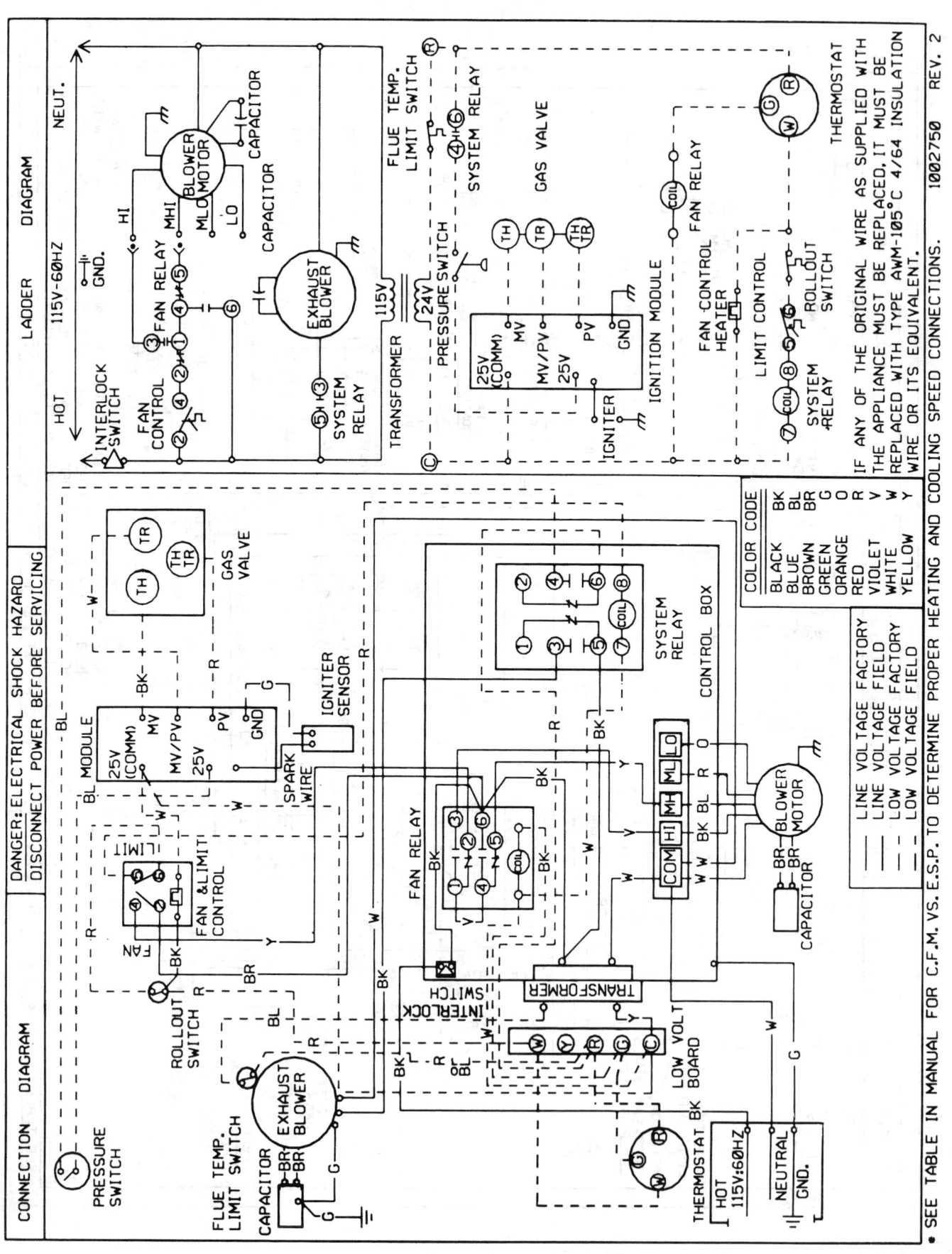

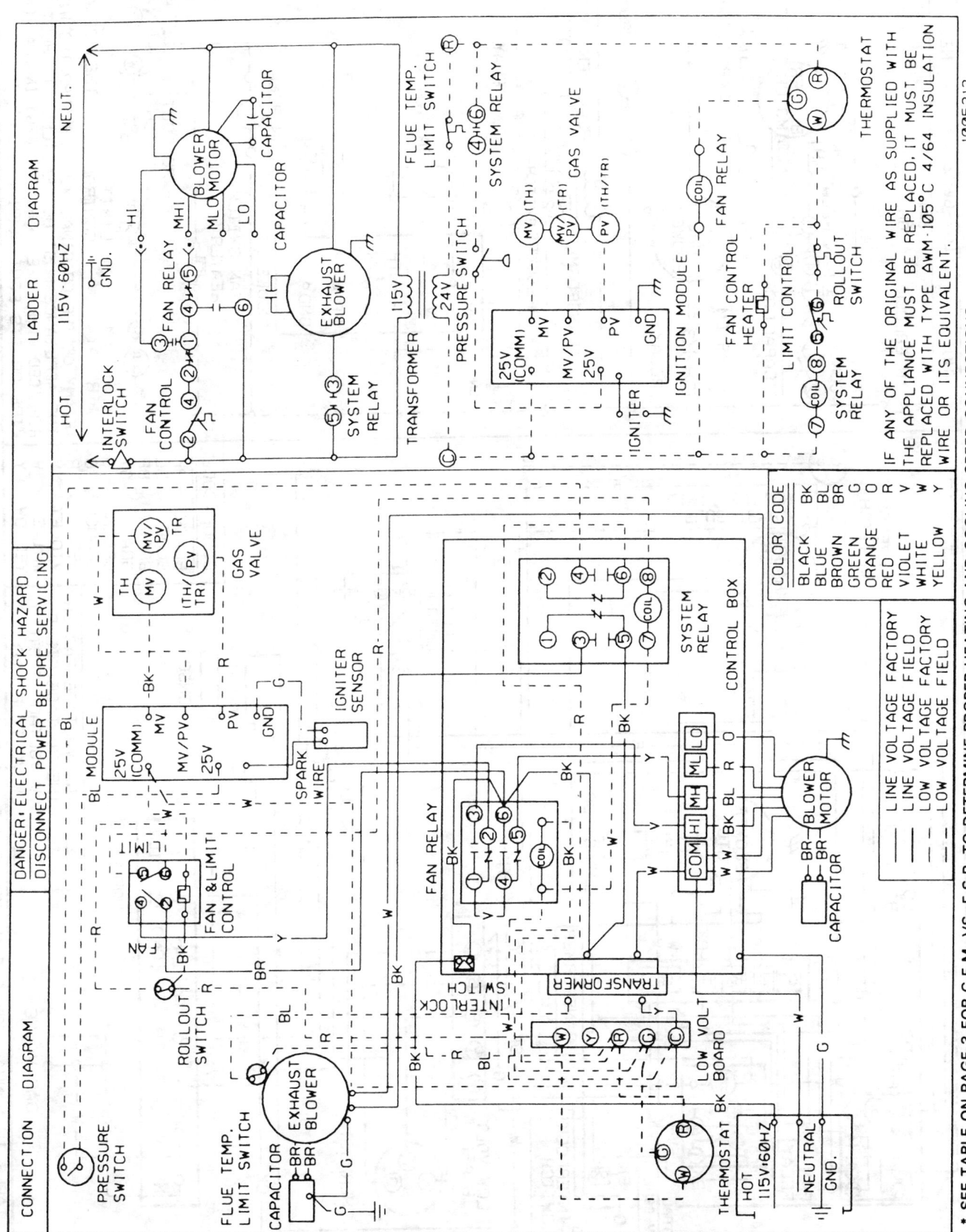

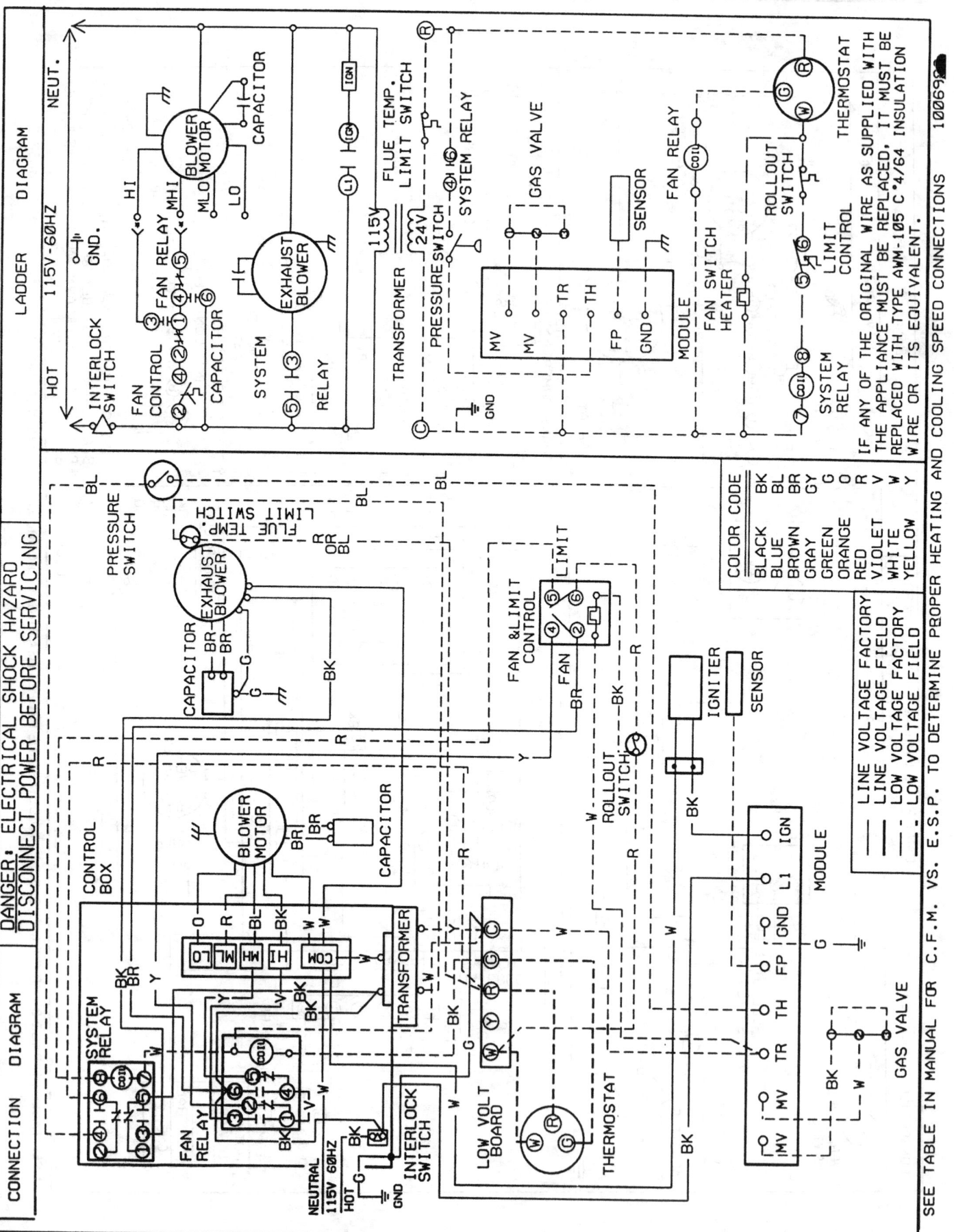

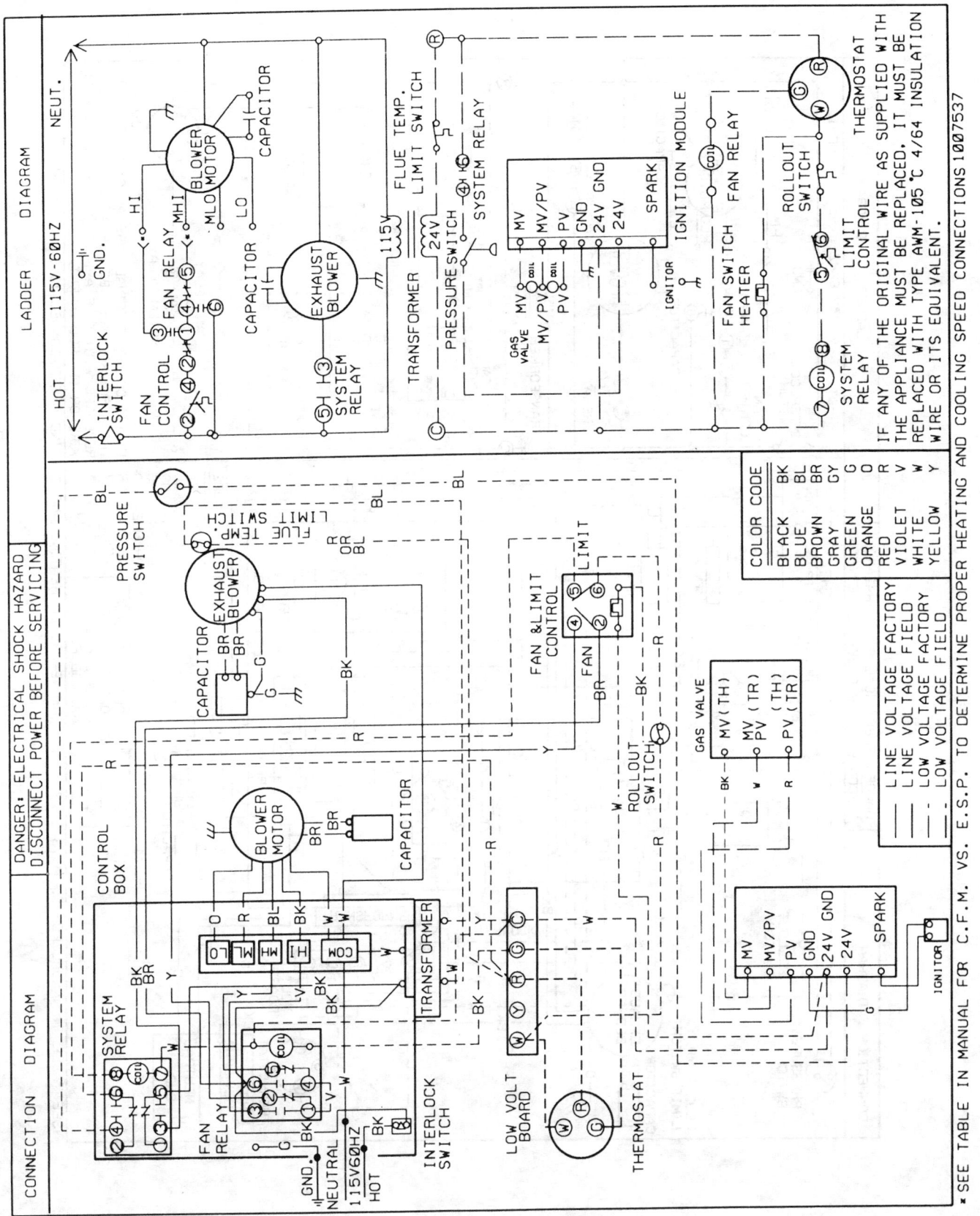

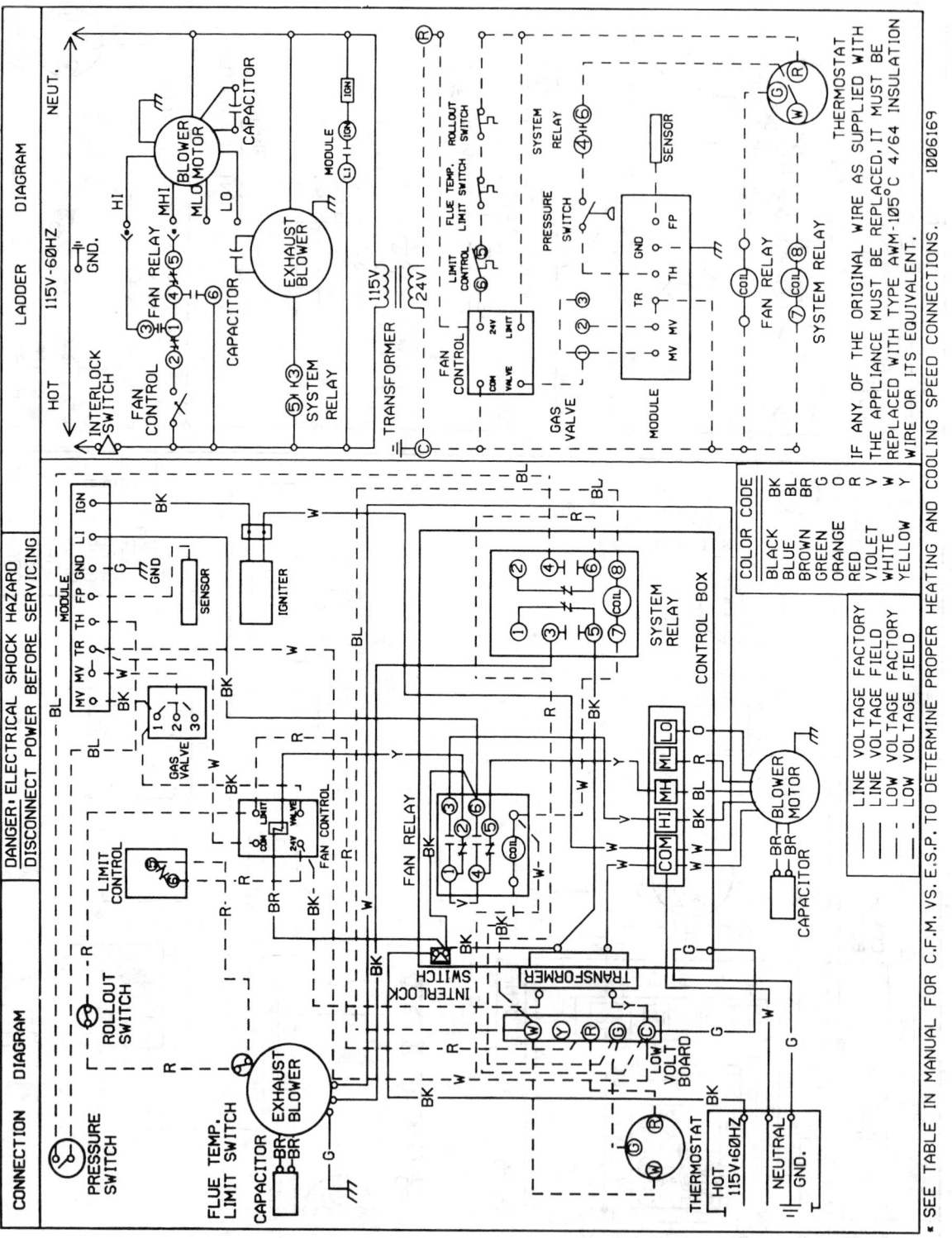

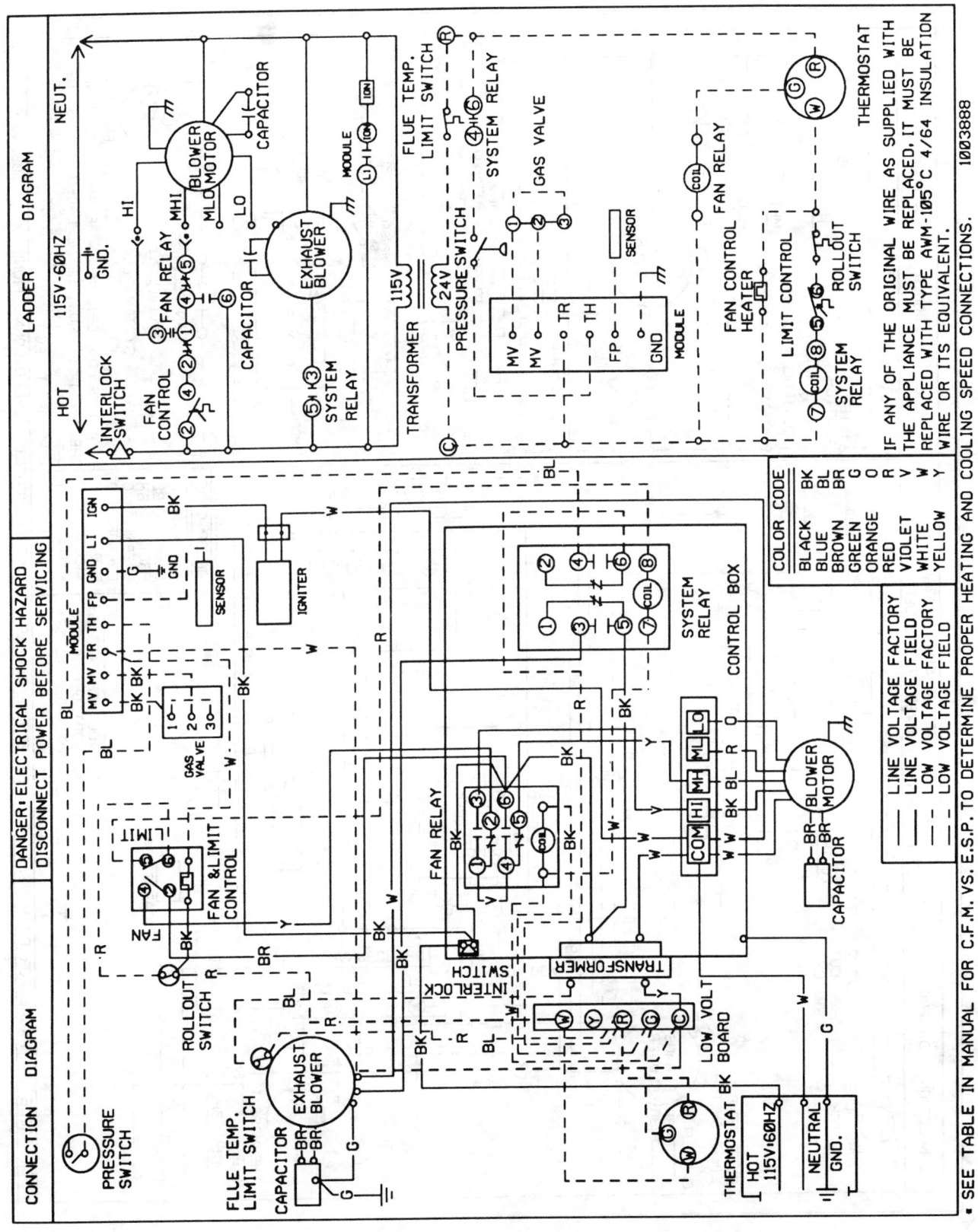

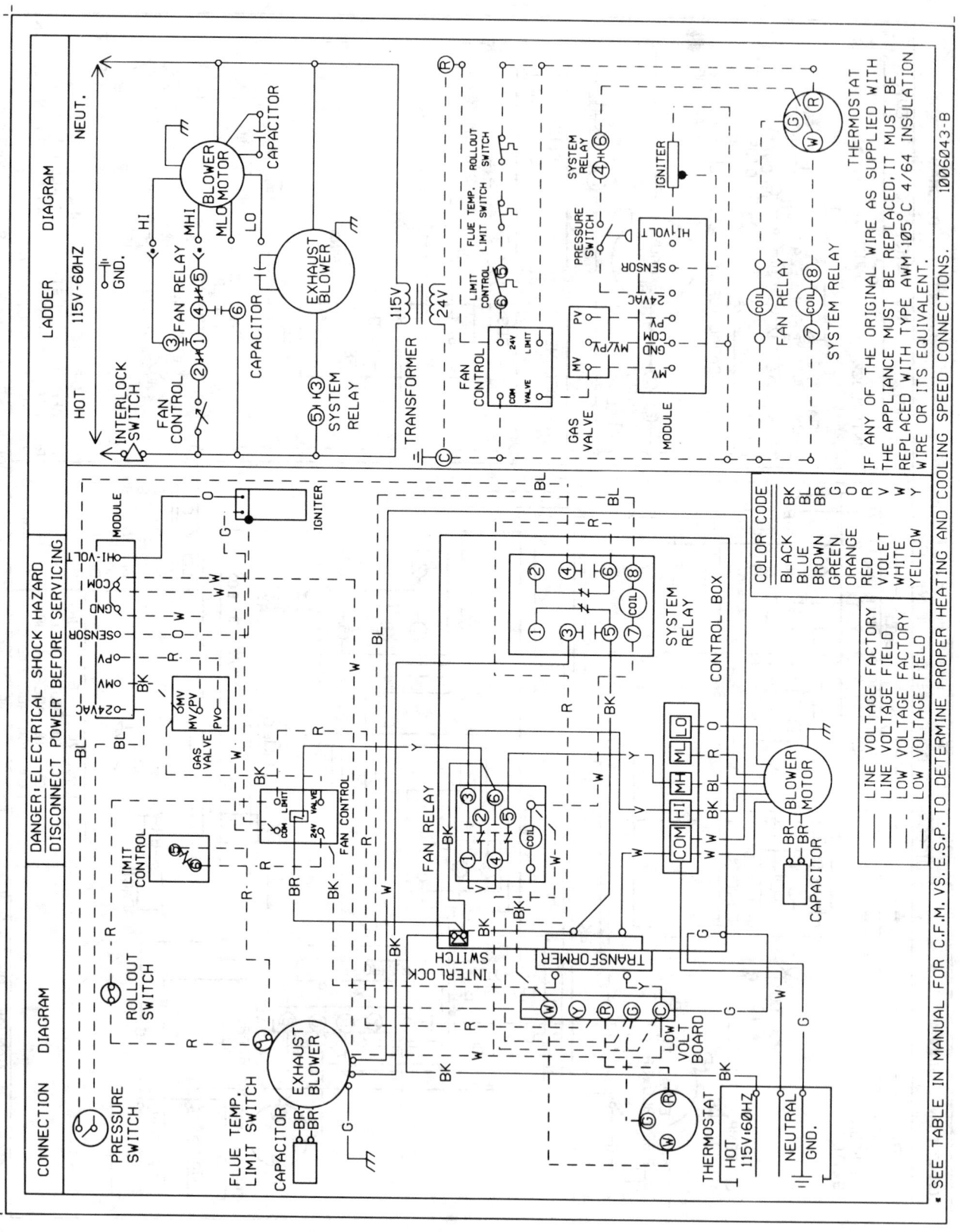

INDEX TECHNICAL SUPPORT MANUALS

Manual Part Number 1005147

Natural Gas Models Upflow

NUGK040KF03	867.769452
NUGK050MF03	867.769412
NUGK050NF03	867.769462
NUGK075DG03	867.769422
NUGK075DG04	867.769423
NUGK075KG03	867.769472
NUGK100DH03	876.769432
NUGK100DH04	876.769433
NUGK100DH05	876.769434
NUGK100KH03	867.769482
NUGK100KH04	867.769483
NUGK100KH05	867.769484
NUGK125DK03	867.769442
NUGK125KK03	867.769492

LP Models Upflow

NULK050MF03	867.779412
NULK075DG03	867.779422
NULK075DG04	867.779423
NULK100DH03	867.779432
NULK100DH04	867.779433
NULK100DH05	867.779434
NULK125DK03	867.779442

INDEX TECHNICAL SUPPORT MANUALS

Manual Part Number 1005218

Natural Gas Models Upflow

NUGK040KF04	867.769453
NUGK040KF05	867.769454
NUGK050MF04	867.769465
NUGK050MF05	867.769426
NUGK050NF04	867.769475
NUGK050NF05	867.769437
NUGK075DG05	867.769487
NUGK075DG06	867.769445
NUGK075KG04	867.769495
NUGK075KG05	867.769474
NUGK100DH06	867.769435
NUGK100DH07	867.769436
NUGK100KH06	867.769485
NUGK100KH07	867.769486
NUGK125DK04	867.769443
NUGK125DK05	867.769444
NUGK125KK04	867.769493
NUGK125KK05	867.769494

LP Models Upflow

NULK050MF03	867.779412
NULK050MF04	867.779413
NULK075DG03	867.779422
NULK075DG04	867.779423
NULK100DH03	867.779432
NULK100DH05	867.779434
NULK125DK03	867.779442
NULK125DK04	867.779443

INDEX TECHNICAL SUPPORT MANUALS

Manual Part Number 1005997

Natural Gas Models Upflow

NUGK040KF06	867.769455
NUGK050MF06	867.769415
NUGK050NF06	867.769465
NUGK075DG07	867.769426
NUGK075KG06	867.769475
NUGK100DH08	867.769437
NUGK100KH08	867.769487
NUGK125DK06	867.769445
NUGK125KK06	867.769495

LP Models Upflow

NULK050MF05	867.779414
NULK075DG05	867.779424
NULK100DH06	867.779435
NULK125DK05	867.779444

Manual Part Number 1006034

Natural Gas Models Upflow

NUGS050AF01	867.769050
NUGS075AG01	867.769060
NUGS100AH01	867.769070
NUGS125AK01	867.769080

LP Models Upflow

NULS050AF01	867.779050
NULS075AG01	867.779060
NULS100AH01	867.779070
NULS125AK01	867.779080

INDEX TECHNICAL SUPPORT MANUALS

Manual Part Number 1006412

Natural Gas Models Upflow

NUGS050AF01 .... 867.769050
NUGS075BG01 .... 867.769061
NUGS100BH01 .... 876.769071
NUGS125AK01 .... 867.769080

LP Models Upflow

NULS050AF01 .... 867.779050
NULS075BG01 .... 867.779061
NULS100BH01 .... 867.779071
NULS125AK01 .... 867.779080

Manual Part Number 1007200

Natural Gas Models Upflow

NUGK040KF06 .... 867.769455
NUGK050MF06 .... 867.769415
NUGK050NF06 .... 867.769465
NUGK075DG07 .... 867.769426
NUGK075KG06 .... 867.769475
NUGK100DH09 .... 867.769438
NUGK100KH09 .... 867.769488
NUGK125DK07 .... 867.769446
NUGK125KK07 .... 867.769496

LP Models Upflow

NULK050MF05 .... 867.779414
NULK075DG05 .... 867.779424
NULK100DH06 .... 867.779435
NULK125DK06 .... 867.779445

Manual Part Number 1007213

Natural Gas Models Upflow

NUGS050AF02 .... 867.769051
NUGS075BG02 .... 867.769062
NUGS100BH02 .... 867.769072
NUGS125AK02 .... 867.769081

LP Models Upflow

NULS050AF02 .... 867.779051
NULS075BG02 .... 867.779062
NULS100BH02 .... 867.779072
NULS125AK02 .... 867.779081

INDEX TECHNICAL SUPPORT MANUALS

Manual Part Number 1007516

Natural Gas Models Upflow

NUGK040KF07
NUGK050MF07
NUGK050NF07
NUGK075DG08
NUGK075KG07
NUGK100DH11
NUGK100KH11
NUGK125DK08
NUGK125KK08

LP Models Upflow

NULK050MF06
NULK075DG06
NULK100DH07
NULK125DK07

Manual Part Number 1007567

Natural Gas Models Upflow

NUGS050AF03
NUGS075BG03
NUGS100BH03
NUGS125AK03

LP Models Upflow

NULS050AF03
NULS075BG03
NULS100BH03
NULS125AK03

INDEX TECHNICAL SUPPORT MANUALS

Manual Part Number 1004542

Natural Gas Models Counterflow

NDGK040KF03 . . . . 867.769502
NDGK050DF03 . . . . 867.769167
NDGK050KF03 . . . . 867.769512
NDGK075DF03 . . . . 867.769177
NDGK075KF03 . . . . 867.769522
NDGK100DG03 . . . . 867.769182
NDGK100KG03 . . . . 867.769532
NDGK125DK03 . . . . 867.769192
NDGK125KK03 . . . . 867.769542

LP Models Counterflow

NDLK050DF03 . . . . 867.779512
NDLK075DF03 . . . . 867.779522
NDLK100DG03 . . . . 867.779532
NDLK125DK03 . . . . 867.779542

Manual Part Number 1005332

Natural Gas Models Counterflow

NDGK040KF04 . . . . 867.769503
NDGK050DF04 . . . . 867.769168
NDGK050KF04 . . . . 867.769513
NDGK075DF04 . . . . 867.769178
NDGK075KF04 . . . . 867.769523
NDGK100DG04 . . . . 867.769183
NDGK100KG04 . . . . 867.769533
NDGK125DK04 . . . . 867.769193
NDGK125KK04 . . . . 867.769543

LP Models Counterflow

NDLK075DF04 . . . . 867.779513
NDLK075DF04 . . . . 867.779523
NDLK100DG04 . . . . 867.779533
NDLK125DK04 . . . . 867.779543

INDEX TECHNICAL SUPPORT MANUALS

Manual Part Number 1005998

Natural Gas Models Counterflow

NDGK040KF05 . . . . 867.769504
NDGK050DF05 . . . . 867.769169
NDGK050KF05 . . . . 867.769514
NDGK075DF05 . . . . 867.769179
NDGK075KF05 . . . . 867.769524
NDGK100DG05 . . . . 867.769184
NDGK100KG05 . . . . 867.769534
NDGK125DK04 . . . . 867.769193
NDGK125KK04 . . . . 867.769543

LP Models Counterflow

NDLK050DF05 . . . . 867.779514
NDLK075DF05 . . . . 867.779524
NDLK100DG05 . . . . 867.779534
NDLK125DK04 . . . . 867.779543

Manual Part Number 1006984

Natural Gas Models Counterflow

NDGK040KF05 . . . . 867.769504
NDGK050DF05 . . . . 867.769169
NDGK050KF05 . . . . 867.769514
NDGK075DF05 . . . . 867.769179
NDGK075KF05 . . . . 867.769524
NDGK100DG05 . . . . 867.769184
NDGK100KG05 . . . . 867.769534
NDGK125DK04 . . . . 867.769193
NDGK125KK04 . . . . 867.769543

LP Models Counterflow

NDLK050DF05 . . . . 867.779514
NDLK075DF05 . . . . 867.779524
NDLK100DG05 . . . . 867.779534
NDLK125DK04 . . . . 867.779543

INDEX TECHNICAL SUPPORT MANUALS

Manual Part Number 1007190

Natural Gas Models Counterflow

Model	Part Number
NDGK040KF05	867.769504
NDGK050DF05	867.769169
NDGK050KF05	867.769514
NDGK075DF06	867.769670
NDGK075KF06	867.769525
NDGK100DG05	867.769184
NDGK100KG05	867.769534
NDGK125DK04	867.769193
NDGK125KK04	867.769543

LP Models Counterflow

Model	Part Number
NDLK050DF05	867.779514
NDLK075DF06	867.779525
NDLK100DG05	867.779534
NDLK125DK04	867.779543

Manual Part Number 1007542

Natural Gas Models Counterflow

NDGK040KF06
NDGK050DF06
NDGK050KF06
NDGK075DF07
NDGK075KF07
NDGK100DG07
NDGK100KG06
NDGK125DK05
NDGK125KK05

LP Models Counterflow

NDLK050DF06
NDLK075DF07
NDLK100DG07
NDLK125DK05

Manufactured by:

Inter-City Products
Corporation

Lavergne, TN USA 37086
Brantford, ONT. CANADA N3T 5P4

Part No. 2146635

7/92

Installation Instructions

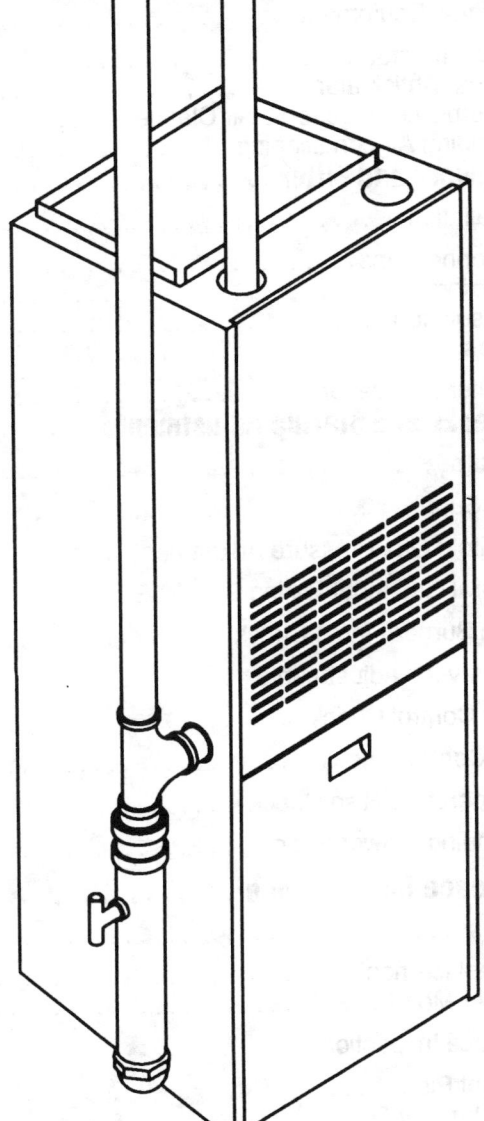

- Safety Rules
- Installation
- Ventilation Air
- Vent & Combustion Piping
- Gas Supply & Piping
- Wiring
- Ductwork Connections
- Start–Up
- Maintenance

WARNING

This furnace is not designed for use in mobile homes, trailers or recreational vehicles. Such use could result in property damage, bodily injury and/or death.

Direct Vent
Condensing Gas Furnace

TABLE OF CONTENTS

1. Safety Labeling and Signal Words

Danger, Warning and Caution

The signal words **DANGER, WARNING** and **CAUTION** are used to identify levels of hazard seriousness. The signal word **DANGER** is only used on product labels to signify an immediate hazard. The signal words **WARNING** and **CAUTION** will be used on product labels and throughout this manual and other manuals that may apply to the product.

Signal Words

DANGER – Immediate hazards which **WILL** result in severe personal injury or death.

WARNING – Hazards or unsafe practices which **COULD** result in severe personal injury or death.

CAUTION – Hazards or unsafe practices which **COULD** result in minor personal injury or product or property damage.

Signal Words in Manuals

The signal word **WARNING** is used throughout this manual in the following manner:

WARNING

The signal word **CAUTION** is used throughout this manual in the following manner:

CAUTION

Product Labeling

Signal words are used in combination with colors and/or pictures on product labels. Following are examples of product labels with explanations of the colors used.

Danger Label

Black printing on a white background except the word **DANGER** which is white with a red background.

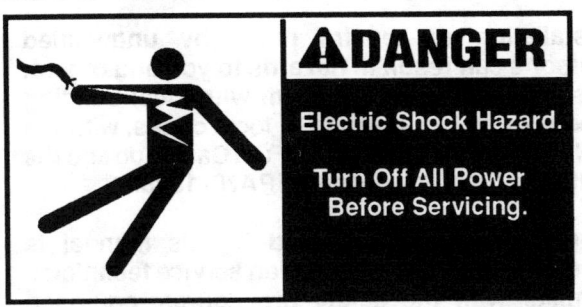

Warning Label

Black printing on a white background except the word **WARNING** which is black with a orange background.

Caution Label

Black printing on a white background except the word **CAUTION** which is black with a yellow background.

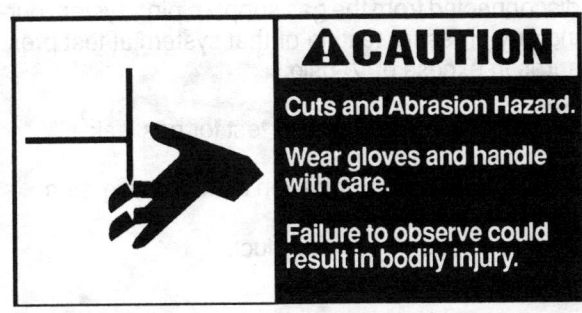

2. Safe Installation Requirements

Safe Installation Requirements

WARNING

Installation or repairs made by unqualified persons can result in hazards to you and others. Insttallation MUST conform with local building codes or, in the absence of local codes, with the ANSIZ223.1–1990 National Fuel Gas Code and the National Electrical Code NFPA70–1988.

The information contained in this manual is intended for use by a qualified service technician familiar with the safety procedures, equipped with the proper tools and test instruments.

Failure to carefully read and follow all instructions in this manual can result in furnace malfunction, property damage, personal injury and/or death.

- Do **NOT** use this furnace as a construction heater.

- This furnace is **NOT** approved for installation in mobile homes, trailers or recreation vehicles.

- Use only the type of gas approved for this furnace (See Rating Plate). Overfiring will result in failure of heat exchanger and cause dangerous operation.

- The furnace and its individual shutoff valve **MUST** be disconnected from the gas supply piping system during any pressure testing of that system at test pressures in excess of $\frac{1}{2}$ psig.

- Do **NOT** use open flame to test for gas leak.

- Provide adequate ventilation air to furnace area.

- Seal supply and return air ducts.

- Furnace must be properly vented to the outside.

It is the personal responsibility and obligation of the customer to contact a qualified installer to assure that the installation conforms to governing codes and ordinances.

WARNING

Carbon Monoxide Poisoning Hazard.

This furnace can NOT be common-vented or connected to any type B, BW or L vent or vent connector, nor to any portion of a factory-built or masonry chimney. If this furnace is replacing a previously common-vented furnace, it may be necessary to resize the existing vent line and chimney to prevent oversizing problems for the other remaining appliances(s). See National Fuel Gas Code, ANSI Z223.1-1990. This furnace MUST be vented to the outside.

Failure to properly vent this furnace or other appliances can result in property damage, personal injury and/or death.

Venting and Combustion Air Check

NOTE: If this installation removes an existing furnace from a common venting system serving other appliances, and to make sure there is adequate combustion air for all appliances, **MAKE THE FOLLOWING CHECK.**

1. Seal any unused openings in the venting system(s).

2. Visually inspect all venting systems for proper size and horizontal pitch to ensure there is no blockage or restriction, leakage, corrosion or other deficiencies which could cause an unsafe condition.

3. Insofar as is practical, close all doors and windows between the space in which the appliances are located and other spaces of the building.

4. Turn on clothes dryers. Turn on any exhaust fans, such as range hoods and bathroom exhausts, so they will operate at maximum speed. Do not operate a summer exhaust fan. Close fireplace dampers.

5. Follow the lighting instructions for each appliance and place it in operation. Adjust thermostat or other control so appliance will operate continuously.

6. Test for spillage at the draft hood relief opening after 5 minutes of main burner operation. Use the flame of a match or candle. Flame should draw towards vent pipe.(**Figure 1**) Repeat this for all appliances that have a draft hood opening.

7. After it has been determined that each appliance vents properly when tested as outlined, return doors, windows, exhaust fans, fireplace dampers and any other gas–burning appliance to their previous condition of use.

8. If improper venting is observed during any of the above tests, the venting system (or other cause) **MUST** be corrected. Refer to the appropriate tables in Appendix G in the National Fuel Gas Code, ANSI Z223.1–1988.

NOTE: If flame pulls towards draft hood, this indicates sufficient infiltration air.

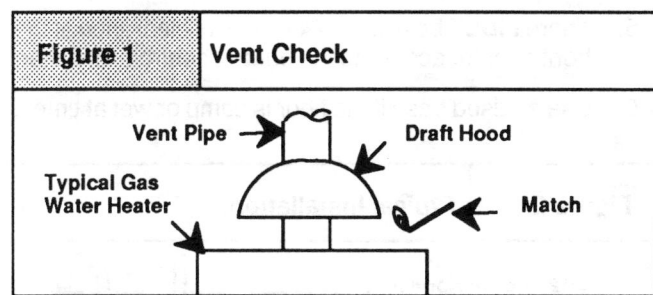

Figure 1 Vent Check

3. Helpful Information

Refer to the following booklets for additional information:

National Fuel Gas Code
NFPA/ANSI Z223.1 – 1988 (or current edition)

National Electrical Code
NFPA 70–1990 (or current edition)

For a nominal charge, these booklets can be ordered from:

American National Standards Institute
1430 Broadway
New York, NY 10018

Basic Tools and Material

Pipe Wrenches	Plenum
Tin Snips	Gas Pipe
Screw Drivers	Thermostat and Wire
Hammer	Electrical Material
Wire Cutting Pliers	Disconnect Switch
Awl	Pipe and Duct Hangers
Drill and Metal Bits	Vent Pipe
Level	Sheet Metal Screws
Folding Rule or Tape	Duct Tape
Sheet Metal Duct	Pipe Joint Compound

4. Installation

NOTE: Installation **MUST** conform with local building codes or in the absence of local codes, with the American National Standards, Z223.1-1988 *National Fuel Gas Code* and the *National Electrical Code*, NFPA70-1990 or current edition.

Location and Clearances

1. Refer to **Figure 2** for typical installation and basic connecting parts required. Supply and return air plenums and duct are also required.

2. If furnace is a replacement, it is usually best to install the furnace where the old one was. Choose the location or evaluate the existing location based upon the minimum clearance requirements and furnace dimensions (**Figure 3**).

CAUTION
Do NOT locate furnace where temperatures may drop below freezing causing improper operation or damage to equipment. Do NOT operate furnace in a corrosive atmosphere containing chlorine, fluorine or any other damaging chemicals. Refer to Ventilation Air Section, *Contaminated Air*.

Installation Requirements

1. Install furnace level.

2. Install furnace as centralized as practical with respect to the heat distribution system.

3. Install the vent and combustion pipes as short as practical, but they **MUST** be at least 5 feet and no more than 40 feet with 4 elbows. (**See Vent and Combustion Piping in this manual**).

4. Do **NOT** install furnace directly on carpeting, tile or other combustible material other than wood flooring.

5. There **MUST** be at least 24 inches of clearance at the front door for access to the burner, controls and filter.

6. Use a raised base if the floor is damp or wet at times.

7. Residential garage installations require:

- Burners and ignition sources installed at least 18 inches above the floor.

- Furnace protected from possible vehicle damage.

Figure 2	Typical Installation

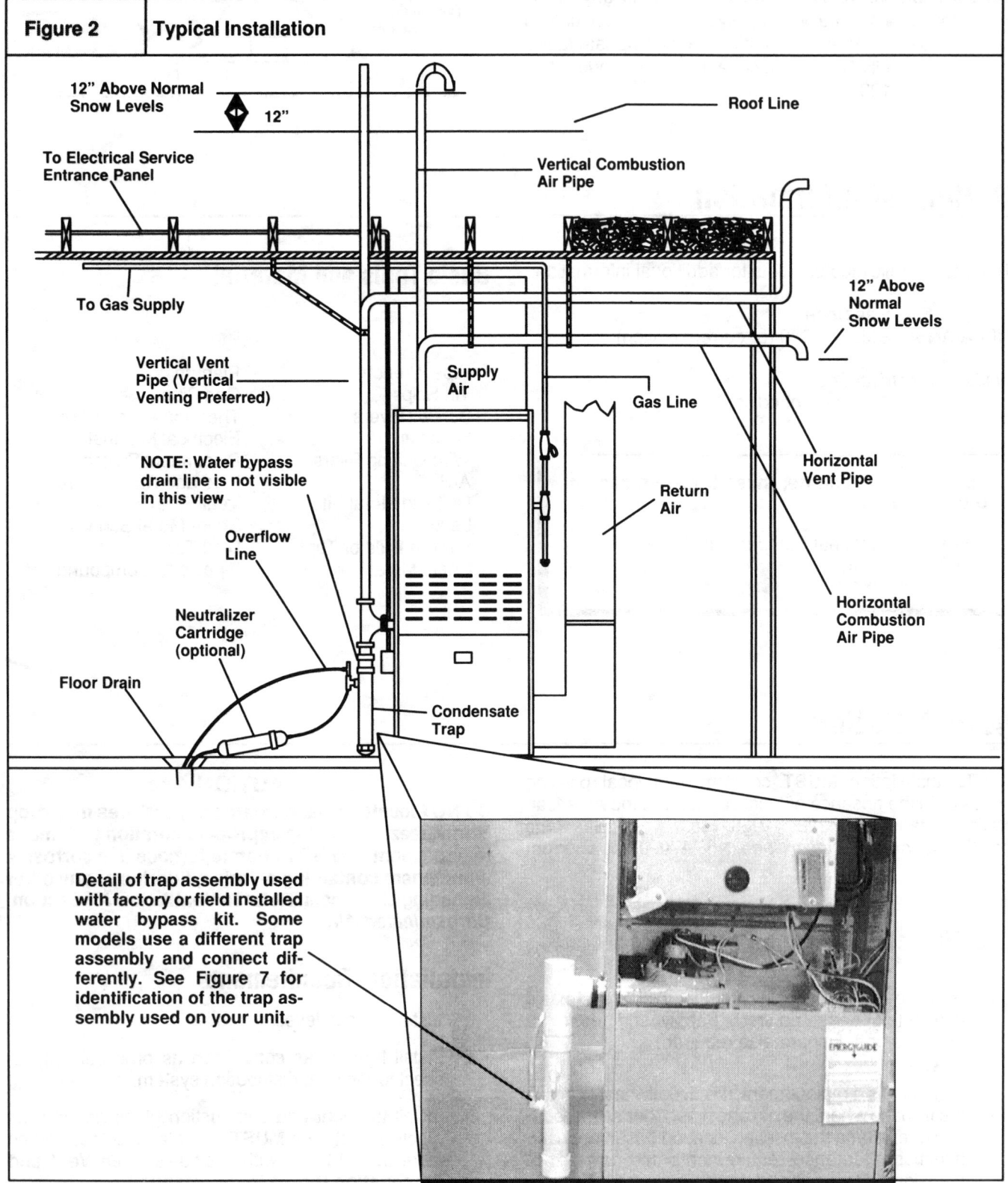

12" Above Normal Snow Levels

12"

Roof Line

To Electrical Service Entrance Panel

Vertical Combustion Air Pipe

To Gas Supply

12" Above Normal Snow Levels

Vertical Vent Pipe (Vertical Venting Preferred)

Supply Air

Gas Line

NOTE: Water bypass drain line is not visible in this view

Return Air

Horizontal Vent Pipe

Overflow Line

Neutralizer Cartridge (optional)

Horizontal Combustion Air Pipe

Floor Drain

Condensate Trap

Detail of trap assembly used with factory or field installed water bypass kit. Some models use a different trap assembly and connect differently. See Figure 7 for identification of the trap assembly used on your unit.

| Figure 3 | Dimensions and Clearances |

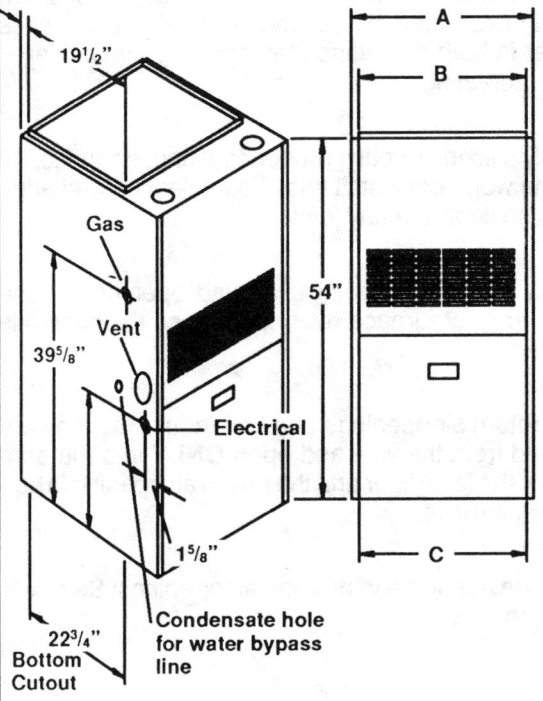

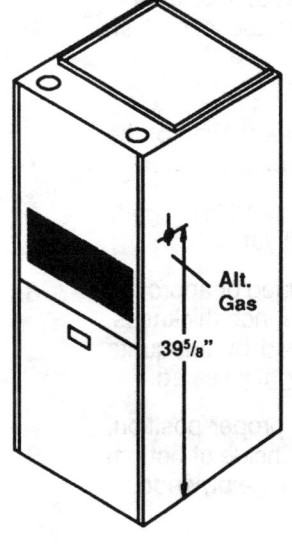

Standard Input BTUH	A	B	C
50 – 75,000	18¼	16¾	14
100,000	21¾	20¼	14
125,000	25½	24	21½

For fire protection clearance, wood and other combustible materials **MUST NOT** be closer than:

0" to flue pipe
1" from top of furnace
3" from front of furnace
0" from rear of furnace
1" from left side of furnace
1" from right side of furnace

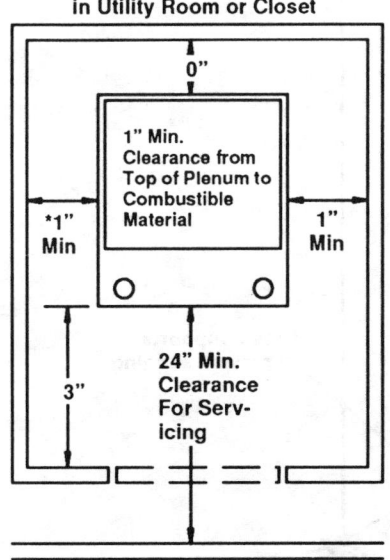

Furnace installation in Utility Room or Closet

*1 inch clearance to combustible material. 4 inch clearance required for venting pipe and drain assembly.

Furnace installation in Basement

24" Min. Clearance For Filter Removal

Outline of Base, If Used

Return Air Cabinet (If Used)

0"

18" Min. Clearance For Access to Rear of Return Air Cabinet

1" Min. Clearance from Top of Plenum to Combustible Material

24" Min. Clearance For Servicing

False Floor Closet Installation

NOTE: This type of installation is in a closet with the furnace placed over an opening in a false floor raised above the regular floor (**Figure 4**). False floor closet installation **MUST** be made in accordance with local building and fire codes.

1. If an existing closet is being reworked, shorten and frame the door so the door bottom is above, or flush with the location of false floor.

2. Construct a level framework of 2 x 4's or 2 x 6's around the inside of the closet. Make the top of the false floor at least 12 inches above existing floor.

3. Cut a piece of plywood (for the false floor) to fit tightly inside the closet, but do **NOT** install yet.

NOTE: Plywood **MUST** be A-C exterior glued or approved for underlayment use, with a minimum ³/₄ inch thickness. Maximum size is 4' x 4' unless supported by adequate bracing with any joints in the false floor tightly sealed.

4. Set the furnace on the plywood in proper position. Open front access panel and mark inside of bottom opening on plywood. Cut opening in the plywood.

5. Unless existing floor is concrete, place a piece of sheet metal on existing floor under position of opening in false floor. Sheet metal should be 2 inches wider in both directions than opening and centered under opening.

6. Apply silicone or butyl rubber caulking around top of frame work and install false floor. Seal around edge of false floor and any joints.

7. Apply a bead of caulking around opening in false floor and set furnace over opening so the bottom is sealed.

8. Cut return air openings and box them in so they are sealed from the wall and open **ONLY** into the area below the false floor and the return air opening in bottom of furnace.

9. Provide adequate ventilation air openings. **See Ventilation Air.**

Figure 4	Closet False Floor Installation

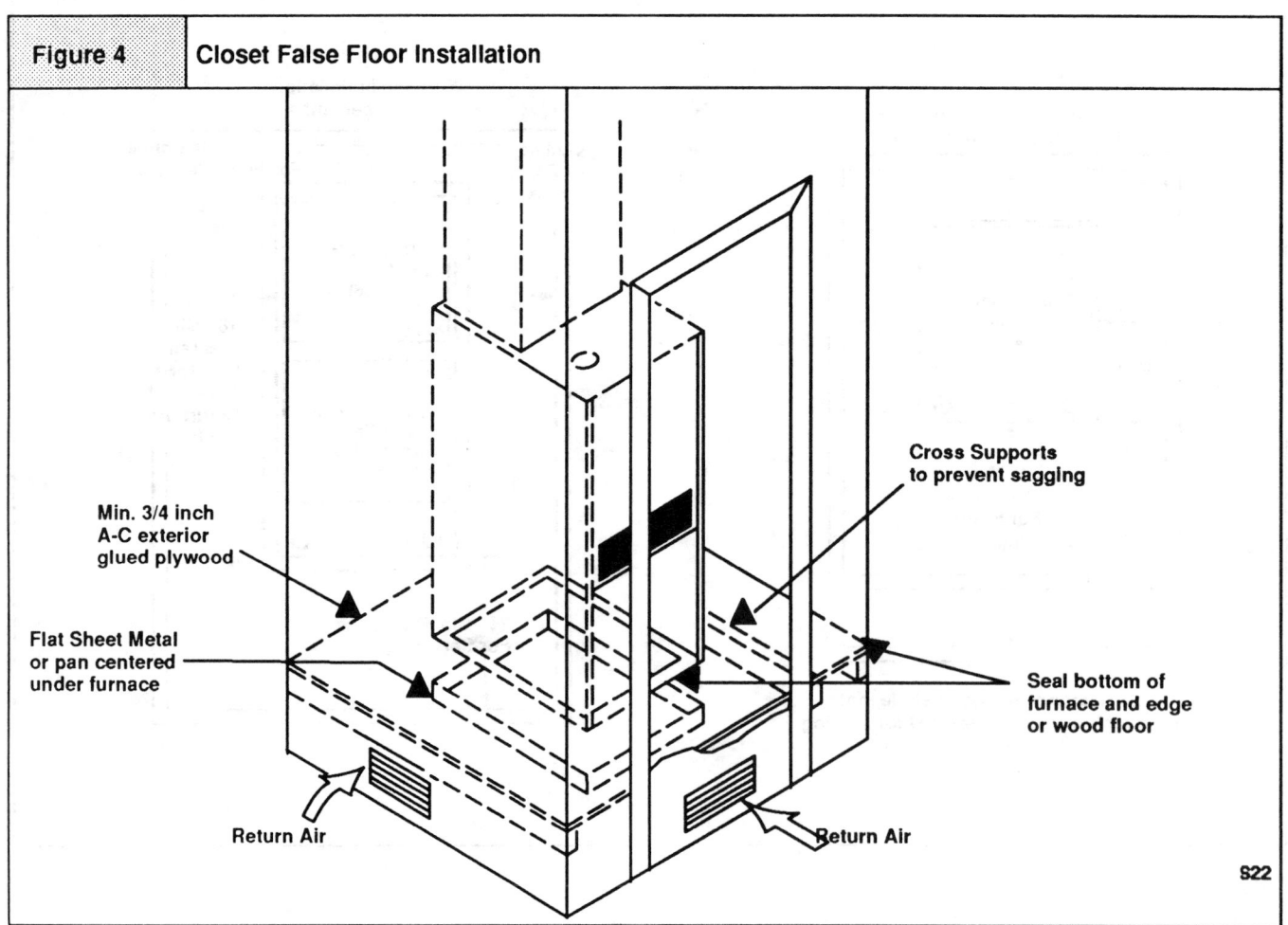

Min. 3/4 inch A-C exterior glued plywood

Flat Sheet Metal or pan centered under furnace

Cross Supports to prevent sagging

Seal bottom of furnace and edge or wood floor

Return Air

Return Air

S22

5. Ventilation Air

Ventilation air is required to maintain furnace components at a safe operating temperature. These instructions are to provide Ventilation Air **ONLY** for the Direct Vent Furnace. These instructions do **NOT** allow for the combustion air requirements for any appliances.

Unconfined space installation

An unconfined space is defined as a space whose volume is not less than 50 cubic feet per 1,000 BTUH of the aggregate (total) input of all appliances installed in that space. Rooms communicating (adjoining) directly with the space in which the appliances are installed, through openings not furnished with doors, are considered a part of the unconfined space.

Normally, no special ventilating means are required for furnaces installed in unconfined spaces.

Confined Space Installation

A confined space is is defined as space with whose volume is less than 50 cubic feet per 1,000 BTUH of the aggregate (total) input rating of all appliances installed in that space Provide confined space with **adequate** air for proper ventilation for furnace.

NOTE: Furnaces that have been installed in accordance with the *False Floor Closet installation* instructions in this manual can also be provided adequate ventilation air as follows:

Requirements

Installation in confined space or closet requires the following:

1. Install a fully louvered door on the closet, or

2. Provide two openings for air, one near the top and one near the bottom of the enclosure or door. Size each opening to accept a grille or louver whose free area is at least one square inch per 1,000 Btuh input of the furnace.

WARNING

Fire and carbon monoxide poisoning hazard.

If other fuel–burning appliances are installed in the same area as the direct vent furnace, they MUST be provided with adequate combustion air as well as ventilation air.

Failure to provide adequate combustion or ventilation air can result in property damage, personal injury and/or death.

---

6. Vent and Combustion Air Piping

WARNING

Carbon monoxide poisoning, fire and explosion hazard.

Read and follow all instructions in this section

Failure to properly vent this furnace can result in property damage, personal injury and/or death.

General Information

NOTE: All instructions **MUST** be followed unless in conflict with type of material being used or local codes.

This furnace removes both sensible and latent heat from the combustion flue gases. Removal of latent heat results in condensation of flue gas water vapor. This condensed water vapor drains from secondary heat exchanger and out of the unit into a PVC drain trap.

Combustion air is obtained through piping running directly from the combustion burner box cover and terminating outdoors in the same atmospheric pressure zone as the vent pipe.

Vent kits are supplied with fittings for 2 inch round PVC (poly–vinyl chloride) or CPVC (chlorinated poly–vinyl chloride) schedule 40 vent pipe. (3 inch round PVC kits are supplied for 75,000, 100,000 and 125,000 BTU models.)

Joining Pipe and Fittings

WARNING

Fire hazard.

Provide adequate ventilation and do NOT assemble near heat source or open flame. Do NOT smoke while using solvent cements and avoid contact with skin or eyes.

Observe all cautions and warnings printed on material containers to prevent possible personal injury and/or death.

NOTE: All PVC and CPVC pipe fittings, solvent cement, primers and procedures **MUST** conform to American National Standard Institute and American Society for Testing and Materials (ANSI/ASTM) standards.

- *Pipe and Fittings* – ASTM D1785, D2466 and D2665
- *PVC Primer and Solvent Cement* – ASTM D2564
- *Procedure for Cementing Joints* – Ref ASTM D2855

CAUTION

Do NOT use solvent cement that has become curdled, lumpy or thickened an do NOT thin. Observe shelf precautions printed on containers. For applications below 32 degrees F., use only low temperature type solvent cement.

NOTE: Cut, prepare and assemble all vent and combustion air pipe before any joint is permanently cemented.

1. Cut pipe end square, remove ragged edges and burrs. Chamfer end of pipe, then clean fitting socket and pipe joint of all dirt, grease or moisture.

NOTE: Stir the solvent cement frequently while using. Use a natural bristle brush or the dauber supplied with the cement. The proper brush size is one inch.

2. After checking pipe and socket for proper fit, wipe socket and pipe with cleaner-primer. Apply a liberal coat of primer to inside surface of socket and outside of pipe. Do **NOT** allow primer to dry before applying cement.

3. Apply a thin coat of cement evenly in the socket. Quickly apply a heavy coat of cement to the pipe end and insert pipe into fittings with a slight twisting movement until it bottoms out.

NOTE: Cement **MUST** be fluid while inserting pipe, if **NOT** recoat pipe.

4. Hold the pipe in the fitting for 30 seconds to prevent the tapered socket from pushing the pipe out of the fitting.

5. Wipe all excess cement from the joint with a rag. Allow 15 minutes before handling. Cure times varies according to fit, temperature and humidity.

Vent and Combustion Air Piping

WARNING

Carbon monoxide poisoning hazard.

Cement or mechanically seal all joints, fittings, etc. to prevent leakage of flue gases.

Failure to properly seal vent piping can result in personal injury and/or death.

Installation Guidelines

NOTE: All vent piping **MUST** be installed in compliance with Part 7, *Venting of Equipment, National Fuel Gas Code* NFPA54/ANSI Z223.1–1988, (or current edition) local codes or ordinance, these instructions, and good trade practices.

1. Vertical piping is preferred because there will be some moisture in the flue gases that may condense as it leaves the vent pipe (See *Special Instruction For Horizontal Vents*).

2. The vent **MUST** exit the furnace at the lower left side.

3. Piping diameter **MUST NOT** be reduced.

4. All piping from the furnace to termination **MUST** slope upwards a minimum of 1/4 inch per foot of run.

5. Use DWV type long radius elbows to change from a vertical run to a horizontal run. If other type elbows are used, then use two, 45 degree elbows in place of one 90 degree elbow with the elbows slightly misaligned to provide slope in the horizontal run.

6. All horizontal pipe runs **MUST** be supported at least every four feet with metal pipe strapping. **NO** sags or dips are permitted.

7. All vertical pipe runs **MUST** be supported every six feet where accessible.

8. The maximum pipe length is 40 total feet. Up to four, 90 degree elbows can be used.

NOTE: Do **NOT** count the take off tee on a horizontal run.

9. The minimum pipe run length is 5 feet.

10. Do **NOT** install the piping in the same chase with a metal or high temperature plastic vent, such as Ultravent from another gas or other fuel burning appliance.

11. Do **NOT** install the piping within 6 inches of vent pipe from another gas or other fuel burning appliance.

12. The piping can be run in the same chase or adjacent to supply or vent pipe for water supply or waste plumbing. It can also be run in the same chase with a vent from another 90+ furnace.

13. The vent outlet **MUST** be installed so as to terminate in the same atmospheric pressure zone as the combustion air intake.

Insulation Guidelines

1. Armaflex or equivalent closed cell foam insulation is recommended. Use one inch thickness or multiple layers if required for extreme climate conditions.

2. If Fiberglas or equivalent insulation is used it must have a vapor barrier. Use R values of 7 up to 10 feet, R–11 if exposure exceeds 10 feet.

3. If it is necessary to insulate piping when a chimney is used as a chase. The top of the chimney **MUST** be sealed flush or crowned up so **ONLY** the piping protrudes.

NOTE: If situations require pipe to be run on the exterior wall to reach a suitable termination point, it **MUST** be properly insulated. It **MUST** be boxed in and sealed against moisture if Fiberglas insulation is used.

4. When the vent or combustion air pipe height above the roof exceeds 30 inches, or if an exterior vertical riser is used on a horizontal vent to get above snow levels, the exterior portion **MUST** be insulated. Use **ONLY** moisture resistant insulation such as Armaflex or other equivalent type of insulation.

5. When combustion air inlet piping is installed above a suspended ceiling, the pipe **MUST** be insulated with moisture resistant insulation such as Armaflex or other equivalent type of insulation.

6. Insulate combustion air inlet piping when run in warm, humid spaces such as basements.

Furnace Connections

WARNING

Carbon monoxide poisoning hazard.

Do NOT install furnace so that indoor air is used for combustion.

Failure to provide outside combustion air can result in personal injury and/or death.

Combustion Air

50,000 BTU MODELS ONLY

1. Connect 2″ PVC pipe for combustion air (except 75,000, 100,000 and 125,000 BTU) to the adapter on top of the Combustion Air Box, **Figure 5.**

75,000, 100,000 and 125, 000 BTU MODELS ONLY

2. A 22 inch x 2½ inch diameter pipe and 3 inch to 2½ inch reducer is furnished to connect the combustion air pipe to furnace. Install as shown in **Figure 5**, and connect 3″ PVC pipe for combustion air to the adapter on top of the furnace.

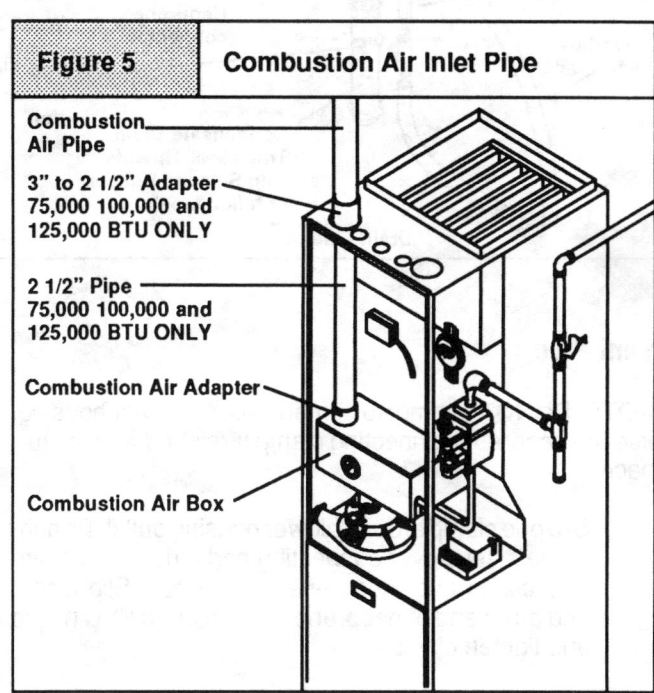

| Figure 5 | Combustion Air Inlet Pipe |

Combustion Air Pipe

3″ to 2 1/2″ Adapter 75,000 100,000 and 125,000 BTU ONLY

2 1/2″ Pipe 75,000 100,000 and 125,000 BTU ONLY

Combustion Air Adapter

Combustion Air Box

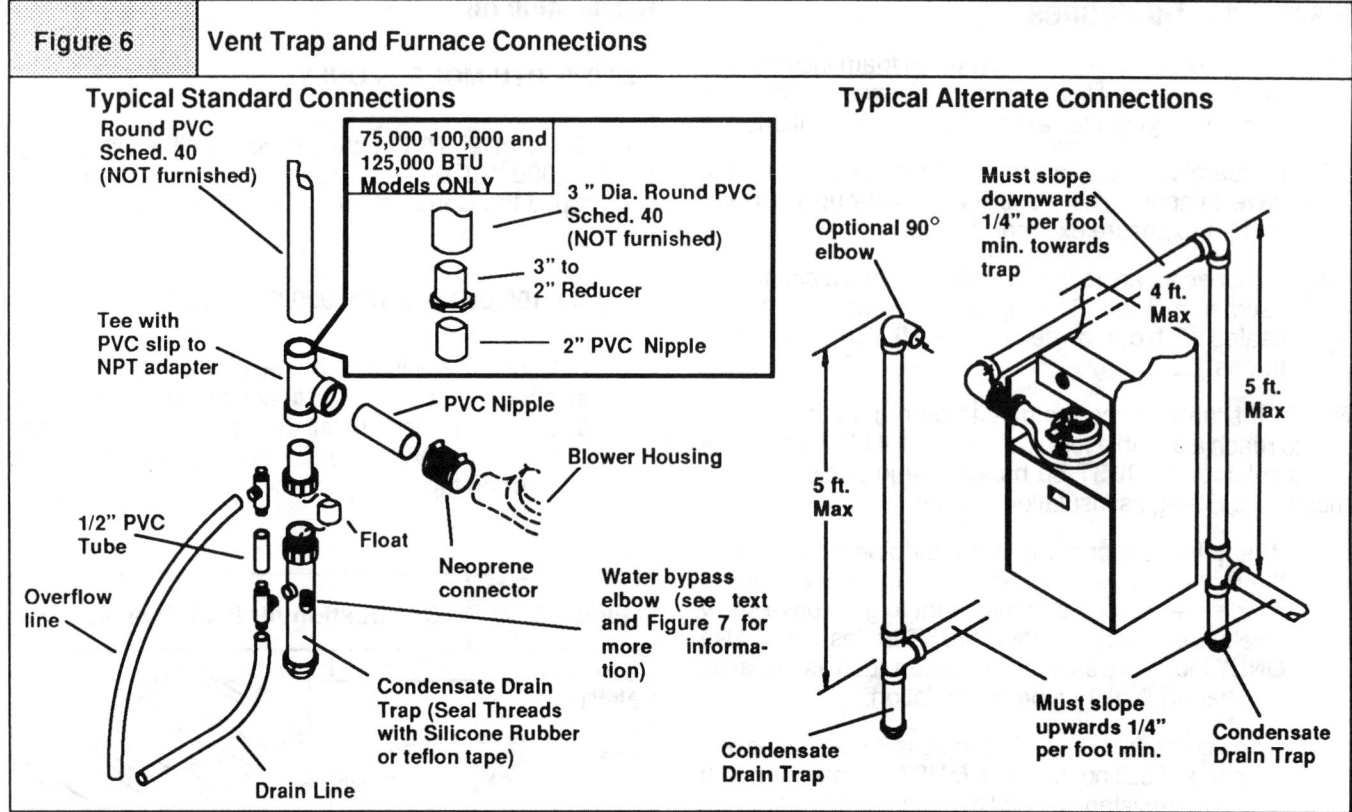

Figure 6 | Vent Trap and Furnace Connections

Vent Pipe

NOTE: Do **NOT** attempt to cement pipe to blower housing, use the neoprene connecting clamp furnished with the furnace.

1. Slip one clamp around blower housing outlet. Fit neoprene sleeve over connecting parts then slide stainless clamp over neoprene and tighten. Slip clamp and other end of neoprene connector on PVC nipple and tighten clamp.

2. If vertical rise exceeds six feet, secure the tee and trap assembly to the furnace with a pipe clamp (field supplied) to remove excessive weight from clamp connection.

3. The optional 90 degree elbow (PVC DWV socket X SLIP/SPIGOT, field siuupplied) may be fastened to the combination blower outlet coupling using the 2" x 2" PVC nipple. Refer to **Figure 6** *Typical Alternate Connections* for drain trap assembly.

Drain Trap Assembly

1. Install the drain trap assembly to provide the necessary 5 inches water column against vent pressure. Ensure all parts fit properly and are correctly oriented before cementing.

2. Install drain trap assembly within 4 feet horizontally and 5 feet vertically (**lower only**) of the furnace blower housing. Refer to typical examples in **Figure 6**.

3. The drain trap **MUST** be reasonably accessible for the homeowner to check.

Condensate Drain and Neutralizer

NOTE: Drain line and overflow line can be ¹/₂ inch PVC flex tube or schedule 40 with a disconnect union so the trap can be removed. Trap assembly provides 5 inches water column so no additional trap is required. Drains must terminate at an inside drain to prevent freezing of condensate and possible property damage.

A separate water bypass elbow is located between the secondary heat exchanger and blower inlet. Tubing is routed out of the cabinet through a grommetted hole in cabinet and connected to fitting at side or bottom of trap. See **Figure 7** to determine which type of trap is used on your unit. Tubing **MUST** route directly to water bypass connection without excess sags or loops in line. Secure tubing with clamps.

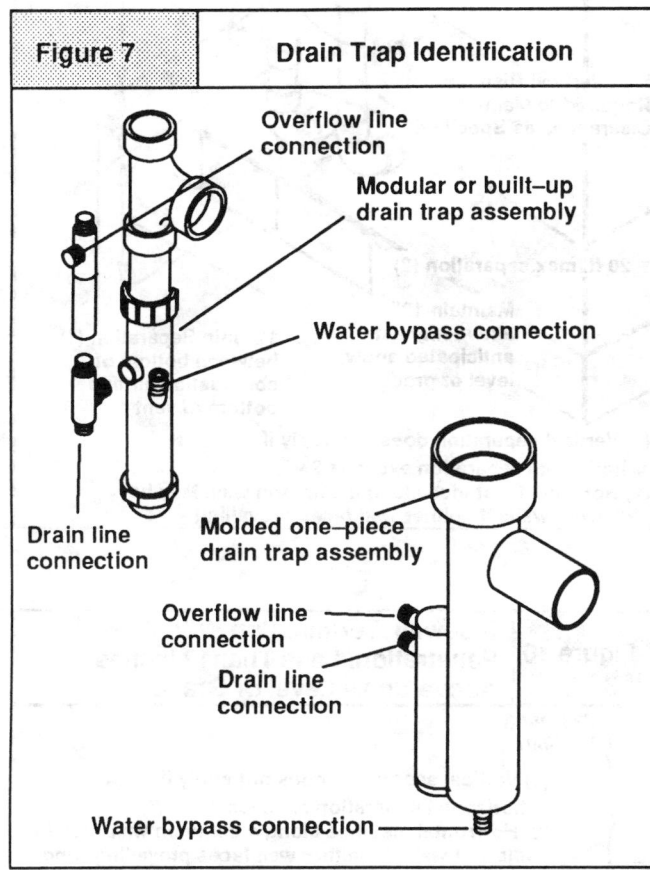

Figure 7	Drain Trap Identification

Overflow line connection

Modular or built–up drain trap assembly

Water bypass connection

Drain line connection

Molded one–piece drain trap assembly

Overflow line connection

Drain line connection

Water bypass connection

1. A condensate or sump pump **MUST** be used if local codes require, or if no inside floor drain is available. A condensate neutralizer cartridge **MUST** be installed in the drain line in a horizontal position **ONLY** unless pump is approved for use with acidic condensate.

2. Install an overflow line if routing to floor drain or sump pump.

3. A condensate pump **MUST** have an auxiliary safety switch to prevent operation of furnace and resulting overlow of condensate in the event of pump failure. The safety switch **MUST** be wired through the R circuit **ONLY** (low voltage) to provide operation in either heating or cooling modes.

Horizontal Piping Runs

NOTE: vent and combustion air piping can be installed horizontally through an outside wall using all of the applicable instructions under *Vent and Combustion Air Installation Guidelines* and these additional requirements. The requirements and limitations for horizontal venting are very strict. **ALL** horizontal vent installations **MUST** be made in accordance with these instructions.

Location and Clearance Requirements

CAUTION
The combustion products and moist flue gases can condense as they leave the terminal elbow. The condensate may freeze on the exterior wall, under the eaves and on surrounding objects. Some discoloration to the exterior of the building could occur.

NOTE: The vent and combustion air pipe location **MUST** meet the requirements listed in the following instructions or applicable codes, whichever specifies the most clearance or strictest limitations.

1. Maintain a 12 inch clearance above ground level, above highest anticipated snow levels and 6 inches out from wall.

NOTE: Ice or snow may cause the furnace to shut down if the vent or combustion air inlet becomes obstructed. If required, use a vertical riser (**Figures 8, 9** and **10**) or shield vent to prevent blockage from drifting snow.

2. Do **NOT** install vent outlet above any walk way or area that may create a hazard or nuisance or be detrimental to the operation of other equipment.

3. Install vent outlet at least four feet from and **NOT** above or below any door, window, gravity inlet or forced air inlet for the building.

4. Install vent outlet at least 4 feet away from any soffit or under eave vent.

5. Do **NOT** install vent outlet under any kind of patio or deck.

6. Locate piping on the side of the building away from prevailing winter winds when practical, but taking into consideration other limitations to determine the best overall location.

Combustion Air and Vent Termination

NOTE: Refer to **Figures 8, 9** and **10** for installation of vent and combustion air termination elbow. Combustion air intake and vent **MUST** terminate in the same atmospheric pressure zone.

CAUTION
Maintain a minimum of 36 inches between combustion air inlet and clothes dryer vent. Locate combustion air inlet as far as possible from swimming pool and swimming pool pump house.

1. Cut two, 2½ inch (3⅝ inch for 75,000 100,000 and 125,000 BTU models) diameter holes through the exterior wall. Do **NOT** make the holes oversized, or it will be necessary to add a sheet metal or plywood plate on the outside with the correct size hole in it.

2. Check hole sizes by making sure it is smaller than the couplings or elbows that will be installed on the out-side. The couplings or elbows **MUST** prevent the pipe from being pushed back through the wall.

3. Extend vent pipe and combustion air pipe through the wall ¾ to 1 inch and seal area between pipe and wall.

Figure 8	**Rooftop Termination**

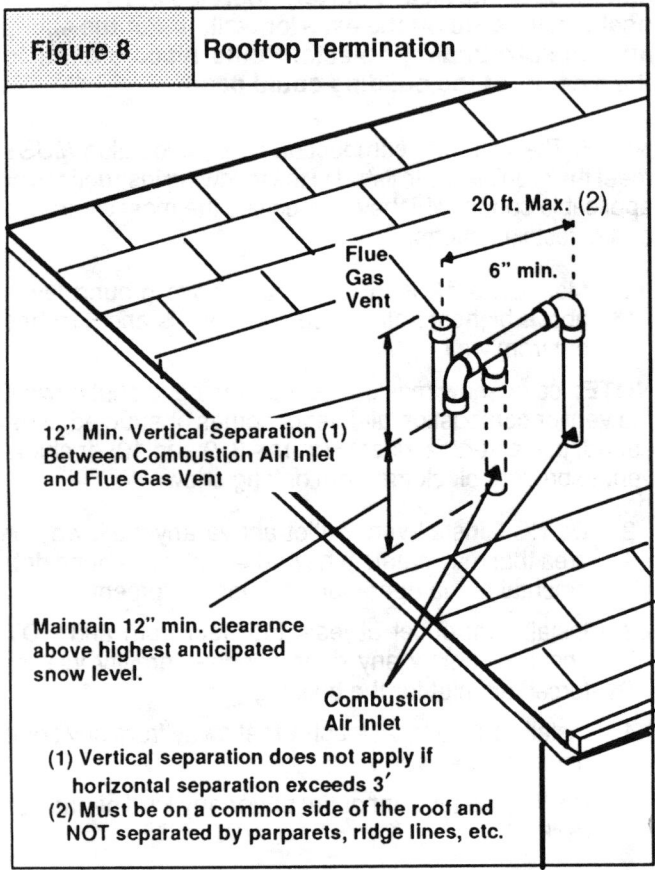

20 ft. Max. (2)

Flue Gas Vent

6" min.

12" Min. Vertical Separation (1) Between Combustion Air Inlet and Flue Gas Vent

Maintain 12" min. clearance above highest anticipated snow level.

Combustion Air Inlet

(1) Vertical separation does not apply if horizontal separation exceeds 3′
(2) Must be on a common side of the roof and NOT separated by parparets, ridge lines, etc.

4. Install the couplings, nipple and termination elbows as shown and maintain spacing between vent and combustion air piping as indicated in **Figures 8, 9, and 10**.

Vent Termination with Exterior Risers

1. Install elbows and pipe to form riser as shown in **Figure 9**.

2. Secure vent pipe to wall with galvanized strap or other rust resistant material to restrain pipe from moving.

3. Insulate pipe with Armaflex or equivalent moisture resistant closed cell foam insulation or Fiberglass in-sulation if boxed in and sealed against moisture.

Figure 9	**Sidewall Termination With Wall Penetration. Less Than 12 Inches Above Snow Level or Grade**

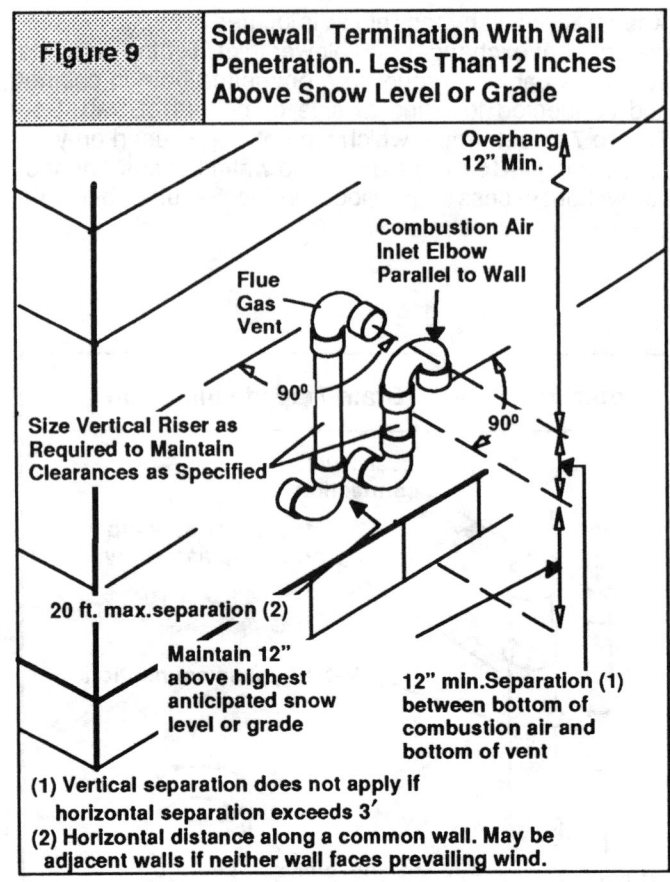

Overhang 12" Min.

Combustion Air Inlet Elbow Parallel to Wall

Flue Gas Vent

90°

90°

Size Vertical Riser as Required to Maintain Clearances as Specified

20 ft. max. separation (2)

Maintain 12" above highest anticipated snow level or grade

12" min. Separation (1) between bottom of combustion air and bottom of vent

(1) Vertical separation does not apply if horizontal separation exceeds 3′
(2) Horizontal distance along a common wall. May be adjacent walls if neither wall faces prevailing wind.

Figure 10	**Sidewall Termination With Wall Penetration. Less Than 12 Inches Above Snow Level or Grade**

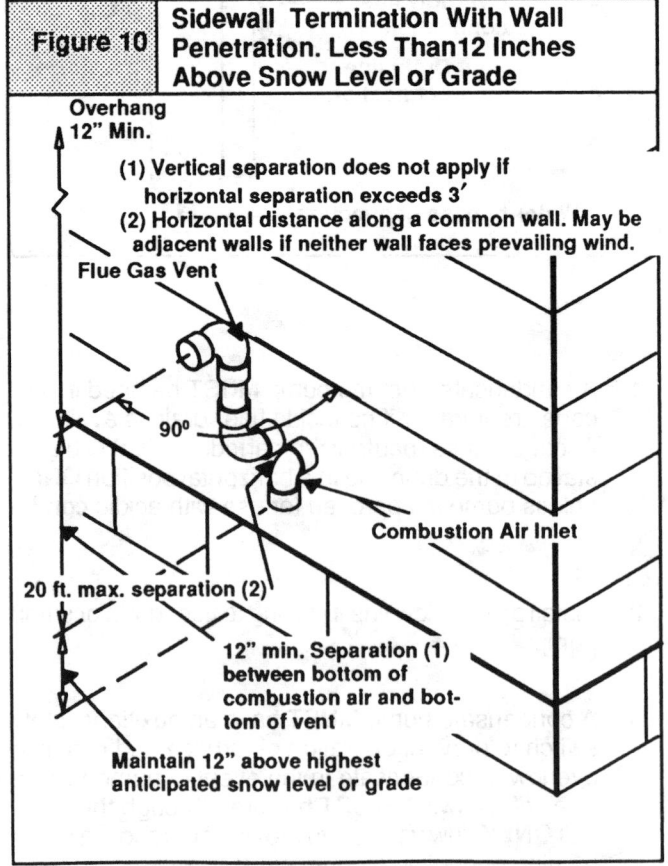

Overhang 12" Min.

(1) Vertical separation does not apply if horizontal separation exceeds 3′
(2) Horizontal distance along a common wall. May be adjacent walls if neither wall faces prevailing wind.

Flue Gas Vent

90°

Combustion Air Inlet

20 ft. max. separation (2)

12" min. Separation (1) between bottom of combustion air and bot-tom of vent

Maintain 12" above highest anticipated snow level or grade

7. Gas Supply and Piping

WARNING

Fire and explosion hazard.

Natural Gas

Models designated for Natural Gas are to be used with Natural Gas ONLY.

LP Gas

LP gas models have orifices sized for commercially pure propane gas. Furnace MUST NOT be used with butane or a mixture of butane unless properly sized orifices are installed by a licensed LP installer.

Failure to follow these instructions can result in property damage, personal injury and/or death.

NOTE: The rating plate is stamped with the model number, gas type and gas input rating. See Section 10 (*Checks and Start–Up Adjustments*) for proper procedures for adjusting manifold pressure.

Gas Pressures

1. Gas input to burners MUST NOT exceed the rated input shown on rating plate.

2. Do NOT allow minimum gas supply pressure to vary downward. Doing so will decrease input to furnace. Refer to Figure 11 for gas supply and manifold pressures.

3. For operation above 2,000 feet, orifice change or manifold pressure adjustment may be required to suit gas supplied. Check with gas supplier. If orifice sizing is needed, it should be based on reducing the input rating by 4 percent for each 1,000 feet above sea level.

Figure 11	Gas Pressures			
Gas Type	Supply Pressure			Manifold Pressure
	Recommended	Max.	Min.	
Natural	7 inches	14 inches	4.5 inches	3.5 inches
LP	11 inches	14 inches	11 inches	10 inches

Figure 12	Orifice Sizes			
Gas Type	Manifold Pressure	Specific Gravity	Heating Value (BTU per Cubic Ft.)	Orifice Size (Drill #)
Natural	3.5 " W.C.	0.6	800	40
			900	41
			1000	42
			1100	43
Propane	10 " W.C.	1.53	2500	54

Orifice Sizing

WARNING

Fire hazard.

Change orifices to obtain desired manifold pressure. Do NOT adjust manifold pressure more than ± 0.3 inches water column to obtain rated input. Do NOT set input rating above that shown on rating plate.

Failure to properly set input pressure can result in property damage, personal injury and/or death.

NOTE: The furnace is supplied with standard orifices for gas shown on rating plate. Factory sized orifices for natural and LP gas are listed in the furnace Technical Support manual and on the rating plate.

1. Ensure furnace is equipped with the correct main burner orifices.

2. Refer to Figure 12 for correct orifice size for a given heating valve and specific gravity for natural and propane gas.

Changing Orifices

WARNING

Electrical shock, fire or explosion hazard.

Turn OFF electric power (at disconnect) and gas supply (at manual valve in gas line) when installing orifices. Installation of orifices requires a qualified service technician.

Failure to properly install orifices can result in property damage, personal injury and/or death.

NOTE: Main burner orifices can be changed for high altitudes.

1. Disconnect gas line from gas valve.

2. Remove Combustion Box front cover and manifold from furnace.

3. Remove the orifices from the manifold and replace them with properly sized orifices.

4. Tighten orifices so there is 1.65 (1–$^{21}/_{32}$) inch from the face of the orifice to the back side of the manifold, **Figure 13**.

5. Reinstall manifold and Combustion Air Box Cover. Ensure burners do **NOT** bind on new orifices.

Figure 13	Changing Orifices

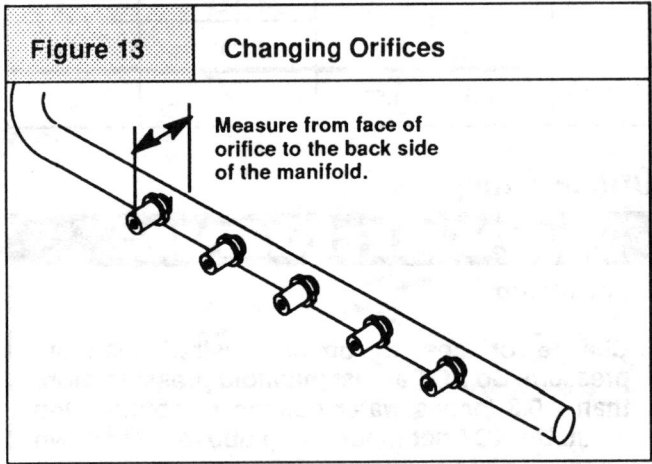

Measure from face of orifice to the back side of the manifold.

Gas Piping Requirements

1. Determine the minimum pipe size from the tables in **Figure 14** and **Figure 15**, basing the length of the run from the main line, gas meter or source to the furnace.

2. Properly size gas pipe to handle combined appliance load or run gas pipe directly from gas meter or LP gas regulator.

3. Install correct pipe size for run length and furnace rating.

4. Measure pipe length from gas meter or LP second stage regulator.

Connections

NOTE: Refer to **Figure 17** for the general layout at the furnace. The rules listed apply to natural and LP gas pipe installations.

1. Use black iron or steel pipe and fittings or other pipe approved by local code.

3. Use ground joint unions and install a drip leg no less than 3 inches long to trap dirt and moisture before it can enter gas valve.

4. Use two pipe wrenches when making connections to prevent gas valve from turning.

5. Provide a $^{1}/_{8}$ inch National Pipe Thread (NPT) plug for test gauge connection immediately up stream of gas supply connection to furnace.

6. Install a manual shut–off valve and tighten all joints securely.

Additional LP Connection Requirements

1. Have a licensed LP gas dealer make all connections at storage tank and check all connections from tank to furnace.

2. If copper tubing is used, it **MUST** comply with limitation set in National Fuel Gas Code.

3. Two–stage regulation of LP gas is recommended.

Figure 14	Natural Gas Pipe Sizes/Capacity				
Natural Gas Capacity **BTU Per Hour Input** (In Thousands)					
Pipe Length (Feet)	Pipe Size I.D.				
	$^{3}/_{8}$"	$^{1}/_{2}$"	$^{3}/_{4}$"	1"	1$^{1}/_{4}$"	1$^{1}/_{2}$"
10	72	132	278	520	1,050	1.600
20	49	92	190	350	730	1,100
30	40	73	152	285	590	890
40	34	63	130	245	500	760
50	30	56	115	215	440	670
60	27	50	105	195	400	610

NOTE: Piping that is too small will prevent the proper amount of gas from reaching the furnace.

Figure 15	LP Gas Pipe Sizes/Capacity					
LP Gas Capacity **BTU Per Hour Input** (In Thousands)						
Pipe Length (Feet)	Copper Tubing O.D.			Iron Pipe O.D.		
	³/₈"	¹/₂"	³/₄"	¹/₂"	³/₄"	1"
10	39	92	329	275	567	1,071
20	26	62	216	189	393	732
30	21	50	181	152	315	590
40	19	41	145	129	267	504
50	—	37	131	114	237	448
60	—	35	121	103	217	409

NOTE: Copper tubing for gas supply MUST comply with limitations in National Fuel Gas code, reference 2.6.3 Metallic Tubing

NOTE: If a gas connector is used, it **MUST** be acceptable to local authority. Connector may **NOT** be used inside the furnace or be secured or supported by the furnace or ductwork. Connectors MUST comply with one of the following standards or a superseding standard.

- ANSI Z21.24a-1983, *Metal Connectors for Gas Appliances*
- ANSI Z21.45b-1983, *Flexible Connectors of Other Than All-Metal Construction for Gas Appliances.*

2. Use pipe joint compound on external (male) threads **ONLY**. Joint compound **MUST** be resistant to any chemical action of LP gases (**Figure 16**).

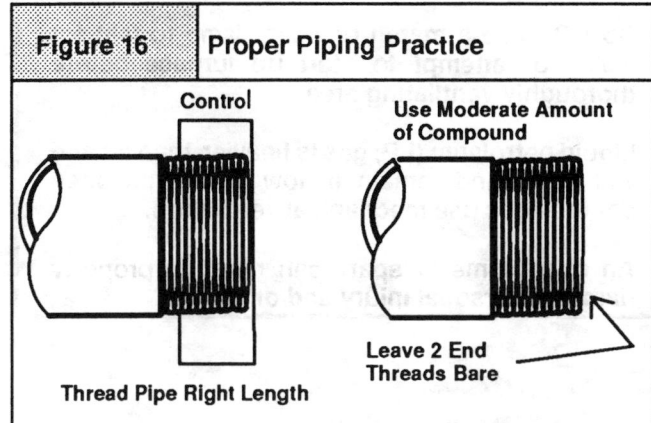

Figure 16	Proper Piping Practice

Control

Use Moderate Amount of Compound

Leave 2 End Threads Bare

Thread Pipe Right Length

Figure 17	Typical Gas Piping

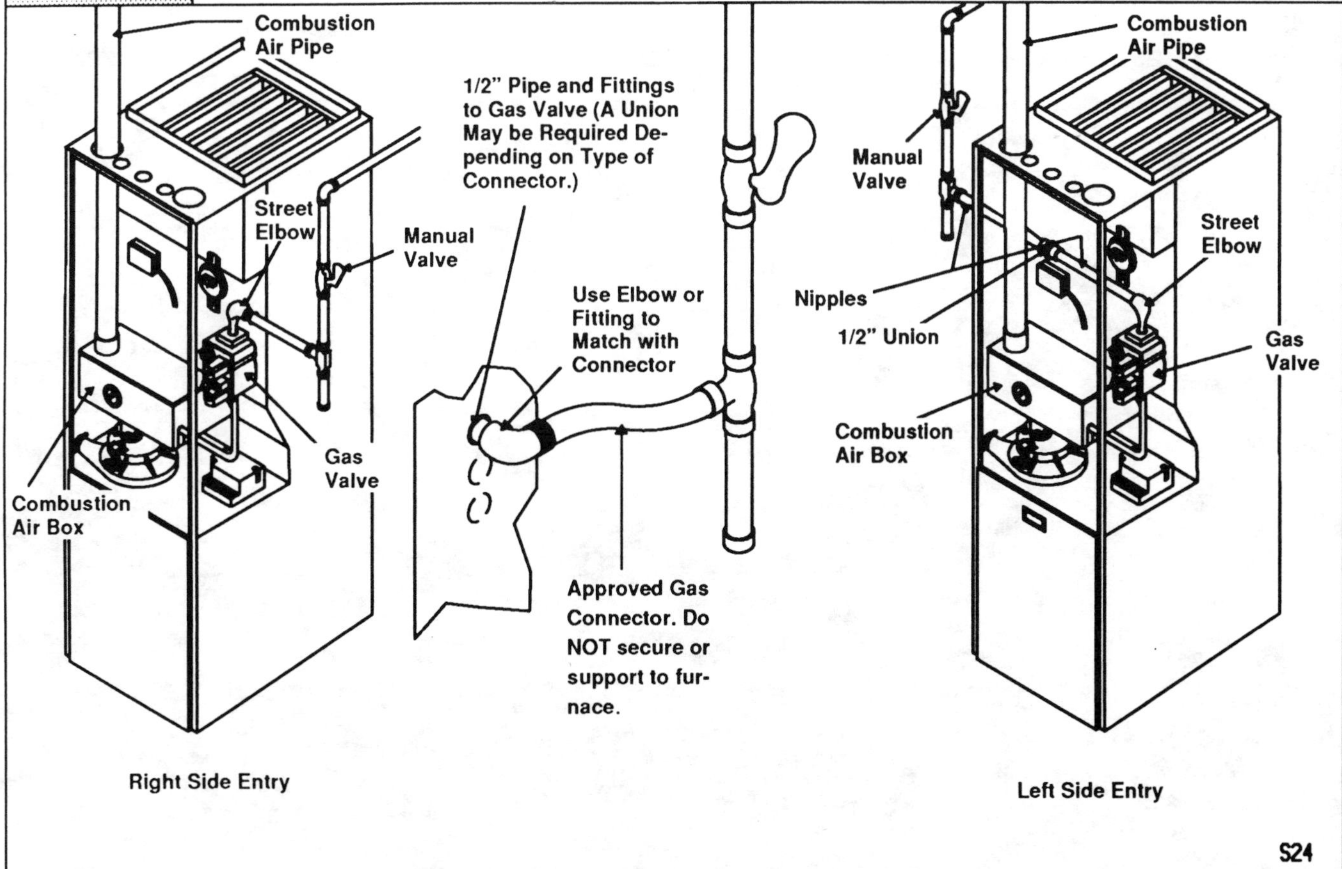

Combustion Air Pipe

1/2" Pipe and Fittings to Gas Valve (A Union May be Required Depending on Type of Connector.)

Street Elbow

Manual Valve

Gas Valve

Combustion Air Box

Use Elbow or Fitting to Match with Connector

Approved Gas Connector. Do NOT secure or support to furnace.

Right Side Entry

Combustion Air Pipe

Manual Valve

Nipples

1/2" Union

Combustion Air Box

Street Elbow

Gas Valve

Left Side Entry

S24

Final Check

1. Test all pipes for leaks.

2. When checking gas piping to furnace, shut **OFF** manual gas valve to furnace.

3. Apply soap suds (or a liquid detergent) to each joint. Bubbles forming indicate a leak.

4. Correct even the smallest leak at once.

5. If orifices were changed, make sure they are checked for leakage.

6. Gas pressure **MUST NOT** exceed ½ PSIG. Checking gas piping above ½ PSIG requires the furnace and manual shut-off valve to be disconnected during testing.

WARNING

Fire or explosion hazard.

Do NOT use a match or open flame to test for leaks, or attempt to start up furnace before thoroughly ventilating area.

Liquid petroleum (LP) gas is heavier than air and will settle and remain in low areas and open depressions use mechanical ventilation.

An open flame or spark can result in property damage, personal injury and/or death.

WARNING

Fire or explosion hazard.

Do NOT exceed specified gas pressures for testing.

Exceeding these pressures can result in property damage, personal injury and/or death

8. Electrical Wiring

WARNING

Electrical shock hazard.

Turn OFF electric power at fuse box or service panel before making any electrical connections and ensure a proper ground connection is made before connecting line voltage.

Failure to do so can result in property damage, personal injury and/or death.

NOTE: A large washer is provided in the parts bag to use with the strain relief or conduit connector. Install connectors through hole in side of furnace then washer and conduit nut.

Grounding

NOTE: A ground lug wire is provided for ground connection. Use #14 AWG insulated copper connection from furnace to service panel or properly driven ground rod.

Power Supply

NOTE: Line voltage circuit is completely factory wired. Make all line voltages connections inside furnace connection or main junction box. **ALL** electrical work **MUST** conform with local codes, ordinances and the national electrical code, NFPA 70–1990 or current edition.

1. Run #14 AWG hot, neutral and ground wires to furnace from a 15 amp power supply circuit through disconnect switch.

NOTE: The power supply to the furnace **MUST** be polarized for the hot surface ignition to work correctly.

2. Do **NOT** connect furnace to existing lighting or other circuit.

3. Do **NOT** complete line voltage connections until furnace is permanently grounded.

4. Make all line voltage and ground connections with copper wires.

5. Complete line voltage connections inside junction box (**Figure 18**).

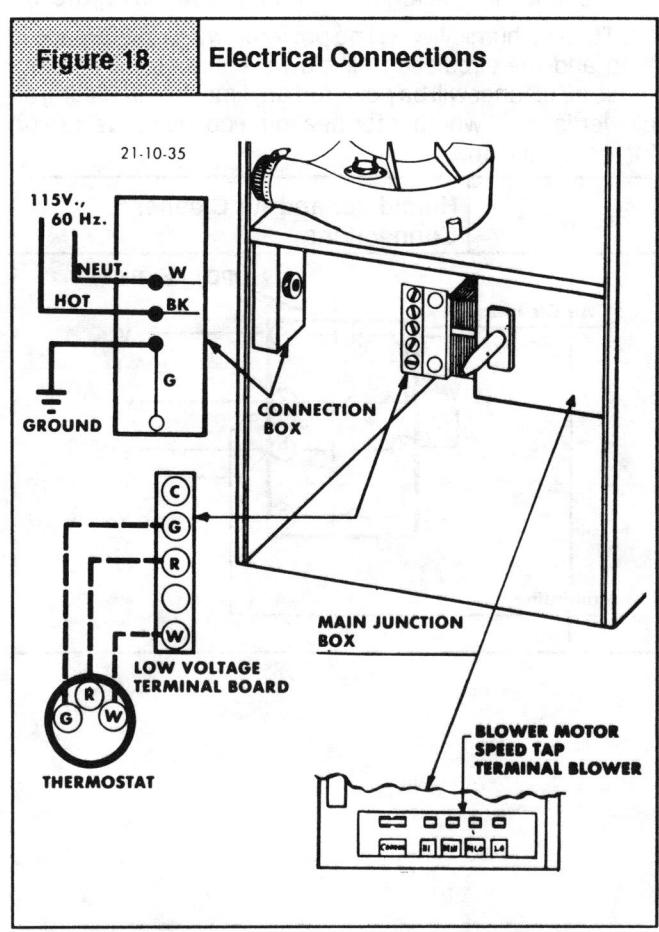

Figure 18	Electrical Connections

Optional Equipment

NOTE: All wiring (except thermostat) from furnace to optional equipment **MUST** conform to the temperature limitations for type **T** wire. Install wiring in accordance with manufacturer's instructions.

Thermostat

NOTE: Thermostat location has an important effect on the operation of the unit. Follow instructions included with thermostat for correct mounting and wiring.

Heat Anticipator

1. Set thermostat heat anticipator in accordance with thermostat instructions to value shown in furnace *Technical Support Manual.*

Humidifier/Electronic Air Cleaner

1. Make connections through use of a sail switch installed in ductwork if the furnace has a SPDT fan relay with only three terminals.

NOTE: If manufacturer does **NOT** supply sail switch, consult place of purchase.

2. Make power connection to furnace fan relay as shown in **Figure 19**. (Furnace is equipped with a DPDT fan relay with six terminals.)

3. Make connections for neutral and ground wires with connections to furnace at wiring shown in **Figure 18**.

NOTE: The humidifier will be powered when the furnace is fired and the circulating air blower comes on. The electronic air cleaner will be powered anytime the circulating air blower is on— whether for heating, cooling or just fan on for air circulation.

Figure 19	Humidifier and Air Cleaner Connections

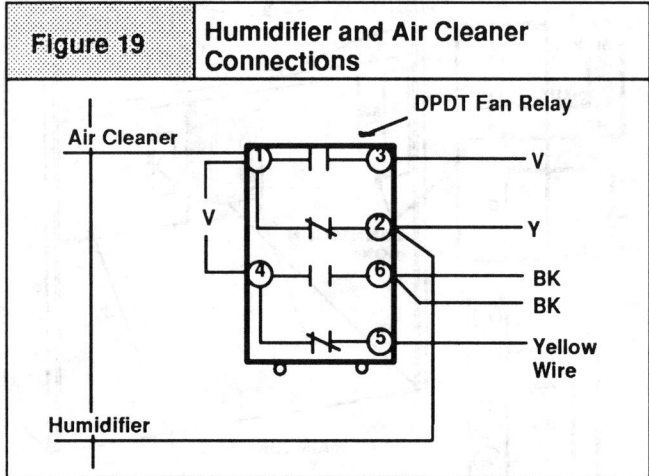

Adding Air conditioning

1. Replace heating only thermostat and cable with heat/cooling thermostat and 4-wire thermostat cable if required.

2. Connect from **W**, **G**, **Y** and **R** on thermostat to **W**, **G**, **Y** and **R** on furnace low voltage terminal board.

3. Connect wires from contactor on condensing unit to **Y** and **C** on low voltage terminal board at furnace.

4. Follow all instructions with condensing unit and evaporator coil.

NOTE: The furnace fan relay will change fan speeds automatically as heat and cool are selected at the thermostat.

9. Ductwork and Filter

WARNING

Carbon monoxide poisoning hazard.

Do NOT draw return air from inside a closet or utility room. Return air duct MUST be sealed to furnace casing.

Failure to properly seal duct can result in personal injury and/or death,

Installation

NOTE: Design and install air distribution system to comply with Air Conditioning Contractors of America manuals or other approved methods that conform to local codes and good trade practices.

1. When furnace supply ducts carry air outside furnace area, seal return air duct to furnace casing and terminate duct outside furnace space .

2. Install air conditioning cooling coil (evaporator) on outlet side of furnace.

3. If separate evaporator and blower units are used, install good sealing dampers for air flow control. Chilled air going through the furnace could cause condensation and shorten the furnace life.

NOTE: Dampers (field supplied) can be either automatic or manual. Manually operated dampers MUST be equipped with a means to prevent furnace or air conditioning operation unless damper is in the full heat or cool position.

Connections

NOTE: Return air can enter through either, or both sides, the bottom, and (by using the optional rear return air cabinet), the back side.

1. For side connections using a 14" x 25" filter, cut out the embossed area ONLY—up to the nibs on each side. This will provide a 12³/₄" x 23³/₄" approximate opening. For side connections using a 16" x 25" filter, cut out the embossed area shown in **Figure 21**. This will provide a 14³/₄" x 23³/₄" approximate opening.

2. Bottom returns can be made by removing the knockout panel in the furnace base. Do **NOT** remove knock-out except for a bottom return.

3. If furnace is equipped with a 20" x 25" or 24" x 25" filter, it must use a bottom return or two side returns using 14" x 25" filters (**NOT** supplied).

4. Installation of locking-type dampers are recommended in all branches, or in individual ducts to balance system's air flow.

5. Non-combustible, flexible duct connectors are recommended for return and supply connections to furnace.

6. If air return grill is located close to the fan inlet, install at least one, 90 degree air turn between fan and inlet grill to reduce noise.

NOTE: To further reduce noise, install acoustical air turning vanes and/or line inside of duct with acoustical material.

Sizing

1. Existing or new ductwork **MUST** be sized to handle the correct amount of airflow for either heating only or heating and air conditioning.

Insulation

1. Insulate ductwork installed in attics or other areas exposed to outside temperatures with a minimum of 2 inch insulation and vapor barrier.

2. Insulate ductwork in indoor unconditioned areas with a minimum of 1 inch insulation with indoor type vapor barrier.

Filters

NOTE: The furnace is provided with 1 filter, either the disposable low velocity type or the washable high velocity type. The size and type of filter supplied with the furnace will handle the airflow required if central air conditioning is used with the furnace.

1. Use either filter type:
   - Washable, high velocity filter based on a maximum air flow rating of 500 FPM.
   - Disposable, low velocity filter based on a maximum air flow of 300 FPM when used with filter grill.

NOTE: Disposable, low velocity filters may be replaced with washable, high velocity filter providing they meet the minimum size areas listed in **Figure 20**. Washable, high velocity filters can be replaced **ONLY** with same type and size.

CAUTION
If filters are suitable for heating application, advise homeowner that filter size will need to be increased if air conditioning is added.

Figure 20	Recommended Filter Area					
CFM Airflow	Disposable Low Velocity (300 FPM Air Flow)			Washable High Velocity (500 FPM Air Flow)		
	Minimum Surface Area (Sq, In.)	Recommended Nominal Size	Qty.	Minimum Surface Area (Sq, In.)	Recommended Nominal Size	Qty.
800	384	20 x 25	1	231	14 x 20	1
900	432	20 x 25	1	260	15 x 20	1
1000	480	20 x 30	1	288	14 x 25	1
1100	528	20 x 30	1	317	15 x 25	1
1200	576	14 x 25	2	346	16 x 25	1
1300	624	14 x 25	2	375	20 x 25	1
1400	672	16 x 25	2	404	20 x 25	1
1500	720	16 x 25	2	432	20 x 25	1
1600	768	20 x 25	2	461	20 x 25	1
1700	816	20 x 25	2	490	20 x 25	1
1800	864	20 x 25	2	519	20 x 30	1
1900	912	20 x 30	2	548	24 x 25	1
2000	960	20 x 30	2	576	24 x 25	1

Figure 21	Side Return Air Cutout

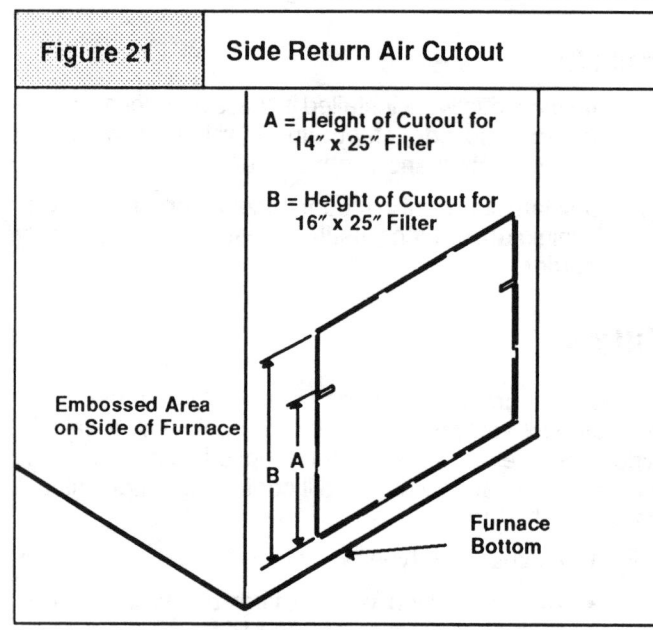

A = Height of Cutout for 14" x 25" Filter

B = Height of Cutout for 16" x 25" Filter

Embossed Area on Side of Furnace

Furnace Bottom

CAUTION

Use ONLY the required knockout that matches return air connections.

1. Insert end of filter strap through slot so that notched area of strap rest in slot. Use pliers to twist end of strap about 30 to 45 degrees to lock strap in place.

2. Use pliers to bend hook in the front edge of the side panel 90 degrees towards center so the filter strap will rest in the notched area.

3. For a bottom return, the filter strap fits under the flange on the front edge of the furnace base.

4. Install filter and secure with filter strap.

Figure 22	Filter Straps

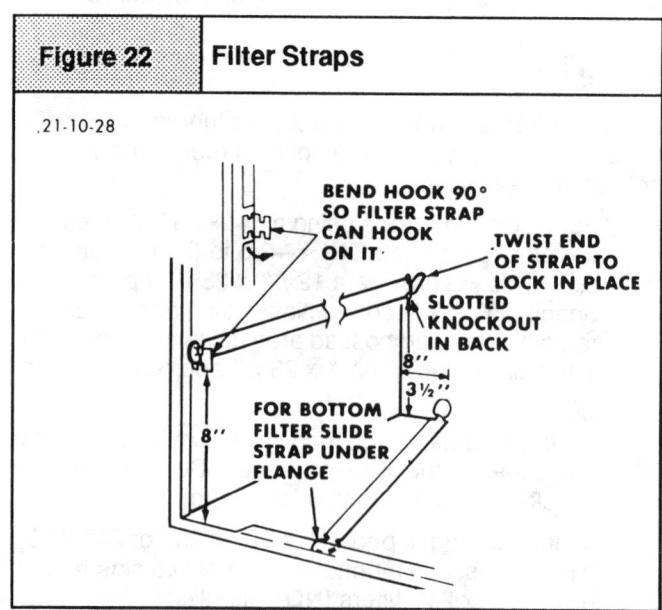

.21-10-28

BEND HOOK 90° SO FILTER STRAP CAN HOOK ON IT

TWIST END OF STRAP TO LOCK IN PLACE

SLOTTED KNOCKOUT IN BACK

FOR BOTTOM FILTER SLIDE STRAP UNDER FLANGE

8"

3½"

8"

Filter Installation

NOTE: There are slotted knockouts approximately $\frac{1}{16}$" x $\frac{1}{2}$" on the back of the furnace (**Figure 21**) that are used to secure the filter strap. Knockouts for side filters are 8 inches up from the bottom and 1 inch in from the side. For a bottom filter the knockouts (one on each side) are 3½ inches in and 1 inch up from bottom. If a combination side and bottom return is used, **ONLY** use one strap on the bottom filter and the other strap for the side filter.

10. Checks and Startup Adjustments

WARNING

Fire or explosion hazard

Thoroughly ventilate area before attempting to start furnace to eliminate any gas accumulation.

Failure to do so can result in property damage, personal injury or death.

Startup

NOTE: Refer to startup procedures in the *Owners Manual*.

CAUTION

If any sparks, odors or unusual noises occur, immediately shut OFF power to furnace. Check for wiring errors or obstruction to blower.

Gas Supply

1. Gas supply pressure should be within minimum and maximum values listed on rating plate. Pressures are usually set by gas suppliers.

Figure 23	Gas Pressure Settings
Gas Type	Manifold Pressure Inches Water Column
Natural	3.5
LP	10

IMPORTANT NOTES
Combustion Air Box Cover MUST be removed when adjusting manifold pressure.

- On LP gas, the rated input is obtained when the BTU content is 2,500 BTU per cubic foot and manifold pressure set at 10 inches.

- If LP gas has a different BTU content, orifices MUST be changed by licensed LP installer.

- Measured input MUST NOT exceed rated input.

- On installations above 2,000 feet, reduce rated input by 4 percent per 1,000 feet above sea level. (Consult your gas supplier)

- Any major change in gas flow requires changing burner orifice size.

Manifold Gas Pressure Adjustment

NOTE: Make adjustment to manifold pressure with burners operating and **combustion air box cover removed**.

1. Connect U-Tube manometer to the tapped opening on the outlet side of gas valve. Use manometer with a 0 to 12 inches water column range.

WARNING

Fire or explosion hazard.

Turn OFF gas at manual shut off valve in gas line before connecting U-Tube manometer.

Failure to do so can result in property damage, personal injury and/or death.

2. Remove adjustment screw cover on gas valve. Turn counterclockwise to decrease pressure and clockwise to increase.

3. Set pressure to value shown in **Figure 23**, ± 0.3 inches water column. (Refer to **Important Notes**). Pressure is also listed on furnace rating plate.

Natural Gas Input Rating Check

NOTE: The gas meter can be used to measure input to furnace. Rating is based on a natural gas BTU content of 1,000 BTU's per cubic foot. Check with gas supplier for actual BTU content.

1. Turn **OFF** gas supply to all other appliances and start furnace, **combustion air box cover MUST be in place**.

2. Time how many seconds it takes the one cubic foot (normally the smallest) dial on the gas meter to make one complete revolution. Refer to **Example**.

Example			
Natural Gas BTU Content	No. of Seconds Per Hour	Time In Seconds	BTU Per Hour
1,000	3,600	48	75,000
1,000 x 3,600 ÷ 48 = 75,000 BTUH			

NOTE: If meter uses a 2 cubic foot dial, divide results (seconds) by two.

3. Relight all appliances and ensure all pilots are operating.

Main Burner Flame Check

NOTE: The main burner flames can be observed through the sight glass in the combustion box in most instances. If a more detailed observation is required, tape a piece of Plexiglas in place of front cover when making flame adjustments.

1. Check for the following (**Figure 24**):

   • Stable, soft and blue flames

   • Flames extending directly upward without curling, floating or lifting off burner

   • Flames do **NOT** touch sides of heat exchanger

NOTE: Dust may cause orange tips or wisps of yellow, but **MUST NOT** have solid, yellow tips.

2. Check main burner flames monthly.

3. If any problems with main burner flames are noted, it may be necessary to adjust to gas pressures, check for drafts or adjust primary air.

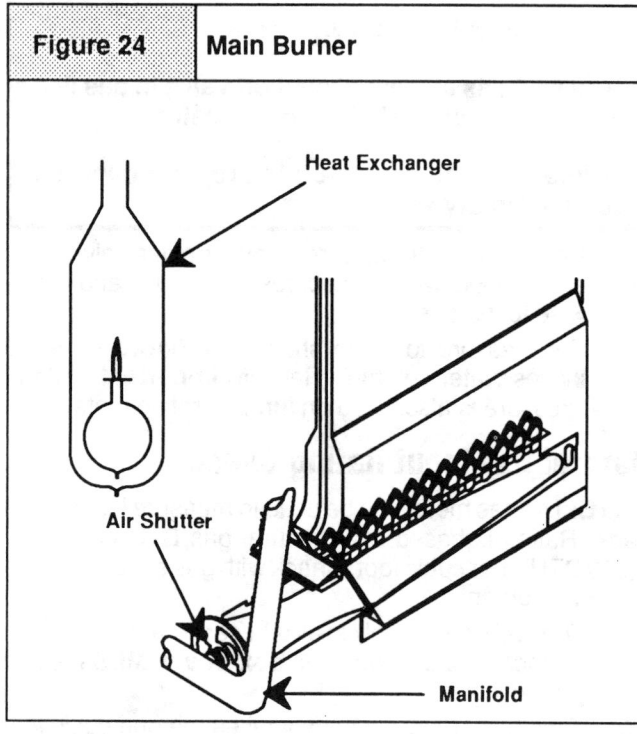

Figure 24	Main Burner

Primary Air Adjustments

NOTE: If the burners are **NOT** equipped with air shutters, **NO** adjust is necessary. Adjustment of the air shutter may be necessary to obtain the correct flame characteristics and/or to minimize resonance heat exchanger noise generated by the burner flame.

1. Place air shutters in the full, open position.

2. Start the furnace. (Refer to lighting instructions on furnace or in the *Owners Information Manual*.)

3. Allow furnace to run for 10 minutes with burner box closed and then check flame characteristics.

4. Loosen shutter locking screw and close shutter until flame has a yellow tip then open just enough to eliminate yellow tip. Tighten locking screw.

5. If resonance noise occurs, close the air shutters just enough to permit the slightest amount possible of yellow tip in the flame.

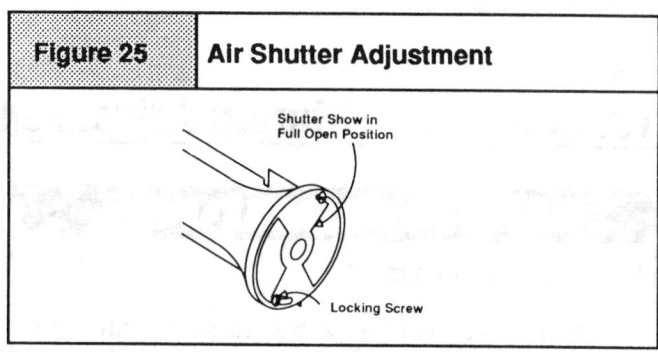

Figure 25	Air Shutter Adjustment

Limit Control Check

WARNING

Fire hazard.

Limit control is factory preset and MUST NOT be adjusted. Use ONLY manufacturer's authorized replacement parts.

Failure to do so can result in personal injury and/or death can result.

1. Operate furnace continuously for 15 minutes.

2. Block return air grille(s) and check that main burners go out and blower continues to run.

3. Remove material blocking return air grille(s). Check that main burners relight after a short cool down period.

Fan Control Check

NOTE: The fan control can be adjusted to turn **ON**, 15 to 90 seconds after the burner lights. It can be adjusted to turn **OFF** 30 to 120 seconds after the burner shuts **OFF**. Refer to **Figure 26**.

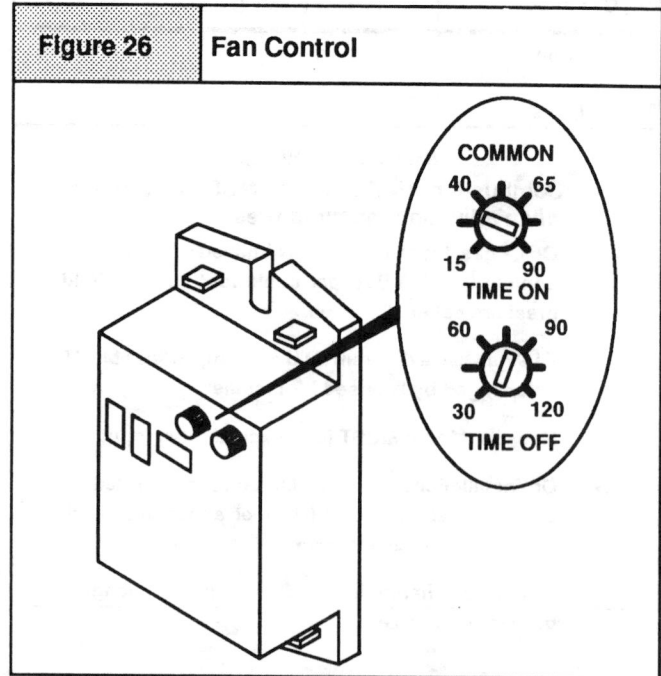

Figure 26	Fan Control

1. The fan control switch is preset at the factory, but, if necessary can be adjusted to obtain a satisfactory comfort level.

2. Operate the furnace and ensure that the blower turns **ON** and **OFF** at the appropriate time to provide the desired comfort level.

Temperature Rise Check

1. Place thermometers in supply and return air registers as close to furnace as possible.

2. Operate furnace continuously for 15 minutes with all registers and duct dampers open.

NOTE: Temperature rise is the difference between supply and return air temperature.

3. Take reading and compare with range specified on rating plate.

NOTE: Correct temperature rise will normally be obtained with a properly designed system operating at rated input and recommended blower speed. If correct temperature rise is **NOT** obtained, it may be necessary to increase blower speed to lower temperature rise or decrease blower speed to increase temperature rise.

Changing Blower Speed

WARNING

Electrical shock hazard.

Turn OFF power to furnace before changing speed taps.

Failure to do so can result in personal injury and/or death.

1. Connect the yellow wire to desired speed tap for heating at the blower motor speed tap terminal block located in furnace junction box. (**Figure 27**).

2. Connect the violet wire to the desired speed tap for cooling.

NOTE: If both heating and cooling are the same speed, connect a spade terminal adapter on the motor speed tap and connect both yellow and violet wires.

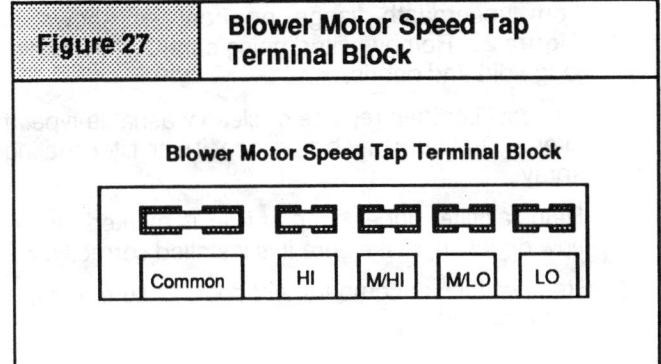

| Figure 27 | Blower Motor Speed Tap Terminal Block |

11. Furnace Maintenance

WARNING

Electrical shock hazard.

Turn OFF power to furnace before performing any maintenance or removing panels.

Failure to do so can result in personal injury and/or death.

NOTE: It is recommended that you have your furnace inspected and serviced on an annual basis (before the heating season) by a qualified service technician.

If you perform maintenance on the furnace yourself, remember that certain mechanical and electrical skills, plus specialized tools are required to properly perform maintenance. Personal injury or death may result if you are **NOT** properly trained. You should call your installing dealer or place of purchase if you are uncertain about your ability to perform maintenance.

Filters

NOTE: Dirty filters are the most common cause of inadequate heating or cooling performance.

- Replace filters monthly
- Replace disposable type filters before they become clogged.
- Use water and mild detergent to clean washable type filters.
- Replace filters with same size and type

NOTE: Some filters are marked with an arrow to indicate the proper direction of air flow through the filter. The arrow **MUST** point in the direction of air flow. Refer to section on filters in this manual.

Replacement

NOTE: The filter can be located in a remote filter rack attached to the outside of the furnace, a separate return air cabinet attached to furnace or a remote filter grille.

- Remote filter grilles and return air cabinets will usually have a hinged door or removable panel to access filter.
- Filter racks attached to the furnace will usually be made so the filter simply slides out one side. Use **ONLY** the same size filter.
- Filter type must be same unless replacing a disposable low-velocity type with a washable high-velocity type.

1. Turn **OFF** electric power at circuit breaker or disconnect switch.

2. Remove blower compartment door.

3. Lift filter strap from hook on side or slide strap to side from underneath flange on front as shown in **Figure 28**. Remove filter, being careful **NOT** to dislodge dirt and debris.

4. Inspect filter then replace or clean washable type. If filter is aluminum mesh type, coat it with filter coating spray.

5. Reinstall filter under strap. If filter is marked for airflow direction, make sure it is installed correctly.

6. Replace blower compartment door. Ensure that it is tightly closed.

Figure 28	Filter Replacement

21-10-28

Blower Motor

- Requires lubrication every five years of normal operation.

- Lubricate by adding ½ teaspoon (2cc) of SAE #10W30 motor oil to each motor bearing through oil tubes or by removing cap plugs in motor end bells.

CAUTION
Do NOT over oil or use 3 in 1 oil, penetrating oil, WD40 or similar oils. Use of these may damage motors.

Furnace Inspection

NOTE: A properly adjusted gas furnace will **NOT** require frequent cleaning. Inspect furnace regularly to ensure safe and efficient operation. A brief monthly inspection that does **NOT** require disassembly is recommended. Have furnace inspected and cleaned (if required) by a qualified service technical annually.

Vent Pipe

- ensure vent pipe and combustion air intake are clear and free of obstructions.
- Check vent pipe for tight joints, secure attachment to furnace and sagging pipe.
- Horizontal sections of vent pipe **MUST** slope upward ¼ inch per foot.

Return Air Duct

- Check that return air ducts are sealed to furnace casing and that duct is in good physical condition.
- Duct **MUST** terminate outside the space with **NO** holes or inlets in furnace space.

Furnace Base

- The floor or furnace base must be in good physical condition.

Furnace Interior

- Remove the combustion air burner cover and use a flashlight to inspect the visible part of the heat exchanger, burners and ignitor.
- Check for loose soot and give particular attention to deterioration from corrosion or other sources.
- If soot or deterioration is found inside furnace do **NOT** operate furnace and call a qualified service technician.

Main Burners

- Allow the furnace to run approximately 10 minutes then inspect the main burner flames (**Figure 24**).
- Contact a qualified service agency at once if an abnormal flame develops.
- Clean burners by gently striking orifice end on a block of wood. This will remove dirt or lint build-up in the tube.
- Replace burners if extremely rusted, crushed, or if burner ports have collapsed.

Condensate Disposal

NOTE: The condensate trap is part of the vent system. A condensate Neutralizer cartridge, if used in the drain line will require some maintenance.

- Disassemble and clean trap and cartridge prior to each heating season, or if drain line become plugged.

- Inspect the drain line and overflow line at least monthly. If the condensate neutralizer cartridge becomes plugged, the condensate will flow through the overflow line. If this happens, clean both cartridge and trap.

- To clean, disconnect the drain line cartridge and unscrew end cap from cartridge. Pour the neutralizer out and thoroughly flush neutralizer and inside of cartridge with water. Pour neutralizer back into cartridge, adding neutralizer if cartridge is less than ³/₄ full. Unscrew trap from vent connecting tee and flush thoroughly with water. Use soap if necessary to clean. Do **NOT** use any kind of solvents. Ensure float is reinstalled in trap (**Figure 6**).

- Reassemble and seal threaded connections with silicone rubber (bathtub caulk) or pipe dope approved for plastic pipe. See repair parts sections in the *Furnace Technical Support Manual*, to order replacement neutralizer. Condensate is acidic, do **NOT** use for any purpose.

12. Cleaning Heat Exchangers

WARNING

Electrical shock, fire or explosion hazard.

Turn OFF electric power at disconnect and gas supply at manual shutoff valve. Have qualified service technician perform the following procedures.

Failure to do so can result in property damage, personal injury and/or death .

NOTE: If filters are inadequate or **NOT** maintained, it may be necessary to clean the exterior surface of the secondary heat exchanger to obtain proper airflow. If the primary heat exchanger requires cleaning, it can be completed without removing or cleaning of the secondary heat exchanger.

The **ONLY** time it should be necessary to disassemble and clean the interior of both the primary and secondary heat exchangers would be due to a sooting condition caused by abnormal combustion.

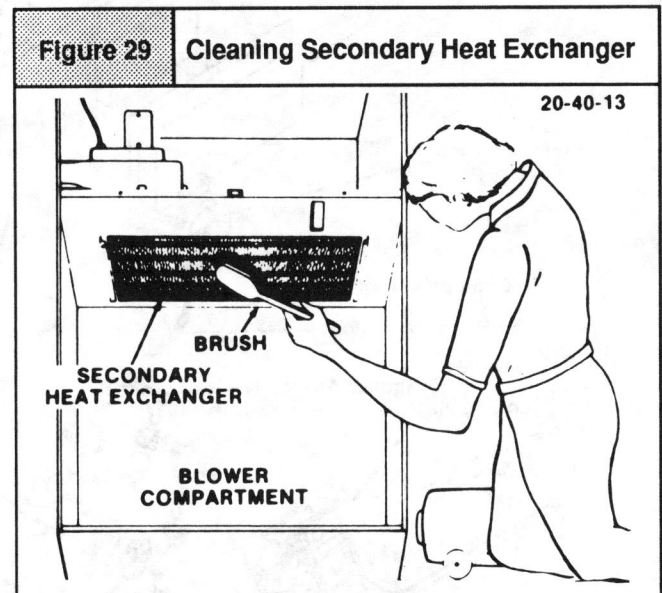

| Figure 29 | Cleaning Secondary Heat Exchanger |

20-40-13

Secondary Heat Exchanger

Exterior Cleaning

1. Turn **OFF** electric power.
2. Remove the two retaining screws holding control box from the top of the blower deck.
3. Lift control box up and outward from furnace. Control box will have enough slack in the wiring to allow the box to be held out of the way to remove blower assembly.
4. Remove the two retaining screws holding blower in position in the slide rails.

5. Blower can now be removed by pulling assembly from furnace. Support blower next to furnace to avoid having to disconnect wiring.

6. Using a stiff bristle brush and a vacuum cleaner, clean dirt and lint build-up from bottom side of secondary heat exchanger. Brush strokes must be with the fin surface to avoid damage to the fins. Use a fin comb to straighten fins (**Figure 29**).

7. Inspect and clean blower wheel using brush and vacuum. Be careful **NOT** to dislodge balance weights (clips) that may be on the blower wheel.

CAUTION

An unbalanced wheel can cause undesirable noise, vibration or blower damage.

Figure 30	Cleaning Heat Exchanger

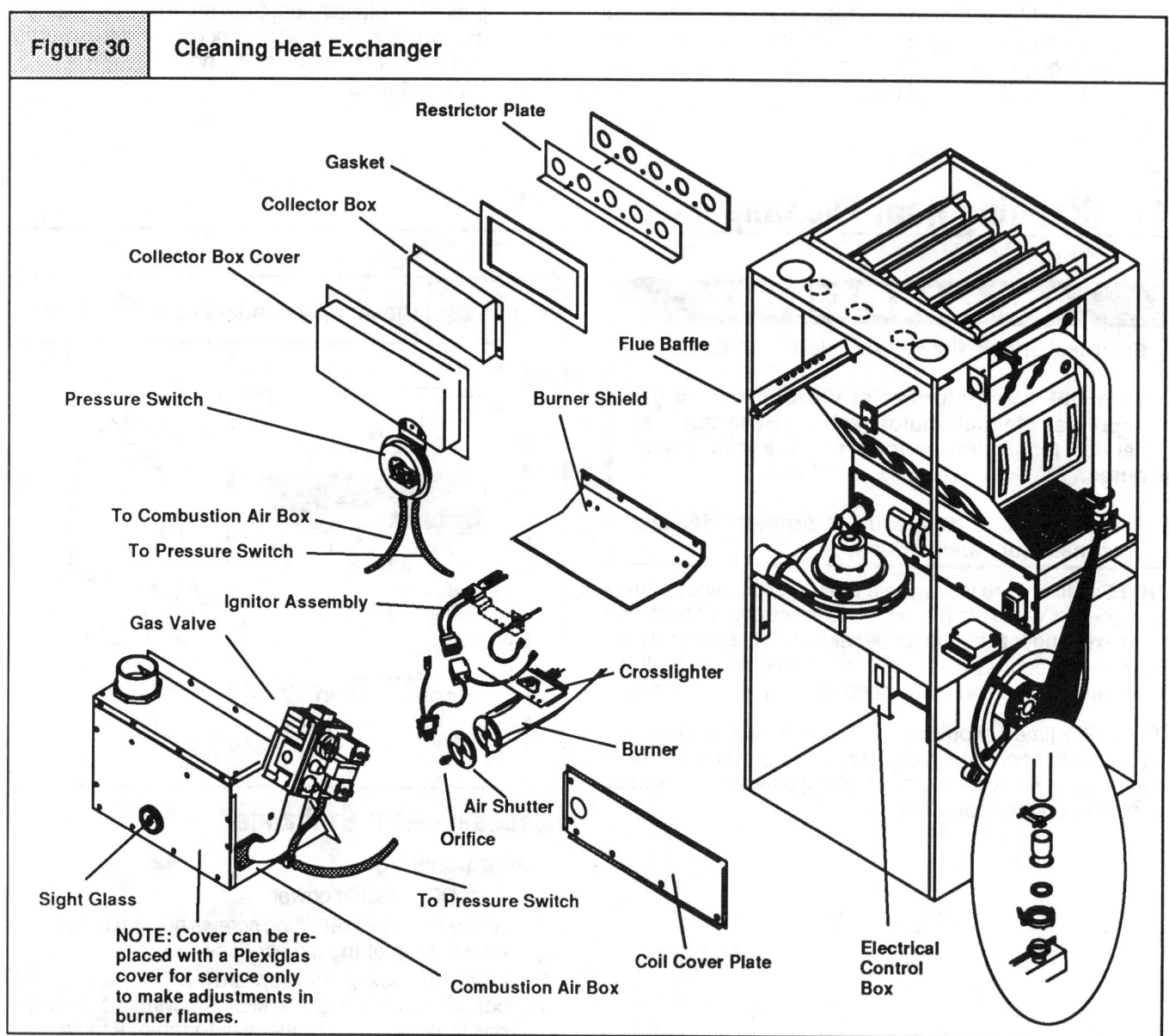

Primary Heat Exchanger

Disassembly

NOTE: The following parts and assemblies **MUST** be removed before the heat exchanger can be cleaned.

1. Disconnect electric power.

2. Remove screws for the collector box cover, collector box and restrictor plate. Handle collector box gasket with care to avoid damage.

3. Remove screws that hold flue baffles in position and carefully pull baffles out.

4. Disconnect gas supply line at union and at gas valve if necessary for removal from furnace and Disconnect electrical leads at gas valve.

5. Remove combustion air burner box, front cover and top.

6. Remove retaining screws for manifold and remove gas valve/manifold assembly. Use caution to prevent damage to hot surface ignition assembly.

7. Pull burners and crosslighter from the heat exchanger. Remove crosslighter from burners by sliding crosslighter toward the orifice end of burner.

Cleaning

1. Use a long flexible handle brush and vacuum cleaner to clean interior of heat exchanger .

2. Gently strike orifice end of burners on a block of wood to remove any dirt or lint build up in the tube.

3. Use a brush and stiff wire to clean the crosslighter.

4. Reassemble in reverse order. See *Reassembly Instructions.*

Secondary Heat Exchanger

Interior Cleaning

NOTE: The following parts and assemblies **MUST** be removed before the heat exchanger can be cleaned.

1. Loosen clamps on the inlet and outlet connectors on the combustion air blower.

2. Remove screws holding combustion air blower to blower deck.

3. Gently wiggle and pull blower outward to disengage from inlet and outlet connectors.

NOTE: Blower can be pulled out so it just clears the furnace casing on the left side without disconnect any wiring. Support blower so it does not hang by the wire.

4. Remove screws holding the electronic hot surface module to the blower deck.

5. Remove screws holding the secondary heat exchanger panel and remove panel.

6. Remove screw holding **Z** bracket to division panel and remove bracket.

7. Loosen screw on the secondary heat exchanger inlet coupling by reaching in on the right side.

CAUTION

Fins on secondary heat exchanger have sharp edges. Wear gloves or cover secondary heat exchanger fins with rags to prevent possible cuts.

Failure to do so can result in personal injury.

8. Remove secondary heat exchanger by pulling straight out and reposition electronic module and wiring, etc. to get past them.

Cleaning

1. Use a $1^{5}/_{8}$ inch plastic cap plug or the palm of your hand to plug either the inlet or outlet port and fill heat exchanger coil with approximately $1^{1}/_{2}$ quarts of hot water.

2. Plug other port and shake coil vigorously. Drain and flush with a hard stream of water from a garden hose. Repeat **Steps 1** and **2**, if required.

3. Thoroughly wash exterior. Do **NOT** use a hard stream of water on the exterior to prevent damage to fins.

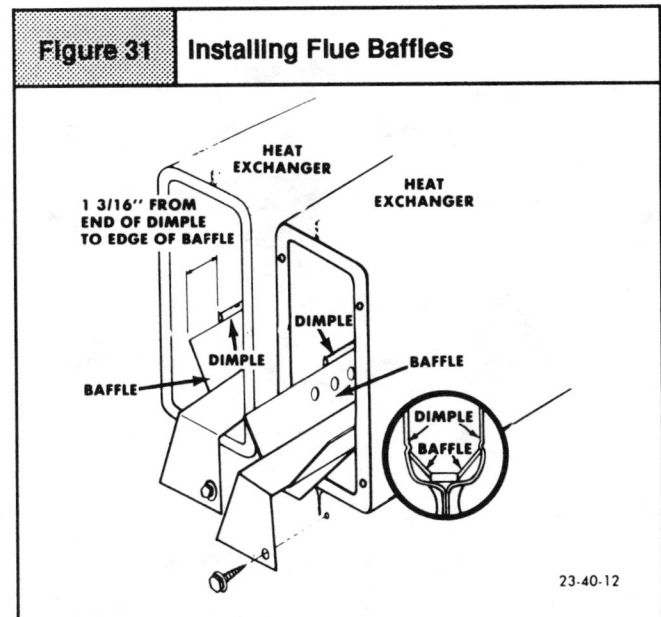

| Figure 31 | Installing Flue Baffles |

23-40-12

Reassembly

NOTE: Reassemble all parts in reverse order as removed, with the following instructions.

1. Install burner into slots of crosslighter and press into position. Back of crosslighter **MUST** be seated firmly against the burner ports.

2. Insert burners into heat exchanger. Burners **MUST** be inserted into the slots at the back of the heat exchanger and level.

3. Install flue baffles as shown in **Figure 31**. Baffles **MUST** be located below dimple in heat exchanger and firm against bottom of flue outlet.

4. Replace any torn or defective insulation.

5. Replace all gaskets and parts that are broken or deteriorated.

6. Test for gas leaks after reassembly. Use a soapy solution on all joints. **ALL** leaks **MUST** be repaired immediately.

7. Perform an operational check of the furnace.

TECHNICAL SUPPORT MANUAL

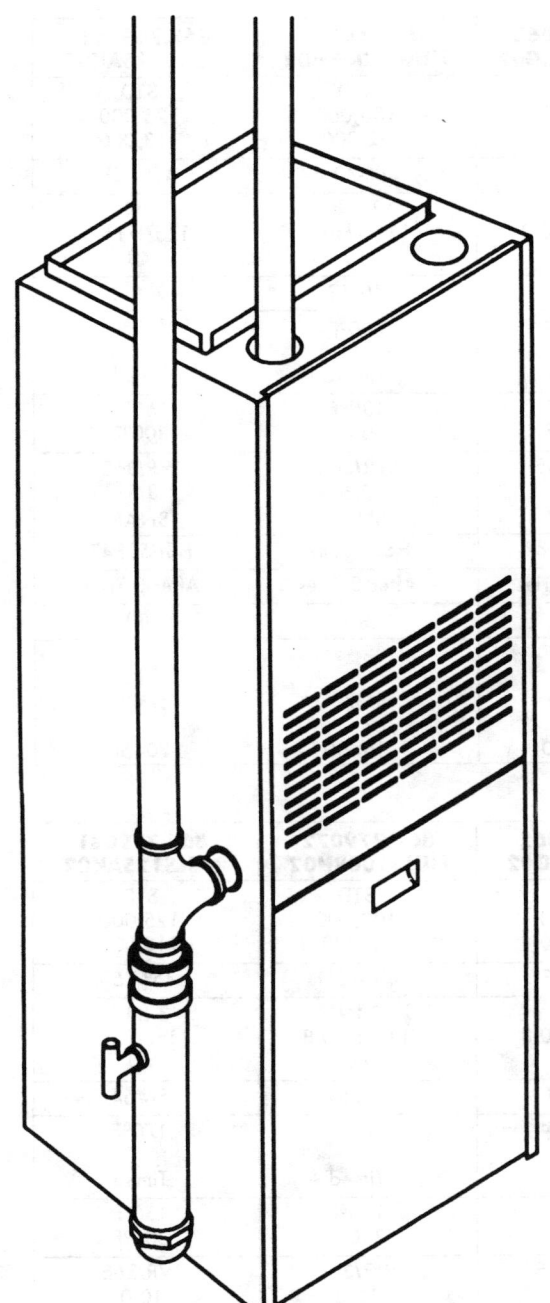

- **Specifications**
- **Wiring Diagrams**
- **Repair Parts**

FOR
NATURAL GAS MODELS

867.769051	NUGS050AF02
867.769062	NUGS075BG02
867.769072	NUGS100BH02
867.769081	NUGS125AK02

L.P. GAS MODELS

867.779051	NULS050AF02
867.779062	NULS075BG02
867.779072	NULS100BH02
867.779081	NULS125AK02

Design Certified by A.G.A.

CONDENSING UPFLOW GAS FURNACES
Save This Manual For Future Reference.

F664-

PRINTED IN U.S.A.

1007213
11-5-91
L.P. 1

TECHNICAL-SERVICE-DATA

Specifications: NUGS

MODEL NO.	867.769051 NUGS050AF02	867.769062 NUGS075BG02	867.769072 NUGS100BH02	867.769081 NUGS125AK02
GAS TYPE-NAT INPUT (BTUH) STD/ALT OUTPUT (BTUH)	STD 50,000 48,000	STD 75,000 70,000	STD 100,000 92,000	STD 125,000 113,000
TEMPERATURE RISE	15°-45°F	35°-65°F	40°-70°F	45°-75°
FLUE SIZE ELEC. VOLTS/PH./F.L.A. TRANSFORMER SIZE (VA)	2 IN. 115/1/8.0 40	3 IN. 115/1/10.8 40	3 IN. 115/1/10.8 40	3 IN. 115/1/11.8 40
ORIFICE NO. REQ'/SIZE	2/#42	3/#42	4/#42	5/#42
LIMIT SETTING FAN SW SETTING OFF	250°F Timed	250°F Timed	200°F Timed	170°F Timed
EXHAUST LIMIT THERMAL SENSOR	120°F 300°F	130°F 300°F	130°F 300°F	150°F 300°F
GAS VALVE MFG/TYPE MANIFOLD PRESSURE REGULATION TYPE	WR/36E 3.5 SNAP	WR/36E 3.5 SNAP	WR/36E 3.5 SNAP	WR/36E 3.5 SNAP
IGNITION TYPE/MODEL	HSI/50E47	HSI/50E47	HSI/50E47	HSI/50E47
LOCK-OUT TIMING	After 3 Tries	After 3 Tries	After 3 Tries	After 3 Tries
ANTICIPATOR SETTING	.80	.80	.80	.80
CAP. RATING MFD/VOLTS COMBUSTION AIR BLOWER) ‡FURNACE BLOWER RATED EXT. STATIC PRESS	5/370 .10-.50	5/370 .12-.50	5/370 .15-.50	5/370 .20-.50

Specifications: NULS

MODEL NO.	867.779051 NULS050AF02	867.779062 NULS075BG02	867.779072 NULS100BH02	867.779081 NULS125AK02
GAS TYPE-L.P. INPUT (BTUH) STD/ALT OUTPUT (BTUH)	STD 50,000 48,000	STD 75,000 70,000	STD 100,000 92,000	STD 125,000 113,000
TEMPERATURE RISE	15°-45°F	35°-65°F	40°-70°F	45°-75°
FLUE SIZE ELEC. VOLTS/PH./F.L.A. TRANSFORMER SIZE (VA)	2 IN. 115/1/8.0 40	3 IN. 115/1/10.8 40	3 IN. 115/1/10.8 40	3 IN. 115/1/11.8 40
ORIFICE NO. REQ'/SIZE	2/#54	3/#54	4/#54	5/#54
LIMIT SETTING FAN SW SETTING OFF	250°F Timed	250°F Timed	200°F Timed	170°F Timed
EXHAUST LIMIT THERMAL SENSOR	120°F 300°F	130°F 300°F	130°F 300°F	150°F 300°F
GAS VALVE MFG/TYPE MANIFOLD PRESSURE REGULATION TYPE	WR/36E 10.0 SNAP	WR/36E 10.0 SNAP	WR/36E 10.0 SNAP	WR/36E 10.0 SNAP
IGNITION TYPE/MODEL	HSI/50E47	HSI/50E47	HSI/50E47	HSI/50E47
LOCK-OUT TIMING	After 3 Tries	After 3 Tries	After 3 Tries	After 3 Tries
ANTICIPATOR SETTING	.80	.80	.80	.80
CAP. RATING MFD/VOLTS COMBUSTION AIR BLOWER) ‡FURNACE BLOWER RATED EXT. STATIC PRESS	5/370 .10-.50	5/370 .12-.50	5/370 .15-.50	5/370 .20-.50

NOTES - ALL MODELS

For elevations above 2000 ft., the standard input rating should be reduced by 4% for each 1000 ft. above sea level by changing orifice size, refer to NFPA54 National Fuel Gas Code. Do not use alternate input rating as a basis to determine firing rate above 2000 ft., use the listed standard input rating. Orifice sizes must not be changed to cause derating EXCEPT for altitude.

The total heat loss from the house or structure as expressed in total BTU/HR must be calculated by manufacturers method or in accordance with the "A.S.H.R.A.E. Guide" or "Manual J-Load Calculations" published by the Air Conditioning Contractors of America. The total heat loss calculated should be equal to or less than the heating capacity. Output based on D.O.E. test procedures, steady state efficiency times input.

BLOWER PERFORMANCE DATA - ALL MODELS

MODEL NUMBER		867.769051 867.779051 NUGS050AF02 NULS050AF02	867.769062 867.779062 NUGS075BG02 NULS075BG02	867.769072 867.779072 NUGS100BH02 NULS100BH02	867.769081 867.779081 NUGS125AK02 NULS125AK02
BLOWER DATA TYPE & SIZE		DD10-8A	DD10-9A	DD10-9A	DD12-11A-T
MOTOR DATA	AMPS @ RPM TYPE/HP	8.0 @ 1050 PSC-1/2 HP	10.8 @ 1050 PSC-3/4 HP	10.8 @ 1050 PSC-3/4 HP	11.8 @ 1050 PSC-1 HP
CAPACITOR MFD/VOLTS		7.5 MFD/370	10 MFD/370	10 MFD/370	15 MFD/370
FILTER DATA #REQ'D/SIZE TYPE		1 - 16x25x1 PERMANENT	1 - 16x25x1 PERMANENT	1 - 16x25x1* PERMANENT	1 - 24x25x1** PERMANENT

	ESP	SPEED				
AIR DELIVERY IN CFM VARYING EXT. STATIC PRESS. (IN. WC.)	0.10	LO MED. LO MED. HI HI	945 1190 1375 1625	1105 1360 1605 1775	1100 1370 1660 1900	1480 1565 1835 2135
	0.20	LO MED. LO MED. HI HI	1325 1180 1360 1570	1090 1335 1560 1705	1076 1350 1608 1830	1475 1560 1820 2090
	0.30	LO MED. LO MED. HI HI	940 1160 1340 1500	1070 1360 1500 1630	1075 1325 1559 1752	1460 1545 1785 2050
	0.40	LO MED. LO MED. HI HI	935 1140 1305 1445	1045 1255 1435 1560	1035 1280 1505 1675	1440 1525 1750 2005
	0.50	LO MED. LO MED. HI HI	920 1110 1265 1385	1010 1205 1375 1485	1000 1245 1445 1605	1425 1500 1715 1965
	0.60	LO MED. LO MED. HI HI	905 1080 1205 1320	960 1140 1295 1395	965 1195 1370 1520	1385 1475 1680 1900
	0.70	LO MED. LO MED. HI HI	870 1030 1145 1245	885 1050 1205 1305	905 1125 1290 1430	1340 1430 1625 1855
	0.80	LO MED. LO MED. HI HI	820 965 1070 1160	815 940 1080 1200	830 1040 1200 1335	1275 1360 1575 1790

** For 5 Ton Cooling - Use a 16'' x 25'' x 1'' Filter (not supplied) on each side of furnace.

WIRING DIAGRAM — ALL MODELS

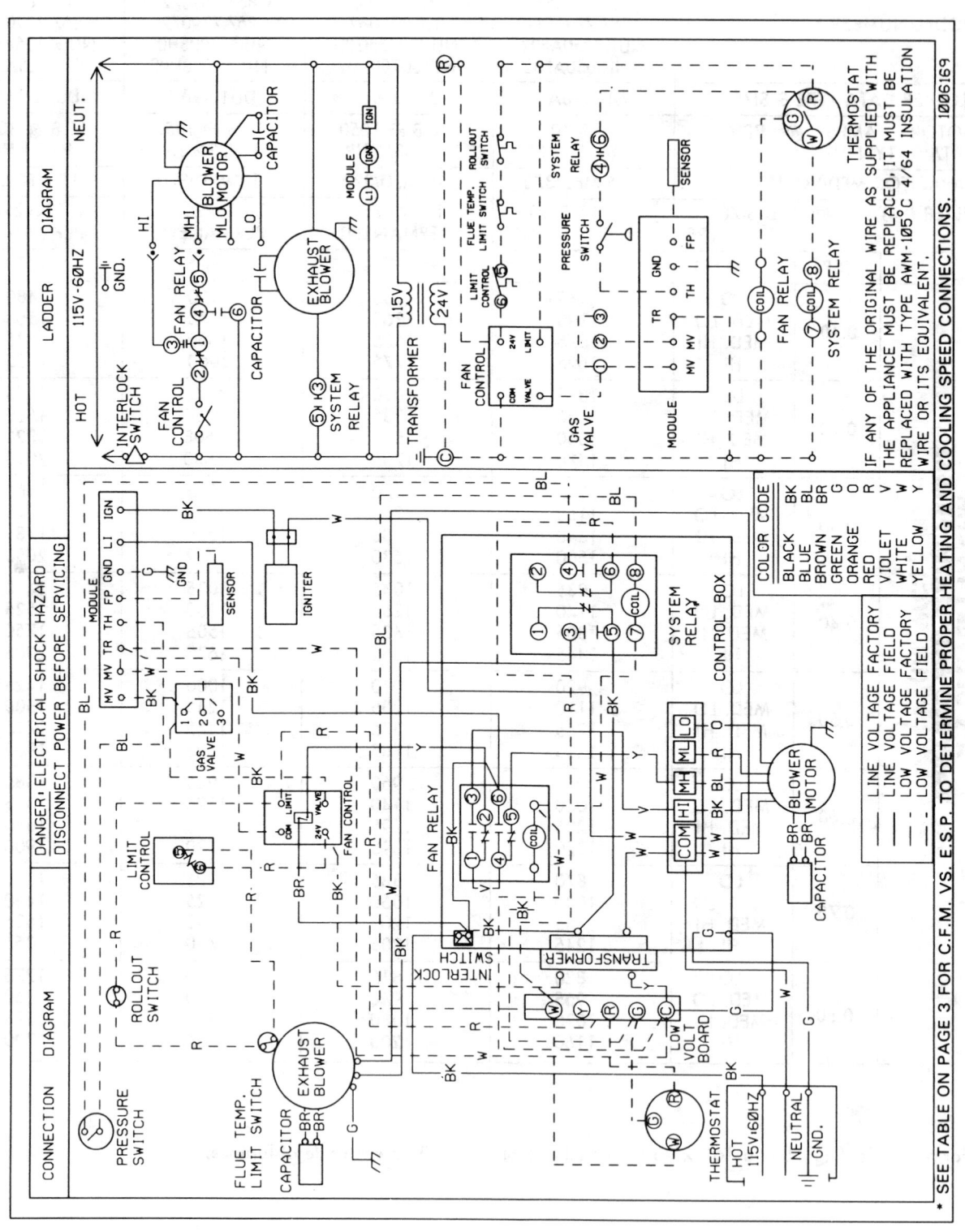

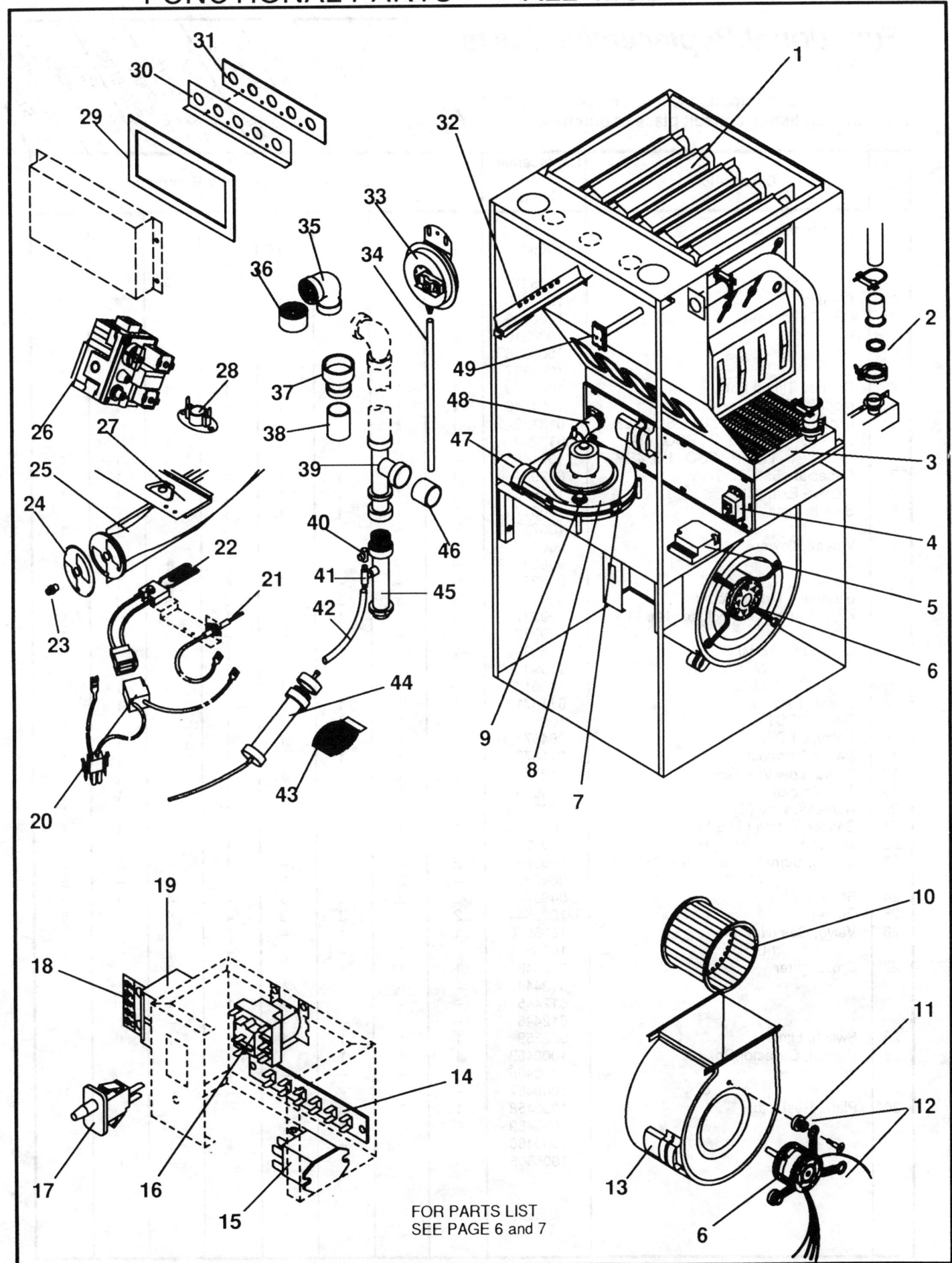

FOR PARTS LIST
SEE PAGE 6 and 7

Functional Replacement Parts

Replacement part supplied will be current active part.
For parts not listed, consult place of purchase.

Key No.	Description	Replacement Part Number	867.769051 NUGS050AF02	867.779051 NULS050AF02	867.769062 NUGS075BG02	867.779062 NULS075BG02	867.769072 NUGS100BH02	867.779072 NULS100BH02	867.769081 NUGS125AK02	867.779081 NULS125AK02
1	Exchanger, Heat	1005812	1	1	–	–	–	–	–	–
		1005813	–	–	1	1	–	–	–	–
		1005814	–	–	–	–	1	1	–	–
		1005815	–	–	–	–	–	–	1	1
2	O–Ring	1000599	1	1	1	1	1	1	1	1
3	Coil, Condenser	1000309	1	1	–	–	–	–	–	–
		1000310	–	–	1	1	–	–	–	–
		1000356	–	–	–	–	1	1	–	–
		1000357	–	–	–	–	–	–	1	1
4	Control, Timed Fan	1005229	1	1	1	1	1	1	1	1
5	Module, Control	1001346	1	1	1	1	1	1	1	1
6	Motor, Blower 1/2 H. P., PSC., 4SP.	613136	1	1	–	–	–	–	–	–
	3/4 H. P., PSC., 4SP.	613209	–	–	1	1	1	1	–	–
	1 H. P., PSC., LG. 4SP.	1001480	–	–	–	–	–	–	1	1
7	Capacitor, 5MFD., 370V.	602974	1	1	1	1	1	1	1	1
8	Blower, Exhaust	1005568	1	1	1	1	1	1	1	1
9	Switch, Limit	1003175	1	1	1	1	1	1	–	–
		1007375	–	–	–	–	–	–	1	1
10	Wheel, Blower	600587	1	1	–	–	–	–	–	–
		96839	–	–	1	1	1	1	–	–
		89726	–	–	–	–	–	–	1	1
11	Grommet, Motor Mount	609194	3	3	3	3	3	3	4	4
12	Kit, Motor Mount (Includes No. 11)	609227	1	1	1	1	1	1	–	–
		609228	–	–	–	–	–	–	1	1
13	Capacitor, 7.5 MFD., 370V.	706338	1	1	–	–	–	–	–	–
	10 MFD., 370V.	91593	–	–	1	1	1	1	–	–
	15 MFD., 370V.	1006176	–	–	–	1	–	–	1	1
14	Board, Terminal	611871	1	1	1	1	1	1	1	1
15	Relay, DPST.	1000742	1	1	1	1	1	1	1	1
16	Relay, DPDT.	96467	1	1	1	1	1	1	1	1
17	Switch, Interlock	611872	1	1	1	1	1	1	1	1
18	Board, Low Volt Terminal	612254	1	1	1	1	1	1	1	1
19	Transformer	613039	1	1	1	1	1	1	1	1
20	Harness, Wire (H S I)	1003360	1	1	1	1	1	1	1	1
21	Senser, Flame (H S I)	1001358	1	1	1	1	1	1	1	1
22	Ignitor, Hot Surface (H S I)	1001344	1	1	1	1	1	1	1	1
23	Orifice, Burner (Nat.) (No. 42)	609563	2	–	3	–	4	–	5	–
	(L. P.) (No. 54)	96466	–	2	–	3	–	4	–	5
24	Shutter, Air	94927	2	2	3	3	4	4	5	5
25	Burner	1005644	2	2	3	3	4	4	5	5
26	Valve, Gas (Nat.)	1005570	1	–	1	–	1	–	1	–
	(L. P.)	1005571	–	1	–	1	–	1	–	1
27	Crosslighter	612443	1	1	–	–	–	–	–	–
		612444	–	–	1	1	–	–	–	–
		612445	–	–	–	–	1	1	–	–
		612446	–	–	–	–	–	–	1	1
28	Switch, Limit	522559	1	1	1	1	1	1	1	1
29	Gasket, Collector Box	1000463	1	1	1	1	–	–	–	–
		1000465	–	–	–	–	1	1	–	–
		1000467	–	–	–	–	–	–	1	1
30	Plate, Restrictor	1000458	1	1	–	–	–	–	–	–
		1000459	–	–	1	1	–	–	–	–
		1000460	–	–	–	–	1	1	–	–
		1005505	–	–	–	–	–	–	1	1

CONTINUED ON PAGE 7

6

Functional Replacement Parts

Replacement part supplied will be current active part.
For parts not listed, consult place of purchase.

Key No.	Description	Replacement Part Number	867.769051 NUGS050AF02	867.779051 NULS050AF02	867.769062 NUGS075BG02	867.779062 NULS075BG02	867.769072 NUGS100BH02	867.779072 NULS100BH02	867.769081 NUGS125AK02	867.779081 NULS125AK02
					Model Number/Quantity Required					
31	Gasket, Restrictor Plate	1000462	1	1						
		1000464	–	–	1	1				
		1000466	–	–	–	–	1	1		
		1000468	–	–	–	–	–	–	1	1
32	Baffle, Flue	613083	2	2	3	3	4	4	5	5
33	Switch, Pressure	1005602	1	1						
		1006405	–	–	1	1				
		1006406	–	–	–	–	1	1		
		1005605	–	–	–	–	–	–	1	1
34	Tube, Sensor (Valve to Comb. Box)	1005751	1	1	1	1	1	1	1	1
	(Pressure Switch to Comb. Box)	1004318	1	1	1	1	1	1	1	1
	(Pressure Switch to Comb. Blower)	1005752	1	1						
		1005753	–	–	1	1				
		1005754	–	–	–	–	1	1		
		1005755	–	–	–	–	–	–	1	1
35	Cap, Vent 90° Elbow	612906	1	1						
		1005777 *	–	–	1	1	1	1	1	1
36	Cap, Vertical Vent	1005774 *	1	1						
		1005795 *	–	–	1	1	1	1	1	1
37	Adapter, 2" to 3"	1005779 *	–	–	1	1	1	1	1	1
38	Nipple, 2" x 3" Lg.	1005775 *	–	–	1	1	1	1	1	1
39	Tee, 2" (Includes Float No. 1000887)	1000892	1	1	1	1	1	1	1	1
40	Tee, 1/2" Union	613051	1	1	1	1	1	1	1	1
41	Tee, 1/2" Male	612696	1	1	1	1	1	1	1	1
42	Tubing, 5/8" O. D.x1/2" I. D. (Optional)	612829	1	1	1	1	1	1	1	1
43	Neutralizer (Optional)	612827	1	1	1	1	1	1	1	1
44	Kit, Neutralizer Tubing (Optional)	612833	1	1	1	1	1	1	1	1
45	Trap (Multiple Piece)	1000891	1	1	1	1	1	1	1	1
)(	Trap (Molded One Piece, Replaces Nos. 39, 40, 41 and 45)	1007356	1	1						
		1007344	–	–	1	1	1	1	1	1
46	Nipple, 2' x 2' Lg.	612689	1	1	1	1	1	1	1	1
47	Coupling, Discharge	1002522	1	1	1	1	1	1	1	1
48	Coupling	1002532	1	1	1	1	1	1	1	1
49	Control, Limit	1005762	1	1	1	1				
		1005761	–	–	–	–	1	1		
		1005763	–	–	–	–	–	–	1	1
)(	Filter, Air (16" X 25" X 1")	613487	1	1	1	1	1	1		
	(24" X 25" X 1")	612706	–	–	–	–	–	–	1	1
)(	Manual, User's Information	1006174	1	1	1	1	1	1	1	1
)(	Manual, Installation	1007217	1	1	1	1	1	1	1	1
)(	Manual, Technical Support	1007213	1	1	1	1	1	1	1	1

)(PART NOT ILLUSTRATED

* STANDARD HARDWARE ITEM, PURCHASE LOCALLY.

FOR PARTS ILLUSTRATION SEE PAGE 5

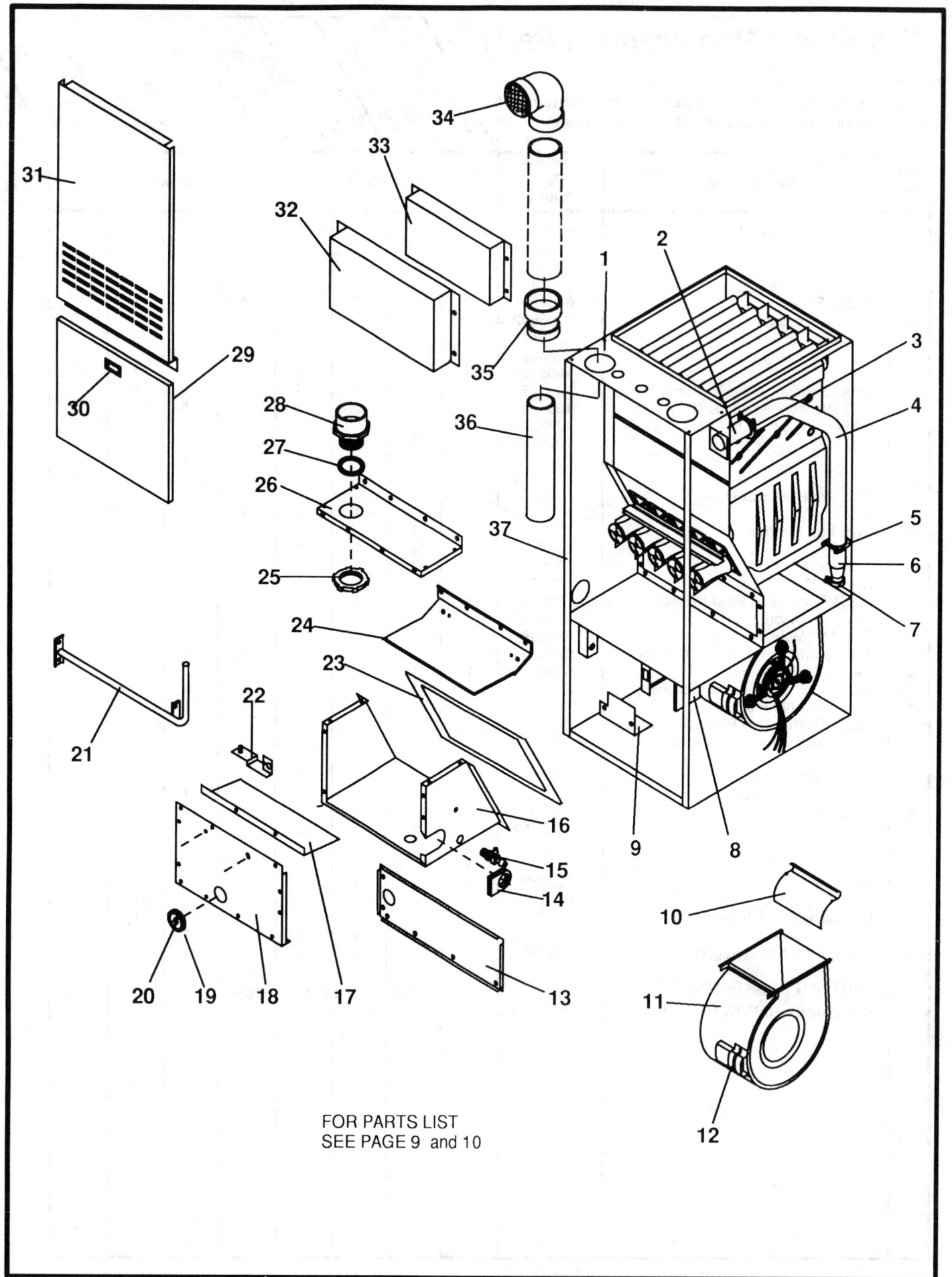

FOR PARTS LIST
SEE PAGE 9 and 10

Replacement Parts

Key No.	Description	Replacement Part Number	867.769051 NUGS050AF02	867.779051 NULS050AF02	867.769062 NUGS075BG02	867.779062 NULS075BG02	867.769072 NUGS100BH02	867.779072 NULS100BH02	867.769081 NUGS125AK02	867.779081 NULS125AK02
1	Panel, Top	1005517	1	1	1	1	–	–		
		1005515	–	–	–	–	1	1		
		1005513	–	–	–	–	–	–	1	1
2	Adapter, Transfer	1000471	1	1	1	1	1	1	1	1
3	Clamp, 2"	1001328	1	1	1	1	1	1	1	1
4	Tube, Transfer	1007098	1	1	1	1	1	1	1	1
5	Clamp, 1–7/8"	1001327	1	1	1	1	1	1	1	1
6	Tube, Adapter	1001553	1	1	1	1	1	1	1	1
7	Coupling, V Retainer	1001187	1	1	1	1	1	1	1	1
8	Box, Junction	611515	1	1	1	1	1	1	1	1
9	Cover, Junction Box	611541	1	1	1	1	1	1	1	1
10	Panel, Cut–Off	95143	1	1						
		95145	–	–	1	1	1	1		
		612992	–	–	–	–	–	–	1	1
11	Housing, Blower	1005114	1	1						
		1005440	–	–	1	1	1	1		
		1001840	–	–	–	–	–	–	1	1
12	Clamp, Capacitor	94364	2	2	2	2	2	2	1	1
	(Blower Motor)	706439	–	–	–	–	–	–	1	1
13	Panel, Coil	1005572	1	1	1	1				
		1005573	–	–	–	–	1	1		
		1005574	–	–	–	–	–	–	1	1
14	Grommet, Manifold	1005554	1	1	1	1	1	1	1	1
15	Tap, Pressure	1005606	1	1	1	1	1	1	1	1
16	Bottom, Combustion Box	1005478	1	1						
		1005477	–	–	1	1				
		1005476	–	–	–	–	1	1		
		1005475	–	–	–	–	–	–	1	1
17	Baffle, Air Difusser	1006617	1	1						
		1006616	–	–	1	1				
		1006615	–	–	–	–	1	1		
		1006614	–	–	–	–	–	–	1	1
18	Front, Combustion Box	1006613	1	1						
		1006612	–	–	1	1				
		1006611	–	–	–	–	1	1		
		1006610	–	–	–	–	–	–	1	1
19	Grommet, Sight Glass	1005522	1	1	1	1	1	1	1	1
20	Glass, Sight	1005508	1	1	1	1	1	1	1	1
21	Manifold	1005496	1	1						
		1005495	–	–	1	1				
		1005494	–	–	–	–	1	1		
		1005493	–	–	–	–	–	–	1	1
22	Bracket, Ignitor	1003363	1	1	1	1	1	1	1	1
23	Gasket, Combustion Box	1005504	1	1						
		1005503	–	–	1	1				
		1005502	–	–	–	–	1	1		
		1005501	–	–	–	–	–	–	1	1
24	Shield, Burner	1005521	1	1						
		1005520	–	–	1	1				
		1005519	–	–	–	–	1	1		
		1005518	–	–	–	–	–	–	1	1
25	Locknut	1005616	1	1						
		1005615	–	–	1	1	1	1	1	1
26	Top, Combustion Box	1005482	1	1						
		1006404	–	–	1	1				
		1006403	–	–	–	–	1	1		
		1005479	–	–	–	–	–	–	1	1

CONTINUED ON PAGE 10

Replacement Parts

Replacement part supplied will be current active part.
For parts not listed, consult place of purchase.

Key No.	Description	Replacement Part Number	867.769051 NUGS050AF02	867.779051 NULS050AF02	867.769062 NUGS075BG02	867.779062 NULS075BG02	867.769072 NUGS100BH02	867.779072 NULS100BH02	867.769081 NUGS125AK02	867.779081 NULS125AK02
			Model Number/Quantity Required							
27	O–Ring	1005600	1	1	–	–	–	–	–	–
		1005601	–	–	1	1	1	1	1	1
28	Adapter, Combustion Air	1005818	1	1	–	–	–	–	–	–
		1005523	–	–	1	1	1	1	1	1
29	Door, Blower	1004488	1	1	1	1	–	–	–	–
		1004493	–	–	–	–	1	1	–	–
		1004498	–	–	–	1	–	–	1	1
30	Handle, Door	1004335	1	1	1	1	1	1	1	1
31	Door, Front	1005560	1	1	1	1	–	–	–	–
		1005562	–	–	–	–	1	1	–	–
		1005564	–	–	–	–	–	–	1	1
32	Cover, Collector Box	1005511	1	1	1	1	–	–	–	–
		1005510	–	–	–	–	1	1	–	–
		1005509	–	–	–	–	–	–	1	1
33	Box, Collector	1000275	1	1	1	1	–	–	–	–
		1000276	–	–	–	–	1	1	–	–
		1000277	–	–	–	–	–	–	1	1
34	Cap, Combustion Air	612906	1	1	–	–	–	–	–	–
		1005777 *	–	–	1	1	1	1	1	1
35	Reducer, Combustion Air (3" to 2–1/2")	1005778 *	–	1	1	1	–	1	1	1
36	Pipe, Combustion Air (2–1/2" x 22" Lg.)	1005760 **	–	1	1	1	–	1	1	1
37	Casing	1007074	1	1	1	1	–	–	–	–
		1007076	–	–	–	–	1	1	–	–
		1007078	–	–	–	–	–	–	1	1

*   STANDARD HARDWARE ITEM, PURCHASE LOCALLY.
** 2–1 / 2" I. D. SCH40DWV / PVC PIPE.

FOR PARTS ILLUSTRATION SEE PAGE 8

Notes

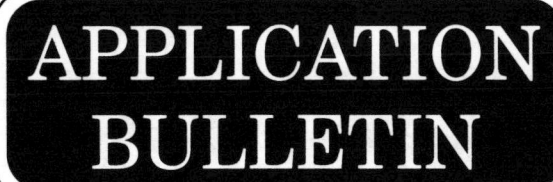

APPLICATION BULLETIN

DATE: February 10, 1992

BULLETIN: ABR008

SUBJECT: FURNACE VENT SIZING

PRODUCTS AFFECTED: NUGS/LS075 & NUGS/LS100 DIRECT VENTED FURNACES

```
****************************************************************
***************** RESIDENTIAL PRODUCT INFORMATION ******************
****************************************************************
```

In September 1990, we increased the size of the vent diameter on the 75,000 and 100,000 Btuh (input) furnaces from 2" to 3". This was done to avoid nuisance lock-outs of the pressure switch under certain operating conditions. We have continued our investigations and can now permit the use of 2" vent piping with specified lengths and a specified maximum number of 90° elbows, per the tables below. Please note that the elbows **must** be **"long radius"** or **"long sweep"** type bends and that two (2) 45° elbows can be substituted for one (1) 90° elbow.

FURNACE MODEL NUGS/NULS075					
INLET VENT PIPE			**OUTLET VENT PIPE**		
LENGTH	DIAMETER	# OF 90° ELLS	LENGTH	DIAMETER	# OF 90° ELLS
5'	2"	5	5'	2"	5
10'	2"	4	10'	2"	4
15'	2"	3	15'	2"	3
20'	2"	2	20'	2"	2
25'	2"	1	25'	2"	1
30'	3"	5	30'	3"	5
35'	3"	5	35'	3"	5
40'	3"	4	40'	3"	4

Cont'd Page 2 .....

Inter-City Products
≡ Heating & Air Conditioning ≡

INTER-CITY PRODUCTS CORPORATION (USA) 1136 HEIL-QUAKER BOULEVARD POST OFFICE BOX 3005 LAVERGNE TENNESSEE 37086-1985 (615) 793-0450

FURNACE MODEL NUGS/NULS100					
INLET VENT PIPE			**OUTLET VENT PIPE**		
LENGTH	DIAMETER	# OF 90° ELLS	LENGTH	DIAMETER	# OF 90° ELLS
5'	2"	5	5'	2"	5
10'	2"	4	10'	2"	4
15'	2"	3	15'	2"	3
20'	3"	5	20'	2"	2
25'	3"	5	25'	2"	1
30'	3"	5	30'	3"	5
35'	3"	5	35'	3"	5
40'	3"	4	40'	3"	4

We intend to further refine this information in terms of maximum **TOTAL EQUIVALENT FEET** and assign equivalent feet values to the elbows. Until this information becomes available, please use the above tables and contact us for assistance, as required.

Axel C. Dietrichson
Manager, Application Engineering

USER'S INFORMATION MANUAL

FOR YOUR SAFETY
Do not store or use gasoline or other flammable vapors and liquids in the vicinity of this or any other appliance.

WARNING
Improper installation, adjustment, alteration, service or maintenance can cause injury or property damage. Refer to this manual. For assistance or additional information consult a qualified installer, service agency or the gas supplier.

FOR YOUR SAFETY WHAT TO DO IF YOU SMELL GAS
- Do not try to light any appliance.
- Do not touch any electrical switch; do not use any phone in your building.
- Immediately call your gas supplier from a neighbor's phone. Follow the gas supplier's instructions.
- If you cannot reach your gas supplier, call the fire department.

TYPICAL NATURAL GAS UPFLOW INSTALLED IN BASEMENT

CONDENSING GAS FURNACES

SAVE THIS MANUAL FOR FUTURE REFERENCE

1006147
11-1-89
LT. 1
PRINTED IN U.S.A.

SAFETY REQUIREMENTS

Your furnace is built to provide many years of safe and dependable service, providing it is properly installed and maintained. However, abuse and/or improper use can shorten the life of the furnace and create hazards for you, the homeowner.

WARNING

> IMPROPER INSTALLATION, ADJUSTMENT, OPERATION, SERVICE, REPAIR, MAIN — TENANCE, OR ALTERATION OF THIS PRODUCT MAY RESULT IN PROPERTY DAMAGE, BODILY INJURY OR DEATH FROM HAZARDS SUCH AS FIRE, EXPLOSION, SMOKE, SOOT, CONDENSATION, ELECTRIC SHOCK OR CARBON MONOXIDE.

The following rules and recommendations, should be followed to insure safe and efficient operation of your furnace.

1. Thoroughly read this manual and all labels on the furnace to help you understand how your furnace operates and the hazards involved with gas and electricity.

2. Do not use this furnace if any part has been under water. Immediately call a qualified service technician to inspect the furnace and to replace any part of the control system and any gas control which has been under water.

3. Make sure the furnace is always connected to an approved vent, in good condition, to carry combustion products outdoors.

4. Never obstruct the vent grilles, or other means, that provide air to the furnace for proper combustion and ventilation of flue gases. If any structural changes are made, such as enclosing the furnace area, or if you add weather stripping, storm windows or another fuel burning appliance in the same area, have a qualified service agency check the combustion air supply.

5. Familiarize yourself with the possible air starvation signals outlined in the Combustion Air and Indoor Humidity section and perform the checks to determine if combustion air is adequate.

6. Maintain safety and service clearances from the furnace as listed on Furnace Clearance Label inside the cabinet. Keep the furnace room or area clean and free of combustible materials at all times. Never store gasoline, paint, aerosol cans, waxes, bleaches, dry cleaning fluid or items such as papers, rags, brooms or dust mops near the furnace.

7. If your furnace is installed in an area with loose fill or exposed insulation, the insulating material must be kept free and clear of furnace as some insulation is combustible. if additional insulation is added make sure the furnace area is checked.

8. Familiarize yourself with the controls that shut off the gas and electrical power to the furace. If the furnace is to be shutdown, for any length of time, turn off both the gas and electrical power. For safety always turn them off before performing service or maintenance on the furnace.

9. Establish a regular service and maintenance schedule to insure efficient and safe operation of the furnace. It is recommended that you have a qualified service agency perform a complete check on the furnace, before each heating season. See Service Technician Checks.

SERVICE TECHNICIAN CHECKS

When the furnace is being inspected for condition and operation have the Service Technician check the following items.

For additional information the Service Technician can consult the installation instructions and applicable service manual for the furnace.

1. Check for adequate combustion air being supplied to the furnace area and all air openings into or from the furnace.

2. Check all flue gas passages including main and pilot burners, heat exchanger, and vent pipe.

3. Check gas pipe and all connections inside and leading to the furnace for leaks.

4. Check electrical wiring and connections.

5. Check supply and return air ducts for leakage, blockage and connections to furnace.

6. Check circulating air blower wheel and motor, clean and lubricate if required.

7. Perform an operational checkout on the furnace to be sure safety controls function and that furnace operates properly.

FREEZING TEMPERATURES AND YOUR HOME

Your furnace is equipped with safety devices that may keep it from operating if there are any abnormal conditions affecting the furnace and L.P. models may not operate if there is even a brief or slight interruption in the electric power or gas supply.

If your furnace remains shut down long enough during freezing temperatures, for water pipes to freeze it could result in serious water damage.

If your home will be unattended during this time you should take these precautions.

1. **Shut the water off at the main inlet into your home and drain the water lines if possible.**

2. **Have someone check your home as often as necessary for temperature conditions that could cause water damage. Suggest they call a qualified service agency if required.**

2

COMBUSTION AIR/INDOOR HUMIDITY

(How They Affect Your Safety and Comfort)

Your home needs to breathe and the different temperatures and humidity ranges in your home during the year makes it necessary to be sensitive to air requirements and potential ventilation problems.

Because of high energy costs for home heating, new materials and methods are being used in construction and remodeling. The improved construction and additional insulation has made these homes much tighter around windows and doors so that air leakage is minimal. This may create a problem in supplying enough combustion and ventilation air for gas-fired or other fuel burning appliances. Fresh air is needed for combustion and ventilation of flue gases.

An energy efficient home or a home using exhaust fans, fireplaces, clothes dryers, and gas appliances increases this problem and your appliances could be starving for air, which is unsafe.

This may result in more and more air being drawn from the house until fresh air is sucked in through an appliance flue or fireplace chimney. **Carbon monoxide can be the result.**

Carbon monoxide or "CO" is a colorless and odorless gas produced when fuel is not burned completely or when the flame does not receive sufficient oxygen.

CARBON MONOXIDE CAN RESULT IN ASPHYXIATION.

Be aware of these air starvation signals:

1. Headaches-Nausea-Dizziness
2. Excessive humidity-Heavily frosted windows or a moist "clammy" feeling in the home.
3. Smoke from the fireplace won't draw up the chimney.
4. Flue gases won't draw up the appliance flue pipe.

1. How do I know if my furnace or other appliances are receiving enough air for proper combustion and ventilation of flue gases?

Use the following checkout procedure to determine if the air leakage into your home is adequate to supply the needs of your appliances and fireplace. If you are uncertain about your ability to perform these checks contact your installing dealer or place of purchase.

A. Make the inspection as follows:

1. Close all doors and windows. If you have a fireplace, start a fire and wait until flames are buring vigorously.
2. Turn on all exhausting devices, such as: kitchen and bathroom exhaust fans - dryers (gas or electric)
3. Turn on all vented gas appliances, such as: heating equipment (includes any room heaters) water heaters.
4. Wait ten (10) minutes for drafts to stabilize.

5. Check for draft hood spillage at each appliance. (Hold a lighted match 2" from draft opening as shown for the typical gas water heater.)

FIG. 1 CHECK FOR DRAFT HOOD SPILLAGE

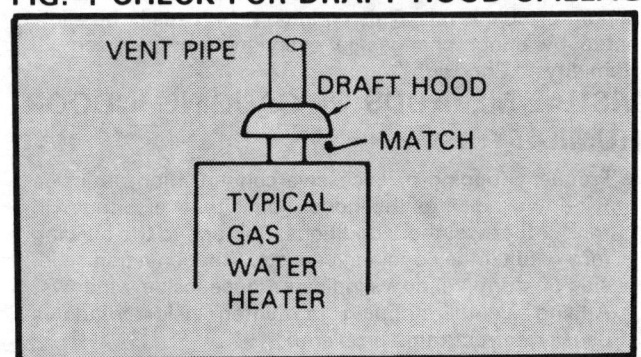

B. No Spillage -

If match flame pulls toward draft hood, this indicates sufficient infiltration air. Return exhausting devices and appliances to the condition you found them.

3

Combustion Air and Indoor Humidity (Cont.)

C. Drafthood Spillage -

If there is spillage at a draft hood - (match goes out or flame wavers away from draft hood):

1. Check for plugged flue connectors and chimneys. Check and repair stoppage and test again.

2. If you have a fireplace, open a window or door near the fireplace and then check for spillage.
   a. If spillage stops, do not use the fireplace until you can supply fresh air by a permanent duct.

3. If you have kitchen and bathroom exhaust fans turn them off and check for spillage.
   a. If spillage stops, do not use exhaust fans until you can supply fresh air by a permanent duct. Circuit breakers for fans should be turned off.

> ## WARNING
>
> DRAFT HOOD SPILLAGE, CAN CAUSE BODILY INJURY OR DEATH FROM ASPHYXIATION. SPILLAGE INDICATES THAT ADDITIONAL AIR MUST BE BROUGHT INTO THE STRUCTURE FROM THE OUTSIDE. KEEP A WINDOW OPEN (MINIMUM 2'') NEAR THE APPLIANCE UNTIL A PERMANENT AIR DUCT IS INSTALLED. CONTACT A QUALIFIED SERVICE AGENCY.

4. Spillage means air starvation and a fresh air duct or air intakes must be installed to provide air directly to the furnace or other gas appliance. These must comply with local and state building codes or in their absence with the National Fuel Gas Code NFPA 54/ANSI Z223.1, current edition.

2. What is Relative Humidity and how do I know if it is too high or low?

Relative humidity is the amount of water vapor in the air relative to the amount the air can hold at the same temperature. Example: At 40% relative humidity, the air can hold 60% more moisture before it is saturated.

The colder the air, the less moisture it can hold. As air is warmed, its ability to hold moisture is increased. Example: A winter day, outdoor temperature 10°F, and relative humidity of 70%. If that air enters a home and is warmed to 72°F the relative humidity will drop to 6% (very dry) if no more moisture is added.

Relative humidity is important to your health and home as proper humidification helps cut down on incidences of respiratory illness and helps keep air cleaner and fresher.

HAS YOUR FURNACE BEEN REPLACED?

This furnace is very efficient and has a much smaller vent (flue) pipe with a blower in the venting system.

The smaller vent and blower increases the efficiency of the furnace but they will also decrease the amount of natural air infiltration into the house. This is because less air will escape up the vent system during the off cycle so less cold dry air will enter the house by infilteration. This in conjunction with other items may cause the humidity to raise to uncomfortable levels. This condition can usually be eliminated by minor changes in everyday routines, see "If Humidity Is Too High".

VISUAL METHODS OF GAUGING INDOOR HUMIDITY:

● Frequent fogging or excessive condensation on inside windows indicates the indoor humidity level is too high for outdoor weather conditions. Damage to the building may result if the condition persists. (Condensation on inside of storm windows indicates loose inside windows. Adding weatherstripping to tighten inside windows usually corrects this problem.)

● Drop three ice cubes into a glass of water and stir. If, within three minutes, moisture does not form on the glass, the air is too dry and a humidifier would be beneficial. (Do not perform this test in the kitchen, as cooking vapors may produce inaccurate results.)

A good relative humidity is one just high enough to barely start condensation along the lower edges or lower corners of the windows. More than that can be damaging.

IF HUMIDITY IS TOO HIGH....

Condensation occurs when warm, moist air contacts a cool surface (window or outside wall, for example), and drops of water or a coating of frost form. The condensation problem increases as the outside temperature decreases.

A high humidity level usually results from bathing or cooking, etc. Suggestions for correcting this problem:

1. Turn down or discontinue use of humidifier.

2. Use range and bathroom exhaust fans while cooking and bathing or open a door or window for a few minutes to bring in cool dryer air.

3. Cook with pans covered.

4. Take shorter baths or showers with cooler water.

5. Install a fresh air intake duct. Cold, dry air brought in from outside to the furnace area lowers the indoor humidity level.

6. If the above measures do not correct the problem, have appliances checked. A malfunctioning appliance can contribute water vapor to the house.

7. If the above items do not correct the problem consult a heating contractor about adding a heat recovery ventilator or air to air heat exchanger.

RECOMMENDED INDOOR HUMIDITY:

Use the following table as a guide. It shows the recommended maximum indoor humidity in relationship to the outdoor temperature.

TEMPERATURE	HUMIDITY
+20°F and above	35%
+10°F	30%
0°F	25%
-10°F	20%
-20°F	15%

YOUR FURNACE

Condensing furnaces have a higher efficiency rating than conventional gas furnaces, but are basically the same in design and operation. The major difference is the addition of a secondary heat exchanger which captures heat that would normally go out the furnace vent pipe.

By capturing this heat the flue gases are cooled to a point where most of the vapor (mainly water) condenses out and must be drained away. Because the gases are cooled down plastic pipe is used for the vent pipe in place of metal.

The following paragraphs and Illustration will help you to understand the main parts of your furnace and how they operate.

DOOR INTERLOCK SWITCH

All of the electrical power for the furnace goes through the Door Interlock Switch and the furnace will not operate if the Blower Door is not properly in place.

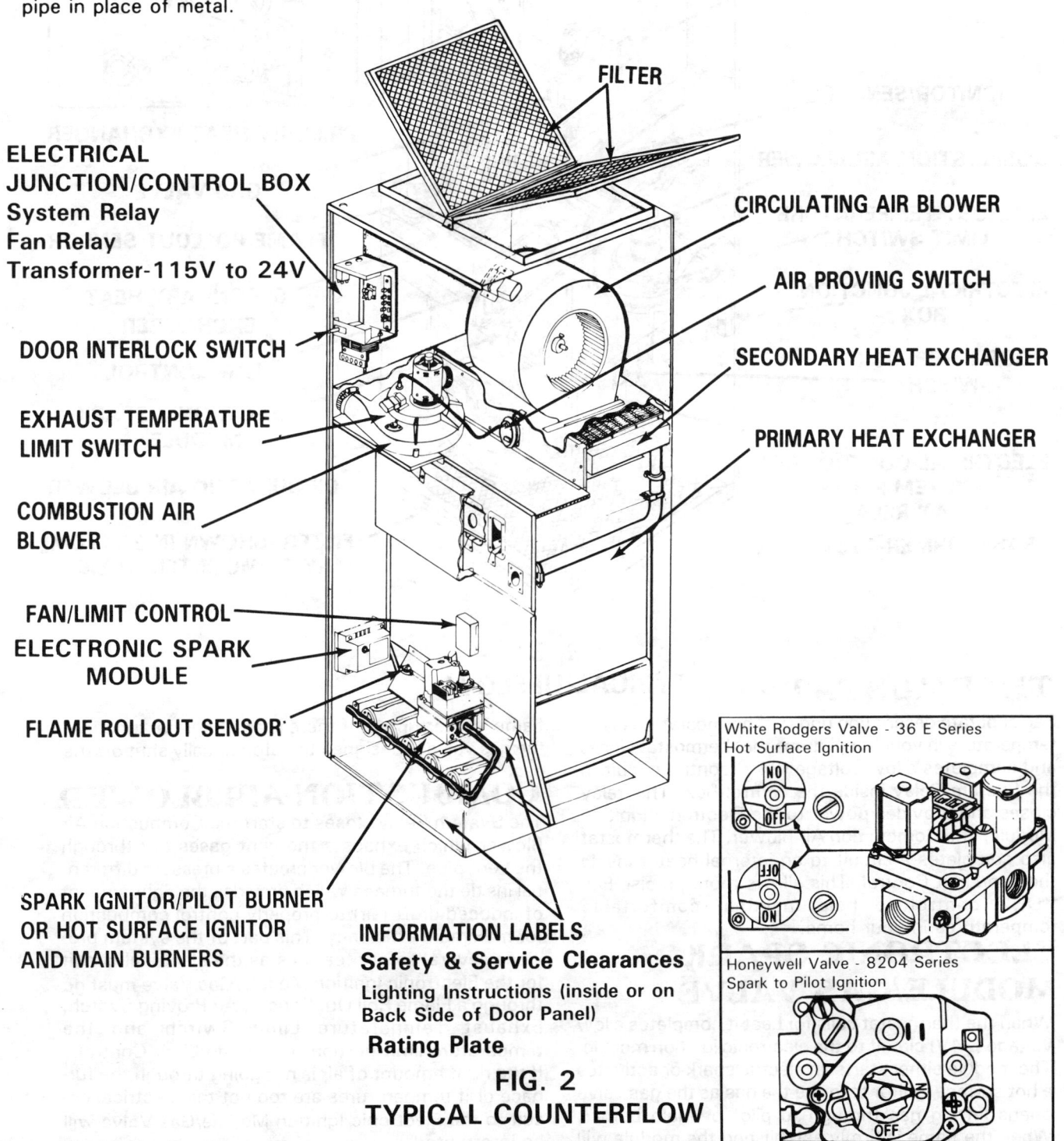

FILTER

ELECTRICAL
JUNCTION/CONTROL BOX
System Relay
Fan Relay
Transformer-115V to 24V

DOOR INTERLOCK SWITCH

EXHAUST TEMPERATURE
LIMIT SWITCH

COMBUSTION AIR
BLOWER

FAN/LIMIT CONTROL
ELECTRONIC SPARK
MODULE

FLAME ROLLOUT SENSOR

SPARK IGNITOR/PILOT BURNER
OR HOT SURFACE IGNITOR
AND MAIN BURNERS

CIRCULATING AIR BLOWER

AIR PROVING SWITCH

SECONDARY HEAT EXCHANGER

PRIMARY HEAT EXCHANGER

INFORMATION LABELS
Safety & Service Clearances
Lighting Instructions (inside or on
Back Side of Door Panel)
Rating Plate

White Rodgers Valve - 36 E Series
Hot Surface Ignition

Honeywell Valve - 8204 Series
Spark to Pilot Ignition

FIG. 2
TYPICAL COUNTERFLOW

Your Furnace (Cont.)

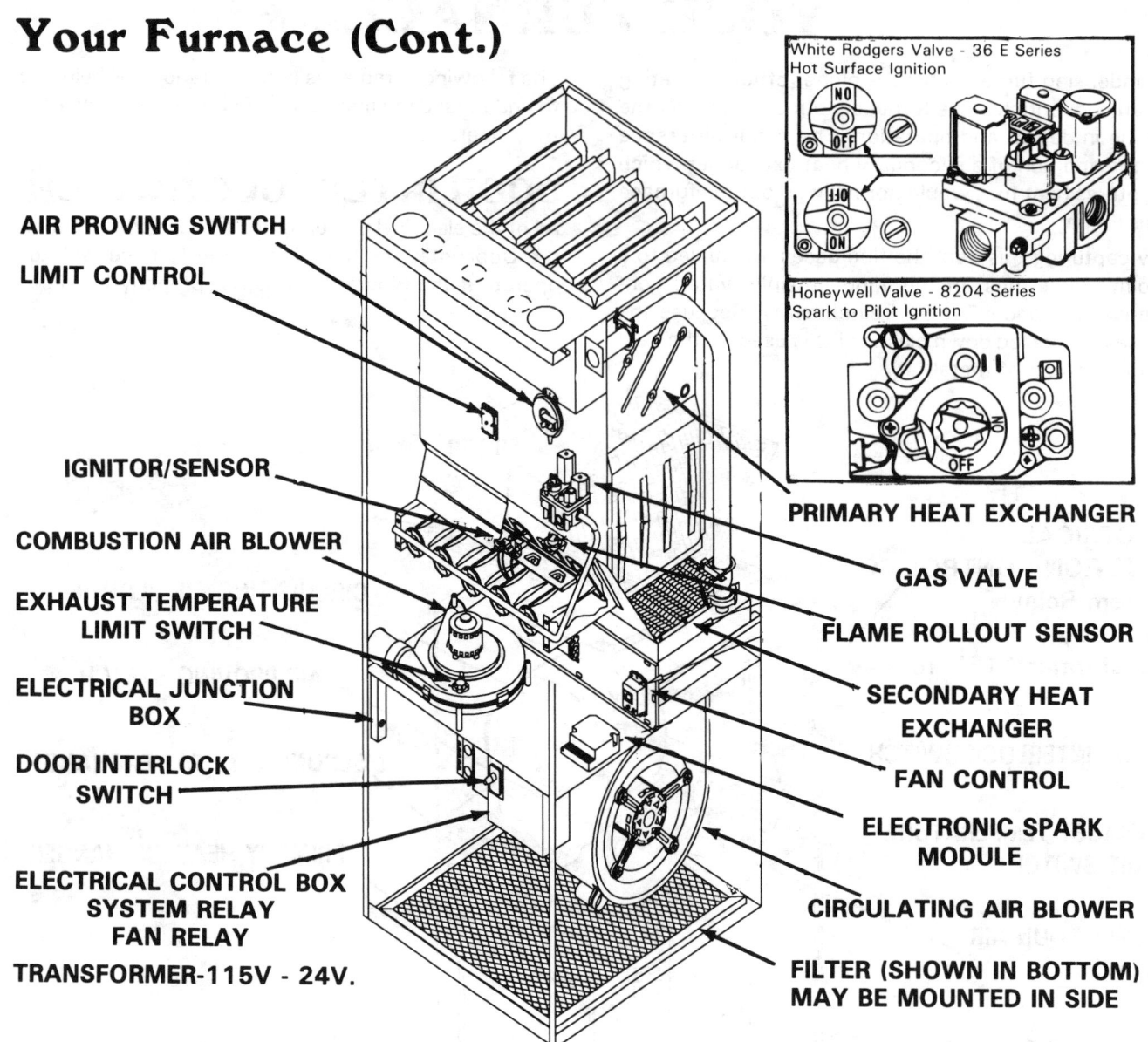

AIR PROVING SWITCH

LIMIT CONTROL

IGNITOR/SENSOR

COMBUSTION AIR BLOWER

EXHAUST TEMPERATURE LIMIT SWITCH

ELECTRICAL JUNCTION BOX

DOOR INTERLOCK SWITCH

ELECTRICAL CONTROL BOX
SYSTEM RELAY
FAN RELAY
TRANSFORMER-115V - 24V.

White Rodgers Valve - 36 E Series
Hot Surface Ignition

Honeywell Valve - 8204 Series
Spark to Pilot Ignition

PRIMARY HEAT EXCHANGER

GAS VALVE

FLAME ROLLOUT SENSOR

SECONDARY HEAT EXCHANGER

FAN CONTROL

ELECTRONIC SPARK MODULE

CIRCULATING AIR BLOWER

FILTER (SHOWN IN BOTTOM) MAY BE MOUNTED IN SIDE

FIG. 3
TYPICAL UPFLOW

THERMOSTAT

See separate description under "Thermostat". As the temperature in your home drops the thermostat closes and completes a low voltage (24V) control circuit to the System Relay inside the control box. This relay closes and provides power to the Electronic Ignition Module, and Combustion Air Blower. The thermostat also completes a circuit to an internal heater inside the Fan/Limit Control. This allows more precise Fan "ON" control to help maintain comfortable temperatures in your home.

ELECTRONIC SPARK MODULE/GAS VALVE

When the thermostat calls for heat it completes a low voltage (24V) circuit to the electronic ignition module. The module either starts an electric spark or activates a hot surface ignitor to ignite the gas as the gas valve opens letting gas through to pilot or main burner. When the flame is firmly established the module will shut off. For safety it will automatically restart if the

flame is interrupted. If there is an abnormal function, the gas valve is designed to automatically shut off the gas.

COMBUSTION AIR BLOWER

The System Relay closes to start the Combustion Air Blower which exhausts the vent gases out through the vent pipe. The blower creates a pressure differential inside the furnace which provides the right amount of induced draft (air) to properly control combustion for maximum efficiency. This part of the system provides several safety features as the electrical circuit for the Electronic Ignition Module/Gas Valve must go through a Flame Roll Out Sensor, Air Proving Switch, Exhaust Temperature Limit Switch and the temperature limit portion of the Fan/Limit Control. If the right amount of air is not going through the furnace or if temperatures are too hot the electrical circuit to the Electronic Ignition Module/Gas Valve will be interrupted.

Your Furnace (Cont.)

CIRCULATING AIR BLOWER

The blower to circulate room air through the furnace starts as soon as the internal heater in the Fan/Limit control causes the fan switch portion to close completing the electrical circuit.

It shuts off when the Fan/Limit Control opens the electrical circuit. This means the blower comes on approximately 30 seconds after the furnace first starts and it continues to run until the furnace cools down. This helps to achieve the maximum comfort and efficiency from your furnace.

FAN RELAY

This relay provides electric power to the Circulating Air Blower for continuous blower operation (Thermostat Fan Switch set to "FAN ON") and for the blower to run if central air conditioning is installed.

FAN/LIMIT CONTROL

The Fan/Limit Control provides power to the Circulating Air Blower to keep it on, until the furnace cools down. The limit portion provides safety because it will open the low voltage control circuit, shutting the furnace down if it gets too hot.

The fan off setting can be adjusted if the fan remains on long enough that cool drafts are felt in the room after the furnace shuts off. Contact a Qualified Service Technician.

DRAIN TRAP ASSEMBLY

The drain trap has a float in it, that will cause the furnace to shut down if the drain lines become obstructed. If this happens the trap and lines must be taken apart and cleaned. See "Condensate Disposal" page 10.

Thermostat

Your furnace will not operate properly without a good quality, correctly installed thermostat. The thermostat location is very important as it must be sensing average room temperatures. It must not be exposed to hot or cold drafts or hot or cold spots on the wall, such as outside walls or a wall with pipes inside or openings into attic.

There are many types and styles of thermostats but the operation is usually similar. **BE SURE TO BECOME FAMILIAR WITH YOUR THERMOSTAT**. The simplest type of thermostat only starts and stops the furnace to maintain the proper room temperature. The most widely used types will control both heating and cooling functions and will have a Fan Switch with Auto and ON settings. On Auto, the Circulating Air Blower will cycle on/off with the furnace but if switched to ON it will run constantly whether or not the furnace is on.

In addition there are thermostats that automatically switch from Heating to Cooling and with night setbacks. The night set-back, or multiple set-back type, will lower the temperature at night or during the day when no one is at home.

HEATING:

Position the Thermostat System Switch on HEAT. Set the Fan Switch to AUTO for the blower to cycle ON/OFF. Select desired temperature setting and furnace will automatically start up and shut off as required to maintain that setting. Fan Switch may be positioned to ON for continuous air circulation only if that position maintains the same or a higher blower speed so you have the same or greater airflow.

CAUTION

Continuous Fan ON at too slow of a blower speed may cause improper furnace operation and possible damage to the heat exchanger.

HEATING ANTICIPATOR:

For more precise comfort control your thermostat may have an adjustable Heat Anticipator. (Some are not adjustable). For most homes the anticipator should be set on the value listed in the Tech Data Sheet which is found in the Furnace Technical Support Manual.

If your furnace cycles ON/OFF with very short ON intervals or if the OFF cycle is so long that room temperatures become uncomfortable the anticipator setting may need to be adjusted.

To Adjust:

Remove thermostat cover and locate anticipator. Check the setting. If not on the setting recommended for your furnace, move the lever to the proper setting and try it for a day or two.

To lengthen burner-on time move the indicator towards "Longer". To decrease burner-on time move in opposite direction.

```
─────────────── NOTE ───────────────
To accurately determine the effect do not adjust
more than half a scale marking at a time and
allow a day between adjustments.
```

COOLING:

For cooling simply position the system switch to COOL Instead of HEAT and thermostat will function in the same manner to control cooling.

Operating Your Furnace

Keep the blower access door and upper access panel in place except for inspection or maintenance. An automatic switch prevents furnace operation if the blower door is not in place.

Before starting your furnace be sure you read and understand all of the procedures in this manual. Check to make sure the furnace filter is clean and correctly installed.

STARTING THE FURNACE

WARNING

DO NOT ATTEMPT TO LIGHT THE BURNER WITH A MATCH OR FLAME OF ANY KIND. YOU COULD BE INJURED.

See page 5 or 6 for location and illustration of gas valve.

1. Turn the thermostat to its lowest temperature setting or to OFF if equipped with a System Select Switch.
2. Turn Manual Shutoff Valve, in the gas line, to OFF position, should be a right angle or 90° to gas line.
3. Turn OFF electric power to furnace at disconnect switch or circuit breaker.
4. Remove furnace access panel, exposing gas controls and burner compartment.
5. Locate gas valve and identify settings. The valve is marked with ON — OFF positions.
6. Turn the knob all the way to OFF. Make sure it has been in OFF for at least 5 minutes before proceeding.
7. Turn the control knob to ON (Depress knob to turn on some Honeywell valves).
8. Replace access panels and doors.
9. Turn ON electric power for furnace.
10. Open the Manual Shutoff Valve in the gas-line.
11. Set thermostat to desired temperature and System Select Switch to HEAT if equipped.

The furnace will begin sparking to ignite the pilot flame or the hot surface ignitor will heat up to ignite the gas. When the system verifies a steady flame the ignition system will shut off.

With Spark to Pilot Ignition (System Retries)

The ignition system will spark to light the pilot for 90 seconds. If the pilot does not light within 90 seconds the system will shut off for 5 minutes and then it will try again for 90 seconds. It will continue to cycle until the pilot lights, (90 seconds ON/5 minutes OFF).

If the unit does not start after 3 tries or approximately 20 minutes, shut the unit off at the thermostat and call a qualified Service Agency.

WITH HOT SURFACE IGNITION SYSTEM

Furnace will make several attempts to lite before going into lockout. Each attempt requires a longer time with the total cycle taking approximately 8-10 minutes.

If air remaining in the lines on a new installation prevents the furnace from lighting **ONE** additional cycle may be tried. If furnace does not light, turn the thermostat to its lowest setting. Wait one minute, then turn it back up above the temperature shown on the thermometer. This starts the ignition cycle over again. **DO NOT REPEAT MORE THAN ONCE.** If furnace will not light, call a qualified Service Agency.

TURNING OFF THE FURNACE

1. Set the thermostat to the lowest setting or set System Select Switch to OFF if equipped.

Should overheating occur or the gas supply fail to shut off, shut off the manual gas valve to the furnace before shutting off the electrical supply.

EXTENDED SHUTDOWN

1. Set thermostat to lowest setting or set System Select Switch to OFF if equipped.
2. Turn Manual Shutoff Valve to off position, right angle or 90° to gas line.
3. Turn electric power off. (May be left "ON" for set-back or chronograph type thermostat with batteries, provided thermostat has a system select switch to place in the "OFF" position.)
4. Turn the gas valve control knob to "OFF".

Winter Shutdown

If there is the possibility of freezing temperatures, remove the condensate drain trap and empty to prevent freezing.

L P Model Furnaces

WARNING

If your L.P. (liquified petroleum) gas furnace is installed in a basement, an excavated area or a confined space, we recommend that you contact your L.P. supplier about installing a warning device that would alert you to a gas leak. We recommend this because L.P. gas is heavier than air and any leaking gas can settle in any low areas or confined spaces. This L.P. gas would create a DANGER OF EXPLOSION OR FIRE. If you suspect the presence of gas, follow the instructions on the cover of this manual.

FURNACE MAINTENANCE

Filter Replacement

It is recommended that you have your furnace inspected and serviced on an annual basis (before the heating season) by a qualified service technician.

You may perform maintenance on the furnace yourself, but remember that certain mechanical and electrical skills and tools are required to properly perform maintenance on the furnace. Personal injury or death may result if you are not properly trained. You should call your installing dealer or place of purchase if you are uncertain about your ability to perform maintenance.

> ### WARNING
> TURN OFF ELECTRIC POWER TO FURNACE BEFORE PERFORMING ANY MAINTENANCE OR REMOVING PANELS, BECAUSE OF THE DANGER OF ELECTRICAL SHOCK.

AIR FILTERS — Monthly

The air filter(s) should be inspected at least monthly and cleaned or replaced as required. There are two types of filters most commonly used. The most widely used is the fiberglass disposable type which should be replaced before it becomes clogged. The other type commonly in use is the washable type constructed of aluminum mesh, foam, or reinforced fibers. Washable filters may be cleaned by soaking in mild detergent and rinsing with water.

> ### NOTE
> Some filters are marked with an arrow to indicate the proper direction of air flow through the filter. When installing the arrow must point in the direction of the air flow. Remember that dirty filters are the most common cause of inadequate heating or cooling performance.

The table on page 10 lists recommended sizes and types of filters that may be used with your furnace, based on the input rating and nominal tons of air conditioning that may be used with the furnace.

However, the furnace installer may have used a larger filter for additional air volume or if the furnace was installed for Heating Only with a remote filter cabinet or central return he may have installed a smaller filter. If air conditioning has been added since your furnace was installed, make sure the filter size is adequate.

Replacement filters should be of the same type and size to ensure adequate air flow and filtering, unless a disposable low velocity filter is replaced with a washable high velocity type.

The filter will normally be found inside the furnace blower compartment, see pages 5 and 6, but alternate locations may be a remote filter rack attached to the outside of the furnace, a separate return air cabinet attached to furnace or a remote filter grille.

Remote filter grilles and return air cabinets will usually have a hinged door or removable panel to be able to remove filter. Filter racks attached to the furnace will usually be made so the filter simply slides out one side for removal. Use only the same size filter. The type must be the same unless replacing a disposable low velocity type, with a washable high velocity type.

> ### WARNING
> NEVER OPERATE FURNACE WITHOUT A FILTER INSTALLED AS DUST AND LINT WILL BUILD UP ON INTERNAL PARTS RESULTING IN LOSS OF EFFICIENCY, EQUIPMENT DAMAGE AND POSSIBLE FIRE.

Filter Replacement/Upflow (Hi-Boy) See Figure 4

1. Turn off electric power for furnace at circuit breaker or disconnect switch.
2. Remove blower compartment door.
3. Pull back on filter clip and remove filter being careful not to dislodge dirt and debris from filter.
4. Inspect filter and replace or clean washable type. If filter is aluminum mesh it should be recoated with filter coating spray.
5. Reinstall filter under clips. If filter is marked for air flow direction make sure it's installed correctly.
6. Replace blower compartment door making sure that it's tightly closed.
7. Turn on electric power for the furnace.

FIGURE 4 FILTER REPLACEMENT

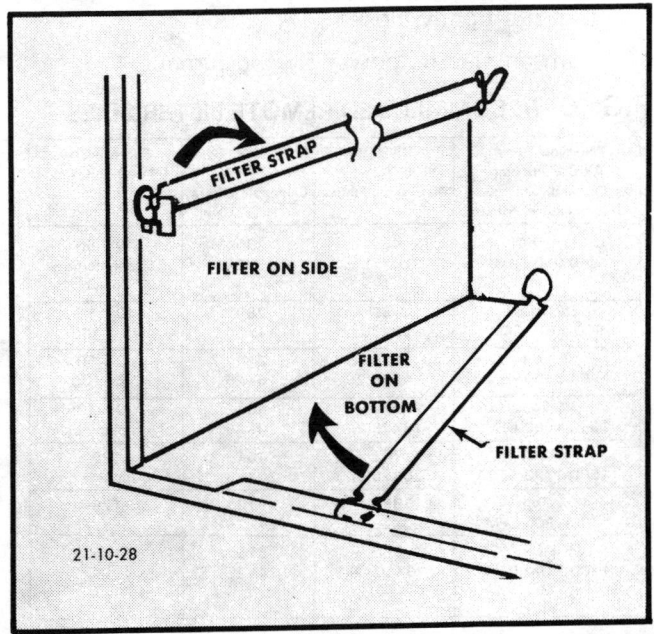

FILTER STRAP

FILTER ON SIDE

FILTER ON BOTTOM

FILTER STRAP

21-10-28

Furnace Maintenance (Cont.)

Filter Replacement/Counterflow

See Figure 5

1. Turn off electric power for furnace at circuit breaker or disconnect switch.
2. Remove blower compartment door.
3. Reach up through right side and lift upward and swing top of filter towards center of furnace then pull filter down and remove being careful not to dislodge dirt and debris from filter.
4. Inspect filter and replace or clean washable type. If filter is aluminum mesh it should be recoated with filter coating spray.

FIG. 5 COUNTERFLOW FILTER REPLACEMENT

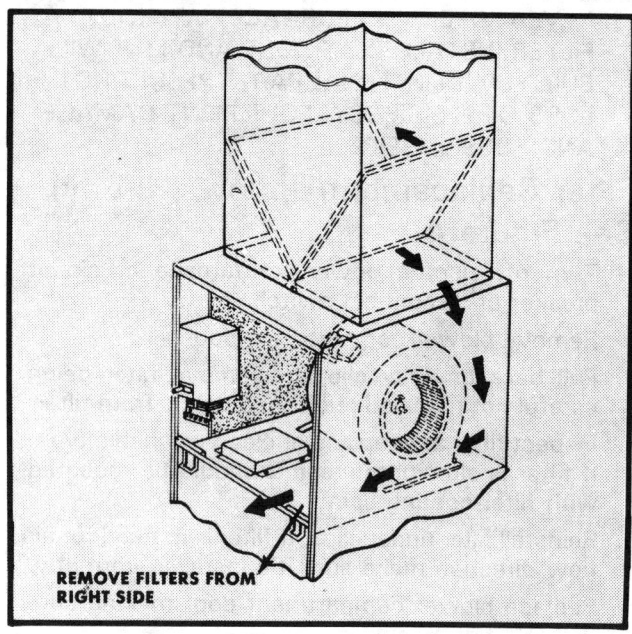

REMOVE FILTERS FROM RIGHT SIDE

5. Reinstall left filter first. If filter is marked for air flow direction make sure it's installed correctly.

6. Replace blower compartment door making sure that it's tightly closed.

7. Turn on electric power for the furnace.

FIG. 6 RECOMMENDED REMOTE FILTER SIZES

Nominal Tons Air Conditioning Nominal Air Flow Cubic Feet per Minute	Recommended Filter Sizes Sq. Inch Surface Area/Nominal Size		Furnace Size 1000 X BTUH
	Disposable Filters	Cleanable Filters	
Up Thru 2 Tons 800-900 CFM	432 20X25	260 15X20	40 50 75 (counterflow) (only)
2½ Tons 900-1100 CFM	480 20X30	288 14X25	40 50 75 100
3 Tons 1100-1300 CFM	576 14X25(2)	346 16X25	40 50 75 100
3½ Tons 1300-1500 CFM	672 16X25(2)	404 20X25	100 125 75 (upflow only)
4 Tons 1500-1700 CFM	768 20X25(2)	461 20X25	125 100 (upflow only)
4½ Tons 1700-1900 CFM	864 20X25(2)	519 24X25	125
5 Tons 1900-2100 CFM	960 20X30 (2)	576 24X25	125

CONDENSATE DISPOSAL Monthly/Annually

Your furnace has a condensate trap as part of the vent system. The moisture in the flue gases will condense and collect in the trap to go to an inside drain or be pumped to a sewer line using a condensate pump.

The Condensate Trap and Condensate Neutralizer Cartridge (if used) in the drain line leading from the trap will require some maintenance. Disassemble and clean trap and cartridge prior to each heating season or if drain line becomes plugged.

Inspect the drain line and overflow line at least monthly. If the Condensate Neutralizer Cartridge becomes plugged the condensate will flow through the overflow line. If this happens clean both cartridge and trap.

TO CLEAN: Disconnect the drain line cartridge and unscrew end cap from cartridge. Pour the neutralizer out and thoroughly flush neutralizer and inside of cartridge with water. Pour neutralizer back into cartridge, adding neutralizer if cartridge is less than ¾ full. Unscrew trap from Vent Connecting Tee and flush throughly with water, use soap if necessary to clean, DO NOT USE any kind of solvents. Make sure float is reinstalled in trap, Ref. Fig. 7.

Reassemble and seal threaded connections with silicone rubber (bathtub caulk) or pipe dope approved for plastic pipe.

See repair parts section in the Furnace Technical Support Manual, to order replacement neutralizer.

Do not use Condensate for any reason as it is acidic.

FURNACE CONDITION AND FLUE GAS PASSAGES/Monthly

A properly adjusted gas furnace should not require cleaning at frequent intervals, but it should be inspected regularly to ensure safe and efficient operation. A brief monthly inspection is recommended that does not require disassembly. In addition you should have the furnace inspected, and cleaned if required, by a qualified service technician annually.

During the monthly inspection check the vent pipe and fresh air intake (if installed), to be sure they are clear and free of obstructions. Check vent pipe for evidence of condensate leakage, tight joints, secure attachment to furnace and sagging pipe.

Horizontal sections of pipe must slope upward 1/4'' per foot except sections between furnace and drain trap when trap is not mounted directly on furnace. Any horizontal section (max. 4' long) must slope down a minimum of 1/4'' per foot to trap. Vertical sections to trap can be a maximum of 5 feet.

Check return air duct to make sure it is sealed to furnace casing and that it is in good physical condition. It must terminate outside the space containing the furnace with no holes or inlets in furnace space.

Furnace Maintenance (Cont.)

FIG. 7 TYPICAL INSTALLATION

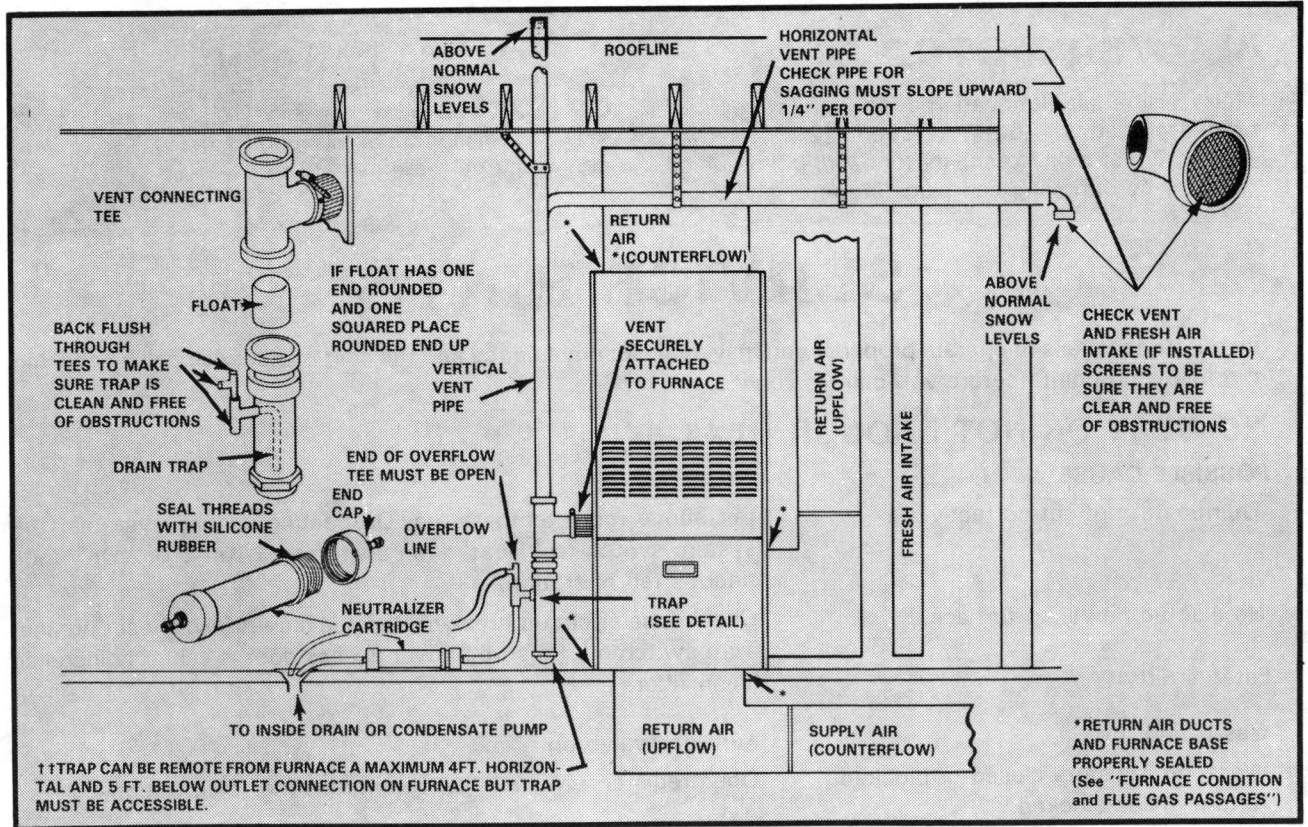

VENT CONNECTING TEE

FLOAT

IF FLOAT HAS ONE END ROUNDED AND ONE SQUARED PLACE ROUNDED END UP

BACK FLUSH THROUGH TEES TO MAKE SURE TRAP IS CLEAN AND FREE OF OBSTRUCTIONS

DRAIN TRAP

SEAL THREADS WITH SILICONE RUBBER

END OF OVERFLOW TEE MUST BE OPEN

END CAP

OVERFLOW LINE

NEUTRALIZER CARTRIDGE

TO INSIDE DRAIN OR CONDENSATE PUMP

††TRAP CAN BE REMOTE FROM FURNACE A MAXIMUM 4FT. HORIZONTAL AND 5 FT. BELOW OUTLET CONNECTION ON FURNACE BUT TRAP MUST BE ACCESSIBLE.

ABOVE NORMAL SNOW LEVELS

ROOFLINE

HORIZONTAL VENT PIPE CHECK PIPE FOR SAGGING MUST SLOPE UPWARD 1/4" PER FOOT

RETURN AIR *(COUNTERFLOW)

VENT SECURELY ATTACHED TO FURNACE

VERTICAL VENT PIPE

RETURN AIR (UPFLOW)

FRESH AIR INTAKE

ABOVE NORMAL SNOW LEVELS

CHECK VENT AND FRESH AIR INTAKE (IF INSTALLED) SCREENS TO BE SURE THEY ARE CLEAR AND FREE OF OBSTRUCTIONS

TRAP (SEE DETAIL)

RETURN AIR (UPFLOW)

SUPPLY AIR (COUNTERFLOW)

*RETURN AIR DUCTS AND FURNACE BASE PROPERLY SEALED (See "FURNACE CONDITION and FLUE GAS PASSAGES")

The floor or furnace base must be in good physical condition. For Upflow Furnace with a bottom return the floor or base area around the furnace must form a seal (no sagging, cracks defects etc.) to prevent air from being pulled in from furnace area, or any defect area must be sealed between floor or base and furnace.

Remove the front panel and use a flashlight to inspect the visible part of the heat exchanger, burners and ignitor. Check for loose soot and give particular attention to obvious deterioration from corrosion or other sources. Check for any signs of condensate leakage inside furnace cabinet.
If soot or deterioration is found or if there is evidence of condensate leakage inside furnace **DO NOT OPERATE FURNACE.**

Call a qualified service technician.

MAIN BURNER & PILOT FLAMES/Monthly

Allow furnace to run approximately 10 minutes then inspect the main burner flames and pilot flame (except furnaces with Hot Surface Ignition).

MAIN BURNER FLAMES should be stable, soft and blue, (dust may cause orange tips or they may have wisps of yellow but they must not have solid yellow tips). They should extend directly upward from burner without curling, floating or lifting off. They must not touch the sides of the heat exchanger.

Contact a qualified service agency at once if an abnormal flame appearance should develop.

FIG. 8 MAIN BURNER FLAME

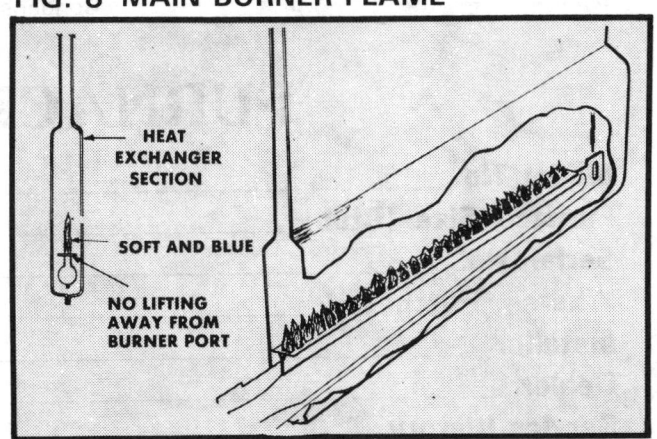

HEAT EXCHANGER SECTION

SOFT AND BLUE

NO LIFTING AWAY FROM BURNER PORT

PILOT FLAME should surround 3/8" to 1/2" of the ignitor/sensor tip.

FIG. 9 PILOT FLAME

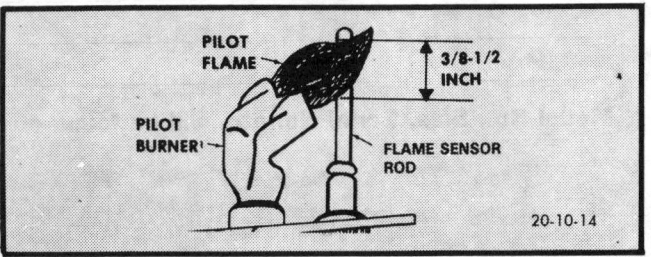

PILOT FLAME

PILOT BURNER

3/8-1/2 INCH

FLAME SENSOR ROD

20-10-14

11

BLOWER MOTOR

Motor will require lubrication every five (5) years of normal operation. Add ½ teaspoon (2 cc) of SAE #10W30 motor oil to each motor bearing through oil tubes or by removing cap plugs in motor end bells.

CAUTION

DO NOT over oil or use 3 in 1 oil, penetrating oil, WD40 or similar oils or oil motor bearings. Use of these may damage these motors.

SERVICE HINTS

If your furnace fails to operate properly, you may avoid inconvenience and the cost of a service call by checking the following points before you call for service.

NO HEAT OR NOT ENOUGH HEAT...

POSSIBLE CAUSE	WHAT TO DO
Thermostat not set correctly.	Set above room temperature. On heating/cooling systems, turn system switch to ''Heat''; fan switch to ''Auto'' or ''On'' (continuous fan operation).
No electric power to furnace.	Check fuse or circuit breaker. Replace blown fuse. Reset breaker. Turn switch on. Be sure blower access door is securely installed.
Filter is dirty.	Clean filter.
Gas is shut off.	Turn gas valve on.
Warm air registers closed or blocked Return grilles blocked.	Open registers. Move rugs, furniture, other obstructions.
Vent Pipe is obstructed or Drain lines on Condensate Trap are obstructed causing float to rise and block vent pipe.	Furnace has a pressure switch in the vent system which interrupts the electrical circuit. If the furnace is vented horizontally check the outlet area to make sure it is not obstructed. Check drain lines and Condensate Trap.

FURNACE RECORD

Model No. _____

Furnace Size (Btuh) _____

Serial No. _____

Where Purchased_____ Date_____

Installer _____ Date _____

Dealer _____ Phone _____

Service History _____

_____

_____

_____

_____

_____

Model No., Size, Serial No. etc. will be found on the Furnace Rating Plate, See Page 5.

Lennox

SERVICE

LENNOX *Industries Inc.*

UNIT INFORMATION

Corp. 826-L7

Litho U.S.A.

Supersedes Corp. 8113-L12

G14 SERIES UNITS

I - INTRODUCTION

The G14 unit is a condensing furnace utilizing the pulse combustion process. Initially, combustion takes place in an enclosed chamber. Then, as combustion products pass through the heat exchange system into a coil, the latent heat of combustion is extracted by condensing water from the exhaust gas.

The unit uses a redundant gas valve to assure safety shut-off as required by A.G.A.

Electronic Direct Spark Ignition is used to initiate combustion. A flame rectification sensor verifies ignition with a protection circuit that permits five trials for ignition before locking out the gas valve and control circuit. The sensor also verifies loss of combustion during a cycle, closing the gas valve and locking out the system. Obstructions to the air intake or exhaust outlet also shut down the unit immediately.

A small blower is used to purge the combustion chamber before and after each heating cycle to provide proper air mixture for start-up.

The units are manufactured for natural gas application. L.P. kits are available for field changeover.

PULSE COMBUSTION PROCESS

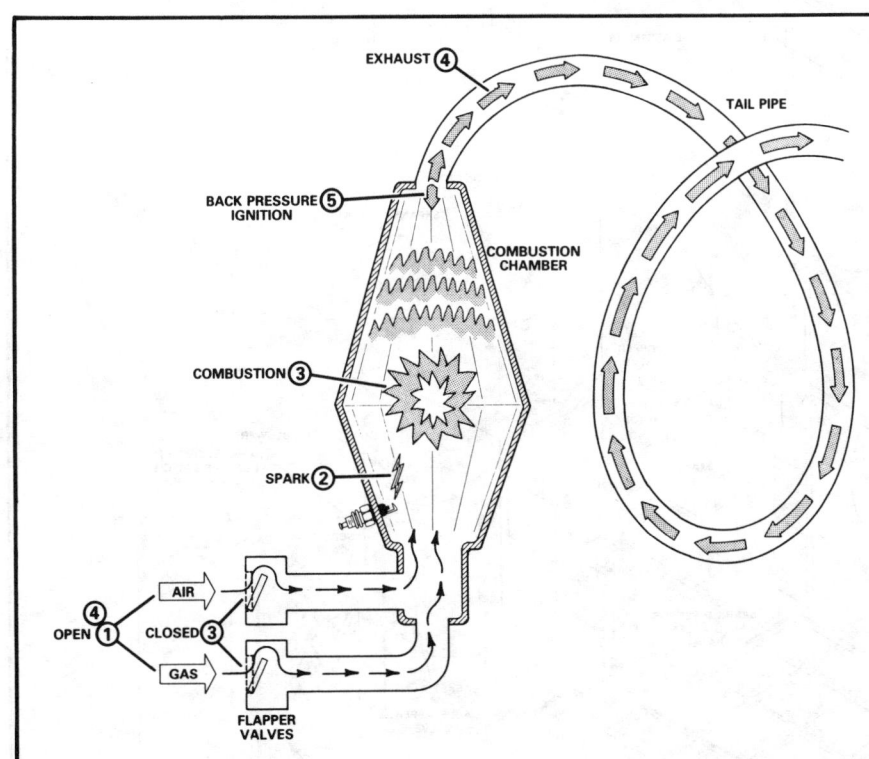

1 - Gas and air enter and mix in combustion chamber.
2 - To start the cycle a spark is used to ignite the gas and air mixture. (This is one 'pulse').
3 - Positive pressure from combustion closes flapper valves and forces exhaust gases down a tailpipe.
4 - Exhaust gases leaving the chamber create a negative pressure. This opens the flapper valves drawing in gas and air.
5 - At the same instant part of the pressure pulse is reflected back from the tailpipe causing the new gas and air mixture to ignite. No spark is needed. (This is another 'pulse').
6 - Steps 4 and 5 repeat 60 to 70 times per second forming consecutive 'pulses' of 1/4 to 1/2 Btu each.

II - PARTS ARRANGEMENT

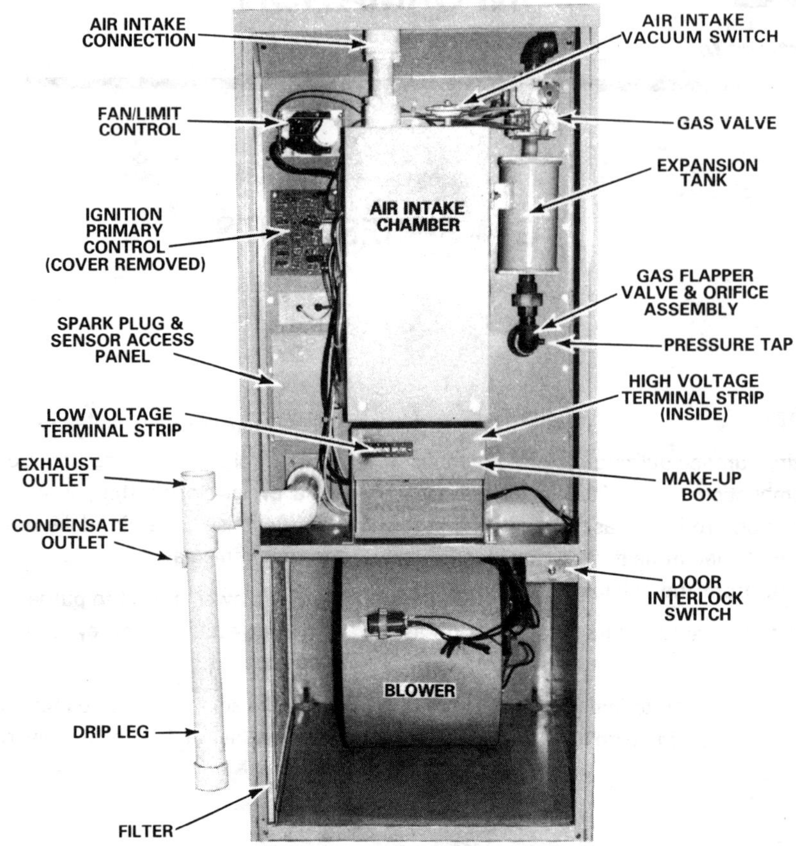

AIR INTAKE CONNECTION
AIR INTAKE VACUUM SWITCH
FAN/LIMIT CONTROL
GAS VALVE
EXPANSION TANK
IGNITION PRIMARY CONTROL (COVER REMOVED)
AIR INTAKE CHAMBER
GAS FLAPPER VALVE & ORIFICE ASSEMBLY
PRESSURE TAP
SPARK PLUG & SENSOR ACCESS PANEL
LOW VOLTAGE TERMINAL STRIP
HIGH VOLTAGE TERMINAL STRIP (INSIDE)
EXHAUST OUTLET
MAKE-UP BOX
CONDENSATE OUTLET
DOOR INTERLOCK SWITCH
DRIP LEG
BLOWER
FILTER

NOTE 40, 60, AND 80 MODELS SHOWN

A - Expanded View

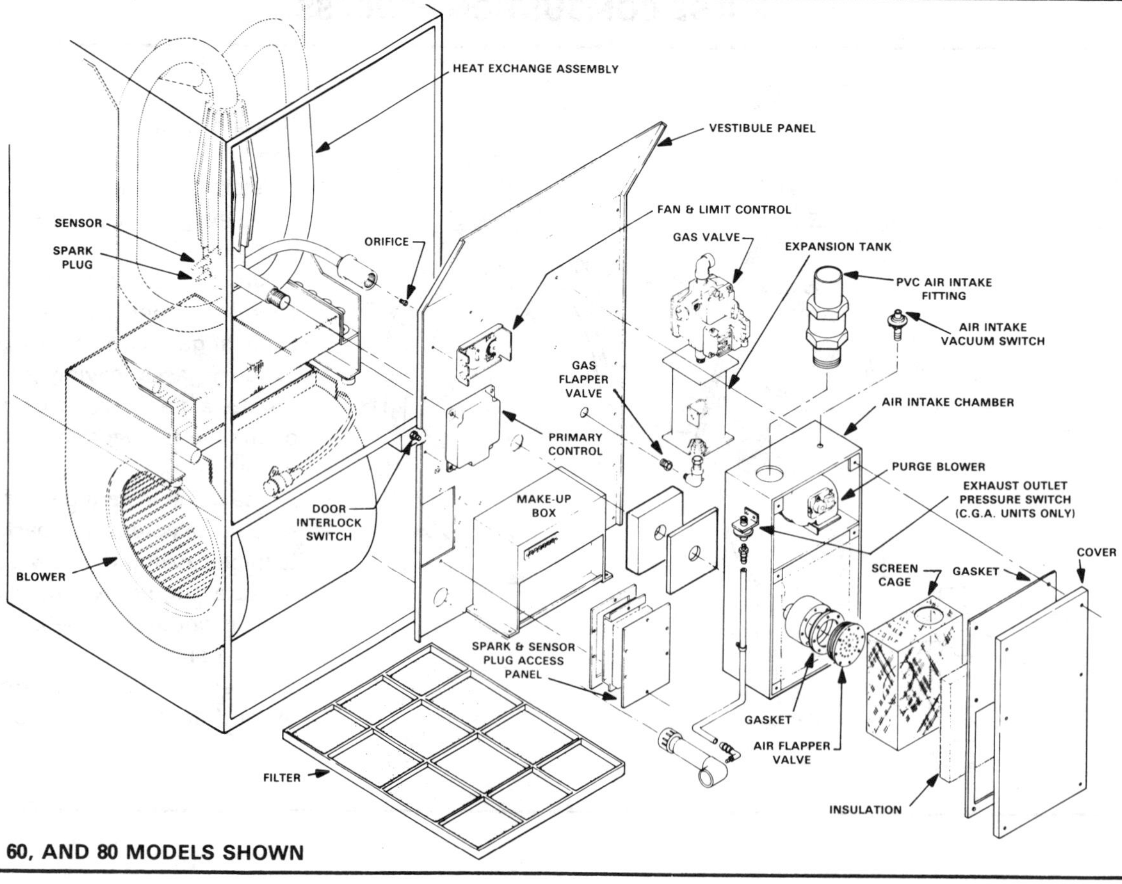

HEAT EXCHANGE ASSEMBLY
VESTIBULE PANEL
SENSOR
SPARK PLUG
ORIFICE
FAN & LIMIT CONTROL
GAS VALVE
EXPANSION TANK
PVC AIR INTAKE FITTING
AIR INTAKE VACUUM SWITCH
GAS FLAPPER VALVE
AIR INTAKE CHAMBER
PRIMARY CONTROL
DOOR INTERLOCK SWITCH
MAKE-UP BOX
PURGE BLOWER
EXHAUST OUTLET PRESSURE SWITCH (C.G.A. UNITS ONLY)
BLOWER
SCREEN CAGE
GASKET
COVER
SPARK & SENSOR PLUG ACCESS PANEL
GASKET
AIR FLAPPER VALVE
FILTER
INSULATION

NOTE 40, 60, AND 80 MODELS SHOWN

III - SPECIFICATIONS & DIMENSIONS

The G14 unit input range covers 40,000 through 130,000 Btuh. See specifications table.

The '40', '60' and '80' models use the same cabinet size as a '110' in the existing G8 through G12 furnace line. The '100' and '130'

models use the same cabinet size as a '137' in the G8 through G12 line. All units in the G14 series use direct drive blowers and accept cooling coils in nominal tonnages up to 5 tons for the '130'.

Slab filters are used for either bottom or side return air for all the unit sizes.

SPECIFICATIONS

Model No.		G14Q3-40	G14Q3-60	G14Q4-60	G14Q3-80	G14Q4-80	G14Q3-100	G14Q5-100	G14Q5-130
Input Btuh		40,000	60,000	60,000	80,000	80,000	100,000	100,000	130,000
High static Certified by A.G.A. (in. wg.)		.50	.50	.50	.50	.50	.45	.50	.50
Temperature rise range (°F)		35 - 65	40 - 70	35 - 65	45 - 75	40 - 70	45 - 75	45 - 75	45 - 75
Vent/Intake air size (in.)		2	2	2	2	2	2	2	2
Gas piping size I.P.S. (in.)	Natural	1/2	1/2	1/2	1/2	1/2	1/2	1/2	1/2
	*LPG	1/2	1/2	1/2	1/2	1/2	1/2	1/2	1/2
Condensate drain connection (SDR11)		1/2	1/2	1/2	1/2	1/2	1/2	1/2	1/2
Blower wheel nom. diam. x width (in.)		10 x 8	10 x 8	11 x 9	10 x 8	11 x 9	10 x 8	12 x 12	12 x 12
Blower motor hp		1/3	1/3	1/2	1/3	1/2	1/2	3/4	3/4
Size of filters (in.) 1 Per Unit		16/25/1	16/25/1	16/25/1	16/25/1	16/25/1	20/25/1	20/25/1	20/25/1
Tons of cooling that can be added		1.5 — 3	1.5 — 3	2.5 — 4	1.5 — 3	2.5 — 4	3 — 3.5	4 — 5	4 — 5
Shippng weight (lbs.)		250	250	255	250	255	286	314	314
Electrical characteristics		120 volts — 60 hertz — 1 phase (All Units)							

*For LPG units a field changeover kit is required and must be ordered extra.

IV - UNIT COMPONENTS

A - Make-up Box (Figure 1)

The make-up box is located below the air intake chamber and is designed to open over the exhaust PVC line when the unit is set up for right hand discharge of exhaust.

1 - Low voltage terminal strip with thermostat markings.
2 - 30VA transformer, 120 volt primary/24 volt secondary.
3 - Double-pole, double-throw indoor blower relay - 24 volt coil.
4 - Power supply and accessory terminal strip.

B - Fan/Limit Control (Figure 2)

A Honeywell fan/limit control is used with a sure start heater. The heater is energized with the gas valve to close the fan contacts after 30 to 45 seconds. The fan 'off' setting is factory adjusted to 90°F. It should not be necessary to change this setting. Fan 'off' settings above 90°F will cause the blower to recycle frequently (after a heating cycle) due to the additional heat that will be retained in the heat exchange assembly.

Do not change the limit factory setting. It is fixed in position for a maximum discharge air of 175°F.

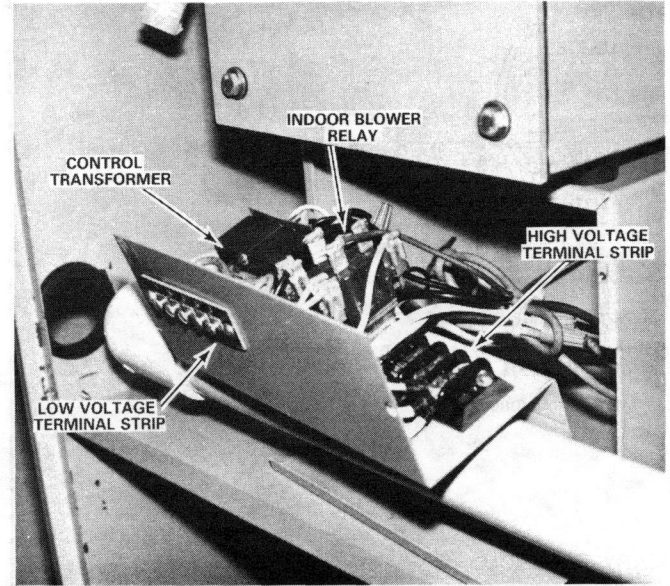

FIGURE 1

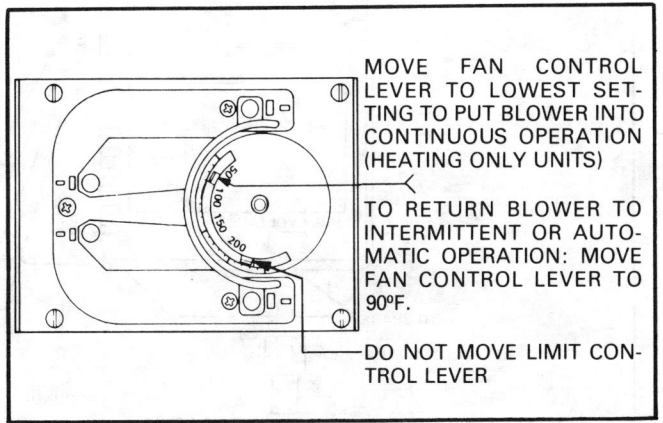

MOVE FAN CONTROL LEVER TO LOWEST SETTING TO PUT BLOWER INTO CONTINUOUS OPERATION (HEATING ONLY UNITS)

TO RETURN BLOWER TO INTERMITTENT OR AUTOMATIC OPERATION: MOVE FAN CONTROL LEVER TO 90°F.

DO NOT MOVE LIMIT CONTROL LEVER

FIGURE 2

C - Ignition Control

1 - An electronic direct spark ignition control with flame rectification sensing is used on all G14 units. The first production units (1981-1982) use ignition controls manufactured by 'Gas Energy' and 'Prestolite'. Other makes of controls may be used on later production units.

2 - The ignition controls are fully interchangeable and connect directly to a unit wiring harness. The wiring harness has a six prong plug (identified as JP1) that mates to a jack connected to the control. Either control will plug into and operate any unit without rewiring.

3 - The 'Prestolite' control is illustrated in Figure 3. The unit wiring harness plugs directly into the jack at the lower right hand corner of the control. Each of the six jack terminals are identified by number and function. The spark and sensor wire connections are made to quick connect terminals on the control as shown.

4 - The 'Gas Energy' control is illustrated in Figure 4. This control has an interconnecting harness used to connect it to the unit wiring harness plug. The terminals and connections are identified by function and number.

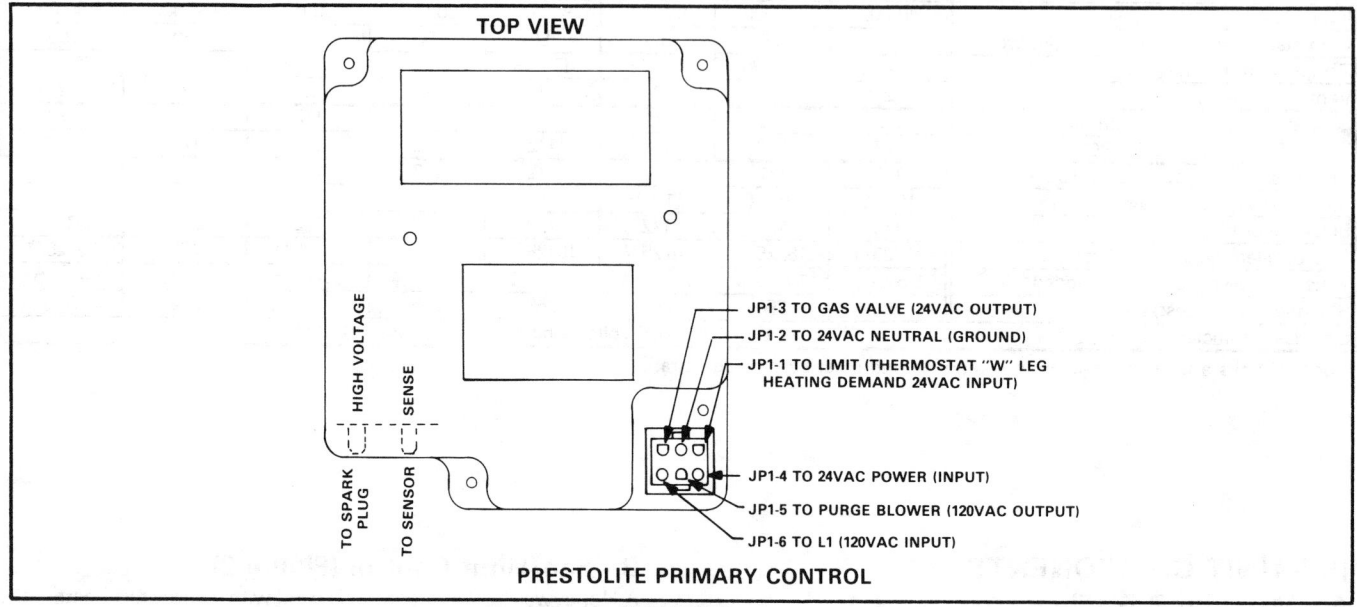

FIGURE 3

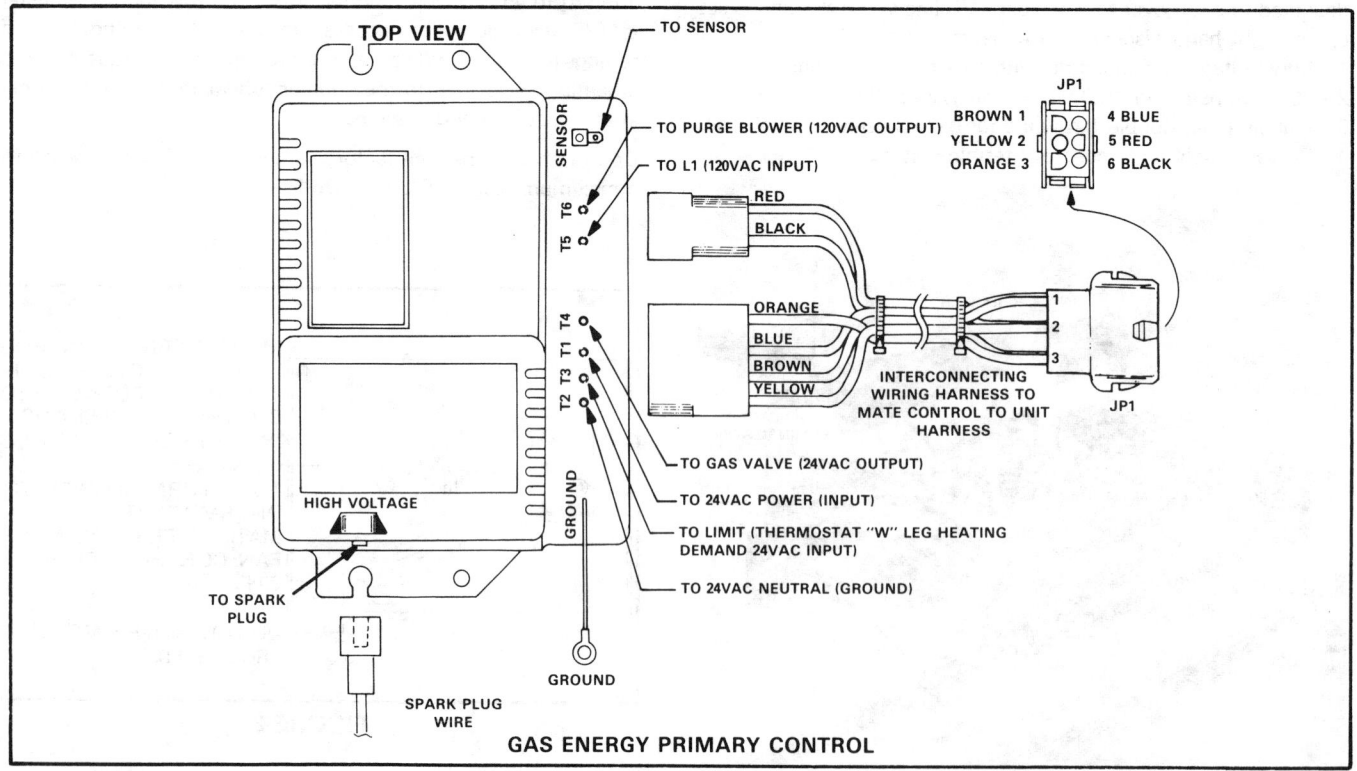

FIGURE 4

5 - The primary control provides four main functions: pre-purge, ignition, flame sensing and post-purge. The ignition attempt sequence of the control provides 5 trials for ignition before locking out. The unit will usually ignite on the first attempt. See Figure 5 for a normal ignition sequence with nominal timings for simplicity.

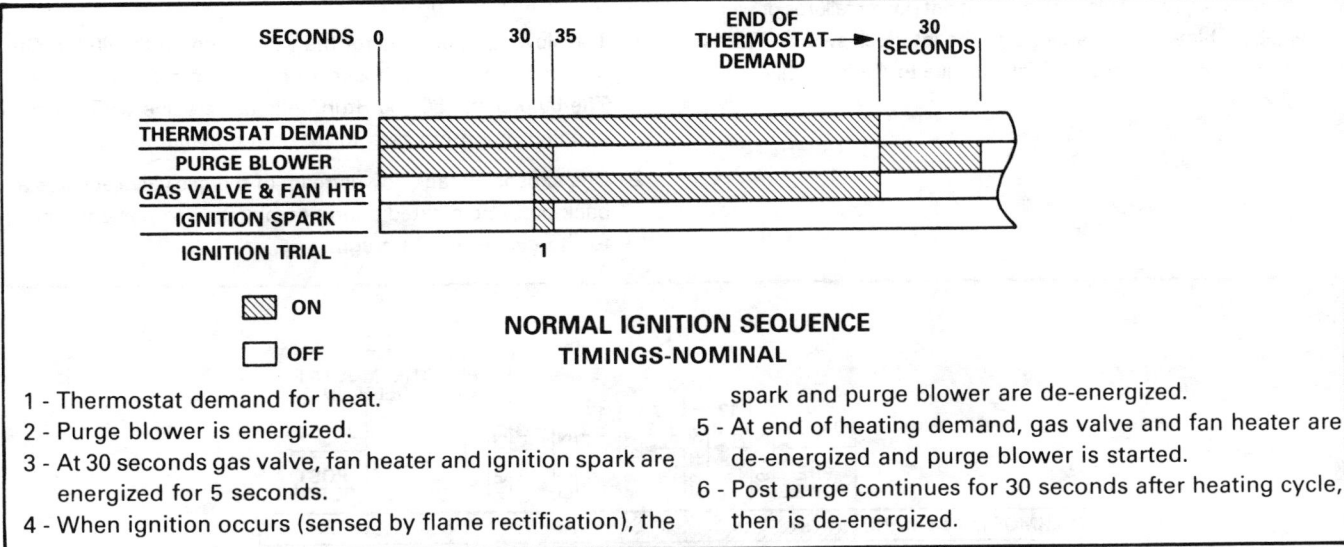

NORMAL IGNITION SEQUENCE
TIMINGS-NOMINAL

1 - Thermostat demand for heat.
2 - Purge blower is energized.
3 - At 30 seconds gas valve, fan heater and ignition spark are energized for 5 seconds.
4 - When ignition occurs (sensed by flame rectification), the spark and purge blower are de-energized.
5 - At end of heating demand, gas valve and fan heater are de-energized and purge blower is started.
6 - Post purge continues for 30 seconds after heating cycle, then is de-energized.

FIGURE 5

6 - Proper gas/air mixture is required for ignition on the first attempt. If there is slight deviation, within tolerance of the unit, a second or third trial may be necessary for ignition. The control will lockout the system if ignition is not obtained within 5 trials. Reset after lockout requires only breaking and re-making the thermostat demand. See Figure 6 for the ignition attempt sequence with retrials (nominal timings given for simplicity). Loss of combustion during a heating cycle is sensed through absense of flame signal causing the control to lockout after five ignition retrials.

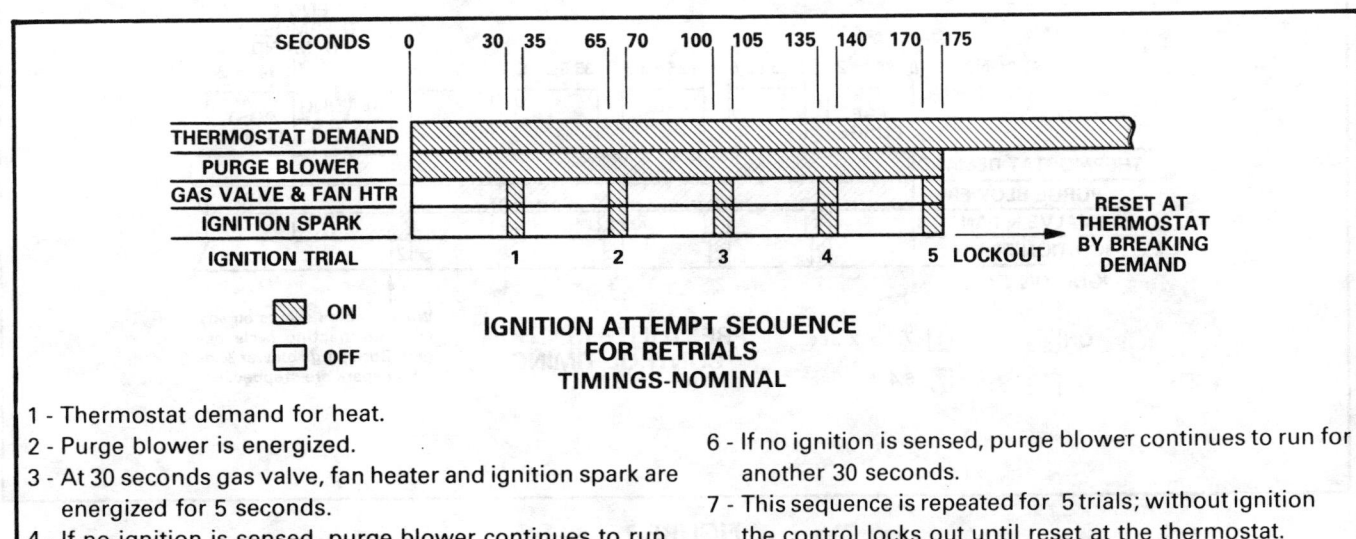

IGNITION ATTEMPT SEQUENCE
FOR RETRIALS
TIMINGS-NOMINAL

1 - Thermostat demand for heat.
2 - Purge blower is energized.
3 - At 30 seconds gas valve, fan heater and ignition spark are energized for 5 seconds.
4 - If no ignition is sensed, purge blower continues to run.
5 - After additional 30 seconds, ignition retrial takes place for another 5 seconds.
6 - If no ignition is sensed, purge blower continues to run for another 30 seconds.
7 - This sequence is repeated for 5 trials; without ignition the control locks out until reset at the thermostat.
8 - With ignition occuring at any trial, the sequence continues as a normal ignition sequence.

FIGURE 6

7 - The specific timings for 'Gas Energy' and 'Prestolite' ignition controls vary, but do not affect unit operation. Both will make 5 trials for ignition before lockout. The specific timing sequences for each are given in Figure 7.

The 'Prestolite' control runs through a post-purge cycle each time power is interrupted to the unit; if main switch is turned off and on again or following intermittent power failures or when replacing blower door energizing the interlock switch. This is a normal operating characteristic unique to the 'Prestolite' control only.

D - Gas Valve & Expansion Tank (Figure 8)

1 - The gas valve is slow opening on '40', '60' and '80' units and immediate opening on '100' and '130' units. It is internally redundant to assure safety shut-off. There are two top terminals and two side terminals that energize the internal redundant solenoids. They are wired in parallel. A manual on-off knob is provided on the valve.

The slow opening time for the valve used on the 40,000 Btuh unit is 7 seconds from zero to maximum manifold pressure. The 60,000 and 80,000 Btuh units use a valve with 4 second opening time from zero to maximum manifold pressure.

2 - The expansion tank downstream of the gas valve absorbs any back-pressure created during combustion to prevent damage to the gas valve diaphragm.

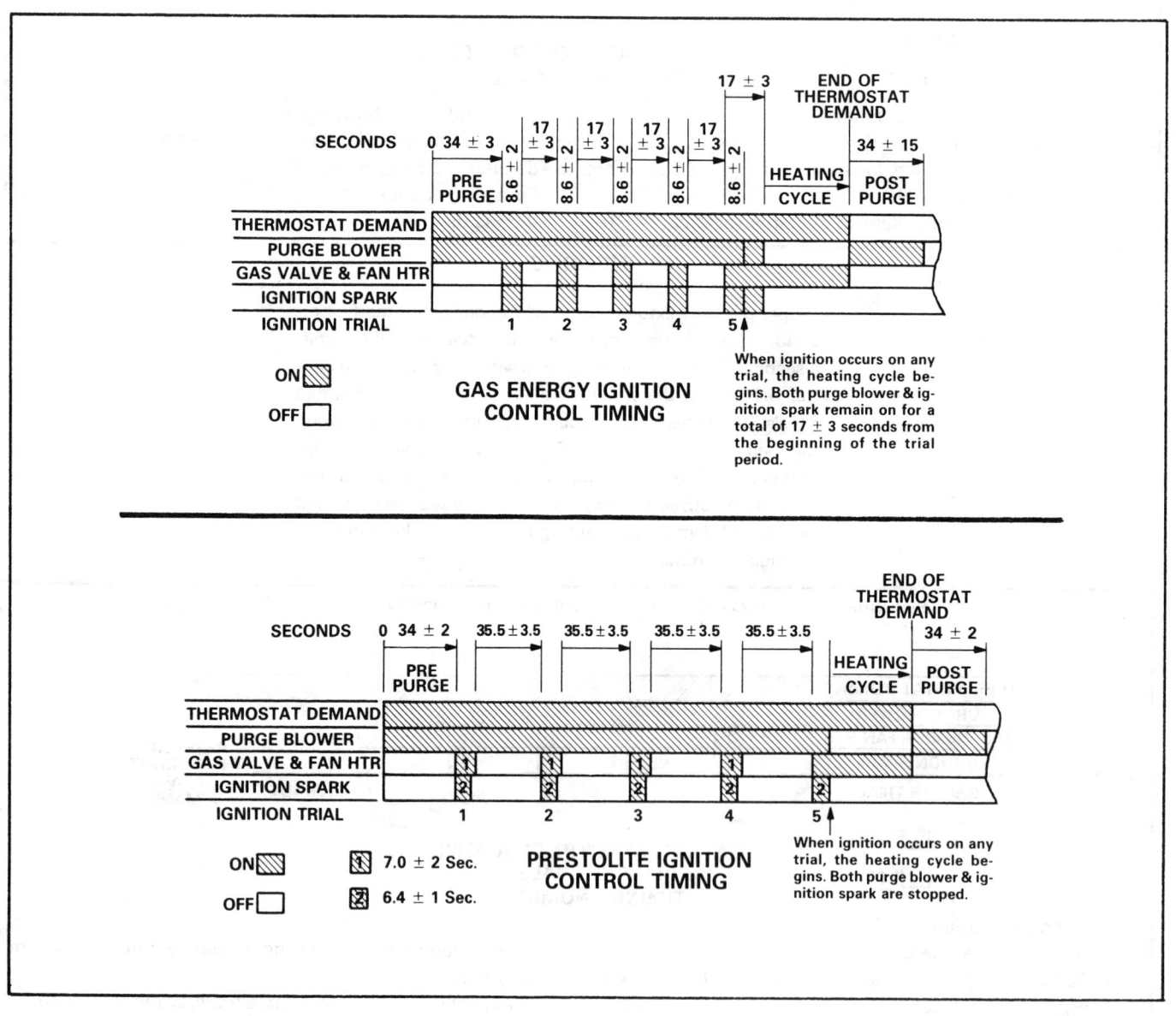

FIGURE 7

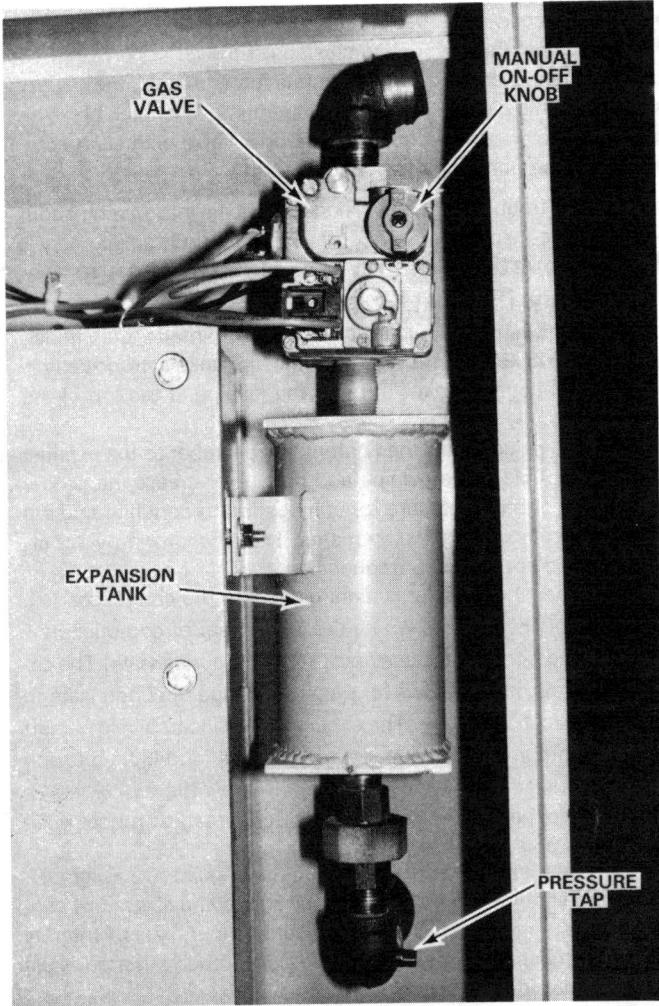

FIGURE 8

E - Air Intake Vacuum Switch

A vacuum switch is mounted on top of the Air Intake Chamber, see Parts Arrangement and Expanded View on Page 2. (Used on both A.G.A. and C.G.A. units.)

The switch is normally closed and remains closed under normal operating conditions. Obstructions or close-off of the air intake pipe causes the switch to open at a 2.0 ± 0.2 in. W.C. vaccum on '40', '60', and '80' models and 3.0 ± 0.2 in. W.C. on '100' and '130' units. When the switch opens it breaks the heat demand circuit to shut down the unit. This is a safety shut down function. The switch automatically resets when the restriction is removed from the air intake.

F - Exhaust Outlet Pressure Switch (C.G.A. Units Only)

This pressure switch is mounted on the side of the Air Intake Chamber and is connected to the exhaust outlet PVC elbow by a length of plastic tubing. Refer to the Expanded View on Page 2.

The switch is normally closed and remains closed under normal operating conditions. Obstruction or close-off of the exhaust outlet pipe causes the switch to open. On the '40' units it opens at 2.75 ± 0.2 in. W.C. pressure; '60' and '80' units it opens at 4.0 ± 0.2

in. W.C. pressure; '100' and '130' units it opens at 5.0 ± 0.2 in. W.C. pressure. When the switch opens it breaks the heat demand circuit to shut down the unit. This is a safety shutdown function. The switch automatically resets when the restriction is removed from the exhaust outlet.

G - Gas Intake Flapper Valve & Orifice (Figure 9)

1 - A union at the bottom of the expansion tank provides for removal of the gas flapper valve assembly and access to the orifice.

The flapper floats freely over the spacer and is opened against the clearance plate by incoming gas pressure. Back pressure from each combustion pulse forces the flapper against the valve body closing off the gas supply.

Refer to the troubleshooting section for specific information about flapper valve inspection and conditions requiring replacement.

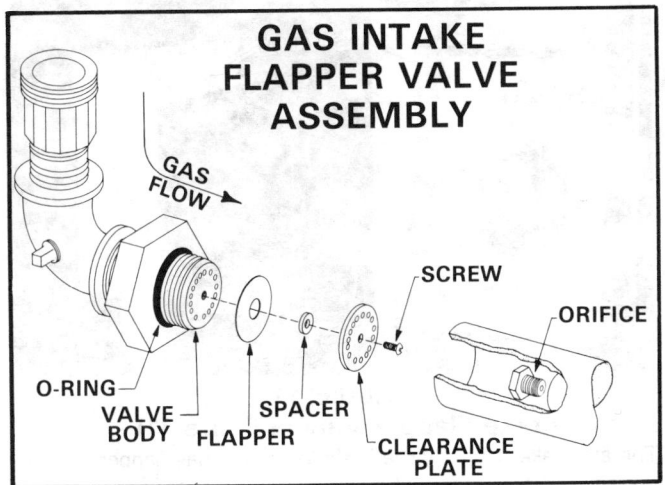

FIGURE 9

2 - Each G14 unit uses only one orifice located downstream of the flapper valve and is sized specifically for each unit. Table 1 references the nominal orifice drill sizes for each unit.

Note: Standard atmospheric burner orifices or orifice blanks cannot be used as replacements in G14 units.

TABLE 1

UNIT	NOMINAL ORIFICE DRILL SIZE			
	NATURAL GAS	L.P. GAS	HIGH ALTITUDE	
			NATURAL GAS	L.P. GAS
G14-40 A.G.A.	#14	#30	*	*
G14-60 A.G.A.	#13	#27	*	*
G14-80 A.G.A.	D	#19	*	*
G14-100 A.G.A.	#9	#21	*	*
G14-130 A.G.A.	#1	#15	*	*
G14-40 C.G.A.	#15	#30	#18	#31
G14-60 C.G.A.	#13	#24	#17	#27
G14-80 C.G.A.	H	#17	1/4"	#20
G14-100 C.G.A.	#9	#21	#9	#23
G14-130 C.G.A.	#1	#15	#1	#17

*High altitude derating is not necessary due to the self compensating effect of the pulse combustion process. If special conditions exist where derating is required, contact the Lennox Division Headquarters in your area.

H - Air Intake Chamber & Purge Blower (Figure 10)

1 - The air intake chamber houses the purge blower and air intake flapper valve assemblies. Air enters through the top inlet, passes through the purge blower and through the flapper valve to the combustion chamber. The entire air intake chamber is mounted on rubber isolators to eliminate vibration.

2 - The purge blower has a 120 volt motor and is permanently lubricated. It is powered only during pre and post purge. During combustion the blower is not powered, but the air is drawn through the blower by negative pressure.

FIGURE 10

I - Air Intake Flapper Valve (Figure 11)

The air intake flapper valve is similar to the gas flapper valve in operation. A flapper floats freely over a spacer between two plates. In actual operation, initially, the flapper is forced against the clearance plate by the purge blower allowing air to enter the com-

bustion chamber. Next, back pressure from combustion forces the flapper against the cover plate closing off the air supply. Finally, as a negative is created in the combustion chamber, the flapper is drawn to the clearance plate and air enters. The back pressure and negative pressures control the flapper valve with each combustion pulse once ignition has occured.

Refer to the troubleshooting section for specific information about flapper valve inspection and conditions requiring replacement.

J - Combustion Chamber & Heat Exchange Assembly (Figure 12)

1 - The combustion chamber has gas and air intake 'manifolds', the gas intake is on the right and the air intake is front-center. The exhaust gas leaves through the tailpipe at the top of the chamber.

2 - The tailpipe connects the combustion chamber to the exhaust gas decoupler. The tailpipe and decoupler create the proper amount of back pressure for combustion to continue and are major heat exchange components. The resonator provides attenuation for acoustic frequencies.

3 - The exhaust decoupler is manifolded into the condenser coil. The condenser coil is where the latent heat of combustion is extracted from the exhaust gas producing condensate. The circuiting of the coil allows for proper drainage of condensate to the exhaust outlet line. The exhaust line is located in lower left hand corner of the vest panel on '40', '60', and '80' units and in the lower right corner of the '100' and '130' units.

4 - The entire heat exchange assembly is mounted on rubber isolation mounts to eliminate vibration.

5 - Each unit input size uses a specific heat exchange assembly. Externally they are the same physical size and shape but cannot be interchanged between unit input sizes. Internal characteristics related to unit input properly match each assembly for the unit input rating.

K - Spark Plug & Sensor

1 - The spark plug and sensor are located on the lower left side of the combustion chamber, see Figure 13. The sensor is the top plug and is longer than the spark plug. The spark plug is

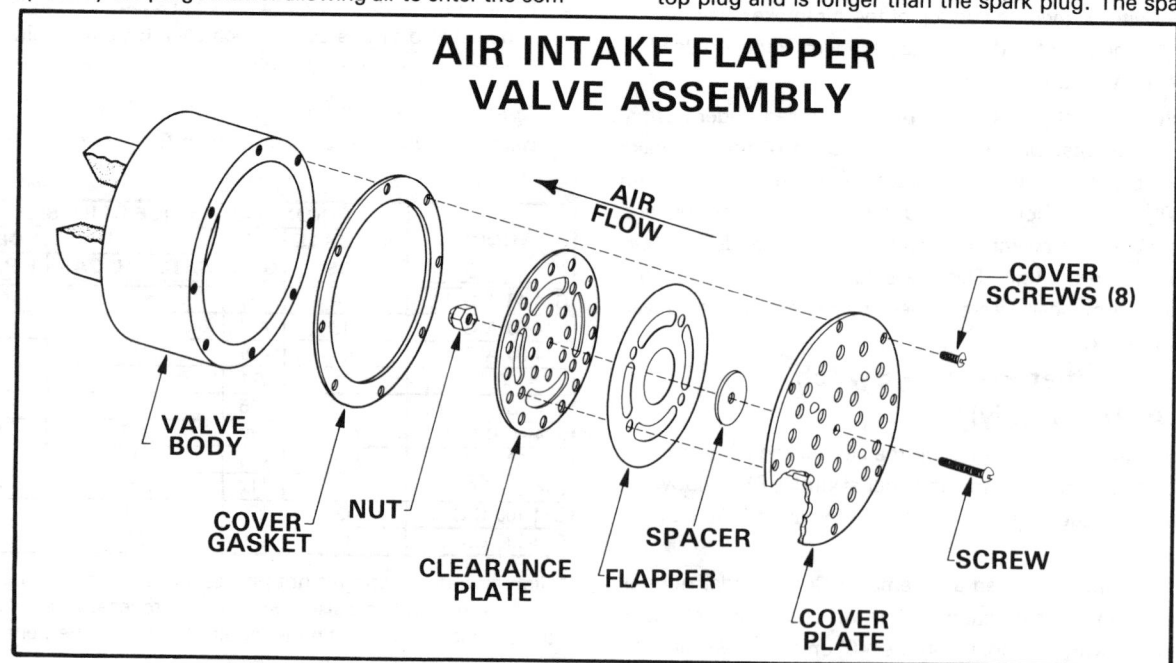

FIGURE 11

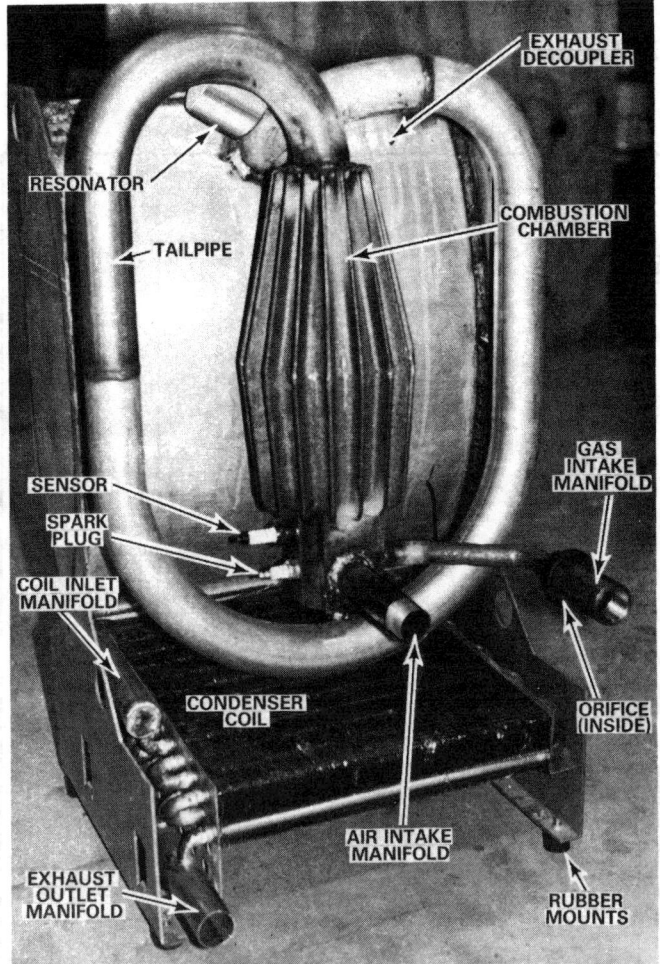

FIGURE 12

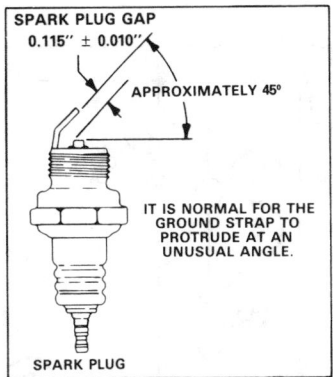

FIGURE 14

5 - The sensor is a spark plug type with a single center electrode (no ground strap) and compression rings to form a seal to the chamber. It, also, should not need regular maintenance.

L - Gas & Air Components Applied to Heat Exchange Assembly (Figure 15)

This photo identifies all the components that make up the basic heating assembly.

1 - Gas flows through the valve, expansion tank, flapper valve and orifice into the combustion chamber.
2 - Air flows through the air flapper valve and directly into the combustion chamber.
3 - Combustion takes place and exhaust gas flows through the tailpipe, exhaust decoupler and condenser coil to the exhaust outlet.

in the lower position. The plugs cannot be interchanged due to different thread diameters.

2 - Spark plug socket wrench size = 3/4 inch. Sensor plug socket wrench size = 11/16 inch.

3 - The spark plug is used in conjunction with the primary control for igniting the initial gas and air mixture. The temperatures in the combustion chamber keep the plug free from oxides and it should not need regular maintenance. Compression rings are used to form the seal to the chamber.

FIGURE 13

4 - Figure 14 gives the proper spark gap setting. Note that proper gapping for use in the G14 produces a ground strap angle unusual in comparison to other spark plug applications. A feeler gauge can be used to check the gap.

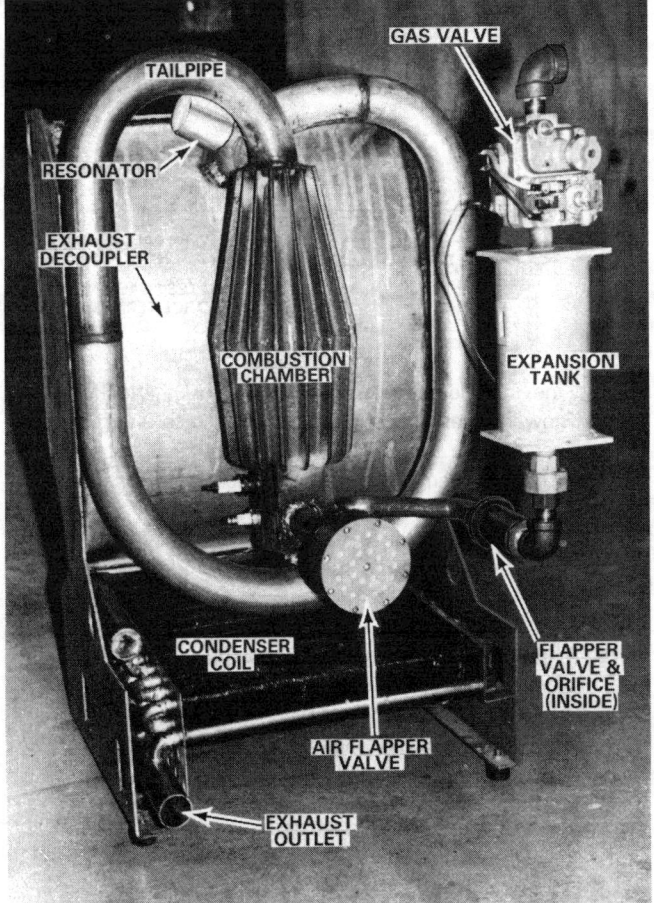

FIGURE 15

V - SEQUENCE OF OPERATION

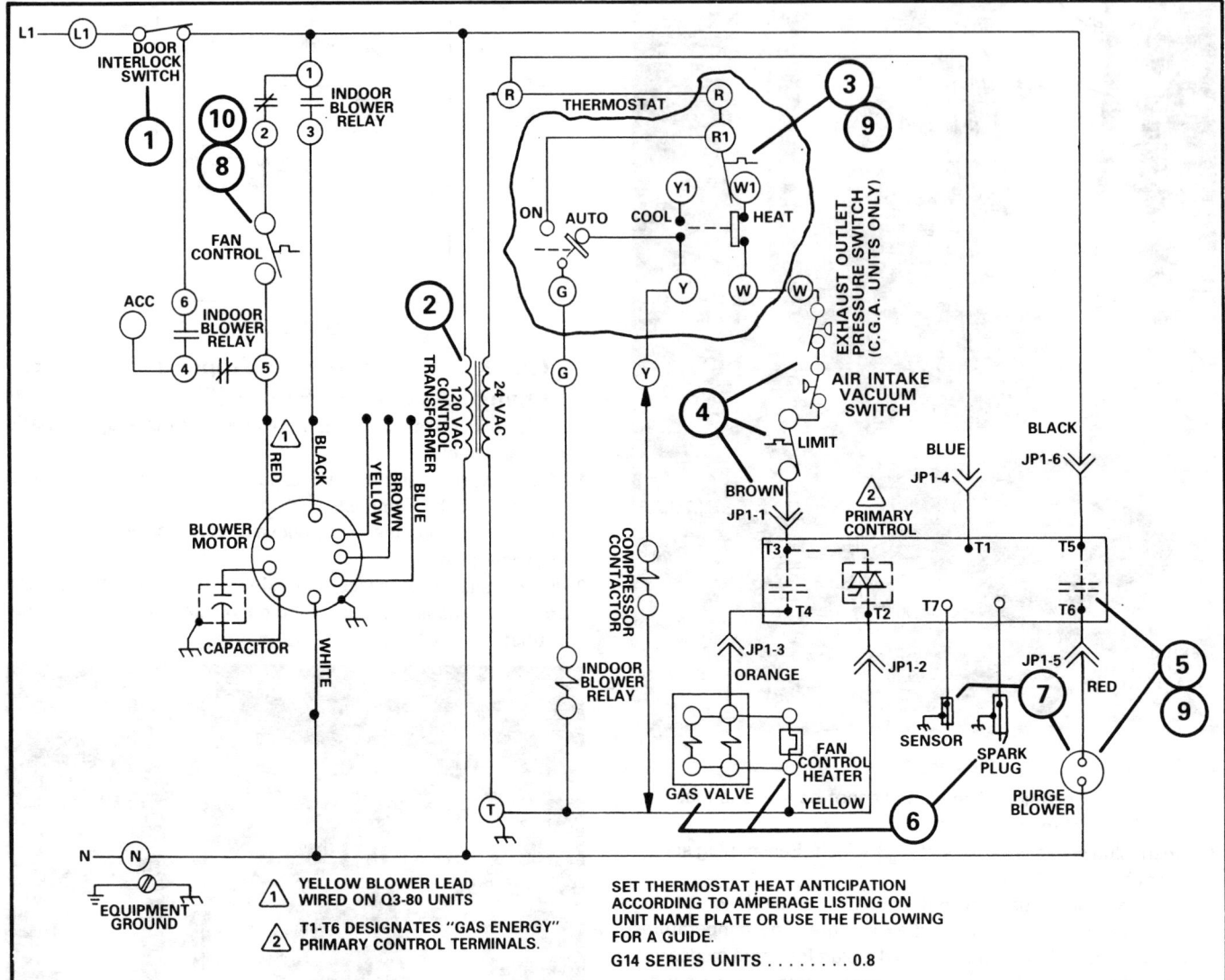

YELLOW BLOWER LEAD WIRED ON Q3-80 UNITS

T1-T6 DESIGNATES "GAS ENERGY" PRIMARY CONTROL TERMINALS.

SET THERMOSTAT HEAT ANTICIPATION ACCORDING TO AMPERAGE LISTING ON UNIT NAME PLATE OR USE THE FOLLOWING FOR A GUIDE.

G14 SERIES UNITS 0.8

1 - Line voltage feeds through the door interlock switch. Blower access panel must be in place to energize unit.

2 - Transformer provides 24 volt control circuit power.

3 - A heating demand closes the thermostat heating bulb contacts.

4 - The control circuit feeds from "W" leg through the exhaust outlet pressure switch (C.G.A. units only), the air intake vacuum switch (A.G.A. & C.G.A. units) and the limit control to energize the primary control.

5 - Through the primary control the purge blower is energized for approx. 30 sec. prepurge.

6 - At the end of prepurge the purge blower continues to run and the gas valve, fan control heater & spark plug are energized for approx. 8 seconds.

7 - The sensor determines ignition by flame rectification and de-energizes the spark plug and purge blower. Combustion continues.

8 - After approximately 30 to 45 seconds the fan control contacts close & energize the indoor blower motor on low speed.

9 - When heating demand is satisfied the thermostat heating bulb contacts open. The primary control is de-energized removing power from the gas valve & fan control heater. At this time the purge blower is energized for a 30 second post purge. The indoor blower motor remains on.

10 - When the air temperature reaches 90°F the fan control contacts open — shutting off the indoor blower.

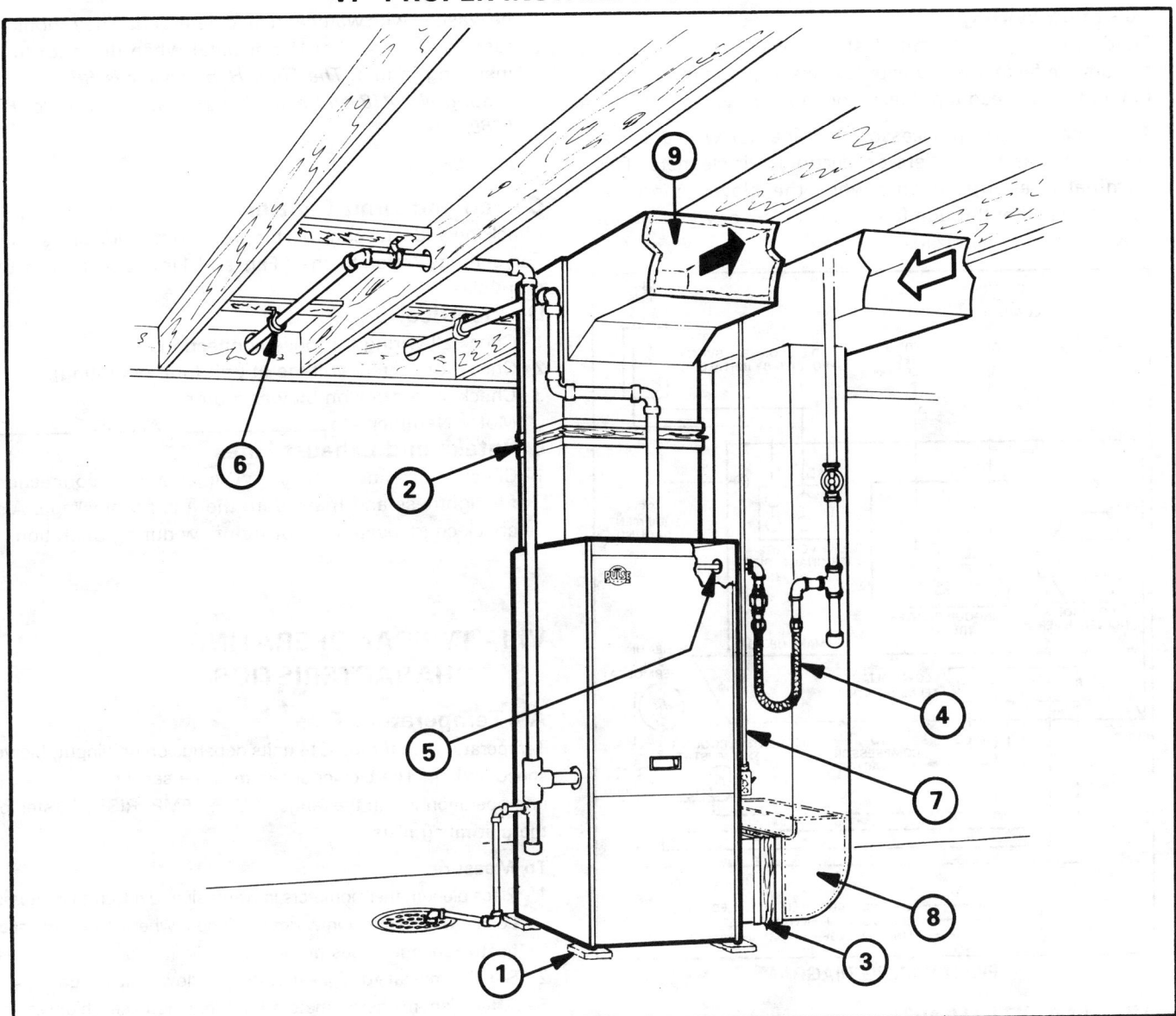

1 - Isolation Mounting Pads
If leveling bolts are not used, vibration isolating pads should be used especially when the unit is installed on wood flooring. "Isomode" pads or equivalent should be used.

2 - Flexible Boot Supply Air Plenum
A flexible canvas boot or equivalent should be used in the supply air plenum, above cooling coil or future coil location.

3 - Flexible Boot Return Air Plenum
A flexible canvas boot or equivalent should be used and located as close to the furnace as possible. Preferably between the furnace and external electronic air cleaner, if used.

4 - Gas Connector (A.G.A. Units Only)

5 - Gas Supply Piping Centered In Inlet Hole
The gas supply pipe should not rest on the unit cabinet.

6 - Isolation Hangers
PVC piping for intake and exhaust lines should be suspended from hangers. A suitable hanger can be fabricated from a 1 inch wide strip of 26 ga. covered with "Armaflex" or equivalent.

7 - Electrical Conduit Isolated From Ductwork and Joists
The electrical conduit can transmit vibration from the unit cabinet to ductwork or joists if clamped to either. It may be clamped tightly to unit cabinet but should not touch ductwork or joists.

8 - Return Air Plenum Insulated Past First Elbow
A 1.5 to 3 lb. density, matt face, 1 inch thick insulation should be used and all exposed edges should be protected from air flow.

9 - Supply Air Plenum Insulated Past First Elbow
Same requirements as return air plenum.

10 - Field Wiring

Field wiring is to terminal strips. Multi-speed blower motors are factory wired with low speed (red) tap for heating and high speed tap (black) for cooling.

The units include an accessory terminal for wiring accessories such as humidifiers or electronic air cleaners. This terminal is energized only when the blower motor is operating (either through fan control circuit or when indoor blower relay is energized.

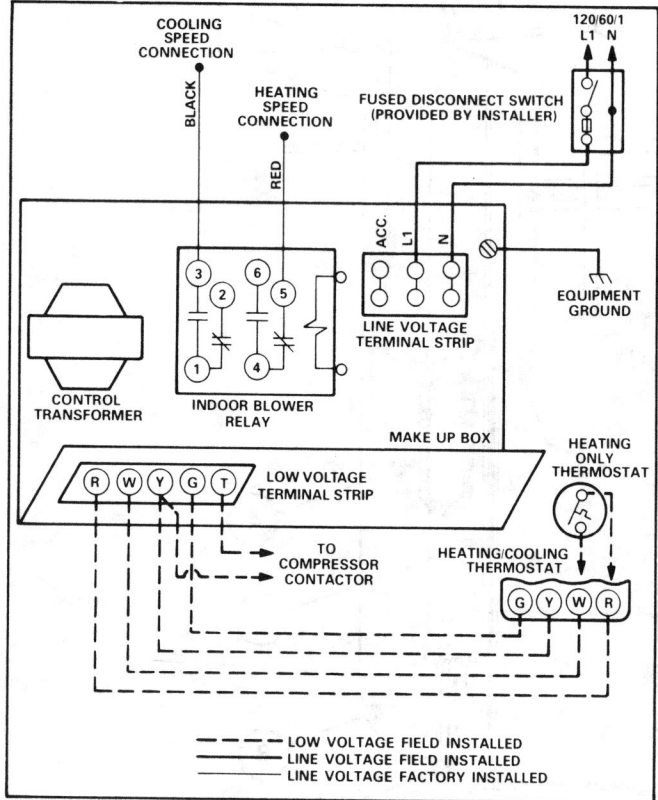

FIELD WIRING DIAGRAM

VII - MAINTENANCE

At the beginning of each heating season, the system should be checked as follows:

A - Blower

1 - Check and clean blower wheel.
2 - Motor Lubrication
Always relubricate motor according to the manufacturer's lubrication instructions on each motor. If no instructions are provided, use the following as a guide.

a - *Motors Without Oiling Ports* - Prelubricated and sealed. No further lubrication required.

b - *Direct Drive Motors With Oiling Ports* - Prelubricated for an extended period of operation. For extended bearing life, relubricate with a few drops of SAE No. 10 non-detergent oil once every two years. It may be necessary to remove blower assembly for access to oiling ports.

B - Filters

1 - Filters must be cleaned or replaced when dirty to assure proper furnace operation.
2 - The washable Scottfoam filters supplied with the G14 can be washed with water and mild detergent. They should be resprayed with Filter Handicoater when dry prior to re-installing in unit. *The Filter Handicoater is RP products coating No. 418 and available as Lennox Part No. P-8-5069.*

C - Fan and Limit Control

Check fan and limit controls for proper operation and setting. For settings, refer to the "Fan and Limit" section in this manual.

D - Electrical

1 - Check all wiring for loose connections.
2 - Check for correct voltage at unit (unit operating).
3 - Check amp-draw on blower motor.
   Motor Nampleplate _____ Actual _____

E - Intake and Exhaust Lines

1 - Check intake and exhaust PVC lines and all connections for tightness and make sure there is no blockage. Also check condensate line for free flow during operation.

VIII - TYPICAL OPERATING CHARACTERISTICS

A - Temperature Rise

Temperature rise for the G14 units depends on unit input, blower speed and hp. The blower speed must be set for unit operation within the range of 'AIR TEMP. RISE °F' listed on the unit rating plate.

To Measure:

1 - Place plenum thermometers in warm air and return air plenums. Locate thermometer in warm air plenum where it will not "see" heat exchanger, thus picking up radiant heat.
2 - Set thermostat to highest setting. Allow unit to run.
3 - After plenum thermometers have reached their highest and steadiest readings, subtract the two readings. The difference should be in the range listed on unit rating plate. If this temperature is low, decrease blower speed; if high, increase blower speed. To change blower motor speed taps see table 2.

TABLE 2

BLOWER SPEED SELECTION

IMPORTANT – TO PREVENT MOTOR BURNOUT, NEVER CONNECT MORE THAN ONE MOTOR LEAD TO ANY ONE CONNECTION. TAPE UNUSED MOTOR LEADS SEPARATELY.

SPEED	BLOWER MOTOR LEAD		
	Q3	Q4	Q5
LOW	RED	RED	RED
MEDIUM LOW	----	----	YELLOW
MEDIUM	YELLOW	YELLOW	BLUE
MEDIUM HI	----	----	BROWN
HIGH	BLACK	BLACK	BLACK

B - Manifold Pressure

Manifold pressure for the G14 units falls into two categories; running pressure and no fire pressure.

Running pressure occurs with the unit operating normally and is 3.8 in. W.C. for natural gas and 9.0 in W.C. for L.P. gases. Checks of running pressure are made as verification of proper regulator adjustment.

No fire pressure is normally higher than the running pressure and occurs when the gas valve is open without the unit ignited. The difference between the running pressure and no fire pressure is unique to the pulse combustion process. For natural gas the no fire pressure is 0.4 in W.C. higher than the running pressure. For L.P. gases the no fire pressure varies less and may be up to 0.2 in. W.C. higher than the running pressure. Checks of no fire pressure are needed only during troubleshooting to verify regulator adjustment.

To Measure:

1 - Remove the 1/8 inch pipe plug from the pressure tap on elbow below expansion tank, see Figure 8 for location of tap. Insert hose barb in tap and connect gauge.

2 - Set thermostat for heating demand (the demand can be started at the unit by jumpering 'R' to 'W' on the low voltage terminal strip).

3 - Check the running pressure after the unit has ignited and is operating normally.

4 - The no fire pressure check during troubleshooting can be made only within the approximate 7 second ignition trials, prior to primary control lockout of the gas valve. The action of the slow opening gas valve regulates the time from zero to maximum manifold pressure; 7 seconds on the '40' units and 4 seconds on the '60' and '80' units. Therefore, maximum manifold no fire pressure can be measured for only a few seconds of each ignition trial following each purge cycle.

C - Flame Signal

A 50 microamp DC meter range is needed to check the flame signal on the ignition primary controls. ('Simpson' models 250, 255 & 260 have a DC microamp range of 0-50 and are suggested for use.)

To Measure:

1 - Place the meter in series between ignition control and sensor wire; positive (+) lead of meter to ignition control (sensor terminal) and negative (−) lead of meter to sensor wire.

2 - Set the thermostat for a heating demand and check the flame signal with the unit operating.

3 - The flame signals are as follows:

   Prestolite Primary Control - 3 to 5 Microamps DC
   Gas Energy Primary Control - 25 to 35 Microamps DC

   The flame signal may rise above these values for the first few seconds after ignition and then levels off within the ranges given.

D - Exhaust Temperature Range

The exhaust temperature should not exceed 130°F for any of the G14's. If it does the unit will trip out on limit indicating a problem.

Most units run with maximum exhaust temperatures of 110°F to 125°F from lower to higher unit inputs. For example, the '40' units maximum of 110°F, '60' units - 120°F and '80' units - 125°F. Exhaust temperatures lower than these values are possible and normal.

E - Exhaust CO_2 Content

When the unit is properly installed and operating normally the CO_2 content of the exhaust gas is 8.5 to 10% for natural gas and 11 to 12% for L.P. gases.

The CO_2 content can be measured using the Bacharach CO_2 test with a Fyrite CO_2 indicator. Other testers are available and instructions packaged with tester should be used.

To Measure:

1 - Drill size "R" or 11/32 in. hole on top of the exhaust outlet PVC elbow (inside unit cabinet) and tap 1/8 - 27 NPT as shown in Figure 16. This hole is used as the CO_2 test port.

2 - Install a hose barb connector into the CO_2 test port. Refer to Figure 16.

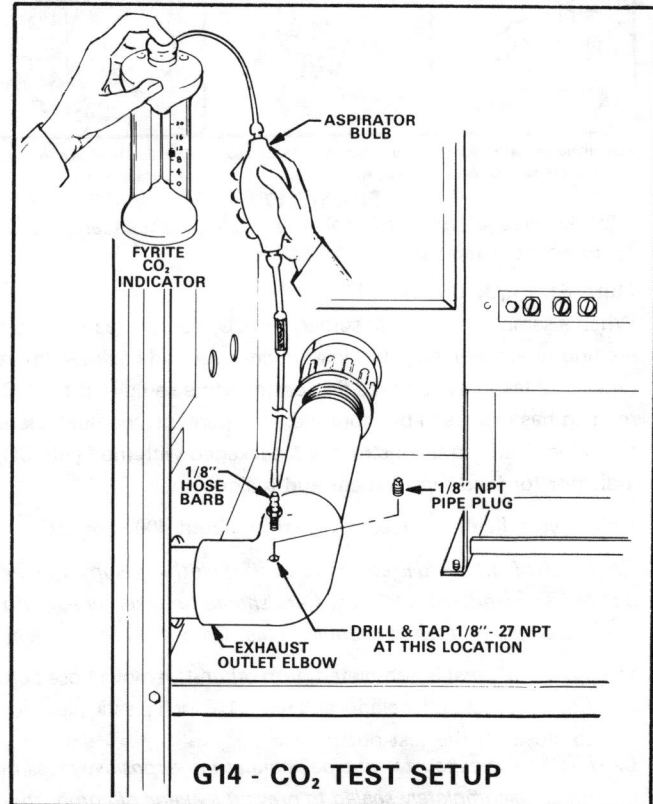

G14 - CO_2 TEST SETUP

FIGURE 16

3 - Attach end of Fyrite sampling tube to hose barb on exhaust outlet elbow.

4 - Set thermostat to highest setting and allow unit to run for 15 minutes.

5 - Adjust the Fyrite indicator scale to zero:
   a. Invert Fyrite until all fluid has run into top reservoir.
   b. Turn Fyrite upright until all fluid has run into bottom reservoir.
   c. Hold Fyrite upright at a 45° angle for five seconds to drain excess fluid droplets from inside surfaces.
   d. Hold Fyrite upright and depress plunger valve and release.

e. Loosen lock nut in rear of scale. Slide scale until zero percent CO_2 scale division lines up with top of fluid column. Tighten scale lock nut.

6 - Holding instrument upright, lay rubber connector to tip on plunger valve at top of instrument.

7 - Depress plunger valve with connector tip and hold down. See Figure 17.

8 - Squeeze and release aspirator bulb 18 times. On the 18th squeeze, the depressed plunger valve is released before releasing aspirator bulb. See Figure 17.

9 - Turn over the CO_2 indicator twice, permitting the test fluid to run back and forth, forcing it to absorb the flue gas sample. See Figure 17.

10 - Hold instrument at a 45° angle for five seconds. Hold upright for several seconds and immediately read percent CO_2 on scale in line with top of fluid column. See Figure 17.

BACHARACH CO_2 TEST

  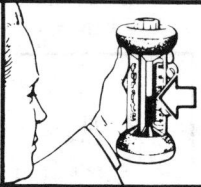

Push rubber connector down - then squeeze bulb 18 times.

Turn bottom side up and back again.

Read fluid level on scale for percentage of gas.

FIGURE 17

If test fluid is cold, run 2 to 3 trial tests to bring the fluid temperature up to ambient temperature.

Fluid Strength Check

When a sample has been absorbed and CO_2 percent read, without venting to atmosphere, turn instrument over and back two more times and take a second reading on the same sample. If the CO_2 reading has increased by more than 1/2 percent, the fluid needs replacing. Refer to the instructions packaged with the Fyrite CO_2 indicator for fluid replacement and maintenance.

Fresh Fyrite fluid will absorb CO_2 from about 400 samples.

CAUTION: Fyrite fluid is corrosive to skin, clothing, some metals, painted and laquered surfaces. Care should be taken when turning instrument and when changing fluid. Do not get fluid in eyes.

11 - When CO_2 test is completed, turn off unit, remove hose barb from exhaust outlet elbow and use a 1/8 inch plastic pipe plug to close off the test port.

CAUTION: The exhaust vent pipe operates under positive pressure and must be completely sealed to prevent leakage of combustion products into the living space.

F - Safety Shutdown

Safety shutdown occurs when any of the following probems are encountered.

1 - Loss of combustion during heating cycle caused by:
   a. Obstruction to air intake piping
   b. Obstruction to exhaust outlet piping
   c. Low gas pressure
   d. Failure of gas flapper valve
   e. Failure of air flapper valve
   f. Failure of main gas valve
   g. Loose spark plug or sensor creating pressure loss.
   h. Loose sensor wire

2 - High limit cutout
   a. Blower failure
   b. Temperature rise too high
   c. Restricted filter or return air
   d. Restricted supply air plenum.

G - Internal Component Temperatures

During operation, temperature at the top of the combustion chamber and tailpipe is 1000-1200°F. At the tailpipe entrance to the exhaust decoupler, temperature has dropped to approximately 600°F. From the exhaust decoupler outlet to coil intake manifold - 350°F. At the coil exhaust outlet manifold, temperatures range approximately 100 - 110°F. These are average temperatures and vary with blower speed and input.

H - Condensate pH Range

The condensate is mildly acidic and can be measured with pH indicators. The pH scale is a measurement of acidity or alkalinity.

	pH Scale		
Acid Range	0		Increasing in acidity
	1		
	2		
	3		
	4	G14	
	5	Condensate	
	6	pH Range	
Neutral	7		
Alkaline Range	8		Increasing in alkalinity
	9		
	10		
	11		
	12		
	13		
	14		

The condensate pH range is 4 to 6. For comparison, pH levels of common liquids are: vinegar 2.4 - 3.4; wine 2.8 - 3.8; orange juice 3 - 4; soft drinks 2 - 4; beer 4 - 5; and tomato juice 4.0 - 4.4. The concentration of the acidity of all these fluids including the condensate is very low and harmless.

I - Acceptable Operating Input

Field adjustments to the unit input are not normally needed due to the specifically sized components for each unit input rating.

The unit may run up to ± 3 to 4 percent of rated input (listed on unit nameplate) due to installation variables such as temperature rise, external static pressure and return air temperature combined with allowable tolerances of components within the unit. This is an acceptable operating range.

Operation of the G14 above or below this acceptable operating range may cause continuity, startup and lockout problems; in general erratic operation.

To achieve accurate input measurements requires time for the unit to 'run in'. The run in time allows the flapper valves to 'seat' and combustion to clean the protective layer of oil residue that may be present from inside surfaces of the heat exchange assembly. This process stabilizes combustion rate and may take one to two hours of continuous operation. Since it is impratical to operate an installed unit for one to two hours continuously, the unit should be allowed to operate normally (cycle on demand) for a period to accumulate several total hours of run time. Overnight operation should provide enough total run time to obtain accurate measurement of input. Just prior to any input check the unit should be run continuously for 15 minutes.

Checking Gas Input - Determine Gas Flow At Meter:

1 - Turn off all other gas appliances, including pilot lights on appliances, if used.

2 - Set the thermostat to highest setting and allow unit to run continuous for 15 minutes. The 15 minute run time is essential to allow unit to stabilize operating rate.

3 - At the gas supply meter using either the one, two or five-foot dial on the meter, time one full revolution (in seconds) with a watch. See Figure 18.

4 - Find the number of seconds for one revolution on the Gas Rate chart, see Table 3. Read the cubic feet for the matching one, two or five-foot dial size from Table 3, multiply this times the Btu per cubic foot content of the gas. The result is the total gas Btuh input.

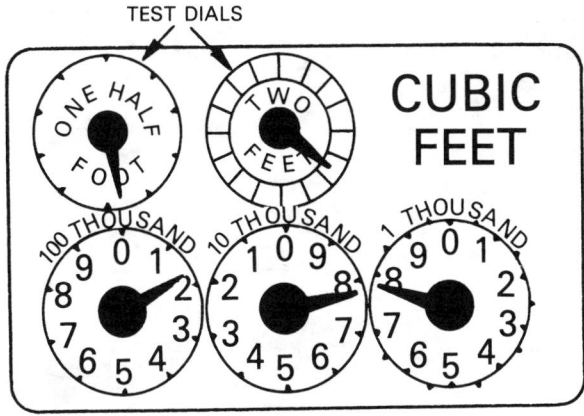

FIGURE 18

EXAMPLE:

a. One revolution on the two-foot dial = 90 seconds.

b. Using the gas rate chart, Table 3, note that 90 seconds = 80 cubic feet of gas per hour.

c. Normally there are 1,000 Btu in each cubic foot of gas. Make adjustment to this figure where gas heating value is other than 1,000 Btu/cu. ft.; contact gas supplier for local Btu/cu. ft. of gas.

d. 80 cu. ft./hr. x 1,000 Btu/cu. ft. = 80,000 Btuh input.

5 - Check the Btuh input figure against the Btuh listed on the unit nameplate.

TABLE 3

METER FLOW RATE															
GAS RATE — CUBIC FEET PER HOUR															
Secs. for One Rev.	Size of Test Dial			Secs. for One Rev.	Size of Test Dial			Secs. for One Rev.	Size of Test Dial		Secs. for One Rev.	Size of Test Dial	
	1 cu. ft.	2 cu. ft.	3 cu. ft.		1 cu. ft.	2 cu. ft.	3 cu. ft.		2 cu. ft.	5 cu. ft.		2 cu. ft.	5 cu. ft.
10	360	720	1800	34	106	212	529	62	116	290	110	----	164
11	327	655	1636	35	103	206	514	64	112	281	112	64	161
12	300	600	1500	36	100	200	500	66	109	273	116	62	155
13	277	555	1385	37	97	195	486	68	106	265	120	60	150
14	257	514	1286	38	95	189	479	70	103	257	125	----	144
15	240	480	1200	39	92	185	462	72	100	250	130	----	138
16	225	450	1125	40	90	180	450	74	97	243	135	----	132
17	212	424	1059	41	----	176	439	76	95	237	140	----	129
18	200	400	1000	42	86	172	429	78	92	231	145	----	124
19	189	379	947	43	----	167	419	80	90	225	150	----	120
20	180	360	900	44	82	164	409	82	88	220	155	----	116
21	171	343	857	45	80	160	400	84	86	214	160	----	113
22	164	327	818	46	78	157	391	86	84	209	165	----	109
23	157	313	783	47	----	153	383	88	82	205	170	----	106
24	150	300	750	48	75	150	375	90	80	200	175	----	103
25	144	288	720	49	----	147	367	92	78	196	180	----	100
26	138	277	692	50	72	144	360	94	----	192	----	----	----
27	133	267	667	----	----	----	----	96	75	188	51	141	353
28	129	257	643	52	69	138	346	98	----	184	53	136	340
29	124	248	621	54	67	133	333	100	72	180	55	131	327
30	120	240	600	56	64	129	321	102	----	176	57	126	316
31	116	232	581	58	62	124	310	104	69	173	59	122	305
32	113	225	563	60	60	120	300	106	----	170	----	----	----
33	109	218	545	----	----	----	----	108	67	167	----	----	----

IX - NEW UNIT STARTUP

Normal setup conditions of a new unit installation require running the unit through several trys for ignition before the unit will run continuously. Initially the unit may start and die several times until air bleeds from the gas piping and residues of oil and water are purged from the heat exchange assembly. Break and remake the thermostat demand to restart ignition sequence at 2 to 3 minute intervals until continuous operation is obtained.

X - TROUBLESHOOTING

Effective troubleshooting of the G14 depends on a thorough understanding of all unit components and their functions as described in this manual. Symptoms of unit operation break down problems into four main categories:

1 - Unit will not run.
2 - Unit starts clean but runs less than 10 seconds.
3 - Unit runs but shuts off before thermostat is satisfied - insufficient heat.
4 - Unit sputter starts and dies.

Each of the four problem categories above are broken down into troubleshooting sequence flow charts on the following pages with additional information provided to explain certain checks. Methods refered to in the flow charts for measurement of manifold running and no fire pressure, flame signal, exhaust CO_2 content and operating input are explained in the previous section on Typical Operating Characteristics.

When troubleshooting a unit be sure that all of the basic checks are covered carefully and double check indications before replacing components. DO AS LITTLE DISASSEMBLY AS POSSIBLE during troubleshooting to prevent introducing additional problems, such as gas or air leaks or damage to components.

Troubleshooting should proceed through the appropriate flow chart following the steps in order after definition of the symptoms. At any point a "NO" answer is encountered and a repair is made, reassemble unit and retest for operation. If the unit does not operate, recheck up to that point and then continue through the chart. Occasionally more than one specific problem may exist.

A - Additional Problems

1 - Blower Runs Continuously

Step 1 - Is thermostat fan switch set to 'on'? Switch to 'auto'.
Step 2 - Fan control 'off' setting below ambient air temperature. Readjust to 90°F.
Step 3 - Defective fan control contacts? Replace if necessary.
Step 4 - Defective blower relay contacts? Replace if necessary.

2 - Frequent Recycling of Blower after Heat Cycle

Fan control 'off' setting may be too high. Readjust to 90°F. Settings above 90°F off do not allow the heat exchange assembly to cool down enough before the blower is stopped. The additional heat retained in the unit when air flow is stopped builds up causing the recycle problem.

3 - Supply Air Blower Does Not Run

Step 1 - Check for loose wiring.
Step 2 - Defective fan control?
Step 3 - Defective blower relay?
Step 4 - Defective blower motor and/or capacitor?

4 - Unit Does Not Shut Off

Step 1 - Defective thermostat?
Step 2 - Shorted 24 VAC control circuit wiring — checkout and repair.
Step 3 - Gas valve stuck open?

CAUTION:

1 - Before servicing spark plug and sensor after unit has been operating, the unit should be allowed to cool down at least 15 minutes before placing hands into the heat section access opening.

2 - To cool completely to room temperature, the blower should be run continuous for approximately 40 minutes.

3 - When servicing the air intake flapper valve, keep in mind that it is only moderatley warm during unit operation. After the unit cycles off, the residual heat in the combustion chamber will transfer back to the valve causing it to become very hot. Allow it to cool for 10 to 15 minutes before handling or run blower as in 2 above.

B - Unit Will Not Run

This flow chart breaks down into four sections; Electrical checks, Gas Checks, Air Checks and Spark Checks. If this chart is com-pleted and the unit remains inoperative, the chart is continued on Page 20.

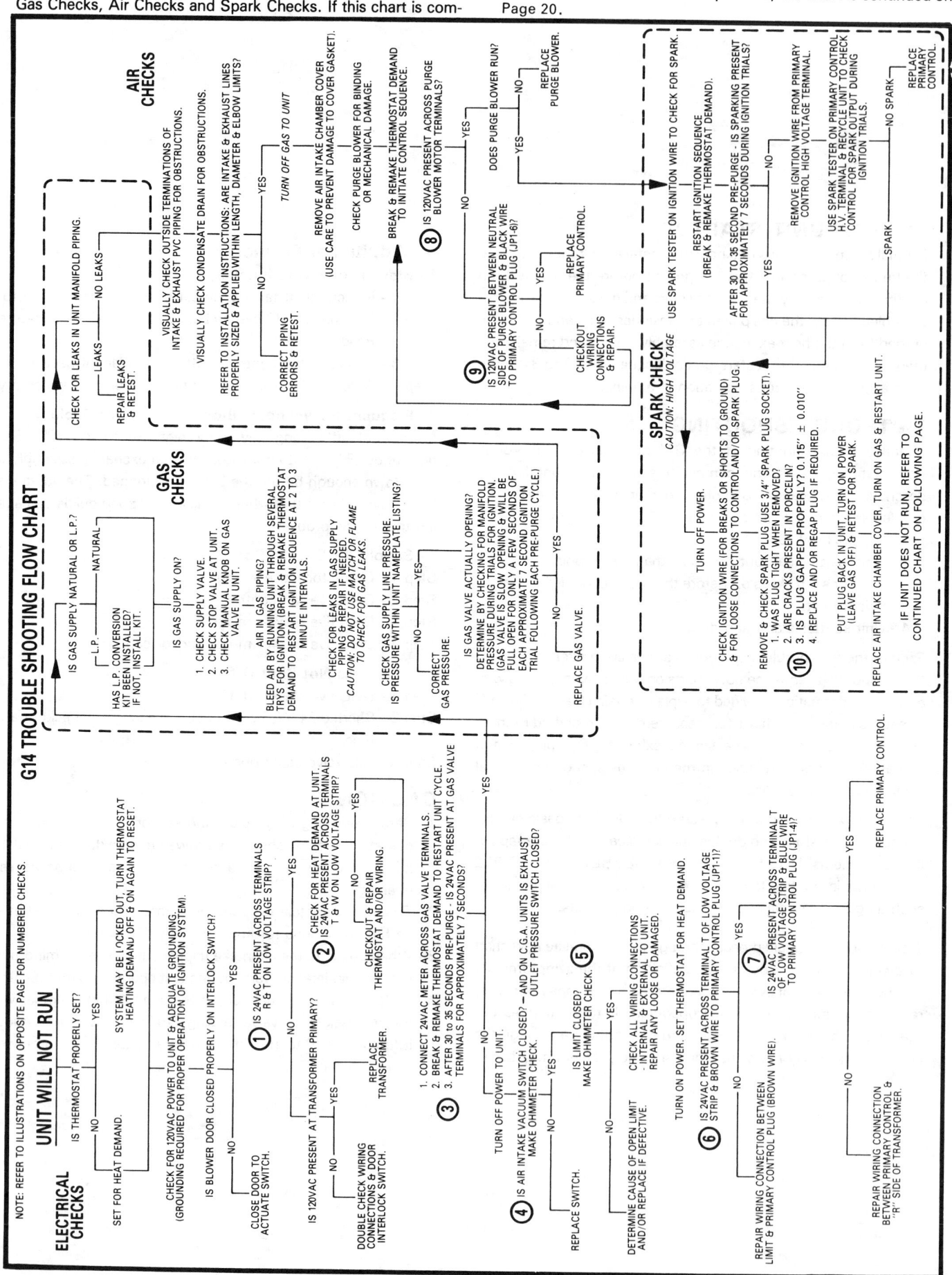

G14 TROUBLE SHOOTING FLOW CHART

NOTE: REFER TO ILLUSTRATIONS ON OPPOSITE PAGE FOR NUMBERED CHECKS.

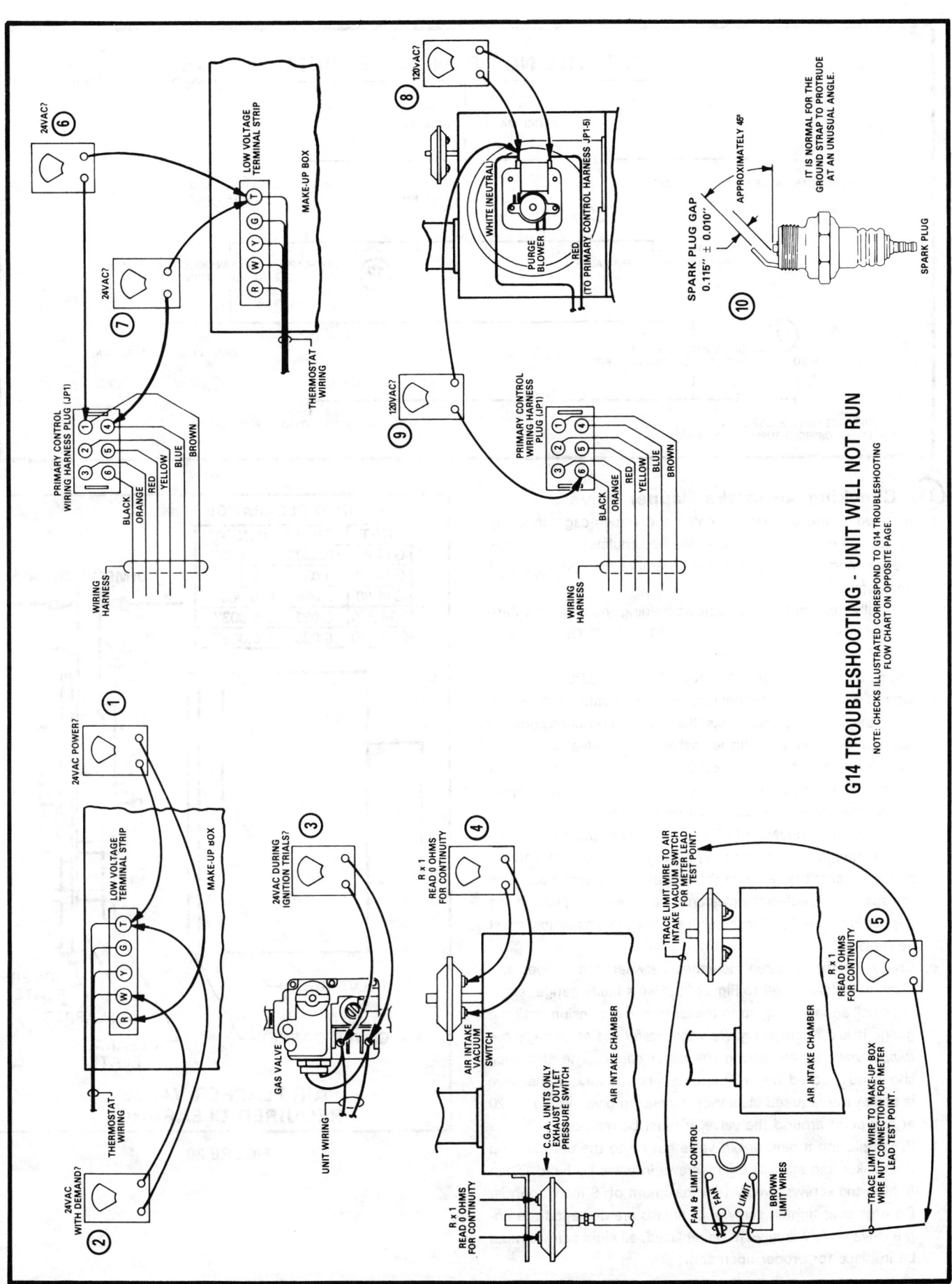

G14 TROUBLESHOOTING - UNIT WILL NOT RUN

NOTE: CHECKS ILLUSTRATED CORRESPOND TO G14 TROUBLESHOOTING FLOW CHART ON OPPOSITE PAGE.

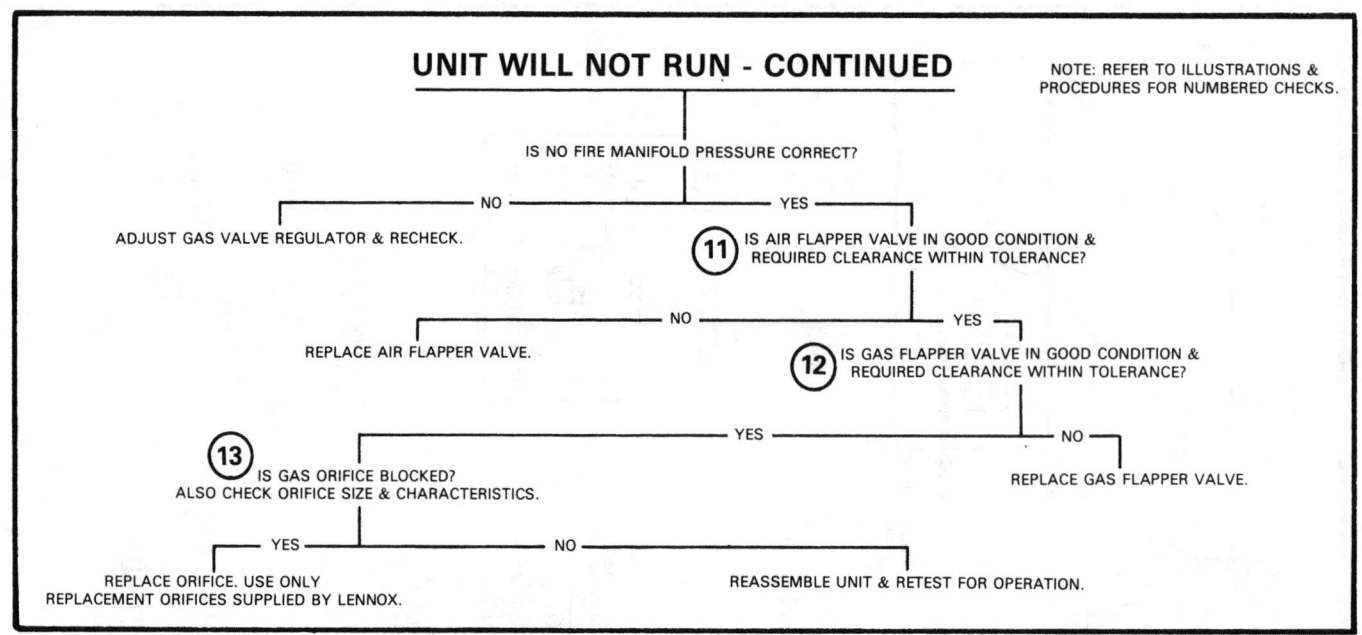

UNIT WILL NOT RUN - CONTINUED

NOTE: REFER TO ILLUSTRATIONS & PROCEDURES FOR NUMBERED CHECKS.

IS NO FIRE MANIFOLD PRESSURE CORRECT?

— NO — ADJUST GAS VALVE REGULATOR & RECHECK.

— YES — (11) IS AIR FLAPPER VALVE IN GOOD CONDITION & REQUIRED CLEARANCE WITHIN TOLERANCE?

— NO — REPLACE AIR FLAPPER VALVE.

— YES — (12) IS GAS FLAPPER VALVE IN GOOD CONDITION & REQUIRED CLEARANCE WITHIN TOLERANCE?

— YES — (13) IS GAS ORIFICE BLOCKED? ALSO CHECK ORIFICE SIZE & CHARACTERISTICS.

— NO — REPLACE GAS FLAPPER VALVE.

— YES — REPLACE ORIFICE. USE ONLY REPLACEMENT ORIFICES SUPPLIED BY LENNOX.

— NO — REASSEMBLE UNIT & RETEST FOR OPERATION.

(11)- Checking Air Intake Flapper Valve

a. Remove air intake chamber cover and screen cage checking for foreign materials that may have accumulated, clean purge blower and upper and lower chamber compartments if necessary.

b. Carefully remove the eight screws holding the air intake flapper valve to the valve body. DO NOT TURN OR REMOVE CENTER SCREW. Remove the valve from the unit being careful not to damage gasket. CAUTION: DO NOT DROP.

c. Do not disassemble internal components of valve. Visually inspect the flapper. On new units, the flapper may not be perfectly flat, it may be curved or dished between the plates - this is normal. On units that have had sufficient 'run in' time, the flapper will be flat. If the flapper is torn, creased or has uneven (frayed) edges, the valve assembly must be replaced.

d. Check for free movement of the flapper over the spacer. Use a feeler gauge blade to carefully move the flapper between the plates. Be sure the center of the flapper is not trapped between the spacer and either the clearance plate or cover plate. If the flapper is trapped under the spacer, the valve assembly must be replaced.

e. Check for the required clearance between the flapper and clearance plate. Refer to Figure 20. Use a feeler gauge, starting small and working up to the clearance dimension until the gauge is just about snug. *Be very careful not to damage the flapper material by forcing the feeler gauge.* The clearance should be checked in 6 or 8 places around the valve. If the valve is out of the required clearance dimension given in Figure 20 at any point around the valve, it must be replaced.

f. When placing a new or old valve back into the unit, line up the gasket and start all eight screws in place by hand. Then tighten the screws evenly to a maximum of 9 inch pounds. Do not over tighten screws, if threads are damaged the entire valve body will have to be replaced; all eight screws must be in place for proper operation.

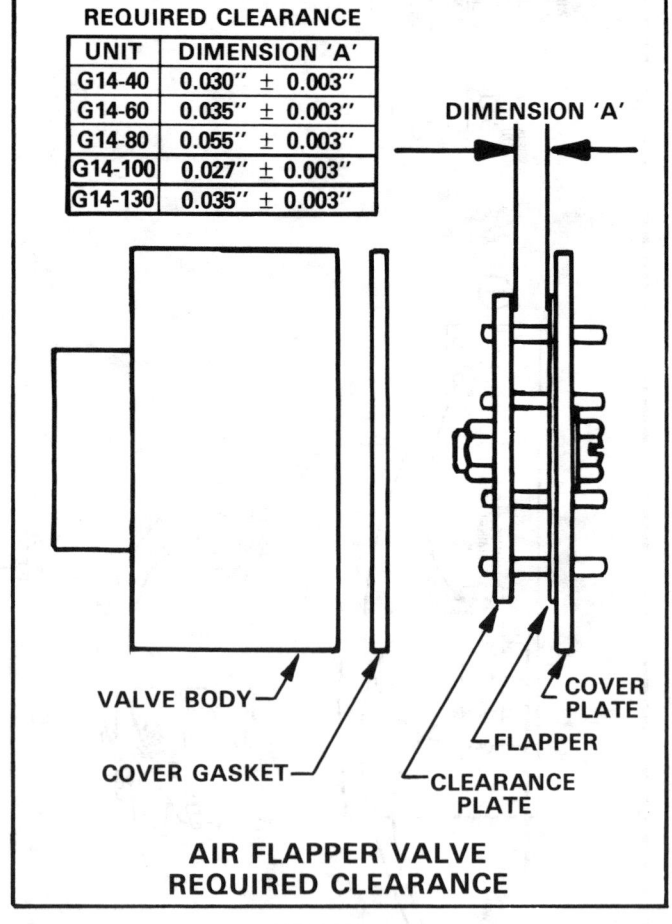

REQUIRED CLEARANCE

UNIT	DIMENSION 'A'
G14-40	0.030" ± 0.003"
G14-60	0.035" ± 0.003"
G14-80	0.055" ± 0.003"
G14-100	0.027" ± 0.003"
G14-130	0.035" ± 0.003"

DIMENSION 'A'

VALVE BODY

COVER GASKET

COVER PLATE

FLAPPER

CLEARANCE PLATE

AIR FLAPPER VALVE REQUIRED CLEARANCE

FIGURE 20

⑫ - Checking Gas Flapper Valve

a. Disconnect union at bottom of expansion tank and remove entire gas flapper valve, nipple and elbow assembly as one piece. It is not recommended to remove elbow and nipple from flapper valve unless the valve is being replaced. Use care not to damage O-Ring when handling valve out of unit. DO NOT DROP.

b. Do not turn or remove center screw of valve assembly. Visually inspect the flapper. The flapper may be dished or curved on new units, this is normal. Units that have 'run in', the flapper will be flat. If the flapper is torn, creased or has uneven (frayed) edges, the valve assembly must be replaced.

c. Check for free movement of the flapper over the spacer. Use a feeler gauge blade to carefully move the flapper between the plates. Be sure the flapper is not trapped between the spacer and either the valve body or clearance plate. If the flapper does not move freely or is trapped under the spacer, the valve assembly must be replaced.

d. Check for the required clearance between the flapper and valve body. Refer to Figure 21. Use a feeler gauge, starting small and working up to the clearance dimensions until the gauge is just about snug. *Be very careful not to damage the flapper material by forcing the gauge.* The clearance should be checked around the valve in several places. If the valve is out of the required clearance dimension given in Figure 21 at any point around the valve, it must be replaced.

e. When placing a new or old valve back into the unit, use care not to damage the O-Ring. DO NOT USE PIPE SEALERS ON THE FLAPPER VALVE THREADS.

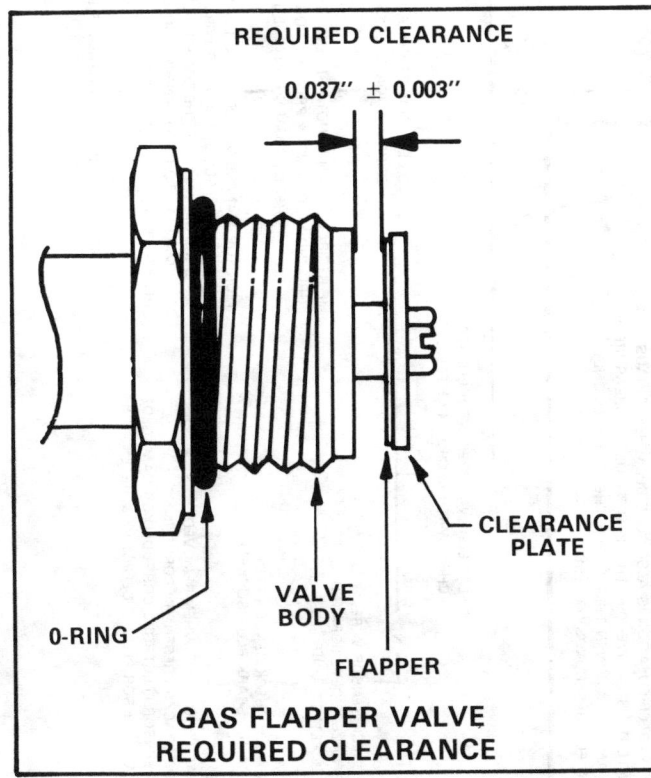

REQUIRED CLEARANCE

0.037" ± 0.003"

O-RING

VALVE BODY

FLAPPER

CLEARANCE PLATE

GAS FLAPPER VALVE REQUIRED CLEARANCE

FIGURE 21

⑬ - Checking Gas Orifice

a. With gas flapper valve assembly removed, use a flashlight to check for blockage of the orifice in the manifold. Use a 1/2 inch shallow socket on 40', '60' and '80' units or 9/16 inch shallow socket on '100' and '130' units with extension to remove the orifice.

b. Check the orifice drill size for the unit as given in Table 1 on page 7. If the orifice is incorrect it must be replaced.

c. Refer to Figures 22 and 23 and the physical characteristics of the orifice. The surface must be flat and the orifice opening must not be chamfered. The orifice taper must be centered and not recessed. If any defects are found the orifice must be replaced.

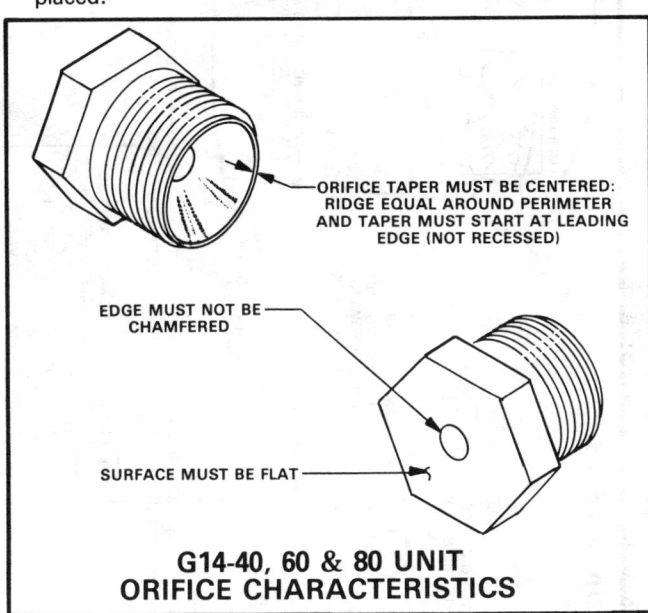

ORIFICE TAPER MUST BE CENTERED: RIDGE EQUAL AROUND PERIMETER AND TAPER MUST START AT LEADING EDGE (NOT RECESSED)

EDGE MUST NOT BE CHAMFERED

SURFACE MUST BE FLAT

G14-40, 60 & 80 UNIT ORIFICE CHARACTERISTICS

FIGURE 22

d. Standard atmospheric burner orifices or orifice blanks cannot be used as replacements for the G14. Only replacement orifices supplied through Lennox should be used.

e. When threading the orifice into the manifold use a rubber or foam insert in the socket to hold threads beyond end of socket. *Carefully align threads by hand turning the socket extension until orifice is in place. Then tighten with driver until snug.*

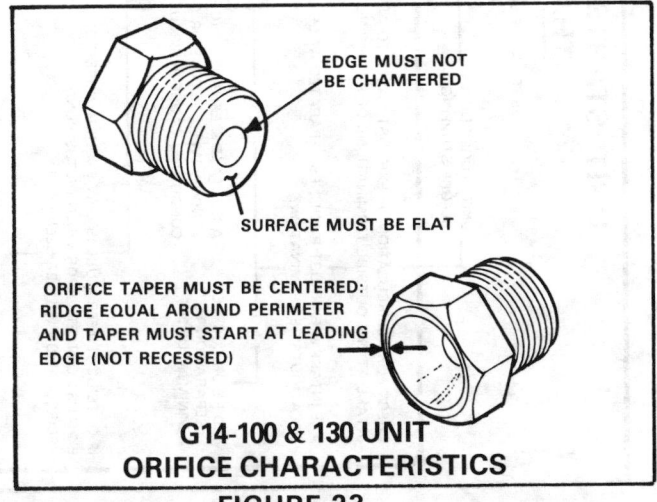

EDGE MUST NOT BE CHAMFERED

SURFACE MUST BE FLAT

ORIFICE TAPER MUST BE CENTERED: RIDGE EQUAL AROUND PERIMETER AND TAPER MUST START AT LEADING EDGE (NOT RECESSED)

G14-100 & 130 UNIT ORIFICE CHARACTERISTICS

FIGURE 23

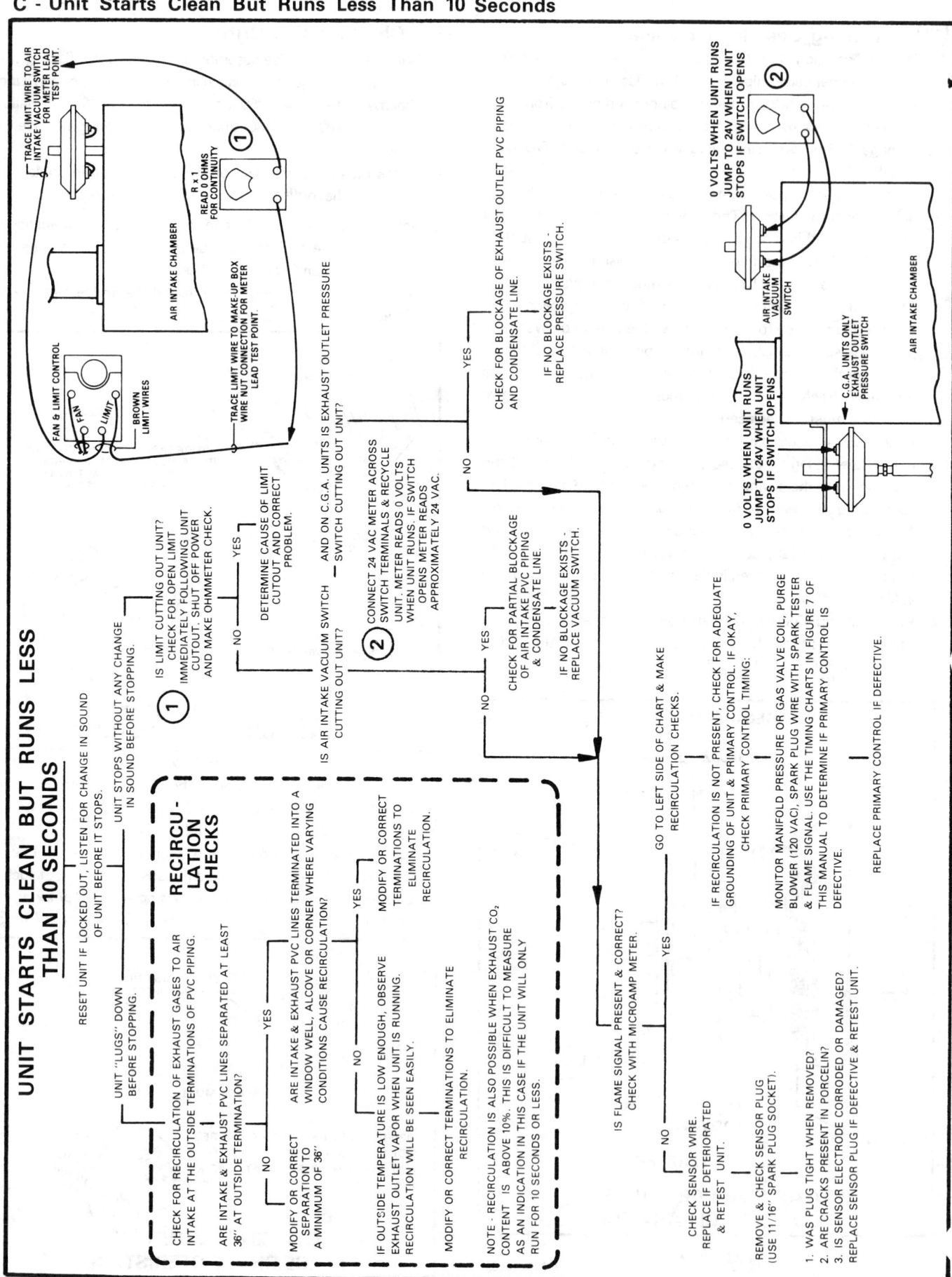

UNIT RUNS BUT SHUTS OFF BEFORE THERMOSTAT IS SATISFIED - INSUFFICIENT HEAT

(1) DOES LIMIT CUTOUT CAUSING UNIT TO SHUT OFF?
TURN OFF POWER & MAKE OHMMETER CHECK IMMEDIATELY FOLLOWING CUTOUT OF UNIT.

NO → IS GAS PRESSURE LOW?
CHECK FOR INTERMITTENT LOW GAS PRESSURE. (MONITOR MANIFOLD PRESSURE)

YES → CORRECT GAS PRESSURE.

NO → IS EXHAUST OUTLET PVC PIPING PARTIALLY BLOCKED OR RESTRICTED?

YES → ELIMINATE BLOCKAGE & RETEST.

NO → IS MANIFOLD RUNNING PRESSURE TOO HIGH?

YES → ADJUST GAS VALVE REGULATOR

NO → IS INPUT HIGH? OUT OF ACCEPTABLE RANGE?

YES → CHECK GAS BTUH INPUT.

NO → IS TEMPERATURE RISE WITHIN PROPER RANGE?

YES → REPLACE LIMIT.

NO → ADJUST BLOWER SPEED.

YES → IS FILTER CLEAN & PROPERLY INSTALLED? CORRECT IF NECESSARY & RETEST UNIT.

ARE SUPPLY & RETURN AIR DUCTS UNRESTRICTED? CORRECT IF NECESSARY & RETEST.

RECIRCULATION CHECKS

CHECK FOR RECIRCULATION OF EXHAUST GASES TO AIR INTAKE AT THE OUTSIDE TERMINATION OF PVC EXHAUST PIPING.

ARE INTAKE & EXHAUST PVC LINES SEPARATED AT LEAST 36'' AT OUTSIDE TERMINATION?

NO → MODIFY OR CORRECT SEPARATION TO A MINIMUM OF 36''.

YES → ARE INTAKE & EXHAUST PVC LINES TERMINATED INTO A WINDOW WELL, ALCOVE OR CORNER WHERE VARYING CONDITIONS CAUSE RECIRCULATION?

YES

NO → IF OUTSIDE TEMPERATURE IS LOW ENOUGH, OBSERVE EXHAUST OUTLET VAPOR WHEN UNIT IS RUNNING. RECIRCULATION WILL BE SEEN EASILY.

IF OUTSIDE TEMPERATURE IS TOO HIGH TO SEE EXHAUST VAPOR OR RECIRCULATION CANNOT BE DETERMINED, CHECK CO_2 CONTENT OF EXHAUST GAS. IF CO_2 CONTENT IS ABOVE 10% RECIRCULATION IS POSSIBLE.

TRACE LIMIT WIRE TO AIR INTAKE VACUUM SWITCH FOR METER LEAD TEST POINT.

(1) R x 1 READ 0 OHMS FOR CONTINUITY

AIR INTAKE CHAMBER

FAN & LIMIT CONTROL
FAN
LIMIT
BROWN LIMIT WIRES

TRACE LIMIT WIRE TO MAKE-UP BOX WIRE NUT CONNECTION FOR METER LEAD TEST POINT.

UNIT SPUTTER STARTS & DIES

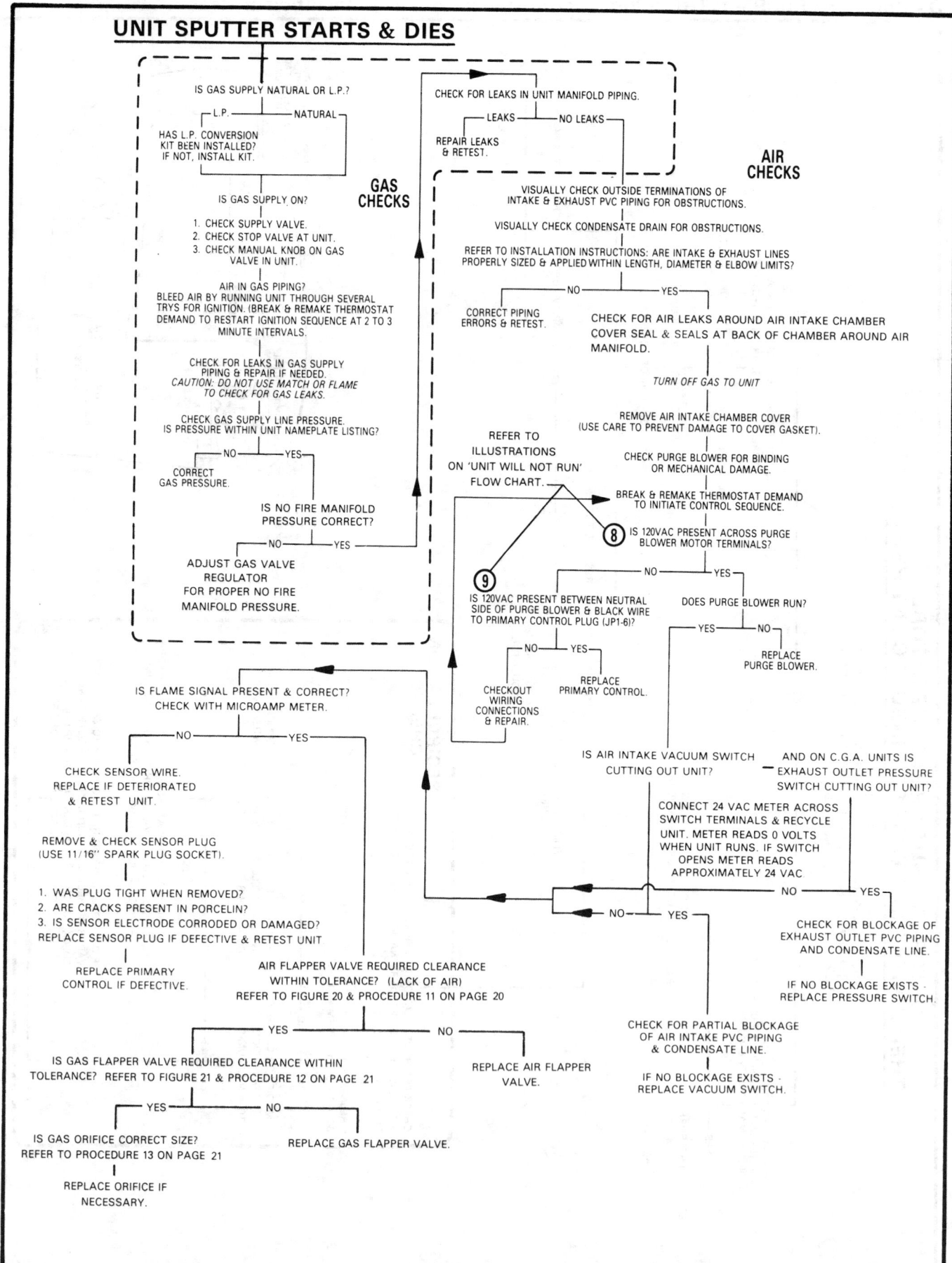

installation
operation
and
service
instructions

G21Q Series Units

GAS UNITS
502,690M
9/91

RETAIN THESE INSTRUCTIONS FOR FUTURE REFERENCE

⚠ WARNING

If the information in this manual is not followed exactly, a fire or explosion may result causing property damage, personal injury or loss of life.

Do not store or use gasoline or other flammable vapors and liquids in the vicinity of this or any other appliance.

Installation and service must be performed by a qualified installer, service agency or the gas supplier.

WHAT TO DO IF YOU SMELL GAS:
- Do not try to light any appliance.
- Open windows.
- Do not touch any electrical switch; Do not use the phone in your building.
- Immediately call your gas supplier from a neighbor's phone. Follow the gas supplier's instructions.
- If you cannot reach your gas supplier, call the fire department.

Technical Literature
Litho U.S.A

Dallas, Texas

TABLE OF CONTENTS

UNIT DIMENSIONS

***GAS PIPING INLETS (Both Sides)
26-1/8" (664mm)
GAS INLETS
2-3/4" (70mm)
7-1/4" (184mm)
I
HIGH VOLTAGE INLET (Both Sides)
LOW VOLTAGE INLET (Both Sides)
FLUE PIPE OUTLET
1-3/4" (44mm)
**18-1/2" (470mm)
D RETURN AIR OPENING (Both Sides)
1-1/2" (38mm)
G
F
*23-1/2" (597mm)

TOP VIEW
17-3/4" (451mm)
C SUPPLY AIR OPENING
H
J
AIR INTAKE

B
E*

*BOTTOM RETURN AIR OPENING
**23-1/2" (597mm) ON UNITS WITH Q4 OR Q5 BLOWERS.
***100 MODELS ONLY
†4-5/8" ON G21Q5-100

Model No.	A	B	C	D	E	F	G	H†	I	J
G21Q3 or Q4 40/60/80 Series	49" (1245mm)	21-1/4" (540mm)	19-1/8" (486mm)	14-1/2" (368mm)	14-1/2" (79mm)	23-5/8" (600mm)	20-1/4" (514mm)	4-1/8" (105mm)	5-1/4" (133mm)	8-1/2" (216mm)
G21Q5-80 and G21Q3 or Q5-100	53" (1346mm)	26-1/4" (667mm)	24-1/8" (613mm)	18-1/2" (470mm)	18-1/2" (470mm)	27-5/8" (702mm)	24-1/4" (616mm)	1-15/16" (49mm)	4-1/2" (114mm)	11" (279mm)

START-UP AND PERFORMANCE CHECK LIST

Job Name _____ Job No. _____ Date _____

Job Location _____ City _____ State _____

Installer _____ City _____ State _____

Unit Model No. _____ Serial No. _____ Serviceman _____

HEAT SECTION

Electrical Connections Tight? ☐

Supply Voltage _____

Condensate Drain In Unconditioned Space (If applicable)

Heat Tape Applied? ☐   Heat Tape Electrical Supply On? ☐

Gas Piping Connections Tight & Leak Tested? ☐

Fuel Type: Natural Gas ☐   LP Gas ☐

Furnace Btu Input _____

Line Pressure (7" Natural Gas; 11" LP Gas)

Regulator Pressure (Refer to unit nameplate) _____

Exhaust Connections Tight? ☐

Intake Connections Tight? ☐

Fan Timer Control Fan Off Setting (180 sec. factory setting) _____

Temperature Rise _____   External Static Pressure _____

Filters Clean & Secure? ☐

THERMOSTAT

Calibrated? ☐   Heat Anticipator Properly Set? ☐   Level? ☐

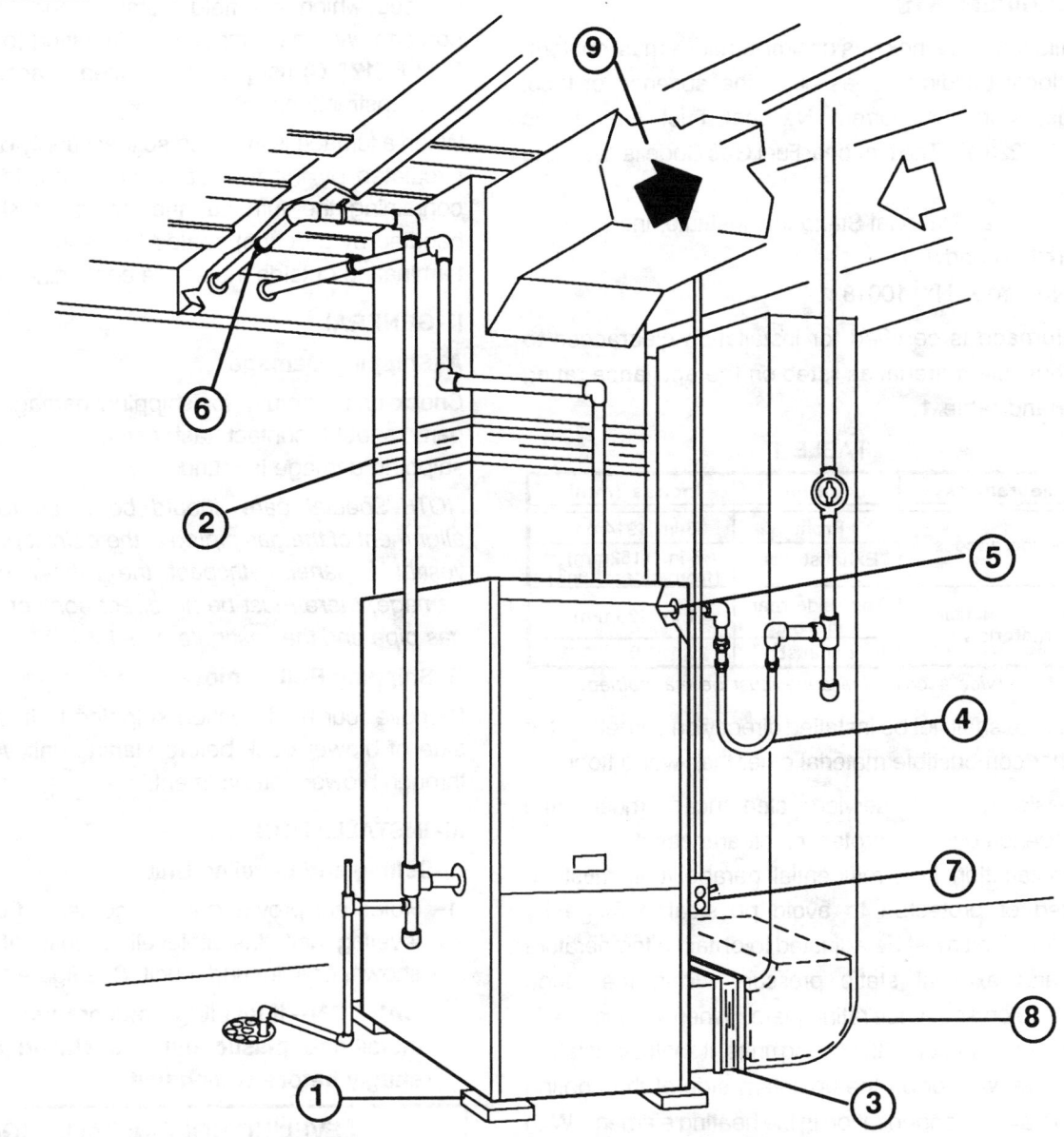

MAJOR INSTALLATION REQUIREMENTS

1– ISOLATION MOUNTING PADS

2– FLEXIBLE BOOT–SUPPLY AIR PLENUM

3– FLEXIBLE BOOT – RETURN AIR PLENUM

4– GAS CONNECTOR

5– GAS SUPPLY PIPING CENTERED IN INLET HOLE

6– ISOLATION HANGERS

7– ELECTRICAL CONDUIT ISOLATED FROM UNIT AND DUCT WORK

8– RETURN AIR PLENUM AND DUCT INSULATED PAST FIRST ELBOW

9– SUPPLY AIR PLENUM AND DUCT INSULATED PAST FIRST ELBOW

THE ABOVE REQUIREMENTS ARE ESSENTIAL TO THE INSTALLATION IN ORDER TO ISOLATE THE UNIT.

I-REQUIREMENTS

Installation of Lennox gas central furnaces must conform with local building codes or, in the absence of local codes, with the current National Fuel Gas Code (ANSI-Z223.1). The National Fuel Gas Code is available from:

American National Standards Institute, Inc.
1430 Broadway
New York, NY 10018

The furnace is certified for installation clearances to combustible material as listed on the appliance rating plate and table 1.

TABLE 1

Clearances	Location	Inches (mm)
Service access	Front	36 in. (914mm)
	Exhaust side	6 in. (152mm) (from side of unit)
To combustible materials	Top, side rear and front	1 in. (25mm)
	Exhaust	0

NOTE-Service access clearance must be maintained.

Appliance shall not be installed directly on carpeting, tile or other combustible material other than wood flooring.

Accessibility and service clearances must take precedence over fire protection clearances.

For installation in a residential garage, unit must be located or protected to avoid physical damage by vehicles. Unit must be adjusted to obtain a temperature rise and external static pressure within the range specified on appliance rating plate. When this furnace is used in conjunction with cooling units, it shall be installed in parallel with or on the upstream side of the cooling units to avoid condensation in the heating element. With a parallel flow arrangement, damper (or other means to control flow of air) shall be adequate to prevent chilled air from entering furnace and, if manually operated, must be equipped with means to prevent operation of either unit unless damper is in full "heat" or "cool" position.

When installed, furnace must be electrically grounded in accordance with current National Electric Code, ANSI/NFPA No. 70, if an external electrical source is utilized. The National Electric Code is available from:

National Fire Protection Association
470 Atlantic Avenue
Boston, MA 02210

Wiring to be done in the field, between the furnace and devices not attached to the furnace or between separate devices which are field-installed and located, shall conform with the temperature limitation for type T wire [63° F (17° C) rise] when installed in accordance with these instructions.

When a furnace is installed so that supply ducts carry air circulated by the furnace to areas outside the space containing the furnace, the return air shall also be handled by a duct(s) sealed to the furnace casing and terminating outside the space containing the furnace.

II-GENERAL

A-Shipping Damage

Check unit carefully for shipping damage. Receiving party should contact last carrier immediately if any shipping damage is found.

NOTE-Special care should be taken to check the alignment of the gas piping at the point it penetrates the vestibule panel. Inspect the rubber grommet for damage; there must be no direct contact between the gas pipe and the vestibule panel.

B-Shipping Bolt Removal

Remove four heat section shipping bolts from bottom side of blower deck before starting unit. Access bolts through blower compartment.

III-INSTALLATION

A-Setting and Leveling Unit

1- Holes are provided in the corners of unit base for leveling unit. Install leveling bolts (if desired as shown, or shim under unit. See figure 1.

CAUTION – If leveling bolts are used, be sure to install the plastic nuts as shown and tighten snugly before setting unit.

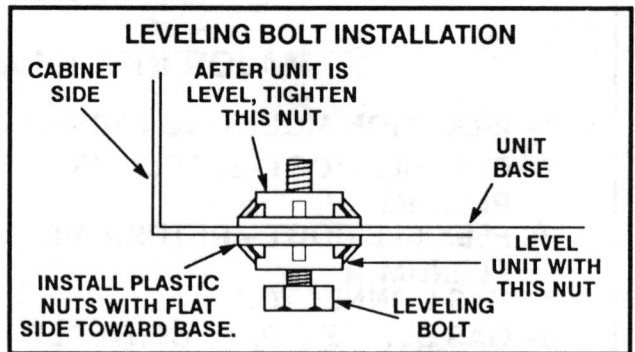

LEVELING BOLT INSTALLATION

FIGURE 1

2- Set unit in desired location keeping in mind clearances listed on unit rating plate. Also keep in mind gas supply connection, electrical supply, vent connections and clearances for installing and servicing unit.

B-Return Air Opening (Figure 2)

Return air can be brought in either side or bottom of unit. Scribe lines show the outline of each return air opening. Remove remaining insulation from around return air opening.

NOTE-Insulation adhesive is only used inside of scribe lines.

NOTE-Units with Q4 and Q5 blowers use larger opening.

1- Cut opening in floor or platform.

2- Flange return air plenum and lower into opening.

3- Place fibered glass insulation strips around opening. Position isolation mounting pads at corners of insulation. Insulation should not overlap the mounting pads. Trim away any excess insulation from strips.

4- Set unit. Make sure unit is sitting on isolation pads.

NOTE-Be careful not to damage fibered glass. Check for tight seal.

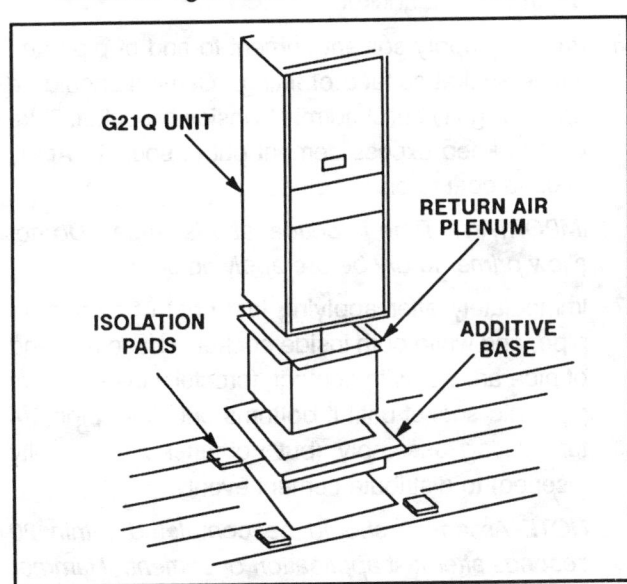

FIGURE 2

C-Filters

G21 series units are equipped with a reusable foam filter. Filter must be in place any time unit is in operation.

For bottom return air opening applications, install filter mounting clips provided and secure with sheet metal screws.

For side return air openings, use U-channel on blower deck and supplied filter rack to install filter. To install filter rack, remove two screws from side of cabinet. Place flange of filter rack inside bottom panel and side panel. Secure with previously removed cabinet/base bottom screws. See figure 3.

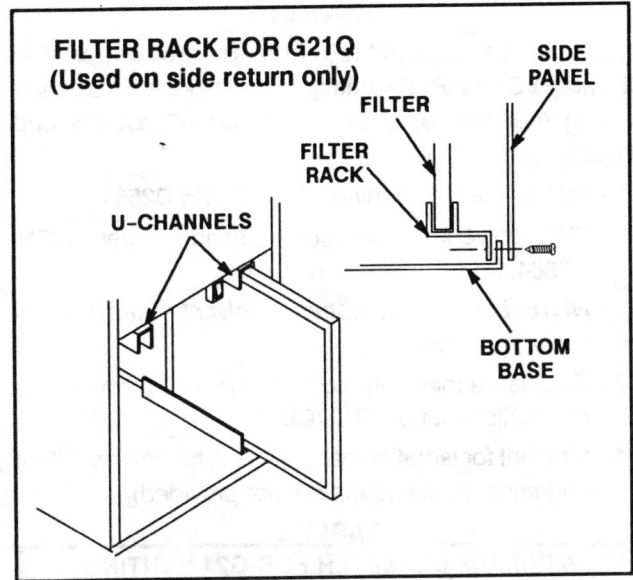

FIGURE 3

D-Duct System

1- Install flexible canvas boots or equivalent on both supply and return air plenums. Boots should be placed as close as possible to unit.

2- Insulate supply air plenum and duct system at least through the first elbow. Use 1-1/2 to 3 lb. density, matt face, 1" thick insulation. Provisions must be made to keep insulation in place and to protect edges from airflow deterioration.

3- Size and install supply and return system using industry-approved standards that result in a quiet and low-static system with uniform distribution.

IV-EXHAUST, CONDENSATE AND INTAKE PIPING

A-Exhaust and Intake Piping Requirements

All schedule 40 PVC pipe, fittings, primer and solvent cement must conform with American National Standard Institute and the American Society for Testing and Materials (ANSI/ASTM) standards. The solvent shall be free flowing and contain no lumps, undissolved particles or any foreign matter that adversely affects the joint strength or chemical resistance of the cement. The cement shall show no gelation, stratification, or separation that cannot be removed by stirring.

CAUTION-Solvent cements for plastic pipe are flammable liquids and should be kept away from all sources of ignition. Do not use excessive amounts of solvent cement when making joints. Good ventilation should be maintained to reduce fire hazard and to minimize breathing of solvent vapors. Avoid contact of cement with skin and eyes.

Materials

Schedule 40 PVC (type 1120 or 1220) pipe sized per table 2 and PVC1 or PVC12 fittings for intake and exhaust piping (not provided) per ASTM D1785, D2466 and D2665.

1- PVC primer (not provided) per ASTM D2564.

2- PVC solvent cement (not provided) per ASTM D2564.

   NOTE–Low temperature solvent cement is recommended.

3- Exhaust termination isolation material (provided in installation kit LB-49107C).

4- Material for isolation hangers –– Armaflex insulation and sheet metal strapping (not provided).

TABLE 2

MINIMUM DIAMETER FOR G21 VENTING					
Pipe Length (Max. Feet)	Number of 90° Elbows				
	0	2	4	6	8
5	2	2	2	2	2
10	2	2	2	2	2
20	2	2	2	2	2-1/2
30	2	2	2	2-1/2	2-1/2
40	2	2	2-1/2	2-1/2	2-1/2
50	2	2-1/2	2-1/2	2-1/2	2-1/2
60	2-1/2	2-1/2	2-1/2	2-1/2	3
70	2-1/2	2-1/2	2-1/2	3	3
80	2-1/2	2-1/2	3	3	3
90	2-1/2	3	3	3	3

Schedule 40 PVC pipe used for exhaust and intake lines should be sized per table 2. Each 90° elbow is equivalent to 5 ft. of vent pipe. Two 45° elbows are equivalent to one 90° elbow. One 45° elbow is equal to 2.5 ft. of vent pipe.

If intake and exhaust piping runs are not equal in length and combination, the larger diameter pipe (as sized per table 2) must be used for both runs. Regardless of the diameter of pipe used, the standard roof and wall terminations described in section D-Intake and Exhaust Piping Terminations should be used. Exhaust piping must terminate with 1-1/2" pipe.

Muffler lengths should be excluded when measuring vent pipe runs for sizing. Vent pipe must be sized at 2 in. between unit and mufflers. PVC drain, waste and vent (DWV) type fittings may be used for intake runs. Exhaust fittings, however, must be schedule 40 PVC.

Procedure for Cementing Joints Per ASTM D2855

WARNING–DANGER OF EXPLOSION! FUMES FROM PVC GLUE MAY IGNITE DURING SYSTEM CHECK. REMOVE SPARK PLUG WIRE FROM IGNITION CONTROL BEFORE 115V POWER IS APPLIED. RECONNECT WIRE AFTER TWO MINUTES.

1- Measure and cut PVC pipe to desired length.

2- Debur and chamfer end of pipe, removing any ridges or rough edges. If end is not chamfered, edge of pipe may remove cement from fitting socket and result in a leaking joint.

3- Clean and dry surfaces to be joined.

4- Test fit joint and mark depth of fitting on outside of pipe.

5- Uniformly apply liberal coat of primer to inside socket surface of fitting and male end of pipe to depth of fitting socket.

6- Promptly apply solvent cement to end of pipe and inside socket surface of fitting. Cement should be applied lightly but uniformly to inside of socket. Take care to keep excess cement out of socket. Apply second coat to end of pipe.

   IMPORTANT–Time is critical at this stage. Do not allow primer to dry before applying cement.

7- Immediately after applying last coat of cement to pipe, and while both inside socket surface and end of pipe are wet with cement, forcefully insert end of pipe into socket until it bottoms out. Turn pipe 1/4 turn during assembly (but not after pipe is fully inserted) to distribute cement evenly.

   NOTE–Assembly should be completed within 20 seconds after last application of cement. Hammer blows should not be used when inserting pipe.

8- After assembly, wipe excess cement from pipe at end of fitting socket. A properly made joint will show a bead around its entire perimeter. Any gaps may indicate a defective assembly due to insufficient solvent.

9- Handle joints carefully until completely set.

B-Exhaust and Condensate Line Piping

If a G21 furnace replaces a furnace which was commonly vented with another gas appliance, the size of the existing vent pipe for that gas appliance must be checked. Without the heat of the original furnace flue products, the existing vent pipe is probably oversized for the single water heater or other appliance. The vent should be checked for proper draw with the remaining appliance.

Removal of Unit from Common Venting System

In the event that an existing furnace is removed from a venting system commonly run with separate gas appliances, the venting system is likely to be too large to properly vent the remaining attached appliances. The following test should be conducted while all appliances (both in operation and those not in operation) are connected to the common venting system. If the venting system has been installed improperly, corrections must be made as outlined in the previous section.

1- Seal any unused openings in the common venting system.

2- Visually inspect the venting system from proper size and horizontal pitch and determine there is no blockage or restriction, leakage, corrosion and other deficiencies which could cause an unsafe condition.

3- Insofar as is practical, close all building doors and windows and all doors between the space in which the appliances remaining connected to the common venting system are located and other spaces of the building. Turn on clothes dryers and any appliances not connected to the common venting system. Turn on any exhaust fans, such as range hoods and bathroom exhausts, so they will operate at maximum speed. Do not operate a summer exhaust fan. Close fireplace dampers.

4- Follow the lighting instruction. Place the appliance being inspected in operation. Adjust thermostat so appliance will operate continuously.

5- Test for spillage at the draft hood relief opening after 5 minutes of main burner operation. Use the flame of a match or candle, or smoke from a cigarette, cigar or pipe.

6- After it has been determined that each appliance remaining connected to the common venting system properly vents when tested as outlined above, return doors, windows, exhaust fans, fireplace dampers and any other gas-burning appliance to their previous condition of use.

7- If improper venting is observed during any of the above tests, the common venting system must be corrected. The common venting system should be resized to approach the minimum size as determined by using the appropriate tables in appendix G in the current standards of the National Fuel Gas Code.

C-Intake and Exhaust Piping Terminations

Intake and exhaust pipes may be routed either horizontally through an outside wall or vertically through the roof. In attic or closet installations, vertical termination through the roof is preferred. Figures 4 through 11 show typical terminations.

1- Use schedule 40 PVC pipe for both intake and exhaust piping

2- Secure all joints, including drain leg, gas tight using approved PVC solvent.

3- Piping diameters should be determined according to length of pipe run. See table 2. Locate intake piping upwind (prevailing wind) from exhaust piping. To avoid recirculation of exhaust gas on roof terminations, end of exhaust pipe must be higher than intake pipe.

Exhaust and intake exits must be in same pressure zone. Do not exit one through the roof and one on the side. Also, do not exit the intake on one side and the exhaust on another side of the house or structure.

4- Intake and exhaust pipes should be placed as close together as possible at termination end (refer to illustrations). Maximum separation is 3 in. on roof terminations and 6 in. on side wall terminations.

5- Exhaust piping must terminate straight out or up as shown. On roof terminations, the intake piping should terminate straight down using two 90° elbows (See figure 4). In rooftop applications, a 2" X 1-1/2" reducer must be used on the exhaust piping at the point where it exits the structure to improve the velocity of exhaust away from the intake piping.

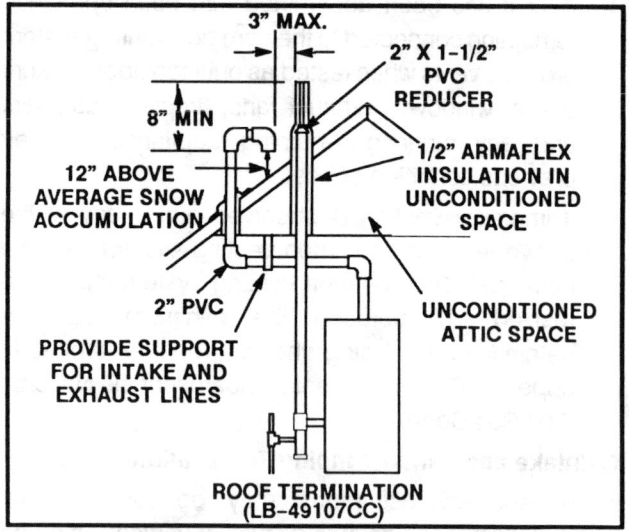

FIGURE 4

ROOF TERMINATION
(LB-49107CC)

3" MAX.

2" X 1-1/2"
PVC
REDUCER

8" MIN

1/2" ARMAFLEX
INSULATION IN
UNCONDITIONED
SPACE

12" ABOVE
AVERAGE SNOW
ACCUMULATION

2" PVC

UNCONDITIONED
ATTIC SPACE

PROVIDE SUPPORT
FOR INTAKE AND
EXHAUST LINES

1/2" ARMAFLEX
INSULATION IN
UNCONDITIONED
SPACE

12" MAX.

2" X 1-1/2" PVC
REDUCER

2" PVC

1-1/2" PVC

1/2" ARMAFLEX
INSULATION

6"
MAX.

2" PVC
COUPLING

OUTSIDE
WALL

8" MIN.

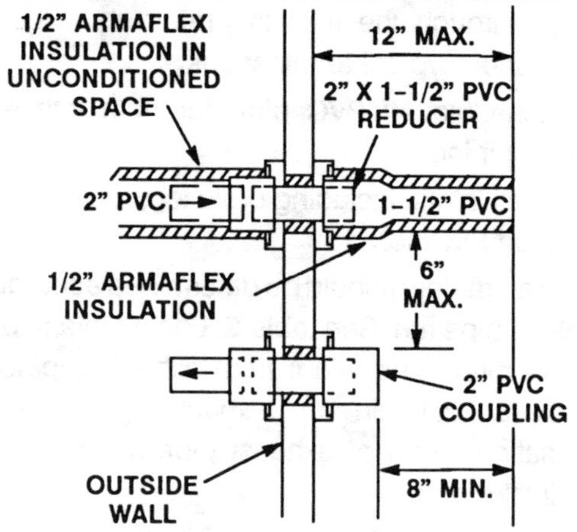

TOP VIEW
WALL TERMINATION
(LB-49107CB)

FIGURE 5

NOTE–If winter design temperature is below 32° F, exhaust piping must be insulated with 1/2" Armaflex or equivalent when run through unheated space. Do not leave any surface area of exhaust pipe open to outside air; exterior exhaust pipe must be insulated with 1/2" Armaflex or equivalent. In extreme cold climate areas, 3/4" Armaflex or equivalent is recommended. Insulation on outside runs of exhaust pipe must be painted or wrapped to protect insulation from deterioration.

IMPORTANT–Care must be taken to avoid recirculation of exhaust back into intake pipe.

6– On side wall exits, exhaust piping should extend a maximum of 12 inches beyond the outside wall. Intake piping should be as short as possible. See figure 5.

7– Minimum separation distance between the end of the exhaust pipe and the end of the intake pipe is 8 inches.

8– If intake and exhaust piping must be run up a side wall to position above snow accumulation or other obstructions, refer to figures 10 and 11 for proper piping method. Piping must be supported every 3 ft. as shown. When exhaust and intake piping must be run up an outside wall, the exhaust piping is reduced to 1-1/2 in. after the final elbow.

9– Position termination ends so they are free from any obstructions and above the level of snow accumulation (where applicable). Termination ends must be a minimum of 12 in. above grade level. Do not point into window wells, stairwells, alcoves, courtyard areas or other recessed areas. Do not position termination ends directly below roof eaves.

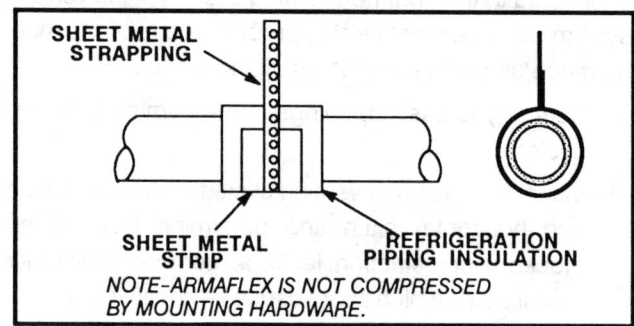

SHEET METAL
STRAPPING

SHEET METAL
STRIP

REFRIGERATION
PIPING INSULATION

NOTE–ARMAFLEX IS NOT COMPRESSED BY MOUNTING HARDWARE.

FIGURE 6

10– Suspend piping at a minimum of every 5 feet using isolation hangers. A suitable hanger can be fabricated by putting a sleeve of Armaflex refrigeration piping insulation around the pipe and suspending it using metal strapping as shown in figure 6. Place a small sheet metal strip between the Armaflex and the metal strapping to prevent crimping. Do not secure piping directly to joist or flooring.

11– In areas where piping penetrates joists or interior walls, hole must be large enough to allow clearance on all sides of pipe through center of hole using an isolation hanger.

12– Isolate piping at the point where it exits the outside wall or roof. Use termination kit LB-49107C.

13– Unit should not be installed in areas normally subject to freezing temperatures.

14– When furnace is installed in a residence where unit is shut down for an extended period of time, such as a vacation home, make provisions for draining drip leg on exhaust line.

D-Intake Piping

1- Cement intake piping in slip connector located at top of unit.

2- Route piping to outside of structure. Continue with installation following instructions given in exhaust and intake piping termination section.

   IMPORTANT-Combustion air intake inlet should not be located within 6 feet of dryer vent, condensing unit, or combustion air inlet or outlet of another appliance. Piping should not exit less than 3 feet from opening into another building.

3- Intake muffler is required for use with G21-80/100 units. Muffler use is optional on G21-40, -60 units. Install intake muffler as outlined in installation instructions packaged with muffler.

E-Exhaust and Condensate Piping

This unit is designed for either right or left side exit of exhaust piping.

NOTE—If unit is equipped with a Q4 or Q5 blower and side return air is used, the exhaust piping must be routed out the side opposite the return air duct.

1- Cut PVC pipe (provided) to the desired length for exit from the unit.

2- Slide PVC pipe through rubber grommet in cabinet. Care must be taken to center pipe hole.

3- Compression elbow is mounted for left side exhaust pipe exit on G21-40/60/80 units and right side exit on G21-100 units. If piping must exit on other side, disconnect exhaust pressure tubing, rotate and tighten compression elbow and reconnect tubing.

   NOTE—Differential pressure switch will not operate properly if tubing is kinked.

4- Cement drain leg assembly to PVC pipe as shown in figure 12.

   IMPORTANT—Bottom portion of drain leg assembly with condensate connection is not cemented to tee. Rotate until condensate connection is in suitable position. Cement into bottom of drain leg assembly tee.

   IMPORTANT—Stand pipe must remain open at the top to vent drain. Open end of pipe must not be used to connect drain hoses or other condensate hoses.

5- Cement exhaust pipe into top of drain leg assembly and route to outside of structure using intake and piping requirements. All horizontal runs of exhaust pipe must slope back toward unit. A minimum of 1/4" drop for each 12" of horizontal run is mandatory for drainage. Horizontal runs of exhaust piping must be supported every 5 ft. using isolation hangers.

   NOTE—Exhaust piping must be insulated with 1/2" Armaflex or equivalent when run through unheated space. Do not leave any area of exhaust pipe open to outside air; exterior exhaust must be insulated with 1/2" Armaflex or equivalent.

   CAUTION—Do not discharge exhaust into an existing stack or stack that also serves another gas appliance. If vertical discharge through an existing unused stack is required, insert PVC pipe inside the stack until the end is even with the top or outlet end of the metal stack.

6- One exhaust muffler is required for use with G21-80/100. Use of a second exhaust muffler on G21-80/100 is optional. Use of one exhaust muffler is optional on G21-40/60 units. Install exhaust line muffler(s) as outlined in installation instructions packaged with muffler.

7- Cement PVC pipe to compression elbow which is already in place.

   IMPORTANT—Care must be taken to assure a secure, tight seal between compression elbow assembly and manifold outlet.

8- Connect condensate drain line (1/2" SDR 11 plastic pipe or tubing) to condensate connection on drip leg assembly and route to open drain. Condensate drain must be sloped downward away from drip leg to drain. If drain level is above drip leg, condensate pump must be used to condensate line. Condensate drain line should be routed only within the conditioned space to avoid freezing of condensate and blockage of drain line.

   CAUTION—Do not use copper tubing or existing copper condensate lines for drain line.

9- Seal unused exhaust line piping hole with snap-plug provided.

   CAUTION—The exhaust vent pipe operates under positive pressure and must be completely sealed to prevent leakage of combustion products into the living space.

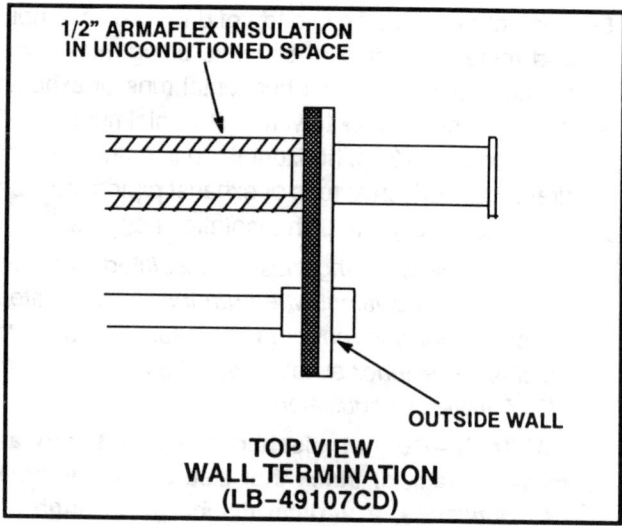

TOP VIEW
WALL TERMINATION
(LB-49107CD)
FIGURE 7

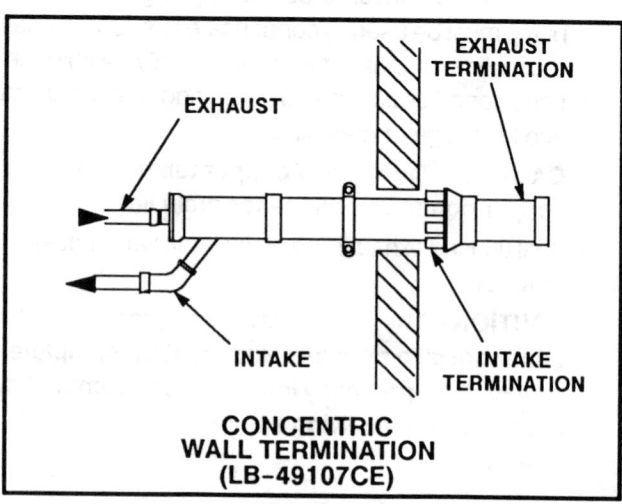

CONCENTRIC
ROOFTOP TERMINATION
(LB-49107CE)
FIGURE 8

EXHAUST TERMINATION

EXHAUST

INTAKE

INTAKE TERMINATION

CONCENTRIC
WALL TERMINATION
(LB-49107CE)
FIGURE 9

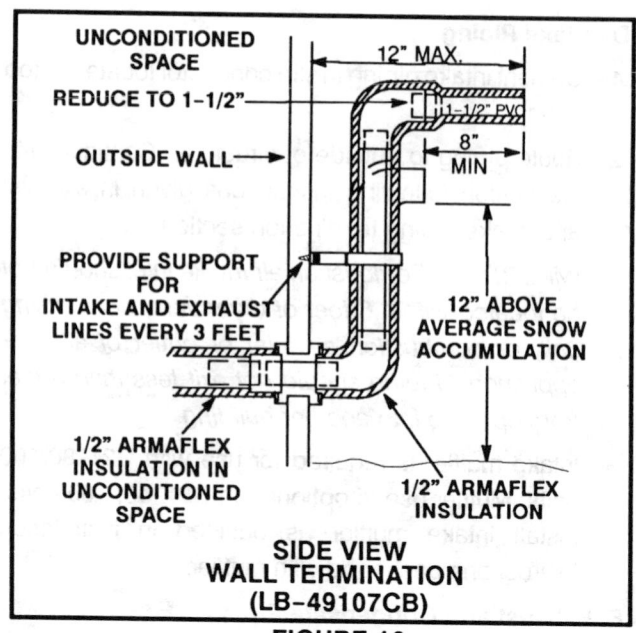

SIDE VIEW
WALL TERMINATION
(LB-49107CB)
FIGURE 10

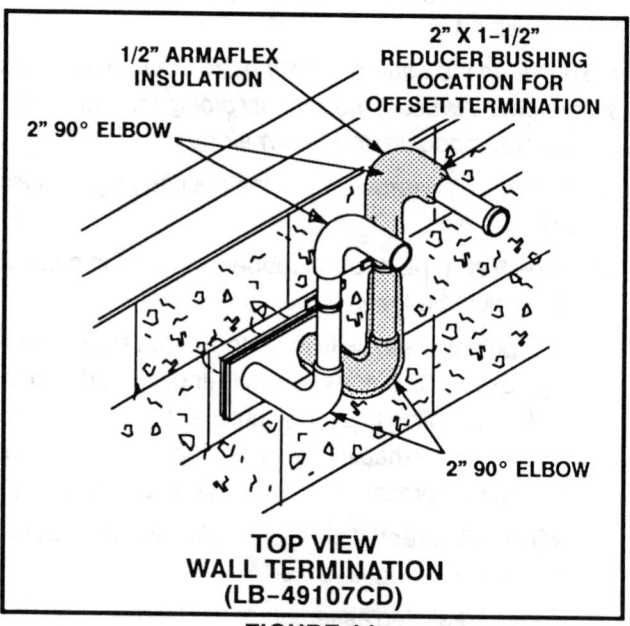

TOP VIEW
WALL TERMINATION
(LB-49107CD)
FIGURE 11

DRAIN LEG ASSEMBLY

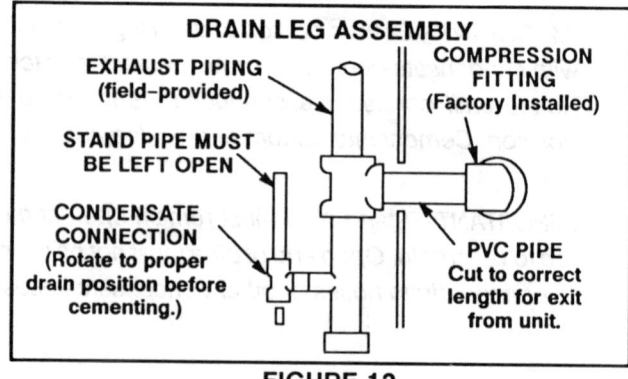

FIGURE 12

V-GAS PIPING

A-Gas Supply

The unit is shipped standard for right-side installation of gas piping. A piping hole is also fabricated in the left side for an alternate piping arrangement.

1- When connecting the gas supply, the length of run from the meter must be considered in determining the pipe size to avoid excessive pressure drop. For correct sizing of gas delivering piping, consult the utility having jurisdiction. A drip leg should be installed in the pipe run to the unit. In some localities, codes may require a manual main shut-off valve and union (furnished by installer) installed external to unit. Union must be of ground joint type.

NOTE-Compounds used on threaded joints of gas piping must be resistant to the actions of liquefied petroleum gases.

2- The use of one of the following gas connectors is recommended:

   * ANS Z21.24 Appliance Connectors of Corrugated Metal Tubing and Fittings.

   * ANS Z21.45 Assembled Flexible Appliance Connectors of Other than All-Metal Construction.

The above connectors may be used if acceptable by the authority having jurisdiction. A gas connector is provided and, if used, should be installed between the manual main shut-off valve and ground joint union. See figure 13 for downflow applications and horizontal applications.

CAUTION-Flexible gas connector must not be used to exit the unit. Flex connector must be installed in U-shaped fashion in order to achieve its purpose (See figure 13). Do not secure to unit ducting or structure.

3- Center gas line through piping hole. Gas line should not touch side of unit. See figure 13 for downflow and horizontal applications.

4- Connect gas supply line.

NOTE – Installer must provide a 1/8" N.P.T. plugged tap in field piping upstream of gas supply connections to unit. Tap must be accessible to test gauge connection. (See figure 14).

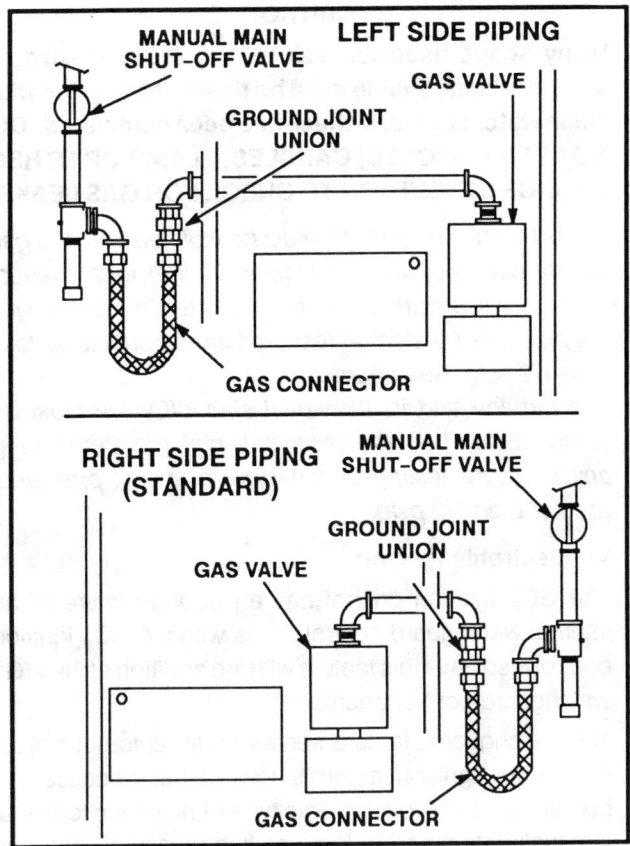

FIGURE 13

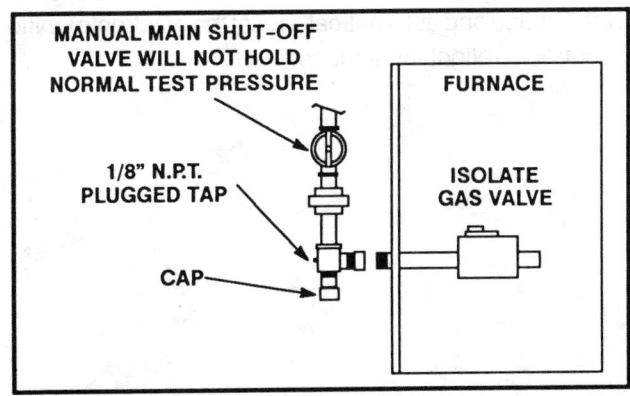

FIGURE 14

B-Leak Check

After gas piping has been completed, carefully check all piping connections (factory and field) for gas leaks. Use a leak detecting solution or other preferred means.

CAUTION

Many soaps used for leak testing are corrosive to certain metals. Piping must be rinsed thoroughly with clean water after leak check has been completed. DO NOT USE MATCHES, CANDLES, FLAME OR OTHER SOURCE OF IGNITION TO CHECK FOR GAS LEAKS.

IMPORTANT-The furnace must be isolated from the gas supply piping system by closing its individual manual shut-off valve during any pressure testing of the gas supply piping system at test pressures equal to or less than 1/2 psig. See figure 14.

The furnace and its individual shut-off valve must be disconnected from the gas supply piping system during any pressure testing of the system at test pressures greater than 1/2 psig.

VI-Electronic Ignition

The GC3 ignition control has an added feature of an internal watchguard control. Units with the GC1 ignition control also have this feature with the addition of the WG1 watchguard control board.

The watchguard feature serves as an automatic reset device for ignition controls locked-out because the burner has failed to light. This type of nuisance lock-out is usually attributed to low gas line pressure. After one hour of continuous thermostat demand for heat, the watchguard will break and remake thermostat demand to the furnace and automatically reset the electronic ignition control to relight the furnace.

VII-ELECTRICAL

1- Select fuse and wire size according to blower motor amps.

2- Access openings are provided on both sides of cabinet to facilitate wiring.

3- Install room thermostat according to instructions provided with thermostat. See figure 15 for new Lennox thermostat nomenclature versus old style nomenclature.

4- Install a separate fused disconnect switch near the unit so that power can be turned off for servicing.

5- Complete wiring connections to equipment using provided wiring diagrams.

6- Electrically ground unit in accordance with local codes or, in the absence of local codes, in accordance with the National Electric Code.

7- Install an auxiliary receptacle near unit.

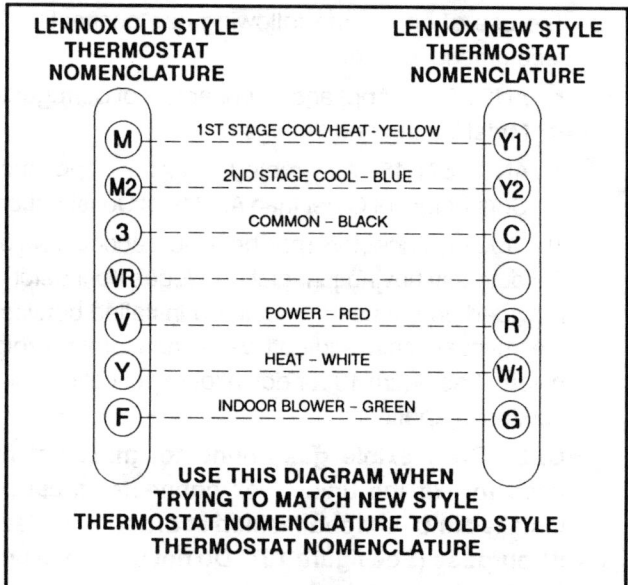

LENNOX OLD STYLE THERMOSTAT NOMENCLATURE		LENNOX NEW STYLE THERMOSTAT NOMENCLATURE
(M)	1ST STAGE COOL/HEAT - YELLOW	(Y1)
(M2)	2ND STAGE COOL - BLUE	(Y2)
(3)	COMMON - BLACK	(C)
(VR)		( )
(V)	POWER - RED	(R)
(Y)	HEAT - WHITE	(W1)
(F)	INDOOR BLOWER - GREEN	(G)

USE THIS DIAGRAM WHEN TRYING TO MATCH NEW STYLE THERMOSTAT NOMENCLATURE TO OLD STYLE THERMOSTAT NOMENCLATURE

FIGURE 15

TYPICAL G21Q WIRING DIAGRAM
WITH GC3 ELECTRONIC IGNITION CONTROL

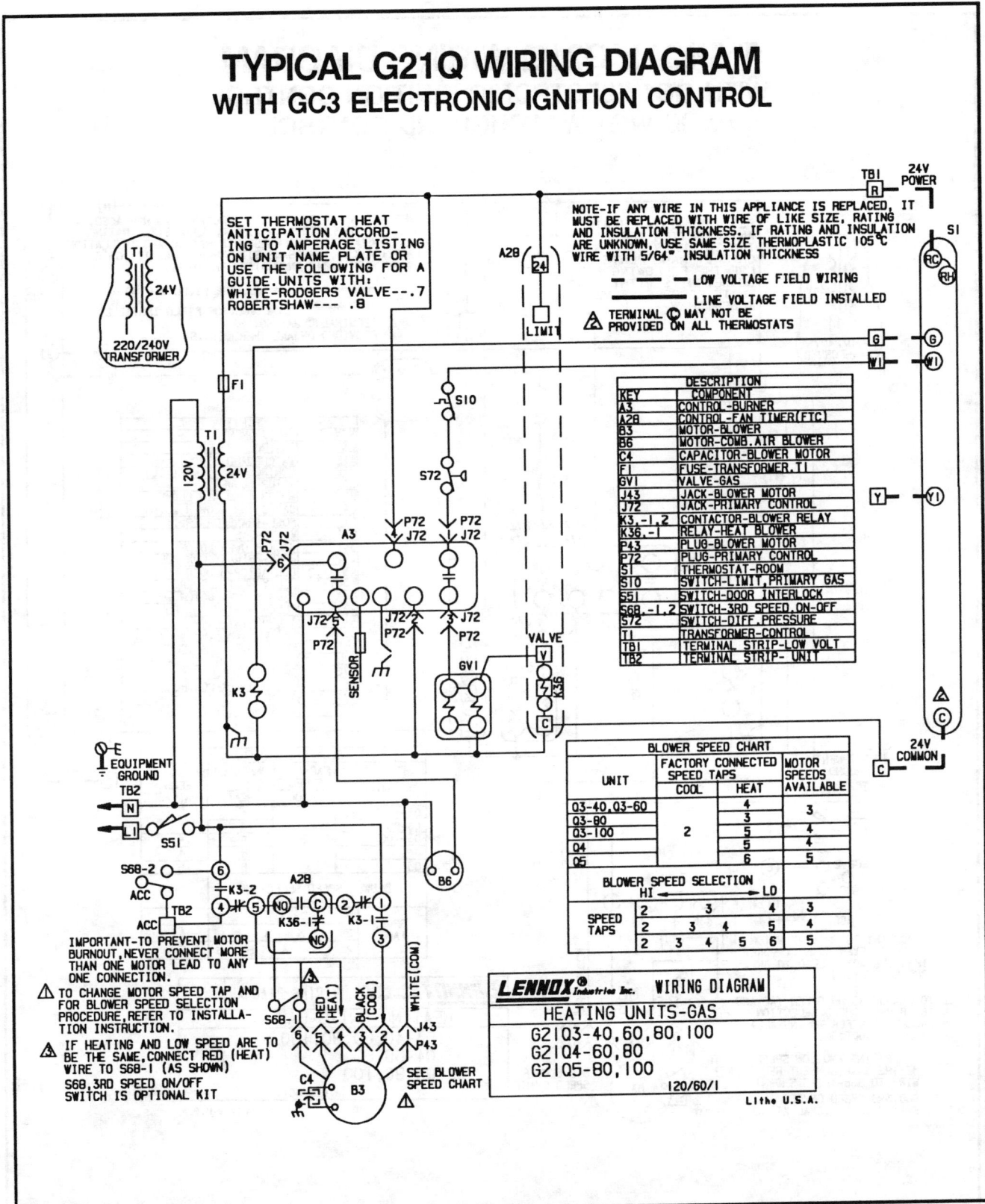

FIGURE 16

TYPICAL G21Q WIRING DIAGRAM
WITH GC1 ELECTRONIC IGNITION CONTROL AND WG1 WATCHGUARD CONTROL

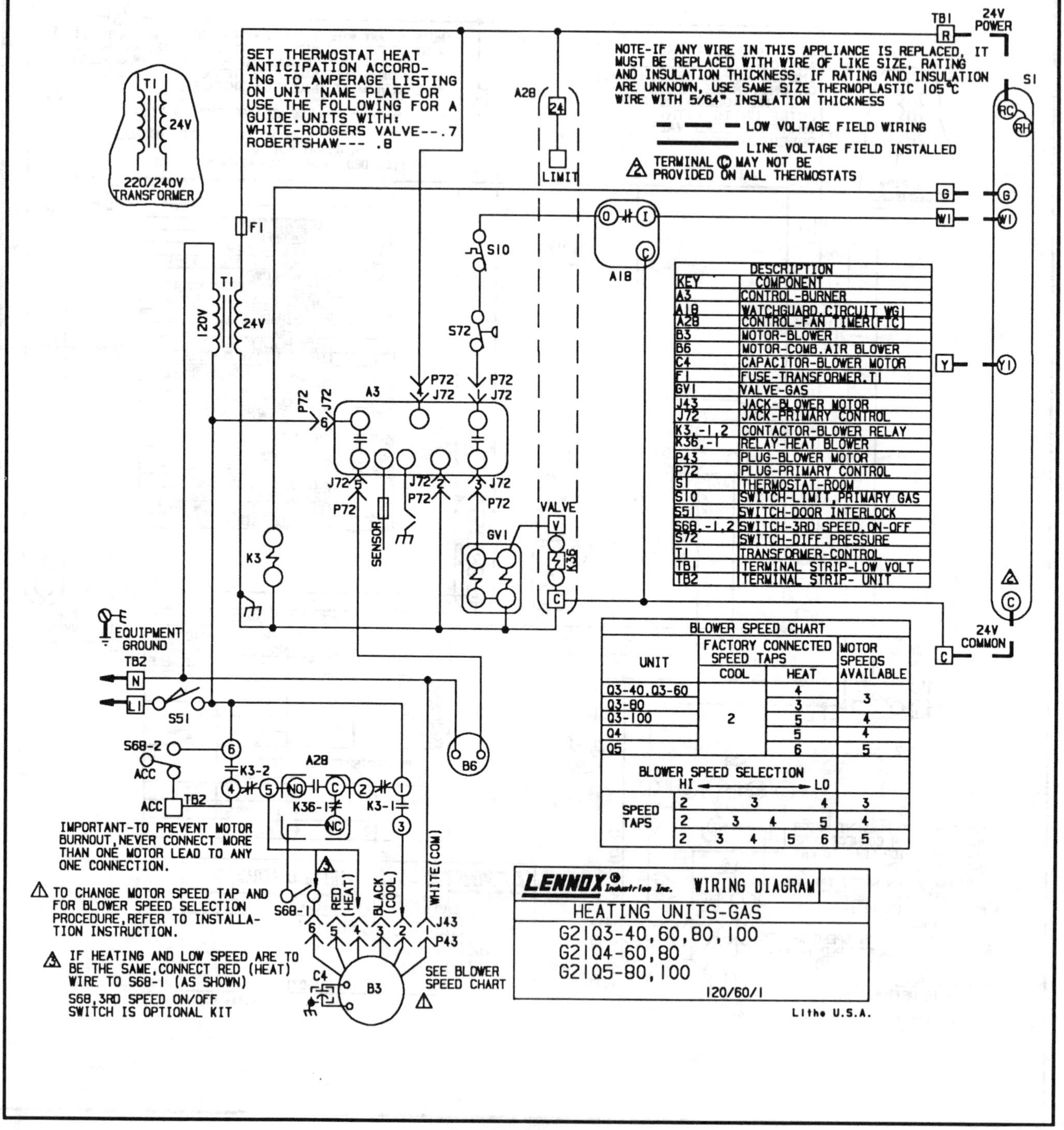

TB1
R
24V POWER
NOTE-IF ANY WIRE IN THIS APPLIANCE IS REPLACED, IT MUST BE REPLACED WITH WIRE OF LIKE SIZE, RATING AND INSULATION THICKNESS. IF RATING AND INSULATION ARE UNKNOWN, USE SAME SIZE THERMOPLASTIC 105°C WIRE WITH 5/64" INSULATION THICKNESS
LOW VOLTAGE FIELD WIRING
LINE VOLTAGE FIELD INSTALLED
TERMINAL C MAY NOT BE PROVIDED ON ALL THERMOSTATS
SET THERMOSTAT HEAT ANTICIPATION ACCORDING TO AMPERAGE LISTING ON UNIT NAME PLATE OR USE THE FOLLOWING FOR A GUIDE.UNITS WITH: WHITE-RODGERS VALVE--.7 ROBERTSHAW--- .8
T1
24V
220/240V TRANSFORMER
A28
24
LIMIT
S1
RC
RH
G
W1
Y
Y1
C
24V COMMON
A18
S10
S72
P72
J72
A3
F1
T1
120V
24V
K3
SENSOR
GV1
VALVE
V
K36
EQUIPMENT GROUND
TB2
N
L1
S51
S68-2
ACC
K3-2
A28
NO
K36-1
NC
K3-1
B6
IMPORTANT-TO PREVENT MOTOR BURNOUT,NEVER CONNECT MORE THAN ONE MOTOR LEAD TO ANY ONE CONNECTION.
TO CHANGE MOTOR SPEED TAP AND FOR BLOWER SPEED SELECTION PROCEDURE,REFER TO INSTALLATION INSTRUCTION.
IF HEATING AND LOW SPEED ARE TO BE THE SAME,CONNECT RED (HEAT) WIRE TO S68-1 (AS SHOWN)
S68,3RD SPEED ON/OFF SWITCH IS OPTIONAL KIT
S68-1
RED (HEAT)
BLACK (COOL)
WHITE(COM)
J43
P43
C4
B3
SEE BLOWER SPEED CHART
LENNOX® Industries Inc.
WIRING DIAGRAM
HEATING UNITS-GAS
G21Q3-40,60,80,100
G21Q4-60,80
G21Q5-80,100
120/60/1
Litho U.S.A.

KEY	DESCRIPTION COMPONENT
A3	CONTROL-BURNER
A18	WATCHGUARD CIRCUIT WG1
A28	CONTROL-FAN TIMER(FTC)
B3	MOTOR-BLOWER
B6	MOTOR-COMB.AIR BLOWER
C4	CAPACITOR-BLOWER MOTOR
F1	FUSE-TRANSFORMER,T1
GV1	VALVE-GAS
J43	JACK-BLOWER MOTOR
J72	JACK-PRIMARY CONTROL
K3,-1,2	CONTACTOR-BLOWER RELAY
K36,-1	RELAY-HEAT BLOWER
P43	PLUG-BLOWER MOTOR
P72	PLUG-PRIMARY CONTROL
S1	THERMOSTAT-ROOM
S10	SWITCH-LIMIT,PRIMARY GAS
S51	SWITCH-DOOR INTERLOCK
S68,-1,2	SWITCH-3RD SPEED,ON-OFF
S72	SWITCH-DIFF.PRESSURE
T1	TRANSFORMER-CONTROL
TB1	TERMINAL STRIP-LOW VOLT
TB2	TERMINAL STRIP- UNIT

BLOWER SPEED CHART

UNIT	FACTORY CONNECTED SPEED TAPS COOL	HEAT	MOTOR SPEEDS AVAILABLE
Q3-40,Q3-60	2	4	3
Q3-80		3	
Q3-100		5	4
Q4		5	4
Q5		6	5

BLOWER SPEED SELECTION HI ← → LO

SPEED TAPS	2	3		4		3
	2	3	4	5		4
	2	3	4	5	6	5

FIGURE 17

VIII-START-UP/ADJUSTMENTS

FOR YOUR SAFETY READ BEFORE LIGHTING

WARNING: Do not use this furnace if any part has been under water. Immediately call a qualified service technician to inspect the furnace and to replace any part of the control system and any gas control which has been under water.

WARNING: If overheating occurs or if gas supply fails to shut off, shut off the manual gas valve to the appliance before shutting off electrical supply.

CAUTION: Before attempting to perform any service or maintenance turn the electrical power to unit OFF at disconnect switch.

BEFORE LIGHTING smell all around the appliance area for gas. Be sure to smell next to the floor because some gas is heavier than air and will settle on the floor.

Use only your hand to push in or turn the gas control knob. Never use tools. If the knob will not push in or turn by hand, do not try to repair it, call a qualified service technician. Force or attempted repair may result in a fire or explosion.

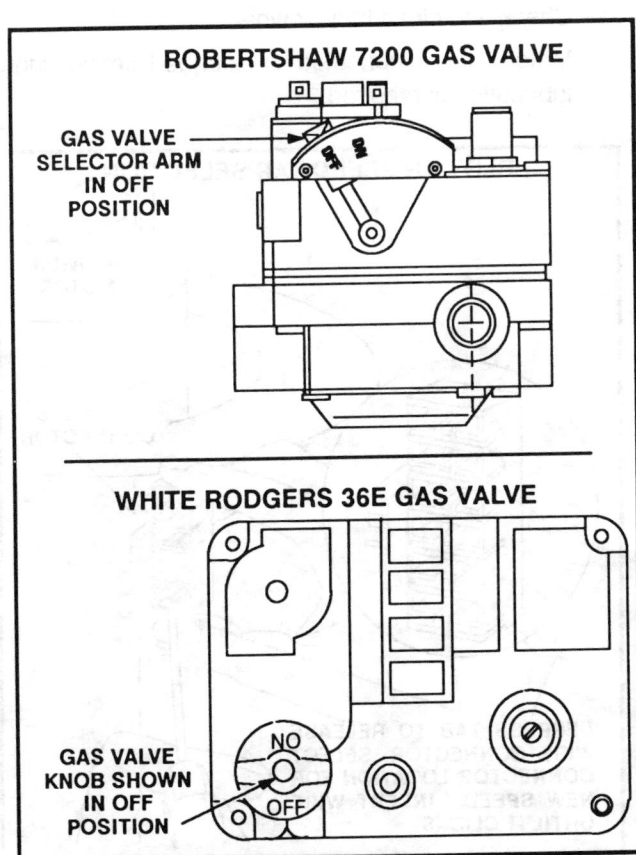

ROBERTSHAW 7200 GAS VALVE

GAS VALVE SELECTOR ARM IN OFF POSITION

WHITE RODGERS 36E GAS VALVE

GAS VALVE KNOB SHOWN IN OFF POSITION

NO

OFF

FIGURE 18

This unit is equipped with an automatic spark ignition system with flame rectification. Once combustion has started, the purge blower and spark ignitor are turned off. Do **not** try to light by hand.

A-Gas Valve Operation

WARNING: If you do not follow these instructions exactly, a fire or explosion may result causing property damage, personal injury or loss of life.

Gas Valve Operation for Robertshaw and White Rodgers Valves (Figure 18)

1– Set thermostat to lowest setting.

2– Turn off all electrical power to furnace.

3– This appliance Is equipped with an ignition device which automatically lights the burner. Do **not** try to light the burner by hand.

4– Remove unit access panel.

5– Turn knob on gas valve clockwise ⭢ to **OFF**. On Robertshaw 7200 gas valve, depress lever on gas control and move to **OFF** and release. Do not force.

6– Wait five (5) minutes to clear out any gas. If you then smell gas, STOP! Immediately call your gas supplier from a neighbor's phone. Follow the gas supplier's instructions. If you do not smell gas go to next step.

7– Turn knob on gas valve counterclockwise ⭠ to **ON**. On Robertshaw 7200 gas valve, depress lever on gas control and move to **ON** and release.

8– Replace unit access panel.

9– Turn on all electrical power to unit.

10– Set thermostat to desired setting.

11– If the furnace will not operate, follow the instructions "To Turn Off Gas To Unit" and call your service technician or gas supplier.

B-To Turn Off Gas To Unit

1– Set thermostat to lowest setting.

2– Turn off all electrical power to unit if service is to be performed.

3– Remove heat section access panel.

4– Turn knob on gas valve clockwise ⭢ to **OFF**. Do not force.

NOTE-On Robertshaw 7200 gas valve, depress lever on gas control and move to OFF and release.

5– Replace unit access panel.

C-Gas Flow

To check proper gas flow to combustion chamber, determine Btu input from the appliance rating plate. Divide this input rating by the Btu per cubic foot of available gas. Result is the number of cubic feet per hour required. Determine the flow of gas through gas meter for 2 minutes and multiply by 30 to get the hourly flow of gas to burner.

D-Gas Pressure

1- Check gas line pressure with unit firing at maximum rate. Normal natural gas inlet line pressure should be 7.0 in. w.c. Normal line pressure for LP gas is 11.0 in. w.c.

   IMPORTANT-Minimum gas supply pressure is listed on unit rating plate for normal input. Operation below minimum pressure may cause nuisance lockouts.

2- After line pressure is checked and adjusted, check regulator pressure. Correct manifold pressure (unit running) is specified on nameplate. To measure, connect gauge to pressure tap in elbow below expansion tank.

E-Heat Anticipation Settings

Units with White Rodgers gas valves -- 0.7
Units with Robertshaw gas valves -- 0.8

F-Limit Control

Limit Control -- Factory set: No adjustment necessary.

G-Fan Timer Control

Fan On Delay Timing -- Fixed at 45 seconds.
Fan Off Delay Timing -- Factory set at 180 seconds and adjustable from 120 to 240 seconds.

H-Temperature Rise and External Static Pressure

Check temperature rise and external static pressure. If necessary, adjust blower speed to maintain temperature rise and external static pressure within range shown on unit rating plate.

I-Electrical

1- Check all wiring for loose connections.

2- Check fuse located on unit control box. Fuse should be a 2 amp AGC fast blow.

3- Check for correct voltage at unit (unit operating).

4- Check amp-draw on blower motor.
   Motor Nameplate _____ Actual _____.

NOTE-Do not secure electrical conduit directly to ducting or structure.

J-Blower Speeds

Blower speed selection is accomplished by changing the taps at the harness connector at the blower motor. See figure 19.

Refer to speed selection chart on unit wiring diagram.

NOTE-CFM readings are taken external to unit with a dry evaporator coil and without accessories.

IX-SERVICE

A-Annual Service

At the beginning of each heating season, system should be checked as follows:

Electrical

1- Check all wiring for loose connections.

2- Check fuse located on unit control box. Fuse should be a 2 amp AGC fast blow.

3- Check for correct voltage at unit (unit operating).

4- Check amp-draw on blower motor.
   Motor Nameplate _____ Actual _____

5- Check to see that heat (if applicable) is operating.

Blower

1- Check and clean blower wheel.

2- Motors are prelubricated for extended life; no further lubrication is required.

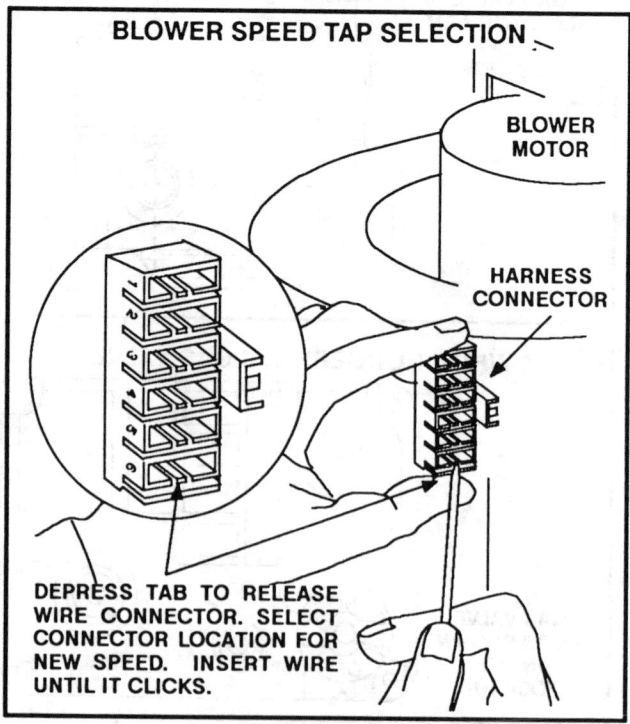

BLOWER SPEED TAP SELECTION

BLOWER MOTOR

HARNESS CONNECTOR

DEPRESS TAB TO RELEASE WIRE CONNECTOR. SELECT CONNECTOR LOCATION FOR NEW SPEED. INSERT WIRE UNTIL IT CLICKS.

FIGURE 19

Filters

1- Filters must be cleaned or replaced when dirty to assure proper furnace operation.

2- Reusable foam filters supplied with G21 can be washed with water and mild detergent. When dry, they should be sprayed with filter handicoater prior to reinstallation. Filter handicoater is RP Products coating no. 481 and is available as Lennox part no. P-8-5069.

3- If replacement is necessary, order Lennox part no. P-9-7831 for 20 X 25 inch filter and P-8-7822 for 16 X 25 inch filter.

Intake and Exhaust Lines

Check intake and exhaust PVC lines and all connections for tightness and make sure there is no blockage. Also check condensate line for free flow during operation.

Insulation

Outdoor piping insulation should be inspected yearly for deterioration. If necessary, replace with same materials.

B-Cleaning Heat Exchanger/Burner Assembly

NOTE-Use papers or protective covering in front of furnace while removing heat exchanger assembly.

CAUTION-Before removing spark plug and sensor wires after unit has been operating, unit should be allowed to cool down at least 15 minutes before placing hands into heat chamber access opening. Residual heat in combustion chamber transfers back to air intake valve causing it to become very hot when unit is first shut down. To cool completely to room temperature, blower should be run continuously for approximately 40 minutes.

1- Turn off both electrical and gas power supplies to furnace.

2- Remove upper and lower furnace access panels.

3- Remove cover or air decoupler box in vestibule panel.

4- Remove insulation pieces from lower section of air decoupler box.

5- Unscrew air valve housing, using either a strap or basin wrench.

6- Disconnect wiring to purge blower.

7- Remover nut from PVC air inlet fitting.

8- Remove nuts from air decoupler box mounting bolts and gas decoupler bracket.

9- Remove air decoupler box from unit.

10- Remove rubber pad(s) from air pipe.

11- Detach PVC exhaust pipe from coil manifold outlet (located in lower corner of vestibule panel).

12- Disconnect gas to unit.

13- Disconnect wiring to gas valve.

14- Break union in gas line just below gas decoiupler. Remove gas valve / gas decoupler / piping assembly.

15- Remove remaining gas piping from fitting at vestibule panel.

IMPORTANT – Hex head fitting contains gas diaphragm valve so care must be taken when handling this portion of piping assembly.

16- Disconnect blower motor wires form control box.

17- Disconnect spark plug and sensor wires form plugs in combustion chamber. (Access plate is located to the left of the air decoupler box.)

18- Remove vest panel.

19- From underside of blower deck, remove four nuts holding rubber heat train mounts.

20- Lift heat train from unit.

21- Backflush heat train with a soapy water solution or steam clean.

IMPORTANT-If unit is backflushed with water, make sure all water is drained from heat train before replacing.

22- Reverse above steps to replace heat exchanger assembly. Be sure rubber seal pad is in place on air pipe and that ground wire on gas valve is put back on the upper-right air decoupler box mounting stud.

G21Q
PARTS ARRANGEMENT

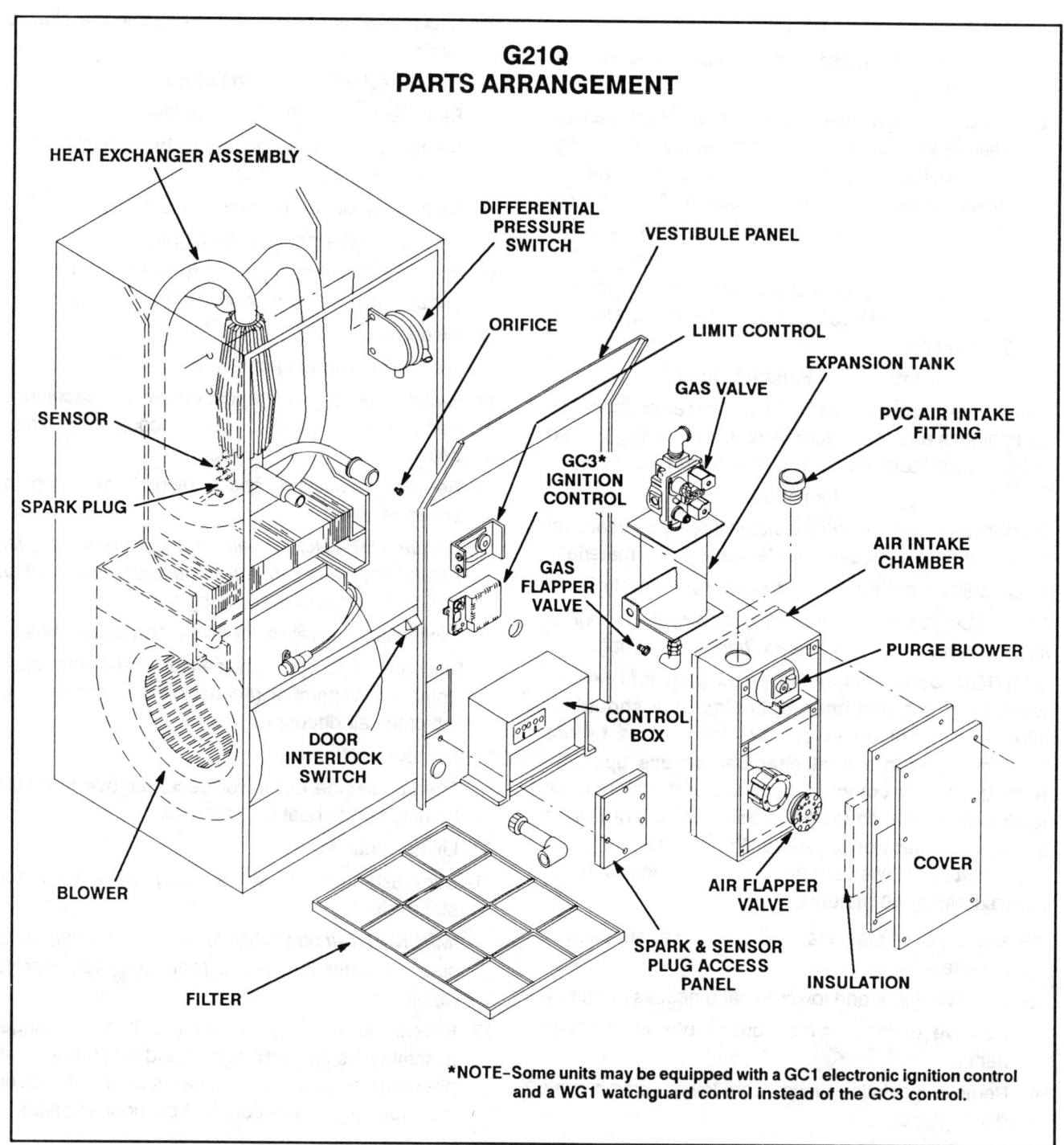

HEAT EXCHANGER ASSEMBLY

DIFFERENTIAL PRESSURE SWITCH

VESTIBULE PANEL

ORIFICE

LIMIT CONTROL

EXPANSION TANK

GAS VALVE

PVC AIR INTAKE FITTING

SENSOR

GC3* IGNITION CONTROL

SPARK PLUG

GAS FLAPPER VALVE

AIR INTAKE CHAMBER

PURGE BLOWER

DOOR INTERLOCK SWITCH

CONTROL BOX

BLOWER

AIR FLAPPER VALVE

COVER

SPARK & SENSOR PLUG ACCESS PANEL

INSULATION

FILTER

*NOTE–Some units may be equipped with a GC1 electronic ignition control and a WG1 watchguard control instead of the GC3 control.

FIGURE 20

X-REPAIR PARTS LIST

The following repair parts are available through independent Lennox dealers. When ordering parts, include the complete furnace model number listed on the unit rating plate. Example: G21Q3-60-1.

CABINET PARTS
Top access panel
Blower panel
Vestibule panel
Control box cover

CONTROL PANEL PARTS
Transformer
Indoor blower relay
Fan Timer Control

BLOWER PARTS
Blower wheel
Motor
Motor mounting frame
Motor capacitor
Blower housing cut-off plate
Blower housing

HEATING PARTS
Heat exchanger assembly
Gas orifice
Gas valve
Gas decoupler
Gas flapper valve
Purge blower
Air intake flapper valve
Primary control
Ignition lead
Spark plug ignitor
Flame sensor lead
Flame sensor

XI-TROUBLESHOOTING

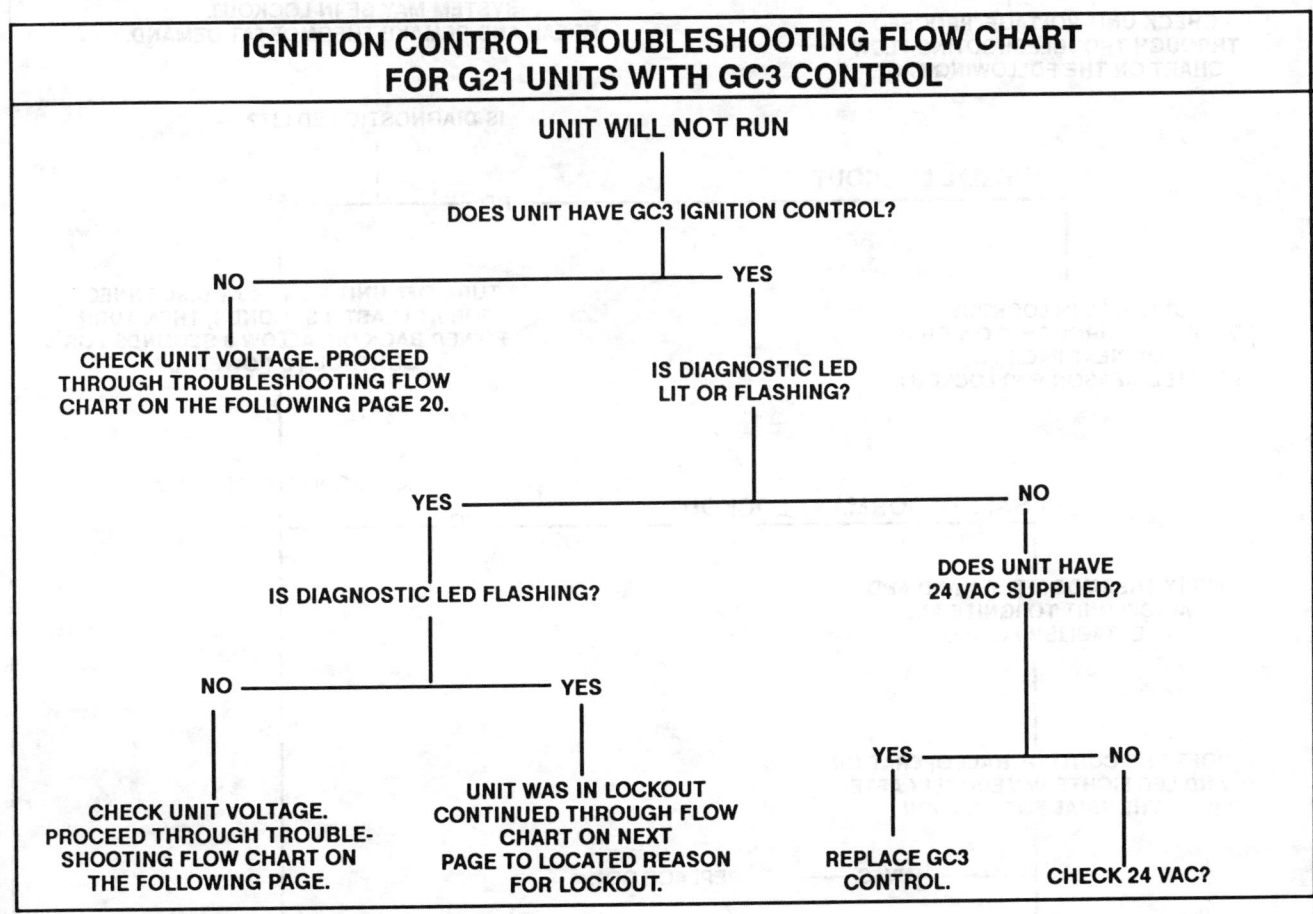

**IGNITION CONTROL TROUBLESHOOTING FLOW CHART
FOR G21 UNITS WITH GC3 CONTROL**

UNIT WILL NOT RUN

DOES UNIT HAVE GC3 IGNITION CONTROL?

NO — CHECK UNIT VOLTAGE. PROCEED THROUGH TROUBLESHOOTING FLOW CHART ON THE FOLLOWING PAGE 20.

YES — IS DIAGNOSTIC LED LIT OR FLASHING?

YES — IS DIAGNOSTIC LED FLASHING?

NO — CHECK UNIT VOLTAGE. PROCEED THROUGH TROUBLE-SHOOTING FLOW CHART ON THE FOLLOWING PAGE.

YES — UNIT WAS IN LOCKOUT CONTINUED THROUGH FLOW CHART ON NEXT PAGE TO LOCATED REASON FOR LOCKOUT.

NO — DOES UNIT HAVE 24 VAC SUPPLIED?

YES — REPLACE GC3 CONTROL.

NO — CHECK 24 VAC?

IGNITION CONTROL TROUBLESHOOTING FLOW CHART
FOR G21 UNITS WITH GC1 CONTROL AND WG1 WATCHGUARD

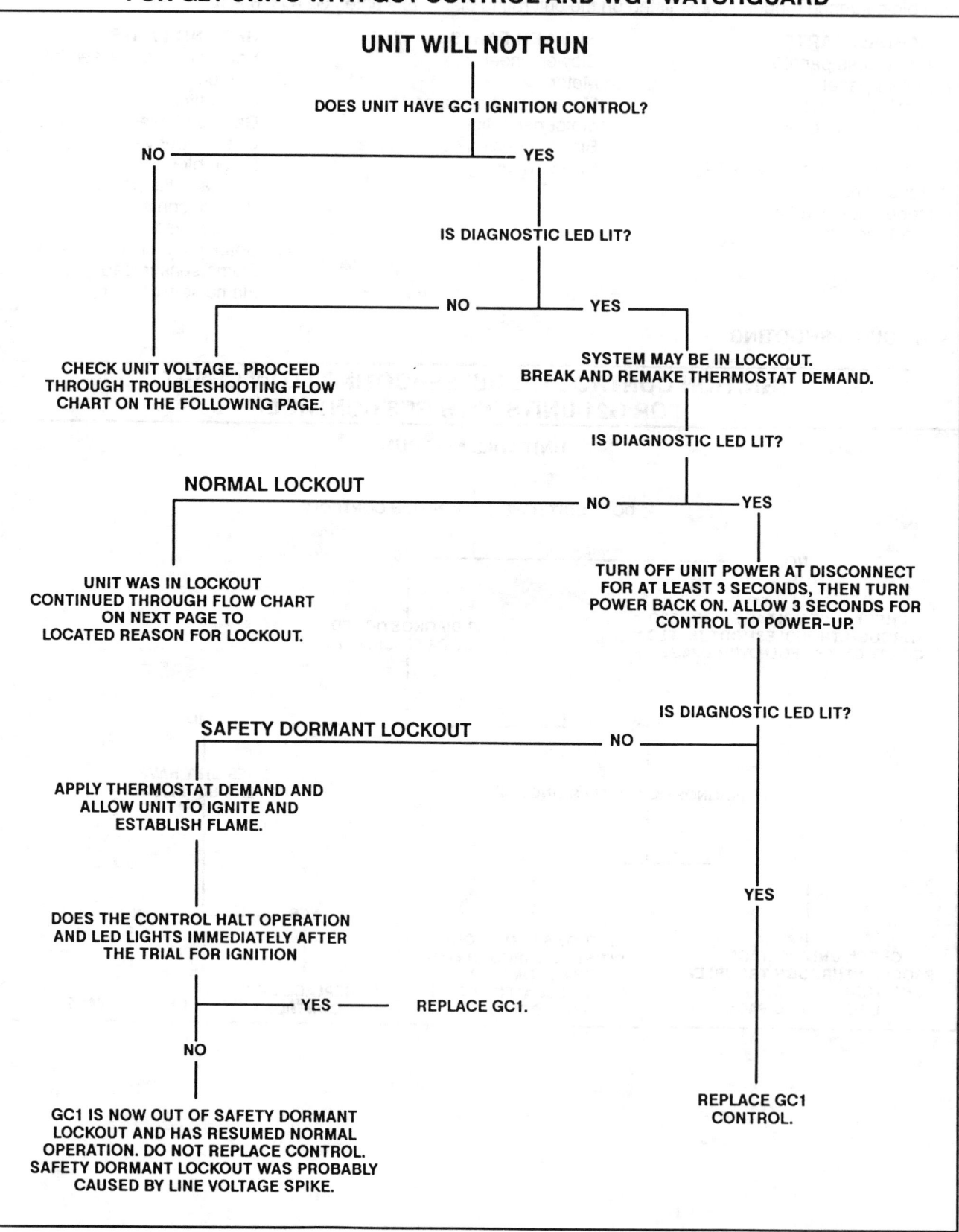

UNIT WILL NOT RUN

DOES UNIT HAVE GC1 IGNITION CONTROL?

NO — YES

IS DIAGNOSTIC LED LIT?

NO — YES

CHECK UNIT VOLTAGE. PROCEED THROUGH TROUBLESHOOTING FLOW CHART ON THE FOLLOWING PAGE.

SYSTEM MAY BE IN LOCKOUT. BREAK AND REMAKE THERMOSTAT DEMAND.

IS DIAGNOSTIC LED LIT?

NORMAL LOCKOUT

NO — YES

UNIT WAS IN LOCKOUT CONTINUED THROUGH FLOW CHART ON NEXT PAGE TO LOCATED REASON FOR LOCKOUT.

TURN OFF UNIT POWER AT DISCONNECT FOR AT LEAST 3 SECONDS, THEN TURN POWER BACK ON. ALLOW 3 SECONDS FOR CONTROL TO POWER-UP.

IS DIAGNOSTIC LED LIT?

SAFETY DORMANT LOCKOUT

NO

APPLY THERMOSTAT DEMAND AND ALLOW UNIT TO IGNITE AND ESTABLISH FLAME.

DOES THE CONTROL HALT OPERATION AND LED LIGHTS IMMEDIATELY AFTER THE TRIAL FOR IGNITION

YES — REPLACE GC1.

NO

YES

GC1 IS NOW OUT OF SAFETY DORMANT LOCKOUT AND HAS RESUMED NORMAL OPERATION. DO NOT REPLACE CONTROL. SAFETY DORMANT LOCKOUT WAS PROBABLY CAUSED BY LINE VOLTAGE SPIKE.

REPLACE GC1 CONTROL.

G21 TROUBLSHOOTING FLOW CHART
(continued form flow chart on previous page)
NOTE—Numbered steps refer to illustrations on last page.

UNIT WILL NOT RUN

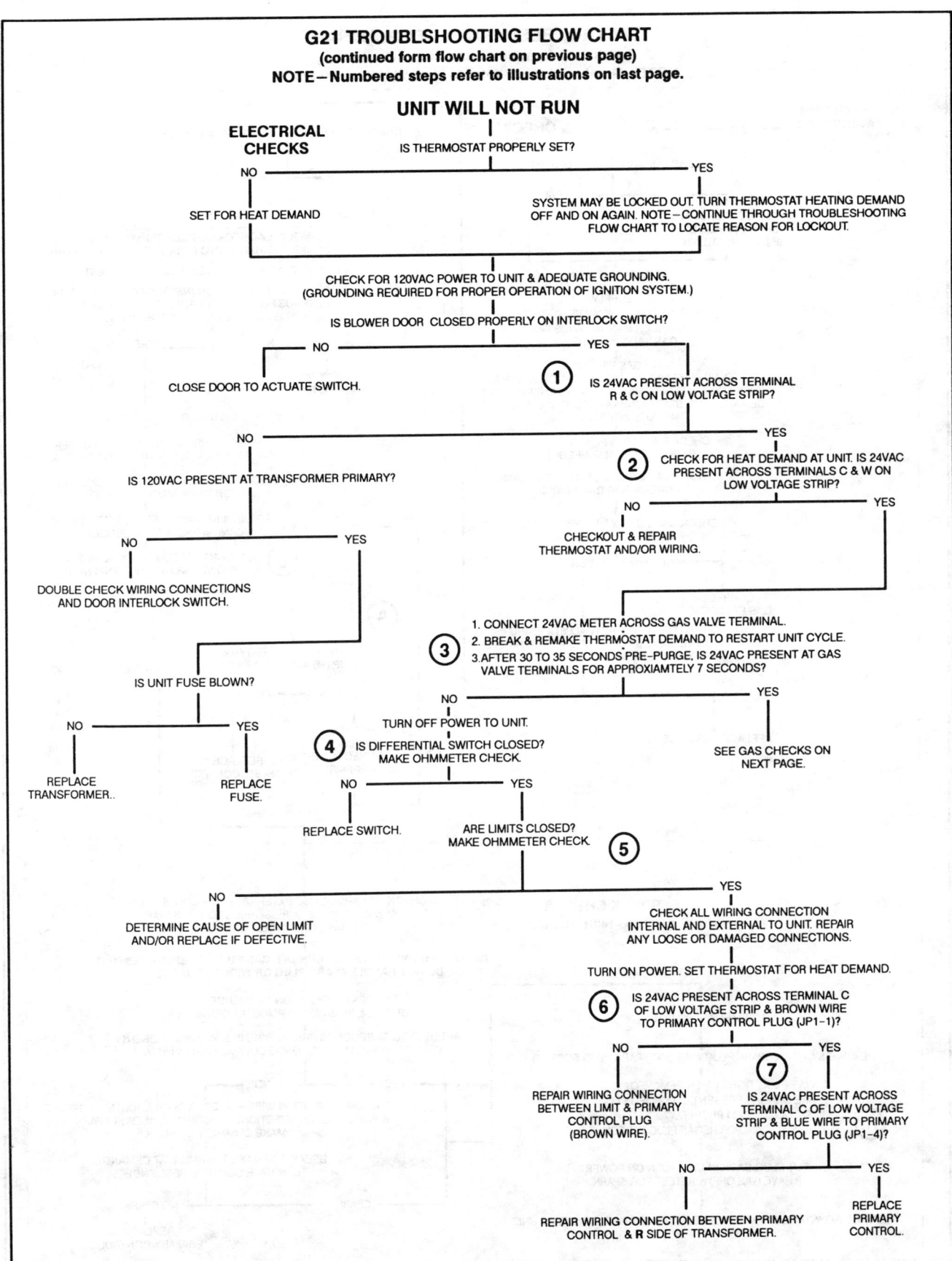

ELECTRICAL CHECKS

IS THERMOSTAT PROPERLY SET?

NO — SET FOR HEAT DEMAND

YES — SYSTEM MAY BE LOCKED OUT. TURN THERMOSTAT HEATING DEMAND OFF AND ON AGAIN. NOTE—CONTINUE THROUGH TROUBLESHOOTING FLOW CHART TO LOCATE REASON FOR LOCKOUT.

CHECK FOR 120VAC POWER TO UNIT & ADEQUATE GROUNDING. (GROUNDING REQUIRED FOR PROPER OPERATION OF IGNITION SYSTEM.)

IS BLOWER DOOR CLOSED PROPERLY ON INTERLOCK SWITCH?

NO — CLOSE DOOR TO ACTUATE SWITCH.

YES — (1) IS 24VAC PRESENT ACROSS TERMINAL R & C ON LOW VOLTAGE STRIP?

NO — IS 120VAC PRESENT AT TRANSFORMER PRIMARY?

YES — (2) CHECK FOR HEAT DEMAND AT UNIT. IS 24VAC PRESENT ACROSS TERMINALS C & W ON LOW VOLTAGE STRIP?

NO — CHECKOUT & REPAIR THERMOSTAT AND/OR WIRING.

YES —

NO — DOUBLE CHECK WIRING CONNECTIONS AND DOOR INTERLOCK SWITCH.

YES — IS UNIT FUSE BLOWN?

(3) 1. CONNECT 24VAC METER ACROSS GAS VALVE TERMINAL.
2. BREAK & REMAKE THERMOSTAT DEMAND TO RESTART UNIT CYCLE.
3. AFTER 30 TO 35 SECONDS PRE-PURGE, IS 24VAC PRESENT AT GAS VALVE TERMINALS FOR APPROXIAMTELY 7 SECONDS?

NO — REPLACE TRANSFORMER..

YES — REPLACE FUSE.

NO — TURN OFF POWER TO UNIT. (4) IS DIFFERENTIAL SWITCH CLOSED? MAKE OHMMETER CHECK.

YES — SEE GAS CHECKS ON NEXT PAGE.

NO — REPLACE SWITCH.

YES — ARE LIMITS CLOSED? MAKE OHMMETER CHECK. (5)

NO — DETERMINE CAUSE OF OPEN LIMIT AND/OR REPLACE IF DEFECTIVE.

YES — CHECK ALL WIRING CONNECTION INTERNAL AND EXTERNAL TO UNIT. REPAIR ANY LOOSE OR DAMAGED CONNECTIONS.

TURN ON POWER. SET THERMOSTAT FOR HEAT DEMAND.

(6) IS 24VAC PRESENT ACROSS TERMINAL C OF LOW VOLTAGE STRIP & BROWN WIRE TO PRIMARY CONTROL PLUG (JP1-1)?

NO — REPAIR WIRING CONNECTION BETWEEN LIMIT & PRIMARY CONTROL PLUG (BROWN WIRE).

YES — (7) IS 24VAC PRESENT ACROSS TERMINAL C OF LOW VOLTAGE STRIP & BLUE WIRE TO PRIMARY CONTROL PLUG (JP1-4)?

NO — REPAIR WIRING CONNECTION BETWEEN PRIMARY CONTROL & R SIDE OF TRANSFORMER.

YES — REPLACE PRIMARY CONTROL.

G21 TROUBLSHOOTING FLOW CHART
(continued form flow chart on previous page)
NOTE—Numbered steps refer to illustrations on last page.

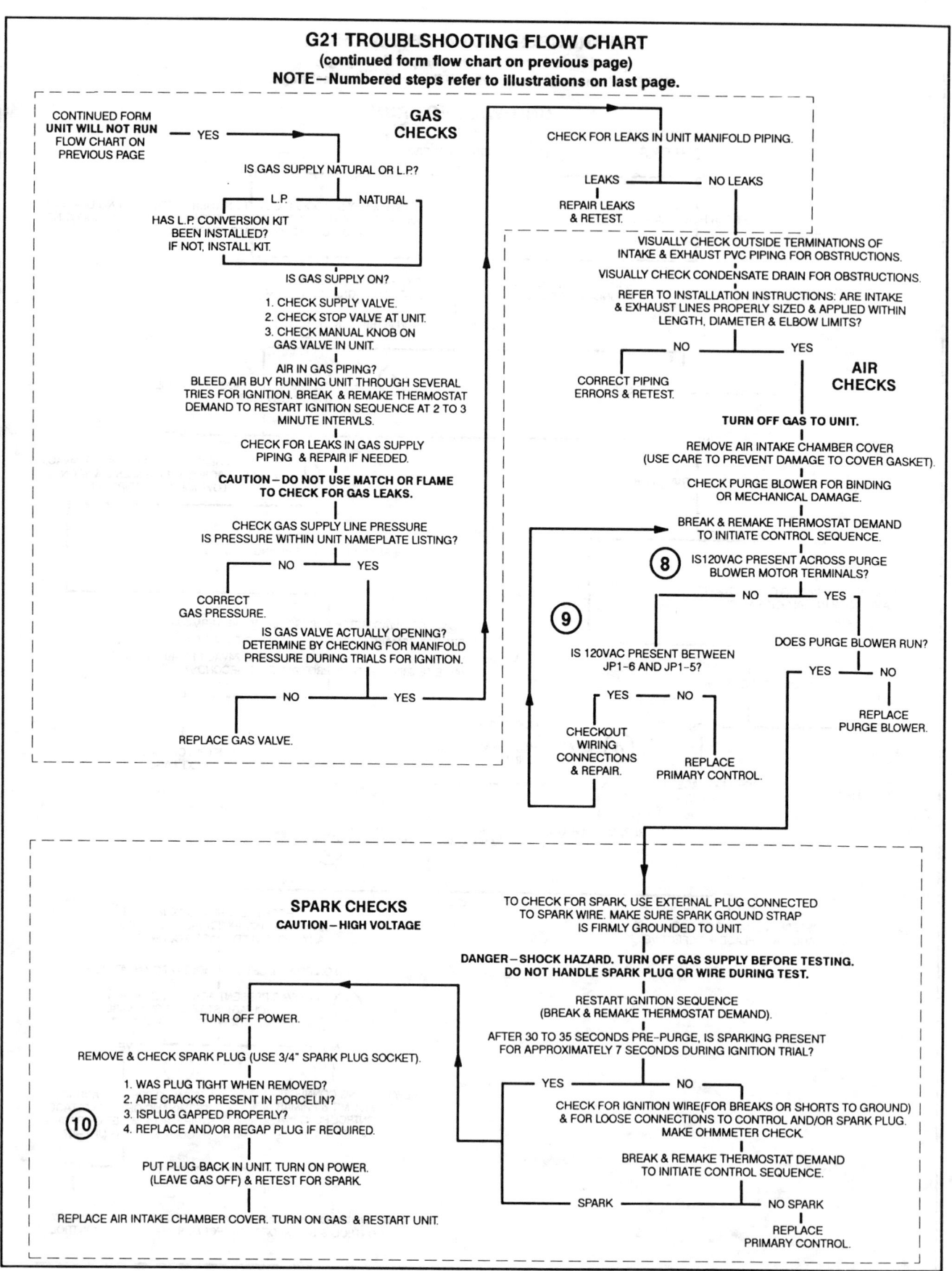

CONTINUED FORM **UNIT WILL NOT RUN** FLOW CHART ON PREVIOUS PAGE — YES →

GAS CHECKS

IS GAS SUPPLY NATURAL OR L.P.?

L.P. / NATURAL

HAS L.P. CONVERSION KIT BEEN INSTALLED? IF NOT, INSTALL KIT.

IS GAS SUPPLY ON?
1. CHECK SUPPLY VALVE.
2. CHECK STOP VALVE AT UNIT.
3. CHECK MANUAL KNOB ON GAS VALVE IN UNIT.

AIR IN GAS PIPING? BLEED AIR BUY RUNNING UNIT THROUGH SEVERAL TRIES FOR IGNITION. BREAK & REMAKE THERMOSTAT DEMAND TO RESTART IGNITION SEQUENCE AT 2 TO 3 MINUTE INTERVLS.

CHECK FOR LEAKS IN GAS SUPPLY PIPING & REPAIR IF NEEDED.

CAUTION—DO NOT USE MATCH OR FLAME TO CHECK FOR GAS LEAKS.

CHECK GAS SUPPLY LINE PRESSURE IS PRESSURE WITHIN UNIT NAMEPLATE LISTING?

NO / YES

CORRECT GAS PRESSURE.

IS GAS VALVE ACTUALLY OPENING? DETERMINE BY CHECKING FOR MANIFOLD PRESSURE DURING TRIALS FOR IGNITION.

NO / YES

REPLACE GAS VALVE.

CHECK FOR LEAKS IN UNIT MANIFOLD PIPING.

LEAKS / NO LEAKS

REPAIR LEAKS & RETEST.

VISUALLY CHECK OUTSIDE TERMINATIONS OF INTAKE & EXHAUST PVC PIPING FOR OBSTRUCTIONS.

VISUALLY CHECK CONDENSATE DRAIN FOR OBSTRUCTIONS.

REFER TO INSTALLATION INSTRUCTIONS: ARE INTAKE & EXHAUST LINES PROPERLY SIZED & APPLIED WITHIN LENGTH, DIAMETER & ELBOW LIMITS?

NO / YES

CORRECT PIPING ERRORS & RETEST.

AIR CHECKS

TURN OFF GAS TO UNIT.

REMOVE AIR INTAKE CHAMBER COVER (USE CARE TO PREVENT DAMAGE TO COVER GASKET).

CHECK PURGE BLOWER FOR BINDING OR MECHANICAL DAMAGE.

BREAK & REMAKE THERMOSTAT DEMAND TO INITIATE CONTROL SEQUENCE.

⑧ IS120VAC PRESENT ACROSS PURGE BLOWER MOTOR TERMINALS?

NO / YES

⑨ IS 120VAC PRESENT BETWEEN JP1-6 AND JP1-5?

YES / NO

CHECKOUT WIRING CONNECTIONS & REPAIR.

REPLACE PRIMARY CONTROL.

DOES PURGE BLOWER RUN?

YES / NO

REPLACE PURGE BLOWER.

SPARK CHECKS
CAUTION—HIGH VOLTAGE

TO CHECK FOR SPARK, USE EXTERNAL PLUG CONNECTED TO SPARK WIRE. MAKE SURE SPARK GROUND STRAP IS FIRMLY GROUNDED TO UNIT.

DANGER—SHOCK HAZARD. TURN OFF GAS SUPPLY BEFORE TESTING. DO NOT HANDLE SPARK PLUG OR WIRE DURING TEST.

RESTART IGNITION SEQUENCE (BREAK & REMAKE THERMOSTAT DEMAND).

AFTER 30 TO 35 SECONDS PRE-PURGE, IS SPARKING PRESENT FOR APPROXIMATELY 7 SECONDS DURING IGNITION TRIAL?

YES / NO

TUNR OFF POWER.

REMOVE & CHECK SPARK PLUG (USE 3/4" SPARK PLUG SOCKET).

⑩
1. WAS PLUG TIGHT WHEN REMOVED?
2. ARE CRACKS PRESENT IN PORCELIN?
3. ISPLUG GAPPED PROPERLY?
4. REPLACE AND/OR REGAP PLUG IF REQUIRED.

PUT PLUG BACK IN UNIT. TURN ON POWER. (LEAVE GAS OFF) & RETEST FOR SPARK.

REPLACE AIR INTAKE CHAMBER COVER. TURN ON GAS & RESTART UNIT.

CHECK FOR IGNITION WIRE(FOR BREAKS OR SHORTS TO GROUND) & FOR LOOSE CONNECTIONS TO CONTROL AND/OR SPARK PLUG. MAKE OHMMETER CHECK.

BREAK & REMAKE THERMOSTAT DEMAND TO INITIATE CONTROL SEQUENCE.

SPARK / NO SPARK

REPLACE PRIMARY CONTROL.

G21 TROUBLESHOOTING FLOW CHART — UNIT WILL NOT RUN

CHECK VOLTAGE AT TERMINAL AND CHECKING THERMOSTAT DEMAND

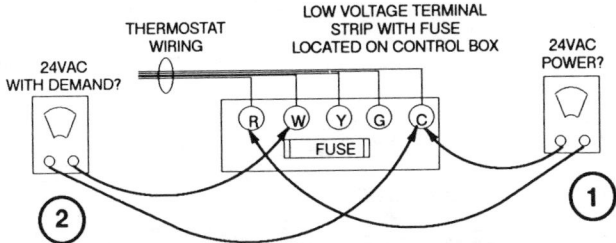

THERMOSTAT WIRING

LOW VOLTAGE TERMINAL STRIP WITH FUSE LOCATED ON CONTROL BOX

24VAC WITH DEMAND?

24VAC POWER?

R W Y G C

FUSE

② ①

CHECKING VOLTAGE AT GAS VALVE

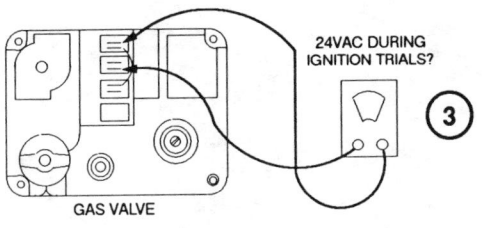

24VAC DURING IGNITION TRIALS?

③

GAS VALVE

CHECKING FOR OPEN SWITCH

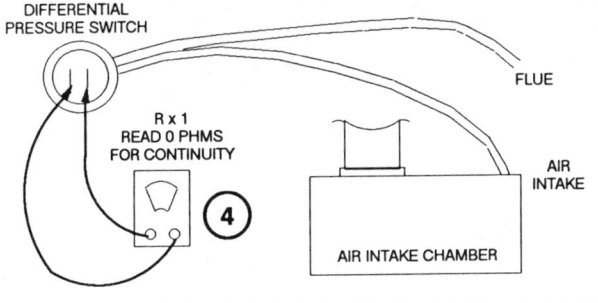

DIFFERENTIAL PRESSURE SWITCH

FLUE

R x 1 READ 0 PHMS FOR CONTINUITY

AIR INTAKE

④

AIR INTAKE CHAMBER

CHECKING FOR OPEN SWITCH IN LIMIT CONTROL

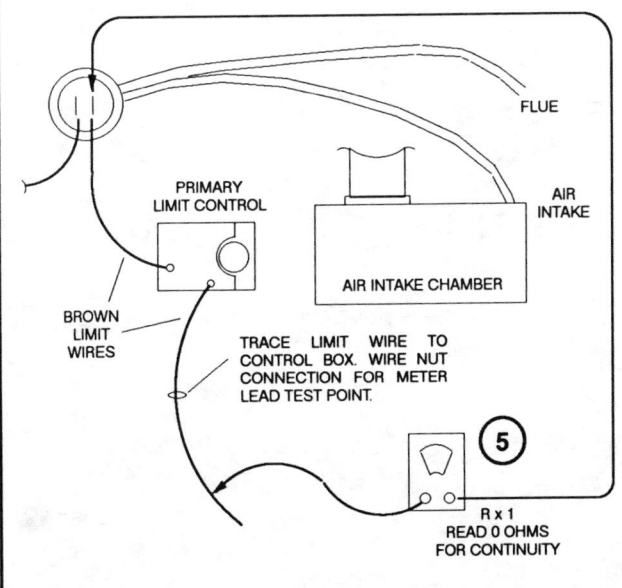

FLUE

PRIMARY LIMIT CONTROL

AIR INTAKE

AIR INTAKE CHAMBER

BROWN LIMIT WIRES

TRACE LIMIT WIRE TO CONTROL BOX. WIRE NUT CONNECTION FOR METER LEAD TEST POINT.

⑤

R x 1 READ 0 OHMS FOR CONTINUITY

CHECKING VOLTAGE AT PRIMARY CONTROL

PRIMARY CONTROL WIRING HARNESS PLUG (JP1)

24VAC?

⑥

WIRING HARNESS

③ ② ①
⑥ ⑤ ④

24VAC?

⑦

LOW VOLTAGE TERMINAL STRIP WITH FUSE LOCATED ON CONTROL BOX

R W Y G C

THERMOSTAT WIRING

CHECKING VOLTAGE AT PURGE BLOWER

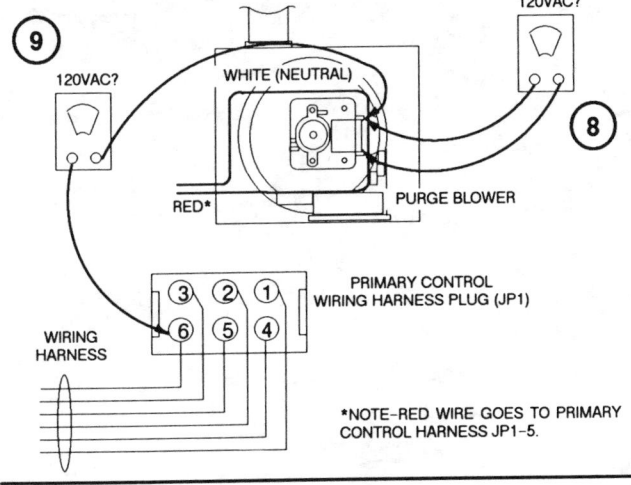

⑨

120VAC?

WHITE (NEUTRAL)

120VAC?

⑧

RED*

PURGE BLOWER

③ ② ①
⑥ ⑤ ④

PRIMARY CONTROL WIRING HARNESS PLUG (JP1)

WIRING HARNESS

*NOTE-RED WIRE GOES TO PRIMARY CONTROL HARNESS JP1-5.

SPARK PLUG

IT IS NORMAL FOR THE ELECTRODE TO PROTRUDE AT AN UNUSUAL ANGLE

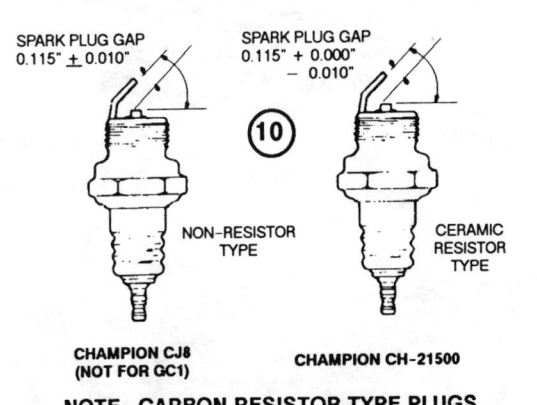

SPARK PLUG GAP 0.115" ± 0.010"

SPARK PLUG GAP 0.115" + 0.000" − 0.010"

⑩

NON-RESISTOR TYPE

CERAMIC RESISTOR TYPE

CHAMPION CJ8 (NOT FOR GC1)

CHAMPION CH-21500

NOTE- CARBON RESISTOR TYPE PLUGS SHOULD NOT BE USED.

ENGINEERING DATA

GAS FURNACES

G21V "PULSE21™" SERIES
UP-FLO GAS FURNACES
*91.7% to 93.5% A.F.U.E.
60,000 to 100,000 Btuh Input
Add-On Cooling — 1-1/2 thru 5 Nominal Tons
*Isolated Combustion System Rating for Non-Weatherized Furnaces

G21V

Bulletin #480176
November 1991

Typical Applications

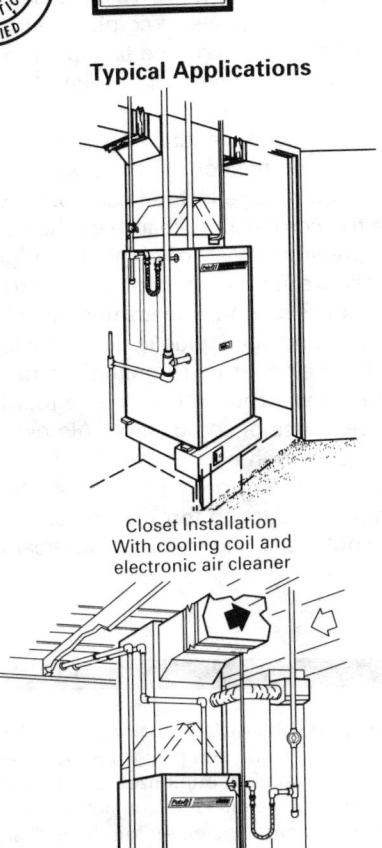

Closet Installation
With cooling coil and
electronic air cleaner

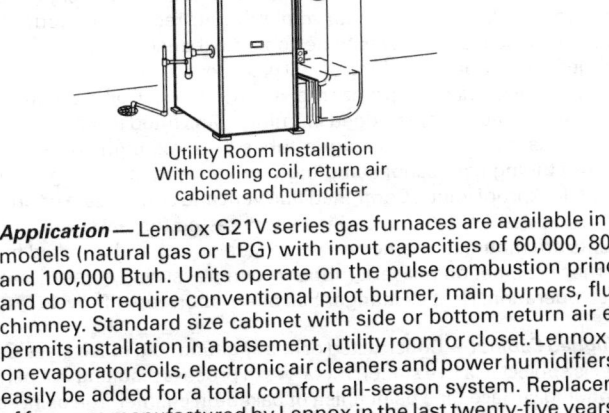

Utility Room Installation
With cooling coil, return air
cabinet and humidifier

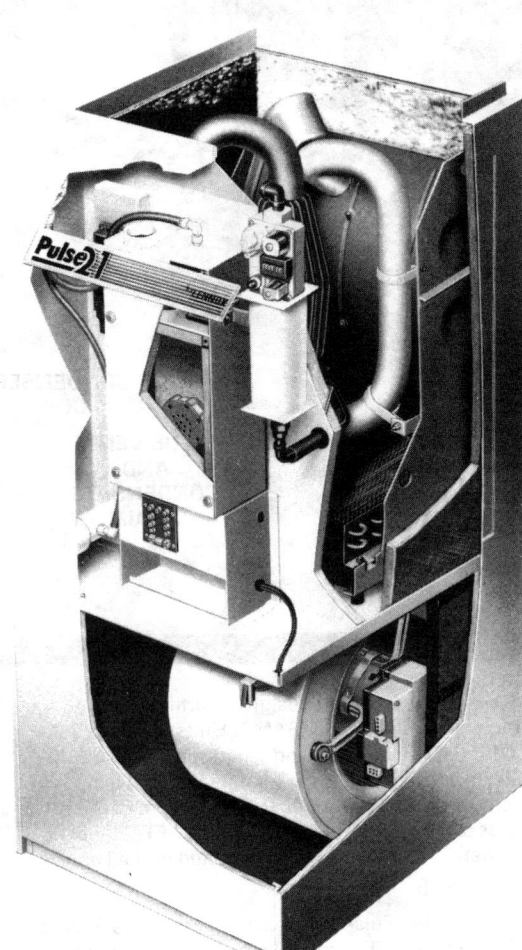

Application — Lennox G21V series gas furnaces are available in four models (natural gas or LPG) with input capacities of 60,000, 80,000 and 100,000 Btuh. Units operate on the pulse combustion principle and do not require conventional pilot burner, main burners, flue or chimney. Standard size cabinet with side or bottom return air entry permits installation in a basement, utility room or closet. Lennox add-on evaporator coils, electronic air cleaners and power humidifiers can easily be added for a total comfort all-season system. Replacement of furnaces manufactured by Lennox in the last twenty-five years can be done with only minor modification to duct work or add-on coils.

Electronically variable speed (VSM) blower motor maintains a specified air volume (cfm) throughout the entire external static range. G21V furnaces also feature a variable heat output in direct proportion to to amount of air delivered by the blower. Burner control is completely automatic. Units are also applicable to the Lennox Harmony Zone Control System and the Lennox Efficiency Plus Humidity Control System.

High efficiency of the G21V series is achieved with a unique heat exchanger design which features: finned cast iron combustion chamber, temperature resistant steel tailpipe, aluminized steel exhaust decoupler section and a finned stainless steel tube condenser coil. Moisture, during the process of combustion, is condensed in the coil, extracting almost every usable Btu out of the gas. Most of the combustion heat is utilized in the heat transfer from the coil, producing flue vent temperatures as low as 100F to 130F which allows the use of PVC (polyvinyl chloride) pipe for venting. Furnace can be vented through a side wall, roof or to the top of an existing chimney with up

to 35 ft. of PVC pipe and up to four 90 degree elbows. Condensate created in the coil (PH ranges from 4.0 to 6.0) is not harmful to standard household plumbing and can be drained into city sewers and septic tanks without damage.

The G21V furnace has no pilot light or burners. An automotive type spark plug is used for ignition on the initial cycle only, saving gas and electrical energy. In the pulse combustion process, the use of atmospheric burners is eliminated, with combustion confined to heat exchanger combustion chamber. Sealed combustion system virtually eliminates the loss of conditioned air due to combustion and stack dilution. Combustion air is piped to the furnace with same type PVC pipe as used for exhaust gases.

Furnace is equipped with a standard type redundant two-stage gas valve in series with a gas expansion tank and gas intake flapper valve. Also factory installed are a air intake flapper valve, purge blower, spark plug igniter, flame sensor with solid-state control, solid-state blower control, limit/modulation control, 50VA transformer, high and low voltage terminal strip and cleanable air filter. Furnished for field installation are a flexible gas line connector, (4) isolation mounting pads, base insulation pad, external side return air filter mounting kit and condensate drip leg.

Optional equipment available are: flue vent/air intake line roof or wall termination installation kits, LPG conversion kits, mufflers and thermostat.

G21V units are shipped completely factory assembled with all controls installed and wired. Units are test fired at the factory before shipment.

©1991 Lennox Industries Inc.

NOTE — Specifications, Ratings and Dimensions subject to change without notice.

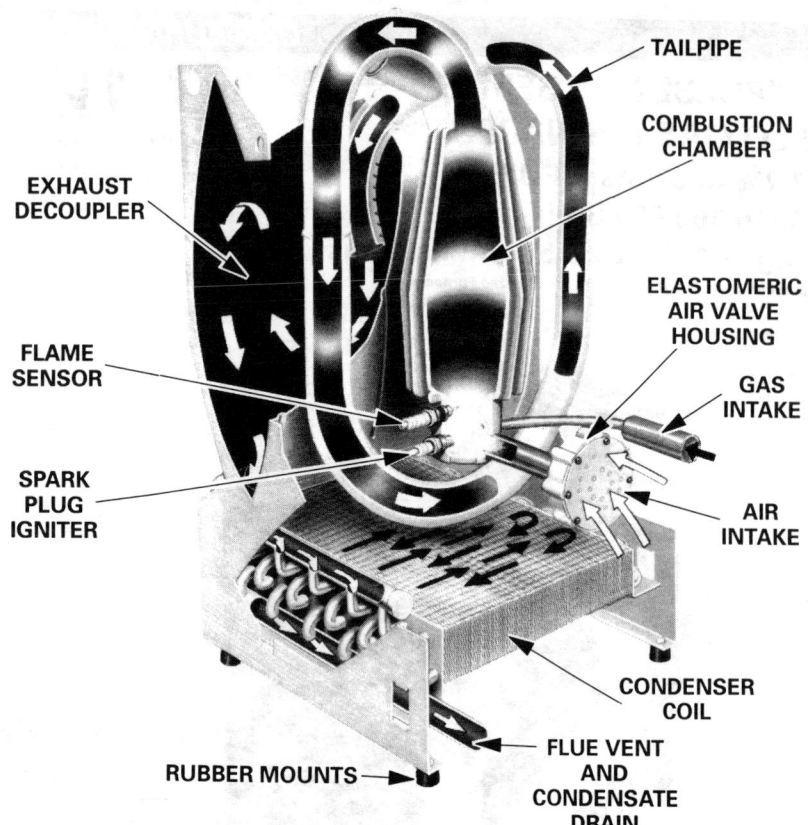

TAILPIPE

COMBUSTION CHAMBER

EXHAUST DECOUPLER

ELASTOMERIC AIR VALVE HOUSING

FLAME SENSOR

GAS INTAKE

SPARK PLUG IGNITER

AIR INTAKE

CONDENSER COIL

RUBBER MOUNTS

FLUE VENT AND CONDENSATE DRAIN

The process of combustion begins as gas and air are introduced into the sealed combustion chamber with the spark plug igniter. Spark from the plug ignites the gas/air mixture, which in turn causes a positive pressure buildup that closes the gas and air inlets. This pressure relieves itself by forcing the products of combustion out of the combustion chamber through the tailpipe into the heat exchanger exhaust decoupler and on into the heat exchanger coil. As the combustion chamber empties, its pressure becomes negative, drawing in air and gas for the next pulse of combustion. At the same instant, part of the pressure pulse is reflected back from the tailpipe at the top of the combustion chamber. The flame remnants of the previous pulse of combustion ignites the new gas/air mixture in the chamber, continuing the cycle. Once combustion is started, it feeds upon itself allowing the purge blower and spark plug igniter to be turned off. Each pulse of gas/air mixture is ignited at a rate of 60 to 70 times per second, producing from one-fourth to one-half of a Btu per pulse of combustion. Almost complete combustion occurs with each pulse. The force of these series of ignitions creates great turbulence which forces the products of combustion through the entire heat exchanger assembly resulting in maximum heat transfer.

FEATURES

Approvals — G21V series furnaces are designed certified by A.G.A. Laboratories and ratings are certified by GAMA. Units meet the California Nitrogen Oxides (NO$_x$) standards and California Seasonal Efficiency requirements. In addition, units have been rated and tested in the Lennox Research Laboratory according to Department of Energy (DOE) test procedures and Federal Trade Commission (FTC) labeling regulations. Blower data is from unit tests conducted in the Lennox Laboratory air test chamber.

Equipment Warranty — G21V "Pulse" heat exchangers have a limited lifetime warranty. Solid-state ignition modules have a limited warranty for three years. All other components have a limited warranty for one year. Refer to Lennox Limited Equipment Warranty certificate included with the equipment for details.

Sequence of Operation — Room thermostat, on a demand for heat, will initiate purge blower operation for a pre-purge cycle (30 seconds) followed by energizing and opening of the gas valve. As ignition occurs, the flame sensor reacts to proof of ignition and de-energizes the spark plug igniter and purge blower. Furnace blower operation is initiated 45 seconds after combustion ignition. When thermostat is satisfied, gas valve is closed and purge blower is re-energized for a post-purge cycle (34 seconds). Furnace blower will remain in operation until "fan off" factory setting of 330 seconds (adjustable from 90 to 330 seconds) is reached. Should loss of flame occur before thermostat is satisfied, flame sensor controls will initiate 5 attempts at re-ignition before locking out unit operation. Additionally, loss of either combustion intake air or flue exhaust will automatically terminate system operation. If unit becomes locked out, Watchguard control automatically resets ignition controls after one hour.

Heat Exchanger Assembly — Lennox developed heat exchanger assembly consists of combustion chamber, tailpipe, exhaust decoupler section and condenser coil. Combustion chamber contains the spark plug igniter, flame sensor and combustion air and gas intake manifolds. Cast iron construction provides excellent radiation of heat over entire surface area. Finned "teardrop" shape design permits total air coverage of all surfaces with low resistance. Tailpipe connects the combustion chamber to the exhaust decoupler section. Precisely

sized and shaped tailpipe is constructed of combination stainless and aluminized steel for superior resistance to high temperatures. Aluminized steel resonator on tailpipe minimizes combustion sound. Heavy gauge aluminized steel exhaust decoupler section has large surface area for maximum heat transfer. Air foil shape design results in complete air coverage with minimum air resistance. Condenser coil intake header connects to bottom of exhaust decoupler section. Large face area and circuiting of coil provides high heat transfer, minimum air resistance and proper moisture drainage. Coil is constructed of exactly spaced ripple-edged aluminum fins fitted to stainless steel tubes. Flared collars on fins grip tubes for maximum contact area. Flared tubing connections and high temperature soldering provide tight, leakproof joints. Combined flue vent and condensate drain outlet is located on the coil. Coil is factory tested for leaks. All components are mounted in a heavy gauge steel frame and installed in the furnace cabinet on resilient rubber mounts assuring quiet, vibration free operation. Heat exchanger has been laboratory life cycle tested.

Rugged Cabinet — Constructed of heavy gauge cold rolled steel. Cabinet is subject to a five station metal wash process resulting in a perfect bonding surface for a paint finish of baked-on enamel. The paint solution and metal are given opposite electrical charges resulting in positive adhesion and even coverage of the paint to the metal surfaces. Heat exchanger section is completely lined with thick (1-1/2 lb./ft.^{3} density) foil faced fiberglass insulation. Blower compartment is completely lined with thick (1-1/2 lb./ft.^{3} density) black mat faced fiberglass insulation. This results in quiet and efficient operation due to the excellent acoustical and insulating properties of fiberglass. Complete service access is accomplished by removing heating section and blower access panels. Removable panel is provided in vestibule panel for access to the spark plug and flame sensor. Holes are located in the base for cabinet leveling. Leveling bolts and nuts are not provided and must be ordered extra. Safety interlock switch automatically shuts power off to unit when blower access panel is removed. Blower assembly may be completely removed from unit for servicing. Electrical inlets, gas line inlets, air intake and exhaust air outlets are provided in both sides of cabinet. Combustion air inlet opening is located in cabinet cap. Return air duct connection can be made on either side or bottom of cabinet.

Powerful Blowers — Units are equipped with quiet multi-speed direct drive blowers. Each blower assembly is statically and dynamically balanced. Change in blower speed is easily accomplished by simple wiring change on VSM motor.

Variable Speed (VSM) Blower Motor — Electronically variable speed (VSM) motor is resiliently mounted. Electronic control on motor allows blower to operate at three of the eleven speeds or air volumes available. The three speeds or air volumes may be field selected depending on size of application and air volume required. See blower performance tables. When units are used with the Harmony Zone Control System, blower motor operates between low and high speed settings depending on number of zones operating.

VSP-1 Solid-State Indoor Blower Control — Circuit board located in wiring junction box contains all necessary controls to automatically operate the furnace. Contains blower timed-on control (45 seconds fixed) and adjustable blower timed-off control (90 to 330 seconds). Blower operation is automatic if limit is tripped. Board also contains a 110v accessory terminal to operate accessories. Three service LED's on board indicate proper system operation.

Wiring Junction Box — Power supply and thermostat connections are made at the wiring junction box located on the vestibule panel. Box contains 50 VA transformer, high and low voltage terminal strips and blower cooling relay. Low voltage terminal strip has a fuse to protect the transformer. Terminal strip permits easy connections for optional power humidifiers and electronic air cleaners. Blower cooling relay activates blower operation for add-on air conditioning cooling.

Combustion Air Intake Box — Contains the purge blower, air intake flapper valve and air valve housing. The 40, 60 and 80 units have a single differential pressure switch mounted inside the unit cabinet. The 100 models have a single differential pressure switch mounted on the vestibule panel. Box is located on vestibule panel. Purge blower is equipped with a permanently lubricated motor. Blower operates only during pre-purge and post-purge cycles. Air is drawn through the blower during the combustion cycle by negative pressure in the combustion chamber. Pressure switches terminate unit operation in case of air intake or flue exhaust blockage. Flapper valve air housing is constructed of an elastomeric non-metallic polymer which reduces operating sound levels. Flapper valve section of the box is completely lined with 1 inch thick (6 lb./ft.^{3} density) duct liner board, black neoprene coated fiberglass. Valve opening and closing is actuated by back pressure and negative pressure in combustion chamber during the heating cycle.

Automatic Two-Stage Gas Valve, Expansion Tank and Gas Intake Flapper Valve — 24 volt redundant two-stage gas control valve combines gas pressure regulation and manual main shutoff valve into one compact combination control. Dual valve design provides double assurance of 100% close off of gas on each heating cycle. Expansion tank is located downstream from the gas valve and absorbs any pressure pulsations. Gas intake flapper valve is installed in the combustion chamber intake manifold between the orifice and expansion tank. Valve is opened by entering gas pressure and closed by back pressure from combustion pulse during the heating cycle. Two-stage gas valve provides a variable heating output for zoned applications or two-stage heating applications.

Ignition Control — Solid-state control provides power for spark plug igniter. Also controls pre-purge and post-purge cycles and re-ignition sequence if loss of flame occurs. Ignition control is factory installed on the vestibule panel.

WatchGuard Control — Solid-state control provides automatic reset of ignition controls after 1 hour of continuous thermostat demand after unit lockout. Factory installed on the vestibule panel.

Limit/Modulation Control — Factory installed and accurately located on vestibule. Fixed limit control provides positive protection from abnormal operating conditions. Automatic reset. Modulation control allows combustion process to cycle on and off to maintain even supply air temperature, while saving energy by utilizing residual heat in the heat exchanger.

Cleanable Air Filters — Washable or vacuum cleanable frame type filter is furnished as standard. Polyurethane media is coated with oil for maximum efficiency. Filter is readily accessible in unit for quick and easy removal for servicing.

External Side Return Air Filter Cabinet (Furnished) — External filter cabinet is furnished for installing air filter external to unit cabinet on side return air applications. Heavy gauge cold rolled steel filter rack assembly, with baked-on enamel finish, field installs on either side of unit cabinet with existing screws. Cabinet utilizes existing filter supplied with unit. Rack has flanges for ease of duct connection.

Installation Recommendations — Lennox recommends the following installation procedures to minimize any vibration transmitted from furnace during operation. Place (4) neoprene rubber isolation mounting pads (furnished) and/or base insulation pad (furnished), 1 inch thick (1-1/2 lb./ft.^{3} density) fiberglass, under the unit. Install flexible duct connectors in the supply air plenum and return air plenum or duct connection. Insulate (1 inch thick, 1-1/2 to 3 lb./ft.^{3} density, mat faced fiberglass) supply and return air plenums through take-off or duct elbow. Use flexible gas connector (furnished) in gas supply piping where allowed by local codes. Insulate (refrigerant piping insulation or equivalent) all straps and hangers used in suspending ducts, electrical conduit, gas piping, combustion air intake piping and flue exhaust piping. In addition, use plastic pipe or tubing for drain line from the heat coil condensate drain leg (furnished) to the drain, do not use copper tubing.

OPTIONAL ACCESSORIES (Must Be Ordered Extra)

Thermostat (Optional) — Heating thermostat is not furnished and must be ordered extra. See Thermostats bulletin in Accessories Section. For non-zoned applications, a two-stage heating thermostat may be used for dual levels of air volume and heating output. For all-season applications, heating-cooling thermostat is available with the condensing unit.

In-Line Mufflers (Optional) — Two mufflers (LB-52057CA) are optional and must be ordered extra. Mufflers field install, vertical or horizontal, one in the intake line and one in the exhaust line.

LPG Conversion Kits (Optional) — For LPG models a conversion kit is available for field changeover from natural gas. Kit is not furnished and must be ordered extra. See Specifications tables.

Concentric Vent/Intake Air Roof/Wall Termination Kit (Optional) — Facilitates installation of combustion air intake pipe and flue exhaust pipe. Kit (LB-49107CE) contains concentric termination assembly, mounting clamp, roof flashing, reducer bushing and 45 degree elbow. Kit requires single hole penetration of roof or wall for installation. Kit is A.G.A. certified and must be ordered extra for field installation.

Vent/Intake Air Roof Termination Kit (Optional) — Facilitates installation of combustion air intake pipe and flue exhaust pipe. Kit contains two neoprene rubber roof flashings and 18 inch insulation sleeve for sealing and isolating intake and exhaust piping penetration in roof. Kit (LB-49107CC) must be ordered extra for field installation.

Vent/Intake Air Wall Termination Kit (Optional) — Facilitates installation of combustion air intake pipe and flue exhaust pipe. Kit must be ordered extra. Select one of the following:

1 — Kit (LB-49107CB) contains 2 stainless steel outside seal caps, 2 galvanized steel inside seal caps, 4 seal rings for the caps and 18 inch insulation sleeve for sealing and isolating intake and exhaust piping penetration of wall. Maintain a maximum of 6 inches between the inlet and outlet openings in the installation of the pipes.

2 — Kit (LB-49107CD) consists of close-couple side-by-side PVC piping with galvanized steel wall cover plate for sealing and isolating piping penetration of the wall. Piping spacing and length is sized for proper wall installations. A.G.A. certified.

SPECIFICATIONS

Model No.		G21V3-60	G21V3-80	G21V5-80	G21V5-100
Input Btuh		60,000	80,000	80,000	100,000
Output Btuh		55,000	73,000	74,000	95,000
*A.F.U.E.		92.4%	92.7%	91.7%	93.5%
California Seasonal Efficiency		88.1%	89.4%	87.2%	89.8%
Temperature rise range (°F)		40 — 70	45 — 75	35 — 65	40 — 70
High static certified by A.G.A. (in wg.)		.80	.80	.80	.80
Gas Piping Size	Natural	1/2	1/2	1/2	1/2
I.P.S. (in.)	**LPG	1/2	1/2	1/2	1/2
Vent/Intake air pipe size connection (in.)		2	2	2	2
Condensate drain connection (in.) SDR11		1/2	1/2	1/2	1/2
Blower wheel nominal diameter x width (in.)		10 x 8	10 x 8	11-1/2 x 9	11-1/2 x 9
Blower motor hp		1/2	1/2	1	1
Number and size of filters (in.)		(1) 16 x 25 x 1	(1) 16 x 25 x 1	(1) 20 x 25 x 1	(1) 20 x 25 x 1
Tons of cooling that can be added		1-1/2, 2, 2-1/2 or 3	2, 2-1/2 or 3	3-1/2, 4 or 5	3-1/2, 4 or 5
Shipping weight (lbs.)		250	250	297	297
Number of packages in shipment		1	1	1	1
Electrical characteristics		120 volts — 60 hertz — 1 phase (less than 12 amps) All models			
External Filter Cabinet (furnished) •Filter size (in.)		(1) 16 x 25 x 1	(1) 16 x 25 x 1	(1) 20 x 25 x 1	(1) 20 x 25 x 1
**LPG kit (optional)		LB-83176CE	LB-83176CF	LB-83176CF	LB-83176CP
Furnace Twinning Kit (optional)		LB-63093CA (All models)			

•Filter is not furnished with cabinet. Filter cabinet utilizes existing filter supplied with G21V unit.
*Annual Fuel Utilization Efficiency based on D.O.E. test procedures and according to F.T.C. labeling regulations. Isolated combustion system rating for non-weatherized furnaces.
**LPG kit must be ordered extra for field changeover.

BLOWER DATA

G21V3-60, G21V3-80 BLOWER PERFORMANCE
FACTORY BLOWER SPEED SETTINGS

G21V3-60		G21V3-80	
Low Speed Cooling	— 2	Low Speed Cooling	— 2
High Speed Cooling	— 11	High Speed Cooling	— 11
Heating Speed	— 6	Heating Speed	— 7

External Static Pressure (in. wg.)	Air Volume (cfm) @ Various Speeds										
	Speed 1	Speed 2	Speed 3	Speed 4	Speed 5	Speed 6	Speed 7	Speed 8	Speed 9	Speed 10	Speed 11
0 thru .80	- - - -	490	635	760	880	1030	1140	1220	1345	1420	1420

NOTE — All air data is measured external to the unit with the air filter in place.

G21V5-80, G21V5-100 BLOWER PERFORMANCE
FACTORY BLOWER SPEED SETTINGS

G21V5-80		G21V5-100	
Low Speed Cooling	— 2	Low Speed Cooling	— 2
High Speed Cooling	— 11	High Speed Cooling	— 11
Heating Speed	— 6	Heating Speed	— 7

External Static Pressure (in. wg.)	Air Volume (cfm) @ Various Speeds										
	Speed 1	Speed 2	Speed 3	Speed 4	Speed 5	Speed 6	Speed 7	Speed 8	Speed 9	Speed 10	Speed 11
0 thru .80	- - - -	770	1015	1305	1510	1685	1820	2010	2050	2100	2100

NOTE — All air data is measured external to the unit with the air filter in place.

A.G.A. INSTALLATION CLEARANCES

Sides	1 inch
Rear	1 inch
Top	1 inch
Front	1 inch
Floor	Combustible
Exhaust Pipe	0 inches
Exhaust Pipe Side	6 inches (service only)

DIMENSIONS (inches)

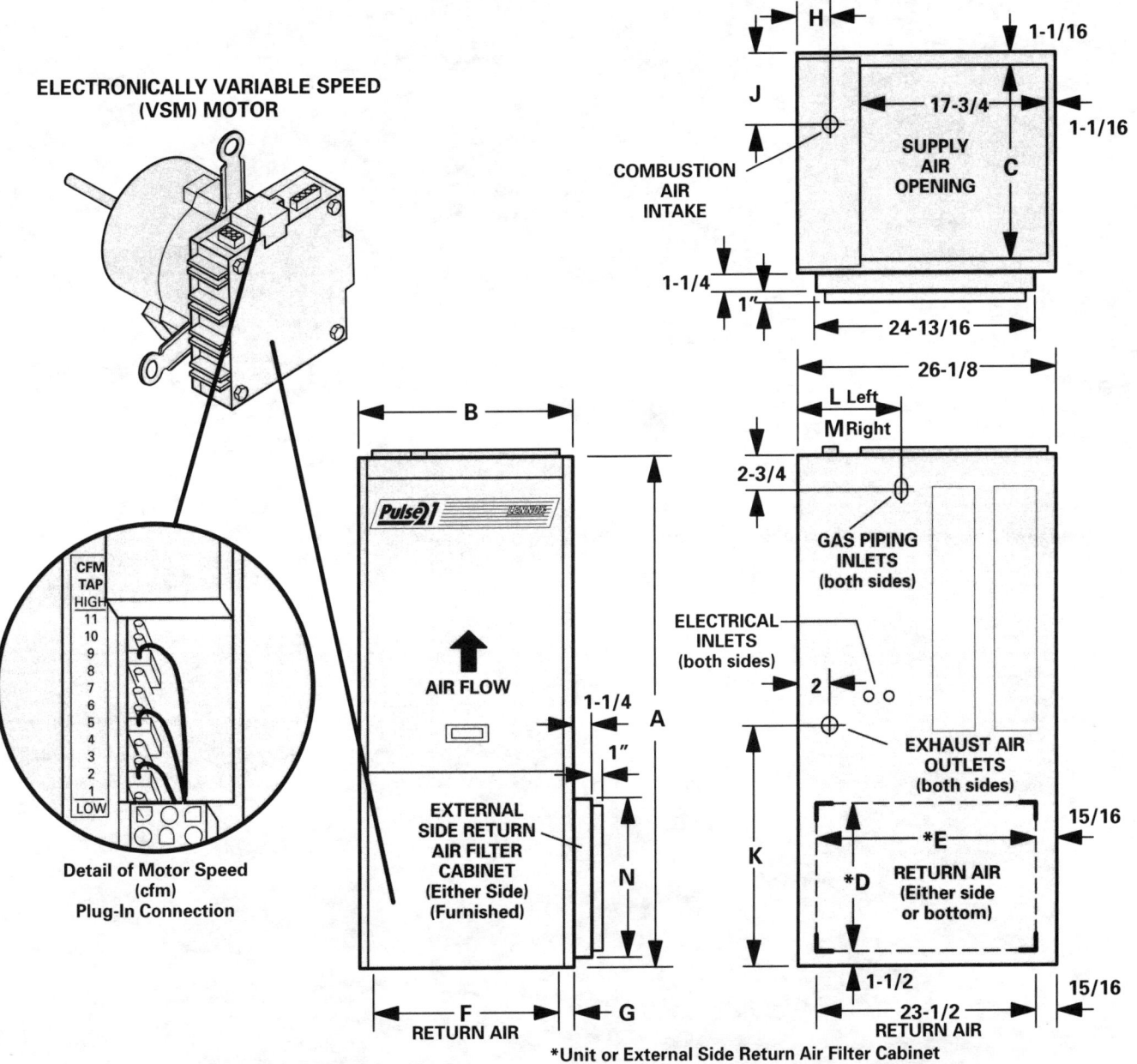

ELECTRONICALLY VARIABLE SPEED
(VSM) MOTOR

Detail of Motor Speed
(cfm)
Plug-In Connection

COMBUSTION AIR INTAKE

SUPPLY AIR OPENING

AIR FLOW

EXTERNAL SIDE RETURN AIR FILTER CABINET (Either Side) (Furnished)

GAS PIPING INLETS (both sides)

ELECTRICAL INLETS (both sides)

EXHAUST AIR OUTLETS (both sides)

RETURN AIR (Either side or bottom)

*Unit or External Side Return Air Filter Cabinet

Model No.	A	B	C	D	E	F	G	H	J	K	L	M	N
G21V3-60 G21V3-80	49	21-1/4	19-1/8	14-1/2	18-1/2	14-1/2	3-3/8	4-1/2	8-1/2	20-1/4	7-1/4	5-1/4	16
G21V5-80 G21V5-100	53	26-1/4	24-1/8	18-1/2	23-1/2	18-1/2	3-7/8	2-1/2	11	24-1/4	4-5/8	4-5/8	20

—5—

LENNOX®
ENGINEERING DATA

ACCESSORIES
HARMONY

Bulletin #480182
February 1992
Supersedes
November 1991

HARMONY™ ZONE CONTROL SYSTEM FOR VARIABLE AIR VOLUME HEATING/COOLING SYSTEMS

Harmony Zone Control System — Harmony system is designed to provide up to four separate heating/cooling zones utilizing a single heating unit (G21V or GSR21V) and single condensing unit (HS14 two-speed). A single speed condensing unit may be used for two zone applications. Two-speed condensing unit may be used for two, three or four zone applications. The system consists of the Harmony control center, Harmony control panel, master thermostat and duct mounted zone dampers with a thermostat in each zone. The zone dampers, in response to a demand from the zone thermostats, automatically open to supply air flow to each zone. At the same time, the variable speed motor (VSM) in the G21V/GSR21V furnace automatically adjusts the air volume to the zones as required. Because of the VSM motors' ability to vary system air volume as required, no bypass damper is required. Individual air volumes for heating or cooling are available to each zone. Dampers are available in either round or rectangular configuration. Each zone is sized for the heating/cooling load. Damper operation and blower air volume is controlled by the control center. Control panel is furnished with the control center. All other components must be ordered separately. The Harmony system saves energy by allowing temperature setback in unoccupied areas while maintaining comfort in occupied areas. System also results in lower equipment costs by eliminating the need for two separate heating/cooling systems.

Control Center — Harmony control center (83H48) contains all necessary relays and controls to operate the system. Control center cabinet is constructed of heavy gauge steel with a enamel paint finish and removeable latching cover. Solid-state circuit board features: low voltage output terminals for four damper zones, furnace and condensing unit connections and low voltage input terminals for four zone thermostats, control panel, furnace, zone damper transformer and pressure switches. Board also has jumper selectors for "Zone Air Volume Selection" (25-95%), "Heating Air Reduction" (0%, 20% or 40%) and "VSM Motor Size" (1/2 or 1 hp). LED's on circuit board indicate "Heating" (Red) or "Cooling" (Green) for each zone. Additional LED's (Red) on board indicate "Zone Damper Operation", "Furnace Operation", "Blower Operation" and "Condensing Unit Operation". A series of four diagnostic LED's are furnished as an aid in servicing system. Up to five dampers may be connected in parallel on each zone without any additional controls (maximum system total of seven). Built-in time delay function prevents short cycling of system. Holes for mounting are furnished and electrical inlets are provided in top and bottom of panel. Dimensions: 13-1/4" x 10" x 1-3/4". Power requirements: 24VAC. Control center is powered by furnace transformer. Shipping weight: 6 lbs.

Control Panel — Furnished as standard with control center. Touch sensitive panel features: "Central" or "Zone" control mode selection, "Heating" or "Cooling" mode selection (only available during "Zone" mode operation), "Auto" or "On" fan control for continuous or intermittent blower operation and system "Off" switch. LED indicators show selected features at a glance. Panel is constructed of high impact Cycolac. Dimensions: 3-1/2" x 4-3/4" x 1-5/16".

System Equipment Data — See flow chart on page 4 for system equipment selection. For G21V/GSR21V series furnace data, see section Heating Units — Gas. For HS14 series two-speed or other single speed condensing units data, see Cooling Units — Condensing Units section. For add-on evaporator coil unit data, see section, Cooling Units — Coils-Blower Coil Units. For EMD14-65 or EMD14M-65 Economizer dampers, see Accessories section.

Sequence of Operation — Two modes of operation are available at the control panel — a "Central" control mode and a "Zone" control mode.

In the "central" mode, heating or cooling selection and temperature demand are controlled by the zone 1 (master) thermostat, all dampers remain open at mechanically preset openings and blower is controlled by the central control board (total zone air volume), delivering air to all zones.

In the "zone" mode, heating or cooling selection is controlled by the control panel, temperature demand is controlled by the individual zone thermostats, zone dampers route air to appropriate zones and blower air volume is adjusted according to central control jumper setting and number of zones operating. Dampers close if there is no demand to that zone.

Individual air volume for each zone is preset at the control center and may be adjusted from 25% to 95% actual air volume. During heating mode, Heating Air Reduction jumpers allow a lower heating air volume to be selected (0%, 20% or 40% reduction of cooling air volume). During continuous ("On") blower operation, all dampers remain open if there is no demand from thermostats.

ZONE SYSTEM
CONTROL
CENTER
(cover removed)

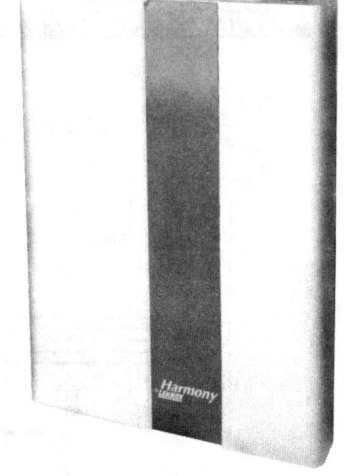

ZONE SYSTEM
CONTROL
CENTER
(cover in place)

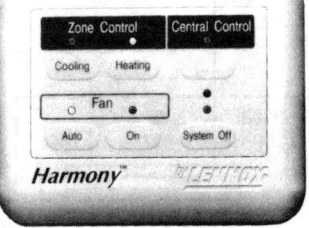

ZONE SYSTEM
CONTROL
PANEL

NOTE — Specifications, Ratings and Dimensions subject to change without notice.

©1992 Lennox Industries Inc.

ROUND
ZONE DAMPER

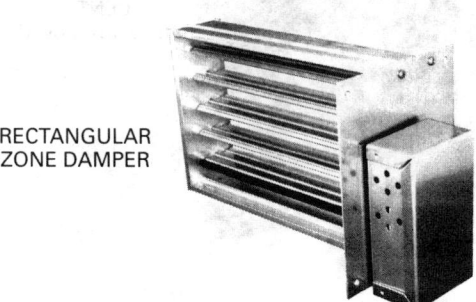

RECTANGULAR
ZONE DAMPER

Master (Zone 1) Thermostat (Optional) — Lennox recommends the use of a programmable thermostat for zone 1 master thermostat. See flowchart on page 4 for recommended thermostats. Also see Thermostats bulletin in Accessories section.

Zone Thermostats (Optional) — Lennox recommends the use of a single stage heating/single stage cooling thermostat for each zone. See flowchart on page 4 for recommended thermostats. Also see Thermostats bulletin in Accessories section.

Round Zone Damper (Optional) — Round damper is constructed of heavy gauge galvanized steel. Damper shell is furnished with one straight end and one crimped end for ease of duct connection. Damper blade rotates smoothly in nylon bearings. Adjustable blade stop is furnished on damper blade for system balancing. Damper features factory installed, heavy duty, synchronous motor with spring return open. Heavy duty steel gearing provides long motor life. Damper springs open in case of power failure. See damper specifications table for sizes, air resistance and shipping weights. Power requirements: 24 VAC.

Rectangular Zone Damper (Optional) — Rectangular damper is constructed of heavy gauge aluminum and stainless steel. Damper is a slip-in, opposed blade type with duct mounting plate on one end for ease of duct connection. Damper rotates smoothly in nylon bearings. A rubber blade stop is furnished for installation on damper blade if system balancing is required. Damper features factory installed, heavy duty, synchronous motor with spring return open. Heavy duty steel gearing provides long motor life. Damper springs open in case of power failure. See damper specifications table for sizes, air resistance and shipping weights. Power requirements: 24 VAC.

Pressure Switches (Optional) — Low Pressure Switch (72H50) and High Pressure Switch (72H51) are required for proper HS14 two-speed condensing unit operation. Switches determine low or high speed compressor operation, based on evaporator load.

Freezestat (Optional) — Freezestat (93G35) is required with single speed condensing units. Prevents condensing unit operation in case of abnormal operating conditions.

Transformer (Optional) — Transformer is required for operation of zone dampers. See flowchart on page 4.

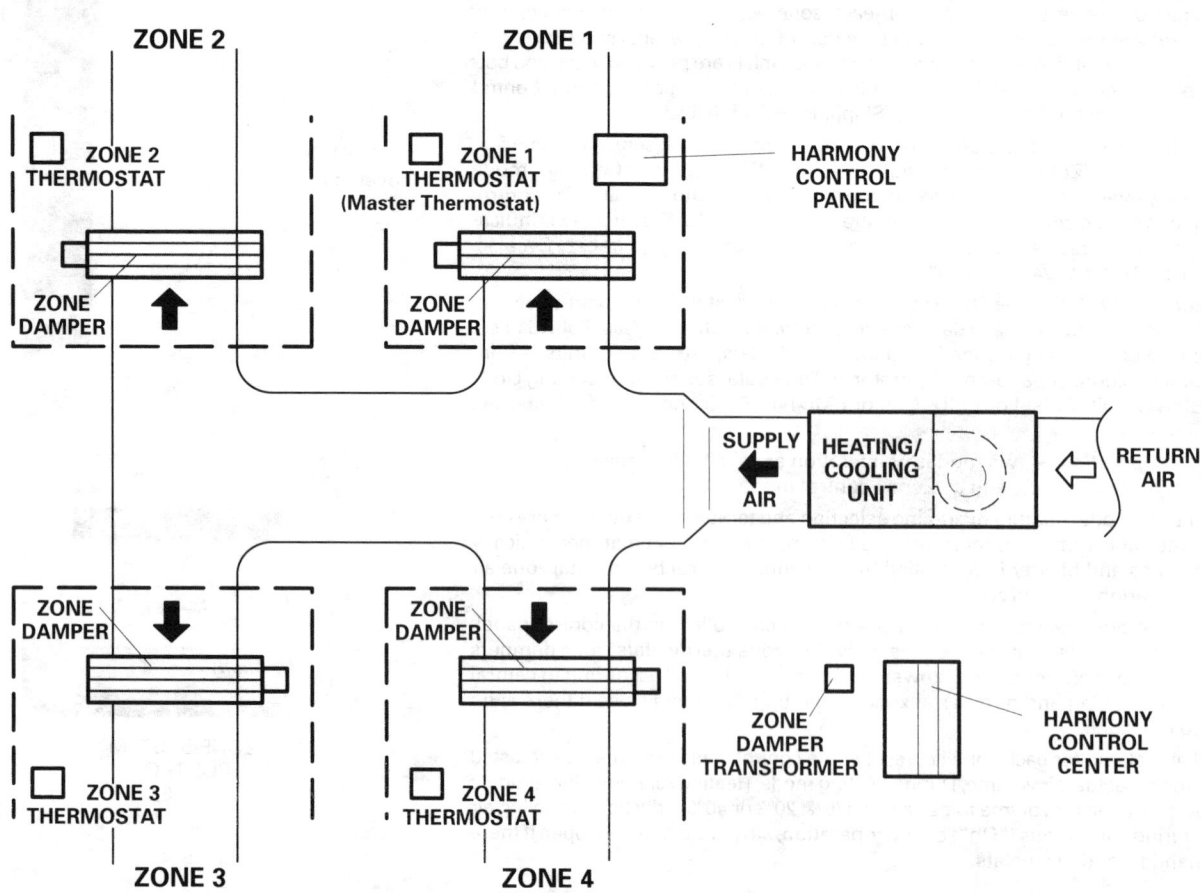

DAMPER SPECIFICATIONS

ROUND ZONE DAMPERS

Model No.	Dimensions Dia. (in.)	Air Volume (cfm)	Total Resistance (In. wg.)	Shipping Weight (lbs.)
DR-06	6	50	.01	2
		100	.04	
		110	.05	
DR-08	8	100	.02	4
		150	.03	
		210	.05	
DR-10	10	100	.01	6
		200	.02	
		325	.05	
DR-12	12	200	.02	8
		350	.03	
		460	.05	
DR-14	14	200	.01	10
		400	.02	
		640	.06	

RECTANGULAR ZONE DAMPERS

Model No.	Dimensions W x H (in.)	Air Volume (cfm)	Total Resistance (In. wg.)	Shipping Weight (lbs.)
DS-1008	10 X 8	100	.01	4
		200	.02	
		325	.05	
DS-1208	12 X 8	120	.01	3
		240	.02	
		395	.05	
DS-1408	14 X 8	200	.02	3
		350	.03	
		460	.05	
DS-1608	16 X 8	225	.02	4
		395	.03	
		520	.05	
DS-1808	18 X 8	185	.01	4
		375	.02	
		600	.06	
DS-2008	20 X 8	210	.01	4
		415	.02	
		665	.06	

FIELD WIRING

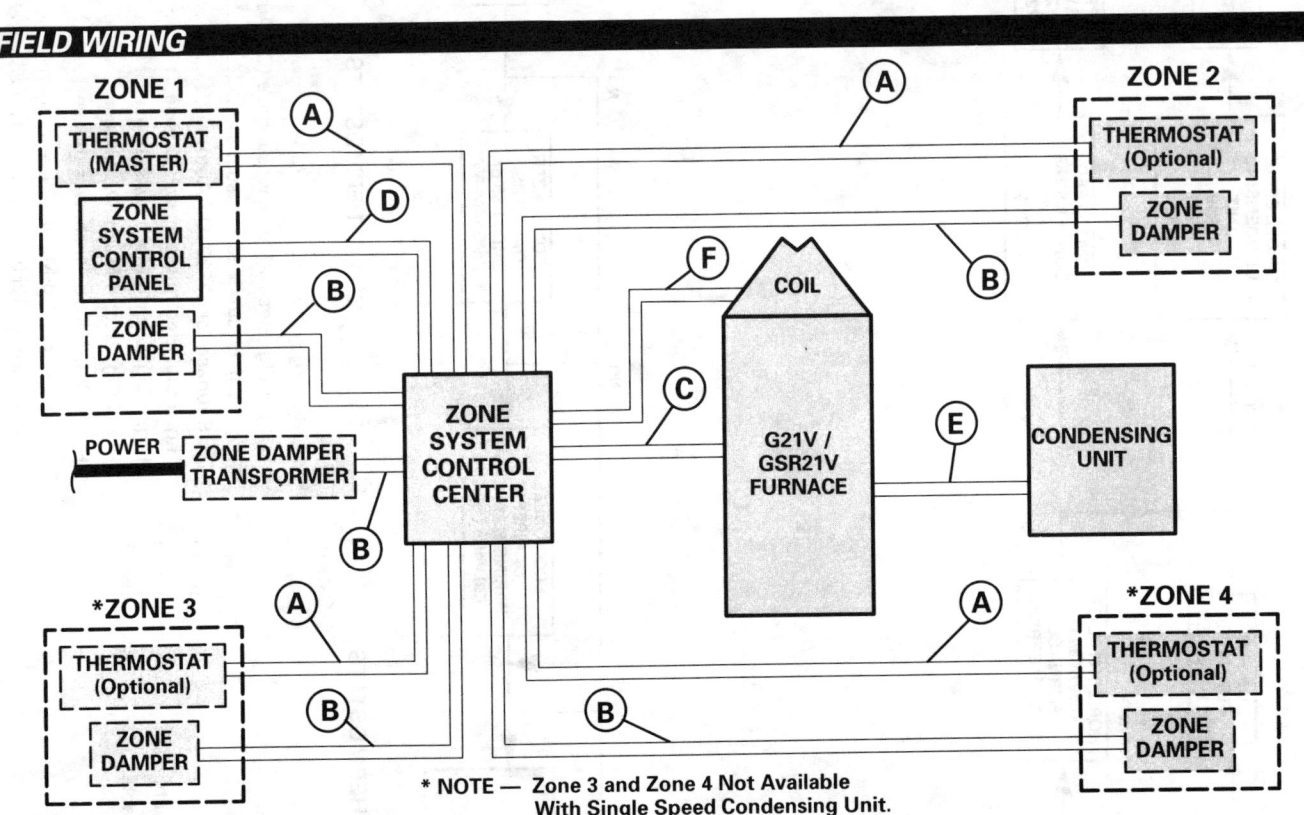

* NOTE — Zone 3 and Zone 4 Not Available With Single Speed Condensing Unit.

A — Three wire low voltage — 18 ga. minimum

B — Two wire low voltage — 18 ga. minimum

C — Eight wire low voltage — 18 ga. minimum

D — Five wire low voltage — 18 ga. minimum

E — Two wire low voltage — single speed condensing unit — 18 ga. minimum

Three wire low voltage — two speed condensing unit — 18 ga. minimum

F — Four wire low voltage — (pressure switches) two speed condensing unit only — 18 ga. minimum

— Field Wiring Not Furnished —

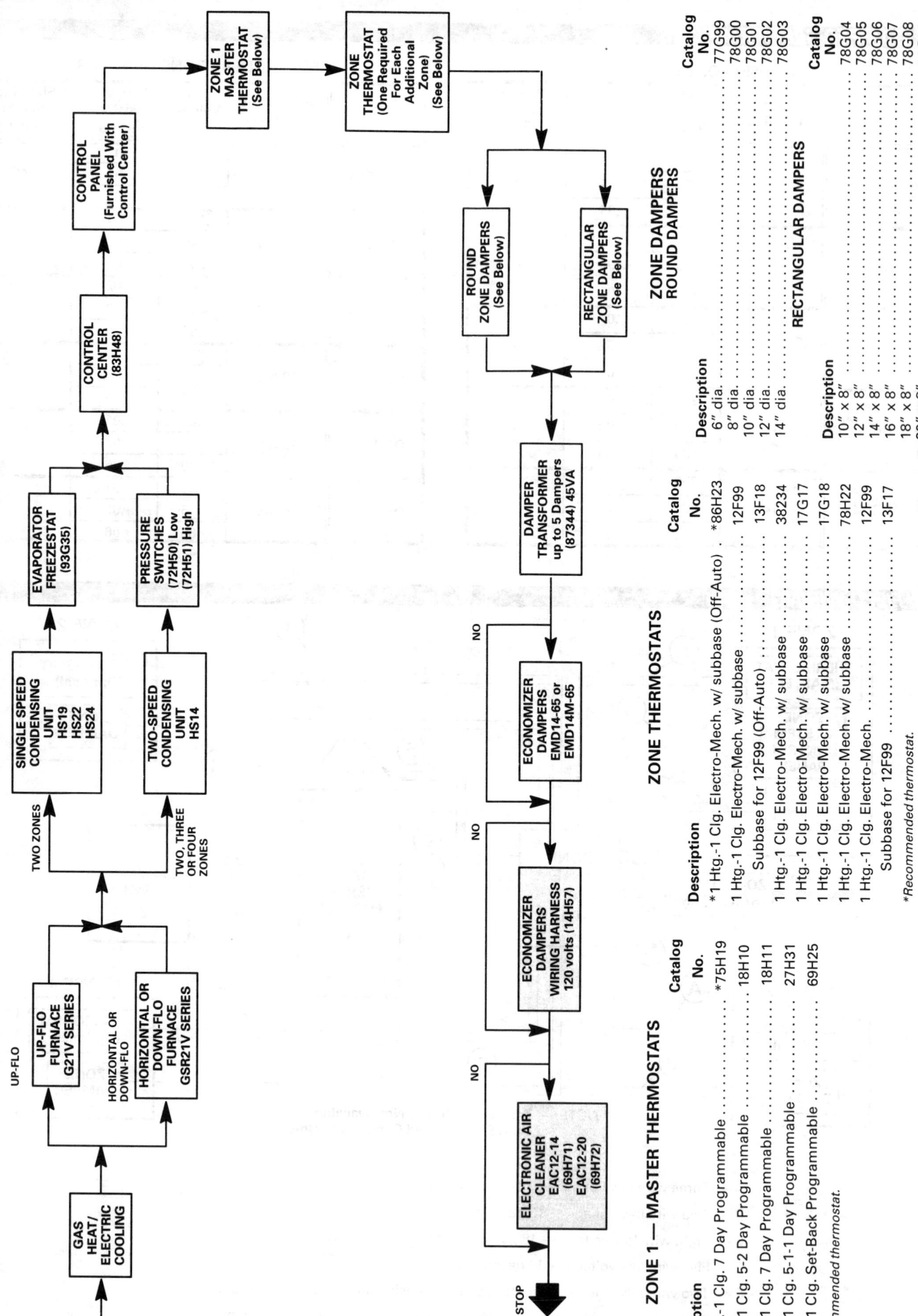

HARMONY ZONE CONTROL SYSTEM SELECTION FLOWCHART

Flowchart boxes:

START → GAS HEAT/ELECTRIC COOLING → UP-FLO FURNACE G21V SERIES (UP-FLO) / HORIZONTAL OR DOWN-FLO FURNACE GSR21V SERIES (HORIZONTAL OR DOWN-FLO)

SINGLE SPEED CONDENSING UNIT HS19 HS22 HS24 (TWO ZONES) → EVAPORATOR FREEZESTAT (93G35)

TWO-SPEED CONDENSING UNIT HS14 (TWO, THREE OR FOUR ZONES) → PRESSURE SWITCHES (72H50) Low (72H51) High

→ CONTROL CENTER (83H48) → CONTROL PANEL (Furnished With Control Center) → ZONE 1 MASTER THERMOSTAT (See Below) → ZONE THERMOSTAT (One Required For Each Additional Zone) (See Below)

→ ROUND ZONE DAMPERS (See Below) / RECTANGULAR ZONE DAMPERS (See Below)

→ DAMPER TRANSFORMER up to 5 Dampers (87344) 45VA → ECONOMIZER DAMPERS EMD14-65 or EMD14M-65 (NO) → ECONOMIZER DAMPERS WIRING HARNESS 120 volts (14H57) (NO) → ELECTRONIC AIR CLEANER EAC12-14 (69H71) EAC12-20 (69H72) (NO) → STOP

ZONE 1 — MASTER THERMOSTATS

Description	Catalog No.
*1 Htg.-1 Clg. 7 Day Programmable	*75H19
1 Htg.-1 Clg. 5-2 Day Programmable	18H10
1 Htg.-1 Clg. 7 Day Programmable	18H11
1 Htg.-1 Clg. 5-1-1 Day Programmable	27H31
1 Htg.-1 Clg. Set-Back Programmable	69H25

*Recommended thermostat.

ZONE THERMOSTATS

Description	Catalog No.
*1 Htg.-1 Clg. Electro-Mech. w/ subbase (Off-Auto)	*86H23
1 Htg.-1 Clg. Electro-Mech. w/ subbase	12F99
Subbase for 12F99 (Off-Auto)	13F18
1 Htg.-1 Clg. Electro-Mech. w/ subbase	38234
1 Htg.-1 Clg. Electro-Mech. w/ subbase	17G17
1 Htg.-1 Clg. Electro-Mech. w/ subbase	17G18
1 Htg.-1 Clg. Electro-Mech. w/ subbase	78H22
1 Htg.-1 Clg. Electro-Mech.	12F99
Subbase for 12F99	13F17

*Recommended thermostat.

ZONE DAMPERS

ROUND DAMPERS

Description	Catalog No.
6" dia.	77G99
8" dia.	78G00
10" dia.	78G01
12" dia.	78G02
14" dia.	78G03

RECTANGULAR DAMPERS

Description	Catalog No.
10" x 8"	78G04
12" x 8"	78G05
14" x 8"	78G06
16" x 8"	78G07
18" x 8"	78G08
20" x 8"	78G09

Rheem

INSTALLATION INSTRUCTIONS FOR UPFLOW HIGH EFFICIENCY CONDENSING GAS FURNACES

APPROVED

WARNING:	FOR YOUR SAFETY	FOR YOUR SAFETY
Improper installation, adjustment, alteration, service or maintenance can cause injury or property damage. Refer to this manual. For assistance or additional information consult a qualified installer, service agency or the gas supplier.	Do not store or use gasoline or other flammable vapors and liquids, or other combustible materials in the vicinity of this or any other appliance.	**WHAT TO DO IF YOU SMELL GAS** • Do not try to light any appliance. • Do not touch any electrical switch; do not use any phone in your building. • Immediately call your gas supplier from a neighbor's phone. Follow the gas supplier's instructions. • If you cannot reach your gas supplier, call the fire department.

DO NOT DESTROY. PLEASE READ CAREFULLY AND KEEP IN A SAFE PLACE FOR FUTURE REFERENCE

CONTENTS

CAUTION: *TO ENSURE PROPER INSTALLATION AND OPERATION OF THIS PRODUCT, COMPLETELY READ ALL INSTRUCTIONS PRIOR TO ATTEMPTING TO ASSEMBLE, INSTALL, OPERATE, MAINTAIN OR REPAIR THE PRODUCT.*

SAFETY RULES

1. Do not install this furnace in a mobile home, trailer or recreational vehicle.

2. Use only the type of gas approved for this furnace. (See Rating Plate.) Overfiring or underfiring will result in failure of heat exchanger and cause dangerous operation.

3. This furnace must be connected only to the approved vent system to carry combustion products outdoors as described in the Vent Pipe Installation section.

4. This furnace must be connected to the approved combustion air supply system to carry combustion air from the outdoors directly to the furnace burner.

5. Never test for gas leaks with an open flame. Use soap suds to check all connections. This will avoid any possibility of fire or explosion.

6. Provide adequate ventilation air to the furnace area.

7. Make sure supply and return air ducts are sealed to furnace casing and entirely separate from area supplying ventilation air.

HELPFUL INFORMATION

The following booklets will help you in making the installation. Obtain latest editions from:

American National Standards Institute
1430 Broadway
New York, NY 10018

ANSI Z223.1 - National Fuel Gas Code (NFPA 54)

ANSI/NFPA No. 70 - National Electric Code or current editions

National Fire Protection Association, Inc.
Batterymarch Park
Quincy, MA 02269

NFPA-90A - Installation of Air Conditioning and Ventilating Systems

NFPA-90B - Warm Air Heating and Air Conditioning Systems

Canadian Gas Association
55 Scarsdale Road
Don Mills, Ontario, Canada M3B 2R3

CAN/CGA-B149.1-M86 - Natural Gas Installation Code.
CAN/CGA-B149.2-M86 - Propane Gas Installation Code.

INSTALLATION REQUIREMENTS

WARNING: IMPROPER INSTALLATION, ADJUSTMENT, ALTERATION, SERVICE OR MAINTENANCE CAN CAUSE INJURY OR PROPERTY DAMAGE. CONSULT A QUALIFIED INSTALLER, SERVICE AGENCY OR THE GAS SUPPLIER FOR INFORMATION OR ASSISTANCE.

INSTALLATION

This furnace should be installed in accordance with the latest edition American National Standard Z223.1 - Booklet entitled "National Fuel Gas Code" (NFPA 54) (in Canada, CAN/CGA B149.1 and .2 Installation Codes for gas burning appliances), and the requirements or codes of the local utility or other authority having jurisdiction including local plumbing or waste water codes.

SHIPPING WEIGHTS

FURNACE INPUT (BTUH)	LBS.
45,000	158
60,000	160
75,000	166
90,000	188
105,000	188
125,000	194

LOCATION

WARNING: THIS FURNACE IS NOT APPROVED FOR INSTALLATION IN A MOBILE HOME. DO NOT INSTALL THIS FURNACE IN A MOBILE HOME. INSTALLATION IN A MOBILE HOME COULD CAUSE FIRE, PROPERTY DAMAGE AND PERSONAL INJURY.

This furnace is suitable for installation in buildings constructed on-site. This heating unit should be located for a short vent run (minimum 5 feet) and should be centralized with respect to the heat distribution system as much as practical. When installed in a utility room, the door of the room should be wide enough to allow the largest part of the furnace to enter, or to permit the replacement of another appliance, such as a water heater.

A gas-fired furnace for installation in a residential garage must be installed so that the burner and the ignition source are located not less than 18" above the floor and the furnace is located or protected to avoid physical damage by vehicles.

This furnace is approved for installation indoors only. Do not install outdoors. Do not install in a location that is exposed to low ambient temperatures (below 40°F).

WARNING COMBUSTIBLE MATERIAL MUST NOT BE PLACED ON OR AGAINST THE FURNACE JACKET. THE AREA AROUND THE FURNACE MUST BE KEPT CLEAR AND FREE OF ALL COM-

BUSTIBLE MATERIALS INCLUDING GASOLINE AND OTHER FLAMMABLE VAPORS AND LIQUIDS. THE HOMEOWNER SHOULD BE CAUTIONED THAT THE FURNACE AREA MUST NOT BE USED AS A BROOM CLOSET OR FOR ANY OTHER STORAGE PURPOSES.

The furnace must be level to assure proper operation. If the furnace is not level, condensate may accumulate in the secondary coil reducing efficiency or causing the furnace to shut down.

The installation should be such that the physical support of the furnace is sound without sagging, cracks, gaps, etc., around the base so as to provide a seal between the support and the base.

This is a direct vent forced air furnace. All of the combustion air is supplied directly from the outdoors and the products of combustion are discharged to the outdoors through a special air intake and vent system made up of factory and field supplied PVC or SDR 26 parts. The minimum length is 5 feet and the maximum length is 40 feet of straight pipe, four 90° sweep elbows and the terminal fittings. See combustion air and venting instructions for details.

All ventilation air openings in the furnace must be in the same room and not separated by partitions or other means that would affect the air flows or pressures. These openings must not be obstructed.

CLEARANCE ACCESSIBILITY

The design of forced air furnaces with input ratings as listed in table below is certified by A.G.A. and C.G.A. Laboratories for the clearances to combustible materials shown in inches.

See rating plate and clearance label on the furnace for specific information on model numbers and minimum clearances to combustible material.

It is recommended that at least 24″ clearance be allowed in front of all furnaces for proper servicing.

ACCESSIBILITY CLEARANCES WHERE GREATER MUST TAKE PRECEDENCE OVER FIRE PROTECTION CLEARANCES.

These models are certified for use on combustible wood floors.

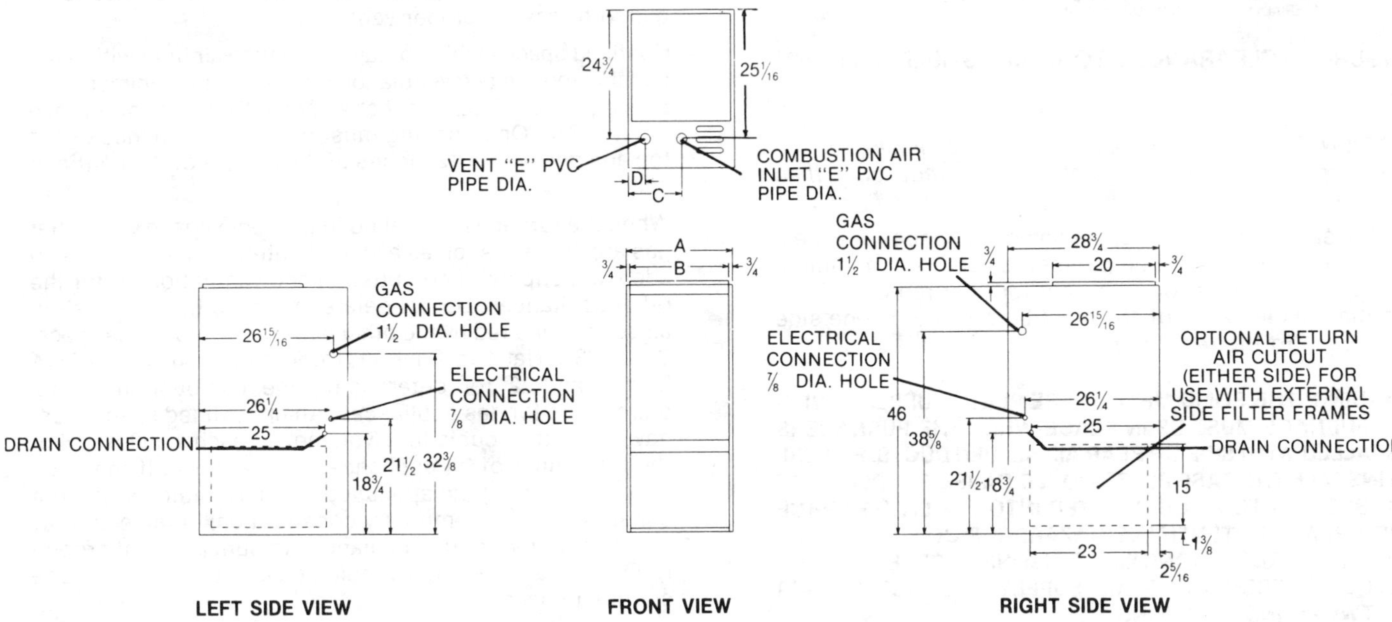

FIGURE 1. DIMENSIONS

TABLE 1. DIMENSIONS AND CLEARANCE TO COMBUSTIBLE MATERIAL (INCHES)

MODEL INPUT	A	B	C	D	E	F	REDUCED CLEARANCE (IN.)					
							LEFT SIDE	RIGHT SIDE	BACK	TOP	FRONT†	VENT
45,000	21	19½	10½	3	2	18½	0	0	0	1	2	0
60,000	21	19½	10½	3	2	18½	0	0	0	1	2	0
75,000	21	19½	10½	3	2	18½	0	0	0	1	2	0
90,000	24½	23	12¼	3¾	3	22	0	0	0	1	2	0
105,000	24½	23	12¼	3¾	3	22	0	0	0	1	2	0
120,000	24½	23	12¼	3¾	3	22	0	0	0	1	2	0

The 2″ front clearance is required for ventilation.

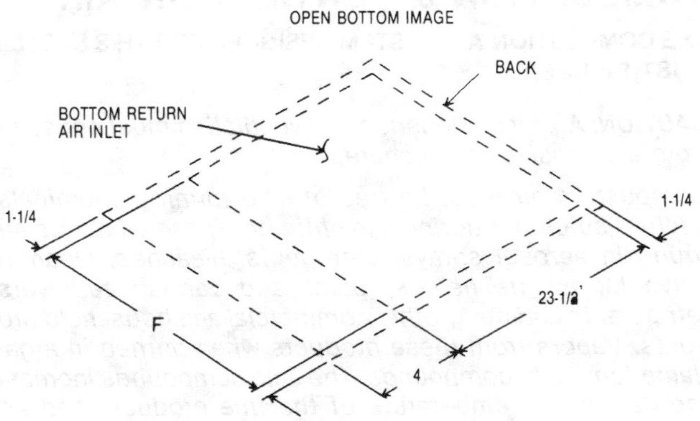

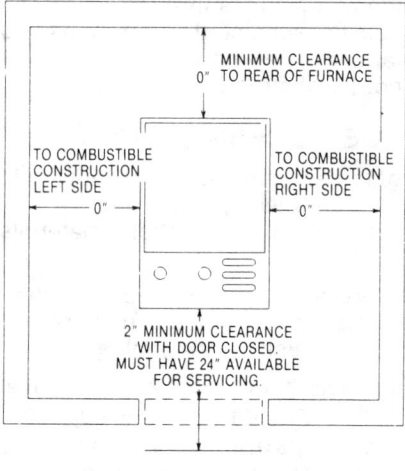

MINIMUM CLEARANCE
0" TO REAR OF FURNACE

TO COMBUSTIBLE
CONSTRUCTION
LEFT SIDE
0"

TO COMBUSTIBLE
CONSTRUCTION
RIGHT SIDE
0"

2" MINIMUM CLEARANCE
WITH DOOR CLOSED.
MUST HAVE 24" AVAILABLE
FOR SERVICING.

FOR FIRE PROTECTION CLEARANCE, WOOD AND OTHER
COMBUSTIBLE MATERIALS MUST NOT BE CLOSER THAN:

0" FROM REAR OF FURNACE   0" TO VENT PIPE
0" FROM LEFT SIDE OF FURNACE   1" FROM TOP OF FURNACE PLENUM
0" FROM RIGHT SIDE OF FURNACE   2" FROM FRONT OF FURNACE*
*MINIMUM 2" FRONT CLEARANCE FOR VENTILATION
(MINIMUM 24" FRONT CLEARANCE FOR SERVICING)
MAY BE INSTALLED ON COMBUSTIBLE WOOD FLOORS.
DO NOT INSTALL DIRECTLY ON CARPETING, TILE OR OTHER COMBUSTIBLE
MATERIAL OTHER THAN WOOD FLOORING.

FIGURE 2. CLEARANCES TO COMBUSTIBLE MATERIAL

CAUTION: This gas-fired furnace must not be installed directly on carpeting, tile or other combustible material other than wood flooring.

For basement installations, the same clearances apply except when a rear return air cabinet is used, 24" minimum clearance must be allowed at rear for filter removal. Also, a minimum clearance of 18" must be allowed on one side for access to the rear filter.

WARNING: THE SOLID METAL BASE PLATE SUPPLIED WITH THE FURNACE MUST BE IN PLACE WHEN THE FURNACE IS INSTALLED WITH SIDE OR REAR AIR RETURN DUCTS. FAILURE TO INSTALL THE BASE PLATE COULD CAUSE PRODUCTS OF COMBUSTION TO BE CIRCULATED INTO THE LIVING SPACE AND CREATE POTENTIALLY HAZARDOUS CONDITIONS, INCLUDING CARBON MONOXIDE POISONING. REFER TO SECTION ON "CIRCULATING AIR SUPPLY" FOR RETURN AIR DUCTWORK INSTRUCTIONS.

FURNACE WIDTH	BASE PLATE NO.	BASE PLATE SIZE
21"	AE-61229-04	$19\frac{3}{4} \times 24\frac{11}{16}$
$24\frac{1}{2}$"	AE-61229-05	$23\frac{1}{4} \times 24\frac{11}{16}$

COMBUSTION & VENTILATION AIR

THE COMBUSTION AIR SYSTEM DESIGNED FOR THIS SYSTEM MUST BE USED.

CAUTION: Air for combustion and ventilation must not come from a corrosive atmosphere.

Combustion air must be free of acid forming chemicals; such as sulphur, fluorine and chlorine. These elements are found in aerosol sprays, detergents, bleaches, cleaning solvents, air fresheners, paint and varnish removers, refrigerants and many other commercial and household products. Vapors from these products when burned in a gas flame form acid compounds. The acid compounds increase the dew point temperature of the flue products and are highly corrosive after they condense.

This is a direct vent forced air furnace and all of the air for combustion is supplied directly to the furnace burner through a special air intake system outlined in these installa-

tion instructions. This system consists of an AS-61457-01 horizontal combustion air terminal and field supplied Schedule 40 or SDR26 PVC pipe.

The maximum air intake pipe length is 40 feet of straight pipe and four 90° sweep elbows plus the termination. The minimum pipe diameter between the furnace and the terminal is 2 inches for the 45,000, 60,000 and 75,000 BTUH models. The minimum pipe diameter between the furnace and the terminal is 3 inches for the 90,000, 105,000 and 120,000 BTUH models. The horizontal combustion air inlet terminal is composed of a screened 90° elbow spaced 6 inches from the wall with up to 18 inches of 2 inch diameter pipe through the wall for all furnace sizes. Vertical, through the roof applications use two 90° elbows to keep the combustion air inlet pointed downward to prevent the entry of rain water.

The minimum air intake pipe length is 5 feet.

IT IS REQUIRED THAT 3 INCH PVC PIPE BE USED FOR ALL INSTALLATIONS AT ALTITUDES ABOVE 4,000 FEET.

The combustion air for this furnace is supplied directly from the outdoors through the combustion air inlet system. When it is installed in a confined space, such as a utility room or closet, 2 inches of space to the front of the furnace is required to assure proper ventilation.

Confined spaces which obtain ventilation air from within the building require two ventilation openings with minimum free area equal to 0.5 square inches per 1,000 BTUH of furnace input rating. One opening must be within 12 inches of the top and one within 12 inches of the bottom of the confined space.

When the furnace is installed in the same space with other gas appliances, such as a water heater, be sure there is an adequate supply of combustion and ventilation air for the other appliances. Do not delete or reduce the combustion air supply required by the other gas appliances in this space. See Z223.1, National Fuel Gas Code (NFPA-54) or CAN/CGA B149.1 and .2 for determining the combustion air requirements for gas appliances. An unconfined space must have at least 50 cubic feet (volume) for each 1,000 BTUH of the total input of all appliances in the space. If the open space containing the appliances is in a building with tight construction (contemporary construction) outside air may still be required for the appliances to burn and vent properly. Outside air openings should be sized the same as for a confined space.

VENT PIPE INSTALLATION

WARNING: PROPER VENT PIPE INSTALLATION IS CRITICAL TO THE SAFE OPERATION OF THE FURNACE, THEREFORE, CAREFULLY READ AND FOLLOW ALL THE INSTRUCTIONS GIVEN IN THIS SECTION.

DEVICES ATTACHED TO THE FLUE OR VENT FOR THE PURPOSE OF REDUCING HEAT LOSS UP THE CHIMNEY, INCLUDING FIELD-INSTALLED DRAFT INDUCERS, HAVE NOT BEEN TESTED AND HAVE NOT BEEN INCLUDED IN THE DESIGN CERTIFICATION OF THIS FURNACE. WE, THE MANUFACTURER, CANNOT AND WILL NOT BE RESPONSIBLE FOR INJURY OR DAMAGE CAUSED BY THE USE OF SUCH UNTESTED AND/OR UNCERTIFIED DEVICES, ACCESSORIES OR COMPONENTS.

This furnace removed both sensible and latent heat from the combustion flue gases. Removal of latent heat results in condensation of flue gas water vapor. This condensed water vapor drains from the secondary heat exchanger and out of the vent into a drain trap. See Figures 3A and 3B.

This furnace must be vented to the outdoors with PVC (polyvinyl chloride) Schedule 40 or SDR26 pipe. The vent pipe is approved for 0 inches clearance from combustible materials.

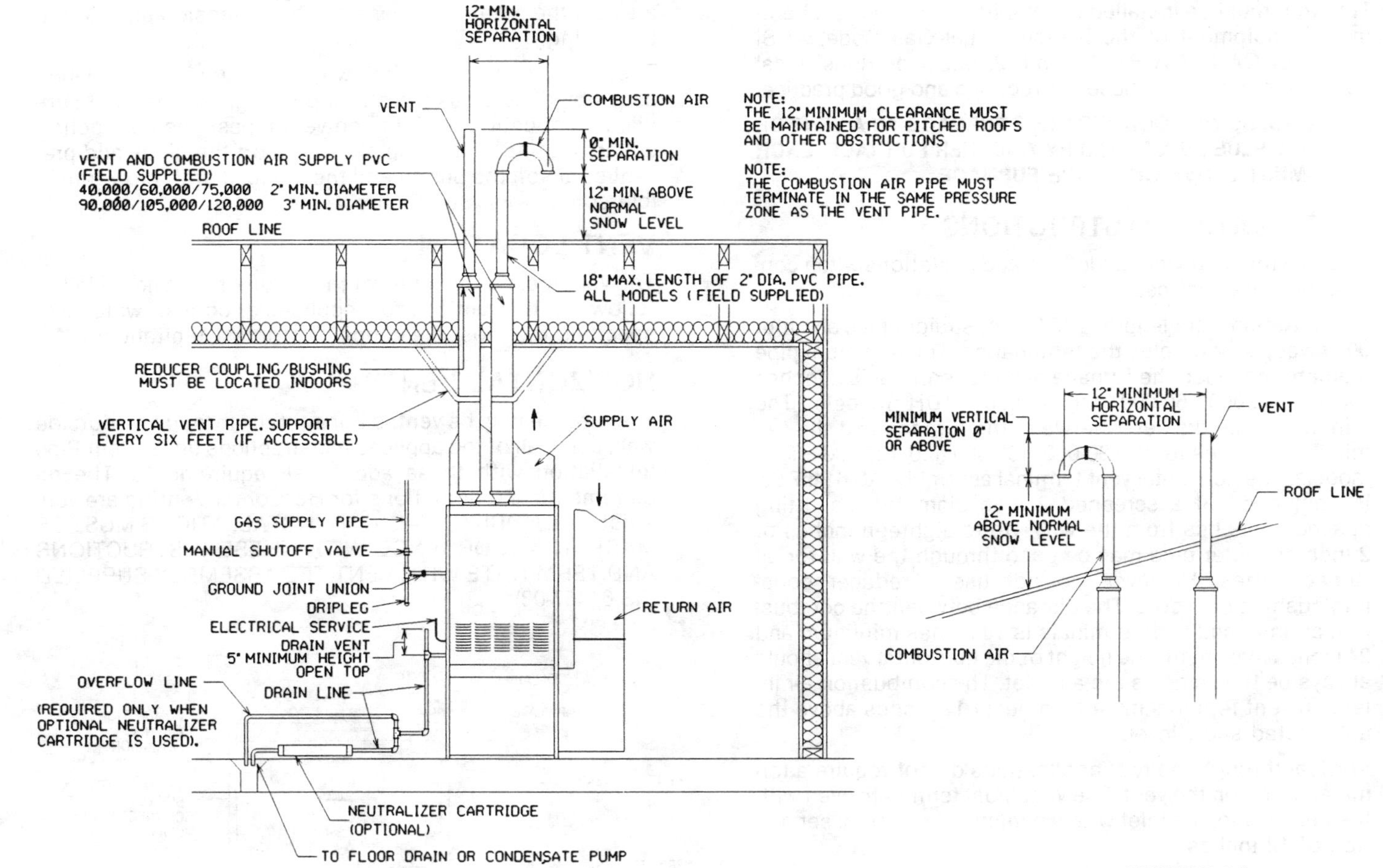

12" MIN.
HORIZONTAL
SEPARATION

VENT

COMBUSTION AIR

0" MIN.
SEPARATION

12" MIN. ABOVE
NORMAL
SNOW LEVEL

NOTE:
THE 12" MINIMUM CLEARANCE MUST
BE MAINTAINED FOR PITCHED ROOFS
AND OTHER OBSTRUCTIONS

NOTE:
THE COMBUSTION AIR PIPE MUST
TERMINATE IN THE SAME PRESSURE
ZONE AS THE VENT PIPE.

VENT AND COMBUSTION AIR SUPPLY PVC
(FIELD SUPPLIED)
40,000/60,000/75,000   2" MIN. DIAMETER
90,000/105,000/120,000   3" MIN. DIAMETER

ROOF LINE

18" MAX. LENGTH OF 2" DIA. PVC PIPE.
ALL MODELS (FIELD SUPPLIED)

REDUCER COUPLING/BUSHING
MUST BE LOCATED INDOORS

VERTICAL VENT PIPE. SUPPORT
EVERY SIX FEET (IF ACCESSIBLE)

SUPPLY AIR

GAS SUPPLY PIPE

MANUAL SHUTOFF VALVE

GROUND JOINT UNION

DRIPLEG

ELECTRICAL SERVICE

DRAIN VENT
5" MINIMUM HEIGHT
OPEN TOP

RETURN AIR

OVERFLOW LINE
(REQUIRED ONLY WHEN
OPTIONAL NEUTRALIZER
CARTRIDGE IS USED).

DRAIN LINE

NEUTRALIZER CARTRIDGE
(OPTIONAL)

TO FLOOR DRAIN OR CONDENSATE PUMP

12" MINIMUM
HORIZONTAL
SEPARATION

VENT

MINIMUM VERTICAL
SEPARATION 0"
OR ABOVE

12" MINIMUM
ABOVE NORMAL
SNOW LEVEL

ROOF LINE

COMBUSTION AIR

FIGURE 3A.
VERTICAL VENTING

REDUCER COUPLING/BUSHING
MUST BE LOCATED INDOORS.

FIGURE 3B.
HORIZONTAL VENTING

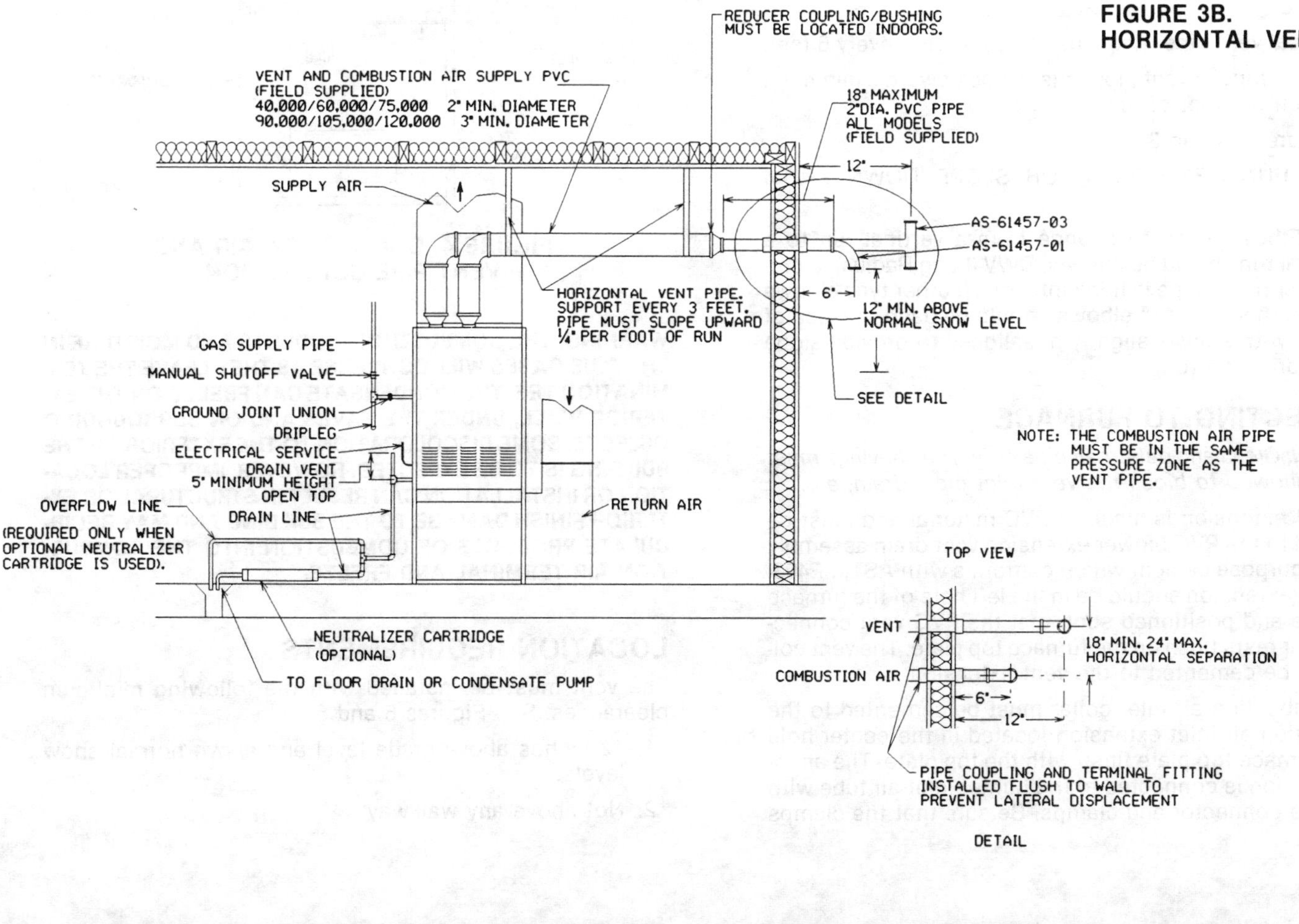

VENT AND COMBUSTION AIR SUPPLY PVC
(FIELD SUPPLIED)
40,000/60,000/75,000   2" MIN. DIAMETER
90,000/105,000/120,000   3" MIN. DIAMETER

18" MAXIMUM
2" DIA. PVC PIPE
ALL MODELS
(FIELD SUPPLIED)

SUPPLY AIR

12"

AS-61457-03

AS-61457-01

HORIZONTAL VENT PIPE.
SUPPORT EVERY 3 FEET.
PIPE MUST SLOPE UPWARD
1/4" PER FOOT OF RUN

6"

12" MIN. ABOVE
NORMAL SNOW LEVEL

GAS SUPPLY PIPE

MANUAL SHUTOFF VALVE

GROUND JOINT UNION

DRIPLEG

ELECTRICAL SERVICE

DRAIN VENT
5" MINIMUM HEIGHT
OPEN TOP

SEE DETAIL

NOTE: THE COMBUSTION AIR PIPE
MUST BE IN THE SAME
PRESSURE ZONE AS THE
VENT PIPE.

RETURN AIR

OVERFLOW LINE
(REQUIRED ONLY WHEN
OPTIONAL NEUTRALIZER
CARTRIDGE IS USED).

DRAIN LINE

NEUTRALIZER CARTRIDGE
(OPTIONAL)

TO FLOOR DRAIN OR CONDENSATE PUMP

TOP VIEW

VENT

COMBUSTION AIR

6"

12"

18" MIN. 24" MAX.
HORIZONTAL SEPARATION

PIPE COUPLING AND TERMINAL FITTING
INSTALLED FLUSH TO WALL TO
PREVENT LATERAL DISPLACEMENT

DETAIL

5

This vent must be installed in compliance with Part 7, Venting of Equipment of the National Fuel Gas Code, ANSI Z223.1 or CAN/CGA B149.1 and .2, latest editions, local codes or ordinances, these instructions and good practice.

WARNING: DO NOT CONNECT THIS FURNACE TO A VENT OR CHIMNEY FLUE BEING USED BY ANOTHER FURNACE. EACH VENT MUST SERVE ONLY ONE FURNACE.

INSTALLATION INSTRUCTIONS

Observe the following guidelines and limitations when constructing the vent assembly.

The maximum vent length is 40 feet of straight pipe and four 90° sweep elbows plus the termination. The minimum pipe diameter between the furnace and the terminal is 2 inches for the 45,000, 60,000 and 75,000 BTUH models. The minimum pipe diameter between the furnace and the terminal is 3 inches for the 90,000, 105,000 and 120,000 BTUH models. The horizontal vent terminal assembly, AS-61457-03, is composed of a screened 2 inch diameter tee fitting spaced 12 inches from the wall. Up to eighteen inches of 2 inch diameter pipe may be used through the wall for all furnace sizes. However, do not use a reducer coupling/bushing outdoors. The distance between the combustion air inlet and vent terminals is 18 inches minimum and 24 inches maximum. The height of the horizontal vent should always be the same as the air inlet. The combustion air inlet and vent terminal must be at least 12 inches above the anticipated snow level.

Vertical through the roof applications do not require a terminal fitting on the vent. The vent must terminate even with the combustion air inlet with a minimum horizontal separation of 12 inches.

The minimum vent pipe length is 5 feet.

IT IS REQUIRED THAT 3 INCH PVC PIPE BE USED FOR ALL INSTALLATION AT ALTITUDES ABOVE 4,000 FEET.

All horizontal vent pipe runs must be supported at least every 3 feet with metal pipe strapping. No sags or dips are permitted.

All vertical vent pipe runs must be supported every 6 feet.

Horizontal runs of vent pipe must slope upward a minimum of ¼ inch per foot of run.

See Figures 3A and 3B.

NEVER RUN VENT DOWN OR SLOPE DOWN FROM FURNACE.

NOTE: Elbows used to change from a vertical run to a horizontal run should be the type DWV (Long Radius) to provide the correct slope in horizontal run. If other type elbows are used, then two 45° elbows should be used in place of one 90°, with elbows slightly misaligned to provide slope in the horizontal runs.

CONNECTING TO FURNACE

CAUTION: Clean and deburr all pipe cuts. The shavings must not be allowed to block the vent inlet pipe, drain, etc.

The vent extension is made of PVC material and must be attached to the PVC blower extension vent drain assembly with all purpose cement which conforms to ASTM F493. The vent extension should be in the left hole of the furnace top plate and positioned such that the PVC vent connection collar rests flush on the furnace top plate. The vent collar must be cemented to the vent extension.

The combustion air inlet collar must be cemented to the combustion air inlet extension located in the center hole of the furnace top plate flush with the top plate. The air inlet extension is connected to the burner inlet air tube with a flexible connector and clamps. Be sure that the clamps

are over the burner inlet pipe and the extension pipe before tightening.

These parts are supplied with the furnace and when assembled, serve as the starting fittings for the field supplied combustion air inlet and vent pipes. The connection collars resting on the top plate support the pipes and prevent strain on the burner and the induced draft blower flexible connectors. See Figure 4.

VENT LOCATION

The vent location must meet the requirements listed in the following instructions or applicable codes, whichever specifies the most clearance or strictest limitations.

HORIZONTAL VENTS

The furnace may be vented horizontally through an outside wall, using all of the applicable instructions under Vent Pipe Installation with these additional requirements. The requirements and limitations for Horizontal Venting are very strict. ALL HORIZONTAL VENT INSTALLATIONS MUST BE MADE IN ACCORDANCE WITH THESE INSTRUCTIONS AND TERMINATE WITH VENT TEE ASSEMBLY SUPPLIED (AS-61457-03).

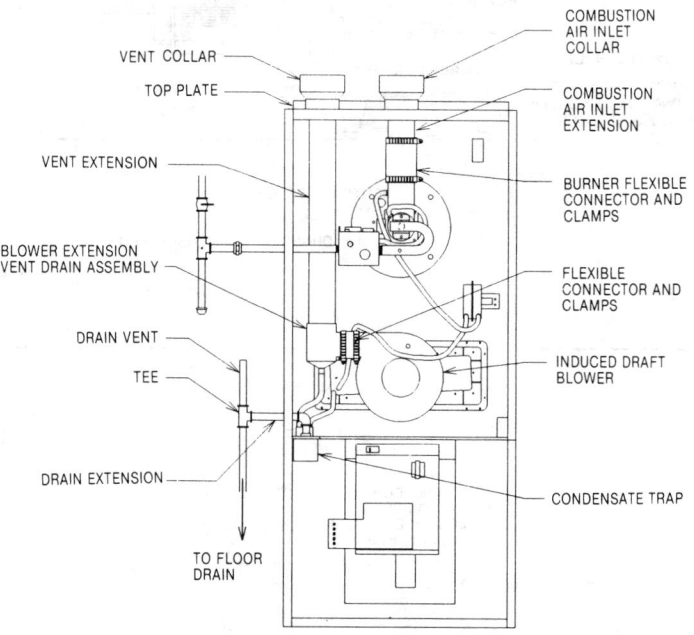

FIGURE 4. COMBUSTION AIR AND VENT PIPE CONNECTION

WARNING: THE COMBUSTION PRODUCTS AND MOISTURE IN THE FLUE GASES WILL CONDENSE AS THEY LEAVE THE TERMINATION TEE. THE CONDENSATE CAN FREEZE ON THE EXTERIOR WALL, UNDER THE EAVES AND ON SURROUNDING OBJECTS. SOME DISCOLORATION TO THE EXTERIOR OF THE BUILDING IS TO BE EXPECTED. HOWEVER, IMPROPER LOCATION OR INSTALLATION CAN RESULT IN STRUCTURAL OR EXTERIOR FINISH DAMAGE TO THE BUILDING AND MAY RECIRCULATE PRODUCTS OF COMBUSTION INTO THE COMBUSTION AIR TERMINAL AND FREEZE.

LOCATION REQUIREMENTS

The vent must be installed with the following minimum clearances. See Figures 5 and 6.

1. 12 inches above grade level and above normal snow levels.

2. Not above any walkway.

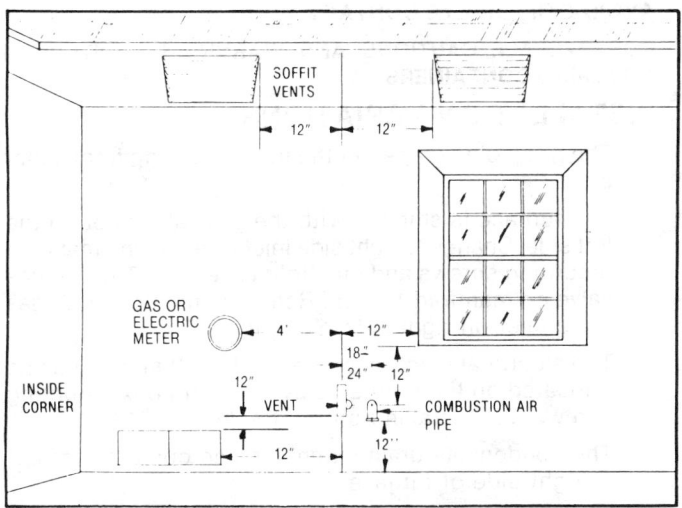

FIGURE 5. MINIMUM CLEARANCES

3. One foot below, horizontally from, or above any door window or gravity air inlet to the building.

4. At least 3 feet above any forced air inlet located within 10 feet, except the combustion air inlet of a direct vent appliance.

5. Minimum horizontal clearance of 4 feet from electric meters, gas meters, regulators, and relief equipment.

6. A minimum vertical clearance of 5 feet below soffit or overhangs.

In addition to the minimum horizontal clearances, the vent location is also governed by the following requirements and guidelines.

1. Do not install under any kind of patio or deck.

2. Do not locate on the side of a building with prevailing winter winds. This will help prevent moisture from freezing on walls and overhangs (under eaves).

3. Do not extend vent directly through brick or masonry surfaces. Use a rust resistant sheet metal or plastic backing plate behind vent. See Figure 6.

4. Do not locate too close to shrubs as condensate may stunt or kill them. See Figure 6.

5. An installation under roof overhangs should be avoided when possible. The underneath side of the eave will collect heavy amounts of moisture and icicles may form.

6. Caulk all cracks, seams and joints within 6 feet horizontally and above and below vent. See Figure 6.

7. Painted surfaces must be sound and in good condition with no cracking, peeling, etc. Painted surfaces will require maintenance.

8. Do not expose 3″ x 2″ reducer/bushing to outdoor ambient temperatures.

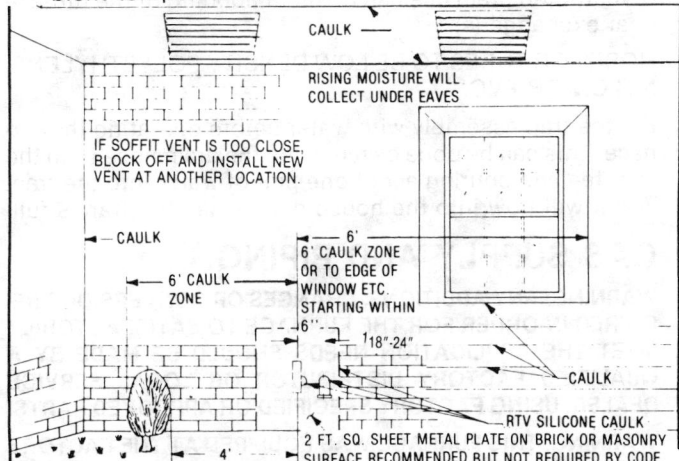

FIGURE 6. MOISTURE ZONES

NOTICE: When the installation of this furnace replaces an existing furnace that is removed from a venting system serving other appliances, the venting system is likely to be too large to properly vent the remaining attached appliances. The performance of the existing venting system must be tested with use of the remaining appliances as set forth below:

The following steps shall be followed with each appliance remaining connected to the common venting system placed in operation, while the other appliances remaining connected to the common venting system are not in operation.

(a) Seal any unused openings in the common venting system.

(b) Visually inspect the venting system for proper size and horizontal pitch and determine there is no blockage or restriction, leakage, corrosion or other deficiencies which could cause an unsafe condition.

(c) Insofar as is practical, close all building doors and windows and all doors between the space in which the appliances remaining connected to the common venting system are located and other spaces of the building. Turn on clothes dryers and any appliance not connected to the common venting system. Turn on any exhaust fans, such as range hoods and bathroom exhausts, so they will operate at maximum speed. Do not operate a summer exhaust fan. Close fireplace dampers.

(d) Follow the lighting instructions. Place the appliance being inspected in operation. Adjust thermostat so appliance will operate continuously.

(e) Test for spillage at the draft hood relief opening after 5 minutes of main burner operation. Use the flame of a match or candle, or smoke from a cigarette, cigar or pipe.

(f) After it has been determined that each appliance remaining connected to the common venting system properly vents when tested as outlined above, return doors, windows, exhaust fans, fireplace dampers and any other gas-burning appliance to their previous conditions of use.

(g) If improper venting is observed during any of the above tests, the common venting system must be corrected.

A reduction of draft in the vent system and possibly the condensing of combustion products could be caused by excessive heat loss through the surface area of the existing oversized vent system. The heat loss may be reduced by replacing single wall vent connectors with Type B vent material, reducing the size of the common vent or using a liner in an oversized masonry chimney. Any steps taken to revise the vent system should be in accordance with the National Fuel Gas Code, ANSI Z223.1 (latest edition).

When resizing any portion of the remaining common vent system, it should be resized to approach the minimum size as determined using the appropriate tables in Appendix G in the National Fuel Gas Code ANSI Z223.1 (latest edition).

OUTDOOR COMBUSTION AIR SUPPLY

NOTE: Refer to section on Combustion and Ventilation Air on page 4 for proper ventilation air supply in confined spaces.

COMBUSTION AIR PIPE INSTALLATION

1. The combustion air pipe must be terminated close to the vent pipe termination. See Figures 3A and 3B.

2. All horizontal runs must be supported every three feet.

3. All vertical runs must be supported every six feet.

4. Elbows used to change from a vertical run to a horizontal run should be type DWV (long radius).

5. Caution should be taken to make sure no water or moisture enters the combustion air pipe to prevent damage to the furnace.

6. Schedule 40 or SDR26 PVC pipe must be used. The maximum length is 40' with a maximum of 4 elbows plus termination.

7. A termination elbow, AS-61457-01, must be used for horizontal installations. A return bend is required for vertical termination.

8. The combustion air pipe must be sealed and adequately supported.

9. A through the wall combustion air inlet terminal should be located 36 inches horizontally from a gas appliance vent.

JOINING PIPE AND FITTINGS

All pipe, fittings, solvent cement, primers and procedures must conform to American National Standard Institute and American Society for Testing and Materials (ANSI/ASTM) standards in the U.S.

Pipe and Fittings - ASTM D1785, D2466 and D2665

PVC Primer and Solvent Cement - ASTM D2564

Procedure for Cementing Joints Ref ASTM D2855

In Canada, all products must be Canadian Standards Association listed types.

CEMENTING JOINTS

All joints in the PVC vent must be properly sealed using the following material and procedure.

PVC CLEANER-PRIMER AND PVC MEDIUM BODY SOLVENT CEMENT.

1. Cut pipe end square, remove ragged edges and burrs. Chamfer edge of pipe, then clean fitting socket and pipe joint area of all dirt, grease and moisture.

2. After checking pipe and socket for proper fit, wipe socket and pipe with cleaner-primer. Apply a liberal coat of primer to inside surface of socket and outside of pipe. DO NOT ALLOW PRIMER TO DRY BEFORE APPLYING CEMENT.

3. Apply a thin coat of cement evenly in the socket. Quickly apply a heavy coat of cement to the pipe end and insert pipe into fitting with a slight twisting movement until it bottoms out.

   NOTE: Cement must be fluid; if not, recoat.

4. Hold the pipe in the fitting for 30 seconds to prevent the tapered socket from pushing the pipe out of the fitting.

5. Wipe all excess cement from the joint with a rag. Allow 15 minutes before handling. Cure time varies according to fit, temperature and humidity.

NOTE: Stir the solvent cement frequently while using. Use a natural bristle brush or the dauber supplied with the can. The proper brush size is one inch.

CAUTION: For Proper Installation:

DO NOT use solvent cement that has become curdled, lumpy or thickened.

DO NOT thin. Observe shelf precautions printed on containers.

For application below 32°F, use only low temperature type solvent cement.

WARNING: DANGER OF FIRE OR BODILY INJURY.

PVC SOLVENT CEMENTS AND PRIMERS ARE HIGHLY FLAMMABLE. PROVIDE ADEQUATE VENTILATION AND DO NOT ASSEMBLE NEAR HEAT SOURCE OR OPEN FLAME. DO NOT SMOKE.

AVOID SKIN OR EYE CONTACT.

OBSERVE ALL CAUTIONS AND WARNINGS PRINTED ON MATERIAL CONTAINERS.

INSTALLATION VARIATIONS

1. The gas piping may enter the furnace casing from either side.

   The furnace is shipped with the gas valve inlet on the left side. Change to right side inlet by removing manifold mounting screws and manifold assembly. Tighten gas valve on manifold ½ turn. Reinstall manifold with gas valve inlet on right side. See Figure 7.

2. The electrical junction box is on the left side. It can be relocated on the right side by removing box mounting screws and installing on right side.

3. The condensate drain extension can come out of left or right side of furnace.

CONDENSATE DRAIN/NEUTRALIZER

Drain line and overflow line can be ½" PVC flex tube or Schedule 40 with a disconnect union so the trap can be removed. Drains should terminate at an inside drain.

WARNING: DO NOT RUN DRAIN OUTDOORS. FREEZING OF CONDENSATE COULD CAUSE PROPERTY DAMAGE.

If no floor drain is available, a condensate pump that is resistant to acidic water must be installed. Pumps are available from your local distributor. If a pump is used that is not resistant to acidic water, a condensate neutralizer must be used ahead of the pump.

Condensate pump must have an auxiliary safety switch to prevent operation of furnace and resulting overflow of condensate in the event of pump failure. The safety switch must be wired through the "R" circuit ONLY (Low Voltage) to provide operation in either heating or cooling modes.

If local codes require, install a Condensate Neutralizer Cartridge in the drain line.

NOTE: Install cartridge in a horizontal position only.

Install overflow line if routing to floor drain. See Figures 3A and 3B.

Connect field supplied drain material to the furnace drain trap. Cement the factory supplied 1/2" ABS street ell in the top opening of the ABS drain trap facing to either the right or left side of the furnace. Stub out of the furnace casing with a short section of PVC pipe. Install a tee fitting with an end upward, fitted with a 5" riser pipe to serve as an antisiphon vent. Run drain tube floor drain or condensate pump. Use an all purpose solvent cement that is compatible for use with ABS and PVC material.

DO NOT connect a common drain line with an air conditioner evaporator coil drain located above the furnace. A blocked or restricted drain line can result in flooding of the furnace heat exchanger(s).

HOSES OR TUBES TO BE LOW DENSITY POLYETHYLENE, NYLON OR PVC.

Fill the trap assembly with water before operating the furnace. This can be done by removing the drain hose from the vent tee and pouring about one pint of water into the trap. Water will flow into the house drain when the trap is full.

GAS SUPPLY AND PIPING

WARNING: ANY ADDITIONS, CHANGES OR CONVERSIONS REQUIRED IN ORDER FOR THE FURNACE TO SATISFACTORILY MEET THE APPLICATION NEEDS SHOULD BE MADE BY A QUALIFIED FACTORY DISTRIBUTOR OR LOCAL SERVICE DEALER, USING FACTORY SPECIFIED OR APPROVED PARTS.

WARNING: THIS FURNACE WAS EQUIPPED AT THE FACTORY FOR USE ON NATURAL GAS ONLY. CONVERSION TO LP OR

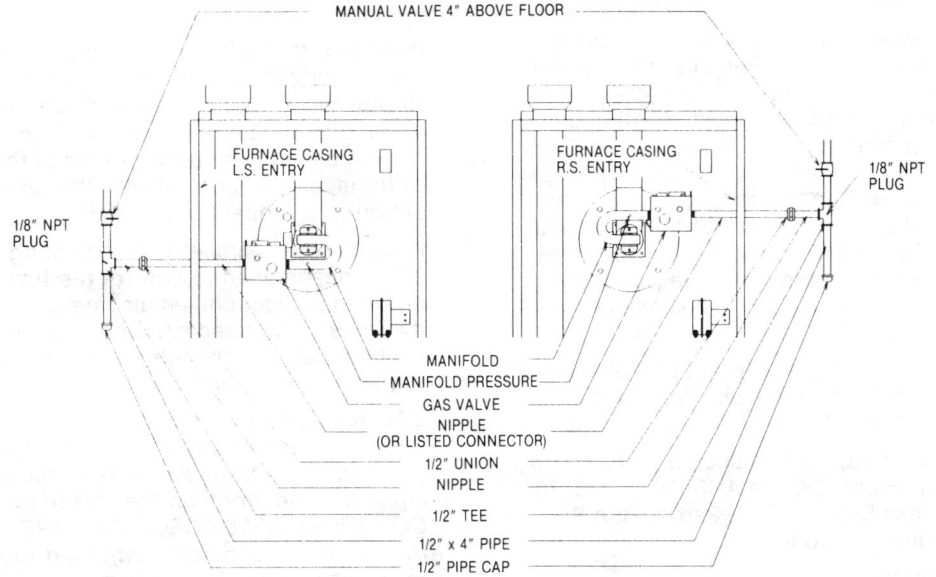

MANUAL VALVE 4" ABOVE FLOOR

FURNACE CASING L.S. ENTRY

FURNACE CASING R.S. ENTRY

1/8" NPT PLUG

1/8" NPT PLUG

MANIFOLD
MANIFOLD PRESSURE
GAS VALVE
NIPPLE (OR LISTED CONNECTOR)
1/2" UNION
NIPPLE
1/2" TEE
1/2" x 4" PIPE
1/2" PIPE CAP

FIGURE 7. GAS PIPING

PROPANE GAS REQUIRES A SPECIAL KIT SUPPLIED BY THE DISTRIBUTOR OR MANUFACTURER. MAILING ADDRESSES ARE LISTED ON THE FURNACE RATING PLATE, PARTS LIST AND WARRANTY. FAILURE TO USE THE PROPER CONVERSION KIT CAN CAUSE FIRE, CARBON MONOXIDE POISONING, EXPLOSION, PERSONAL INJURY OR PROPERTY DAMAGE.

Care should be taken after installation of this equipment that gas control valve not be subjected to high gas supply line pressure. The furnace and its individual shut off valve must be disconnected from the gas supply piping system during any pressure testing of that system at test pressure in excess of 1/2 PSIG (3.48 kPa). The furnace must be isolated from the gas supply piping system by closing its individual manual shut off valve during any pressure testing of the gas supply piping system at test pressure equal to or less than 1/2 PSIG (3.48 kPa).

In making gas connections, avoid strains as they may cause noise and damage controls. Use back-up wrench to prevent twisting control assembly when connecting supply piping.

To check for leaks in piping use a soap and water solution or other approved method.

WARNING: DO NOT USE AN OPEN FLAME.

CONNECT GAS SERVICE from meter to control assembly. See Figure 7 for typical hook-up. A ground-joint union must be installed so the control assembly may be easily removed. A 1/8" NPT plug must be provided on the supply pipe to the valve for the purpose of making measurements of the inlet gas pressure. A manual shutoff valve or plugcock should be installed in the gas line outside of the furnace casing. Valve should be readily accessible for turning on or off. A drip leg should be installed in the gas supply line as close to the furnace as possible. A pipe compound resistant to the action of liquefied petroleum gases must be used at all threaded pipe connections.

Gas piping should be installed in accordance with local codes and regulations of the utility company. Consult local gas company for location of manual main valve. The gas line should be of adequate size to prevent undue pressure drop and never smaller than the pipe size to the combination gas valve. It is recommended that the size of pipe selected be in accordance with TABLE 2 for the length of pipe required and connected to the furnace as illustrated.

GAS SUPPLY

Recommended gas supply pressures are 5" to 7" water column pressure for natural gas and 11" to 14" water column pressure for LP gas. A maximum gas supply pressure of 14" water column should not be exceeded on either gas. The minimum supply pressures for purposes of input adjustment must be 5" W.C. for natural gas and 11" for LP gas.

TABLE 2. GAS PIPE CAPACITY TABLE (CU. FT./HR.)

Capacity of gas pipe of different diameters and lengths in cu. ft. per hr. with pressure drop of 0.3 in. and specific gravity of 0.60 (natural gas).

Nominal Iron Pipe Size, Inches	Length of Pipe, Feet							
	10	20	30	40	50	60	70	80
1/2	132	92	73	63	56	50	46	43
3/4	278	190	152	130	115	105	96	90
1	520	350	285	245	215	195	180	170
1 1/4	1,050	730	590	500	440	400	370	350
1 1/2	1,600	1,100	890	760	670	610	560	530

After the length of pipe has been determined, select the pipe size which will provide the minimum cubic feet per hour required for the gas input rating of the furnace. By formula:

$$\text{Cu. Ft. Per Hr. Required} = \frac{\text{Gas Input of Furnace (BTU/HR)}}{\text{Heating Value of Gas (BTU/FT}^3\text{)}}$$

The gas input of the furnace is marked on the furnace rating plate. The heating value of the gas (BTU/FT^3) may be determined by consulting the local natural gas utility or the LP gas supplier.

ELECTRICAL WIRING

WARNING: TURN OFF ELECTRIC POWER AT FUSE BOX OR SERVICE PANEL BEFORE MAKING ANY ELECTRICAL CONNECTIONS.

WARNING: CABINET MUST BE PERMANENTLY GROUNDED. A GROUND SCREW IS PROVIDED IN THE JUNCTION BOX FOR THIS PURPOSE.

GROUND CONNECTION MUST BE COMPLETED BEFORE MAKING LINE VOLTAGE CONNECTIONS.

ELECTRICAL CONNECTIONS

The electrical supply requirements are listed on the furnace rating plate.

Use a separate fused branch electrical circuit containing a properly sized fuse or circuit breaker. Run this circuit directly from the main switch box to an electrical disconnect which must be readily accessible and located within sight of the furnace. Connect from the disconnect to the black and white furnace connection leads inside the junction box on the left side of the furnace, in the control compartment. This junction box and internal wiring may be field relocated to the right side of the furnace for convenience. For proper connection, refer to the appropriate wiring diagram located on the inside cover of the furnace control box and in these instructions.

NOTE: L1 (hot) and L2 (neutral) polarity must be observed when making field connections to the furnace. The ignition module will not sense flame if L1 and L2 are reversed.

Installation of the electric supply line should be in accordance with the National Electric Code ANSI/NFPA No. 70, latest edition, or Canadian Electric Code Part 1 - CSA Standard C22.1 and local building codes.

This can be obtained from:

National Fire Protection Association
Batterymarch Park
Quincy, MA 02269

Canadian Standards Association
178 Roxdale Boulevard
Roxdale, Ontario, Canada M9W 1R3

THERMOSTAT

Install room thermostat in accordance with instruction sheet in box with the thermostat. Run thermostat lead wires inside blower compartment. Connect thermostat as shown on wiring diagram. Never install thermostat on an outside wall or where it will be influenced by drafts, concealed hot or cold water pipe or ducts, lighting fixtures, radiation from fireplace, rays of sun, lamps, television, radios or air streams from registers. Refer to instructions packed with thermostat for "heater" adjustment or selection.

HEAT ANTICIPATOR SETTINGS

The thermostat heat anticipator setting is .60 amps. To field check setting: (a) add the current draw of the various components in the system or (b) measure the current flow on either the R or W thermostat circuits and set the thermostat heat anticipator according to the current flow measured.

CIRCULATING AIR SUPPLY

The circulating air supply may be taken either: (1) from outside the building, (2) from return air ducts from several rooms, or (3) any combination of the two. When outside air is utilized, the system must be designed and adjusted such that the temperature of the supply air to the furnace will not be below 50°F during the heating season. When using a combination of outside air and return air, be sure the ducts are so designed and a diverting damper so installed that the volume of circulating air entering the furnace cannot be reduced or restricted below that which would normally enter through the circulating air intake of the furnace.

Plenum chambers and air ducts shall be installed in accordance with the Standard for the Installation of Air Conditioning and Ventilating Systems, NFPA No. 90A, or the Standard for the Installation of Warm Air Heating and Air Conditioning Systems, NFPA No. 90B.

When the furnace is installed so that the supply ducts carry air circulated by the furnace to areas outside the space containing the furnace, the return air shall be handled by a duct or ducts sealed to the furnace casing and terminating outside the space containing the furnace. If there is no complete return air duct system, the return air connection must be sealed to the furnace casing and run full size to a location outside the utility room or space housing the furnace to prevent a negative pressure on the venting systems. When a cooling coil is used in connection with a furnace, it must be installed downstream of the furnace (outlet end of furnace) or in parallel with the furnace to avoid condensation in the heating element.

If installed in parallel with a cooling unit, the damper, or other means used to control the flow of air, must be adequate to prevent chilled air from entering the furnace, and if manually operated must be equipped with means to prevent operation of the other unit unless the damper is in the full heat or cool position.

CAUTION: One of the most common causes of trouble in forced air heating systems is insufficient return air to the furnace. The return air system should be approximately equal to or greater than the area of the warm air discharge. CONSULT LOCAL CODES FOR SPECIAL REQUIREMENT. Blower speed should be adjusted to maintain the air rise range shown on the rating plate.

WARNING: THE SOLID METAL BASE PLATE SUPPLIED WITH THE FURNACE MUST BE IN PLACE WHEN THE FURNACE IS INSTALLED WITH SIDE OR REAR AIR RETURN DUCTS. FAILURE TO INSTALL THE BASE PLATE COULD CAUSE PRODUCTS OF COMBUSTION TO BE CIRCULATED INTO THE LIVING SPACE AND CREATE POTENTIALLY HAZARDOUS CONDITIONS, INCLUDING CARBON MONOXIDE POISONING.

RETURN AIR

WARNING: NEVER ALLOW THE PRODUCTS OF COMBUSTION OR THE FLUE PRODUCTS TO ENTER THE RETURN AIR DUCTWORK OR THE CIRCULATED AIR SUPPLY. ALL RETURN DUCTWORK MUST BE ADEQUATELY SEALED AND SECURED TO THE FURNACE WITH SHEET METAL SCREWS, AND JOINTS TAPED. ALL OTHER DUCT JOINTS MUST BE SECURED WITH APPROVED CONNECTIONS AND SEALED AIRTIGHT. WHEN A FURNACE IS MOUNTED ON A PLATFORM WITH RETURN THROUGH THE BOTTOM, IT MUST BE SEALED AIRTIGHT BETWEEN THE FURNACE AND THE RETURN AIR PLENUM. THE FLOOR OR PLATFORM MUST PROVIDE SOUND PHYSICAL SUPPORT OF THE FURNACE WITHOUT SAGGING, CRACKS, GAPS, ETC., AROUND THE BASE AS TO PROVIDE A SEAL BETWEEN THE SUPPORT AND THE BASE.

FAILURE TO PREVENT PRODUCTS OF COMBUSTION FROM BEING CIRCULATED INTO THE LIVING SPACE CAN CREATE POTENTIALLY HAZARDOUS CONDITIONS, INCLUDING CARBON MONOXIDE POISONING THAT COULD RESULT IN PERSONAL INJURY OR DEATH.

DO NOT, UNDER ANY CIRCUMSTANCES, CONNECT RETURN OR SUPPLY DUCTWORK TO OR FROM ANY OTHER HEAT PRODUCING DEVICE SUCH AS A FIREPLACE INSERT, STOVE, ETC. DOING SO MAY RESULT IN FIRE, CARBON MONOXIDE POISONING, EXPLOSION, PERSONAL INJURY OR PROPERTY DAMAGE.

Install the cold air return to terminate through the floor under the furnace. A direct connection should be made to the bottom of the furnace. For installations where return air ducts cannot be run under the floor, return air may be taken from the side or rear as required. A side or rear return air cabinet is available from the manufacturer.

When side air return is used, determine the size opening required, and scribe a line between the knockout squares, cut out opening along these lines.

WARNING: BLOWER AND BURNERS MUST NEVER BE OPERATED WITHOUT BLOWER DOOR IN PLACE. THIS IS TO PREVENT DRAWING GAS FUMES (WHICH COULD CONTAIN HAZARDOUS CARBON MONOXIDE) INTO THE HOME THAT COULD RESULT IN PERSONAL INJURY OR DEATH.

START-UP PROCEDURE

TO PUT FURNACE IN OPERATION

Carefully read and follow the instruction label on how to put the furnace in operation located on the furnace. This label will have complete and detailed instructions for the gas control system on the furnace.

FOR YOUR SAFETY READ BEFORE OPERATING.

WARNING: IF YOU DO NOT FOLLOW THESE INSTRUCTIONS EXACTLY, A FIRE OR EXPLOSION MAY RESULT CAUSING PROPERTY DAMAGE, PERSONAL INJURY, OR LOSS OF LIFE.

A. This appliance does not have a pilot. It is equipped with an ignition device which automatically lights the burners. Do <u>not</u> try to light the burner by hand.

B. **BEFORE OPERATING** smell around the appliance area for gas. Be sure to smell next to the floor because some gas is heavier than air and will settle to the floor.

WHAT TO DO IF YOU SMELL GAS

- Do not try to light any appliance.
- Do not touch any electric switch; do not use any phone in your building.
- Immediately call your gas supplier from a neighbor's phone. Follow the gas supplier's instructions.
- If you cannot reach your gas supplier, call the fire department.

C. Use only your hand to push in or turn the gas control knob. Never use tools. If the knob will not push in or turn by hand, don't try to repair it, call a qualified service technician. Force or attempted repair may result in a fire or explosion.

D. Do not use this appliance if any part has been under water. Immediately call a qualified service technician to inspect the appliance and replace any part of the control system and any gas control that has been under water.

OPERATING INSTRUCTIONS

1. **STOP!** Read all safety information above.
2. Set the thermostat to lowest setting.
3. Turn off all electric power to the appliance.
4. This appliance does not have a pilot. It is equipped with an ignition device which automatically lights the burner. Do <u>not</u> try to light the burner by hand.
5. Remove control door.

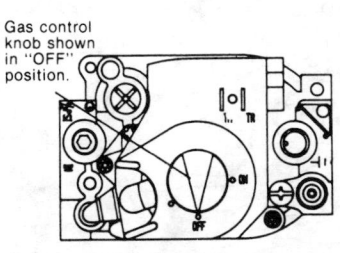
Gas control knob shown in "OFF" position.

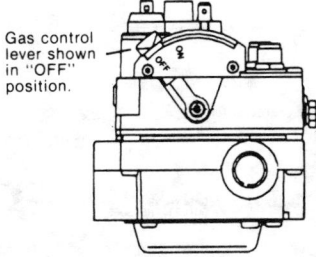
Gas control lever shown in "OFF" position.

HONEYWELL VALVE   **ROBERTSHAW 7200 VALVE**

FIGURE 8. GAS CONTROL VALVES

6. HONEYWELL VALVE: Turn manual gas control knob clockwise to "Off" position.

   ROBERTSHAW 7200 VALVE: Depress control lever and move to "Off" position.

7. Wait five (5) minutes to clear out any gas. If you then smell gas, **STOP!** Follow "B" in the safety information above. If you don't smell gas, go on to the next step.

8. HONEYWELL VALVE: Turn on gas to main burners by turning the manual gas control knob counterclockwise to "On" position.

   ROBERTSHAW 7200 VALVE: Move gas control lever from "Off" to "On" position.

9. Replace control door.

10. Turn on all electric power to the furnace.

11. Set thermostat to desired setting.

12. If the appliance will not operate, follow the instructions "To Turn Off Gas Appliance" and call your service technician or gas appliance.

TO TURN OFF GAS APPLIANCE

1. Set thermostat to lowest setting.

2. Turn off all electrical power to the appliance if service is to be performed.

3. Remove control door.

4. HONEYWELL VALVE: Turn manual gas control knob to "Off" position.

   ROBERTSHAW 7200 VALVE: Depress control lever and move to "Off" position.

5. Replace control door.

WARNING: SHOULD OVERHEATING OCCUR OR THE GAS SUPPLY FAIL TO SHUT OFF, SHUT OFF THE MANUAL GAS VALVE TO THE APPLIANCE BEFORE SHUTTING OFF THE ELECTRICAL SUPPLY.

SEQUENCE OF OPERATION (HONEYWELL INTEGRATED CONTROL)

This furnace is equipped with a Honeywell S9201A integrated ignition and blower control board. This control combines functions of the hot surface ignition 100% lockout safety control and fixed time off blower controls. It also provides a low voltage heat-cool thermostat control terminal board and connection points for field installed humidifier and electronic air cleaner optional accessories. Two indicator lights are also provided to aid the service technician.

Upon heat call from the thermostat (connection of R and W terminals), the induced draft blower starts and runs through a 30 second prepurge cycle. After the induced draft blower starts, the air proving differential negative pressure switch closes and starts the main burner ignition cycle. The hot surface ignitor is energized for 36 seconds to heat up, then the gas valve is energized to start gas flow to the main burner for ignition. The main gas burner flame is sensed by the de-energized hot surface ignitor within 0.8 seconds. If main burner flame is not sensed within the six second maximum trial for ignition time, the control will repeat the prepurge and ignition cycle for four additional retries. After a total of five cycles without sensing main burner flame, the system will then go into a 100% lockout mode. During the lockout mode neither the hot surface ignitor or the gas valve will be energized until the system is reset by opening the thermostat (disconnecting the R and W terminals) or interrupting the electrical power for ten seconds or longer. The induced draft blower and main burner will shut off when the thermostat is satisfied (R and W open).

The fixed time blower control will start the circulating air blower on heat speed thirty seconds after the main burner is ignited. The circulating air blower will continue to run during burner operation then shut down 120 seconds after the burner shuts off. The circulating air blower will start and run on heating speed if the thermostat fan switch is in "ON" position and the thermostat mode switch is in the "HEAT" position. Upon heat call while in this mode, the blower will stop and go through delay until 30 seconds after the burner lights. The blower control will automatically energize the circulating air blower if any of the high temperature limit controls are actuated (open) to shut off the main burner due to abnormal temperature conditions.

When the thermostat is in the cooling mode, the blower control will start and stop the circulating air blower on cooling speed when the thermostat contacts close or open respectively. The circulating air blower will start and run continuously at heating speed with the thermostat fan switch in the "ON" position and the thermostat in the cooling mode. The blower will step up to cooling speed when both terminals G and Y are energized.

The heating mode blower time off delay can be changed to 80, 160 or 200 seconds by moving the blower time delay switch screws, SW1 and 2. Both screws are tightened "down" in contact with the base wires to give the factory setting of 120 second blower off delay. To raise the screw "up" and break contact with the base wires, rotate screw counterclockwise one half turn with a screwdriver. Blower off delay can be changed by positioning the SW screws as follows:

SW1	SW2	Off Delay
Down	Down	120 Seconds
Down	Up	160 Seconds
Up	Down	80 Seconds
Up	Up	200 Seconds

FIELD INSTALLED OPTION ACCESSORIES

1. Electronic air cleaner line voltage power can be supplied from the screw terminal "EAC" and a line voltage neutral screw terminal on the control board. This will power the electronic air cleaner whenever the circulating air blower is in operation.

2. Humidifier line voltage power can be supplied from screw terminal "HUM" to a line voltage neutral screw terminal on the control board. This will power the humidifier whenever the burner is on and the circulating air blower is operating in the heating mode.

TWINNING INSTRUCTIONS

Twinning or parallel operation of two furnaces, side by side, connected to a common duct system controlled by a single thermostat can be done with the Honeywell S9201A-1002 integrated furnace control. The interconnection of the controls and instructions shown in wiring diagram 90-22932-01 will assure simultaneous operation of both indoor air blowers independent of the status of both furnaces. These applications occur when two furnaces are used to supply air for a 7-1/2 or 10 ton cooling coil in a common duct system.

1. Install the furnaces side by side and make connections to the common supply and return air duct system. It may be convenient to convert to right side gas and electrical supply to the furnace on the right side.

2. Identify the furnaces number 1 and 2 for wiring purposes. The wiring diagram arbitrarily shows furnace 1 on the left.

3. Be sure the electrical supply to both furnaces is turned off while making these electrical connections.

4. Disconnect and remove the control transformer from furnace 2.

5. Connect control power (24VAC) from terminal board "R" in furnace 1 to the "24VAC" terminal (1/4" push on) on the control board in furnace 2.

6. Connect control power (24VAC) from terminal board "C" in furnace 1 to terminal board "C" in furnace 2. ·

7. Connect control (24VAC) jumper wires between terminal board "W," "Y" and "G" in furnace 1 and furnace 2.

8. Connect a line voltage (115VAC) 12 gauge jumper wire between "EAC" screw terminal on board in furnace 1 to "EAC" screw terminal on board in furnace 2. The wire must be enclosed in the furnace cabinets, secured in place and protected with bushings where it goes through cabinet or control box sides. This wire should have 4/64 inch insulation (or equivalent) with 105 degree C temperature rating similar to other line voltage conductors within the furnace.

9. Install a control thermostat (not supplied) in accordance with its instructions. Control power (24VAC) to the thermostat must be supplied from terminal board "R" in furnace 2. Connect the thermostat to "W" (heat), "Y" (cool) and "G" (fan) common to both furnaces as shown.

HELP GUIDES

A service trouble shooting guide is located on the outside of the control box cover. A wiring diagram is located on the inside of the control box cover. *CAUTION: The control box contains exposed line voltage connections; disconnect power before servicing. Service must be by a trained qualified service technician.*

Two green LED help lights are located on the control board. Both lights should be on when the board is operating. The first light marked PWR is lit when 24 volt power is available on the control board. The second light marked OK is lit when the board control circuit is operational.

TO RATE FURNACE

Manifold pressure should be set at 3.5" W.C. for natural gas. The inlet pressure to the gas valve must be between 5" and 7" W.C. Only small variations in the gas flow should be made by means of the pressure regulator adjustment. In no case should the final manifold pressure vary more than plus or minus 0.3" W.C. from the above specified pressures.

The gas control pressure regulator references the pressure in the burner tube rather than atmosphere or room pressure. The manifold pressure may be measured as the differential between the manifold, pressure tap or gas valve outlet and the burner tube pressure, with the regulator cap in place.

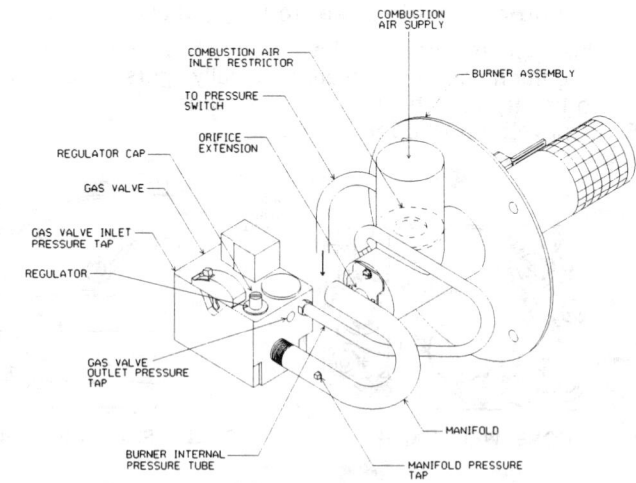

FIGURE 9. BURNER, MANIFOLD & GAS VALVE

The manifold pressure can also be measured by removing the burner internal pressure tube where it connects to the barbed fitting on the gas valve and measuring the positive manifold gas pressure at the manifold tap or the gas valve outlet. After measuring the manifold pressure, reconnect the burner internal pressure tube to the barbed fitting on the gas valve to maintain the proper pressure differential between the manifold and the burner when measuring input rate. *CAUTION: If the burner internal pressure tube is not reconnected, the furnace will be overfired due to the negative pressure in the burner and heat exchanger. See Figure 9 for location of connections.*

To adjust pressure regulator, remove regulator cap and turn adjustment screw clockwise to increase pressure or counterclockwise to decrease pressure, then replace regulator cap securely. Any necessary major changes in the gas flow rate should be made by changing the size of the burner orifices. To change orifice spud, shut off manual main gas valve and remove the manifold. For furnaces for use on LP gas, the LP gas supply pressure must be set between 11.0" and 14.0" W.C. by means of the tank or branch supply regulators. The furnace manifold pressure should be set at 10" W.C. at the combination gas control valve. For elevations up to 2,000 feet, rating plate input ratings apply. For elevations over 2,000 feet, reduce input 4% for each 1,000 feet above sea level. In Canada, conversion for elevations above 4,500 feet is required to be done by a certified conversion station.

Check of input is important to prevent overfiring of the furnace beyond its design-rated input. NEVER SET INPUT ABOVE THAT SHOWN ON THE RATING PLATE. Use TABLE 3 or formula to determine input rate. Start furnace and measure time required to burn one cubic foot of gas. Time the meter with only the furnace in operation.

Prior to checking furnace input, make certain that all other gas appliances are shut off, with the exception of pilot burners.

TABLE 3. METER TIME

METER TIME IN MINUTES AND SECONDS FOR NORMAL INPUT RATING OF FURNACES EQUIPPED FOR NATURAL OR LP GAS											
INPUT BTU/HR	METER SIZE CU. FT.	HEATING VALUE OF GAS BTU PER CU. FT.									
		900		1000		1040		1100		2500	
		MIN.	SEC.	MIN.	SEC.	MIN.	SEC.	MIN.	SEC.	MIN.	SEC.
45,000	ONE	1	12	1	20	1	23	1	28	3	20
	TEN	12	0	13	20	13	50	14	40	33	20
60,000	ONE	0	54	1	0	1	3	1	6	2	30
	TEN	9	0	10	0	10	24	11	0	25	0
75,000	ONE	0	44	0	48	0	50	0	53	2	0
	TEN	7	12	8	0	8	19	8	48	20	0
90,000	ONE	0	36	0	40	0	42	0	44	1	40
	TEN	6	0	6	40	7	0	7	20	16	40
105,000	ONE	0	31	0	34	0	36	0	38	1	26
	TEN	5	10	5	40	6	0	6	20	14	20
120,000	ONE	0	27	0	30	0	31	0	33	1	15
	TEN	4	30	5	0	5	10	5	30	12	30

$$\text{Input BTU/HR} = \frac{\text{Heating Value of Gas (BTU/FT}^3) \times 3600}{\text{Time in Seconds (for 1 cu. ft.) of Gas}}$$

LIMIT CONTROL

The high limit cut-off is set at the factory and cannot be adjusted. It is calibrated to prevent the air temperature leaving the furnace from exceeding the maximum outlet air temperature.

A manual reset type limit control is installed on the induced draft blower housing for sensing flue gas temperature. This control will shut off the main burners if abnormal flue gas temperatures occur.

FAN CONTROL

The furnace is equipped with an integrated circuit board which controls the on-off sequence of the blower motor. The on-off timings are factory fixed for consistent blower operation. This control also includes a relay to control the induced draft blower and a low voltage terminal board for thermostat connections. The low voltage terminals are located on the lower left side of the control box in the blower compartment.

PRESSURE SWITCH

The furnace has a differential pressure switch to sense adequate flow of combustion air/products through the furnace and vent system. The pressure switch will also shut the furnace down if a blocked drain condition exists.

HOT SURFACE IGNITER

This furnace has a hot surface ignition system which features direct ignition of the main burners. The igniter also acts as a sensor for proving main burner ignition.

CAUTION: The hot surface igniter is fragile. Care must be taken to avoid direct contact with the dark silicon carbide elements.

Leads to the igniter are line voltage when there is a call for heat. Keep hands or tools away to prevent hazard of electrical shock. Shut off electrical power before servicing any of the controls.

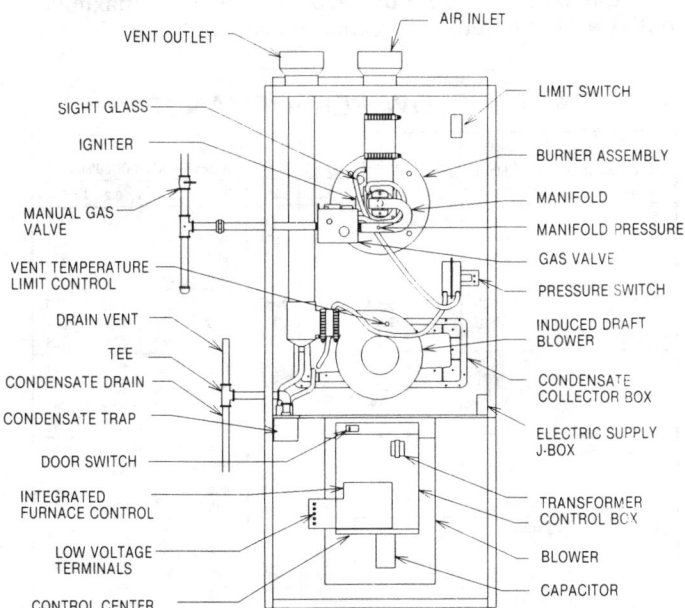

FIGURE 10. COMPONENT LOCATION (FURNACE SHOWN WITH CONTROL AND BLOWER DOORS REMOVED)

CONTROL CIRCUIT SAFETY CHECK AND SEQUENCE OF OPERATION

The safety features of the control system should be checked to assure proper operation of the furnace.

1. Thermostat contacts close and supply power through safety limit switches to the control circuit which starts the draft inducer. When required, combustion air is established through the pressure switch, power is then supplied to the ignition control. A prepurge period of approximately 30 seconds followed by an igniter heat up of 36 seconds. After the igniter heats up, the main valve opens and the burner is lit. The igniter then

becomes the flame sensor. To check the sense function of the igniter, turn off the gas supply at the gas valve and allow the furnace to attempt to relight. Since no gas will flow, the igniter should lockout the system after 5 ignition attempts.

2. The high limit control located on the vestibule control compartment prevents excessive outlet air temperature by shutting off the automatic main gas valve when the maximum outlet air temperature has been reached. To check the control operation, block the air flow through the unit temporarily. The limit switch should function to turn off the automatic gas valve within a few minutes. Remove blockage immediately after switch operates.

GAS FURNACE (DIRECT DRIVE) INSTRUCTIONS FOR CHANGING BLOWER SPEED

CAUTION: Disconnect electrical supply to furnace before attempting to change blower speed.

Blower motor is wired for blower speeds required for normal operation as shown in Table 4.

If additional blower motor speed taps are available (leads connected to M1 and M2 on fan control), speeds may be changed if necessary to fit requirements of particular installation. Reconnect the unused motor leads to M1 or M2. Check motor lead color for speed designation.

Heating speeds should not be reduced where it could cause the furnace air temperature rise to exceed the maximum outlet air temperature specified for the unit.

TABLE 4. AIR FLOW PERFORMANCE

Input Rating	Blower Size	Motor H.P.	Blower Speed	CFM AIR DELIVERY EXTERNAL STATIC PRESSURE INCHES WATER COLUMN							
				0.8	0.7	0.6	0.5	0.4	0.3	0.2	0.1
45.000	11x7	1 4 HP 4-Speed	High	—	1178	1206	1234	1263	1290	1315	1338
			Med-Hi	—	870	891	911	930	946	959	969
			Med-Lo	—	753	778	794	807	822	843	873
			Low *	—	527	561	585	600	611	621	632
60.000	11x7	1 3 HP 3 Speed	High	—	1403	1450	1488	1518	1539	1552	1559
			Med *	—	1011	1033	1041	1040	1031	1020	1008
			Low	—	751	759	765	768	766	756	738
75.000	11x7	1 2 HP 4-Speed	High	—	1447	1491	1530	1563	1591	1617	1641
			Med-Hi	—	1253	1285	1309	1324	1330	1326	1313
			Med-Lo *	—	1006	1027	1041	1049	1054	1058	1063
			Low	—	858	884	900	906	905	900	893
90.000	11x10	3/4 HP 4-Speed	High	—	1840	1915	1980	2045	2105	2160	2210
			Med-Hi	—	1620	1675	1720	1755	1790	1820	1845
			Med-Lo	—	1380	1420	1440	1460	1465	1475	1480
			Low *	—	1090	1140	1170	1185	1200	1210	1215
105.000	11x10	3/4 HP 4-Speed	High	—	1771	1850	1927	1998	2060	2110	2145
			Med-Hi	—	1556	1611	1678	1748	1810	1853	1868
			Med-Lo	—	1308	1362	1399	1425	1446	1469	1500
			Low *	—	1047	1098	1136	1163	1183	1198	1212
120.000	11x10	3/4 HP 4-Speed	High	—	1870	1980	2040	2085	2120	2145	2160
			Med-Hi	—	1660	1770	1840	1895	1935	1960	1980
			Med-Low *	—	1440	1520	1580	1640	1680	1725	1760
			Low	—	1190	1240	1280	1320	1360	1390	1410

Air conditioning models are A.G.A. or C.G.A. certified at a maximum E.S.P. of 0.5 inches.
Furnaces are certified as follows:
Input of 45.000 BTU/HR at .10 E.S.P. (in.) (.12 E.S.P. CGA) Canadian
Input of 60, 75.000 BTU/HR at .12 E.S.P. (in.)
Input of 90.000 BTU/HR at .15 E.S.P. (in.)
Input of 105 & 120.000 and larger at .20 E.S.P. (in.)
* Heating Speed
Motor Wire Colors
Common - White
High Speed - Black
Medium High or Medium Speed - Blue
Medium Low Speed - Yellow
Low Speed - Red

MAINTENANCE

WARNING: DISCONNECT MAIN ELECTRICAL POWER TO THE UNIT BEFORE ATTEMPTING ANY MAINTENANCE.

1. Keep the air filters clean. A new home may require more frequent attention to the filters until dust from construction is removed.

2. How to Clean Filters:

Plastic Impregnated Fiber - Vacuum clean; wash with detergent and water. Air dry thoroughly and reinstall.

CAUTION: Do not operate your system for extended periods without filters. A portion of the dust entrained in the air may temporarily lodge in the air duct runs and at the supply registers. Any recirculated dust particles will be heated and charred by contact with the furnace heat exchanger. This residue will soil ceilings, walls, drapes, carpets, and other household articles.

MINIMUM FILTER SIZES (IN.)

Width	Filter Size	Type
21	20 x 25 x 1	Cleanable
24½	24 x 25 x 1	Cleanable

3. With horizontal venting, the screen in the vent terminal should be inspected and cleaned annually. Also, inspect and clean the screen in the inlet air terminal.

4. Be sure the condensate drain line does not become blocked or plugged. Visual inspection of the condensate flow can easily be made while the furnace is in operation. A flashlight can be used to illuminate the discharge end placed into the drain.

5. Clean the condensate trap and flush the condensate tube to make sure condensate flows freely while the furnace is in operation.

6. If a means to neutralize the condensate has been provided for the furnace, be sure to follow the maintenance schedule furnished by the manufacturer.

7. Start the furnace and observe its operation. Watch the burner flames to see if they are bright blue. If not remove, inspect and clean burner.

LUBRICATION

The blower motor sleeve bearings are prelubricated by the motor manufacturer and may not require attention for an indefinite period of time. However, our recommendations are as follows:

1. Motors without oiling ports -

Prelubricated and sealed. No further lubrication should be required, but in case of bearing noise problems, the blower and the motor end bells can be disassembled and the bearings relubricated by a qualified service person.

2. Motors with oiling ports -

Add from 10 to 20 drops of Electric Motor Oil or an SE grade of non-detergent SAE-10 or 20 motor oil to each bearing every two years for somewhat continuous duty, or at least every five years for light duty. DO NOT over oil, because excessive lubrication can damage the motor. Lubrication to be done by a qualified service person.

3. To access oiling ports or to relubricate sealed indoor blower motor bearings, removal of the blower assembly and motor is required.

Remove the four screws securing the control center box to the blower shelf and move the box aside. The thermostat and/or motor leads may have to be disconnected. Remove the two screws at the front of the blower housing, securing the assembly to the blower shelf flange. Slide the blower assembly forward until it is disengaged from the shelf track. To remove the motor, loosen the set screw(s) that secure the blower wheel to the motor shaft, remove the screw securing the motor ground wire to the housing, remove the three screws from the motor mounting bracket legs and slide the bracket assembly from the housing.

Lubricate the blower motor bearings and reassemble by reversing the procedure outlined above. Exercise care when sliding the blower assembly into the shelf track and be sure that the rear flange engages properly. This procedure should be done by a qualified service person.

In any event, clean motor periodically to prevent the possibility of overheating due to an accumulation of dust and dirt on the windings or on the motor exterior. And, as suggested elsewhere in these instructions, the air filters should be kept clean because dirty filters can restrict airflow and the motor depends upon sufficient air flowing across and through it to keep from overheating.

The induced draft blower is prelubricated and sealed. No further lubrication is required.

CLEANING PRIMARY HEAT EXCHANGER AND BURNER ASSEMBLY

The furnace should operate for many years without excessive scale buildup in the flue passageways. However, it is recommended that a qualified technician familiar with this furnace inspect the flue passageways, the vent system and the main burners for correct operation, paying particular attention to deterioration from corrosion or other sources. The flue passageways and vent system should be inspected and cleaned (if required) by a qualified serviceman after the second year of service and annually thereafter using the following procedure.

1. Turn off all power to the furnace and set the thermostat lever to the lowest temperature.

2. Shut off the gas supply to the furnace either at the meter or at a manual valve in the supply piping.

3. Remove the control door from the furnace.

4. Turn the gas control knob to the "Off" position.

5. Disconnect tubing at burner barbed connections.

6. Loosen clamps on air inlet coupling and slide up on air inlet pipe.

7. Remove manifold and gas valve assembly.

8. Disconnect igniter leads at connector.

9. Remove burner mounting nuts and slide burner assembly out using care not to touch or bump the igniter. It is very fragile.

10. The furnace can now be cleaned by the use of a wire brush with a flexible handle. Slide the brush in burner opening. Sweeping back and forth will loosen any scale, allowing it to fall to the bottom of the combustion chamber. The debris can now be cleaned with the nozzle of a vacuum cleaner.

11. Check the ports on the gas burners to make certain that they are clean. Brushing or jarring may loosen any accumulation.

12. With a light and a mirror, the condition of the inside of the element can be observed.

13. Reassemble steps 1 through 9 in reverse order.

REPLACEMENT PARTS

Replacement parts for service are listed in the PARTS LIST included with these instructions.

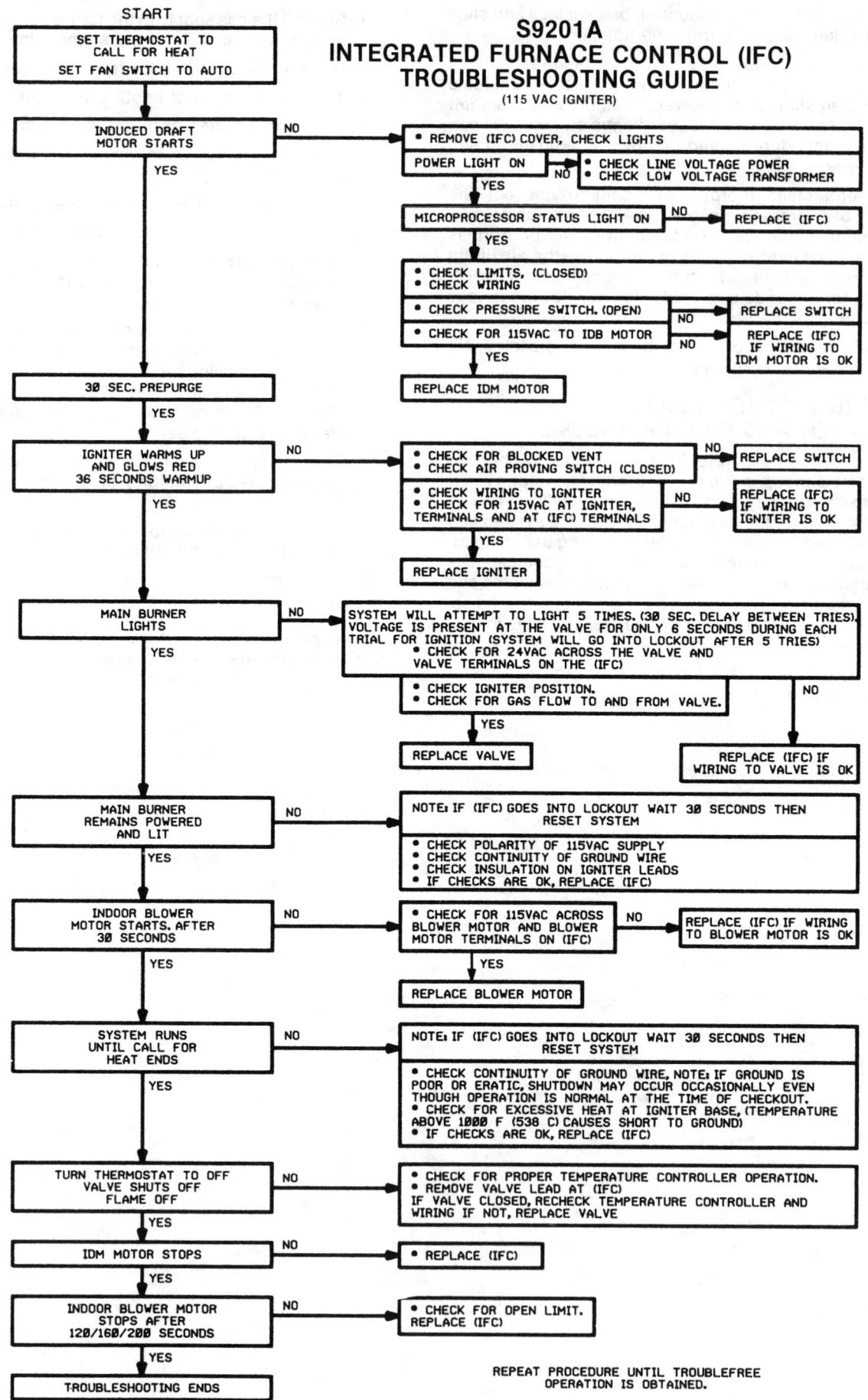

S9201A
INTEGRATED FURNACE CONTROL (IFC)
TROUBLESHOOTING GUIDE
(115 VAC IGNITER)

START

SET THERMOSTAT TO CALL FOR HEAT
SET FAN SWITCH TO AUTO

INDUCED DRAFT MOTOR STARTS — NO →
- REMOVE (IFC) COVER, CHECK LIGHTS

POWER LIGHT ON — NO →
- CHECK LINE VOLTAGE POWER
- CHECK LOW VOLTAGE TRANSFORMER

YES

MICROPROCESSOR STATUS LIGHT ON — NO → REPLACE (IFC)

YES

- CHECK LIMITS, (CLOSED)
- CHECK WIRING

- CHECK PRESSURE SWITCH. (OPEN) — NO → REPLACE SWITCH
- CHECK FOR 115VAC TO IDB MOTOR — NO → REPLACE (IFC) IF WIRING TO IDM MOTOR IS OK

YES

REPLACE IDM MOTOR

YES

30 SEC. PREPURGE

YES

IGNITER WARMS UP AND GLOWS RED 36 SECONDS WARMUP — NO →
- CHECK FOR BLOCKED VENT
- CHECK AIR PROVING SWITCH (CLOSED) — NO → REPLACE SWITCH

- CHECK WIRING TO IGNITER
- CHECK FOR 115VAC AT IGNITER, TERMINALS AND AT (IFC) TERMINALS — NO → REPLACE (IFC) IF WIRING TO IGNITER IS OK

YES

REPLACE IGNITER

YES

MAIN BURNER LIGHTS — NO →
SYSTEM WILL ATTEMPT TO LIGHT 5 TIMES. (30 SEC. DELAY BETWEEN TRIES). VOLTAGE IS PRESENT AT THE VALVE FOR ONLY 6 SECONDS DURING EACH TRIAL FOR IGNITION (SYSTEM WILL GO INTO LOCKOUT AFTER 5 TRIES)
- CHECK FOR 24VAC ACROSS THE VALVE AND VALVE TERMINALS ON THE (IFC)

- CHECK IGNITER POSITION.
- CHECK FOR GAS FLOW TO AND FROM VALVE. — NO →

YES

REPLACE VALVE

REPLACE (IFC) IF WIRING TO VALVE IS OK

YES

MAIN BURNER REMAINS POWERED AND LIT — NO →
NOTE: IF (IFC) GOES INTO LOCKOUT WAIT 30 SECONDS THEN RESET SYSTEM

- CHECK POLARITY OF 115VAC SUPPLY
- CHECK CONTINUITY OF GROUND WIRE
- CHECK INSULATION ON IGNITER LEADS
- IF CHECKS ARE OK, REPLACE (IFC)

YES

INDOOR BLOWER MOTOR STARTS. AFTER 30 SECONDS — NO →
- CHECK FOR 115VAC ACROSS BLOWER MOTOR AND BLOWER MOTOR TERMINALS ON (IFC) — NO → REPLACE (IFC) IF WIRING TO BLOWER MOTOR IS OK

YES

REPLACE BLOWER MOTOR

YES

SYSTEM RUNS UNTIL CALL FOR HEAT ENDS — NO →
NOTE: IF (IFC) GOES INTO LOCKOUT WAIT 30 SECONDS THEN RESET SYSTEM

- CHECK CONTINUITY OF GROUND WIRE. NOTE: IF GROUND IS POOR OR ERATIC, SHUTDOWN MAY OCCUR OCCASIONALLY EVEN THOUGH OPERATION IS NORMAL AT THE TIME OF CHECKOUT.
- CHECK FOR EXCESSIVE HEAT AT IGNITER BASE, (TEMPERATURE ABOVE 1000 F (538 C) CAUSES SHORT TO GROUND)
- IF CHECKS ARE OK, REPLACE (IFC)

YES

TURN THERMOSTAT TO OFF VALVE SHUTS OFF FLAME OFF — NO →
- CHECK FOR PROPER TEMPERATURE CONTROLLER OPERATION.
- REMOVE VALVE LEAD AT (IFC)
IF VALVE CLOSED, RECHECK TEMPERATURE CONTROLLER AND WIRING IF NOT, REPLACE VALVE

YES

IDM MOTOR STOPS — NO →
- REPLACE (IFC)

YES

INDOOR BLOWER MOTOR STOPS AFTER 120/160/200 SECONDS — NO →
- CHECK FOR OPEN LIMIT. REPLACE (IFC)

YES

TROUBLESHOOTING ENDS

REPEAT PROCEDURE UNTIL TROUBLEFREE OPERATION IS OBTAINED.

NOTE: STATIC DISCHARGE CAN DAMAGE INTEGRATED FURNACE CONTROL (IFC)

92-22744-02-01

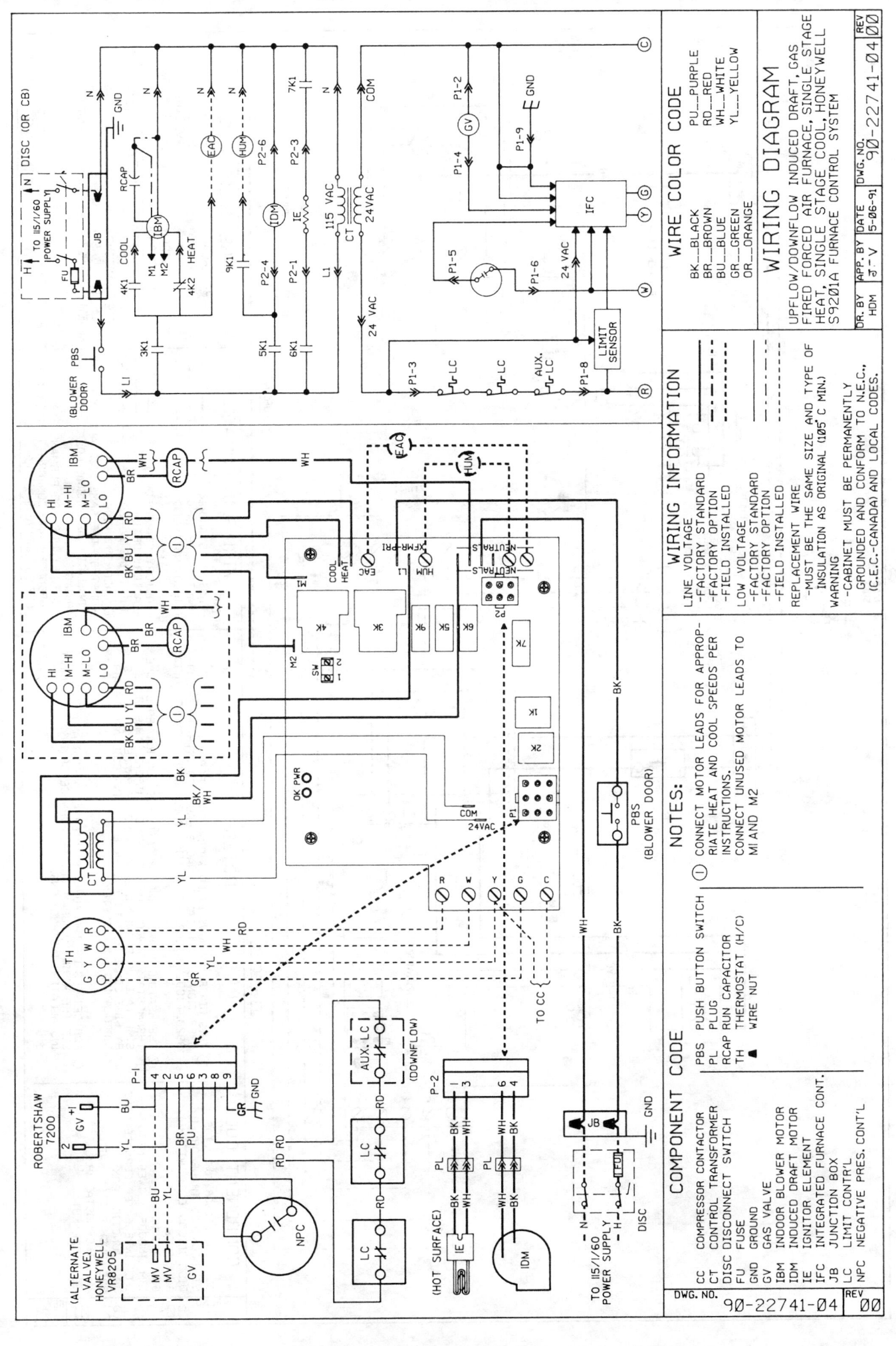

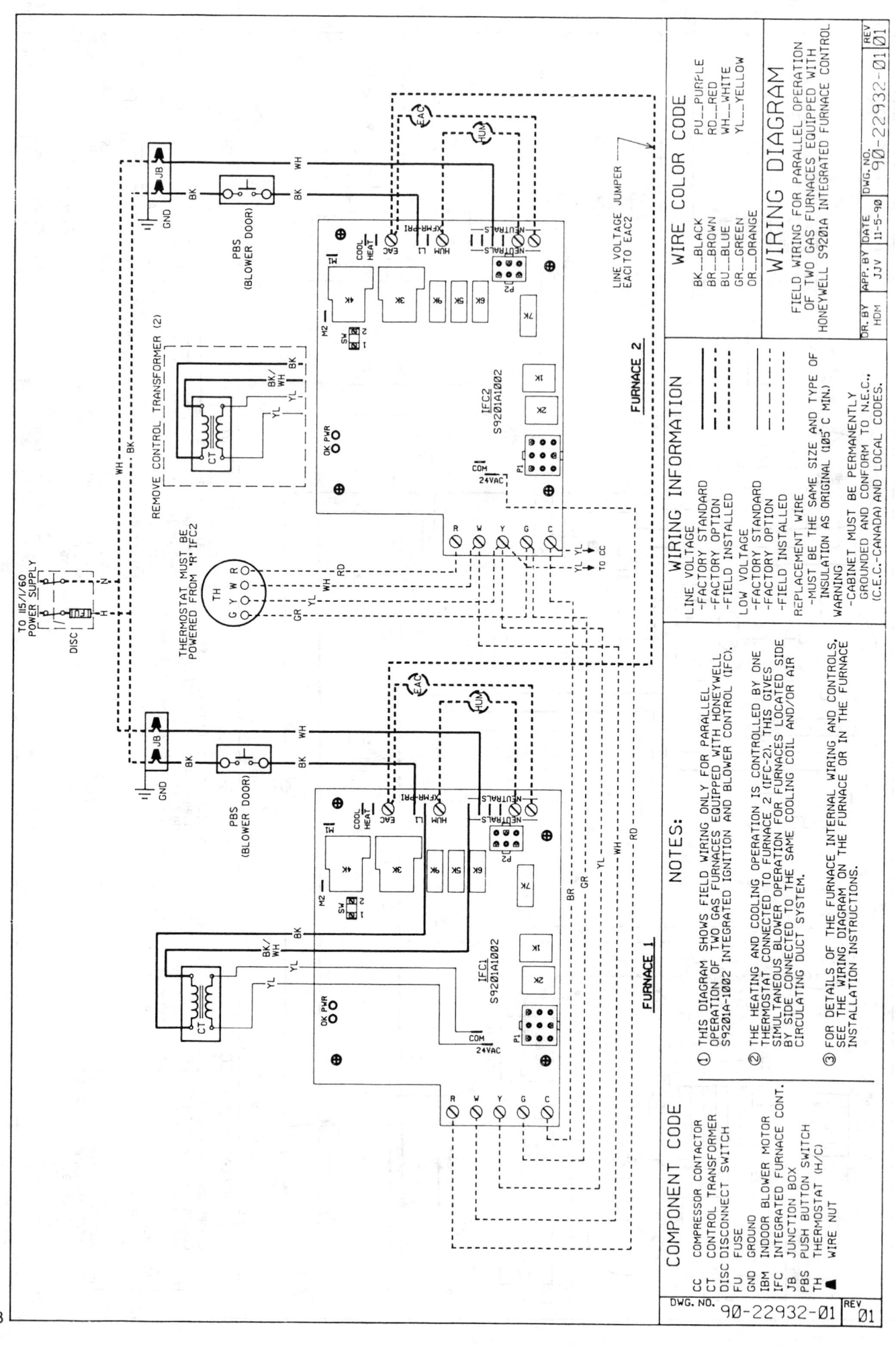

WIRING DIAGRAM

FIELD WIRING FOR PARALLEL OPERATION OF TWO GAS FURNACES EQUIPPED WITH HONEYWELL S9201A INTEGRATED FURNACE CONTROL

| DWG. NO. | 90-22932-01 | REV 01 |

| DR. BY HDM | APP. BY JJV | DATE 11-5-90 |

WIRE COLOR CODE

BK—BLACK
BR—BROWN
BU—BLUE
GR—GREEN
OR—ORANGE

PU—PURPLE
RD—RED
WH—WHITE
YL—YELLOW

WIRING INFORMATION

LINE VOLTAGE
  -FACTORY STANDARD
  -FACTORY OPTION
  -FIELD INSTALLED

LOW VOLTAGE
  -FACTORY STANDARD
  -FACTORY OPTION
  -FIELD INSTALLED

REPLACEMENT WIRE
  -MUST BE THE SAME SIZE AND TYPE OF INSULATION AS ORIGINAL (105°C MIN.)

WARNING
  -CABINET MUST BE PERMANENTLY GROUNDED AND CONFORM TO N.E.C.. (C.E.C.-CANADA) AND LOCAL CODES.

NOTES:

① THIS DIAGRAM SHOWS FIELD WIRING ONLY FOR PARALLEL OPERATION OF TWO GAS FURNACES EQUIPPED WITH HONEYWELL S9201A-1002 INTEGRATED IGNITION AND BLOWER CONTROL (IFC).

② THE HEATING AND COOLING OPERATION IS CONTROLLED BY ONE THERMOSTAT CONNECTED TO FURNACE 2 (IFC-2). THIS GIVES SIMULTANEOUS BLOWER OPERATION FOR FURNACES LOCATED SIDE BY SIDE CONNECTED TO THE SAME COOLING COIL AND/OR AIR CIRCULATING DUCT SYSTEM.

③ FOR DETAILS OF THE FURNACE INTERNAL WIRING AND CONTROLS, SEE THE WIRING DIAGRAM ON THE FURNACE OR IN THE FURNACE INSTALLATION INSTRUCTIONS.

COMPONENT CODE

CC COMPRESSOR CONTACTOR
CT CONTROL TRANSFORMER
DISC DISCONNECT SWITCH
FU FUSE
GND GROUND
IBM INDOOR BLOWER MOTOR
IFC INTEGRATED FURNACE CONT.
JB JUNCTION BOX
PBS PUSH BUTTON SWITCH
TH THERMOSTAT (H/C)
◢ WIRE NUT

| DWG. NO. | 90-22932-01 | REV 01 |

FURNACE 2

FURNACE 1

IFC2
S9201A1002

IFC1
S9201A1002

REMOVE CONTROL TRANSFORMER (2)

THERMOSTAT MUST BE POWERED FROM "R" IFC2

LINE VOLTAGE JUMPER
EAC1 TO EAC2

TO 115/1/60 POWER SUPPLY

USER'S INFORMATION MANUAL
CONDENSING FURNACES

WARNING:
Improper installation, adjustment, alteration, service or maintenance can cause injury or property damage. Refer to this manual. For assistance or additional information consult a qualified installer, service agency or the gas supplier.

FOR YOUR SAFETY

Do not store or use gasoline or other flammable vapors and liquids, or other combustible materials in the vicinity of this or any other appliance.

FOR YOUR SAFETY
WHAT TO DO IF YOU SMELL GAS
- Do not try to light any appliance.
- Do not touch any electrical switch; do not use any phone in your building.
- Immediately call your gas supplier from a neighbor's phone. Follow the gas supplier's instructions.
- If you cannot reach your gas supplier, call the fire department.

DO NOT DESTROY. PLEASE READ CAREFULLY AND KEEP IN A SAFE PLACE FOR FUTURE REFERENCE

CAUTION: READ THESE INSTRUCTIONS THOROUGHLY BEFORE ATTEMPTING TO OPERATE OR MAINTAIN THIS FURNACE.

This furnace has been designed to give you many years of efficient, dependable home comfort. With regular maintenance, this furnace will operate satisfactorily year after year. Please read this manual to familiarize yourself with operation, maintenance and safety procedures.

DEVICES ATTACHED TO THE FLUE OR VENT FOR THE PURPOSE OF REDUCING HEAT LOSS UP THE CHIMNEY, INCLUDING FIELD-INSTALLED DRAFT INDUCERS, HAVE NOT BEEN TESTED AND HAVE NOT BEEN INCLUDED IN THE DESIGN CERTIFICATION OF THIS FURNACE. WE, THE MANUFACTURER, CANNOT AND WILL NOT BE RESPONSIBLE FOR INJURY OR DAMAGE CAUSED BY THE USE OF SUCH UNTESTED AND/OR UNCERTIFIED DEVICES, ACCESSORIES OR COMPONENTS.

SAFETY

WARNING: IMPROPER INSTALLATION, ADJUSTMENT, ALTERATION, SERVICE OR MAINTENANCE CAN CAUSE INJURY OR PROPERTY DAMAGE. REFER TO THE INSTALLATION INSTRUCTIONS PROVIDED WITH THE FURNACE AND THIS MANUAL. FOR ASSISTANCE OR ADDITIONAL INFORMATION, CONSULT A QUALIFIED INSTALLER, SERVICE AGENCY OR THE GAS SUPPLIER.

CAREFULLY FOLLOW THESE SAFETY RULES:

1. Combustible material must not be placed on or against the furnace jacket. The area around the furnace must be kept clear and free of all combustible materials including gasoline and other flammable vapors and liquids.

2. A furnace installed in an attic or other insulated space must be kept free and clear of insulating material. Examine the furnace area when installing the furnace or adding more insulation. Some materials may be combustible.

3. To prevent carbon monoxide poisoning, replace all blower doors and compartment covers after servicing the furnace. Do not operate the unit without all panels and doors securely in place.

4. Should overheating occur, or the gas valve fail to shut off the gas supply, turn off the manual gas valve to the furnace before turning off the electrical supply.

5. Any additions, changes or conversions required in order for the furnace to satisfactorily meet the application needs should be made by a qualified factory distributor or local service dealer, using factory specified or approved parts. Read your WARRANTY. Contact the WARRANTOR for conversion information. This furnace was equipped at the factory for use on NATURAL GAS ONLY. Conversion to LP (propane) gas requires a special kit supplied by the WARRANTOR.

WARNING: OBSTRUCTION OF THE AIR VENT ON AN LP (PROPANE) TANK REGULATOR CAN CAUSE EXPLOSION OR FIRE RESULTING IN SERIOUS PERSONAL INJURY OR PROPERTY DAMAGE. PERIODICALLY INSPECT AND CLEAN THE AIR VENT SCREEN TO PREVENT ANY OBSTRUCTION. KEEP PROTECTIVE REGULATOR COVER IN PLACE, AS EXPOSURE TO THE ELEMENTS CAN CAUSE ICE BUILDUP AND REGULATOR FAILURE.

6. A furnace needs an adequate supply of combustion and ventilation air for proper and safe operation. Do not block or obstruct air openings on the furnace or air openings supplying the area where the furnace is installed. Do not store anything around the furnace that could block the flow of fresh air to the unit. Your installation may receive air from the inside heated space, from the outside, from the attic or crawl space. Whenever adding insulation, be sure the air supply openings are not covered.

7. Do not use this furnace if any part has been under water. Immediately call a qualified service technician to inspect the furnace and to replace any part of the control system and any gas control which has been under water.

92-20802-61-00

SYSTEM OPERATION INFORMATION

1. Keep the air filters clean. Your heating system will operate more efficiently and provide better heating, more economically.

2. Arrange your furniture and drapes so that the supply air registers and the return air grilles are unobstructed.

3. Close doors and windows. This will reduce the heating load on your system.

4. Avoid excessive use of kitchen exhaust fans.

5. Do not permit the heat generated by television, lamps, or radios to influence the thermostat operation.

6. If you desire to operate your system with constant air circulation, please ask advice from your servicing contractor.

During the heating season the operation of the warm air furnace is automatic. Your installing dealer has provided a wall mounted thermostat which is sensitive to the change in temperature of the air moving around the thermostat. Your thermostat will have switches to select some or all of the following instructions:

HEAT - Turns heating on when temperature drops below setpoint.

COOL - Turns cooling on when temperature rises above setpoint.

AUTO - Turns cooling or heating on as required to maintain setpoint.

OFF - Turns heating and cooling modes off. (Fan may still run in FAN-ON.)

FAN-ON - Turns fan on for continuous operation.

FAN-AUTO - Fan cycles on and off with cooling or heating operation.

STARTING OR SHUTTING DOWN YOUR FURNACE

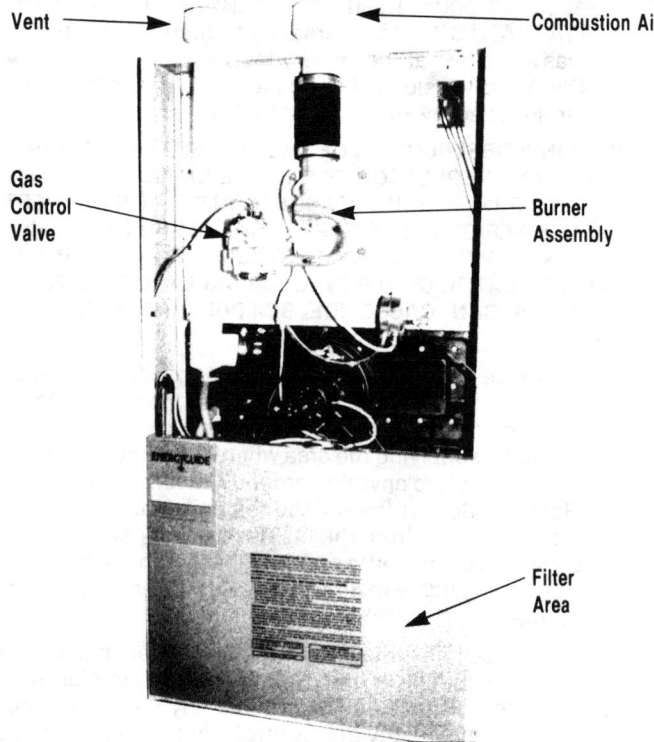

Figure 1. Gas Control Valve - Upflow Model
(Furnace shown with control door removed.)

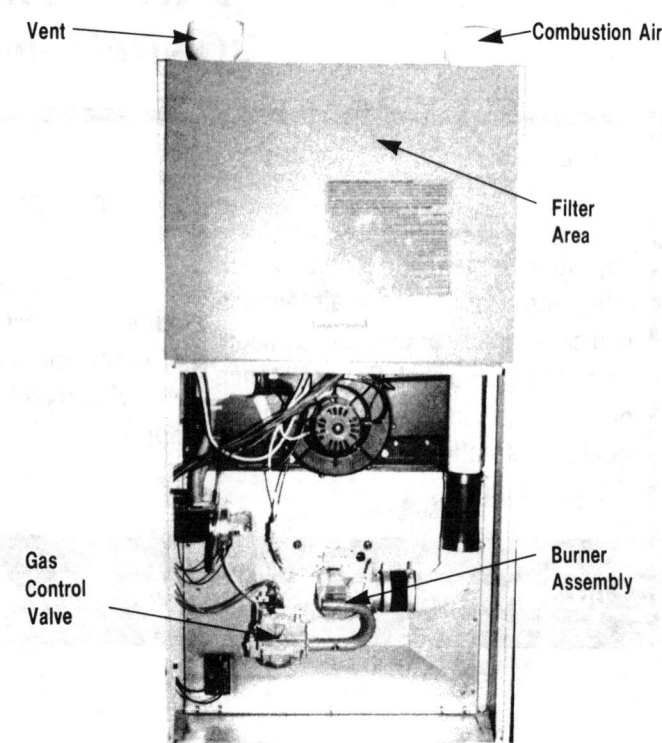

Figure 2. Gas Control Valve - Downflow Model
(Furnace shown with control door removed.)

START-UP PROCEDURE

FOR YOUR SAFETY READ BEFORE OPERATING

WARNING: IF YOU DO NOT FOLLOW THESE INSTRUCTIONS EXACTLY, A FIRE OR EXPLOSION MAY RESULT CAUSING PROPERTY DAMAGE, PERSONAL INJURY OR LOSS OF LIFE.

A. This appliance does not have a pilot. It is equipped with an ignition device which automatically lights the burner. Do not try to light the burner by hand.

B. BEFORE OPERATING smell all around the appliance area for gas. Be sure to smell next to the floor because some gas is heavier than air and will settle on the floor.

   WHAT TO DO IF YOU SMELL GAS
   • Do not try to light any appliance.
   • Do not touch any electric switch; do not use any phone in your building.
   • Immediately call your gas supplier from a neighbor's phone. Follow the gas supplier's instructions.
   • If you cannot reach your gas supplier, call the fire department.

C. Use only your hand to push in or turn the gas control knob. Never use tools. If the knob will not push in or turn by hand, don't try to repair it, call a qualified service technician. Force or attempted repair may result in a fire or explosion.

D. Do not use this appliance if any part has been under water. Immediately call a qualified service technician to inspect the appliance and to replace any part of the control system and any gas control which has been under water.

OPERATING INSTRUCTIONS

1. STOP! Read all safety information above.
2. Set the thermostat to lowest setting.
3. Turn off all electric power to the appliance.
4. This appliance does not have a pilot. It is equipped with an ignition device which automatically lights the burner. Do not try to light the burner by hand.
5. Remove control door.

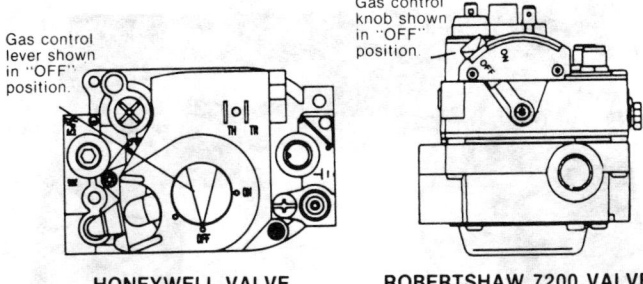

Gas control lever shown in "OFF" position.

Gas control knob shown in "OFF" position.

HONEYWELL VALVE   **ROBERTSHAW 7200 VALVE**

6. HONEYWELL VALVE: Turn manual gas control knob clockwise to "Off" position.

   ROBERTSHAW 7200 VALVE: Depress control lever and move to "Off" position.

7. Wait five (5) minutes to clear out any gas. If you then smell gas, STOP! Follow 'B' in the safety information above. If you don't smell gas, go to next step.

8. HONEYWELL VALVE: Turn on gas to main burners by turning the manual gas control knob counterclockwise to "On" position.

   ROBERTSHAW 7200 VALVE: Move gas control lever from "Off" to "On" position.

9. Replace control door.
10. Turn on all electric power to the appliance.
11. Set thermostat to desired setting.
12. If the appliance will not operate, follow the instructions "To Turn Off Gas To Appliance" and call your service technician or gas supplier.

TO TURN OFF GAS TO APPLIANCE

1. Set the thermostat to the lowest setting.
2. Turn off all electric power to the appliance if service is to be performed.
3. Remove control door.
4. HONEYWELL VALVE: Turn manual gas control knob clockwise to "Off" position.

   ROBERTSHAW 7200 VALVE: Depress control lever and move to "Off" position.

5. Replace control door.

MODE D'EMPLOI

1. Mettez le thermostat à la position la plus basse.
2. Coupez tout courant électrique vers l'appareil.
3. Cet appareil n'a pas de veilleuse. Il est équipé d'un système d'allumage qui allume le brûleur automatiquement. N'essayez pas d'allumer le brûleur manuellement.
4. Enlevez la porte de la commande.
5. SOUPAPE HONEYWELL: Tournez la commande manuelle contrôlant le gaz dans le sens des aiguilles d'une montre vers la position "Off".

   SOUPAPE ROBERTSHAW 7200: Pressez le levier de contrôle et placez-le à la position "Off".

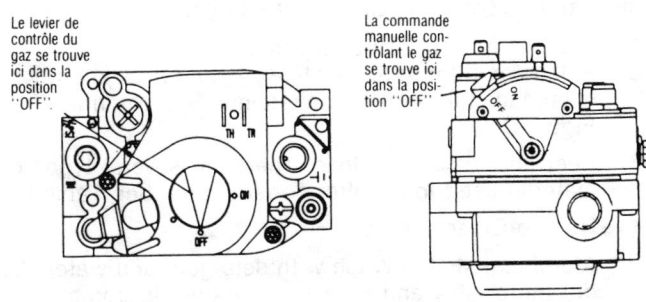

Le levier de contrôle du gaz se trouve ici dans la position "OFF".

La commande manuelle contrôlant le gaz se trouve ici dans la position "OFF".

SOUPAPE HONEYWELL   **SOUPAPE ROBERTSHAW 7200**

6. *PRUDENCE—Assurez-vous que la commande manuelle contrôlant le gaz se trouve dans la position "OFF" depuis au moins cinq minutes.*

7. SOUPAPE HONEYWELL: Faites circuler le gaz vers les brûleurs principaux en tournant la commande manuelle contrôlant le gaz dans le sens contraire des aiguilles d'une montre vers la position "On".

   SOUPAPE ROBERTSHAW 7200: Déplacez le levier de contrôle du gaz de la position "Off" vers la position "On".

8. Replacez la porte de commande.
9. Rallumez le courant électrique vers l'appareil.
10. Placez le thermostat dans la position voulue.
11. Si l'appareil ne fonctionne pas, suivez les instructions "Comment couper le débit de gaz vers l'appareil" et appelez votre réparateur ou votre fournisseur de gaz.

COMMENT COUPER LE DÉBIT DE GAZ VERS L'APPAREIL

1. Placez le thermostat à la position la plus basse.
2. Coupez tout courant électrique vers l'appareil si son entretien est requis.
3. Enlevez la porte de commande.
4. SOUPAPE HONEYWELL: Tournez la commande manuelle contrôlant le gaz dans le sens des aiguilles d'une montre vers la position "Off".

   SOUPAPE ROBERTSHAW 7200: Pressez le levier de contrôle et placez-le sur la position "Off".

5. Replacez la porte de la commande.

MAINTENANCE

CAUTION: *DO NOT OPERATE YOUR SYSTEM FOR EX-TENDED PERIODS WITHOUT FILTERS. A PORTION OF THE DUST ENTRAINED IN THE AIR MAY TEMPORARILY LODGE IN THE AIR DUCT RUNS AND AT THE SUPPLY REGISTERS. ANY RECIRCULATED DUST PARTICLES WILL BE HEATED AND CHARRED BY CONTACT WITH THE FURNACE HEAT EXCHANGER. THIS RESIDUE WILL SOIL CEILINGS, WALLS, DRAPES, CARPETS, AND OTHER HOUSEHOLD ARTICLES.*

UPFLOW FILTER MAINTENANCE

WARNING: TURN OFF ELECTRICAL POWER TO FURNACE BEFORE REMOVING FRONT ACCESS DOOR.

1. Remove bottom front access door.
2. Pull up on filter retainer while pushing back slightly to disengage the front of the retainer. Remove filter. See Figure 3.
3. Keep the air filters clean. There are several types of material used in air filter construction. See Table I.
4. How to Clean Filters:

   Aluminum Mesh - Wash with detergent and water. Air dry thoroughly and renew the coating in compliance with the manufacturer's instructions.

   Plastic Impregnated Fiber - Vacuum clean and reinstall.
5. Reinstall clean filter.
6. Replace front access door.
7. Restore electrical power to furnace.

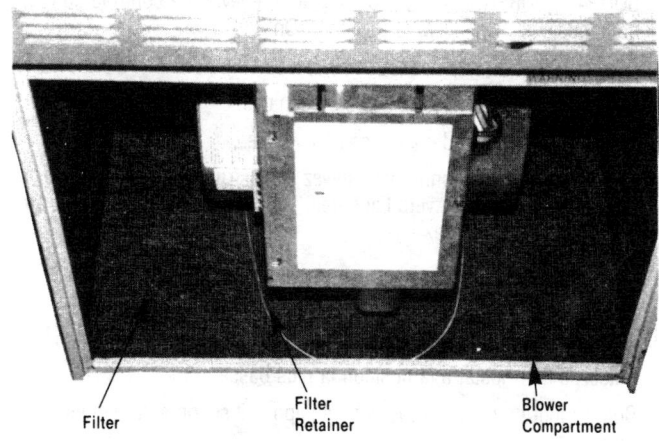

Figure 3. Upflow Filter Maintenance

MINIMUM FILTER SIZES			
Width	Upflow Filter Size	Downflow Filter Size	Type
21	20 x 25 x 1	20 x 22 x 1	Cleanable
24½	24 x 25 x 1	24 x 22 x 1	Cleanable

TABLE I

DOWNFLOW FILTER MAINTENANCE

WARNING: TURN OFF ELECTRICAL POWER TO FURNACE BEFORE REMOVING FRONT ACCESS DOOR.

1. Remove top front access door.
2. Disengage filter retaining wire. See Figure 4.
3. Bend corners of filter down and pull filter out of blower compartment.
4. Clean filter with a vacuum and reinstall.

   NOTE: Make sure filter's grid support is toward the blower.
5. Engage filter retaining wire to hold filter in place.
6. Replace top front access door.
7. Restore electrical power to furnace.
8. Inspect filter monthly and replace when necessary.

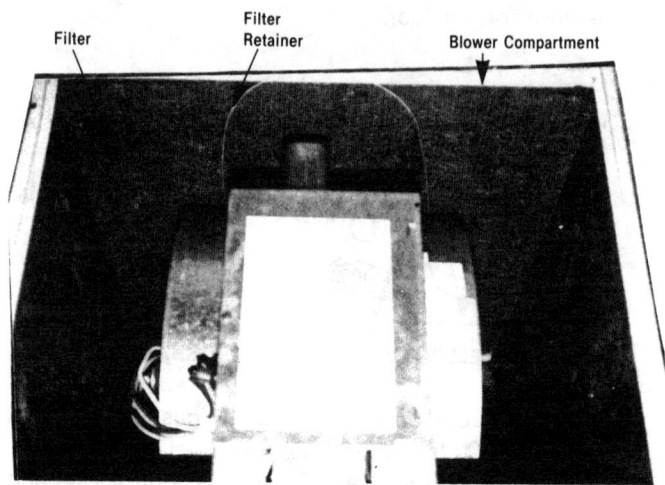

Figure 4. Downflow Filter Maintenance

COMBUSTION AREA AND VENT SYSTEM

The furnace should operate for many years without excessive scale buildup in the flue passageways. However it is recommended that the homeowner inspect the flue passageways, the vent system and the main burner for continued safe operation paying particular attention to deterioration from corrosion or other sources. The flue passageways and vent system should be inspected and cleaned (if required) by a qualified serviceman after the second year of service and annually thereafter.

To inspect the combustion area and vent system, you will need a flashlight.

1. Turn off the electrical supply to your furnace and remove the access doors.
2. Inspect the gas burner area for dirt, rust, or scale.

WARNING: IF DIRT, RUST, SOOT OR SCALE ACCUMULATIONS ARE FOUND, DO NOT OPERATE YOUR FURNACE. CALL YOUR DEALER TO INSPECT THE HEAT EXCHANGERS FOR LEAKS. LEAKS CAN CAUSE TOXIC FUMES TO ENTER YOUR HOME.

3. Inspect the vent-air intake system. Be sure that the vent and air intake pipes are connected to the furnace, slope upward without sagging, and are physically sound, without holes or excessive corrosion. Refer to the installation instructions for details on vent-air intake installation.

WARNING: IF HOLES ARE FOUND IN THE VENT PIPE, OR IF IT HAS BECOME DISCONNECTED, TOXIC FUMES CAN ESCAPE INTO YOUR HOME. DO NOT OPERATE YOUR FURNACE. CALL YOUR DEALER FOR SERVICE.

4. Be sure that the return air duct connections are physically sound, are sealed to the furnace casing and terminate outside the space containing the furnace.
5. Be sure the physical support of the furnace is sound, without sags, cracks, etc., around the base so as to provide a seal between the support and the base.
6. Look for obvious signs of deterioration of the furnace.

4

7. With horizontal venting, inspect and clean the screen inside the horizontal vent terminal.

8. With horizontal air intake, inspect and clean the screen inside the termination elbow.

9. If your furnace is free of the above conditions, replace the access doors and restore electrical power to your furnace.

10. Start your furnace and observe its operation. Watch the burner flames to see if they are bright blue. If you observe a suspected malfunction, or that the burner flames are not bright blue, call your dealer.

CONDENSATE

NOTE: This furnace is equipped with a differential pressure switch which will shut off the main burners if the condensate drain line becomes blocked.

Be sure this condensate drain line does not become blocked or plugged. Visual inspection of condensate flow can easily be made while the furnace is in operation. A flashlight can be used to illuminate the discharge end placed in the sewer opening.

Clean and flush the condensate tube to make sure condensate flows freely while the furnace is in operation.

If a means to neutralize the condensate has been provided for the furnace, be sure to follow the maintenance schedule furnished by the manufacturer.

LUBRICATION

The blower motor sleeve bearings are prelubricated by the motor manufacturer and may not require attention for an indefinite period of time. However, our recommendations are as follows:

A. Motors without oiling ports —
   Prelubricated and sealed. No further lubrication should be required, but in case of bearing problems, the blower and the motor end bells can be disassembled and the bearing relubricated by a qualified service person.

B. Motors with oiling ports —
   Add from 10 to 20 drops of Electric Motor Oil or an SE grade of non-detergent SAE-10 or 20 motor oil to each bearing every two years for somewhat continuous duty, or at least every five years for light duty. Take care not to over oil, because excessive lubrication can damage the motor. Lubrication should only be done by a qualified service person.

C. Blower motor lubrication should only be done by a qualified service person. Refer to "Lubrication" in the Installation and Operating Instructions that came with this furnace for detailed instructions.

The induced draft blower is prelubricated and sealed. No further lubrication is required.

In any event, clean motor periodically to prevent the possibility of overheating due to an accumulation of dust and dirt on the windings or on the motor exterior. And, as suggested elsewhere in these instructions, the air filters should be kept clean because dirty filters can restrict airflow and the motor depends upon sufficient air flowing across and through it to keep from overheating.

OWNER SERVICE HINTS

PROBLEM

Insufficient Heating
Burner is operating
Blower is operating

REMEDY

1. Increase temperature setting on thermostat.

2. Check return air filters and clean or change, if necessary.

3. Recheck to assure that all supply registers and diffusers are open.

4. Check closing of all doors and windows.

5. Call your servicing contractor.

PROBLEM

Main Burner is off

REMEDY

1. Check for "On" position of gas valve.

2. Refer to your electric ignition instructions provided with unit.

3. Call your service contractor.

PROBLEM

Blower operates
Main Burner on & off fast
(Cycling)

REMEDY

1. Check cleanliness of the air filters.

2. Recheck to assure that all supply registers and diffusers are open.

3. Call your servicing contractor.

Trane

SERVICE FACTS

Gas Furnace — Induced Draft — 2 Stage Heat

Model: TUD100R948A

Dwg. No. 21X340078 P01

Library	
Product Section	
Product	
Model	
Literature Type	
Sequence	
Date	
File No.	
Supercedes	

B - FURNACES

IMPORTANT — This document contains a wiring diagram, a parts list, and service information. This is customer property and is to remain with this unit. Please return to service information pack upon completion of work.

WARNING: HAZARDOUS VOLTAGE - DISCONNECT POWER BEFORE SERVICING

PRODUCT SPECIFICATIONS ①

MODEL	TUD100R948A
TYPE	Hot Surface Ignition
RATINGS ②	
Input BTUH - 2nd Stage	100,000
1st Stage	65,000
Temp. rise (Min.-Max.) °F.	35 - 65
BLOWER DRIVE	DIRECT
Diameter-Width (In.)	10 x 8
No. Used	1
Speeds (No.)	4
CFM vs. in. w.g.	See Fan Performance Table
Motor HP	1/2
R.P.M.	1075
Volts/Ph/Hz	115/1/60
FILTER — Furnished?	Yes
Type Recommended	
Lo. Vel. (No.-Size-Thk.)	
Hi Vel. (No.-Size-Thk.)	1 - 20x25 - 1in.
VENT — Size (in.)	4 Round
HEAT EXCHANGER	
Type-Fired	Alum. Steel
-Unfired	
Gauge (Fired)	20
ORIFICES — Main	
Nat. Gas. Qty. — Drill Size	5 — 44
L.P. Gas Qty. — Drill Size	5 — 55
GAS VALVE	Redundant - Two Stage
PILOT SAFETY DEVICE	
Type	Hot Surface Ignition
BURNERS — Type	Multiport Inshot
Number	5
POWER CONN. — V/Ph/Hz ③	115/1/60
Ampacity (In Amps)	12
Fuse Size — Max. (Amps)	15
PIPE CONN. SIZE (IN.)	1/2
DIMENSIONS	H x W x D
Uncrated (In.)	40 x 21 x 28
WEIGHT	
Shipping (Lbs.) / Net (Lbs)	162 / 151

① Central Furnace heating designs are certified by the American Gas Association Inc. Laboratories.

② Ratings shown are for elevations up to 2000 feet. For elevations above 2000 feet; Ratings should be reduced at the rate of 4% for each 1000 feet above sea level.

③ The above wiring specifications are in accordance with National Electrical Code; however, installations must comply with local codes.

EMERGENCY SHUT-OFF INSTRUCTIONS

IF IT IS SUSPECTED THAT A FAILURE OF THE ELECTRICAL, FUEL, OR MECHANICAL SYSTEMS WITHIN THIS FURNACE HAS OCCURRED, THE GAS SUPPLY SHOULD IMMEDIATELY BE TURNED OFF AT THE MANUAL GAS VALVE, LOCATED IN THE BURNER COMPARTMENT AND/OR AT LEVER-HANDLED COCK, AND ELECTRICAL POWER TO THE FURNACE SHOULD BE DISCONNECTED. THE FAILURE MUST BE CORRECTED BY A QUALIFIED SERVICER BEFORE OPERATING THE FURNACE.

ABNORMAL CONDITIONS

1. EXCESSIVE COMBUSTION PRESSURE VENT OR FLUE BLOCKAGE

If pressure against induced draft blower outlet becomes excessive, the pressure switch will react and shut off the gas valve until acceptable combustion pressure is again available.

2. LOSS OF FLAME

If loss of flame occurs during a heating cycle, when flame is not present at the sensor, the flame control module will close the gas valve. The flame control module will then recycle the ignition sequence, then if ignition is not achieved, it will shut off the gas valve and lock out the system.

3. POWER FAILURE

If there is a power failure during a heating cycle, the system will restart the ignition sequence automatically when power is restored, if the thermostat still calls for heat.

4. GAS SUPPLY FAILURE

If loss of flame occurs during a heating cycle, the system flame control module will re-cycle the ignition sequence. If ignition is not achieved, the flame control module will shut off the gas valve and lock out the system.

5. INDUCED DRAFT BLOWER FAILURE

If pressure is not sensed by the pressure switch, the contacts will remain open and not allow the gas valve to open, therefore the unit will not start. If failure occurs during a running cycle, the pressure switch contacts will open and the gas valve will close to shut the unit down.

SAFETY NOTICE

THIS INFORMATION IS INTENDED FOR USE BY INDIVIDUALS POSSESSING ADEQUATE BACKGROUNDS OF ELECTRICAL AND MECHANICAL EXPERIENCE. ANY ATTEMPT TO REPAIR A CENTRAL AIR CONDITIONING PRODUCT MAY RESULT IN PERSONAL INJURY AND OR PROPERTY DAMAGE. THE MANUFACTURER OR SELLER CANNOT BE RESPONSIBLE FOR THE INTERPRETATION OF THIS INFORMATION, NOR CAN IT ASSUME ANY LIABILITY IN CONNECTION WITH ITS USE.

SAFETY WARNING

ALL PARTS OF THIS PRODUCT CAPABLE OF CONDUCTING ELECTRICAL CURRENT ARE GROUNDED. IF GROUNDING WIRES, SCREWS, STRAPS, CLIPS, NUTS OR WASHERS USED TO COMPLETE A PATH TO GROUND ARE REMOVED FOR SERVICE, THEY MUST BE RETURNED TO THEIR ORIGINAL POSITION AND PROPERLY FASTENED.

SEQUENCE OF OPERATION

Thermostat call for heat (2-stage thermostat)

Call for 1st stage only:

R and W1 thermostat contacts close signaling the control module to run its self-check routine. After the control module has verified that the 1st stage pressure switch contacts are open and the limit switch(es) contacts are closed, the draft blower will be energized.

As the induced draft blower comes up to speed, the pressure switch contacts will close and the ignitor warm up period will begin. The ignitor will heat for approx. 17 seconds, then the gas valve is energized in 1st stage to permit gas flow to the burners. The flame sensor confirms that ignition has been achieved within the 6 second ignition trial period.

As the flame sensor confirms that ignition has been achieved, the delay to fan ON period begins timing and after approx. 45 seconds the I.D. blower motor will be energized at low speed and will continue to run during the heating cycle.

Call for 2nd stage after 1st stage:

R and W2 thermostat contacts close signaling a call for 2nd stage heat. After a 30 second delay, the induced draft blower will be energized on high speed and the 2nd stage pressure switch contacts will close allowing the gas valve to be energized in 2nd stage and the I.D. blower motor in high speed.

2nd stage satisfied, 1st stage still called:

R and W2 thermostat contacts open signaling that 2nd stage heating requirements are satified. The induced draft blower is reduced to low speed allowing the 2nd stage pressure switch contacts to open and the gas valve is reduced to 1st stage. After aprox. 30 seconds the I.D. blower motor is reduced to low speed.

1st stage satisfied:

R and W1 thermostat contacts open signaling that 1st stage heating requirements are satified. The gas valve will close and the induced draft blower will be de-energized. The I.D. blower motor will continue to run for the fan off period (Field selectable at 90, 120, 150 or 210 seconds), then will be de-energized by the control module.

Thermostat call for heat (1-stage Thermostat)

R and W1/W2 (jumpered) thermostat contacts close signaling a call for heat. 1st stage sequence of operation remains the same as above. 2nd stage heat has a 10 minute delay from the time of 1st stage ignition.

Thermostat satisfied:

R and W1/W2 (jumpered) contacts open signaling the control module to close the gas valve and de-energize the induced draft blower. The I.D. blower motor will continue to operate at high heat speed for approx. 30 seconds after the flames are extinguished and then is switched to low heat speed for the remaining FAN-OFF period.

LIMIT SWITCH CHECK OUT

To check for proper operation of the limit switches, set the thermostat to a temperature higher than the indicated temperature to bring on the gas valve. Restrict the airflow by blocking the return air or by disconnecting the blower. When the furnace reaches the maximum outlet temperature as shown on the rating plate, the burners must shut off. If they do not shut off after a reasonable time and over-heating is evident, a faulty limit switch is probable and the limit switch must be replaced. After checking the operation of the limit control, be sure to remove the paper or cardboard from the return air inlet, or reconnect the blower.

AIRFLOW ADJUSTMENT

Check inlet and outlet air temperatures to make sure they are within the ranges specified on the furnace rating nameplate. If the airflow needs to be increased or decreased, see the wiring diagram for information on changing the speed of the blower motor.

WARNING: Disconnect power to the unit before removing the blower door.

This unit is equipped with a blower door switch which cuts power to the blower and gas valve causing shutdown when the door is removed. Operation with the door removed or ajar can permit the escape of dangerous fumes. All panels must be securely closed at all times for safe operation of the furnace.

INDOOR BLOWER TIMING

Heating: The control module controls the indoor blower. The blower start is fixed at 45 seconds after ignition. The FAN-OFF period is field selectable by dip switches at 90, 120, 150, or 210 seconds. The factory setting is 150 seconds, (See unit wiring diagram above).

Cooling: The fan delay off period is factory set at 0 seconds. The option for 80 second delay off is field selectable, (See unit wiring diagram above).

ROOM AIR THERMOSTAT HEAT ANTICIPATOR ADJUSTMENT

Set the thermostat heat anticipator according to the current flow measured, or the settings found in the notes on the furnace wiring diagram.

FLAME ROLL-OUT DEVICE

All models are equipped with a fusible link on the burner cover. In case of flame roll-out, the link will open (melt) and cause the circuit to open which shuts off all flow of gas.

COMBUSTION AND INPUT CHECK

1. Make sure all gas appliances are off except the furnace.

2. Clock the gas meter with the furnace operating (determine the dial rating of the meter) for one revolution.

3. Match the "Sec" column in the gas flow (in cfh) Table 3 (on back page) with the time clocked.

4. Read the "Flow" column opposite the number of seconds clocked.

5. Use the following factors if necessary.

For 1 Cu. Ft. Dial Gas Flow CFH = Chart Flow Reading % 2
For 1/2 Cu Ft. Dial Gas Flow CFH = Chart Flow Reading % 4
For 5 Cu. Ft. Dial Gas Flow CFH = 10X Chart Flow Reading % 4

6. Multiply the final figure by the heating value of the gas obtained from the utility company and compare to the nameplate rating. This must not exceed the nameplate rating.

7. Changes can be made by adjusting the manifold pressure or changing orifices (orifice change may not always be required).

a. Attach a manifold pressure gauge.

b. Remove the slot screw on top of the gas valve for 2nd stage manifold pressure adjustment. Remove slot screw on outlet side for 1st stage adjustment.

c. Turn the adjustment nut in to increase the gas flow rate, and out to decrease the gas flow rate using a 3/32" hex wrench.

d. The 2nd stage final manifold pressure setting shall be no less than 3.0" W.C. and no more than 3.7" W.C. with an input of no more than nameplate rating and no less than 93 % of the nameplate rating, unless the unit is derated for high altitude.

For LP gases, the 2nd stage final manifold pressure shall be no less than 10.0" W.C. and no more than 10.5" W.C. with an input of no more than the nameplate rating and no less than 93 % of the nameplate rating, unless the unit is derated for altitude.

TABLE 1 - FINAL MANIFOLD PRESSURE SETTINGS

FUEL	1st STAGE		2nd STAGE	
	MINIMUM	MAXIMUM	MINIMUM	MAXIMUM
Natural Gas	1.4"W.C.	1.7" W.C.	3.0" W.C.	3.7" W.C.
LP Gas	4.0" W.C.	4.5" W.C.	10.0" W.C.	10.5" W.C.

TABLE 2

Input Rating BTUH (000)	No. of Burners	Main Burner Orifice	
		Drill Size	
		Nat. Gas	Propane
100	5	44	55

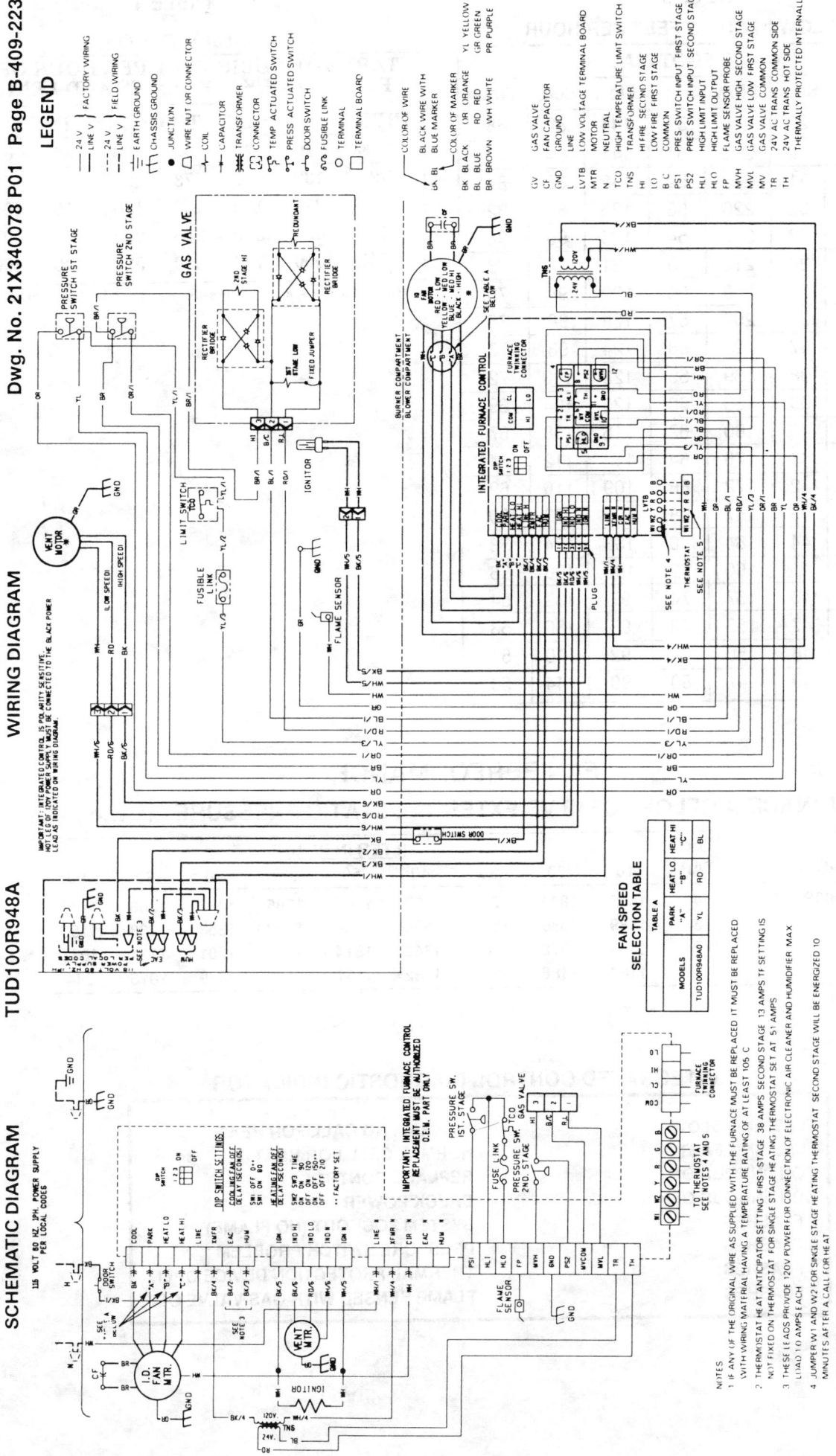

SCHEMATIC DIAGRAM   WIRING DIAGRAM

TUD100R948A   Dwg. No. 21X340078 P01   Page B-409-223

From Dwg. 21D340043 Rev 5

TABLE 3
GAS FLOW IN CUBIC FEET PER HOUR

2 CUBIC FOOT DIAL							
Sec.	Flow	Sec.	Flow	Sec.	Flow	Sec.	Flow
8	900	29	248	50	144	82	88
9	800	30	240	51	141	84	86
10	720	31	232	52	138	86	84
11	655	32	225	53	136	88	82
12	600	33	218	54	133	90	80
13	555	34	212	55	131	92	78
14	514	35	206	56	129	94	76
15	480	36	200	57	126	96	75
16	450	37	195	58	124	98	73
17	424	38	189	59	122	100	72
18	400	39	185	60	120	104	69
19	379	40	180	62	116	108	67
20	360	41	176	64	112	112	64
21	343	42	172	66	109	116	62
22	327	43	167	68	106	120	60
23	313	44	164	70	103	124	58
24	300	45	160	72	100	128	56
25	288	46	157	74	97	132	54
26	277	47	153	76	95	136	53
27	267	48	150	78	92	140	51
28	257	49	147	80	90	144	50

TABLE 4
Natural Gas Only
TABLE OF CUBIC FEET PER HOUR OF GAS FOR VARIOUS PIPE SIZES AND LENGTHS

PIPE SIZE	LENGTH OF PIPE IN FEET						
	10	20	30	40	50	60	70
1/2	132	92	73	63	56	50	46
3/4	278	190	152	130	115	105	96
1	520	350	285	245	215	195	180
1-1/4	1050	730	590	520	440	400	370

This Table is based on pressure drop of 0.3 inch W.C. and 0.6 SP. GR. gas.

FAN PERFORMANCE
FURNACE AIRFLOW (CFM) VS EXTERNAL STATIC PRESSURE (in. w.g.)

MODEL	SPEED TAP	.10	.20	.30	FILTER IN PLACE .40	.50	.60	.70	.80	.90
TUD100R948A	4	1880	1844	1796	1737	1666	1585	1492	1388	1273
	3	1589	1586	1568	1536	1489	1828	1353	1263	1158
	2	1359	1370	1367	1348	1314	1265	1201	1121	1027
	1	1167	1186	1191	1182	1161	1127	1079	1018	944

INTEGRATED CONTROL DIAGNOSTIC INDICATOR

FLASHING SLOW	NORMAL. NO CALL FOR HEAT
FLASHING FAST	NORMAL CALL FOR HEAT
CONTINUOUS ON	REPLACE CONTROL
CONTINUOUS OFF	CHECK POWER
2 FLASHES	SYSTEM LOCKOUT (NO FLAME)
3 FLASHES	PRESSURE SWITCH PROBLEM
4 FLASHES	THERMAL PROTECTION DEVICE OPEN
5 FLASHES	FLAME SENSED WITH GAS VALVE OFF

Gas Furnace
Model: TUC100B948A0

IMPORTANT — This document is customer property and is to remain with this unit. Please return to service information pack upon completion of work.

Library	
Product Section	
Product	
Model	
Literature Type	
Sequence	
Date	
File No.	
Supersedes	

PRODUCT SPECIFICATIONS

MODEL	TUC100B948A0
TYPE	UPFLOW, INTERMITTENT ELECTRONIC IGNITION
RATINGS ① Input, BTUH ② Temp. Rise (Min. — Max.) °F.	100,000 40 — 70
BLOWER DRIVE Dia. — Width (in.) No. Used Speeds (No.) CFM vs. in. w.g. Motor HP R.P.M. Volts/Ph/Hz	DIRECT 10 x 10 1 4 SEE FAN PERFORMANCE TABLE 3/4 1075 115/1/60
FILTER — Furnished? Type Recommended Lo Vel. (No. - Size - Thk.) Hi Vel. (No. - Size - Thk.)	YES 1 - 20 x 25 - 1 in.
VENT — Size (in.)	3.0 ROUND
HEAT EXCHANGER Type — Fired — Unfired Gauge (Fired)	ALUMINIZED STEEL TYPE 1 29-4C 20
ORIFICES — Main Nat. Gas Qty. — Drill Size L.P. Gas Qty. — Drill Size	3 — 33 3 — 52
GAS VALVE	REDUNDANT — SINGLE STAGE
DIRECT IGNITION DEVICE Type	HOT SURFACE
BURNERS — Type Number	LINEAR 3
POWER CONN. — V/Ph/Hz ④ Ampacity (In Amps) Fuse Size — Max. (Amps.)	115/1/60 16 20
PIPE CONN. SIZE (IN.)	1/2
DIMENSIONS Crated (in.)	H x W x D 50-1/2 x 26 x 33-1/2
WEIGHT Shipping (lbs.) / Net (lbs.)	228 / 220

¹ Central furnace heating designs are certified by the American Gas Association Inc. Laboratories.
² Ratings shown are for elevations up to 2000 feet. For elevations above 2000 feet; ratings should be reduced at the rate of 4% for each 1000 feet above sea level.
³ Based on U.S. Government Standard Tests.
⁴ The above wiring specifications are in accordance with the National Electrical Code; however installations must comply with local codes.

OPTIONAL EQUIPMENT

ELECTRONIC AIR CLEANER BEF140C100A
AIR CLEANER RELAY KIT BAY24X043
LP CONVERSION KIT BAYLPKT208
HIGH ALTITUDE PRESSURE SWITCH BAYHALT203

EMERGENCY SHUT-OFF INSTRUCTIONS

IF IT IS SUSPECTED THAT A FAILURE OF THE ELECTRICAL, FUEL, OR MECHANICAL SYSTEMS WITHIN THIS FURNACE HAS OCCURRED, THE GAS SUPPLY SHOULD IMMEDIATELY BE TURNED OFF AT THE MANUAL GAS VALVE, LOCATED IN THE BURNER COMPARTMENT AND/OR AT LEVER-HANDLED COCK, AND ELECTRICAL POWER TO THE FURNACE SHOULD BE DISCONNECTED. THE FAILURE MUST BE CORRECTED BY A QUALIFIED SERVICER BEFORE OPERATING THE FURNACE.

WARNING — DO NOT ATTEMPT TO MANUALLY LIGHT THE GAS BURNERS.

Lighting instructions appear on each unit. Each installation must be checked out at the time of initial start up to insure proper operation of all components. Check out should include putting the unit through one complete cycle as outlined below.

Turn on main electrical supply and set thermostat above indicated temperature. The ignitor will automatically heat for approx. 15 seconds, then the gas valve is energized to permit gas flow to burners. After ignition and flame is established, flame control module monitors flame and supplies power to gas valve until thermostat is satisfied. Ignition sequence will cycle 3 times before lockout.

If burner fails to ignite, lower thermostat setting or disconnect electrical supply, wait 5 minutes, raise thermostat setting above indicated temperature, turn electrical supply on. Unit will repeat lighting sequence.

SAFETY NOTICE

THIS INFORMATION IS INTENDED FOR USE BY INDIVIDUALS POSSESSING ADEQUATE BACKGROUNDS OF ELECTRICAL AND MECHANICAL EXPERIENCE. ANY ATTEMPT TO REPAIR A CENTRAL AIR CONDITIONING PRODUCT MAY RESULT IN PERSONAL INJURY AND OR PROPERTY DAMAGE. THE MANUFACTURER OR SELLER CANNOT BE RESPONSIBLE FOR THE INTERPRETATION OF THIS INFORMATION, NOR CAN IT ASSUME ANY LIABILITY IN CONNECTION WITH ITS USE.

RECONNECT ALL GROUNDING DEVICES

ALL PARTS OF THIS PRODUCT CAPABLE OF CONDUCTING ELECTRICAL CURRENT ARE GROUNDED. IF GROUNDING WIRES, SCREWS, STRAPS, CLIPS, NUTS OR WASHERS USED TO COMPLETE A PATH TO GROUND ARE REMOVED FOR SERVICE, THEY MUST BE RETURNED TO THEIR ORIGINAL POSITION AND PROPERLY FASTENED.

DISCONNECT POWER BEFORE SERVICING

REDDI PARTS

COMPONENT	QTY.	DESCRIPTION	CAT. #
Blower Wheel	1	10" D x 10" W, CW, Convex	WG74X0035
Burner Asm.	3	Replacement	WW54X0112
Capacitor	1	20 MFD, 440V	WW20X0127
Fan - Induced Draft	1	Replacement	WW73X0128
Fan & Limit Control (Fan/Limit)	1	Limit 180°, 8" probe, Honeywell #L4064A2265	WW29X0902
Filter	1	25" x 20"	WG85X0051
Flame Sensor	1	Fenwal # 22-100000-6H	WW37X0058
Fusible Link (Flame Roll-Out)	1	333°F./167°C Rating, Micro Devices # 404333A	WG09X0033
Gas Valve	1	White Rodgers # 36E01-221	WG19X0231
Heat Exchanger Asm.	1	Replacement	WW93X0081
Ignitor Asm.	1	Norton # 271N	WW37X0057
Ignitor Control	1	24 VAC, 60 Hz., 3-Try Board White Rodgers # 50E47-160	WW37X0056
Motor, Blower	1	3/4 HP, 1075 RPM, 115V, 60 Hz., 1 Phase 4 Speed, 11.5 FLA, PSC, Sleeve Bearings	WG94X0158
Motor, Comb. Air/I.D. Blower (Vent)	1	Replacement	WG94X0059
Orifice	3	# 35 Drill	WG16X0186
Recup Cell Asm.	1	Replacement	WW93X0096
Relay - Blower	1	DPDT, 16 FLA, 48 LRA @ 120V, 24V Coil, P & B # KUM1078	WW24X0231
Relay - Combustion Blower	1	SPST, 12 FLA, 60 LRA @ 125V, 24V Coil, P & B # S87R1A2B1D-24V	WG24X0119
Relay - Time Delay	1	13.8 RLA, 82.8 LRA @ 120V, 40-80 Sec. Time Delay, Therm-O-Disc	WW24X0232
Switch - Aux. Limit	1	Open 155 ± 5°F., Close 125 ± 7°F., T.I. # NT01L-0388	WG23X0090
Switch - Blower Door	1	SPST, N.O., 3/4 HP @ 125 VAC	WG23X0073
Switch - Pressure	1	1.08" ± .06" Neg. W.C. 28VA Pilot Duty @ 24V, Tridelta # FS6399-615	WW26X0110
Transformer	1	115V Pri., 24V Sec., 35VA Jard # TF-351124-B11A	WW32X0092
Wheel - Combustion Blower	1	Vernco # AL-30-101	WG92X0162

SEQUENCE OF OPERATION

With gas and electrical power "OFF"
1. Duct connections properly sealed
2. Filters in place
3. Venting properly assembled
4. Condensate drains properly installed
5. Blower door in place on upflow models

Adjust Heat Anticipator on Thermostat, for Natural or Propane Gas, as follows:
0.85 — White Rodgers Gas Valve
0.75 — Robertshaw Gas Valve
0.65 — Johnson Gas Valve

Turn knob on main gas valve within unit to "OFF". Turn external gas valve to "ON". Purge air from gas lines. After purging, check all gas connections with soapy solution-**DO NOT CHECK WITH AN OPEN FLAME**. Allow 5 minutes for any gas that might have escaped to dissipate. LP Gas being heavier than air may require forced ventilation. Turn knob on gas valve in unit to "ON".

THERMOSTAT CALLS FOR HEAT

R and W Thermostat contacts close to supply power, through safety limit switches, to control circuit which starts the induced draft blower. When the required combustion air is established, the pressure switch allows power to flow through the safety controls to the flame control module and the ignitor.

The ignitor will heat for approx. 15 seconds, then the gas valve is energized to permit gas flow to the burners. The flame sensor confirms that ignition has been achieved within the 7 second ignition trial period. All models utilize a remote sensor.

If the sensor does not confirm ignition in the 7 second time period, the ignition control will go into a re-try mode. During the second trial for ignition there is a 60 second prepurge time for the induced draft motor, then the ignitor will energize for approx. 45 seconds before the gas valve is energized. If the flame sensor does not confirm ignition during the second trial, the ignition control will go into a re-try mode again. The sequence of operation will be repeated. If the ignition has not been confirmed at the end of the 3rd trial, the ignition control will go into lock out mode.

When ignition and flame is established, the flame control module monitors the flame and supplies power to the gas valve until the thermostat is satisfied. When the thermostat is satisfied, R and W contacts open to de-energize the control circuit.

MAIN BURNER ADJUSTMENT

The regulator on the unit's gas valve is set for a manifold pressure of 3.5" W.C. for natural gas. This can be checked by turning off the gas valve and removing the plug on the gas valve labeled "Outlet Pressure" and installing a water manometer. Turn the gas valve to ON. Put furnace in operation and read the manometer. If it is necessary to adjust the gas consumption, turn the adjustment screw (beneath cap on pressure regulator) clockwise or counterclockwise to 3.5" W.C. After checking pressure, turn gas valve OFF. Remove manometer. Replace plug. Put furnace into operation and leak check plug with soapy solution for leaks.

For LP Models, the regulator is set at 10.5" W.C. If the gas consumption needs to be increased, do not exceed 11.0" W.C.

FLAME CHARACTERISTICS

On LP (propane) units, some light yellow tipping of the outer mantle is normal. Inner mantle should be bright blue. Natural gas units should not have any yellow tipped flames.

Another method for checking input is to clock the gas meter with **all** other gas appliances turned OFF.

HIGH ALTITUDE INSTALLATIONS:

IMPORTANT: The sea level rated input of the furnace installed at elevations above 2,000 feet should be reduced 4% for each 1,000 feet above sea level. For example, for elevations of 4,350 feet, the correct input would be the sea level rated input less 17%. Check orifice size. The drill size is stamped on the gas manifold or see Table below.

Input Rating BTUH (000)	No. of Burners	Main Burner Orifice	
		Drill Size	
		Nat. Gas	Propane
100	3	33	52

CONTROLS AND SAFETY SWITCH ADJUSTMENT

The fan switch is factory set for the blower to come on at 120°F.* and to shut off at 100°F. These settings are satisfactory in most applications. If necessary to field adjust, disconnect power and proceed as follows:

1. Adjust fan-off pointer to shut blower off when air at the farthest register begins to feel cool, or to setting that will not cause blower to recycle.
2. Fan-on differential is adjustable from 25°F. to 45°F. above fan-off setting.

*NOTE: Do not exceed 125°F. fan-on setting for counterflow units.

BLOWER

All models are factory equipped with a transformer and relay. Additional relay is not required when adding air conditioning.

REMOVAL OF BLOWER:

If the blower should have to be removed for servicing, **TURN OFF ELECTRICITY TO UNIT** and disconnect electrical leads at control box.

AIRFLOW ADJUSTMENT

Check inlet and outlet air temperatures to make sure they are within the ranges specified on the furnace rating plate. If airflow needs to be increased or decreased, see wiring diagram for information on changing the wiring to blower motor. Speed changes can be made at molex plug in blower compartment.

LIMIT SWITCH CHECK OUT

The limit switch is a safety device designed to electrically de-energize the gas valve should the furnace become overheated.

To check for proper operation of the limit switches, set thermostat to temperature higher than room temperature to energize the gas valve. Restrict airflow by blocking the return air or by interrupting power to the blower. When the furnace reaches the maximum outlet temperature as shown on the rating plate, the burners must shut off. If they do not shut off after a reasonable time and over-heating is evident, a faulty limit is probable and the limit switch must be replaced. After checking the operation of the fan and limit control, be sure to remove the paper or cardboard restriction from the filter.

FLAME ROLL-OUT DEVICE

All models are equipped with a fusible link on burner cover. In case of flame roll-out, link will open and close off flow of all gas. See instruction label on front panel of heat exchanger.

FAN PERFORMANCE
FURNACE AIRFLOW (CFM) VS EXTERNAL STATIC PRESSURE (m. w.g.)

MODEL	SPEED TAP	— FILTER IN PLACE —								
		.10	.20	.30	.40	.50	.60	.70	.80	.90
TUC100B948A0	4	1869	1790	1706	1620	1517	1430	1322	1210	1079
	3	1666	1603	1532	1461	1370	1300	1214	1096	953
	2	1569	1515	1451	1379	1302	1225	1130	1032	899
	1	1268	1232	1195	1145	1090	1032	951	—	—

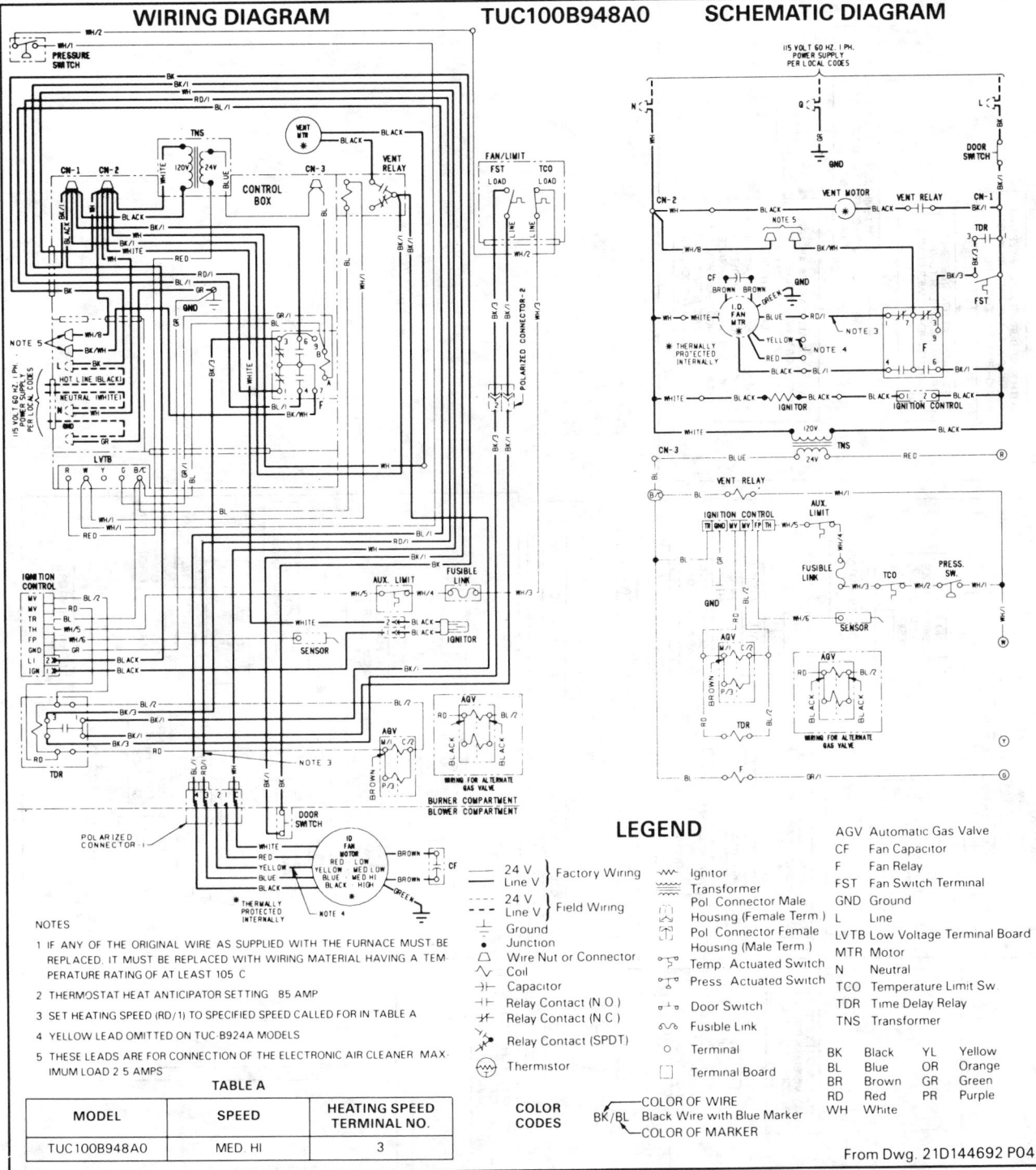

WIRING DIAGRAM — TUC100B948A0 — SCHEMATIC DIAGRAM

NOTES

1. IF ANY OF THE ORIGINAL WIRE AS SUPPLIED WITH THE FURNACE MUST BE REPLACED, IT MUST BE REPLACED WITH WIRING MATERIAL HAVING A TEMPERATURE RATING OF AT LEAST 105 C

2. THERMOSTAT HEAT ANTICIPATOR SETTING .85 AMP

3. SET HEATING SPEED (RD/1) TO SPECIFIED SPEED CALLED FOR IN TABLE A

4. YELLOW LEAD OMITTED ON TUC-B924A MODELS

5. THESE LEADS ARE FOR CONNECTION OF THE ELECTRONIC AIR CLEANER. MAXIMUM LOAD 2.5 AMPS

TABLE A

MODEL	SPEED	HEATING SPEED TERMINAL NO.
TUC100B948A0	MED. HI	3

LEGEND

AGV	Automatic Gas Valve		
CF	Fan Capacitor		
F	Fan Relay		
FST	Fan Switch Terminal		
GND	Ground		
L	Line		
LVTB	Low Voltage Terminal Board		
MTR	Motor		
N	Neutral		
TCO	Temperature Limit Sw.		
TDR	Time Delay Relay		
TNS	Transformer		

BK	Black	YL	Yellow
BL	Blue	OR	Orange
BR	Brown	GR	Green
RD	Red	PR	Purple
WH	White		

24 V } Factory Wiring
Line V

24 V } Field Wiring
Line V

Ignitor
Transformer
Pol. Connector Male Housing (Female Term.)
Pol. Connector Female Housing (Male Term.)
Ground
Junction
Wire Nut or Connector
Coil
Capacitor
Relay Contact (N O)
Relay Contact (N C)
Relay Contact (SPDT)
Thermistor
Temp. Actuated Switch
Press. Actuated Switch
Door Switch
Fusible Link
Terminal
Terminal Board

COLOR CODES
BK/BL ── COLOR OF WIRE
Black Wire with Blue Marker
── COLOR OF MARKER

From Dwg. 21D144692 P04

WARNING

The furnace must be isolated from the gas supply piping system by closing its individual manual shutoff valve during any pressure testing of the gas supply piping system at test pressures equal to or greater than 1/2 psig.

All gas fittings must be checked for leaks using a soapy solution before lighting furnace. **DO NOT CHECK WITH AN OPEN FLAME.**

WARNING: DISCONNECT POWER TO UNIT BEFORE REMOVING BLOWER DOOR.

Unit is equipped with a blower door switch which cuts power to blower and gas valve causing shutdown when door is removed. Unit must not be altered to allow operation with the blower door removed. Operation with doors removed or ajar can permit the escape of dangerous fumes. All panels must be securely closed at all times for safe operation of the furnace.

PRINCIPLE OF OPERATION

The system utilizes a silicone carbide element for ignition. The ignitor is an electrically heated resistance element which thermally ignites the gas. The flame detector circuit utilizes flame rectification for monitoring the gas flame.

Upon a call for heat, the element is powered from the 120 VAC line and allowed to heat for 15 seconds (typical). Then the main valve is powered, permitting gas flow to the burner for the trial-for-ignition period. At the end of this period, the flame sensor checks for the presence of flame. If flame is present, the system will monitor it and hold the main valve open. If flame is not established within the trial-for-ignition period, the system will recycle through the complete ignition sequence. If after three cycles combustion has not been established, the system will lockout. If lockout does occur the combustion vent motor will continue to run as long as the thermostat is calling for heat.

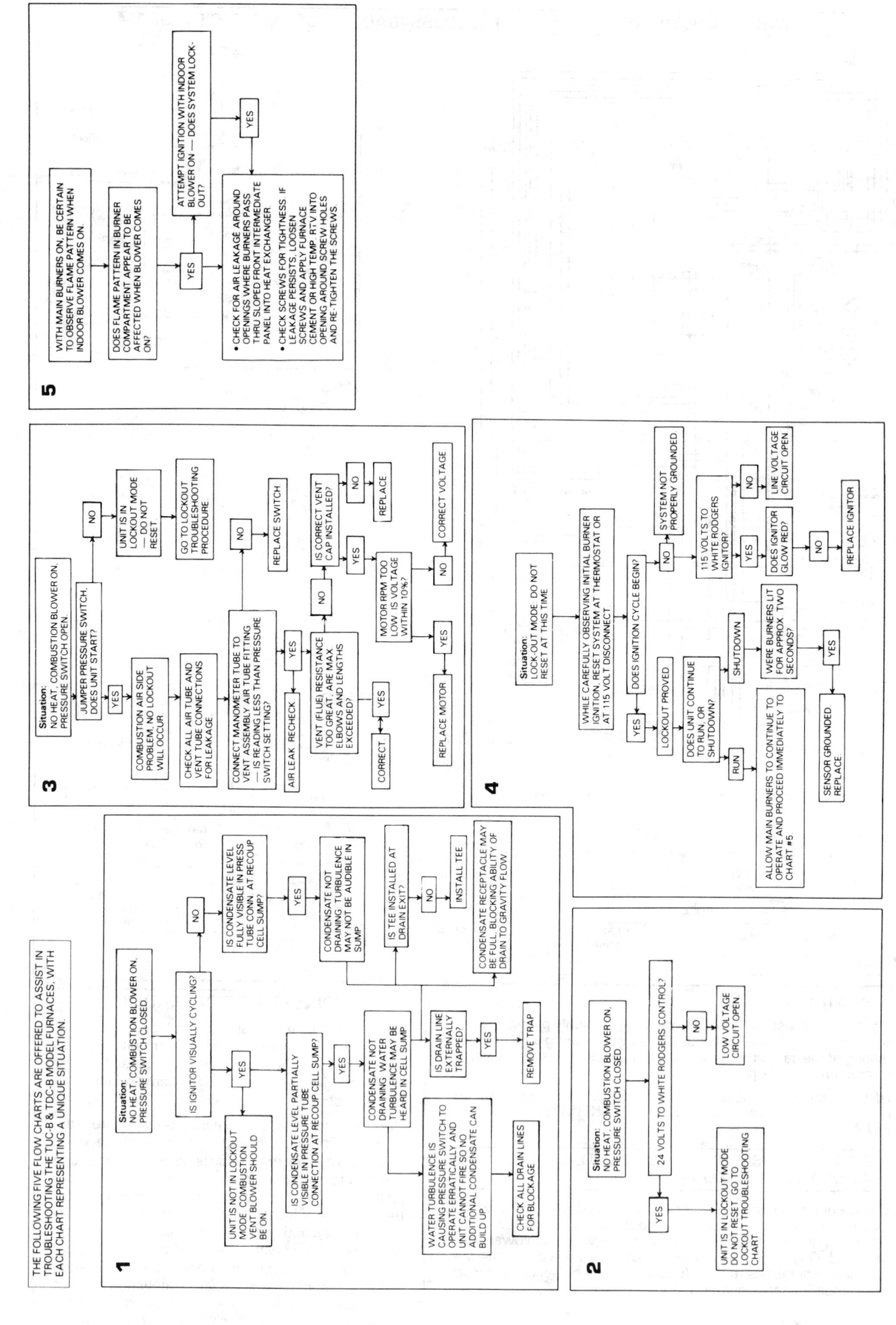

THE FOLLOWING FIVE FLOW CHARTS ARE OFFERED TO ASSIST IN TROUBLESHOOTING THE TUC-B & TDC-B MODEL FURNACES, WITH EACH CHART REPRESENTING A UNIQUE SITUATION.

1

Situation: NO HEAT. COMBUSTION BLOWER ON, PRESSURE SWITCH CLOSED

- IS IGNITOR VISUALLY CYCLING?
  - NO → IS CONDENSATE LEVEL FULLY VISIBLE IN PRESS TUBE CONN AT RECOUP CELL SUMP? → YES → CONDENSATE NOT DRAINING. TURBULENCE MAY NOT BE AUDIBLE IN SUMP
  - YES → UNIT IS NOT IN LOCKOUT MODE. COMBUSTION VENT BLOWER SHOULD BE ON
- IS CONDENSATE LEVEL PARTIALLY VISIBLE IN PRESSURE TUBE CONNECTION AT RECOUP CELL SUMP? → YES → CONDENSATE NOT DRAINING. WATER TURBULENCE MAY BE HEARD IN CELL SUMP
- IS DRAIN LINE EXTERNALLY TRAPPED? → YES → REMOVE TRAP
- IS TEE INSTALLED AT DRAIN EXIT? → NO → INSTALL TEE
- CONDENSATE RECEPTACLE MAY BE FULL. BLOCKING ABILITY OF DRAIN TO GRAVITY FLOW
- WATER TURBULENCE IS CAUSING PRESSURE SWITCH TO OPERATE ERRATICALLY AND UNIT CANNOT FIRE SO NO ADDITIONAL CONDENSATE CAN BUILD UP
- CHECK ALL DRAIN LINES FOR BLOCKAGE

2

Situation: NO HEAT. COMBUSTION BLOWER ON, PRESSURE SWITCH CLOSED

- 24 VOLTS TO WHITE RODGERS CONTROL?
  - NO → LOW VOLTAGE CIRCUIT OPEN
  - YES → UNIT IS IN LOCKOUT MODE. DO NOT RESET GO TO LOCKOUT TROUBLESHOOTING CHART

3

Situation: NO HEAT, COMBUSTION BLOWER ON, PRESSURE SWITCH OPEN

- JUMPER PRESSURE SWITCH. DOES UNIT START?
  - NO → UNIT IS IN LOCKOUT MODE — DO NOT RESET → GO TO LOCKOUT TROUBLESHOOTING PROCEDURE
  - YES → COMBUSTION AIR SIDE PROBLEM. NO LOCKOUT WILL OCCUR
- CHECK ALL AIR TUBE AND VENT TUBE CONNECTIONS FOR LEAKAGE
- CONNECT MANOMETER TUBE TO VENT ASSEMBLY AIR TUBE FITTING — IS READING LESS THAN PRESSURE SWITCH SETTING?
  - NO → REPLACE SWITCH
  - YES → AIR LEAK RECHECK
- VENT (FLUE) RESISTANCE TOO GREAT. ARE MAX ELBOWS AND LENGTHS EXCEEDED?
  - CORRECT / YES → IS CORRECT VENT CAP INSTALLED?
    - NO → REPLACE
    - YES → MOTOR RPM TOO LOW. IS VOLTAGE WITHIN 10%?
      - NO → CORRECT VOLTAGE
      - YES → REPLACE MOTOR

4

Situation: LOCK-OUT MODE. DO NOT RESET AT THIS TIME

- WHILE CAREFULLY OBSERVING INITIAL BURNER IGNITION, RESET SYSTEM AT THERMOSTAT OR AT 115 VOLT DISCONNECT
- DOES IGNITION CYCLE BEGIN?
  - NO → SYSTEM NOT PROPERLY GROUNDED → 115 VOLTS TO WHITE RODGERS IGNITOR?
    - NO → LINE VOLTAGE CIRCUIT OPEN
    - YES → DOES IGNITOR GLOW RED? → NO → REPLACE IGNITOR
  - YES → LOCKOUT PROVED
- DOES UNIT CONTINUE TO RUN, OR SHUTDOWN?
  - SHUTDOWN → WERE BURNERS LIT FOR APPROX. TWO SECONDS? → YES → SENSOR GROUNDED REPLACE
  - RUN → ALLOW MAIN BURNERS TO CONTINUE TO OPERATE AND PROCEED IMMEDIATELY TO CHART #5

5

WITH MAIN BURNERS ON, BE CERTAIN TO OBSERVE FLAME PATTERN WHEN INDOOR BLOWER COMES ON

- DOES FLAME PATTERN IN BURNER COMPARTMENT APPEAR TO BE AFFECTED WHEN BLOWER COMES ON?
  - YES → ATTEMPT IGNITION WITH INDOOR BLOWER ON — DOES SYSTEM LOCK-OUT?
    - YES →
      - CHECK FOR AIR LEAKAGE AROUND OPENINGS WHERE BURNERS PASS THRU SLOPED FRONT INTERMEDIATE PANEL INTO HEAT EXCHANGER
      - CHECK SCREWS FOR TIGHTNESS IF LEAKAGE PERSISTS, LOOSEN SCREWS AND APPLY FURNACE CEMENT OR HIGH TEMP RTV INTO OPENING AROUND SCREW HOLES AND RE-TIGHTEN THE SCREWS

User's Information Manual
Super efficiency condensing gas furnace BLU-K, BLD-K and TUC Models

FOR YOUR SAFETY: WHAT TO DO IF YOU SMELL GAS
- Do not try to light any appliance.
- Do not touch any electrical switch; do not use any phone in your building
- Immediately call your gas supplier from a neighbor's phone. Follow the gas supplier's instructions.
- If you cannot reach your gas supplier, call the fire department.

WARNING:
Improper installation, adjustment, alteration, service, or maintenance can cause injury or property damage. Refer to this manual. For assistance or additional information, consult a qualified installer, service agency, or the gas supplier.

FOR YOUR SAFETY
Do not store or use gasoline or other flammable vapors and liquids in the vicinity of this or any other appliance.

Safety notice.

WARNING: Improper installation, adjustment, alteration, service or maintenance can cause injury or property damage. Refer to the installation instructions provided with the furnace and this manual. For assistance or additional information consult a qualified installer, service agency or the gas supplier.

This information is intended for use by individuals possessing adequate backgrounds of electrical and mechanical experience. Any attempt to repair a central air conditioning product may result in personal injury and/or property damage. American Standard Inc. or seller cannot be responsible for the information, nor can it assume any liability in connection with its use.

Do not use this furnace if any part has been under water. Immediately call a qualified service technician to inspect the furnace, and to replace any part of the control system and any gas control which has been under water.

Filter maintenance reduces energy use.

A clean filter saves money.

When the furnace circulates and filters the air in your home, dust and dirt particles build up on the filter. Excessive accumulation can block the airflow, forcing the unit to work harder to maintain desired temperatures.

And the harder your unit has to work, the more energy it uses. So you pay more any time your system is running with a dirty filter.

CAUTION: Never operate your unit for either heating or cooling with filters removed.

Help ensure top efficiency by cleaning the filter once a month.

Clean it twice a month during seasons when the unit runs more often.

You can leave the filter in the frame and vacuum it. Or you can take it out of the furnace and wash it with a household detergent.

Both methods are quick and easy — and guaranteed to improve the performance of your system.

Replacing your filter.

When replacing your furnace filters, always use the same size and type that was originally supplied. Filters are available from your dealer.

Where disposable filters are used, they must be replaced every month with the same size as originally supplied.

How to remove your filter.

WARNING: Disconnect power to unit before removing blower door.

Downflow furnace.

Downflow furnaces are factory supplied with two or four 10" x 20" x 1" disposable air filters which are located in a filter cabinet atop the furnace.

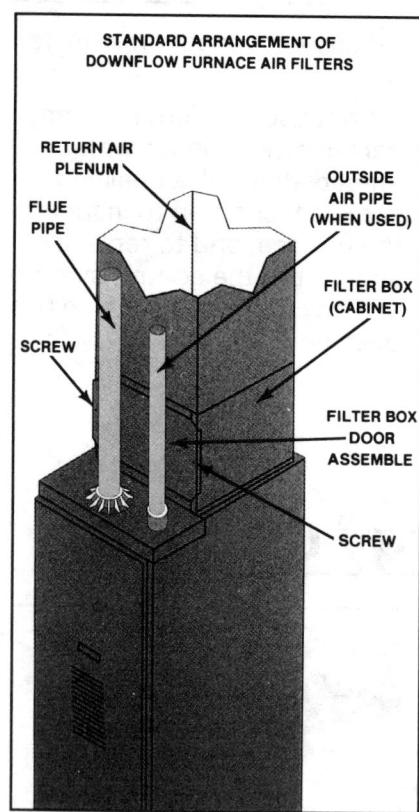

STANDARD ARRANGEMENT OF
DOWNFLOW FURNACE AIR FILTERS

RETURN AIR PLENUM

FLUE PIPE

OUTSIDE AIR PIPE (WHEN USED)

FILTER BOX (CABINET)

SCREW

FILTER BOX DOOR ASSEMBLE

SCREW

Access to filters requires removal of two screws and filter box door assembly (see illustration). Filters are removed by pulling out of box, flexing as required to pass pipes. Replacement filters are inserted in a reverse order making sure they are properly positioned in retaining clips on back panel. Then re-install filter box door assembly and secure with two screws previously removed.

Upflow furnace.

Upflow furnaces are factory supplied with a standard size permanent type air filter which may be located within the furnace blower compartment in either a BOTTOM or SIDE (left or right) return air inlet.

To replace filters, remove blower access door, loosen the filter retaining wire at the front of the unit. Replace the filter in the same manner, making sure that filter retaining wire is secured in place in both front and back of the unit. Replace blower access door.

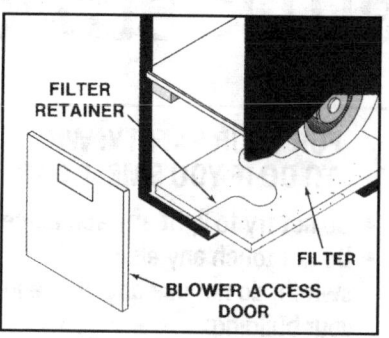

FILTER RETAINER

FILTER

BLOWER ACCESS DOOR

A bottom return air inlet as above features a 16" x 25" x 1" filter in the 18" wide furnaces, a 20" x 25" x 1" filter in the 24" wide furnace cabinets.

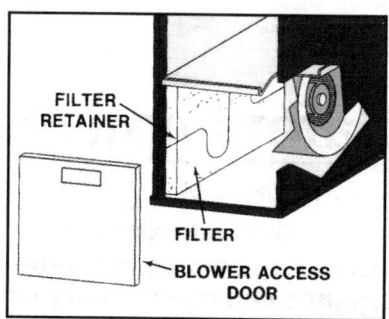

FILTER RETAINER

FILTER

BLOWER ACCESS DOOR

A left or right return air inlet as above (left side shown) requires trimming of factory supplied filter to 14½" x 25" x 1" for both the 18" and 24" wide furnaces.

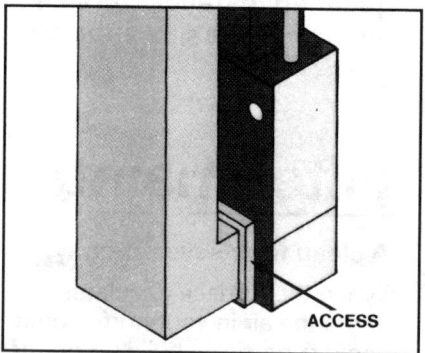

ACCESS

Air filters may also be located outside of the furnace using a SIDE FILTER FRAME.

The Problem Solver.

Save time and money.
Before calling for service, check the following:

Problem	Possible Trouble	Possible Remedy
No Heating - Blower Does Not Operate	1. Thermostat set incorrectly.	1. Adjust thermostat - See operating instructions.
	2. Blown fuse or tripped circuit breaker.	2. Replace or reset protective device or call for servicer.
	3. Defective component.	3. Most controls are automatic and will recycle. If your unit still does not operate, call for servicer.
	4. Burner may not ignite.	4. Call servicer.
	5. Main gas line turned off.	5. Have gas company check.
	6. Blower door removed or ajar.	6. Close door securely to restore power to blower and gas valve.
Insufficient Heating - Blower Operates Continuously	1. Dirty air filters. 2. Blocked supply or return registers.	1. Clean or replace filters. 2. Make sure registers are open and no obstacles blocking off the air.
Unusual Noise		**Call your servicer.**

A furnace is not a household appliance. It is complex and requires professional maintenance and repair.

That's why attempts at "do-it-yourself" repairs on an in-warranty unit may void the remainder of your warranty.

Other than performing the simple maintenance recommended in this manual, you should not attempt to make any adjustments to your furnace. Your dealer will be able to take care of any questions or problems you may have. A periodic inspection of your furnace should be made by a qualified service agency at the start of each heating season.

Keep your furnace looking like new for years.

Clean the enamel finish of your furnace with ordinary soap and water. For stubborn grease spots, use a household detergent. Lacquer thinner or other synthetic solvents may damage the finish.

Do away with surprise repair bills with Service Agreement.

Service Agreements may be available from your Dealer or installer. The agreement has the following advantages:
1. Established cost for service resulting from normal usage . . . no need for an unexpected service cost to "upset" budgeted expenses.
2. Includes both parts and labor for the duration of the Agreement. **Be certain you read the Agreement for complete details and exclusions.**
3. Service is performed by servicers knowledgeable of the operation of this equipment.

Maintenance information for the owner.

Never stop the system by shutting off the main power.

If the main power to your air conditioner is ever disconnected for more than three hours, turn off the thermostat. Then wait for at least three more hours after the power has been restored before turning the thermostat back on. Failure to follow this procedure could result in damage to your air conditioning system.

1. **General Inspection —** Examine the furnace installation for the following items:
   a. All flue product carrying areas external to furnace (i.e. chimney, vent connector) are clear and free of obstruction.
   b. The vent connector is in place, slopes upward and is physically sound without holes or excessive corrosion.

   c. The return air duct connection(s) is physically sound, is sealed to furnace and terminates outside space containing the furnace.
   d. The physical support of the furnace should be sound without sagging, cracks, gaps, etc., around the base so as to provide a seal between the support and the base.
   e. There are no obvious signs of deterioration of furnace.

2. **Blowers** — The blower size and speed determine the air volume delivered by the furnace. The blower motor bearings are factory lubricated and under normal operating conditions usually do not require servicing. Your Servicer can advise you if oiling is required. Annual cleaning of the blower wheel and housing is recommended for maximum air output and combustion air requirements.

3. **Igniter** — This unit has a special hot surface direct ignition device that automatically lights the burners. Please note that it is very fragile and should be handled with care.

   CAUTION: Do not touch igniter. It is extremely hot.

* 4. **Burner** — Gas burners do not normally require scheduled servicing, however, accumulation of lint may cause a yellowing flame or delayed ignition. Either condition indicates that a service call is required. For best operation burners must be cleaned annually using brushes and vacuum cleaner.

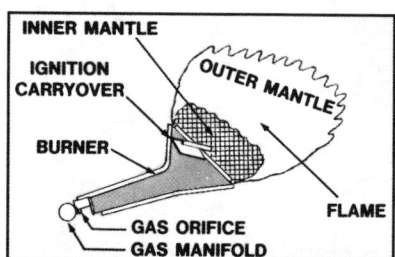

Turn off gas and electric power supply. Remove burners by first removing burner door and burner retention bracket, then lift front burner from the spud. Slide burner out of the chamber. Use a soft brush that will pass through the venturi to clean the inside of the burner. Reassemble parts by reversal of the above procedure.

NOTE: On LP (propane) units, some light yellow tipping of the outer mantle is normal. Inner mantle should be bright blue.

Natural gas units should not have any yellow tipped flames. This condition indicates that a service call is required. For best operation, burners must be cleaned annually using brushes and vacuum cleaner.

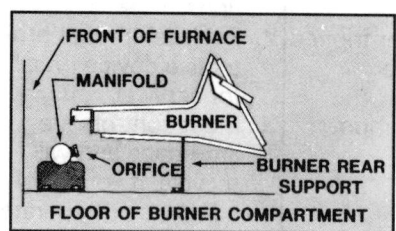

*NOTE: On LP (propane) units, due to variations in BTU content and altitude, servicing may be required at shorter intervals.

5. **Heat Exchanger/Flue Pipe** — These items must be inspected for signs of corrosion, and/or deterioration at the beginnning of each heating season by a qualified service technician and cleaned annually for best operation. To clean flue gas passages, follow these recommendations.

   a. Inspect flue pipe exterior for cracks, leaks, holes or leaky joints. Some discoloration of PVC pipe is normal.
   b. Turn off gas and electrical power to furnace.
   c. Remove burner compartment door from furnace.
   d. Inspect around insulation covering the flue collector box. Inspect induced draft blower connections from recuperative cell and to the flue pipe connection.
   e. Remove burner retention bracket and burner cover, but do not remove burners. Use a mirror and flashlight to inspect interior of heat exchanger by looking past burners. Be careful not to damage the igniter, flame sensor or other components.
   f. If any corrosion is present, contact a service agency. Heat exchangers should be cleaned by a qualified service technician.
   g. After inspection is completed replace the burner cover, retention bracket and

furnace door.
   h. Restore gas and electrical power to furnace. Check unit for normal operation.

6. **Circuit Protection** — If blower or gas valve fail to operate, the cause could be the circuit breaker or a loose or blown fuse. Replace fuse or reset circuit breaker.

7. **Operation** — Your warm air furnace should not be operated in a corrosive atmosphere. Paint solvents, cleaning chemicals, spray propellants, and bleaches should not be used in the vicinity of the furnace during normal operation.

8. **Furnace Condensate Drain Tubes** — Condensate drain tubes must be checked periodically to assure that condensate can flow freely from unit to drain. Drain tubes should not be installed where freeze up is a possibility.

9. **Cooling Coil Condensate Drain** — If you have a cooling coil installed with your furnace, condensate drains should be checked and cleaned periodically to assure that condensate can drain freely from coil to drain. If condensate cannot drain freely water damage could occur.

A cap is installed in drain system inside cabinet for inspection and cleaning as needed. When cap is removed, threads must be re-sealed, with pipe thread sealant when reinstalled. If a drain problem cannot be corrected, call your service company.

Warning.

Unit is equipped with a blower door switch which cuts power to blower and gas valve causing shutdown when door is removed. Unit must not be altered to allow operation with the blower door removed. Operation with doors removed or ajar can permit the escape of dangerous fumes. All panels must be securely closed at all times for safe operation of furnace.

WARNING: Should over-heating occur, or gas supply fail to shut off, shut off the manual gas valve to the furnace before shutting off the electrical supply.

In the event that electrical, fuel or mechanical failures occur, the owner should immediately turn off the gas supply at the manual gas valve located in the burner compartment and electrical power to the furnace and contact your servicer.

For your safety.

Furnace area must be kept clear and free of combustible materials, gasoline, and other flammable vapors and liquids.

Air for combustion and ventilation.

The flow of combustion and ventilating air must not be obstructed from reaching the furnace. Air openings provided in the casing of furnace must be kept free of obstructions which would restrict airflow, thereby affecting efficiency and safe operation of your furnace. Also, air openings provided to the area in which the furnace is installed and the space around the furnace shall not be blocked or obstructed. Keep this in mind should you choose to remodel the area which contains your furnace.

Furnaces must have air for proper performance. If this furnace is installed in an attic or other insulated space it must be kept free and clear of all insulating materials as some insulating materials are combustible.

If additional insulation is added after the furnace is installed, the area around furnace must be inspected to ensure it is free and clear of insulation.

Condensate drain.

Provisions must be made to prevent winter freeze-up of condensate drain tubes from furnace. Frozen condensate tubes will result in furnace shutdown. Condensate drain tubes should not be installed where freeze up is a possibility.

To light furnace.

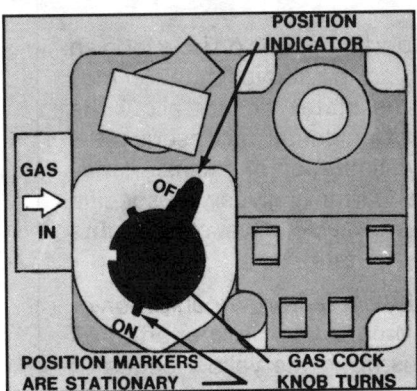

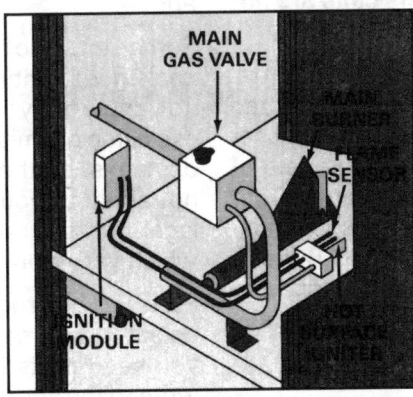

Lighting instructions.

Your furnace is equipped with a hot surface direct ignition device.

WARNING: Do not attempt to manually light the furnace.

1. **Please read all safety information in this book before operating furnace.**
2. Set thermostat to lowest setting. Turn off all electric power to furnace.
3. Remove control access panel.
4. Turn gas cock knob on main gas valve within unit clockwise to "OFF" position. If external gas cock is used, turn to "OFF" position. Allow 5 minutes for any gas within unit to escape. LP gas being heavier than air may require forced ventilation. If you smell gas STOP! Follow the "What To Do If You Smell Gas" instructions on the front cover of this book. If you don't smell gas, go to next step.
5. Turn gas cock knob counterclockwise to "ON" marker.
6. Replace control access panel.
7. Turn on main electrical supply and set thermostat to desired setting. Combustion blower will start and ignition device will start to heat up. After approximately 45 seconds main gas valve will open and burners will ignite.
8. When thermostat is satisfied, main burners will extinguish.
9. If main burners fail to ignite lower thermostat setting or disconnect electrical supply, wait 5 minutes, raise thermostat setting above indicated temperature.
10. If furnace will not light, turn "OFF" all gas and electricity to unit and call Servicer or gas supplier.

For complete shut-down.

Turn gas cock knob on main gas valve to "OFF" position. Disconnect electrical supply to unit.

CAUTION: If this is done during the cold weather months, provisions must be taken to prevent freeze-up of all water pipes and water receptacles.

Whenever your house is to be vacant, arrange to have someone inspect your house for proper temperature. If your furnace should fail to operate, damage could result, such as frozen water pipes.

Safety cutoff device. (Thermal limit)

All models are equipped with a fusible link behind gas valve. In case of main gas valve malfunction and consequent overheating, link will open and close off flow of all gas. See instruction label on front panel of heat exchanger.

Limited Lifetime Warranty
High Efficiency Condensing Gas Furnace
BLU-K-B, BLD-K-B, TUC and TDC Models (Parts Only)

This warranty is extended by American Standard Inc., to the original purchaser for lifetime under the conditions as defined below or for a period of twenty years from the date of installation for any succeeding owner.

If any part of your Gas-fired Furnace fails because of a manufacturing defect within one year from the date of original purchase, Warrantor will furnish without charge the required replacement part. **Any local transportation, related service labor, air filters and diagnosis calls are not included.**

Lifetime Warranty

The heat exchanger of the Gas Furnace when installed to serve a single family residence or single condominium is warranted to the original purchaser for use during his or her lifetime, provided the dwelling is the original purchaser's primary, uninterrupted residence from the date of purchase until a defect in the heat exchanger or recouperative cell is discovered. Warrantor will furnish without charge a replacement heat exchanger F.O.B. nearest Parts Distribution point. **Any local transportation related service labor, and diagnosis calls are not included.**

Twenty Year Warranty

The heat exchanger of the Gas Furnace is warranted to any purchaser other than the original purchaser for a period of twenty (20) years from the date of original installation of the furnace, subject to proof of original purchase. Warrantor will furnish without charge a replacement heat exchanger F.O.B. nearest Parts Distribution point. **Any local transportation, related service labor, and diagnosis calls are not included.**

Extended Warranties

First Thru Tenth Year — type 29-4C Vent Pipe — (Applies only to installations requiring type 29-4C Vent Pipe). If the external type 29-4C vent pipe fails due to perforation caused by corrosion within ten years from the date of original installation of the gas-fired Furnace and vent pipe, Warrantor will furnish without charge a replacement external vent pipe F.O.B. nearest Parts Distribution point. **Any local transportation, related service labor and diagnosis calls are not included. This war**ranty covers only the type 29-4C external vent pipe specified by Warrantor for use with your Gas Furnace.

During the first year following installation, Warrantor will furnish a replacement for a covered failed component, F.O.B. nearest Parts Distribution point. Thereafter in order to fill any Extended Warranty, **WARRANTOR WILL AT ITS OPTION** provide a heat exchanger including recouperative cell without charge F.O.B. nearest Parts Distribution point or allow a credit (in the amount of the then current wholesale price) of an equivalent heat exchanger toward the purchase price of a comparable heating unit. **Any local transportation, related service labor and diagnosis calls are not included.**

This warranty does not cover failure of your Gas Furnace if it is damaged while in your possession or if the failure is caused by unreasonable use. In no event shall Warrantor be liable for incidental or consequential damages. **In no event shall any implied warranty of merchantability or fitness for use exceed the term of the limited warranty stated above.**

Some states do not allow limitations on how long an implied warranty lasts, so the above limitation may not apply to you. Some states do not allow the exclusion or limitation of incidental or consequential damages, so the above limitation or exclusion may not apply to you. This warranty gives you specific legal rights, and you may also have other rights which vary from state to state.

Parts will be provided by our factory organization or an authorized service organization in your area. All you need do is look us up in the yellow pages or write to the address given below. If you wish further help or information concerning this warranty, contact:

Manager — Product Service
American Standard Inc.
Troup Highway
Tyler, Texas 75711-9010

American Standard Inc.,
Troup Highway, Tyler, Texas 75711-9010
Warrantor

PW-204-2688

All model designs are certified by the American Gas Association Laboratories to comply with national standards for safety performance and durability.

American Standard Inc.
6200 Troup Highway
Tyler, TX 75711-9010

Pub No 32-1001-01
© American Standard Inc. 1988

User's Information Manual
High efficiency induced draft gas furnace
TUD-R 2 Stage Heat Models

Safety notice.

WARNING: Improper installation, adjustment, alteration, service or maintenance can cause injury or property damage. Refer to the installation instructions provided with the furnace and this manual. For assistance or additional information consult a qualified installer, service agency or the gas supplier. This information is intended for use by individuals possessing adequate backgrounds of electrical and mechanical experience. Any attempt to repair a central air conditioning product may result in personal injury and/or property damage. American Standard Inc. or seller cannot be responsible for the interpretation of this information, nor can it assume any liability in connection with its use.

Do not use this furnace if any part has been under water. Immediately call a qualified service technician to inspect the furnace and to replace any part of the control system and any gas control which has been under water.

Understand the signal words DANGER, WARNING, AND CAUTION. These words are safety alert words. DANGER indicates the most serious hazards which <u>will</u> result in severe personal injury or death. WARNING indicates hazards which <u>could</u> result in personal injury or death. CAUTION is used to indicate unsafe practices which could result in minor injury or property damage.

Important Information

There must be a free flow of fresh air sufficient for efficient combustion and safe ventilation of your furnace. Do not allow the louvers on the front panels of your furnace to become blocked as this will restrict the flow of fresh air.

The combustion air for your furnace must be fresh uncontaminated air. Paints, varnishes, laundry bleaches, detergents, many household cleaners, water softening salts, adhesives, and all such products release fumes containing compounds which could lead to early heat exchanger and vent system deterioration. Do not store these type of products near your furnace and consider fresh air for your furnace during construction or remodeling.

Never store gasoline, combustible materials, or other flammable liquids or vapors near your furnace.

If you have a problem, check the "Problem Solver" section of this manual before you call for a possibly unneeded service call.

Parts and controls of this furnace are unique. Should service or modification be required, be sure your servicer uses only factory authorized parts, kits, or accessories for this furnace.

Filter maintenance reduces energy use.

A clean filter saves money.

When the furnace circulates and filters the air in your home, dust and dirt particles build up on the filter. Excessive accumulation can block the airflow, forcing the unit to work harder to maintain desired temperatures.

And the harder your unit has to work, the more energy it uses. So you pay more any time your system is running with a dirty filter.

CAUTION: Never operate your unit for either heating or cooling with filters removed.

Help ensure top efficiency by cleaning the filter once a month. Clean it twice a month during seasons when the unit runs more often.

You can clean the filter with a vacuum, OR you can wash it with a household detergent.

Both methods are quick and easy, and guaranteed to improve the performance of your system.

Your filter may or may not be framed.

Replacing your filter.

When replacing your furnace filters, always use the same size and high velocity type that was originally supplied. Filters are available from your dealer. Where disposable filters are used, they must be replaced every month with the same size as originally supplied.

How to remove your filter.

WARNING: Disconnect power to unit before removing blower door.

Upflow furnaces are factory supplied with a standard size permanent type air filter which may be located within the furnace blower compartment in either a

BOTTOM or SIDE (left or right) return air inlet. To replace filters, remove blower access door, loosen the filter retaining wire at the front of the unit. After cleaning, replace the filter in the same manner making sure that filter retaining wire is secured in place in both front and back of the unit. Replace blower access door.

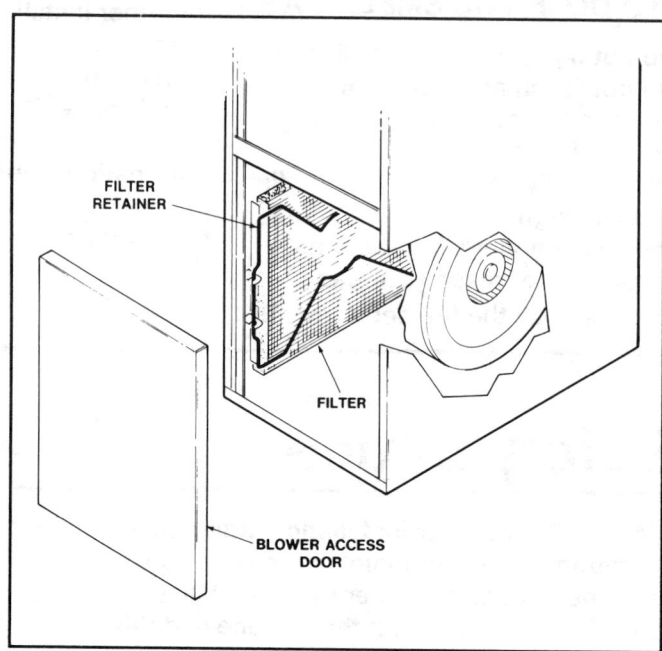

A bottom return air inlet as shown features a 14" x 25" x 1" filter in the 14-1/2" wide furnace cabinets; a 17" x 25" x 1" filter in the 17-1/2" wide models; a 21" x 25" x 1" filter in the 21" wide furnace cabinets; and a 24" x 25" x 1" filter in the 24-1/2" wide cabinet.

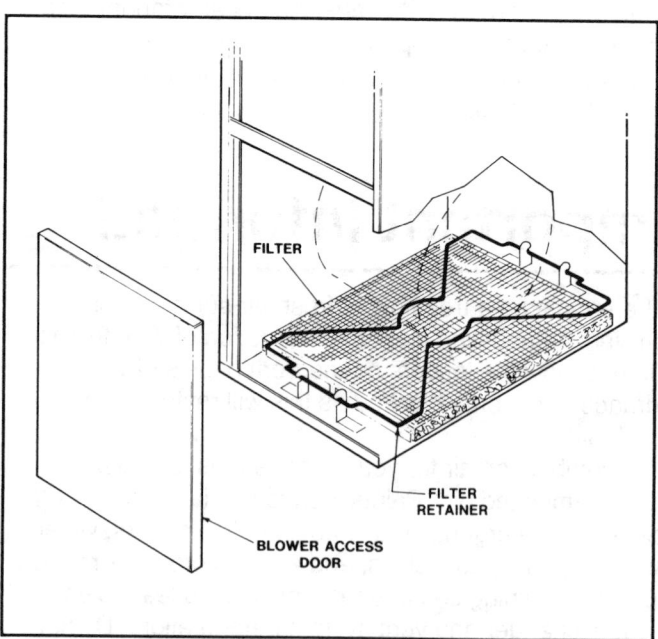

A left or right return air inlet as above (left side shown) requires trimming of factory supplied filter to 17" x 25" x 1" for both the 21" and 24-1/2" wide furnaces.

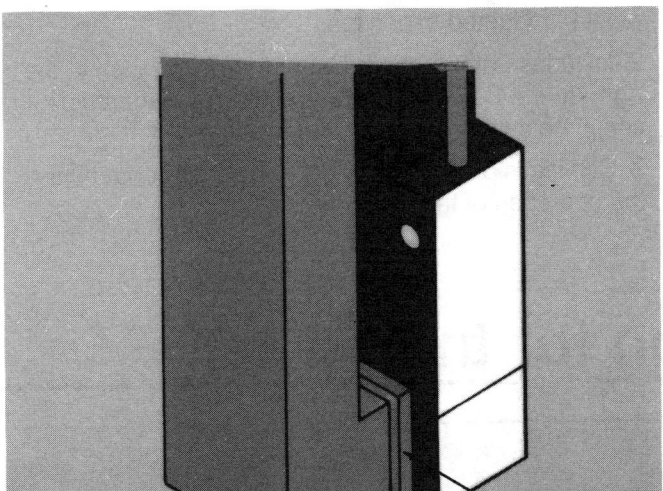

ACCESS

Air filters may also be located outside of the furnace using a **SIDE FILTER FRAME.**

FILTER TABLE

MODELS	CABINET WIDTH	FILTER QTY & SIZE
TUD040R924A	14-1/2	1 - 16 X 25 X 1
TUD060R936A	14-1/2	1 - 16 X 25 X 1
TUD080R936A	17-1/2	1 - 17 X 25 X 1
TUD100R936A	17-1/2	1 - 17 X 25 X 1
TUD100R948A	21	1 - 21 X 25 X 1
TUD120R954A	21	1 - 21 X 25 X 1
TUD100R960A	24-1/2	1 - 24 X 25 X 1
TUD120R960A	24-1/2	1 - 24 X 25 X 1
TUD140R960A	24-1/2	1 - 24 X 25 X 1

The problem solver.

A furnace is not a household appliance. It is complex and requires professional maintenance and repair. That's why attempts at "do-it-yourself" repairs on an in-warranty unit may void the remainder of your warranty. Other than performing the simple maintenance recommended in this manual, you should not attempt to make any adjustments to your furnace. Your dealer will be able to take care of any questions or problems you may have. A periodic inspection of your furnace should be made by a qualified service agency at the start of each heating season.

Keep your furnace looking like new for years.

Clean the enamel finish of your furnace with ordinary soap and water. For stubborn grease spots, use a household detergent. Lacquer thinner or other synthetic solvents may damage the finish.

Save time and money. Before calling for service, check the following:

Problem	Possible Trouble	Possible Remedy
No Heating - Blower Does not operate	1. Thermostat set incorrectly. 2. Blown fuse or tripped circuit breaker. 3. Defective component. 4. Burner may not ignite. 5. Main gas line turned off. 6. Blower door removed or ajar.	1. Adjust thermostat See operating instructions 2. Replace or reset protective device or call for servicer. 3. Most controls are automatic and will recycle. If your unit still does not operate call for servicer. 4. Call Servicer. 5. Have gas company check. 6. Close door securely to restore power to blower and gas valve.
Insufficient Heating - Blower operates continuously	1. Dirty air filters. 2. Blocked supply or return registers.	1. Clean or replace filters. 2. Make sure registers are open and no obstacles blocking off the air.
Unusual Noise		Call your servicer

3

Do away with surprise repair bills with a Service Agreement.

Service Agreements may be available from your Dealer or Installer. The agreement has the following advantages:

1. Established cost for service resulting from normal usage ... no need for an unexpected service cost to

"upset" budgeted expenses.

2. Includes both parts and labor for the duration of the Agreement. Be certain you read the Agreement for complete details and exclusions.

3. Service is performed by servicers knowledgeable of the operation of this equipment.

Maintenance information for the owner.

Never stop the cooling system by shutting off the main power.

If the main power to your air conditioner is ever disconnected for more than three hours, turn off the thermostat. Then wait for at least three more hours after the power has been restored before turning the thermostat back on. Failure to follow this procedure could result in damage to your air conditioning system.

1. GENERAL INSPECTION - Examine the furnace installation for the following items:

 a. All flue product carrying areas external to the furnace (i.e. chimney, vent connector) are clear and free of obstruction.

 b. The vent connector is in place, slopes upward and is physically sound without holes or excessive corrosion.

 c. The return air duct connection(s) is physically sound, is sealed to the furnace and terminates outside the space containing the furnace.

 d. The physical support of the furnace should be sound without sagging, cracks, gaps, etc., around the base so as to provide a seal between the support and the base.

 e. There are no obvious signs of deterioration of the furnace.

2. BLOWERS - The blower size and speed determine the air volume delivered by the furnace. The blower motor bearings are factory lubricated and under normal operating conditions usually do not require servicing. Your Servicer can advise you if oiling is required. Annual cleaning of the blower wheel and housing is recommended for maximum air output and combustion air requirements.

3. IGNITER - This unit has a special hot surface direct ignition device that automatically lights the burners. Please note that it is very fragile and should be handled with care.

CAUTION: Do not touch igniter. It is extremely hot.

4. BURNER - Gas burners do not normally require scheduled servicing, however, accumulation of foreign material may cause a yellowing flame or delayed ignition. Either condition indicates that a service call is required. For best operation burners must be cleaned annually using brushes and vacuum cleaner.

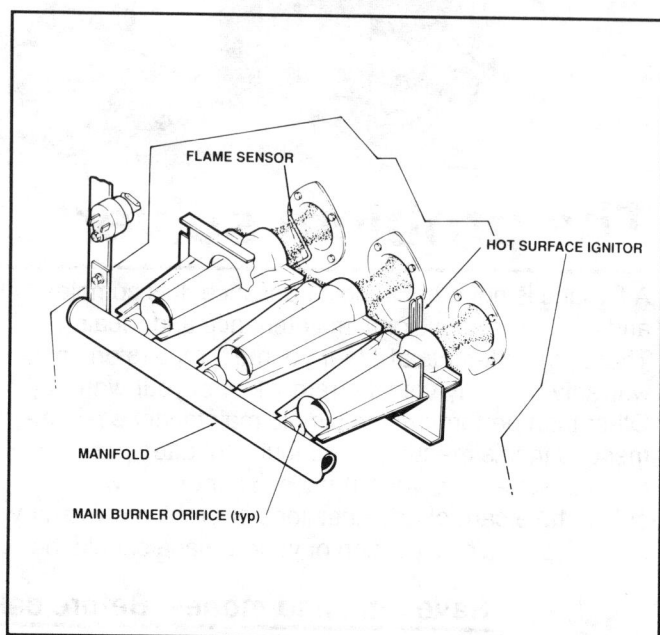

Turn off gas and electric power supply. To clean burners, remove burner cover, remove top burner bracket and lift burner from orifice. **NOTE: Be careful not to break igniter when removing burner assembly.** Clean burners with brush and/or vacuum cleaner. Reassemble parts by reversal of the above procedure.

NOTE: On LP (propane) units, some light yellow tipping of the outer mantle is normal. Inner mantle should be bright blue.

Natural gas units should not have any yellow tipped flames. This condition indicates that a service call is required. For best operation, burners must be cleaned annually using brushes and vacuum cleaner.

NOTE: On LP (propane) units, due to variations in BTU content and altitude, servicing may be required at shorter intervals.

5. HEAT EXCHANGER/FLUE PIPE - These items must be inspected for signs of corrosion, and/or deterioration at the beginning of each heating season by a qualified service technician and cleaned annually for best operation. To clean flue gas passages, follow recommendations below:

 a. Turn off gas and electric power supply.

4

b. Disconnect flue pipe from induced draft blower outlet.

c. Remove screws fastening flue collector box; remove collector box and flue restriction plate.

d. Remove burners.

e. Clean heat exchangers using wire brush and vacuum cleaner. Check flue pipe and clean if required.

f. Inspect and clean induced draft blower wheel.

g. Replace parts removed in reverse order. Inspect burners. Clean if necessary.

h. Restore gas supply. Check for leaks using a soap solution. Restore electrical supply. Check unit for normal operation.

6. CIRCUIT PROTECTION - If blower or gas valve fail to operate, the cause could be the circuit breaker or a loose or blown fuse. Replace fuse or reset circuit breaker.

7. OPERATION - Your warm air furnace should not be operated in a corrosive atmosphere. Paint solvents, cleaning chemicals, spray propellants, and bleaches should not be used in the vicinity of the furnace during normal operation.

8. AIR CIRCULATION - To ensure increased comfort, blower on this unit may be operated continuously for both heating and cooling. This will result in constantly filtered air and aid in maintaining more even temperatures by avoiding temperature stratification throughout conditioned area. To accomplish constant air circulation set thermostat fan switch to "ON".

9. COOLING COIL CONDENSATE DRAIN - If you have a cooling coil installed with your furnace, condensate drains should be checked and cleaned periodically to assure that condensate can drain freely from coil to drain. If condensate cannot drain freely water damage could occur.

WARNING:

Unit is equipped with a blower door switch which cuts power to blower and gas valve causing shutdown when door is removed. Unit must not be altered to allow operation with the blower door removed. Operation with doors removed or ajar can permit the escape of dangerous fumes. All panels must be securely closed at all times for safe operation of the furnace.

WARNING.

Should overheating occur, or the gas supply fail to shut off, shut off the manual gas valve to the furnace before shutting off the electrical supply.

In the event that electrical, fuel or mechanical failures occur, the owner should immediately turn off the gas supply at the manual gas valve located in the burner compartment and electrical power to the furnace and contact servicer.

For your safety.

Furnace area must be kept clear and free of combustible materials, gasoline, and other flammable vapors and liquids.

Air for combustion and ventilation.

The flow of combustion and ventilating air must not be obstructed from reaching the furnace. Air openings provided in the casing of furnace must be kept free of obstructions which would restrict airflow, thereby affecting efficiency and safe operation of your furnace. Also, air openings provided to the area in which the furnace is installed and the space around the furnace shall not be clocked or obstructed. Keep this in mind should you choose to remodel the area which contains your furnace. Furnaces must have air for proper performance.

If additional insulation is added after the furnace is installed, the area around the furnace must be inspected to ensure it is free and clear of insulation. If this furnace is installed in an attic or other insulated space it must be kept free and clear of all insulating materials as some insulating materials are combustible.

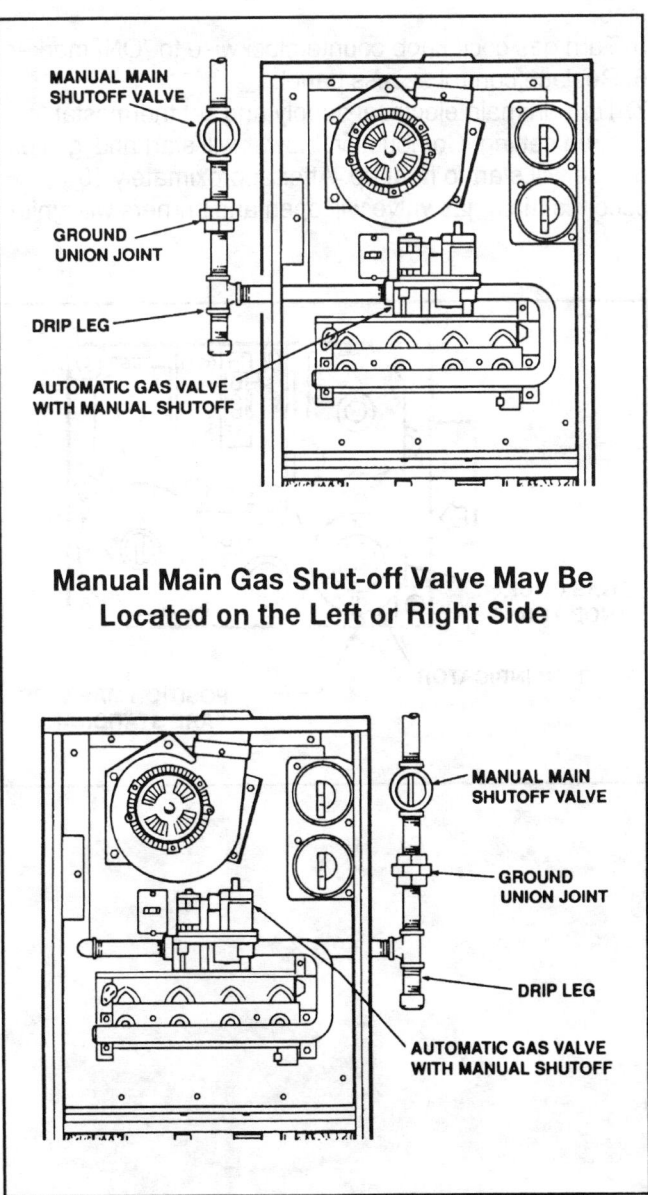

Manual Main Gas Shut-off Valve May Be Located on the Left or Right Side

To light furnace.

Lighting instructions.

Your furnace is equipped with a hot surface direct ignition device.

WARNING.

Do not attempt to manually light the furnace.

1. Please read all safety information in this book before operating furnace.

2. Set thermostat to lowest setting. Turn off all electric power to furnace.

3. Remove control access panel.

4. Turn gas cock knob on main gas valve with unit clockwise to "OFF" position. If external gas cock is used, turn to "OFF" position. All 5 minutes for any gas within unit to escape. LP gas being heavier than air may require forced ventilation. If you smell gas STOP! Follow the "What To Do If You Smell Gas" instructions on the front cover of this book. If you don't smell gas, go to next step.

5. Turn gas cock knob counterclockwise to "ON" marker.

6. Replace control access panel.

7. Turn on main electrical supply and set thermostat to desired setting. Combustion blower will start and ignition device will start to heat up. After approximately 15 seconds main gas valve will open and burners will ignite.

8. When thermostat is satisfied, main burners will extinguish.

9. If main burners fail to ignite lower thermostat setting or disconnect electrical supply, wait 5 minutes, raise thermostat setting above indicated temperature.

10. If furnace will not light, turn "OFF" all gas and electricity to unit and call Servicer or gas supplier

For complete shut-down:

Turn gas cock knob on main gas valve to "OFF" position. Disconnect electrical supply to unit.

CAUTION: If this is done during the cold weather months, provisions must be taken to prevent freeze-up of all water pipes and water receptacles.

Whenever your house is to be vacant, arrange to have someone inspect your house for proper temperature. If your furnace should fail to operate, damage could result, such as frozen water pipes.

Safety cutoff device (thermal limit):

All models are equipped with a fusible link behind gas valve. In case of main gas valve malfunction and consequent overheating, link will open and close off flow of all gas. See instruction label on from panel of heat exchanger.

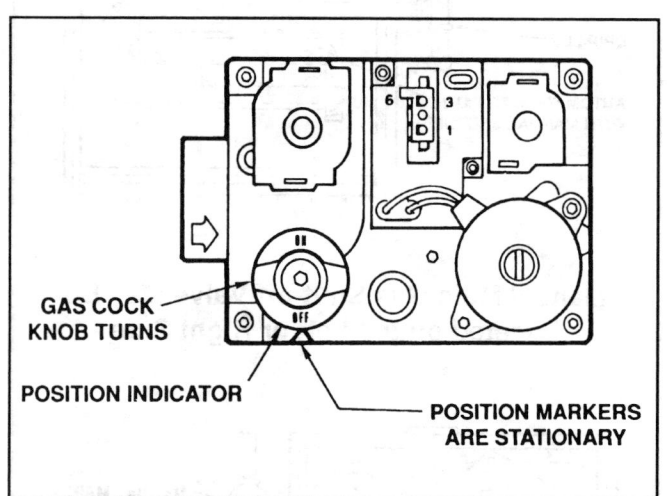

GAS COCK
KNOB TURNS

POSITION INDICATOR

POSITION MARKERS
ARE STATIONARY

Important warranty and service information.

LIMITED WARRANTY
HIGH EFFICIENCY INDUCED DRAFT GAS FURNACE
TUD and TDD-R Models
(Parts Only)

This warranty is extended by American Standard Inc., to the original purchaser and to any succeeding owner of the real property of which the Gas Furnace is originally affixed, and applies to products purchased and retained for use within the U.S.A.

If any part of your Gas Furnace fails because of a manufacturing defect within one year from the date of original purchase, Warrantor will furnish without charge the required replacement part. **Any local transportation, related service labor, air filters and diagnosis calls are not included.**

In addition, if the steel heat exchanger fails because of a manufacturing defect within the second through twentieth year from the date of original purchase, Warrantor will furnish without charge a replacement heat exchanger. **Any local transportation, related service labor and diagnosis calls are not included.**

This warranty does not cover failure of your Gas Furnace if it is damaged while in your possession or if the failure is caused by unreasonable use. In no event shall Warrantor be liable for incidental or consequential damages. **In no event shall any implied warranty of merchantability or fitness for use exceed the term of the limited warranty stated above.**

Some states do not allow limitations on how long an implied warranty lasts, so the above limitation may not apply to you. Some states do not allow the exclusion or limitation of incidental or consequential damages, so the above limitation or exclusion may not apply to you. This warranty gives you specific legal rights, and you may also have other rights which vary from state to state.

Parts will be provided by our factory organization or an authorized service organization in your area. All you need do is look us up in the yellow pages or write to the address given below. If you wish further help or information concerning this warranty, contact:

Manager — Product Service
American Standard Inc.
Troup Highway
Tyler, Texas 75711-9010

American Standard Inc.,
Troup Highway, Tyler, Texas 75711-9010
Warrantor

GW-512-2791

Warranty Information
It's always a good idea to keep records which will save you time and money. If it's necessary to have your FURNACE repaired, the service man will want to know if your unit is still under Warranty. To save time, take a few minutes to record the following information here:
Model Number: _____
Serial Number: _____
Date of Purchase _____

Service Information
Call your installing dealer if the unit is inoperative. Before you call, always check the following to be sure service is really required:

a. Be sure the main switch that supplies power to the unit is in the "ON" position.

b. Replace any burned-out fuses or reset circuit breakers.

c. Be sure the thermostat is properly set.

American Gas
Association
Design certified

American Standard Inc.
6200 Troup Highway
Tyler, TX 75711-9010

Pub. No. 32-5009-01
©American Standard Inc. 1991

21X340073 PO1
8/91

York

GAS-FIRED FURNACES
STELLAR™ HIGH-EFFICIENCY
DOWNFLOW CONDENSING MODELS

MODELS: **P1CD, (STYLE A)**   **57 THRU 114 MBH OUTPUT**

DOWNFLOW MODELS
TYPE FSP
DIRECT VENT FURNACE

FOR YOUR SAFETY

WHAT TO DO IF YOU SMELL GAS

- **Do not try to light any appliance.**

- **Open windows.**

- **Do not touch any electrical switch; do not use any phone in your building.**

- **Extinguish any open flames.**

- **Immediately call your gas supplier from a neighbor's phone. Follow the gas supplier's instructions.**

- **If you cannot reach your gas supplier, call the fire department.**

FOR YOUR SAFETY

Do not store or use gasoline or other flammable vapors and liquids in the vicinity of this or any other appliance.

WARNING: Improper installation, adjustment, alteration, service or maintenance can cause injury or property damage. Refer to this manual. For assistance or additional information, consult a qualified installer, service agency or the gas supplier.

CAUTION
THIS PRODUCT MUST BE INSTALLED IN STRICT COMPLIANCE WITH THE ENCLOSED INSTALLATION INSTRUCTIONS AND ANY APPLICABLE LOCAL, STATE, AND NATIONAL CODES INCLUDING, BUT NOT LIMITED TO, BUILDING, ELECTRICAL, AND MECHANICAL CODES.

WARNING
IMPROPER INSTALLATION MAY CREATE A CONDITION WHERE THE OPERATION OF THE PRODUCT COULD CAUSE PERSONAL INJURY OR PROPERTY DAMAGE.

TABLE OF CONTENTS

GENERAL INFORMATION

DESCRIPTION

This Category IV, sealed combustion furnace is designed for residential installation, provided space temperature is 32°F or higher.

Model P1CD units may be converted to propane (LP) gas if factory supplied components are used. Conversions required in order for the appliance to satisfactorily meet the application must be made by a CES distributor, conversion station or other qualified agency, using factory specified and/or approved parts.

INSPECTION

As soon as a unit is received, it should be inspected for possible damage during transit. If damage is evident, the extent of the damage should be noted on the carrier's freight bill. A separate request for inspection by the carrier's agent should be made in writing. Also, before installation the unit should be checked for screws or bolts which may have loosened in transit.

NOTES, CAUTIONS, & WARNINGS

The installer should pay particular attention to the words:

NOTE, CAUTION and **WARNING. NOTES** are intended to clarify or make the installation easier. **CAUTIONS** are given to prevent equipment damage. **WARNINGS** are given to alert the installer that personal injury and/or equipment or property damage may occur if installation procedures are not handled properly.

CAUTION: The cooling coil must be installed in the supply air duct, downstream of the furnace.

The furnace room must not be used as a broom closet or for any other storage purposes as a fire hazard may be created. Never store items such as the following on, near, or in contact with the furnace.

1. Spray or aerosol cans, rags, brooms, dust mops, vacuum cleaners or other cleaning tools.

2. Soap powders, bleaches, waxes or other cleaning compounds; plastic items or containers; gasoline, kerosene, cigarette lighter fluid; dry-cleaning fluids or other volatile fluid.

3. Paint thinners and other painting compounds.

4. Paper bags or other paper products.

WARNING: Never operate the furnace with the blower door removed. To do so could result in serious personal injury and/or equipment damage.

WARNING: This furnace may not be common vented with any other appliance since it requires separate, properly-sized air intake and vent lines. The furnace shall not be connected to any type of B, BW or L vent or vent connector, and not connected to any portion of a factory-built or masonry chimney.

If this furnace is replacing a common-vented furnace, it may be necessary to resize the existing vent line and chimney to prevent oversizing problems for the new combination of units. Refer to the National Gas Code (ANSI Z223.1-) or CANI-B149.1 or .2 Installation Code (latest editions).

The following steps shall be followed with each appliance remaining connected to the common venting system placed in operation, while the other appliances remaining connected to the common venting system are not in operation:

1. Seal any unused openings in the common venting system.

2. Visually inspect venting system for proper size and horizontal pitch and determine there is no blockage or restriction, leakage, corrosion or other deficiencies which could cause an unsafe condition.

Central Environmental Systems

3. Insofar as is practical, close all building doors and windows and all doors between the space in which the appliances remaining connected to the common venting system are located and other spaces of the building. Turn on clothes dryers and any appliance not connected to the common venting system. Turn on any exhaust fans, such as range hoods and bathroom exhausts, so they will operate at maximum speed. Do not operate a summer exhaust fan. Close fireplace dampers.

4. Follow the lighting instructions. Place the appliance being inspected in operation. Adjust the thermostat so the appliance will operate continuously.

5. Check for spillage at the draft hood relief opening after 5 minutes of main burner operation. Use the flame of a match or candle.

6. After it has been determined that each appliance remaining connected to the common venting system properly vents when tested as outlined above, return doors, windows, exhaust fans, fireplace dampers and any other gas burning appliance to their previous conditions of use.

7. If improper venting is observed during any of the previous tests, the common venting system must be corrected.

8. Any corrections to the common venting system must be in accordance with the National Fuel Gas Code Z223.1 or CAN1-B149.1 or .2 Installation Code (latest editions). If the common vent system must be resized, it should be resized to approach the minimum size as determined using the appropriate tables in Appendix G of the above codes.

LIMITATIONS & LOCATION

This furnace should be installed in accordance with all national and local building/safety codes and requirements, or in the absence of local codes, with the National Fuel Gas Code ANSI Z223.1 or CAN1-B149.1 or .2 Installation Code (latest editions), local plumbing or waste water codes, and other applicable codes.

CAUTION: Do not install the furnace in an unconditioned space or garage that could experience ambient temperatures of 32°F (0°C) or lower.

CAUTION: This unit must be installed in a level (1/4") position side-to-side and front-to-back to provide proper condensate drainage.

CAUTION: Do not allow return air temperature to be below 55°F for extended periods. To do so may cause condensate to occur in the main fired heat exchanger.

WARNING: Furnaces shall not be installed directly on carpeting, tile, wood or other combustible material. An accessory combustible floor base is available to allow installation on combustible flooring and must be used.

The size of the unit should be based on an acceptable heat loss calculation for the structure.

Check the rating plate to make certain the unit is equipped for the type of gas supplied, and proper electrical characteristics are available.

For installations in U.S. above 2,000 feet, reduce input 4% for each 1,000 feet above sea level.

Do not install this unit in a mobile home.

A furnace installed in a residential garage shall be located so that all burners and burner ignition devices are located not less that 18" above the garage floor, and located or protected to prevent damage by vehicles.

Allow clearances from combustible materials as listed under "Clearances to Combustibles", ensuring that service access is allowed for both the burners and blower.

SPECIFIC UNIT INFORMATION

CLEARANCES TO COMBUSTIBLES

Minimum clearances from combustible construction are in inches:

```
Top ...........................................................1
Front .........................................................3
Vent Piping ...............................................0
Rear ..........................................................0
Sides .........................................................0
```

CLEARANCES FOR ACCESS

Ample clearances should be provided to permit easy access to the unit. The following minimum clearances are recommended:

1. Twenty-four (24) inches between the front of the furnace and an adjacent wall or another appliance, when access is required for servicing and cleaning.

2. Eighteen (18) inches at the side where access is required for passage to the front when servicing or for inspection or replacement of flue/vent connections.

NOTE: In all cases, accessibility clearances shall take precedence over clearances for combustible materials where accessibility clearances are greater.

NOMENCLATURE

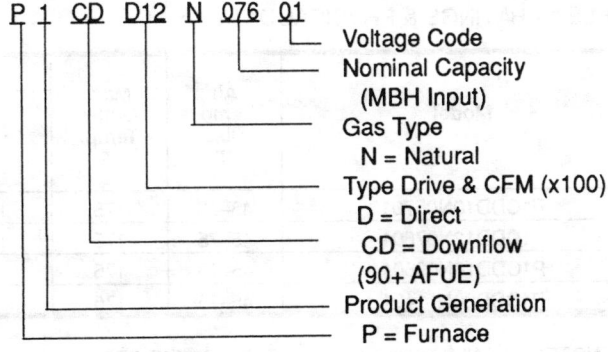

```
P  1  CD  D12  N  076  01
                            └── Voltage Code
                        └────── Nominal Capacity
                                 (MBH Input)
                    └────────── Gas Type
                                 N = Natural
              └──────────────── Type Drive & CFM (x100)
                                 D = Direct
          └────────────────── CD = Downflow
                                 (90+ AFUE)
       └───────────────────── Product Generation
    └──────────────────────── P = Furnace
```

DIMENSIONS

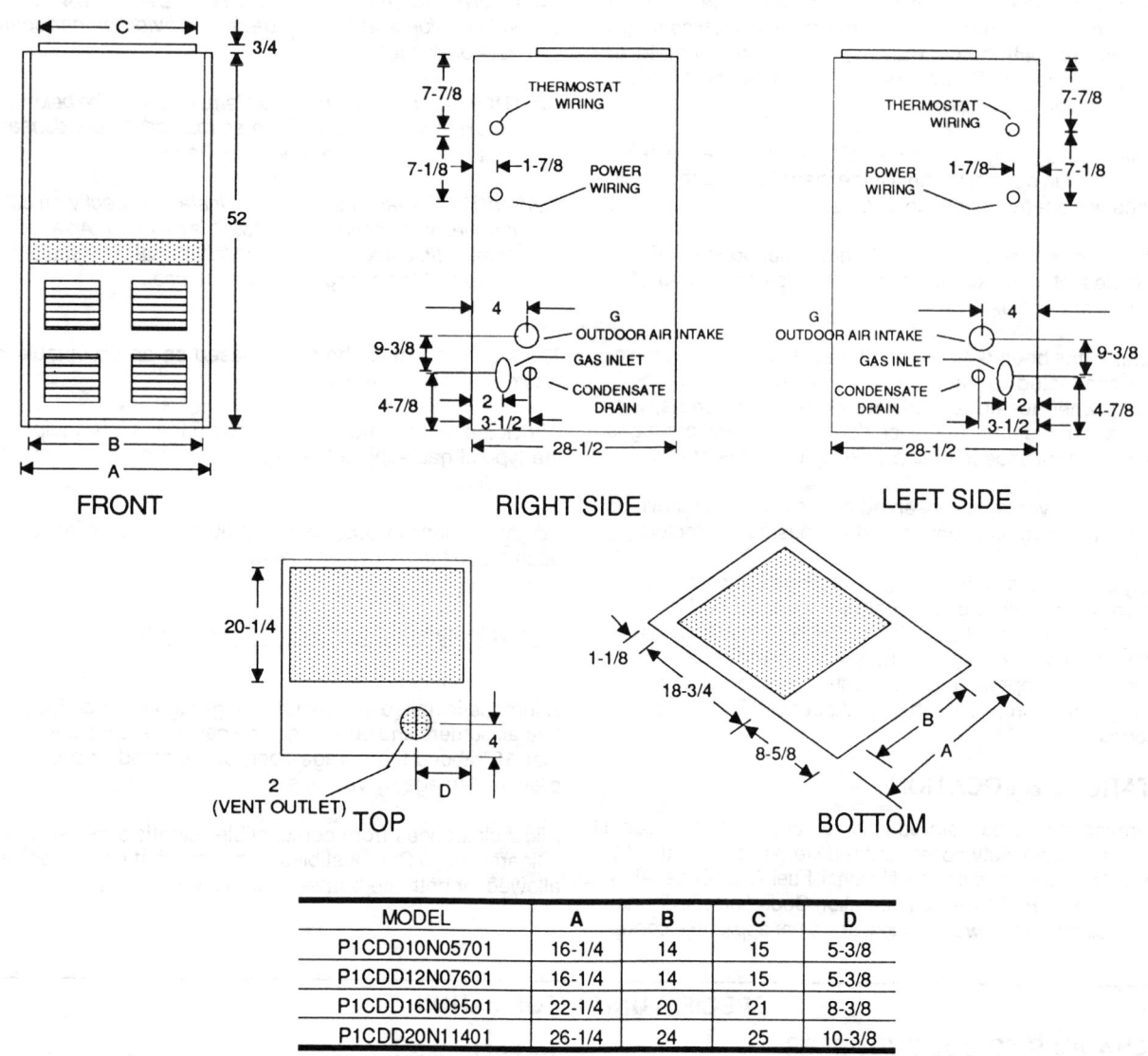

FRONT

RIGHT SIDE

LEFT SIDE

TOP

BOTTOM

MODEL	A	B	C	D
P1CDD10N05701	16-1/4	14	15	5-3/8
P1CDD12N07601	16-1/4	14	15	5-3/8
P1CDD16N09501	22-1/4	20	21	8-3/8
P1CDD20N11401	26-1/4	24	25	10-3/8

All dimensions are in inches, and are approximate. Certified dimensions are available upon request.

TABLE 1 - RATINGS & PHYSICAL DATA

Model	Air Temp. Rise °F	Max. Outlet Temp. °F	Blower		Filter Size	Unit Amps	Max. Over-Current Protect.	Min. Wire Size (AWG) @ 75 ft. One Way
			HP	Size	Supplied (2)			
P1CDD10N05701	45–75	175	1/2	10–6	14 x 20	15	20	14
P1CDD12N07601	45–75	175	1/2	10–8	14 x 20	15	20	14
P1CDD16N09501	45–75	175	1	10–10	14 x 20	17	20	12
P1CDD20N11401	45–75	175	1	11–10	14 x 20	17	20	12

NOTES: 1. All furnaces are factory wired for 115-1-60 operation.
2. All Filters supplied with the furnace are high-velocity, cleanable type.

When the furnace is used in conjunction with a cooling coil, the furnace must be installed parallel with or on the upstream side of the cooling unit to avoid condensation in the primary heat exchanger. When a parallel flow arrangement is used, the dampers or other means used to control air flow shall be adequate to prevent chilled air from entering the furnace, and if manually operated, must be equipped with means to prevent operation of either unit unless the damper is in the full heat or cool position.

The furnace shall be located:

1. Where a minimum amount of air intake/vent piping and elbows will be required.

2. As centralized with the air distribution as possible.

3. In an area where ventilation facilities provide for safe limits of ambient temperature under normal operating conditions. Ambient temperatures must not fall below 32°F (0° C).

4. Where it will not interfere with proper air circulation in the confined space.

5. Where the outdoor combustion air/vent terminal will not be blocked or restricted.

6. Where it will not interfere with the cleaning, servicing or removal of other appliances.

COMBUSTION AIR AND VENT SYSTEM

This furnace requires outdoor combustion air. Two separate, properly-sized pipes must be used; one bringing outdoor air from the accessory terminal kit outdoors to the furnace combustion air intake, and one from the furnace vent connection back to the terminal kit located outdoors.

The vent terminal kit should be located either through the wall (horizontal or side vent) or through the roof (vertical vent). Care should be taken to locate side vented systems where trees or shrubs will not block or restrict supply air from entering or combustion products from leaving the terminal. Also, the terminal assembly should be located as far as possible from a swimming pool or a location where swimming pool chemicals might be stored. Care must be taken such that the terminal assembly outdoors follows the clearances listed in the following table:

LOCATION CLEARANCE

Dryer Vent .... 4 feet
Plumbing Vent Stack .... 3 feet
Gas Appliance Vent Terminal .... 1 foot
From any opening where vent
  gas could enter the building .... 1 foot
Above grade and anticipated
  snow depth .... 1 foot
Above grade when adjacent to
  a public walkway .... 7 feet
From electric, gas meters, regulators and relief
  equipment - min. horizontal distance .... 4 feet

NOTE: *Consideration must be given for degradation of building materials by flue gases.*

For proper vent/combustion air intake sizing and installation, see the section of this instruction "Combustion Air/Vent Pipe Sizing."

UNIT INSTALLATION

DUCTWORK

The duct system's design and installation must:

1. Handle an air volume appropriate for the served space and within the operating parameters of the furnace specifications.

2. Be installed in accordance with standards of NFPA (National Fire Protection Association) as outlined in NFPA pamphlets 90A and 90B or applicable national, provincial, local fire and safety codes.

3. Create a closed duct system. The supply duct system must be connected to the furnace outlet and the return duct system must be connected to the furnace inlet. Both supply and return duct systems must terminate outside the space containing the furnace.

4. Generally complete a path for heated or cooled air to circulate through the heating and air conditioning equipment and to and from the conditioned space.

NOTE: *The supply air temperature differential between the side discharge versus front to back discharge is substantial. It is recommended that whenever possible position the furnace so left and right side supply air is dominate.*

After the unit is in the desired position, fasten the supply ductwork to the furnace duct flanges. A removable access panel should be provided in the outlet duct such that smoke or reflected light would be observable inside the casing to indicate the presence of leaks in the heat exchanger. This access cover shall be attached in such a manner as to prevent leaks. Flexible duct connectors are recommended to connect both the supply and return ducts to the furnace.

Return air ductwork flanges are provided at the top of the unit. Before connecting the return air ductwork to the furnace, refer to the "Filters" section of this instruction. The supply air ductwork connects to the bottom of the furnace.

When the return duct system is not complete, the return connection must be run full size from the furnace to a location outside the utility room or basement. For further details, consult Section 5.3, Air Combustion and Ventilation of the National Fuel Gas Code, ANSI Z223.1 or CAN1-B149.1 or .2 Installation Code (latest editions).

FILTERS

Two 14" x 20" x 1" permanent washable filters are supplied with each unit. Downflow furnace filters are installed above the furnace, extending into the ductwork as shown in Figure 1. Branch ducts must enter above the height of dimension FH.

The filter rack should be secured to the center of the front and rear flanges at the furnace's return air opening. Drill a hole through the rear duct flange into the filter rack and secure it with a sheet metal screw.

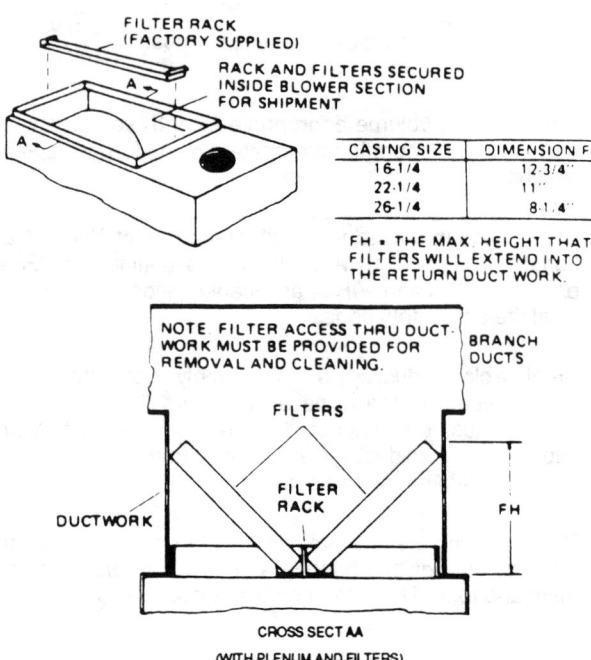

CASING SIZE	DIMENSION FH
16-1/4	12-3/4"
22-1/4	11"
26-1/4	8-1/4"

FH = THE MAX. HEIGHT THAT FILTERS WILL EXTEND INTO THE RETURN DUCT WORK.

FIGURE 1 - DOWNFLOW FILTERS

COMBUSTION AIR/VENT PIPE SIZING

Refer to Table 2 to select the proper size piping for combustion air intake and venting. The size will be determined by a combination of furnace model, total length of run, and the number of elbows required. The following rules must also be observed.

1. Long radius elbows are recommended for all units.

2. Elbows are assumed to be 90 degrees. Two 45 degree elbows count as one 90 degree elbow.

3. Elbow count refers to combustion air piping and vent piping separately. For example, if the table allows for 5 elbows, this will allow a maximum of 5 elbows in the combustion air piping and a maximum of 5 elbows in the vent piping.

4. The inlet air elbow and accessory vent terminal kit parts are already accounted for, and should not be counted in the allowable total indicated in the table.

5. Combustion air and vent piping must be of the same diameter.

6. All piping and fittings are to be Schedule 40 PVC, PVC-DWV, ABS-DWV, SDR-21 PVC, or SDR-26 PVC.

COMBUSTION AIR INTAKE

All models are provided with a 2" diameter intake elbow.

VENT PIPE CONNECTIONS

All models are provided with a 2" diameter vent pipe.

With reference to the piping sizing table, where units are installed with a vent pipe of a different diameter than the inlet and exhaust, reduction or expansion fittings must be incorporated by the installer at the time of installation, external to the furnace.

PIPING ASSEMBLY

The final assembly procedure for the vent/combustion air piping is as follows:

1. Cut piping to the proper length, beginning at the furnace.

2. Remove burrs from inside and outside the piping.

3. Chamfer the outer edges of the piping.

4. Dry-fit the entire vent/combustion air piping assembly.

5. Disassemble the piping and apply cement primer and cement per the cement manufacturer's instructions. Primer and cement must conform to ASTM D2564 for PVC, or ASTM D2235 for ABS piping.

WARNING: *Solvent cements are flammable and must be used in well-ventilated areas only. Keep them away from heat, sparks and open flames (including pilots). Do not breathe vapors and avoid contact with skin and eyes.*

6. All joints must be made to provide a permanent, air-tight, water-tight seal. The only exception to this means of joining is where the exhaust is joined to the venter assembly. This joint should be made with RTV sealant. A stainless steel screw must be used in the hole in the venter housing to keep the street elbow in proper alignment.

WARNING: *Never operate furnace without the exhaust joined to the venter assembly with a stainless steel screw and sealed with RTV sealant.*

If the intake elbow is used for right side exit, carefully remove the plastic cap which is factory installed on the upper left vent connection.

TABLE 2 - COMBUSTION AIR/VENT PIPE SIZES

MODEL	Pipe Size	MAXIMUM ELBOWS PER TOTAL RUN						
		0–5 feet	5-10 feet	10-15 feet	15-20 feet	20-25 feet	25-30 feet	30-35 feet
P1CDD10N05701	1-1/2"	3	2	Note 1	Note 1	Note 1	Note 1	Note 1
	2"	5	5	5	5	5	5	5
P1CDD12N07601	1-1/2"	3	2	Note 1	Note 1	Note 1	Note 1	Note 1
	2"	5	5	5	5	5	5	5
P1CDD16N09501	3"	5	5	5	5	5	5	5"
P1CDD20N11401	3"	5	5	5	5	5	5	5

NOTE: 1. Must use the larger pipe size indicated

Central Environmental Systems

Move the tubing and clamp from the upper right connection to the upper left connection. Place the plastic cap onto the upper right connection.

7. Support the combustion air and vent piping such that it is angled at least 1/4" per linear foot upward from the furnace. Piping must be supported with pipe hangers to prevent sagging. Maximum spacing between hangers is five (5) feet, except SDR-PVC piping, where maximum spacing is three (3) feet.

8. Seal around the openings where the combustion air and vent piping pass through the roof or side wall.

CAUTION: Vent piping must be insulated with 1/2" Armaflex insulation if it will be subjected to freezing temperatures such as routing through unheated areas or through an unused chimney.

The combustion air piping must be insulated with 1/2" thick Armaflex insulation if it is installed above a suspended ceiling or in a warm, humid space such as a laundry room to prevent possible condensation from forming on the outside of the pipe.

VENT TERMINAL ASSEMBLY

The combustion air and vent piping must terminate outdoors using an accessory Vent Terminal Assembly Kit.

Two vent terminal kits are available. The 1VK0307 kit is to be used with 1-1/2" and 2" piping, and the 1VK0308 is to be used for 3" piping.

NOTE: The 3" vent terminal kit contains one elbow having a "splitter baffle" in one opening. This elbow must be used for the combustion air intake.

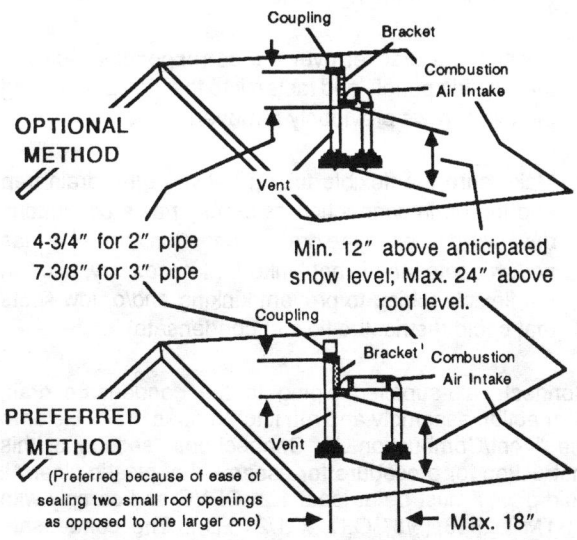

FIGURE 2 - ROOFTOP TERMINATION

Each kit contains the following components:

1. One terminal bracket with 2 - 90 degree PVC elbows.
2. One 90 degree PVC street elbow.
3. One 90 degree PVC elbow.
4. One PVC pipe coupling.
5. Installation instructions.

Central Environmental Systems

This terminal kit may be used for rooftop or side wall installation. Rooftop termination is the recommended means, and should be arranged according to one of the methods shown in Figure 2.

NOTE: The vent terminal kit and exposed piping may be painted the same color as the building to make them less noticeable.

If optional side wall venting is to be used, installation of the terminal kit should be as shown in Figure 3.

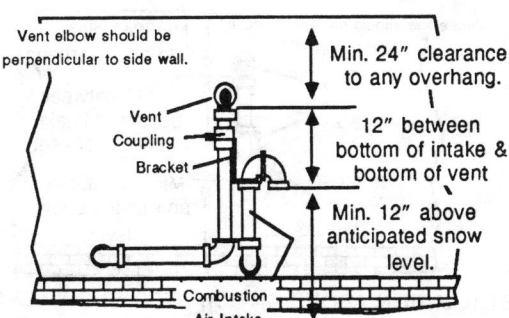

INSTALLATION BELOW ANTICIPATED SNOW LEVEL

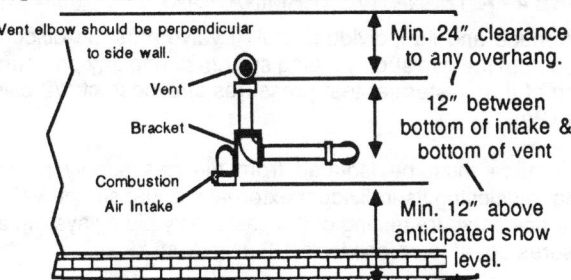

INSTALLATION ABOVE ANTICIPATED SNOW LEVEL

FIGURE 3 - SIDE WALL TERMINATION

If for some reason, combustion air and vent piping can not be routed together, the optional side wall termination may be carried out as illustrated in Figure 4.

The installation procedure for the terminal kit is as follows:

1. Cut all combustion air and vent piping so the vent termination fittings and brackets can be dry-fitted together.

2. All piping should be free of burrs inside and out, and the outside edge should be chamfered.

3. Reassembly all piping and fittings using cement primer and cement per the cement manufacturer's instructions. Primer and cement must conform to ASTM 2564 for PVC or ASTM D2235 for ABS piping.

4. Reattach and tighten the vent termination bracket.

GAS PIPING

NOTE: An accessible manual shutoff valve must be installed upstream of the furnace gas controls and within 6 feet of the furnace. A 1/8" NPT plugged tapping, accessible for test gauge connection must be installed immediately upstream of the gas supply connection to the furnace.

7

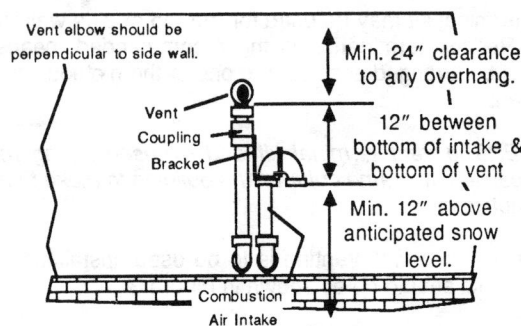

INSTALLATION BELOW ANTICIPATED SNOW LEVEL

Vent elbow should be perpendicular to side wall.

Vent
Coupling
Bracket

Min. 24" clearance to any overhang.

12" between bottom of intake & bottom of vent

Min. 12" above anticipated snow level.

Combustion Air Intake

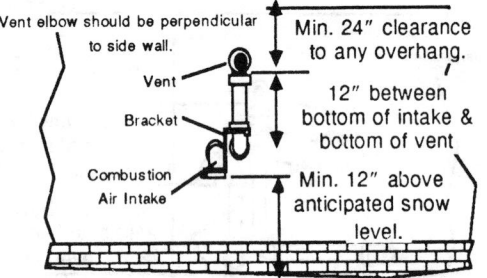

INSTALLATION ABOVE ANTICIPATED SNOW LEVEL

Vent elbow should be perpendicular to side wall.

Vent

Bracket

Combustion Air Intake

Min. 24" clearance to any overhang.

12" between bottom of intake & bottom of vent

Min. 12" above anticipated snow level.

FIGURE 4 - ALTERNATIVE TERMINATIONS

The furnace and its individual shutoff valve must be disconnected from the gas supply piping system during any pressure testing of that system at test pressures in excess of 1/2 psig (3.48 kPa).

The furnace must be isolated from the gas supply piping system by closing its individual external manual shutoff valve during any pressure testing of the gas supply piping system at pressures equal to or less than 1/2 psig (3.48 kPa).

Gas piping may be connected from either side of the furnace. Sizing and installation of the supply gas line should comply with the local utility requirements. The gas supply should be a separate line, installed in accordance with the National Fuel Gas Code, ANSI Z223.1 or CAN1-B149.1 or .2 Installation Codes (latest editions).

Some utility companies require pipe sizes larger than the maximum sizes listed. Using the properly sized wrought iron, steel or approved flexible pipe, make gas connections to the

unit. Installation of a drop leg and ground union joint is required (see Figure 5).

WARNING: *Compounds used on threaded joints of gas piping must be resistent to the action of liquified petroleum gases. After connections are made, leak-test all pipe connections.*

WARNING: *Do not use an open flame or other source of ignition for leak testing. Set the manual gas valve to the OFF position.*

For the purpose of input adjustment, the minimum inlet gas pressure must be equal to or greater than that shown on the rating label.

CONDENSATE PIPING

The condensate drain connection assembly is located in the lower right front corner of the burner compartment. It consists of a mounting bracket and 1/2" CPVC coupling (ASTM D2846) with flexible tubing connected to it.

The installation procedure for condensate piping is as follows:

1. Determine whether the condensate drain line will be installed through the right or left side of the furnace.

2. Carefully remove the 3/4" diameter knockout from the appropriate side of the furnace. Knockouts are provided on each side.

3. For left side or alternate drain connection, it will be necessary to relocate the condensate drain connection assembly.

   a. Remove the two screws that secure the drain connection bracket to the right side panel .

   b. Position the bracket over the corresponding holes in the left side panel , and fasten it to the side panel using the two screws previously removed.

   c. Make sure the flexible tubing between the drain trap and the drain connection assembly has a continuous downward slope to the drain connection assembly, has no low spots, and is not kinked. (If necessary, shorten the flexible tubing to prevent kinking and/or low spots that could restrict the flow of condensate).

4. Connect field-supplied piping to the condensate drain connection assembly and run it to an open drain. Refer to the "Vent/Combustion Air Connections" section of this instruction for procedure for assembly of plastic pipe. All field piping must be at least 1/2" CPVC and comply with ASTM D2846 (5/8" O.D. x 1/2" I.D.). The condensate piping may be tied together with the air conditioning condensate drain if the air conditioning condensate drain line is trapped upstream of the tie-in and the combined drains are constructed of CPVC piping. Where necessary, an accessory condensate pump may be used.

5. All pipe joints must be properly cemented using CPVC primer and cement that conforms to ASTM F493.

6. The furnace contains an internal trap. Therefore, no external trap should be used.

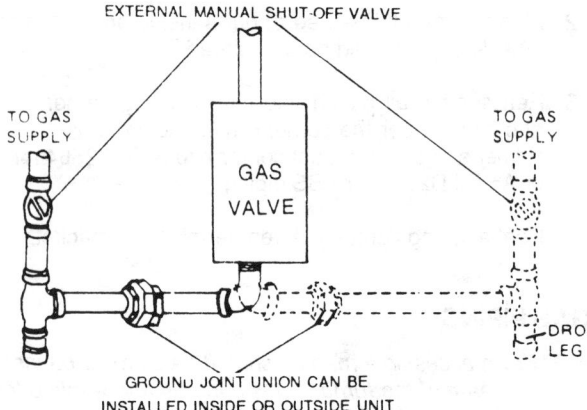

EXTERNAL MANUAL SHUT-OFF VALVE

TO GAS SUPPLY

GAS VALVE

TO GAS SUPPLY

DROP LEG

GROUND JOINT UNION CAN BE INSTALLED INSIDE OR OUTSIDE UNIT

FIGURE 5 - GAS PIPING

Central Environmental Systems

7. If a condensate pump is used, it must be suitable for use with acidic water.

8. Where required, an accessory neutralizer can be installed in the drain line, external to the furnace.

VENT PIPE CONDENSATE DRAIN

In some applications, condensate may form in the vent pipe as it passes through spaces cooler than the flue gas temperature. Figure 6 shows a recommended installation of a condensate tee trap to provide for removal of condensate from the venting system.

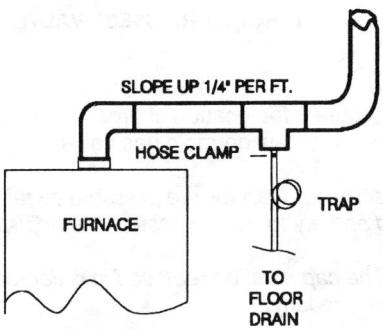

FIGURE 6 - VENT CONDENSATE DRAIN

ELECTRICAL DATA

Use copper conductors only!

Field wiring to the unit must conform to and be grounded in accordance with the provisions of the National Electrical Code ANSI/NFPA No. 70-1987, Canadian Electric Code C22.1, Part 1 and/or local codes.

Electric wires which are field installed shall conform with the temperature limitation for 63°F/35°C rise wire when installed in accordance with instructions. Specific electrical data is given for the furnace on its rating plate and in Table 1 of this instruction.

ELECTRICAL CONNECTIONS

The furnace's control system depends on correct polarity of the power supply. Connect the power supply as shown on the unit wiring label on the inside of the blower compartment door.

Provide a power supply separate from all other circuits. Install overcurrent protection and disconnect switch per local/national electrical codes.

The switch should be reasonably close to the unit for convenience in servicing. With the disconnect switch in the OFF position, check all wiring against the unit wiring label. Also, see the wiring diagram in this instruction.

Install the field-supplied thermostat. The thermostat instructions for wiring are packed with the thermostat. With the thermostat in the OFF position and the main electrical source disconnected, complete the low voltage wiring from the thermostat to the terminal. Set the heat anticipator on the thermostat to .36 amps.

NOTE: *Some thermostats do not have adjustable anticipators. On such thermostats, adjust cycle rate to prevent possible short on/off cycles.*

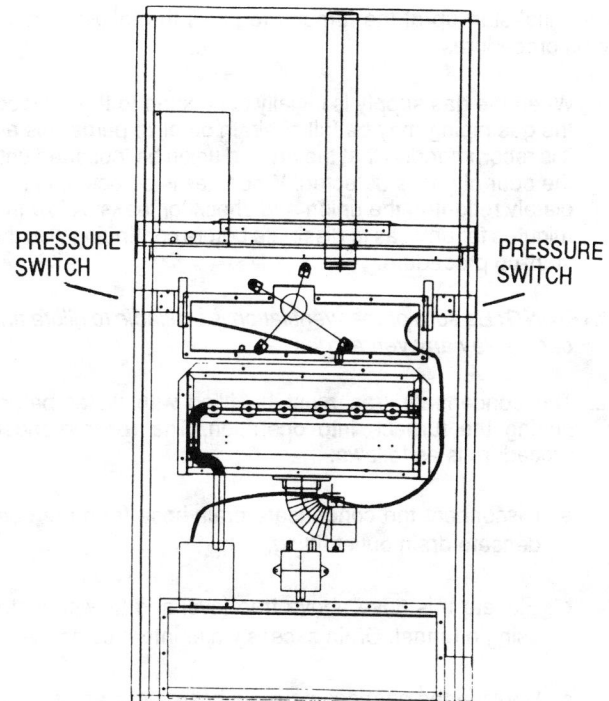

FIGURE 7 - PRESURE SWITCH TUBING

SAFETY CONTROLS

Interlock Switch

This unit is equipped with an electrical interlock switch mounted in the blower compartment.

This switch interrupts power at the unit when the panel covering the blower compartment is removed. This prevents operation of the automatic gas control valve and the blower.

WARNING: *Blower and burner must never be operated without the blower panel in place.*

Electrical supply to this unit is dependent upon the panel that covers the blower compartment being in place and properly positioned.

CAUTION: *Main power to the unit must still be interrupted at the main power disconnect switch before any service or repair work is to be done to the unit. Do not rely upon the interlock switch as a main power disconnect.*

Pressure Switches

This furnace is supplied with pressure switches which monitor the flow through the combustion air/vent piping system. The tubing and switch configuration is shown in Figure 7. These switches de-energize the ignition control module and the gas valve if any of the following conditions are present.

1. Blockage of combustion air piping or terminal.

2. Blockage of vent piping or terminal.

3. Failure of combustion air blower motor.

4. Blockage of condensate drain piping.

START-UP AND ADJUSTMENTS

The initial start-up of the furnace requires the following additional procedures:

1. When the gas supply is initially connected to the furnace, the gas piping may be full of air. In order to purge this air, it is recommended that the ground union be loosened until the odor of gas is detected. When gas is detected, immediately retighten the union and check for leaks. Allow five minutes for any gas to dissipate before continuing with the start-up procedure.

WARNING: Be sure proper ventilation is available to dilute and carry away any vented gas.

2. The condensate trap must be filled with water before putting the furnace into operation. The recommended procedure is as follows:

   a. Disconnect the condensate drain hose from the condensate drain outlet fitting.

   b. Elevate this hose above trap level and fill with water using a funnel. Drain excess water into a container.

   c. Replace the condensate drain hose and clamps.

3. All electrical connections made in the field and in the factory should be checked for proper tightness.

IGNITION SYSTEM CHECKOUT/ADJUSTMENT

1. Turn the gas supply ON at external valve and main gas valve.

2. Set the thermostat above room temperature to call for heat.

3. System start-up will occur as follows:

   a. Venter motor will start and come up to speed. Shortly after venter start-up, the hot surface igniter will glow for about 17 seconds.

   b. After warm-up cycle, ignition module will energize (open) the main gas valve for seven seconds.

NOTE: Burner ignition may not be satisfactory on first start-up due to residual air in gas line, or until gas pressure (manifold) is adjusted.

6. With furnace in operation, paint the pipe joints and valve gasket lines with a rich soap and water solution. Bubbles indicate a gas leak. Take appropriate steps to stop the leaks. If the leak persists, replace the component.

WARNING: DO NOT omit this test! NEVER use a flame to check for gas leaks.

ADJUSTMENT OF MANIFOLD GAS PRESSURE

1. Turn gas off at main gas valve. Remove 1/8" plug in the main gas valve body and install proper manometer tube adapter fitting. Connect line from gas valve tap to manometer

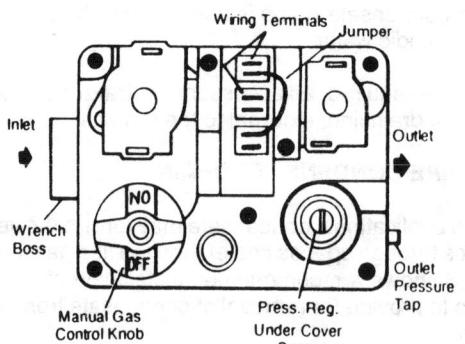

WHITE-RODGERS 36E01 VALVE

FIGURE 8 - GAS VALVE

2. Refer to Figure 8 for location of pressure regulator adjustment cap and screw on main gas valve.

NOTE: The screw-off cap for the pressure regulator must be removed entirely to gain access to the adjustment screw.

WARNING: The cap must be replaced in order for the furnace to operate properly.

3. Turn gas and electrical supplies ON. Start furnace and observe manifold pressure on manometer.

4. Adjust manifold pressure by adjusting gas valve regulator screw: for natural gas, set at 3.5" W.C.; for propane (LP) gas, set at 10.0" W.C.

WARNING: The manifold pressure must be checked with the screw-off cap in place on the pressure regulator.

WARNING: If manifold pressure is too high, an over-fire condition exists which could cause heat exchanger failure. If the manifold pressure is too low, sooting and eventual clogging of the heat exchanger could occur.

If gas valve regulator is turned in, or clockwise, manifold pressure is increased. If screw is turned out, or counterclockwise, manifold pressure will decrease.

WARNING: Once the correct gas pressure to the burners has been established, turn the gas valve knob to OFF and turn the electrical supply switch to OFF; then remove the pressure tap at the gas valve and re-install the plug, using a compound (on the threads) resistant to the action of LP gases.

Turn the electrical and gas supplies back on, and with the burners in operation, check for gas leakage around the plug with a soap and water solution.

ELECTRONIC AIR CLEANER CONNECTION

Two 1/4" spade terminals (AC and AC N) are located on the 50A50 control for electronic air cleaner connections. The terminals provide 115 VAC (1.0 amp maximum) during circulating blower operation.

HUMIDIFIER CONNECTION

Two 1/4" spade terminals (HUM and HUM N) are located on the 50A50 control for humidifier connection. The terminals provide 115 VAC (1.0 amp maximum) during heat speed operation of the circulating blower.

ADJUSTMENT OF FAN-OFF CONTROL SETTINGS

The fan-off setting must be long enough to adequately cool the furnace, but not so long that cold air is blown into the heated space. The fan-off timing may be adjusted by setting the option switches located (see Figure 9) on the 50A50 module as follows:

OPTION SWITCH POSITIONS		
To Delay Fan-Off By:	Set Switch	
	1	2
60 Sec.	On	On
90 Sec.	Off	On
120 Sec.	On	Off
180 Sec.	Off	Off

Factory set at 30 sec. on and 60 sec. off (adjustable)

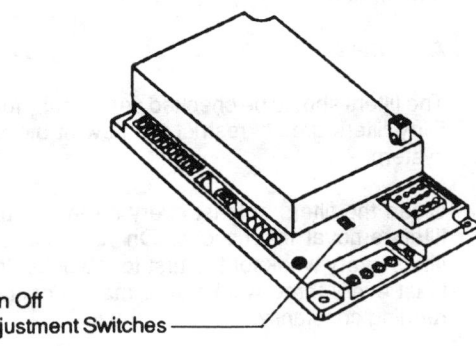

Fan Off
Adjustment Switches

FIGURE 9 - LOCATION OF FAN OFF ADJUSTMENT

CHECKING GAS INPUT (NATURAL GAS)

NOTE: Front door of burner box must be secured when checking gas input.

1. Turn off all other gas appliances connected to the gas meter.

2. With the furnace turned on, measure the time needed for one revolution of the hand on the smallest dial on the meter. A typical domestic gas meter usually has a 1/2 or 1 cubic foot test dial.

3. Using the number of seconds for each revolution and the size of the test dial increment, find the cubic feet of gas consumed per hour from Table 3.

NOTE: To find the Btuh input, multiply the number of cubic feet of gas consumed per hour by the BTU content of the gas in your particular locality. Contact your gas company for this information, as it varies widely from city to city.

EXAMPLE: It is found by measurement that it takes 26 seconds for the hand on the 1 cubic foot dial to make a revolution with only a 120,000 Btuh furnace running. Using this information, locate 26 seconds in the first column of Table 3. Read across to the column headed "1 cubic foot" where you will see that 138 cubic feet of gas per hour are consumed by the furnace at that rate. Multiply 138 by 850 (the BTU rating of the gas obtained from the local gas company). The result is 117,300 Btuh, which is close to the 120,000 Btuh rating of the furnace.

If the actual input is not within 5% of the furnace rating, with allowance being made for the permissible range of the regulator setting (0.3 inches W.C.), replace the orifice spuds with spuds of the proper size.

Central Environmental Systems

TABLE 3 - GAS RATE (CUBIC FEET PER HOUR)

Seconds for one Revolution	Size of Test Dial	
	1/2 Cubic Ft.	1 cubic ft.
10	180	360
12	150	300
14	129	257
16	113	225
18	100	200
20	90	180
22	82	164
24	75	150
26	69	138
28	64	129
30	60	120
32	56	113
34	53	106
36	50	100
38	47	95
40	45	90
42	43	86
44	41	82
46	39	78
48	37	75
50	36	72
52	35	69
54	34	67
56	32	64
58	31	62
60	30	60

CAUTION: Be sure to relight any gas appliances that were turned off at the start of this input check.

ADJUSTMENT OF TEMPERATURE RISE

The temperature rise, or temperature difference between the return air and the heated air from the furnace, must be within the range shown on the furnace rating plate and within the application limitations shown in Table 1. After the temperature rise has been determined, the cfm can be calculated.

After about 20 minutes of operation, determine the furnace temperature rise. Take readings of both the return air and the heated air in the ducts, about six feet from the furnace where they will not be affected by radiant heat. Increase the blower speed to decrease the temperature rise; decrease the blower speed to increase the rise.

All direct-drive blowers have multi-speed blowers. Refer to the unit wiring diagram and connect the blower motor for the desired speed. The blower motor speed taps are located in the control box in the blower compartment.

OPERATION & MAINTENANCE

SEQUENCE OF OPERATION

Hot Surface Ignition System

WARNING: Do not attempt to light this furnace by hand (with a match or any other means). There may be a potential shock hazard from the components of the hot surface ignition system. The furnace can only be lit automatically by its hot surface ignition system.

The following describes the sequence of operation of the furnace. Refer to the schematic wiring diagram (page 14) for component location.

Continuous Blower

On the cooling/heating thermostats with fan switch, when the fan switch is set in the ON position, a circuit is completed between terminals R and G of the thermostat. The motor is energized through the black, high-speed tap. The blower then operates on high speed.

Intermittent Blower

When the system switch is set on HEAT and the fan switch is set on AUTO, and the room thermostat calls for heat, a circuit is completed between terminals R and W of the thermostat. When the proper amount of combustion air is being provided, a pressure switch activates the 50A50 ignition control. A second pressure switch (2LP-1 or 2LP-2) and the limit are in this circuit, and must be in the closed position for the ignition control to be activated.

The 50A50 ignition control provides a 17-second warm-up period. The gas valve then opens for seven seconds.

As gas starts to flow and ignition occurs, the flame sensor begins its sensing function. If a flame is detected within seven seconds after ignition, normal furnace operation continues until the thermostat circuit between R and W is opened. After flame is present for 30 seconds, the circulating blower is energized.

When the thermostat circuit opens, the ignition control is deenergized. With the ignition control deenergized, the gas flow stops and the burner flames are extinguished. The venter continues to operate for 15 seconds after the gas flow stops.

The blower motor continues to operate for the amount of time set by the fan-off delay dip switches on the ignition control module. The heating cycle is then complete, and the unit is ready for the start of the next heating cycle.

If flame is not detected within the seven second sensing period, the gas valve is de-energized. The 50A50 control is equipped with a re-try option. This provides a 60-second wait following an unsuccessful ignition attempt (flame not detected).

After the 60 second wait, the ignition sequence is restarted with an additional 10 seconds of igniter warm-up time. If this ignition attempt is unsuccessful, one more re-try will be made before lockout.

50A50 HOT SURFACE IGNITION CONTROL

All 50A50 controls will repeat the ignition sequence for a total of four recycles if flame is lost within the first 10 seconds of establishment.

If flame is established for more than 10 seconds after ignition, the controller will clear the ignition attempt (re-try) counter. If flame is lost after 10 seconds, it will restart the ignition sequence. This can occur a maximum of five times.

During burner operation, a momentary loss of power of 50 milliseconds or longer will drop out the main gas valve. When the power is restored, the gas valve will remain de-energized, and a restart of the ignition sequence will begin immediately.

A momentary loss of gas supply, flame blowout, or a shorted or open condition in the flame probe circuit will be sensed within 0.8 seconds. The gas valve will de-energize and the control will restart the ignition sequence after waiting 60 seconds. Recycles will begin and the burner will operate normally if the gas supply returns, or the fault condition is corrected prior to the last ignition attempt. Otherwise, the control will lockout.

If the control is locked out, it may be reset by momentary power interruption of 1/20 second or longer. Either the 24v thermostat or line voltage may be interrupted.

MAINTENANCE

The manufacturer recommends that maintenance is performed by a qualified service agency for cleaning vent/air intakes, condensate drains and neutralizers, burners, primary and secondary heat exchangers and the blower motor and wheel assembly. An annual inspection of these components is recommended.

Air Filters

The filters should be checked periodically for dirt accumulation. Dirty filters greatly restrict the flow of air and overburden the system.

Clean the filters at least every three months. See page 5 for filter removal instructions. On new construction, check the filters every week for the first four weeks. Inspect the filters at least every three weeks after that, especially if the system is running constantly.

Air filters supplied with the furnace are the high-velocity, cleanable type. Clean these filters by washing in warm water. Make sure to shake all the water out of the filter and have it reasonable dry before installing it in the furnace. When replacing filters, be sure to use the same size and type as originally supplied.

Lubrication

Blower motors in these furnaces are permanently lubricated and do not require periodic oiling.

Burner Removal/Cleaning

The main burners should be checked periodically for dirt accumulation.

If cleaning is required, follow the steps listed below:

1. Turn off the electrical power to the unit.

2. Remove the lower access door.

3. Remove the front cover of the burner box.

4. Vacuum the burner assembly to remove dirt or dust.

Cleaning the Primary Heat Exchanger

1. Turn off the main manual gas valve external to the furnace.

2. Turn off the electrical power to the furnace.

3. Remove the access door.

4. Disconnect the gas supply piping and control wiring from the gas valve.

5. Remove the front cover of the burner box.

6. Remove the screws holding the burner box assembly to the vestibule panel.

7. Remove the burner box assembly.

8. To reach the lower portion of the heat exchanger, remove the flue box cover and flue baffles.

9. With a stiff wire brush, brush out loose scale or soot.

10. Vacuum the burner assembly and heat exchanger.

11. Replace all parts removed for cleaning by reversing the order of disassembly.

12. Reconnect all wiring and gas piping.

13. Restore electrical power and gas supply to the furnace.

Cleaning the Secondary Heat Exchanger

1. Remove the screw in the venter outlet. Disconnect the drain lines from the venter and from the condensate drain pan. Remove the venter blower and the condensate pan.

2. With a stiff wire brush, brush out loose scale or soot.

3. Vacuum the secondary heat exchanger.

4. Replace the condensate pan and drain hose. Replace venter and use RTV sealant to seal vent pipe to venter outlet. Secure vent pipe to venter outlet with stainless steel screw.

WARNING: Never operate furnace without the exhaust joined to the venter assembly with a stainless steel screw and sealed with RTV sealant.

BLOWER CARE

Even with good filters properly in place, blower wheels and motors will become dust laden after long months of operation. The entire blower assembly should be inspected annually. If the motor and wheel are heavily coated with dust, they can be brushed and cleaned with a vacuum cleaner.

The procedure for removing the blower assembly for cleaning is as follows:

1. Disconnect the electrical supply to the furnace.

2. Remove the access panels.

3. Disconnect the two wire harness plugs from the top of the control box.

4. Remove the four screws holding the control box and position it out of the way.

5. Remove screw from outlet of venter housing which holds vent pipe in place. Remove screw securing vent pipe to blower deck and remove vent pipe. Disconnect from field supplied venting.

6. Remove screws which retain blower to the blower deck.

7. Vacuum the motor and the blower wheel using a soft brush attachment. Care must be used not to disturb the balance weights (clips) on the blower wheel vanes.

8. Before reinstalling the blower assembly, inspect the secondary heat exchanger which is visible directly below the blower opening in the blower deck. If it requires cleaning, vacuum it with a soft brush attachment and follow the direction of the fins.

9. Reinstall the blower assembly. Replace the mounting screws that hold the blower assembly to the front portion of the blower deck. The two mounting screws used on the sides of the blower are used for shipping purposes only, and are not necessary after the furnace has been installed.

10. Reinstall the control box and reconnect the wiring harness plugs.

11. Replace vent pipe in venter outlet and seal joint using RTV sealant. Replace stainless steel screw securing vent pipe to venter outlet. Replace stainless steel screw securing vent pipe to blower deck. Reconnect to field supplied vent pipe

WARNING: Never operate furnace without the exhaust vent in place or joined to the venter assembly with a stainless steel screw and sealed with RTV sealant.

12. Replace the access doors and restore the electrical supply to the unit.

TROUBLESHOOTING

The following visual checks should be made before troubleshooting:

1. Check to see that the power to the furnace and the 50A50 control module is ON.

2. The manual shutoff valves in the gas line to the furnace must be open.

3. Make sure all wiring connections are secure.

4. Review the sequence of operation.

Start the system by setting the thermostat above the room temperature. Observe the system's response. Then use the Troubleshooting Table to check the system's operation.

To use the troubleshooting table, begin by reading the upper left-hand box and then following the instructions in each box. If the condition described in the box is true (yes answer), go down to the next box. If the condition is not true (no answer), go to the box to the right. Continue checking and answering the questions in the boxes until the problem is explained and corrective action is described. After any maintenance or repair, the troubleshooting sequence should be repeated until normal system operation is obtained.

WARNING: Do not try to repair controls. Replace defective controls with CES Source 1 Parts.

WIRING DIAGRAM -ALL MODELS

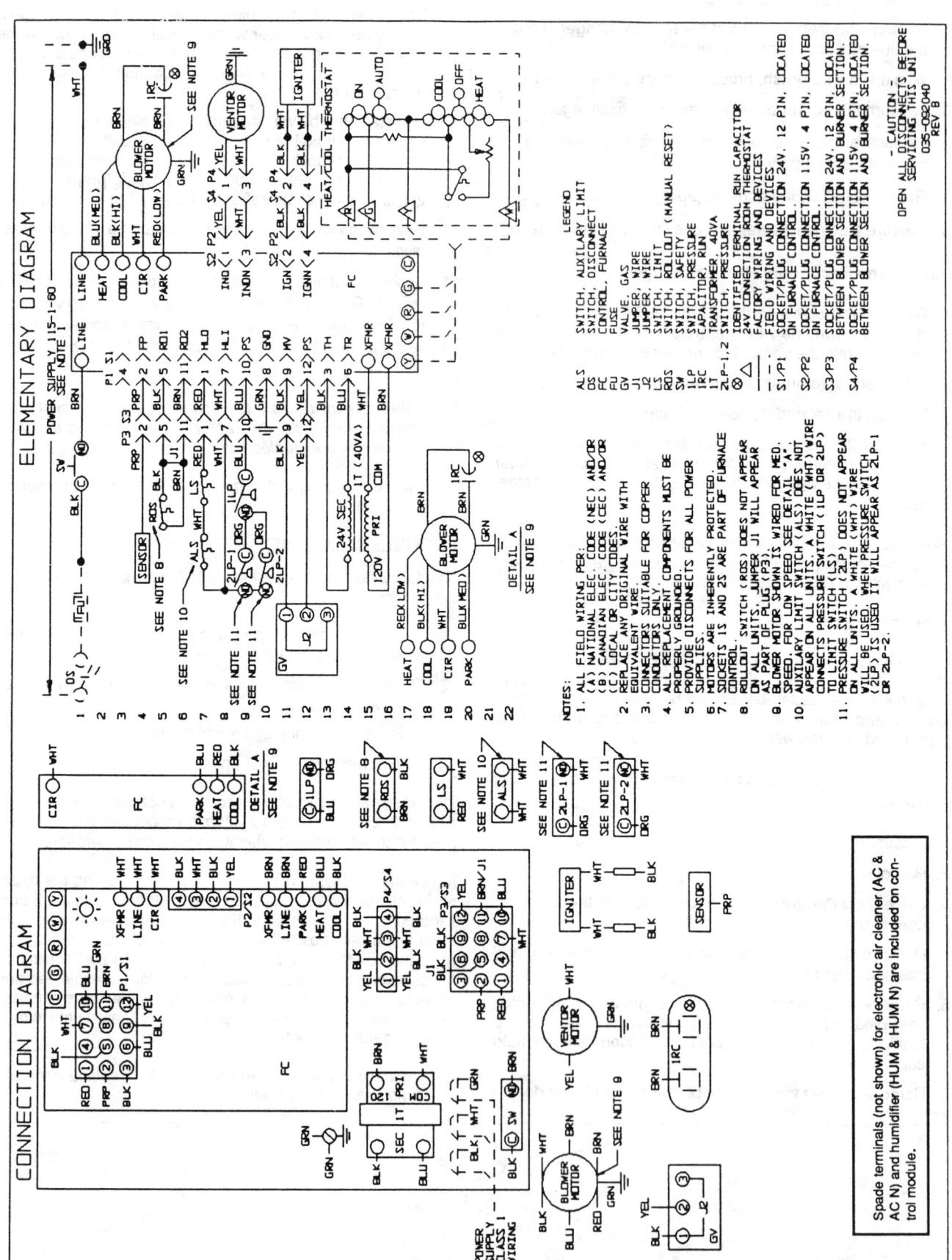

WHITE-RODGERS 50A50 TROUBLESHOOTING TABLE

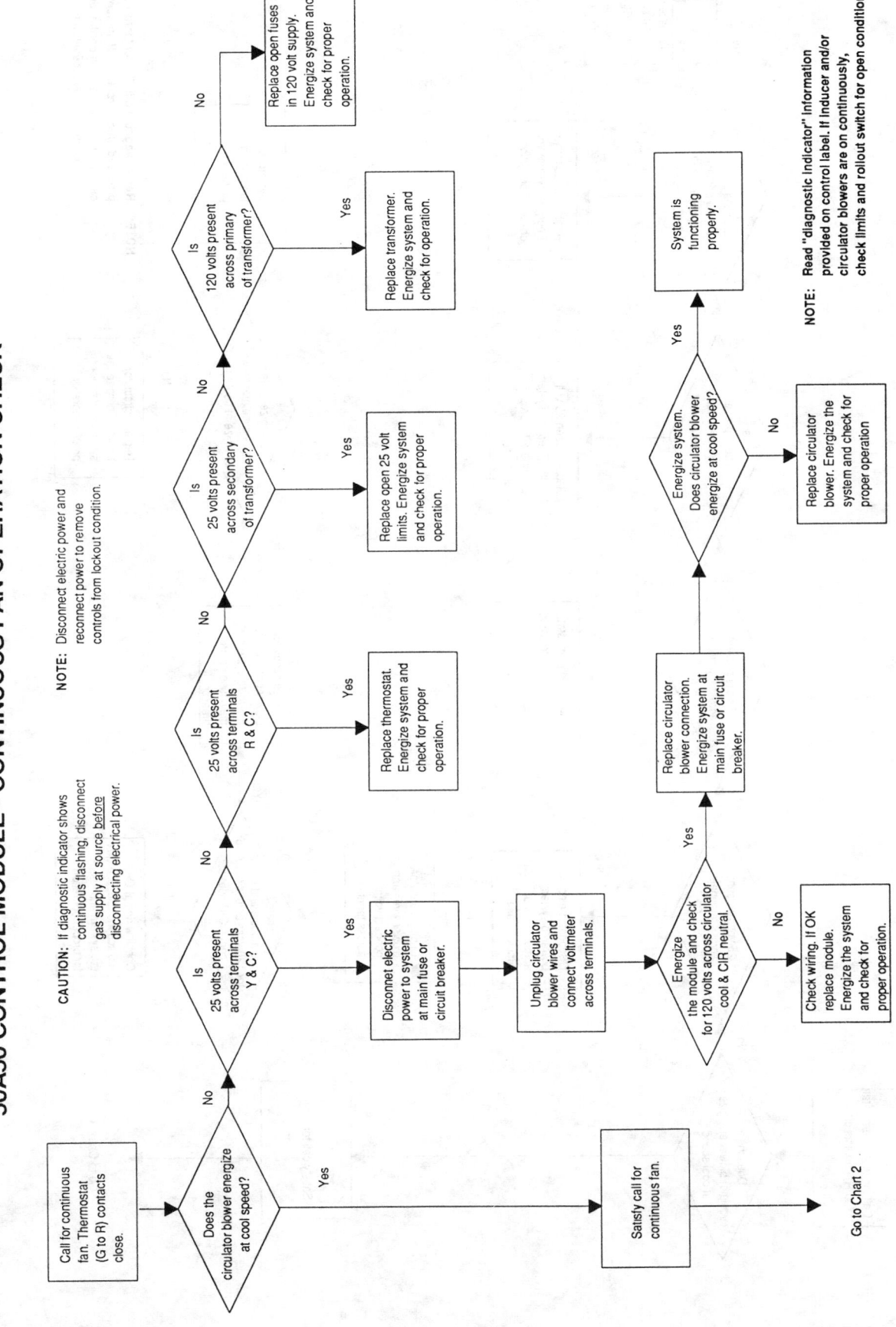

50A50 CONTROL MODULE - CONTINUOUS FAN OPERATION CHECK

50A50 CONTROL MODULE - COOL OPERATION CHECK

Continued from Chart 1

→ Call for cool. Thermostat (Y to R) contacts close.

→ **Does the circulator blower energize at cool speed?**

No ↑

Is 25 volts present across terminals Y & C?

- **No** → **Is 25 volts present across terminals R & C?**
  - **No** → **Is 25 volts present across secondary of transformer?**
    - **No** → **Is 120 volts present across primary of transformer?**
      - **No** → Replace open fuses in 120 volt supply. Energize system and check for proper operation
      - **Yes** → Replace transformer. Energize system and check for operation.
    - **Yes** → Replace open 25 volt limits. Energize system and check for proper operation.
  - **Yes** → Replace thermostat. Energize system and check for proper operation.
- **Yes** → Disconnect electric power to system at main fuse or circuit breaker.

→ Unplug circulator blower wires and connect voltmeter across terminals.

→ **Energize the module and check for 120 volts across circulator cool & CIR neutral?**

- **No** → Check wiring. If OK, replace module, energize the system and check for proper operation.
- **Yes** → Replace circulator blower connection. Energize system at main fuse or circuit breaker.

→ **Energize system. Does circulator blower energize at cool speed?**

- **No** → Replace circulator blower. Energize the system and check for proper operation
- **Yes** → System is functioning properly.

Yes (from "Does the circulator blower energize at cool speed?") → Satisfy call for cool. → Go to Chart 3

NOTE: Read "diagnostic Indicator" information provided on control label. If Inducer and/or circulator blowers are on continuously, check limits and rollout switch for open condition.

50A50 CONTROL MODULE - HEAT CYCLE PREPURGE CHECK

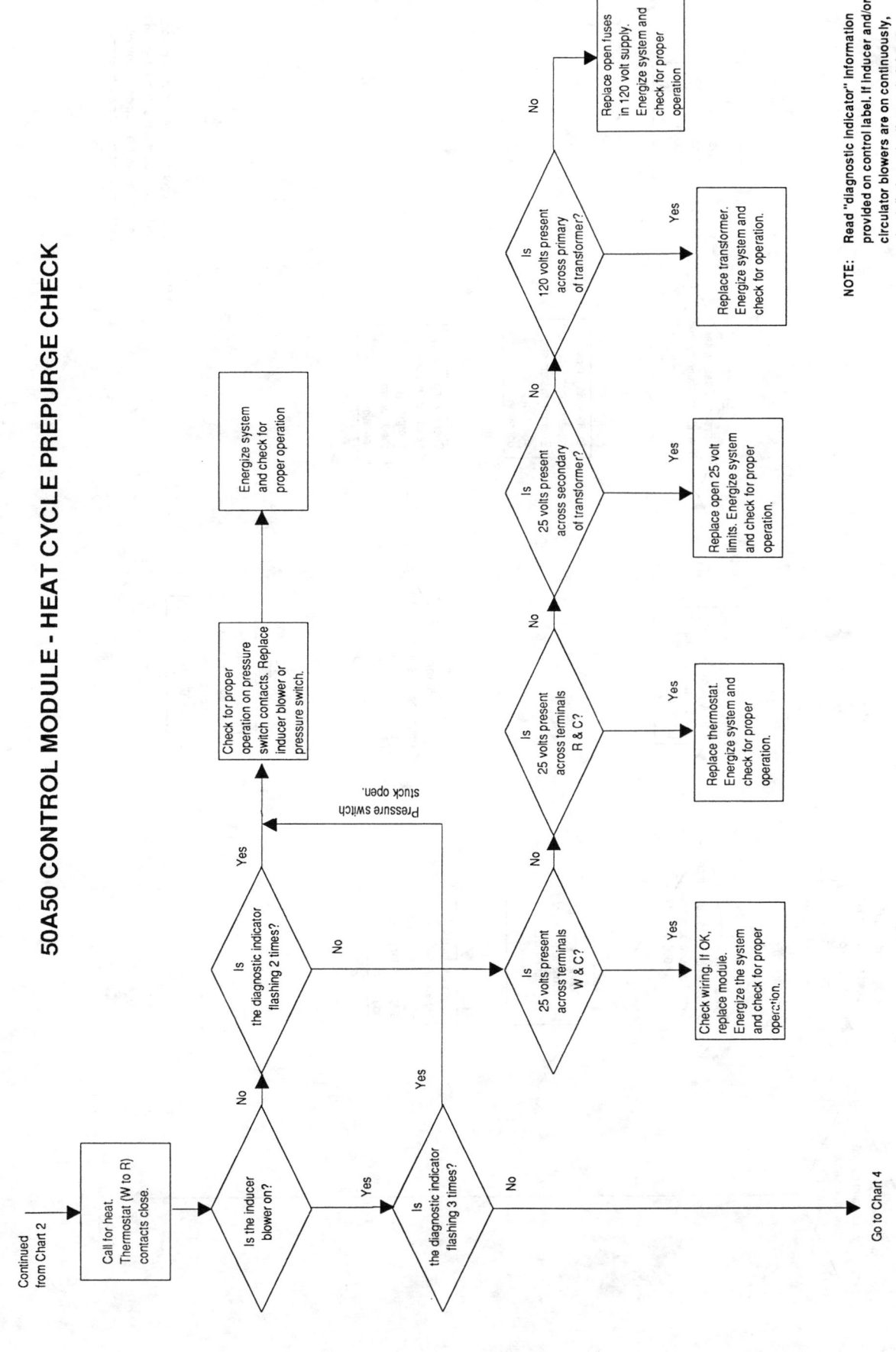

NOTE: Read "diagnostic indicator" information provided on control label. If inducer and/or circulator blowers are on continuously, check limits and rollout switch for open condition.

50A50 CONTROL MODULE - HEAT CYCLE IGNITER WARM-UP CHECK

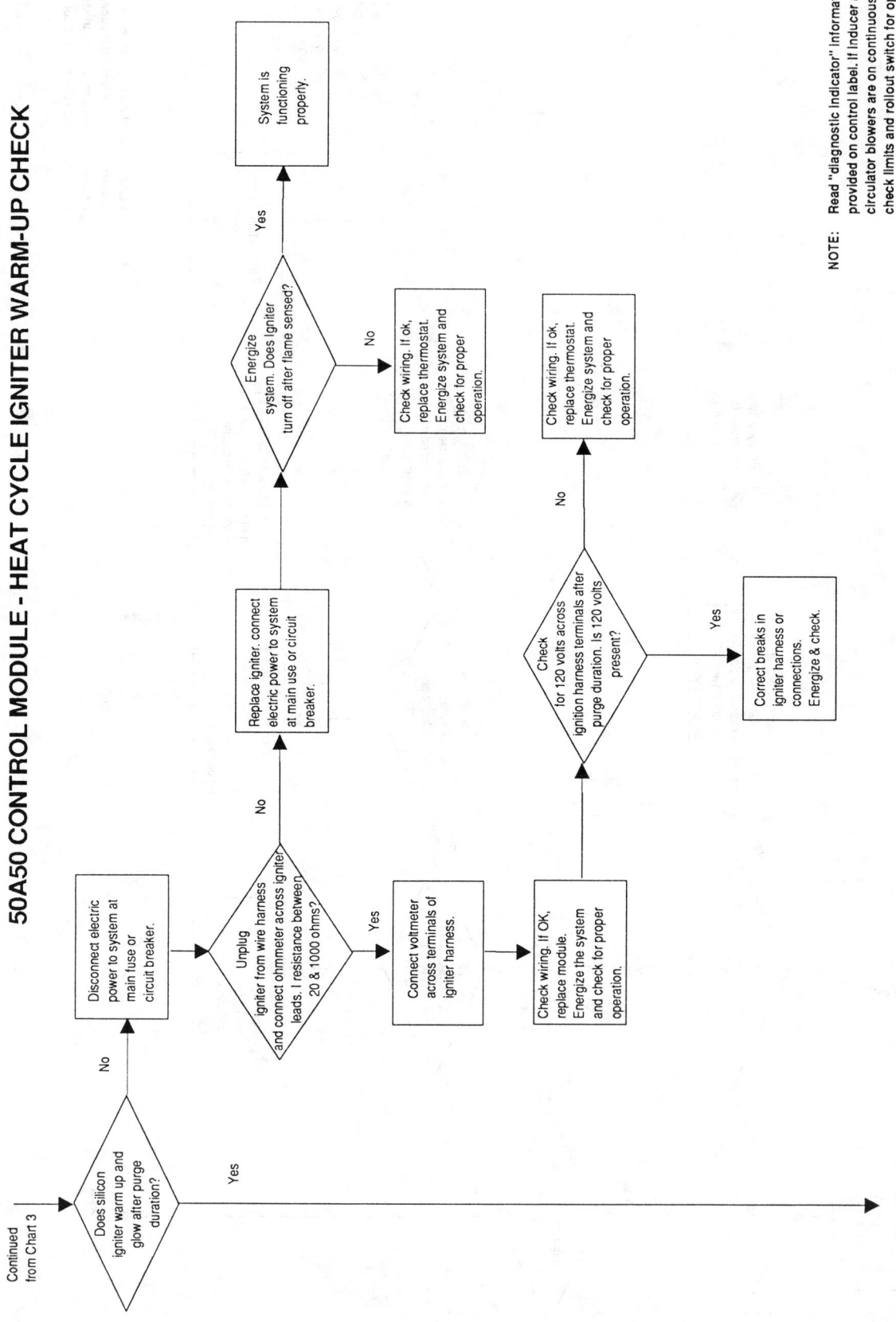

Continued from Chart 3

Does silicon igniter warm up and glow after purge duration?

No → **Disconnect electric power to system at main fuse or circuit breaker.**

→ **Unplug igniter from wire harness and connect ohmmeter across igniter leads. I resistance between 20 & 1000 ohms?**

No → **Replace igniter. connect electric power to system at main use or circuit breaker.**

→ **Energize system. Does Igniter turn off after flame sensed?**

Yes → **System is functioning properly.**

No → **Check wiring. If ok, replace thermostat. Energize system and check for proper operation.**

Yes → **Connect voltmeter across terminals of igniter harness.**

→ **Check wiring. If OK, replace module. Energize the system and check for proper operation.**

→ **Check for 120 volts across ignition harness terminals after purge duration. Is 120 volts present?**

No → **Check wiring. If ok, replace thermostat. Energize system and check for proper operation.**

Yes → **Correct breaks in igniter harness or connections. Energize & check.**

Yes → Go to Chart 5

NOTE: Read "diagnostic indicator" Information provided on control label. If Inducer and/or circulator blowers are on continuously, check limits and rollout switch for open condition.

Central Environmental Systems

50A50 CONTROL MODULE - HEAT CYCLE MAIN BURNER IGNITION CHECK

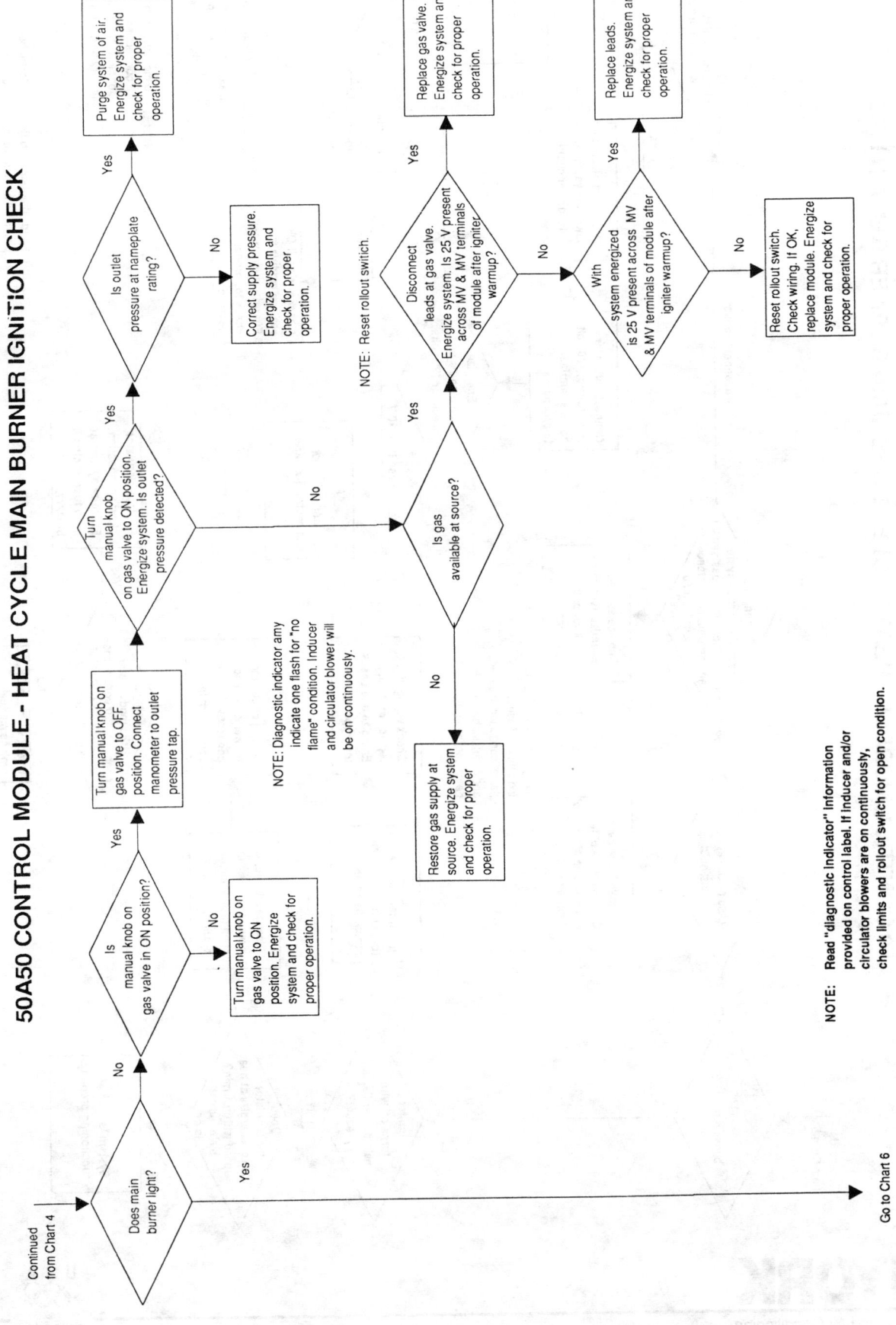

Continued from Chart 4

Does main burner light?

Yes → Go to Chart 6

No ↓

Is manual knob on gas valve in ON position?

No → Turn manual knob on gas valve to ON position. Energize system and check for proper operation.

Yes ↓

Turn manual knob on gas valve to OFF position. Connect manometer to outlet pressure tap.

Turn manual knob on gas valve to ON position. Energize system. Is outlet pressure detected?

Yes → **Is outlet pressure at nameplate rating?**

Yes → Purge system of air. Energize system and check for proper operation.

No → Correct supply pressure. Energize system and check for proper operation.

No → **Is gas available at source?**

Yes → **Disconnect leads at gas valve. Energize system. Is 25 V present across MV & MV terminals of module after igniter warmup?**

NOTE: Reset rollout switch.

Yes → Replace gas valve. Energize system and check for proper operation.

No → **With system energized is 25 V present across MV & MV terminals of module after igniter warmup?**

Yes → Replace leads. Energize system and check for proper operation.

No → Reset rollout switch. Check wiring. If OK, replace module. Energize system and check for proper operation.

No → Restore gas supply at source. Energize system and check for proper operation.

NOTE: Diagnostic indicator amy indicate one flash for "no flame" condition. Inducer and circulator blower will be on continuously.

NOTE: Read "diagnostic indicator" Information provided on control label. If inducer and/or circulator blowers are on continuously, check limits and rollout switch for open condition.

50A50 CONTROL MODULE - HEAT CYCLE - DOES MAIN BURNER REMAIN LIT?

Continued from Chart 5

Does main burner remain lit?
- No →
- Yes →

Does flame sense probe have carbon or dust build-up?
- No →
- Yes → Remove flame sensor. Clean surface of flame rod with fine steel wool then reinstall.

NOTE: Diagnostic indicator amy indicate one flash for "no flame" condition. Inducer and circulator blower will be on continuously.

Disconnect Sensor. Deenergize system. Connect ohmmeter across sensor and ground. Is resistance less than 50 ohms?
- No →
- Yes → Replace sensor. Energize system and check for proper operation.

Reconnect sensor

Is ground wire connected to furnace and control box GND terminal wired to burner ground?
- No → Correct ground wiring.
- Yes → Disconnect electric power to system at main fuse or circuit breaker.

Connect voltmeter from control 120 volt hot line terminal to ground.

Energize the module and check for 120 volts.
- Less than 30 V → Reverse 120 V hot and neutral line wires.

Is flame probe located properly in flame? Proper location will provide adequate flame current.
- Yes → Check wiring. If OK, replace module. Energize system and check for proper operation.
- No → Reposition flame probe to achieve 1.0 μ amp or make flame current.

NOTE: Flame probe must be in flame to sense.

NOTE: To measure flame current, disconnect sensor lead. Connect microamp meter in series with lead and sensor.

NOTE: Read "diagnostic Indicator" information provided on control label. If Inducer and/or circulator blowers are on continuously, check limits and rollout switch for open condition.

Igniter turns off with main burner flame present?
- No → Igniter remains heated (bright red) with main burner flame present. → Check wiring. If OK, replace module. Energize and check for proper operation.
- Yes →

Does the circulator blower energize at heat speed within 1 min? Main burner is on.
- No → Disconnect electric power to system at main fuse or circuit breaker. → Unplug circulator blower wires and connect voltmeter across terminals.
- Yes → System is functioning properly.

Energize the module and check for 120 volts across terminals circulator heat & Cir netural?
- Yes → Replace circulator blower connection. Energize system at main fuse or circuit breaker.
- No → Check wiring. If OK, replace module. Energize the system and check for proper operation.

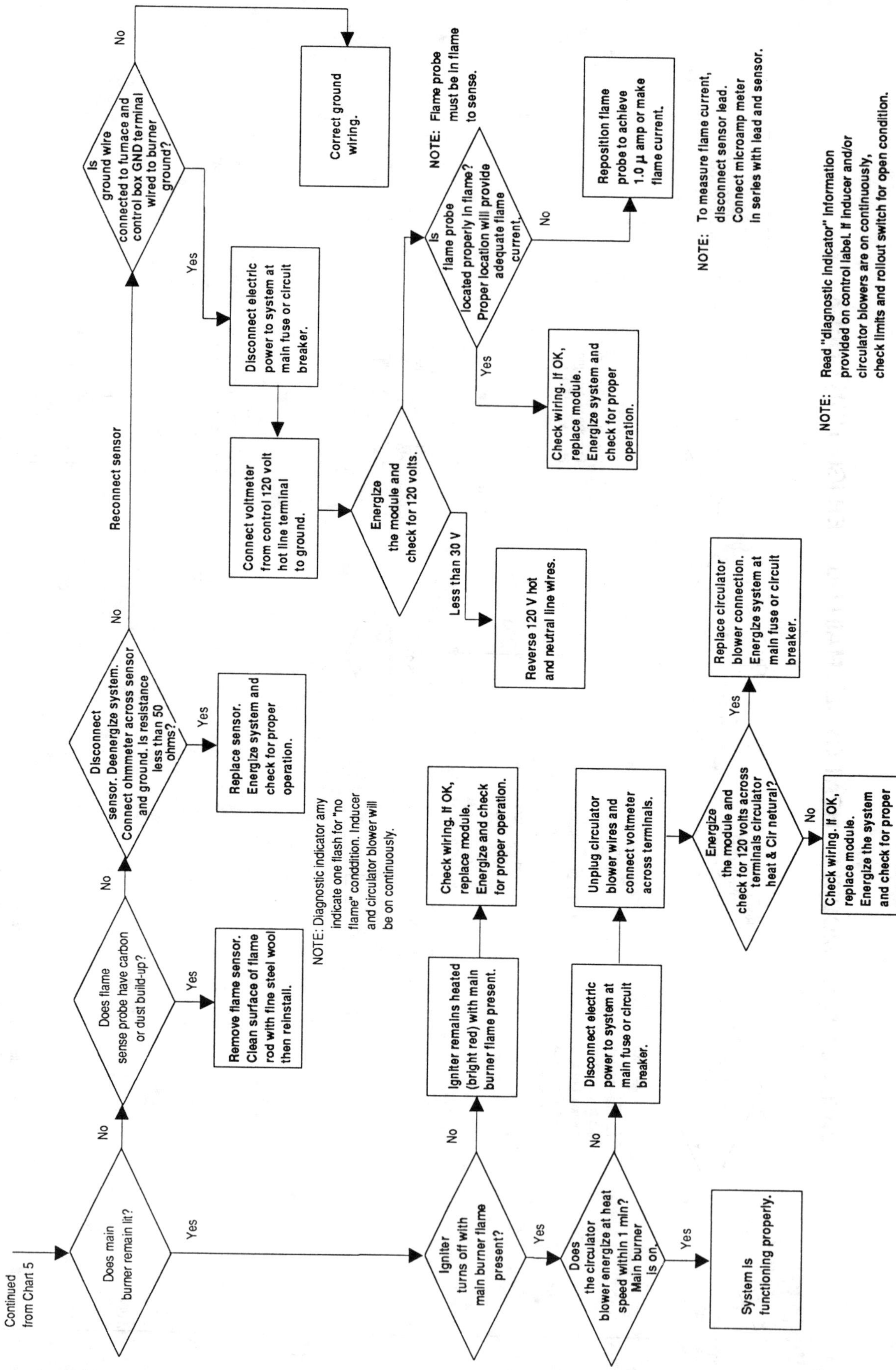

YORK®

Heating and Air Conditioning

USA
Official Sponsor of the 1992 U.S. Olympic Team
36USC380

P.O. Box 1592, York, Pennsylvania USA 17405-1592
Subject to change without notice.   Printed in U.S.A.
Copyright © by York International Corporation 1992.   All rights reserved.   RPC 8M 492 1.10   Code: SBY

Supersedes: 650.65-N3Y (991)
650.65-N3Y

GAS-FIRED FURNACES
Stellar– PLUS HIGH-EFFICIENCY
UPFLOW CONDENSING MODELS

035-09037

MODELS: P2UD, (STYLE C) - PROPANE (LP), 37 THRU 131 MBH OUTPUT
P2UD, (STYLE C) - NATURAL GAS, 37 THRU 131 MBH OUTPUT

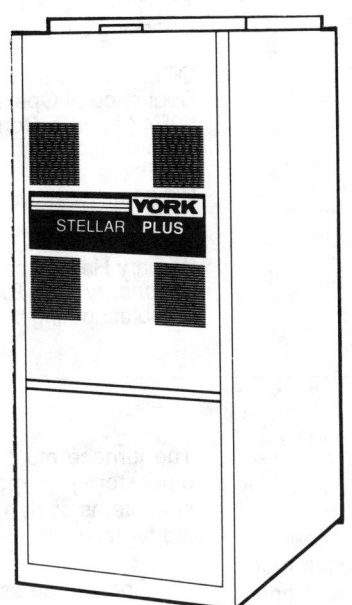

UPFLOW MODELS
TYPE FSP
DIRECT VENT FURNACE

FOR YOUR SAFETY

WHAT TO DO IF YOU SMELL GAS

- **Do not try to light any appliance.**

- **Open windows.**

- **Do not touch any electrical switch; do not use any phone in your building.**

- **Extinguish any open flames**.

- **Immediately call your gas supplier from a neighbor's phone. Follow the gas supplier's instructions.**

- **If you cannot reach your gas supplier, call the fire department.**

FOR YOUR SAFETY

Do not store or use gasoline or other flammable vapors and liquids in the vicinity of this or any other appliance.

WARNING: Improper installation, adjustment, alteration, service or maintenance can cause injury or property damage. Refer to this manual. For assistance or additional information, consult a qualified installer, service agency or the gas supplier.

TABLE OF CONTENTS

GENERAL INFORMATION

DESCRIPTION

This Category IV, sealed combustion furnace is designed for residential installation in a basement, closet, recreation room or garage, provided space temperature is 32°F or higher. Model P2UD Style C units are available for direct installation using natural gas (N) or propane (P). Ensure the proper model is installed for the available fuel gas.

Model P2UD (P) units may be converted to natural gas if factory supplied components are used. High altitude conversions required in order for the appliance to satisfactorily meet the application must be made by a CES distributor, conversion station or other qualified agency, using factory specified and/or approved parts.

INSPECTION

As soon as a unit is received, it should be inspected for possible damage during transit. If damage is evident, the extent of the damage should be noted on the carrier's freight bill. A separate request for inspection by the carrier's agent should be made in writing. Also, before installation the unit should be checked for screws or bolts which may have loosened in transit. There are no shipping or spacer brackets which need to be removed.

NOTES, CAUTIONS, & WARNINGS

The installer should pay particular attention to the words:

NOTE, CAUTION and **WARNING. NOTES** are intended to clarify or make the installation easier. **CAUTIONS** are given to prevent equipment damage. **WARNINGS** are given to alert the installer that personal injury and/or equipment or property damage may occur if installation procedures are not handled properly.

CAUTION: The cooling coil must be installed in the supply air duct, downstream of the furnace.

The furnace must not be used as a broom closet or for any other storage purposes as a fire hazard may be created. Never store items such as the following on, near, or in contract with the furnace.

1. Spray or aerosol cans, rags, brooms, dust mops, vacuum cleaners or other cleaning tools.

2. Soap powders, bleaches, waxes or other cleaning compounds; plastic items or containers; gasoline, kerosene, cigarette lighter fluid; dry-cleaning fluids or other volatile fluid.

3. Paint thinners and other painting compounds.

4. Paper bags or other paper products.

WARNING: Never operate the furnace with the blower door removed. To do so could result in serious personal injury and/or equipment damage.

WARNING: This furnace may not be common vented with any other appliance, since it requires separate, properly-sized air intake and vent lines. The furnace shall not be connected to any type of B, BW or L vent or vent connector, and not connected to any portion of a factory-built or masonry chimney.

If this furnace is replacing a common-vented furnace, it may be necessary to resize the existing vent line and chimney to prevent oversizing problems for the new combination of units. Refer to the National Gas Code (ANSI Z223.1-) or CANI-B149.1 or .2 Installation Code (latest editions).

The following steps shall be followed with each appliance remaining connected to the common venting system placed in operation, while the other appliances remaining connected to the common venting system are not in operation:

1. Seal any unused openings in the common venting system.

2

2. Visually inspect venting system for proper size and horizontal pitch and determine there is no blockage or restriction, leakage, corrosion or other deficiencies which could cause an unsafe conditon.

3. Insofar as is practical, close all building doors and windows and all door between the space in which the applicances remaining connected to the common venting system are located and other spaces of the building. Turn on clothes dryers and any appliance not connected to the common venting system. Turn on any exhaust fans, such as range hoods and bathroom exhausts, so they will operate at maximum speed. Do not operate a summer exhaust fan. Close fireplace dampers.

4. Follow the lighting instructions. Place the appliance being inspected in operation. Adjust the thermostat so the appliance will operate continuously.

5. Check for spillage at the draft hood relief opening after 5 minutes of main burner operation. Use the flame of a match or candle.

6. After it has been determined that each appliance remaining connected to the common venting system properly vents when tested as outlined above, return doors, windows, exhaust fans, fireplace dampers and any other gas burning appliance to their previous conditions of use.

7. If improper venting is observed during any of the previous tests, the common venting system must be corrected.

8. Any corrections to the common venting system must be in accordance with the National Fuel Gas Code Z223.1 or CAN1-B149.1 or .2 Installation Code (latest editions). If the common vent system must be resized, it chould be resized to approach the minimum size as determined using the appropriate tables in Appendix G of the above codes.

LIMITATIONS & LOCATION

This furnace should be installed in accordance with all national and local building/safety codes and requirements, or in the absence of local codes, with the National Fuel Gas Code ANSI Z223.1 or CAN1-B149.1 or .2 Installation Code (latest editions), local plumbing or waste water codes, and other applicable codes.

CAUTION: Do not install the furnace in an unconditioned space or garage that could experience ambient temperatures of 32°F (0°C) or lower.

CAUTION: This unit must be installed in a level (1/4") position side-to-side and front-to-back to provide proper condensate drainage.

CAUTION: Do not allow return air temperature to be below 55°F for extended periods. To do so may cause condensate to occur in the main fired heat exchanger.

WARNING: Furnaces shall not be installed directly on carpeting, tile or other combustible material other than wood flooring.

The size of the unit should be based on an acceptable heat loss calculation for the structure.

Check the rating plate to make certain the unit is equipped for the type of gas supplied, and proper electrical characteristics are available.

For installations above 2,000 feet, reduce input 4% for each 1,000 feet above sea level. In Canada, derate input 10% for elevations at 2,000-4,500 feet.

Do not install this unit in a mobile home.

A furnace installed in a residential garage shall be located so that all burners and burner ignition devices are located not less that 18" above the garage floor, and located or protected to prevent damage by vehicles.

SPECIFIC UNIT INFORMATION

CLEARANCES TO COMBUSTIBLES

Minimum clearances from combustible construction are in inches:

```
Top ............................................................. 1
Front ........................................................... 3
Vent Piping ................................................... 0
Rear ............................................................ 0
Sides ........................................................... 0
```

CLEARANCES FOR ACCESS

Ample clearances should be provided to permit easy access to the unit. The following minimum clearances are recommended:

1. Twenty-four (24) inches between the front of the furnace and an adjacent wall or another appliance, when access is required for servicing and cleaning.

2. Eighteen (18) inches at the side where access is required for passage to the front when servicing or for inspection or replacement of flue/vent connections.

NOTE: In all cases, accessibility clearances shall take precedence over clearances for combustible materials where accessibility clearances are greater.

NOMENCLATURE

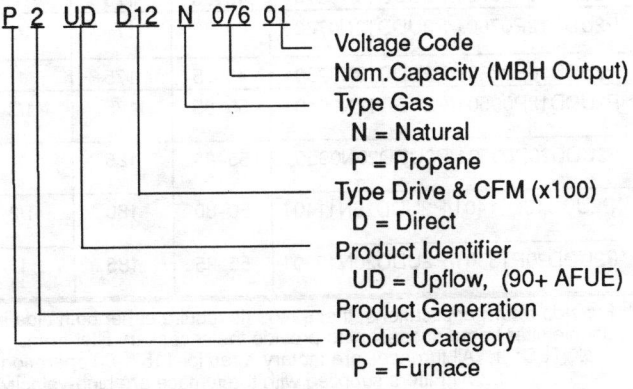

DIMENSIONS

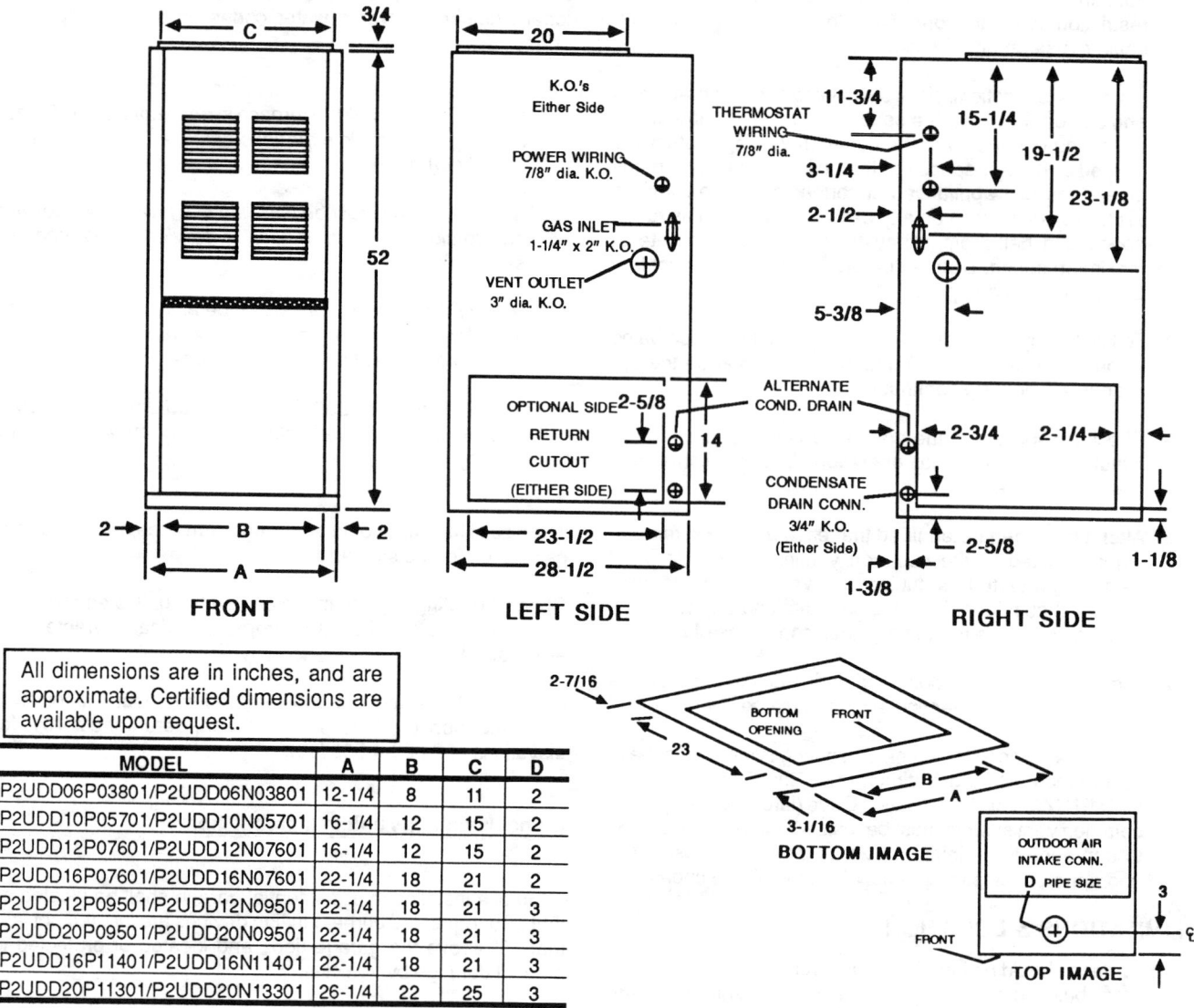

FRONT

LEFT SIDE

RIGHT SIDE

BOTTOM IMAGE

TOP IMAGE

All dimensions are in inches, and are approximate. Certified dimensions are available upon request.

MODEL	A	B	C	D
P2UDD06P03801/P2UDD06N03801	12-1/4	8	11	2
P2UDD10P05701/P2UDD10N05701	16-1/4	12	15	2
P2UDD12P07601/P2UDD12N07601	16-1/4	12	15	2
P2UDD16P07601/P2UDD16N07601	22-1/4	18	21	2
P2UDD12P09501/P2UDD12N09501	22-1/4	18	21	3
P2UDD20P09501/P2UDD20N09501	22-1/4	18	21	3
P2UDD16P11401/P2UDD16N11401	22-1/4	18	21	3
P2UDD20P11301/P2UDD20N13301	26-1/4	22	25	3

TABLE 1 - RATINGS & PHYSICAL DATA

Model	Air Temp. Rise °F	Max. Outlet Temp. °F	Blower HP	Blower Size	Filter Size Supplied	Filter Size Suggested Bottom	Unit Full Load Amps	Max. Over-Current Protect.	Min. Wire Size (AWG) @ 75 ft. One Way
P2UDD06P03801/P2UDD06N03801	45–75	175	1/5	9–4	16x26 Side	12x26	↑Less Than 12 amps↓	15	14
P2UDD10P05701/P2UDD10N05701	45–75	175	1/3	10–6	16x26 Side	16x26		15	14
P2UDD12P07601/P2UDD12N07601	45–75	175	1/2	11–8	16x26 Side	16x26		15	14
P2UDD16P07601/P2UDD16N07601	45–75	175	1	11–10	16X26 Side	20x26	14.74	20	12
P2UDD12P09501/P2UDD12N09501	55–85	185	1/3	10–8	20x26 Btm See Note 3	20x26	<12	15	14
P2UDD20P09501/P2UDD20N09501	55–85	185	1	11–10	20x26 Btm See Note 3	20x26	14.74	15	12
*P2UDD16P11401/P2UDD16N11401	50–80	180	1/2	10–10	20x26 Btm See Note 3	20x26	< 12	15	14
*P2UDD20P13301/P2UDD20N13301	55–85	185	1	11–10	24x26 Btm 16x26 Side*	24x26	14.74	20	12

* For side return applications, these models require either both side inlets, bottom and one side or the use of the optional single-side return double filter frame accessory to provide the necessary filter area.

NOTES: 1. All furnaces are factory wired for 115-1-60 operation.
2. All Filters supplied with the furnace are high-velocity, cleanable type.
3. Recommended filter size for side return is 16x26.

Allow clearances from combustible materials as listed under "Clearances to Combustibles", ensuring that service access is allowed for both the burners and blower.

When the furnace is used in conjunction with a cooling coil, the furnace must be installed parallel with or on the upstream side of the cooling unit to avoid condensation in the primary heat exchanger. When a parallel flow arrangement is used, the dampers or other means used to control air flow shall be adequate to prevent chilled air from entering the furnace, and if manually operated, must be equipped with means to prevent operation of either unit unless the damper is in the full heat or cool position.

The furnace shall be located:

1. Where a minimum amount of air intake/vent piping and elbows will be required.

2. As centralized with the air distribution as possible.

3. In an area where ventilation facilities provide for safe limits of ambient temperature under normal operating conditions. Ambient temperatures must not fall below 32°F (0° C).

4. Where it will not interfere with proper air circulation in the confined space.

5. Where the outdoor combustion air/vent terminal will not be blocked or restricted.

6. Where it will not interfere with the cleaning, servicing or removal of other appliances.

COMBUSTION AIR AND VENT SYSTEM

This furnace requires outdoor combustion air. Two separate, properly-sized pipes must be used; one bringing outdoor air from the accessory terminal kit outdoors to the furnace combustion air intake (top of unit), and one from the furnace vent connection (left or right side of unit) back to the terminal kit located outdoors.

The vent terminal kit should be located either through the wall (horizontal or side vent) or through the roof (vertical vent). Care should be taken to locate side vented systems where trees or shrubs will not block or restrict supply air from entering or combustion products from leaving the terminal. Also, the terminal assembly should be located as far as possible from a swimming pool or a location where swimming pool chemicals might be stored. Care must be taken such that the terminal assembly outdoors follows the clearances listed in the following table:

LOCATION CLEARANCE

Dryer Vent .... 4 feet
Plumbing Vent Stack .... 3 feet
Gas Appliance Vent Terminal .... 4 feet
From any opening where vent
gas could enter the builiding .... 1 foot
Above grade and anticipated
snow depth .... 1 foot
Above grade when adjacent to
a public walkway .... 7 feet
From electric, gas meters, regulators and relief
equipment - min. horizontal distance .... 4 feet

NOTE: Consideration must be given for degradation of building materials by flue gases.

Central Environmental Systems

For proper vent/combustion air intake sizing and installation, see the section of this instruction "Combustion Air/Vent Pipe Sizing."

UNIT INSTALLATION

DUCTWORK

The duct system's design and installation must:

1. Handle an air volume appropriate for the served space and within the operating parameters of the furnace specifications.

2. Be installed in accordance with standards of NFPA (National Fire Protection Association) as outlined in NFPA pamphlets 90A and 90B or applicable national, provincial, local fire and safety codes.

3. Create a closed duct system. The supply duct system must be connected to the furnace outlet and the return duct system must be connected to the furnace inlet. Both supply and return duct systems must terminate outside the space containing the furnace.

4. Generally complete a path for heated or cooled air to circulate through the air conditioning and heating equipment and to and from the conditioned space.

NOTE: The supply air temperature differential between the side discharge versus front to back discharge is substantial. It is recommended that whenever possible position the furnace so left and right side supply air is dominate.

After the unit is in the desired position, fasten the supply ductwork to the furnace duct flanges. A removable access panel should be provided in the outlet duct such that smoke or reflected light would be observable inside the casing to indicate the presence of leaks in the heat exchanger. This access cover shall be attached in such a manner as to prevent leaks. Flexible duct connectors are recommended to connect both the supply and return ducts to the furnace.

Return air ductwork may be connected to an upflow furnace in one of the following two ways:

1. Bottom Return - Before attaching the ductwork to the furnace bottom, see the "Filters" section of this instruction.

2. Side Return - Cut a hole in the side panel of the furnace using the right-angle markings (See Figure 1) as a guide for position and size of the opening. Install a single side return filter frame accessory if one is required. If this accessory is not needed, the ductwork can be fastened directly to the furnace opening.

NOTE: Some accessory side filter packages require cutting a slightly larger opening.

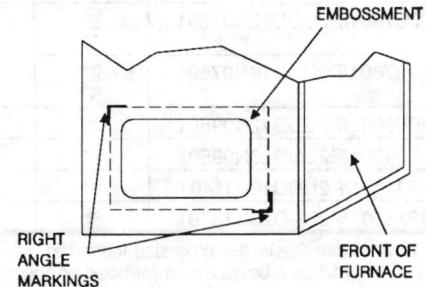

FIGURE 1 - SIDE RETURN CUTOUT MARKINGS

When the return duct system is not complete, the return connection must be run full size from the furnace to a location outside the utility room or basement. For further details, consult Section 5.3, Air Combustion and Ventilation of the National Fuel Gas Code, ANSI Z223.1 or CAN1-B149.1 or .2 Installation Code (latest editions).

FILTERS

The type and size of filter(s) to be used are shown in Table 1.

12-1/4 & 16-1/4 Inch Width Furnaces

All 12-1/4 & 16-1/4 inch wide furnaces are shipped with filters mounted on the left side. Filters may be relocated to the bottom or right side as follows:

The wire filter retainer must be moved if the return air application requires moving of the filter from the side to the bottom location or vice versa. When relocating filters, it may be necessary to trim the filter to the proper size.

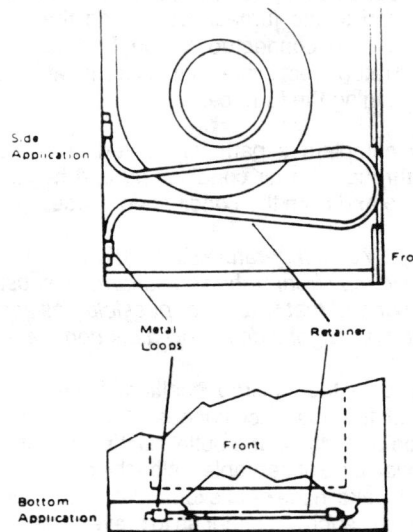

FIGURE 2 -FILTER RETAINER

The ends of the retainer are attached to the rear panel in two metal loops (See Figure 2). The ends must be squeezed together to free them from the loops. The retainer may then be moved to the new location, and the ends inserted in the loops on the rear panel at the new location. Loops are provided for retainer location to accomodate filter application on the bottom or either side of the furnace.

To remove a filter from the bottom location, push the closed end of the filter retainer to the left until it clears the lip on the front of the furnace base, which acts as a catch for the retainer. When the retainer is clear of the lip, lift up. The retainer will pivot in the loops. This will expose the filter to allow removal. To reinstall the filter, simply reverse this procedure.

To remove a filter from the side location, push the closed end of the filter retainer down until it clears the flange on the side panel, which acts as a catch for the retainer. When the retainer is clear of the flange, it will pivot in the loops. Swing the retainer toward the center of the furnace. This will expose the filter to allow removal. To reinstall the filter, simply reverse this procedure.

22-1/4 & 26-1/4 Inch Width Furnaces

22-1/4 & 26-1/4 inch wide furnaces are shipped with filters mounted in the bottom and left side.

For filter removal in these models, follow the same procedure as with the small width furnaces.

COMBUSTION AIR/VENT PIPE SIZING

Refer to Table 2 to select the proper size piping for combustion air intake and venting. The size will be determined by a combination of furnace model, total length of run, and the number of elbows required. The following rules must also be observed.

1. Long radius elbows are recommended for all units except for the 133 MBH unit for which long radius elbows are mandatory.

2. Elbows are assumed to be 90 degrees. Two 45 degree elbows count as one 90 degree elbow.

3. Elbow count refers to combustion air piping and vent piping separately. For example, if the table allows for 5 elbows, this will allow a maximum of 5 elbows in the combustion air piping and a maximum of 5 elbows in the vent piping.

4. The combustion air blower elbow and vent terminal kit parts are already accounted for, and should not be counted in the allowable total indicated in the table.

5. Combustion air and vent piping must be of the same diameter.

6. All piping and fittings are to be Schedule 40 PVC, PVC-DWV, ABS-DWV, SDR-21 PVC, or SDR-26 PVC.

TABLE 2 - COMBUSTION AIR/VENT PIPE SIZES

MODEL	Pipe Size	MAXIMUM ELBOWS PER TOTAL RUN (See Note 2)						
		0–5 feet	5-10 feet	10-15 feet	15-20 feet	20-25 feet	25-30 feet	30-35 feet
P2UDD06P03801/P2UDD06N03801	1-1/2" 2"	5 5	4 5	3 5	2 5	Note 1 5	Note 1 5	Note 1 5
P2UDD10P05701/P2UDD10N05701	1-1/2" 2"	3 5	2 5	Note 1 5	Note 1 5	Note 1 5	Note 1 5	Note 1 5
P2UDD12P07601/P2UDD12N07601	2" 3"	5 5	4 5	3 5	2 5	1 5	Note 1 5	Note 1 5
P2UDD16P07601/P2UDD16N07601	2" 3"	5 5	4 5	3 5	2 5	1 5	Note 1 5	Note 1 5
P2UDD12P09501/P2UDD12N09501	2"	5	5	5	5	5	5	30' Max
P2UDD20P09501/P2UDD20N09501	2"	5	5	5	5	5	5	30' Max
P2UDD16P11401/P2UDD16N11401	3"	5	5	5	5	5	5	5
P2UDD20P13301/P2UDD20N13301	3"	5	5	5	5	5	5	30' Max.

NOTES: 1. Must use the larger pipe size indicated
2. Long radius elbows recommended on these units, except 133 MBH, when over 25 feet of piping is required. On 133 MBH units, use of long radius elbows is mandatory.

Central Environmental Systems

COMBUSTION AIR INTAKE

Once the piping size is determined, the combustion air intake collar must be connected to that size piping. Units rated from 40 MBH input (38 MBH ouput) to 80 MBH input (76 MBH output) are equipped with intake collars which accomodate 2" diameter piping. The 100 MBH input (95 MBH output) through 144 MBH input (133 MBH output) units have collars for 3" diameter piping. When the pipe size listed in the table is different than the size of the intake collar provided with the unit, transition should be made as illustrated in Figure 3.

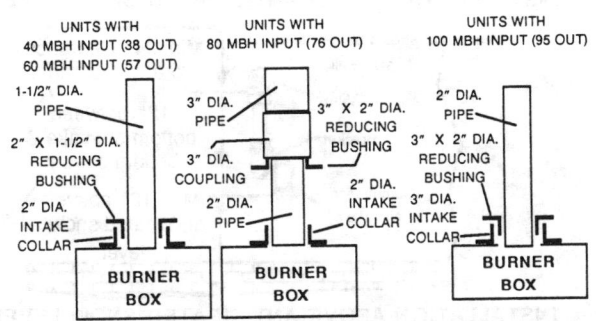

FIGURE 3 - COMBUSTION AIR PIPING ADAPTATION

VENT PIPE CONNECTIONS

All models are provided with a combustion blower elbow which accomodates 2" diameter vent pipe. This elbow may be positioned for right or left exit from the unit as required for the installation conditions.

With reference to the piping sizing table, where units are installed with a vent pipe of a different diameter than the exhaust elbow, reduction or expansion fittings must be incorporated by the installer at the time of installation. Examples of such adaptation are shown in Figure 4.

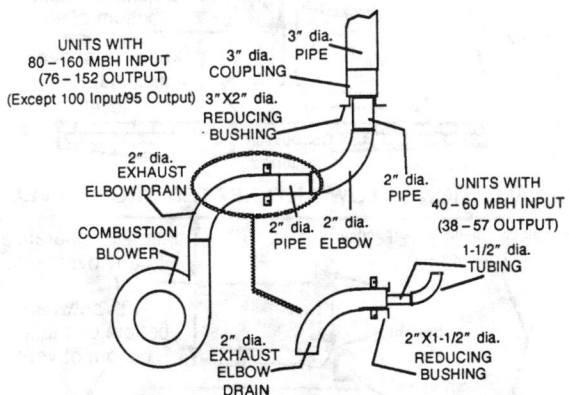

FIGURE 4 - VENT PIPING ADAPTATION

PIPING ASSEMBLY

The final assembly procedure for the vent/combustion air piping is as follows:

1. Cut piping to the proper length, beginning at the furnace.

2. Deburr the piping inside and outside.

3. Chamfer the outer edges of the piping.

4. Dry-fit the entire vent/combustion air piping assembly.

Central Environmental Systems

5. Disassemble the piping and apply cement primer and cement per the cement manufacturer's instructions. Primer and cement must conform to ASTM D2564 for PVC, or ASTM D2235 for ABS piping.

WARNING: *Solvent cements are flammable and must be used in well-ventilated areas only. Keep them away from heat, sparks and open flames (including pilots). Do not breathe vapors and avoid contact with skin and eyes.*

6. All joints must be made to provide a permanent, air-tight, water-tight seal. The only exception to this means of joining is where the street elbow is joined to the venter assembly. This joint should be made with RTV sealant. A stainless steel screw should be used in the hole in the venter housing to keep the street elbow in proper alignment. If the street eblow is changed for a left side exit, the stainless steel screw must be reused in the new location. If the elbow/drain is used for right side exit, carefully remove the plastic cap which is factory installed on the lower left drain connection (see Figure 5). Move the drain tubing and clamp from the lower right drain connection to the lower left drain connection. Place the plastic cap onto the lower right drain connection.

NOTE: *The upper drain connections are not used in upflow installations and are internally sealed.*

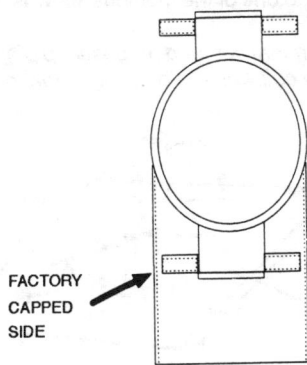

FIGURE 5 - DRAIN CAP LOCATION

Drain tubing from the drain fitting must be routed to prevent kinks and low spots. This tubing may have to be cut and shortened when the flue piping is routed out of the left side of the furnace.

7. Support the combustion air and vent piping such that it is angled 1/4" per linear foot upward from the furnace. Piping should be supported with pipe hangers to prevent sagging. Maximum spacing between hangers is five (5) feet, except SDR-PVC piping, where maximum spacing is three (3) feet.

8. Seal around the openings where the combustion air and vent piping pass through the roof of side wall.

CAUTION: *Vent piping must be insulated with 1/2" Armaflex insulation if it will be subjected to freezing temperatures such as routing through unheated areas or through an unused chimney.*

The combustion air piping must be insulated with 1/2" thick Armaflex insulation if it is installed above a suspended ceiling or in a warm. humid space such as a laundry room to prevent possible condensation from forming on the outside of the pipe.

VENT TERMINAL ASSEMBLY

The combustion air and vent piping must terminate outdoors using the Vent Terminal Assembly Kit.

Two vent terminal kits are available. The 1VK0307 kit is to be used with 1-1/2" and 2" piping, and the 1VK0308 is to be used for 3" piping.

NOTE: The 3" vent terminal kit contains one elbow having a "splitter baffle" in one opening. This elbow must be used for the combustion air intake.

Each kit contains the following components:

1. One terminal bracket with 2 - 90 degree PVC elbows.

2. One 90 degree PVC street elbow.

3. One 90 degree PVC elbow.

4. One PVC pipe coupling.

5. Installation instructions.

This terminal kit may be used for rooftop or side wall installation. Rooftop termination is the recommended means, and should be arranged according to one of the methods shown in Figure 6.

NOTE: The vent terminal kit and exposed piping may be painted the same color as the building to make them less noticable.

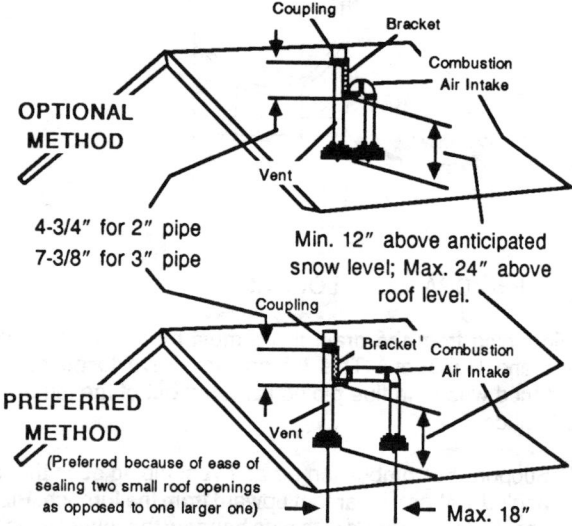

FIGURE 6 - ROOFTOP TERMINATION

If optional side wall venting is to be used, installation of the teminal kit should be as shown in Figure 7.

If for some reason, combustion air and vent piping can not be routed together, the optional side wall termination may be carried out as illustrated in Figure 8.

The installation procedure for the terminal kit is as follows:

1. Cut all combustion air and vent piping so the vent termination fittings and brackets can be dry-fitted together.

2. All piping should be deburred inside and out, and the outside edge should be chamfered.

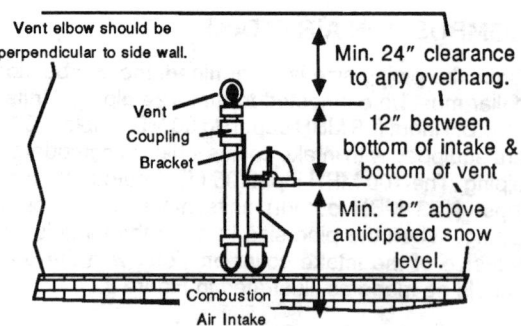

INSTALLATION BELOW ANTICIPATED SNOW LEVEL

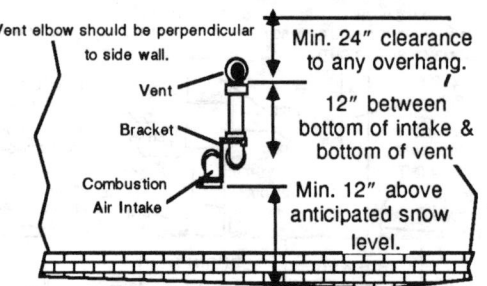

INSTALLATION ABOVE ANTICIPATED SNOW LEVEL

FIGURE 7 - SIDE WALL TERMINATION

3. Reassembly all piping and fittings using cement primer and cement per the cement manufacturer's instructions. Primer and cement must conform to ASTM 2564 for PVC or ASTM D2235 for ABS piping.

4. Reattach and tighten the vent termination bracket.

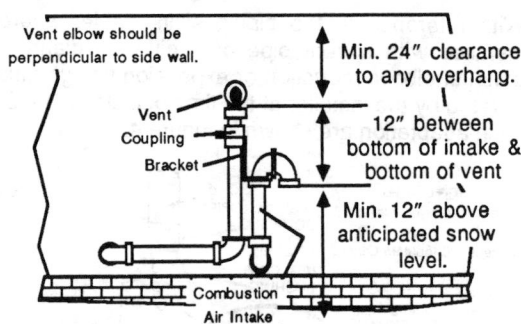

INSTALLATION BELOW ANTICIPATED SNOW LEVEL

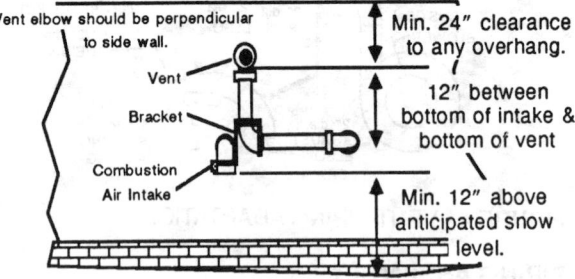

INSTALLATION ABOVE ANTICIPATED SNOW LEVEL

FIGURE 8 - ALTERNATIVE TERMINATIONS

GAS PIPING

NOTE: An accessible manual shutoff valve must be installed upstream of the furnace gas controls and within 6 feet of the furnace. A 1/8" NPT plugged tapping, accessible for test guage connection must be installed immediately upstream of the gas supply connection to the furnace.

Central Environmental Systems

The furnace and its individual shutoff valve must be disconnected from the gas supply piping system during any pressure testing of that system at test pressures in excess of 1/2 psig (3.48 kPa).

The furnace must be isolated from the gas supply piping system by closing its individual external manual shutoff valve during any pressure testing of the gas supply piping system at pressures equal to or less than 1/2 psig (3.48 kPa).

Gas piping may be connected from either side of the furnace. Sizing and installation of the supply gas line should comply with the local utility requirements. The gas supply should be a separate line, installed in accordance with the National Fuel Gas Code, ANSI Z223.1 or CAN1-B149.1 or .2 Installation Codes (latest editions).

Some utility companies require pipe sizes larger than the maximum sizes listed. Using the properly sized wrought iron, steel or approved flexible pipe, make gas connections to the unit. Installation of a drop leg and ground union joint is required (see Figure 9).

WARNING: *Compounds used on threaded joints of gas piping must be resistent to the action of liquified petroleum gases. After connections are made, leak-test all pipe connections.*

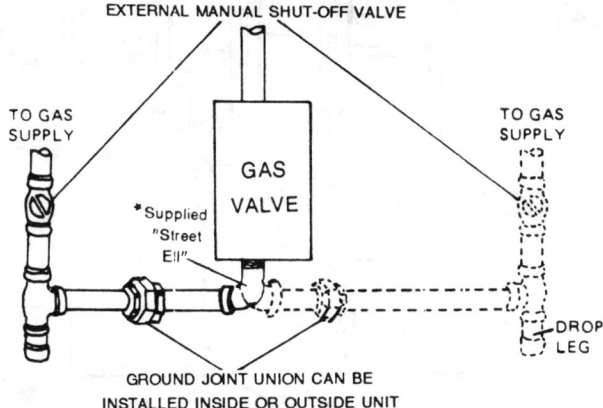

*NOTE: Unit is supplied with a 1/2" "street ell." Where "street ell's" are not allowed, a pipe nipple and elbow may be used. Use of nipple and elbow may require field extending side opening of casing for proper clearance.

FIGURE 9 - GAS PIPING

WARNING: *Do not use an open flame or other source of ignition for leak testing. Set the manual gas valve to the OFF position.*

For the purpose of input adjustment, the minimum inlet gas pressure must be equal to or greater than that shown on the rating label.

CONDENSATE PIPING

The condensate drain connection assembly is located in the lower right front corner of the blower compartment. It consists of a mounting bracket and 1/2" CPVC coupling (ASTM D2846) with flexible tubing connected to it (see Figure 10). Another knockout is provided above the original location if the added height is necessary. New holes must be drilled for the screws if this alternate location is used.

The installation procedure for condensate piping is as follows:

1. Determine whether the condensate drain line will be installed through the right or left side of the furnace.

2. Carefully remove the 3/4" diameter knockout from the appropriate side of the furnace. Two knockouts are provided on each side, one located toward the lower right front of the blower compartment, the other is located above the first.

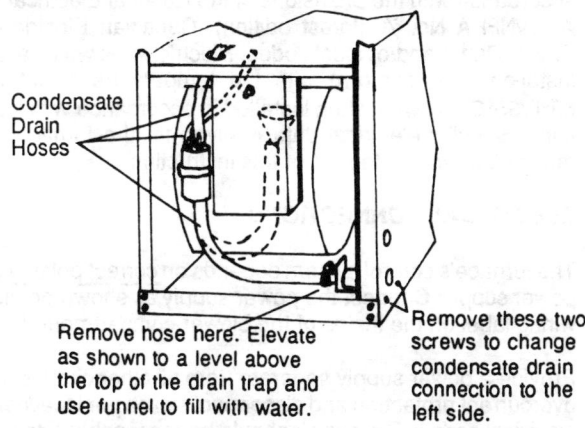

NOTE: Trap is shown exposed but may be behind controls box.

FIGURE 10 - CONDENSATE DRAIN CONNECTION

3. For left side or alternate drain connection, it will be necessary to relocate the condensate drain connection assembly.

   a. Remove the two screws that secure the drain connection bracket to the right side panel (see Figure 10).

   b. Position the bracket over the corresponding holes in the left side panel (new holes must be drilled for alternate locations), and fasten it to the side panel using the two screws previously removed.

   c. Make sure the flexible tubing between the drain trap mounted on the blower and the drain connection assembly has a continuous downward slope to the drain connection assembly, has no low spots, and is not kinked. (If necessary, shorten the flexible tubing to prevent kinking and/or low spots that could restrict the flow of condensate.

4. Connect field-supplied piping to the condensate drain connection assembly and run it to an open drain. Refer to the "Vent/Combustion Air Connections" section of this instruction for procedure for assembly of plastic pipe. The manufacturer recommends 1/2" I.D. CPVC or equivalent field installed drain pipe. The condensate piping may be tied together with the air conditioning condensate drain if the air conditioning condensate drain line is trapped upstream of the tie-in and the combined drains are constructed of the same material.

5. All pipe joints must be cleaned, de-burred and cemented using CPVC primer and cement.

6. The furnace contains an internal trap. Therefore, no external trap should be used.

7. If a condensate pump is used, it must be suitable for use with acidic water.

8. Where required, a field-supplied neutralizer can be installed in the drain line, external to the furnace.

ELECTRICAL DATA

Use copper conductors only!

Field wiring to the unit must conform to and be grounded in accordance with the provisions of the National Electrical Code ANSI/NFPA No. 70-(latest edition), Canadian Electric Code C22.1, Part 1 and/or local codes. Electric wires which are field installed shall conform with the temperature limitation for 63°F/35°C rise wire when installed in accordance with instructions. Specific electrical data is given for the furnace on its rating plate and in Table 1 of this instruction.

ELECTRICAL CONNECTIONS

The furnace's control system depends on <u>correct polarity</u> of the power supply. Connect the power supply as shown on the unit wiring label on the inside of the blower compartment door.

Provide a power supply separate from all other circuits. Install overcurrent protection and disconnect switch per local/national electrical codes. The switch should be reasonably close to the unit for convenience in servicing. With the disconnect switch in the OFF position, check all wiring against the unit wiring label. Also, see the wiring diagram in this instruction.

Install the field-supplied thermostat. The thermostat instructions for wiring are packed with the thermostat. With the thermostat in the OFF position and the main electrical source disconnected, complete the low voltage wiring from the thermostat to the terminal board on the low voltage transformer. Set the heat anticipator on the thermostat to 1.16 amps when the White-Rodgers 36E01 gas valve is used,

NOTE: *Some thermostats do not have adjustable anticipators. On such thermostats, adjust cycle rate to prevent possible short on/off cycles.*

The 24-volt transformer is sized for the furnace components only, and should not be connected to auxiliary devices such as humidifiers, air cleaners, etc.

SAFETY CONTROLS

Interlock Switch

This unit is equipped with an Electrical Interlock Switch mounted in the blower compartment. This switch interrupts power at the unit when the panel covering the blower compartment is removed. This prevents operation of the automatic gas control valve and the blower.

WARNING: *Blower and burner must never be operated without the blower panel in place.*

Electrical supply to this unit is dependent upon the panel that covers the blower compartment being in place and properly positioned.

CAUTION: *Main power to the unit must still be interrupted at the main power disconnect switch before any service or repair work is to be done to the unit. Do not rely upon the interlock switch as a main power disconnect.*

Flusible Link Control

This control is mounted in the burner compartment. If the temperature in the burner compartment exceeds 306°F, the circuit to the igniter control and the gas valve is de-energized.

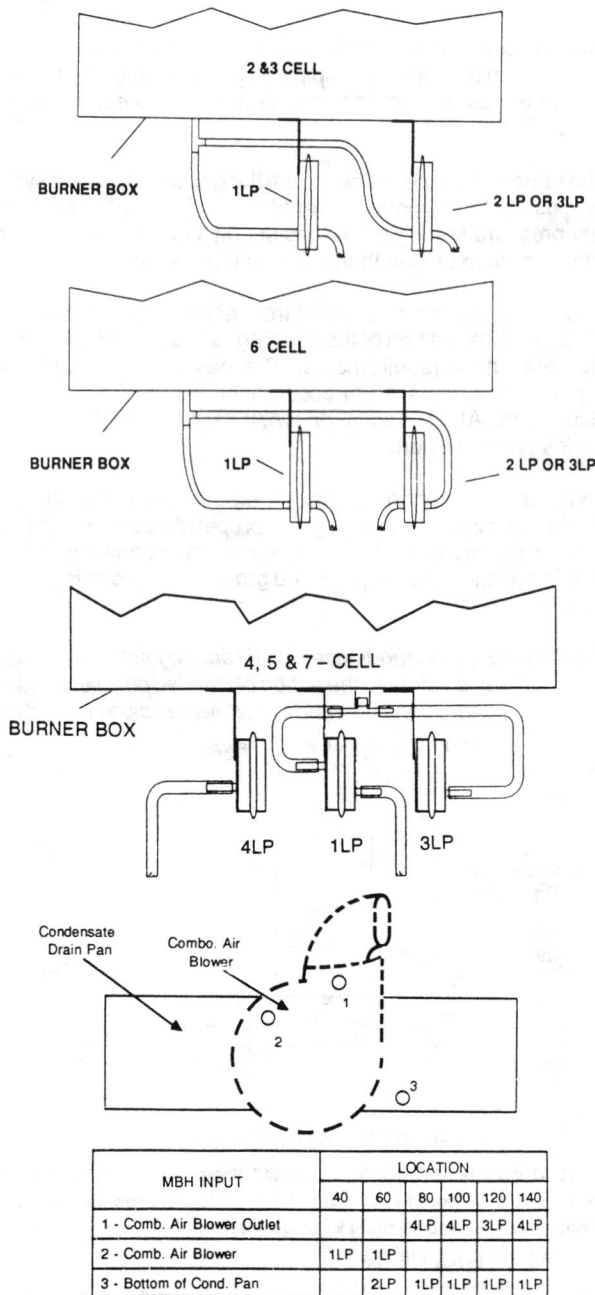

MBH INPUT	LOCATION					
	40	60	80	100	120	140
1 - Comb. Air Blower Outlet			4LP	4LP	3LP	4LP
2 - Comb. Air Blower	1LP	1LP				
3 - Bottom of Cond. Pan		2LP	1LP	1LP	1LP	1LP

FIGURE 11 - PRESSURE SWITCH TUBING

This is a one-time fusible link type control. Before the furnace can operate again, the control must be replaced. The operation of this control indicates a possible burner flame rollout condition. The combustion air blower and the combustion air pressure switch should be checked for proper operation.

Pressure Switches

This furnace is supplied with pressure switches which monitor the flow through the combustion air/vent piping system. These switches de-energize the ignition control module and the gas valve if any of the following conditions are present.

1. Blockage of combustion air piping or terminal.

2. Blockage of vent piping or terminal.

3. Failure of combustion air blower motor.

Central Environmental Systems

4. Blockage of condensate drain piping.

START-UP AND ADJUSTMENTS

The initial start-up of the furnace requires the following additional procedures:

1. When the gas supply is initially connected to the furnace, the gas piping may be full of air. In order to purge this air, it is recommended that the ground union be loosened until the odor of gas is detected. When gas is detected, immediately retighten the union and check for leaks. Allow five minutes for any gas to dissipate before continuing with the start-up procedure.

WARNING: Be sure proper ventilation is available to dilute and carry away any vented gas.

2. The condensate trap must be filled with water before putting the furnace into operation. The recommended procedure is as follows:

   a Disconnect the condensate drain hose from the condensate drain outlet fitting (see Figure 10 for location of this hose connection).

   b. Elevate this hose above trap level and fill with water using a funnel. Drain excess water into a container.

   c. Replace the condensate drain hose and clamps.

3. All electrial connections made in the field and in the factory chould be checked for proper tightness.

IGNITION SYSTEM CHECKOUT/ADJUSTMENT

1. Turn gas supply OFF at external gas valve or main gas valve (see Figure 12). Turn electrical power ON. Check that 24 volts is available across terminal C and R on the transformer.

2. Check ignition module operation as follows:

   a. Set the thermostat to "heat". Turn thermostat setting above room temperature to call for heat.

   b. Through burner box observation ports watch for the hot surface igniter to begin glowing.

   c. If glow is seen, turn the thermostat down (below room temperature). Hot surface igniter should stop glowing.

3. Turn the gas supply ON at external valve and main gas valve.

4. Set the thermostat above room temperature to call for heat.

5. System start-up will occur as follows:

   a. Venter motor will start and come up to speed. Shortly after venter start-up, the hot surface igniter will glow for about 45 seconds.

   b. After warm-up cycle, ignition module will energize (open) the main gas valve for four seconds. At the same time power to the igniter is shut off.

NOTE: Burner ignition may not be satisfactory on first start-up due to residual air in gas line, or until gas pressure (manifold) or burner air shutters are adjusted.

6. With furnace in operation, paint the pipe joints and valve gasket lines with a rich soap and water solution. Bubble indicate a gas leak. Take appropriate steps to stop the leaks. If the leak persists, replace the component.

*WARNING: DO NOT omit this test! **NEVER** use a flame to check for gas leaks.*

ADJUSTMENT OF MANIFOLD GAS PRESSURE

1. Turn gas off at main gas valve. Remove 1/8" plug in the main gas valve body and install proper manometer tube adapter fitting. Connect line from gas valve tap to manometer

2. Refer to Figure 12 for location of pressure regulator adjustment cap and screw on main gas valve.

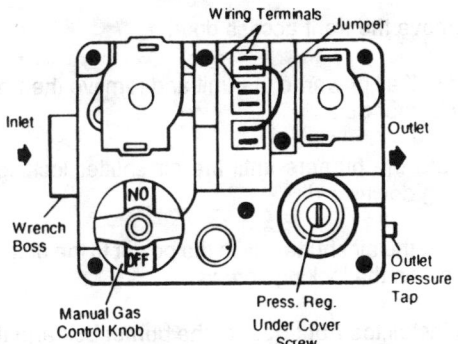

WHITE-RODGERS 36E01 VALVE

FIGURE 12 – GAS VALVE

NOTE: The screw-off cap for the pressure regulator must be removed entirely to gain access to the adjustment screw. Loosening or tightening the cap does not adjust the flow of gas.

3. Turn gas and electrical supplies ON. Start furnace and observe manifold pressure on manometer.

4. Adjust manifold pressure by adjusting gas valve regulator screw: for natural gas, set at 3.5" W.C.; for propane (LP) gas, set at 10.0" W.C.

WARNING: If manifold pressure is too high, an over-fire condition exists which could cause heat exchanger failure. If the manifold pressure is too low, sooting and eventual clogging of the heat exchanger could occur.

If gas valve regulator is turned in, or clockwise, manifold pressure is increased. If screw is turned out, or counterclockwise, manifold pressure will decrease.

WARNING: Once the correct gas pressure to the burners has been established, turn the gas valve knob to OFF and turn the electrical supply switch to OFF; then remove the pressure tap at the gas valve and re-install the plug, using a compound (on the threads) resistant to the action of LP gases.

Turn the electrical and gas supplies back on, and with the burners in operation, check for gas leakage around the plug with a soap and water solution.

ADJUSTMENT OF PRIMARY AIR

The main burners should be in operation for 15 minutes before making the primary air adjustment. The burner flame should not contain any yellow color. With the furnace operating at full input, adjust the primary air of the burners as follows:

CAUTION: The front panel of the burner enclosure must be in place and secured with screws during normal operation and when observing the proper burner flame.

Natural Gas or Propane (LP)

All models are shipped with the air shutters wide open. Local variations in the gas supply may require changes in the settings described above. To change the air shutter settings, use the following procedure:

1. Remove the front access door.

2. Turn off all power to the unit and remove the front door of the burner box.

3. Rotate the burners until the air shutter locking screw is facing downward.

4. Adjust the air shutters with the power to the unit "OFF" and retighten the locking screws.

5. Re-install the front door of the burner box and tighten the screws which secure the door.

6. Restore power and start the unit. Observe through the observation port to see if all flames are now blue in color. If yellow flames are still visible, repeat steps 2 thru 6.

7. Cycle the burners on an off a few times to verify the burners are lighting promptly and properly

8. Replace front access door.

CHECKING GAS INPUT (NATURAL GAS)

NOTE: Front door of burner box must be secured when checking gas input.

1. Turn off all other gas appliances connected to the gas meter.

2. With the furnace turned on, measure the time needed for one revolution of the hand on the smallest dial on the meter. A typical domestic gas meter usually has a 1/2 or 1 cubic foot test dial.

3. Using the number of seconds for each revolution and the size of the test dial increment, find the cubic feet of gas consumed per hour from Table 3.

TABLE 3 – GAS RATE (CUBIC FEET PER HOUR)

Seconds for one Revol.	Size of Test Dial	
	1/2 cu. ft.	1 cu. ft.
10	180	360
12	150	300
14	129	257
16	113	225
18	100	200
20	90	180
22	82	164
24	75	150
26	69	138
28	64	129
30	60	120
32	56	113
34	53	106
36	50	100
38	47	95
40	45	90
42	43	86
44	41	82
46	39	78
48	37	75
50	36	72
52	35	69
54	34	67
56	32	64
58	31	62
60	30	60

NOTE: To find the Btuh input, multiply the number of cubic feet of gas consumed per hour by the BTU content of the gas in your particular locality. Contact your gas company for this information, as it varies widely from city to city.

EXAMPLE: It is found by measurement that it takes 26 seconds for the hand on the 1 cubic foot dial to make a revolution with only a 120,000 Btuh furnace running. Using this information, locate 26 seconds in the first column of Table 3. Read across to the column headed "1 cubic foot" where you will see that 138 cubic feet of gas per hour are consumed by the furnace at that rate. Multiply 138 by 850 (the BTU rating of the gas obtained from the local gas company). The result is 117,300 Btuh, which is close to the 120,000 Btuh rating of the furnace.

If the actual input is not within 5% of the furnace rating, with allowance being made for the permissible range of the regulator setting (0.3 inches W.C.), replace the orifice spuds with spuds of the proper size.

CAUTION: Be sure to relight any gas appliances that were turned off at the start of this input check.

ADJUSTMENT OF TEMPERATURE RISE

The temperature rise, or temperature difference between the return air and the heated air from the furnace, must be within the range shown on the furnace rating plate and within the application limitations shown in Table 1. After the temperature rise has been determined, the cfm can be calculated.

After about 20 minutes of operation, determine the furnace temperature rise. Take readings of both the reutrn air and the heated air in the ducts, about six feet from the furnace where

Central Environmental Systems

they will not be affected by radiant heat. Increase the blower speed to decrease the temperature rise; decrease the blower speed to increase the rise.

All direct-drive blowers have multi-speed blowers. Refer to the unit wiring diagram and connect the blower motor for the desired speed. The blower motor speed taps are located in the control box in the blower compartment.

ADJUSTMENT OF FAN CONTROL SETTINGS

NOTE: Fan-On time is preset and not adjustable. When the main gas valve is energized (ON), two wires carry 24v to a small heater which controls fan-on timing.

To adjust the Fan Off setting:

1. Turn the furnace on.

2. Let the furnace operate for 20 minutes.

3. Turn the furnace off.

4. Read the thermometer when the blower stops.

5. If this temperature is too high when the blower stops, lower the fan off setting; if the temperature is too low, raise the setting.

6. If adjustments are made to the fan off setting, check the operation of the furnace by repeating the previous steps.

CAUTION: When fan on and fan off adjustments are made, be careful not to rotate the fan control dial. If the dial is allowed to rotate, the control could be damaged and operate erratically.

CHECKING LIMIT CONTROL

With the main burners operating, cover all return air grilles with paper to restrict the flow of return air. In a few minutes, the burners should be shut off by the limit control. Remove the return air restrictions. The main blower will cool the unit. The limit switch should reclose and main burners relight after the proper re-start period.

OPERATION & MAINTENANCE
SEQUENCE OF OPERATION
Hot Surface Ignition System

WARNING: Do not attempt to light this furnace by hand (with a match or any other means). There may be a potential shock hazard from the components of the hot surface ignition system. The furnace can only be lit automatically by its hot surface ignition system.

The following describes the sequence of operation of the furnace. Refer to the schematic wiring diagram (pages 16 and 17) for component location.

Continous Blower

On the cooling/heating units with fan switch, when the fan switch is set in the ON position, a circuit is completed between terminals R and G of the thermostat. This energizes the 1R relay. Contact 1R-1 closes and contact 1R-2 opens. The motor is energized through the black, high-speed tap. The blower then operates on high speed.

Central Environmental Systems

Intermittent Blower

When the system switch is set on HEAT and the fan switch is set on AUTO, and the room thermostat calls for heat, a circuit is completed between terminals R and W of the thermostat. This energizes the venter relay 3R, which energizes the venter. When the proper amount of combustion air is being provided, a pressure switch activates the 50E47 ignition control. A second and third pressure switch (2LP or 3LP and 4LP)(not used in all models), the fusible link (FL) and the limit are in this circuit, and must be in the closed position for the ignition control to be activated.

The 50E47 ignition control provides a 45-second warm-up period. The gas valve then opens for four seconds.

As gas starts to flow and ignition occurs, the flame sensor begins its sensing function. If a flame is detected within four seconds after ignition, normal furnace operation continues until the thermostat circuit between R and W is opened. When the supply air temperature reaches 115 to 125°F, the fan switch closes.

When the thermostat circuit opens, the venter is deenergized, along with the ignition control. With the ignition control deenergized, the gas flow stops and the burner flames are extinguished.

The blower motor continues to operate until the supply air temperature drops to between 85 and 100°F. When this occurs, the fan switch opens, de-energizing the blower motor. The heating cycle is then complete, and the unit is ready for the start of the next heating cycle.

If flame is not detected within the four second sensing period, the gas valve is de-energized. The 50E47 control is equipped with a re-try option. This provides a 60-second wait following an unsuccessful ignition attempt (flame not detected). After the 60 second wait, the ignition sequence is restarted with an additional 10 seconds of igniter warm-up time. If this ignition attempt is unsuccessful, one more re-try will be made before lockout.

50E47 HOT SURFACE IGNITION CONTROL

All 50E47 controls will repeat the ignition sequence for a total of five recycles if flame is lost within the first 10 seconds of establishment.

If flame is established for more than 10 seconds after ignition, the controller will clear the ignition attempt (re-try) counter. If flame is lost after 10 seconds, it will restart the ignition sequence. This can occur a maximum of five times.

During burner operation, a momentary loss of power of 50 milliseconds or longer will drop out the main gas valve. When the power is restored, the gas valve will remain de-energized, and a restart of the ignition sequence will begin immediately.

A momentary loss of gas supply, flame blowout, or a shorted or open condition in the flame probe circuit will be sensed within 0.8 seconds. The gas valve will de-energize and the control will restart the ignition sequence after waiting 60 seconds. Recycles will begin and the burner will operate normally if the gas supply returns, or the fault condition is corrected prior to the last ignition attempt. Otherwise, the control will lockout.

If the control is locked out, it may be reset by momentary power interruption of 1/20 second or longer. Either the 24v thermostat or line voltage may be interrupted.

MAINTENANCE

The manufacturer recommends that maintainance is performed by a qualified service agency for cleaning vent/air intakes, condensate drains and neutralizers, burners, primary and secondary heat exchangers and the blower motor and wheel assembly. An annual inspection of these components is recommended.

Air Filters

The filters should be checked periodically for dirt accumulation. Dirty filters greatly restrict the flow of air and overburden the system.

Clean the filters at least every three months. See page 4 for filter removal instructions. On new construction, check the filters every week for the first four weeks. Inspect the filters at least every three weeks after that, especially if the system is running constantly.

Air filters supplied with the furnace are the high-velocity, cleanable type. Clean these filters by washing in warm water. Make sure to shake all the water out of the filter and have it reasonable dry before installing it in the furnace. When replacing filters, be sure to use the same size and type as originally supplied.

Lubrication

Blower motors in these furnaces are permanently lubricated and do not require periodic oiling.

Burner Removal/Cleaning

The main burners should be checked periodically for dirt accumulation.

If cleaning is required, follow the steps listed below:

1. Turn off the electrical power to the unit.

2. Remove the upper access door.

3. Remove the front cover of the burner box.

4. Remove burners by applying pressure to burner retention spring until burner can be removed from the hole in burner support plate.

5. Slide burner down and off the orifice, leaving the retention spring in place.

6. Burners may be cleaned by rinsing in hot water.

7. Reassemble the burners in the reverse order.

Cleaning the Primary Heat Exchanger

1. Turn off the main manual gas valve external to the furnace.

2. Turn off the electrical power to the furnace.

3. Remove the access door.

4. Disconnect the gas supply piping and control wiring from the gas valve.

5. Remove the screws holding the burner box assembly to the vestibule panel.

6. Remove the burner box assembly.

7. Remove the burner restrictor plate to gain access to the upper portion of the heat exchanger.

8. To reach the lower portion of the heat exchanger, remove the flue box cover and flue baffles.

9. With a stiff wire brush, brush out loose scale or soot.

10. Vacuum the burner assembly and heat exchanger.

11. Replace all parts removed for cleaning by reversing the order of disassembly.

12. Reconnect all wiring and gas piping.

13. Restore electrical power and gas supply to the furnace.

Cleaning the Secondary Heat Exchanger

1. Follow steps 1 thru 10 under "Cleaning the Primary Heat Exchanger."

2. Remove the vent piping from the venter housing. Disconnect the drain lines from the venter and from the condensate drain pan. Remove the venter blower and the condensate pan. The turbulators can then be gently removed from the secondary heat exchanger.

3. With a stiff wire brush, brush out loose scale or soot.

4. Vacuum the secondary heat exchanger.

5. Finish the cleaning procedure by following steps 11 thru 13 under "Cleaning the Primary Heat Exchanger."

BLOWER CARE

Even with good filters properly in place, blower wheels and motors will become dust laden after long months of operation. The entire blower assembly should be inspected annually. If the motor and wheel are heavily coated with dust, they can be brushed and cleaned with a vacuum cleaner.

The procedure for removing the blower assembly for cleaning is as follows:

1. Disconnect the electrical supply to the furnace.

2. Remove the access panels.

3. Disconnect the two wire harness plugs from the top of the control box.

4. Remove the four screws holding the control box and position it out of the way.

5. Remove the hoses from the top of the condensate drain trap, and one at the front of the trap.

6. Remove screws which retain blower to the blower deck.

Central Environmental Systems

7. Remove the blower assembly with the condensate drain trap and the control wiring still attached.

8. Vacuum the motor and the blower wheel using a soft brush attachment. Care must be used not to disturb the balance weights (clips) on the blower wheel vanes.

9. Before reinstalling the blower assembly, inspect the lower portion of the secondary heat exchanger which is visible directly above the blower opening in the blower deck. If it requires cleaning, vacuum it with a soft brush attachment and follow the direction of the fins.

10. Reinstall the blower assembly and the condensate drain hoses. Replace the mounting screws that hold the blower assembly to the front portion of the blower deck. The two mounting screws used on the sides of the blower are used for shipping purposes only, and are not necessary after the furnace has been installed.

11. Reinstall the control box and reconnect the wiring harness plugs.

12. Replace the access doors and restore the electrical supply to the unit.

WARNING: *If the condensate drain trap has been emptied during this cleaning procedure, refill it by following the instructions found in the "Start-Up and Adjustment" section on this instruction.*

13. Operate the unit for a minimum of five minutes and carefully check for any leaks in the condensate drain connections.

TROUBLESHOOTING

The following visual checks should be made before troubleshooting:

1. Check to see that the power to the furnace and the 50E47 control module is ON.

2. The manual shutoff valves in the gas line to the furnace must be open.

3. Make sure all wiring connections are secure.

4 Review the sequence of operation.

Start the system by setting the thermostat above the room temperature. Observe the system's response. Then use the Troubleshooting Table to check the system's operation.

Start the system by reading the upper left-hand box and then following the instructions in each box. If the condition described in the box is true (yes answer), go down to the next box. If the condition is not true (no answer), go to the box to the right. Continue checking and answering the questions in the boxes until the problem is explained and corrective action is described. After any maintenance or repair, the troubleshooting sequence should be repeated until normal system operation is obtained.

WARNING: *Do not try to repair controls. Replace defective controls with CES Source 1 Parts.*

WIRING DIAGRAM - 37 thru 56 MBH OUTPUT

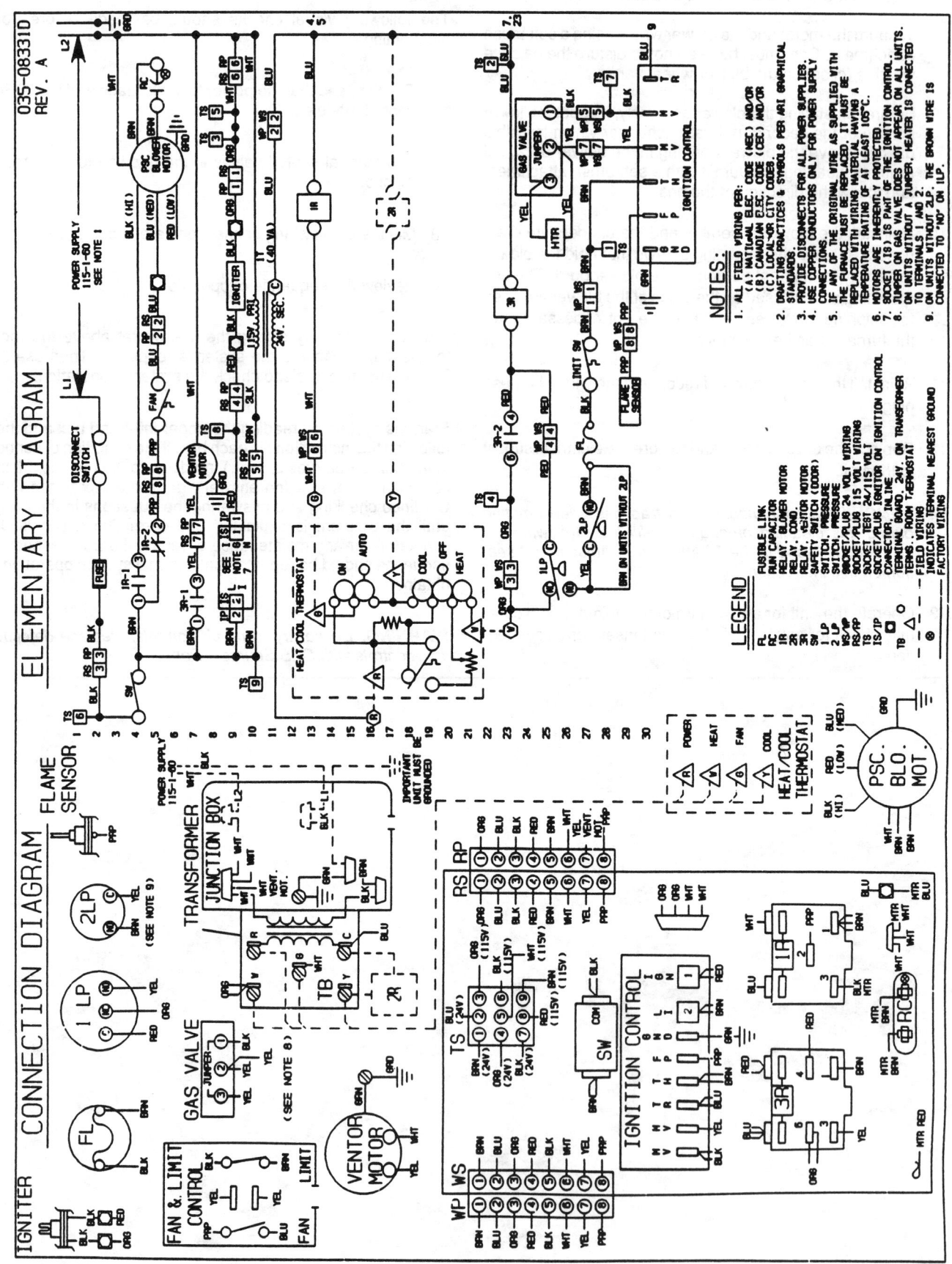

WIRING DIAGRAM - 75 thru 131 MBH OUTPUT

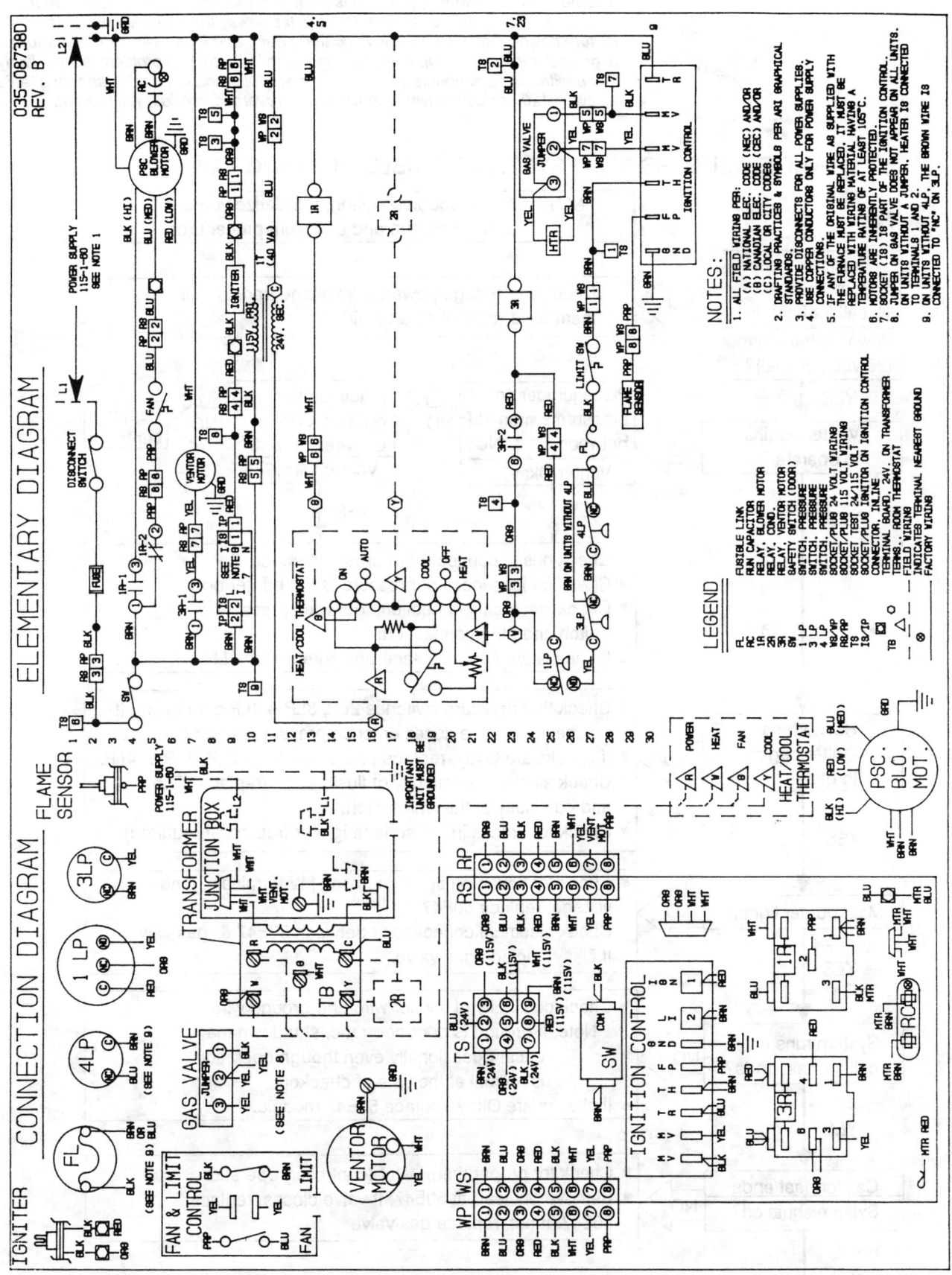

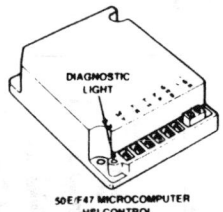

NOTE: *Models with diagnostic light have self-diagnostic capabilities. If the light on the module is on continuously, the fault is likely to be internal to the module. To make sure, interrupt the line or 25 volt thermostat power for a few seconds and then restore power. If internal fault is indicated again, and flame sensor is not shorted to ground, replace control. A flashing light also indicates the problem is most likely in the external components or wiring. Proceed following the troublshooting table below. If 120 vac power wiring to furance is reversed, module will lockout.*

WHITE-RODGERS 50E47 SYSTEM TROUBLESHOOTING TABLE

NOTE: Before troubleshooting, familiarize yourself with the startup and checkout procedures.

Turn thermostat (controller) to call for heat
Power at transformer (25 vac nominal)?

— NO → Check line voltage power, low voltage transformer thermostat (controller) and wiring

YES ↓

Venter begins to operate — NO →

Place jumper on pressure switch 1LP between "C" & "NC" Venter Okay? — NO → Place jumper on relay 3R between "1" & "3" Venter Okay? — NO → Replace venter

YES ↓ (1LP)   YES ↓ (3R)

- Check hoses connected to pressure switch 1LP.
- Check for looseness of pressure tap in burner box.
- Check electrical connections between pressure switch and plug connections.
- If checks are Okay, replace pressure switch 1LP.

YES ↓

Hot surface igniter begins to glow? — NO →

- Check that pressure switches 2LP, 3LP & 4LP are closed. If no check for loose hoses or loose pressure tap on burner box.
- If checks are Okay, replace pressure switch 2LP, 3LP or 4LP.
- Check electrical continuity of flusible link (replace if open) and continuity of the limit switch.
- Check for cracks in hot surface igniter (replace if required)

YES ↓

Main burner lights? — NO →

- Check for 25vac on 50E47 acrosss MV terminals. If no voltage, replace 50E47.
- Check electrical connections between 50E47 & gas valve. If Okay, replace gas valve.

YES ↓

System runs until call for heat ends? — NO →

- Check continuity of sensor wire and ground wire.
  Note: If ground is poor or erratic, shutdown may occur occasionally even though operation is normal at the time of checkout.
- If checks are Okay, replace 50E47 module.

YES ↓

Call for heat ends
System shuts off? — NO →

- Check for proper thermostat (controller) operation.
- Remove MV lead at 50E47. If valve closes, replace 50E47. If not, replace gas valve

YES ↓

Troubleshooting ends

Repeat procedure until troublefree operation is obtained.

Official Sponsor
of the 1992
U.S. Olympic Team

Heating and Air Conditioning

U S A

Official Sponsor
of the 1992
U.S. Olympic Team

36USC380

Stellar™ HIGH-EFFICIENCY
GAS-FIRED FURNACES
INDUCED-DRAFT UPFLOW, DOWNFLOW

035-09106

MODELS :

P3UC (STYLE A) UPFLOW 45 thru 106 MBH OUTPUT
P3CC (STYLE A) DOWNFLOW 45 thru 106 OUTPUT
P9UC (STYLE A) UPFLOW 45 thru106 MBH OUTPUT
P9CC (STYLE A) DOWNFLOW 45 thru 106 MBH OUTPUT

UPFLOW MODELS

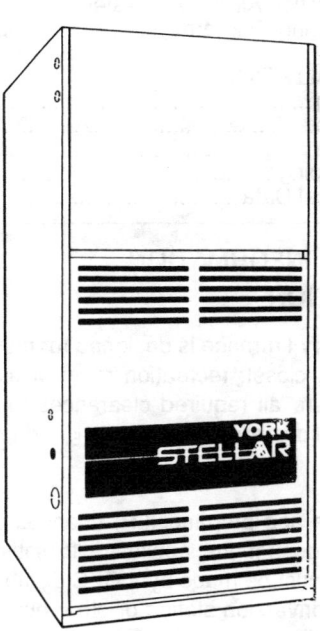

DOWNFLOW MODELS

FOR YOUR SAFETY

WHAT TO DO IF YOU SMELL GAS

- Do not try to light any appliance.

- Open windows.

- Do not touch any electrical switch;
  do not use any phone in your building.

- Extinguish any open flame.

- Immediately call your gas supplier from
  a neighbor's phone. Follow the gas
  supplier's instructions.

- If you cannot reach your gas supplier,
  call the fire department.

FOR YOUR SAFETY

Do not store or use gasoline or other flammable
vapors and liquids in the vicinity of this or any
other appliance.

WARNING: Improper installation, adjustment, alteration, service or maintenance can cause injury or property damage. Refer to this manual. For assistance or additional information, consult a qualified installer, service agency or the gas supplier.

TABLE OF CONTENTS

GENERAL INFORMATION

DESCRIPTION

This Category I furnace is designed for residential installation in a garage, closet, recreation room or alcove or any other location where all required clearances to combustibles and other restrictions are met. It is designed for natural gas-fired operation.

High altitude and propane (LP) changes or conversions required in order for the appliance to satisfactorily meet the application must be made by a CES distributor, or in Canada a certified conversion station or other qualified agency, using factory specified and/or approved parts.

The appliance shall be installed so that all electrical components are protected from water.

INSPECTION

As soon as a unit is received, it should be inspected for possible damage during transit. If the damage is evident, the extent of the damage should be noted on the carrier's freight bill. A separate request for inspection by the carrier's agent should be made in writing. Also, before installation, the unit should be checked for screws or bolts which may have loosened in transit. A sheet metal bracket, fastened between the blower assembly and the blower housing is used to support the blower assembly during shipment. Loosen two screws to remove bracket, then replace screws.

NOTES, CAUTIONS & WARNINGS

The installer should pay particular attention to the words: NOTE, CAUTION and WARNING. NOTES are intended to clarify or make the installation easier. CAUTIONS are given to prevent equipment damage. WARNINGS are given to alert the installer that personal injury and/or equipment or property damage may occur if installation procedures are not handled properly.

CAUTION: The cooling coil must be installed in the supply air duct downstream of the furnace. The furnace area must not be used as a broom closet or for any other storage

purposes, as a fire hazard may be created. Never store items such as the following on, near or in contact with the furnace.

1. *Spray or aerosol cans, rags, brooms, dust mops, vacuum cleaners or other cleaning tools.*

2. *Soap powders, bleaches, waxes or other cleaning compounds; plastic items or containers; gasoline, kerosene, cigarette lighter fluid, dry-cleaning fluids or other volatile fluid.*

3. *Paint thinners and other painting compounds.*

4. *Paper bags or other paper products.*

WARNING: Never operate the furnace with the blower door removed. To do so could result in serious personal injury and/or equipment damage.

WARNING: Each furnace in this series is a Category I furnace, suitable for common venting with another non-induced, gas-fired appliance. Each furnace in this series is a Category III furnace when vented horizontally using the high temperature plastic pipe specified in Application Date 650.64-AD2V- "Venting Guide for High Efficiency Induced Draft Upflow/Downflow Furnaces."

If this furnace is replacing a common-vented furnace, it may be necessary to resize the existing vent line and chimney to prevent oversizing problems for the new combination of units. See National Fuel Gas Code (ANSI Z223.1-), or CAN1-B149.1 or .2, Installation Code (latest editions).

The following steps shall be followed with each appliance remaining connected to the common venting system placed in operation, while the other appliances remaining connected to the common venting system are not in operation.

1. Seal any unused openings in the common venting system.

2. Visually inspect venting system for proper size and horizontal pitch and determine there is no blockage or restriction, leakage, corrosion or other deficiencies which could cause an unsafe condition.

Central Environmental Systems

3. Insofar as is practical, close all building doors and windows and all doors between the space in which the appliances remaining connected to the common venting system are located and other spaces of the building. Turn on clothes dryers and any other appliances not connected to the common venting system. Turn on any exhaust fans, such as range hoods and bathroom exhausts so they will operate at maximum speed. Do not operate a summer exhaust fan. Close fireplace dampers.

4. Follow the lighting instructions. Place the appliance being operated in operation. Adjust thermostat so appliance will operate continuously.

5. After it has been determined that each appliance remaining connected to the common venting system properly vents when tested as outlined above, return doors, windows, exhaust fans, fireplace dampers and any other gas-burning appliance to their previously conditions of use.

6. If improper venting is observed during any of the previous tests, the common venting system must be corrected.

7. Any corrections to the common venting system must be in accordance with the National Fuel Gas Code Z223.1, or CAN1 B149.1 or .2 Installation Code (latest editions). If the common vent system must be resized, it should be resized to approach the minimum size as determined using the appropriate tables in Appendix G of the above codes.

LIMITATIONS AND LOCATION

This furnace should be installed in accordance with all national and local building/safety codes and requirements, or, in the absence of local codes, with the National Fuel Gas Code, ANSI Z223.1, or CAN1 B149.1 or .2 Installation Code, local plumbing or waste water codes, and other applicable codes.

WARNING: *Upflow furnaces shall not be installed directly on carpeting, tile or other combustible material other than wood flooring. For downflow furnaces, an accessory combustible floor base is available to allow installation on combustible flooring.*

The size of the unit should be based on an acceptable heat loss calculation for the structure.

Check the rating plate to make certain the unit is equipped for the type of gas supplied, and proper electrical characteristics are available.

For installations above 2,000 feet, reduce input 4% for each 1,000 feet above sea level.

DO NOT INSTALL THIS UNIT IN A MOBILE HOME.

A furnace installed in a residential garage shall be located so that all burners and burner ignition devices are located no less than 18" above the garage floor, and located or protected to prevent damage by vehicles.

Allow clearances from combustible materials as listed under "CLEARANCES TO COMBUSTIBLES" ensuring that service access is allowed for both the burners and blower.

When the furnace is used in conjunction with a cooling coil, the furnace must be installed parallel with or on the upstream side of the cooling unit to avoid condensation in the primary heat exchanger. When a parallel flow arrangement is used, the dampers or other means used to control air flow shall be adequate to prevent chilled air from entering the furnace, and if manually operated, must be equipped with means to prevent operating of either unit unless the damper is in the full heat or cool position.

SPECIFIC UNIT INFORMATION

CLEARANCES TO COMBUSTIBLES

Minimum clearances from combustible construction are in inches:

```
Top.................................................1
Front ..............................................6
Vent Piping .....................................6*
Rear ..............................................0
Sides ............................................0
  *May be 1" for Type B-1/BH vent
```

CLEARANCES FOR ACCESS

Ample clearances should be provided to permit easy access to the unit. The following minimum clearances are recommended:

1. Twenty-four (24) inches between the front of the furnace and an adjacent wall or another appliance, when access is required for servicing and cleaning.

2. Eighteen (18) inches at the side where access is required for passage to the front when servicing or for inspection or replacement of flue/vent connections.

NOTE: *In all cases, accessibility clearances shall take precedence over clearances for combustible materials where accessibility clearances are greater.*

NOMENCLATURE

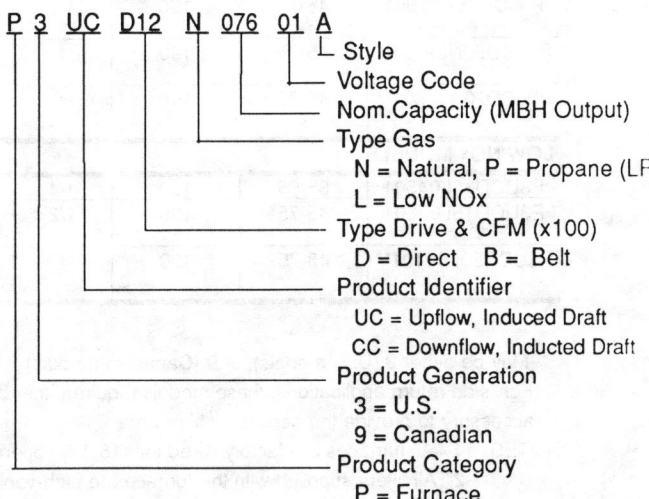

P 3 UC D12 N 076 01 A

- └ Style
- Voltage Code
- Nom.Capacity (MBH Output)
- Type Gas
  - N = Natural, P = Propane (LP)
  - L = Low NOx
- Type Drive & CFM (x100)
  - D = Direct   B = Belt
- Product Identifier
  - UC = Upflow, Induced Draft
  - CC = Downflow, Inducted Draft
- Product Generation
  - 3 = U.S.
  - 9 = Canadian
- Product Category
  - P = Furnace

DIMENSIONS–UPFLOW MODELS

FRONT   LEFT SIDE   RIGHT SIDE

TOP IMAGE

All dimensions are in inches, and are approximate.
Certified dimensions are available upon request.

MODEL	A	B	C	D (U.S.)	D (Canada)	E
P*UCD12N04501	16-1/4	12	15	3	4²	5-5/32
P*UCD08N06101	16-1/4	12	15	4¹	4²	5-5/32
P*UCD12N06101	16-1/4	12	15	4¹	4²	5-5/32
P*UCD12N07601	22-1/4	18	21	4¹	4²	8-5/32
P*UCD16N07601	22-1/4	18	21	4¹	4²	8-5/32
P*UCD20N09101	22-1/4	18	21	4¹	5²	8-5/32
P*UCD20N10601	26-1/4	22	25	4¹	5²	10-5/32
P3UCD12L04501	16-1/4	12	15	3	–	5-5/32
P3UCD16L07601	22-1/4	18	21	4¹	–	8-5/32
P3UCD20L09101	22-1/4	18	21	4¹	–	8-5/32

* May be either 3 (U.S. models), or 9 (Canadian models)
1 A 3" to 4" adapter is supplied for U.S. units and must be field installed.
2 A flue adapter is supplied for Canadian models when vertically vented, and must be field installed.

TABLE 1 –RATINGS & PHYSICAL DATA–UPFLOW MODELS

Model	Air Temp Rise °F	Maximum Outlet Temp. °F	Blower		Filter Size		Unit Amps	Max. Over-Current Protect.	Min. Wire Size (AWG) @ 75 Ft. One Way
			HP	Size	Supplied	Suggested Bottom			
P*UCD12N04501	35-65	190	1/4	10–7	16X26 Side	16x26	<15	15	14
P*UCD08N06101	45-75	190	1/4	9–6	16X26 Side	16x26	<15	15	14
P*UCD12N06101	45-75	190	1/3	10–7	16X26 Side	16x26	<15	15	14
P*UCD12N07601	45-75	190	1/3	10–9	20X26 Btm See Note 3	20x26	<15	15	14
P*UCD16N07601	45-75	190	1/2	10–10	20X26 Btm See Note 3	20x26	<15	15	14
P*UCD20N09101**	45-75	190	1	10–10	20X26 Btm 16x26 Side	20x26	<17	20	12
P*UCD20N10601**	45-75	190	1	10–10	24X26 Btm 16x26 Side	24x26	<17	20	12
LOW NOx MODELS									
P3UCD12L04501	35-65	190	1/4	10–7	16x26 Side	16x26	<15	15	14
P3UCD16L07601	45-75	190	1/2	10–10	20x26 Btm See Note 3	20x26	<15	15	14
P3UCD20L09101**	45-75	190	1	10–10	20x26 Btm 16x26 Side	20x26	<17	20	12

* May be either 3 (U.S. models), or 9 (Canadian models)
** For side return applications, these models require either both side inlets or the use of the optional dual filter side return frame accessory to provide the necessary filter area
NOTES: 1. All furnaces are factory wired for 115-1-60 operation.
2. All filters supplied with the furnace are high-veolocty, cleanable type
3. Recommended filter size for side return is 16x26.

Central Environmental Systems

DIMENSIONS–DOWNFLOW MODELS

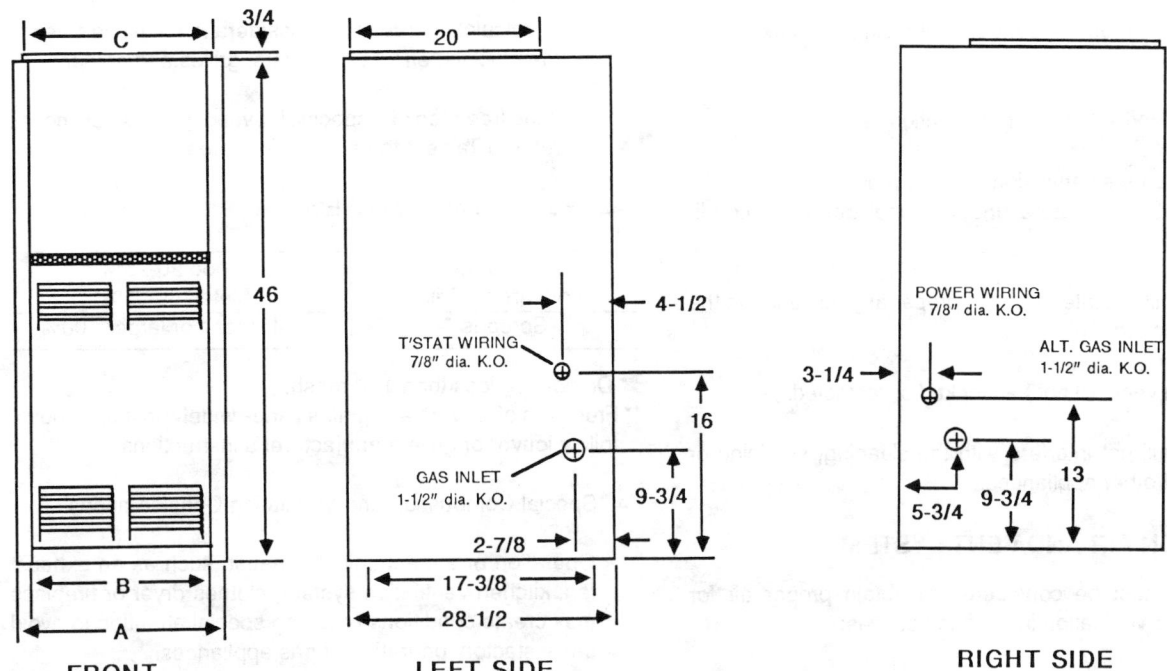

FRONT   LEFT SIDE   RIGHT SIDE

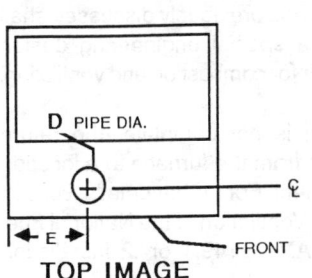

TOP IMAGE

MODEL	A	B	C	D (U.S.)	D (Canada)	E
P*CCD08N04501	16-1/4	13-1/2	15	3	4[2]	5-5/32
P*CCD12N06101	16-1/4	13-1/2	15	4[1]	4[2]	5-5/32
P*CCD12N07601	22-1/4	19-1/2	21	4[1]	4[2]	8-5/32
P*CCD16N07601	22-1/4	19-1/2	21	4[1]	4[2]	8-5/32
P*CCD16N09101	22-1/4	19-1/2	21	4[1]	4[2]	8-5/32
P*CCD20N10601	26-1/4	24-1.2	25	4[1]	5[2]	10-5/32
P3CCD08L04501	16-1/4	13-1/2	15	3	–	5-5/32
P3CCD12L06101	16-1/4	13-1/2	15	4[1]	–	5-5/32
P3CCD16L07601	22-1/4	19-1/2	21	4[1]	–	8-5/32
P3CCD20L10601	26-1/4	24-1/2	25	4[1]	–	10-5/32

* May be either 3 (U.S. models), or 9 (Canadian models)
[1] A 3" to 4" adapter is factory supplied for U.S. units
[2] A flue adapter is supplied for Canadian models when vertically vented, and must be field installed.

All dimensions are in inches, and are approximate. Certified dimensions are available upon request.

TABLE 2 – RATINGS & PHYSICAL DATA–DOWNFLOW MODELS

Model	Air Temp Rise °F	Maximum Outlet Temp °F	Blower HP	Blower Size	Filter Size	Total Unit Amps	Max. Over-Current Protect.	Min. Wire Size (AWG) @ 75'
P*CCD08N04501	45-75	190	1/5	10–6	(2) 14X20	<15	15	14
P*CCD12N06101	45-75	190	1/4	10–8	(2) 14X20	<15	15	14
P*CCD12N07601	45-75	190	1/3	10–9	(2) 14X20	<15	15	14
P*CCD16N07601	45-75	190	1/2	10–10	(2) 14X20	<15	15	14
P*CCD16N09101	45-75	190	1/2	10–10	(2) 14X20	<15	15	14
P*CCD20N10601	45-75	190	1	11–10	(2) 14X20	<17	20	12
LOW NOx MODELS								
P3CCD08L04501	45-75	190	1/5	10–6	(2) 14X20	<15	15	14
P3CCD12L06101	45-75	190	1/4	10–8	(2) 14X20	<15	15	14
P3CCD16L07601	45-75	190	1/2	10–10	(2) 14X20	<15	15	14
P3CCD20L10601	45–75	190	1	11–10	(2) 14X20	<17	20	12

* May be either 3 (U.S. models), or 9 (Canadian models).
NOTES: 1. Wire size based on copper conductors, 60°C, 3% voltage drop.
2. Continous return air temperature should not be below 55°F.

The furnace should be located:

1. Where a minimum amount of vent piping and elbows will be required.

2. As centralized with the air distribution as possible.

3. In an area where ventilation facilities provide for safe limits of ambient temperature under normal operating conditions.

4. Where it will not interfere with proper air circulation in the confined space.

5. Where the vent will not be blocked or restricted.

6. Where it will not interfere with the cleaning, servicing or removal of other appliances.

COMBUSTION AIR AND VENT SYSTEM

The following must be considered to obtain proper air for combustion and ventilation in confined spaces:

1. Air Source from Inside the Building -

   Two permanent openings, one within 12 inches of the top of the confined space and one within 12 inches of the bottom, shall each have a free area of not less than one square inch per 1,000 Btuh of total input rating of all appliances located in the space. The openings shall communicate freely with interior areas having adequate infiltration from the outside.

NOTE: *At least 100 square inches free area shall be used for each opening.*

2. Air Source from Outdoors -

   Two permanent openings, one within 12 inches of the top of the confined space and one within 12 inches of the bottom, shall communicate directly, or by means of ducts, with the outdoors or to such crawl or attic spaces that freely communicate with the outdoors.

   a. Vertical Ducts - Each opening shall have a free area of not less than one square inch per 4,000 Btuh of total input of all appliances located in the space.

EXAMPLE:

$$\frac{\text{Total Input of All Appliances}}{4000} = \text{Square Inches Free Area}$$

   b. Horizontal Ducts - Each opening shall have a free area of not less than one square inch per 2,000 Btuh of total input of all appliances located in the space.

NOTE: *Ducts shall have the same cross-sectional area as the free area in the opening to which they are connected. The minimum dimension of rectangular ducts shall be three inches.*

3. Louvers, Grilles and Screens

   a. In calculating free area, consideration must be given to the blocking effects of louvers, grilles and screens.

   b. If the free area of a specific louver or grille is not known, refer to Table 3 to estimate free area.

TABLE 3 - ESTIMATED FREE AREA

Wood or Metal Louvers or Grilles	Wood 20-25%* Metal 60-75%*
Screens**	1/4" mesh or larger 100%

* Do not use less than 1/4" mesh.
** Free area of louvers and grilles varies widely; installer should follow louver or grille manufacturer's instructions.

4. Special Combustion and Ventilation Considerations

   Operation of a mechanical exhaust, such as an exhaust fan, kitchen ventilation system, clothes dryer or fireplace may create conditions requiring special attention to avoid unsatisfactory operation of gas appliances.

The size of combustion air openings previously discussed shall not necessarily govern when a special engineering design ensures an adequate supply of air for combustion and ventilation.

Where the return duct system is not complete, the return connection must be run full size from the furnace to a location outside the utility room or basement. For further details, consult Section 5.3, Air combustion and Ventilation of the National Fuel Gas Code, ANSI Z223.1 or CAN1 B149.1 or .2 Installation Code-latest editions.

UNIT INSTALLATION

VENTING

The furnace should be vented in accordance with the National Gas Code, ANSI Z223.1 or CAN1 B149.1 or .2 Installation Code (latest editions), and requirements or codes of the local utility or other authority having jurisdiction.

This furnace may be vertically or horizontally vented. Refer to Application Data (Form 650.64-AD2V) for information on venting procedures and requirements for this furnace.

DUCTWORK - GENERAL

The duct system's design and installation must:

1. Handle an air volume appropriate for the served space and within the operating parameters of the furnace specifications.

2. Be installed in accordance with standards of NFPA (National Fire Protection Association) as outlined in NFPA pamphlets 90A and 90B.

3. Create a closed duct system. The supply system must be connected to the furnace outlet and the return duct system must be connected to the furnace inlet. Both supply and

Central Environmental Systems

return duct systems must terminate outside the space containing the furnace.

4. Generally complete a path for heated or cooled air to circulate through the air conditioning and heating equipment and to and from the conditioned space

After the unit is in the desired position, fasten the supply ductwork to the furnace duct flanges. A removable access panel should be provided in the outlet duct such that smoke or reflected light would be observable inside the casing to indicate the presence of leaks in the heat exchanger. This access cover shall be attached in such a manner as to prevent leaks. Flexible duct connectors are recommended to connect both the supply and return ducts to the furnace.

Ductwork - Upflow Models

Return air ductwork may be connected to an upflow furnace in one of the following two ways:

1. Bottom Return - Before attaching the ductwork to the furnace bottom, see the "Filters" section of this instruction.

2. Side return - Cut a hole in the side panel of the furnace using the right-angle markings (See Figure 1) as a guide for position and size of the opening. Install a single side return filter frame accessory if one is required. If this accessory is not needed, the ductwork can be fastened directly to the furnace opening.

NOTE: Some accessory side filter packages require cutting a slightly larger opening.

Where the return duct system is not complete, the return connection must run full size to a location outside the utility room or basement. For further details, consult Section 5.3 (Air for Combustion and Ventilation) of the National Fuel Gas Code, ASNI Z223.1, or Can1 B149.1 or .2, Installation Code-latest editions.

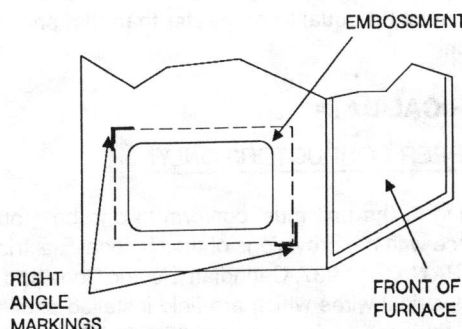

FIGURE 1 - SIDE RETURN CUTOFF MARKINGS

Ductwork - Downflow Models

Return air ductwork flanges are provided at the top of the unit. Before connecting the return air ductwork to the furnace, refer to the "Filters" section of this instruction. The supply air ductwork connects to the bottom of the furnace.

Central Environmental Systems

FILTERS

Upflow Models

16-1/4" Width Furnaces

All 16-1/4" wide furnaces are shipped with filters mounted on the left side. Filters may be located to the bottom or right side as follows:

The wire filter retainer must be moved if the return air application requires moving of the filter from the side to the bottom location, or vice versa. When relocating filters, it may be necessary to trim the filter to the proper size.

The ends of the retainer are attached to the rear panel in two metal loops (See Figure 2). The ends must be squeezed together to free them from the loops. The retainer may then be moved to the new location, and the ends inserted in the loops on the rear panel at the new location. Loops are provided for retainer location to accommodate filter application on the bottom or either side of the furnace.

To remove a filter from the bottom location, push the closed end of the filter retainer to the left until it clears the lip on the front of the furnace base, which acts as a catch for the retainer. When the retainer is clear of the lip, lift up. The retainer will pivot in the loops. This will expose the filter to allow removal. To re-install the filter, simply reverse this procedure.

To remove a filter from the side location, push the closed end of the filter down until it clears the flange on the side panel, which acts as a catch for the retainer. When the retainer is clear of the flange, it will pivot in the loops. Swing the retainer toward the center of the furnace. This will expose the filter to allow removal. To re-install the filter, simply reverse this procedure.

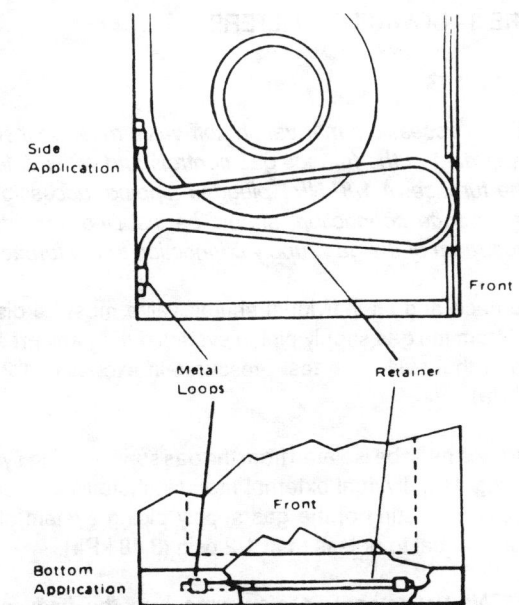

FIGURE 2 - FILTER RETAINER

22-1/4" and 26-1/4" Width Furnaces

22-1/4" and 26-1/4" wide furnaces are shipped with filters mounted in the bottom and left side.

Downflow Models

Two 14" x 20" x 1" permanent washable filters are supplied with each unit. Downflow furnace filters are installed above the furnace, extending into the ductwork as shown in Figure 3. Branch ducts must enter above the height of dimension FH.

The filter rack should be secured to the center of the front and rear flanges at the furnace's return air opening. Drill a hole through the rear duct flange into the filter rack and secure it with a sheet metal screw.

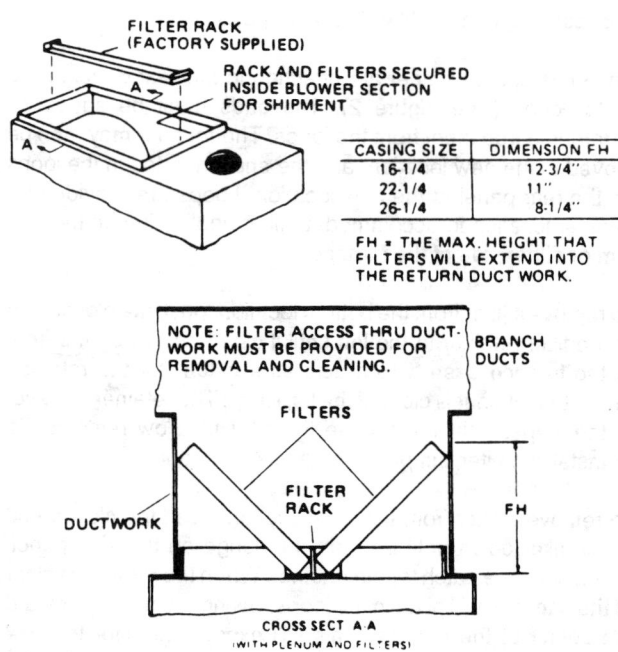

CASING SIZE	DIMENSION FH
16-1/4	12-3/4"
22-1/4	11"
26-1/4	8-1/4"

FH = THE MAX. HEIGHT THAT FILTERS WILL EXTEND INTO THE RETURN DUCT WORK.

FIGURE 3 - DOWNFLOW FILTERS

GAS PIPING

NOTE: *An accessible manual shutoff valve must be installed upstream of the furnace gas controls and within 6 feet of the furnace. A 1/8" NPT plugged tapping, accessible for test gauge connection, should be installed immediately upstream of the gas supply connection to the furnace.*

The furnace and its individual shutoff valve must be disconnected from the gas supply piping system during any pressure testing of that system at test pressures in excess of 1/2 psig (3.48 kPa).

The furnace must be isolated from the gas supply piping system by closing its individual external manual shutoff valve during any pressure testing of the gas supply piping system at test pressures equal to or less than 1/2 psig (3.48 kPa).

CAUTION: *Never apply a pipe wrench to the body of the combination automatic gas valve. A wrench must be placed on the projection or wrench boss of the valve when installing piping to it.*

Gas piping may be connected from either side of the furnace. Sizing and installation of the supply gas line should comply with the local utility requirements. The gas supply should be a separate line, installed in accordance with the National Fuel Gas Code, ANSI Z223.1, or CAN1 B149.1 or .2 Installation Codes (latest editions).

Some utility companies, or local codes, require pipe sizes larger than the maximum sizes listed. Using the properly sized wrought iron, approved flexible or steel pipe, make gas connections to the unit. Installation of a drop leg and ground union is required (See Figure 4).

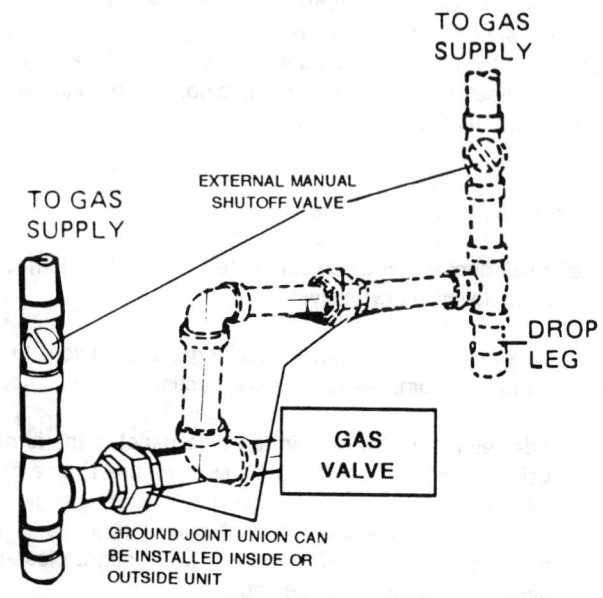

FIGURE 4 - GAS PIPING

WARNING: *Compounds used on threaded joints of gas piping must be resistant to the action of liquified petroleum gases. After connections are made, leak-test all pipe connections.*

WARNING: *Do not use an open flame or other source of ignition for leak testing. Set the manual gas valve to the off position.*

For the purpose of input adjustment, the minimum inlet gas pressure must be equal to or greater than that shown on the rating label.

ELECTRICAL DATA

USE COPPER CONDUCTORS ONLY!

Field wiring to the unit must conform to and be grounded in accordance with the provisions of the National Electrical Code ANSI/NFPA No. 70-1987, Canadian Electric Code and /or local codes. Electrical wires which are field installed shall conform with the temperature limitation for 63°F/35°C rise wire when installed in accordance with instructions. Specific electrical data is given on the furnace rating plate and Tables 1 and 2 of this instruction.

ELECTRICAL CONNECTIONS

The furnace's control system depends on correct polarity of the power supply. Connect the power supply as shown on the unit wiring label on the inside of the blower compartment door.

Central Environmental Systems

Provide a power supply separate from all other circuits. Install overcurrent protection and disconnect switch per local/national electrical codes. The switch should be reasonable close to the unit for convenience in servicing. With the disconnect switch in the OFF position, check all wiring against the unit wiring label. Also, see the wiring diagram in this instruction.

Install the field-supplied thermostat. The thermostat instructions for wiring are packed with the thermostat. With the thermostat set in the OFF position and the main electrical source disconnected, complete the low-voltage wiring from the thermostat to the terminal board on the low-voltage transformer. Set the heat anticipator on the thermostat to 1.24 amps.

The 24-volt, 40 VA transformer is sized for the furnace components only, and should not be connected to auxiliary devices such as humidifiers, air cleaners, etc.

NOTE: Some thermostats do not have adjustable anticipators. On such thermostats, adjust cycle rate to prevent possible short on/off cycles.

SAFETY CONTROLS

Blower Door Safety Switch
This unit is equipped with an electrical interlock switch mounted in the blower compartment. This switch interrupts all power at the unit when the panel covering the blower compartment is removed.

WARNING: Blower and burner must never be operated without the blower panel in place.

Electrical supply to the unit is dependent upon the panel that covers the blower compartment being in place and properly positioned.

CAUTION: Main power supply to the unit must still be interrupted at the main power disconnect switch before any service or repair work is to be done to the unit. Do not rely upon the interlock switch as a main power disconnect.

Rollout Switch Control
This control is mounted on the burner assembly. If the temperature in the burner compartment exceeds its set point, the igniter control and the gas valve are de-energized. This is manual reset control and must be reset before operation can continue. The operation of this control indicates a malfunction in the combustion air blower or a blocked vent pipe connection.

Pressure Switch
This furnace is supplied with a differential pressure switch which monitors the flow through the furnace and venting system. This switch de-energizes the ignition control module and the gas valve if any of the following conditions are present:

1. Blockage of internal flue gas passageways.

2. Blockage of vent piping.

3. Failure of combustion air blower/motor.

Vent Safety Switch (Canadian Models Only)
This unit is equipped with a vent safety system. This vent safety switch de-energizes the pressure switch when a blocked vent condition is present. Refer to venting instructions for location and installation.

Auxiliary Limit (Downflow models only)
The auxiliary limit is used to provide protection from excessive temperatures under reversed air flow conditions, or blower or motor failure.

START-UP AND ADJUSTMENTS

The initial start-up of the furnace requires the following additional procedures.

1. When the gas supply is initially connected to the furnace, the gas piping may be full of air. In order to purge this air, it is recommended that the ground joint union be loosened until the odor of gas is detected. When gas is detected, immediately retighten the union and check for leaks. Allow five (5) minutes for any gas to dissipate before continuing with the start-up procedure.

2. All electrical connections made in the field and in the factory should be checked for proper tightness.

IGNITION SYSTEM CHECKOUT/ADJUSTMENT

1. Turn control system power ON, and turn gas supply OFF.

2. Check the control module operation as follows:

   a. Set thermostat above room temperature to call for heat.

   b. Watch for the hot surface igniter to glow in burner compartment.

   c. Turn the thermostat down to end the call for heat.

3. Turn the gas supply ON.

4. Set the thermostat above room temperature to call for heat.

5. Start the system as follows:

   a. The venter comes up to proper speed and the hot surface igniter begins to glow for 45 seconds.

   b. After this warm-up period, the main valve is energized for four (4) seconds as the hot surface igniter is de-energized.

   c. The burners light and are sensed by the remote sensor.

NOTE: Burner ignition may not be satisfactory until the gas input and combustion air have been adjusted.

6. With the main burner in operation, paint the pipe joints and valve gasket lines with a rich soap and water solution. Bubbles indicate gas leakage. To stop the leaks, tighten all joints and screws. If the leak persists, replace the component.

WARNING: DO NOT omit this test! NEVER use a flame to check for gas leaks!

ADJUSTMENT OF MANIFOLD GAS PRESSURE

Measure the manifold gas pressure. The manifold pressure should be set at 3.5" W.C. for natural gas.

This manifold pressure should easily be attainable when the line pressure is between the maximum and minimum supply pressures stated on the rating plate. The minimum supply pressure listed on the rating plate should only be used if a higher pressure within the allowable range is not attainable.

Refer to Figure 5 (Gas Valve) for plug location for installing pressure tap to check gas pressure to burners.

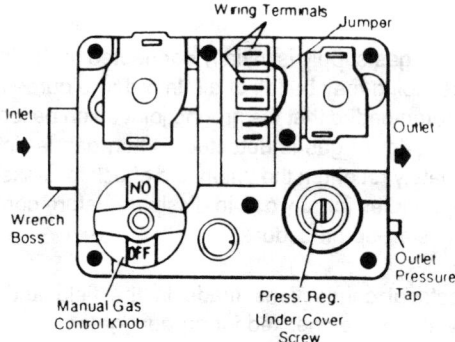

FIGURE 5 - GAS VALVE (White-Rodgers 36E01)

WARNING: The stated gas pressure should always be used. If the pressure is too high, overfiring and premature failure of the heat exchanger could occur. If the pressure is too low, sooting and eventual clogging of the heat exchanger could occur.

NOTE: The screw-off cap for the pressure regulator must be removed entirely to gain access to the adjustment screw. Loosening or tightening the cap does not adjust gas flow.

The gas flow may be adjusted by turning the pressure regulator adjustment screw clockwise to increase the pressure or counter-clockwise to decrease the pressure.

WARNING: Once the correct gas pressure to the burners has been established, turn the gas valve knob to OFF and turn the electrical supply switch OFF; then, remove the pressure tap at the gas valve and re-install the plug using a compound (on the threads) resistant to the action of LP gases. Replace cap on regulator adjusting screw.

Turn the electrical and gas supplies back on, and with the burners in operation, check for gas leakage around the plug with a soap and water solution.

ADJUSTMENT OF PRIMARY AIR

The main burners should be in operation for 15 minutes before making the primary air adjustment. The burner flame should not contain any yellow color. With the furnace operating at full input, adjust the primary air of the burners as follows:

Natural Gas - All models are shipped with the air shutters partially open.

Local variations in the gas supply may require changes in the settings described above. To change the air shutter settings, use the following procedure:

1. Remove the front access door.

2. Loosen shutter locking screw.

3. Adjust the air shutters with power to the unit "OFF", and retighten the locking screws.

4. Observe through the observation port to see if all flames are now blue in color. See Figure 6 for proper flame pattern. If yellow flames ar still visible, repeat steps 2 thru 4.

5. Cycle the burners on and off a few times to verify the burners are lighting promptly and properly.

6. Replace the front access door.

CHECKING GAS INPUT

1. Turn off all other gas appliances connected to gas meter.

2. With the furnace turned on, measure the time needed for one revolution of the hand on the smallest dial on the meter. A typical domestic gas meter usually has a 1/2 or 1 cubic foot test dial.

3. Using the number of seconds for each revolution and the size of the test dial increment, find the cubic feet of gas consumed per hour from Table 4.

TABLE 4 – GAS RATE (CUBIC FEET PER HOUR)

Seconds for One Revolution	Size of Test Dial	
	1/2 cubic foot	1 cubic foot
10	180	360
12	150	300
14	129	257
16	113	225
18	100	200
20	90	180
22	82	164
24	75	150
26	69	138
28	64	129
30	60	120
32	56	113
34	53	106
36	50	100
38	47	95
40	45	90
42	43	86
44	41	82
46	39	78
48	37	75
50	36	72
52	35	69
54	34	67
56	32	64
58	31	62
60	30	60

NOTE: To find the Btuh input, multiply the number of cubic feet of gas consumed per hour by the BTU content of the gas in your particular locality. Contract your gas company for this information, as it varies widely from city to city.

EXAMPLE: It is found by measurement that it takes 26 seconds for the hand to turn on the 2 cubic foot dial to make a revolution with only a 120,000 Btuh furnace running. Using this information, locate 26 seconds in the first column of Table 4. Read across to the column headed "1 Cubic Foot" where you will see that 138 cubic feet of gas per hour ar consumed by the furnace at that rate. Multiply 138 by 850 (the BTU rating of the gas obtained from the local gas company). The result is 117,300 Btuh, which is close to the 120,000 Btuh rating of the furnace.

CAUTION: Be sure to relight any gas appliances that were turned off at the start of this input check.

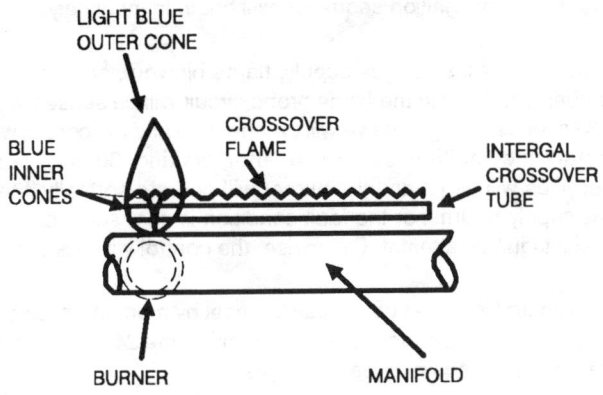

FIGURE 6 - PROPER FLAME APPEARANCE

ADJUSTMENT OF TEMPERATURE RISE

The temperature rise, or temperature difference between the return air and the heated air from the furnace, must be within the range shown on the furnace rating plate and within the application limitations shown in Tables 1 and 2. After the temperature rise has been determined, the cfm can be calculated.

After about 20 minutes of operation, determine the furnace temperature rise. Take readings of both the return air and the heated air in the ducts, about six feet from the furnace where they will not be affected by radiant heat. Increase the blower speed to decrease the temperature rise; decrease the blower speed to increase the rise.

All direct-drive blowers have multi-speed motors. Refer to the unit wiring diagram and connect the blower motor for the desired sped. The blower motor speed taps are located in the control box in the blower compartment.

ADJUSTMENT OF FAN CONTROL SETTINGS

Place a thermometer in the heated air duct, about six feet from

the furnace, where it won't be affected by radiant heat. Usually the fan control is set so that the thermometer reads about 125 F when the blower starts, and about 85 F when it stops.

The Fan On setting of the fan control must be high enough to allow the air in the furnace to be heated enough so that no cold air is blown into the heated space, but no so high that the furnace might be damaged by excessive heat.

To adjust the Fan On setting (Upflow Models Only):

1. Turn the furnace on.

2. Read the thermometer when the blower starts.

3. If this temperature is too high when the blower starts, lower the fan on setting. If the temperature is too low, raise the setting.

4. If adjustments are made to the fan on setting, check the operation of the furnace by repeating the previous steps.

The Fan Off setting must be low enough to adequately cool the furnace, but not so low that cold air is blown into the heated space.

To adjust the Fan Off setting:

1. Turn the furnace on.

2. Let the furnace operate for 20 minutes.

3. Turn the furnace off.

4. Read the thermometer when the blower stops.

5. If this temperature is too high when the blower stops, lower the fan off setting. If the temperature is too low, raise the setting.

6. If adjustments are made to the fan off setting, check the operation of the furnace by repeating the previous steps.

CAUTION: When fan on and fan off adjustments are made, be careful not to rotate the fan control dial. If the dial is allowed to rotate, the control could be damaged and operate erratically.

CHECKING LIMIT CONTROL

With the main burners operating, cover all return air grilles with paper to restrict the flow of return air. In a few minutes the burners should be shut off by the limit control. Remove the return air restrictions. The main blower will cool the unit. The limit switch should reclose and the main burners relight after the proper re-start period.

Central Environmental Systems

OPERATION AND MAINTENANCE

SEQUENCE OF OPERATION

Hot Surface Ignition System

WARNING: Do not attempt to relight this furnace by hand(with a match or any other means). There may be a potential shock hazard from the components of the hot surface ignition system. The furnace can only be lit automatically by its hot surface ignition system.

The following describes the sequence of operation of the furnace. Refer to the schematic wiring diagram (pages 15 thru 18) for component locations.

CONTINUOUS BLOWER

On cooling/heating units with fan switch, when the fan switch is set in the ON position, a circuit is completed between terminals R and G of the thermostat. This energizes the 1R relay. Contact 1R-1 closes and contact 1R-2 opens. The motor is energized through the black, high speed tap. The blower then operates on high speed.

INTERMITTENT BLOWER

When the system is set on HEAT and the fan switch is set on AUTO, and the room thermostat calls for heat, a circuit is completed between terminals R and W of the thermostat. This energizes the venter relay 3R, which energizes the venter. When the proper amount of combustion air is being provided, a pressure switch (1LP) activates the 50E47 ignition control. The rollout switch control, primary limit and auxiliary limit are also in this circuit and must be in the closed position for the ignition control to be activated.

The 50E47 ignition control provides a 45-second warm-up period. The gas valve then opens for four seconds.

As gas starts to flow and ignition occurs, the flame sensor begins its sensing function. If a flame is detected within four seconds after ignition, normal furnace operation continues until the thermostat circuit between terminals R and W is opened. After approximately 60 seconds (or the supply air temperature reaches 155° to 125°F), the fan switch closes.

When the thermostat opens, the venter is de-energized, along with the ignition control. With the ignition control de-energized, the gas flow stops and the burner flames are extinguished.

The blower motor continues to operate until the supply air temperature drops to between 85°and 100°F. When this occurs, the fan switch opens, de-energizing the blower motor. The heating cycle is then complete, and the unit is ready for the start of the next heating cycle.

If flame is not detected in the four second sensing period, the gas valve is de-energized. The 50E47 control is equipped with a re-try option. This provides a 60-second wait following an unsuccessful ignition attempt (flame not detected). After the 60 second wait, the ignition sequence is restarted with an additional 10 seconds of igniter warm-up time. If this ignition attempt is unsuccessful, one more re-try will be made before lockout.

50E47 SERIES HOT SURFACE IGNITION CONTROL

All White-Rodgers 50E47 controls will repeat the ignition sequence for a total of five recycles is flame is lost within the first 10 seconds of establishment.

If flame is established for more than 10 seconds after ignition, the controller will clear the ignition attempt (re-try) counter. If flame is lost after 10 seconds, it will restart the ignition sequence. This can occur a maximum of five times.

During burner operation, a momentary loss of power of 50 milliseconds or longer will drop out the main gas valve. When power is restored, the gas valve will remain de-energized, and a restart of the ignition sequence will begin immediately.

A momentary loss of gas supply, flame blowout, or a shorted or open condition in the flame probe circuit will be sensed with 0.8 seconds. The gas valve will de-energize and the control will restart the ignition sequence after waiting 60 seconds. Recycles will begin and the burner will operate normally if the gas supply returns, or the fault condition is connected prior to the last ignition attempt. Otherwise, the control will lock out.

If the control is locked out, it may be reset by momentary power interruption of 1/20 second or longer. Either the 24-volt thermostat or line voltage may be interrupted.

MAINTENANCE

Air Filters

The filters should be checked periodically for dirt accumulation. Dirty filters greatly restrict the flow of air and overburden the system.

Clean the filters at least every three months. See the section titled "Filters" for filter removal instructions. On new construction, check the filters every week for the first four weeks. Inspect the filters every three weeks after that, especially if the system is running constantly.

All filters supplied with the furnace are the high-velocity, cleanable type. Clean these filters by washing in warm water. Make sure to shake all the water out of the filter and have it reasonably dry before installing it in the furnace. When replacing filters, be sure to use the same size and type as originally supplied.

Lubrication

Blower motors in these furnaces are permanently lubricated and do not require periodic oiling.

Burner Removal/Cleaning

The main burners should be checked periodically for dirt accumulation.

If cleaning is required, follow this procedure:

1. Turn off the electrical power to the unit.

2. Remove the access door.

3. Remove the igniter.

4. Turn off the gas supply at the external manual shutoff valve and loosen the ground union joint.

5. Remove the airshield.

6. Remove the four screws that hold the burner assembly to the vest panel and remove the assembly.

7. Remove burners from the burner assembly.

8. Burners may be cleaned by rinsing in hot water.

9. Reassemble the burners in the reverse order, making sure the burner shield is tightened securely in place.

Cleaning the Heat Exchanger

1. Turn off the main manual gas valve external to the furnace.

2. Turn off electrical power to the furnace.

3. Remove the upper access door.

4. Disconnect wires from HSI sensor, rollout switch and HSI igniter. Remove igniter **carefully**, as it is easily broken.

5. Remove the airshield. Remove the four screws that hold the burner assembly to the vestibule panel and remove the assembly. The lower portion of the heat exchanger will now be exposed.

6. Remove inducer blower and motor at the top of the furnace. Remove upper plate.

7. With upper exchanger opening exposed, remove restrictor baffle from each cell. Upper portion of the heat exchanger is now exposed.

8. With a stiff wire brush, brush inside cells at top and bottom. Vacuum loose scales and dirt from each cell.

9. Clean - vacuum all burners.

10. Replace all components in reverse order. Reconnect all wiring.

11. Restore electrical power and gas supply to the furnace.

12. Check furnace operation.

Central Environmental Systems

BLOWER CARE

Even with good filters properly in place, blower wheels and motors will become dust laden after long months of operation. The entire blower assembly should be inspected annually. If the motor and wheel are heavily coated with dust, they can be brushed and cleaned with a vacuum cleaner.

The procedure for removing the blower assembly for cleaning is as follows:

1. Disconnect the electrical supply to the furnace.

2. Remove the access panels.

3. Disconnect the two wire harness plugs from the top of the control box.

4. Remove the four screws holding the control box and position it out of the way.

NOTE: Steps 5, 6, and 7 apply to downflow models only.

5. Disconnect the flue pipe at the top of the unit.

6. Loosen the screws holding the top panel and move the panel aside.

7. Lift out the flue passage pipe an inner flue pipe as an assembly and set aside.

8. Remove the screws which retain blower to blower deck.

9. Remove the blower assembly with the control wiring still attached.

10. Vacuum the motor and the blower using a soft brush attachment. Care must be used not to disturb any balance weights (clips) on the blower wheel vanes.

11. Before reinstalling the blower assembly, inspect the heat exchanger which is visible in the blower opening of the blower deck. If it requires cleaning, vacuum it with a soft brush attachment.

12. Reinstall the blower assembly. Replace mounting screws that hold the blower assembly to the front portion of the blower deck. Two mounting screws used on the sides of the blower are used for shipping purposes only, and are not necessary after the furnace has been installed.

NOTE: Steps 13 and 14 apply to downflow models only.

13. Replace the flue passage pipe/inner flue pipe assembly and top panel.

14. Connect the flue pipe at the top of the unit.

15. Reinstall the control box and reconnect the wiring harness plugs.

16. Replace the access doors and restore the electrical supply to the unit.

TROUBLESHOOTING

The following visual checks should be made before troubleshooting:

1. Check to see that the power to the furnace and the 50E47 control module is ON.

2. The manual shutoff valves in the gas line to the furnace must be open.

3. Make sure all wiring connections are secure.

4. Review the sequence of operation.

Start the system by setting the thermostat above room temperature. Observe the system's response. Then use the Troubleshooting Table on page 19 to check the system's operation.

Use the table by reading the upper left-hand box and then following the instructions in each box. If the condition described in the box is true (yes answer), go down to the next box. If the condition is not true (no answer), go to the box to the right.

Continue checking and answering the questions in the boxes until the problem is explained and corrective action is described. After any maintenance or repair, the troubleshooting sequence can be repeated until normal system operation is obtained.

WARNING: Do not try to repair controls. Replace defective controls with CES Source 1 Parts.

UNIT WIRING DIAGRAM - ALL MODELS

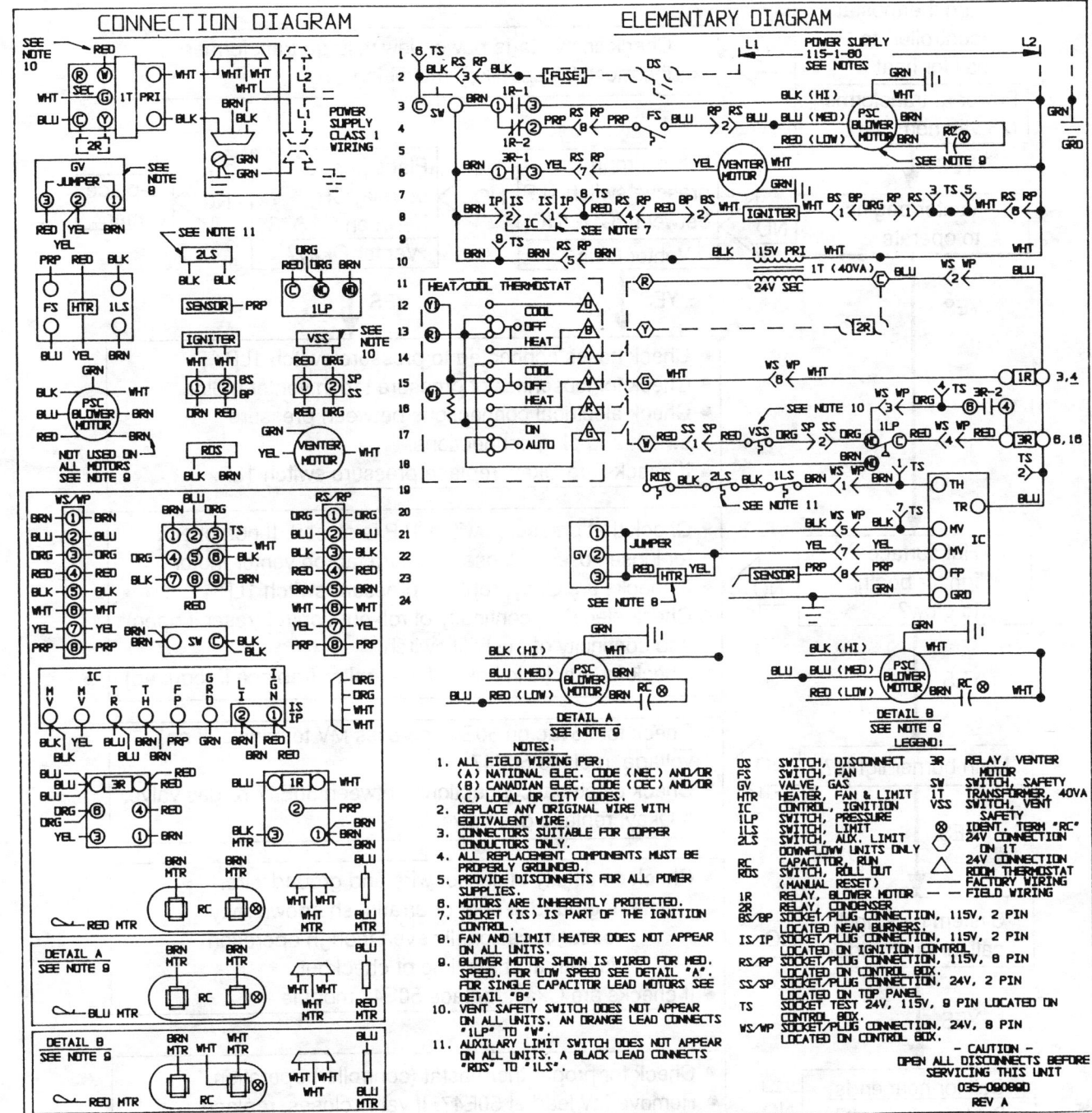

WHITE-RODGERS 50E47 SYSTEM TROUBLESHOOTING TABLE

NOTE: Before troubleshooting, familiarize yourself
with the startup and checkout procedures.

Turn thermostat (controller) to call for heat — **NO** → Check line voltage power, low voltage transformer thermostat (controller) and wiring

Power at transformer (25 vac nominal)?

YES

Venter begins to operate — **NO** → Place jumper on pressue switch 1LP between "C" & "NC" — Venter Okay? — **NO** → Place jumper on relay 3R between "1" & "3" — Venter Okay? — **NO** → Replace venter

YES

YES (from pressure switch 1LP Venter Okay?) **YES** (from relay 3R Venter Okay?)

- Check hoses connected to pressure switch 1LP.
- Check for looseness of pressure tap in venter motor.
- Check electrical connections between pressure switch and plug connections.
- If checks are Okay, replace pressure switch 1LP.

Hot surface igniter begins to glow? — **NO** →
- Check that pressure switch 1LP is closed. If not, check for loose hoses or loose pressure tap on venter motor
- If checks are Okay, replace pressure switch 1LP.
- Check electrical continuity of roll-out control (reset if open) and continuity of the limit switch.
- Check for cracks in hot surface igniter (replace if required)

YES

Main burner lights? — **NO** →
- Check for 25vac on 50E47 acrosss MV terminals. If no voltage, replace 50E47.
- Check electrical connections between 50E47 & gas valve. If Okay, replace gas valve.

YES

System runs until call for heat ends? — **NO** →
- Check continuity of sensor wire and ground wire.
  Note: If ground is poor or erratic, shutdown may occur occasionally even though operation is normal at the time of checkout.
- If checks are Okay, replace 50E47 module.

YES

Call for heat ends System shuts off? — **NO** →
- Check for proper thermostat (controller) operation.
- Remove MV lead at 50E47. If valve closes, replace 50E47. If not, replace gas valve

YES

Troubleshooting ends   Repeat procedure until troublefree operation is obtained.

YORK

Heating and Air Conditioning

USA
Official Sponsor
of the 1992
U S Olympic Team
3M/SC380

P.O. Box 1592, York, Pennsylvania USA 17405-1592
Subject to change without notice.   Printed in U.S.A.
Copyright © by York International Corporation 1992.   All rights reserved.   CPC 25M 292 .88   Codes: SBY   650.64-N3Y

FORM 650.65-TB13Y (890)

TRAINING COURSE

Sales/Application
and
Installation/Service

STELLAR PLUS™ P2UD HIGH EFFICIENCY
UPFLOW GAS FURNACES
(USE WITH A/V PROGRAM 650.65-TS13Y [890])

YORK SILVER STAR TRAINING COURSE

STELLAR 2000 SERIES

FEATURING THE STELLAR PLUS MODELS OF HIGH-EFFICIENCY, UPFLOW GAS FURNACES

Part 1 of this presentation describes the important SALES and APPLICATION features that will help you increase your sales.

Part 2 covers all aspects of INSTALLATION and SERVICE procedures including components, venting, controls, required adjustments and electrical circuitry. This information will enable the technician to systematically install, adjust and diagnose service problems on this product.

ADDITIONAL RECOMMENDED LITERATURE

Installation Instructions	650.65-N2Y
Vent Kit Instructions	650.65-N1.1V
Propane Kit Instructions	650.65-N1.2V

TIME REQUIRED

1. Slides & Script .... 3 hrs.
2. Quiz .... 30 min.
3. Correcting quiz and discussion .... 30 min.

SLIDE 1
This is a Silver Star presentation . . . .

ANOTHER

Silver Star

TRAINING COURSE
from **YORK**
Heating and Air Conditioning

SLIDE 2
Introducing another star in York's galaxy of fine products . . The Stellar 2000 series. . . .

SLIDE 3
Featuring the STELLAR PLUS Model P2UD line of deluxe High Efficiency upflow Condensing furnaces.

P2UD
HIGH EFFICIENCY
UPFLOW
GAS FURNACES

SLIDE 4

In this part of the program we will describe the important Sales and Application features that will help you increase your sales of high efficiency furnaces.

SALES
AND
APPLICATION
FEATURES

SLIDE 5

Before we go any further, lets take a moment to look at the gas furnace market. For the past several years approximately 2,000,000 furnaces have been sold annually. On the average, upflow furnaces account for approximately 70.8% of these sales or about 1.42 million furnaces per year. During this presentation we will be concerned only with a segment of this upflow furnace market.

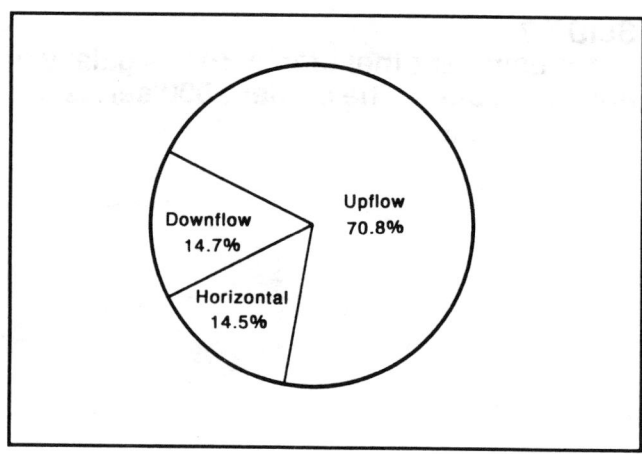

SLIDE 6

This slide illustrates upflow furnace sales in the various, isolated combustion air, AFUE ranges. Note that 21.8% of the upflow furnaces sold last year had an AFUE rating of 85% or greater. This 21.8% market share is the segment we are concerned with and it translates to 309,560 furnaces nationally.

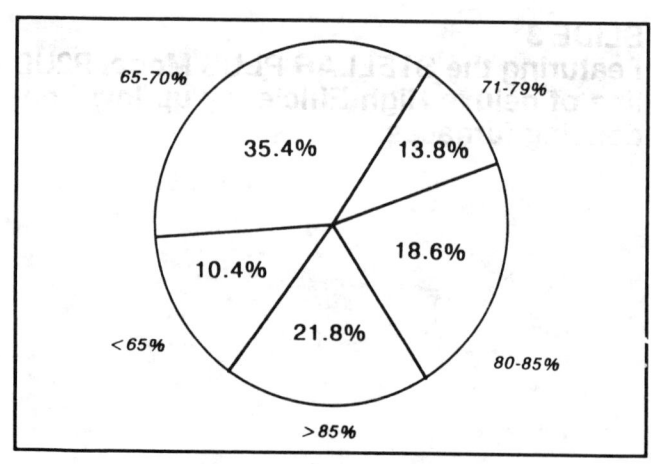

2

SLIDE 7

Locally, however, increased sales are dependent on the market needs or requirements in your specific area. For example, these local needs may be one or more of the following:

- Nitrous Oxide (NOx) requirements
- Intermittent Ignition Devices
- Annual Fuel Utilization Efficiency (AFUE) requirements
- Propane operation
- Low BTU input/high airflow

The P2UD line of furnaces will satisfy your local needs for a High-Efficiency Upflow furnace.

> **Local Needs**
> **or**
> **Requirements**

SLIDE 8

An finally, to satisfy the application needs of a specific installation, these compact furnaces can be installed in a basement, closet, alcove, recreation room or garage where the ambient temperature does not fall below 32°F and the continuous return air temperature is not below 55°F.

> ## APPLICATIONS
> - **Basement**
> - **Closet**
> - **Alcove**
> - **Recreation Room**
> - **Garage**

SLIDE 9

Intended primarily for residential use in the replacement market or for installation in new homes. They can also be installed in a non-residential application such as commercial buildings. The warranty, which we will cover later, is different for residential and non-residential use.

> **Replacement Installations**
> **New Homes**
> **Commercial Buildings**

SLIDE 10
All models may be either side wall or thru-roof vented using approved plastic type combustion air and vent piping. A Vent Termination Kit is required and will be discussed later in this program. Now that we have had an overview of market requirements and applications, let's look at the units themselves.

**Side Vent
Roof Vent**

SLIDE 11
This series of High-Efficiency upflow furnaces have numerous, improved design features such as heat exchangers, burners, igniter, and collector box, as you will soon see. They are completely factory assembled, wired and tested to assure dependable operation. Additionally, all models have heavy duty transformers and fan relays, making them suitable for heating only or heating/cooling applications.

- **Factory Assembled**
- **Factory Tested**
- **Heating Only**

  or

  Heating/Cooling

SLIDE 12
To assist you in selecting the proper model for a particular application let's take a moment to review the product nomenclature for a typical unit.

Reading from left to right, the "P" stands for furnaces. The "2" designates second generation. The "UD" relates to the Product Identifier and stands for High-Efficiency, Condensing Upflow. The "D06" means Direct Drive Blower, 600 CFM. In the next position, a "N" would mean natural gas while a "P" would mean propane gas. The next three digits, in this case "038" indicates the OUTPUT heating capacity in thousands of BTU/Hour. The last two digits "01" is the voltage code and means 115 volts, single phase, 60 Hertz power supply.

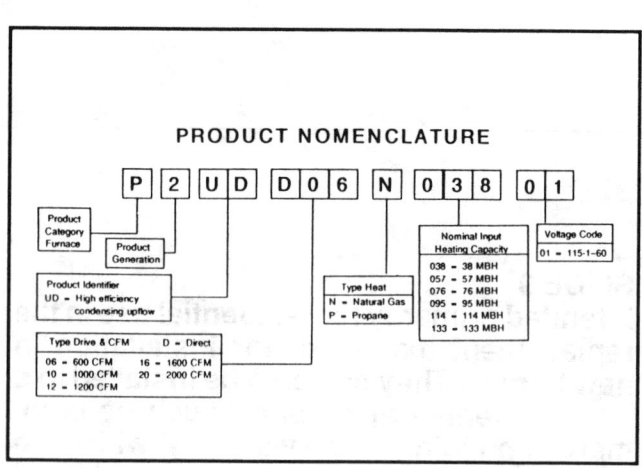

SLIDE 13
The P2UD series of High-Efficiency furnaces are available in six models with OUTPUT capacities of 38, 57, 76, 95, 114, and 133 MBH.

P2UD
Heating Capacities
(MBH Output)

38, 57, 76, 95, 114, 133

SLIDE 14
To assure safe operation, the units are Design Certified by the American Gas Association (A.G.A.).

A.G.A. tests for compliance with requirements of the American National Standards Institute (ANSI).

These stringent ANSI tests are designed to simulate adverse operation conditions and are concerned with such items as:

1. Combustion and ignition characteristics.

2. Limit and safety control operation.

3. Heat exchanger and casing temperature.

4. Wiring and component temperatures.

SLIDE 15
These High-Efficiency Condensing Upflow furnaces have steady-state efficiencies of 95% and Annual Fuel Utilization Efficiency (AFUE) ratings of at least 92%. AFUE is determined in accordance with D.O.E. test procedures by the isolated combustion method.

92%

A.F.U.E.

SLIDE 16
Depending upon the model selected, the California Seasonal Efficiency is 86% to 89% as determined in accordance with test procedures as specified by the State of California Energy Commission.

California Seasonal

Efficiency

86 - 89%

SLIDE 17
These furnaces have not been designed or built specifically for any particular market area. However, their low Nox emission level (35 nanograms/joule) complies with ALL local requirements without the use of additional parts or accessories.

Low NOx

SLIDE 18
All sizes of P2UD's are available in Natural Gas or Propane Gas models. Natural gas models CANNOT be field converted for propane operation. Propane models however, can be converted for use with natural gas using an accessory conversion kit.

Factory Built
for
Natural or
Propane Gas

Central Environmental Systems

SLIDE 19

For replacement installations, the wide range of BTU outputs permit selection of a unit that will closely match the heating requirements of the dwelling. This will result in a lower installed cost and more efficient operation of the unit for the consumer.

<u>Replacement Installations</u>

SLIDE 20

This chart taken from the consumer brochure compares different furnace efficiencies and shows the yearly savings. If the current furnace is more than 10 years old the efficiency is likely to be about 60%. If the old furnace had an annual heating cost of $700 the cost of operation of a new 92% AFUE P2UD furnace will be between $425 and $450. In this instance, the annual savings amounts to more than $250.

Furnace Efficiency	Approximate Annual Operating Cost						
60%	$400	$500	$600	$700	$800	$900	$1000
65	365	460	550	640	735	825	915
70	340	425	510	595	575	760	845
75	315	395	470	550	630	710	785
80	295	365	440	515	585	660	735
90	255	320	385	450	515	580	640
95	240	305	365	425	485	545	605

To use this chart:
Estimate the approximate efficiency of your current furnace. If it's more than 10 years old the efficiency is likely to be about 60%. Then locate your current operating cost, based on your gas heating bill, eliminating the cost of operating other gas appliances. Use the chart to estimate how much you can save with a more efficient furnace.

Example:
If your current furnace is 60% AFUE, and your annual operating cost is $700, the cost to operate a new 90% AFUE furnace will be between $425 and $450, an annual savings of more than $250.

SLIDE 21

The newer, energy-efficient homes require less BTU's to maintain comfort conditions at winter design temperatures. These homes therefore require smaller capacity furnaces. However, in heating/cooling applications, because of the cooling load, they require high airflow. All furnaces in this series have direct-drive blowers with multi-speed motors. This permits blower operation at one cfm for cooling and another cfm for heating.

New Homes

SLIDE 22

These compact units have been designed so that they fit and match perfectly with companion air conditioning coils. The low height of 52 inches enable them to be installed with or without air conditioning coils in areas with low ceilings. All models are 28-1/2" from front to rear. The widths vary from 12-1/4 inches to 26-1/4 inches depending on capacity.

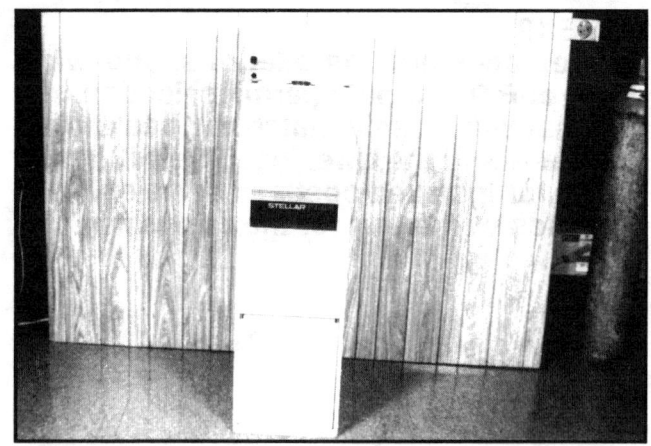

SLIDE 23

To provide many years of satisfactory service the one piece casing is fabricated from heavy-gauge steel. For additional rigidity and to prevent noise complaints, the casings are formed with surface indentations. The indentations also helps to prevent distortion or damage during shipping and handling. To provide a durable, smooth finish, all panels are degreased, bonderized, and finished with baked enamel.

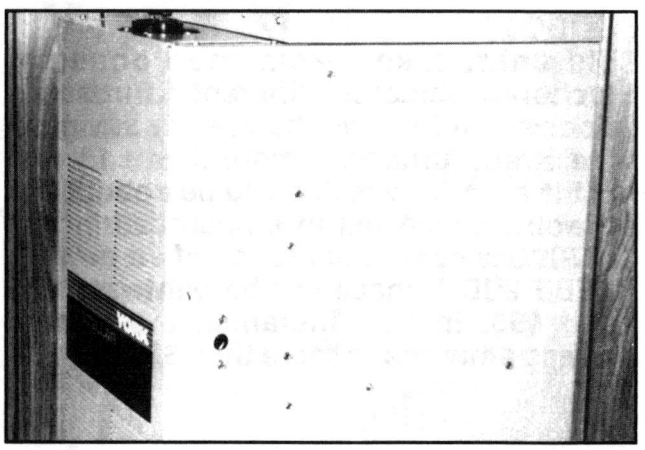

SLIDE 24

For ease of installation, maintenance, and/or service the front panels may be lifted up and removed allowing full access to all component parts.

SLIDE 25

All units are furnished with permanent-type, high-velocity filters. The filters are shipped mounted on the side or bottom and can be relocated if necessary. On the 133 MBH models, 2 filters are supplied. The filters are secured in position with a wire retaining loop which is easily removable for filter maintenance. The control box in front of the blower houses the blower access door interlock switch.

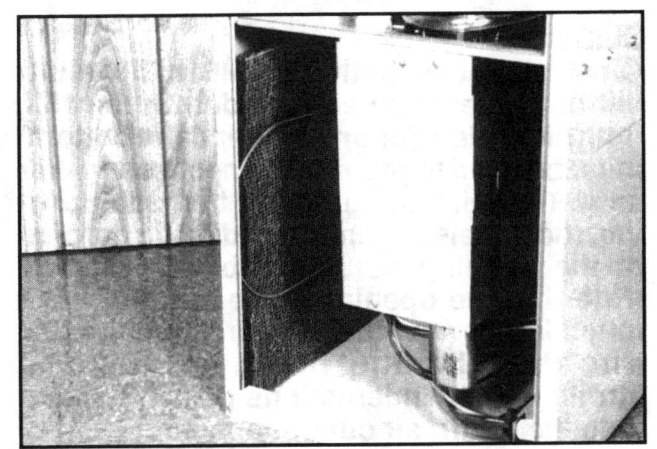

SLIDE 26

The blower access door interlock switch is factory-installed on all models. When the blower access door is removed this switch will automatically de-energize the electrical power, preventing operation of the unit.

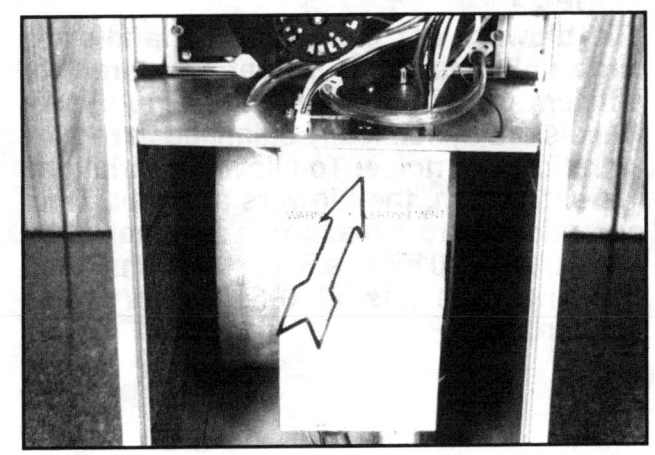

SLIDE 27

All models have a solid bottom panel, permitting installation on wooden floors. On all sizes the return air may enter from the bottom of the unit. Should the installation require bottom return air, the bottom closure panel is easily knocked out.

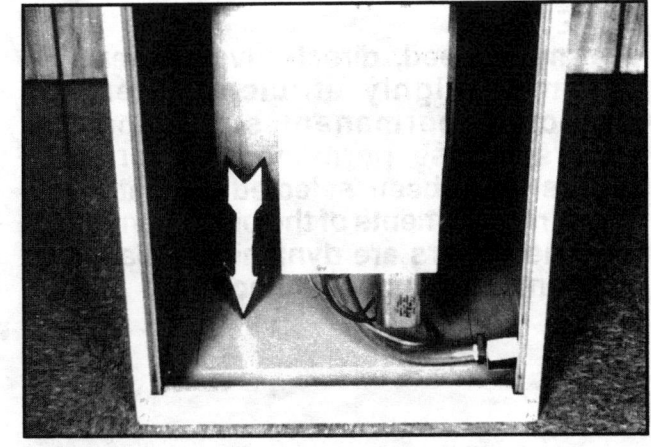

SLIDE 28
On all sizes except the 133 MBH, the return air may enter from either side. On the 133, both side inlets (or one side and the bottom) must be used to provide the necessary filter area. To facilitate cutting of the side opening, the panels are embossed at the corners of the opening as shown by the arrows. Note that the opening to be cut out has a much larger area than the area outlined by the stregthening indentations. The strengthening indentations DO NOT designate the return air cut out opening.

SLIDE 29
The blowers are oversized to handle the large volumes of air that are required for efficient operation. In addition, the blower wheels are dynamically balanced to reduce vibration and noise. To allow for variations in duct design, the blowers are capable of operating up to a maximum external static pressure of 1.0 inches water column. Rated CFM however, is at 0.5 inches water column.

SLIDE 30
The multi-speed, direct-drive blowers are driven by highly efficient, internally protected, permanent split-capacitor motors. These permanently lubricated motors have been selected to match the power requirements of the blower and ducting. The motors are dynamically balanced to minimize vibration and noise.

SLIDE 31
To further reduce any noise or vibration, the motor/blower assembly is mounted to the blower housing using neoprene rubber isolation pads.

SLIDE 32
The low voltage thermostat control wiring is connected to the conveniently located terminal board mounted on the transformer. The power supply wiring is connected to color coded lead wires in the factory mounted junction box located on the right side panel. All units have been tested and approved for use with either circuit breakers or fuses. The ampacity of every P2UD is lower than a comparable P1UD. The lower ampacity of the new P2UD's means smaller wire and fuse/circuit breakers are needed which result in a lower installed cost.

SLIDE 33
All models are furnished with a factory installed condensate trap. No additional traps in the condensate drain line are required.

SLIDE 34

The condensate drain connection is factory installed on the right side as seen here. If proper slope of the drain is not possible, there is an additional knockout 2-5/8" higher that may be used. For ease of installation two similar knockouts are also on the left panel for field conversion to left side drainage.

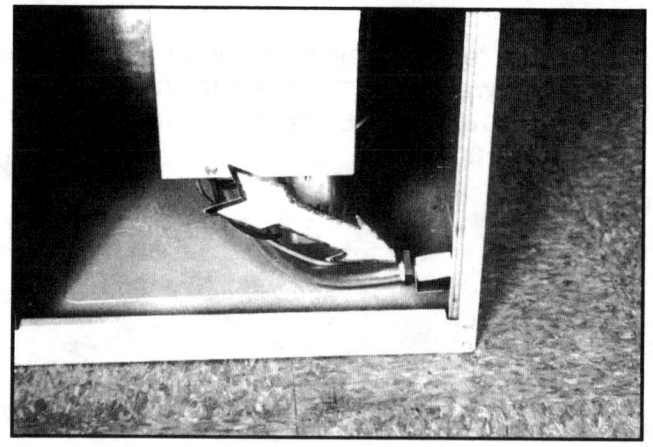

SLIDE 35

The interior casing surfaces of the heat exchanger area are covered with reinforced, heavy-gauge, foil-faced insulation. The addition of this type of insulation is multi-purpose. First of all, it reduces casing surface temperatures permitting the units to be installed at reduced clearances from combustible construction materials. Secondly, it helps to reduce air velocity noise. And lastly, the reinforcement prevents shredding of the foil facing which could be carried by the air stream into the living quarters or restrict air flow through an evaporator coil. Note the crimped seam on the heat exchanger sections. These units have two heat exchangers, a primary and a secondary.

SLIDE 36

The primary heat exchanger is comprised of individual half sections fabricated in one piece, from heavy-gauge aluminized sheet steel. Aluminized steel is used to assure long, dependable life. For maximum heat transfer each half section is die formed into an efficient convoluted surface.

SLIDE 37

The two half sections are folded and joined together by a crimped seam on the remaining three sides. This seam ensures a leak proof assembly without destroying the integrity of the aluminized coating as is common with a welded seam. An additional expansion/sizing operation, in the burner area, has been incorporated into the design of the P2UD heat exchangers. The expansion/sizing assures dimensional tolerances and increases the distance between the burner flame and the heat exchanger surface to prevent hot spots.

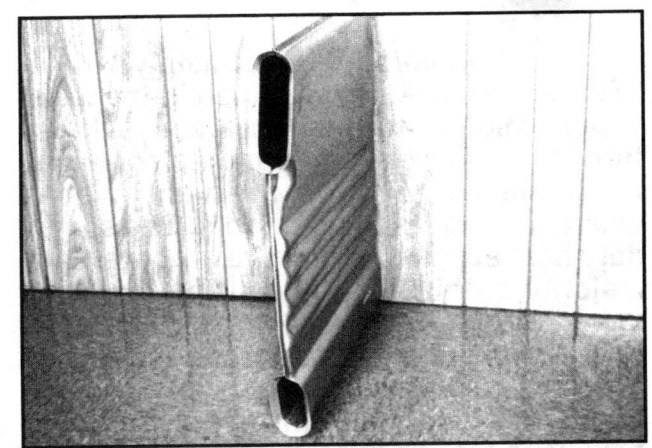

SLIDE 38

The individual sections are secured to the furnace division panel at the burner area and flue gas outlet area with high-torque, non-strippable screws. The resultant heat exchanger assembly is corrosion resistant, leakproof, highly efficient and extremely reliable.

SLIDE 39

To prevent damage during shipping or handling, the individual heat exchanger sections are supported at the rear by a factory installed retainer. The retainer prevents excessive movement of the sections yet permits them to move freely. Allowing this normal movement prevents expansion/contraction noises during operation of the unit. The retainer may be mounted at the top as we see here or it may be mounted much lower. In either case the bracket does not have to be removed during installation.

SLIDE 40
During combustion approximately 10% of the available heat is contained in the water vapor which is formed. In a conventional furnace the water vapor, and its contained heat, passes through the vent system where it is lost to the outdoors. The secondary heat exchanger however, recovers the majority of this heat by causing the water vapor to condense and liberate its latent heat. The resultant condensate, because of absorbed carbon dioxide, is a weak carbonic acid solution with about the same acidity as orange juice.

Secondary
Heat
Exchanger

SLIDE 41
Here we see the secondary heat exchanger assembly. It is factory installed beneath the bottom of the primary heat exchanger directly above the room air blower and will condense .57 gallons of water per hour on the 95 MBH (100 MBH input) model. The combustion gases enter the secondary heat exchanger, at the rear collector box, through the round transfer tubes.

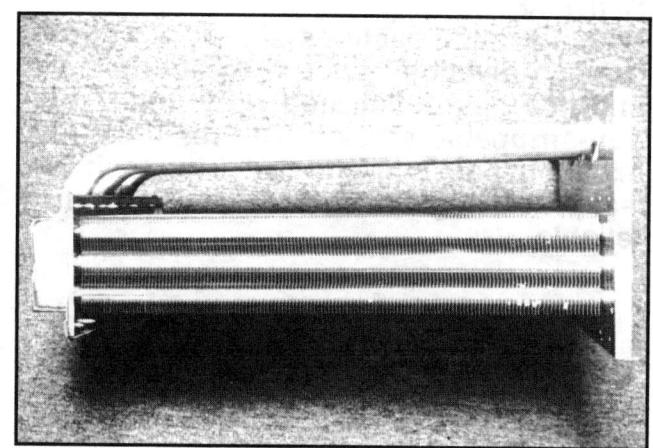

SLIDE 42
The transfer tubes and mounting flanges are constructed of aluminized steel. The flanges, to prevent destruction of the aluminized coating, are mechanically secured to the transfer tubes by an expansion operation. Also, to prevent leaks or corrosion, the transfer tubes are installed using ceramic fiber gaskets and stainless steel screws.

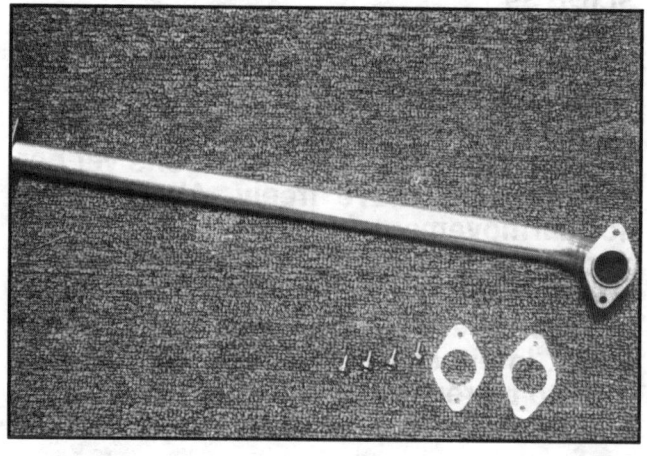

SLIDE 43

All parts of the secondary heat exchanger are constructed of AL29-4C stainless steel, the fins however, are aluminum for superior heat transfer. The flue gases make one pass through the coil from rear to front. Each tube has an internal turbulator baffle (AL29-4C) that swirls the flue gases to improve heat conduction. AL29-4C stainless steel is used because of its superior resistance to acid corrosion.

SLIDE 44

The P2UD series of furnaces now use a collector box that is constructed of high-density, corrosion-proof, non-metallic material to cover the front area of the secondary heat exchanger. This new high-density material prevents hair line cracking that, in time can enlarge, causing condensate leakage. In addition, an extra-thick, closed cell neoprene gasket is used to ensure a water tight, air tight seal between the collector box and the secondary heat exchanger.

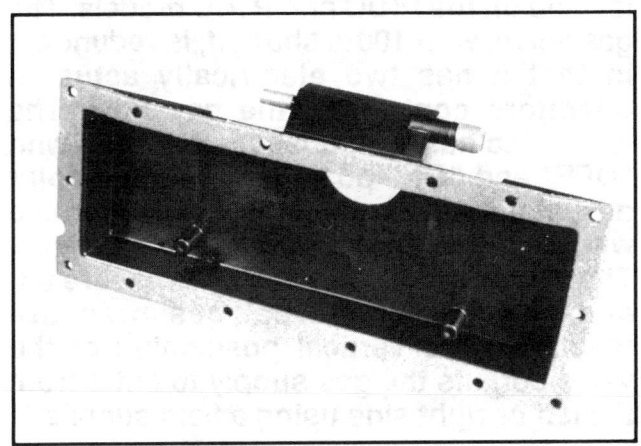

SLIDE 45

The lower portion of this slide shows the installed collector box. Mounted on the collector box is the permanently lubricated, venter blower that draws the combustion gases through the heat exchangers and discharges them outdoors through the vent system. These furnaces are classified Category IV and must be vented using approved plastic type vent piping. For disposal of condensate within the vent system a drain, is factory installed at the bottom of the discharge elbow. The discharge elbow is factory positioned for left venting and may be turned for right venting if required. The casings are provided with knockouts on both right and left side panels for ease of vent installation. The venter blower hous-

ing and wheel are also constructed of non-metallic material for corrosion protection. The motor, wheel, and housing can be removed as an entire assembly.

SLIDE 46

Exhaustive testing has determined that normal household products such as detergents, softening agents and bleach, to name a few, have contributed to corrosion in heat exchangers. With condensing type furnaces this corrosion rate is accelerated. By using 100% outdoor air for combustion, the effects of indoor contaminates are eliminated.

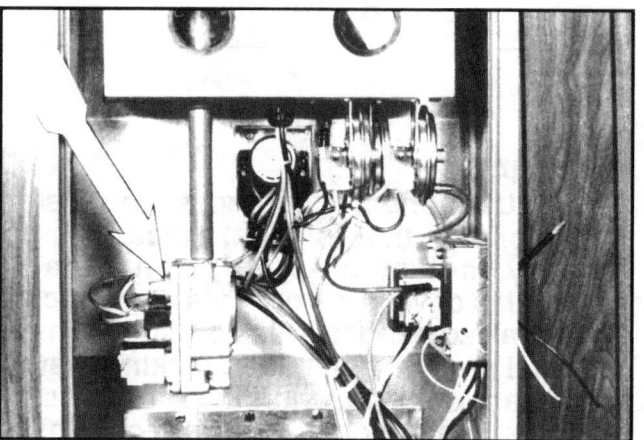

100%
Outdoor
Combustion Air

SLIDE 47

To standardize on controls, all sizes of P2UD furnaces use the same model White-Rodgers ignition control and gas valve that is used on the P*UC and P1CC models. The gas valve, with 100% shut off, is redundant in that it has two electrically actuated operators controlling the gas flow. The valve also has a manual control knob with "OFF" and "ON" position. The knob color designates the type of gas. Blue is for use with natural gas and Red for propane gas. The valve also has a manifold pressure tap and adjustable internal gas pressure regulator. The vertical positioning of the valve permits the gas supply to enter from the left or right side using a field supplied

street-ell. The casings are provided with elongated knockouts on both right and left side panels for ease of piping installation.

SLIDE 48

On all models the combination fan and limit control is conveniently located. Note that the control does not require a cover. The electrical connections are recessed within the body to prevent the possibility of an electrical hazard.

The limit setting is factory preset to prevent excessive discharge temperature in the event that there is a loss or reduction in air flow due to blower or blower motor failure, dirty filters, or restricted ductwork. Fan operation is "Time On, Temperature Off". Inside the fan portion of the control there is a 24 volt heater and bimetal actuated switch. The heater terminals, indicated by the arrow, receive power from the gas valve control circuit. During a call for heat the gas

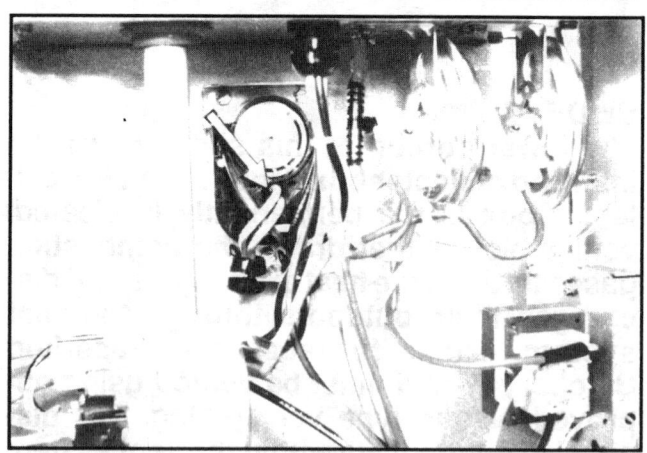

valve opens immediately, and after approximately 40 to 50 seconds, or the supply air temperature reaches 115 to 125° F, the blower motor is energized. The blower off setting is field adjustable.

SLIDE 49

The furnaces are supplied with non-adjustable pressure switches which monitor the flow through the combustion air/vent piping system. The switches de-energize the ignition control module and the gas valve if any of the following conditions are present.

1. Restriction or blockage of combustion air piping.
2. Restriction or blockage of vent piping.
3. Failure of the combustion air blower or motor.
4. Restriction or blockage of condensate drain piping.
5. Restriction or blockage in the primary or secondary heat exchanger.

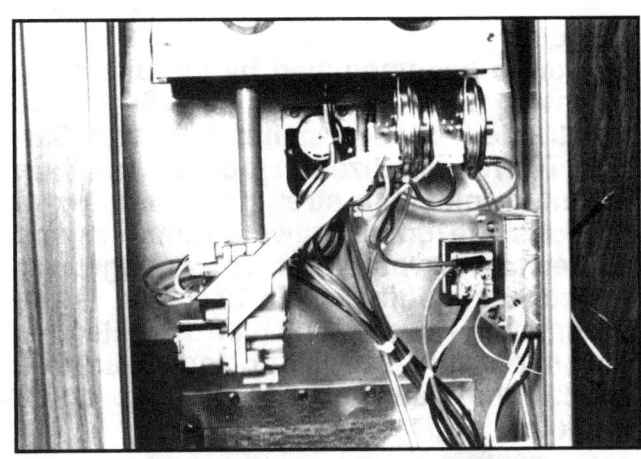

For ease of service, the pressure settings are stamped on the switches. During the second portion of this program, we will discuss the operation of these switches and how to check them in the field.

SLIDE 50

The burners used on the P2UD series of furnaces are newer versions of the type originally used. The original burners (left) were rated at 40,000 BTU/Hour and produced a lazy flame that tended to be noisy and difficult to control. Also, minute changes in the combustion air flow patterns and very small changes in burner alignment could cause the flame to impinge upon the heat exchanger. The improved burner (right) is rated at 20,000 BTU/Hr. and has a more defined venuri which creates a high velocity and greater turbulence of the gas/air mixture. The resultant flame is very tight and controlled. The positioning ring on the front of the burner fits into a factory installed index plate which assures precise

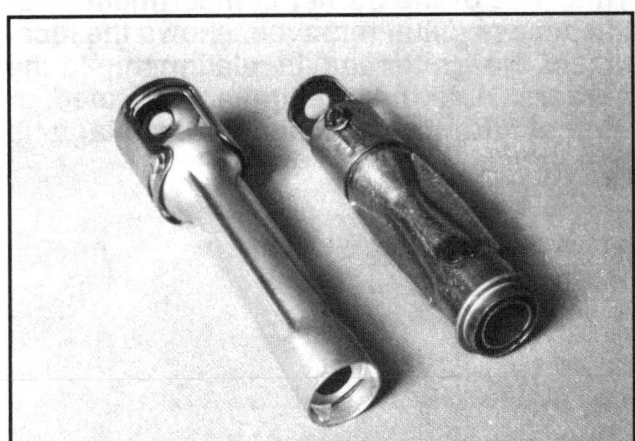

burner/heat exchanger alignment. The burners are designed to be "universal" in that they will operate equally well on natural or propane gas.

SLIDE 51

The burners are installed in this compartment at the top of the furnace. Locating the burners at the top contributes to the compact size of these units. The combustion gases are routed through the primary heat exchanger directly into the secondary heat exchanger without the use of a "dead cell" or large diameter transfer tubes that require additional room. The two round observation ports permit visual inspection of the burners during operation. The combustion air inlet flange is indicated by the small arrow.

SLIDE 52

Ignition is accomplished by the use of an improved hot surface igniter (HSI) that is constructed of recrystallized Silicon Carbide. The recrystalization process results in maximum possible strength. Operating at 120 volts, the igniter draws 4.5 amperes and is rated at 540 watts/hour. The wattage consumed per ignition cycle however, is approximately 7 watts. This type of igniter has been used successfully for many years and has but one drawback. "If you bump it, you break it."

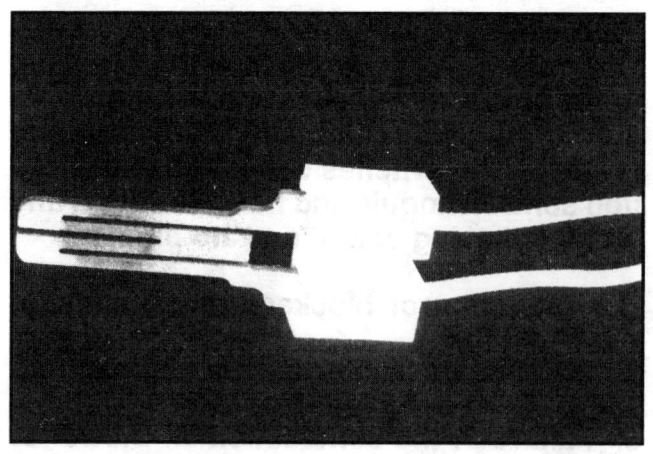

SLIDE 53

This slide of the burner compartment, with the access panel removed, shows the location of the igniter and its relationship to the burners. Note that the igniter is located so that accidental contact or breakage is prevented.

SLIDE 54

Looking at the burner heads you can see the stainless steel perforated runner tube. The igniter is factory positioned so that its 2500° temperature can light the gas/air mixture issuing from the burner and runner tube simultaneously. The flame, burning on the surface of the runner tube, then ignites each of the remaining burners. The small arrow is pointing to the orifice which meters the correct amount of gas for proper operation of the runner tube. The burner compartment sealing surfaces on P2UD furnaces now have a high temperature silicone rubber gasket. The addition of this gasket prevents ambient air from being drawn into the burner box which may affect the burner flame pattern.

SLIDE 55

This slide shows another component of the HSI system, the flame sensor probe. The flame sensor probe is merely a rod of high temperature alloy secured into a glazed ceramic insulator.

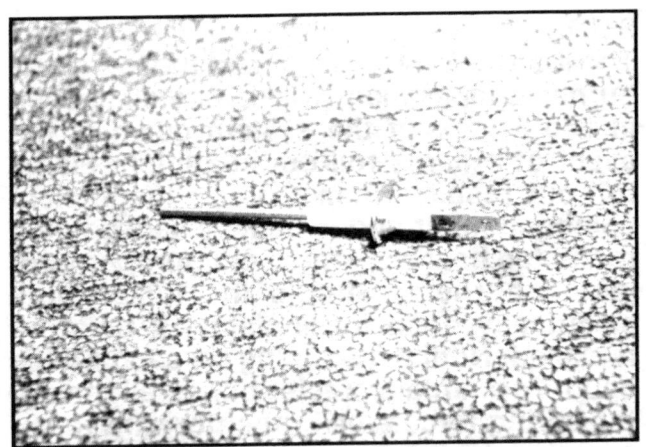

SLIDE 56

Here we see the importance of the high temperature rating of the rod. The probe is factory positioned so that it is directly in the flame path of the extreme right burner. The flame sense probe, in conjunction with an electronic ignition control module, proves the presence of flame. If flame is not detected the ignition control will de-energize the gas valve.

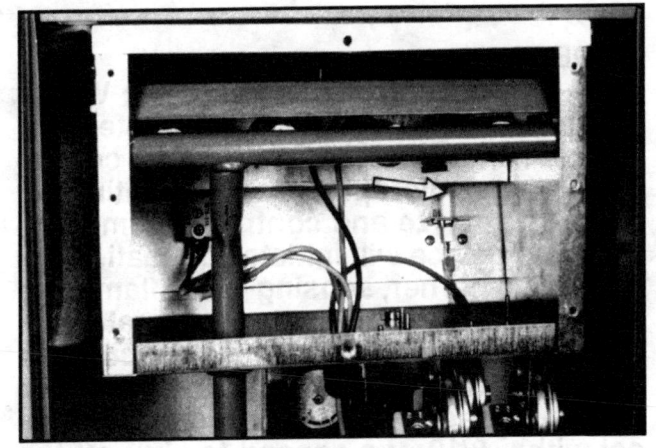

SLIDE 57

An additional safety device that is factory installed on all models is a fusible link.

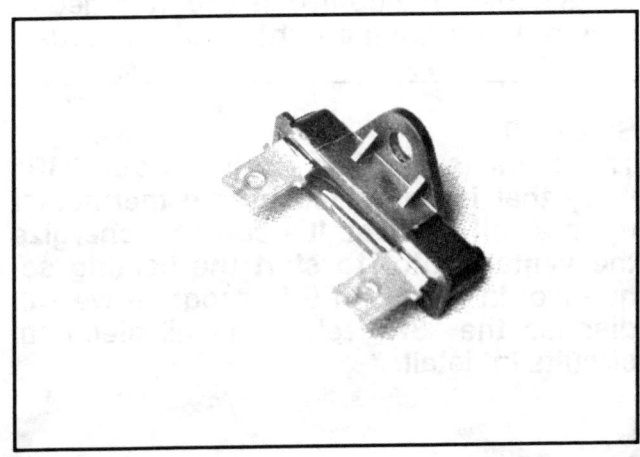

SLIDE 58

The fusible link is mounted at the top of the burner compartment. It is calibrated to open at 306°F to shut the system down in the event of a flame roll out condition. Flame roll out results when there is insufficient combustion air and is usually caused by a restricted heat exchanger or improperly adjusted burners. Should the fusible link electrically open its contacts, it must be replaced. Notice the spring on each orifice spud. To maintain precise burner alignment the spring forces the burner forward to ensure that the burner positioning ring is held tightly against the burner indexing plate. Burner or orifice removal is easily accomplished. Simply pull back on the burner until it is free of the hole in the index

plate, lower the front of the burner then move it forward until it clears the orifice spud and remove the burner.

SLIDE 59

Here in the control box we see the White-Rodgers control module. It is a self-testing, automatic gas interrupted ignition control employing a microprocessor to continually monitor, analyze and control system function. The device will initiate automatic ignition of the burner, sensing of the flame and system shutoff during normal operation. In the event of abnormal operation such as an internal fault or loss of flame the microprocessor electronic circuitry will react within 8/10 of a second to shut off the gas flow. It is programmed for two re-trys in the event of no flame during initial operation and 5 recycle's if flame is lost after 4 seconds of operation. If either of the above is exceeded, the control is programmed to lock out and flash a light emitting diode

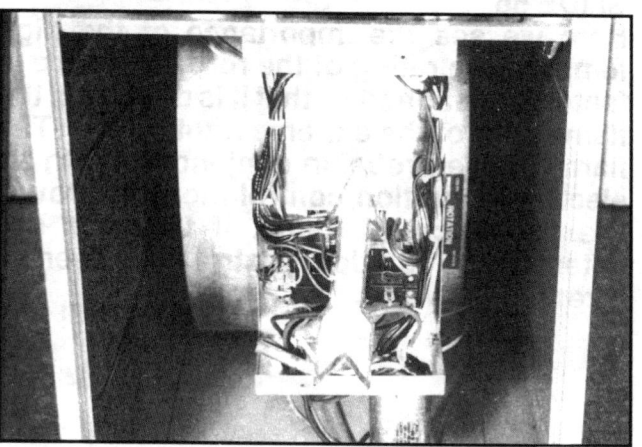

(LED). If the LED is on continuously, the fault is likely to be internal to the module. The control is universal in that it can be applied to natural or propane gas units.

SLIDE 60

The arrow is pointing to the 24 volt "3R" relay that is energized by the thermostat upon a call for heat. It's contacts energize the venter motor to start the heating sequence. In Part 2 of this program we will discuss the "3R" relay and all electrical circuits in detail.

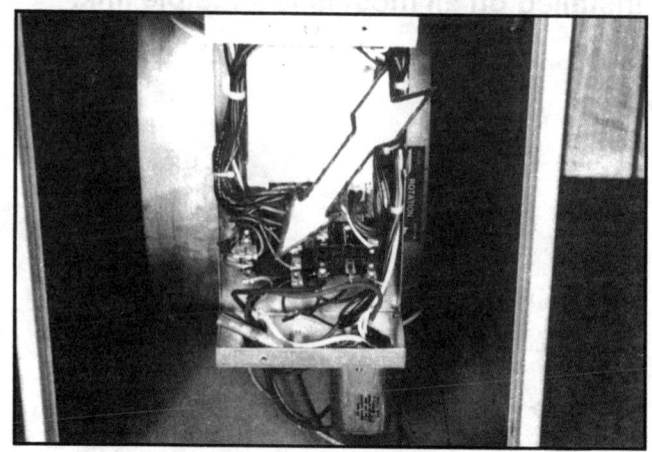

20

SLIDE 61

Also in the control compartment is the blower relay that permits blower operation at a higher CFM for cooling and lower CFM for heating. This blower relay is energized automatically by the thermostat control circuit during a call for cooling.

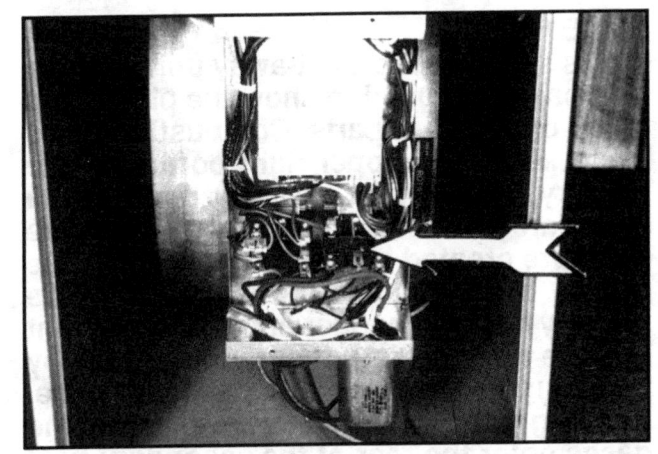

SLIDE 62

Here we see the plug-in electrical connectors for changing the blower speed. This simple adjustment permits the blower to deliver the required amount of air for proper operation.

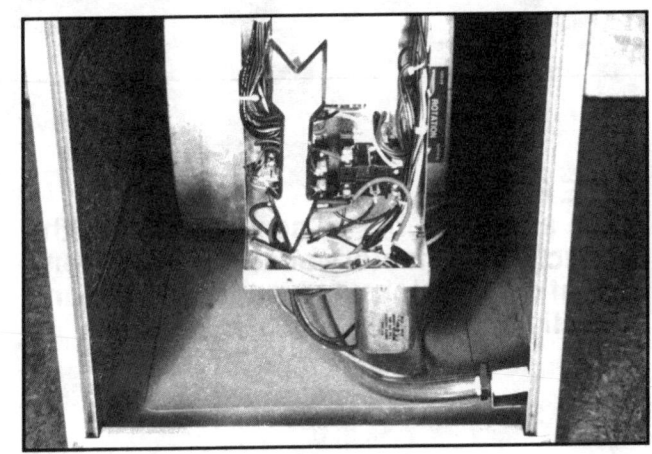

SLIDE 63

For ease of maintenance all blower assemblies are easily removed. The arrows are pointing to the two plugs that permit the blower and control box to be removed as an assembly. The black cap between the plugs covers a socket which is used for factory testing of the unit during the final assembly. These sockets and plugs will prove invaluable for troubleshooting the system and will be covered in Part 2 of this program.

To standardize on parts, the control box, socket, plugs, and components we have pointed out are used on all sizes and models of P*UD, P*UC and P1CC furnaces.

Central Environmental Systems

SLIDE 64

In this slide we see a cut-away unit with the side panel removed to show the placement of the component parts. Combustion takes place within the upper right portion of the primary heat exchanger. The combustion gases are drawn downward by the negative pressure created by the venter blower. The gases, at approximately 450°F, exit the primary heat exchanger at the bottom right and are directed into the transfer tubes by the insulated, aluminized deflector plate. Passing through the transfer tubes the gases enter the rear of the secondary heat exchanger. As the gases are drawn through the secondary heat exchanger additional heat is extracted causing the water vapor to condense which liberates additional heat.

The gases, at approximately 130°F enter the front collector box where they are drawn into the venter blower and then discharged outdoors through the vent system.

SLIDE 65

There are a number of accessories for the product line. A factory accessory is a component or kit which can be added to the unit in the field during the initial installation.

Accessories

SLIDE 66

All sizes of P2UD factory-built propane units can be converted for use with natural gas using the appropriate accessory conversion kit. Consult the Tech Guide or price list for the conversion kit numbers. Each kit contains main burner orifice spuds, runner tube orifice spud, pressure regulator conversion spring and heat exchanger orifice plate. Detailed instructions and conversion labels are also provided with the kits.

Propane Gas to Natural Gas Conversion Kit (Factory Built P2UD)

SLIDE 67

To properly vent the unit you MUST install a vent terminal kit. There are two vent kits available, a 2" and a 3". The selection of the proper kit is dependent upon the furnace size and the length of intake or vent piping and number of elbows. Proper selection will be covered later in this presentation. To permit the kit to be universal for all applications the kit is shipped un-assembled. The vent kit does not have a screen to prevent leaves, birds, bugs and debris from entering either vent. The use of a screen may cause frost to form and cause the furnace to shutdown.

SLIDE 68

A Side Filter Rack, that fits all models, is available that permits moving the factory supplied filter to the outside of the furnace. To supply the required filter area for the 133 MBH model two filter racks, one on each side, must be used.

SLIDE 69

A Side Filter Cabinet that is applicable to all sizes of furnaces is also available. It is shipped knocked down and is assembled to the side of the furnace. The use of this kit on the 133 MBH model permits the return air to enter from a single side.

SLIDE 70

Rear return air plenums are also available. Four sizes are available to permit the plenums to match the furnace cabinet size. The plenum comes knocked down and is assembled to the rear of the furnace. The location and dimensions of the rear opening to be cut in the furnace are given in the installation instructions.

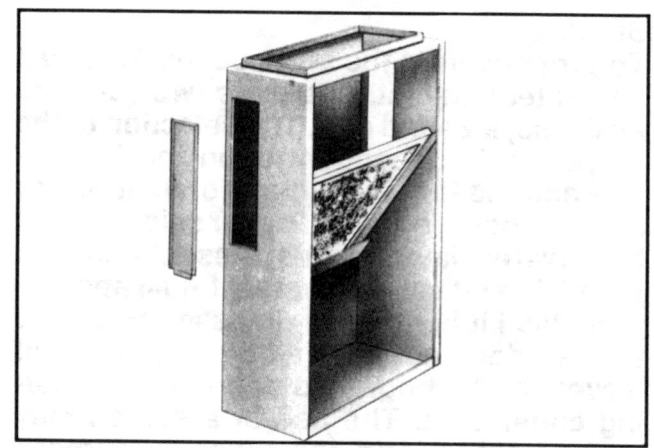

SLIDE 71

If your codes prohibit you from draining the furnace condensate into the sewer system a neutralizer kit is available. The neutralizer medium disintegrates as it neutralizes the condensate and should be renewed yearly. The neutralizer medium is available from the service parts department.

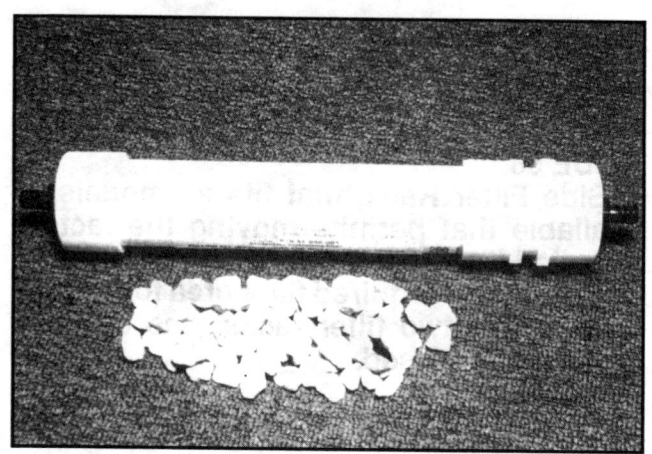

SLIDE 72

For those instances where a floor drain is not available, the accessory condensate pump must be used. This pump has a vertical lift capacity of 17 feet and can drain the condensate from both the furnace and evaporator coil. The pump has an auxiliary safety switch that must be field wired into the low voltage control circuit to prevent overflow damage or operation of the furnace or air conditioner in the event of pump failure. The pump is manufactured of corrosion resistant materials and does not require the installation of a neutralizer kit ahead of it.

SLIDE 73

With the design features and the component parts that comprise the unit, it should last a lifetime. We're so sure that a properly installed STELLAR PLUS furnace will last that you can offer a limited life time warranty on both heat exchangers to the original purchase for as long as he owns his home, and for twenty years from original installation date for a second owner. For non-residential installations however, the limited warranty is 20 years from date of installation.

> # Lifetime Warranty
>
> ## On Heat Exchanger

SLIDE 74

PLUS, a two year limited parts warranty on all other internal operating components PLUS, if on a properly installed STELLAR PLUS furnace, either heat exchanger fails during the first two years, YORK will replace the furnace, provided the owner pays the replacement labor costs.

> ## Two Year
>
> ## Parts Warranty

SLIDE 75

During this portion of the presentation we have discussed the Sales and Application features of the STELLAR PLUS P2UD series furnaces.

Let's quickly review several key features:

* Full Heating Capacity Range - from 38 to 133 MBH Output.

* Wide Air Conditioning Range - from 1 ton (400 cfm) to 5 tons (2000 cfm).

* Broad Application Range - Basement, Closet, Alcove, Recreation Room or Garage.

* Compact - Only 52 inches tall, 28-1/2" front to rear and, depending on the model selected, 12-1/4" to 26-1/4" wide.

* Consistent Design - Coil line-up matches furnace line-up to achieve size for size integration.

> ## 38 - 133 MBH
> ## 400 - 2,000 CFM
> ## Broad Application
> ## Compact
> ## Consistent Design
> ## Natural or Propane
> ## Improved

* Factory-built for Natural or Propane gas.

* Improved inshot type burners and latest technology heat exchangers.

SLIDE 76

The overview of these features will permit you to answer questions a consumer may have regarding the furnace. If you have any questions on sales and application still un-answered, we will take the time to answer them now.

Questions

and

Discussion

SLIDE 77

In Part 2, we will discuss specific installation and service procedures for the P2UD series of furnaces. Basic topics such as venting and installation procedures will not be discussed as they are covered in detail in the Ladders of Learning training books.

Part Two

P2UD
Installation and Service

SLIDE 78

The first point we must stress is, that even though these furnaces use 100% outside air for combustion, they have NOT been designed, tested or certified for mobile home installation. They MUST NOT BE IN-STALLED IN A MOBILE HOME.

MOBILE HOMES

Central Environmental Systems

SLIDE 79
Choose a level location for the furnace with the following in mind:

1. Where a minimum amount of air intake/vent piping and elbows will be required.
2. As centralized with the air distribution system as possible.
3. In an area where ventilation facilities provide for safe limits of ambient temperature under normal operating conditions.
4. Where ambient temperatures will not fall below 32°F.
5. Where it will not interfere with proper air circulation in the confined space.
6. Where it will not interfere with the cleaning, servicing or removal of other

Furnace

Location

appliances.
7. Where the continuous return air temperature is not below 55°.

SLIDE 80
To provide proper condensate drainage the unit MUST be installed in a LEVEL position (plus or minus 1/4") side-to-side and front-to-back. An unlevel unit may prevent proper drainage and cause the pressure switch to shut the furnace off.

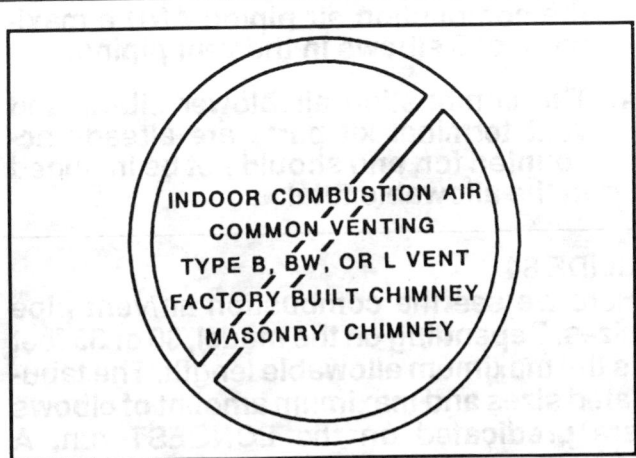

SLIDE 81
This furnace requires outdoor combustion air. Two separate, properly-size pipes must be used; one bringing outdoor air from the accessory terminal kit outdoors to the furnace combustion air intake (top of unit), and one from the furnace vent connection (left or right side of unit) back to the terminal kit located outdoors. This series of furnaces MAY NOT BE COMMON-VENTED with any other appliance, since they require separate, properly sized air intake and vent lines. In addition, they SHALL NOT BE CONNECTED to any type of B, BW, or L vent or vent connector, and not be connected to any portion of a factory-built or masonry chimney. If this furnace is replacing a previously common-vented furnace consult the National Fuel Gas Code to determine if

INDOOR COMBUSTION AIR
COMMON VENTING
TYPE B, BW, OR L VENT
FACTORY BUILT CHIMNEY
MASONRY CHIMNEY

it is necessary to resize the existing vent line and chimney to prevent oversizing problems for the other remaining appliance(s).

SLIDE 82

The furnace may be either side wall or roof vented. The location of the vent terminal outdoors must follow the clearances listed here.

LOCATION	CLEARANCE
Dryer Vent	4 feet
Plumbing Vent Stack	3 feet
Gas Appliance Vent Terminal	4 feet
From any openings where vent gases could enter the building	1 foot
Above grade and anticipated snow depth	1 foot
Above grade when adjacent to a public walkway	7 feet

LOCATION	CLEARANCE
Dryer Vent	4 feet
Plumbing Vent Stack	3 feet
Gas Appliance Vent Terminal	4 feet
From any opening where vent gases could enter the building	1 foot
Above grade and anticipated snow depth	1 foot
Above grade when adjacent to a public walkway	7 feet

Care should be taken to locate side vented systems away from swimming pools, swimming pool chemical storage areas, trees, shrubs, and at least 4 feet from an inside corner of the building.

SLIDE 83

The combustion air/vent pipe size is determined by the furnace model, total length of run, and the number of elbows required. The following rules must also be observed:

1. Long radius elbows are recommended.

2. Elbows are assumed to be 90°. Two 45° elbows count as one 90° elbow.

3. Elbow count refers to combustion air piping AND vent piping separately. For example, if the table allows for 5 elbows, this will allow a maximum of 5 elbows in the combustion air piping AND a maximum of 5 elbows in the vent piping.

4. The combustion air blower elbow and vent terminal kit parts are already accounted for, and should not be included in the allowable total.

Combustion Air/Vent Pipe Size

- Long radius elbows
- Elbow count for both pipes
- Vent terminal & combustion elbows NOT to be included
- Combustion Air & Vent same diameter
- Schedule 40 PVC, DWV ABS-DWV, SDR-21 PVC, SDR-26 PVC

5. Combustion air and vent piping MUST BE the same diameter.

6. All piping and fittings are to be Schedule 40 PVC, PVC-DWV, ABS-DWV, SDR-21 PVC or SDR-26 PVC.

SLIDE 84

Here we see the combustion air/vent pipe sizes. Depending on the model, 30 or 35 feet is the maximum allowable length. The tabulated sizes and maximum amount of elbows are predicated on the LONGEST run. A reduction in the amount of elbows does NOT permit longer lengths. If the allowable length or amount of elbows is exceeded the unit may have poor combustion, pulsate, or shut off on the pressure switch.

VENT SIZING TABLE

MODEL	Pipe Size	MAX. ELBOWS PER TOTAL RUN (See Note 2)						
		0-5 Feet	5-10 Feet	10-15 Feet	15-20 Feet	20-25 Feet	25-30 Feet	30-35 Feet
P2UDD06†03801	1-1/2"	5	4	3	2	Note 1	Note 1	Note 1
	2"	5	5	5	5	5	5	5
P2UDD10†05701	1-1/2"	3	2	Note 1	Note 1	Note 1	Note 1	Note 1
	2"	5	5	5	5	5	5	5
P2UDD12†07601	2"	5	4	3	2	1	Note 1	Note 1
	3"	5	5	5	5	5	5	5
P2UDD12†09501	2"	5	5	5	5	5	5	30' Max.
P2UDD16†11401	3"	5	5	5	5	5	5	5
P2UDD20†13301	3"	5	5	5	5	5	5	30' Max.

† May Be P = Propane (LP) or N = Natural Gas

Notes:
1. Must use the larger pipe size indicated.
2. Long radius elbows shown on this unit when over 25 feet of piping is required.

SLIDE 85

Once the piping size is determined, the combustion air intake collar must be connected to that size piping. Units rated from 40 MBH input (38 MBH output) to 80 MBH input (76 MBH output) are equipped with intake collars which accommodate 2" diameter vent pipe. If the determined vent pipe size is of a different diameter than the exhaust street-elbow, reduction or expansion fittings must be incorporated at the time of installation. Examples of such adaption are seen here.

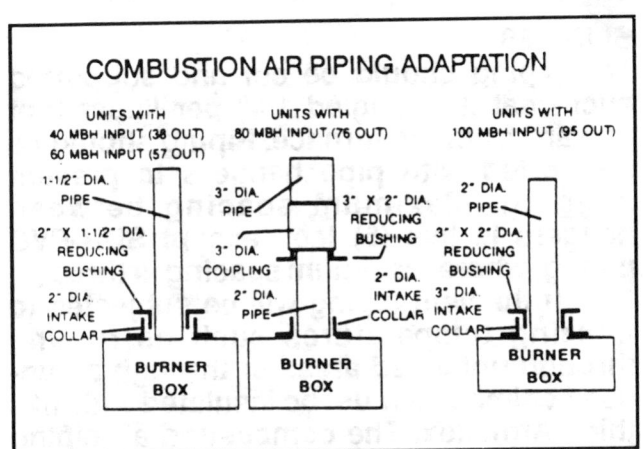

COMBUSTION AIR PIPING ADAPTATION

SLIDE 86

All models are provided with a combustion blower street-elbow which accommodates 2" diameter vent pipe. If the determined vent pipe size is of a different diameter than the exhaust street-elbow, reduction or expansion fittings must be incorporated at the time of installation. Examples of such adaption are seen here.

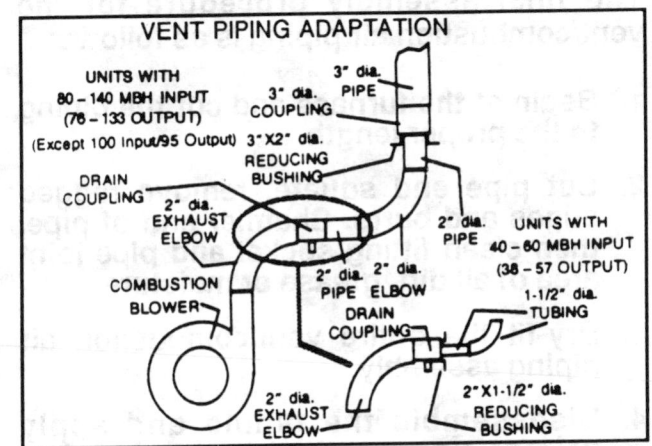

VENT PIPING ADAPTATION

SLIDE 87

When installing the piping all joints must be made so that they are permanent, air-tight, and water-tight. DO NOT ATTEMPT TO CEMENT THE STREET-ELBOW INTO THE VENTER HOUSING. A stainless steel screw is provided to mechanically secure the street-elbow into the venter housing. Also, run a small bead of RTV sealant around the street-elbow, about 1/4" from the end, before inserting into the venter outlet. In addition, it is suggested that a removable joint (RTV/stainless screws) be installed at a strategic location should disassembly be required at a later time.

Piping Must be:

- Permanent
- Air-Tight
- Insulated

SLIDE 88

The piping should be cut and supported such that it is angled 1/4" per linear foot upward from the furnace. Piping should be supported with pipe hangers to prevent sagging. Maximum spacing between hangers is five (5) feet, except SDR-PVC piping, where maximum spacing is three (3) feet. If the vent piping will be subjected to freezing temperatures such as routing through unheated areas or through an unused chimney it must be insulated with 1/2" thick Armaflex. The combustion air piping must be insulated with 1/2" thick Armaflex insulation if installed above a suspended ceiling or in a warm, humid space, such as a laundry room.

Piping Must be:

- Angled Upward
- Supported
- Insulated

SLIDE 89

The final assembly procedure for the vent/combustion air piping is as follows:

1. Begin at the furnace and cut the piping to the proper length.

2. Cut pipe end square, remove ragged edges and burrs. Chamfer end of pipe, then clean fitting socket and pipe joint area of all dirt, grease or moisture.

3. Dry-fit the entire vent/combustion air piping assembly.

4. Disassemble the piping and apply primer and cement following the manufactures instructions.

Piping Assembly

- Cut to length
- Cut square & de-burr
- Chamfer & clean
- Check for proper fit
- Apply cleaner/primer
- Apply cement

Connect field-supplied piping to the condensate drain connection and run it to an open drain following the above procedure.

SLIDE 90

The combustion air and vent piping MUST terminate outdoors using the proper size Vent-Terminal Assembly Kit. The vent-terminal kit is an integral part of the design certified furnace and must be installed without alteration or modification. The 1VK0307 kit is to be used with 1-1/2" and 2" piping, and the 1VK0308 kit is to be used for 3" piping. The 3" vent-terminal kit contains one elbow having a "splitter baffle" in one opening. This elbow must be used for the combustion air intake. The terminal kits may be used for rooftop or side wall installation. Rooftop termination is the recommended means and should be arranged according to one of the methods shown here.

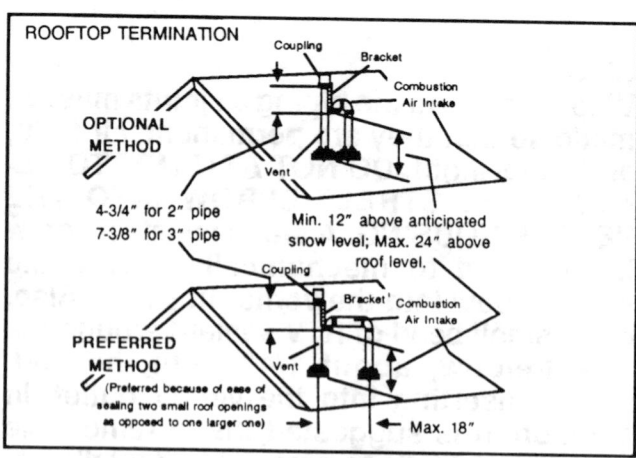

30

SLIDE 91
These examples of side wall venting show:

1. The vent kit terminating both above and below anticipated snow levels.

2. When combustion air and vent piping can not be routed together, the position of the pipes may be reversed so that the combustion air inlet is on the opposite side of the vent. In all instances, roof-top included, the intake elbow is BELOW the vent.

Once the method of termination has been determined, the illustration showing that method must be followed such that all dimensional requirements are fulfilled.

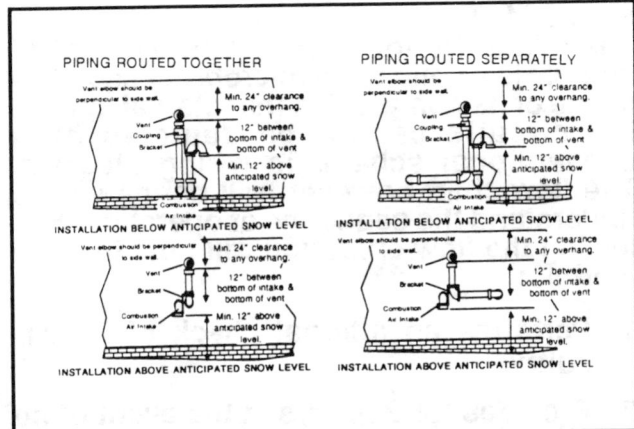

When properly installed, any wind, whether of steady or blowing velocity, blowing toward or across the vent-terminal hits the discharge and inlet at the same instant canceling each other out.

SLIDE 92
The thermostat heat anticipator setting is 1.16 amps for all P2UD models. Some thermostats have a "cycle rate" rather than an adjustable anticipator. On these thermostats adjust the cycle rate down to prevent possible short on/off cycles.

Thermostat

Anticipator Setting

1.16

SLIDE 93
The next subject for discussion is the Hot Surface Ignition system. The Hot Surface Ignition (HSI) system proves the presence of flame through the principle of flame rectification. Flame rectification is a process which converts alternating current (AC) into direct current (DC).

An AC voltage is applied to the flame sensor electrode. When flame is present, the fuel gas molecules between the flame sense electrode and the burner (ground) become conductive and allow a small current to flow. Because of the surface area difference between the flame sensor electrode and the burner (ground electrode) this current flow takes place mostly in one direction only.

The detector circuit in the HSI control module accepts ONLY this pulsating DC

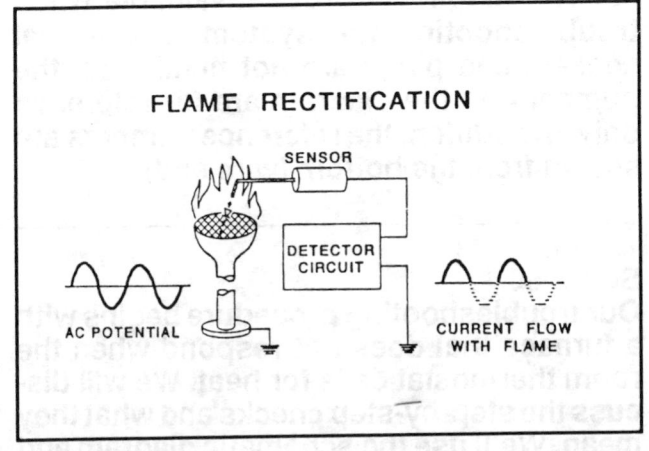

FLAME RECTIFICATION

signal, of at least 2 micro-amps, as proof that flame is present. If a short circuit exists (between the electrodes), an AC signal is detected. The HSI control module does NOT accept this as proof of flame. If flame is not present current cannot flow and NO signal is detected.

SLIDE 94

The White-Rodgers HSI (50E47) control module is a solid state microprocessor. Because of its complexity, the internal operating circuits cannot be depicted by a conventional schematic wiring diagram. The flow chart however, will permit you to understand the sequence of operation. Following the flow chart you will note that the module:

1. Performs an internal check of its circuitry.

2. Provides for 2 re-trys in the event of no flame during initial operation.

3. Provides for 5 recycles if flame is lost after 4 seconds of operation.

The flow chart does not show it but, a momentary loss of gas supply, flame blowout, or a shorted or open condition in the flame sense circuit will be detected in 8/10 of a second and the system will shut down. After a 60 second delay recycles will

---

HSI FLOW CHART

(See Appendix)

---

begin. If the fault is corrected prior to the last ignition attempt, the unit will operate normally. Otherwise, the control will lock-out. A Light Emitting Diode (LED) provides visual indication as to the likely cause of lockout. To reset the control interrupt either the 24 volt or line voltage circuit for 1/20 second or longer.

SLIDE 95

Now that we have seen the sequence of operation we are prepared to troubleshoot a "no heat" service problem using the connection and elementary diagrams. The connection diagram shows the electrical components and their relative location within the unit. The legend tells us that the white plug and socket (WP/WS) is 24 volts, the rust colored (RP/RS) is 115 volts and the test socket (TS) is 24/115 volts. These sockets and plugs will prove invaluable when troubleshooting the system. The actual sockets and plugs are not numbered, the numbers on the diagram are for reference only. In addition, the reference numbers are shown from the bottom (wire end).

CONNECTION DIAGRAM

(See Appendix)

TO PREVENT A WRONG DIAGNOSIS WHEN TROUBLESHOOTING A SYSTEM, MAKE SURE YOU ARE TESTING THE PROPER CIRCUIT.

SLIDE 96

Our troubleshooting procedure begins with a furnace that does not respond when the room thermostat calls for heat. We will discuss the step-by-step checks and what they mean. We'll use the schematic diagram and a voltmeter to check for AC voltage at various locations until the problem is located. If the problem is found to be a defective control do not try to repair it. REPLACE DEFECTIVE CONTROLS WITH CES SOURCE 1 Parts.

ELEMENTARY DIAGRAM

(See Appendix)

Central Environmental Systems

SLIDE 97

A schematic wiring diagram is drawn to simplify the circuit for ease of understanding. It separates the various load circuits and identifies the controls which cause each load to function. The terminals on a load or control may not be in exactly the same location or sequence as shown in the diagram. To simplify this wiring diagram it has been divided at the transformer to show only the 24 volt or 115 volt circuits.

1ST - Check for 24 volts at terminal "R" and "C" on the control transformer terminal board. If it's not present we'll continue to the 2ND step.

```
1ST CHECK

(See Appendix)
```

SLIDE 98

2ND - Check for 115 volts at TS9 and TS5. If voltage is present it indicates either an open circuit or the transformer is defective. Check all lead wires and plugs/sockets for continuity before replacing any component. Do NOT insert the meter probe into the female electrical contact in the plug. (You might enlarge the contact and cause an open circuit when the plug is mated into the socket and further compound your problem.) If you find that 115 volts is not available at TS9 and TS5, make the 3RD check.

3RD - Check for 115 volts at TS6 and TS5. No voltage here indicates an open fuse or broken wire in the power supply. If voltage is present check the door switch. (Remember that power is shut off when the blower access door is removed.)

```
2ND & 3RD CHECKS

(See Appendix)
```

SLIDE 99

If 24 volts was originally present at terminals "R" and "C" on the transformer terminal board, make the 4TH check.

4TH - Check for 24 volts at terminals "W" and "C" on the transformer terminal board. If it's not present, check the room thermostat, sub-base and thermostat wiring. If power is available at "W" and "C", check for 24 volts at TS4 and TS2. If 24 volts is not available, check the orange and blue wires and also the WP/WS terminals 2 & 3. With 24 volts available at TS4 and TS2, make the 5TH check.

```
4TH CHECK

(See Appendix)
```

SLIDE 100

5TH - Check for 24 volts at the coil terminals of the 3R (venter) relay. If 24 volts is not available the 1LP pressure switch is checked next. Put one lead of the voltmeter on the 3R coil terminal (blue wire) and the other lead on the pressure switch common terminal (red wire). If 24 volts is not available there but is available at the NC terminal, the pressure switch is open and must be replaced. If 24 volts was originally present at the 3R coil terminals the 3R relay contacts, on lines 7 and 23 should be closed. To determine if the relay or venter motor is defective we must continue to the next check.

5TH CHECK

(See Appendix)

SLIDE 101

6TH - Check for 115 volts between terminal 3 of the 3R relay and TS5. No voltage tells us that the 3R relay is not closing its contacts and must be replaced. If 115 volts is available the motor must be checked next. Carefully feel the motor. If it is at ambient temperature, the windings have an open circuit and the motor/venter must be replaced. If the motor is hot, the internal thermostat may have opened because of a defective run capacitor. Using a capacitor tester, check the run capacitor. (The run capacitor is located in the venter/motor junction box.) If the capacitor is okay, the motor has one open winding and the motor/venter assembly must be replaced.

6TH CHECK

(See Appendix)

If the venter motor runs but the hot surface igniter does not glow, we must check to see if the 1LP pressure switch has closed its contacts.

SLIDE 102

7TH - Check for 24 volts between TS2 and the normally open (NO) contact of the 1LP pressure switch. If 24 volts is not available, check the pressure switch hose connections to make sure they are tight and not leaking. Consult the installation instructions to ensure that the vent system length or amount of elbows has not been exceeded and all hoses are connected to the proper pressure taps. Make sure the pressure tap in the burner box is tight and is not leaking. Check the electrical connections on the pressure switch and plug connectors. If, after making these checks, the pressure switch contacts still do not close, we will check the pressure at the switch.

7TH CHECK

(see Appendix)

SLIDE 103

To check the pressure you must have an inclined manometer, or slope gauge as it is also known. This inclined manometer is capable of measuring 1" water column pressure and the fluid level is indicating 0.68" WC on the scale. The .68 reading could be positive or it can be negative. If the "B" connection is open to the atmosphere and we apply a POSITIVE pressure to the "A" connection, the liquid level will rise and indicate the pressure. However, if the "A" connection is open to the atmosphere and we apply a negative pressure to the "B" connection, atmospheric pressure being greater, forces the liquid level upward to indicate the pressure. The inclined manometer can also measure the difference between two positive or two negative pressures and we call this the differential pressure. Lets consider two NEGATIVE pres-

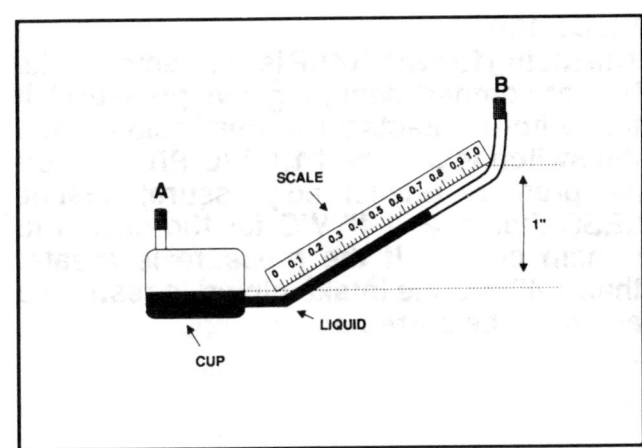

sures, a -1" WC and a -1.68" WC. If the greater pressure (-1.68) is connected to "B" and the lesser pressure (-1") is connected to "A", the difference between the two pressures (.68") is indicated on the scale.

SLIDE 104

The pressure switches are now marked with the calibration point (in inches of W.C.), and eiher a PR or PF to designate Pressure Rise or Pressure Fall. If specific tolerances are not listed, the following may be used as a guide: ±.05" W.C. up to .50" and 10% for set points higher. To properly use the inclined manometer(0-2" WC), it must be connected, using tees, at the pressure switch. If you are not sure whether the measured pressure is positive or negative, start the furnace and momentarily connect the tube from the pressure switch to each side of the inclined manometer until the fluid rises and indicates a pressure reading on the scale.

Presssure Switch Calibration

- **Inches Water Column**

- **PR = Pressue Rise**

- **PF = Pressure Fall**

- **±.05" W.C. up to .50" 10% for Higher Set Points**

SLIDE 105

Since the pressure in the burner compartment and collector box are less than atmospheric pressure, the 1LP switch is measuring a NEGATIVE differential pressure. If the switch is stamped .78" WC PF, the normal differential pressure (measured pressure) must be GREATER than .78" WC to close the normally open (NO) contact and permit normal operation. If the volume of gases through the furnace is gradually reduced because of a restruction or operating problem, it will also cause the differential pressure to gradually fall until it reaches the set point of .78" WC PF and open the (now closed) NO contact to de-energize the ignition control module and gas valve. If the measured pressure is less than the set

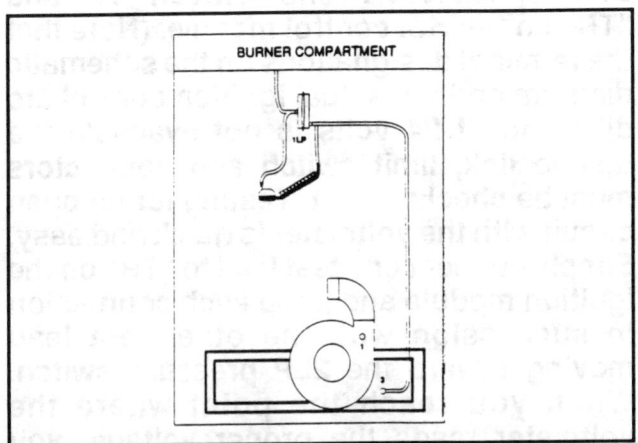

point, the combustion blower, heat exchangers, drain and intake/vent piping must be inspected.

SLIDE 106

The tubing to switch 4LP is connected to the burner compartment (negative pressure). If the switch contacts are normally closed and the switch is stamped 1.4"WC PR, the normal pressure (measured pressure) must be LESS than the -1.4" WC for the switch to remain closed. If the pressure is greater than 1.4" WC the intake piping is restricted and must be corrected.

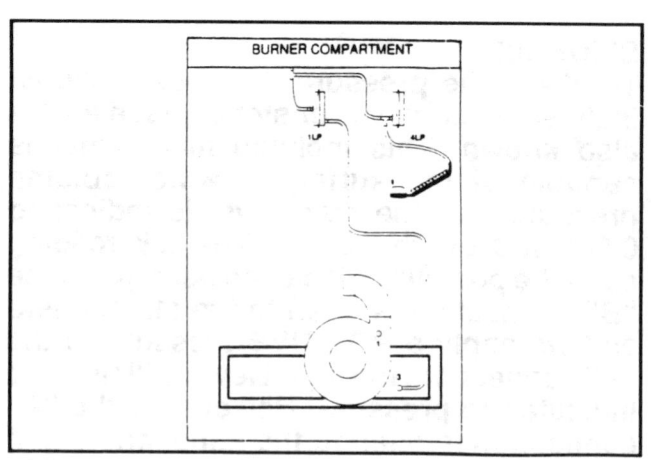

SLIDE 107

If 3LP has NC contacts and is stamped 1.4" WC PR, the contacts open on pressure rise. Since the tubing is connected to the outlet of the venter housing the pressure is positive. The measured pressure must be LESS than 1.4" WC. If the pressure is greater than 1.4" the vent piping is restricted. If the pressure is less than 1.4" and the NC switch contacts are open, the switch is defective. After correcting the problem the hot surface igniter should glow. If not, go on to the next check.

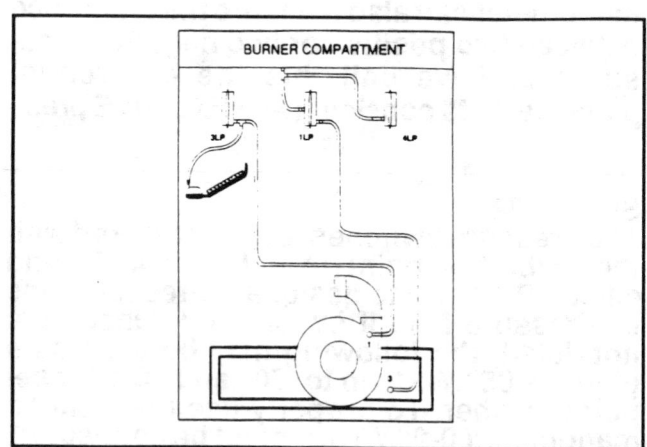

SLIDE 108

8TH - Check for 24 volts between "TH" and "TR" on the HSI control module. (Note that the terminal designations on the schematic diagram and the actual ignition control are different.) If 24 volts is not available the fusible link, limit switch and connectors must be checked next. Testing for an open circuit with the voltmeter is quick and easy. Simply connect one test lead to "TR" on the ignition module and jump each connection in succession with the other test lead moving toward the 2LP pressure switch. When you reach the point where the voltmeter reads the proper voltage, you have just jumped the open switch or broken wire, whichever the case may be. If 24 volts

8TH CHECK

(See Appendix)

is available at "TR" and "TH" and the igniter still does not glow, proceed to the next check.

SLIDE 109

9TH - DISCONNECT ELECTRIC POWER TO THE SYSTEM AT THE MAIN FUSE OR CIRCUIT BREAKER. Connect an ohmmeter across TS3 and TS8. The resistance of the igniter should be between 20 and 50 ohms. If you are unable to read the proper resistance, the igniter must be replaced. Prior to replacement, check the igniter leads and plugs/sockets to the ignition module and recheck the resistance of the igniter to verify your findings. If the igniter and wiring are okay, the ignition module is defective and must be replaced. If the LED is on continuously, interrupt the power for a few seconds. If an internal fault is indicated again, replace the control. At this point the igniter should glow and the main burners

9TH CHECK

(See Appendix)

ignite. If the main burners do not light, we must check further.

SLIDE 110

10TH - Check for 24 volts between "MV"/"MV" at the ignition control module. (After the ignition warm up period, you only have 4 seconds in which to perform these tests). If there is no voltage at these terminals replace the module. If voltage is available at the module, check to see if it is available at terminals 1 & 2 of the gas valve and also between terminals 2 & 3. If voltage is at 1 & 2 and NOT at 2 & 3 the jumper wire is open. If 24 volts is present at the gas valve check the following: the gas supply shut-off valve; the supply gas pressure (must be greater than 4" WC, but not exceeding 14" WC) and the gas valve manual control knob which should be turned to the ON position. If after these checks the gas valve does not

10TH CHECK

(See Appendix)

open, replace the gas valve. Note the heater connected to the gas valve terminals 2 & 3. This is the fan heater in the combination fan and limit control.

SLIDE 111

If the furnace operation is erratic or if the main burners do not stay on until the call for heat ends, check the ground wire and sensor wire. If the ground is poor or erratic, shut down may occur occasionally even though operation is normal at the time of checkout. To test the sensor wire, you must disconnect it from the "FP terminal of the ignition control. Check the wire for continuity. If you do not measure zero ohms, replace the wire. Next, test the wire for infinity between the (removed) terminal and the furnace casing. If you do not read infinity, either the wire or the flame sense electrode is defective and must be replaced.

**Check Sensor
& Ground Wires**

If the flame sense probe is in the burner flame and all checks are okay, replace the ignition control module.

SLIDE 112

The next subject we will cover is maintenance. (When performing any maintenance, the electrical power must be turned off.) The most neglected maintenance item is filters. The filters should be checked periodically for dirt accumulation. Clean the filters at least every three months. On new construction, check the filters every week for the first four weeks. Inspect the filters at least every three weeks after that, especially is the system is running constantly. All filters supplied with the furnace are the high-velocity, cleanable type. Clean these filters by washing in warm water. When replacing filters, be sure to use the same size and type as originally supplied.

SLIDE 113

Blower motors in these furnaces are of the permanently lubricated type and do not require periodic oiling. Even with good filters properly in place, blower wheels and motors will become dust laden after long months of operation and will require cleaning.

Blower Servicing

SLIDE 114

Squeeze the locking tabs on the electrical plugs and pull straight up to remove them from the control box sockets. As shown here, the next step is to remove the four retaining screws that secure the control box. Hold the box when removing it to prevent it from falling and position it out of the way.

SLIDE 115

Remove the hoses from the condensate trap. After removing the blower retaining screws the blower/motor assembly, control box and interconnecting wires are removed from the furnace for maintenance or repair. Reverse the above steps to reinstall the components. When replacing the control box retaining screws remember to reconnect the venter motor ground wire.

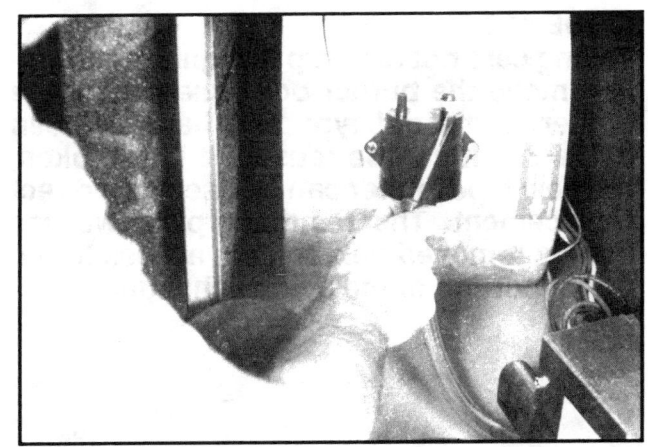

SLIDE 116

With proper adjustments the heat exchangers should not require cleaning for a prolonged period of time. However, when cleaning is required, follow these steps: Turn off the main manual gas valve external to the furnace and the electrical power. Remove the screws which secure the combustion air inlet flange to the burner box. To prevent air leaks and hasten reassembly, replace one of the screws back into the burner box. Because of the different type of screws required for proper assembly, this "replacement" procedure should be followed when performing any type of service or repair.

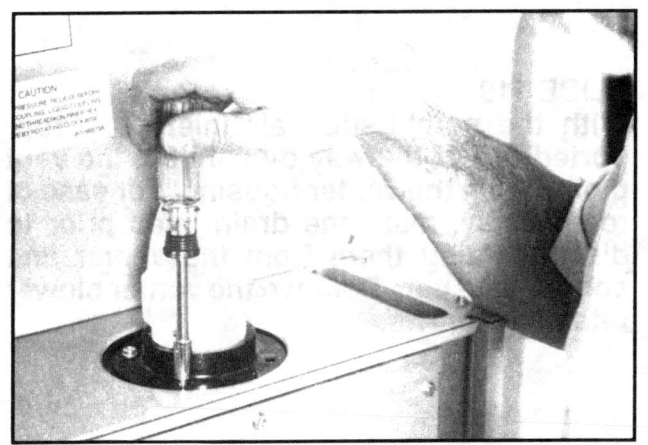

SLIDE 117

If the evaporator condensate drain line is in the way, remove it. With the burner access door removed, disconnect the igniter, sensor, and fusible link wiring. After removing the harness connector nut, pull the harness connector and wires from the burner box. Disconnect the gas supply piping and control wiring from the gas valve. Remove the screws that hold the burner box assembly to the vestibule as shown here.

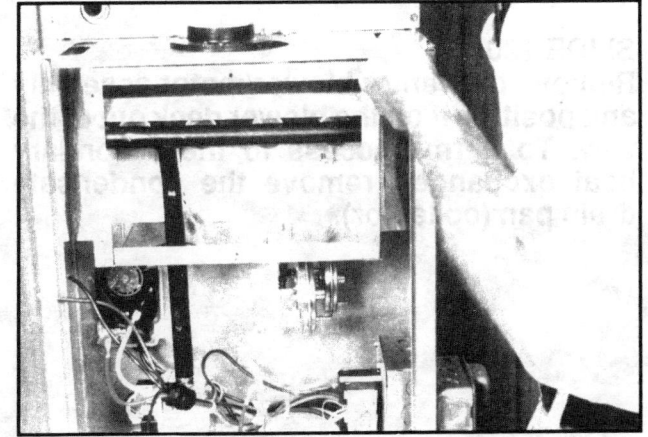

SLIDE 118
Taking care not to bump the igniter, carefully remove the burner box assembly. If the igniter is the old type and has not been damaged, it can be reused. If it is broken, the new type igniter can be used as a direct replacement. The restricter plate we see here is removed next to gain access to the upper portion of the heat exchanger.

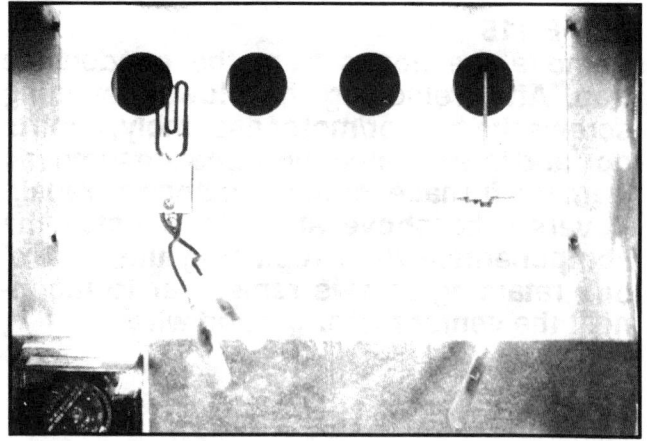

SLIDE 119
With the combustion air inlet pipe supported out of the way disconnect the vent piping from the venter housing. For ease of reassembly, mark the drain lines prior to disconnecting them from the venter and condensate pan. Remove the venter blower retaining screws.

SLIDE 120
Remove the venter blower/motor assembly and position it on the blower deck out of the way. To permit access to the secondary heat exchanger, remove the condensate drain pan (collector).

Central Environmental Systems

SLIDE 121
To reach the lower portion of the primary heat exchanger the insulated, aluminized deflector plate must be removed. With the deflector plate off, remove the screws holding the finger baffles and pull the baffles out of the unit.

SLIDE 122
After removing the turbulators from the secondary heat exchanger, brush out loose scale or soot from both heat exchangers with a stiff wire brush. Use a vacuum cleaner to clean the debris from the heat exchangers. Visually inspect the heat exchanger sections for cracks, openings or excessive corrosion. If any of these conditions are found, the heat exchanger must be replaced. To save time during reassembly, all parts removed for cleaning MUST BE replaced by reversing the order of disassembly. When reinstalling the finger baffles make sure they are positioned horizontally and are resting on the internal indentations in the heat exchanger. Reconnect all wiring and gas piping prior to restoring electrical power and gas supply to the unit.

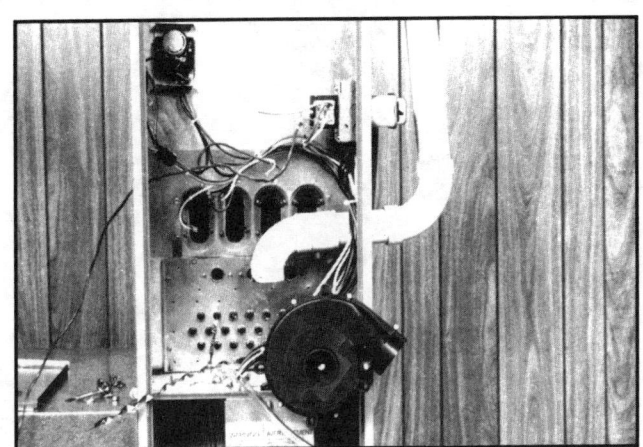

SLIDE 123
Questions and answers . . . . before this or any installation is complete there are some questions that must be answered. Some questions you must ask yourself, some questions will be asked by the equipment owner, and there are also questions that will not be asked. Some questions to ask yourself are: Is the main power turned on? Is the gas supply on? Are all system function operable from the thermostat? Have you cleaned your work area and disposed of any trash? Have you presented the equipment owner with all the instructions and paperwork? Have you explained the warranty and warranty procedures? If at this time the owner of the equipment does not have any questions, it may be that he does not know what to ask. Show him where the filters are, explain how often to change them and show him how. Explain the thermostat operation and how to set it. Explain how the system works and the sequence of operation. Show him where the disconnect switch, fuses and gas supply shutoff valve is located. Explain how to shut the system down.

Questions

and

Answers

Having the answers to these questions may very well prevent you from having to come back at a later time.

Hopefully, during this presentation we have answered some of the questions you may have wanted to ask. If you have any unanswered questions, we will answer them now.

NOTES

APPENDIX - SEQUENCE OF OPERATION - HSI CONTROL

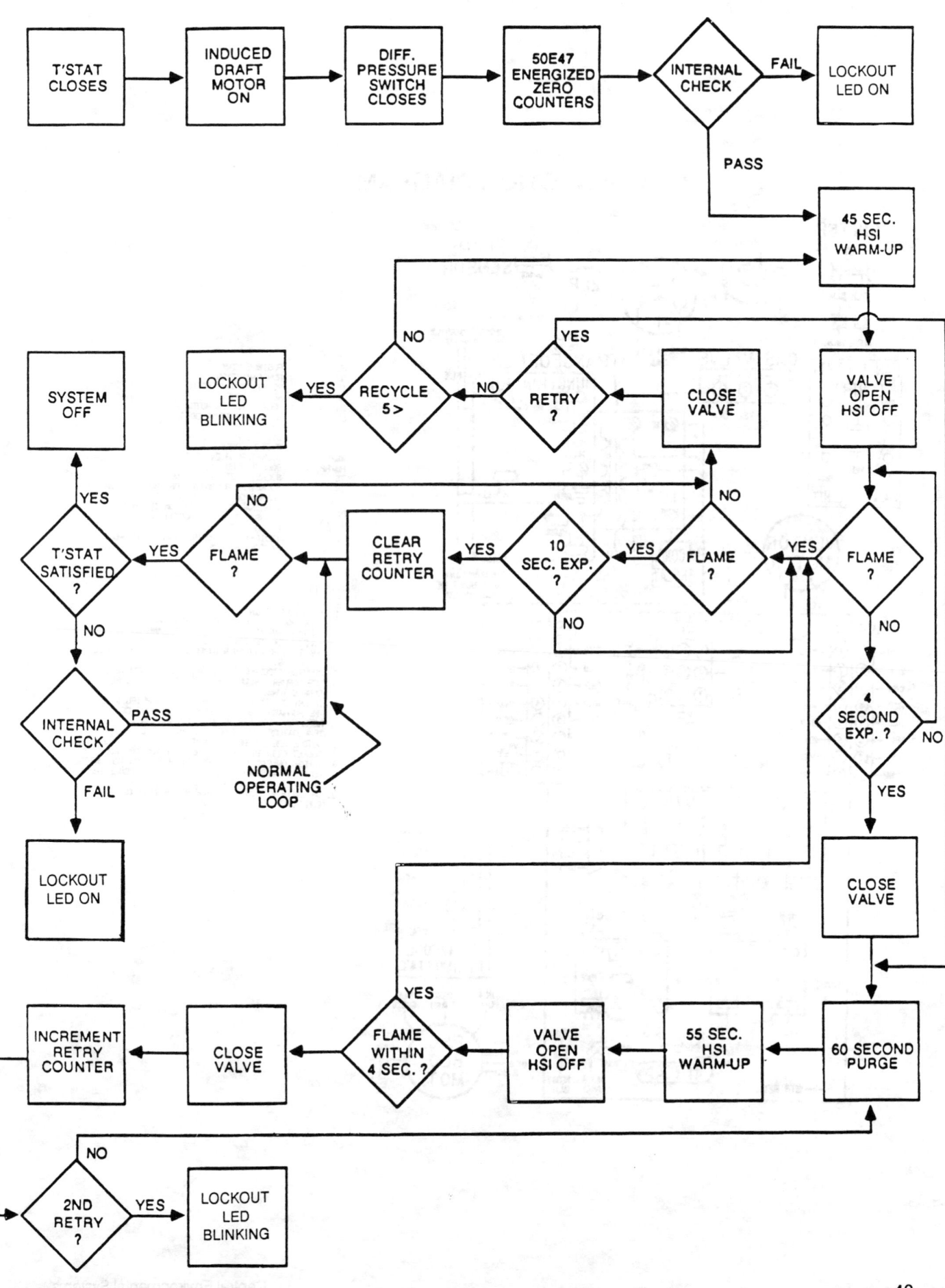

CONNECTION DIAGRAM

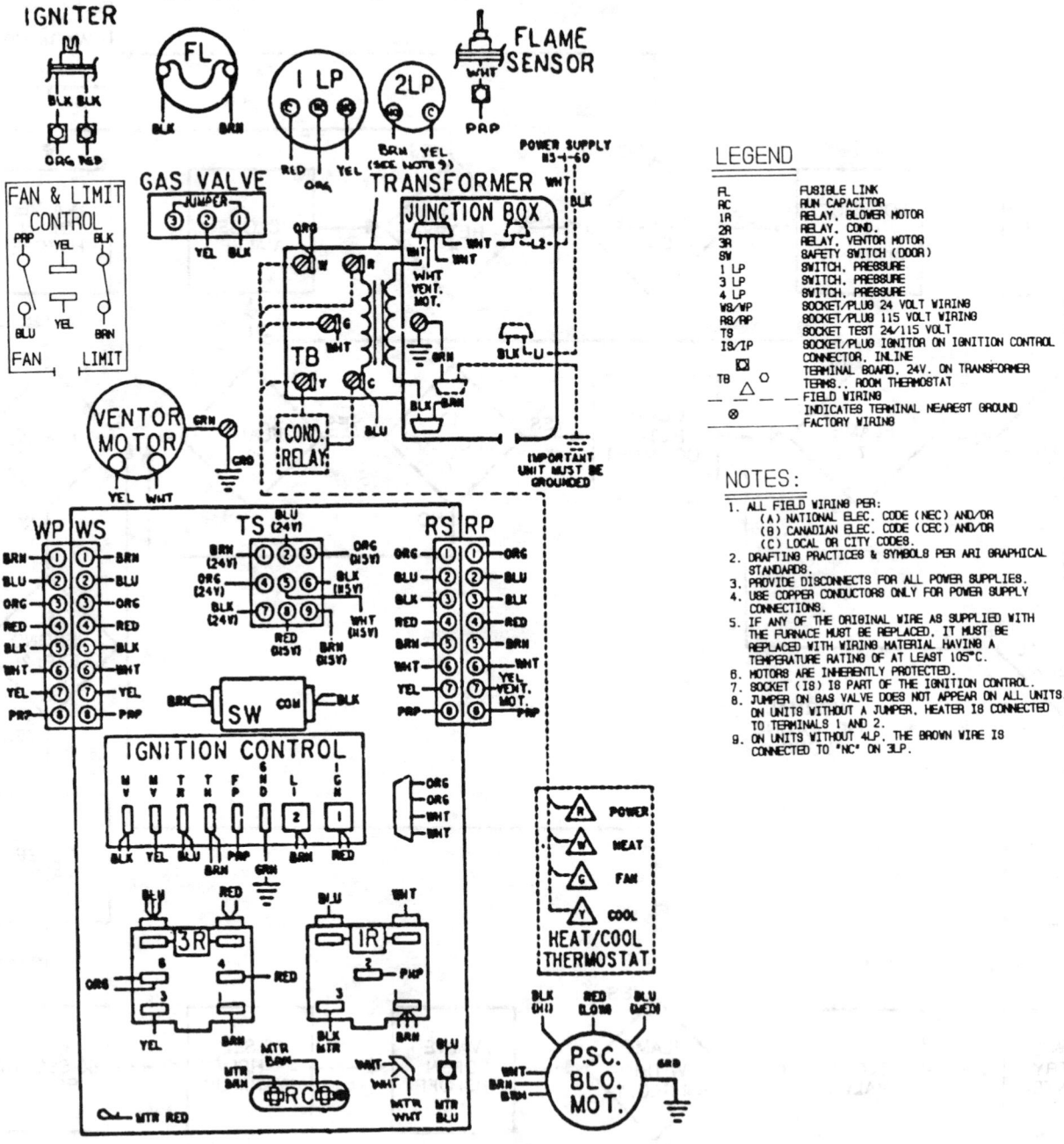

APPENDIX - ELEMENTARY DIAGRAM

ELEMENTARY DIAGRAM

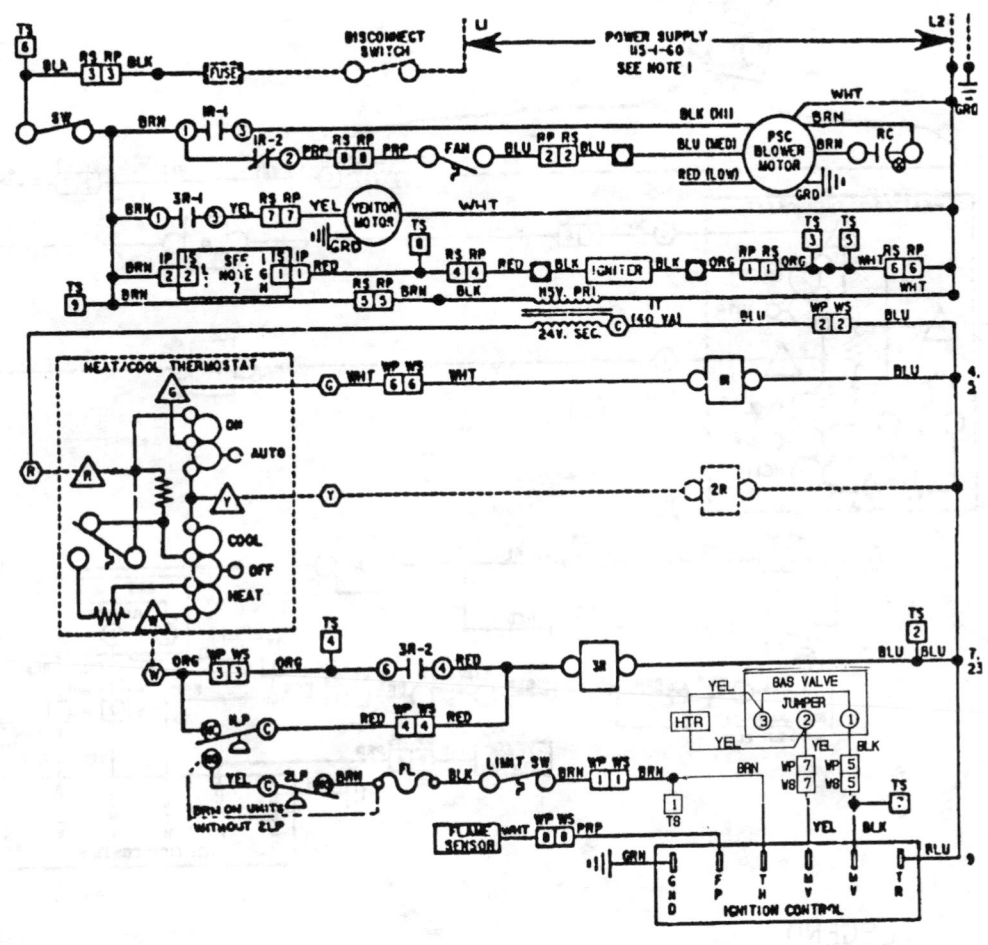

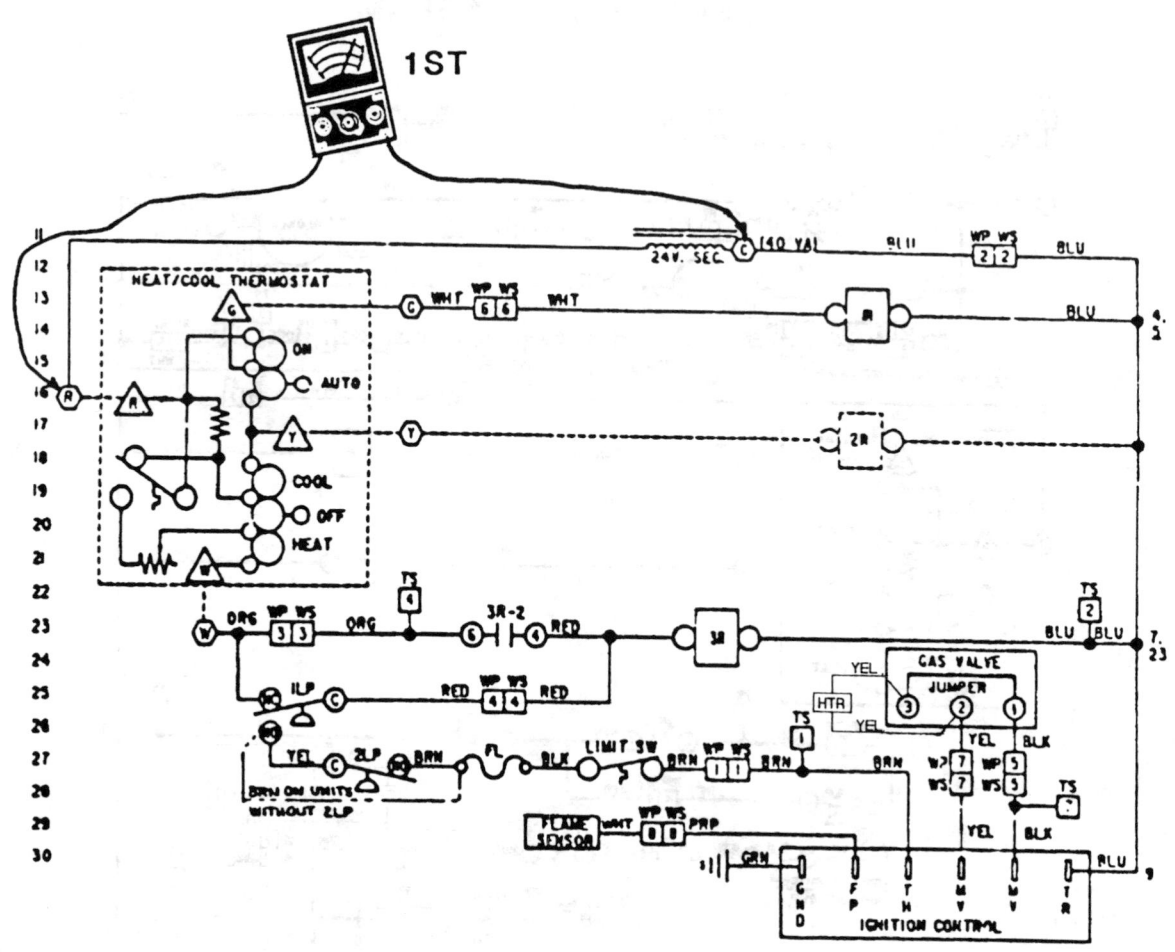

LEGEND

FL	FUSIBLE LINK
RC	RUN CAPACITOR -
IR	RELAY, BLOWER MOTOR
2R	RELAY, COND.
3R	RELAY, VENTOR MOTOR
SW	SAFETY SWITCH (DOOR)
I LP	SWITCH, PRESSURE
2 LP	SWITCH, PRESSURE
WS/WP	SOCKET/PLUG 24 VOLT WIRING
RS/RP	SOCKET/PLUG 115 VOLT WIRING
TS	SOCKET TEST 24/115 VOLT
IS/IP	SOCKET/PLUG IGNITOR ON IGNITION CONTROL
	CONNECTOR, INLINE
TB	TERMINAL BOARD, 24V. ON TRANSFORMER
	TERMS.. ROOM THERMOSTAT
	FIELD WIRING
●	INDICATES TERMINAL NEAREST GROUND
	FACTORY WIRING

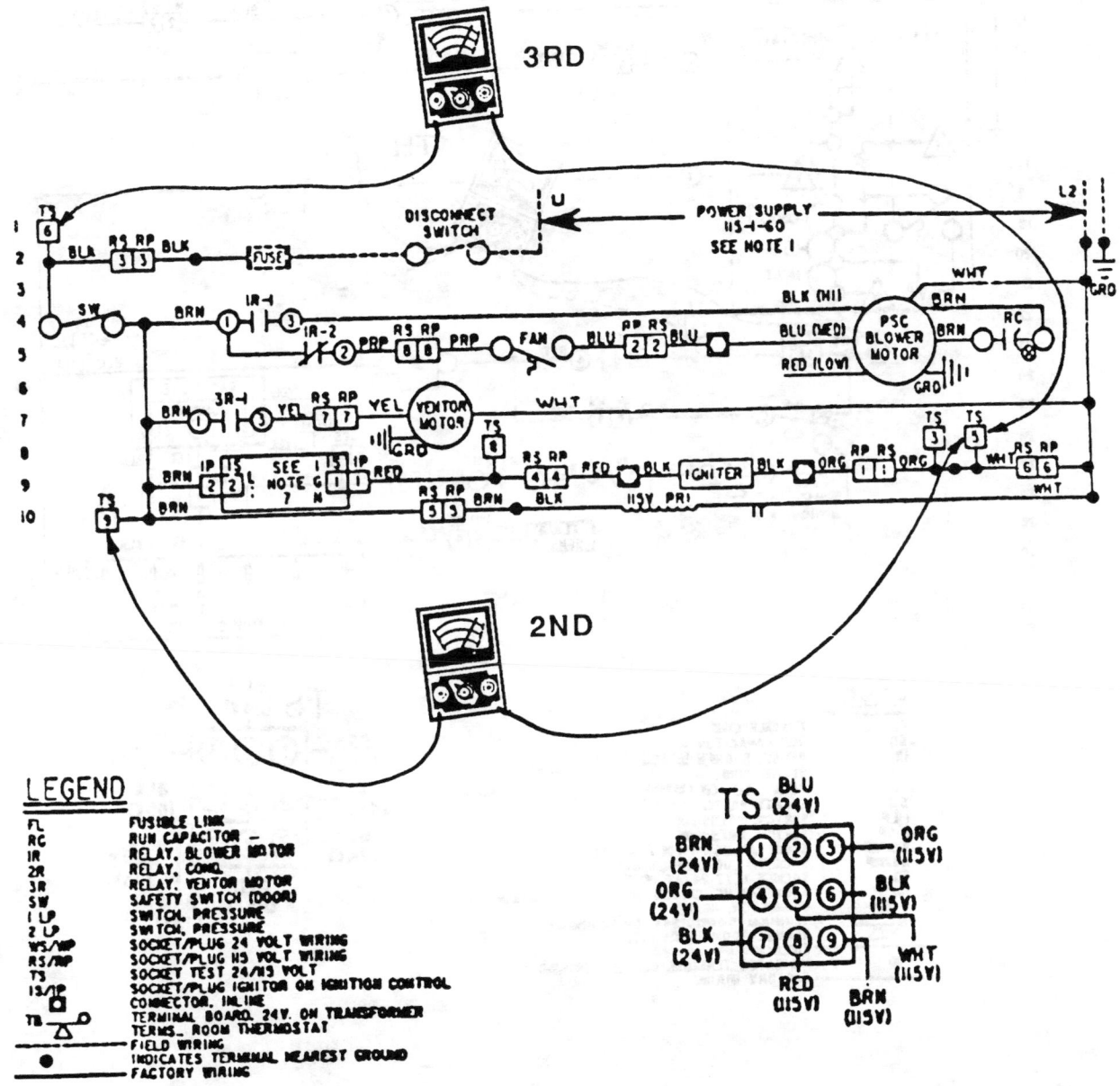

LEGEND

FL	FUSIBLE LINK
RC	RUN CAPACITOR —
IR	RELAY, BLOWER MOTOR
2R	RELAY, COND.
3R	RELAY, VENTOR MOTOR
SW	SAFETY SWITCH (DOOR)
1 LP	SWITCH, PRESSURE
2 LP	SWITCH, PRESSURE
WS/WP	SOCKET/PLUG 24 VOLT WIRING
RS/RP	SOCKET/PLUG 115 VOLT WIRING
T3	SOCKET TEST 24/115 VOLT
IS/IP	SOCKET/PLUG IGNITOR ON IGNITION CONTROL
	CONNECTOR, INLINE
TB	TERMINAL BOARD, 24V. ON TRANSFORMER
	TERMS.. ROOM THERMOSTAT
	FIELD WIRING
●	INDICATES TERMINAL NEAREST GROUND
	FACTORY WIRING

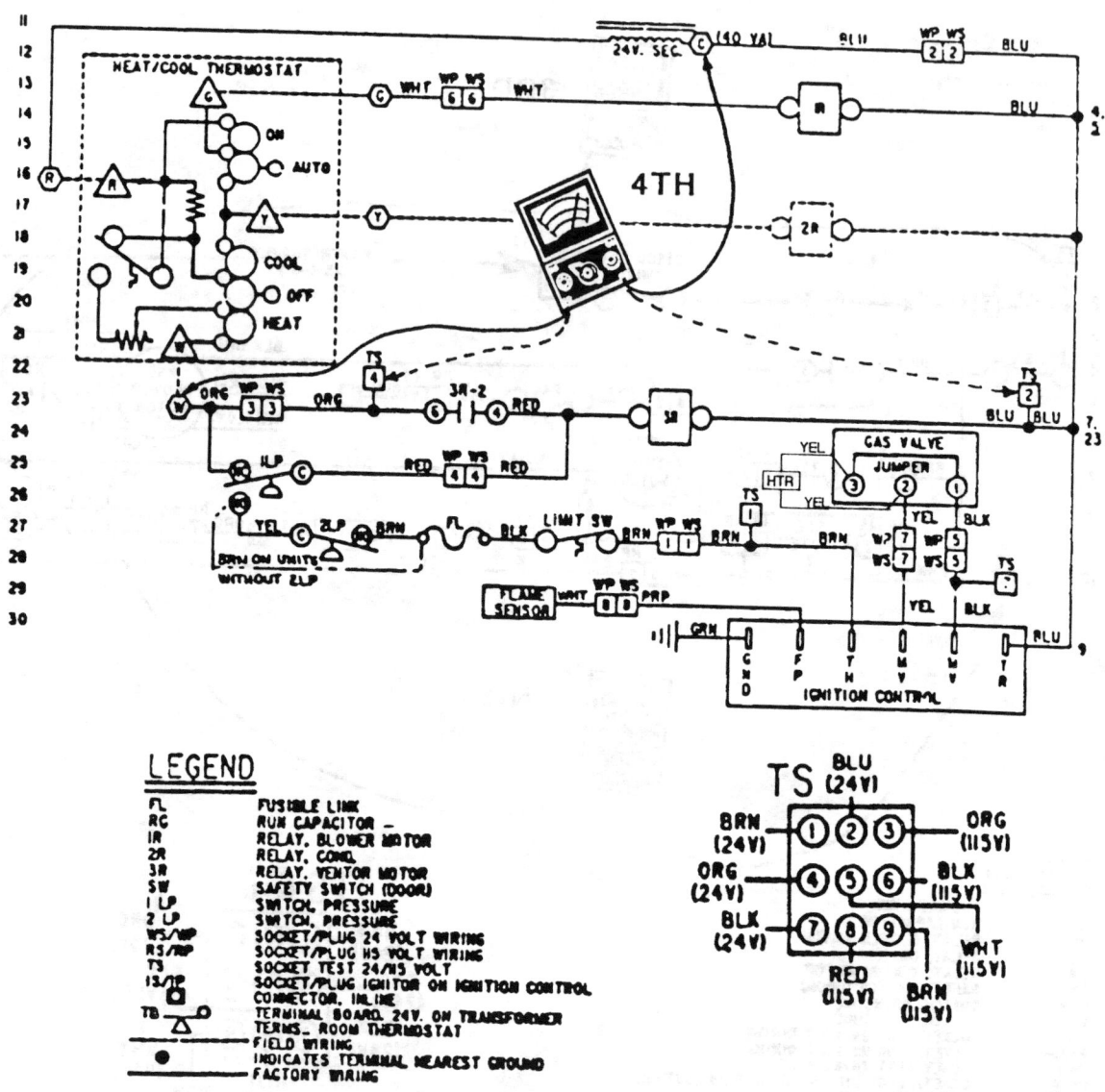

LEGEND

FL	FUSIBLE LINK
RC	RUN CAPACITOR
IR	RELAY, BLOWER MOTOR
2R	RELAY, COND.
3R	RELAY, VENTOR MOTOR
SW	SAFETY SWITCH (DOOR)
1 LP	SWITCH, PRESSURE
2 LP	SWITCH, PRESSURE
WS/WP	SOCKET/PLUG 24 VOLT WIRING
RS/RP	SOCKET/PLUG 115 VOLT WIRING
TS	SOCKET TEST 24/115 VOLT
IS/IP	SOCKET/PLUG IGNITOR ON IGNITION CONTROL
	CONNECTOR, INLINE
TB	TERMINAL BOARD 24V. ON TRANSFORMER
TERMS.	ROOM THERMOSTAT
	FIELD WIRING
●	INDICATES TERMINAL NEAREST GROUND
	FACTORY WIRING

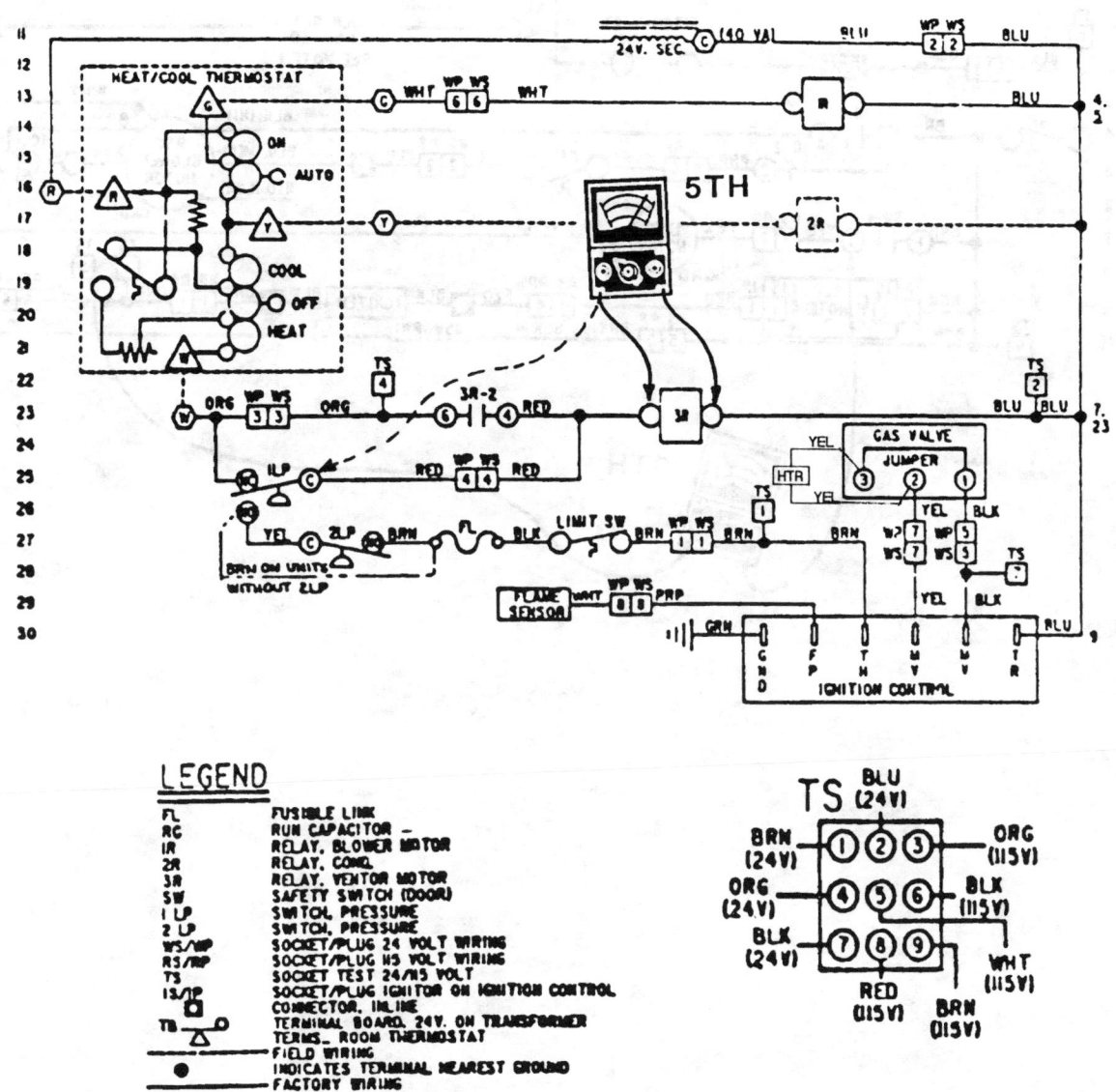

LEGEND

FL	FUSIBLE LINK
RC	RUN CAPACITOR -
IR	RELAY, BLOWER MOTOR
2R	RELAY, COOL
3R	RELAY, VENTOR MOTOR
SW	SAFETY SWITCH (DOOR)
1 LP	SWITCH, PRESSURE
2 LP	SWITCH, PRESSURE
WS/WP	SOCKET/PLUG 24 VOLT WIRING
RS/RP	SOCKET/PLUG 115 VOLT WIRING
TS	SOCKET TEST 24/115 VOLT
IS/IP	SOCKET/PLUG IGNITOR ON IGNITION CONTROL
	CONNECTOR, INLINE
TB	TERMINAL BOARD, 24V. ON TRANSFORMER
	TERMS. ROOM THERMOSTAT
— — —	FIELD WIRING
●	INDICATES TERMINAL NEAREST GROUND
——	FACTORY WIRING

APPENDIX - TROUBLESHOOTING - STEP 6

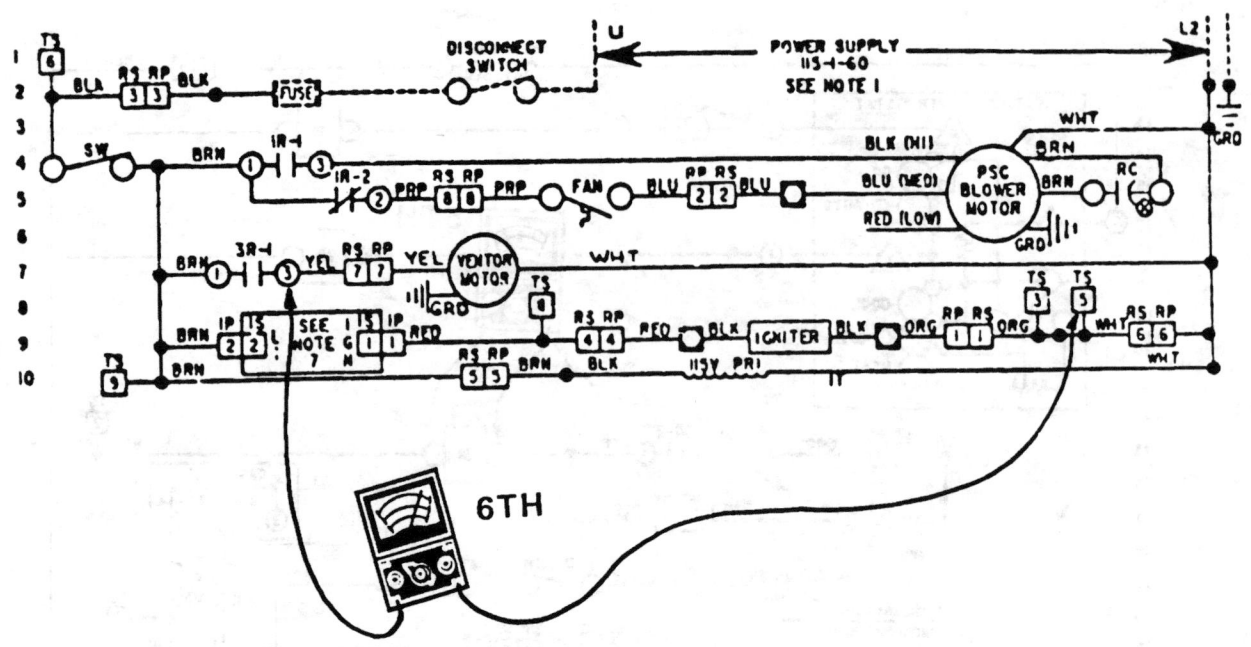

6TH

Central Environmental Systems

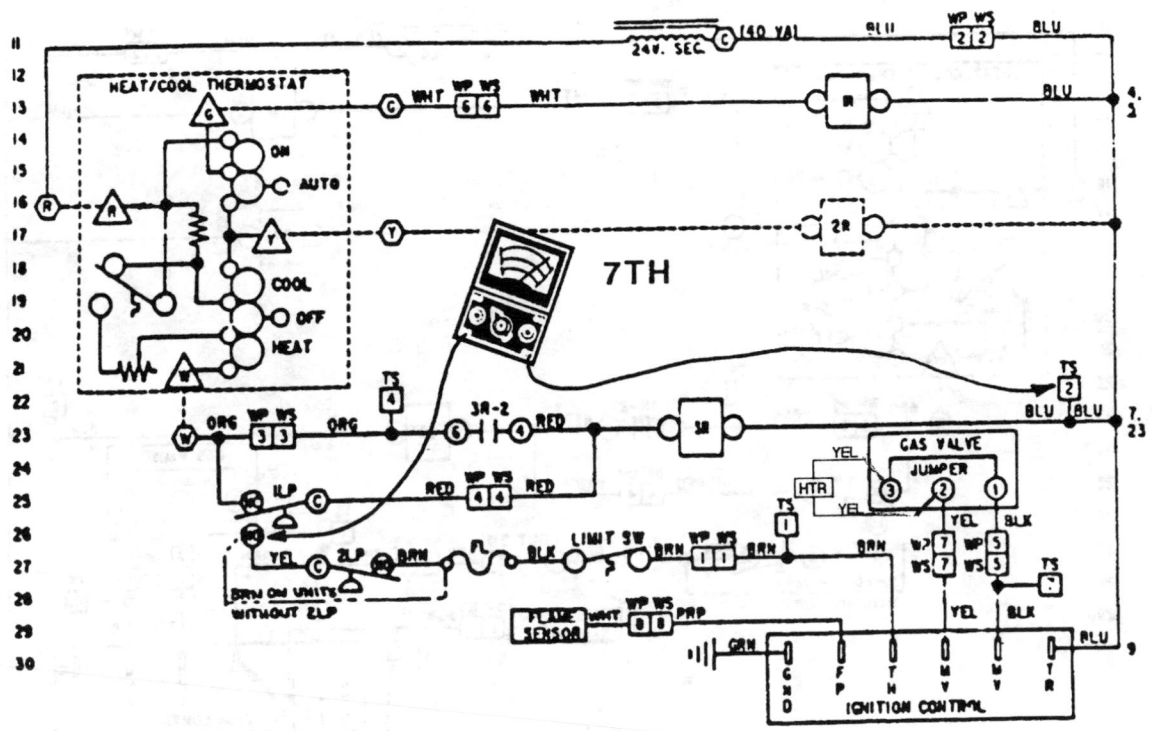

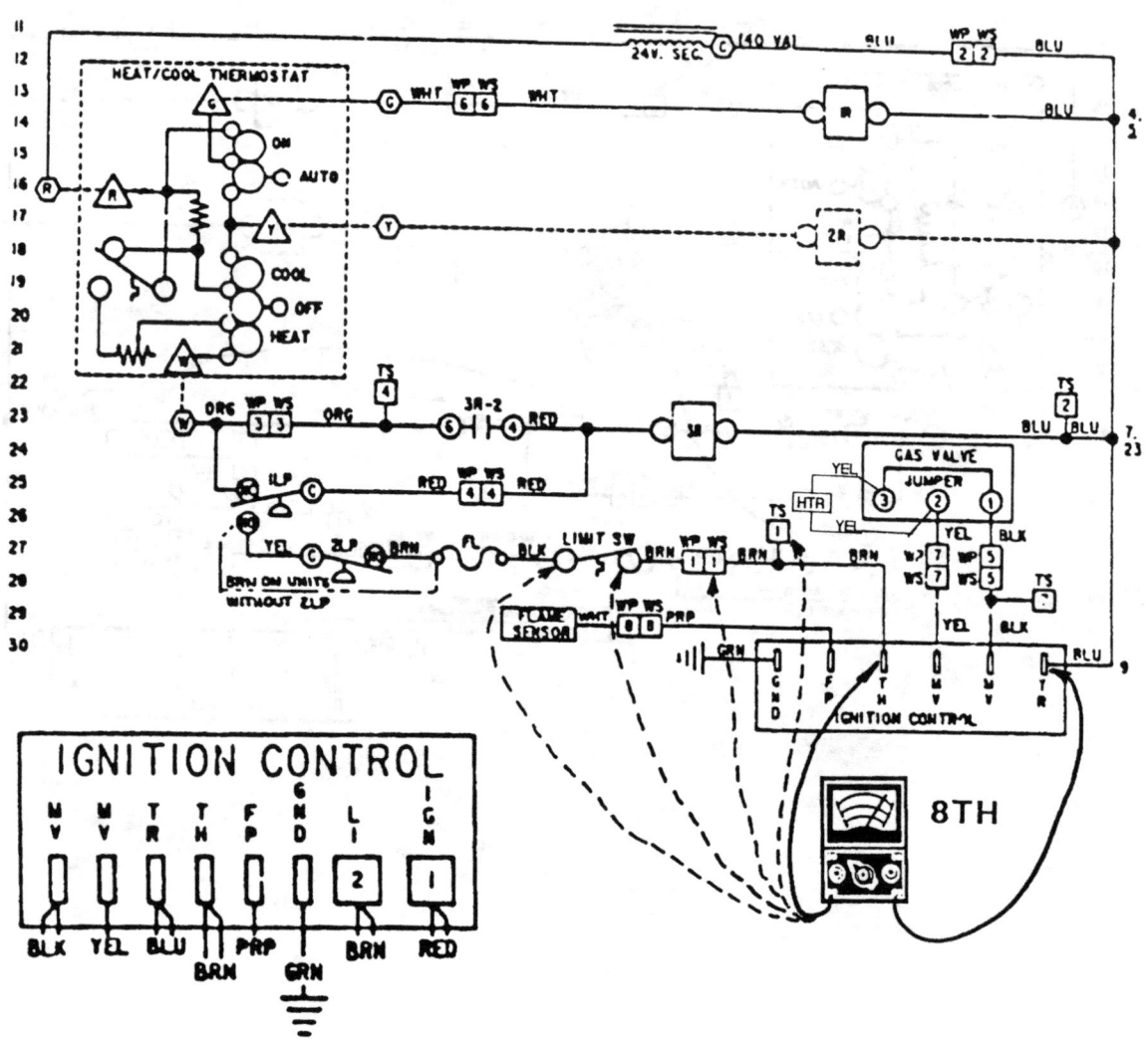

POWER <u>OFF</u>

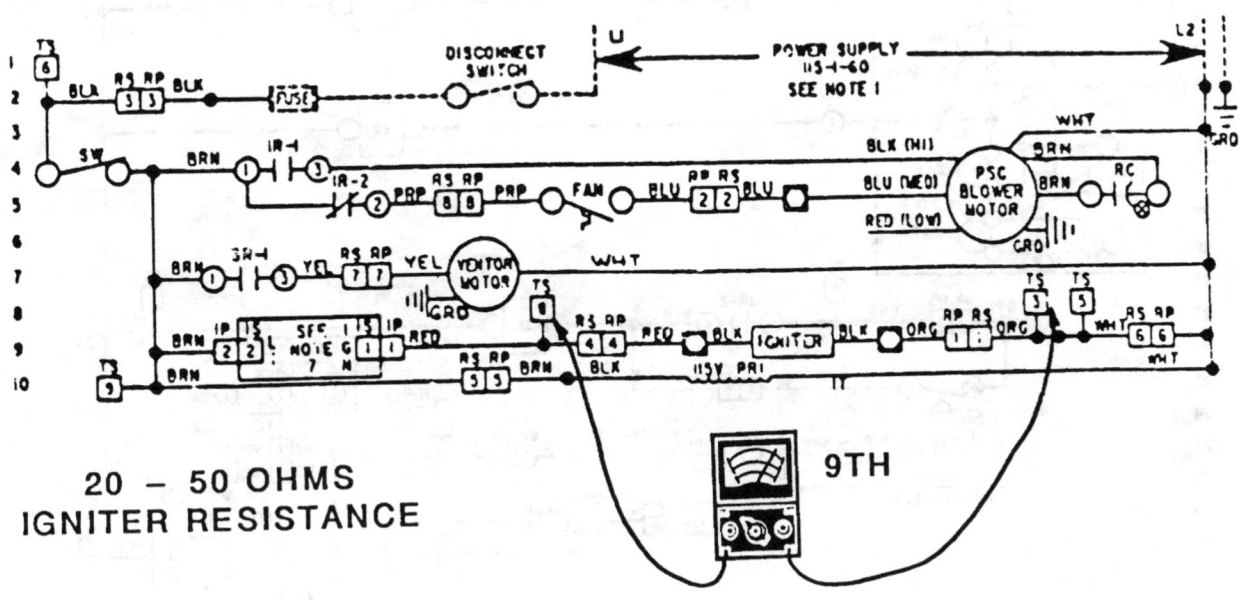

**20 — 50 OHMS
IGNITER RESISTANCE**

9TH

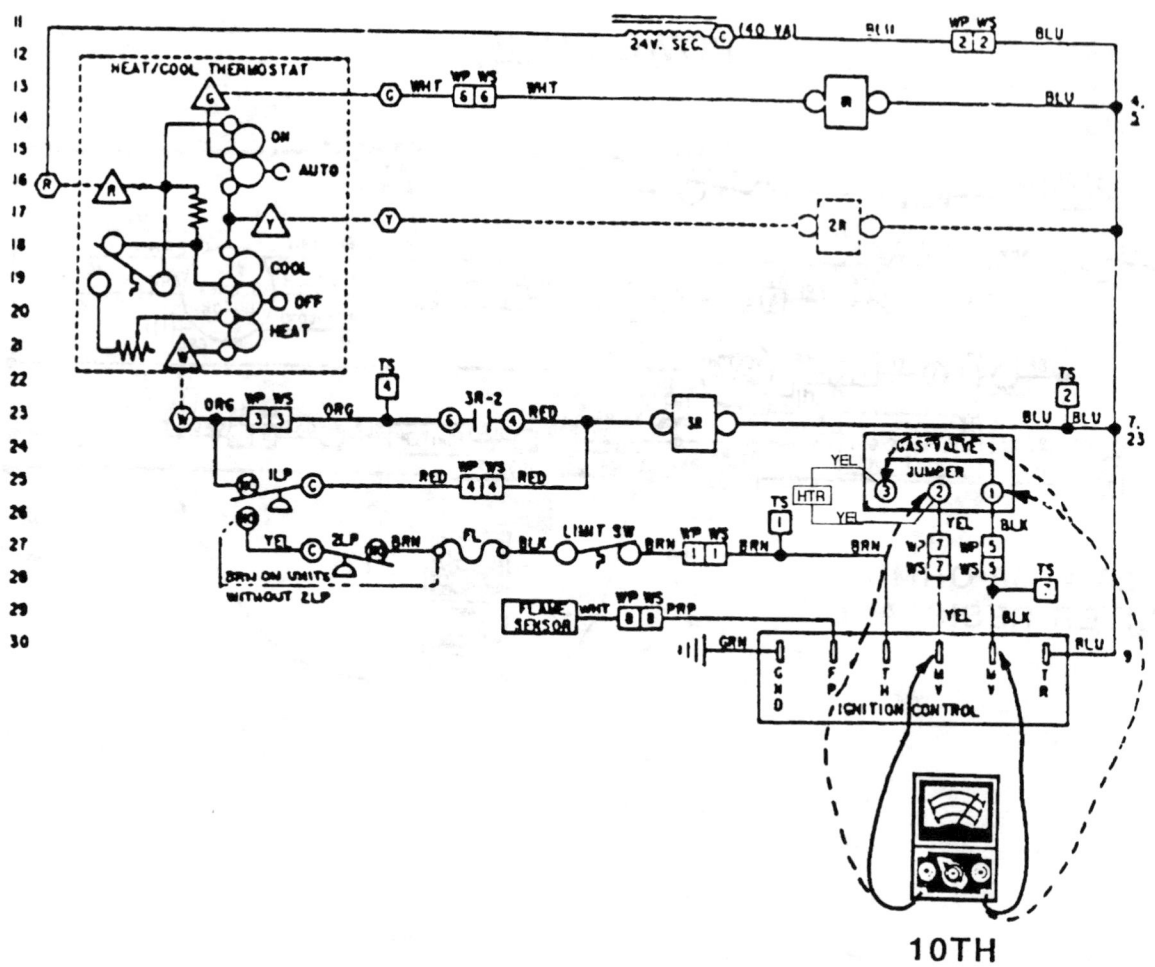

10TH

HIGH-EFFICIENCY CONDENSING GAS FURNACE QUIZ

NAME _____   DATE_____

TRUE (T) or FALSE (F) Circle correct answer

T   F   1. These furnaces have AFUE ratings of 85%.

T   F   2. Conversion from natural gas to propane is not permitted.

T   F   3. For heat/cool applications, the accessory relay kit is required.

T   F   4. Because of the closed base, these units may be installed directly on carpeting or floor tiles.

T   F   5. The fan off setting is not adjustable.

T   F   6. On all sizes of units the gas supply piping may enter from side.

T   F   7. The primary heat exchanger has a 10-year warranty.

T   F   8. The furnaces are classified category IV and must be vented using approved plastic vent pipe.

T   F   9. The differential pressure switches are non-adjustable.

T   F   10. When converting from propane to natural gas the ignition module does not have to be changed.

T   F   11. Since these units utilize 100% outside combustion air, they may be installed in mobile homes.

T   F   12. These units may be roof vented or side vented.

T   F   13. The maximum amount of elbows permitted for either combustion air or vent piping is 10.

T   F   14. Combustion air and vent piping may be of different diameter.

T   F   15. The HSI control module is programmed for 2-retries in the event of no flame during initial operation.

T   F   16. If the HSI control module is locked out and the LED is flashing, the control module must be replaced.

T   F   17. The venter motor must be oiled annually.

T   F   18. The secondary heat exchanger has a 1-year warranty.

T   F   19. The fan on setting is field adjustable.

T   F   20. The hot surface igniter consumes approximately 7 watts per ignition cycle.

T   F   21. Rear return air is not permitted on these units.

T   F   22. The maximum length of combustion air/vent pipe is 25 feet.

T   F   23. The vent terminal kit is not required so long as proper dimensions are maintained.

T   F   24. To test the pressure switches requires an inclined manometer capable of measuring 1.5" WC.

T   F   25. The heat exchanger sections must be cleaned annually.

YORK®

Heating and Air Conditioning

U S A
Official Sponsor
of the 1992
U.S. Olympic Team
36USC380

P.O. Box 1592, York, Pennsylvania USA 17405-1592
Subject to change without notice.   Printed in U.S.A.
Copyright © by York International Corporation 1991.   All rights reserved.   RPC 1M 1191 1.50

Supersedes: 650.65-TB11Y (988)
650.65-TB13Y

YORK®

Silver Star

TRAINING COURSE

Sales/Application
and
Installation/Service

GAS-FIRED FURNACES
MODELS P3UC/P3CC (UPFLOW AND DOWNFLOW)
(USE WITH A/V PROGRAM 650.64-TS13Y [890])

SILVER STAR TRAINING PROGRAM
GAS FIRED FURNACES
P3UC/P3CC (UPFLOW AND DOWNFLOW)

NOTE: Leaders should read this preface prior to conducting a meeting.

SCOPE

Part 1 covers sales and application information. Part 2 explains installation and service techniques for the P3UC/P3CC Models. The text may be divided depending on the interests of the audience.

OBJECTIVE

To communicate, in laymans terms, all aspects of this product.

Recommended Literature for 80 + Furnaces

Sales Brochure	650.64-CM3Y
Technical Guide	650.64-TG3Y
Installation Instructions	650.64-N3Y
Propane Conversion Kit Instructions	650.64-N3.1V
Vent Application Guide	650.64-AD2V
Fundamentals of Venting	503.81-TB21V
Gas Furnace Application/Installation	504.48-TB21V

SLIDE 1
This is a Silver Star presentation . . .

SLIDE 2
Introducing another star in York's galaxy of fine products . . . The Stellar 2000 Series . . .

SLIDE 3
Featuring the P3UC/P3CC line of upflow and downflow fuel gas furnaces.

P3UC/P3CC
UPFLOW
DOWNFLOW
GAS FURNACES

SLIDE 4
In this part of the program we will describe the important sales and application features that will help you to increase your sales of high efficiency furnaces.

SALES
AND
APPLICATION
FEATURES

SLIDE 5
Let's take a moment to talk about the gas furnace market. For the past several years, approximately 2 million furnaces have been sold annually. On the average, the models sold are 70.8% upflow, 14.7% downflow and 14.5% horizontal. During this presentation we will be concerned only with the upflow and downflow furnaces which represent 85.5% of sales.

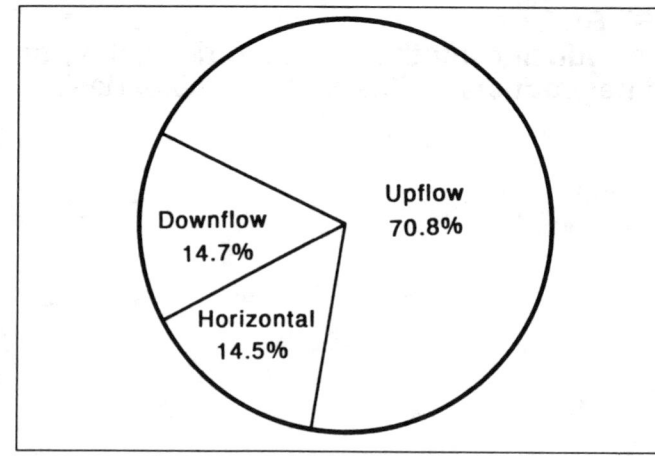

SLIDE 6
This slide illustrates upflow furnace sales in the various AFUE ranges. note that 13.8% of sales were in the 71-79 AFUE range. Three years ago this segment was only 5%. As you can see, the trend is toward higher efficiency furnaces.

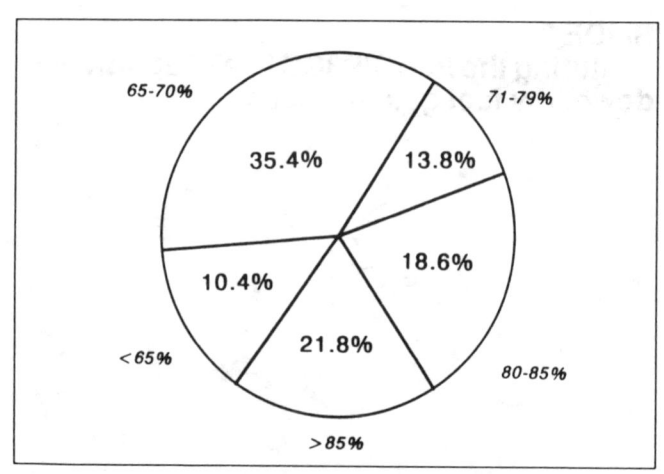

2

SLIDE 7
Here we see a similar chart for the downflow furnaces. Last year furnaces below 71% AFUE accounted for 58.4% of total sales. Three years ago this same segment accounted for 85.4%. These charts tell us that nationally, there is a strong and growing market for high efficiency upflow and downflow furnaces.

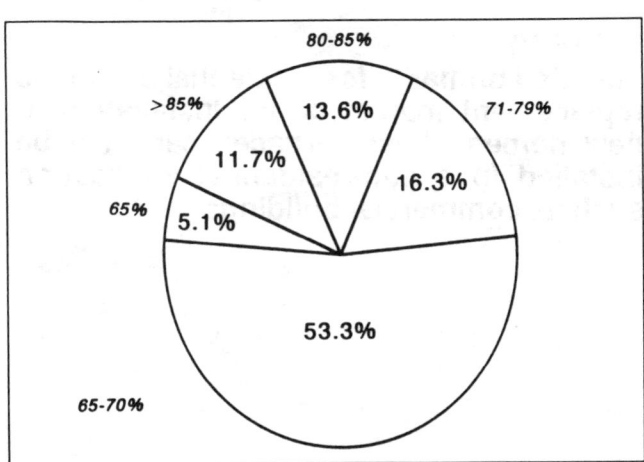

SLIDE 8
Locally, however, increased sales are dependent on the market needs or requirements in your specific area. For example, these local needs may include one or more of the following.

- Nitrous Oxide (NOx) requirement
- Intermittent Ignition Devices.
- Annual Fuel Utilization Efficiency (AFUE) requirements
- Propane operation
- Low BTU input/high airflow

In all except extreme instances, this line of furnaces will satisfy your local needs.

LOCAL NEEDS

or

REQUIREMENTS

SLIDE 9
And finally, to satisfy the application needs of a specific installation, these compact furnaces can be installed in a basement, closet, alcove, recreation room or garage where the continuous return air temperature is not below 55°F.

APPLICATIONS

- BASEMENT
- CLOSET
- ALCOVE
- RECREATION ROOM
- GARAGE

SLIDE 10

Intended primarily for residential use in the replacement market or for installation in new homes, these furnaces can also be installed in a non-residential application such as commercial buildings.

> **Replacement Installations**
>
> **New Homes**
>
> **Commercial Buildings**

SLIDE 11

This series of furnaces have numerous improved design features such as heat exchangers, burners, igniter and pressure switch sensing method. They are completely factory assembled, wired and tested to assure dependable operation. Additionally, all models have heavy duty 40VA transformers and fan relays, making them suitable for heating only or heating/cooling applications.

> **Factory Assembled**
> **Factory Tested**
> **Heating Only**
> **or**
> **Heating/Cooling**

SLIDE 12

To assure safe operation, the units are Design Certified by the American Gas Association (A.G.A.).

A.G.A. tests for compliance with requirements of the American National Standards Institute (ANSI).

These stringent ANSI tests are designed to simulate adverse operating conditions and are concerned with such items are:

1. Combustion and ignition characteristics.

2. Limit and safety control operation.

3. Heat exchanger and casing temperature.

4. Wiring and component temperatures.

Central Environmental Systems

SLIDE 13

To assist you in selecting the proper model for a particular application, let's take a moment to review the product nomenclature for a typical unit.

Reading from left to right, the "P" stands for furnaces. The "3" designates third generation. The "CC" relates to the product identifier and stands for High Efficiency Downflow. Another identifier, "UC" denotes High Efficiency Upflow. The "D08" means direct drive blower, 800 CFM. The letter "N" means Natural Gas, (Low Nox models would have an "L" in this position). The next three digits, in this case "045", indicate the Nominal Output Heating Capacity in thousands of BTU/Hr. The last two digits "01" is the voltage code and means 115 volts, single phase, 60 Hertz power supply.

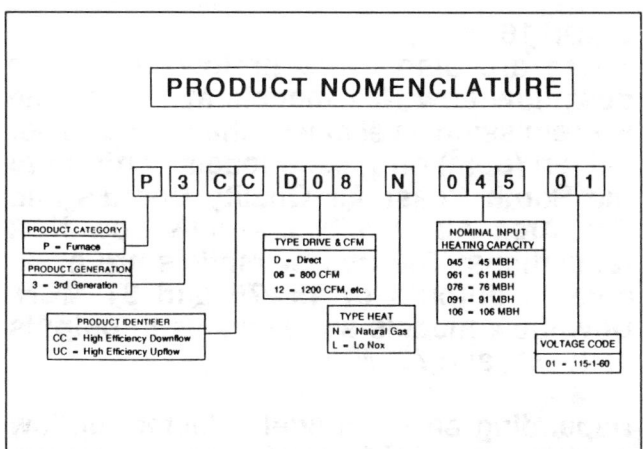

SLIDE 14

The P3UC upflow models are currently available with OUTPUT capacities of 45, 61, 76, 91 and 106 MBH. Depending upon the model, these units have annual Fuel Utilization Efficiency (AFUE) ratings between 78% and 79.4%. AFUE is determined in accordance with D.O.E. test procedures by the isolated combustion air method.

P3UC Upflow

Heating Capacities
(MBH Output)

45, 61, 76, 91, 106

SLIDE 15

The P3CC downflow models are available with OUTPUT capacities of 45, 61, 76, 91 and 106 MBH. Depending upon the model, these units have AFUE ratings between 79.3% and 80.2%.

P3CC Downflow

Heating Capacities
(MBH Output)

45, 61, 76, 91, 106

SLIDE 16

There are also three upflow and three downflow Low Nox models available. The Nox emission level of less than 40 parts per million (ppm) on these models conform to the South Coast Air Quality Board's and Bay Area Air Quality District's Low NOx regulations. The upflow models are available in outputs of 45, 76 and 91 MBH. Downflow models are available in outputs of 45, 61, and 76 MBH.

Depending on the model selected, upflow models have California Seasonal Efficiencies between 78.3% and 79.4% as determined in accordance with test procedures as specified by the state of California Energy Commission. The California Seasonal Efficiencies of downflow models range between 79.5% and 80.0%.

Low NOx Models
(MBH Output)

- Upflow 45, 76, 91
- Downflow 45, 61, 76

SLIDE 17

For replacement installations, the range of BTU outputs permit selection of a unit that will closely match the heating requirements of the dwelling. This will result in more efficient operation of the unit for the consumer.

Replacement Installations

SLIDE 18

Factory built for use with natural gas, all sizes and model, except Low NOx models, can be field converted for propane operation. An accessory natural gas to propane conversion kit is required and will be covered later.

Natural Gas
or
Propane

(Except Low NOx Models)

SLIDE 19

This chart taken from the consumer brochure compares different furnace efficiencies and shows the yearly savings. If the current unit is more than 10 years old, the efficiency is likely to be about 60%. If the annual operating cost is $700, the cost to operate a new 80% AFUE furnace will be $515, an annual savings of $185.

Approximate Annual Operating Cost

Furnace Efficiency							
60%	$400	$500	$600	$700	$800	$900	$1000
65%	$365	$460	$550	$640	$735	$825	$915
70%	$340	$425	$510	$595	$675	$760	$845
75%	$315	$395	$470	$550	$630	$710	$785
80%	$295	$365	$440	$515	$585	$660	$735
85%	$255	$320	$385	$450	$515	$580	$640
90%	$240	$305	$365	$425	$485	$545	$605

To use this chart:

Estimate the approximate efficiency of your current furnace. If it's more than 10 years old, the efficiency is likely to be about 60%. Then locate your current annual operating costs, based on your gas heating bill, eliminating the cost of operating other gas appliances. Use the chart to estimate how much you can save with a more efficient furnace.

Example:

If your current furnace is 60% AFUE, and your annual operating cost is $700, the cost to operate a new 80% AFUE furnace will be about $515, an annual savings of about $185.

SLIDE 20

The newer, energy-efficient homes require less BTU's to maintain comfort conditions at winter design temperatures. These homes require smaller capacity furnaces. However, in heating/cooling applications, because of the cooling load, they require high airflow. All furnaces in this series have direct-drive blowers with multi-speed motors. This permits blower operation at one CFM for cooling and another CFM for heating.

New Homes

SLIDE 21

Although these units are considered residential furnaces, they may be used for commercial applications as long as the capacity of the furnace is sufficient to handle the heat loss of the building. Of course, a word of caution is in order . . . in most areas, heating codes vary between residential and commercial buildings, with commercial codes being more stringent. Always be sure the product you choose will meet the specs for the job.

Building Codes — Residential Codes — Commercial Codes

SLIDE 22

These compact units have been designed so that they will fit and match perfectly with companion air conditioning coils. All upflow and downflow models are only 46" high. These low heights allow their installation with or without air conditioning coils in areas with low ceilings. All models are 28-1/2 inches from front to rear. The widths vary from 16-1/4" to 26-1/4", depending on capacity.

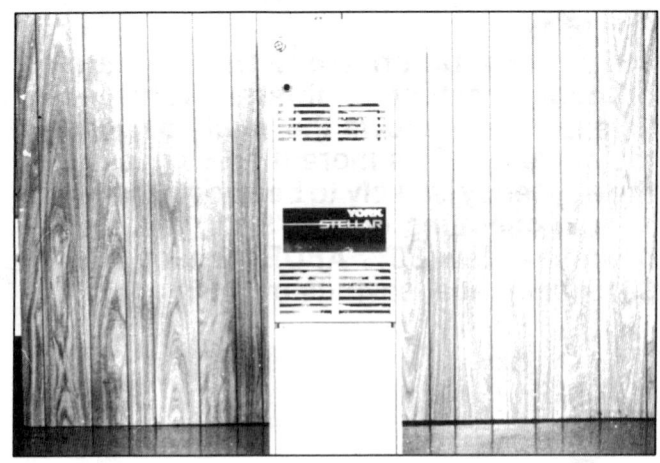

SLIDE 23

To provide many years of satisfactory service, the one piece casing is fabricated from heavy-gauge steel. For additional rigidity and to prevent noise complaints, the casing panels are formed with surface indentations. This also helps to prevent distortion or damage during shipping and handling. To provide a durable, smooth finish, all panels are degreased, bonderized, and finished with baked enamel. For ease of installation and maintenance, the front panels may be lifted up and removed, allowing full access to all component parts.

SLIDE 24

All units are furnished with permanent-type, high-velocity filters. On upflow models the filters are shipped mounted on the side or bottom and can be relocated if necessary. On the 91 and 106 MBH model two filters are supplied. The filters are secured in position with a wire retaining loop which is easily removable for filter maintenance. The control box in front of the blower houses the blower access door interlock switch.

SLIDE 25
The blower access door interlock switch is factory-installed on all models. When the blower access door is removed this switch will automatically de-energize the electrical power, preventing operation of the unit.

SLIDE 26
All the upflow units have a solid bottom panel, permitting installation on wooden floors. On all sizes the return air may enter from the bottom of the unit. Should the installation require bottom return air, the bottom closure panel is easily knocked out.

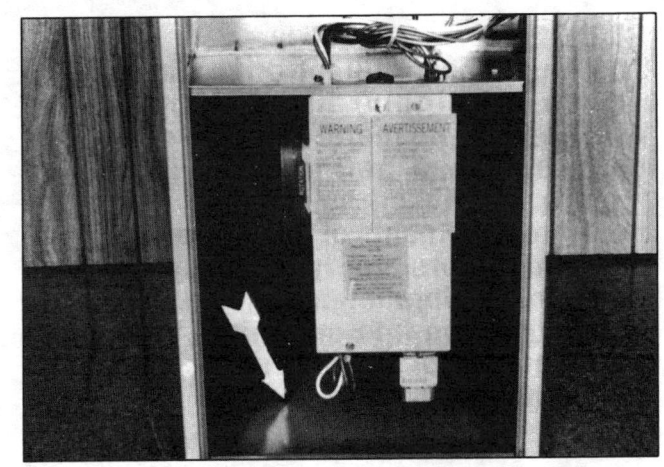

SLIDE 27
On all sizes, except the 91 and 106 MBH models, the return air may enter from either side. On the 91 and 106 MBH model both side inlets (or one side and the bottom) must be used to provide the necessary filter area. To facilitate cutting of the side opening, the panels are embossed at the corners of the opening as shown by the arrows. Note that the area to be cut off has a much larger area than the area outlined by the strengthening indentations. The strengthening indentations DO NOT designate the return air cut out opening.

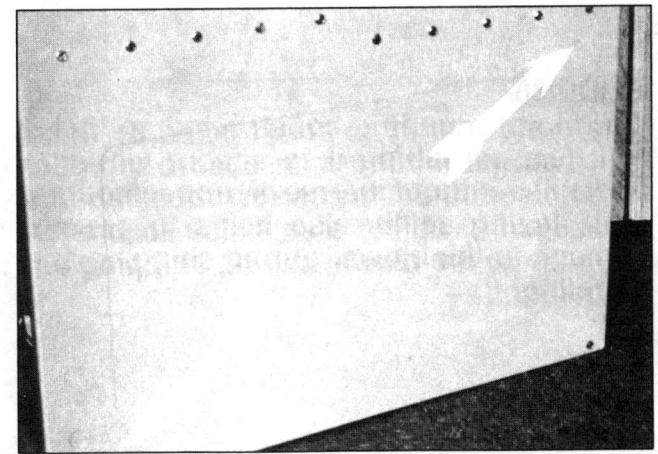

SLIDE 28

The blowers are oversized to handle the large volumes of air that are required for efficient operation. In addition, the blower wheels are dynamically balanced to reduce vibration and noise. To allow for variations in duct design, the blowers are capable of operating up to a maximum external static pressure of 0.75 inches water column. Rated CFM, however, is at 0.5 inches water column.

SLIDE 29

The multi-speed, direct drive blowers are driven by highly efficient, internally protected, permanent split-capacitor motors. These permanently lubricated motors have been selected to match the power requirements of the blower and ducting. The motors are dynamically balanced to minimize vibration and noise. To assure quiet, vibration-free operation, the motors are secured to the blower by an unique motor mount.

SLIDE 30

The motor mount is constructed so that it will flex, permitting it to absorb vibration and noise without the use of rubber mounts. This flexing ability also helps to prevent damage to the blower during shipping and handling.

Central Environmental Systems

SLIDE 31

The low voltage thermostat control wiring is connected to the conveniently located terminal board mounted on the transformer. The power supply wiring is connected to color coded lead wires in the factory mounted junction box. All units have been tested and approved for use with either circuit breakers or fuses.

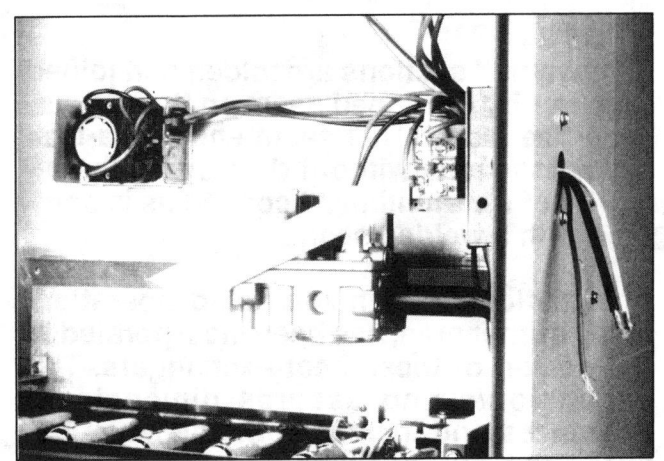

SLIDE 32

The interior casing surfaces of the heat exchanger area are covered with reinforced, heavy-gauge, foil-faced insulation. The addition of this type of insulation is multi-purpose. First of all, it reduces casing surface temperatures permitting the units to be installed at reduced clearances from combustible construction materials. Secondly, it helps reduce air velocity noise. And, lastly, the reinforcement prevents shredding of the foil facing which could be carried by the air stream into the living quarters or restrict air flow through an evaporator coil. Note the crimped seam on the heat exchanger sections.

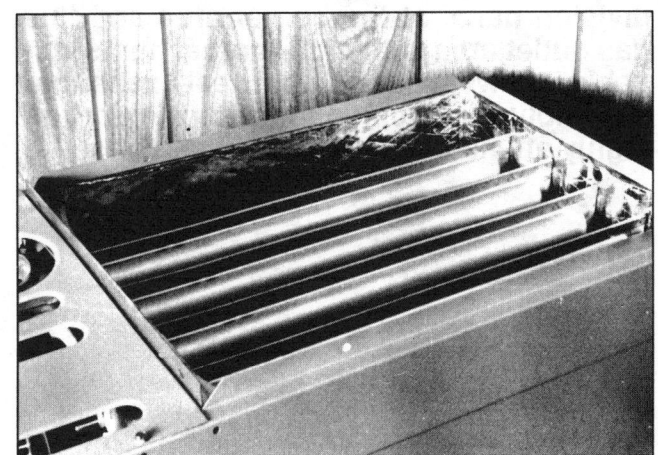

SLIDE 33

The heat exchanger is comprised of individual half-sections fabricated in one piece from heavy-gauge aluminized sheet steel. Aluminized steel is used to assure long, dependable life. For maximum heat transfer, each half section is die formed into an efficient convoluted surface.

SLIDE 34

The two half sections are folded and joined together by a crimped seam on the remaining three sides. This seam ensures a leak proof assembly without destroying the integrity of the aluminized coating as is common with a welded seam.

An additional expansion/sizing operation, in the burner area, has been incorporated in the design of these heat exchangers. The expansion/sizing assures dimensional tolerances and increases the distance between the burner flame and the heat exchanger surface to prevent hot spots.

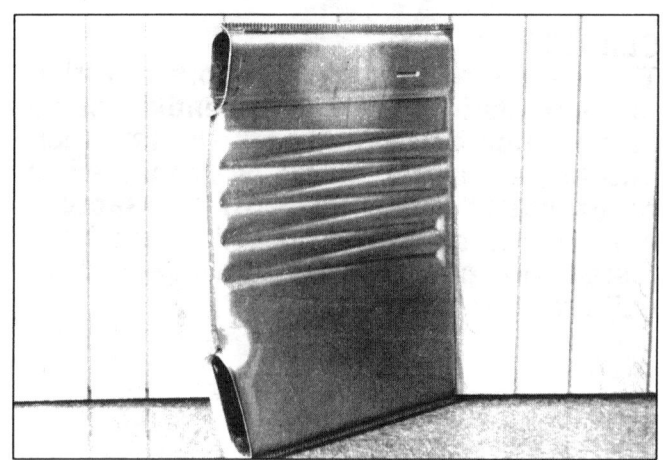

SLIDE 35

The individual sections are secured to the division panel at the burner area and flue gas outlet area with high-torque, non-strippable screws. The resultant heat exchanger assembly is corrosion resistant leak-proof, highly efficient and extremely reliable.

SLIDE 36

To prevent damage during shipment or handling, the individual heat exchanger sections are supported at the rear by a factory installed retainer. The retainer prevents excessive movement of the sections, yet permits them to move freely. Allowing normal movement prevents expansion and contraction noises during operation of the unit. The retainer may be mounted at the top, as we see here, or it may be mounted much lower. In either case the bracket does not have to be removed during installation.

Central Environmental Systems

SLIDE 37

Here we see the induced combustion blower that draws the combustion gases through the heat exchanger sections with controlled flow for maximum efficiency. When using metal vents in vertical applications only, these furnaces are Category I listed and may be common vented with another non-induced, gas-fired appliance. In addition, each furnace in this series is a Category III furnace when horizontally using the high temperature plastic pipe as specified in the Application Data. The flue collar (combustion air blower outlet) on all sizes and models is 3". For those models that require venting with a 4" vent pipe a 3" to 4" adapter is factory supplied. The adapter is factory installed on downflow units; up flow models require field installation.

SLIDE 38

The induced combustion blower housing is fabricated of steel and dipped into a protective coating bath before being sealed with high temperature silicone sealant. The motor and galvanized blower wheel are dynamically balanced to minimize operating sound levels. On the front of the motor is an oversize auxiliary fan which draws ambient air through the motor to maintain low motor operating temperatures. The motor is factory lubricated for an estimated 10 year period. However, it does have oiling provisions and may be oiled periodically to prolong its life.

SLIDE 39

All sizes of furnaces use a White-Rodgers hot surface ignition system and gas valve with 100% shutoff. The gas valve is redundant in that it has two electrically actuated operators controlling the gas flow. The valve also has a manual control knob with "OFF" and "ON" positions and an internal gas pressure regulator. In addition, it has a manifold pressure tap and gas pressure adjustment. For ease of installation, knockouts in the casing side panels permit gas piping connection from either side.

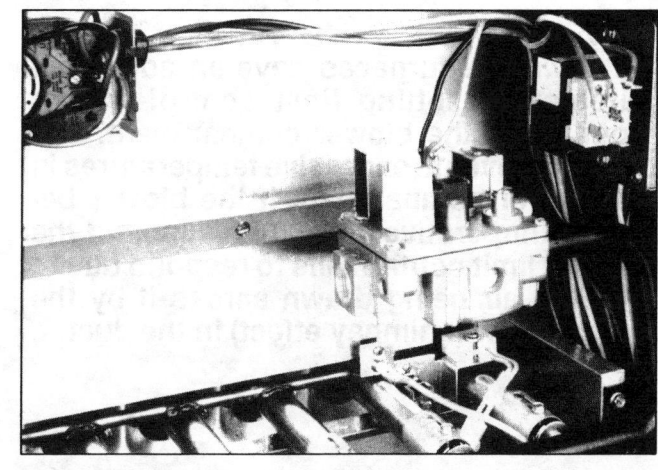

SLIDE 40

On all models the combination fan and limit control is conveniently located. Note that the control does not have a cover. The electrical connections are recessed within the body to prevent the possibility of an electrical hazard.

The limit setting is factory preset to prevent excessive discharge temperatures in the event there is a loss or reduction in air flow due to blower or blower motor failure, dirty filters, or restricted ductwork or grills. On upflow models the fan ON and OFF temperature settings may be adjusted to achieve optimum performance for each particular installation.

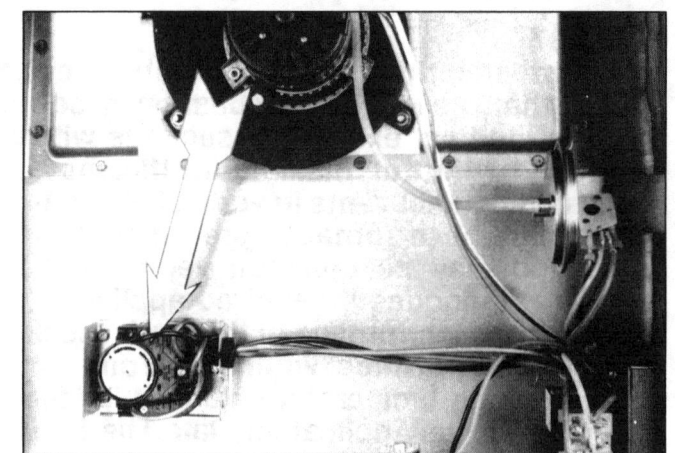

SLIDE 41

On downflow models fan operation is "Time On, Temperature Off." Inside the fan portion of the combination fan and limit control there is a bimetal actuated switch and 24 volt heater. The arrow is pointing to the heater electrical connections that receive power from the 24 volt gas valve control circuit. When energized, the gas valve opens immediately, and the main burners ignite after approximately 40-50 seconds (or the supply air temperature reaches 115 to 125°F) the fan switch closes and energizes the blower motor.

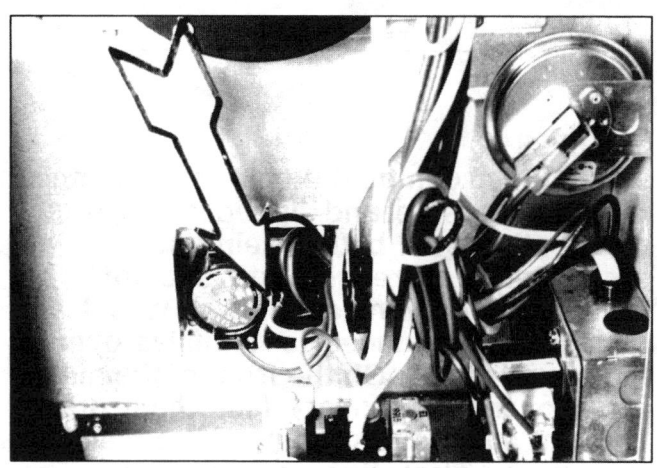

SLIDE 42

All downflow furnaces have an additional automatic-resetting limit control that is mounted in the blower compartment. The control prevents excessive temperatures in the blower compartment if the blower becomes inoperative for any reason, and the primary limit control fails to respond due to the cold air being drawn across it by the reverse-flow (chimney effect) in the duct.

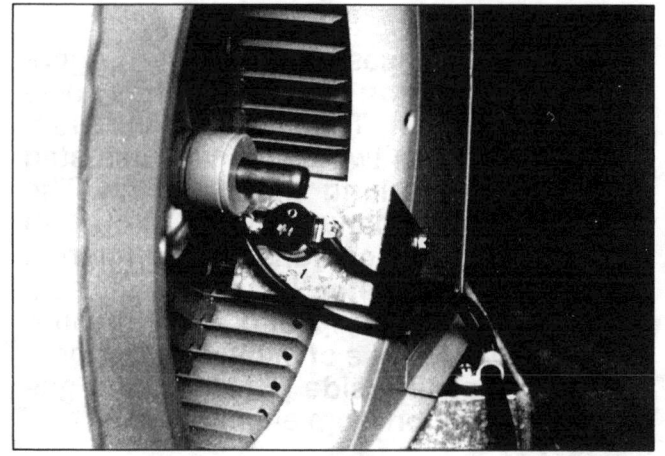

Central Environmental Systems

SLIDE 43

In addition to the limit control we have just described, all sizes and models have flame roll-out protection. The roll-out control is a bi-metal actuated switch that may be manually reset.

Roll Out

Protected

SLIDE 44

The arrow shows the location of the roll-out control. If the temperature in the burner compartment exceeds the control set point, the contracts open the 24 volt circuit to de-energize the ignition control module and gas valve. Operation of the roll-out control indicates a malfunction in the combustion air blower, restricted or blocked heat exchanger section(s) or vent system.

SLIDE 45

All models are supplied with a non-adjustable pressure switch which monitors the flow of gases through the furnace. This switch de-energizes the ignition control module and gas valve if there is a blockage of internal flue gas passageways or failure of the combustion air blower/motor. The method of sensing flow has been changed so that it recognizes an operating problem without causing nuisance shutdown. For ease of service, the pressure settings are stamped on the switches. During the second portion of this program we will discuss how to check them in the field.

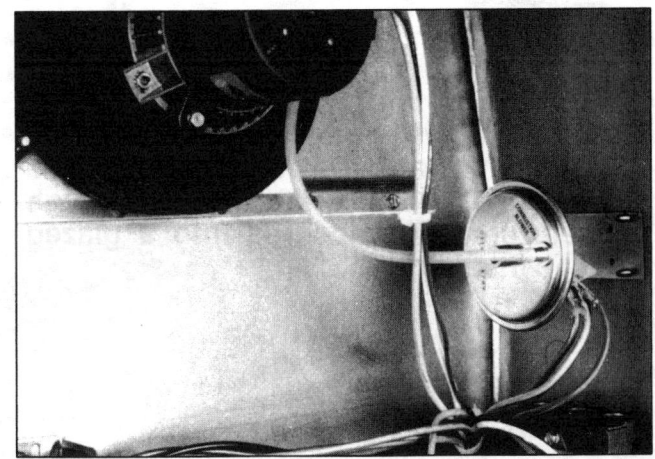

SLIDE 46
Ignition is accomplished by the use of an improved Hot Surface Igniter (HSI) that is constructed of recrystallized Silicon Carbide. The recrystallization process results in maximum possible strength. Operating at 120 volts, the igniter draws 4.5 amperes and is rated at 540 watts/hour. The wattage consumed per ignition cycle, however, is approximately 7 watts. This type of igniter has been used successfully for many years and has but one drawback: "If you bump it, you break it."

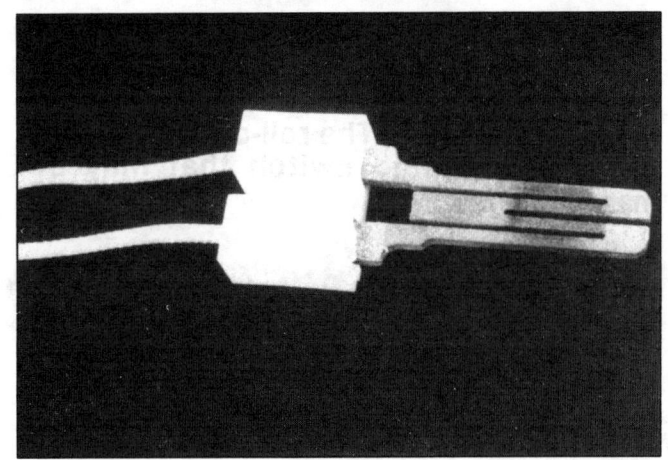

SLIDE 47
The igniter is positioned just above the burner so that accidental contact or breakage is prevented. The location of the igniter permits its 2500°F temperature to light the gas/air mixture issuing from the lanced ports of the burner. The carryover then ignites each of the remaining burners. The cut-out portion just above the igniter permits observation of the igniter during operation.

SLIDE 48
This slide shows another component of the HSI system, the flame sensor probe. The flame sensor probe is merely a rod of high temperature alloy secured into a glazed ceramic insulator.

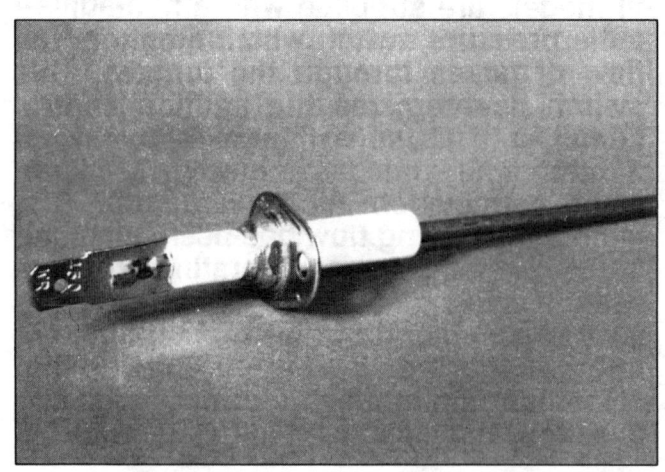

SLIDE 49
Here we see the importance of the high temperature rating of the rod. The probe is factory positioned so that it is directly in the burner flame path. The flame sense probe, in conjunction with an electronic control module, proves the presence of flame. If flame is not detected the gas valve will close.

SLIDE 50
This series of furnaces feature a new design of extra quiet, aluminized lanced port burners with adjustable air shutters. All burners are shipped with the air shutters partially open and are easily adjusted. The burners are universal in that they will operate equally well on natural gas or propane gas.

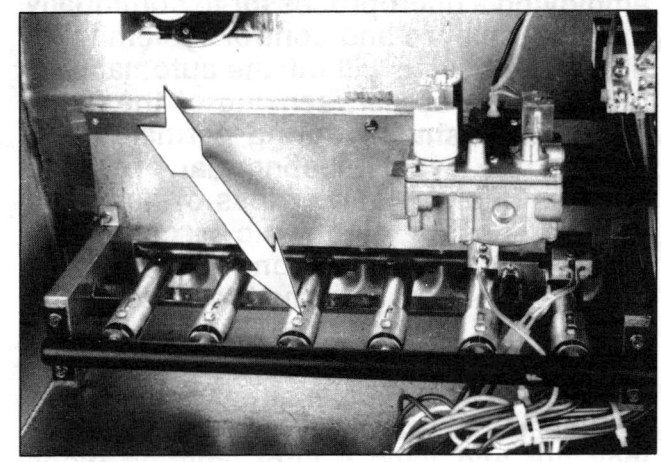

SLIDE 51
To prevent ignition or resonance problems caused by burner misalignment, the burners and carry-over are welded into a one-piece assembly. This design feature prevents the burners from shifting or twisting out of position.

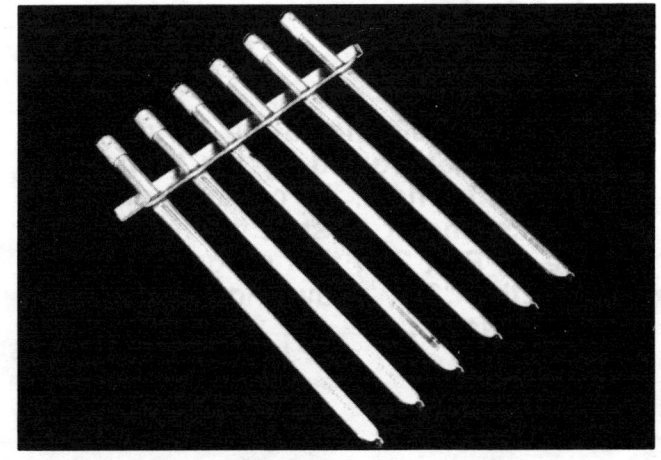

SLIDE 52

To help prevent lint build-up beneath the burner ports, they are lanced and angled in the direction of the incoming gas/air mixture. This design permits airborne dust and lint to be carried directly to the flame where they are incinerated. This slide also shows the positioning of the igniter, flame sensor probe and roll-out control in relationship to the burners. The roll-out control reset button is between the two terminals.

SLIDE 53

Here in the control box we see the White-Rodgers control module. It is a self-testing automatic gas interrupted ignition control employing a microprocessor to continually monitor, analyze and control system functions. The device will initiate automatic ignition of the burner, sensing of the flame and system shutoff during normal operation. In the event of abnormal operation such as internal fault or loss of flame the microprocessor electronic circuitry will react within 8/10 of a second to shut off the gas flow. It is programmed for 2 retries in the event of no flame during initial operation, and 5 recycles if flame is lost after 4 seconds of operation. If either of the above is exceeded, the control is programmed to lock out and flash a light emitting diode (LED). If the LED is on continuously, the fault is likely to be internal to the module. The control is universal in that it can be applied to natural or propane gas units.

SLIDE 54

The arrow is pointing to the 24 volt "3R" relay that is energized by the thermostat upon a call for heat. Its contacts energize the inducer motor to start the heating sequence. In Part 2 of this program, we will discuss the "3R" relay and all electrical circuits in detail.

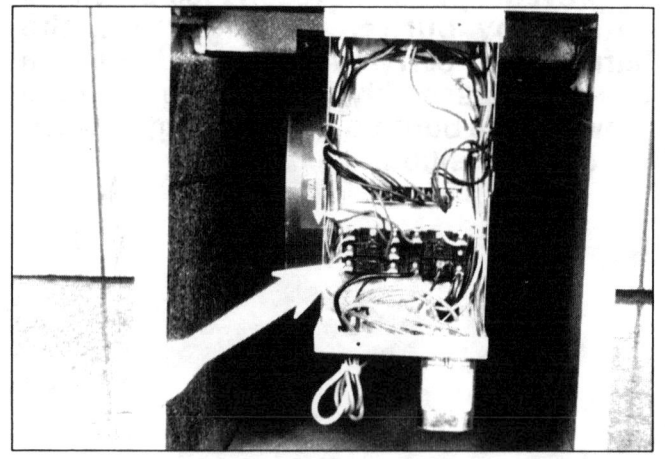

SLIDE 55
Also in the control box is the blower relay that permits blower operation at a higher CFM for cooling and lower CFM for heating. The blower relay is energized automatically by the thermostat control circuit during a call for cooling.

SLIDE 56
Here we see the plug-in electrical connectors for changing the blower speed. This simple adjustment permits the blower to deliver the required amount of air for proper operation.

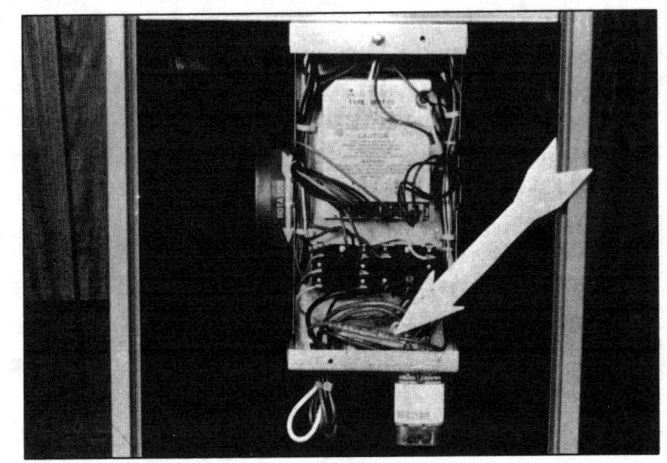

SLIDE 57
For ease of maintenance all blower assemblies are easily removed. The two plugs permit the control box and blower to be removed as an assembly. The black cap between the plugs covers a socket which is used for factory testing of the unit during the final assembly. These sockets and plugs will prove invaluable for troubleshooting the system and will be covered in Part 2 of this program.

SLIDE 58
There are a number of accessories for the product line. A factory accessory is a component or kit which can be added to the unit in the field during the initial installation.

Accessories

SLIDE 59
If propane gas is the supplied fuel, an accessory Propane Conversion Kit is required. The kit is universal in that it will convert all sizes of P3UC/P3CC units, except Low NOx units, for LP operation. The kit contains main burner orifice spuds and a pressure regulator conversion spring. Detailed instructions and conversion labels are also provided in the kit.

Propane Conversion Kit
(Not for Low NOx Models)
1. Main Burner Spuds
2. Gas Valve Conversion Spring
3. Instructions
4. Conversion Labels

SLIDE 60
A side filter rack is available for all sizes of P3UC upflow units. This rack permits moving the factory supplied filter to the outside of the furnace. To supply the required filter area for the 91 and 106 MBH models, two filter racks one on each side, must be used.

SLIDE 61
A side filter cabinet that is applicable to all sizes of P3UC upflow furnaces is also available. It is shipped knocked down and is assembled to the side of the furnace. The use of this kit on the 91 and 106 models permits the return air to enter from a single side.

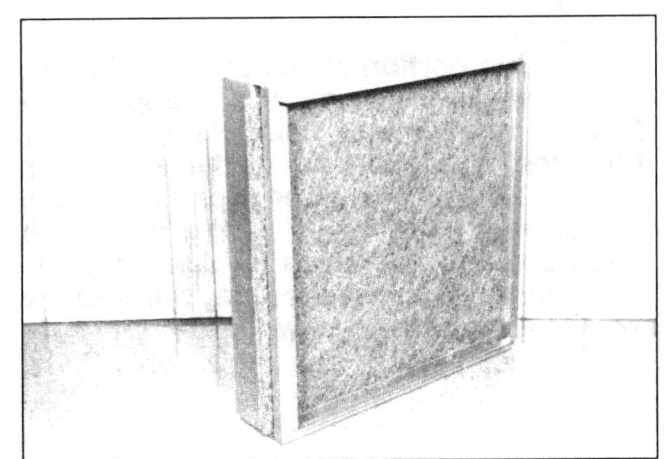

SLIDE 62
Rear return air plenums are also available. Three sizes are available to permit the plenums to match the furnace cabinet size. The plenum comes knocked down and is assembled to the rear of the furnace. The location and dimensions of the rear opening to be cut in the furnace are given in the installation instructions.

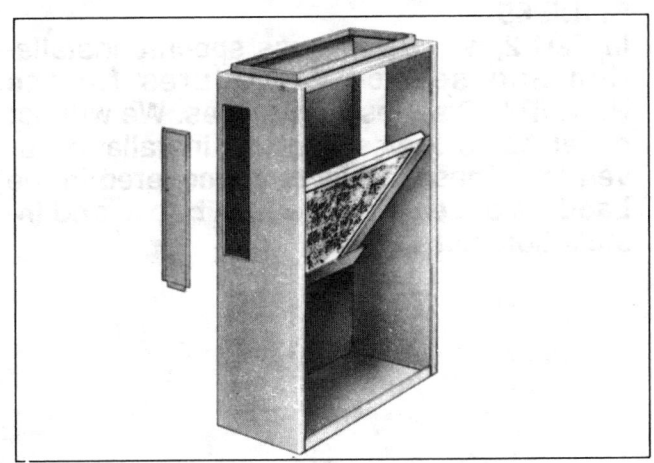

SLIDE 63
For downflow P3CC models three sizes of accessory combustible floor bases are available. One size will fit the 45 and 61 MBH models. Another size is used for the 76 and 91 MBH models and another size will fit the 106 MBH model.

Downflow Models
Combustible Floor Base

(Three Sizes Available)

SLIDE 64

During this portion of the presentation we have discussed the sales and application features of the Stellar P3UC/P3CC series of furnaces. The overview of these features will permit you to answer questions a consumer may have regarding the furnace. If you have any questions on sales and application still unanswered, we will take the time to answer them now.

*Questions
and
Discussion*

SLIDE 65

In Part 2, we will discuss specific installation and service procedures for the P3UC/P3CC series of furnaces. We will not cover basic topics such as installation or venting. These subjects are covered in the Ladders of Learning training books and installation instructions.

Part Two
P3UC/P3CC
Installation
And Service

SLIDE 66

However, in addition to basic venting procedures and limitations, there are additional venting limitations that pertain to this series of furnaces. The venting tables in the National Fuel Gas Code ANSI Z223.1-1988 (NFPA 54) or the Fundamentals of Venting (Form 503.81-TB21V) must NOT be used to size the venting system. For vertical vent installations the vent pipe may be B-1, single-wall metal, or high temperature plastic pipe. Common venting with another appliance (non-induced) is permitted only on vertical vent installations using metal vents. High temperature plastic pipe may be used in dedicated vertical vent installations. For horizontal venting high temperature plastic pipe MUST be used.

Venting

Liimitations

Central Environmental Systems

SLIDE 67
Type B-1 or single-wall metal vent for vertical installation:

1. Must NOT be smaller that the vent connection size on the furnace.

2. Must NOT be more than 45 feet in total length.

3. Must NOT have more than 5 elbows. A reduction in the amount of elbows used does NOT permit a longer vent length.

The Vent System Must Not:

• Be Smaller Than Furnace Connection

• Be more than 45 ft. in Length

• Have More Than 5 Elbows

SLIDE 68
A chart such as this is in the installation instructions furnished with the furnace. It is to be used when venting these furnaces with one water heater. The table shows the correct sizes for the furnace vent connector "F", the heat connector "H" and the common vent "C".

COMMON VENTING FOR ONE FURNACE AND ONE WATER HEATER

FURNACE INPUT (MBH)	"F" FURNACE VENT DIA.	"H" WATER HTR. VENT DIA.	"C" COMMON VENT DIA.
57	3"	3" or 4"	4"
76	4"	3" or 4"	5"
95	4"	3" or 4"	5"
114	4"	3" or 4"	6"

SLIDE 69
These furnaces may also be vertical or horizontal vented using vent material made from GE's ULTEM resin which will withstand flue gas temperatures up to 460°F. The material components are Hart & Cooley's ULTRAVENT™ material or Plexco PLEXVENT® material.

When using high temperature plastic pipe:

1. Follow the manufacturer's instructions.

2. Maximum vent length is 30 ft. plus 5 sweep elbows.

3. Refer to Vent Application Guide (650.64-AD2V).

High Temperature Plastic Pipe

1. Follow Manufacturer's Instructions

2. Maximum Length 30 ft. Plus 5 Sweep Elbows

3 Refer to Venting Guide 650.65-AD2V.

SLIDE 70

The first point we should stress it that these furnaces have NOT been designed, tested or certified for mobile home installation. THEY MUST NOT BE INSTALLED IN A MOBILE HOME.

MOBILE HOMES

SLIDE 71

The thermostat heat anticipator settings are 1.16 amps for the upflow models and 1.21 amps for the downflow models. Downflow models require a higher setting because of the heater in the combination fan and limit control. Some thermostats have a "cycle rate" rather that an adjustable anticipator. On these thermostats adjust the cycle rate down to prevent possible short on/off cycles.

**Thermostat
Anticipator Settings**

Upflow Models 1.16

Downflow 1.21

SLIDE 72

The next subject for discussion is the Hot Surface Ignition system. The Hot Surface Ignition (HSI) system proves the presence of flame through the principle of flame rectification. Flame rectification is a process which converts alternating current (AC) into direct current (DC).

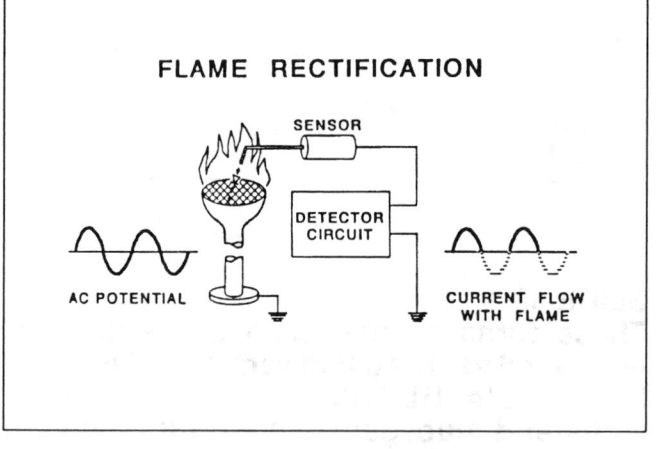

FLAME RECTIFICATION

An AC voltage is applied to the flame sensor electrode. When flame is present, the fuel gas molecules between the flame sense electrode and the burner (ground) become conductive and allow a small current to flow. Because of the surface area difference between the flame sensor electrode and the burner (ground electrode) this current flow takes place mostly in one direction only.

The detector circuit in the HSI control module accepts ONLY this pulsating DC signal, of at least 2 micro-amps, as proof that flame is present. If a short circuit exists (between the electrodes), an AC signal is detected. The HSI control module does not accept this as proof of flame. If flame is not present current cannot flow and NO signal is detected.

SLIDE 73

The White-Rodgers HSI (50E47) control module is a solid state microprocessor. Because of its complexity, the internal operating circuits cannot be depicted by the conventional schematic wiring diagram. The flow chart however, will permit you to understand the sequence of operation. Following the flow chart you will note that the module:

1. Performs an internal check of its circuitry.

2. Provides for 2 retries in the event of no flame during initial operation.

3. Provides for 5 recycles if flame is lost after 4 seconds of operations.

The flow chart does not show it, but a momentary loss of gas supply, flame blowout, or a shorted or open condition in the flame sense circuit will be detected in 8/10 of a second and the system will shut

HSI FLOW CHART
(See Appendix)

down. After a 60 second delay recycles will begin. If the fault is corrected prior to the last ignition attempt, the unit will operate normally. Otherwise, the control will lockout. A Light Emitting Diode (LED) provides visual indication as to likely cause of lockout. To reset the control, interrupt either the 24 volt or line voltage circuit for 1/20 second or longer.

SLIDE 74

Now that we have seen the sequence of operation we are prepared to troubleshoot a "no heat" service problem using the connection and elementary diagrams. The connection diagram shows the electrical components and their relative location within the unit. The legend tells us that the white plug and socket (WP/WS) is 24 volts, the rust colored (RP/RS) is 115 volts and the test socket (TS) is 24/115 volts. These sockets and plugs will prove invaluable when troubleshooting the system. The actual sockets and plugs are not numbered, the numbers on the diagram are for reference only. In addition, the reference numbers are shown from the bottom (wire end). TO PREVENT A WRONG DIAGNOSIS WHEN

CONNECTION DIAGRAM
(See Appendix)

TROUBLESHOOTING A SYSTEM, MAKE SURE YOU ARE TESTING THE PROPER CIRCUIT.

SLIDE 75

The troubleshooting procedure begins with a furnace that does not respond when the room thermostat calls for heat. We will discuss the step-by-step checks and what they mean. We'll use the schematic diagram and a voltmeter to check for AC voltage at various locations until the problem is located. If the problem is found to be a defective control, do not try to repair it. REPLACE DEFECTIVE CONTROLS WITH CES SOURCE 1 PARTS.

ELEMENTARY DIAGRAM
(See Appendix)

SLIDE 76

A schematic wiring diagram is drawn to simplify the circuit for ease of understanding. It separates the various load circuits and identifies the controls which cause each load to function. The terminals on a load or control may not be in exactly the same location or sequence as shown in the diagram. To further simplify this wiring diagram, it has been divided at the transformer to show only the 24 volt or 115 volt circuits.

1ST - Check for 24 volts at terminals "R" and "C" on the control transformer terminal board. If it's not present we'll continue to the 2ND step.

```
1ST CHECK
(See Appendix)
```

SLIDE 77

2ND - Check for 115 volts at TS9 and TS5. If voltage is present it indicates either an open circuit or the transformer is defective. Check all lead wires and plugs/sockets for continuity before replacing any component. Do NOT force or insert the meter probe into the female electrical contact in the plug (you might enlarge the contact and compound your problem by causing an open circuit when the plug is mated into the socket). If you find that 115 volts is not available at TS9 and TS5, do the 3RD check.

3RD - Check for 115 volts at TS6 and TS5. No voltage here indicates an open fuse or broken wire in the power

```
2ND & 3RD CHECKS
(See Appendix)
```

supply. If voltage is present, check the door switch. (Remember that power is shut off when the blower access door is removed.

SLIDE 78

If 24 volts was originally present at terminals "R" and "C" on the transformer terminal board, do the 4TH check.

4TH - Check for 24 volts at terminals "W" and "C" on the transformer terminal board. If it's not present check the room thermostat,, subbase and thermostat "R" and "W" wiring. If power is available at "W" and "C" check for 24 volts at TS4 and TS2. If 24 volts is not available check the orange and blue wires and also the WP/WS terminals 2 & 3. With 24 volts available at TS4 and TS2 make the 5TH check.

```
4TH CHECK
(See Appendix)
```

SLIDE 79

5TH - Check for 24 volts at the coil terminals of the 3R (venter) relay. If 24 volts is not available the 1LP pressure switch is checked next. Put one lead of the voltmeter on the 3R coil terminal (blue wire) and the other lead on the pressure switch common terminal (red wire). If 24 volts is not available there but is available at the NC terminal, the pressure switch is open and must be replaced. If 24 volts was originally present at the 3R coil terminals the 3R relay contacts on lines 7 and 23 should be closed. To determine if the relay or venter motor is defective we must continue to the next check.

5TH CHECK
(See Appendix)

SLIDE 80

6TH - Check for 115 volts between terminal 3 of the 3R relay and TS5. No voltage tells us that the 3R relay is not closing its contacts and must be replaced. If 115 volts is available the motor must be checked next. Carefully feel the motor. If it is at ambient temperature, the windings have an open circuit and the motor/venter must be replaced. If the motor is hot the internal thermostat may have opened. The motor/venter must be removed to ensure that no debris is preventing the blower wheel from turning. If the bearings are tight, lubricating oil may help free them, If this doesn't work, the motor/venter

6TH CHECK
(See Appendix)

must be replaced. If the venter motor runs but the hot surface igniter does not glow we must check to see if the 1LP pressure switch is closing its contacts.

SLIDE 81

7TH - Check for 24 volts between TS2 and the normally open (NO) contact of the 1LP pressure switch. If 24 volts is not available check the pressure switch hose connections to make sure they are tight and not leaking. Consult the installation instructions to ensure that the vent system length, or amount of elbows has not been exceeded. Check the electrical connections on the pressure switch and plug connectors. If after making these checks the pressure switch contacts still do not close, we will check the pressure at the switch.

7TH CHECK
(See Appendix)

Central Environmental Systems

SLIDE 82

To check the pressure you must have an inclined manometer. This inclined manometer is capable of measuring 1" water column pressure and the fluid level is indicating 0.68" WC on the scale. The .68 reading could be positive or it could be negative. If the "B" connection is open to the atmosphere and we apply a POSITIVE pressure to the "A" connection, the pressure forces the liquid level to rise, indicating the applied pressure. However, if the "A" connection is open to the atmosphere and we apply a NEGATIVE pressure to the "B" connection, atmospheric pressure being greater, forces the liquid level to rise indicating the amount of negative pressure.

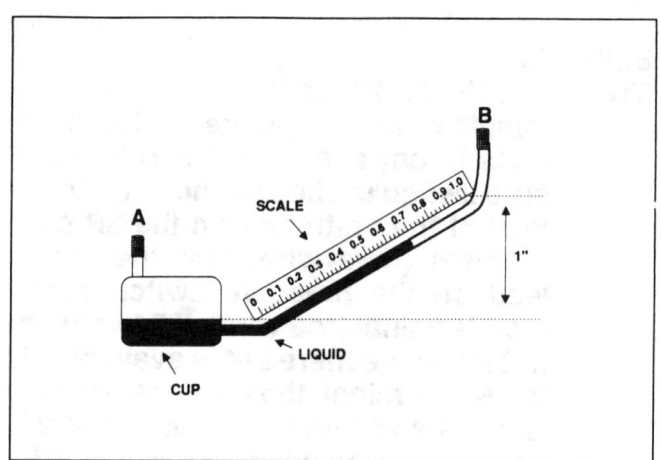

SLIDE 83

The pressure switches are now marked with the calibration point (in inches of W.C.) and either a PR or PF to designate Pressure Rise or Pressure Fall. If specific tolerances are not listed, the following may be used as a guide: ±.05" W.C. up to .50" and 10% for set points higher. To properly use the inclined manometer it must be connected, using tees at the pressure switch. If you are not sure whether the measured pressure is positive or negative, start the furnace and momentarily connect the tube from the pressure switch to each side of the inclined manometer until the fluid rises and indicates a pressure reading on the scale.

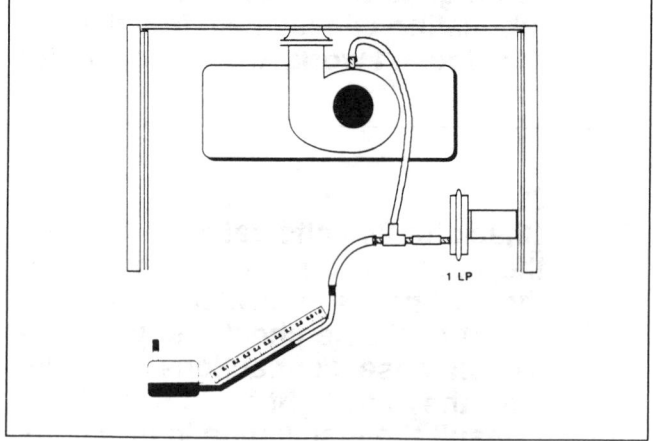

Pressure Switch Calibration

- **Inches Water Column**
- **PR = Pressure Rise**
- **PF = Pressure Fall**
- **±.05" W.C. up to .50" - 10% for Higher Set Points**

SLIDE 84

Here we see that 1LP is measuring a NEGATIVE pressure. If the switch is stamped .78" W.C. PF, the normal pressure (measured pressure) must be GREATER than .78" W.C. to close the normally open (NO) contact. If, for example, the pressure is .98" W.C. (greater than .78"), and the NO switch contacts do not close, the switch is defective and must be replaced. However, assuming normal operation (closed NO contact) and because of restriction or operating problem, the volume of gases is gradually reduced, it will cause the negative pressure to fall until it reaches the set point (.78" W.C. PF) and open the (now closed) NO contact closing the gas valve. If the measured pressure is less than the set point, the combustion blower, vent and heat exchangers must be inspected. After correcting the problem the hot surface igniter should glow. If not, go to the next check.

28

SLIDE 85

8TH - Check for 24 volts between "TH" and "TR" on the HSI control module. (Note that the terminal designations on the schematic diagram and the actual ignition control are different. Remember also to depress the door switch.) If 24 volts is not available, the limit switch, auxiliary limit switch, roll out control and connectors must be checked next. Testing for an open circuit with the voltmeter is quick and easy. Simply connect one test lead to "TR" on the ignition module and jump each connection in succession with the other test lead moving toward the 1LP pressure switch. When you reach the point where the voltmeter reads the proper voltage, you have just jumped the open switch or broken wire, whichever the case may be. If 24 volts is available at "TR" and "TH" and the igniter still does not glow, proceed to the next check.

8TH CHECK
(See Appendix)

SLIDE 86

9TH - DISCONNECT ELECTRIC POWER TO THE SYSTEM AT THE MAIN FUSE OR CIRCUIT BREAKER. Connect an ohmmeter across TS3 and TS8. The resistance of the igniter should be between 20 and 50 ohms. If you are unable to read the proper resistance, the igniter must be replaced. Prior to replacement, check the igniter leads and plugs/sockets back to the ignition module and recheck the resistance of the igniter to verify your findings. If the igniter and wiring are okay the ignition module is defective and must be replaced. If the LED is on continuously, interrupt the power for a few seconds. If an internal fault is indicated again, replace the control. At this point the igniter should glow and the main burners ignite. If the main burners do not light we must check further.

9TH CHECK
(See Appendix)

SLIDE 87

10TH - Check for 24 volts between "MV/MV" at the ignition control module. (After the igniter warm-up period, you have four seconds in which to perform these tests). If there is no voltage at these terminals, replace the module. If voltage is available at the module, check if it is available at terminals 1 and 3 of the gas valve and also between terminals 2 and 3. If voltage is at 1 & 2 and NOT at 2 & 3 the jumper wire is open. If 24 volts is present at the gas valve check the following: the gas supply shutoff valve; the supply gas pressure (must be greater than 4" W.C. but not exceeding 14" W.C.), and the gas valve manual control knob is turned to the ON position. If after these checks the gas valve does not open, replace the gas valve. Note the heater connected to the gas valve terminals 2 & 3. This is the fan heater in the combination fan and limit control.

**10TH CHECK
(See Appendix)**

SLIDE 88

If the furnace operation is erratic or if the main burners do not stay on until the call for heat ends, check the sensor wire and ground wire. If the ground is poor or erratic, shutdown may occur occasionally even though operation is normal at the time of checkout. To test the sensor wire, you must disconnect it from the "FP" terminal of the ignition module. Check the wire for continuity. If you do not measure zero ohms, replace the wire. Next, test the wire for infinity between the (removed) terminal and the furnace casing. If you do not read infinity, either the wire or flame sense electrode is defective and must be replaced. If the flame sense probe is in the burner flame and all checks are okay, replace the ignition control module.

**Check Sensor
& Ground Wires
For Continuity**

SLIDE 89
The next subject we will cover is maintenance. (When performing any maintenance the electrical power must be turned off.) The most neglected maintenance item is filters. The filters must be checked periodically for dirt accumulation. Clean the filters at least every three months. On new construction check the filters every week for the first four weeks. Inspect the filters at least every three weeks after that, especially if the system is running constantly. All filters supplied with the furnace are the high velocity, cleanable type. Clean these filters by washing in warm water. When replacing filters, be sure to use the same size and type as originally supplied.

SLIDE 90
Blower motors in these furnaces are permanently lubricated and do not require periodic oiling. Even with good filters properly in place, blower wheels and motors will become dust laden after long months of operation.

Blower Servicing

SLIDE 91
Squeeze the locking tabs on the electrical plugs and pull straight up to remove them from the control box sockets. As shown here, the next step is to remove the four retaining screws that secure the control box. Hold the box when removing it to prevent it from falling and position it out of the way.

SLIDE 92
After removing the blower retaining screws the blower/motor assembly, control box and inter-connecting wires are removed from the furnace for maintenance or repair. Reverse the above steps to reinstall the components.

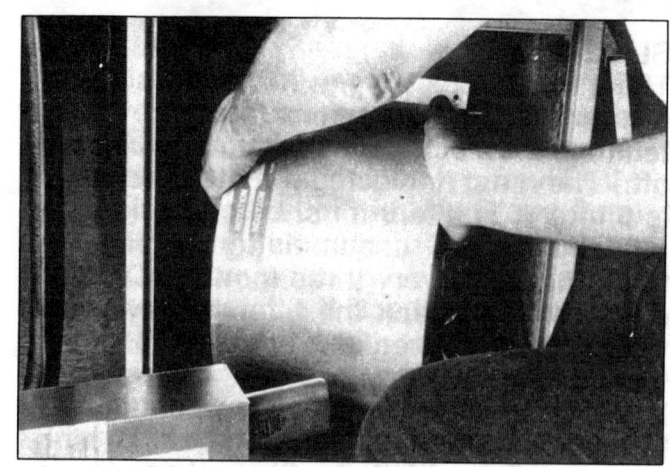

SLIDE 93
The main burners should be checked periodically for dirt accumulation. Turn off the main manual gas valve external to the furnace and electrical power. With the upper access door removed, disconnect the igniter, sensor and roll out control wiring. After removing the bracket retaining screw, taking care not to touch or bump the igniter, carefully remove the entire assembly.

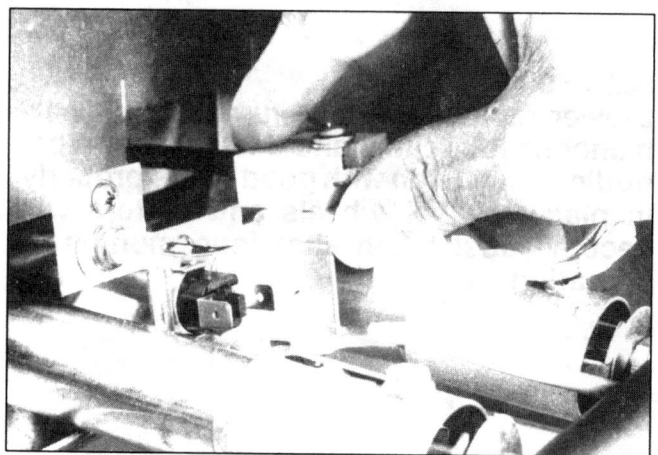

SLIDE 94
The burners are secured to the manifold brackets by two screws. Removing these screws, one on each side of the manifold bracket, permits the entire burner assembly to be easily removed.

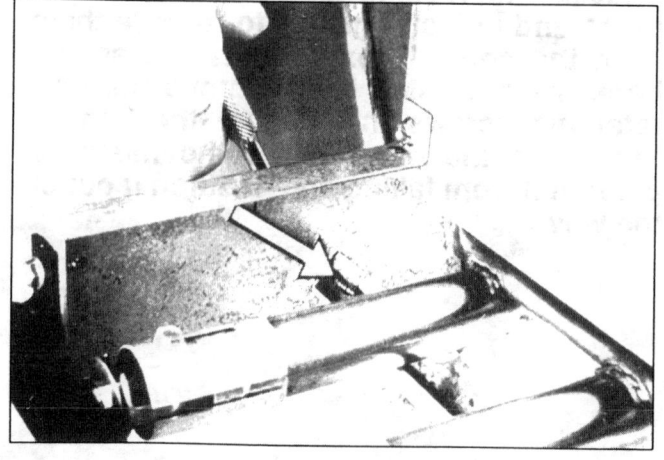

SLIDE 95

With the retaining screws removed, the burner assembly is moved forward, into the heat exchanger, until the burners clear the orifice spuds. Push down on the burners until they clear the bottom on the manifold and pull the entire burner assembly out of the furnace. The burners may be cleaned by rinsing them, from the top down, in hot water paying particular attention to the carry-over.

SLIDE 96

With proper adjustments the heat exchanger sections should not require cleaning for prolonged period of time. However, when cleaning is required, remove the burners and the burner shield to provide access to the bottom of the heat exchanger sections. Note that to remove the burner assembly or burner shield does not require disconnection of the gas piping or removal of the manifold.

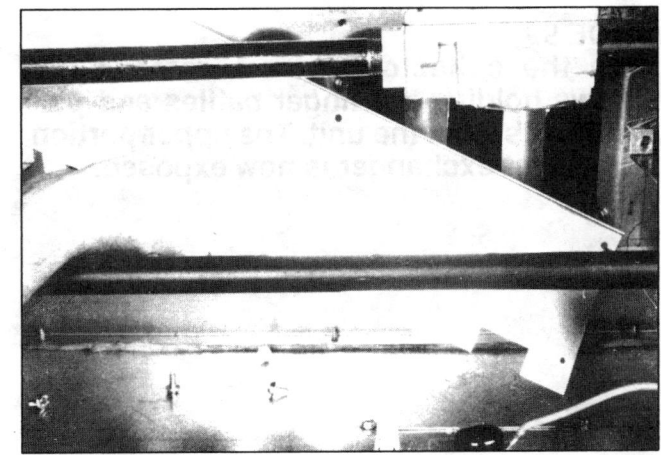

SLIDE 97

To permit access to the upper portion of the heat exchanger sections follow these steps:

1. Remove the vent pipe. If the evaporator condensate drain line is in the way, remove it.

2. Remove the tubing from the venter. (If there is more than one tube, mark or identify them so you will be able to reinstall them properly).

3. Remove the venter blower retaining screws.

SLIDE 98
Remove the venter blower and support it out of the way. Remove the screws securing the collector box and remove it.

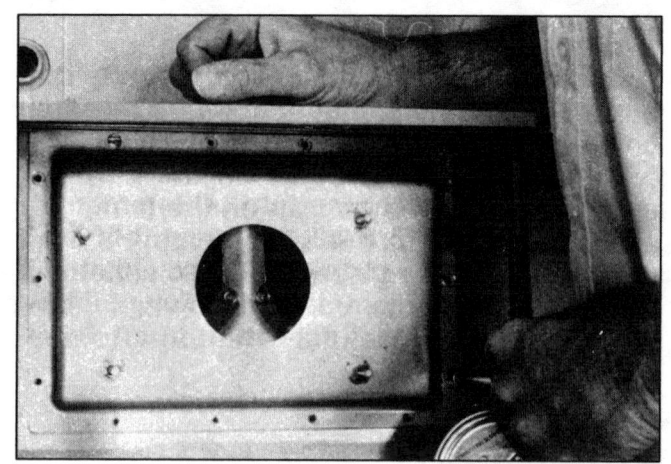

SLIDE 99
With the collector box off, remove the screws holding the finger baffles and pull the baffles from the unit. The upper portion of the heat exchanger is now exposed.

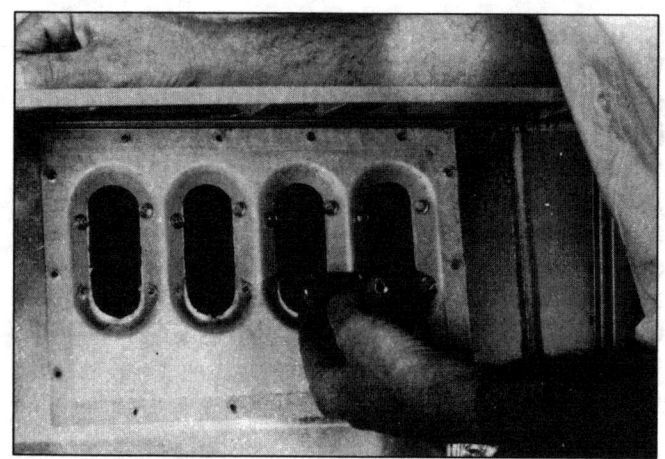

SLIDE 100
Brush out loose scale or soot from the top and bottom of the heat exchanger with a stiff wire brush. Use a vacuum cleaner to clean debris from the heat exchanger. Visually inspect heat exchanger sections for cracks, openings, or excessive corrosion. (If any of these conditions are found the heat exchanger must be replaced). To save time during reassembly, all parts removed for cleaning must be replaced by reversing the order of disassembly. When reinstalling the finger baffles make sure that they are positioned horizontally and are resting on the internal indentations in the heat exchanger. Reconnect all wiring and gas piping prior to restoring electrical power and gas supply to the unit.

1. Brush Out Heat Exchangers
2. Vacuum Debris
3. Inspect Heat Exchangers
4. Reassemble
5. Reconnect Piping and Power

Central Environmental Systems

SLIDE 101

Questions and answers . . . before this or any installation is complete, there are some questions that must be answered. Some are questions you must ask yourself, some questions will be asked by the equipment owner, and there are some questions that will not be asked. Some questions to ask yourself are: Is the main power turned on? Is the gas supply on? Are all systems operable from the thermostat? Have you cleaned your work area and disposed of any trash? Have you presented the equipment owner will all the instructions and paperwork? Have you explained the warranty and warranty procedures?

If at this time the owner of the equipment does not have any questions it may be he does not know what to ask. Show him where the filters are, explain how often to change them and show him how. Explain the thermostat operation and how to set it. Explain how the system works and the sequence of operation. Show him where the disconnect switch, fuses and gas supply shut off valve is located. Explain how to shut the system down.

Having the answers to these questions may very well prevent you from having to come back at a later time.

Hopefully, during this presentation we have answered some of the questions you may have wanted to ask. If you still have any unanswered questions, we will answer them now.

QUESTIONS
AND
ANSWERS

Central Environmental Systems

APPENDIX - SEQUENCE OF OPERATION - HSI CONTROL

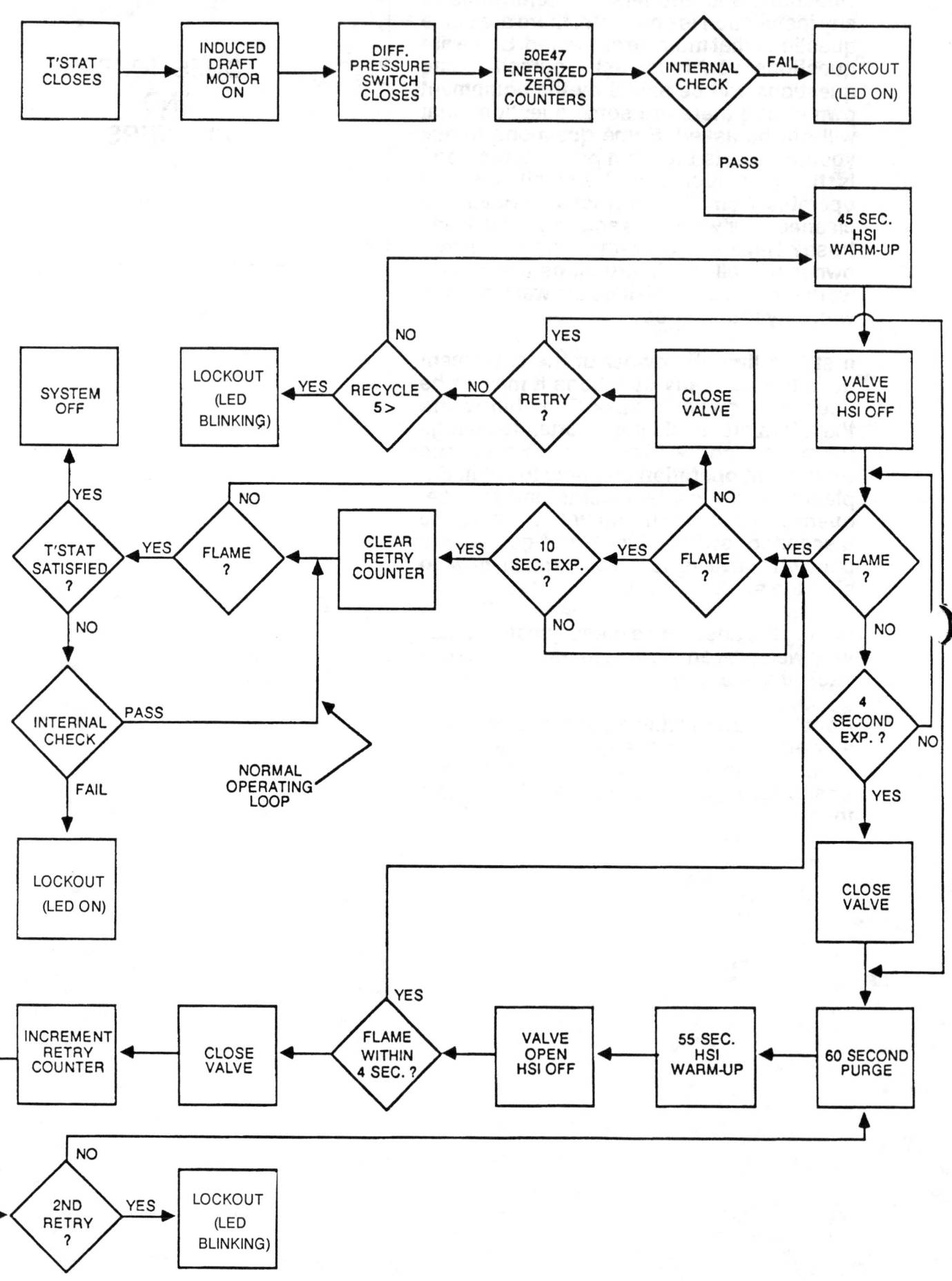

APPENDIX - CONNECTION DIAGRAM

CONNECTION DIAGRAM

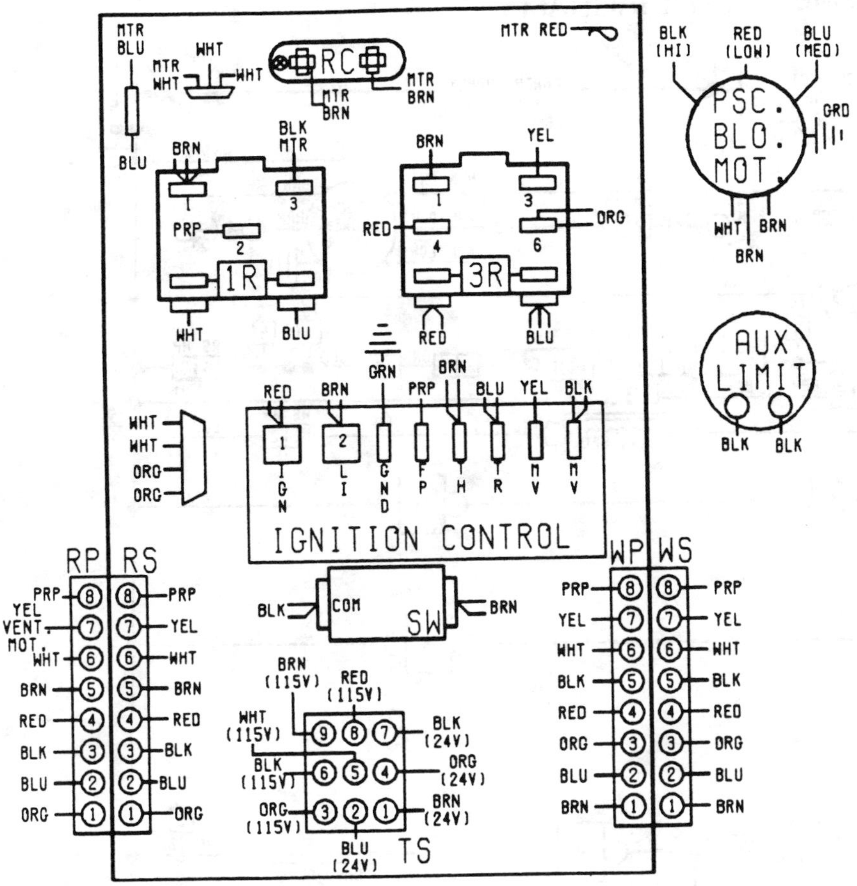

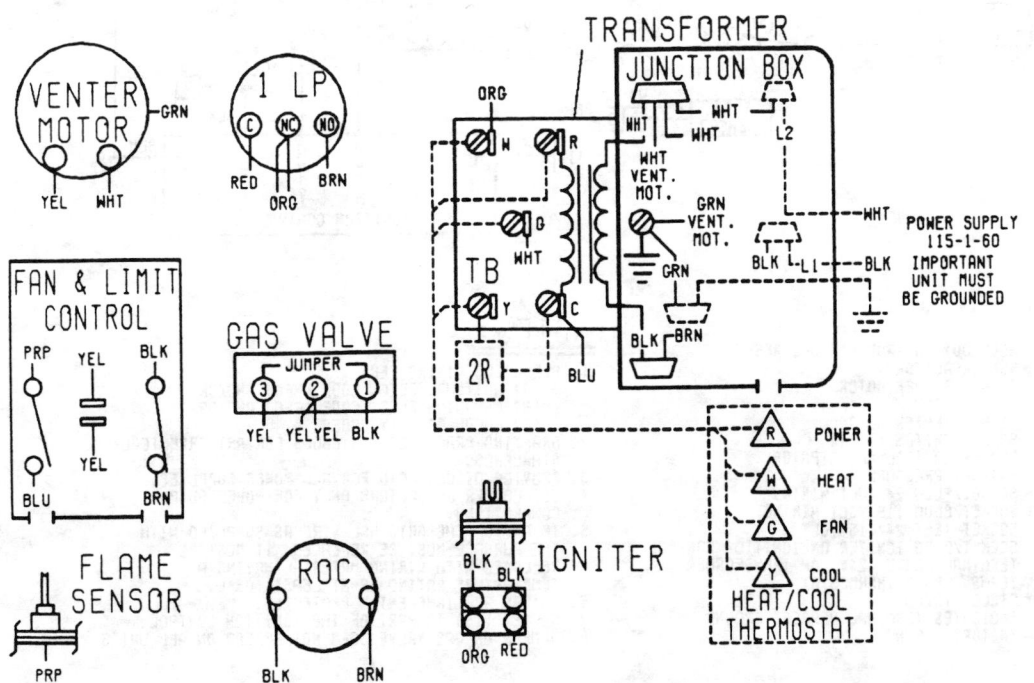

LEGEND

ROC	ROLL OUT CONTROL (MANUAL RESET)
RC	RUN CAPACITOR
1R	RELAY, BLOWER MOTOR
2R	RELAY, COND.
3R	RELAY, VENTER MOTOR
SW	SAFETY SWITCH (DOOR)
HTR	BLOWER CONTROL ANTICIPATOR
1 LP	SWITCH, PRESSURE
WS/WP	SOCKET/PLUG 24 VOLT WIRING
RS/RP	SOCKET/PLUG 115 VOLT WIRING
TS	SOCKET TEST 24/115 VOLT
IS/IP	SOCKET/PLUG IGNITER ON IGNITION CONTROL
TB ○	TERMINAL BOARD, 24V. ON TRANSFORMER
△	TERMS., ROOM THERMOSTAT
-------	FIELD WIRING
⊗	INDICATES TERMINAL NEAREST GROUND
———	FACTORY WIRING

NOTES:

1. ALL FIELD WIRING PER:
   (A) NATIONAL ELEC. CODE (NEC) AND/OR
   (B) CANADIAN ELEC. CODE (CEC) AND/OR
   (C) LOCAL OR CITY CODES.
2. DRAFTING PRACTICES & SYMBOLS PER ARI GRAPHICAL STANDARDS.
3. PROVIDE DISCONNECTS FOR ALL POWER SUPPLIES.
4. USE COPPER CONDUCTORS ONLY FOR POWER SUPPLY CONNECTIONS.
5. IF ANY OF THE ORIGINAL WIRE AS SUPPLIED WITH THE FURNACE MUST BE REPLACED, IT MUST BE REPLACED WITH WIRING MATERIAL HAVING A TEMPERATURE RATING OF AT LEAST 105°C.
6. MOTORS ARE INHERENTLY PROTECTED.
7. SOCKET (IS) IS PART OF THE IGNITION CONTROL.
8. JUMPER ON GAS VALVE DOES NOT APPEAR ON ALL UNITS.

ELEMENTARY DIAGRAM

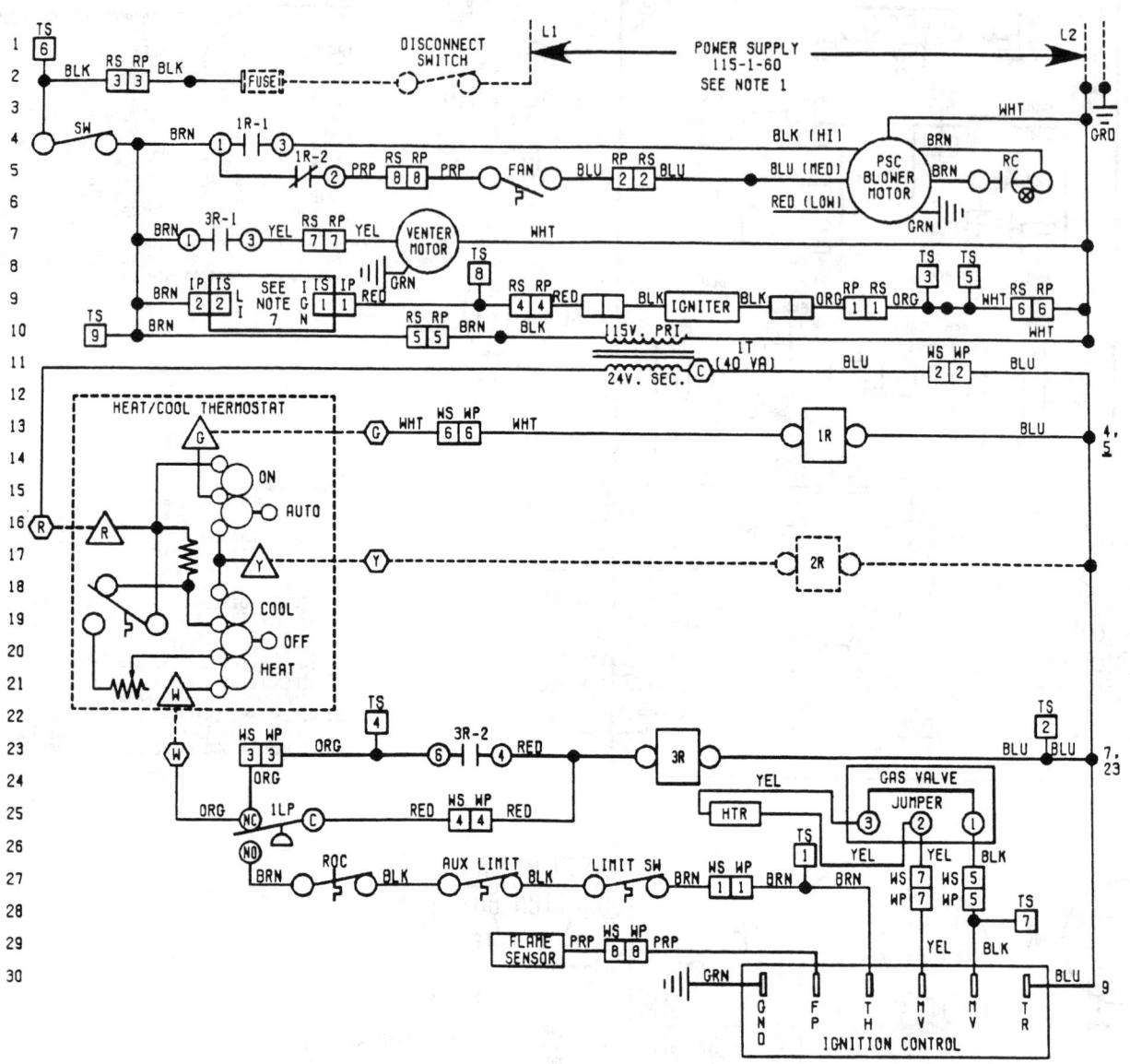

LEGEND

ROC	ROLL OUT CONTROL (MANUAL RESET)
RC	RUN CAPACITOR
1R	RELAY, BLOWER MOTOR
2R	RELAY, COND.
3R	RELAY, VENTER MOTOR
SW	SAFETY SWITCH (DOOR)
HTR	BLOWER CONTROL ANTICIPATOR
1 LP	SWITCH, PRESSURE
WS/WP	SOCKET/PLUG 24 VOLT WIRING
RS/RP	SOCKET/PLUG 115 VOLT WIRING
TS	SOCKET TEST 24/115 VOLT
IS/IP	SOCKET/PLUG IGNITER ON IGNITION CONTROL
TB	TERMINAL BOARD, 24V. ON TRANSFORMER
	TERMS., ROOM THERMOSTAT

------------ FIELD WIRING

⊗ INDICATES TERMINAL NEAREST GROUND

――――― FACTORY WIRING

NOTES:

1. ALL FIELD WIRING PER:
   (A) NATIONAL ELEC. CODE (NEC) AND/OR
   (B) CANADIAN ELEC. CODE (CEC) AND/OR
   (C) LOCAL OR CITY CODES.
2. DRAFTING PRACTICES & SYMBOLS PER ARI GRAPHICAL STANDARDS.
3. PROVIDE DISCONNECTS FOR ALL POWER SUPPLIES.
4. USE COPPER CONDUCTORS ONLY FOR POWER SUPPLY CONNECTIONS.
5. IF ANY OF THE ORIGINAL WIRE AS SUPPLIED WITH THE FURNACE MUST BE REPLACED. IT MUST BE REPLACED WITH WIRING MATERIAL HAVING A TEMPERATURE RATING OF AT LEAST 105°C.
6. MOTORS ARE INHERENTLY PROTECTED.
7. SOCKET (IS) IS PART OF THE IGNITION CONTROL.
8. JUMPER ON GAS VALVE DOES NOT APPEAR ON ALL UNITS.

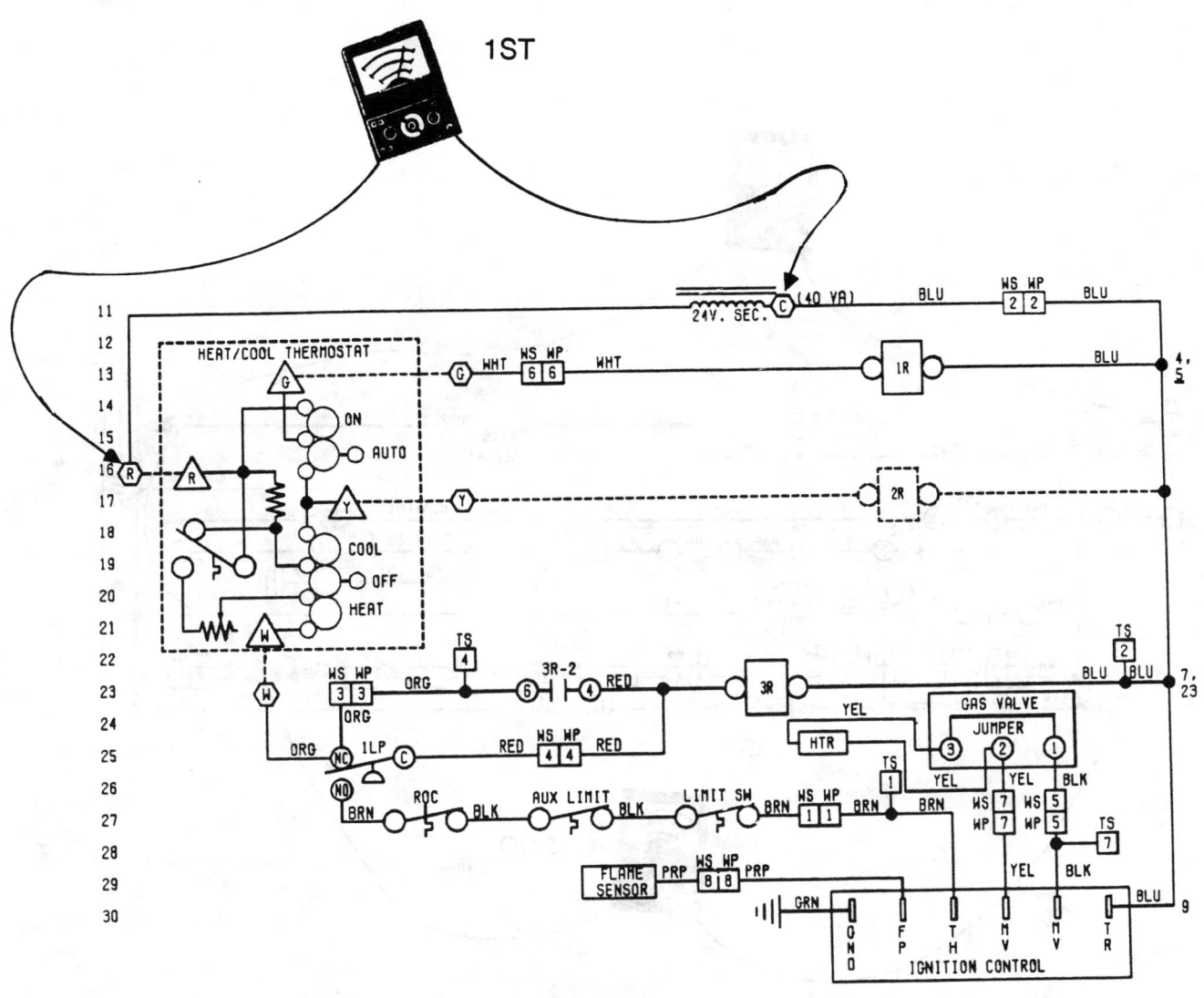

1ST

LEGEND

ROC	ROLL OUT CONTROL (MANUAL RESET)
RC	RUN CAPACITOR
1R	RELAY, BLOWER MOTOR
2R	RELAY, COND.
3R	RELAY, VENTER MOTOR
SW	SAFETY SWITCH (DOOR)
HTR	BLOWER CONTROL ANTICIPATOR
1 LP	SWITCH, PRESSURE
WS/WP	SOCKET/PLUG 24 VOLT WIRING
RS/RP	SOCKET/PLUG 115 VOLT WIRING
TS	SOCKET TEST 24/115 VOLT
IS/IP	SOCKET/PLUG IGNITER ON IGNITION CONTROL
TB	TERMINAL BOARD, 24V. ON TRANSFORMER
△ ○	TERMS., ROOM THERMOSTAT
---------	FIELD WIRING
⊗	INDICATES TERMINAL NEAREST GROUND
————	FACTORY WIRING

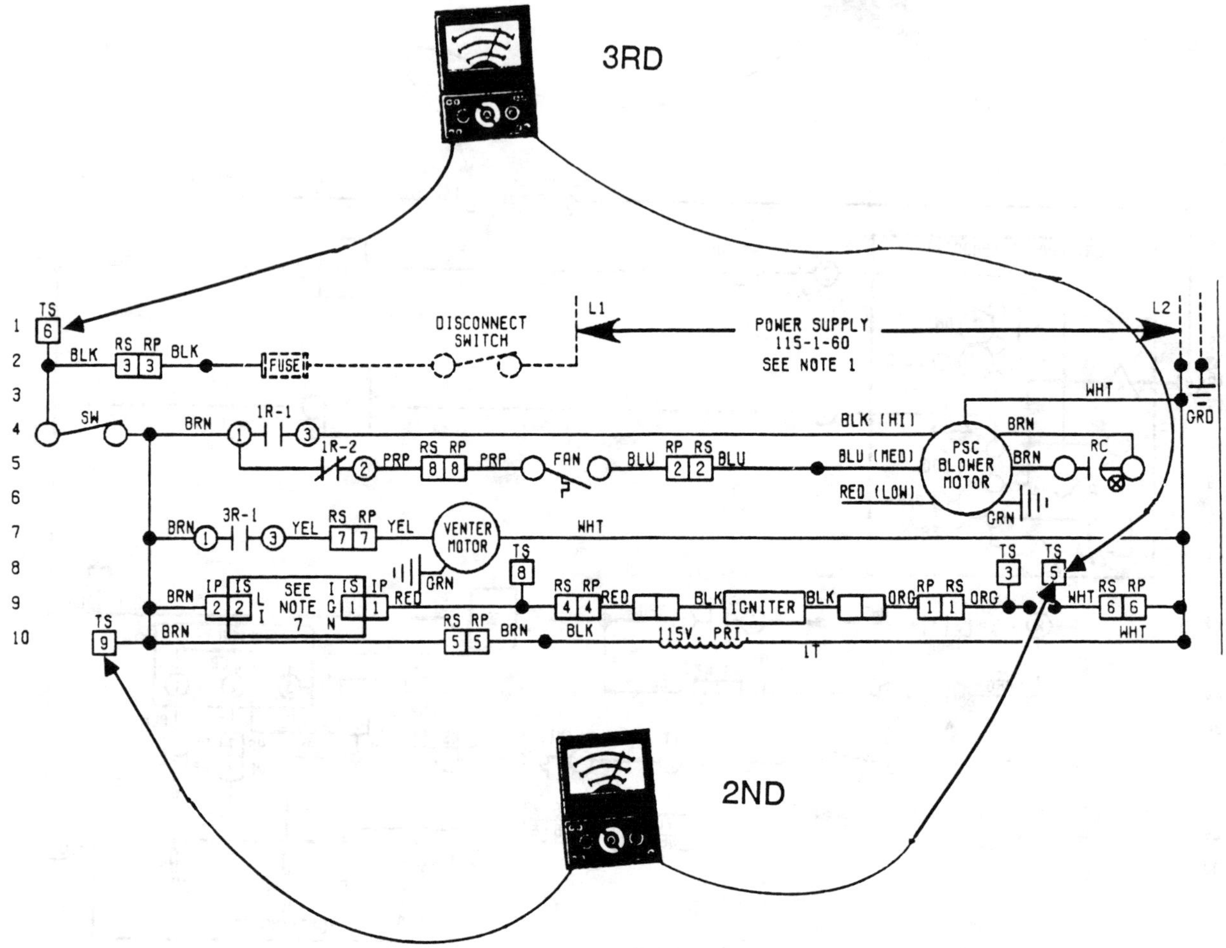

3RD

2ND

TS 6

BLK RS RP BLK [FUSE] DISCONNECT SWITCH L1 POWER SUPPLY 115-1-60 SEE NOTE 1 L2
3 3

SW BRN 1R-1 1R-2 PRP RS RP PRP FAN BLU RP RS BLU BLK (HI) WHT GRD
1 3 2 8 8 2 2 BLU (MED) PSC BLOWER MOTOR BRN RC
RED (LOW) BRN GRN

BRN 3R-1 YEL RS RP YEL VENTER MOTOR WHT TS 8 TS 3 TS 5
1 3 7 7 GRN

BRN IP IS SEE IS IP RED RS RP RED BLK IGNITER BLK ORG RP RS ORG WHT RS RP
2 2 NOTE 7 1 1 4 4 1 1 6 6
WHT

TS 9 BRN RS RP BRN BLK 115V. PRI WHT
5 5 1T

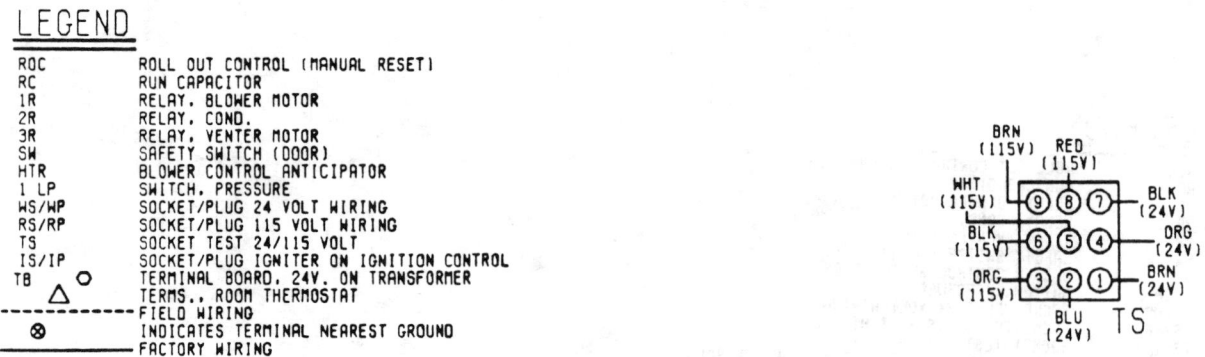

LEGEND

ROC	ROLL OUT CONTROL (MANUAL RESET)
RC	RUN CAPACITOR
1R	RELAY, BLOWER MOTOR
2R	RELAY, COND.
3R	RELAY, VENTER MOTOR
SW	SAFETY SWITCH (DOOR)
HTR	BLOWER CONTROL ANTICIPATOR
1 LP	SWITCH, PRESSURE
WS/WP	SOCKET/PLUG 24 VOLT WIRING
RS/RP	SOCKET/PLUG 115 VOLT WIRING
TS	SOCKET TEST 24/115 VOLT
IS/IP	SOCKET/PLUG IGNITER ON IGNITION CONTROL
TB ○	TERMINAL BOARD, 24V. ON TRANSFORMER
△	TERMS., ROOM THERMOSTAT
--------	FIELD WIRING
⊗	INDICATES TERMINAL NEAREST GROUND
——	FACTORY WIRING

BRN (115V) RED (115V)
WHT (115V) 9 8 7 BLK (24V)
BLK (115V) 6 5 4 ORG (24V)
ORG (115V) 3 2 1 BRN (24V)
BLU (24V) TS

40

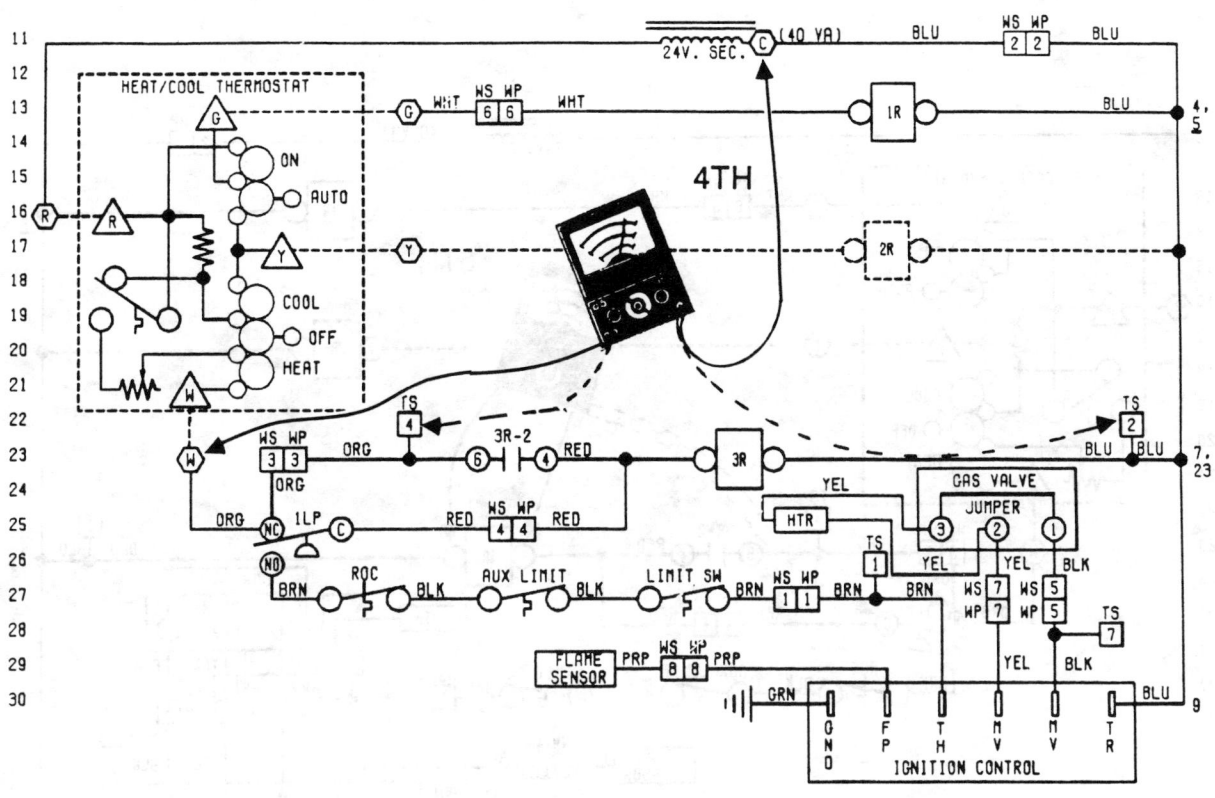

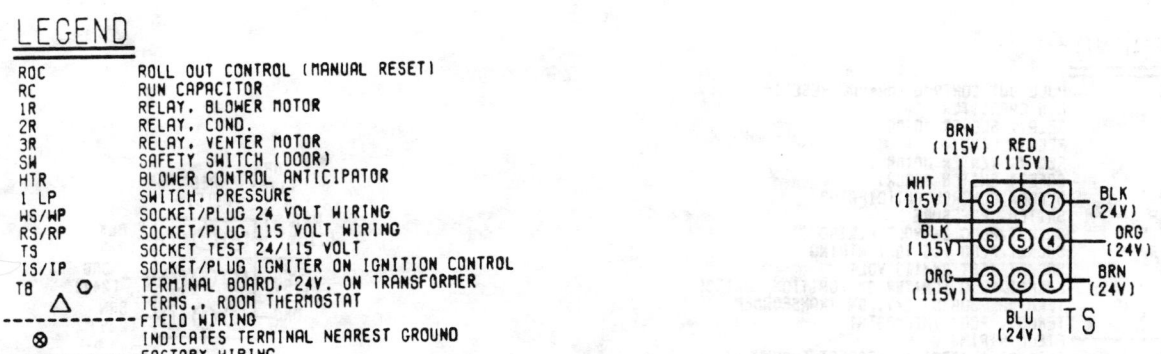

LEGEND

ROC	ROLL OUT CONTROL (MANUAL RESET)
RC	RUN CAPACITOR
1R	RELAY, BLOWER MOTOR
2R	RELAY, COND.
3R	RELAY, VENTER MOTOR
SW	SAFETY SWITCH (DOOR)
HTR	BLOWER CONTROL ANTICIPATOR
1 LP	SWITCH, PRESSURE
WS/WP	SOCKET/PLUG 24 VOLT WIRING
RS/RP	SOCKET/PLUG 115 VOLT WIRING
TS	SOCKET TEST 24/115 VOLT
IS/IP	SOCKET/PLUG IGNITER ON IGNITION CONTROL
TB	TERMINAL BOARD, 24V. ON TRANSFORMER
△	TERMS., ROOM THERMOSTAT
- - - - -	FIELD WIRING
⊗	INDICATES TERMINAL NEAREST GROUND
———	FACTORY WIRING

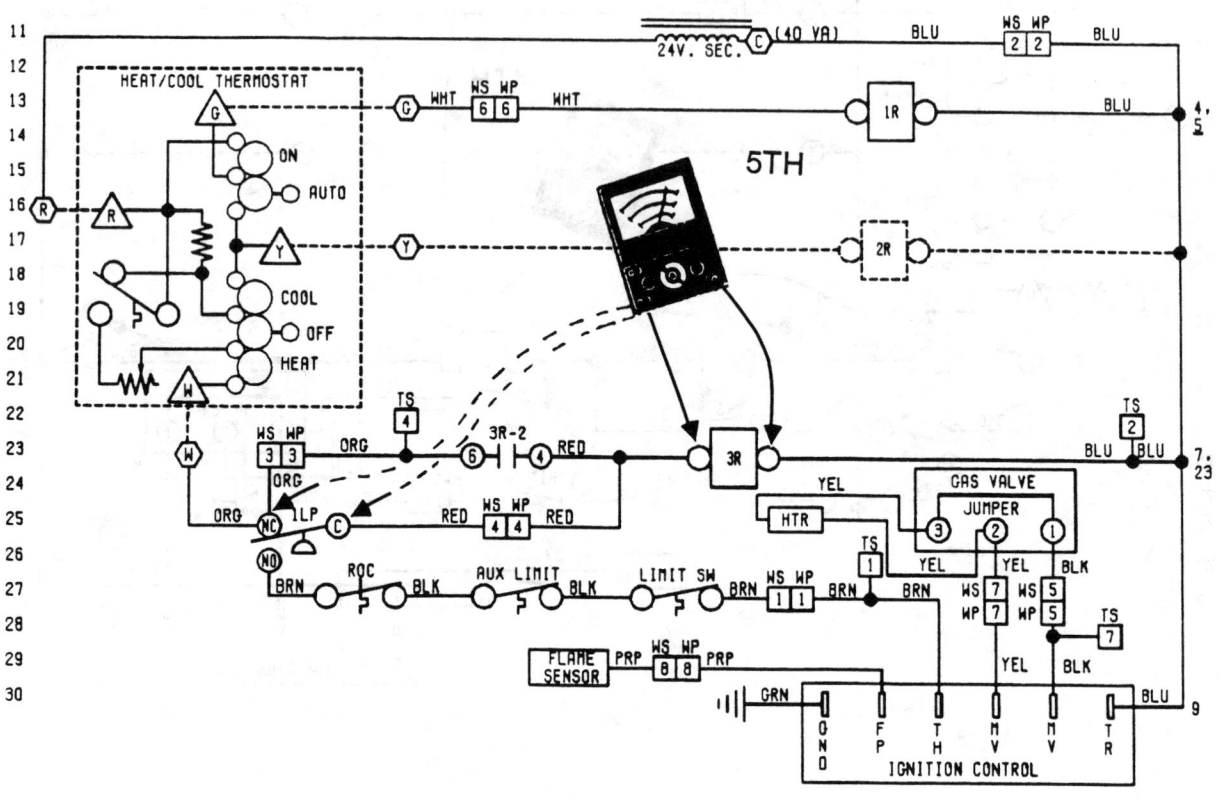

LEGEND

ROC	ROLL OUT CONTROL (MANUAL RESET)
RC	RUN CAPACITOR
1R	RELAY, BLOWER MOTOR
2R	RELAY, COND.
3R	RELAY, VENTER MOTOR
SW	SAFETY SWITCH (DOOR)
HTR	BLOWER CONTROL ANTICIPATOR
1 LP	SWITCH, PRESSURE
WS/WP	SOCKET/PLUG 24 VOLT WIRING
RS/RP	SOCKET/PLUG 115 VOLT WIRING
TS	SOCKET TEST 24/115 VOLT
IS/IP	SOCKET/PLUG IGNITER ON IGNITION CONTROL
TB ○	TERMINAL BOARD, 24V. ON TRANSFORMER
△	TERMS., ROOM THERMOSTAT
- - - - -	FIELD WIRING
⊗	INDICATES TERMINAL NEAREST GROUND
———	FACTORY WIRING

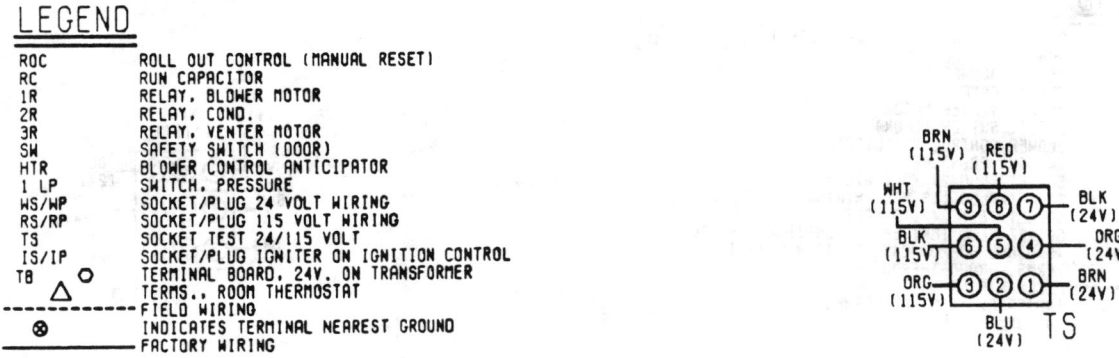

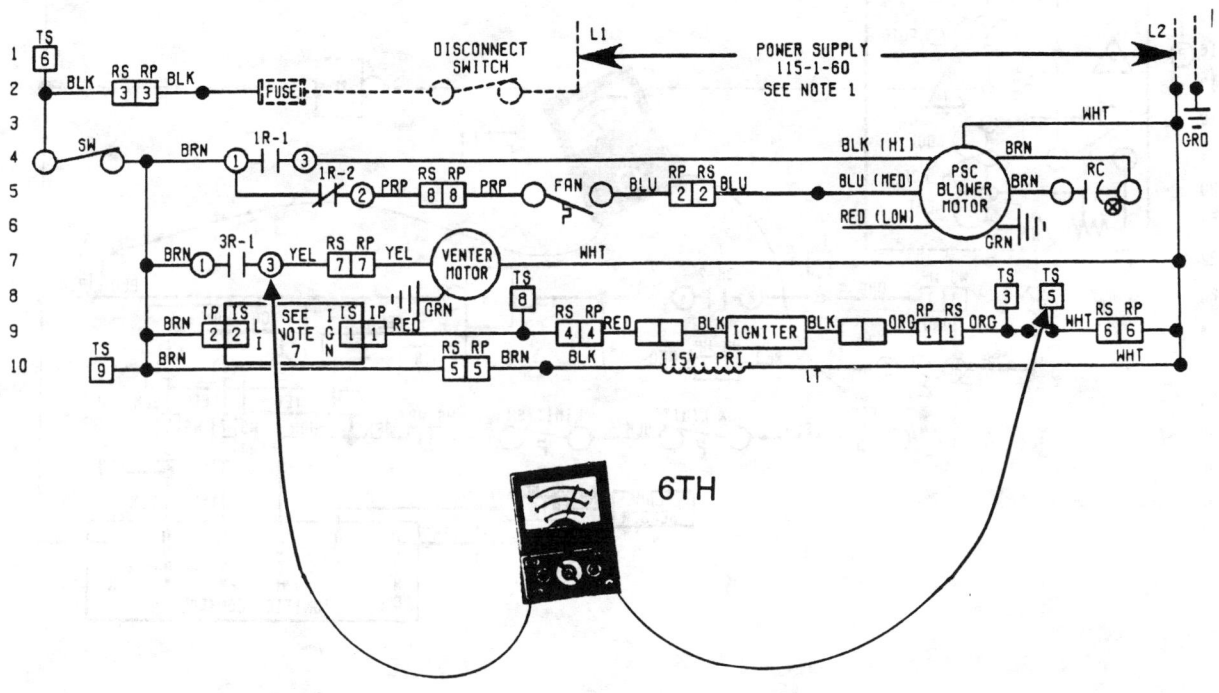

6TH

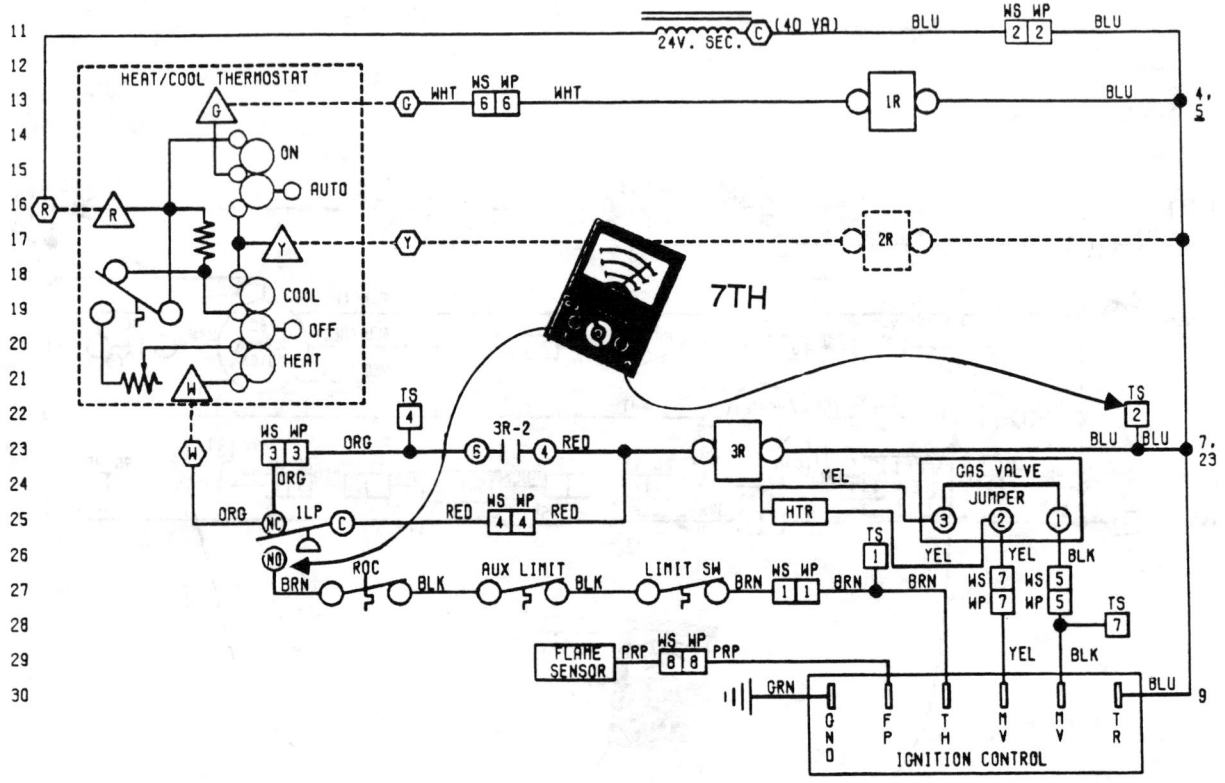

Central Environmental Systems

24V. SEC.
(40 VA)
BLU
WS WP
HEAT/COOL THERMOSTAT
ON
AUTO
COOL
OFF
HEAT
WHT
1R
2R
TS
3R-2
ORG
RED
3R
YEL
GAS VALVE
JUMPER
HTR
1LP
ROC
BLK
AUX LIMIT
LIMIT SW
BRN
FLAME SENSOR
PRP
GRN
IGNITION CONTROL
8TH

GRN
BRN
RED
PRP
BLU
YEL
BLK
IGNITION CONTROL

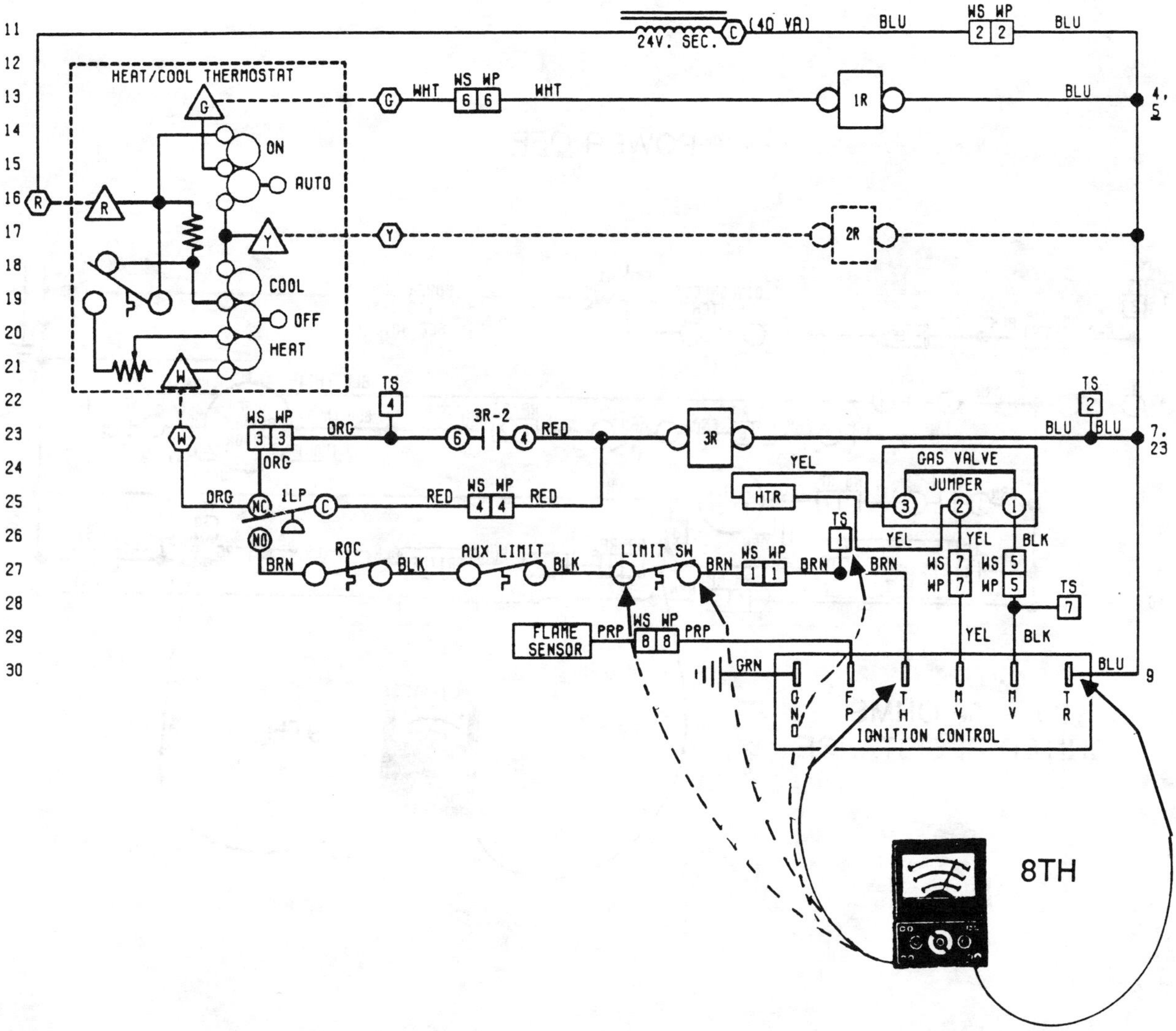

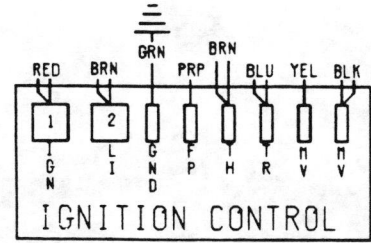

POWER <u>OFF</u>

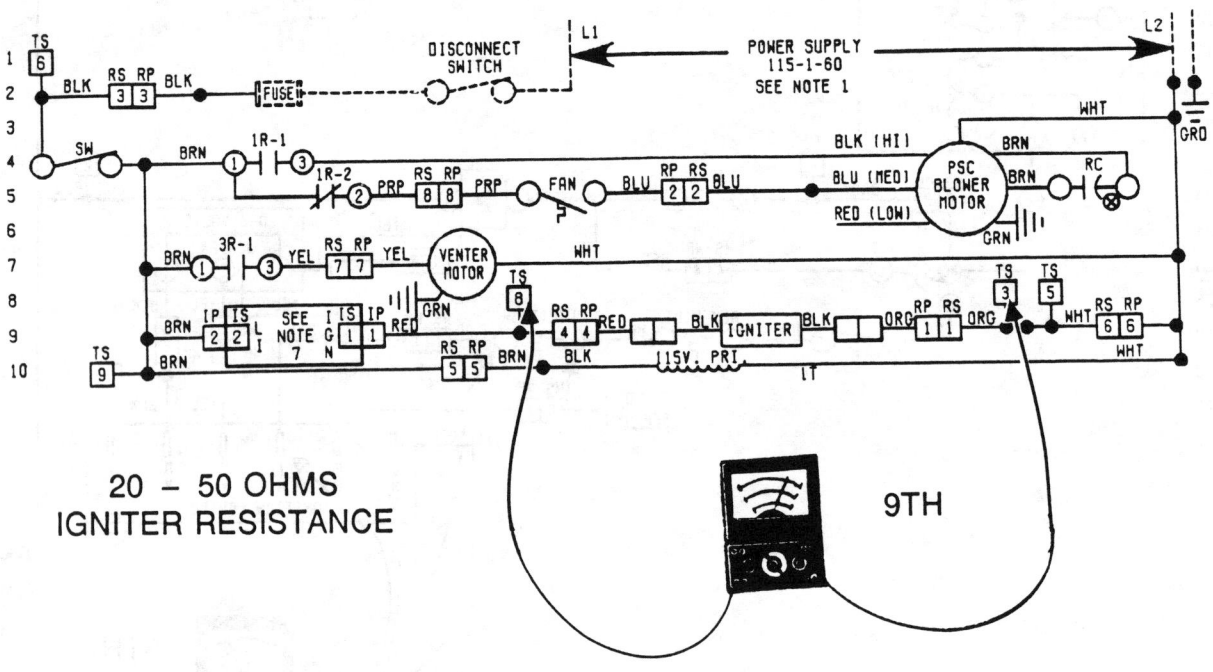

20 — 50 OHMS
IGNITER RESISTANCE

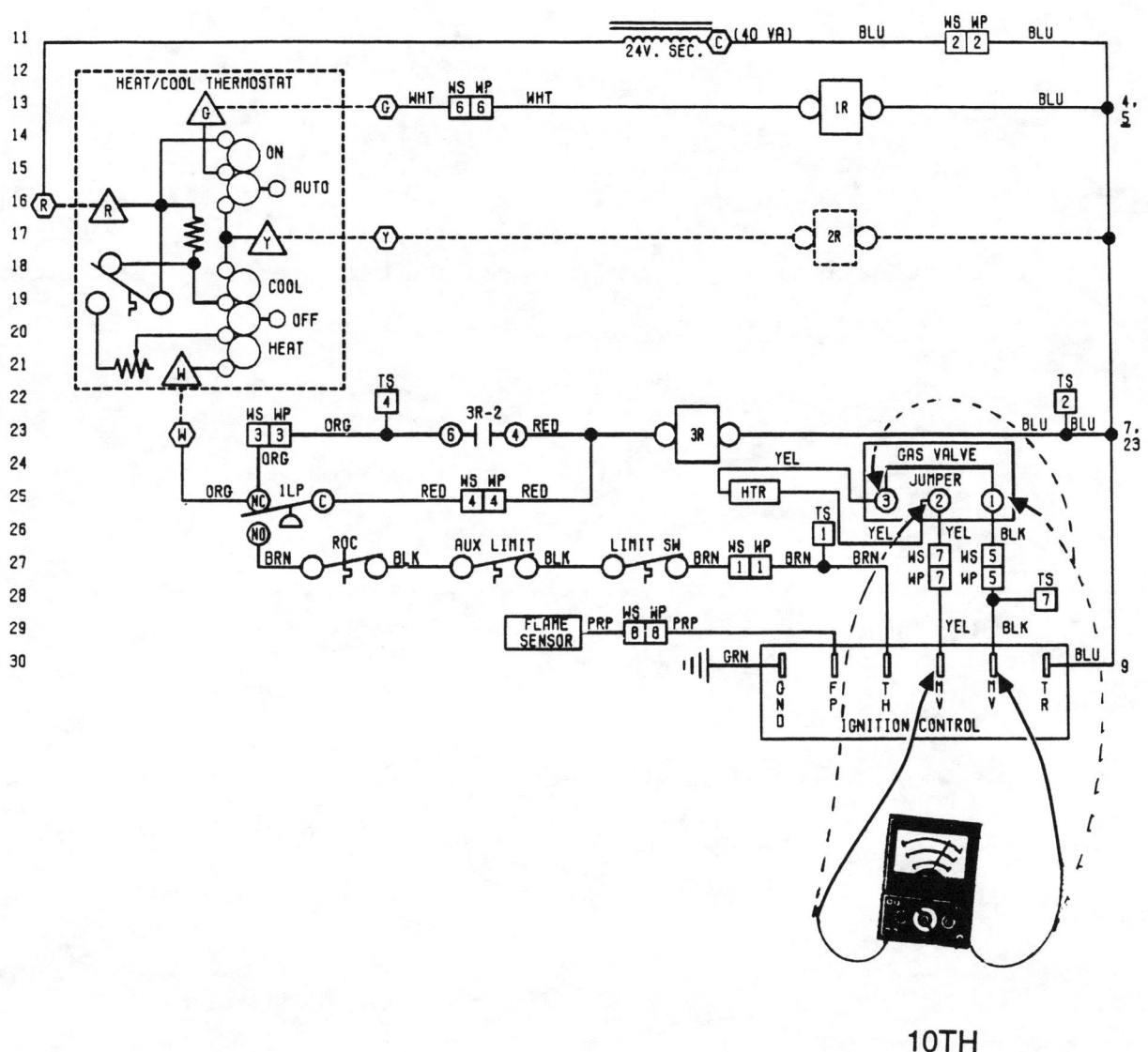

10TH

NOTES

NAME _____ DATE _____

FUNDAMENTALS OF HIGH EFFICIENCY GAS FURNACES QUIZ

TRUE (T) or FALSE (F) Circle correct answer.

T F   1. These furnaces have AFUE ratings greater than 78%.

T F   2. Low NOx units can not be converted to operate on propane gas.

T F   3. For heat/cool applications an accessory relay kit is required.

T F   4. Because of the closed base pan, upflow models may be installed directly on wooden floors.

T F   5. On all sizes and models the gas supply piping may enter from either side.

T F   6. The units may be side wall vented.

T F   7. The transformer is sized for heating/cooling operation.

T F   8. These models are classified Category I and may be common vented.

T F   9. The HSI control module is programmed for two re-trys in the event of no flame during initial operation.

T F 10. If the flame roll-out protector is "open" the venter motor will not operate.

T F 11. When converting to propane operation the HSI ignition module does not have to be changed.

T F 12. On counterflow models the blower is energized 40-50 seconds after the gas valve opens.

T F 13. On upflow models bottom return air is not permitted.

T F 14. All sizes and models are tested and approved for installation in a mobile home.

T F 15. If the ignition control module is locked out and the LED is flashing, the control module must be changed.

T F 16. The test socket holes are numbered from the top.

T F 17. During a call for heat the thermostat completes the R to W circuit.

T F 18. If 24 volts is not available at the 3R relay coil, the relay is defective.

T F 19. If voltage is available at the transformer primary terminals but not at the secondary, the roll-out protector is "open".

T F 20. For proper operation the thermostat heat anticipator must be set at .4 amps.

T F 21. These units are approved for use with either fuses or circuit breakers.

T F 22. When converting to propane operation, the burners must be changed.

T F 23. Accessory rear return air plenums are available for upflow models.

T F 24. The differential pressure switches are field adjustable.

T F 25. After five years of operation the blower motors must be oiled annually.

Heating and Air Conditioning

Official Sponsor
of the 1992
U.S. Olympic Team
36USC380

P.O. Box 1592, York, Pennsylvania USA 17405-1592
Subject to change without notice. Printed in U.S.A.
Copyright © by York International Corporation 1991. All rights reserved. RPC 2M 1191 1.35

Supersedes: 650.64-TB11Y (1289)
650.64-TB13Y

Other Books in the Pro Series

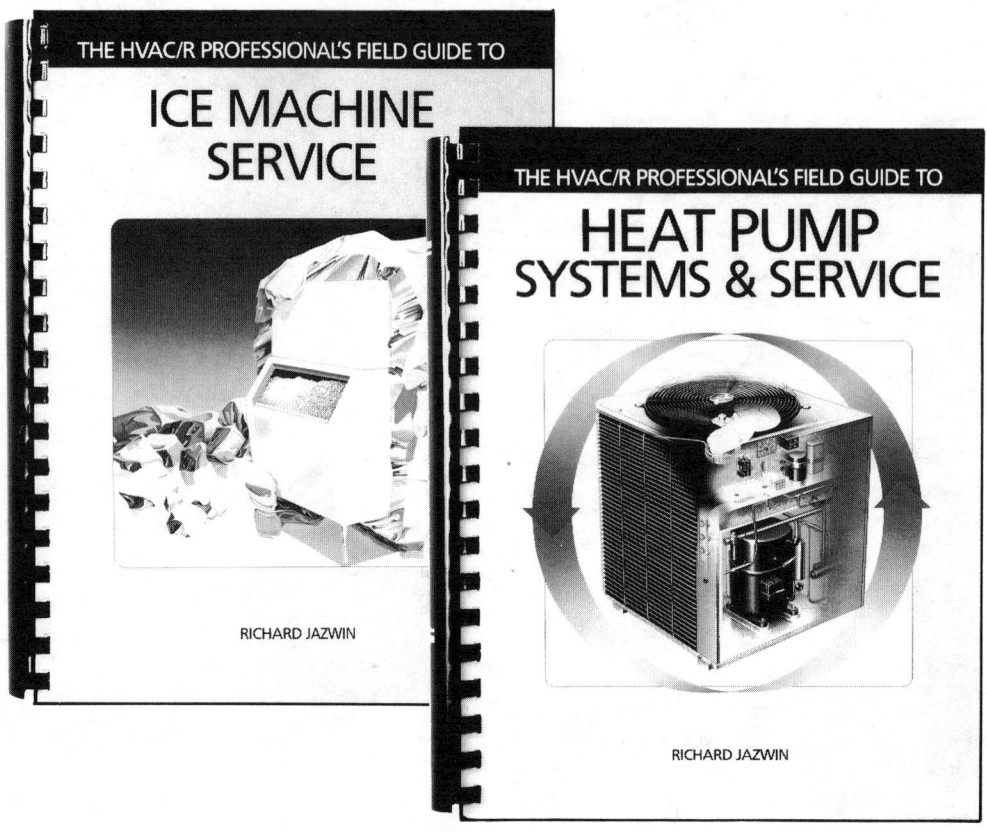